01

THIRD EDITION

Clinical Virology Manual

THIRD EDITION

Clinical Virology Manual

EDITORS

Steven Specter
Department of Medical Microbiology and Immunology
College of Medicine
University of South Florida
Tampa, Florida

Richard L. Hodinka
Clinical Virology Laboratory
Children's Hospital of Philadelphia
and
Departments of Pediatrics and Pathology
University of Pennsylvania School of Medicine
Philadelphia, Pennsylvania

Stephen A. Young
TriCore Reference Laboratories
and
Department of Molecular Genetics and Microbiology
University of New Mexico School of Medicine
Albuquerque, New Mexico

ASM
PRESS

WASHINGTON, D.C.

Address editorial correspondence to ASM Press, 1752 N St., NW, Washington, DC 20036-2904, USA

Send orders to ASM Press, P.O. Box 605, Herndon, VA 20172, USA
Phone: 800-546-2416; 703-661-1593
Fax: 703-661-1501
E-mail: books@asmusa.org
Online: www.asmpress.org

Library of Congress Cataloging-in-Publication Data
Clinical virology manual/edited by Steven Specter, Richard L. Hodinka, and Stephen A.
Young—3rd ed.
    p. cm.
  Includes bibliographical references and index.
  ISBN 1-55581-173-6
    1. Diagnostic virology—Handbooks, manuals, etc. I. Specter, Steven. II. Hodinka, Richard L.
III. Young, Stephen A.
  QR387.C48 2000
  616'.0194—dc21   00-040163

We dedicate this book to our colleague and friend Jerry Lancz, who served as an editor of the first two editions but has since retired. His foresight and insights created a publication that has brought greater understanding and unity to a field that continues to expand. We also dedicate it to our wives, Randie, Kitty, and Linda, and to our children, Ross, Rachel, Ryan, Tyler, Brett, and Jesse, whose patience and support sustain us through all our endeavors.

Contents

Contributors

DAVID A. ANDERSON
Hepatitis Research Unit and Australian Centre for Hepatitis Virology, Macfarlane Burnet Centre for Medical Research, Fairfield, Victoria 3078, Australia

JAMES G. ANTHONY
Research and Development., Digene Corporation, Silver Spring, MD 20904

LAURE AURELIAN
Virology/Immunology Laboratories, Departments of Pharmacology and Experimental Therapeutics and of Microbiology and Immunology, School of Medicine, University of Maryland, Baltimore, MD 21201

HAROLD L. BALLEW
Centers for Disease Control and Prevention, Atlanta, Ga.

WILLIAM J. BELLINI
Measles Virus Section, Centers for Disease Control and Prevention, Atlanta, GA 30333

MAURO BENDINELLI
Department of Biomedicine and Retrovirus Center, University of Pisa, Pisa, Italy

THOMAS A. BRAWNER
1355 Ravean Ct., Encinitas, CA 92024

R. MARK L. BULLER
Department of Molecular Microbiology and Immunology, Saint Louis University Health Sciences Center, St. Louis, MO 63104

JOHN J. DOCHERTY
Department of Microbiology/Immunology, Northeastern Ohio Universities College of Medicine, Rootstown, OH 44272

SHEILA C. DOLLARD
Herpesvirus Section, Centers for Disease Control and Prevention, Atlanta, GA 30333

DEAN D. ERDMAN
Respiratory and Enterovirus Branch, Centers for Disease Control and Prevention, Atlanta, GA 30333

ROBERT C. GALLO
Institute of Human Virology, University of Maryland, Baltimore, MD 21201

HELEN GAY
Department of Microbiology/Immunology, Northeastern Ohio Universities College of Medicine, Rootstown, OH 44272

CURT A. GLEAVES
Infectious Diseases and Molecular Diagnostics, Providence Portland Medical Center, Portland, OR 97213

HANS-PETER GRUNERT
Department of Virology, Institute for Infectious Diseases Medicine, University Hospital Benjamin Franklin, Free University of Berlin, 12203 Berlin, Germany

BRIAN HJELLE
Department of Pathology, Molecular Genetics, and Microbiology and Department of Biology, University of New Mexico School of Medicine, Albuquerque, NM 87131

RICHARD L. HODINKA
Clinical Virology Laboratory, Children's Hospital of Philadelphia, and Departments of Pediatrics and Pathology, University of Pennsylvania School of Medicine, Philadelphia, PA 19104

G. D. HSIUNG
Virology Reference Laboratory, VA Connecticut Healthcare System, West Haven, CT 06516

XI JIANG
Center for Pediatric Research, Children's Hospital of the King's Daughters and Eastern Virginia Medical School, Norfolk, VA 23510

MARIE L. LANDRY
Department of Laboratory Medicine, Yale University School of Medicine, New Haven, CT 06520

DIANE S. LELAND
Department of Pathology and Laboratory Medicine, Riley Hospital for Children, Indianapolis, IN 46202

DAVID A. LENNETTE
Virolab, Inc., 1204 Tenth St., Berkeley, CA 94710

EVELYNE T. LENNETTE
Virolab, Inc., 1204 Tenth St., Berkeley, CA 94710

ARTHUR L. LEWIS
Epidemiology Research Center, Office of Laboratory Services, Florida Department of Health and Rehabilitative Services, Tampa, Fla.

LISA LINSKE-O'CONNELL
Advanced Bioscience Laboratories, Inc., Kensington, MD 20895-1078

ATTILA T. LÖRINCZ
Research and Development, Digene Corporation, Silver Spring, MD 20904

FABRIZIO MAGGI
Department of Biomedicine and Retrovirus Center, University of Pisa, Pisa, Italy

BRIAN W. J. MAHY
National Center for Infectious Diseases, Centers for Disease Control and Prevention, Atlanta, GA 30333

DAVID O. MATSON
Center for Pediatric Research, Children's Hospital of the King's Daughters and Eastern Virginia Medical School, Norfolk, VA 23510

ROBYN McGUIRE
Research and Development, Chemicon International, Inc., Temecula, CA 92590

LEROY C. McLAREN
Department of Molecular Genetics and Microbiology, School of Medicine, University of New Mexico, Albuquerque, NM 87107

JAMES McSHARRY
Department of Microbiology, Immunology and Molecular Genetics, Albany Medical College, Albany, NY 12208

MARK B. MEADS
Department of Medical Microbiology and Immunology, University of South Florida, Tampa, FL 33612-4799

PETER G. MEDVECZKY
Department of Medical Microbiology and Immunology, University of South Florida, Tampa, FL 33612-4799

DOUGLAS K. MITCHELL
Center for Pediatric Research, Children's Hospital of the King's Daughters and Eastern Virginia Medical School, Norfolk, VA 23510

MAURICE A. MUFSON
Department of Medicine, Marshall University School of Medicine, Huntington, WV 25701-3655

ISA K. MUSHAHWAR
Experimental Biology Research, Abbott Laboratories, 1401 Sheridan Rd., North Chicago, IL 60064-6269

STANLEY J. NAIDES
Division of Rheumatology, Milton S. Hershey Medical Center, Pennsylvania State University, Hershey, PA 17033

FREDERICK S. NOLTE
Department of Pathology and Laboratory Medicine, Emory University School of Medicine, Atlanta, GA 30332

MIGUEL L. O'RYAN
Department of Microbiology, University of Chile, Santiago, Chile

PHILIP E. PELLETT
Herpesvirus Section, Centers for Disease Control and Prevention, Atlanta, GA 30333

MARTIN PETRIC
Virology Laboratory, Department of Pediatric Laboratory Medicine, The Hospital for Sick Children, Toronto, Ontario M5G 1X8, Canada

MAURO PISTELLO
Department of Biomedicine and Retrovirus Center, University of Pisa, Pisa, Italy

CHRISTOPHER M. POKABLA
Department of Microbiology/Immunology, Northeastern Ohio Universities College of Medicine, Rootstown, OH 44272

CHARLES A. REED
Department of Pediatrics, Washington University School of Medicine, St. Louis, Mo.

CHARLES R. RINALDO, JR.
Clinical Virology Laboratory, University of Pittsburgh Medical Center, Pittsburgh, PA 15213, and Department of Infectious Diseases and Microbiology, University of Pittsburgh Graduate School of Public Health, Pittsburgh, PA 15261

JOHN T. ROEHRIG
Arbovirus Disease Branch, Division of Vector-Borne Infectious Diseases, National Center for Infectious Diseases, U.S. Public Health Service, U.S. Department of Health and Human Services, Fort Collins, CO 80522

DAVID T. ROWE
Department of Infectious Diseases and Microbiology, University of Pittsburgh Graduate School of Public Health, Pittsburgh, PA 15261

JULIUS SCHACHTER
Chlamydia Research Laboratory, Department of Laboratory Medicine, University of California—San Francisco, San Francisco, CA 94110

JÖRG SCHÜPBACH
Swiss National Center for Retroviruses, University of Zurich, CH-8028 Zurich, Switzerland

TED E. SCHUTZBANK
Infectious Disease Laboratory, Research and Development, Diagnostic Products Corporation, Los Angeles, CA 91381

JOHN L. SEVER
Department of Pediatrics, Children's National Medical Center, Washington, DC 20010-2970

KEERTI V. SHAH
Department of Molecular Microbiology and Immunology, Johns Hopkins University School of Public Health, Baltimore, MD 21205

ROGER D. SMITH
Department of Pathology and Laboratory Medicine, University of Cincinnati College of Medicine, Cincinnati, OH 45267-0529

THOMAS F. SMITH
Section of Clinical Microbiology, Department of Laboratory Medicine, Mayo Clinic, Rochester, MN 55905

LILLIAN M. STARK
Epidemiology Research Center, Office of Laboratory Services, Florida Department of Health and Rehabilitative Services, Tampa, Fla.

KIRSTEN ST. GEORGE
Clinical Virology Laboratory, University of Pittsburgh Medical Center, Pittsburgh, PA 15213

ELLA M. SWIERKOSZ
Departments of Pathology and Pediatrics, Saint Louis University School of Medicine, St. Louis, MO 63104

MARIA SZYMANSKI
Virology Laboratory, Department of Pediatric Laboratory Medicine, The Hospital for Sick Children, Toronto, Ontario M5G 1X8, Canada

CHARLES V. TRIMARCHI
Rabies Laboratory, New York State Department of Health, Albany NY 12201

MARIALINDA VATTERONI
Department of Biomedicine and Retrovirus Center, University of Pisa, Pisa, Italy

ANN WARFORD
SRA Life Sciences, Rockville, MD 20850

DANNY L. WIEDBRAUK
Departments of Clinical Pathology and Pediatrics, William Beaumont Hospital, Royal Oak, MI 48073-6769

RAWIA S. YASSIN
Department of Pathology and Laboratory Medicine, University of Cincinnati College of Medicine, Cincinnati, OH 45267-0529

STEPHEN A. YOUNG
TriCore Reference Laboratories and Department of Molecular Genetics and Microbiology, University of New Mexico School of Medicine, Albuquerque, NM 87107

HEINZ ZEICHHARDT
Department of Virology, Institute for Infectious Diseases Medicine, University Hospital Benjamin Franklin, Free University of Berlin, 12203 Berlin, Germany

Preface to the Third Edition

The aims of the *Clinical Virology Manual* remain the same as those of the first edition; thus, the original preface is included to describe those goals. The third edition is updated and expanded from the second edition. It has been expanded from 36 chapters to 39 chapters and 3 appendixes. The original section on reference laboratories now comprises the appendixes. Many of the chapters have been updated and expanded, while some of the more standard virology techniques of the past have been retained from the second edition (chapters 6, 8, 11, and 12). Four new chapters have been added to the Laboratory Procedures section; these include replacing the chapter on PCR with one chapter on molecular diagnostics and one on quantitative molecular technologies, as well as chapters on the use of flow cytometry in viral diagnostics and automation in the virology laboratory. These are intended to address much of the modernization that has occurred in the past several years. In the Viral Pathogens section, we have separated the coverage of viral hepatitis into two chapters along the lines of route(s) of infection; separated cytomegalovirus, Epstein-Barr virus,

and varicella-zoster virus from human herpesvirus 6; added human herpesviruses 7 and 8 to the latter chapter; and added a chapter on rodent-borne viruses. The information in the appendixes has been updated.

This edition also brings several major changes, including a new publisher, the retirement of one of the original editors, and the addition of two new editors. We are pleased that ASM Press is now publishing this edition and hope that ASM members as well as nonmembers will find this manual a useful adjunct to the *Manual of Clinical Microbiology* and *Manual of Clinical Laboratory Immunology*. There are a number of chapters for which the authors have changed as a result of change of professional focus, retirement, or death. We hope that this edition is a credit to those who preceded this effort, especially Jerry Lancz, to whom this edition is dedicated.

STEVEN SPECTER
RICHARD L. HODINKA
STEPHEN A. YOUNG

Preface to the First Edition

Clinical virology is an area that is undergoing rapid expansion. As a service for patient care, the utility of the clinical virology laboratory has increased significantly in the past decade. Due to the availability of commercial test kits, sophisticated yet simple diagnostic reagents, and the standardization of laboratory assays, accurate, reliable and, in many instances, rapid protocols are currently available for the diagnosis of a variety of viral agents producing human infections. Thus, the demands (on both the physician and the clinical laboratory virologist) for the diagnosis of viral infections will continue to increase. With this in mind, this volume is written as both an aid to the clinician and as a guide for the clinical laboratory.

This manual has three sections. The first describes laboratory procedures to detect viruses. The initial chapters deal with quality control in the laboratory and specimen handling, areas that are critical for an effective diagnostic laboratory. This is followed by individual chapters that provide information or a detailed protocol on how to set up and test samples for viral diagnosis using this technique. Both classical and the newer, more experimental techniques are described in detail.

The second section focuses on the viral agents. Viruses are grouped into chapters based on a target organ-system categorization. In this way, viruses producing infection in a particular organ or tissue are discussed and compared in a single chapter. This approach more accurately reflects the problems and choices faced by the attending physician and clinical technician for the diagnosis of a viral infection. Each chapter includes information relating basic, pathogenic, immunologic, and protective measures concerning each virus group, as well as information on its isolation, propagation, and diagnosis. This section also includes a chapter on *Chlamydia*. There are two reasons for including this family: the clinical laboratory often isolates and diagnoses *Chlamydia*, and the techniques used in its isolation and diagnosis are used in other instances.

The third section is designed to be used for reference. Here we supply information about Federal Reference Laboratories at the Centers for Disease Control and their role in the diagnosis of viral infection. The diagnostic and regulatory activities of state health laboratories and services available at individual hospital laboratories are provided in survey form. This listing is somewhat incomplete in that it contains information provided in response to an initial questionnaire and follow-up.

The aim and scope of this volume are service: to the physician, as a source of basic and clinical information regarding viruses and viral diseases, and to the laboratories, as a reference source to aid in the diagnosis of virus infection by providing detailed information on individual techniques and the impetus to expand services offered.

STEVEN SPECTER
GERALD LANCZ

LABORATORY PROCEDURES FOR DETECTING VIRUSES

I

Quality Assurance in Clinical Virology

ANN WARFORD

1

Quality assurance in clinical virology requires a comprehensive program for surveillance and improvement of all aspects of laboratory services. Laboratory testing for health assessment, disease diagnosis, or treatment begins with patient preparation and sampling and continues through testing, reporting of results to patient care providers, and appropriate notification of results and test interpretation. In a 1996 report of a prospective study of the type and frequency of laboratory testing problems in primary care physicians' offices during a 6-month period, a rate of 1.1 problems per 1,000 visits was found (Nutting et al., 1996). Twenty-seven percent of these test problems had an impact on patient care, including serious effects such as unnecessary hospitalization, prolonged hospital stay, more invasive diagnostic procedures, and delays in treatment. However, only 25% of the laboratory problems involved test analysis or inconsistent results; 75% of errors occurred in specimen collection and transport (43%) or timely provider notification of results (32%). This and other studies (Boone et al., 1982; Bartlett et al., 1994) confirm the need for laboratory involvement in improving the total testing process, including preanalytical and postanalytical steps, if laboratory services are to be meaningful and beneficial in patient health care.

REGULATORY REQUIREMENTS

Effective September 1992, with the implementation of the federal Clinical Laboratory Improvement Amendments of 1998 (CLIA-88), all clinical laboratories in the United States are regulated by the Health Care Financing Administration (HCFA) unless state health department regulations exceed and are approved by HCFA (Centers for Disease Control and Prevention [CDC], 1992; HCFA, 1992). The provisions of CLIA-88 include licensure, inspections conducted by HCFA or HCFA-approved organizations such as the College of American Pathologists (CAP) and the Joint Commission for Accreditation of Healthcare Organizations (JCAHO), and sanctions for failure to meet mandated standards. The stated purpose of CLIA-88 regulation of laboratories is to improve laboratory quality and achieve accurate and reliable laboratory results. The main quality standards of the regulatory and accrediting organizations can be categorized as personnel qualifications, responsibilities and competency assessment, proficiency

testing for all analytes and staff, written and approved procedures, method verification and validation, test reagent and equipment quality control and preventive maintenance and, lastly, patient test management, which includes ongoing assessment and improvement of all laboratory services. In the references, those references that are marked with an asterisk provide in-depth information regarding U.S. clinical laboratory regulations and accreditation requirements as well as useful quality assurance resources.

PROCEDURE MANUAL

An essential tool for the laboratory staff is a complete and current procedure manual available at the bench. The manual should contain a detailed, stepwise procedure for all tasks performed in the laboratory written according to guidelines established by the National Committee for Clinical Laboratory Standards (NCCLS) (1996). The required elements of the procedure manual are title, test principle, patient preparation, specimen collection, transport and storage, reagents, standards and controls, supplies, instrumentation (including calibration and maintenance), quality control frequency and acceptable limits, corrective action for unacceptable quality control, test steps, calculations, expected values, reference range, critical values, linearity and detection limits, method limitations and interfering substances, method validation, references, implementation and review dates, and author(s).

A copy of a manufacturer's kit package insert does not meet the requirements for the laboratory's written procedures. In addition to formal procedures for each type of patient test performed, written procedures are required for care provider sample collection and handling and must specify specimen rejection criteria, such as shown in Table 1. Specimen collection information must also be provided to medical and nursing staffs and as part of periodic laboratory hospital inservice education programs to be effective. No procedure in the laboratory can compensate for erroneous specimen collection and handling. Written protocols are also required for proficiency testing, safety, and the quality assurance and improvement program. Each written procedure must be reviewed and approved by the laboratory director and updated when method improvements are implemented.

TABLE 1 Examples of specimen rejection criteria[a]

Problem	Specimen	Test	Action	
			Reject (phone for new sample)	Process and test with disclaimer
Delay in transit	Clotted blood	Serology		>24 h
	Whole blood (unspun)	Culture/PCR	>12 h (whole blood)	6–12 h (whole blood)
	Serum or plasma (RT)	PCR	>72 h (RT)	25–72 h (RT)
	Serum or plasma (cold)	PCR		>72 h (cold pack, refrigerated)
	PPT tube (unspun)	PCR	>12 h (unspun/whole blood)	6–12 h (unspun/whole blood)
	PPT tube (spun)	PCR	>72 h (RT)	25–72 h (RT)
	PPT tube (spun)	PCR		>72 h (cold pack, refrigerated)
	Nonblood	Clostridium difficile toxin		>4 h (C. difficile toxin in stool)
		Viral culture		>48 h (refrigerated) for viral cultures
Heparin (green top)	Whole blood	PCR	Any (cannot use for PCR)	
Hemolysis	Serum	Serology	Looks like whole blood	Mild/moderate hemolysis (note serum appearance in computer)
Lipemia/icterus	Serum	Serology		Note appearance in computer
Mislabeled or unidentified	Any (except surgery)	Any	Reject/recollect	Tissue/CSF (have physician identify and sign, add disclaimer)
Dry swab, wood, calcium alginate, or charcoal swab	Swab	Culture		Note unsatisfactory swabs in computer with disclaimer
Container gross external contamination	Any (except surgery)	Any	Reject and recollect	Tissue/CSF (have submitter or supervisor disinfect with bleach)
Duplicate (<24 h)	Any except surgery (BAL, biopsy, CSF)	Any	Reject duplicate blood, urine, or stool	Process if requested by physician
Fixative (Formalin)	Any	Any	Reject and recollect	
Non-VTM (Bacti culturette)	Swab	Culture/DFA	Cannot use culturette for DFA/EIA or Chlamydia	Can use culturette for viral culture (transfer to VTM as soon as possible)
Nonstandard source or collection method	Sputum or stool for respiratory viruses	Culture/DFA	Reject, recollect NP/Tht/BAL	Add disclaimer
QNS	Any	Any		Call for physician's test priority list
Inadequate cellular material	Lesions, swabs	DFA	Call for recollection	

[a]Abbreviations: RT, room temperature; PPT, plasma processing tube (BD); BAL, bronchoalveolar lavage; CSF, cerebrospinal fluid; VTM, viral transport medium (SP buffer); DFA, direct fluorescent-antibody assay; EIA, enzyme immunoassay; NP, nasopharyngeal swab; Tht, throat swab; QNS, quantity not sufficient.

STAFF

The key to a quality viral diagnostic service is the laboratory staff. Staff qualifications for education, experience, training, and licensure or certification vary greatly among regulatory and accrediting agencies, with CLIA-88 having the minimum requirements (August et al., 1990; HCFA, 1992). Virology testing is categorized in CLIA-88 as moderate- and high-complexity testing, with only a few infectious mononucleosis serology kits listed as "waived," i.e., exempt from many CLIA-88 regulations. Virology laboratories can offer level-one testing, which consists of immunoassays for antigen detection without microscopy, or level-two high-complexity testing for viral isolation and identification and all other viral diagnostics. Because most virology methods are complex and subjective, requiring independent analysis and decisions, adequate education and training in theory and methods are essential for quality results. Several studies have correlated the level of education, training, and certification or licensure with laboratory performance quality as measured in proficiency surveys (Gerber et al., 1991; Hancock et al, 1993; Woods and Bryan, 1994; CDC, 1996; Shahangian, 1998). Continuing education is certainly desirable for all virologists in this rapidly changing field and is required in some states, particularly those with licensure requirements for laboratory personnel. Among the laboratory director's responsibilities are written qualifications, duties, and responsibilities for all staff and assurance that staffing levels are adequate for the type and volume of testing performed. Excessive workloads are not consistent with quality, particularly with subjective tasks requiring judgment, such as microscopy.

PROFICIENCY TESTING

CLIA-88 has adopted an external, graded proficiency test program(s) (PT) as the main indicator of the quality of laboratory testing performance. All laboratories must participate in PT for each analyte or test for which patient testing is performed; laboratories that fail consecutive challenges or two of the three annual testing events are subject to severe sanctions. Proficiency testing must be performed in the same manner and with the same staff as are routine patient samples. Known proficiency samples are an imperfect measure of a laboratory's performance accuracy and reliability because (i) they are recognized challenges which have penalties for failure and are prone to special attention; (ii) they test only the analytical phase of testing, not specimen collection, transport, or usual result reporting; (iii) they consist of a laboratory adapted virus(es) or pooled, processed body fluids spiked with analyte, which may have a matrix effect which renders them inaccurate with certain methods; and (iv) they cannot test analyte concentrations near the assay cutoff due to nonconsensus results with borderline levels. However, PT samples do still detect staff human errors and some poorly performing methods. PT unknown sample testing and analysis of results provided by programs such as CAP PT surveys also provide an educational resource for the laboratory. If no graded proficiency samples are available for tests performed, the laboratory must validate these methods for accuracy and reliability

TABLE 2 Top HCFA inspection CLIA-88 deficiencies cited 1996 to 1998[a]

Proficiency testing program for each specialty and subspecialty inadequate

Quality assurance plan; lack of comprehensive written plan for maintaining quality of overall testing process, identifying problems, and implementing corrective action

Quality control not documented with at least two levels of controls for each day of testing

Preventive maintenance and function checks of instrumentation inadequate

Competency assessment program of staff performance inadequate

Daily supervisory review of quality control, preventive maintenance, and patient test results not performed

Procedure manual and job descriptions without lab director's written designation of responsibilities and duties of staff

Correlation of multiple test methods for same analytes not documented

[a]Sources: Chapin and Baron, 1996; Belanger, 1998.

TABLE 3 Troubleshooting unacceptable patient or proficiency test results

Procedure or method
- Equipment, reagents, standards, quality control materials
- Limitations of methodology: sensitivity, specificity, precision, linear range
- Written procedure erroneous

Technical factors
- Incubation time, temperature, humidity, carbon dioxide
- Pipetting, dilutions, calculations
- Misinterpretation, not following written protocol

Staff or staffing
- Training, experience, continuing education
- Use of overtime, per diem, rotating staff
- Workload-to-staff ratio

Clerical error(s)
- Mislabeling, transcription, units, computer entry

Sample or sampling
- Transport time and/or temperature
- Interfering substances, contamination
- Organism or analyte not present or not viable on receipt

Obtain input on preventive measures from lab staff and others

twice annually by other means, such as samples split with a reference laboratory, known samples, and patient clinical correlations including chart review. Blind quality control has been reported to offer the best measure of routine laboratory performance and can be accomplished with samples split and relabeled prior to receipt in the laboratory to assess reproducibility (Boone et al., 1982; Farrington, et al., 1995; Gray et al., 1995a, 1995b; Shahangian, 1998). Inadequate PT performance is the most common post-CLIA-88 inspection citation (Table 2) (Chapin and Baron, 1996; Belanger, 1998). Any type of PT assessment is useless without investigation and efforts to improve system problems. PT failures provide an opportunity for evaluation of factors contributing to test performance problems (Table 3), and use of total quality management methods with staff input from all sections and levels is recommended and outlined by NCCLS (1997) and others (Engebretson and Cembrowski, 1992). Investigations by CDC and CAP showed that approximately 20% of repeated PT failures had no cause identified by the laboratory and that on-site technical consultation was required for performance improvement (Boone et al., 1982; Hoeltge and Duckworth, 1987).

STAFF COMPETENCY ASSESSMENT PROGRAM

Annual competency assessment and training verification of laboratory staff is also mandated by CLIA-88 and is another of the main HCFA inspection deficiencies cited (Table 2). Competency assessment is even more critical to the quality of laboratory testing since it requires evaluation of testing personnel in all of the routine patient testing procedures, including preanalytical and postanalytical steps, quality control, and instrument methods, as well as analysis. The mandated competency assessment procedures (Table 4) are (i) direct observation of test performance, instrument main-

TABLE 4 Staff training verification and competency assessment documentation

---

Technical supervisor must assess and verify staff performance of procedures promptly, accurately, and proficiently at least annually by use of the following:

- Direct observation of routine test performance, instrument maintenance and function checks, and microscopy and interpretation
- Monitoring worksheets, result recording, and reporting
- Testing proficiency samples, previously analyzed specimens, blind controls, and/or reference samples
- Daily review of quality control records and preventive maintenance records
- Additional procedures such as written or verbal tests, continuing education, problem solving of test failures, and evaluation of critical incidents, error reports, or complaints
- Reevaluation required with each change in methods

---

tenance and function checks, and microscopy and interpretation; (ii) monitoring worksheets, result recording, and reporting; (iii) testing of proficiency samples, previously analyzed samples, and blind controls or reference samples; and (iv) daily review of quality control and preventive maintenance records. Competency assessment can also consist of continuing education, written or oral tests, and evaluation of critical incidents, error reports, or complaints and should include evaluation of problem-solving ability, particularly concerning test failures. Evaluation of testing staff for troubleshooting ability can be aided by use of the form shown in Fig. 1, which is recommended for document-

ing reports of laboratory problems and complaints. By incorporating many of the documents normally used in laboratory operations, competency verification does not have to be an onerous process. Some assessment documentation items might include a training checklist created from the major and critical steps of the procedure manual, daily worksheet and results review checks, repeat testing of positive and equivocal results that are normally performed, confirmatory test results, and review of microscopy, quality control, and preventive maintenance results. As with proficiency testing, poor staff competency indicates the need for evaluation of laboratory systems for recruiting, staffing patterns, training, continuing education, and retention of a qualified staff.

METHOD PERFORMANCE VERIFICATION AND VALIDATION

Method performance verification is required for all U.S. Food and Drug Administration (FDA)-approved instruments, kits, and test systems to demonstrate that accuracy, precision, and reportable range are comparable to those established by the manufacturers. This verification usually consists of parallel testing of the new product with a standard method of known performance characteristics. A minimum of 20 known positive specimens and 50 negative samples has been recommended for this evaluation by McCurdy and colleagues (Elder et al., 1997). For non-FDA-approved methods, establishment of the performance characteristics of accuracy, precision, analytical sensitivity, specificity, interfering substances, and reportable and references ranges is CLIA-88 mandated. Recommendations for in-house developed molecular assay validation are specified by

Problem Description (include test, date, sample ID)
Steps taken to evaluate and solve problem:
Problem reported to: Date:
Corrective Action:
Further Preventive Measures:
Comments:
Prepared by: Date:
Reviewed by: Date:

FIGURE 1 Report of laboratory problem, complaint, or error (adapted from August et al., 1990).

NCCLS (1995a), the Association for Molecular Pathology (AMP) (1999), and the American Society for Microbiology (Elder et al., 1997). These authors suggest that home-brew microbiology assays should be validated with 50 samples known to contain microbes and 100 analyte negative specimens. For some viruses, particularly those that are difficult to culture, 50 positive patient samples may not be available within the institution, and collaboration with a reference or large public health laboratory is recommended in addition to obtaining reference standards from commercial or government sources.

QUALITY CONTROL

Monitoring of equipment, reagents and kits, and environmental conditions is required at least each day of use for all sections of clinical virology. All quality control procedures and results must be documented appropriately and actively reviewed, and corrective action must be implemented in a timely manner (Table 5). Common requirements include (i) instrument function tests each day of use, equipment calibration, preventive maintenance, and service according to manufacturer's instructions; (ii) temperature, humidity, and/or carbon dioxide measurements daily, with measurement verified by external standards periodically; and (iii)

labeling of all reagents, media, solutions, stains, antisera, and kits with identity, concentration, reactivity, purity (or sterility), storage conditions, source, safety hazard information, and dates of preparation, receipt, use, and expiration. Each new reagent or kit lot must be tested in parallel with prior product of satisfactory performance. At least two levels of control, positive and negative, must be tested on each day of use for all test methods, and three control levels—reactive, weakly reactive, and negative—are required for quantitative assays in addition to standards or calibrators, as specified, by verified methods. In addition to the controls that are part of test kits, use of external controls from a different source, such as CAP or commercial suppliers, has been reported to be valuable in detecting random and systematic errors in testing (Gray et al., 1995a, 1995b; Yen-Lieberman et al., 1996). Failure to document positive and negative control results and instrument function tests and preventive maintenance are among the top HCFA inspection deficiencies cited (Table 2).

Cell Culture Quality Control

Quality control requirements specific for viral isolation are listed briefly in Table 6 (Miller and Wentworth, 1985; Lennette, 1995; Clarke, 1998) and include maintenance of

---

TABLE 5 Mandated quality assurance documentation

Proficiency testing performed for all assays in the same manner and by the same staff as patient sample testing in routine work flow with lab director's review, investigation, and corrective action for failures
- Alternate quality assurance program for assays and analytes without formal, graded proficiency samples to verify test accuracy and reliability, e.g., use of samples split with reference lab, standard reference materials, and/or evaluation of patient clinical outcomes

Quality control: monitoring, weekly supervisory review, and corrective action
- Equipment and instrument preventive maintenance and service according to manufacturer's instructions, function tests, and calibrations
- Temperature-dependent equipment monitored each day of use and thermometers verified with standard; carbon dioxide levels in incubators monitored daily and internal monitor verified weekly with Fyrite analyzer
- Reagents, media, stains, solutions, and antisera labeled with identity, concentration, storage conditions, source, safety hazard information, and dates of preparation, receipt, use, and expiration
- Reagent and medium reactivity, analysis, and/or purity and sterility, parallel testing (each lot), positive and negative control results (each day of use)
- Antiserum and antigen titers, parallel testing (each lot), positive and negative controls (each day of use)
- Quantitative test controls of three levels: negative, reactive, and weakly reactive
- External (non-kit) control limits established (if used) to detect random and systematic error(s)

Procedure manual written in NCCLS format reviewed annually by director including:
- Preanalytical steps of patient identification, sample collection, transport, and storage and specimen rejection policy
- Analytical methods with verification of performance claims or validation
  - Verify assay performance for accuracy, precision, reportable range, and reference range
  - Validate in-house assay(s) performance by establishing accuracy, precision, analytical sensitivity and specificity, reportable range, and reference range with known positive and negative samples
- Postanalytic processes including prompt and accurate reporting with provider notification of critical values, interpretation and charting of test results, and amended reports
- Quality assurance program for total testing process with active evaluation and improvement to detect and correct significant errors

Specimen log and test requisitions with complete patient information including diagnosis, infection type, source, and collection date and time

Staff: laboratory director, technical supervisor, testing personnel
- Written duties, responsibilities, and qualifications (education, experience, and licensure/certification)
- Ongoing training verification, performance evaluations, continuing education
- Annual competency and proficiency testing, color discrimination ability, and consistent microscopy detection and interpretation

---

TABLE 6 Cell culture-specific quality control documentation

Cell culture records with cell type, passage number, source, age (<10 days postseed), receipt date, and maintenance media (each lot and date)

Uninoculated cells of each lot maintained as negative controls for mycoplasma contamination, cytopathology (CPE), and nonreactivity in immunofluorescence, hemagglutination, or hemadsorption identification assays

Cell cultures available and incubated appropriately for isolation or detection of entire range of viruses for which services are offered

Inoculated cell cultures checked at least every other day for initial 14 days of incubation

Culture media and additives checked for sterility, growth promotion, and absence of toxicity

Antisera and antigens checked for positive and negative reactivity each day of use

Red blood cell suspensions checked periodically by spectrophotometer

Buffers and diluents checked for pH, sterility, and absence of cytotoxicity

Water quality: pyrogen- and bacterium-free type I or superior water

Biological safety cabinet appropriate for classification of viruses propagated and certified at least annually

Daily decontamination of hoods and benches with high level disinfectant

cell culture records with cell type, source, passage number, and age of cells; cells should be inoculated within 8 to 10 days of seeding for optimal performance. The maintenance medium lot and date of use for each cell culture should also be recorded. Incubation and observation of uninoculated cells, maintained in the same manner as inoculated cells, is recommended as a negative control for cytopathology (cytopathic effect [CPE]), mycoplasma contamination, and identification methods such as hemadsorption and/or hemagglutination and immunofluorescence. Buffers, media, and additives should be checked for pH, sterility, growth promotion, and absence of toxicity in cell culture. Water that is used with cell cultures should be pyrogen- and bacteria-free type I or superior water as defined by NCCLS (1997). Primary cell culture endogenous agents such as simian virus 5 (SV5), SV40, foamy retrovirus, herpesvirus B, or contaminants from media or additives such as serum can produce CPE which may mimic or interfere with viral CPE and may cause hemadsorption with red blood cells or nonspecific fluorescence.

The laboratory must have cells available to provide optimal performance, and they must be incubated appropriately to permit isolation or detection of all viruses for which services are offered. Some accrediting agencies, such as CAP, specify types of cells and incubation periods for isolation of representative viruses (CAP, 1999). For cell culture contamination prevention and safety of staff, virology benches and safety cabinets should be disinfected at least daily with a high-level disinfectant such as 10% sodium hypochlorite (bleach). A class II or higher biological safety cabinet with HEPA filters and external venting, certified annually, plus facilities and procedures that are appropriate to the biohazard level of the viruses tested, as defined by the CDC (1993), must be used for virus isolation.

Molecular Testing Quality Control

Quality control procedures which are specific for molecular testing are listed in Table 7 (NCCLS, 1995a; AMP, 1999). In addition to the usual reactive, weakly reactive, and negative controls for each day of testing, use of a reagent blank(s) and an internal control in each specimen for detection of inhibition is recommended by NCCLS (1995a) for molecular assays. Optimal quality control is achieved with use of the sample internal controls as well as other controls from the initial steps of specimen processing and is continued through all subsequent steps—amplification, separation, hybridization, and detection. Prevention of contamination is also critical to quality assurance in molecular testing, and NCCLS suggests the use of chemicals such as uracil-N-glycosylase and isopsoralens for contamination control. Environmental controls such as plugged pipette tips, frequent glove changes, and separate equipment and rooms for pre- and postamplification steps are also suggested. Reagents, particularly oligonucleotide primers and probes, must be of known sequence, concentration, and purity and must be prepared using good manufacturing practices. The quality control requirements for instruments such as thermal cyclers, pipettes, spectrophotometers, and luminometers must have functional tests and temperature monitoring each day of use. Preventive maintenance and calibrations must be in compliance with manufacturers' instructions. Pipette calibration and temperature verification are critical to accuracy and reliability in molec-

TABLE 7 Molecular testing-specific quality control documentation

Amplified DNA testing procedures must prevent nucleic acid cross-contamination by:
- Separate reagent preparation, specimen preparation, and amplification/detection areas
- Dedicated equipment for each area including lab coats, pipettes, aerosol barrier pipet tips, and powder-free disposable gloves
- Unidirectional work flow from clean (DNA free) to dirty (amplified DNA) areas
- Incorporation of contamination prevention chemicals such as psoralens and/or uracil-N-glycosylase parallel tested to evaluate DNA inactivation

Primers and probes: known sequence, concentration, and purity and parallel testing of each lot

Positive and negative controls for all phases of testing: specimen processing, amplification, hybridization, and detection
- Additional recommended controls include reagent blank and internal control for inhibition
- Molecular weight standards for electrophoresis (each run)

Thermal cyclers, water baths, heat blocks, and incubators: follow manufacturer's instructions for use, preventive maintenance, and service; perform calibration at least every 6 months; monitor temperature and function (each day of use)

Spectrophotometers, luminometers, and pipettes: follow manufacturer's instructions for use, preventive maintenance, and service; perform calibration at least every 6 months; monitor function (each day of use)

ular testing; NCCLS recommends thermal cycler monitoring monthly with a thermocouple and pipette calibration at least twice per year.

LABORATORY DESIGN, SPACE, AND SAFETY

One of the laboratory director's responsibilities is the provision of adequate space to ensure quality of testing and safety of the laboratory staff. Design and space requirements vary greatly depending on the level and types of viral diagnostic services offered and the biohazard level of the viruses handled. An approved NCCLS document (1998a) is available for design guidance. For both viral isolation and viral molecular testing, it is recommended that separate areas for reagent preparation and specimen processing, which are not contaminated with infectious viral agents or nucleic acids, be maintained. A separate biological safety cabinet is desirable for "clean work" with uninfected cell cultures, reagent and medium preparation, and/or sample processing prior to culture or nucleic acid amplification. The "dirty" areas of positive culture identification and amplified nucleic acid detection are ideally located in separate rooms and/or safety cabinets, and the work flow should be unidirectional, starting each day in the clean areas and proceeding to potentially contaminated areas without backtracking. Universal safety precautions are required in all clinical laboratories and are particularly important in virology laboratories which handle infectious pathogens in high concentrations. Use of disposable gloves and gowns and frequent disinfection of work surfaces and biological safety cabinets with 10% bleach followed by 70% alcohol are important both for safety and to prevent contamination of the work areas.

EVALUATION OF REFERENCE LABORATORIES

Referral of esoteric and low volume tests to an off-site laboratory does not relieve the referring laboratory of quality monitoring of the testing. It is a laboratory director's responsibility to select a reference laboratory that meets state and federal regulations and provides quality services. NCCLS (1995b, 1998b) suggests the following criteria for reference laboratory evaluation: proof of licensure and accreditation; use of FDA or validated assays; copies of validation studies and written protocols; PT results; frequent internal and external quality control use; appropriate test turnaround time; qualified laboratory staff; high-quality methods, instrumentation, and facilities; specimen stability during transport; good result-reporting systems; policies for resolving questionable results; and cost. Parallel testing with split samples of new assays critical to patient management may also be advisable.

COMPREHENSIVE QUALITY ASSURANCE AND IMPROVEMENT

Although quality assurance and improvement of the overall testing process are mandated and critical to accurate and reliable laboratory services, requirements for the preanalytical and postanalytical phases are limited in CLIA-88 to (i) assessing patient management, (ii) comparing test results performed by different methods, and (iii) correlating test results with patient information. Interventions to improve the quality of specimens tested and the time to reporting of test results yield patient and physician satisfaction and address the majority of the problems affecting patient care reported by physicians (Nutting et al., 1996). Comparing

results of different methods for identifying viral infections can be rewarding in clinical virology; diverse methods are available and can be improved with test data obtained in conjunction with patient clinical information. CAP has a program, called Q-probes, for comparing preanalytical and postanalytical indicators of quality among laboratories. Past indicators have included measures of turnaround time, reference laboratory quality, test order accuracy, and other critical issues but only one survey specific to virology—hepatitis test utilization. Quality assurance has been challenged as too costly and of unproven benefit, but several studies have demonstrated that the costs of quality control and quality assurance are low compared to the costs of quality failures—erroneous test results, repeated tests, recollected samples, and delays in test reporting (Westgard et al., 1984; Bartlett et al., 1994; Farrington et al., 1995). Based on a review and analysis of many published quality control procedures and studies, Bartlett et al. emphasize the need for continuous quality improvement processes, as detailed in NCCLS quality management guidelines (1997, 1998c), with particular emphasis on error prevention to increase accuracy through effective training, communication, and continuing education of staff (Bartlett et al., 1994). JCAHO requires demonstration of programs for monitoring, intervention, and improvement of quality indicators affecting patient outcomes (JCAHO, 1998). Although quality indicators that are clearly linked with improved patient outcomes are still to be established, such patient outcome studies will be the ideal performance measures of laboratory quality.

REFERENCES

*Association for Molecular Pathology. 1999. Recommendations for in-house development and operation of molecular diagnostic tests. *Am. J. Clin. Pathol.* 111:449–463.

*August, M. J., J. A. Hindler, T. W. Huber, and D. L. Sewell. 1990. Cumitech 3A. Quality control and quality assurance practices in clinical microbiology. Coordinating ed., A. S. Weissfeld. American Society for Microbiology, Washington, D.C.

Bartlett, R. C., M. Mazens-Sullivan, J. Z. Tetreault, S. Lobel, and J. Nivard. 1994. Evolving approaches to management of quality in clinical microbiology. *Clin. Microbiol. Rev.* 7:55–87.

Belanger, A. C. 1998. The Joint Commission and CLIA: a five year retrospective. *Med. Lab. Observ.* Feb. 1998:45–48.

Boone, D. J., H. J. Hansen, T. L. Hearn, D. S. Lewis, and D. Dudley. 1982. Laboratory evaluation and assistance efforts: mailed, onsite and blind proficiency testing surveys conducted by the Centers for Disease Control. *Am. J. Public Health* 72:1364–1368.

*Centers for Disease Control and Prevention. 1992. Regulations for implementing the Clinical Laboratory Improvement Amendments of 1988: a summary. *Morbid. Mortal. Weekly Rep.* 41:RR–2.

Centers for Disease Control and Prevention. 1996. Clinical laboratory performance on proficiency testing samples—United States, 1994. *Morbid. Mortal. Weekly Rep.* 45:193–196.

*Centers for Disease Control and Prevention and National Institutes of Health. 1993. *Biosafety in Microbiological and Biomedical Laboratories,* 3rd ed. U.S. Government Printing Office, Washington, D.C.

Chapin, K., and E. J. Baron. 1996. Impact of CLIA 88 on the clinical microbiology laboratory. *Diagn. Microbiol. Infect. Dis.* 23:35–43.

*Clarke, L. M. 1998. Viral culture: selection, assessment, quality control, and maintenance of uninoculated cell cultures, p. 463–471. In H. D. Isenberg (ed.), Essential Procedures for Clinical Microbiology. ASM Press, Washington, D.C.

*College of American Pathologists. 1999. Laboratory Improvement: Laboratory Accreditation Program Checklists. College American Pathologists, Chicago, Ill.

*Elder, B. L., S. A. Hansen, J. A. Kellogg, F. J. Marsik, and R. J. Zabransky. 1997. Cumitech 31. Verification and validation of procedures in the clinical microbiology laboratory. Coordinating ed., B. W. McCurdy. American Society for Microbiology, Washington, D.C.

Engebretson, M. J., and G. S. Cembrowski. 1992. Achieving the Health Care Financing Administration limits by quality improvement and quality control. Arch. Pathol. Lab. Med. 116:781–787.

Farrington, M., M. Amphlett, D. F. Brown, and S. Messer. 1995. Fifteen percent of microbiology reports are wrong: further experience with an internal quality assessment and audit scheme. J. Hosp. Infect. 30(Suppl.):364–371.

Gerber, A. R., R. O. Valdieserri, C. A. Johnson, R. E. Schwartz, J. S. Hancock, and T. L. Hearn. 1991. Quality of laboratory performance in testing for HIV type 1 antibody. Arch. Pathol. Lab. Med. 115:1091–1096.

Gray, J. J., T. G. Wreghitt, T. A. McKee, P. McIntyre, C. E. Roth, D. J. Smith, G. Sutehall, G. Higgins, R. Geraghty, R. Whetstone, et al. 1995a. Internal quality assurance in a clinical virology laboratory. I. Internal quality assessment. J. Clin. Pathol. 48:168–173.

Gray, J. J., T. G. Wreghitt, T. A. McKee, P. McIntyre, C. E. Roth, D. J. Smith, G. Sutehall, G. Higgins, R. Geraghty, R. Whetstone, et al. 1995b. Internal quality assurance in a clinical virology laboratory. II. Internal quality control. J. Clin. Pathol. 48:198–202.

Hancock, J. S., R. N. Taylor, C. A. Johnson, A. R. Gerber, and W. O. Schalla. 1993. Quality of laboratory performance in testing for human immunodeficiency virus type 1 antibody. Arch. Pathol. Lab. Med. 117:1148–1155.

*Health Care Financing Administration, Department of Health and Human Services. 1992. Clinical Laboratory Improvement Amendments of 1988: final rule. Federal Register 28 Feb. 1992:7146.

Hoeltge, G. A., and J. K. Duckworth. 1987. Review of proficiency testing performance of laboratories accredited by the College of American Pathologists. Arch. Pathol. Lab. Med. 111:1011–1014.

*Joint Commission on Accreditation of Health Care Organizations. 1998. 1998–99 Comprehensive Accreditation Manual for Pathology and Clinical Laboratory Services. Joint Commission on Accreditation of Health Care Organizations, Oakbrook Terrace, Ill.

*Lennette, D. A. 1995. General principles for laboratory diagnosis of viral, rickettsial and chlamydial infections. In E. H. Lennette, D. A. Lennette, and E. T. Lennette (ed.), Diagnostic Procedures for Viral, Rickettsial and Chlamydial Infections. American Public Health Association, Washington, D.C.

*Miller, J. M., and B. B. Wentworth. 1985. Methods for Quality Control in Diagnostic Microbiology. American Public Health Association, Washington, D.C.

*National Committee for Clinical Laboratory Standards. 1991. Preparation and Testing of Reagent Water in the Clinical Laboratory, 2nd ed. C3-A2. 11(13). NCCLS, Wayne, Pa.

*National Committee for Clinical Laboratory Standards. 1995a. Molecular Diagnostic Methods for Infectious Diseases. Approved Guideline. MM3-A. NCCLS, Wayne, Pa.

*National Committee for Clinical Laboratory Standards. 1995b. Training Verification for Laboratory Personnel. GP21-A. NCCLS, Wayne, Pa.

*National Committee for Clinical Laboratory Standards. 1996. Clinical Laboratory Technical Procedure Manuals. 3rd ed. GP2-A2. NCCLS, Wayne, Pa.

*National Committee for Clinical Laboratory Standards. 1997. Continuous Quality Improvement: Essential Management Approaches and Their Use in Proficiency Testing. Proposed Guideline. GP22-P. NCCLS, Wayne, Pa.

*National Committee for Clinical Laboratory Standards. 1998a. Laboratory Design. Approved Guideline. GP18-A. NCCLS, Wayne, Pa.

*National Committee for Clinical Laboratory Standards. 1998b. Selecting and Evaluating a Referral Laboratory. Approved Guideline. GP9-A. NCCLS, Wayne, Pa.

*National Committee for Clinical Laboratory Standards. 1998c. A Quality System Model for Healthcare, Proposed Guideline. GP26-P. NCCLS, Wayne, Pa.

Nutting, P. A., D. S. Main, P. M. Fischer, T. M. Stull, M. Pontious, M. Seifert, Jr., D. J. Boone, and S. Holcomb. 1996. Problems in laboratory testing in primary care. JAMA 275:635–639.

Shahangian, S. 1998. Proficiency testing in laboratory medicine. Uses and limitations. Arch. Pathol. Lab. Med. 122:15–27.

Westgard, J. O., P. Hyltoft Peterson, and T. Groth. 1984. The quality costs of an analytical process. Scand. J. Clin. Lab. Investig. 44(Suppl. 172):228–236.

Woods, G. L., and J. A. Bryan. 1994. Detection of Chlamydia trachomatis by direct fluorescent antibody staining. Arch. Pathol. Lab. Med. 118:483–488.

Yen-Lieberman, B., D. Brambilla, B. Jackson, J. Bremer, R. Coombs, M. Cronin, S. Herman, D. Katzenstein, S. Leung, H. J. Lin, P. Palumbo, S. Rasheed, J. Todd, M. Vahey, and P. Reichelderfer. 1996. Evaluation of a quality assurance program for quantitation of human immunodeficiency virus type 1 RNA in plasma by the AIDS Clinical Trials Group virology laboratories. J. Clin. Microbiol. 34:2695–2701.

References marked with an asterisk provide information about U.S. clinical laboratory regulations, accredition requirements, and quality assurance resources.

Specimen Requirements: Selection, Collection, Transport, and Processing

THOMAS F. SMITH

2

In the past 15 years, there has been a rapid expansion in the implementation of rapid and clinically useful tests in diagnostic virology. In the early 1980s, the availability of specific reagents, particularly monoclonal antibodies, was essential in the development of tests for rapid (24 h) cell culture detection of viruses (particularly herpesvirus) and for the formatting of single unit membrane enzyme immunoassays (EIAs) for 15- to 30 min detection of both antigens (respiratory syncytial virus [RSV], rotavirus, influenza A virus) and antibodies (rubella virus, human immunodeficiency virus [HIV]). Monoclonal antibodies were also essential for the design of EIAs for sensitive and specific measurement of antibodies in serologic tests (hepatitis viruses, HIV, Epstein-Barr virus [EBV]) that are essential as first line diagnostic tests for detecting viral infection.

The development and utilization of molecular nucleic acid techniques for production of synthetic peptides for use in serologic tests and genetically engineered cell lines for the rapid detection of viral infections, particularly herpes simplex virus (HSV), and the creation of gene libraries for the recognition of new viruses (hepatitis C virus) have demonstrated the practical implications of this technology for modern diagnostic laboratories. Of course, the most remarkable innovation has been the amplification of target nucleic acids of viruses in specimens by PCR. The transition of PCR from a research assay to the routine clinical laboratory has been facilitated by many technical advances that now include automated instruments capable of yielding quantitative results for specific virus nucleic acids in 30 to 60 min. One of the most outstanding applications of PCR in clinical virology is the detection of HSV DNA in cerebrospinal fluid (CSF) specimens from patients with central nervous system (CNS) disease. Neither conventional cell culture recovery of the virus from CSF nor early detection of intrathecal antibodies is adequate for laboratory diagnosis of CNS disease caused by this virus. PCR has clearly added a new dimension for the diagnosis of HSV and many other viral infections.

New technology and novel assays have made the appropriate selection of specimens and interpretation of test results more complex than in the past. For example, initial presentation of a patient with symptoms of HIV infection should prompt submission of serum specimens for serologic testing. If specific antibodies to this virus are detected, a plasma specimen should be tested by a quantitative assay to measure viral load, i.e., the copy number of HIV RNA, to assess the need for prompt antiviral therapy of the patient. In addition, serial monitoring of the HIV load in plasma allows assessment of therapeutic efficacy. Importantly, it should be noted that the Western blot (WB) is a supplementary test to confirm the specificity of detection of HIV antibodies by EIA. Although the WB can be an individual orderable test in many reference laboratories, EIA should always be performed (with repeatedly reactive results) prior to this supplemental test. This testing algorithm should always be followed, since approximately 15 to 20% of serum specimens from uninfected individuals produce indeterminate (nondiagnostic) bands in WB assays. These false-positive results trigger unnecessary laboratory tests and clinical follow-up assessments involved with counseling of the patient.

Overall, in recent years, the development and implementation of these new test technologies, together with specific specimen requirements and interpretation of results for these assays, have demonstrated the need for frequent communication between the clinical service and diagnostic laboratory. No technique, regardless of how rapid, is useful unless the selection and quality of the specimen is adequate regarding the source, method of transport, and means of processing. Finally, the ever-increasing cost of medical practice obligates the laboratory director to constantly update written specimen requirements to help clinicians obtain a cost-effective and definitive viral diagnosis.

SPECIMEN SELECTION

Comprehensive tables listing general disease categories and associated viruses are essential for the clinician as a guide for submitting viral specimens. Tables should include optimal specimen types, methods of collection, volumes required, and conditions and containers for transport (Table 1). This information should be updated for institutional procedure guides at least on an annual basis; ideally, electronic ordering of laboratory tests requires that a specimen be submitted only according to the appropriate conditions.

The modern virology laboratory has a mixture of conventional assay methods (cell culture, serology) together with specific antigen detection or detection of target nucleic acids by amplification methods. For example, cytomegalovirus (CMV) detected in urine during the first week

TABLE 1. Specimen information for diagnostic virology services

General disease category	Virus	Serology[a]	Specimen(s) to be submitted for culture	Culturette swab	Sterile screw-cap container	Vacutainer (serology)	Vol	Other considerations
Respiratory pharyngitis, croup, bronchitis, pneumonia	Influenza virus	Yes	Throat swab	X		X	5 ml of sterile clotted blood for serology	
			Sputum		X			
			BAL		X			
	Parainfluenza virus	No	Throat swab	X				
			BAL		X			
	RSV	NA	Nasopharyngeal wash or aspirate; nasopharyngeal swab (Calgiswab)		X	X		Nasopharyngeal wash or aspirate preferred to nasopharyngeal swab; laboratory diagnosis by culture, immunofluorescence, or EIA
	Enterovirus	No	Throat swab	X				
	Herpes simplex virus	NA	Throat swab	X		X		
			Sputum		X			
	Adenovirus	Yes	Throat swab	X		X	5 ml of sterile clotted blood for serology	
	CMV	NA	Throat swab	X				
			Sputum		X			
			BAL		X			
			Blood (5 ml of heparinized blood in green capped tube)			X		
			Urine		X		5 ml of urine	
Exanthem (maculopapular)	Adenovirus	Yes	Throat swab	X		X	5 ml of sterile clotted blood for serology	
	Enterovirus	No	Throat swab	X				
			Rectal swab	X				
	Rubella virus	Yes				X	5 ml of sterile clotted blood for serology	Single serum only required for determination of immune status to rubella virus; for serologic evidence of acute-phase infections (rubella or other agents listed), two or more serum specimens are needed taken 2–3 weeks apart unless IgM antibody can be demonstrated with the first serum specimen
	Measles virus (rubeola)	Yes				X	5 ml of sterile clotted blood for serology	

TABLE 1. (*Continued*)

	Virus	Culture	Specimen			Serology / amount	Other considerations
	Parainfluenza virus[b]	No	Throat swab	X	X		
	RSV[b]	Yes	Nasopharyngeal wash or aspirate; nasopharyngeal swab (Calgiswab)	X	X	5 ml of sterile clotted blood for serology	
Exanthem (vesicular)	HSV	Yes	Vesicle swab; vesicle scrapings on slide for HSV DFA test (see "Other considerations")	X	X	5 ml of sterile clotted blood for serology	Vesicle scrapings for DFA test: place small drop of saline in each of two separate areas of a glass slide 5 to 10 mm apart; transfer skin scrapings from a scalpel blade to the saline and spread the cells over a small circular area (5 to 10 mm in diameter); dry the slide at room temperature and transport the slides to the laboratory in a cardboard mailer
	VZV	Yes	Vesicle swab; vesicle scrapings on slide for VZV DFA test	X	X	5 ml of sterile clotted blood for serology	
CNS (aseptic meningitis and encephalitis)	HSV	Yes (see "Other considerations")	CSF	X	X	5 ml of sterile clotted blood for serology; 1–2 ml of CSF for serology	PCR for detection of HSV DNA in CSF
			Vesicle	X			
			Brain biopsy				
	Enterovirus	No	CSF	X	X	1–2 ml	
			Throat swab	X	X		
			Rectal swab	X	X		
			Serum (infants)			1–2 ml	
			Brain biopsy, if available				
	LaCrosse (California virus)	Yes			X	5 ml of sterile clotted blood for serology	Viral isolation not attempted
	EBV	Yes	CSF			1–2 ml	PCR for detection of EBV DNA in CSF
	VZV	NA	CSF			1–2 ml	PCR for detection of VZV DNA in CSF
	CMV	NA	CSF			1–2 ml	PCR for detection of CMV DNA in CSF

(*Continued on next page*)

TABLE 1. Specimen information for diagnostic virology services (*Continued*)

General disease category	Virus	Serology[a]	Specimen(s) to be submitted for culture	Container or transport device for specimen type			Vol	Other considerations
				Culturette swab	Sterile screw-cap container	Vacutainer (serology)		
	Measles virus (rubeola)	Yes				X	5 ml of sterile clotted blood for serology	Viral isolation generally not successful
	St. Louis encephalitis	Yes				X	5 ml of sterile clotted blood for serology	Viral isolation not attempted
	Western equine encephalitis	Yes				X	5 ml of sterile clotted blood for serology	Viral isolation not attempted
	Rabies	Yes				X	5 ml of sterile clotted blood for serology	Antibody to rabies virus determined by Veterinary Diagnostic Laboratory, Kansas State University, Manhattan, with sera from individuals exposed to a possibly rabid animal
	Mumps	Yes	CSF Urine Throat swab (Stensen's duct)	X	X X	X	5–10 ml of urine 5 ml of sterile clotted blood for serology	
Infectious mononucleosis	EBV	Yes				X	5 ml of sterile clotted blood for serology	Viral isolation not attempted; immunofluorescence test for antibodies to EBV indicated in those patients with heterophile-negative (rapid test method) determinations
	CMV	Yes	Urine Throat swab	X	X	X	5–10 ml of urine 5 ml of sterile clotted blood for serology	
Hepatitis	HAV, HBV, HCV	Antigen and/or antibody test				X	5 ml of sterile clotted blood for antigen or antibody determinations	EBV and CMV occasionally cause hepatitis, especially in immunocompromised patients
Gastroenteritis	Rotavirus	Antigen detection	Stool (5 g)		X			Rotavirus antigen in stool specimens detected by EIA
	Norwalk-like agents	Yes				X	5 ml of sterile clotted blood for serology	Assays for antibodies to Norwalk-like agents are performed in only a few research laboratories

TABLE 1. (Continued)

	Agent	Serology	Specimen				Amount of specimen	Other considerations
	Adenovirus	Yes	Stool (5 mg) or rectal swab	X		X	5 ml of sterile clotted blood for serology	Few "high numbered" serotypes of adenoviruses have been associated with gastroenteritis
Genital infections	HSV	NA (see "Other considerations")	Vesicle swab	X		X	5 ml of sterile clotted blood for serology	Serology generally is not applicable; rarely, it may be useful for primary genital infections due to HSV; specimens for culture are recommended
			Vesicle scrapings on slide for HSV DFA test (see "Other considerations")					Vesicle scrapings for DFA test: place small drop of saline in each of two separate areas of a glass slide 5 to 10 mm apart; transfer skin scrapings from a scalpel blade to the saline and spread the cells over a small circular area (5 to 10 mm in diameter); dry the slide at room temperature and transport the slides to the laboratory in a cardboard mailer
	HPV	Yes	Endocervical swab, biopsy tissue, paraffin-embedded tissue	Specific collection kits required				HPV can be detected in Papanicolaou-stained cells; alternatively, certain biotypes (6/11/42/43/44, low risk; 16/18/31/33/35/45/51/52/56, high risk) can be detected by commercial kits using nucleic acid detection methods
Congenital	CMV	Yes	Urine		X		5–10 ml of urine	Detection of CMV in urine from neonates at 10 days old indicates congenital infection due to CMV
			Throat swab			X	5 ml of sterile clotted blood for serology	
	HSV	Yes	Vesicle swab	X		X	5 ml of sterile clotted blood for serology	Presence of IgM to CMV or HSV in cord or neonatal blood indicates congenital infection
			Throat swab			X	5 ml of sterile clotted blood for serology	
	Rubella virus	Yes				X	5 ml of sterile clotted blood for serology	Viral isolation not attempted; IgM antibody to rubella virus should be assayed using serum from babies up to 6 months of age; IgG antibody should not be determined since its presence reflects only passive transfer from the mother

(Continued on next page)

TABLE 1. Specimen information for diagnostic virology services (*Continued*)

General disease category	Virus	Serology[a]	Specimen(s) to be submitted for culture	Container or transport device for specimen type			Vol	Other considerations
				Culturette swab	Sterile screw-cap container	Vacutainer (serology)		
	Enterovirus	No	Throat swab	X				Cord blood is a useful and productive specimen for recovering enteroviruses in cell culture in congenital infections
			Rectal swab	X				
			Serum (see "Other considerations")		X			
Ocular infections	HSV	Yes (see "Other considerations")	Swab (eye)	X		X	5 ml of sterile clotted blood for serology	Serology may be useful only for primary ocular infections due to HSV; specimens for culture are recommended
			Vesicle swab	X				Vesicle scrapings for DFA test: place small drop of saline in each of two separate areas of a glass slide 5 to 10 mm apart; transfer skin scrapings from a scalpel blade to the saline and spread the cells over a small circular area (5 to 10 mm in diameter); dry the slide at room temperature and transport the slides to the laboratory in a cardboard mailer
	Adenovirus	Yes	Swab (eye)	X		X	5 ml of sterile clotted blood for serology	

[a]Availability of serology for laboratory diagnosis of infection. NA, not applicable.
[b]Occurs less frequently.

of life in a neonate connotes interuterine infection of the mother with this virus. Conversely, CMV recovery from urine of an immunocompromised patient indicates that the patient is infected with this agent but does not imply that it represents the etiology of an associated systemic disease. Similarly, early detection of CMV DNA by PCR in the blood of organ transplant patients can trigger preemptive antiviral therapy, whereas nucleic acid amplification is a sensitive assay for infection but is nonspecific as a marker for disease once symptoms have developed.

The availability of test methods in a diagnostic virology laboratory may dictate the most appropriate assay for achieving optimal results, but selection of the appropriate specimen is critical to successful detection of viral antigens. For rapid testing of influenza A virus, throat washings or nasopharyngeal aspirates are superior to throat swabs (90% versus 60%) using single-unit membrane EIAs. Conversely, both influenza A and B virus can be detected with >90% sensitivity by direct immunofluorescence using cells obtained from a throat swab compared to conventional cell culture recovery of the virus.

SPECIMEN COLLECTION

Timing

Specimens should be collected early in the acute phase of infection. However, the duration of viral shedding depends on the type of virus and systemic involvement, as well as other factors. For example, the duration of RSV shedding is usually 3 to 7 days, but with a range of 1 to 36 days (Hall et al., 1976). Further, viral excretion was of greater duration in patients with lower respiratory tract disease than in those with clinical manifestations limited to the upper respiratory tract.

The presence of a virus in a specimen may depend on the source of the sample. Generally, varicella-zoster virus (VZV) can be recovered from lesions for up to 7 days after the initial vesicles appear but from blood by culture or nucleic acid amplification only during the late incubation period or days 1 to 4 of the acute illness (Mainka et al., 1998).

Interpretation of the significance of enterovirus isolates from stool specimens is confounded by the prolonged shedding (6 to 8 weeks) of these viruses, especially in children (Modlin, 1986). Since virus is commonly present for a shorter time in the oropharynx and often yields virus from specimens collected during the first 5 to 7 days of illness, interpretation of a positive virus isolation is less ambiguous.

The immunocompetence of a patient with a viral illness has a significant effect on the time and duration of virus excretion. For example, high concentrations of HSV ($>10^4$ PFU) are detected in specimens from lesions of immunocompromised patients for more than 3 weeks, while the mean duration of HSV shedding from immunocompetent men and women with genital infection is 11.4 days (Corey et al., 1983). Adenovirus, especially in bone marrow transplant patients, can disseminate to the CNS, respiratory system, liver, and intestine with fatality rates greater than 50% (Hierholzer, 1992). Echoviruses have been recovered from the CSF of children with agammaglobulinemia for periods varying from 2 months to 3 years (McKinney et al., 1987). VZV could be recovered from the buffy coat of blood specimens from patients with malignant disease for at least 8 days after the onset of cutaneous lesions but could not be isolated from patients with typical varicella and no underlying malignancy (Myers, 1979). New lesions due to VZV infection in immunologically normal children develop over a 4-day period, while new lesions form in most immunocompromised children for more than 5 days. Immunocompromised adults with zoster shed virus for longer (7.0 days) than do otherwise normal adults (5.3 days). In addition, zoster is much more likely to disseminate cutaneously in immunocompromised than in immunocompetent hosts (Balfour, 1988).

Some viruses, such as CMV, rubella virus, adenoviruses, and enteroviruses can be excreted from various sites in the asymptomatic individual for months or years. Therefore, the timing of specimen collection is critical with regard to being able to associate the virus with the current disease. For example, a CMV-positive urine specimen collected during the first 2 weeks of life from an infant with congenital disease is strong evidence supporting viral etiology of the congenital anomalies, presumably since the agent was acquired in utero and active viral infection and replication of the virus are present. Conversely, isolation of CMV from the same source after 2 weeks would not discriminate between congenitally and postnatally acquired CMV infection. Thus, CMV infection in an infant obtained as a result of passage of the fetus through the birth canal (postnatal infection) would likely require at least 2 weeks for viral replication to reach detectable levels in the urine.

Adeno- and enteroviruses can be excreted in stool specimens for several days to weeks, even though the ability of these viruses to cause gastroenteritis is not established. Nevertheless, both viruses are recognized pathogens, causing acute upper respiratory tract disease (Atmar and Georghiou, 1993; Miller et al., 1996). The isolation of either agent from a stool specimen, but not from other sources (respiratory tract), would not have etiologic significance. Importantly, disseminated adenovirus infections can occur in immunocompromised hosts with the virus being recovered from the same body sites as CMV, and even cytopathic effects in cell cultures can be similar to those of this virus (Landry et al., 1987). Interestingly, both enteroviruses and adenoviruses were recovered over a period of several weeks from a patient with bare lymphocyte syndrome, a form of partial combined immunodeficiency (Arens et al., 1987). Thus, persistent or chronic excretion of these viruses may merely reflect the compromised ability of the host to eliminate viral replication; on the other hand, adeno- and enteroviruses may produce disease in patients with limited antiviral defenses.

Volume and Number of Specimens

Ideally, nasal washes and nasopharyngeal aspirates are required for rapid identification of influenza virus type A antigen by membrane EIA. Extracts of throat swabs from patients with influenza virus infection are inferior to nasal washes and aspirates. For example, 139 specimens from the respiratory tract processed at the Mayo Clinic (Rochester, Minn.) (swabs, 68; nasal aspirate, 44; tracheal secretion, 14; sputum, 11; bronchoalveolar lavage, 2) were tested by immunofluorescence, EIA, and inoculation into cell cultures. Collectively, with all techniques, 61 specimens (44%) (nasal aspirates, 50%; swabs, 43%; sputum, 64%; tracheal aspirates, 21%) were positive for influenza type A virus. Of the 61 total positive specimens, 57 (93%) were positive by immunofluorescence; only 30 (49%) were detected by EIA. Immunofluorescence was positive for influenza virus type A in 40 of 61 specimens (66%) in which other techniques (EIA, cell culture) were positive for the virus. Influenza virus type A was detected exclusively by immunofluorescence in 17 (28%) and by EIA in 1 (2%) of the specimens yielding positive results. Therefore, immunofluorescence is preferable to membrane EIA for influenza virus regarding sensitivity; in addition, both influenza virus types (A and B) can be detected by immunofluorescence.

Laboratory diagnosis of CMV viremia by shell vial provides valuable information for determining symptomatic infection in immunocompromised patients, particularly organ transplant recipients. A leukocyte inoculum concentration of 4×10^5 cells, compared to 2×10^5 cells, produced a 36% increase in the sensitivity of CMV detection by the shell vial assay (Lipson et al., 1996). Further, since viremia with CMV is intermittent (especially when small volumes of blood [5 ml] are processed from leukopenic patients), recovery of the virus can be enhanced by both increasing the volume of anticoagulase treated blood to 10 ml and collecting two samples (within 3 h) (Patel et al., 1995a).

Source

Throat, Nasopharyngeal Swab, and Nasal Washing

Specimens from the respiratory tract can represent almost one-half of the source material and one-third of the total viruses diagnosed in the clinical laboratory. Throat or nasal washings may be more productive for viral isolation than throat or nasal swabs, but few comparisons have been reported. Generally, nasopharyngeal aspirates are more productive than nasopharyngeal swabs for the diagnosis of RSV (Ahluwalia et al., 1987; Mackie et al., 1991). The convenience of using a swab for medical personnel and the willingness of the patient to allow collection of this specimen (compared with washings) are important factors in this choice. Swabs are considered superior to nasal wash because several problems are associated with the latter. Nasal wash specimens submitted for fluorescent antibody detection of viral antigen often contain debris such as mucus, squamous cells, leukocytes, and erythrocytes. In addition, the number of cells obtained by nasal wash has been found to be smaller than that obtained by nasal and oropharyngeal swabs combined (Kim et al., 1983; Blumenfeld et al., 1984). Frayha et al. (1989) demonstrated that nasopharyngeal swabs and nasopharyngeal aspirates were equally effective for the diagnosis of RSV, parainfluenza virus, and influenza virus in children.

Sputum

Recovery of a virus from sputum does not necessarily reflect lower respiratory tract infection but may be due to "contamination" of the specimen with the agent present in the throat. Alternatively, recovery or cytologic detection of a virus from a transtracheal aspirate or transbronchial biopsy indicates lower respiratory tract involvement (Blumenfeld et al., 1984). Although Kimball et al. (1983) obtained an overall isolation rate of 20% from patients diagnosed as having radiologically confirmed pneumonia, they suggested that not all of the viruses recovered from this source may be of clinical importance as the etiologic agents of lower respiratory tract disease in their study population. Thus, the type and frequency of viral isolates was as follows: influenza virus (H_3N_2), six; RSV, two; HSV, nine; rhinovirus, three. The results caused the investigators to conclude that only the influenza virus and RSV (total, 8%) caused lower respiratory tract disease in their patients. This conclusion was based on the recognition of these viruses as lower respiratory tract pathogens and the presence of radiographic evidence of pneumonia in these eight patients but not in those from

whom other viruses (HSV, rhinoviruses) were detected. On this basis, they speculated that sputum specimens may be of particular value in the laboratory diagnosis of lower respiratory tract disease due to viruses. Unfortunately, because of the severity of illness in their study population, these investigators were unable to obtain throat washings for comparison with the sputum specimens.

Interestingly, HSV and CMV are capable of growth in human alveolar macrophages; thus, bronchopulmonary lavage specimens yielding these cells may be useful as a relatively simple technique for laboratory diagnosis of lower respiratory tract infection due to these viruses without resorting to an open-lung biopsy procedure (Kahn and Jones, 1988).

BAL

Bronchoscopy, transbronchial biopsy, and bronchoalveolar lavage (BAL) have been useful procedures for obtaining specimens, especially for the diagnosis of *Pneumocystis carinii* infections. Of these, BAL has provided an alternative to open lung biopsy for the rapid diagnosis of viral infections, especially CMV (Martin and Smith, 1986). With this procedure, the entire tracheobronchial tree is inspected with a fiber-optic bronchoscope. After removing the scope, cleaning it of secretions, and reinserting it into the involved segment of the lung, saline is instilled and then removed by vacuum to obtain the lavage specimen of suspended cells. The cells are washed and inoculated onto cell cultures or stained and assayed directly for virus using immunologic or nucleic acid detection methods. In 1 year at the Mayo Clinic, 80 viral isolates were obtained from BAL specimens. Seventy-one (89%) were CMV, four were parainfluenza virus, three were influenza virus, and two were enterovirus. Woods et al. (1990) indicated that the presence of alveolar lymphopenia in a patient and laboratory diagnosis of CMV using antibodies to early antigens of the virus were highly suggestive of disease, although correlation of these results with positive cytologic findings (inclusion bodies) would increase the specificity of tests regarding CMV etiology. Respiratory infections are common after lung transplantation. Of 18 patients tested, BAL specimens yielded parainfluenza virus more commonly (15) than did nasopharyngeal swab (3) or sputum (1) in symptomatic individuals posttransplant (Wendt et al., 1995).

Rectal Swabs and Stool Specimens

The use of feces as a specimen for virus isolation has been reduced for cases of gastroenteritis with the realization that viruses that are noncultivable in cell culture (rotavirus, Norwalk-like agent [caliciviruses], astroviruses, and perhaps some adenoviruses) are responsible for most cases of viral gastroenteritis (Christensen, 1989; Mautner et al., 1995; Mitchell et al., 1995). It is estimated that rotaviruses are the most prevalent cause of gastroenteritis worldwide. Approximately 3 million cases of symptomatic rotavirus infection occur in the United States each year. Fortunately, diagnostic virology laboratories can diagnose this infection by rapid membrane enzyme immunoassay kits using stool (preferably) or rectal swabs (Dennehy et al., 1988). Enteroviruses can be isolated as commonly from the feces of individuals without disease as from patients with gastroenteritis. However, it is frequently useful to submit stool or rectal swabs in addition to other specimens (particularly CSF) for laboratory diagnosis of CNS disease presumed to be of enteroviral etiology. In this situation, an enterovirus may be excreted in the gastrointestinal tract, but the agent may not be recovered from CSF. Importantly, the etiologic association of such isolates from the gastrointestinal tract becomes somewhat more tenuous than from CSF, owing to the common excretion of these agents in stool subsequent to respiratory tract infection (Bowen et al., 1983). There is a higher rate of virus isolation from stool specimens, but these are less convenient than rectal swabs for the diagnosis of viral gastroenteritis (Mintz and Drew, 1980). Detection of the fastidious adenovirus types (40 and 41) by nucleic acid probes may help to clarify the role of these agents in gastrointestinal diseases (Krajden et al., 1990; Hussain et al., 1996; Vizzi and Ferraro, 1996).

Viruses surrounded by a lipid envelope generally are not found in an active form in stool specimens, although CMV is an important cause of intestinal disease and has been recovered from that source (Drew, 1988).

Urine

Laboratory diagnosis of CMV from immunosuppressed patients and detection of HSV and VZV from dermal specimens have been important for the medical management of patients with antiviral drugs in recent years. Over a 4-year period (1994 to 1998), urine samples represented 21% of the specimens submitted to the Mayo Clinic Virology Laboratory compared with genital (38%), blood (24%), dermal (15%), and lung tissue (2%) samples (Fig. 1). (Other sources, such as respiratory, CSF, eye, rectal/stool, and other tissue specimens, were not tabulated.)

Many viruses are excreted in urine during the incubation period. Mumps, adenovirus, and CMV are commonly recovered from urine after symptoms develop. Although very rarely found in the diagnostic laboratory, mumps virus can be isolated from urine when specimens from other sites are negative, as in the case of CNS disease. Papillomavirus also is excreted in urine, but detection requires highly sensitive techniques such as nucleic acid hybridization (Melchers et al., 1989; Shen et al., 1996). Similarly, the polyomaviruses, JC and BK, can be detected by immunologic methods in shell vial cell cultures, but detection by in situ hybridization or nucleic acid amplification is more sensitive than isolation of the virus in cell culture (Marshall et al., 1990). The clinical significance of polyomavirus excretion in urine needs to be determined.

Urine specimens submitted for laboratory diagnosis of CMV infection can be inoculated directly into cell cultures because the virus is not concentrated in urine sediment after low-speed centrifugation. However, in one study, low speed centrifugation (500 × g for 10 min) of urine specimens prior to inoculation presumably removed toxic materials and thus increased viral detection in shell vial cell cultures (Lipson et al., 1990). Because of the relatively low titer, CMV in urine must be concentrated first to be detectable by nucleic acid hybridization (Landini et al., 1990; Bennion et al., 1998). Urine is the best single specimen for recovery of CMV, but in some cases the virus has been isolated using only a throat swab (Dominguez et al., 1993; Visseren et al., 1997). At the Mayo Clinic, CMV was detected in 6% of 14,000 urine specimens; CMV accounted for all but 4 (adenovirus, 2; HSV, 2) of 874 isolates (99.5%) (Fig. 1). This demonstrated that CMV excretion could be sporadic from body sites and therefore that multiple specimens should be processed. In third-trimester pregnant women, cervical excretions appreciably exceeded urinary shedding of virus, the rates being 11.6% (48 of 404) and 6.3% (29 of 463), respectively (Reynolds et al., 1973). It must be recognized that recovery of CMV from urine should be interpreted only as evidence of infection, not necessarily as an etiologic agent of disease with the exception of cases of very early detection (0 to 14 days postnatal) in congenitally infected neonates. As expected, HSV was recovered more frequently from specimens obtained from the cervical canal than from urine from pregnant women with acute genital infections (Kawana et al., 1982).

Dermal Lesions

HSV (70%), VZV (29%), coxsackievirus group A (1%), and perhaps some echoviruses are the principal agents recovered on a routine basis from dermal lesions (Smith, 1983; Smith and Wold, 1991). Dermal lesions represented only 15% of

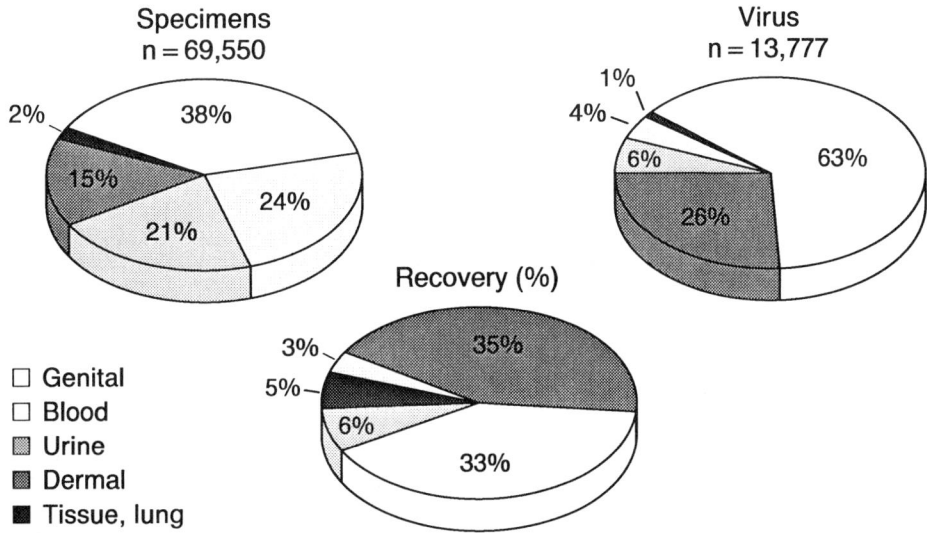

FIGURE 1 Recovery of viruses from several specimen sources at the Mayo Clinic, 1994 to 1998

almost 70,000 specimens but accounted for over one-fourth of the viral isolates, with an isolation rate of 35% (Fig. 1). Almost equal numbers of HSV-1, HSV-2, and VZV were detected from dermal specimens. The ability to detect or isolate HSV varies with the stage of the lesion. For example, HSV was recovered from 94% of vesicular lesions, 87% of pustular lesions, 70% of ulcers, and 27% of crusted lesions (Moseby et al., 1981). Similarly, smears prepared with cells obtained from vesicles for Papanicolaou, crystal violet, or immunofluorescence staining were superior to cells obtained from ulcers for the diagnosis of HSV infection. Skin biopsy of cutaneous lesions may be important in the diagnosis of systemic CMV and human herpesvirus 8 infections of immunocompromised patients (PCR detection) (Swanson and Feldman, 1987; Mendez et al., 1998c). Importantly, the laboratory is necessary to distinguish HSV from VZV in that the clinical presentation of the patient may not be typical for these specific viral infections (Rubben and Baron, 1997). Nucleic acid amplification has clearly given a new dimension to the detection and management of viral CNS diseases (Gutierrez and Prober, 1998).

CSF

Nonpolio enteroviruses are the most common isolates from CSF. HSV, an important cause of CNS disease, rarely has been isolated from CSF except when in association with meningitis caused by HSV-2 (Rubin, 1983). Alternatively, detection of amplified DNA sequences in CSF of patients with HSV has replaced viral culture methods as the "gold standard" diagnostic test for CNS infections due to this virus, with detection rates of about 7% (Mitchell et al., 1997). Although isolation rates of viruses from CSF specimens generally are low (<4%), this source has been almost exclusively productive for the recovery of enteroviruses. Enterovirus RNA can be reverse transcribed to DNA and detected by PCR with primers selected from the 5'-nontranslated (conserved) portion of the viral genome (Rotbart, 1995). Evaluations and clinical experience have indicated these amplification assays to be superior to culture detection of enteroviruses (Gorgievoki-Hrishoho et al., 1998; Pozo et al., 1998; van Vliet et al., 1998). The test is not commercially available. CMV, VZV, and adenoviruses are rarely isolated from CSF in immunocompromised hosts. Togaviruses, although present in CSF, usually are not collected for testing because they do not replicate well in the cell cultures commonly used for routine viral diagnosis.

Eyes

HSV (66%) and adenovirus (34%) are the viruses commonly associated with eye infections (Smith, 1984); however, enteroviruses have been associated with hemorrhagic conjunctivitis (Shulman et al., 1997). Cytomegalovirus retinitis is recognized as an initial manifestation of AIDS (Centers for Disease Control and Prevention, 1992). Nucleic acid based amplification tests have been described experimentally for detection of HIV, HSV adenovirus, and other agents from eye specimens (Shimazaki et al., 1994; Jackson et al., 1996). Neonates infected with other agents, such as rubella virus, HSV, VZV, and human papillomavirus (HPV), may exhibit ocular involvement as disseminated infection (Kimberlin et al., 1996; Jacobs, 1998).

Blood

Isolation of viruses from the blood provides evidence of acute-phase infection and symptomatic disease (Ljungman et al., 1993). For example, CMV viremia in patients prior to bone marrow transplantation is predictive of both CMV pneumonia and gastrointestinal disease. For detection of CMV viremia within the same episode of disease, viremia was 90% sensitive and 80% specific for predicting concurrent organ involvement in liver transplant recipients (Badley et al., 1996). Early detection of CMV DNA in blood specimens from organ recipients posttransplantation can prompt preemptive therapy for CMV viremia (Patel et al., 1994; 1995b; Patel and Paya, 1997; Mendez et al., 1998b). In addition, quantitative application of nucleic acid target amplification may identify individuals who will develop symptomatic CMV infection (Mendez et al., 1998a; Mazzulli et al., 1999). Similarly, quantitative CMV DNA amplification may be useful for monitoring the efficacy of antiviral therapy (Gerna et al., 1998). Specific separation procedures that allow the collection of different leukocyte fractions of the blood will yield higher isolation rates of viruses such as CMV and VZV compared with buffy coat preparations, especially with specimens from neutropenic patients (Paya et al., 1988; Reina et al., 1998). Inoculation of three, rather than two, shell vial cell cultures provides the most sensitive cell culture detection of CMV (Paya et al., 1987; Reina et al., 1987). Localization of CMV target DNA in the cellular fraction of blood is consistent with culture results (Mendez et al., 1998b). Of 16,700 total specimens processed during a 5-year period, almost one-fourth were from blood, with an isolation rate of 3%. Only 2 (HSV) of 523 isolates were other than CMV. Enteroviruses were isolated from 14 of 31 frozen serum specimens obtained from hospitalized patients with enterovirus infection that had been documented previously by the recovery of virus from stool, throat, or CSF (Prather et al., 1984). Amplification of enterovirus nucleic acids by PCR will hopefully be common soon in most laboratories (Ahmed et al., 1997). Similarly, whole blood was used to document congenital infection due to echovirus type 11 (Jones et al., 1980). Viremia with HIV may predispose patients to coinfection with CMV in the blood (Shibata et al., 1988).

Detection of CMV infection in blood leukocytes (pp65 antigenemia test) is less sensitive than DNA amplification by PCR. More importantly, the antigenemia test is poorly controlled because of lack of standardization among laboratories, the limited time regarding cellular separation from whole blood, and variable results with stored specimens (Shinkai et al., 1997; Mazzulli et al., 1999).

Tissue

Generally, lung and other tissue from the respiratory tract and brain tissue are the only specimens from organs that yield viruses in cell cultures. Occasionally liver and, rarely, spleen tissue have yielded CMV or HSV. Of 95 viral isolates from tissue over an 8-year period (isolation rate, 3.6%), 82 (86%) of these were CMV, HSV, parainfluenza virus, influenza virus, rhinovirus, and adenovirus from respiratory tissues (Smith, 1983). Interestingly, an updated analysis of isolates from tissue (1994 to 1998) yielded almost identical results (Fig. 1). Obviously, selection of particular tissue specimens will improve the recovery rate of viruses from this source. For example, of 105 open-lung biopsy specimens, obtained mostly from immunosuppressed adults, CMV was recovered from 20 patients (19%) and influenza virus type A was recovered from 1 patient. Generally, however, the recovery rate is less than 10% and the majority of isolates are CMV from lung specimens. The low rate of isolation may stem from the release of viral inhibitors in tissue after homogenization. Alternatively, enzymatic digestion of tissue fragments has provided higher rates of viral isolation compared with homogenized specimens. The ability to detect viruses such as CMV, HPV, and JC virus from tissue fixed in formaldehyde and embedded in paraffin adds a new dimension to the laboratory diagnosis of viral infection (Telenti et al., 1990).

TRANSPORT

Swabs

Swabs are used for the collection of specimens from throat, dermal, rectal, and ocular sites. Of a total of almost 19,000 specimens submitted for viral culture during 1988 at the Mayo Clinic, 12,000 (63%) were collected with swabs (Smith and Wold, 1991). A variety of fibers have been used for the tip of the shaft of commercially available swabs, including rayon, cotton, dacron, polyester, and calcium alginate. All materials, with the notable exception of calcium alginate-tipped swabs, have been acceptable for general use for collection of viral specimens.

Stability of Viruses

In general, viruses that are enveloped, such as the herpesviruses, HIV, and the myxo- and paramyxoviruses (especially RSV), are relatively labile compared with those without envelopes (Tjotta et al., 1991). Similarly, survival of viruses is favored by cool temperatures (4°C) and an approximately neutral pH (Abad et al., 1994). Nevertheless, these viruses survive transit for at least 1 to 3 days if maintained at 4°C (Shinkai et al., 1997). HSV can survive for as long as 2 h on the surface of skin, 3 h on cloth, and 4 h on plastic (Turner et al., 1982).

Transport System

Comprehensive studies of viral transport media are difficult to find in the scientific literature. The reason is likely the substantial expense of comparing two or more types of media for transport of swabs from the clinician's office to the laboratory. The most comprehensive review of transport systems for viruses appeared approximately 10 years ago. Swab-tube combinations that are commercially available (Culturette, Virocult) have been demonstrated to be adaptable to medical practices because of acceptable recovery rates of viruses, based on product evaluations in clinical practices (Huntoon et al., 1981; Perez et al., 1984; Johnson, 1990). Importantly, these products have replaced traditional "home-brew" transport solutions that ranged from skimmed milk to bacteriologic broth and buffered gelatin. Generally, these in-house media were evaluated for efficacy only with laboratory strains of viruses; study results likely had little relevance to survival of viruses from fresh specimens from clinical sources.

At the Mayo Clinic, Culturettes (Becton-Dickinson Microbiology Systems, Cockeysville, Md.) have been used for transport of viral specimens (swabs) for over 25 years. The swab consists of a plastic tube containing a rayon-tipped applicator and an ampule of modified Stuart's transport medium. These swabs have allowed recovery of commonly isolated viruses in clinical laboratories at rates ranging up to 50%. An advantage for our institution is that one swab, the Culturette, may be used for specimen collection and transport of organisms appropriate for diagnostic use in bacteriology, mycology, parasitology, and virology.

Since the composition of the Culturettes has not changed since its initial development over 30 years ago, a clinical evaluation with these swabs performed almost 20 years ago is still relevant today. In summary, 80 viruses from throat specimens were recovered from 200 children (40% isolation rate). Swabs with Stuart's transport medium were as effective as Culturettes that contained either Hanks' balanced salt solution (HBSS) or Liebowitz-Emory medium (LEM) with charcoal (Huntoon et al., 1981). Nevertheless, to be able to assess statistically significant data for differences between enveloped and nonenveloped viruses and among total viruses recovered from each of the three transport systems (Culturettes and Stuart's, Culturettes and HBSS, or Culturettes and LEM), several hundred more patients would need to be entered in the study. This is the obvious reason for the lack of clinically relevant comparisons and the dependence on testing laboratory strains for efficacy of transport.

A few other major viral transport media are available. Micro Test Multi-Microbe transport medium (M4-3) contains 3 ml of HBSS supplemented with bovine serum albumin, gelatin, sucrose, and glutamic acid. Phenol red is incorporated to indicate the pH. Vancomycin, amphotericin B, and colistin are used to inhibit contamination. The manufacturer (MicroTest, Inc., Lilburn, Ga.) claims equal or superior stability of high-titered laboratory strains of CMV, HSV-1, or influenza virus type A in M4-3 medium compared with Bartels' viral transport medium (VTM). Only one publication reports an evaluation of M4-3 viral transport performance with fresh clinical specimens. In that study, M4-3 and VTM were equivalent for the detection of HSV from oral and genital sources.

ViraTrans, ChlamTrans, and FlexTrans (Bartels, Issaquah, Wash.) are available for transport of specimens for virus and chlamydia infection with both agents. All three transport media (2 ml) are placed in 15-ml conical centrifuge tubes containing glass beads. The medium compositions are all similar; the basic composition includes a buffer or other medium containing antibiotics and phenol red. The efficacies of these transport media were compared to those of other systems using laboratory strains of virus to demonstrate equivalency.

The incorporation of protein, such as serum, albumin, or gelatin, into transport medium has been advocated as a means of stabilizing viruses during transit to the laboratory. We participated in a four-site evaluation of a laboratory test kit for rapid diagnosis of genital HSV infections. Two laboratories used a transport medium containing 0.5% gelatin and another incorporated 2% fetal bovine serum as a stabilization agent. Our laboratory used the Culturette during the study, and almost all specimens were submitted from another reference laboratory that required a minimum of 24 h of transit time between collection and receipt at the Mayo Clinic laboratory. Transport times for specimens submitted to the other three laboratories were routinely less than 24 h. The addition of protein to stabilize HSV during transit provided no apparent positive effect on the overall isolation rate of virus, since isolation rates of the virus ranged from 27% (HBSS plus 0.5% gelatin) to 34% (Culturette). Admittedly, this is a comparison of apples with oranges, because the specimens are not common; however, the methods used at all investigational sites were similar.

Storage of Specimens

Viruses such as adeno- and enteroviruses that do not have structurally labile lipid envelopes survive freeze-thaw procedures with relatively little loss of viral titer. Conversely, a single freeze-thaw cycle may decrease the titer of HSV by 100-fold. Even storage at room temperature for 1 to 30 days significantly reduces the infectivity of this virus. For short-term (5 days) transit or storage of most viral suspensions, therefore, the specimen should be held at 4°C rather than frozen.

Preservation of the infectivity of labile viruses such as RSV and VZV for reference purposes can be accomplished without refrigeration by initially freeze-drying the preparation in the presence of a stabilizer containing sucrose, phosphate compounds, sodium glutamate, and bovine albumin. More practically, solutions of dimethyl sulfoxide in concentrations of 5%, sucrose-phosphate-glutamate containing 1% bovine albumin, and other formulations containing serum, glycerol, skim milk, or other proteins have been shown to be useful cryoprotectants for viruses (Tannock et al., 1987; Gallo et al., 1989; Johnson, 1990).

PROCESSING OF SPECIMENS

Inoculation

Generally, 0.2 to 0.3 ml of specimen is inoculated onto human diploid fibroblast (HDF) cells, primary monkey kidney cells, and a continuous cell line, such as HeLa or HEp-2 cells. Primary rabbit kidney cells may be used in place of HDF cells for the recovery of HSV (Moore, 1984), and rhabdosarcoma and buffalo green monkey kidney cells have been found to yield more enteroviruses than do the usual cell systems.

Removal of liquid medium from cell monolayers before inoculation of a specimen, in order to allow for adsorption of viral particles to the cells (1 h), probably enhances the rate of recovery of these viruses. Based on studies that have demonstrated that low-speed centrifugation (2,000 to 3,000 × g) of the specimen inoculum onto cell culture monolayers increases the efficiency of chlamydia, we applied this technique to urine specimens for the diagnosis of human strains of CMV (Gleaves et al., 1984). Rapid detection of herpesviruses, especially, has been performed for several years using the shell vial assay; conventional tube cell cultures are also inoculated in order to detect (after several days of incubation) low-titered viruses in specimens. In this procedure, 0.2 ml of urine is inoculated onto monolayers of MRC-5 cells seeded in 1-dram shell vials containing a circular cover slip. The inoculated shell vials are then centrifuged at 700 × g for 1 h, medium is added back, and the vials are incubated at 36°C for 16 to 48 h. At this time, the cover slip containing the cell monolayer is tested for the presence of early antigen of CMV by immunofluorescence. This rapid test reduces the detection time for CMV from an average of 8 days in conventional tube cell cultures to 1 day (16 h). The sensitivity of the rapid test was 100% compared with conventional cell culture isolation. If the specimens were not centrifuged, however, the sensitivity was only 37.5%. This technique has been applied in the laboratory for the rapid diagnosis of HSV infections and as applicable to other viruses, including adenovirus, influenza A, and VZV (Gleaves et al., 1985; Espy et al., 1986, 1987). The only limitation of the methodology is the availability of specific monoclonal or polyclonal antibodies for use as a probe to detect early antigens synthesized by a virus.

Inoculation of specimens directly into cell cultures at the bedside has yielded 500-fold-higher titers of certain agents, such as RSV (Hall and Douglas, 1975; Skinner et

al., 1997). However, a study of 135 samples inoculated at bedside or held for 3 h at 4°C before transport to the laboratory showed no difference in the rate of RSV recovery. Of 51 positive specimens, 44 (86%) were positive by both inoculation procedures, three (6%) were recovered only from specimens inoculated at bedside, and four (8%) were positive for RSV only when the specimen was inoculated in the laboratory (Bromberg et al., 1984).

SEROLOGIC DETERMINATIONS

Although techniques and high-quality reagents have reduced the time required for the diagnosis of many viral infections to just a few hours, the need for some serologic tests is still apparent. Agents such as EBV, rubella virus, measles virus, parvovirus, arboviruses (especially tovaviruses), hepatitis viruses, and HIV do not replicate in the battery of cell cultures generally used in the clinical laboratory. As such, serology remains the test of choice for the laboratory diagnosis of these virus infections. Further, assessment of the immune status of individuals to rubella virus, CMV (renal transplant recipients), and VZV (children with neoplastic disease) has remained an important diagnostic function of viral laboratories. The detection of virus-specific immunoglobulin M (IgM) antibodies in the acute phase serum of patients, in contrast to IgG antibodies, usually indicates a recent primary infection (with a few exceptions, e.g., HSV) with a particular agent. Inclusion of the proper controls and separation methods is necessary to provide specific, reliable results. Thus, sensitive techniques for IgM determination using a single specimen provide a necessary complement to rapid techniques for demonstrating viral antigens directly in specimens or after amplification in cell culture. (See chapter 15.)

On a broader basis, specimens for serologic testing for viral, bacterial, fungal, and parasitic infections are submitted to the infectious disease serology laboratory within our division. Collectively, this laboratory accounts for over 500,000 tests, 45% of the specimens submitted for serology, culture, and molecular diagnosis of microbial infection at our institution. Consolidation of serologic tests into one large physical area allows specimen processing (liquid transfer, dilution, addition to microtiter plates or tubes), analysis, spectrophotometric determinations, and print-out of results to be carried out by automation with instruments designed to provide maximum output and objective results.

FUTURE CONSIDERATIONS

The need for conventional tube cell culture detection of viruses will likely be phased out in the coming years. Viruses causing respiratory tract infections (mainly RSV, influenza virus, enterovirus, adenovirus, and parainfluenza virus) can be detected in immunofluorescence assays after initial replication in shell vial cell cultures for 16 to 48 h or by single-unit EIAs in 15 min of "hands-on" time.

Amplification of viral nucleic acids in CSF, blood, and tissue specimens has already lifted diagnostic virology to a higher level. New viral etiologic agents will be recognized with PCR formats to include amplification, detection, melting point analysis, and nucleic acid products for sequence analysis, antiviral susceptibility, pathogenicity, and epidemiologic information critical for patient management. From a technical standpoint, these data are presently available in a single workday.

Most importantly, with laboratory reimbursement predicated on a "capitated" basis, it is necessary to control hospital costs by obtaining specific diagnostic results with a minimum amount of laboratory testing. Clinicians need to know which specimens to submit for specific viral diagnosis. The laboratory must become closely involved with the medical practice to provide relevant, rapid test results that have a beneficial effect on patient management. This may involve indications for specific antiviral therapy or ultimate reduction of hospital stays.

Diagnostic virology laboratories with commercial partners in development have led the way in clinical microbiology regarding the formatting, application, and implementation of rapid, relevant assays. Finally, with expanding technical innovations, readily available electronic test information and ordering has become a necessity.

REFERENCES

Abad, F. X., R. M. Pinto, and A. Bosch. 1994. Survival of enteric viruses on environmental fomites. *Appl. Environ. Microbiol.* **60:**3704–3710.

Ahluwalia, G., J. Embree, P. McNicol, B. Law, and G. W. Hammond. 1987. Comparison of nasopharyngeal aspirate and nasopharyngeal swab specimens for respiratory syncytial virus diagnosis by cell culture, indirect immunofluorescence assay, and enzyme-linked immunosorbent assay. *J. Clin. Microbiol.* **25:**763–767.

Ahmed, A., F. Brito, C. Goto, S. M. Hickey, K. D. Olsen, M. Trujillo, and G. H. McCracken, Jr. 1997. Clinical utility of the polymerase chain reaction for diagnosis of enteroviral meningitis in infancy. *J. Pediatr.* **131:**393–397.

Arens, M. Q., A. P. Knutsen, K. B. Schwarz, S. T. Roodman, and E. M. Swierkosz. 1987. Multiple and persistent viral infections in a patient with Bare Lymphocyte Syndrome. *J. Infect. Dis.* **156:**837–841.

Atmar, R. L., and P. R. Georghiou. 1993. Classification of respiratory tract picornavirus isolates as enterovirus or rhinovirus by using reverse transcription-polymerase chain reaction. *J. Clin. Microbiol.* **31:**1544–1546.

Badley, A. D., R. Patel, D. F. Portela, W. S. Harmsen, T. F. Smith, D. M. Ilstrup, J. L. Steers, R. H. Wiesner, and C. V. Paya. 1996. Prognostic significance and risk factors of untreated cytomegalovirus viremia in liver transplant recipients. *J. Infect. Dis.* **173:**446–449.

Balfour, H. H., Jr. 1988. Varicella-zoster virus infections in immunocompromised hosts: a review of the natural history and management. *Am. J. Med.* **85:**68–73.

Bennion, D. W., L. J. Wright, R. A. Watt, A. A. Whiting, and J. F. Carlquist. 1998. Optimal recovery of cytomegalovirus from urine as a function of specimen preparation. *Diagn. Microbiol. Infect. Dis.* **31:**337–342.

Blumenfeld, W., E. Wager, and W. K. Hadley. 1984. Use of transbronchial biopsy for diagnosis of opportunistic pulmonary infections in acquired immunodeficiency syndrome (AIDS). *Am. J. Clin. Pathol.* **81:**1–5.

Bowen, G. S., M. C. Fisher, A. Deforest, C. M. Thompson, Jr., B. Kleger, and H. Friedman. 1983. Epidemic of meningitis and febrile illness in neonates caused by ECHO type 11 virus in Philadelphia. *Pediatr. Infect. Dis.* **2:**359–363.

Bromberg, K., B. Daidone, L. Clarke, and M. F. Sierra. 1984. Comparison of immediate and delayed inoculation of HEp-2 cells for isolation of respiratory syncytial virus. *J. Clin. Microbiol.* **20:**123–124.

Centers for Disease Control and Prevention. 1992. Revised classification system for HIV infection and expanded surveillance case definition for AIDS among adolescents and adults. *Morbid. Mortal. Weekly Rep.* **41:**1–19.

Christensen, M. L. 1989. Human viral gastroenteritis. *Clin. Microbiol. Rev.* **2:**51–89.

Corey, L., H. G. Adams, Z. A. Brown, and K. K. Holmes. 1983. Genital herpes simplex virus infections: clinical manifestations, course, and complications. *Ann. Intern. Med.* **98:**958–972.

Dennehy, P. H., D. R. Gauntlett, and W. E. Tente. 1988. Comparison of nine commercial immunoassays for the detection of rotavirus in fecal specimens. *J. Clin. Microbiol.* **26:**1630–1634.

Dominguez, E. A., L. H. Taber, and R. B. Couch. 1993. Comparison of rapid diagnostic techniques for respiratory syncytial and influenza A virus respiratory infections in young children. *J. Clin. Microbiol.* **31:**2286–2290.

Drew, W. L. 1988. Cytomegalovirus infection in patients with AIDS. *J. Infect. Dis.* **158:**449–456.

Espy, M. J., T. F. Smith, M. W. Harmon, and A. P. Kendal. 1986. Rapid detection of influenza virus by shell vial assay with monoclonal antibodies. *J. Clin. Microbiol.* **24:**677–679.

Espy, M. J., J. C. Hierholzer, and T. F. Smith. 1987. The effect of centrifugation on the rapid detection of adenovirus in shell vials. *Am. J. Clin. Pathol.* **88:**358–360.

Frayha, H., S. Castriciano, J. Mahony, and M. Chernesky. 1989. Nasopharyngeal swabs and nasopharyngeal aspirates equally effective for the diagnosis of viral respiratory disease in hospitalized children. *J. Clin. Microbiol.* **27:**1387–1389.

Gallo, D., J. S. Kimpton, and P. J. Johnson. 1989. Isolation of human immunodeficiency virus from peripheral blood lymphocytes stored in various transport media and frozen at −60°C. *J. Clin. Microbiol.* **27:**88–90.

Gerna, G., E. Percivalle, F. Baldanti, A. Sarasini, M. Zavattoni, M. Furione, M. Torsellini, and M. G. Revello. 1998. Diagnostic significance and clinical impact of quantitative assays for diagnosis of human cytomegalovirus infection/disease in immunocompromised patients. *New Microbiol.* **21:**293–308.

Gleaves, C. A., D. J. Wilson, A. D. Wold, and T. F. Smith. 1985. Detection and serotyping of herpes simplex virus in MRC-5 cells using centrifugation and monoclonal antibodies 16 h postinoculation. *J. Clin. Microbiol.* **21:**29–32.

Gleaves, C. A., C. F. Lee, C. I. Bustamante, and J. D. Meyers. 1988. Use of urine monoclonal antibodies for laboratory diagnosis of varicella-zoster virus infection. *J. Clin. Microbiol.* **26:**1623–1625.

Gorgievski-Hrisoho, M., J.-D. Schumacher, N. Vilimonovic, D. Germann, and L. Matter. 1998. Detection by PCR of enteroviruses in cerebrospinal fluid during a summer outbreak of aseptic meningitis in Switzerland. *J. Clin. Microbiol.* **36:**2408–2412.

Gutierrez, K. M., and C. G. Prober. 1998. Encephalitis. Identifying the specific cause is key to effective management. *Postgrad. Med.* **103:**129–130.

Hall, C. B., and R. G. Douglas, Jr. 1975. Clinically useful method for the isolation of respiratory syncytial virus. *J. Infect. Dis.* **131:**1–5.

Hall, C. B., R. G. Douglas, Jr., and J. M. Geiman. 1976. Respiratory syncytial virus infections in infants: quantitation and duration of shedding. *J. Pediatr.* **89:**11–15.

Hierholzer, J. C. 1992. Adenoviruses in the immunocompromised host. *Clin. Microbiol. Rev.* **5:**262–274.

Huntoon, C. J., R. F. House, Jr., and T. F. Smith. 1981. Recovery of viruses from three transport media incorporated into Culturettes. *Arch. Pathol. Lab. Med.* **105:**436–437.

Hussain, M. A., P. Costello, D. J. Morris, A. S. Bailey, G. Corbitt, R. J. Cooper, and A. B. Tullo. 1996. Comparison of primer sets for detection of fecal and ocular adenovirus infection using the polymerase chain reaction. *J. Med. Virol.* **49:**187–194.

Jackson, R., D. J. Morris, R. J. Cooper, A. S. Bailey, P. E. Klapper, G. M. Cleator, and A. B. Tullo. 1996. Multiplex polymerase chain reaction for adenovirus and herpes simplex virus in eye swabs. *J. Virol. Methods* **56:**41–48.

Jacobs, R. F. 1998. Neonatal herpes simplex virus infections. *Semin. Perinatol.* **22:**64–71.

Johnson, F. B. 1990. Transport of viral specimens. *Clin. Microbiol. Rev.* **3:**120–131.

Jones, M. J., M. Kolb, H. J. Votava, R. L. Johnson, and T. F. Smith. 1980. Intrauterine echovirus type 11 infection. *Mayo Clin. Proc.* **55:**509–512.

Kahn, F. W., and J. M. Jones. 1988. Analysis of bronchoalveolar lavage specimens from immunocompromised patients with a protocol applicable in the microbiology laboratory. *J. Clin. Microbiol.* **26:**1150–1155.

Kawana, T., K. Kawogoe, K. Takizawa, J. T. Chen, T. Kawaguchi, and S. Sakamoto. 1982. Clinical and virologic studies on female genital herpes. *Obstet. Gynecol.* **60:**456–461.

Kim, H. W., R. G. Wyatt, B. F. Fernie, C. D. Brandt, J. O. Arrobio, B. C. Jeffries, and R. H. Parrott. 1983. Respiratory syncytial virus detection by immunofluorescence in nasal secretions with monoclonal antibodies against selected surface and internal proteins. *J. Clin. Microbiol.* **18:**1399–1404.

Kimball, A. M., H. M. Foy, M. K. Cooney, I. D. Allan, M. Mattock, and J. J. Plorde. 1983. Isolation of respiratory syncytial and influenza viruses from the sputum of patients hospitalized with pneumonia. *J. Infect. Dis.* **147:**181–184.

Kimberlin, D. W., F. D. Lakeman, A. M. Arvin, C. G. Prober, L. Corey, D. A. Powell, S. K. Burchett, R. F. Jacobs, S. E. Starr, and R. J. Whitley. 1996. Application of the polymerase chain reaction to the diagnosis and management of neonatal herpes simplex virus disease. *J. Infect. Dis.* **174:**1162–1167.

Krajden, M., M. Brown, A. Petrasek, and P. J. Middleton. 1990. Clinical features of adenovirus enteritis: a review of 127 cases. *Pediatr. Infect. Dis. J.* **9:**636–641.

Landini, M. P., B. Trevisani, M. X. Guan, A. Ripalti, T. Lazzarotto, and M. La Placa. 1990. A simple and rapid procedure for the direct detection of cytomegalovirus in urine samples. *J. Clin. Lab. Anal.* **4:**161–164.

Landry, M. L., C. K. Y. Fong, K. Neddermann, L. Solomon, and G. D. Hsiung. 1987. Disseminated adenovirus infection in an immunocompromised host: pitfalls in diagnosis. *Am. J. Med.* **83:**555–559.

Lipson, S. M., P. Costello, S. Forlenza, B. Agins, and K. Szabo. 1990. Enhanced detection of cytomegalovirus in shell vial cell culture monolayers by preinoculation treatment of urine with low-speed centrifugation. *Curr. Microbiol.* **20:**39–42.

Lipson, S. M., L. H. Folk, and S. H. Lee. 1996. Effect of leukocyte concentration and inoculum volume on the laboratory identification of cytomegalovirus in peripheral blood by the centrifugation culture-antigen detection methodology. *Arch. Pathol. Lab. Med.* **120:**53–56.

Ljungman, P., R. DeBock, C. Cordonnier, H. Einsele, D. Engelhard, J. Grundy, A. Locasciulli, P. Reusser, and P. Ribaud. 1993. Practices for cytomegalovirus diagnosis, prophylaxis and treatment in allogeneic bone marrow transplant recipients: a report from the working party for infectious diseases of the EBMT. *Bone Marrow Transplant.* **12:**399–403.

Mackie, P. L., P. J. Madge, S. Getty, and J. Y. Paton. 1991. Rapid diagnosis of respiratory syncytial virus infection by using

pernasal swabs. *J. Clin. Microbiol.* **29:**2653–2655.

Mainka, C., B. Fuss, H. Geiger, H. Hofelmayr, and M. H. Wolff. 1998. Characterization of viremia at different stages of varicella-zoster virus infection. *J. Med. Virol.* **56:**91–98.

Marshall, W. F., A. Telenti, J. Proper, A. J. Aksamit, and T. F. Smith. 1990. Rapid detection of polyomavirus BK by a shall vial cell culture assay. *J. Clin. Microbiol.* **28:**1613–1615.

Martin, W. J., II, and T. F. Smith. 1986. Rapid detection of cytomegalovirus in bronchoalveolar lavage specimens by a monoclonal antibody method. *J. Clin. Microbiol.* **23:**1006–1008.

Mautner, V., V. Steinthorsdòttir, and A. Bailey. 1995. Enteric adenoviruses. *Curr. Top. Microbiol. Immunol.* **199:**229–282.

Mazzulli, T., L. W. Drew, B. Yen-Lieberman, D. Jekic-McMullen, D. J. Kohn, C. Isada, G. Moussa, R. Chua, and S. Walmsley. 1999. Multicenter comparison of the Digene Hybrid Capture CMV DNA assay (version 2.0), the pp65 antigenemia assay, and cell culture for detection of cytomegalovirus viremia. *J. Clin. Microbiol.* **37:**958–963.

McKinney, R. E., Jr., S. L. Katz, and C. M. Wilfert. 1987. Chronic enteroviral meningoencephalitis in agammaglobulinemic patients. *Rev. Infect. Dis.* **9:**334–356.

Melchers, W. J. G., R. Schift, E. Stolz, J. Lindeman, and W. G. V. Quint. 1989. Human papillomavirus detection in urine samples from male patients by the polymerase chain reaction. *J. Clin. Microbiol.* **27:**1711–1714.

Mendez, J., M. Espy, T. F. Smith, J. Wilson, R. Wiesner, and C. V. Paya. 1998a. Clinical significance of viral load in the diagnosis of cytomegalovirus disease after liver transplantation. *Transplantation* **65:**1477–1481.

Mendez, J. C., M. J. Espy, T. F. Smith, J. A. Wilson, and C. V. Paya. 1998b. Evaluation of PCR primers for early diagnosis of cytomegalovirus infection following liver transplantation. *J. Clin. Microbiol.* **36:**526–530.

Mendez, J. C., G. W. Procop, M. J. Espy, C. V. Paya, and T. F. Smith. 1998c. Detection and semiquantitative analysis of human herpesvirus 8 DNA in specimens from patients with Kaposi's sarcoma. *J. Clin. Microbiol.* **36:**2220–2222.

Miller, R. F., C. Loveday, J. Holton, Y. Shorvell, G. Patel, and N. S. Brink. 1996. Community-based respiratory viral infections in HIV positive patients with lower respiratory tract disease: a prospective bronchoscopic study. *Genitourin. Med.* **72:**9–11.

Mintz, L., and W. L. Drew. 1980. Relation of culture site to the recovery of nonpolio enteroviruses. *Am. J. Clin. Pathol.* **74:**324–326.

Mitchell, D. K., S. S. Monroe, X. Jiang, D. O. Matson, R. I. Glass, and L. K. Pickering. 1995. Virologic features of an astrovirus diarrhea outbreak in a day care center revealed by reverse transcriptase-polymerase chain reaction. *J. Infect. Dis.* **172:**1437–1444.

Mitchell, P. S., M. J. Espy, T. F. Smith, D. R. Toal, P. N. Rys, E. F. Berbari, D. R. Osmon, and D. H. Persing. 1997. Laboratory diagnosis of central nervous system infections with herpes simplex virus by PCR performed with cerebrospinal fluid specimens. *J. Clin. Microbiol.* **35:**2873–2877.

Modlin, J. F. 1986. Perinatal echovirus infection: insights from a literature review of 61 cases of serious infection and 16 outbreaks in nurseries. *Rev. Infect. Dis.* **8:**918–926.

Moore, D. F. 1984. Comparison of human fibroblast cells and primary rabbit kidney cells for isolation of herpes simplex virus. *J. Clin. Microbiol.* **19:**548–549.

Moseby, R. C., L. Corey, D. Benjamin, C. Winter, and M. L. Remington. 1981. Comparison of viral isolation, direct immunofluorescence, and indirect immunoperoxidase techniques for detection of genital herpes simplex virus infection. *J. Clin. Microbiol.* **13**:913–918.

Myers, M. G. 1979. Viremia caused by varicella-zoster virus: association with malignant progressive varicella. *J. Infect. Dis.* **140**:229–233.

Patel, R., and C. V. Paya. 1997. Infections in solid-organ transplant recipients. *Clin. Microbiol. Rev.* **10**:86–124.

Patel, R., D. W. Klein, M. J. Espy, W. S. Harmsen, D. M. Ilstrup, C. V. Paya, and T. F. Smith. 1995a. Optimization and detection of cytomegalovirus viremia in transplantation recipients by shell vial assay. *J. Clin. Microbiol.* **33**:2984–2986.

Patel, R., T. F. Smith, M. Espy, D. Portela, R. H. Wiesner, R. A. F. Krom, and C. V. Paya. 1995b. A prospective comparison of molecular diagnostic techniques for the early detection of cytomegalovirus in liver transplant recipients. *J. Infect. Dis.* **171**:1010–1014.

Patel, R., T. F. Smith, M. Espy, R. H. Wiesner, R. A. F. Krom, D. Portela, and C. V. Paya. 1994. Detection of cytomegalovirus DNA in sera of liver transplant recipients. *J. Clin. Microbiol.* **32**:1431–1434.

Paya, C. V., A. D. Wold, and T. F. Smith. 1987. Detection of cytomegalovirus infections in specimens other than urine by shell vial assay and conventional tube cell cultures. *J. Clin. Microbiol.* **25**:755–757.

Paya, C. V., A. D. Wold, and T. F. Smith. 1988. Detection of cytomegalovirus from blood leukocytes separated by Sepracell-MN and Ficoll-Paque/Macrodex methods. *J. Clin. Microbiol.* **26**:2031–2033.

Perez, T. R., P. L. Mosman, and S. V. Jachau. 1984. Experience with Virocult as a viral collection and transportation system. *Diagn. Microbiol. Infect. Dis.* **2**:7–9.

Pozo, F., I. Casas, A. Tenorio, G. Trallero, and J. M. Echevarria. 1998. Evaluation of a commercially available reverse transcription-PCR assay for diagnosis of enteroviral infection in archival and prospectively collected cerebrospinal fluid specimens. *J. Clin. Microbiol.* **36**:1741–1745.

Prather, S. L., J. A. Jenista, and M. A. Menegus. 1984. The isolation of nonpolio enteroviruses from serum. *Diagn. Microbiol. Infect. Dis.* **2**:353–357.

Reina, J., I. Blanco, and M. Munar. 1997. Determination of the number of blood samples needed for optimal detection of cytomegalovirus viremia in immunocompromised patients using a shell-vial assay. *Eur. J. Clin. Microbiol. Infect. Dis.* **16**:318–321.

Reina, J., J. Saurina, V. Fernandez-Baca, I. Blanco, and M. Munar. 1998. An increase in the number of polymorphonuclear leukocytes inoculated on shell-vial culture increases the sensitivity of this assay in the detection of cytomegalovirus in the blood of immunocompromised patients. *Diagn. Microbiol. Infect. Dis.* **31**:425–428.

Reynolds, D. W., S. Stagno, T. S. Hasty, M. Tiller, and C. A. Alford, Jr. 1973. Maternal cytomegalovirus excretion and perinatal infeciton. *N. Engl. J. Med.* **289**:1–5.

Rotbart, H. A. 1995. Enteroviral infections of the central nervous system. *Clin. Infect. Dis.* **20**:971–981.

Rubben, A., and J. M. Baron. 1997. Routine detection of herpes simplex virus and varicella-zoster virus by polymerase chain reaction reveals that initial herpes zoster is frequently misdiagnosed as herpes simplex. *Br. J. Dermatol.* **137**:259–261.

Rubin, S. J. 1983. Detection of viruses in spinal fluid. *Am. J. Med.* **75**:124–128.

Shen, J., J. E. Tate, C. P. Crum, and M. L. Goodman. 1996. Prevalence of human papillomavirus (HPV) in benign and malignant tumors of the upper respiratory tract. *Hum. Pathol.* **9**:15–20.

Shibata, D., W. J. Martin, M. D. Appleman, D. M. Causey, J. M. Leedom, and N. Arnheim. 1988. Detection of cytomegalovirus DNA in peripheral blood of patients infected with human immunodeficiency virus. *J. Infect. Dis.* **158**:1185–1192.

Shimazaki, J., K. Tsubota, M. Sawa, S. Kinoshita, T. Ohkura, and M. Honda. 1994. Detection of human immunodeficiency virus, hepatitis B virus, and hepatitis C virus in donor eyes using polymerase chain reaction. *Br. J. Ophthalmol.* **78**:859–862.

Shinkai, M., S. A. Bozzetti, W. Powderly, P. Frame, and S. A. Spector. 1997. Utility of urine and leukocyte cultures and plasma DNA polymerase chain reaction for identification of AIDS patients at risk for developing human cytomegalovirus disease. *J. Infect. Dis.* **175**:302–308.

Shulman, L. M., Y. Manor, R. Azar, R. Handsher, A. Vonsover, E. Mendelson, S. Rothman, D. Hassin, T. Halmut, B. Abramovitz, and N. Varsano. 1997. Identification of a new strain of fastidious enterovirus 70 as a causative agent of an outbreak of hemorrhagic conjunctivitis. *J. Clin. Microbiol.* **35**:2145–2149.

Skinner, G. R., M. A. Billstrom, S. Randall, A. Ahmad, S. Patel, J. Davies, and A. Deane. 1997. A system for isolation, transport and storage of herpes simplex viruses. *J. Virol. Methods* **65**:1–8.

Smith, T. F. 1983. Clinical uses of the diagnostic virology laboratory. *Med. Clin. North Am.* **67**:935–951.

Smith, T. F. 1984. Diagnostic virology in the community hospital. *Postgrad. Med.* **75**:215–223.

Smith, T. F., and A. D. Wold. 1991. Changing trends of diagnostic virology in a tertiary care medical center, p. 1–16. *In* L. M. de la Maza and E. M. Peterson (ed.), *Medical Virology X.* Elsevier, New York, N.Y.

Swanson, S., and P. S. Feldman. 1987. Cytomegalovirus infection initially diagnosed by skin biopsy. *Am. J. Clin. Pathol.* **87**:113–116.

Tannock, G. A., J. C. Hierholzer, D. A. Bryce, C.-F. Chee, and J. A. Paul. 1987. Freeze-drying of respiratory syncytial viruses for transportation and storage. *J. Clin. Microbiol.* **25**:1769–1771.

Telenti, A., A. J. Aksamit, Jr., J. Proper, and T. F. Smith. 1990. Detection of JC virus DNA by polymerase chain reaction in patients with progressive multifocal leukoencephalopathy. *J. Infect. Dis.* **162**:858–861.

Tjotta, E., O. Hungess, and B. Grinde. 1991. Survival of HIV-1 activity after disinfection, temperature and pH changes, or drying. *J. Med. Virol.* **35**:223–227.

Turner, R., Z. Shehab, K. Osborne, and J. O. Hendley. 1982. Shedding and survival of herpes simplex virus from "fever blisters." *Pediatrics* **70**:547–549.

van Vliet, K. E., M. Glimaker, P. Lebon, P. E. Klapper, C. E. Taylor, M. Ciardi, H. G. A. M. van der Avoort, R. J. A. Dierpersloot, J. Kurtz, M. F. Peeters, G. M. Cleator, and A. M. van Loon for The European Union Concerted Action on Viral Meningitis and Encephalitis. 1998. Multicenter evaluation of the Amplicor Enterovirus PCR test with cerebrospinal fluid from patients with aseptic meningitis. *J. Clin. Microbiol.* **36**:2652–2657.

Visseren, F. L., K. P. Bouter, M. J. Pon, J. B. Hoekstra, D. W. Erkelens, and R. J. Diepersloot. 1997. Patients with diabetes mellitus and atherosclerosis: a role for cytomegalovirus? *Diabetes Res. Clin. Pract.* **36**:49–55.

Vizzi, E., and D. Ferraro. 1996. Detection of enteric adenoviruses 40 and 41 in stool specimens by monoclonal antibody-based enzyme immunoassays. *Res. Virol.* **147**:333–339.

Wendt, C. H., J. M. Fox, and M. I. Hertz. 1995. Paramyxovirus infection in lung transplant recipients. *J. Heart Lung Transplant.* **14:**479–485.

Woods, G. L., A. B. Thompson, S. L. Rennard, and J. Lin- der. 1990. Detection of cytomegalovirus in bronchoalveolar lavage specimens: spin amplification and staining with a monoclonal antibody to the early nuclear antigen for diagnosis of cytomegalovirus pneumonia. *Chest* **98:**568–575.

Primary Isolation of Viruses

MARIE L. LANDRY AND G. D. HSIUNG

3

Viruses are obligate intracellular parasites and therefore require living cells in which to replicate. This is very different from the cultivation of bacteria, for which nutrient broth or agar plates suffice. The living cells essential for virus isolation and assay can be in the form of cultured cells, embryonated eggs, or laboratory animals, such as newborn mice (Fig. 1). The variety of methods and host systems employed for the isolation of viruses from clinical specimens reflects the fact that the optimum growth conditions for different viruses differ tremendously. If an insensitive host system is inoculated with a specimen containing a particular virus, or if suboptimal growth conditions exist, the virus will likely not be isolated and a false-negative result will be obtained. Thus, each laboratory must select the host systems and methods needed to optimize the isolation of those viruses causing the most morbidity in the patient populations it serves.

Although embryonated eggs and laboratory animals are useful for the isolation of certain viruses, cell cultures in monolayers are the sole isolation system utilized in most clinical virology laboratories. Thus, this chapter will focus primarily on cell culture techniques, with virus isolation methods using other host systems included in the appendix at the end of the chapter.

VIRUS ISOLATION IN CELL CULTURES

Background

The discovery by Enders, Weller, and Robbins in the late 1940s that poliovirus replicates in cultivated mammalian cells derived from non-nervous system tissues revolutionized and simplified procedures for the isolation of viruses (Enders et al., 1949). Until that time, intact animals or embryonated eggs were the common systems used. After that landmark discovery, cell cultures were prepared for virus studies from a wide variety of animal and human tissues, and as a result, in the years following, most of the common viruses we are familiar with today were discovered.

Types of Cell Culture

Cell cultures are generally separated into three types (Table 1): primary cells, which are prepared directly from animal or human tissues and usually can be subcultured for only one or two passages; diploid cell cultures, which are usually

derived from human tissues, either fetal or newborn, and can be subcultured 20 to 50 times before senescence; and continuous cell lines, which can be established from human or animal tissues, from tumors, or following the spontaneous transformation of normal tissues. The last type has a heteroploid karyotype and can be subcultured an indefinite number of times. However, sensitivity to virus infection may change after serial passage and after passage in different laboratories.

Recent innovations in monolayer cell culture include the use of mixtures of two different cell lines and genetically altered cells (see below). In addition, suspension cultures of human lymphocytes have allowed isolation of several human retroviruses and herpesviruses (see below).

Variation in Sensitivity to Different Viruses

Cell cultures vary greatly in their sensitivities to different viruses (Table 2). If a virus is inoculated into an insensitive cell culture, the virus will not be able to replicate and a negative result will be obtained. When small amounts of virus are present in a clinical sample, a positive result may be obtained only when the most sensitive systems are used. Therefore, it is critical that those caring for the patients inform the laboratory of the clinical syndrome and/or virus(es) suspected, so that the most sensitive cell cultures can be used and appropriate detection methods can be employed. Laboratories must also reassess the sensitivity characteristics of cell cultures intermittently, since significant changes can occur over time (Landry et al., 1982; Coffin and Hodinka, 1995; McCarter and Robinson, 1997).

Supplies and Equipment Needed

The materials needed for the isolation of viruses in cell culture (Table 3) are those necessary for the safe handling and inoculation of cell cultures, maintenance and observation of cell cultures, and preservation and storage of clinical specimens and virus isolates. Although the clinical virologist is primarily interested in virus isolation, maintaining different cell cultures in healthy condition is absolutely necessary in order to ensure good results. A wide variety of cell cultures are available commercially and can be purchased and delivered weekly according to the needs of the laboratory. Both the quantity and types of cell cultures used will vary with the seasonal variations in virus activity. Some laboratories may elect to prepare some of their cell cultures in the laboratory from available animal or human tissues (e.g., rabbit kidney or human newborn foreskin) or to passage certain cell lines (e.g., HEp-2 cell or human diploid fibroblast [HDF] strains) for reasons of availability, economy, or quality. Instructions for the preparation and maintenance of cell cultures can be found in several reference books (Hsiung et al., 1994; Schmidt and Emmons, 1989).

Obtaining and Processing Specimens

Although obtaining and processing specimens have been reviewed in the previous chapter, it should be reiterated that without appropriate specimens that are properly collected early in illness and promptly transported to the laboratory, the subsequent time and effort spent in isolation attempts will be wasted. Accomplishing this is an important task of the clinical virology laboratory and requires continuing communication with and education of clinicians.

Virus Isolation Methods

Conventional Cell Culture

■ Inoculation and Incubation

To isolate a spectrum of viruses, cell cultures of several different types are inoculated, such as HDF, a human heteroploid cell line (e.g., A549), and a primary monkey kidney (MK) cell culture. Alternatively, for specific indications (e.g., herpes simplex virus infection), limited cultures intended to detect only one or two virus types can be performed.

The cell type(s) most sensitive to the suspected viruses in the clinical specimen should be included. Ideally, only healthy, freshly prepared, young cell cultures should be used, because aged cells are less sensitive to virus infection. All cell cultures should be examined under the microscope before inoculation to ensure that the cells are in good con-

TABLE 1 Cell cultures commonly used in clinical virology laboratories

Cell culture	Examples	No. of subpassages
Primary	Kidney tissues from monkeys, rabbits, etc.; embryos from chickens, guinea pigs, etc.	1 or 2
Diploid (limited passage)	Human embryonic lung or human newborn foreskin	20 to 50
Heteroploid	Human epidermoid carcinoma of larynx (HEp-2) or of lung (A549)	Indefinite

dition. Monolayer cultures in screw-cap tubes are most commonly used in clinical laboratories, as described below. However, when large numbers of samples are processed, 24-well plates can be more efficient. Although techniques may vary somewhat for different viruses, in general, the following procedures apply.

1. Pour off the culture medium and inoculate specimens (0.1 to 0.3 ml) into each culture tube. Uninoculated cultures should be kept in parallel for comparison.

2. Allow the specimen to adsorb in the incubator at 35°C for 30 to 60 min. Then 1.0 to 1.5 ml of maintenance medium should be added, and the inoculated cultures should be returned to the incubator. Inoculated cultures can be placed in a rotating drum if available, which is optimal for the isolation of respiratory viruses, especially rhinoviruses, and results in earlier appearance of cytopathic effect (CPE) for many viruses. More rapid rotation of cultures also can enhance the speed and sensitivity of virus recovery (Mavromoustakis et al., 1988). If stationary racks are used to conserve space, it is critical that culture tubes be positioned so that the cell monolayer is bathed in nutrient medium; otherwise, the cells will degenerate, especially at the edge of the monolayer.

TABLE 2 Variation in sensitivity of cell cultures to infection by viruses commonly isolated in clinical virology laboratories

Virus	Cell culture[a]			
	PMK	HDF	HEp-2	A549
RNA virus				
Enterovirus	+++	++	+/−	+/−
Rhinovirus	+	+++	+	−
Influenza virus	+++	+	−	−
RSV	++	+	+++	+/−
DNA virus				
Adenovirus	+	++	+++	+++
HSV	+	++	++	+++
VZV	+	+++	−	+++
CMV	−	+++	−	−

[a] PMK, primary MK. Degree of sensitivity: +++, highly sensitive; ++, moderately sensitive; +, low sensitivity; +/−, variable; −, not sensitive.

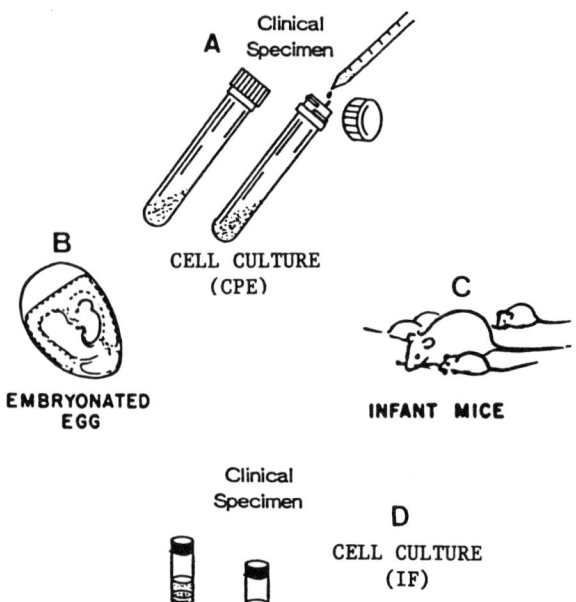

FIGURE 1 Host systems for virus isolation. (A) tissue culture method (CPE); (B) embryonated eggs; (C) newborn mice; (D) tissue culture (shell vial centrifugation). (Reprinted from Hsiung et al., 1994, with permission of the publisher.)

TABLE 3 Supplies and equipment needed for isolation of viruses in cell culture

Procedure	Supplies and equipment needed
Inoculation of cell cultures	Laminar flow hood, centrifuge, pipettes, automatic pipetting device, pipette jar and discard can, disposable gloves, disinfectant, and sterile glass- and plasticware
Maintenance of cell cultures	Culture media, serum, antibiotics, 4°C refrigerator, test tube racks and/or rotating drum, 35°C incubators, CO_2 incubator, water bath, and upright and inverted microscopes
Staining of shell vials and identification of isolates	Centrifuge, centrifuge tubes, PBS, monoclonal antibodies and reagents, and fluorescence microscope
Preservation and storage of viruses	Freezer vials, ultralow-temperature freezer (-70°C), and DMSO as stabilizer

3. Check inoculated culture tubes daily for the first week, then every other day, for virus-induced CPE. Compare the inoculated tubes with uninoculated control tubes from the same lot of cell cultures.

4. Certain specimens, such as urine and stool, frequently will be toxic to the cell cultures, and this toxicity can be confused with virus-induced cytopathology. With such specimens, it is a good practice to check inoculated tubes, either after adsorption or within 24 h of inoculation, and refeed them with fresh medium if necessary. If toxic effects are extensive, it may be necessary to subpassage the inoculated cells in order to dilute the toxic factors and provide viable cells for virus growth.

5. Bacterial or fungal contamination will require filtration, using a 450 μm-pore-size filter, of either the inoculated culture supernatant or the original specimen, followed by inoculation of fresh cultures.

6. Inoculated cultures and the uninoculated controls are generally kept for observation for virus-induced effects for 10 to 14 days. Exceptions include cultures for herpes simplex virus (HSV) only, which may be terminated at 7 days, and for cytomegalovirus (CMV), which are commonly kept for 3 to 4 weeks. During this time, cell cultures may need to be refed to maintain the cells in good condition. Some cultures, such as HEp-2 cells, may require refeeding or subculturing every few days. Great care must be taken when refeeding cultures that cross contamination from one specimen to another does not occur. Separate pipettes should be used for separate specimens. Aerosols, spatter, and contamination of test tube caps and gloves should be avoided.

7. When virus-induced effects occur and progress to include 50% of the monolayer, specific identification can usually be obtained by immunofluorescence (IF) of infected cells (see "Identification of Virus Isolates" below). Passage of infected cultures, especially in doubtful cases, into a fresh

culture of the same cell type may be necessary to ensure recovery of virus for further identification of the isolate. For certain cell-associated viruses, such as CMV or varicella-zoster virus (VZV), it is necessary to trypsinize and passage intact infected cells. Adenovirus can be subcultured after freezing and thawing infected cells, which disrupts the cells and releases intracellular virus.

8. For certain fastidious viruses, when the amount of infectious virus in the specimen is low or when the patient has received antiviral therapy, blind passage (i.e., subculture of the inoculated culture in the absence of virus-induced effects) into a set of fresh culture tubes may be necessary before virus growth can be detected.

In order to enhance viral replication and provide a more rapid and sensitive diagnosis, herpesvirus-infected cell cultures have been treated with dimethyl sulfoxide (DMSO) and/or dexamethasone. The results reported have been mixed, with some authors reporting enhancement (West et al., 1989) and others reporting no effect (Espy et al., 1988). This discrepancy may be explained by the age of the treated cells. DMSO and dexamethasone appear to result in more rapid and extensive viral CPE when confluent, static cells are used but not when actively growing subconfluent cell cultures are inoculated.

■ Detection of Virus-Induced Effects
CPE

Many viruses can be identified by the characteristic cellular changes they induce in susceptible cell cultures. These can be visualized under the light microscope. Examples of CPEs characteristic of a number of common viruses are shown in Fig. 2 and described in greater detail in the sections on individual viruses below. The degree of CPE is usually graded from + to + + + + as it progresses from involving less than 25% of the cell monolayer (+) to 50% (+ +), 75% (+ + +), and finally 100% (+ + + +). There are two important points that should be emphasized regarding CPE induced by virus:

1. The rate at which the CPE progresses may help to distinguish similar viruses; for example, HSV progresses rapidly to involve entire monolayers of several cell systems (Fig. 3), whereas two other herpesviruses, CMV and VZV, grow primarily in HDFs and progress slowly over a number of days or weeks.

2. The type of cell culture(s) in which the virus replicates is important; that is, although the CPEs may be similar within a virus group, the susceptibility of different cell types to different viruses may differ greatly. For example, both polioviruses and echoviruses induce similar CPEs in primary rhesus MK (RhMK) cells; however, echoviruses do not induce CPE in HEp-2 cells, thus allowing presumptive identification (Fig. 4).

It should be cautioned that virus-induced CPE must be distinguished from "nonspecific" CPE caused by toxicity of specimens, contamination with bacteria or fungi, or old cells. A subculture onto fresh cells should amplify virus effects and dilute toxic effects. On occasion, foci of cells inoculated onto the culture monolayer from the original specimen or from another cell culture can be mistaken for viral CPE. With experience, the appearance of the cellular changes, taken together with the susceptible cell systems, the specimen source, and clinical disease, usually allows a presumptive diagnosis to be made as soon as the virus-induced cellular changes occur.

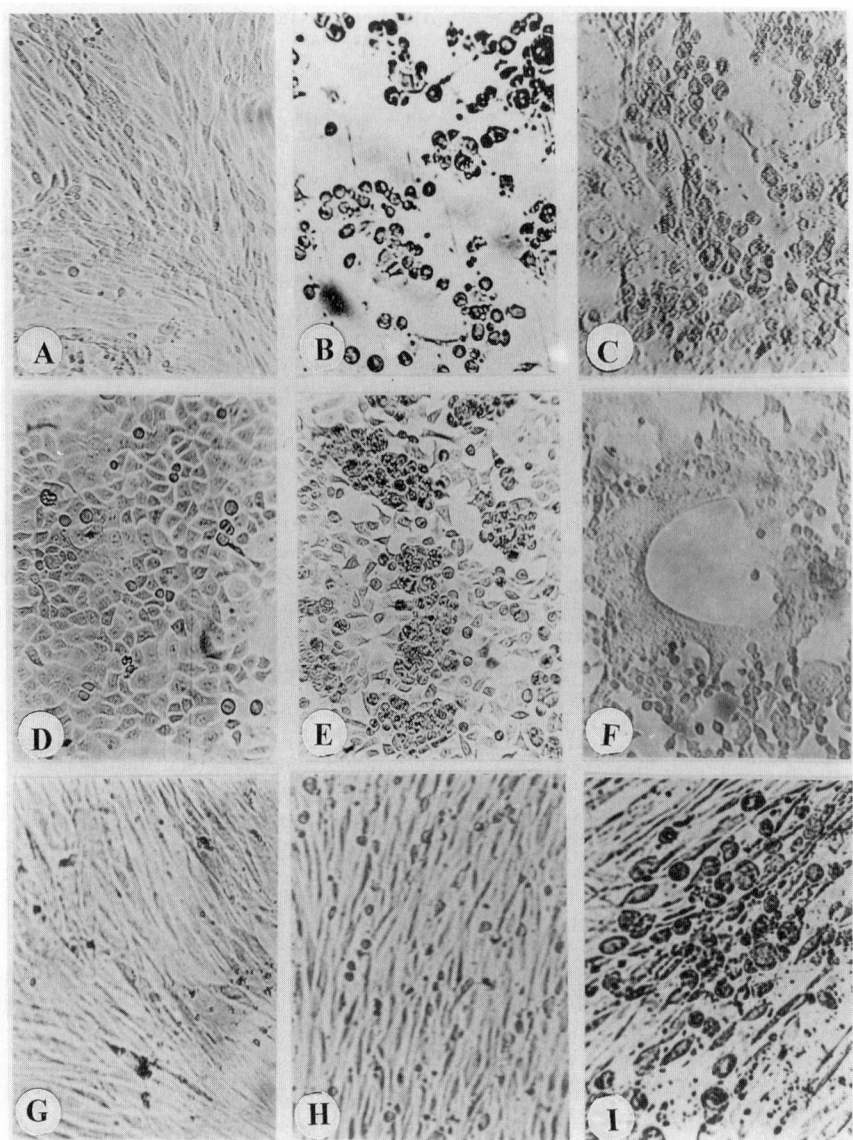

FIGURE 2 Examples of characteristic CPEs of different viruses. (A) Uninfected RhMK cells; (B) poliovirus CPE in RhMK cells; (C) influenza B virus CPE in RhMK cells; (D) uninfected HEp-2 cells; (E) adenovirus CPE in HEp-2 cells; (F) RSV CPE in HEp-2 cells; (G) uninfected HDFs; (H) rhinovirus CPE in HDFs; and (I) CMV CPE in HDFs.

Hemadsorption

Parainfluenza and sometimes influenza viruses may not induce distinctive cellular changes; however, these viruses possess hemagglutinins, which have an affinity for red blood cells (RBC) and are expressed on the surfaces of infected cells. When a freshly obtained guinea pig RBC suspension is added to the infected cultures, the RBC adsorb onto the infected cells, resulting in a hemadsorption phenomenon, as shown in Fig. 5B. Caution should be exercised when aged guinea pig RBC are used, however, because nonspecific hemadsorption often occurs in an uninoculated culture (Fig. 5C) and should be distinguished from that resulting from a specific viral infection. Furthermore, the hemadsorption test is usually performed at 4 or 22°C because the RBC will elute when incubated at 37°C. When a culture shows positive hemadsorption, infected cells are transferred to a slide and stained by IF to identify the causative virus. Alternatively, the culture fluid can be subcultured into a fresh culture to confirm the virus isolation and to permit further identification.

It should be noted that not all viruses that agglutinate RBC can adsorb them onto infected cell monolayers, because hemadsorption is a property of those viruses that bud from the host cell membrane during maturation and thus express viral hemagglutinin on the surface of the infected cell. The technique is further described in chapter 13.

Immunostaining

To more rapidly detect virus growth, shorten the observation period, and avoid repeated examinations for CPE, intact cell culture monolayers can be stained with fluorescein- or horseradish peroxidase-labeled antibodies to viral

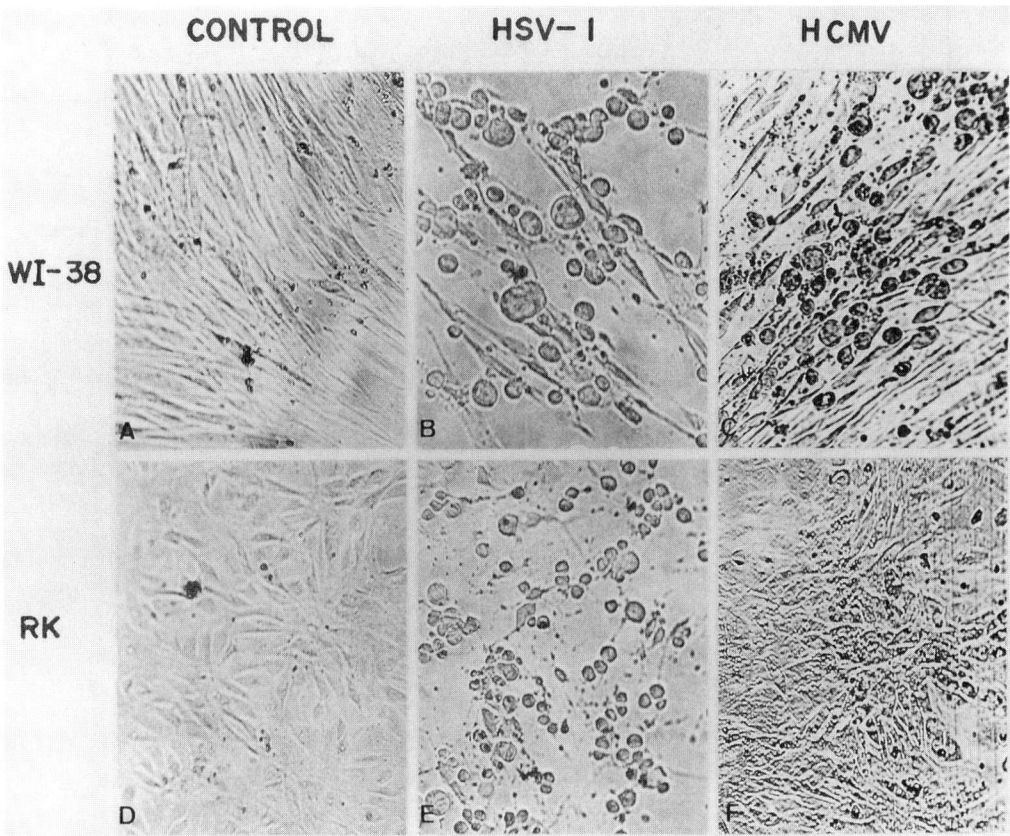

FIGURE 3 Cell sensitivity and rate of progression of CPE of two herpesviruses: HSV-1 and human CMV (HCMV). (A) Uninfected WI-38 cells; (B) extensive HSV-1 CPE in WI-38 cells 2 days postinoculation; (C) CMV CPE in WI-38 cells 1 week postinoculation; (D) uninfected (RK) cells; (E) extensive HSV-1 CPE in RK cells 1 day postinoculation; (F) absence of CMV CPE in RK cells 2 weeks postinoculation. (Reprinted from Hsiung [1982], with permission.)

antigens, usually within 1 to 6 days after inoculation. Immunostaining is most readily accomplished when cell cultures are grown in plates or shell vials rather than roller tubes and is usually performed when only one virus is being sought (Miller and Howell, 1983). Alternatively, at the end of the observation period and in the absence of positive CPE or hemadsorption, cells can be scraped or trypsinized from roller tubes and blindly stained for detection of viral antigens prior to being discarded. With the availability of pooled monoclonal antibodies, multiple viral pathogens can now be detected with a single immunostain (Rabalais et al., 1992).

Interference

In addition to detection by immunostaining, certain viruses that do not readily induce CPE in infected cultures can also be detected by their ability to interfere with the growth of a second virus inoculated into the same culture. This test has been used most frequently in the detection of rubella virus and is referred to as the "interference phenomenon." African green monkey kidney cell cultures infected with rubella virus commonly do not show CPE. After a 10-day incubation, a standard dose of a challenge virus, such as echovirus 11, is inoculated into the same tube, as well as into a control tube without rubella virus, and incubated for two or more additional days. The infection of these cells by

rubella virus will inhibit the replication of a superinfection by echovirus 11. Thus, echovirus CPE will be observed in the control tube but will not be observed in rubella virus-infected cultures, due to the presence of an interfering agent.

Centrifugation Culture (Shell Vial Technique)

Conventional virus isolation requires observation of roller tube cultures for CPE, which can take weeks to appear. However, the rapid diagnosis of viral infections is increasingly important in patient management. One of the most significant contributions to rapid diagnosis in the clinical laboratory has been the application of centrifugation cultures to viral diagnosis.

For a number of years, it has been recognized that low-speed centrifugation of monolayers enhances infectivity of viruses (Hudson et al., 1976) as well as chlamydia. The mechanism for this effect is unclear and may involve centrifugation of virus aggregates or of virus attached to cell debris or an effect on cell membranes to enhance virus entry.

In 1984 the use of centrifugation cultures followed by staining with a monoclonal antibody at 24 h postinoculation was first reported for CMV (Gleaves et al., 1984) (Fig. 6). Subsequent reports have documented its usefulness in rapid diagnosis of other viruses, including HSV, VZV,

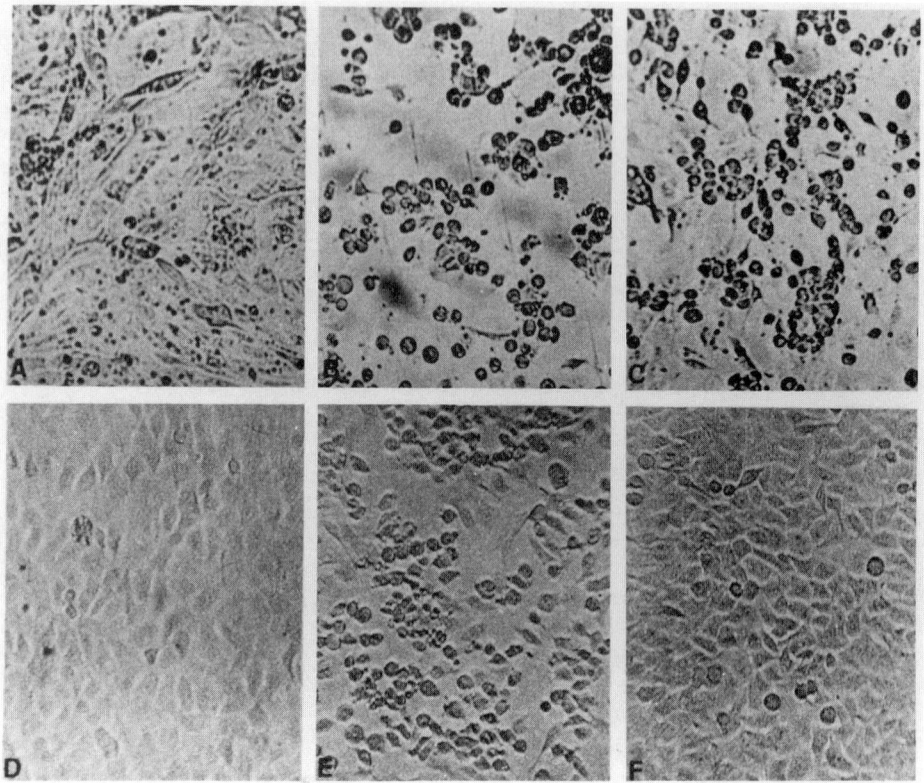

FIGURE 4 Differential sensitivity of cell cultures to enteroviruses. (A) Uninfected RhMK cells; (B) poliovirus-infected RhMK cells showing advanced CPE; (C) echovirus-infected RhMK cells showing advanced CPE; (D) uninfected HEp-2 cells; (E) poliovirus-infected HEp-2 cells showing advanced CPE; (F) echovirus-infected HEp-2 cells showing absence of CPE. (Reprinted from Hsiung et al., 1994, with permission of the publisher.)

adenovirus, respiratory viruses, and polyomavirus BK (Gleaves et al., 1985, 1988; Espy et al., 1987; Marshall et al., 1990; Olsen et al., 1993). In addition, when the inoculum is standardized, semiquantitative results can be obtained by counting the number of virus-positive cells (Buller et al., 1992; Slavin et al., 1992).

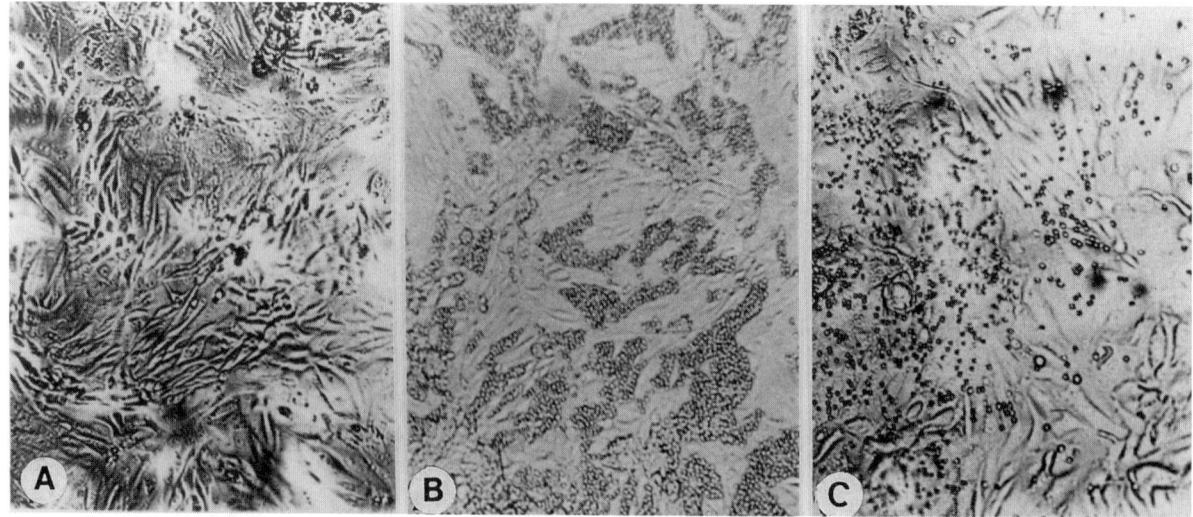

FIGURE 5 Hemadsorption of guinea pig RBC by parainfluenza virus in MK cells. (A) Uninfected MK cells; (B) specific hemadsorption in parainfluenza virus-infected MK cells; (C) nonspecific hemadsorption seen with aged RBC in uninfected cell cultures. (Modified from Hsiung et al., 1994.)

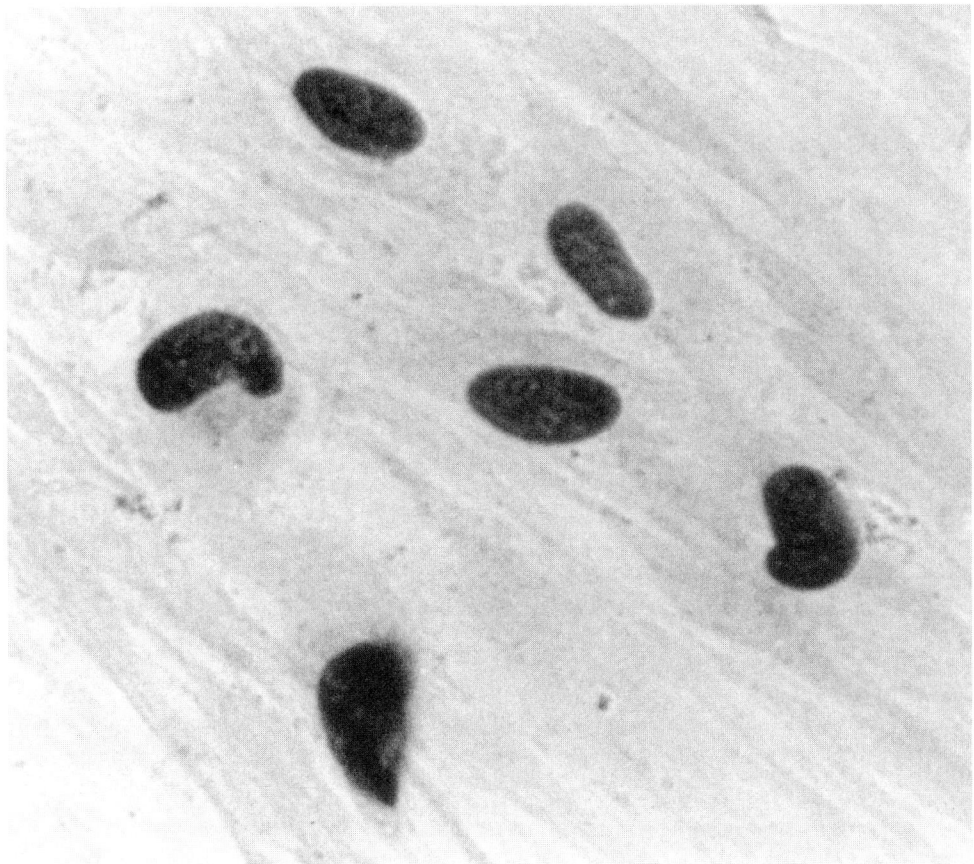

FIGURE 6   Centrifugation culture; detection of CMV early antigens in infected nuclei at 16 to 24 h postinoculation (immunoperoxidase stain).

The shell vial technique combines (i) cell culture to amplify virus in the specimen, (ii) centrifugation to enhance viral infectivity, and (iii) early detection of virus-induced antigen (before CPE) by the use of high-quality specific antibodies. It can be used for any virus that replicates in cell culture and for which a specific antibody is available. For viruses with a long replication cycle, such as CMV, viral antigens produced early in the replication cycle can be detected many days before CPE is apparent by light microscopy. For viruses that replicate faster, e.g., HSV, or if the antibodies available are directed toward late rather than early replication products, less time is gained by using the shell vial technique (Table 4).

Although centrifugation culture enhances the rapidity of diagnosis of viral infections, numerous studies have demonstrated that for maximal sensitivity, both conventional culture and centrifugation cultures should be performed in parallel (Rabella and Drew, 1990; Espy et al., 1991; Rabalais et al., 1992; Olsen et al., 1993). The overall sensitivity of the shell vial technique varies with the type of specimen (Paya et al., 1987), the length and temperature of centrifugation (Shuster et al., 1985), the virus, the cell cultures, the antibody employed, and the time of fixation and staining. The use of young cell monolayers (Fedorko et al., 1989) with inoculation of multiple shell vials enhances the recovery rate (Paya et al., 1988). Toxicity, particularly problematic with blood and urine specimens, can lead to cell death and the loss of the monolayer, necessitating blind pas-

sage of the specimen or specimen reinoculation. Furthermore, with all rapid techniques that use specific antibodies

TABLE 4   Time to virus detection by conventional and centrifugation cultures[a]

Virus	Time (days) to virus detection	
	Conventional culture [avg (range)]	Shell vial centrifugation culture[b]
Respiratory viruses		
RSV	6 (2–14)	1–2
Influenza A virus	2 (1–7)	1–2
Influenza B virus	2 (1–7)	1–2
Parainfluenza virus 1–4	6 (1–14)	1–2
Adenovirus	6 (1–14)	2–5
Rhinovirus	5 (2–14)	Not available
Herpesviruses[c]		
HSV	2 (1–7)	1–2
VZV	6 (3–14)	2–5
CMV	8 (1–28)	1–2

[a] Modified from Landry, 1997.
[b] Incubation time prior to fixation and immunostaining with monoclonal antibodies.
[c] Isolation of EBV and HHV-6 requires cocultivation with umbilical cord lymphocytes. Routine diagnosis is by antibody response.

or nucleic acid probes to detect viruses, only the virus sought will be detected. Conventional isolation using a spectrum of cell cultures can detect a variety of virus types, including unanticipated agents (Blanding et al., 1989). However, the recent use of antibody pools, which can detect between two and seven different viruses, has facilitated the rapid detection of multiple agents (Brumback et al., 1993; Olsen et al., 1993; Engler and Preuss, 1997) and increased the versatility of shell vial cultures.

■ Reagents and Equipment

Antibodies to specific viral types, either fluorescein or peroxidase labeled

Cold acetone

Shell vials; 1 dram, 15 by 45 mm, with caps or stoppers

Coverslips; 12-mm diameter

Cell cultures grown on coverslips in shell vials and sensitive to the suspected viruses

Low-speed centrifuge with adapters for shell vials

Humidified chamber

Rotator or rocker

Suction flask and vacuum source

■ Test procedure

Inoculation of Shell Vials

1. Prepare three shell vials for blood specimens suspected of containing CMV; for all other specimens and viruses, prepare two shell vials.
2. Remove cap and aspirate medium from shell vial.
3. Inoculate prepared specimen onto monolayer; 0.2 to 0.3 ml per vial.
4. Replace cap and centrifuge (30 to 60 min at 700 × g).
5. Aspirate inoculum for blood, urine, and stool samples and then rinse with 1 ml of medium to reduce toxicity.
6. Add 1.0 ml of maintenance medium to each shell vial and incubate at 35°C for 1 to 2 days.

Fixation of Coverslips in Shell Vials

1. Before fixation, inspect the coverslips for toxicity, contamination, etc. If necessary, passage the cell suspension to a new vial and repeat incubation before staining.
2. If the monolayer is intact, aspirate the medium from the shell vials and rinse once with 1.0 ml of phosphate-buffered saline (PBS). If the monolayer appears fragile, do not rinse with PBS.
3. Aspirate medium completely, add 1.0 ml of cold acetone to each shell vial and allow cells to fix for 10 min.
4. Aspirate the acetone and allow the coverslips in the shell vial to dry completely.

Staining of Coverslips

1. Add 1.0 ml of PBS to each coverslip and then aspirate the PBS.
2. Pipette 150 μl of antiserum (appropriately titrated and diluted) into the shell vial. Replace the cap.
3. Rock the tray holding the shell vials to distribute the reagent and then check to see that no coverslips are floating above the reagent.
4. Place the rack holding the shell vials in a humidified chamber in the 35°C incubator.
5. Incubate for 30 min.
6. Add 1.0 ml of PBS to the shell vial and then aspirate the PBS. Pipette a second 1.0 ml of PBS into the shell vial, allow the monolayer to soak for 5 min, and then aspirate the PBS.

For direct assays (primary antibody is labeled), go directly to step 9. For indirect assays (primary antibody is not labeled), continue with steps 7 through 10.
7. Pipette 150 μl of labeled conjugate onto the monolayer.
8. Repeat steps 3 through 6.
9. Add distilled water to the shell vial. Using forceps and a wire probe, remove the coverslip and blot it on tissue or absorbent paper, e.g., a Kimwipe. (Exposure of the slide to distilled water should be kept to less than 30 to 45 s.)
10. Add 1 drop of mounting fluid to a properly labeled slide and place the coverslip on the mounting fluid with the cell side down, being careful not to trap air bubbles.

Reading Procedure

Coverslips are examined using a 20× objective with a fluorescence microscope equipped with the appropriate filters to maximize detection of the fluorescein isothiocyanate label (or a light microscope if a peroxidase label is used). A known positive control is run for each viral antigen with each assay. Noninfected monolayers are fixed and stained as negative antigen controls. For indirect IF assay, normal goat serum, or PBS plus fluorescein isothiocyanate conjugate, is used as a negative serum control.

The pattern of fluorescence varies depending upon the virus sought, the antibody used, the cell culture, and the stage of virus replication. Even a single cell, characteristically stained, is considered a positive result. The test should be repeated if (i) the staining pattern is not typical for the virus sought, (ii) nonspecific staining is observed, or (iii) the staining color is more yellow than green.

Virus Assay and Identification

Virus Infectivity Assay by the Endpoint of CPE

At times it is necessary to quantitate the amount of infectious virus present in a specimen or a cell culture. The specimen can be assayed by determining the highest dilution of the fluid that produces CPE (or hemadsorption or interference) in 50% of the cell cultures inoculated; this endpoint is the 50% tissue culture infectious dose ($TCID_{50}$), determined as follows.
1. Add 0.9 ml (each) of Hanks' balanced salt solution (HBSS) to six sterile test tubes.
2. Add 0.1 ml of virus suspension to the first dilution tube, mix it thoroughly, and transfer 0.1 ml of the mixture to the next tube.
3. With a separate 1-ml pipette, mix the suspension and transfer 0.1 ml to the next tube; continue this process with all six tubes, using separate pipettes.
4. Inoculate 0.1 ml of each dilution of the virus suspension into a tube of a sensitive cell culture, four tubes per dilution (use a separate pipette for each dilution).
5. After the appropriate incubation period, record the number of tubes at each dilution with CPE.
6. Determine the 50% endpoint by the method of Reed and Muench (1938) (Table 5). The calculation is as follows: 50% endpoint = (percent with CPE of >50% − 50%)/(percent with CPE of >50% − percent with CPE of <50%) = (83 − 50)/(83 − 40) = 0.7. Therefore, at a dilution of $10^{-4.7}$, 0.1 ml of this virus will contain 1 $TCID_{50}$. At a dilution of $10^{-2.7}$, 0.1 ml will contain 100 $TCID_{50}$.

Plaque Formation

Many viruses that produce CPE, and also certain viruses that do not produce detectable CPE under fluid medium in cell cultures, may be detected by their ability to form plaques in cell monolayers under a solid or semisolid medium, such as agar, agarose, or methyl cellulose (Hsiung et

TABLE 5 Example of determination of TCID$_{50}$

Virus dilution	No. of cultures[a]		Cumulative no. of cultures[b]		CPE ratio	% Cultures with CPE[d]
	With CPE	No CPE	With CPE[c]	No CPE		
10^{-3}	4	0	9	0	9/9	100
10^{-4}	3	1	5	1	5/6	83
10^{-5}	2	2	2	3	2/5	40
10^{-6}	0	4	0	7	0/7	0

[a]Four cultures inoculated per virus dilution.
[b]Cumulative number of cultures is obtained by adding the numbers in the direction indicated by the arrows in the corresponding columns at left.
[c]It is assumed that a culture showing CPE at a given dilution of virus also shows CPE at a higher virus concentration.
[d]In the example shown, the 50% endpoint lies between the virus dilutions 10^{-4} and 10^{-5}. For calculation of the endpoint, see the text.

al., 1994). For example, monolayers can be fixed and stained with crystal violet after the medium is removed to visualize plaques (Fig. 7). Virus plaques can be manifest as clear areas where infected cells have detached (e.g., HSV) or as darkly stained foci of infected cells (e.g., CMV). Virus particles from a focus of infection are localized by the solid overlay medium, and the virus spreads from infected cells only to adjacent cells, resulting in discrete foci of infection. Because one infectious unit is capable of initiating one plaque, this technique can be used both for accurate quantitative assay of virus infectivity and for purification of virus strains. Plaque assays are not commonly used in a clinical laboratory at the present time; therefore, detailed procedures are not included in this chapter and the reader is referred to the reference list at the end of the chapter.

It should be noted, however, that as demand for antiviral sensitivity testing of virus isolates such as HSV has increased, plaque reduction assays are more frequently performed (Safrin et al., 1994). Figure 7 shows the reduction in HSV-2-induced plaques in the presence of increasing concentrations of acyclovir (ACV). The cell system used in antiviral assays has a significant effect on the amount of drug required to inhibit the virus (Hu and Hsiung, 1989). Highly sensitive cell systems, such as guinea pig embryo (GPE) cells, require more ACV to effect a 50% reduction in plaques than less sensitive chicken embryo (CE) and Vero cells. For HSV antiviral assays, Vero cells are the most commonly used.

Identification of Virus Isolates

Presumptive identification can usually be made on the basis of characteristic virus-induced effects (e.g., the type of CPE or hemadsorption) and selective cell sensitivity. For final identification, staining of infected cells with either fluorescein- or peroxidase-labeled antibodies is most commonly used. Monolayers showing CPE or hemadsorption are trypsinized or dislodged, and the cells are transferred to a welled slide and then stained with specific antibody. Nucleic acid hybridization, enzyme-linked immunosorbent assay (ELISA), latex agglutination, or more recently PCR with sequencing, can also be used for virus identification. When determination of the specific type is requested for enteroviruses or adenoviruses, neutralization of virus-induced cytopathology in cell culture can be performed. These tests will be discussed in greater detail in subsequent chapters.

Occasionally, a new isolate cannot be identified by the standard tests. The morphological properties of the infecting virus can be determined by electron microscopy (EM), if available, with subsequent identification by molecular techniques. Alternatively, the more basic properties of the new agent, such as the nucleic acid type, the size of the isolate, and the presence or absence of a lipid envelope, can be

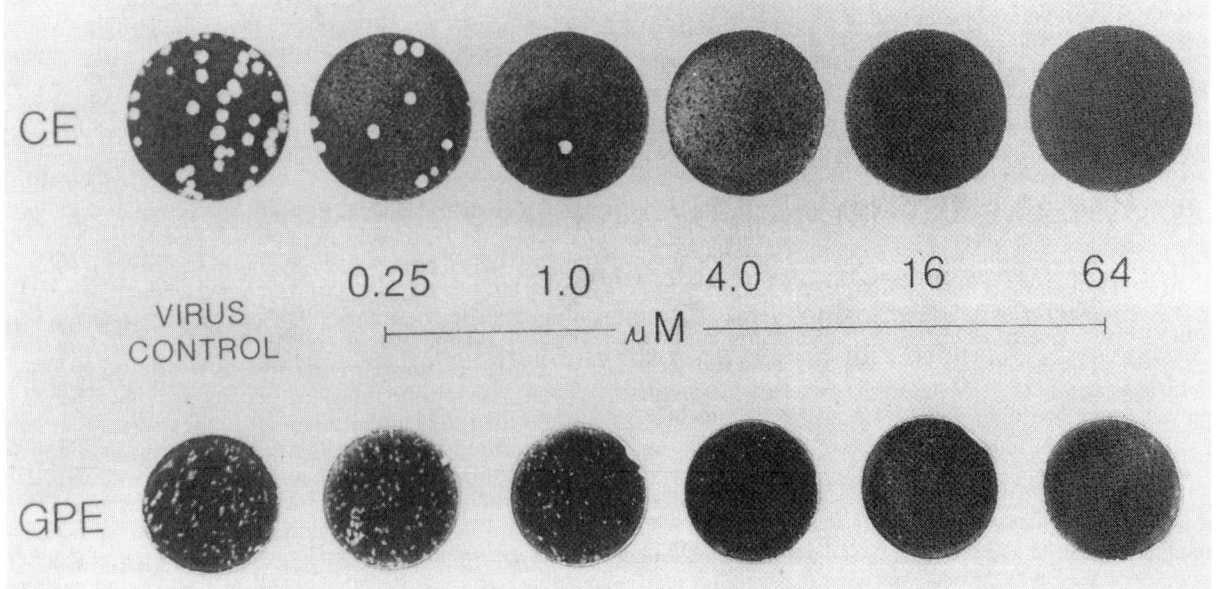

FIGURE 7 ACV inhibition of HSV-2-induced plaque formation in CE cells and GPE cells. Note that in more sensitive cell systems (GPE cells), a higher concentration of ACV is required to inhibit viral plaques than in the less sensitive CE cells. (Reprinted from Landry et al., 1989, with permission of the publisher.)

determined (Hsiung et al., 1994). The nucleic acid type can be determined by exposure to 5'-bromodeoxyuridine, an inhibitor of DNA viruses, followed by assay of virus infectivity. The virus size can be estimated by passing infected culture fluids through a series of filters, and the presence of a lipid envelope can be determined by exposure of the virus to ether followed by reinoculation into culture to determine if infectivity has been lost.

Newer Culture Methods

Mixed-Cell Cultures

In order to detect more viruses with fewer cell cultures and to eliminate the need for primary MK cells, mixed-cell cultures have been developed (Y. Huang and B. Turchek, 14th Annu. Clin. Virol. Sympo, poster M15, 1998; Y. Huang, S. Hite, and V. Duane, 15th Annu. Clin. Virol. Symp., poster T29, 1999). There is at present a variety of combinations of cultures to choose from, depending upon the viruses sought. These include a mixture for HSV and VZV (MRC-5 and CV-1 cells), a mixture for respiratory viruses (mink lung [ML] and A549 cells), and two mixtures (human rhabdomyosarcoma [RD] and H292 cells and Buffalo green monkey kidney [BGMK] and A549 cells) for enteroviruses. Mixed cells can be used either in conventional culture or in centrifugation culture with early immunostaining using pooled antibodies. Results to date have been very promising.

Genetically Modified Cell Lines

Genetic modification of cell lines is an emerging technology with potential for the diagnostic laboratory (Olivo, 1996). Genetic elements derived from viral, bacterial, or cellular sources are stably introduced into a cell, and when the target virus enters the cell, an event in the viral replication cycle triggers the production of a measurable enzyme. In a simple histochemical assay, infected cells stain a characteristic color. This approach has been given the acronym enzyme-linked virus-inducible system (ELVIS) and has been shown to be feasible for both DNA and RNA viruses, although different strategies are necessary for enzyme induction. In contrast to CPE, ELVIS can be read by an untrained observer, and the earliest stages of infection can be reliably detected.

The ELVIS method for detection of HSV appears to be simple, sensitive, and rapid, and it is an attractive assay, especially for laboratories processing large numbers of specimens. HSV ELVIS can be used for the simultaneous detection, identification, and typing of HSV isolates from clinical specimens (M. A. Lewinski, S. M. Mendoza, D. A. Bruckner, and D. R. Scholl, 37th ICAAC, poster H-96, 1997). However, ELVIS remains somewhat less sensitive than the most sensitive of conventional cell culture systems when specimens contain only a few infectious HSV particles.

ELVIS culture methodology is also amenable to automation, with some advantages over standard immunological methods, such as ELISA: it can provide more sensitive results and can detect infectious virus. Lastly, genetically modified cell lines can also facilitate antiviral susceptibility testing.

Lymphocyte Cultures

Although not available in routine diagnostic laboratories, human lymphocyte cultures are used in research and in some specialized reference laboratories to isolate several human herpesviruses (Epstein-Barr virus [EBV], human herpesvirus type 6 [HHV-6], and HHV-7), and retroviruses (human immunodeficiency virus types 1 and 2 [HIV-1 and HIV-2] and human T-cell leukemia virus types 1 and 2 [HTLV-1 and HTLV-2]). Human leukocytes for cocultivation can be obtained from seronegative donors or from cord blood of uninfected newborn infants.

EBV is able to transform uninfected human leukocytes into continuous cell lines, and EBV nuclear antigen can be detected in the nuclei of transformed cells. Unfortunately, in vitro transformation of cord blood lymphocytes by EBV in clinical specimens usually requires 30 to 90 days of incubation.

HTLV-1 and -2 and HHV-6 and -7 can be isolated by cocultivating samples with cord blood lymphocytes stimulated with phytohemagglutinin or interleukin 2. Identification is done by PCR or with monoclonal antibodies, as well as by reverse transcriptase assay of supernatant fluid for HTLV.

Since HIV lyses CD4 lymphocytes, successful isolation necessitates repeated additions of fresh, stimulated, uninfected peripheral blood lymphocytes to cultures. Virus growth is detected by periodically assaying supernatant fluid for reverse transcriptase or viral antigen or by EM. Continuous mature T-cell lines from leukemic patients are useful in propagating HIV laboratory strains but are not as sensitive for primary isolation of wild-type viruses.

Advantages and Limitations of Cell Culture

The advantages of cell culture for virus diagnosis include relative ease compared with animal inoculation, a broad spectrum and sensitivity compared with other available diagnostic methods, and the recovery of unknown or unexpected infectious virus(es) that may be present in the specimen (Mackenzie, 1999). It is limited by the inherent time delay required for virus growth, by the difficulty in maintaining cell cultures, by the sometimes variable quality of cultures, and by the decreased sensitivity of cell lines at higher passage levels. Contamination with adventitious agents occurs, including bacteria, fungi, and mycoplasma, which can inhibit the growth of viruses in clinical specimens (Hsiung, 1968; Smith, 1970; Stanbridge, 1971; Chu et al., 1973). Endogenous viruses that are latent in the tissue culture or the calf serum can be reactivated during cultivation and cause CPE or hemadsorption and thus can be confused with virus isolated from the patient's specimen (Fong and Landry, 1992). Viruses potentially contaminating primary MK cell cultures, such as herpes B virus, can pose a serious health risk as well. Furthermore, some common viruses do not as yet produce identifiable effects in readily available cell cultures—for example, hepatitis viruses, rotavirus, Norwalk virus, some group A coxsackieviruses, and togaviruses—so that other methods of detection are necessary.

To get the best results from primary isolation in cell culture, it is most important to maintain healthy cell cultures and to have a spectrum of cell types available. In general, a primary MK cell culture, an HDF strain, and a human heteroploid cell line (e.g., A549) constitute a satisfactory combination. If the isolation of a particular virus is a high priority, the most sensitive system available should be selected.

Another problem in isolating certain viruses, especially of the orthomyxo- and paramyxovirus groups, is the presence of inhibitory substances and/or antibodies in calf serum used in the cell culture medium (Krizanova and Rathova, 1969). Ideally, maintenance medium for inoculated cultures should be serum free; however, serum is required for long-term maintenance of cells. Using fetal or agamma calf serum reduces this problem but adds to expense. To date, no completely satisfactory, chemically defined medium is available.

In recent years, reducing the time to obtain results has become increasingly important, in order to implement infection control measures and to initiate antiviral therapy.

In addition, hospital stays are shorter, and if patient management is to be affected, results must be available quickly. The need to reduce costs and do more with fewer personnel has created additional pressures. Innovations in virus culture methods that address these concerns have included rapid shell vial centrifugation cultures, use of pooled antibodies for detection of multiple viruses, and, most recently, mixtures of two cell systems in one culture. Thus, virus isolation in cell culture continues to play a major role in viral diagnostic laboratories.

Viruses Commonly isolated in Clinical Laboratories

HSV-1 and HSV-2

HSV-1 and HSV-2 infect a wide variety of cell cultures and animals. However, differences in the sensitivities of cell cultures are evident, particularly when specimens contain low titers of virus (Zhao et al., 1987). Four continuous cell lines, ML, RD, A549, and NCI-H292 cells, have been found to be highly sensitive, more so than many HDF strains (Woods and Young, 1988; Johnston et al., 1990; Hierholzer et al., 1993). Early studies found primary cell cultures, such as rabbit kidney (RK), human embryonic kidney (HEK), and GPE cells, to be very sensitive to HSV infection (Landry et al., 1982). A more recent evaluation, however, found MRC-5 to be more sensitive than RK cells obtained from two commercial suppliers (McCarter and Robinson, 1997), underscoring the need for intermittent re-evaluation of cell susceptibilities in each laboratory. The recently introduced mixtures of two sensitive cell lines in one culture may help address this problem (Huang et al., 15th Annu. Clin. Virol. Symp.).

Vesicular fluids, throat swabs, and genital lesions are the most common sources for virus isolation. HSV produces a rapid degeneration of cells, often appearing within 24 h of inoculation of the cell culture (Fig. 3). The CPE begins as clusters of enlarged, rounded, refractile cells and spreads to involve the entire monolayer, usually within 48 h. The formation of multinucleated giant cells can also be seen with HSV-2 and is more apparent in epithelial than in fibroblast cells. Subcultures are performed by passaging 0.2 ml of supernatant fluid to a fresh culture tube. Over 90% of positives will be identified within 3 to 5 days. Occasionally, CPE develops later and, rarely, will be detected after blind passage.

Centrifugation cultures in shell vials or 24-well plates can be stained from 1 to 3 days after inoculation; however, staining at 1 day may miss some low-titered samples (Espy et al., 1991).

Identification of virus as HSV and differentiation as type 1 or 2 is most readily done by IF or immunoperoxidase stains using monoclonal antibodies (Balkovic and Hsiung, 1985; Miller and Howell, 1983). Genetically engineered cell lines can also be used to isolate and identify HSV (Lewinski et al., 37th ICAAC).

VZV

HDF and A549 cells are the most sensitive for the isolation of VZV, although the virus has also been isolated in other human epithelial cells, primary MK cells, CV-1 cells and occasionally GPE cells. Vesicle fluid and lesion swabs are the usual sources for VZV isolation. The virus is quite labile; therefore, prompt inoculation into cell culture is desirable.

Cytopathology starts as foci of rounded enlarged cells, as seen with HSV; however, the onset and progression are much slower and the foci of CPE tend to progress linearly along the axes of the cells, similar to CMV. However, VZV-infected foci degenerate more rapidly than those infected by CMV. CPE first appears 3 to 7 days after inoculation but may take 2 or 3 weeks. The virus is cell associated, and subpassages are performed by trypsinization and passage of infected intact cells to fresh monolayers of cells. Final identification is by IF with monoclonal antibodies.

Centrifugation cultures are significantly more sensitive than conventional roller tubes (Gleaves et al., 1988; Coffin and Hodinka, 1995). For best results, staining of monolayers at 2 and again at 4 to 5 days after inoculation is recommended. Use of mixed-cell cultures (Huang et al., 15th Annu. Clin. Virol. Symp.) and antibody pools with different fluorescent labels may optimize detection of both HSV and VZV in a single culture (Brumback et al., 1993).

CMV

HDFs are the single most successful conventional culture system for the isolation of CMV. The source of the fibroblasts can be either human embryonic tissues or newborn foreskin cells. The latter, however, lose their sensitivity after the 10th to 15th passage. ML cells have proved very useful for CMV centrifugation cultures, especially for blood samples (Gleaves et al., 1992). Virus can be isolated from a variety of specimens, including urine, saliva, tears, milk, semen, stools, vaginal or cervical secretions, peripheral blood leukocytes, bronchoalveolar lavage fluid, lung, liver, and gastrointestinal biopsy tissues.

CPE may develop within a few days to many weeks, depending on the amount of virus in the specimen. Characteristic CPE consists of foci of enlarged, refractile cells that slowly enlarge over weeks and often do not involve the entire monolayer (Fig. 3C). Thus, it is important that the monolayers be maintained in good condition for at least 3 weeks. On the other hand, when a high titer of CMV, such as is contained in urine samples from congenitally infected babies, is inoculated, one may see generalized rounding at 24 h that can be confused with HSV. Incubation of cultures at 36 instead of 33°C results in more rapid onset of CPE and higher isolation rates (Gregory and Menegus, 1983). For subculture, early passage of intact infected cells is essential. Monolayers should be trypsinized and then dispensed onto fresh uninfected cells. Identification of isolates can be accomplished by IF. Since CPE is slow to advance, passage of trypsinized monolayers into centrifugation cultures followed by staining at 24 or 48 h can provide a more rapid confirmation.

As described above, centrifugation cultures have had a major impact on the rapid diagnosis of CMV infections. However, CMV antigenemia has replaced centrifugation cultures in many clinical laboratories for rapid detection and quantitation of CMV in blood (van der Bij et al., 1988; Landry and Ferguson, 1993). For optimal recovery rates, it is generally recommended that conventional cultures be performed in parallel (Rabella and Drew, 1990).

Adenovirus

In general, human adenoviruses produce CPE in continuous human cell lines, such as A549, and in HDF and HEK cell cultures. Each of these cell systems has disadvantages: the continuous cell lines may be difficult to maintain; HEK cells are often not readily available and are expensive; HDFs are less sensitive, and the changes produced are not characteristic (Mahafzah and Landry, 1989). Nonhuman cells, such as RhMK cells, are of variable sensitivity, and virus growth is slower. Centrifugation cultures can provide a more rapid diagnosis, but staining at 2 days and again at 4 to 5 days may be needed for optimum sensitivity (Espy et al., 1987; Mahafzah and Landry, 1989). Throat swabs, nasopharyngeal swabs, eye swabs, and stool are good sources of virus; the choice depends on the clinical syndrome.

Characteristic CPE consists of grapelike clusters of rounded cells (Fig. 2E), which appear in 2 to 7 days with types 1, 2, 3, 5, 6, and 7. Other adenovirus types may require 3 to 4 weeks or blind passage. Adenovirus remains cell associated, similar to VZV and CMV; however, adenovirus is nonenveloped and stable to freezing and thawing. Therefore, two to three cycles of freezing at 70°C and thawing disrupts the cells and releases intranuclear infectious virus. Enteric adenovirus types 40 and 41, associated with gastroenteritis, do not grow readily in A549 cells or HDFs but can be isolated in H292 cells (Hierholzer et al., 1993).

Identification of isolates as adenoviruses can be done by IF with antihexon antibody. Neutralization tests with type-specific antiserum or molecular analysis will identify virus types.

Enteroviruses

In general, enteroviruses grow best in epithelial cells of primate origin. Polio- and coxsackie B viruses grow well in primary MK, HEp-2, and BGMK cells, and echovirus grows well in primary MK and RD (a rhabdomyosarcoma cell line) cells but not in HEp-2 cells (Table 6). The universal host for coxsackie-virus group A is the newborn mouse; however, some strains grow in HDF, HEK, MK, or RD cells. Since inoculation of multiple cell types optimizes enterovirus detection (Dragan and Menegus, 1986), the use of mixed-cell cultures may result in greater yield while conserving time and resources. Enteroviruses can be recovered from feces, throat swabs, cerebrospinal fluid, blood, vesicle fluid, conjunctival swabs, and urine.

Characteristically, infected monolayer cells round up, become refractile, shrink, degenerate, and then detach from the surface of the culture vessel (Fig. 2B). Virus in the supernatant fluid can be subpassaged. Preliminary identification can be determined by characteristic CPE and differential cell susceptibility (Hsiung et al., 1994; Johnston and Siegel, 1990). Fluorescein-labeled monoclonal antibodies are available for rapid preliminary identification of commonly isolated enteroviruses (Rigonan et al., 1998). Final identification and serotyping by microneutralization tests in cell culture using antiserum pools is expensive and time-consuming and is reserved for reference laboratories. In the future, identification will be performed by molecular techniques (Oberste et al., 1999; Muir et al., 1998).

Rhinovirus

Rhinoviruses are classified as picornaviruses along with the enteroviruses but can be separated from the enteroviruses by their sensitivity to low pH. Many rhinovirus types were originally isolated in organ cultures of the human embryonic trachea. However, they can be isolated in cells of human origin (usually HDFs), certain strains of HeLa cells, and human fetal tonsil cells. The varying sensitivities of different lots of cells can be a problem. Recent work has identified WI-38 and HeLa-I cells as the most sensitive (Arruda et al., 1996). Sources of virus include nasal swabs or washes and throat swabs. Cultivation at 33°C in a roller drum apparatus is optimal, and CPE may occur from the first to the third week of incubation. The CPE is similar to that of the enteroviruses, starts as foci of rounded cells, and spreads gradually (Fig. 2H). However, CPE may not progress and may even disappear; if it is not progressing, subpassage of supernatant fluids from infected cells should be performed. Isolates are identified by characteristic CPE and inactivation at pH 3. Typing by neutralization tests is reserved for research laboratories.

Influenza virus

Primary MK cells are the most widely used cell culture for isolation of influenza virus, although the host range may be increased by addition of trypsin to the medium (Frank et al., 1979). Madin-Darby canine kidney (MDCK), MRC-5, and ML cells (Schultz-Cherry et al., 1998) have all been used successfully, especially in centrifugation cultures (Reina et al., 1997). Influenza virus is also reliably isolated in eggs (Smith and Reichrath, 1974). Nasopharyngeal aspirates and swabs, nasal washings, and throat swabs are good sources for virus and should be collected early in illness, preferably in the first 24 to 48 h. Serum components may inhibit influenza virus from replicating. Therefore, serum should be removed from cell cultures by rinsing them with HBSS before inoculation, and cultures should be maintained in serum-free medium after inoculation. Incubation at 33°C in a roller drum is optimal for isolation. The presence of virus is generally detected by hemadsorption of guinea pig RBC onto infected monolayers (Fig. 5B). CPE is seen with influenza (Fig. 2C), but it usually occurs later than the detection of virus by hemadsorption. Subcultures can be performed by passaging the supernatant fluids. Isolates can be identified as influenza virus A or B by IF using monoclonal antibodies. Subtype and strain identification are determined by hemagglutination inhibition, although subtyping by monoclonal antibody staining has recently been reported (Tkacova et al., 1997).

Parainfluenza Virus

Primary MK cells are the most sensitive system for isolation of these viruses. Parainfluenza virus grows poorly or not at all in hens' eggs. HEK, HDF, and HEp-2 cells are less sensitive. Some success has recently been reported with H292 cells (Hierholzer et al., 1993).

Nasopharyngeal aspirates and swabs, nasal washings, and throat swabs are good sources for virus. Cell cultures should be washed with HBSS before inoculation and refed with medium without serum. Incubation at 33 to 36°C in a roller drum is optimal. The presence of virus is detected by hemadsorption (Fig. 5B), which occurs before CPE. Parainfluenza virus type 2 may produce syncytia, especially in HEp-2 cells. On subculture, parainfluenza virus type 3 may also induce syncytium formation. In those instances when high levels of virus are present, hemadsorption may be detected in the infected cultures within a few days; with specimens containing less virus, 10 days or more of incubation may be necessary. Identification is by IF or by hemadsorption inhibition.

TABLE 6 Host susceptibility to enteroviruses

Virus	No. of serotypes	Sensitivity[a]				
		RhMK	HDF	HEp-2	RD	Newborn mouse
Poliovirus	3	++	++	++	++	−
Coxsackie B virus	6	++	++	++	−	++
Coxsackie A virus	23	+/−	+/−	−	++	++
Echovirus	30	++	+	−	++	−

[a] Degree of sensitivity: + +, sensitive; +, less sensitive; +/−, variable; −, not sensitive.

RSV

Respiratory syncytial virus (RSV) grows best in continuous cell lines, such as HEp-2, in which it produces characteristic syncytia (Fig. 2F). However, syncytium formation is variable and viral replication may be missed. If HEp-2 cells are confluent and 5 to 7 days old when inoculated, syncytia may not form. Rather, nonspecific rounding may occur. Syncytium formation is also dependent upon the presence of adequate levels of glutamine and calcium in the medium (Marquez and Hsiung, 1967; Shahrabadi and Lee, 1988). Primary MK cells show CPE, and HDF cells support RSV growth but are less sensitive and the cytopathology is not characteristic (Arens et al., 1986); however, centrifugation culture can improve detection.

HEp-2 cells have become so difficult to work with that many laboratories rely on rapid diagnostic methods, such as ELISA and IF, to detect most of the positive samples. Virus culture is still useful for optimum detection of RSV and for the recovery of other viral respiratory pathogens (Blanding et al., 1989). RSV is found in respiratory secretions from the nose and oropharynx. Sample collection is important, and RSV is more reliably detected from nasopharyngeal aspirates than from swabs in children (Ahluwalia et al., 1987). Identification is by IF.

APPENDIX
Additional Methods for Virus Isolation

Explant culture or cocultivation

Clinical specimens, such as fresh tissue cells, can be grown out (explanted) or cocultivated with cells susceptible to the suspected agent in the specimen and observed for the development of CPE. The procedure is as follows.
1. Mince tissues finely in 0.25% trypsin.
2. Centrifuge at low speed to pellet the cells.
3. Remove the trypsin.
4. Resuspend in HBSS with antibiotics to a 10 or 20% suspension.
5. For cocultivation, inoculate 0.1 to 0.2 ml onto cell monolayers.
6. For explant culture, resuspend in growth medium with 20% calf serum and disperse into petri dishes or flasks.

Organ culture

For certain fastidious viruses, such as rhinovirus, coronavirus, or rotavirus, isolation can be accomplished using whole organs or part of an organ in vitro, such as human embryonic trachea or intestine, allowing preservation of architecture and/or function. Virus growth may be detected by subculture of nutrient fluids onto monolayers or by observation of cessation of ciliary activity and eventual degeneration of epithelial cells. This technique is tedious and is not routinely used in clinical laboratories. Detailed procedures can be found in the textbooks in the reference list.

Virus isolation in embryonated eggs
Background

The chicken embryo is a highly sensitive host for the primary isolation of several virus types. Some influenza A virus strains are more readily isolated in embryonated eggs than in cell culture, although the reverse is true for influenza B virus strains. The chicken embryo is also a sensitive host for the isolation of mumps virus; however, MK cells are equally sensitive for the rapid isolation of this agent. For primary isolation of influenza and mumps viruses, inoculation of the amniotic sac is necessary. Subsequently, virus isolates can be adapted to grow in the allantoic sac.

The chorioallantoic membrane (CAM) is highly susceptible to pock formation by HSV and poxviruses. With the availability of numerous sensitive cell cultures for the isolation of HSV and vaccinia virus, and with the eradication of smallpox, CAM inoculation is now rarely employed. HSV and poxviruses can be differentiated by the morphological characteristics of the pocks that they produce in this system.

Maintenance and source of eggs

Fertile hens' eggs for virus isolation should be obtained from flocks that are free of infection (Newcastle disease virus and mycoplasma can be particularly troublesome). The eggs should be incubated at 37°C in an atmosphere of 40 to 70% humidity to ensure proper development of the air sac. To prevent adhesion of the embryonic membranes and to keep the embryo centralized, the eggs should be turned two to four times each day. After 4 to 5 days of incubation, the eggs are candled to determine which ones are fertile and contain developing embryos.

Inoculation and incubation

This section is modified from Hsiung et al., 1994.

Amniotic or allantoic cavity inoculation

1. Use 7- to 13-day-old embryonated eggs (optimum for mumps virus, 7 to 8 days; optimum for influenza virus, 10 to 13 days).
2. Candle the eggs to locate the embryo and detect movement; make a puncture through the shell over the air sac.
3. Inoculate 0.1 to 0.2 ml of the specimen into the amniotic sac, which is entered with a quick stabbing motion, and/or into the allantoic cavity (as the needle is withdrawn) using a 1 3/4 in.-long, 23-gauge needle; use three or four eggs per specimen. When the needle is in the amniotic cavity, gentle pressure will move the embryo (Fig. A1A). (These procedures should be performed while the egg is illuminated on a candler.)
4. Seal the hole in the shell with Scotch tape.
5. Incubate the eggs at 35 to 37°C with the air sac uppermost.
6. Candle inoculated eggs daily. Discard those that die within 24 h after inoculation.

CAM inoculation

1. Candle eggs containing 9- to 12-day-old developing chicken embryos; mark an area free from large blood vessels on the side where the embryo is located and the area over the air sac.
2. Drill two slits in the eggshell, one on the side and the other over the air sac.
3. Puncture the shell membrane under the slits with a sterile needle; care should be taken not to damage the CAM.

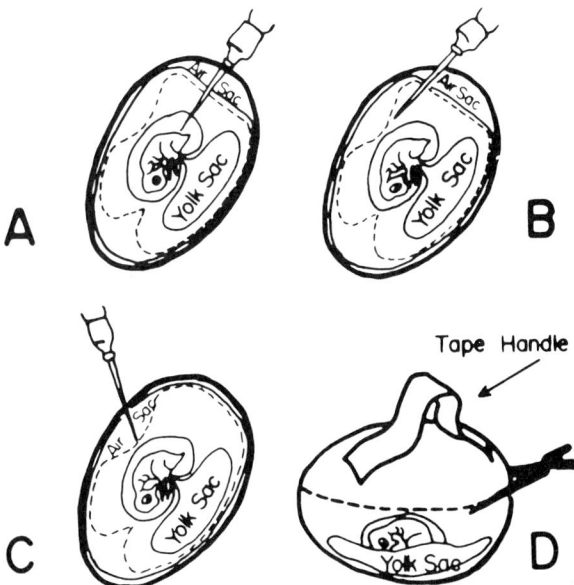

FIGURE A1 Embryonated-egg inoculation and harvesting. (A) Amniotic cavity inoculation; (B) allantoic cavity inoculation. (C) CAM inoculation (note artificial air sac); (D) CAM harvesting. (Reprinted from Hsiung et al., 1994, with permission of the publisher.)

4. With a rubber bulb, gently apply suction at the hole over the air sac. If this procedure is carried out while the egg is being candled, one can see the CAM drop and a new, artificial air sac form.

5. Place 0.05 ml of inoculum on the dropped membrane by inserting the needle through the side slit to a depth of about 5 mm (Fig. A1C). Withdraw the needle very slowly. Seal the opening with Scotch tape or wax and incubate the egg at 35 to 37°C with the artificial air sac uppermost.

6. Candle the inoculated eggs daily; discard those that die within 24 h after inoculation.

Harvesting, assay, and identification of isolates
Amniotic and allantoic fluids

1. Harvest amniotic fluids and allantoic fluids separately, 2 to 4 days after inoculation for influenza virus and 5 to 7 days after inoculation for mumps virus. The procedure for harvesting egg fluids is as follows.
   a. Chill the eggs at 4°C overnight or for 30 min at −20°C and then 2 to 4 h in the refrigerator in order to clot the blood.
   b. Open the eggshell over the air sac.
   c. Cut out and remove the overlying shell membrane and CAM with scissors.
   d. Aspirate amniotic and allantoic fluids separately with sterile capillary pipettes.
2. Carry out a spot hemagglutination test by mixing 0.025 ml of allantoic or amniotic fluid, both undiluted and a 1:5 dilution in PBS with 0.025 ml of a 0.5% suspension of guinea pig or chick RBC.
3. Allow this mixture to stand at room temperature for 35 to 45 min before reading it. If a hemagglutinating virus is present, the RBC will not settle to the bottom but will remain in suspension. Nonspecific hemagglutination can be caused by bacterial contamination.
4. Blind passage of egg fluids in eggs may be necessary before virus is detected.
5. Isolates are identified by the inhibition of hemagglutination using specific antisera (hemagglutination inhibition test).

CAM

1. Place the inoculated egg on a holder so that the slit through which the inoculum was delivered faces upward.
2. Make a tape handle over the area of inoculation (Fig. A1D).
3. Cut off the top half of the eggshell, including the infected area, and gently remove the CAM, which is attached to the shell.
4. Place the infected CAM in a petri dish with a few millimeters of PBS; spread the membrane flat against the bottom of the dish; place the dish on a dark surface to facilitate counting of pocks.
5. Variola, vaccinia, monkeypox, and cowpox viruses and HSV-1 and -2 can be differentiated by the morphological characteristics of the pocks they produce on the CAM. The pocks can be ground and examined by EM or used as antigens for complement fixation or agar gel precipitation tests or for molecular analysis.

Virus Isolation in Mice
Background

Although suckling mice are the definitive hosts for certain viruses, isolation of virus in mice is rarely performed in clinical laboratories today. Group A and B coxsakieviruses were originally isolated in suckling mice and were differentiated by the pathological lesions and illnesses that they produce. Coxsackie B viruses, however, are readily isolated in cell culture. Although no universal cell system for the isolation of group A viruses has been found, several cell systems, such as RD cells and HDFs, support the growth of a number of the coxsackie A virus serotypes.

The suckling mouse is also the universal host for togaviruses, many of which are etiologic agents of encephalitis. Several cell lines and cell cultures derived from insects are now available for many of these viruses.

Inoculation and observation for illness

Pregnant mice are obtained, and the entire litter of mice is inoculated with the clinical specimen within 24 to 48 h of birth, as

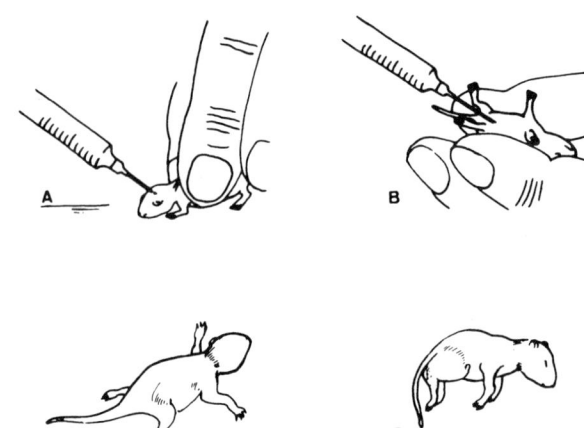

FIGURE A2 Newborn mouse inoculation. (A) Intracerebral inoculation; (B) intraperitoneal inoculation; (C) flaccid hindlimb paralysis as seen with coxsackie A virus infection; (D) spastic paralysis secondary to coxsackie B virus infection. (Modified from Hsiung et al., 1994.)

described above (Fig. A2). Older mice are less susceptible to infection. Mice should be obtained from pretested virus-free mouse colonies (Sendai virus, coronavirus, and many others can be troublesome). Materials needed include animal facilities for housing noninfected and infected animals, needles, syringes, scissors, and forceps.

1. Inoculate newborn mice with the specimen within 24 to 48 h of birth with 0.01 to 0.02 ml/mouse intracerebrally and/or with 0.03 to 0.05 ml/mouse intraperitoneally, using a 27-gauge 3/8-in-long needle and a 1/2-ml syringe (Fig. A2).
2. Check the inoculated mice twice daily for signs of illness, paralysis, or death (Fig. A2)

Harvesting and processing infected tissues

This section is partially excerpted from Hsiung et al., 1994.
1. Harvest mouse brain or skeletal muscle when the animals are paralyzed or when other symptoms appear.
2. Make a 10% tissue suspension and inoculate it into additional mice and/or appropriate cell cultures, if the latter are available, for further study.

Identification of isolates

Mouse protection neutralization test or inhibition of CPE in cell culture can be performed for final identification. It is important to recognize that the mice may also become ill due to their own endogenous viruses and that cannibalism may occur.

When one works with togaviruses, it should be appreciated that the isolation and identification of some of these viruses can be associated with risks to laboratory personnel. Unless appropriate precautions can be taken, no attempts should be made to isolate these viruses outside of the appropriate reference facilities.

REFERENCES

Ahluwalia, G., J. Embree, P. McNicol, B. Law, and G. Hammond. 1987. Comparison of nasopharyngeal aspirate and nasopharyngeal swab specimens for respiratory syncytial virus diagnosis by cell culture, indirect immunofluorescence assay, and enzyme-linked immunosorbent assay. *J. Clin. Microbiol.* 25:763–767.

Arens, M. Q., E. M. Swierkosz, R. R. Schmidt, P. Armstrong, and K. A. Rivetna. 1986. Enhanced isolation of respiratory syncytial virus in cell culture. *J. Clin. Microbiol.* 23:800–802.

Arruda, E., C. E. Crump, B. S. Rollins, A. Ohlin, and F. G. Hayden. 1996. Comparative susceptibilities of human embryonic fibroblasts and HeLa cells for isolation of human rhinoviruses. *J. Clin. Microbiol.* **34:**1277–1279.

Balkovic, E. S., and G. D. Hsiung. 1985. Comparison of immunofluorescence with commercial monoclonal antibodies to biochemical and biologic techniques for typing clinical herpes simplex virus isolates. *J. Clin. Microbiol.* **22:**870–872.

Blanding, J. G., M. G. Hoshiko, and H. R. Stutman. 1989. Routine viral culture for pediatric respiratory specimens submitted for direct immunofluorescence testing. *J. Clin. Microbiol.* **27:**1438–1440.

Brumback, B. G., P. G. Farthing, and S. N. Castellino. 1993. Simultaneous detection of and differentiation between herpes simplex virus and varicella-zoster viruses with two fluorescent probes in the same test system. *J. Clin. Microbiol.* **31:**3260–3263.

Buller, R. S., T. C. Bailey, N. A. Ettinger, M. Keener, T. Langlois, J. P. Miller, and G. A. Storch. 1992. Use of a modified shell vial technique to quantitate cytomegalovirus viremia in a population of solid-organ transplant recipients. *J. Clin. Microbiol.* **30:**2620–2624.

Chu, F. C., J. B. Johnson, H. C. Orr, P. G. Probst, and J. C. Petricciani. 1973. Bacterial virus contamination of fetal bovine sera. *In Vitro* **9:**31–34.

Coffin, S. E., and R. L. Hodinka. 1995. Utility of direct immunofluorescence and virus culture for detection of varicella-zoster virus in skin lesions. *J. Clin. Microbiol.* **33:**2792–2795.

Dragan R., and M. A. Menegus. 1986. A combination of four cell types for rapid detection of enteroviruses in clinical specimens. *J. Med. Virol.* **19:**219–228.

Enders, J. F., T. H. Weller, and F. C. Robbins. 1949. Cultivation of the Lansing strain of poliomyelitis virus in cultures of various human embryonic tissues. *Science* **109:**8587.

Engler, H. D., and J. Pruess. 1997. Laboratory diagnosis of respiratory virus infections in 24 hours by utilizing shell vial cultures. *J. Clin. Microbiol.* **35:**2165–2167.

Espy, M. J., J. C. Hierholzer, and T. F. Smith. 1987. The effect of centrifugation on the rapid detection of adenovirus in shell vial. *Am. J. Clin. Pathol.* **88:**358–360.

Espy, M. J., A. D. Wold, D. M. Ilstrup, and T. F. Smith. 1988. Effect of treatment of shell vial cell cultures with dimethylsulfoxide and dexamethasone for detection of cytomegalovirus. *J. Clin. Microbiol.* **26:**1091–1093.

Espy, M. J., A. D. Wold, D. J. Jespersen, M. F. Jones, and T. F. Smith. 1991. Comparison of shell vials and conventional tubes seeded with rhabdomyosarcoma and MRC-5 cells for the rapid detection of herpes simplex virus. *J. Clin. Microbiol.* **29:**2751–2753.

Fedorko, D. P., D. M. Ilstrup, and T. F. Smith. 1989. Effect of age of shell vial monolayers on detection of cytomegalovirus from urine specimens. *J. Clin. Microbiol.* **27:**2107–2109.

Fong, C. K. Y., and M. L. Landry. 1992. An adventitious viral contaminant in commercially supplied A549 cells: identification of infectious bovine rhinotracheitis virus and its impact on diagnosis of infection in clinical specimens. *J. Clin. Microbiol.* **30:**1611–1613.

Frank, A. L., R. B. Couch, C. A. Griffis, and B. D. Baxter. 1979. Comparison of different tissue cultures for isolation and quantitation of influenza and parainfluenza viruses. *J. Clin. Microbiol.* **10:**32–36.

Gleaves, C. A., T. F. Smith, E. A. Shuster, and G. R. Pearson. 1984. Rapid detection of cytomegalovirus in MRC-5 cells inoculated with urine specimens by using low-speed centrifugation and monoclonal antibody to an early antigen. *J. Clin. Microbiol.* **19:**917–919.

Gleaves, C. A., D. J. Wilson, A. D. Wold, and T. F. Smith. 1985. Detection and serotyping of herpes simplex virus in MRC-5 cells by use of centrifugation and monoclonal antibodies 16 h postinoculation. *J. Clin. Microbiol.* **21:**29–32.

Gleaves, C. A., C. F. Lee, C. I. Bustamante, and J. D. Meyers. 1988. Use of murine monoclonal antibodies for laboratory diagnosis of varicella-zoster virus infection. *J. Clin. Microbiol.* **26:**1623–1625.

Gleaves C. A., D. A. Hursh, and J. D. Meyers. 1992. Detection of human cytomegalovirus in clinical specimens by centrifugation culture with a nonhuman cell line. *J. Clin. Microbiol.* **30:**1045–1048.

Gregory, W. W., and M. A. Menegus. 1983. Effect of incubation temperature on isolation of cytomegalovirus from fresh clinical specimens. *J. Clin. Microbiol.* **18:**1003–1005.

Hierholzer, J. C., E. Castells, G. G. Banks, J. A. Bryan, and C. T. McEwen. 1993. Sensitivity of NCI-H292 human lung mucoepidermoid cells for respiratory and other human viruses. *J. Clin. Microbiol.* **31:**1504–1510.

Hsiung, G. D. 1968. Latent virus infections in primate tissues with special reference to simian viruses. *Bacteriol. Rev.* **32:**185–205.

Hsiung, G. D., C. K. Y. Fong,. and M. L. Landry. 1994. *Hsiung's Diagnostic Virology,* 4th ed. Yale University Press, New Haven, Conn.

Hu, J. M., and G. D. Hsiung. 1989. Evaluation of antiviral agents. I. *In vitro* perspectives. *Antivir. Res.* **11:**217–232.

Hudson, J. B., V. Misra, and T. R. Mosmann. 1976. Cytomegalovirus infectivity: analysis of the phenomenon of centrifugal enhancement of infectivity. *Virology* **72:**235–243.

Johnston, S .L. G., and C. S. Siegel. 1990. Presumptive identification of enterovirus with RD, HEP-2, and RMK cell lines. *J. Clin. Microbiol.* **28:**1049–1050.

Johnston, S. L. G., K. Wellens, and C. S. Siegel. 1990. Rapid isolation of herpes simplex virus by using mink lung and rhabdomyosarcoma cell cultures. *J. Clin. Microbiol.* **28:**2806–2807.

Krizanova, O., and V. Rathova. 1969. Serum inhibitors of myxoviruses. *Curr. Top. Microbiol. Immunol.* **47:**125–151.

Landry, M. L. 1997. Rapid viral diagnosis, p. 608-617. *In* N. R. Rose, E. Conway de Macario, J. D. Folds, H. C. Lane, and R. M. Nakamura (ed.), *Manual of Clinical Laboratory Immunology,* 5th ed. American Society for Microbiology, Washington, D.C.

Landry, M. L., and D. Ferguson. 1993. Comparison of quantitative cytomegalovirus antigenemia assay with culture methods and correlation with clinical disease. *J. Clin. Microbiol.* **31:**2851–2856.

Landry, M. L., D. R. Mayo, and G. D. Hsiung. 1982. Comparison of guinea pig embryo cells, rabbit kidney cells and human embryonic fibroblast cell strains for the isolation of herpes simplex virus. *J. Clin. Microbiol.* **15:**842–847.

Landry, M. L., D. R. Mayo, and G. D. Hsiung. 1989. Rapid and accurate viral diagnosis. *Pharmacol. Ther.* **40:**287–328.

Mackenzie, J. S. 1999. Emerging viral diseases: an Australian perspective. *Emerg. Infect. Dis.* **5:**1–8.

Mahafzah, A. M., and M. L. Landry. 1989. Evaluation of conventional culture, centrifugation culture, and immunofluorescence reagents for the rapid diagnosis of adenovirus infections. *Diagn. Microbiol. Infect. Dis.* **12:**407–411.

Marquez, A., and G. D. Hsiung. 1967. Influence of glutamine on multiplication and cytopathic effect of respiratory syncytial virus. *Proc. Soc. Exp. Biol. Med.* **124:**95–99.

Marshall, W. F., A. Telenti, J. Proper, A. J. Aksamit, and T. F. Smith. 1990. Rapid detection of polyomavirus BK by a shell vial cell culture assay. *J. Clin. Microbiol.* **28:**1613–1615.

Mavromoustakis, C. T., D. T. Witiak, and J. H. Hughes. 1988. Effect of high-speed rolling on herpes simplex virus detection and replication. *J. Clin. Microbiol.* **26:**2328–2331.

McCarter, Y. S., and A. Robinson. 1997. Comparison of MRC-5 and primary rabbit kidney cells for the detection of herpes simplex virus. *Arch. Pathol. Lab. Med.* **121:**122–124.

Miller, M. J., and C. L. Howell. 1983. Rapid detection and identification of herpes simplex virus in cell culture by a direct immunoperoxidase staining procedure. *J. Clin. Microbiol.* **18:**550–553.

Muir, P., U. Kammerer, K. Korn, M. N. Mulders, and T. Poyry. 1998. Molecular typing of enteroviruses: current status and future requirements. *Clin. Microbiol. Rev.* **11:**202–227.

Oberste, S. M., K. Maher, D. R. Kilpatrick, M. R. Flemister, B. A. Brown, and M. A. Pallansch. 1999. Typing of human enteroviruses by partial sequencing of VP1. *J. Clin. Microbiol.* **37:**1288–1293.

Olivo, P. D. 1996. Transgenic cell lines for detection of animal viruses. *Clin. Microbiol. Rev.* **9:**321–334.

Olsen, M. A., K. M. Shuck, A. R. Sambol, S. M. Flor, J. O'Brien, and B. J. Cabera. 1993. Isolation of seven respiratory viruses in shell vials: a practical and highly sensitive method. *J. Clin. Microbiol.* **31:**422–425.

Paya, C., A. D. Wold, and T. F. Smith. 1987. Detection of cytomegalovirus infections in specimens other than urine by shell vial assay and conventional tube cell cultures. *J. Clin. Microbiol.* **25:**755–757.

Paya, C., A. D. Wold, D. M. Ilstrup, and T. F. Smith. 1988. Evaluation of number of shell vial cultures per clinical specimen for rapid diagnosis of cytomegalovirus infection. *J. Clin. Microbiol.* **26:**198–200.

Rabalais, G. P., G. G. Stout, K. L. Ladd, and K. M. Cost. 1992. Rapid diagnosis of respiratory viral infections by using a shell vial assay and monoclonal antibody pool. *J. Clin. Microbiol.* **30:**1505–1508.

Rabella, N., and W. L. Drew. 1990. Comparison of conventional and shell vial cultures for detecting cytomegalovirus infection. *J. Clin. Microbiol.* **28:**806–807.

Reed, L. J., and H. A. Muench. 1938. A simple method of estimating fifty percent end points. *Am. J. Hyg.* **27:**493–497.

Reina, J., V. Fernandez-Baca, L. Blanco, and M. Munar. 1997. Comparison of Madin-Darby canine kidney cells (MDCK) with a green monkey continuous cell line (Vero) and human lung embryonated cells (MRC-5) in the isolation of influenza A virus from nasopharyngeal aspirates by shell vial culture. *J. Clin. Microbiol.* **35:**1900–1901.

Rigonan, A. S., L. Mann, and T. Chonmaitree. 1998. Use of monoclonal antibodies to identify serotypes of enterovirus isolates. *J. Clin. Microbiol.* **36:**1877–1881.

Safrin, S., T. Elbeik, and J. Mills. 1994. A rapid screen test for in vitro susceptibility of clinical herpes simplex virus isolates. *J. Infect. Dis.* **169:**879–882.

Schmidt, N. J., and R. W. Emmons (ed.). 1989. *Diagnostic Procedures for Viral, Rickettsial and Chlamydial Infections.* American Public Health Association, Washington, D.C.

Schultz-Cherry, S., N. Dybdahl-Sissoko, M. McGregor, and V. S. Hinshaw. 1998. Mink lung epithelial cells: unique cell line that supports influenza A and B virus replication. *J. Clin. Microbiol.* **36:**3718–3720.

Shahrabadi, M. S., and P. W. K. Lee. 1988. Calcium requirement for syncytium formation in HEp-2 cells by respiratory syncytial virus. *J. Clin. Microbiol.* **26:**139–141.

Shuster, E. A., J. S. Beneke, G. E. Tegtmeier, G. E. Pearson, C. A. Gleaves, A. D. Wold, and T. F. Smith. 1985. Monoclonal antibody for rapid laboratory detection of cytomegalovirus infections: characterization and diagnostic application. *Mayo Clin. Proc.* **60:**577–585.

Slavin, M. A., C. A. Gleaves, H. G. Schock, and R. A. Bowden. 1992. Quantification of cytomegalovirus in bronchoalveolar lavage fluid after allogeneic marrow transplantation by centrifugation culture. *J. Clin. Microbiol.* **30:**2776–2779.

Smith, K. O. 1970. Adventitious viruses in cell cultures. *Prog. Med. Virol.* **12:**302–336.

Smith, T. F., and L. Reichrath. 1974. Comparative recovery of 1972–1973 influenza virus isolates in embryonated eggs and primary rhesus monkey kidney cell cultures after one freeze-thaw cycle. *Am. J. Clin. Pathol.* **61:**579–584.

Stanbridge, E. 1971. Mycoplasmas and cell cultures. *Bacteriol. Rev.* **35:**206–227.

Tkacova, M., E. Vareckova, I. C. Baker, J. M. Love, and T. Ziegler. 1997. Evaluation of monoclonal antibodies for subtyping of currently circulating human type A influenza viruses. *J. Clin. Microbiol.* **35:**1196–1198.

van der Bij, W., J. Schirm, R. Torensma, W. J. van Son, A. M. Tegzess, and T. H. The. 1988. Comparison between viremia and antigenemia for detection of cytomegalovirus in blood. *J. Clin. Microbiol.* **26:**2531–2535.

West, P. G., B. A. Aldrich, R. Hartwig, and G. J. Haller. 1989. Increased detection of herpes simplex virus in MRC-5 cells treated with dimethyl sulfoxide and dexamethasone. *J. Clin. Microbiol.* **27:**770–772.

Woods, G. L., and A. Young. 1988. Use of A-549 cells in a clinical virology laboratory. *J. Clin. Microbiol.* **26:**1026–1028.

Zhao, L., M. S. Landry, E. S. Balkovic, and G. D. Hsiung. 1987. Impact of cell culture sensitivity and virus concentration on rapid detection of herpes simplex virus by cytopathic effects and immunoperoxidase staining. *J. Clin. Microbiol.* **25:**1401–1405.

The Cytopathology of Virus Infections

ROGER D. SMITH AND RAWIA S. YASSIN

4

Virus-infected cells that exfoliate or are scraped from the skin or mucous membranes may exhibit readily identifiable morphological changes that permit rapid diagnosis. In many instances, the cytologic alterations may be so distinctive as to be pathognomonic for infection with a specific agent. In others, the changes may point to a virus group or merely raise a suspicion of infection to be confirmed by other means.

The purpose of this chapter is to describe the methods used to obtain and prepare cells for cytologic examination and to illustrate characteristic changes encountered in common virus infections. Most of the methods described are used routinely by diagnostic cytology laboratories and are best applied by the cytotechnologist or pathologist who analyzes cytologic material on a daily basis.

PREPARATION AND STAINING

The proper collection, fixation, preparation, and staining of specimens for cytology is essential. Rapid fixation in 95% alcohol or with cytology spray fixative before the smear dries is imperative for accurate interpretation. The purpose of fixation is to maintain the existing form and structure of the cellular elements and to achieve consistent staining characteristics and identifiable structures. Improper specimen preparation will decrease the diagnostic accuracy and may lead to false-positive or false-negative results. Air drying causes nuclear swelling and distortion, cytoplasmic vacuolization, and atypical staining. These changes can mimic nuclear and/or cytoplasmic alterations seen with some virus infections or can distort the characteristic details to such an extent that viral cytopathology cannot be identified. For example, the ground-glass nuclear appearance seen in early herpes simplex virus (HSV) infection can be confused with the smudgy nuclear detail seen in air-dried specimens. Aqueous fixatives (e.g., formalin) result in poor staining and irregular condensation of chromatin, which can be mistaken for a nuclear inclusion.

The preparatory methods utilized for the microscopic examination of cytologic specimens can be divided into four categories: direct smears, preparation by cytocentrifugation, membrane filter preparation, and cell block preparation (Bales and Durfee, 1979).

Direct Smears

Direct scraping of vesicular or bullous lesions of the skin and mucous membranes is the simplest method of cell collection for the identification of viral changes. The scrapings from the base and edges of the suspected lesion should be smeared evenly onto a clean glass slide and immediately fixed with 95% alcohol or cytology spray fixative. Once fixed, the smears are almost indefinitely stable at room temperature and can be stored, mailed, or transported to a cytology laboratory for further processing. The slides are then stained with a modified Papanicolaou staining technique. Proper sampling is important because the crusted or eczematous areas often fail to show the diagnostic cellular features. If a specific lesion, such as an ulcer or unroofed vesicle, is present, the base and edges of the lesion should be thoroughly scraped with a spatula, tongue blade, or endoscopic brush to insure proper sampling.

■ Modified Papanicolaou Stain

The modified Papanicolaou stain technique is used to detect cytologic changes due to viral infection. The procedure for staining is as follows:

1. Ten dips in 95% ethyl alcohol (EtOH)
2. Ten dips in 70% EtOH
3. Ten dips in 50% EtOH
4. Ten dips in distilled H_2O
5. Two minutes in hematoxylin (Gill hematoxylin, consisting of 2,190 ml of distilled H_2O, 750 ml of ethylene glycol, 6 g of hematoxylin (C.I. no. 75290), 0.6 g of sodium iodate, 528 g of aluminum sulfate, and 60 ml of glacial acetic acid)
6. One minute under running tap water
7. One minute in Scott's tap water substitute (consisting of 10 g of anhydrous magnesium sulfate, 2 g of sodium bicarbonate, and 1 liter of distilled water)
8. Ten dips in tap water
9. Ten dips in 50% EtOH
10. Ten dips in 95% EtOH
11. One and one-half minutes in OG-6′
12. Ten dips in 95% EtOH
13. Ten dips in 95% EtOH
14. Three minutes in EA-65′
15. Ten dips in 95% EtOH

16. Ten dips in 95% EtOH
17. Ten dips in 95% EtOH
18. Ten dips in 100% EtOH
19. Ten dips in 100% EtOH
20. Ten dips in 100% EtOH–Hemo-De2 (equal amounts)
21. Ten dips in Hemo-De2, in three consecutive dishes
22. Coverslipping the slide with Permount medium

Cytocentrifugation and Filtration

Cytocentrifugation and filtration techniques have been developed which concentrate small numbers of cells suspended in fluids, and they are the preferred methods for preparation of samples from urine, cerebrospinal fluid, and bronchial wash, bronchial alveolar lavage (BAL), and body cavity fluids. To determine which of the two techniques should be used, one needs to consider the expected number of cells in the fluid. Fluids containing many cells should be centrifuged. The specimen is placed in a centrifuge tube and spun at 1,500 rpm for 10 min. The supernatant fluid is decanted, leaving a volume of 2 ml in the tube, and the sediment is resuspended. The specimen is then prepared using the standard cytocentrifugation technique (Barrett, 1976). If little or no sediment is present, then the filtration technique (Gill, 1976) is the method of choice. The use of membrane filters should be limited to cases where the number of cells is low and where additional cell sampling presents a problem.

Cytospin preparations offer several advantages. The cells are evenly dispersed on the slide (monolayer), little or no background artifact is present, and the cell preparation can be utilized easily for other diagnostic procedures, such as special stains, immunofluorescence, immunoperoxidase, and electron microscopy. Cell block preparations require a histopathology laboratory and are used when there is an abundance of cellular material that can be embedded in paraffin and sectioned for histologic examination.

VIRUS CYTOPATHOLOGY

The skin and eye, as well as the respiratory, genital, and urinary tracts, are locations that readily yield cytologic material for rapid diagnosis of virus. Characteristic cytologic changes depend on the cytopathic effect of a virus in infected cells, which need not include all cells of the involved organ. In practice, however, it is most useful to consider cytologic alterations in the context of the organ system affected and the clinical presentation. Therefore, the following discussion and illustrations are organized according to organ system.

Viral Infections of the Respiratory Tract

Smears of cells obtained from nasal and throat swabs, tracheal aspirates, sputum, bronchial washings and brushings, and pulmonary lavage (i.e., BAL) may exhibit cytologic alterations that are diagnostic of virus infection (Table 1).

In adults, the most frequently encountered virus that is readily detectable by cytology is cytomegalovirus (CMV) (Warner et al., 1964). Particularly in patients with AIDS or those who are receiving immunosuppressive therapy for transplantation or cancer chemotherapy, the rapid cytologic identification of characteristic inclusions may be a great asset in patient management because CMV isolation in tissue culture may take many days and often weeks.

TABLE 1 Cytopathology of respiratory viral infections

Virus	Clinical presentation	Cytologic findings[a]
Adenovirus	Upper respiratory infection, pneumonia	Small multiple eosinophilic IN, inclusions (early); large, single, dense basophilic IN inclusion (late)
CMV	Pneumonia	Cytomegaly; large, single, amphophilic IN inclusions; small PAS-positive IC inclusions
HSV	Tracheobronchitis	Large ground-glass nucleus, (early); eosinophilic IN, inclusions (late); multinuclearity with nuclear molding
Parainfluenza virus	Bronchitis, pneumonia	Cytomegaly; single or multiple nuclei; small eosinophilic IC inclusions
RSV	Tracheobronchitis, pneumonia	Large multinucleated cells; IC basophilic inclusions with prominent halos
Measles virus	Prodromal	Mulberrylike clusters of lymphocytic nuclei in nasal secretions
	Pneumonia	Multinucleated giant cells with IN and IC eosinophilic inclusions
Nonspecific (many viruses)	Bronchitis, pneumonia	Ciliocytophoria

[a] IN, intranuclear; IC, intracytoplasmic; PAS, periodic acid-Schiff reaction.

Because CMV infection involves the lungs, a deep specimen containing pulmonary macrophages is needed. Patients with CMV pneumonia rarely produce abundant sputum, thus requiring bronchial washings and brushings or pulmonary lavage to obtain an adequate specimen. The characteristic cytologic changes are seen in pulmonary macrophages or in cells lining the alveoli. They most often exhibit a single nucleus, but occasionally two or more nuclei that are four to six times their normal size are present (Fig. 1). Early in the infection these nuclei contain amphoteric or basophilic inclusions that are granular, and the inclusions become condensed and surrounded by a halo in the later stages of infection. Smaller, more eosinophilic oval cytoplasmic inclusions that are periodic acid-Schiff reaction positive are often, but not invariably, present. At later stages of the infection, cytoplasmic inclusions may predominate, with an empty or collapsed nucleus (Fig. 2).

Characteristic inclusion-bearing cells may be observed in sputum and bronchial washings in HSV tracheobronchitis, which is often encountered in immunosuppressed and burn patients (Vernon, 1982). Inclusion-bearing cells tend to be multinucleated and contain either eosinophilic intranuclear inclusions that are centrally located and surrounded by a halo or, at an earlier stage of infection, ground-glass inclusions that stain poorly (Fig. 3). The chro-

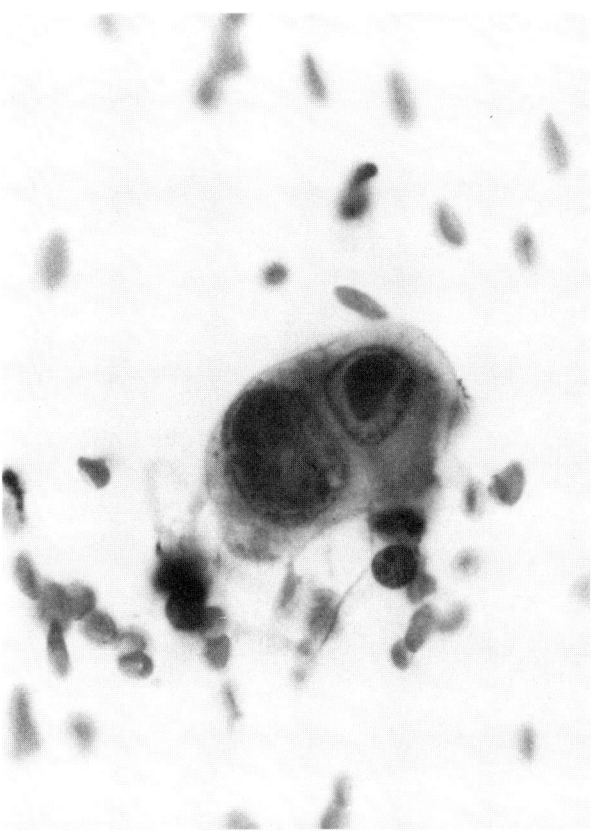

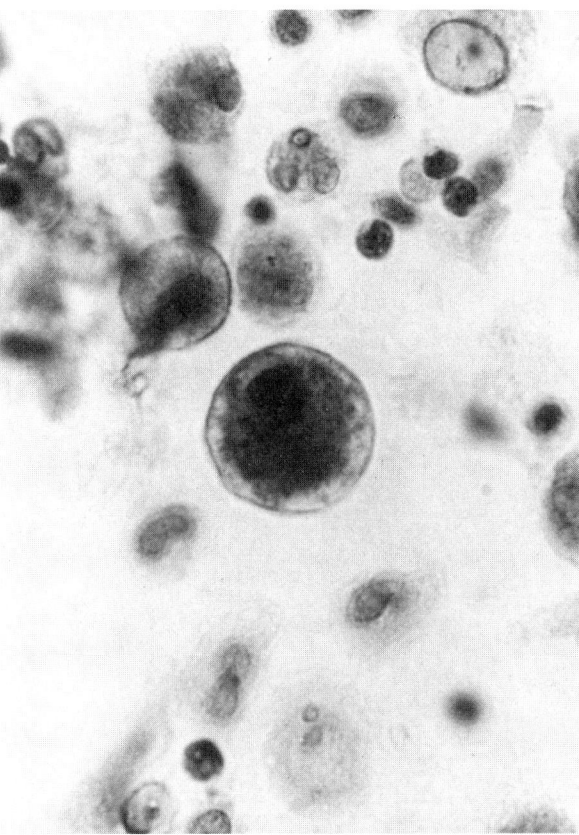

FIGURE 1   CMV in bronchial brushing; large basophilic intranuclear inclusions in binucleated cell and small cytoplasmic inclusions; pap stain. Magnification, ×800.

FIGURE 2   CMV in urine; multiple intracytoplasmic inclusions with small degenerating intranuclear inclusion in late stage of infection; pap stain. Magnification, ×800.

matin often appears as a basophilic ring condensed at the periphery of the nuclear membrane. Cytoplasmic inclusions are not present. When there are multiple nuclei, they are frequently molded or indented by each other (Fig. 4). Because HSV tends to produce cellular necrosis, the backgrounds of these smears usually contain an abundance of cellular debris.

Adenovirus infections of the upper respiratory tract may be identified by smears of secretions from the nasopharynx. Adenovirus pneumonia may be diagnosed by finding typical intranuclear inclusions in bronchial or epithelial cells obtained by bronchoscopy. At an early stage (Fig. 5), the nucleus contains multiple small, rounded eosinophilic inclusions, each surrounded by a halo. At a later stage, a single larger, dense, basophilic intranuclear inclusion is seen (Fig. 6).

Respiratory syncytial virus (RSV) and parainfluenza viruses frequently cause bronchitis and pneumonia in infants and young children. They can be rapidly diagnosed by identifying characteristic cytologic changes in respiratory epithelial cells obtained by nasopharyngeal swabs and from tracheal aspirates (Naib et al., 1968). Parainfluenza virus can be differentiated from RSV by the presence of large cells containing a single nucleus and multiple small eosinophilic inclusions. In RSV infection, the epithelial cells are large and multinucleate and the cytoplasm contains multiple basophilic inclusions with prominent halos.

Measles can be detected during the prodrome by finding mulberrylike clusters of lymphocytes having up to 50 nuclei in smears of nasal secretions (Tomkins and Macanlay, 1955). Measles giant cells (Fig. 7) are multinucleate respiratory epithelial cells with intranuclear and cytoplasmic inclusions. These cells may appear in the sputa of patients with measles pneumonia.

A nonspecific change referred to as ciliocytophoria is found in various inflammatory diseases of the respiratory tract and, in particular, virus infections (Pierce and Hirsch, 1958). The ciliated bronchial epithelial cells undergo a degenerative process in which a pinching off occurs between the cytoplasm and nucleus, resulting in detached tufts of cilia and a degenerating nucleus and cytoplasm (Fig. 8). The degenerated cytoplasm may contain small, round eosinophilic inclusion bodies (Takahashi, 1981). Ciliocytophoria occurs most frequently with influenza, parainfluenza, and adenovirus infection but may also occur in bronchiectasis and other nonviral inflammatory conditions. It is more frequently encountered in sputum specimens than in those obtained by bronchoscopy.

Virus Infections of the Urinary Tract

Although many viruses that cause systemic infections have been isolated from the urine, those most readily diagnosed by urine cytology are CMV, HSV, and a member of the papovavirus group, designated the BK virus (BKV) (Table 2).

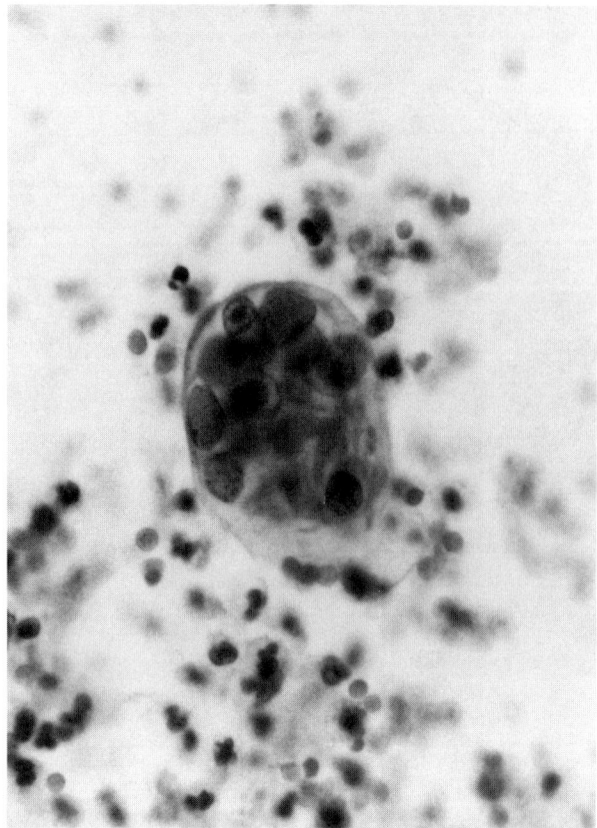

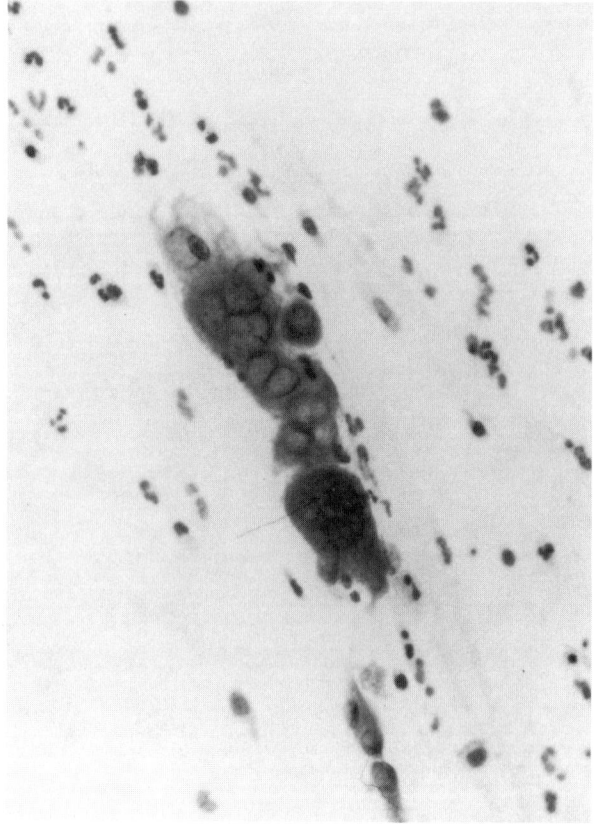

FIGURE 3 HSV in sputum; epithelial cells showing the ground-glass nuclear appearance of the early inclusions; pap stain. Magnification, ×600.

FIGURE 4 HSV in bronchial brushing; multinucleated cell with intranuclear inclusions in different stages of development and molding of the nuclei; pap stain. Magnification, ×800.

TABLE 2 Virus infections of the urinary tract

Virus	Clinical presentation	Cytologic findings[a]
Adenovirus	Hemorrhagic cystitis	Dense basophilic IN inclusions in transitional cells
BKV	Urethral stenosis in renal transplant and asymptomatic, interstitial nephritis in immunosuppressed patients	Large full mucoid IN inclusions (early); dense full basophilic inclusions bulging from cytoplasm (late)
CMV	Asymptomatic in immmunosuppressed patients	Large basophilic IN inclusions surrounded by halo; multiple eosinophilic IC inclusions
HSV	Generalized infection or local cystitis; may be contaminant from herpes genitalis	Ground-glass nuclei (early); eosinophilic IN inclusions (late); multinuclearity; may be part of tubular cast
Measles virus	Measles with exanthema	Multinucleated giant cells with IC inclusions

[a] IN, intranuclear; IC, intracytoplasmic.

In infections by each of these viruses, epithelial cells of the urinary tract from the renal tubules to the bladder and urethra detach and enter the urine. These cells often contain characteristic inclusions. Because of the relatively small number of cells in a large fluid volume, filtration or cytocentrifugation is necessary to obtain a suitable preparation (Schumann et al., 1977). Each of these infections occurs most often, but not exclusively, in immunosuppressed patients (commonly renal transplant recipients), and they may coexist as mixed infections. CMV-infected urothelial cells were first described in the urine of newborn infants with cytomegalic inclusion disease (Fetterman, 1952). Cytologic examination of the urinary sediment is indicated with any infant suspected of having neonatal CMV infection. Positive cytology is seen in approximately 50% of neonates that will subsequently have CMV-positive cultures (Hanshaw et al., 1968). CMV is the most frequently encountered viral infection in renal transplant recipients and in patients with AIDS, complicating the human immunodeficiency virus infection. In one study it was found in 31% of 2,354 cytologically examined routine urine samples obtained from 91 patients (Traystman et al., 1980). Cellular changes include cytomegaly of the urothelial cells with typical large round and smooth intranuclear inclusions surrounded by a clear halo and smaller eosinophilic cytoplasmic inclusions. This cytopathology most often involves single cells. Although the classical large, dense intranuclear inclusion is the

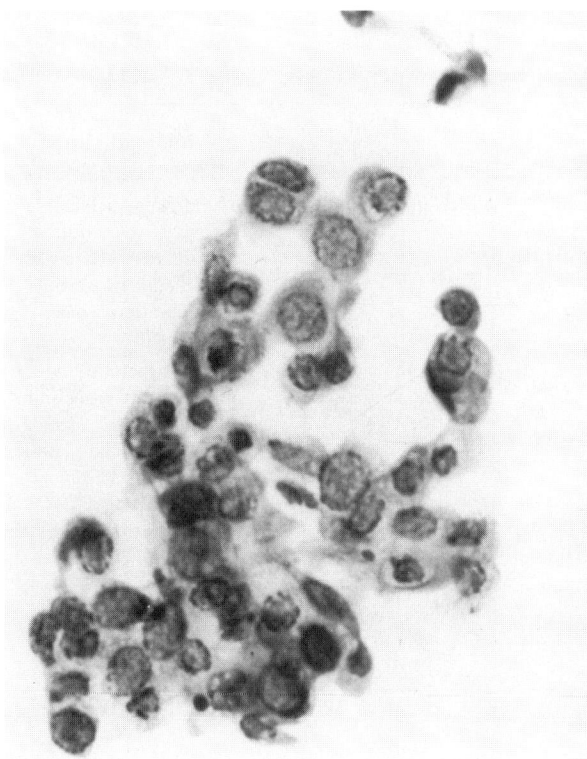

FIGURE 5 Adenovirus in conjunctival scraping; granular intranuclear inclusions with condensed chromatin at the nuclear membrane are characteristic of the early stage of infection; pap stain. Magnification, ×600.

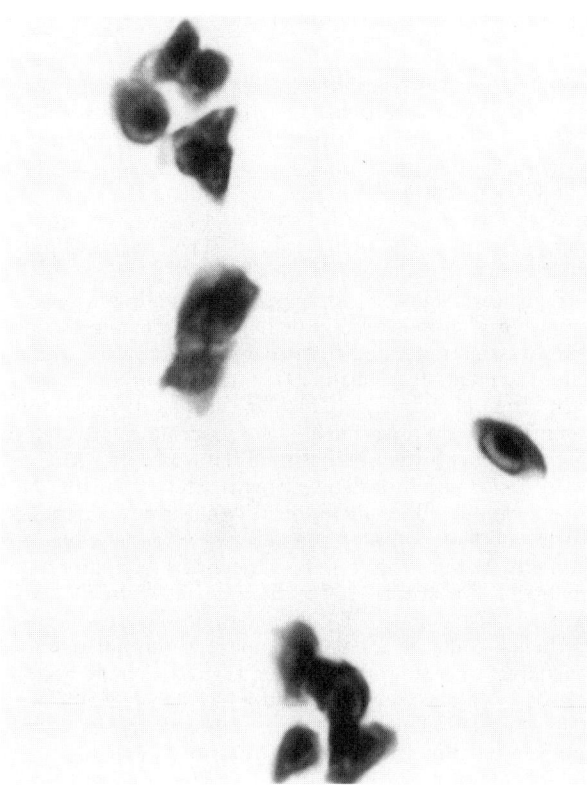

FIGURE 6 Adenovirus in conjunctival scraping; conjunctival cells contain a dense, sometimes teardrop-shaped intranuclear inclusion surrounded by a clear halo in the late stage of infection; pap smear. Magnification, ×800.

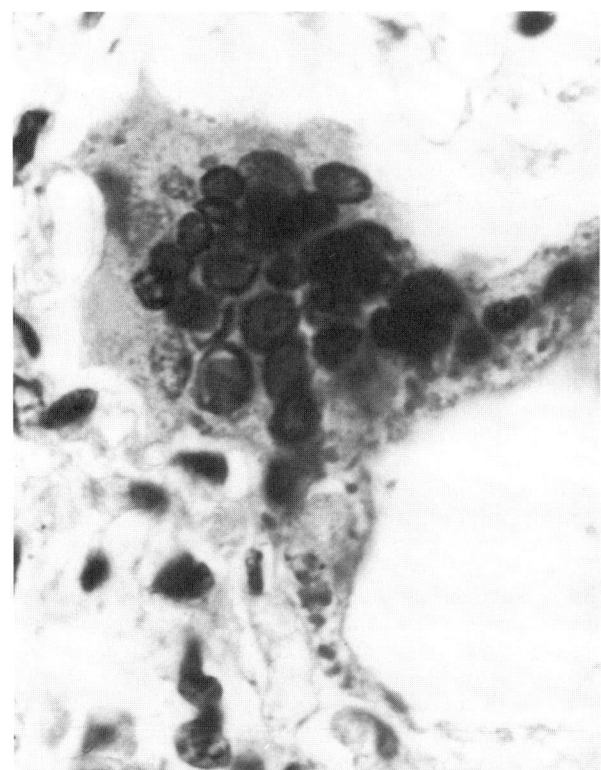

FIGURE 7 Measles pneumonia; paraffin-embedded lung tissue showing a multinucleated giant cell with multiple intranuclear and cytoplasmic inclusions; hematoxylin and eosin stain. Magnification, ×800.

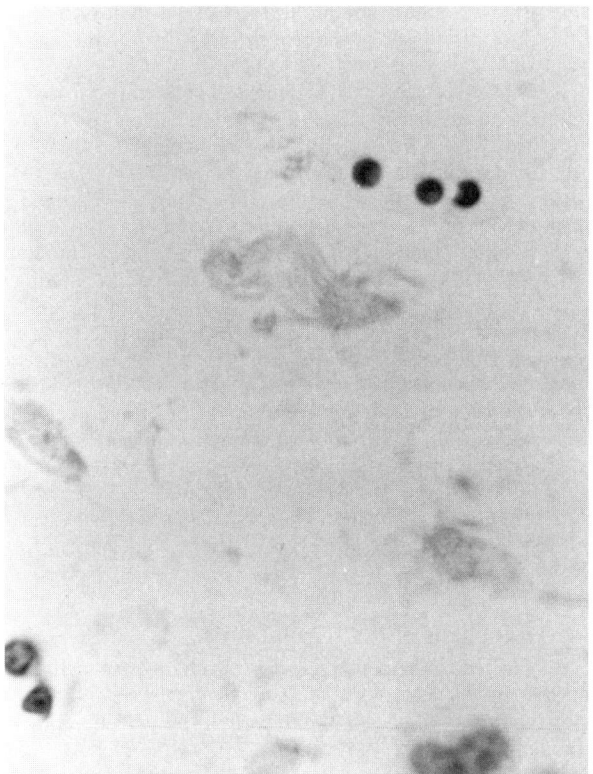

FIGURE 8 Ciliocytophoria in sputum; a tuft of detached cilia and nuclear debris; pap stain. Magnification, ×600.

easiest to identify, occasional cells may be binucleated (Fig. 1), and some cells may have large dense eosinophilic cytoplasmic inclusions with little or no evidence of an intranuclear inclusion (Fig. 2).

BKV was first isolated from the urine of a 39-year-old man who developed ureteral stenosis 4 months after renal transplantation (Gardner et al., 1971). The urine sediment contained abnormal transitional cells with dense intranuclear inclusions composed of crystalline arrays of papovavirus virions as revealed by electron microscopy. BKV was later isolated from many asymptomatic transplant recipients (Coleman, 1975) and two other patients with ureteral stenosis (Coleman et al., 1978). More recently, BKV in the urine has been associated with a viral interstitial nephritis in renal transplant patients (Mathur et al., 1997) and in individuals with AIDS (Smith et al., 1998). Finding BKV inclusions in the urine in patients with transplants indicates the necessity for renal biopsy to differentiate rejection from BKV nephritis for appropriate treatment. At an early stage, the most recognizable cytologic change due to BKV infection is the presence of epithelial cells with enlarged nuclei, each filled by a mucoid, grayish-staining inclusion (Fig. 9). A more homogeneous, densely basophilic inclusion is detached at a later stage (Fig. 10). This is sometimes referred to as a "decoy cell." Frequently, the nucleus appears to be bulging from the cytoplasm or thrusting from it, giving a cometlike or half-moon effect (Fig. 11). Although most involved cells have single nuclei, occasional binucleated forms are seen with both stages of inclusions present. At a later stage, the inclusion shrinks from the nuclear membrane, leaving an incomplete, thin halo. The intranuclear inclusions of BKV can be distinguished from those of CMV by the complete and consistent halo around the CMV inclusion and the lack of cytoplasmic inclusions in BKV (compare Fig. 1 and 10).

Cytologic changes of HSV-infected cells in the urinary sediment are similar to those described in cells of the respiratory tract. They include multinuclear syncytial cells with enlarged ground-glass nuclei, seen at an early stage of infection, and typical eosinophilic, often angulated intranuclear inclusions surrounded by a halo at a later stage. Elongated clumps of infected epithelial cells, probably of tubular origin, may contain inclusions in varying stages of development (Fig. 12). Urinary sediment cells characteristic of HSV may occur in a generalized HSV infection involving the kidney or in a localized cystitis; they may also result from herpes genitalis, particularly when there is vaginal involvement during the infection (Masukawa et al., 1972). Other cytologic changes that may be observed in urinary sediment cells include intranuclear inclusions of adenovirus associated with acute hemorrhagic cystitis in children (Numazaki et al., 1973) and inclusion-bearing cells in the urine of patients with measles (Bolande, 1961).

Virus Infections of the Genital Tract

Cytologic recognition of typical viral changes in cells of routine "pap" smears is the most readily available and cost-effective method of detecting genital herpes infections (Table 3).

Herpes genitalis is now the second most prevalent sexually transmitted disease, and detection of infections by cytology can be followed by successful treatment with acyclovir. Cytologic recognition of HSV is of critical importance in directing the management of pregnancy near term.

TABLE 3 Virus infections of the genital tract

Virus	Clinical presentation	Cytologic findings[a]
HSV (types I and II)	Herpes genitalis	Ground-glass nuclei (early); eosinophilic IN inclusions (late with peripheral chromatin condensation; multinuclearity
HPV	Condyloma acuminatum; cervical dysplasia	Enlarged hyperchromatic nucleus; rare basophilic IN inclusions; perivascular cytoplasmic clearing and vacuolar degeneration (koilocytotic change)
Molluscum contagiosum virus	Vaginal, penile, or perineal papule with central umbilication	Large densely staining IC inclusions displacing nucleus; squamous cell often bean shaped

[a]IN, intranuclear; IC, intracytoplasmic.

The overall incidence of HSV in routine vaginal smears has been reported as approximately 0.3% (Naib, 1980), although this figure varies greatly depending on the patient population. The sensitivity of cytology for detecting HSV infection depends somewhat on the location of the herpetic lesions and the adequacy of the sample. In one study (Vontver et al., 1979), 41% of 69 cases with external lesions that were HSV culture positive had positive smears, but the rate was 23% with women that had only cervical lesions. Similar results were reported in a study of 76 patients with genital HSV comparing virus isolation, immunofluorescence, immunoperoxidase, and cytology as means of making a diagnosis (Moseley et al., 1981). The overall positive cytology rate was 37.6%, but it was 47.9% for cases with vaginal or cutaneous lesions. Significantly, there were a number of cases in which the cytology was positive but virus isolation in tissue culture from the same sample was negative. This discrepancy is repeatedly encountered in reported studies comparing virus isolation and cytopathology with various viruses at different sites, and it indicates that the greatest diagnostic yield with virus amenable to cytologic diagnosis is from the combination of cytology and virus isolation or identification by immunohistology or molecular-probe techniques.

The identifiable cytologic changes in genital HSV infection are identical to those described with infections of the respiratory and urinary tracts. At an early stage the enlarged nuclei have a bland ground-glass appearance with the chromatin displaced to the periphery, resulting in an apparent thickening of the nuclear membrane (Fig. 13). At a later stage, the nucleus contains an eosinophilic inclusion surrounded by a clear halo. Multinuclear cells are common, with up to 10 nuclei which often exhibit molding. Inclusions in multinucleated cells may all be at the same stage (Fig. 14) or may exhibit different stages of development (Fig. 15). HSV-1 and HSV-2 produce identical morphological changes and cannot be differentiated on the basis of cytology. Although it has been reported that primary HSV infection can be differentiated from recurrent or secondary infection by a predominance of

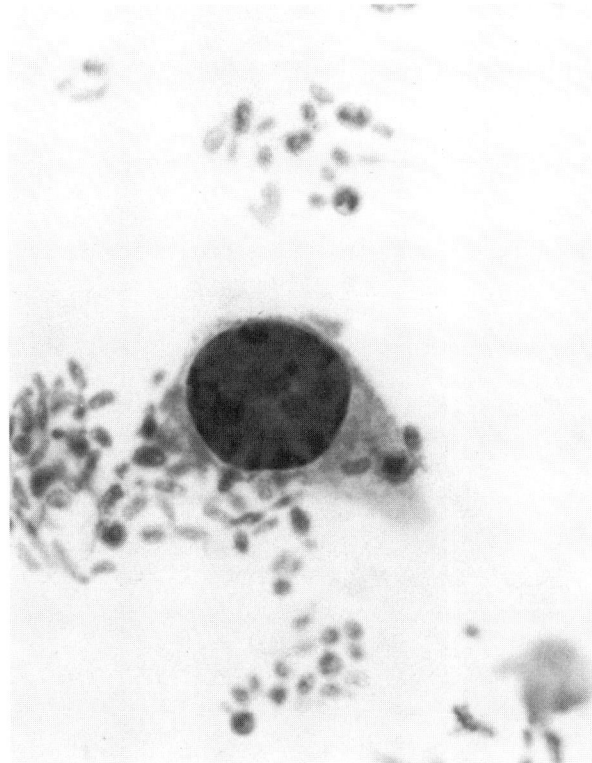

FIGURE 9 BKV in urine; uroepithelial cell with early changes of BKV infection. The enlarged nucleus is filled with a grayish-staining inclusion having a mucoid appearance. Note the *Candida* yeast in the background. Pap stain. Magnification, ×800.

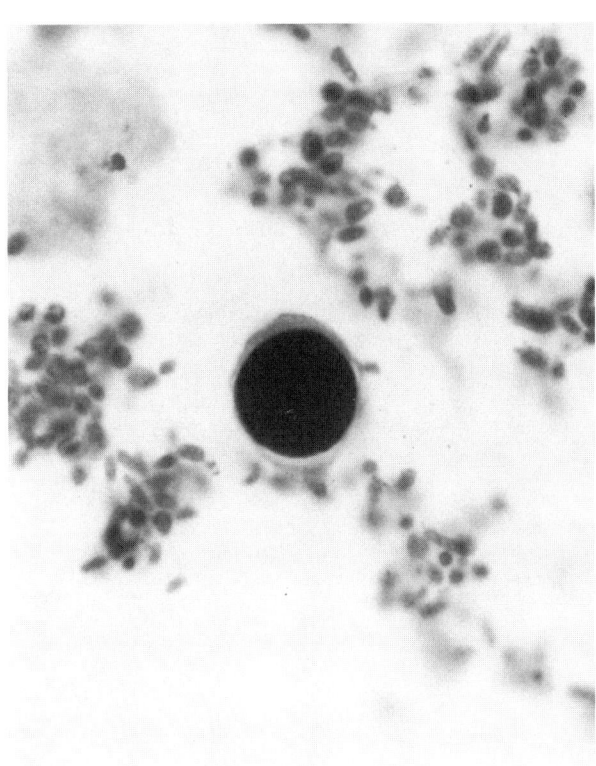

FIGURE 10 BKV in urine. A transitional cell contains a large dense inclusion filling and expanding the nucleus. The background contains yeast. Pap stain. Magnification, ×800.

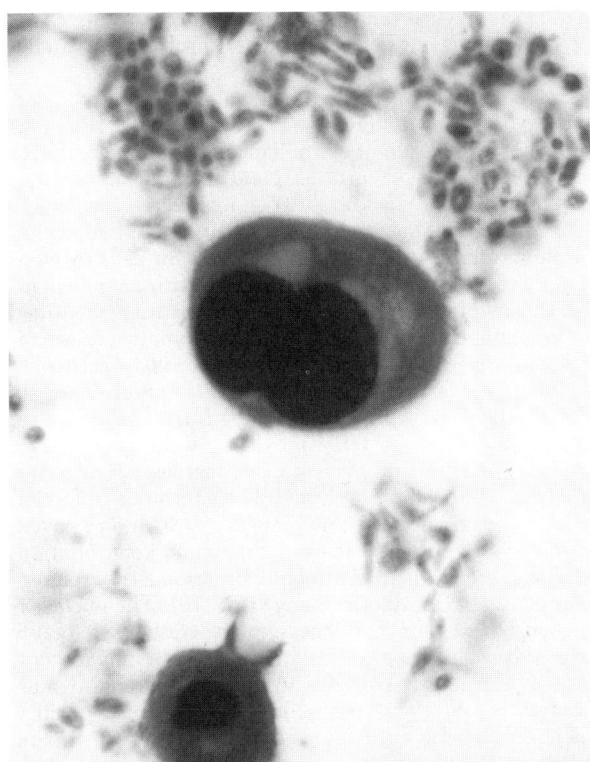

FIGURE 11 BKV in urine. In the late stage, there is cytoplasmic degeneration, leaving dense intranuclear inclusions appearing to bud from the nucleus. Pap stain. Magnification, ×800.

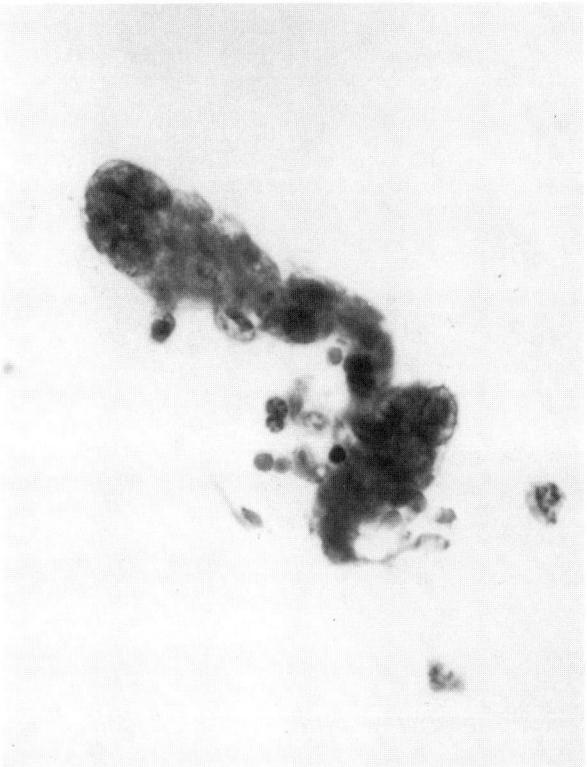

FIGURE 12 HSV in urine; a renal tubular cast showing characteristic HSV inclusions and nuclear molding; pap stain. Magnification, ×800.

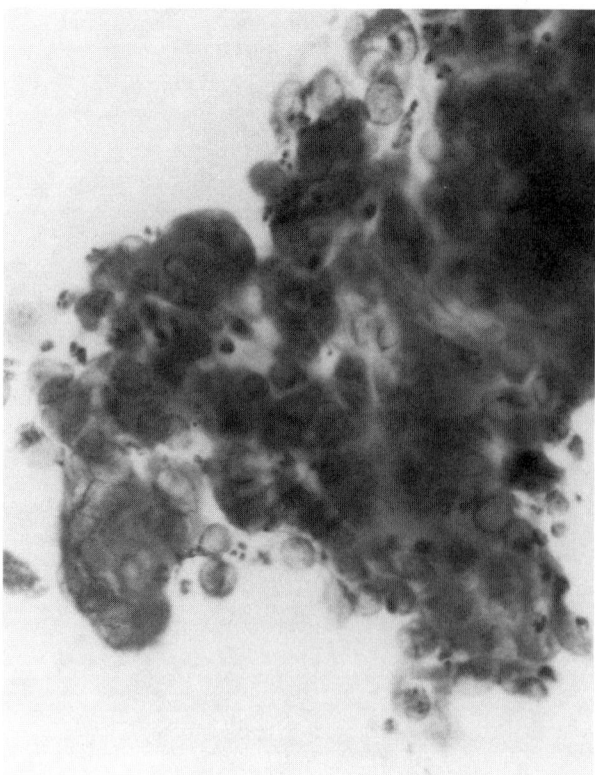

FIGURE 13 HSV in genital smear. An early stage of the infection shows ground-glass-appearing inclusions in a multinucleated mass of cells. Pap stain. Magnification, ×600.

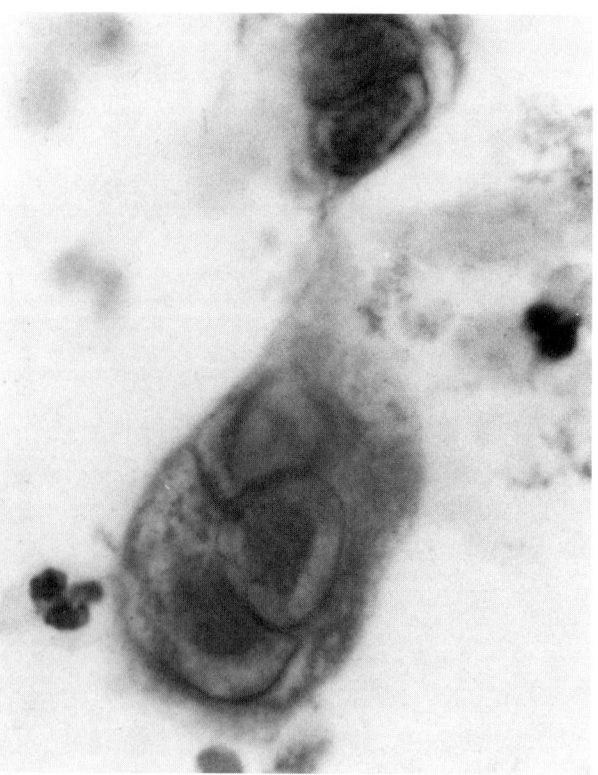

FIGURE 14 HSV in genital smear. A multinucleated cell contains angulated intranuclear inclusions surrounded by halos. There is condensation of chromatin at the nuclear membrane and nuclear molding. Pap stain. Magnification, ×800.

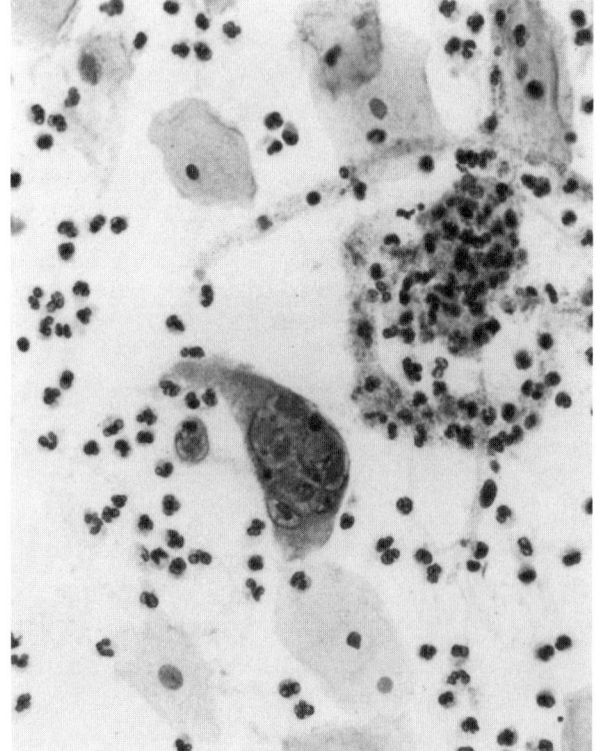

FIGURE 15 HSV in genital smear. Note inclusions at different stages of development within one multinucleated epithelial cell. Pap stain. Magnification, ×600.

bland ground-glass nuclei in primary infection (Ng et al., 1970), this has not been confirmed in subsequent studies (Naib, 1980).

Infection of the cervical or vaginal mucosa and the skin of the perineum with a human papilloma virus (HPV) may result in proliferation of epithelial cells forming vegetative papillary growths known as condyloma acuminata, or venereal warts. Atypical cellular changes often accompany the proliferative process and may result in dysplastic cellular alterations similar to those of malignant cells (Meisels et al., 1981). HPV has been associated with premalignant cervical dysplasia and carcinoma and has been identified by immunohistology and/or molecular probes in association with squamous cell carcinoma of the vulva (Zachow et al., 1982; Pilotti et al., 1984) and cervix (Shroyer et al., 1993). At the present time, cytology is the only practical way of identifying an HPV infection of the genital tract when a characteristic gross condylomatous lesion is not observed. The changes attributable to virus infection involve squamous cells, which appear swollen and have a perinuclear halo with poor cytoplasmic keratinization, resulting in irregular staining. This produces a picture referred to as koilocytotic change (Fig. 16). The nucleus is frequently enlarged and occasionally contains a poorly defined basophilic inclusion which, by electronmicroscopy, is composed of virus particles and fibrillar material (Caras-Cordero et al., 1981). Although the incidence of HPV infection and the accuracy of diagnosis based on finding koilocytotic change by cytology have not yet been fully documented, the identification of HPV infection of the genital tract is likely to be important in preventing sexual transmission of HPV. Women who have genital HPV infection should have follow-up with periodic cytology to

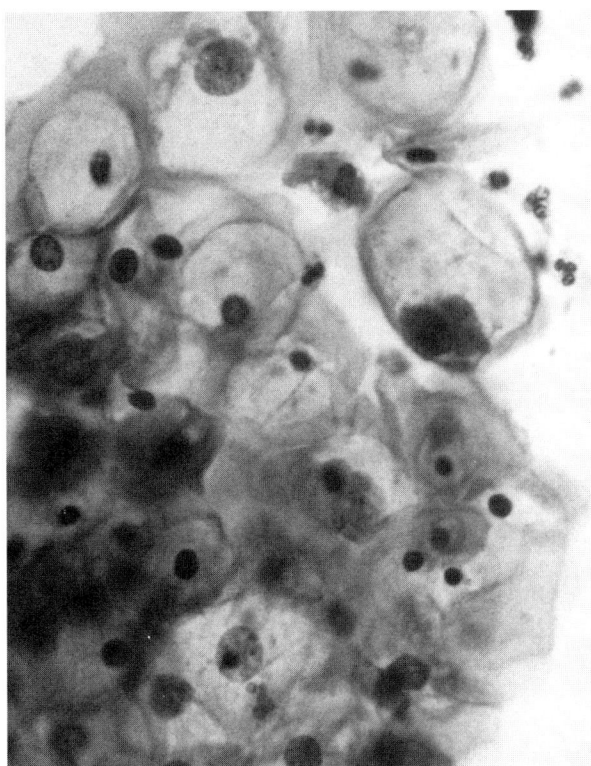

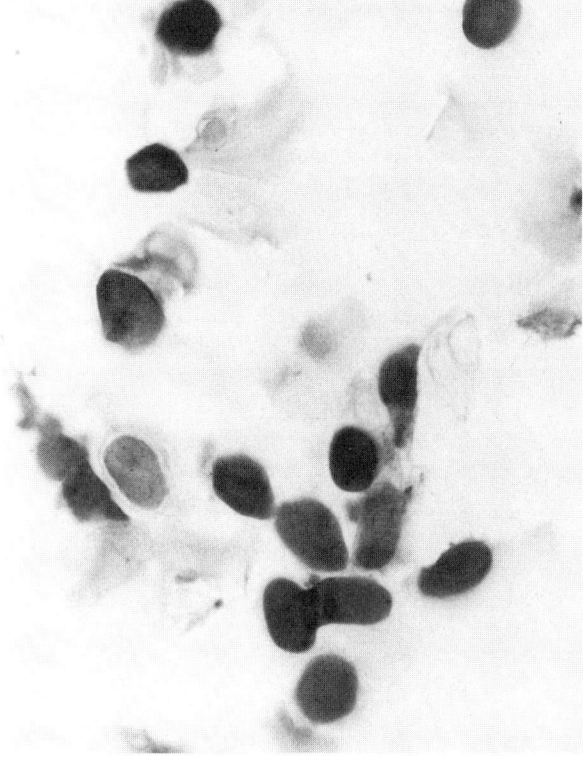

FIGURE 16 Papillomavirus in genital smear. Enlarged squamous cells exhibit koilocytotic changes characteristic of a condylomatous lesion. Pap smear. Magnification, ×800.

FIGURE 17 Molluscum contagiosum in genital or skin lesion smear; molluscum bodies from a vaginal smear showing large intracytoplasmic inclusions displacing and compressing the nucleus; pap stain. Magnification, ×800.

detect early atypical changes that may indicate malignant transformation. Other viruses that have been identified by characteristic cytologic findings in vaginal smears include CMV and the poxvirus that causes molluscum contagiosum (Brown et al., 1981). Molluscum contagiosum is a benign cutaneous infection, most often observed in children and young adults, which is easily transmitted by direct contact. Circumstantial evidence suggests the infection is transmitted between young adults during sexual intercourse. Although the lesions have a characteristic appearance consisting of a small, firm papule with a centralized umbilication on the skin or vaginal mucosa, virus isolation techniques to confirm the diagnosis are not yet available. However, the histopathology necessitating biopsy and the morphological changes in individual cells as observed in a pap-stained smear are diagnostic. The cytologic changes consist of large, densely staining cytoplasmic inclusions occupying the entire squamous cell and resulting in peripheral displacement of a flattened nucleus (Fig. 17). The cells frequently assume a bean shape, with the nucleus displaced to the concave aspect.

Viruses Infecting the Eye

In many common ocular lesions, and particularly in those involving the cornea, a rapid diagnosis is essential in order to initiate therapy and avoid progressive corneal damage. Although biopsy and virus isolation are the usual definitive procedures for the diagnosis of virus infections involving the cornea and conjunctiva, exfoliative cytology offers a simple, inexpensive, and rapid means of diagnosing aden-

ovirus keratoconjunctivitis and keratitis due to herpesviruses (Naib et al., 1967; Schumann et al., 1980) (Table 4). In addition, characteristic cytologic changes consisting of multinucleated giant cells in measles keratoconjunctivitis and the large dense cytoplasmic inclusions of molluscum contagiosum in conjunctival and eyelid lesions may also yield a definitive diagnosis, although these infections are rarely encountered as isolated ophthalmic diseases. Finally, cytology is most useful for the diagnosis of chlamydial conjunctivitis, which may be clinically difficult to differentiate from the disease caused by adenovirus but can be diagnosed by the finding of characteristic cytoplasmic inclusions caused by the chlamydial infection (Gupta et al., 1979).

Specimens containing conjunctival or corneal cells should be collected by a physician, preferably an ophthalmologist, by swab or superficial conjunctival or corneal scraping. Corneal scraping requires examination with a slit lamp microscope for localization of the lesion. The collected material is immediately spread on an alcohol-moistened slide, and after partial evaporation, the slide is placed in 95% EtOH for proper fixation. Because there are usually very few cells and little fluid substrate, air drying of the smears is a frequent problem, but it can be avoided by immediate fixation of the specimen on the slide.

In adenovirus infections, the conjunctival or corneal cells are mixed with lymphocytes and plasma cells and contain distinctive intranuclear inclusions. In the early stages of infection (Fig. 5), the intranuclear inclusions are multi-

TABLE 4 Virus and bacterial infections of the eye

Virus or bacterium	Clinical presentation	Cytologic findings[a]
Adenovirus	Acute (epidemic) keratoconjunctivitis and conjunctivitis with pharyngitis	Multiple eosinophilic IN inclusions (early); dense central basophilic inclusions surrounded by halo (late)
HSV	Corneal vesicle or ulcer; may be isolated ophthalmic lesion or with other HSV vesicles	Multinucleated cells with eosinophilic IN inclusions surrounded by halo; nucleus has ground-glass appearance (early stage)
Molluscum contagiosum virus	Reddish papular 5-mm diameter lesions of eyelid or conjunctiva	Large dense basophilic IC inclusions displacing nucleus
Varicella-zoster virus	Vesicular eruptions in dermatome involving eye (shingles) or accompanying chicken pox	Multinucleated cells with IN eosinophilic inclusions
Chlamydia	Granular conjunctiva with corneal ulcerations (trachoma) or conjunctivitis only (inclusion conjunctivitis)	Enlarged corneal (trachoma only) and conjunctival cells with numerous IC basophilic inclusions surrounded by individual halos

[a]IN, intranuclear; IC, intracytoplasmic.

FIGURE 18 Chlamydia in conjunctival scraping. Epithelial cells contain intracytoplasmic inclusions surrounded by clear halos. Pap stain. Magnification, ×800.

ple, small, and sometimes granular and eosinophilic with accentuated nuclear membranes due to condensation of chromatin. At a later stage, the small inclusions coalesce to form a single dense basophilic body, usually centrally located and surrounded by a clear halo (Fig. 6).

Keratoconjunctivitis caused by HSV can occur as a primary lesion or as part of a systemic infection. Superficial scrapings from the margin of the ulcerated area will usually contain multinucleated cells with characteristic large eosinophilic intranuclear inclusions surrounded by prominent halos. Early in the infection, as with HSV at other sites, scrapings will reveal enlarged nuclei having a ground-glass appearance.

Herpes zoster keratitis is well recognized clinically and usually does not require additional diagnostic confirmation, such as cytology. However, scrapings of the lesions will yield cells similar to the ones found in HSV infection, although it is reported that syncytia and intranuclear inclusions are less prominent than with HSV (Naib et al., 1967). The acute conjunctivitis that occurs, usually during the prodromal of measles, is also associated with characteristic cytologic findings seen in conjunctival smears or scrapings. The characteristic cells may contain up to 100 round nuclei surrounded by an abundant cytoplasm in which there are numerous eosinophilic inclusions. Occasionally, similar eosinophilic inclusions can be found within the nuclei. These findings can precede the appearance of the typical exanthem by 2 to 3 days.

The morphological changes in conjunctival and corneal cells due to virus infection must be differentiated from the cytologic changes due to chlamydial infections causing trachoma and inclusion conjunctivitis. In both these diseases, the epithelial cells generally are enlarged and have abundant cytoplasm containing clusters of basophilic intracytoplasmic inclusions, each surrounded by a large individual halo (Fig. 18). In trachoma, the corneal cells are involved, whereas only conjunctival cells show the changes in the more benign chlamydial conjunctivitis. The presence of cells containing cytoplasmic inclusions in cytologic examination of specimens from the eye suggests chlamydial, rather than viral, infection.

REFERENCES

Bales, C., and G. Durfee. 1979. Cytologic techniques, p. 1187–1266. *In* L. Koss (ed.), *Diagnostic Cytology*, 3rd ed., vol. 2. J. B. Lippincott Co., Philadelphia, Pa.

Barrett, D. 1976. Cytocentrifugation technique, p. 80–83. *In* C. Keebler, J. Reagan, and G. L. Wied (ed.), *Compendium on Cytopreparatory Techniques*, 4th ed. Tutorials of Cytology, Chicago, Ill.

Bolande, R. P. 1961. Significance and nature of inclusion-bearing cells in the urine of patients with measles. *N. Engl. J. Med.* **265:**919–923.

Brown, S. T., J. F. Nalley, and S. N. Kraus. 1981. Molluscum contagiosum. *Sex. Transm. Dis.* **8:**227–234.

Caras-Cordero, M., C. Morin, M. Roy, M. Fortier, and A. Meisels. 1981. Origin of the koilocytes in condylomata of the human cervix. Ultrastructural study. *Acta Cytol.* **25:**383–392.

Coleman, D. V. 1975. The cytodiagnosis of human polyoma infection. *Acta Cytol.* **9:**93–96.

Coleman, D. V., E. F. D. MacKenzie, S. D. Gardner, J. M. Poulding, B. Amer, and W. J. Russell. 1978. Human polyomavirus (BK) infection and ureteric stenosis in renal allograft recipients. *J. Clin. Pathol.* **31:**338–347.

Fetterman, G. H. 1952. New laboratory aid in clinical diagnosis of inclusion disease of infancy. *Am. J. Clin. Pathol.* **22:**424–427.

Gardner, S. D., A. M. Field, D. V. Coleman, and B. Hulme. 1971. New human papovavirus (BK) isolated from urine after renal transplantation. *Lancet* **i:**1253–1257.

Gill, G. 1976. Methods of cell collection on membrane filters, p. 34–44. *In* C. Keebler, J. Reagan, and G. L. Wied (ed.), *Compendium on Cytopreparatory Techniques*, 4th ed. Tutorials of Cytology, Chicago, Ill.

Gupta, P. A., E. F. Lee, Y. S. Erozan, J. K. Frost, S. T. Geddes, and P. A. Donovan. 1979. Cytologic investigations in chlamydia infection. *Acta Cytol.* **23:**315–320.

Hanshaw, J. B., H. J. Steinfeld, and C. J. White. 1968. Fluorescent-antibody test for cytomegalovirus macroglobulin. *N. Engl. J. Med.* **279:**566–570.

Masukawa, T., J. C. Jarancis, M. Rytel, and R. F. Mattingly. 1972. Herpes genitalis virus isolation from human bladder urine. *Acta Cytol.* **16:**416–428.

Mathur, V. S., J. L. Olson, T. M. Darragh, and T. S. B. Yen. 1997. Polyomavirus-induced interstitial nephritis in two renal transplant recipients: case reports and review of the literature. *Am. J. Kidney Dis.* **29:**754–758.

Meisels, A., M. Ray, M. Fortier, C. Morin, M. Caras-Cordero, K. V. Shah, and H. Turgeon. 1981. Human papillomavirus infection of the cervix. *Acta Cytol.* **25:**7–16.

Moseley, R. C., L. Corey, D. Benjamin, C. Winter, and M. L. Remington. 1981. Comparison of viral isolation, direct immunofluorescence, and direct immunoperoxidase techniques for detection of genital herpes simplex virus infection. *J. Clin. Microbiol.* **13:**913–918,

Naib, Z. M., A. S. Clepper, and S. R. Elliott. 1967. Exfoliative cytology as an aid in diagnosis of ophthalmic lesions. *Acta Cytol.* **11:**295–303.

Naib, Z., J. Stewart, W. Dowdle, H. Casey, W. Marine, and A. Nahmias. 1968. Cytologic features of viral respiratory tract infections. *Acta Cytol.* **12:162-171.**

Naib, Z. M. 1980. Exfoliative cytology in the rapid diagnosis of herpes simplex infection, p. 381–386. *In* A. J. Nahmias, W. R. Dowdle, and R. F. Schinazi (ed.), *The Human Herpes Virus: an Interdisciplinary Perspective.* Elsevier/North Holland Biomedical Press, Amsterdam, The Netherlands.

Ng, A. B. P., J. W. Reagan, and S. S. Yen. 1970. Herpes genitalis: clinical and cytopathological experience with 256 patients. *Obstet. Gynecol.* **36:**645–651.

Numazaki, Y., T. Kumasaka, N. Yano, M. Yamanaka, T. Miyazawe, S. Takai, and N. Ishida. 1973. Further study on acute hemorrhagic cystitis due to adenovirus type 11. *N. Engl. J. Med.* **289:**344–347.

Pierce, C. H., and J. G. Hirsch. 1958. Ciliocytophoria relationship to viral respiratory infections in humans. *Proc. Soc. Exp. Biol. Med.* **98:**489–492.

Pilotti, S., F. Rilke, K. V. Shah, G. D. Torre, and G. DePalo. 1984. Immunohistochemical and ultrastructural evidence of papilloma virus infection associated with in situ and microinvasive squamous cell carcinoma of the vulva. *Am. J. Surg. Pathol.* **8:**751–761.

Schumann, G. B., S. Beng, and R. Hill. 1977. Use of the cytocentrifuge for the detection of cytomegalovirus inclusions in the urine of renal allograft patients. *Acta Cytol.* **21:**168–172.

Schumann, G. B., G. J. O'Dowd, and P. A. Spinnler. 1980. Eye cytology. *Lab. Med.* **11:**533–540.

Shroyer, K. R., G. S. Lovelace, M. L. Abarca, R. H. Fennell, M. E. Corkill, W. D. Woodard, and G. H. Davilla. 1993. Detection of human papillomavirus DNA by in situ hybridization and polymerase chain reaction in human papillomavirus equivocal and dysplastic cervical biopsies. *Hum. Pathol.* **24:**1012–1016.

Smith, R. D., J. H. Galla, K. Skahan, P. Anderson, C. C. Linnemann, G. S. Ault, C. F. Ryschkewitsch, and G. L. Stoner. 1998. Tubulointerstitial nephritis due to a mutant polyomavirus BK virus strain, BKV(Cin), causing end-stage renal disease. *J. Clin. Microbiol.* **36:**1660–1665.

Takahashi, M. 1981. *Color Atlas of Cancer Cytology*, p. 291. Igaku-Shoin, Tokyo, Japan.

Tomkins, V., and I. C. Macanlay. 1955. A characteristic cell in nasal secretions during prodromal measles. *J. Am. Med. Assoc.* **157:**711–712.

Traystman, M. D., P. K. Gupta, K. V. Shah, M. Reissig, L. Y. Cowles, W. D. Hillis, and J. K. Frost. 1980. Identification of viruses in the urine of renal transplant recipients by cytomorphology. *Acta Cytol.* **24:**501–510.

Vernon, S. E. 1982. Cytologic features on non-fatal herpes virus tracheobronchitis. *Acta Cytol.* **26:**237–242.

Vontver, L. A., W. C. Reeves, M. Rathay, L. Corey, M. A. Remington, E. Tolentino, A. Schweid, and K. K. Holmes. 1979. Clinical course and diagnosis of genital herpes virus infection and evaluation of topical surfactant therapy. *Am. J. Obstet. Gynecol.* **133:**548–554.

Warner, N. E., E. A. McGrew, and S. Nanos. 1964. Cytologic study of sputum in cytomegalic inclusion disease. *Acta Cytol.* **8:**311–315.

Zachow, K. R., R. S. Ostrow, M. Bender, S. Watts, T. Okagaki, F. Pass, and A. J. Faras. 1982. Detection of human papillomavirus DNA in anogenital neoplasms. *Nature* **300:**771–773.

Electron Microscopy and Immunoelectron Microscopy

MARTIN PETRIC AND MARIA SZYMANSKI

5

In clinical virology, electron microscopy (EM) has achieved a role equivalent to that of conventional light microscopy in clinical microbiology. In both cases, the microorganism is visually detected and its features are enhanced by staining. In a clinical specimen, EM allows the rapid detection of the virus, at least at the level of the family into which it is classified, with a very high degree of specificity. However, depending on the concentration and stability of the virus, EM has limited sensitivity compared to other methods of virus diagnosis. While diagnosis at the level of the virus family is frequently clinically meaningful, further identification of the virus at the level of the genus must be accomplished by other methods, such as isolation in cell culture or immunospecific or molecular approaches.

HISTORY

The ability to visualize viruses had a major impact on the development of virology. The particle one visualizes may only be the relatively inert, metabolically inactive portion of the virus life cycle, but it remains a fundamental feature in the classification and recognition of viruses. Since viruses range from 20 to over 300 nm in diameter, their features can only be identified at the nanometer level of resolution, and the optimal instrument to accomplish this is the electron microscope.

Prior to the development of the electron microscope, viruses were considered invisible or ultramicroscopic. Early applications of EM to clinical virology involved distinguishing poxvirus from varicella-zoster virus (van Rooyen and Scott, 1948). The major advances came with the development of simple negative-staining technology, which allowed the viral ultrastructure to be visualized (Brenner and Horne, 1959). This opened the door to the application of EM to virus diagnosis and led to unexpected findings, especially in the case of gastroenteritis viruses, as shown in Fig. 1 and 2 (Middleton et al., 1977; Petric, 1999).

Immunoelectron microscopy arose from the combination of EM with the immunospecific interaction of viruses with their respective antibodies (Almeida et al., 1963). Initially, this involved the detection of virus clumps formed when a virus preparation was reacted with its specific antibody. Further developments have resulted in the use of colloidal-gold-labeled antibodies to identify virus particles or morphological subviral units (Stannard et al., 1982). Specific antibodies have also been applied to coated grids to selectively enhance the binding of virus particles and thereby enhance the sensitivity and specificity of the procedure (Gerna et al., 1988).

METHODS

In clinical virology laboratories, the main application of EM is in the direct visualization of the virus in negatively stained specimens. Conversely, the examination of viruses in thin-sectioned specimens has been largely assumed by pathology laboratories, where viruslike particles are one of a number of markers that are examined in tissues. Accordingly, the methods discussed below will concern only the direct visualization of viruses after negative staining.

Basic Principles

Diagnostic applications of EM involve a number of discrete steps (Doane and Anderson, 1987). These include the adsorption of viruses to the coated EM grid, the interaction of the viruses with the negative stain, and the recognition of specific morphological features on examination. The copper grid, approximately 3 mm in diameter, is coated with a thin film of Formvar or Parlodion, which is further stabilized by coating it with a layer of carbon. This surface is able to adsorb virus particles or their subunits, as well as other particulate matter in the specimen. It has been shown that viruses present in a drop of water will concentrate over time on the surface of the drop, and this inherently enriches the quantity of virus in contact with the grid (Johnson and Gregory, 1993).

To be amenable to visualization, the virus must be present at 10^6 to 10^7 particles per ml (Doane and Anderson, 1987). The need for such a high concentration can be rationalized if one considers that after accounting for the thickness of the grid bars and the peripheral ring of the grid, the effective area to which viruses can bind and be observed is approximately 3 mm^2. If it is assumed that only the particles present within 10 to 100 μm (100 to 1,000 diameters of a 100-nm-diameter virus) of the grid surface are available to interact through diffusion with the grid coating, then the effective volume that is sampled for examination is between 0.03 and 0.3 μl. If the above assumptions are correct, this represents 30 to 3,000 particles in a specimen containing

10^6 to 10^7 particles per ml. Only a fraction of these available particles will actually bind to the grid coating and survive the staining process to be visualized. In this context, it can therefore be readily understood why such high concentrations of virus in the specimen are needed for diagnosis by EM to be feasible.

Negative staining is based on the principle that the virus particles on the grid will remain unstained whereas the

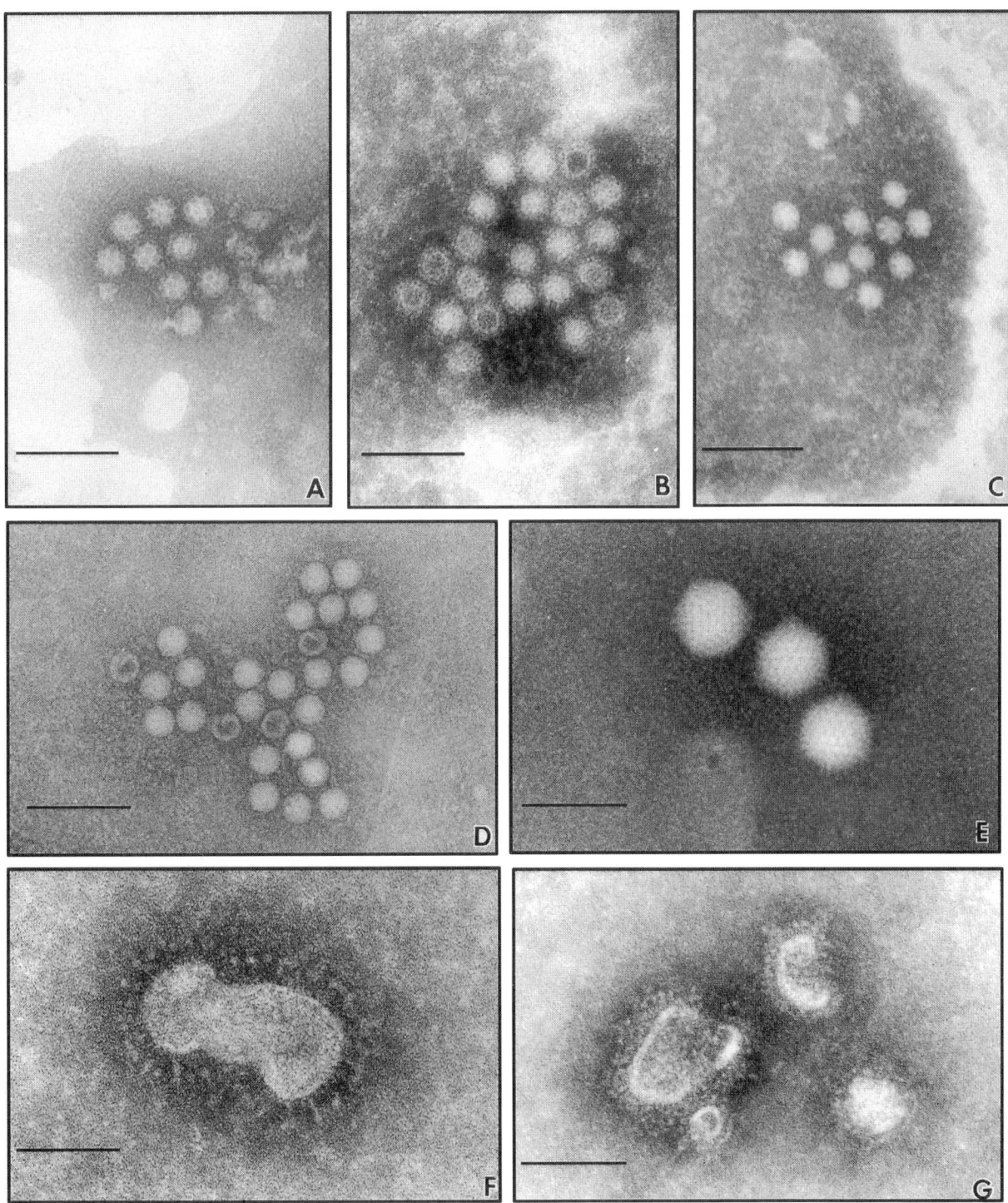

FIGURE 1 Gastroenteritis viruses detected in stool specimens by EM using the direct-application method. (A) Calicivirus; (B) Norwalk-like virus; (C) astrovirus; (D) small round virus; (E) adenovirus; (F) coronavirus; (G) torovirus. Bars, 100 nm. (Reprinted from Petric, 1999, with permission.)

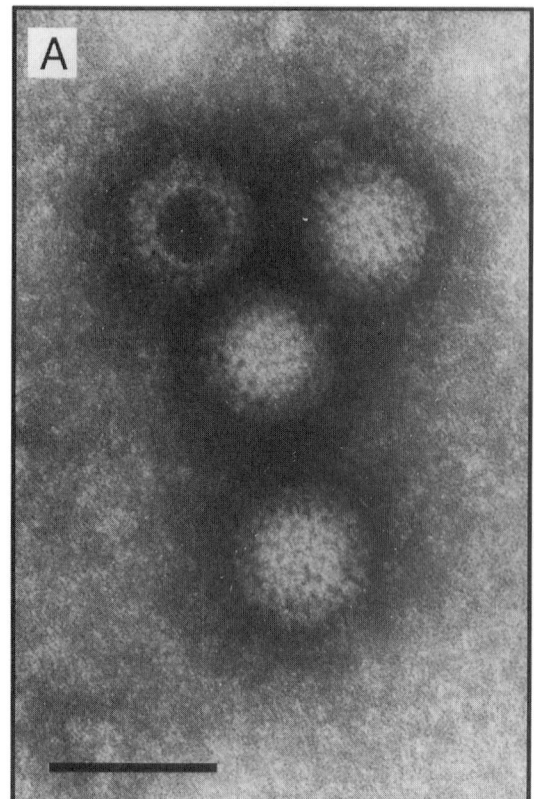

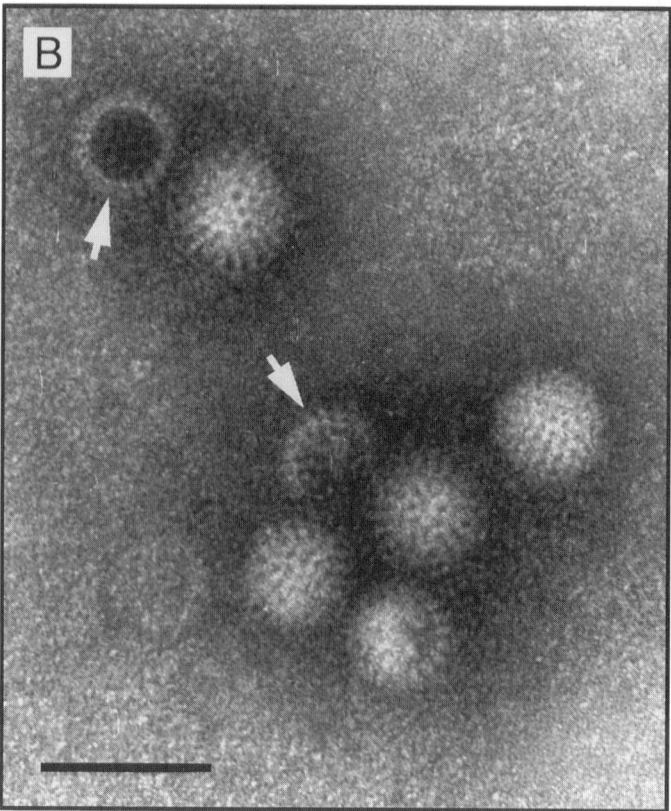

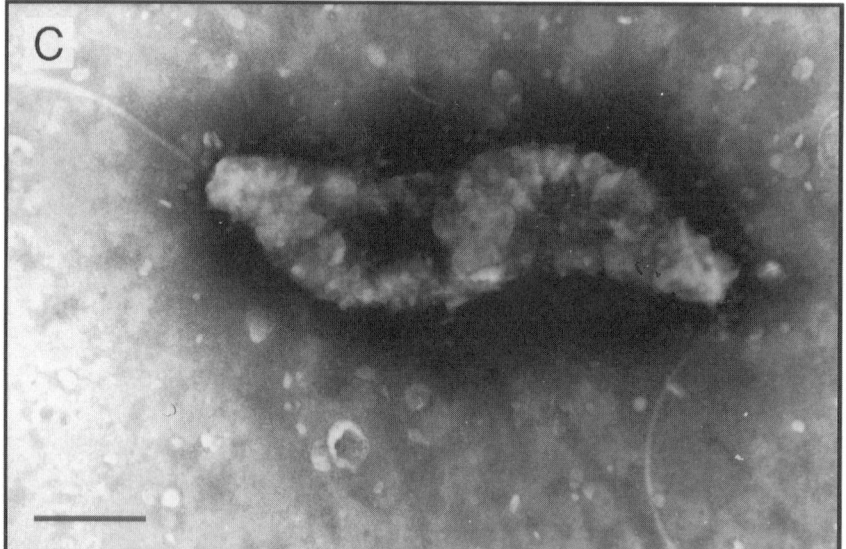

FIGURE 2 Microorganisms diagnosed by EM using the direct-application method. (A) Reovirus; (B) rotavirus. Note the differences between the arrangements of capsomeres in rotavirus and reovirus. The arrows indicate single capsid particles. (C) Campylobacter. Note the sinusoidal appearance with bipolar flagella. Bars, 100 nm.

background will be dark due to the heavy metal deposit of the stain. Negative contrast is therefore achieved, with areas of the grid having limited stain deposits (virus particles) being electron lucent and areas with more extensive stain deposits (background) being electron dense. In reality, the virus particles do take up stain into their interiors as well as into crevices on their surfaces. This is important, since it allows the visualization of viral capsomeres and peplomers in addition to the size and shape of the virus particles. The stains most commonly used include phosphotungstic acid and uranyl acetate (Hayat and Miller, 1990). The former has the formula $PW_{12}O_{40}{}^{3-}$ and an ionic diameter of 0.8 nm. Uranyl acetate has an ionic diameter of 0.5 nm and is also a weak fixative that stabilizes lipid membranes and is reactive with phosphate and carbonate groups. When uranyl acetate is used, the specimens must not contain phosphate buffer. Because of their charged nature, the ions of the stain may not interact with some specimens, and a wetting agent, such as bovine serum albumin or bacitracin may be added to the stain solution to overcome this problem (Gregory and Pirie, 1973).

Negatively stained preparations must be examined at a relatively high magnification and at a high level of resolution and contrast. A magnification of ×50,000 is generally sufficient to allow the recognition of most clinically relevant viruses, and the further 10-fold magnification achieved by examining the image through the binocular magnifier allows for the definitive recognition of most viruses or their subunits. With the application of image capture technologies, the specimen can be examined visually on a high-resolution CRT screen. This can be used to obtain a better resolution of the image at a higher magnification for more precise identification of the virus. For virus diagnosis, the microscope must be aligned for very high resolution and contrast. Peplomers and capsids require an effective resolution of at least 2 nm to be identified. Similarly, the contrast must be very high in negatively stained specimens, and this is generally achieved with objective lenses of short focal length and appropriate objective apertures. Accordingly, microscopes optimized for thin-section examination may prove suboptimal for virus diagnosis in negatively stained specimens.

Negative-Staining Methods

A number of approaches, all using very simple materials, have been described for negative staining of the specimen (reviewed by Doane and Anderson, 1987, and Hayat and Miller, 1990). The copper specimen grids generally used are 200 to 400 mesh. These are coated with Formvar (polyvinyl formal) and stabilized with carbon film. Coated grids can be obtained commercially, but they are best coated on the premises and can be stored for a prolonged period under vacuum. The negative stain may be a 2% solution of phosphotungstic acid adjusted to pH 6.5 with KOH, a 1% solution of uranyl acetate at pH 4.0, or a 3% solution of ammonium molybdate, pH 7.2. The stains are made up in distilled water and sterilized by filtration. To promote the wetting properties of the phosphotungstic acid preparation, bovine serum albumin may be added up to a concentration of 0.05%. The manipulations of the grid are achieved by using fine electron microscopy forceps adapted to lock in a closed position, a supply of Whatman 1 filter paper for blotting, and a short-wavelength UV lamp for sterilization of the specimens.

Direct-Application Method

The direct-application method is the simplest and most commonly used approach, especially for laboratories where the specimen volume is high. A drop of the specimen is placed on a grid that is held in the forceps. Alternatively, a drop of the specimen may be placed on a glass slide or other disposable surface, such as parafilm, and the coated surface of the grid is touched to the specimen. The drop is wicked off with the edge of a strip of filter paper, and a drop of the negative stain is added. The stain is then wicked off, and the grid is air dried.

A useful variation of this method has recently been reported in which the coated surface of the grid is altered to bind the virus more effectively (MacRae and Srivastava, 1998). Prior to the application of the sample, the grid is floated on a drop of 0.1% poly-L-lysine (MW, 35,000) and washed with a drop of distilled water. We found that this approach enhanced the detection and recognition of gastroenteritis viruses from stools but provided no advantage in the detection of herpesviruses from lesion aspirates.

Water Drop Method

The direct-application method is ideal for clinical specimens that have been resuspended in water or a 1% solution of ammonium acetate (Doane et al., 1969). For specimens with a relatively high salt content at physiological or higher levels, the salts tend to precipitate on the grid surface and obscure the viruses present. The water drop method is a rapid approach to remove the excess salt. A drop of sterile distilled water or 1% ammonium acetate solution is placed on a waxed surface such as parafilm. A small drop of the specimen is then placed on top of the original drop. The coated surface of the grid is touched to the surface of the drop, and the grid is then processed for negative staining as described above. Because of the dilution effect inherent in this method, only specimens expected to have a high virus concentration, such as those obtained by density gradient centrifugation or amplification in cell culture, are best processed by this approach.

Agar Diffusion Method

For specimens containing high salt in which the virus is present in a lower concentration, the agar diffusion method has advantages (Anderson and Doane, 1972). The wells of a flexible microtiter plate are filled with approximately 300 μl of 1% agar dissolved in distilled water and can then be stored at 4°C. A coated specimen grid is placed on the surface of the agar, and a drop of the specimen is applied to the grid surface. The specimen drop is allowed to be absorbed into the agar at room temperature for 30 to 60 min. The grid is removed from the agar and processed for negative staining as described above. This approach is more time-consuming than the water drop method. However, it serves to concentrate the virus in the specimen while removing the excess salt present.

Airfuge Ultracentrifugation

The Airfuge ultracentrifuge has been developed expressly for the centrifugal deposition of viruses in the specimen on the EM grid (Hammond et al., 1981; Hayat and Miller, 1990). Coated specimen grids are placed at the periphery of each sector of the EM-90 rotor. Ninety microliters of the specimen clarified by centrifugation at at least 10,000 rpm for 10 min are placed in each respective rotor sector. The specimens are centrifuged at 90,000 rpm for 30 min. The grids are removed with forceps, and each grid is inverted briefly onto a drop of negative stain and dried. Decontamination of the rotor is achieved by immersing it in a solution of 2.5% glutaraldehyde or 10% formalin and rinsing it with ethanol (Doane and Anderson, 1987). Alternatively, a fixed-angle rotor can be used in the Airfuge ultracentrifuge to concentrate the virus. Up to 1,000-fold-increased sensitivity has been reported, although in our experience this is difficult to achieve. Specimens with a greater volume, such as urine, may be processed by differential centrifugation, namely, low-speed centrifugation at 10,000 × g for 10 min followed by ultracentrifugation at 100,000 × g for 1 h. The pellet is then resuspended in a small volume of 1% ammonium acetate solution and processed for negative staining.

Other methods of processing specimens for EM include the pseudoreplica approach and the use of polyethylene glycol to concentrate the virus in the specimen. The psuedoreplica approach involves the application of a Formvar solution to the surface of agar on which a specimen has been dried (Lee et al., 1978). The Formvar is then stained with phosphotungstic acid and mounted on a grid. Concentration of virus with polyethylene glycol (MW, 6,000) has also been used effectively (Hebert, T. T., 1963). While these are elegant approaches for specific viruses and specimens, they do not lend themselves to routine use in most diagnostic laboratories.

MORPHOLOGICAL FEATURES

Viruses are identified at the family or genus level by the distinct size and structure of the virion, the appearance of the nucleocapsid, or the peplomers of the envelope. Icosahedral viruses ranging from herpesviruses to parvoviruses have nucleocapsids of a defined diameter, and they may also have well-defined capsomeres, as shown in Fig. 3. Viruses with helical symmetry may also have well-defined nucleocapsids, such as those of the paramyxoviruses seen in Fig. 4B. These viruses may also have unique peplomers or spike proteins by which they can be identified, such as those of respiratory syncytial virus or influenza virus (Fig. 4A and C). For most electron microscopists, the identification of the large viruses, such as herpesviruses or poxviruses, shown in Fig. 3A, C, and D, is considered trivial. To identify smaller viruses, the electron microscope must be optimally configured for resolution and contrast. Hence, the differentiation of rotaviruses from reoviruses (Fig. 2A and B) or Norwalk viruses from astroviruses (Fig. 1B and C) requires not only knowledge of the respective structures but also a suitably configured electron microscope.

Despite the minimal manipulations of negative staining, some very sensitive viruses do not survive the process intact. For viruses with a distinct and stable nucleocapsid, such as the herringbone structures of the nucleocapsid of paramyxoviruses (Fig. 4B), this in itself serves to identify the virus. Conversely, viruses such as rubella virus (Fig. 4F) need to be intact to be identified. Such viruses can be stabilized by treating the specimen with 0.5% glutaraldehyde for 30 min at 4°C prior to applying it to the grid (Muller et al., 1983). This stabilizes the virus envelope sufficiently to allow for visualization of the intact particles after negative staining.

APPLICATIONS

Skin Lesions

Vesicular skin eruptions are generally caused by viruses such as herpes simplex virus or varicella-zoster virus (Fig. 3A). While infections with these viruses can be diagnosed with confidence by most health care providers, a laboratory diagnosis is called for in unusual presentations, in immunocompromised patients, or in cases where significant exposure of contacts has occurred which necessitates broad-range intervention, such as the administration of immune serum globulin. Direct examination of a specimen from such lesions can provide a rapid, specific, and relatively sensitive diagnosis. If the lesion is vesicular, the fluid is best aspirated

with a 25-gauge needle on a 1 ml syringe. The fluid is drawn into the needle without having it enter the syringe barrel. It is then expelled onto a glass slide and allowed to dry. Where it is ambiguous whether the lesion is a vesicle or a papule, scraping the surface with the beveled edge of the needle is an effective procedure because vesicles break and fluid will be visualized whereas papules remain intact. The fluid is then collected in the needle or picked up by pressing a slide to the droplet. The volumes of fluid obtained by this process range from 1 to 10 μl and prove adequate for diagnosis by EM. The needle may be rinsed by aspiration of medium that is then used for isolation or testing by PCR. Crusted lesions can be lifted with a syringe needle and transferred to the glass slide with the wet surface touched to the slide. Enough virus is generally present in the wet part of the crust to achieve a successful EM diagnosis. Ulcers can be sampled by scraping them with the beveled edge of the syringe needle, although such specimens may not be as readily productive. Lesions on mucous membranes, such as oral and genital lesions, can be collected by aspiration of the vesicle or scraping the ulcer. Such specimens often have less virus, and a diagnosis by EM is more challenging and time-consuming. Suspected herpetic whitlow occurring on the palms may present as a painful whitish discoloration. In collecting such specimens, one should take account of the fact that the skin on the palmar surface is substantially thicker, and more aggressive sampling may be required to reach the vesicle contents.

Other skin lesions for which a virus diagnosis can be established by EM include the pearly lesions of molluscum contagiosum and vesicles such as those of hand, foot, and mouth disease. The scraping from mulluscum lesions may be minimally productive, but the rate of success at visualizing orthopoxviruses (Fig. 3C) in such specimens is very high. A presumptive diagnosis of hand, foot, and mouth disease can be made by visualizing small round enteroviruses in aspirates of hand lesions. Lesions of contagious pustular dermatitis are generally large and vesicular. Parapox virus can be readily identified by EM in aspirates of these lesions (Fig. 3D).

In preparing lesion aspirate specimens for negative staining, it is ideal if the specimen is dried on the glass slide and then resuspended in a minimal volume of fluid. This can be accomplished by resuspending the specimen, regardless of its initial volume, in as little as 1 μl of 1% ammonium acetate and touching the preparation with a coated grid. This procedure frequently results in a more concentrated preparation of virus on the grid surface.

Stool Specimens

Viruses are the most common etiologic agents of gastroenteritis, and these include the rotaviruses, caliciviruses (including the Norwalk group), astroviruses, adenoviruses, coronaviruses, and toroviruses (Fig. 1 and 2). With this broad spectrum of agents that grow poorly if at all in cell culture, comprehensive immunospecific assays have proven difficult to develop. Accordingly, EM continues to have an important role in the diagnosis of these agents (reviewed by Petric, 1999). This is in part also due to the well-characterized structures of these agents, which are relatively stable under processing, and to the relatively high concentrations of these viruses in stool specimens, owing to the large surface area of the gut, where they replicate.

Stool specimens generally need to be resuspended in either water or a 1% ammonium acetate solution. They may

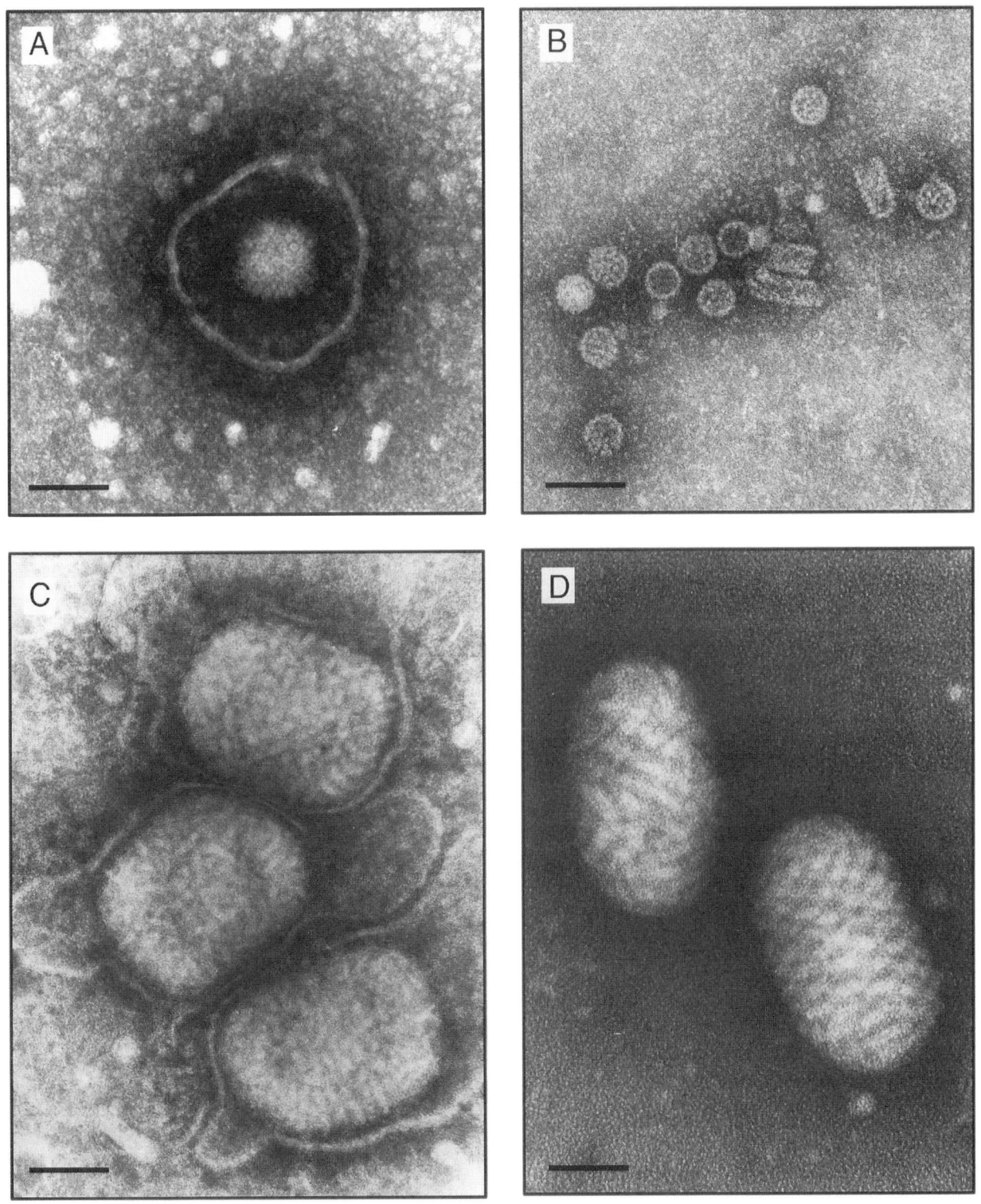

FIGURE 3 Viruses seen by EM in lesion specimens from skin or mucous membrane secretions. (A) Herpes group virus from herpes, varicella, or shingles. (B) Papovavirus from respiratory tract secretions of an immunocompromised patient. (C) Orthopoxvirus from molluscum contagiosum. (D) Parapoxvirus from orf. Bars, 100 nm.

be made as a 10 to 20% suspension that is then subjected to low-speed centrifugation to clarify the specimen. However, in the event of a high specimen load, a small amount of the stool can be suspended in approximately 10 volumes of a 1% ammonium acetate solution and applied directly to the grid. Centrifugation of the specimen poses the risk that it

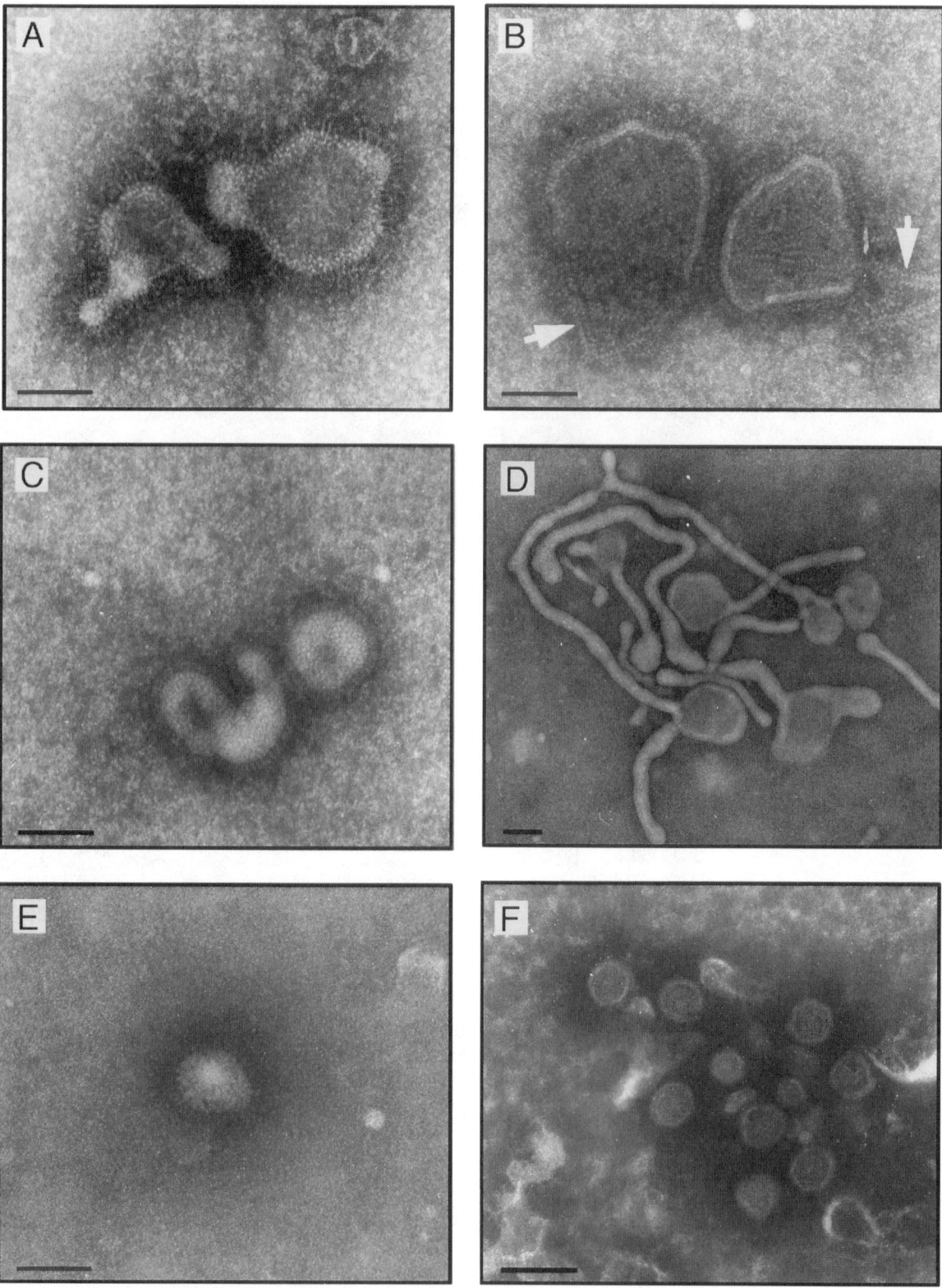

FIGURE 4 Agents seen in EM examination of cell cultures showing cytopathic effects. (A) Respiratory syncytial virus. The virus particles are generally intact with well-defined fringes of spikes. (B) Parainfluenza virus. Note that many particles are broken, with the nucleocapsid visible as herringbone rods (arrows). (C) Influenza virus. Note the well-defined spike proteins. (D) Mycoplasma-like organisms seen in cell cultures. (E) Foamy agent seen in contaminated cell cultures. (F) Rubella virus from infected cell culture. The preparations were treated with glutaraldehyde to stabilize the viruses. Bars, 100 nm.

may remove the immune complexes that form between the virus and the coproantibody, and it is more time-consuming due to additional manipulations of the specimen.

Since the numbers of etiologic agents involved are often small, the operator must have a thorough knowledge of the virus morphologies both under ideal conditions and under partial disintegration. For example, rotaviruses can be present as intact smooth particles or as partly degraded particles whose outer shells have been lost (Fig. 2B). Likewise, it is essential that the electron microscope be set at maximum resolution and contrast to recognize and differentiate between the astroviruses and the Norwalk group of viruses (Fig. 1). Lastly, we have found that *Campylobacter* has a unique sinusoidal morphology with bipolar flagella in characteristic sockets, which allows for its presumptive identification by EM (Fig. 2C). Specimens containing these structures are then referred for bacteriological workup.

Urine Specimens

Viruses chronically shed in the urine include adenovirus, cytomegalovirus, and human papovaviruses (Howell et al., 1998). These may be present in asymptomatic subjects, and their detection in the urine is difficult to interpret (Lecatsas and Boes, 1980). However, in immunocompromised patients the diagnosis of these viruses may be relevant. Since adenovirus can cause hemorrhagic cystitis, its detection in the urine is of diagnostic significance. To detect viruses in the urine, it is generally important to concentrate the specimen. Ultracentrifugation at 100,000 × g for 1 h will deposit most viruses. However, the resuspended pellets can contain extensive other debris that may make identification of the viruses difficult.

Other Specimens

Respiratory viruses, such as influenza virus, have been detected by EM from respiratory secretions (reviewed by Hayat and Miller, 1990). However, with the ready availability of immunofluorence microscopy and antigen-based enzyme immunoassays, detection by EM has become suboptimal. We have detected papovaviruses from respiratory tract secretions of patients who suffered from severe combined immunodeficiency (Fig. 3B). Likewise, detection of herpesviruses and paramyxoviruses has been reported in cerebrospinal fluid specimens from patients with complications of varicella or mumps, respectively. Prior to the development of sensitive immunospecific assays for hepatitis B virus, the presence of the hepatitis B virus surface antigen in the blood could be confirmed by immunoelectron microscopy (Fig. 5C). Tissues can be examined for viruses by direct EM. Papillomaviruses from warts and rotaviruses from the autopsy specimens of fatal cases of gastroenteritis can be detected by this approach (Carlson et al., 1978). Examination of other tissues for viruses generally requires fixation and sectioning, and these processes are more effectively performed in pathology laboratories. It must be remembered that viruses in thin-sectioned preparations do not have the same characteristic morphology as those visualized by negative staining.

Identification of Viruses in Cell Culture

EM represents an effective broad-spectrum method for examining infected cell cultures for the presence of a virus (Fig. 4). This is feasible, since the limited numbers of viruses present in the clinical specimens with which the cell culture was inoculated have been amplified to the level at which they can readily be detected by EM. Although identifying the virus present in the inoculated cell culture can

be done very effectively by fluorescence microscopy after staining the culture with specific monoclonal antibody, EM continues to offer some advantages. With specimens such as stools, either the appropriate antibodies may not be available or the causes of cytopathic effects may be diverse, and therefore the presence of a virus may be readily determined by EM. In other instances, the cytopathic effects may be relatively nonspecific and a broad-spectrum process such as EM can identify the virus. A further advantage of examining cell cultures by EM is that contaminating viruses that cause cytopathic effect in primary cell cultures, such as the simian paramyxovviruses, mycoplasmalike structures, and foamy agents (Fig. 4D and E) can be readily recognized (Doane and Anderson, 1987).

Cell culture preparations showing cytopathic effects are processed for EM by removing and storing the medium and adding 2 to 3 drops of 1% ammonium acetate to the monolayer to resuspend the cells. After a few minutes, the cell preparation is collected by scraping the monolayer, and a portion is applied to the coated grid and processed for negative staining. The cell culture medium may also be examined by applying it directly to the coated grid, or virus suspected to be present may be concentrated by ultracentrifugation as described above. Examination of the medium is preferred in cases where the cell culture has undergone extensive cytopathic effect with complete disruption of the monolayer. To reduce the residual salt in the medium, a drop of 1% ammonium acetate is placed on the grid after the cell lysate has been wicked off and removed prior to the application of the negative stain.

IMMUNOELECTRON MICROSCOPY

When a virus preparation is mixed with its specific antibody, the viruses or their subunits can be visualized by negative-contrast EM as aggregated forms of recognizable virus units. This approach has had a major role in the definitive identification and morphological characterization of viruses including rubella virus, hepatitis B virus, Norwalk virus, and, most recently, the human torovirus, as shown in Fig. 5A (Best et al., 1967; Bayer et al., 1968; Kapikian et al., 1972; Duckmanton et al., 1997).

Immunoelectron microscopy has been used effectively to increase the sensitivity of virus detection and to serotype a number of viruses (Anderson and Doane, 1973; Edwards et al., 1975; Petrovicova and Juck, 1977). With a defined viral antigen, this approach has also been used to detect seroconversion (Kapikian et al., 1975; Duckmanton et al., 1997).

The detection of virus-antibody interaction by immunoelectron microscopy has been greatly improved by the use of antibodies labeled with colloidal gold, as shown in Fig. 5D (Stannard et al., 1982; Beesley and Betts, 1985). This approach can include antibody directly labeled with the gold particles, labeled anti-species antibody, or gold-labeled protein G or protein A, which are available commercially (Geohegan and Ackerman, 1977). These procedures are sensitive to concentrations of the reagents and require careful titration and adequate controls.

Direct Immunoelectron Microscopy

Various conditions for the reaction of the virus preparation with the antibody have been reported. For stool specimens, a 1:50 dilution of the serum is mixed with a virus preparation, incubated for 2 h at 37°C, and centrifuged at 14,000 rpm for 15 min. The pellet is resuspended in 1% ammonium acetate, applied to the coated grid, and

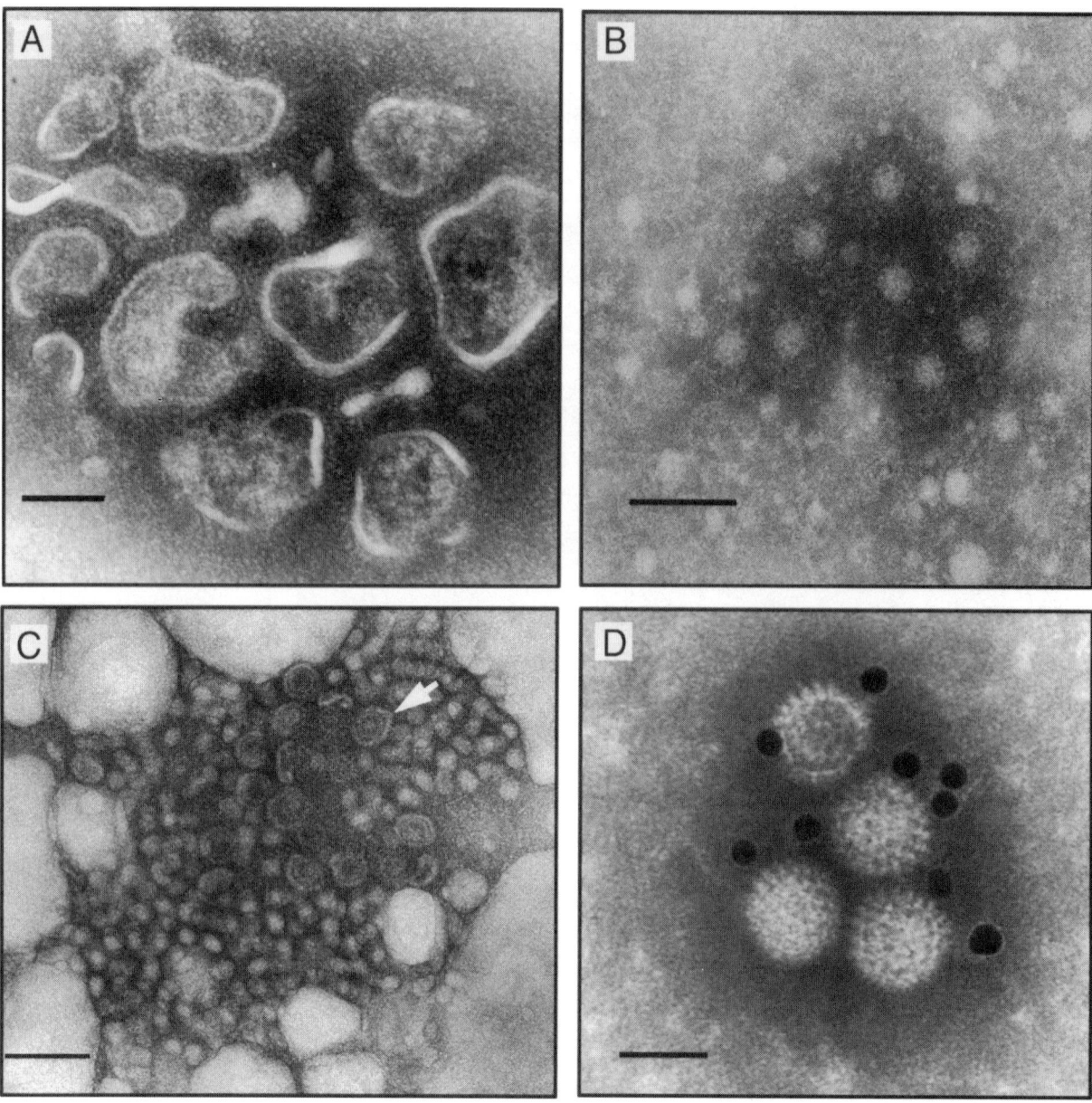

FIGURE 5 Immunoelectron microscopy of viruses. (A) Human torovirus reacted with patient convalescent serum. (B) Astrovirus from a stool specimen reacted with antiserum produced in guinea pig. (C) Serum from a hepatitis B virus-infected patient reacted with reference antiserum to hepatitis B surface antigen. Note the intact hepatitis B virions (arrow) and 22-nm diameter surface antigen spheres. (D) Rotavirus reacted with reference antibody and protein A labeled with colloidal gold. Note the association of gold granules with the virus particles. Bars, 100 nm. (Reprinted from Hopley and Doane, 1985, with permission.)

processed as outlined above. This approach was used in the initial detection of the Norwalk virus and the recent characterization of the human torovirus (Kapikian et al., 1972; Duckmanton et al., 1997). Other variations on this method include incubating the virus with the antiserum for 1 h at 37°C followed by overnight incubation at 4°C. The mixture is applied to the grid using the agar diffusion method described above (10). This is an important step, since the direct application of serum to the grid may result in excess deposits of material, making the recognition of the virus and its subunits difficult.

Serum-in-Agar Method

The serum-in-agar approach is based on the principle that antibody in an agar block on which a coated grid has been placed will diffuse and react with the specimen to form immune complexes on the grid surface (Anderson and Doane, 1973). A 1% molten-agar preparation is cooled to 45 to 50°C, mixed with the desired antibody preparation (serum or monoclonal antibody), and placed in the wells of a microtiter plate as described above for the agar diffusion method. After the agar has solidified, a coated grid is placed on its surface and a drop of the virus specimen is applied to the grid. The drop is allowed to dry at room temperature, and the grid is stained as described above for the direct-application method. This has been reported to be a very sensitive immunoelectron microscopy approach. For optimal performance, the antisera must be used in the optimal concentration range, so titration of these reagents is necessary.

Immunogold EM

A very important development in the diagnosis of viruses by EM is the ability to use antibody conjugated to colloidal gold as an immunospecific marker. This has allowed the definitive identification of virus particles or their subunits based on morphology and immune reactivity, as shown in Fig. 5D. Protein A or protein G labeled with colloidal gold will react with a wide spectrum of antibodies, and it can therefore be used to identify viruses that have reacted with the antibodies. There are different approaches for carrying out the reaction of the virus with the antibody-colloidal gold conjugate. One common approach involves the adaptation of the serum-in-agar method described above, in which the antibody is incorporated into the agar (Hopley and Doane, 1985; Doane and Anderson, 1987). The virus preparation is mixed with an equal volume of a colloidal gold conjugate of protein A, which is suspended in a buffer of 0.5 M Tris-HCl, pH 7.0, 0.15 M NaCl, and 0.5 M polyethylene glycol (MW, 20,000), and the mixture is applied to the grid surface. The virus-conjugate mixture is allowed to dry on the agar surface, and the grid is rinsed in 3 sequential drops of protein A-conjugate buffer, followed by 2 drops of distilled water. The preparation is then subjected to negative staining as described above. The association of the colloidal gold beads of uniform size with the virus particles or their subunits is indicative of an immunospecific reaction.

Solid-Phase Immunoelectron Microscopy

A fundamental principle of negative-contrast EM is the attachment of the virus to the surface of the coated grid. This attachment can be selectively enhanced by coating the grid surface with an antibody to the virus. This approach has been reported to enhance virus attachment by 10- to 100-fold, thereby improving the sensitivity of the procedure for virus detection by that degree. A further improvement to this approach has been to initially coat the grid with a solution of protein A followed by the specific antibody.

The approach consists of applying a drop of protein A solution to a coated grid (Doane and Anderson, 1987). The grid may be floated on a drop of protein A solution (1 mg/ml). The grid is then drained and washed by touching it to 3 drops of 0.05 M Tris-HCl buffer, pH 7.2, for a total wash time of 1 to 2 min. The grid is then inverted over a drop of a 1:100 dilution of antiserum for 10 min and again washed by passing it over 3 drops of Tris-HCl buffer. The grid is then inverted on a drop of the specimen for 30 min at room temperature, rinsed in Tris-HCl buffer, and stained as described above. This method is useful for detecting a suspected virus present in small quantities in the specimen.

ADVANTAGES OF EM AS A DIAGNOSTIC APPROACH

EM is the most direct approach for virus diagnosis. It has one of the highest degrees of specificity of any diagnostic method, at the level of the family to which the virus belongs, since the virus or its subunits have unique and definitive morphologies. The method provides the most rapid diagnosis of a virus infection for certain specimens, such as lesion aspirates and stools, as well as identification of viruses in inoculated cell cultures showing cytopathic effects. Since it is based on virus morphology, it can detect a very wide spectrum of viruses in a single examination, which overcomes the limitations of other assays that are based on a positive or negative response with specific probes or cell lines. Immunoelectron microscopy allows for increased sensitivity and specificity. It is also allows for the investigation of the immune response to viruses that cannot be cultivated and for which other tests have not yet been developed. From a functional perspective, it has the advantage that it does not require infectious particles and can be effectively performed on appropriately fixed specimens, provided that the virus morphology has remained intact. Finally, in the case of asymptomatic shedding of viruses, such as adenoviruses in stool specimens, the lower sensitivity of EM is an advantage in that it can only detect virus at the high concentrations that are more consistent with a disease process.

LIMITATIONS

Perhaps the major limitation of EM is the cost, which includes the instrument and its maintenance. The costs represent a major investment for most institutions, and an electron microscope is often acquired for shared use with pathology or research laboratories. This arrangement may result in the instrument not being readily available for rapid virus diagnosis. It can also lead to the acquisition of an instrument in which the resolution and contrast are compromised and therefore suboptimal for observing viruses in negatively stained specimens. The second limitation is the availability of technical staff with a good knowledge of virus morphology who have a sound understanding of the process and can attend to minor maintenance matters. The third limitation is the need for an adequate virus concentration in the specimen. Hence, stools and lesion aspirates are acceptable specimens whereas cerebrospinal fluids and respiratory secretions are generally not acceptable for diagnosis by EM. Specimens such as urine require concentration, which makes the procedure more labor intensive. The final limitation of EM rests in the fact that the specimens must be examined individually and hence the procedure is not open to automation. A qualified operator can only prepare and screen between 30 and 40 specimens per day. The process is therefore ideal for screening relatively few specimens and is not amenable to commercial or other laboratories that process a high volume of specimens.

TIPS

Ideally, a virology laboratory should have its own dedicated electron microscope. While the newest models have excellent features of image processing and alignment, models dating from 30 years ago are still functional and produce excellent images. Hence, an electron microscope is an investment that will last for a very long time. To ensure that a microscope is optimally aligned and operating with the appropriate apertures and providing optimal resolution and contrast to clearly see the viruses and their features, it is valuable to obtain grids with negatively stained preparations of the viruses of interest from a reference laboratory. With these grids the microscope can be tuned to provide clear images of viruses that can be readily recognized.

For routine diagnostic work, the microscope should have

the capacity to readily switch from a low magnification (×500 to ×1,000) to a high magnification (×50,000) suitable for virus identification. At low magnification, the grid is scanned to select grid areas that possess an adequate amount of appropriately stained specimen. The instrument is then switched to high magnification to search for virus particles in the selected areas.

It is important that specimens be appropriately collected. Stool specimens in containers are far preferable to rectal swabs, since one is assured that an adequate quantity of material has been collected (Howell et al., 1998). Aspirates of skin lesions are best collected by the virologist or the designated technologist(s) trained to operate the electron microscope. This ensures consistency in the collection of the specimen and its expeditious processing. It also allows for a decision to be made by experienced professionals as to whether the lesion is suitable for sampling.

Since the above procedures involve work with infectious specimens, adequate safety precautions must be employed. The handling of the specimen is preferably done in a laminar flow hood. This may prove challenging when handling stool specimens if the hood is not vented to the outside. A potential safety risk arises when the stained specimen on the grid is removed from the hood and transferred to the electron microscope. Exposure of the grid to a UV light source of 700 to 1,000 μW per cm^2 at a distance of 10 to 15 cm for approximately 10 min is considered adequate to inactivate the viruses on its surface (Doane and Anderson, 1987). Similarly, forceps used to handle grids should be dipped in ethanol and flamed immediately after use. With repeat use, metal forceps will become warped. They can be reconditioned by regrinding them. Ceramic forceps are longer lasting in this respect. Grids to be discarded should be either autoclaved or immersed in a 2.5% glutaraldehyde solution.

FUTURE PERSPECTIVES

Computerization has had a major positive impact on the operation of electron microscopes. The developments in image manipulation allow for the display of images on a CRT monitor, which further simplifies the recognition of virus particles or their subunits and is a major advantage for teaching. Such images can be readily printed or sent to a reference laboratory electronically or stored for future reference, which greatly facilitates consultation and interpretation. These developments are expected to make EM more user friendly and reduce the amount of training required to become proficient in this technology.

CONCLUSIONS

The developments in immunoelectron microscopy and improvements in instrumentation have greatly enhanced the sensitivity and specificity of diagnostic EM. However, these manipulations are time-consuming; hence, routine diagnostic EM continues to use the simple direct-application method with refinements being used only on more challenging or unorthodox specimens. With the development of enzyme immunoassays and molecular approaches, such as PCR and reverse transcription-PCR, EM is now mainly used in reference laboratories or laboratories in tertiary health care centers. However, in these centers it remains an indispensable process in the delivery of a comprehensive virus diagnosis.

REFERENCES

Almeida, J., B. Cinader, and A. Howatson. 1963. The structure of antigen antibody complexes. *J. Exp. Med.* **118:**327–340.

Anderson, N., and F. W. Doane. 1972. Agar diffusion method for negative staining of microbial suspensions in salt solutions. *Appl. Microbiol.* **24:**495–496.

Anderson, N., and F. W. Doane. 1973. Specific identification of enteroviruses by immuno-electron microscopy using a serum in agar diffusion method. *Can. J. Microbiol.* **19:**585–589.

Bayer, M. E., B. S. Blumberg, and B. Werner. 1968. Particles associated with Australia antigen in the sera of patients with leukemia, Down's syndrome and hepatitis. *Nature* **218:**1057–1059.

Beesley, J. E., and M. P. Betts. 1985. Virus diagnosis: a novel use for the protein A-gold probe. *Med. Lab. Sci.* **42:**161–165.

Best, J. M., J. E. Banatvala, J. D. Almeida, and A. P. Waterson. 1967. Morphological characteristics of rubella virus. *Lancet* **ii:**237–239.

Brenner, S., and R. W. Horne. 1959. A negative stain method for high resolution electron microscopy of viruses. *Biochim. Biophys. Acta.* **34:**103–110.

Carlson, J., P. J. Middleton, M. T. Szymanski, J. Huber, and M. Petric. 1978. Fatal rotavirus gastroenteritis. *Am. J. Dis. Child.* **132:**477–479.

Doane, F. W., and N. Anderson. 1987. *Electron Microscopy in Diagnostic Virology: a Practical Guide and Atlas.* Cambridge University Press, New York, N.Y.

Doane, F. W., N. Anderson, A. Zbitnew, and A. J. Rhodes. 1969. Application of electron microscopy to the diagnosis of virus infections. *Can. Med. Assoc. J.* **100:**1043–1049.

Duckmanton, L., B. Luan, J. Devenish, R. Tellier, and M. Petric. 1997. Characterization of torovirus from human fecal specimens. *Virology* **239:**158–168.

Edwards, E. A., W. A. Valters, L. G. Boehm, and M. J. Rosenbaum. 1975. Visualization by immune electron microscopy of viruses associated with acute respiratory diseases. *J. Immunol. Methods* **8:**159–167.

Geohegan, W. D., and G. A. Ackerman. 1977. Adsorption of horseradish peroxidase, ovomucoid and anti-immunoglobulin to colloidal gold for the direct detection of concanavalin A, wheat germ agglutinin, and goat anti-human immunoglobulin G on cell surfaces at the electron microscope level. *J. Histochem. Cytochem.* **25:**1187–1200.

Gerna, G., A. Sarasini, B. S. Coulson, M. Parea, M. Torsellini, E. Arbustini, and M. Battaglia. 1988. Comprehensive sensitivities of solid phase immune electron microscopy and enzyme linked immunosorbent assay for subtyping human rotavirus strains with neutralizing monoclonal antibodies. *J. Clin. Microbiol.* **26:**1143–1151.

Gregory, D. W., and B. J. S. Pirie. 1973. Wetting agents for biological electron microscopy. *J. Microsc.* **99:**261–265.

Hammond, G., P. R. Hazleton, I. Chuang, and B. Klisko. 1981. Improved detection of viruses by electron microscopy after direct ultracentrifuge preparation of specimens. *J. Clin. Microbiol.* **14:**210–221.

Hayat, M. A., and S. Miller. 1990. *Negative Staining.* McGraw Hill, New York, N.Y.

Hebert, T. T. 1963. Precipitation of plant viruses by polyethylene glycol. *Phytopathology* **53:**362.

Hopley, J., and F. W. Doane. 1985. Development of a sensitive protein A-gold immunoelectron microscopy method for detecting viral antigens. *J. Virol. Methods* **12:**135–147.

Howell, D. N., C. M. Payne, S. A. Miller, and J. D. Shelburne. 1998. Special techniques in diagnostic electron microscopy. *Hum. Pathol.* **29:**1339–1346.

Johnson, R. P. C., and D. W. Gregory. 1993. Viruses accumulate spontaneously near droplet surfaces: a method to concentrate viruses for electron microscopy. *J. Microsc.* **171:**125–136.

Kapikian, A. Z., R. G. Wyatt, R. Dolin, T. S. Thornhill, A. R. Kalica, and R. M. Chanock. 1972. Visualization by immune electron microscopy of a 27 nm particle associated with acute infectious nonbacterial gastroenteritis. *J. Virol.* **10:**1075–1081.

Kapikian, A., Z., S. M. Feinstone, R. H. Purcell, R. G. Wyatt, T. S. Thornhill, A. R. Kalica, and R. M. Chanock. 1975. Detection and identification by immune electron microscopy of fastidious agents associated with respiratory illness, acute non-bacterial gastroenteritis and hepatitis A. *Perspect. Virol.* **9:**9–47.

Lecatsas, G., and E. G. Boes. 1980. Urinary virus excretion in pregnancy. *S. Afr. Med. J.* **57:**988–990.

Lee, F. K., A. J. Nahmias, and S. Stagno. 1978. Rapid diagnosis of cytomegalovirus infection in infants by electron microscopy. *N. Engl. J. Med.* **299:**1266–1270.

MacRae, J., and M. Srivastava. 1998. Detection of viruses by electron microscopy: an efficient approach. *J. Virol. Methods* **72:**105–108.

Middleton, P. J., M. T. Szymanski, and M. Petric. 1977. Viruses associated with acute gastroenteritis in young children. *Am. J. Dis. Child.* **131:**733–737.

Muller, G., M. Bruns, L. M. Peralta, and F. Lehman-Grube. 1983. Lymphocytic choriomeningitis virus. IV. Electron microscopic investigation of the virion. *Arch. Virol.* **74:**229

Petric, M. 1999. Caliciviruses, adenoviruses and other diarrheic viruses, p. 1005–1013. *In* P. R. Murray, E. J. Baron, M. A. Pfaller, F. C. Tennover, and R. H. Yolken (ed.), *Manual of Clinical Microbiology*, 7th ed. ASM Press, Washington, D.C.

Petrovicova, A., and A. S. Juck. 1977. Serotyping of Coxsackie viruses by immune electron microscopy. *Acta Virol.* **21:**165–167.

Stannard, L. M., M. Lennon, M. Hodgekiss, and H. Smuts. 1982. An electron microscopic demonstration of immune complexes of hepatitis B e-antigen using colloidal gold as a marker. *J. Med. Virol.* **9:**165–175.

van Rooyen, C. E., and G. D. Scott. 1948. Smallpox diagnosis with special reference to electron microscopy. *Can. J. Public Health* **39:**467–477.

The Interference Assay

CHARLES A. REED

6

HISTORY

For many decades, rubella was looked upon as a rather unimportant and mild disease of childhood. It was commonly referred to as German measles or the three-day measles and seemed to cause few complications. Rubella had first been described in the German medical literature by DeBergen in 1752 and Orlow in 1758 (Emminghaus, 1870), but it was not until 1941 that the teratogenic effects of this virus, when acquired during pregnancy, became well documented. It was then that an Australian ophthalmologist, Gregg (1941), reported a high incidence of cataracts and other anomalies among the offspring of mothers who had contracted rubella early in pregnancy during a rubella epidemic in Australia. A few years later, Swan et al. (1943; 1946) reported more cases of congenital defects, with the later observations related especially to rubella. Since then, many reports have appeared in the literature confirming these findings and reporting many other congenital defects and complications of pregnancy following maternal rubella. Through the years, the pattern of events with their resulting effects have become known as "the rubella syndrome." Many retrospective studies and other reports in the literature are in agreement with the earlier findings (Skinner, 1961; also see chapter 36).

With the advent of newer virologic techniques, more precisely defined chemical media, and a multiplicity of cell lines developed in the late 1940s and early 1950s, the world of virology exploded with the isolation and identification of many viruses (Melnick et al., 1979). However, it was not until 1962 that two groups of investigators succeeded in isolating the rubella virus in two different systems. Weller and Neva were able to isolate the virus in primary human amnion cell cultures, which demonstrated a visible cytopathic effect (CPE) with time (Weller and Neva, 1962). Parkman et al. (1962) were able to demonstrate the presence of the virus in primary African green monkey kidney (AGMK) cell cultures, which produced no rubella virus-associated CPE. However, it was shown that rubella virus infection interfered with the CPE production of a challenge virus (e.g., ECHO-11). This was the first demonstration of the interference assay that could be used to identify rubella virus subsequent to its isolation in cell culture. At about the same time, Veronelli et al. reported similar findings using a continuous rhesus monkey cell line, LLC-MK2, which had been shown to be useful for virus research (Hull et al., 1962; Veronelli et al., 1962).

Parkman et al. continued their research with the rubella virus interference phenomenon, and it was shown that infection in cell culture interfered with CPE production by many enteroviruses, myxoviruses, and arboviruses in AGMK cell cultures (Parkman et al., 1964a). Thirteen different cell lines were tested, and it was found that virus replication occurred in 11 of these cell lines, but the interference with the CPE by a challenge virus varied with the different lines. In addition, their studies showed that rubella virus is heat labile and that specimens should be chilled or frozen after collection. In continuing studies, Parkman et al. demonstrated that rubella virus replication and neutralization tests showed similar results when the AGMK and LLC-MK2 cell lines were compared (Parkman et al., 1964b). All of the techniques and methods were put to the test when the United States was swept by a major epidemic of rubella in 1964–1965, and most virology laboratories started to culture for the rubella virus. With the licensure of a rubella vaccine in 1969, however, the demands on the clinical virus laboratory for culturing rubella virus have been significantly reduced.

SPECIMEN COLLECTION, TRANSPORT, AND STORAGE

As with all virus isolation, the process starts with the proper collection and handling of the specimen (see chapter 2). Once a sample is inoculated onto tissue culture, the remainder of the specimen should be frozen and stored at −70°C.

For rubella virus identification by the interference assay, the following specimens are appropriate (Sanders, 1978; Herrmann, 1979).

1. Respiratory. Nasal and pharyngeal swabs from children and adults. Throat washings may be suitable from adults. These should be collected no later than 5 days after the onset of the rash.

2. Blood. Heparinized blood should be collected as near the onset of the rash as possible and as early as 7 days after possible exposure.

3. Cerebrospinal fluid. Cerebrospinal fluid is most useful in cases involving suspected congenital rubella.

4. Urine. Urine should be collected as aseptically as possible. Urine has been shown to contain virus for months

after the birth of a congenitally infected infant (Cooper and Krugman, 1966).

5. Tissues. All tissues and body fluids may be collected and tested when appropriate. These include ocular tissues and fluids, autopsy or biopsy tissues, placenta, amniotic fluid, and any fetal tissues that are accessible.

SPECIMEN PREPARATION AND INOCULATION

Specimens collected on swabs should be expressed into 2 ml of maintenance medium if not collected directly in a transport broth. All specimens—with the exception of cerebrospinal fluid—should be treated with the appropriate antimicrobial agents according to each laboratory's preference. Tissue material should be ground and prepared as a 10% suspension in the same maintenance medium that is to be added to the final cell culture. Our laboratory utilizes Earle's basic salt solution, Eagle's minimum essential medium (EMEM), and 2% fetal calf serum as a maintenance medium. Adjust the pH of all specimens to 7.0 to 7.2.

Select the cell line to be inoculated. The primary AGMK cell line is the most widely used, but it has drawbacks that will be discussed later. LLC-MK2 is used in our laboratory if problems arise with the AGMK cells. Remove the medium from the selected tube cultures, using four tubes for each specimen. Inoculate 0.25 ml of the specimen directly onto each cell culture followed by a 1-h adsorption period at 36°C (stationary rack). Add 1.5 ml of fresh maintenance medium to each tube, and incubate in a roller drum. Observe each tube daily, and record any changes noted.

ECHO-11 VIRUS CHALLENGE AND NEUTRALIZATION

Rubella virus may not produce visible CPE but may be detectable by the property of viral interference. For example, if rubella virus is growing in AGMK cells, these cells become resistant to ECHO-11 virus, to which they are ordinarily highly susceptible. As a result, no CPE is produced following infection by ECHO-11 virus. The following procedure is our adaptation of the procedures of Parkman et al. (1964b).

■ Challenge

1. Prepare a stock pool of ECHO-11 virus in LLC-MK2 cells, aliquot it into tubes (1 ml per tube), and store it at −70°C.
2. Titer this challenge virus (ECHO-11) in LLC-MK2 cells to an endpoint by the Reed-Muench (1938) method.
3. Inoculate submitted specimens that are suspected to contain rubella virus (including all specimens from newborns with congenital malformations) into four tubes of AGMK cells of the same lot.
4. At 10 days postinoculation, remove the medium from two inoculated tubes and two control tubes of AGMK cells.
   a. Add 100 50% tissue culture infective doses ($TCID_{50}$s) in 0.1 ml of ECHO-11 challenge virus each to one inoculated tube and one control tube. Add 10 $TCID_{50}$s in 0.1 ml of ECHO-11 challenge virus to each of the remaining inoculated and control tubes. These dilutions of virus stock are made in EMEM supplemented with 2% fetal calf serum.
   b. Incubate the tubes in a stationary rack for 1 h at 36°C; manually rotate the tubes every 10 to 15 min.
   c. Add 1.5 ml of EMEM with 2% fetal calf serum to each tube at the end of this incubation period.
   d. Incubate the tubes in a stationary rack at 36°C.
5. Examine the tubes daily; interference is present if extensive CPE develops in the control cell cultures but not in the cell cultures inoculated with the clinical specimen.

Note: This method allows for the detection of low-titer virus replication. If the 10-$TCID_{50}$ challenge tube is negative for ECHO-11, the unchallenged tubes should be harvested and passed to four new tubes. Repeat steps 3 and 4 above.

■ Interference-Neutralization Test

1. Harvest material (by one freeze-thaw cycle) from the two inoculated tubes that were not subjected to challenge virus.
2. Place 0.15 ml of the harvested material in a sterile screw-cap tube (16 by 75 mm), and add 0.15 ml of rubella antiserum at a dilution to represent 50 neutralizing-antibody units per 0.1 ml.
3. As a control, incubate the harvest material without antiserum.
4. Incubate samples in a stationary rack for 1 h at 36°C.
5. Inoculate 0.2 ml of the virus-antiserum mixture and 0.1 ml of the untreated virus into one tube each of AGMK cells; incubate the tubes at 36°C in a stationary rack. Duplicate tubes may be used.
6. After 5 days, inoculate each tube with 100 $TCID_{50}$ of ECHO-11 virus; if duplicate tubes were prepared, one set may be challenged with 10 $TCID_{50}$ of ECHO-11 virus; the virus is confirmed as rubella virus if virus interference is prevented by antiserum treatment and is again demonstrated in the control tube.

Note: Several different log dilutions of virus can be employed at step 6 if the virus titer is too high for the neutralizing capacity of the antibody.

PROS AND CONS

The "rubella challenge-interference" system is probably the most rapid system for isolating the virus from a patient specimen in a single passage. It can be achieved within 14 days on primary AGMK cells. Other cell lines that are capable of replicating rubella virus usually require more than one blind passage before a visible CPE is noted. During the period of time required for virus replication, the physical condition of the cell sheet is critical and must be visually compared very carefully with uninfected control tubes. Most laboratories today do not pass the specimen from tubes of a set that has shown a positive ECHO-11 virus CPE. The specimen is reported negative after 14 days.

However, because laboratories may experience contamination with indigenous resident simian viruses in purchased AGMK cell lines from time to time, these cell lines should be examined daily for simian-virus-associated CPE and tested for the presence of these simian viruses. If the cell cultures are positive for contaminating agents, they should not be used because misleading results may occur. If specimens have already been inoculated in tubes of that lot, reinoculate a new lot of tested cells from the specimen sample that has been held in frozen storage. This problem is not encountered if the LLC-MK2 line is used; however, LLC-MK2 cells are not as sensitive for the primary isolation of rubella virus. The

LLC-MK2 line can be used for the neutralization test subsequent to the primary isolation of the virus.

Because of the expense of the cell cultures and the laboratory personnel time that is required to isolate and challenge rubella virus and then confirm the isolation, this protocol has been questioned, especially if a serologic test will give the physician the necessary information. However, virus isolation and identification by interference assay is still the standard for proof of current rubella virus infection.

REFERENCES

Cooper, L. Z., and S. Krugman. 1966. Diagnosis and management: congenital rubella. *Pediatrics* **37**:335–338.

Emminghaus, H. 1870. Uber rubeolen. *Jahrb. Kinderheilkd.* **4**:47–59.

Gregg, N. M. 1941. Congenital cataract following German measles in the mother. *Trans. Ophthalmol. Soc. Aust.* **3**:35–46.

Herrmann, K. L. 1979. Rubella virus, p. 725–766. *In* E. H. Lennette and N. J. Schmidt (ed.), *Diagnostic Procedures for Viral, Rickettsial and Chlamydial Infections*, 5th ed. American Public Health Association, Washington, D.C.

Hull, R. H., W. R. Cherry, and O. J. Tritch. 1962. Growth characteristics of monkey kidney cell strains LLC-MK1, LLC-MK2 and LLC-MK2 (NCTC-3 196) and their utility in virus research. *J. Exp. Med.* **115**:903–917.

Melnick, J. L., H. A. Wenner, and C. A. Phillips. 1979. Enteroviruses, p. 471–534. *In* E. H. Lennette and N. J. Schmidt (ed.), *Diagnostic Procedures for Viral, Rickettsial and Chlamydial Infections*, 5th ed. American Public Health Association, Washington, D.C.

Parkman, P. D., E. L. Buescher, and M. S. Artenstein. 1962. Recovery of rubella virus from army recruits. *Proc. Soc. Exp. Biol. Med.* **111**:225–230.

Parkman, P. D., E. L. Buescher, M. S. Artenstein, J. M. McCown, F. K. Mundon, and A. D. Druzd. 1964a. Studies of rubella. I. Properties of the virus. *J. Immunol.* **93**:595–607.

Parkman, P. D., F. K. Mundon, J. M. McCown, and E. L. Buescher. 1964b. Studies of rubella. II. Neutralization of the virus. *J. Immunol.* **93**:608–617.

Reed, L. J., and H. Muench. 1938. A simple method of estimating fifty percent end points. *Am. J. Hyg.* **27**:493–497.

Sanders, C. V., Jr. 1978. Diagnostic virology, p. 259–281. *In* H. Rothschild, F. Allison, Jr., and C. Howe (ed.), *Human Diseases Caused by Viruses*. Oxford University Press, New York, N.Y.

Skinner, C. W. 1961. The rubella problem. *Am. J. Dis. Child.* **101**:78–86.

Swan, C., A. L. Tostevin, B. Moore, H. Mayo, and G. H. B. Black. 1943. Congenital defects in infants following infectious disease during pregnancy. *Med. J. Aust.* **2**:201–210.

Swan, C., A. L. Tostevin, and G. H. B. Black. 1946. Final observations on congenital defects in infants following infectious diseases during pregnancy with special reference to rubella. *Med. J. Aust.* **2**:889–908.

Veronelli, J. A., H. F. Maassab, and A. V. Hennessy. 1962. Isolation in tissue culture of an interfering agent from patients with rubella. *Proc. Soc. Exp. Biol. Med.* **111**:472–476.

Weller, T. H., and F. A. Neva. 1962. Propagation in tissue culture of cytopathic agents from patients with rubella-like illness. *Proc. Soc. Exp. Biol. Med.* **111**:215–225.

Immunofluorescence

TED E. SCHUTZBANK AND ROBYN McGUIRE

7

HISTORY

The use of fluorochrome-labeled antibody reagents to visualize microbial antigens was first described by Coons et al. (1942). By the 1950s, this technique was used routinely to detect viruses and other infectious agents (Arnaud et al., 1976; Litwin and Grose, 1992) in patient specimens (Biegeleisen et al., 1959; Nichols and McCoumb, 1962; Kirsh and Kissling, 1963; Schaap et al., 1963; Fedova and Zelenkova, 1965; Uchida et al., 1965; Haire, 1969; Baratta et al., 1975; Fulton and Middleton, 1975; Lennette et al., 1976; Olding-Stenkvist et al., 1976; Bikbulatov et al., 1978; Anestad et al., 1983; Grandien and Olding-Stenkvist, 1984) and tissue culture (Gardner and McQuillin, 1968; Lennette et al., 1976; Minnich and Ray, 1980). The early immunofluorescence assays (FA) used noncommercial preparations of polyclonal antisera directed against the target virus and a secondary reagent labeled with either rhodamine or fluorescein. These reagents had significant cross-reactivity due to the presence of antibodies to host cell proteins. Interpretation of results required a great deal of training, skill, and experience. A detailed description of these early methods is provided by Gardner and McQuillin (1974).

In the early 1980s, reports indicated an improvement in the quality and consistency of immunofluorescence, as the use of purified monoclonal antibodies resulted in the elimination of many of the nonspecific staining problems (Gardner and McQuillin, 1968; Athanasiu, 1985). Over the next few years, reagents containing monoclonal antibodies directed against herpesviruses and respiratory viruses were described (Showalter et al., 1981; Balachandran et al., 1982; Volpi et al., 1983; Balkovic and Hsiung, 1985; Routledge et al., 1985; Shalit et al., 1985; Swierkosz et al., 1985; Waner et al., 1985; Walls et al., 1986). Although monoclonal antibodies are more difficult technically to produce than polyclonal antibodies, the availability of commercial reagents helped to expand the range of testing offered by clinical virology laboratories. Moreover, many of these reagents were labeled directly with fluorescein, which shortened the assay to a single step, with either a 15- or 30-min incubation, rather than the usual two-step, 60-min assay. Now, more than 50 years after the first report, immunofluorescence has become one of the primary technologies used by diagnostic virology laboratories (Gallo, 1983).

FLUORESCENCE AND THE FLUORESCENCE MICROSCOPE

Fluorescence

Fluorescence is defined as the adsorption of a specific wavelength of light by a specific molecule, followed by the nearly instantaneous reemission of light at a different wavelength. Both the excitation and emission wavelengths are defined by the molecular structure of the fluorescent compound, or fluorochrome. When the molecule absorbs light at the excitation frequency, its electrons move to a higher energy state. As the electrons lose energy and return to their ground state, energy is emitted in the form of light. The wavelength of the emitted light is always longer than the excitation wavelength (referred to as "Stokes fluorescence").

A wide variety of fluorochromes have been chemically modified so that they can be directly coupled to proteins. The excitation-emission spectra of one of the most commonly used fluorochromes, fluorescein isothiocyanate (FITC), are shown in Fig. 1. By coupling fluorochromes directly to antibody molecules, these "fluorescent antibodies" can be used as a stain to detect the presence of specific antigens in clinical specimens and virus-infected cell cultures. For this procedure, a fluorescence microscope is required.

Fluorescence Microscopy

A fluorescence microscope is used to permit irradiation of the specimen with light at the desired excitation wavelength and to allow discrimination between that and the weaker emitted light emanating from the specimen. Historically, there are two types of fluorescence microscopes, transmitted-light and incident-light (epifluorescence) microscopes. Transmitted-fluorescence microscopes, which are relatively obsolete, are not discussed in this chapter. The three most critical components of the epifluorescence microscope are the light source, the filters, and the objective lenses.

The most often used light sources are high-pressure mercury or halogen vapor lamps, with the former being the most common. Mercury lamps give off high-intensity light in the near-ultraviolet-to-blue range (peak intensities, 313 to 578 nm), the range used to excite the most commonly used fluorochromes in clinical virology, such as fluorescein

(490 nm), which gives a green fluorescence, and rhodamine (540 nm), which fluoresces red. In fluorescence microscopy, light from the lamp passes through a barrier filter, which allows only the excitation wavelength to pass (Fig. 2). This light is directed by a dichroic mirror through the objective lens, where it is focused on the specimen. The light emitted from the specimen is collected by the objective lens, travels through the mirror and the second barrier filter (both of which allow light at the emission frequency only to pass) and on to the eyepieces of the microscope. The stained specimen glows brightly against a dark background, because only the emitted wavelength is allowed to reach the eyepieces.

APPLICATIONS

Since the early studies, immunofluorescence has been used to detect viruses in patient specimens and tissue culture.

Direct Specimen Testing

A large number of viral agents can be detected directly in a variety of clinical specimens submitted to virology laboratories. Among the viruses that infect the respiratory tract and which can readily be detected in nasopharyngeal exudates are respiratory syncytial virus (RSV), influenza viruses A and B, parainfluenza viruses 1 through 4, and adenovirus. Measles, mumps, and herpesviruses (herpes simplex virus

[HSV] and varicella-zoster virus [VZV]) are also readily detected in patient specimens (Taber et al., 1976; Kumamoto et al., 1988; Costello et al., 1993; Waner, 1994). Rabies virus can be detected by immunofluorescence in the brain tissue of infected animals (Gardner and McQuillin, 1980), in skin biopsy specimens taken from the neck and legs of infected individuals, and in corneal smears (Warrell et al., 1988). Direct detection of viruses in specimens has an advantage over culture isolation due to more timely reporting of results; immunofluorescence has an advantage over enzyme immunoassays, as specimen adequacy can be determined (Balachandran et al., 1982; Lawrence et al., 1984; Ahluwalia et al., 1987; Chomel et al., 1991; Njayou et al., 1991; Takimoto et al., 1991; Ramirez de Arellano et al., 1992; Olsen et al., 1993) and more than one virus may be detected (Grandien et al., 1984; Subbarao et al., 1989; Waner, 1994). One report has suggested that timely reports of viral infections, as determined by direct specimen testing, will minimize the inappropriate use of antibiotics (Woo et al., 1997).

While the specificity of direct specimen detection is generally excellent, sensitivity is often low compared to culture, except for labile viruses such as RSV and VZV (Minnich et al., 1980; Schmidt et al., 1980; Douglas, 1985; Solomon, 1988; Takimoto et al., 1991; Thomas et al., 1991; Dahl et al., 1997; Freymuth et al., 1997). The use of cytocentrifugation, rather than drop suspension, to prepare slides has been shown to improve sensitivity for the detection of HSV (Landry et al., 1997), and many laboratories are now using cytocentrifuged slides for all direct specimen testing (Landry et al., 1997; Doing et al., 1998). However, direct testing of patient specimens should not be relied on as the only method of virus detection; culture is recommended for all specimens that are negative by direct specimen testing.

Immunofluorescence is the preferred method for the detection of the 65-kDa lower matrix protein of cytomegalovirus (CMV), phosphoprotein 65 (pp65), in peripheral blood leukocytes of patients at high risk for severe CMV disease. It is not yet clear if the presence of the antigen reflects infection of leukocytes, phagocytosis, or both. However, an increase in the number of positive leukocytes has been seen in patients with active clinical disease. The antigenemia assay has been shown to be a rapid, sensitive, and specific method to predict and differentiate CMV disease from asymptomatic infection in organ transplant patients and persons with AIDS. Immunofluorescence is not the first test of choice for detection of viruses in tissue samples, as the structural morphology of the tissue cannot be readily discerned.

To detect CMV pp65 antigen, slides are prepared from peripheral blood leukocytes isolated from anticoagulated blood (sodium citrate, sodium heparin, EDTA, or acid citrate dextrose are acceptable) by cytocentrifugation (Storch et al., 1994; Landry et al., 1997). The leukocytes may be isolated by using dextran sedimentation or other separation methods or by lysis of red cells by ammonium chloride (Ho et al., 1998). For optimum sensitivity, studies have shown that the samples should be processed within 8 h of collection, as there can be a loss of positive cells beyond this time (Landry et al., 1995). Positive cells can still be detected in anticoagulated blood that is several days old (Brumback et al., 1997), but quantitation is not accurate. Formalin is superior to acetone for fixation (Perez et al., 1995); the fixed cells are permeabilized with a non-ionic detergent such as NP-40 and stained. Positive cells show typical stain-

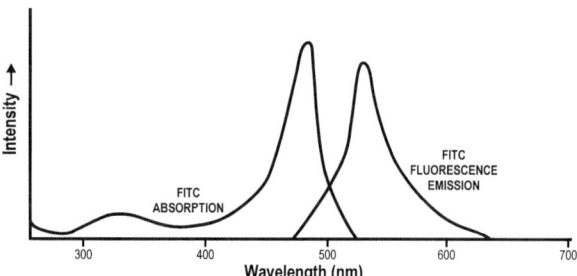

FIGURE 1 Excitation and emission spectra of FITC.

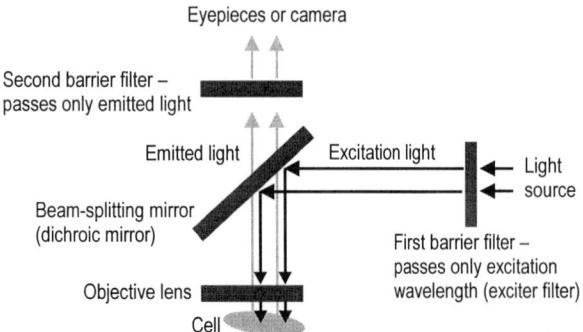

FIGURE 2 Optical system of an epifluorescence upright microscope. See text for details. (Courtesy of Douglas Kline, Department of Biological Sciences, Kent State University, Kent, Ohio.)

ing of the nucleus (Color plate 1E [see color insert]). Results are expressed as the number of positive nuclei per 2×10^5 leukocytes. The significance of the result depends on clinical diagnosis, patient history, symptoms, and results of other tests, including previous CMV pp65 results. Commercial kits containing the reagents needed to perform the antigenemia assay are available.

Standard Culture

Cell culture is the "gold standard" for virus detection and diagnosis. As virus is amplified by growth in cells, it offers the advantage of sensitivity over direct specimen testing, except for fragile viruses as previously indicated. The major disadvantage of culture, on the other hand, is the length of time required before reporting of results: anywhere from 7 to 28 days depending on the virus.

Each virus, or family of viruses, has a distinct cytopathic effect (CPE) in culture that can be used for presumptive identification. Presumptive identification of ortho- and paramyxoviruses can also be made by adsorption of blood cells (hemadsorption) to infected cells. Definitive identification is then confirmed by immunofluorescence.

Commercial FA reagents are available for the majority of cultivable viruses, including RSV, parainfluenza virus types 1, 2, 3, and 4, influenza viruses A and B, measles and mumps viruses, HSV-1 and HSV-2, CMV, VZV, adenovirus, and many of the enteroviruses. These reagents include those that identify individual viruses (CMV, HSV, and RSV, etc.), antibody pools directed against several viruses (as for the common respiratory viruses), or those with broad group identification (pan-enterovirus reagents). Both indirect (IFA) and direct (DFA) reagents are available.

Shell Vials

The spin-amplified shell vial assay has significantly shortened the turnaround time for virus identification in cell culture, particularly for those viruses, such as CMV, that take a long time to grow (Gleaves et al., 1984). It is also the standard method for culturing *Chlamydia* (a bacterium that is cultured in virology when isolation is necessary).

Shell vial cultures of MRC-5 cells are used for CMV, and McCoy cells are used for *Chlamydia;* however, other cells, such as RMK and A549 cells, etc., have been used successfully to isolate other viruses. Many viruses, such as CMV and herpesviruses, can be detected within 24 to 48 h, before the development of CPE; others may require up to 5 days before detection (Engler et al., 1997). Depending on the virus, different cell lines and combinations of multiple cell lines can be used in shell vial culture systems. Respiratory viruses, CMV, VZV, rubella virus, enteroviruses, HSV, human herpesvirus 6 (HHV-6), and HHV-7 have all been detected in shell vial cultures.

Shell vials are 1-dram glass vials containing a coverslip, upon which cell monolayers are grown. Vials with subconfluent monolayers (particularly for CMV) are preferred. Specimens are inoculated directly into the vial and centrifuged at 25 to 37°C at $700 \times g$ for 40 min. Higher *g* forces for shorter time periods have also been shown to be effective (Engler et al., 1994). Maintenance medium is added to the vials, and they are incubated at 37°C for 18 h to 4 or 5 days.

After the incubation period, the maintenance medium is removed; the coverslips are washed with phosphate-buffered saline (PBS) and fixed with acetone. The cover-slips can then be stained directly in the shell vial or removed and mounted on a microscope slide for staining (Clarke, 1998).

Specimen Collection and Processing

Proper collection and transport of specimens is critical for successful detection of viral antigens by FA. Specimens that are considered most appropriate for analysis by immunofluorescence include nasopharyngeal swabs, nasopharyngeal aspirates or washes, bronchoalveolar lavage samples, swabs or scrapings from vesicular lesions, tissue biopsies (e.g., lung, liver, and brain), blood leukocytes, conjunctival cells, corneal scrapings, and urine sediment. Specimens that should be considered inappropriate are throat swabs and gargles, which contain very small amounts of virus-infected cells. A comprehensive discussion of the selection, collection, transport, and processing of specimens can be found in chapter 2. A brief discussion of how various types of specimens are processed for FA testing is found below.

Nasopharyngeal Aspirates/Washes, Tissue Aspirates, and Swabs Submitted in Viral Transport Medium

These specimens require centrifugation to collect cells for FA analysis. When swab specimens are submitted in viral transport medium, thoroughly mix the sample on a vortex mixer and firmly press the swab against the inside of the transport tube to express all collected cells, virus, and fluid. Centrifuge the specimen at $600 \times g$ for 5 min to pellet the cells. The cell pellet is suspended with 5 ml of sterile PBS at a pH of 7.0 to 7.6 and centrifuged as above. All but 100 to 200 µl of the PBS is removed; a uniform suspension of cells in the remaining fluid is made by gently pipetting up and down. A drop of the cell suspension is placed into one or more wells of a Teflon-coated multiwell glass slide and air dried. The slide may be placed on a slide warmer to facilitate drying, but care must be taken to avoid excessively high temperatures that may denature the viral antigens, resulting in loss of reactivity with the antibody. Alternatively, methods for preparing slides from cell suspensions by cytocentrifugation have been reported (Landry et al., 1997; Doing et al., 1998). Cell spots prepared with a cytocentrifuge were found to be more uniform than those prepared by the method described above and had less nonspecific staining, as mucous and other debris were centrifuged out of the well; also, they were easier to interpret by personnel with less training and experience. An additional benefit is that the smaller cell spot required less reagent to stain the cells, resulting in cost savings for the laboratory.

Preparation of Slides Directly from Swabs

Often it is expedient to make a slide from a swabbed clinical specimen immediately after the sample is taken. This is usually the case with swabs obtained from vesicular lesions caused by HSV, VZV, or enterovirus infections. To prepare these slides, a fresh vesicular lesion should be chosen and opened with a sterile scalpel. The base of the vesicle is swabbed firmly with a cotton or Dacron (*not* calcium alginate) swab. The swab is then rolled (*not* rubbed) onto the appropriate number of wells of a Teflon-coated glass slide. (Rubbing the swab onto the slide may result in damage to the cells, making the slide difficult to read.) Slides prepared in this manner should be air dried completely before the next step.

Fixation

Prior to staining, the specimen must be properly fixed. The most common fixative used is cold acetone. Acetone absorbs moisture if it is not properly stored. Moisture causes a nonspecific hazing during the fixation process; therefore, it is important to ensure that fresh acetone is used for fixing the specimens. Acetone that appears to be cloudy should be discarded. The time of fixation of specimens does not appear to be overly important; 2 to 10 min is more than adequate for this purpose. It is also possible to flood the slide with acetone and allow it to evaporate. However, the majority of commercial kits have validated their reagents with slides fixed for 10 min, and changing the fixation procedure may adversely affect staining. Many high-volume laboratories have replaced the traditional 1-dram vial used for shell vial assays with plastic multiple-well cell culture plates. In this case, a mixture of methanol-acetone or ethanol-acetone must be used as a fixative, since polystyrene is not compatible with 100% acetone. Methanol alone as a fixative is not compatible with many FA reagents; water is not recommended as a diluent in place of methanol or ethanol.

PRACTICAL DETAILS FOR DIRECT AND INDIRECT IMMUNOFLUORESCENCE STAINING

Historically, most immunofluorescence staining procedures have been indirect; that is, the specimen is initially reacted with a primary antigen-specific antibody reagent, binding of which is detected using a secondary antibody-fluorochrome conjugate directed against the primary antibody (Fig. 3A). This procedure is referred to as the indirect fluorescent antibody method, or IFA. More recently, the use of directly labeled primary antibodies has gained in popularity and availability. A significant advantage of using directly labeled antibodies for antigen detection is speed; by eliminating the second staining step, the assay time is virtually cut in half or better. The use of directly coupled antibodies is referred to as the direct fluorescent antibody assay, or DFA (Fig. 3B). A downside of the DFA procedure is that the intensity of fluorescence may be lower than with IFA. Commercial reagents may compensate for the lower intensity by using a cocktail of two or more antibodies that react with different antigens or different epitopes on the same antigen; this results in a greater number of labeled antibodies binding to each infected cell and thus in increased intensity of fluorescence. Most commercially available DFA reagents or labeled secondary antibodies contain a counterstain such as Evans' blue to allow nonfluorescing cells to be visualized. Evans' blue-stained cells fluoresce red under the microscope, making it easier for the operator to focus on the specimen and, in the case of slides made from patients' specimens, to determine the adequacy of the specimen by the number and morphology of cells present in the well.

The following procedures should be used as a guide for performing both DFA or IFA assays:

1. During the staining process, slides should be placed in a humidified chamber to prevent evaporation of reagents. Several types of humidified chambers are available com-

A

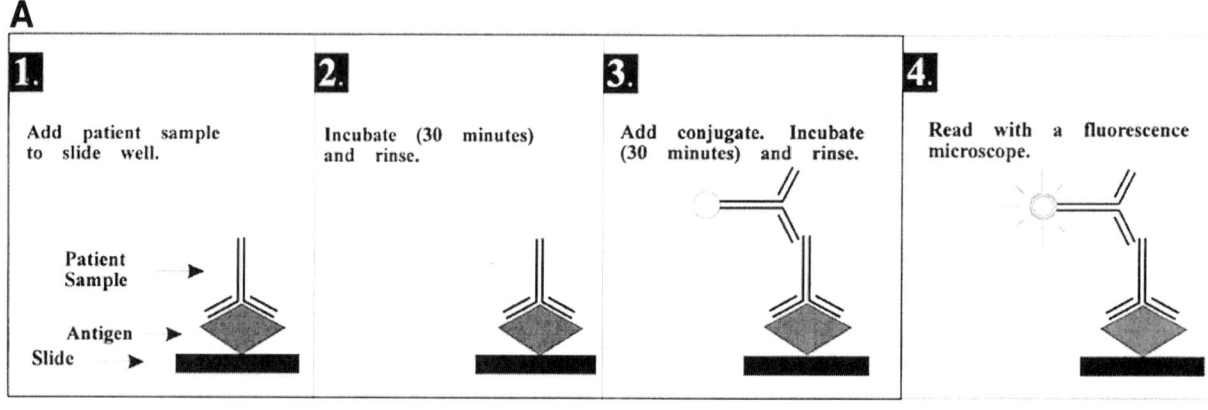

B

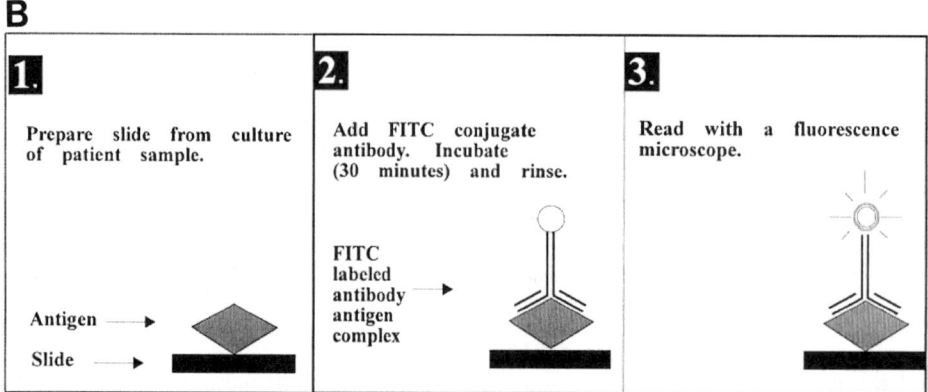

FIGURE 3 Schematic representations of IFA (A) and DFA (B).

mercially; alternatively, a box with a wet sponge or moistened paper towels will suffice.

2. The specimen is completely overlaid with the primary antibody, after which the slide is placed in the chamber and incubated at 37°C. The length of incubation is variable, based on the antibody concentration and its affinity for the antigen. Most commercial immunofluorescence kits suggest a 15- to 30-min incubation.

3. After staining, the primary antibody must be removed by washing. Immersion in a Coplin jar containing fresh PBS is the most gentle and efficient method. The addition of Tween 20 to the buffer can help to remove any nonspecific binding of antibody. A series of one or two washes of 1-min durations is sufficient if the primary antibody is a monoclonal antibody. Polyclonal antibodies may require longer washing. Overwashing is undesirable, since it may lead to diminished staining and false-negative results. Washing conditions need to be established for each different antibody being used. If the primary antibody is directly coupled to the fluorochrome (DFA), skip to step 7.

4. Remove excess PBS from the slide, being careful not to touch the areas of the slide containing the specimen.

5. Overlay the specimen with the labeled secondary antibody and incubate as above.

6. Wash the slide as described in step 3.

7. If PBS is used for washing, the slides should be mounted immediately. Excess PBS should be blotted or shaken from the slide, again taking care not to disturb the specimen. If the slides are to be allowed to air dry before mounting, rinse in distilled water and then dry completely. A drop of mounting medium is placed on the specimen area followed by a glass coverslip. Mounting medium, usually supplied with commercial kits, contains buffered glycerol, pH 9.0, with photobleach inhibitors to prolong fluorescence.

Reading and interpretation of results for DFA and IFA require critical evaluation to ensure reliable results. Staining patterns are highly dependent on the type of virus being detected as well as on the specificity of the primary antibody being used. The control slides supplied with commercial kits may be used as an indicator of expected results. Direct smear preparations can contain fragments of disrupted cells as well as leukocytes, both of which can trap fluorescein and result in staining artifacts, which makes the slide difficult to interpret. Contaminating bacteria, such as Staphylococcus aureus, can bind the Fc portion of antibodies and may cause false-positive results. Specimen quality can be verified by the number and type of cells present on the slide. For example, good nasopharyngeal specimens should have at least three columnar epithelial cells per high-power field (×400). Antigen-negative specimens that contain fewer cells or that contain a large number of squamous epithelial cells should be considered poor quality, with suspect results. Each laboratory must develop its own criteria for interpreting slides as positive and negative, based on staining patterns and intensity of fluorescence.

TROUBLESHOOTING

False-positive results with FA can be due to several causes, such as nonspecific staining due to binding between the Fc region of the antibody and protein A or G of S. aureus, a contaminant in some respiratory specimens. Inflammatory cells such as lymphocytes and polymorphonuclear leukocytes contain Fc receptors, and herpesvirus-infected cells (including CMV, HSV, and VZV) (Keller et al., 1976; Litwin and Grose, 1992; MacCormac and Grundy, 1996)

also express Fc receptors that might cause nonspecific binding of antibodies. False-negative results can arise from a variety of causes; the most common are described in Table 1. The troubleshooting guide in Table 1 describes commonly encountered problems with FA and suggests ways of resolving them (Clarke, 1998).

QUALITY ASSURANCE/QUALITY CONTROL

The majority of FA reagents currently used by clinical laboratories are purchased from commercial sources. These reagents fall under the jurisdiction of the U.S. Food and Drug Administration (FDA), which, at minimum, requires that they be manufactured under "current good manufacturing practices" to ensure consistent quality. Many kits also require premarket approval [510(k) clearance] before they can be used for diagnosis. Consequently, performance characteristics, such as sensitivity, specificity, cross-reactivity, and prevalence, etc., have been established for these reagents and are included in package inserts. Validation of commercial reagents usually requires two to three clinical sites with a minimum of 200 specimens per site, depending on the prevalence of the virus.

Home-brew assays, in contrast, are developed by individual laboratories. The critical components for home-brew assays are now regulated by the FDA and must be labeled "Analyte Specific Reagent" by the manufacturer. Their sale is restricted to CLIA high-complexity labs, public health labs, Veterans Administration hospitals, and manufacturers. These assays also must be validated before they can be used by the laboratory. NCCLS standards should be followed for reagent and assay validation.

Quality Assurance

Quality assurance establishes standard operating procedures (SOPs) for all aspects of testing, including specimen collection and processing, assay protocols, validation requirements, and quality control (QC) for all tests performed in a laboratory. Critical components, such as swabs, buffers, and transport media, etc., should also be included, along with package inserts from the commercial kits used in the laboratory. If home-brew assays are used, recipes, procedures, and QC specifications should be included. Package inserts should be read carefully, with close attention paid to the intended use of the kit, assay procedures, QC, performance characteristics, and limitations of the assay. Any deviation from manufacturers' recommended procedures must be validated by the laboratory.

SOPs are living documents and should be reviewed and updated regularly to ensure that they reflect actual practices in the laboratory.

Quality Control

QC refers to those controls that are run at established intervals to ensure that procedures and reagents (including cells) are behaving appropriately. Commercial kits usually include control slides to be used along with direct specimen testing and to act as a control for the reagent. For culture confirmation, known positive and negative isolates should be inoculated into cell culture to ensure proper cell culture technique, slide preparation, and staining. Previously isolated viruses can be used as controls, or known strains may be purchased from commercial sources such as the American Type Culture Collection. QC specimens should always be tested first to confirm that the assay has worked correctly.

TABLE 1 Problems most frequently encountered with FA[a]

Problem	Cause(s)	Solution
Weak or no specific fluorescence (including positive controls)	Wrong (too weak) concentration of immunoreagent	Retitrate reagent
	Deterioration of reagent	Retitrate and/or replace; store properly, and aliquot if necessary
	Counterstain too strong	Review concentration and counterstaining time
	Rapid fading of fluorescence	pH of mounting medium must be greater than that of wash buffer; use mounting medium that contains photobleaching inhibitor
	Wrong filters or inadequate light source	Review use of maintenance of equipment; replace light bulb
	Cannot achieve sharp focus because of dirty or improperly focused optics	Review procedures for maintenance and use of optics
	Nonspecific glare or haze masking specific fluorescence	Allow cell spots to air dry completely before mounting; use fresh acetone
	Antigens destroyed by fixative	Do not use fixatives other than acetone without determining effect on antigen stability
Weak or no fluorescence with test slides but positive controls acceptable	Controls do not adequately reflect actual test conditions	Review procedure; determine whether same or different cells, reagents, or methods were used for test and control slides
	Improper slide cleaning, preparation, or fixation	Review procedures
Nonspecific fluorescence and/or false positives with test slides and negative controls	Wrong concentration (too strong) of immunoreagent used	Retitrate reagent
	Cross-reactive immunoreagent	Identify non-cross-reactive concentration of reagent or obtain a better product
	Binding of antibody via Fc receptors	Use conjugated $F(ab')_2$ fragments
	Autofluorescence of cells	Use alternative cell line, or incorporate counterstain for better resolution
	Trapping of immunoreagent on heavy or raised preparations	Avoid heavy cell preparations which obscure cellular morphology and complicate reading; increase number of washes during staining
	Inappropriate immersion oil	Use immersion oil designated for fluorescence microscopy
	Inadequate removal of fixative	Air dry cell spots completely after fixing; rinse slides with PBS and/or water just prior to staining
	Water in acetone	Replace acetone
	Slides dried during staining	Use properly sealed moist chamber during incubation; apply adequate reagent volumes to cell spots
	Mounting medium not appropriate for immunofluorescence, or pH too low	Obtain proper reagents; pH of mounting medium must be greater than that of wash buffer

(Continued on next page)

TABLE 1 *(Continued)*

	Mounting medium applied to wet cells	Allow cell spots to air dry completely before mounting
	Wrong filters or equipment	Review use and maintenance of equipment
	Reading error	Must be able to differentiate specific patterns of fluorescence from nonspecific staining
	Precipitated material in immunoreagent	Microcentrifuge reagent for 1 min, or filter through a 0.45-μm-pore-size filter
Nonspecific fluorescence and/or false positives with test slides but negative controls acceptable	Controls do not adequately reflect test conditions	Review procedure; determine whether same or different cells, reagents, or methods were used for test and control slides
	Reading error	More reading experience needed, since controls may be easier to read and interpret than test slides
	Reagent cross-contamination	Take care to confine reagents to wells
	Inoculum debris or microorganisms on the test slide	Increase number of washes performed on monolayers prior to harvesting

[a] Reprinted from Isenberg (1998) with permission of the publisher.

If the controls fail, the assay is considered to have failed as well.

Care must be taken with passing and maintaining clinical isolates used as controls, particularly RNA viruses. Passing an RNA virus at a high multiplicity of infection may cause an increase in the production of defective interfering particles (hence a lower infectivity of the virus stock) and reading errors during RNA replication, resulting in mutations in the RNA and leading to poor or absent staining due to the loss of target epitopes. Similarly, at high passage numbers, some cell lines are no longer susceptible to infection by certain viruses, e.g., HSV infection of A549 cells.

RECENT ADVANCES IN FLUORESCENT ANTIBODY ASSAY DEVELOPMENT

In the past few years, there has been increased interest in the use of fluorescence immunoassays in screening of direct specimens for the presence of more than one virus by using pools of antibodies directed against multiple viruses that cause similar diseases. The most common application of the use of pooled antibodies is for the detection of respiratory viral pathogens (Balkovic et al., 1985; Johnson et al., 1993; McDonald and Quennec, 1993). Several kits that are now commercially available contain a screening reagent comprised of monoclonal antibodies directed against the seven most common respiratory viruses: influenza virus types A and B, parainfluenza virus types 1, 2, and 3, RSV, and adenovirus. Depending on the manufacturer, these respiratory virus panels are configured as either IFA or DFA. Ciliated epithelial cells infected with any of the above viruses react with one or more (in the case of multiple infections) of the antibodies in the pool, giving a positive result. The antibodies in these pools may be directly labeled with one or more fluorochromes (DFA) or have a common fluorescein-conjugated secondary antibody (IFA); typing of individual viruses is not possible. To determine with which virus (or occasionally multiple viruses) the patient is infected, the commercial kits typically contain individual typing reagents specific for each of the above viruses.

A difference in staining patterns between two viruses can be used to detect and differentiate these viruses in a pooled reagent. An example of such an application is demonstrated in Color plate 1A and B (see color insert). These cells were infected with either HSV-1 (A) or HSV-2 (B) and stained with a single reagent. The reagent was prepared by mixing the HSV-1 and HSV-2 DFA reagents together to detect the cytoplasmic (HSV-1) or nuclear (HSV-2) viral antigens. Because the morphological staining patterns are so clearly different, the two viruses can be easily discriminated with a combined reagent.

A significant advance in the use of pooled monoclonal antibodies has been the application of multiple fluorescence (Brumback et al., 1993; Brumback and Wade, 1994, 1996). This was first described as a home-brew assay using different reagents for each virus, with more than one staining step. The reagents were labeled with fluorochromes that emitted blue, green, yellow, and red fluorescence and allowed for the detection of multiple viruses in one well. This type of staining requires several different filter sets in order to visualize the fluorescence emitted from 400 to 600 μm. Color plate 1C shows influenza A and B virus-infected cells stained with a single commercial reagent, as seen with a standard FITC filter set. This and similar reagents contain monoclonal antibodies directed against two or more viruses, with each antibody directly coupled to a different fluorochrome. As shown in Color plate 1C, cells infected with influenza A virus are stained apple-green with the primary component in the reagent, while the influenza B virus-infected cells show a yellow-gold-to-orange fluorescence when stained with the secondary component of the reagent. If a rhodamine/TRITC filter set is available, the apple-green fluorescence will disappear and the yellow-gold-staining cells will fluoresce a bright, or "hot," pink.

The ability to detect two or more different viruses in the same well minimizes the number of slides that need to be prepared and read, can result in significant cost benefits to the laboratory, and improves the turnaround times for reporting of direct specimen results. Some training is necessary for acclimation to reading the yellow-gold-orange stain of dually fluorescent reagents or to differentiate staining patterns of viruses stained with pooled reagents; however, most laboratory personnel become proficient very quickly. These reagents can be used for direct specimen detection or for culture confirmation.

REFERENCES

Ahluwalia, G., J. Embree, P. McNicol, B. Law, and G. W. Hammond. 1987. Comparison of nasopharyngeal aspirate and nasopharyngeal swab specimens for respiratory syncytial virus diagnosis by cell culture, indirect immunofluorescence assay, and enzyme-linked immunosorbent assay. *J. Clin. Microbiol.* **25:**763–767.

Anestad, G., N. Breivik, and T. Thoresen. 1983. Rapid diagnosis of respiratory syncytial virus and influenza A virus infections by immunofluorescence: experience with simplified procedure for the preparation of cell smears from nasopharyngeal secretions. *Acta Pathol. Microbiol. Immunol. Scand. B* **91:**267–271.

Arnaud, J. P., J. J. Prat, C. Griscelli, M. Gentilini, and C. Nezelof. 1976. Diagnosis of *Pneumoncystis carinii* pneumonia by indirect immunofluorescence. Technic, value of limitations of the method. *Nouv. Presse Med.* **5:**2607–2610.

Athanasiu, P. 1985. Rapid viral diagnosis by immunofluorescence reactions. *Virologie* **36:**295–301.

Balachandran, N., B. Frame, M. Chernesky, E. Kraiselburd, Y. Kouri, D. Garcia, C. Lavery, and W. E. Rawls. 1982. Identification and typing of herpes simplex viruses with monoclonal antibodies. *J. Clin. Microbiol.* **16:**205–208.

Balkovic, E. S., and G. D. Hsiung. 1985. Comparison of immunofluorescence with commercial monoclonal antibodies to biochemical and biological techniques for typing clinical herpes simplex virus isolates. *J. Clin. Microbiol.* **22:**870–872.

Baratta, L., M. Valeriani, and G. Andreoni. 1975. Rapid diagnosis of respiratory viral disease by means of indirect immunofluorescence test. *Ann. Sclavo.* **17:**40–49.

Biegeleisen, J. Z., Jr., L. V. Scott, and V. Lewis, Jr. 1959. Rapid diagnosis of herpes simplex virus infection with fluorescent antibody. *Science* **129:**640.

Bikbulatov, R. M., T. Liarskaia, and S. A. Demidova. 1978. Use of the immunofluorescence method for the diagnosis of cytomegalovirus infection. *Vopr. Virusol.* **2:**241–323.

Brumback, B. G., and C. D. Wade. 1994. Simultaneous culture for adenovirus, cytomegalovirus, and herpes simplex virus in same shell vial by three-color fluorescence. *J. Clin. Microbiol.* **32:**2289–2290.

Brumback, B. G., and C. D. Wade. 1996. Simultaneous rapid culture for four respiratory viruses in the same cell monolayer using a differential multicolored fluorescent confirmatory stain. *J. Clin. Microbiol.* **34:**798–801.

Brumback, B. G., S. N. Bolejack, M. V. Morris, C. Mohla, and T. E. Schutzbank. 1997. Comparison of culture and the antigenemia assay for detection of cytomegalovirus in blood specimens submitted to a reference laboratory. *J. Clin. Microbiol.* **35:**1819–1821.

Brumback, B. G., P. G. Farthing, and S. N. Castellino. 1993. Simultaneous detection of and differentiation between herpes simplex and varicella-zoster viruses with two fluorescent probes in the same test system. *J. Clin. Microbiol.* **31:**3260–3263.

Chomel, J. J., D. Pardon, D. Thouvenot, J. P. Allard, and M. Aymard. 1991. Comparison between three rapid methods for the direct diagnosis of influenza and the conventional isolation procedure. *Biologicals* **19:**287–292.

Clarke, L. 1998. Viruses, rickettsiae, chlamydiae, and mycoplasmas, p. 451–559. *In* H. D. Isenberg (ed.), *Essential Procedures for Clinical Microbiology.* ASM Press, Washington, D.C.

Coons, A. H., H. J. Creech, R. N. Jones, and E. Berliner. 1942. The demonstration of pneumococcal antigen in tissues by the use of fluorescent antibody. *J. Immunol.* **45:**159.

Costello, M. J., N. T. Smernoff, and M. Yungbluth. 1993. Laboratory diagnosis of viral respiratory tract infections. *Lab. Med.* **24:**150–157.

Dahl, H., J. Marcia, and A. Linde. 1997. Antigen detection: the method of choice in comparison with virus isolation and serology for laboratory diagnosis of herpes zoster in human immunodeficiency virus-infected patients. *J. Clin. Microbiol.* **35:**347–349.

Doing, K. M., M. A. Jerkofsky, E. G. Dow, and J. Jellison. 1998. Use of fluorescent-antibody staining of cytocentrifuge-prepared smears in combination with cell culture for direct detection of respiratory viruses. *J. Clin. Microbiol.* **36:**2112–2114.

Douglas, R. G., Jr. 1985. Chemotherapy of respiratory virus infections. *In* L. M. de la Maza and E. M. Peterson (ed.), *Medical Virology IV.* Lawrence Erlbaur Associates, London, United Kingdom.

Engler, H. D., and J. Preuss. 1997. Laboratory diagnosis of respiratory virus infections in 24 hours by utilizing shell vial cultures. *J. Clin. Microbiol.* **35:**2165–2167.

Engler, H. D., and S. T. Selepak. 1994. Effect of centrifuging shell vials at 3,500 × *g* on detection of viruses in clinical specimens. *J. Clin. Microbiol.* **32:**1580–1582.

Fedova, D., and L. Zelenkova. 1965. The use of the fluorescent antibody method for the rapid diagnosis of the A2 influenza virus. II. The identification of influenza virus in nasal smears by the fluorescent antibody technique. *J. Hyg. Epidemiol.* (Prague) **9:**135.

Freymuth, R., A. Vabret, F. Galateau-Salle, J. Ferey, G. Eugene, J. Petitjean, E. Gennatay, J. Brouard, M. Jokik, J. F. Duhamel, and B. Guillois. 1997. Detection of respiratory syncytial virus, parainfluenza virus 3, adenovirus and rhinovirus sequences in respiratory tract of infants by polymerase chain reaction and hybridization. *Clin. Diagn. Virol.* **8:**31–40.

Fulton, R. E., and P. J. Middleton. 1975. Immunofluorescence in diagnosis of measles infections in children. *J. Pediatr.* **86:**17–22.

Gallo, D. 1983. Uses of immunofluorescence in diagnostic virology. *Am. J. Med. Technol.* **49:**157–162.

Gardner, P. S., and J. McQuillin. 1968. Application of the immunofluorescent antibody technique in the rapid diagnosis of respiratory syncytial virus infection. *Br. Med. J.* **2:**340.

Gardner, P. S., and J. McQuillin. 1974. *Rapid Virus Diagnosis: Application of Immunofluorescence.* Butterworths, London, United Kingdom.

Gardner, P. S., and J. McQuillin. 1980. *Rapid Virus Diagnosis: Application of Immunofluorescence,* 2nd ed. Butterworths, London, United Kingdom.

Gleaves, C. A., T. F. Smith, E. A. Shuster, and G. R. Pearson. 1984. Rapid detection of cytomegalovirus in MRC5 cells inoculated with urine specimens by using low-speed centrifugation and monoclonal antibody to an early antigen. *J. Clin. Microbiol.* **19:**917–919.

Grandien, M., and E. Olding-Stenkvist. 1984. Rapid diagnosis of viral infections in the central nervous system. *Scand. J. Infect. Dis.* **16:**1–8.

Haire, M. 1969. Rapid identification of rubella-virus antigen from throat swabs. *Lancet* **i:**920.

Ho, S. K. N., C.-Y. Lo, I. K. P. Cheng, and T.-M. Chan. 1998. Rapid cytomegalovirus pp65 antigenemia assay by direct erythrocyte lysis and immunofluorescence staining. *J. Clin. Microbiol.* **36:**638–640.

Isenberg, H. D. (ed.). 1998. *Essential Procedures for Clinical Microbiology.* ASM Press, Washington, D.C.

Johnson, B. R., K. Osinusi, W. I. Aderele, and O. Tomori. 1993. Viral pathogens of acute lower respiratory infections in pre-school Nigerian children and clinical implications of multiple microbial identifications. *West Afr. J. Med.* **12:**11–20.

Keller, R., R. Peitchel, J. N. Goldman, and M. Goldman. 1976. An IgG-Fc receptor induced in cytomegalovirus-infected human fibroblasts. *J. Immunol.* **116:**772–777.

Kirsh, D., and R. Kissling. 1963. The use of immunofluorescence in the rapid presumptive diagnosis of variola. *Bull. W. H. O.* **29:**126.

Kumamoto, Y., T. Hirose, S. Ikegaki, T. Inoke, T. Gouro, S. Tabata, H. Yoshio, and H. Sakaoka. 1988. A clinical study of genital herpes and the clinical efficacy of acyclovir tablets. *Hinyokika Kiyo* **34:**383–393.

Landry, M. L., S. Cohen, and K. Huber. 1997. Comparison of EDTA and acid-citrate-dextrose collection tubes for the detection of cytomegalovirus antigenemia and infectivity in leukocytes before and after storage. *J. Clin. Microbiol.* **35:**305–306.

Landry, M. L., D. Ferguson, and J. Wlochowski. 1997. Detection of herpes simplex virus in clinical specimens by cytospin-enhanced direct immunofluorescence. *J. Clin. Microbiol.* **35:**302–304.

Landry, M. L., E. Ferguson, S. Cohen, K. Huber, and P. Wetherill. 1995. Effect of delayed specimen processing on cytomegalovirus antigenemia test results. *J. Clin. Microbiol.* **33:**257–259.

Lawrence, T. G., D. B. Budzko, and B. W. Wilcke, Jr. 1984. Detection of herpes simplex virus in clinical specimens by enzyme-linked immunosorbent assay. *Am. J. Clin. Pathol.* **81:**339–341.

Lennette, D. A., R. W. Emmons, and E. H. Lennette. 1976. Rapid diagnosis of mumps virus infections by immunofluorescence methods. *J. Clin. Microbiol.* **2:**81–84.

Litwin, V., and C. Grose. 1992. Herpesviral Fc receptors and their relationship to the human Fc receptors. *Immunol. Res.* **11:**226–238.

Llanes-Rodas, R., and C. Liu. 1956. Rapid diagnosis of measles from urinary sediments stained with fluorescent antibody. *N. Engl. J. Med.* **275:**516.

MacCormac, L. P., and J. E. Grundy. 1996. Human cytomegalovirus induces an Fc γ receptor (Fc γR) in endothelial cells and fibroblasts that is distinct from the human cellular Fc γRs. *J. Infect. Dis.* **174:**1151–1161.

Matthey, S., D. Nicholson, S. Ruhs, B. Alden, M. Knock, K. Schultz, and A. Schmuecker. 1992. Rapid detection of respiratory viruses by shell vial culture and direct staining by using pooled and individual monoclonal antibodies. *J. Clin. Microbiol.* **30:**540–544.

McDonald, J. C., and P. Quennec. 1993. Utility of a respiratory virus panel containing a monoclonal antibody pool for screening of respiratory specimens in nonpeak respiratory syncytial virus season. *J. Clin. Microbiol.* **31:**2809–2811.

Minnich, L., and C. G. Ray. 1980. Comparison of direct immunofluorescent staining of clinical specimens for respiratory virus antigens with conventional isolation techniques. *J. Clin. Microbiol.* **12:**391–394.

Nichols, R. L., and E. E. McComb. 1962. Immunofluorescent studies of trachoma and related antigens. *J. Immunol.* **89:**545.

Njayou, M., A. Balla, and E. Kapo. 1991. Comparison of four techniques of measles diagnosis: virus isolation, immunofluorescence, immunoperoxidase and ELISA. *Indian J. Med. Res.* **93:**430–434.

Olding-Stenkvist, E., and M. Grandien. 1976. Rapid diagnosis of influenza A infection by immunofluorescence. Methodological problems and clinical material. *Acta Pathol. Microbiol. Scand. B* **85:**296–302.

Olsen, M. A., K. M. Shuck, and A. R. Sambol. 1993. Evaluation of Abbott TestPack RSV for the diagnosis of respiratory syncytial virus infections. *Diagn. Microbiol. Infect. Dis.* **16:**105–109.

Pérez, J. L., M. De Oña, J. Niubò, H. Villar, S. Melón, A. García, and R. Martín. 1995. Comparison of several fixation methods for cytomegalovirus antigenemia assay. *J. Clin. Microbiol.* **33:**1646–1649.

Ramirez de Arellano, E., J. Aznar, and A. Pascual. 1992. Evaluation of 5 methods for the diagnosis of infections by respiratory syncytial virus. *Enferm. Infect. Microbiol. Clin.* **10:**103–106.

Routledge, E. G., J. McQuillin, A. C. Samson, and G. L. Toms. 1985. The development of monoclonal antibodies to respiratory syncytial virus and their use in diagnosis by indirect immunofluorescence. *J. Med. Virol.* **15:**305–320.

Schaap, G. J. P., and R. A. Velthoen. 1963. Technical remarks on the application of the fluorescent antibody test (F.A. test) in the diagnosis of rabies. *Antonie Leeuwenhoek* **29:**216.

Schmidt, J. J., D. Gallo, V. Devlin, J. D. Woodie, and R. W. Emmons. 1980. Direct immunofluorescence staining for detection of herpes simplex and varicella-zoster virus antigens in vesicular lesions and certain tissue specimens. *J. Clin. Microbiol.* **12:**651–655.

Shalit, I., P. A. McKee, H. Beauchamp, and J. L. Waner. 1985. Comparison of polyclonal antiserum versus monoclonal antibodies for the rapid diagnosis of influenza A virus infections by immunofluorescence in clinical specimens. *J. Clin. Microbiol.* **22:**877–879.

Showalter, S. D., M. Zweig, and B. Hampar. 1981. Monoclonal antibodies to herpes simplex virus type 1 proteins, including the immediate-early protein ICP4. *Infect. Immun.* **34:**684–692.

Solomon, A. R. 1988. New diagnostic tests for herpes simplex and varicella zoster infections. *J. Am. Acad. Dermatol.* **18:**218–221.

Storch, G. A., M. Gaudreault-Keener, and P. C. Welby. 1994. Comparison of heparin and EDTA transport tubes for detection of cytomegalovirus in leukocytes by shell vial assay, pp65 antigenemia assay, and PCR. *J. Clin. Microbiol.* **32:**2581–2583.

Subbarao, E. K., J. Griffis, and J. L. Waner. 1989. Detection of multiple viral agents in nasopharyngeal specimens yielding respiratory syncytial virus (RSV). An assessment of diagnostic strategy and clinical significance. *Diagn. Microbiol. Infect. Dis.* **12:**327–332.

Swierkosz, E. M., M. Q. Arens, R. R. Schmidt, and T. Armstrong. 1985. Evaluation of two immunofluorescence assays with monoclonal antibodies for typing of herpes simplex virus. *J. Clin. Microbiol.* **21:**643–644.

Taber, L. H., F. Brasier, R. B. Couch, S. B. Greenberg, D. Jones, and V. Knight. 1976. Diagnosis of herpes simplex virus infection by immunofluorescence. *J. Clin. Microbiol.* **3:**309–312.

Takimoto, S., M. Grandien, M. A. Ishida, M. S. Pereira, T. M. Paiva, T. Ishimaru, E. M. Makita, and C. H. Martinez. 1991. Comparison of enzyme-linked immunosorbent assay, indirect immunofluorescence assay, and virus isolation for detection of respiratory viruses in nasopharyngeal secretions. *J. Clin. Microbiol.* **29:**470–474.

Thomas, E. E., and L. E. Book. 1991. Comparison of two rapid methods for detection of respiratory syncytial virus (RSV) (TestPack RSV and Ortho RSV ELISA) with direct immunofluorescence and virus isolation for the diagnosis of pediatric RSV infection. *J. Clin. Microbiol.* **29:**632–635.

Uchida, Y., and S. J. Kimura. 1965. Fluorescent antibody studies with agents of varicella and herpes zoster propagated in vitro. *Proc. Soc. Exp. Biol. Med.* **86:**789.

Volpi, A., A. D. Lakeman, L. Pereira, and S. Stagno. 1983. Monoclonal antibodies for rapid diagnosis and typing of genital herpes infections during pregnancy. *Am. J. Obstet. Gynecol.* **146:**813–815.

Walls, H. H., M. W. Harmon, J. J. Slagle, C. Stocksdale, and A. P. Kendal. 1986. Characterization and evaluation of monoclonal antibodies developed by typing influenza A and B viruses. *J. Clin. Microbiol.* **23:**240–245.

Waner, J. L. 1994. Mixed viral infections: detection and management. *Clin. Microbiol. Rev.* **7:**143–151.

Waner, J. L., N. J. Whitehurst, T. Downs, and D. G. Graves. 1985. Production of monoclonal antibodies against parainfluenza 3 virus and their use in diagnosis by immunofluorescence. *J. Clin. Microbiol.* **22:**535–538.

Warrell, M., S. Looareesuwan, S. Manatsathit, N. White, P. Phaupradit, A. Vehhahiva, C. Hoke, D. Burke, and D. Warrell. 1988. Rapid diagnosis of rabies and post-vaccinal encephalitides. *Clin. Exp. Immunol.* **71:**229–234.

Woo, P. C. Y., S. S. Chiu, W.-H. Seto, and M. Peiris. 1997. Cost-effectiveness of rapid diagnosis of viral respiratory tract infections in pediatric patients. *J. Clin. Microbiol.* **35:**1579–1581.

Radioimmunoassay

ISA K. MUSHAHWAR AND THOMAS A. BRAWNER

8

Since the introduction of radioimmunoassay (RIA) techniques (Yalow and Berson, 1960) for the determination of endogenous human plasma insulin, RIA has become the essential key analytical method in many sciences. Besides endocrinology, RIA has been a valuable tool in such fields as enzymology (Kolb and Grodsky, 1970), hematology (Rutland, 1984), toxicology (Shimada et al., 1983), pharmacology (Robinson and Smith, 1983), parasitology (Avraham et al., 1982), neurochemistry (Dowse et al., 1983), microbiology (Zollinger et al., 1976), plant pathology (Ghabrial and Shepherd, 1980), diagnostic medicine (Matsui et al., 1982), mycology (Poor and Cutler, 1979), and virology (Wiktor et al., 1972). Thus, RIA has contributed to the revolution in biology during the past 40 years in many other disciplines besides endocrinology. This trend will continue to have a positive impact on public health in the future through diagnostic and epidemiological studies in the areas of infectious diseases and cancer.

COMPETITIVE BINDING RIA

The original RIA described by Yalow and Berson (1960) and subsequently by other investigators (Faiman and Ryan, 1967; Odell et al., 1967; Goodfriend and Ball, 1969) was a competitive binding RIA, where the competition between an unlabeled antigen and a radiolabeled counterpart for a limited amount of antibody is monitored. This is illustrated as follows: $Ag^* + Ab + Ag^0 \rightarrow Ag^*Ab + Ag^0Ab$, where Ag^* represents radiolabeled antigen, Ag^0 is the unlabeled antigen, Ab is the antibody, Ag^*Ab is a complex of radiolabeled antigen with antibody, and Ag^0Ab is a complex of unlabeled antigen with antibody. The higher the concentration of Ag^0, the lower the concentration of radioactive Ag^*Ab and the higher the concentration of free Ag^*. The reaction is designed to take place in a solution giving a mixture of bound and free radiolabeled antigens. Separation of antibody-bound antigen from the free antigen is achieved by electrophoretic techniques or a variety of immunoprecipitation techniques (Morgan and Lazarow, 1962; Hales and Randle, 1963).

SOLID-PHASE RADIOIMMUNOASSAYS

A simpler and widely used variation of the competitive binding RIA is a solid-phase RIA, which is also based on competitive inhibition (Catt and Tregear, 1967) utilizing radiolabeled antigen and solid-phase antibody for the separation of bound and free antigen as follows:

$$SP\text{-}Ab + Ag^* \rightarrow \frac{SP\text{-}Ab\text{-}Ag^0}{SP\text{-}Ab\text{-}Ag^*} + Ag^* + Ag^0$$

where the count rate of radioactive antigen (Ag^*), bound through antibody linkage to the surface of the solid phase (SP-Ab) to form the radiolabeled antigen-antibody complex (SP-Ab-Ag^*) is in direct proportion to the quantity of competitive unlabeled antigen (Ag^0) in the reaction mixture. Removal of the reaction mixture and washing the solid phase (polystyrene tube) serves as an effective and highly reproducible method for separating bound and free antigen. Soon after the introduction of solid-phase assays, it became apparent that RIA procedures based on principles other than competitive inhibition could be developed. This was the direct solid-phase sandwich RIA (Wide et al., 1971). These methods were found to be applicable to all biological substances that have a minimum of two binding sites and thus can be utilized to assay for both antigen and antibodies. Soon, a direct solid-phase sandwich RIA utilizing, for the first time, radiolabeled specific immunoglobulin as a probe for macromolecular viral antigens was introduced (Ling and Overby, 1972). The potential advantages of using radiolabeled antibody over radiolabeled antigen (e.g., viruses) are the greater availability, ease of labeling, and immobilization of immunoglobulins.

There has been a rapid increase in the variety of solid-phase RIAs developed for the detection and quantification of a multitude of antigens and their corresponding antibodies (Overby and Mushahwar, 1979; Mushahwar and Overby, 1983). For the most part, these assays have been used extensively in highly specialized research environments. Most importantly, the vast majority of these tests never made the transition from research procedures to commercially available and standardized procedures (Hill and Matsen, 1983). Because the hepatitis viruses have been widely characterized with highly specific and commercially available RIAs, we will use these models to describe the various direct and indirect (competitive) methods that employ solid-phase sandwich RIA to detect the antigens and antibodies of viral hepatitis (Table 1).

TABLE 1 Hepatitis

Hepatitis marker	Reference(s)
HBV	
HBsAg	Hollinger et al., 1971; Ling and Overby, 1972; Purcell et al., 1973
Anti-HBc IgG	Neurath et al., 1978; Overby and Ling, 1976; Purcell et al., 1973; Vyas and Roberts, 1977
Anti-HBs	Ginsberg et al., 1973
Anti-HBe	Blum et al., 1979; Frosner et al., 1978; Mushahwar et al., 1978; Neurath et al., 1979
HBeAg	Blum et al., 1979; Frosner et al., 1978; Mushahwar et al., 1978; Neurath et al., 1979
Anti-HBc IgM	Chau et al., 1983
HBcAg	Overby and Ling, 1976; Purcell et al., 1973; Vyas and Roberts, 1977
HDV[a]	
HDAg	Rizzetto et al., 1980; Mushahwar and Decker, 1983
Anti-HD IgG	Mushahwar and Decker, 1983; Rizzetto et al., 1980
Anti-HD IgM	Smedile et al., 1982
HAV	
HAV	Decker et al., 1979; Hollinger et al., 1975; Hollinger and Maynard, 1976; Purcell et al., 1976
Anti-HAV IgG	Decker et al., 1979; Hollinger and Maynard, 1976; Purcell et al., 1976; Safford et al., 1980
Anti-HAV IgM	Bradley et al., 1977; Decker et al., 1981; Devine et al., 1979; Flehmig et al., 1979; Lemon et al., 1980; Roggendorf et al., 1980
Anti-HAV IgA	Overby et al., 1981
HCV	
Anti-HCV IgG	Kuo et al., 1989

[a] HDV, hepatitis D virus.

HEPATITIS B SEROLOGIC MARKERS

Three distinct viral antigens and their respective antibodies are used for the diagnosis of hepatitis B virus (HBV) infections: hepatitis B surface antigen (HBsAg) and its antibody (anti-HBs), hepatitis B core antigen (HBcAg) and its antibody (anti-HBc); and hepatitis B e antigen (HBeAg) and its antibody (anti-HBe). These antigens and antibodies occur sequentially in serum during the course of disease and recovery (Mushahwar et al., 1981). Because HBsAg is a defective particle without nucleic acid and is produced in large quantities, it is an easily detectable marker of an ongoing HBV infection. Seroconversion to anti-HBs is an indication of recovery and eventual immunity. HBcAg is found in the internal core of the DNA-containing virus and has not been found to occur free in serum. Anti-HBc rises with the onset of viremia and persists through recovery and immunity. This antibody is sometimes the only detectable marker for exposure to the virus. HBeAg rises in serum at the same time as HBsAg, and its presence is associated with acute disease, chronic disease, and the presence of infec-

tious virus. Seroconversion to anti-HBe signifies a better clinical prognosis and a lower level of viremia.

DIRECT SOLID-PHASE SANDWICH ASSAYS

Radiometric Assays for HBV Antigens and Antibodies

The direct solid-phase sandwich RIA procedure for antigen detection is divided into three main operational stages as follows: adsorption of unlabeled antibody onto the solid phase, binding of the antigen by the adsorbed antibody, and detection of bound antigen by reaction with radiolabeled and highly specific antibody. The first successful direct solid-phase RIA for the detection of viral antigens was the AUSRIA prototype (Ling and Overby, 1972) for the detection of HBsAg. In this system, highly specific human anti-HBs was adsorbed to polystyrene tubes. The serum to be tested for HBsAg was incubated with the anti-HBs-coated tube, giving a complex of anti-HBs and HBsAg in the solid phase. The radioactive probe in the next reaction was affinity purified guinea pig anti-HBs labeled with radioactive iodine (^{125}I).

A direct solid-phase RIA for HBeAg similar to AUSRIA was developed (Mushahwar et al., 1978), employing radiolabeled specific antibodies as illustrated in Fig. 1. Briefly, 6-mm-diameter polystyrene beads are coated with anti-HBe by incubating a dilution of antiserum with the beads at pH 9.0 for 24 h at room temperature. These antibody-coated beads are then used as the solid-phase antibody. Anti-HBe immunoglobulin G (IgG) preparations are radiolabeled with ^{125}I by the chloramine-T method (Greenwood et al., 1963), often resulting in a specific radioactivity of 18 to 25 μCi/μg of IgG. The ^{125}I labeled anti-HBe solution is diluted to approximately 3 μCi/ml in a diluent containing 50% fetal calf serum. For HBeAg detection, 0.2-ml serum samples are incubated overnight at room temperature with the antibody-coated beads. The beads are then washed with distilled water and further incubated in 0.2 ml of ^{125}I-labeled anti-HBe solution at 45°C for 4 h. The beads are then washed and measured for ^{125}I-labeled anti-HBe uptake. The resulting count rate of the multiple-layered beads is directly proportional to the HBeAg concentration in the sample. Human convalescent antisera containing anti-HBe are a source of reagents for the diagnostic RIA described, so that animal immunizations with purified preparations of HBeAg are not necessary. The manufacturer's package insert for HBeAg determination describes the procedures in a stepwise manner.

The presence or absence of HBeAg is determined by comparing the counts per minute of the unknown specimen to a predetermined cutoff value (Mushahwar et al., 1978). Specimens whose count rates are equal to or greater than this cutoff value are considered to be reactive for HBeAg. For a test to be valid, the mean value for the positive control should be at least four times the negative-control mean. It is recommended by the manufacturer that three negative and three positive HBeAg controls should be assayed with each run. A similar direct solid-phase sandwich RIA for anti-HBs detection has been developed (Ginsberg et al., 1973). The principle is illustrated in Fig. 2.

Highly purified HBsAg is bound to a polystyrene bead by adsorption to produce the solid-phase antigen. The serum specimen to be assayed for anti-HBs is incubated with the solid-phase antigen. If anti-HBs is present, it will complex with the antigen. ^{125}I-HBsAg is added in the next step; it will bind to the antibody already trapped on the bead, form-

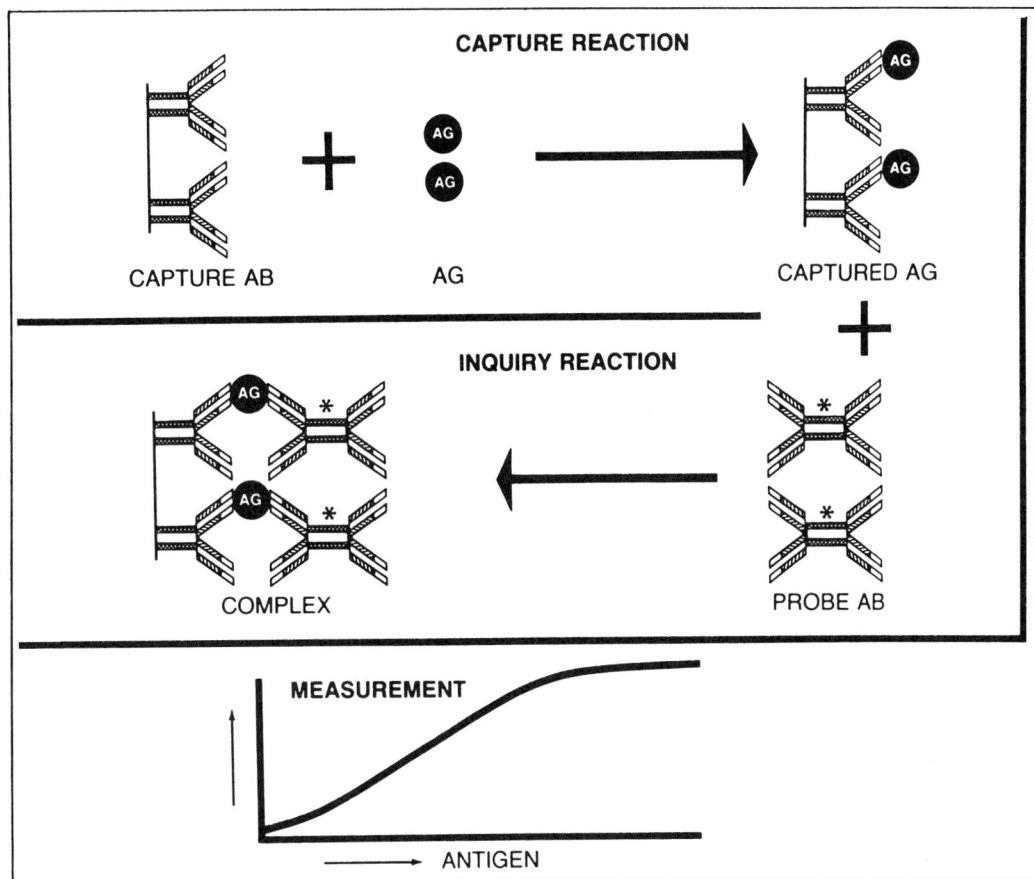

FIGURE 1 Schematic representation of a direct solid-phase assay designed to detect HBeAg. The procedure involves capture of the antigen by antibody attached to the solid phase (top) and subsequent binding of labeled antibody to exposed sites on the captured antigen (middle). The amount of antigen measured is in direct proportion to the count rate of the washed solid phase (bottom).

ing an antigen–antibody–^{125}I-HBsAg complex. The amount of radioactivity on the bead is in proportion to the concentration of anti-HBs in the serum. The step-by-step procedure is described in the manufacturer's package insert. An example of this is the AUSAB test (Abbott Laboratories) for anti-HBs. It is recommended that seven negative and three positive controls be assayed with each run of unknowns.

Again, as in the case of HBeAg detection, reactive and nonreactive specimens are determined by relating the net counts per minute of unknowns to a cutoff value calculated by multiplying the negative-control mean by the factor 2.1. Specimens with values greater than the cutoff are considered reactive for anti-HBs.

INDIRECT (COMPETITIVE) RADIOMETRIC ASSAYS FOR HBV ANTIBODIES

In these assays, radiolabeled antibodies are utilized as probes for the detection of serum antibodies in either a one-step or a two-step procedure. The solid phase is an appropriate antigen-coated polystyrene tube or bead. The radiolabeled probe will have fewer antigen binding sites for the reaction when serum antibodies are present. Hence, the final count rate of the solid phase will be inversely proportional to the

amount of antibody in the serum specimen being tested. These techniques have been applied to the detection of anti-HBc (Overby and Ling, 1976) and anti-HBe (Mushahwar et al., 1978).

One-Step Procedure

The anti-HBc assay is illustrated in Fig. 3. The solid-phase reagent is a polystyrene bead coated with purified HBcAg. The bead is first reacted with the serum specimen to be assayed for anti-HBc. The resulting HBcAg–anti-HBc complex is challenged with purified human ^{125}I–anti-HBc IgG in a second step. The level of radioactivity bound to the bead is inversely proportional to the amount of anti-HBc in the test specimen. A competitive binding RIA is available commercially (Table 2). The step-by-step procedure is described in the manufacturer's package insert.

Five positive and five negative controls are tested with each run of unknown specimens. The presence or absence of anti-HBc is determined by comparing the net counts per minute of the specimen to a cutoff value, calculated as the sum of the negative-control and positive-control means divided by two. Specimens whose count rates are equal to or lower than the cutoff value are considered reactive for anti-HBc.

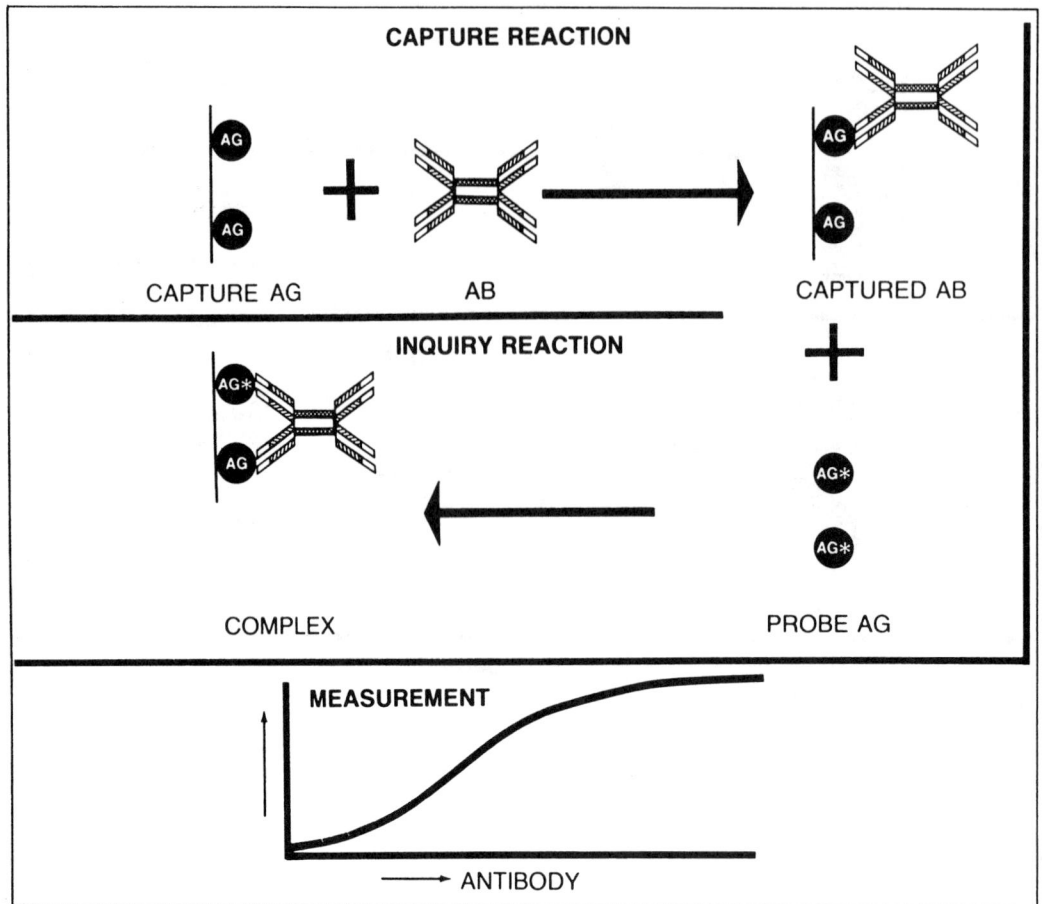

FIGURE 2 Schematic representation of direct solid-phase assay to detect anti-HBs. The two-step procedure involves capture of specific antibody by antigen bound to the solid phase (top) and binding of labeled antigen to specific available sites on the captured antibody (middle). The amount of antibody measured is in direct proportion to the count rate of the washed solid phase (bottom).

TABLE 2 Commercial hepatitis RIA kits

Test	Serologic marker	Product name	Solid-phase configuration	Manufacturer
Antigen	HBsAg	AUSRIA II	6-mm-diameter bead	Abbott Laboratories
		Heparia[a]	6-mm-diameter bead	North American Biologicals, Inc.
		Rialyze[a]	Gel column	Ames Co.; Division of Miles Laboratories
		AUK-3	6-mm-diameter bead	Sorin Biomedica
		Riasure[a]	Microbeads	Electro-Nucleonics Laboratories, Inc.
		Clinical assays[a]	Plastic tube	Connought Laboratories (Travenol Laboratories)
		NML HBsAg RIA	6-mm-diameter bead	Nuclear-Medical Laboratories
	HBeAg	ABBOTT-HBe	6-mm-diameter bead	Abbott Laboratories
		EBK	6-mm-diameter bead	Sorin Biomedica
Antibody	Anti-HBs	AUSAB	6-mm-diameter bead	Abbott Laboratories
		AB-AUK	6-mm-diameter bead	Sorin Biomedica
		Hepab[a]	6-mm-diameter bead	North American Biologicals, Inc.
		Clinical assays[a]	Plastic tube	Connought Laboratories (Travenol Laboratories)
	Anti-HBc (total)	CORAB	6-mm-diameter bead	Abbott Laboratories
		AB-COREK	6-mm-diameter bead	Sorin Biomedica
	Anti-HBc (IgM)	CORAB-M	6-mm-diameter bead	Abbott Laboratories
		CORE-IGMK	6-mm-diameter bead	Sorin Biomedica
	Anti-HAV (IgM)	HAVAB-M	6-mm-diameter bead	Abbott Laboratories
	Anti-HBe	ABBOTT-HBe	6-mm-diameter bead	Abbott Laboratories
		EBK	6-mm-diameter bead	Sorin Biomedica
	Anti-HAV (total)	HA VAR	6-mm-diameter bead	Abbott Laboratories
	Anti-delta (total)	ABBOTT-ANTI-DELTA	6-mm-diameter bead	Abbott Laboratories

[a] No longer commercially available.

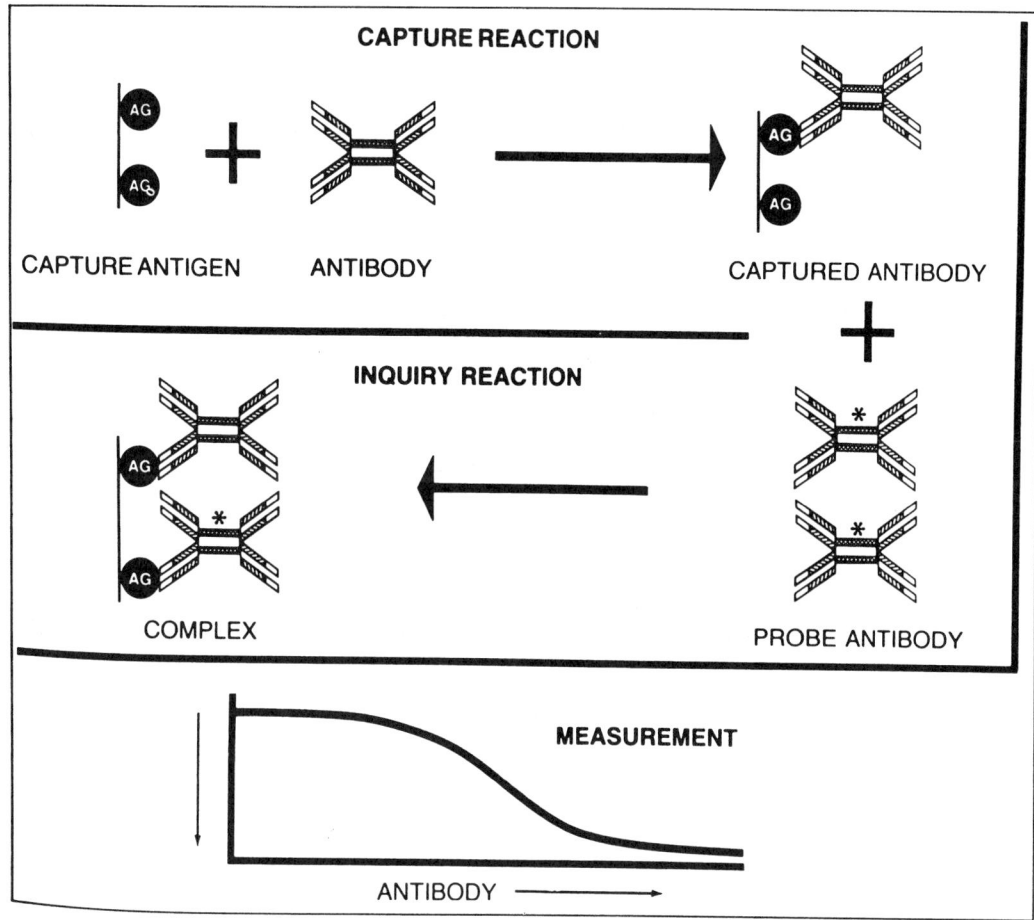

FIGURE 3 Schematic representation of an indirect competitive solid-phase assay for anti-HBc utilizing labeled antibodies. The two-step procedure involves capture of antibody by an antigen (HBcAg)-coated solid phase (top) and binding of labeled antibody to available sites on the bound antigen (middle). The count rate of the washed solid phase is inversely proportional to the amount of antibody in the assay specimen (bottom).

Two-Step Procedure

The anti-HBe assay has been described in detail (Mushahwar et al., 1978). This was the first successful indirect solid-phase RIA for the detection of serum anti-HBe utilizing a two-step neutralization procedure. The principle of the assay is illustrated in Fig. 4. In this assay, 0.1 ml of the patient's serum is mixed with an equal volume of standardized HBeAg-positive serum (neutralizing reagent). The mixture is incubated overnight at room temperature with the solid phase, a polystyrene bead coated with high-titer human anti-HBe serum. After being washed, the bead is incubated with ^{125}I-labeled anti-HBe IgG for 4 h at 45°C, as described above for the direct solid-phase HBeAg assay. The quantity of neutralizing reagent in the initial incubation has been selected to give over 10,000 cpm on the bead in the presence of a serum that is negative for anti-HBe. A 50% or more reduction in the count rate indicates the presence of anti-HBe in the test sample. This assay is available commercially (Table 2). The step-by-step procedure is described in the manufacturer's package insert. Three negative and three positive anti-HBe controls are used each time the test is run for anti-HBe analysis.

The presence or absence of anti-HBe is determined by comparing the count rate of the specimen to a cutoff value calculated from the sum of the negative control and positive control divided by two. Specimens whose count rates are equal to or less than the cutoff value are considered reactive for anti-HBe.

ASSAY FOR IgM-CLASS ANTIBODIES

Virus-specific IgM antibody has been confirmed as a prominent early immune response in many viral infections. Because it is relatively short-lived, it is a good marker for acute disease (Chau et al., 1983). A reliable and reproducible RIA for the detection of hepatitis A virus (HAV) IgM antibody (anti-HAV IgM) has been described (Decker et al., 1979). The anti-HAV IgM assay is based on the following reactions and is illustrated in Fig. 5.

1. SP-Abμ + IgM → Sp-Abμ–IgM
2. SP-Abμ-IgM + HAV → SP-Abμ–IgM–HAV
3. SP-Abμ–IgM–HAV + Ab*-HAV → SP-Abμ–IgM–HAV–Ab*-HAV

SP-Abμ is a polystyrene surface coated with μ-chain-specific goat anti-human antibody. In step 1, if anti-HAV IgM is present in a patient's serum, it will be bound by the

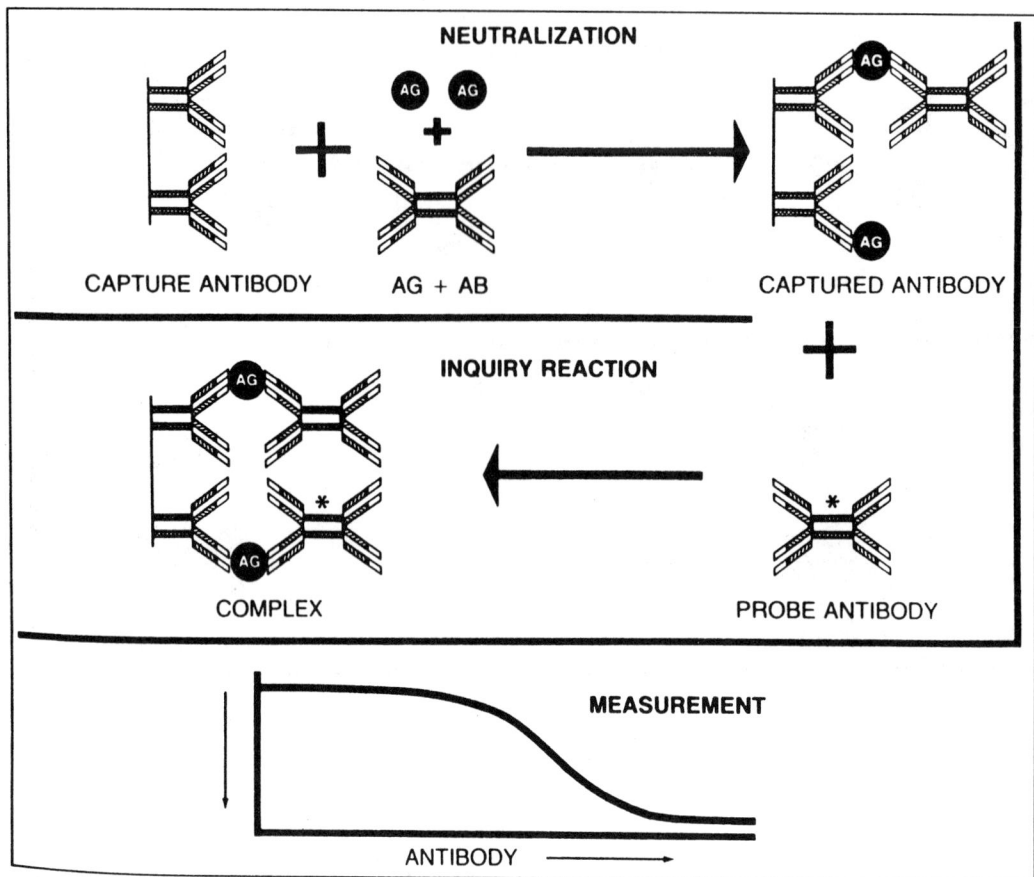

FIGURE 4 Schematic representation of a competitive solid-phase sandwich assay for anti-HBe antibodies. The solid phase, coated with anti-HBe, is incubated with a standardized amount of HBeAg and a sample of the patient's serum. The anti-HBe in the patient's serum combines with the antigen and reduces the amount of antigen binding to the solid-phase-bound anti-HBe (top). The sandwich is formed by the addition of labeled anti-HBe (middle). The amount of labeled anti-HBe bound is inversely proportional to the amount of specific antibody present in the patient's serum (bottom).

μ-chain-specific solid-phase antibody. In step 2, HAV will be attached to this complex to form the SP-Abμ–IgM–HAV. This complex is then detected in step 3 by incubation with the probe antibody, Ab*-HAV, an ^{125}I-labeled human anti-HAV IgG. The resulting count rate of the multiple-layer product SP-Abu–IgM–HAV–Ab*-HAV is in proportion to the anti-HAV IgM concentration in the patient's serum. This test was shown to be highly specific. No cross-reactions were observed in the presence of high-titer anti-HAV IgG when hepatitis A convalescent-serum samples were tested. This assay is commercially available. The step-by-step procedure is described in the manufacturer's package insert.

Two negative and three positive controls are used each time the test is run for anti-HAV IgM analysis. The presence or absence of anti-HAV IgM is determined by comparing the count rate of the specimen to a predetermined cutoff value (Decker et al., 1979).

OTHER VIRAL MARKERS

Several reviews (Overby and Mushahwar, 1979; Mushahwar and Overby, 1983) have surveyed the application of RIA for rapid diagnosis of viral, bacterial, and fungal diseases. The following review describes the use of RIA to identify viral antigens or antiviral antibodies appearing in the literature. Tables 3 and 4 summarize the information presented in this review.

Enteroviruses

Enterovirus Antibody Detection

A solid-phase IgM capture RIA designed to detect antibody specifically directed against coxsackievirus B4 and BS has been described (Morgan-Capner and McSorley, 1983). Positive results were obtained with sera from patients with heterologous enterovirus infection. These results may be due to anamnestic IgM response or shared, common antigenic determinants. The authors note that specificity may be improved by the development of monoclonal detector antibodies.

IgM-class as well as IgG-class antibodies to coxsackievirus A7, A9, A16, B2, B4, or B5 or echovirus 4, 17, or 25 could be detected using RIA procedures (Torfason et al., 1984). IgG titers were demonstrated to be higher in the convalescent-phase sera. Modification of the indirect IgM test resulted in an antibody capture assay that demonstrated type specificity. The authors note that the antigen must be carefully standardized in order to maintain the type specificity.

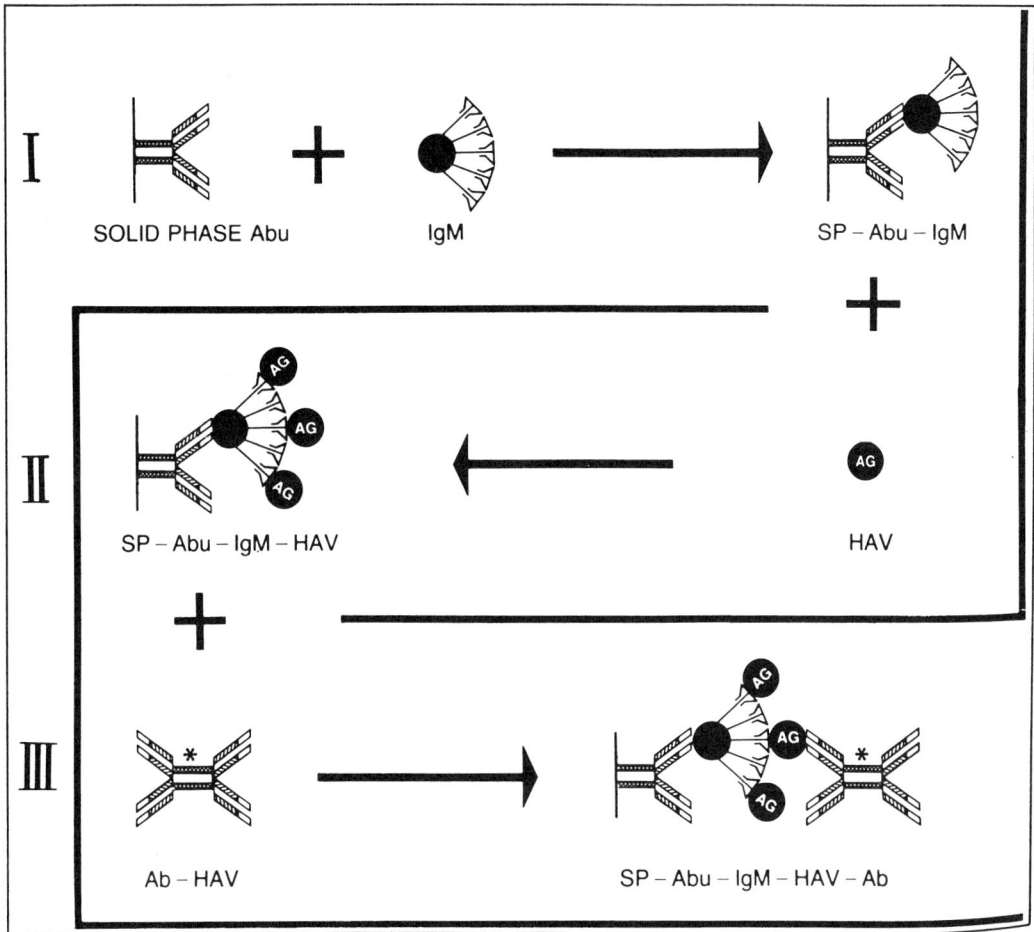

FIGURE 5 Schematic representation of a direct assay for IgM-class anti-HAV antibody. The solid phase is coated with anti-human antibody specific for the μ chain. The three-step procedure involves capture of IgM antibodies by the solid-phase μ chain-specific antibody (I), subsequent binding of HAV to the captured antibody (II), and binding of radiolabeled antibody to the solid-phase-bound complex (III). The amount of antibody in the serum sample is directly proportional to the count rate of the washed solid phase.

Togavirus

Antialphavirus Antibody

A previous review (Mushahwar and Overby, 1983) has detailed the sensitive and specific procedures used to identify antibody to Venezuelan, western, and eastern equine encephalitis viruses (Jahrling et al., 1978).

Antiflavivirus Antibody

Trent et al. (1976) have described a solid-phase assay for the detection of IgG and IgM antibodies to specific St. Louis encephalitis (SLE) virus structural protein. Other investigators (Wolff et al., 1981) have extended the solid-phase system to use the crude antigens now commonly used to diagnose SLE virus infections. Their results indicated that the procedure was as sensitive as conventional serologic tests but not as specific. The procedure was capable of differentiating SLE from similar clinical infections with alphaviruses. However, infections caused by related flaviviruses could not be accurately differentiated.

Anti-Rubella Virus Antibody

The ability to diagnose acute infections by serologic methods becomes more important when the clinical signs and symptoms are subtle. The principle of diagnosing rubella virus infections by the detection of virus-specific IgM antibody in a serum sample is now well accepted. Solid-phase RIA has been shown to detect IgM, as well as IgG, antibodies to rubella virus. The adsorption of purified rubella virus to polystyrene beads resulted in a highly sensitive test capable of detecting IgM antibodies (Meurman et al., 1977; Meurman, 1978; Mortimer et al., 1981) or IgM or IgG antibodies (Kalimo et al., 1976). The development of a solid-phase IgM capture procedure (Mortimer et al., 1981; Tedder et al., 1982) resulted in a high degree of sensitivity and a decreased probability of interference by rheumatoid factor. The use of monoclonal antibody directed against the hemagglutinin protein of rubella virus resulted in a strong and specific reaction when used in the MACRIA assay (Tedder et al., 1982). Evaluation of clinical samples with this system may result in greater correlation with more

TABLE 3 Antigen detection

Organism	Radiolabeled reagent	Sample[a]	Separation technique	Reference
Rotavirus	Ab	Stool	Polystyrene bead	Sarkkinen et al., 1979b
	Ab	Stool	Polystyrene bead	Haikala et al., 1983
	Ab	Stool	Microtiter plate	Cukor et al., 1978
	Ab	Stool	Microtiter plate	Cukor et al., 1984
	Ab	Stool	Polystyrene tube	Middleton et al., 1977
	Ab	Stool	Polystyrene bead	Sarkkinen et al., 1980
Norwalk agent	Ab	Stool	Microtiter plate	Greenberg et al., 1978
Adenovirus	Ab	Stool	Polystyrene bead	Halonen et al., 1980
	Ab	NP	Polystyrene bead	Sarkkinen et al., 1981
	Ab	NP	Polystyrene bead	Vesikari et al., 1982
HSV	Protein A	Virus	Microtiter plate	Marsden et al., 1984
	Protein A	Cell	Filter paper disc	Cleveland et al., 1979
	Ab	Virus	Polystyrene tube	Enlander et al., 1976
	Ab	Cell	Glass vials	Forghani et al., 1974
Respiratory viruses (RSV)	Ab	NP	Polystyrene bead	Sarkkinen et al., 1981
	Ab	NP	Polystyrene bead	Meurman et al., 1984
	Ab	NP	Polystyrene bead	Vesikari et al., 1982
Parainfluenza virus type 2	Ab	NP	Polystyrene bead	Sarkkinen et al., 1981
Parainfluenza virus type 3	Ab	NP	Polystyrene bead	Vesikari et al., 1982
Influenza virus	AMP	NP	Polystyrene bead	Coonrod et al., 1984
	Ab	NP	Polystyrene bead	Sarkkinen et al., 1981
	Ab	NP	Polystyrene bead	Versikari et al., 1982

[a] NP, nasopharyngeal sample; virus, virus from cell culture; cell, antigens in cell culture.

TABLE 4 Antibody detection

Organism	Ab species identified	Radiolabed reagent	Separation technique	Reference
Enteroviruses	IgM	Ab	Polystyrene bead	Morgan-Capner and McSorley, 1983
Togaviruses	IgG, IgM	Ab, Ag	Polystyrene bead	Torfason et al., 1984
Rubella virus	IgM	Ab	Microtiter plate	Kangro et al., 1981
	IgM	Ab	Polystyrene bead	Mortimer et al., 1981
	IgG, IgM	Ab	Polystyrene bead	Meurman, 1978
	IgG, IgM	Ab	Polystyrene bead	Kalimo et al., 1976
	IgG, IgM	Ab	Polystyrene bead	Meurman and Granfors, 1977
	IgM	Ab	Polystyrene bead	Tedder et al., 1982
SLE virus	IgM	Ab	Red cells	Sexton et al., 1982
	IgG, IgM	Ab	Polyvinyl microtiter plate	Wolff et al., 1981
Rotavirus	IgG	Ab	Bead	Trent et al., 1976
	IgG, IgA, IgM	Ab	Polystyrene bead	Sarkkinen et al., 1979
	IgG, IgA, IgM	Ab	Filter paper	Watanabe and Holmes, 1977
Herpesviruses	IgA	Ab	Microtiter plate	Cukor et al., 1979
CMV	IgG	Ab	Polyvinyl microtiter plate	Kimmel et al., 1980
	IgM	Ab	Microtiter plate, polystyrene well	Kangro et al., 1984
HSV	IgG	Ab	Microtiter plate	Dreesman et al., 1979
	IgG	Ab	Infected cells	Smith et al., 1974
	IgG, IgA, IgM	Ab	Plastic-coated bead	Patterson et al., 1978
	IgG, IgM	Ab	Polystyrene bead	Kalimo et al., 1977a
	IgG, IgM	Ab	Polystyrene bead	Kalimo et al., 1977a
	IgG	Ab	Polyvinyl chloride plate	Matson et al., 1983
	IgG	Ab	Polyvinyl chloride plate	Adler-Storthz et al., 1983
VZV	IgG	Ab	Polyvinyl microtiter plate	Friedman et al., 1979
	IgG, IgM	Ab	Polyvinyl microtiter plate	Arvin and Koropchak, 1980
	IgG	Ab	Microtiter plate	Benzie-Campbell et al., 1981
	IgM	Ab	Polystyrene tube and beads	Tedder et al., 1981
Adenovirus	IgA	Ab	Polystyrene bead	Halonen et al., 1979b

traditional hemagglutination assays used to evaluate immune status. Other solid-phase systems, such as microtiter plates (Kangro et al., 1981) and the use of red cells (Sexton et al., 1982), have been successfully employed to detect IgM antibodies to rubella virus.

Rotavirus

Antigen Detection

The use of solid-phase detection systems for the identification of rotavirus antigens in stool specimens has become increasingly common. The assays have taken three general forms, as follows: the capture of antigen on microtiter plates (Kalica et al., 1977; Cukor et al., 1978, 1984; Greenberg et al., 1978), on polystyrene beads (Sarkkinen et al., 1979a; Haikala et al., 1983), or using polystyrene tubes (Middleton et al., 1977).

Comparison of RIA procedures (Sarkkinen et al., 1979a) to latex agglutination (Haikala et al., 1983) indicates that RIA is more sensitive and less subject to interference by particulate material. The key to a sensitive and specific test for antigen identification is the selection of immune reagents of the highest quality. Cukor et al. (1978) noted that screening of antiserum by conventional procedures, such as complement fixation (CF), is not suitable or appropriate for selecting antibody capable of identifying rotavirus in stool samples with great sensitivity. The use of a monoclonal antibody directed against the common group-specific antigen has resulted in a sensitive and specific test (Cukor et al., 1984).

Antibody Detection

Sarkkinen et al. (1979b) described a solid-phase RIA for the detection of human rotavirus-specific IgG, IgM, and IgA antibodies. This system, using viral antigens to capture specific antibodies, was found to be more sensitive than the reference CF test. Cukor et al. (1979) have described a procedure for the detection of IgA in human milk. The RIA was capable of detecting antibody for longer periods of time than enzyme-linked immunosorbent assay (ELISA), immunofluorescence assay (IFA), or neutralization assays.

Norwalk Virus

Antigen Detection

A second viral agent has been linked to acute episodes of diarrhea or vomiting (Cukor and Blacklow, 1984). The Norwalk virus is a small, 27-nm-diameter nonenveloped particle. Although the particle was visualized in 1972, major advances were not made until an RIA capable of detecting the agent was reported (Greenberg et al., 1978). The use of a solid-phase system employing microtiter plates allows a large number of patient specimens to be screened.

Adenovirus

Antigen Detection

Adenovirus infections may result in virus shedding from two different sites. Virus shed in feces has been identified in a manner similar to that described for rotavirus. Halonen et al. (1980) described a simple yet sensitive method for identifying adenovirus antigen in stool suspensions. Confirmatory tests have shown the procedure to be sensitive while maintaining high specificity.

Identification of adenovirus antigens in nasopharyngeal secretions by RIA has been demonstrated (Sarkkinen et al., 1981). The investigators demonstrated complete correlation with reference to the IFA.

CMV

Antibody Detection

The presence or absence of IgG antibodies to cytomegalovirus (CMV) was determined (Kimmel et al., 1980) by using an indirect RIA detection system. The procedure was capable of accurately detecting anti-CMV antibody in the presence of antibodies to other members of the herpesvirus family. A modification of the original RIA incorporating a primary 1:100 dilution resulted in a rapid screening procedure with excellent correlation with serum titration experiments (Kimmel et al., 1980).

Kangro et al. (1984) compared RIA procedures with ELISA and found comparable sensitivity with sera from adults. The RIA procedures were more sensitive, however, when cord serum was used. The results indicate that IFA procedures were less sensitive than RIA or ELISA. The identification of IgM antibodies with an indirect solid-phase RIA has been described (Griffiths and Kangro, 1984). The procedure was shown to be highly specific for detecting congenital infection and sensitive for identifying primary CMV infection in pregnant women.

HSV

Antibody Detection

Early application of RIA procedures for the detection of specific antibodies to herpes simplex virus (HSV) used virus-infected fixed monolayers to measure naturally occurring immunoglobulins (Smith et al., 1974). Extension of this procedure involved adsorption of viral antigens to the surface of a solid phase with results comparable to those with monolayers (Smith et al., 1974). Refinement of serum dilution procedures and incubation times resulted in greater sensitivity (Patterson and Smith, 1973).

Improvements in the procedures, such as attachment of viral antigens to polystyrene beads (Kalimo et al., 1977c), use of specific viral antigens (Kalimo et al., 1977a), or removal of cross-reacting antigens by sample adsorption (Patterson et al., 1978), resulted in lower background and greater sensitivity and specificity.

Cross-reactivity between members of the HSV group was observed to be a problem during the early phases of assay development (Kalimo et al., 1977b). Adsorption of patient serum with potential cross-reacting organisms or using a solid phase coated with specific antigens resolves major problems of cross-reactivity with other herpesviruses (Patterson et al., 1978; Smith and Kennell, 1981). The identification and use of specific HSV type 1 (HSV-1) and HSV-2 glycoproteins attached to the solid phase (Matson et al., 1983) allowed identification of an immune response to either HSV-1 or HSV-2 with high levels of sensitivity and specificity (Forghani et al., 1975; Adler-Storthz et al., 1983).

Antigen Detection

Generally, the identification of HSV in clinical samples has required virus isolation in cell culture and subsequent identification by reaction with fluorescent- or enzyme-tagged antibody. RIA technology has been applied to the detection of HSV antigens in cell culture (Forghani et al., 1974;

Cleveland et al., 1979), virus-containing fluids (Enlander et al., 1976), or clinical specimens (Forghani et al., 1974). Identification of viral antigens was shown to be highly sensitive and specific when clinical specimens were examined (Forghani et al., 1974).

Commonly, labeled antibodies have been used to detect specific antigens. Staphylococcal protein A labeled with ^{125}I has been used to detect immune complexes (Cleveland et al., 1979). This generic approach lends itself to modification for the identification of other viral antigens.

VZV

Anti-VZV Antibody

Several investigators have described methods for the detection of IgM (Tedder et al., 1981), IgG (Friedman et al., 1979; Benzie-Campbell et al., 1981; Richman et al., 1981), or IgG and IgM antibodies to varicella-zoster virus (VZV) (Arvin and Koropchak, 1980). All of the assays were shown to be more specific and sensitive than the standard CF procedures.

Respiratory Viruses

Influenza Virus Antigen Detection

Coonrod et al. (1984) observed that viral-antigen detection was less sensitive during the first 3 days after the onset of infection and that antigen could be detected longer than infectious virus. A comparison of two antigen detection systems (Sarkkinen et al., 1981) indicates that the sensitivities and specificities of RIA and IFA are equivalent.

Solid-phase assays for the detection of influenza A virus antigens from clinical samples have been described (Sarkkinen et al., 1981; Coonrod et al., 1984).

Other Respiratory Viruses

Identification of viral antigens in the appropriate clinical specimen is the most direct method of identifying a causative agent. Sarkkinen et al. (1981) reported the use of a solid-phase antigen capture RIA procedure to identify the presence of respiratory syncytial virus (RSV), parainfluenza virus type 2, or adenovirus in nasopharyngeal secretions. Agreement between RIA and IFA was very good for all viruses. Correlation between IFA and RIA was independent of the sample fraction, either mucus or cells, assayed for all viruses except RSV. Lower sensitivity was observed when only the mucus fraction (devoid of cells) was tested in the case of RSV.

The ability of RIA procedures to identify viral antigens in nasopharyngeal samples by RIA and a comparison with diagnosis by other serologic methods has been described (Meurman et al., 1984). The results indicate that the ability to identify viral antigen in the sample was dependent on the time after the onset of symptoms and the age of the patient. Antigen detection was more sensitive in specimens taken from children under 6 months of age compared with those from older children.

Other RIAs

Since the publication of the first edition of this book in 1986, several RIAs have been described for the detection of a variety of viral antigens and antibodies. The following summary describes these RIAs and covers the literature between 1984 and 1989.

Adenovirus Antigen

Ruuskanen et al. (1984) have described an RIA for the detection of adenovirus antigen in the nasopharyngeal specimens of children with febrile exudative tonsillitis. The rapid detection of adenovirus antigen permitted withdrawal of unnecessary and ineffective antibiotic treatment for most of these children.

Astrovirus Antibody

The development and evaluation of RIAs for the detection of IgM and IgG antibodies to astrovirus have been described (Wilson and Cubitt, 1988). The tests were shown to be sensitive and specific and suitable for screening large numbers of sera. The use of the assays has established that astrovirus type 1 is prevalent in the United Kingdom and that not only infants but also school-age children and elderly patients are affected by this virus.

Coxsackie B Virus Antibody

IgM antibody capture RIAs (Pugh, 1984) were developed to detect coxsackie virus B1- to B5-specific IgM. This specific antibody was detected in sera from all patients with coxsackie B virus infections that were confirmed by virus isolation; however, sera from some patients with rising neutralizing antibody titers were negative in the assay. Frequent heterotypic responses were seen among the positive sera. When sera from patients with enterovirus infections other than coxsackie B virus were studied, they were also found to be reactive to the coxsackie B virus antigen used for this test.

Dengue Virus

A monoclonal RIA was developed for detection of dengue virus (Monath et al., 1986a) in infected cell culture fluids and blood samples from dengue patients. The assay was found to be 10-fold more sensitive for dengue type 2 than for dengue type 1 and 3 viruses and 100-fold more sensitive than for dengue type 4 virus. Virus was more frequently detected in cases of primary infection than in cases of superinfection.

HCV Antibody

The isolation of a clone of nucleic acid from the genome of the non-A, non-B hepatitis agent (Choo et al., 1989) allowed for the subsequent expression of an encoded protein (Choo et al., 1989; Kuo et al., 1989) and offered the first opportunity to develop specific tests for a new RNA virus subsequently named hepatitis C virus (HCV). A usable polypeptide (C-100) derived from about 11% of the viral genome has been found to contain at least two antigenic epitopes that are recognized by sera from many patients with chronic HCV infection. A very sensitive solid-phase RIA for anti-HCV has been developed (Kuo et al., 1989) utilizing this polypeptide.

HSV Antibodies

A solid-phase RIA for the detection of IgG antibodies to HSV using a mouse monoclonal antibody specific for the Fc portion of human IgG as the radiolabeled detecting antibody has been developed (Berry et al., 1987). Compared to virus neutralization assays, the RIA had a sensitivity of 100% and a specificity of 93%.

VZV Antibody

Kangro et al. (1988) described a simple and sensitive IgM antibody capture RIA that utilizes VZV antigen and a single monoclonal anti-VZV antibody for the detection of anti-VZV IgM. This assay was also compared to IFA to detect IgM responses in patients with varicella (chicken pox) and patients with zoster (shingles). IgM antibodies were detected in all patients with varicella. The IgM antibodies appeared shortly after the onset of the rash, reached peak levels 1 to 4 weeks later, and then declined to low or undetectable levels in most patients. IgM antibodies were also detected in 98.2% of patients with zoster, but these IgM levels were significantly lower than the levels of IgM produced in response to varicella. There was wider individual variation both in magnitude and duration of the IgM responses.

Yellow Fever Virus

A solid-phase RIA was developed (Monath et al., 1986b) for the detection of yellow fever virus in infected cell culture supernatant fluid and clinical samples. The test employed a flavivirus group-reactive monoclonal antibody attached to a polystyrene bead as the solid support and a radiolabeled type-specific antibody probe in a simultaneous sandwich RIA format. The sensitivity of the assay was found to be 100 pg of gradient-purified virion protein per 100 ml. Specificity was approximately 99.4%.

SUMMARY

Since its introduction 40 years ago as a highly sensitive, specific, and reproducible method for the determination of insulin levels in plasma (Yalow and Berson, 1960), RIA has been a valuable tool for the measurement of many molecules. Over the years, RIA has proven to be one of the key serologic procedures in diagnostic medicine and has been utilized in many clinical and epidemiological investigations. The wide use of RIA in the field of infectious diseases has considerably improved our diagnostic accuracy and efficiency and enabled us to avoid diagnostic errors based on the use of less sensitive immunologic procedures.

REFERENCES

Adler-Storthz, K., D. O. Matson, E. Adam, and G. R. Dreesman. 1983. A micro solid-phase radioimmunoassay for detection of herpesvirus type-specific antibody: specificity and sensitivity. *J. Virol. Methods* **6:**85–97.

Arvin, A. M., and C. M. Koropchak. 1980. Immunoglobulins M and G to varicella-zoster virus measured by solid-phase radioimmunoassay: antibody responses to varicella and herpes zoster infection. *J. Clin. Microbiol.* **12:**367–374.

Avraham, H., J. Golenser, Y. Fazitt, D. T. Spira, and D. Sulitzeanu. 1982. A highly sensitive solid-phase radioimmunoassay for the assay of Plasmodium falciparum antigens and antibodies. *J. Immunol. Methods* **53:**61–68.

Benzie-Campbell, A., H. O. Kangro, and R. B. Heath. 1981. The development and evaluation of a solid-phase radioimmunoassay (RIA) procedure for the determination of susceptibility to varicella. *J. Virol. Methods* **2:**149–158.

Berry, N. J., J. E. Grundy, and P. D. Griffiths. 1987. Radioimmunoassay for the detection of IgG antibodies to herpes simplex virus and its use as a prognostic indicator of HSV excretion in transplant recipients. *J. Med. Virol.* **21:**147–154.

Blum, H. E., G. Dolken, and W. Gerok. 1979. Solid-phase radioimmunoassay for hepatitis Be antigen (HBeAg). *Klin. Wochenschr.* **57:**1129–1132.

Bradley, D. W., J. E. Maynard, S. H. Hindman, C. L. Hornbeck, H. A. Fields, K. A. McCaustland, and E. H. Cook, Jr. 1977. Serodiagnosis of viral hepatitis A: detection of acutephase immunoglobulin M antihepatitis A virus by radioimmunoassay. *J. Clin. Microbiol.* **5:**521–530.

Catt, K. J., and G. W. Tregear. 1967. Solid-phase radioimmunoassay in antibody coated tubes. *Science* **158:**1570–1572.

Chau, K. H., M. P. Hargie, R. H. Decker, I. K. Mushahwar, and L. R. Overby. 1983. Serodiagnosis of recent hepatitis B infection by IgM class anti-HBc. *Hepatology* **3:**142–149.

Choo, Q. L., G. Kuo, A. J. Weiner, L. R. Overby, D. W. Bradley, and M. Houghton. 1989. Isolation of a cDNA clone derived from a blood-borne non-A, non-B viral hepatitis genome. *Science* **244:**359–361.

Cleveland, P. H., D. D. Richman, M. N. Oxman, M. G. Wickham, P. S. Binder, and D. M. Worthem. 1979. Immobilization of viral antigens on filter paper for a ^{125}I staphylococcal protein A immunoassay: a rapid and sensitive technique for detection of herpes simplex virus antigens and antiviral antibodies. *J. Immunol. Methods* **29:**369–386.

Coonrod, J. D., R. F. Betts, C. C. Linnemann, Jr., and L. C. Hsu. 1984. Etiological diagnosis of influenza A virus by enzymatic radioimmunoassay. *J. Clin. Microbiol.* **19:**361–365.

Cukor, G., and N. R. Blacklow. 1984. Human viral gastroenteritis. *Microbiol. Rev.* **48:**157–179.

Cukor, G., M. K. Berry, and N. R. Blacklow. 1978. Simplified radioimmunoassay for detection of human rotavirus in stools. *J. Infect. Dis.* **138:**906–910.

Cukor, G., N. R. Blacklow, F. E. Capozza, Z. F. K. Panjvani, and F. Bednarek. 1979. Persistence of antibodies to rotavirus in human milk. *J. Clin. Microbiol.* **9:**93–96.

Cukor, G., D. M. Perron, R. Hudson, and N. R. Blacklow. 1984. Detection of rotavirus in human stools by using monoclonal antibody. *J. Clin. Microbiol.* **19:**888–892.

Decker, R. H., L. R. Overby, C. M. Ling, G. Frosner, F. Deinhardt, and J. Boggs. 1979. Serology of transmission of hepatitis A in humans. *J. Infect. Dis.* **139:**74–82.

Decker, R. H., S. M. Kosakowski, A. S. Vanderbilt, C. M. Ling, and L. R. Overby. 1981. Diagnosis of acute hepatitis A by HAVAB-M, a direct radioimmunoassay for IgM anti-HAV. *Am. J. Clin. Pathol.* **76:**140–147.

Devine, R. E., F. Sit, and R. Larke. 1979. Laboratory diagnosis of acute hepatitis A virus (HAV) infection by detection of HAV-specific IgM antibody using radioimmunoassay. *Can. J. Public Health* **70:**58.

Dowse, C. A., P. R. Carnegie, D. S. Linthium, and C. C. A. Bernard. 1983. Solid-phase radioimmunoassay for human myelin basic protein using a monoclonal antibody. *J. Neuroimmunol.* **5:**135–144.

Dreesman, G. R., D. O. Matson, R. J. Courtney, E. Adam, and J. L. Melnick. 1979. Detection of herpes virus type-specific antibody by a micro solid-phase radioimmunometric assay. *Intervirology* **12:**115–119.

Enlander, D., L. V. D. Remedios, P. M. Weber, and L. Drew. 1976. Radioimmunoassay for herpes simplex virus. *J. Immunol. Methods* **10:**357–362.

Faiman, C., and R. J. Ryan. 1967. Radioimmunoassay for human follicle-stimulating hormone. *J. Clin. Endocrinol.* **27:**444–447.

Flehmig, B., M. Ranke, H. Berthold, and H. J. Gerth. 1979. A solid-phase radioimmunoassay for detection of IgM antibodies to hepatitis A virus. *J. Infect. Dis.* **140:**169–175.

Forghani, B., N. J. Schmidt, and E. H. Lennette. 1974. Solidphase radioimmunoassay for identification of Herpesvirus hominis types 1 and 2 from clinical materials. *Appl. Microbiol.* **28:**661–667.

Forghani, B., N. J. Schmidt, and E. H. Lennette. 1975. Solid-phase radioimmunoassay for typing herpes simplex viral antibodies in human sera. *J. Clin. Microbiol.* **2:**410–418.

Friedman, M. G., S. Leventon-Kriss, and I. Saroy. 1979. Sensitive solid-phase radioimmunoassay for detection of human immunoglobulin G antibodies to varicella-zoster virus. *J. Clin. Microbiol.* **9:**1–10.

Frosner, G. G., M. Brodersen, G. Papaevangelou, Y. Sugg, H. Hass, I. K. Mushahwar, C. M. Ling, L. R. Overby, and F. Deinhardt. 1978. Detection of HBeAg and anti-HBe in acute hepatitis B by a sensitive radioimmunoassay. *J. Med. Virol.* **3:**67–76.

Ghabrial., S. A., and R. J. Shepherd. 1980. A sensitive radioimmunosorbent assay for the detection of plant viruses. *J. Gen. Virol.* **48:**311–317.

Ginsberg, A. L., M. E. Conrad, W. H. Bancroft, C.-M. Ling, and L. R. Overby. 1973. Antibody to Australia antigen: detection with a simple radioimmune assay, incidence in military populations, and role in the prevention of hepatitis B with gammaglobulin. *J. Lab. Clin. Med.* **82:**317–325.

Goodfriend, T. L., and D. Ball. 1969. Radioimmunoassay of bradykinin: chemical modification to enable use of radioactive iodine. *J. Lab. Clin. Med.* **73:**501–511.

Greenberg, H. B., R. G. Wyatt, J. Valdesco, A. R. Kalica, W. T. London, R. M. Chanock, and A. Z. Kapikian. 1978. Solid-phase microtiter radioimmunoassay for detection of the Norwalk strain of acute nonbacterial epidemic gastroenteritis virus and its antibodies. *J. Med. Virol.* **2:**97–108.

Greenwood, F. C., W. M. Hunter, and J. S. Glover. 1963. The preparation of ^{131}I-labeled human growth hormone of high specific radioactivity. *Biochem. J.* **89:**114–123.

Griffiths, P. O., and H. O. Kangro. 1984. A user's guide to the indirect solid-phase radioimmunoassay for the detection of cytomegalovirus-specific IgM antibodies. *J. Virol. Methods* **8:**271–282.

Haikala, O., J. O. Kokkonen, M. K. Leinonen, T. Nurmi, R. Mantyjarvi, and H. K. Sarkkinen. 1983. Rapid detection of rotavirus in stool by latex agglutination: comparison with radioimmunoassay and electron microscopy and clinical evaluation of the test. *J. Med. Virol.* **11:**91–97.

Hales, C. N., and P. J. Randle. 1963. Immunoassay of insulin with insulin-antibody precipitate. *Biochem. J.* **88:**137–146.

Halonen, P., H. Bennich, E. Torfason, T. Karlsson, B. Ziola, M.-T. Matikainen, E. Hjertsson, and T. Wesslen. 1979a. Solid-phase radioimmunoassay of serum immunoglobulin A antibodies to respiratory syncytial virus and adenovirus. *J. Clin. Microbiol.* **10:**192–197.

Halonen, P., O. Meurman, M. T. Matikainen, E. Torfason, and H. Bennich. 1979b. Ig-A antibody response in acute rubella determined by solid-phase radioimmunoassay. *J. Hyg.* (London) **83:**69–75.

Halonen, P., H. Sarkkinen, P. Arstila, E. Hjertsson, and E. Torfason. 1980. Four-layer radioimmunoassay for detection of adenovirus in stool. *J. Clin. Microbiol.* **11:**614–617.

Hill, H. R., and J. M. Matsen. 1983. Enzyme-linked immunosorbent assay and radioimmunoassay in the serologic diagnosis of infectious diseases. *J. Infect. Dis.* **147:**258–263.

Hollinger, F. B., and J. E. Maynard. 1976. Recent diagnostic techniques for detecting hepatitis A virus and antibody. *Rush Presbyterian St. Luke's Med. Bull.* **15:**93–103.

Hollinger, F. B., V. Vorndam, and G. R. Dreesman. 1971. Assay of Australia antigen and antibody employing double-antibody and solid-phase radioimmunoassay techniques and comparison with the passive hemagglutination methods. *J. Immunol.* **107:**1099–1111.

Hollinger, F. B., D. W. Bradley, J. E. Maynard, G. R. Dreesman, and J. L. Melnick. 1975. Detection of hepatitis A viral antigen by radioimmunoassay. *J. Immunol.* **115:**1464–1466.

Jahrling, P. B., R. A. Hesse, and J. F. Metzger. 1978. Radioimmunoassay for quantitation of antibodies to alphaviruses with staphylococcal protein A. *J. Clin. Microbiol.* **8:**54–60.

Kalica, A. R., R. H. Purcell, M. M. Sereno, R. G. Wyatt, H. W. Kim, R. M. Chanock, and A. Z. Kapikian. 1977. A microtiter solid-phase radioimmunoassay for detection of the human reovirus-like agent in stools. *J. Immunol.* **118:**1275–1279.

Kalimo, K. O. K., O. H. Meurman, P. E.. Halonen, B. R. Ziola, M. K. Viljanen, K. Granfors, and P. Toivanen. 1976. Solid-phase radioimmunoassay of rubella virus immunoglobulin G and immunoglobulin M antibodies. *J. Clin. Microbiol.* **4:**117–123.

Kalimo, K. O. K., R. J. Martilla, K. Granfors, and M. K. Viljanen. 1977a. Solid-phase radioimmunoassay of human immunoglobulin M and immunoglobulin G antibodies against herpes simplex virus type 1 capsid, envelope, and excreted antigens. *Infect. Immun.* **15:**883–889.

Kalimo, K. O. K., R. J. Martilla, B. R. Ziola, M. T. Matikainen, and M. Panelius. 1977b. Radioimmunoassay of herpes simplex and measles virus antibodies in serum and CSF of patients without infectious or demyelinating diseases of the central nervous system. *J. Med. Virol.* **10:**431–438.

Kalimo, K. O. K., B. R. Ziola, M. K. Viljanen, K. Granfors, and P. Toivanen. 1977c. Solid-phase radioimmunoassay of herpes simplex virus IgG and IgM antibodies. *J. Immunol. Methods* **14:**183–195.

Kangro, H. O., C. Jackson, and R. B. Heath. 1981. Comparison of radioimmunoassay and the gel filtration technique for routine diagnosis of rubella during pregnancy. *J. Hyg.* (London) **87:**249–255.

Kangro, H. O., J. C. Booth, T. M. F. Bakir, Y. Tryhorn, and S. Sutherland. 1984. Detection of IgM antibodies against cytomegalovirus: comparison of two radioimmunoassays, enzyme-linked immunosorbent assay and immunofluorescent antibody test. *J. Med. Virol.* **14:**73–80.

Kangro, H. O., A. Ward, H. Argent, R. B. Heath, J. E. Cradock-Watson, and M. K. Ridehalgh. 1988. Detection of specific IgM in varicella and herpes zoster by antibody-capture radioimmunoassay. *Epidemiol. Infect.* **101:**187–195.

Kimmel, N., M. G. Friedman, and I. Sarov. 1980. Detection of human cytomegalovirus-specific IgG antibodies by a sensitive solid-phase radioimmunoassay and by the rapid screening test. *J. Med. Virol.* **5:**195–203.

Kolb, H. J., and G. M. Grodsky. 1970. Biological and immunological activity of fructose 1,6-di-phosphatase. Application of a quantitative displacement radioimmunoassay. *Biochemistry* **9:**4900–4906.

Kuo, G., Q. L. Choo, H. J. Alter, G. L. Gitnick, A. G. Redeker, R. H. Purcell, T. Miyamura, J. L. Dienstag, M. J. Alter, C. E. Stevens, G. E. Tegtmeier, F. Bonino, M. Colombo, W. S. Lee, C. Kuo, K. Berger, J. R. Shuster, L. R. Overby, D. W. Bradley, and M. Houghton. 1989. An assay for circulating antibodies to a major etiologic virus of human non-A, non-B hepatitis. *Science* **244:**362–364.

Lemon, S. M., C. D. Brown, D. S. Brooks, T. E. Simms, and W. H. Bancroft. 1980. Specific immunoglobulin M response to hepatitis A virus determined by solid-phase radioimmunoassay. *Infect. Immun.* **28:**927–936.

Ling, C.-M., and L. R. Overby. 1972. Prevalence of hepatitis B virus antigens as revealed by direct radioimmune assay with ^{125}I-antibody. *J. Immunol.* **109**:834–841.

Marsden, H. S., A. Buckmaster, J. W. Palfreyman, R. G. Hope, and A. C. Minson. 1984. Characterization of the 92,000-dalton glycoprotein induced by herpes simplex type 2. *J. Virol.* **50**:547–554.

Matson, D. O., K. Alder-Storthz, E. Adam, and G. R. Dreesman. 1983. A micro solid-phase radioimmunoassay for detection of herpesvirus type-specific antibody: parameters involved in standardization. *J. Virol. Methods* **6**:71–83.

Matsui, A., H. T. Psacharopoulos, and M. P. Mowat. 1982. Radioimmunoassay of serum glycocholic acid, standard laboratory tests of liver function and liver biopsy findings: comparative study of children with liver disease. *J. Clin. Pathol.* **35**:1011–1017.

Meurman, O., and K. Granfors. 1977. Completion of hemagglutination inhibition test by solid-phase radioimmunoassay test in routine diagnostic rubella serology. *Med. Biol.* **55**:241–244.

Meurman, O. H. Sarkkinen, O. Ruuskanen, P. Hänninen, and P. Halonen. 1984. Diagnosis of respiratory syncytial virus infection in children: comparison of viral antigen detection and serology. *J. Med. Virol.* **14**:61–65.

Meurman, O. H. 1978. Antibody responses in patients with rubella infection determined by passive hemagglutination, hemagglutination inhibition, complement fixation, and solid-phase radioimmunoassay tests. *Infect. Immun.* **19**:369–372.

Meurman, O. H., M. K. Viljanen, and K. Granfors. 1977. Solid-phase radioimmunoassay of rubella virus immunoglobulin M antibodies: comparison with sucrose density gradient centrifugation test. *J. Clin. Microbiol.* **5**:257–262.

Middleton, P. J., M. D. Holdway, M. Petric, M. T. Szymanski, and J. S. Tam. 1977. Solid-phase radioimmunoassay for the detection of rotavirus. *Infect. Immun.* **16**:439–444.

Monath, T. P., J. R. Wands, L. J. R. Hill, M. K. Gentry, and D. J. Gubler. 1986a. Multisite monoclonal immunoassay for dengue viruses: detection of viraemic human sera and interference by heterologous antibody. *J. Gen. Virol.* **67**:639–650.

Monath, T. P., L. J. Hill, N. V. Brown, C. B. Cropp, J. J. Schlesinger, J. F. Saluzzo, and J. R. Wands. 1986b. Sensitive and specific monoclonal immunoassay for detecting yellow fever virus in laboratory and clinical specimens. *J. Clin. Microbiol.* **23**:129–134.

Morgan, C. R., and A. Lazarow. 1962. Immunoassay of insulin using a two-antibody system. *Proc. Soc. Exp. Biol. Med.* **110**:29–32.

Morgan-Capner, P., and C. McSorley. 1983. Antibody capture radioimmunoassay (MACRIA) for coxsackievirus B4 and B5-specific IgM. *J. Hyg.* (London) **90**:333–349.

Mortimer, P. P., R. S. Tedder, M. H. Hambling, M. S. Shafi, F. Burkhardt, and U. Schilt. 1981. Antibody capture radioimmunoassay for anti-rubella IgM. *J. Hyg.* (London) **86**:139–153.

Mushahwar, I. K., and R. H. Decker. 1983. Prevalence of anti-delta in various HBsAg positive populations, p. 269. In G. Verme, F. Bonino, and M. Rizzetto (ed.), *Viral Hepatitis and Delta Infection.* Alan R. Liss, New York, N.Y.

Mushahwar, I. K., and L. R. Overby. 1983. Radioimmune assays for diagnosis of infectious diseases, p. 167–194. In F. S. Ashkar (ed.), *Radiobioassays.* CRC Press, Boca Raton, Fla.

Mushahwar, I. K., L. R. Overby, G. Frosner, F. Deinhardt, and C. M. Ling. 1978. Prevalence of hepatitis B e-antigen and its antibody as detected by radioimmunoassays. *J. Med. Virol.* **2**:77–87.

Mushahwar, I. K., J. L. Dienstag, H. F. Polesky, L. C. McGrath, R. H. Decker, and L. R. Overby. 1981. Interpretation of various serological profiles of hepatitis B virus infection. *Am. J. Clin. Pathol.* **76**:773–777.

Neurath, A. R., W. Szmuness, C. E. Stevens, N. Strick, and E. J. Harley. 1978. Radioimmunoassay and some properties of human antibodies to hepatitis B core antigen. *J. Gen. Virol.* **38**:549–559.

Neurath, A. R., N. Strick, W. Szmuness, C. E. Stevens, and E. J. Harley. 1979. Radioimmunoassay of hepatitis B e-antigen (HBeAg); identification for HBeAg not associated with immunoglobulins. *J. Gen. Virol.* **42**:493.

Odell, W. D., G. T. Ross, and P. L. Rayford. 1967. Radioimmunoassay for luteinizing hormone in human plasma or serum: physiological studies. *J. Clin. Investig.* **46**:248–255.

Overby, L. R., and C.-M. Ling. 1976. Radioimmune assay for anti-core as evidence for exposure to hepatitis B virus. *Rush Presbyterian St. Luke's Med. Bull.* **15**:83–92.

Overby, L. R., and I. K. Mushahwar. 1979. Radioimmune assays, p. 39–69. In M. W. Rytel (ed.), *Rapid Diagnosis in Infectious Disease.* CRC Press, Boca Raton, Fla.

Overby, L. R., C.-M. Ling, R. H. Decker, I. K. Mushahwar, and K. Chau. 1981. Serodiagnostic profiles of viral hepatitis, p. 169–182. In W. Szmuness, H. J. Alter, and J. E. Maynard (ed.), *Viral Hepatitis 1981 International Symposium.* Franklin Institute Press, Philadelphia, Pa.

Patterson, W. R., and K. O. Smith. 1973. Improvement of radioimmunoassay for measurement of viral antibody in human sera. *J. Clin. Microbiol.* **2**:130–133.

Patterson, W. R., W. E. Rawls, and K. O. Smith. 1978. Differentiation of serum antibodies to herpesvirus types 1 and 2 by radioimmunoassay. *Proc. Soc. Exp. Biol. Med.* **157**:273–277.

Poor, A. H., and J. E. Cutler. 1979. Partially purified antibodies used in a solid-phase radioimmunoassay for detecting candidal antigenemia. *J. Clin. Microbiol.* **9**:362–368.

Pugh, S. F. 1984. Heterotypic reactions in a radioimmunoassay for coxsackie B virus specific IgM. *J. Clin. Pathol.* **37**:433–439.

Purcell, R. H., D. C. Wong, H. J. Alter, and P. V. Holland. 1973. Microtiter solid-phase radioimmunoassay for hepatitis B antigen. *Appl. Microbiol.* **26**:478–484.

Purcell, R. H., D. C. Wong, Y. Moritsugo, J. L. . Dienstag, J. A. Routenberg, and J. D. Boggs. 1976. A microtiter solid-phase radioimmunoassay for hepatitis A antigen and antibody. *J. Immunol.* **116**:349–356.

Richman, D. D., P. H. Cleveland, M. N. Oxman, and J. A. Zaia. 1981. A rapid radioimmunoassay using ^{125}I-labeled staphylococcal protein A for antibody to varicella-zoster virus. *J. Infect. Dis.* **143**:693–699.

Rizzetto, M., D. J. Gocke, G. Verme, J. W.-K. Shih, R. H. Purcell, and J. L. Gerin. 1979. Incidence and significance of antibodies to delta antigen in hepatitis B virus infection. *Lancet* **ii**:986–990.

Rizzetto, M., J. W.-K Shih, and J. L. Gerin. 1980. The hepatitis B virus-associated delta antigen: isolation from liver, development of solid-phase radioimmunoassays for delta antigen and anti-delta and partial characterization of delta antigen. *J. Immunol.* **125**:318–324.

Robinson, K., and R. N. Smith. 1983. Methadone radioimmunoassay: two simple methods. *J. Pharm. Pharmacol.* **35**:566–569.

Roggendorf, M., G. G. Frosner, F. Deinhardt, and R. Scheidt. 1980. Comparison of solid-phase test systems for demonstrating antibodies against hepatitis A virus (anti-HAV) of the IgM-class. *J. Med. Virol.* **5**:47–62.

Rutland, P. C. 1984. The development of a radioimmunoassay for the measurement of fetal haemoglobin and its use in determining the distribution of HbF in the British population. *Med. Lab. Sci.* **41**:84–85.

Ruuskanen, O., H. Sarkkinen, O. Meurman, P. Hurme, T. Rossi, P. Halonen, and P. Hanninen. 1984. Rapid diagnosis of adenoviral tonsillitis: a prospective clinical study. *J. Pediatr.* **104**:725–728.

Safford, S. E. S., S. B. Needleman, and R. H. Decker. 1980. Radioimmunoassay for detection of antibody to hepatitis A virus. *Am. J. Clin. Pathol.* **74**:25–31.

Sarkkinen, H. K., P. E. Halonen, and P. P. Arstila. 1979a. Comparison of four-layer radioimmunoassay and electron microscopy for detection of human rotavirus. *J. Med. Virol.* **4**:255–260.

Sarkkinen, H. K., O. H. Meurman, and P. E. Halonen. 1979b. Solid-phase radioimmunoassay of IgA and IgM antibodies to human rotavirus. *J. Med. Virol.* **3**:281–289.

Sarkkinen, H. K., H. Tuokko, and P. E. Halonen. 1980. Comparison of enzyme-immunoassay and radioimmunoassay for detection of human rotaviruses and adenoviruses from stool specimens. *J. Virol. Methods* **1**:331–341.

Sarkkinen, H. K., P. E. Halonen, and A. A. Salmi. 1981. Detection of influenza A virus by radioimmunoassay and enzyme immunoassay from nasopharyngeal specimens. *J. Med. Virol.* **7**:213–220.

Sexton, S. A., J. Hodgsen, and P. Morgan-Capner. 1982. The detection of rubella-specific IgM by an immunosorbent assay with solid-phase attachment of red cells (SPARC). *J. Hyg.* (London) **88**:453–461.

Shimada, N., K. Ushioda, S. Nagatsuka, T. Ueda, and T. Yokoshima. 1983. Comparison between radioreceptor assay and RIA for the determination of dihydroergotoxine in rabbit plasma samples. *J. Immunol. Methods* **65**:191–198.

Smedile, A., C. Lavarini, O. Crivelli, G. Raimondo, M. Fassone, and M. Rizzetto. 1982. Radioimmunoassay detection of IgM antibodies to the HBV-associated delta antigen: clinical significance in delta infection. *J. Med. Virol.* **9**:131–138.

Smith, K. O., and W. Kennell. 1981. Differentiation of members of the human herpesviridae family by radioimmunoassay. *Infect. Immun.* **33**:491–497.

Smith, K. O., W. D. Gehle, and A. W. McCracken. 1974. Radioimmunoassay techniques for detecting naturally occurring viral antibody in human sera. *J. Immunol. Methods* **5**:337–344.

Tedder, R. S., P. P. Mortimer, and R. B. Lord. 1981. Detection of antibody to varicella-zoster virus by competitive and IgM-antibody capture immunoassay. *J. Med. Virol.* **8**:89–101.

Tedder, R. S., J. L. Yao, and M. J. Anderson. 1982. The production of monoclonal antibodies to rubella hemagglutinin and their use in antibody-capture assays for rubella-specific IgM. *J. Hyg.* (London) **88**:335–350.

Torfason, E. G., C. Kallander, and P. Halonen. 1981. Solid-phase radioimmunoassay of serum IgG, IgM and IgA antibodies to cytomegalovirus. *J. Med. Virol.* **7**:85–96.

Torfason, E. G., G. Frisk, and H. Diderholm. 1984. Indirect and reverse radioimmunoassays and their apparent specificities in the detection of antibodies to enteroviruses in human sera. *J. Med. Virol.* **13**:13–31.

Trent, D. W., C. L. Harvey, A. Quereshi, and D. LeStourgeon. 1976. Solid-phase radioimmunoassay for antibodies to flavivirus structural and nonstructural proteins. *Infect. Immun.* **13**:1325–1333.

Vesikari, T., A.-L. Kuusela, H. K. Sarkkinen, and P. E. Halonen. 1982. Clinical evaluation of radioimmunoassay of nasopharyngeal secretions and serology for diagnosis of viral infections in children hospitalized for respiratory infections. *Pediatr. Infect. Dis.* **1**:391–394.

Vyas, G. N., and I. M. Roberts. 1977. Radioimmunoassay of hepatitis B core antigen and antibody with autologous reagents. *Vox Sang.* **33**:369.

Watanabe, H., and I. H. Holmes. 1977. Filter-paper solid-phase RIA for human rotavirus surface immunoglobulins. *J. Clin. Microbiol.* **6**:319–324.

Wide, L., K. E. Kirkham, and W. M. Hunger. 1971. Solid-phase antigen-antibody systems, p. 405. *In* L. Wide, K. E. Kirkham, and W. M. Hunger (ed.), *Radioimmunoassay Methods.* Churchill-Livingstone, Edinburgh, United Kingdom.

Wiktor, R. J., H. Koprowski, and F. Dixon. 1972. Radioimmunoassay procedure for rabies binding antibodies. *J. Immunol.* **109**:464–470.

Wilson, S. A., and W. D. Cubitt. 1988. The development and evaluation of radioimmune assays for the detection of immune globulins M and G against astrovirus. *J. Virol. Methods* **19**:151–159.

Wolff, K. L., D. J. Muth, B. W. Hudson, and D. W. Trent. 1981. Evaluation of the solid-phase radioimmunoassay for diagnosis of St. Louis encephalitis infection in humans. *J. Clin. Microbiol.* **14**:135–140.

Yalow, R. S., and S. A. Berson. 1960. Immunoassay of endogenous plasma insulin in man. *J. Clin. Investig.* **39**:1157–1175.

Zollinger, W. D., J. M. Dalrymple, and M. S. Artenstein. 1976. Analysis of parameters affecting the solid-phase radioimmunoassay quantitation of antibody to meningococcal antigens. *J. Immunol.* **117**:1788–1798.

Enzyme Immunoassay

DIANE S. LELAND

9

Identification of viruses isolated in cell culture, detection of viral antigens in clinical samples, and identification and quantitation of viral antibodies in serum, all functions routinely performed in the diagnostic-virology laboratory, can be accomplished using assays based on the principle of enzyme immunoassay (EIA). EIA is used here to describe a variety of assays that are based on the binding of antibodies with their antigens and detection of this reaction using a component conjugated with an active enzyme; this enzyme subsequently acts on its substrate to produce a color change. The test result is determined by observing or measuring color. In heterogeneous EIAs, one of the components is bound to a solid phase, facilitating separating antigen-antibody complexes from the solution.

HISTORY

EIAs were first described in the early 1970s (Avrameas and Guilbert, 1971; Engvall and Perlmann, 1971; Van Weeman and Schuurs, 1971), when it was discovered that antibodies could be conjugated to enzymes without losing their reactivity. EIA systems use antibody labeled with enzymes rather than radionuclides, with sensitivities and specificities equivalent to those of the radioimmunoassay (RIA) systems. RIAs are described in chapter 8. During the 1970s and 1980s, EIAs for detecting a wide variety of viral antibodies were developed. Several reviews of EIA were published during the 1980s (Yolken, 1980; Voller et al., 1982; Wardley and Crowther, 1982; Avrameas, 1983), and a listing of EIAs developed for viral antibody detection and identification during the 1970s and 1980s has been published (O'Beirne and Sever, 1992). O'Beirne and Sever (1992) also included practical information on preparing EIA reagents and performing EIA testing. At present, most EIAs for viral antibody and antigen detection are commercial kits that include all reagents and provide exact performance specifications. Commercial kits also provide detailed instructions for performing the assay. The laboratory technician can perform many EIAs without dealing with the development of methods or with extensive and laborious reagent preparation steps. This chapter deals with principles of EIA and contemporary applications of EIA in viral antibody and antigen detection, primarily focusing on the diagnostic-virology laboratory rather than research applications.

METHODS USED

Immunoperoxidase Staining (Histochemical EIA and Immunohistologic Staining)

One very simple qualitative application of EIA is immunoperoxidase staining, also called histochemical EIA or immunohistologic staining. Most of these assays use the enzyme horseradish peroxidase, and they are often called "immunoperoxidase staining." Tissue sections or cells collected from lesions, the respiratory tract, the genital tract, or from virus-infected cell cultures are fixed on a microscope slide. Antibodies of known specificity are added. In direct staining, the antibodies are labeled with peroxidase enzymes. After an incubation period and rinsing, a substrate solution is added; in areas where peroxidase antibodies have bound, the peroxidase enzymes act on the substrate to produce an insoluble colored reaction product. Direct-staining procedures usually require only about 30 min. In indirect staining, the primary antibodies are unlabeled. After incubation and rinsing, a preparation of peroxidase-labeled antispecies antibodies (directed against the species in which the primary antibody was raised) is added. These "detection" antibodies bind to primary antibody molecules bound in the first step. After incubation and rinsing, a substrate is added, and color development occurs. Indirect staining requires approximately 90 min. Results for both direct and indirect staining are evaluated by viewing the preparation through a standard light microscope. The intensity, distribution, and pattern of the staining are evaluated.

Immunoperoxidase staining has been applied in the clinical laboratory for detection of many viral antigens and antibodies, although this staining is not used as frequently as immunofluorescence. The advantage of immunoperoxidase over immunofluorescence is that a fluorescence microscope is not required for evaluation of immunoperoxidase stains. The disadvantages of immunoperoxidase methods include the additional time required for color development during the staining process and nonspecific staining that may be due to indigenous peroxidases in some clinical specimens. Applications and procedures for immunoperoxidase staining are described in detail in chapter 10.

Immunoblotting

Immunoblotting is another application of EIA used in the clinical virology laboratory. Blotting, which refers to the transfer of DNA, RNA, or protein from electrophoretic gels to a membrane, is used to prepare the antigen. An EIA method can be used to detect antibodies reacting with the blotted antigen to identify or characterize either the blotted antigen or the antibodies (Lee et al., 1992). In immunoblotting, the antigen used in the test system is usually a protein antigen that has been transferred using the Western blotting technique. In this technique, the electrophoretically separated antigen is transferred by an electric current onto nitrocellulose paper, and the paper is dried and cut into strips. The EIA is performed by flooding the antigen strip with antibodies. Then, following incubation and rinsing of the strip, the appropriate enzyme-labeled antibody is added, followed by substrate. Reactivity is signaled by the presence of colored bands at appropriate positions and of sufficient intensity on the strip. Immunoblotting is described in chapter 18.

Noncompetitive Solid-Phase EIAs

Noncompetitive solid-phase EIAs are the type used most commonly in clinical-virology laboratories. In this configuration, either the antigens or antibodies are bound to a solid phase, such as the wall of a test tube or microwell or the surface of a plastic bead. A known viral antigen(s) is bound to the solid phase if antibodies are to be identified; a known viral antibody, often monoclonal, is bound to the solid phase if antigens are to be identified. If antibodies are to be identified, the test material is usually the patient's serum. If antigens are to be identified, the test material is a clinical sample (i.e., material collected from the patient's throat, lesion, or genital area) or a suspension of virus-infected cells from a cell culture. After the test material is allowed to react with the solid phase and the unbound reactants are rinsed away, a preparation of enzyme-labeled antibodies is added. Following incubation and rinsing, a substrate solution is added; the enzyme (attached to the antibodies) acts on this substrate to produce a color change. The color change may be measured visually but is most often quantitated spectrophotometrically.

A tube-based EIA for detection of human immunoglobulin G (IgG) is diagrammed in Fig. 1. In this assay system, a known viral antigen(s) is bound to the wall of the test tube. The patient's serum is added to the test tube. After incubation and washing of the test tube, a detection preparation of enzyme-labeled anti-human IgG is added. If the patient's IgG reacted with the antigens on the wall of the tube during the first incubation, the bound IgG is recognized by the detection antibodies. If the patient's IgG did not bind during the initial incubation, the detection antibodies will not bind. After the test tube is washed to remove any unbound detection antibodies, a substrate solution is added to the reactants, incubated, and observed for a change in color. A color change indicates that the patient's IgG bound to antigen in the first step; detection antibody then bound in the second step and produced a color change. A lack of color change indicates that none of the patient's IgG bound in the first step; hence, there was no binding of detection antibody and no subsequent color change.

An antibody detection EIA system utilizing a plastic bead as the solid-phase surface is shown in Fig. 2. Known viral antigen is bound to the surface of the bead. The antigen-coated bead is placed in a dilution of the patient's serum. If the patient's serum contains virus-specific IgG antibodies, they will bind to the antigens on the bead. If virus-specific IgG is absent, no binding of the patient's antibodies will take place. Following an incubation period and

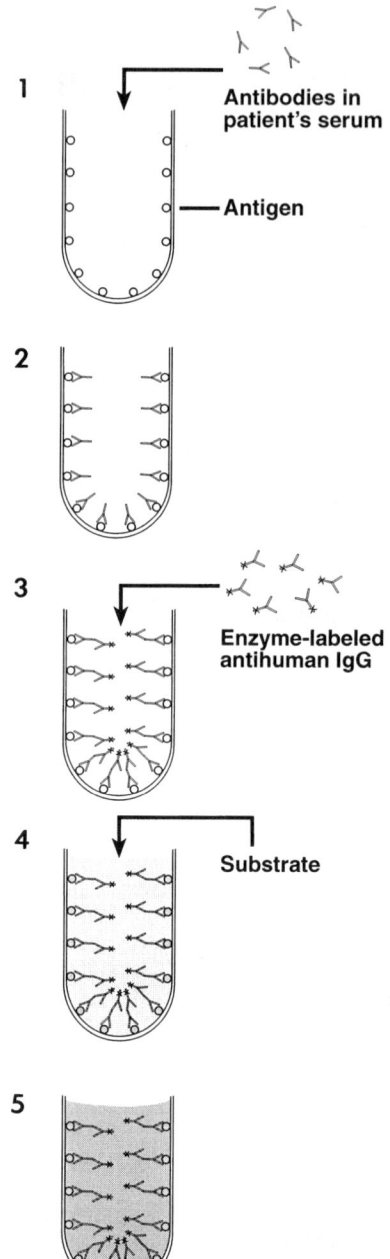

FIGURE 1 Tube-based noncompetitive EIA for antibody detection. (Step 1) The patient's serum is added to a tube or microwell that is coated with known antigen. The mixture is incubated. (Step 2) If the patient's antibodies recognize the antigen, they will bind. Thorough rinsing follows this incubation period. (Step 3) Enzyme-labeled anti-human IgG is added. This recognizes the bound antibodies and binds to them. Thorough rinsing follows this incubation period. (Step 4) A substrate solution is added, and the mixture is incubated. (Step 5) The enzymes that are part of the enzyme-labeled anti-human IgG act on the substrate to produce a color change.

rinsing, the bead is transferred to a solution containing detection antibodies (enzyme-labeled anti-human IgG). If the patient's IgG bound to antigens on the bead in step 1,

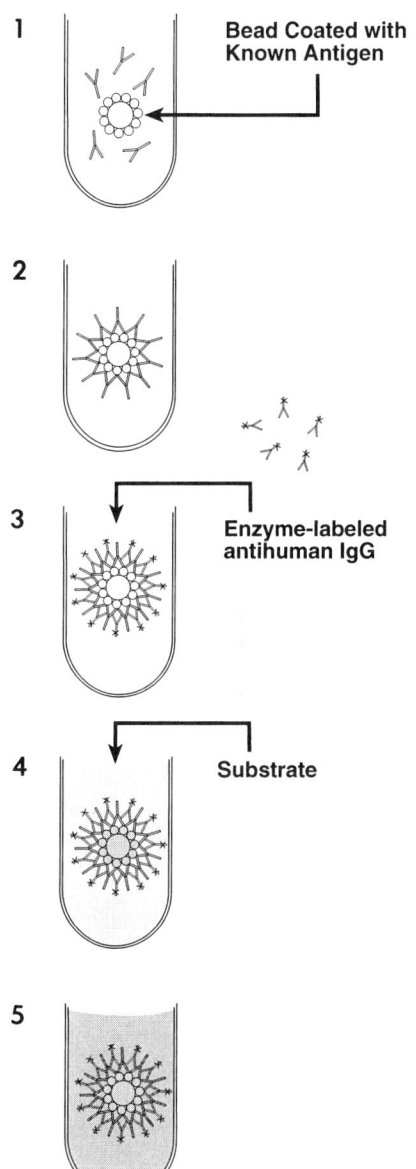

1 Bead Coated with Known Antigen

2

3 Enzyme-labeled antihuman IgG

4 Substrate

5

FIGURE 2 Bead-based noncompetitive EIA for antibody detection. (Step 1) An antigen-coated bead is incubated in a dilution of the patient's serum. The mixture is incubated. (Step 2) If the antibodies in the patient's serum recognize the antigen, they will bind. Thorough rinsing follows this incubation period. (Step 3) Enzyme-labeled anti-human IgG is added. This recognizes the bound antibodies and binds to them. Thorough rinsing follows this incubation period. (Step 4) A substrate solution is added, and the mixture is incubated. (Step 5) The enzymes that are part of the bound enzyme-labeled antihuman IgG act on the substrate to produce a color change.

the detection antibodies will bind. If the patient's IgG did not bind in step 1, the detection antibodies will not be able to bind. Following incubation and rinsing, the bead is transferred to a substrate solution. Color change will be observed if the patient's IgG antibody bound in the initial incubation, and no color change will occur if the patient's IgG antibodies did not bind in the initial step.

A variation used in the detection system of one antibody detection assay EIA features the addition of enzyme-labeled antigen (rather than enzyme-labeled anti-human IgG) in the detection step. A patient's antibodies that bound to antigens on the bead during the initial incubation will then bind the enzyme-labeled antigen. This binding is a specific reaction between the antibodies and the enzyme-labeled antigen. This is in contrast to the binding of enzyme-labeled anti-human IgG (used in most EIA systems), which binds to any human IgG rather than targeting only antibodies of the desired specificity.

In antigen detection EIAs, the patient's sample (feces, throat swab, lesion swab, or serum) or a suspension of virus-infected cells from a cell culture is incubated with an antibody-coated solid phase. The antibodies on the solid phase function by "capturing" target viral antigen in the sample (Fig. 3). Detection antibodies (enzyme-conjugated antiviral antibodies) are added subsequently or concurrently and bind to captured viral antigen. After incubation and rinsing, a substrate solution is added, and a color change is observed.

Competitive Solid-Phase EIA

In competitive solid-phase EIA systems, as in noncompetitive solid-phase systems, either an antigen or an antibody is bound to the solid phase. In contrast, in the competitive system a measured amount of known enzyme-labeled component (of the same specificity as the component being detected in the assay) is added along with the patient's sample (Fig. 4). Hence, the labeled component "competes" against the unlabeled component in the patient's sample for binding to the components on the solid phase. After incubation and rinsing, the solid phase is exposed to a substrate solution. If the patient's sample contained a large amount of the component in question, it would bind, and a minimal amount of the labeled component (added in the EIA system) would bind to the solid phase; therefore, slight or no color change would result. If the patient's sample contained little or none of the component in question, the labeled component (added in the EIA system) would bind, and color change would be observed. In the competitive EIAs, an absence of color is interpreted as positivity of the desired analyte while the presence of color change indicates absence or negativity.

Membrane EIAs

Rapid (10- to 20-min) EIA systems are available in which the test system is a self-contained unit assembled in individual modules or cassettes. In one such system, viral antibody is attached to a membrane in the test packet. After extraction or filtration of the patient's clinical sample, the sample is poured onto the membrane, and viral antigen in the sample is captured by the viral antibodies bound to the membrane. Then enzyme-labeled viral antibodies are added; these bind to the bound viral antigens. After the membrane is rinsed to remove unattached antibodies, a substrate solution is added, and color develops on the membrane in the area, usually triangular or X shaped, in which the original viral antibodies were attached.

In a second type of packet EIA, pretreatment of the patient's clinical sample is required. This includes filtering and then mixing the sample in a tube containing biotin-labeled viral antibodies and microparticles coated with viral antibodies. This mixture is poured through the test packet, and enzyme-labeled anti-biotin antibodies are added, fol-

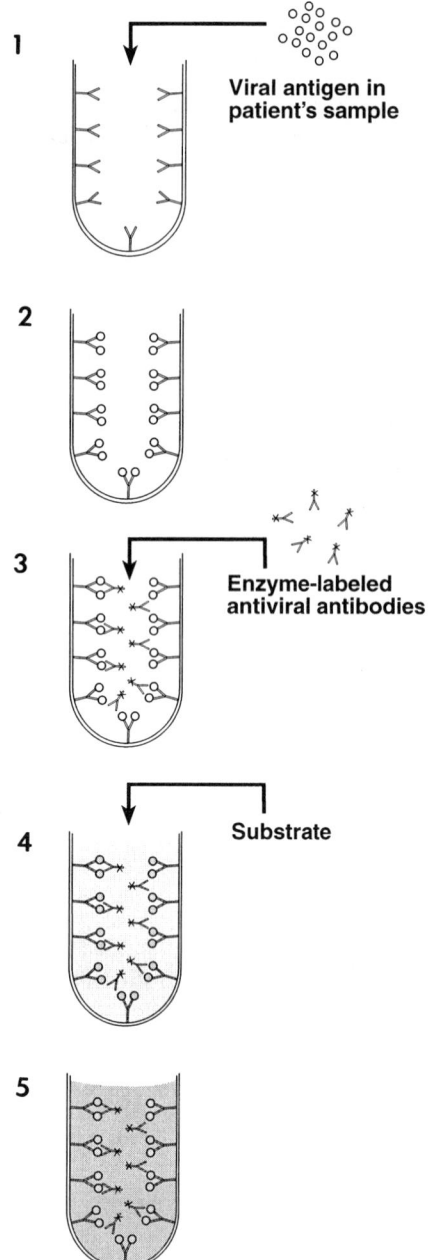

FIGURE 3 Noncompetitive tube-based EIA for antigen detection. (Step 1) The patient's sample (feces, throat or nasal wash, etc.) is added to a tube or microwell that is coated with antiviral antibodies of known specificities. The mixture is incubated. (Step 2) If the antibodies coating the tube recognize the antigen in the patient's sample, they will bind, or capture, the antigen. Thorough rinsing follows this incubation period. (Step 3) Enzyme-labeled antiviral antibodies are added. They will bind to the viral antigen that was captured in step 2. Thorough rinsing follows this incubation period. (Step 4) A substrate solution is added, and the mixture is incubated. (Step 5) The enzymes that are part of the bound enzyme-labeled antiviral antibodies act on the substrate to produce a color change.

lowed by substrate. Color development occurs on the pad in the packet. These systems are expensive but are useful

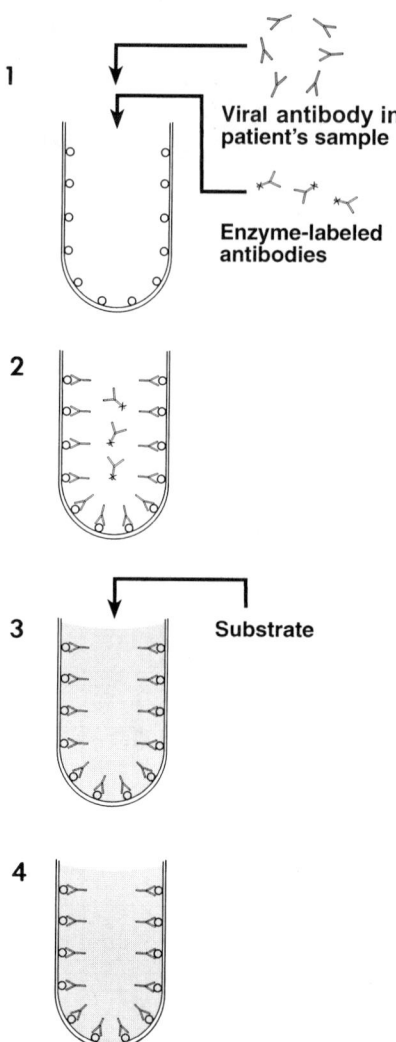

FIGURE 4 Competitive tube-based EIA for antibody detection. (Step 1) The patient's serum and enzyme-labeled antibodies are added to an antigen-coated tube or microwell. The mixture is incubated. (Step 2) The patient's antibodies compete with the enzyme-labeled antibodies for binding the antigen. If the patient's antibodies are present in sufficient quantities, they will bind, and few, if any, of the enzyme-labeled antibodies succeed in binding. Thorough rinsing follows this incubation period. (Step 3) A substrate solution is added. (Step 4) There is no color change because unlabeled antibodies from the patient's serum have bound and few, if any, enzyme-labeled antibodies are present.

when a rapid result is needed and trained virologists are not available.

Chemiluminescence and Biotin-Avidin EIAs

Chemiluminescence and biotin-avidin EIAs involve traditional EIA-type steps. In chemiluminescence, the enzyme label may be horseradish peroxidase. The enzyme acts on pyrogallol, isoluminol, acridinium ester, or oxalate ester, oxidizing these substrates to produce a burst of light (luminescence). This burst can be measured by a luminometer.

The amount of luminescence, like the color change in traditional solid-phase EIAs, is proportional to the amount of antigen-antibody binding.

In biotin-avidin EIAs, biotin-conjugated primary antibodies are incubated with the test sample. Unbound antibodies are washed away. Then enzyme-conjugated avidin or streptavidin is added; one molecule of avidin can react with four biotin molecules and bind strongly in a reaction that is nonimmunological and nearly irreversible. After addition of a substrate solution, a color change is observed. This system avoids the nonspecificity of other EIAs involving antispecies antibody (Forghani and Hagens, 1995).

OIAs

Another variation on the principle of EIA is now available in a format called optical immunoassay (OIA). In OIA systems, the surface of a silicon wafer covered with an optical coating serves as the solid-phase support matrix. In OIA systems for antigen detection, a capture antibody is attached to the solid phase. White light reflected through this surface appears golden unless the thickness of the molecular thin film is changed. When a clinical sample containing the desired antigen is placed on the test device, the capture antibody binds the antigen. Enzyme-conjugated antibodies are added that bind to the bound antigen. A substrate solution is then added. The binding of components increases the thickness of the molecular film. By a process known as mass enhancement, white light reflected through this film appears purple or blue, indicating a positive result. If a clinical sample does not contain the desired antigen, the capture antibody will not bind antigen from the sample, conjugate will not bind, there will be no change in the original molecular thickness, and reflected white light will appear gold. This testing takes only 15 to 20 min to complete and requires no special expertise or laboratory equipment.

General Considerations in EIA Testing

In all EIA systems, the quality of the reagents and the basic design of the system are critical if the assay is to function reliably. The components and formats of the various assays may vary widely. Most EIA components are available commercially, and each component must be analyzed carefully.

Solid-Phase Support

Commonly used solid-phase supports include polystyrene or polyvinyl tubes, beads, and microtiter plates. Antigen(s) or antibodies are usually bound to the solid phase by passive adsorption, although covalent binding has been used. Blocking of any additional binding sites on the solid phase is necessary in order to reduce nonspecific binding of reagents or of patients' sample material. This blocking is accomplished by using bovine serum albumin, antibody-negative serum, gelatin, or nonionic detergent, such as Tween 20. The amount of antigen or antibody adsorbed to the solid phase is critical for accuracy of the assay. Oversensitization or uneven binding of adsorbed components may adversely affect the quality of the results.

Some EIAs designed for antigen detection use a membrane, such as nitrocellulose or nylon, as the solid-phase support. Large amounts of antibody can be bound to such membranes, which is important for assay sensitivity and speed (Forghani and Hagens, 1995). Membranes may be

contained in a cassette-type plastic packet or attached to a plastic stick.

Antigens

EIA systems often use antigens that are not highly purified. Culture lysates or extracts are used in many systems. At present, cloned antigens produced by recombinant DNA technology are used in some systems. The use of the cloned antigens allows optimal amounts of antigens to be attached to the beads. False-positive results due to nonspecific binding or reactions with tissue-related antigens obtained in systems that use culture lysate antigens have been largely eliminated through the use of the cloned antigens. However, antibodies may bind to the carrier portion of the cloned antigen. For example, human superoxide dismutase can be used in the production of some hepatitis C virus (HCV) proteins. Certain testing systems for HCV antibodies include a control that detects antibodies that bind human superoxide dismutase and are not specific for the HCV-encoded portions of the recombinant HCV antigens.

Antibodies

Antibodies used in EIA systems for detection of viral antigens must be of high quality, including high titer, high affinity, and high avidity, if the system is to function reliably. Monoclonal antibodies or pools of monoclonal antibodies directed against related epitopes are now used in many systems to ensure specificity. This is in contrast to the reactions sometimes encountered with polyclonal antibodies that may react with a variety of related antigens. The specificity of a monoclonal antibody may also be its greatest disadvantage because other epitopes are missed.

Conjugates

Antibodies chemically linked or conjugated with an enzyme serve in EIAs to detect or mark the presence of antigen-antibody complexes. These substances go through a conjugation process involving application of a substance that cross-links the enzyme to the antibody while allowing the antibody to retain its binding and the enzyme to retain its hydrolytic activity. Glutaraldehyde has been widely used for this purpose. The aldehyde group reacts with the amino groups of lysine on the antibody and enzyme. Sodium metaperiodate has also been used to combine the amino group of an antibody with the carbohydrates of an enzyme (Forghani and Hagens, 1995). The antibody to which the enzyme is conjugated must be of high quality, just like any antibody used in any EIA system.

Enzymes

Horseradish peroxidase, coupled to antibody by the use of periodate, is often used in EIA procedures. This enzyme can produce insoluble reaction products, completely soluble colored products that can be measured spectrophotometrically, or a burst of light that is measured as chemiluminescence. Alkaline phosphatase, coupled by the use of glutaraldehyde, is also used widely due to its reproducible reactivity.

Substrates

In EIAs, either a chromogenic substrate—one that produces a colored product—or a chemiluminescent substrate—one that produces a burst of light—is used. The quality of the assay depends on the capacity of the substrate to reliably yield a stable or readily measurable product. Some of the substrates for horseradish peroxidase have been reported to be carcinogenic or mutagenic, but all substrates, whether potentially carcinogenic or not, should be handled carefully. Inappropriate handling, exposure to light, etc., can adversely affect the reactivity of the substrate.

Stop Solutions

Most EIAs include as the final step the addition of a solution designed to stop the reaction and stabilize the color of the reactants for a limited time (approximately 2 h) so that the reactions will not change before or while they are being read. Most stop solutions are composed of acid, usually sulfuric, phosphoric, or combinations of the two.

Detection

EIAs, such as immunoperoxidase staining or immunoblotting, depend on the visual examination of the distribution of color change reaction. In immunoperoxidase staining, this is done with a light microscope. With immunoblotting, a visual inspection is sufficient to determine the intensities and positions of reactive bands. For most solid-phase EIAs, a spectrophotometric readout is used rather than visual assessment by the technologist. For tube EIAs (or bead-based EIAs with the final step performed in a test tube), a spectrophotometer is used. Routinely, the spectrophotometer is interfaced to a computer which is programmed to evaluate the spectrophotometric values, calculate ranges, determine cutoff values, and interpret results for individual patient samples. For EIAs performed in microwell plates, spectrophotometric plate readers are available. They have the capacity to determine values for multiwell plates in a few seconds. These plate readers, like the spectrophotometers described above, can be interfaced with computers that are programmed to perform all mathematical calculations and interpret results.

Reporting Results

Because most EIA tests have a spectrophotometric readout that is converted to a numerical value (continuous scale), it is possible to report results in either a qualitative (positive versus negative) or quantitative format.

Qualitative Reporting

Qualitative reporting is the most basic type of assessment. The reactivity level of the test sample is compared with a predetermined cutoff point. If the sample meets or exceeds the cutoff value, the report of "reactive," "positive," or perhaps "antibody (or antigen) detected" is made. If the reactivity level of the test sample fails to reach the cutoff value, the report for the sample is "nonreactive," "negative," or perhaps "antibody (or antigen) not detected." For some assays, manufacturers define a "gray zone" for patients' values that fall very near the cutoff value. Such results are reported as "equivocal" or "indeterminate."

Quantitation for Single Samples

Quantitation for most EIAs remains unique to the assay as specified by the manufacturer. There is little standardization. The actual absorbance reading (or appropriate numeri-

cal score) for the test samples may be reported, along with the absorbance value or score that represents the "cutoff value" for the test method. By comparing the magnitude of the sample value to that of the cutoff value, the physician can estimate the level of antibody. If the sample value falls very near the cutoff value, the sample has a low level of antibody. Because actual absorbance readings are presented, the numbers are largely meaningless unless the cutoff value is reported. In systems in which the absorbance value is converted to some type of unit (e.g., EIA unit), this value may be reported rather than the absorbance value. However, like the absorbance value, it is of little value quantitatively unless the cutoff value is provided for comparison. Conversion of EIA values to serial-dilution titers is discouraged.

When the absorbance reading of the test sample is divided by the cutoff value of the assay, a quotient termed the "index value" is generated. Because the index value represents a comparison of the test sample's value and the cutoff value, all values can be interpreted by the same criteria: an index value of <1.0 indicates a negative test result (antibody not detected), and an index value of ≥1.0 indicates a positive test result (antibody detected). The magnitude of the index value correlates with the quantity of antibody. Some manufacturers use terms other than index value for this quotient, but the same criteria apply.

Comparison of Levels in Paired (Acute versus Convalescent) Samples

Physicians may sometimes rely on determining differences in antibody levels between two samples collected from the patient: one sample is collected as early as possible in the illness (acute), and the second is collected 2 to 3 weeks later (convalescent). In traditional antibody detection tests performed on twofold serial dilutions, endpoints that show a fourfold or greater difference in antibody level between the acute and convalescent samples indicate current or very recent infection. This is an accepted measurement when comparing titers from serial dilutions. Because EIAs are reported in continuous-scale numbers rather than in dilutions, new rules must be supplied by manufacturers of EIA test products to allow discrimination between significant and insignificant differences in acute and convalescent antibody levels.

One manufacturer suggests that a quotient (the convalescent index value divided by the acute index value) of 2.1 indicates a significant difference in antibody levels for some types of antibodies (Diamedix Corporation, package insert for measles IgG EIA kit, 1998). However, the necessary quotient for significance, even with products produced by the same manufacturer, may vary according to the type of antibody being measured (e.g., the criteria for a significant difference in levels of rubella antibodies may be different from those for measles antibody). Many manufacturers do not define the parameters for evaluating the magnitude of difference between two samples.

Quality Control

Unfortunately, serum samples containing standardized levels of antibodies or antigens are not widely available. This makes nationwide or worldwide standardization of assays difficult, if not impossible. However, most commercially available EIA products are marketed with calibrator samples that control calibration; for some products, several calibrators are included in order to generate a curve. Most EIA systems also include both positive and negative serum samples, intended for testing in each run of patient samples. All of these control materials must produce results that fall within acceptable limits with each run of testing if the run is to be considered valid and the patients' results reported. One additional type of control that must be incorporated into EIA testing is a lot-to-lot control that is tested when reagents with new lot numbers are put into service. The purpose of this type of control is to ensure that each new lot of reagents produces results comparable to those produced by the previous lot. This type of control is especially important in tests that provide a quantitative value. Lot-to-lot control materials are not routinely provided by manufacturers and must be purchased or otherwise obtained by the laboratory. Patient samples (or pooled patient materials) previously tested may be used for this purpose. Prior to being put into service, all lot-to-lot control material must be tested in duplicate or replicate in several runs of testing, including reagents with more than one lot number, in order to define an acceptable range. Routinely, a total of 20 to 30 values is generated, from which a mean value is calculated; then a range that includes ±2 or ±3 standard deviations from the mean is defined. The value obtained when the lot-to-lot control is tested with a new lot of reagents is expected to fall within this range.

APPLICATIONS

General Comments

Solid-phase EIAs are routine in clinical laboratories because they are suitable for the testing of large numbers of samples, require little technical expertise to perform, and are generally cost-effective. Most color change reactions are objectively measured by a spectrophotometer, which eliminates subjective assessments by technologists and provides results in the form of continuous-scale numerical readings. Many EIAs require little, if any, expensive instrumentation.

Manual tube and microplate EIA methods require approximately 2 to 4 h to complete and are relatively inexpensive to run, while the test membrane-type EIAs require only 10 to 15 min to run and are more expensive. One of the major drawbacks of the solid-phase EIAs is that patterns of reactivity cannot be evaluated to differentiate specific from nonspecific antibody binding, a differentiation that is possible in immunofluorescence and immunoperoxidase assays, in which results are examined microscopically. The test volume and technologists' time and expertise will often be the determining factors in making the selection of the best EIA.

EIAs for Viral-Antigen Detection

Tests for viral-antigen detection, especially for respiratory-virus antigens, have been shown to be more useful in patient management than either traditional virus isolation (Woo et al., 1997) or viral detection in "rapid" culture using centrifugation-enhanced inoculation (Adcock et al., 1997). Children with respiratory syncytial virus (RSV) confirmed by EIA received antibiotic therapy for fewer days than RSV-negative children, and physicians felt that the RSV EIA results influenced their antibiotic decisions (Adcock et al., 1997). Rapid diagnosis of respiratory-virus infections

allowed significantly reduced hospital stays, antibiotic use, and laboratory utilization, resulting in cost savings and improved patient management (Woo et al., 1997). Discontinuation of unnecessary antibiotic therapy is not the only benefit of a rapid virus diagnosis. With the availability of antivirals, proper treatment of some viral illnesses can be implemented as soon as the infection is confirmed. Because antivirals are virus specific (rather than broad spectrum like some antibiotics), rapid and definitive identification of viruses is needed.

EIAs for the detection of certain viral antigens (hepatitis B virus and rotavirus) represent the only practical approach for diagnosis because these agents do not proliferate in standard cell cultures. Although other viruses, such as herpes simplex virus (HSV), influenza A and B viruses, and RSV, all proliferate in standard cell cultures, antigen detection by EIA provides a more timely result. In general, for these culturable viruses, the rapid antigen detection methods are less sensitive than virus isolation, but the results are readily available.

EIA for viral-antigen detection, in contrast to the immunofluorescence assays for viral-antigen detection, do not depend on visual evaluation of intact infected cells. EIA methods can detect free virus found in mucus or watery secretions that would be missed by immunofluorescence. In addition, EIAs successfully detect noninfectious viruses and viral-antigen fragments. These will fail to propagate in cell cultures, yielding a false-negative culture result.

As with all of the rapid detection methods, a single type of viral antigen is detected with each assay, so if viruses other than the target virus are present in the sample, they will remain undetected. In contrast, virus isolation detects most culturable viruses present in a sample.

EIA methods for detection of rotavirus antigen in fecal and other types of samples are especially popular because this virus will not proliferate in standard cell cultures and immunofluorescence techniques are not useful in the identification of rotavirus. The EIAs for rotavirus detection have been shown to produce results which are, in many cases, comparable in sensitivity to those of electron microscopy (Gerna et al., 1987) and latex agglutination (Lipson and Zelinsky-Papez, 1989). Similarly, antigens of the nonculturable enteric adenoviruses 40 and 41 in fecal samples are routinely detected by EIA.

Another virus whose presence is usually confirmed through antigen detection is the hepatitis B virus; this virus does not proliferate in standard cell cultures and cannot be detected through immunofluorescence. Although the original antigen detection methods for hepatitis B antigens were RIAs, most have been converted to EIAs. Hepatitis B virus antigens are found in high titer in peripheral blood, which is the specimen of choice for hepatitis B virus antigen testing. Human immunodeficiency virus (HIV) does not proliferate in standard cell cultures and is not detectable directly through immunofluorescence, but EIA may be used for HIV type 1 (HIV-1) antigen detection. An EIA for HIV-1 antigen detection is a component of the blood-borne-pathogen testing performed on all blood transfused in the United States.

RSV is also frequently identified through antigen detection EIAs. Although RSV proliferates well in standard cell cultures and can be identified by immunofluorescence, the rapid antigen EIAs are popular because they easily accommodate the large volume of RSV tests performed during winter months when RSV is prevalent, and they require little technologist's time or expertise. Many RSV antigen detection EIAs are available commercially, and the sensitivity and specificity vary from product to product (Table 1). It is important to note that all specimens are not equally satisfactory for RSV EIA testing. As shown in Table 1, RSV antigen detection varies considerably between nasal washes and bronchoalveolar lavage samples (Englund et al., 1996). Also of importance is the age of the patient being tested. RSV EIA determinations performed on 54 older adults failed to detect RSV in any of the 11 patients diagnosed by serology as RSV infected; RSV was isolated from 6 of the 9 patients cultured for RSV (Falsey et al., 1996).

EIA reagents are available for detection of influenza A virus. This virus will usually proliferate in standard cultures, although there is variability among strains, and influenza A virus cytopathic effect is often very subtle and difficult to detect. Influenza A virus antigen can also be detected by immunofluorescence. The EIAs for detection of influenza A virus antigen have been compared with influenza A virus isolation in cell cultures (Table 1). Because the incidence of influenza A virus is low in most populations, the use of an EIA for antigen detection may be costly and may have low yield in terms of numbers of positives identified. However, when selected populations with high incidences of influenza A virus are tested, the use of an antigen detection method becomes more cost-effective.

HSV types 1 and 2, which proliferate readily in standard cell cultures, can be identified by either immunofluorescence or EIA. The sensitivities and specificities reported in several evaluations of HSV EIAs are shown in Table 1.

OIA has recently been studied for direct detection of chlamydia antigen in clinical specimens. In studies of 152 stored and 37 fresh ocular specimens from infants, the sensitivities of OIA were 94.2 and 100%, respectively, and the specificities were 97 and 92.6%, respectively, compared to those of chlamydia isolation in culture (Roblin et al., 1997). In contrast, the sensitivity of OIA in detection of chlamydia antigen was 73.8% in a study of 306 genital samples collected from women at a sexually transmitted disease clinic (Pate et al., 1998). OIA has also been used in influenza A and B virus antigen detection. OIA detects a nucleoprotein found in both influenza A and B viruses. OIA has been reported to detect influenza A and B viruses as effectively as 14-day viral culture for nasal aspirates, nasopharyngeal

TABLE 1 Comparison of the sensitivities and specificities of various EIA methods for viral-antigen detection

Virus	Method (manufacturer)/description	% Sensitivity[a]	% Specificity[a]	No. of samples tested	Reference
HSV	Herpcheck (DuPont)/manual microwell EIA	93	92	346	Johnston et al., 1992
		85	96	377	Ogburn et al., 1994
		84	98	473	Ogburn et al., 1994
		72	100	80[b]	Ogburn et al., 1994
		84	97	168[b]	Ogburn et al., 1994
		74	98	21,522	Verano and Michalski, 1995
	Premier (Meridian)/manual microwell EIA	64	99	136	Thomas and Book, 1991
	SureCell (Kodak)/membrane EIA	76	99	440	Kodak, package insert, 1992
		70	99	365	Kodak, package insert, 1992
	VIDAS (Vitek)/automated EIA	92	89	356	Johnston et al., 1992
RSV	Directigen (Becton Dickinson)/ membrane EIA	61	95	86	Dominguez et al., 1993
		76	73	100	Rothbarth et al., 1991
		15, 71, 89[c]	≥97	539	Englund et al., 1996
	Test Pack (Abbott)/membrane EIA	57	98	86	Dominguez et al., 1993
		93[d]	90[d]	104	Miller et al., 1993
		92	97	100	Rothbarth et al., 1991
		91	96	152	Thomas and Book, 1991
		83	87	65	Todd et al., 1995
	VIDAS (Vitek)/automated EIA	83[d]	93[d]	140	Miller et al., 1993
		96[d]	94[d]	87	Miller et al., 1993
Influenza A virus	Directigen FLU-A (Becton Dickinson)/membrane EIA	100	92	190	Waner et al., 1991
		84	100	160	Chomel et al., 1992
		100	100	51	Todd et al., 1995
	AB FLU (Biostar)/OIA[e]	88	69[f]	79	Covalciuc et al., 1998

[a] Sensitivity and specificity compared to virus isolation.
[b] Genital specimens only.
[c] 15, adult nasal wash; 71, adult endotracheal tube; 89, adult bronchoalveolar lavage.
[d] Sensitivity and specificity relative to direct immunofluorescence.
[e] Detects both influenza A and influenza B viruses.
[f] Most of the OIA-positive, culture-negative samples were positive by reverse transcription-PCR.

swabs, and throat swabs and is only slightly less sensitive with sputum specimens (Covalciuc et al., 1999).

EIAs for Viral Antibody Detection

The assortment of commercially available EIA diagnostic test kits designed for detection of various viral antibodies is enormous. These require minimal technical expertise and provide timely, cost-effective results. EIAs have replaced many of the older, more labor-intensive antibody detection methods. The decision to select one EIA over another may be difficult, with the decision being based on availability of personnel, cost per test, testing volume, and expertise required. Most commercially available viral-antibody detection kits and systems are distributed along with specific testing protocols. These protocols should be followed without modification to ensure accurate test results. No EIA testing procedures are presented in this text.

Manual bead-based EIAs are widely available for measurement of antibodies and antigens associated with the blood-borne pathogens hepatitis B virus, HCV, HIV, and human T-lymphotrophic viruses I and II. Each of the bead-based EIAs follows similar steps and requires 2.5 to 4 h to complete. Instrumentation is available to aid with some steps of the bead-based EIAs, but total automation has not yet been achieved.

Manual EIA systems with testing performed in microwell plates are available for detecting antibodies against cytomegalovirus (CMV), HSV, hepatitis A, B, and C viruses, HIV, Epstein Barr virus, measles virus, mumps virus, rubella virus, varicella-zoster virus, and perhaps others. The manual microwell EIAs usually require minimal expertise from the technologist and are completed in 3 to 4 h.

Several manufacturers have developed automated EIA systems. Such systems can perform a variety of viral-antibody detection immunoassays, most commonly CMV IgG and IgM, Epstein-Barr virus capsid antibody IgG and IgM, HSV types 1 and 2 IgG and IgM, measles virus IgG, mumps virus IgG, rubella virus IgG and IgM, and varicella-zoster virus IgG. This list is expanding rapidly as the technology is refined. The basic principle of these procedures is the same as that for the comparable manual EIAs, except that in the automated system, an instrument dilutes, transfers, mixes, rinses, and evaluates results for all of the assays. Routinely, the technologist "builds" a microplate by assembling microwells coated with the antigen that will be used in the test system. This microwell plate is positioned on the instrument along with the necessary reagents (diluent, conjugate, and substrate) and tubes containing undiluted patients' serum samples that will be assayed for the presence of antibodies.

After the technologist positions all reagents and samples as required, the instrument is activated, and its microcomputer directs it through the appropriate timed cycles of diluting, reagent addition, and washing. When the entire test protocol is completed, the microcomputer may signal the technologist to transfer the microtiter plate to a spectrophotometric plate reader that measures the color of each microtiter well, or the instrument may have the capacity to complete the spectrophotometric reading without transfer of the plate. The microcomputer processes spectrophotometric color measurements to calculate the final values and interpret the results for each sample.

With these instruments, a single immunoassay (e.g., rubella virus IgG) can be performed on all of the samples or two or more compatible assays (e.g., rubella virus IgG and CMV IgG) can be performed simultaneously on all of the samples. Likewise, three to eight separate compatible assays may be performed on limited numbers of samples in the same run of testing.

Membrane EIAs have been developed for antibody detection. A membrane EIA in a cassette format is available for detection of HIV-1 antibodies. In testing of 1,466 samples from a sexually transmitted disease clinic, an HIV-1 membrane EIA yielded 100% sensitivity and 99.5% specificity compared to an EIA with Western blot confirmation (Kassler et al., 1995). Recently, a rapid test using a nitrocellulose membrane strip and EIA steps for HIV-1 and HIV-2 antibody detection was described (Hiroyasu et al., 1999). The nitrocellulose strip has a conjugate site (containing HIV-1 and HIV-2 recombinant antigens conjugated to selenium colloid) and a capture site (containing HIV-1 and HIV-2 antigens). If a patient's sample contains anti-HIV-1 or anti-HIV-2 antibodies, the antibodies first react with the antigen-selenium colloid conjugates. As these complexes flow past the capture site, the antibodies react with the antigens at the site, forming a visible red line within 15 min. This method was equal in sensitivity and specificity to traditional EIAs. This test may be an excellent approach for HIV antibody testing at remote sites as well as in traditional laboratories.

Other formats for membrane EIAs are being examined. A dot immunobinding assay has been used to simultaneously detect antibodies to measles, mumps, and rubella viruses. Filter paper soaked with serum or whole blood was applied to nitrocellulose sheets containing dots of the three viral antigens. Following sequential incubation, rinsing, addition of conjugate, and so on, the results obtained compared favorably with those obtained by standard EIAs (Condorelli and Ziegler, 1993).

IgM- and IgA-Specific EIAs

By changing the specificity of the enzyme-labeled component of the EIA test, an IgG-specific assay can be modified for detection of IgM or IgA. Anti-human IgM conjugate will detect IgM and anti-human IgA will detect IgA. However, this modification alone is usually not sufficient for IgM detection. IgM is present in serum in small quantities compared to IgG, and IgG has been shown to interfere with IgM detection. This interference, as well as IgM-specific testing methods, is discussed in chapter 15.

Many IgM-specific EIA procedures have an additional step at the beginning of the EIA—the "capture" step. Anti-IgM antibodies attached to a solid phase bind, or capture, any IgM in the test serum. The assay is completed by adding reagents that will determine the specificity of the captured IgM.

TIPS

Each EIA assay has unique performance features. However, there are some aspects of EIA testing that are largely universal. A listing of some of these universally applicable reminders follows.

1. EIA techniques are very sensitive. To obtain consistently high-quality results, maintain consistent pipetting technique, incubation times, and temperature conditions. The room temperature, as well as the test incubation tem-

perature, is important. Bring all reagents to room temperature before testing.

2. Washing steps are very important. Improperly washed wells may give erroneous results, and splashing and splattering of reagents may affect test results. Do not scratch coated wells during washing and aspiration. Avoid microbial contamination of the reagents and the water used for washing.

3. Do not allow wells to dry out between assay steps.

4. Store unused microwells in sealed pouches containing desiccant strips. If the desiccant has a color change indicator, check the indicator each time the pouch is opened. If the color indicator has changed color, do not use the microwells. Do not reuse microwells.

5. Avoid contamination of the substrate solution with conjugate or other oxidants which will cause the solution to change color prematurely. Protect most substrates from light. Check the color of the substrate solution. No color change or precipitate should be observed prior to use.

6. Stop solutions usually contain acid. Handle them with care because they may cause burns or irritation to the skin. If they are turbid, do not use them. Avoid contact of substrate and stop solutions with oxidizing agents and metal parts.

7. Do not mix reagent bottle caps.

8. Do not mix reagents from different lots.

9. Many components of EIA assay kits contain sodium azide as a preservative. Sodium azide is known to form lead or copper azide in laboratory plumbing. These azides may explode on percussion, such as hammering. To prevent formation of lead or copper azide, flush drains thoroughly with water after disposing of solutions containing sodium azide. Sodium azide may also inhibit conjugate activity. Use clean pipette tips for conjugate addition so that azide is not carried from other reagents.

10. Observe all instructions for the spectrophotometer. Check the adjustment of the proper wavelength and optional reference wavelength.

11. Before reading microwells on a spectrophotometric plate reader, clean the undersides of the wells to remove condensation.

12. Calibrate dispensing equipment to ensure that the delivery volume is correct and equipment is free of contamination.

13. Be sure to observe the manufacturer's instructions concerning the use of heat-inactivated serum. Some EIAs should not be performed on serum that has been heat inactivated.

14. Specimens containing precipitate may give inconsistent test results. Such specimens should be clarified by centrifugation.

15. If comparisons with other methods are required, always perform both tests simultaneously.

16. If comparisons of the reactivities of two samples are to be made, test both samples in the same run of testing. Do not compare values obtained for samples tested in different runs. Comparing antibody levels determined in runs of testing performed days or weeks apart may allow insignificant random variations in testing to appear as significant differences.

FUTURE

Although EIA has enjoyed enormous popularity in many clinical diagnostic laboratories, there are several applications of EIA that hold great promise for the future. EIAs are

being used to replace older, more cumbersome assays to accomplish tasks previously considered nonroutine and laborious.

Epidemiology and Serotyping

Novel applications of EIA have proven effective in aiding epidemiological investigations and in serotyping. Three EIAs, each incorporating monoclonal antibodies against one of three different serotypes (P1A, P1B, and P2A) of rotavirus, although not as sensitive as reverse transcription PCR or a hybridization assay, successfully assigned P types to human fecal rotavirus strains (Masendycz et al., 1997). In another study, an avian influenza A H5 antibody-specific indirect EIA was established as part of an epidemiological investigation to determine the extent of human-to-human transmission and to establish risk factors associated with infection. The H5-specific EIA proved 100% sensitive and 92% specific in detecting avian influenza antibodies in the sera of children younger than 15 years of age and was more sensitive for avian influenza A antibody detection than standard hemagglutination inhibition assays (Rowe et al., 1999).

Recently, EIA products became available commercially that accurately differentiate HSV-1 and HSV-2 antibodies, a differentiation that was not possible previously with commercial EIAs. The antigens in the new EIA are type-specific glycoproteins G from HSV-1 and HSV-2 which have been immunoaffinity purified with monoclonal antibodies. Of 193 sera tested by this new EIA, 21 had equivocal results; however, for the 172 samples with unequivocal results, the sensitivity and specificity were 95 and 96%, respectively, for HSV-1 and 98 and 97% for HSV-2 compared to immunoblotting (Ashley et al., 1998).

Assays Performed on Samples Other Than Serum

Body fluids other than serum are now being assayed for various types of antibodies. Urine and oral fluids (gingival-crevicular transudate and oral mucosal transudate) are readily obtained by noninvasive means, not requiring a venipuncture for sample collection. They are available in sufficient quantities from most patients. Compared to serum antibody determinations, oral-fluid samples tested by EIA to determine immune status for measles, mumps, and rubella virus antibodies showed sensitivities of 97, 94, and 93%, respectively, and specificities of 100, 94, and 98%, respectively (Thieme et al., 1994).

Both gingival-crevicular transudate and urine samples tested by an enzyme-linked fluorescent assay for HIV-1 and HIV-2 antibodies yielded promising results. In testing of 187 HIV-infected individuals and 115 noninfected individuals, oral fluids yielded 100% sensitivity and specificity compared to serum results and urine samples yielded 95.2% sensitivity and 97.4% specificity (Martinez et al., 1999).

Urine samples assayed by an EIA procedure for RSV and influenza A virus antibodies yielded results that were comparable to those obtained with serum samples for these patients (Ireland and Nicholson, 1996). Likewise, saliva samples tested for influenza A virus IgG antibodies in a fluorescein isothiocyanate–anti-fluorescuin isothiocyanate amplification EIA had results that correlated better with serum IgG results than those generated by RIA (Vyse et al., 1999).

Automation

EIA automation is the direction that many immunoserology diagnostic laboratories favor. Newer instrumentation allows many EIAs to be performed simultaneously, often during off-hours when laboratory personnel are not present. Once the instrument is programmed and supplied with all of the necessary samples, the instrument completes all phases of testing and holds the data for the technologist to harvest during the next regular work shift. The more sophisticated instruments are capable of performing a wider variety of assays and have "on-board" reagent storage, and many can be interfaced with laboratory computer systems to allow data to go directly from the instrument into the data management program. All of these advances are helping to lower laboratory costs while providing high-quality assay results. Automation is discussed extensively in chapter 22.

CONCLUSIONS

EIA technology, with its extreme versatility, offers an avenue for performing many of the assessments required in the diagnostic-virology laboratory. EIA is currently a mainstay in diagnostic testing, and continuing development will undoubtedly produce additional EIA-based assays to replace antiquated technologies and solve contemporary problems.

REFERENCES

Adcock, P. M., G. G. Stout, M. A. Hauck, and G. S. Marshall. 1997. Effect of rapid viral diagnosis on the management of children hospitalized with lower respiratory tract infection. *Pediatr. Infect. Dis. J.* **16:**842–846.

Ashley, R. L., L. Wu, J. Pickering, M. Tu, and L. Schnorenberg. 1998. Premarket evaluation of a commercial glycoprotein G-based enzyme immunoassay for herpes simplex virus type-specific antibodies. *J. Clin. Microbiol.* **36:**294–295.

Avrameas, S. 1983. Enzyme immunoassays and related techniques: development and limitations. *Curr. Top. Microbiol. Immunol.* **104:**93–99.

Avrameas. S., and B. Guilbert. 1971. A method for quantitative determination of cellular immunoglobulins by enzyme-labeled antibodies. *Eur. J. Immunol.* **1:**394–396.

Chomel, J. J., M. F. Remilleux, P. Marchand, and M. Aymard. 1992. Rapid diagnosis of influenza A. Comparison with ELISA immunocapture and culture. *J. Virol. Methods* **37:**337–343.

Condorelli, F., and T. Ziegler. 1993. Dot immunobinding assay for simultaneous detection of specific immunoglobulin G antibodies to measles virus, mumps virus, and rubella virus. *J. Clin. Microbiol.* **31:**717–719.

Covalciuc, K. A., K. H. Webb, and C. A. Carlson. 1999. Comparison of four clinical specimen types for detection of influenza A and B viruses by optical immunoassay (FLUE OIA Test) and cell culture methods. *J. Clin. Microbiol.* **37:**3971–3974.

Dominguez, E. A., L. H. Taber, and R. B. Couch. 1993. Comparison of rapid diagnostic techniques for respiratory syncytial and influenza A virus respiratory infections in young children. *J. Clin. Microbiol.* **31:**2286–2290.

Englund, J. A., P. A. Piedra, A. Jewell, K. Patel, B. B. Baxter, and E. Whimbey. 1996. Rapid diagnosis of respiratory syncytial virus infections in immunocompromised adults. *J. Clin. Microbiol.* **34:**1649–1653.

Engvall, E., and P. Perlman. 1971. Enzyme-linked immunosorbent assay (ELISA). Quantitative assay of immunoglobulin G. *Immunochemistry* **8:**871–874.

Falsey, A. R., R. M. McCann, W. J. Hall, and M. M. Criddle. 1996. Evaluation of four methods for the diagnosis of respiratory syncytial virus infection in older adults. *J. Am. Geriatrics Soc.* **44:**71–73.

Forghani, B., and S. Hagens. 1995. Diagnosis of viral infections by antigen detection, p. 79–96. In E. H. Lennette, D. A. Lennette, and E. T. Lennette (ed.), *Diagnostic Procedures for Viral, Rickettsial, and Chlamydial Infections.* American Public Health Association, Washington, D.C.

Gerna, G., A. Sarasini, N. Passarani, M. Torsellini, M. Parea, and M. Battaglia. 1987. Comparative evaluation of a commercial enzyme-linked immunoassay and solid-phase immune electron microscopy for rotavirus detection in stool specimens. *J. Clin. Microbiol.* **25:**1137–1139.

Hiroyasu, A., B. Petchclai, K. Khupulsup, T. Kurimura, and K. Takeda. 1999. Evaluation of a rapid immunochromatographic test for detection of antibodies to human immunodeficiency virus. *J. Clin. Microbiol.* **37:**367–370.

Ireland, D. C., and K. G. Nicholson. 1996. Diagnosis of respiratory virus infections using GACELISA of urinary antibodies. *J. Immunol. Methods* **195:**73–80.

Johnston, S. L. G., S. Hamilton, P. Bindra, D. A. Hursh, and C. A. Gleaves. 1992. Evaluation of an automated immunodiagnostic assay system for direct detection of herpes simplex virus antigen in clinical specimens. *J. Clin. Microbiol.* **30:**1042–1044.

Kassler, W. J., C. Haley, W. K. Jones, A. R. Gerber, E. J. Kennedy, and J. R. George. 1995. Performance of a rapid, on-site human immunodeficiency virus antibody assay in a public health setting. *J. Clin. Microbiol.* **33:**2899–2902.

Lee, H. H., M. Canavaggio, and J. D. Burczak. 1992. Immunoblotting, p. 195-210. In E. H. Lennette (ed.), *Laboratory Diagnosis of Viral Infections*, 2nd ed. Marcel Dekker, Inc., New York, N.Y.

Lipson, S. M., and K. A. Zelinsky-Papez. 1989. Comparison of four latex agglutination (LA) and three enzyme-linked immunosorbent assays (ELISA) for the detection of rotavirus in fecal specimens. *Am. J. Clin. Pathol.* **92:**637–643.

Martinez, P. M., A. R. Torres, R. O. de Lejarazu, A. Montoya, J. F. Martin, and J. M. Eiros. 1999. Human immunodeficiency virus antibody testing by enzyme-linked fluorescent and Western blot assays using serum, gingival-crevicular transudate, and urine samples. *J. Clin. Microbiol.* **37:**1100–1106.

Masendycz, P. J., E. A. Palombo, R. J. Gorrell, and R. F. Bishop. 1997. Comparison of enzyme immunoassay, PCR, and type-specific cDNA probe techniques for identification of group A rotavirus gene 4 types (P types). *J. Clin. Microbiol.* **35:**3104–3108.

Miller, H., R. Milk, and F. Diaz-Mitoma. 1993. Comparison of the VIDAS RSV assay and the Abbott Testpack RSV with direct immunofluorescence for detection of respiratory syncytial virus in nasopharyngeal aspirates. *J. Clin. Microbiol.* **31:**1336–1338.

O'Beirne, A. J., and J. L. Sever. 1992. Enzyme immunoassay, p. 153–188. In S. Specter and G. Lancz (eds.), *Clinical Virology Manual*, 2nd ed. Elsevier, New York, N.Y.

Ogburn, J. R., J. T. Hoffpauir, E. Cole, K. Hood, D. Michael, T. Nguyen, S. Raden, B. Raju, V. Reisinger, and P. E. Oefinger. 1994. Evaluation of new transport medium for detection of herpes simplex by culture and direct enzyme-linked immunosorbent assay. *J. Clin. Microbiol.* **32:**3082–3084.

Pate, M. S., P. B. Dixon, K. Hardy, M. Crosby, and E. W. Hook III. 1998. Evaluation of the Biostar Chlamydia OIA Assay with specimens from women attending a sexually transmitted disease clinic. *J. Clin. Microbiol.* **36:**2183–2186.

Roblin, P. M., M. Gelling, A. Kutlin, N. Tsumura, and M. R. Hammerschlag. 1997. Evaluation of a new optical immunoassay for diagnosis of neonatal chlamydial conjunctivitis. *J. Clin. Microbiol.* **35:**515–516.

Rothbarth, P. H., M.-C. Hermus, and P. Schrijnemakers. 1991. Reliability of two new test kits for rapid diagnosis of respiratory syncytial virus infection. *J. Clin. Microbiol.* **29:**824–826.

Rowe, T., R. A. Abernathy, J. Hu-Primmer, W. W. Thompson, X. Lu, W. Lim, K. Fukuda, N. J. Cox, and J. M. Katz. 1999. Detection of antibody to avian influenza A (H5N1) virus in human serum by using a combination of serologic assays. *J. Clin. Microbiol.* **37:**937–943.

Thieme, T., S. Piacentini, S. Davidson, and K. Steingart. 1994. Determination of measles, mumps, and rubella immunization status using oral fluid samples. *JAMA* **272:**219–221.

Thomas, E. E., and L. E. Book. 1991. Comparison of two rapid methods for detection of respiratory syncytial virus (RSV) (TestPack RSV and Ortho RSV ELISA) with direct immunofluorescence and virus isolation for the diagnosis of pediatric RSV infection. *J. Clin. Microbiol.* **29:**632–635.

Todd, S. J., L. Minnich, and J. L. Waner. 1995. Comparison of rapid immunofluorescence procedure with TestPack RSV and Directigen FLU-A for diagnosis of respiratory syncytial virus and influenza A virus. *J. Clin. Microbiol.* **33:**1650–1651.

Van Weeman, B. K., and A. H. W. M. Schuurs. 1971. Immunoassay using haptoenzyme conjugates. *FEBS Lett.* **24:**77–81.

Verano, L., and F. J. Michalski. 1995. Comparison of a direct antigen enzyme immunoassay, Herpchek, with cell culture for detection of herpes simplex virus from clinical specimens. *J. Clin. Microbiol.* **33:**1378–1379.

Voller, A., D. E. Bidwell, and A. Bartlett. 1982. ELISA techniques in virology, p. 59–81. *In* C. R. Howard (ed.), *New Developments in Practical Virology.* Alan R. Liss, New York, N.Y.

Vyse, A. J., D. W. G. Brown, B. J. Cohen, J. R. Samuel, and D. J. Nokes. 1999. Detection of rubella virus-specific immunoglobulin G in saliva by an amplification-based enzyme-linked immunosorbent assay using monoclonal antibody to fluorescein isothiocyanate. *J. Clin. Microbiol.* **37:**391–395.

Waner, J. L., S. J. Todd, H. Shalaby, P. Murphy, and L. V. Wall. 1991. Comparison of Directigen FLU-A with viral isolation and direct immunofluorescence for the rapid detection and identification of influenza A virus. *J. Clin. Microbiol.* **29:**479–482.

Wardley, R. C., and J. R. Crowther (ed.). 1982. *The ELISA: Enzyme-Linked Immunosorbent Assay in Veterinary Research and Diagnosis.* (*Current Topics in Veterinary Medicine and Animal Science,* vol 22). Martinus Nijhoff Publishers, The Hague, The Netherlands.

Woo, P. C., S. S. Chiu, W. H. Seto, and M. Peiris. 1997. Cost-effectiveness of rapid diagnosis of viral respiratory tract infections in pediatric patients. *J. Clin. Microbiol.* **35:**1579–1581.

Yolken, R. H. 1980. Enzyme-linked immunosorbent assay (ELISA): a practical tool for rapid diagnosis of viruses and other infectious agents. *Yale J. Biol. Med.* **53:**85–92.

Peroxidase-Antiperoxidase Detection of Viral Antigens in Cells

JOHN J. DOCHERTY, CHRISTOPHER M. POKABLA, AND HELEN GAY

10

Immunocytochemical staining, a sensitive and specific method to detect localized antigens with labeled antibodies, has been used extensively to identify viral antigens. Coons introduced immunocytochemistry in 1942 by using fluorescein-labeled anti-pneumococcal III antibody to find pneumococcal antigens in the livers and the spleens of experimentally infected mice (Coons et al., 1941, 1942). In the following years, immunofluorescence (IF) was used to detect many different bacterial and viral antigens in vivo and in vitro (Coons et al., 1950; Kaplan et al., 1950; Kurstak, 1971). Information gathered from such studies advanced our knowledge of the structures and functions of viral proteins, the transforming mechanisms of viruses, and viral replication cycles. Indeed, IF continues to be used successfully in a variety of ways, in both the research and the clinical laboratories.

As with any procedure, there has been a continuing quest to improve on the original procedures of Coons. By the mid 1960s, a procedure that used the basic methodology of previously described fluorescence methods began to emerge, but the fluor was replaced with an enzyme. When the procedure was used to detect virus antigens, a precipitate formed, marking the location of the antibody-antigen reaction. This immunoenzymatic procedure, which was reported by Ram et al. (1966), used acid phosphatase-conjugated antibodies to localize tissue antigens. However, because of rapid loss of enzymatic activity by the antibody-enzyme conjugate (Nakane and Pierce, 1967), acid phosphatase was replaced with the more stable enzyme horseradish peroxidase (HRP). Introduced by Nakane and Pierce, immunoperoxidase staining involves reacting the antigen with HRP-conjugated antibody and then visualizing the complex by exposing the tissue to a solution containing appropriate electron transfer substrates, such as 3,3'-diaminobenzidine tetrahydrochloride (DAB) and hydrogen peroxide (Nakane and Pierce, 1967). The bound enzyme first reduces the hydrogen peroxide to water and then oxidizes DAB. The resulting oxidized DAB polymerizes to form a nondiffusible, insoluble dark-brown precipitate that localizes at the site of the antigen (Pearse, 1972). This immunoenzymatic staining method overcame many of the weaknesses of IF. The peroxidase-stained preparations, once mounted, are essentially permanent, can be viewed with an ordinary light microscope, and avoid cellular autofluorescence (Spendlove, 1967). The peroxidase enzyme is available in pure form and is stable after chemical conjugation, and because it is small, it readily penetrates tissue (Sternberger, 1974). The practical advantages offered by the immunoperoxidase staining method have made it an attractive procedure for the detection of a wide variety of viral antigens. The use of peroxidase does have one major drawback. The enzyme is endogenous to some mammalian tissues (Straus, 1971; Streefkerk, 1972; Weir et al., 1974; Blain et al., 1975), which results in nonspecific staining that can interfere with the interpretation of the immunoperoxidase staining test results. This problem can be solved by eliminating endogenous peroxidase by pretreatment of tissues with peroxidase-inactivating reagents (Straus, 1971, 1979; Streefkerk, 1972; Weir et al., 1974). Most of the endogenous enzyme activity is eliminated by this treatment, but some investigators have reported that some of these treatments also damage antigens (Fink et al., 1979; Straus, 1979).

During the years of refinement of both the IF and immunoperoxidase staining techniques, improvements were introduced that greatly increased the sensitivity of these staining methods. Coons had first introduced IF as a direct stain in which the fluorochrome was directly attached to the specific or primary antibody (Coons et al., 1941, 1942). Twelve years later, Weller and Coons showed that IF staining sensitivity could be increased 10-fold if the tissue was first treated with unlabeled specific antibody and then treated with fluorescein-labeled anti-immunoglobulin (e.g., anti-immunoglobulin G [anti-IgG]) antibody (Weller and Coons, 1954). Through this indirect staining, several anti-IgG molecules could attach to a single bound primary antibody, thus greatly amplifying the fluorescent signal and thereby increasing detection of the antigen (Sternberger, 1974). In addition, indirect staining made the technique more general because a single preparation of fluorochrome-labeled anti-IgG could be used to detect antibodies raised in the same species to a variety of different antigens, avoiding the necessity of labeling each specific antibody.

The indirect staining method has been applied to immunoperoxidase staining to detect a wide variety of viruses, including canine distemper virus (Higgins et al., 1982), enteric adenovirus (Cevenini et al., 1984), enteroviruses (Herrmann et al., 1974), hepatitis B virus-associated delta antigen (Recchia et al., 1981), hepatitis B virus surface antigen (Hsu et al., 1983), human cytomegalovirus (Gerna et al., 1976), mouse mammary tumor virus (Keydar et al., 1978), papillomavirus antigen (Kurman et al., 1981; Syrjanen and Pyrhonen, 1982), rotavirus (Chasey, 1980; Cevenini et al., 1984; Grom and Bernard, 1985), yellow fever virus (de la Monte et al., 1983), Epstein-Barr virus (Musiani et al., 1986), influenza virus type A (Gardner et al., 1978), and respiratory syncytial virus (Gardner et al., 1978). Two technical problems have come to light regarding this method. First, the chemical conjugation process inactivates some of the HRP that is coupled to the antibody; second, some of the antibody remains unconjugated. Both the unconjugated antibody and the antibody conjugated to inactive enzyme compete with antibody molecules labeled with active enzyme for the antigenic sites and thus decrease the sensitivity of the assay.

In 1970, Sternberger et al. (1970) introduced an immunoperoxidase procedure that required no chemical conjugation of enzyme to antibody. Peroxidase was bound to an anti-peroxidase antibody via specific antibody-antigen interactions. Viral antigens were localized by this unlabeled-antibody-enzyme technique, schematically presented in Fig. 1, when the specimen was treated sequentially with the following reagents.

1. Antibody to a specific antigen raised in species A (primary antibody)
2. Anti-species A serum raised in species B (bridge antibody) applied in sufficient excess so that one binding site of the antibody attaches to the primary antibody and the other binding site remains free
3. A purified antibody-enzyme complex made up of anti-HRP raised in species A that had been combined with HRP
4. DAB and H_2O_2

This method, known as the unlabeled-peroxidase–antiperoxidase (PXAPX) technique, is 20 times more sensitive

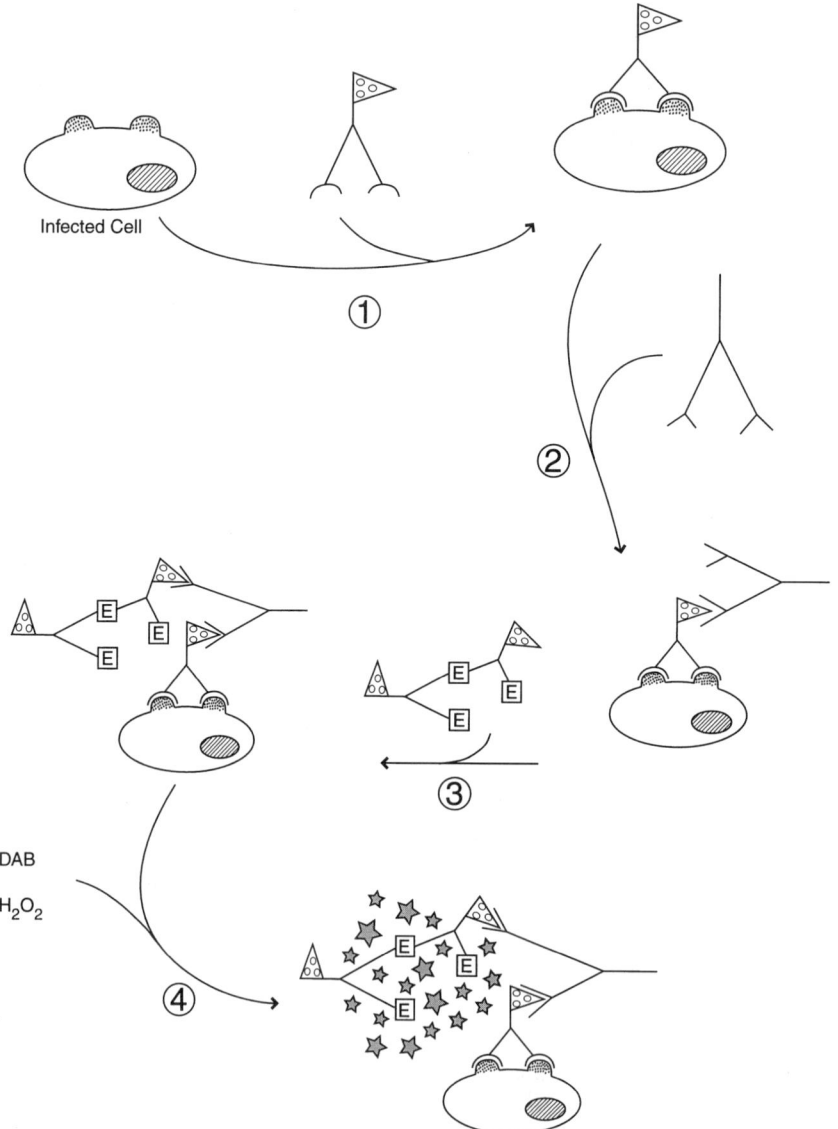

FIGURE 1 Schematic representation of the unlabeled PXAPX method of antigen detection. 1, Primary antibody; 2, bridge antibody; 3, PXAPX complexes; 4, H_2O_2; E, enzyme.

than the indirect immunoperoxidase staining method and 100 to 1,000 times more sensitive than IF (Pearson et al., 1979). Sensitivity is greatly increased because each viral antigen is attached to one or more active enzyme molecules.

The PXAPX technique has been used to detect a wide variety of viruses, including canine distemper virus, cytomegalovirus, dengue virus, hepatitis B virus, herpes simplex virus (HSV type 1 [HSV-1] and HSV-2), mumps virus, papillomavirus, paramyxovirus, rabies virus, retroviruses, and transmissible gastroenteritis virus (Table 1).

We have used the PXAPX technique routinely to detect HSV in either tissue culture or tissue scrapings from human genitalia. In this chapter, we will provide the reader with our insights and experience using immunoenzymatic PXAPX staining to detect viral antigens. It is not our intent to belabor the pros and cons of one procedure over another but rather to describe the PXAPX method that is routinely used in our laboratory. The procedure outlines can serve as a foundation and may be adapted and modified to the specifications of any laboratory or virus, as conditions dictate.

STAINING CONSIDERATIONS

Successful staining by the PXAPX method requires consideration of several critical factors. One of the most important requirements is the choice of appropriate tissue fixative. Detection of antigens cannot occur by any immunocytochemical method if they have been destroyed or significantly altered by the fixative. An appropriate fixative for the antigen(s) to be examined must first be determined, and this may vary for different viruses. Secondly, optimum concentrations of the three different antisera used must be established, regardless of whether the sera were prepared in the research laboratory or obtained from a commercial source. The establishment of appropriate dilutions is carried out by block titration of the three antisera and recording those combinations that produce optimum results. Once approximate values are obtained, we frequently adjust to final con-

centrations by titrating one of the three (frequently the primary antibody) while holding the other two constant and subjectively evaluating the quality of the stain for optimum results. Optimum results are defined as maximum staining intensity of the antigen with minimal staining of tissue that does not contain the antigen. This may be repeated for the bridging antibody and the anti-peroxidase antibody. The dilutions and incubations used in the procedure detailed below were optimized for the detection of HSV-2 in acetone (Pearson et al., 1979)- or 70% ethanol (Steinkamp and Crissman, 1974)-fixed tissue culture cells or specimens obtained from patients. The staining procedure is easily adaptable to other virus systems, and the specific dilutions and incubation times described here can be used as a starting point for optimizing the procedure for other viruses.

MATERIALS FOR IMMUNOPEROXIDASE STAINING

The materials listed below are for the PXAPX staining system, in which the antiviral serum and anti-peroxidase serum were produced in rabbits and the bridge anti-rabbit antibody was produced in a goat. Other combinations of hosts can be used, but it is essential that consistency be maintained, i.e., that the two hosts are not too closely related and that the antiviral and anti-peroxidase antibodies have been raised in the same host. Normal, nonimmune serum of the host in which the bridge antibody was raised (in this case, a goat) is used to block nonspecific binding of antibodies and is layered on the samples before the addition of each of the three different antibodies. The nonimmune serum contains an insignificant amount of cross-reacting antibody that does not interfere with the specific binding of any of the antibodies.

Sera

    Nonimmune goat serum (NGS); prepared in our laboratory but can be commercially obtained

    Antiviral serum raised in rabbits; prepared in our laboratory but can be commercially obtained

    Preimmune serum from rabbits; prepared in our laboratory but can be commercially obtained

    Anti-rabbit serum raised in goats; commercially obtained; antibody protein concentration, 15 mg/ml

    Rabbit PXAPX; commercially obtained; antibody protein concentration, 3 mg/ml

Phosphate-buffered saline (PBS; 0.01 M sodium phosphate [pH 7.2], 0.15 M NaCl)

Absolute methanol

Hydrogen peroxide (fresh)

DAB

Absolute ethanol (EtOH); 90% EtOH and 70% EtOH

Xylene

Permount

METHODS

Phosphate-buffered saline is used throughout for all dilutions and washes. All procedures are performed at room temperature (~20°C).

  1. If the samples are immersed in fixative, rinse them with PBS (5 min) before beginning to stain; this is not necessary if the samples have been previously fixed and air dried. We

TABLE 1 Applications of unlabeled-PXAPX technique to detect viral antigens

Virus	Reference(s)
Canine distemper virus	Ducatelle et al., 1980
Cytomegalovirus	Jiwa et al., 1990
Dengue virus	Churdboonchart et al., 1984; Makino et al., 1991; Okuno et al., 1977
Hepatitis B virus (surface antigen and core antigen)	Machado-Vieira and Sarno, 1982
HSV-1 or -2	Adams et al., 1984; Flaitz and Hammond, 1988; Gay et al., 1984; Kurchak et al., 1977; Pearson et al., 1979
Mumps virus	Yasuda et al., 1983
Papillomavirus	Shu and Yang, 1987
Paramyxovirus	Schwendemann et al., 1982
Rabies virus	Bourgon and Charlton, 1987; Last et al., 1994; Palmer et al., 1985; Zimmer et al., 1990
Retroviruses	Suni et al., 1981
Transmissible gastroenteritis virus	Chu et al., 1982

routinely fix samples in acetone at 4°C for 5 to 10 min, air dry them, and store them desiccated at 4°C for up to 21 days before staining them.

2. Cover the samples with a few drops of a 1:30 dilution of NGS for 5 min; if the samples are dry at the beginning of the staining procedure, prewet them by dipping them into PBS before adding NGS.

3. Drain the NGS by tipping the sample onto a paper towel at about a 90° angle; this removes the liquid that has accumulated at the edges of the coverslip or slide.

4. Overlay the samples with a few drops of an appropriate dilution of antiviral serum and incubate them for 1 h in a humid chamber; the optimum dilution for anti-HSV-2 serum raised in rabbits in this laboratory has ranged from 1:200 to 1:2,000. (Controls include the following: positive control, virus-infected cells plus immune serum; negative control, virus-infected cells plus preimmune serum, virus-infected cells treated with PBS in place of the primary antibody, or uninfected cells treated with immune serum.)

5. Wash the samples twice in PBS (5 min each time).

6. Cover the samples with a few drops of diluted NGS for 5 min at room temperature and then drain them.

7. Overlay each sample with a few drops of a 1:10 dilution of goat anti-rabbit serum; incubate the samples for 20 min and drain them.

8. Wash the samples twice in PBS (5 min each time).

9. Immerse the samples in absolute methanol for 15 min.

10. Wash the samples three times in PBS (5 min each time).

11. Cover the samples with a few drops of diluted NGS for 5 min and then drain them.

12. Apply a few drops of rabbit PXAPX diluted 1:100 in 1% NGS–PBS (1 ml of NGS plus 99 ml of PBS) for 20 min and then drain them.

13. Wash the samples twice in PBS (5 min each time).

14. Stain the samples by immersing them in the disclosing reaction mixture for 7 min while they are protected from light; the disclosing reaction mixture consists of 0.025% DAB in PBS, freshly prepared and filtered (0.45-μm-pore-size filter), to which hydrogen peroxide is added to a final concentration of 0.005%. (Note: stir the DAB rapidly for at least 30 min [in the dark] before filtering it and adding hydrogen peroxide.)

15. Wash the stained samples twice in PBS (5 min each time).

16. If the samples are not to be counterstained, they are dehydrated by immersing them sequentially in a series of graded alcohols (i.e., 70% EtOH for 1 min, 90% EtOH for 1 min, and absolute EtOH for 1 min) and then in xylene and mounted with Permount and coverslips for light microscopic studies.

17. Counterstain (optional). An added attraction of the PXAPX procedure is that it can easily be interfaced with other staining procedures that may assist in viewing specific virus-induced cellular changes. In this regard, we have successfully interfaced the PXAPX procedure with the Papanicolaou stain (PAP) (Papanicolaou, 1963) by merely taking the sample through a standard PAP procedure after it was stained by PXAPX (Pearson et al., 1979).

RESULTS

Immunoperoxidase-stained samples contain dark-brown (if DAB has been used as a substrate) deposits at the site of the antigen and very little staining elsewhere unless the specimen has been counterstained with a stain such as PAP.

Control specimens, in which preimmune serum or PBS is used in place of the primary serum, will often show some diffuse, light-brown, nonspecific staining throughout the cell or tissue specimen, particularly if the sample has not been counterstained. When Vero cells infected with HSV-2 were stained with PXAPX-PAP and examined by light microscopy, infected cells with dark-brown-stained cytoplasm and nuclei were readily seen (Color plate 2a [see color insert]). A closer examination of Color plate 2a reveals typical HSV syncytia recruiting adjacent cells. Uninfected cells surrounding the infected cells contained light-blue-stained nuclei with a clear cytoplasm (Color plate 2a). Color plate 2b shows cells with various degrees of brown stain in the cytoplasm, from light tan to dark brown, which presumably reflect the stage of infection by the virus (i.e., early versus late) in that cell. Cells with lightly stained tan cytoplasm (Color plate 2b) were structurally indistinguishable from uninfected cells with a clear cytoplasm and likely represent cells early in the viral replication cycle. However, as the infection progressed, the cells became more darkly stained and began to develop characteristic HSV cytopathic effects (i.e., rounding and syncytium formation) (Color plate 2b).

The intensity of the stain can be regulated to various degrees by the length of time the sample is in the disclosing reaction mixture. As the staining time is increased, however, DAB can be oxidized by atmospheric oxygen, resulting in progressive deposition of product over the entire sample. This reduces the contrast between stained and unstained areas in the sample and may make interpretation of the results difficult. Manipulation of the DAB and H_2O_2 concentrations can result in faster or slower substrate conversion and is frequently adjusted in a subjective manner to meet the demands or constraints of a particular system. It is possible, however, to overstain the sample so that normal tissue is intensely stained and cannot be distinguished from infected tissue.

We used PXAPX and PXAPX-PAP to stain exfoliated cells from a female patient with a confirmed HSV-2 genital infection. The HSV-2-infected cells stained only by PXAPX were brown (Color plate 2c), while uninfected cells remained colorless or light tan. When this sample was counterstained with PAP, the infected cells retained their distinct brown color (Color plate 2d) and uninfected cells stained blue or pink with dark blue or dark red nuclei. Although we chose to use a common stain (i.e., PAP), many other histological stains could be used in conjunction with the PXAPX method should the staining of specific cellular elements, viral inclusions, or pathological changes be required.

DISCUSSION

It has been our experience that PXAPX staining produces consistent, sensitive, and reliable results. Some of the advantages of the procedure are that it is straightforward and uncomplicated, the reagents needed can be easily prepared in the laboratory or are commercially available, no extraordinary equipment is required to carry out the procedure or view the results, and the stained preparation is permanent. We have found the last characteristic to be particularly valuable for reviewing samples months to years after they are stained.

As with most procedures, the PXAPX method described has inherent weaknesses. The disadvantages include the rather long period of time required to complete the PXAPX procedure, which precludes its use when rapid results are required. Indirect immunoperoxidase staining is more rapid than the PXAPX procedure described, and it should be considered when speed of obtaining results is more important than test sensitivity (Benjamin, 1975). Additionally, DAB is a suspected carcinogen, and care must be taken when handling and disposing of this material. Because of the carcinogenic properties of DAB, it may be wiser to use 4-chloro-1-napthol (Hawkes et al., 1982) or *p*-phenylenediamine dihydrochloride–pyrocatechol (Hanker et al., 1977).

Perhaps the biggest drawback to immunoperoxidase staining that has occasionally complicated its use is the background staining caused by endogenous peroxidase in mammalian mucosecretions (Blain et al., 1975) and inflammatory cells (Straus, 1971). This is particularly troublesome for us because most of our studies are of HSV infections of the female genital tract. This anatomical site is rich in mucosecretions and, during an active herpetic infection, inflammatory cells. Because of these two sources of endogenous peroxidase, nonspecific staining can make the detection of a positive reaction extremely difficult or impossible. Consequently, the endogenous enzyme must be inactivated in order to obtain a satisfactory immunoperoxidase stain. Fortunately, there are several methods to inactivate endogenous peroxidase, such as pretreatment with methanol (Straus, 1971, 1979; Streefkerk, 1972; Weir et al., 1974). However, some reports suggest that such treatment may alter antigenic structure (Fink et al., 1979; Straus, 1979), and we have found this to be true for HSV antigens. Indeed, if an HSV-positive sample is treated with methanol prior to the PXAPX procedure, the HSV antigens are altered to such an extent that they are not recognized by the primary antibody. This results in a false negative, a highly undesirable result in either the clinical or research laboratory.

When we were confronted with this apparent impasse, we reasoned that while the methanol adversely affected the viral antigens, it may not affect the antibodies after they have reacted with HSV antigens. Therefore, we ran a series of experiments in which methanol treatment was attempted at each stage of the procedure after the reaction with primary antibody (Fig. 1). These studies eventually demonstrated that methanol treatment after goat anti-rabbit IgG, but before rabbit PXAPX, preserved the viral antigen primary antibody reaction, inactivated endogenous enzyme, and did not interfere with the detection of viral antigen (Pearson et al., 1979).

Nonetheless, because the possibility for antigenic alteration by such pretreatment exists, we have used an immunoenzymatic staining method that uses a nonmammalian enzyme, glucose oxidase (Gay et al., 1984). Essentially, the procedure is the same as the PXAPX method, but glucose oxidase-coupled antibody is used in place of peroxidase-coupled antibody. Because glucose oxidase is not present in mammalian cells, this modification has allowed us to eliminate methanol yet preserve the sensitivity and permanency of the PXAPX method (Campbell and Bhatnager, 1976; Suffin et al., 1979; Clark et al., 1982; Gay et al., 1984). An example of this method is seen in Color plate 2e and f. Vero cells infected with HSV-2 were processed using glucose oxidase-coupled antibody instead of peroxidase-coupled antibodies. In Color plate 2e, 2-*p*-iodophenyl-3-*p*-nitrophenyl-5-phenyl-2H-tetrazolium chloride (INT) was used as a substrate and infected cells stained dark red. In Fig. 2f, 3,3'-(3,3'-dimethoxy-4,4'-biphenylylene)-bis-(2-*p*-nitrophenyl-5-phenyl-2H-tetrazolium chloride) (NBT) was used as a substrate and the infected cells stained dark blue-black. In both cases, uninfected cells were colorless.

The PXAPX method has proven to be a valuable tool in virological studies. It is simple, very sensitive, and inexpensive, and it provides a permanent preparation. It can be used for most viral antigens and can be configured to meet the requirements of an individual laboratory. It has been successfully used with a large number of different viruses and can be easily modified to detect any antigen under investigation.

REFERENCES

Adams, R. L, D. R. Springall, and M. M. Levene. 1984. The immunocytochemical detection of herpes simplex virus in cervical smears—a valuable technique for routine use. *J. Pathol.* 143:241–247.

Benjamin, D. R. 1975. Use of immunoperoxidase for rapid viral diagnosis, p. 89–96. In D. Schlessinger (ed.), *Microbiology—1975*. American Society for Microbiology, Washington, D.C.

Blain, J. A., P. J. Heald, A. E. Mack, and C. E. Shaw. 1975. Peroxidase in human cervical mucus during the menstrual cycle. *Contraception* 11:677–680.

Bourgon, A. R., and K. M. Charlton. 1987. The demonstration of rabies antigen in paraffin-embedded-tissues using the peroxidase-antiperoxidase method: a comparative study. *Can. J. Vet. Res.* 51:117–120.

Campbell, G. T., and A. S. Bhatnager. 1976. Simultaneous visualization by light microscopy of two pituitary hormones in a single tissue section using a combination of indirect immunohistochemical methods. *J. Histochem. Cytochem.* 24:448–452.

Cevenini, R., F. Rumpianesi, M. Mazzaracchio, M. Donati, E. Falcieri, and I. Sarov. 1984. A simple immunoperoxidase method for detecting enteric adenovirus and rotavirus in cell culture. *J. Infect.* 8:22–27.

Chasey, D. 1980. Investigation of immunoperoxidase-labelled rotavirus in tissue culture by light and electron microscopy. *J. Gen. Virol.* 50:195–200.

Chu, R. M., N. Li, R. D. Glock, and R. F. Ross. 1982. Applications of peroxidase-antiperoxidase staining technique for detection of transmissible gastroenteritis virus in pigs. *Am. J. Vet. Res.* 43:77–81.

Churdboonchart, V., K. Kamsattaya, S. Yoksan, P. Sinarachatanant, and N. Vhamarapravati. 1984. Application of peroxidase-antiperoxidase (PAP) staining for detection and localization of dengue-2 antigen. I. In an endogenous peroxidase containing cell systems. *Southeast Asian J. Trop. Med. Public Health* 15:547–553.

Clark, C. A., E. C. Downs, and F. J. Primus. 1982. An unlabeled antibody method using glucose oxidase-antiglucose oxidase complexes (GAG): a sensitive alternative to immunoperoxidase for the detection of tissue antigens. *J. Histochem. Cytochem.* 30:27–34.

Coons, A. H., H. J. Creech, and R. N. Jones. 1941. Immunological properties of an antibody containing a fluorescent group. *Proc. Soc. Exp. Biol. Med.* 47:200–202.

Coons, A. H., H. J. Creech, R. N. Jones, and E. Berliner. 1942. The demonstration of pneumococcal antigens in tissues by the use of the fluorescent antibody. *J. Immunol.* 45:159–170.

Coons, A. H., J. C. Snyder, F. S. Cheever, and E. S. Murray. 1950. Localization of antigen in tissue cells. IV. Antigens of rickettsiae and mumps virus. *J. Exp. Med.* 91:31–38.

de la Monte, S. M., A. L. Linhares, A. P. A. Travassos Da Rosa, and F. P. Pinheiro. 1983. Immunoperoxidase detection of yellow fever virus after natural and experimental infections. *Trop. Geogr. Med.* **35:**234–241.

Ducatelle, R., W. Coussement, and J. Hoorens. 1980. Demonstration of canine distemper viral antigen in paraffin sections, using unlabeled antibody-enzyme method. *Am. J. Vet. Res.* **41:**1860–1862.

Fink, B., E. Loepfe, and R. Wyler. 1979. Demonstration of viral antigens in cryostat sections by a new immunoperoxidase procedure eliminating endogenous peroxidase activity. *J. Histochem. Cytochem.* **27:**686–688.

Flaitz, C. M., and H. L. Hammond. 1988. The immunoperoxidase method for the rapid diagnosis of intraoral herpes simplex virus infection in patients receiving bone marrow transplants. *Spec. Care Dentist.* **8:**82–85.

Gardner, P. S., M. Grandien, and J. McQuillin. 1978. Comparison of immunofluorescence and immunoperoxidase methods for viral diagnosis at a distance: a WHO collaborative study. *Bull. W.H.O.* **56:**105–110.

Gay, H., W. R. Clark, and J. J. Docherty. 1984. Detection of herpes simplex virus infection using glucose oxidase-antiglucose oxidase immunoenzymatic stain. *J. Histochem. Cytochem.* **32:**447–451.

Gerna, G., A. Vasquez, C. J. McCloud, and R. W. Chambers. 1976. The immunoperoxidase technique for rapid human cytomegalovirus identification. *Arch. Virol.* **50:**311–321.

Grom, J., and S. Bernard. 1985. Virus enzyme-linked cell immunoassay (VELCIA): detection and titration of rotavirus antigen and demonstration of rotavirus neutralizing and total antibodies. *J. Virol. Methods* **10:**135–144.

Hanker, J. S., P. E. Yates, C. B. Metz, and A. Rustioni. 1977. A new specific, sensitive and non-carcinogenic reagent for the demonstration of horseradish peroxidase. *Histochem. J.* **9:**789–792.

Hawkes, R., E. Niday, and J. Gordon. 1982. A dot-immunoblotting assay for monoclonal and other antibodies. *Anal. Biochem.* **119:**142–147.

Herrmann, J. E., S. A. Morse, and F. Collins. 1974. Comparison of techniques and immunoreagents used for indirect immunofluorescence and immunoperoxidase identification of enteroviruses. *Infect. Immun.* **10:**220–226.

Higgins, R. J., S. Krakowka, A. E. Metzler, and A. Koestner. 1982. Immunoperoxidase labeling of canine distemper virus replication cycle in VERO cells. *Am. J. Vet. Res.* **43:**1820–1824.

Hsu, H., M. Kao, Y. Lin, D. Chen, and C. Lee. 1983. Detection of hepatitis B surface antigen in liver tissue: a comparative study on the sensitivity of histochemical and immunoperoxidase techniques. *J. Formosan Med. Assoc.* **82:**657–666.

Jiwa, M., R. D. M. Steebergen, F .E. Zwaah, P. M. Kluin, A. K. Raap, and M. van der Ploeg. 1990. Three sensitive methods for the detection of cytomegalovirus in lung tissue of patients with interstitial pneumonitis. *Am. J. Clin. Pathol.* **93:**491–494.

Kaplan, M. H., A. H. Coons, and H. W. Deane. 1950. Localization of antigen in tissue cells. III. Cellular distribution of pneumococcal polysaccharides types II and III in the mouse. *J. Exp. Med.* **91:**15–30.

Keydar, I., R. Mesa-Tejada, M. Ramanarayanan, T. Ohno, C. Fenoglio, R. Hu, and S. Spiegelman. 1978. Detection of viral proteins in mouse mammary tumors by immunoperoxidase staining of paraffin sections. *Proc. Natl. Acad. Sci. USA* **75:**1524–1528.

Kurchak, M., D. R. Dubbs, and S. Kit. 1977. Detection of herpes simplex virus-related antigens in the nuclei and cytoplasm of biochemically transformed cells with peroxidase/anti-peroxidase immunological staining and direct immunofluorescence. *Int. J. Cancer* **20:**371–380.

Kurman, R. J., K. H. Shah, W. D. Lancaster, and A. B. Jensen. 1981. Immunoperoxidase localization of papillomavirus antigens in cervical dysplasia nad vulvar condylomas. *Gynecology* **140:**931–935.

Kurstak, E. 1971. The immunoperoxidase technique: localization of viral antigens in cells, p. 423–444. *In* K. Maramorosch and H. Koprowski (ed.), *Methods in Virology*, vol. 5. Academic Press, New York, N.Y.

Last, R. D., J. E. Jardine, M. M. E. Smit, and J. J. Van Der Lugt. 1994. Application of immunoperoxidase techniques to formalin-fixed brain tissue for the diagnosis of rabies in southern Africa. *Onderstepoort J. Vet. Res.* **61:**183–187.

Machado-Vieira, L. M., and E. N. Sarno. 1982. The efficacy of the immunoperoxidase technique for the detection of several antigens after destaining of stored stained sections. *Braz. J. Med. Biol. Res.* **15:**265–268.

Makino, Y., M. Tadano, S. Arakaki, and T. Fukunaga. 1991. Potential use of a baculovirus-expressed Dengue-4E protein as a diagnostic antigen in regions endemic for Dengue and Japanese Encephalitis. *Am. J. Trop. Med. Hyg.* **45:**636–643.

Musiani, M., M. Zerbini, M. Plazzi, and M. LaPlaca. 1986. Double immunoenzymatic staining for the simultaneous detection of Epstein-Barr virus induced antigens. *Histochemistry* **84:**15–17.

Nakane, P. K., and G. B. Pierce. 1967. Enzyme-labeled antibodies: preparation and application for the localization of antigens. *J. Histochem. Cytochem.* **14:**929–931.

Okuno, Y., F. Sasao, T. Fukunaga, and K. Fukai. 1977. An application of PAP (peroxidase-anti-peroxidase) staining technique for the rapid titration of Dengue virus type 4 infectivity. *Biken J.* **20:**29–33.

Palmer, D. G., P. Ossent, M. M. Suter, and E. Ferrari. 1985. Demonstration of rabies viral antigen in paraffin tissue sections: comparison of the immunofluorescence technique with the unlabeled antibody enzyme method. *Am. J. Vet. Res.* **46:**283–286.

Papanicolaou, G. N. 1963. Technique, p. 3–12. *In Atlas of Exfoliative Cytology.* Harvard University Press, Cambridge, Mass.

Pearse, A. G. E. 1972. *Histochemistry: Theoretical and Applied,* vol. 2. Churchill Livingston, Edinburgh, United Kingdom.

Pearson, N. S., G. Fleagle, and J .J. Docherty. 1979. Detection of herpes simplex virus infection of female genitalia by the peroxidase-antiperoxidase method alone or in conjunction with the Papanicolaou stain. *J. Clin. Microbiol.* **10:**737–746.

Ram, J. S., P. K. Nakane, D. G. Rawlinson, and G. B. Pierce. 1966. Enzyme-labeled antibodies for ultrastructural studies. *Fed. Proc.* **25:**732.

Recchia, S., R. Rizzi, F. Acquauiva, M. Rizzetto, V. Tison, F. Bonino, and G. Verme. 1981. Immunoperoxidase staining of the HBV-associated delta antigen in paraffinated liver specimens. *Pathologica* **73:**773–777.

Schwendemann, G., J. S. Wolinsky, G. Hatzidimitrou, D. C. Merz, and M. N. Waxham. 1982. Postembedding immunocytochemical localization of paramyxovirus antigens by light and electron microscopy. *J. Histochem. Cytochem.* **30:**1313–1319.

Shu, L., and X. Yang. 1987. Human papillomavirus infection and carcinogenesis of cancer of uterine cervix. II. Immunoperoxidase localization of papillomavirus antigen in tissues of the uterine cervix. *Chung Hua Chung Liu Tsa Chih* **9:**430–431.

Spendlove, R. S. 1967. Microscopic techniques, p. 482–520. *In* K. Maramorosch and H. Koprowski (ed.), *Methods in Virology*, vol. 3. Academic Press, New York, N.Y.

Steinkamp, J. A., and H. A. Crissman. 1974. Automated analysis of deoxyribonucleic acid, protein and nuclear to cytoplasmic relationships in tumor cells and gynecologic specimens. *J. Histochem. Cytochem.* **22:**616–621.

Sternberger, L. A. 1974. *Immunocytochemistry.* Prentice-Hall, Englewood Cliffs, N.J.

Sternberger, L. A., P. H. Hardy, J. J. Cuculis, and H. G. Meyer. 1970. The unlabeled antibody enzyme method of immunohistochemistry. Preparation and properties of soluble antigen-antibody complex (horseradish peroxidase-antihorseradish peroxidase) and its use in identification of spirochetes. *J. Histochem. Cytochem.* **18:**315–333.

Straus, W. 1971. Inhibition of peroxidase by methanol and by methanol-nitroferricyanide for use in immunoperoxidase procedures. *J. Histochem. Cytochem.* **19:**682–688.

Straus, W. 1979. Peroxidase procedures. Technical problems encountered during their application. *J. Histochem. Cytochem.* **27:**1349–1351.

Streefkerk, J. G. 1972. Inhibition of erythrocyte pseudoperoxidase activity by treatment with hydrogen peroxide following methanol. *J. Histochem. Cytochem.* **20:**829–831.

Suffin, S. C., K. B. Muck, J. C. Young, K. Lewin, and D. D. Porter. 1979. Improvement of the glucose oxidase immunoenzymatic technic. Use of a tetrazolium whose formazan is stable without heavy metal chelation. *Am. J. Clin. Pathol.* **71:**492–496.

Suni, J., T. Wahlstrom, and A. Vaheri. 1981. Retrovirus p30-related antigen in human syncytiotrophoblasts and IgG antibodies in cord-blood sera. *Int. J. Cancer* **28:**559–566.

Syrjanen, K. J., and S. Pyrhonen. 1982. Demonstration of human papilloma virus antigen in the condylomatous lesions of the uterine cervix by immunoperoxidase technique. *Gynecol. Obstet. Investig.* **14:**90–96.

Weir, E. E., T. G. Pretlow, A. Pitts, and E. E. Williams. 1974. Destruction of endogenous peroxidase activity in order to locate cellular antigens by peroxidase-labeled antibodies. *J. Histochem. Cytochem.* **22:**51–54.

Weller, T. H., and A. H. Coons. 1954. Fluorescent antibody studies with agents of varicella and herpes zoster propagated in vitro. *Proc. Soc. Exp. Biol. Med.* **86:**789–794.

Yasuda, Y., Y. Hosaka, T. Fukunaga, Y. Okuno, and K. Fukai. 1983. Application of the PAP (peroxidase-anti-peroxidase) staining technique for the rapid titration of mumps virus infectivity. *Biken J.* **26:**93–97.

Zimmer, K., D. Wiegand, D. Manz, J. W. Frost, M. Reinacher, and K. Frese. 1990. Evaluation of five different methods for routine diagnosis of rabies. *J. Vet. Med.* **37:**392–400.

Complement Fixation Test*

LILLIAN M. STARK AND ARTHUR L. LEWIS

11

INTRODUCTION

The introduction of new or improved techniques often results in the rejection of older methodologies as inadequate. In the early years of clinical virology, the neutralization test served as the standard against which the specificities and sensitivities of alternative methods were evaluated. However, the refinement of the complement fixation (CF) test and the purification of its reagents subsequently made that test the standard. Presently, enzyme-linked immunosorbent assay (ELISA) and radioimmunoassay predominate because of their increased sensitivity, speed, and ability to be automated. Nevertheless, the CF test is still valuable for antibody studies, as well as virus identification and typing. Automation of the dilution and pipetting steps are possible, but not necessary, and the test does not require expensive equipment, such as a microplate spectrophotometer. The same protocol is used for all CF tests, regardless of which antigen or antibody is assayed, allowing convenient screening against a large battery of agents simultaneously.

Test Principles

The CF test comprises two antigen-antibody reactions. The first is the specific reaction between either a known antigen and unknown (patient) serum or an unknown antigen and specific antiserum. This occurs in the presence of a predetermined amount of complement. The specific antigen-antibody reaction binds (fixes) the complement, preventing it from reacting in the second stage of the test. If the antigen and antibody used in the test do not react with each other, the complement is not fixed and thus is free to react in the next step.

A hemolytic system is used to detect a lack of CF. The second antigen-antibody reaction occurs when sheep red blood cells (SRBC) are reacted with rabbit antibody to SRBC (hemolysin). This sensitizes the cells, causing them to lyse in the presence of free complement, but not when the complement has been previously fixed by a specific antigen-antibody reaction. Thus, hemolysis occurs when the test antigen and serum lacking antibody do not react specifically with each other (negative test result), and conversely, a lack of hemolysis indicates an antigen-antibody reaction (positive test result). The reactions are diagrammed in Fig. 1.

Background

The CF test was introduced in 1909 by Wasserman et al. for syphilis serology. Thereafter, the test was applied in veterinary studies: vaccinia in calves, foot and mouth disease, psittacosis, and the arthropod-borne encephalitides (Ciuca, 1929; Rice, 1948a, 1948b; Bankowski et al., 1953). There were, however, difficulties in adapting the test to routine diagnosis of human diseases of viral etiology.

Early work used crude antigens of low titer, containing extraneous material, which were nonspecific and anticomplementary (Howitt, 1937; Sosa-Martinez and Lennette, 1955). Antigens prepared from infected mouse brain contained interfering substances which could be removed by high-speed centrifugation (Kidd and Friedewald, 1942; Havens et al., 1943; Lennette et al., 1956a). Seitz filtration removed these substances but lowered the virus titer (Casals and Palacios, 1941), whereas repeated freeze-thaw cycles prior to centrifugation increased specificity without titer loss. A cephalinlike substance in tissue extracts was responsible for some nonspecific reactions with normal serum (Maltaner, 1946).

Lipid extraction was used for the preparation of CF antigens for the equine arthropod-borne encephalitides, lymphocytic choriomeningitis (LCM), and St. Louis encephalitis viruses (Howitt, 1937; De Boer and Cox, 1947; Espana and Hammon, 1948; Casals and Palacios, 1949). Virus antigens were prepared from the brains of 1-week-old mice which had been inoculated intracerebrally when less than 3 days old, using a sucrose-acetone extraction method (Clark and Casals, 1958). This procedure provided an antigen suitable for both CF and hemagglutination (HA) testing (Hammon and Sather, 1969). A lymphocytic choriomeningitis antigen was prepared from guinea pig spleen (Smadel et al., 1939). The CF test has been used extensively in arbovirology; studies indicate the California encephalitis group of arthropod-borne viruses to be large and complex. Cross-antigenic comparisons among isolates by CF provided data indicating differences which were not apparent by other examinations, such as HA inhibition (HI) and neutralization tests. CF findings indicated the California encephalitis group was composed of 13 serotypes, 11 present in the Western hemisphere and 2 present in the Eastern hemisphere (Sather and Hammon, 1967).

*Reprinted from the second edition.

Step 1. Primary test reaction

ANTIGEN-X + ANTISERUM-A + COMPLEMENT → ANTIGEN-X + ANTISERUM-A
+ COMPLEMENT

ANTIGEN-X + ANTISERUM-X + COMPLEMENT → (ANTIGEN-X − ANTIBODY-X
− COMPLEMENT)

Step 2. Detection of primary test reaction

[S-RBC = sheep red blood cells sensitized with hemolysin]

ANTIGEN-X + ANTISERUM-A + COMPLEMENT + S-RBC → HEMOLYSIS OF RBC

(ANTIGEN-X − ANTIBODY-X − COMPLEMENT) + S-RBC → NO HEMOLYSIS
(RBC BUTTON)

FIGURE 1 Test principles.

Virus propagation in embryonated chicken eggs (Woodruff and Goodpasture, 1931; Goodpasture et al., 1931, 1933; Stevenson and Butler, 1933) allowed for the production of high-titer antigens for various orthomyxoviruses and paramyxoviruses (Beveridge and Burnet, 1946; French, 1952; Whitney et al., 1953; Sosa-Martinez and Lennette, 1955; Lennette et al., 1956a, 1956b).

Virus propagation in cell cultures (Enders et al., 1949; Robbins et al., 1950, 1951; Dulbecco, 1952; Rowe et al., 1953; Enders and Peebles, 1954; Hilleman and Werner, 1954; Huebner et al., 1954; Henle et al., 1955) allowed the utilization of cell culture fluids as a source of CF antigens. Problems of low antigen titer (Ruckle and Rogers, 1957; Schmidt, 1957; Girardi et al., 1958; Weller and Witton, 1958; Taylor-Robinson and Downie, 1959) and nonspecific and anticomplementary reactions (Svedmyr et al., 1952; Black and Melnick, 1954) occurred. Attempts to concentrate the antigens or remove nonspecific inhibitors by heat were unsuccessful (Svedmyr et al., 1953; LeBouvier et al., 1954; Black and Melnick, 1955; Schmidt and Lennette, 1956). Other attempts to improve antigen preparations included reducing the volume of culture medium (Schmidt et al., 1957), use of a roller bottle (Churcher et al., 1959) or suspension cultures (Westwood et al., 1960; Suggs et al., 1961; Halonen et al., 1967), antigen purification on sucrose density gradients (Julkunen et al., 1984), and varying the composition of the culture medium and the multiplicity of infection (Schmidt, 1969).

In addition to identification of unknown agents in infected host systems, such as cell cultures or mouse brain, the CF test has been used to detect rotavirus antigens directly in stool specimens. Anticomplementary activity present in some specimens was removed by absorption of the clarified stool suspension supernatant with calf serum (Pauri et al., 1981).

Some nonspecific reactions in the CF test were also attributed to the presence of natural antibody in normal serum, which could be removed by heating the serum for 30 min at 60 to 65°C (Kidd and Friedewald, 1942). Nonspecific complement fixation also occurred with sera that were positive in the Wasserman test (Lichter, 1953), as well as when autoantibodies were present (Thorn et al., 1988).

The early CF tests were performed in test tubes, required large quantities of reagents, and used various protocols. The Communicable Disease Center (now the Centers for Disease Control and Prevention, U.S. Department of Health and Human Services, Atlanta, Ga.) developed a standardized CF test in the late 1950s, which with the development of microtechniques (Takatsy, 1950; Sever, 1962) became the laboratory branch CF test (Casey, 1965). Revisions to the protocol were made in 1974 and 1981. Because it requires minimal quantities of reagents, microtiter CF has become the standard methodology for CF.

New variations on the basic CF test protocol have included an ELISA-CF, where peroxidase-labeled complement rather than the hemolytic system was used to detect CF antibodies (Taguchi, 1988). Rapid-rate kinetic turbidometric assays (Fulton and Dininno, 1985) and complement-mediated neutralization tests (Hishiyama et al., 1988), as well as a single radial CF test in agarose plates (Sato et al., 1983, 1988; Ochiai et al., 1987), have also been developed. A monoclonal antibody to substitute for the polyclonal hemolysin has been investigated (Ossewaarde and deBooij, 1989).

Comparison with Other Methods

CF has been compared with various immunoassay procedures to detect antibodies to cytomegalovirus (CMV). The specific clinical population tested influenced the correlation of positive results between test types (Ravaoarino et al., 1984). Although the CF test generally gave lower titers than the ELISA, both tests detected similar antibody prevalence rates in the group of healthy blood donors tested (Dzierzanowska et al., 1986). Miller et al. (1989) compared six serological techniques, including CF, with virus isolation in transplant patients and concluded that no single serological test could reliably and rapidly detect primary CMV infection. They suggested that a combination of different methods be used, e.g., CF and ELISA procedures, which do not necessarily test for the same category of specific antibodies. Thorn et al. (1988) indicated that preferential elevation of CF antibodies against CMV relative to the CMV immunoglobulins detected with ELISA may be reflective of the pathogenic process in Sjögren's syndrome.

ELISA and CF, and HI tests were compared for measles antibody determination (Ferrante et al., 1987). The range of titers (sensitivity) was greater with ELISA than with CF; both were considered superior to HI. For mycoplasma infections, good correlations were seen between CF and ELISA (Fischer et al., 1986; Wreghitt and Sillis, 1987). Inouye et al. (1981) compared CF with immune adherence HA and felt the ratio of antibody titer detected in each test was of value in diagnosing recent infection for rotavirus and Japanese encephalitis. In a study monitoring seroconversion after administration of a rotavirus vaccine, CF was not as sensitive as the immunoassay used when the entire population was evaluated but was better in the older than in the younger recipients (Midthun et al., 1989).

Antigens produced from influenza virus may be either the 5, soluble (internal nucleoprotein), or V, virion (hemagglutinin or neuraminidase), type. The S antigen is type specific for influenza A, B, or C; the V antigen is strain specific and stimulates the formation of protective antibody (Hoyle, 1952; Lief and Henle, 1956a, 1956b). The antibody response to the S antigen is a rapid development of immunoglobulin, which is more readily measured with the CF than with the HI test (Duca et al., 1979). V antigen neuraminidases have also been distinguished by using monoclonal antibodies in a CF test (Holmes et al., 1982). ELISA was shown to be more sensitive than CF (Julkunen et al., 1984), but the authors also state that during nonepidemic periods CF might be preferable because it allows routine simultaneous testing against several viral antigens.

The ability to screen against a battery of antigens (or antisera) with a single test is one of the major advantages of the CF procedure. It has been used that way in studies to test for possible etiologic agents of a specific syndrome (Hudson et al., 1981) as well as in retrospect to determine clinical conditions associated with infection (Puolakkainen et al., 1987). In our laboratory, sera are tested routinely against a battery of respiratory or central nervous system agents using CF to evaluate the association between potential etiologic agents and the clinical picture.

GENERAL CONSIDERATIONS

Technique

The CF test requires a capable, well-trained technologist for accurate results. Since it is performed by a microtiter procedure, minute amounts of reagents are used and precision in delivery is crucial. Calibrated micropipettes with disposble tips are beneficial. When dropper pipettes and loops are used for delivery and dilution, they must be properly cleaned and calibrated and used carefully to ensure reliability. If large numbers of specimens are to be examined, automatic microdilution and reagent delivery systems are of value.

All reagents must be very carefully prepared according to established laboratory protocol; the test is very sensitive to errors in reagent composition and concentration. All reagents should be properly stored according to the supplier's directions. Multiple freeze-thaw cycles are to be avoided; thus, it is best to aliquot sensitive reagents in small volumes for storage. New lots of hemolysin, antisera, and antigens must be titrated before they are used in routine testing. The SRBC should be handled gently and should not be used if there is any hemolysis during washing or storage.

Specimens

Good-quality specimens are essential for a meaningful test. Serum should be collected in a sterile tube without preservatives or anticoagulants. Hemolysis must be avoided. Sufficient blood must be drawn to provide at least 1 to 2 ml of serum. If any red blood cells remain after the serum is decanted from the clot, they should be removed by centrifugation prior to freezing the specimen. As bacterial contamination of the serum may produce erroneous test results, the specimen should be handled aseptically.

Sera may sometimes react with complement in the test. This "anticomplementary" behavior may be present in the serum when collected or may develop during storage. Heat inactivation of the serum (by incubating the serum aliquot in a 56°C water bath for 30 min) may alleviate this; otherwise, the serum should be treated with complement to remove anticomplement components. If this is not effective, another serum specimen should be collected. Unfortunately, the sera of some individuals may persistently contain anticomplementary factors. These factors are usually present at lower serum dilutions, so a high specific titer may still be recognizable in the specimen.

The CF test may be used to identify an unknown antigen, such as a virus isolate made in cell culture. The culture should be rapidly frozen (in a dry ice-alcohol bath) and thawed three times and clarified by centrifugation. An uninoculated culture should be similarly processed for use as the tissue control in the test. If nonspecific reactions occur, the antigen may be clarified by extraction with an equal volume of Freon. The test antiserum may be absorbed with noninfected (control) tissue to eliminate nonspecific reactions.

Reagents

Complement from different animal sources is commercially available; however, it differs in sensitivity in the CF test (Sethi et al., 1981). Most frequently, guinea pig complement is used at five 50% hemolytic complement units (CH_{50}) (laboratory branch CF test). Some test protocols call for complement to be used at two full hemolytic units, in which case it should be titrated in the presence of each test antigen, as antigens vary in the ability to bind complement (Lennette, 1969).

Antigens may be whole virions (V antigens) or produced by the fragmentation of the virus (S antigens). The latter are a mixture, including nucleoprotein (Craigie and Wishart, 1936a, 1936b; Hoyle, 1952; Ada and Perry, 1954; Schafer and Zillig, 1954; Paucker et al., 1956; Westwood et al., 1965; Cohen and Wilcox, 1966). S antigens have been demonstrated for a number of viruses, and they vary in stability and specificity (Chambers et al., 1950; Black and Melnick, 1955; Ende et al., 1957). In mumps (Henle et al., 1948) and influenza (Duca et al., 1979) infections, anti-S antibodies rise more rapidly than anti-V antibodies and can thus be used to distinguish recent from previous infections.

Reagents for the CF test can be prepared in-house, but this may lead to difficulties in quality control and test comparisons. Commercial suppliers known to us at the time of writing are listed in Table 1.

Test Interpretation

For a test to be considered valid, all controls must be within the limits specified in the test procedure. These include sensitivity of the red blood cells to lysis, concentration of complement used and its reaction with the red blood cells, sensitivity of the antiserum-antigen reaction, and absence of reaction with the tissue (antigen) or serum control. If the controls fail, corrective action and a repeat test must be performed before the results are reported.

CF serology can only identify an agent responsible for a recent infection when both acute and convalescent sera are tested simultaneously. The acute specimen should be collected as close to the onset of illness as possible. The convalescent specimen should be collected 2 to 3 weeks later. In immunologically impaired patients or in the presence of an

TABLE 1 Commercial suppliers of reagents for CF test[a]

Supplier and contact information	Antigens	Antisera	Complement	Hemolysin	RBC
Accurate Chemical & Scientific Corp. 300 Shames Dr. Westbury, NY 11590 (516) 333-2221; (800) 645-6264	X	X	X	X	
Diamedix 2140 N. Miami Ave. Miami, FL 33217 (305) 324-2300; (800) 327-4565			X	X	X
Hillcrest Biologicals 10703 Progress Way Cypress, CA 90630-4714 (213) 420-2657; (800) 445-0185	X				
Microbix 341 Bering Ave. Toronto, Ontario M8Z 3A8 Canada (416) 234-1624	X	X	X		
Rockland, Inc. Box 316 Gilbertsville, PA 19525 (215) 369-1008			X	X	X
Virion (U.S.), Inc. 4 Upperfield Rd. Morristown, NJ 07960 (201) 993-8219; (800) 524-2689	X	X	X	X	
Bio Whittaker Inc. 8830 Biggs Ford Rd. Walkersville, MD 21793-0127 (301) 898-7025; (800) 638-3976	X	X	X	X	X

[a] Reagents available from each supplier are indicated by X. RBC, red blood cells.

agent sensitive to the presence of administered antibiotics, where antibody production may be delayed, a third serum collected 4 to 6 weeks after onset may be of value. It is generally accepted that a fourfold rise in titer to a specific agent is indicative of recent infection. Nevertheless, the entire patient history must be considered, since heterologous anamnestic reactions may occur. In such cases, the individual experienced a previous infection or immunization, and a later infection with another virus spurred a sharp, rapid rise in immunoglobulin G antibodies to the first agent. This is not uncommon with the group A and B arboviruses or paramyxoviruses. In addition, stable elevated titers to multiple agents may indicate a generalized stimulation of immunological activity unrelated to those agents.

TEST SETUP

Equipment
Freezer, $-20°C$ ($-70°C$ freezer preferable)
Refrigerator, 4°C
Incubator, 37°C
Water baths, 37 and 56°C
Calibrated centrifuge
Centrifuge microtiter plate carriers
pH meter
Balance
Liquid aspiration device
Microtiter plate vibrator
Timer
Test tube racks
Test tubes (13 by 100 and 15 by 125 mm)
Microtiter plates (U style; 96-well format)
Diluter (automatic, calibrated loops, or pipettor type)
Microtiter dropper pipettes (0.025 and 0.05 ml)
Serological pipettes (1, 2, 5, and 10 ml)
Pipettors and disposable tips (0.025 and 0.05 ml)
Glassware (Erlenmeyer and volumetric flasks, graduated cylinders, beakers, and centrifuge tubes)
Mirror reading device (optional)
Reagents
pH buffer standards
Hydrochloric acid (1 N)
Stock 1 M $MgCl_2$–0.3 M $CaCl_2$ solution
Veronal-buffered diluent (VBD) (stock)
Gelatin
SRBC (2.8% suspension)

Hemolysin

Complement

Antigens

Antigen controls (tissue controls)

Sera (known positive antibody status)

Sera (known negative antibody status)

Sera (test [unknown])

Reagent Preparation

Label all reagents with name, concentration, date prepared, and technician's initials.

■ 1 M MgCl₂–0.3 M CaCl₂ Solution

1. Add 20.3 g of $MgCl_2 \cdot 6H_2O$ and 4.4 g of $CaCl_2 \cdot 2H_2O$ to 70 ml of reagent-grade water in a 100-ml volumetric flask.
2. Mix by swirling.
3. Fill to the 100-ml mark with reagent-grade water.
4. This reagent may be filter sterilized for extended storage. Store in the refrigerator.

■ VBD Stock Solution (5× VBD)

1. Add 83.0 g of NaCl to 1,500 ml of reagent-grade water in a 2-liter volumetric flask.
2. Add 10.19 g of *Na*-5,5-diethyl barbiturate (sodium barbital, or Veronal).
3. Mix by swirling until the chemicals are completely dissolved.
4. Add 34.0 ml of 1 N HCl and mix by swirling.
5. Add 5.0 ml of stock $MgCl_2 \cdot CaCl_2$.
6. Bring the total solution volume up to 2 liters with reagent-grade water.
7. Check the pH of a 1:5 dilution of the buffer (1 ml of 5× VBD plus 4 ml of reagent-grade water); if the pH is below 7.3 or above 7.4, discard the buffer and prepare fresh stock.
8. This reagent may be filter sterilized for extended storage. Store in the refrigerator.

■ Gelatin Water

1. Add 1 g of gelatin to 200 ml of reagent-grade water in a 2-liter flask.
2. Bring the solution to a boil, swirling to dissolve the gelatin.
3. Remove from heat and allow to cool.
4. Add 600 ml of sterile reagent-grade water.
5. Cover tightly.
6. Store in the refrigerator for no longer than 1 week.

■ VBD Working Solution

The solution should be prepared on the day of use.
1. Add 200 ml of stock 5× VBD to 800 ml of gelatin water.
2. Check the pH; if the pH is below 7.3 or above 7.4, discard the buffer and prepare fresh stock 5× VBD.

■ Stock Hemolysin (1:100)

1. Dissolve 0.85 g of NaCl in 100 ml of reagent-grade water; add 5.0 g of phenol and swirl to dissolve.
2. Aliquot 4.0 ml of 5% phenol saline into a 125-ml Erlenmeyer flask.
3. Add 94.0 ml of cold VBD to the flask and mix by swirling.
4. Add 2.0 ml of glycerinized hemolysin to the flask and mix by swirling.
5. Store at 4°C.

Test Outline

The steps required for the performance of a CF test, according to the protocols used our laboratory, are as follows:
1. Prepare the reagents. The gelatin water, VBD, color standards, sensitized SRBC, diluted complement, antisera, and antigen must be freshly prepared for each test.
2. Wash and standardize the SRBC (see "Preparation of Erythrocytes").
3. If a new lot of hemolysin is to be used, perform a hemolysin titration (see the appropriate section).
4. Prepare color standards (see the appropriate section).
5. Perform a complement titration (see the appropriate section). Complement must be titrated each time an antigen titration or diagnostic test is performed; the minimal acceptable hemolytic titer is 250 CH_{50}/ml.
6. If a new lot of antigen is to be used, perform an antigen titration (see the appropriate section). This procedure may also be used for the identification of an unknown antigen by testing against antisera to suspected agents.
7. Perform the diagnostic serology test (see the appropriate section).

TEST PERFORMANCE

Preparation of Sheep Erythrocytes

■ Cell Washing

1. Place 10 ml of the SRBC suspension in a 250-ml centrifuge bottle.
2. Add approximately 100-ml of cold VBD and mix gently.
3. Centrifuge the mixture at 600 × g at 4°C for 10 min.
4. Aspirate the supernatant.
5. Gently resuspend the cells in 100 ml of fresh cold VBD; centrifuge and aspirate the suspension as described above.
6. Repeat this procedure for a total of three washes.
7. Resuspend the cells in approximately 20 ml of cold VBD; transfer to a 40-ml conical, graduated centrifuge tube and fill it to the 40-ml mark with VBD.
8. Centrifuge the suspension at 600 × g at 4°C for 10 min.
9. Note the volume of the packed cells, and aspirate the supernatant without disturbing the cells. If the supernatant is colorless, proceed with cell standardization. If hemolysis is observed, the cells are not suitable for use, and a new lot should be obtained.

■ Cell Standardization Centrifugation Method

1. To calculate the volume of VBD required for a 2.8% cell suspension, multiply the volume of packed cells obtained after the final wash centrifugation (see the appropriate section) by 34.7.
2. Suspend the cells in the calculated volume of VBD in an appropriate-size Erlenmeyer flask by gentle swirling.
3. To check the accuracy of the dilution, pipette 7.0 ml of the suspension into a 10-ml graduated centrifuge tube having an accuracy of ±0.025 ml in the 0- to 1-ml range.
4. Centrifuge the suspension at 600 × g at 4°C for 10 min.
5. If the volume of the packed cells is 0.2 ml, the suspension is accurate, and the 7-ml sample may be resuspended and returned to the original flask for use in the test.
6. If the volume of the packed cells is not 0.2 ml, calculate the correction factor, which is the actual volume of packed cells read from the centrifuge tube divided by 0.2. This factor is multiplied by the original volume of the cell suspension minus 7.0 ml to determine the cor-

rected cell suspension volume. Compute the difference between the corrected volume and the existing volume; this is the volume of VBD that is used to adjust the cell suspension (steps 7 and 8).

7. If the corrected volume is less than the existing volume, aliquot a portion of the cell suspension from the flask sufficient to include the volume of VBD to be removed to a centrifuge tube. Centrifuge the suspension at 600 × g at 4°C for 5 min. Pipette and discard the calculated excess VBD from the supernatant. Resuspend the cells, returning them to the flask. Recheck the dilution accuracy (step 3).

8. If the corrected volume is greater than the existing volume, add the required amount of VBD to the flask, resuspend the cells, and recheck the dilution accuracy (step 3).

Hemolysin Titration

■ Preparation of a 1:1,000 Hemolysin Dilution
1. Place 9.0 ml of cold VBD in a 15- by 125-mm test tube labeled 1:1,000 hemolysin.
2. Add 1.0 ml of the 1:100 stock hemolysin and mix well with the pipette.

■ Preparation of Additional Hemolysin Dilutions
1. See Table 2 for the dilution protocol.
2. Label six 15- by 125-mm test tubes with the dilutions listed.
3. Using a 5.0-ml pipette, place the designated volumes of VBD into the appropriately labeled tubes.
4. Using a 5.0-ml pipette, add 1.0 ml of the 1:1,000 hemolysin dilution to each tube.

■ Preparation of a 1:400 Dilution of Complement
1. Place undiluted complement in an ice bath.
2. Measure 100 ml of cold VBD in a 100-ml graduated cylinder. Use a 1-ml pipette to remove 0.25 ml, and transfer the remainder to a 125-ml Erlenmeyer flask.
3. Draw up the undiluted complement in a 1.0-ml pipette to the 0.6-ml mark; wipe the pipette tip. Holding the pipette vertically, deliver 0.25 ml of complement dropwise into the VBD in the flask.
4. Mix by swirling it gently.
5. Place the 1:400 dilution of complement at 4°C for at least 20 min before using it; do not use it after 2 h.

■ Preparation of Hemolysin-Sensitized Cells
1. Place seven 13- by 100-mm test tubes in a rack.
2. Label the first 1:1,000; label the remaining six tubes with the dilutions listed in Table 2.
3. Add 1.0 ml of the standardized 2.8% SRBC to each of the seven tubes.
4. Thoroughly mix the 1:1,000 hemolysin dilution (see the appropriate section), and add 1.0 ml slowly, with constant swirling, to the tube labeled 1:1,000.
5. To each of the labeled tubes, add 1.0 ml of the appropriate hemolysin dilution (see the appropriate section) in a like manner.
6. Shake the rack and place it in a 37°C water bath for 15 min.

■ Hemolysin Titration
1. Place seven 12- by 75-mm test tubes in a rack. Label the first 1:1,000; label the remaining six tubes with the dilutions listed in Table 2.
2. Add 0.4 ml of VBD to each tube.

3. Add 0.4-ml of the 1:400 complement dilution (see the appropriate section) to each tube; shake the rack.
4. Add 0.2 ml of the SRBC sensitized with each of the hemolysin dilutions (see the appropriate section) to the appropriately labeled tube containing VBD and complement; mix each tube by shaking.
5. Incubate the tubes in a 37°C water bath for 1 h; shake the rack after 30 min.
6. Prepare color standards (see "Color Standards" below).
7. Centrifuge the tubes at 600 × g for 5 min.
8. Compare each tube with the prepared color standards. If the tube matches a color standard, record the percent hemolysis; if not, interpolate to the nearest 5% and record.
9. Plot on arithmetic (linear) graph paper the percent hemolysis obtained at each hemolysin dilution (Fig. 2). Draw a line through the plotted points and determine the plateau region. This is the area where additional hemolysin produces no marked increase in hemolysis.
10. The second dilution on the plateau is the optimal dilution for the tested lot of hemolysin to be used for cell sensitization (in Fig. 2, this is a 1:2,000 dilution). The minimal acceptable titer (dilution) is ≥1:2,000.

Note: commercial complement diluted 1:400 will generally yield 30 to 80% hemolysis for cells optimally sensitized as described above. To obtain the correct percent hemolysis with a less active complement sample, it may be necessary to use a 1:300 dilution. With a very potent complement preparation, a 1:500 dilution may be needed.

Color Standards

■ Preparation of Hemoglobin Solution
1. Pipette 1.0 ml of the thoroughly mixed 2.8% SRBC suspension into a 15- by 125-mm test tube.
2. Add 7.0 ml of distilled water, and shake the tube until all the cells are lysed.
3. Add 2.0 ml of stock buffer (5× VBD) to the tube, and mix well.

■ Preparation of 0.28% Red Blood Cell Suspension
1. Pipette 1.0 ml of the thoroughly mixed 2.8% SRBC suspension into a 15- by 125-mm test tube.
2. Add 9.0 ml of cold VBD and mix well.

■ Preparation of Color Standards
1. Label 11 12- by 75-mm test tubes with the percent hemolysis given in Table 3 (0 to 100), and place in a rack in ascending order.
2. Using a 2.0-ml pipette, deliver the appropriate volume of hemoglobin solution listed in Table 3 to each tube.

TABLE 2 Preparation of additional hemolysin dilutions

Final dilution	VBD (ml)	1:1,000 hemolysin (ml)
1:1,500	0.5	1.0
1:2,000	1.0	1.0
1:2,500	1.5	1.0
1:3,000	2.0	1.0
1:4,000	3.0	1.0
1:8,000	7.0	1.0

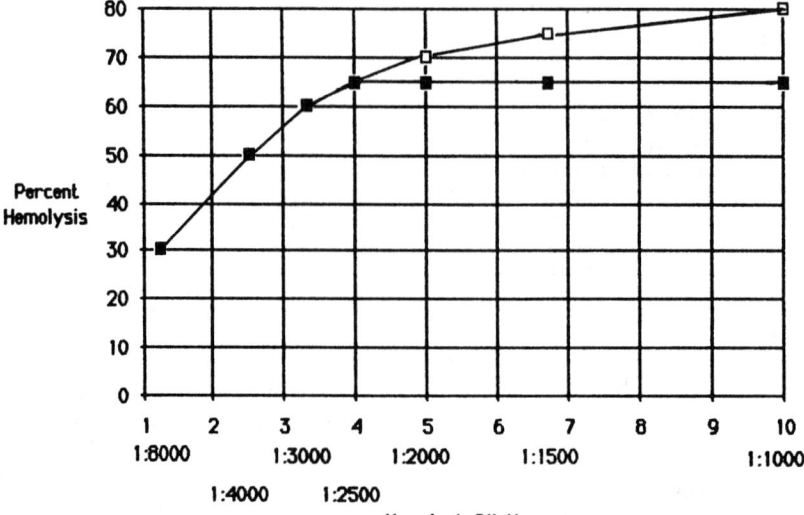

FIGURE 2 Determination of the optimal hemolysin dilution. The location for each of the hemolysin dilutions on the *x* axis is determined by dividing the lowest dilution by the next higher dilution and multiplying by the number of blocks allotted on the *x* axis. For example, 1,000/1,500 × 10 blocks = 6.7 (hemolysin dilution, 1:1,500); 1,000/2,000 × 10 blocks = 5.0 (hemolysin dilution, 1:2,000); 1,000/2,500 × 10 blocks = 4.0 (hemolysin dilution, 1:2,500). The solid squares indicate acceptable titrations, and the open squares indicate unacceptable titrations. The optimal dilution in this titration is 1:2,000 (i.e., the second dilution on the plateau).

TABLE 3 Preparation of color standards

Reagent	Vol of reagent for % hemolysis of:										
	0	10	20	30	40	50	60	70	80	90	100
Hemoglobin solution	0.0	0.1	0.2	0.3	0.4	0.5	0.6	0.7	0.8	0.9	1.0
0.28% Cell suspension	1.0	0.9	0.8	0.7	0.6	0.5	0.4	0.3	0.2	0.1	0.0

3. Using a 2.0-ml pipette, deliver the appropriate volume of 0.28% SRBC suspension listed in Table 3 to each tube.
4. Shake the rack vigorously to mix the contents of the tubes well.
5. Centrifuge the tubes at 600 × *g* for 5 min at ambient temperature.
6. Remove the tubes from the centrifuge without agitation, and store them in a rack in the refrigerator until they are needed.

Complement Titration
■ Preparation of Sensitized Cells
1. Add 8.0 ml of the standardized 2.8% SRBC suspension to a 125-ml Erlenmeyer flask.
2. Prepare 10 ml of the optimal hemolysin dilution from the 1:100 stock hemolysin solution, and add 8.0 ml of this dilution to the SRBC with rapid stirring.
3. Incubate the mixture for 15 min in a 37°C water bath, mixing it at 7 min.

■ Preparation of a 1:400 dilution of complement
1. See the appropriate section above.
2. A less potent complement may be tested at a 1:300 dilution. Use exactly 99.67 ml of VBD, adding 0.33 ml of undiluted complement.

■ Complement Titration
1. Label two sets of five 12- by 75-mm test tubes with numbers 0 through 4, and place them in a rack; the titration is performed in duplicate.
2. Add VBD to each tube in the amounts given in Table 4.
3. Add the 1:400-diluted complement to each tube in the amounts given in Table 4.
4. Shake the rack to mix the reagents.
5. Add 0.2 ml of sensitized SRBC to each tube.
6. Shake the rack, and place it in a 37°C water bath; after 15 min of incubation, shake the rack; continue incubation an additional 15 min.

TABLE 4 Complement titration

Tube no.	Amt of reagent (ml)		
	VBD	Complement (1:400)	Sensitized RBC[a]
0 (cell control)	0.80	0	0.20
1	0.60	0.20	0.20
2	0.55	0.25	0.20
3	0.50	0.30	0.20
4	0.40	0.40	0.20

[a] RBC, red blood cells.

7. Prepare color standards (see the appropriate section above) while the titration tubes are incubating.
8. Remove the rack from the water bath after the total 30-min incubation; centrifuge the titration tubes and color standard tubes at 600 × g for 5 min.
9. Compare each tube with the color standards; record the percent hemolysis, interpolating to the nearest 5% if an exact match does not occur.
10. Calculate the average percent hemolysis for each pair of duplicate tubes.
11. If lysis is greater than 90% or less than 10%, repeat the procedure using complement that has been diluted more or less than 1:400, respectively.

■ Computation of the Complement Volume Producing 50% Hemolysis (CH$_{50}$)

1. Use Table 5 to determine the ratio ($x/100 - x$) for the average percent hemolysis calculated for each pair of tubes in "complement titration," step 10.
2. For each of the four tubes containing complement, plot on log graph paper the ratio value ($x/100 - x$) on the x axis versus the volume of 1:400 complement in ml on the y axis (Fig. 3).
3. Examine the graph to see whether two of the plotted points are on the left and two are on the right of the vertical 1.0 line. If so, proceed with the next step; otherwise, the complement titration must be repeated. If more than two points are to the right of the line, too much complement was used; repeat the process using a more dilute complement (1:500). If more than two points occur to the left of the line, there was insufficient complement in the test; repeat the process using a less dilute complement (1:300).
4. Draw a line between the two plotted points for tubes 1 and 2, and find its midpoint.
5. Draw a line between the two plotted points for tubes 3 and 4, and find its midpoint.
6. Draw a line between the two midpoints, and determine the slope. From any point near the left end of the line, measure horizontally to a point 10 cm to the right; measure the vertical distance, in centimeters, from that point upward to the midpoint line, and divide the vertical distance by 10 cm to obtain the slope.
7. Test reproducibility requires that the slope be between 0.18 and 0.22; if the computed slope is not within this range, the complement titration must be repeated.
8. From the intersection of the vertical 1.0 line with the line joining the two midpoints, draw a dotted horizontal line to the y axis. Read the volume in milliliters of the 1:400 complement dilution at the point where the dotted line intersects the y axis. This volume contains 1 CH$_{50}$; this volume is multiplied by 5 to obtain the volume of 5 CH$_{50}$.

TABLE 5 Percent ratio

x	$x/100 - x$
10	0.111
15	0.176
20	0.250
25	0.330
30	0.430
35	0.540
40	0.67
45	0.82
50	1.00
55	1.22
60	1.50
65	1.86
70	2.33
75	3.00
80	4.00
85	5.70
90	9.00

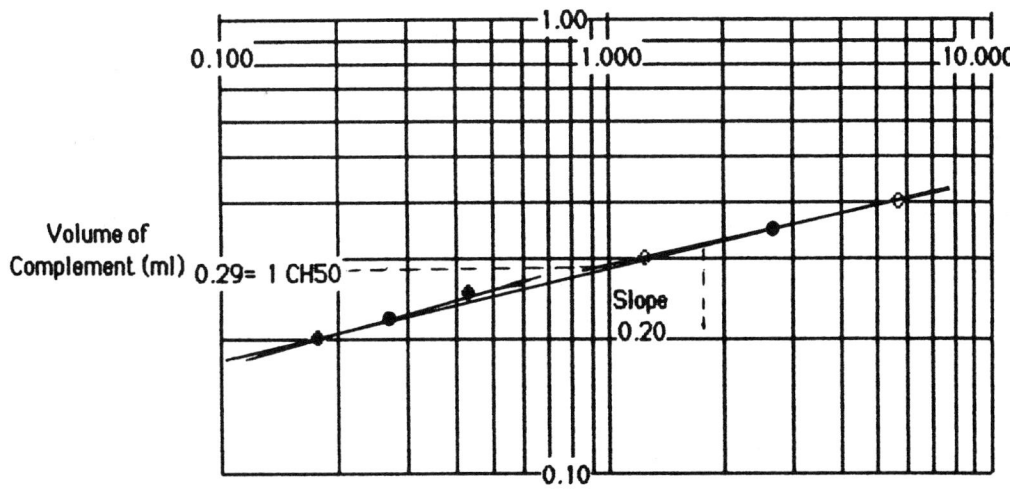

FIGURE 3 Determination of the optimal volume of complement.

9. The test requires 5 CH$_{50}$ in 0.4 ml. The dilution of complement which will provide this (x^*) is calculated by the following equation:

$$\frac{\text{Dilution of complement used in titration}}{\text{Volume containing 5 CH}_{50}} = \frac{x^*}{0.4}$$

For example, if the volume of 1:400 complement dilution containing 5 CH$_{50}$ is 1.3 ml, then the dilution is given by the following equation: $400/1.3 = x^*/0.4$, and $x^* = 123$. Thus, in this example, complement should be used in the test at a dilution of 1:123.

Antigen Titration

■ **Setup**

1. Label a 96-well microtiter plate as shown in Fig. 4; each square represents a well, and the diagram indicates which reactants will be added to each well.
2. Prepare the 1:8 starting antiserum dilution by adding 0.2 ml of specific antiserum to a 12- by 75-mm test tube; add 1.4 ml of VBD and mix. Inactivate the mixture in a 56°C water bath for 30 min; cool it to room temperature.
3. Label tubes for the known positive antigen at its optimal dilution and for the antigen tissue control.
4. Label 13- by 100-mm test tubes for serial twofold dilutions of the test antigen. Add 1.0 ml of cold VBD to each of the antigen dilution tubes; if the test is to identify an unknown antigen, use the antigen undiluted and at 1:2, 1:4, and 1:8 dilutions. Include a tube of undiluted tissue control for this antigen. If the test is to determine the optimal dilution of a new lot of a previously used known CF antigen, dilute the antigen to include two dilutions above and two below the optimal dilution of the previous lot in the titration. The same tissue control used for the known positive antigen (step 3) may be used.

■ **Addition of Reactants to the Microtiter Plate**

1. Deliver 0.05 ml of heat-inactivated antiserum diluted 1:8 to each of the wells indicated with an "S" in Fig. 4.

ANTISERUM DILUTION

	8	16	32	64	128	256	512	1024		
A-1	S									antigen dilutions
A-2	S									
A-3	S									
A-4	S									
A-5	S									
+A	S									reference antigen
+TAC	S									tissue control
SC	S									serum control
A-1	5u	2.5u	1.25u	V 5u		5u	2.5u	1.25u	A-5	complement controls
A-2	5u	2.5u	1.25u	V 2.5u		5u	2.5u	1.25u	+A	
A-3	5u	2.5u	1.25u	V 1.25u		5u	2.5u	1.25u	+TAC	
A-4	5u	2.5u	1.25u	CC	CC	S 5u	S 2.5u	S 1.25u	SC	
	5	2.5	1.25			5	2.5	1.25		

UNITS OF COMPLEMENT

FIGURE 4 Antigen titration microtiter plate.

2. Using a calibrated dropper pipette, deliver 0.025 ml of VBD to each of the remaining serum dilution wells.
3. Prepare serial twofold serum dilutions (an automatic programmable dilutor, microtiter pipetting devices with disposable tips, or pretested metal loops may be used). Transfer 0.025 ml of serum from the well labeled 8 into the well labeled 16, mix it with the VBD, and continue to transfer and mix across the plate.
4. All reagents should be kept cold, preferably in an ice bath, prior to delivery to the microtiter plate.
5. Prepare the optimal dilution of the known positive antigen with VBD. A tissue antigen control (TAC) is prepared from an uninfected host (e.g., cell culture or mouse brain) of the same type in which the antigen was produced. A tissue control must be used for each type of antigen tested.
6. Prepare serial twofold dilutions of test antigen in the labeled test tubes. Add 1.0 ml of antigen to the tube labeled 1:2 and mix it well with the pipette. Prepare the 1:4 dilution by adding 1.0 ml of the 1:2 dilution to the VBD in the tube labeled 1:4. Mix it well with the pipette, and continue serial transfer through the last required dilution.
7. Using a 0.025-ml dropper pipette, add 0.025 ml of antigen dilution 1 to the eight wells labeled A-1 and to the three wells of the complement controls labeled A-1. Repeat for each of the remaining antigen dilutions. Use row A-5 for the test TAC if it differs from the tissue of the known positive antigen.
8. Using a 0.025-ml dropper pipette, add 0.025 ml of the optimal dilution of the known positive antigen to the eight +A wells and to the three wells of the complement control labeled +A; similarly, add the positive TAC to the eight +TAC wells and the three +TAC complement control wells.
9. Using a 0.025-ml dropper pipette, add 0.025 ml of VBD to the eight serum control wells (SC) and to each of the complement control wells. Add 0.05 ml of VBD to the three VBD control wells (V). Add 0.1 ml of VBD to each cell control well (CC).
10. Shake the plates on an orbital shaker for 20 to 30 s; cover the plates and hold them in the refrigerator while preparing the diluted complement.

■ Preparation of Complement

1. Prepare 5 ml of complement per plate plus an additional 5 ml for the complement controls at the dilution previously determined by complement titration. To calculate the volume of undiluted complement required, multiply the total volume needed by the reciprocal of the dilution determined in the complement titration. For example, if 40 ml of a 1:133 dilution of complement are needed, the volume of undiluted complement necessary is 40/133, or 0.3 ml; therefore, use 0.3 ml of undiluted complement plus 39.7 ml of VBD.
2. Add the calculated volume of cold VBD to a flask. Add the calculated volume of complement dropwise to the VBD, mix it by swirling, and allow the mixture to stand for 20 min on ice or in the refrigerator.
3. Label a 15- by 125-mm test tube "2.5"; add 2.0 ml of VBD and 2.0 ml of the 5-CH_{50} complement dilution, and mix gently. Label a second tube "1.25"; add 3.0 ml of VBD and 1.0 ml of the 5-CH_{50} complement dilution, and mix gently.
4. Using a 0.05-ml dropper pipette, add 0.05 ml of the 5-CR_{50} complement dilution to all wells in the antigen titration and to the wells labeled 5u (u stands for units) in the complement control.
5. Using a 0.05-ml dropper pipette, add 0.05 ml of the 2.5-CH_{50} complement dilution to all wells labeled 2.5 in the complement control; do the same with the 1.25-CH_{50} dilution.
6. Shake the plates on an orbital mixer for 20 to 30 s; cover the plates and incubate them at 4°C for 15 to 18 h (refrigerate overnight).

■ Preparation of Hemolysin-Sensitized SRBC

1. Determine the volume of sensitized SRBC needed for the test by multiplying the total number of plates plus one times 3 ml per plate.
2. Prepare a volume of optimally diluted hemolysin equal to one-half of the total volume calculated in step 1 plus about 10%.
3. Swirl the 2.8% standardized SRBC suspension to achieve a homogeneous suspension. Into a small flask, aliquot a volume of suspended SRBC equal to one-half the total volume calculated in step 1.
4. Place the flask of SRBC on a rotating platform, and while the flask is rotating, use a 10-ml pipette to add a volume of the diluted hemolysin equal to the volume of the SRBC suspension and mix it well.
5. Incubate the cells in a 37°C water bath for 15 min, swirling it to mix it after 7 min.
6. Meanwhile, remove the prepared test plates from the refrigerator and allow them to warm to room temperature.
7. Using a 0.025-ml dropper pipette, add 0.025 ml of sensitized SRBC to each well on the plate. Tap the plate edge to mix the contents, and seal the plate with transparent sealing tape. Vibrate the plate for 20 s on an orbital shaker.
8. Place the plates on a level shelf in a 37°C incubator for 30 min; do not stack the plates. Tap the plate edges at 15 min to resuspend the SRBC.

■ Reading the Test

1. If color standards less than 24 h old are available, they may be used; otherwise, prepare fresh color standards (see the appropriate section).
2. Centrifuge the plates at 300 × g for 5 min.
3. Read and record the percent hemolysis of the complement controls using the color standards; interpolate to the nearest 5%.
4. Compare the readings with those in Table 6 to determine if they are acceptable. Any values less than those shown are considered to be anticomplementary; therefore, the test is not acceptable and should be repeated.

TABLE 6 Acceptable percent hemolysis in control test reactions

Type of control	% Hemolysis		
	5 CH_{50}	2.5 CH_{50}	1.25 CH_{50}
Antigen	100	85–100	0–75
VBD	100	90–100	40–75
Serum	100	90–100	0–75
Tissue	100	85–100	0–75

5. Read and record the percent hemolysis for each titration well; an example is given in Table 7.
6. Draw a line through the approximate 30% hemolysis endpoints of each antigen dilution, interpolating where necessary; do not include dilutions showing anticomplementary activity.
7. The optimal antigen dilution is the one at which the greatest amount of complement is fixed, based on the following criteria.
   a. All dilutions showing anticomplementary activity plus the next highest dilution must be excluded from the curve.
   b. If two antigen dilutions give identical reactions within the optimal dilution curve, select the one which gives greater fixation to the right of the curve.
   c. If two antigen dilutions give identical fixation reactions, select the lower dilution.
   d. The titer of the reference antiserum with the test antigen must be within one twofold dilution of the titer obtained with the reference antigen.
8. The antigen titration procedure may be used to identify an unknown antigen by testing against known antisera; the degree of fixation by the unknown antigen and specific antiserum is compared with that of the reference antigen with its specific antiserum.

Diagnostic Serology Test

■ Setup

1. Each serum specimen may be tested against any number of antigens; however, for each antigen both known positive and known negative antisera must be run concurrently in the test.
2. Label one 12- by 75-mm test tube for each unknown, known positive, and known negative serum to be tested.
3. Aliquot 0.2 ml of each serum into the appropriate tube; to each tube, add 1.4 ml of cold VBD, resulting in a 1:8 starting serum dilution.
4. Inactivate the sera by placing the tubes in a 56°C water bath for 30 min.
5. Label microtiter plates; an example is given in Fig. 5. If numerous sera are to be run, it is advantageous to arrange the plates so the unknown sera are batched for each antigen tested, all the test serum controls are on

one plate, the antigen controls (known positive and negative sera) are on another plate, and the complement controls are on a third plate; the exact arrangement will be dictated in part by the equipment available to dilute the reagents and samples.
6. Prepare the test antigen at its previously determined optimal dilution. To determine the total volume required, multiply the number of wells receiving the test antigen by 0.025 ml and add 0.5 to 1 ml to allow excess for some loss of sample during pipetting. Prepare the tissue control for each antigen at the same dilution as the test antigen; compute the total volume needed in the same manner. Store all antigens at 4°C until they are needed.

■ Addition of Reactants to the Microtiter Plate

1. Deliver 0.05 ml of each cooled, inactivated serum to the appropriately labeled wells (1:8) on the test and control plates.
2. Using a calibrated dropper pipette, deliver 0.025 ml of VBD to each of the remaining serum dilution wells.
3. Prepare serial twofold serum dilutions; an automatic programmable dilutor, microtiter pipetting devices with disposable tips, or pretested metal loops may be used. Transfer 0.025 ml of serum from the well labeled 8 into the well labeled 16, mix it with the VBD, and continue to transfer and mix across the plate.
4. Using a calibrated dropper pipette, deliver 0.025 ml of the optimal dilution of test antigen to each of the wells in the antigen portion of the test; add antigen to the positive and negative control serum and the complement control wells.
5. Using a calibrated dropper pipette, deliver 0.025 ml of the diluted tissue control to each of the wells in the tissue control portion of the test plate; also add this volume to the positive and negative control serum and the complement control wells.
6. Using a calibrated dropper pipette, deliver 0.025 ml of VBD to the complement control wells, 0.05 ml to the VBD control wells, and 0.1 ml to the cell control wells.
7. Shake the plates on an orbital shaker for 20 to 30 s; cover the plates and hold them in the refrigerator while preparing the diluted complement.

TABLE 7 Determination of optimal antigen dilution[a]

Antigen	Antigen dilution	% Hemolysis										
		1:8[b]	1:16	1:32	1:64	1:128	1:256	1:512	1:1,024	5u[c]	2.5u	1.25u
Test antigen	1:2	0	0	0	10	40	70	100	100	70	50	0
	1:4	0	0	0	30	45	80	100	100	100	85	0
	1:8	0	0	0	0	20	60	90	100	100	95	10
	1:16	0	0	0	0	2	70	100	100	100	100	40
	1:32	0	0	0	20	0	100	100	100	100	100	40
Reference antigen	Optimal	0	0	0	0	10	50	100	100	100	100	35
Tissue control	Optimal	100	100	100	100	100	100	100	100	100	100	40
Serum control	None	100	100	100	100	100	100	100	100	100	100	40
VBD control	None									100	90	60

[a] The 1:2 dilution of the test antigen is anticomplementary. The optimal dilution is 1:8, based on the criteria described in the text.
[b] Reference antiserum dilution.
[c] Complement control.

		Components (ml)	Serum Dilution					
			1:8	1:16	1:32	1:64	1:128	1:256
Unknown Serum Assay	Test	Unknown Serum Test Antigen 5 CH_{50}	0.025 0.025 → 0.050					
	Serum Controls	Unknown Serum VBD 5 CH_{50}	0.025 0.025 → 0.050					
	Tissue Controls	Unknown Serum Normal Tissue Antigen* 5 CH_{50}	0.025 0.025 → 0.050					
Known Positive Serum Assay	Test	Known Positive Serum Test Antigen 5 CH_{50}	0.025 0.025 → 0.050					
	Serum Controls	Known Positive Serum VBD 5 CH_{50}	0.025 0.025 → 0.050					
	Tissue Controls	Known Positive Serum Normal Tissue Antigen[a] 5 CH_{50}	0.025 0.025 → 0.050					
Known Negative Serum Assay	Test	Known Negative Serum Test Antigen 5 CH_{50}	0.025 0.025 → 0.050					
	Serum Controls	Known Negative Serum VBD 5 CH_{50}	0.025 0.025 → 0.050					
	Tissue Controls	Known Negative Serum Normal Tissue Antigen[a] 5 CH_{50}	0.025 0.025 → 0.050					
	Cell Control	VBD	0.1					

Complement Controls for Test

Complement Controls	Components (ml)	5-Unit CH_{50}	2.5-Unit[b] CH_{50}	1.25-Unit[b] CH_{50}
Antigen	Test Antigen VBD Complement	0.025 0.025 0.050	0.025 0.025 0.050	0.025 0.025 0.050
Tissue[a]	Normal Tissue Antigen VBD Complement	0.025 0.025 0.050	0.025 0.025 0.050	0.025 0.025 0.050
VBD	VBD Complement	0.050 0.050	0.050 0.050	0.050 0.050

FIGURE 5 Diagnostic serology microtiter test. Notes: *a*, normal tissue antigen used at same dilution as test antigen; *b*, the 2.5- and 1.25-CH_{50} are 1:2 and 1:4 dilutions, respectively, of the 5-CH_{50} in 0.05 ml of complement.

■ Preparation of Complement
See under "Antigen Titration" above.

■ Preparation of Hemolysin-Sensitized SRBC
See under "Antigen Titration" above.

■ Reading the Test

1. If color standards less than 24 h old are available, they may be used; otherwise, prepare fresh color standards (see the appropriate section above).
2. Centrifuge the plates at $300 \times g$ for 5 min.
3. Read and record the percent hemolysis of the complement controls using the color standards; interpolate to the nearest 5%.
4. Compare the readings with those in Table 6 to determine if they are acceptable. Any values less than those shown are considered to be anticomplementary; therefore, the test is not acceptable and should be repeated.
5. Read and record the percent hemolysis in each test well. The serum titer is the highest serum dilution resulting in less than 30% hemolysis.
6. Diagnostic sera that do not react with the tissue control and are not anticomplementary (i.e., do not have titer in the serum control plate) may have their results reported; if the serum is anticomplementary or reacts with tissue at the 1:8 dilution but not at higher dilutions, reactions observed at the higher dilutions are considered valid.
7. If the serum reacts with the tissue control, it must be tissue treated and retested. Mix 0.2 ml of undiluted serum with 0.2 ml of the tissue control in a 12- by 75-mm test tube, making a 1:2 dilution of the serum, and incubate it in a 37°C water bath for 1 h. Add 1.2 ml of VBD, making a 1:8 dilution of the serum. Inactivate the serum in a 56°C water bath for 30 min. The serum is now ready for retesting.
8. If the serum is anticomplementary, it is treated in the same manner, substituting complement for the tissue control.
9. If the serum is both anticomplementary and tissue reactive, treat 0.2 ml of serum with 0.2 ml of tissue and 0.4 ml of complement. After the 1-h incubation at 37°C, add 0.8 ml of VBD and inactivate the mixture in a 56°C water bath for 30 min.
10. Serum lipids, if excessive, may interfere with the test; treat 0.5 ml of serum with 0.5 ml of Freon and mix them well in a vortex mixer. Centrifuge the mixture at $600 \times g$ for 15 min, and carefully aspirate the supernatant fluid containing the serum. Proceed with reading the test.

REFERENCES

Ada, G. L., and B. T. Perry. 1954. Studies on the soluble complement-fixing antigens of influenza virus. III. The nature of the antigens. *Aust. J. Exp. Biol. Med. Sci.* **32**:177–186.

Bankowski, R. A., R. W. Wichmann, and M. Kummer. 1953. A complement-fixation test for identification of immunological types of the virus of vesicular exanthema of swine. *Am. J. Vet. Res.* **14**:145–149.

Beveridge, W. I. B., and F. M. Burnet. 1946. *The Cultivation of Viruses and Rickettsiae in the Chick Embryo. Medical Research Council Special Report Series No. 256*, p. 1–92. His Majesty's Statistics Office, London, England.

Black, F. L., and J. L. Melnick. 1954. The specificity of the complement fixation test in poliomyelitis. *Yale J. Biol. Med.* **26**:385–393.

Black, F. L., and J. L. Melnick. 1955. Appearance of soluble and cross-reactive complement-fixing antigens on treatment of poliovirus with formalin. *Proc. Soc. Exp. Biol. Med.* **89**:353–355.

Casals, J., and R. Palacios. 1941. The complement fixation test in the diagnosis of virus infections of the central nervous system. *J. Exp. Med.* **99**:429–449.

Casals, J., and R. Palacios. 1949. Acetone-ether extracted antigens for complement fixation with certain neurotropic viruses. *Proc. Soc. Exp. Biol. Med.* **70**:339–343.

Casey, H. L. 1965. Adaptation of LBCF method of microtechnique. In *Standard Diagnostic Complement-Fixation Method and Adaptation to Microtest. Public Health Monograph No. 74, Public Health Service Publication No. 1228.* U.S. Government Printing Office, Washington, D.C.

Chambers, L. A., S. S. Cohen, and J. R. Clawson. 1950. Studies on commercial typhus vaccines. II. The antigenic fractions of disrupted epidemic typhus rickettsiae. *J. Immunol.* **65**:459–463.

Churcher, G. M., F. W. Sheffield, and W. Smith. 1959. Poliomyelitis virus flocculation: the reactivity of unconcentrated cell-culture fluids. *Br. J. Exp. Pathol.* **40**:87–95.

Cuica, A. 1929. The reaction of complement-fixation in foot-and-mouth disease as a means of identifying the different types of virus. *J. Hyg.* **28**:325–339.

Clark, D. H., and J. Casals. 1958. Techniques for hemagglutination and hemagglutination-inhibition with arthropod-borne viruses. *Am. J. Trop. Med. Hyg.* **7**:561–573.

Cohen, G. H., and W. C. Wilcox. 1966. Soluble antigens of vaccinia infected mammalian cells I. Separation of virus-induced soluble antigens into two classes on the basis of physical characteristics. *J. Bacteriol.* **92**:676–686.

Craigie, J., and F. O. Wishart. 1936a. The complement fixation reaction in variola. *Can. Public Health J.* **27**:371–379.

Craigie, J., and F. O. Wishart. 1936b. Studies on the soluble precipitable substances of vaccinia. II. The precipitable substances of dermal vaccine. *J. Exp. Med.* **64**:819–830.

DeBoer, C. J., and H. R. Cox. 1947. Specific complement-fixing diagnostic antigens for neurotropic virus diseases. *J. Immunol.* **55**:193–204.

Duca, M., V. Moroisanu, L. Handrache, V. Ciubitaru, L. Ionescu, and M. Liscia. 1979. Efficiency of complement fixation (CF) test with internal nucleoprotein (NP) antigen and hemagglutination-inhibition (HI) test in the serodiagnosis of A and B influenza infections. *Arch. Roum. Pathol. Exp. Microbiol.* **38**:331–337.

Dulbecco, R. 1952. Production of plaques in monolayer tissue cultures by single particles of an animal virus. *Proc. Natl. Acad. Sci.* **38**:747–752.

Dzierzanowska, D., J. Kolodziejczyk, and I. Lourie. 1986. Determination of cytomegalovirus antibody titers in sera of volunteer blood donors by use of complement fixation and two immunoenzymatic tests. *Arch. Immunol. Ther. Exp.* (Warsaw) **34**:397–402.

Ende, M., A. Van den Polson, and G. S. Turner. 1957. Experiments with the soluble antigen of rabies in suckling mouse brain. *J. Hyg.* **55**:361–373.

Enders, J. F., and T. C. Peebles. 1954. Propagation in tissue cultures of cytopathogenic agents from patients with measles. *Proc. Soc. Exp. Biol. Med.* **80**:277–286.

Enders, J. F., T. H. Weller, and F. C. Robbins. 1949. Cultivation of the Lansing strain of poliomyelitis virus in cultures of various human embryonic tissues. *Science* **109**:85–87.

Espana, C., and W. M. Hammon. 1948. An improved benzene-extracted complement-fixing antigen applied to the diagnosis of the arthropod-borne virus encephalitides. *J. Immunol.* **59:**31–44.

Ferrante, P., G. Achilli, G. Gerna, and F. Bergamini. 1987. Subacute sclerosing panencephalitis: detection of measles antibody in serum and cerebrospinal fluid by enzyme-linked immunosorbent assay, complement fixation, and hemagglutination inhibition. *Microbiologica* **10:**111–118.

Fischer, G. S., W. I. Sweimler, and B. Kleger. 1986. Comparison of Mycoplasmelisa with complement fixation test for measurement of antibodies to *Mycoplasma pneumoniae. Diagn. Microbiol. Infect. Dis.* **4:**139–145.

French, E. L. 1952. Murray Valley encephalitis. Isolation and characterization of the aetiological agent. *Med. J. Aust.* **1:**100–103.

Fulton, R. E., and V. L. Dininno. 1985. Rapid rate-kinetic turbidometric assay for quantitation of viral complement fixing antibodies. *J. Virol. Methods* **12:**13–24.

Girardi, A. J., J. Warren, C. Goldman, and B. Jeffries. 1958. Growth and CF antigenicity of measles virus in cells deriving from human heart. *Proc. Soc. Exp. Biol. Med.* **98:**18–22.

Goodpasture, E. W., A. M. Woodruff, and G. J. Buddingh. 1931. The cultivation of vaccine and other viruses in the chorioallantoic membrane of chick embryos. *Science* **74:**371–372.

Goodpasture, E. W., A. M. Woodruff, and G. J. Buddingh. 1933. Use of embryo chick in investigation of certain pathological problems. *South. Med. J.* **26:**418–420.

Halonen, P. E., H. L. Casey, J. A. Stewart, and A. D. Hall. 1967. Rubella complement fixing antigen prepared by alkaline extraction of virus grown in suspension culture of BHK-21 cells. *Proc. Soc. Exp. Biol. Med.* **125:**167–172.

Hammon, W. M., and G. E. Sather. 1969. Arboviruses, p. 227–280. *In* E. H. Lennette and N. J. Schmidt (ed.), *Diagnostic Procedures for Viral and Rickettsial Infections*, 4th ed. American Public Health Association, New York, N.Y.

Havens, W. P., Jr., D. W. Watson, R. H. Green, G. L. Lavin, and I. E. Smadel. 1943. Complement fixation with neurotropic viruses. *J. Exp. Med.* **77:**139–153.

Henle, G., S. Harris, and W. Henle. 1948. The reactivity of various human sera with mumps complement fixation antigens. *J. Exp. Med.* **88:**133–147.

Henle, G., S. Harris, W. Henle, and F. Deinhardt. 1955. Propagation and primary isolation of mumps virus in tissue culture. *Proc. Soc. Exp. Biol. Med.* **89:**556–560.

Hilleman, M. R., and J. H. Werner. 1954. Recovery of new agent from patients with acute respiratory illness. *Proc. Soc. Exp. Biol. Med.* **85:**183–188.

Hishiyama, M., M. Tsurudome, Y. Ito, A. Yamada, and A. Sugiura. 1988. Complement mediated neutralization test for determination of mumps vaccine induced antibody. *Vaccine* **6:**423–427.

Holmes, K. T., A. W. Hampson, R. I. Raison, R. G. Webster, W. J. O'Sullivan, and C. E. Mountford. 1982. A comparison of two antineuraminidase monoclonal antibodies by complement activation. *Eur. J. Immunol.* **12:**523–526.

Howitt, B. F. 1937. The complement fixation reaction in experimental equine encephalomyelitis, lymphocytic choriomeningitis and St. Louis type of encephalitis. *J. Immunol.* **33:**235–250.

Hoyle, L. 1952. Structure of the influenza virus. The relation between biological activity and chemical structure of virus fractions. *J. Hyg.* **50:**229–245.

Hudson, L. D., S. Adelman, and C. W. Lewis. 1981. Pityriasis rosea. Viral complement fixation studies. *J. Am. Acad. Dermatol.* **4:**544–546.

Huebner, R. J., W. P. Rowe, T. G. Ward, R. H. Parrott, and J. A. Bell. 1954. Adenoidalpharyngeal conjunctival agents. A newly recognized group of common viruses of the respiratory system. *N. Engl. J. Med.* **251:**1077–1086.

Inouye, S., S. Matsuno, and R. Kono. 1981. Difference in antibody reactivity between complement fixation and immune adherence hemagglutination tests with virus antigens. *J. Clin. Microbiol.* **14:**241–246.

Julkunen, I., M. Kleemola, and T. Hovi. 1984. Serological diagnosis of influenza A and B infections by enzyme imunoassay. Comparison with the complement fixation test. *J. Virol. Methods* **9:**7–14.

Kidd, J. G., and W. F. Friedewald. 1942. A natural antibody that reacts in vitro with a sedimentable constituent of normal tissue cells. I. Demonstration of the phenomenon. *J. Exp. Med.* **76:**543–556.

LeBouvier, G. L., G. D. Laurence, E. M. Parfitt, M. G. Jennens, and A. Goffe. 1954. Typing of poliomyelitis virus by complement fixation. *Lancet* **ii:**531–532.

Lennette, E. H. 1969. General principles underlying laboratory diagnosis of viral and rickettsial infections p. 1–65. *In* E. H. Lennette and N. J. Schmidt (ed.), *Diagnostic Procedures for Viral and Rickettsial Diseases*, 4th ed. American Public Health Association, New York, N.Y.

Lennette, E. H., A. Wiener, B. J. Neff, and M. N. Hoffman. 1956a. A chick embryo-derived complement-fixing antigen for western equine encephalomyelitis. *Proc. Soc. Exp. Biol. Med.* **92:**575–577.

Lennette, E. H., A. Wiener, M. I. Ota, F. Y. Fujimoto, and M. N. Hoffman. 1956b. Rapid identification of isolates of western equine encephalomyelitis virus by the complement fixation technique. *Am. J. Hyg.* **64:**270–275.

Lichter, A. G. 1953. Observation on anticomplementary reactions. *Am. Med. Assoc. Arch. Dermatol. Syphilis* **67:**362–368.

Lief, F. S., and W. Henle. 1956a. Studies on the soluble antigen of influenza virus. I. The release of S antigen from elementary bodies by treatment with ether. *Virology* **2:**753–771.

Lief, F. S., and W. Henle. 1956b. Studies on the soluble antigen of influenza virus. III. The decreased incorporation of S antigen into elementary bodies of increasing incompleteness. *Virology* **2:**772–797.

Maltaner, F. 1946. Significance of thromboplastic activity of antigens used in complement fixation tests. *Proc. Soc. Exp. Biol. Med.* **62:**302–304.

Midthun, K., L. I. Pang, J. Flores, and A. Z. Kapikian. 1989. Comparison of immunoglobulin A (IgA), IgG, and IgM enzyme-linked immunosorbent assays, plaque reduction neutralization assay, and complement fixation in detecting seroresponses to rotavirus vaccine candidates. *J. Clin. Microbiol.* **27:**2799–2804.

Miller, H., B. McCulloch, M. P. Landini, and F. Rossier. 1989. Comparison of immunoblotting with other serological methods and virus isolation for the early detection of primary cytomegalovirus infection in allograft recipients. *J. Clin. Microbiol.* **27:**2672–2677.

Ochiai, H., M. Shibata, S. Sato, and S. Niwayama. 1987. Single radial complement fixation test using NP containing plates: a simple and sensitive method for the detection of influenza infection. *J. Virol. Methods* **15:**151–158.

Ossewaarde, J. M., and A. D. deBooij. 1989. Development of a monoclonal antibody for use as an amboceptor in complement fixation tests. *J. Virol. Methods* **25:**13–20.

Paucker, K., F. S. Lief, and W. Henle. 1956. Studies on the soluble antigen of influenza virus. IV. Fractionation of elementary bodies labeled with radio-active phosphorus. *Virology* **2:**798–810.

Pauri, P., P. Bagnarelli, and M. Clementi. 1981. Complement fixation test for rotavirus detection: comparison and analysis of different methods to reduce anticomplementary activity of some specimens. *J. Virol. Methods* **3:**329–335.

Puolakkainen, M., M. Kousa, and P. Saikku. 1987. Clinical conditions associated with positive complement fixation serology for *Chlamydiae. Epidemiol. Infect.* **98:**101–108.

Ravaoarinoro, M., M. Reginster, A. Doraal, and D. Sondag-Thull. 1984. Comparison between an enzyme-linked immunosorbent assay and a complement fixation test in assessing the age related acquisition of cytomegalovirus antibodies in groups of the population living in Belgium. *Acta Clin. Belg.* **39:**6–12.

Rice, C. E. 1948a. Inhibitory effects of certain avian and mammalian antisera in specific complement fixation systems. *J. Immunol.* **59:**365–378.

Rice, C. E. 1948b. Some factors influencing the selection of a complement-fixation method. II. Parallel use of the direct and indirect techniques. *J. Immunol.* **60:**11–21.

Robbins, F. C., J. F. Enders, and T. H. Weller. 1950. Cytopathogenic effect of poliomyelitis viruses in vitro on human embryonic tissues. *Proc. Soc. Exp. Biol. Med.* **75:**370–374.

Robbins, F. C., J. F. Enders, T. H. Weller, and G. L. Florentino. 1951. Studies on the cultivation of poliomyelitis viruses in tissue culture. V. The direct isolation and serologic identification of virus strains in tissue culture from patients with nonparalytic and paralytic poliomyelitis. *Am. J. Hyg.* **54:**286–293.

Rowe, W. P., R. J. Huebner, L. K. Gilmore, R. H. Parrott, and T. G. Ward. 1953. Isolation of a cytopathogenic agent from human adenoids undergoing spontaneous degeneration in tissue culture. *Proc. Soc. Exp. Biol. Med.* **84:**570–573.

Ruckle, G., and K. D. Rogers. 1957. Studies with measles virus. II. Isolation of virus and immunological studies in persons who have had the natural disease. *J. Immunol.* **78:**341–355.

Sather, G. E., and W. M. Hammon. 1967. Antigenic patterns within the California-encephalitis-virus group. *Am. J. Trop. Med. Hyg.* **16:**548–557.

Sato, S., S. Motoda, I. Iwase, and K. Jo. 1983. Single radial complement fixation test using complement film. Assay of the antibody response to strain and type specific antigens of influenza. *J. Virol. Methods* **7:**57–64.

Sato, S., H. Ochiai, and S. Niwayama. 1988. Application of the single radial complement fixation test for serodiagnosis of influenza, respiratory synctial, mumps, adeno type 3 and herpes simplex type 1 virus infections. *J. Med. Virol.* **24:**395–404.

Schafer, W., and W. Zillig. 1954. Über den Aufbau des Virus-Elementarteilchens der klassischen Geflügelpest. I. Gewinnung, physikalisch-chemische und biologische Eigenschaften einiger Spaltprodukte. *Z. Naturforsch.* **9b:**779–788.

Schmidt, N. J. 1957. An inquiry into the use of the complement fixation test for the typing of poliomyelitis viruses. *Am. J. Hyg.* **66:**119–130.

Schmidt, N. J. 1969. Tissue culture methods and procedures for diagnostic virology, p. 79–178. *In* E. H. Lennette and N. J. Schmidt (ed.), *Diagnostic Procedures for Viral and Rickettsial Disease,* 4th ed. American Public Health Association, New York, N.Y.

Schmidt, N. J., and E. H. Lennette. 1956. Modification of the homotypic specificity of poliomyelitis complement fixing antigens by heat. *J. Exp. Med.* **104:**99–102.

Schmidt, N. J., E. H. Lennette, J. H., Doleman, and S. J. Hagens. 1957. Factors influencing the potency of poliomyelitis complement-fixing antigens produced in tissue culture systems. *Am. J. Hyg.* **66:**1–9.

Sethi, J., D. Pei, and Y. Hirshaut. 1981. Choice and specificity of complement in complement fixation assay. *J. Clin. Microbiol.* **13:**888–890.

Sever, J. L. 1962. Application of a microtechnique to viral serological investigations. *J. Immunol.* **88:**320–329.

Smadel, J. E., R. D. Baird, and M. J. Wall. 1939. Complement fixation in infections with the virus of lymphocytic choriomeningitis. *Proc. Soc. Exp. Biol. Med.* **40:**71–73.

Sosa-Martinez, J., and E. H. Lennette. 1955. Studies on a complement fixation test for herpes simplex. *J. Bacteriol.* **70:**205–215.

Stevenson, W. D. H., and M. B. E. Butler. 1933. Dermal strains of vaccina virus grown on the chorio-allantoic membrane of chick embryos. *Lancet* **i:**228–230.

Suggs, M. R., Jr., H. L. Casey, D. D. Sligh, A. R. Fodor, and W. F. McLimmans. 1961. A batch-type concentration and purification procedure for poliovirus complement fixing antigen. *J. Bacteriol.* **82:**789–791.

Svedmyr, A., J. R. Enders, and A. Holloway. 1952. Complement fixation with Brunhilde and Lansing poliomyelitis viruses propagated in tissue culture. *Proc. Soc. Exp. Biol. Med.* **79:**296–309.

Svedmyr, A., J. R. Enders, and A. Holloway. 1953. Complement fixation with the three types of poliomyelitis viruses propagated in tissue culture. *Am. J. Hyg.* **57:**60–70.

Taguchi, F. 1988. New complement fixation test with peroxidase labeled complement C1q for direct and quantitative determination of antibodies to herpes simplex virus. *Microbiol. Immunol.* **32:**1167–1173.

Takatsy, G. 1950. A new method for the preparation of serial dilutions in a quick and accurate way. *Kiserletes Orvostudomany* **2:**293–296.

Taylor-Robinson, D., and A. W. Downie. 1959. Chickenpox and herpes zoster. I. Complement fixation studies. *Br. J. Exp. Pathol.* **40:**398–409.

Thorn, J. J., P. Oxholm, and H. K. Andersen. 1988. High levels of complement fixing antibodies against cytomegalovirus in patients with primary Sjogren's syndrome. *Clin. Exp. Rheumatol.* **6:**71–74.

Weller, T. H., and H. M. Witton. 1958. The etiologic agents of varicella and herpes zoster. Serologic studies with the viruses as propagated in vitro. *J. Exp. Med.* **108:**869–890.

Westwood, J. C. N., G. Appleyard, D. Taylor-Robinson, and H. T. Zwartouw. 1960. The production of high titre poliovirus in concentrated suspensions of tissue culture cells. *Br. J. Exp. Pathol.* **41:**105–111.

Westwood, J. C. N., H. T. Zwartouw, G. Appleyard, and D. H. J. Titmuss. 1965. Comparison of the soluble antigens and virus particle antigens of vaccinia virus. *J. Gen. Microbiol.* **38:**47–53.

Whitney, E., L. M. Kraft, W. B. Lawson, and L. Gordon. 1953. Noninfectious complement-fixing antigen from embryonated hens' eggs infected with lymphocytic choriomeningitis virus. *Proc. Soc. Am. Bacteriol.* **1953:**50.

Woodruff, A. M., and E. W. Goodpasture. 1931. The susceptibility of the chorioallantoic membrane of chick embryos to infection with the fowl-pox virus. *Am. J. Pathol.* **7:**209–222.

Wreghitt, T. G., and M. Sillis. 1987. An investigation of the *Mycoplasma pneumoniae* infections in Cambridge in 1983 using a capture enzyme-linked immunosorbent assay (ELISA), indirect immunofluorescence (IF) and complement fixation (CF) tests. *Isr. J. Med. Sci.* **23:**704–708.

Neutralization*

HAROLD C. BALLEW

12

Neutralization of a virus is defined as loss of infectivity through reaction of the virus with specific antibody. To test for virus neutralization, virus and serum are mixed under appropriate conditions, incubated, and then used to inoculate a susceptible living host for detection of unneutralized virus by reactions such as cytopathic effect (CPE), plaque formation, and metabolic inhibition in cell cultures; death of or paralysis in animals; and pock formation on the chorioallantoic membrane (CAM) of embryonated hens' eggs.

The neutralization test has been used in virology longer than any other serologic procedure. Despite its relative antiquity, neutralization is one of the most specific and widely used serologic procedures in diagnostic virology. Neutralization techniques can be used to identify a virus isolate or to measure the antibody response of an individual to a virus. Because of its high immunologic specificity, the neutralization test is often the standard against which the specificity of other serologic procedures is evaluated.

One of the first principles formulated in virus serologic testing was the so-called "percentage law" (Andrewes and Elford, 1933), which states that when virus is added to excess antibody, the percentage of virus not neutralized is the same regardless of the amount of virus added. With the advent of plaquing procedures for viruses, the neutralization reaction could be accurately evaluated (Dulbecco et al., 1956). In their studies on the kinetics of the reaction, Dulbecco et al. found that there is a linear interdependence between the rate of neutralization and the concentration of antibody and that there is always a fraction of virus that is not inactivated. Other investigators who helped to establish the basic properties of the neutralization test are Tyrrell and Horsfall (1953), Salk et al. (1954), and Mandel (1960). The contributions of these and other investigators to the neutralization test are not described in this chapter but are adequately reviewed by Horsfall and Tamm (1960) and Maramorosch and Koprowski (1967).

STANDARDIZATION OF TEST MATERIALS

Before the neutralization test is performed, the known components that are to be used must be standardized. To identify a virus isolate, a known pretitered antiserum or standardized serum pool is used. Conversely, to measure the

antibody response of an individual to a virus, a known pretitered virus is employed.

Virus

The known virus consists of extracts from infected tissues or fluids from cell cultures and embryonated hens' eggs. This virus can be purchased commercially or prepared by inoculating a susceptible host system with a stock virus and harvesting it at the optimal time. The known virus should always be titrated in the host system to be used for the neutralization test, whether purchased commercially or prepared by host system inoculation.

To titrate a known virus or virus isolate, prepare serial 10-fold dilutions in a maintenance medium and inoculate a susceptible host system with fixed volumes of each dilution. Observe the host for signs of infection, which indicates virus that is not neutralized. The virus endpoint titer is the reciprocal of the highest dilution of virus that infects 50% of the host systems. This endpoint dilution contains one 50% tissue culture infective dose ($TCID_{50}$) of virus per unit of volume, or the amount of virus that will infect 50% of the cell cultures inoculated. In animals, if death is the criterion employed, then this endpoint dilution is called the 50% lethal dose (LD_{50}). The concentration of virus generally used in the neutralization test is 100 $TCID_{50}$ or 100 LD_{50} per unit of volume (Fig. 1).

Serum

A specific immune serum can be purchased commercially or prepared by immunizing susceptible animals and harvesting the serum at the optimal time. The antiserum should always be titrated in the neutralization test against its homologous virus, whether it is purchased commercially or prepared by immunization.

To titrate a specific immune serum or test serum, prepare serial twofold dilutions of the serum and mix each dilution with an equal volume of standardized virus (usually 100 $TCID_{50}$). The virus and serum mixtures are usually incubated for 1 h at room temperature or at 37°C. The time and temperature for incubating the virus and serum mixtures varies with different viruses. Inoculate a susceptible host system with each virus-serum mixture. The serum antibody titer is the reciprocal of the highest dilution of antiserum protecting against the virus. The endpoint dilution contains 1 antibody (Ab) unit per unit

---

*Reprinted from the second edition.

127

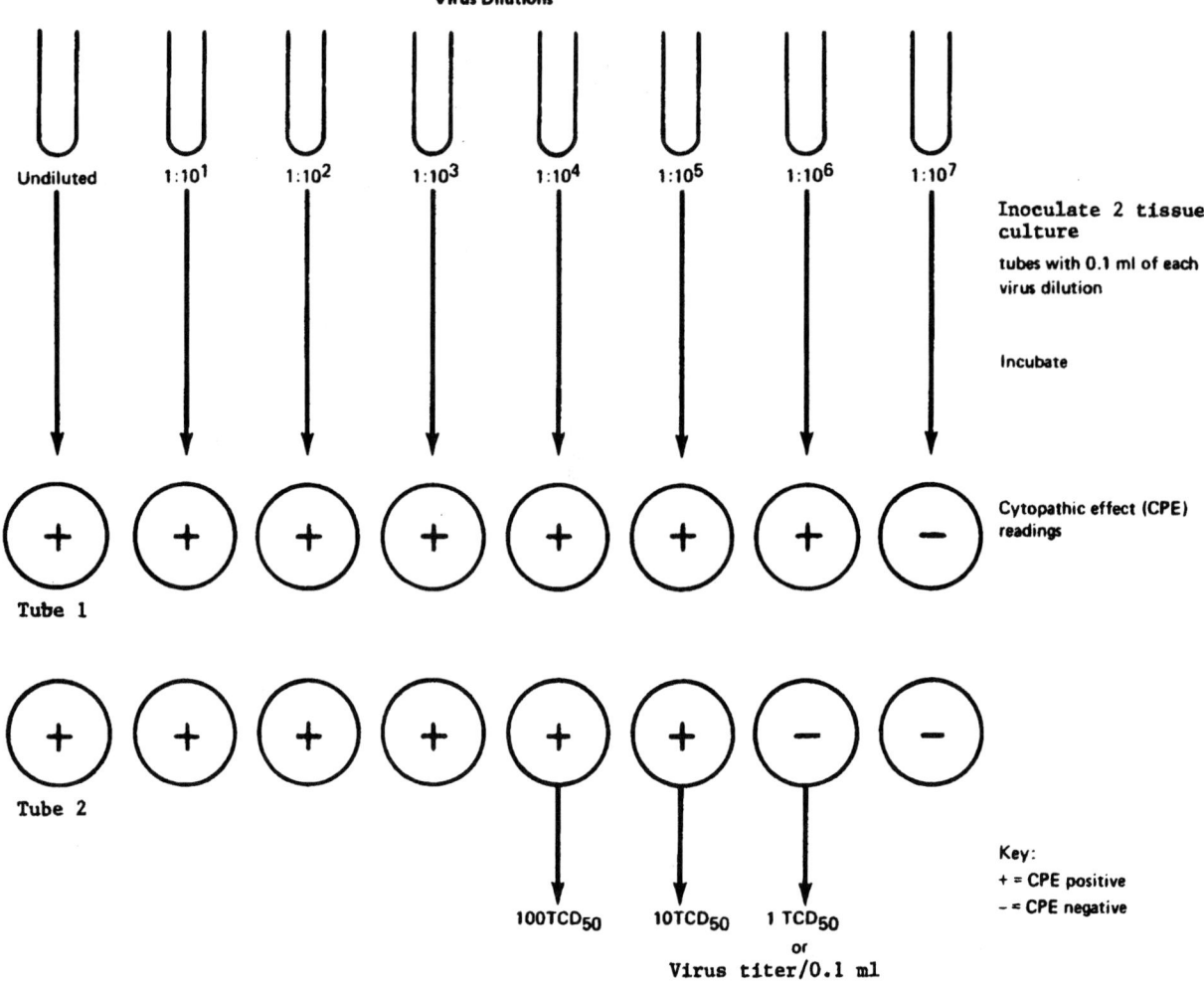

FIGURE 1 This virus titration demonstrates that the endpoint titer is the highest dilution of virus that infects 50% of the inoculated cell cultures.

of volume. The standardized concentration of antiserum generally used in the neutralization test is 20 Ab units per unit of volume (Fig. 2). In Fig. 2, 1 Ab unit/0.1 ml is used as an example to demonstrate how antibody units are calculated.

Earlier studies have shown that the neutralization titer of serum is reduced in certain circumstances by heat, dilution, repeated freezing-thawing, and prolonged storage at 4°C. Part of the neutralization titer may be restored by adding "accessory factors," undefined substances that often are present in freshly collected serum or serum that has been maintained in a frozen state, to the virus-serum mixtures when they are prepared. However, it is usually possible to detect seroconversion in the absence of accessory factors. In addition, these factors may be eliminated by heating the sera to be tested at 56°C for 30 min. Many laboratories do nothing about these factors. The sera may be used without inactivation or without the addition of fresh serum. There is no general agreement on the importance of accessory factors, but accessory factors and nonspecific neutralizing substances have been discussed in detail by Ginsberg and Horsfall (1949), Sabin (1950), Tyrrell and Horsfall (1953), and Mandel (1960).

Host System

One of the basic requirements for all neutralization tests is a living host system to demonstrate unneutralized virus. In the neutralization procedure, virus and serum are mixed, incubated, and then injected into or exposed to a susceptible host system in which the presence of surviving virus may be detected. The host used is primarily determined by the infectious and lethal capacity of the virus, as well as by host availability. The three host systems commonly used for the neutralization test are cell cultures, embryonated hens' eggs, and mice.

Cell Cultures

Where feasible, cell cultures are the preferred host. Cell cultures are the most important hosts for performing the neutralization test because they are susceptible to a wide range of viruses, are readily available, and have no immune system to influence the test. The two types of cell cultures used in the neutralization test are suspension and monolayer cultures. In the suspension culture, the virus-infected cells grow floating in the medium. After the virus

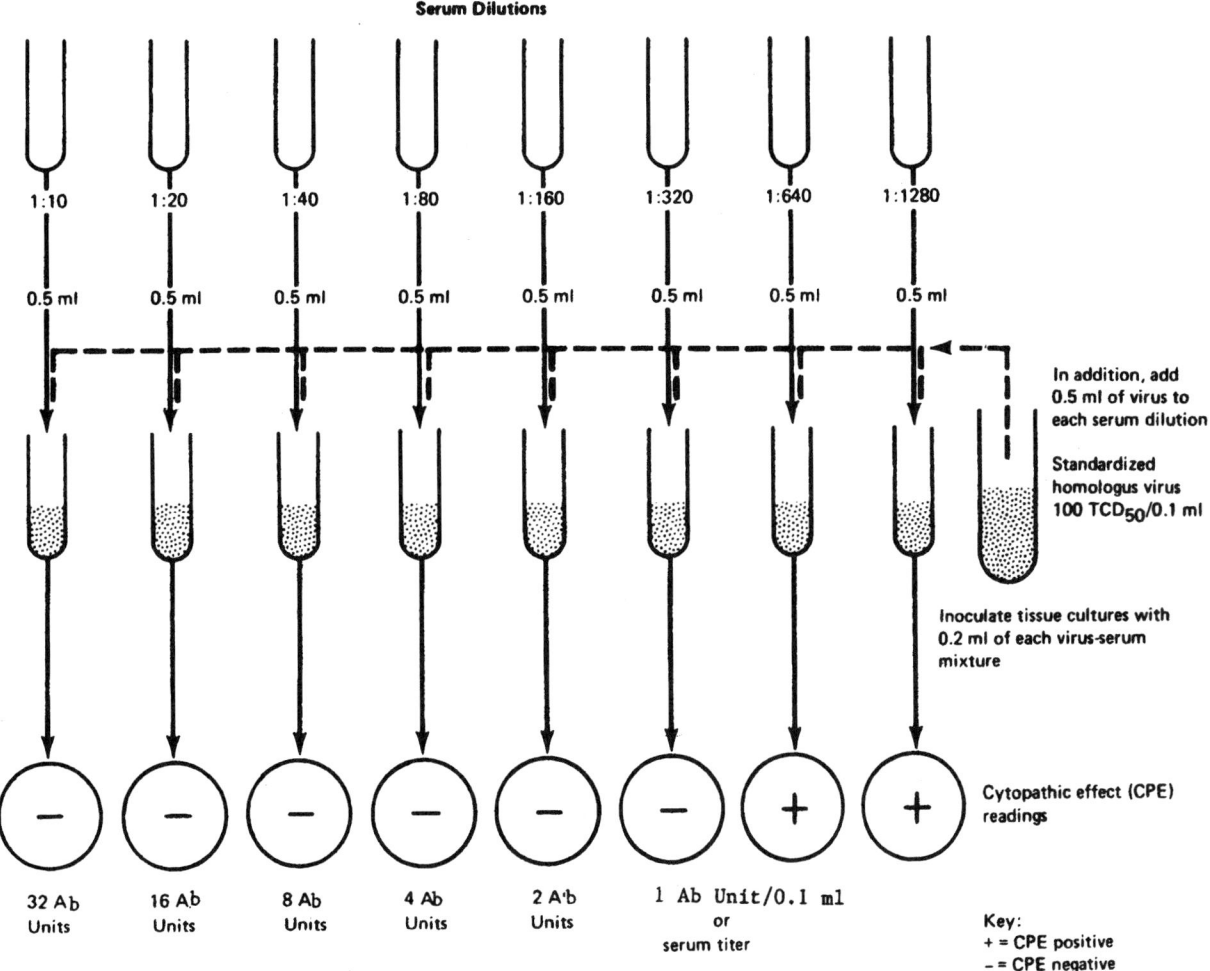

FIGURE 2 This serum titration demonstrates that the antibody titer is the highest dilution of antiserum that protects against the standardized virus.

replicates in the cells, progeny are released into the medium and the unneutralized virus may be detected by reactions such as CPE, change in pH of the medium, or hemadsorption. In monolayer cell cultures, the virus replicates in cell cultures that have been overlaid with agar (Dulbecco et al., 1956). This "agar cover" prevents the virus from establishing secondary foci of infection and keeps the initial infection localized, giving rise to plaques. One plaque is produced by each infectious unit, or PFU, in the original virus suspension. The counting of plaques is a very accurate method for quantitating virus. The prevention of PFU with specific antiserum in the neutralization test is called "plaque reduction."

Embryonated Hens' Eggs

The inoculation of various tissues or cavities of the developing embryo for attempted virus isolation and identification is standard procedure in many laboratories. The route of inoculation depends on the virus to be isolated. For example, the amniotic cavity is used for isolating influenza viruses and mumps virus, and the CAM is often used for isolating variola virus, vaccinia virus, and herpes simplex virus. Unneutralized virus replicates in the embryonated hens' egg and can be detected by the death of the embryo, by pock formation on the CAM, or by agglutination of erythrocytes using amniotic or allantoic fluids containing hemagglutinating virus.

Mice

Adult and suckling mice are the animals most frequently used for virus isolation and identification. The white Swiss mouse is used extensively for isolating and determining unneutralized arboviruses, certain enteroviruses, and rabiesvirus. The age of the mouse has a great influence on its susceptibility to disease. Thus, suckling mice may be more susceptible to viruses than adult mice. Unneutralized virus is usually determined by death of or paralysis in infected mice.

TEST PROCEDURES

Cell culture is the most frequently used host system for performing the neutralization test. In the following types of neutralization tests, suspension cell cultures primarily are used as the host system.

Constant Virus, Varying Serum

Mix a selected dilution of virus (usually 100 $TCID_{50}$, as determined from a previous titration) with various dilutions of acute- and convalescent-phase sera. Incubate the virus-serum mixture, and then inoculate a susceptible host system with the mixture. The reciprocal of the highest dilution of acute- and convalescent-phase sera protecting the host against the virus is the serum titer.

■ Materials

Susceptible host system such as cell cultures, embryonated hens' eggs, or mice

Maintenance medium

Known positive virus (pretitered and standardized to contain 100 $TCID_{50}$ or LD_{50}/0.1 ml)

Acute- and convalescent-phase sera (usually inactivated at 56°C for 30 min)

■ Procedures

1. Prepare serial twofold dilutions of acute- and convalescent-phase sera (1:10 through 1:5,120) in 0.5-ml volumes.
2. Mix 0.5 ml of standardized known positive virus with each serum dilution.
3. Dilute the known standardized virus (1:10 through 1:1,000).
4. Incubate the virus-serum mixtures at about 37°C for 1 h.
5. After incubation, inoculate each of three cell culture tubes with 0.2 ml of each virus-serum mixture (when inoculating mice, use 0.02 to 0.04 ml of mixture).
6. Inoculate three cell cultures with 0.1 ml of known positive virus dilutions (undiluted through 1:1,000); these dilutions are used as a back titration to confirm the test potency of 100 $TCID_{50}$ (Fig. 1).
7. Include three uninoculated cell cultures for controls.
8. Incubate the cell cultures at 33°C to 35°C in a slanted position.
9. Observe the cell cultures daily for CPE.

Interpretation

A fourfold rise in antibody titer between acute- and convalescent-phase sera is considered diagnostically significant. In this test, CPE in cell cultures is used to detect unneutralized virus. This unneutralized virus may be detected in embryonated hens' eggs by pocks on the CAM and in mice by death or paralysis.

Constant Antiserum, Varying Virus

For virus identification, various dilutions of virus are mixed with a constant antiserum dilution. These dilutions are incubated to allow the virus and antiserum to react. Each virus and serum mixture is then inoculated into a susceptible host system. The dilution of virus that infects 50% of host systems is considered the endpoint dilution.

■ Materials

Susceptible host system such as cell cultures, embryonated hens' eggs, or mice

Maintenance medium, such as Eagle's modified essential medium

Known positive antiserum (inactivated at 56°C for 30 min and standardized to contain 20 Ab units/0.1 ml)

Known negative serum (inactivated at 56°C for 30 min)

Virus isolate

■ Procedures

1. Prepare serial 10-fold dilutions (1:10 through 1:108) of the virus isolate in maintenance medium.
2. Mix 0.5 ml of each isolate dilution (undiluted through 1:108) with 0.5 ml of known standardized antiserum.
3. Mix 0.5 ml of each isolate dilution (undiluted through 1:108) of the virus with 0.5 ml of known negative serum.
4. Incubate the virus-serum mixtures at 37°C for 1 h.
5. Inoculate each of three cell culture tubes with 0.2 ml of each virus-serum mixture (when inoculating mice, use 0.02 to 0.04 ml of the mixture).
6. Include three uninoculated cell cultures for controls.
7. Incubate the cell cultures at 33 to 35°C in a slanted position.
8. Observe the cell cultures daily for CPE.

Interpretation

The dilution of virus that infects 50% of the host system is considered the endpoint dilution. A difference of at least 2 logs or two tubes must be demonstrated between the normal and the immune antiserum to show significant neutralization. Virus dilutions are generally made using 10-fold dilutions. The standardized antiserum dilution in this test is selected based on its ability to neutralize 100 $TCID_{50}$ or 100 LD_{50} of virus, which gives the greatest amount of sensitivity and specificity for the test.

Constant Virus, Constant Antiserum

A selected dilution of virus (usually 100 $TCID_{50}$ as determined from prior titration) is mixed with a selected dilution of known antiserum (usually 20 Ab units/0.1 ml). The mixture is incubated and then injected into a susceptible host system for observation of unneutralized virus. The virus is identified if the antiserum neutralizes the infectivity of the virus.

■ Materials

Susceptible host system such as cell cultures, embryonated hens' eggs, or mice

Maintenance medium

Known positive antiserum (inactivated at 56°C for 30 min and standardized to contain 20 Ab units/0.1 ml)

Unknown virus isolated (pretitered and standardized to contain 100 $TCID_{50}$/0.1 mL)

■ Procedures

1. Mix 0.5 ml of the standardized unknown virus isolate with 0.5 ml of the known standardized positive antiserum.
2. Dilute the standardized unknown virus isolate (1:10 through 1:1,000).
3. Incubate the virus-serum mixture at about 37°C for 1 h.
4. Inoculate each of three susceptible cell culture tubes with 0.2 ml of the virus-serum mixture (when inoculating mice, use 0.02 to 0.04 ml of the mixture).
5. Inoculate three susceptible cell culture tubes with 0.1 ml of the standardized unknown virus dilutions (undiluted through 1:1,000); these dilutions are used as a back titration to confirm the test potency of 100 $TCID_{50}$.
6. Include three uninoculated cell culture tubes for controls.
7. Incubate the cell cultures at 33 to 35°C in a slanted position.
8. Observe the cell cultures daily for CPE.

TABLE 1 Calculation of 50% mortality (virus titration) with sample data

| Virus dilution | No. of deaths/no. inoculated | Cumulative no. of: | | Mortality ratio | % Mortality |
		Deaths	Survivors		
10^{-4} (1:10^4)	5/5	10	0	10/10	100
10^{-5} (1:10^5)	4/5	5	1	5/6	83
10^{-6} (1:10^6)	1/5	1	5	1/6	17
10^{-7} (1:10^7)	0/5	0	10	0/10	0

Interpolation formula: $\dfrac{\% \text{ mortality greater than } 50\% - 50\%}{\% \text{ mortality greater than } 50\% - \% \text{ mortality less than } 50\%}$

Substituting: $\dfrac{83 - 50}{83 - 17} = \dfrac{33}{66} = 0.5$

    a. Multiply the interpolative value by the negative $\log_{10}$ of the dilution ratio.
        Negative $\log_{10}$ of the dilution ratio = −1
        Interpolative value   = 0.5
        Corrected interpolative value   = −0.5
    b. The endpoint dilution associated with 50% mortality is located between the 10^{-5} and 10^{-6} dilutions.
    c. The $\log_{10}$ of the 50% endpoint dilution is estimated by adding the corrected interpolative value to the $\log_{10}$ of the dilution above 50%: −5 + (−0.5) = −5.5.
    d. The 50% endpoint dilution is estimated at $10^{-5.5}$.
    e. The 50% titer is estimated at $10^{5.5}$.

Interpretation

The absence of CPE in the cell cultures inoculated with the virus-serum mixtures identifies the virus isolate. Antibody in serum can be determined also by running a single dilution of a test serum against a standard dose of known virus. Neutralization indicates the presence of specific antibody.

Varying Virus, Varying Antiserum

This type of neutralization test should be used with caution. The titers of the unknown virus and the known antiserum have not been predetermined. Various dilutions of the virus and antiserum are made, combined, incubated, and then inoculated into a susceptible host system in which the presence of unneutralized virus may be detected. The test can actually give maximum information about both the virus and the antiserum in relation to one another. Although this test design can be used for neutralization results, it is mainly used as a routine procedure in the block titration of antigen in the complement fixation test.

CALCULATIONS OF 50% ENDPOINTS

The 50% endpoint can be used in several different reactions. A 50% mortality ratio is expressed as LD_{50}, a 50% infective dose is expressed as as ID_{50} in cell cultures, and a CPE in 50% of the cultures is expressed as $TCID_{50}$.

Reed-Muench Method

Tables 1 and 2 illustrate the Reed-Muench method.

Karber Method for Calculating 50% Mortality

The Karber formula holds that the negative logarithm of the 50% endpoint titer is equal to the negative logarithm of the highest virus concentration used, as follows:

TABLE 2 Calculation of 50% endpoint dilution of virus plus constant immune serum with sample data

| Virus dilution | + | constant serum | No. of deaths/no. inoculated | Cumulative no. of: | | Mortality ratio | % Mortality |
				Deaths	Survivors		
10^{-2} (1:10^2)	+	IS	5/5	10	0	10/10	100
10^{-3} (1:10^3)	+	IS	3/5	5	2	5/7	71
10^{-4} (1:10^4)	+	IS	2/5	2	5	2/7	29
10^{-5} (1:10^5)	+	IS	0/5	0	10	0/10	0

Interpolation formula: $\dfrac{\% \text{ mortality greater than } 50\% - 50\%}{\% \text{ mortality greater than } 50\% - \% \text{ mortality less than } 50\%}$

Substituting: $\dfrac{71 - 50}{71 - 29} = \dfrac{21}{42} = 0.5$

    a. Multiply the interpolative value by the negative $\log^{10}$ of the dilution ratio.
        Negative $\log_{10}$ of the dilution ratio = −1
        Interpolative value   = 0.5
        Corrected interpolative value   = −0.5
    b. The endpoint dilution associated with 50% mortality is located between the 10^{-3} and 10^{-4} dilutions.
    c. The $\log_{10}$ of the 50% endpoint dilution is estimated by adding the corrected interpolative value to the $\log_{10}$ of the dilution above 50%: −3 + (−0.5) = −3.5.
    d. The 50% endpoint dilution is estimated at $10^{-3.5}$.
    e. The 50% titer is estimated at $10^{3.5}$.

The logarithmic difference between the 50% titers of the virus titration in Table 1 and the virus plus immune serum titration in Table 2 is 2.0 (5.5 − 3.5). Generally, a reduction of at least 2 in $\log_{10}$ (titer) of the virus must be demonstrated by a test serum to show significant neutralization.

$$\left[\frac{\text{sum of \% mortality at each dilution}}{100} - 0.5 \times \text{logarithm of dilution} \right]$$

Example:

Virus dilution	Mortality ratio	% Mortality
10^{-1} (1:10^1)	8/8	100
10^{-2} (1:10^2)	8/8	100
10^{-3} (1:10^3)	8/8	100
10^{-4} (1:10^4)	4/8	50
10^{-5} (1:10^5)	0/8	0

$$-1.0 - \left[\left(\frac{100 + 100 + 100 + 50}{100} - 0.5 \right) \times (\log_{10}) \right]$$

$$= -1.0 - \left[\left(\frac{350}{100} - 0.5 \right) \times \log_{10} \right]$$

$$= 1.0[(3.5 - 0.5) \times 1]$$

$$= -1.0 - 3.0$$

$$= 4.0$$

Thus, the 50% endpoint dilution is $10^{-4.0}$ and the 50% endpoint titer is $10^{4.0}$.

PREPARATION OF SERUM POOLS

To avoid the time and expense required for typing each enterovirus isolate with individual antiserum, the use of serum pools is recommended (Lim and Benyesh-Melnick, 1960; Schmidt et al., 1961). By one method, several antisera that react against different enterovirus serotypes are combined into a pool. A number of pools containing different antisera are set up. Neutralization of the isolate by one of the serum pools indicates that the isolate is one of the enteroviruses whose antiserum is included in the pool. For final identification, the isolate is run in a neutralization test against each antiserum in the pool. By a second method, each antiserum is present in two different intersecting pools. For example, antibody to poliovirus type 1 is the only antiserum that is included in pools C and F, and poliovirus type 1 is the only virus that is neutralized by both of these pools. The method of incorporating antisera into intersecting pool schemes is not included in this chapter but is adequately reviewed by Lennette and Schmidt (1979) and Melnick et al. (1977).

Serum pools should be prepared with only known pretitered antisera. The procedure for preparing enterovirus serum pools follows.

High-titered antiserum must be used. The titers of the sera must be high enough so that their combination in a pool will result in a total serum concentration not greater than 10%. Titer each serum individually against 100 $TCID_{50}$ of its prototype virus per 0.1 ml (Fig. 2).

1. Prepare serial twofold dilutions (1:10 through 1:1,280) of serum and mix each dilution with an equal volume containing 100 $TCID_{50}$ of virus.
2. After incubation at 37°C for 1 h, inoculate three cell culture tubes (animals may be used) with 0.2 ml of each virus-serum mixture.
3. Determine the titer of each serum by examining the cell cultures for CPE; this endpoint dilution or antibody titer is the last dilution of serum demonstrating complete neutralization (no CPE) and contains 1 Ab unit/0.1 ml.
4. 20 Ab units/0.1 ml are required in the pool (to calculate 20 Ab units/0.1 ml, divide 20 into the denominator of

the dilution containing 1 Ab unit/0.1 ml, e.g., 320/20 = 16; thus, a 1/16 dilution contains 20 Ab units/0.1 ml). With the sera titered and the dilution containing 20 Ab units of serum/0.1 ml calculated, incorporate the sera into the pool at the dilution that will give 20 Ab units/0.1 ml.

Example: To make a pool of polioviruses 1, 2, and 3, immune sera (IS) have the following titers:

Poliovirus 1 IS, 1:5,120
Poliovirus 2 IS, 1:20,480
Poliovirus 3 IS, 1:10,240

Divide the titers by 20 to get 20 Ab units/0.1 ml.

$$\text{Polio 1 IS: } \frac{5,120}{20} = 256$$

$$\text{Polio 2 IS: } \frac{20,480}{20} = 1,024$$

$$\text{Polio 3 IS: } \frac{10,240}{20} = 512$$

Decide on the desired volume of the pool. To make a 1,000 ml poliovirus pool using the above figures, divide the denominator of the dilution containing 20 Ab units/0.1 ml into the volume of the pool desired to obtain the amount of each undilute serum to be added to the pool.

$$\text{Polio 1 IS: } \frac{1,000}{256} = 3.90 \text{ ml}$$

$$\text{Polio 2 IS: } \frac{1,000}{1,024} = 0.98 \text{ ml}$$

$$\text{Polio 3 IS: } \frac{1,000}{512} = 1.95 \text{ ml}$$

Total the volumes of sera to be added:

$$\begin{array}{r} 3.90 \text{ ml} \\ 0.98 \text{ ml} \\ + 1.95 \text{ ml} \\ \hline 6.83 \text{ ml} \end{array}$$

Subtract this from the total volume of the pool to determine the amount of diluent to be used:

$$\begin{array}{r} 1,000.00 \text{ ml} \\ - 6.83 \text{ ml} \\ \hline 993.17 \text{ ml} \end{array}$$

Combine the diluent (maintenance medium) with the immune sera in the following volumes:

Maintenance medium . . . . 993.17 ml
Polio 1 IS . . . . 3.90 ml
Polio 2 IS . . . . 0.98 ml
Polio 3 IS . . . . 1.95 ml

Aliquot the pools and store them frozen at −20°C or lower. This results in a pool containing 20 Ab units/0.1 ml of each serum.

INTERPRETATION OF SEROLOGIC TEST RESULTS

The neutralization test is frequently used to identify a virus isolate, and it is also used to indicate a recent virus infection by demonstrating a significant rise in antibody titer to a specific virus using paired sera. Usually, a fourfold rise in antibody titer between acute- and convalescent-phase sera run in the same test is considered to be diagnostically significant. Recent infections cannot be distinguished from past infections by examining a single serum specimen, as a high antibody titer with a single serum specimen is no guarantee of recent infection. However, a single serum specimen may be of value in determining whether an individual has been exposed to a viral agent at some time in the past. In some diseases, the presence of antibody against the causative agent indicates immunity to the disease. Information gained from a single serum specimen may also be of epidemiologic value in determining the number of individuals in a population exposed to certain agents and, where applicable, in assessing herd immunity. The serum collection time is extremely important in all serologic tests. The acute-phase serum should be collected as soon as possible after the onset of disease. The convalescent-phase serum is usually collected 2 to 3 weeks later. If the acute-phase serum is drawn too late, antibodies may be approaching maximal levels; therefore, a significant rise in antibody titer may not be demonstrated.

Interpretation of Tests for Viral Antibodies

The examples shown in Fig. 3 demonstrate the basic principles for interpreting serologic test results; they do not represent tests for any particular virus.

Test 1

Serum sample	Dilution									Controls
	Undiluted	1:10	1:20	1:40	1:80	1:160	1:320	1:640	1:1,280	
Acute phase	+	+	+	+	+	+	+	+	+	−
Convalescent phase	+	+	+	+	+	+	+	+	+	−

The results of this test indicate that the individual has not been exposed to the tested virus.

Test 2

Serum sample	Dilution									Controls
	Undiluted	1:10	1:20	1:40	1:80	1:160	1:320	1:640	1:1,280	
Acute phase	−	−	+	+	+	+	+	+	+	−
Convalescent phase	−	−	−	−	−	+	+	+	+	−

The results of this test demonstrate an eightfold rise in antibody titer between acute- and convalescent-phase sera, suggesting recent infection with the tested virus.

Test 3

Serum sample	Dilution									Controls
	Undiluted	1:10	1:20	1:40	1:80	1:160	1:320	1:640	1:1,280	
Single serum	−	−	−	−	−	−	+	+	+	−

The antibody titer of 160 for a single serum sample indicates past infection with the tested virus. A high antibody titer is no guarantee of recent infection.

Test 4

Serum sample	Dilution									Controls
	Undiluted	1:10	1:20	1:40	1:80	1:160	1:320	1:640	1:1,280	
Acute phase	−	−	−	−	−	+	+	+	+	−
Convalescent phase	−	−	−	−	−	+	+	+	+	−

The antibody titers of 80 for the paired sera indicate infection with the tested virus at some time in the past. It is very important to collect the sera at the proper time: if the acute-phase serum is drawn too late, the antibody titer may have already risen to the point that, when compared with the convalescent-phase serum titer, a rise in antibody titer may not be apparent.

FIGURE 3 Examples of serologic test results. +, virus infectivity; −, no virus infectivity.

REFERENCES

Andrewes, C. H., and W. J. Elford. 1933. Observation on anti page 1: the percentage law. *Br. J. Exp. Pathol.* **14:**367–374.

Dulbecco, R., M. Voit, and A. G. R. Strickland. 1956. A study on the basic aspects of neutralization of two animal viruses, western equine encephalitis virus and poliomyelitis virus. *Virology* **2:**162–205.

Ginsberg, H. S., and F. L. Horsfall, Jr. 1949. A labile component of normal serum which combines with various viruses. Neutralization of infectivity and inhibition of hemagglutination by the component. *J. Exp. Med.* **90:**475–495.

Horsfall, F. L., Jr., and I. Tamm (ed.). 1960. *Viral and Rickettsial Infections of Man*, 4th ed. J. B. Lippincott, Philadelphia, Pa.

Lennette, E. H., and N. J. Schmidt (ed.). 1979. *Diagnostic Procedures for Viral, Rickettsial and Chlamydia Infections*, 5th ed. American Public Health Association, Washington, D.C.

Lim, K. A., and M. Benyesh-Melnick. 1960. Typing of viruses by combination of antiserum pools. Application of typing of enteroviruses (coxsackie and ECHO). *J. Immunol.* **84:**309–317.

Mandel, B. 1960. Neutralization of viral infectivity. Characterization of virus-antibody complex, including association, disassociation and host cell interaction. *Ann. N. Y. Acad. Sci.* **83:**515–527.

Maramorosch, K., and H. Koprowski. (ed.). 1967. *Methods in Virology*, vol. III. Academic Press, New York, N.Y.

Melnick, J. L., N. J. Schmidt, B. Hampil, and H. H. Ho. 1977. Lyophilized combination pools of enterovirus equine antisera: preparation and test procedures for the identification of field strains of 19 group A coxsackievirus serotypes. *Intervirology* **8:**172–181.

Sabin, A. B. 1950. The dengue group of viruses and its family relationship. *Bacteriol. Rev.* **14:**225–232.

Salk, J. E., J. S. Younger, and E. N. Ward. 1954. Use of color change of phenol red as the indicator in titrating poliomyelitis virus or its antibody in a tissue-culture system. *Am. J. Hyg.* **60:**214–230.

Schmidt, N. J., R. W. Guenther, and E. H. Lennette. 1961. Typing of ECHO virus isolates of immune serum pools, the intersecting serum scheme. *J. Immunol.* **87:**623–626.

Tyrrell, D. A. J., and F. L. Horsfall, Jr. 1953. Neutralization properties. *J. Exp. Med.* **97:**845–862.

BIBLIOGRAPHY

Ballew, H. C., F. T. Forrester, H. C. Lyerla, W. M. Velleca, and B. R. Bird. 1977. *Laboratory Diagnosis of Viral Diseases.* Centers for Disease Control, Atlanta, Ga.

Ballew, H. C., F. T. Forrester, H. C. Lyerla, W. M. Velleca, B. R. Bird, and J. D. Roberts. 1979. *Basic Laboratory Methods in Virology.* Centers for Disease Control, Atlanta, Ga.

Ballew, H. C., F. T. Forrester, and H. C. Lyerla. 1983. *Laboratory Methods for Diagnosing Respiratory Virus Infections.* Centers for Disease Control, Atlanta, Ga.

Ballew, H. C., H. C. Lyerla, and F. T. Forrester. 1979. *Laboratory Methods for Diagnosing Herpesvirus Infections.* Centers for Disease Control, Atlanta, Ga.

Habel, K., and N. P. Salzman. 1969. *Fundamental Techniques in Virology.* Academic Press, New York, N.Y.

Hatch, M. H., and G. E. Marchetti. 1971. Isolation of echoviruses with human embryonic lung fibroblast cells. *Appl. Microbiol.* **22:**736–737.

Hsiung, G. D. 1982. *Diagnostic Virology*, 3rd ed. Yale University Press, New Haven, Conn.

Lennette, E. H., A. Balows, W. J. Hausler, Jr., and J. P. Truant (ed.). 1980. *Manual of Clinical Microbiology*, 3rd ed. American Society for Microbiology, Washington, D.C.

Lennette, D. A., S. Specter, and K. D. Thompson (ed.). 1979. *Diagnosis of Viral Infections.* University Park Press, Baltimore, Md.

Melnick, J. L., and B. Hampil. 1965. *WHO Collaborative Studies on Enterovirus Reference Antisera.* (Bulletin WHO 33.) World Health Organization, Geneva, Switzerland.

Wallis, C., and J. F. Melnick. 1967. Virus aggregation as the cause of the non-neutralizable persistent fraction. *J. Virol.* **1:**478–488.

Hemadsorption and Hemagglutination Inhibition

STEPHEN A. YOUNG AND LEROY C. McLAREN

13

HEMADSORPTION

Two genera of common respiratory viruses, the influenza and parainfluenza viruses, and several other viruses, such as mumps virus and Newcastle disease virus, replicate in cell cultures but frequently do not produce cytopathic effects. The cell lines routinely used for isolation of these viruses include primary monkey kidney cells, continuous cultures of Madin-Darby canine kidney cells, and rhesus monkey kidney (LLC-MK2) cells. Although the replication of these viruses may not produce cytopathic effects, the presence of replicating virus can be detected by hemadsorption (Hsiung, 1982; Swenson, 1992a; Leland, 1996).

The final step in the maturation phase of these viruses is the budding of the virion through the plasma membrane of the cell. The insertion of viral proteins necessary for the budding of virus from the cell membrane facilitates the binding of red blood cells (RBCs) to the infected cells. This process is referred to as hemadsorption, and the RBCs from several species (human O, chicken, or guinea pig) can be used in this method. The process is relatively simple; a suspension of RBCs is added to the cell monolayer, and the mixture is incubated at the required temperature and then observed microscopically for adsorption of RBCs to the cells in the monolayer.

Materials

Guinea pig RBCs

A 10% (vol/vol) suspension of RBCs in Alsever's solution is used. This suspension is stable for 1 week when stored at 4°C.

Uninoculated cell culture controls

Cell cultures inoculated with a patient sample(s)

Cell cultures inoculated with known hemadsorbing strains of virus

Method

1. Transfer 5.0 ml of the RBCs in Alsever's solution to a graduated centrifuge tube and add 5.0 ml of phosphate-buffered saline (PBS); mix well, and centrifuge the mixture at 900 × g for 5 min.
2. Discard the supernatant, and resuspend the RBC pellet in 10 ml of PBS.
3. Repeat the pelleting and resuspension steps two more times.
4. After discarding the supernatant fluid from the final centrifugation, determine the packed cell volume of the RBCs.
5. Resuspend the RBCs in PBS to a final concentration of 10% (vol/vol). The stock suspension of RBCs should not be stored for more than 1 week at 4°C.
6. On each day of testing, prepare a 0.4% (vol/vol) suspension of guinea pig RBCs in PBS from the 10% (vol/vol) stock solution of guinea pig RBCs.
7. Remove the cell culture fluid from each monolayer in a biological safety cabinet. The cell culture fluid can be saved in a sterile tube or discarded, depending on the method used to identify hemadsorbing virus.
8. Add 0.2 ml of the 0.4% RBC solution to each cell culture.
9. Incubate the cell cultures at either 4°C or room temperature for 30 min.
10. Rock each tube gently, and then observe each monolayer microscopically. The uninoculated cell monolayer generally does not have cells displaying hemadsorption (Fig. 1A). However, as the cells in the monolayer age they become sticky and can appear to have RBCs that are hemadsorbing. The cell cultures inoculated with a known hemadsorbing virus should have RBCs adsorbed to the cells (Fig. 1B). The monolayers can show various degrees of hemadsorption, and at times the supernatant will also have hemagglutination of the RBCs.
11. For hemadsorption-negative cultures, the RBC suspension can be removed and fresh cell culture medium can be added; the cultures are then reincubated.
12. The identification of the hemadsorbing virus can be done with either the supernatant or the infected cells. The monolayer can be removed from the surface of the tube, and the cells are then attached to a glass slide and stained in either a direct or indirect immunofluorescent assay. The binding of virus-specific antibody is detected by direct observation of the stained cells with a fluorescent microscope. The supernatant can be tested by hemagglutination inhibition (HAI).

HAI TEST

A wide variety of viruses have the ability to combine with and agglutinate RBCs (hemagglutination). Adenoviruses, arboviruses, some enteroviruses, influenza virus, parainfluenza viruses, mumps virus, measles virus, and reoviruses

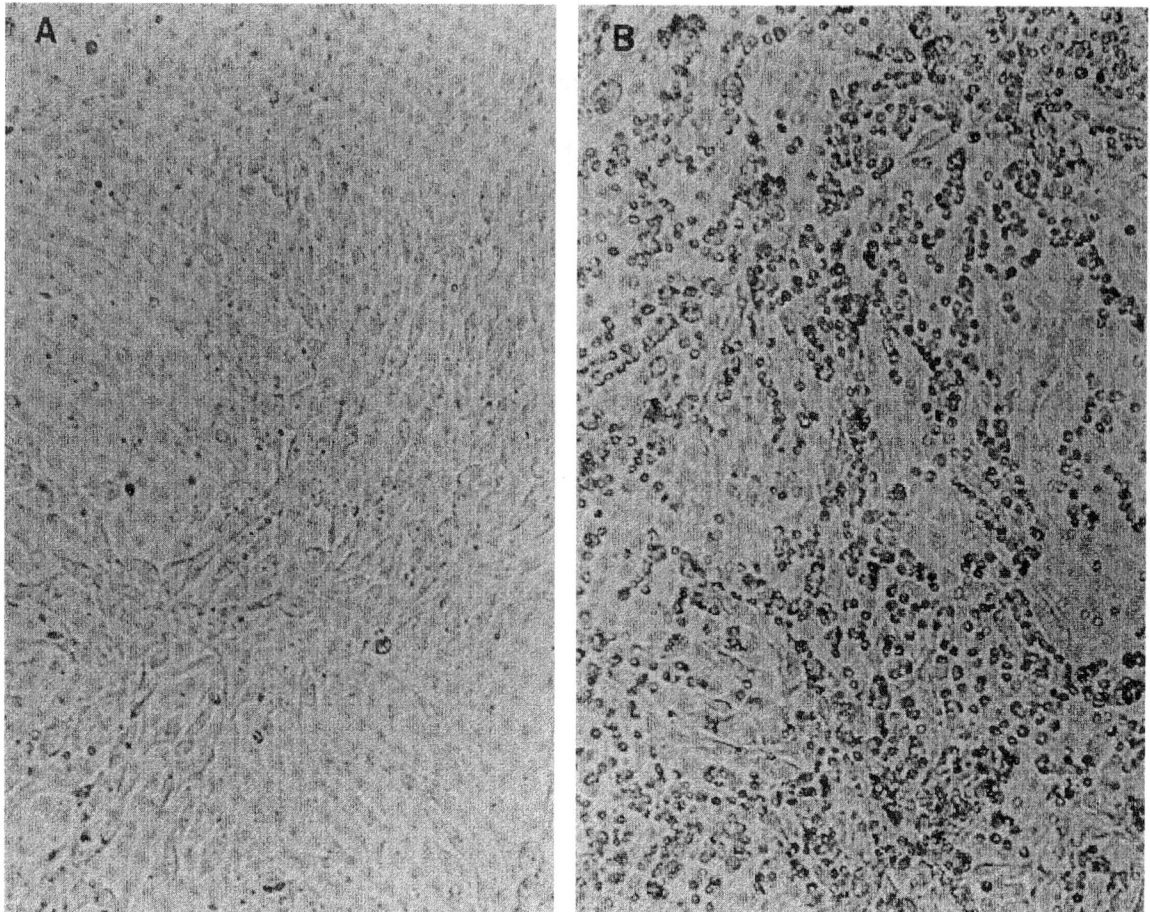

FIGURE 1 Hemadsorption. (A) Uninoculated primary monkey kidney cell; (B) primary monkey kidney cells infected with influenza A virus.

have this property (Hsiung, 1982; Schmidt and Emmons, 1989; Swenson, 1992b; Hodinka, 1999). The hemagglutination test can be used to detect either the virus or the amount of antigen in a sample. Antibodies that have the ability to react with either the virus or the antigen can prevent hemagglutination. This is the principle of the HAI test. In the past, the HAI test was routinely used in clinical virology laboratories to detect antibody to several of these viruses. However, the HAI test is no longer a routine procedure in most clinical laboratories for either serological diagnosis or virus identification. The movement away from the HAI test is primarily due to the time required to perform the test, even though it is relatively easy to perform and requires only inexpensive equipment and reagents (Mahony and Chernesky, 1999).

Serological diagnosis by the HAI test is accomplished by making serial dilutions of a patient's serum, mixing each dilution with a fixed amount of viral hemagglutinin (HA), and then adding an RBC suspension. The animal species from which the indicator RBCs are collected is dependent on the viral HA used in the assay. If the serum contains antibody, the virus will be bound and unable to agglutinate the RBCs. If there are no antibodies in the patient's serum, the virus will agglutinate the RBCs. The reciprocal of the highest dilution of the patient's serum that will completely inhibit agglutination is termed the antibody titer (Hsiung, 1982; Schmidt and Emmons, 1989; Swenson, 1992b).

Serological diagnosis by the HAI test can be complicated by the presence of nonspecific viral inhibitors (nonantibody) and red cell agglutinins in the patient's serum (Hsiung, 1982; Swenson, 1992b; Hodinka, 1999). The presence of such inhibitors can give rise to false-positive results in the HAI test. Therefore, several procedures were developed to remove the inhibitors or RBC agglutinins. These procedures include RBC adsorption, kaolin adsorption, receptor-destroying enzyme treatment, and heat inactivation or potassium periodate (KIO) treatment. The detailed procedures for serum pretreatment have been described elsewhere (Hsiung, 1982; Swenson, 1992b).

The most common use of the HAI test in laboratories today is for the subtyping of influenza virus isolates by state health department or World Health Organization collaborating influenza surveillance laboratories. A detailed procedure for the typing of influenza virus isolates is presented by Swenson (1992b).

The specificity of the HAI test varies with the virus. The reaction can be highly specific for certain viruses (influenza-parainfluenza groups) and less specific for other viruses (arboviruses). A procedure for measuring antibody to influenza viruses is presented below.

Materials

Blood from a guinea pig or other mammalian or avian species in Alsever's solution at a 10% (vol/vol) concentration (This suspension is stable for 2 weeks at 4°C.)

Amniotic, allantoic, or tissue culture fluid containing influenza virus

Alsever's solution

PBS, 0.01 M, pH 7.2

Disposable microtiter plates, "U" type

Calibrated diluting loops, 0.025 and 0.050 ml

Dropping pipettes, 0.025 and 0.050 ml per drop

Blotter paper with calibrating circles

Microtiter plate reading mirror

37 and 56°C water baths

Receptor destroying enzyme, 100 U/ml

Preparation of RBC Suspension

To prepare the RBC suspension, see steps 1 through 5 of the hemadsorption method, above.

Titration of HA

1. Prepare a 0.5% (vol/vol) solution of RBCs in PBS from the stock suspension (1 ml of 10% [vol/vol] RBCs added to 19 ml of PBS).
2. Prepare stocks of the influenza A and B viruses currently circulating in one of the following formats: cell culture fluid, allantoic fluid, or amniotic fluid.
3. Dilute the influenza A and B stock solution 1:10 in PBS, and add 0.1 ml to the first well of each row on the microtiter plate. Add 50 μl of PBS to wells 2 through 9 in that row.
4. Place a calibrated 0.05-ml diluter in well 1, and transfer the contents to well 2.
5. Move the diluter from well to well, resulting in a dilution series from 1:10 to 1:2,560.
6. Prepare an RBC control well by adding 0.05 ml of PBS to a well on the microtiter plate.
7. Add 50 μl of the 0.5% (vol/vol) RBC suspension to each well using the calibrated dropping pipettes.
8. Mix the contents of the plate gently, cover the plate, and incubate it at room temperature until the RBC controls form a tight button (1 to 2 h).

9. The HA titer is the highest dilution of virus capable of causing agglutination; agglutinated RBCs form a lattice on the bottom of the well, while nonagglutinated RBCs will form a discrete button at the bottom of the well (Fig. 2). The HA titer is the reciprocal of the highest dilution of virus showing agglutination and represents 1 HA unit/0.05 ml of virus. Dilute the virus suspension to contain 4 HA units/0.025 ml (or 8 HA units/0.05 ml) for the HAI test. If the HA titer is 160, then the original virus stock will be diluted 1:20 for the HAI test.

HAI Test for Influenza Virus

1. Acute and convalescent sera, pretreated to remove nonspecific inhibitors, should be tested for antibodies to one or more of the circulating strains of influenza virus. The HA titer for each virus used in the assay should be 4 HA units/0.025 ml.
2. Reference antiserum for each virus should be included to confirm the identities of virus strains used in the test.
3. Prepare a 1:10 dilution of the treated acute and convalescent serum in PBS, and add 0.05 ml of this dilution to well 1 of a dedicated row.
4. Add 0.025 ml of PBS to wells 2 through 9 in each row.
5. Prepare twofold dilutions of each treated serum using a calibrated 0.025-ml microtiter diluter. This is accomplished by placing the diluter in the first well of each row and sequentially transferring 0.025 ml from well 1 to well 9. The serum dilutions span the range from 1:10 to 1:2,560. This same dilution series is used with the control antiserum.
6. A control reaction in which only the serum and cells are added is set up for each serum.
7. Add 0.025 ml of each virus suspension (4 HA units/0.025 ml) to each serum dilution and the serum control.

   A back titration of the virus suspension is included in each HAI assay. Five wells in a row are used for each back titration. Fifty microliters of PBS is added to wells 1 through 5 and three additional wells. Fifty microliters of the virus suspension is added to well 1. Twofold dilutions of the virus suspension are prepared using a precalibrated 0.05-ml microtiter diluter.

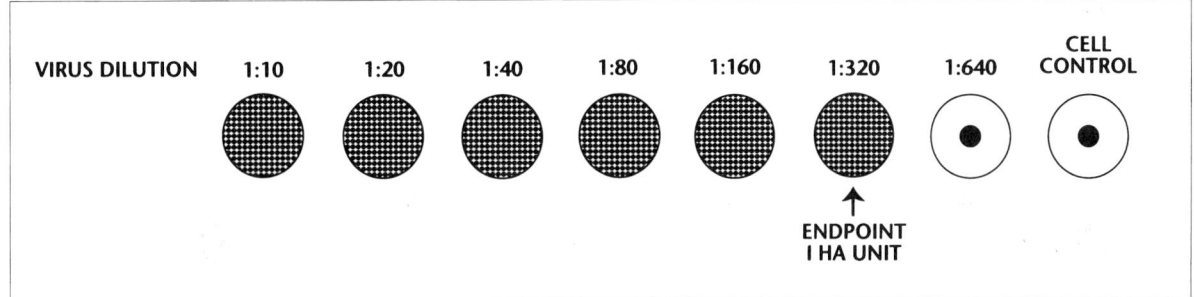

FIGURE 2 HA endpoint determination.

The diluter is placed in the first well, and then 0.05 ml is transferred sequentially from wells 1 to 4.

8. Gently shake the microtiter plate after the serum dilutions and HA have been added, cover the plate, and incubate it at room temperature for 30 min.

9. Add 0.05 ml of the 0.5% RBC suspension to each well using a calibrated dropping pipette. This suspension should be added to all the serum dilutions, the back titration, serum controls, and RBC control wells.

10. Gently shake the microtiter plate and incubate it at room temperature until the RBC control has a button at the bottom of the well (30 to 60 min).

11. The back titration of the virus suspension should show hemagglutination in the first two to three wells, indicating the dilution of the virus suspension was correct. The serum and RBC controls should show the absence of agglutination.

12. The HAI titer of each serum is defined as the highest dilution of serum that completely inhibits hemagglutination. The HAI titer of the acute-phase serum is 10, and for the convalescent serum the titer is 160. A fourfold or greater rise in HAI titers is interpreted as significant and is indicative of recent influenza virus infection or vaccination (Hsiung, 1982; Swenson, 1992b). An example of an acute and a convalescent serum titrated against a single influenza virus strain is presented in Fig. 3.

Other Viruses to Which the HAI Test Is Applicable

The procedure described above is applicable to the parainfluenza viruses, rubella virus, measles virus, reovirus, adenovirus, and togavirus groups with some modifications. The appropriate chapters in this text should be consulted for details. Modifications that are necessary for these viruses are summarized below (Hsiung, 1982; Swenson, 1992b).

1. Modifications for parainfluenza viruses include the use of human O, guinea pig, or chicken RBCs; pretreatment of the serum with RDE heat treatment; and, in a majority of cases, adsorption with guinea pig RBCs.

2. Modifications for rubella virus include the following: 1-day-old chick or goose RBCs are used, the diluent is HEPES-saline-albumen-gelatin, and sera must be treated with $MnCl_2$-heparin and adsorbed with chick RBCs to remove nonspecific inhibitors and agglutinins.

3. Modifications for the adenovirus group include the following: rhesus monkey RBCs are used for group 1 adenovirus serotypes, and rat RBCs are used for adenoviruses in group II, and the sera must be heat inactivated at 56°C for 30 min and adsorbed with the type of RBCs used in the HAI test.

4. Modifications for the reoviruses include the use of human O RBCs and pretreatment of the sera by both heat inactivation and kaolin adsorption.

5. Modifications for members of the arbovirus group include the use of buffers with pH values ranging from 6.0 to 7.4 and pretreatment of the sera by heat inactivation and kaolin adsorption.

6. Modifications for the serotypes of enteroviruses that are capable of HAI include using human O RBCs, changing the incubation temperature depending upon the serotype, and pretreatment of the sera by heat inactivation and kaolin adsorption. Coxsackievirus A-20, A-21, and A-24 and echovirus types 3, 11, 13, and 19 agglutinate at 4°C. Coxsackievirus B-1, B-3, and B-5 and echovirus types 6, 7, 12, 20, and 21 hemagglutinate at 37°C.

REFERENCES

Hodinka, R. L. 1999. Serological tests in clinical virology, p. 195–211. In E. H. Lennette and T. F. Smith (ed.), Laboratory Diagnosis of Viral Infections, 3rd ed. Marcel Dekker, Inc., New York, N.Y.

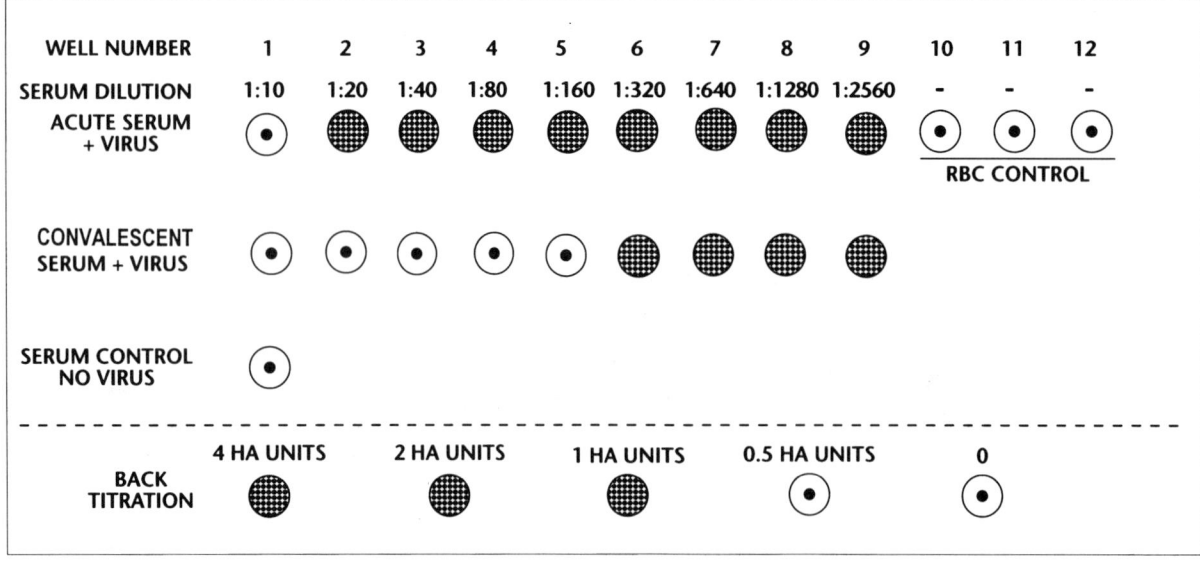

FIGURE 3   Determination of HAI titers for acute- and convalescent-phase sera.

Hsiung, G. D. 1982. Hemagglutination and the hemagglutination—inhibition test, p. 35–41. *In* G. D. Hsiung (ed.), *Diagnostic Virology*, 3rd ed. Yale University Press, New Haven, Conn.

Leland, D. S. 1996. Hemadsorption testing, p. 195–197. *In* D. S. Leland (ed.), *Clinical Virology*. W. B. Saunders Company, Philadelphia, Pa.

Mahony, J. B., and M. A. Chernesky. 1999. Immunoassay for the diagnosis of infectious diseases, p. 202–214. *In* P. R. Murray, E. J. Baron, M. A. Pfaller, F. C. Tenover, and R. H. Yolken (ed.), *Manual of Clinical Microbiology*, 7th ed. Amercian Society for Microbiology, Washington, D.C.

Schmidt, N. J., and R. W. Emmons. 1989. General principles of laboratory diagnostic methods for viral, rickettsial and chlamydial infections, p. 1–35. *In* N. J. Schmidt and R. W. Emmons (ed.), *Diagnostic Procedures for Viral, Rickettsial and Chlamydial Infections*, 6th ed. American Public Health Association, Washington, D.C.

Swenson, P. D. 1992a. Detection of viruses by hemadsorption, p. 8.8.1–8.8.5. *In* H. D. Isenberg (ed.), *Clinical Microbiology Procedures Handbook*. American Society for Microbiology, Washington, D.C.

Swenson, P. D. 1992b. Hemagglutination inhibition test for the identification of influenza viruses, p. 8.12.1–8.12.11. *In* H. D. Isenberg (ed.), *Clinical Microbiology Procedures Handbook*. American Society for Microbiology, Washington, D.C.

Immune Adherence Hemagglutination

EVELYNE T. LENNETTE AND DAVID A. LENNETTE

14

Each of the various serologic test procedures described in this volume has its advantages and disadvantages. Their relative utilities depend on the specific requirements of the laboratory worker, and selection of an appropriate assay requires careful analysis, as no assay is ideal for all situations. The following discussion is oriented to the needs of a clinical diagnostic laboratory. Clinical laboratories need serologic test procedures of high specificity and with sensitivity suitable both for determination of immunity and for diagnosis of current, or recent, infections. Procedures should be convenient and rapid to perform on a single sample or on a hundred serum samples. The demand on equipment and reagents should be within the reach of smaller laboratories with limited resources. Test antigens should be varied and readily available or easily made. Of significance to a reference laboratory is the ease with which assays may be adapted for use with new or additional test antigens.

Some years ago, when we reviewed many serologic procedures applicable for use in virology, two procedures appeared to meet most of the requirements just outlined. One was the complement fixation (CF) test, and the other was immune adherence hemagglutination (IAHA), a test virtually unknown in the United States although well developed in Japan. In the intervening years, enzyme immunoassays (EIA) have displaced CF testing for most routine serology. Interestingly, CF tests remain in use in some reference virology laboratories, insensitive as they are compared to other serologic techniques: radioimmunoassay (RIA), EIA, hemagglutination inhibition, and indirect fluorescence assay (IFA). Continued use of the CF test is encouraged by the wider range of available antigens suitable for this assay, as well as a large amount of experience with the CF procedure for clinical diagnosis, as shown by publications in the medical literature. In contrast, the limited inventory of commercially available antigens for enzyme-linked immunosorbent assays, users' reluctance to deal with the hazards and waste disposal problems of RIA procedures, and the unsuitability of IFAs for examining large numbers of specimens have all acted to limit uniform adoption of these assays. Yet, despite the progress made with EIA in recent years, IAHA remains a viable alternative for many reference laboratories.

In 1976, immune adherence assays were known in the United States only to a few researchers studying the hepatitis A virus and varicella-zoster virus (VZV) and to a number of Japanese workers who were instrumental in developing the procedure for use in clinical laboratories. As indicated below, IAHA is as specific as CF but with sensitivity comparable to that of EIA and IFA. Its main advantages are quantitation, simplicity, and the potential for automation. Additional uses for the IAHA test have developed since publication of the first edition of this book, and they are referenced in Table 1.

HISTORICAL BACKGROUND

The IAHA assay is based on a phenomenon known as serologic adhesion, observed by Levaditi (1901). After injecting antibody-coated cholera vibrios into guinea pigs, he noted platelet aggregation (or adhesion) to the vibrios. In the same year, similar observations were reported by Laveran and Mesnil (1901) with *Trypanosoma lewisi* and immune rats. Kritchewski later showed that this adhesion reaction required complement. Other French and Russian workers then improved and applied the adhesion test for the in vitro assay of antibodies to trypanosomes, leishmania, leptospira, and spirochetes (Lamanna, 1957). In 1952, Nelson showed that similar adhesion occurs if human erythrocytes are substituted for platelets. The agglutination was examined with a microscope after antigen, serum, complement, and erythrocytes were mixed on a slide. Nelson (1963) coined the name immune adherence for the reaction. Ito and Tagaya (1966) adapted the procedure for use with microtest plates. Although the potential utility of the IAHA procedure was recognized early, it failed to come into widespread use due to a shortcoming: the agglutination reaction was reversible and was not stable enough for the reactions to be read reliably. This problem was finally overcome when Mayumi and coworkers (1971) introduced the use of dithiothreitol (DTT) as a stabilizer while developing an IAHA procedure to detect hepatitis B surface antigen serum. With that single improvement, the repertoire of antigens found to be suitable for use in the IAHA assay grew quickly to include enteroviruses, adenoviruses, and hepatitis B virus. In the United States, the IAHA test was first used in hepatitis A seroepidemiology studies (Hilleman et al., 1975; Miller et al., 1975). Table 1 provides a list of many of the successful applications of IAHA assays, showing that it has been used with many viral, bacterial, fungal, and mycoplasmal antigens.

TABLE 1　Applications of IAHA

Organism class	Antigen(s)	Reference(s)
Bacterial	*L. pneumophila*	Lawrence and Wentworth, 1985
Chlamydial	*Chlamydia psittaci; Chlamydia trachomatis*	Lennette and Lennette, 1978
Fungal	*Candida albicans*	Nagayoshi et al., 1980
	Blastomyces	Lennette, unpublished
	Histoplasma capsulatum (mycelial)	Lennette, unpublished
Rickettsial	*Coxiella burnetii*	Lennette, unpublished
Viral	Adenoviruses	Ito and Tagaya, 1966
	BK papovavirus	E. T. Lennette and K. Shah, unpublished data
	Cytomegalovirus	Dienstag et al., 1976
	Dengue viruses	Inouye et al., 1980
	Enteroviruses	Ito and Tagaya, 1966
	EBV	Lennette et al., 1982
	Herpes simplex virus	Ito and Tagaya, 1966
	Hantaviruses	Suntharee et al., 1981
	Hepatitis A virus	Miller et al., 1975
	Hepatitis B virus	Tsuda et al., 1975
	Japanese encephalitis virus	Inouye et al., 1981
	Mouse mammary tumor virus	Nagayoshi et al., 1981
	Measles virus (rubeola)	Lennette et al., 1979
	Norwalk virus	Kapikian et al., 1978
	Rabies virus	Budzko et al., 1983; Zanetti et al., 1989
	Rotavirus	Kapikian et al., 1981; Nagayoshi et al., 1980
	Simian virus 40	Ichikawa et al., 1987
	VZV	Gershon et al., 1976
Heterophile	Paul-Bunnell heterophile	Kapikian et al., 1981

IAHA MICROTITER PROCEDURE

In the IAHA test, antibodies and antigens form complexes during the first incubation, which is followed by the addition of complement. The resulting antibody-antigen-complement complexes then react with C3b receptors on human erythrocytes to cause hemagglutination.

The equipment, reagents, and manipulations required for the performance of the IAHA test are similar to those used for the CF test. The following is a step-by-step description of the procedure. The compositions of the solutions used for the IAHA test are found in the appendix to this chapter.

1. Dilute the test sera in diluting buffer, Veronal-buffered saline (VBS; pH 3), at 1:4, and then inactivate the sera at 56°C for 30 min.
2. Rinse V-well microtiter plates prior to use with VBS containing gelatin (GVB). Discard the rinse solution by inverting the plates over a sink, and then rap the inverted plates against a hard surface to remove residual buffer.
3. Add 1 drop (25 µl) of VBS containing bovine serum albumin (BVB) to every well that will be used.
4. Add 1 drop of each inactivated serum to the first and eighth well of one row on the test plate. Prepare serial twofold dilutions of the added serum (two sets of seven and five wells), using microdilutors.
5. Add 1 drop of positive test antigen (previously titrated to contain 1 to 2 U of reactivity) to the first seven wells of each row. Add identically diluted negative (control) antigen to wells 8 through 12. The correct antigen concentrations are determined by block titrations with control reference sera.
6. Shake the plates for 10 s with a vibrating mixer, and then incubate them at 37°C for 30 min. It is also acceptable to incubate the plates in a refrigerator overnight

(useful with heat-labile antigens). All plates should be covered during the incubations to minimize losses due to evaporation.

7. Add 1 drop of diluted guinea pig complement (1:100 in BVB) to all wells on the test plates. Again, mix and incubate the plates at 37°C for 40 min. Determine the exact dilution of complement by a prior block titration; ordinarily no excess complement is used.
8. At the end of the incubation, stop the reaction by adding 1 drop of DTT-EDTA to each well, and then add 1 drop of a 0.8% suspension of human erythrocytes, type O, to each well.
9. Hemagglutination should be complete and readable within 1 h. Positive reactions are those showing >50% agglutination. The agglutination pattern is usually stable at room temperature; for best results, we chill the test plates (overnight) if they are not read within an hour or so.

Either U- or V-bottom microtest plates can be used for IAHA tests, although the hemagglutination patterns are easier to read on the V plates. We find that plates made of polystyrene plastic may give nonspecific binding of erythrocytes, which can be eliminated by prewetting the plates with a buffer containing carrier protein. If gelatin is used in the prewetting buffer, each lot of gelatin should be screened for suitability, as some lots contain heat-labile substances that interfere with the IAHA test. The interference is eliminated by autoclaving the gelatin stock solution for 15 min at 121°C.

Many acute-phase serum samples appear to contain immune complexes, which persist even after the diluted serum has been inactivated at 56°C. These complexes give positive IAHA reactions, even in the absence of added test antigens. Sera containing these immune complexes are not found as frequently as are sera that are reported as "anti-complementary" in the CF test; nevertheless, the immune complexes are an annoyance. We have been able to eliminate the nonspecific reactions due to these immune complexes by diluting test sera in barbital-saline adjusted to pH 3 (essentially unbuffered) and then heating the diluted sera at 56°C for 30 min. It appears that this modified inactivation procedure removes interference from the complexes. If the serum is not diluted more than 1:5, no subsequent adjustment of pH is necessary.

Commercially available antigens for use with the CF test have been satisfactory for use in the IAHA test, provided the antigen CF test titer was at least 1:8. Antigen concentrations needed for the IAHA test are generally only one-fourth or one-eighth those required for the CF test, which may result in significant savings of cost and material. We have prepared many different viral antigens and compared them to commercially available materials. In nearly all cases, antigens prepared from cell culture material in house have been of better quality and higher titer than the commercially prepared antigens. The following procedure is satisfactory for the preparation of many viral antigens for use in the IAHA test.

Infected and uninfected cell cultures are separately suspended in saline equal in volume to one-fifth their original growth medium. The cell suspensions are then disrupted by sonication and clarified by centrifugation at 10,000 × g for 10 min. Further purification or concentration of the antigen preparation is usually unnecessary. The preparation can be dispensed in small volumes and stored at −70°C for indefinite periods. When proper conditions for virus replication are used, antigen titers of 32 or greater should be expected. Photochemical inactivation of virus infectivity in the antigen preparation with psoralens and long-wave UV irradiation has given us excellent results, using the method described by Hanson et al. (1978).

We have prepared bacterial antigens using the confluent growth obtained on appropriate agar media, which can be scraped, suspended in saline, and inactivated by heating or the addition of formalin (at 0.1%). The suspension is washed twice by centrifugation at 10,000 × g for 20 min. The washed sediment is then adjusted to a 10% suspension and sonicated to disrupt the cells. For disruption of 1 to 2 ml of suspension, we use three cycles of 15 s each at 10-W output of a microprobe-type sonicator. The preparation is clarified at 10,000 × g and stored at −70°C until it is used. The antigens can be prepared very efficiently in this manner, and the antigen titer achievable is usually between 500 and 2,000.

Egg-derived antigens have not been satisfactory in the IAHA test. Such antigens include rickettsiae and chlamydiae grown in yolk sacs. Extraction of lipids from these antigens with Freon 113 improves their performance in the IAHA test, but at the cost of decreased antigen titers. Marmoset liver-derived hepatitis A virus is purified by density gradient centrifugation to prepare antigen suitable for IAHA tests (Miller et al., 1975). The use of Freon 113 and other chlorofluorocarbons has undesirable environmental effects and is being eliminated in most laboratories.

The stability of viral and bacterial antigens for the IAHA test is very good, provided they do not become microbially contaminated. Addition of 0.01% thimerosal preservative reduces, but will not eliminate, such contamination. A minor problem encountered with frozen antigen preparations is precipitation of antigen that has been stored

and thawed. This can usually be counteracted by brief sonication of the freshly thawed antigen.

RED BLOOD CELLS

Primate erythrocytes and nonprimate platelets are reported to be suitable indicator cells for the IAHA test (Nelson, 1963). The only widely used cell type is human type O erythrocytes. The nature of the reactive site on the cells is unknown—the presence of C3b receptors on the cells is necessary but not sufficient to provide IAHA reactivity, as only about one out of three O-positive donors has suitable erythrocytes (Klopstock et al., 1963). The receptors are sensitive to proteolytic enzymes and neuraminidase, consistent with a glycoprotein composition (E. T. Lennette, unpublished observations). The ability of cells to react in the IAHA procedure in the presence of immune complexes is apparently a permanent property of the donor: erythrocytes obtained from one of the authors are known to have been reactive over a span of >20 years. Hence, the suitability of any particular donor need only be checked once. Obtaining a supply of reactive erythrocytes is the obstacle preventing some laboratories from adopting IAHA procedures. There is no commercial supply of pretested blood, so each laboratory needs to find its own supply. Our laboratory uses two sources of blood: laboratory personnel and local blood banks. Type O blood from a blood bank can be used for 2 to 3 weeks; the sample segments (which are often discarded) from blood bags may supply enough cells for a useful number of IAHA tests; they are used at a rate of one segment per day in our laboratory. Often there are as many as eight segments from each bag of blood that is used for blood derivatives; these segments are not used by the blood bank and could be made available to the clinical laboratory. Blood collected in EDTA is not suitable for use in the IAHA test.

In our laboratory, a reference antigen and a set of reference sera are used to screen blood samples for suitable erythrocytes. Only erythrocytes that give acceptable serum titers are selected for use. Donated blood segments can be stored for up to 2 weeks at 4°C before use. Although this means of obtaining erythrocytes is less convenient than having access to a panel of preselected donors, we find it presents the fewest problems. We do not recommend pooling cells from randomly selected donors; using pooled erythrocytes with the IAHA procedure gives substantially lower sensitivity. Limited efforts have been made to increase the shelf life of erythrocytes for use in the IAHA test. Cells fixed with glutaraldehyde or formaldehyde can be used in IAHA procedures, but fixed cells are less satisfactory than fresh cells. A method for freezing human erythrocytes for later use in the IAHA test has been reported (Lawrence and Wentworth, 1985), which eliminates the need to screen blood donors.

HEMAGGLUTINATION PATTERNS

The agglutination pattern seen in the IAHA test is uniform and does not vary appreciably with different test antigens. However, one should become familiar with the agglutination patterns produced by individual sera. The patterns in microtest plates are best examined with the aid of a magnifying mirror. A specific positive reaction should appear as uniformly granular agglutination, and a negative reaction should appear as a cell button with a smooth outline. While reading test results, one should compare the patterns obtained with test sera to those obtained with the control or reference sera. With practice, it is usually possible to differentiate specific agglutination from nonspecific reactions. In addition to the agglutination sometimes seen with negative control antigens, some sera give an uncharacteristic agglutination with the appropriate test antigen. Specific reactions give a slightly granular or coarse agglutination; nonspecific reactions often produce a fine-textured or matte agglutination. Even when freshly collected sera are inactivated in low-pH diluent, about 1% of the specimens tested have equal titers with positive and control antigens. It is usually necessary to resort to another test system to evaluate such sera.

DETECTION OF IgM ANTIBODIES

Immunoglobulin M (IgM) antibodies are not efficiently detected by the CF assay. As both CF and IAHA tests detect complement-activating immune complexes, it is somewhat surprising to discover that IgM antibodies are detectable by IAHA assay. The first evidence that IgM antibodies react well in IAHA tests came from the application of IAHA testing to the detection of the Paul-Bunnell antibodies associated with infectious mononucleosis caused by Epstein-Barr virus (EBV). The IAHA test was both more sensitive and more specific than the standard differential heterophile agglutination test and ox-cell hemolysis test (Lennette and Lennette, 1978; Evans and Niederman, 1982). Paul-Bunnell antibodies are exclusively of the IgM class. Other evidence that IgM antibodies are reactive in IAHA came from work employing sera that were fractionated by sucrose gradient density centrifugation in studies using sera obtained from varicella and zoster patients (Gershon et al., 1981). Fractions containing either VZV-specific IgM or IgG were both reactive in the IAHA test, provided that the IAHA test was performed promptly after fractionation of the sera. In our experience, concentrated sucrose has a detrimental effect on IAHA-reactive IgM even after brief storage: we recommend other methods for physical separation of immunoglobulin classes, e.g., Quik-sep IgM columns (Isolab, Inc., Akron, Ohio).

We have observed that the detection of IgM antibodies by IAHA depends on the quality of the test antigens used. Using sera containing Paul-Bunnell heterophile antibody, it was possible to show that antigen lots varied greatly in reactivity. Results obtained by box titration of different lots of antigen can differ by as much as 10-fold. This observation has been extended to include other commercially available viral antigens for use in CF tests, e.g., cytomegalovirus and VZV antigens. The lot-to-lot variation of an antigen preparation provided by any one supplier is seen as a shift in the optimal titer of each lot of antigen when IgM-containing reference sera are tested. Although the optimal titer of antigen varies from lot to lot, the reference serum titers do not vary. In contrast, the variations in antigen preparations from supplier to supplier affect the titers of the test sera. That is, the serum titer obtained can vary greatly depending upon the source of antigen. This variation is seen mainly with IgM-containing sera and is not appreciable with sera containing IgG only, e.g., from immune donors, suggesting different reactivities of these immunoglobulin classes to viral antigens. At present, there is no standardization among suppliers as to the composition of viral test antigens used in serologic reactions to contain any particular specificity. The differences are noticeable in the CF test but are more pronounced in the IAHA test, especially when sera containing IgM antibodies are tested. We recommend that serum panels for antigen titration and evaluation include known IgM-positive sera when possible.

SENSITIVITY AND SPECIFICITY

IAHA is most often compared to CF, due to their similarities in manipulations and reagent requirements. Also, both procedures measure complement activation by immune complexes: the IAHA test measures it directly, and the CF test measures it by complement depletion. There are, however, a few differences between the two tests that explain the advantages found with IAHA.

First, the tests differ in the nature of the indicator system used to measure the immune complexes formed in vitro. In the CF test, a measured amount of complement is added to the test wells; any complement not bound in the test reactions is then indicated by the addition of sheep red blood cell (SRBC)–anti-SRBC complex. The remaining complement reacts to lyse the SRBCs. This procedure is inherently insensitive to small differences in depletion of the excess complement added at the beginning of the test. In the IAHA procedure, only the complement reacted in the test is measured; the test is insensitive to excess unreacted complement. Any agglutination above a very low background is significant. This difference in mechanism between the two tests accounts for the four- to eightfold increase in sensitivity shown by the IAHA method.

The increased sensitivity of the IAHA test has been shown to be adequate for its reliable use for determination of immune status, even of immunosuppressed patients (Gershon et al., 1976). Agents for which this has been shown include VZV, cytomegalovirus, EBV, and others (Table 1). This increased sensitivity, together with the ability to detect IgM antibodies, allows the IAHA test to demonstrate a more rapid and pronounced titer increase during an infection than can be seen with the CF test. With the CF test, a minimum of 2 weeks is advised for the reliable detection of antibody titer changes between acute-phase and convalescent-phase serum samples. With the IAHA test, 8- to 16-fold titer changes can frequently be found with sera collected 3 to 5 days apart during the acute or early convalescent phase of illness.

IAHA is comparable to IFA, rather than CF, in its sensitivity, as shown by parallel serologic testing of numerous sera with varied antigens. Although IFA titers are often slightly higher than those obtained with IAHA, the ability of the IFA to differentiate positive from negative sera is the same as that of the IAHA test. In our laboratory, IFA and the IAHA test can be used almost interchangeably, providing useful and complementary test systems for a wide range of antigens.

IAHA is comparable in reactivity to both fluorescent antibody to membrane antigens and neutralization tests (Gershon et al., 1976) for determination of varicella-zoster immune status. The data of Baba et al. (1987) showed that the IAHA test clearly differentiates infections with VZV from those with herpes simplex virus. For determination of postimmunization immunity to rabies, IAHA was found to be as suitable as the "rapid" fluorescent-focus inhibition test (RFFIT) which is commonly used (Budzko et al., 1983; Bota et al., 1987; Zanetti et al., 1989) and more convenient to perform: the RFFIT takes several days to complete and requires the use of live rabies virus.

Limited comparisons of IAHA to RIA and EIA indicate that IAHA is slightly less sensitive than these methods. The sensitivity difference does not appear to be significant in routine clinical applications. The high specificity of IAHA, as well as its less demanding and less arbitrary standardization requirements, is a distinct advantage over EIA or RIA.

IAHA tests are useful to detect antigenic differences among related viruses, and they offer very good type specificity. For example, IAHA has been used to subgroup rotaviruses in a system different from that developed by neutralization tests (Kapikian et al., 1981). IAHA is also reported to be useful for serotyping dengue virus isolates (Suntharee et al., 1981), and it has been used to classify hemorrhagic fever with renal syndrome viruses into serotypes (Sugiyama et al., 1984), where it proved superior to IFA for classification. In our own laboratory, IAHA has been used to subtype *Legionella pneumophila* with hyperimmune rabbit sera. The ability to detect antigenic differences against a background of cross-reacting specificities allows the IAHA test to be used in monitoring antigen purification, e.g., IAHA purification of rotavirus extracted from stools (Kapikian et al., 1981). In the purification of EBV antigens, IAHA was found to detect specific EBV antigens and was unaffected by the presence of other antigens (impurities) in the antigen preparations.

ADVANTAGES AND DISADVANTAGES

The principal limitation of the IAHA test lies in its inability to differentiate antibodies of different immunoglobulin subclasses. Although the IAHA assay detects IgM antibodies as well as IgG antibodies, the two classes cannot be measured separately unless they are physically separated, as by column chromatography or density gradient centrifugation. In addition, IAHA will not react with antibodies that do not fix complement, e.g., IgA. Another disadvantage of the IAHA test is that it is difficult to use with hemagglutinating viruses. In theory, it is possible to obtain antigens free of hemagglutinins. In practice, commercial CF antigens for viruses, such as influenzas and mumps viruses, often do contain hemagglutinin activity and are not suitable for use in the IAHA test.

We find the IAHA test to be satisfactory for routine use for reasons other than sensitivity and specificity. The procedure is quite economical in the use of reagents and supplies. Due to its increased sensitivity compared to the CF test, the amounts of antigen and complement used may be reduced about fourfold, on average. Although no special equipment is required to perform IAHA tests, partial to full automation of the test procedure is possible. Titer endpoints are usually very sharp, with 4+ agglutination in one well and no agglutination in the next well. Thus, there is little uncertainty or subjectivity in obtaining an accurate titer—often a problem with reading IFA, for example. Titers are reproducible, both within test runs and between runs, which makes it easy to detect significant titer changes.

Another advantage of the IAHA assay is the simplicity of pretest preparations. Every test component of the CF procedure has to be monitored carefully and titered for each test run. With the IAHA procedure, every component is added in excess and needs to be titered only once for each lot of reagent, as long as the reagent is stable during storage. Also, the CF test usually requires overnight incubation, whereas the IAHA test is completed within 3 to 4 h.

APPENDIX
Veronal buffer (VB) 5× stock

Dissolve 43.0 g of NaCl and 4.6 g of diethyl barbituric acid in 950 ml of warm deionized water. Adjust the pH with NaOH to 7.4. Add 2.5 ml of $MgCl_2$-$CaCl_2$ solution, and adjust the volume to 1 liter. The $MgCl_2$-$CaCl_2$ solution contains 20.33 g of $MgCl_2$·$6H_2O$ and 4.4 g of $CaCl_2$ in 100 ml of water.

Serum dilution buffer

VB (1×) is prepared by diluting the above-mentioned stock with 4 volumes of deionized water. The pH is adjusted to 3.0 with 2 N HCl.

BVB

Add bovine serum albumin, fraction V, to VBS (1X) to a final concentration of 1 mg/ml.

GVB

Add 2.5% autoclaved gelatin to VBS (1X) to a final concentration of 0.125%.

EDTA-DTT-VBS buffer

Add 2 parts 0.1 M EDTA, pH 7.5 (disodium EDTA), to 3 parts VBS (1X). Add DTT to a final concentration of 3 mg/ml before using the buffer; after DTT is added, the buffer should not be used for more than 3 to 4 weeks, stored at 5°C.

REFERENCES

Baba, K., K. Shiraki, T. Kanasaki, K. Yamanishi, P. L. Ogra, H. Yabuuchi, and M. Takahashi. 1987. Specificity of skin test with varicella-zoster virus antigen in varicella-zoster and herpes simplex virus infections. *J. Clin. Microbiol.* **25:**2193–2196.

Bota, C., R. Anderson, S. Goyal, L. Charamella, D. Howard, and D. Briggs. 1987. Comparative prevalence of rabies antibodies among household and unclaimed/stray dogs as determined by the immune adherence hemagglutination assay. *Int. J. Epidemiol.* **16:**472–476.

Budzko, D. B., L. J. Charamella, D. Jelinek, and G. R. Anderson. 1983. Rapid test for detection of rabies antibodies in human serum. *J. Clin. Microbiol.* **17:**481–484.

Dienstag, J. L., W. L. Cline, and R. H. Purcell. 1976. Detection of cytomegalovirus antibodies by immune adherence hemagglutination. *Proc. Soc. Exp. Biol. Med.* **153:**543–548.

Evans, A. S., and J. C. Niederman. 1982. EBV-IgA and new heterophile antibody tests in diagnosis of infectious mononucleosis. *Am. J. Clin. Pathol.* **77:**555–560.

Gershon, A. A., Z. G. Kalter, and S. P. Steinberg. 1976. Detection of antibody to varicella-zoster virus by immune adherence hemagglutination. *Proc. Soc. Exp. Biol. Med.* **151:**762–765.

Gershon, A. A., S. P. Steinberg, W. Borkowsky, D. A. Lennette, and E. T. Lennette. 1981. IgM to varicella-zoster virus: demonstration in patients with and without clinical zoster. *Pediatr. Infect. Dis.* **1:**164–166.

Hanson, C. V., J. L. Riggs, and E. H. Lennette. 1978. Photochemical inactivation of DNA and RNA viruses by psoralen derivatives. *J. Gen. Virol.* **40:**345–358.

Hilleman, M. R., P. J. Provost, W. J. Miller, V. M. Villarejos, O. L. Ittensohn, and W. J. McAleer. 1975. Development and utilization of complement-fixation and immune adherence tests for human hepatitis A virus and antibody. *Am. J. Med. Sci.* **270:**93–98.

Ichikawa, T., N. Minamoto, T. Kinjo, K. Matsubayashi, and I. Narama. 1987. A serological survey of simian virus 40 in monkeys. *Microbiol. Immunol.* **31:**1001–1008.

Inouye, S., S. Matsuno, A. Hasegawa, K. Miyamura, R. Kono, and L. Rosen. 1980. Serotyping of dengue viruses by an immune adherence hemagglutination test. *Am. J. Trop. Med. Hyg.* **29:**1389–1393.

Inouye, S., S. Matsuno, and R. Kono. 1981. Difference in antibody reactivity between complement fixation and immune adherence hemagglutination tests with virus antigens. *J. Clin. Microbiol.* **14:**241–246.

Ito, M., and I. Tagaya. 1966. Immune adherence hemagglutination test as a new sensitive method for titration of animal virus antigens and antibodies. *Jpn. J. Med. Sci. Biol.* **19:**109–126.

Kapikian, A. Z., H. B. Greenberg, W. L. Cline, A. R. Kalica, R. G. Wyatt, H. James, Jr., N. L. Lloyd, R. M. Chanock, R. W. Ryder, and H. W. Kim. 1978. Prevalence of antibody to the Norwalk agent by a newly developed immune adherence hemagglutination assay. *J. Med. Virol.* **2:**281–294.

Kapikian, A. Z., W. L. Cline, H. B. Greenberg, R. G. Wyatt, A. R. Kalica, C. E. Banks, H. James, Jr., J. Flores, and R. M. Chanock. 1981. Antigenic characterization of human and animal rotaviruses by immune adherence hemagglutination assay (IAHA): evidence for distinctness of IAHA and neutralization antigens. *Infect. Immun.* **33:**415–425.

Klopstock, A., J. Schwartz, and N. Zipkis. 1963. Individual difference of the reactivity of human erythrocytes in the immune adherence hemagglutination test. *Vox Sang.* **8:**382–383.

Lamanna, C. 1957. Adhesion of foreign particles to particulate antigens in the presence of antibody and complement (serological adhesion). *Bacteriol. Rev.* **21:**30–45.

Laveran, A., and F. Mesnil. 1901. Recherches morphologiques et experimentales sur le trypanosome des rats (Tr. lewisi Kent). *Ann. Inst. Pasteur* **15:**673–714.

Lawrence, T., and B. Wentworth. 1985. Freezing and rejuvenation of human O erythrocytes for use in the immune hemagglutination test. *J. Clin. Microbiol.* **22:**654–655.

Lennette, D. A., E. T. Lennette, B. B. Wentworth, M. L. V. French, and G. L. Lattimer. 1979. Serology of Legionnaires' disease: comparison of indirect immunofluorescent antibody, immune adherence hemagglutination, and indirect hemagglutination tests. *J. Clin. Microbiol.* **10:**876–879.

Lennette, E. T., and D. A. Lennette. 1978. Immune adherence hemagglutination: alternative to complement-fixation serology. *J. Clin. Microbiol.* **7:**282–285.

Lennette, E. T., E. Ward, G. Henle, and W. Henle. 1982. Detection of antibodies to Epstein-Barr virus capsid antigen by immune adherence hemagglutination. *J. Clin. Microbiol.* **15:**69–73.

Levaditi, C. 1901. Sur l'etat de la cytose dans la plasma des animaux normaux et des organismes vaccine contre le vibrion cholerique. *Ann. Inst. Pasteur* **15:**894–927.

Mayumi, M. K., K. Okochi, and K. Nishioka. 1971. Detection of Australian antigen by means of immune adherence hemagglutination test. *Vox Sang.* **20:**178–181.

Miller, W. J., P. J. Provost, W. J. McAleer, O. L. Ittensohn, V. M. Villarejos, and M. R. Hilleman. 1975. Specific immune adherence assay for human hepatitis A antibody application to diagnostic and epidemiologic investigations. *Proc. Soc. Exp. Biol. Med.* **149:**254–261.

Nagayoshi, S., H. Yamaguchi, T. Ichikawa, M. Miyazu, T. Morishima, T. Ozaki, S. Isomura, S. Suzuki, and M. Hoshino. 1980. Changes of the rotavirus concentration in faeces during the course of acute gastroenteritis as determined by the immune adherence hemagglutination test. *Eur. J. Pediatr.* **134:**99–102.

Nagayoshi, S., M. Imai, Y. Tsutsui, S. Saga, M. Takahashi, and M. Hoshino. 1981. Use of the immune adherence hemagglutination test for titration of breast cancer patients' sera cross-reacting with purified mouse mammary tumor virus. *Gann* **72:**98–103.

Nelson, D. S. 1963. Immune adherence. *Immunology* **3:**131–180.

Segal, E., N. Vardinon, J. Foldes, J. Schwartz, and E. Eylan. 1977. Serum anti-*Candida albicans* antibodies in candidal and non-candidal vaginitis. *Zentbl. Bakteriol. Orig. A* **239:**548–553.

Sugiyama, K., Y. Matsuura, C. Morita, S. Shiga, Y. Akao, T. Komatsu, and T. Kitamura. 1984. An immune adherence assay for discrimination between etiologic agents of hemorrhagic fever with renal syndrome. *J. Infect. Dis.* **149:**67–73.

Suntharee, R., C. Charnchudhi, A. Sompop, C. Kanai, A. Igarashi, and S. Inouye. 1981. Isolation and identification of dengue viruses combined use of C6/36 cells and the immune adherence hemagglutination test. *Jpn. J. Med. Sci. Biol.* **34:**375–379.

Tsuda, F., T. Takahasi, K. Takahashi, Y. Miyakawa, and M. Mayumi. 1975. Determination of antibody to hepatitis B core antigen by means of immune adherence hemagglutination. *J. Immunol.* **115:**834–838.

Zanetti, C. R., M. S. Tino, E. L. Chamelet, M. M. Ishizuka, and O. A. Pereira. 1989. Simplification of immune adherence hemagglutination test for detection of rabies antibodies in human serum. *Rev. Inst. Med. Trop. Sao Paulo* **31:**341–345.

Immunoglobulin M Determinations

DEAN D. ERDMAN

15

The presence of specific antibody activity due to immunoglobulins in serum was reported as early as the 1930s (Heidelberger and Pederson, 1937). Subsequent studies demonstrated that the first immunoglobulins to appear after a primary antigenic stimulus were of the immunoglobulin M (IgM) class. These IgM antibodies reportedly disappeared rapidly, usually within a few weeks, and were replaced by IgG antibodies that persisted for a longer period. Decades later, Schluederberg (1965) suggested for the first time that detection of virus-specific IgM antibodies could be of value in recognizing recent virus infections. Today, IgM determinations have become routine procedures in many diagnostic virology laboratories.

The transient nature of the IgM antibody response appears to hold true for most primary virus infections; virus-specific IgM antibodies typically appear between 7 and 10 days after primary infection, reach maximal levels within 2 to 3 weeks, and then decline to undetectable levels after about 3 months, whereas IgG antibodies typically persist for years, if not indefinitely (Fig. 1). Consequently, detection of virus-specific IgM antibodies is now well established as a potentially valuable method for the rapid diagnosis of recent or current virus infections. This approach provides a considerable advantage over classical serological testing, which requires the demonstration of a significant rise in antibody titer between acute- and convalescent-phase serum specimens. For this approach to be successful, however, the IgM antibody response must be virus specific, transient (i.e., present only with recent infection by the particular virus), and measurable with adequate reliability and sensitivity.

METHODS USED FOR IgM ANTIBODY DETERMINATION

Since the introduction of the first applications of IgM determination in diagnostic virology, a variety of methods have been developed for this purpose (for a more complete review, see Meurman [1983]). These methods can generally be separated into three groups (Table 1): (i) those based on comparing IgM titers before and after chemical inactivation of serum IgM, (ii) those based on the physicochemical separation of IgM from other serum immunoglobulin classes, and (iii) those based on solid-phase immunologic detection of IgM antibodies. This chapter discusses the relative merits of each of these approaches.

Methods Based on Chemical Inactivation of IgM

One of the earliest methods for determination of virus-specific IgM antibodies involved pretreatment of serum with mercaptans, such as 2-mercaptoethanol and dithiothreitol, which have the capacity to selectively split IgM molecules into immunologically inactive forms by breaking the disulfide bonds between the polypeptide chains (Banatvala et al., 1967). A fourfold or greater decrease in antibody titer following mercaptan treatment was considered indicative of the presence of virus-specific IgM antibodies. For this method to be effective, however, at least 75% of the total virus-specific antibody must be of the IgM class. Because this is the case only during the very early stages of most virus infections, the diagnostic value of this approach is quite limited. Furthermore, pretreatment with mercaptans can give variable results: with insufficient treatment, IgM monomers can reassociate and regain immune reactivity, resulting in false-negative reactions, while too-rigorous treatment can reduce IgG molecules, producing false-positive results. For these reasons, this method is no longer acceptable for detection of virus-specific IgM, but it can be used to confirm successful separation of IgM antibodies by other methods.

Methods Based on Physicochemical Separation of IgM

Physicochemical separation methods were originally developed to separate IgM antibodies from other serum immunoglobulins in order to facilitate assay by conventional serological tests, e.g., complement fixation (CF) and hemagglutination inhibition (HI) assays. It was later recognized the IgM separation could also benefit some solid-phase immunoassays (see below).

Sucrose Density Gradient Centrifugation

One of the earliest methods for recovering IgM antibodies employs high-speed centrifugation of serum on sucrose gradients. Because IgM proteins have a higher sedimentation coefficient (19S) than other immunoglobulins (7S to 11S), IgM antibodies can be physically separated from other antibodies by rate-zonal centrifugation on sucrose gradients. Lipoprotein molecules, including most of the nonspecific inhibitors of rubella virus hemagglutination, have low densities and therefore remain near the top of the gradient following centrifugation. This technique was introduced in

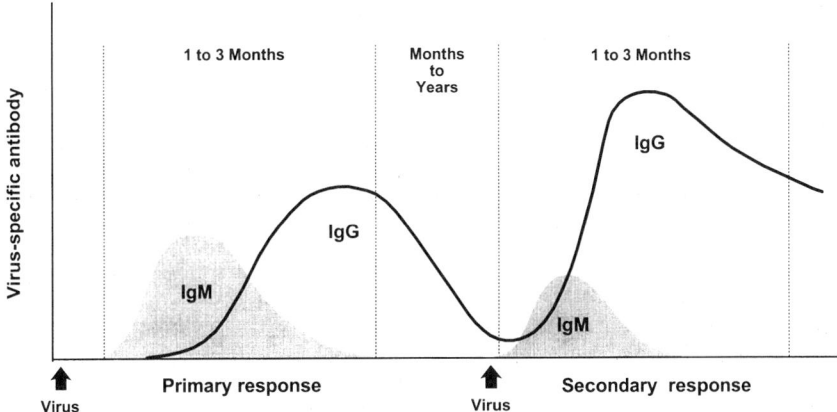

FIGURE 1 Primary and secondary immune responses to a hypothetical virus infection. After primary exposure to viral antigen, there is usually an early and pronounced rise in IgM antibodies, followed by a gradual rise in IgG. IgM antibodies then typically decline to undetectable levels within 1 to 3 months, whereas IgG antibodies can persist for years. Because immunological memory develops, reexposure to the same or similar virus usually results in an early and enhanced rise in IgG antibodies and an attenuated IgM response.

1968 for rapid diagnosis of recent rubella virus infection by demonstration of the presence of IgM antibodies (Vesikari and Vaheri, 1968). Since that time, various modifications of the method have been published (Forghani et al., 1973; Caul et al., 1978) and the test has been applied to the diagnosis of many virus infections (Frösner et al., 1979; Al-Nakib, 1980; Hawkes et al., 1980).

To perform sucrose density gradient centrifugation, a density gradient is prepared by layering 1.4-ml amounts of 37, 23, and 10% (wt/vol) solutions of sucrose in 0.01 M phosphate-buffered saline (PBS), pH 7.2, on a 0.2-ml cushion of 50% sucrose in a 5-ml ultracentrifuge tube. The sucrose layers are allowed to equilibrate for 4 to 6 h at 4°C, and then 0.2 ml of a 1:2 dilution of the test serum in PBS, pretreated, if necessary, to remove nonspecific serum components that would interfere with the assay of IgM antibody, is carefully layered on top of the gradient. The gradient is then centrifuged at 157,000 × g for 16 h in a swinging-bucket rotor or for 2 h using vertical rotors with reorienting gradients (Frösner et al., 1979). Ten to 12 fractions of about 0.4 ml each are collected by puncturing the bottom of the tube. The IgM antibodies concentrate in the bottom three or four fractions, IgG antibodies separate primarily in frac-

tions six to eight, and the lipoproteins (nonspecific inhibitors) remain near the top of the gradient. The first four fractions (i.e., those presumed to contain IgM) must be checked for both IgM antibody activity and the presence of contaminating human IgG (Caul et al., 1974). The isolated IgM fractions can then be tested for antiviral activity by any suitable serological tests.

Sucrose density gradient centrifugation combined with classical serological assays (e.g., CF and HI) are not as sensitive as some of the more recently developed solid-phase IgM immunoassays. Moreover, sucrose gradient centrifugation is a rather laborious procedure, and the high cost of the equipment places it out of the financial reach of most clinical laboratories. However, because of its high degree of specificity and overall reliability, this method is generally considered to be the standard for comparison when new IgM antibody tests are developed.

Column Chromatography

Column chromatographic methods based on gel filtration (Frisch-Niggemeyer, 1982), ion-exchange chromatography (Johnson and Libby, 1980), and, to a lesser extent, affinity chromatography (Barros and Lebon, 1975) have been used for many years to separate and isolate serum IgM antibodies. These methods offer simple and cost-effective alternatives to sucrose density gradient centrifugation while providing similar results if performed properly.

IgM separation on appropriately sized gels takes advantage of the differences in size between IgM molecules (molecular weight [MW], 900,000) and other immunoglobulins (MW, 150,000 to 400,000). Sephacryl S-300 and Sephadex G-200 (Amersham Pharmacia Biotech, Piscataway, N.J.) are commercially available products that have been used for this purpose. Sephacryl S-300 is preferred because it does not need to be rehydrated and it allows high-flow-rate filtration of serum under pressure for rapid specimen processing without over-packing problems or deformation of the column (Morgan-Capner et al., 1980). Since serum lipoproteins and nonspecific cell agglutinins

TABLE 1 Methods of IgM antibody determination

Methods based on chemical inactivation of IgM
  Alkylation reduction by mercaptans

Methods based on physicochemical separation of IgM
  Sucrose density gradient centrifugation
  Column chromatography
  IgG absorption

Methods based on solid-phase immunologic detection of IgM
  Indirect IgM immunoassays
  Reverse, or "capture," IgM immunoassays

can elute from these columns along with the IgM, these substances must be removed prior to serum fractionation if they interfere with the assay. Serum, pretreated with heparin and $MnCl_2$ to remove lipoproteins and cell adsorbed to remove nonspecific agglutinins, is layered on the top of the gel column and eluted through the column with Tris-buffered saline (0.02 M Tris in 0.15 M NaCl, pH 7.5). Discrete fractions are collected for titration of antibody activity. Each new column should be standardized with known specific IgM-positive and IgM-negative serum samples. With both Sephacryl S-300 and Sephadex G-200 columns, IgM is eluted in the first protein peak and IgG in the second; IgA may also be present in the first peak eluted from the Sephadex G-200 column but not from the Sephacryl S-300 column.

The specificity of gel separation of IgM for diagnosis of virus infections is very high, provided a number of factors that can cause false-positive results are taken into consideration (Pattison et al., 1976). Prolonged storage of serum at −20°C or bacterial contamination of serum may make pretreatment of it ineffective, resulting in false-positive results. Also, if the serum has been preheated at 56°C or higher, IgG may aggregate and therefore could elute into the IgM fractions after gel filtration. To minimize misinterpretation of the gel fractionation test, any presumptive IgM antibody activity in the first peak should be confirmed by an IgM-specific assay.

Ion-exchange chromatography, based on the differential binding of IgM and IgG antibodies to anion-exchange resins such as quaternary aminoethyl-Sephadex A50, has also been used for separation of IgM antibodies (Johnson and Libby, 1980). Serum loaded onto columns is washed, and most IgG and lipoproteins are eluted. IgM is recovered by elution at an acidic pH. As with sucrose density gradient fractionation and gel filtration, ion-exchange chromatography does not entirely eliminate IgG or IgA antibodies from the IgM fraction, and some loss of IgM antibodies can be expected (Elder and Smith, 1987; Elder et al., 1987).

IgG Absorption

A significant improvement in the speed and convenience of IgM separation has been achieved with the development of methods for "absorption" of IgG antibodies. These methods are based on the capacity of certain bacterial cell wall proteins to bind and remove IgG in serum specimens. Briefly, the IgG absorbent is added to the serum specimen according to the manufacturer's instructions, incubated to allow binding between the absorbent and the IgG, and then centrifuged briefly to pellet the IgG-absorbent complex (some procedures do not require centrifugation). The supernatant is then ready for assay for IgM antibodies.

Ankerst et al. (1974) first applied this concept for detection of rubella virus IgM antibodies after absorption of serum with *Staphylococcus aureus* bacteria. Some strains of *S. aureus* possess cell wall protein A (SPA), which binds to the Fc receptor of IgG and can be used to absorb and remove the majority of the serum IgG. However, SPA binds strongly only to IgG subclasses IgG1, IgG2, and IgG4—not IgG3 (Kronvall and Williams, 1969), which constitutes up to 5% of the original IgG antibody and has been shown to possess a disproportionately high level of antiviral activity (Beck, 1981). SPA can also absorb significant amounts of IgM, with potential loss of assay sensitivity; consequently, SPA is no longer considered an acceptable method of IgM detection.

Streptococcal protein G (SPG), a cell wall protein from group G streptococci, offers improvements over SPA for IgG absorption. SPG binds all four IgG subclasses and does not bind to IgM antibodies (Bjork and Kronvall, 1984). A recombinant form of SPG was successfully used to remove IgG antibodies in serum specimens, permitting detection of IgM and IgA antibodies to human immunodeficiency virus type 1 (Weiblen et al., 1990). In this study, plasma specimens were incubated with a 50% suspension of SPG-agarose and centrifuged briefly, and the supernatant was recovered for assay. Up to a 99.95% reduction in the IgG concentration was reported, although repeated treatments were necessary with specimens containing high-titer IgG. However, unlike SPA, SPG does not bind to IgA antibodies, which could possibly cause false-positive or false-negative results depending on the method employed for IgM detection. To address this possibility, Fuccillo et al. (1992) pretreated serum with a combination of SPA and streptococci by the method of Kronvall et al. (1979) to remove most serum IgG and IgA antibodies. For use with classical CF and HI assays, absorbed serum must be carefully evaluated to ensure that residual antibody activity is exclusively IgM.

IgG absorption methods are now routinely used to complement indirect solid-phase immunoassays for virus-specific IgM detection (see below). Preadsorption has been shown to reduce nonspecific IgM activity in serum and to significantly increase the sensitivity of indirect IgM assays by removing most competing IgG. In addition to SPA and SPG, anti-human IgG antibody preparations have been used successfully to bind and block serum IgG activity. Commercial preparations of SPA and SPG (from, e.g., Pierce, Rockford, Ill., and Amersham Pharmacia Biotech) and of anti-human IgG (e.g., GullSORB [Meridian Diagnostics, Cincinnati, Ohio] and IgG Removal Reagent [Stellar Bio Systems, Columbia, Md.]) are available separately and as part of commercial antiviral IgM diagnostic kits.

Methods Based on Solid-Phase Immunologic Detection of IgM

The availability of class-specific antiglobulins, specifically anti-human IgM, and solid-phase supports used to bind and separate immunoreactants has revolutionized IgM detection and led to rapid development of the wide range of assay formats currently used in diagnostic virology (Meurman, 1983). The major distinguishing features of solid-phase immunoassays are the choices of indicator label (e.g., fluorescein isothiocyanate, enzymes, radioisotopes, or erythrocytes) and solid phase (e.g., plastic 96-well microtiter plates, beads, tubes, or nitrocellulose/nylon sheets). Solid-phase immunoassays can be differentiated further into indirect and reverse, or "capture," forms based on the orientation of the immunoreactants on the solid phase. The indirect and capture formats have advantages and disadvantages that are described further below.

Indirect IgM Immunoassays

Indirect IgM immunoassays are characterized by the binding of viral antigen to a solid-phase surface followed by incubation with the serum specimen (Fig. 2). Specific IgM antibodies present in the specimen bind to the antigen and are subsequently detected with anti-human IgM antibody labeled with a suitable marker. Because of their technical simplicity

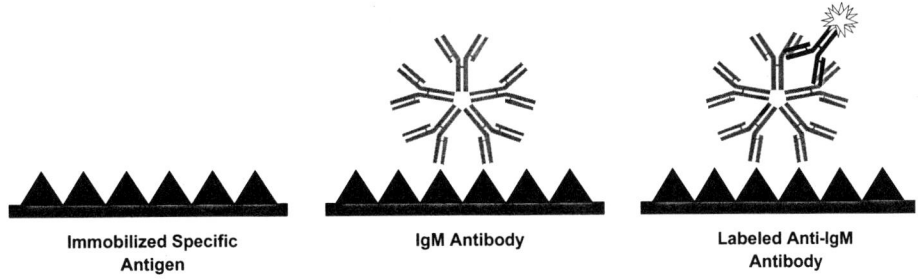

FIGURE 2 Indirect solid-phase immunoassay for IgM antibody.

and limited reagent requirements, indirect IgM immunoassays are particularly popular among virologists and commercial test kit vendors. However, two major problems often encountered with these assays can limit their sensitivity and specificity: (i) interference by IgM class rheumatoid factor (RF) and (ii) competition between specific IgG and IgM antibodies in patient serum specimens for available epitopes on the antigen bound to the solid phase. Each of these factors can play an important role in the reliability of an indirect IgM antibody assay system.

False-positive results can occur in indirect IgM assays when RF antibodies present in the serum attach to complexes of IgG antibody bound to the antigen on the solid phase (Fig. 3). RF is an antibody primarily of the IgM class that reacts with the Fc portion of bound IgG. RF is found in a high percentage of persons with rheumatoid arthritis and related connective tissue diseases and can also be found to varying degrees in people with subacute bacterial endocarditis, chronic liver disease, parasitic infections, and tuberculosis, as well as during pregnancy and among apparently normal healthy persons, particularly neonates and the elderly (Meurman, 1983; Fuccillo et al., 1992). Transient appearance of RF has also been associated with acute infection with parvovirus B19, measles virus, rubella virus, and cytomegalovirus and following heavy prophylactic vaccination. In practice, the effects of RF can vary with assays for different viruses without obvious explanation (Salonen et al., 1980).

Because indirect IgM immunoassays fix antigen directly to the solid phase, components of the cell line in which the

virus was grown or expressed (i.e., cellular proteins and nucleic acids) may be present on the solid-phase surface and serve as targets for serum antibodies with specificity for these components, e.g., antinuclear antibodies that cross-react with DNA-histone complexes (Fuccillo et al., 1992). If these antibodies are of the IgM class or are IgG antibodies associated with RF, false-positive results can occur unless each specimen is also tested against a negative-control antigen. Interference by this mechanism can be minimized, but not always eliminated, by extensive purification of the antigen or by using Western blot assays, in which the antigen components are separated by gel electrophoresis and blotted prior to IgM testing.

False-negative results can occur in indirect IgM assays if virus-specific IgG antibodies (and to a lesser extent IgA antibodies) that can compete with IgM for available epitopes on the antigen bound to the solid phase are present in the serum (Heinz et al., 1981). This can be a problem particularly if the specimen is collected late after the onset of symptoms, when IgG antibodies have begun to rise, or after repeat infections (as with respiratory syncytial virus, parainfluenza viruses, and enteroviruses) or reactivations (as with herpesviruses), where an early and enhanced IgG response can obscure the presence of low-level IgM antibodies (Fig. 1).

To address these concerns, serum to be tested by indirect IgM assays should always be pretreated by one of the methods described previously to remove interfering IgG antibodies. IgG absorption methods have the advantage of also removing RF by removing RF-IgG complexes, although the efficiency of these methods needs to be determined on a

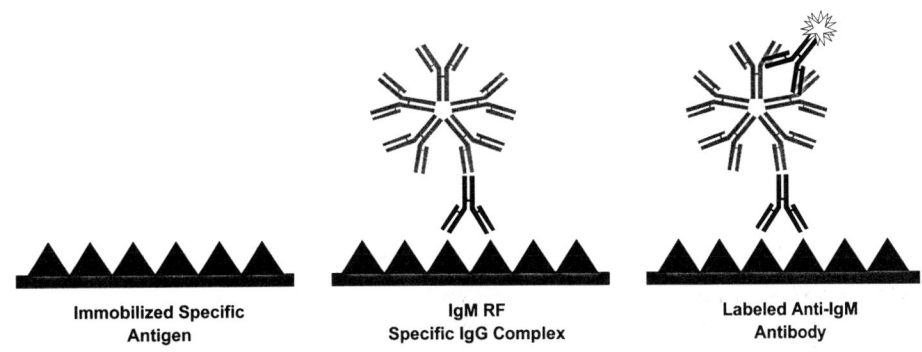

FIGURE 3 False-positive IgM result caused by RF interference in indirect solid-phase immunoassay.

case-by-case basis (Fuccillo et al., 1992). If RF is of particular concern, additional steps can be taken to reduce its influence (Meurman, 1983), i.e., (i) by blocking the binding sites of RF by adding aggregated IgG, which has higher affinity for RF than native IgG, to the serum diluent; (ii) by using labeled F(ab')₂ fragments as detector antibody, thereby eliminating the Fc portion of the IgG molecule that binds RF; and (iii) by determining the optimal dilution of serum that retains assay sensitivity while minimizing background RF binding. Independently of methods used to pretreat serum specimens, all specimens must be run against negative-control antigen (e.g., noninfected tissue culture lysate) to reveal nonspecific activity.

The first of the indirect assays to be applied for the detection of virus-specific IgM antibodies was the immunofluorescence assay (IFA) (Baublis and Brown, 1968). In this test, the antigen most often used is virus-infected cells fixed to glass slides. The method is essentially identical to indirect IFA for IgG antibodies, except that fluorescein isothiocyanate-labeled anti-human IgM detector antibody is used. The reading of IFA-IgM tests requires considerable skill and experience. Nonspecific staining may cause false-positive readings, but an experienced IFA microscopist can minimize false-positive results by differentiating patterns of specific and nonspecific fluorescence, a possibility that does not exist for radioimmunoassays (RIAs) or enzyme immunoassays (EIAs). In experienced hands and with the use of high-quality reagents, the IFA-IgM test can be both sensitive and reliable.

Solid-phase RIAs have been used to detect viral antibodies since the early 1970s and, despite diminished popularity, are still used by some laboratories for virus-specific IgM detection. The major advantages of the RIA are high sensitivity and potential for automation. The major disadvantages of RIAs have been the relatively short shelf-life of ¹²⁵I-labeled conjugates and the desire of most diagnostic laboratories to avoid the problems associated with the use and disposal of radioisotopes. Hence, IFAs and EIAs have largely replaced RIAs for IgM detection in the diagnostic virology laboratory.

EIAs were first reported for the detection of virus-specific IgM antibodies by Voller and Bidwell (1976). In principle, this method is identical to the RIA: it uses similar antigen preparations and solid-phase supports. However, radiolabeled conjugates are replaced with enzyme labels, typically alkaline phosphatase or horseradish peroxidase. The enzyme conjugates for EIAs have a long shelf-life, compared with RIA conjugates, and are more readily adapted to commercial development. EIAs are also remarkably flexible as to form and test results and can be read subjectively by eye or objectively by using automated multichannel spectrophotometers linked to computers to aid data analysis.

Reverse, or "Capture," IgM Immunoassays

Another approach for avoiding the problems of competitive interference from IgG and nonspecific reactivity seen with the traditional indirect immunoassays described previously is the reverse, or capture, IgM design (Fig. 4). This method, first reported for detecting virus-specific IgM antibodies by Flehmig (1978) and more fully described by Duermeyer et al. (1979), employs a solid-phase surface coated with anti-human IgM antibodies to "capture" and bind IgM antibodies in the clinical specimen. The washing process removes IgG and any immune complexes in the specimen. The addition of specific viral antigen, followed by a second, usually enzyme-labeled, antiviral antibody, completes the test. This approach has attracted considerable support among virologists, and capture IgM immunoassays have now been described for most of the viruses of public health importance.

Capture immunoassays have proven to be very sensitive and specific for detection of virus-specific IgM antibodies and, where comparisons have been made, are generally superior to indirect assay designs (Heinz et al., 1981; Roggendorf et al., 1981; Forghani et al., 1973; McCartney et al., 1986; Gerna et al., 1987; Besselaar et al., 1989; Re and Landini, 1989). Because the first step in the capture assay leads to separation of IgM antibodies, competition between IgM and other immunoglobulins (primarily IgG and IgA) does not occur and interference with RF is greatly reduced. Nevertheless, there are potential problems with the capture assay design that require consideration. Because virus-specific IgM antibodies must compete with nonspecific IgM for available sites on the capture phase, assay sensitivity can vary with the relative proportion of specific to nonspecific IgM antibodies in the specimen. RF interference still exists with IgM capture assays, although to a lesser extent than with indirect designs. For example, IgM-RF may be captured and bound to the solid phase and then bound to the labeled antiviral antibodies. RF interference by this mechanism can be minimized for capture immunoassays by using the methods described for indirect assays and by (i) substituting monoclonal capture and detector antibodies for animal polyclonal reagents, which appears to reduce unwanted RF activity (Wielaard et al., 1985,

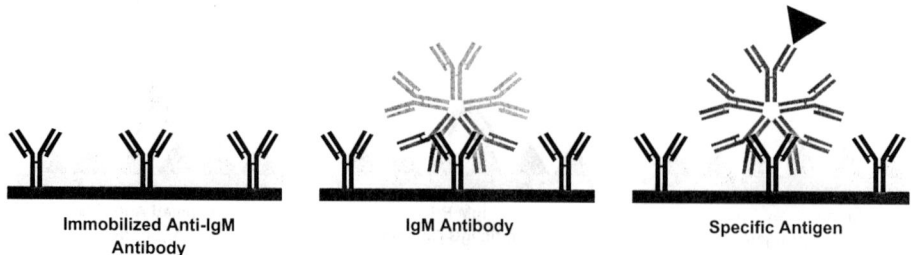

Immobilized Anti-IgM Antibody **IgM Antibody** **Specific Antigen**

FIGURE 4 Reverse, or "capture," solid-phase immunoassay for IgM antibody.

Chantler and Evans, 1986), and (ii) eliminating detector antibody entirely through direct labeling of antigen with enzyme (Schmitz et al., 1980; Nielsen et al., 1987; Tuokko, 1988; Morinet et al., 1991) or using erythrocytes if the antigen has hemagglutinating properties (Krech and Wilhelm, 1979; Van der Logt et al., 1981, 1985; Hilfenhaus et al., 1993). A more significant, but less common, mechanism of RF interference in capture immunoassays is the binding of RF-antiviral IgG complexes to the capture phase, which can mimic specific IgM antibody (Fig. 5). These specimens generally have very high levels of both RF and antiviral IgG antibodies, although they are less commonly encountered. The main limitations preventing broader commercial availability of capture IgM immunoassays for viral diagnosis have been their complexity and cost of manufacture, although this situation is changing.

INTERPRETATION OF IgM ASSAY RESULTS

The diagnostic value of specific IgM antibody assays is variable and dependent on the virus and the infection in question. Generally, transient IgM responses are characteristic of acute virus infections caused by viruses that elicit long-lasting immunity, such as with rubella virus, measles virus, mumps virus, parvovirus B19, and hepatitis A virus. In these infections, a reliable diagnosis can usually be made by specific IgM antibody testing of a single serum specimen taken early in the illness. In infections with viruses belonging to groups of closely related strains or serotypes (e.g., herpesviruses, adenoviruses, enteroviruses, parainfluenza viruses, and togaviruses), IgM serodiagnosis may be complicated by the possible absence of a specific IgM response, as well as by possible false-positive reactions to related viruses.

In addition to considerations of assay design, false-negative IgM antibody results may occur in a number of clinical settings. The IgM response is typically poor in children under 3 months old, as has been shown in studies of respiratory syncytial virus infection (Welliver et al., 1980), and tends to be weak generally for many respiratory virus infections (e.g., respiratory syncytial virus, parainfluenza viruses, adenoviruses, and influenza viruses), possibly because these infections are localized and less immunogenic or because they represent repeated infections that often yield a diminished IgM response. Immunocompromised persons may fail to generate detectable IgM antibodies or may present a delayed response. In rare cases, e.g., aplastic crisis caused by parvovirus B19 infection, onset of symptoms may precede development of IgM antibodies and therefore require a different diagnostic method, e.g., PCR assay. There can also be considerable individual variation in the appearance of IgM antibodies after onset of symptoms among heterogeneous populations, as has been demonstrated for rubella virus infection (Meurman, 1983).

False-positive IgM antibody results may occur due to cross-reactions between closely related viruses. Such cross-reactions have been reported for togavirus (Wolff et al., 1981) and coxsackie B virus (Schmidt et al., 1968) infections. Moreover, evidence of the occurrence of true polyclonal IgM production in cases of acute infectious mononucleosis has also been reported (Morgan-Capner et al., 1983); these data suggest that production of various IgM antibodies may result from Epstein-Barr virus-induced stimulation of B lymphocytes already committed by prior antigenic stimulation. Cross-reactions by this or other mechanisms have also been reported to occur between IgM tests for rubella virus, measles virus, and parvovirus B19 (Kurtz and Anderson, 1985; Jenkerson et al., 1995). In general, however, heterologous IgM antibody responses are low compared with homologous titers. Nevertheless, these observations emphasize the importance of careful interpretation of positive virus-specific IgM tests together with the complete clinical picture.

Finally, the expected duration of the IgM response must be considered in interpreting the significance of a positive test result. Variations in the temporal appearance of IgM antibodies, including the occurrence of prolonged IgM antibody responses, can result in difficulties in interpreting the significance of test results in relation to the clinical illness in question. Generally, the IgM antibody response following an acute virus infection is of limited duration. However, persistent IgM antibodies have been observed in complicated infections (e.g., postmeasles encephalitis), chronic infections (e.g., hepatitis B virus and parvovirus B19), and congenital infections (e.g., rubella virus and parvovirus B19) and in some immunosuppressed patients (e.g., cytomegalovirus). The persistence of specific IgM in these cases appears to be related to the persistence of viral antigen (or even replicating virus) in the patient. Occasionally, prolonged IgM antibody responses have been observed without any apparent reason. Also, as more sensitive methods for

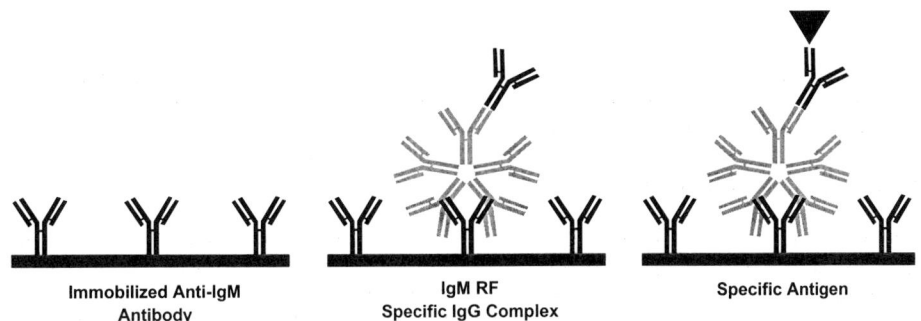

FIGURE 5 False-positive IgM result caused by RF interference in reverse, or "capture," solid-phase immunoassay.

TABLE 2 A selection of commercial vendors of diagnostic kits for virus-specific IgM antibodies

Virus	Vendor(s)[a]
Cytomegalovirus	2, 3, 6, 8, 9, 10, 11, 12, 13, 14
Dengue and other arboviruses	5, 7, 11
Epstein-Barr virus	2, 6, 8, 9, 10, 11, 12, 13, 14
Hepatitis A virus	1, 6
Hepatitis B virus	1, 6
Herpes simplex virus types 1 and 2	2, 3, 8, 9, 10, 11, 12, 13, 14
Herpes simplex virus types 6 and 7	10, 11
Measles virus	2, 5, 9, 10, 11, 12, 13
Mumps virus	2, 10, 11, 12
Parvovirus B19	4, 11
Rubella virus	2, 3, 4, 8, 9, 10, 11, 12, 13, 14
Varicella-zoster virus	2, 11, 12, 13

[a] 1, Abbott Laboratories, Abbott Park, Ill.; 2, Accurate Chemical and Scientific Corp., Westbury, N.Y.; 3, Alpha Diagnostics International, San Antonio, Tex.; 4, Biotrin International, Boston, Mass.; 5, Chemicon International, Temecula, Calif.; 6, DiaSorin, Stillwater, Minn.; 7, Integrated Diagnostics, Baltimore, Md.; 8, Intracel, Issaquah, Wash.; 9, Meridian Diagnostics, Cincinnati, Ohio; 10, Stellar Bio Systems, Columbia, Md.; 11, Tri-Delta Diagnostics, Morris Plains, N.J.; 12, Trinity Biotech, Jamestown, N.Y.; 13, Wampole Laboratories, Cranbury, N.J.; 14, Zeus Scientific, Raritan, N.J.

IgM determination are developed, the time following an acute infection during which specific IgM is detectable will increase. For diagnosis of an acute infection, the ideal maximum duration of specific IgM antibodies should be 1 to 3 months. It may therefore be necessary to limit the sensitivity of some assays to retain the optimal diagnostic usefulness of the methods.

Sensitive and reliable methods for the determination of virus-specific IgM antibodies have been developed for most human viral pathogens, and commercial reagents and complete diagnostic kits are available for many virus infections (Table 2). Substantial improvements in assay design and reagent quality have occurred during the past two decades, and continued improvements can be expected with the development and wider availability of well-defined recombinant and synthetic viral antigens. These methods, when adopted for routine use in clinical laboratories, should bring considerable improvement to viral diagnostic services.

REFERENCES

Al-Nakib, W. 1980. A modified passive-haemagglutination technique for the detection of cytomegalovirus and herpes simplex virus antibodies: application in virus-specific IgM diagnosis. *J. Med. Virol.* **5:**287–293.

Ankerst, J., P. Christensen, L. Kjellén, and G. Kronvall. 1974. A routine diagnostic test for IgA and IgM antibodies to rubella virus: absorption of IgG with *Staphylococcus aureus*. *J. Infect. Dis.* **130:**268–273.

Banatvala, J. E., J. M. Best, E. A. Kennedy, E. E. Smith, and M. E. Spence. 1967. A serological method for demonstrating recent infection by rubella virus. *Br. Med. J.* **3:**285–286.

Barros, M. F., and P. Lebon. 1975. Séparation des anticorps IgM anti-rubéole par chromatographie d'affinité. *Biomed. Express* **23:**184–188.

Baublis, J. V., and G. C. Brown. 1968. Specific response of the immunoglobulins to rubella infection. *Proc. Soc. Exp. Biol. Med.* **128:**206–210.

Beck, O. E. 1981. Distribution of virus antibody activity among human IgG subclasses. *Clin. Exp. Immunol.* **43:**626–632.

Besselaar, T. G., N. K. Blackburn, and N. Aldridge. 1989. Comparison of an antibody-capture IgM enzyme-linked immunosorbent assay with IgM-indirect immunofluorescence for the diagnosis of acute Sindbis and West Nile infections. *J. Virol. Methods* **25:**337–346.

Bjorck, L., and B. Kronvall. 1984. Purification and some properties of streptococcal protein G, a novel IgG-binding reagent. *J. Immunol.* **133:**969–974.

Caul, E. O., S. J. Hobbs, P. C. Roberts, and S. K. R. Clarke. 1978. Evaluation of a simplified sucrose gradient method for the detection of rubella-specific IgM in routine diagnostic practice. *J. Med. Virol.* **2:**153–163.

Caul, E. O., G. W. Smyth, and S. K. R. Clarke. 1974. A simplified method for the detection of rubella-specific IgM employing sucrose density fractionation and 2-mercaptoethanol. *J. Hyg.* **73:**329–340.

Chantler, S., and C. J. Evans. 1986. Selection and performance of monoclonal and polyclonal antibodies in an IgM antibody capture enzyme immunoassay for rubella. *J. Immunol. Methods* **87:**109–117.

Duermeyer, W., F. Wielaard, and J. van der Veen. 1979. A new principle for the detection of specific IgM antibodies applied in an ELISA for hepatitis A. *J. Med. Virol.* **4:**25–32.

Elder, B. L., and T. F. Smith. 1987. Evaluation of cytomegelisa immunoglobulin M assay and comparison with indirect fluorescent antibody testing of QAE-Sephadex A50-treated sera. *Am. J. Clin. Pathol.* **87:**230–235.

Elder, B. L., C. D. Shelley, and T. F. Smith. 1987. Evaluation of quaternary aminoethyl-Sephadex A50 column chromatography for detection of anti-cytomegalovirus immunoglobulin M. *Mayo Clin. Proc.* **62:**345–350.

Flehmig, B. 1978. Laboratoriumsdiagnose der Hepatitis A-Infektion. *Bundesgesundheitsblatt* **21:**277–283.

Forghani, B., N. J. Schmidt, and E. H. Lennette. 1973. Demonstration of rubella IgM antibody by indirect fluorescent antibody staining, sucrose density gradient centrifugation and mercaptoethanol reduction. *Intervirology* **1:**48–59.

Frisch-Niggemeyer, W. 1982. Simple and rapid chromatographic separation of IgM using microcolumns and stained sera. *J. Virol. Methods* **5:**135–142.

Frösner, G. G., R. Scheid, H. Wolf, and F. Deinhardt. 1979. Immunoglobulin M anti-hepatitis A virus determination by reorienting gradient centrifugation for diagnosis of acute hepatitis A. *J. Clin. Microbiol.* **9:**476–478.

Fuccillo, D. A., D. A. Vacante, and J. L. Sever. 1992. Rapid viral diagnosis, p. 545–553. *In* N. R. Rose, E. C. deMacario, J. L. Fahey, H. Friedman, and G. M. Penn (ed.), *Manual of Clinical Laboratory Immunology*, 4th ed. American Society for Microbiology, Washington, D.C.

Gerna, I., M. Zannino, M. G. Revello, E. Petruzzelli, and M. Dovis. 1987. Development and evaluation of a capture enzyme-linked immunosorbent assay for determination of rubella immunoglobulin M using monoclonal antibodies. *J. Clin. Microbiol.* **25:**1033–1038.

Hawkes, R. A., C. R. Boughton, F. Ferguson, N. I. Lehmann. 1980. Use of immunoglobulin M antibody to hepatitis B core antigen in diagnosis of viral hepatitis. *J. Clin. Microbiol.* **11:**581–583.

Heidelberger, M., and K. O. Pedersen. 1937. The molecular weight of antibodies. *J. Exp. Med.* **65:**393–414.

Heinz, F. X., M. Roggendorf, H. Hofmann, C. Kunz, and F. Dienhardt. 1981. Comparison of two different enzyme immunoassays for detection of immunoglobulin M antibodies against tick-borne encephalitis virus in serum and cerebrospinal fluid. *J. Clin. Microbiol.* **14:**141–146.

Hilfenhaus, S., B. J. Cohen, C. Bates, S. Kajigaya, N. S. Young, M. Zambon, and P. P. Mortimer. 1993. Antibody capture haemadherence tests for parvovirus B19-specific IgM and IgG. *J. Virol. Methods* **45:**27–37.

Jenkerson, S. A., M. Beller, J. P. Middaugh, and D. D. Erdman. 1995. False positive rubeola IgM tests. *N. Engl. J. Med.* **332:**1103–1104.

Johnson, R. B., Jr., and R. Libby. 1980. Separation of immunoglobulin M (IgM) essentially free of IgG from serum for use in systems requiring assay of IgM-type antibodies without interference from rheumatoid factor. *J. Clin. Microbiol.* **12:**451–454.

Krech, U., and J. A. Wilhelm. 1979. A solid-phase immunosorbent technique for the rapid detection of rubella IgM by haemagglutination inhibition. *J. Gen. Virol.* **44:**281–286.

Kronvall, G., and R. C. Williams, Jr. 1969. Differences in anti-protein A activity among IgG subgroups. *J. Immunol.* **103:**828–833.

Kronvall, G., A. Simmons, E. B. Myhre, and S. Jonsson. 1979. Specific absorption of human serum albumin, immunoglobulin A, and immunoglobulin G with selected strains of group A and G streptococci. *Infect. Immun.* **25:**1–10.

Kurtz, J. B., and M. J. Anderson. 1985. Cross-reactions in rubella and parvovirus specific IgM tests. *Lancet* **ii:**1356.

McCartney, R. A., J. E. Banatvala, and E. J. Bell. 1986. Routine use of mu-antibody-capture ELISA for the serological diagnosis of Coxsackie B virus infections. *J. Med. Virol.* **19:**205–212.

Meurman, O. 1983. Detection of antiviral IgM antibodies and its problems—a review. *Curr. Top. Microbiol. Immunol.* **104:**101–131.

Morgan-Capner, P., E. Davies, and J. R. Pattison. 1980. Rubella-specific IgM detection using Sephacryl S-300 gel filtration. *J. Clin. Pathol.* **33:**1082–1085.

Morgan-Capner, P., R. S. Tedder, and J. E. Mace. 1983. Rubella-specific IgM reactivity in sera from cases of infectious mononucleosis. *J. Hyg.* **90:**407–413.

Morinet, F., A. M. Courouce, F. Galibert, and Y. Perol. 1991. The use of labeled fusion protein for detection of B19 parvovirus IgM antibodies by an immunocapture test. *J. Virol. Methods* **32:**21–30.

Nielsen, C. M., K. Hansen, H. M. Andersen, J. Gerstoft, and B. F. Vestergaard. 1987. An enzyme labelled nuclear antigen immunoassay for detection of cytomegalovirus IgM antibodies in human serum: specific and non-specific reactions. *J. Med. Virol.* **22:**67–76.

Pattison, J. R., J. E. Mace, and D. S. Dane. 1976. The detection and avoidance of false-positive reactions in tests for rubella-specific IgM. *J. Med. Microbiol.* **9:**355–357.

Re, M. C., and M. P. Landini. 1989. IgM to human cytomegalovirus: comparison to two enzyme immunoassays and IgM reactivity to viral polypeptides detected by immunoblotting. *J. Clin. Lab. Anal.* **3:**169–173.

Roggendorf, M., F. Heinz, F. Deinhardt, and C. Kunz. 1981. Serological diagnosis of acute tick-borne encephalitis by demonstration of antibodies of the IgM class. *J. Med. Virol.* **7:**41–50.

Salonen E.-M., A. Vaheri, J. Suni, and O. Wager. 1980. Rheumatoid factor in acute viral infections: interference with determination of IgM, IgG, and IgA antibodies in an enzyme immunoassay. *J. Infect. Dis.* **142:**250–255.

Schluederberg, A. 1965. Immune globulins in human viral infections. *Nature* **205:**1232–1233.

Schmidt, N. J., E. H. Lennette, and J. Dennis. 1968. Characterization of antibodies produced in natural and experimental coxsackievirus infections. *J. Immunol.* **100:**99–106.

Schmitz, H., U. von Deimling, and B. Flehmig. 1980. Detection of IgM antibodies to cytomegalovirus (CMV) using an enzyme-labelled antigen (ELA). *J. Gen. Virol.* **50:**59–68.

Tuokko, H. 1988. The detection of measles specific immunoglobulin M antibodies using biotinylated antigens. *Acta Pathol. Microbiol. Immunol. Scand.* **96:**491–496.

van der Logt, J. T. M., A. M. van Loon, and J. van der Veen. 1981. Hemadsorption immunosorbent technique for determination of rubella immunoglobulin M antibody. *J. Clin. Microbiol.* **13:**410–415.

van der Logt, J. T. M., A. M. van Loon, and J. van der Veen. 1982. Detection of parainfluenza IgM antibody by hemadsorption immunosorbent technique. *J. Med. Virol.* **10:**213–221.

Vesikari, T., and A. Vaheri. 1968. Rubella: a method for rapid diagnosis of a recent infection by demonstration of the IgM antibodies. *Br. Med. J.* **1:**221–223.

Voller, A., and D. E. Bidwell. 1976. Enzyme-immunoassays for antibodies in measles, cytomegalovirus infections and after rubella vaccination. *Br. J. Exp. Pathol.* **57:**243–247.

Weiblen, B. J., R. T. Schumacher, and R. Hoff. 1990. Detection of IgM and IgA HIV antibodies after removal of IgG with recombinant protein G. *J. Immunol. Methods* **126:**199–204.

Welliver, R. C., T. N. Kaul, T. I. Putnam, M. Sun, K. Riddlesberger, and P. L. Ogra. 1980. The antibody response to primary and secondary infection with respiratory syncytial virus: kinetics of class-specific responses. *J. Pediatr.* **96:**808–813.

Wielaard, F., A. Denissen, J. van Elleswijk-v. d. Berg, and G. Van Gemert. 1985. Clinical validation of an antibody-capture anti-rubella IgM-ELISA. *J. Virol. Methods* **10:**349–354.

Wolff, K. L., D. J. Muth, B. W. Hudson, and D. W. Trent. 1981. Evaluation of the solidphase radioimmunoassay for diagnosis of St. Louis encephalitis infection in humans. *J. Clin. Microbiol.* **14:**135–140.

Antiviral Drug Susceptibility Testing*

ELLA M. SWIERKOSZ

16

Within the past decade, antiviral agents have become available for a wide variety of viral infections (Table 1). While virus isolates from untreated patients are usually susceptible to antiviral agents, therapy with certain agents has led to the emergence of drug-resistant variants. Increasingly, diagnostic virology laboratories are requested to perform in vitro antiviral susceptibility testing. This chapter will briefly summarize the antiviral agents in use, the mechanisms of resistance to these agents, clinical situations where antiviral resistance has emerged, methods employed for susceptibility testing, and guidelines for interpretation of antiviral susceptibility testing. Immune globulin preparations for prophylaxis of viral infections will not be discussed. The major antiviral agents in use and the mechanisms of resistance are listed in Tables 1 to 3 and have been reviewed in detail (Bean, 1992; Beutner, 1995; De Clercq, 1995; Hayden, 1995; Cirelli et al., 1996; Hirsch et al., 1996; Hodinka, 1997; Erice, 1999). Continuous updates of human immunodeficiency virus (HIV) treatment guidelines are available from the Worldwide Web site of the HIV/AIDS Treatment Information Service at http://www.hivatis.org.

ANTIVIRAL AGENTS AND MECHANISMS OF ACTION AND RESISTANCE

Herpesviruses

Agents active against various herpesviruses include acyclovir, cidofovir, famciclovir, foscarnet, ganciclovir, idoxuridine, trifluridine, valacyclovir, and vidarabine. Vidarabine (9-beta-D-arabinofuranosyladenine; adenine arabinoside), an adenosine nucleoside analog, was the first systemic antiviral agent licensed for use in the United States. It was shown to be effective for treatment of herpes simplex virus (HSV) encephalitis, and neonatal infection (Bean, 1992). An ophthalmic preparation is used topically for HSV keratitis. Vidarabine's toxicity and poor solubility in water contributed to its replacement by acyclovir. Vidarabine-resistant HSV can be generated in vitro but is not a problem clinically (Coen et al., 1985). Despite its in vitro activity

against acyclovir-resistant HSV, vidarabine has not been effective for treatment of HSV in HIV-infected individuals with HSV infection (Safrin et al., 1991b).

Acyclovir [9-(2-hydroxyethoxymethyl)guanine; acycloguanosine], a guanosine analog, has in vitro activity against a number of herpesviruses but is used clinically for HSV and varicella-zoster virus (VZV) infections. Acyclovir has largely supplanted vidarabine because of its ease of administration, low toxicity, and efficacy (Bean, 1992; Hayden, 1995). Acyclovir is phosphorylated by a virus-specific thymidine kinase (TK) to its monophosphate form, which is subsequently phosphorylated to acyclovir triphosphate by cellular enzymes. Acyclovir triphosphate competes with the natural substrate dGTP for viral DNA polymerase. Because it has a higher affinity for the viral polymerase than does dGTP, it is preferentially incorporated into newly synthesized viral DNA. This results in termination of DNA synthesis, because acyclovir lacks the 3' hydroxyl group necessary to form phosphodiester linkages with incoming nucleotides (Bean, 1992). Inactivation of the HSV DNA polymerase also occurs as a consequence of the tight binding of the enzyme to the template-primer-acyclovir monophosphate complex that forms during incorporation of acyclovir monophosphate (Reardon and Spector, 1989).

Acyclovir resistance arises as a result of mutations in either the viral TK or DNA polymerase genes. TK mutations arise at a high frequency (10^{-3} to 10^{-4}) (Hirsch et al., 1996). Mutation in the TK gene can produce strains deficient in TK (TK$^-$), which results in reduced or absent phosphorylation of acyclovir. A second type of TK mutation results in altered substrate binding properties for acyclovir (TK^A). The majority of acyclovir-resistant clinical isolates of HSV and VZV are TK$^-$, although rare TK^A or DNA polymerase mutants have been recovered (Chatis and Crumpacker, 1992; Field and Biron, 1992). TK mutations render these viruses less pathogenic in animals (Field and Darby, 1980). However, in immunocompromised hosts, such as bone marrow transplant and HIV-infected patients, TK$^-$ virus can cause progressive disease (Bean et al., 1987; Englund et al., 1990). TK$^-$ viruses are, however, also restricted in their ability to reactivate from latency (Coen et al., 1989; Hill et al., 1991b). Therefore, subsequent recurrences in the same patient could again be responsive to acyclovir (TK$^+$) (Safrin et al., 1990, 1991a). Reactivation of acyclovir-resistant HSV with an altered TK has been docu-

* This chapter contains information presented in the *Manual of Clinical Microbiology*, 7th ed. (Swierkosz and Hodinka, 1999).

TABLE 1 Antiviral agents for treatment of herpesviruses

Drug class/mechanism of action	Antiviral agent(s)	Clinical indications for use[a]	Mechanism(s) of resistance
Nucleoside analog/inhibitor of viral DNA polymerase	Acyclovir	Localized and systemic HSV and VZV infections; prophylaxis of transplant recipients	Mutations in viral TK and viral DNA polymerase (*pol*) genes
	Cidofovir	CMV retinitis in patients with AIDS	Mutations in DNA *pol* gene
	Famciclovir (active metabolite, penciclovir)	Recurrent genital HSV in adults; zoster in adults	Mutations in viral TK and viral DNA *pol* genes
	Ganciclovir	Treatment of CMV retinitis in immunocompromised patients; prophylaxis of transplant recipients and patients with AIDS	Mutations in viral UL97 (phosphotransferase) and UL54 (*pol*) genes
	Valacyclovir (active metabolite, acyclovir)	Recurrent genital herpes in immunocompetent adults; zoster in immunocompetent adults	Mutations in viral TK and viral DNA *pol* genes
Nucleoside analog/inhibitor of DNA synthesis (topical only)	Trifluridine	HSV keratitis	Unknown
	Vidarabine	HSV keratitis	Unknown
Pyrophosphate analog/inhibitor of viral DNA polymerase	Foscarnet	Treatment of CMV retinitis in patients with AIDS and mucocutaneous acyclovir-resistant HSV infections in immunocompromised patients	Mutations in viral DNA *pol* gene

[a] FDA-approved uses.

mented in an immunocompetent individual, but it is a rare occurrence (Kost et al., 1993).

Acyclovir-resistant HSV has rarely been recovered from patients with normal immunity (Lehrman et al., 1986). It is not uncommon to recover resistant HSV from transplant and HIV-infected patients after prolonged treatment with acyclovir (Wade et al., 1983; Englund et al., 1990; Safrin et al., 1990, 1991b). Acyclovir-resistant VZV has been recovered from AIDS patients after prolonged acyclovir therapy (Pahwa et al., 1988; Linnemann et al., 1990; Safrin et al., 1991a).

Recently, the l-valyl ester of acyclovir, valacyclovir, has been licensed for oral treatment of genital HSV infection and herpes zoster in immunocompetent adults. Valacyclovir is rapidly converted to acyclovir after oral administration and results in a three- to fivefold increase in acyclovir bioavailability compared to oral acyclovir. Because of its increased bioavailability, valacyclovir can be administered in a less frequent oral dosage regimen than is required for acyclovir (Beutner, 1995).

Famciclovir, 2-[2-(2-amino-9*H*-purin-9-yl)ethyl]-1,3-propanediol diacetate, is a synthetic acyclic guanine derivative. Famciclovir is the orally administered prodrug of the active antiviral compound penciclovir, which is structurally similar to acyclovir. Following oral administration, famciclovir undergoes rapid biotransformation to penciclovir, which has inhibitory activity against HSV type 1 (HSV-1), HSV-2, and VZV (Cirelli et al., 1996). The bioavailability of penciclovir after an oral dose of 500 mg of famciclovir is approximately 77%. In contrast, the bioavailability of acyclovir is approximately 10 to 20% (Hayden, 1995). Famciclovir has been approved for use in immunocompetent adults for the treatment of acute zoster (shingles) and recurrent episodes of genital herpes. The long intracellular half-life of penciclovir triphosphate in HSV- and VZV-infected cells permits less frequent administration of famciclovir than acyclovir (Cirelli et al., 1996). Like acyclovir, penci-

clovir requires phosphorylation by viral TK to a monophosphate form that is converted to penciclovir triphosphate by cellular enzymes. Penciclovir triphosphate inhibits HSV and VZV DNA polymerases competitively with dGTP, thus inhibiting viral DNA synthesis and ultimately viral replication. Unlike acyclovir triphosphate, penciclovir triphosphate is not an obligate chain terminator (Earnshaw et al., 1992). Penciclovir-resistant mutants of HSV and VZV can result from mutations in viral TK or DNA polymerase genes. Acyclovir-resistant HSV and VZV mutants deficient in TK can be cross-resistant to penciclovir, but cross-resistance does not occur with all acyclovir-resistant isolates (Safrin and Phan, 1993; Talarico et al., 1993).

A topical agent, trifluridine (trifluorothymidine), a fluorinated thymidine analog, has been used extensively for treatment of HSV keratitis. The triphosphate form of trifluridine is a competitive inhibitor of DNA polymerases. Resistance in clinical isolates of HSV has not been described, although resistant isolates can be generated in vitro.

Ganciclovir [9-(1,3-dihydroxy-2-propoxy)methylguanine] is another guanosine analog that is similar in structure to acyclovir. Its spectrum of activity includes HSV and cytomegalovirus (CMV). However, it is more active than acyclovir against CMV. Because it is more toxic than acyclovir, it is used only to treat serious CMV infections. In addition to intravenous ganciclovir, oral ganciclovir and ganciclovir intraocular implants have become available. Ganciclovir is used clinically for treatment of CMV diseases, including gastroenteritis, pneumonitis, and retinitis. Prophylactic ganciclovir is used for prevention of retinitis in HIV-infected patients and of CMV disease in transplant patients. Ganciclovir is activated to its monophosphate form by virus-specific TK in HSV-infected cells and by a CMV-specific phosphotransferase in CMV-infected cells (Littler et al., 1992). It is subsequently phosphorylated to ganciclovir triphosphate by cellular kinases. Ganciclovir

TABLE 2 Antiviral agents for treatment of HIV-1

Drug class/mechanism of action	Antiviral agent(s)	Clinical indications for use[a]	Mechanism(s) of resistance
Nucleoside analog/NRTI	ABC	Treatment of HIV-1 in combination with ZDV and lamivudine 3TC	Mutations in viral RT gene
	ddI	Treatment of HIV infection in combination with ZDV, d4T, or 3TC and a PI	Mutations in viral RT gene
	3TC	Treatment of HIV infection in combination with ZDV, ddI, or d4T and a PI	Mutations in viral RT gene
	d4T	Treatment of HIV infection in combination with ddI or 3TC and a PI	Mutations in viral RT gene
	ddC	Treatment of HIV infection in combination with ZDV and a PI	Mutations in viral RT gene
	ZDV	Treatment of HIV infection in combination with ddI, ddC, or 3TC and a PI; monotherapy for prophylactic use in pregnant women	Mutations in viral RT gene
NNRTI	Delavirdine	Treatment of HIV infection in combination with two NRTI	Mutations in viral RT gene
	Efavirenz Nevirapine		
PI	Amprenavir	Treatment of HIV-1 infection in combination with other antiretrovirals	Mutations in viral protease gene
	Indinavir	Treatment of HIV-1 infection in combination with two NRTI	Mutations in viral protease gene
	Nelfinavir	Treatment of HIV-1 infection in combination with two NRTI	Mutations in viral protease gene
	Ritonavir	Treatment of HIV-1 infection in combination with two NRTI	Mutations in viral protease gene
	Saquinavir	Treatment of HIV-1 infection in combination with two NRTI	Mutations in viral protease gene

[a] Source, http://www.hivatis.org.

triphosphate acts as a competitive inhibitor of viral DNA polymerase but, unlike acyclovir, does not cause DNA chain termination (Bean,1992). Substitution or deletion mutations within the putative catalytic domain of the viral phosphotransferase gene, UL97, have accounted for most of the low-level ganciclovir resistance described in clinical isolates (50% inhibitory concentrations $IC_{50}s$, ≥ 8 and <30 μM) (Biron et al., 1986; Biron, 1991; Stanat et al., 1991; Chou et al., 1995a, 1995b; Smith et al., 1997; Baldanti et al., 1998b; Erice, 1999). The most commonly observed UL97 mutations are M460V, A594V, and L595S (Erice, 1999). Isolates with high-level ganciclovir resistance (IC_{50}, ≥ 30 μM) have mutations in the viral DNA polymerase gene (UL54-pol) in addition to UL97 mutations (Smith et al., 1997). The UL54 mutants can be cross-resistant to cidofovir or foscarnet (Smith et al., 1997; Erice, 1999). Multidrug-resistant isolates, resistant to ganciclovir, cidofovir, and foscarnet, contain additional UL54 substitution mutations (Smith et al., 1997). All ganciclovir-resistant CMV isolates recovered to date have been from immunocompromised patients, the majority of whom had AIDS (Erice, 1999).

Foscarnet (trisodium phosphonoformate) is a pyrophosphate analog with in vitro activity against all herpesviruses, hepatitis B virus, and HIV type 1 (HIV-1). It serves as a noncompetitive inhibitor of viral DNA polymerases and of reverse transcriptase (RT) of HIV-1. Foscarnet has been used for prophylaxis and treatment of CMV retinitis and treatment of acyclovir-resistant HSV and VZV infections (Safrin et al., 1990, 1991a, 1991b). Clinical isolates of foscarnet-resistant HSV, CMV, and VZV have been recovered almost exclusively from immunocompromised patients (Safrin et al., 1991b; Fillet et al., 1995; Baldanti et al., 1996; Smith et al., 1997). Foscarnet-resistant HSV isolates are usually susceptible or borderline susceptible to acyclovir, suggesting that the binding site of acyclovir triphosphate on the viral DNA polymerase may differ from that for foscarnet (Safrin et al., 1991b). Resistance in herpesviruses is due to mutations of the viral DNA polymerase gene. Foscarnet-resistant CMV isolates cross-resistant to ganciclovir and cidofovir have also been recovered (Smith et al., 1997).

Cidofovir [(S)-1-(3-hydroxy-2-phosphonylmethoxypropyl)cytosine; HPMPC] has broad-spectrum activity against adenoviruses, herpesviruses, iridoviruses, papovaviruses,

TABLE 3 Miscellaneous antiviral agents[a]

Drug class/mechanism of action	Antiviral agent(s)	Clinical indications for use[b]	Mechanism(s) of resistance
Primary symmetrical amine; prevents virus uncoating	Amantadine	Prophylaxis and treatment of influenza A virus	Mutations in the transmembrane domain of the viral M2 gene of influenza A virus
	Rimantidine	Prophylaxis for influenza A virus in children and adults; treatment of influenza A in adults	Mutations in the transmembrane domain of the viral M2 gene of influenza A virus
IFN; inhibition of virus replication by stimulating a variety of cellular responses	IFN-α-n3; human leukocyte derived	Intralesional treatment of condylomata acuminata	Not described
	IFN-α-2a; recombinant	Treatment of chronic hepatitis C	Not described
	IFN-α-2b; recombinant	Treatment of external condylomata acuminata and chronic hepatitis B and C; treatment of chronic hepatitis C in combination with ribavirin	Not described
Nucleoside analog; inhibition of multiple virus replication processes	Ribavirin	Treatment of hospitalized infants and young children with severe RSV infections; treatment of chronic hepatitis C in combination with IFN-α-2b (recombinant)	Development of resistance has not been evaluated in vitro or in clinical trials

[a] Immunoglobulin preparations, available for a variety of viral infections, are not discussed.
[b] FDA-approved uses.

and poxviruses. It has been approved in the United States for prophylaxis and treatment of CMV retinitis. Clinically, it has also been used for treatment of acyclovir- and foscarnet-resistant HSV (Snoeck et al., 1994). Cidofovir is converted intracellularly to its active metabolite, the diphosphorylated form of HPMPC, HPMPCpp, by successive phosphorylation steps, presumably by cellular kinases. Unlike acyclovir and related nucleoside analogs, initial monophosphorylation by viral thymidine kinase or other viral kinases is not required. HPMPCpp acts as an alternative substrate of dCTP and an inhibitor of viral DNA polymerase. Incorporation of two consecutive HPMPCpp molecules into DNA is required for chain termination. Because cidofovir does not require phosphorylation to a monophosphate form by viral thymidine kinase, it is active against TK$^-$ HSV and VZV. Resistance to cidofovir is due to mutations in the CMV UL54 gene. As described above, multidrug resistance to cidofovir, ganciclovir, and foscarnet can occur as a result of *pol* gene mutations (Smith et al., 1997; Erice, 1999). It is noteworthy that the level of cidofovir resistance described to date is low, resulting in IC$_{50}$s of 4 to 8 μM, while the peak plasma levels achievable are in the range of 40 to 70 μM. Thus, reduced susceptibility to cidofovir in vitro may not be clinically significant (Cherrington et al., 1998).

RESPIRATORY VIRUSES

Ribavirin (1-β-D-ribofurnosyl-1,2,4-trizole-3-carboxamide) is a synthetic nucleoside analog of guanosine and inosine with in vitro activity against a wide variety of viruses, including respiratory syncytial virus (RSV), influenza A and B viruses, parainfluenza virus, Lassa fever virus, and HIV-1. Ribavirin is currently licensed in the United States for treatment of RSV by small-particle aerosol. The exact

mechanisms of action of ribavirin are unclear, but several theories have been proposed. Ribavirin monophosphate inhibits cellular inosine monophosphate dehydrogenase activity, thereby reducing GTP levels. Ribavirin triphosphate inhibits translation of viral transcripts by interfering with 5′ capping of viral mRNA. Also, ribavirin triphosphate inhibits the RNA-dependent RNA polymerase activity of influenza and LaCrosse viruses (Hirsch et al., 1996). To date, no clinically significant resistance to ribavirin has been encountered.

Amantadine (1-adamantanamine hydrochloride), a primary symmetrical amine with a cagelike structure, and rimantadine, the methyl derivative of amantadine, are active against influenza A virus. Neither drug inhibits influenza B virus at clinically achievable drug concentrations. These drugs have been approved in the United States for prophylaxis and treatment of influenza A virus. Amantadine and rimantadine, which have identical spectra and mechanisms of activity, act by inhibiting the ion channel activity of the influenza A M2 protein, thus preventing viral uncoating and release of viral RNA into the cytoplasm (Bean, 1992). Resistance to these drugs is due to a single-amino-acid change in the transmembrane portion of the M2 protein (Belshe et al., 1988). Clinical isolates of influenza A virus resistant to amantadine and rimantadine have been readily recovered during therapy, which may limit the widespread use of these two drugs (Hayden et al., 1991).

HIV-1

NRTI

Zidovudine (azidothymidine; 3′-azido-3′-deoxythymidine; AZT; ZDV) is a thymidine analog with activity against HIV-1, HIV-2, and other human retroviruses. It was the first antiviral agent approved for primary therapy of HIV-1 infection. It belongs to a class of antiretroviral drugs labeled nucleoside analog RT inhibitors (NRTI). Other NRTI approved subsequently, in order of approval, include didanosine (2′,3′-dideoxyinosine; ddI), zalcitabine (2′,3′-dideoxycytidine; ddC), stavudine (2′,3′-didehydro-2′-deoxythymidine; d4T), lamivudine (2′,3′-dideoxy, 3′-thiacytidine; 3TC), and abacavir {(1S,cis)-4-[2-amino-6-(cyclopropylamino)-9H-purin-9-yl]-2-cyclopentene-1-methanol; ABC}. NRTI are phosphorylated by cellular enzymes to 5′-triphosphate forms that serve as competitive inhibitors of HIV RT. Incorporation of the triphosphate forms leads to chain termination because these agents lack a free 3′ hydroxyl group to form a phosphodiester bond with the incoming nucleotide (De Clercq, 1995). Because of the rapid emergence of NRTI-resistant HIV-1 isolates in patients treated with a single drug, combination therapy of NRTI with non-nucleoside analog RT inhibitors (NNRTI) and protease inhibitors (PI) is currently recommended by the Panel of Clinical Practices for Treatment of HIV Infection (Carpenter et al., 1997). The single exception is the prophylactic use of ZDV in pregnant women to reduce the risk of perinatal HIV transmission (Centers for Disease Control and Prevention, 1998).

ZDV resistance was first described in isolates from patients receiving prolonged monotherapy with ZDV. Larder et al. (1989) measured the ZDV susceptibilities of HIV-1 isolates from HIV-infected individuals on therapy for various times by a plaque assay with CD4$^+$ HeLa cells. The IC$_{50}$s for isolates from untreated patients were 0.01 to 0.05 μM, and those for isolates obtained after 6 or more months of therapy were 0.04 to 6 μM. Genotypic characterization of these isolates demonstrated five amino acid substitutions in the reverse transcriptase gene: M41L, D67N, K70R, T215Y/F, and K219Q (Larder and Kemp, 1989; Boucher et al., 1992; Kellam et al., 1992). Resistance to ddI and cross-resistance to ddC and ABC are induced by the RT mutation L74V, which also resulted in a reversal of the ZDV resistance due to the T215Y mutation (St. Clair et al., 1991; Abravaya et al., 1995). The M184V/I mutation is also responsible for causing resistance to ddI and cross-resistance to ddC, 3TC, and ABC (De Clercq, 1995; Antiviral Resistance Website [http://www.viral-resistance.com]). The K65R mutation confers resistance to ddI, ddC, 3TC, and ABC, and the V75T mutation induces resistance to d4T (De Clercq,1995; Arts and Wainberg, 1996; http://www.viral-resistance.com). Mutation Q151M conferred partial resistance to ZDV, ddI, ddC, and d4T, while a combination of Q151M and three or four additional mutations (V75I, F77L, F116Y or A62V, V75I, F77L, and F116Y) conferred high-level resistance to these agents and low-level cross-resistance to 3TC (Eron et al., 1993; Iversen et al., 1996).

NNRTI

The NNRTI class includes agents with diverse chemical structures that are unrelated to nucleosides. The currently licensed NNRTI include nevirapine, delavirdine, and efavirenz. NNRTI bind to a non-substrate-binding site that is located in the vicinity of the substrate-binding site of RT, causing a disruption of the enzyme's catalytic site. Because resistant viruses emerge rapidly when NNRTI are administered as monotherapy, NNRTI must always be administered with antiretroviral agents with other mechanisms of activity. The mutations associated to date with resistance to delavirdine or nevirapine are A98G, L100I, K103N, K103T, V106A, V108I, Y181C/I, Y188C, G190A, and P236L (Schinazi et al., 1999; http://

www.viral-resistance.com). While mutations associated with efavirenz resistance have been generated in vitro, naturally occurring mutants have not been described (Schinazi et al., 1999; http://www.viral-resistance.com).

PI

HIV-1 protease is essential for the posttranslational cleavage of precursor polyproteins encoded by the *gag* and *pol* genes. These nonfunctional polyproteins must be cleaved into smaller functional proteins to produce infectious virions (Boden and Markowitz, 1998). PI competitively bind to the active site of the enzyme, thus preventing insertion of precursor polyproteins into the protease active site. This inhibition prevents cleavage of the viral polyproteins, resulting in the production of noninfectious virus particles (Boden and Markowitz, 1998). Five PI have been approved by the Food and Drug Administration (FDA) for treatment of HIV-1 infection: amprenavir, indinavir, nelfinavir, ritonavir, and saquinavir. Because of the rapid emergence of drug-resistant viruses, monotherapy with PI is not recommended (Carpenter et al., 1997; http://www.hivatis.org). Mutations within the HIV-1 protease gene conferring resistance to PI have been described. Mutations at codons 48 and 90 confer saquinavir resistance in vivo. The primary mechanism of ritonavir resistance in vivo is an initial mutation at position 82 of the protease gene followed by sequential accumulation of mutations at eight additional codons (Boden and Markowitz, 1998). Mutations in at least 11 protease gene codons were associated with in vivo indinavir resistance. A mutation in codon 30 has been identified in nelfinavir-resistant virus in vivo; other mutations occasionally occurred, but always in combination with the codon 30 mutation (Boden and Markowitz, 1998). Elucidation of the mechanism of in vivo emergence of resistance to amprenavir is ongoing; however, a mutation in codon 50 of the protease gene appears critical (Boden and Markowitz, 1998). Cross-resistance to all PI currently licensed has been associated with a minimum of four substitutions in the protease gene at codons 46, 63, 82, and 84 (Boden and Markowitz, 1998). Additional mutations can accumulate sequentially, conferring cross-resistance to other PI (Boden and Markowitz, 1998; Schinazi et al., 1999). An additional mechanism of resistance to PI includes mutations in any of the nine cleavage sites of the HIV-1 protease (Boden and Markowitz, 1998). An updatable database listing mutations of retroviral genes associated with resistance to antiviral agents active against HIV-1 is available online at http://www.viral-resistance.com.

IFN

Interferons (IFN) are a group of cytokines with complex antiviral, immunomodulating, and antiproliferative activities. IFN are proteins produced by eukaryotic cells in response to various inducers, including viruses that act on uninfected cells to render them resistant to viral infection. There are three types of IFN based on the cell type from which they were derived: IFN-α, IFN-β, and IFN-γ. IFN themselves are not antiviral but induce proteins in exposed cells that inhibit specific viral functions, such as penetration, uncoating, synthesis of viral mRNA, viral protein synthesis, and virus assembly and release. A review of IFN activity has been published (Hayden, 1995). Two recombinant IFN, IFN-α-2a and IFN-α-2b, are approved for use in the United States. IFN-α-2a is approved for treatment of chronic hepatitis C. IFN-α-2b is approved for treatment of chronic hepatitis B and C and external condylomata acuminata. Combination therapy of IFN-α-2b with ribavirin has recently been approved for treatment of chronic hepatitis C. IFN-α-n3 (human leucokyte derived) has been approved for treatment of external condylomata acuminata (Table 3).

INDICATIONS FOR ANTIVIRAL SUSCEPTIBILITY TESTING

Antiviral susceptibility testing is essential for defining mechanisms of antiviral resistance, for determining the frequency with which drug-resistant virus mutants emerge in clinical practice, to test for cross-resistance to alternative agents, and when evaluating new antiviral agents. With the exception of antiviral susceptibility testing for HSV, results of in vitro phenotypic assays are usually not available quickly enough to be relevant for patient management. Nevertheless, the clinical deterioration of patients undergoing antiviral therapy can be associated with resistant virus. The results of antiviral susceptibility testing may be helpful in certain clinical situations. Persistent or worsening HSV or VZV infection during treatment with acyclovir may indicate drug resistance. Alternative therapies, such as foscarnet and cidofovir, are available. Furthermore, HSV or VZV causing recurrent infection is often TK competent and therefore again susceptible to acyclovir, which has minimal toxicity compared to foscarnet (Bean et al., 1987). Persistent or worsening CMV retinitis, pneumonitis, or colitis that is unresponsive to ganciclovir may also indicate drug-resistant virus. Cidofovir and foscarnet can be used as alternative agents. Influenza A virus isolates resistant to amantadine and rimantadine readily emerged during clinical trials of these drugs; therefore, continuous shedding or transmission of influenza A virus in a population on prophylaxis or treatment with these agents may be due to drug resistance. Monitoring HIV-1 isolates is essential for assessing the net effect of multiple-resistance mutations of isolates from patients receiving combination therapy and to test for cross-resistance to alternative antiretroviral drugs. The impact of susceptibility testing of HIV-1 on individual patient management decisions is unclear.

DEFINITION OF ANTIVIRAL RESISTANCE

Antiviral resistance is a decrease in susceptibility to an antiviral drug that can be clearly established by in vitro testing and can be confirmed by genetic analysis of the virus and biochemical study of the altered enzymes. In vitro drug resistance must be distinguished from clinical resistance, in which the viral infection fails to respond to therapy. Clinical failures may or may not be due to the presence of a drug-resistant virus. Failure to achieve clinical response also hinges on other factors, such as the patient's immunological status and the pharmacokinetics of the drug in the individual patient. For example, limited penetration of a drug into the central nervous system may allow the escape of HIV-1 despite suppression of the virus at other sites. Poor oral absorption and binding to plasma proteins may limit the bioavailability of certain drugs. Furthermore, administration of certain antiretroviral drugs in combination may interfere with absorption or stimulate elimination of one or more of the coadministered drugs. Patient-specific factors, such as intolerance to an antiretroviral drug or an intercurrent infection, can also lead to increases in HIV-1 plasma viremia despite in vitro susceptibility (Carpenter et al., 1997; Rodriguez-Rosado et al., 1999).

Variables of Antiviral Susceptibility Testing

To date, no standards exist for antiviral susceptibility testing. NCCLS has established a subcommittee to develop a standard for susceptibility testing of HSV as a first step in developing consensus protocols for antiviral susceptibility testing. The major obstacle to standardization of antiviral susceptibility testing is that many variables influence the final result. These include (i) the cell line, (ii) the viral inoculum titer, (iii) the incubation time, (iv) the concentration range of the antiviral agent tested, (v) the reference strains, (vi) the assay method, (vii) the endpoint criteria, (viii) the calculation of the endpoint, and (ix) the interpretation of the endpoint. Testing of a single virus isolate may lead to greatly different endpoints depending on the type of cell culture used (Harmenberg et al., 1980; De Clercq, 1982; R. Wittrock, R. T. Sarisky, R. L. Hodiuka, M. Levin, A. Weinberg, and J. J. Leary, unpublished data). The titer of the virus inoculum is also critical; too large an inoculum can make a susceptible isolate appear resistant; too small an inoculum can make all isolates appear susceptible (Harmenberg et al., 1980). The length of time that virus incubates in the presence of a drug must be sufficient to allow detection of small plaques, in the case of a plaque reduction assay (PRA), or to allow growth of a subpopulation of resistant virus which may replicate at a lower rate than wild-type virus (Baldanti et al., 1996). The prolonged incubation time of peripheral blood mononuclear cell (PBMC)-based assays for HIV-1 susceptibility testing has been shown to select for subpopulations of HIV-1 variants not present in the starting inoculum (Kusumi et al., 1992). Moreover, different assay methods can produce different results. For example, the dye uptake (DU) assay for HSV susceptibility testing produces higher IC_{50}s than the PRA (Hill et al., 1991a). The concentration range of the drug tested affects the quality of the dose-response curve and therefore the validity of endpoint calculations. Susceptibility results are usually expressed as IC_{50}s because of the greater mathematical precision of the 50% endpoint than of a 90 or 99% endpoint. (Synonyms for IC_{50} are 50% inhibitory dose and 50% effective dose.) However, debate continues concerning the appropriate endpoint, i.e., 50 versus 90 or 99% inhibition. IC_{50}s are more precise and reproducible, but IC_{90}s may correlate better with clinical response and are better at detecting subpopulations of drug-resistant strains among sensitive ones (Baldanti et al., 1998a). Moreover, few studies have correlated in vitro results with clinical response (Lehrman et al., 1986; Bean et al., 1987; Jacobson et al., 1990; St. Clair et al., 1993; Richman, 1995; Durant et al., 1999).

Another critical variable is the heterogeneity within the population of a virus "isolate." A single clinical isolate actually represents a mixture of drug-susceptible and drug-resistant phenotypes (Hill et al., 1991; Richman et al., 1991; St. Clair et al., 1991; Baldanti et al., 1998a). A virus population that has never encountered an antiviral agent is predominantly drug susceptible; resistant virus may be present at low levels. The presence of low levels of resistant virus in a population that is predominantly drug susceptible might not be reflected in the IC_{50} but would manifest its presence in higher IC_{90}s or IC_{99}s. At this time it is unknown whether a small fraction of drug-resistant virus is important to the behavior of the virus in vivo or how such a fraction might affect the response of the infection to therapy in an otherwise healthy host. However, in immunocompromised patients, under the continued selective pressure of antiviral therapy, resistant virus can emerge, and its presence can often be correlated with progressive viral disease (Sibrack et al., 1982; Bean, 1987; Pahwa et al., 1988; Safrin et al., 1990, 1991a, 1991b; Englund et al., 1990; Jacobson et al., 1990; Chatis and Crumpacker, 1992; Erice, 1999).

The genetic locus at which a mutation occurs also affects susceptibility testing endpoints. For example, mutations in the UL97 (phosphotransferase) gene confer low-level ganciclovir resistance on CMV while mutations in both the UL97 and UL54 genes confer high-level resistance to ganciclovir (Smith et al., 1997).

TESTING METHODS

Phenotypic versus Genotypic Assays

Phenotypic assays are in vitro susceptibility assays that measure the inhibitory effect of antiviral agents on the entire virus population in a patient isolate. A variety of endpoint measurements have been utilized, including a reduction in the number of plaques; inhibition of viral DNA synthesis; reduction in the yield of a viral structural protein, i.e., hemagglutinin of influenza virus or the p24 antigen of HIV; and reduction in the enzymatic activity of a functional protein, i.e., HIV-1 RT. Phenotypic assays in use include PRA, DU, DNA hybridization, enzyme immunoassay (EIA), and yield reduction assays for herpesviruses and influenza A viruses and PBMC cocultivation and recombinant-virus assays for HIV-1. Genotypic assays analyze viral nucleic acid to detect specific mutations that cause antiviral-drug resistance. Genotyping has been applied primarily to CMV and HIV-1. Both assay types have unique features that complement each other. Phenotypic assays are better suited to assess the combined effect of multiple-resistance mutations on drug susceptibility. This is especially important for viruses such as CMV and HIV-1 that acquire resistance-associated mutations in multiple genes that may be manifested as new patterns of resistance, cross-resistance, multidrug resistance, or even reversal of resistance (Jacobson et al., 1990; Hill et al., 1991b; St. Clair et al., 1991; Smith et al., 1997). However, most phenotypic assays are labor-intensive and expensive and have long turnaround times. Genotypic assays are relatively inexpensive and have shorter turnaround times but cannot detect mutations outside the selected target. Interpretation of genotypic assays is problematic due to the complex interactions of resistance mutations that result in a particular drug resistance phenotype. The utility of either type of assay in the management of a particular patient appearing unresponsive to therapy is under investigation. The major phenotypic and genotypic assays in use will be discussed. Detailed protocols for the most commonly used phenotypic assays have been published elsewhere (Swierkosz and Biron, 1994) and are not presented here. However, the principle of each test and guidelines for interpretation of results are discussed.

Control Strains

Simultaneous testing of control strains is crucial when antiviral susceptibility testing is being done. Reference strains should include genetically and phenotypically well-characterized drug-susceptible and drug-resistant isolates. Drug-resistant strains chosen for reference should include those with drug resistance phenotypes relevant to the mode of action of the drug to be tested. For example, for testing

nucleoside analogs, which require phosphorylation by viral TK, i.e., acyclovir versus HSV and VZV, TK-negative or -deficient strains should be included. Susceptibility testing of CMV should include both UL97 and UL54 mutants. For HIV-1 testing, mutants resistant to both NRTI and NNRTI and mutants resistant to PI should be included. The National Institute of Allergy and Infectious Diseases AIDS Research and Reference Reagent Program (Bethesda, Md.) provides upon request a number of reference strains of HSV, VZV, and CMV, including the drug-resistant strains mentioned above and the laboratory control strains CMV AD169 and VZV Oka. For susceptibility testing of HIV-1, the AIDS Research and Reference Reagent Program has a repository of strains with various resistance phenotypes. Pharmaceutical companies and the American Type Culture Collection are also sources of control strains.

PRAs for CMV, HSV, and VZV

The PRA has classically been the standard method of antiviral susceptibility testing to which new methods are compared (Biron and Elion, 1980; Hayden et al., 1980; McLaren et al., 1983; Biron et al., 1985; Hill et al., 1991a). Because many variations of the PRA have been reported, the NCCLS has formed a subcommittee to develop a standard for PRA testing of HSV. The principle of the PRA is the inhibition of viral plaque formation in the presence of an antiviral agent. The concentration of antiviral agent inhibiting plaque formation by 50% is considered the IC_{50}. Although the PRA is tedious and consumes more reagents than other methods, for small-scale testing of isolates, it is appropriate. Prior to performing the antiviral susceptibility assay per se, titers of HSV isolates must be determined to ensure an inoculum appropriate for the surface area of the assay wells or plates (i.e., approximately 100 PFU/60-mm-wide tissue culture plate). Because clinical strains of CMV and VZV are cell associated and because low-titer cell-free stocks are less stable during storage, infected cell suspensions (obtained by trypsin treatment of the infected monolayer) can be conveniently used for these viruses. Two to three passages of clinical strains of CMV and VZV in cell culture are usually necessary to obtain a sufficient titer of virus. Low-passage isolates should be used because they are more likely to be representative of the original mixed population of the clinical isolate than a higher-passage stock would be. Well-characterized drug-susceptible and drug-resistant strains of HSV, CMV AD169, and VZV Oka or Ellen serve as reference strains. Stepwise instructions for performance of the PRA for HSV, VZV, and CMV have been published elsewhere (Hill et al., 1991a; Swierkosz and Biron, 1994, 1995). A modified PRA has been described that utilizes Vero or CV-1 cells that have been stably transformed with the *Escherichia coli lacZ* gene under the control of an HSV-1 early promoter which express beta-galactosidase only after infection with HSV. Plaques were visualized after histochemical staining for beta-galactocidase (Tebas et al., 1995; 1998). Proposed susceptibility breakpoints determined by the PRA are listed in Table 4.

DU

The DU assay has been used for many years for susceptibility testing of HSV (McLaren et al., 1983; Hill, 1991a; Swierkosz and Biron, 1995). The assay is based on the preferential uptake of a vital dye (neutral red) by viable cells but not by nonviable cells. The extent of viral lytic activity is determined by the relative amount of dye bound to viable cells after infection with HSV compared with the amount

TABLE 4 Proposed guidelines for antiviral susceptibility results of herpes group and influenza A viruses

Virus	Antiviral agent	Method	IC$_{50}$ Denoting resistance	Reference
HSV	Acyclovir	PRA	≥2 μg/ml	Field and Darby, 1980; Collins et al., 1982; McLaren et al., 1983; Swierkosz et al., 1987
		DNA hybridization	≥2 μg/ml	Englund et al., 1990
		DU	≥3 μg/ml	Hill et al., 1991a
	Famciclovir (active metabolite, penciclovir)	PRA and DNA hybridization	Definitive breakpoints cannot be established	Coen et al., 1985
	Foscarnet	PRA	>100 μg/ml	Coen et al., 1985
	Vidarabine	PRA	≥2-fold increase of IC$_{50}$ compared to control or pretherapy isolate	
VZV	Acyclovir	PRA and DNA hybridization	≥3- to 4-fold increase of IC$_{50}$ compared to pretherapy isolate or control strain	Biron and Elion, 1980; Jacobson et al., 1990; Linnemann et al., 1990; Safrin et al., 1991a
	Famciclovir	PRA and DNA hybridization	Definitive breakpoints cannot be established	Standring-Cox et al., 1996
	Foscarnet	Late antigen reduction assay	300 μM	Fillet et al., 1995
Human CMV	Cidofovir	PRA and DNA hybridization	≥4 μM	Cherrington et al., 1998, Erice, 1999
	Foscarnet	PRA and DNA hybridization	>400 μM	Chou et al., 1997; Erice, 1999
	Ganciclovir	PRA and DNA hybridization	≥12 μM	Erice, 1999
Influenza A virus	Amantadine or rimantadine	EIA	>0.1 μg/ml	Belshe et al., 1988, 1989

bound to uninfected cells. The dye bound by viable cells is eluted by ethanol and measured colorimetrically. The drug concentration inhibiting viral lytic activity by 50% is considered the IC$_{50}$.

The DU assay consistently gives IC$_{50}$s of acyclovir that are three to five times greater than those given by PRA. This difference is most likely due to the higher inoculum used in the DU assay (500 PFU/ml) and to the use of a liquid overlay which allows drug-resistant virus to "amplify," thus resulting in a more sensitive detection of small amounts of drug-resistant virus. Therefore, the DU assay uses a cutoff IC$_{50}$ of ≥3 μg/ml to denote acyclovir resistance (Table 4).

DNA Hybridization

Commercially available DNA-DNA hybridization test kits (Hybriwix Probe Systems, Athens, Ohio) have been used for susceptibility testing of HSV, CMV, and VZV (Swierkosz et al., 1987; Dankner et al., 1990; Englund et al., 1990; Jacobson et al., 1990; Safrin et al., 1990, 1991a, 1991b; Chou et al., 1995a, 1995b, 1997; Smith et al., 1997). Each kit contains cell culture plates, culture medium, lysing reagent, nylon "wicks," ^{125}I-labeled probe, hybridization solution, wash reagent, and positive and negative control wicks. Cell culture wells are infected with virus, overlaid with dilutions of antiviral agent, and incubated to allow virus growth. After the lysis of cells, lysates containing viral and cellular DNA are transferred to the wicks by capillary action. Following hybridization with virus-specific probe and washing of the wicks, the radioactivity of the dried wicks is counted in a gamma counter. The concentration of drug that causes a 50% reduction in counts compared with that of untreated virus control is the IC$_{50}$. The turnaround

time of the commercial hybridization reactions, excluding infection and incubation of cell culture wells, is approximately 4 h. Disadvantages of the Hybriwix assays include the use of a radioisotope, the short shelf life of the probe (2 months), and the relatively high cost of the kit. It is more cost-effective in a laboratory where large numbers of isolates are analyzed.

Yield Reduction Assay

The yield reduction assay reflects the ability of an antiviral agent to inhibit the production of infectious virus rather than the formation of a plaque. Cell monolayers are infected with virus, incubated in the presence of the antiviral compound, and then lysed. Cell-free virus titers are subsequently determined by plaque assay. The endpoint is defined as the concentration of antiviral which reduces virus yield by 50% in comparison with that of untreated control cultures. When used for susceptibility testing of HSV against penciclovir and acyclovir, the IC$_{50}$s for penciclovir were equivalent to or lower than those for acyclovir. The greater activity of penciclovir is postulated to be the result of the extended half-life of penciclovir-triphosphate (Wittrock et al., unpublished).

EIAs

EIAs have been developed for the susceptibility testing of HSV, VZV, and influenza A virus (Berkowitz and Levin, 1985; Rabalais et al., 1987; Belshe et al., 1989; Safrin et al., 1996). EIA permits quantitative measurement of viral activity by spectrophotometric analysis; IC$_{50}$s are calculated as the concentrations of antiviral agent that reduce the absorbance to 50% of that of the virus control. EIA is more suited to the routine diagnostic laboratory than the PRA and the DU assay. Susceptibility results for HSV and VZV determined by this method have correlated well with those obtained by the PRA. Susceptibility testing of influenza A virus by EIA utilizes antibodies to influenza A virus hemagglutinins (H1 or H3); viral hemagglutinin expression correlates with virus growth. Amantadine and rimantadine activities are measured by inhibition of hemagglutinin expression. Amantadine- and rimantadine-susceptible and -resistant isolates, whose M2 gene sequences are known, serve as controls and must be tested in parallel with patient isolates. A protocol for the EIA for susceptibility testing of influenza A virus has been published (Swierkosz and Biron, 1995).

Flow Cytometry

Flow cytometry has been applied to susceptibility testing of HSV and CMV (Lipson et al., 1997; Pavic et al., 1997; McSharry et al., 1998). While the IC_{50}s measured by flow cytometry were numerically different from those determined by plaque reduction, drug-susceptible isolates could be readily distinguished from drug-resistant isolates. Advantages of antiviral susceptibility testing by flow cytometry include the potential for automation, the objectivity of the assay, and a shorter turnaround time than the PRA.

Measurement of HSV and VZV TK Activity by Plaque Autoradiography

Functional viral TK is required for the initial phosphorylation of acyclovir. To determine whether resistance to acyclovir is due to diminished or altered viral TK, two plaque autoradiograph methods are used (Martin et al., 1985). Incorporation of [^{125}I]iodododeoxycytidine (IdC), a pyrimidine analog selectively phosphorylated by the VZV- or HSV-specific TK, correlates well with the acyclovir-phosphorylating potential of HSV and VZV isolates. Incorporation of [^{14}C]thymidine (dT) specifically assesses the thymidine-phosphorylating activities of these isolates and is useful for analyzing resistance to pyrimidine nucleoside analogs (acyclovir is a purine nucleoside analog). Most acyclovir-resistant HSV and VZV fail to incorporate both substrates due to diminished TK activity (TK$^-$); occasionally, strains with altered substrate (TK^A) activity are seen that fail to incorporate IdC but are able to incorporate dT. For IdC incorporation, Vero cells (HSV) and MRC-5 cells (VZV) are used. For dT incorporation, LMTK$^-$ TK$^-$ mouse LM cells (Roswell Park Memorial Institute, Buffalo, N.Y.) are used for HSV; the TK$^-$ cell line 143B is used for VZV dT incorporation. These assays provide both quantitative and qualitative evaluations of the TK status of a mixed population of TK$^+$ and TK$^-$ strains. The IdC and dT plaque autoradiograph methods have been detailed elsewhere (Martin et al., 1985; Swierkosz and Biron, 1995).

HIV

A number of phenotypic assays are in use for testing the susceptibility of HIV-1 isolates to NRTI (Larder et al., 1989; Linnemann et al., 1990; St. Clair et al., 1991). A serious limitation of these procedures is the fact that not all clinical isolates grow in the cell culture lines used in these assays. The AIDS Clinical Trials Group developed an assay performed in PBMCs that allowed the growth of almost all clinical isolates of HIV-1 (Japour et al., 1993). Virus activity is quantitated by measurement of the p24 antigen of HIV-1. The PBMC assay, however, is labor-intensive, costly, and difficult to control because of the many variables, and it has a long turnaround time (weeks). Moreover, this

assay requires cocultivation of infected PBMCs with uninfected donor PBMCs to produce a stock of the clinical isolate being tested, which has been shown to select for subpopulations of HIV-1 not present in the original isolate (Kusumi et al., 1992).

A new generation of phenotypic assays, recombinant virus assays (RVA), have been developed to circumvent these problems. RVA involve PCR amplification of complete RT or protease gene (PR) coding sequences either directly from the patient's PBMCs or after isolation of virus by cocultivation in PBMCs. The amplified RT or PR gene sequences from the patient strain are then cotransfected into CD4$^+$ T lymphocytes in parallel with an HIV-1 proviral molecular clone in which the original RT or RT and PR coding sequences had been deleted. Thus, CD4$^+$ cells containing a chimeric virus are generated, which contain the patient's RT or RT and PR sequences in a background of an HIV-1 strain from which the original RT or RT and PR sequences had been deleted (Kellam and Larder, 1994; Shi and Mellors, 1997; Hertogs et al., 1998). The susceptibilities of the chimeric viruses to all clinically available RT and PR inhibitors are subsequently determined in a single assay either by PRA in CD4$^+$ HeLa cells or by MT4 cell viability-based assay. RVA thus allow determination of the phenotypic resistance patterns of circulating virus in vivo and circumvent the problem of selection of nonrepresentative variants during cultivation. The RVA can be completed in approximately 10 days from the time of cotransfection. It is not known, however, what proportion of the total population of virus a subpopulation of resistant virus must achieve to be detectable by this assay (Hertogs et al., 1998).

In vitro susceptibility assays which test combinations of antiretroviral agents have been performed to determine drug synergy or antagonism (Prichard et al., 1993; Deminie et al., 1996; Chong and Pagano, 1997). Synergy and antagonism have been determined by the method of Chou and Talay (1984) and Prichard et al. (1993). A computer software program called Calcusyn (Biosoft, Cambridge, United Kingdom) has also been used (Chong and Pagano, 1997). In vitro testing of drug combinations is a necessary first step in evaluating possible drug combinations for use in vivo.

Genotypic Assays

The genetic basis for antiviral resistance has been extensively studied for CMV and HIV-1. As new antiviral agents become available, resistance-associated mutations must be elucidated. The major advantage of genotypic assays is the relatively rapid turnaround time compared to phenotypic assays, which may translate to earlier and better management of patients (Rodriguez-Rosado et al., 1999).

CMV

Genotypic assays have been used to screen CMV isolates for mutations associated with ganciclovir resistance. Both UL97 and UL54 mutations can be detected. A number of approaches have been successfully applied to genotyping CMV. PCR amplification and sequencing of the entire UL54 gene and the fragment of the UL97 gene spanning the conserved domains of phosphotransferase detected the mutations currently known to confer resistance to ganciclovir (Chou et al., 1995a; Baldanti et al., 1996; Smith et al., 1997; Erice, 1999). Alternatively, PCR amplification of short fragments of the UL97 gene followed by restriction endonuclease digestion has been used to detect mutations at positions 460, 594, and 595 (Chou et al., 1995a, 1995b). UL97 mutations could be detected in 89% of ganciclovir-resistant isolates, while UL54 mutations were present in all high-level ganciclovir-resistant isolates (Smith et al., 1997). PCR amplification followed by restriction endonuclease digestion could recognize mutant virus representing 10% of the total virus population (Smith et al., 1997). UL97-associated ganciclovir resistance mutations have also been detected directly in patient blood and cerebrospinal fluid (Spector et al., 1995; Wolf et al., 1995; Boivin et al., 1996). Marker transfer experiments have been used to definitively determine that a particular mutation is associated with drug resistance. For this purpose, PCR-amplified UL97 and UL54 fragments containing resistance-associated mutations were cotransfected with CMV strain AD169 (drug susceptible). The resulting recombinant plaques were assayed for antiviral susceptibility by PRA. To further verify transfer of the mutations in question, sequencing was performed across the transfected fragment (Baldanti et al., 1996; Chou et al., 1997).

HIV

Genotypic analysis of HIV-1 RT and PR genes has been employed for direct detection of mutations associated with HIV drug resistance (Rabalais et al., 1987; Larder et al., 1991; Richman et al., 1991; Eron et al., 1992; Shafer et al., 1996). Direct amplification of HIV gene regions containing resistance mutations followed by automated sequencing has been successfully used for detection of mutations associated with HIV antiviral resistance (Larder et al., 1993). Automated high-speed DNA sequencing has recently been developed by Visible Genetics Inc. (Toronto, Ontario, Canada) and Applied Biosystems (Foster City, Calif.) and should allow analysis of both the RT and PR genes within hours (Schuurman et al., 1999). DNA sequencing by using matrix hybridization (GeneChip) has been developed by Affymetrix Inc. (Santa Clara, Calif.); it permits resolution of sequence variants of the gag-pol-pro regions of HIV-1 (Chee et al., 1996; Kozal et al., 1996; Persing et al., 1996). The "line probe assay" analyzed PCR-amplified RT gene regions by hybridization to probes immobilized on a membrane strip. The probes were designed to cover different RT gene polymorphisms and drug-associated mutations (Stuyver et al., 1997). Other methods applied to direct detection of HIV drug resistance include ligase chain reaction (Abravaya et al., 1995), RNase A mismatch cleavage assay (López-Galíndez et al., 1991), self-sustained sequence replication amplification reaction (Gingeras et al., 1991), and a point mutation assay (Kaye et al., 1992).

A concern with each of the genotypic methods is that mutant variants present at low frequency may not be detectable and that mixtures of HIV-1 strains with minor sequence variations may not be distinguishable (Shafer et al., 1996). Genotypic assays do allow more rapid and efficient detection of resistance than phenotypic assays and may allow earlier detection of emerging resistance. However, the complexity of these tests makes them impractical for most routine diagnostic virology laboratories, and wide interlaboratory variability in testing exists (Schuurman et al., 1999). Moreover, genotypic assays can detect only known resistance-associated mutations. Finally, the applicability of genotypic analysis of HIV-1 isolates to individual patient management is unclear because of the complex interactions among different combinations of resistance mutations (Richman, 1995; Boyer et al., 1998). Recently, Durant et al. (1999) documented the benefit of HIV-1 genotyping in choosing alternative antiretroviral regimens in pretreated individuals with therapeutic failure.

Measurement of plasma HIV-1 RNA levels (viral load) reflects the extent of virus replication in an infected individual and has been shown to be the strongest predictor of clinical outcome (Carpenter et al., 1997; Marschner et al., 1998). The three methods currently in use include reverse transcription-PCR (Monitor HIV; Roche Molecular Systems), nucleic acid sequence-based amplification (NASBA; Organon Teknika), and branched-chain DNA signal amplification (Quantiplex; Bayer). The monitoring of HIV-1 RNA levels is used to assess the relative risk of disease progression and the efficacy of antiretroviral therapy and is the predominant test used by clinicians to monitor HIV-infected patients. Because treatment failure may be due to a variety of pharmacological factors and poor patient compliance in addition to emergence of resistant virus, viral-load monitoring is essential for formulating patient management decisions regarding initiation and alteration of therapy. Declining HIV-1 RNA levels during treatment indicates response to therapy, while a significant rise in RNA levels indicates treatment failure (Carpenter et al., 1997; Marschner et al., 1998).

INTERPRETATION OF ANTIVIRAL SUSCEPTIBILITY RESULTS

Table 4 lists breakpoint $IC_{50}s$ proposed by various investigators for herpesviruses and influenza A virus. At present, there are no definitive interpretive standards for antiviral susceptibility testing. The concentration of antiviral at which virus is considered susceptible to the antiviral agent has generally been based on median susceptibilities of large numbers of clinical isolates from patients prior to, during, and after antiviral therapy. Because of the variables that affect antiviral susceptibility results, the absolute IC_{50} can vary from assay to assay and from laboratory to laboratory. Moreover, in vitro results indicating susceptibility or resistance may not correlate with the response of the infection to therapy in vivo. As discussed previously, the clinical response of the patient also depends upon host-specific factors and drug pharmacokinetics. A poor clinical response may occur even though the antiviral susceptibility testing indicates in vitro susceptibility (Dekker et al., 1983). Patients with HSV infections who are immunocompromised may fail to respond to therapy despite in vitro $IC_{50}s$ indicating susceptibility to vidarabine or acyclovir (Englund et al., 1990; Safrin et al., 1990, 1991b). Conversely, HSV isolates for which $IC_{50}s$ of acyclovir are >2 μg/ml can occasionally be recovered from otherwise healthy hosts who have responded to acyclovir therapy (Lehrman et al., 1986). Thus, a high IC_{50} derived by in vitro susceptibility testing is not sufficient to designate a viral strain as resistant. Neither can in vitro susceptibility to a drug a priori predict successful clinical outcome. Whenever possible, evidence of genetic alteration of the virus should be considered as well.

Interpretation of antiviral susceptibility test results is further complicated by the variability in endpoint due to testing methodologies (McLaren et al., 1983; Cole and Balfour, 1987; Hill et al., 1991a; Pepin et al., 1992; Hodinka, 1997). Because endpoints are dependent on the test method, each new method and antiviral agent must be correlated with a historic standard that has been used to test large numbers of isolates. Also, the absolute IC_{50} may vary from assay to assay and laboratory to laboratory. Moreover, because small subpopulations of resistant virus may not be reflected in $IC_{50}s$, $IC_{90}s$ may be more predictive of clinical response. One approach to interpreting susceptibility endpoints is to compare the IC_{50}s of an isolate obtained prior to therapy (or of a well-characterized reference control strain) with that of an isolate obtained during therapy; a significant increase in the ratio of such IC_{50}s denotes resistance. However, pretherapy isolates are often unavailable, and the IC_{50} ratio considered clinically significant is unclear. Large-scale collaborative comparisons of methods with the same virus isolates are necessary to standardize antiviral susceptibility testing and to establish definitive interpretive guidelines. Only when a standardized assay is adopted can prospective studies be performed to correlate in vivo response with in vitro susceptibility. Such studies are essential before definitive interpretive breakpoints are established.

Susceptibility testing of penciclovir illustrates the effect that cell line and testing method have on endpoint. When acyclovir, and penciclovir, which is structurally similar to acyclovir, were tested with HSV isolates in Vero cells by PRA, penciclovir appeared less active than acyclovir and HSV-2 isolates appeared to be resistant to penciclovir. In contrast, penciclovir appeared more active than acyclovir against some HSV isolates when SCC25 cells were used. The drugs appeared to have comparable activities when tested in MRC-5 and A549 cells (Wittrock et al., unpublished). When clinical isolates of VZV were tested by PRA and DNA hybridization, IC_{50}s obtained by DNA hybridization were significantly lower than those obtained by the PRA. Variability in endpoint was also seen with VZV, depending on the composition of the inoculum (cell free versus cell associated) (Standring-Cox et al., 1996). Therefore, breakpoints for susceptibility testing of penciclovir with HSV and VZV cannot be established at this time.

Absolute IC_{50}s denoting resistance of HIV-1 to antiretroviral drugs cannot be assigned. HIV-1 in vitro susceptibility results are usually expressed as the "fold increase" in the IC_{50} of an isolate obtained during therapy compared to that of a pretreatment isolate or a drug-susceptible isolate (Hertogs et al., 1998; Schinazi et al., 1999). In general, single drug resistance mutations are associated with 2- to 10-fold increases in IC_{50} while multiple mutations in the same gene can confer a >100-fold increase in IC_{50} (Schinazi et al., 1999).

There is controversy surrounding the significance and clinical utility of in vitro susceptibility testing and genotypic analysis of HIV-1 (Richman, 1995). Antiretroviral therapy may fail for reasons other than the emergence of drug-resistant virus, such as drug antagonism, noncompliance, increased clearance of one antiretroviral drug when coadministered with another drug, inadequate penetration of a drug into a sequestered site (i.e., the central nervous system), or malabsorption of a drug from the gastrointestinal tract (Hertogs et al., 1998). The relative proportion of resistant virus in the total virus population is also an important determinant of clinical response (Baldanti et al., 1998a). The presence of the syncytium-inducing phenotype has also been associated with an increased risk of disease progression (Richman and Bozzette, 1994). Both phenotypic and genotypic assays have drawbacks that limit their applicability to patient management. At present, alteration in patient therapy in based on monitoring of the HIV-1 load. Nevertheless, in vitro susceptibility testing of HIV-1 is important for determining the genetic sites of drug resistance mutations, the effects of sequential mutations on drug efficacy, and the rate of emergence of drug resistance mutations when various drug combinations are used and when new drugs are tested.

SUMMARY

No consensus protocol exists for antiviral susceptibility testing. Standardization is hampered by the many variables that affect susceptibility test results. No single assay method, cell line, or inoculum composition (cell free versus cell associated) appears sufficient for testing all viruses. A major problem with culture-based susceptibility testing of viruses other than HSV is that assays may require weeks to complete, a fact that limits their utility in the management of acute cases (Drew et al., 1993). Genotypic assays for UL97 and UL54 mutations of CMV may replace culture-based assays as more resistance-associated mutations are identified. The clinical significance of susceptibility testing of HIV-1 isolates remains questionable. Additional studies are required to correlate drug resistance mutations to antiretroviral drugs with patient outcome. Finally, combination therapy for HIV-1 has necessitated the development of assays capable of evaluating drug combinations for synergy, indifference, or antagonism.

REFERENCES

Abravaya, K., J. J. Carrino, S. Muldoon, and H. H. Lee. 1995. Detection of point mutations with a modified ligase chain reaction (Gap-LCR). *Nucleic Acids Res.* **23:**675–682.

Arts, E. J., and M. A. Wainberg. 1996. Mechanisms of nucleoside analog antiviral activity and resistance during human immunodeficiency virus reverse transcription. *Antimicrob. Agents Chemother.* **40:**527–540.

Baldanti, F., M. R. Underwood, S. C. Stanat, K. K. Biron, S. Chou. A. Sarasini, E. Silini, and G. Gerna. 1996. Single amino acid changes in the DNA polymerase confer foscarnet resistance and slow-growth phenotype, while mutations in the UL97-encoded phosphotransferase confer ganciclovir resistance in three double-resistant human cytomegalovirus strains recovered from patients with AIDS. *J. Virol.* **70:**1390–1395.

Baldanti, F., K. K. Biron, and G. Gerna. 1998a. Interpreting human cytomegalovirus antiviral drug susceptibility testing: the role of mixed virus populations. *J. Infect. Dis.* **177:**823.

Baldanti, F., M. R. Underwood, C. L. Talarico, L. Simoncini, A. Sarasini, K. K. Biron, and G. Gerna. 1998b. The Cys 607→Tyr change in the UL97 phosphotransferase confers ganciclovir resistance to two human cytomegalovirus strains recovered from two immunocompromised patients. *Antimicrob. Agents Chemother.* **42:**444–446.

Bean, B. 1992. Antiviral therapy: current concepts and practices. *Clin. Microbiol. Rev.* **5:**146–182.

Bean, B., C. Fletcher, J. Englund, S. N. Lehrman, and M. N. Ellis. 1987. Progressive mucocutaneous herpes simplex infection due to acyclovir-resistant virus in an immunocompromised patient: correlation of viral susceptibilities and plasma levels with response to therapy. *Diagn. Microbiol. Infect. Dis.* **7:**199–204.

Belshe, R. B., M. H. Smith, C. B. Hall, R. Betts, and A. J. Hay. 1988. Genetic basis of resistance to rimantadine emerging during treatment of influenza virus infection. *J. Virol.* **62:**1508–1512.

Belshe, R. B., B. Burk, F. Newman, R. L. Cerruti, and I. S. Sim. 1989. Resistance of influenza A virus to amantadine and rimantadine: results of one decade of surveillance. *J. Infect. Dis.* **159:**430–435.

Berkowitz, F. E., and M. J. Levin. 1985. Use of an enzyme-linked immunosorbent assay performed directly on fixed infected cell monolayers for evaluating drugs against varicella-zoster virus. *Antimicrob. Agents Chemother.* **28:**207–210.

Beutner, K. R. 1995. Valacyclovir: a review of its antiviral activity, pharmacokinetic properties, and clinical efficacy. *Antiviral Res.* **28**:281–290.

Biron, K. K. 1991. Ganciclovir-resistant human cytomegalovirus isolates; resistance mechanisms and in vitro susceptibility to antiviral agents. *Transplant. Proc.* **23**(Suppl. 3):162–167.

Biron, K. K., and G. B. Elion. 1980. In vitro susceptibility of varicella-zoster virus to acyclovir. *Antimicrob. Agents Chemother.* **18**:443–447.

Biron, K. K., S. C. Stanat, J. B. Sorrell, J. A. Fyfe, P. M. Keller, C. U. Lambe, and D. J. Nelson. 1985. Metabolic activation of the nucleoside analog q-[2-hydroxy-1-(hydroxymethyl)ethoxymethyl] guanine in human diploid fibroblasts infected with human cytomegalovirus. *Proc. Natl. Acad. Sci. USA* **82**:2473–2477.

Biron, K. K., J. A. Fyfe, S. C. Stanat, L. K. Leslie, J. B. Sorrell, C. U. Lambe, and D. M. Coen. 1986. A human cytomegalovirus mutant resistant to the nucleoside analog 9-{[2-hydroxy-1-(hydroxymethyl)ethoxy]methyl}guanine (BW B759U) induces reduced levels of BW B759U triphosphate. *Proc. Natl. Acad. Sci. USA* **83**:8769–8773.

Boden, D., and M. Markowitz. 1998. Resistance to human immunodeficiency virus type 1 protease inhibitors. *Antimicrob. Agents Chemother.* **42**:2775–2783.

Boivin, G., S. Chou, M. R. Quirk, A. Erice, and M. C. Jordon. 1996. Detection of ganciclovir resistance mutations and quantitation of cytomegalovirus (CMV) DNA in leukocytes of patients with fatal disseminated CMV disease. *J. Infect. Dis.* **173**:523–528.

Boucher, C. A. B., E. O'Sullivan, J. W. Mulder, C. Ramautarsing, P. Kellam, G. Darby, J. M. A. Lange, J. Goudsmit, and B. A. Larder. 1992. Ordered appearance of zidovudine resistance mutations during treatment of 18 human immunodeficiency virus-positive subjects. *J. Infect. Dis.* **165**:105–110.

Boyer, P. L., H.-Q. Gao, and S. H. Hughes. 1998. A mutation at position 190 of human immunodeficiency virus type 1 reverse transcriptase interacts with mutations at positions 74 and 75 via the template primer. *Antimicrob. Agents Chemother.* **42**:447–452.

Carpenter, C. C. J., M. A. Fischl, S. M. Hammer, M. S. Hirsch, D. M. Jacobsen, D. A. Katzenstein. J. S. G. Montaner, D. D. Richman, M. S. Saag, R. T. Schooley, M. A. Thompson, S. Vella, P. G. Yeni, and P. A. Volberding. 1997. Antiretroviral therapy for HIV infection in 1997. Updated recommendations of the International AIDS Society-USA panel. *JAMA* **277**:1962–1969.

Centers for Disease Control and Prevention. 1998. Public Health Service task force recommendations for the use of antiretroviral drugs in pregnant women infected with HIV-1 for maternal health and for reducing perinatal HIV-1 transmission in the United States. *Morb. Mortal. Wkly. Rep.* **47**(RR-2):16–17.

Chatis, P. A., and C. S. Crumpacker. 1992. Resistance of herpesviruses to antiviral drugs. *Antimicrob. Agents Chemother.* **36**:1589–1595.

Chee, M., M. Yang, E. Hubbell, A. Berno, X. C. Huang, D. Stern, J. Winkler, D. J. Lockart, M. S. Morris, and S. P. A. Fodor. 1996. Accessing genetic information with high-density DNA arrays. *Science* **274**:610–614.

Cherrington, J. M., M. D. Fuller, P. D. Lamy, R. Miner, J. P. Lalezari, S. Nuessle, and W. L. Drew. 1998. In vitro antiviral susceptibilities of isolates from cytomegalovirus retinitis patients receiving first- or second-line cidofovir therapy: relationship to clinical outcome. *J. Infect. Dis.* **178**:1821–1825.

Chong, K.-T., and P. J. Pagano. 1997. In vitro combination of

PNU-140690, a human immunodeficiency virus type 1 protease inhibitor, with ritonavir against ritonavir-sensitive and -resistant clinical isolates. *Antimicrob. Agents Chemother.* **41**:2367–2373.

Chou, S., A. Erice, M. C. Jordon, G. M. Vercellotti, K. R. Michels, C. L. Talarico, S. C. Stanat, and K. K. Biron. 1995a. Analysis of the UL97 phosphotransferase coding sequence in clinical cytomegalovirus isolates and identification of mutations conferring ganciclovir resistance. *J. Infect. Dis.* **171**:576–583.

Chou, S., S. Guentzel, K. R. Michels, R. C. Miner, and W. L. Drew. 1995b. Frequency of UL97 phosphotransferase mutations related to ganciclovir resistance in clinical cytomegalovirus isolates. *J. Infect. Dis.* **171**:239–242.

Chou, S., G. Marousek, S. Guentzel, S. E. Follansbee, M. E. Poscher, J. P. Lalezari, R. C. Miner, and W. L. Drew. 1997. Evolution of mutations conferring multidrug resistance during prophylaxis and therapy for cytomegalovirus disease. *J. Infect. Dis.* **176**:786–789.

Chou, T.-C., and P. J. Talay. 1984. Quantitative analysis of dose effect relationships: the combined effects of multiple drugs of enzyme inhibitors. *Adv. Enzyme Regul.* **22**:27–55.

Cirelli, R., K. Herne, M. McCrary, P. Lee, and S. K. Tyring. 1996. Famciclovir: review of clinical efficacy and safety. *Antiviral Res.* **29**:141–151.

Coen, D. M., H. E. Fleming, Jr., L. K. Leslie, and M. J. Retondo. 1985. Sensitivity of arabinosyladenine-resistant mutants of herpes simplex virus to other antiviral drugs and mapping of drug hypersensitivity mutations to the DNA polymerase locus. *J. Virol.* **53**:477–488.

Coen, D. M., M. Kosz-Vnenchak, J. G. Jacobson, D. A. Leib, C. L. Bogard, P. A. Schaffer, K. L. Tyler, and D. M. Knipe. 1989. Thymidine kinase-negative herpes simplex virus mutants establish latency in mouse trigeminal ganglia but do not reactivate. *Proc. Natl. Acad. Sci. USA* **86**:4736–4740.

Cole, N. L., and H. H. Balfour, Jr. 1987. In vitro susceptibility of cytomegalovirus isolates from immunocompromised patients to acyclovir and ganciclovir. *Diagn. Microbiol. Infect. Dis.* **6**:255–261.

Collins, P., G. Appleyard, and N. M. Oliver. 1982. Sensitivity of herpes virus isolates from acyclovir clinical trials. *Am. J. Med.* **73**(Suppl.):380–382.

Dankner, W. M., D. Scholl, S. C. Stanat, M. Martin, R. L. Sonke, and S. A. Spector. 1990. Rapid antiviral DNA-DNA hybridization assay for human cytomegalovirus. *J. Virol. Methods* **28**:293–298.

De Clercq, E. 1982. Comparative efficacy of antiherpes drugs in different cell lines. *Antimicrob. Agents Chemother.* **21**:661–663.

De Clercq, E. 1995. Antiviral therapy for human immunodeficiency virus infections. *Clin. Microbiol. Rev.* **8**:200–239.

Dekker, C., M. N. Ellis, C. McLaren, G. Hunter, J. Rogers, and D. W. Barry. 1983. Virus resistance in clinical practice. *J. Antimicrob. Chemother.* **12**:137–152.

Deminie, C. A., C. M. Bechtold, D. Stock, M. Alam, F. Djang, A. H. Balch, T.-C. Chou, M. Prichard, R. J. Colonno, and P. F. Lin. 1996. Evaluation of reverse transcriptase and protease inhibitors in two-drug combinations against human immunodeficiency virus replication. *Antimicrob. Agents Chemother.* **40**:1346–1351.

Drew, W. L., R. Miner, and E. Saleh. 1993. Antiviral susceptibility testing of cytomegalovirus: criteria for detecting resistance to antivirals. *Clin. Diagn. Virol.* **1**:179–185.

Durant, J., P. Clevenbergh, P. Halfon, P. Delgiudice, S. Porsin, P. Simonet, N. Montagne, C. A. Boucher, J. M.

Schapiro, and P. Delamonica. 1999. Drug-resistance genotyping in HIV-1 therapy: the VIRADAPT randomized controlled trial. *Lancet* **353**:2195–2199.

Earnshaw, D. L., T. H. Bacon, S. J. Darlison, K. Edmonds, R. M. Perkins, and R. A. Vere Hodge. 1992. Mode of antiviral action of penciclovir in MRC-5 cells infected with herpes simplex virus type 1 (HSV-1), HSV-2, and varicella-zoster virus. *Antimicrob. Agents Chemother.* **36**:2747–2757.

Englund, J. A., M. E. Zimmerman, E. M. Swierkosz, J. L. Goodman, D. R. Scholl, and H. H. Balfour, Jr. 1990. Herpes simplex virus resistant to acyclovir. A study in a tertiary care center. *Ann. Intern. Med.* **112**:416–422.

Erice, A. 1999. Resistance of human cytomegalovirus to antiviral drugs. *Clin. Microbiol. Rev.* **12**:286–297.

Eron, J. J., P. Gorczyca, J. C. Kaplan, and R. T. D'Aquila. 1992. Susceptibility testing by polymerase chain reaction DNA quantitation: a method to measure drug resistance of human immunodeficiency virus type 1 isolates. *Proc. Natl. Acad. Sci. USA* **89**:3241–3245.

Eron, J. J., Y.-K. Chow, A. M. Caliendo, J. Videler, K. M. Devore, T. P. Cooley, H. A. Liebman, J. D. Kaplan, M. S. Hirsch, and R. T. D'Aquila. 1993. *pol* mutations conferring zidovudine and didanosine resistance with different effects in vitro yield multiply resistant human immunodeficiency virus type 1 in vivo. *Antimicrob. Agents Chemother.* **37**:1480–1487.

Field, A. K., and K. K. Biron. 1992. "The end of innocence" revisted: resistance of herpesviruses to antiviral drugs. *Clin. Microbiol. Rev.* **7**:1–13.

Field, H. J., and G. Darby. 1980. Pathogenicity in mice of strains of herpes simplex virus which are resistant to acyclovir in vitro and in vivo. *Antimicrob. Agents Chemother.* **17**:209–216.

Fillet, A.-M., B. Visse, E. Caumes, B. Dumont, M. Gentillini, and J.-M. Huraux. 1995. Foscarnet-resistant multidermal zoster in a patient with AIDS. *Clin. Infect. Dis.* **21**:1348–1349.

Gingeras, T. R., P. Prodanovich, T. Latimer, J. C. Guatelli, D. D. Richman, and K. J. Barringer. 1991. Use of self-sustained sequence replication amplification reaction to analyze and detect mutations in zidovudine-resistant human immunodeficiency virus. *J. Infect. Dis.* **164**:1066–1074.

Harmenberg, J., B. Wahren, and B. Oberg. 1980. Influence of cells and virus multiplicity on the inhibition of herpesviruses with acycloguanosine. *Intervirology* **14**:239–244.

Hayden, F. G. 1995. Antiviral agents, p. 411–450. *In* G. L. Mandell, J. E. Bennett, and R. Dolin (ed.), *Principles and Practice of Infectious Diseases*, 4th ed. Churchill Livingstone, New York, N.Y.

Hayden, F. G., K. M. Cote, and G. D. Douglas, Jr. 1980. Plaque inhibition assay for drug susceptibility testing of influenza viruses. *Antimicrob. Agents Chemother.* **17**:865–870.

Hayden, F. G., S. J. Sperber, R. B. Belshe, R. D. Clover, A. J. Hay, and S. Pyke. 1991. Recovery of drug-resistant influenza A virus during therapeutic use of rimantadine. *Antimicrob. Agents Chemother.* **35**:1741–1747.

Hertogs, K., M.-P. De Bethune, V. Miller, T. Ivens, P. Schel, A. Van Cauwenberge, C. Van Den Eynde, V. Van Gerwen, H. Azijn, M. Van Houtte, F. Peeters, S. Staszewski, M. Conant, S. Bloor, S. Kemp, B. Larder, and R. Pauwels. 1998. A rapid method for simultaneous detection of phenotypic resistance to inhibitors of protease and reverse transcriptase in recombinant human immunodeficiency virus type 1 isolates from patients treated with antiretroviral drugs. *Antimicrob. Agents Chemother.* **42**:269–276.

Hill, E. L., M. N. Ellis, and P. Nguyen-Dinh. 1991a. Antiviral and antiparasitic susceptibility testing, p.1184–1188. *In* A. Balows, W. J. Hausler, Jr., K. L. Herrmann, H. D. Isenberg, and

H. J. Shadomy (ed.), *Manual of Clinical Microbiology*, 5th ed. American Society for Microbiology, Washington, D.C.

Hill, E. L., G. A. Hunter, and M. N. Ellis. 1991b. In vitro and in vivo characterization of herpes simplex virus clinical isolates recovered from patients infected with human immunodeficiency virus. *Antimicrob. Agents Chemother.* **35**:2322–2328.

Hirsch, M. S., J. C. Kaplan, and R. T. D'Aquila. 1996. Antiviral agents, p. 431–466. *In* B. N. Fields, D. M. Knipe, P. M. Howley, R. M. Chanock, J. L. Melnick, T. P. Monath, B. Roizman, and S. E. Straus (ed.), *Virology*, 3rd ed., Lippincott-Raven, Philadelphia, Pa.

Hodinka, R. L. 1997. What clinicians need to know about antiviral drugs and viral resistance. *Infect. Dis. Clin. N. Am.* **11**:945–967.

Iversen, A. K., R. W. Shafer, K. Wehrly, M. A. Winters, J. I. Mullins, B. Chesebro, and T. C. Merigan. 1996. Multidrug-resistant human immunodeficiency virus type 1 strains resulting from combination antiretroviral therapy. *J. Virol.* **70**:1086–1090.

Jacobson, M. A., T. G. Berger, S. Fikrig, P. Cecherer, J. W. Moohr, S. C. Stanat, and K.K. Biron. 1990. Acyclovir-resistant varicella zoster virus infection after chronic oral acyclovir therapy in patients with the acquired immunodeficiency syndrome (AIDS). *Ann. Intern. Med.* **112**:187–191.

Japour, A. J., D. L. Mayers, V. A. Johnson, D. R. Kuritzkes, L. A. Beckett, J. M. Arduino, J. Lane, R. J. Black, P. S. Reichelderfer, R. T. D'Aquila, C. S. Crumpacker, the RV-43 Study Group, and the AIDS Clinical Trials Group Virology Committee Resistance Working Group. 1993. Standarized peripheral blood mononuclear cell culture assay for determination of drug susceptibilities of clinical human immunodeficiency virus type 1 isolates. *Antimicrob. Agents Chemother.* **37**:1095–1101.

Kaye, S., C. Loveday, and R. S. Tedder. 1992. A microtitre format point mutation assay: application to the detection of drug resistance in human immunodeficiency virus type-1 infected patients treated with zidovudine. *J. Med. Virol.* **37**:241–246.

Kellam, P., and B. A. Larder. 1994. A recombinant virus assay: a rapid, phenotypic assay for assessment of drug susceptibility of human immunodeficiency virus type 1 isolates. *Antimicrob. Agents. Chemother.* **38**:23–30.

Kellam, P., C. A. B. Boucher, and B. A. Larder. 1992. Fifth mutation in human immunodeficiency virus type 1 reverse transcriptase contributes to the development of high-level resistance to zidovudine. *Proc. Natl. Acad. Sci. USA* **89**:1934–1938.

Kost, R. G., E. L. Hill, M. Tigges, and S. E. Straus. 1993. Recurrent acyclovir resistant genital herpes in an immunocompetent patient. *N. Engl. J. Med.* **329**:1777–1782.

Kozal, M. J., N. Shah, N. Shen, R. Yang, R. Fucini, T. C. Merigan, D. D. Richman, D. Morris, E. Hubbell, M. Chee, and T. R. Gingeras. 1996. Extensive polymorphisms observed in HIV-1 clade B protease gene using high-density oligonucleotide arrays. *Nat. Med.* **7**:753–759.

Kusumi, K., B. Conway, S. Cunningham, A. Berson, C. Evans, A. K. N. Iversen, D. Colvin, M. V. Gallo, S. Coutre, E. G. Shpaer, D. V. Faulkner, A. DeRonde, S. Volkman, C. Williams, M. S. Hirsch, and J. I. Mullins. 1992. Human immunodeficiency virus type 1 envelope gene structure and diversity in vivo and after cocultivation in vitro. *J. Virol.* **66**:875–885.

Larder, B. A., and S. D. Kemp. 1989. Multiple mutations in HIV-1 reverse transcriptase confer high-level resistance to zidovudine (AZT). *Science* **46**:1155–1158.

Larder, B. A., G. Darby, and D. D. Richman. 1989. HIV with reduced sensitivity to zidovudine (AZT) isolated during prolonged therapy. *Science* **243:**1731–1734.

Larder, B. A., P. Kellam, and S. D. Kemp. 1991. Zidovudine resistance predicted by direct detection of mutations in DNA from HIV-infected lymphocytes. *AIDS* **5:**137–144.

Larder, B. A., A. Kohli, P. Kellam, S. D. Kemp, M. Kronick, and R. D. Henfrey. 1993. Quantitative detection of HIV-1 drug resistance mutations by automated DNA sequencing. *Nature* **365:**671–673.

Lehrman, S. N., J. M. Douglas, L. Corey, and D. W. Barry. 1986. Recurrent genital herpes and suppressive oral acyclovir therapy. Relation between clinical outcome and in-vitro drug sensitivity. *Ann. Intern. Med.* **104:**786–790.

Linnemann, C. C., Jr., K. K. Biron, W. G. Hoppenjans, and A. M. Solinger. 1990. Emergence of acyclovir-resistant varicella zoster virus in an AIDS patient on prolonged acyclovir therapy. *AIDS* **4:**577–579.

Lipson, S. M., M. Soni, F. X. Biondo, D. H. Shepp, M. H. Kaplan, and T. Sun. 1997. Antiviral susceptibility testing-flow cytometric analysis (AST-FCA) for the detection of cytomegalovirus drug resistance. *Diagn. Microbiol. Infect. Dis.* **28:**123–129.

Littler, E., A. D. Stuart, and M. S. Chee. 1992. Human cytomegalovirus UL97 open reading frame encodes a protein that phosphorylates the antiviral nucleoside analogue ganciclovir. *Nature* **358:**160–162.

López-Galíndez, C., J. M. Rojas, R. Nájera, D. D. Richman, and M. Perucho. 1991. Characterization of genetic variation and 3′-azido-3′-deoxythymidine-resistance mutations of human immunodeficiency virus by the RNase A mismatch cleavage method. *Proc. Natl. Acad. Sci. USA* **88:**4280–4284.

Marschner, I. C., A. C. Collier, R. W. Combs, R. T. D'Aquila, V. DeGruttola, M. A. Fischl, S. M. Hammer, M. D. Hughes, V. A. Johnson, D. A. Katzenstein. D. D. Richman, L. M. Smeaton, S. A. Spector, and M. S. Saag. 1998. Uses of changes in plasma levels of human immunodeficiency virus type 1 RNA to assess the clinical benefit of antiretroviral therapy. *J. Infect. Dis.* **177:**40–47.

Martin, J. L., M. N. Ellis, P. M. Keller, K. K. Biron, S. N. Lehrman, D. Barry, and P. A. Furman. 1985. Plaque autoradiography assay for the detection and quantitation of thymidine kinase-deficient and thymidine kinase-altered mutants of herpes simplex virus in clinical isolates. *Antimicrob. Agents Chemother.* **28:**181–187.

McLaren, C., M. N. Ellis, and G. A. Hunter. 1983. A colorimetric assay for the measurement of the sensitivity of herpes simplex viruses to antiviral agents. *Antiviral Res.* **3:**223–234.

McSharry, J. M., N. S. Lurain, G. L. Drusano, A. Landay, J. Manischewitz, M. Nokta, M. O'Gorman, H. M. Shapiro, A. Weinberg, P. Reichelderfer, and C. Crumpacker. 1998. Flow cytometric determination of ganciclovir susceptibilities of human cytomegalovirus clinical isolates. *J. Clin. Microbiol.* **36:**958–964.

Pahwa, S., K. Biron, W. Lim, P. Swenson, M. H. Kaplan, N. Sadick, and R. Pahwa. 1988. Continuous varicella-zoster infection associated with acyclovir resistance in a child with AIDS. *JAMA* **260:**2879–2882.

Pavic, I., A. Hartmann, A. Zimmermann, D. Michel, W. Hampl, I. Schleyer, and T. Mertens. 1997. Flow cytometric analysis of herpes simplex virus type 1 suceptibility to acyclovir, ganciclovir, and foscarnet. *Antimicrob. Agents Chemother.* **41:**2686–2692.

Pepin, J.-M., F. Simon, M. C. Dazza, and F. Brun-Vezinet. 1992. The clinical significance of in vitro cytomegalovirus susceptibility to antiviral drugs. *Res. Virol.* **143:**126–128.

Persing, D. H., D. A. Relman, and F. C. Tenover. 1996. Genotypic detection of antimicrobial resistance, p. 43–47. *In* D. H. Persing (ed.), *PCR Protocols for Emerging Infectious Diseases.* American Society for Microbiology, Washington, D.C.

Prichard, M. N., L. E. Prichard, and C. Shipman, Jr. 1993. Strategic design and three-dimensional analysis of antiviral drug combinations. *Antimicrob. Agents Chemother.* **37:**540–545.

Rabalais, G. P., M. J. Levin, and F. E. Berkowitz. 1987. Rapid herpes simplex virus susceptibility testing using an enzyme-linked immunosorbent assay performed in situ on fixed virus-infected monolayers. *Antimicrob. Agents Chemother.* **31:**946–948.

Reardon, J. E., and T. Spector. 1989. Herpes simplex virus type 1 DNA polymerase. Mechanism of inhibition by acyclovir triphosphate. *J. Biol. Chem.* **264:**7405–7411.

Richman, D. D. 1995. Clinical significance of drug resistance in human immunodeficiency virus. *Clin. Infect. Dis.* **21**(Suppl.2):S166–S169.

Richman, D. D., and S. A. Bozzette. 1994. The impact of the syncytium-inducing phenotype of human immunodeficiency virus on disease progression. *J. Infect. Dis.* **169:**968–974.

Richman, D. D., J. C. Guatelli, J. Grimes, A. Tsiatis, and T. R. Gingeras. 1991. Detection of mutations associated with zidovudine resistance in human immunodeficiency virus utilizing the polymerase chain reaction. *J. Infect. Dis.* **164:**1075–1081.

Rodriguez-Rosado, R., C. Briones, and V. Soriano. 1999. Introduction of HIV drug resistance testing in clinical practice. *AIDS* **13:**1007–1014.

Safrin, S., and L. Phan. 1993. In vitro activity of penciclovir against clinical isolates of acyclovir-resistant and foscarnet-resistant herpes simplex virus. *Antimicrob. Agents Chemother.* **37:**2241–2243.

Safrin, S., T. Assaykeen, S. Follansbee, and J. Mills. 1990. Foscarnet therapy for acyclovir-resistant mucocutaneous herpes simplex virus infection in 26 AIDS patients: preliminary data. *J. Infect. Dis.* **161:**1078–1084.

Safrin, S., T. G. Berger, I. Gilson, P. R. Wolfe, C. B. Wofsy, J. Mills, and K. K. Biron. 1991a. Foscarnet therapy in five patients with AIDS and acyclovir-resistant varicella-zoster virus infection. *Ann. Intern. Med.* **115:**19–21.

Safrin, S., C. Crumpacker, P. Chatis, R. Davis, R. Hafner, J. Rush, H. A. Kessler, B. Landry, J. Mills, and the AIDS Clinical Trials Group. 1991b. A controlled trial comparing foscarnet with vidarabine for acyclovir-resistant mucocutaneous herpes simplex in the acquired immunodeficiency syndrome. *N. Engl. J. Med.* **325:**551–555.

Safrin, S., E. Palacios, and B. J. Leahy. 1996. Comparative evaluation of microplate enzyme-linked immunosorbent assay versus plaque reduction assay for antiviral susceptibility testing of herpes simplex virus isolates. *Antimicrob. Agents Chemother.* **40:**1017–1019.

Schinazi, R. F., B. A. Larder, and J. W. Mellors. 1999. Mutations in retroviral genes associated with drug resistance: 1999–2000 update. *Int. Antivir. News* **7:**46–69.

Schuurman, R., L. Demeter, P. Reichelderfer, J. Tijnagel, T. de Groot, and C. Boucher. 1999. Worldwide evaluation of DNA sequencing approaches for identification of drug resistance mutations in the human immunodeficiency virus type 1 reverse transcriptase. *J. Clin. Microbiol.* **37:**2291–2296.

Shafer, R. W., M. A. Winters, D. L. Mayers, A. J. Japour, D. R. Kuritzkes, O. S. Weislow, F. White, A. Erice, K. J. Sannerud, A. Iversen, F. Pena, D. Dimitrov, L. M. Frenkel, and P. S. Reichelderfer. 1996. Interlaboratory comparison of sequence-specific PCR and ligase detection reaction to detect a human immunodeficiency virus type 1 drug resistance mutation. The AIDS Clinical Trials Group Virology Committee Drug Resistance Working Group. *J. Clin. Microbiol.* **34:**1849–1853.

Shi, C., and J. W. Mellors. 1997. A recombinant retroviral system for rapid in vivo analysis of human immunodeficiency virus type 1 susceptibility to reverse transcriptase inhibitors. *Antimicrob. Agents Chemother.* **41:**2781–2785.

Sibrack, C. D., L. T. Gutman, C. M. Wilfert, C. McLaren, M. H. St. Clair, P. M. Keller, and D. W. Barry. 1982. Pathogenicity of acyclovir-resistant herpes simplex virus type 1 from an immunodeficient child. *J. Infect. Dis.* **146:**673–682.

Smith, I. L., J. M. Cherrington, R. E. Jiles, M. D. Fuller, W. R. Freeman, and S. A. Spector. 1997. High-level resistance of cytomegalovirus to ganciclovir is associated with alterations in both the UL97 and DNA polymerase genes. *J. Infect. Dis.* **176:**69–77.

Snoeck, R., G. Andrei. M. Gerard, A. Silverman, A. Hedderman, J. Balzarini, C. Sadzot-Delvaux, G. Tricot, N. Clumeck, and E. De Clercq. 1994. Successful treatment of progressive mucocutaneous infection due to acyclovir- and foscarnet-resistant herpes simplex virus with (S)-1-(3-hydroxy-2-phosphonylmethoxypropyl)cytosine (HPMPC). *Clin. Infect. Dis.* **18:**570–580.

Spector, S. A., K. Hsia, D. Wolf, M. Shinkai, and I. Smith. 1995. Molecular detection of human cytomegalovirus and determination of genotypic ganciclovir resistance in clinical specimens. *Clin. Infect. Dis.* **21**(Suppl. 2)**:**S170–S173.

Stanat, S. C., J. E. Reardon, A. Erice, M. C. Jordan, W. L. Drew, and K. K. Biron. 1991. Ganciclovir-resistant cytomegalovirus clinical isolates: mode of resistance to ganciclovir. *Antimicrob. Agents Chemother.* **35:**2191–2197.

Standring-Cox, R., T. H. Bacon, and B. A. Howard. 1996. Comparison of a DNA probe assay with the plaque reduction assay for measuring the sensitivity of herpes simplex virus and varicella-zoster virus to penciclovir and acyclovir. *J. Virol. Methods* **56:**3–11.

St. Clair, M. H., J. L. Martin, G. Tudor-Williams, M. C. Bach, C. L. Vavro, D. M. King, P. Kellam, S. D. Kemp, and B. A. Larder. 1991. Resistance to ddI and sensitivity to AZT induced by a mutation in HIV-1 reverse transcriptase. *Science* **253:**1557–1559.

St. Clair, M. H., P. M. Hartigan, J. C. Andrews, C. L. Vavro, M. S. Simberkoff, J. D. Hamilton, and the VA Cooperative Study Group. 1993. Zidovudine resistance, syncytium-induc-ing phenotype, and HIV disease progression in a case-control study. *J. Acquir. Immune Defic. Syndr.* **6:**891–897.

Stuyver, L., A. Wyseur, A. Rombout, J. Louwagie, T. Scarcez, C. Verhofstede, D. Rimland, R. F. Schinazi, and R. Rossau. 1997. Line probe assay for rapid detection of drug-selected mutations in the human immunodeficiency virus type 1 reverse transcriptase gene. *Antimicrob. Agents Chemother.* **41:**284–291.

Swierkosz, E. M., and K. K. Biron. 1994. Antiviral susceptibility testing, p. 8.26.2–8.26.21. *In* H. D. Isenberg (ed.), *Clinical Microbiology Procedures Manual, Supplement 1.* American Society for Microbiology, Washington, D.C.

Swierkosz, E. M., and K. K. Biron. 1995. Antiviral susceptibility testing, p. 139–154. *In* E. H. Lennette, D. A. Lennette, and E. T. Lennette (ed.), *Diagnostic Procedures for Viral, Rickettsial and Chlamydial Infections,* 7th ed. American Public Health Association, Washington, D.C.

Swierkosz, E. M., and R. L. Hodinka. 1999. Antiviral agents and susceptibility tests, p. 1624–1639. *In* P. R. Murray, E. J. Baron, M. A. Pfaller, F. C. Tenover, and R. H. Yolken (ed.), *Manual of Clinical Microbiology,* 7th ed. ASM Press, Washington, D.C.

Swierkosz, E. M., D. R. Scholl, J. L. Brown, J. D. Jollick, and C. A. Gleaves. 1987. Improved DNA hybridization method for detection of acyclovir-resistant herpes simplex virus. *Antimicrob. Agents Chemother.* **31:**1465–1469.

Talarico, C. L., W. C. Phelps, and K. K. Biron. 1993. Analysis of the thymidine kinase genes from acyclovir-resistant mutants of varicella-zoster virus isolated from patients with AIDS. *J. Virol.* **67:**1024–1033.

Tebas, P., E. C. Stabel, and P. D. Olivo. 1995. Antiviral susceptibility testing with a cell line which expresses β-galactosidase after infection with herpes simplex virus. *Antimicrob. Agents Chemother.* **39:**1287–1291.

Tebas, P., D. Scholl, J. Jollick, K. McHarg, M. Arens, and P. D. Olivo. 1998. A rapid assay to screen for drug-resistant herpes simplex virus. *J. Infect. Dis.* **177:**217–220.

Wade, J. C., C. McClaren, and J. D. Meyers. 1983. Frequency and significance of acyclovir resistant herpes simplex virus isolated from marrow transplant patients receiving multiple courses of treatment with acyclovir. *J. Infect. Dis.* **148:**1077–1082.

Wolf, D. G., I. L. Smith, D. J. Lee, W. R. Freeman, M. Flores-Aguilar, and S. A. Spector. 1995. Mutations in human cytomegalovirus UL97 gene confer clinical resistance to ganciclovir and can be detected directly in patient plasma. *J. Clin. Investig.* **95:**257–263.

Nucleic Acid Hybridization

JAMES G. ANTHONY, LISA LINSKE-O'CONNELL, AND ATTILA T. LÖRINCZ

17

Virus detection is accomplished by many different and often unrelated methods. Choosing the most appropriate test for any particular situation is a complicated process. Presently, there is no universally superior procedure, although that day may be approaching. The traditional mainstay of viral diagnosis is virus isolation in cell culture. However, this approach is time consuming, requires considerable expertise, and is especially difficult with fastidious viruses. This makes alternative methods such as immunoassays, hybridization, and PCR much more attractive in a clinical setting. For a number of viral agents that cannot be grown easily, such as the hepatitis and gastroenteritis viruses and papillomavirus, direct detection of viral components, especially the genetic material, offers the most logical and accessible method for diagnosis. Targets for direct immunoassays include protein or carbohydrate moieties. In specific cases, it is possible to measure enzymatic activities associated with viral infections, such as the reverse transcriptases of retroviruses or the DNA polymerase of hepatitis B virus (HBV). PCR and other nucleic acid amplification methods (TMA, NASBA, SDA) have revolutionized molecular biology and have begun to filter into the realm of routine molecular diagnostics. The sensitivity of these methods is unsurpassed. Nevertheless, technical barriers, including issues of contamination, accurate quantitation, sample preparation, and high cost, have slowed the routine application of PCR and other amplification methods for standard clinical testing. In recent years, significant progress has been made in the development and application of signal amplification-based hybridization methods for viral diagnostics.

Hybridization is a powerful method for detecting viruses even without the use of PCR amplification. The genetic material of the virus forms the basis of hybridization tests. Often there is sufficient viral DNA or RNA in clinical specimens to perform direct testing and quantitation. In situ hybridization (ISH) is well suited for detection of viruses, including cytomegalovirus (CMV), herpes simplex virus (HSV), human papillomavirus (HPV), and others, in human tissues. Dot blot and Southern blot hybridizations require extracted nucleic acids, with subsequent loss of information on cell or tissue localization. Nevertheless, these methods may also be useful. More recently, sandwich-based signal amplification (SSA) methods using microtiter plates have been developed. These methods offer the advantages of the traditional DNA dot blot test but with substantially improved sensitivity, specificity, and reproducibility, and they include Hybrid Capture (Lorincz, 1996), branched DNA (Hendricks et al., 1995), and the DNA enzyme immunoassay (Prati et al., 1995). The SSA formats greatly ease sample handling and increase sample throughput. Homogenous assays that are performed without wash steps have also been developed; they require no extraction steps and minimal sample processing steps. Advances at the cutting edge of nucleic acid diagnostics include the use of microslides or microchip hybridization formats, which will likely be among the diagnostic platforms of the future. Sample processing and data analysis are two critical elements that must be addressed before microarray formats can be applied for routine clinical analysis. Simplification of sample processing and data analysis to increase throughput, the development of ultrasensitive and highly specific tests, and cost containment are key challenges facing the wide dissemination of new viral diagnostic methods. This chapter is restricted to SSA probe tests and hybridization methods not employing amplification.

HISTORY

The modern era of nucleic acid hybridization testing was ushered in by E. M. Southern's discovery that gel-electrophoretically separated DNA could be transferred efficiently to a solid support (nitrocellulose) to generate a replica of the gel. The gel replica could then be probed with radioactively labeled DNA to detect the DNA target of interest. This elucidation of the fundamentals of blot hybridization was quickly followed by the use of similar methods for the detection of RNA ("Northern" blots). A simplified version of the filter-based hybridization method is known as the dot blot. This method omits the electrophoretic gel separation steps and immobilizes the nucleic acid of interest directly on the solid support. Dot blotting has gained widespread acceptance as a more rapid, but less specific, method of nucleic acid detection. Another variation of the blot hybridization method is ISH, which detects nucleic acids directly on thin tissue sections. ISH requires an even higher level of manipulation but gives histological detail not available by the other methods. Before the invention of PCR and signal amplification methods, nucleic acid targets were detected directly using radioactively labeled probes. The primary hybridization detection methods for viral samples were Southern/dot blotting, RNA (Northern) blotting, and ISH. The development of PCR and SSA techniques has led to substantial improvement in sensitivity and ease of use (see "Specific Hybridization Methods," below, for detailed descriptions). Additionally, SSA techniques have decreased the amount of necessary labor and reduced the percentage of false-positive results. As such, SSA methods have become widely accepted in the clinical laboratory.

Since the time of Southern's original technique, there have been great improvements in the generation of nucleic acid probes. The establishment of inexpensive and rapid oligonucleotide syntheses, the use of phage-based transcription vectors coupled with the newer SSA techniques, and the use of PCR and other target amplification-based methods have simplified and reduced the time and effort required for probe production. The gain in sensitivity in target amplification is a result of production of multiple copies of the target nucleic acid that can then be detected by conventional means. Signal amplification does not result in an increase in target concentration but essentially uses the target as a bridge to amplify the signal through either target-specific accumulation of labeled molecules or target-specific immobilization of enzymes that can activate multiple substrate molecules.

There have also been many advances in the development of nucleic acid labels, especially in the area of fluorescent labels. Fluorescent labels have been particularly useful for ISH and microarray applications. Hybridization methods utilizing 96-well microplate formats have become routine. The next logical step, use of 384- and 1,536-well microplates, has already begun. More recently, the use of addressable arrays of DNA probes on slides and silicon chips has taken a foothold in the research laboratory, paving the way for the future of molecular diagnostics (Marshall and Hodgson, 1998).

BASIC PRINCIPLES

Base Pairing and Denaturation

Although there have been many advances in the sensitivity, speed, and reproducibility of nucleic acid hybridization assays, the basic principles of hybridization remain relatively unchanged from the first applications of blot hybridization. The intrinsic affinity of complementary nucleic acid strands accounts for the specificity of the method. This affinity can be affected by a variety of factors including temperature, pH, ion concentration, probe size and guanosine-cytidine (GC) content, denaturing agents (e.g., formamide), microscopic viscosity (e.g., glycerol), and macroscopic viscosity (e.g., dextran sulfate and other hybridization accelerants) (Hames and Higgins, 1985; Kohne, 1997).

DNA or RNA molecules can exist in either the single-stranded or the double-stranded state. Under the proper conditions, the thermodynamic driving force of complementary single strands of DNA or RNA favors duplex formation. These double-stranded forms can be converted to their single-stranded forms by elevated temperatures or by denaturation at highly alkaline pH (RNA is degraded at over pH 10). Following denaturation, the probe and target molecules are renatured by neutralization and/or a decrease in temperature. The optimal temperature for hybridization is determined by the melting temperature (T_m) of the complementary regions. The T_m is highly influenced by the ion

concentration and the presence of denaturing agents, such as formamide, in the hybridization solution.

The hybridization of two complementary DNA strands is a highly specific process in which only complementary or highly homologous strands can pair. This pairing process forms the basis of DNA-testing methodology. After hybridization, the specific duplexes must be distinguished from unreacted molecules and nonspecific hybrids. By one procedure, a first reactant (target) is immobilized on a solid support. After hybridization, the excess liquid-phase probe is washed away. Nitrocellulose and nylon filters, polystyrene microtiter plates, and glass slides are the most commonly used solid supports for DNA tests (Meinkoth and Wahl, 1984; Hames and Higgins, 1985; Lorincz, 1989). Immobilization may be by passive adsorption (Nikiforov and Rogers, 1995) or, preferably, by covalent attachment (Rasmussen et al., 1991).

Assay Range

The useful detection range of an assay must encompass clinically significant values for the method to be worthwhile. Typically, there is a trade-off between sensitivity improvements and specificity. Less-sensitive tests are more prone to false-negative results, while very sensitive tests are likely to have a higher clinical and, possibly, analytical false-positive rate. The nature of the viral target and the clinical significance of the virus concentration usually determine the degree of sensitivity that is required. In some cases, it would be detrimental to have an overly sensitive assay yielding unnecessary positive results for clinically insignificant virus levels. In other cases, accurate quantitation is a requirement for test utility (e.g., monitoring of HBV or human immunodeficiency virus [HIV] load).

Cross-reactivity Concerns

Natural divergence has produced more than 80 types of HPV, distinguishable by differences in their nucleotide base sequences (de Villiers, 1997). Nevertheless, many areas of high homology exist between different HPV types. Similarly, there are many variants of the hepatitis C virus (HCV) and HIV. In fact, all viruses consist of families of more or less related members clustered about consensus sequences. Hybridization tests can be cross-reactive (related but divergent nucleotide sequences can be detected by stabilization of poorly matched hybrids) or highly specific (very little reactivity with divergent virus types). The degree of cross-reactivity depends on the relatedness of the nucleic acid strands and the "stringency" under which the reaction is conducted.

Qualitative versus Quantitative Tests

Most tests are qualitative, that is, they are able to distinguish only the presence or absence of a virus above the specified lowest detectable level. Recently, the virus levels for CMV, HBV, HCV, and HIV were found to be predictive of patient prognosis. Such viral load tests have gained widespread acceptance for monitoring the effectiveness of drug therapy. These quantitative tests include a series of known concentrations of virus or other suitable controls (calibrators) to produce a standard curve over the clinically meaningful range. This curve is then used to calculate the absolute amount of virus in the patient sample. Great care must be taken to include positive and negative calibrators to ensure the accuracy of the standard curve. Samples must be taken, stored, and processed in such a way as to ensure that they are representative of the levels present in the patient. This extra care in sample handling plus the additional controls and calibrators increase the cost and difficulty of producing and running these types of assays.

Controlling the Hybridization Test

Hybridization test specificity and sensitivity can be regulated by varying stringency via manipulation of the concentration of salt or other chemicals and by changing the temperature of the hybridization and wash solution (Table 1). The filter hybridization reaction is performed by incubating the filter in a defined solution containing the probes. Then, unbound and nonspecifically bound probes are removed by extensive washing of the filter. Reviews by Meinkoth and Wahl (1984), Hames and Higgins (1985), and Lörincz (1989) contain detailed treatments of this subject and of specific hybridization and wash solutions. Hybridization on microplates and glass slides may be accomplished by means similar to filter hybridization (Hendricks et al., 1995). In addition, the hybrids may be formed in solution followed by capture of hybrids using specific binding reactions, such as biotin-streptavidin bonding, or antibody recognition of new structures on the hybrids (Lörincz, 1996). Immobilization of the target and target-bound probes is followed by wash steps to remove unbound probe and/or other reactants. Hybridization specificity using oligonucleotide probes can be controlled by the length and complexity of the probe, hybridization temperature, salt concentration, and formamide. The T_m (also called the T_d) for a given RNA-DNA hybrid or DNA duplex is the temperature at which 50% of the molecules are in a double-stranded form. The T_m for oligonucleotides between 14 and 70 bases can be estimated by the formula T_m (in degrees Celsius) $= 81.5 + 16.6 (\log M) + 0.41 (\% GC) - (600/N) - 0.72 (\%$ formamide), where M is the molarity of the sodium concentration, N is the chain length (in nucleotides), and % GC is the percentage of dG plus dC bases in the probe (Sambrook et al., 1989). DNA-RNA duplexes usually have a T_m of about 10°C higher than that of DNA-DNA; however, this relationship is variable depending on the concentration of formamide (Lorincz and Reid, 1989). As a general rule, the T_m goes up as the length of the duplex and percent GC

increases. For probes larger than 70 bases, the specificity of the hybridization is usually determined by the stringency of the wash conditions following hybridization. Typically, this is done by varying the salt concentration and temperature of the wash buffer until the desired balance of sensitivity and specificity for the probe-target pair is achieved. Enzymatic and chemical elimination (e.g., RNase and base treatment) of unhybridized and partially hybridized probes is an additional method of controlling hybridization specificity.

GENERAL CONSIDERATIONS

Specimen Handling

Regarding hybridization as a method of virus detection in clinical samples, there are several points to consider. The first is the nature of the sample target nucleic acid: whether the target is single-stranded DNA, double-stranded DNA, single-stranded RNA, or double-stranded RNA determines the types of precautions needed for handling. DNA is a stable molecule that can survive relatively unfavorable conditions for long periods and can be easily extracted from most clinical specimens (Sambrook et al., 1989). In contrast, RNA is a labile molecule that requires careful handling; however, even RNA can be preserved intact more easily than can the viability of some fastidious viruses and bacteria. The composition of clinical specimens is heterogeneous and may consist of large amounts of mucus, blood, proteins, and microorganisms, etc. Thus, the storage and extraction of each type of specimen must be optimized to yield the highest-quality nucleic acid. Freezing on dry ice is a universally acceptable but inconvenient state of transport and storage. Most specimens yield suitable DNA and often even acceptable RNA after storage at −70°C for several months. Transport and storage at 4°C for a few days is often adequate for the recovery of DNA but not RNA, except in situations such as HIV or HCV detection in plasma where the genomic RNA is partially protected by the viral capsid. Several different specialized specimen transport media are commercially available for testing for chlamydia, gonorrhea, and HPV in crude clinical specimens. These media can stabilize DNA for several weeks at room temperature (Warford and Levy, 1989; Ferris et al., 1998). Fixation of specimens in buffered formalin is suitable for subsequent analysis by ISH or PCR (Moench, 1987; Shibata et al., 1988). PreservCyt, a medium for preservation of cellular morphology of cytology specimens, is also a suitable (alcoholic) fixative that is conducive to DNA testing (Peyton et al., 1998; Manos et al., 1999), but it is suboptimal for preservation of RNA (unpublished data).

Extraction of Nucleic Acids

To perform dot blotting or Southern blotting, DNA must be extracted from specimens. Large samples of solid tissue must be pulverized. All extracted samples are treated with a digestion solution containing protease and detergent (Sambrook et al., 1989). The sample, whether blood, Pap smear, spinal fluid, or sputum, contains various levels of extraneous organic material such as cells, bacteria, and mucus. Thus, the source and the hybridization method determine the extent of extraction required. In some instances, the virus or bacteria can be concentrated by centrifugation before extraction of the pelleted material. Alternatively, the nucleic acid can be extracted, concentrated by a chromatographic column (e.g., Qiagen [see Krajden et al., 1996]) or by precipitation with salt and ethanol or isopropanol, and

TABLE 1 Principal controlling factors of hybridization tests[a]

Temperature	Increase gives higher stringency; optimal hybridization is at $T_d - 25°C^b$
Cations (Na$^+$)	Increased concentration gives lower stringency, favors hybridization rate (up to about 1.5 M), and enhances duplex stability
Formamide	Increased concentration raises stringency (lowers T_d) independently of temperature or salt concentration
Dextran sulfate or polyethylene glycol . .	Accelerates hybridization rate

[a] Reprinted with permission (Lorincz, 1989).
[b] T_d, denaturation temperature (temperature at which 50% of nucleic acid duplexes are denatured). T_d can be estimated for any defined, totally matched DNA-DNA duplex by the formula T_d (in degrees Celsius) $= 81.5 + 16.6 (\log M) + 0.41 (\% GC) - 0.72 (\%$ formamide), where M is the molarity of the sodium concentration and % GC is the percentage of dG plus dC bases in the DNA. The T_d of DNA-RNA duplexes is usually about 10°C higher than that of DNA-DNA; however, this relationship is variable depending on the formamide concentration.

then resuspended in a buffered solution. The liberated DNA can be analyzed directly by dot blot, microplate, or microslide methods or purified further with organic solvents before use in Southern blot analysis (Southern, 1975; Meinkoth and Wahl, 1984; Lörincz, 1989).

Probe Preparation

Construction of good probes is a prerequisite for any successful hybridization test. Probes can be either DNA or RNA, short (oligonucleotides) or long, and they can be coupled to a wide array of adducts for use in subsequent detection. Kits for making DNA and RNA probes are available from several companies, including Amersham (Arlington Heights, Ill.), Life Technologies (Gaithersburg, Md.), and Promega Biotech (Madison, Wis.). A commonly used procedure for preparing labeled DNA probes is called random primer labeling (Feinberg and Vogelstein, 1983). In this method, the input DNA is denatured and small random DNA oligonucleotide primers hybridize to complementary regions on the template DNA and act as origins of synthesis for a DNA polymerizing enzyme. The result is a labeled (e.g., ^{32}P, biotin, and digoxigenin-deoxynucleoside triphosphate labels) complementary copy of the template DNA strand. DNA of biological origin to be used as a probe for clinical testing must always be pure and have the plasmid vector sequences removed by gel electrophoresis or by some other method. This is especially true in specimens that frequently have large numbers of bacteria harboring plasmids that may be cross-reactive with these vector sequences (e.g., genital swabs). Labeled probes can also be generated by PCR by incorporating the labeled nucleotides during the amplification reaction. Methods for preparation of single-stranded, labeled DNA probes, either during PCR or postamplification, serve as a convenient source of DNA probe. Single-stranded DNA can be prepared during PCR by the asymmetric PCR method, which uses an excess of one primer, resulting in the accumulation of single-stranded DNA of the same strand as the abundant primer (Scully et al., 1990). Another method uses postamplification production of single-stranded probes through one nuclease-protected oligonuleotide, e.g., thioester protection (Nikiforov et al., 1994) (Fig. 1). Following amplification, the double-stranded DNA contains one protected strand and one unprotected strand. The DNA is reacted with 5' exonuclease to digest the unprotected strand, leaving the protected strand intact. In both methods, a label can be incorporated either during or after amplification.

RNA probes are made in vitro with RNA polymerase encoded by certain bacteriophages (Little and Jackson, 1987). Plasmids that contain specific phage RNA promoter sequences (e.g., T7 phage promoters [Davanloo et al., 1984]) upstream of an inserted DNA sequence that is to be copied into the probe have been constructed. In the presence of the phage RNA polymerase and appropriate nucleotides, substrates, and salts, the plasmids direct the synthesis of single-stranded RNA probes complementary to these DNA regions. Carefully prepared RNA probes are practically free of sequences that are cross-reactive with bacterial plasmids. Transcription probes can also be generated by incorporation of phage promoter sequences in oligonucleotides used for PCR (Sitzmann and LeMotte, 1993) (Fig. 2). Following amplification, the promoter bearing PCR products can be used for transcription without the need for linearization or removal of plasmid sequences. The use of RNase A to digest unhybridized regions of RNA improves specificity and makes RNA probes equal or superior to DNA probes.

Signal Detection

Following washing, the immobilized target is ready for signal detection. The method used depends on the type of probe label. Radioactive signals are usually detected by autoradiography for ISH and filter-based tests or by scintillation counting in microtiter plate-based tests. For filter tests, the most popular radioactive label is ^{32}P. For ISH tests, the most common radioactive labels used are ^{35}S and ^{3}H. Microtiter plate-based tests may make use of a variety of radioactive or nonradioactive labels. Recently, nonradioactive chemiluminesence- or fluorescence-based systems with

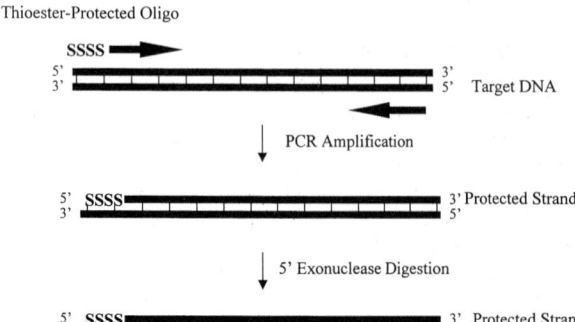

FIGURE 1 Postamplification production of single-stranded DNA probes using thioester protection and 5' exonuclease digestion.

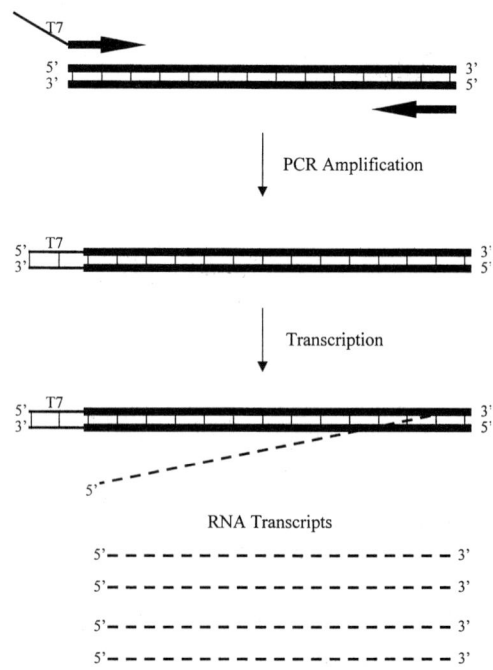

FIGURE 2 Production of transcription templates for RNA production by PCR.

superior performance have become available (e.g., ECL, etc. [see Kenten et al., 1992; Nelson, 1998; Szollosi et al., 1998]).

Although frequently used for sensitive detection assays, radioactive probes have the disadvantages of short useful life, radiation hazards, and disposal problems. Nonradioactive detection systems have essentially replaced the use of radioisotopes. Examples include antibodies to haptenated probes, streptavidin and biotin, and enzymes directly conjugated to the DNA or RNA (Matthews and Kricka, 1988). When probes are labeled with biotin, the presence of probe-target hybrids is detected with an enzyme such as alkaline phosphatase conjugated to streptavidin. Streptavidin binds tightly to biotin and thus localizes the enzymatic activity to the position of the hybrids on the solid phase (filter, microtiter well, array spot). Many different substrates are available for these enzymes, and the substrates generally fall into two categories: colorimetric and chemiluminescent. For alkaline phosphatase, a commonly used colorimetric substrate is a mixture of BCIP (5-bromo-4-chloro-3-indolylphosphate) and NBT (nitroblue tetrazolium). Chemiluminescent substrates have been developed for alkaline phosphatase (e.g., CDP-Star [Tropix], Lumi-Phos530 [Lumigen Inc.], galactosidase, glucuronidase, and a number of other enzymes). Enzymatic activation of the chemiluminescent substrates is achieved by removing a functional group (a phosphate group in the case of alkaline phosphatase). This results in the formation of an excited form of the substrate molecule, which decays and emits light detectable by a luminometer. In addition, chemiluminescent substrates that are chemically activated and used directly coupled to the probe molecule are quite popular (Kling, 1997). Chemically activated substrates, such as acridinium esters (AEs), are treated with base and oxidizing agents (typically hydrogen peroxide) to form an excited state, which releases a photon during decay to the ground state (Kling, 1997). Bioluminescent proteins have also been adapted for nucleic acid detection. The bioluminescent aequorin protein has been used in a novel PCR detection system. In this method, the aequorin protein is chemically coupled to anti-Digoxin antibodies that bind to Digoxin-incorporated PCR products (Xiao et al., 1996). With signal enhancement, chemiluminescent methods provide superior sensitivity to colorimetric detection and often perform more rapidly than radioactive labels (Matthews et al., 1985; Nelson and Kacian, 1990). Probe labels can also be used in homogeneous or near-homogeneous formats. The probe molecules remain in solution, and the signal is generated either by a close association of two probe molecules brought together by virtue of their association with the target (such as in fluorescent resonance energy transfer [Szollosi et al., 1998]) or where probe molecules, immobilized on a surface, can be distinguished from probe in solution. An example is electrochemiluminescence, in which only probe near the (electrode) surface generates a signal (Kenten et al., 1992). Another homogeneous method is the hybridization protection assay (HPA), which utilizes oligonucleotide probes conjugated with a highly chemiluminescent AE (Fig. 3). After hybridization of probes to target molecules, the sample is treated with weak base destroying the AE on the unhybridized probes and rendering the AE permanently nonchemiluminescent. Hybridization of AE-labeled probes to their exact complements protects the AE from hydrolysis (Nelson, 1998). Subsequent exposure to hydrogen peroxide under basic conditions causes the AE to decompose, forming N-methylacridone, which emits light (Kling, 1997).

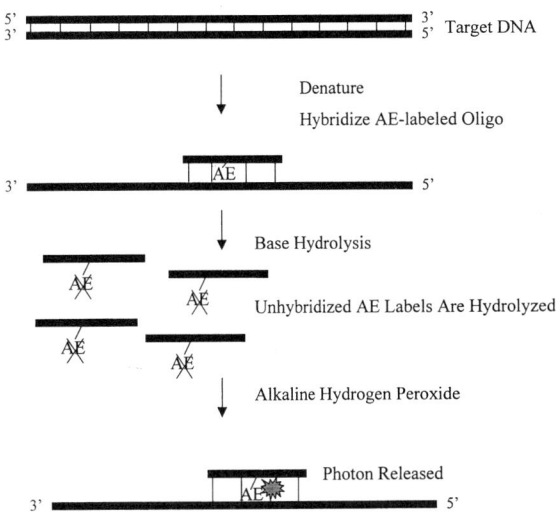

FIGURE 3 Detection of a DNA target by HPA.

Because of background problems, the available nonradioactive detection systems perform poorly with dot blots of clinical specimens but are as sensitive and reproducible as ^{32}P-labeled probes in Southern blots. Microtiter-based nonradioactive hybridization tests have also been developed. Typically, hybridization is performed in solution, followed by specific capture of the hybrids onto a solid support using antibodies or other avid intermolecular attractions. Alternatively, the probe may be immobilized on magnetic particles and used to remove the target from solution by hybridization. Such methods are less prone to background problems and do not require the use of radioactive probes. Our laboratory has applied a solution hybridization method, Hybrid Capture II (HC II), using 1,2-dioxetane-based chemiluminescence (Pollard-Knight et al., 1990) to the detection of HPV, HBV, CMV, HSV and chlamydiae in crude clinical specimens. Results show that HC II is much more sensitive and specific than dot blot hybridization employing ^{32}P-labeled probes. In fact, HC II approaches PCR levels of sensitivity (Peyton et al., 1998; Clavel et al., 1999; Girdner et al., 1999). Another area where radiolabeled probes offer little or no advantage is in ISH. Nonradioactive (fluorescent) probe-based in situ tests are as good as or better than most radioactive versions (Seelig and Tibedo, 1997). In addition, microslide array methods have primarily utilized fluorescent probes.

SPECIFIC HYBRIDIZATION METHODS

Four basic DNA tests—the dot blot, the Southern blot, ISH, and solid-surface sandwich hybridization (plates, beads, slides)—are particularly important, especially in combination with PCR amplification, for detecting pathogenic microorganisms. As PCR and tests based on this technique are described in chapter 19, they are not dealt with in detail here.

Dot Blotting

Dot blot hybridization is well suited for analysis of large numbers of specimens (Kimpton et al., 1988; Agha et al., 1989; Lorincz, 1989; Warford and Levy, 1989). Dot blotting can be performed with crude clinical samples or with nucleic acids purified from specimens. To avoid false positive results with DNA probes, specimens must be extracted with phenol and chloroform prior to application to the filter. In contrast, ^{32}P-labeled RNA probes perform well with crude specimens and give results comparable to the Southern blot (Lorincz, 1989; Warford and Levy, 1989).

A limitation of the dot blot method is that it lacks the characteristic DNA banding patterns provided by Southern blots, which are responsible for the latter test's excellent specificity. Dot blots also are unsuitable for low-stringency hybridization. But if adequate care is taken to optimize the specimen preparation, hybridization, and washing conditions and to avoid overloading with specimen DNA, the dot blot can reliably provide specific and sensitive detection.

Southern Blotting

In 1975, Southern's report of a procedure for faithfully reproducing images of DNA banding patterns in electrophoretic gels on filters revolutionized the study of nucleic acids (Southern, 1975). Southern demonstrated that DNA fragments separated according to size in gels could be denatured and transferred to nitrocellulose filters. This DNA could then be hybridized with probes, thus allowing the detection of specific fragments of DNA. In trained hands, the Southern blot has been an invaluable research tool. However, because of the skill required to perform the test and the several days required to obtain results, this method is not easily adapted to clinical laboratories. Today, SSA tests, PCR, and other amplification methods have largely replaced the Southern blot for routine diagnostic use.

In Situ Hybridization

ISH is used to detect DNA or RNA targets within cells (Stoler and Broker, 1986; Moench, 1987; Eversole et al., 1988; Schmidbauer et al., 1988). Thin sections (3 to 5 μm) of formalin-fixed and paraffin-embedded biopsy specimens placed on specially treated glass slides can be analyzed microscopically for target nucleic acids with the retention of good histological detail. ISH uses labeled probes on thin tissue sections. This requires a high level of manipulation but gives histological detail not available by the other methods.

Prior to hybridization, the sections are deparaffinized in xylene and rehydrated through graded concentrations of ethanol. The sections are usually treated with protease to make the cells more permeable. Cellular DNA is then denatured by alkali or heat (RNA targets do not require denaturation) and allowed to hybridize with DNA or RNA probes. The slides are then washed to remove the unhybridized probes, and probe-target hybrids are detected as appropriate, depending on whether the probes were labeled with a radioisotope (detection with photographic emulsion), a fluorescent adduct (detection by UV microscope), or some other adduct that permits the localization of alkaline phosphatase (detection with BCIP-NBT). New developments that have increased the sensitivity of ISH include target amplification methods (e.g., in situ PCR [see chapter 19 for a discussion of these methods]) and signal amplification methods such as tyramide signal amplification (TSA) (Wiedorn et al., 1999). TSA can be used to enhance sensitivity in either direct or indirect formats. Typically, both formats use biotinylated nucleic acid probes, and after wash steps the slides are reacted with streptavidin-labeled horse-

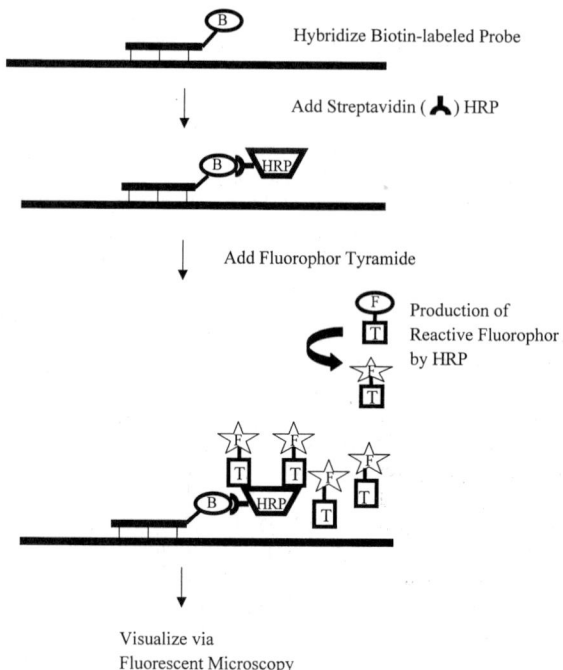

FIGURE 4 ISH by the direct TSA method.

radish peroxidase (HRP) enzyme. In the direct TSA assay (Fig. 4), fluorescence-labeled tyramide molecules (e.g., tyramide-fluorescein, tyramide-cyanine 3) are added and the HRP catalyzes accumulation of the fluorescence-labeled tyramide molecules on the surface, where they can be viewed by fluorescence microscopy. In the indirect method (Fig. 5), biotin-labeled tyramide molecules are added and immobilized to the probe site by HRP catalysis. The slide is washed and incubated with streptavidin-HRP molecules; after removal of the unbound streptavidin-HRP, a chromogenic HRP substrate is added and the slide is viewed by light microscopy. Both methods result in a substantial improvement in sensitivity, and because signal amplification is localized to the probe region there is little or no loss in signal resolution (Wiedorn et al., 1999). Robotic handling devices are now available for performing ISH (Brigati, 1988; Riben et al., 1997). Further developments along these lines should make ISH attractive for use in routine clinical testing laboratories.

Sandwich Hybridization

Sandwich hybridization methods typically use target DNA or RNA as a bridge between two specific probes (Fig. 6). One probe immobilizes the target and, indirectly through the target sequence, the second probe is immobilized on the solid support. The second probe generates the detection signal either directly or indirectly. The method enables a large number of patient samples to be quantitatively or qualitatively measured with a single probe set. Sandwich hybridization methods generally use microplates, magnetic beads, glass slides, or microarrays as solid supports. Some procedures include special signal amplification steps and are thus termed SSA tests.

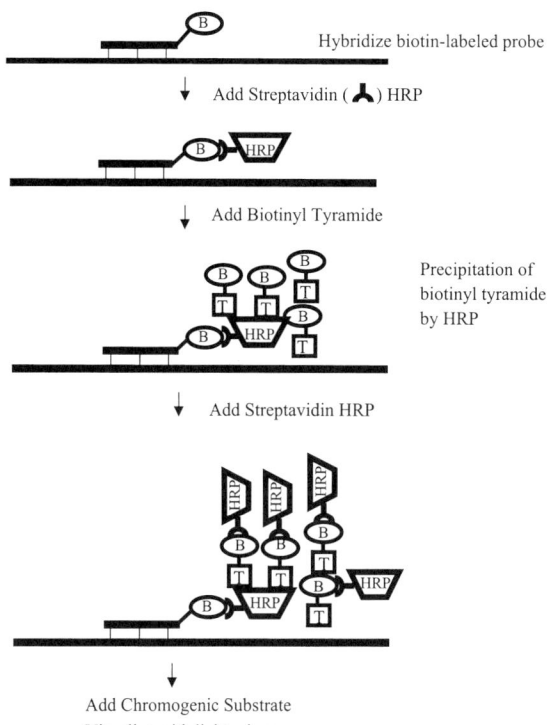

FIGURE 5 ISH by the indirect TSA method.

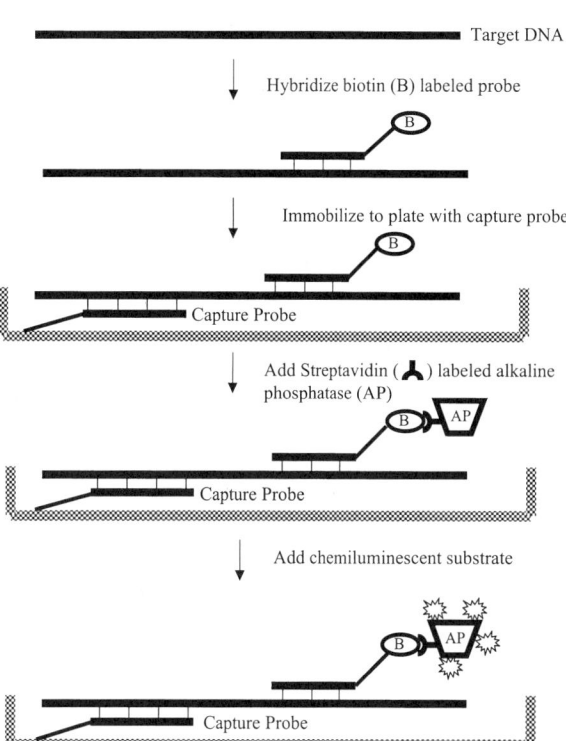

FIGURE 6 Sandwich hybridization detection of a nucleic acid target with an immobilized oligonucleotide capture probe and a biotinylated detection probe.

Plate-Based Hybridization Methods

There have been a variety of plate-based methods for detection of nucleic acids. The vast majority of these methods are for detection of postamplification products and rely on immobilized DNA capture probes to facilitate detection of the target-amplified products. The most widespread method of immobilization is to label the capture probe (de Beenhouwer et al., 1995) or the amplification product (Erhardt et al., 1996) with biotin and capture on streptavidin-coated microplates. Alternatives for microplate immobilization of capture probes include the use of end-modified capture probes for covalent linkage to the plate (Rasmussen et al., 1991) or adsorption of the DNA capture probes in the presence of salt or cationic detergents (Nikiforov and Rogers, 1995). A second target-specific signal-generating probe can be hybridized prior to, during, or subsequent to target immobilization. If the capture probes have been covalently coupled or adsorbed to the plate surface, typically a biotin-labeled signal probe is used, followed by the addition of streptavidin-labeled enzymes (e.g., alkaline phosphatase) for substrate activation. Another SSA microplate-based method, HC II (Fig. 7), utilizes RNA probes to facilitate both capture and detection of the target molecule and thus differs from the classic sandwich hybridization format. Microplates containing immobilized antibody specific for RNA-DNA hybrids in combination with target-specific RNA probes are used to capture the target DNA. Following hybridization, RNA-DNA hybrids are captured on the plates and, after the nonhybridized nucleic acid has been removed, an alkaline phosphatase-labeled anti-RNA-DNA monoclonal antibody is used to detect the immobilized hybrids. After incubation with the chemiluminescent substrate CDP Star (Tropix PE, Bedford, Mass.), the number of hybrids immobilized can be quantified (Lorincz, 1996).

Bead-Based Formats

Magnetic bead-based capture methods have gained popularity due to their ready adaptation to automation and ability to concentrate the sample. These methods are similar to the microplate format, utilizing separate capture and signal-generating probes. Target immobilization can be achieved during or after capture probe hybridization. This immobilization can be carried out by several means: noncovalent adsorption, covalent binding of capture oligonucleotides via functionalized groups, or the addition of binding agents such as biotin to the capture probe, followed by immobilization on streptavidin-coated beads (Zammatteo et al., 1997; O'Meara et al., 1998). Typically, posthybridization capture formats utilize streptavidin-coated beads in combination with biotinylated capture probes and labeled signal-generating probes.

Slide and Microchip Arrays

Slide and microchip arrays have evolved in the past few years (for a good general review, see the January 1999 supplement to *Nature Genetics* [Cohen, 1999]). They basically fall into two categories, cDNA and oligonucleotide arrays. The oligonucleotide arrays are generally either chemically coupled to the array surface or synthesized directly on arrays. The cDNA arrays are generally adsorbed to the slide surface, which has been pretreated with polylysine, amino silanes, or aminoreactive silanes to enhance adherence. The slides are then air dried and treated with UV light to cross-link and covalently couple the DNA to the slide. For virus testing, the arrays can potentially be used for multiple-virus screening tests and detection of mutations that effect drug resistance. At present, this technology is still in its infancy and is used mainly for research or pharmaceutical screening.

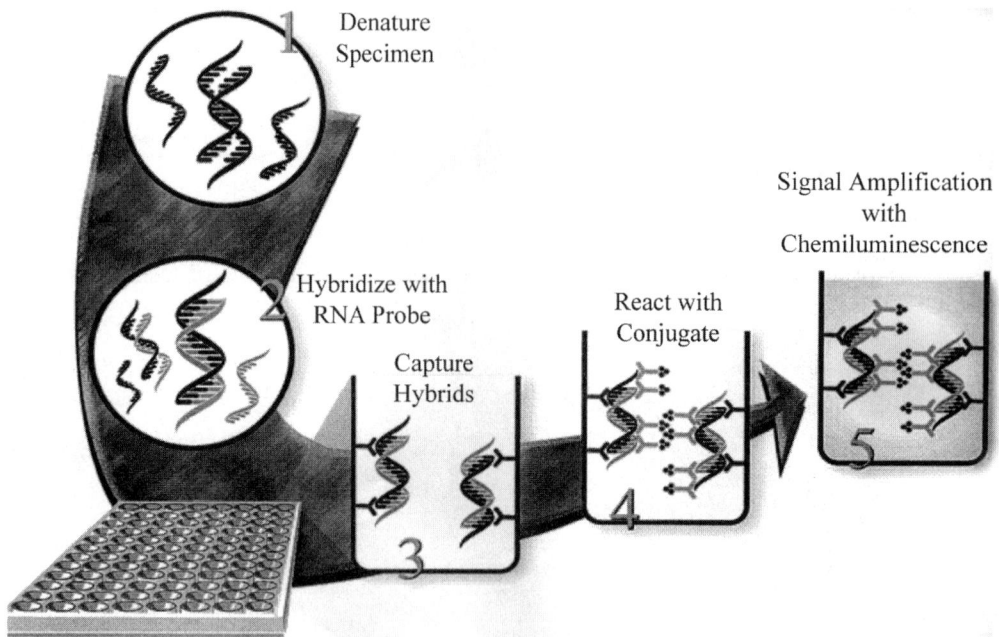

FIGURE 7 HC II detection of a DNA target.

APPLICATION OF HYBRIDIZATION TO VIRUS TESTING

In principle, hybridization tests can be used to detect any organism with very high sensitivity and specificity (Zwadyk and Cooksey, 1987; Yolken, 1988). This is especially true of tests based on PCR or other target amplification methods, which can detect one or a few organisms in a specimen with a vast excess of other biomass. Although SSA tests may not achieve the analytical sensitivity possible with PCR, they are highly sensitive and often attain or exceed PCR-like levels of clinical sensitivity because they are easier to perform and are not affected by inhibitors. Nucleic acid-based tests work equally well on dead and viable organisms and sometimes work even with partially degraded specimens. Due to the wide diversity of tests and applications, it is not possible to present a comprehensive list of all the variations of hybridization testing. Instead, a few relevant examples are discussed.

Cytomegalovirus

CMV has a large DNA genome, some regions of which cross-react strongly with the human genome. It is essential that these regions be eliminated from probes. Spector et al. (1984) used dot blot hybridization (with the *Eco*RI B and D fragments of the CMV AD169 strain as probes) to detect CMV DNA in urine and buffy coat specimens from neonates and bone marrow transplant recipients. DNA testing gave results faster than culture, and in some cases positive DNA tests were obtained with immunosuppressed patients who had negative culture tests. Subsequently, these patients were shown to have CMV disease. Mazzulli et al. (1999) performed a multicenter comparison of the CMV Hybrid Capture test with antigenemia, shell vial culture, and tube culture, with sensitivities of 95, 94, 43, and 46%, respectively. In addition, the specificity of the CMV Hybrid Capture assay was better than 99% in a group of healthy, nonimmunosuppressed volunteers. Thus, the rapid DNA test was more sensitive than culture and was as sensitive as the more cumbersome, microscope-based antigenemia test.

ISH testing for CMV may be helpful in arriving at a definitive diagnosis when characteristic viral inclusions are not present in histological specimens of transplant patients suspected of active CMV infection. Weiss et al. (1990) used ISH to detect CMV DNA in 26 (5%) of 477 lung or heart biopsies from a group of heart-lung recipients. ISH had a sensitivity of 85% and a specificity of 99% compared with histopathology. In contrast, culture was positive in only 68% of the patients diagnosed as CMV positive by both ISH and histopathology. ISH was positive for CMV DNA in four cases that were negative by histopathology. These cases probably represented true CMV infection, because biopsies from other sites showed evidence of CMV infection. Thus, from the above studies it appears that DNA testing for CMV has advantages in speed and sensitivity over culture and immunofluorescence and may be the test of choice for early diagnosis of CMV in immunosuppressed patients.

Human Papillomavirus

Hybridization testing is particularly well suited for detecting HPV. Peyton et al. (1998) compared the use of PCR and HC II for detection of HPV from cervical specimens by a variety of specimen collection strategies. The relative sensitivity of the HC II method is greater than 90% compared with the "gold standard," PCR. However, some samples were positive by HC II and negative by PCR. Subsequently, many of these specimens were shown to contain HPV DNA. The sensitivity of the HC II method for detection of HPV in women with cervical intraepithelial neoplasia (CIN), grade 2/3, or with invasive cancer has been reported to be 90 to 98% in recent studies. HPV DNA testing has proven to be a very valuable tool when used in a triage format following the Pap test for the management of women with Pap test results showing atypical squamous cells of undetermined significance (ASCUS). Use of HPV testing for women with ASCUS results in a substantial (60%) reduction in expensive and invasive procedures such as colposcopy (Lorincz et al., 1992; Clavel et al., 1999; Manos et al., 1999). Importantly, HC II was reported to have far superior ease of use. Richart and Nuovo (1990) used a nonradioactive in situ test to analyze 70 specimens of low-grade CIN. They found that 91% of these CIN specimens were positive for one or more of 14 different HPV types, leading to the suggestion that ISH may be a good method for correctly assigning histopathologically borderline specimens to the disease or normal category. Wiedorn et al. compared TSA-enhanced ISH with in situ PCR for the detection of HPV in 100 paraffin-embedded cervical biopsy specimens. They found the two methods to be nearly equivalent in terms of sensitivity (Wiedorn et al., 1999). The TSA-ISH method has been shown to detect a single HPV copy in formalin-fixed SiHa cells (Zehbe et al., 1997). The utility of hybridization for detecting and typing HPV is compelling. There is no culture system for HPV. Direct immunoassays are insensitive and cannot detect clinically important HPVs in a type-specific manner. Serology assays also do not appear to be particularly useful at present (Lorincz, 1996).

Hepatitis B Virus

Most studies have shown a close correlation between the presence of HBV e antigen (HBeAg) and the presence of HBV DNA in serum (Krogsgaard et al., 1985; Lok et al., 1985; Krogsgaard, 1988). Detection of HBV DNA in serum may be a marker of active virus replication in the liver. Krogsgaard et al. (1985) reported that detection of HBV DNA in serum by dot blotting upon admission to the hospital was positively correlated with clinical symptoms of short duration in individuals prior to hospital admission. During follow-up of patients, HBV DNA was always cleared before HBsAg (surface antigen) and generally before HBeAg (Krogsgaard, 1988). However, a few patients have been found to remain HBV DNA positive after becoming HBeAg negative, indicating that under certain circumstances HBV DNA tests may be more sensitive than tests for HBeAg. To some extent, these data were affected by the sensitivity and particular cutoffs of the tests employed. It is now known that the standard dot blot test for HBV DNA is of low sensitivity. More recently, it has become obvious that the best way of monitoring HBV-infected patients is by quantitative tests for serum HBV DNA. Pontisso et al. (1986) and Bortolotti et al. (1986) reported that conversion from HBV DNA positive to negative (i.e., detectable to undetectable DNA) within 1 or 2 weeks after onset of symptoms showed a strong correlation with recovery. Conversely, persistence of circulating HBV DNA for more than 8 weeks after onset of symptoms indicated the development of chronic HBV infection (Krogsgaard et al., 1985). A recent study by Niesters, comparing PCR (Roche Monitor), branched chain (bDNA) (Chiron) (Fig. 8), and direct hybridization using signal amplification (Digene's second-generation HBV HC II ultrasensitive), showed that the PCR and HC II methods had similar dynamic ranges, were more sensitive than the bDNA test, and were quite useful in monitoring the quantitative levels of HBV in patients undergoing lamivudine treatment (Niesters et al., 1998). These studies also indicated that HBV DNA testing is a particularly efficacious monitoring modality. Rijntjes et al. used ISH to detect HBV DNA in formalin-fixed liver spec-

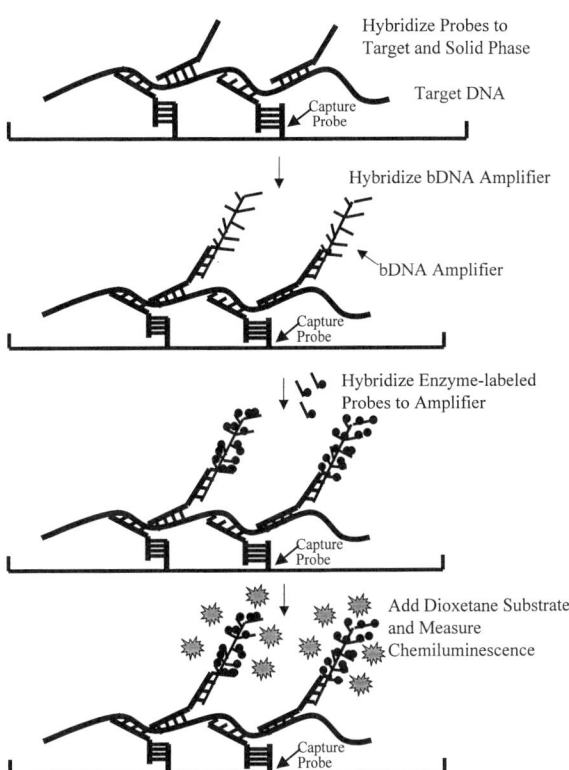

FIGURE 8 The bDNA assay utilizes target-specific oligonucleotide probes for capture and for directing the hybridization of an amplifier bDNA. The bDNA is then hybridized with enzyme-labeled oligonucleotides which activate the chemiluminescent substrate.

imens, including three cases in which the patients were not seropositive for HBV (Rijntjes et al., 1985). Southern blotting has also been used to demonstrate that HBV genomes are integrated into human DNA in hepatomas (Shafritz et al., 1981; Lie-Injo et al., 1983).

In conclusion, HBV DNA analysis is useful in certain situations, such as monitoring antiviral therapy. Conventional serological methods remain the principal method for primary diagnosis of HBV-related conditions.

Herpes Simplex Virus

When using nucleic acid probes to detect HSV, it is important to use specific clones of HSV DNA from which sequences that are cross-reactive to human DNA have been eliminated. Redfield et al. (1983) utilized the HSV-1 *Bam*HI A fragment to detect HSV-2 DNA in swabs of genital ulcers. Even though the HSV-1 probe was two- to four-fold less sensitive for detection of HSV-2 DNA than for the fully complementary HSV-1 DNA, the sensitivity of the DNA test was 78% and the specificity was 100% compared to isolation in cell culture. Schmidbauer et al. (1988) compared ISH with immunocytochemistry for diagnosis of HSV encephalitis. The two methods had comparable sensitivity (72%). However, in cases with a disease course longer than 1 month, ISH was more sensitive. Recent work comparing the Digene HSV HC II test and viral culture tests to PCR demonstrated that HC II had a sensitivity of 93.2% and a specificity of 100%, while culture had a sensitivity of 84.1% and a specificity of 100% (Cullen et al., 1997). The rapid turnaround of the hybridization method makes it more amenable for use in clinical applications, where the speed of results is a concern.

CONSIDERATIONS AND TIPS

Problems in Test Performance and Interpretation: Choosing the Correct Test

The various detection tests each have specific useful characteristics. For example, the speed and simplicity of the dot blot test make it useful for routine testing of large numbers of specimens. Southern blots can be used to verify dot blot results, to search for new viruses, and to track the spread of pathogenic variants in a population. An investigator wishing to localize viral DNA or RNA to specific cells or tissue regions can opt for ISH. However, this method cannot easily detect target DNA in low copy numbers unless it is coupled with signal amplification methods such as TSA, PCR, or NASBA. ISH can also be used to analyze paraffin-embedded tissues, specimens that are poorly suited for dot blot or Southern blot analysis.

While microarray methods have made rapid advances in the research laboratory, microplate formats have become the method of choice for the development of new nucleic acid-based tests for the clinical laboratory. Microplates offer the advantages of superior sensitivity with proven, relatively cheap technology that greatly improves assay throughput, labor, and cost compared to tube- and filter-based methods. Microplates are also easily adaptable to automation, which is the wave of the future for most large-scale testing in clinical laboratories. Currently, 96-well formats are the best developed, but work is underway to develop 384- and even 1,536-well plates, with the prospect of further improvements in sample throughput, reduced reagent costs, and reduced sample volume requirements.

All tests must be carefully monitored to detect erroneous results. False negatives can occur in several ways, e.g., the specimen is taken or stored improperly or an error occurs in the technical performance of the tests, etc. Each set of analyses should be accompanied by positive and negative controls. Positive controls help to assess that signal generation is working properly. They consist of specimens containing known amounts of specific target DNAs at various concentrations, including concentrations close to the lower limit of detection desired. Negative controls are specimens known to be free of target DNA or that mimic as closely as possible true negative specimens. Several of these should be analyzed with each batch of tests. If any are positive, the entire set of test results is suspect. Quantitative tests need even more care and should be based on a robust standard curve with well-spaced calibrators spanning the useful range of the test. There are detailed statistical techniques for tracking the performance of diagnostic tests; however, they are outside the scope of this review.

Pros and Cons of Nucleic Acid Testing

The full potential of nucleic acid testing is being realized slowly. Hybridization tests can be applied to the detection of virtually any organism, the major technical limitation being the identification of useful type-specific probes. Sensitivity is no longer an issue since the development of DNA and RNA target and signal amplification technologies. Culture is often too slow to help acute-stage management of the patient. Also, cell culture can detect the presence of only viable organisms and in some situations is not even an option (e.g., HPV detection). Thus, there are obvious advantages of hybridization testing. This approach is generally highly sensitive, specific, and relatively inexpensive. Furthermore, hybridization tests do not require viable (and potentially infectious) organisms. Often, hybridization tests are superior to serology. Antibodies to disease agents cannot be detected until well into the course of infection. Also, seropositivity persists well beyond the resolution of the acute infection. Measurement of specific immunoglobulin classes can provide information on the timing of acute infections. However, such tests can be error prone. In contrast, hybridization tests are extremely useful for quickly detecting acute-phase infection. ISH is particularly useful in this regard in directly identifying the infectious agent in diseased tissues. Hybridization tests can be more sensitive and specific than immunodetection of viral antigens. The specimens can be subjected to relatively harsh treatment and still be acceptable for DNA analysis but not for immunoassays.

Unfortunately, hybridization tests have been encumbered by many problems. In particular, specimen preparation procedures, test formats, and the time required to run the tests need improvement. Many of the tests employed in the past used radioactively labeled probes, which are clearly not practical for most routine laboratories. Nonradioactive probes have been available for years (Matthews and Kricka, 1988) and have recently improved to the point where they offer high sensitivity and specificity.

Detection of RNA by slot blot procedures can be problematic, as RNA is often rapidly degraded in clinical specimens. Rotbart et al. (1987) observed that the immediate addition of RNase inhibitors to specimens minimized the loss of RNA. Thus, improvements in technology should permit RNA detection as easily as DNA detection.

FUTURE DEVELOPMENTS

Historically, nonradioactive nucleic acid detection assays suitable for clinical use have suffered from a lack of sensitivity and specificity. This has led to the development of both target amplification methods (e.g., PCR and NASBA) and signal amplification methods (e.g., TSA and HC II). When first developed, target amplification methods required electrophoretic gel analysis of the amplification products. Subsequently, less cumbersome methods that eliminated the need for gel analysis were developed. Currently, one of the more popular target amplification methods is the Taqman PCR system, which couples the amplification and detection steps through the use of oligonucleotide probes labeled with both fluorescent reporter and quencher dyes that must be within a predetermined distance of each other. During PCR, the oligonucleotide probe is cleaved by the 5′ exonuclease activity of *Taq* DNA polymerase when it hybridizes to the target being amplified. Cleavage of the probe releases the two dye molecules and generates an increase in the fluorescence intensity of the reporter dye due to separation from the quencher. The Taqman assay measures amplification indirectly through this process of hydrolysis of hybridized probe. The labeled probe must be present in great excess for the signal generated by hydrolysis to be quantitative. However, this results in high background, because the quenching is never quantitative and the background from unhybridized probe can be quite high; nevertheless, this is a very powerful method for viral nucleic acid detection. Still, it does not address the ultimate goal of the optimal clinical protocol, because of the complexity of the sample preparation steps and the requirement for multiple internal controls. Signal amplification methods may address this need. Signal amplification methods that can process samples in a single tube without the need for extraction or precipitation steps have been developed; specimens can be treated (lysed) directly in the collection medium and then transferred to microplates for detection analysis. Since the signal amplification steps are applied after target immobilization, there is less need for internal controls and quantification of the results is more easily achieved.

In large reference laboratories, the need for cheaper and faster ways to produce a large volume of reliable patient results with less highly trained technicians leads to the need for automation and miniaturization. Currently, several systems have portions of the sample preparation, assay, or data analysis steps automated. The ultimate goal is a fully automated system that loads the sample at one end and produces a result at the other. There is a trade-off with these systems. The time and labor saved with an automated system are in part absorbed by the efforts needed to set up and maintain the automated equipment. The cost of these fully automated systems can also be prohibitive for small and middle-sized laboratories.

The newest methodology, chip arrays, uses basic hybridization techniques coupled with extreme miniaturization and fluorescent detection to enable a single sample to be probed for a large number of possible sequences. It is at the other extreme from sandwich techniques that measure multiple samples with one probe. Other hybridization-based technologies, such as dendrimers, optical or surface plasma resonance-based tests, and molecular beacons, have also been developed. They offer the promise of continued advancement in sensitivity, ease of use, and throughput of nucleic acid detection (Tyagi and Kramer, 1996; Nilsen et al., 1997; Schneider et al., 1997).

CONCLUSIONS

Hybridization testing has a clear place in the routine detection of viruses and other disease agents. Researchers in academia and in the commercial sector are actively working on the remaining problems. Several companies are now selling signal amplification-based nucleic acid probe assays for detecting a variety of viral agents; they include Bayer (Norwood, Mass.), Third Wave (Madison, Wis.), Integrated Genetics (Framingham, Mass.), and Digene Corporation (Beltsville, Md.), to name but a few (see also Zwadyk and Cooksey, 1987). The impact of hybridization testing on patient management is likely to grow considerably in the coming years.

REFERENCES

Agha, S. A., L. A. Mahmoud, L. C. Archard, A. M. Abd-Elaal, S. Selwyn, A. D. Mee, and J. C. Coleman. 1989. Early diagnosis of cytomegalovirus infection in renal transplant and dialysis patients by DNA-DNA hybridisation assay. *J. Med. Virol.* **27:**252–257.

Bortolotti, F., A. Bertaggia, C. Crivellaro, M. Armigliato, A. Alberti, P. Pontisso, C. Chermello, and G. Realdi. 1986. Chronic evolution of acute hepatitis type B: prevalence and predictive markers. *Infection* **14:**64–67.

Brigati, D. J., L. R. Budgeon, E. R. Unger, D. Koebler, C. Cuomo, T. Kennedy, and J. M. Perdomo. 1988. Immunocyto-

chemistry is automated: development of a robotic workstation based upon the capillary action principle. *J. Histotechnol.* **11:**165–183.

Clavel, C., M. Masure, J.-P. Bory, I. Putaud, C. Mangeonjean, M. Lorenzato, R. Gabriel, C. Quereux, and P. Birembaut. 1999. Hybrid Capture II-based human papillomavirus detection, a sensitive test to detect in routine high-grade cervical lesions: a preliminary study on 1518 women. *Br. J. Cancer* **80:**1306–1311.

Cohen, B. (ed.). 1999. *Nature Genetics* (supplement), vol. 21. *The Chipping Forecast.* Nature America Inc., New York, N.Y.

Cullen, A. P., C. D. Long, and A. T. Lorincz. 1997. Rapid detection and typing of herpes simplex virus DNA in clinical specimens by the Hybrid Capture II signal amplification probe test. *J. Clin. Microbiol.* **35:**2275–2278.

Davanloo, P., A. H. Rosenberg, J. J. Dunn, and F. W. Studier. 1984. Cloning and expression of the gene for bacteriophage T7 RNA polymerase. *Proc. Natl. Acad. Sci. USA* **81:**2035–2039.

de Beenhouwer, H., Z. Liang, P. de Rijk, C. van Eekeren, and F. Portaels. 1995. Detection and identification of mycobacteria by DNA amplification and oligonucleotide-specific capture plate hybridization. *J. Clin. Microbiol.* **33:**2994–2998.

de Villiers, E.-M. 1997. Papillomavirus and HPV typing. *Clin. Dermatol.* **15:**199–206.

Erhardt, A., S. Schaefer, N. Athanassiou, M. Kann, and W. H. Gerlich. 1996. Quantitative assay of PCR-amplified hepatitis B virus DNA using a peroxidase-labelled DNA probe and enhanced chemiluminescence. *J. Clin. Microbiol.* **34:**1885–1891.

Eversole, L. R., C. E. Stone, and A. M. Beckman. 1988. Detection of EBV and HPV DNA sequences in oral "hairy" leukoplakia by in situ hybridization. *J. Med. Virol.* **26:**271–277.

Feinberg, A. P., and B. Vogelstein. 1983. A technique for radiolabeling DNA restriction endonuclease fragments to high specific activity. *Anal. Biochem.* **132:**6–13.

Ferris, D. G., T. C. Wright, Jr., M. S. Litaker, R. M. Richart, A. T. Lorincz, X.-W. Sun, and C. D. Woodworth. 1998. Comparison of two tests for detecting carcinogenic HPV in women with Papanicolaou smear reports of ASCUS and LSIL. *J. Fam. Pract.* **46:**136–141.

Girdner, J. L., A. P. Cullen, T. G. Salama, L. He, A. Lorincz, and T. C. Quinn. 1999. Evaluation of the Digene Hybrid Capture II CT-ID test for detection of *Chlamydia trachomatis* in endocervical specimens. *J. Clin. Microbiol.* **37:**1579–1581.

Hames, B. D., and S. J. Higgins (ed.). 1985. *Nucleic Acid Hybridisation: a Practical Approach.* IRL Press, Oxford, England.

Hendricks, D. A., B. J. Stowe, B. S. Hoo, J. Kolberg, B. D. Irvine, P. D. Neuwald, M. S. Urdea, and R. P. Perrillo. 1995. Quantitation of HBV DNA in human serum using a branched DNA (bDNA) signal amplification assay. *Am. J. Clin. Pathol.* **104:**537–546.

Kenten, J. H., S. Gudibande, J. Link, J. J. Willey, B. Curfman, E. O. Major, and R. J. Massey. 1992. Improved electrochemiluminescent label for DNA probe assays: rapid quantitative assays of HIV-1 polymerase chain reaction products. *Clin. Chem.* **38:**873–879.

Kimpton, C. P., G. Corbitt, and D. J. Morris. 1988. Detection of cytomegalovirus by dot-blot DNA hybridization using probes labelled with ^{32}P by nick translation or random hexanucleotide priming. *Mol. Cell. Probes* **2:**181–188.

Kling, J. 1997. Luminescence developments help scientists see the light. *Scientist* **11:**16–17.

Kohne, D. E. 23 April 1997. Accelerated nucleic acid reassociation method. Gen-Probe Incorporated, San Diego, Calif.

U.S. patent 85,304,635. [European patent 7(EP 0 167 366 B1), 28 June 1985.]

Krajden, M., P. Shankaran, C. Bourke, and W. Lau. 1996. Detection of cytomegalovirus in blood donors by PCR using the Digene SHARP Signal System assay: effects of sample preparation and detection methodology. *J. Clin. Microbiol.* **34:**29–33.

Krogsgaard, K. 1988. Hepatitis B virus DNA in serum. Applied molecular biology in the evaluation of hepatitis B infection. *Liver* **8:**257–283.

Krogsgaard, K., P. Kryger, J. Aldershvile, P. Andersson, C. Brechot, and the Copenhagen Hepatitis Acuta Programme. 1985. Hepatitis B virus DNA in serum from patients with acute hepatitis B. *Hepatology* **5:**10–13.

Lie-Injo, L. E., M. Balasegaram, C. G. Lopez, and A. R. Herrera. 1983. Hepatitis B virus DNA in liver and white blood cells of patients with hepatoma. *DNA* **2:**301–308.

Little, P. F. R., and I. J. Jackson. 1987. Application of plasmids containing promoters specific for phage-encoded RNA polymerases, p. 1–18. *In* D. M. Glover (ed.), *DNA Cloning III. A Practical Approach*. IRL Press, Washington, D.C.

Lok, A. S. F., P. Karayiannis, T. P. Jowett, M. J. F. Fowler, P. Farci, J. Monjardino, and H. C. Thomas. 1985. Studies of HBV replication during acute hepatitis followed by recovery and acute hepatitis progressing to chronic disease. *J. Hepatol.* **1:**671–679.

Lorincz, A. T. 1989. Human papillomavirus testing. *Diagn. Clin. Testing* **27:**28–37.

Lorincz, A. T. 1996. Molecular methods for the detection of human papillomavirus infection. *Obstet. Gynecol. Clin. N. Am.* **23:**707–730.

Lorincz, A. T., and R. Reid. 1989. Association of human papillomavirus with gynecologic cancer. *Curr. Opin. Oncol.* **1:**123–132.

Lorincz, A. T., R. Reid, A. B. Jenson, M. D. Greenberg, W. Lancaster, and R. J. Kurman. 1992. Human papillomavirus infection of the cervix: relative risk associations of 15 common anogenital types. *Obstet. Gynecol.* **79:**328–337.

Manos, M. M., W. K. Kinney, L. B. Hurley, M. E. Sherman, J. Shieh-Ngai, R. J. Kurman, J. E. Ransley, B. J. Fetterman, J. S. Hartinger, K. M. McIntosh, G. F. Pawlick, and R. A. Hiatt. 1999. Identifying women with cervical neoplasia: using human papillomavirus DNA testing for equivocal Papanicolaou results. *JAMA* **281:**1605–1610.

Marshall, A., and J. Hodgson. 1998. DNA chips: an array of possibilities. *Nat. Biotechnol.* **16:**27–31.

Matthews, J. A., and L. J. Kricka. 1988. Analytical strategies for the use of DNA probes. *Anal. Biochem.* **169:**1–25.

Matthews, J. A., A. Batki, C. Hynds, and L. J. Kricka. 1985. Enhanced chemiluminescent method for the detection of DNA dot-hybridization assays. *Anal. Biochem.* **151:**205–209.

Mazzulli, T., L. W. Drew, B. Yen-Lieberman, D. Jekic-McMullen, D. J. Kohn, C. Isada, G. Moussa, R. Chua, and S. Walmsley. 1999. Multicenter comparison of the Digene Hybrid Capture CMV DNA assay (version 2.0), the pp65 antigenemia assay, and cell culture for detection of cytomegalovirus viremia. *J. Clin. Microbiol.* **37:**958–963.

Meinkoth, J., and G. Wahl. 1984. Hybridization of nucleic acids immobilized on solid supports. *Anal. Biochem.* **138:**267–284.

Moench, T. R. 1987. *In situ* hybridization. *Mol. Cell. Probes* **1:**195–205.

Nelson, N. C. 1998. Rapid detection of genetic mutations using the chemiluminescent hybridization protection assay (HPA): overview and comparison with other methods. *Crit. Rev. Clin. Lab. Sci.* **35:**369–414.

Nelson, N. C., and D. L. Kacian. 1990. Chemiluminescent DNA probes: a comparison of the acridinium ester and dioxetane detection systems and their use in clinical diagnostic assays. *Clin. Chim. Acta* **194:**73–90.

Nikiforov, T. T., and Y.-H. Rogers. 1995. The use of 96-well polystyrene plates for DNA hybridization-based assays: an evaluation of different approaches to oligonucleotide immobilization. *Anal. Biochem.* **227:**201–209.

Nikiforov, T. T., R. B. Rendle, M. L. Kotewicz, and Y.-H. Rogers. 1994. The use of phosphorothioate primers and exonuclease hydrolysis for the preparation of single-stranded PCR products and their detection by solid-phase hybridization. *PCR Methods Appl.* **3:**285–291.

Nilsen, T. W., J. Grayzel, and W. Prensky. 1997. Dendritic nucleic acid structures. *J. Theor. Biol.* **187:**273–284.

O'Meara, D., Z. Yun, A. Sonnerborg, and J. Lundeberg. 1998. Cooperative oligonucleotides mediating direct capture of hepatitis C virus RNA from serum. *J. Clin. Microbiol.* **36:**2454–2459.

Peyton, C. L., M. Schiffman, A. T. Lorincz, W. C. Hunt, I. Mielzynska, C. Bratti, S. Eaton, A. Hildesheim, L. A. Morera, A. C. Rodriguez, M. E. Sherman, and C. M. Wheeler. 1998. Comparison of PCR- and Hybrid Capture-based HPV detection systems using multiple cervical specimen collection strategies. *J. Clin. Microbiol.* **36:**3248–3254.

Pollard-Knight, D., A. C. Simmonds, A. P. Schaap, H. Akhavan, and M. A. W. Brady. 1990. Nonradioactive DNA detection on Southern blots by enzymatically triggered chemiluminescence. *Anal. Biochem.* **185:**353–358.

Pontisso, P., F. Bortolotti, E. Schiavon, L. Chemello, A. Alberti, and G. Realdi. 1986. Serum hepatitis B virus DNA in acute hepatitis type B. *Digestion* **34:**46–50.

Prati, D., B. D. Rawal, C. Dang, C. Capelli, and G. N. Vyas. 1995. DNA enzyme immunoassay of the PCR-amplified HLA-DQ alpha gene for estimating residual leukocytes in filtered blood. *Clin. Diagn. Lab. Immunol.* **2:**182–185.

Rasmussen, S. R., M. R. Larsen, and S. E. Rasmussen. 1991. Covalent immobilization of DNA onto polystyrene microwells: the molecules are only bound at the 5′ end. *Anal. Biochem.* **198:**138–142.

Redfield, D. C., D. D. Richman, S. Albanil, M. N. Oxman, and G. M. Wahl. 1983. Detection of herpes simplex virus in clinical specimens by DNA hybridization. *Diagn. Microbiol. Infect. Dis.* **1:**117–128.

Riben, M. W., J. H. Malfetano, T. Nazeer, P. J. Muraca, R. A. Ambros, and J. S. Ross. 1997. Identification of HER-2/*neu* oncogene amplification by fluorescence *in situ* hybridization in stage I endometrial carcinoma. *Mod. Pathol.* **10:**823–831.

Richart, R. M., and G. J. Nuovo. 1990. Human papillomavirus DNA in situ hybridization may be used for the quality control of genital tract biopsies. *Obstet. Gynecol.* **75:**223–226.

Rijntjes, P. J. M., T. J. M. van Ditzhuijsen, A. M. van Loon, U. J. G. M. van Haelst, F. B. Bronkhorst, and S. H. Yap. 1985. Hepatitis B virus DNA detected in formalin-fixed liver specimens and its relation to serologic markers and histopathologic features in chronic liver disease. *Am. J. Pathol.* **120:**411–418.

Rotbart, H. A., M. J. Levin, N. L. Murphy, and M. J. Abzug. 1987. RNA target loss during solid phase hybridization of body fluids—a quantitative study. *Mol. Cell. Probes* **1:**347–358.

Sambrook, J., E. F. Fritsch, and T. Maniatis. 1989. *Molecular Cloning: a Laboratory Manual*, 2nd ed. Cold Spring Harbor Laboratory Press, Cold Spring Harbor, N.Y.

Schmidbauer, M., H. Budka, and P. Ambros. 1988. Comparison of in situ DNA hybridization (ISH) and immunocytochemistry for diagnosis of herpes simplex virus (HSV) encephalitis in tissue. *Virchows Arch. A Pathol. Anat. Histopathol.* **414**:39–43.

Schneider, B. H., J. G. Edwards, and N. F. Hartman. 1997. Hartman interferometer: versatile integrated optic sensor for label-free, real-time quantification of nucleic acids, proteins, and pathogens. *Clin. Chem.* **43**:1757–1763.

Scully, S. P., M. E. Joyce, N. Abidi, and M. E. Bolander. 1990. The use of polymerase chain reaction generated nucleotide sequences as probes for hybridization. *Mol. Cell. Probes* **4**:485–495.

Seelig, S., and S. E. Tibedo. 1997. Fluorescence in situ hybridization. *IVD Technol.* **3**:26–36.

Shafritz, D. A., D. Shouval, H. I. Sherman, S. J. Hadziyannis, and M. C. Kew. 1981. Integration of hepatitis B virus DNA into the genome of liver cells in chronic liver disease and hepatocellular carcinoma: studies in percutaneous liver biopsies and post-mortem tissue specimens. *N. Engl. J. Med.* **305**:1067–1073.

Shibata, D., Y. S. Fu, J. W. Gupta, K. V. Shah, N. Arnheim, and W. J. Martin. 1988. Detection of human papillomavirus in normal and dysplastic tissue by the polymerase chain reaction. *Lab. Investig.* **59**:555–559.

Sitzmann, J. H., and P. K. LeMotte. 1993. Rapid and efficient generation of PCR-derived riboprobe templates for in situ hybridization histochemistry. *J. Histochem. Cytochem.* **41**:773–776.

Southern, E. M. 1975. Detection of specific sequences among DNA fragments separated by gel electrophoresis. *J. Mol. Biol.* **98**:503–517.

Spector, S. A., J. A. Rua, D. H. Spector, and R. McMillan. 1984. Detection of human cytomegalovirus in clinical specimens by DNA-DNA hybridization. *J. Infect. Dis.* **150**:121–126.

Stoler, M. H., and T. R. Broker. 1986. In situ hybridization detection of human papillomavirus DNAs and messenger RNAs in genital condylomas and a cervical carcinoma. *Hum. Pathol.* **17**:1250–1258.

Szollosi, J., S. Damjanovich, and L. Matyus. 1998. Application of fluorescence resonance energy transfer in the clinical laboratory: routine and research. *Cytom. Commun. Clin. Cytom.* **34**:159–179.

Tyagi, S., and F. R. Kramer. 1996. Molecular beacons: probes that fluoresce upon hybridization. *Nat. Biotechnol.* **14**:303–308.

Warford, A. L., and R. A. Levy. 1989. Use of commercial DNA probes. *Clin. Lab. Sci.* **2**:105–108.

Weiss, L. M., L. A. Movahed, G. J. Berry, and M. E. Billingham. 1990. In situ hybridization studies for viral nucleic acids in heart and lung allograft biopsies. *Am. J. Clin. Pathol.* **93**:675–679.

Wiedorn, K. H., H. Kuhl, J. Galle, J. Caselitz, and E. Vollmer. 1999. Comparison of in-situ hybridization, direct and indirect in-situ PCR as well as tyramide signal amplification for the detection of HPV. *Histochem. Cell Biol.* **111**:89–95.

Xiao, L., C. Yang, C. O. Nelson, B. P. Holloway, V. Udhayakumar, and A. A. Lal. 1996. Quantitation of RT-PCR amplified cytokine mRNA by aequorin-based bioluminescence immunoassay. *J. Immunol. Methods* **199**:139–147.

Yolken, R. H. 1988. Nucleic acids or immunoglobulins: which are the molecular probes of the future? *Mol. Cell. Probes* **2**:87–96.

Zammatteo, N., I. Alexandre, I. Ernest, L. Le, F. Brancart, and J. Remacle. 1997. Comparison between microwell and bead supports for the detection of human cytomegalovirus amplicons by sandwich hybridization. *Anal. Biochem.* **253**:180–189.

Zehbe, I., G. W. Hacker, H. Su, C. Hauser-Kronberger, J. F. Hainfeld, and R. Tubbs. 1997. Sensitive *in situ* hybridization with catalyzed reporter deposition, streptavidin-nanogold, and silver acetate autometallography: detection of single-copy human papillomavirus. *Am. J. Pathol.* **150**:1553–1561.

Zwadyk, P., Jr. and R. C. Cooksey. 1987. Nucleic acid probes in clinical microbiology. *Crit. Rev. Clin. Lab. Sci.* **25**:71–103.

Application of Western Blotting to Diagnosis of Viral Infections

MARK B. MEADS AND PETER G. MEDVECZKY

18

Molecular biological techniques have an increasing role in the laboratory diagnosis of viral infections as a consequence of their being more sensitive and specific than methods developed in the past. These modern techniques are especially important for the diagnosis of infections caused by agents that are difficult to propagate in tissue or cell culture. For example, techniques such as Western blotting (also referred to as immunoblotting) and PCR are part of the growing panel of diagnostic procedures for human immunodeficiency virus (HIV) and human T-cell leukemia virus type 1 (HTLV-1) and HTLV-2 infections (Gallo et al., 1986; Mullis et al., 1986; Carlson et al., 1987, Centers for Disease Control, 1988a, 1988b, 1989; Saiki et al., 1988; Consortium for Retrovirus Serology Standardization, 1988; Healey and Howard, 1989; Hirsch and Curren, 1990). This chapter discusses the principle, describes the methodology, and provides some practical clinical applications of Western blotting.

HISTORY AND PRINCIPLE OF WESTERN BLOTTING

The diagnosis of viral infection is often based on the detection of specific circulating antibodies to viral antigens in serum samples. Enzyme immunoassays (EIA) often are used for diagnosis of viral diseases as well as for screening of blood and blood products for viruses. Although EIA are very sensitive and highly specific, false-positive reactions occur. Given the medical and social significance of particular virus infections, e.g., HIV, HTLV, and the hepatitis viruses, it is important that diagnostic tests for these virus infections be as specific, accurate, and sensitive as possible. Although the sensitivities and specificities of some EIA can be greater than 95%, e.g., the licensed EIA for HIV (Petricciani, 1985; Centers for Disease Control, 1988b), it is a standard laboratory practice to repeat a positive EIA test. According to guidelines adopted by the U.S. Public Health Service, if a second positive EIA result is obtained upon repeat testing, the diagnosis must be confirmed by another assay, most often Western blotting (Centers for Disease Control, 1988a, 1988b, 1989; Consortium for Retrovirus Serology Standardization, 1988).

Several "blotting" techniques have been developed. The initially described technique is referred to as "Southern" blotting, after its originator, E. M. Southern (1975). It is a fundamental tool for the analysis of DNA fragments. An analogous method for analyzing RNAs was dubbed "Northern" blotting (Alwine et al., 1977) as a molecular biologist's joke. The humor continued when a modification of the nucleic acid blotting methods for the study of proteins was developed and referred to as "Western" blotting, more properly called "immunoblotting" (Towbin et al., 1979; Burnette, 1981).

These different blotting techniques share a common principle. Complex mixtures of macromolecules (DNA, RNA, or protein) are first separated by size in rectangular "slab" gels by using electrophoresis. After separation, the molecules are transferred ("blotted") onto the surface of a membrane, and the separated and immobilized nucleic acid fragments or proteins are detected and/or identified on the membrane by using specific molecular probes. For example, RNA blots can be probed with radioactive cDNA probes. The bound RNA hybridizes in situ with the labeled probe, and the reaction is then visualized by autoradiography (Alwine et al., 1977).

The basic approach to the use of Western blotting for the diagnosis of viral infections begins with purified virions that are disrupted by ionic detergent treatment, releasing viral proteins. As shown schematically in Fig. 1, these virion proteins are then separated on gels by sodium dodecyl sulfate-polyacrylamide gel electrophoresis (SDS-PAGE). These gels, approximately 1 mm thick, are formed between two square glass plates; 14 by 14 cm is a typical size. Viral proteins are denatured by boiling in SDS–2-mercaptoethanol buffer, and a sample containing a few hundred micrograms of protein is loaded across the top of the gel. After electrophoresis, the gel is placed on a membrane (nitrocellulose or nylon, etc.) of the same size and this gel-nitrocellulose unit is placed on top of several layers of wet filter paper. Filter papers are then laid on the gel and proteins are transferred ("blotted") to the membrane by electrophoresis. Transfer may be done by "semidry" electroblotters when the gel-nitrocellulose-paper sandwich is placed between two rectangular metal or carbon electrodes (apparatus available from Hoefer Scientific Instruments Inc., San Francisco, Calif.; Fisher Inc., Springfield, N.J.; or other distributors and manufacturers). In an older version of electroblotting, which is still preferred by some investigators, the sandwich is held together by a plastic device and is submerged in buffer in a large electrophoresis tank (Towbin et

182

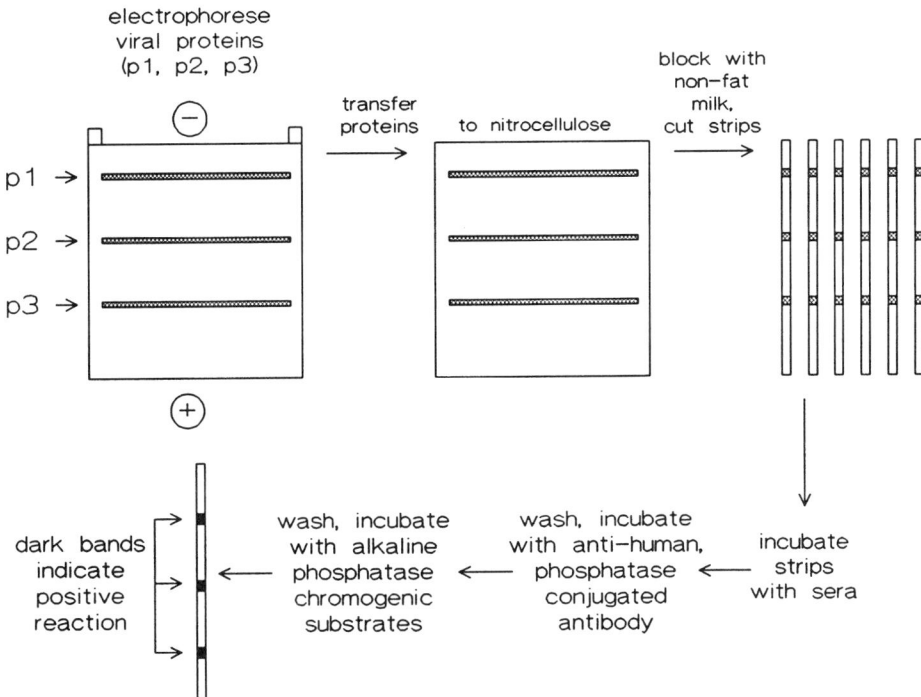

FIGURE 1 Schematic representation of Western blotting.

al., 1979; Burnette, 1981). In either case, the transfer of proteins from the gel to the membrane is mediated by electrical current so that the pattern of proteins obtained by SDS-PAGE is preserved during transfer. After transfer, the membrane is incubated with a buffer containing nonspecific proteins such as milk casein or serum albumin to block all unoccupied areas that could serve as binding sites. This blocking step prevents nonspecific adsorption of immunoglobulin (Ig) proteins to the nitrocellulose filter during subsequent steps. The nitrocellulose sheet is then cut into several strips and each strip is ready to use for the detection of antiviral antibodies.

If a patient serum containing anti-HIV antibodies is incubated with a Western blot strip, the antibodies specific to the individual viral proteins form stable complexes with the transferred protein species and the antibodies remain bound to those antigens even after extensive washing. Typically, patients who have seroconverted have antibodies to several proteins of the viral agent (Centers for Disease Control, 1988a, 1988b, 1989; Consortium for Retrovirus Serology Standardization, 1988).

After washing to remove unbound antibodies, the last step in the procedure is to visualize the patients' bound Ig. This can be achieved by using radiolabeled *Staphylococcus aureus* protein A or a second antibody that is labeled with a radioisotope or an enzyme. Clinical laboratories usually use safer, nonradioactive methods such as alkaline phosphatase- or horseradish peroxidase enzyme-coupled anti-human antibodies. The anti-human Ig and the enzyme are coupled covalently in a manner that allows the activity of both molecules to remain intact.

Detection of patient antibody is accomplished by monitoring enzyme activity linked to the anti-human antibody (see chapter 9). The enzyme activity can be demonstrated

in situ by incubating the membrane with appropriate chromogenic or luminogenic substrates. The products of chromogenic reactions are insoluble in water and develop a dark color at the site of enzyme activity. During the last step of the Western blot procedure, dark bands (blue for alkaline phosphatase, brown for peroxidase) corresponding to the physical location of the reactive viral proteins appear on the nitrocellulose filter. Luminogenic substrates produce light at the site of enzyme activity. Luminol releases blue-green light when it is oxidized by a reaction involving hydrogen peroxide and horseradish peroxidase. The reaction is enhanced in the presence of phenols, which increase light output and duration. A permanent record of results is obtained by exposing membranes to blue light-sensitive X-ray film a few minutes after addition of substrates. Film is exposed for up to 1 h, although exposures from 10 s to 10 min are usually all that are necessary.

In summary, the cascade of steps in performing Western blotting includes electrophoretic separation of viral proteins, nitrocellulose immobilization of viral proteins, viral protein capture of patient antibodies, patient antibody binding to enzyme-labeled anti-human antibody, and detection of the presence of human antibody by an enzyme reaction that produces visible dark bands or chemiluminescence at the site of these molecular complexes.

THE WESTERN BLOT PROCEDURE

The various aspects of purification of virions and viral proteins and the details of SDS-PAGE are not discussed here, since clinical laboratories rarely have sufficient resources to perform these steps. We recommend the purchase of blotted viral proteins or Western blot kits available from several commercial sources.

List of Materials Needed

Instruments and Materials

Rocking platform

Vacuum aspirator with flasks

pH meter

Adjustable micropipettes (1 to 20 μl, 10 to 200 μl, and 100 to 1,000 μl) and sterile tips

Pasteur and serological pipettes

Western blot incubation trays.

> The nitrocellulose strips are usually shipped in Western blot trays. If you wish to purchase separate trays, two types are available: disposable and reusable. Reusable trays require thorough cleaning after exposure to reagents; disposable trays are more convenient but more costly in the long run. Before ordering trays, contact the manufacturer of the Western blot nitrocellulose strips you plan to purchase, since dimensions of the strips and trays vary among different commercial sources.

Polaroid or other camera to photograph results for permanent record

Reagents

Tris base (Trizma [Sigma product T-8524 or equivalent]; Sigma Chemical Co., St. Louis, Mo.)

Hydrochloric acid (5 N)

Sodium chloride (Sigma product S9625 or equivalent)

Magnesium chloride, $6H_2O$ (Sigma product M9272 or equivalent)

EDTA (Sigma product E5134 or equivalent)

Milli-Q or equivalent highly purified or double-distilled H_2O, autoclaved

Nonfat dry milk (from any supermarket)

Nitrocellulose strips with blotted proteins (see below)

Anti-human IgG, alkaline phosphatase conjugated (product S3821 or equivalent; Promega, Madison, Wis.)

Tween 20 (polyoxyethylenesorbitan monolaurate [Sigma product P1379 or equivalent])

Nitroblue tetrazolium (NBT) powder or NBT tablets (Sigma product N 6876 or N 5514 [for tablets] or equivalent)

N,N-Dimethylformamide (Sigma product D8654)

5-bromo-4-chloro-3-indolyl phosphate (BCIP), p-toluidine salt, or BCIP tablets (Sigma product B 8503 or B 0274 [for tablets] or equivalent)

Solutions

TST buffer (10 mM Tris-HCl [pH 8.0], 150 mM NaCl, 0.05% Tween 20) (1 liter)

> Dissolve 1.21 g of Tris base and 8.76 g of NaCl in about 900 ml of sterile Milli-Q H_2O, and adjust pH to 8.0 with 5 N HCl. Add 0.5 ml of Tween 20 and fill with H_2O to 1 liter. Store at 4°C.

TST buffer with 5% dry milk (10 mM Tris-HCl [pH 8.0], 150 mM NaCl, 0.05% Tween 20) (50 ml)

> Dissolve 2.5 g of nonfat dry milk in 50 ml of TST buffer; store at −20°C.

Alkaline phosphatase buffer (100 mM NaCl, 5 mM $MgCl_2$, 100 mM Tris-HCl [pH 9.5]) (100 ml)

> Dissolve 1.21 g of Tris base, 0.10 g of $MgCl_2 \cdot 6H_2O$, and 0.58 g of NaCl in about 90 ml of sterile Milli-Q H_2O, and adjust pH to 9.5 with 5 N HCl. Fill with H_2O to 100 ml. Store at 4°C.

NBT solution (2 ml)

> Dissolve 100 mg of NBT powder or tablets in 2 ml of 70% N,N-dimethylformamide (mixture of 0.7 ml of Dimethylformamide and 0.3 ml of H_2O). Store in the dark at −20°C.

BCIP solution (1 ml)

> Dissolve 50 mg of BCIP in 1 ml of 100% N,N-dimethylformamide. Store in the dark at −20°C.

Stop solution (5 mM EDTA, 50 mM Tris-HCl [pH 7.5]) (100 ml)

> Dissolve 1.21 g of Tris base, 0.186 g of EDTA, and 0.58 g of NaCl in about 90 ml of sterile Milli-Q H_2O, and adjust pH to 7.5 with 5 N HCl. Fill with H_2O to 100 ml. Store at 4°C.

Step-By-Step Procedure

This description is for a procedure that uses an alkaline phosphatase-labeled second antibody for detection. Some minor modifications may be needed if another system is used (Sambrook et al., 1989).

1. Check condition of equipment required, collect reagents, check expiration date of Western blot strips and sera (do not use beyond expiration date), and prepare solutions.

2. Incubate with primary antibody.
   a. Dilute serum 1:50 in TST buffer–5% dry milk. Prepare enough to allow use of 0.05 ml per cm^2 of Western blot filter strip. For example, if the area of the test strip is 5 cm^2, then a minimum of 0.25 ml of diluted test serum should be prepared. Appropriate controls are essential in Western blotting; positive and a negative control sera appropriately diluted must be included in all assays (manufacturers of Western blot kits offer such controls).
   b. Place Western blot strips in a multiwell tray using forceps, and mark each well for future identification of each reaction.
   c. Tilt tray at about a 30° angle and add 0.05 ml of TST buffer–5% dry milk per cm^2 of filter strip to the bottom of each well. Slowly lower the tray to a horizontal position so that the strips adsorb buffer gradually and evenly. Add diluted serum to the appropriate wells, close the lid, and incubate at room temperature for 2 h on a rocking platform with gentle agitation.

3. Wash filters five times with TST.
   a. Remove lid, tilt tray, and aspirate the liquid by Pasteur pipette connected to a vacuum flask. Do not allow filters to dry out!
   b. Fill wells about halfway with TST and wash with gentle agitation (rocking) for 7 min.
   c. Remove TST and repeat this wash procedure four times for a total of five washes. Make sure that the buffer is completely removed after each wash.

4. Incubation with secondary antibody.
   a. Dilute the anti-human IgG alkaline phosphatase conjugate according to the manufacturer's specifications in TST buffer–5% dry milk.
   b. Add the diluted serum to each well (0.1 ml of diluted serum per cm^2 of filter strip).
   c. Close the lid and incubate at room temperature for 1 h on a rocking platform with gentle agitation.

5. Wash filters five times with TST, as described in step 3.

6. Incubate with chromogenic substrates.

During the final period, prepare the chromogenic substrates; mix 66 µl of NBT stock with 10 ml of alkaline phosphatase buffer in a test tube, and then add 33 µl of BCIP and mix well. Remove the washing buffer and add 0.1 ml of chromogenic substrate solution per cm² of filter strip. Incubate the tray with gentle agitation as before, and monitor the development of dark blue bands.

7. Stop reaction.

When strong bands are visible with the positive control serum and test samples but the negative control is still clear, the reaction has stopped. Optimal incubation time with substrates varies from a few minutes to half an hour and depends on the reagents and the titer of antibodies. To stop the reaction, quickly aspirate the substrate solution and add 2 ml of stop solution. Some experimentation is necessary to determine optimal conditions for each new set of reagents and to stop the reaction at the right moment. A timer should be used to monitor the incubation time with the chromogenic substrates in order to note the optimal time for future reference. Overincubation usually results in high background and/or appearance of bands in the negative control serum. If this should occur, the Western blot must be repeated with a shortened incubation period with the chromogenic substrate.

8. Interpretation of the assay, with special hints.

In general, a Western blot test is considered positive when the patient's serum is reactive with more than one viral antigen. For example, a licensed HIV test is interpreted as positive when multiple bands are present, e.g., p24, p31, gp41, and gp160 (Centers for Disease Control, 1988a, 1988b, 1989; Consortium for Retrovirus Serology Standardization, 1988). The question is how to identify these proteins. Most manufacturers offer sequentially numbered Western blot strips that have originated from a single gel. The strips are numbered such that adjacent numbers refer to originally adjacent strips. The key to accurate results is to use a pair of adjacent nitrocellulose Western blot strips that originated from the same area of the gel used in SDS-PAGE. One of these strips should be reacted with the positive control serum, and the adjacent strip should be reacted with the patient sample. When the test is complete, strips should be aligned numerically according to the manufacturer's numbering and should be compared side by side and photographed. This method of analysis can help in the interpretation of dubious and nonspecific reactions.

Alternative Procedure: Chemiluminescence

Chemiluminescence has two main advantages compared to chromogenic techniques: first, it enables a greater-than-10-fold increase in sensitivity without the use of isotopes (Schneppenheim et al., 1991; Constantine et al., 1994), and second, exposure times can be varied to increase or decrease sensitivity. A disadvantage is that it requires the use of a darkroom and developing equipment. Although alkaline phosphatase can be used in this technique, systems based on this enzyme are more complex and less convenient than those based on peroxidase. For these reasons, the luminol-peroxidase system is recommended. This procedure is identical to that described in the Western blot step-by-step sec-

tion, with a few exceptions. The secondary antibody used in step 4 must be conjugated to horseradish peroxidase, and steps 6 and 7 should be replaced with the following (Schneppenheim et al., 1991).

6. Add just enough visualization solution to cover each strip, and allow the reaction to proceed for 1 min at room temperature. Remove excess solution from the strips and place them face down on a sheet of clear plastic wrap, and fold the plastic wrap around the strips so that they are completely covered on both sides. Next, place the strips in plastic wrap face up in an X-ray film cassette and make sure the plastic wrap is dry and smooth (wrinkles in plastic wrap between film and strips create high background). Close the cassette until the film is exposed in a darkroom. The film can be exposed immediately; do not wait longer than 20 min, as luminescence will begin to decline.

7. In a darkroom, place the film on top of the strips and close the cassette. The film should first be exposed for only 10 s. If more sensitivity is required, the exposure can be as long as 1 h; longer exposures, however, increase background. If less sensitivity is required, leave the strips in the cassette for an hour or more and repeat exposure.

Materials and Reagents
Visualization solution (prepare immediately before use) (Schneppenheim et al., 1991)

0.5 ml of 10× luminol stock (40 mg of luminol [Sigma product A 8511] in 10 ml of DMSO). Store at $-20°C$.

0.5 ml of 10× p-iodophenol stock (10 mg [Aldrich product I-1,020-1] in 10 ml of DMSO). Store at $-20°C$.

2.5 ml of 100 mM Tris-Cl, pH 7.5

25 µl of 3% H_2O_2

H_2O (to 5 ml)

Anti-human IgG, horseradish peroxidase conjugated (Promega product W4031 or equivalent)

Clear plastic wrap

Blue light-sensitive X-ray film and cassette

Darkroom and X-ray film developing equipment

ADVANTAGES AND DISADVANTAGES OF THE WESTERN BLOT ASSAY

It is obvious that the Western blot assay is more specific than EIA, since in the Western blot assay, antibodies to several antigens are detected simultaneously by using a group of electrophoretically separated viral proteins. At present, Western blotting offers a very reliable confirmatory assay for HIV-1 infection (Centers for Disease Control, 1988b, 1989). A potential competitor for the Western blot assay has been reported for the diagnosis of HIV infection. This new technique, called "recombinant-antigen immunoblot assay" (RIBA-HIV216), utilizes a set of purified antigens produced by recombinant technology (Oroszlan and Copeland, 1985; Steimer et al., 1986a; Truett et al., 1989; Lillehoy et al., 1990; Busch et al., 1991). It remains to be determined whether this recombinant-protein assay is more specific and/or more sensitive than the standard Western blot assay.

Western blot assays can produce nonspecific, or so-called "indeterminate," reactions in which often only one band is seen. These results are sometimes attributed to, among other causes, an underlying autoimmune disease that results in production of antibodies to cellular antigens (Healey and Howard, 1989). However, indeterminate status may precede a truly positive status and thus be indicative of infection. Such a result requires a specimen to be collected from the patient at a later date for retesting. As demonstrated in Fig. 2, an indeterminate reaction is observed at an early stage of HIV infection (day 14), with the antibody response directed to only a single HIV protein, p24. Gradually, the patient develops antibodies to several other viral proteins and by day 35 is considered HIV positive.

Similarly, false-negative reactions have been described (Kissler et al., 1987), but the occurrence of false-negative reactions is relatively rare. Perhaps the main disadvantage of Western blotting is that it requires experienced personnel, which limits its use to laboratories capable of performing more-specialized services. Therefore, training of laboratory personnel to perform this technique is highly recommended.

COMMERCIAL KITS AND NITROCELLULOSE STRIPS WITH BLOTTED PROTEINS

Nitrocellulose strips with blotted viral proteins or complete kits with all necessary reagents are available from commercial sources for HIV-1, HTLV-1, and hepatitis C virus Western blot assays. A current list of all licensed kits and manufacturers can be found at the website of the U.S. Food and Drug Administration (www.fda.gov/cber/products/testkits.htm). Chemiluminescence kits containing premixed visualization solutions and secondary antibody are available from Amersham Pharmacia Biotech (Piscataway, N.J.) and NEN (Boston, Mass.).

REFERENCES

Alwine, J. C., P. J. Kemp, and G. R. Stark. 1977. Method for the detection of specific RNAs in agarose gels by transfer to diazobenzylmethyl-paper and hybridization with DNA probes. *Proc. Natl. Acad. Sci. USA* **74:**5350–5354.

Burnette, W. H. 1981. Western blotting: electrophoretic transfer of proteins from SDS-polyacrylamide gels to unmodified nitrocellulose and radiographic detection with antibody and radioiodinated protein A. *Anal. Biochem.* **112:**195–203.

Busch, M. P., Z. El Amad, T. M. McHugh, D. Chien, and A. J. Polito. 1991. Reliable confirmation and quantitation of human immunodeficiency virus type 1 antibody using a recombinant-antigen immunoblot assay. *Transfusion* **31:**129–137.

Carlson, J. R., J. Yee, S. H. Hinrichs, M. L. Bryant, M. B. Gardner, and N. C. Pedersen. 1987. Comparison of indirect immunofluorescence and Western blot for the detection of anti-human immunodeficiency virus antibodies. *J. Clin. Microbiol.* **25:**494–497.

Centers for Disease Control. 1988a. Licensure of screening tests for antibody to human T-lymphotropic virus type I. *Morb. Mortal. Wkly. Rep.* **37:**736–747.

Centers for Disease Control. 1988b. Update: serological testing for antibody to human immunodeficiency virus. *Morb. Mortal. Wkly. Rep.* **36:**833–840.

Centers for Disease Control. 1989. Interpretation and use of the Western blot assay for serodiagnosis of human immunodeficiency virus type 1 infection. *Morb. Mortal. Wkly. Rep.* **38**(Suppl. 7):1–7.

Consortium for Retrovirus Serology Standardization. 1988. Serological diagnosis of human immunodeficiency virus infection by Western blotting. *JAMA* **260:**674–679.

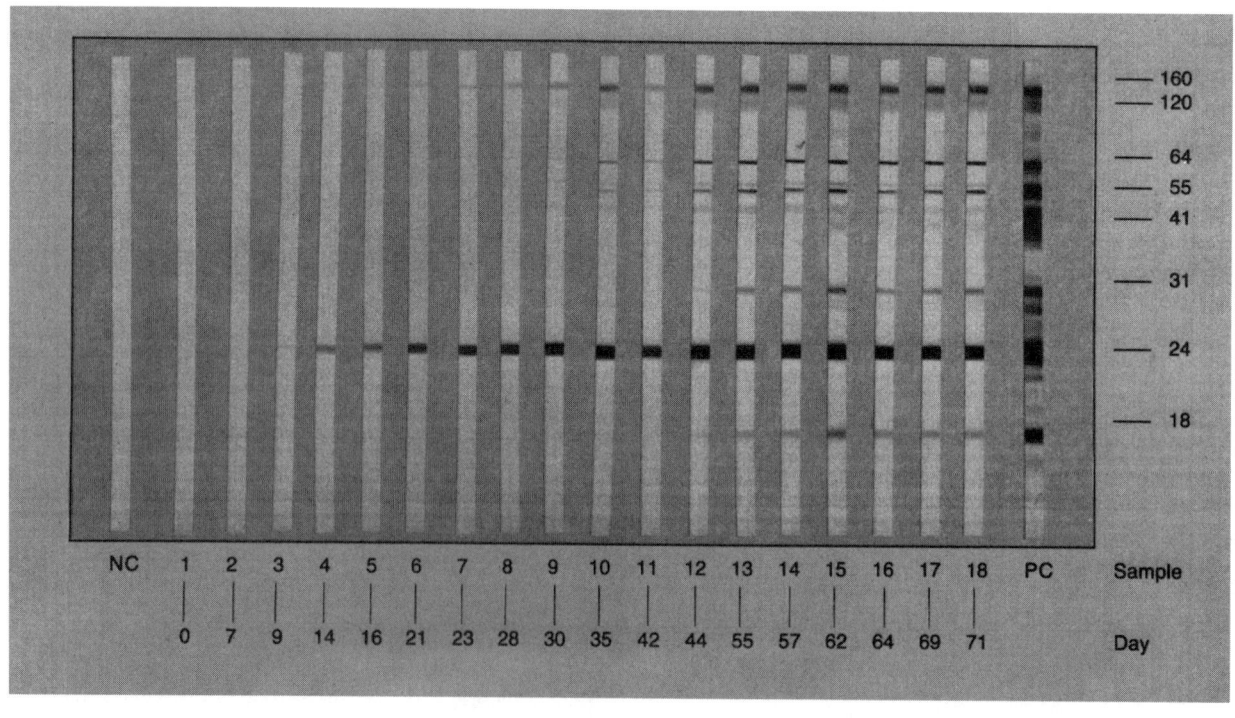

FIGURE 2 Demonstration of seroconversion after exposure to HIV by Western blotting (kindly provided by Boston Biomedica, Inc., Bridgewater, Mass.). Western blot assay was performed with a Biotech Western blot kit. Serum samples were taken from an individual at intervals, as indicated. NC, negative control serum; PC, positive control serum.

Constantine, N. T., J. Bansal, X. Zhang, K. C. Hyams, and C. Hayes. 1994. Enhanced chemiluminescence as a means of increasing the sensitivity of Western blot assays for HIV antibody. *J. Virol. Methods* **47**:153–164.

Gallo, D., J. L. Diggs, G. R. Shell, P. J. Dailey, M. N. Hoffman, and J. L. Riggs. 1986. Comparison of detection of antibody to the acquired immune deficiency syndrome virus by enzyme immunoassay, immunofluorescence, and Western blot methods. *J. Clin. Microbiol.* **23**:1049–1051.

Healey, D. S., and T. S. Howard. 1989. Activity to non-viral proteins on Western blot mistaken for reactivity to HIV glycoproteins. *AIDS* **3**:545–546.

Hirsch, M. S., and J. Curren. 1990. Human immunodeficiency viruses. Biology and medical aspects, p. 1545–1570. *In* B. N. Fields and D. M. Knipe (ed.), *Virology*. Raven Press, New York, N.Y.

Kissler, H. A., B. Blauw, J. Spear, D. A. Paul, L. A. Falk, and A. Landay. 1987. Diagnosis of human immunodeficiency virus infection in seronegative homosexuals presenting with an acute viral syndrome. *JAMA* **258**:1196–1199.

Lillehoy, E. P., S. S. Alexander, C. J. Dubrule, S. Wiktor, R. Adams, C. Tai, A. Manns, and W. A. Blattner. 1990. Development and evaluation of a human T-cell leukemia virus type I serological confirmatory assay incorporating a recombinant envelope polypeptide. *J. Clin. Microbiol.* **28**:2653–2658.

Mullis, K. B., F. Faloona, S. J. Scharf, R. K. Saiki, G. T. Horn, and H. A. Erlich. 1986. Specific enzymatic amplification of DNA in vitro: the polymerase chain reaction. *Cold Spring Harbor Symp. Quant. Biol.* **51**:263–273.

Oroszlan, S., and T. D. Copeland. 1985. Primary structure and processing of gag and env gene products of human T-cell leukemia viruses HTLV-I$_{CR}$ and HTLV-II$_{ATK}$. *Curr. Top. Microbiol. Immunol.* **115**:221–233.

Petricciani, J. C. 1985. Licensed tests for antibody to human T-lymphotropic virus type III. Sensitivity and specificity. *Ann. Intern. Med.* **103**:726–729.

Saiki, R. K., D. H. Gelfand, F. Stoffel, S. J. Scharf, R. Higuchi, G. T. Horn, K. B. Mullis, and H. A. Erlich. 1988. Primer-directed enzymatic amplification of DNA with a thermostable DNA polymerase. *Science* **239**:487–491.

Sambrook, J., E. F. Fritsch, and T. Maniatis. 1989. *Molecular Cloning: a Laboratory Manual*, 2nd ed., p. 18.60–18.75. Cold Spring Harbor Laboratory Press, Cold Spring Harbor, N.Y.

Schneppenheim, R., U. Budde, N. Dahllmann, and P. Rautenberg. 1991. Luminography—a new, highly sensitive visualization method for electrophoresis. *Electrophoresis* **12**:367–372.

Southern, E. M. 1975. Detection of specific sequences among DNA fragments separated by gel electrophoresis. *J. Mol. Biol.* **98**:503–517.

Steimer, K. S., K. W. Higgins, M. A. Powers, J. C. Stephans, A. Gyenes, C. George-Nascimento, P. Luciw, P. J. Barr, R. A. Hallewell, and R. Sanchez-Pescador. 1986a. Recombinant polypeptide from the endonuclease region of the acquired immune deficiency syndrome retrovirus polymerase (*pol*) gene detects serum antibodies in most infected individuals. *J. Virol.* **58**:9–16.

Steimer, K. S., J. P. Puma, M. A. Powers, C. George-Nascimento, J. C. Stephens, J. A. Levy, R. Sanchez-Pescador, P. Luciw, P. J. Barr, and R. A. Hallewell. 1986b. Differential antibody responses of individuals infected with the AIDS-associated retroviruses surveyed using the viral core antigen p26gag expressed in bacteria. *Virology* **150**:283–290.

Towbin, H., T. Staehelin, and J. Gordon. 1979. Electrophoretic transfer of proteins from polyacrylamide gels to nitrocellulose sheets: procedure and some applications. *Proc. Natl. Acad. Sci. USA* **76**:4350–4354.

Truett, M. A., D. Y. Chien, T. L. Calarco, R. K. DiNello, and A. J. Polito. 1989. Recombinant immunoblot assay for the detection of antibodies to HIV, p. 121–141. *In* P. A. Luciw and K. S. Steimer (ed.), *HIV Detection by Genetic Engineering Methods*. Marcel Dekker, New York, N.Y.

Nucleic Acid Amplification Methods

DANNY L. WIEDBRAUK

19

Nucleic acid detection methods play an increasingly important role in the detection of viral illnesses. Once the province of esoteric, "boutique," and university research laboratories, nucleic acid methods are now important and necessary procedures in many hospital laboratories. However, the rapid evolution of nucleic acid detection technologies and products means that clinical laboratories must choose products and services from a bewildering array of vendors and technologies. Because each technology has its own testing characteristics, equipment requirements, and sensitivity levels (Table 1), matching test characteristics and diagnostic utility for an individual institution can be difficult. This chapter will attempt to demystify nucleic acid testing methods by describing the leading target, probe, and signal amplification technologies for detecting viral nucleic acids and summarizing how these technologies work.

NUCLEIC ACID AMPLIFICATION

Nucleic acid amplification methods are usually classified as target or probe amplification methods, depending upon the source of the nucleic acid that is amplified. Target amplification methods are among the oldest and best characterized of the nucleic acid amplification methodologies. Target amplification methods use enzymatic tools to increase the concentration of the target nucleic acids in the sample. In contrast with target amplification methods, probe amplification methods increase the concentrations of certain probe species when the target nucleic acid is present in the sample. Probe amplification methods, such as the ligase chain reaction (LCR), are at least as sensitive and specific as target amplification procedures for the detection of infectious agents in clinical specimens (Davis et al., 1998; Puolakkainen et al., 1998). However, probe amplification methods have not been used in quantitative procedures.

One of the greatest strengths and a major weakness of nucleic acid amplification methods is their exquisitely high sensitivity. These procedures can generate millions of DNA or RNA copies from template sequences. Product carryover and cross contamination of specimens or common reagents with amplified products (amplicons) can cause false-positive results. Therefore, stringent amplicon control measures must be utilized in all nucleic acid amplification technologies in order to limit these problems.

False-negative results can occur when the specimen contains chemical or biological substances that inhibit the enzymatic amplification process. False-negative reactions can occur through a number of mechanisms. Specimens containing EDTA can chelate divalent cations, such as Mg^{2+}, that are necessary for the enzymatic reaction. Proteases can degrade the amplification enzymes, and the presence of ribo- or deoxyribonucleases can degrade nucleic acid targets and/or primers. In addition, a wide variety of biological and chemical substances can directly inhibit the enzymes responsible for nucleic acid amplification. Nucleic acid amplification methods that employ multiple enzymes are generally more sensitive to the presence of inhibitory substances than methods utilizing one enzyme. Therefore, stringent sample preparation methods and specimen inhibition controls are necessary in order to minimize false-negative results (National Committee for Clinical Laboratory Standards, 1995).

Although nucleic acid amplification methods are somewhat more complicated to use than signal amplification methods, target and probe amplification procedures are used more frequently than signal amplification technologies. This increased usage has been fostered by the increasing availability of commercial test kits and high-quality reagents, primers, and controls.

PCR

Developed by researchers at the Cetus Corporation (Saiki et al., 1985; Mullis and Faloona, 1987), PCR is one of the oldest and best-known methods for replicating specific DNA sequences in vitro. In its simplest form, the PCR procedure utilizes two 15- to 30-base oligonucleotide primers that are complementary to nucleic acid sequences that flank the target DNA sequence to be amplified. These primers are included in a reaction mixture containing the target nucleic acid, a heat-stable DNA polymerase, a defined solution of salts, and excess amounts of each of the four deoxynucleoside triphosphates (dNTPs). The mixture is then subjected to repeated cycles of defined temperature changes. These thermal changes facilitate the denaturation (94 to 97°C) of the template DNA, the annealing (55 to 72°C) of the primers to the target DNA, and the extension (72°C) of the primers so that the target DNA sequence is replicated (Fig. 1). During the next heating cycle, the strands separate, and the original DNA and the newly synthesized DNA strand both serve as

TABLE 1 Summary of nucleic acid amplification testing methods

Method	Target	Thermal cycling requirement	Sensitivity (organisms/ml)
PCR	DNA	Yes	5–50
LCR	DNA	Yes	5–50
NASBA	RNA	No	5–50
TMA	RNA	No	5–50
SDA	DNA	No	5–50
Fluorogenic-probe systems	DNA	Yes	50–500
Cleavase invader	DNA	No	500–1,000
Electron transfer	DNA	No	100,000
HPA	RNA	No	100,000
Hybrid capture	DNA RNA	No	500–1,000

templates for another round of DNA replication. Thus, the number of target DNA strands doubles with each thermal cycle. PCR procedures for infectious agents typically consist of 20 to 40 thermal cycles. These procedures produce a 10^5- to 10^6-fold increase in target nucleic acid concentrations within 3 to 4 h. The amplified DNA can be detected by a variety of methods, including capillary electrophoresis, solid-phase or solution hybridization, high-performance liquid chromatography, agarose gel electrophoresis with direct visualization of stained nucleic acids, and Southern blotting. PCR procedures can amplify only DNA targets. For RNA detection, the RNA must be converted to cDNA with a reverse transcriptase enzyme prior to amplification. Some of the newer thermostable DNA polymerases, such as *Tth*, have both reverse transcriptase and DNA polymerase activities when employed in specialized reaction buffers containing manganese rather than magnesium (Myers and Gelfand, 1991). Such bifunctional polymerases greatly simplify RNA amplification procedures in the clinical laboratory.

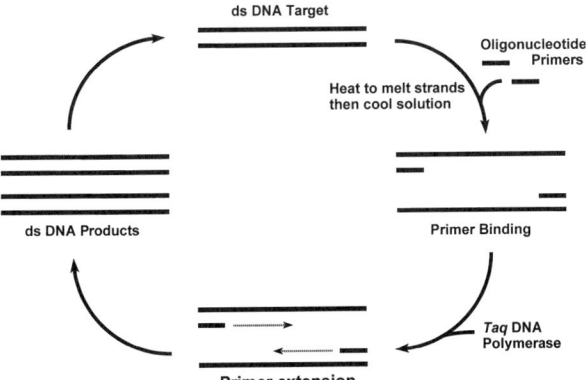

FIGURE 1 PCR. The double-stranded DNA (ds DNA) target (top) is heated to separate the strands. As the solution cools, the two oligonucleotide primers bind to opposite strands on the target DNA. The thermostable *Taq* DNA polymerase extends the primers according to the nucleotide sequence of the target DNA strand to produce double-stranded DNA products. The old and new strands both serve as templates for further DNA synthesis during the next cycle of heating and cooling.

Because PCR was the first target amplification technology to find widespread usage in the clinical laboratory, a wide variety of monoplex, multiplex (Edwards and Gibbs, 1994), and quantitative PCR procedures are available for clinical use. In addition, a portable, battery-operated PCR detection system which utilizes fluorescence resonance energy transfer (FRET) technology has been described for field use (Belgrader et al., 1998, 1999).

PCR is an extremely sensitive and specific procedure. However, the reaction can be inhibited by a variety of substances, including heme (Mercier et al., 1990; Ruano et al., 1992), heparin (Beutler et al., 1990; Holodniy et al., 1991), phenol (Katcher and Schwartz, 1994), polyamines (Ahokas and Erkkila, 1993), plant polysaccharides (Demeke and Adams, 1992), urine (Khan et al., 1991; Chernesky et al., 1997; Berg et al., 1997; Mahony et al., 1998; Toye et al., 1998), vitreous fluids (Wiedbrauk et al., 1995), and calcium alginate (Wadowski et al., 1994). The inhibition profile of the PCR procedure depends largely upon the type of polymerase used in the reaction (Wiedbrauk et al., 1995) and the purity of the nucleic acid to be amplified.

NASBA AND TMA

Nucleic acid sequence-based amplification (NASBA) and transcription-mediated amplification (TMA) are functionally identical isothermal target amplification procedures that are based upon the replication events that occur during retroviral transcription (Guatelli, 1990). In these procedures, reverse transcriptase, RNase H, and T7 RNA polymerase are used to generate new RNA targets via double-stranded DNA intermediates (Fig. 2). Like PCR, NASBA and TMA utilize oligonucleotide primers that are complementary to the target nucleic acid sequences. However, at least one of these primers also contains a promoter sequence for T7 RNA polymerase. When the primer anneals with the target, the promoter end of the primer does not anneal because it does not contain complementary sequences (Fig. 2). The annealed end of the primer is extended by reverse transcriptase. RNase H degrades the RNA in the RNA-DNA hybrid, allowing the second primer to bind to the cDNA. Reverse transcriptase then extends the 5′ end of the primer. The resulting double-stranded DNA contains a complete, transcriptionally competent T7 RNA polymerase promoter. T7 RNA polymerase binds to this promoter and produces 50 to 1,000 antisense RNA copies of the original target. These antisense transcripts are in turn converted to T7 promoter-containing double-stranded cDNA copies and used as transcription templates. This process continues in a self-sustained, cyclic fashion under isothermal conditions until components of the reaction mixture become limited or inactivated. In this procedure, each DNA template generates not one but many RNA copies, and transcription takes place continuously, without thermocycling. The resulting products can be detected by a variety of methods. NASBA and TMA can produce a 10^7-fold increase in the nucleic acid target in 60 to 90 min.

While NASBA and TMA can theoretically utilize both DNA and RNA targets, RNA targets are preferred in both methods. NASBA and TMA differ only in the number of enzymes used to catalyze the reaction. NASBA utilizes three separate enzymes, while TMA uses two. The TMA procedure employs a native reverse transcriptase enzyme that also has RNase H activity. Both procedures produce qualitative and quantitative results, and multiplex testing procedures have been developed. Instrumentation is now available for both methods, and a real-time detection sys-

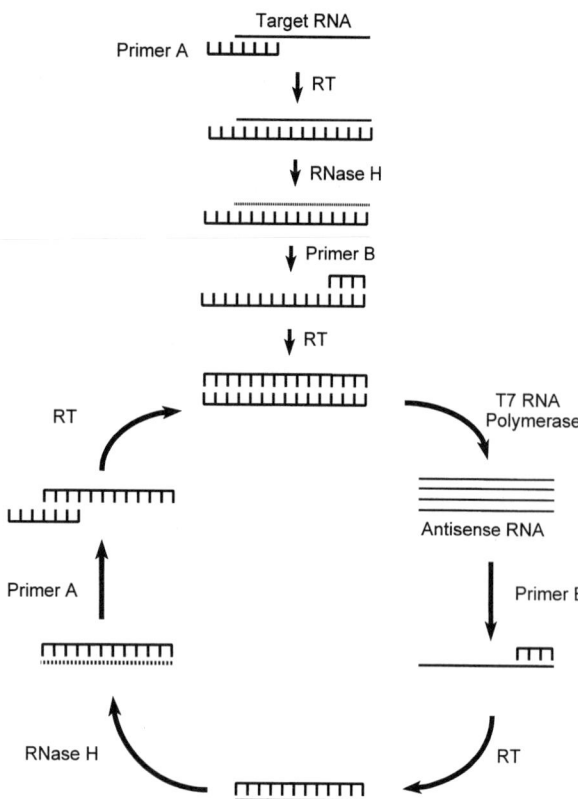

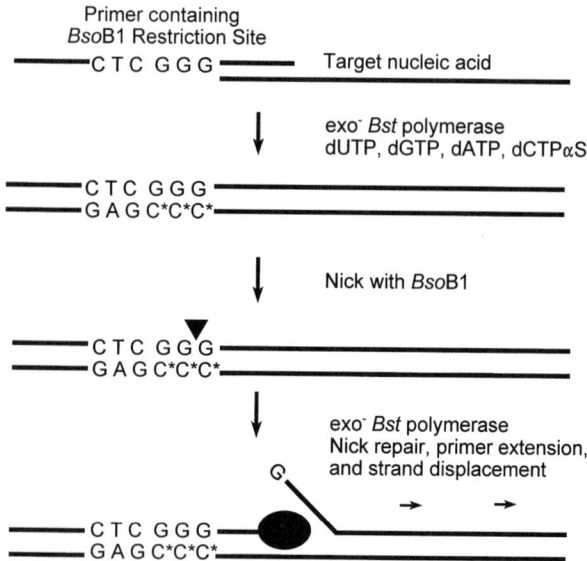

FIGURE 2   NASBA and TMA. Primer A, which contains the promoter sequence for the T7 RNA polymerase and sequences complementary to the target RNA, binds to the target RNA strand (top). Reverse transcriptase (RT) extends the primer according to the genetic sequence of the target strand, and RNase H degrades the RNA portion of the DNA-RNA hybrid molecule. Primer B binds to the complementary DNA, and reverse transcriptase extends the primer to make a complete, transcription-competent double-stranded DNA intermediate. The T7 RNA polymerase generates 50 to 1,000 antisense RNA transcripts (from the original RNA), each of which can be converted to transcription-competent double-stranded DNA as before.

FIGURE 3   SDA. An oligonucleotide primer containing a *Bso*BI restriction site (5'-CTCGGG) binds to the complementary target nucleic acid. The primer and target are extended by a thermostable exo⁻, *Bst* DNA polymerase (large oval) in the presence of dGTP, dATP, dUTP, and a dCTP that contains an alpha-thiol group (dCTPαS). The resulting DNA synthesis generates a double-stranded *Bso*BI recognition site, with one strand containing 5' phosphorothiolate linkages (asterisks). *Bso*BI nicks the strand (arrowhead) without cutting the complementary thiolated strand, and the exo⁻ *Bst* polymerase extends the nucleic acid strand from the nick. The original nucleic acid is displaced rather than degraded because the DNA polymerase does not have 5' exonucleolytic activity. The restriction site is regenerated by the polymerase. This amplification scheme becomes exponential when an antisense primer containing a *Bso*BI site is added to the reaction.

tem has recently been described for the NASBA assay that significantly shortens the assay time (Leone et al., 1998).

NASBA and TMA reactions can be inhibited by a variety of substances (Witt and Kemper, 1999), including urine (Mahony et al., 1998) and a variety of proteins and polysaccharides (personal observations). For this reason, most of the newer NASBA and TMA procedures utilize extensive nucleic acid purification methods prior to amplification. TMA procedures currently use target capture methods on paramagnetic beads, while the NASBA methods utilize the guanidinium extraction procedures described by Boom et al. (1990, 1999) to prevent inhibition.

SDA

The strand displacement amplification (SDA) method is an isothermal DNA amplification procedure developed by Walker et al. (1992). In its current configuration (Fig. 3), a primer containing a *Bso*BI restriction site (5' GGGCTC) hybridizes with complementary sequences on the target

nucleic acid. The primer is extended by an exonuclease-deficient (exo⁻) *Bst* DNA polymerase in the presence of dGTP, dATP, dTTP, and a derivitized dCTP containing an alpha-thiol group (dCTPαS) (Spargo et al., 1996). The resulting DNA synthesis generates a double-stranded *Bso*BI recognition site, one strand of which contains phosphorothiolate linkages located 5' to each deoxycytosine residue. The presence of the phosphorothiolate bond causes the *Bso*BI restriction enzyme to nick the recognition site without cutting the complementary thiolated strand. The exo⁻ *Bst* polymerase fragment initiates another round of DNA synthesis at the nick. However, the DNA located downstream of the nick is not degraded because the exo⁻ Klenow fragment lacks 5' exonucleolytic activity. Instead, the downstream DNA fragment is displaced as the new DNA molecule is synthesized. The displacement step regenerates the *Bso*BI site. Nicking and strand displacement steps cycle continuously until the reaction mixture components become limited. This linear amplification process can produce a new DNA target every 3 min (Walker et al., 1995). The amplification scheme becomes exponential when sense and antisense primers containing *Bso*BI sites are used. The resulting procedure doubles the number of target sequences every 3 min until the reaction mixture components become rate limiting. With the exception of an initial boiling step to denature the nucleic acids, SDA reactions are isothermal and are carried out at 50 to 60°C for 2 h. Performing SDA at stringent operating temperatures decreases the background amplification due to mispriming (Walker and Linn, 1996). Mispriming and background levels can be further reduced by using the single-stranded-DNA binding protein from gene 32 of bacteriophage T4 (Walker et al., 1996). This protein also enhances the ability to amplify longer (200- to 1,000-bp) target sequences (Walker et al., 1996). Real-time detection of SDA products by fluorescence polarization has recently been described; this method allows the detection of the cryptic plasmid of *Chlamydia trachomatis* in just 30 min (Little et al., 1999). Little is known about the inhibition profile of this procedure.

LCR

Like PCR, LCR utilizes a thermocycler and target-specific oligonucleotides which are complementary to specific sequences on the target DNA (Barany, 1991a, 1991b). However, LCR utilizes four oligonucleotides that completely cover the target sequences rather than flank them. The LCR procedure also employs a thermostable ligase rather than a DNA polymerase.

In this procedure, the target DNA is denatured by heating it to 94°C, and the oligonucleotides anneal to the target sequences as the reaction mixture cools to 65°C. Because the oligonucleotides are constructed so that they hybridize adjacent to each other, DNA ligase interprets the gap between the ends of these oligomers as a nick in need of repair and covalently links them. The oligomers remain joined during subsequent target denaturation cycles and serve as templates for the hybridization and ligation of other oligonucleotides.

LCR methods are extremely sensitive, but the specificity of the original procedure was relatively low due to false-positive reactions caused by target-independent ligation. Gap junction LCR methods (Fig. 4) have corrected this problem. In the gap-junction procedure, the oligonucleotide primers anneal to the target so that a 3- to 7-base gap is formed between the oligomers. A thermostable DNA polymerase and a single dNTP are included in the reaction mixture to fill the gap. Thus, target-independent ligation products cannot serve as ligation templates and are not amplified. PCR-like extension of the 5′ primer does not occur because only one dNTP is present. Once the gap has been filled, the thermostable DNA ligase joins the two oligonucleotides and the cycle continues as described above.

The presence of the target sequence can be determined electrophoretically or chemically. In the Abbott Diagnostics (Abbott Park, Ill.) system, one side of the oligomeric pair is biotinylated while the other oligomer has a fluorescent label (Fig. 4). After amplification, the reaction mixture is mixed with streptavidin-coated microparticles. The microparticles are captured, and the free primers are removed by washing. The presence of full-length oligonucleotides generates a fluorescent signal in the instrument. As one of the first nucleic acid detection methods to incorporate an automated detection system, the Abbott LCx system proved to be a very successful method for detecting *C. trachomatis* and *Neisseria gonorrhoeae* in clinical specimens. The LCx system is as sensitive and specific as PCR (Davis et al., 1998; Puolakkainen et al., 1998), but the detection system makes it difficult to test a large number of specimens. Various degrees of inhibition have been reported when urine specimens from pregnant women were tested (Chernesky et al., 1997; Mahony et al., 1998).

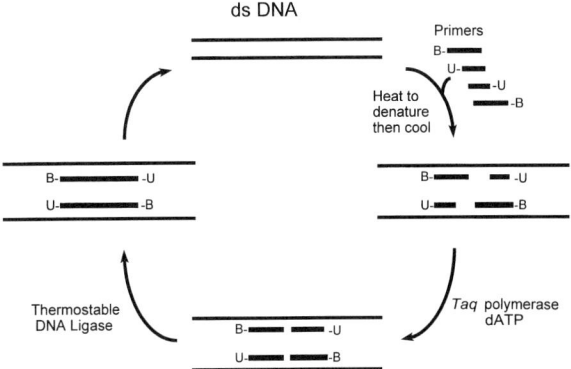

FIGURE 4 Gap junction LCR. Double-stranded DNA (top) is heated to separate the strands, and the oligonucleotide primers anneal to the target so that a gap of 3 to 7 nucleotides is formed between the primers. The primers are chosen so that filling the gap requires a single oligonucleotide. A thermostable DNA polymerase and a single dNTP are included in the reaction mixture to fill the gap. Once the gap is filled, the thermostable DNA ligase joins the two oligonucleotides and the cycle continues as described above. Detection of the ligated oligonucleotides can be done through the interaction of biotin (B) with a streptavidin solid phase. The presence of label (U) on the solid phase after washing signifies the presence of full-length oligonucleotides.

CPT

Cycling probe technology (CPT) (Duck et al., 1990; Beggs et al., 1996; Bekkaoui et al., 1996) is a rapid isothermal-probe amplification method developed by ID Biomedical Corporation in Burnaby, British Columbia, Canada. The CPT system is a linear probe amplification system that employs a chimeric DNA-RNA-DNA probe (Fig. 5). The internal RNA portion of the probe is complementary to, and hybridizes with, the target DNA. RNase H, which is present in the reaction mixture, degrades the RNA portion of the DNA-RNA hybrid, and the noncomplementary DNA portions of the probe dissociate from the target. Another intact probe segment binds to the target DNA, and the cycle repeats. Detection of the probe fragments can be accomplished by gel electrophoresis, chemiluminescence, or fluorescence. In its current configuration, the chimeric probe is labeled with fluorescein at the 5′ terminus and with biotin at the 3′ terminus. The reaction occurs at a constant temperature in streptavidin-coated microwells (see "EIA-Based Detection" below). A horseradish peroxidase-conjugated antifluorescein antibody is used to detect the presence of full-length probes (Bekkaoui et al., 1999). Earlier versions of this methodology had high background levels and were partially or completely inhibited by the presence of nonhomologous DNA (Modrusan et al., 1998). However, the addition of spermadine and EGTA to the reaction mixture significantly reduced the background levels and the inhibitory effects of extraneous DNA (Modrusan et al, 1998). While less sensitive than other amplified methods, CPT is especially interesting because the probe fragments are not amplifiable, thus minimizing some of the product carryover problems associated with other amplification methods.

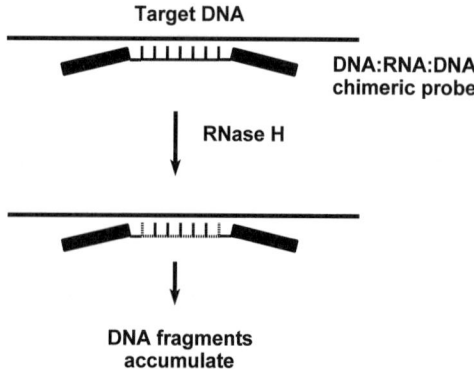

FIGURE 5 CPT. CPT is an isothermal method which utilizes a chimeric DNA-RNA-DNA probe. The internal RNA portion of the probe is complementary to, and hybridizes with, the target DNA. RNase H degrades the RNA of the DNA-RNA hybrid, and the noncomplementary DNA portions of the probe dissociate from the target. Once the DNA fragments dissociate, another intact probe segment binds to the target DNA and repeats the cycle.

CLEAVASE INVADER ASSAY

The Cleavase invader assay is a probe amplification system that utilizes two probes and a flap endonuclease derived from an archaebacterial species. In this assay, the invader probe is fully complementary and binds tightly to the target nucleic acid (Fig. 6). The signal probe, which has two domains, is provided in vast excess. The 3' end of the signal probe is complementary to the target sequence and binds immediately downstream from the invader probe. The 5' end of the signal probe does not hybridize with the target. Once the invader and signal probes bind to the target, the flap endonuclease cleaves the noncomplementary portion of the signal probe. Because the hybridization reaction is

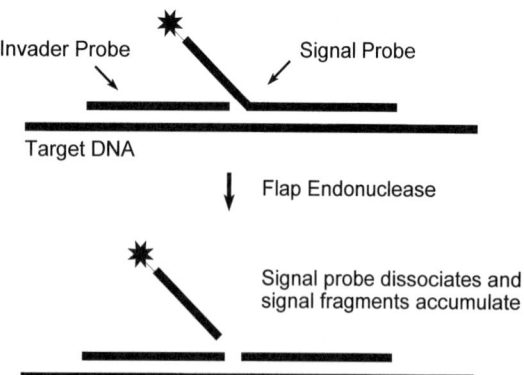

FIGURE 6 Cleavase invader assay. The invader probe and the labeled (asterisk) signal probe hybridize adjacent to each other so that a portion of the signal probe does not hybridize. The archaebacterial flap endonuclease cleaves the unhybridized portion of the signal probe. The signal probe dissociates because the reaction is performed at or near the melting temperature for the probe. Another signal probe hybridizes, and the cycle repeats. Signal probe fragments accumulate when the target nucleic acids are present.

performed at or near the melting temperature of the signal probe, the cleaved signal probe is replaced by an uncleaved signal probe 50 to 500 times per min. This procedure will produce a 3,000- to 30,000-fold increase in signal in 60 min. The cleaved signal probe can be detected by electrophoresis, FRET (Ryan et al., 1999), or enzyme immunoassay (EIA). While less sensitive than other amplified methods, the Cleavase invader assay has a broad dynamic range and appears to be well suited for nucleic acid quantitations. Little is known about the inhibition profile of this system.

SIGNAL AMPLIFICATION

In contrast to target and probe amplification systems, signal amplification methods are designed to increase the signal strength of a detection system without increasing the number of target molecules. Signal amplification methods have several advantages over target amplification procedures. Signal amplification methods are generally simpler to perform and are not as susceptible to carryover contamination, and quantitative signal amplification methods usually have significantly broader dynamic (linear) ranges. The first-generation signal amplification methods were less sensitive than target amplification procedures. However, the most sensitive signal amplification assays (e.g., branched DNA [bDNA]) are at least as sensitive as NASBA and PCR.

EIA-BASED DETECTION

Early attempts to automate nucleic acid assays and reduce the turnaround time and labor requirement of these procedures relied heavily upon enzyme immunoassay (EIA) technologies. EIA methods are often used to detect amplified DNA products because automation is readily available and familiar to those who work in the clinical laboratory. The most common EIA method for detecting amplified target DNA is shown in Fig. 7. In this procedure, biotinylated capture probes and enzyme-labeled detection probes are allowed to hybridize (sequentially or simultaneously) to the target DNA in a streptavidin-coated microtiter plate well. PCR products can be captured directly by using biotinylated primers. After hybridization is complete, the unbound probes are removed by several high-stringency washes using a standard microtiter plate washer. An appropriate chromogen or substrate is added to the wells, and the absorbance of the solution is measured using a standard microtiter plate spectrophotometer. These procedures are usually qualitative, but the concentration of target DNA is usually proportional to the final absorbance of the chromogen or substrate solution.

EIA detection systems are sensitive, relatively fast, and easy to perform. EIA systems utilizing nucleic acid probes are very specific, and primer dimers and other nonspecific amplification products are not detected. The multiple washing steps tend to minimize any inhibition caused by the specimen. A number of procedural modifications have been described for different solid supports (paper, latex, etc.) and detection systems (e.g., fluorescence or chemiluminescence).

An interesting modification of the EIA procedure has been introduced by Digene Diagnostics Inc. (Silver Spring, Md.) for use with their RNA probe systems. In this system the target DNA is denatured and allowed to hybridize to an unlabeled RNA probe (Fig. 8). The hybridization mixture is then transferred to an antibody-coated microtiter plate, and the DNA-RNA hybrids are captured. The unbound materials are removed by washing, and the DNA-RNA hybrids are detected with a unique alkaline phosphatase-labeled mono-

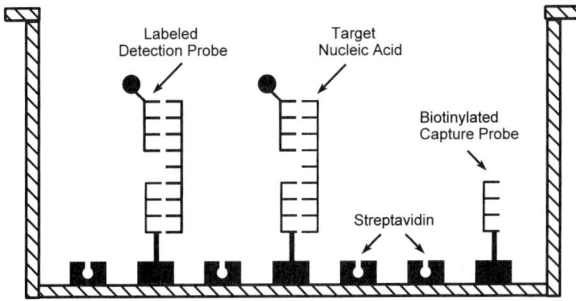

FIGURE 7 Typical EIA detection system. The target nucleic acids are heated and allowed to hybridize with biotinylated capture probes. The resulting mixture is placed in a microtiter plate well containing immobilized strepavidin. The unhybridized nucleic acids are washed away, and a labeled detection probe is allowed to hybridize. After another wash, an appropriate substrate is added. A signal (color, fluorescence, light, etc.) is generated when the target nucleic acid is present in the sample.

clonal antibody that is specific for DNA-RNA hybrids. After several washes, a dioxitane substrate is added to the well, and the chemiluminescent signal is measured with a luminometer. The use of a chemiluminescent substrate and the ability to bind multiple antibodies to each hybrid significantly enhance the signal generation properties of this assay. Using 10-fold dilutions of two Eurohep HBV reference plasma specimens, Kessler et al. (1998) reported that the detection limit of the Digene hybridization assay was

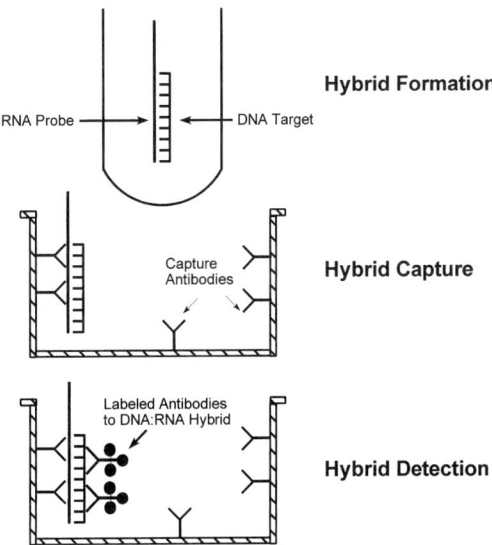

FIGURE 8 Hybrid capture. The target DNA is denatured and allowed to hybridize to a large unlabeled RNA probe. The DNA-RNA hybrids are captured in a microtiter plate well by immobilized, hybrid-specific antibodies. Labeled monoclonal antibodies to the DNA-RNA hybrid are added and bind to the length of the hybrid. After another wash, the chemiluminescent substrate is added. Light is generated if the target DNA is present in the sample.

10^6 to 10^7 copies/ml compared with 10^3 copies/ml with the Amplicor PCR assay. Cullen et al. (1997) reported that the analytical sensitivity of the Digene hybrid capture II test for herpes simplex virus was 5×10^3 to 1×10^4 copies per assay.

bDNA ASSAY

The bDNA system from Chiron Corporation (Emeryville, Calif.) (Urdea et al., 1987) is one of the most powerful signal amplification systems described to date. This procedure (see chapter 20) utilizes an intricate network of oligonucleotide capture probes, target probes, novel branched secondary probes, and short, enzyme-labeled tertiary probes to capture the target and produce a signal. The target probes are complementary to several areas on the target nucleic acid and attach to the target in several places. The distal end of the some of the target probes are complementary to the oligonucleotides (capture probes) that are immobilized on the microplate. The distal ends of other target probes are complementary to one arm of the branched secondary probe. The branched secondary probe possesses 15 or more branches that are complementary to the labeled tertiary probes. Although this multitiered probe system appears complex, it is relatively easy to use and can attach from 60 to 300 enzyme molecules to each target nucleic acid strand. After the hybridization and stringency washing steps are complete, dioxitane, a chemiluminescent substrate, is added, and the resulting light is measured in a luminometer.

bDNA methods can be used to detect viral nucleic acids at their naturally occurring concentrations without nucleic acid replication. This method has an extremely broad dynamic range that appears to be well suited to nucleic acid quantitation. The newest bDNA methods are almost as sensitive as PCR, detecting as few as 50 human immunodeficiency virus (HIV) copies/ml of plasma (Collins et al., 1997).

HPA

The hybridization protection assay (HPA) (Gen-Probe, San Diego, Calif.) was one of the first probe assays to receive Food and Drug Administration (FDA) clearance for clinical diagnostic use. The HPA utilizes a novel chemilumescent technology to amplify the signal while minimizing background noise (Fig. 9). The HPA employs a chemiluminescent acridinium ester label that is covalently coupled to the oligonucleotide probes via an acid-sensitive ether bond (Arnold et al., 1989). Once the ether bond is hydrolyzed, the label becomes permanently nonluminescent. Probes that are bound to target nucleic acids are protected from hydrolysis and retain their chemiluminescence. While this is a qualitative assay, the amount of chemiluminescence produced in the HPA is proportional to the amount of probe-target hybrid formed. The HPA by itself has significantly less sensitivity than PCR tests. The commercial HPA procedure for *C. trachomatis*, for instance, is very specific, but it has no more sensitivity than commercially available EIAs (Schachter, 1997).

REAL-TIME DETECTION OF NUCLEIC ACIDS

FRET System

Real-time detection systems for detecting amplification or hybridization events are being used by an increasing number of molecular diagnostics laboratories. The advantages of these methods are their rapid throughput and reduced labor

Target Present   **Target Absent**

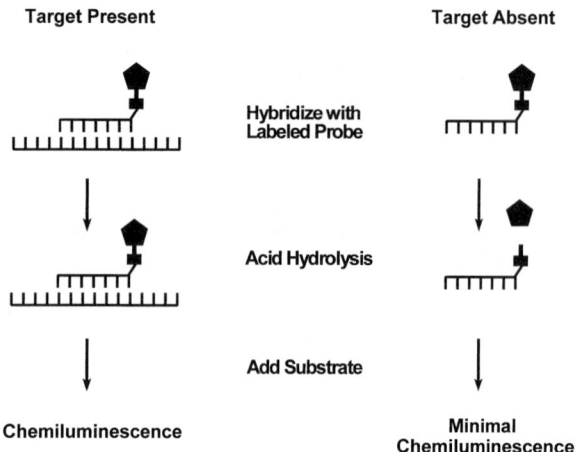

Hybridize with
Labeled Probe

Acid Hydrolysis

Add Substrate

Chemiluminescence   Minimal
Chemiluminescence

Primer extension

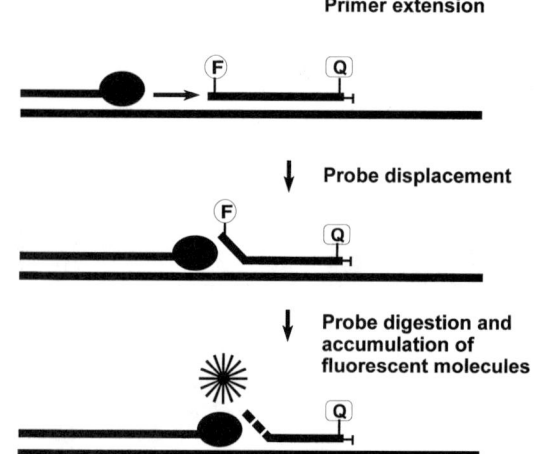

Probe displacement

Probe digestion and
accumulation of
fluorescent molecules

FIGURE 9 Hybridization protection assay. The oligonucleotide detector probe contains a chemiluminescent acridinium ester that is covalently attached to the probe via an acid-sensitive ether bond (top right). Complementary base pairing of the probe to the target protects the ether bond from acid hydrolysis. In the absence of base pairing, the ether bond is hydrolyzed, and the label becomes permanently nonluminescent.

FIGURE 10 FRET system. The FRET system utilizes a reporter oligonucleotide that has a fluorescent dye (F) covalently coupled to the 5′ end and a quencher dye (Q) on the 3′ end. For PCR procedures, the reporter probe hybridizes internally to the flanking PCR primers. As the upstream primer is extended, the reporter oligonucleotide is digested by the 5′→3′ nuclease activity of the polymerase (large oval) Digestion separates the reporter and quencher molecules and allows the reporter molecule to fluoresce strongly. Reporter molecules accumulate with each amplification cycle.

costs compared to more traditional methods. One such method is the TaqMan detection system from Roche and Perkin-Elmer. This method, which is used in conjunction with PCR, employs two standard PCR primers and an internal reporter oligonucleotide. The oligonucleotide has a fluorescent reporter molecule coupled to the 5′ end (Fig. 10) and a quencher molecule at the 3′ end (Livak et al., 1995). When the probe is intact, the proximity of the quencher molecule suppresses the fluorescence of the reporter molecule (Holland et al., 1991). During PCR, the reporter probe is digested by the 5′→3′ nuclease activity of the polymerase. This digestion separates the reporter and quencher molecules, causing a significant increase in the background fluorescence in the tube (Lee et al., 1993; Livak et al., 1995). Cleavage also removes the probe from the target strand, thus allowing primer extension to continue normally. Additional reporter dye molecules are cleaved from their respective probes with each thermal cycle, causing a cumulative increase in fluorescence intensity that is proportional to the amount of amplicon produced. The ability to monitor the real-time progress of the PCR allows the laboratory to quantify nucleic acids based upon the time it takes for the fluorescence to reach an arbitrary threshold. The higher the starting copy number of the nucleic acid target, the sooner the arbitrary threshold is reached.

The FRET procedure is somewhat less sensitive than standard PCR (Ryncarz et al., 1999), principally due to the extremely small sample size. Kawai et al. (1999) and Martell et al. (1999) reported that the TaqMan system had a lower limit of detection of 1×10^3 to 2×10^3 copies/ml for hepatitis C virus. However, both reports state that the TaqMan method has a dynamic (linear) range of 5 log units within a single reaction tube, which is significantly better than traditional PCR methods.

Instrumentation is available which can measure the fluorescence intensity after each thermal cycle. In addition, some instruments can detect and quantify several fluorescent targets simultaneously. These molecular beacon assays (Piatek et al., 1998) can simultaneously detect multiple targets through the release and accumulation of different colored fluorescent dyes in the reaction mixture.

Most of the FRET systems are closed-tube procedures, which can reduce amplicon carryover problems that can plague other nucleic acid amplification methods. FRET technology has been used with a variety of nucleic acid detection methodologies, including PCR (Livak et al., 1995), SDA (Walker and Linn, 1996), and Cleavase invader assays (Ryan et al., 1999). The rapid throughput of this system, the ability to quantitate nucleic acids in a single tube, and the applicability to several amplification systems make the FRET system a valuable tool for the diagnosis of viral infections despite its slightly lower sensitivity levels.

ELECTRON TRANSFER AND BIOELECTRIC DETECTION

A new method for detecting hybridization events which uses ruthinium-based electron donor and acceptor pairs attached to short nucleic acid probes has been described by Meade and Kayyem (1995). These investigators have demonstrated that the rate of electron transfer from the donor to the acceptor molecule is significantly shorter after hybridization than it is for a single-stranded probe (Meade and Kayyem, 1995; Meade, 1996). This difference is presumably due to the longer transfer time required for electrons to move down the phosphate backbone of a single-stranded molecule versus electron transfer down the middle of the helical DNA duplex. Newer versions of this technology are based upon the specific generation of electrical current through the reversible oxidation and reduction of metal complex (ferrocene) labels on nucleic acid targets (Yu et al., 1999). In this procedure, the target nucleic acid is allowed to simultaneously hybridize to gold electrode-bound capture probes and to ferrocene-labeled signaling probes. The probes are constructed so that they are complementary to adjacent regions of the target. Prior nucleic acid amplification is necessary to detect most clinically relevant infectious agents. However, current research by Clinical Micro Sensors, Inc. (Pasadena, Calif.), suggests that amplification-independent detection may be a realistic possibility (D. Farkas, personal communication). These methods are especially interesting because they are adaptable to existing biosensor technologies and they reportedly have fewer inhibition problems than enzyme-based amplification systems (D. Farkas, personal communication).

CHOOSING AN AMPLIFICATION SYSTEM

Choosing an appropriate amplification system is difficult at this time because (i) only a few tests have been approved by the FDA, (ii) an increasing number of vendors are entering the market, and (iii) no manufacturer has a broad range of tests. FDA approval is a critical parameter for some laboratories, because they do not have the staff and/or expertise to develop new assays and perform quality assurance monitoring of reagents and intermediate solutions. Other laboratories prefer FDA-approved products because some insurance companies classify in-house-developed assays as "experimental" and refuse to pay for them.

Test system choices can be difficult even when choosing among FDA-approved procedures. A number of factors influence test system choices, including the expected test menu, patient population, anticipated test volumes, technical expertise, space constraints, and whether the laboratory has previously purchased nucleic acid amplification and detection equipment. For example, some laboratories that perform the Roche HIV Monitor test for HIV RNA quantitation (viral load testing) may prefer to use the Amplicor PCR test from Roche Diagnostics because the test procedures are similar and the same equipment can be used for both assays. However, it is well known that the *N. gonorrhoeae* test from Roche cross-reacts with certain strains of *Neisseria subflava* and *Neisseria cinerea* (Farrell, 1999). While these organisms are normally considered to be commensal organisms of the upper respiratory tract, transient carriage in the genital tract cannot be overlooked. Indeed, Farrell (1999) reported that 15 of 96 (15.9%) Amplicor-positive and 14 of 17 (82.3%) Amplicor-equivocal specimens collected from the local indigenous population of Queensland, Australia, could not be confirmed by a more sensitive PCR method. Thus, the performance of a particular assay may work well in one patient population and have significant limitations in another.

Anticipated test volumes also have an important impact upon test selection. Low-volume laboratories (<30 specimens/day) may elect to use LCR methods because they are sensitive, specific, and easy to use. However, the LCR detection system can only detect 19 samples/batch, and the detection time for a single batch is approximately 1 h. Higher-volume laboratories may need more than one thermocycler and/or detection system in order to test all their specimens on one shift. The COBAS Amplicor system also provides a significant level of automation for low-volume laboratories. The COBAS system can test up to 24 individual samples and provide 48 results in multiplex testing for *C. trachomatis* and *N. gonorrhoeae*. The cycle time for the COBAS system is approximately 6 h. The BDProbeTecS-DA system from Becton Dickinson is designed to handle larger test volumes. The BDProbeTec system can test up to 94 specimens per batch, and as many as six batches can be performed in an 8-h shift. In contrast to PCR and LCR assays, BDProbeTec testing can be done in a single room, and each specimen has an amplification control.

Laboratories that develop in-house assays are currently limited to PCR and NASBA methodologies. Other amplification methods described in this chapter are part of closed assay systems whose manufacturers discourage or prohibit in-house development. The most popular in-house method is PCR. PCR and its derivatives (e.g., reverse transcriptase PCR, in situ PCR, and FRET detection) have been used by an increasing number of clinical laboratories to detect and quantify nucleic acid targets. Several manufacturers are assisting in the development of in-house PCR tests by making matched reagent sets and amplicon detection systems. The use of these systems can significantly shorten development time. Laboratories that use PCR methods will need separate work areas for reagent preparation, specimen extraction, and amplicon detection. The laboratory will also need at least one thermocycler and amplicon detection equipment, such as hybridization incubators, a microplate washer, a microplate reader, gel electrophoresis equipment, and a UV light box. Laboratories performing PCR for clinical purposes must also obtain a license from Roche Diagnostics and a pay a 9 to 15% royalty on all clinical tests. Laboratories using the TaqMan procedure must pay TaqMan and PCR royalties for clinical assays.

NASBA has also been used for in-house test development. It is of interest to note that Organon-Teknika is actively supporting in-house test development by providing a "basic kit" and technical support for primer selection, hybridization conditions, and troubleshooting. Unlike PCR tests, in-house NASBA tests may be used royalty free as long as the laboratory uses the basic kit. NASBA procedures require less space and equipment than PCR methods, but laboratories using NASBA will need heating blocks and a dedicated detection system from Organon-Teknika. Higher-volume laboratories may also want to purchase the automated nucleic acid extraction system. NASBA procedures have some limitations, however. In its current form, the NASBA procedure can detect only RNA targets. Extraction and handling of RNA is more demanding than that of DNA due to the presence of hardy RNases in the environment which can degrade the target. Thus, manual RNA extraction procedures for NASBA are more time intensive than standard PCR procedures.

Nucleic acid detection is still in its infancy, and choosing among test methodologies will be difficult until more tests are available commercially. Until then, molecular virology laboratories must utilize multiple methods to detect viral agents in clinical specimens.

CONCLUSIONS

Virus detection is increasingly important for the diagnosis of viral disease, clinical management of patients on antiviral therapies, detecting the emergence of drug resistance, and reducing the length of hospitalization. To be useful, however, virus detection information must be delivered within a clinically relevant time frame. Many of the nucleic acid detection methods mentioned in this chapter have significantly reduced the time to result for virus detection, thereby contributing to improved patient management. Nucleic acid detection methods have also made significant improvements in our ability to detect fastidious and slow-growing viruses (e.g., human parvovirus, Epstein-Barr virus, and certain enteroviruses), viruses that are dangerous to amplify in culture (e.g., HIV and certain hemorrhagic fever viruses), and viruses that are present in low concentrations (Wiedbrauk and Hodinka, 1998). More importantly, nucleic acid methods have expanded the role of the clinical virology laboratory by allowing laboratories to detect viruses that do not grow in culture (e.g., hepatitis B virus, hepatitis C virus, and human papillomaviruses). While nucleic acid detection methods will never completely replace culture and direct fluorescent antibody assay methods, they will continue to play an important role in the detection and monitoring of viral diseases.

REFERENCES

Ahokas, H., and M. J. Erkkila. 1993. Interference of PCR amplification by the polyamines, spermine and spermidine. *PCR Methods Appl.* **3**:65–68.

Arnold, L. J., P. W. Hammond, W. A. Wiese, and N. C. Nelson. 1989. Assay formats involving acridinium-ester-labeled DNA probes. *Clin. Chem.* **35**:1588–1594.

Barany, F. 1991a. Genetic disease detection and DNA amplification using cloned thermostable ligase. *Proc. Natl. Acad. Sci. USA* **88**:189–193.

Barany, F. 1991b. The ligase chain reaction in a PCR world. *PCR Methods Appl.* **1**:5–16.

Beggs, M. L., M. D. Cave, C. Marlowe, L. Cloney, P. Duck, and K. D. Eisenach. 1996. Characterization of *Mycobacterium tuberculosis* complex direct repeat sequence for use in cycling probe reaction *J. Clin. Microbiol.* **34**:2985–2989.

Bekkaoui, F., I. Poisson, W. Crosby, L. Cloney, and P. Duck. 1996. Cycling probe technology with RNase H attached to an oligonucleotide. *BioTechniques* **20**:240–248.

Bekkaoui, F., J. P. McNevin, C. H. Leung, G. L. Peterson, A. Patel, R. S. Bhatt, and R. N. Bryan. 1999. Rapid detection of the *mecA* gene in methicillin resistant staphylococci using colorimetric cycling probe technology. *Diagn. Microbiol. Infect. Dis.* **34**:83–90.

Belgrader, P., W. Benett, D. Hadley, G. Long, R. Mariella, Jr., F. Milanovich, S. Nasarabadi, W. Nelson, J. Richards, and P. Stratton. 1998. Rapid pathogen detection using a microchip PCR array instrument. *Clin. Chem.* **44**:2191–2194.

Belgrader, P., W. Benett, D. Hadley, J. Richards, P. Stratton, R. Mariella, Jr., and F. Milanovich. 1999. PCR detection of bacteria in seven minutes. *Science* **284**:449–450.

Berg, E. S., G. Anestad, H. Mo, G. Storvold, and K. Skaug. 1997. False-negative results of a ligase chain reaction assay to detect *Chlamydia trachomatis* due to inhibitors in urine. *Eur. J. Clin. Microbiol. Infect. Dis.* **16**:727–731.

Beutler, E., T. Gelbart, and W. Kuhl. 1990. Interference of heparin with the polymerase chain reaction. *BioTechniques* **9**:166.

Boom, R., C. J. Sol, M. M. Salimans, C. L. Jansen, P. M. Wertheim-van Dillen, and J. van der Noordaa. 1990. Rapid and simple method for purification of nucleic acids. *J. Clin. Microbiol.* **28**:495–503.

Boom, R., C. Sol., M. Beld, J. Weel, J. Goudsmit, and P. Wertheim-van Dillen. 1999. Improved silica-guanidiniumthiocyanate DNA isolation procedure based on selective binding of bovine alpha-casein to silica particles. *J. Clin. Microbiol.* **37**:615–619.

Chernesky, M. A., D. Jang, J. Sellors, K. Luinstra, S. Chong, C. Castriciano, and J. B. Mahony. 1997. Urinary inhibitors of polymerase chain reaction and ligase chain reaction and testing of multiple specimens may contribute to lower assay sensitivities for diagnosing *Chlamydia trachomatis* infected women. *Mol. Cell. Probes* **11**:243–249.

Collins, M. L., B. Irvine, D. Tyner, E. Fine, C. Zayati, C. Chang, T. Horn, D. Ahle, J. Detmer, L.-P. Shen, J. Kolberg,

S. Bushnell, M. S. Urdea, and D. D. Ho. 1997. A branched DNA signal amplification assay for quantification of nucleic acid targets below 100 molecules/ml. *Nucleic Acids Res.* **25**:2979–2984.

Cullen, A. P., C. D. Long, and A. T. Lorincz. 1997. Rapid detection and typing of herpes simplex virus DNA in clinical specimens by the hybrid capture II signal amplification probe test *J. Clin. Microbiol.* **35**:2275–2278.

Davis, J. D., P. K. Riley, C. W. Peters, and K. H. Rand. 1998. A comparison of ligase chain reaction to polymerase chain reaction in the detection of *Chlamydia trachomatis* endocervical infections. *Infect. Dis. Obstet. Gynecol.* **6**:57–60

Demeke, T., and R. Adams. 1992. The effects of plant polysaccharides and buffer additives on PCR. *BioTechniques* **12**:332–334.

Duck, P., G. Alvarado-Urbina, B. Burdick, and B. Collier. 1990. Probe amplifier system based on chimeric cycling oligonucleotides. *BioTechniques* **9**:142–148.

Edwards, M. C., and R. A. Gibbs. 1994. Multiplex PCR: advantages, development, and applications. *PCR Methods Appl.* **3**:S65–S75.

Farrell, D. J. 1999. Evaluation of AMPLICOR *Neisseria gonorrhoeae* PCR using *cppB* nested PCR and 16S rRNA PCR. *J. Clin. Microbiol.* **37**:386–390.

Guatelli, J. C., K. M. Whitfield, D. Y. Kwoh, K. J. Barringer, D. D. Richman, and T. R. Gingeras. 1990. Isothermal, *in vitro* amplification of nucleic acids by a multienzyme reaction modeled after retroviral replication. *Proc. Natl. Acad. Sci. USA* **87**:1874–1878.

Holland, P. M., R. D. Abramson, R. Watson, and D. H. Gelfand. 1991. Detection of specific polymerase chain reaction product by utilizing the 5′ to 3′ exonuclease activity of *Thermus aquaticus* DNA polymerase. *Proc. Natl. Acad. Sci. USA* **88**:7276–7280.

Holodniy, M., S. Kim, D. Katzenstein, M. Konrad, E. Groves, and T. C. Merigan. 1991. Inhibition of human immunodeficiency virus gene amplification by heparin. *J. Clin. Microbiol.* **29**:676–679.

Katcher, H. L., and I. Schwartz. 1994. A distinctive property of *Tth* DNA polymerase: enzymatic amplification in the presence of phenol. *BioTechniques* **16**:84–92.

Kawai, S., O. Yokosuka, T. Kanda, F. Imazeki, Y. Maru, and H. Saisho. 1999. Quantification of hepatitis C virus by TaqMan PCR: comparison with HCV Amplicor Monitor assay. *J. Med. Virol.* **58**:121–126.

Kessler, H. H., K. Pierer, E. Dragon, H. Lackner, B. Santner, D. Stunzner, E. Stelzl, B. Waitzl, and E. Marth. 1998. Evaluation of a new assay for HBV DNA quantitation in patients with chronic hepatitis B. *Clin. Diagn. Virol.* **9**:37–43.

Khan, G., H. O. Kangro, P. J. Coates, and R. B. Heath. 1991. Inhibitory effects of urine on the polymerase chain reaction for cytomegalovirus DNA. *J. Clin. Pathol.* **44**:360–365.

Lee, L. G., C. R. Connell, and W. Bloch. 1993. Allelic discrimination by nick-translation PCR with fluorogenic probes. *Nucleic Acids Res.* **21**:3761–3766.

Leone, G., H. van Schijndel, B. van Gemen, F. R. Kramer, and C. D. Schoen. 1998. Molecular beacon probes combined with amplification by NASBA enable homogeneous, real-time detection of RNA. *Nucleic Acids Res.* **26**:2150–2155.

Little, M. C., J. Andrews, R. Moore, S. Bustos, L. Jones, C. Embres, G. Durmowicz, J. Harris, D. Berger, K. Yanson, C. Rostkowski, D. Yursis, J. Price, T. Fort, A. Walters, M. Collis, O. Llorin, J. Wood, F. Failing, C. O'Keefe, B. Scrivens, B. Pope, T. Hansen, K. Marino, K. Williams, and M. Boenisch. 1999. Strand displacement amplification and homogeneous real-time detection incorporated in a second-generation DNA probe system, BDProbeTecET. *Clin. Chem.* **45**:777–784.

Livak, K. J., S. J. A. Flood, J. Marmaro, W. Giusti, and K. Deetz. 1995. Oligonucleotides with fluorescent dyes at opposite ends provide a quenched probe system useful for detecting PCR product and nucleic acid hybridization. *PCR Methods Appl.* **4**:357–362.

Mahony, J., S. Chong, D. Jang, K. Luinstra, M. Faught, D. Dalby, J. Sellors, and M. Chernesky. 1998. Urine specimens from pregnant and nonpregnant women inhibitory to amplification of *Chlamydia trachomatis* nucleic acid by PCR, ligase chain reaction, and transcription-mediated amplification: identification of urinary substances associated with inhibition and removal of inhibitory activity. *J. Clin. Microbiol.* **36**:3122–3126.

Martell, M., J. Gomez, J. I. Esteban, S. Sauleda, J. Quer, B. Cabot, R. Esteban, and J. Guardia. 1999. High-throughput real-time reverse transcription-PCR quantitation of hepatitis C virus RNA. *J. Clin. Microbiol.* **37**:327–332.

Meade, T. J. 1996. Electron transfer reactions through the DNA double helix. *Met. Ions Biol. Syst.* **32**:453–478.

Meade, T. J., and J. F. Kayyem. 1995. Electron transfer through DNA: site-specific modification of duplex DNA with ruthenium donors and acceptors. *Angew. Chem. Int. Ed. Engl.* **34**:352–354.

Mercier, B., C. Gaucher, O. Feugeas, and C. Mazurier. 1990. Direct PCR from whole blood, without DNA extraction. *Nucleic Acids Res.* **18**:5908.

Modrusan, Z., F. Bekkaoui, and P. Duck. 1998. Spermine-mediated improvement of cycling probe reaction. *Mol. Cell. Probes* **12**:107–116.

Mullis, K. B., and F. A. Faloona. 1987. Specific synthesis of DNA *in vitro* via a polymerase-catalyzed chain reaction. *Methods Enzymol.* **155**:335–350.

Myers, T. W., and D. H. Gelfand. 1991. Reverse transcription and DNA amplification by a *Thermus thermophilus* DNA polymerase. *Biochemistry* **30**:7661–7666.

National Committee for Clinical Laboratory Standards. 1995. *Molecular Diagnostic Methods for Infectious Diseases. Approved Guideline.* NCCLS document MM3-A. National Committee for Clinical Laboratory Standards, Wayne, Pa.

Piatek, A. S., S. Tyagi, A. C. Pol, A. Telenti, L. P. Miller, F. R. Kramer, and D. Alland. 1998. Molecular beacon sequence analysis for detecting drug resistance in *Mycobacterium tuberculosis. Nat. Biotechnol.* **16**:359–363.

Puolakkainen, M., E. Hiltunen-Back, T. Reunala, S. Suhonen, P. Lahteenmaki, M. Lehtinen, and J. Paavonen. 1998. Comparison of performances of two commercially available tests, a PCR assay and a ligase chain reaction test, in detection of urogenital *Chlamydia trachomatis* infection. *J. Clin. Microbiol.* **36**:1489–1493.

Ruano, G., E. M. Pagliaro, T. R. Schwartz, K. Lamy, D. Messina, R. E. Gaensslen, and H. C. Lee. 1992. Heat-soaked PCR: an efficient method for DNA amplification with applications to forensic analysis. *BioTechniques* **13**:266–274.

Ryan, D., B. Nuccie, and D. Arvan. 1999. Non-PCR-dependent detection of the factor V Leiden mutation from genomic DNA using a homogeneous invader microtiter plate assay. *Mol. Diagn.* **4**:135–144.

Ryncarz, A. J., J. Goddard, A. Wald, M. L. Huang, B. Roizman, and L. Corey. 1999. Development of a high-throughput quantitative assay for detecting herpes simplex virus DNA in clinical samples. *J. Clin. Microbiol.* **37**:1941–1947.

Saiki, R. K., S. Scharf, F. Faloona, K. B. Mullis, G. T. Horn, H. A. Erlich, and N. Arnheim. 1985. Enzymatic amplification of β-globin genomic sequences and restriction site analysis for diagnosis of sickle cell anemia. *Science* **230**:1350–1354.

Schachter, J. 1997. DFA, EIA, PCR, LCR, and other technologies: what tests should be used for diagnosis of chlamydia infections? *Immunol. Investig.* **26**:157–161.

Spargo, C. A., M. S. Fraiser, M. van Cleve, D. J. Wright, C. M. Nycz, P. A. Spears, and G. T. Walker. 1996. Detection of M. *tuberculosis* DNA using thermophilic strand displacement amplification. *Mol. Cell. Probes* **10**:247–256.

Toye, B., W. Woods, M. Bobrowska, and K. Ramotar. 1998. Inhibition of PCR in genital and urine specimens submitted for *Chlamydia trachomatis* testing. *J. Clin. Microbiol.* **36**:2356–2358.

Urdea, M. S., J. A. Running, T. Horn, J. Clyne, L. Ku, and B. D. Warner. 1987. A novel method for the rapid detection of specific nucleotide sequences in crude biologic samples without blotting or radioactivity: application to the analysis of hepatitis B in serum. *Gene* **61**:253–264.

Wadowsky, R. M., S. Laus, T. Libert, S. J. States, and G. D. Ehrlich. 1994. Inhibition of PCR-based assay for *Bordetella pertussis* by using calcium alginate fiber and aluminum shaft components of a nasopharyngeal swab. *J. Clin. Microbiol.* **32**:1054–1057.

Walker, G. T., and C. P. Linn. 1996. Detection of *Mycobacterium tuberculosis* DNA with thermophilic strand displacement amplification and fluorescence polarization. *Clin. Chem.* **42**:1604–1608.

Walker, G. T., M. C. Little, J. G. Nadeau, and D. D. Shank. 1992. Isothermal *in vitro* amplification of DNA by a restriction enzyme/DNA polymerase system. *Proc. Natl. Acad. Sci. USA* **89**:392–396.

Walker, G. T., C. A. Spargo, C. M. Nycz, J. A. Down, M. S. Dey, A. H. Walters, D. R. Howard, W. E. Keating, M. C. Little, J. G. Nadeau, S. R. Jurgensen, V. R. Neece, and P. Zwadyk, Jr. 1995. A chemiluminescent DNA probe test based on strand displacement amplification, p. 329–349. *In* D. L. Wiedbrauk and D. H. Farkas (ed.), *Molecular Methods for Virus Detection.* Academic Press, San Diego, Calif.

Walker, G. T., C. P. Linn, and J. G. Nadeau. 1996. DNA detection by strand displacement amplification and fluorescence polarization with signal amplification using a DNA binding protein. *Nucleic Acids Res.* **24**:348–353.

Wiedbrauk, D. L., and R. L. Hodinka. 1998. Applications of the polymerase chain reaction, p. 98–116. *In* S. Specter, M. Bendinelli, and H. Friedman (ed.), *Rapid Detection of Infectious Agents.* Plenum Press, New York, N.Y.

Wiedbrauk, D. L., J. C. Werner, and A. M. Drevon. 1995. Inhibition of PCR by aqueous and vitreous fluids. *J. Clin. Microbiol.* **33**:2643–2646.

Witt, D. J., and M. Kemper. 1999. Techniques for the evaluation of nucleic acid amplification technology performance with specimens containing interfering substances: efficacy of boom methodology for extraction of HIV-1 RNA. *J. Virol. Methods* **79**:97–111.

Yu, C. J., Y. Chong, J. F. Kayyem, and M. Gozin. 1999. Soluble ferrocene conjugates for incorporation into self-assembled monolayers. *J. Org. Chem.* **64**:2070–2079.

Quantitative Molecular Techniques

FREDERICK S. NOLTE

20

The development of quantitative nucleic acid amplification techniques has created new opportunities for the clinical laboratory to influence the diagnosis and management of patients with viral diseases. These techniques provide important information that can be used to predict disease progression, distinguish symptomatic from asymptomatic infection, monitor the development of antiviral resistance, and assess the efficacy of antiviral therapy. Prior to development of these techniques, clinicians were limited to either laborious culture methods available only in research laboratories or insensitive antigen detection assays to measure viral load. Viral load is a relatively new term; it first appeared in the scientific literature in 1987 in a paper by Jonas Salk (Salk, 1987) proposing that viral load in human immunodeficiency virus type 1 (HIV-1)-infected individuals could be reduced by boosting the immune response, leading to reduced morbidity, mortality, and disease transmission. Viral-load assays assess the overall virus replicative activity that reflects the underlying disease process, usually by quantCitation of the viral nucleic acid in the blood (Dailey and Hayden, 1999).

Viral-load testing in HIV-1 infection is the best example of how quantitative molecular techniques have increased our understanding of a disease process and improved patient care. The HIV-1 viral-load test is powerfully predictive for disease progression and is currently used to start, monitor, and change antiretroviral therapy (Saag et al., 1996). Viral-load assays may have similar impact on the care of patients with other chronic viral infections for which the therapy is currently not optimal, such as hepatitis B and hepatitis C. Viral-load assays also provide important information in distinguishing active from latent herpesvirus infections in immunocompromised patients.

In this chapter the quantitative molecular techniques that serve as the bases of viral-load assays will be described. The key issues and important variables that affect assay performance will be reviewed. The reader is directed to the National Committee for Clinical Laboratory Standards guideline *Quantitative Molecular Method for Infectious Diseases* (1999) for more information on implementing these assays in the clinical laboratory.

HISTORY

The first description of PCR by Mullis and colleagues was a milestone in biotechnology and heralded the beginning of molecular diagnostics (Saiki et al., 1988). The first applications of PCR in clinical virology were qualitative and provided very sensitive methods for the detection of viral nucleic acids in clinical specimens. However, semiquantitative and quantitative applications of PCR for viral-gene analysis followed, as it was realized that a linear relationship existed between the amounts of input template and amplification product.

Initial attempts at quantitative PCR involved simple quantitation of the amplification product (Oka et al., 1990; Warren et al., 1991). However, since PCR results in an exponential amplification of the initial target copy number, small differences in amplification efficiency from sample to sample lead to large and unpredictable differences in the amount of final reaction product. The unpredictable variability of individual amplification reactions prevents reliable quantitation of nucleic acids by this approach.

Other early attempts at quantitative PCR used limiting dilutions of samples (Simmonds et al., 1990; Brillanti et al., 1991). In limiting-dilution PCR, a series of dilutions of the sample are made and PCR is performed on multiple aliquots of each dilution; the mean number of target molecules can be calculated based on the percentage of aliquots that are PCR positive at a given dilution. The precision of this approach is limited by the precision with which the dilution series is made. Furthermore, limiting-dilution PCR is impractical for routine use owing to the large number of PCRs that must be performed to quantitate each sample.

The first attempts to control for sample-to-sample variation in amplification efficiency used coamplification of internal reference templates. Single-copy cellular genes (Kellog et al., 1990) or ubiquitously expressed transcripts (Noonan et al., 1990) have been used as internal reference templates. In concept, variables influencing the amplification efficiency should affect both templates similarly. In practice, however, different PCR templates may have very different thermodynamics and amplification efficiencies. The different template sequences may influence the amounts of both products in an unpredictable manner. Also, this approach is limited in virological applications in that it cannot be used to quantitate extracellular templates, such as viruses, in serum or plasma samples.

Quantitative competitive PCR (cPCR) was the first truly quantitative PCR method developed and is used widely for quantitation of viral nucleic acids (Gilliland et al.,

1990; Piatak et al., 1993). cPCR relies on the inclusion of an internal control competitor in each reaction. The competitor molecules have the same primer binding sequences as the target molecule and are similar in size and base composition to the target but are distinguishable from it. Therefore, the efficiency of each reaction can be normalized to the internal competitor. A known amount of competitor is added to each sample, and the ratio of PCR products reflects the ratio between the initial amounts of competitor and target. Several lines of evidence show that cPCR is presently the most reliable approach for quantitative PCR and reverse transcriptase (RT)-PCR.

Real-time quantitative PCR assays have recently been developed (Heid et al., 1996). The term real-time PCR refers to methods in which the target amplification and detection steps occur simultaneously in the same tube. These methods employ novel fluorogenic probes or fluorescent dyes to monitor the PCR product as it accumulates, and they require special thermal cyclers with precision optics that can monitor the fluorescence emission from the sample wells. The use of real-time methods in cPCR assays enhances their performance characteristics and represents the future of quantitative PCR assays.

Although PCR is the best-developed and most widely used nucleic acid amplification strategy, other strategies have been developed, and several serve as the bases of quantitative assays for viral nucleic acids. Quantitative assays based on nucleic acid sequence-based amplification (NASBA), branched DNA (bDNA), and hybrid capture are commercially available. Quantitative assays based on transcription-mediated amplification (TMA), strand displacement amplification, and ligase chain reaction are currently in commercial development.

METHODS

PCR

PCR is a simple, in vitro chemical reaction that permits the synthesis of essentially limitless quantities of a targeted nucleic acid sequence. This is accomplished through the action of a DNA polymerase that, under the right conditions, can copy a strand of DNA. At its simplest, PCR consists of target DNA, a molar excess of two oligonucleotide primers, a heat-stable DNA polymerase, an equimolar mixture of deoxyribonucleotide triphosphates (dATP, dCTP, dGTP, and dTTP), $MgCl_2$, KCl, and a Tris-HCl buffer. The two primers flank the sequence to be amplified, typically <100 to several hundred bases, and are complementary to opposite strands of the target.

To initiate a PCR, the reaction mixture is heated to separate the two strands of target DNA and then cooled to permit the primers to anneal to the target DNA in a sequence-specific manner. The DNA polymerase then initiates extension of the primers toward one another at their 3' ends. The primer extension products are dissociated from the target DNA by heating. Each extension product, as well as the original target, can serve as a template for subsequent rounds of primer annealing and extension.

A PCR cycle consists of three steps, denaturation, extension, and annealing. At the end of each cycle, the PCR products are theoretically doubled. Thus, after n PCR cycles, the target sequence can be amplified 2^n-fold. The whole procedure is carried out in a programmable thermal cycler that precisely controls the temperatures at which the steps occur, the length of time that the reaction is held at

the different temperatures, and the number of cycles. Ideally, after 20 cycles of PCR, a millionfold amplification is achieved, and after 30 cycles, a billionfold amplification results. In practice, the amplification may not be completely efficient due to failure to optimize the reaction conditions or the presence of inhibitors of the DNA polymerase. In such cases, the total amplification is best described by the expression $(1 + e)^n$, where e is the amplification efficiency ($0 \leq e \leq 1$) and n is the total number of cycles.

A variety of PCR-based strategies have been developed to accurately quantitate DNA and RNA targets in clinical specimens, including simple quantitation of the amplification product (Oka et al., 1990; Warren et al., 1991), limiting dilutions of samples (Simmonds et al., 1990; Brillanti et al., 1991), coamplification of internal reference templates (Kellog et al., 1990; Noonan et al., 1990), and cPCR (Gilliland et al., 1990; Piatak et al., 1993). However, it is generally accepted that cPCR is the most reliable and robust approach to gene quantitation (Clementi et al., 1993).

The basic concept behind cPCR is the coamplification in the same reaction tube of two different templates of equal or similar lengths with the same primer binding sequences. Since both templates are amplified with the same primer pair, identical thermodynamics and amplification efficiencies are ensured. The amount of one of the templates must be known, and after amplification, the products from the two templates must be distinguishable. Different types of competitors have been used in cPCR, but in general those competitors similar to the target in size and base composition work most effectively. RNA competitors should be used in quantitative RT-PCRs to address the problem of variable RT efficiency.

The yield of PCR product is described by the equation $Y = I(1 + e)^n$, where Y is the quantity of PCR product, I is the quantity of template at the beginning of the reaction, e is the efficiency of the reaction, and n is the number of cycles. In cPCR, this equation is written for the two templates as follows: competitor, $Y_c = I_c(1 + e)^n$; target, $Y_t = I_t(1 + e)^n$. Since e and n are the same for both the competitor and target, the relative product ratio, Y_c/Y_t depends directly on their initial concentration ratio, I_c/I_t, and the function $Y_c/Y_t = I_c/I_t$ is linear.

A single concentration of competitor is sufficient, in theory, to quantitate an unknown amount of target without the use of a standard curve. However, because analysis of two template species present in a sample at widely different amounts was imprecise in practice, cPCR using several concentrations of competitor within the expected concentration range of the target was generally performed. However, recently it was shown in a study of different approaches to the standardization of cPCR (Haberhausen et al., 1998) that this approach provided no more accurate results than the use of a single concentration of competitor. The commercially available quantitative PCR and RT-PCR assays for cytomegalovirus (CMV), HIV, and hepatitis C virus (HCV) (Roche Molecular Systems, Pleasanton, Calif.) all use a single concentration of a competitor (quantitation standard [QS]) to determine the initial concentration of the target.

The commercially available quantitative PCR assays incorporate several technological innovations that have made both qualitative and quantitative PCR feasible for clinical laboratories. The RT-PCR assays for quantitation of HIV-1 and HCV RNAs use a single enzyme, a recombinant DNA polymerase from *Thermus thermophilus*, to carry out both reverse transcription of the target RNA and amplification of the cDNA (Myers and Gelfand, 1991). Prior to the development of this polymerase, RT-PCR involved cumbersome two-enzyme systems. The assays incorporate dUTP and uracil-N-glycosylase for the control of carryover contamination (Longo et al., 1990). A colorimetric microtiter plate system is used for detection of the amplified products. In this system, biotinylated primers are used to amplify the target, and the biotin-containing PCR product is denatured and added to the microtiter well coated with specific capture probes. After hybridization with the capture probe, bound product is detected with a streptavidin-enzyme conjugate and a chromogenic substate (Loeffelholz et al., 1992). This system for PCR product detection resembles an enzyme immunoassay and uses microtiter plate washers and readers commonly found in clinical laboratories. Finally, all of the manual quantitative PCR assays are being adapted for the COBAS system, which automates the amplification and detection steps.

A schematic of the Amplicor Monitor system (Roche) for the quantitation of HIV-1 RNA is shown in Fig. 1. After amplification, the samples are denatured and aliquots are diluted in microtiter wells coated with probes specific for either the virus or the QS. After hybridization and color development, the ratio of HIV-1 RNA signal to that of the QS is used to calculate the HIV-1 RNA copies in the original sample by the following formula: virus copy number = (optical density × dilution factor)$_{virus}$/(optical density × dilution factor)$_{QS}$ × input QS copies.

Real-Time PCR

Real-time PCR methods, in which the target amplification and detection steps occur simultaneously in the same tube, require special thermal cyclers with precision optics that can monitor the fluorescence emissions from the sample wells. The computer software supporting the thermal cycler monitors the data throughout the PCR at every cycle and generates an amplification plot for each reaction.

In the simplest format, the PCR product is detected as it is produced by using fluorescent dyes that preferentially bind to double-stranded DNA. SYBR Green I is one such dye that has been used in this application (Morrison et al., 1998). In the unbound state, the fluorescence is relatively low, but when the dye is bound to double-stranded DNA, the fluorescence is greatly enhanced. The dye will bind to both the specific and nonspecific PCR products. The specificity of the detection can be improved through melting curve analysis. The specific amplified product will have a characteristic melting peak at its predicted melting temperature (T_m), whereas the primer dimers and other nonspecific products should have different T_ms or give broader peaks (Ririe et al., 1997).

The specificity of real-time PCR can also be increased by including hybridization probes in the reaction mixture. These probes are labeled with fluorescent dyes or with combinations of fluorescent and quencher dyes. In the 5′ nuclease PCR assay (TaqMan), the 5′-to-3′ exonuclease activity of Taq DNA polymerase is used to cleave a nonextendable hybridization probe during the primer extension phase of PCR (Holland et al., 1991). This approach uses dual-labeled fluorogenic hybridization probes. One fluorescent dye serves as the reporter, and its emission spectrum is quenched by the second fluorescent dye. The nuclease degradation of the hybridization probe releases the reporter dye, resulting in an increase in its peak fluorescent emission. The increase in fluorescent emission indicates that specific PCR product has been made, and the intensity of the fluorescence is related to the amount of product (Heid et al., 1996).

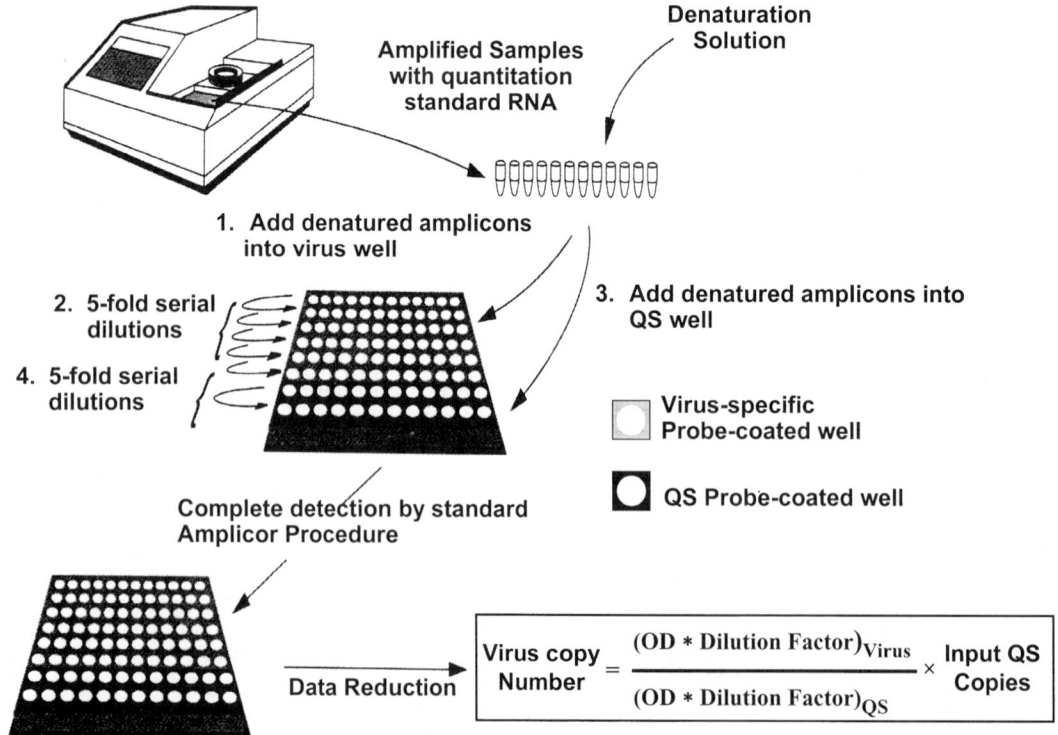

FIGURE 1 Schematic of the Roche Amplicor monitor system for the quantitation of HIV-1 RNA. OD, optical density. (Provided by Roche Molecular Systems.)

Fluorescence resonance energy transfer (FRET) is the basis of another approach to real-time PCR (Lay and Wittwer, 1997). This method requires two specially designed sequence-specific oligonucleotide probes. These hybridization probes are designed to hybridize next to each other on the product molecule. The 3' end of one probe is labeled with a donor dye, and the 5' end of the other probe is labeled with an acceptor dye. The donor dye is excited by an external light source and, instead of emitting light, transfers its energy to the acceptor dye by a process called FRET. The excited acceptor dye emits light at a longer wavelength than the unbound donor dye, and the intensity of the acceptor dye light emission is proportional to the amount of PCR product.

Real-time detection and quantitation of PCR product can also be accomplished by using molecular beacons (Tyagi et al., 1998). Molecular beacons are hairpin-shaped oligonucleotide probes with internally quenched fluorophores whose fluorescence is restored when they bind to a target nucleic acid. They are designed in such a way that the loop portion of the molecule is a probe sequence complementary to the target sequence. The stem is formed by the annealing of complementary arm sequences on the ends of the probe. A fluorescent dye is attached to the end of one arm, and a quenching molecule is attached to the end of the other arm. The stem keeps the fluorophore and quencher in close proximity, and no light emission occurs. When the probe encounters a target molecule, it forms a hybrid that is longer and more stable than the stem and undergoes a conformational change that forces the stem apart and causes the fluorophore and quencher to move away from each other, restoring the fluorescence.

Real-time PCR methods decrease the time required to perform nucleic acid assays because there is no post-PCR processing time. Also, since amplification and detection occur in the same closed tube, these methods eliminate the postamplification manipulations that can lead to laboratory contamination with amplicon. Real-time PCR methods provide accurate and reproducible quantitation of target nucleic acids because analysis is performed early in the log phase of product accumulation. These methods also have very large dynamic ranges of starting target molecule determination: at least 5 orders of magnitude. To date, no real-time quantitative PCR assays for viral nucleic acids are commercially available.

NASBA

NASBA is one of several transcription-based amplification methods that amplify RNA targets (Compton, 1991). In contrast to PCR, NASBA is an isothermal process that employs three enzymes, avian myeloblastosis virus RT, RNase H, and T7 RNA polymerase (Fig. 2). The amplification steps involve the formation of cDNAs from the target RNA with oligonucleotide primers containing a T7 RNA polymerase binding site. RNase H then degrades the initial strand of target RNA in the RNA-DNA hybrid after it has served as a template for the first primer. The second primer binds to the newly formed cDNA and is extended, resulting in the formation of double-stranded cDNAs with an intact T7 promoter. This DNA molecule serves as a substrate for the T7 RNA polymerase, which transcribes multiple copies of antisense RNA. The antisense RNA molecules can also bind the second primer and, through the combined enzymatic activities, new double-stranded cDNA molecules are synthesized, which in turn serve as transcription templates for the RNA poly-

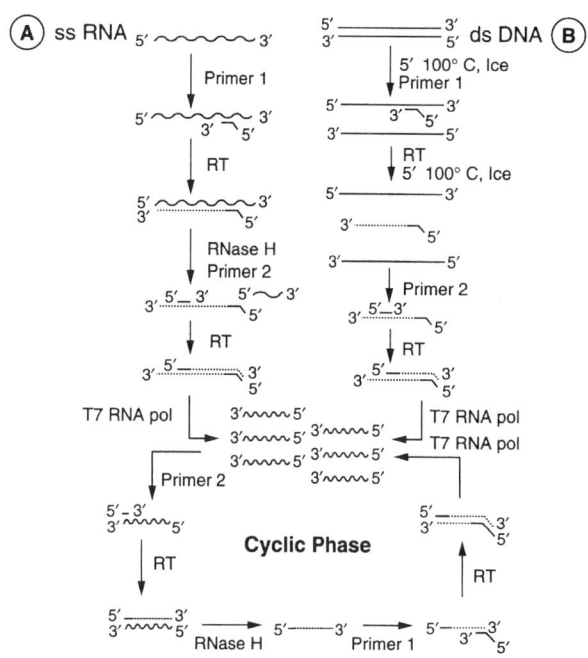

FIGURE 2 NASBA amplification pathways for single-stranded (ss) RNA (A) and double-stranded (ds) DNA (B). RT, avian myeloblastosis virus RT; pol, T7 RNA polymerase; wavy lines, RNA; dashed lines, newly synthesized DNA; solid lines, primers. (Provided by Organon-Teknika.)

merase. In this way exponential amplification of the targeted RNA sequence occurs.

TMA is similar in concept to NASBA but employs only two enzymes, an RT and T7 RNA polymerase. In TMA, the RT degrades the initial RNA template as it synthesizes the cDNA.

The NucliSens HIV-1 QT assay (Organon Teknika, Boxtel, The Netherlands) for quantitation of HIV-1 RNA in plasma and other specimens is based on NASBA (Van Gemen et al., 1993). The nucleic acids are extracted from specimens by using guanidinium thiocyanate and then purified by adherence to acidified silica. The extraction procedure is very robust and can be used to extract and purify nucleic acids from a variety of clinical specimens (Boom et al., 1990). An instrument that automates the sample preparation steps is available.

The primers in this assay target the *gag* region and amplify the target RNA, as well as the three internal calibrator RNA molecules that are included in each reaction. The internal calibrators are present at low, medium, and high copy numbers and are added to the sample prior to nucleic acid extraction to control for sample-to-sample variation of nucleic acid extraction and amplification. Each calibrator has a unique internal sequence that allows for discrimination of the calibrator and the target amplicons with specific probes that are labeled with ruthenium. The amplicons from the target and the three calibrators, Qa, Qb, and Qc, are hybridized to capture oligonucleotides bound to magnetic beads and are detected by hybridization to specific ruthenium-labeled probes in four separate microtubes (Fig. 3). The magnetic beads with the amplicon-probe complex are attached to the surface of an electrode with a magnet, and a voltage applied to this electrode triggers an electrochemiluminescent reaction. The light emitted by the

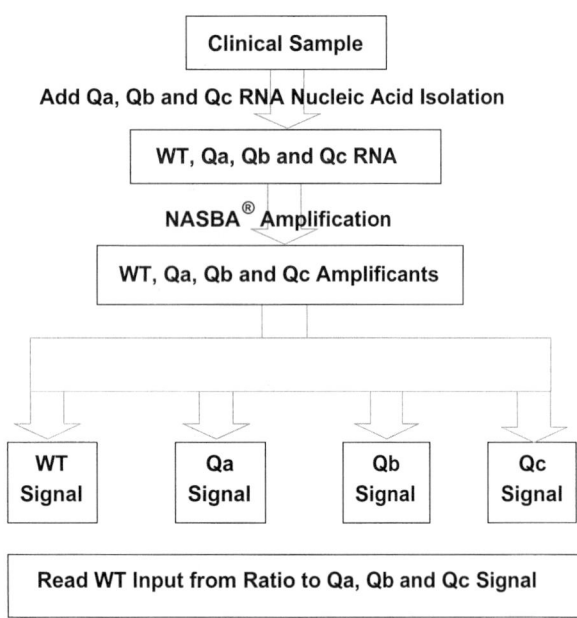

FIGURE 3 Principle of viral nucleic acid quantitation with NASBA. WT, wild-type sequence; Q, calibrator; ECL, electrochemiluminescence. (Provided by Organon-Teknika.)

ruthenium label is proportional to the amount of amplicon. The original starting concentration of the target is calculated from the relative amounts of the four amplicons.

bDNA

The bDNA signal amplification system is a solid-phase sandwich hybridization assay incorporating multiple sets of synthetic oligonucleotide probes and several simultaneous hybridization steps (Nolte, 1999). Multiple target-specific

probes (five to nine, depending upon the assay), termed capture extenders, are used to capture the target nucleic acid (DNA or RNA) onto the surface of a microtiter well plate (Fig. 4). A second set of target-specific probes (18 to 39, depending upon the assay), termed label extenders, hybridize to the target and, in the first-generation assays, also serve as binding sites for the synthetic bDNA amplifier molecules. The amplifier molecules each have 15 identical arms, each of which can bind three alkaline phosphatase-labeled probes. As many as 3,000 enzyme-labeled probes can be hybridized to each target molecule in this manner. Detection of the bound labeled probes is achieved by incubating the complex with an enzyme-triggerable chemiluminescent substrate, dioxetane, and measuring the light emission. Since the number of target molecules is not altered, the resulting signal is directly proportional to the concentration of the target nucleic acid. The quantity of the target in the sample is determined from a standard curve.

In the second- and third-generation bDNA assays, a preamplifier molecule is used to further increase the number of labeled probes that can be bound to the target (Fig. 4). The label extender probes are designed such that two probes must be bound to adjacent regions of the target for efficient hybridization to the preamplifier molecule to occur. Each preamplifier molecule can bind multiple amplifier molecules. In the second-generation bDNA assay for HIV-1 RNA, each captured RNA molecule may bind as many as 10,080 separate alkaline phosphatase-labeled probes (Kern et al., 1996). The lowest concentration of HIV-1 RNA that could be distinguished from the negative control in this assay was 390 copies/ml.

The first and second generations of the bDNA assays were limited by nonspecific hybridization between the amplification probes and other nucleic acids. Short regions of hybridization between any of the probes composing the amplification system (preamplifier, amplifier, and labeled probe) and any nontarget nucleic acid sequence lead to amplification of the background signal. Capture probes, capture extenders, and the sample nucleic acid are all

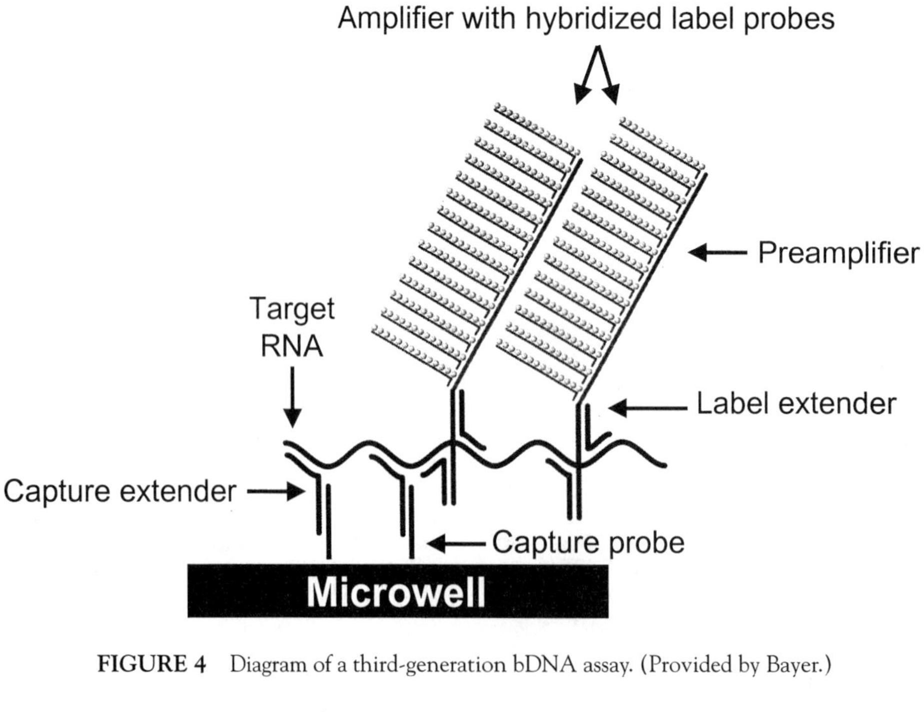

Amplifier with hybridized label probes

← Preamplifier

Target RNA

← Label extender

Capture extender →

← Capture probe

Microwell

FIGURE 4 Diagram of a third-generation bDNA assay. (Provided by Bayer.)

sources of this background hybridization (Collins et al., 1997).

In order to reduce the potential for hybridization to all nontarget nucleic acids, the nonnatural bases isocytidine (isoC) and isoguanosine (isoG) were incorporated into the amplification probes in the third-generation assays. IsoC and isoG are among the 6 base pairs capable of forming Watson and Crick base pairs joined by mutually exclusive hydrogen-bonding schemes incorporating three hydrogen bonds (Piccirilli et al., 1990; Switzer et al., 1993). Although isoC and isoG can be incorporated into duplex DNA and RNA by DNA and RNA polymerases, these potential extra letters in the genetic alphabet are not found in natural oligonucleotides. The isoC and isoG bases pair with each other but not with any of the four naturally occurring bases. Sequences containing isoC-isoG base pairs are more thermally stable than their C-G congeners. Procedures have been described for using an automated DNA synthesizer to prepare derivatives of these isobases suitable for incorporation into DNA (Switzer et al., 1993).

In the third-generation assay for HIV-1, approximately every fourth nucleotide of the preamplifier, amplifier, and alkaline phosphatase-labeled probe was either isoC or isoG, which attenuates the hybridization of these molecules to natural sequences (Collins et al., 1997). The use of isoC- and isoG-containing probes increased the target-specific amplification without a concomitant increase in the background from nontarget sequences, thereby greatly enhancing the sensitivity to a detection limit of approximately 50 molecules/ml.

Recently, Bayer introduced the System 340 platform that automates incubation, washing, reading, data processing, and report generation. It is anticipated that automation will dramatically reduce operator-to-operator differences while decreasing labor requirements. In one laboratory, the average coefficient of variation was reduced 43% and the hands-on labor was reduced 39% with the automated platform for the HIV-1 RNA 2.0 assay (Bayer, unpublished data). Future improvements in reproducibility and operational efficiency await the development of a system for automated sample preparation.

Hybrid Capture Assay

Hybrid capture assays employ a signal amplification technology that can be applied to the detection and quantitation of DNA or RNA target molecules without amplification (Cope et al., 1997c; Mazzulli et al., 1999). This technology depends upon the formation of DNA-RNA hybrid molecules and uses antibodies specific for these hybrids to capture and detect them. A large, single-stranded RNA probe is added to the specimen containing the target DNA. The DNA-RNA hybrids are captured by antibodies specific for these hybrids coating a solid support. Bound hybrids are then reacted with alkaline phosphatase-conjugated antibodies specific for DNA-RNA hybrids. The enzyme-conjugated antibodies are detected by the addition of a chemiluminescent substrate (Fig. 5). Because each DNA-RNA hybrid binds approximately 1,000 antibody conjugate molecules, each of which is labeled with three alkaline phosphatase molecules, the resulting signal is amplified at least 3,000-fold. The light output is measured with a luminometer, and the intensity of the emitted light is proportional to the concentration of target DNA in the specimen. For quantitative determinations, a series of standards are run in parallel, allowing the generation of a standard curve that is used to quantitate the amount of target nucleic acid in the specimen.

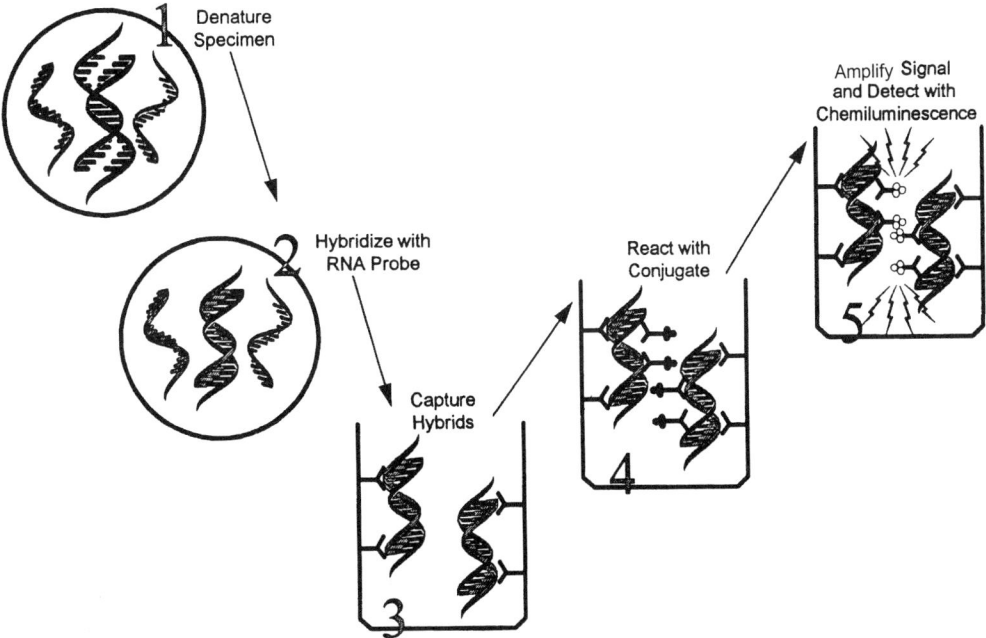

FIGURE 5 Diagram of a hybrid capture assay. The specimen is denatured (1) and then hybridized with an RNA probe (2). If the target nucleic acid is present, the resulting DNA-RNA hybrid is captured by an antibody specific for the hybrid (3). Multiple alkaline-phosphatase-conjugated antibodies bind to the captured hybrids (4). A chemiluminescent substrate emits light that is measured in a luminometer (5). (Provided by Digene.)

Key Assay Performance Issues

There are a number of key performance characteristics that affect the results, use, and interpretation of quantitative nucleic acid assays (Dailey and Hayden, 1999). All of these characteristics should be considered when comparing different assays and technologies.

Linearity

The extent to which assay results are proportional to the concentration of analyte is termed assay linearity. Linearity is determined by testing samples containing multiple levels of analyte and plotting the assay results against the expected results. The extent to which this plot approximates a straight line is a measure of the assay linearity. The linear range of an assay is the concentration range over which the assay quantitates with acceptable accuracy, precision, and linearity. The linear range should include levels of viral load commonly seen in patients.

Accuracy

The accuracy of an assay is the extent to which it agrees with the true value or an accepted reference standard. Unfortunately, accepted reference standards for viral nucleic acids are not widely available. One important exception is HCV RNA. A World Health Organization international standard for nucleic acid amplification assays has been established (Saldanha et al., 1999). The standard has been assigned a value of 10^5 IU/ml and is available from the National Institute for Biological Standards and Control, South Mims, United Kingdom.

The calibrators used in different quantitative nucleic acid assays are characterized by using different techniques, and as a result, the number of virus copies reported by one assay may not agree with that reported by another assay. The availability of accepted reference standards should help establish the accuracies of the various assays and permit more comparison between methods. Because different viral-load assays currently use different standards, the same assay should be used for longitudinal monitoring of patients. If clinical decisions are based on specific viral-load levels published in the literature, remember that those numbers may only apply to the particular assay used in that study.

Precision

Agreement between replicate measurements of the same sample is termed precision. Precision is a gauge of the variability inherent in an assay due to error or noise. The precisions of viral-load assays have been reported using a variety of different statistical measures, including standard deviation, coefficient of variation, range, and fold difference (Dailey and Hayden, 1999). Regardless of which measure is used, viral-load assay variability is best analyzed after $\log_{10}$ transformation of the data. Log transformation of viral-load data is important because viral replication is a multiplicative, not an additive, process. In addition, the parametric statistical tests often used to describe and analyze viral-load data cannot be used with untransformed data because without log transformation the data are not normally distributed and variability is not homogenous (Dailey and Hayden, 1999).

Tolerance Limit

The tolerance limit of an assay is the difference between two sequential patient samples that can be considered to be significantly different within a certain confidence interval, usually 95% (Patel, 1986). The tolerance limit should be a function of both the biological variability of the viral load and the precision of the assay. The tolerance limit for HIV-1 RNA assays is generally accepted to be 0.5 $\log_{10}$ copies of HIV-1 RNA. As a result, only changes greater than 0.5 $\log_{10}$ (3.2-fold) in the HIV-1 viral load are considered significant.

Lower Limit of Quantitation

The lower limit of quantitation is the lowest concentration of analyte that can be quantitated with acceptable accuracy and precision. Viral-load assays tend to be less precise at or near the limit of quantitation. As a result, the significance of changes in the viral load at the lowest levels of quantitation should be evaluated carefully.

The limit of detection is the lowest concentration of analyte that can be discriminated for background but not necessarily quantitated accurately. Limit of detection and analytical sensitivity are terms that are used interchangeably. A viral-load assay may accurately quantitate levels as low as 500 copies/ml but may detect samples as positive with as few as 50 copies/ml.

Specificity

The specificity of an assay is the extent to which it measures only the analyte it was intended to measure. The specificities of viral-load assays are evaluated by testing specimens from patients without evidence of infection with that virus, usually seronegative individuals. Viral-load assays are typically used to evaluate patients for whom the diagnosis is already established. These tests have not been designed as diagnostic tests and have not been tested extensively on seronegative populations. The specificities of HIV-1 RNA assays range from 95 to 98% (Dailey and Hayden, 1999). False positives in signal amplification assays can result from samples with high background, and in target amplification assays, they can result from amplicon product contamination. Inappropriate use of these assays has led to misdiagnoses (Rich et al., 1999).

Subtype Genetic Variation

A key component of all viral-load assays is the hybridization of oligonucleotide probes or primers to the target nucleic acid. Two of the important targets for viral-load assays, HIV-1 and HCV, are RNA viruses that display impressive genetic heterogeneity. Probe and primer design must take into account the inherent genetic variability of the target if the assay is to provide equal quantitation of all subtypes (Kwok et al., 1990). Primer mismatches at critical positions in the target may not completely prevent amplification in a PCR assay, but they can substantially reduce the efficiency of amplification, leading to results that are much lower than anticipated (Vandamme et al., 1996; Dunne and Crowe, 1997; Hawkins et al., 1997).

APPLICATIONS

HIV-1

When HIV-1 enters a new host, there is typically a burst of viremia that is then controlled by the onset of immune responses. The subsequent level of virus in the plasma is a reflection of the equilibrium reached between the virus and the host after the initial battle and is generally maintained for years. This steady-state level varies from individual to individual and is predictive of long-term clinical outcome in the absence of effective antiretroviral therapy. This was first demonstrated in 1996 as part of a multicenter AIDS cohort study (Mellors et al., 1996). In that study, investigators measured the viral load in stored plasma samples collected from 180 untreated HIV-1-infected men and traced the fates of those patients. The risk of AIDS and death in study subjects was directly related to the viral load at study entry. Plasma viral load was a better predictor of progression to AIDS and death than the number of CD4$^+$ lymphocytes. These results established the fact that viral load critically influences disease progression and suggested that lowering virus levels as much as possible and as early as possible with therapy may be essential to prolonging life. Subsequent studies have both confirmed that viral load influences disease progression and changed the way new therapies are evaluated (Coombs et al., 1996; Katzenstein et al., 1996; O'Brien et al., 1996).

In addition to determining prognosis, viral-load data are also useful in monitoring therapeutic response in patients receiving antiretroviral therapy. Multiple analyses of more than 5,000 patients who participated in 18 clinical drug trials with viral-load monitoring demonstrated a reproducible dose-response association between decreases in the magnitude of plasma viremia and improved clinical outcome, based on standard endpoints of new AIDS-defining diagnoses, and survival (Centers for Disease Control and Prevention, 1998) This relationship occurred over a range of patient baseline characteristics, including pretreatment viral load, CD4$^+$-cell count, and prior antiretroviral drug experience.

Viral-load testing has become the essential parameter in guiding decisions to begin or change antiretroviral therapy (Carpenter et al., 1997). Plasma HIV-1 RNA levels should be determined at the time of diagnosis and every 3 to 4 months thereafter in untreated patients. Viral load should also be assessed immediately prior to and again 4 to 8 weeks after beginning antiretroviral therapy. The initial efficacy of the therapy can be assessed at the second time point because in most patients a potent antiretroviral therapy should result in a large decrease (0.5 to 0.75 log$_{10}$ units) in the viral load by 4 to 8 weeks. In most patients, the viral load will continue to decline and should become undetectable (<500 copies/ml) by 12 to 16 weeks of therapy. Failure to achieve a significant decrease in the viral load as described above should prompt the clinician to reassess the patient and/or consider a change in the drug regimen.

Once the patient is on therapy, the viral load should be measured every 3 to 4 months to evaluate the continuing efficacy of the therapy. If the viral load remains above 500 copies/ml after 6 months of therapy, the result should be confirmed, and if confirmed, a change in therapy should be considered. More-sensitive viral-load assays are now available that can quantitate HIV RNA down to a level of 50 copies/ml. Data from clinical trials strongly suggest that lowering the plasma viral load to below 50 copies/ml is associated with more complete and durable virus suppression than that produced by reducing the viral load to between 50 and 500 copies/ml (Gulick et al., 1997).

HCV

Quantitative determinations of HCV RNA in sera have been used to increase our understanding of the natural history of chronic HCV infection, predict which patients are likely to respond to therapy, and determine whether a virological response to therapy has occurred. HCV viral load is relatively stable without significant fluctuation in untreated patients with chronic hepatitis C (Nguyen et al., 1996). However, unlike in HIV-1 infection, the HCV load does not correlate with the severity of disease or the prognosis (National Institutes of Health, 1997). High HCV load is an important risk factor in mother-to-child transmission of the virus. Although the overall risk of transmission of HCV from infected mother to child is low (approximately 5%), the risk of transmission increases with high levels (≥10^6 copies/ml) of maternal viremia (Ohto et al., 1994).

Rates of response to a course of standard therapy with interferon and ribavirin are higher in patients with low levels of HCV RNA (Davis et al., 1998; McHutchison et al., 1998; Poynard et al., 1998). Low-level HCV viremia is usually defined as $<2 \times 10^6$/ml. Although viral load is an independent predictor of response to therapy, viral genotype is a stronger predictor than viral load, and the level of viremia should not be used as a reason to deny treatment (Lok and Gunaratnam, 1997; National Institutes of Health, 1997; Anonymous, 1999).

Response to therapy in hepatitis C is defined biochemically as the normalization of alanine aminotransferase levels and virologically as absence of detectable HCV RNA in the blood. Typically, end-of-treatment responses are assessed by a sensitive qualitative rather than quantitative assay. Although there is no consensus on the use of quantitative HCV tests to monitor patients on therapy, viral-load monitoring may provide early information on the likelihood of therapeutic response.

HBV

Approximately 5% of individuals infected with HBV become chronic carriers of the HBV surface antigen (HBsAg). Chronic HBsAg carriers can be divided into two groups: those with low-level viral replication and normal liver function tests and those with active viral replication and progressive liver disease. The secretory version of the HBV core protein, the e antigen (HBeAg), has traditionally served as a marker for active viral replication. However, the level of HBV DNA in serum or plasma better reflects the replicative activity of HBV than the presence or absence of HBeAg.

Monitoring levels of HBV DNA in serum may be useful for identifying individuals most likely to respond to antiviral therapy, evaluating the efficacy of therapy, and following viral load after therapy (Kuhn et al., 1989; Perrillo et al., 1990; Zoulim et al., 1992; Hendricks et al., 1995; Ranki et al., 1995). During therapy, the clinician can assess its efficacy by measuring either the absolute reduction or the kinetics of decrease in the viral load.

CMV

CMV load determinations have been used to predict the subsequent development or relapse of CMV disease before the onset of symptoms in solid-organ transplant recipients, bone marrow transplant recipients, and AIDS patients. There is substantial evidence that weekly monitoring of CMV load during the first 3 months after transplant is useful to predict CMV disease in solid-organ transplant recipients (Cope et al., 1997a, 1997b; Toyoda et al., 1997). However, the breakpoints for DNA copy number that predict the development of CMV disease in different patient populations are not well defined. Available data on DNA copy numbers from different laboratories are difficult to compare because of differences in the methods used to quantitate CMV DNA (Boeckh and Boivin, 1998).

There is only a moderate association between high systemic viral load and CMV disease in allogeneic bone marrow transplant recipients (Zaia et al., 1997; Gor et al.,

1998). A significant portion of patients may develop CMV disease with low viral load or progress rapidly from low DNA levels to overt disease. Severe acute graft-versus-host disease requiring treatment with steroids is the clinical setting in which rapid progression to CMV disease may occur. Quantitative molecular methods are probably not needed to monitor CMV-seropositive autograft recipients due to the low incidence of CMV disease in these patients.

In HIV-1-infected individuals there is an association between a high CMV load and CMV disease (Boivin et al., 1997b; Rasmussen et al., 1995). The CMV load also predicts the development of CMV disease (Bowen et al., 1997; Shinkai et al., 1997) and response to antiviral therapy (Spector et al., 1996). In addition, a high CMV load is an independent predictor of poor survival in most studies (Spector et al., 1998). A variety of samples, including whole blood, plasma, and leukocytes, as well as a variety of molecular methods, have been used for the viral-load determinations, and it is not known which sample or method is optimal.

A decline in systemic CMV load occurs after initiation of effective antiviral therapy in solid-organ transplant recipients, bone marrow transplant recipients, and HIV-1-infected individuals (Boeckh and Boivin, 1998). Viral-load testing may also be used to monitor the emergence of drug-resistant CMV strains and the progression of CMV disease (Boivin et al., 1996, 1997a; Smith et al., 1996).

There has also been considerable interest in using quantitative nucleic acid tests to assess CMV load in the central nervous system, lungs, and eyes. CMV quantitation has been used to study the viral load in cerebrospinal fluid in the different central nervous system manifestations of CMV infection in AIDS patients. Patients with CMV polyradiculopathy have higher viral loads in the cerebrospinal fluid than patients with CMV encephalitis (Shinkai et al., 1997). AIDS patients with CMV encephalitis had 10- to 1,000-fold-higher CMV loads in autopsy brain specimens than did patients without histologic evidence of encephalitis (Kuhn et al., 1995). Quantitative nucleic acid tests can distinguish asymptomatic pulmonary shedding of CMV from CMV pneumonia in immunocompromised patients (Boivin et al., 1996). Testing for CMV DNA in the eyes may provide a virological measurement of the response to therapy in patients with low or undetectable systemic viral load and may provide a means to study the pathogenesis of CMV retinitis. The aqueous humor provides an accessible sample for DNA quantitation and may reflect viral load in the eyes of AIDS patients (Gerna et al., 1994; S. Mitchell, Editorial, *Br. J. Ophthalmol.* **80:**195–196, 1996).

Other Herpesviruses

Epstein Barr virus (EBV) load may be a useful prognostic marker for development of EBV-related posttransplant lymphoproliferative disorders (PTLD) and to monitor the effects of appropriate interventions (Riddler et al., 1994; Rowe et al., 1997). EBV-related PTLD occurs in 1 to 10% of transplant recipients and involves expansions of B or T cells ranging from reactive hyperplasias to large-cell lymphomas. PTLD can be rapidly fatal if not accurately diagnosed and often responds to decreased immunosuppression and reconstitution of the immune system.

Quantitative nucleic acid assays may also help establish disease associations for other ubiquitous herpesviruses, including human herpesviruses 6 and 7. Similar to CMV, the high rates of reactivation and asymptomatic excretion of these viruses make it difficult to relate infection to disease without using viral load as a measure of active virus replication.

ADVANTAGES AND DISADVANTAGES

Prior to the development of quantitative molecular techniques, laboratories were limited to cumbersome culture methods or insensitive antigen detection methods to determine viral load. The molecular methods are more rapid and cost-effective than conventional virological methods and have created new opportunities for the clinical laboratory to influence patient care. The molecular approach is the only approach available for the detection and quantitation of medically important viruses like HCV and is the only practical approach for quantitation of the other medically important viruses.

Each of the quantitative molecular techniques has particular strengths and limitations that are inherent in the underlying nucleic acid amplification strategies. The major commercial viral-load assays employ either target or signal amplification. Target amplication systems typically employ a single pair of primers, each of which is usually 20 to 40 bases in length. Mismatches between the primer and target sequence can lead to failure to amplify the target or to inefficient amplification, depending on the number and positions of the mismatches. Primers must be carefully selected from highly conserved regions to ensure equal amplification of all genotypes. Signal amplification assays employ either numerous probes or large probes that cover large regions of the target genome. In practice, signal amplification assays are less prone to errors resulting from target sequence heterogeneity.

Target amplification methods are more sensitive than signal amplification methods, with limits of detection in the 20- to 50-copy/ml range for the most sensitive PCR-based assays. These methods achieve this level of sensitivity through producing billions of copies of the target nucleic acid. A major challenge in developing quantitative target amplification assays has been in establishing the relationship between the initial amount of the target sequence in the specimens and the amount of amplified product. As a result, target amplification methods tend to be less precise than signal amplification methods. The number of target molecules is not altered with signal amplification methods. Therefore, the amount of signal is directly proportional to the amount of target sequence present in the clinical sample.

Target amplification methods are enzymatic processes that are prone to sample inhibition and, as a consequence, employ multistep sample preparation protocols to partially purify nucleic acids. Sample preparation is less cumbersome for signal amplification assays, since they are less prone to sample inhibition.

False-positive reactions are a concern with both target and signal amplification assays but for very different reasons. The tremendous numbers of product molecules produced in target amplification assays can be difficult to contain. Physical separation of pre- and postamplification activities, unidirectional work flow, plugged pipette tips, physical containment, and enzymatic and chemical methods for amplicon inactivation are all used to limit the occurrence of false positives due to amplicon cross contamination (Kwoh et al., 1989). None of these practices is necessary with signal amplification systems because the target molecules are not amplified. False positives occur in signal amplification assays due to sample matrix effects that can lead to high background counts. The reported false-positive rates for the various bDNA assays are 2 to 5%.

TIPS

The commercially available viral-load assays all have their particular strengths and limitations. No single viral-load assay is ideal for every laboratory setting. When choosing among the available assays, many issues should be considered, including the types of patients and specimens to be tested, available work space, technical skills of the laboratory staff, work flow, volume of testing, and time to results. Table 1 compares the features of the major commercially available quantitative HIV-1 RNA assays.

FUTURE

Viral-load testing in HIV-1 infection is the best example of how quantitative molecular methods can increase our understanding of a disease process and improve patient care. Viral-load assays are likely to become the standard of care in the management of patients with other chronic viral infections for which therapy is not optimal. The best candidates on the near horizon are HCV and HBV.

Data are accumulating on the use of CMV load testing in the diagnosis and monitoring of solid-organ transplant recipients and AIDS patients. Viral-load testing will also be used more to distinguish active from latent infections with other members of the herpesvirus family in immunocompromised patients.

On the technology front, there will be increased use of automation in viral-load assays (see chapter 22). Significant progress has already been made in automating the amplification and detection steps. Sample preparation remains the greatest challenge for automation. With the development of real-time PCR and other strategies that combine amplification and detection in a single closed vessel, quantitative molecular methods will become less technically demanding and faster. These methods also eliminate many of the concerns about false-positive results due to amplicon cross contamination.

CONCLUSIONS

Viral-load assays have created new opportunities for the clinical laboratory to influence patient care. The clinical applications of these assays are increasing and are no longer limited to HIV-1 infections. A thorough understanding of the key performance issues and features of the available quantitative molecular techniques is essential to the practice of modern clinical virology.

REFERENCES

Boeckh, M., and G. Boivin. 1998. Quantitation of cytomegalovirus: methodologic aspects and clinical applications. *Clin. Microbiol. Rev.* **11:**533–554.

Boivin, G., S. Chgou, M. R. Quirk, A. Erice, and M. C. Jordan. 1996. Detection of ganciclovir resistance mutations quantitation of cytomegalovirus (CMV) DNA in leukocytes of patients with fatal disseminated CMV disease. *J. Infect. Dis.* **173:**523–528.

Boivin, G., C. Gilbert, M. Morissette, J. Handfield, N. Goyette, and M. G. Bergeron. 1997a. A case of ganciclovir-resistant cytomegalovirus (CMV) retinitis in a patient with AIDS: longitudinal molecular analysis of the CMV viral load and viral mutations in blood compartments. *AIDS* **11:**867–873.

Boivin, G., J. Handfield, G. Murray, E. Toma, R. Lalonde, J. G. Lazar, and T. G. Bergeron. 1997b. Quantitation of cytomegalovirus (CMV) DNA in leukocytes of human immunodeficiency virus-infected subjects with and without CMV disease by using PCR and the SHARP signal detection system. *J. Clin. Microbiol.* **35:**522–526.

Boom, R., C. Sol, M. Salimans, C. Jansen, P. M. Wertheim-van Dillen, and J. van derNoordaa. 1990. Rapid and simple method for purification of nucleic acids. *J. Clin. Microbiol.* **28:**495–503.

Bowen, E. F., C. A. Sabin, P. Wilson, P. D. Griffiths, C. C. Davey, M. A. Johnson, and V. C. Emery. 1997. Cytomegalovirus (CMV) viraemia detected by polymerase chain reaction identifies a group of HIV-positive paitents at high risk of CMV disease. *AIDS* **11:**889–893.

Brillanti, C., J. A. Garson, P. W. Tuke, C. Ring, M. Briggs, C. Masci, M. Miglioli, L. Barbara, and R. S. Tedder. 1991. Effect of α-interferon therapy on hepatitis C viraemia in community acquired chronic non-A, non-B hepatitis: a quantitative polymerase chain reaction study. *J. Med. Virol.* **34:**136–141.

Carpenter, C. C. J., M. A. Fischl, S. M. Hammer, M. S. Hirsch, D. M. Jacobsen, D. A. Katzenstein, J. S. G. Montaner, D. D. Richman, M. S. Saag, R. T. Schooley, M. A. Thompson, S. Vella, P. G. Yen, and P. A. Volberding. 1997. Antiretroviral therapy for HIV infection in 1997. (Updated recommendations of the International AIDS Society-USA Panel). *JAMA* **277:**1962–1969.

TABLE 1   Comparison of the major commercially available quantitative HIV-1 RNA assays

Assay feature	Roche	Bayer	Organon-Teknika
Amplification method	RT-PCR	bDNA	NASBA
Detection method	Colorimetric	Chemiluminescence	Electrochemiluminescence
Gene target	*gag*	*pol*	*gag*
Specimen(s)	Plasma	Plasma	Plasma and other fluids
Sample vol (ml)	0.2–0.5	1–2	0.1–1
Anticoagulant(s)	EDTA; ACD	EDTA	All
No. of standards	1 internal	5 external	3 internal
Dynamic range (copies/ml)	400–750,000 (standard) 50–50,000 (ultrasensitive)	500–1,000,000 (version 2.0) 50–50,000 (version 3.0)	80–10,000,000
Intra-assay SD ($\log_{10}$)	0.15–0.33	0.12–0.32	0.13–0.23
Kit (batch) size (no. of tests)	24 (≥12)	96 (≥18)	50 (≥10)
Turnaround time	1 or 2 days	2 days	1 or 2 days

Centers for Disease Control and Prevention. 1998. Report of the NIH panel to define principles of therapy of HIV infection and guidelines for the use of antiretroviral agents in HIV-infected adults and adolescents. *Morb. Mortal. Wkly. Rep.* **47(RR-5):**1–41.

Clementi, M., S. Menzo, P. Bagnarelli, A. Manzin, A. Valenza, and P. E. Varaldo. 1993. Quantitative PCR and RT-PCR in virology. *PCR Methods Appl.* **2:**191–196.

Collins, M. L., B. Irvine, D. Tyner, E. Fine, C. Zayati, C. Chang, T. Horn, D. Ahle, J. Detmer, L.-P. Shen, J. Kolbert, S. Bushnell, M. S. Urdea, and D. D. Ho. 1997. A branched DNA signal amplification assay for quantification of nucleic acid targets below 100 molecules/ml. *Nucleic Acids Res.* **25:**2979–2984.

Compton, J. 1991. Nucleic acid sequence-based amplification. *Nature* **350:**91–92.

Anonymous. 1999. EASL International Consensus Conference on Hepatitis C. Paris, 26–28 February 1999, Cousensus statement. European Association for the study of the Liver. *J. Hepatol.* **30:**956–961.

Coombs, R. W., S. L. Welles, C. Hooper, P. S. Reichelderfer, R. T. D'Aquila, A. J. Japour, V. A. Johnson, D. R. Kuritzkes, D. D. Richman, S. Kwok, J. Todd, J. B. Jackson, V. DeGruttola, C. S. Crumpacker, and J. Kahn. 1996. Association of plasma human immunodeficiency virus type-1 RNA level with risk of clinical progression in patients with advanced infection. *J. Infect. Dis.* **174:**704–714.

Cope, A. V., C. Sabin, A. Burroughs, K. Rolles, P. D. Griffiths, and V. C. Emery. 1997a. Interrelationships among quantity of human cytomegalovirus (HCMV) DNA in blood, donor-recipient serostatus, and administration of methylprednisolone as risk factors for HCMV disease following liver transplantation. *J. Infect. Dis.* **176:**1484–1490.

Cope, A. V., P. Sweny, C. Sabin, L. Rees, P. D. Griffiths, and V. C. Emery. 1997b. Quality of cytomegalovirus viruria is a major risk factor for cytomegalovirus disease after renal transplantation. *J. Med. Virol.* **52:**200–205.

Cope, J. J., A. Hildesheim, M. H. Schiffman, M. M. Manos, A. T. Lorincz, R. D. Burk, A. G. Glass, C. Greer, J. Burkland, K. Helgesen, D. R. Scott, M. E. Sherman, R. J. Kurman, and K.-L. Liaw. 1997c. Comparison of the hybrid capture tube test and PCR for detection of human papillomavirus DNA in cervical specimens. *J. Clin. Microbiol.* **35:**2262–2265.

Dailey, P. J., and D. Hayden. 1999. Viral load assays: methodologies, variables, and interpretations, p. 119–129. *In* P. T. Cohan, M. A. Sande, and P. A Volberding (ed.), *The AIDS Knowledge Base*, 3rd ed. Lippincott-Raven Publishers, Philadelphia, Pa.

Davis, G. L., R. Esteban-Mur, V. Rustgi, J. Hoefs, S. C. Gordon, C. Trepo, M. L. Shiffman, S. Zeuzem, A. Craxi, M.-H. Ling, and J. Albrecht. 1998. Interferon alpha-2b alone or in combination with ribavirin for the treatment of relapse of chronic hepatitis C. International Hepatitis Interventional Therapy Group. *N. Engl. J. Med.* **339:**1493–1499.

Dunne, A. L., and S. M. Crowe. 1997. Comparison of branched DNA and reverse transcriptase polymerase chain reaction for quantifying six different HIV-1 subtypes in plasma. *AIDS* **11:**2–3.

Gerna, G., F. Baldanti, A. Sarasini, M. Furione, E. Percivalle, M. G. Revello, D. Zipeto, and D. Zella. 1994. Effect of foscarnet induction treatment on quantitation of human cytomegalovirus DNA in bone marrow transplant recipients. *Br. J. Haematol.* **91:**674–683.

Gilliland, G., S. Perrin, K. Blanchard, and H. F. Bunn. 1990. Analysis of cytokine mRNA and DNA: detection and quantitation by competitive polymerase chain reaction. *Proc. Natl. Acad. Sci. USA* **87:**2725–2729.

Gor, D., C. Sabin, H. G. Prentice, N. Vyas, S. Man, P. D. Griffiths, and V. C. Emery. 1998. Longitudinal fluctuations in cytomegalovirus load in bone marrow transplant patients: relationship between peak virus load, donor/recipient serostatus, acute GVHD and CMV disease. *Bone Marrow Transplant.* **21:**597–605.

Gulick, R. M., J. W. Mellors, D. Havlir, J. J. Eron, C. Gonzalez, D. McMahon, D. D. Richman, F. T. Valentine, L. Jonas, A. Meibohm, E. A. Emini, and J. A. Chodakewitz. 1997. Treatment with indinavir, zidovudine, and lamivudine in adults with human immunodeficiency virus infection and prior antiretroviral therapy. *N. Engl. J. Med.* **337:**734–739.

Haberhausen, G., J. Pinsl, C.-C. Kuhn, and C. Markert-Hahn. 1998. Comparative study of different standardization concepts in quantitative competitive reverse transcription-PCR assays. *J. Clin. Microbiol.* **36:**628–633.

Hawkins, A., F. Davidson, and P. Simmonds. 1997. Comparison of plasma virus loads among individuals infected with hepatitis C virus (HCV) genotypes 1, 2, and 3 by quantiplex HCV RNA assay versions 1 and 2, Roche monitor assay, and an in-house limiting dilution method. *J. Clin. Microbiol.* **35:**187–192.

Heid, C., J. Stevens, K. J. Livak, and P. Mickey Williams. 1996. Real time quantitative PCR. *Genome Res.* **6:**986–994.

Hendricks, D. A., B. J. Stowe, B. S. Hoo, J. Kolberg, B. D. Irvine, P. D. Neuwald, M. S. Urdea, and R. P. Perrillo. 1995. Quantitation of HBV DNA in human serum using a branched DNA (bDNA) signal amplification assay. *Am. J. Clin. Pathol.* **104:**537–546.

Holland, P. M., R. D. Abramson, R. Watson, and D. H. Gelfand. 1991. Detection of specific polymerase chain reaction product by utilizing the 5′ → 3′ exonuclease activity of *Thermus aquaticus* DNA polymerase. *Proc. Natl. Acad. Sci. USA.* **88:**7276–7280.

Katzenstein, D. A., S. M. Hammer, M. D. Hughes, M. Gundacker, J. B. Jackson, S. Fiscus, S. Rasheed, T. Elbeik, R. Reichman, A. Japour, T. C. Merigan, and M. S. Hirsch. 1996. The relation of virologic and immunologic markers to clinical outcomes after nucleoside therapy in HIV-infected adults with 200 to 500 CD4 cells per cubic millimeter. AIDS Clinical Trials Group Study 175 Virology Study Team. *N. Engl. J. Med.* **335:**1091–1098.

Kellog, D. E., J. J. Sninsky, and S. Kwok. 1990. Quantitation of HIV-1 proviral DNA relative to cellular DNA by the polymerase chain reaction. *Anal. Biochem.* **189:**202–208.

Kern, D., M. Collins, T. Fultz, J. Detmer, S. Hamren, J. J. Peterkin, P. Sheridan, M. Urdea, R. White, T. Yeghiazarian, and J. Todd. 1996. An enhanced-sensitivity branched-DNA assay for quantification of human immunodeficiency virus type 1 RNA in plasma. *J. Clin. Microbiol.* **34:**3196–3202.

Kuhn, J. E., T. Wendland, H. J. Eggers, E. Lorentzen, U. Wieland, B. Eing, M. Kiessling, and P. Gass. 1995. Quantitation of human cytomegalovirus genomes in the brain of AIDS patients. *J. Med. Virol.* **47:**70–82.

Kuhn, M. C., A. L. McNamara, R. P. Perrillo, C. M. Cabal, and C. R. Campbell. 1989. Quantitation of hepatitis B viral DNA by solution hybridization: comparison with DNA polymerase and hepatitis B e antigen during antiviral therapy. *J. Med. Virol.* **27:**274–281.

Kwoh, D. Y., G. R. David, K. M. Whitfield, H. L. Chapelle, L. J. DiMichele, and T. R. Gingeras. 1989. Transcription-based amplification system and detection of amplified human immunodeficiency virus type 1 with a bead-based sandwich hybridization format. *Proc. Natl. Acad. Sci. USA* **86:**1173–1177.

Kwok, S., D. E. Kellogg, N. McKinney, D. Spasic, L. Goda, C. Levenson, and J. J. Sninsky. 1990. Effects of primer-template mismatches on the polymerase chain reaction: human immunodeficiency virus type 1 model studies. *Nucleic Acids Res.* **18**:999–1005.

Lay, M. J., and C. T. Wittwer. 1997. Real-time fluorescence genotyping of factor V leiden during rapid cycle PCR. *Clin. Chem.* **43**:2262–2267.

Loeffelholz, M. J., C. A. Lewinski, S. R. Silver, A. Purohit, S. A. Herman, D. A. Buonagurio, and E. A. Dragon. 1992. Detection of *Chlamydia trachomatis* in endocervical specimens by polymerase chain reaction. *J. Clin. Microbiol.* **30**:2847–2851.

Lok, A. S., and N. T. Gunaratnam. 1997. Diagnosis of hepatitis C. *Hepatology* **26**:48S–56S.

Longo, M. C., M. S. Berninger, and J. L. Hartley. 1990. Use of uracil DNA glycosylase to control carry-over contamination in polymerase chain reactions. *Gene* **93**:125–128.

Mazzulli, T., L. W. Drew, B. Yen-Lieberman, D. Jekic-McMullen, D. J. Kohn, C. Isada, G. Moussa, R. Chua, and S. Walmsley. 1999. Multicenter comparison of the Digene hybrid capture CMV DNA assay (version 2.0), the pp65 antigenemia assay, and cell culture for detection of cytomegalovirus viremia. *J. Clin. Microbiol.* **37**:958–963.

McHutchison, J. G., S. C. Gordon, E. R. Schiff, M. L. Shiffman, W. M. Lee, V. K. Rustgi, Z. D. Goodman, M.-H. Ling, S. Cort, and J. K. Albrecht. 1998. Interferon alfa-2b alone or in combination with ribavirin as initial treatment for chronic hepatitis C. Hepatitis Interventional Therapy Group. *N. Engl. J. Med.* **339**:1485–1492.

Mellors, J. W., C. R. Rinaldo, Jr., P. Gupta, R. M. White, J. A. Todd, and L. A. Kingsley. 1996. Prognosis in HIV-1 infection predicted by the quantity of virus in plasma. *Science* **272**:1167–1170.

Morrison, T., J. J. Weiss, and C. T. Wittwer. 1998. Quantification of low copy transcripts by continuous SYBR green I dye monitoring during amplification. *BioTechniques* **24**:954–958.

Myers, T. W., and D. H. Gelfand. 1991. Reverse transcription and DNA amplification by a *Thermus thermophilus* DNA polymerase. *Biochemistry* **30**:7661–7666.

National Committee for Clinical Laboratory Standards. 1999. *Quantitative Molecular Method for Infectious Diseases. Proposed guideline.* NCCLS document MM6-P. National Committee for Clinical Laboratory Standards; Wayne, Pa.

National Institutes of Health. 1997. Management of Hepatitis C. *NIH Consens. Statement* **15**:1–41.

Nguyen, T., A. Sedghi-Vaziri, L. Wilkes, T. Mondala, P. Pockros, J. Lindsay, and J. McHutchison. 1996. Fluctuations in viral load (HCV RNA) are relatively insignificant in untreated patients with chronic HCV infection. *J. Viral Hepatol.* **3**:75–78.

Nolte, F. S. 1999. Branched DNA signal amplification for direct quantitation of nucleic acid sequences in clinical specimens. *Adv. Clin. Chem.* **33**:201–235.

Noonan, K. E., C. Beck, T. A. Holzmayer, J. E. Chin, J. S. Wunder, I. L. Andrulis, A. F. Gazdar, C. L. Wilman, B. Griffith, D. D. von Hoff, and I. B. Robinson. 1990. Quantitative analysis of MDR1 (multidrug resistance) gene expression in human tumors by polymerase chain reaction. *Proc. Natl. Acad. Sci. USA* **87**:7160–7164.

O'Brien, W. A., P. M. Hartigan, D. Martin, J. Esinhart, A. Hill, S. Benoit, M. Rubin, M. S. Simberkoff, and J. D. Hamilton. 1996. Changes in plasma HIV-1 RNA and CD4+ lymphocyte counts and the risk of progression to AIDS. *N. Engl. J. Med.* **334**:426–431.

Ohto, H., S. Terazawa, N. Sasaki, N. Sasaki, K. Hino, C. Ishiwata, M. Kako, N. Ukue, C. Endo, A. Matsui, H. Okamoto, and S. Mishiro. 1994. Transmission of hepatitis C virus from mothers to infants. The Vertical Transmission of Hepatitis C Virus Collaborative Study Group. *N. Engl. J. Med.* **330**:744–750.

Oka, S., K. Urayama, Y. Hirabayashi, O. Kiyokata, H. Goto, K. Mitamura, S. Kimura, and K. Shimata. 1990. Quantitative analysis of human immunodeficiency virus type 1 DNA in asymptomatic carriers using the polymerase chain reaction. *Biochem. Biophys. Res. Commun.* **167**:1–8.

Patel, J. K. 1986. Tolerance limits—a review. *Common Stat. Theory Method* **15**:2917–2962.

Perrillo, R. P., E. R. Schiff, G. L. Davis, H. C. Bodenheimer, K. Linsay, J. Payne, J. L. Dienstag, C. O'Brien, C. Tamburro, I. M. Jacobson, R. Sampliner, D. Feit, J. Lefkowitch, M. Kuhns, C. Meschievitz, B. Sanghvi, J. Albrecht, and A. Bigas. 1990. A randomized, controlled trial of interferon alfa-2b alone and after prednisone withdrawal for the treatment of chronic hepatitis B. The Hepatitis Interventional Therapy Group. *N. Engl. J. Med.* **321**:295–301.

Piatak, M., Jr., K.-C. Luk, B. Williams, and J. D. Lifson. 1993. Quantitative competitive polymerase chain reaction for accurate quantitation of HIV DNA and RNA species. *Bio Techniques* **14**:70–80.

Piccirilli, J. A., T. Krauch, S. E. Moroney, and S. A. Benner. 1990. Enzymatic incorporation of a new base pair into DNA and RNA extends the genetic alphabet. *Nature* **343**:33–37.

Poynard, T., P. Marcellin, S. Lee, C. Niederau, G. S. Minuk, G. Ideo, V. Bain, J. Heathcote, S. Zeuzem, C. Trepo, and J. Albrecht. 1998. Randomised trial of interferon a2b plus ribavirin for 48 weeks or for 24 weeks versus interferon a2b plus placebo for 48 weeks for treatment of chronic infection with hepatitis C virus. International Hepatitis Interventional Therapy Group. *Lancet* **352**:1426–1432.

Ranki, M., H. M. Schatzl, R. Zachoval, M. Uusi-Oukari, and P. Lehtovaara. 1995. Quantification of hepatitis B virus DNA over a wide range from serum for studying viral replicative activity in response to treatment and the recurrent infection. *Hepatology* **21**:1492–1499.

Rasmussen, L., S. Morris, D. Zipeto, J. Fessel, R. Wolitz, A. Dowling, and T. C. Merigan. 1995. Quantitation of human cytomegalovirus DNA from peripheral blood cells of human immunodeficiency virus-infected patient could predict cytomegalovirus retinitis. *J. Infect. Dis.* **171**:177–182.

Rich, J. D., N. A. Merriman, E. Mylonakis, T. C. Greenough, T. P. Flanigan, B. J. Mady, and C. C. J. Carpenter. 1999. Misdiagnosis of HIV infection by HIV-1 plasma viral load testing: a case series. *Ann. Intern. Med.* **130**:37–39.

Riddler, S. A., M. C. Breinig, and J. L. C. McKnight. 1994. Increased levels of circulating Epstein-Barr virus (EBV)-induced lymphocytes and decreased EBV nuclear antigen antibody responses are associated with the development of post-transplant lymphoproliferative disease in solid-organ transplant recipients. *Blood* **84**:972–984.

Ririe, K., R. P. Rasmussen, and C. T. Wittwer. 1997. Product differentiation by analysis of DNA melting curves during the polymerase chain reaction. *Anal. Biochem.* **245**:154–160.

Rowe, D. T., L. Qu, J. Reyes, N. Jabbour, E. Yunis, P. Putnam, S. Todo, and M. Green. 1997. Use of quantitative competitive PCR to measure Epstein-Barr virus genome load in the peripheral blood of pediatric transplant patients with lymphoproliferative disorders. *J. Clin. Microbiol.* **35**:1612–1615.

Saag, M. S., M. Holodniy, D. R. Kuritzkes, W. A. O'Brien, R. Coombs, M. E. Poscher, D. M. Jacobsen, G. M. Shaw,

D. D. Richman, and P. A. Volberding. 1996. HIV viral load markers in clinical practice. *Nat. Med.* **2:**625–629.

Saiki, R. K., D. H. Gelfand, S. Stoffel, S. J. Scharf, R. Higuchi, K. B. Mullis, G. Horn, and H. A. Ehrlich. 1988. Primer-directed enzymatic amplification of DNA with a thermostable DNA polymerase. *Science* **239:**487–491.

Saldanha, J., N. Lelie, and A. Heath. 1999. Establishment of the first international standard for nucleic acid amplification technology (NAT) assays for HCV RNA. WHO Collaborative Study Group. *Vox Sang.* **76:**149–158.

Salk, J. 1987. Prospects for the control of AIDS by immunizing seropositive individuals. *Nature* **327:**473–476.

Shinkai, M., S. A. Bozzette, W. Powderly, P. Frame, and S. A. Spector. 1997. Utility of urine and leukocyte cultures and plasma DNA polymerase chain reaction for identification of AIDS patients at risk for developing human cytomegalovirus disease. *J. Infect. Dis.* **175:**302–308.

Simmonds, P., P. Balfe, J. Peutherer, C. A. Ludlam, J. O. Bishop, and A. J. Leigh Brown. 1990. Human immunodeficiency virus-infected individuals contain provirus in small numbers of peripheral mononuclear cells at low copy numbers. *J. Virol.* **64:**864–872.

Smith, I. L., M. Shinkai, W. R. Freeman, and S. A. Spector. 1996. Polyradiculopathy associated with ganciclovir-resistant cytomegalovirus in an AIDS patient: phenotypic and genotypic characterization of sequential virus isolates. *J. Infect. Dis.* **173:**1481–1484.

Spector, S. A., G. F. McKinley, J. P. Lalezari, T. Samo, R. Andruczk, S. Follansbee, P. D. Sparti, D. V. Havlir, G. Simpson, W. Buhles, R. Wong, and M. Stempien. 1996. Oral ganciclovir for the prevention of cytomegalovirus disease in persons with AIDS. Roche Cooperative Oral Ganciclovir Study Group. *N. Engl. J. Med.* **334:**1491–1497.

Spector, S. A., R. Wong, K. Hsia, M. Pilcher, and M. J. Stempien. 1998. Plasma cytomegalovirus (CMV) DNA load predicts CMV disease and survival in AIDS patients. *J. Clin. Investig.* **101:**497–502.

Switzer, C. Y., S. E. Moroney, and S. Benner. 1993. Enzymatic recognition of the base pair between isocytidine and isoguanosine. *Biochemistry* **32:**10489–10496.

Toyoda, M., J. B. Carlos, O. A. Galera, K. Galfayan, X. M. Zhang, Z. L. Sun, L. S. C. Czer, and S. C. Jordon. 1997. Correlation of cytomegalovirus DNA levels with response to antiviral therapy in cardiac and renal allograft recipients. *Transplantation* **63:**957–963.

Tyagi, S., D. P. Bratu, and F. R. Kramer. 1998. Multicolor molecular beacons for allele discrimination. *Nat. Biotech.* **16:**49–53.

Vandamme, A.-M., J.-C. Schmit, S. Van Dooren, K. Van Laethem, E. Gobbers, W. Kok, P. Goubau, M. Witvrouw, W. Peetermans, E. De Clercq, and J. Desmyter. 1996. Quantification of HIV-1 RNA in plasma: comparable results with the NASBA HIV-1 RNA QT and the AMPLICOR HIV monitor test. *J. Acquir. Immune Defic. Syndr. Hum. Retrovirol.* **13:**127–139.

Van Gemen, B., T. Kievits, P. Nara, H. G. Huisman, S. Jurriaans, J. Goudsmit, and P. Lens. 1993. Qualitative and quantitative detection of HIV-1 RNA by nucleic acid sequence-based amplification. *AIDS* **7**(Suppl. 2):S107–S110.

Warren, W., T. Wheat, and P. Knudsen. 1991. Rapid analysis and quantitation of PCR products by high performance liquid chromatography. *BioTechniques* **11:**250–255.

Zaia, J. A., G. M. Gallex-Hawkins, B. R. Tegtmeier, A. ter Veer, X. Li, J. C. Niland, and S. J. Forman. 1997. Late cytomegalovirus disease in marrow transplantation is predicted by virus load in plasma. *J. Infect. Dis.* **176:**782–785.

Zoulim, F., L. Mimms, M. Floreani, C. Pichoud, I. Chemin, A. Kay, L. Vitvitski, and C. Trepo. 1992. New assays for quantitative determination of viral markers in management of chronic hepatitis B virus infection. *J. Clin. Microbiol.* **30:**1111–1119.

Flow Cytometry

JAMES McSHARRY

21

INTRODUCTION

Virus-cell interactions have been analyzed for more than 25 years by using fluorochrome-labeled antibodies in conjunction with flow cytometry. In vitro studies of virus-cell interactions include the detection and quantification of (i) the binding of fluorochrome-labeled viruses to their receptors on cell surfaces; (ii) fluorochrome-labeled viral antigens on the surfaces, in the cytoplasm, and in the nuclei of virus-infected cells; (iii) virus-induced apoptosis; and (iv) the effects of virus infection on the expression of cellular antigens and nucleic acid synthesis. In vivo studies have used fluorochrome-labeled monoclonal antibodies and flow cytometry to detect and quantify virus-infected cells directly in clinical specimens. The published literature before 1994 has been reviewed (McSharry, 1994). Since that time, the number of publications reporting on the use of flow cytometry in virology has exploded to the point that a review of the extant literature is beyond the scope of this chapter. Therefore, this chapter will not be inclusive but will describe some of the more recent uses of flow cytometry in virology, including (i) a brief history of flow cytometry; (ii) a description of the principles of flow cytometry and its use in detecting and quantifying human cytomegalovirus (HCMV)-infected tissue culture cells; (iii) its use in studying apoptosis of virus-infected cells; (iv) its use in measuring the effect of virus infection on the cell cycle; (v) an updated review and critique of the published reports of the use of flow cytometry for the direct detection of virus-infected cells in clinical specimens and for the detection of human immunodeficiency virus (HIV), herpes simplex virus types 1 and 2 (HSV-1 and HSV-2), and HCMV in vitro; and (vi) its use in drug susceptibility testing of clinical HCMV isolates.

History

The fluorescence-activated cell sorter was developed in the early 1970s to study isolated populations of viable cells for immunology. The original instruments were large, contained powerful dual lasers that were water cooled, and were able to analyze and sort cells. In the modern era (1990 to the present), the flow cytometer has been simplified to a more compact, air-cooled, single-laser analytical instrument for use in immunology, cell biology, molecular biology, pathology, and diagnostic microbiology. For a detailed history of flow cytometry, see Shapiro (1995).

FLOW CYTOMETRY

Definition of Flow Cytometry

Flow cytometry can be defined as the measurement of the physical and/or chemical characteristics of cells while they pass single file in a fluid stream through a measuring apparatus (Watson, 1991; Shapiro, 1995; Robinson et al., 1999). Simple flow cytometers use a single argon ion laser to simultaneously measure the light scatter properties of cells and any fluorescence associated with the cells. The light scatter properties of each cell are measured when cells pass through the laser beam and scatter the light. The scattered light passes through lenses that separate it into forward angle light scatter (FW-SC) and right angle light scatter (RT-SC). The light scatter properties of cells are used to count the cells and determine their size and granularity. If the cells passing through the laser beam are labeled with one or more fluorochromes directed against specific cellular and/or viral components, the laser light excites the fluorochrome(s), causing each fluorochrome to emit light at a higher wavelength. The emitted light from light scatter and/or fluorochromes is separated into individual components by lenses, captured by photomultiplier tubes, and digitized, and the data are displayed on a monitor screen, printed on paper to yield a hard copy, and stored in a computer for a permanent file that can be used for further data analysis at a future time. Thus, the flow cytometric analysis of fluorochrome-labeled cells yields information on the number, size, and granularity of the cells, the number of fluorochrome-labeled cells, and the amount of fluorochrome associated with each cell. The simultaneous measurement of a number of physical and biochemical characteristics of each cell is known as multiparametric analysis. The ability of flow cytometry to perform multiparametric analyses gives this technology the analytical power to identify populations of cells (phenotypes) that are undergoing particular biological events (apoptosis, viral infection, etc.). The analysis of the light scatter properties of cells is often used to separate different cell populations in a sample containing cells of different sizes and complexities, such as peripheral blood cells. Fluorochrome-labeled antibodies to cellular and/or viral antigens and nucleic acids are used to further identify cells within each of the separated populations. Several thousand cells can be analyzed per second, yielding statistically significant data in a short time. In this manner, the multiparametric analysis of the physical, biological, and chemical properties of each cell passing through the laser beam can be performed, yielding the information required to characterize the properties of the cells under study.

The Flow Cytometer

Figure 1 illustrates a generic flow cytometer that has the ability to use a single argon ion laser to distinguish the light scatter properties of cells and up to three different fluorochromes associated with each cell. More sophisticated flow cytometers use two or more lasers to examine additional fluorochromes. Some flow cytometers are cell sorters that have the capacity to physically separate cells out of a population and collect that specific cell population. However, the studies described in this chapter only require an instrument with a single argon ion laser that has the capacity to simultaneously analyze the light scatter properties of cells and two or three cell-associated fluorochromes.

Flow Cytometric Analysis of Virus-Infected Cell Cultures

Flow cytometric analysis of virus-infected tissue culture cells is becoming an extremely useful tool for studying virus-cell interactions, viral pathogenesis, and drug susceptibility (McSharry, 1994, 1995, 1998, 1999). A typical flow cytometric analysis of uninfected and HCMV-infected human foreskin fibroblasts (HFF) is shown in Fig. 2. Uninfected and HCMV-infected cells were permeabilized with methanol, treated with fluorescein isothiocyanate (FITC)-labeled monoclonal antibody to the HCMV immediate-early (IE) antigens (MAB810; Chemicon International, Inc., Temecula, Calif.), and analyzed for the percentage of antigen-positive cells by flow cytometry. This monoclonal antibody recognizes an epitope that is shared by both HCMV IE-1 and IE-2 antigens and has been widely used in diagnostic virology for confirmation of HCMV infection in tissue cultures (Mazeron et al., 1992). The left-hand panels in Fig. 2 are dot blots of the flow

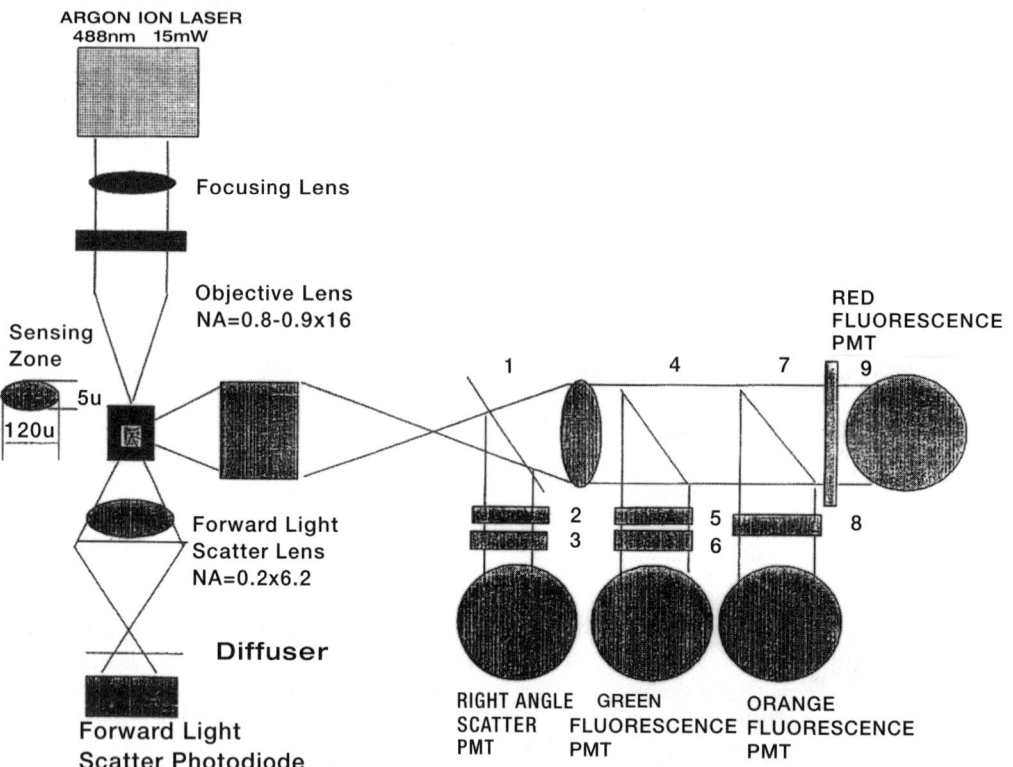

FIGURE 1 Diagram of a flow cell with attached optical systems. As the cells pass through the flow cell (the small black square box at center left), a laser beam intersects the stream of cells and scatters light. FW-SC passes through the forward light scatter lens, and the energy is collected by the forward light scatter photodiode. RT-SC passes through the objective lens, the beam splitter, the laser line filter, and the diffuser, and the energy is collected by the right angle scatter photomultiplier tube (PMT). If the cells are labeled with fluorescent molecules, the laser will excite these molecules, which emit light of higher energies. The emitted energies of different wavelengths pass through the objective lens and various filters and are collected and amplified by the various PMTs. The amplified signals are converted into digital information and stored in a computer for further analysis. 1, beam splitter; 2, laser line filter (396- to 496-nm band pass); 3, diffuser; 4, dichroic mirror 1 (570-nm long-pass filter); 5, laser cut filter (490-nm short cut); 6, green filter (515- to 530-nm band pass); 7, dichroic mirror 2 (610-nm long pass); 8, orange filter (565- to 592-nm band pass); 9, red filter (660-nm long pass). Reprinted from McSharry (1994).

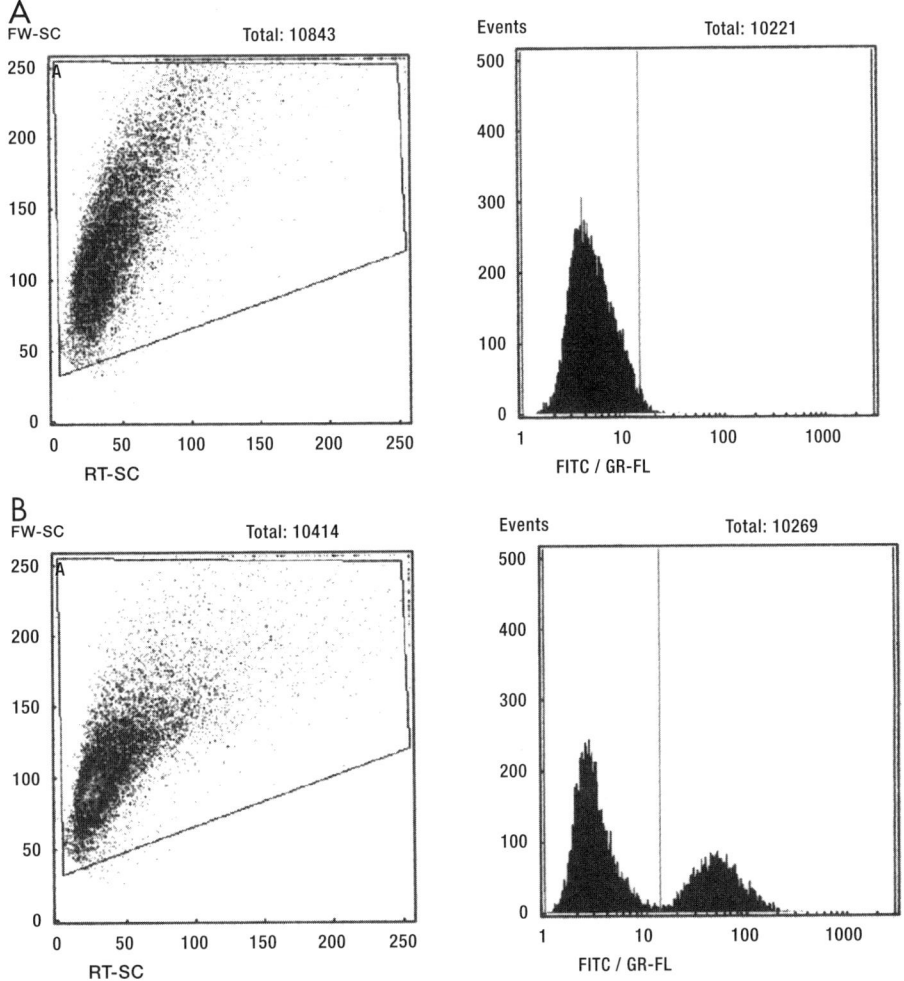

FIGURE 2 Flow cytometric analysis of uninfected and HCMV-infected HFF treated with FITC-labeled monoclonal antibody to the HCMV IE antigens. Uninfected (A) and HCMV-infected (B) HFF were permeabilized with methanol, treated with an FITC-labeled monoclonal antibody to the HCMV IE antigens, and analyzed for the percentage of antigen-positive cells by flow cytometry. The left-hand panels represent the FW-SC and RT-SC analysis of the cells. Cells of the correct size to be intact cells are gated. The right-hand panels represent the analysis of the fluorescence intensities of uninfected and HCMV-infected cells.

cytometric analysis of the light scatter properties (FW-SC versus RT-SC) of uninfected (Fig. 2A) and HCMV-infected (Fig. 2B) cells. Flow cytometric analysis of the light scatter properties of the cells is used to separate intact cells from debris. The cells are collected, and a gate (defined region) is placed around approximately 10,000 events that possess the light scatter properties characteristic of intact cells. The right-hand panels in Fig. 2 are histograms of the flow cytometric analysis of the fluorescence intensities of the uninfected HFF (Fig. 2A) and HCMV-infected cells (Fig. 2B) within the gates illustrated in the respective left-hand panels. The histogram for uninfected cells (Fig. 2A) shows a homogenous population of cells with low FITC fluorescence intensity (FITC/GR-FL) (between 1 and 10 on the x axis), indicating that uninfected cells do not contain the HCMV IE antigens. The histogram for HCMV-infected cells (Fig. 2B) shows two peaks, one with low FITC/GR-FL (between 1 and 10 on the x axis), similar to that for uninfected cells (Fig. 2A), and a second peak of

higher FITC/GR-FL (between 10 and 110 on the x axis), representing HCMV-infected cells that are synthesizing the IE antigens. These results show that the use of this monoclonal antibody to the HCMV IE antigens and flow cytometry can easily distinguish between the uninfected and HCMV-infected cell populations. Furthermore, the analysis will determine the percentages of cells in the uninfected and HCMV-infected cell populations, information that can be used to determine the effect of drugs on virus replication (Fig. 3 through 5). The flow cytometric analysis of virus-infected cells is rapid, quantitative, objective, and easily performed.

APOPTOSIS

Infection of cells with a number of viruses induces apoptosis (programmed cell death) (O'Brien, 1998; Everett and McFadden, 1999). Some viruses produce viral proteins in infected cells that delay or inhibit apoptosis (Teodoro and

Branton, 1997; Griffin and Hardwick, 1999). An early event in apoptosis is the redistribution of phosphoserine from the inner leaflet to the outer leaflet of the plasma membrane (Zhang et al., 1997). Late events in apoptosis result in nuclear condensation, fragmented DNA, and changes in cell size. Each of these characteristics of apoptotic cells can be detected and quantified by flow cytometry. Fragmented DNA in apoptotic cells binds less propidium iodide (PI) than chromosome length DNA, and this property of apoptotic cells can be used to measure apoptosis by flow cytometry. Early studies used the flow cytometric analysis of the binding of PI to DNA to demonstrate the presence of a PI binding peak with less fluorescence intensity than the normal G_0/G_1 peak in cells undergoing apoptosis (McSharry, 1994). In another procedure, fluorochrome-labeled nucleotides can be added to the free ends of the fragmented DNA by terminal deoxynucleotidyl transferase and the apoptotic cells containing fragmented DNA can be detected and quantified by flow cytometry (Bromidge et al., 1995; Chapman et al., 1995). Interactions between the Fas receptor and the Fas ligand lead to apoptosis in activated T cells (Kaplan and Sieg, 1998). The increased expression of the Fas receptor on the cell surface can be detected and quantified with fluorochrome-labeled monoclonal antibody to Fas antigen and flow cytometry. Finally, treatment of cells undergoing apoptosis with fluorochrome-labeled annexin V, a protein that binds to phosphoserine, followed by flow cytometry can detect and quantify the apoptotic cells with phosphoserine on their surfaces (Zhang et al., 1997). Flow cytometric analysis of these four properties of apoptotic cells have been used to detect and quantify the number of HIV-infected cells undergoing apoptosis. Furthermore, it is often possible to label cell surface markers with a fluorochrome-labeled antibody so that the phenotype of the apoptotic cells can be determined or to use fluorochrome-labeled monoclonal antibodies to viral antigens to determine if virus-infected cells or uninfected bystander cells undergo apoptosis. The power of multiparametric flow cytometry has been used to attempt to answer questions concerning HIV infection and apoptosis.

The role of HIV infection of CD4$^+$ T cells in apoptosis is controversial. Most studies suggest that infectious virus is required to induce apoptosis in T cells and that the infected cells undergo apoptosis. However, other studies suggest that noninfectious virus particles or purified or recombinant viral components, such as gp120, can induce apoptosis in CD4$^+$ T cells. Furthermore, some studies suggest that uninfected bystander cells undergo apoptosis, not the virus-infected cells. My purpose in the discussion below is not to resolve this controversy but to illustrate the various ways that flow cytometry has been used to detect and quantify apoptotic HIV-infected and uninfected cells.

In one study, a protease-defective, gp120-containing virus particle induced apoptosis, as measured by PI binding to DNA, in 40 to 50% of T-cell-enriched, mitogen-stimulated peripheral blood mononuclear cells (PBMC) obtained from HIV-seronegative individuals (Kameoka et al., 1997). Under the same conditions of infection, infectious virus (HIV$_{LAI}$) or recombinant gp120 induced apoptosis in less than 15% of these T cells. These results suggest that protease-defective HIV particles containing large amounts of gp120 can induce a substantial proportion of uninfected T cells to undergo apoptosis.

However, Noraz et al. (1997) showed that only HIV-infected cells undergo apoptosis. In this study, PBMC were obtained from HIV-seronegative donors and purified CD4$^+$ T cells were obtained by using magnetic beads with anti-

CD4 antibody followed by elution of the CD4$^+$-T-cell population. A multiparametric flow cytometric analysis of virus infection and apoptosis was performed. Phycoerythrin (PE)-labeled monoclonal antibody to gp120 was used to detect HIV-infected cells, FITC-labeled annexin V was used to detect apoptotic cells, and 7-AAD was used to measure cell viability. The effect of HIV infection on Fas antigen expression was separately measured using FITC-labeled monoclonal antibody to Fas antigen and flow cytometry. The results showed that only HIV-infected cells undergo apoptosis. Furthermore, the percentage of Fas antigen-positive cells did not differ in HIV-infected and uninfected cells, suggesting that Fas antigen expression is not modulated by HIV infection.

In a third study, FITC-labeled monoclonal antibody to Fas antigen, PI staining of fragmented DNA, and flow cytometry were used to measure spontaneous apoptosis and Fas expression in PBMC obtained from HIV-seropositive and HIV-seronegative individuals (Patki et al., 1997). In HIV-infected subjects with less than 100 CD4$^+$ T cells/μl, 22.4% of the PBMC were apoptotic on the basis of PI staining and 64.4% of the cells expressed the Fas antigen. This contrasted with 6.3% apoptotic cells and 14.5% of cells expressing the Fas antigen in PBMC obtained from HIV-seronegative individuals. Most of the Fas antigen-positive cells were not apoptotic, suggesting that Fas antigen expression may not be associated with apoptosis in this virus-cell system and confirming a previous report (Noraz et al., 1997). In that study, there was no attempt to separate or enrich for CD4$^+$ cells, and both CD8$^+$ and CD4$^+$ T cells were probably present in the PBMC. Furthermore, there was no attempt to phenotype the cells undergoing apoptosis. The only conclusion that can be reached is that PBMC from HIV-seropositive patients have increased apoptosis compared to PBMC from HIV-seronegative individuals.

To determine the effect of maternal HIV infection on apoptosis in T cells of newborns, cord blood lymphocytes (CBL) were obtained from uninfected or infected babies born to HIV-infected mothers (Economides et al., 1998). Fluorochrome-labeled monoclonal antibodies to CD5 (a T-cell marker), CD4, and CD8 cells were used to phenotype the apoptotic cells, and 7-AAD was used to distinguish among viable cells, early apoptotic cells, and necrotic cells (Schmid et al., 1994). The results of the flow cytometric analysis of these cells showed that both CD4$^+$ and CD8$^+$ CBL T cells obtained from uninfected babies born to HIV-infected mothers had an increased percentage of apoptotic cells compared to CBL of uninfected babies born to HIV-seronegative mothers. Since the increased apoptosis was found in CBL of uninfected babies, it suggests that exposure to HIV or its products can induce apoptosis and that actual infection with HIV may not be required, at least in vivo. These results support the bystander hypothesis of apoptosis in HIV-infected individuals.

Finally, Herbein et al. (1998a, 1998b) used a recombinant HIV that expresses green fluorescent protein (GFP) to infect either nonadherent lymphocytes or a mixture of adherent monocytes and nonadherent lymphocytes from HIV-seronegative donors. Apoptosis was determined by flow cytometric analysis of cells with the terminal deoxynucleotidyl transferase or annexin V assay. When the culture contained only lymphocytes (CD4$^+$ and CD8$^+$ T cells), the vast majority of apoptotic cells were HIV-infected CD4$^+$ T cells. However, when monocytes and lymphocytes were present in the culture, both HIV-infected CD4$^+$ T cells and uninfected CD8$^+$ T cells were apoptotic. These results show that HIV-infected CD4$^+$ lymphocytes undergo apoptosis and that in the presence of monocytes/macrophages, uninfected CD8[+] T cells also undergo apoptosis, suggesting that the monocytes present HIV antigens to CD8[+] T cells to induce apoptosis (Ameisen, 1998).

CELL CYCLE ANALYSIS

Normal resting cells have a 2N DNA content before DNA synthesis and a 4N DNA content after DNA synthesis. The binding of PI to nucleic acids (DNA and RNA) is directly proportional to the amount of nucleic acid present in the cell. Thus, the flow cytometric analysis of the fluorescence intensities of cells treated with PI can be used to determine their DNA content. Flow cytometric analysis of normal resting cells treated with RNase and PI yields a histogram with a peak of low fluorescence intensity representing cells in the G_0/G_1 phase of the cell cycle, followed by a smaller peak with twice the fluorescence intensity of the first peak representing cells in the G_2/M phase of the cell cycle. The space between the two peaks represents cells in the S phase of the cell cycle. Flow cytometric analysis of PI binding to DNA in methanol-permeabilized, RNase-treated cells has been used to determine the ploidies of normal and cancer cells (Robinson et al., 1999); it can also be used to determine the effect of virus replication on the cell cycle. Some examples of the use of flow cytometry to measure the effect of virus infection on the cell cycle are presented below.

Infection of a variety of human cells with HIV results in cell cycle arrest in the G_2 phase of the cell cycle. Analysis of mutant viruses suggests that arrest is associated with the Vpr accessory protein of HIV. Transfection of cells with plasmids containing the wild-type Vpr gene results in G_2 arrest followed by apoptosis (Stewart et al., 1997; Hrimech et al., 1999). Furthermore, virus particles containing Vpr are able to induce G_2 arrest in the presence of inhibitory concentrations of reverse transcriptase and protease inhibitors, suggesting that virus replication is not required to arrest cells in the G_2 phase of the cell cycle (Poon et al., 1998).

HCMV infection of fibroblasts arrests cells in the G_1/S phase of the cell cycle (Bresnahan et al., 1996; Dittmer and Mocarski, 1997; Salvant et al., 1998). The UL69 gene product, a component of the virus particle, is at least partially responsible for arresting infected cells in this phase (Lu and Shenk, 1999). Similar results have been reported for HSV (de Bruyn Kops and Knipe, 1988) and Epstein-Barr virus (Cayrol and Flemington, 1996).

When a virus infects a cell, it attempts to take over the cell to its own advantage. In the studies reported above, infection of cells with HIV blocked the cell cycle in G_2, a phase of the cell cycle in which large amounts of transactivating factors required for HIV growth have accumulated in the cell, enabling HIV to replicate more efficiently (Goh et al., 1998). A similar situation exists in HCMV-infected cells that are arrested in the G_1 phase of the cell cycle. Since herpesviruses encode many of the genes required for viral DNA replication, the presence of large amounts of the building blocks for DNA synthesis in the cell is sufficient. Thus, the virus need only use the large pool of nucleotides for the synthesis of viral DNA.

DETECTION OF VIRUS-INFECTED CELLS IN CLINICAL SPECIMENS

HIV

In the early 1990s, a number of groups demonstrated that fluorochrome-labeled monoclonal antibodies in conjunction with flow cytometry could be used to directly detect and quantify HIV-infected cells in peripheral blood (reviewed in McSharry, 1994). These studies showed that the percentage of p24 antigen-positive cells increased with disease progression and that the number of CD4[+] T helper cells was inversely proportional to the number of p24 antigen-positive cells. The percentage of antigen-positive cells decreased to background levels when AIDS patients were treated with zidovudine. These results suggested that fluorochrome-labeled monoclonal antibodies to HIV antigens and flow cytometry could be used both for the rapid diagnosis of HIV infection and to monitor antiretroviral therapy in AIDS patients. However, under the experimental conditions used, this technique failed to consistently detect HIV-infected cells in asymptomatic HIV-seropositive patients. Thus, the use of monoclonal antibodies to HIV p24 antigen combined with flow cytometry has not become part of the battery of diagnostic techniques used to monitor patients with HIV infection.

A number of procedures involving PCR and flow cytometry were developed to improve sensitivity for detection of HIV-infected blood cells obtained from HIV-seropositive patients at all stages of the disease (Yang et al., 1994, 1995; Dorenbaum et al., 1997). In these studies, proviral DNA was isolated from lymphocytes obtained from HIV-infected individuals, amplified by heminested PCR using biotinilated primers, hybridized to digoxigenin-labeled dUTP probes, and then bound to beads via a strepavidin linkage. The HIV DNA-containing beads were incubated with an FITC-labeled antibody to digoxigenin, and the fluorescent beads were detected and counted by flow cytometry. This assay could detect as few as three copies of proviral DNA in a cell line persistently infected with HIV and could detect amplified DNA from HIV-infected lymphocytes in HIV-seropositive patients and in HIV-infected babies born to HIV-infected mothers. This level of detection was comparable to the detection of HIV in plasma by the PCR techniques available at the time.

Others have used in situ PCR or in situ reverse transcription (RT)-PCR and flow cytometry to detect HIV-infected cells in peripheral blood obtained from HIV-seropositive patients (Patterson et al., 1993, 1995, 1998; Re et al., 1994, Gibellini et al., 1995). Lymphocytes were first phenotyped by using PE-labeled monoclonal antibodies to CD4 to label the CD4[+] T helper cells. Then the cells were permeabilized with Permafix (Ortho Diagnostic Systems, Inc., Raritan, N.J.) to perform the in situ PCRs. Digoxigenin-labeled dUTP-containing primers were used in the PCRs or RT-PCRs, and FITC-labeled antibody to digoxigenin was used to detect cells with amplified proviral DNA or viral RNA. The results showed that between 17.3 and 55.5% of the CD4[+] T cells carried proviral DNA but very few, if any, CD4[+] T cells contained viral RNA (Patterson et al., 1995).

In a more recent study, lymphocytes were labeled with PE-labeled monoclonal antibodies to CD4 to label CD4[+] T helper cells or with PE-labeled monoclonal antibodies to CD14 to label CD14[+] monocytes. The cells were permeabilized with Permafix. Proviral DNA was detected by PCR as described above. Viral RNA was detected by a fluorescent in situ hybridization method. The phenotyped, permeabilized cells were incubated with a cocktail of 5′-carboxyfluorescein-labeled probes that cover a large proportion of the expressed viral RNAs. Flow cytometric analysis of both the CD4[+] and CD14[+] cells showed that the CD14[+] monocytes were expressing viral RNA and that the CD4[+] T cells containing proviral DNA did not express viral RNA (Patterson et al., 1998). These results suggest that the HIV-infected monocytes, not the HIV-infected lymphocytes, are producing the virus found in the circulation. This conclusion may be biased, because CD4[+] T cells that express viral RNA and infectious virus turn over more rapidly than the CD14[+] monocytes or CD4[+] T cells with inactive proviral DNA. Therefore, the majority of cells available for analysis by flow cytometry would be the latently infected CD4[+] T cells and the longer-lived infected monocytes.

There is ample evidence that flow cytometry can be used to directly detect and quantify HIV-infected cells in clinical samples. With improvements in methodology, including the use of fluorochrome-labeled monoclonal antibodies to combinations of viral antigens, flow cytometry may find its place in rapid diagnosis of virus infection in HIV-seropositive patients. Validation of these techniques for diagnostic use in HIV disease will require direct comparisons with current procedures for detecting HIV in clinical specimens, such as RT-PCR of virus in plasma, p24 antigenemia assays, and coculture assays (Hammer et al., 1993).

HCMV

HCMV-infected cells in bronchoalveolar lavage (BAL) specimens and PBMC have been detected and quantified by fluorochrome-labeled antibodies and flow cytometry (Elmendorf et al., 1988; McSharry, 1994; Imbert-Marcille et al., 1997). In an earlier study of the direct detection of HCMV-infected cells in peripheral blood, PBMC were analyzed for the presence of IE-antigen-positive cells by indirect immunofluorescence and flow cytometry (McSharry, 1994). The results showed that 0.6 to 8.1% of the PBMC obtained from AIDS patients or kidney transplant patients with acute HCMV disease were positive for the HCMV IE antigen. Less than 1% of the PBMC of healthy controls were positive for the IE antigen. In the most recent study, Imbert-Marcille et al. (1997) used monoclonal antibodies to IE, pp65, or late HCMV antigens to detect antigen-positive leukocytes and used flow cytometry to quantify the antigen-positive cells. The pp65 monoclonal antibody detected between 0.04 and 1.15% of antigen-positive leukocytes, whereas monoclonal antibodies to IE and late antigens detected lower percentages. Leukocytes obtained from normal controls had 0.01 to 0.08% antigen-positive cells. The discrepancy in the percentage of antigen-positive blood cells between these two studies is due to the different populations of cells that were analyzed and the different monoclonal antibodies used. In the former study, lymphocytes and monocytes were analyzed; the latter study included an analysis of lymphocytes, monocytes, and granulocytes. Since granulocytes make up a high percentage of the leukocytes, the percentage of antigen-positive cells should be lower in the second study. In addition, the use of monoclonal antibody to the pp65 antigen was unique to that study. Since the pp65 antigen is found in granulocytes in HCMV-infected patients, a different percentage of antigen-positive cells may be expected.

Taken together, these publications suggest that monoclonal antibodies to HCMV antigens and flow cytometry can be used to detect HCMV in BAL specimens and peripheral blood cells obtained from patients with acute HCMV disease. However, as presently performed, this technology has not made its mark in this field. Further study using combinations of antibodies to IE, early (pp65), and late antigens may make it a more useful assay. The development of technologies combining in situ PCR and flow cytometry for detection of HCMV-infected BAL cells and peripheral blood cells may also enhance the feasibility of this technology for the rapid diagnosis of HCMV disease. Validation of any of these flow cytometry-based technologies will require comparisons with the current procedures for detecting HCMV in clinical specimens (Boeckh and Boivin, 1998; Mazzulli et al., 1999).

DETECTION OF VIRUS-INFECTED CULTURED CELLS

HSV-1 and HSV-2

HSVs replicate rapidly in tissue culture cells (Roizman and Sears, 1995). Laboratory diagnosis of HSV infection involves inoculation of a clinical specimen into tissue culture cells, production of cytopathic effect (CPE), and confirmation with monoclonal antibodies or bead agglutination assays that HSV is causing the CPE. The assay can be performed in 1 day to 1 week. As a rapid alternative for the laboratory diagnosis of HSV infection, fluorochrome-labeled monoclonal antibodies to type-specific HSV antigens and flow cytometry were used to type, detect, and quantify HSV clinical specimens after amplification in tissue culture (McSharry et al., 1990). The technique requires only overnight incubation of the clinical specimen with tissue culture cells and 2 h of preparation before flow cytometric analysis. Therefore, a quantitative laboratory diagnosis, including typing, is available in less than 24 h after obtaining the clinical isolate. This rapid, quantitative flow cytometric technique has the advantage of automation and speed and is less labor-intensive and subjective than microscopic observation to determine the presence of CPE and fluorescence microscopy to confirm the laboratory diagnosis of HSV infection. The ability to detect HSV-infected cells by using monoclonal antibodies to HSV antigens and flow cytometry set the stage for using this technology to screen antiviral drugs against HSV (C. Chutkowski, J. Mahony, and J. J. McSharry, Abstr. 95th Gen. Meet. Am. Soc. Microbiol. 1995, abstr. S-20, 1995) and the development of drug susceptibility assays for HSV.

Antiviral Drug Susceptibility Testing of HSV

A flow cytometry-based drug susceptibility assay for HSV-1 was recently reported (Pavic et al., 1997). These investigators used a monoclonal antibody to the HSV glycoprotein, gC, and flow cytometry to determine the effective concentrations for 50% inhibition (IC$_{50}$s) of acyclovir, ganciclovir, and foscarnet for a number of HSV-1 laboratory strains and clinical isolates. The analysis was performed after only 18 h of incubation of virus with cells in the presence of various concentrations of drug. The results showed that flow cytometry could be used to determine 50% inhibitory concentrations (IC$_{50}$s) and that the IC$_{50}$s were similar to those obtained with more time-consuming and labor-intensive infectivity assays. Drug-susceptible and drug-resistant virus strains were easily distinguished with the flow cytometry assay. The IC$_{50}$s could be determined over a wide range of viral multiplicities of infection (MOI), making it possible to analyze a larger sample of virus than is possible in a plaque reduction assay (PRA). Of particular interest was the ability to detect 1 to 2% drug-resistant viruses in a population of drug-susceptible viruses. These results suggest that flow cytometry is a rapid, quantitative, and easy method for determining the drug susceptibilities of HSV clinical isolates.

HIV

The earlier reports of the use of monoclonal antibodies to HIV antigens and flow cytometry to detect HIV-infected cultured cells have been reviewed (McSharry, 1994). A number of more recent studies have used monoclonal antibodies to HIV antigens and flow cytometry to detect HIV-infected cells in vitro (Zolla-Pazner et al., 1995; Chen et al., 1997; Cecilia et al., 1998). These authors used either monoclonal antibodies to viral antigens or reporter genes to detect HIV infection and used flow cytometry to quantify the virus-infected cells. Cecilia et al. (1998) constructed human osteosarcoma cells that expressed CD4, CXCR4, and/or CCR5 coreceptors and a Tat-dependent GFP. They used these cells to determine the different neutralization sensitivities of primary HIV isolates in the context of coreceptor use.

Drug Susceptibility Testing of HIV

One recent study used flow cytometry to determine the drug susceptibility of HIV (Gervaix et al., 1997). A stable reporter T-cell line (CEM-GFP) that expresses CD4, CXCR4 coreceptor, and GFP under the control of the HIV long terminal repeat was established and used to determine the effects of antiretroviral drugs on the percentage of HIV-infected cells expressing GFP. Two-color immunofluorescence involving PE-labeled monoclonal antibody to gp120 and GFP expression showed that CEM-GFP cells infected with HIV express both gp120 and GFP in a dose- and time-dependent manner. Furthermore, zidovudine, nevirapine, and saquinavir reduced the percentage of HIV-infected cells that express GFP. The only drawback to this very exciting finding is that the CEM-GFP cells express CD4 and the CXCR4 coreceptor and are only susceptible to infection with lymphotropic virus. They are not susceptible to infection with macrophage-tropic viruses that use the CCR5 coreceptor. The use of permanent cell lines that express CD4, both coreceptors, and HIV long terminal repeat-driven GFP should make it possible to establish drug susceptibility testing for both lymphotropic and macrophage-tropic HIV clinical isolates (Cecilia et al., 1998).

HCMV

The use of indirect immunofluorescence and flow cytometry for the detection and quantification of HCMV-infected tissue culture cells has been reported (Elmendorf et al., 1988; Schols et al., 1989; McSharry, 1998). The results clearly showed that HCMV-infected cells treated with monoclonal antibodies to viral antigens followed by FITC-labeled goat antimouse antibody and PI can be detected and quantified by flow cytometry and that uninfected cells were easily distinguished from HCMV-infected cells.

Using monoclonal antibodies to IE, early, and late HCMV antigens and an FITC-labeled second antibody and flow cytometry, the time course of the synthesis of these antigens in cells infected with HCMV at an MOI of 1 was determined (Schols et al., 1989; McSharry et al., 1998b). The data showed that IE antigens were synthesized within 2 h postinfection and their synthesis continued throughout the infection. Late antigen synthesis began between 72 and 96 h postinfection and reached maximum production between 96 and 120 h postinfection. This information has been used to develop procedures for screening antiviral drugs for potential activity against HCMV, for determining the IC_{50}s and IC_{90}s of various antiviral compounds that inhibit HCMV synthesis, and for determining the drug susceptibilities of HCMV clinical isolates (Kesson et al., 1998; Lipson et al., 1997; McSharry et al., 1998a, 1998b; J. J. McSharry, N. S. Luarain, A. C. McDonough, B. A. Olson, M. Czernewski, A. Landay, and C. Crumpacker, Abstr. 38th Intersci. Conf., Antimicrob. Agents Chemother., abstr. H-106, 1998; J. J. McSharry, C. Talarico, M. Davis, and K. K. Birou, Abstr. 38th Intersci. Conf. Antimicrob. Agents Chemother., abstr. H-107, 1998).

HCMV Drug Susceptibility Assay

The procedures currently used in our laboratory for determining the drug susceptibilities of HCMV laboratory strains and clinical isolates are described in some detail below. It should be evident to the reader that the same procedures can be used to screen antiviral drugs for activity against HCMV and for determining the drug susceptibility of any virus that replicates in cultured cells and for which antibodies are available to detect the virus-infected cells by flow cytometry.

Materials and Methods

Cells and medium. Monolayer cultures of MRC-5 cells (ATCC CCL 171) at passages 20 to 25 or HFF at passages 9 to 15 were used throughout these experiments. Minimum essential medium (MEM) supplemented with 10% fetal bovine serum (FBS), penicillin, streptomycin, and amphotericin B was used to grow cells and virus.

Viruses and virus-infected cells. The ganciclovir-sensitive AD169 laboratory strain of HCMV was obtained from David Anders, New York State Health Department; its ganciclovir-resistant derivative, D6/3/1, was obtained from Nell Lurain (Lurain et al., 1994), and susceptible and resistant HCMV clinical isolates were obtained from The Albany Medical Center Hospital, Rush Medical Center, and the Virology Quality Assurance Laboratory of the Division of AIDS, National Institutes of Health.

For the production of cell-free virus, HFF monolayers are inoculated with HCMV at an MOI of 0.1 to 1 PFU/cell in the presence of MEM supplemented with 10% FBS and antibiotics. After a 2-h adsorption period at 37°C, the inoculum is removed, MEM supplemented with 3% FBS and antibiotics is added, and the infected monolayers are incubated at 37°C until the whole monolayer shows CPE. The infected cells are incubated for three more days, and released virus is harvested from the medium. Cell debris is removed from the medium containing released virus by centrifugation at $400 \times g$ for 10 min, and the supernatant is divided into 1-ml portions and frozen at $-70°C$.

For cell-associated virus, virus-infected cells at an MOI of 0.1 virus-infected cell/uninfected cell are added directly to the medium overlying HFF monolayers. The infected monolayers are incubated at 37°C until at least 90% of the cells in the monolayer exhibit CPE consistent with HCMV infection. The infected cells are removed from the flask with trypsin-EDTA, resuspended in MEM supplemented with 10% FBS, and pelleted at $400 \times g$ for 10 min. The pellet is resuspended in MEM supplemented with 20% FBS and 10% dimethyl sulfoxide to a final concentration of 10^6/ml, and the cells are frozen at $-70°C$ in a Styrofoam box overnight and then placed in liquid nitrogen.

Antiviral compounds. Ganciclovir, cidofovir, 1263W94, RPR CMV423, and RPR127025 were generous gifts from Roche/Syntex Laboratories, Nutley, N.J.; Gilead Sciences, Foster City, Calif.; Glaxo Wellcome, Triangle Park, N.C.; and Rhone Poulenc Rorer, Paris, France, respectively.

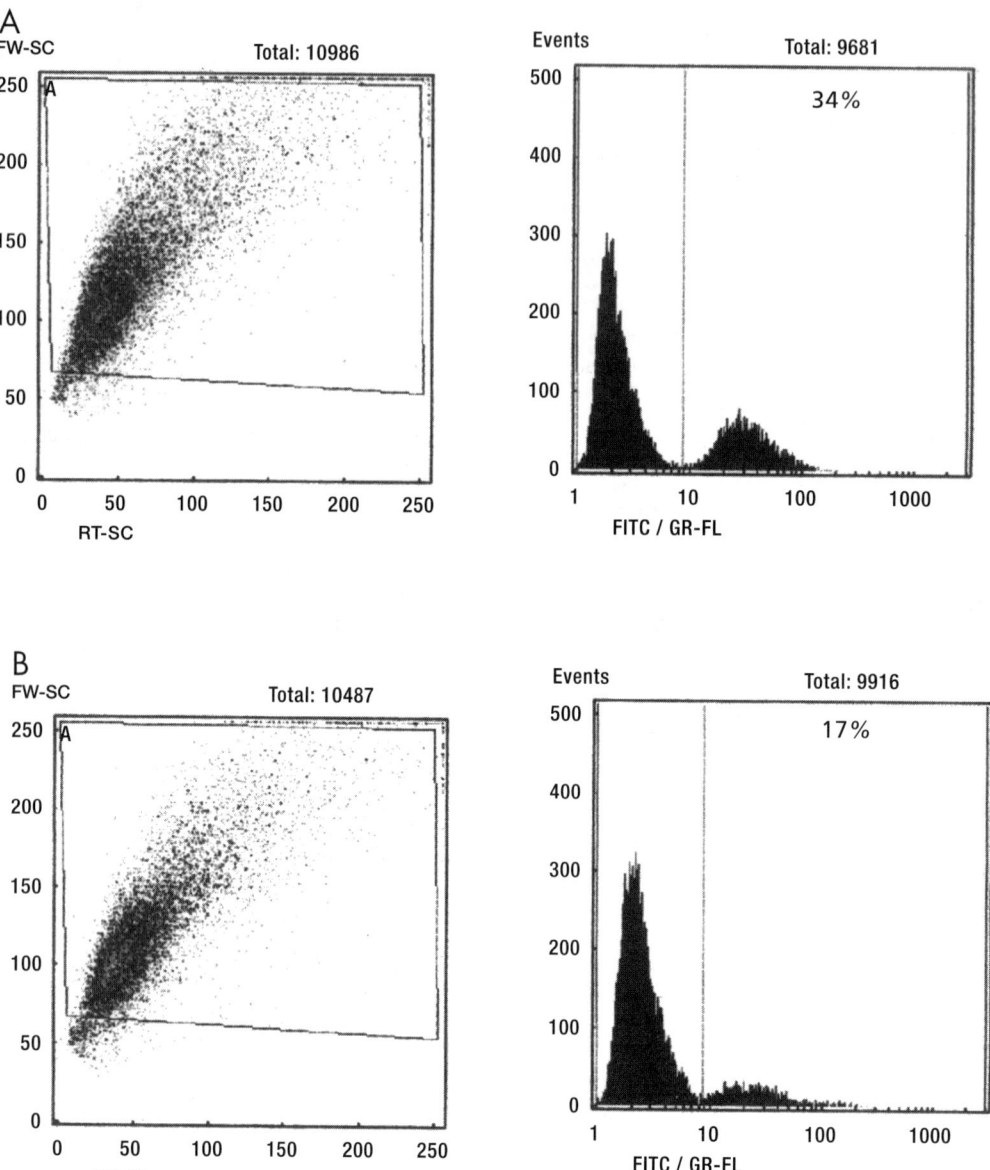

FIGURE 3 Effect of foscarnet on IE antigen synthesis in cells infected with a foscarnet-suscepti-ble clinical sample. Medium containing 0 (A) or 200 (B) μM foscarnet was added to HFF monolay-ers. Virus-infected cells were added to the flask at an MOI of 0.01. After incubation at 37°C for 144 h, the cells were harvested with trypsin-EDTA, permeabilized with methanol, treated with a monoclonal antibody to the HCMV IE antigens, and analyzed by flow cytometry for the percentage of antigen-positive cells. Initially, the cells were analyzed for FW-SC versus RT-SC to identify intact cells and exclude debris. The intact cells were gated and analyzed for the percentage of antigen-pos-itive cells in the population. In the absence of foscarnet, 34% of the cells synthesized the IE anti-gens. In the presence of 200 μM foscarnet, only 17% of the cells expressed the IE antigens.

Foscarnet was purchased from Sigma Chemical Co., St. Louis, Mo.

PRA. To determine the drug susceptibility of cell-free or cell-associated virus, PRAs were performed. For cell-free virus, the medium was removed from the wells in a 24-well plate and approximately 50 PFU of virus in 0.2 ml was added to monolayer cultures of HFF. After a 2-h adsorption period, the inoculum was removed, an agarose overlay containing various concentrations of drug was added, and

the plates were incubated at 37°C for 10 days. The monolayers were fixed with paraformaldehyde, the agarose was removed, the monolayers were stained with crystal violet, and the plaques were counted with the aid of a light microscope.

For cell-associated virus, medium containing various concentrations of drugs was added to the wells in a 24-well plate, approximately 50 virus-infected cells were added directly to the medium, and the plates were incubated at 37°C for 7 days. The medium was removed, the monolayers

A

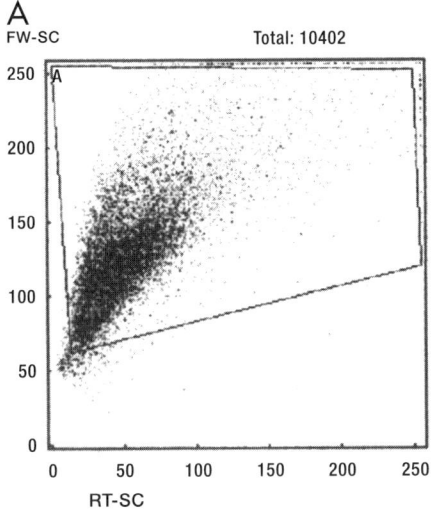

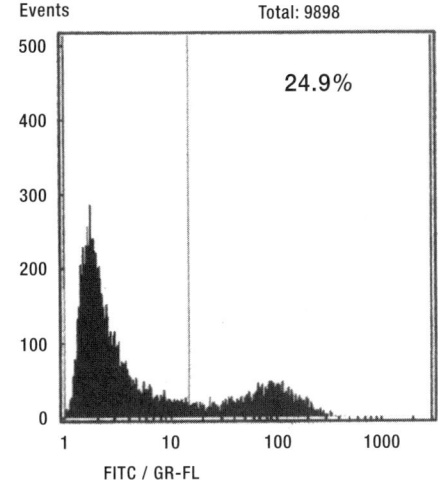

B

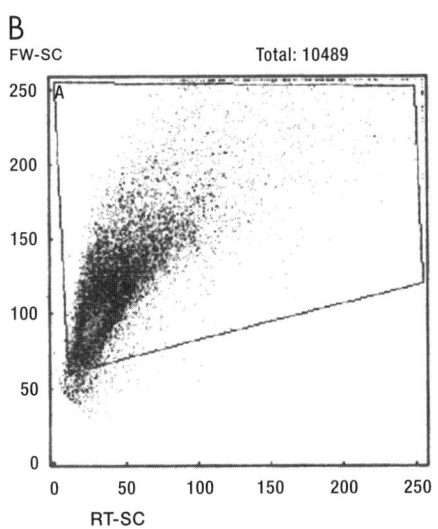

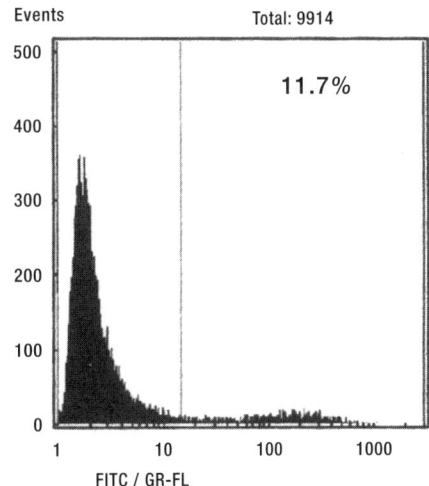

FIGURE 4 Effect of foscarnet on late-antigen synthesis in cells infected with a foscarnet-susceptible clinical isolate. As in Fig. 3 except that a monoclonal antibody to an HCMV late antigen was used to identify the antigen-positive cells. In the absence of foscarnet (A), 24.9% of the cells were positive for the late antigen. In the presence of 200 μM foscarnet (B), only 11.7% of the cells were positive for the late antigen.

were fixed with paraformaldehyde and stained with crystal violet, and the plaques were counted with the aid of a light microscope.

In each case, the reduction in the number of PFU was calculated as a percentage of the control, and the IC_{50}s and IC_{90}s were calculated by plotting the percent reduction against the drug concentration with SlideWright Plus software.

Flow cytometry drug susceptibility assay. To determine the drug susceptibilities of cell-associated HCMV clinical isolates, medium containing various concentrations of drug is added to HFF monolayers in 25-cm^2 flasks. Then the monolayers are infected with HCMV-infected cells at an MOI of 0.001 to 0.01 by adding the infected cells directly to the medium. After incubation at 37°C for 144 h, the

medium is removed and the mock-infected and virus-infected cells are removed from the culture flask with trypsin-EDTA, resuspended in MEM supplemented with 10% FBS, pelleted at 400 × g for 10 min, washed once with phosphate-buffered saline (PBS) without Ca^{2+} and Mg^{2+}, and put on ice for 60 min. The cells are made permeable by the addition of ice-cold absolute methanol to the residual PBS in the tube to yield a final concentration of 90% methanol. The permeable cells are stored at −70°C until they are prepared for analysis by flow cytometry.

To treat the permeable cells with monoclonal antibodies, the cells are centrifuged at low speed, the methanol is removed, and the cells are washed once with PBS without Ca^{2+} and Mg^{2+}. Monoclonal antibodies (FITC-labeled monoclonal antibody to the HCMV IE antigens [MAB810] or FITC-labeled monoclonal antibody to the HCMV gB

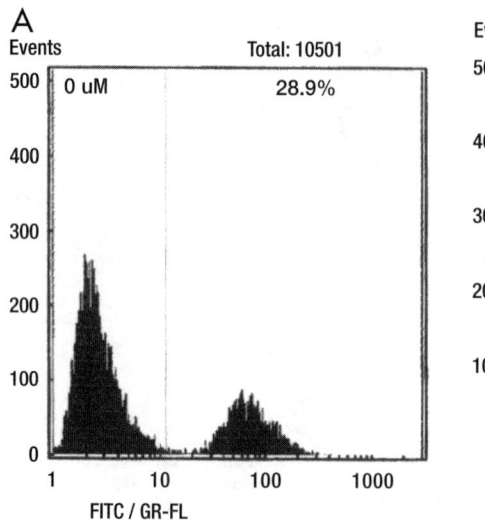
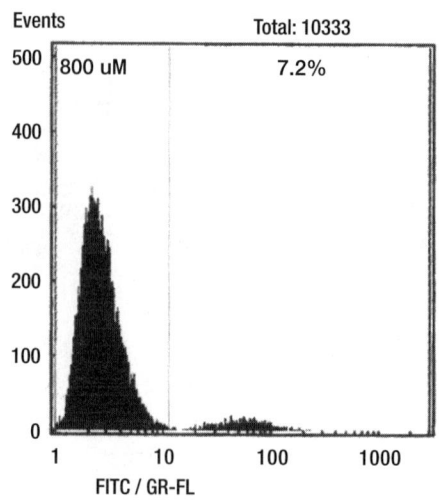

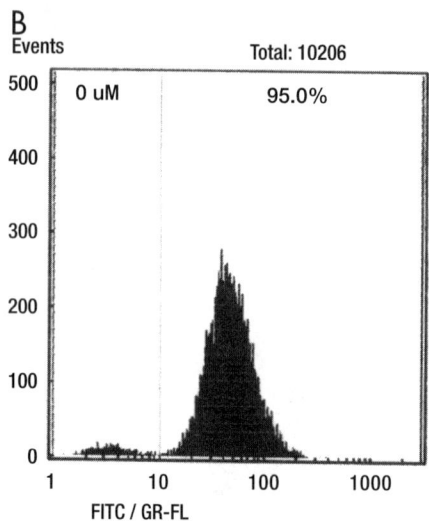
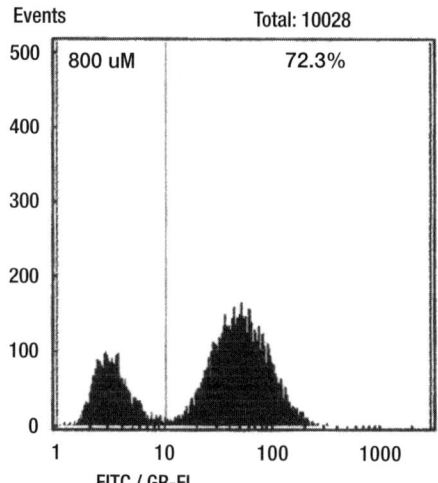

FIGURE 5 Effect of foscarnet on the synthesis of IE antigen in cells infected with a foscarnet-susceptible clinical isolate (A) or a foscarnet-resistant clinical isolate (B). Infection was performed as for Fig. 3. Of the cells infected with the foscarnet-susceptible clinical isolate, 28.9% synthesize the IE antigen in the absence of foscarnet whereas only 7.2% of the cells synthesize the IE antigen in the presence of 800 μM foscarnet. Of the cells infected with the foscarnet-resistant clinical isolate, 95.0% synthesize the IE antigen in the absence of foscarnet whereas 72.3% synthesize the IE antigen in the presence of 800 μM foscarnet.

late antigen [MAB8126] obtained from Chemicon International) diluted to the appropriate concentration in 0.007% Evans Blue containing 0.1% bovine serum albumin and sodium azide (Chemicon International) are added to the permeabilized cells, and they are incubated for 60 min at 37°C. After the incubation period, the cells are washed three times in PBS-Tween 20 (Chemicon International) and analyzed by flow cytometry.

A Cytoronabsolute analytical flow cytometer (Ortho Diagnostic Systems, Inc.) with a 15-mW argon ion laser set at 488 nm for excitation was used for these studies. The initial analysis was FW-SC versus RT-SC to distinguish intact cells from smaller debris. A gate was drawn around the events representing intact cells, and the events within the gate were analyzed for FITC fluorescence intensity. The results are presented as the percent of antigen-positive cells in the analyzed population. Controls consisted of uninfected cells treated with the FITC-labeled monoclonal antibodies to IE or late antigens and HCMV-infected cells treated with FITC-labeled isotype control antibodies.

Results

Figures 3 and 4 illustrate the analysis of the effect of foscarnet on the percentage of cells synthesizing the IE (Fig. 3) or late (Fig. 4) antigen. In the absence of foscarnet, 34% of the cells synthesized the IE antigen and 24.9% of the cells synthesized the late antigen. In the presence of 200 μM foscarnet, only 17% of the cells synthesized the IE antigen and 11.7% of the cells synthesized the late antigen. On the basis of this type of analysis for each of six drug concentrations, the IC_{50}s for this clinical isolate were calculated to be 99.58 μM foscarnet based on the analysis of the percentage of cells synthesizing the IE antigen and 102.06 μM foscarnet based on the percentage of cells synthesizing the late antigen. The two IC_{50}s are in excellent agreement and demonstrate that, under conditions of low MOI for HCMV-infected cells, IC_{50}s can be determined using the effect of drugs on the synthesis of either the IE antigen or the late antigen. Analysis of the effect of drugs on IE antigen synthesis can be performed on virus-infected cells after 72 to 96 h of incubation, making it a more rapid assay. The IC_{50} for this clinical isolate by a PRA was 91.27 μM foscarnet. These results show that under conditions of low MOI (0.01) the IC_{50}s derived from the flow cytometry drug susceptibility assay using monoclonal antibodies to either the IE or late antigen are in excellent agreement with the IC_{50}s obtained with the PRA. One drawback of the flow cytometry drug susceptibility assay is that it does not always yield an IC_{90} or IC_{95} because of the presence of input virus-infected cells. Despite this drawback, the speed, accuracy, and ease of the assay far outweigh its disadvantages.

Figure 5 illustrates the flow cytometric analysis of the effect of 0 or 800 μM foscarnet on the percentage of cells synthesizing the IE antigen following infection with a foscarnet-susceptible (Fig. 5A) and a foscarnet-resistant (Fig. 5B) HCMV clinical isolate. The analysis shows that in the presence of 0 μM foscarnet, 28.9% of the cells infected with a foscarnet-sensitive clinical isolate and 95% of the cells infected with a foscarnet-resistant clinical isolate synthesized the IE antigen. In the presence of 800 μM foscarnet, only 7.2% of the cells synthesized the IE antigen for the drug-susceptible isolate while 72.3% of the infected cells were antigen positive for the foscarnet-resistant strain. For the foscarnet-resistant clinical isolate, there was less than a 50% reduction in the percentage of antigen-positive cells, showing that this isolate is resistant to the highest concentration of foscarnet used. Therefore, the flow cytometry assay can clearly distinguish between sensitive and resistant clinical isolates.

The broad applicability of this assay for determining IC_{50}s for HCMV clinical isolates was demonstrated when it was used to determine the susceptibilities of a set of 35 HCMV clinical isolates to six different antiviral compounds (McSharry et al., 38th ICAAC, abstr. H-106 and H-107). The data comparing the average IC_{50}s obtained by the flow cytometry assay using the IE or late antigen and by PRA for these six compounds are presented in Table 1. On a micromolar basis, the antiviral activity of these compounds is RPRCMV423 > RPR127025 > 1263W94 > cidofovir > ganciclovir > foscarnet. There is excellent correlation between the IC_{50}s obtained using the effect of drug on the reduction in the percentage of cells synthesizing either the IE or late antigens and between those obtained with the flow cytometry assay and the PRA for the individual drugs.

In summary, the flow cytometry drug susceptibility assay for HCMV clinical isolates is accurate, rapid, and quantitative and can be automated. It can be used to measure the susceptibilities of HCMV clinical isolates to several antiviral compounds that may have different modes of action. The assay readily distinguished between drug-susceptible and drug-resistant clinical isolates. Finally, the assay yields IC_{50}s that are equivalent to those obtained with the PRA.

ADVANTAGES AND DISADVANTAGES OF DRUG SUSCEPTIBILITY ASSAYS

The current procedures for determining the susceptibilities of HCMV clinical isolates to antiviral drugs were recently reviewed (Erice, 1999). They include both phenotypic and genotypic assays. Genotypic assays are often rapid and can be automated. However, they detect only known or previously characterized genetic changes associated with drug resistance. Unknown mutations associated with drug resistance may be overlooked. Phenotypic assays measure drug susceptibility irrespective of the mutation and may be more useful for detecting drug susceptibility. The currently used phenotypic assays are very time-consuming, labor-intensive, and subjective. In contrast, the flow cytometric analysis of viral drug susceptibility is rapid, objective, and simple to perform and can be automated. The assay for HCMV drug susceptibility using the monoclonal antibody to the IE antigen and flow cytometry can be completed after only 72 to 96 h of incubation of virus with drugs, making it at least as rapid as any of the phenotypic assays currently in use (Erice, 1999). Another major advantage is the fact that the flow cytometer does all of the counting, not a microscopist who may be prone to subjective observations. For HSV, the flow cytometry-based drug susceptibility test takes less than 24 h and, if type-specific antibodies are used, can type the isolate at the same time. Modern flow cytometers are easy to use, and any certified medical technologist can be trained to perform these assays in a short time. The real skill is matching the appropriate monoclonal antibodies with the task at hand.

TABLE 1 Comparison of average IC_{50} of six drugs for HCMV clinical isolates

| Drug[a] | IC_{50} (±SD) (μM) | | |
| | Flow cytometry | | PRA |
	IE antigen	Late antigen	
CMV423	0.031 ± 0.016	0.018 ± 0.012	0.028 ± 0.022
127025	0.0497 ± 0.039	ND[b]	ND[b]
1263W94	0.379 ± 0.203	0.384 ± 0.138	0.312 ± 0.094
CID	0.33 ± 0.09	0.706 ± 0.28	0.704 ± 0.29
GCV	4.93 ± 1.43	3.76 ± 1.65	2.77 ± 1.47
FOS	241.00 ± 90.92	191.08 ± 82.50	140.67 ± 42.24

[a]CMV423 and 127025 are experimental drugs from Rhone Poulenc Rorer; 1263W94 is an experimental drug from GlaxoWellcome. CID, cidofovir; GCV, ganciclovir; FOS, foscarnet.
[b]ND, not determined.

The major disadvantage of the flow cytometry-based drug susceptibility assays is the need for a flow cytometer to perform the assay. However, as more medical centers obtain flow cytometers, their availability for use in these assays will increase. Once the manufacturers become more aware of the multiple uses of their instruments, they will market them for the clinical microbiology laboratory. Another concern is the cost of the instruments and the fluorochrome-labeled monoclonal antibodies used to detect virus-infected cells. Analytical instruments from the two major companies that sell them cost approximately $80,000. Some instruments are available for as little as $30,000. With time, the cost of the instruments will be reduced. Furthermore, when the time and labor savings associated with this technology are calculated, the costs of the instrument and the reagents are no longer prohibitive.

TIPS

Since most of the procedures for treating virus-infected cells with antibodies outlined in this chapter use permeabilized cells, some guidance on permeabilization methods is appropriate. The research literature is full of different techniques for permeabilizing cells. However, many of these techniques work only for the cell system used. When dealing with tissue culture cells, we have found that methanol permeabilization, as outlined above, is the best. It puts holes in both the plasma membrane and the nuclear envelope, allowing antibodies to enter the cytoplasm and the nucleus and bind to specific antigens and any unbound materials to be washed out. However, methanol destroys the light scatter properties of cells and some of the cell surface antigens used to identify cells in clinical samples. For treating blood cells and labeling the cell surface as well as internal antigens, any of the blood cell-permeabilizing reagents available from flow cytometry companies can be used. These new fixatives, such as Permeafix (Ortho Diagnostic Systems), Fix and Perm (Caltag Laboratories, San Francisco, Calif.), and FACS Lysing Reagent and FACS Permeabilizing Solution (Becton Dickinson Co., San Jose, Calif.), can permeabilize blood cells without destroying light scatter properties and surface antigens. These reagents were designed to work with blood cells and do not always work with other types of cells.

The final tips concern the quality of the viral-antigen-specific monoclonal antibodies used for determining the presence of virus-infected cells. Since reagents for use in flow cytometry in viral diagnosis are not commercially available at this time, the laboratory investigator must identify usable monoclonal antibodies for the task at hand and make sure that they are diluted to the proper working dilution. Depending on the virus-cell system, a number of monoclonal antibodies may need to be screened to determine which ones are useful for the task. As more flow cytometry procedures are used in the clinical laboratory, the commercial availability of standardized reagents will increase.

FUTURE USES OF FLOW CYTOMETRY IN CLINICAL VIROLOGY

There are at least five areas where flow cytometry can have an impact in the clinical virology laboratory. First, flow cytometry could be used for the rapid detection of virus-infected cells directly in clinical samples. With the availability of permeabilization reagents for blood cells, and possibly BAL specimens, and as more fluorochrome-labeled monoclonal antibodies to viral antigens become available, the flow cytometric detection and quantification of virus-infected cells in clinical samples will become the method of choice for rapid laboratory diagnosis. This change will help eliminate the use of the labor-intensive and subjective methods involving fluorescence microscopy. With automation, the detection of virus-infected cells directly in clinical samples will add speed and efficiency to the clinical virology laboratory so that diagnostic virology will become a practical adjunct to clinical diagnosis. Second, the use of flow cytometry for detection and quantification of virus-infected cells in culture will add speed and automation to the clinical laboratory. In the future, technologists will not spend many hours at the microscope looking for CPE or fluorescence. As a screen for antiviral drugs, flow cytometry will replace the standard PRA. Third, as has been demonstrated in this review, the role of apoptosis in viral infection is easily studied using multiparametric flow cytometry. Flow cytometry will add speed and the ability to quantify the cells undergoing apoptosis. Fourth, flow cytometry will be used to quantify virus-infected cells that have had either viral RNA or DNA amplified by PCR procedures. Fifth, the multiparametric flow cytometric analysis of cytokine-producing cells and of the effects of virus infection on cytokine production has recently been published (Maino and Picker, 1998). As these flow cytometric techniques that are increasingly being used in the research laboratory with great success become adapted to the clinical laboratory in the near future, this promising technology will become practical in the clinical virology laboratory, resulting in greater efficiency and lower costs.

CONCLUSIONS

This review has demonstrated that fluorochrome-labeled antibodies and flow cytometry can be used to detect virus-infected cells in culture and directly in clinical samples. Direct detection of virus-infected cells in clinical samples is currently problematic. However, with further development of the use of combinations of monoclonal antibodies to viral antigens and the continued development of the use of PCR and RT-PCR to amplify viral nucleic acids in virus-infected cells, flow cytometry may become a useful method for the direct detection of virus-infected cells in clinical specimens. The use of flow cytometry to measure the effect of virus infection on apoptosis and the cell cycle is already well established. The use of this technology for screening antiviral drugs and for drug susceptibility assays for HCMV has been presented in enough detail to demonstrate its great potential for the rapid, quantitative analysis of virus-drug interactions. Similar analyses could be performed for the detection of any virus that grows in tissue culture cells and for which antibodies to viral antigens are available for the detection of virus-infected cells. Expanded use of fluorochrome-labeled monoclonal antibodies to viral antigens and flow cytometry for the detection and quantification of virus-infected tissue culture cells will save time and effort and make the diagnostic laboratory more efficient and productive.

I acknowledge Ann Ogden-McDonough and Betty Olson for their excellent technical assistance.

This work was supported in part by grants AI32367 and AI41690 from the National Institutes of Health.

REFERENCES

Ameisen, J. C. 1998. Setting death in motion. *Nature* **395:**117–119.

Boeckh, M., and G. Boivin. 1998. Quantitation of cytomegalovirus: methodologic aspects and clinical applications. *Clin. Microbiol. Rev.* **11:**533–554.

Bresnahan, W. A., I. Boldogh, E. A. Thompson, and T. Albrecht. 1996. Human cytomegalovirus inhibits cellular DNA synthesis and arrests productively infected cells in G1. *Virology* **224:**150–160.

Bromidge, T. J., D. J. Howe, S. A. Johnson, and M. J. Phillips. 1995. Adaptation of the TdT assay for semi-quantitative flow cytometric detection of DNA strand breaks. *Cytometry* **20:**257–260.

Cayrol, C., and E. K. Flemington. 1996. The Epstein Barr virus bZIP transcription factor Zta causes G_0/G_1 cell cycle arrest through induction of cyclin dependent kinase inhibitors. *EMBO J.* **15:**2748–2759.

Cecilia, D., V. K. Kewalramani, J. O'Leary, B. Volsky, P. Nyambi, S. Burda, S. Xu, D. R. Littman, and S. Zolla-Pazner. 1998. Neutralization profiles of primary human immunodeficiency virus type 1 isolates in the context of coreceptor usage. *J. Virol.* **72:**6988–6996.

Chapman, R.S., C. M. Chresta, A. A. Herberg, H. M. Beere, S. Heer, A. D. Whetton, J. A. Hickman, and C. Dive. 1995. Further characterization of the in situ terminal deoxynucleotidyl transferase (TdT) assay for the flow cytometric analysis of apoptosis in drug resistant and drug sensitive leukaemic cells. *Cytometry* **20:**245–256.

Chen, B. K., M. B. Feinberg, and D. Baltimore. 1997. The κB sites in the human immunodeficiency virus type 1 long terminal repeat enhance virus replication yet are not absolutely required for viral growth. *J. Virol.* **71:**5495–5504.

de Bruyn Kops, A., and D. M. Knipe. 1988. Formation of DNA replication structures in herpes virus-infected cells requires a viral DNA binding protein. *Cell* **55:**857–868.

Dittmer, D., and E. S. Mocarski. 1997. Human cytomegalovirus infection inhibits G1/S transition. *J. Virol.* **71:**1629–1634.

Dorenbaum, A., K. S. Venkateswaran, G. Yang, A. M. Comeau, D. Wara, and G. N. Vyas. 1997. Transmission of HIV-1 in infants born to seropositive mothers: PCR amplified proviral DNA detected by flow cytometric analysis of immunoreactive beads. *J. Acquir. Immune Defic. Syndr. Hum. Retrovirol.* **15:**35–42.

Economides, A., I. Schmid, D. J. Anisman-Posner, S. Plaeger, Y. J. Bryson, and C. H. Uttenbogaart. 1998. Apoptosis in cord blood T lymphocytes from infants of human immunodeficiency virus-infected mothers. *Clin. Diagn. Lab. Immunol.* **5:**230–234.

Elmendorf, S., J. J. McSharry, J. Laffin, D. Fogleman, and J. M. Lehman. 1988. Detection of an early cytomegalovirus antigen with two color quantitative flow cytometry. *Cytometry* **9:**254–260.

Erice, A. 1999. Resistance of human cytomegalovirus to antiviral drugs. *Clin. Microbiol. Rev.* **12:**286–297.

Everett, H., and G. McFadden. 1999. Apoptosis: an innate immune response to virus infection. *Trends Microbiol.* **7:**160–165.

Gervaix, A., D. West, L. M. Leoni, D. D. Richman, F. Wong-Staal, and J. Corbeil. 1997. A new reporter cell line to monitor HIV infection and drug susceptibility in vitro. *Proc. Natl. Acad. Sci. USA* **94:**4653–4658.

Gibellini, D., G. Zauli, M. C. Re, G. Furlini, S. Lolli, A. Bassini, C. Celeghini, and M. La Placa. 1995. In situ polymerase chain reaction technique revealed by flow cytometry as a tool for gene detection. *Anal. Biochem.* **228:**252–258.

Goh, W. C., M. E. Rogel, C. M. Kinsey, S. F. Michael, P. N. Fultz, M. A. Nowak, B. H. Hahn, and M. Emerman. 1998. HIV-1 Vpr increases viral expression by manipulation of the cell cycle: a mechanism for selection of Vpr in vivo. *Nat. Med.* **4:**65–71.

Griffin, D. E., and J. M. Hardwick. 1999. Perspective: virus infections and the death of neurons. *Trends Microbiol.* **7:**155–160.

Hammer, S., C. Crumpacker, R. D'Aquila, B. Jackson, J. Lathey, D. Livnat, and P. Reichelderfer. 1993. Use of virologic assays for detection of human immunodeficiency virus in clinical trials: recommendations of the AIDS Clinical Trials Group Virology Committee. *J. Clin. Microbiol.* **31:**2557–2564.

Herbein, G., U. Mahlknecht, F. Batliwalla, P. Gregersen, T. Pappas, J. Butler, W. A. O'Brien, and E. Verden. 1998a. Apoptosis of CD8[+] T cells is mediated by macrophages through interactions with chemokine receptor CXCR4. *Nature* **395:**189–194.

Herbein, G., C. Van Lint, J. L. Lovett, and E. Verdin. 1998b. Distinct mechanisms trigger apoptosis in human immunodeficiency virus type 1 infected and in uninfected bystander T lymphocytes. *J. Virol.* **72:**660–670.

Hrimech, M., X.-J. Yao, F. Bachand, N. Rougeau, and E. A. Cohen. 1999. Human immunodeficiency virus type 1 (HIV-1) Vpr functions as an immediate early protein during HIV-1 infection. *J. Virol.* **73:**4101–4109.

Imbert-Marcille, B. M., N. Robillard, A. S. Poirier, M. Coste-Burel, D. Cantarovich, N. Milpied, and S. Billaudel. 1997. Development of a method for direct quantitation of cytomegalovirus antigenemia by flow cytometry. *J. Clin. Microbiol.* **35:**2665–2669.

Kameoka, M., T. Kimura, Y.-H. Zheng, S. Suzuki, F. Fujinaga, R. B. Luftig, and K. Ikuta. 1997. Protease-defective, gp120-containing human immunodeficiency virus type 1 particles induce apoptosis more efficiently than does wild-type or recombinant gp120 protein in healthy-donor-derived peripheral blood T cells. *J. Clin. Microbiol.* **35:**41–47.

Kaplan, D., and S. Seig. 1998. Role of the Fas/Fas ligand apoptotic pathway in human immunodeficiency virus type 1 disease. *J. Virol.* **72:**6279–6282.

Kesson, A. M., F. Zeng, A. L. Cunningham, and W. D. Rawlinson. 1998. The use of flow cytometry to detect antiviral resistance in human cytomegalovirus. *J. Virol. Method.* **71:**177–186.

Lipson, S. M., M. Soni, F. X. Biondo, D. H. Shepp, M. H. Kaplan, and T. Sun. 1997. Antiviral susceptibility testing-flow cytometric analysis (AST-FCA) for the detection of cytomegalovirus drug resistance. *Diagn. Microbiol. Infect. Dis.* **28:**123–129.

Lu, M., and T. Shenk. 1999. Human cytomegalovirus UL69 protein induces cells to accumulate in G1 phase of the cell cycle. *J. Virol.* **73:**676–683.

Lurain, N. S., L. E. Spafford, and K. D. Thompson. 1994. Mutation in the UL97 open reading frame of human cytomegalovirus strains resistant to ganciclovir. *J. Virol.* **68:**4427–4431.

Maino, V. C., and L. J. Picker. 1998. Identification of functional subsets by flow cytometry: intracellular detection of cytokine expression. *Cytometry* **34:**207–215.

Mazeron, M. C., G. Jahn, and B. Plachter. 1992. Monoclonal antibody E-13 (M-810) to human cytomegalovirus recognizes an epitope encoded by exon 2 of the major immediate early gene. *J. Gen. Virol.* **73:**2699–2703.

Mazzulli, T., W. L. Drew, B. Yen-Liberman, D. Jekic-McMullen, D. J. Kohn, C. Isada, G. Moussa, R. Chua, and S. Wamsley. 1999. Multicenter comparison of the Digene hybrid capture CMV DNA assay (version 2.0), the pp65 anti-genemia assay, and cell culture for detection of cytomegalovirus viremia. *J. Clin. Microbiol.* **37**:958–963.

McSharry, J. J. 1994. Uses of flow cytometry in virology. *Clin. Microbiol. Rev.* **7**:576–604.

McSharry, J. J. 1995. Flow cytometry based antiviral resistance assays. *Clin. Immunol. Newsl.* **9**:113–119.

McSharry, J. J. 1998. Flow cytometric analysis of virally infected cells: in vitro and in vivo studies, p. 39–56. *In* S. Specter, M. Bendinelli, and H. Friedman (ed.), *Rapid Detection of Infectious Agents*, Plenum Press, New York, N.Y.

McSharry, J. J. 1999. Flow cytometric antiviral drug suscepti-bility assays. *Clin. Immunol. Newsl.* **19**:1–14.

McSharry, J. J., R. Costantino, M. B. McSharry, R. A. Venezia, and J. M. Lehman. 1990. Rapid detection of herpes simplex virus in clinical samples by flow cytometry after ampli-fication in tissue culture. *J. Clin. Microbiol.* **28**:1864–1866.

McSharry, J. J, N. S. Lurain, G. L. Drusano, A. Landay, J. Manischewitz, M. Nokta, M. O'Gorman, H. M. Shapiro, A. Weinberg, P. S. Reichelderfer, and C. Crumpacker. 1998a. Flow cytometric determination of ganciclovir susceptibilities of human cytomegalovirus clinical isolates. *J. Clin. Microbiol.* **36**:958–964.

McSharry, J. J., N. S. Lurain, G. L. Drusano, A. Landay, M. Notka, M. O'Gorman, A. Weinberg, H. M. Shapiro, P. S. Reichelderfer, and C. Crumpacker. 1998b. Rapid ganciclovir susceptibility assay using flow cytometry for human cytomegalovirus clinical isolates. *Antimicrob. Agents Chemoth-er.* **42**:2326–2331.

Noraz, N., J. Gozlan, J. Corbeil, T. Brunner, and S. A. Spector. 1997. HIV-induced apoptosis of activated primary CD4+ T lymphocytes is not mediated by Fas-Fas ligand. *AIDS* **11**:1671–1680.

O'Brien, V. 1998. Viruses and apoptosis. *J. Gen. Virol.* **79**:1833–1845.

Patki, A. H., D. L. Georges, and M. M. Lederman. 1997. CD4+-T-cell counts, spontaneous apoptosis, and Fas expression in peripheral blood mononuclear cells obtained from human immunodeficiency virus type 1-infected subjects. *Clin. Diagn. Lab. Immunol.* **4**:736–741.

Patterson, B. K., M. Till, P. Otto, C. Goolsby, M. R. Furtado, L. J. McBride, and S. M. Wolinsky. 1993. Detection of HIV-1 DNA and messenger RNA in individual cells by PCR-driven in situ hybridization and flow cytometry. *Science* **260**:976–979.

Patterson, B. K., C. Goolsby, V. Hodara, K. L. Lohman, and S. M. Wolinsky. 1995. Detection of CD4+ T cells harboring human immunodeficiency virus type 1 DNA by flow cytometry using simultaneous immunophenotyping and PCR-driven in situ hybridization: evidence of epitope masking of the CD4 cell surface molecule in vivo. *J. Virol.* **69**:4316–4322.

Patterson, B. K., V. L. Mosiman, L. Cantarero, M. Furta-do, M. Bhattacharya, and C. Goolsby. 1998. Detection of HIV-RNA-positive monocytes in peripheral blood of HIV-positive patients by simultaneous flow cytometric analysis of intracellular HIV RNA and cellular phentotype. *Cytometry* **31**:265–274.

Pavic, I., A. Hartmen, A. Zimmermann, D. Michel, W. Hampl, I. Schleyer, and T. Mertens. 1997. Flow cytometric analysis of herpes simplex virus type 1 susceptibility to acy-clovir, ganciclovir, and foscarnet. *Antimicrob. Agents Chemoth-er.* **41**:2686–2692.

Poon, B., K. Grovit-Ferbas, S. A. Stewart, and I. S. Y. Chen. 1998. Cell cycle arrest by Vpr in HIV-1 virions and insensitivi-ty to antiviral agents. *Science* **281**:266–269.

Re, M. C., G. Furlini, D. Gibellini, M. Vignoli, E. Ramazzot-ti, S. Lolli, S. Ranieri, and M. La Placa. 1994. Quantification of human immunodeficiency virus type 1-infected mononu-clear cells in peripheral blood of seropositive subjects by newly developed flow cytometry analysis of the product of an in situ PCR assay. *J. Clin. Microbiol.* **32**:2152–2157.

Robinson, J. P., Z. Darzynkiewicz, P. Dean, A. Orfao, P. Rabinowitch, C. Stewart, H. Tanke, and L. Wheeles. 1999. *Current Protocols in Cytometry.* John Wiley and Sons, Inc., New York, N.Y.

Roizman, B., and A. Sears. 1995. Herpes simplex viruses and their replication, p. 2231–2296. *In* B. N. Fields, D. M. Knipe, and P. M. Howley (ed.), *Fields Virology.* Lippincott-Raven Press, New York, N.Y.

Salvant, B. S., E. A. Fortunato, and D. H. Spector. 1998. Cell cycle dysregulation by human cytomegalovirus: influence of the cell cycle phase at the time of infection and effects on cyclin transcription. *J. Virol.* **72**:3729–3741.

Schmid, I., C. H. Uittenbogaart, B. Keld, and J. V. Giorgi. 1994. A rapid method for measuring apoptosis and dual color immunofluorescence by single laser flow cytometry. *J. Immunol. Methods* **170**:145–157.

Schols, D., R. Snoeck, J. Neyts, and E. De Clercq. 1989. Detection of immediate early, early and late antigens of human cytomegalovirus by flow cytometry. *J. Virol. Methods* **26**:247–254.

Shapiro, H. M. 1995. *Practical Flow Cytometry*, 3rd ed. Wiley-Liss, Inc., New York, N.Y.

Stewart, S. A., B. Poon, J. B. M. Jowett, and I. S. Y. Chen. 1997. Human immunodeficiency virus type 1 Vpr induces apoptosis following cell cycle arrest. *J. Virol.* **71**:5579–5592.

Teodoro, J. G., and P. E. Branton. 1997. Regulation of apop-tosis by viral gene products. *J. Virol.* **71**:1739–1746.

Watson, J. V. 1991. *Introduction to Flow Cytometry.* Cambridge University Press, New York, N.Y.

Yang, G., S. Garhwal, J. C. Olson, and G. N. Vyas. 1994. Flow cytometric immunodetection of human immunodefi-ciency virus type 1 proviral DNA by heminested PCR and digoxigenin-labeled probes. *Clin. Diagn. Lab. Immunol.* **1**:26–31.

Yang, G., J. C. Olson, R. Pu, and G. N. Vyas. 1995. Flow cytometric detection of human immunodeficiency virus type 1 proviral DNA by the polymerase chain reaction incor-porating digoxigenin- or fluorescein-labeled dUTP. *Cytometry* **21**:197–202.

Zhang, G., V. Gurtu, S. R. Kain, and G. Yan. 1997. Early detection of apoptosis using a fluorescent conjugate of annexin V. *BioTechniques* **23**:525–531.

Zolla-Pazner, S., J. O'Leary, S. Burda, M. K. Gorny, M. Kim, J. Mascola, and F. McCutchan. 1995. Serotyping of pri-mary human immunodeficiency virus type 1 isolates from diverse geographic locations by flow cytometry. *J. Virol.* **69**:3807–3815.

Automation in Diagnostic Virology

CURT A. GLEAVES

22

Until recently, automation has not played a large role in the diagnostic virology laboratory. However, increased workloads have led to a demand for automated procedures. This increase in workload is due in part to the rapid growth of molecular diagnostics, effective antiviral therapy and monitoring, laboratory consolidation in hospitals, and health care reform (Chan, 1996; Felder and Kost, 1998; Bauer, 1999; B. Dale, presented at Roche workshop, "The Future of Molecular Diagnostics," 1999). Additionally, many of the test procedures in the diagnostic virology laboratory are labor intensive, and it can take several hours to several days before the test can be completed. Automation would alleviate much of this labor requirement as well as decrease turnaround time (TAT) (Chan, 1996; Tang et al., 1997; Felder and Kost, 1998).

HISTORY

Historically, automation in the diagnostic virology laboratory has been primarily associated with viral antibody testing (serology) rather than direct detection of the virus itself or one of many viral components (i.e., viral antigen[s] or viral nucleic acid). Until the late 1980s and early 1990s, with few exceptions (the IMx and VIDAS), most instruments used in the virology laboratory (e.g., enzyme immunoassay [EIA] plate readers and washers) consumed a lot of technologists' time and at best were considered semiautomated if they included a software component for result calculations. Compared to other clinical laboratory sections, where automation has been a mainstay for a number of years, the diagnostic virology laboratory is just now learning to embrace automation.

GENERAL CONSIDERATIONS

Several questions and considerations need to be addressed before automated procedures are implemented in the virology laboratory.

1. What test(s) can and should be automated? Labor-intensive tests, such as molecular testing, and high-volume testing, such as prenatal viral serologies and *Chlamydia trachomatis*, hepatitis, and human immunodeficiency virus (HIV) testing, which require high throughput with rapid TAT, are good examples.

2. What is the performance of the test procedures with the automated system? Have studies been done to show performance, such as sensitivity, specificity, and reproducibility, comparable to that of current nonautomated assays? Have comparisons with other automated systems been performed? (Morré et al., 1999; Waites et al., 1999; C. Gleaves, M. Nishimoto, M. Campbell, and R. Dworkin, *Abstr. 99th Gen. Meet. Am. Soc. Microbiol. 1999*, abstr. V-11, p. 658, 1999).

3. What are the cost and capital expense? What does it cost to bring the system into the work flow of the diagnostic virology laboratory? This will include reagent cost; labor cost, which includes the testing procedure; and the maintenance required. Most companies would prefer to sell the automated instrument, a maintenance contract, and test kits needed to perform the assay. However, with the ever-changing health care environment and with technology advancing rapidly, maintaining flexibility by renting a system for a specific reagent may be more cost-effective. Additionally, one needs to consider that (particularly in the United States) not all of the automated systems available have large test menus (this is truer for molecular testing than for viral-serology testing). Therefore, it is likely that two or three different automated systems will be needed to perform the testing that is required.

Other considerations include the space requirements for the system or systems to be used, staffing needs, and the skill level required to operate the equipment and perform the automated procedures. How will automation fit into the work flow of an already labor-intensive, very complex laboratory (Kiehl, 1986; Chan, 1996; Tang and Persing, 1999)?

AUTOMATED INSTRUMENTS AND TESTS AVAILABLE FOR THE DIAGNOSTIC VIROLOGY LABORATORY

Major advances have been achieved in adapting older instruments and designing new instruments for automating the diagnostic virology process. The emergence of molecular techniques for diagnostic virology has helped to advance this process, and there are several instruments and procedures available for the virology laboratory (Table 1). Examples are the Abbott IMx and LCx systems. Both of these systems were adapted from the Abbott TDx platform. The IMx has been used for several years for viral serologies, such as TORCH testing and HIV antibody and hepatitis panels, and for the detection of *C. trachomatis* (Pezzlo and Tilton, 1994). The LCx system was specifically developed for the detection of ligase chain reaction (LCR)-amplified products from both *C. trachomatis* and *Neisseria gonorrhoeae* in clinical samples. New instruments, such as the TIGRIS and QUANTIPLEX 340 systems, have been designed specifically for high-throughput molecular diagnostics and for monitoring antiviral therapy, primarily for HIV- and hepatitis C virus (HCV)-infected individuals (Yen-Lieberman, 1999; L. Merrill and M. Longiaru, presented at Gen-Probe technical workshop, 1998).

Viral Serology

As mentioned above, automated systems for viral serologies have been used in the laboratory for several years. Several automated systems are commercially available for routine use and offer adequate test menus for the laboratory to perform TORCH testing, hepatitis serologies, and HIV testing (Table 1). Basically, the methodologies for most of the systems are EIA based, involving capture molecules (viral antigens or antibodies) coating a solid phase, such as polystyrene beads (AxSYM and Immulite) or microparticles (IMx), or other capture devices, such as the SPR used in the VIDAS system (Rogers et al., 1989). Some systems, such as the AxSYM, IMx, and VIDAS analyzers, are considered closed systems. Once the serum samples are loaded into a specimen cup, tray, or strip along with detection reagents and wash buffer, the instrument performs all of the necessary functions, including adding the appropriate reagents, incubations, washes, and detection of the specific analyte being tested for. Once these systems are closed, they cannot be opened to add additional tests until the tests being performed are completed.

The Immulite 1000 and 2000 are examples of random-access systems for viral serologies. At any time during the process, additional tests and reagents can be added to the system without disrupting any current testing being performed (Babson, 1991; Gleaves et al., *Abstr. 99th Gen. Meet. Am. Soc. Microbiol. 1999*).

Open automated systems can be beneficial for laboratories that perform several different manual plate EIAs from various manufacturers; the open systems allow these manual EIAs to be performed in an automated format. Some of these systems have their own specially designed plate assays but still allow other manufacturers' assays to be used (Table 1).

Virus Detection

There has not been much effort to automate "classic" virus detection. There are manual EIAs for hepatitis B virus surface antigen (HBsAg), herpes simplex virus (HSV), rotavirus, enteric adenoviruses, and influenza A virus, and these tests can be performed with the open automated systems. IMx, AxSYM, VIDAS, and Immulite have automated tests available for HbsAg and various systems have tried to add testing for HSV, respiratory syncytial virus, rotavirus, and *Clostridium difficile* toxins with various degrees of success (Rogers et al., 1989; Johnston et al., 1992; Pezzlo and Tilton, 1994; VIDAS system technical update, BioMerieux, 1999). Since the late 1980s, several systems have been used for the direct detection of *C. trachomatis*; however, with few exceptions, much of the current testing for *C. trachomatis* is performed by molecular techniques (Tang and Persing, 1999).

Molecular Virology

While automation has been important for various tests in virology, it is molecular diagnostics that has catalyzed the recent expansion of automation in diagnostic virology, with more molecular techniques being developed and used diagnostically (see chapter 19) (Winn-Deen, 1996). The Gen-Probe nucleic acid hybridization assays initially paved the way for molecular diagnostics in the clinical laboratory (Kohne, 1990).

As mentioned earlier, the Abbott LCX system was the first automated system for the detection of amplified product for both *C. trachomatis* and *N. gonorrhoeae* in clinical samples following specimen processing and amplification (Barany, 1991; Buimer et al., 1996). The LCX system has great potential for automating several procedures in the virology laboratory, and the system is being evaluated and tested for use with PCR-amplified products for viral testing of HIV and HCV.

Another chemistry analyzer, the COBAS system, has been converted for molecular diagnostics and can automatically perform the amplification and detection steps of PCR. The COBAS Amplicor system automates both the amplification and product detection steps of PCR once the specimens are extracted and placed in the instrument (DiDomenico et al., 1996; Jungkind et al., 1996). COBAS is currently being used for both qualitative and quantitative PCR for HIV and HCV and for detection of *Mycobacterium tuberculosis*, cytomegalovirus (CMV), and both *C. trachomatis* and *N. gonorrhoeae* in clinical samples (Dale, Roche Workshop, 1999).

The major disadvantages of the above-mentioned automated systems for molecular diagnostics are the specimen-processing and nucleic acid extraction steps. In most laboratories, these are still performed manually; however, the TECAN GENESIS RSP100, an automated pipetting instrument, has shown promise for accelerating sample processing for the LCx assays (K. L. Hanson and C. P. Cartwright, *Proc. Fifteenth Annu. Clin. Virol. Symp.*, abstr. S30, 1999).

Three automated systems are available for specimen processing and nucleic acid isolation, the ROBOT 9604, the NucliSens Extractor, and the COBAS AmpliPrep. The ROBOT 9604 from Qiagen is a fully automated workstation for extraction of both DNA and RNA from clinical specimens (technical bulletin, Qiagen, 1998). The purified nucleic acids are suitable for use with any of the molecular amplification methods currently available (i.e., PCR, LCR, nucleic acid sequence-based amplification, and transcription-mediated amplification [TMA]). It can process 96 samples in 2 h and can be used with any of the current manual extraction procedures offered by Qiagen. The NucliSens Extractor is a walk-away automated system which uses the Boom method, a solid-phase-based extraction technology for the isolation of nucleic acids from a variety of specimen types (Boom et al., 1990; van Buul et al., 1998). The systems allows 10 samples per h to be processed, and the final nucleic acid product is suitable for a wide variety of molecular applications, including sequencing and cloning. The COBAS AmpliPrep is an automated nucleic acid extraction instrument that can be used for the quantitative recovery of nucleic acids (Dale, Roche workshop, 1999). The procedure uses a specific nucleic acid probe, bound to a magnetic bead, which binds the released RNA or DNA following cell lysis. Currently, the COBAS AmpliPrep is a dedicated extraction instrument. It has been developed for specific COBAS Amplicor assays only, such as those for HIV and HCV, and cannot be used as an "open" nucleic acid extraction system like the other systems described above.

TABLE 1 Automated instruments and methods for diagnostic virology

Manufacturer	Product	Methodology	Menu	Features	Limitations
Abbott Laboratories	IMx, AxSYM	Both systems are antigen-antibody EIAs using fluorogenic enzyme substrates (MEIA) and fluorescence polarization	All hepatitis markers, TORCH panels, immunoglobulin G (IgG) and IgM, HIV-1 and -2, other retrovirus markers	IMx: bench unit, 20 tests/h, onboard dilution of sample; AxSYM: free-standing unit, 110 samples/h, onboard dilution of sample; both have data management	IMx: sample vol (150 μl/sample), run size, no random access; AxSYM: space requirements, most menu items not available in the United States
	LCx	Molecular detection using LCR for target amplification; amplified products bind to microparticles (MEIA) and are detected with a fluorogenic enzyme substrate	*Chlamydia* and GC; MTB and HIV-1 viral load	Bench unit, 22 tests/h, onboard sample dilution, data management	Throughput, no random access, space requirements, limited menu, some menu items not available in the United States
Bayer Corp. Diagnostics Division (Chiron)	Immuno 1	EIA using colorimetry	Rubella virus IgG and IgM, toxoplasma IgG, IgM, and IgE	Sample vol (5 μl), 120 tests/h, onboard washing and refrigeration, no daily or weekly maintenance, walk-away for up to 7 h	Space requirements, limited test menu for viral serologies
	QUANTIPLEX bDNA System 340	Single-amplification probe assay using a chemiluminescent substrate for detection	CMV, HIV-1, HCV, HBV	High throughput; run 84 tests/batch with version 2.0 or 178 tests/batch with version 3.0, version 3.0 can get up to 50 copies/ml of sample	Specimen vol (2 ml for version 2.0 and 1 ml for version 3.0), costs high for low-throughput laboratories, not Food and Drug Administration approved, addition of reagents manual
Beckman Instruments Inc.	ACCESS	Paramagnetic particle, chemiluminescent immunoassay	Rubella virus IgG, toxoplasma IgG, *C. trachomatis*	Sample vol (20 μl), 100 tests/h, bench unit, open reagent system, minimal maintenance, data management system, onboard dilution	Limited test menu
BioChem Immuno Systems Inc.	LABOTECH	Fully automated microplate analyzer; can perform 300 viral serologies by enzyme-linked immunosorbent assay-based methods; colorimetric measurement of endpoint	System can be programmed to completely process most manufacturers' viral serology kits; this includes HIV and hepatitis assays	Bench unit, open reagent system, data management system, onboard dilution	Dependent on other manufacturers' kits

(Continued on next page)

TABLE 1 Automated instruments and methods for diagnostic virology (*Continued*)

BioMerieux Vitek Inc.	VIDAS, Mini-VIDAS	Enzyme-linked fluorescent assay (ELFA); VIDAS PROBE test strips contain all of the reagents to perform amplification (TMA) and detection by ELFA	Serology: Rubella virus, toxoplasma, CMV IgG and IgM; varicella-zoster virus (VZV), rubeola, and mumps viruses; antigen detection: respiratory syncytial virus, rotavirus, *C. difficile*, *Chlamydia*; probe: *Chlamydia*, GC, MTB, and HIV-1 RNA test	Bench unit, fully automated, low maintenance, one-point calibration every 2 weeks, can perform several different tests per run	Limited run size, can only handle up to 28 samples/run, sample vol of 100 μl, bench space for the instrument and computer station; molecular tests are in development
Bio-Rad	CODA Complete Microplate Analysis System	Open-format EIA using colorimetry; can be used for back-end detection of amplified products	Epstein-Barr virus EBV serologies, IgG and IgM TORCH panels, HSV type 1 (HSV-1) and -2 IgG, other identification serologies; AmpliTek LCR kit can be used with the system	Bench unit, fully automated, low maintenance, can perform semiautomated assays, complete data management software	Space requirements, limited test menu, cost
Diagnostic Products Corp.	Immulite 1000, Immulite 2000	Chemiluminescent bead immunoassay	Full IgG TORCH panels, hepatitis testing	Random-access instrument, can complete 60 tests /h, samples can be added to the instrument on a continuous basis	Limited test menu and space requirements
Diamedix	MAGO, Automated EIA Processor	Basic EIA colorimetric measurement of endpoint; now have an open system	Measles virus IgG and TORCH serologies	Batch or random-access instrument, ready-to-use reagents, capable of performing several different assays per run	Limited test menu and space requirements; however, other assays can be added to the system
DiaSorin	COPALIS	High-resolution light-scattering analysis to measure particle coupling; uses laser technology, multiplex testing	TORCH serologies	Maximum of 72 tests/run, results in 12 min, bench unit with real-time status monitoring, detection of four distinct tests/cuvette plus an internal reference standard	Limited menu, space requirements
Gen-Probe	Leader 450i	Nucleic acid hybridization with chemiluminescent detection; TMA detection	*Chlamydia*, GC, and MTB probes; TMA for *Chlamydia* and MTB	Automated luminometer capable of processing 250 tubes/run	Labor-intensive before adding tubes to the 450i
	TIGRIS	Fully automated amplification system for TMA	Proposed menu: HIV-1, HCV, *Chlamydia*, CT-GC, MTB, HBV, viral-load testing for HIV-1	Fully integrated and automated; performs sample processing, amplification, and detection; can run 500 samples/8-h shift	Availability of test menu, space requirements, cost prohibitive for low-throughput laboratories

(*Continued on next page*)

TABLE 1 *(Continued)*

Meridian Diagnostics (Gull Laboratories)	DUET Workstation	Basic EIA colorimetric measurement of endpoint	Runs all of Gull's viral IgG and IgM serologies (EBV, CMV, HSV-1 and -2, VZV, measles, rubella, and rubeola viruses, and *Toxoplasma*); it is also an open system, and other manufacturer's kits can be formatted for it	Can perform 2 EIA plates/run (192 tests), can run IgG and IgM assays at the same time, fully automated, handles all sample and reagent additions, can run 1 to 8 different tests/sample	Cannot be formatted for HIV or hepatitis immunoassays
INTRACEL (Bartels)	Alpha 4 Immunoassay System	Fully automated open immunoassay system; EIA using colorimetry detection	Several ready-to-use infectious disease serology kits; can be programmed to perform other manufacturer's kits as well	Can process 384 wells/run; automatically pipettes all reagents, controls, and samples; can store 180 protocols in memory	Space requirements, cost per test for a low-vol laboratory
Organon-Teknika	Xtractor	Extraction and isolation of nucleic acids from clinical specimens	Isolation of both DNA and RNA	Sample versatility (serum, plasma, whole blood, tissues, cells, etc.), inactivates microbial agents, flexible sample vol (0.01 to 2 ml)	Space requirements, cost, can extract only 10 samples at a time
Packard	MultiPROBE II Automated Liquid Handling System	Computer-controlled X-Y-Z robotic liquid-handling system; equipped with a 4-channel fluid design that produces consistent and reproducible high-quality results	Open system designed for automation of sample preparation procedures	Liquid transfers can be performed in a multitipped mode from any combination of laboratory containers	Space requirements, cost; best for high-throughput laboratories
Qiagen	ROBOT 9604	Integrated modular workstation for nucleic acid preparation	Isolation of both DNA and RNA	High throughput (up to 96 samples processed in <2 h), completely automated; all of the current QIAamp viral extraction procedures can be performed	Space requirements, cost; for high-throughput laboratories
Roche	COBAS Amplicor	PCR system that automates both amplification and product detection by colorimetric reading	*Chlamydia*, HCV (HIV-1, MTB, and GC not available in the United States)	Parallel amplification and detection, unattended overnight operation, up to 100 detections without operator intervention	Limited menu, up-front specimen processing, not an open system for other non-Amplicor PCR methods, cost
	COBAS AmpliPrep	Uses a specific nucleic acid probe, bound to a magnetic bead, which binds the released RNA or DNA after cell lysis	Specific AmpliPrep systems are being developed for HIV and HCV, with HBV and CMV to follow	Automated nucleic acid extraction, quantitative recovery of viral nucleic acid, throughput of 24 to 30 specimens/h, contamination control; when coupled with the COBAS TaqMan system, a fully automated extraction, amplification, and detection system will be available	Availability, space requirements, expense

Other Molecular Virology Automation and Testing

Several other diagnostics companies have developed or are developing new automated molecular testing as well as working on fully automated molecular diagnostic systems. Bayer Diagnostics has acquired the rights to the Chiron Quantiplex single-amplification assays (for branched DNA) and is currently marketing the Q340 semiautomated system for the viral-load quantitation of HIV, HCV, and hepatitis B virus (HBV) (Yen-Lieberman, 1999; M. A. Martin, C. A. Vitkauskas, B. Niesters, D. A. Tabaczynski, P. Jorgensen, G. Bhandari, S. Mayer, R. L. Hodinka, and Y. Yoshi, *Proc. Fifteenth Annu. Clin. Virol. Symp.*, abstr. M5, 1999). Bio-Merieux Vitek has joined with Gen-Probe to develop TMA amplification assays called VIDAS PROBE tests to be used with the VIDAS automated system (VIDAS technical update, BioMerieux, 1999). The VIDAS PROBE test strip will contain all of the necessary reagents to perform both the amplification and product detection reactions in the VIDAS instrument. This will allow the amplification reaction to be automated and closed inside the disposable test strip. The SPR component of the VIDAS system will be coated with a specific capture probe for product detection, and each test will include an internal control for recognition of false-positive results. The BDProbeTec ET System from Becton Dickinson uses strand displacement amplification technology (see chapter 19). Specimen processing is performed initially, and the amplification and detection steps are run concurrently in the instrument. The results are ready in an hour, and an amplification control is run with each sample to reduce any false-negative results (R. R. Kendrick, R. L. Sautter, W. D. LeBar, and T. Rudolph, *Proc. Fifteenth Annu. Clin. Virol. Symp.*, abstr. S34, 1999).

Another automated molecular technique is called matrix hybridization or chip technology (see chapter 19) (Fodor et al., 1993). The advantage of this system is its ability to resolve mixtures of amplified sequences, allowing several viral-DNA probes to be attached to a chip and an amplified specimen product to be tested for a variety of viral infections or antivirus mutations. The first system available is for the detection of HIV type 1 (HIV-1) genome mutations for HIV-1 drug resistance (Kozal et al., 1996).

Total Automated Systems for Molecular Virology

One of the earliest automated systems for molecular virology was the VITROS system developed by Johnson and Johnson in the early 1990s. It consisted of a closed-vessel system based on PCR, and infectious agents could be detected within 2 h (Findlay et al., 1993). Since then, this automated system, called VITROS ECi, has been developed into a random access system for immunodiagnostics (G. Colebrooke, A. Stalham, E. Effendowicz, C. Thomas, J. Wright, and S. Edwards, *Proc. Fifteenth Annu. Clin. Virol. Symp.*, abstr. M2, 1999).

Two fully automated instrument systems are being developed for molecular diagnostics. TIGRIS, from Gen-Probe, is the first instrument system to completely automate nucleic acid amplification testing. The system performs sample processing and amplification (TMA) and utilizes dual kinetic assay technology for amplified-product detection. The system is reported to be able to process 500 tests in 8 h or 1,000 samples in 12 h. The current proposed menu includes dual detection of *C. trachomatis* and *N. gonorrhoeae* and dual detection of HIV and HCV for blood bank screening and HIV-1 viral-load testing (McDonough et al., 1998; Merrill and Longiaru, Gen-Probe technical workshop, 1998). Another fully automated extraction, amplification

(PCR), and detection system for molecular virology will combine COBAS AmpliPrep and COBAS TaqMan from Roche Biomolecular Systems. One major advantage of this system is that COBAS TaqMan is a homogenous real-time detection system, allowing the detection of the targeted product to occur simultaneously with the amplification process. Since there will be no post-PCR detection assay, the total assay time could be decreased from 6 to 2.5 h (Martell et al., 1999; Mercier et al., 1999; Dale, Roche workshop, 1999; W. Ping, Roche workshop, 1999).

CONCLUSIONS

Although instrumentation and automation in diagnostic virology have lagged behind other disciplines in the clinical laboratory, including microbiology, new technology and instrumentation have resulted in many new approaches. These approaches have contributed to faster testing TAT and result reporting for physicians; in some instances the need for primary virus isolation prior to detection has been eliminated, and new techniques have led to greater accuracy and sensitivity in detection of low concentrations of viral agents in clinical specimens. These advantages both help the laboratory and are beneficial to the patient, allowing for better patient management, decreased empirical therapy, and more cost-effective use of antiviral and antimicrobial agents. Additional advantages include computer interfacing with other departments, optimizing the workflow in the laboratory, and a reduction in labor and labor-intensive procedures. However, reduction in labor due to automation has not always been realized in the general diagnostic laboratory and may prove more difficult to achieve in diagnostic virology (Pezzlo and Tilton, 1994).

In the last decade, the diagnostic virology laboratory has become more integrated with the general diagnostic laboratory. With this integration, however, have come many internal and external challenges. Internally, availability of qualified technologists and decreased hospital space and hospital resources for the laboratory are continual problems. Managed-care competition, cost compression, and increased regulation of the testing laboratory are but a few of the external problems that the laboratory faces. Automation and instrumentation for diagnostic virology are major expenditures for the hospital, and justifying them to directors and administrators in our current health care climate is difficult. This is made more difficult when, as mentioned earlier (Table 1), many of the automated instruments currently available for diagnostic virology are only capable of performing one or a handful of virology procedures.

It should not be long before the majority of viral diagnostic procedures become automated. Several automated systems currently have the capability of being used for this purpose, particularly for viral serology and molecular virology (Chan, 1996; Tang and Persing, 1999; Dale, Roche workshop, 1999). It is only a matter of time before manufacturers begin to develop such systems and make them available to the laboratory. As diagnostic virologists, we must be involved in the development, evaluation, and application of these new systems and procedures. It is our responsibility to determine which automated systems and tests will best suit our laboratory and physicians' and patients' needs.

REFERENCES

Babson, A. L. 1991. The Cirrus IMMULITE: a new random access automated immunoassay system. *J. Clin. Immunoassay* **14:**83–88.

Barany, F. 1991. The ligase chain reaction in a PCR world. *PCR Methods Appl.* **1:**5–16.

Bauer, S. 1999. Lab automation fantasies and realities. *Adv. Lab.* **8:**24–28.

Boom, R., C. J. A. Sol, M. M. M. Salimans, C. L. Jansen, P. M. E. Wertheim-van Diilen, and J. van der Noordaa. 1990. Rapid and simple method for purification of nucleic acids. *J. Clin. Microbiol.* **28:**495–503.

Buimer, M., G. J. J. Van Doornum, S. Ching, P. G. H. Peerbooms, P. K. Plier, D. Ram, and H. H. Lee. 1996. Detection of *Chlamydia trachomatis* and *Neisseria gonorrhoeae* by ligase chain reaction-based assays with clinical specimens from various sites: implications for diagnostic testing and screening. *J. Clin. Microbiol.* **34:**2395–2400.

Chan, D. W. 1996. General introduction, p. 1–8. *In* D. W. Chan (ed.), *Immunoassay Automation: An Update Guide to Systems.* Academic Press, San Diego, Calif.

DiDomenico, N., H. Link, R. Knobel, T. Caratsch, W. Weschier, Z. G. Loewy, and M. Rosenstraus. 1996. COBAS AMPLICOR: fully automated RNA and DNA amplification and detection system for routine diagnostic PCR. *Clin. Chem.* **42:**1915–1923.

Felder, R. A., and G. J. Kost. 1998. Automation of diagnostic testing. *Med. Lab. Obs.* **30:**22–27.

Findlay, J. B., S. M. Atwood, L. Bergmeyer, J. Chemelli, K. Christy, T. Cummins, W. Donish, T. Ekeze, J. Falvo, and D. Patterson. 1993. Automated closed-vessel system for in vitro diagnostics based on polymerase chain reaction. *Clin. Chem.* **39:**1927–1933.

Fodor, S. P., R. P. Rava, X. C. Huang, A. C. Pease, C. P. Holmes, and C. L. Adams. 1993. Multiplex biochemical assays with biological chips. *Nature* **364:**555–556.

Johnston, S. L. G., S. Hamilton, R. Bindra, D. A. Hursh, and C. A. Gleaves. 1992. Evaluation of an automated immunodiagnostic assay system for direct detection of herpes simplex virus antigen in clinical specimens. *J. Clin. Microbiol.* **30:**1042–1044.

Jungkind, D., S. DiRenzo, K. G. Beavis, and N. S. Silverman. 1996. Evaluation of automated COBAS AMPLICOR PCR system for detection of several infectious agents and its impact on laboratory management. *J. Clin. Microbiol.* **34:**2778–2783.

Kiehl, B. L. 1986. Considerations for automation of virology testing, p. 239–244. *In* S. Specter and G. J. Lancz (ed.), *Clinical Virology Manual.* Elsevier, New York, N.Y.

Kohne, D. E. 1990. The use of DNA probes to detect and identify microorganisms, p. 11–35. *In* B. Kleger (ed.), *Rapid Methods in Clinical Microbiology.* Plenum Press, New York, N.Y.

Kozal, M. J., N. Shah, N. Shen, R. Yang, R. Fucini, T. C. Merigan, D. D. Richman, D. Morris, E. Hubbell, M. Chee, and T. R. Gingeras. 1996. Extensive polymorphisms observed in HIV-1 clade B protease gene using high-density oligonucleotide arrays. *Nat. Med.* **2:**753–759.

Martell, M., J. Gomez, J. I. Esteban, S. Sauleda, J. Quer, B. Cabot, R. Esteban, and J. Guardia. 1999. High-throughput real-time reverse transcription-PCR quantitation of hepatitis C virus RNA. *J. Clin. Microbiol.* **37:**327–332.

McDonough, S. H., C. Giachetti, Y. Yang, D. P. Kolk, E. Billyard, and L. Mimms. 1998. High throughput assay for the simultaneous or separate detection of human immunodeficiency virus (HIV) and hepatitis type C virus (HCV). *Infusion Ther. Transfus. Med.* **25:**164–169.

Mercier, B., L. Burlot, and C. Ferec. 1999. Simultaneous screening for HBV DNA and HCV RNA genomes in blood donations using a novel TaqMan PCR assay. *J. Virol. Methods* **77:**1–9.

Morré, S. A., I. G. M. Van Valkengoed, R. M. Moes, A. J. P. Boeke, C. J. L. M. Meijer, and A. J. C. Van den Brule. 1999. Determination of *Chlamydia trachomatis* prevalence in an asymptomatic screening population: performance of the LCx and COBAS Amplicor tests with urine specimens. *J. Clin. Microbiol.* **37:**3092–3096.

Pezzlo, M. T., and R. C. Tilton. 1994. Automation and instrumentation in clinical microbiology, p. 197–210. *In* B. J. Howard, J. F. Keiser, T. F. Smith, A. S. Weissfeld, and R. C. Tilton (ed.), *Clinical and Pathogenic Microbiology,* 2nd ed. Mosby, St. Louis, Mo.

Rogers, C. H., K. L. Hoffman, and R. M. Juris. 1989. An automated system for infectious disease diagnostics. *Am. Clin. Lab.* **8:**35–37.

Tang, Y. W., and D. H. Persing. 1999. Molecular detection and identification of microorganisms, p. 215–248. *In* P. R. Murray, E. J. Baron, M. A. Pfaller, F. C. Tenover, and R. H. Yolken (ed.), *Manual of Clinical Microbiology,* 7th ed. American Society for Microbiology, Washington, D.C.

Tang, Y. W., G. W. Procop, and D. H. Persing. 1997. Molecular diagnostics of infectious diseases. *Clin. Chem.* **43:**2021–2038.

van Buul, C., H. Cuypers, P. Lelie, M. Chudy, M. Nubling, R. Melseri, A. Nabble, and P. Oudshoorn. 1998. The NucliSen extractor for automated nucleic acid isolation. *Infusion Ther. Transfus. Med.* **25:**147–151.

Waites, K. B., K. R. Smith, M. A. Crum, R. D. Hockett, A. H. Wells, and E. W. Hook III. 1999. Detection of *Chlamydia trachomatis* endocervical infections by ligase chain reaction versus ACCESS *Chlamydia trachomatis* antigen assay. *J. Clin. Microbiol.* **37:**3072–3073.

Winn-Deen, E. S. 1996. Automation of molecular genetic methods—Part 2. DNA amplification techniques. *J. Clin. Ligand Assay* **19:**21–26.

Yen-Lieberman, B. 1999. Quantitative HIV-1 RNA evaluation of methods and quality assurance issues, p. 67–81. *In Molecular Virology Workshop.* Pan American Society for Clinical Virology, Clearwater, Fla.

VIRAL PATHOGENS

Respiratory Viruses

MAURICE A. MUFSON

23

The viruses that primarily infect the respiratory tract, designated "respiratory viruses," include the influenza viruses, adenoviruses, parainfluenza viruses, respiratory syncytial virus (RSV), coronaviruses, and rhinoviruses (Table 1). Viral infections of the respiratory tract as a group annually account for substantial morbidity and mortality; influenza viruses account for the most deaths, mainly among high-risk persons. These infections also represent an important cause of days lost from work and school, accounting for significant losses both financially and socially.

Minor upper respiratory infections, namely colds, bronchitis, flu, and flu-like illnesses, rank high among the most common diagnoses of persons treated as outpatients. Rhinoviruses are most often associated with these minor respiratory illnesses, although the other respiratory viruses cause these infections also (Table 2) (Pitkaranta and Hayden, 1998). Viral pneumonia and bronchiolitis account for about 1/10 of community-acquired lower respiratory tract infections. All of the respiratory viruses can cause pneumonia in children and adults, although the rhinoviruses do so infrequently and usually only in persons with serious underlying pulmonary disease; normally they are not considered to be a lower respiratory tract pathogen. RSV and the parainfluenza viruses are the main causes of bronchiolitis among infants and children. During epidemics of influenza virus and RSV, many of the pneumonias among adults are a result of secondary bacterial infections (Cate, 1998).

Viral respiratory tract infections occur worldwide. Influenza virus, parainfluenza viruses, and RSV occur epidemically. Adenoviruses, coronaviruses, and rhinoviruses occur endemically. Persons of all age groups are susceptible to viral respiratory tract infections, although children suffer about twice as many infections as adults (Table 3). Viruses that infect the respiratory tract are transmitted usually by direct contact and less often by aerosols. Many factors contribute to the severity of illness, including the characteristics of the virus, inoculum size, and host factors, such as age, general health and underlying diseases, immune status, socioeconomic status, and nutritional state. Viruses that principally target other organs but also infect the respiratory tract include the coxsackie viruses, echoviruses, herpes simplex virus, and varicella zoster virus (they are discussed in other chapters).

An etiologic diagnosis of viral respiratory tract illnesses requires laboratory confirmation, because the symptoms and signs of these illnesses lack the specificity that permits etiologic recognition on clinical grounds alone with any confidence. Influenza virus infection usually produces a respiratory tract infection more serious than those of other viruses, with systemic symptoms and signs, but these clinical findings do not unequivocally indicate influenza. Its epidemic occurrence in the winter in the northern hemisphere and the identification of influenza virus in a community lend some degree of certainty to calling a flu-like illness the flu. A positive laboratory test is required for the highest degree of certainty, and a rapidly done test is especially important now that effective chemotherapeutic drugs for the treatment of influenza virus infection are available. Similarly, during the annual winter epidemic of RSV, an infant or child with a serious bronchiolitis illness likely is infected with this virus. A rapid diagnostic test positive for the virus can provide timely etiologic confirmation and thus allow antiviral treatment with ribavirin to be considered in seriously ill infants with underlying cardiopulmonary diseases (Edell et al., 1998; Rodriguez, 1999; Rodriguez et al., 1999).

Several new diagnostic tests for respiratory viruses are commercially available or under development, obviating the need to attempt virus isolation in cell culture (Table 4). These tests include PCR and multiplex PCR, utilizing gene technology; enzyme immunoassay (EIA); other immunoassays; immunofluorescence; and an optical immunoassay for specific and rapid identification of individual virus infections (Kellogg, 1991; Chomel et al., 1992; Dominguez et al., 1993; Olsen et al., 1993; Leonardi et al., 1994; Covalciuc et al., 1999). Such rapid diagnostic procedures provide speed and exhibit a high degree of sensitivity and specificity when compared with the "gold standard" of virus isolation. Commercially available diagnostic tests for the rapid identification of influenza virus or RSV include the antigen detection optical immunoassay test, which uses a silicon wafer for influenza virus (FLU OIA); immunoassay kits for influenza A virus (Directigen Flu-A; Zstatflu; Primea EIA; Enzygnost Influenza A[Ag], and Enzygnost Influenza B[Ag]); a direct EIA for RSV (Abbott TestPack RSV and Enzygnost RSV [Ag]); and a monoclonal antibody-based immunofluorescence test for both of these viruses (Imagen for RSV and FluA for influenza virus) (Chomel et al., 1992; Dominguez et al., 1993; Olsen et al., 1993; Leonardi et al., 1994; Covalciuc et al., 1999).

TABLE 1 Viruses that primarily infect the respiratory tract

Family	Genus	Virus and serotypes
Orthomyxoviridae	*Influenzavirus*	Influenza virus, types A, B, and C
Paramyxoviridae	*Respirovirus*	Parainfluenza virus, types 1, 2, 3, 4A, and 4B
	Pneumovirus	RSV, subgroups A and B
Adenoviridae	*Adenovirus*	Adenovirus serotypes 1–49
Picornaviridae	*Rhinovirus*	Major group, 89 types; minor group, 10 types
Coronaviridae	*Coronavirus*	Group 1, HCV 229E; group 2, HCV OC43

Respiratory virus infections represent a major public health problem because of their worldwide occurrence, rapid spread in the community, and considerable morbidity and mortality. It is not surprising that the development of effective and safe chemotherapeutic drugs and of highly immunogenic and protective vaccines for each of the individual respiratory viruses is a continuing priority. Except for the influenza viruses, it remains an elusive goal.

The influenza virus vaccine and several anti-influenza virus drugs provide means for preventing and treating this most serious of respiratory virus infections. Influenza vaccine, first developed in the early 1940s, still produced in embryonated eggs, and recently manufactured as a purified preparation of neuraminidase and hemagglutinin, elicits a satisfactory antibody response and provides a high measure of protection. Safe and inexpensive, it represents a cost-effective means to prevent most influenza illnesses and to reduce mortality. Several influenza virus-specific drugs are also available for the treatment of influenza; these include the adamantanes, amantadine and rimantadine, and the newly introduced competitive inhibitors of neuraminidase, zanamavir and oseltamivir (Monto et al., 1999). The neuraminidase inhibitors represent uniquely specific and effective treatment for both influenza A and B viruses, and if they are applied in a timely manner to the treatment of proven influenza virus infection only, these drugs could significantly reduce the serious outcomes of this infection.

Two different approaches are available for the management of RSV infection. These include a humanized monoclonal antibody for prophylaxis (palivizumab [Snyagis]) and ribavirin (Virazole), an anti-RSV drug. Aerosolized ribavirin offers some therapeutic benefit in the treatment of RSV illness among infants with underlying cardiopulmonary diseases; however, it seems that its benefit in the treatment of otherwise-well children lies in reduced reactive-airway disease later in life (Edell et al., 1998; Rodriguez, 1999; Rodriguez et al., 1999). Palivizumab, a humanized monoclonal antibody directed against the fusion (F) protein of RSV, appears to be an effective immunoprophylaxis for serious infections in premature infants and children suffering bronchopulmonary dysplasia (The IMpact-RSV Study Group, 1998).

A live attenuated orally administered adenovirus vaccine incorporating only serotypes 4 and 7, applicable only for military recruits, is widely used in this special population. Other respiratory virus vaccines are under development, including vaccines for RSV and the parainfluenza viruses, but none seems near approval for general use. The need demands effective vaccines and safe and cost-effective drugs specific for these respiratory virus infections.

INFLUENZA VIRUS

Of all viruses that infect the respiratory tract, influenza viruses cause the predominant number of serious acute respiratory tract illnesses, typically the flu and pneumonia (Cox and Subbarao, 1999). Influenza A virus was isolated in 1933, influenza B virus was isolated in 1940, and influenza C virus was isolated in 1951. The estimated number of deaths in the United States during the pandemics of 1957 and 1968 was 100,000, and attack rates approached 50% (Nicholson, 1992). Influenza and pneumonia deaths increased among elderly persons during recent decades, underscoring the need for prompt diagnosis and treatment and assiduous use of preventive measures, including avoiding infected persons and widespread use of the influenza vaccine.

Influenza A and B viruses belong to the family *Orthomyxoviridae* and the genus *Influenzavirus*; influenza C virus belongs to a separate genus. The virions are enveloped particles containing a single-stranded, negative-sense segmented-RNA genome that is surrounded by a helical capsid; influenza A and B viruses contain eight segments of RNA, and influenza C contains seven segments. The enveloped virion contains several structural proteins, including the matrix (M) protein, the nucleocapsid protein, and three large proteins that function in RNA replication and transcription. The outer peplomers, or spikes, consist of two glycoproteins, the hemagglutinin (HA), which is involved in the attachment of virus to cells and the initiation of infection, and the neuraminidase (NA), which facilitates release of the virus from cells. These two glycoproteins exhibit sub-

TABLE 2 Relative importance of viruses in upper and lower respiratory tract disease

Virus	Importance in the following respiratory tract disease[a]:				
	Common cold	Flu/flu-like illness	Otitis media	Bronchiolitis	Pneumonia
Influenza virus	++++	++++	−	+	++++
Parainfluenza virus	+++	+	−	++++	++++
RSV	+++	+	−	++++	++++
Adenovirus	+++	++	−	−	++++
Rhinovirus	++++	+	+	−	−
Coronavirus	++	+	+	−	+

[a] + to ++++, minimal to major importance; − no or negligible importance.

TABLE 3 Relative importance of respiratory viruses among different age groups

Virus	Importance in the following age group[a]:			
	Infants and children	Adolescents	Adults	Aged
Influenza virus	+++	+++	++++	++++
Parainfluenza virus	++++	+	++	+
RSV	++++	+	+	++
Adenovirus	+++	++++	++	
Rhinovirus	++++	+++	++++	+++
Coronavirus	+++	+	+	−

[a]+ to ++++, minimal to major importance; − no or negligible importance.

stantial antigenic variation, which is especially well documented among influenza A viruses; 15 HA subtypes and 9 NA subtypes are recognized (Table 5). The high rate of antigenic variation observed among the HA and NA glycoproteins of influenza A virus derives from its segmented genome and is the result of antigenic drift (due to stepwise mutations of the genes that encode the HA and NA), reflected in variations in the antigenic characteristics of the HA and NA glycoproteins, and antigenic shift (due to reassortment of genome segments from two different influenza viruses), with major changes in the HA or possibly NA or both. This variation holds the potential for the development of a pandemic (Table 5). Such antigenic variation also enables the virus to infect and produce disease among populations of persons who would otherwise possess immunity, because their antibody fails to recognize the new antigenic components of the virus (Gerhard et al., 1983).

Influenza A and B viruses account for more morbidity and mortality than any of the other respiratory viruses. During epidemic years, influenza viruses produce substantially more morbidity and mortality than in nonepidemic years. Elderly persons and persons of any age who have underlying cardiac and pulmonary diseases, diabetes, and immunosuppressive diseases are at increased risk of life-threatening respiratory tract illness from influenza virus infection. Usually 20,000 excess deaths and more than 100,000 excess instances of hospitalization occur during epidemics of influenza virus infection in the United States. None of the other respiratory viruses manifests an impact on the health of the population equal to that of the influenza viruses.

Influenza viruses comprise three types, A, B, and C. Only influenza A and B viruses are clinically important. Influenza C virus infection occurs uncommonly, and it usually is associated with mild upper respiratory tract illness; however, rarely it can be isolated from persons with bronchitis or pneumonia (Katagiri et al., 1987; Moriuchi et al., 1991; Greenbaum et al., 1998). Influenza A and B viruses

produce epidemics during the winter months in temperate climates, on a 1- to 3-year cycle for influenza A virus and on a 4- to 7-year cycle for influenza B virus. Pandemics of influenza A virus occur at approximately 10-year intervals (Table 5). Recently, two major influenza pandemics in the United States occurred: the "Asian flu" in 1956 to 1957 and the "Hong Kong flu" in 1968. The "swine flu" outbreak was limited to a single military base in 1968; however, that year a special swine flu vaccine was administered widely as part of the swine flu immunization program, but that virus never spread (Webster et al., 1992; Dowdle, 1997). The influenza viruses possessing the H1 HA continually circulate in swine (Chambers et al., 1991). Rarely, they infect humans and produce serious and fatal disease (Wentworth et al., 1994).

During the past three years, two new influenza viruses isolated in Hong Kong, an H5N1 and an H9N2, spread from chickens and poultry to a small number of persons; of 18 persons infected with the H5N1 virus, 6 died (Table 5) (Peiris et al., 1999; Zhou et al., 1999). These infections signal their epidemic potential and the peril that such natural recombinants of influenza virus hold for humans. Birds and fowl are the natural hosts of influenza viruses, and the influenza viruses that infect humans likely derived from avian reservoirs by reassortment (Webster et al., 1992). Also, avian-human reassortant viruses circulate in swine over long periods of time and possibly are transmitted to humans by this route (Claas et al., 1994).

The incubation period of influenza viruses is 1 to 2 days. The onset of disease is abrupt, with sudden high fever, headache, myalgias, rhinitis, tracheobronchitis, pharyngitis, and a cough. The intensity of the signs and symptoms characterizes the illness as the flu. In uncomplicated cases, the illness abates in about 1 week. Primary influenza virus pneumonia can complicate the course of illness, especially in elderly and other high-risk persons who suffer underlying cardiovascular or pulmonary diseases, diabetes, or immune disorders (Yousuf et al., 1997). Secondary bacterial infec-

TABLE 4 Procedures for laboratory diagnosis of respiratory viruses

Virus	Assays for diagnosis of virus infection[a]		
	Virus isolation (identification)	Antigen response	Antibody
Influenza virus	Cell cultures, eggs (HI, HAdI, IF)	IF, ELISA, RT-PCR	HI, HAdI, ELISA-IgM
Parainfluenza virus	Cell cultures (HI, HAdI, IF)	IF, ELISA, RT-PCR	CF, HI, ELISA-IgM
RSV	Cell cultures (IF, ELISA)	IF, ELISA, RT-PCR	CF, ELISA-IgM
Adenovirus	Cell cultures (HI, IF, NT)	IF, ELISA, RT-PCR	CF, ELISA, HI, NT
Rhinovirus	Cell cultures (NT)	RT-PCR	ELISA, NT
Coronavirus	Cell cultures (NT, HI)	IF, RT-PCR	CF, ELISA, IF

[a]CF, complement fixation; HAdI, hemadsorption inhibition; HI, hemagglutination inhibition; IF, immunofluorescence; NT, neutralization test.

TABLE 5 Major influenza A strains during the past 60 years

Common name of strain	Subtype[a]	Year(s) of occurrence
"Spanish flu"	H1N1	Pandemic, 1918–1920, and until about 1956
"Asian flu"	H2N2	Pandemic, 1956–1957, and until 1967
"Hong Kong flu"	H3N2	Pandemic, 1968, and continues to be present
"Swine flu"	H1N1	1976; limited to one U.S. military base
"Russian flu"	H1N1	Prevalent 1977 and continues to be present
Chicken flu in Hong Kong	H5N1	1997; limited to 18 cases in humans (Zhou et al., 1999)
Poultry flu in Hong Kong	H9N2	1999; limited to two cases in humans (Peiris et al., 1999)

[a]H, hemagglutinin; N, neuraminidase.

tion can also develop during serious influenza virus infection in the elderly and persons with underlying cardiovascular or pulmonary diseases and diabetes. The onset of productive cough, with purulent sputum, new fever spikes, chest pain, and shortness of breath portends the beginning of bacterial superinfection and pneumonia. *Streptococcus pneumoniae*, *Staphylococcus*, or *Haemophilus influenzae* are the most-common causes for these superinfections. Appropriate antibiotic treatment must be initiated promptly. Fatalities occur commonly in influenza virus pneumonia, especially in persons in the high-risk groups.

The laboratory diagnosis of influenza virus infection by isolation of the virus in cell cultures, the gold standard, is no longer satisfactory for routine diagnosis, because the physician needs to know rapidly whether a patient is infected with influenza virus if influenza virus-specific chemotherapeutic drugs are to be used in its treatment. However, virus isolation remains an important tool in epidemiologic studies and for the recognition of the specific type of influenza circulating in the community (Table 4). Nasal aspirates, nasopharyngeal swabs, throat swabs, and sputum specimens collected for this purpose should be kept at 4°C until processed. As influenza viruses do not produce cytopathic effects in susceptible cultures of monkey kidney cells, growth of the virus must be detected by hemadsorption procedures, and isolates can be typed by hemadsorption inhibition procedures (Table 4). Most clinical isolates of influenza virus grow in primary rhesus or cynomologous monkey cell cultures or continuous rhesus kidney (LLC-MK2) cell cultures or in Madin-Darby canine kidney (MDCK) cells in about 2 to 5 days. Serotyping isolates by using hemagglutination inhibition procedures can be done the same day, or hemadsorption inhibition, a more sensitive technique which requires subpassage of the isolate and an additional 2 to 3 days of culture, can be used. A new rapid culture assay involves overnight growth of influenza virus from clinical specimens in MDCK cells and staining of viral antigen by an immunoperoxidase technique employing pools of monoclonal antibodies specific for 10 influenza A virus HA types and influenza B viruses (Ziegler et al., 1995).

During the past decade, rapid diagnostic procedures replaced virus isolation for confirming the diagnosis of influenza virus infection; these tests detect minute amounts of viral antigen in infected cells present in respiratory secretions and take about 15 to 20 min to complete. The common rapid diagnostic methods involve EIA, other immunoassays such as the optical immunoassay, and immunofluorescence. Commercially available diagnostic tests for the rapid identification of influenza viruses include the antigen detection optical immunoassay test, which uses a silicon wafer for influenza virus (FLU OIA); several immunoassays, including Directigen Flu-A, Zstatflu, Primea EIA, Enzygnost Influenza A[Ag], and Enzygnost Influenza

B[Ag]; and a monoclonal antibody-based immunofluorescence test (Chomel et al., 1992; Dominguez et al., 1993; Olsen et al., 1993; Leonardi et al., 1994; Covalciuc et al., 1999).

An immunologic response, with detection of a fourfold or greater rise in serum antibody titers during convalescence, also provides evidence of acute infection with influenza virus. The tests employed for serologic diagnosis are hemagglutination inhibition, hemadsorption inhibition, and an immunoglobulin M (IgM) enzyme-linked immunosorbent assay (ELISA). As these tests require acute-phase and convalescent-phase serum specimens tested together, they have limited use in establishing the diagnosis in individual cases within a time frame that permits judicious use of anti-influenza virus drugs, but they are useful for tracking epidemics in the community. IgM-specific and IgA-specific antibodies against influenza virus reach peak levels about 14 days after infection, and IgG-specific antibody reaches peak levels about 4 to 7 weeks later. Serologic diagnosis and typing of influenza virus can be complicated by the fact that the anamnestic response to infection is highest to the strain causing the primary infection, even when there is subsequent infection by other strains. This has been termed the "doctrine of original antigenic sin."

Recently, a new class of drugs, competitive NA inhibitors, became available for the treatment of influenza A and B virus infections. Currently these include zanamavir and oseltamivir (Vellayappan et al., 1982; Hayden et al., 1999; Monto et al., 1999). These drugs exhibit specific anti-NA activity against both influenza A and B viruses. Based on derivatives of the sialic acid analogue 2-deoxy-2,3-dehydro-N-acetylneuraminic acid, they are potent inhibitors of influenza virus NA because they exhibit very low binding affinities for influenza NA. The dose of zanamivir is 10 mg twice daily administered by inhalation for 5 days. When administered very soon after the onset of laboratory-proven influenza virus infection, the duration of symptoms and signs, including fever, was shortened significantly by at least 1 day (Monto et al., 1999). The dose of oseltamivir is 75 mg twice daily for 5 days. Both NA inhibitors can also be used for prophylaxis of influenza; they are given either once or twice per day during the period that influenza virus is in the community. Properly used in cases of proven influenza virus infection, these drugs can significantly reduce the serious outcomes of this infection. Moreover, laboratory determination of influenza virus infection by rapid techniques not only provides a basis for using zanamivir or oseltamivir, but also precludes unnecessary use of antibiotics in these cases.

Amantadine (l-adamantanamine hydrochloride [Symmetrel]) and rimantadine (Flumadine) are effective prophylactic drugs for influenza A virus infections only (Bricaire et al., 1990). These drugs interfere with replication by blocking the action of the M2 protein. Ideally, they should be started before the person becomes infected; these drugs are 70 to 90% effective as prophylaxis. The dose is one 100-mg tablet twice daily and should be taken for the time necessary to protect the person from infection. Amantadine and rimantadine also can be used to treat the flu, apparently with some success, especially in the first 3 days of illness, because they tend to shorten the duration of illness slightly (Wintermeyer and Nahata, 1995). However, these drugs produce side effects involving the central nervous system that include delirium, hallucinations, and behavioral disorders and involving the gastrointestinal tract. These symptoms and signs are less likely to occur at lower dosages of the drugs. Resistant strains of amantadine and rimantadine develop, but they rarely circulate (Ziegler et al., 1999).

The influenza virus vaccine serves an important role in protecting high-risk persons from the flu and pneumonia, and it should be widely used. It shows at least 70 to 90% effectiveness in young adults and 30 to 50% effectiveness in elderly adults in preventing disease, depending upon the year and the vaccine formulation. The vaccine comprises two recent type A strains containing the newest HA and NA surface antigens and one type B strain (Table 6). The current vaccine, designated a "subvirion" or "purified" vaccine, consists of only these surface antigens (and not whole virus). Unlike the whole-virus vaccines still available (but not as widely used), the purified or subvirion vaccine can be given safely to children and adults. Annual influenza virus vaccination is recommended for persons at high risk of serious disease, namely persons 65 years of age or older, persons of any age with underlying cardiovascular and pulmonary diseases, persons with chronic metabolic diseases, residents of nursing homes and chronic care facilities, children, and adolescents receiving chronic aspirin therapy. It is recommended also for persons who have the opportunity to transmit influenza virus infection to persons in these high-risk groups (Table 7). Because influenza viruses exhibit antigenic drift almost annually and antigenic shift at several-year intervals, the vaccine needs to be given each year. The trivalent influenza vaccine for 1999 to 2000 includes the strains A/Beijing/262/95-like (H1N1), A/Sydney/5/97-like (H3N2), and B/Beijing/184/93-like (although in the United States the vaccine contains the antigenic equivalent B strain B/Yamanashi/166/98) (Table 6). Influenza vaccine is contraindicated in persons with chicken egg allergy.

Influenza virus epidemics impose an overwhelming burden of illness with substantial morbidity and mortality. None of the other respiratory viruses shows the same impact. The development and use of effective anti-influenza virus drugs to treat illness and of effective vaccines to prevent illness remain no less important today than in the past decade. One strategy for a new influenza virus vaccine is a live attenuated, cold-adapted vaccine that can be administered intranasally (Belshe et al., 1998). When tested in

TABLE 6 Influenza vaccine composition for 1999 to 2000

Influenza virus	Subtype[a]	Quantity of virus per dose (μg)
A/Beijing/262/95-like	H1N1	15
A/Sydney/5/97-like	H3N2	15
B/Beijing/184/93-like[b]		15

[a]H, hemagglutinin; N, neuraminidase.
[b]In the United States, the vaccine will contain the antigenically equivalent strain B/Yamanashi/166/98 because of its better growth.

TABLE 7 Recommendations for dosage of influenza vaccine for 1999 to 2000

Age group of recipients	Type of vaccine[a]	Dosage (ml)	No. of doses
6–35 mo	Split virus only	0.25	1 or 2[b]
3–8 yr	Split virus only	0.50	1 or 2[b]
9–12 yr	Split virus only	0.50	1
12 yr or more	Whole or split virus	0.50	1

[a]Children must receive only a split-virus vaccine (also called a "subvirion" or "purified-surface-antigen" vaccine). All vaccines are administered intramuscularly.
[b]Children 9 years of age or younger receiving vaccine for the first time should be given two doses of vaccine 1 month apart.

children 15 to 71 months of age, it proved safe and efficacious against the influenza A (H3N2) and B strains of 1996 to 1997. More treatment options that are effective, easy to use, and inexpensive need to be developed, especially if they are to be used worldwide in both developing and developed countries.

PARAINFLUENZA VIRUS

Parainfluenza viruses rank among the most important groups of respiratory viruses that cause acute upper and lower respiratory tract disease in infants and children (Table 2). First isolated from humans in 1956, parainfluenza viruses belong to the genus *Respirovirus* in the family *Paramyxoviridae* and contain a negative single-stranded negative-sense RNA genome that encodes six mRNAs, with each mRNA coding for one protein, namely, HA-NA (HN), F protein, nucleocapsid protein, phosphoprotein, polymerase protein, and M protein. Five serotypes comprise the group of parainfluenza viruses: types 1, 2, 3, 4A, and 4B. Parainfluenza viruses possess at least two envelope glycoproteins, the HN and the F protein. HN mediates adsorption of the virion to cell receptors and the enzymatic cleavage of sialic acid (neuraminic acid) residues of the virus that must be removed to prevent self-aggregation of virus particles during release from cells. The F glycoprotein facilitates penetration of virus into cells and mediates immunity. Parainfluenza virus infections are transmitted by direct contact and by aerosol droplets.

Parainfluenza virus types 1, 2, and 3 occur worldwide and among persons of all age groups (Table 3). Parainfluenza viruses 4A and B are much less prevalent. Parainfluenza virus type 1 occurs in epidemics during the fall season in alternate years, usually the odd-numbered years, while parainfluenza virus type 2 occurs sporadically; parainfluenza virus type 3 occurs yearly in endemic areas. The incubation period is usually between 2 and 6 days. Parainfluenza virus types 1, 2, and 3 are the main viruses that cause most croup illnesses (laryngotracheobronchitis) in infants and young children less than 5 years of age (Table 8) (Mufson et al., 1970; Marx et al., 1997). Parainfluenza virus type 3 causes pneumonia and bronchiolitis in infants and small children and is second only to RSV in causing these conditions. Parainfluenza virus type 4 occurs only infrequently, and it is usually associated with mild symptoms of upper respiratory tract illness (rhinorrhea, sore throat, and cough). Among adults, parainfluenza viruses usually cause minor upper respiratory tract illnesses, and very infrequently types 1 and 3 cause pneumonia (Marx et al., 1999).

TABLE 8 Parainfluenza viruses and respiratory tract syndromes in infants and children

Respiratory tract syndrome	No. of patients	% of illnesses associated with indicated parainfluenza virus serotype[a]		
		Type 1	Type 2	Type 3
Pneumonia	169	5.9	4.7	31.9
Bronchiolitis	120	6.7	4.2	23.4
Laryngotracheobronchitis	24	8.3	20.8	41.6

[a]All patients were tested both by virus isolation and by antibody response. (Data from Mufson et al., 1970.)

Parainfluenza virus infections of the upper respiratory tract usually present as rhinitis, pharyngitis, and bronchitis, often associated with fever. Rhinorrhea, cough, and pharyngeal erythema are the usual findings. The infection can progress to croup, with the typical bark-like cough. Although uncommon, cyanosis develops in the presence of marked airway obstruction. Persistent fever and productive cough signal lower respiratory tract involvement. Infants and children with bronchiolitis and pneumonia manifest wheezing, tachypnea, retractions, and sometimes cyanosis.

Primary infection with parainfluenza viruses provides some measure of immunity, but it is neither complete nor long lasting. Reinfections occur commonly (van Wyke Coelingh et al., 1990). Primary infection in seronegative infants and children induces ample hemagglutination inhibition and neutralizing antibody responses, with a robust response to HN sites and a weak response to F protein sites (van Wyke Coelingh et al., 1990). Reinfection in children produced much higher hemagglutination inhibition and neutralizing antibody levels than primary infection and substantial responses to the HN and F protein sites. Second infections rarely cause illness as severe as that seen with primary infection. Immunity also correlates well with the amount of secretory IgA present in the mucous membranes of the nose.

For the laboratory diagnosis of parainfluenza virus infections, virus isolation in tissue culture remains the gold standard. Respiratory-secretion specimens must be kept at 4°C until processed. Parainfluenza viruses grow slowly and usually require 3 to 5 days until they can be detected by hemadsorption procedures, because they do not form lytic cytopathic effects in cell cultures. Isolates can be typed by using hemadsorption inhibition (Table 4).

Rapid diagnostic tests that detect parainfluenza virus antigens in respiratory secretions within minutes to hours after collection of the specimen are preferred and are now widely available. These tests include direct immunofluorescence, EIA, and multiplex reverse transcription-PCR (RT-PCR) (Echevarría et al., 1998; Fan et al., 1998; Osiowy, 1998; Grondahl et al., 1999). Direct immunofluorescence and EIAs are specific for each parainfluenza virus type and show high sensitivity and specificity. The new RT-PCR procedures rapidly detect all parainfluenza virus types in a single nucleic acid amplification assay, with similarly high sensitivity and specificity (Corne et al., 1999). RT-PCR procedures can detect and differentiate multiple respiratory viruses in a single amplification procedure by utilizing several pairs of primers that are specific not only to the genes of parainfluenza viruses types 1, 2, and 3, but also to genes of other respiratory viruses commonly associated with the same serious respiratory tract diseases as the parainfluenza viruses (for example, RSV A and B, influenza viruses A and B, and adenoviruses types 1 to 7)

(Echevarría et al., 1998; Fan et al., 1998; Osiowy, 1998; Grondahl et al., 1999).

An immunologic response with detection of a diagnostic (fourfold or greater) rise in serum antibody levels during convalescence also provides evidence of infection. Antibody measurement procedures require obtaining a convalescent-phase serum sample about 3 weeks after the onset of illness. They can be done when virus isolation and direct detection of viral antigen are negative or not available. Antibody can be measured by complement fixation, hemagglutination inhibition, or neutralization. Interpretation of serologic tests relative to the parainfluenza virus type producing infection may be complicated by the heterotypic antibody response seen in some infected individuals.

Vaccines for the parainfluenza viruses, as with RSV, are still being developed, but hold some promise. These include two candidate vaccines of parainfluenza virus type 3, namely, a live, cold-passaged, temperature-sensitive attenuated vaccine (cp-45) derived from the JS wild-type virus and a replication-defective vaccinia virus recombinant expressing the HN or F proteins of a human parainfluenza virus type 3 (Karron et al., 1995; Ray et al., 1996; Durbin et al., 1999; Skiadopoulos et al., 1999). No antiviral drugs are available for the treatment of parainfluenza virus infection.

RSV

Human RSV, the most important etiological agent of respiratory tract disease in infants and children, was first isolated in 1956 from a symptomatic laboratory chimpanzee during an outbreak of illness resembling the common cold. Shortly afterward, the virus was shown to be a human pathogen. Since that time, epidemiologic studies have shown that RSV represents the single most important cause of serious lower respiratory tract disease, especially bronchiolitis and pneumonia, in infants and children, accounting for more than 90,000 hospital admissions annually in infants (Table 2) (Hall, 1999).

RSV belongs in the family *Paramyxoviridae* and genus *Pneumovirus*. The virus lacks both an HA and an NA and its nucleocapsid diameter is 12 to 15 nm, compared with 18 nm of the parainfluenza viruses (Richman et al., 1971). RSV is an enveloped virus with a single-stranded, negative-polarity, nonsegmented RNA genome that encodes at least 10 viral proteins (Sullender, 2000). These include three transmembrane glycoproteins (namely, the G protein, a heavily glycosylated glycoprotein which mediates attachment to the cell; the F protein, which promotes membrane fusion; and the small hydrophobic protein), five structural proteins, and two nonstructural proteins (NS1 and NS2). The structural proteins include the two matrix proteins (M and M2 or 22K) and three nucleocapsid proteins (nucleocapsid, polymerase, and phosphoproteins). By employing monoclonal antibody technology to antigenic analysis of RSV, two subgroups, A and B, were recognized on the basis of their reactivity with a panel of monoclonal antibodies (Anderson et al., 1985; Mufson et al., 1985; Gimenez et al., 1986). The major difference between strains of subgroups A and B was found in their G transmembrane glycoproteins. They differed in a single subgroup-specific linear epitope represented by 15 amino acids, residues 174 to 188, of the G protein; the activity of this peptide appeared to depend upon intrapeptide disulfide bonds (Akerlind-Stopner et al., 1990). Heterogeneity also exists within the A and B subgroups, manifested by genetic diversity and immunologic reactivity (Akerlind et al., 1988; Sullender et al., 1991; Mufson and Stanek, 1996). On the basis of reactivity with monoclonal antibodies directed against the G protein, subgroup B strains could be differentiated into two variants, subgroups B1 and B2 (Akerlind et al., 1988).

Epidemics of RSV occur yearly during the winter in temperate climates. The two subgroups can cocirculate in the same geographic area simultaneously, usually with a predominance of subgroup A (Mufson et al., 1988, 1991; Thomas et al., 1994). In the northern temperate zones, the peak of the epidemic is usually January to April, and in the tropics, it is during the rainy seasons; in alternate years the epidemic starts early, late in November or December, and late, in January (Fig. 1) (Mufson et al., 1973, 1991). The virus causes disease mainly among infants and children,

with infants between 6 weeks and 6 months of age being the ones most often affected (Table 9). Among adults, RSV infections are less common; they usually result in mild upper respiratory illnesses, but uncommonly they can cause community-acquired pneumonia (Dowell et al., 1996). Elderly persons, especially institutionalized adults and elderly persons with underlying heart and lung disease, are at high risk for developing serious RSV infections (Mlinaric-Galinovic et al., 1996; Falsey, 1998; Walsh et al., 1999). Although the rate of occurrence of serious RSV infections in elderly adults is not as high as that in infants and children, such RSV infections account for 2 to 9% of all hospitalizations for pneumonia among the elderly (Han et al., 1999). RSV also causes severe pneumonia and death in persons with leukemia and bone marrow transplants (Whimbey et al., 1995).

RSV infections are transmitted during close contact. The incubation period is about 4 to 5 days. RSV causes upper and lower respiratory tract illnesses (Table 2). In children, when the lower respiratory tract becomes involved (about 25 to 40% of infections), the child can develop a cough, fever, sneezing, wheezing, hyperventilation, rhonchi, fine rales, and otitis media. The chest roentgenogram is usually normal, but in more-severe cases, it often reveals hyperinflation and pulmonary infiltrates. Dyspnea, hyperexpansion, and tachypnea can develop, and apnea may occur. Children with laryngotracheobronchitis,

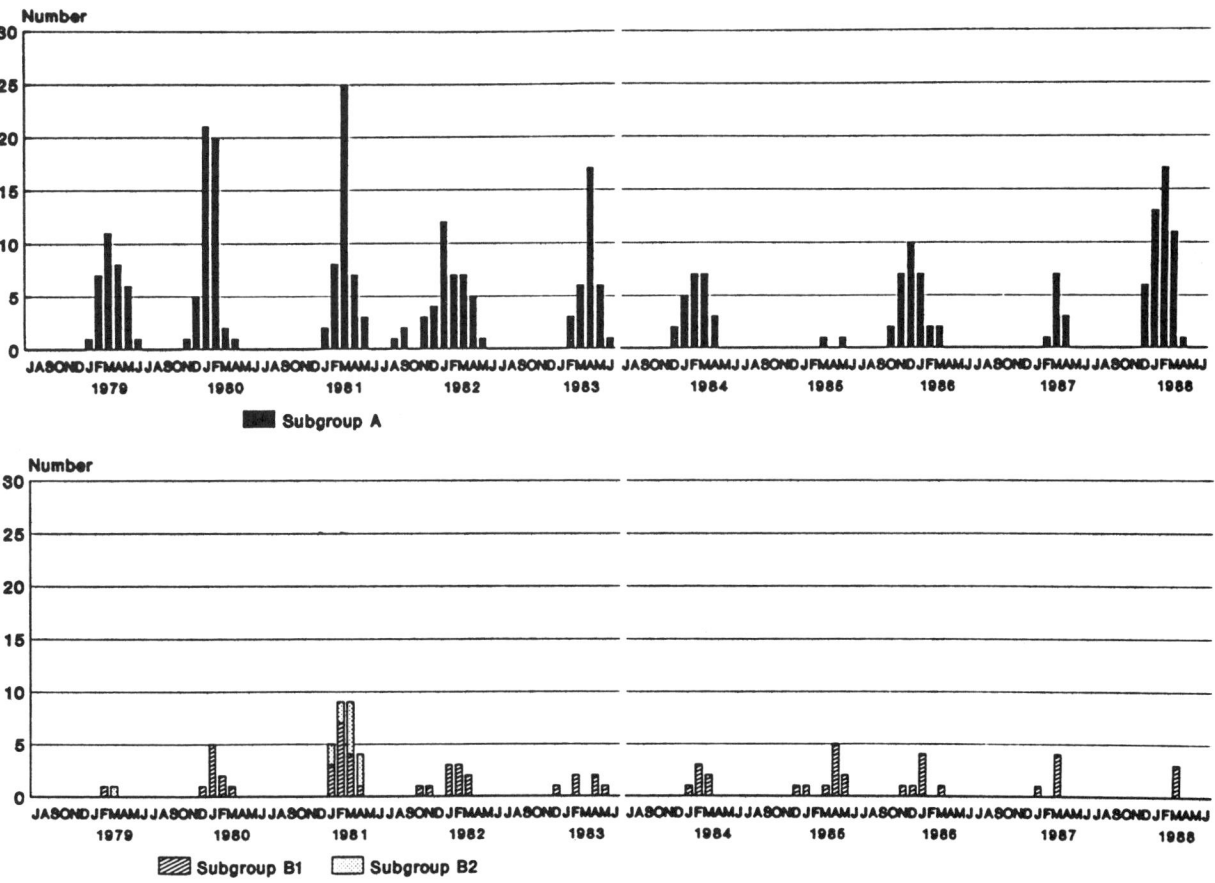

FIGURE 1 Temporal distribution of RSV subgroups A, B1, and B2 from July 1978 through June 1988 in Huntington, W.Va. RSV epidemics occur each winter in temperate climates starting early (November or December) and late (January or February) in alternating years. (Reprinted from Mufson et al., 1991.)

TABLE 9 Subgroup of RSV and acute respiratory tract illness in infants and children

Clinical diagnosis	No. (%) of illnesses associated with indicated subgroup of virus[a]		
	A	B1	B2
Pneumonia	127 (39.8)	25 (34.2)	6 (46.1)
Bronchiolitis	97 (30.4)	11 (15.1)	2 (15.4)
Laryngotracheobronchitis	30 (9.4)	12 (6.4)	2 (15.4)
Upper respiratory tract illness or bronchitis	65 (20.4)	25 (34.3)	3 (23.1)

[a]Significantly fewer cases of bronchiolitis were associated with subgroup B strains (chi-square test, 14.41; df, 6; P, 0.025). (Data from Mufson et al., 1991.)

bronchiolitis, or pneumonia often require hospitalization due to the need for intubation and ventilation. In an uncomplicated RSV infection, recovery occurs after 7 to 12 days. Generally, the mortality is less than 2% among children admitted to hospitals. Nevertheless, when a child has underlying cyanotic congenital heart disease, the case-fatality rate may be as high as 37%. Other preexisting conditions that are associated with more-severe disease or deaths include bronchopulmonary disease, renal disease, immunosuppression, prematurity, and age of 3 months or younger (Walsh et al., 1997). Depending upon the population and time period, either subgroup A or B infection, or neither, may be associated with more-severe respiratory tract illness (Mufson et al., 1991; Walsh et al., 1997; Hornsleth et al., 1998).

The role of RSV, if any, in sudden infant death syndrome (SIDS) remains to be elucidated. Epidemiologic studies showed no statistically significant association between SIDS and the occurrence of RSV infection among infants younger than 18 months of age admitted to hospitals during four annual epidemics (Nelson et al., 1975). In postmortem examinations, RSV nucleic acid was not detected significantly more often in lungs of infants who died from SIDS than from lungs of infants in a control group with known causes of death (Cubie et al., 1997).

Reinfections with RSV occur commonly. Among schoolchildren, the risk for RSV infection during each epidemic is about 20 to 40%. First infection with subgroup A strains provides some protection from a second infection with the homologous, but not the heterologous, subgroup of the virus (Mufson et al., 1987). High levels of preexisting circulating neutralizing, anti-F and anti-G antibodies in adults provided some measure of protection against reinfection (Hall et al., 1991).

The diagnosis of RSV can be made by isolation of virus in cell cultures, by direct examination of respiratory secretions using immunofluorescence or ELISA, or by demonstration of a fourfold rise in antibody between acute- and convalescent-phase serum samples (Table 4). However, isolation of RSV in cell culture or direct detection of viral antigen is preferred to antibody detection for diagnosis of infection, since it yields information sooner. Since RSV replicates in the respiratory epithelia of the nasopharynx, nasopharyngeal swab and nasal aspirate specimens are the best sources for virus. As RSV is very labile, specimens should be transported on wet ice to the laboratory and kept at 4°C until processed. Freeze-thawing can cause substantial loss of infectivity. HEp-2 cells are most commonly employed in cell cultures for isolation of RSV. The virus

produces multinucleated giant cells, syncytia, in these cultures. The formation of cytopathic effects indicative of virus growth in tissue cultures usually takes about 5 to 9 days in HEp-2 cells (or other susceptible cell cultures).

A number of commercial rapid diagnostic EIA kits which identify RSV antigen in respiratory secretions are available. These EIA kits provide high degrees of sensitivity and specificity, although they vary with the commercial kit (Table 4). The results are available in minutes to hours. The commercially available tests for the rapid identification of RSV include direct EIAs for RSV (Abbott TestPack RSV and Enzygnost RSV [Ag]) and a monoclonal antibody-based immunofluorescence test for both RSV and influenza viruses (Imagen for RSV and FluA for influenza virus) (Dominguez et al., 1993; Olsen et al., 1993).

Antibody response during convalescence detected by using complement fixation or ELISA methods can also be employed to diagnose an RSV infection (Table 4). However, as these procedures require a serum specimen obtained 18 to 21 days after the onset of illness, this approach is more likely to be used in epidemiologic and research studies. Children 4 to 9 months of age possess maternal antibodies that may complicate interpretation of the data.

Rapid diagnosis of RSV infection is important because of the availability of two means of management of these infections, namely a chemotherapeutic drug, ribavirin, and a humanized monoclonal antibody for prophylaxis, palivizumab. A synthetic nucleoside, ribavirin is appropriate for the treatment of infants with serious RSV illness or infants with underlying diseases that place them at high risk for severe RSV illness; however, its benefit in the treatment of otherwise-well children lies in reduced reactive-airway disease later in life (Edell et al., 1998; Rodriguez, 1999; Rodriguez et al., 1999). It is administered as a continuous aerosol at 20 mg/ml for 18 to 20 h each day during the first several days of illness and can hasten recovery.

A second approach to the management of RSV includes administration of palivizumab for prophylaxis (American Academy of Pediatrics Committee on Infectious Diseases and Committee of Fetus and Newborn, 1998; The IMpact-RSV Study Group, 1998). Palivizumab appears to be effective for immunoprophylaxis of serious infections in premature children and children suffering bronchopulmonary dysplasia (The IMpact-RSV Study Group, 1998). Administered intramuscularly into the thigh, the dose of palivizumab is 15 mg/kg of body weight per month for the 5 months during the RSV epidemic season, usually December through April. High-titered immune globulin (RSVIG) also provides satisfactory prophylaxis of RSV among high-risk infants; however, it must be administered intravenously.

Administration of dexamethasone during the first 3 days of illness showed no effect in the treatment of infants with bronchiolitis (De Boeck et al., 1997). These patients were given 0.6 mg/kg intravenously in two doses on the first day of illness and 0.015 mg/kg on each of the next two days.

A vaccine for RSV is of the highest priority because of the common occurrence of the virus in infants and children and the severity of disease. The development of a vaccine for this virus started in the early 1960s and began with an inactivated vaccine that induced specific circulating antibody that failed to protect the children who received it and enhanced the severity of respiratory disease in some of them (Kim et al., 1969). Since that time, several novel approaches for the development of an RSV vaccine, including live attenuated virus vaccines (such as the cold-passaged virus vaccine cp52) and various subunit vaccines (such as purified F protein and recombinant vaccines using virus vectors and the F protein), have been investigated (Murphy et al., 1989; Collins et al., 1995; Crowe et al., 1996; Groothuis et al., 1998; Collins et al., 1999). These vaccines are experimental, and although some of them are being tested preliminarily in adults and children, including live attenuated virus vaccines and a purified fusion protein (PFP-2) vaccine, none is nearly ready for larger trials (Groothuis et al., 1998).

ADENOVIRUS

Adenoviruses were isolated from primary cell cultures of adenoids from children and were named by Rowe, Huebner, and colleagues in the early 1950s (Ginsberg, 1999). Currently, 49 human adenovirus serotypes have been identified; they can be grouped into six subgenera, designated subgenuses A through F, on the basis of differing classification schemes, including a nonnested PCR, hemagglutination characteristics, or the percent G+C content of their DNA (Table 10) (Kidd et al., 1996).

Adenovirus particles contain a linear, double-stranded DNA molecule of about 35 kb, which is surrounded by an icosahedral capsid with one (two in some enteric types) fiber-like projection extending from each of the 12 vertices, which possess knob-like ends that are the attachment sites for target cells. The adenovirus virion is nonenveloped (naked). It is not affected by the low pH in the stomach, the bile from the gallbladder, or the proteolytic enzymes secreted by the pancreas and replicates well in the gastrointestinal tract.

Adenovirus infections occur worldwide. The transmission of adenovirus infection varies from sporadic to epidemic. Epidemics usually occur among infants and children and among military recruits. The incubation period ranges from 5 to 8 days. The fecal-oral route and, to a lesser extent, aerosols account for the spread of infection in infants and children. Since adenoviruses are very stable, they can be transmitted in water and by fomites, for example, ophthalmological instruments. The initial site of infection can be the conjunctivae, oropharynx, or the intestine, and the infection subsequently spreads to the regional adenoid tissues. Rarely, adenovirus can be isolated from the lungs, livers, kidneys, and brains of patients with fatal infections.

Adenoviruses are important causes of acute upper and lower respiratory tract illnesses in infants and children (Table 2) (Yamadera et al., 1995). In industrialized countries, adenoviruses cause 2 to 5% of all acute respiratory tract diseases. Types 1, 2, 3, 5, and 7 account for 87% of adenovirus infections (Table 11). Military recruits who acquire adenovirus infection usually develop either upper respiratory tract disease, pharyngitis, or pneumonia (Ylikoski and Karjalainen, 1989; Barraza et al., 1999). The types that commonly infect military recruits include 4, 7, 14, and 21. Genetic variation occurs among types 4 and 7a (Crawford-Miksza et al., 1999).

Subgenus A comprises types 12, 18, and 31, which infect the gastrointestinal tract only. Subgenus B cluster B:1 contains types 3, 7, 16, and 21 and cluster B:2 contains types 11, 14, 34, and 35 (Table 10) (Kidd et al., 1996). Of the types in subgenus B, types 3, 7, 14, and 21 cause epidemics of acute respiratory tract infection among children and mil-

TABLE 10 Classification of adenoviruses by subgenus characteristics

Subgenus	Serotype(s)	Hemagglutination pattern with indicated erythrocytes		% G+C of DNA
		Rhesus	Rat	
A	12, 18, 31	Negative	None	48–49
B:1	3, 7, 16, 21	Complete	Negative	50–52
B:2	11, 14, 34, 35	Complete	Negative	50–52
C	1, 2, 5, 6	Negative	Partial[a]	57–59
D	8–10, 13, 15, 17, 19, 20, 22–30, 32, 33, 36–39, 42–49	Negative	Complete	58
E	4	Negative	Partial[a]	57–61
F	40, 41	Negative	Partial	57–59

[a]The hemagglutination test requires adding a heterotypic antiserum to the diluent (type 4 antiserum for subgenus C and type 6 antiserum for subgenus E).

TABLE 11 Common adenovirus serotypes associated with respiratory tract diseases among children, adolescents, and adults

Respiratory tract disease	Serotypes associated with illness[a]		
	Children	Adolescents[b]	Adults
Pneumonia	1–3, 7	3, 4, 7	Rare
Acute upper respiratory tract disease	1–3, 5–7	3, 4, 7, 14, 21	3, 7
Pharyngoconjunctival fever	3, 4, 7, 14	3, 4, 7, 14	3, 4, 7, 14
Epidemic keratoconjunctivitis			8, 19, 37

[a]The most common types are shown; other types may be associated with these syndromes very infrequently or on a sporadic basis.
[b]Adolescents are usually represented by military recruits; common types in military recruits are 3, 4, 7, 14, and 21.

itary recruits (Table 11). Type 3 commonly causes conjunctivitis of much less serious consequences than type 8. Type 11 causes acute hemorrhagic cystitis; rarely, type 35 is shed in the urine of the patients with cystitis (Li et al., 1991). Type 7 causes pneumonia among young children, with the potential for permanent lung damage. Subgenus C includes types 1, 2, 5, and 6, which are associated with acute respiratory tract diseases and fever among children younger than 5 years old. Subgenus D, the largest subgenus, contains 31 serotypes, namely, 8 to 10, 13, 15, 17, 19, 20, 22 to 30, 32, 33, 36 to 39, and 42 to 49, most of which infect the eye, causing conjunctivitis and epidemic keratoconjunctivitis (especially type 8), and the gastrointestinal tract (Table 10) (Colon, 1991). Subgenus E comprises only type 4, which commonly causes disease, mainly acute upper respiratory tract disease, pharyngitis, and pneumonia, predominantly among military recruits (Barraza et al., 1999). Subgenus F includes types 40 and 41, which are second only to rotaviruses as the most frequent causes of infantile viral gastroenteritis. Types B:1 and B:2 of subgenus A are oncogenic in rodents but not in humans.

Adenoviruses can be isolated in cell cultures from the respiratory tract, conjunctivae, feces, and urine. Isolates can be presumptively recognized in cell cultures by their characteristic cytopathic effect of grape-like clusters of infected cells that develop about 2 to 5 days after inoculation with a virus-positive specimen (Table 4). Immunologic identification of adenovirus serotypes can be done by using PCR or immunofluorescence; individual isolates can be serotyped by using hemagglutination inhibition, neutralization, immunofluorescence, or DNA restriction analysis procedures. Adenoviruses also can be detected directly in respiratory secretions by using PCR, EIA, immunofluorescence, and immunochromatography (Hietala et al., 1988; Bruckova et al., 1989; Kidd et al., 1996; Akhtar et al., 1999; Tsutsumi et al., 1999). Several PCR methods are available for the detection and identification of adenoviruses. A multiplex RT-PCR procedure carried out directly on respiratory secretions identifies adenovirus types 1 through 7; parainfluenza virus types 1, 2, and 3; and RSV subtypes A and B with higher sensitivity than immunofluorescence (Osiowy, 1998). Immunochromatography is a new, fairly sensitive, and highly specific rapid test for identifying adenovirus types 1, 2, 3, 5, and 7 in pharyngeal swab specimens (Tsutsumi et al., 1999).

Adenovirus infection can be established serologically by the detection of a fourfold or greater rise in antibody titers during convalescence. Type-specific antibody responses can be assayed by hemagglutination inhibition (for hemagglutinating strains), EIA, and neutralization assays. However, except for epidemiologic investigations, antibody determination is a much less efficient means for determining adenovirus infection than direct detection of adenovirus by PCR or other rapid methods.

No drugs are available for adenovirus infections of the respiratory tract. In a single case report, an AIDS patient with adenovirus pneumonia improved after administration of ribavirin intravenously and then orally (Maslo et al., 1997). After the drug was stopped because of anemia, the patient had a relapse of the pneumonia.

Live attenuated vaccines for types 4 and 7, consisting of these types in an enteric coated capsule, are recommended for and routinely administered only to military recruits (Ludwig et al., 1998; Barraza et al., 1999). Starting about 1970, they were administered to military recruits entering training. The vaccine strains cause a subclinical intestinal infection, resulting in the production of local antibody in both the respiratory and the intestinal tracts, and their administration results in a significant reduction in illness among exposed recruits. About two-thirds of recruits new to military service lacked antibody to these types, and nearly 90% lacked antibody to one of the two types (Ludwig et al., 1998). At one military facility, administration of vaccine to all recruit trainees prevented any outbreaks of acute respiratory disease associated with adenovirus infection (Barraza et al., 1999). However, when adenovirus vaccine was not given to any recruit during an interval of almost one year, a major outbreak of type 4 respiratory tract disease occurred, with hospitalization rates as high as 11.6% (Barraza et al., 1999). These findings demonstrate the efficacy and importance of the use of live attenuated adenovirus vaccines in military recruits.

Human recombinant adenoviruses can be constructed that express recombinant proteins of differing viruses, including rabies glycoprotein, hepatitis B surface antigen, the spike protein of coronaviruses, the core protein of hepatitis C virus, RSV F and G proteins, and human immunodeficiency virus type 1 envelope protein (Natuk et al., 1994; Yarosh et al., 1996; Brader et al., 1997; Bruce et al., 1999). Such recombinant adenovirus-vectored vaccines provide a novel mechanism for developing vaccines for viruses that fail conventional vaccine development means.

Adenoviruses comprise a large group of diverse types that account for substantial morbidity from acute viral diseases of the respiratory tract, eye, gastrointestinal tract, and urinary tract of children and adults. They also cause opportunistic infections in severely immunosuppressed persons. An effective vaccine for the main types that infect children would have wide use in the pediatric populations.

RHINOVIRUS

Rhinoviruses comprise a group of more than 100 serotypes and cause more common-cold (minor upper respiratory tract) illnesses than any other virus that infects the respiratory tract. They account for about one-half of common colds occurring in children and adults (Hendley, 1999). These infections account for substantial absenteeism from schools and work, amounting annually to about 26 million days lost from schools and about 23 million days lost from work. Persons with asthma can suffer exacerbations associated with rhinovirus infection, and in persons with underlying lung disease and immune disorders, rhinovirus can cause pneumonia.

About 5 decades ago, the search for the common-cold virus resulted in the isolation of the first candidate rhinoviruses. The studies conducted in the subsequent 30 years revealed that the rhinoviruses comprise more than 100 antigenically distinct serotypes (Cooney et al., 1982; Hamparian et al., 1987). Rhinoviruses belong to the family *Picornaviridae*. They are small (15 to 30 nm), naked viruses, containing a single-stranded, positive-sense RNA genome. Rhinoviruses encode four structural proteins, three of which are capsid proteins (VP2, VP3, and VP1) and one that is internal (VP4), and several nonstructural proteins, including two proteases, an RNA-dependent RNA polymerase, and a small protein. Unlike the enteroviruses, rhinoviruses lose infectivity upon exposure to mild acids (pH of <5), accounting for their failure to infect the gut. Rhinoviruses can be grouped into two receptor families, designated major and minor, on the basis of utilization of cell attachment surface receptors. The major group, comprising about 90% of the more than 100 serotypes, utilizes intercellular adhesion molecule 1 (ICAM-1, CD54), a class 1 membrane protein and member of the immunoglobulin superfamily (Greve et al., 1989; Staunton et al., 1989; Uncapher et al., 1991). The minor group, comprising 10 serotypes, utilizes low-density lipoprotein receptors (Marlovits et al., 1998; Schober et al., 1998). A final group comprises a few miscellaneous serotypes, including serotype 87, which requires sialic acid for attachment. The surface of the virus capsid contains a surface depression or canyon encircling the fivefold axes of the pentamer that accommodates the sites of receptor attachment (Rossmann et al., 1994).

Rhinovirus infections occur year-round, with peaks in the early spring and fall months in the temperate zones, and they occur during the rainy period in the tropics. They are spread person to person by two mechanisms: by direct contact with ill persons whose hands and fingers have become contaminated with infectious secretions, the main mode of transmission, and by aerosol.

Rhinoviruses replicate in the cells of mucous membranes of the nose. A direct correlation between the quantity of virus in secretions and the occurrence and severity of illness exists. Peak concentrations of virus are shed on the second and third days of the illness. As optimal replication of rhinoviruses takes place at 33 to 34°C, the concentration of rhinovirus in the nose is 10 to 100 times higher than in pharyngeal secretions.

The usual incubation period of rhinovirus illness is 1 to 4 days, although it may be as long as 7 days. Symptoms of the common cold include rhinorrhea, nasal obstruction, sneezing, sore throat, cough, and headache (Table 2). Fever is uncommon. The illness usually lasts about one week. The symptoms of rhinovirus colds reflect the host response and are not direct effects of the virus on the host (Hendley, 1998). Despite the severity of symptoms, little tissue damage occurs in the mucosa. The inflammatory response derives from the release of cytokines, including IL-8, to start an inflammatory cascade (Hendley, 1998; Turner et al., 1998). Secondary complications due to bacterial superinfection rarely occur.

Rhinoviruses also cause otitis media in infants and children; these viruses can be recovered from middle ear fluid in 10% of subacute or chronic cases that are negative for bacterial pathogens (Arola et al., 1990a, 1990b). However, rhinovirus RNA was detected in the middle ear effusions of 19 of 100 infants with otitis media by RT-PCR compared with virus isolation from only 5 of the infants positive by RT-PCR (Pitkaranta et al., 1998). Virus RNA was detected as late as 8 weeks after the onset of infection, suggesting that rhinovirus is a common cause of otitis media with effusion in infants and that RT-PCR procedures provide the best method for diagnosing this infection.

Although rhinoviruses predominantly infect the upper respiratory tract, recent studies associate rhinovirus infections with exacerbations of asthma and with acute lower respiratory tract disease, especially in persons with chronic obstructive pulmonary disease or cystic fibrosis, immunocompromised adults, infants, and elderly persons (Abzug et al., 1990; Pitkaranta and Hayden, 1998; Gern and Busse, 1999). Rhinoviruses grow less well at the higher temperature of the lungs than they do in the nose, but they reach sufficiently high titers to establish infection there (Papadopoulos et al., 1999). Rhinoviruses, as well as other respiratory viruses (less frequently), play a major role in provoking exacerbations of asthma in children and adults. Rhinoviruses possibly do this through several mechanisms involving production of proinflammatory chemokines and cytokines, increased number of inflammatory cells, heightened eosinophilia, and their effects on airway smooth muscle (Grunberg and Sterk, 1999; Hakonarson et al., 1999; Sanders, 1999). The symptoms and signs of asthma exacerbations do not distinguish rhinovirus exacerbations from those caused by other pathogens.

Type-specific immunity develops following infection and is characterized by IgG neutralizing antibodies in serum and secretions. However, long-lasting immunity best correlates with the level and secretion of IgA antibodies from the nasal mucous membranes. Patients who experience a common cold caused by rhinovirus usually will not get another cold caused by a different rhinovirus within at least 4 weeks, due to viral interference and local production of interferon.

The diagnosis of rhinovirus infection requires laboratory tests infrequently, as the overwhelming number of illnesses associated with rhinovirus infection are common colds and diagnosis is based on clinical presentation alone (Table 4). Nonetheless, when laboratory confirmation of rhinovirus infection seems indicated, it can be approached by utilizing one of several procedures, including virus isolation, detection of viral RNA by RT-PCR, antigen detection by direct immunofluorescence in cells from respiratory secretions, or detection of a fourfold rise in antibody titers by neutralization test or EIA.

Virus isolation in cell culture from nasopharyngeal or oropharyngeal swab specimens or nasal aspirates is very sensitive, especially when the specimens are collected early after the onset of illness, mainly during the first day or two of illness. However, virus isolation is labor-intensive and requires several days until virus growth occurs, so its use as a routine diagnostic test needs substantial justification. Rhinoviruses grow well in fetal human fibroblast cells and primary human embryonic kidney cells and produce changes in the cell cultures characteristic of the group of picornaviruses. Confirmation of the individual rhinovirus serotype requires further antigenic characterization with type-specific antibody.

Rhinovirus infections associated with otitis media can be confirmed in the laboratory by virus isolation and RNA detection in middle ear effusions by RT-PCR (Arola et al., 1990a, 1990b; Pitkaranta et al., 1998). RT-PCR is unlikely to be available except as a research procedure. In persons with exacerbations of asthma, rhinovirus can be isolated from respiratory secretions. It can be isolated from bronchoalveolar lavage fluid and endotracheal aspirate of adults with underlying chronic pulmonary disease and from infants (Abzug et al., 1990; Ghosh et al., 1999). As rhinoviruses infrequently cause pneumonia and as the yield from virus isolation attempts is low, carrying out this procedure on a routine basis is not justified.

Specific antibody to rhinovirus can be measured by macro- and microneutralization and plaque-reduction tests, and an ELISA using antibody capture of rhinovirus antigen can detect IgA antibodies (Barclay et al., 1988). None of these tests is employed for routine diagnosis.

The investigation of rhinovirus infections in the past two decades widened the spectrum of diseases potentially associated with this group of viruses. The data on rhinovirus infection in otitis media seem especially cogent because of the detection of rhinovirus RNA in middle ear effusions by

RT-PCR as long as several weeks after the onset of illness. Quite possibly, patients with these illnesses might benefit by specific antiviral treatment. However, no antiviral drugs are currently available for the treatment of rhinovirus infections. The development of drugs against rhinoviruses holds more interest and value now that these viruses apparently cause substantial numbers of exacerbations of asthma and otitis media.

CORONAVIRUS

Coronaviruses, also an important cause of common colds, belong to the family *Coronaviridae*, genus *Coronavirus*. They are enveloped, spherical particles with helical nucleocapsids of 80 to 160 nm in diameter, possessing a single-stranded, positive-sense RNA genome. Nonhemagglutinating coronaviruses possess three major proteins, namely, the surface or spike (S) peplomeric glycoprotein of 150 kDa in unglycosylated form, with 1,173 amino acids and an M_r of 128,600; the M glycoprotein, which is highly hydrophobic, mediates incorporation of the S glycoprotein into the envelope, contains 230 amino acids, and has an M_r of 26,216; and the nucleocapsid phosphoprotein of 55 kDa (Raabe et al., 1990; Mounir and Talbot, 1992; Kunkel and Herrier, 1993). Hemagglutinating coronaviruses possess a fourth protein, the HA esterase glycoprotein. Coronaviruses comprise four serogroups; the prototype human coronavirus strains belong in serogroups 1 and 2. Serogroup 1 includes the human prototype strain HCV 229E and coronaviruses of several other species, including feline enteric coronavirus and porcine transmissible gastroenteritis virus. The cell surface receptor of HCV 229E is human aminopeptidase N (hAPN[CD123]) (Yeager et al., 1992). Serogroup 2 includes the human prototype strain HCV OC43, a hemagglutinating coronavirus, and coronaviruses of several other species including mouse hepatitis virus (de Haan et al., 1999). HLA class 1 antigen is the cell surface receptor of HCV OC43 (Collins, 1993, 1994). The S protein of HCV OC43 appears to be the major HA acting on the receptor *N*-acetyl-9-*O*-acetylneuraminic acid of the erythrocyte surface (Kunkel and Herrier, 1993). The two human prototype strains, HCV 229E and HCV OC43, were isolated and characterized in the mid-1960s (Hamre and Procknow, 1966; Kapikian, 1975). Serogroup 3 includes only avian bronchitis virus and serogroup 4 includes only bluecomb disease virus of turkeys.

Coronavirus infections occur worldwide, usually appearing sporadically throughout the winter and spring. The virus is spread by large droplets via the respiratory route. Among children, coronavirus infection occurs commonly, one infection per child per year, which is about three times as often as in adults (Schmidt et al., 1986). These infections are associated with about 5 to 15% of upper respiratory tract illnesses.

Coronaviruses mainly cause common colds and acute upper respiratory tract illness (Table 2) (Pohl-Koppe et al., 1995; Makela et al., 1998). Infrequently, they are associated with febrile respiratory tract illness and otitis media (Arola et al., 1990a; Pitkaranta et al., 1998). Coronavirus RNA can be detected by RT-PCR in a very small proportion of middle ear effusions in children with otitis media requiring tympanostomy tubes (Pitkaranta et al., 1998). The clinical features of acute upper respiratory tract illnesses consist of coryza, rhinorrhea and nasal congestion, sore throat, and pharyngeal edema. Fever, headache, and cough are uncommon in mild upper respiratory tract illness. The illness lasts about one week. Coronavirus infections in persons with chronic chest disease can cause exacerbations, with symptoms of lower respiratory tract illness (Buscho et al., 1978; Wiselka et al., 1993). The mean duration of these exacerbations lasted about 11 days.

Pneumonia rarely occurs with coronavirus infection (Riski and Hovi, 1980). Coronavirus was reported as the cause of pneumonia in an adult with breast cancer who had received high-dose chemotherapy and autologous bone marrow transplant (Folz and Elkordy, 1999). The virus was detected by electron microscopic examination of bronchoalveolar lavage fluid. The clinical findings in coronavirus pneumonia do not distinguish it from other viral pneumonias.

Whether coronavirus causes central nervous system disease is moot, and additional investigations continue. HCV OC43 and HCV 229E can infect primary cultures of neural cells, including fetal astrocytes and adult microglia and astrocytes (Bonavia et al., 1997; Arbour and Talbot, 1998; Arbour et al., 1999). However, only fetal astrocytes released infectious virus. Myelin basic protein and virus-reactive T-cell lines, mostly CD4 positive, established in culture from persons with multiple sclerosis showed immunologic cross-reactivity between myelin and coronavirus antigens (Talbot et al., 1996). By using RT-PCR, coronavirus RNA was demonstrated in a small proportion of the brains from persons with multiple sclerosis (Stewart et al., 1992; Talbot et al., 1993). Coronavirus RNA can be detected in a very small number of cerebrospinal fluid specimens obtained from persons with multiple sclerosis and from persons suffering acute monosymptomatic optic neuritis (Cristallo et al., 1997; Dessau et al., 1999). However, the proportion of positive cerebrospinal fluid specimens from these ill patients was about the same as that from control patients.

Type-specific antibody to a single coronavirus prototype develops in response to infection and persists for long periods, with a mean duration of about 4 months (Macnaughton, 1982). Circulating and mucosal antibodies confer protection from infection and illness, and most commonly these antibodies are to the virion peplomers. The prevalence of coronavirus antibody increases with age, so that most persons older than 6 years possess circulating antibody to at least one prototype coronavirus strain (Furuuchi et al., 1978). Nonetheless, reinfections are common. Presence of or change in antibody titer to one coronavirus prototype is not influenced by the level of antibody to the other prototype (Schmidt, 1984).

The procedures for laboratory diagnosis of coronavirus infections associated with acute respiratory tract illness include virus isolation in cell cultures (only serogroup 1 HCV 229E), detection of viral antigen by immunofluorescence, detection of viral RNA by RT-PCR, and demonstration of a fourfold or greater rise in antibody titer during convalescence (Table 4). Virus isolation is complicated, and it is not recommended as a routine diagnostic test, especially since most illnesses associated with coronavirus infections are common colds. Virus isolation and detection of viral RNA by RT-PCR remain research tools. The procedures for antibody measurement include EIA and neutralization assays and, for serogroup 2 HCV OC43, hemagglutination inhibition.

Acute respiratory tract illnesses associated with coronavirus infection are treated symptomatically. Neither vaccine (none is under development) nor specific antiviral drugs is available for coronavirus infections.

The role, if any, coronavirus infection plays in multiple sclerosis or other chronic central nervous system disease of undetermined etiology continues to be investigated. Without gene technology, the detection of coronavirus would not be possible, except in respiratory secretions of persons with acute respiratory tract illnesses. RT-PCR detection of coronavirus RNA in brains of persons with multiple sclerosis opens new research prospects and poses questions about its pathogenesis that deserve additional investigations considering the results of preliminary studies. Does the detection of coronavirus RNA in otitis media effusions provide adequate reason for following through with the development of anticoronavirus drugs for the treatment of this serious illness of infants and children? We need to determine the frequency of coronavirus infection in otitis media and effusion and the consequences of these infections first by additional studies.

REFERENCES

Abzug, M. J., A. C. Beam, E. A. Gyorkos, and M. J. Levin. 1990. Viral pneumonia in the first month of life. *Pediatr. Infect. Dis. J.* **9:**881–885.

Akerlind, B., E. Norrby, C. Orvell, and M. A. Mufson. 1988. Respiratory syncytial virus: heterogeneity of subgroup B strains. *J. Gen. Virol.* **69:**2145–2154.

Akerlind-Stopner, B., G. Utter, M. A. Mufson, C. Orvell, R. A. Lerner, and E. Norrby. 1990. A subgroup-specific antigenic site in the G protein of respiratory syncytial virus forms a disulfide-bonded loop. *J. Virol.* **64:**5143–5148.

Akhtar, N., J. Ni, D. Stromberg, G. L. Rosenthal, N. E. Bowles, and J. A. Towbin. 1999. Tracheal aspirate as a substrate for polymerase chain reaction detection of viral genome in childhood pneumonia and myocarditis. *Circulation* **99:**2011–2018.

American Academy of Pediatrics Committee on Infectious Diseases and Committee of Fetus and Newborn. 1998. Prevention of respiratory syncytial virus infections: indications for the use of palivizumab and update on the use of RSV-IGIV. *Pediatrics* **102:**1211–1216.

Anderson, L. J., J. C. Hierholzer, C. Tsou, R. M. Hendry, B. F. Fernie, Y. Stone, and K. McIntosh. 1985. Antigenic characterization of respiratory syncytial virus strains with monoclonal antibodies. *J. Infect. Dis.* **151:**626–633.

Arbour, N., G. C. Lachance, M. Tardieu, N. R. Cashman, and P. J. Talbot. 1999. Acute and persistent infection of human neural cell lines by human coronavirus OC43. *J. Virol.* **73:**3338–3350.

Arbour, N., and P. J. Talbot. 1998. Persistent infection of neural cell lines by human coronaviruses. *Adv. Exp. Med. Biol.* **440:**575–581.

Arola, M., O. Ruuskanen, T. Ziegler, J. Mertsola, K. Nanto-Salonen, A. Putto-Laurila, M. K. Viljanen, and P. Halonen. 1990a. Clinical role of respiratory virus infection in acute otitis media. *Pediatrics* **86:**848–855.

Arola, M., T. Ziegler, H. Puhakka, O. P. Lehtonen, and O. Ruuskanen. 1990b. Rhinovirus in otitis media with effusion. *Ann. Otol. Rhinol. Laryngol.* **99:**451–453.

Barclay, W. S., K. A. Callow, M. Sergeant, and W. Al-Nakib. 1988. Evaluation of an enzyme-linked immunosorbent assay that measures rhinovirus-specific antibodies in human sera and nasal secretions. *J. Med. Virol.* **25:**475–482.

Barraza, E. M., S. L. Ludwig, J. C. Gaydos, and J. F. Brundage. 1999. Reemergence of adenovirus type 4 acute respiratory disease in military trainees: report of an outbreak during a lapse in vaccination. *J. Infect. Dis.* **179:**1531–1533.

Belshe, R. B., P. M. Mendelman, J. Treanor, J. King, W. C. Gruber, P. Piedra, D. I. Bernstein, F. G. Hayden, K. Kotloff, K. Zangwill, D. Iacuzio, and M. Wolff. 1998. The efficacy of live attenuated, cold-adapted, trivalent, intranasal influenzavirus vaccine in children. *N. Engl. J. Med.* **338:**1405–1412.

Bonavia, A., N. Arbour, V. W. Yong, and P. J. Talbot. 1997. Infection of primary cultures of human neural cells by human coronaviruses 229E and OC43. *J. Virol.* **71:**800–806.

Brader, K. R., J. K. Wolf, M. C. Hung, D. Yu, M. A. Crispens, K. L. van Golen, and J. E. Price. 1997. Adenovirus E1A expression enhances the sensitivity of an ovarian cancer cell line to multiple cytotoxic agents through an apoptotic mechanism. *Clin. Cancer Res.* **3:**2017–2024.

Bricaire, F., C. Hannoun, and J. P. Boissel. 1990. Prevention of influenza A. Effectiveness and tolerance of rimantadine hydrochloride. *Presse Med.* **19:**69–72.

Bruce, C. B., A. Akrigg, S. A. Sharpe, T. Hanke, G. W. Wilkinson, and M. P. Cranage. 1999. Replication-deficient recombinant adenoviruses expressing the human immunodeficiency virus Env antigen can induce both humoral and CTL immune responses in mice. *J. Gen. Virol.* **80:**2621–2628.

Bruckova, M., M. Grandien, C. A. Pettersson, and L. Kunzova. 1989. Use of nasal and pharyngeal swabs for rapid detection of respiratory syncytial virus and adenovirus antigens by enzyme-linked immunosorbent assay. *J. Clin. Microbiol.* **27:**1867–1869.

Buscho, R. O., D. Saxtan, P. S. Shultz, E. Finch, and M. A. Mufson. 1978. Infections with viruses and Mycoplasma pneumoniae during exacerbations of chronic bronchitis. *J. Infect. Dis.* **137:**377–383.

Cate, T. R. 1998. Impact of influenza and other community-acquired viruses. *Semin. Respir. Infect.* **13:**17–23.

Chambers, T. M., V. S. Hinshaw, Y. Kawaoka, B. C. Easter-day, and R. G. Webster. 1991. Influenza viral infection of swine in the United States 1988–1989. *Arch. Virol.* **116:**261–265.

Chomel, J. J., M. F. Remilleux, P. Marchand, and M. Aymard. 1992. Rapid diagnosis of influenza A. Comparison with ELISA immunocapture and culture. *J. Virol. Methods* **37:**337–343.

Claas, E. C., Y. Kawaoka, J. C. de Jong, N. Masurel, and R. G. Webster. 1994. Infection of children with avian-human reassortant influenza virus from pigs in Europe. *Virology* **204:**453–457.

Collins, A. R. 1993. HLA class I antigen serves as a receptor for human coronavirus OC43. *Immunol. Invest.* **22:**95–103.

Collins, A. R. 1994. Human coronavirus OC43 interacts with major histocompatibility complex class I molecules at the cell surface to establish infection. *Immunol. Invest.* **23:**313–321.

Collins, P. L., M. G. Hill, E. Camargo, H. Grosfeld, R. M. Chanock, and B. R. Murphy. 1995. Production of infectious human respiratory syncytial virus from cloned cDNA confirms an essential role for the transcription elongation factor from the 5' proximal open reading frame of the M2 mRNA in gene expression and provides a capability for vaccine development. *Proc. Natl. Acad. Sci. USA* **92:**11563–11567.

Collins, P. L., S. S. Whitehead, A. Bukreyev, R. Fearns, M. N. Teng, K. Juhasz, R. M. Chanock, and B. R. Murphy. 1999. Rational design of live-attenuated recombinant vaccine virus for human respiratory syncytial virus by reverse genetics. *Adv. Virus Res.* **54:**423–451.

Colon, L. E. 1991. Keratoconjunctivitis due to adenovirus type 8: report on a large outbreak. *Ann. Ophthalmol.* **23:**63–65.

Cooney, M. K., J. P. Fox, and G. E. Kenny. 1982. Antigenic groupings of 90 rhinovirus serotypes. *Infect. Immun.* **37:**642–647.

Corne, J. M., S. Green, G. Sanderson, E. O. Caul, and S. L. Johnston. 1999. A multiplex RT-PCR for the detection of parainfluenza viruses 1-3 in clinical samples. *J. Virol. Methods* **82:**9–18.

Covalciuc, K. A., K. H. Webb, and C. A. Carlson. 1999. Comparison of four clinical specimen types for detection of influenza A and B viruses by optical immunoassay (FLU OIA test) and cell culture methods. *J. Clin. Microbiol.* **37:**3971–3974.

Cox, N. J., and K. Subbarao. 1999. Influenza. *Lancet* **354:**1277–1282.

Crawford-Miksza, L. K., R. N. Nang, and D. P. Schnurr. 1999. Strain variation in adenovirus serotypes 4 and 7a causing acute respiratory disease. *J. Clin. Microbiol.* **37:**1107–1112.

Cristallo, A., F. Gambaro, G. Biamonti, P. Ferrante, M. Battaglia, and P. M. Cereda. 1997. Human coronavirus polyadenylated RNA sequences in cerebrospinal fluid from multiple sclerosis patients. *New Microbiol.* **20:**105–114.

Crowe, J. E., Jr., P. T. Bui, C. Y. Firestone, M. Connors, W. R. Elkins, R. M. Chanock, and B. R. Murphy. 1996. Live subgroup B respiratory syncytial virus vaccines that are attenuated, genetically stable, and immunogenic in rodents and non-human primates. *J. Infect. Dis.* **173:**829–839.

Cubie, H. A., L. A. Duncan, L. A. Marshall, and N. M. Smith. 1997. Detection of respiratory syncytial virus nucleic acid in archival postmortem tissue from infants. *Pediatr. Pathol. Lab. Med.* **17:**927–938.

De Boeck, K., N. Van der Aa, S. Van Lierde, L. Corbeel, and R. Eeckels. 1997. Respiratory syncytial virus bronchiolitis: a double-blind dexamethasone efficacy study. *J. Pediatr.* **131:**919–921.

de Haan, C. A., M. Smeets, F. Vernooij, H. Vennema, and P. J. Rottier. 1999. Mapping of the coronavirus membrane protein domains involved in interaction with the spike protein. *J. Virol.* **73:**7441–7452.

Dessau, R. B., G. Lisby, and J. L. Frederiksen. 1999. Coronaviruses in spinal fluid of patients with acute monosymptomatic optic neuritis. *Acta Neurol. Scand.* **100:**88–91.

Dominguez, E. A., L. H. Taber, and R. B. Couch. 1993. Comparison of rapid diagnostic techniques for respiratory syncytial and influenza A virus respiratory infections in young children. *J. Clin. Microbiol.* **31:**2286–2290.

Dowdle, W. R. 1997. Pandemic influenza: confronting a re-emergent threat. The 1976 experience. *J. Infect. Dis.* **176**(Suppl. 1)**:**S69–S72.

Dowell, S. F., L. J. Anderson, H. E. Gary, Jr., D. D. Erdman, J. F. Plouffe, T. M. File, Jr., B. J. Marston, and R. F. Breiman. 1996. Respiratory syncytial virus is an important cause of community-acquired lower respiratory infection among hospitalized adults. *J. Infect. Dis.* **174:**456–462.

Durbin, A. P., C. J. Cho, W. R. Elkins, L. S. Wyatt, B. Moss, and B. R. Murphy. 1999. Comparison of the immunogenicity and efficacy of a replication-defective vaccinia virus expressing antigens of human parainfluenza virus type 3 (HPIV3) with those of a live attenuated HPIV3 vaccine candidate in rhesus monkeys passively immunized with PIV3 antibodies. *J. Infect. Dis.* **179:**1345–1351.

Echevarría, J. E., D. D. Erdman, E. M. Swierkosz, B. P. Holloway, and L. J. Anderson. 1998. Simultaneous detection and identification of human parainfluenza viruses 1, 2, and 3 from clinical samples by multiplex PCR. *J. Clin. Microbiol.* **36:**1388–1391.

Edell, D., E. Bruce, K. Hale, and V. Khoshoo. 1998. Reduced long-term respiratory morbidity after treatment of respiratory syncytial virus bronchiolitis with ribavirin in previously healthy infants: a preliminary report. *Pediatr. Pulmonol.* **25:**154–158.

Falsey, A. R. 1998. Respiratory syncytial virus infection in older persons. *Vaccine* **16:**1775–1778.

Fan, J., K. J. Henrickson, and L. L. Savatski. 1998. Rapid simultaneous diagnosis of infections with respiratory syncytial viruses A and B, influenza viruses A and B, and human parainfluenza virus types 1, 2, and 3 by multiplex quantitative reverse transcription- polymerase chain reaction-enzyme hybridization assay (Hexaplex). *Clin. Infect. Dis.* **26:**1397–1402.

Folz, R. J., and M. A. Elkordy. 1999. Coronavirus pneumonia following autologous bone marrow transplantation for breast cancer. *Chest* **115:**901–905.

Furuuchi, S., M. Shimizu, and Y. Shimizu. 1978. Field trials on transmissible gastroenteritis live virus vaccine in newborn piglets. *Natl. Inst. Anim. Health Q.* (Tokyo) **18:**135–142.

Gerhard, W., C. Hackett, and F. Melchers. 1983. The recognition specificity of a murine helper T cell for hemagglutinin of influenza virus A/PR/8/34. *J. Immunol.* **130:**2379–2385.

Gern, J. E., and W. W. Busse. 1999. Association of rhinovirus infections with asthma. *Clin. Microbiol. Rev.* **12:**9–18.

Ghosh, S., R. Champlin, R. Couch, J. Englund, I. Raad, S. Malik, M. Luna, and E. Whimbey. 1999. Rhinovirus infections in myelosuppressed adult blood and marrow transplant recipients. *Clin. Infect. Dis.* **29:**528–532.

Gimenez, H. B., N. Hardman, H. M. Keir, and P. Cash. 1986. Antigenic variation between human respiratory syncytial virus isolates. *J. Gen. Virol.* **67:**863–870.

Ginsberg, H. S. 1999. The life and times of adenoviruses. *Adv. Virus Res.* **54:**1–13.

Greenbaum, E., A. Morag, and Z. Zakay-Rones. 1998. Isolation of influenza C virus during an outbreak of influenza A and B viruses. *J. Clin. Microbiol.* **36:**1441–1442.

Greve, J. M., G. Davis, A. M. Meyer, C. P. Forte, S. C. Yost, C. W. Marlor, M. E. Kamarck, and A. McClelland. 1989. The major human rhinovirus receptor is ICAM-1. *Cell* **56:**839–847.

Grondahl, B., W. Puppe, A. Hoppe, I. Kuhne, J. A. Weigl, and H. J. Schmitt. 1999. Rapid identification of nine microorganisms causing acute respiratory tract infections by single-tube multiplex reverse transcription-PCR: feasibility study. *J. Clin. Microbiol.* **37:**1–7.

Groothuis, J. R., S. J. King, D. A. Hogerman, P. R. Paradiso, and E. A. Simoes. 1998. Safety and immunogenicity of a purified F protein respiratory syncytial virus (PFP-2) vaccine in seropositive children with bronchopulmonary dysplasia. *J. Infect. Dis.* **177:**467–469.

Grunberg, K., and P. J. Sterk. 1999. Rhinovirus infections: induction and modulation of airways inflammation in asthma. *Clin. Exp. Allergy* **29**(Suppl. 2)**:**65–73.

Hakonarson, H., C. Carter, N. Maskeri, R. Hodinka, and M. M. Grunstein. 1999. Rhinovirus-mediated changes in airway smooth muscle responsiveness: induced autocrine role of interleukin-1 beta. *Am. J. Physiol.* **277:**L13–L21.

Hall, C. B. 1999. Respiratory syncytial virus: a continuing culprit and conundrum. *J. Pediatr.* **135:**2–7.

Hall, C. B., E. E. Walsh, C. E. Long, and K. C. Schnabel. 1991. Immunity to and frequency of reinfection with respiratory syncytial virus. *J. Infect. Dis.* **163:**693–698.

Hamparian, V. V., R. J. Colonno, M. K. Cooney, E. C. Dick, J. M. Gwaltney, Jr., J. H. Hughes, W. S. Jordan, Jr., A. Z. Kapikian, W. J. Mogabgab, and A. Monto. 1987. A collaborative report: rhinoviruses—extension of the numbering system from 89 to 100. *Virology* **159**:191–192.

Hamre, D., and J. J. Procknow. 1966. A new virus isolated from the human respiratory tract. *Proc. Soc. Exp. Biol. Med.* **121**:190–193.

Han, L. L., J. P. Alexander, and L. J. Anderson. 1999. Respiratory syncytial virus pneumonia among the elderly: an assessment of disease burden. *J. Infect. Dis.* **179**:25–30.

Hayden, F. G., J. J. Treanor, R. S. Fritz, M. Lobo, R. F. Betts, M. Miller, N. Kinnersley, R. G. Mills, P. Ward, and S. E. Straus. 1999. Use of the oral neuraminidase inhibitor oseltamivir in experimental human influenza: randomized controlled trials for prevention and treatment. *JAMA* **282**:1240–1246.

Hendley, J. O. 1998. The host response, not the virus, causes the symptoms of the common cold. *Clin. Infect. Dis.* **26**:847–848.

Hendley, J. O. 1999. Clinical virology of rhinoviruses. *Adv. Virus Res.* **54**:453–466.

Hietala, J., M. Uhari, and H. Tuokko. 1988. Antigen detection in the diagnosis of viral infections. *Scand. J. Infect. Dis.* **20**:595–599.

Hornsleth, A., B. Klug, M. Nir, J. Johansen, K. S. Hansen, L. S. Christensen, and L. B. Larsen. 1998. Severity of respiratory syncytial virus disease related to type and genotype of virus and to cytokine values in nasopharyngeal secretions. *Pediatr. Infect. Dis. J.* **17**:1114–1121.

The IMpact-RSV Study Group. 1998. Palivizumab, a humanized respiratory syncytial virus monoclonal antibody, reduces hospitalization from respiratory syncytial virus infection in high-risk infants. *Pediatrics* **102**:531–537.

Kapikian, A. Z. 1975. The coronaviruses. *Dev. Biol. Stand.* **28**:42–64.

Karron, R. A., P. F. Wright, F. K. Newman, M. Makhene, J. Thompson, R. Samorodin, M. H. Wilson, E. L. Anderson, M. L. Clements, and B. R. Murphy. 1995. A live human parainfluenza type 3 virus vaccine is attenuated and immunogenic in healthy infants and children. *J. Infect. Dis.* **172**:1445–1450.

Katagiri, S., A. Ohizumi, S. Ohyama, and M. Homma. 1987. Follow-up study of type C influenza outbreak in a children's home. *Microbiol. Immunol.* **31**:337–343.

Kellogg, J. A. 1991. Culture vs direct antigen assays for detection of microbial pathogens from lower respiratory tract specimens suspected of containing the respiratory syncytial virus. *Arch. Pathol. Lab. Med.* **115**:451–458.

Kidd, A. H., M. Jonsson, D. Garwicz, A. E. Kajon, A. G. Wermenbol, M. W. Verweij, and J. C. de Jong. 1996. Rapid subgenus identification of human adenovirus isolates by a general PCR. *J. Clin. Microbiol.* **34**:622–627.

Kim, H. W., J. G. Canchola, C. D. Brandt, G. Pyles, R. M. Chanock, K. Jensen, and R. H. Parrott. 1969. Respiratory syncytial virus disease in infants despite prior administration of antigenic inactivated vaccine. *Am. J. Epidemiol.* **89**:422–434.

Kunkel, F., and G. Herrler. 1993. Structural and functional analysis of the surface protein of human coronavirus OC43. *Virology* **195**:195–202.

Leonardi, G. P., H. Leib, G. S. Birkhead, C. Smith, P. Costello, and W. Conron. 1994. Comparison of rapid detection methods for influenza A virus and their value in health-care management of institutionalized geriatric patients. *J. Clin. Microbiol.* **32**:70–74.

Li, Q. G., J. Hambraeus, and G. Wadell. 1991. Genetic relationship between thirteen genome types of adenovirus 11, 34, and 35 with different tropisms. *Intervirology* **32**:338–350.

Ludwig, S. L., J. F. Brundage, P. W. Kelley, R. Nang, C. Towle, D. P. Schnurr, L. Crawford-Miksza, and J. C. Gaydos. 1998. Prevalence of antibodies to adenovirus serotypes 4 and 7 among unimmunized US Army trainees: results of a retrospective nationwide seroprevalence survey. *J. Infect. Dis.* **178**:1776–1778.

Macnaughton, M. R. 1982. Occurrence and frequency of coronavirus infections in humans as determined by enzyme-linked immunosorbent assay. *Infect. Immun.* **38**:419–423.

Makela, M. J., T. Puhakka, O. Ruuskanen, M. Leinonen, P. Saikku, M. Kimpimaki, S. Blomqvist, T. Hyypia, and P. Arstila. 1998. Viruses and bacteria in the etiology of the common cold. *J. Clin. Microbiol.* **36**:539–542.

Marlovits, T. C., C. Abrahamsberg, and D. Blaas. 1998. Very-low-density lipoprotein receptor fragment shed from HeLa cells inhibits human rhinovirus infection. *J. Virol.* **72**:10246–10250.

Marx, A., H. E. Gary, Jr., B. J. Marston, D. D. Erdman, R. F. Breiman, T. J. Torok, J. F. Plouffe, T. M. File, Jr., and L. J. Anderson. 1999. Parainfluenza virus infection among adults hospitalized for lower respiratory tract infection. *Clin. Infect. Dis.* **29**:134–140.

Marx, A., T. J. Torok, R. C. Holman, M. J. Clarke, and L. J. Anderson. 1997. Pediatric hospitalizations for croup (laryngotracheobronchitis): biennial increases associated with human parainfluenza virus 1 epidemics. *J. Infect. Dis.* **176**:1423–1427.

Maslo, C., P. M. Girard, T. Urban, S. Guessant, and W. Rozenbaum. 1997. Ribavirin therapy for adenovirus pneumonia in an AIDS patient. *Am. J. Respir. Crit. Care Med.* **156**:1263–1264.

Mlinaric-Galinovic, G., A. R. Falsey, and E. E. Walsh. 1996. Respiratory syncytial virus infection in the elderly. *Eur. J. Clin. Microbiol. Infect. Dis.* **15**:777–781.

Monto, A. S., D. M. Fleming, D. Henry, R. de Groot, M. Makela, T. Klein, M. Elliott, O. N. Keene, and C. Y. Man. 1999. Efficacy and safety of the neuraminidase inhibitor zanamivir in the treatment of influenza A and B virus infections. *J. Infect. Dis.* **180**:254–261.

Moriuchi, H., N. Katsushima, H. Nishimura, K. Nakamura, and Y. Numazaki. 1991. Community-acquired influenza C virus infection in children. *J. Pediatr.* **118**:235–238.

Mounir, S., and P. J. Talbot. 1992. Sequence analysis of the membrane protein gene of human coronavirus OC43 and evidence for O-glycosylation. *J. Gen. Virol.* **73**:2731–2736.

Mufson, M. A., B. Åkerlind-Stopner, C. Örvell, R. B. Belshe, and E. Norrby. 1991. A single-season epidemic with respiratory syncytial virus subgroup B2 during 10 epidemic years, 1978 to 1988. *J. Clin. Microbiol.* **29**:162–165.

Mufson, M. A., R. B. Belshe, C. Orvell, and E. Norrby. 1987. Subgroup characteristics of respiratory syncytial virus strains recovered from children with two consecutive infections. *J. Clin. Microbiol.* **25**:1535–1539.

Mufson, M. A., R. B. Belshe, C. Orvell, and E. Norrby. 1988. Respiratory syncytial virus epidemics: variable dominance of subgroups A and B strains among children, 1981–1986. *J. Infect. Dis.* **157**:143–148.

Mufson, M. A., H. E. Krause, H. E. Mocega, and F. W. Dawson. 1970. Viruses, Mycoplasma pneumoniae and bacteria associated with lower respiratory tract disease among infants. *Am. J. Epidemiol.* **91:**192–202.

Mufson, M. A., H. D. Levine, R. E. Wasil, H. E. Mocega-Gonzalez, and H. E. Krause. 1973. Epidemiology of respiratory syncytial virus infection among infants and children in Chicago. *Am. J. Epidemiol.* **98:**88–95.

Mufson, M. A., C. Orvell, B. Rafnar, and E. Norrby. 1985. Two distinct subtypes of human respiratory syncytial virus. *J. Gen. Virol.* **66:**2111–2124.

Mufson, M. A., and R. J. Stanek. 1996. Identification of a variant subgroup A strain of respiratory syncytial virus. *J. Clin. Microbiol.* **34:**2493–2496.

Murphy, B. R., P. L. Collins, L. Lawrence, J. Zubak, R. M. Chanock, and G. A. Prince. 1989. Immunosuppression of the antibody response to respiratory syncytial virus (RSV) by pre-existing serum antibodies: partial prevention by topical infection of the respiratory tract with vaccinia virus-RSV recombinants. *J. Gen. Virol.* **70:**2185–2190.

Natuk, R. J., A. R. Davis, P. K. Chanda, M. D. Lubeck, M. Chengalvala, S. C. Murthy, M. S. Wade, S. K. Dheer, B. M. Bhat, and K. K. Murthy. 1994. Adenovirus vectored vaccines. *Dev. Biol. Stand.* **82:**71–77.

Nelson, K. E., M. A. Greenberg, M. A. Mufson, and V. K. Moses. 1975. The sudden infant death syndrome and epidemic viral disease. *Am. J. Epidemiol.* **101:**423–430.

Nicholson, K. G. 1992. Clinical features of influenza. *Semin. Respir. Infect.* **7:**26–37.

Olsen, M. A., K. M. Shuck, and A. R. Sambol. 1993. Evaluation of Abbott TestPack RSV for the diagnosis of respiratory syncytial virus infections. *Diagn. Microbiol. Infect. Dis.* **16:**105–109.

Osiowy, C. 1998. Direct detection of respiratory syncytial virus, parainfluenza virus, and adenovirus in clinical respiratory specimens by a multiplex reverse transcription-PCR assay. *J. Clin. Microbiol.* **36:**3149–3154.

Papadopoulos, N. G., G. Sanderson, J. Hunter, and S. L. Johnston. 1999. Rhinoviruses replicate effectively at lower airway temperatures. *J. Med. Virol.* **58:**100–104.

Peiris, M., K. Y. Yuen, C. W. Leung, K. H. Chan, P. L. Ip, R. W. Lai, W. K. Orr, and K. F. Shortridge. 1999. Human infection with influenza H9N2. *Lancet* **354:**916–917.

Pitkaranta, A., and F. G. Hayden. 1998. Rhinoviruses: important respiratory pathogens. *Ann. Med.* **30:**529–537.

Pitkaranta, A., J. Jero, E. Arruda, A. Virolainen, and F. G. Hayden. 1998. Polymerase chain reaction-based detection of rhinovirus, respiratory syncytial virus, and coronavirus in otitis media with effusion. *J. Pediatr.* **133:**390–394.

Pohl-Koppe, A., T. Raabe, S. G. Siddell, and V. ter Meulen. 1995. Detection of human coronavirus 229E-specific antibodies using recombinant fusion proteins. *J. Virol. Methods* **55:**175–183.

Raabe, T., B. Schelle-Prinz, and S. G. Siddell. 1990. Nucleotide sequence of the gene encoding the spike glycoprotein of human coronavirus HCV 229E. *J. Gen. Virol.* **71:**1065–1073.

Ray, R., M. S. Galinski, B. R. Heminway, K. Meyer, F. K. Newman, and R. B. Belshe. 1996. Temperature-sensitive phenotype of the human parainfluenza virus type 3 candidate vaccine strain (cp45) correlates with a defect in the L gene. *J. Virol.* **70:**580–584.

Richman, A. V., F. A. Pedreira, and N. M. Tauraso. 1971. Attempts to demonstrate hemagglutination and hemadsorption by respiratory syncytial virus. *Appl. Microbiol.* **21:**1099–1100.

Riski, H., and T. Hovi. 1980. Coronavirus infections of man associated with diseases other than the common cold. *J. Med. Virol.* **6:**259–265.

Rodriguez, W. J. 1999. Management strategies for respiratory syncytial virus infections in infants. *J. Pediatr.* **135:**45–50.

Rodriguez, W. J., J. Arrobio, R. Fink, H. W. Kim, and C. Milburn. 1999. Prospective follow-up and pulmonary functions from a placebo-controlled randomized trial of ribavirin therapy in respiratory syncytial virus bronchiolitis. Ribavirin Study Group. *Arch. Pediatr. Adolesc. Med.* **153:**469–474.

Rossmann, M. G., N. H. Olson, P. R. Kolatkar, M. A. Oliveira, R. H. Cheng, J. M. Greve, A. McClelland, and T. S. Baker. 1994. Crystallographic and cryo EM analysis of virion-receptor interactions. *Arch. Virol. Suppl.* **9:**531–541.

Sanders, S. P. 1999. Asthma, viruses, and nitric oxide. *Proc. Soc. Exp. Biol. Med.* **220:**123–132.

Schmidt, O. W. 1984. Antigenic characterization of human coronaviruses 229E and OC43 by enzyme-linked immunosorbent assay. *J. Clin. Microbiol.* **20:**175–180.

Schmidt, O. W., I. D. Allan, M. K. Cooney, H. M. Foy, and J. P. Fox. 1986. Rises in titers of antibody to human coronaviruses OC43 and 229E in Seattle families during 1975–1979. *Am. J. Epidemiol.* **123:**862–868.

Schober, D., P. Kronenberger, E. Prchla, D. Blaas, and R. Fuchs. 1998. Major and minor receptor group human rhinoviruses penetrate from endosomes by different mechanisms. *J. Virol.* **72:**1354–1364.

Skiadopoulos, M. H., S. Surman, J. M. Tatem, M. Paschalis, S. L. Wu, S. A. Udem, A. P. Durbin, P. L. Collins, and B. R. Murphy. 1999. Identification of mutations contributing to the temperature-sensitive, cold-adapted, and attenuation phenotypes of the live-attenuated cold-passage 45 (cp45) human parainfluenza virus 3 candidate vaccine. *J. Virol.* **73:**1374–1381.

Staunton, D. E., V. J. Merluzzi, R. Rothlein, R. Barton, S. D. Marlin, and T. A. Springer. 1989. A cell adhesion molecule, ICAM-1, is the major surface receptor for rhinoviruses. *Cell* **56:**849–853.

Stewart, J. N., S. Mounir, and P. J. Talbot. 1992. Human coronavirus gene expression in the brains of multiple sclerosis patients. *Virology* **191:**502–505.

Sullender, W. M. 2000. Respiratory syncytial virus genetic and antigenic diversity. *Clin. Microbiol. Rev.* **13:**1–15.

Sullender, W. M., M. A. Mufson, L. J. Anderson, and G. W. Wertz. 1991. Genetic diversity of the attachment protein of subgroup B respiratory syncytial viruses. *J. Virol.* **65:**5425–5434.

Talbot, P. J., S. Ekande, N. R. Cashman, S. Mounir, and J. N. Stewart. 1993. Neurotropism of human coronavirus 229E. *Adv. Exp. Med. Biol.* **342:**339–346.

Talbot, P. J., J. S. Paquette, C. Ciurli, J. P. Antel, and F. Ouellet. 1996. Myelin basic protein and human coronavirus 229E cross-reactive T cells in multiple sclerosis. *Ann. Neurol.* **39:**233–240.

Thomas, E., M. J. Margach, C. Orvell, B. Morrison, and E. Wilson. 1994. Respiratory syncytial virus subgroup B dominance during one winter season between 1987 and 1992 in Vancouver, Canada. *J. Clin. Microbiol.* **32:**238–242.

Tsutsumi, H., K. Ouchi, M. Ohsaki, T. Yamanaka, Y. Kuniya, Y. Takeuchi, C. Nakai, H. Meguro, and S. Chiba. 1999. Immunochromatography test for rapid diagnosis of adenovirus respiratory tract infections: comparison with virus isolation in tissue culture. *J. Clin. Microbiol.* **37:**2007–2009.

Turner, R. B., K. W. Weingand, C. H. Yeh, and D. W. Leedy. 1998. Association between interleukin-8 concentration in nasal secretions and severity of symptoms of experimental rhinovirus colds. *Clin. Infect. Dis.* **26:**840–846.

Uncapher, C. R., C. M. DeWitt, and R. J. Colonno. 1991. The major and minor group receptor families contain all but one human rhinovirus serotype. *Virology* **180:**814–817.

van Wyke Coelingh, K. L., C. C. Winter, E. L. Tierney, S. L. Hall, W. T. London, H. W. Kim, R. M. Chanock, and B. R. Murphy. 1990. Antibody responses of humans and nonhuman primates to individual antigenic sites of the hemagglutinin-neuraminidase and fusion glycoproteins after primary infection or reinfection with parainfluenza type 3 virus. *J. Virol.* **64:**3833–3843.

Vellayappan, K., J. Teo, and S. Doraisingham. 1982. Respiratory syncytial virus infections in children. *J. Singapore Paediatr. Soc.* **24:**69–77.

Walsh, E. E., A. R. Falsey, and P. A. Hennessey. 1999. Respiratory syncytial and other virus infections in persons with chronic cardiopulmonary disease. *Am. J. Respir. Crit. Care Med.* **160:**791–795.

Walsh, E. E., K. M. McConnochie, C. E. Long, and C. B. Hall. 1997. Severity of respiratory syncytial virus infection is related to virus strain. *J. Infect. Dis.* **175:**814–820.

Webster, R. G., W. J. Bean, O. T. Gorman, T. M. Chambers, and Y. Kawaoka. 1992. Evolution and ecology of influenza A viruses. *Microbiol. Rev.* **56:**152–179.

Wentworth, D. E., B. L. Thompson, X. Xu, H. L. Regnery, A. J. Cooley, M. W. McGregor, N. J. Cox, and V. S. Hinshaw. 1994. An influenza A (H1N1) virus, closely related to swine influenza virus, responsible for a fatal case of human influenza. *J. Virol.* **68:**2051–2058.

Whimbey, E., R. B. Couch, J. A. Englund, M. Andreeff, J. M. Goodrich, I. I. Raad, V. Lewis, N. Mirza, M. A. Luna, and B. Baxter. 1995. Respiratory syncytial virus pneumonia in hospitalized adult patients with leukemia. *Clin. Infect. Dis.* **21:**376–379.

Wintermeyer, S. M., and M. C. Nahata. 1995. Rimantadine: a clinical perspective. *Ann. Pharmacother.* **29:**299–310.

Wiselka, M. J., J. Kent, J. B. Cookson, and K. G. Nicholson. 1993. Impact of respiratory virus infection in patients with chronic chest disease. *Epidemiol. Infect.* **111:**337–346.

Yamadera, S., K. Yamashita, M. Akatsuka, N. Kato, M. Hashido, S. Inouye, and S. Yamazaki. 1995. Adenovirus surveillance, 1982-1993, Japan. A report of the National Epidemiological Surveillance of Infectious Agents in Japan. *Jpn. J. Med. Sci. Biol.* **48:**199–210.

Yarosh, O. K., A. I. Wandeler, F. L. Graham, J. B. Campbell, and L. Prevec. 1996. Human adenovirus type 5 vectors expressing rabies glycoprotein. *Vaccine* **14:**1257–1264.

Yeager, C. L., R. A. Ashmun, R. K. Williams, C. B. Cardellichio, L. H. Shapiro, A. T. Look, and K. V. Holmes. 1992. Human aminopeptidase N is a receptor for human coronavirus 229E. *Nature* **357:**420–422.

Ylikoski, J., and J. Karjalainen. 1989. Acute tonsillitis in young men: etiological agents and their differentiation. *Scand. J. Infect. Dis.* **21:**169–174.

Yousuf, H. M., J. Englund, R. Couch, K. Rolston, M. Luna, J. Goodrich, V. Lewis, N. Q. Mirza, M. Andreeff, C. Koller, L. Elting, G. P. Bodey, and E. Whimbey. 1997. Influenza among hospitalized adults with leukemia. *Clin. Infect. Dis.* **24:**1095–1099.

Zhou, N. N., K. F. Shortridge, E. C. J. Claas, S. L. Krauss, and R. G. Webster. 1999. Rapid evolution of H5N1 influenza viruses in chickens in Hong Kong. *J. Virol.* **73:**3366–3374.

Ziegler, T., H. Hall, A. Sanchez-Fauquier, W. C. Gamble, and N. J. Cox. 1995. Type- and subtype-specific detection of influenza viruses in clinical specimens by rapid culture assay. *J. Clin. Microbiol.* **33:**318–321.

Ziegler, T., M. L. Hemphill, M. L. Ziegler, G. Perez-Oronoz, A. I. Klimov, A. W. Hampson, H. L. Regnery, and N. J. Cox. 1999. Low incidence of rimantadine resistance in field isolates of influenza A viruses. *J. Infect. Dis.* **180:**935–939.

Enteroviruses

HEINZ ZEICHHARDT AND HANS-PETER GRUNERT

24

INTRODUCTION

Human enteroviruses are small RNA viruses that belong to the family *Picornaviridae* ("pico" [Greek], small). The *Picornaviridae* comprise the following genera of human and animal pathogenic viruses: *Enterovirus, Rhinovirus, Cardiovirus, Aphthovirus, Hepatovirus,* and *Parechovirus.* (For classification and nomenclature of the viruses, see Minor et al. [1995] and Pringle [1997]; updated information can also be found on the website of the International Committee on Taxonomy of Viruses [http://www.ncbi.nlm.nih.gov/ICTV/].) The members of the *Enterovirus* genus that infect humans are grouped together because of similar physicochemical properties; they include the polioviruses, coxsackievirus groups A and B, echoviruses, and enteroviruses 68 to 71 (Table 1). The formerly named enterovirus type 72 has been reclassified as hepatitis A virus in its own genus, *Hepatovirus.* Echovirus types 22 and 23 have been reclassified in the new genus *Parechovirus.* All enteroviruses inhabit the human alimentary tract, and most of them are able to infect the central nervous system (CNS). In addition, these viruses induce a variety of clinical syndromes. The reviews of Melnick (1996), Rueckert (1996), and Zeichhardt and Grunert (1999) are recommended for extensive additional readings on enteroviruses.

History of Virus Discovery

Crippling paralytic disease was already recorded in ancient times (Melnick, 1982). The disease was characterized for the first time as poliomyelitis with flaccid paralysis by the German orthopedist Heine and the Swedish pediatrician Medin and therefore was also called Heine-Medin disease. Poliomyelitis was established as a viral disease in 1909 when Landsteiner and Popper (1909) transmitted paralytic disease to monkeys by inoculating them with filtered stool from a patient with paralytic disease. During the next 40 years, animal inoculation was the method of choice for virus inoculation and study. Thus, in 1948, in Coxsackie, N.Y., a virus was isolated in suckling mice that had been inoculated with a cell-free filtrate of stools obtained from two children suffering from paralysis (Dalldorf and Sickles, 1948). This virus, which could not be neutralized by antiserum against any of the three polioviruses, became the first member of the group A coxsackieviruses. The first of the group B coxsackieviruses was isolated in 1949 (Melnick et al., 1949).

The major breakthrough for diagnosing and controlling poliomyelitis was the observation that poliovirus could be propagated in human embryonic tissues in culture (Enders et al., 1949). These tissue cultures allowed easy isolation of the viruses and were prerequisite for the development of vaccines, including both the inactivated vaccine of Salk and the attenuated (oral) vaccine of Sabin. After the introduction of tissue culture, the isolation of many other enteroviruses was possible. The first echoviruses ("echo" stands for "enteric cytopathic human orphan") were discovered in 1951 by Robbins et al. (1951). These viruses often were isolated from stools of healthy children and therefore could not be related to a disease; hence, these "enteric viruses" were called "orphan viruses." Poliovirus type 1 and the related human rhinovirus type 14 were the first human pathogenic viruses of which the three-dimensional structure was shown by X-ray analysis (Hogle et al., 1985; Rossmann et al., 1985). Molla and coworkers (1991) established for the first time the cell-free synthesis of infective poliovirus particles from isolated viral RNA in a cytoplasmic extract from noninfected cells.

BIOLOGY

Structure

Enteroviruses are small RNA viruses consisting of a spherical, nonenveloped protein shell which encapsidates one molecule of single-stranded positive-sense RNA (Fig. 1). The structural and functional properties of poliovirus are described below, as this virus is the best-characterized enterovirus. The poliovirus particle has a diameter of 27 to 30 nm, a buoyant density in CsCl of 1.34 g/ml, a molecular mass of 8.25×10^6 Da, and a sedimentation coefficient of 156S to 160S (for reviews, see Mirzayan and Wimmer [1994], Rueckert [1996], and Zeichhardt and Grunert [1999]).

The virus capsid has an icosahedral symmetry and consists of 60 protomers. Each protomer comprises the four nonglycosylated virus proteins VP1 (34 kDa), VP2 (30 kDa), VP3 (27 kDa), and VP4 (7.5 kDa). The virus protein VP4 is myristoylated at its N terminus. As shown by X-ray analysis, the virus proteins VP1, VP2, and VP3 are located at the capsid surface, whereas VP4 is located in the interior of the capsid in close contact with the viral RNA (Hogle et al., 1985). The capsid proteins VP1, VP2, and VP3 reveal a pseudoequivalent packing arrangement which is also shown for some spherical plant viruses. This suggests that picornaviruses are old viruses, in evolutionary terms.

The capsid surface of all three poliovirus serotypes shows a depression, also called a "canyon," around each of the 12 fivefold symmetry axes. The canyon is formed at the junction of VP1 and VP3 and is involved in the binding of the virus-specific receptor (see "Replication in Cell Culture," below). This virus-receptor interaction was first described as the "canyon hypothesis" for the closely related human rhinovirus 14 by Rossmann et al. (1985).

The single-stranded positive-sense RNA genome of poliovirus type 1 is approximately 7,500 nucleotides long (molecular mass, 2.6×10^6 Da) and codes in a single open reading frame for all viral capsid and functional proteins (Fig. 2). The genomes of several enteroviruses have been sequenced completely. For updated information, see "The Picornavirus Home Page" (http://www.iah.bbsrc.ac.uk/virus/picornaviridae/). A small virus protein (VPg, for "virus protein genome-linked") is covalently bound to the 5′ end of the RNA molecule, and the 3′ end is polyadenylated (approximately 60 nucleotides). The 5′ and 3′ termini of the viral RNA cover nontranslated regions (NTRs) of different lengths (5′ NTR, 743 nucleotides; 3′ NTR, 68 nucleotides) showing a high degree of secondary structures. A clover leaf-like secondary structure in the 5′ NTR was identified as the internal ribosomal entry site (IRES) for the initiation of translation at the AUG codon in position 743.

The positive-sense viral RNA serves as mRNA. The translation of this RNA yields a single large precursor polyprotein. The coding region is subdivided into three regions (Fig. 2). Region P1 codes for the capsid proteins VP0 (precursor of VP4 and VP2), VP3, and VP1. Regions P2 and P3 code for functional proteins such as the viral proteases (2A, 3C, and 3CD), the genome-linked virus protein VPg, and the viral RNA polymerase (3D). The P1 capsid precursor protein is released from the nascent polyprotein by protease 2A cleavage. In additional steps, P1 and the precursor for the functional proteins (P2 and P3) are processed by proteases 3C and 3CD. The final cleavage of VP0 into VP4 and VP2 takes place when the newly synthesized viral RNA is encapsidated at the end of the virus maturation process.

TABLE 1 Serotypes of the genus *Enterovirus*

Virus(es)	Serotypes[a]
Poliovirus	1, 2, 3
Coxsackievirus group A	A1–A22, A24
Coxsackievirus group B	B1–B6
Echovirus[b]	1–7, 9, 11–21, 24–27, 29–33
Other enteroviruses	68–71

[a] Echovirus 8 has been deleted because it was identical to echovirus 1. The following viruses have been reclassified: coxsackievirus A23 as echovirus 9, echovirus 10 as reovirus 1, echoviruses 22 and 23 as parechoviruses 1 and 2 (respectively), echovirus 28 as human rhinovirus 1A, echovirus 34 as coxsackievirus A24, enterovirus 72 as hepatitis A virus. Further information can be found at the website of the International Committee on Taxonomy of Viruses (http://www.ncbi.nlm.nih.gov/ICTV/).
[b] "Echo" stands for "enteric cytopathic human orphan."

Antigenicity and Neutralization

Preparations of poliovirus contain two antigens that can be detected in complement fixation (CF) and precipitation tests: infective (or "native") virus, called D (or N) antigen, and noninfective virus, called C antigen or occasionally H (heated) antigen. The antigenic sites at the capsid surface determine the type-specific antigenicity of the enteroviruses that is best investigated in neutralization tests. At the surface of poliovirus, the capsid proteins VP1, VP2, and VP3 present four type-specific immunodominant epitopes, the "neutralization antigenic sites" N-Ag I, II, IIIA, and IIIB (Fig. 1B) (for reviews, see Mosser et al. [1989], Mirzayan and Wimmer [1994], and Rueckert [1996]). In contrast, for coxsackievirus type B3, the major neutralizing antigenic sites are located on capsid protein VP2 (Beatrice et al., 1980).

Several mechanisms have been proposed for neutralization (for reviews, see Dimmock [1984], Mosser et al. [1989], and Stewart and Nemerow [1997]), each interfering with the entry of enteroviruses into their host cells (see "Replication in Cell Culture," below). Neutralizing antibodies can (i) reduce the number of infective units merely by aggregation, (ii) create large immune complexes incapable of adsorbing to the cell surface, (iii) induce a conformational change in the capsid, or (iv) stabilize the capsid and thereby prevent virus uncoating. Stoichiometric analysis of the neutralization reaction has revealed that binding of one polyclonal antibody or four monoclonal antibodies at a single virion is sufficient to neutralize infectivity (Wetz et al., 1986; Icenogle et al., 1983).

Several antigenic relationships between enteroviruses have been observed. In neutralization tests, poliovirus types 1 and 2 partially cross-react. Furthermore, coxsackievirus types A3 and A8, A11 and A15, and A13 and A18, and echovirus types 1 and 8, 6 and 30, and 12 and 29, are antigenically related. The cross-reactivity between several enteroviruses observed in the CF test may be due to common antigenic sites of the virus proteins that are located in the interior of the capsid. These sites are accessible only when a soluble antigen is used. Such immunologic cross-reactivity was confirmed by the immunoblot technique (Mertens et al., 1983). In addition, these antigenic relationships are reflected by limited homologies among the genomic RNAs and viral proteins of the different enteroviruses, as studied by sequence alignments (Palmenberg, 1989; Rueckert, 1996). The homology of the complete genomic sequences for different enteroviruses is greater than 50% and for different strains within a species is greater than 75%. The homologies of the amino acid sequences of the capsid proteins are about 70% among the three serotypes of the poliovirus Sabin strains and >50% between polioviruses and coxsackieviruses.

Reactivity to Chemical and Physical Agents and Virus Storage

All enteroviruses are resistant to low pH (pH 3) and several proteolytic enzymes, which is the prerequisite for virus passage through the stomach and duodenum. The viruses are resistant to several disinfectants, such as 70% alcohol, 5% lysol, and 1% quaternary ammonium compounds, and to ether, deoxycholate, and various other detergents that destroy lipid-containing viruses. In general, enteroviruses are inactivated by the following chemicals (see Melnick [1982] and Moore and Morens [1984]): formaldehyde (0.3%), HCl (0.1 N), free residual chlorine (0.3 to

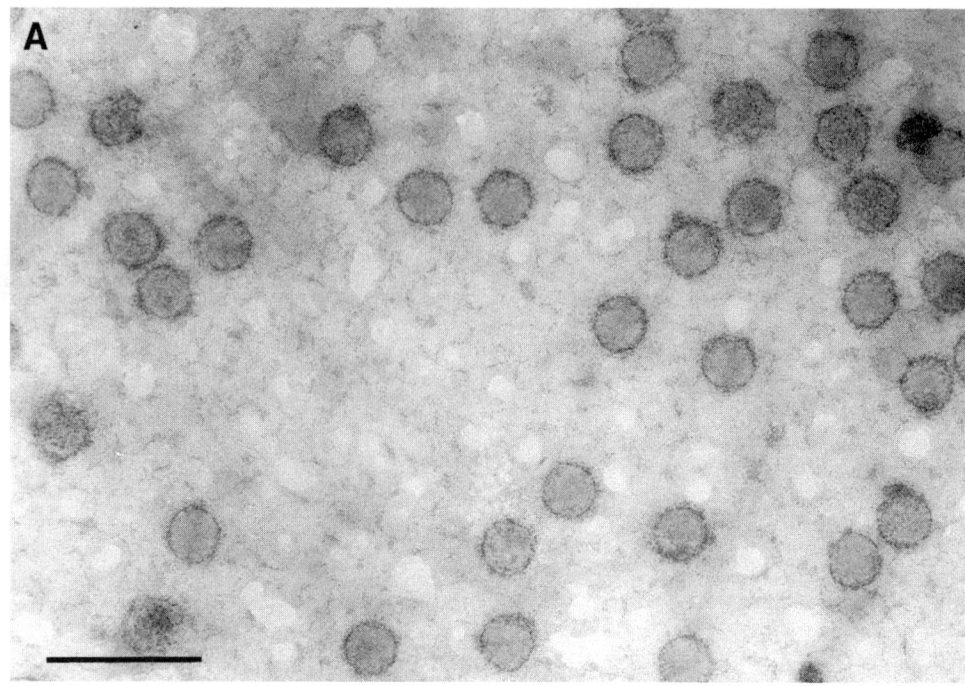

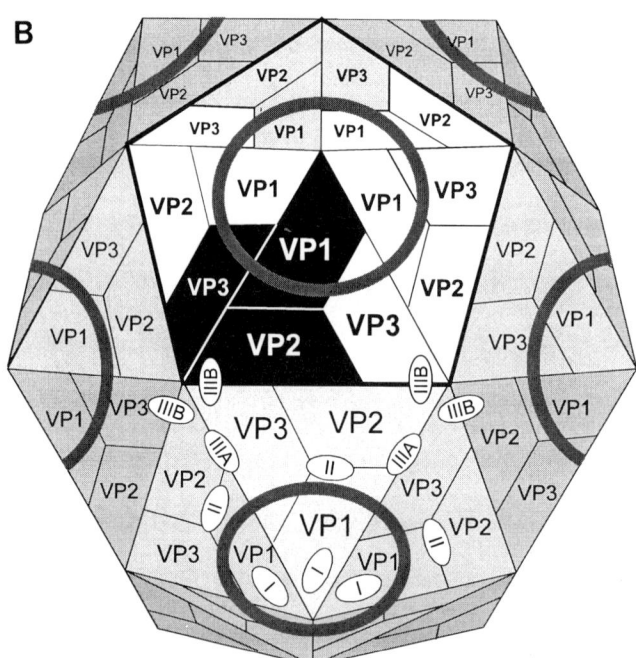

FIGURE 1 Morphology of poliovirus. (A) Transmission electron micrograph of poliovirus type 1 particles, negatively stained with 0.5% uranyl acetate. Bar, 100 nm. (B) Schematic representation of the three-dimensional structure of a poliovirus particle and the four neutralizing antigenic (N-Ag) sites. X-ray crystallographic structure analysis of poliovirus type 1 (Hogle et al., 1985) has revealed an icosahedral capsid structure typical of enteroviruses. The capsid surface is composed of 60 protomers, each consisting of the capsid proteins VP1, VP2, and VP3 (black areas). Each of the 12 fivefold symmetry axes is surrounded by five protomers, forming a pentamer (surrounded by a bold black line). The attachment site for the virus-specific receptor is a depression around the five-fold symmetry axis, also called the canyon (dark grey circles). Each of the three surface-exposed capsid proteins contains immunodominant antigenic sites at which neutralizing antibodies bind, resulting in virus neutralization. Four N-Ag sites (white ellipses) have been mapped to surface loop extensions (for a review, see Mirzayan and Wimmer [1994]). N-Ag I is a continuous sequence in VP1 mapping to amino acids 95 to 105. N-Ag II is a discontinuous site mapping to amino acids 221 to 226 of VP1 and amino acids 164 to 172 and 270 in VP2. N-Ag III consists of two independent discontinuous sites. N-Ag IIIA is composed of amino acids 58 to 60 and 71 to 73 of VP3, and N-Ag IIIB consists of amino acid 72 of VP2 and amino acids 76 to 79 of VP3. The smallest capsid protein, VP4, lies buried in the capsid shell in close association with the single molecule of viral RNA.

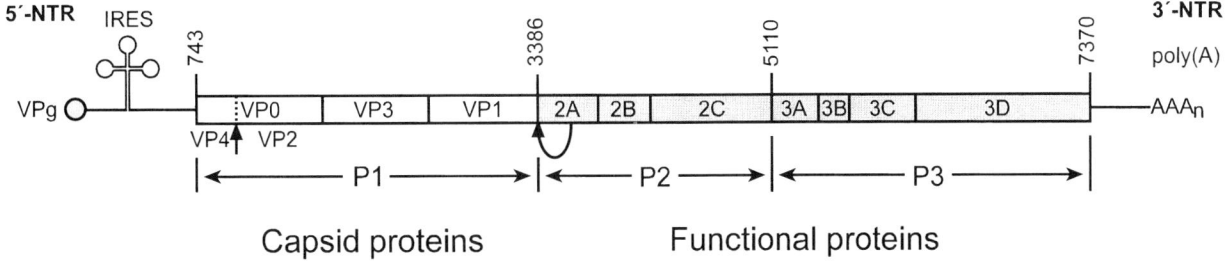

FIGURE 2 Genome organization of poliovirus type 1. The poliovirus genome is a single-stranded positive-sense RNA of approximately 7,500 nucleotides. Nucleotides 743 to 7370 code in a single open reading frame for the capsid proteins (white boxes in coding region P1) and functional proteins (grey boxes in coding regions P2 and P3). The 5' and 3' NTRs are shown as lines. The IRES is shown schematically with the two-dimensional structure. The virus protein VPg is covalently linked to the terminal uracil of the 5' NTR. For further details, see "Structure" and "Replication in Cell Culture," in the text.

0.5 ppm), and other halogens (free residual bromine or iodine, ~0.5 ppm at 10 min of contact time). The presence of organic matter with the virus may result in protection against inactivation. For this reason, 3% formaldehyde is recommended for disinfection. The following physical conditions are inactivating: drying, heat (50°C for 1 h in the absence of magnesium chloride), and light (in the presence of vital dyes such as neutral red, acridine orange, and proflavine).

Enteroviruses are stable for years when stored at −70°C. Storage at −20°C results in some loss in titer over periods of months, while enteroviruses in suspension stored at 4°C usually will stay viable for weeks.

Replication in Cell Culture

The cell tropism of enteroviruses determines their pathogenicity mechanism. The tropism mainly depends on the binding of the virus to a virus-specific receptor at the surface of a susceptible host cell. In addition, virus reproduction in the host cell is influenced by different cellular factors (for a review, see Andino et al. [1999]). The viral replication cycle can be divided into two phases. The early phase comprises virus adsorption to the host cell, virus penetration into the cell, and virus uncoating with release of the genomic RNA from the capsid. The later phase comprises synthesis of viral proteins and RNA, virus assembly and maturation, and finally release of virus progeny from the infected host cell (Fig. 3). The reproduction of poliovirus is the best characterized of that of all enteroviruses and is described in detail below (for reviews, see Mirzayan and Wimmer [1994], Rueckert [1996], and Zeichhardt and Grunert [1999]).

Virus Adsorption

The adsorption of the virus to a virus-specific receptor at the surface of a host cell is the first step in the virus reproduction cycle. Several of the enterovirus receptors belong to receptor families such as the immunoglobulin superfamily or the integrins. The receptors for the three types of polioviruses, the six types of group B coxsackieviruses, and the three types of group A coxsackieviruses belong to the immunoglobulin superfamily.

The poliovirus receptor (CD155) is a glycosylated three-domain membrane protein (67 to 80 kDa) that appears in three isoforms (Mendelsohn et al., 1989; Koike et al., 1990). During virus adsorption to the host cell, at least

domain 1 of the poliovirus receptor binds in the canyon of the virus capsid and induces conformational alterations as a prerequisite for virus uncoating (Freistadt and Racaniello, 1991; Bernhardt et al., 1994; Morrison et al., 1994). The group B coxsackieviruses use the coxsackievirus adenovirus receptor (CAR), a membrane glycoprotein (46 kDa) that is also used by adenoviruses 2 and 5 (Bergelson et al., 1997). The receptor for coxsackieviruses A13, A18, and A21 is intercellular adhesion molecule 1 (ICAM-1) (CD54) (Colonno et al., 1986). Some receptors for enteroviruses belong to a second superfamily of proteins, the integrins. Vitronectin ($\alpha_v\beta_3$) is the receptor for coxsackievirus A9 and for echovirus 9 strain Barty (Roivainen et al., 1994; Nelsen-Salz et al., 1999). Echoviruses 1 and 8 use very late antigen 2 (VLA-2) ($\alpha_2\beta_1$) as their receptor (Bergelson et al., 1993). Another membrane protein, the decay-accelerating factor (DAF) (CD55), is the receptor for echoviruses 6, 7, 12, and 21 and for enterovirus 70 (Bergelson et al., 1994; Karnauchow et al., 1996).

For some enteroviruses, additional binding proteins (accessory factors) that are not essential for cell tropism but support attachment of the virus to the host cell surface have been reported. Binding of poliovirus to the host cell surface is supported by accessory factors such as the lymphocyte homing receptor (CD44) or membrane glycoproteins (50 and 23 to 25 kDa) (Barnert et al., 1992; Shepley and Racaniello, 1994; Bouchard and Racaniello, 1997). Additionally, coxsackieviruses B1, B3, and B5 (Shafren et al., 1995) and coxsackievirus A21 (Shafren et al., 1997) bind to DAF.

Virus Penetration and Uncoating

After attachment to the virus-specific receptor, poliovirus enters the host cell by a mechanism of receptor-mediated endocytosis (penetration) (Zeichhardt et al., 1985; Willingmann et al., 1989). The virus is internalized within 1 to 5 min after adsorption and reaches the acid compartments (endosomes) via "clathrin-coated vesicles" within 15 to 20 min after adsorption. Virus uncoating, with the release of genomic RNA from the capsid, takes place in the acidic endosomes (pH 5.5). During entry, the virus capsid undergoes conformational changes (for a review, see Rueckert [1996]). The internal virus protein VP4 is released and the N terminus of VP1 becomes accessible at the surface of the capsid. These virus particles, called A-particles, have a reduced sedimentation coefficient (from 156S to 135S) and

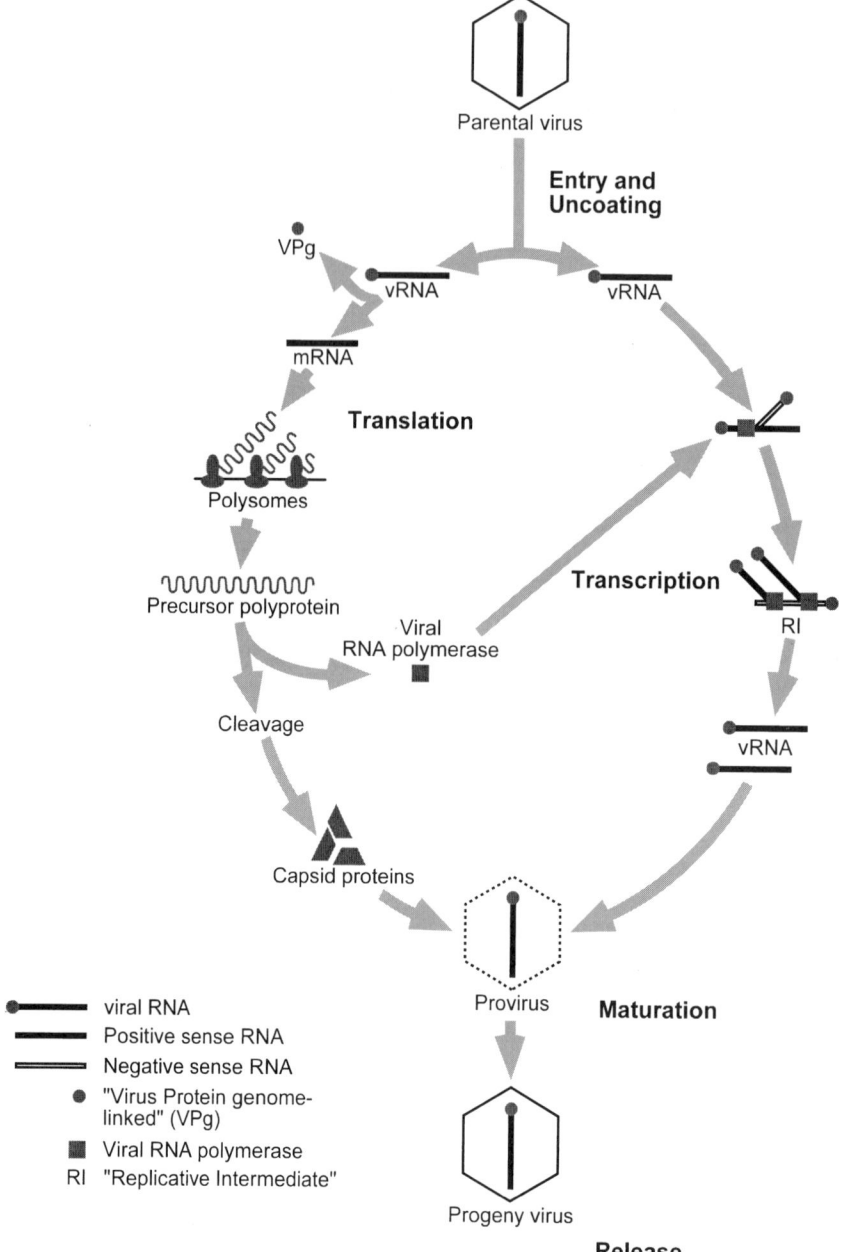

FIGURE 3 Replication cycle of poliovirus. Parental poliovirus enters its host cell by receptor-mediated endocytosis and releases its viral RNA from the virus capsid (uncoating) in acidic organelles (endosomes). After release of VPg from the parental viral RNA, protein synthesis of poliovirus starts at the rough endoplasmic reticulum. The viral precursor polyprotein is autocatalytically cleaved by viral proteases, resulting in the viral RNA polymerase, and via several precursor proteins in the virus capsid proteins. At the smooth endoplasmic reticulum, the viral RNA polymerase synthesizes new viral RNA. Positive-sense RNA serves as a template for negative-sense RNA molecules, which themselves are templates for new positive-sense RNA. This positive-sense RNA is released from multistranded replicative intermediates (RI) and used for either further viral transcription, translation, or encapsidation into assembling provirus particles. After the capsid protein precursor VP0 has finally been cleaved into VP2 and VP4, the maturation of progeny virus is completed. One cycle of poliovirus reproduction takes about 6 h.

altered antigenic properties. Finally the viral RNA is released into the cytoplasm for subsequent syntheses of viral proteins and RNA. Recently, an alternative entry mecha- nism in which the viral RNA is released from the virus capsid independently of low pH was reported (Tosteson and Chow, 1997).

Viral Protein and RNA Synthesis

After the uncoating step, the virus protein VPg is released from the 5′ terminus of the viral genome (Fig. 3). The parental viral positive-sense RNA serves as a template for the translation which takes place at the rough endoplasmic reticulum. In contrast to all cellular mRNAs, the viral RNA lacks an m^7G cap group. Viral translation is initiated in the 5′ NTR at the IRES. Translation starts at the AUG condon in position 743 and results in a single precursor polyprotein, which is autoproteolytically cleaved by protease 2A into the capsid precursor protein (P1) and the precursor for the functional proteins (P2 and P3). Further proteolytic processing is mediated by proteases 3C and 3CD, which yields, among others, the protein VPg and the viral RNA polymerase 3D. Initiation of transcription of the viral RNA depends on secondary structures in the 5′ and 3′ NTRs and is performed by the viral RNA polymerase and cellular factors. The positive-stranded genomic RNA serves as a template for the synthesis of negative-stranded RNA. During RNA synthesis, multistranded "replicative intermediates" are formed, which release newly synthesized positive-stranded RNA serving as a template for further translation, transcription, or subsequent virus assembly.

Virus Assembly and Maturation

During membrane-associated assembly of new virus particles, different intermediates are generated. At first, protomers are formed by aggregation of proteins VP0, VP3, and VP1. Five protomers assemble to 1 pentamer, of which 12 form an empty protein shell of 80S, the procapsid. One molecule of VPg containing positive-stranded genomic RNA is encapsidated in one procapsid. Virus maturation is completed by a final cleavage of VP0 into VP4 and VP2 (maturation cleavage). The resulting new infective virus particles (156S to 160S) are often condensed in crystal-like areas near cytoplasmic membranes.

Virus Release

One reproduction cycle lasts 6 to 8 h for poliovirus but can be some hours longer for other enteroviruses. At the end of the first cycle, the infected host cells are lysed. Up to 10^4 to 10^5 newly synthesized virus particles are released from a single infected cell (Rueckert, 1996). The mechanism of virus release from the cell is yet not clearly understood. It has been shown for poliovirus that progeny virus is found in the supernatant prior to cytolysis. Further, only a small portion of the virus particles are mature infective virions. Ratios of infective virus to total virus particles of $1:10^1$ to $1:10^3$ have been observed.

Shortly after infection (1 to 3 h postinfection), most enteroviruses induce a pronounced inhibition of cellular RNA, protein, and thereby cellular DNA synthesis (Diefenthal et al., 1973; for reviews, see Mirzayan and Wimmer [1994], Rueckert [1996], and Zeichhardt and Grunert [1999]). Poliovirus induces a "shutoff" of host cell protein synthesis, which is caused by proteolytic cleavage of the cellular protein p220, the eukaryotic initiation factor of translation eIF-4G. This initiation factor is part of the cap-binding complex eIF-4F, which initiates translation of host cell proteins by recognition of the capped mRNA. This proteolytic cleavage is mediated by viral protease 2A. Viral protease 3C seems to be involved in inhibition of host cell transcription.

Most enteroviruses are strongly cytolytic; that is, they induce a cytopathic effect (CPE), resulting in destruction of the cell by lysis. A typical example of such a CPE can be seen in Fig. 4, which shows poliovirus type 1 infecting a monolayer of HEp-2 cells (Zeichhardt et al., 1982). Most of the infected cells detach from the surface of the culture vessel. The remaining cells withdraw from adjacent cells, round up, and are attached to the substratum by long filopodia (Diefenthal and Habermehl, 1967). The microvilli at the cell surface merge and disappear. Ultrathin sections of poliovirus-infected cells show drastic changes in the interior of the cell, such as vesicles arranged in clusters in the cytoplasm and a lobed nucleus with irregular distribution of condensed chromatin (Dales et al., 1965; Zeichhardt et al., 1982). Poliovirus induces characteristic mitotic changes and chromosomal aberrations. The early stage of replication induces enhancement of mitosis; later stages result in an arrest of mitosis in the metaphase (colchicine-like effect). Chromosomal damage is characterized by single chromatid breaks and pulverization (Habermehl et al., 1966; Bartsch et al., 1969).

PATHOGENESIS AND CLINICAL SYNDROMES

The mechanism of pathogenicity of enterovirus infections is the lytic infection of host cells resulting in severe CPEs (see "Replication in Cell Culture," above). Enteroviruses can lead to cyclic infections in their hosts with a viremia and subsequent transport of the virus to target organs (spinal cord and brain, meninges, myocardium, skin, and liver, etc.) (for reviews, see Moore and Morens [1984], Melnick [1996], and Zeichhardt and Grunert [1999]). Clinical syndromes associated with enterovirus infections are summarized in Table 2.

Incubation Times

All polioviruses, coxsackieviruses of groups A and B, and echoviruses have incubation times ranging from 2 to 35 days, with an average of 7 to 14 days. The shortest incubation period, 12 to 72 h, has been reported for local infections of the eye by enterovirus type 70.

Asymptomatic Infections

Infections with enteroviruses are very common; however, it should be reemphasized that the most common forms of infection are silent, mild, or subclinical, although they nevertheless induce immunity. It has been reported that 90 to 95% of poliovirus infections are asymptomatic, while only 0.1 to 1.0% of infections cause paralytic poliomyelitis. Asymptomatic infections are most common for polioviruses, followed by echoviruses and coxsackieviruses (50%) (reviewed by Moore and Morens [1984]). The high incidence of inapparent infections with enteroviruses may be due to virus passage through the gut. The cells of the epithelia of the gut normally have a high rate of turnover. Although 10^4 or even more infective virus particles can be reproduced in one infected cell of the gut, and consequently 10^6 to 10^9 viruses can be detected in 1 g of feces of enterovirus-infected persons, virus-induced lysis of the cells might be without clinical consequence. Clinical symptoms occur only after massive infection of the gut epithelia.

Polioviruses

The course of infection of polioviruses is the best understood of all enteroviruses and therefore is presented as a typical example (Fig. 5). The portal of entry of polioviruses is the alimentary tract via the mouth. During the incubation

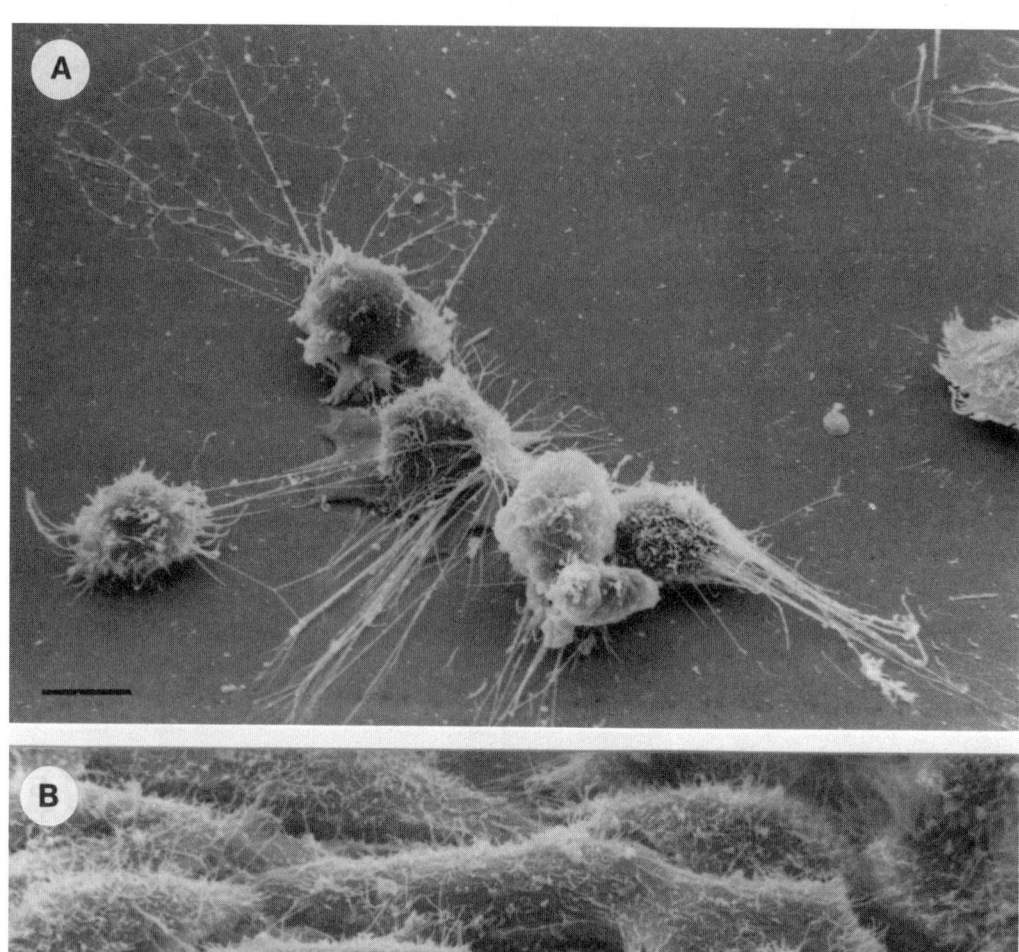

FIGURE 4 Scanning electron micrograph of HEp-2 cells infected with poliovirus type 1. (A) Infected cells show a severe CPE, characterized by rounded-up cells that are attached to the substratum only by long filopodia. (B) Mock-infected HEp-2 cells are characterized as a monolayer of confluent cells with evenly distributed microvilli at the plasma membrane. Bars, 10 μm.

period (6 to 20 days), poliovirus multiplies in the mucosal tissues of the pharynx, the lymphoid tissue (tonsils and Peyer's patches), and the gut. For this reason, virus is spread via oral and fecal routes beginning shortly after infection. In most cases (90 to 95%), virus infection is inapparent (i.e., the patient is asymptomatic).

TABLE 2 Enterovirus infections and their clinical syndromes[a]

Virus(es)	Type(s)	Clinical syndrome(s)
Poliovirus	1, 2, 3	Abortive poliomyelitis (minor illness, undifferentiated febrile illness)
		Nonparalytic poliomyelitis (aseptic meningitis)
		Paralytic poliomyelitis (major illness), encephalitis (infrequently)
		Postpolio syndrome
Coxsackievirus group A	2, 3, 4, 5, 6, 8, 10	Herpangina (vesicular pharyngitis)
	10	Acute lymphatic or nodular pharyngitis
	2, 4, 7, 9, 10	Aseptic meningitis
	7, 9	Paralysis (infrequently)
	4, 14, 16	Myocarditis, pericarditis
	4, 5, 6, 9, 16	Exanthema
	4, 5, 9, 10, 16	Hand, foot, and mouth disease
	9, 16	Pneumonitis of children
	21, 24	Common cold
	4, 9	Hepatitis
	18, 20, 21, 22, 24	Infantile diarrhea
	24	Acute hemorrhagic conjunctivitis
	Different types	Undifferentiated febrile illness
Coxsackievirus group B	1, 2, 3, 4, 5	Pleurodynia
	1, 2, 3, 4, 5	Bornholm disease (epidemic pleurodynia or acute epidemic myalgia)
	1, 2, 3, 4, 5, 6	Aseptic meningitis
	2, 3, 4, 5	Paralysis (infrequently)
	1, 2, 3, 4, 5	Severe systemic infection in infants, meningoencephalitis, myocarditis
	1, 2, 3, 4, 5	Myocarditis, pericarditis, chronic cardiovascular disease
	4, 5	Upper respiratory illness and pneumonia
	5	Exanthema
	2, 5	Hand, foot, and mouth disease
	5	Hepatitis
	1, 2, 4	Pancreatitis
	4	Diabetes
	1, 2, 3, 4, 5, 6	Undifferentiated febrile illness
Echovirus	1–7, 9, 11, 13–23, 25, 27, 30, 31	Aseptic meningitis
	4, 6, 9, 11, 30; possibly 1, 7, 13, 14, 16, 18, 31	Paralysis (infrequently)
	2, 6, 9, 19; possibly 3, 4, 7, 11, 14, 18, 22	Encephalitis, ataxia, or Guillain-Barré syndrome
	2, 4, 6, 9, 11, 16, 18; possibly 1, 3, 5, 7, 12, 14, 19, 20	Exanthema, Boston exanthema disease (echovirus 16)
	4, 9, 11, 20, 25; possibly 1–3, 6–8, 16, 19, 22	Respiratory illness
	7, 11	Conjunctivitis
	1, 6, 9	Epidemic myalgia (infrequently)
	1, 6, 9, 19	Myocarditis and pericarditis (infrequently)
	4, 9	Hepatitis
	Different types	Diarrhea
	Different types	Undifferentiated febrile illness
Other enteroviruses	68	Pneumonia and bronchiolitis
	70	Acute hemorrhagic conjunctivitis
	71	Aseptic meningitis
	70, 71	Paralysis
	70, 71	Meningoencephalitis
	71	Hand, foot, and mouth disease

[a]Modified from Melnick (1996).

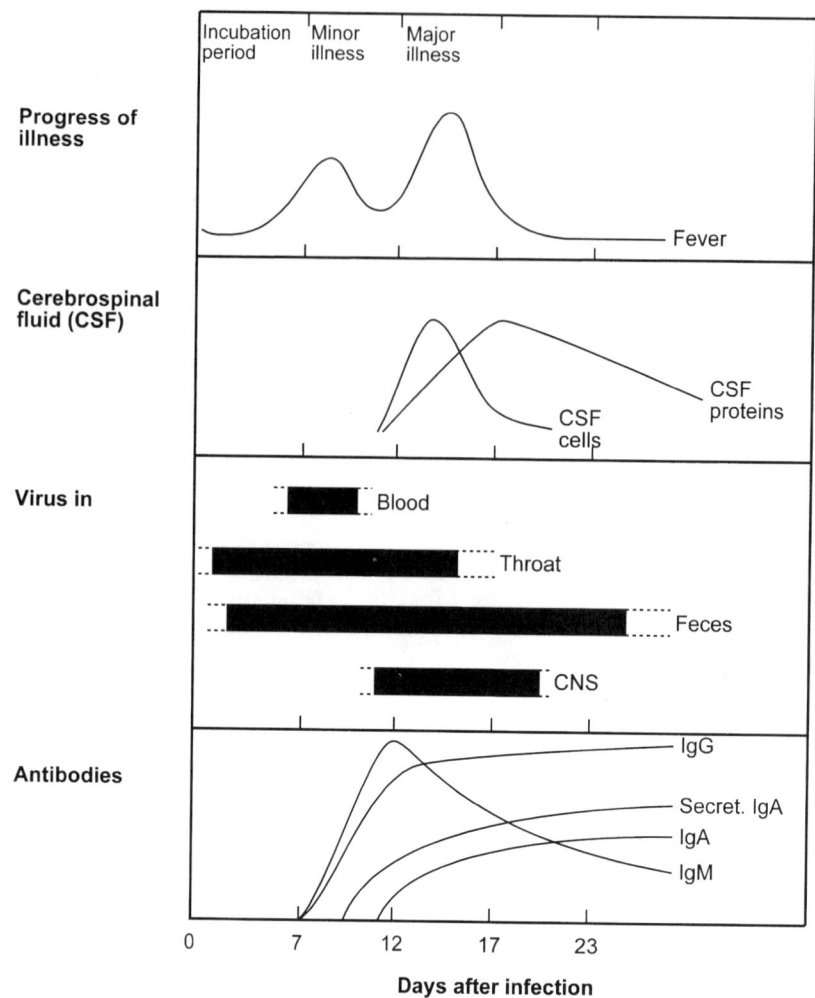

FIGURE 5 Course of poliovirus infection.

Abortive Poliomyelitis (Minor Illness)

The virus can be spread to the draining lymph nodes, leading to a viremia characterized by recovery of virus from the blood stream for a few days (6 to 9 days after infection). During this time, the first nonspecific clinical symptoms (e.g., fever, malaise, sore throat, and sometimes headache and vomiting) are observed. In about 4 to 8% of poliovirus infections, illness does not proceed and takes the form of only a minor illness ("abortive poliomyelitis") (Table 2).

Nonparalytic Poliomyelitis (Aseptic Meningitis)

If poliovirus crosses the blood-brain barrier (by an as-yet-unknown mechanism), it can infect its target cells in the CNS, causing a nonparalytic poliomyelitis (1 to 2%) and paralytic poliomyelitis (0.1 to 1.0%). This is accompanied by increased levels of cells and proteins in the cerebrospinal fluid (CSF). In nonparalytic poliomyelitis, patients have the same prodromal illness as those with minor illness, followed after 3 to 7 days by an illness similar to aseptic meningitis that is commonly accompanied by high fever, back pain, and muscle spasm. In general, patients rapidly recover from the disease after 2 to 10 days.

Paralytic Poliomyelitis (Major Illness)

In rare cases (0.1 to 1%), infection of the CNS with poliovirus causes paralytic poliomyelitis, also called "major illness." Paralytic poliomyelitis additionally comprises flaccid paralysis (involvement of the whole muscle) or paresis (involvement of only some muscle groups), which is due to spinal and/or bulbar damage. The bulbar poliomyelitis (ascendant infection) is less common than the spinal form and has a poor prognosis due to damage of cerebral nerve or vegetative centers. In the case of the spinal illness, recovery of motor function may occur to some degree after some months; however, any remaining paralysis is permanent. Encephalitis occurs in rare cases.

The source of these severe clinical symptoms is the very specific extraintestinal target cell range of polioviruses: especially the anterior horn cells of the spinal cord, but also the dorsal root ganglia, certain brain stem centers, cerebellum, spinal sensory columns, and occasionally the cerebral motor cortex. Histologic changes first observed are vascular engorgement, accompanied by perivascular infiltration with lymphocytes and also polymorphonuclear neutrophils, plasma cells, and microglia. As experimental infections of rhesus monkeys by Bodian in the 1950s showed, the infected anterior horn cells are severely damaged by the poliovirus-specific CPE, characterized by a decrease in the size of Nissl bodies in the cytoplasm (chromatolysis) together with nuclear damages such as condensation and breakage of chromatin (Bodian et al., 1959).

Certain factors have been reported to increase the severity of a poliomyelitis infection, including very young and very old age, male sex, chronic undernutrition, corticosteroid treatment, physical exertion, hypoxia, cold, irradiation, tonsillectomy, pregnancy, adrenal-related endocrine changes, and possibly hypercholesterolemia (reviewed by Moore and Morens [1984] and Melnick [1996]).

Postpolio Syndrome (Progressive Postpoliomyelitis Muscle Atrophy)

Several decades after experience with paralytic poliomyelitis, a small number of patients show recrudescence of paralysis and muscle atrophy. The pathogenesis of this "progressive postpoliomyelitis muscle atrophy" is yet not well understood. It has been suggested that the muscle atrophy is caused by an additive effect of physiological aging and the long-lasting loss of neuromuscular functions (Johnson, 1984). Whether the postpolio syndrome is a consequence of a poliovirus infection persisting over decades is controversial (as discussed by Melchers et al. [1992] and Julien et al. [1999]).

Coxsackieviruses, Echoviruses, and Enteroviruses 68 to 71

Coxsackieviruses, echoviruses, and enteroviruses 68 to 71 have a less-specific extraintestinal target organ range than does poliovirus and therefore can lead to a wider range of illnesses (Table 2). As described for polioviruses, coxsackieviruses and echoviruses multiply primarily in the pharynx and small intestine and are shed in the feces for up to 1 month and in respiratory secretions for several days. Generally speaking, coxsackieviruses and echoviruses, besides the alimentary tract, can infect the meninges, CNS, myocardium and pericardium, striated muscles, respiratory tract, eye, and skin. No disease has been assigned yet to enterovirus 69.

Meningitis and CNS Disease

Most coxsackieviruses of groups A and B, as well as echoviruses, can induce meningitis and in some cases paresis and paralysis. The early symptoms, such as fever, malaise, headache, nausea, and abdominal pain, resemble the minor illness of poliomyelitis and are followed by irritation of the meninges with neck and back stiffness and vomiting. Very often, aseptic meningitis and mild paresis is accompanied by rashes. In general, the manifestation in the CNS is less severe than in poliovirus infections and recovery is very often complete. Cases of severe paralysis, which may be confused with paralytic poliomyelitis without a precise laboratory diagnosis, have been described for infections with coxsackieviruses A7, A9, and B2 to B5 and some echoviruses. Meningoencephalitis can be induced by group B coxsackieviruses, especially in children. Infections with enterovirus 70 can induce hemorrhagic conjunctivitis, which can be accompanied by a poliomyelitis-like disease in rare cases. Encephalitis, ataxia, and Guillain-Barré syndrome are also associated with echovirus infections. Epidemic outbreaks can be attributed to enterovirus infections. Echovirus 9 caused a pandemic outbreak of aseptic menin-

gitis from 1955 to 1960. During an epidemic outbreak of hand, foot, and mouth disease caused by enterovirus 71 in California from 1969 to 1973, several cases of aseptic meningitis, meningoencephalitis, and paralysis occurred in parallel.

Herpangina

Herpangina (vesicular pharyngitis) is induced only by coxsackieviruses of group A (mainly A2 to A6, A8, and A10). The disease is characterized by discrete vesicles (sometimes very small) on the tongue, anterior pillars of the fauces, posterior pharynx, palate, uvula, or tonsils. Herpangina preferably attacks infants, who may suffer from abrupt onset of fever, sore throat, vomiting, and abdominal pain. Coxsackievirus A10 may also cause lymphatic pharyngitis.

Respiratory and Acute Febrile Illnesses

Infections of the upper and lower respiratory tract with acute febrile illnesses of a few days' duration and without distinct features are caused by nearly all coxsackieviruses and echoviruses. These infections are similar to the common cold and mainly occur during summer and autumn. In children and adults, pneumonia may be caused by coxsackieviruses, which are also responsible for pneumonitis of infants. Children may additionally suffer from pneumonia and bronchiolitis due to enterovirus 68 and an influenza-like illness caused by enterovirus 71.

Exanthemas and Hand, Foot, and Mouth Disease

Several coxsackieviruses are responsible for rubelliform rashes of young children. Infants often show exanthemas (maculopapular, sometimes morbilli or rubelliform) accompanied by febrile illnesses and/or pharyngitis after infection with echoviruses. An epidemic outbreak of a maculopapular rash that was due to echovirus 16 occurred in Boston, Mass., in the early 1950s and was hence called "Boston exanthema disease."

Vesicles on the hands and feet are characteristic of hand, foot, and mouth disease, which is mainly caused by coxsackievirus A16 but is also associated with coxsackieviruses A4, A5, A9, A10, B2, and B5. Herpangina with generalized oral vesicular lesions may occur. Enterovirus 71 is responsible for hand, foot, and mouth disease in parallel to aseptic meningitis, encephalitis, and acute flaccid paralysis. Fatal cases, especially in the young, may occur due to pulmonary edema and pulmonary hemorrhage.

Conjunctivitis

Coxsackievirus A24 and enterovirus 70 are the main causes of enteroviral conjunctivitis. An epidemic outbreak of acute hemorrhagic conjunctivitis originated from a variant of coxsackievirus A24. This virus, first discovered in Singapore and Hong Kong in 1970, was spread throughout Southeast Asia and occurred in American Samoa in 1986 with an incidence of 47%. Also, enterovirus 70 was responsible for several million cases of acute hemorrhagic conjunctivitis. These outbreaks occurred in Africa, Southeast Asia, Japan, and India from 1969 to 1971, and enterovirus 70 was later introduced via French Polynesia into the United States. A local infection of the eye is characteristic of both viruses; enterovirus 70, however, may additionally cause a CNS disease with poliomyelitis-like paralysis in rare cases. Enterovirus 70 is the only enterovirus with a very short incubation time (24 h; range 12 to 72 h). Infections of the eye resulting in conjunctivitis without hemorrhage can be caused by echoviruses 7 and 11. It should be emphasized that conjunctival fluids of patients with the above viruses are highly infective and may therefore be responsible for frequent transmission.

Pleurodynia

Pleurodynia, also called epidemic myalgia or Bornholm disease, is caused by coxsackieviruses B1 to B5. After malaise, headache, and anorexia, an abrupt onset of fever occurs, followed by severe chest pain (devil's grip) and abdominal pain. This disease, lasting from 2 days to 2 weeks, most often attacks children and adolescents and may be accompanied by generalized muscle hypotonia. Epidemic outbreaks, the first of which was observed on the Baltic island of Bornholm (Denmark) from 1930 to 1932, typically occur in late summer and autumn. Muscle pain in the lower extremities can be due to echovirus infections. Signs of pleurodynia may occur if this myalgia affects the intercostal muscles. Sporadic cases of pleurodynia have been reported for coxsackieviruses A4, A6, A9, and A10 and for echoviruses 1, 6, and 9.

Myocarditis, Pericarditis, and Chronic Cardiovascular Disease

Among the enteroviruses, coxsackieviruses of group B are the main cause of myocarditis, pericarditis, and dilated cardiomyopathy. In some cases, coxsackieviruses A4, A14, and A16, as well as echoviruses 1, 6, 9, and 19, lead to myocarditis and pericarditis. The myocardium, endocardium, and pericardium are the targets of coxsackieviruses and echoviruses. Edema, diffuse focal necrosis, and signs of acute inflammation occur in the infected myocardium. Meningism and convulsions can occur in parallel to the cardiac disease. Fatal cases of myocarditis with a death rate of 50% may be observed in newborns. Pericarditis more often occurs in older children and adolescents and in general is less severe than neonatal myocarditis. Coxsackieviruses B2 to B6 can be the cause of chronic cardiovascular disease, with recurrent pericarditis accompanied by persisting virus-specific IgM. In situ hybridization reveals that myocytes are persistently infected with these coxsackieviruses. Virus persistence may be the reason for long-lasting necrosis in the myocardium.

Gastrointestinal Illness

Infections with several coxsackieviruses and echoviruses can lead to nonspecific clinical symptoms sometimes accompanied by diarrhea. Fatal cases can occur in newborn babies. Whereas generalized infections leading to hepatitis are reported for coxsackieviruses and echoviruses, pancreatitis has been associated with coxsackieviruses of group B.

Diabetes Mellitus

Juvenile-onset, insulin-dependent diabetes mellitus is suspected to be caused by infections with coxsackieviruses of group B. An autoimmunity mechanism may be involved in diabetes, as reported for experimental infections of animals.

Intrauterine and Neonatal Infections

Pregnancy and infection with enteroviruses is an unsolved problem. In spite of several negative reports, maternal infections during the first trimester of pregnancy are suspected to result in the following anomalies in the infected fetus: coxsackievirus types B2 or B4, urogenital anomalies;

coxsackievirus types B3 or B4, cardiovascular anomalies; coxsackievirus type A9, digestive system malformations (reviewed in Moore and Morens [1984]; see also Melnick [1996]). Because the teratogenic risk of intrauterine infections with enteroviruses is lower by several orders of magnitude than that of infections with rubella virus, a recommendation of abortion is not generally accepted for women with proven enterovirus infection in the first trimester of pregnancy.

Fatal disease in newborn babies, often leading to rapid death, may be due to nosocomial transmission of some coxsackieviruses and echoviruses. Fulminant systemic infections ("viral sepsis") with acute myocarditis or pericarditis, encephalitis, and hepatitis (often hemorrhagic with fatal kidney disorders) are reported for nursery outbreaks and sporadic infections with coxsackieviruses of group B and echovirus 11. Severe diarrhea in young children may lead to dramatic disorders of ion and water balance.

IMMUNE RESPONSE

Humoral and secretory antibodies play a major role in immunity to enterovirus infections (Fig. 5). The role of cellular immunity in these infections is not yet well defined. Humoral immunity is mediated by type-specific neutralizing immunoglobulin G (IgG), IgM, and IgA, which prevent hematogenous spread of virus to the target organs. IgM appears first (7 to 10 days after infection).

Virus-specific IgM persists for at least 4 weeks in 90% of infections. Virus-specific IgG and IgA appear a few days after IgM. IgG persists for years and therefore mediates acquired humoral immunity. Production of antibodies in the CNS after poliovirus infection has been reported. Serum antibodies also may reach the CNS by crossing the blood-brain barrier, due to breakdown of the integrity of the meninges. Secretory IgA is induced 2 to 4 weeks after poliovirus infection and is located mainly in nasopharyngeal and gut tissues. Secretory IgA prevents or limits the excretion of polioviruses in the alimentary tract. A baby born to a mother who is immune to an enterovirus is protected against corresponding virus infection within its first months of life due to the transplacental transmission of virus-specific IgGs from mother to fetus.

EPIDEMIOLOGY

Mode of Transmission

Human enteroviruses have humans as their only reservoir (for growth and pathogenicity in animals, see Table 3). Enteroviruses can be isolated from the lower and/or upper alimentary tract and therefore can be spread by both the fecal-oral and respiratory routes (Melnick, 1982, 1996). In areas with poor sanitary conditions, fecal-oral transmission is predominant. Fecal contamination of fingers, objects of a normal household (e.g., toys and towels), and food may be responsible for enterovirus transmission. Transmission by respiratory routes can occur early in infection because the virus replicates in the upper respiratory tract. Sexual transmission of enteroviruses can occur by anal-oral routes. Blood transfusions and insect bites are not responsible for virus transmission. Enteroviruses can be isolated from sewage, therefore, a fecal-water-oral route of transmission is possible. Food-borne acquisition of enteroviruses has been noted. Nosocomial transmission of enteroviruses typically takes place in newborn nurseries and has been reported for

TABLE 3 Growth of enteroviruses in human and monkey kidney cell lines and pathogenicity for animals[g]

Viruses	Growth in:		Pathogenicity	
	Human cells	Monkey kidney cells	Mice	Monkeys
Polioviruses	+	+	Some[a]	+
Coxsackieviruses A	Some[b]	Some[c]	+[c]	Some[d]
Coxsackieviruses B	+	+	+[c]	Some
Echoviruses	Some[b]	+[e]	Some[f]	Some
Enteroviruses 68–71	+	+/−	+/−	Some

[a]Some strains of each type have been adapted to mice.
[b]Some strains grow preferentially in, or have been adapted to, human cell cultures. Coxsackievirus types A11, A13, A15, A16, A18, A20, and A21 may be isolated directly in human cells.
[c]Coxsackievirus types A7, A9, and B strains grow readily in monkey kidney cells; some strains grow poorly in mice and fail to produce disease in these animals.
[d]Especially coxsackievirus type A7.
[e]Echovirus type 21 is cytopathogenic for human epithelial cells but not for monkey kidney cells.
[f]Whereas the prototype and other strains of echovirus 9 are not pathogenic for mice, a number of other strains, especially after passage in monkey kidney cells, produce paralysis in mice (severe coxsackievirus-type myositis).
[g]Modified from Melnick et al. (1979).

several coxsackieviruses of group A and group B and echoviruses. Virus-containing samples, e.g., conjunctival fluids of patients infected with coxsackievirus A24 or enterovirus 70, are highly infective and can lead to a rapid spread of hemorrhagic conjunctivitis.

Geography, Season, Socioeconomy, Sex, Age, and Risk Groups

Enteroviruses are found worldwide. Enterovirus infections characteristically take place during the summertime in areas of the north temperate zone. In tropical and semitropical areas, enterovirus infections occur throughout the year. Persons of low socioeconomic status living in urban areas receive greater exposure to enteroviruses and have higher incidence of subclinical infections than do persons of higher socioeconomic status. This led to the paradox that poliomyelitis in the prevaccine era was a disease of "development" (Melnick, 1982); in other words, improvement in hygienic and socioeconomic conditions was associated with lower subclinical exposure and an increase in incidence of severe illnesses. The improved hygiene might also have led to a decrease in mixed infections with other enteroviruses of patients infected with polioviruses. Mixed infections with more than one enterovirus at the same time can result in interference, leading to a suppression of replication of one of the viruses.

Diseases due to enteroviruses occur more frequently in males than in females (male-to-female ratio, 1.5:1 to 2.5:1) (Moore and Morens, 1984). Generally, young children are the main transmitters of enteroviruses. Echovirus type 9 was found in 50 to 70% of children, compared with 17 to 33% of adults. The age of initial infection decreases with poor hygienic conditions and low socioeconomic status. Severity of disease is also related to age. Poliovirus infection in adults is more likely to lead to paralysis than in children, and infections with coxsackieviruses of group A and echoviruses are usually milder in children than in adults. In contrast, coxsackieviruses of group B can more likely induce fulminant "viral sepsis," myocarditis, encephalitis, and death in newborns than in older children and adults. A reason for the different susceptibilities between children and adults for enterovirus infections might be found in a changing pattern of virus-specific receptors during development and aging.

In general, risk groups for enterovirus infections with poor prognosis are patients with acquired (AIDS), iatrogenic (immune suppressive therapy), and physiological (newborn from a nonimmune mother) immune deficiencies. As noted above, a baby born to a mother immune to an enterovirus is protected against the corresponding enterovirus infection within its first months of life.

Several enteroviruses can be responsible for epidemic or even pandemic outbreaks of disease. Echovirus 16 was responsible for an epidemic outbreak of the "Boston exanthema disease" in Massachusetts in the early 1950s. A variant of coxsackievirus A24 and enterovirus 70 caused independent large epidemics of hemorrhagic conjunctivitis originating in Southeast Asia from 1969 to 1971 and spreading to industrialized countries in the 1980s (for details, see "Coxsackieviruses, Echoviruses, and Enteroviruses 68 to 71," above). Starting in 1998, enterovirus 71 has led to a severe outbreak of meningitis, encephalitis, or acute flaccid paralysis with at least 55 deaths in Taiwan. This virus may be an accidental import from Malaysia, as sequence similarities between different isolates from both countries show. For updated reports of enterovirus outbreaks, see the World Health Organization (WHO) website "Disease Outbreak News" (http://www.who.int/emc/outbreak_news/).

Before the era of vaccination against polioviruses, frequent epidemic outbreaks of poliomyelitis occurred worldwide, with a high incidence of fatal cases. The introduction of the inactivated polio vaccine (IPV) by Jonas Salk in 1954 and the oral polio vaccine (OPV) by Albert Sabin in 1962 led to a drastic reduction of poliomyelitis all over the world. In the prevaccine era, approximately 10 cases of acute flaccid paralysis (AFP) per 100,000 people occurred in the United States. In the Federal Republic of Germany, approximately 5,000 cases of AFP per year were reported before introduction of the oral Sabin vaccine. The enormous benefit of global vaccination programs led by the WHO and local health authorities is documented by the reduction of worldwide poliomyelitis cases to only 35,251 in 1988. By the end of 1998, worldwide cases of poliomyelitis had decreased to 6,349. For updated information, see the WHO website "The Global Polio Eradication Initiative" (http://www.who.int/vaccines-polio/). Poliovirus has already been eliminated from the Americas and from the industrialized countries of Europe (Fig. 6). WHO anticipates worldwide eradication of poliovirus in the first decade of the 21st century. Prerequisite for such success, which will be the second eradication of an infectious disease, after smallpox, is continuous vaccination efforts in order to ensure a high prevalence of antibodies against poliovirus in the population for interruption of virus transmission. This is especially important for individuals travelling to and from regions of endemicity.

Before 1993, frequent outbreaks were observed in the United States, Canada, and The Netherlands among enclaves of religious groups refusing vaccination. Ten cases of poliomyelitis (9 AFPs) were reported in Finland from 1984 to 1985. During this period, a genetically altered wild-type poliovirus type 3, against which the administered inactivated polio vaccine induced only partial immunity, was spread. In Albania, the former Yugoslavia, and Greece, 167 cases of poliomyelitis with 17 deaths occurred in 1996.

LABORATORY DIAGNOSIS

Methods used for laboratory diagnosis of enterovirus infections have been summarized elsewhere (Melnick et al., 1979; Melnick, 1996; Zeichhardt and Grunert, 1999).

Virus Isolation and Identification

Virus Propagation in Cell Culture and Animals

Specimens for virus isolation are usually stools and rectal swabs, throat swabs and washings, and CSF. Successful isolation of virus from clinical specimens depends on the time postinfection, which relates to the pathogenesis of enterovirus infections. For example, virus is most readily isolated from the throat shortly after infection up to 15 days or more, from stools and rectal swabs up to 4 weeks after infection, and from CSF during the time of manifestation of symptoms involving the CNS, usually 2 to 3 weeks after infection (Fig. 5). Isolation from stools is most promising, as virus concentrations in feces are higher than in other specimens (up to 10^6 to 10^9 virus particles per gram of feces). Viruses inducing vesicular rashes, such as coxsackievirus types A4, A5, A9, A10, A16, B2, and B5 and enterovirus 71, can also be isolated from the lesions. Virus isolation from blood is successful during viremia (days 6 to 9 after infection); however, due to the short period of this phase, virus isolation is generally not attempted from this source. In addition, all specimens from target organs generally yield virus if a biopsy or autopsy specimen is taken during the clinical manifestation of disease.

Originally, pathologic lesions produced in mice were used for distinction between coxsackieviruses of groups A and B. More recently, cell cultures are most commonly used for isolating most of the enteroviruses (Melnick et al., 1979; Melnick, 1996) (Table 3). Cells usually used for growth of polioviruses, most coxsackieviruses of group A, all coxsackieviruses of group B, and echoviruses are human embryonic fibroblasts of the skin or lung, permanent human amnion cells, transformed human cell lines such as HeLa and HEp-2 cells, and primary monkey cell lines such as rhesus and African green monkey kidney cells. An exception is the growth of several types of coxsackieviruses of group A. Some of these viruses replicate only in cells derived from a human rhabdomyosarcoma cell line or in newborn mice. Coxsackievirus types A1, A19, and A22 remain as types

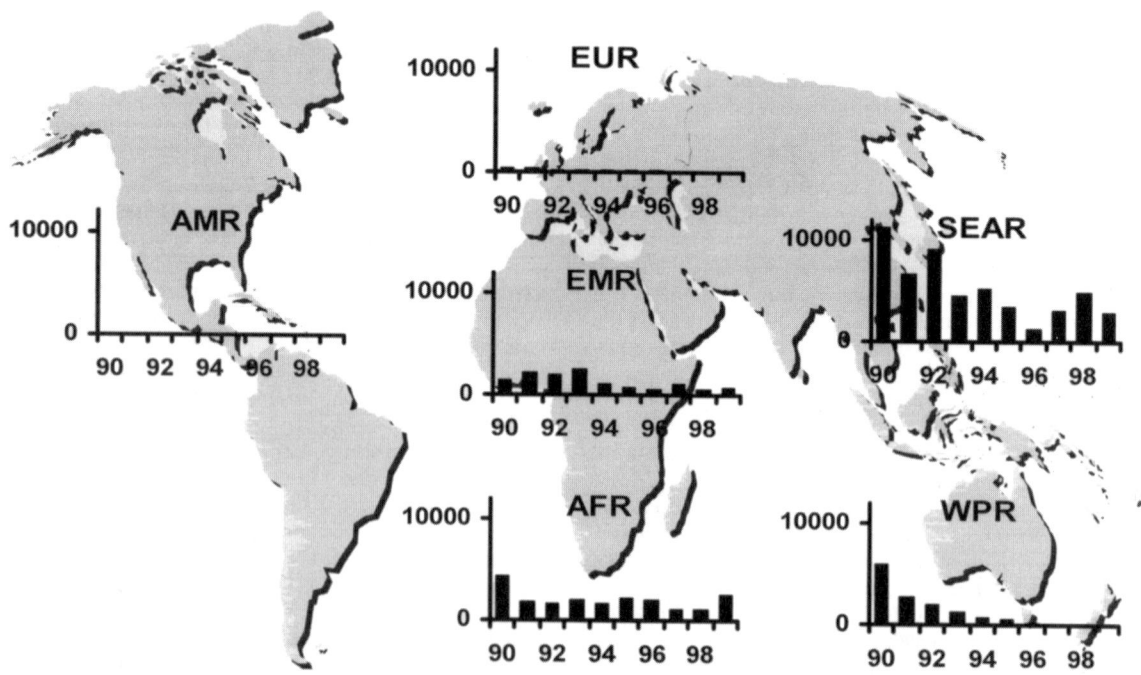

FIGURE 6 Global annual polio cases reported by the WHO from 1990 to 1999. The map shows the number of polio cases per year for each WHO region (AFR, African region; AMR, American region; EMR, eastern Mediterranean region; EUR, European region; SEAR, southeast Asian region; WPR, western Pacific region). For reference, see "Geography, Season, Socioeconomy, Sex, Age, and Risk Factors," in the text.

that require newborn mice for cultivation (Melnick, 1982). Virus isolation in mice, however, is restricted to a few reference laboratories worldwide.

Internationally Standardized Antiserum Pools for Virus Neutralization

The appearance of the typical CPE in cell cultures (see "Replication in Cell Culture," above) is proof of the presence of virus in a specimen. Neutralization tests in which virus infectivity is inhibited by corresponding specific antibodies are used for identification of virus isolates. Most commonly, pools of internationally standardized hyperimmune equine antisera introduced by Lim and Benyesh-Melnick (1960) are used for typing of isolates. The serum pools can be obtained from the WHO Collaborating Centre for Virus Reference and Research, Copenhagen, Denmark. Typing is performed with eight pools of Lim-Benyesh-Melnick (LBM) serum (pools A to H) according to a pattern of "intersecting sera" (Fig. 7). Neutralization of an isolate with pools C and F proves that the isolate consists of poliovirus type 1. For typing of enteroviruses which can be propagated only in animals, LBM antiserum pools J to P are available. In contrast to neutralization tests, enzyme or radioimmunoassays and CF tests do not allow type-specific identifi-

cation, due to immunologic cross-reactions (see "Serologic Diagnosis," below).

Virus Detection by Microscopy

Direct virus visualization by electron microscopy is performed in specialized laboratories. Virus can be detected in preparations from stools due to the high concentration of virus in feces. This is achieved by the technique of negative staining (Fig. 1A). Immunoelectron microscopy allows direct virus identification. Specimens containing virus are incubated with virus-specific antiserum, and the resulting virus-antibody complexes are visualized.

Indirect immunofluorescence may be useful for directly typing enteroviruses in specimens obtained at biopsy or autopsy. However, this technique has not been used extensively.

Detection of Virus Genome

Virus genome detection is useful only for selected diagnostic purposes (Rotbart, 1990; Mulders et al., 1995; van der Avoort et al., 1995), as the pronounced sequence homology between the different enteroviruses does not allow virus differentiation down to serotypes. This is why PCR and other nucleic acid detection techniques cannot replace the immunological neutralization test.

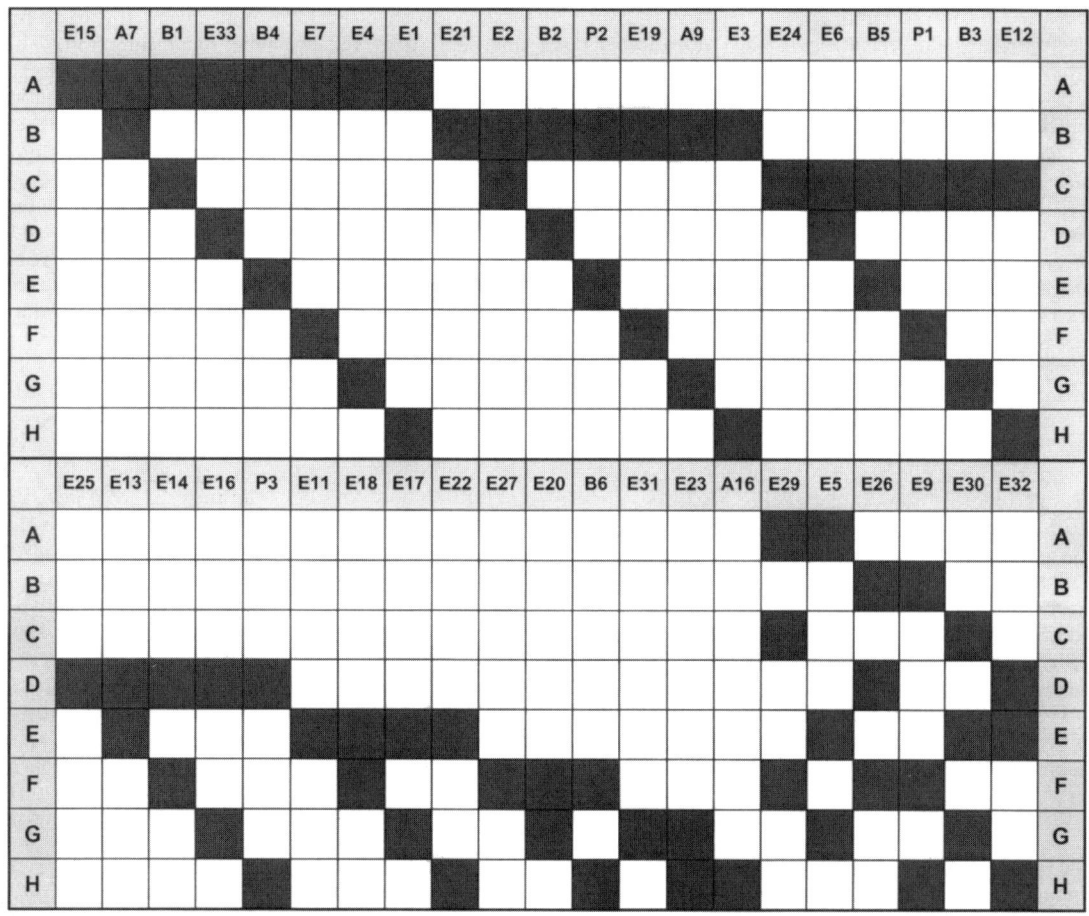

FIGURE 7 LBM antiserum pools for typing of enteroviruses. Filled boxes represent neutralization of enterovirus by antiserum pools A to H. The typing of a virus is based on combined neutralization of different antiserum pools. (Adapted from Melnick et al. [1979].)

Reverse transcriptase PCR is combined with restriction fragment length polymorphism (RFLP) analysis or sequencing for virus typing and differentiation between wild-type and Sabin vaccine-like poliovirus strains. Genome variation between enterovirus isolates and strain-specific oligonucleotide patterns are detected by oligonucleotide analysis (fingerprinting) and sequencing. Fingerprints analyzed by two-dimensional gel electrophoresis are useful tools for epidemiologic surveillance. In situ hybridization with virus-specific gene probes is used for the detection of enterovirus genomes in biopsy or autopsy samples. The use of coxsackievirus B3-specific cDNA for hybridization in cardiac specimens showed for the first time that dilated myocarditis is associated with persistent coxsackievirus B3 infection of the heart (Kandolf et al., 1993).

Differentiation between Wild-Type and Sabin Vaccine-Like Poliovirus Strains

Poliovirus strain characterization for differentiation of wild and vaccine-like viruses is performed in specialized laboratories usually by the following techniques (for a review, see WHO [1992]): (i) determination of growth "markers" such as reproductive capacity at elevated temperatures and plaquing capacities, (ii) intratypic serodifferentiation of virus strains by polyclonal strain-specific antibodies (van Wezel and Hazendonk, 1979; van Loon et al., 1993), (iii) use of monoclonal antibodies (Osterhaus et al., 1983; Ferguson et al., 1986; van Loon et al., 1993), and (iv) oligonucleotide analysis ("fingerprinting") and nucleic acid sequencing (Schweiger et al., 1994; Mulders et al., 1995).

Serologic Diagnosis

Serologic diagnosis in combination with virus identification is the most favorable method of confirming an infection with enteroviruses.

Documentation of a virus-specific serologic response (fourfold or greater rise in antibody titer or a single high titer of virus-specific IgM) confirms a recent enterovirus infection. Ideally, serum from a patient should be collected at the beginning of the illness and 7 to 10 days later, and both samples should be tested at the same time (Fig. 5). Neutralization tests with tissue cultures are most commonly used for serologic diagnosis of infections with polioviruses, coxsackieviruses, and echoviruses, as neutralizing antibodies react with the virus in a type-specific manner (see "Virus Isolation and Identification," above). For those coxsackieviruses of group A that do not grow in cell culture, neutralization tests in mice can be performed.

The CF test is another serologic procedure used to diagnose enteroviruses. It is of limited value because it allows detection of only group-specific antibodies. The same holds true for enzyme and radioimmunoassays. Enzyme immunoassays for coxsackieviruses have been reported (Katze and Crowell, 1980). Theoretically, hemagglutination inhibition tests can be performed with some hemagglutinating strains of coxsackieviruses of groups A and B, some echoviruses, and enterovirus type 68.

PREVENTION AND THERAPY

Hygienic Prevention

As enteroviruses are predominantly transmitted by fecal-oral routes, proper sanitation is of major importance for the prevention of infections. Obeying hygiene rules also prevents nosocomial transmission. Several cases in which clinical personnel transmitted enteroviruses due to lack of normal hygienic conditions have been reported. It should be emphasized that changing of coats and gloves between patient visits as well as hygienic hand rub and hand washing are effective for the prevention of transmission. In particular, improper removal of feces, including diapers but also other virus-containing specimens, is a major source for transmission of enteroviruses.

Although poliovirus-infected patients are not considered highly contagious, these patients should be isolated when hospitalized due to the drastic clinical consequences these infections might have. Isolation is especially appropriate during the final phase of poliovirus eradication, as virus shedding should be limited under all circumstances.

Immunization

Of all enteroviruses, only infections with polioviruses can presently be prevented by vaccination. Prerequisite for developing antipoliomyelitis vaccines was the introduction of tissue culture for the production of virus in large quantity (Enders et al., 1949). Two vaccines against the three serotypes of poliovirus are available. (i) The formalin-inactivated vaccine (IPV) developed by Salk and coworkers and first licensed in 1954 in the United States, administered intramuscularly, prevents poliovirus infections by inducing humoral neutralizing antibodies. (ii) The live attenuated vaccines (OPV) developed by Sabin, Koprowski, Melnick, Cox, and others in 1962 multiply in the gastrointestinal tract and give rise to subclinical infection. The OPV induces humoral IgG, as does the IPV, and additionally elicits secretory IgA in the gut. Therefore, vaccination with OPV not only prevents virus spread through the bloodstream to the CNS but also blocks primary virus multiplication in the gut. There is a possibility that one of the attenuated types of trivalent vaccine will not multiply due to interference from one of the other types or from other enteroviruses during mixed infections. For this reason, the vaccine is administered three times several weeks apart, in the north temperate zone, and preferably not during the summer months, when there is high frequency of other enterovirus infections (see "Geography, Season, Socioeconomy, Sex, Age, and Risk Groups," above). Vaccination with OPV carries a small risk of development of clinical poliomyelitis, including paralysis; there is a risk of one case of vaccine-associated paralytic poliomyelitis (VAPP) per 1.2 million doses of OPV administered (Melnick, 1996).

The reasons for OPV-associated poliomyelitis are changes in the genome of the attenuated virus, often point mutations, that result in reoccurrence of neurovirulence (for reviews, see Nomoto et al. [1989] and Racaniello et al. [1989]). Interestingly, these mutations occur in the 5' NTR, which is necessary for the initiation of viral protein synthesis, in addition to regions coding for the viral capsid proteins and RNA polymerase. The importance of the 5' NTR for an increase in neurovirulence was observed in a single base change from uracil to cytosine at position 472 in the IRES (Evans et al., 1985). The risk of a nonimmune contact person acquiring VAPP by infection with a neurovirulent Sabin virus excreted from a vaccinee is 1 per 5 million doses.

According to national recommendations, all infants and young children should be vaccinated with OPV or IPV (in general three times for basic immunization). Nonimmune children and adults who were not vaccinated as infants should first be immunized with IPV, followed by OPV. Primary administration of IPV reduces the risk of acquiring

VAPP, which is higher in children and adults than in infants. At the edge of poliovirus eradication, booster vaccinations (in general 10 years after the last vaccination) are of major importance and should be administered to everyone, including clinical personnel. In order to prevent shedding of attenuated poliovirus in clinical surroundings, it is recommended that health care workers be vaccinated only with IPV. In some countries, where poliomyelitis has already been eliminated, health authorities have decided to administer IPV instead of OPV, as the risk of acquiring VAPP by OPV is higher than that of developing clinical symptoms after infection with wild-type poliovirus.

Under defined conditions, a passive immunization can be advisable for interruption of the routes of virus transmission and prophylaxis, especially to prevent fulminant infections with "viral sepsis," myocarditis, encephalitis, and death in newborn babies after infection with coxsackieviruses and echoviruses. Passive immunization of nonimmune patients with immunoglobulins obtained from sera of convalescing patients can be effective only when administered as soon as possible after contact and not later than 72 h after exposure.

Therapy

Effective therapy against enterovirus infections cannot as yet be utilized for treatment of humans. A promising new drug candidate, pleconaril, is in its first clinical trials and shows inhibition of coxsackievirus A21 shedding in respiratory infections (Schiff and Sherwood, 2000). All other antiviral compounds developed so far are effective only against some enteroviruses in cell culture. Antiviral compounds that either affect the virus-specific RNA polymerase, such as guanidine and some benzimidazoles, or prevent virus uncoating by capsid stabilization (e.g., WIN compounds) have been proven to inhibit virus reproduction only under in vitro conditions (Zeichhardt et al., 1987; Eggers, 1988; Willingmann et al., 1989; Mosser and Rueckert, 1996).

REFERENCES

Andino, R., D. Boddeker, D. Silvera, and A. V. Gamarnik. 1999. Intracellular determinants of picornavirus replication. *Trends Microbiol.* **7:**76–82.

Barnert, H., H. Zeichhardt, and K.-O. Habermehl. 1992. Identification of 50- and 23-/25-kDa HeLa cell membrane glycoproteins involved in poliovirus infection: occurrence of poliovirus-specific binding sites on susceptible and nonsusceptible cells. *Virology* **186:**533–542.

Bartsch, H. D., K.-O. Habermehl, and W. Diefenthal. 1969. Correlation between poliomyelitisvirus-reproduction-cycle, chromosomal alterations and lysosomal enzymes. *Arch. Gesamte Virusforsch.* **27:**115–127.

Beatrice, S. T., M. G. Katze, B. A. Zajac, and R. L. Crowell. 1980. Induction of neutralizing antibodies by the coxsackievirus B3 virion polypeptide, VP2. *Virology* **104:**426–438.

Bergelson, J. M., N. St. John, S. Kawaguchi, M. Chan, H. Stubdal, J. Modlin, and R. W. Finberg. 1993. Infection by echoviruses 1 and 8 depends on the α 2 subunit of human VLA-2. *J. Virol.* **67:**6847–6852.

Bergelson, J. M., M. Chan, K. R. Solomon, N. F. St. John, H. Lin, and R. W. Finberg. 1994. Decay-accelerating factor (CD55), a glycosylphosphatidylinositol-anchored complement regulatory protein, is a receptor for several echoviruses. *Proc. Natl. Acad. Sci. USA* **91:**6245–6249.

Bergelson, J. M., J. A. Cunningham, G. Droguett, E. A. Kurt Jones, A. Krithivas, J. S. Hong, M. S. Horwitz, R. L. Crowell, and R. W. Finberg. 1997. Isolation of a common receptor for Coxsackie B viruses and adenoviruses 2 and 5. *Science* **275:**1320–1323.

Bernhardt, G., J. Harber, A. Zibert, M. deCrombrugghe, and E. Wimmer. 1994. The poliovirus receptor: identification of domains and amino acid residues critical for virus binding. *Virology.* **203:**344–356.

Bodian, D. 1959. Poliomyelitis: pathogenesis and histopathology, p. 479–518. *In* T. M. Rivers and F. L. Horsefall (ed.), *Viral and Rickettsial Infections of Man* J. B. Lippincott, Philadelphia, Pa.

Bouchard, M. J., and V. R. Racaniello. 1997. CD44 is not required for poliovirus replication. *J. Virol.* **71:**2793–2798.

Colonno, R. J., P. L. Callahan, and W. J. Long. 1986. Isolation of a monoclonal antibody that blocks attachment of the major group of human rhinoviruses. *J. Virol.* **57:**7–12.

Dales, S., H. J. Eggers, I. Tamm, and G. E. Palade. 1965. Electron microscopic study of the formation of poliovirus. *Virology* **26:**379–389.

Dalldorf, G., and G. M. Sickles. 1948. An unidentified, filterable agent isolated from the feces of children with paralysis. *Science* **108:**61–63.

Diefenthal, W., and K.-O. Habermehl. 1967. Die Bedeutung mikrokinematographischer Methoden in der Virologie. *Res. Film.* **6:**22–30.

Diefenthal, W., K.-O. Habermehl, P. R. Lorenz, and T. Beneke. 1973. Virus-induced inhibition of host cell synthesis. *Adv. Biosci.* **11:**127–148.

Dimmock, N. J. 1984. Mechanisms of neutralization of animal viruses. *J. Gen. Virol.* **65:**1015–1022.

Eggers, H. J. 1988. Assay systems: testing of antiviral drugs in cell culture (in vitro), p. 139–148. *In* E. De Clercq and R. T. Walker (ed.), *Antiviral Drug Development*, Series A: Life Sciences, vol. 143. Plenum Press, New York, N.Y.

Enders, J. F., T. H. Weller, and F. C. Robbins. 1949. Cultivation of the Lansing strain of poliomyelitis virus in cultures of various human embryonic tissue. *Science* **109:**85–87.

Evans, D. M. A., G. Dunn, P. D. Minor, G. C. Schild, A. J. Cann, G. Stanway, J. W. Almond, K. Currey, and J. V. Maizel. 1985. Increased neurovirulence associated with a single nucleotide change in a non-coding region of the Sabin type 3 poliovaccine genome. *Nature* **314:**548–550.

Ferguson, M., D. I. Magrath, P. D. Minor, and G. C. Schild. 1986. WHO collaborative study on the use of monoclonal antibodies for the intratypic differentiation of poliovirus strains. *Bull. W.H.O.* **64:**239–246.

Freistadt, M. S., and V. R. Racaniello. 1991. Mutational analysis of the cellular receptor for poliovirus. *J. Virol.* **65:**3873–3876.

Habermehl, K. O., R. Diefenthal, and W. Diefenthal. 1966. Der Einfluß von Virusinfektionen auf den Ablauf der Zellteilung. *Zentbl. Bakteriol. Hyg. 1 Orig.* **199:**273–314.

Hogle, J. M., M. Chow, and D. J. Filman. 1985. Three-dimensional structure of poliovirus at 2.9 Å resolution. *Science* **229:**1358–1365.

Icenogle, J., H. Shiwen, G. Duke, S. Gilbert, R. R. Rueckert, and J. Anderegg. 1983. Neutralization of poliovirus by a monoclonal antibody: kinetics and stoichiometry. *Virology* **127:**412–425.

Johnson, R. T. 1984. Late progression of poliomyelitis paralysis: discussion of pathogenesis. *Rev. Infect. Dis.* **6**(Suppl. 2)**:**568–570.

Julien, J., I. Leparc-Goffart, B. Lina, F. Fuchs, S. Foray, I. Janatova, M. Aymard, and H. Kopecka. 1999. Postpolio syndrome: poliovirus persistence is involved in the pathogenesis. *J. Neurol.* **246:**472–476.

Kandolf, R., K. Klingel, R. Zell, H. C. Selinka, U. Raab, W. Schneider Brachert, and B. Bultmann. 1993. Molecular pathogenesis of enterovirus-induced myocarditis: virus persistence and chronic inflammation. *Intervirology* **35:**140–51.

Karnauchow, T. M., D. L. Tolson, B. A. Harrison, E. Altman, D. M. Lublin and K. Dimock. 1996. The HeLa cell receptor for enterovirus 70 is decay-accelerating factor (CD55). *J. Virol.* **70:**5143–5152.

Katze, M. G., and R. L. Crowell. 1980. Immunological studies of the group B coxsackieviruses by the sandwich enzyme-linked immunosorbent assay (ELISA) and immunoprecipitation. *J. Gen. Virol.* **50:**357–367.

Koike, S., H. Horie, I. Ise, A. Okitsu, M. Yoshida, N. Iizuka, K. Takeuchi, T. Takegami, and A. Nomoto. 1990. The poliovirus receptor protein is produced both as membrane-bound and secreted forms. *EMBO J.* **9:**3217–3224.

Landsteiner, K., and E. Popper. 1909. Übertragung der Poliomyelitis acuta auf Affen. *Z. Immunitaetsforsch. Orig.* **2:**377–390.

Lim, K. A., and M. Benyesh-Melnick. 1960. Typing of viruses by combinations of antiserum pools: application to typing of enteroviruses (Coxsackie and echo). *J. Immunol.* **84:**309–317.

Melchers, W., M. de Visser, P. Jongen, A. van Loon, R. Nibbeling, P. Oostvogel, D. Willemse, and J. Galama. 1992. The postpolio syndrome: no evidence for poliovirus persistence. *Ann. Neurol.* **32:**728–732.

Melnick, J. L. 1982. Enteroviruses, p. 187-251. *In* A. S. Evans (ed.), *Viral Infections of Humans: Epidemiology and Control.* Plenum Medical Books, New York, N.Y.

Melnick, J. L. 1996. Enteroviruses: polioviruses, coxsackieviruses, echoviruses, and newer enteroviruses, p. 655–712. *In* B. N. Fields, D. M. Knipe, P. M. Howley, R. M. Chanock, J. L. Melnick, T. P. Monath, B. Roizman, and S. E. Straus (ed.), *Virology,* 3rd ed., vol. 1. Lippincott-Raven Press, New York, N.Y.

Melnick, J. L., E. W. Shaw, and E. C. Curnen. 1949. A virus isolated from patients diagnosed as nonparalytic poliomyelitis or aseptic meningitis. *Proc. Soc. Exp. Biol. Med.* **71:**344–349.

Melnick, J. L., H. A. Wenner, and C. A. Phillips. 1979. Enteroviruses, p. 471–534. *In* E. H. Lennette and N. J. Schmidt (ed.), *Diagnostic Procedures for Viral, Rickettsial and Chlamydial Infections.* American Public Health Association, Washington, D.C.

Mendelsohn, C. L., E. Wimmer, and V. R. Racaniello. 1989. Cellular receptor for poliovirus: molecular cloning, nucleotide sequence, and expression of a new member of the immunoglobulin superfamily. *Cell* **56:**855–865.

Mertens, T., U. Pika, and H. J. Eggers. 1983. Cross antigenicity among enteroviruses as revealed by immunoblot technique. *Virology* **129:**431–442.

Minor, P. D., F. Brown, E. Domingo, E. Hoey, A. King, N. Knowles, S. Lemon, A. Palmenberg, R. R. Rueckert, G. Stanway, E. Wimmer, and M. Yin-Murphy. 1995. Picornaviridae. Taxonomic structure of the family, p. 329-336. *In* F. A. Murphy, C. M. Fauquet, D. H. L. Bishop, S. A. Ghabrial, A. W. Jarvis, G. P. Martelli, M. A. Mayo, and M. D. Summers (ed.), *Virus Taxonomy: Sixth Report of the International Committee on Taxonomy of Viruses.* Springer Verlag, New York, N.Y.

Mirzayan, C., and E. Wimmer. 1994. Polioviruses: molecular biology, p. 1119-1132. *In* R. G. Webster and A. Granoff (ed.), *Encyclopedia of Virology,* vol. 3. Academic Press, London, United Kingdom.

Molla, A., A. V. Paul, and E. Wimmer. 1991. Cell-free, de novo synthesis of poliovirus. *Science* **254:**1647–1651.

Moore, M., and D. Morens. 1984. Enteroviruses, including polioviruses, p. 407–483. *In* R. B. Belshe (ed.), *Textbook of Human Virology.* PSG Publishing Company, Littleton, Mass.

Morrison, M. E., Y. J. He, M. W. Wien, J. M. Hogle, and V. R. Racaniello. 1994. Homolog-scanning mutagenesis reveals poliovirus receptor residues important for virus binding and replication. *J. Virol.* **68:**2578–2588.

Mosser, A. G., D. M. Leippe, and R. R. Rueckert. 1989. Neutralization of picornaviruses: support for the pentamer bridging hypothesis, p. 155-167. *In* B. L. Semler and E. Ehrenfeld (ed.), *Molecular Aspects of Picornavirus Infection and Detection.* American Society for Microbiology, Washington, D.C.

Mosser, A. G., and R. R. Rueckert. 1996. Picornaviruses: capsid-binding agents, p. 13-40. *In* D. D. Richman (ed.), *Antiviral drug resistance.* John Wiley & Sons, Chichester, United Kingdom.

Mulders, M. N., M. P. G. Koopmans, H. G. A. M. van der Avoort, and A. M. van Loon. 1995. Detection and characterization of poliovirus—the molecular approach, p. 137–156. *In* Y. Becker and G. Darai (ed.), *PCR: Protocols for Diagnosis of Human and Animal Virus Diseases.* Springer-Verlag, Berlin, Germany.

Nelsen-Salz, B., H. J. Eggers, and H. Zimmermann. 1999. Integrin $\alpha_v\beta_3$ (vitronectin receptor) is a candidate receptor for the virulent echovirus 9 strain Barty. *J. Gen. Virol.* **80:**2311–2313.

Nomoto, A., N. Kawamura, M. Kohara, and M. Arita. 1989. Expression of the attenuation phenotype of poliovirus type 1, p. 297-306. *In* B. L. Semler and E. Ehrenfeld (ed.), *Molecular Aspects of Picornavirus Infection and Detection.* American Society for Microbiology, Washington, D.C.

Osterhaus, A. D., A. L. van Wezel, T. G. Hazendonk, F. G. UytdeHaag, J. A. van Asten, and B. van Steenis. 1983. Monoclonal antibodies to polioviruses. Comparison of intratypic strain differentiation of poliovirus type 1 using monoclonal antibodies versus cross-absorbed antisera. *Intervirology* **20:**129–136.

Palmenberg, A. C. 1989. Sequence alignments of picornaviral capsid proteins, p. 211-241. *In* B. L. Semler and E. Ehrenfeld (ed.), *Molecular Aspects of Picornavirus Infection and Detection.* American Society for Microbiology, Washington, D.C.

Pringle, C. R. 1997. Virus taxonomy 1997. *Arch. Virol.* **142:**1727–1733.

Racaniello, V. R., N. La Monica, E. G. Moss, and R. O'Neill. 1989. Genetic analysis of neurovirulence, using a mouse model for poliomyelitis, p. 281–296. *In* B. L. Semler and E. Ehrenfeld (ed.), *Molecular Aspects of Picornavirus Infection and Detection.* American Society for Microbiology, Washington, D.C.

Robbins, F. C., J. F. Enders, T. H. Weller, and G. L. Florentino. 1951. Studies on the cultivation of poliomyelitis viruses in tissue culture. V. The direct isolation and serologic identification of virus strains in tissue culture from patients with nonparalytic and paralytic poliomyelitis. *Am. J. Hyg.* **54:**286–293.

Roivainen, M., L. Piirainen, T. Hovi, I. Virtanen, T. Riikonen, J. Heino and T. Hyypia. 1994. Entry of coxsackievirus A9 into host cells: specific interactions with $\alpha_v\beta_3$ integrin, the vitronectin receptor. *Virology* **203:**357–365.

Rossmann, M. G., E. Arnold, J. W. Erickson, E. A. Frankenberger, J. P. Griffith, H. J. Hecht, J. R. Johnson, G. Kramer,

M. Luo, A. G. Mosser, R. R. Rueckert, B. Sherry, and G. Vriend. 1985. Structure of a human common cold virus and functional relationships to other picornaviruses. *Nature* 317:145–153.

Rotbart, H. A. 1990. PCR amplification of enteroviruses, p. 372–377. *In* M. A. Innis, D. H. Gelfand, J. J. Sninsky, and T. J. White (ed.), *PCR Protocols: a Guide to Methods and Applications.* Academic Press, San Diego, Calif.

Rueckert, R. R. 1996. Picornaviridae and their replication, p. 609-654. *In* B. N. Fields, D. M. Knipe, P. M. Howley, R. M. Chanock, J. L. Melnick, T. P. Monath, B. Roizman, and S. E. Straus (ed.), *Virology*, 3rd ed., vol. 1. Lippincott-Raven Press, New York, N.Y.

Schiff, G. M., and J. R. Sherwood. 2000. Clinical activity of pleconaril in an experimentally induced coxsackievirus A21 respiratory infection. *J. Infect. Dis.* 181:20–26.

Schweiger, B., E. Schreier, B. Bothig, and J. M. Lopez Pila. 1994. Differentiation of vaccine and wild-type polioviruses using polymerase chain reaction and restriction enzyme analysis. *Arch. Virol.* 134:39–50.

Shafren, D. R., R. C. Bates, M. V. Agrez, R. L. Herd, G. F. Burns, and R. D. Barry. 1995. Coxsackieviruses B1, B3, and B5 use decay accelerating factor as a receptor for cell attachment. *J. Virol.* 69:3873–3877.

Shafren, D. R., D. J. Dorahy, R. A. Ingham, G. F. Burns, and R. D. Barry. 1997. Coxsackievirus A21 binds to decay-accelerating factor but requires intercellular adhesion molecule 1 for cell entry. *J. Virol.* 71:4736–4743.

Shepley, M. P., and V. R. Racaniello. 1994. A monoclonal antibody that blocks poliovirus attachment recognizes the lymphocyte homing receptor CD44. *J. Virol.* 68:1301–1308.

Stewart, P. L., and G. R. Nemerow. 1997. Recent structural solutions for antibody neutralization of viruses. *Trends Microbiol.* 5:229–233.

Tosteson, M. T., and M. Chow. 1997. Characterization of the ion channels formed by poliovirus in planar lipid membranes. *J. Virol.* 71:507–511.

van der Avoort, H. G., B. P. Hull, T. Hovi, M. A. Pallansch, O. M. Kew, R. Crainic, D. J. Wood, M. N. Mulders, and A.

M. van Loon. 1995. Comparative study of five methods for intratypic differentiation of polioviruses *J. Clin. Microbiol.* 33:2562–2566.

van Loon, A. M., A. Ras, P. Poelstra, M. Mulders, and H. van der Avoort. 1993. Intratypic differentiation of polioviruses, p. 359-369. *In* E. Kurstak (ed.), *Measles and Poliomyelitis.* Springer-Verlag, Berlin, Germany.

van Wezel, A. L., and A. G. Hazendonk. 1979. Intratypic serodifferentiation of poliomyelitis virus strains by strain-specific antisera. *Intervirology* 11:2–8.

Wetz, K., P. Willingmann, H. Zeichhardt, and K.-O. Habermehl. 1986. Neutralization of poliovirus by polyclonal antibodies requires binding of a single IgG molecule per virion. *Arch. Virol.* 91:207–220.

Willingmann, P., H. Barnert, H. Zeichhardt, and K.-O. Habermehl. 1989. Receptor-mediated endocytosis of poliovirus type 1: recovery of structurally intact and infectious virus from HeLa cells until viral uncoating. *Virology* 168:417–420.

World Health Organization. 1992. New approaches to poliovirus diagnosis using laboratory techniques: memorandum from a WHO meeting. *Bull. W. H. O.* 70:27–33.

Zeichhardt, H., and H.-P. Grunert. 1999. Enteroviruses, p. 3.1–3.12. *In* D. Armstrong and J. Cohen (ed.), *Infectious Diseases.* Mosby Publishers, London, United Kingdom.

Zeichhardt, H., M. J. Otto, M. A. McKinlay, P. Willingmann, and K.-O. Habermehl. 1987. Inhibition of uncoating of poliovirus by disoxaril (WIN 51711). *Virology* 160:281–285.

Zeichhardt, H., J. R. Schlehofer, K. Wetz, H. Hampl, and K.-O. Habermehl. 1982. Mouse Elberfeld (ME) virus determines the cell surface alterations when mixedly infecting poliovirus-infected cells. *J. Gen. Virol.* 58:417–428.

Zeichhardt, H., K. Wetz, P. Willingmann, and K.-O. Habermehl. 1985. Entry of poliovirus type I and Mouse Elberfeld (ME) virus into HEp-2 cells: receptor-mediated endocytosis and endosomal or lysosomal uncoating. *J. Gen. Virol.* 66:483–492.

Rotavirus, Enteric Adenoviruses, Caliciviruses, Astroviruses, and Other Viruses Causing Gastroenteritis

DAVID O. MATSON, MIGUEL L. O'RYAN, XI JIANG, AND DOUGLAS K. MITCHELL

25

Viruses were first recognized to be etiologic agents of acute diarrhea in the 1930s when a filterable agent that caused diarrhea in rabbits was discovered (Hodes, 1977). After the successful cultivation of enteroviruses in the late 1950s, a number of these viruses were identified in stool specimens from children with gastroenteritis (Ramos-Alvarez and Sabin, 1958; Bell and Grist, 1968). Subsequent studies failed to substantiate a role for enteroviruses as a cause of gastroenteritis, because the rates of enterovirus detection in children with or without diarrhea were similar (Yow et al., 1970). In the same year as this summary of negative findings, immune electron microscopy (IEM) was used to visualize the first proven etiologic agent of acute gastroenteritis (Kapikian et al., 1972). In these experiments, convalescent-phase serum mixed with an acute-phase stool from a patient with gastroenteritis during an outbreak in Norwalk, Ohio, generated virus-antibody complexes that could be visualized. In the following two decades, a large number of viruses associated with gastroenteritis in children and adults were identified by the use of EM and IEM techniques, the discovery of new cell lines for viral culture, the development of antibody-based assays, and genomic analysis. Most of these viruses do not replicate in standard cell lines. Table 1 lists currently known viral gastroenteritis pathogens. The classification scheme and the number of viruses in the list are expected to change as more is learned about the molecular virology and epidemiology of enteric viruses.

ROTAVIRUSES

Rotaviruses are frequent pathogens in human, other mammalian, and avian hosts. Human rotaviruses were first detected by EM observation in biopsy specimens from the duodenums of children with acute diarrhea (Bishop et al., 1973). Rotaviruses are the most common cause of severe infantile gastroenteritis worldwide.

Biology

Rotaviruses are in the family *Reoviridae*, genus *Rotavirus*. The genome of rotaviruses consists of 11 segments of double-stranded RNA; thus, rotaviruses are among the few human pathogens whose genetic material is in pieces. The virus is unenveloped, approximately 75 nm in diameter, and composed of three concentric protein shells: the outer shell, inner shell, and core (Fig. 1) (Estes, 1996). The genome is located within the core, and antigenic epitopes detected in diagnostic assays are on the two shells. Each genome segment encodes at least one viral protein. The viral proteins can be structural or nonstructural proteins and function in replication, viral assembly, budding, determination of host range, and viral pathogenesis (Estes, 1996). Nonstructural protein 4 (NSP4) is potentially a "viral enterotoxin" (Ball et al., 1996).

Six distinct rotavirus groups (A through F) have been identified serologically on the basis of common antigens (Table 2) (Bridger, 1988; Desselberger, 1988). Three of these groups (A, B, and C) have been identified in humans. Group A rotaviruses are the common cause of infantile gastroenteritis worldwide. Group B rotaviruses have caused epidemics of cholera-like diarrhea in China and sporadic cases elsewhere (Hung et al., 1984; Brown et al., 1987; Nakata et al., 1987; Peñaranda et al., 1989; Eiden et al., 1994). Group B rotaviruses have been successfully isolated in swine kidney cells by using pancreatin treatment (Sanekata et al., 1996). Group C rotaviruses have caused outbreaks of diarrhea and sporadic cases in many countries, including the United States (Espejo et al., 1984; Bridger et al., 1986; Matsumoto et al., 1989; Caul et al., 1990; Jiang et al., 1995a). This group may prove to be more common than previously considered once sensitive assays are developed. An interesting association between group C rotaviruses and biliary atresia requires confirmation (Riepenhoff-Talty et al., 1996). Group A rotaviruses represent more than 95% of currently identified strains in humans worldwide, and further discussion, unless otherwise stated, will focus on this group.

Group A rotaviruses were first divided into two antigenic types, subgroups I and II, on the basis of antigenic differences of epitopes on the major group antigen VP6 (Estes and Cohen, 1989). Subgroup classification has been largely replaced by more immunologically relevant classification based upon neutralization epitopes of the outer capsid proteins, VP4 and VP7 (Offit et al., 1986; Hoshino et al., 1988). "G serotype" refers to the antigenic specificity of VP7, the outer-capsid glycoprotein, and "P serotype" refers to the antigenic specificity of VP4, the spike protein that is protease (trypsin) sensitive (Estes, 1996). Neutralization assays measure reactivity predominantly to VP7 proteins (Estes, 1996). Ten G serotypes and nine P serotypes have been detected in humans (Table 3) (Steele et al., 1995;

TABLE 1 Viruses associated with gastroenteritis

Rotavirus
Enteric adenoviruses, types 40 and 41, perhaps types in
  subgenus A
Caliciviruses in the Norwalk-like and Sapporo-like genera
Astroviruses
Other viruses[a]
  Coronaviruses
  Parvoviruses
  Pestiviruses
  Toroviruses

[a]An association with gastroenteritis is suspected but not certain.

Estes, 1996; Masendycz et al., 1997). Sequence and mono-clonal antibody analysis has detected G-type intratypic variation, but the immunological significance of these variants is uncertain (Matson et al., 1990a; Raj et al., 1992; Wen et al., 1997). P types are more difficult to characterize and are less certainly defined (Estes, 1996). Twenty genomic P types (genotypes) have been identified to date. These genotypes correspond to one or more of the described P antigenic types, under a separate numbering system (Estes, 1996). Rotavirus types are designated as follows: P antigenic type [P genetic type] G type. As an example, the human M37 strain is P2A[6]G1 (Estes, 1996; Gentsch et al., 1996).

Pathogenesis

Information on the pathophysiology of rotavirus has been obtained mostly from animal models, supported by studies from biopsy specimens from seriously ill children and adult volunteers. Rotavirus infects primarily mature enterocytes located in the mid- and upper villous epithelium (Davidson et al., 1975). The upper small intestine is most commonly involved, although lesions may extend to the distal ileum and colon (Phillips, 1988). A broad interaction between intestinal cells and rotavirus structural and nonstructural proteins occurs, resulting in death of infected villous enterocytes. NSP1 and NSP2 are possible host range determinants. In the piglet model, NSP3, NSP4, and NSP7 are needed for diarrhea to occur (Hoshino et al., 1995). Once infected, the villous enterocyte is sloughed, resulting in an altered mucosal architecture that becomes stunted and flattened. This is followed by a blastic response in the crypt cells, which tends to involve most of the affected villi. Ischemia also may play a role in the loss and stunting of villi (Stephen, 1988). During the recovery phase, the enteroblastic cells mature and reconstruct the villous structure. Because of the loss of mature enterocytes on the tips of the villi, the surface area of the intestine is reduced. Diarrhea

TABLE 2 Host range of rotavirus antigenic groups

Group	Host(s) or reservoir(s)
A	Human, primate, horse, pig, dog, cat, rabbit, mouse, cow, bird
B	Human, pig, cow, sheep, rat
C	Human, pig, ferret
D	Chicken
E	Pig
F	Chicken
G	Chicken

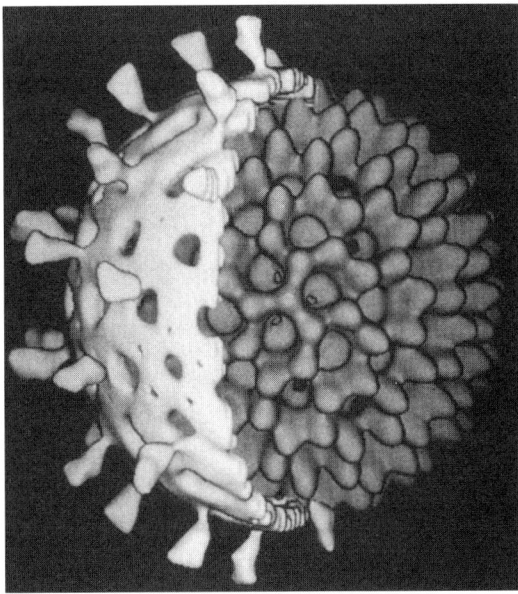

FIGURE 1 Three-dimensional representation of the inner and outer capsids of rotavirus. The outer capsid is displayed in the left portion of the figure. The spikes protruding from the outer capsid are formed by VP4. The matrix of the outer capsid is VP7. The inner capsid, formed by VP6, is displayed in the right portion of the figure, with the outer capsid stripped away. (Courtesy of B. V. V. Prasad, Baylor College of Medicine, Houston, Tex.)

may result from decreased surface area, disruption in epithelial integrity, transient disaccharide deficiency, and/or altered countercurrent mechanisms. The mucosal damage can result in lactose intolerance that may prolong the duration of diarrhea in a small proportion of children. Recently, NSP4 was found to induce age-dependent diarrhea in CD1 mice by triggering calcium-dependent chloride and water secretion, leading to a hypothesis that NSP4 acts as a "viral enterotoxin" (Ball et al., 1996). The potential role of this protein in human disease requires further studies. The clinical spectrum of disease suggests that different pathophysiologic mechanisms may predominate in different individuals, leading to a spectrum of intestinal abnormalities ranging from mild to severe malabsorptive and/or mild to severe secretory diarrhea.

Diverse mechanisms participate in the protective immune response against rotavirus infection and illness, although conflicting studies on protective immunity have been published. Information from in vitro studies, studies in different animal models (mice, rabbits, and piglets), epidemiological studies in children exposed to natural infections, and vaccine-related studies has provided some conclusive evidence on which immune mechanisms protect against rotavirus infection.

Natural rotavirus infections protect against subsequent rotavirus infections (Bishop et al., 1983; Velazquez et al., 1996). Although rotavirus can infect the same child many times, second and subsequent infections are less likely than first infections to be symptomatic (O'Ryan et al., 1990b; Velazquez et al., 1996). Two rotavirus infections confer complete protection against a third severe symptomatic infection (Velazquez et al., 1996). Natural rotavirus infections are associated with rises in serum and/or mucosal antirotavirus immunoglobulin M (IgM), IgG, and IgA anti-

TABLE 3 G and P types of group A rotaviruses[a]

Designation and distribution of G types			Designation and distribution of P types[b]			
Type	Human	Animal species	Genotype	Serotype	Human	Animal species
1	Yes	Cow	1	6		Cow, monkey
2	Yes		2			Monkey
3	Yes	Monkey, dog, cat, rabbit, mouse, pig	3	5	Yes	Monkey, dog, cat
4	Yes	Pig	4	1B	Yes	
5	Yes	Pig, horse	5	7		Cow
6	Yes	Cow	6	2A	Yes	
				2B		Pig
7	No	Chicken, turkey, pigeon, cow	7	9		Pig
8	Yes	Cow	8	1A	Yes	
9	Yes		9	3	Yes	Cat
10	Yes	Cow, sheep	10	4	Yes	
11		Pig	11	8	Yes	Cow
12	Yes		12			Horse
13		Horse	13	3B	Yes	
14		Horse	14		Yes	Pig
			15			Sheep
			16	10		Mouse
			17			Cow, pigeon
			18			Horse
			19			Pig
			20			Mouse

[a]Modified from information provided by Mary Estes.
[b]Biologic correlation between P genotype and serotype is not clearly established.

body titers. The presence and titer of either serum or mucosal antirotavirus antibodies correlate with protection against rota-virus infection and illness (Riepenhoff-Talty et al., 1981; Davidson et al., 1983; Hjelt et al., 1987; Grimwood et al., 1988; Coulson et al., 1990; Hjelt and Grauballe, 1990; Matson et al., 1993; O'Ryan et al., 1994; Offit, 1996). After a primary rotavirus infection, infants and young children develop neutralizing antibodies in serum directed against the G type of the infecting strain (homotypic response) and to a lesser extent against noninfecting G types (heterotypic response) (Chiba et al., 1986; O'Ryan et al., 1994; Offit, 1996). Repeated infections tend to broaden the G-type specificity of the response and protective immunity (O'Ryan et al., 1994). Naturally acquired VP7- and VP4-specific antibodies correlate with protection against infection and illness. A rotavirus reinfection is more likely to be by a G type different from that causing the original infection and, if of the same type, is most commonly asymptomatic (O'Ryan et al., 1990; Velazquez et al., 1996). Breast-feeding protects against rotavirus infections, by decreasing exposure and providing antirotavirus IgA antibodies and the rotavirus-binding glycoprotein lactadherin (Newburg et al., 1998; Matson et al., 1995a). Transplacental serum antirotavirus IgG antibody correlates with protection, in the absence of serum IgA antibody, in the non-breast-feeding infant (Matson et al., 1995a; Velazquez et al., 1996). Virus-host interaction and protective immunity result from an interaction of humoral (isotype-, G-, P-, and other protein-specific) and cell-mediated immune mechanisms, as well as viral load and interaction with other organisms in the intestine (Offit, 1996).

Rotavirus has an incubation period of 1 to 3 days (Davidson et al., 1975; Kapikian et al., 1983). Excretion of rotavirus in stool can precede the onset of symptoms by sev-

eral days and can continue for 8 to 10 days after symptoms end (Pickering et al., 1988).

Rotavirus illness occurs in all age groups, but is most common in infants from 6 months to 2 years of age. In infants and young children, the most common clinical presentation is an abrupt onset of vomiting, followed by explosive watery diarrhea and moderate to high fever (Table 4). The number of vomiting episodes and diarrhea stools determines the severity of resulting dehydration. Dehydration most commonly is isotonic and may lead to metabolic acidosis and death (Rodriguez et al., 1977; Echeverria et al., 1983; Ho et al., 1989). The majority of rotavirus infections are asymptomatic or mild, except among infants 3 to 8 months of age (Pickering et al., 1988; O'Ryan et al., 1990; Velazquez et al., 1996). Asymptomatic rotavirus infections are common in neonates because of passively acquired maternal immunity, breast-feeding, and possible infection with less-virulent strains (Chrystie et al., 1978; Bryden et al., 1982; Rodriguez et al., 1982; Bishop et al., 1983;

TABLE 4 Duration and severity of symptoms of rotavirus gastroenteritis[a]

Symptom	Mean (Mn) or median (Md) value	Range
Duration of diarrhea	4 days (Mn)	0–12 days
Maximum no. of stools/24 h	6 (Md)	0–30
Duration of vomiting	2 days (Md)	0–8 days
Maximum no. of vomiting episodes/24 h	3 (Md)	0–20
Maximum fever	38.9°C (Md)	37.5–40.5°C

[a]Reprinted from Li Pang et al., 1999, with permission.

Gorziglia et al., 1986; Tam et al., 1990; Matson et al., 1995a). The mean duration of illness in immunocompetent hosts is 4 to 5 days, although mild infections can last 3 days or less and severe infections can last 10 days or more (Li Pang et al., 1999). Rotavirus diarrhea is most commonly watery, occasionally with mucus (Pickering et al., 1977; Rodriguez et al., 1977). Bloody stools are rare and suggest the possibility of mixed infections if rotavirus is detected (Prado and O'Ryan, 1994). Mixed rotavirus infections with other viral, bacterial, and parasitic enteropathogens account for up to 15% of diarrhea episodes in developing countries (Rodriguez et al., 1977; Brandt et al., 1983; Prado and O'Ryan, 1994).

Concurrent respiratory tract symptoms have been frequently associated with rotavirus infection, although this may represent concurrent winter respiratory virus infection (Rodriguez et al., 1977; Goldwater et al., 1979; Lewis et al., 1979). Rotavirus has been temporally associated with a wide array of diseases, including intussusception (Konno et al., 1978; Mulcahy et al., 1982; Nicolas et al., 1982), Reye's syndrome and encephalitis (Salmi et al., 1978), aseptic meningitis (Wong et al., 1984), sudden infant death syndrome (Yolken and Murphy, 1982), inflammatory bowel disease (Blacklow and Cukor, 1982), neonatal necrotizing enterocolitis (Rotbart et al., 1983), Kawasaki's syndrome (Matsuno et al., 1983), and myocarditis (Gregorio et al., 1997). The rotavirus genome has been detected by reverse transcriptase-PCR (RT-PCR) in the spinal fluid of children with aseptic meningitis (Nishimura et al., 1993; Pang et al., 1996; Ushijima et al., 1994). A possible association between group C rotavirus and biliary atresia exists (Riepenhoff-Talty et al., 1996). Concerns about the occurrence of intussusception shortly after vaccine administration resulted in the withdrawal of the first licensed rotavirus vaccine (Centers for Disease Control and Prevention [CDC], 1999).

Chronic rotavirus infections can occur in immunodeficient children, although these infections are uncommon. The disease may be more severe in extremely malnourished individuals (Saulsbury et al., 1980; Dagan et al., 1990; Uhnoo et al., 1990a). Liver and/or kidney infection and damage in immunodeficient animals and humans have been reported (Uhnoo et al., 1990b; Gilger et al., 1992).

Infection of adults occurs often among those in close contact with young children and has been reported in travelers, military personnel, elderly persons in institutions, and hospitalized adults (Kapikian et al., 1976; Kim et al., 1977; Hrdy, 1987; Von Bonsdorff et al., 1978; Vollet et al., 1979; Halvorsrud and Orstavik, 1980; Holzel et al., 1980; Pickering et al., 1981; Marrie et al., 1982; Grimwood et al., 1983). Infections in adults tend to be asymptomatic or to cause mild illness (Wenman et al., 1979).

Epidemiology

Rotavirus infection occurs worldwide, affecting nearly every child younger than 5 years of age, regardless of socioeconomic status or environmental conditions. Disease is more often severe and fatal in children in developing countries, a direct consequence of delayed medical care (Bern and Glass, 1994). About 875,000 deaths occur yearly as a consequence of rotavirus infection (World Health Organization, 1997). Rotavirus-associated gastroenteritis occurs most frequently between 6 and 24 months of age, although it can occur in all age groups (Kapikian et al., 1976; Rodriguez et al., 1977; Brandt et al., 1983; Hrdy, 1987). The incidence of

rotavirus diarrhea in community-based studies ranges from 0.2 to 0.8 episodes per child per year (DeZoysa and Feachem, 1985). In the child care setting, the incidence of symptomatic rotavirus infection is about 0.5 episodes per child per year (Bartlett et al., 1988). Rotavirus accounts for 20 to 40% of the acute watery diarrhea episodes that require medical care and 20 to 60% of the hospitalizations for acute diarrhea in children younger than 5 years of age (Matson and Estes, 1990; Ryan et al., 1996; O'Ryan et al., in press). Between 1 in 25 and 1 in 75 children will be hospitalized for rotavirus diarrhea in the United States (Glass et al., 1996). Rotavirus outbreaks are common in closed environments such as child care centers and hospitals (Vial et al., 1988; O'Ryan and Matson, 1990).

Rotavirus infections typically are most common during the autumn and winter (Konno et al., 1983; Bartlett et al., 1988; Matson and Estes, 1990; Glass et al., 1996; Ryan et al., 1996). In North America, the annual rotavirus season begins in late fall in Mexico and moves across the continental United States from southwest to northeast, resulting in a peak of rotavirus illness in March and April in eastern Canada and the northeastern United States (LeBaron et al., 1990). Infections are rare during summer months. A similar west-to-east migration of peak rotavirus activity also appears to occur in South America (O'Ryan et al., in press), but seasonality is less marked than in the United States. The reason for this "migration" of rotavirus activity from west to east is unknown. In tropical areas, rotavirus tends to circulate year-round without a seasonal pattern (Cook et al., 1990).

Information on the epidemiology of rotavirus G and P types is now available for most areas worldwide. Four PG combinations (P1A[8]G1, P1B[4]G2, P1A[8]G3, and P1A[8]G4) occur in more than 95% of typed specimens. Of these combinations, P1A[8]G1 is the most common type (Matson et al., 1990a; O'Ryan and Matson, 1990; Gentsch et al., 1996). Several antigenic types may cocirculate in large urban areas, although one or two types usually predominate (Gentsch et al., 1996; O'Ryan et al., 1997). Changes of predominating types are regional events, probably limited by the extent of exposure and herd immunity. Less-common types continue to circulate in regions where one or two types predominate. In India, P2A[6]G9 types have been detected; in Bangladesh, diverse P genotypes among G9 strains have been detected; and in Brazil, P1A[8]G5, G8, and G10 strains have been detected (Ramachandran et al., 1996; Santos et al., 1998; Unicomb et al., 1999). These findings suggest that reassortment between common and uncommon human and/or animal strains occurs and that in some locations otherwise-uncommon types can or will predominate (Palombo and Bishop, 1995; Gentsch et al., 1996). Isolation of less-common types has been more frequent among neonates with nosocomial rotavirus infections (Tam et al., 1990; Gentsch et al., 1996; Kilgore et al., 1996). Some of these strains also seem to be associated with asymptomatic infections, although whether naturally acquired asymptomatic strains exist is controversial.

As few as 10^2 virus particles are required to cause infection, and up to 10^{11} particles per g of stool are excreted at the peak of infection, shortly after the onset of illness (Nagayoshi et al., 1980; Vesikari et al., 1981a). Rotaviruses are transmitted from person to person by the fecal-oral route. Several studies have reported spread from person to person that was so rapid, and in the absence of a common source such as contaminated water, that respiratory spread was suspected (Foster et al., 1980; Gurwith et al., 1981). Transmission within families, child care centers, and hospitals is common, and spread may be facilitated by contact with asymptomatic or ill attendants, children, or parents (Pickering et al., 1981; Rodriguez et al., 1982; Grimwood et al., 1983; Samadi et al., 1983; Flewett, 1983; O'Ryan et al., 1990). Nosocomial outbreaks may be controlled by closing the affected ward and grouping infected children and exposed staff (Flewett, 1983; O'Ryan et al., 1990). Crowding is a significant risk factor for increased rotavirus transmission in families and child care centers (Gurwith et al., 1983; Keswick et al., 1983; Velazquez et al., 1996). Contaminated water can be a source of rotavirus-associated outbreaks, although these outbreaks are uncommon (Lycke et al., 1978; Hopkins et al., 1984; Tao et al., 1984; Dubois et al., 1997). Houseflies may be vectors for rotavirus transmission (Tan et al., 1997). The importance of other factors, such as temperature, relative humidity, exposure to domestic animals, food products, and excretion by asymptomatic individuals, in transmission of rotavirus remains unclear.

Rotavirus stability has been studied by using simian SA11 rotavirus as a model. SA11 is stable over a pH range of 3.5 to 10.0, and calcium (2 mM) is required for stability of infectious particles. Infectivity is lost by treatment with low concentrations (5 mM) of ethylenediamine tetraacetic acid (EDTA) or by heating at 50°C for 15 min with 2 M $MgCl_2$. The presence of fecal material protects rotaviruses from the action of disinfectants and enhances survival on environmental surfaces (Keswick et al., 1983). A solution of 70% ethanol or an ethanol-containing disinfectant is effective for surface decontamination (Brade et al., 1981). Chlorhexidine gluconate and povidine-iodine are ineffective (Tan and Schnagl, 1981). Lysol, formalin, and iodophor preparations also may be effective. Chlorine has been shown to rapidly inactivate SA11 in water; however, human rotavirus strains may be more resistant to chlorine treatment than animal strains (Butler and Harakeh, 1983). Rotaviruses can survive in treated drinking water, sewage, and marine water and on surfaces in child care centers and probably hospitals.

Diagnosis

Stool samples are the only clinical specimens for rotavirus detection. Stools collected during the first several days of illness have the highest concentration of virus (Vesikari et al., 1981a; Pickering et al., 1988). For optimal detection, stool specimens should be tested immediately, but they may be held at 4°C if the delay is minimal. Otherwise, samples should be frozen at −70°C if viability is to be conserved. For antigen detection methods, a rectal swab is usually sufficient. For viral culture and maximal antigen detection sensitivity, bulk stool is usually required (Wyatt et al., 1983). Stool samples can be stored in a clean receptacle until tested; transport media are not required. Repeat freezing and thawing significantly reduce viability and the possibility of antigenic typing by enzyme-linked immunosorbent assay (ELISA).

Several methods permit rapid, practical detection and characterization of rotaviruses. The most widely used assays include ELISA, latex agglutination, RT-PCR, EM, IEM, and gel electrophoresis of the viral genomic RNA.

Cell Culture

Human rotaviruses are difficult to cultivate from clinical specimens. Rotaviruses may be recovered in roller tube cultures by treating stool specimens with trypsin or pancreatin and by incorporating a small amount of these enzymes into the maintenance medium (Hasegawa et al., 1982; Kutsuzawa et al., 1982; Birch et al., 1983). A variety of cell lines have been tested for rotavirus isolation from clinical samples, including MDBK, PK-15, BSC-1, LLC-MK2, BGM, CV-1, MA104, CaCo-2, and HRT-29 (Estes et al., 1983; Kapikian and Chanock, 1996).

Immunoassays

ELISAs and latex agglutination assays are currently the most widely used techniques for detection of rotavirus. Many commercial kits are available for rotavirus detection and they differ in specificity and sensitivity (Table 5) (Dennehy et al., 1999). Antibody-based detection assays utilize fluorescent or enzyme conjugates (Birch et al., 1979; Yolken and Stopa, 1979). Positive results can be confirmed by blocking with a polyclonal antibody (Pickering et al., 1988). RT-PCR is the "gold standard" for confirming immunoassay results. Detections of rotaviruses in North America in the summer commonly reflect improper assay performance (LeBaron et al., 1992).

Monoclonal antibodies to a wide variety of rotavirus epitopes have been developed. G-typing monoclonal antibodies have been widely used as research tools for epidemiologic studies (O'Ryan et al., 1990; Gentsch et al., 1996). P-typing monoclonal antibodies have limited applicability because of the difficulty of generating P-type-specific antibodies (Masendycz et al., 1999).

Genomic Analysis

Highly conserved sequences exist at the ends of all rotavirus genomic segments and for all types. Partially conserved sequences elsewhere within individual segments also exist. Type-specific regions of rotavirus genes 4 (VP4) and 9 (VP7) have been identified (Green et al., 1988; Gentsch et al., 1992). RT-PCR and hybridization methods for type-common and type-specific detection have been described (Olive and Sethi, 1989; Steele et al., 1995; Kapikian and Chanock, 1996). These tests are highly specific and are 10 to 100 times more sensitive than ELISA is. RT-PCR with type-specific primers amplifies fragments of different lengths and is particularly useful for G or P genotyping (Padilla-Noriega et al., 1998).

EM and IEM

EM was the original diagnostic technique used for identification of rotavirus. Rotaviruses are readily recognized in stool specimens because of their characteristic 70-nm size and morphology (Fig. 2; Hammond et al., 1982). EM also is important for the detection of non-group A rotaviruses that share the typical morphology but not the group antigens of group A rotaviruses. IEM has been employed for detection, identification, and typing of rotaviruses (Brandt et al., 1981; Gerna et al., 1989; Wu et al., 1990). A disadvantage is that particles visualized by IEM are coated with antibody, which masks virus structure. Coating the grid with antibody before applying the sample (solid-phase IEM) avoids this problem; this method is more sensitive than standard IEM (Wu et al., 1990).

Electropherotyping

Electropherotyping is the assessment of the migration pattern of the 11 rotavirus double-stranded RNA genome segments in polyacrylamide gels by electrophoresis (Herring et al., 1982). This was the most widely used diagnostic method before immunoassays were developed. The technique is as sensitive as EM but is less sensitive than immunoassays are. Electro-

TABLE 5 Commercially available diagnostic assays for viral gastroenteritis pathogens[a]

Virus	Assay type	Assay name	Manufacturer
Rotavirus	ELISA	Rotaclone	Cambridge Biotech
		Pathfinder	Sanofi Diagnostics Pasteur
		Rotavirus ELISA	International Diagnostics
		Wellcozyme	Wellcome Diagnositics
		Rotazyme II	Abbott Laboratories
		TestPack	Abbott Laboratories
	Latex agglutination	Rotastat	International Diagnostics
		Meritec Rotavirus	Meridian Diagnostics
		Virogen Rotatest	Wampole Laboratories
		Wellcome RV Latex	Wellcome Diagnositics
Adenovirus	ELISA	Adenoclone	Cambridge Biotech
Astrovirus	ELISA	IDEIA Astrovirus	DAKO Diagnostics

[a]Modified from Mitchell et al., 1999a.
[b]All manufacturers except DAKO Diagnostics are in the United States.

pherotyping can be used in epidemiological studies to monitor specific strains (Kapikian and Chanock, 1996). In large geographic areas, many different electropherotypes cocirculate. A single electropherotype usually predominates in an outbreak (Garbarg-Chenon et al., 1985; O'Ryan et al., 1990).

Electropherotyping has limited utility for typing. Many electropherotypes exist within one G and/or P type, and similar electropherotypes may occur in different antigenic types (Pipittajan et al., 1991). One useful finding has been that G2 strains characteristically display a short electropherotype pattern, in which gene segments 10 and 11 migrate a relatively short distance (Fig. 3). Electropherotyping may provide a clue to a novel strain. For example, supershort electropherotype patterns have been recently reported from Japan; these strains belong to uncommon antigenic types (Nakagomi et al., 1999).

Electropherotyping is the only widely available method for identifying non-group A rotaviruses. Commercial rotavirus ELISAs detect only group A rotaviruses. An EM-positive sample that does not react in a commercial ELISA can be assigned to group B to F by the markedly different electropherotypes of non-group A rotaviruses. There is no established classification based upon electropherotypes, although some classification schemes have been proposed (Desselberger, 1988; Kusuya et al., 1996).

Serology

Neutralization assays have been used to determine antibody responses to specific rotavirus serotypes (Estes and Cohen, 1989). Alternatively, epitope-blocking assays utilizing monoclonal antibodies can be used to measure antibody responses to specific epitopes, such as those that are G type specific (Shaw et al., 1985; Matson et al., 1992). Group-specific IgM, IgG, and IgA antibodies can be measured by ELISA by using perhaps any group A rotavirus as the test antigen. About one-quarter to one-third of rotavirus infections are detected only by seroresponse, because virus excretion is below virus detection assay limits (Matson et al., 1993; Velazquez et al., 1996).

ENTERIC ADENOVIRUSES

Biology

Enteric adenoviruses (EA) are in the family *Adenoviridae*, genus *Mastadenovirus*. The classification of adenoviruses into subgenera A to F is based upon differences in viral hemagglutination or oncogenic properties, the size of structural proteins, and DNA homology (Wadell, 1984; Wadell et al., 1986). Further differentiation into one of the 47 different serotypes is based upon neutralization assays. The term EA commonly is used to refer to types 40 and 41 in subgenus F. Strains in subgenus A (types 12, 18, and 31) also may cause enteric illness, although study results are conflicting on this point. Adenoviruses are nonenveloped viruses that are 70 to 90 nm in diameter (Fig. 2). The viral capsid has icosahedral symmetry and antenna-like projections emerging from each of the 12 vertices. The genome is 33-to-45-kb linear double-stranded DNA (Van Loon et al., 1985).

EA types 40 and 41 are important causes of viral gastroenteritis in children (Uhnoo et al., 1984; Brandt et al., 1985; Madeley, 1986; Kotloff et al., 1989; Cruz et al., 1990). When studies of EAs and gastroenteritis extend to identification of non-EA adenovirus types, adenoviruses from subgenus A (especially type 31) also are usually detected, although much less frequently than 40 and 41. The associations of gastroenteritis with non-40/41 types are confounded by the ability of most adenovirus types to cause respiratory tract infection. Unless otherwise specified, the remaining discussion refers to subgenus F infections only.

Pathogenesis

There is limited information on the pathophysiologic mechanisms involved in EA infections. Observations from a single human case suggest that the nuclei of villous epithelial cells are the site of virus concentration and that microvillus structure is altered (Phillips, 1988). How this affects intestinal function is not known.

The incubation period is 3 to 10 days. EA infections are less likely to be associated with high fever or dehydration but are more likely to cause prolonged illness than rotavirus infection (Grimwood et al., 1995). Diarrhea lasts from 6 to 9 days in most cases, with a range of 4 to 23 days (Uhnoo et al., 1986; Hierholzer et al., 1988; Kotloff et al., 1989; Van et al., 1992). Vomiting and fever may precede or accompany the diarrhea. Persistent lactose intolerance has been reported in a few children infected with EAs (Uhnoo et al., 1986). An association between EAs and celiac disease has been postulated but not proven (Kagnoff et al., 1984; Lahdeaho et al., 1993).

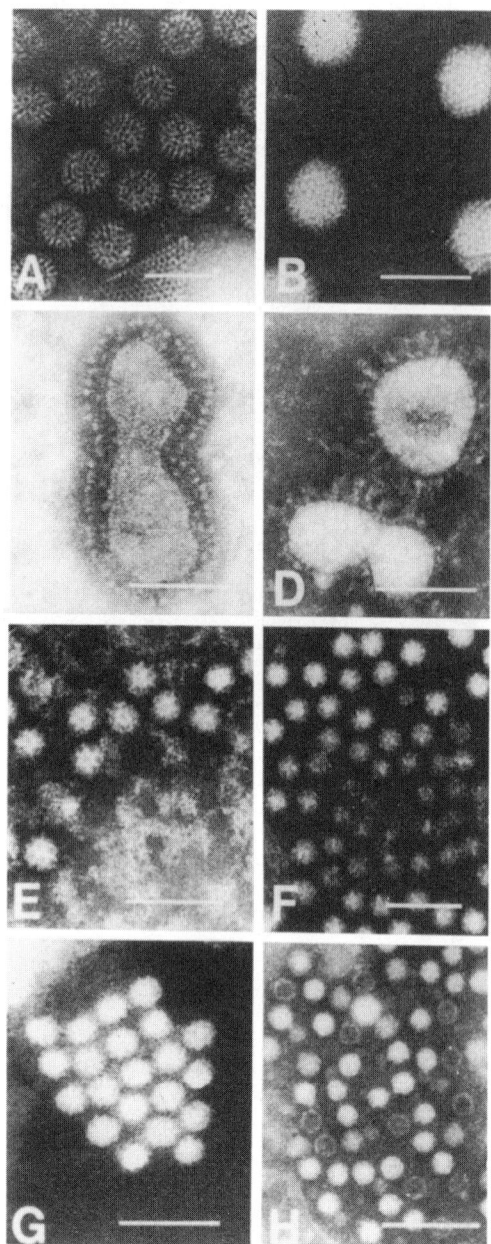

FIGURE 2 EMs of virus particles found in stool specimens from patients with gastroenteritis and visualized by negative staining. Specific viruses and the original magnifications of the micrographs are as follows: (A) rotavirus (×185,000), (B) EA (×234,000), (C) coronavirus (×249,000), (D) Breda virus (×249,000), (E) Sapporo-like calicivirus (×250,000), (F) astrovirus (×196,000), (G) Norwalk-like calicivirus (×249,000), (H) parvovirus (×249,000). The micrographs in panels C, D, and F through H were provided by T. Flewett; the Sapporo-like calicivirus micrograph (E) was originally obtained by C. R. Madeley. The bar represents 100 nm. (Reprinted from Estes and Graham, 1988, with permission.)

Epidemiology

EAs cause 2 to 22% of diarrhea episodes in children in cross-sectional studies (Mickan and Kok, 1994; Noel et al., 1994; Yamashita et al., 1995; Grimwood et al., 1995; Vizzi

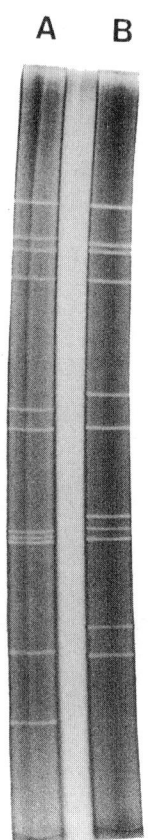

FIGURE 3 RNA electrophoresis of human rotaviruses in a polyacrylamide gel. Lane A contains a VP7 type 3 rotavirus which displays a long RNA electropherotype, named by the migration of segments 10 and 11. Lane B contains a VP7 type 2 rotavirus with a characteristic short RNA electropherotype.

et al., 1996; Durepaire et al., 1996; Wang and Chen, 1997; Steele et al., 1998; Bon et al., 1999; Caeiro et al., 1999), although these findings may be underestimates because of inadequate detection methods (Ahluwalia et al., 1994). Both EA types appear to be widespread and to cause endemic diarrhea. Outbreaks have been reported in orphanages, hospitals, and child care settings (Richmond et al., 1979; Chiba et al., 1983; Paerregard et al., 1990; Van et al., 1992; Taylor et al., 1997b). Seasonal and year-to-year shifts from one serotype to another have been shown at several sites (Wadell et al., 1988; Mickan and Kok, 1994; Grimwood et al., 1995). The seasonality of EA infection is uncertain, although certainly less marked than that exhibited by rotavirus and not found in every study.

EAs are more likely to infect children than adults (Vesikari et al., 1981b). Antibody prevalence to EAs has been shown to increase from 20% during the first 6 months of life to 50% or greater by the third to fourth year of life (Shinozaki et al., 1987). The mode of transmission is most probably fecal-oral (Isaacs et al., 1986).

Diagnosis

EM was the initial detection method for EAs, but commercial assays that use monoclonal antibodies incorporated into an ELISA now are available (Singh-Naz et al., 1988).

Hybridization assays have also been developed for EAs, but they are less sensitive than antigen detection techniques (Hammond et al., 1987). EAs can be cultivated in Graham 293 and CaCo-2 cells (Ahluwalia et al., 1994; Pinto et al., 1994), and such cultivation can increase EA detection by amplification of the virus before detection assays are utilized. Restriction enzyme analysis of DNA extracted from virus isolates is the definitive method for classifying individual isolates (Takiff et al., 1981; Wadell, 1984), although genotyping methods are being developed (Scott-Taylor et al., 1993; Rousell et al., 1993; Allard et al., 1994; Kidd et al., 1996; Li et al., 1999).

CALICIVIRUSES

Biology

Caliciviruses are single-stranded, positive-sense RNA viruses that are closely related to picornaviruses. Caliciviruses are in the family *Caliciviridae* and include four genera: "Norwalk-like viruses," "Sapporo-like viruses," *Vesivirus,* and *Lagovirus* (Berke et al., 1997; Green et al., 1998). The first two genera previously were known as human caliciviruses (HuCVs), because they had been recovered only from cases of acute gastroenteritis in humans. *Vesivirus* and *Lagovirus* caliciviruses were known as animal caliciviruses because they had been recovered only from animal sources. These distinctions have become invalid, because strains falling into the Norwalk-like and Sapporo-like genera have been found in diarrhea specimens from domestic animals (Liu et al., 1999; Dastjerdi et al., 1999) and human infections with *Vesivirus* strains have been proven (Smith et al., 1998). Hepatitis E virus previously was considered to be a calicivirus, but it now is excluded from the *Caliciviridae* (Berke et al., 1997). The Norwalk-like and Sapporo-like caliciviruses are the common causes of human illness, and the remaining discussion will be restricted to these genera.

Caliciviruses are small (30 to 35 nm in diameter), round, and structured when visualized by EM. Two morphologic types have been described: "typical" caliciviruses have a "Star of David" appearance and include the Sapporo-like, *Vesivirus,* and *Lagovirus* caliciviruses. "Small round structured viruses" are Norwalk-like caliciviruses. This morphologic distinction is not always clinically useful, because particle morphology frequently is indistinct in clinical samples. Cryo-EM has proven that typical and small round structured virus morphologies differ in the height of surface arches on the virion (Prasad et al., 1994a, 1994b).

The viral genome is 7.5 kb long, including a poly-A tail (Neill and Mengeling, 1988; Meyers et al., 1991b; Jiang et al., 1993b; Liu et al., 1995). A 2.5-kb subgenomic RNA also is encapsidated and is believed to be the transcript for capsid gene synthesis (Dunham et al., 1998; Meyers et al., 1991a). Caliciviruses encode a single capsid protein of 60 kDa (Jiang et al., 1990; Neill et al., 1991; Jiang, et al., 1992b). A smaller viral antigen (~30 kDa) has been found in stools of patients and in the baculovirus expression culture and likely is a cleavage product from the capsid protein (Jiang et al., 1992b). The calicivirus capsid contains 90 dimers of the capsid protein that form surface arches and a T = 3 icosahedral capsid morphology unique among human viral pathogens (Prasad et al., 1994a). Caliciviruses and picornaviruses are genetically related, in a single superfamily.

Caliciviruses are antigenically and genetically diverse (Ando et al., 1994, 1995; Jiang et al., 1995a, 1996a; Noel et al., 1997). Many prototype strains were identified by EM or

IEM in the 1970s and 1980s, and many more strains were characterized by RT-PCR and sequencing in the 1990s. Because the antigenic relationships among caliciviruses still are not clear, most of them have been named after the location of first discovery. For example, the prototype Norwalk virus was named from an outbreak of acute gastroenteritis in 1968 at an elementary school in Norwalk, Ohio (Adler and Zickl, 1969; Kapikian et al., 1972). Similar Norwalk-like strains include Hawaii virus (Wyatt et al., 1974; Dolin et al., 1975; Thornhill et al., 1977), Snow Mountain virus (Dolin et al., 1982), Montgomery County virus (Wyatt et al., 1974; Thornhill et al., 1977), Taunton agent (Caul et al., 1979; Caul and Appleton, 1982), and Otofuke agent (Taniguchi et al., 1979). Structure revealed by EM has not always been a clue to the correct virus family. For example, the "minireovirus" strain had surface morphology suggesting a small reovirus, but later it was found to be a Norwalk-like virus and named "Toronto virus" (Lew et al., 1994b; Leite et al., 1996).

After cloning of the Norwalk genome in 1990, a large number of strains were characterized by RT-PCR with primers based upon the Norwalk virus genomic sequence. Southampton virus (Lambden et al., 1993), Mexico virus (Jiang et al., 1995d), Desert Storm virus (Lew et al., 1994a), Lordsdale virus (Dingle et al., 1995), Bristol virus (Green et al., 1994), and Grimsby virus (Hale et al., 1999) are among the best characterized strains. Many other strains also have been characterized, and partial sequences of their genome can be found in the GenBank database. Sequence analysis of these new strains showed that they were related to the prototype Norwalk virus. Phylogenetic analysis showed that currently known Norwalk-like viruses can be divided into two genogroups, I and II, represented by the prototypes Norwalk virus and Snow Mountain virus, respectively (Wang et al., 1994; Jiang et al., 1995c). Strains within each genogroup can be further divided into several genetic clades. Clades in genogroup I are represented by Norwalk virus, Desert Storm virus, and Southampton virus, and clades in genogroup II are represented by Snow Mountain virus, Hawaii virus, Mexico virus, and Lordsdale virus(Jiang et al., 1995e). New clades are continually being identified, indicating that Norwalk-like viruses have extensive genetic diversity. The use of genogroups as a term will be supplanted once the biologic significance of different clades is apparent.

Sapporo calicivirus was first identified in 1982 from an outbreak of acute gastroenteritis in an orphanage in Sapporo, Japan. Full-length or partial genomic sequences of several Sapporo-like viruses have been described, including Manchester virus (Liu et al., 1995), Houston/86 (Matson et al., 1995b), Houston/90 (Jiang et al., 1997), Parkville virus (Noel et al., 1997), and London/92 (Jiang et al., 1997). Sapporo-like viruses include at least three genetic clades: one clade contains the prototype Sapporo virus, Manchester virus, and Houston/86; the second clade contains Houston/90 and Parkville virus; and the third clade contains the London/92 strain (Jiang et al., 1997). Sapporo-like viruses likely are more genetically diverse than currently recognized, because reagents for studying the Sapporo-like genus have been more difficult to produce than those for Norwalk-like viruses.

Unlike rotaviruses, EAs, and astroviruses, caliciviruses cannot be cultivated in the laboratory, so antigenic relationships have been determined mostly by binding of antibody (antigenic type) rather than neutralization (serotype). The antigenic relationships among caliciviruses remain unclear. Strains in different genera are antigenically distinct. Strains in different clades are antigenically distinct in antigen-detection ELISAs, but when acute- and convalescent-phase serum pairs are tested in antibody-detection assays, responses across clades occur (Jiang et al., 1996b). These findings indicate that, at least within genera, some (minor) common epitopes exist. A general trend has been observed using a combination of antigen and antibody detection assays: strains positive in an antigen detection assay share >95% sequence identity with each other in the RNA polymerase gene region of the calicivirus genome. Strains inducing antibody responses detected in an antibody detection assay share ≥85% sequence identity in the same genomic region. The biologic significance of these findings is uncertain.

Pathogenesis

Limited information on the pathophysiology of Norwalk-like and Sapporo-like viruses is available. Approximately 50% of volunteers administered Norwalk virus developed illness (Graham et al., 1994). The incubation period has a mean of 24 to 48 h, with a range of 4 to 77 h. Maximal excretion of virus in feces occurs at the onset of illness and shortly thereafter. The longest detection of virus infection is up to 2 weeks after onset of illness. Caliciviruses can be detected in stool specimens, and the amount of excreted virus correlates with the intensity of clinical symptoms. Although only half of volunteers developed illness, 90% of the volunteers excreted Norwalk virus in the stool when tested by RT-PCR and recombinant ELISAs; therefore, about 30 to 40% of the volunteers had subclinical infection.

Pathologic examination of the intestines of adult volunteers infected with Norwalk virus showed abnormal histologic findings in the mucosa of the proximal small bowel, including mucosal inflammation, absorptive cell abnormalities, villous shortening, crypt hypertrophy, and increased epithelial cell mitosis (Agus et al., 1973; Dolin et al., 1975). The gastric and colonic mucosae remain histologically normal. These abnormal findings persist for at least 4 days after clinical symptoms cease; however, the virus has not been detected within abnormal mucosal cells. Pathologic examination of patients following infection with Sapporo-like viruses is lacking. Clarification of these pathophysiologic issues may result from the recent detection of Norwalk-like and Sapporo-like calicivirus infections in domestic animals (Dastjerdi et al., 1999; Guo et al., 1999; Liu et al., 1999).

Immunity following Norwalk-like virus infections is believed to be short-lived. Volunteers can be reinfected by the same strains several months after first challenge with the viruses. Volunteer studies showed that pre-existing antibodies against Norwalk virus were a risk factor for infection; however, an association between the presence of serum antibody and resistance to HuCV gastroenteritis in children living in developing countries and Japan has been reported (Wyatt et al., 1974; Parrino et al., 1977; Baron et al., 1984; Nakata et al., 1985; Johnson et al., 1990; Graham et al., 1994). In one multiple-challenge study, volunteers exposed to Norwalk virus exhibited short-term immunity, and higher antibody levels were associated with protection after repeated exposure. Why fecal or serum antibody is a correlate of protection in some studies and a risk factor for illness in others is not understood. Norwalk-like and Sapporo-like caliciviruses now have been found to contain multiple antigenic types, and shared antigenic epitopes among different antigenic types exist, which may play a role in the immunity observations.

Clinical manifestations of calicivirus infection are known from volunteer studies or studies of large outbreaks of acute gastroenteritis. Caliciviruses cause diarrhea, vomiting, nausea, abdominal cramps, fever, and malaise; diarrhea and vomiting are most common (Adler and Zickl, 1969; Brondum et al., 1985; Graham et al., 1994). Stool specimens generally do not contain blood or mucus. Some studies suggest that vomiting is more frequent among children, while diarrhea is more common among adults (Kaplan et al., 1982a). The illness usually lasts 3 to 4 days and does not lead to chronic infection, except in severely immunocompromised hosts.

Epidemiology

Caliciviruses infect all age groups. Antibody prevalence rises sharply in the first 2 to 3 years of life. Significant additional increases in antibody prevalence occur during school years and in early adulthood. Seroprevalence and age-specific geometric mean titers in developing countries tend to be higher, at earlier ages, than in developed countries.

Caliciviruses have been detected by using RT-PCR detection of viral RNA in stool specimens in up to 20% of diarrhea episodes that occur at home or in patients that are seen in the physician office setting and in about 3% of hospital-associated acute gastroenteritis (X. Jiang, N. Wilton, W. M. Zhong, T. Farkas, E. Barrett, M. Guerrero, G. Ruiz-Palacios, K. Y. Green, A. D. Hale, M. K. Estes, L. K. Pickering, and D. O. Matson, invited presentation, Int. Workshop Hum. Caliciviruses, 1999; Pang et al., 1999). Norwalk-like viruses cause 50 to 60% of food- and water-associated outbreaks of acute gastroenteritis in North America (Curry et al., 1987; Bean et al., 1996; Dalton, 1997; Green, 1997; Fankhauser et al., 1998). Sapporo-like caliciviruses are detected less frequently, but it is uncertain whether these lower detection rates are related to inadequate reagents.

Norwalk-like viruses are transmitted mainly by fecal-oral contamination. Limited evidence suggests airborne transmission may occur. The major public health concern about Norwalk-like viruses has been their ability to cause large outbreaks of gastroenteritis. Such outbreaks usually have high attack rates and have occurred in schools, restaurants, summer camps, hospitals, and nursing homes and on ships (Adler and Zickl, 1969; Baron et al., 1982; Dolin et al., 1982; Kaplan et al., 1982b; Nakata et al., 1983; Brondum et al., 1985; Lew et al., 1994b; Akihara et al., 1996; CDC, 1997; Dalton, 1997; Ang, 1998; Fankhauser et al., 1998; Brugha et al., 1999). Exposure to a common source of virus, such as contaminated food or water, usually can be identified. Outbreaks resulting from consumption of uncooked shellfish are common. Some such outbreaks have been traced to infected food handlers (White et al., 1986; Parashar, 1998). In large outbreaks, initial infections result from a common source of exposure to the viruses. Later infections in the same outbreak result from person-to-person transmission. Large outbreaks frequently occur on ships, such as aircraft carriers, affecting hundreds of on-board personnel. Most shipboard outbreaks probably begin after new exposures (food, water, or person). Intense crowding and sharing of facilities in military ships probably intensify person-to-person spread. Norwalk-like viruses have been detected on environmental surfaces, which also could be an important source in spread of the diseases. Norwalk-like virus infections occur year-round, although they are more common in the winter season.

Epidemiologic patterns for Sapporo-like virus infections are similar to those for Norwalk-like virus infections, but

Sapporo-like viruses usually occur mostly in children, and outbreaks of Sapporo-like acute gastroenteritis in adults are unusual. Antibody prevalence studies indicate that virtually all children have been infected with Sapporo-like viruses by 5 years of age (Nakata et al., 1996). Studies conducted in child care centers have shown that calicivirus-associated diarrhea is widespread, though sporadic, and that there is a high proportion of asymptomatic infections (Matson et al., 1990b). Like Norwalk-like viruses, Sapporo-like virus infections occur year-round, but winter seasonal predominance has been suggested. Outbreaks with a high attack rate involving the elderly in nursing homes have been reported in Japan and England.

Limited information on the stability of caliciviruses is available. Norwalk virus is stable at $-70°C$ and is resistant to acid treatment (pH 2.7 for 3 h) and ether treatment, properties similar to those of other viruses causing diarrhea. The virus also may be resistant to heat treatment for 30 min at 60°C (Dolin et al., 1972). Fresh, unfrozen stool specimens best preserve particle morphology for EM examination. For detection of antigens and RNA by ELISAs and RT-PCR, stool specimens stored at 4°C or in a frozen condition can be used; however, a temperature at $<-20°C$ is recommended for long-term storage for these specimens. The fact that Norwalk-like viruses frequently cause large food- and waterborne outbreaks suggests that the viruses survive in food and water and possibly on environment surfaces.

Diagnosis

Caliciviruses are found in stool specimens. Norwalk-like viruses have been detected in vomitus (Greenberg et al., 1979), but such reports are unusual.

A number of methods for diagnosis of calicivirus infections have been developed. These methods rely upon direct visualization of virions, antibody reactions, or genomic amplification. Cell culture and animal models are not available. Recently developed ELISAs based upon baculovirus-derived capsid proteins are highly sensitive and specific and are useful for serologic studies (Jiang et al., 1992b, 1995b, 1995c; Graham et al., 1994). The most important advantage of these new assays is that unlimited reagents can be produced in the baculovirus expression system.

RT-PCR

RT-PCR is widely used for clinical and epidemiological studies and currently is the most sensitive method for virus detection (Jiang et al., 1992a, 1999; Moe et al., 1994; Wang et al., 1994; Green et al., 1995). Primers based upon conserved regions of calicivirus genomes (e.g., the RNA polymerase region), such as p289 and p290 (Jiang et al., 1999), permit detection of most known calicivirus strains. Sequencing of RT-PCR products allows tracing of individual strains. RT-PCR also has been used to detect the viruses in environmental specimens, including drinking water and contaminated foods (Beller et al., 1997; Gray et al., 1997; Schwab et al., 1998; Sunen and Sobsey, 1999).

Antigen-Detection ELISAs

Recently developed antigen-detection ELISAs utilize hyperimmune antisera to baculovirus-expressed capsid proteins that form virion-like particles (Jiang et al., 1992b). These particles are immunologically indistinguishable from native virions. The sensitivity and specificity of these assays are excellent; however, Norwalk-like viruses and Sapporo-like viruses are antigenically diverse, and multiple strains with different genetic identities cocirculate in all populations studied (Jiang et al., 1999b). Antigen-detection ELISAs detect only closely related strains (≥95% sequence identity in the RNA polymerase genomic region). Currently expressed Norwalk-like viral capsid proteins include those for Norwalk virus, Mexico virus, Toronto virus, Hawaii virus, Lordsdale virus, Southampton virus, and Grimsby virus (Dingle et al., 1995; Jiang et al., 1995c; Leite et al., 1996; Green et al., 1997; Hale et al., 1999; Pelosi et al., 1999). Application of assays for new antigenic types of Norwalk-like viruses has significantly increased detection rates in investigations of outbreaks and sporadic cases of acute gastroenteritis (Jiang et al., Int. Workshop Hum. Caliciviruses, 1999). A continued effort to identify new strains and express their capsid genes will lead to a more useful application of antigen-detection ELISAs in clinical and epidemiological studies of human calicivirus infections.

Monoclonal antibody-based ELISAs for detection of Norwalk viruses have been described (Treanor et al., 1988; Herrmann et al., 1995). One potential advantage of monoclonal antibody-based assays is broad test specificity if type-common epitopes can be found.

Antibody Detection ELISAs

The first method developed from baculovirus-expressed HuCV capsid proteins was to use the proteins to coat the microtiter plate directly for detection of antibody in serum specimens (Jiang et al., 1992b). Because the virion-like particles used for the coating were highly purified, the assay was extremely sensitive. This format initially was developed to measure total Ig in human sera. The same format later was adapted for detection of individual Igs (IgA, -G, and -M) (Monroe et al., 1993b; Treanor et al., 1993; Parker and Cubitt, 1994; Gray et al., 1994) by changing the conjugates specific to individual Ig isotypes. The assays for IgG or total Ig were widely used in serosurveys of HuCV infections in many countries (Numata et al., 1994; Parker et al., 1994; Dimitrov et al., 1997). The IgA and IgM ELISAs are believed to be more specific for diagnosis. The IgM ELISAs are particularly useful when paired acute- and convalescent-phase sera are not available. The Norwalk virus IgA ELISA also has been used to detect secretory IgA in fecal specimens (Okhuysen et al., 1995).

A capture format IgM ELISA was developed for recombinant Norwalk Snow Mountain viruses (Treanor et al., 1993; Gray et al., 1994; Brinker et al., 1998). In this format, an anti-human IgM antiserum was used to capture the IgM molecules from human sera. HuCV-specific IgM then was detected by addition of recombinant capsid antigens, followed by detection with monoclonal or hyperimmune antibodies against the recombinant capsid antigens. This capture IgM ELISA was designed to increase test sensitivity by removing excess IgG in serum samples. The monoclonal antibody-based IgM ELISAs have been reported to be more sensitive than the direct antigen-coating IgM ELISAs. The IgM detection ELISAs may distinguish the type of infecting viruses better than other assays.

EM and IEM

EM still is useful for detection of viruses in clinical specimens when RT-PCR and antigen and antibody detection ELISAs fail. EM also is useful to detect multiple pathogens in a single specimen. The sensitivity of EM is limited to approximately 10^6 particles per g of stool, and caliciviruses usually are excreted at lower concentrations than this during illness. EM detects only a minority of calicivirus infections. IEM permits enhanced detection of calicivirus particles (Kapikian et al., 1972).

ASTROVIRUSES

Biology

Human astroviruses are nonenveloped, single-stranded RNA viruses first identified in 1975 in diarrhea stool specimens from children by EM (Appleton and Higgins, 1975; Madeley and Cosgrove, 1975). The genomic organization is unique among positive-strand RNA viruses and warrants classification of astroviruses as a separate family, the *Astroviridae*, genus *Astrovirus* (Monroe et al., 1993a; Lee and Kurtz, 1994; Willcocks et al., 1995). Astroviruses are 28 nm in diameter by direct EM and have a distinctive five- or six-pointed star on the surface of the particle and a smooth particle edge (Fig. 2).

Eight human astrovirus antigenic types have been identified and adapted to growth in tissue culture (Kurtz and Lee, 1984; Herrmann et al., 1988; Willcocks et al., 1990; Jonassen et al., 1995; Noel et al., 1995). The astrovirus genome has three open reading frames (ORFs) in which ORF1a and ORF1b encode nonstructural proteins and ORF2 encodes the capsid precursor protein (Gibson et al., 1998; Jiang et al., 1993a; Lewis et al., 1994; Willcocks et al., 1994). The complete genomic sequences of types 1 and 2 and the capsid gene sequences of types 4, 5, 6, and 8 have been reported (Jiang et al., 1993a; Monroe et al., 1993a; Lee and Kurtz, 1994; Lewis et al., 1994; Willcocks et al., 1994, 1995).

Pathogenesis

Astroviruses have been detected in the low villous epithelium, in the surface epithelium of a flat biopsy, and in macrophages of the lamina propria (Phillips, 1988). In animals, astrovirus-infected cells showed vacuolation, which was followed by degeneration and cell death, which led to villous atrophy (Kurtz, 1988). The physiologic mechanisms impaired by this cytopathic effect are unknown. Short-term monosaccharide intolerance and prolonged cow's milk intolerance have been reported (Nazer et al., 1982), although these probably are uncommon events.

The incubation period is 1 to 4 days after exposure. Astrovirus illness lasts approximately 4 days and causes diarrhea (82%), vomiting (60%), abdominal distention (25%), and mild dehydration (6 to 20%), but the severity of illness usually is less than that of rotavirus illness (Nazer et al., 1982; Cruz et al., 1992; Guerrero et al., 1998). Illness incidence peaks in the winter (Grohmann, 1985).

Epidemiology

Human astrovirus infections occur worldwide. The incidence of astrovirus diarrhea ranges from 2% of children seeking medical care in Baltimore to 17% of children with persistent (>14 days) diarrhea in Bangladesh (Kotloff et al., 1992; Gaggero et al., 1998; Unicomb et al., 1998). Seroprevalence studies show that 60 to 90% of children are infected with astrovirus by 5 years of age (Kriston et al., 1996; Maldonado et al., 1998). Infection occurs mainly in children younger than 2 years of age and is frequently asymptomatic (Mitchell et al., 1993). Astroviruses are endemic pathogens, but diarrhea outbreaks associated with astrovirus have been reported in communities, schools, child care centers, geriatric care facilities, and hospitals (Kurtz et al., 1977; Ashley et al., 1978; Konno et al., 1982; Gray et al., 1987; Lewis et al., 1989; Herrmann et al., 1991; Leite et al., 1991; Moe et al., 1991; Stewien et al., 1991; Kotloff et al., 1992; Mitchell et al., 1993; Oishi et al., 1994; Utagawa et al., 1994; Mitchell et al., 1995b).

Astroviruses may account for 3 to 5% of hospital admissions for diarrhea in children in both developed and developing countries (Ellis et al., 1984; Shetty et al., 1995; Palombo and Bishop 1996; Unicomb et al., 1998). Other studies report that 1 to 10% of stool samples submitted to hospital laboratories contain astrovirus as detected by EM, ELISA, or RT-PCR (Kurtz et al., 1977; Esahli et al., 1991; Bates et al., 1993; Donelli et al., 1993; Gaggero et al., 1998; Marx et al., 1998; Shastri et al., 1998). These studies intermix community-acquired and nosocomial cases; 25 to 50% of astrovirus infections detected in these studies were from nosocomial infections (Esahli et al., 1991; Kotloff et al., 1992; Shastri et al., 1998). In addition, astrovirus caused a prolonged outbreak of diarrhea among immunocompromised patients in a pediatric bone marrow transplant unit, with excretion occurring for 60 to 90 days despite usual infection control measures (Cubitt et al., 1999).

Outbreaks of diarrhea caused by astrovirus in two geriatric care facilities have been reported. In these outbreaks, 20 to 80% of the patients were infected during the outbreak. These high attack rates suggest that the elderly were susceptible to infection due to waning immunity, high or common exposure in a crowded facility, or a combination of factors. Large outbreaks of foodborne astrovirus infection in Japan have involved thousands of children and adults (Oishi et al., 1994).

A seroprevalence survey conducted in 1978 demonstrated that group-specific antiastrovirus antibody was acquired by 75% of children by 5 to 10 years of age (Kurtz and Lee, 1978). In volunteer studies, 20 of 26 adult volunteers who ingested astrovirus developed group-specific antibody, determined by using IEM for antibody detection (Kurtz et al., 1979; Midthun et al., 1993). Among serum samples collected between 1965 and 1976 from children in Washington, D.C., antibody prevalence to astrovirus type 5 was 13% in children 6 months to 3 years of age and 41% in older children (Midthun et al., 1993). Antiastrovirus type 1 antibodies were identified in 90% of 253 children in London by 5 years of age and antiastrovirus type 6 antibodies were identified in 10 to 30% of all age groups older than 1 year of age by using recombinant astrovirus antigen in an immunofluorescence assay (Kriston et al., 1996). In children and adults in an age-stratified sample in the Netherlands, Koopmans and coworkers (Koopmans et al., 1998) detected neutralizing antibodies to type 1 (91%), type 3 (69%), type 4 (50%), type 5 (36%), and type 2 (31%) astroviruses. Mitchell and co-workers (Mitchell et al., 1999b) studied changes in prevalence of antiastrovirus type 1 and 3 antibodies over a 3-year period in a single community. Increasing prevalence of anti-type 3 antibody and decreasing prevalence of anti-type 1 antibody, especially of transplacental antibody, suggested major shifts in population exposure to the two types.

Transmission in children usually is person to person. Children younger than 36 months of age attending child care centers and homes are at a greater risk of diarrhea than children cared for at home (Bartlett et al., 1985). Infants younger than 1 year of age are at greater risk of acquiring astrovirus and were more likely to develop diarrhea than older children (Lew et al., 1991; Mitchell et al., 1993, 1995b, 1999c). Astrovirus infected 11 to 62% of infants and toddlers in a child care center during several outbreaks of diarrhea, with ELISA as the detection method. Detection by ELISA revealed an attack rate of 50%, whereas RT-PCR detected an attack rate of 89% among children exposed during one large outbreak (Mitchell et al., 1993, 1995b).

Astrovirus type 1 has been the prevalent type in the United Kingdom, accounting for over 65% of cases (Lee and Kurtz, 1994; Noel and Cubitt, 1994). Recent surveys indicate that other types are increasing in the United Kingdom, and this change in types may be associated with an increase in the number of outbreaks of illness (Lee and Kurtz, 1994; Willcocks et al., 1995). Multiple astrovirus types cocirculate wherever studied (Mitchell et al., 1995b; Noel et al., 1995; Guerrero et al., 1998).

Diagnosis

Methods used for detection of astrovirus significantly affect the estimation of the number of astrovirus infections. Initial studies of the incidence of astrovirus infection were based upon virus detection by EM. Currently available detection assays rely upon ELISA and RT-PCR designs.

Cell Culture

Astroviruses grow in human colonic carcinoma (CaCo-2), monkey kidney (LLC-MK2), and PLC/PRF/5 hepatoma cells in the presence of trypsin (Lee and Kurtz, 1981; Jonassen et al., 1995; Noel et al., 1995; Taylor et al., 1997). Astrovirus can be cultivated from fresh or frozen stool samples. Some investigators have used tissue culture to amplify virus prior to detection by other methods (see below) (Pinto et al., 1994).

Immunoassays

Production of a group-specific monoclonal antibody (8E7) and its use in combination with rabbit polyclonal antisera enabled development of ELISAs for detection of astrovirus antigens (Herrmann et al., 1988, 1990; Moe et al., 1991; Mitchell et al., 1993). Availability of the antigen-detection ELISA facilitated epidemiologic studies, demonstrating that astroviruses are a significant cause of diarrhea in many regions of the world, with infection rates significantly higher (2 to 9%) than those determined in surveys utilizing EM as the detection method (≤3%). A commercial antigen detection ELISA which uses the same monoclonal antibody is available in Europe (Table 5). Noel and coworkers (Noel et al., 1995) utilized hyperimmune antisera as the capture antibody and 8E7 as the detector antibody in a typing assay. The ELISAs are suitable for detection of astroviruses in outbreak and sporadic-diarrhea stool specimens. ELISA sensitivity is considerably less than that of RT-PCR for the detection of astrovirus in samples from patients with asymptomatic infections, although the greater sensitivity of RT-PCR has not been quantified (Mitchell et al., 1995b).

RT-PCR

The human astrovirus genome has both type-specific and type-common regions. An RT-PCR method for detection of type-common astrovirus RNA in stool specimens used primers within conserved sequence at the 3′ terminus of the genome (Jonassen et al., 1995). This assay was more sensitive than an ELISA for detection of astrovirus in stool specimens. RT-PCR has been applied for group- or type-specific astrovirus detection (Jonassen et al., 1993, 1995; Willcocks et al., 1994, 1995; Mitchell et al., 1995a, 1995b; Noel et al., 1995; Saito et al., 1995; Matsui et al., 1998). RT-PCR detection of astrovirus in stool samples among children in closed populations, such as child care centers, is an even more sensitive (3-fold) detection method than the ELISAs, but RT-PCR is somewhat cumbersome and has not been applied to large community-based studies. RT-PCR was used by Marx and coworkers (Marx et al., in press) to detect astrovirus in environmental water samples, thereby providing a sensitive method to further define possible reservoirs of infection.

EM and IEM

Astroviruses usually are detected by EM. Only a few virions on an EM grid have the typical star morphology (Madely and Cosgrove, 1975; Kurtz et al., 1977; Madeley and Cosgrove, 1975; Madeley et al., 1978; Madeley, 1979; Oliver and Phillips, 1988; Monroe et al., 1991). IEM enhances astrovirus detection but may require homotypic antisera for optimal application.

RT-PCR Typing

Polyclonal antisera to astrovirus types 1 through 8 raised in rabbits have served as the reference reagents throughout the world (Kurtz and Lee, 1984). In some studies, RT-PCR combined with RNA sequencing has been used to identify types (Noel et al., 1995; Belliot et al., 1997; Mitchell et al., 1999c). Belliot and coworkers (Noel et al., 1995; Belliot et al., 1997) compared sequence information from ORF1a, ORF1b, and the 5′ end of ORF2 and found sufficient sequence variability for type differentiation but probably insufficient variability to distinguish strains within the same type. Matsui and coworkers (Matsui et al., 1998) designed type-specific RT-PCR primers to determine the types of 35 isolates. Both sequityping and genotyping correlated with polyclonal ELISA typing. Preliminary findings indicated that recombination events may occur in human astrovirus strains that result in discrepant results between the nonstructural protein regions and structural protein regions of the genome.

Serology

Assays for antibody to astroviruses are in development. A neutralization assay also has been used to detect type-specific antibodies (Koopmans et al., 1998). The most recent seroprevalence studies have used type-specific capsid proteins produced in a baculovirus system as recombinant antigen in a microimmunofluorescence assay or ELISA to detect human antiastrovirus antibody (Kriston et al., 1996; Mitchell et al., 1999a). Comparable assays for all astrovirus types will be useful for further definition of the epidemiology of astroviruses.

OTHER VIRUSES

A number of different viruses have been identified in stool samples from humans with acute gastroenteritis, but their role in causing illness in humans remains unclear.

Coronaviruses are 180 to 200 nm in diameter and have characterisitic petal-shaped surface projections that give the appearance of a "solar corona" when visualized by EM (Fig. 2) (Caul et al., 1975; Garwes, 1982). These viruses are well documented as a cause of gastroenteritis in animals (Garwes, 1982). In humans, coronaviruses cause the common cold. They have been found in stools of children with gastroenteritis, in newborns with necrotizing enterocolitis, and in the intestinal epithelium of a 13 year-old boy who died of diarrhea (Vaucher et al., 1982; Ettig and Altschuler, 1985; Gerna et al., 1985). They also have been identified in stool specimens from homosexuals and from people without diarhea who live in crowded areas with a water supply that received inadequate treatment (Ashley and Caul, 1988). EM examination of samples suspected to contain human enteric coronaviruses must be interpreted with caution, because pieces of sloughed intestinal epithelium may appear as coronavirus-like particle forms.

Parvoviruses and parvovirus-like agents are small (20 to 30 nm), featureless, round particles (Fig. 2) (Caul and Appleton, 1982). Studies from Norway and England report this virus in stool specimens as the sole agent in a few children with gastroenteritis (Kjeldsberg, 1977; Oliver and Phillips, 1988).

Pestiviruses are 40 to 60 nm in diameter and belong to the *Togaviridae*. A monoclonal antibody-based ELISA detected this virus in stool specimens from 30 of 128 children with diarrhea of unknown etiology, in 1 of 28 children without diarrhea, and in 1 of 31 children with rotavirus diarrhea (Yolken et al., 1989). The diarrhea was mild and accompanied by respiratory tract symptoms.

In 1984, a virus resembling the Breda virus of calf diarrhea was detected in stool specimens from 20 individuals with gastroenteritis (Fig. 2) (Beards et al., 1984). A case of hemorrhagic enterocolitis associated with Breda virus infection has also been reported (Lacombe et al., 1988). Toroviruses now have been detected in sporadic and nosocomial cases of gastroenteritis in children (Koopmans et al., 1997; Jamieson et al., 1998).

UNUSUAL FEATURES, INSIGHTS, AND FUTURE DIRECTIONS

When one assesses the prevalence of viral gastroenteritis pathogens in cross-sectional studies, whether of outpatient or inpatient illness, one finds these viruses to be as common in the developed world as in the developing world. This is in contrast to findings from similar studies for bacterial and parasitic gastroenteritis pathogens, which are far more prevalent in the developing world. These and other findings indicate that the viral gastroenteritis pathogens are resistant to the water, sewage, and food control measures that have so markedly reduced the incidence of bacterial and parasitic gastroenteritis in the developing world. These findings have compelled consideration of vaccines as the most likely successful preventive strategy for viral gastroenteritis.

A feature of immunity against viral gastroenteritis pathogen infection and illness is that it takes multiple infections to induce protection against severe disease. For rotaviruses, this has been determined in longitudinal studies (Velazquez et al., 1996), in which two infections were required to induce complete protection against severe disease. In addition, protective immunity is never complete in adults, even after multiple infections. The same pattern of immunity appears to govern the response to infections with the other viral gastroenteritis pathogens and likely is a common feature of mucosal immunity. The consequences of this immunity pattern are that severe infections can occur over the first several years of life in infants and that contagious infections can occur in all ages.

Movement of viral gastroenteritis pathogens between humans and animals is suspected for all of the known and suspected causes of illness in humans. This movement is proven for rotaviruses and is well supported for caliciviruses and toroviruses. This movement likely is a common feature of the epidemiology of the viral gastroenteritis pathogens, and its existence complicates prevention strategies, for potential reservoirs for human infection likely are very extensive.

TREATMENT AND PREVENTION

Currently, there is no specific treatment recommended for virus-associated gastroenteritis. As for other causes of acute gastroenteritis, the cornerstone of treatment is prevention of dehydration by appropriate use of oral rehydration solutions (Stephen, 1988; CDC, 1992; American Academy of Pediatrics, 1996). Intake of a regular diet should be limited only according to tolerance and for a limited period of time. Use of lactose-free formula should be reserved for the few cases with documented carbohydrate malabsorption. Use of antiviral antibodies (hyperimmune antisera, colostrum, antibody-supplemented formula, or human serum immunoglobulin) has been associated with im-provement of rotavirus gastroenteritis and protection against infection in both animal and human studies in some studies but not in others (Guarino et al., 1991; Brunser et al., 1992; Ebina et al., 1992). The role of artificial passive protection as a potential candidate for wide clinical use seems distant. Rotavirus infections can be prevented by vaccination; protection against severe rotavirus infection has ranged from 70 to more than 90% (Kapikian and Chanock, 1996; Santosham et al., 1997; Pérez-Schael et al., 1997; Joensu et al., 1997). The hope for rotavirus vaccines has been clouded by the recent withdrawal of the first rotavirus vaccine because of a suspected association with intussusception (CDC, 1999). Standard prevention methods, including contact isolation procedures in hospitals, specified diaper changing areas in child care centers, and good hand washing techniques, are required to decrease risk of transmission.

CONCLUSIONS

Enteric viruses currently are recognized as important etiologic agents of acute diarrhea. For some of these agents (rotaviruses, EA, caliciviruses, and astroviruses), the association with disease has been clearly established, while for others (coronaviruses, parvoviruses, pestiviruses, and Breda viruses), the association is less clear and is often based upon single outbreaks. Future epidemiologic studies are required to clarify this issue. Knowledge of the molecular biology of these viruses is increasing and will be helpful in expanding the tools available to detect these enteric viruses. As more knowledge is gained at both the epidemiologic and basic molecular levels, new viruses causing enteric infections likely will be discovered and a clearer understanding of the relationship among the currently recognized viruses will be forthcoming.

This work was supported in part by National Institutes of Health grants HD 13021 and AI 37093.

REFERENCES

Adler, I., and R. Zickl. 1969. Winter vomiting disease. *J. Infect. Dis.* **119:**668–673.

Agus, S. G., R. Dolin, R. G. Wyatt, A. J. Tousimis, and R. S. Northrup. 1973. Acute infectious nonbacterial gastroenteritis: intestinal histopathology. Histologic and enzymatic alterations during illness produced by the Norwalk agent in man. *Ann. Intern. Med.* **79:**18–25.

Ahluwalia, G. S., T. H. Scott-Taylor, B. G. Klisko, and W. Hammond. 1994. Comparison of detection methods for adenovirus from enteric clinical specimens. *Diagn. Microbiol. Infect. Dis.* **18:**161–166.

Akihara, S., M. Nakayama, J. Kakizawa, K. Bosu, K. Izumi, and H. Ushijima. 1996. An outbreak of Norwalk-like virus infection in Tokyo and Saitama in late 1995. *Kansenshogaku Zasshi* **70:**840–841.

Allard, A., A. Kajon, and G. Wadell. 1994. Simple procedure for discrimination and typing of enteric adenoviruses after detection by polymerase chain reaction. *J. Med. Virol.* **44:**250–257.

American Academy of Pediatrics, Provisional Committee on Quality Improvement, Subcommittee on Acute Gastroenteritis. 1996. Practice parameter: the management of acute gastroenteritis in young children. *Pediatrics* **97:**424–435.

Ando, T., S. S. Monroe, J. R. Gentsch, Q. Jin, D. C. Lewis, and R. I. Glass. 1995. Detection and differentiation of antigenically distinct small round-structured viruses (Norwalk-like viruses) by reverse transcription-PCR and Southern hybridization. *J. Clin. Microbiol.* **33:**64–71.

Ando, T., M. N. Mulders, D. C. Lewis, M. K. Estes, S. S. Monroe, and R. I. Glass. 1994. Comparison of the polymerase region of small round structured virus strains previously classified in three antigenic types by solid-phase immune electron microscopy. *Arch. Virol.* **135:**217–226.

Ang, L. H. 1998. An outbreak of viral gastroenteritis associated with eating raw oysters. *Commun. Dis. Public Health* **1:**38–40.

Appleton, H., and P. G. Higgins. 1975. Viruses and gastroenteritis in infants. *Lancet* **i:**1297–1230.

Ashley, C., and O. E. Caul. 1988. Human enteric coronaviruses, p. 91–95. *In* M. Farthing (ed.), *Viruses in the Gut.* Smith Kline & French, Welwyn Garden City, United Kingdom.

Ashley, C. R., E. O. Caul, and W. K. Paver. 1978. Astrovirus-associated gastroenteritis in children. *J. Clin. Pathol.* **31:**939–943.

Ball, J. M., P. Tian, C. Q. Zeng, A. P. Morris, and M. K. Estes. 1996. Age-dependent diarrhea induced by a rotaviral nonstructural glycoprotein. *Science* **272:**101–102.

Baron, R. C., H. B. Greenberg, G. Cukor, and N. R. Blacklow. 1984. Serological responses among teenagers after natural exposure to Norwalk virus. *J. Infect. Dis.* **150:**531–534.

Baron, R. C., F. D. Murphy, H. B. Greenberg, C. E. Davis, D. J. Bregman, G. W. Gary, J. M. Hughes, and L. B. Schonberger. 1982. Norwalk gastrointestinal illness: an outbreak associated with swimming in a recreational lake and secondary person-to-person transmission. *Am. J. Epidemiol.* **115:**163–172.

Bartlett, A. V., M. Moore, and G. W. Gary. 1985. Diarrheal illness among infants and toddlers in day care centers. II. Comparison with day care homes and households. *J. Pediatr.* **107:**503–509.

Bartlett, A. V., R. R. Reeves, and L. K. Pickering. 1988. Rotavirus in infant-toddler day care centers: epidemiology relevant to disease control strategies. *J. Pediatr.* **113:**435–441.

Bates, P. R., A. S. Bailey, D. J. Wood, D. J. Morris, and J. M. Couriel. 1993. Comparative epidemiology of rotavirus, sub-

genus F (types 40 and 41) adenovirus and astrovirus gastroenteritis in children. *J. Med. Virol.* **39:**224–228.

Bean, N. H., J. S. Goulding, C. Lao, and F. J. Angulo. 1996. Surveillance for foodborne-disease outbreaks—United States, 1988–1992. *Morb. Mortal. Wkly. Rep.* **45:**1–73.

Beards, G. M., J. G. Green, C. Hall, and T. H. Flewett. 1984. An enveloped virus in stools of children and adults with gastroenteritis that resembles the Breda virus of calves. *Lancet* **i:**1050–1052.

Bell, E. J., and N. R. Grist. 1968. Viruses in diarrheal disease. *Br. Med. J.* **4:**741–742.

Beller, M., A. Ellis, S. H. Lee, M. A. Drebot, S. A. Jenkerson, E. Funk, M. D. Sobsey, O. D. Simmons, 3rd, S. S. Monroe, T. Ando, J. Noel, M. Petric, J. P. Middaugh, and J. S. Spika. 1997. Outbreak of viral gastroenteritis due to a contaminated well. International consequences. *JAMA* **278:**563–568.

Belliot, G., H. Laveran, and S. S. Monroe. 1997. Detection and genetic differentiation of human astroviruses: phylogenetic grouping varies by coding region. *Arch. Virol.* **142:**1323–1334.

Berke, T., B. Golding, X. Jiang, D. W. Cubitt, M. Wolfaardt, A. W. Smith, and D. O. Matson. 1997. Phylogenetic analysis of the caliciviruses. *J. Med. Virol.* **52:**419–424.

Bern, C., and R. I. Glass. 1994. Impact of diarrheal disease worldwide, p. 1–26. *In* A. Z. Kapikian (ed.), *Viral Infections of the Gastrointestinal Tract,* 2nd ed. Marcel Dekker, New York, N.Y.

Birch, C. J., N. I. Lehmann, A. J. Hawker, J. A. Marshall, and I. D. Gust. 1979. Comparison of electron microscopy, enzyme-linked immunosorbent assay, solid-phase radioimmunoassay, and indirect immunofluorescence for detection of human rotavirus antigen in faeces. *J. Clin. Pathol.* **32:**700–705.

Birch, C. J., S. M. Rodger, J. A. Marshall, and I. D. Gust. 1983. Replication of human rotavirus in cell culture. *J. Med. Virol.* **11:**241–250.

Bishop, R. F., G. L. Barnes, E. Cipriani, and J. S. Lund. 1983. Clinical immunity after neonatal rotavirus infection. A prospective longitudinal study in young children. *N. Engl. J. Med.* **309:**72–76.

Bishop, R. F., G. P. Davidson, I. H. Holmes, and B. J. Ruck. 1973. Virus particles in epithelial cells of duodenal mucosa from children with acute nonbacterial gastroenteritis. *Lancet* **ii:**1281–1283.

Blacklow, N. R., and G. Cukor. 1982. Viruses and gastrointestinal disease, p. 75–87. *In* D. A. J. Tyrrell and A. Z. Kapikian (ed.), *Virus Infections of the Gastrointestinal Tract.* Marcel Dekker, New York, N.Y.

Bon, F., P. Fascia, M. Dauvergne, D. Tenenbaum, H. Planson, A. M. Petion, P. Pothier, and E. Kohli. 1999. Prevalence of group A rotavirus, human calicivirus, astrovirus, and adenovirus type 40 and 41 infections among children with acute gastroenteritis in Dijon, France. *J. Clin. Microbiol.* **37:**3055–3058.

Brade, L., W. A. Schmidt, and I. Gattert. 1981. [Relative effectiveness of disinfectants against rotaviruses (author's translation)]. Zur relativen Wirksamkeit von Desinfektionsmitteln gegenüber Rotaviren. *Zentbl. Bakteriol. Mikrobiol. Hyg.* **174:**151–159.

Brandt, C. D., H. W. Kim, W. J. Rodriguez, J. O. Arrobio, B. C. Jeffries, E. P. Stallings, C. Lewis, A. J. Miles, R. M. Chanock, A. Z. Kapikian, and R. H. Parrott. 1983. Pediatric viral gastroenteritis during eight years of study. *J. Clin. Microbiol.* **18:**71–78.

Brandt, C. D., H. W. Kim, W. J. Rodriguez, J. O. Arrobio, B. C. Jeffries, E. P. Stallings, C. Lewis, A. J. Miles, M. K. Gardner, and R. H. Parrott. 1985. Adenovirus and pediatric gastroenteritis. *J. Infect. Dis.* **151:**437–443.

Brandt, C. D., H. W. Kim, W. J. Rodriguez, L. Thomas, R. H. Yolken, J. O. Arrobio, A. Z. Kapikian, R. H. Parrott, and

R. M. Chanock. 1981. Comparison of direct electron microscopy, immune electron microscopy, and rotavirus enzyme-linked immunosorbent assay for detection of gastroenteritis viruses in children. *J. Clin. Microbiol.* **13:**976–981.

Bridger, J. C. 1988. Non-group A rotavirus, p. 79–82. In M. Farthing (ed.), *Viruses in the Gut.* Smith Kline & French, Welwyn Garden City, United Kingdom.

Bridger, J. C., S. Pedley, and M. A. McCrae. 1986. Group C rotaviruses in humans. *J. Clin. Microbiol.* **23:**760–763.

Brinker, J. P., N. R. Blacklow, M. K. Estes, C. L. Moe, K. J. Schwab, and J. E. Herrmann. 1998. Detection of Norwalk virus and other genogroup 1 human caliciviruses by a monoclonal antibody, recombinant-antigen-based immunoglobulin M capture enzyme immunoassay. *J. Clin. Microbiol.* **36:**1064–1069.

Brondum, J., K. C. Spitalny, R. L. Vogt, K. Godlewski, H. P. Madore, and R. Dolin. 1985. Snow Mountain agent associated with an outbreak of gastroenteritis in Vermont. *J. Infect. Dis.* **152:**834–837.

Brown, D. W., G. M. Beards, G. M. Chen, and T. H. Flewett. 1987. Prevalence of antibody to group B (atypical) rotavirus in humans and animals. *J. Clin. Microbiol.* **25:**316–319.

Brugha, R., I. B. Vipond, M. R. Evans, Q. D. Sandifer, R. J. Roberts, R. L. Salmon, E. O. Caul, and A. K. Mukerjee. 1999. A community outbreak of food-borne small round-structured virus gastroenteritis caused by a contaminated water supply. *Epidemiol. Infect.* **122:**145–154.

Brunser, O., J. Espinoza, G. Figueroa, M. Araya, E. Spencer, H. Hilpert, H. Link-Amster, and H. Brüssow. 1992. Field trial of an infant formula containing anti-rotavirus and anti-*Escherichia coli* milk antibodies from hyperimmunized cows. *J. Pediatr. Gastr. Nutr.* **15:**63–72.

Bryden, A. S., M. E. Thouless, C. J. Hall, T. H. Flewett, B. A. Wharton, P. M. Mathew, and I. Craig. 1982. Rotavirus infections in a special-care baby unit. *J. Infect. Dis.* **4:**43–48.

Butler, M., and M. S. Harakeh. 1983. Inactivation of rotavirus in wastewater effluents by chemical disinfection, p. 282–289. In M. Butler, A. R. Medlen, and R. Morris (ed.), *Viruses and Disinfection of Water and Wastewater.* University of Surrey, Guildford, United Kingdom.

Caeiro, J. P., J. J. Mathewson, M. A. Smith, Z. D. Jiang, M. A. Kaplan, and H. L. DuPont. 1999. Etiology of outpatient pediatric nondysenteric diarrhea: a multicenter study in the United States. *Pediatr. Infect. Dis. J.* **18:**94–97.

Caul, E., J. Ashley, M. Darville, and J. Bridger. 1990. Group C rotavirus associated with fatal enteritis in a family outbreak. *J. Med. Virol.* **30:**201–205.

Caul, E. O., and H. Appleton. 1982. The electron microscopical and physical characteristics of small round human fecal viruses: an interim scheme for classification. *J. Med. Virol.* **9:**257–265.

Caul, E. O., C. Ashley, and J. V. Pether. 1979. "Norwalk"-like particles in the epidemic gastroenteritis in the U.K. *Lancet* **ii:**1292.

Caul, E. O., W. K. Paver, and S. K. R. Clark. 1975. Coronavirus particles in faeces from patients with gastroenteritis. *Lancet* **i:**1972.

Centers for Disease Control and Prevention. 1992. The management of acute diarrhea in children: oral rehydration, maintenance, and nutritional therapy. *Morb. Mortal. Wkly. Rep.* **41:**1–20.

Centers for Disease Control and Prevention. 1997. Outbreak of cyclosporiasis—Northern Virginia-Washington, D.C.-Baltimore, Maryland, metropolitan area, 1997. *Morb. Mortal. Wkly. Rep.* **46:**689–691.

Centers for Disease Control and Prevention. 1999. Intussusception among recipients of rotavirus vaccine—United States 1998–1999. *Morb. Mortal. Wkly. Rep.* **48:**577–581.

Chiba, S., I. Nakamura, S. Urasawa, S. Nakata, K. Taniguchi, K. Funinago, and T. Nakao. 1983. Outbreak of infantile gastroenteritis due to type 40 adenovirus. *Lancet* **i:**954–957.

Chiba, S., S. Nakata, and S. Urasawa. 1986. Protective effect of naturally acquired homotypic and heterotypic antibodies. *Lancet* **i:**417–421.

Chrystie, I. L, B. M. Totterdell, and J. E. Banatvala. 1978. Asymptomatic endemic rotavirus infections in the newborn. *Lancet* **i:**1176–1178.

Cook, S. M., R. I. Glass, C. W. LeBaron, and M. S. Ho. 1990. Global seasonality of rotavirus infections. *Bull. W. H. O.* **68:**171–177.

Coulson, B. S., K. Grimwood, P. J. Masendycz, J. S. Lund, N. Mermelstein, R. F. Bishop, and G. L. Barnes. 1990. Comparison of rotavirus immunoglobulin A coproconversion with other indices of rotavirus infection in a longitudinal study in children. *J. Clin. Microbiol.* **28:**1367–1374.

Cruz, J. R., A. V. Bartlett, J. E. Herrmann, P. Caceres, N. R. Blacklow, and F. Cano. 1992. Astrovirus-associated diarrhea among Guatemalan ambulatory rural children. *J. Clin. Microbiol.* **30:**1140–1144.

Cruz, J. R., P. Caceres, F. Cano, J. Flores, A. Bartlett, and B. Toron. 1990. Adenovirus type 40 and 41 and rotaviruses associated with diarrhea in children from Guatemala. *J. Clin. Microbiol.* **28:**1780–1784.

Cubitt, W. D., D. K. Mitchell, M. J. Carter, M. M. Willcocks, and H. Holzel. 1999. Application of electronmicroscopy, enzyme immunoassay, and RT-PCR to monitor an outbreak of astrovirus type 1 in a paediatric bone marrow transplant unit. *J. Med. Virol.* **57:**313–321.

Curry, A., T. Riordan, J. Craske, and E. O. Caul. 1987. Small round structured viruses and persistence of infectivity in food handlers. *Lancet* **ii:**864–865.

Dagan, R., Y. Bar-David, B. Sarov, M. Katz, I. Kassis, D. Greenberg, R. I. Glass, C. Z. Margolis, and I. Sarov. 1990. Rotavirus diarrhea in Jewish and Bedouin children in the Negev region of Israel: epidemiology, clinical aspects and possible role of malnutrition in severity of illness. *Pediatr. Infect. Dis. J.* **9:**314–321.

Dalton, C. 1997. An outbreak of Norwalk virus gastroenteritis following consumption of oysters. *Commun. Dis. Intell.* **21:**321–322.

Dastjerdi, A. M., J. Green, C. I. Gallimore, D. W. Brown, and J. C. Bridger. 1999. The bovine Newbury agent-2 is genetically more closely related to human SRSVs than to animal caliciviruses. *Virology* **254:**1–5.

Davidson, G. P., R. F. Bishop, R. R. Townley, and I. H. Holmes. 1975. Importance of a new virus in acute sporadic enteritis in children. *Lancet* **i:**242–246.

Davidson, G. P., R. J. Hogg, and C. P. Kirubakaran. 1983. Serum and intestinal immune response to rotavirus enteritis in children. *Infect. Immun.* **40:**447–452.

Dennehy, P. H., M. Hartin, S. M. Nelson, and S. F. Reising. 1999. Evaluation of the ImmunoCardSTAT! rotavirus assay for detection of group A rotavirus in fecal specimens. *J. Clin. Microbiol.* **37:**1977–1979.

Desselberger, U. 1988. Molecular epidemiology of rotavirus, p. 55–64. In M. Farthing (ed.), *Viruses in the Gut.* Smith Kline & French, Welwyn Garden City, United Kingdom.

DeZoysa, I., and R. G. Feachem. 1985. Interventions for the control of diarrhoeal diseases among young children: rotavirus and cholera immunization. *Bull. W. H. O.* **63:**569–583.

Dimitrov, D. H., S. A. Dashti, J. M. Ball, E. Bishbishi, K. Alsaeid, X. Jiang, and M. K. Estes. 1997. Prevalence of antibodies to human caliciviruses (HuCVs) in Kuwait established by ELISA using baculovirus-expressed capsid antigens representing two genogroups of HuCVs. *J. Med. Virol.* **51:**115–118.

Dingle, K. E., P. R. Lambden, E. O. Caul, and I. N. Clarke. 1995. Human enteric Caliciviridae: the complete genome sequence and expression of virus-like particles from a genetic group II small round structured virus. *J. Gen. Virol.* **76:**2349–2355.

Dolin, R., N. R. Blacklow, H. DuPont, R. F. Buscho, R. G. Wyatt, J. A. Kasel, R. Hornick, and R. M. Chanock. 1972. Biological properties of Norwalk agent of acute infectious nonbacterial gastroenteritis. *Proc. Soc. Exp. Biol. Med.* **140:**578–583.

Dolin, R., A. G. Levy, R. G. Wyatt, T. S. Thornhill, and J. D. Gardner. 1975. Viral gastroenteritis induced by the Hawaii agent. Jejunal histopathology and serologic response. *Am. J. Med.* **59:**761–768.

Dolin, R., R. C. Reichman, K. D. Roessner, T. S. Tralka, R. T. Schooley, W. Gary, and D. Morens. 1982. Detection by immune electron microscopy of the Snow Mountain agent of acute viral gastroenteritis. *J. Infect. Dis.* **146:**184–189.

Donelli, G., F. Superti, A. Tinari, M. L. Marziano, D. Caione, C. Concato, and D. Menichella. 1993. Viral childhood diarrhoea in Rome: a diagnostic and epidemiological study. *New Microbiol.* **16:**215–225.

Dubois, E., F. Le Guyader, L. Haugarreau, H. Kopecka, M. Cormier, and M. Pommepuy. 1997. Molecular epidemiological survey of rotaviruses in sewage by reverse transcriptase seminested PCR and restriction fragment length polymorphism assay. *Appl. Environ. Microbiol.* **63:**1794–1800.

Dunham, D. M., X. Jiang, T. Berke, A. W. Smith, and D. O. Matson. 1998. Genomic mapping of a calicivirus VPg. *Arch. Virol.* **143:**2421–2430.

Durepaire, N., S. Ranger-Rogez, and F. Denis. 1996. Evaluation of rapid culture centrifugation method for adenovirus detection in stools. *Diagn. Microbiol. Infect. Dis.* **24:**25–29.

Ebina, T., M. Ohta, Y. Kanamura, Y. Yamamoto-Osumi, and K. Baba. 1992. Passive immunizations of suckling mice and infants with bovine colostrum containing antibodies to human rotavirus. *J. Med. Virol.* **38:**117–123.

Echeverria, P., N. R. Blacklow, G. C. Cukor, S. Vibulband-hitkit, S. Changchawalit, and P. Boonthai. 1983. Rotavirus as a cause of severe gastroenteritis in adults. *J. Clin. Microbiol.* **18:**663–667.

Eiden, J., A. Mouzinho, D. Lindsay, R. Glass, Z. Ying-Fang, and J. Lin Taylor. 1994. Serum antibody response to recombinant major inner capsid protein following human infection with group B rotavirus. *J. Clin. Microbiol.* **32:**1599–1603.

Ellis, M. E., B. Watson, B. K. Mandal, E. M. Dunbar, J. Craske, A. Curry, J. Roberts, and J. Lomax. 1984. Microorganisms in gastroenteritis. *Arch. Dis. Child.* **59:**848–855.

Esahli, H., K. Breback, R. Bennet, A. Ehrnst, M. Eriksson, and K. O. Hedlund. 1991. Astroviruses as a cause of nosocomial outbreaks of infant diarrhea. *Pediatr. Infect. Dis. J.* **10:**511–515.

Espejo, R. T., F. Puerto, C. Soler, and N. Gonzalez. 1984. Characterization of a human pararotavirus. *Infect. Immun.* **44:**112–116.

Estes, M. K. 1996. Rotaviruses and their replication, p. 1625–1673. *In* B. N. Fields, D. M. Knipe, and P. M. Howley (ed.), *Virology*, 3rd ed. Raven Press, New York, N.Y.

Estes, M. K., and J. Cohen. 1989. Rotavirus gene structure and function. *Microbiol. Rev.* **53:**410–449.

Estes, M. K., and D. Y. Graham. 1988. Viral infections of the intestine, p. 566–578. *In* G. Gitnick (ed.), *Principles and Practice of Gastroenterology and Hepatology.* Elsevier Science Publishing, New York, N.Y.

Estes, M. K., E. L. Palmer, and J. F. Obijeski. 1983. Rotaviruses: a review. *Curr. Top. Microbiol. Immunol.* **105:**123–184.

Ettig, P. J., and G. P. Altschuler. 1985. Fatal gastroenteritis associated with coronavirus-like particles. *J. Dis. Child.* **139:**245–248.

Fankhauser, R. L., J. S. Noel, S. S. Monroe, T. Ando, and R. I. Glass. 1998. Molecular epidemiology of "Norwalk-like viruses" in outbreaks of gastroenteritis in the United States. *J. Infect. Dis.* **178:**1571–1578.

Flewett, T. H. 1983. Rotavirus in the home and hospital nursery. *Br. Med. J.* **287:**568–569.

Foster, S. O., E. L. Palmer, G. W. Gary, M. L. Martin, K. L. Herrmann, P. Beasley, and J. Sampson. 1980. Gastroenteritis due to rotavirus in an isolated Pacific island group: an epidemic of 3,439 cases. *J. Infect. Dis.* **141:**32–39.

Gaggero, A., M. O'Ryan, J. S. Noel, R. I. Glass, S. S. Monroe, N. Mamani, V. Prado, and L. F. Avendano. 1998. Prevalence of astrovirus infection among Chilean children with acute gastroenteritis. *J. Clin. Microbiol.* **36:**3691–3693.

Garbarg-Chenon, A., J. Brussieux, A. Boisivon, J. C. Nicolas, and F. Bricout. 1985. Epidemiology of human rotaviruses in a maternity unit as studied by electrophoresis of genomic RNA. *Eur. J. Epidemiol.* **1:**33–36.

Garwes, D. J. 1982. Coronaviruses in animals, p. 315–359. *In* D. A. J. Tyrrell and A. Z. Kapikian (ed.), *Virus Infections of the Gastrointestinal Tract.* Marcel Dekker, New York, N.Y.

Gentsch, J., R. Glass, P. Woods, V. Gouvea, M. Gorziglia, J. Flores, B. Das, and M. Bhan. 1992. Identification of group A rotavirus gene 4 types by polymerase chain reaction. *J. Clin. Microbiol.* **30:**1365–1373.

Gentsch, J. R., M. Ramachandran, B. K. Das, J. P. Leite, A. Alfieri, R. Kumar, M. K. Bhan, and R. I. Glass. 1996. Review of G and P typing results from a global collection of rotavirus strains: implications for vaccine development. *J. Infect. Dis.* **174:**S30–S36.

Gerna, G., N. Passarani, and F. L. Rondanelli. 1985. Human enteric coronaviruses: antigenic relatedness to human coronavirus OC43 and possible etiologic role in viral gastroenteritis. *J. Infect. Dis.* **151:**796–803.

Gerna, G., N. Passarani, L. E. Unicomb, M. Parea, A. Sarasini, M. Battaglia, and R. F. Bishop. 1989. Solid-phase immune electron microscopy and enzyme-linked immunosorbent assay for typing of human rotavirus strains by using polyclonal and monoclonal antibodies: a comparative study. *J. Infect. Dis.* **159:**335–339.

Gibson, C. A., J. Chen, S. A. Monroe, and M. R. Denison. 1998. Expression and processing of nonstructural proteins of the human astroviruses. *Adv. Exp. Med. Biol.* **440:**387–391.

Gilger, M. A., D. O. Matson, M. E. Conner, H. M. Rosenblatt, M. J. Finegold, and M. K. Estes. 1992. Extraintestinal rotavirus infections in children with inmunodeficiency. *J. Pediatr.* **120:**912–917.

Glass, R., P. E. Kilgore, R. C. Holman, S. Jin, J. C. Smith, P. A. Woods, M. J. Clarke, M. S. Ho, and J. R. Gentsch. 1996.

The epidemiology of rotavirus diarrhea in the United States: surveillance and estimates of disease burden. *J. Infect. Dis.* **174:**S5–S11.

Goldwater, P. N., I. L. Chrystie, and J. E. Banatvala. 1979. Rotaviruses and the respiratory tract. *Br. Med. J.* **2:**1551.

Gorziglia, M., Y. Hoshino, A. Buckler-White, I. Blumentals, R. Glass, J. Flores, A. Z. Kapikian, and R. M. Chanock. 1986. Conservation of amino acid sequence of VP8 and cleavage region of 84-kDa outer capsid protein among rotaviruses recovered from asymptomatic neonatal infection. *Proc. Natl. Acad. Sci. USA* **83:**7039–7043.

Graham, D. Y., X. Jiang, T. Tanaka, A. R. Opekun, H. P. Madore, and M. K. Estes. 1994. Norwalk virus infection of volunteers: new insights based on improved assays. *J. Infect. Dis.* **170:**34–43.

Gray, J. J., C. Cunliffe, J. Ball, D. Y. Graham, U. Desselberger, and M. K. Estes. 1994. Detection of immunoglobulin M (IgM), IgA, and IgG Norwalk virus-specific antibodies by indirect enzyme-linked immunosorbent assay with baculovirus-expressed Norwalk virus capsid antigen in adult volunteers challenged with Norwalk virus. *J. Clin. Microbiol.* **32:**3059–3063.

Gray, J. J., J. Green, C. Cunliffe, C. Gallimore, J. V. Lee, K. Neal, and D. W. Brown. 1997. Mixed genogroup SRSV infections among a party of canoeists exposed to contaminated recreational water. *J. Med. Virol.* **52:**425–429.

Gray, J. J., T. G. Wreghitt, W. D. Cubitt, and P. R. Elliot. 1987. An outbreak of gastroenteritis in a home for the elderly associated with astrovirus type 1 and human calicivirus. *J. Med. Virol.* **23:**377–381.

Green, J., C. I. Gallimore, J. P. Norcott, D. Lewis, and D. W. Brown. 1995. Broadly reactive reverse transcriptase polymerase chain reaction for the diagnosis of SRSV-associated gastroenteritis. *J. Med. Virol.* **47:**392–398.

Green, K. Y. 1997. The role of human caliciviruses in epidemic gastroenteritis. *Arch. Virol.* **13:**S153–S165.

Green, K. Y., T. Ando, M. S. Balayan, I. N. Clarke, M. K. Estes, D. O. Matson, S. Nakata, J. D. Neil, M. J. Studdert, and H.-J. Thiel. Caliciviridae. In F. A. Murphy, C. M. Fauquet, D. H. L. Bishop, S. A. Ghabrial, A. W. Jarvis, G. P. Martelli, M. A. Mayo, and M. D. Summers (ed.), *Virus Taxonomy. 7th Report of the International Committee on Taxonomy of Viruses,* in press. Academic Press, Orlando, Fla.

Green, K. Y., A. Z. Kapikian, J. Valdesuso, S. Sosnovtsev, J. J. Treanor, and J. F. Lew. 1997. Expression and self-assembly of recombinant capsid protein from the antigenically distinct Hawaii human calicivirus. *J. Clin. Microbiol.* **35:**1909–1914.

Green, K. Y., J. F. Sears, K. Taniguchi, K. Midthun, Y. Hoshino, M. Gorziglia, K. Nishikawa, S. Urasawa, A. Z. Kapikian, R. M. Chanock, and J. Flores. 1988. Prediction of human rotavirus serotype by nucleotide sequence analysis of the VP7 protein gene. *J. Virol.* **62:**1819–1823.

Green, S. M., K. E. Dingle, P. R. Lambden, E. O. Caul, C. R. Ashley, and I. N. Clarke. 1994. Human enteric *Caliciviridae:* a new prevalent small round-structured virus group defined by RNA-dependent RNA polymerase and capsid diversity. *J. Gen. Virol.* **75:**1883–1888.

Greenberg, H. B., R. G. Wyatt, and A. Z. Kapikian. 1979. Norwalk virus in vomitus. *Lancet* **i:**55.

Gregorio, L., C. L. Sutton, and D. A. Lee. 1997. Central pontine myelinolysis in a previously healthy 4-year-old child with acute rotavirus gastroenteritis. *Pediatrics* **99:**738–743.

Grimwood, K., G. D. Abbott, D. M. Fergusson, L. C. Jennings, and J. M. Allan. 1983. Spread of rotavirus within families: a community based study. *Br. Med. J.* **287:**575–577.

Grimwood, K., R. Carzino, G. L. Barnes, and B. F. Bishop. 1995. Patients with enteric adenovirus gastroenteritis admitted to an Australian pediatric teaching hospital from 1981 to 1992. *J. Clin. Microbiol.* **33:**131–136.

Grimwood, K., J. C. Lund, B. S. Coulson, I. L. Hudson, R. F. Bishop, and G. L. Barnes. 1988. Comparison of serum and mucosal antibody responses following severe acute rotavirus gastroenteritis in young children. *J. Clin. Microbiol.* **26:**732–738.

Grohmann, G. 1985. Viral diarrhoea in Australia, p. 25–28. In S. Tzipori (ed.), *Infectious Diarrhoea in the Young.* Excerpta Medica Elsevier, Amsterdam, The Netherlands.

Guarino, A., S. Guandalini, F. Albano, A. Mascia, G. De Ritis, and A. Rubino. 1991. Enteral inmunoglobulins for treatment of protracted rotaviral diarrhea. *Pediatr. Infect. Dis. J.* **10:**612–614.

Guerrero, M. L., J. S. Noel, D. K. Mitchell, J. J. Calva, A. L. Morrow, J. Martinez, G. Rosales, F. R. Velazquez, S. S. Monroe, R. I. Glass, L. K. Pickering, and G. M. Ruiz-Palacios. 1998. A prospective study of astrovirus diarrhea of infancy in Mexico City. *Pediatr. Infect. Dis. J.* **17:**712–727.

Guo, M., K. O. Chang, M. E. Hardy, Q. Zhang, and L. J. Saif. 1999. Cloning and sequencing of a porcine enteric calicivirus genetically related to Sapporo-like viruses. *International Workshop on Human Caliciviruses.* Centers for Disease Conrol and Prevention, Atlanta, Ga.

Gurwith, M., W. Wenman, D. Gurwith, J. Brunton, S. Feltham, and H. Greenberg. 1983. Diarrhea among infants and young children in Canada: a longitudinal study in three northern communities. *J. Infect. Dis.* **147:**685–692.

Gurwith, M., W. Wenman, D. Hinde, S. Feltham, and H. Greenberg. 1981. A prospective study of rotavirus infection in infants and young children. *J. Infect. Dis.* **144:**218–224.

Hale, A. D., S. E. Crawford, M. Ciarlet, J. Green, C. Gallimore, D. W. Brown, X. Jiang, and M. K. Estes. 1999. Expression and self-assembly of Grimsby virus: antigenic distinction from Norwalk and Mexico viruses. *Clin. Diagn. Lab. Immunol.* **6:**142–145.

Halvorsrud, J., and I. Orstavik. 1980. An epidemic of rotavirus-associated gastroenteritis in a nursing home for the elderly. *Scand. J. Infect. Dis.* **12:**161–164.

Hammond, G. W., G. S. Ahluwalia, F. G. Barker, G. Horsman, and P. R. Hazelton. 1982. Comparison of direct and indirect enzyme inmunoassays with direct ultracentrifugation before electron microscopy for detection of rotaviruses. *J. Clin. Microbiol.* **16:**53–59.

Hammond, G. W., C. Hannan, T. Yeh, K. Fischer, G. Mauthe, and S. E. Strauss. 1987. DNA hybridization for diagnosis of enteric adenovirus infection from directly spotted human fecal specimens. *J. Clin. Microbiol.* **25:**1881–1885.

Hasegawa, A., S. Matsuno, S. Inouye, R. Kono, Y. Tsurukubo, A. Mukoyama, and Y. Saito. 1982. Isolation of human rotaviruses in primary cultures of monkey kidney cells. *J. Clin. Microbiol.* **16:**387–390.

Herring, A. J., N. F. Inglis, C. K. Ojeh, D. R. Snodgrass, and J. D. Menzies. 1982. Rapid diagnosis of rotavirus infection by direct detection of viral nucleic acid in silver-stained polyacrylamide gels. *J. Clin. Microbiol.* **16:**473–477.

Herrmann, J. E., N. R. Blacklow, S. M. Matsui, T. L. Lewis, M. K. Estes, J. M. Ball, and J. P. Brinker. 1995. Monoclonal antibodies for detection of Norwalk virus antigen in stools. *J. Clin. Microbiol.* **33:**2511–2513.

Herrmann, J. E., R. W. Hudson, D. M. Perron-Henry, J. B. Kurtz, and N. R. Blacklow. 1988. Antigenic characterization of cell-cultivated astrovirus serotypes and development of astrovirus-specific monoclonal antibodies. *J. Infect. Dis.* **158:**182–185.

Herrmann, J. E., N. A. Nowak, D. M. Perron-Henry, R. W. Hudson, W. D. Cubitt, and N. R. Blacklow. 1990. Diagnosis of astrovirus gastroenteritis by antigen detection with monoclonal antibodies. *J. Infect. Dis.* **161:**226–229.

Herrmann, J. E., D. N. Taylor, P. Echeverria, and N. R. Blacklow. 1991. Astroviruses as a cause of gastroenteritis in children. *N. Engl. J. Med.* **324:**1757–1760.

Hierholzer, J. C., R. Wigand, L. J. Anderson, T. Adrian, and J. W. Gold. 1988. Adenoviruses from patients with AIDS: a plethora of serotypes and a description of five new serotypes of subgenus D (types 43–47). *J. Infect. Dis.* **157:**804–813.

Hjelt, K., and P. C. Grauballe. 1990. Protective levels of intestinal rotavirus antibodies. *J. Infect. Dis.* **161:**352–353.

Hjelt, K., P. C. Grauballe, A. Paerregaard, O. H. Nielsen, and P. A. Krasilnikoff. 1987. Protective effect of pre-existing rotavirus-specific immunoglobulin A against naturally acquired rotavirus infection in children. *J. Med. Virol.* **21:**39–47.

Ho, M., R. I. Glass, S. S. Monroe, H. P. Madore, S. Stine, P. F. Pinsky, W. D. Cubitt, C. Ashley, and E. O. Caul. 1989. Viral gastroenteritis aboard a cruise ship. *Lancet* **ii:**961–964.

Hodes, H. L. 1977. Viral gastroenteritis. *Am. J. Dis. Child.* **131:**729–731.

Holzel, H., W. D. Cubitt, D. A. McSwiggan, P. J. Sanderson, and J. Church. 1980. An outbreak of rotavirus infection among adults in a cardiology ward. *J. Infect.* **2:**33–37.

Hopkins, R. S., G. B. Gaspard, F. P. Williams, R. J. Karlin, G. Cukor, and N. R. Blacklow. 1984. A community waterborne gastroenteritis outbreak: evidence for rotavirus as the agent. *Am. J. Public Health* **74:**263–265.

Hoshino, Y., L. J. Saif, S. Y. Kang, M. M. Sereno, W. K. Chen, and A. Z. Kapikian. 1995. Identification of group A rotavirus genes associated with virulence of a porcine rotavirus and host range restriction of a human rotavirus in the gnotobiotic piglet model. *Virology* **209:**274–280.

Hoshino, Y., L. J. Saif, and M. M. Sereno. 1988. Infection immunity of piglets to either VP3 or VP7 outer capsid protein confers resistance to challenge with a virulent rotavirus bearing the corresponding antigen. *J. Virol.* **62:**744–748.

Hrdy, D. B. 1987. Epidemiology of rotaviral infection in adults. *Rev. Infect. Dis.* **9:**461–469.

Hung, T., C. G. Wang, Z. Y. Fang, Z. Y. Chou, X. J. Chang, X. G. Lion, G. M. Chen, H. L. Yao, T. X. Chao, W. Ye, S. S. Den, and W. Chang. 1984. Waterborne outbreak of rotavirus diarrhoea in adults in China caused by a novel rotavirus. *Lancet* **i:**1139–1142.

Isaacs, D., D. Day, and S. Crook. 1986. Childhood gastroenteritis: a population study. *Br. Med. J.* **293:**545–546.

Jamieson, F. B., E. E. Wang, C. Bain, J. Good, L. Duckmanton, and M. Petric. 1998. Human torovirus: a new nosocomial gastrointestinal pathogen. *J. Infect. Dis.* **178:**1263–1269.

Jiang, B., P. H. Dennehy, S. Spangenberger, J. R. Gentsch, and R. I. Glass. 1995a. First detection of group C rotavirus in fecal specimens of children with diarrhea in the United States. *J. Infect. Dis.* **172:**45–50.

Jiang, B., S. S. Monroe, E. V. Koonin, S. E. Stine, and R. I. Glass. 1993a. RNA sequence of astrovirus: distinctive genomic organization and a putative retrovirus-like ribosomal frameshifting signal that directs the viral replicase synthesis. *Proc. Natl. Acad. Sci. USA* **90:**10539–10543.

Jiang, X., W. D. Cubitt, T. Berke, W. Zhong, X. Dai, S. Nakata, L. K. Pickering, and D. O. Matson. 1997. Sapporo-like human caliciviruses are genetically and antigenically diverse. *Arch. Virol.* **142:**1813–1827.

Jiang, X., D. Cubitt, J. Hu, X. Dai, J. Treanor, D. O. Matson, and L. K. Pickering. 1995b. Development of an ELISA to detect MX virus, a human calicivirus in the Snow Mountain Agent genogroup. *J. Gen. Virol.* **76:**2739–2747.

Jiang, X., D. Y. Graham, K. Wang, and M. K. Estes. 1990. Norwalk virus genome cloning and characterization. *Science* **250:**1580–1583.

Jiang, X., P. W. Huang, W. M. Zhong, T. Farkas, W. D. Cubitt, and D. O. Matson. 1999. Design and evaluation of a primer pair that detects both Norwalk- and Sapporo-like caliciviruses by RT-PCR. *J. Virol. Methods* **83:**145–154.

Jiang, X., D. O. Matson, W. D. Cubitt, and M. K. Estes. 1996a. Genetic and antigenic diversity of human caliciviruses (HuCVs) using RT-PCR and new EIAs. *Arch. Virol.* **12:**S251–S262.

Jiang, X., D. O. Matson, G. M. Ruiz-Palacios, J. Hu, J. Treanor, and L. K. Pickering. 1995c. Expression, self-assembly, and antigenicity of a Snow Mountain agent-like calicivirus capsid protein. *J. Clin. Microbiol.* **33:**1452–1455.

Jiang, X., D. O. Matson, F. R. Velazquez, J. J. Calva, W. M. Zhong, J. Hu, G. M. Ruiz-Palacios, and L. K. Pickering. 1995d. Study of Norwalk-related viruses in Mexican children. *J. Med. Virol.* **47:**309–316.

Jiang, X., E. Turf, J. Hu, E. Barrett, X. M. Dai, S. Monroe, C. Humphrey, L. K. Pickering, and D. O. Matson. 1996b. Outbreaks of gastroenteritis in elderly nursing homes and retirement facilities associated with human caliciviruses. *J. Med. Virol.* **50:**335–341.

Jiang, X., J. Wang, and M. K. Estes. 1995e. Characterization of SRSVs using RT-PCR and a new antigen ELISA. *Arch. Virol.* **140:**363–374.

Jiang, X., J. Wang, D. Y. Graham, and M. K. Estes. 1992a. Detection of Norwalk virus in stool by polymerase chain reaction. *J. Clin. Microbiol.* **30:**2529–2534.

Jiang, X., M. Wang, D. Y. Graham, and M. K. Estes. 1992b. Expression, self-assembly, and antigenicity of the Norwalk virus capsid protein. *J. Virol.* **66:**6527–6532.

Jiang, X., M. Wang, K. Wang, and M. K. Estes. 1993b. Sequence and genomic organization of Norwalk virus. *Virology* **195:**51–61.

Joensuu, J., E. Koskenniemi, X. Li Pang, and T. Vesikari. 1997. Randomized placebo-controlled trial of rhesus-human reassortant rotavirus vaccine for prevention of severe rotavirus gastroenteritis. *Lancet* **350:**1205–1209.

Johnson, P. C., J. J. Mathewson, H. L. DuPont, and H. B. Greenberg. 1990. Multiple-challenge study of host susceptibility to Norwalk gastroenteritis in US adults. *J. Infect. Dis.* **161:**18–21.

Jonassen, T. O., E. Kjeldsberg, and B. Grinde. 1993. Detection of human astrovirus serotype 1 by the polymerase chain reaction. *J. Virol. Methods* **44:**83–88.

Jonassen, T. O., C. Monceyron, T. W. Lee, J. B. Kurtz, and B. Grinde. 1995. Detection of all serotypes of human astrovirus by the polymerase chain reaction. *J. Virol. Methods* **52:**327–334.

Kagnoff, M. F., R. K. Austin, J. J. Hubert, J. E. Bernardin, and D. O. Kasarda. 1984. Possible role for a human adenovirus in the pathogenesis of celiac disease. *J. Exp. Med.* **160:**1544–1557.

Kapikian, A. Z., and R. M. Chanock. 1996. Rotaviruses, p. 1656–1708. *In* B. N. Fields, D. M. Knipe, and P. M. Howley (ed.), *Virology*, 3rd ed. Raven Press, New York, N.Y.

Kapikian, A. Z., H. W. Kim, R. G. Wyatt, W. L. Cline, J. O. Arrobio, C. D. Brandt, W. J. Rodriguez, S. A. Sack, R. M. Chanock, and R. H. Parrott. 1976. Human reovirus-like agent as the major pathogen associated with winter gastroenteritis in hospitalized infants and young children. *N. Engl. J. Med.* **294:**965–972.

Kapikian, A. Z., R. G. Wyatt, R. Dolin, T. S. Thornhill, A. R. Kalica, and R. M. Chanock. 1972. Visualization by immune electronmicroscopy of a 27-nm particle associated with acute infectious non-bacterial gastroenteritis. *J. Virol.* **10:**1075–1081.

Kapikian, A. Z., R. G. Wyatt, M. M. Levine, R. H. Yolken, D. H. Vankirk, R. Dolin, H. B. Greenberg, and R. M. Chanock. 1983. Oral administration of human rotavirus to volunteers: induction of illness and correlates of resistance. *J. Infect. Dis.* **147:**95–106.

Kaplan, J. E., R. A. Goodman, L. B. Schonberger, E. C. Lippy, and G. W. Gary. 1982a. Gastroenteritis due to Norwalk virus: an outbreak associated with a municipal water system. *J. Infect. Dis.* **146:**190–197.

Kaplan, J. E., L. B. Schonberger, G. Varano, N. Jackman, J. Bied, and G. W. Gary. 1982b. An outbreak of acute nonbacterial gastroenteritis in a nursing home. Demonstration of person-to-person transmission by temporal clustering of cases. *Am. J. Epidemiol.* **116:**940–948.

Keswick, B. H., L. K. Pickering, H. L. DuPont, and W. E. Woodward. 1983. Survival and detection of rotaviruses on environmental surfaces in day care centers. *Appl. Environ. Microbiol.* **46:**813–816.

Kidd, A. H., M. Jonsson, D. Gareicz, A. E. Kajon, A. G. Wermenbol, M. W. Verweij, and J. C. DeJong. 1996. Rapid subgenus identification of human adenovirus isolates by a general PCR. *J. Clin. Microbiol.* **34:**622–627.

Kilgore, P., L. Unicomb, J. R Gentsch, J. Albert, C. Mcelroy, and R. I. Glass. 1996. Neonatal rotavirus infection in Bangladesh: strain characterization and risk factors for nosocomial infection. *Pediatr. Infect. Dis. J.* **15:**672–677.

Kim, H. W., C. D. Brandt, A. Z. Kapikian, R. G. Wyatt, J. P. Arrobio, W. J. Rodriguez, R. M. Chanock, and R. H. Parrott. 1977. Human reovirus-like agent infection: occurrence in adult contacts of pediatric patients with gastroenteritis. *JAMA* **238:**404–407.

Kjeldsberg, E. 1977. Small spherical viruses in faeces from gastroenteritis patients. *Acta Pathol. Microbiol. Immunol. Scand.* **85:**351–354.

Konno, T., H. Suzuki, N. Ishida, R. Chiba, K. Mochizuki, and A. Tsunoda. 1982. Astrovirus-associated epidemic gastroenteritis in Japan. *J. Med. Virol.* **9:**11–17.

Konno, T., H. Suzuki, N. Katsushima, A. Imai, F. Tazawa, T. Kutsuzawa, S. Kitaoka, M. Sakamoto, N. Yazaki, and N. Ishida. 1983. Influence of temperature and relative humidity on human rotavirus infection in Japan. *J. Infect. Dis.* **147:**125–128.

Konno, T., H. Suzuki, T. Kutsuzawa, A. Imai, N. Katsushima, M. Sakamoto, S. Kitaoka, R. Tsuboi, and M. Adachi. 1978. Human rotavirus infection in infants and young children with intussusception. *J. Med. Virol.* **2:**265–269.

Koopmans, M. P., M. H. Bijen, S. S. Monroe, and J. Vinje. 1998. Age-stratified seroprevalence of neutralizing antibodies to astrovirus types 1 to 7 in humans in The Netherlands. *Clin. Diagn. Lab. Immunol.* **5:**33–37.

Koopmans, M. P., E. S. Goosen, A. A. Lima, I. T. McAuliffe, J. P. Nataro, L. J. Barrett, R. I. Glass, and R. L. Guerrant. 1997. Association of torovirus with acute and persistent diarrhea in children. *Pediatr. Infect. Dis. J.* **16:**504–507.

Kotloff, K. L., J. E. Herrmann, N. R. Blacklow, R. W. Hudson, S. S. Wasserman, J. G. Morris, Jr., and M. M. Levine. 1992. The frequency of astrovirus as a cause of diarrhea in Baltimore children. *Pediatr. Infect. Dis. J.* **11:**587–589.

Kotloff, K. L., G. A. Losonsky, J. G. Morris, S. S. Wasserman, H. Singh-Nac, and M. M. Levine. 1989. Enteric adenovirus infection and childhood diarrhea: an epidemiologic study in three clinical settings. *Pediatrics* **84:**219–225.

Kriston, S., M. M. Willcocks, M. J. Carter, and W. D. Cubitt. 1996. Seroprevalence of astrovirus types 1 and 6 in London, determined using recombinant virus antigen. *Epidemiol. Infect.* **117:**159–164.

Kurtz, J., and T. Lee. 1978. Astrovirus gastroenteritis age distribution of antibody. *Med. Microbiol. Immunol.* (Berlin) **166:**227–230.

Kurtz, J. B. 1988. Astroviruses, p. 70–78. *In* M. Farthing (ed.), *Viruses in the Gut.* Smith Kline & French, Welwyn Garden City, United Kingdom.

Kurtz, J. B., and T. W. Lee. 1984. Human astrovirus serotypes. *Lancet* **ii:**1405.

Kurtz, J. B., T. W. Lee, J. W. Craig, and S. E. Reed. 1979. Astrovirus infection in volunteers. *J. Med. Virol.* **3:**221–230.

Kurtz, J. B., T. W. Lee, and D. Pickering. 1977. Astrovirus associated gastroenteritis in a children's ward. *J. Clin. Pathol.* **30:**948–952.

Kusuya, M., R. Fujii, M. Hamano, J. Nakamura, M. Yamada, S. Nii, and T. Mori. 1996. Molecular analysis of outer capsid glycoprotein (VP7) genes from two isolates of human group C rotavirus with different genome electropherotypes. *J. Clin. Microbiol.* **34:**3185–3189.

Kutsuzawa, T., T. Konno, H. Suzuki, A. Z. Kapikian, T. Ebina, and N. Ishida. 1982. Isolation of human rotavirus subgroups 1 and 2 in cell culture. *J. Clin. Microbiol.* **16:**727–730.

Lacombe, D., F. Lamouliatte, C. Billeud, and B. Sandler. 1988. Breda virus and hemorrhagic enteropathy. Reminder apropos of 1 case. *Arch. Fr. Pediatr.* **45:**442.

Lahdeaho, M. L., M. Lehtinen, H. R. Rissa, H. Hyoty, T. Reunala, and M. Maki. 1993. Antibodies to E1b protein-derived peptides of enteric adenovirus type 40 are associated with celiac disease and dermatitis herpetiformis. *Clin. Immunol. Immunopathol.* **69:**300–305.

Lambden, P. R., E. O. Caul, C. R. Ashley, and I. N. Clarke. 1993. Sequence and genome organization of a human small round-structured (Norwalk-like) virus. *Science* **259:**516–519.

LeBaron, C. W., J. R. Allen, M. Hebert, P. Woods, J. Lew, and R. I. Glass. 1992. Outbreaks of summer rotavirus linked to laboratory practices. The National Rotavirus Surveillance System. *Pediatr. Infect. Dis. J.* **11:**860–865.

LeBaron, C. W., J. Lew, R. I. Glass, J. W. Weber, and G. M. Ruiz-Palacios. 1990. The rotavirus study group. Annual rotavirus epidemic patterns in North America. *JAMA* **264:**983–987.

Lee, T. W., and J. B. Kurtz. 1981. Serial propagation of astrovirus in tissue culture with the aid of trypsin. *J. Gen. Virol.* **57:**421–424.

Lee, T. W., and J. B. Kurtz. 1994. Prevalence of human astrovirus serotypes in the Oxford region 1976–92, with evidence for two new serotypes. *Epidemiol. Infect.* **112:**187–193.

Leite, J. P., T. Ando, J. S. Noel, B. Jiang, C. D. Humphrey, J. F. Lew, K. Y. Green, R. I. Glass, and S. S. Monroe. 1996. Characterization of Toronto virus capsid protein expressed in baculovirus. *Arch. Virol.* **141:**865–875.

Leite, J. P., O. M. Barth, and H. G. Schatzmayr. 1991. Astrovirus in faeces of children with acute gastroenteritis in Rio de Janeiro, Brazil. *Mem. Inst. Oswaldo Cruz* **86:**489–490.

Lew, J. F., A. Z. Kapikian, X. Jiang, M. K. Estes, and K. Y. Green. 1994a. Molecular characterization and expression of the capsid protein of a Norwalk-like virus recovered from a Desert Shield troop with gastroenteritis. *Virology* **200:**319–325.

Lew, J. F., C. L. Moe, S. S. Monroe, J. R. Allen, B. M. Harrison, B. D. Forrester, S. E. Stine, P. A. Woods, J. C. Hierholzer, J. E. Herrmann, N. R. Blacklow, A. V. Bartlett, and R. I. Glass. 1991. Astrovirus and adenovirus associated with diarrhea in children in day care settings. *J. Infect. Dis.* 164:673–678.

Lew, J. F., M. Petric, A. Z. Kapikian, X. Jiang, M. K. Estes, and K. Y. Green. 1994b. Identification of minireovirus as a Norwalk-like virus in pediatric patients with gastroenteritis. *J. Virol.* 68:3391–3396.

Lewis, D. C., N. F. Lightfoot, W. D. Cubitt, and S. A. Wilson. 1989. Outbreaks of astrovirus type 1 and rotavirus gastroenteritis in a geriatric in-patient population. *J. Hosp. Infect.* 14:9–14.

Lewis, H. M., J. V. Parry, H. A. Davies, R. P. Parry, A. Mott, R. R. Dourmashkin, P. J. Sanderson, D. A. Tyrrell, and H. B. Valman. 1979. A year's experience of the rotavirus syndrome and its association with respiratory illness. *Arch. Dis. Child.* 54:339–346.

Lewis, T. L., H. B. Greenberg, J. E. Herrmann, L. S. Smith, and S. M. Matsui. 1994. Analysis of astrovirus serotype 1 RNA, identification of the viral RNA-dependent RNA polymerase motif, and expression of a viral structural protein. *J. Virol.* 68:77–83.

Li, Q. G., A. Henningsson, P. Juto, F. Elgh, and G. Wadell. 1999. Use of restriction fragment analysis and sequencing of a serotype-specific region to type adenovirus isolates. *J. Clin. Microbiol.* 37:844–847.

Li Pang, X., J. Joensuu, and T. Vesikari. 1999. Human calicivirus-associated sporadic gastroenteritis in Finnish children less than two years of age followed prospectively during a rotavirus vaccine trial. *Pediatr. Infect. Dis J.* 18:420–426.

Liu, B. L., I. N. Clarke, E. O. Caul, and P. R. Lambden. 1995. Human enteric caliciviruses have a unique genome structure and are distinct from the Norwalk-like viruses. *Arch. Virol.* 140:1345–1356.

Liu, B. L., P. R. Lambden, H. Gunther, P. Otto, M. Elschner, and I. N. Clarke. 1999. Molecular characterization of a bovine enteric calicivirus: relationship to the Norwalk-like viruses. *J. Virol.* 73:819–825.

Lycke, L., J. Bloomberg, G. Berg, A. Ericksson, and L. Madsen. 1978. Epidemic acute diarrhoea in adults associated with infantile gastroenteritis virus. *Lancet* ii:1056–1057.

Madeley, C. R. 1979. Comparison of the features of astroviruses and caliciviruses seen in samples of feces by electron microscopy. *J. Infect. Dis.* 139:519–523.

Madeley, C. R. 1986. The emerging role of adenoviruses as inducers of gastroenteritis. *Pediatr. Infect. Dis.* 5:S63–S74.

Madeley, C. R., and B. P. Cosgrove. 1975. 28nm particles in faeces in infantile gastroenteritis. *Lancet* ii:451–452.

Madeley, C. R., B. P. Cosgrove, and E. J. Bell. 1978. Stool viruses in babies in Glasgow. 2. Investigation of normal newborns in hospital. *J. Hyg.* (London). 81:285–294.

Maldonado, Y., M. Cantwell, M. Old, D. Hill, M. L. Sanchez, L. Logan, F. Millan-Velasco, J. L. Valdespino, J. Sepulveda, and S. Matsui. 1998. Population-based prevalence of symptomatic and asymptomatic astrovirus infection in rural Mayan infants. *J. Infect. Dis.* 178:334–339.

Marrie, T. J., S. H. Lee, R. S. Faulkner, J. Ethier, and C. H. Young. 1982. Rotavirus infection in a geriatric population. *Arch. Intern. Med.* 142:313–316.

Marx, F. E., M. B. Taylor, and W. O. K. Grabow. 1998. The prevalence of human astrovirus and enteric adenovirus infection in South African patients with gastroenteritis. *South Afr. J. Epidemiol. Infect.* 13:5–9.

Marx, F. E., M. B. Taylor, and W. O. K. Grabow. 1998. The application of a reverse transcriptase-polymerase chain reaction-oligonucleotide probe assay for the detection of human astroviruses in environmental water. *Water Res.* 32:2147–2153.

Masendycz, P. J., E. A. Palombo, G. L. Barnes, and R. F. Bishop. 1999. Rotavirus diversity: what surveillance will tell us. *Commun. Dis. Intell.* 23:198–199.

Masendycz, P. J., E. A. Palombo, R. J. Gorrell, and R. F. Bishop. 1997. Comparison of enzyme immunoassay, PCR, and type-specific cDNA probe techniques for identification of group A rotavirus gene 4 types (P types). *J. Clin. Microbiol.* 35:3104–3108.

Matson, D. O., and M. K. Estes. 1990. Impact of rotavirus infection at a large pediatric hospital. *J. Infect. Dis.* 162:598–605.

Matson, D. O., M. K. Estes, J. W. Burns, H. B. Greenberg, K. Taniguchi, and S. Urasawa. 1990a. Serotype variation of human group A rotaviruses in two regions of the United States. *J. Infect. Dis.* 162:605–614.

Matson, D. O., M. K. Estes, T. Tanaka, A. V. Bartlett, and L. K. Pickering. 1990b. Asymptomatic human calicivirus infection in a day care center. *Pediatr. Infect. Dis. J.* 9:190–196.

Matson, D. O., M. L. O'Ryan, I. Herrera, L. K. Pickering, and M. K. Estes. 1993. Fecal antibody responses to symptomatic and asymptomatic rotavirus infections. *J. Infect. Dis.* 167:577–583.

Matson, D. O., M. L. O'Ryan, L. K. Pickering, S. Chiba, S. Nakata, P. Raj, and M. K. Estes. 1992. Characterization of serum antibody responses to natural rotavirus infections in children by VP7-specific epitope-blocking assays. *J. Clin. Microbiol.* 30:1056–1061.

Matson, D. O., R. Velazquez, A. L. Morrow, J. N. Shults, G. M. Ruiz-Palacios, and L. K. Pickering. 1995a. Protective effect of breastfeeding upon first rotavirus infection and illness in a cohort of Mexican children. *Pediatr. Res.* 37:752.

Matson, D. O., W.-M. Zhong, S. Nakata, X. Jiang, L. K. Pickering, and M. K. Estes. 1995b. Molecular characterization of a human calicivirus with sequence relationships closer to animal caliciviruses than other known human caliciviruses. *J. Med. Virol.* 45:215–222.

Matsui, M., H. Ushijima, M. Hachiya, J. Kakizawa, L. Wen, M. Oseto, K. Morooka, and J. B. Kurtz. 1998. Determination of serotypes of astroviruses by reverse transcription-polymerase chain reaction and homologies of the types by the sequencing of Japanese isolates. *Microbiol. Immunol.* 42:539–547.

Matsumoto, K., M. Hatano, K. Kobayashi, A. Hasegawa, S. Yamazaki, S. Nakata, S. Chiba, and Y. Kimura. 1989. An outbreak of gastroenteritis associated with acute rotaviral infection in schoolchildren. *J. Infect. Dis.* 160:611–615.

Matsuno, S., E. Utawaga, and A. Sugiura. 1983. Association of rotavirus infection with Kawasaki syndrome. *J. Infect. Dis.* 148:177.

Meyers, G., C. Wirblich, and H. J. Thiel. 1991a. Genomic and subgenomic RNAs of rabbit hemorrhagic disease virus are both protein-linked and packaged into particles. *Virology* 184:677–686.

Meyers, G., C. Wirblich, and H. J. Thiel. 1991b. Rabbit hemorrhagic disease virus—molecular cloning and nucleotide sequencing of a calicivirus genome. *Virology* 184:664–676.

Mickan, L. D., and T. W. Kok. 1994. Recognition of adenovirus types in faecal samples by southern hybridization in South Australia. *Epidemiol. Infect.* 112:603–613.

Midthun, K., H. B. Greenberg, J. B. Kurtz, G. W. Gary, F. Y. Lin, and A. Z. Kapikian. 1993. Characterization and seroepidemiology of a type 5 astrovirus associated with an outbreak of gastroenteritis in Marin County, California. *J. Clin. Microbiol.* **31:**955–962.

Mitchell, D. K., X. Jiang, and D. O. Matson. 1999a. Gastrointestinal infections, p. 79–92. *In* A. G. Storch (ed.), *Essentials of Diagnostic Virology.* Churchill Livingstone, New York, N.Y.

Mitchell, D. K., D. O. Matson, W. D. Cubitt, L. J. Jackson, M. M. Willcocks, L. K. Pickering, and M. J. Carter. 1999b. Prevalence of antibodies to astrovirus types 1 and 3 in children and adolescents in Norfolk, Virginia. *Pediatr. Infect. Dis. J.* **18:**249–254.

Mitchell, D. K., D. O. Matson, X. Jiang, T. Berke, S. S. Monroe, M. J. Carter, M. M. Willcocks, and L. K. Pickering. 1999c. Molecular epidemiology of childhood astrovirus infection in child care centers. *J. Infect. Dis.* **180:**514–517.

Mitchell, D. K., D. O. Matson, X. Jiang, and L. K. Pickering. 1995a. Use of polymerase chain reaction for determination of astrovirus serotypes associated with diarrheal outbreaks among children attending day care centers (DCCs). *Pediatr. Res.* **37:**184A.

Mitchell, D. K., S. S. Monroe, X. Jiang, D. O. Matson, R. I. Glass, and L. K. Pickering. 1995b. Virologic features of an astrovirus diarrhea outbreak in a day care center revealed by reverse transcriptase-polymerase chain reaction. *J. Infect. Dis.* **172:**1437–1444.

Mitchell, D. K., R. Van, A. L. Morrow, S. S. Monroe, R. I. Glass, and L. K. Pickering. 1993. Outbreaks of astrovirus gastroenteritis in day care centers. *J. Pediatr.* **123:**725–732.

Moe, C. L., J. R. Allen, S. S. Monroe, H. E. Gary, C. D. Humphrey, J. E. Herrmann, N. R. Blacklow, C. Carcamo, M. Koch, K. H. Kim, and R. I. Glass. 1991. Detection of astrovirus in pediatric stool samples by immunoassay and RNA probe. *J. Clin. Microbiol.* **29:**2390–2395.

Moe, C. L., J. Gentsch, T. Ando, G. Grohmann, S. S. Monroe, X. Jiang, J. Wang, M. K. Estes, Y. Seto, and C. Humphrey. 1994. Application of PCR to detect Norwalk virus in fecal specimens from outbreaks of gastroenteritis. *J. Clin. Microbiol.* **32:**642–648.

Monroe, S. S., R. I. Glass, N. Noah, T. H. Flewett, E. O. Caul, C. I. Ashton, A. Curry, A. M. Field, R. Madeley, and P. J. Pead. 1991. Electron microscopic reporting of gastrointestinal viruses in the United Kingdom, 1985-1987. *J. Med. Virol.* **33:**193–198.

Monroe, S. S., B. Jiang, S. E. Stine, M. Koopmans, and R. I. Glass. 1993a. Subgenomic RNA sequence of human astrovirus supports classification of *Astroviridae* as a new family of RNA viruses. *J. Virol.* **67:**3611–3614.

Monroe, S. S., S. E. Stine, X. Jiang, M. K. Estes, and R. I. Glass. 1993b. Detection of antibody to recombinant Norwalk virus antigen in specimens from outbreaks of gastroenteritis. *J. Clin. Microbiol.* **31:**2866–2872.

Mulcahy, D. L., K. R. Kamath, L. M. de Silva, S. Hodges, I. W. Carter, and M. J. Cloonan. 1982. A two-part study of the aetiological role of rotavirus in intussusception. *J. Med. Virol.* **9:**51–55.

Nagayoshi, S., H. Yamaguchi, T. Ichikawa, M. Miyazu, T. Morishima, T. Ozaki, S. Isomura, S. Suzuki, and M. Hoshino. 1980. Changes of the rotavirus concentration in faeces during the course of acute gastroenteritis as determined by the immune adherence hemagglutination test. *Eur. J. Pediatr.* **134:**99–102.

Nakagomi, T., Y. Horie, Y. Koshimura, H. Greenberg, and O. Nakagomi. 1999. Isolation of a human rotavirus strain with a super-short RNA pattern and new P2 subtype. *J. Clin. Microbiol.* **37:**1213–1216.

Nakata, S., S. Chiba, H. Terashima, Y. Sakuma, R. Kogasaka, and T. Nakao. 1983. Microtiter solid-phase radioimmunoassay for detection of human calicivirus in stools. *J. Clin. Microbiol.* **17:**198–201.

Nakata, S., S. Chiba, H. Terashima, T. Yokoyama, and T. Nakao. 1985. Humoral immunity in infants with gastroenteritis caused by human calicivirus. *J. Infect. Dis.* **152:**274–279.

Nakata, S., M. K. Estes, D. Y. Graham, S. S. Wang, G. W. Gary, and J. L. Melnick. 1987. Detection of antibody to group B adult diarrhea rotaviruses in humans. *J. Clin. Microbiol.* **25:**812–818.

Nakata, S., K. Kogawa, K. Numata, S. Ukae, N. Adachi, D. O. Matson, M. K. Estes, and S. Chiba. 1996. The epidemiology of human calicivirus/Sapporo/82/Japan. *Arch. Virol.* **12:**S263–S270.

Nazer, H., S. Rice, and J. A. Walker-Smith. 1982. Clinical associations of stool astrovirus in childhood. *J. Pediatr. Gastroenterol. Nutr.* **1:**555–558.

Neill, J. D., and W. L. Mengeling. 1988. Further characterization of the virus-specific RNAs in feline calicivirus infected cells. *Virus Res.* **11:**59–72.

Neill, J. D., I. M. Reardon, and R. L. Heinrikson. 1991. Nucleotide sequence and expression of the capsid protein gene of feline calicivirus. *J. Virol.* **65:**5440–5447.

Newburg, D. S., J. A. Peterson, G. M. Ruiz-Palacios, D. O. Matson, A. L. Morrow, J. Shults, M. L. Guerrero, P. Chaturvedi, S. O. Newburg, C. D. Scallan, M. R. Taylor, R. L. Ceriani, and L. K. Pickering. 1998. Protection of breast-fed children against symptomatic rotavirus infection by human milk lactadherin. *Lancet* **351:**1150–1154.

Nicolas, J. C., D. Ingrand, B. Fortier, and F. Bricout. 1982. A one-year virological survey of acute intussusception in childhood. *J. Med. Virol.* **9:**267–271.

Nishimura, S., H. Ushijima, S. Nishimura, H. Shiraishi, C. Kanazawa, T. Abe, K. Kaneko, and Y. Fukuyama. 1993. Detection of rotavirus in cerebrospinal fluid and blood of patients with convulsions and gastroenteritis by means of the reverse transcription polymerase chain reaction. *Brain Dev.* **15:**457–459.

Noel, J., and W. D. Cubitt. 1994. Identification of astrovirus serotypes from children treated at the Hospitals for Sick Children, London 1981–93. *Epidemiol. Infect.* **113:**153–159.

Noel, J., A. Mansoor, U. Thaker, J. Hermann, D. Perron-Henry, and W. D. Cubitt. 1994. Identification of adenoviruses in faeces from patients with diarrhoea at the Hospitals for Sick Children, London, 1989–1992. *J. Med. Virol.* **43:**84–90.

Noel, J. S., T. W. Lee, J. B. Kurtz, R. I. Glass, and S. S. Monroe. 1995. Typing of human astroviruses from clinical isolates by enzyme immunoassay and nucleotide sequencing. *J. Clin. Microbiol.* **33:**797–801.

Noel, J. S., B. L. Liu, C. D. Humphrey, E. M. Rodriguez, P. R. Lambden, I. N. Clarke, D. M. Dwyer, T. Ando, R. I. Glass, and S. S. Monroe. 1997. Parkville virus: a novel genetic variant of human calicivirus in the Sapporo virus clade, associated with an outbreak of gastroenteritis in adults. *J. Med. Virol.* **52:**173–178.

Numata, K., S. Nakata, X. Jiang, M. K. Estes, and S. Chiba. 1994. Epidemiological study of Norwalk virus infections in Japan and Southeast Asia by enzyme-linked immunosorbent assays with Norwalk virus capsid protein produced by the baculovirus expression system. *J. Clin. Microbiol.* **32:**121–126.

Offit, P. A. 1996. Host factors associated with protection against rotavirus disease: the skies are clearing. *J. Infect. Dis.* **174:**S59–S64.

Offit, P. A., H. F. Clark, and G. Blavat. 1986. Reassortant rotavirus containing structural proteins VP3 and VP7 from different parents are protective against each parental strain. *J. Virol.* **57:**376–378.

Oishi, I., K. Yamazaki, T. Kimoto, Y. Minekawa, E. Utagawa, S. Yamazaki, S. Inouye, G. S. Grohmann, S. S. Monroe, S. E. Stine, C. Carcamo, T. Ando, and R. I. Glass. 1994. A large outbreak of acute gastroenteritis associated with astrovirus among students and teachers in Osaka Japan. *J. Infect. Dis.* **170:**439–443.

Okhuysen, P. C., X. Jiang, L. Ye, P. C. Johnson, and M. K. Estes. 1995. Viral shedding and fecal IgA response after Norwalk virus infection. *J. Infect. Dis.* **171:**566–569.

Olive, D. M., and S. K. Sethi. 1989. Detection of human rotavirus by using an alkaline phosphatase-conjugated synthetic DNA probe in comparison with enzyme-linked immunoassay and polyacrylamide gel analysis. *J. Clin. Microbiol.* **27:**53–57.

Oliver, A. R., and A. D. Phillips. 1988. An electron microscopical investigation of faecal small round viruses. *J. Med. Virol.* **24:**211–218.

O'Ryan, M., I. Pérez-Schael, N. Mamani, A. Peña, B. Salina, G. González, and J. Gómez. Prospective study of the impact of rotavirus disease in South America. *Pediatr. Res.*, in press.

O'Ryan, M. L., N. Mamani, L. F. Avendaño, J. Cohen, A. Peña, J. Villarroel, A. Chávez, F. Valdivieso, and D. O. Matson. 1997. Molecular epidemiology of human rotaviruses in Santiago, Chile. *Pediatr. Infect. Dis. J.* **16:**305–311.

O'Ryan, M. L., and D. O. Matson. 1990. Viral gastroenteritis pathogens in the day care center setting. *Semin. Pediatr. Infect. Dis.* **1:**252–262.

O'Ryan, M. L., D. O. Matson, M. K. Estes, and L. K. Pickering. 1990. Molecular epidemiology of rotavirus in children attending day care centers in Houston. *J. Infect. Dis.* **162:**810–816.

O'Ryan, M. L., D. O. Matson, M. K. Estes, and L. K. Pickering. 1994. Anti-rotavirus G type-specific and isotype antibodies in children with natural rotavirus infections. *J. Infect. Dis.* **169:**504–511.

Padilla-Noriega, L., M. Méndez-Toss, G. Menchaca, J. F. Contreras, P. Romero-Guido, F. I. Puerto, H. Guiscafré, F. Mota, I. Herrera, R. Cedillo, O. Muñoz, J. Calva, M. de Lourdes Guerrero, B. S. Coulson, H. B. Greenberg, S. López, and C. F. Arias. 1998. Antigenic and genomic diversity of human rotavirus VP4 in two consecutive epidemic seasons in Mexico. *J. Clin. Microbiol.* **36:** 1668–1692.

Paerregard, A., A. Hjelt, J. Genner, J. Moslet, and P. A. Krasilnikoff. 1990. Role of enteric adenoviruses in acute gastroenteritis in children attending day care centers. *Acta Paediatr. Scand.* **79:**370–371.

Palombo, E., and R. Bishop. 1995. Genetic and antigenetic characterization of a serotype G6 human rotavirus isolated in Melbourne, Australia. *J. Med. Virol.* **47:**348–354.

Palombo, E. A., and R. F. Bishop. 1996. Annual incidence, serotype distribution, and genetic diversity of human astrovirus isolates from hospitalized children in Melbourne, Australia. *J. Clin. Microbiol.* **34:**1750–1753.

Pang, X. L., J. Joensuu, and T. Vesikari. 1996. Detection of rotavirus RNA in cerebrospinal fluid in a case of rotavirus gastroenteritis with febrile seizures. *Pediatr. Infect. Dis. J.* **15:**543–545.

Pang, X. L., J. Joensuu, and T. Vesikari. 1999. Human calicivirus-associated sporadic gastroenteritis in Finnish children less than two years of age followed prospectively during a rotavirus vaccine trial. *Pediatr. Infect. Dis. J.* **18:**420–426.

Parashar, U. D., L. Dow, R. L. Fankhauser, C. D. Humphrey, J. Miller, T. Ando, K. S. Williams, C. R. Eddy, J. S. Noel, T. Ingram, J. S. Bresee, S. S. Monroe, and R. I. Glass. 1998. An outbreak of viral gastroenteritis associated with consumption of sandwiches: implications for the control of transmission by food handlers. *Epidemiol. Infect.* **121:**615–621.

Parker, S. P., and W. D. Cubitt. 1994. Measurement of IgA responses following Norwalk virus infection and other human caliciviruses using a recombinant Norwalk virus protein EIA. *Epidemiol. Infect.* **113:**143–151.

Parker, S. P., W. D. Cubitt, X. J. Jiang, and M. K. Estes. 1994. Seroprevalence studies using a recombinant Norwalk virus protein enzyme immunoassay. *J. Med. Virol.* **42:**146–150.

Parrino, T. A., D. S. Schreiber, J. S. Trier, A. Z. Kapikian, and N. R. Blacklow. 1977. Clinical immunity in acute gastroenteritis caused by Norwalk agent. *N. Engl. J. Med.* **297:**86–89.

Pelosi, E., P. R. Lambden, E. O. Caul, B. Liu, K. Dingle, Y. Deng, and I. N. Clarke. 1999. The seroepidemiology of genogroup 1 and genogroup 2 Norwalk-like viruses in Italy. *J. Med. Virol.* **58:**93–99.

Peñaranda, M. E., W. D. Cubitt, P. Sinarachatanant, D. N. Taylor, S. Likanonsakul, L. Saif, and R. I. Glass. 1989. Group C rotavirus infections in patients with diarrhea in Thailand, Nepal, and England. *J. Infect. Dis.* **160:**392–397.

Pérez-Schael, I., M. Guntiñas, M. Pérez, V. Pagone, A. M. Rojas, R. Gonzalez, W. Cunto, Y. Hoshino, and A. Kapikian. 1997. Efficacy of the rhesus-based quadrivalent vaccine in infants and young children in Venezuela. *N. Engl. J. Med.* **337:**1181–1187.

Philipps, A. D. 1988. Mechanisms of mucosal injury: human studies, p. 30–40. *In* M. Farthing (ed.), *Viruses in the Gut.* Smith Kline & French, Welwyn Garden City, United Kingdom.

Pickering, L. K., A. V. Bartlett, R. R. Reeves, and A. Morrow. 1988. Asymptomatic excretion of rotavirus before and after rotavirus diarrhea in children in day care centers. *J. Pediatr.* **112:**361–365.

Pickering, L. K., H. L. DuPont, J. Olarte, R. Conklin, and C. Ericsson. 1977. Fecal leukocytes in enteric infections. *Am. J. Clin. Pathol.* **68:**562–565.

Pickering, L. K., D. G. Evans, H. L. DuPont, J. J. Vollet, and D. J. Evans. 1981. Diarrhea caused by Shigella, rotavirus, and Giardia in day-care centers: prospective study. *J. Pediatr.* **99:**51–56.

Pinto, R. M., J. M. Diez, and A. Bosch. 1994. Use of the colonic carcinoma cell line CaCo-2 for in vivo amplification and detection of enteric viruses. *J. Med. Virol.* **44:**310–315.

Pipittajan, P., S. Kasempimolporn, N. Ikegami, K. Akatani, C. Wasi, and P. Sinarachatanant. 1991. Molecular epidemiology of rotaviruses associated with pediatric diarrhea in Bangkok, Thailand. *J. Clin. Microbiol.* **29:**617–624.

Prado, V., and M. O'Ryan. 1994. Acute gastroenteritis in Latin America. *Infect. Dis. Clin. N. Am.* **1:**77–106.

Prasad, B. V., D. O. Matson, and A. W. Smith. 1994a. Three-dimensional structure of the primate calicivirus. *J. Mol. Biol.* **240:**256–264.

Prasad, B. V., R. Rothnagel, X. Jiang, and M. K. Estes. 1994b. Three-dimensional structure of baculovirus-expressed Norwalk virus capsids. *J. Virol.* **68:**5117–5125.

Raj, P., D. O. Matson, B. Coulson, R. Bishop, K. Taniguchi, S. Urasawa, H. Greenberg, and M. K. Estes. 1992. Comparisons of rotavirus VP7-typing monoclonal antibodies by competition binding assay. *J. Clin. Microbiol.* **30:**704–711.

Ramachandran, M., B. Das, A. Vij, R. Kumar, S. Bhambal, N. Kesari, H. Rawat, L. Bahl, S. Thakur, P. Woods, R. Glass, M. Bhan, and J. Gentsch. 1996. Unusual diversity of human rotavirus G and P genotypes in India. *J. Clin. Microbiol.* **34:**436–439.

Ramos-Alvarez, M., and A. B. Sabin. 1958. Enteropathic viruses and bacteria. *JAMA* **167:**147–156.

Rennels, M., R. Glass, P. Dennehey, D. Bernstein, M. Pichichero, E. Zito, M. Mack, B. Davidson, and A.

Kapikian. 1996. Safety and efficacy of high-dose rhesus-human reassortant rotavirus vaccines—report of the National Multicenter Trial. *Pediatrics* **97:**7–13.

Richmond, S. J., E. O. Caul, S. M. Dunn, C. R. Ashley, S. K. Clarke, and N. R. Seymour. 1979. An outbreak of gastroenteritis in young children caused by adenoviruses. *Lancet* **i:**1178–1181.

Riepenhoff-Talty, M., S. Bogger-Goren, P. Li, P. J. Carmody, H. J. Barrett, and P. L. Ogra. 1981. Development of serum and intestinal antibody response to rotavirus after naturally acquired rotavirus infection in man. *J. Med. Virol.* **8:**215–222.

Riepenhoff-Talty, M., V. Gouvea, M. J. Evans, L. Svensson, E. Hoffenberg, R. J. Sokol, I. Uhnoo, S. J. Greenberg, K. Schakel, G. Zhaori, J. Fitzgerald, S. Chong, M. El-Yousef, A. Nemeth, M. Brown, D. Piccoli, J. Hyams, D. Ruffin, and T. Rossi. 1996. Detection of group C rotavirus in infants with extrahepatic biliary atresia. *J. Infect. Dis.* **174:**8–15.

Rodriguez, W. J., H. W. Kim, J. O. Arrobio, C. D. Brandt, R. M. Chanock, A. Z. Kapikian, R. G. Wyatt, and R. H. Parrott. 1977. Clinical features of acute gastroenteritis associated with human rotavirus-like agent in infants and young children. *J. Pediatr.* **91:**188–193.

Rodriguez, W. J., H. W. Kim, C. D. Brandt, A. B. Fletcher, and R. H. Parrott. 1982. Rotavirus: a cause of nosocomial infection in the nursery. *J. Pediatr.* **101:**274–277.

Rotbart, H. A., M. J. Levin, R. H. Yolken, D. K. Manchester, and J. Jantzen. 1983. An outbreak of rotavirus-associated neonatal necrotizing enterocolitis. *J. Pediatr.* **103:**454–459.

Rousell, J., M. E. Zajdel, P. D. Howdle, and G. E. Blair. 1993. Rapid detection of enteric adenoviruses by means of the polymerase chain reaction. *J. Infect.* **27:**271–275.

Ryan, M. J., M. Ramsay, D. Brown, N. J. Gay, C. P. Farrington, and P. G. Wall. 1996. Hospital admissions attributable to rotavirus infection in England. *J. Infect. Dis.* **174:**12–18.

Saito, K., H. Ushijima, O. Nishio, M. Oseto, H. Motohiro, Y. Ueda, M. Takagi, S. Nakaya, T. Ando, R. Glass, and K. Zaiman. 1995. Detection of astroviruses from stool samples in Japan using reverse transcription and polymerase chain reaction amplification. *Microbiol. Immunol.* **39:**825–828.

Salmi, T. T., P. Arstila, and A. Koivikko. 1978. Central nervous system involvement in patients with rotavirus gastroenteritis. *Scand. J. Infect. Dis.* **10:**29–31.

Samadi, A. R., M. I. Huq, and Q. S. Ahmed. 1983. Detection of rotavirus in handwashings of attendants of children with diarrhoea. *Br. Med. J.* **286:**188.

Sanekata, T., Y. Kuwamoto, S. Akamatsu, N. Sakon, M. Oseto, K. Taniguchi, S. Nakata, and M. K. Estes. 1996. Isolation of group B porcine rotavirus in cell culture. *J. Clin. Microbiol.* **34:**759–761.

Santos, N., R. Lima, C. Pereira, and V. Gouvea. 1998. Detection of rotavirus types G8 and G10 among Brazilian children with diarrhea. *J. Clin. Microbiol.* **36:**2727–2729.

Santosham, M., L. Moulton, R. Raymond, J. Croll, R. Weathersbolt, R. Ward, J. Forro, E. Zito, M. Mack, G. Brenneman, and B. Davidson. 1997. Efficacy and safety of high-dose rhesus-human reassortant rotavirus vaccine in Native American populations. *J. Pediatr.* **131:**632–638.

Saulsbury, F. T., J. A. Winkelstein, and R. H. Yolken. 1980. Chronic rotavirus infection in immunodeficiency. *J. Pediatr.* **97:**61–65.

Schwab, K. J., F. H. Neill, M. K. Estes, T. G. Metcalf, and R. L. Atmar. 1998. Distribution of Norwalk virus within shellfish following bioaccumulation and subsequent depuration by detection using RT-PCR. *J. Food Prot.* **61:**1674–1680.

Scott-Taylor, T. H., G. Ahluwalia, M. Dawood, and G. W. Hammond. 1993. Detection of enteric adenoviruses with synthetic oligonucleotide probes. *J. Med. Virol.* **41:**328–337.

Shastri, S., A. M. Doane, J. Gonzales, U. Upadhyayula, and D. M. Bass. 1998. Prevalence of astroviruses in a children's hospital. *J. Clin. Microbiol.* **36:**2571–2574.

Shaw, R. D., D. L. Stoner-Ma, M. K. Estes, and H. B. Greenberg. 1985. Specific enzyme-linked immunoassay for rotavirus serotypes 1 and 3. *J. Clin. Microbiol.* **22:**286–291.

Shetty, M., T. A. Brown, M. Kotian, and P. G. Shivananda. 1995. Viral diarrhoea in a rural coastal region of Karnataka India. *J. Trop. Pediatr.* **41:**301–303.

Shinozaki, T., K. Araki, H. Ushijima, and R. Fuji. 1987. Antibody response to enteric adenovirus types 40 and 41 in sera from people in various age groups. *J. Clin. Microbiol.* **25:**1679–1682.

Singh-Naz, N., W. J. Rodriguez, A. H. Kidd, and C. D. Brandt. 1988. Monoclonal antibody enzyme-linked immunosorbent assay for specific identification and typing of subgroup F adenoviruses. *J. Clin. Microbiol.* **26:**297–300.

Smith, A. W., E. S. Berry, D. E. Skilling, J. E. Barlough, S. E. Poet, T. Berke, J. Mead, and D. O. Matson. 1998. In vitro isolation and characterization of a calicivirus causing a vesicular disease of the hands and feet. *Clin. Infect. Dis.* **26:**434–439.

Steele, A. D., H. R. Basetse, N. R. Blacklow, and J. E. Herrmann. 1998. Astrovirus infection in South Africa: a pilot study. *Ann. Trop. Paediatr.* **18:**315–319.

Steele, A. D., M. C. van Niekerk, and M. J. Mphahlele. 1995. Geographic distribution of human rotavirus VP4 genotypes and VP7 serotypes in five South African regions. *J. Clin. Microbiol.* **33:**1516–1519.

Stephen, J. 1988. Functional abnormalities in the intestine, p. 41–44. *In* M. Farthing (ed.), *Viruses in the Gut.* Smith Kline & French, Welwyn Garden City, United Kingdom.

Stewien, K. E., E. L. Durigon, H. Tanaka, A. E. Gilio, and E. R. Baldacci. 1991. Occurrence of human astrovirus in Sao Paulo City, Brazil. *Rev. Saude Publica* **25:**157–158. (In Portuguese.)

Sunen, E., and M. D. Sobsey. 1999. Recovery and detection of enterovirus, hepatitis A virus and Norwalk virus in hardshell clams (Mercenaria mercenaria) by RT-PCR methods. *J. Virol. Methods* **77:**179–187.

Takiff, H. E., S. E. Strauss, and C. F. Garon. 1981. Propagation and in vitro studies of previously non-cultivable enteral adenoviruses in 293 cells. *Lancet* **ii:**832–834.

Tam, J. S., B. J. Zeng, S. K. Lo, C. Y. Yeung, M. Lo, and M. H. Ng. 1990. Distinct population of rotaviruses circulating among neonates and older infants. *J. Clin. Microbiol.* **28:**1033–1038.

Tan, J. A., and R. D. Schnagl. 1981. Inactivation of a rotavirus by disinfectants. *Med. J. Aust.* **1:**19–23.

Tan, S. W., K. L. Yap, and H. L. Lee. 1997. Mechanical transport of rotavirus by the legs and wings of *Musca domestica*. *J. Med. Entomol.* **34:**527–531.

Taniguchi, K., S. Urasawa, and T. Urasawa. 1979. Virus-like particle, 35 to 40 nm, associated with an institutional outbreak of acute gastroenteritis in adults. *J. Clin. Microbiol.* **10:**730–736.

Tao, H., C. Guangmu, W. Changan, Y. Aenli, F. Zhaoying, C. Tungxin, C. Zinyi, Y. Weiwe, C. Xuejian, D. Shuasen, L. Xiaoguang, and C. Weicheng. 1984. Waterborne outbreak of rotavirus diarrhea in adults in China caused by a novel rotavirus. *Lancet* **i:**1139–1142.

Taylor, M. B., W. O. Grabow, and W. D. Cubitt. 1997a. Propagation of human astrovirus in the PLC/PRF/5 hepatoma cell line. *J. Virol. Methods* **67**:13–18.

Taylor, M. B., F. E. Marx, and W. O. Grabow. 1997b. Rotavirus, astrovirus and adenovirus associated with an outbreak of gastroenteritis in a South African child care centre. *Epidemiol. Infect.* **119**:227–230.

Thornhill, T. S., R. G. Wyatt, A. R. Kalica, R. Dolin, R. M. Chanock, and A. Z. Kapikian. 1977. Detection by immune electron microscopy of 26- to 27-nm viruslike particles associated with two family outbreaks of gastroenteritis. *J. Infect. Dis.* **135**:20–27.

Treanor, J., R. Dolin, and H. P. Madore. 1988. Production of a monoclonal antibody against the Snow Mountain agent of gastroenteritis by in vitro immunization of murine spleen cells. *Proc. Natl. Acad. Sci. USA* **85**:3613–3617.

Treanor, J. J., X. Jiang, H. P. Madore, and M. K. Estes. 1993. Subclass-specific serum antibody responses to recombinant Norwalk virus capsid antigen (rNV) in adults infected with Norwalk, Snow Mountain, or Hawaii virus. *J. Clin. Microbiol.* **31**:1630–1634.

Uhnoo, I., E. Olding-Stenkvist, and A. Kreuger. 1986. Clinical features of acute gastroenteritis associated with rotavirus, enteric adenoviruses, and bacteria. *Arch. Dis. Child.* **61**:732–738.

Uhnoo, I., M. Riepenhoff-Talty, T. Dharakul, P. Chegas, J. E. Fisher, H. B. Greenberg, and P. L. Ogra. 1990b. Extramucosal spread and development of hepatitis in immunodeficient and normal mice infected with rhesus rotavirus. *J. Virol.* **64**:361–368.

Uhnoo, I., G. Wadell, L. Svensson, and M. E. Johansson. 1984. Importance of enteric adenoviruses 40 and 41 in acute gastroenteritis in infants and children. *J. Clin. Microbiol.* **20**:365–372.

Uhnoo, I. S., J. Freihorst, M. Riepenhoff-Talty, J. E. Fisher, and P. L. Ogra. 1990a. Effect of rotavirus infection and malnutrition on uptake of a dietary antigen in the intestine. *Pediatr. Res.* **27**:152–160.

Unicomb, L. E., N. N. Banu, T. Azim, A. Islam, P. K. Bardhan, A. S. Faruque, A. Hall, C. L. Moe, J. S. Noel, S. S. Monroe, M. J. Albert, and R. I. Glass. 1998. Astrovirus infection in association with acute, persistent and nosocomial diarrhea in Bangladesh. *Pediatr. Infect. Dis. J.* **17**:611–614.

Unicomb, L. E., G. Podder, J. R. Gentsch, P. A. Woods, K. Z. Hasan, A. S. G. Faruque, M. J. Albert, and R. I. Glass. 1999. Evidence of high-frequency genomic reassortment of group A rotavirus strains in Bangladesh: emergence of type G9 in 1995. *J. Clin. Microbiol.* **37**:1885–1891.

Ushijima, H., K. Q. Xin, S. Nishimura, S. Morikawa, and T. Abe. 1994. Detection and sequencing of rotavirus VP7 gene from human materials (stools, sera, cerebrospinal fluids, and throat swabs) by reverse transcription and PCR. *J. Clin. Microbiol.* **32**:2893–2897.

Utagawa, E. T., S. Nishizawa, S. Sekine, Y. Hayashi, Y. Ishihara, I. Oishi, A. Iwasaki, I. Yamashita, K. Miyamura, S. Yamazaki, S. Inouye, and R. I. Glass. 1994. Astrovirus as a cause of gastroenteritis in Japan. *J. Clin. Microbiol.* **32**:1841–1845.

Van, R., C. C. Wun, M. L. O'Ryan, D. O. Matson, L. Jackson, and L. K. Pickering. 1992. Outbreaks of human enteric adenovirus types 40 and 41 in Houston day care centers. *J. Pediatr.* **120**:512–521.

Van Loon, A. E., R. Maas, R. T. Vaessen, A. M. Reemst, J. S. Sussenbach, and T. H. Rozijn. 1985. Cell transformation by the left terminal regions of the adenovirus 40 and 41 genomes. *Virology* **147**:227–230.

Vaucher, Y. E., C. G. Ray, L. L. Minnich, C. M. Payne, D. Beck, and P. Lowe. 1982. Pleomorphic, enveloped, virus-like particles associated with gastrointestinal illness in neonates. *J. Infect. Dis.* **145**:27–36.

Velazquez, R., D. O. Matson, J. J. Calva, M. Lourdes Guerrero, A. L. Morrow, S. Carter-Campbell, R. I. Glass, M. K. Estes, L. K. Pickering, and G. M. Ruiz-Palacios. 1996. Rotavirus infection in infants as protection against subsequent infections. *N. Engl. J. Med.* **335**:1022–1028.

Vesikari, T., M. Maki, H. J. Sarkkinen, P. P. Arstila, and P. E. Halonn. 1981a. Rotavirus, adenovirus, and non-viral enteropathogens in diarrhea. *Arch. Dis. Child.* **56**:264–270.

Vesikari, T., H. K. Sarkkinen, and M. Maki. 1981b. Quantitative aspects of rotavirus excretion in childhood diarrhoea. *Acta Paediatr. Scand.* **70**:717–721.

Vial, P. A., K. L Kotloff, and G. A. Losonsky. 1988. Molecular epidemiology of rotavirus infection in a room for convalescing newborns. *J. Infect. Dis.* **157**:668–673.

Vizzi, E., D. Ferraro, A. Cascio, R. DiStefano, and S. Arista. 1996. Detection of enteric adenoviruses 40 and 41 in stool specimens by monoclonal antibody-based enzyme immunoassays. *Res. Virol.* **147**:333–339.

Vollet, J. J., C. D. Ericsson, G. Gibson, L. K. Pickering, H. L. DuPont, S. Kohl, and R. H. Conklin. 1979. Human rotavirus in an adult population with travelers' diarrhea and its relationship to the location of food consumption. *J. Med. Virol.* **4**:81–87.

Von Bonsdorff, C. H., T. Hovi, P. Makela, and A. Morttinen. 1978. Rotavirus infections in adults in association with acute gastroenteritis. *J. Med. Virol.* **2**:21–28.

Wadell, G. 1984. Molecular epidemiology of human adenoviruses. *Curr. Top. Microbiol.* **110**:191–220.

Wadell, G., A. Allard, L. Svennson, and I. Uhnoo. 1988. Enteric adenoviruses, p. 70–78. In M. Farthing (ed.), *Viruses in the Gut*. Smith Kline & French, Welwyn Garden City, England.

Wadell, G., M. L. Hammarskjold, G. Winberg, T. M. Varsanyi, and G. Sundell. 1986. Genetic variability of adenoviruses. *Ann. N. Y. Acad. Sci.* **354**:16–42.

Wang, B., and X. Chen. 1997. The molecular epidemiology study on enteric adenovirus in stool specimens collected from Wuhan area by using digoxigenin labeled DNA probes. *J. Tongji Med. Univ.* **17**:79–82.

Wang, J., X. Jiang, H. P. Madore, J. Gray, U. Desselberger, T. Ando, Y. Seto, I. Oishi, J. F. Lew, and K. Y. Green. 1994. Sequence diversity of small, round-structured viruses in the Norwalk virus group. *J. Virol.* **68**:5982–5990.

Wen, L., M. Nakayama, Y. Yamanishi, O. Nishio, Z. Y. Fang, O. Nakagomi, and K. Arak. 1997. Genetic variation in the VP7 gene of human rotavirus serotype (G3 type) isolated in China and Japan. *Arch. Virol.* **142**:1481–1489.

Wenman, W. M., D. Hinde, S. Feltham, and M. Gurwith. 1979. Rotavirus infection in adults. Results of a prospective family study. *N. Engl. J. Med.* **301**:303–306.

White, K. E., M. T. Osterholm, J. A. Mariotti, J. A. Korlath, D. H. Lawrence, T. L. Ristinen, and H. B. Greenberg. 1986. A foodborne outbreak of Norwalk virus gastroenteritis. Evidence for post-recovery transmission. *Am. J. Epidemiol.* **124**:120–126.

Willcocks, M. M., T. D. Brown, C. R. Madeley, and M. J. Carter. 1994. The complete sequence of a human astrovirus. *J. Gen. Virol.* **75**:1785–1788.

Willcocks, M. M., M. J. Carter, F. R. Laidler, and C. R. Madeley. 1990. Growth and characterisation of human faecal astrovirus in a continuous cell line. *Arch. Virol.* **113:**73–81.

Willcocks, M. M., J. B. Kurtz, T. W. Lee, and M. J. Carter. 1995. Prevalence of human astrovirus serotype 4: capsid protein sequence and comparison with other strains. *Epidemiol. Infect.* **114:**385–391.

Wong, C. J., Z. Price, and D. A. Bruckner. 1984. Aseptic meningitis in an infant with rotavirus gastroenteritis. *Pediatr. Infect. Dis. J.* **3:**244–246.

World Health Organization. 1997. Vaccine research and development. Rotavirus vaccines for developing countries. *Weekly Epidemiol. Rec.* **72:**35–40.

Wu, B., J. Mahony, G. Simon, and M. Chernesky. 1990. Sensitive solid-phase immune electron microscopy double-antibody technique with gold-immunoglobulin G complexes for detecting rotavirus in cell culture and feces. *J. Clin. Microbiol.* **28:**864–868.

Wyatt, R. G., R. Dolin, N. R. Blacklow, H. L. DuPont, R. F. Buscho, T. S. Thornhill, A. Z. Kapikian, and R. M. Chanock. 1974. Comparison of three agents of acute infectious nonbacterial gastroenteritis by cross-challenge in volunteers. *J. Infect. Dis.* **129:**709–714.

Wyatt, R. G., H. D. James, A. L. Pittman, Y. Hoshino, H. B. Greenberg, A. R. Kalica, J. Flores, and A. Z. Kapikian. 1983. Direct isolation in cell culture of human rotaviruses and their characterization into four serotypes. *J. Clin. Microbiol.* **18:**310–317.

Yamashita, Y., M. Hattori, M. Oseto, M. Mori, H. Inouye, K. Takagi, Y. Ishimaru, and S. Nakano. 1995. Epidemiological studies on enteric adenovirus gastroenteritis in children. *Kansenshogaki Zasshi* **69:**377–382.

Yolken, R., F. Leister, J. Almeido-Hill, E. Dubovi, R. Reid, and M. Santosham. 1989. Infantile gastroenteritis associated with excretion of pestivirus antigens. *Lancet* **i:**517–519.

Yolken, R. H., and M. Murphy. 1982. Sudden infant death syndrome associated with rotavirus infection. *J. Med. Virol.* **10:**291–296.

Yolken, R. H., and P. J. Stopa. 1979. Enzyme-linked fluorescence assay: ultrasensitive solid-phase assay for detection of human rotavirus. *J. Clin. Microbiol.* **10:**317–321.

Yow, M. D., J. L. Melnick, R. J. Blattner, W. B. Stephenson, N. M. Robinson, and M. A. Burkhardt. 1970. The association of viruses and bacteria with infantile diarrhea. *Am. J. Epidemiol.* **92:**33–39.

Waterborne Hepatitis

DAVID A. ANDERSON

26

Viral hepatitis is the general term for inflammatory disease of the liver caused by at least five different viruses; all of these viruses—hepatitis A, B, C, D, and E viruses (HAV, HBV, HCV, HDV, and HEV, respectively)—have a definite association with acute viral hepatitis. The waterborne hepatitis viruses, HAV and HEV, both cause acute and generally self-limiting infections, and although fulminant hepatitis can occur, patients do not progress to long-term-carrier status. At the time of clinical presentation, patients may excrete large amounts of infectious virus; HAV in particular is very easily transmitted from person to person and is often the cause of large-scale, common-source outbreaks which are often traced to contaminated food. This contrasts with the blood-borne hepatitis viruses (described in chapter 27), which progress to chronic infections in many cases, with serious long-term sequelae, but have a lower chance of person-to-person transmission, as they are transmitted only through parenteral contact and, in the case of hepatitis B and D, sexual contact. Appropriate clinical management of patients with HAV or HEV infection is therefore dependent on exclusion of the potentially more serious HBV, HCV, and HDV infections, and is important for detection of outbreaks of hepatitis A in developed countries and both hepatitis A and hepatitis E in developing countries.

Despite great differences between the viruses, the clinical presentation of viral hepatitis is quite uniform in the acute phase, and differential diagnosis is therefore dependent on specific tests for each of the viruses. The selection of appropriate diagnostic tests to be used should be based on an assessment of the most likely risk factors for each infection, but interpretation of test results must also take into account the predictive values of the tests performed, which may vary widely.

This chapter will address the biology and epidemiology of the waterborne viruses, HAV and HEV, with particular emphasis on the value of diagnostic assays in different settings and the current and future prospects for prevention and control of HAV and HEV infections.

BIOLOGY OF HAV AND HEV

Both HAV and HEV are small viruses with genomes of single-stranded, positive-strand RNA (Fig. 1), with icosahedral viral particles that lack a lipid envelope. Much about the replication cycle of HAV has been learned over the past 20 years since its first propagation in cell culture, including the details of its replication, protein processing, and assembly. The intensive study of related members of the picornavirus family, such as poliovirus, has also given insights into the HAV replication cycle. Conversely, HEV has proven to be largely refractory to cell culture, and thus little is known of its replication cycle. Additionally, although HEV shares some common genome features with the calicivirus family (Cubitt et al., 1995), it is considered to be distinct from those viruses, such that it is taxonomically "unclassified" (Pringle, 1998). Little of the replication cycle can therefore be inferred from studies of other viruses.

Biology of HAV

As for all picornaviruses, replication of HAV begins with attachment to specific cellular receptors, internalization of viral particles, and uncoating or disassembly of the capsid to release the viral RNA. The positive-strand RNA genome encodes only a single open reading frame (ORF) that is translated to yield a giant polyprotein; the virus-encoded 3C protease cleaves the polyprotein at a number of sites, liberating eight proteins, including the RNA-dependent RNA polymerase (RDRP) $3D^{pol}$ and the protease $3C^{pro}$. The small VPg protein (part of 3AB) then serves as a primer for the synthesis by the RDRP of a complementary, negative-strand RNA and subsequently for further copies of positive-strand RNA. These identical daughter positive-strand RNAs serve as templates for further transcription to amplify the RNA pool, as mRNA for synthesis of viral proteins, and finally as new genomes for encapsidation within the viral particle (see Rueckert, 1990, for a comprehensive review of picornaviral replication). The HAV particle is initially composed of three proteins, VP0, VP1-2A (also known as PX), and VP3 (Anderson and Ross, 1990), which are liberated from the polyprotein by 3C (Martin et al., 1995) (Fig. 1A). Five copies of each of the three proteins associate to form pentamers, and twelve copies of the pentamer form virions (with viral RNA) or empty capsids with the same antigenicity. Following assembly, 2A is removed from VP1 by cellular enzymes and/or the 3C protease, and in the final maturation step, VP0 is cleaved via an RNA-dependent mechanism to yield VP2 and VP4 (Bishop and Anderson,

1993). Virus is released from cells without cell lysis, but the exact mechanism is unknown. However, recent studies have shown that HAV is preferentially released at the apical surface of polarized epithelial cells (CaCo2) (Blank et al., 2000), consistent with the release of virus into the bile canaliculi and hence the gut.

Diagnosis and immunization for hepatitis A are not complicated by strain differences, despite significant levels of amino acid variation in the capsid proteins, because HAV exists as a single serotype worldwide. This is especially notable in the case of the highly divergent AGM-27, which is considered a true simian strain of HAV that causes disease in African green monkeys and other lower primates (Emerson et al., 1991; Tsarev et al., 1991); this strain shows some potential for the development of a live attenuated hepatitis A vaccine because its antigenic sites are conserved but it has low pathogenicity in chimpanzees (Emerson et al., 1996). The antigenic sites of HAV are formed through the complex interactions of the proteins within and between pentamers (Stapleton et al., 1993); this fact has hampered the production of HAV antigenic material through recombinant DNA techniques, although recent studies using vaccinia vectors have shown promising results (LaBrecque et al., 1998).

Biology of HEV

While our knowledge of HEV replication is currently inadequate, it is clear that it will differ significantly from HAV replication in its mechanism of RNA replication and transcription and its translation scheme for viral proteins (Fig. 1B). HEV proteins are encoded in three separate ORFs, which suggests that multiple mRNA species would be synthesized by the RDRP in addition to the full-length genomic and antigenomic strands. Thus, a more complex scheme of RNA transcription is anticipated. Additionally, multiple RNA species have been detected in the livers of HEV-infected primates (Tam et al., 1991; Yarbough et al., 1991); however, we know nothing of the mechanism used for transcription of these species.

Translation of ORF1 yields a polyprotein (PORF1) containing sequence motifs consistent with RDRP, RNA helicase, methyltransferase, and protease activities (Tam et al., 1991), although none of these activities has been

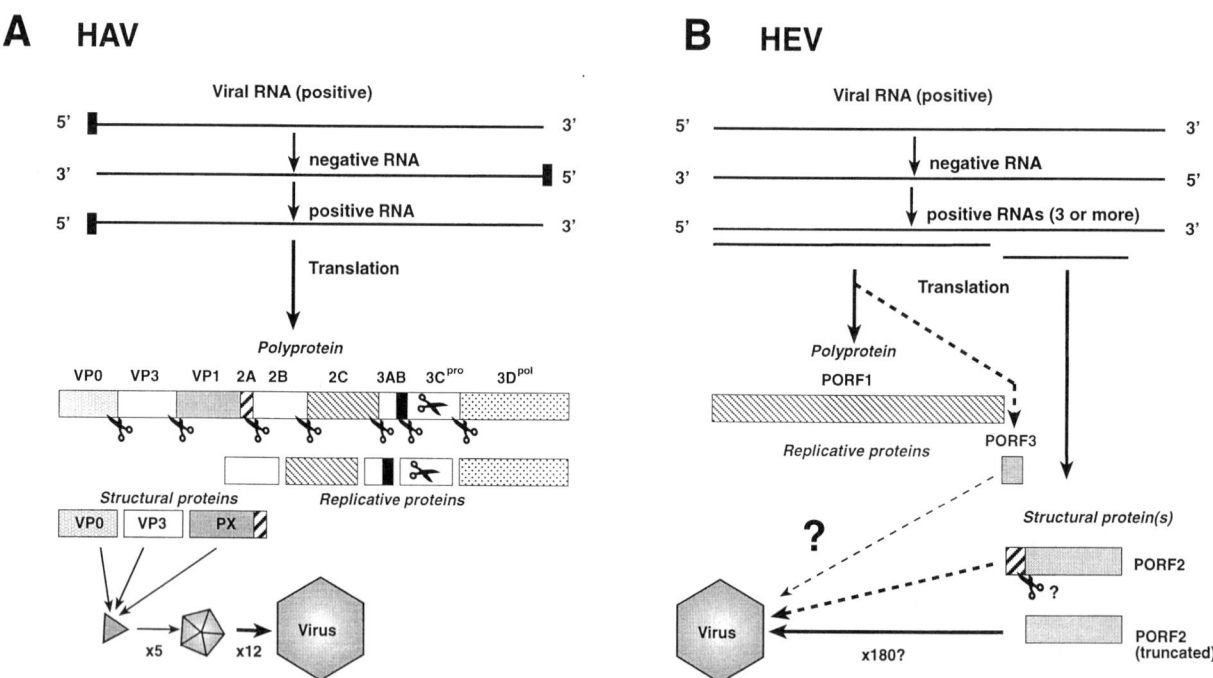

FIGURE 1 Genome replication and the encoded proteins of HAV (A) and HEV (B). Both viruses have positive-strand RNA genomes of around 7,500 nucleotides. (A) HAV replication proceeds via transcription from the genome to give full-length negative- and then positive-strand RNA, which can either be assembled into the virus particle or be used to translate further copies of a single, giant polyprotein which is processed by viral protease to yield the replicative proteins and capsid proteins. Assembly of five copies of each of the three capsid proteins (VP0, VP3, and PX) into pentamers and assembly of the 12 pentamers into capsids are required to form the antigenic sites of the virus. (B) Details of HEV replication are unknown, but it most likely produces a full-length negative-strand RNA and then full-length positive-strand RNA (new viral genomes) as well as subgenomic mRNAs which are used to translate the ORF2 (capsid) and ORF3 proteins, as well as the ORF1 polyprotein, which is cleaved at unknown sites to yield the replicative proteins. Cleavage of full-length PORF2 results in assembly of VLPs from the truncated product, but it is not known whether this or PORF3 is involved in the normal assembly process of the virus.

directly demonstrated and the sizes of mature proteins are unknown. ORF2 encodes the capsid protein, PORF2, and expression of truncated forms of this protein (lacking the first 111 amino acids) in insect cells leads to the assembly of virus-like particles (VLPs, or similar particles described as subviral particles [SVPs]) (He et al., 1993; Tsarev et al., 1993; Li et al., 1997; Zhang et al., 1997; Robinson et al., 1998). Curiously, expression of full-length PORF2 commonly leads to accumulation of insoluble products and large amounts of protein degradation (Torresi et al., 1999), but it is not known whether the authentic virus particle contains a truncated PORF2 (as in VLPs) or the full-length PORF2 in association with other factors, such as the viral RNA, which may promote its proper assembly into viral particles.

Notwithstanding this, VLPs produced using recombinant DNA techniques appear to possess the important antigenic epitopes of the virus (Li et al., 1997; Zhang et al., 1997; Robinson et al., 1998), and they are likely to be instrumental in our further understanding of HEV biology and efforts towards control of HEV infection.

One candidate as a cofactor for HEV assembly is the ORF3-encoded protein (PORF3), a highly basic protein which has been reported to associate with both PORF2 (Jameel et al., 1996) and the cytoskeleton (Zafrullah et al., 1997). Notably, ORF3 partially overlaps both ORF1 and ORF2 (Fig. 1), and it is possible that PORF3 may be translated from a bicistronic RNA, a mechanism which is used by feline calicivirus for its overlapping ORF2 and ORF3 (Herbert et al., 1996). The HEV PORF3 is also immunogenic, and many patients produce a strong but transient antibody response against the protein; however, this does not necessarily imply that the protein is structural, since reactivity to the nonstructural 3C protease in most patients infected with HAV has also been observed (Stewart et al., 1997).

Following assembly, HEV particles must presumably be released from the hepatocyte via the apical surface to reach the gut for transmission, as for HAV (Blank et al., 2000), but this has not been demonstrated.

Genetic and antigenic variations between HEV strains are far more pronounced than for HAV. Early studies identified the prototype Burmese (Reyes et al., 1990) and Mexican (Huang et al., 1992) strains and showed that a number of epitopes were type specific (Yarbough et al., 1991), but more recent studies have greatly expanded our knowledge of the diversity of HEV. The list of known HEV types now includes putative "genotypes" (as defined by Wang and colleagues [Wang et al., 1999] on the basis of relatively conserved ORF1 sequences); the Burmese and related strains (including African, Eastern European, and most Chinese strains) are classified as genotype 1, the Mexican strain is classified as genotype 2, the swine HEV strain (Meng et al., 1997) and closely related strains isolated from patients infected in the United States (Kwo et al., 1997; Schlauder et al., 1998) are classified as genotype 3, and distinct isolates from China (Wang et al., 1999) and Taiwan (Hsieh et al., 1998) are classified as genotype 4. As ORF1 encodes replicative proteins, it is not surprising that even greater variation is seen in the ORF2 and ORF3 proteins, with ORF3 varying by more than 50% between many strains. This genetic and antigenic diversity has important implications for epidemiology, diagnosis, and control, which will be discussed in detail below (see "Epidemiology of HEV," "Diagnosis of HEV Infection," and "Prevention of Waterborne Hepatitis").

PATHOGENESIS

Clinical Characteristics

Acute infections with any of the hepatitis viruses, including HAV and HEV, cannot be distinguished by clinical characteristics or pathological examinations. However, HEV infection is unique among the hepatitis virus infections, in that it is associated with a high mortality during pregnancy, approaching 30% in patients in the third trimester, due to fulminant hepatitis with a very rapid onset. Outside of pregnancy, the rate of fulminant hepatitis for HEV is probably around 10-fold higher than that for HAV (1 versus 0.1%) (Balayan, 1997). The reasons for such severe outcomes in HEV infection are not known, but it is clear that women should take all possible precautions to avoid exposure to HEV during pregnancy. A major risk factor in this regard is travel to areas in which HEV is endemic, such as India and Pakistan.

Infection with HAV or HEV can result in a broad range of clinical outcomes, from subclinical infections (especially in children) to fulminant hepatitis. Clinical and even fulminant hepatitis A infections do occur in children, and there is no reason to exclude the enterically transmitted hepatitis viruses from diagnostic consideration on the basis of patient age.

Clinical presentation of acute viral hepatitis commonly begins with nonspecific, flu-like symptoms, such as fever, headache, anorexia, nausea, and abdominal discomfort. The first distinctive sign of hepatitis is usually dark urine, followed by pale feces and jaundice (yellow discoloration of the skin and sclera); however, some patients will not show visible signs of jaundice despite having other severe symptoms. Physical examination will usually reveal an enlarged, tender liver.

Liver function tests are an important adjunct to diagnosis; raised levels of serum bilirubin, aspartate aminotransferase, and alanine aminotransferase are detected at the time of onset and usually resolve after a period of 3 to 4 weeks. Normalization of liver enzymes usually marks complete recovery; however, many patients report an intolerance to fatty foods which may last for years. When acute hepatitis is strongly suspected on the basis of clinical symptoms, it is appropriate to test for specific hepatitis virus infections (see "Diagnosis of HAV and HEV infections," below) without waiting for the results of liver function tests.

Relapses of hepatitis A are rare and occur in around 7% of patients one month postinfection (Sjogren et al., 1987), and prolonged disease with HEV has also been observed in some areas in which HEV is endemic.

While both HAV and HEV cause fulminant hepatitis infrequently, the onset of encephalopathy can be quite rapid (around 7 days from onset of dark urine for hepatitis A); laboratory examination reveals an increase in prothrombin time and an increase in bilirubin levels (Ross et al., 1991).

Natural History of Virus Infection

The generalized course of infection and serological responses for both HAV and HEV are shown in Fig. 2. Following ingestion of water or food contaminated with either virus, the infection is presumed to be initiated via cells lining the alimentary tract, although direct evidence of this has only recently been obtained, for HAV only (Asher et al., 1995). Virus then spreads to the liver and eventually infects a large proportion of the hepatocyte population but without caus-

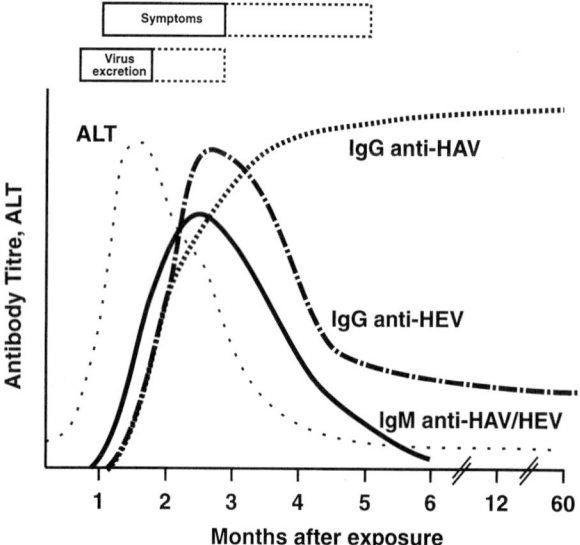

FIGURE 2 Serological and virological courses of infection with HAV or HEV. For HAV, the serological responses shown are typical of those detected with numerous commercially available assays. For HEV, the serological responses shown are those which probably occur in most patients, but the detection of these responses varies widely, depending on the assay used. High levels of HAV-specific IgG provide lifelong protection from reinfection, but HEV-specific IgG declines rapidly during the first six months and may not persist at protective levels for life.

ing direct cytolytic damage. After an incubation period of 4 to 6 weeks after exposure to the virus, liver damage results and is thought to be mediated by the cellular immune response to the viruses; infections in young children generally follow a benign course. Virus is excreted through bile to the gut, and very high titers of infectious HAV are present in most patients at the onset of illness (Coulepis et al., 1980), while titers of HEV are lower. Virus is generally cleared within several weeks, although prolonged excretion has been reported for both viruses. In general, titers of excreted virus will be highest before the onset of obvious symptoms.

Although the vast majority of progeny virus is excreted in the feces, both HAV and HEV produce a viremia around the time of clinical presentation which may last for some weeks. Transmission of HAV via blood products has been demonstrated (Mosley et al., 1994; Soucie et al., 1998), and inactivation procedures to eliminate HAV, as well as the more obvious blood-borne viruses such as HBV, HCV, and human immunodeficiency virus, are clearly required in the manufacture of blood products (Lemon, 1994, 1995). Fortunately, such processes have now been validated for HAV, and further transmission is unlikely (Lemon, 1995; Biesert et al., 1996; Adcock et al., 1998).

Immunity

High titers of immunoglobulin G (IgG) and IgM (and IgA) antibodies are produced by the time of disease onset. For HAV, IgM will often be detectable prior to IgG and IgG will continue to rise for some weeks after onset, whereas HEV-specific IgM and IgG will both peak at around the time of

onset. Although the viruses infect via mucosal surfaces, it is clear that circulating specific IgG is protective against both viruses when present in sufficient titer. Infection with HAV produces lifelong immunity, with high levels of specific IgG (Fig. 2) being maintained. Studies with gamma globulin prophylaxis have allowed minimal protective levels of antibody to be determined (Stapleton et al., 1985). It is also clear from passive immunization studies with high-titer macaque gamma globulin that antibody is sufficient to confer protection from HEV (Tsarev et al., 1994). However, titers of anti-HEV IgG appear to decline by around 90% in the first 6 months after infection before stabilizing, and it is not yet known whether levels of anti-HEV remain high enough to confer protection for life. Indeed, gamma globulin prepared from humans has proven ineffective in conferring protection, presumably due to low titers of antibody (Khuroo and Dar, 1992; Chauhan et al., 1998). Definition of protective antibody levels and antibody specificities will be a major step in the development of effective vaccines against HEV.

EPIDEMIOLOGY OF HEPATITIS A AND E

Despite the fact that the primary mode of transmission of both viruses is via water contaminated with human feces, HAV and HEV have very different distributions of both disease and infection worldwide. These differences appear to be linked at least partly to the efficiency of virus transmission, as HAV is excreted in very high titers, allowing both waterborne and person-to-person spread. However, questions remain regarding the prevalence of HEV infection in areas such as Nepal, where infection with HAV is almost universal and yet less than 50% of the population show evidence of exposure to HEV (Clayson et al., 1997; Anderson et al., 1999).

Epidemiology of HAV

HAV shows three major patterns of infection worldwide, largely reflecting sanitation standards (Ross et al., 1991). In countries or areas with poor sanitation, the high levels of HAV exposure result in most individuals becoming infected at an early age; however, in these circumstances HAV is not a major public health problem, because the symptoms tend to be mild or absent in children and the adult population is no longer susceptible. However, clinical HAV in young children is certainly seen on occasions, most obviously in foodborne outbreaks in schools in developed countries (Reid and Robinson, 1987; Niu et al., 1992; Hutin et al., 1999), but also in populations where HAV is endemic (I. L. Shrestha, personal communication).

Conversely, in countries or populations with very high standards of personal and public sanitation, few individuals are exposed to HAV at an early age, and thus the adult population will largely be susceptible to infection. Because of the high transmissibility of HAV, any large population is likely to have ongoing sporadic HAV infections at a low level, causing significant levels of disease in these susceptible adults. The HAV disease burden in the United States was estimated at more than 140,000 cases and 80 deaths annually in 1990 (Hadler, 1991). The public health problem of HAV in these areas is further compounded by the potential for large outbreaks of HAV spread by contaminated foods. In the past, such outbreaks were largely confined to shellfish, which by virtue of their filter feeding are able to concentrate viruses such as HAV from very large volumes of water and which are generally eaten raw or only lightly cooked. Even slight fecal contamination of waterways can lead to large outbreaks of shellfish-associated HAV infection in the susceptible adult population, such as that seen in Shanghai in 1988 (Halliday et al., 1991) and in the United States in 1991 (Desenclos et al., 1991). However, the increased transport of fresh and frozen foodstuffs between countries now creates additional sources of infection, such as a number of outbreaks in the United States which were traced to frozen strawberries and raspberries imported from Mexico (Reid and Robinson, 1987; Niu et al., 1992; Hutin et al., 1999), where they presumably had been exposed to contaminated water during production or processing. Intravenous and other illicit drug use is also associated with a higher risk of HAV infection and may play a major role in the epidemiology of HAV in some societies (Shaw et al., 1999).

An important aspect in the epidemiology of HAV is the high titer of virus produced, combined with the very great physical stability of the viral particle, which is relatively insensitive to extremes of heat: infectious virus is readily recovered after heating for 10 min at 60°C or at 78°C when in the presence of 1 M $MgCl_2$ (Anderson, 1987). These factors undoubtedly contribute to the high rate of secondary HAV infection among household contacts. Conversely, while relatively little is known of the stability of HEV particles, the low rate of person-to-person spread of HEV is likely to be due to a combination of lower particle stability and/or lower virus titers in excreta than those of HAV.

Epidemiology of HEV

The epidemiology of HEV infection is very poorly understood, especially in countries where the disease is not considered to be endemic (such as the United States). This is largely a consequence of the variable sensitivity and specificity of the assays which have been used to determine seroprevalence (that is, past infection), as IgG responses to different HEV antigens vary widely.

HEV infection is most easily recognized in its epidemic form, occurring every 7 to 10 years in countries in which it is endemic. In general, epidemics are associated with the wet season (summer in many countries in which HEV is endemic), and the highest clinical attack rate is among young adults. For example, HEV was shown retrospectively to be responsible for 16 of 17 epidemics of enterically transmitted hepatitis in India (Arankalle et al., 1994). In these epidemics, the detection of around 70% serological reactivity among patients (using first-generation assays) is sufficient to implicate HEV as a major cause of the outbreak. The role of HEV in sporadic cases of hepatitis is less well defined, but studies in Nepal and India have shown that HEV is responsible for the majority of cases of acute sporadic hepatitis (Arankalle et al., 1993; Clayson et al., 1997). Serological reactivity to HEV among patients with sporadic hepatitis in Nepal during the wet season in 1997 is shown in Fig. 3, implicating HEV in around 70% of cases, even though there was no epidemic.

Early studies detected surprisingly low rates of anti-HEV in countries where epidemics of HEV were known to have occurred in the past and where sporadic infection is ongoing. However, it is now clear that these assays failed to detect many individuals in these countries with past infection due to loss of the antibody against the specific antigens

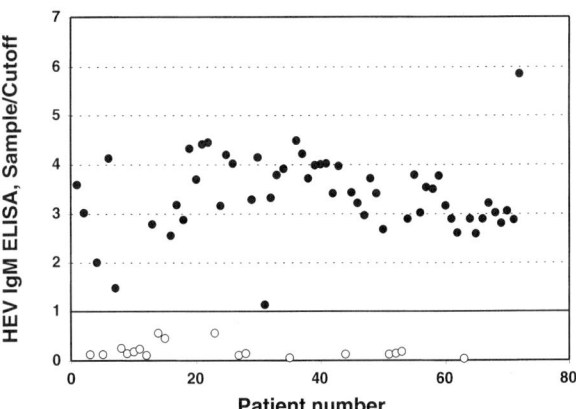

FIGURE 3 HEV IgM reactivities of patients with acute hepatitis in Nepal. Samples from patients with symptoms consistent with acute hepatitis in the month of August 1997 were tested by a prototype HEV IgM ELISA based on the ORF2.1 antigen at the Siddhi Polyclinic, Nepal. Results are shown as the sample-to-cutoff ratio. Closed circles, patients diagnosed as having acute HEV infection; open circles, patients negative for HEV-specific IgM (unpublished data from I. L. Shrestha and the author).

used, which is in fact the property allowing these same assays to be used for diagnosis of acute HEV infection. As shown in Fig. 4, most patients infected during an epidemic of HEV in the Xingiang province of China showed IgG reactivity by the first-generation HEV enzyme-linked immunosorbent assay (ELISA) within the first 46 days after

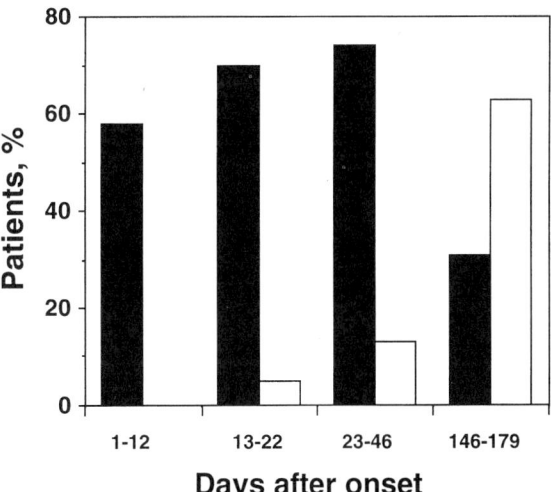

FIGURE 4 Loss of detectable HEV-specific antibody as a function of the assay used. Samples were collected at intervals from individual patients infected with HEV during an epidemic and were tested for HEV-specific IgG using a first-generation ELISA containing recombinant HEV proteins from ORF2 and ORF3. Solid bars indicate patients who were reactive for antibody at the indicated time after illness; open bars indicate patients who were nonreactive after having being reactive at earlier times after illness (negative seroconversion). (Modified from *Virus and Life* [Anderson, 1995], with permission of the publisher.)

illness, but by 6 months more than 60% of these patients had lost their reactivity and would thus be unreactive in seroepidemiological surveys using this assay. Conversely, individual patients may retain antibody against these antigens for long periods (Dawson et al., 1992; Goldsmith et al., 1992), which limits the usefulness of IgG assays for diagnosis of acute infection (see below); however, the proportion of reactive patients is uncertain.

Other assays have now been developed which should eventually help to resolve HEV epidemiology by their ability to detect specific IgG in the majority of patients who have recovered from HEV infection (Li et al., 1997; Zhang et al., 1997; Robinson et al., 1998; Anderson et al., 1999). On a cautionary note, however, there is no consensus regarding the true rate of past HEV infection in developed countries. All assays for HEV-specific IgG that have been developed to date detect at least 1 to 2% prevalence in countries in which HEV is presumably nonendemic, but very poor concordance between HEV assays in a study of a serum panel containing many such reactive specimens has been seen (Mast et al., 1998), and this lack of concordance is reflected in the disparity among recent reports of HEV epidemiology. For example, while ELISAs based on two different preparations of HEV SVPs detected anti-HEV in approximately 20% of serum samples from the United States (Thomas et al., 1997), very similar rates of anti-HEV were reported in a country in which HEV is endemic (Saudi Arabia) using these same assays (Ghabrah et al., 1998). In contrast, an ELISA based on the ORF2.1 protein expressed in *Escherichia coli* appears to be effective for detection of past HEV infection in a country in which HEV is endemic (Nepal); almost 33% of these individuals were reactive, while less than 2% of serum samples from Australia were positive by this assay (Anderson et al., 1999). It is not yet clear whether this reflects a higher specificity of the ORF2.1 assay, which detects low rates in a population in which HEV is not endemic, or whether it reflects a higher sensitivity of the SVP ELISA for the zoonotic strains of HEV which may circulate widely in countries in which it is not endemic.

Is There a Zoonotic Reservoir of HEV?

Further studies are required to determine the true rates of HEV infection in both developing and developed countries; in particular, three recent reports have forced a dramatic change in our assumptions regarding HEV prevalence in the United States and similar countries. In 1997, Kwo and colleagues at the Mayo Clinic detected HEV in a 62-year-old man with no history of recent travel outside the United States. Isolation of partial HEV RNA sequences from this patient revealed that it was a unique HEV genotype, designated US-1, that was different from the known Burmese and Mexican reference strains (Kwo et al., 1997). Subsequently, Meng and colleagues at the National Institutes of Health discovered a swine strain of HEV that infected 100% of pigs in most commercial piggeries studied in the United States but that did not cause obvious disease in the pigs and reported the RNA sequence of this virus, designated swine HEV (Meng et al., 1997). Finally, Schlauder and colleagues demonstrated that the US-1 isolate was a separate isolate of the swine HEV genotype, implicating zoonotic transmission of the virus, and diagnosed acute HEV infection in a second U.S. patient by using an ELISA based on peptides containing the unique US-1 amino acid

sequence (Schlauder et al., 1998). Together, these reports strongly support the conclusion of Kwo that "HEV should be considered an etiologic agent in patients with acute non-ABC hepatitis in the United States" (Kwo et al., 1997). Subsequent reports have revealed that clinical HEV, often associated with novel HEV genotypes, is present at low levels in many countries where it was presumed to be exotic (Hsieh et al., 1998; Worm et al., 1998; Schlauder et al., 1999; Zanetti et al., 1999) and that swine (and perhaps other domestic animals as well as rodents) in different parts of the world harbor HEV infection (Clayson et al., 1995a; Maneerat et al., 1996; Chandler et al., 1999).

There is thus compelling evidence that infection with zoonotic strains of HEV can cause low rates of clinical HEV in countries where conventional spread of human HEV strains does not occur. This reinforces the need for HEV to be considered in the diagnostic criteria for patients in these countries; the ability to detect such strains will be important for diagnostics, but further work and the use of improved serological assays will be required before we gain a clear understanding of the epidemiology of HEV.

Finally, it should be noted that the epidemiologies of HAV and HEV will often not be uniform within a country, illustrated by the very high rates of anti-HAV in Australian Aboriginal children in the Northern Territory compared with those in the general Australian population (Bowden et al., 1994) and the high rates of anti-HEV in indigenous populations of both Malaysia (Seow et al., 1999) and Australia (Riddell and coworkers, unpublished data) compared with those in the respective general populations.

DIAGNOSIS OF HAV AND HEV INFECTION

Correct diagnosis of viral hepatitis depends on reliable assays for each of the viruses, and the choice of which viruses are the most likely causes (and thus which assays should be given priority) must take into account the patient history and the risk factors for exposure to each virus. In the absence of a relevant travel history to developing countries within the past 6 weeks, HEV is much less likely to be implicated than HAV and in most settings would also be considered less likely than HBV and HCV. However, there is no doubt that HEV should now be considered along with HAV (and the blood-borne hepatitis viruses) in the diagnosis of acute hepatitis in developed countries; exclusion of HEV cannot be made on the lack of travel history alone (Kwo et al., 1997; Schlauder et al., 1998).

Diagnosis of HAV Infection

Diagnosis of HAV infection is very straightforward, with the detection of HAV-specific IgM from a single sample of serum or plasma being highly predictive of acute hepatitis A. HAV has a single, immunodominant antigenic site to which all patients react strongly, and the many diagnostic assays available are based on the detection of IgM reacting with intact (but inactivated) particles of HAV. Assays for total anti-HAV and anti-HAV IgG are also available, but their utility in diagnosis is extremely limited (as they require the detection of rising titers of antibody over a 2- to 4-week period). These assays may be useful, however, for determining patient antibody status prior to the administration of inactivated HAV vaccines, but this is generally not cost-effective (Van Doorslaer et al., 1994).

The technology for detection of HAV-specific IgM is sufficiently well established that there are no marked differences between the diagnostic performance of the laboratory-based testing platforms. Choices of diagnostic assays for HAV may reasonably be made on the basis of convenience for the testing laboratory; assays range from single-strip ELISAs for small numbers of specimens to completely automated systems for high-volume laboratories. Rapid, point-of-care assays for diagnosis of acute HAV infection may become available in the future (see "Rapid Diagnosis of Hepatitis A and E," below).

Diagnosis of HEV Infection

Diagnosis of HEV infection is less well established; a number of research and commercial immunoassays are available in various countries, but they have major differences in their sensitivity and specificity (Mast et al., 1998). Prior to the development of comprehensive serological tests, the detection of HEV RNA during viremia provided the only unequivocal evidence for acute HEV infection, with very good sensitivity when performed under ideal conditions (Clayson et al., 1995b). This method has also been instrumental in the detection of divergent HEV strains, in which the serological responses have not been detected by some assays (Hsieh et al., 1998; Schlauder et al., 1999; Wang et al., 1999); however, such detection can only be achieved by sensitive reverse transcription-PCR (RT-PCR), with attendant problems of specificity and specimen transport and handling, which are exacerbated by the conditions in countries where HEV is endemic. In addition, the period of viremia is short, and RT-PCR is therefore unsatisfactory for routine diagnosis of HEV.

The appropriate use and interpretation of serological assays for HEV infection must take into account the widely varying prevalence of HEV infection worldwide.

Diagnosis of HEV Infection in Areas of Low Prevalence

In areas of presumed low prevalence of HEV infection, such as the United States and other countries with high levels of sanitation, test specificity will have a very large impact on the predictive value of test results. The detection of HEV-specific IgG formed the basis of first-generation diagnostic assays for HEV (manufactured by Genelabs Diagnostics and Abbott Diagnostics) and had considerable value for the diagnosis of acute hepatitis in travelers returned from areas in which HEV is endemic (Dawson et al., 1992), among whom the incidence may be much higher than the background rate of reactivity. However, around 2% of the healthy population in areas in which HEV is not endemic are reactive in these assays, and if HEV should be considered in the diagnosis of sporadic acute hepatitis in patients without a travel history (Kwo et al., 1997), the need for more specific tests becomes evident. As an example, if the incidence of HEV infection among patients with acute hepatitis in the United States were 0.2%, then only 1 in 10 patients reactive in a test for HEV-specific IgG would be a true positive. The detection of HEV-specific IgM should therefore become the method of choice for diagnosis of acute HEV infection in areas of low prevalence, such as the United States and western Europe. One such assay, manufactured by Genelabs Diagnostics (Singapore), is currently commercially available in some countries, but the false-positive rate in U.S. blood donors with this assay is 26 of 856 (3%) (Genelabs, 1998), severely limiting its usefulness in

areas in which HEV is not endemic. Studies published by Genelabs scientists have also demonstrated that the antigens present in this assay may fail to detect around 40% of patients with acute HEV infection (Yarbough et al., 1996). Other assays with improved specificity and sensitivity should become available in the near future through the use of more-advanced recombinant antigens which are in use in research laboratories (Favorov et al., 1996; Li et al., 1997; Zhang et al., 1997; Robinson et al., 1998; Anderson et al., 1999). Any such improved assays must also be able to detect infections caused by the diverse HEV strains that appear to be present worldwide.

Diagnosis of HEV Infection in Areas of High Prevalence

While high titers of IgG in patients are suggestive of acute infection, this correlation is imperfect (Ghabrah et al., 1998; Anderson et al., 1999). Thus, the detection of IgG antibody is of little use for diagnosis of acute infection in developing countries where HEV is endemic and where large numbers of patients will have antibody from past infections. The practical limitations of RT-PCR also preclude its widespread use in these settings, and detection of HEV-specific IgM is therefore the method of choice.

As HEV accounts for as much as 70% of the acute sporadic hepatitis in countries in which HEV is endemic, the specificity of assays is less important than in settings in which HEV is not endemic. For example, if the false-positive rate of an assay is 2%, then only 1 in 35 patients with hepatitis in areas in which HEV is endemic might be misdiagnosed with acute HEV. Ideally, assay sensitivity should be sufficiently robust to allow assays to be processed and interpreted manually, where equipment such as ELISA washers and readers are not available. Again, it is likely that the next generation of assays based on improved antigens will be required for use in areas of high prevalence; assays which are currently being developed show promise for HEV diagnosis in these areas (Fig. 3).

Rapid Diagnosis of Hepatitis A and E

Rapid, point-of-care diagnosis of diseases is an admirable goal, with clear advantages for both the clinician and the patient. In view of the common major routes of transmission and relatively good prognosis of HAV and HEV, their rapid diagnosis would also be useful as a first line of investigation. The result could quickly determine whether laboratory investigations of HBV or HCV are necessary, since there are attendant concerns that these viruses can raise in patients.

While automated laboratory diagnostic assays can give results in less than an hour for HAV, no point-of-care assays are currently available for HAV and HEV, although prototype assays for HAV have been demonstrated in the past. However, recent advances in immunochromatographic technologies have been used to develop prototype point-of-care assays for the detection of both HAV- and HEV-specific IgM (Fig. 5). Such assays should prove useful in the clinical settings of developed countries, providing a rapid result to guide the selection of further tests where necessary. In developing countries and refugee settings, these tests would offer the further benefits of reduced collection of venous blood (which is not well tolerated in some cultural settings) and rapid identification of outbreaks remote from laboratory facilities.

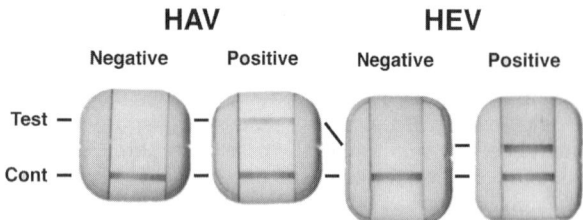

FIGURE 5 Rapid immunochromatographic test for the detection of HAV-specific and HEV-specific IgM. In these examples, 45 μl of patient serum was applied to the sample pad, buffer was applied to the reagent pad containing viral antigens complexed with monoclonal antibody/colloidal gold conjugate, and the test result was visible within 5 min (unpublished data from the author's laboratory [HEV] and from T. Howard, AMRAD Biotech [HAV]). (Prototype test kits were kindly provided by AMRAD ICT, Sydney, Australia.

PREVENTION OF WATERBORNE HEPATITIS

Due to the enteric transmission of waterborne hepatitis, control on a public health basis relies primarily on protection of the water supply from contamination with human feces. However, outbreaks of HAV infection associated with minimally processed foods (Desenclos et al., 1991; Niu et al., 1992; Hutin et al., 1999), together with the relatively high rate of person-to-person transmission, contribute to a sustained level of hepatitis A disease in the United States and other developed countries where general standards of hygiene are very high (Hadler, 1991). Scrupulous personal hygiene is vital in preventing spread from index cases (especially in food handlers or among household members).

Passive immunization with human gamma globulin has been used for many years to provide short-term protection against hepatitis A to travelers and to control outbreaks. The more recently available, highly efficacious, inactivated vaccines against hepatitis A (such as Havrix and VAQTA) have replaced the use of gamma globulin for protection of individuals at high risk of HAV infection, such as travelers, child care workers, homosexual males, and intravenous drug users. These vaccines have protective efficacies of greater than 95% after two doses, with minimal side effects that are largely confined to local reactions (Newcomer et al., 1994; Van Damme et al., 1994). Recent reports suggest that immunization with the inactivated vaccines is also appropriate for intervention in outbreak settings and for household contacts (Sagliocca et al., 1999). The development of combined vaccines for hepatitis A and B will be of particular use for health care workers, travelers, and child care workers, who are at risk for both infections (Ambrosch et al., 1994). Active immunization against hepatitis A is strongly recommended for travelers to countries where standards of hygiene and sanitation are low. A thorough cost-benefit analysis of active immunization, passive immunization, and testing for antibody prior to HAV immunization has been made based on experiences in the United Kingdom (Van Doorslaer et al., 1994), but in general there is no benefit to testing for antibody prior to immunization.

Despite the efficacy of HAV vaccines and the likely availability of HEV vaccines in the future, travelers to areas in which these diseases are endemic should be reminded that all precautions should be taken to avoid other waterborne infections, as widespread infection with these hepatitis viruses is a sure indication that water is regularly subjected to contamination with human feces and other viruses and bacteria are thus likely to be present. Only bottled or freshly boiled water should be consumed, and ice, salads, and raw or partly cooked shellfish should be avoided.

No vaccines are yet available for protection against hepatitis E, but the prospects for these are very good. Studies by Purcell and colleagues at the National Institutes of Health (Tsarev et al., 1994, 1997) have clearly demonstrated that antibody is sufficient to confer protection against hepatitis E and that immunization with the SVP produced following expression of the truncated ORF2 (capsid) protein gives rise to protective levels of antibody. Further development of this recombinant subunit vaccine is expected, but it is likely to be at least several years before a vaccine is available. Further studies are also required to determine the need for HEV vaccination in areas in which HEV is not endemic, in view of the high rate of seroprevalence detected in the United States using some assays and the evidence for zoonotic infection with HEV (see "Is There a Zoonotic Reservoir of HEV?," above).

Although HAV (and soon HEV) may be partly controlled with vaccines, there is evidence that some cases of waterborne hepatitis are due to as-yet-unidentified viral agents. In studies conducted in India and parts of Africa, screening of patients with sporadic hepatitis for serological evidence of HAV, HBV, HCV, and HEV infection failed to detect a specific cause in around half the cases (Arankalle et al., 1993; Arankalle et al., 1995; Coursaget et al., 1995); however, it is probable that many of these cases may be acute HEV infection which cannot be detected by currently available assays. More convincingly, a retrospective study of one outbreak of waterborne hepatitis, on the Indian islands of Andaman in 1987, failed to detect any evidence of HEV; thus, it appears that a novel agent was responsible for that outbreak, which involved more than 300 patients (Arankalle et al., 1994).

UNUSUAL FEATURES AND FUTURE PROSPECTS

The reason for the high mortality associated with acute HEV during pregnancy remains obscure (Tsarev et al., 1995), and the role of zoonotic HEV infection in areas in which HEV is both endemic and nonendemic, such as the United States, requires further study. Serological studies of HEV prevalence must be interpreted with caution because of the widely varying sensitivity and specificity of the assays and because prevalence can vary greatly between populations within countries. The real impact of HEV infection will not be known until testing of patients with sporadic hepatitis for HEV becomes routine, but in order for the results to be meaningful, widely available assays which have very high specificity and sensitivity must be developed.

HAV will continue to be a major public health problem, despite the use of inactivated vaccines in high-risk individuals, and the combined effects of declining antibody prevalence (leading to a larger susceptible population) and high levels of foreign travel may result in an increased incidence of clinical HAV in many developed countries. Foodborne outbreaks will contribute significant morbidity, since a minute amount of infectious feces in waterways may be sufficient to cause widespread HAV contamination of shellfish.

CONCLUSIONS

For the patient with acute hepatitis, the major value in specific diagnosis lies in defining the long-term prognosis once the acute disease has passed and the risk of transmission to partners, family members, and other contacts. In the case of HAV and HEV, a diagnosis provides reassurance that there will be none of the long-term sequelae associated with the blood-borne forms of hepatitis and highlights the importance of hygiene and the use of HAV vaccine or gamma globulin in preventing further transmission. However, the diagnosis of HAV and HEV also has important public health benefits, as efforts to control these infections cannot succeed without thorough evaluation of patients with sporadic hepatitis. Hepatitis A is readily diagnosed with the high-quality tests available, and safe, effective vaccines are already available. Hepatitis E must be suspected in patients with acute hepatitis in whom tests for other hepatitis viruses are negative, as the evidence that zoonotic forms of HEV can cause disease in humans in countries such as the United States is compelling. However, the accurate diagnosis of hepatitis E, a better understanding of HEV disease burden, and appropriate use of the vaccine(s) which is in development will require improved diagnostic immunoassays. These should be available in the near future.

Studies in my laboratory have been supported in part by the WHO Programme for Vaccine Development (HAV), the National Health and Medical Research Council of Australia (HEV), and the Research Fund of the Macfarlane Burnet Centre for Medical Research.

REFERENCES

Adcock, W. L., A. MacGregor, J. R. Davies, M. Hattarki, D. A. Anderson, and N. H. Goss. 1998. Chromatographic removal and heat inactivation of hepatitis A virus during manufacture of human albumin. *Biotechnol. Appl. Biochem.* **28:**85–94.

Ambrosch, F., G. Wiedermann, F. E. Andre, A. Delem, H. Gregor, H. Hofmann, E. D'Hondt, M. Kundi, J. Wynen, and C. Kunz. 1994. Clinical and immunological investigation of a new combined hepatitis A and hepatitis B vaccine. *J. Med. Virol.* **44:**452–456.

Anderson, D. A. 1987. Cytopathology, plaque assay, and heat inactivation of hepatitis A virus strain HM175. *J. Med. Virol.* **22:**35–44.

Anderson, D. A. 1995. Hepatitis E. *Virus Life* **14:**7–9.

Anderson, D. A., F. Li, M. A. Riddell, T. Howard, H.-F. Seow, J. Torresi, G. Perry, D. Sumarsidi, S. M. Shrestha, and I. L. Shrestha. 1999. ELISA for IgG-class antibody to hepatitis E virus based on a highly conserved, conformational epitope expressed in *Escherichia coli. J. Virol. Methods* **81:**131–142.

Anderson, D. A., and B. C. Ross. 1990. Morphogenesis of hepatitis A virus: isolation and characterization of subviral particles. *J. Virol.* **64:**5284–5289.

Arankalle, V. A., M. S. Chadha, S. A. Tsarev, S. U. Emerson, A. R. Risbud, K. Banerjee, and R. H. Purcell. 1994. Seroepidemiology of water-borne hepatitis in India and evidence for a third enterically-transmitted hepatitis agent. *Proc. Natl. Acad. Sci. USA* **91:**3428–3432.

Arankalle, V. A., L. P. Chobe, J. Jha, M. S. Chadha, K. Banerjee, M. O. Favorov, T. Kalinina, and H. Fields. 1993. Etiology of acute sporadic non-A, non-B viral hepatitis in India. *J. Med. Virol.* **40:**121–125.

Arankalle, V. A., J. Jha, M. O. Favorov, A. Chaudhari, H. A. Fields, and K. Banerjee. 1995. Contribution of HEV and

HCV in causing fulminant non-A, non-B hepatitis in western India. *J. Viral Hepat.* **2:**189–193.

Asher, L. V., L. N. Binn, T. L. Mensing, R. H. Marchwicki, R. A. Vassell, and G. D. Young. 1995. Pathogenesis of hepatitis A in orally inoculated owl monkeys (*Aotus trivirgatus*). *J. Med. Virol.* **47:**260–268.

Balayan, M. S. 1997. Epidemiology of hepatitis E virus infection. *J. Viral Hepat.* **4:**155–165.

Biesert, L., S. Lemon, A. Goudeau, H. Suhartono, L. Wang, and H. D. Brede. 1996. Viral safety of a new highly purified factor VIII (OCTATE). *J. Med. Virol.* **48:**360–366.

Bishop, N. E., and D. A. Anderson. 1993. RNA-dependent cleavage of VP0 capsid protein in provirions of hepatitis A virus. *Virology* **197:**616–623.

Blank, C. A., D. A. Anderson, M. Beard, and S. M. Lemon. 2000. Infection of polarized cultures of human intestinal epithelial cells with hepititis A virus: vectorial release of progeny virions through apical cellular membranes. *J. Virol.* **74:**6476–6484.

Bowden, F. J., B. J. Currie, N. C. Miller, S. A. Locarnini, and V. L. Krause. 1994. Should aboriginals in the "top end" of the Northern Territory be vaccinated against hepatitis A? *Med. J. Aust.* **161:**372–373.

Chandler, J. D., M. A. Riddell, F. Li, R. J. Love, and D. A. Anderson. 1999. Serological evidence for swine hepatitis E virus infection in Australian pig herds. *Vet. Microbiol.* **68:**97–107.

Chauhan, A., J. B. Dilawari, R. Sharma, M. Mukesh, and S. R. Saroa. 1998. Role of long-persisting human hepatitis E virus antibodies in protection. *Vaccine* **16:**755–756.

Clayson, E. T., B. L. Innis, K. S. Myint, S. Narupiti, D. W. Vaughn, S. Giri, P. Ranabhat, and M. P. Shrestha. 1995a. Detection of hepatitis E virus infections among domestic swine in the Kathmandu Valley of Nepal. *Am. J. Trop. Med. Hyg.* **53:**228–232.

Clayson, E. T., K. S. Myint, R. Snitbhan, D. W. Vaughn, B. L. Innis, L. Chan, P. Cheung, and M. P. Shrestha. 1995b. Viremia, fecal shedding, and IgM and IgG responses in patients with hepatitis E. *J. Infect. Dis.* **172:**927–933.

Clayson, E. T., M. P. Shrestha, D. W. Vaughn, R. Snitbhan, K. B. Shrestha, C. F. Longer, and B. L. Innis. 1997. Rates of hepatitis E virus infection and disease among adolescents and adults in Kathmandu, Nepal. *J. Infect. Dis.* **176:**763–766.

Coulepis, A. G., S. A. Locarnini, N. I. Lehmann, and I. D. Gust. 1980. Detection of hepatitis A virus in the feces of patients with naturally acquired infections. *J. Infect. Dis.* **141:**151–156.

Coursaget, P., D. Leboulleux, Y. Gharbi, N. Enogat, M. A. Ndao, A. M. Coll-Seck, and R. Kastally. 1995. Etiology of acute sporadic hepatitis in adults in Senegal and Tunisia. *Scand. J. Infect. Dis.* **27:**9–11.

Cubitt, D., D. W. Bradley, M. J. Carter, S. Chiba, M. K. Estes, L. J. Saif, F. L. Schaffer, A. W. Smith, M. J. Studdert, and H. J. Thiel. 1995. Caliciviridae, p. 359–363. *In* F. A. Murphy, C. M. Fauquet, D. L. Bishop, S. A. Ghabrial, A. W. Jarvis, G. P. Martelli, M. A. Mayo, and M. D. Summers (ed.), *Virus Taxonomy: the Classification and Nomenclature of Viruses. The Sixth Report of the ICTV.* Springer-Verlag, Vienna, Austria.

Dawson, G. J., I. K. Mushahwar, K. H. Chau, and G. L. Gitnick. 1992. Detection of long-lasting antibody to hepatitis E virus in a US traveller to Pakistan. *Lancet* **340:**426–427.

Desenclos, J. C., K. C. Klontz, M. H. Wilder, O. V. Nainan, H. S. Margolis, and R. A. Gunn. 1991. A multistate outbreak of hepatitis A caused by the consumption of raw oysters. *Am. J. Public Health* **81**:1268–1272.

Emerson, S. U., S. A. Tsarev, S. Govindarajan, M. Shapiro, and R. H. Purcell. 1996. A simian strain of hepatitis A virus, AGM-27, functions as an attenuated vaccine for chimpanzees. *J. Infect. Dis.* **173**:592–597.

Emerson, S. U., S. A. Tsarev, and R. H. Purcell. 1991. Biological and molecular comparisons of human (HM-175) and simian (AGM-27) hepatitis A viruses. *J. Hepatol.* **13**(Suppl. 4):S144–S145.

Favorov, M. O., Y. E. Khudyakov, E. E. Mast, T. L. Yashina, C. N. Shapiro, N. S. Khudyakova, D. L. Jue, G. G. Onischenko, H. S. Margolis, and H. A. Fields. 1996. IgM and IgG antibodies to hepatitis E virus (HEV) detected by an enzyme immunoassay based on an HEV-specific artificial recombinant mosaic protein. *J. Med. Virol.* **50**:50–58.

Genelabs. 1998. Instruction manual, HEV IgM ELISA, 3-98 ed. Genelabs Diagnostics, Singapore.

Ghabrah, T. M., S. Tsarev, P. O. Yarbough, S. U. Emerson, G. T. Strickland, and R. H. Purcell. 1998. Comparison of tests for antibody to hepatitis E virus. *J. Med. Virol.* **55**:134–137.

Goldsmith, R., P. O. Yarbough, G. R. Reyes, K. E. Fry, K. A. Gabor, M. Kamel, S. Zakaria, S. Amer, and Y. Gaffar. 1992. Enzyme-linked immunosorbent assay for diagnosis of acute sporadic hepatitis E in Egyptian children. *Lancet* **339**:328–331.

Hadler, S. C. 1991. Global impact of hepatitis A virus infection: changing patterns, p. 14–20. *In* F. B. Hollinger, S. M. Lemon, and H. H. Margolis (ed.), *Viral Hepatitis and Liver Disease.* Williams & Wilkins, Baltimore, Md.

Halliday, M. L., L. Y. Kang, T. K. Zhou, M. D. Hu, Q. C. Pan, T. Y. Fu, Y. S. Huang, and S. L. Hu. 1991. An epidemic of hepatitis A attributable to the ingestion of raw clams in Shanghai, China. *J. Infect. Dis.* **164**:852–859.

He, J., A. W. Tam, P. O. Yarbough, G. R. Reyes, and M. Carl. 1993. Expression and diagnostic utility of hepatitis E virus putative structural proteins expressed in insect cells. *J. Clin. Microbiol.* **31**:2167–2173.

Herbert, T. P., I. Brierley, and T. D. Brown. 1996. Detection of the ORF3 polypeptide of feline calicivirus in infected cells and evidence for its expression from a single, functionally bicistronic, subgenomic mRNA. *J. Gen. Virol.* **77**:123–127.

Hsieh, S. Y., P. Y. Yang, Y. P. Ho, C. M. Chu, and Y. F. Liaw. 1998. Identification of a novel strain of hepatitis E virus responsible for sporadic acute hepatitis in Taiwan. *J. Med. Virol.* **55**:300–304.

Huang, C. C., D. Nguyen, J. Fernandez, K. Y. Yun, K. E. Fry, D. W. Bradley, A. W. Tam, and G. R. Reyes. 1992. Molecular cloning and sequencing of the Mexico isolate of hepatitis E virus (HEV). *Virology* **191**:550–558.

Hutin, Y. J., V. Pool, E. H. Cramer, O. V. Nainan, J. Weth, I. T. Williams, S. T. Goldstein, K. F. Gensheimer, B. P. Bell, C. N. Shapiro, M. J. Alter, and H. S. Margolis. 1999. A multistate, foodborne outbreak of hepatitis A. National Hepatitis A Investigation Team. *N. Engl. J. Med.* **340**:595–602.

Jameel, S., M. Zafrullah, M. H. Ozdener, and S. K. Panda. 1996. Expression in animal cells and characterization of the hepatitis E virus structural proteins. *J. Virol.* **70**:207–216.

Khuroo, M. S., and M. Y. Dar. 1992. Hepatitis E: evidence for person-to-person transmission and inability of low dose immune serum globulin from an Indian source to prevent it. *Indian J. Gastroenterol.* **11**:113–116.

Kwo, P. Y., G. G. Schlauder, H. A. Carpenter, P. J. Murphy, J. E. Rosenblatt, G. J. Dawson, E. E. Mast, K. Krawczynski, and V. Balan. 1997. Acute hepatitis E by a new isolate acquired in the United States. *Mayo Clin. Proc.* **72**:1133–1136.

LaBrecque, F. D., D. R. LaBrecque, D. Klinzman, S. Perlman, J. B. Cederna, P. L. Winokur, J. Q. Han, and J. T. Stapleton. 1998. Recombinant hepatitis A virus antigen: improved production and utility in diagnostic immunoassays. *J. Clin. Microbiol.* **36**:2014–2018.

Lemon, S. M. 1994. The natural history of hepatitis A: the potential for transmission by transfusion of blood or blood products. *Vox Sang.* **4**:19–23.

Lemon, S. M. 1995. Hepatitis A virus and blood products: virus validation studies. *Blood Coagul. Fibrinolysis* **6**(Suppl. 2):S20–S22.

Li, T. C., Y. Yamakawa, K. Suzuki, M. Tatsumi, M. A. Razak, T. Uchida, N. Takeda, and T. Miyamura. 1997. Expression and self-assembly of empty virus-like particles of hepatitis E virus. *J. Virol.* **71**:7207–7213.

Maneerat, Y., E. T. Clayson, K. S. Myint, G. D. Young, and B. L. Innis. 1996. Experimental infection of the laboratory rat with the hepatitis E virus. *J. Med. Virol.* **48**:121–128.

Martin, A., N. Escriou, S. F. Chao, M. Girard, S. M. Lemon, and C. Wychowski. 1995. Identification and site-directed mutagenesis of the primary (2A/2B) cleavage site of the hepatitis A virus polyprotein: functional impact on the infectivity of HAV RNA transcripts. *Virology* **213**:213–222.

Mast, E. E., M. J. Alter, P. V. Holland, and R. H. Purcell. 1998. Evaluation of assays for antibody to hepatitis E virus by a serum panel. Hepatitis E Virus Antibody Serum Panel Evaluation Group. *Hepatology* **27**:857–861.

Meng, X.-J., R. H. Purcell, P. G. Halbur, J. R. Lehman, D. M. Webb, T. S. Tsareva, J. S. Haynes, B. J. Thacker, and S. U. Emerson. 1997. A novel virus in swine is closely related to the human hepatitis E virus. *Proc. Natl. Acad. Sci. USA* **94**:9860–9865.

Mosley, J. W., M. J. Nowicki, C. K. Kasper, E. Donegan, L. M. Aledort, M. W. Hilgartner, and E. A. Operskalski. 1994. Hepatitis A virus transmission by blood products in the United States. Transfusion Safety Study Group. *Vox Sang.* **67**(Suppl. 1):24–28.

Newcomer, W., B. Rivin, R. Reid, L. H. Moulton, M. Wolff, J. Croll, C. Johnson, L. Brown, D. Nalin, and M. Santosham. 1994. Immunogenicity, safety and tolerability of varying doses and regimens of inactivated hepatitis A virus vaccine in Navajo children. *Pediatr. Infect. Dis. J.* **13**:640–642.

Niu, M. T., L. B. Polish, B. H. Robertson, B. K. Khanna, B. A. Woodruff, C. N. Shapiro, M. A. Miller, J. D. Smith, J. K. Gedrose, and M. J. Alter. 1992. Multistate outbreak of hepatitis A associated with frozen strawberries. *J. Infect. Dis.* **166**:518–524.

Pringle, C. R. 1998. Virus taxonomy—San Diego 1998. *Arch. Virol.* **143**:1449–1459.

Reid, T. M. S., and H. G. Robinson. 1987. Frozen raspberries and hepatitis A. *Epidemiol. Infect.* **98**:109–112.

Reyes, G. R., M. A. Purdy, J. P. Kim, K. C. Luk, L. M. Young, K. E. Fry, and D. W. Bradley. 1990. Isolation of a cDNA from the virus responsible for enterically transmitted non-A, non-B hepatitis. *Science* **247**:1335–1339.

Robinson, R. A., W. H. Burgess, S. U. Emerson, R. S. Leibowitz, S. A. Sosnovtseva, S. Tsarev, and R. H. Purcell. 1998. Structural characterization of recombinant hepatitis E virus ORF2 proteins in baculovirus-infected insect cells. *Protein Expr. Purif.* **12**:75–84.

Ross, B. C., D. A. Anderson, and I. D. Gust. 1991. Hepatitis A virus and hepatitis A infection. *Adv. Virus Res.* **39**:209–253.

Rueckert, R. R. 1990. Picornaviridae and their replication, p. 507–548. *In* B. N. Fields, D. M. Knipe, R. M. Knipe, R. M.

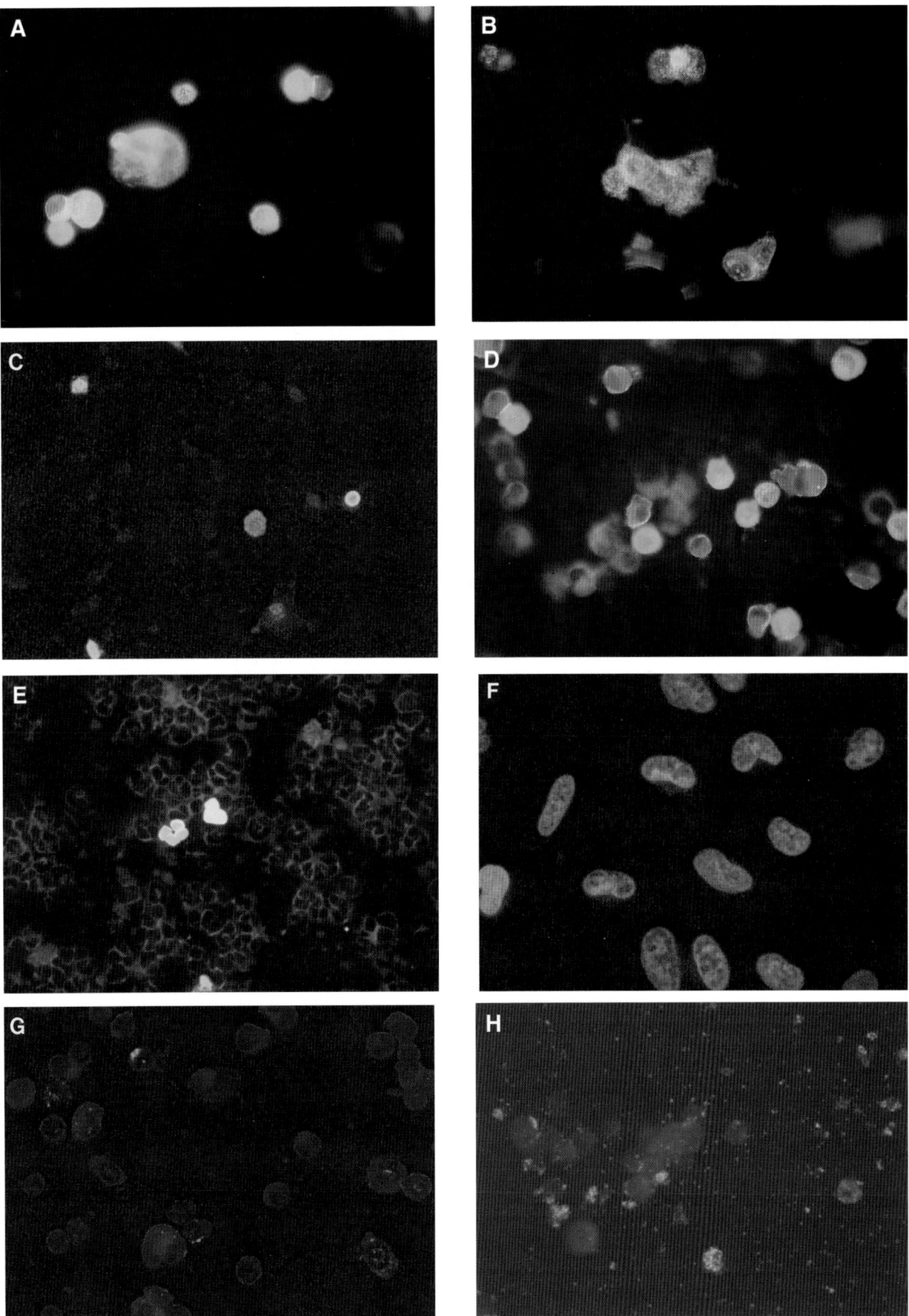

COLOR PLATE 1

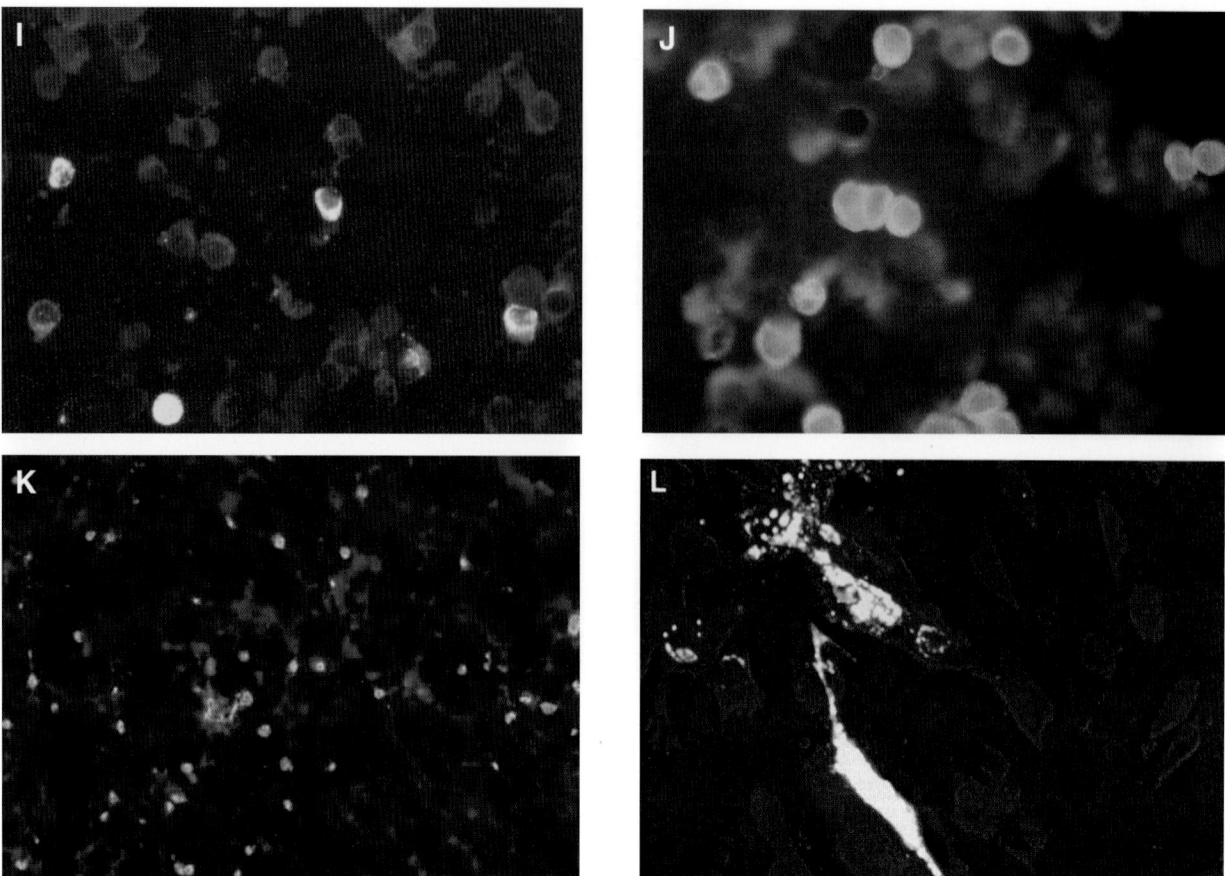

COLOR PLATE 1 (A) Vero cells infected with HSV-1 and stained with the PathoDx HSV-1 typing reagent. Note that staining is limited to the cytoplasm and membrane of the infected cells. Magnification, ×400. (Courtesy of Diagnostic Products Corporation.) (B) HSV-2-infected Vero cells stained with the PathoDx HSV-2 typing reagent. Note the intense speckling pattern limited to the cell nucleus. Magnification, ×400. (Courtesy of Diagnostic Products Corporation.) (C) Dual infection of RMK cells with both influenza A virus and influenza B virus. Cells are stained with the Light Diagnostics SimulFluor FluA/FluB reagent. Influenza A virus-infected cells appear apple-green while influenza B virus-infected cells stain yellow-gold to orange when viewed with an FITC filter set. Magnification, ×200. (Courtesy of Chemicon International, Inc.) (D) Infection of MDCK-1 cells with influenza A virus. Cells are stained with a PathoDx influenza A virus-specific fluoresceinated monoclonal antibody. Magnification, ×400. (Courtesy of Diagnostic Products Corporation.) (E) Peripheral blood leukocytes stained for the presence of the CMV lower matrix protein pp65 with Light Diagnostics CMV pp65 reagents. CMV pp65 is typically found in the nucleus of infected cells. Magnification, ×400. (Courtesy of Chemicon International, Inc.) (F) MRC-5 cells infected with CMV and stained by IFA using monoclonal antibody directed against the immediate-early antigen. Note the staining in the nucleus of the cells only. Magnification, ×400. (Courtesy of Chemicon International, Inc.) (G) Uninfected RMK cells stained with the Light Diagnostics Pan-Enterovirus reagent. Magnification, ×200. (Courtesy of Chemicon International, Inc.) (H) Echovirus type 4-infected RMK cells stained with the Light Diagnostics Pan-Enterovirus reagent. Note the speckled "starry night" staining pattern typical of the Pan-Enterovirus reagent. Magnification, ×200. (Courtesy of Chemicon International, Inc.) (I) RMK cells infected with echovirus type 4 and stained with Echo 4 MAB reagent from Light Diagnostics. Magnification, ×200. (Courtesy of Chemicon International, Inc.) (J) Adenovirus type 5-infected HEp-2 cells stained by DFA using a PathoDx group-specific fluoresceinated monoclonal antibody. Magnification, ×400. (Courtesy of Diagnostic Products Corporation.) (K) Thin section of lung necropsy material from a 6-year-old child with fatal RSV pneumonia. The slide was stained by IFA using primary monoclonal antibody directed against RSV and a fluoresceinated goat anti-mouse secondary antibody. Note the intense fluorescent staining in the alveoli. (Courtesy of Diagnostic Products Corporation.) (L) VZV-infected MRC-5 cells stained with the Light Diagnostics SimulFluor HSV/VZV reagent. Infected cells stain yellow-gold to orange when viewed with an FITC filter set. Magnification, ×200. (Courtesy of Chemicon International, Inc.)

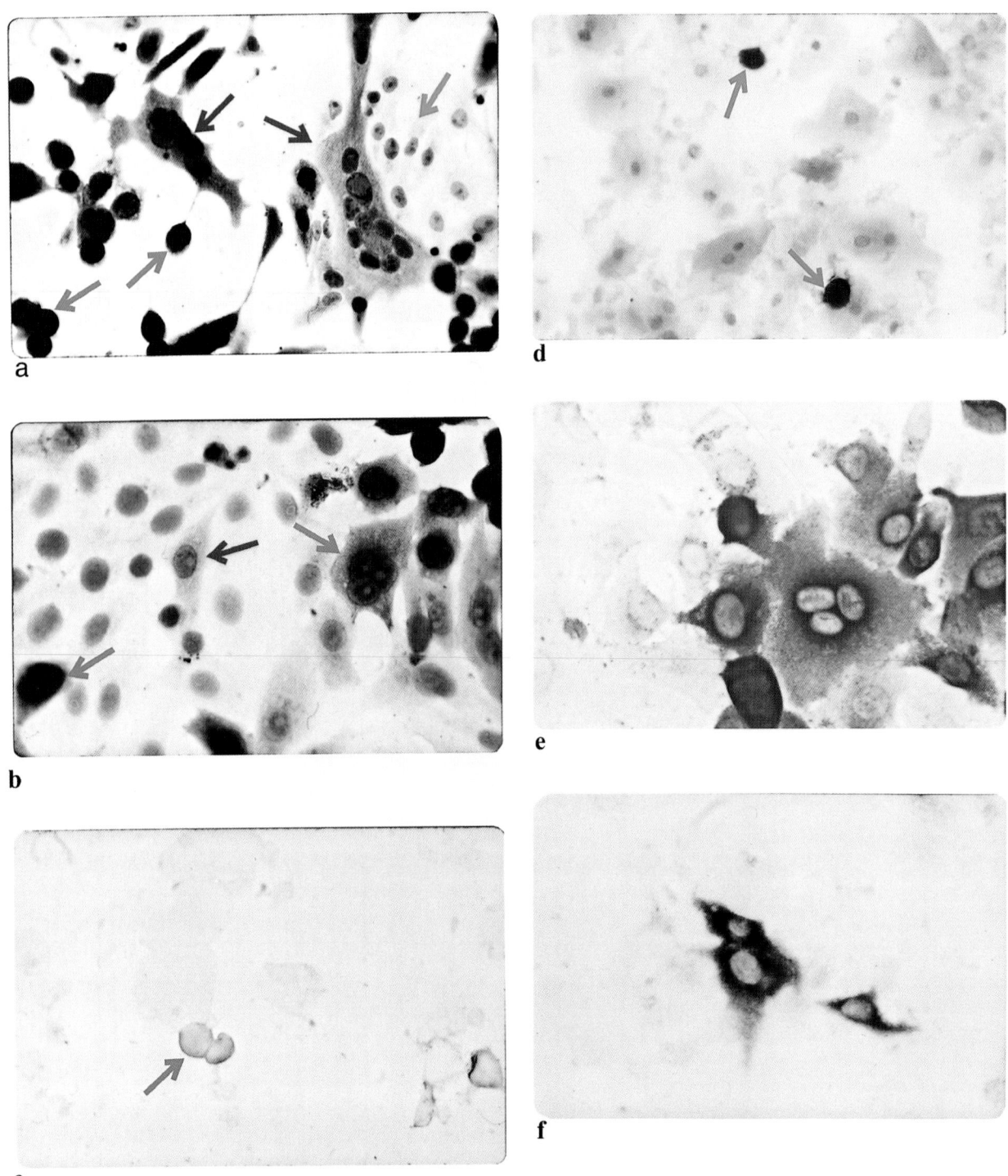

COLOR PLATE 2 (a) HSV-2-infected cells stained by PXAPX-PAP. Green arrow, normal cells, not infected; blue arrows, rounding of infected cells; red arrows, syncytium formation in HSV-infected cells. (b) HSV-2-infected cells stained by PXAPX-PAP. Red arrow, early infection; blue arrows, late infection. (c) Exfoliated cells from a female patient with a confirmed HSV-2 genital infection stained by PXAPX. Blue arrow, infected cell. (d) Exfoliated cells from a female patient with a confirmed HSV-2 genital infection stained by PXAPX-PAP. Blue arrows, HSV-2-infected cells. (e) HSV-2-infected cells stained with glucose oxidase-coupled antibody. INT was used as the substrate. (f) HSV-2-infected cells stained with glucose oxidase-coupled antibody. NBT was used as the substrate.

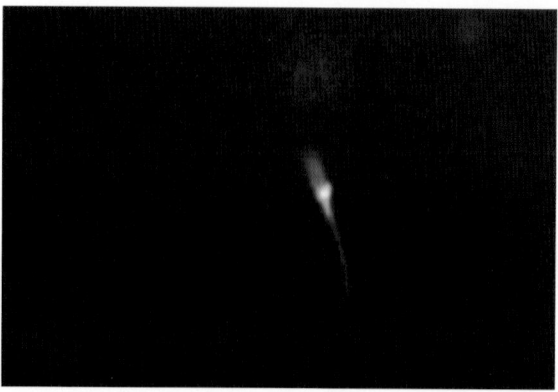

COLOR PLATE 3   Semen from an asymptomatic patient
stained with antibody to HSV-2 by indirect IF reveals the presence of viral antigen in the spermatozoon. Virus was isolated
from seminal fluid from the same patient.

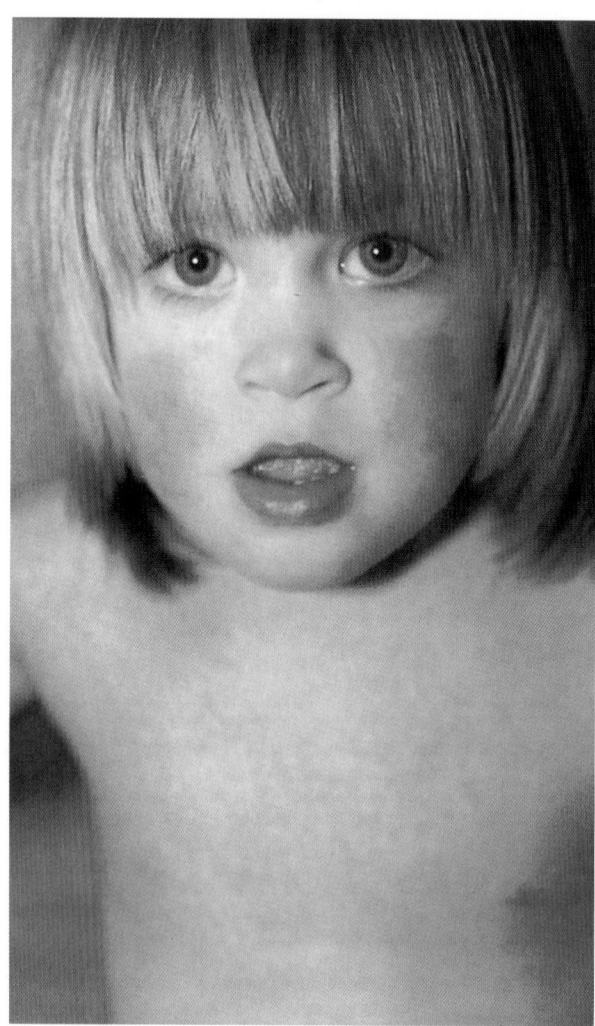

COLOR PLATE 4   "Slapped cheeks" and lacy macular rash
on the torso (not in focus) of a child with erythema infectiosum (from Feder, 1994, with permission).

Chanock, M. S. Hirsch, J. L. Melnick, T. P. Morath, and B. Roizman (ed.), *Virology*. Raven Press, New York, N.Y.

Sagliocca, L., P. Amoroso, T. Stroffolini, B. Adamo, M. E. Tosti, G. Lettieri, C. Esposito, S. Buonocore, P. Pierri, and A. Mele. 1999. Efficacy of hepatitis A vaccine in prevention of secondary hepatitis A infection: a randomised trial. *Lancet* **353:**1136–1139.

Schlauder, G. G., G. J. Dawson, J. C. Erker, P. Y. Kwo, M. F. Knigge, D. L. Smalley, J. E. Rosenblatt, S. M. Desai, and I. K. Mushahwar. 1998. The sequence and phylogenetic analysis of a novel hepatitis E virus isolated from a patient with acute hepatitis reported in the United States. *J. Gen. Virol.* **79:**447–456.

Schlauder, G. G., S. M. Desai, A. R. Zanetti, N. C. Tassopoulos, and I. K. Mushahwar. 1999. Novel hepatitis E virus (HEV) isolates from Europe: evidence for additional genotypes of HEV. *J. Med. Virol.* **57:**243–251.

Seow, H.-F., N. M. B. Mahomed, J.-W. Mak, M. A. Riddell, F. Li, and D. A. Anderson. 1999. Seroprevalence of antibodies to hepatitis E virus in the normal blood donor population and two aboriginal communities in Malaysia. *J. Med. Virol.* **59:**164–168.

Shaw, D. D., D. C. Whiteman, A. D. Merritt, D. M. El-Saadi, R. J. Stafford, K. Heel, and G. A. Smith. 1999. Hepatitis A outbreaks among illicit drug users and their contacts in Queensland, 1997. *Med. J. Austr.* **170:**584–587.

Sjogren, M. H., H. Tanno, O. Fay, S. Sileoni, B. D. Cohen, D. S. Burke, and R. J. Feighny. 1987. Hepatitis A virus in stool during clinical relapse. *Ann. Intern. Med.* **106:**221–226.

Soucie, J. M., B. H. Robertson, B. P. Bell, K. A. McCaustland, and B. L. Evatt. 1998. Hepatitis A virus infections associated with clotting factor concentrate in the United States. *Transfusion* **38:**573–579.

Stapleton, J. T., R. Jansen, and S. M. Lemon. 1985. Neutralizing antibody to hepatitis A virus in immune serum globulin and in the sera of human recipients of immune serum globulin. *Gastroenterology* **89:**637–642.

Stapleton, J. T., V. Raina, P. L. Winokur, K. Walters, D. Klinzman, E. Rosen, and J. H. McLinden. 1993. Antigenic and immunogenic properties of recombinant hepatitis A virus 14S and 70S subviral particles. *J. Virol.* **67:**1080–1085.

Stewart, D. R., T. S. Morris, R. H. Purcell, and S. U. Emerson. 1997. Detection of antibodies to the nonstructural 3C proteinase of hepatitis A virus. *J. Infect. Dis.* **176:**593–601.

Tam, A. W., M. M. Smith, M. E. Guerra, C. C. Huang, D. W. Bradley, K. E. Fry, and G. R. Reyes. 1991. Hepatitis E virus (HEV): molecular cloning and sequencing of the full-length viral genome. *Virology* **185:**120–131.

Thomas, D. L., P. O. Yarbough, D. Vlahov, S. A. Tsarev, K. E. Nelson, A. J. Saah, and R. H. Purcell. 1997. Seroreactivity to hepatitis E virus in areas where the disease is not endemic. *J. Clin. Microbiol.* **35:**1244–1247.

Torresi, J., F. Li, S. A. Locarnini, and D. A. Anderson. 1999. Only the non-glycosylated fraction of hepatitis E virus capsid (open reading frame 2) protein is stable in mammalian cells. *J. Gen. Virol.* **80:**1185–1188.

Tsarev, S. A., S. U. Emerson, M. S. Balayan, J. Ticehurst, and R. H. Purcell. 1991. Simian hepatitis A virus (HAV) strain AGM-27: comparison of genome structure and growth in cell culture with other HAV strains. *J. Gen. Virol.* **72:**1677–1683.

Tsarev, S. A., T. S. Tsareva, S. U. Emerson, S. Govindarajan, M. Shapiro, J. L. Gerin, and R. H. Purcell. 1994. Successful passive and active immunization of cynomolgus monkeys against hepatitis E. *Proc. Natl. Acad. Sci. USA* **91:**10198–10202.

Tsarev, S. A., T. S. Tsareva, S. U. Emerson, S. Govindarajan, M. Shapiro, J. L. Gerin, and R. H. Purcell. 1997. Recombinant vaccine against hepatitis E: dose response and protection against heterologous challenge. *Vaccine* **15:**1834–1838.

Tsarev, S. A., T. S. Tsareva, S. U. Emerson, A. Z. Kapikian, J. Ticehurst, W. London, and R. H. Purcell. 1993. ELISA for antibody to hepatitis E virus (HEV) based on complete open-reading frame-2 protein expressed in insect cells: identification of HEV infection in primates. *J. Infect. Dis.* **168:**369–378.

Tsarev, S. A., T. S. Tsareva, S. U. Emerson, M. K. Rippy, P. Zack, M. Shapiro, and R. H. Purcell. 1995. Experimental hepatitis E in pregnant rhesus monkeys: failure to transmit hepatitis E virus (HEV) to offspring and evidence of naturally acquired antibodies to HEV. *J. Infect. Dis.* **172:**31–37.

Van Damme, P., S. Thoelen, M. Cramm, K. De Groote, A. Safary, and A. Meheus. 1994. Inactivated hepatitis A vaccine: reactogenicity, immunogenicity, and long-term antibody persistence. *J. Med. Virol.* **44:**446–451.

Van Doorslaer, E., G. Tormans, and P. Van Damme. 1994. Cost-effectiveness analysis of vaccination against hepatitis A in travelers. *J. Med. Virol.* **44:**463–469.

Wang, Y., R. Ling, J. C. Erker, H. Zhang, H. Li, S. Desai, I. K. Mushahwar, and T. J. Harrison. 1999. A divergent genotype of hepatitis E virus in Chinese patients with acute hepatitis. *J. Gen. Virol.* **80:**169–177.

Worm, H. C., H. Wurzer, and G. Frosner. 1998. Sporadic hepatitis E in Austria. *N. Engl. J. Med.* **339:**1554–1555.

Yarbough, P. O., E. Garza, A. W. Tam, Y. Zhang, P. McAtee, and T. R. Fuerst. 1996. Assay development of diagnostic tests for IgM and IgG antibody to hepatitis E virus, p. 294–296. *In* Y. Buisson, P. Coursaget, and M. Kane (ed.), *Enterically-Transmitted Hepatitis Viruses*. La Simarre, Joue-les-Tours, France.

Yarbough, P. O., A. W. Tam, K. E. Fry, K. Krawczynski, K. A. McCaustland, D. W. Bradley, and G. R. Reyes. 1991. Hepatitis E virus: identification of type-common epitopes. *J. Virol.* **65:**5790–5797.

Zafrullah, M., M. H. Ozdener, S. K. Panda, and S. Jameel. 1997. The ORF3 protein of hepatitis E virus is a phosphoprotein that associates with the cytoskeleton. *J. Virol.* **71:**9045–9053.

Zanetti, A. R., G. G. Schlauder, L. Romano, E. Tanzi, P. Fabris, G. J. Dawson, and I. K. Mushahwar. 1999. Identification of a novel variant of hepatitis E virus in Italy. *J. Med. Virol.* **57:**356–360.

Zhang, Y., P. McAtee, P. O. Yarbough, A. W. Tam, and T. Fuerst. 1997. Expression, characterization, and immunoreactivities of a soluble hepatitis E virus putative capsid protein species expressed in insect cells. *Clin. Diagn. Lab. Immunol.* **4:**423–428.

Blood-Borne Hepatitis Viruses:
Hepatitis B, C, D, and G Viruses and TT Virus

MAURO BENDINELLI, MAURO PISTELLO, FABRIZIO MAGGI,
AND MARIALINDA VATTERONI

27

Some hepatotropic viruses of humans do not appear to have developed sufficiently efficient means of host-to-host spread. Possibly as compensation for this limitation, these viruses have a remarkable tendency to persist for many years in their hosts, often for life, and to circulate copiously in the peripheral blood. As a consequence, apparent or inapparent percutaneous exposure to blood or blood products represents a frequent occasion of contagion.

Traditionally, blood-borne hepatitis viruses include hepatitis B virus (HBV), hepatitis C virus (HCV), and hepatitis D virus (HDV), which collectively infect and produce disease in several hundred million people worldwide, thus representing major health and diagnostic issues (London and Evans, 1996). In recent years, however, the list of viruses that are commonly transmitted through infected blood and that have been implicated in liver disease has expanded to include hepatitis G virus (HGV), TT virus (TTV), and SEN virus (SEN-V), three agents that have been discovered thanks to recently developed molecular approaches and that are currently the focus of intensive investigation. Although the significance of these novel (and widespread) viruses as agents of liver disease remains to be defined, they are discussed in this chapter because virus-containing blood represents an important source of infection.

HBV, HCV, and HDV have widely diverse genomic properties and replication strategies. Nonetheless, their pathological consequences are similar and offer no reliable clinical clues about the respective viruses. Moreover, many of the infections they produce tend to have an insidious onset and extremely irregular courses of disease. As a result, clinicians depend extensively on the laboratory for their demonstration, differentiation, and staging. Another feature shared by these viruses is that they are not, or are poorly, cultivable in vitro, which renders virus isolation impracticable for diagnostic use. Laboratory diagnosis has, therefore, always had to rely on immunoassays demonstrating the viral antigens or the corresponding antibodies. However, with the recent massive expansion of molecular virology, assays that demonstrate and quantify the viral genomes have come into widespread use, thanks to rapid turnaround times and availability of commercial user-friendly kits as well as to the important information they provide (Specter et al., 1998).

Because a detailed account of the structural, biological, pathogenic, and epidemiological properties of the viruses discussed is beyond the purposes of this book, this chapter will highlight only those features that are pertinent to a correct understanding of laboratory diagnosis.

SPECTRUM OF DISEASES PRODUCED

Primary, acute infections by HBV, HCV, and HDV are often asymptomatic or paucisymptomatic with influenzalike manifestations. When clinically manifest, they produce hepatic pathologies that range from mild, anicteric or icteric self-limiting hepatitis to rare but frequently fatal fulminant hepatitis. Various proportions of acutely infected persons (see individual viruses below) do not resolve the primary infection and become chronically infected. The clinical manifestations of chronic infection by HBV, HCV, and HDV are also essentially indistinguishable. Again, they vary greatly both in pattern and severity: persistently infected patients may remain relatively asymptomatic for extended periods of time or indefinitely (and are sometimes termed "healthy carriers") or develop chronic hepatitis with various degrees of activity and a remittent or stable course, which may eventually lead to hepatic failure and cirrhosis after variable but generally long intervals.

Importantly, the mechanisms of acute and chronic liver damage are only partially understood, but most data converge to indicate that the lesions are not due to cytopathic effects directly produced by the viruses themselves but are, in part at least, mediated by immune attack on viral-antigen-expressing hepatocytes. Indeed, when cell-mediated immune responses are immature or compromised, as in individuals infected perinatally or undergoing immunosuppressive therapies, liver injury is usually less than would be expected from the extent of virus replication. Conversely, in chronically infected patients, an abrupt cessation of immunosuppressive therapies frequently elicits serious deterioration of the clinical and biochemical manifestations of hepatitis, despite the fact that the levels of viremia may concomitantly decline.

Individuals chronically infected with HBV and HCV also have a greatly increased risk (20 to 300 times, depending on the specific populations considered) of developing hepatocellular carcinoma (HCC). The fact that these tumors develop after 20 to 30 years of chronic HBV or HCV infection, mostly on a background of liver cirrhosis, suggests that the tremendous activity of cell death and regeneration, coupled with chronic inflammation, occurring in infected livers is a major driving force in their genesis. However, the fine mechanisms linking HBV and HCV infections to liver cancer and the possible role of cofactors remain unclear (Idilman et al., 1998).

The diagnosis and assessment of the severity of acute and chronic hepatitis rely largely on biochemical exploration of liver integrity and function (Table 1). The levels of aspartate- and especially alanine-aminotransferase (ALT) are of the utmost importance, since they are highly sensitive indicators of hepatocellular damage and provide a simple means of monitoring hepatitis activity. The dynamics of enzyme alterations may even furnish clues about the causing virus, albeit weak ones: e.g., an abrupt, sharp elevation of ALT is considered suggestive of hepatitis A, whereas fluctuating levels are more typical of hepatitis C. While measurement of viremia levels is becoming increasingly important for monitoring the evolution of infection and response to therapies (see below), liver endoscopy, and especially liver biopsy (when indicated), remain essential for staging associated liver pathologies and for prognosis.

Infections with HBV, HCV, and HDV are also associated with a variety of extrahepatic manifestations. These include glomerulonephritis (HBV and HCV infections), arthralgias, the specific form of vasculitis known as polyarteritis nodosa (more frequently seen in HBV infection), cryoglobulinemia (more frequently seen in HCV infection), and possibly Sjögren's syndrome, porphyria cutanea tarda, and others. Some extrahepatic manifestations such as these are explained on the basis of immune complex formation and deposition. Others have a more obscure pathogenesis. Antinuclear, anti-LKM-1 and anti-GOR autoantibodies and other serological markers of autoimmunity are present in many patients with chronic HCV infection.

As already mentioned, the pathogenic potentials of HGV, TTV, and the more recently identified SEN-V are still undefined. The clinical manifestations for which HGV and TTV are presently being considered as possible etiological agents include hepatic and extrahepatic diseases, but no causal relationships have yet been established on solid evidence (see below).

HBV

The incubation period of hepatitis B is usually 1.5 to 3 months but sometimes considerably longer. The clinical course is generally more severe than in other forms of viral hepatitis, although anicteric or even symptomless infections are common, especially in children. The symptoms associated with acute infection (malaise, nausea, anorexia, low-grade fever, jaundice, dark urine, and pale stools) usually last 2 to 3 weeks but may be much more protracted.

The Virus

HBV was first demonstrated in the 1960s, when the so-called Australian antigen, which had been serendipitously detected in the blood of an aborigine, was linked to what was then mainly known as serum or long-incubation hepatitis. Together with a number of similar mammalian and avian viruses, HBV is currently classified within the family *Hepadnaviridae*. As such, it is a DNA virus which uses a reverse transcriptase and an RNA intermediate to replicate itself. Integration of the viral genome in host cell chromosomes occurs but is not required for virus replication and may be a late event in infection. Key structural and replication features of HBV are summarized in Table 2 and Fig. 1. The virus is relatively heat stable but is rapidly destroyed by acids and lipid solvents. In one study, viral infectivity was found to survive drying and storage on contaminated surfaces for at least several days (Bond et al., 1981).

In vivo, the liver is by far the predominant site of replication. However, the hepatotropism of HBV is not absolute. Viral genomes, replicative intermediates, and other footprints of virus replication have repeatedly been detected in extrahepatic sites, including bile duct epithelium, pancreas and kidney cells, lymphocytes, and monocytes. Chimpanzees are readily infected and exhibit an infection course similar to that of humans. In vitro, limited propagation of HBV has been achieved in primary cultures of human hepatocytes. Transfected cell lines are also widely used for the study of viral biosyntheses.

Multiple pieces of evidence have clearly established that antibodies to the pericapsid proteins (HBsAg) are protective and that, based on cross-protection criteria, there is only one serotype of the virus. Isolates have, however, been classified into subtypes based on subtle differences in the antigenic determinants of the HBsAg. The *a* determinant is common to all subtypes and contains an important neutralization domain, while others tend to be present in one of two allelic forms (e.g., *d-y* and *w-r*). This results in subtypes that present different antigenic formulas: those most frequent worldwide are *adw, adr,* and *ayw*. The usefulness of this classification is exclusively epidemiological. More recently, HBV isolates have also been classified into seven genotypes (A to G) based on genetic relatedness, but the implications of this grouping for severity and other parameters of infection are not understood (Chan et al., 1999; Lindh et al., 1999).

Virus mutants are also found in some patients and may result from single mutations in a specific domain of the HBV genome that affect multiple viral functions, due to extensive overlapping of the viral open reading frames (ORFs) (Locarnini, 1998). The importance of these mutants in determining pathogenic features, diagnostic difficulties, resistance to drugs, escape from vaccine-induced

TABLE 1 Some biochemical tests of serum for liver function evaluation

Test	Information provided
ALT and aspartate aminotransferase	Degree of hepatocellular injury and necrosis
Gamma-glutamyl transferase	Marker of cholestasis
Alkaline phosphastase	Confirmation of cholestasis
Direct bilirubin	Degree of cholestasis
Albumin, cholinesterase, total cholesterol, prothrombin time, etc.	Degree of damage to protein-synthetic activity of liver
Ferritin	Degree of damage to storage activity of liver
Immunoglobulin profile	Immune system activation

TABLE 2 Structural, replication, and integration properties of HBV; salient features

Structure

    Spherical 42-nm-diameter virion, also referred to as Dane particle

    Icosahedral 27-nm-diameter capsid, or core, composed of numerous copies of a single phosphoprotein (Table 3)

    Tight envelope, composed of three virus-coded proteins (Table 3) and cell-derived lipids

    Relaxed, circular 3.2-kb double-stranded DNA, with cohesive 5′ ends; the minus strand is nicked and covalently linked to a
        5′-terminal protein; the plus strand is incomplete for a variably long stretch and bears a capped RNA primer at its 5′ terminal

    Genome organized in four extensively overlapping ORFs: pre-S–S (envelope glycoproteins), pre-C–C (core protein and HBeAg),
        P (reverse transcriptase, RNase H, and terminal genomic protein), and X (multifunctional regulatory protein), all coded in the
        minus strand

    Enzymatic activities packaged in virion: reverse transcriptase-DNA polymerase and RNase H

Replication

    Entry into cells probably mediated by attachment of the viral L protein to unknown cellular receptor(s)

    Virion DNA converted to a supercoiled, covalently closed, completely double-stranded circular form and transported to the nucleus

    Genome transcribed to produce multiple copies of a full-length pregenomic RNA intermediate and of subgenomic mRNA

    Pregenomic RNA intermediate packaged into procapsid structure, together with reverse transcriptase, and reverse transcribed into the
        minus DNA strand, on which an incomplete plus DNA strand is then synthesized

    Nucleocapsids mature and bud through the endoplasmic reticulum, acquiring the envelope

    No overt cytopathic effects

Integration

    In many long-term chronically infected individuals and in over 85% of HBV-positive HCCs, the chromosomes of hepatocytes contain
        integrated viral DNA, often in multiple copies per cell

    Mechanism of integration poorly understood; the virus does not code for integration machinery

    Integrated genomes are often defective due to deletions or rearrangments, hence reactivation is very unlikely

    Transcription of integrated DNA restricted to some subgenomic mRNAs

    No specific integration sites in the cell genome

    Role in HCC generation uncertain

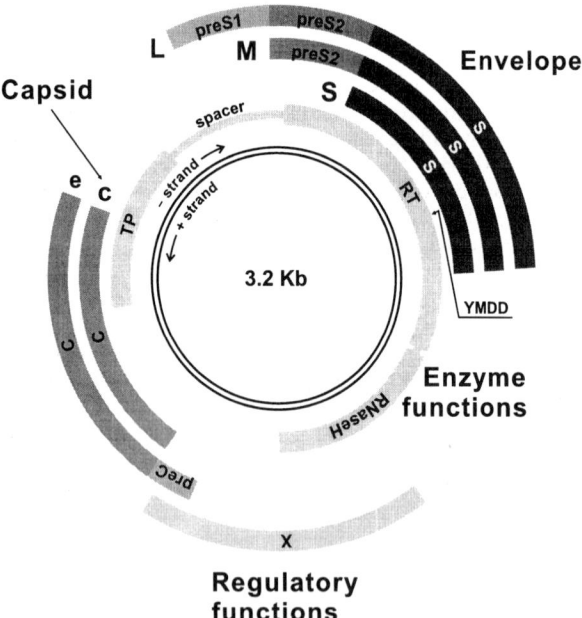

FIGURE 1 Organization of the HBV genome and encoded proteins. The double circle represents the functional covalently closed, completely double-stranded circular viral DNA found in infected cells (Table 2). The arcs represent the protein products. c, capsid (core); e, e antigen; L, M, and S, envelope glycoproteins; RT, reverse transcriptase-DNA polymerase; TP, terminal protein. The approximate location in the RT of the YMDD amino acid motif important for resistance to lamivudine is also shown.

immunity, and other aspects of the infection produced has been the subject of intensive investigations in recent years. Mutations in the pre-S–S ORF may prevent recognition by anti-HBs antibodies, mutations in the pre-C–C ORF may result in loss of HBV e antigen (HBeAg) production and escape from cell-mediated immune mechanisms, and mutations in the P ORF can be associated with resistance to specific chemotherapeutic agents, thus significantly affecting patient management in clinical practice. Additional HBV mutants that have recently attracted great attention are the so-called HBeAg-minus or -defective viral variants, which carry a mutation resulting in a stop codon at position 1896 or in other sites at the end of the pre-C region and are therefore unable to produce the antigen. They have been implicated in particularly severe forms of acute and chronic hepatitis as well as in the relapse of apparently resolved infections (Brunetto et al., 1999; Ngui et al., 1999; Zuckerman and Zuckerman, 1999).

For more comprehensive reviews of the structural and biological properties of HBV, see Ganem, 1996; Hollinger, 1996; and Vyas and Yen, 1999.

Dynamics of HBV Replication and Antiviral Immune Responses in Infected Patients

The bulk of information about the dynamics of HBV infection has been collected by measuring the levels of viral antigens and antibodies in the blood by immunoassays and measuring those of viral DNA by nucleic acid hybridization assays. The general frame that has emerged from these studies is still valuable, but the much more sensitive gene amplification methods now extensively used for DNA detection and quantification have somewhat modified the picture. Thus, for example, it has been seen that patients who by conventional methods appear to have eradicated the infection may continue to harbor the virus for much longer periods than previously believed, although the clinicopathological implications of these recent findings have yet to be fully evaluated.

The possible courses and phases of HBV infection have been variously classified (Lee, 1997). From the clinical virology standpoint, however, it seems useful to limit ourselves to distinguishing between self-limited acute infections and those that instead persist longer and become chronic. By convention, hepatitis is considered chronic when it does not resolve within 6 months from onset.

Self-Limited Acute Infections

Typically, viremia becomes detectable 3 to 5 weeks before the development of clinical symptoms (i.e., 2 to 6 weeks before the onset of ALT abnormalities) and peaks soon thereafter (Fig. 2A). Complete infectious 42-nm-diameter virus particles circulate abundantly in the plasma of infect-

ed individuals, frequently reaching levels of 10^8/ml or greater, a feature that is reflected in the positivity of all the diagnostic assays that detect viral molecules (HBV DNA, viral polymerase, HBsAg, and HBeAg). However, HBV is unusual among human viruses in that the plasma of patients also contains a large excess of HBsAg proteins that do not participate in the formation of complete virions but are instead aggregated in spherical and filamentous lipoprotein structures 17 to 25 nm in diameter. These subvirion particles, which are noninfectious due to the absence of the viral genome, are even more abundant (up to 10^6 times) than complete virions. As a result, the blood contains an extremely high concentration of HBsAg (on the order of 50 to 300 μg/ml), which explains why this antigen is such a sensitive marker for identifying infected persons (see below).

Nonspecific effectors of antiviral resistance, including interferons (IFN), natural killer cells, and cytokines, are believed to represent the first line of resistance against infection. Cytotoxic T lymphocytes (CTL) and other effectors of cell-mediated immune responses, such as the antiviral cytokines, are most likely the second line of defense. Numerous CTL specific for epitopes of the core proteins expressed on infected hepatocytes are recruited to the liver and probably act in concert with the various types of IFN. Thanks to their ability to enhance major histocompatibility complex class I protein expression, these mediators are in fact known to render infected hepatocytes more vulnerable to the cytolytic action of CTL (under normal conditions, hepatocytes display little major histocompatibility complex class I protein on the cell surface). CTL are considered pivotal in containing HBV burdens and, when this occurs, in eliminating the infection, but they are also major contributors to the generation of hepatic damage (Chang and Chisari, 1999). Indeed, a state of partial or split immunotolerance is often invoked to explain why many infections remain asymptomatic with little histological evidence of liver injury. As already noted, subclinical infections are especially numerous when the immune system is poorly reactive (e.g., due to neonatal age or other causes).

Anti-HBV humoral immune responses are believed to be of less importance in the above-mentioned contexts but are certainly of paramount importance for laboratory diagnosis, which is usually performed after the initial preserological phase of infection, characterized by high levels of viremia in the absence of antiviral antibodies. Anti-HBc immunoglobulin M antibodies (IgM) are the first antibodies to appear. They typically develop at the same time as ALT changes, increase rapidly to reach considerable titers, and are then progressively, nearly completely, replaced by high titers of anti-HBc IgG.

Most likely as a result of the mounting immune response, viremia undergoes an initial decline concomitant with the appearance of clinical symptoms. Subsequently, viremia continues to decline and, except in few complicated cases, becomes undetectable, at least by conventional serological (HBsAg) and hybridization (HBV DNA) methods, in a few weeks. Importantly, this phase is usually preceded by the disappearance of HBeAg from the blood and the appearance of anti-HBe antibodies. Concomitant with or subsequent to viremia termination, hepatic histology, biochemical parameters, and clinical conditions normalize. Anti-HBs antibodies, which are particularly important because they are capable of neutralizing the virus, also develop and become measurable just before or, more often, a few weeks after the disappearance of HBsAg from the

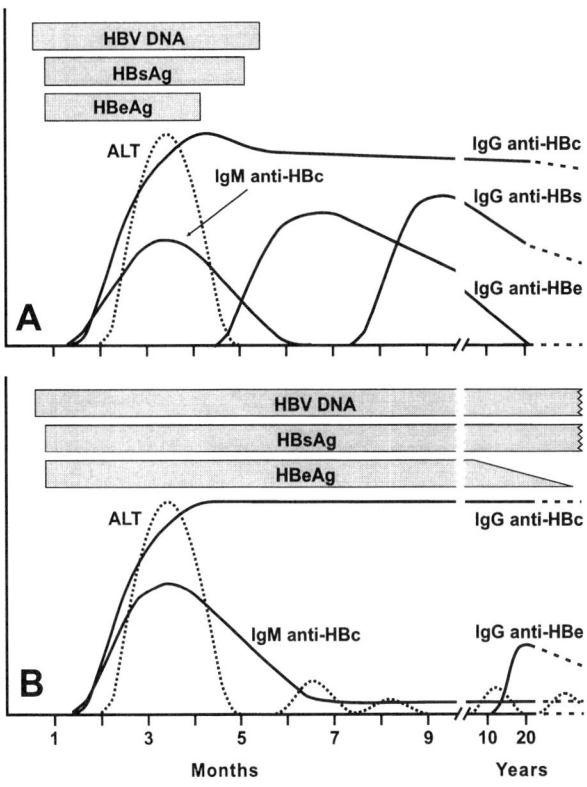

FIGURE 2 Typical courses of HBV infection. (A) Acute self-limited infection. (B) Infection that does not resolve and becomes chronic. While the virological and immunological events depicted are seen in virtually all patients, clinical symptoms develop in a proportion of infected individuals that varies greatly with age at infection and other parameters. The shaded bars represent the periods during which the markers indicated are demonstrable in the circulation. As described in the text, in an undefined proportion of patients minute amounts of HBV DNA are demonstrable by sensitive gene amplification techniques well beyond the periods indicated.

serum. Thus, by all criteria, the hepatitis and the underlying infection appear to be resolved. Anti-HBc antibodies persist for life in the great majority of patients (>90%), while anti-HBs antibodies have a somewhat greater tendency to become undetectable with passing years.

Until recently, the series of events resulting in clinical recovery described above was considered to correspond to a complete eradication of the infecting virus. However, duplicating what was previously observed after the resolution of chronic infections (see below), recent studies using DNA amplification methods have shown that, at least in some patients, recovery from self-limited acute infection is followed by the persistence of low numbers of viral genomes in plasma for months or years (Yotsuyanagi et al., 1998).

Chronic Infections

The proportion of acute infections that progress to chronicity (Fig. 2B) varies markedly depending on a number of variables, including the dose and genetic heterogeneity of the viral inoculum, the route of transmission, and the patient's age and immune responsiveness. Of these, age is by far the most important, likely as a result of differences in the ways the virus is confronted by the host's immune responses. Thus, individuals acquiring the infection perinatally become chronically infected in greater than 90% of cases versus 15 to 30% in childhood and fewer than 5% in adulthood (Chang, 1998).

It has been noted that asymptomatic or mild primary infections tend to become chronic more frequently than others. Practically, however, persistence of the virus is revealed only by the observation that viremia does not tend to subside. In fact, chronically infected patients are characterized by the persistence of markers of active virus replication at levels that remain for years in the same range or slightly lower than in acute infections. They can be asymptomatic ($\approx$40%) or alternate clinical manifestations with variably long periods of remission. Viremia is relatively stable over time or, more often, fluctuates in association with fluctuations in ALT and anti-HBc IgM levels.

Recently, the availability of drugs that effectively inhibit HBV replication has permitted estimates of HBV dynamics in chronically infected patients. Thus, it has been calculated that in chronic HBeAg-positive carriers, total HBV release into the circulation is on the order of 10^{11} or more virions per day. This value is similar to what is reported for human immunodeficiency virus type 1 (HIV-1). In contrast, the minimum half-life of HBV-infected cells was calculated to be considerably longer than that of HIV-1-infected cells (10 to 100 days versus 1 day), most probably as a result of the lower cytopathogenicity of HBV. However, the number of hepatocytes destroyed by HBV and renewed per day was calculated to be around 10^9, i.e., in the same range as the CD4$^+$ T lymphocytes destroyed daily by HIV-1. Clearance of HBV virions from plasma occurs with a half-life of approximately 1 to 3 days and a daily turnover of about 50% (Nowak et al., 1996; Zeuzem et al., 1997).

In any case, with passing years of persistence, the levels of infectious virus present in the blood tend to slowly decline, as revealed by decreasing concentrations of HBeAg and HBV DNA. Eventually, usually after many years of HBV persistence, the HBeAg becomes undetectable, and this is generally followed by seroconversion to anti-HBe. However, in the great majority of cases the levels of circulating HBsAg remain essentially unchanged for decades or for life. In fact, spontaneous recovery from chronic infec-

tion is observed at rates of only 0.5% per year. When resolution occurs, virological and serological markers evolve with a pattern similar to that described for acute infections, albeit at a slower pace.

Recently, through gene amplification techniques, persons who, based on conventional methods, were believed to have completely cleared a chronic infection for a long time (for example, individuals positive for anti-HBc antibody only) have been found to harbor minute amounts of HBV-DNA in the plasma and liver. Although this postconversion phase is often termed nonreplicative or integrative, the residual viral DNA in plasma has been found to be associated with complete virus particles and that in the liver has been found to be extrachromosomal and to undergo low-grade replication (Loriot et al., 1997; Jurinke et al., 1998; Mason et al., 1998). Further studies are needed to define the clinical significance of this immunologically undetectable form of HBV carriage, which also occurs in animal models (Michalak et al., 1999). The phenomenon, however, has already been invoked to explain the relapses of apparently cured HBV infections that are occasionally observed as a result of HIV- or drug-induced immunosuppression (Altfeld et al., 1998; Dhedin et al., 1998).

Epidemiology

The prevalence of HBV carriers in the general population varies widely throughout the world, reaching up to 10 to 20% in certain areas of sub-Saharan Africa, China, and Southeast Asia and being as low as 0.1 to 0.5% in Western Europe and North America. Pockets of comparatively higher endemicity are, however, found within any geographic area in specific ethnic, behavioral, and professional groups.

Infected humans are the only source of contagion. In the past, hepatitis B was a frequent complication of blood transfusions and of the therapeutic use of blood products, but it is now a rare occurrence in these clinical settings as a result of specific screening of donations, virucidal treatment of blood derivatives, and other preventive measures. Due to high virus content, minute traces of infected blood suffice for transmission; thus, improperly shared usage of, and incidental injuries with, syringes, needles, medical and toilet instruments, etc., represent other important occasions of contagion. Lower quantities of HBV are also present in semen, vaginal and menstrual secretions, saliva, urine, tears, and breast milk. Sexual intercourse is also a frequent route of dissemination, as shown by the observation that in low-endemicity areas, the highest rates of infection are found in sexually active young adults, with peaks among male homosexuals.

Vertical transmission from HBV-positive mothers to babies is also frequently observed (in the absence of proper immunoprophylaxis, almost 90% of children born to HBeAg-positive mothers become infected) and represents a major mechanism of virus spread, especially in areas of hyperendemicity. The fact that the babies become HBsAg positive only 1 to 3 months after birth and other evidence indicate that transmission occurs during parturition or perinatally rather than in utero. Intrafamilial spread with no apparent vertical, parenteral, or sexual exposure also occurs and appears to be especially important for young children living in high-endemicity areas or institutions. Other modes of transmission are possible but epidemiologically unimportant. For example, experiments have shown that oral transmission is feasible but requires large inocula. However, in developed countries, over one-fourth of infected individuals do not have, or do not report, high-risk behaviors or events.

Prevention and Treatment

Intervention strategies for limiting HBV dissemination include reduced use of nonautologous blood transfusions, accurate selection of blood and organ donors, proper education of health care workers and other high-risk subjects, decontamination of high-risk environments, administration of HBV-specific immune globulins (HBIG), chemoprophylaxis, and vaccination. Prevention of occupational and nosocomial transmission is essentially based on the presumption that any person might carry the virus and on consequent strict implementation and maintenance of infection control practices. Changes in risk behavior of illegal injecting drug users should also be pursued. In an attempt to prevent less common modes of transmission, HBV-negative sexual partners and household contacts of virus carriers should be informed of how to reduce the risk of becoming infected (vaccination is of primary importance), while infected individuals should be counseled on how they can prevent transmitting the virus to others as well as on secondary prevention practices that can reduce the risk of progressing toward increasingly severe liver pathology. In particular, HBV carriers should be advised to refrain from drinking alcohol, to be vaccinated against hepatitis A, and to seek appropriate medical care.

The administration of high-titer HBIG consistently, albeit transiently, decreases the risk of acquiring the infection and the severity of hepatitis but, given the high efficacy of vaccination for preexposure prophylaxis, should be limited to postexposure prophylaxis. Combined immunoprophylaxis at birth with both HBIG and vaccination for babies born to HBV-positive women (which should be regarded as postexposure prophylaxis) is especially important and highly effective.

The first anti-HBV vaccine was developed in the early 1970s and was composed of highly purified HBsAg harvested from infected human plasma and processed to remove or inactivate any infectivity. This immunogen has now been widely replaced by recombinant HBsAg vaccines. Although both types of vaccine have been extensively used with excellent safety and efficacy records (protection lasts for at least 10 years [Assad and Francis, 1999]), further improvements have been proposed. Among these, the incorporation of highly conserved pre-S peptides has been proposed for protecting those who do not respond adequately to the standard vaccines. In many countries, routine vaccination of infants is recommended; however, it is widely recognized that adolescents should also be targeted due to the importance of sexual contact for virus transmission (Centers for Disease Control and Prevention, 1999). Information is starting to emerge that vaccinated individuals have a reduced incidence of HCC as well as of HBV infections (Chang et al., 1997). It is therefore likely that, in the near future, the HBV vaccine will come to be regarded as the first antitumor vaccine effective in humans.

Whereas there are no specific therapies of proven benefit for acute hepatitis, several have proved effective against chronic infections with compensated liver disease. The long-term risk of cirrhosis and HCC as well as the infectivity of virus carriers largely justifies the clinical use of these therapies. Treatment with recombinant IFN-α at doses ranging from 3 to 10 million U three to seven times weekly for 16 to 32 weeks usually leads to sustained improvement of clinical symptoms, ALT normalization, and disappearance of HBV DNA, followed by clearance of the HBeAg from serum in one-third of patients and, eventually, clearance of HBsAg in 8% of patients. This rate of cure is approximately three to four times higher than that observed in untreated controls followed for similar periods of time (Vajro et al., 1998; Janssen et al., 1999). Low HBV DNA levels (<200 pg/ml), high serum ALT (>100 IU/ml), short duration of infection, and severe inflammation in liver biopsy are considered predictors of sustained responsiveness to IFN therapy. Interestingly, resolution is usually heralded by a flare up in ALT levels, which is assumed to reflect massive immune clearance of infected hepatocytes. Despite the low levels of HBV DNA that frequently remain detectable in liver tissue (see above), relapses are rare.

Several agents that directly affect virus replication have also been examined for anti-HBV activity in vivo (Liaw, 1997). The cytosine analog lamivudine, or (−)3TC, an enantiomer of 3′-thiacytidine that was originally developed as a reverse transcriptase inhibitor for anti-HIV therapy, is probably the most powerful inhibitor of HBV replication in vivo tested so far. At doses of 50 to 600 mg per day, circulating virus is rapidly reduced in virtually 100% of patients and, after a course of several months, is abated below the detection limits of PCR in a significant proportion of cases. However, after withdrawal of the compound, most patients relapse (Dienstag et al., 1999; Pessoa and Wright, 1999). Lamivudine-resistant reverse transcriptase gene mutants are a frequent finding in long-term-treated patients, but virus replication generally remains at moderate levels, probably due to reduced replication competence (Hoofnagle, 1998; Ling and Harrison, 1999). Combination therapies with IFN and lamivudine and other antiviral agents (such as famciclovir, adefovir dipirovil, and other nucleoside analogs) are under evaluation for efficacy and ability to prevent drug resistance (Ono-Nita et al., 1999).

Currently, liver transplantation is the only available treatment for decompensated cirrhosis and HCC. However, in HBV-infected patients this treatment is very commonly followed by the recurrence of infection in the grafted liver and by a particularly aggressive course of the ensuing hepatitis due to immunosuppression (steroids are known to activate HBV expression). The presence of HBV DNA and HBeAg in the prospective recipient are considered highly predictive of posttransplant recurrence (Angus, 1997). The drugs discussed above are also being evaluated in the posttransplant context, alone, in various combinations, and in association with HBIG, with some initial success (Dodson et al., 1999). Therapeutic infusion of autologous CTL expanded in vivo has also been proposed.

Laboratory Diagnosis and Follow-Up

Tests for Viral Antigens and Antibodies

Traditionally, the laboratory diagnosis of HBV infection is performed with immunoassays which demonstrate the presence in the patient's serum of specific viral proteins (Table 3) or of the corresponding antibodies. The radioimmunoassays initially used have been almost completely replaced by commercially available enzyme immunoassays (EIA) in various formats, generally characterized by excellent specificity and sensitivity. An algorithm guiding the use of such assays is provided in Fig. 3, and a key to the interpretation of test results is given in Table 4.

The key marker for diagnosis is the HbsAg, which becomes demonstrable in serum 2 to 6 weeks prior to

TABLE 3 Major viral proteins exploited for the laboratory diagnosis of HBV infection

Function (current name)	Properties
Envelope glycoproteins or surface antigen (HBsAg)	The bulk of the immunoreactive HBsAg found in plasma is the S glycoprotein (24–27 kDa). HBV virions and subviral aggregates also contain small amounts of two larger products of the pre-S–S ORF known as middle (M) and large (L) pre-S proteins (31 and 39 kDa, respectively) that result from initiation of translation at different AUG codons (Fig. 1).
Capsid phosphoprotein or core antigen (HBcAg) Truncated capsid protein or e antigen (HBeAg)	The ORF coding for the capsid protein has two translation initiation codons. While initiation from the first codon leads to production of HBcAg (22 kDa), initiation from the second codon produces HBeAg (Fig. 1). HBeAg is not incorporated into virions but is released by infected cells as a soluble glycosylated or unglycosylated protein. HBeAg detection in serum represents an indirect marker for active viral replication.

biochemical evidence of liver damage and remains positive throughout the course of infection, except in its late integrative stages. The lower limit of sensitivity of the assays currently in use is 0.1 to 2 ng/ml of serum. When this marker is negative, one can exclude an active infection except for a very few instances in which the amounts of antigen produced are too small for consistent detectability, as may occur in fulminant liver failure and in neonatal infections.

Mutations in the *a* determinant may also affect the performance of HBsAg detection immunoassays (Coleman et al., 1999). By contrast, a positive test indicates the existence of an HBV infection that can be either acute or chronic. When the HBsAg test result is positive, one must proceed to test for the presence of antibodies to HBcAg, which are confirmatory for the diagnosis of infection. Usually, anti-HBc IgM appears 2 to 4 weeks after the HBsAg and, unless

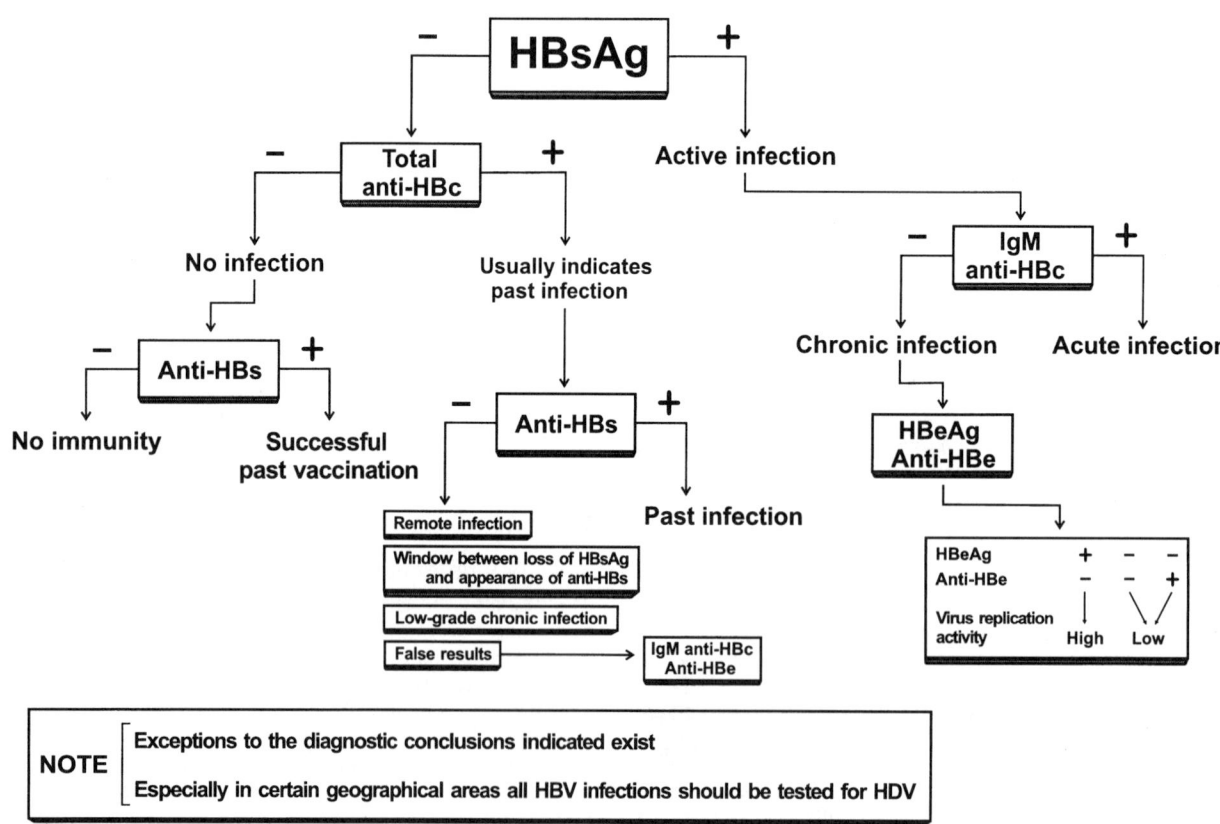

FIGURE 3 Algorithm for serological diagnosis of HBV infection.

TABLE 4 Interpretation of serological test results for HBV infection

Results[a]						Interpretation
HBsAg	HBeAg	Anti-HBe	Anti-HBc IgM	Anti-HBc IgG	Anti-HBs	
+	+	−	−	−	−	Primary infection, very early stages
+	+	−	+	−	−	Primary infection, early stages
+	+	−	−	+	−	Primary infection, late stages
−	−	+	±	+	+	Resolution of primary infection
+	±	−	−	+	−	Chronic infection
±	−	+	−	+	±	Resolution of chronic infection
−	−	±	−	±	±	Past infection
−	−	−	−	−	+	Past vaccination
−	−	−	−	+	−	False positive; passively acquired antibody; resolved infection with loss of detectable anti-HBs
−	−	−	−	−	−	No infection

[a] +, positive; −, negative; ±, may be either positive or negative

the assay method is particularly sensitive, become undetectable in a few months, but it is important to keep in mind that low levels can persist for much longer and can reappear or increase in titer (>10 IU/ml) when viral replication increases. Thus, a reactivation of viral replication can be revealed by quantitative determination of anti-HBc IgM as well as by quantitation of HBV DNA.

Additional markers useful for establishing the activity of infection include HBeAg, anti-HBe, and anti-HBs. With the exclusion of infections produced by HBeAg-minus mutants, the HBeAg is found in the patient's blood when the virus is actively replicating, and its continuous presence therefore correlates with active hepatitis. Conversely, disappearance of this antigen and appearance of the corresponding antibodies is indicative of reduced viral replication. In acute infections, these events usually herald resolution, since they are usually followed, within weeks or months, by the disappearance of the HBsAg from the serum and, with a lag of weeks, by the development of anti-HBs antibodies.

Detection and Measurement of Viral DNA in Plasma

In clinical practice, conventional immunoassays permit an accurate diagnosis of most acute and chronic HBV infections, but there are instances where determining the presence of HBV DNA in plasma is useful or even essential for correct interpretation of immunological data (Table 5). Although HBeAg still serves as an important marker for monitoring infection activity (with rare exceptions, it is detectable only in patients who circulate HBV DNA), a negative assay result and seroconversion to anti-HBe does not preclude viral replication (Maruyama et al., 1993). It should also be noted that in infections sustained by HBeAg-minus variants, anti-HBe can coexist with active viral replication. Tests for the viral reverse transcriptase -DNA polymerase may also be useful, but they are much less sensitive.

In essence, detection of viral DNA must be regarded as the most accurate indicator of HBV replication and infectivity. As would be expected, this is especially true when the viral DNA is detected by methods based on gene amplification rather than by direct nucleic acid hybridization methods. Even though hybridization assays of higher sensitivity have been described (Ranki et al., 1995), the lower detection limit of commercial DNA hybridization assays ranges on the order of 10^6 to 10^7 genomes/ml, whereas amplification methods can reproducibly reveal a few hundred genomes (Zaaijer et al., 1994; Hwang et al., 1996). Thus, in one study HBV DNA was detected by PCR in more than 50% of patients who tested negative for HBeAg (Berger et al., 1998). Detection of HBV DNA also circumvents the problem of demonstrating and staging the infections sustained by genetic variants of HBV, such as the HBeAg-minus mutants.

HBV DNA can now be quantitated with good reproducibility over a wide range of values. The procedure has maximum utility for recognizing those patients who might benefit from IFN treatment and in the follow-up of responses to therapies. For example, persistently elevated ALT levels despite HBeAg clearance in patients undergoing antiviral treatments suggests that an HBeAg-minus mutant has emerged and has replaced the original wild-type virus. This possibility is easily verified by testing for viral DNA. It is most likely that, with better standardization and the development of tests with wider linear dynamic ranges and reduced cost, periodic determination of HBV DNA levels will soon become the key for managing hepatitis B patients, especially those with low viremia levels. Thus, it has recently been proposed that a level of 10^5 genomes/ml can be used as a sort of cutoff to distinguish the asymptomatic carrier and chronic hepatitis states and that treatments reducing the viremia to less than 10^4 genomes/ml should be regarded as successful because they are not accompanied by significant liver injury (Niitsuma et al., 1997). Additional studies are in progress to establish what levels of HBV replication should be considered clinically significant and justify antiviral intervention.

HBV Genotyping and Determination of Drug Resistance

Validated methods for HBV genotyping include sequencing and restriction pattern analysis of amplified segments of the viral genome (Lindh et al., 1998). Since the potential influence, if any, of the viral genotype on the course of HBV infection has yet to be established, the utility of genotyping is presently limited to epidemiological investigations. With the introduction in therapy of effective nucleoside analogs, determination of drug resistance is instead acquiring an

TABLE 5 Usefulness of HBV, HCV, and HDV genome detection and quantitation in different clinical situations

Clinical situation	Utility[a]		
	HBV	HCV	HDV
Diagnosis of acute infection; early stages	+	+++	+
Diagnosis of acute infection; late stages	−	++	++
Evaluation of discordant serological profiles	++	++	++
Diagnosis of vertical infection	+	++	+
Diagnosis of chronic infection	−	++	++
Diagnosis of infection in immunosuppressed patients	++	++	+
Assessment of blood infectivity	+++	+++	+++
Follow-up of clinical course	+	+	+
Prediction of responsiveness to IFN treatment	++	+	?
Evaluation of the efficacy of therapies	+++	+++	++
Assessment of infection eradication	+++	+++	+++

[a]+, useful; ++, very useful; +++, most useful; −, not useful; ?, unknown.

increasing importance in the management of HBV-infected patients as well as in other viral infections. Thus, for example, a classical lamivudine resistance mutation, due to changes in the YMDD motif of the RNA-dependent DNA polymerase (Fig. 1), occurs in 15 to 30% of patients after 1 year of treatment with this compound. It consists of the substitution of either valine or isoleucine for methionine in the highly conserved motif Tyr-Met-Asp-Asp of the C domain of the catalytic site of the viral polymerase. Currently, the method of choice for detecting lamivudine-resistant variants is sequencing of HBV in the P ORF regions known to be involved in mutations towards drug resistance. More recently proposed rapid methods based on restriction fragment length polymorphism and real-time PCR have yet to be validated in the field (Allen et al., 1999).

Liver Biopsy

Liver biopsy results do not directly correlate with the virologic stage of infection (Lee, 1997) but are extremely useful for prognosis and for making decisions regarding patient management. Biopsy is usually performed before starting antiviral treatments in order to exclude nonviral causes of liver damage and after treatments to evaluate their efficacy and possibly should be carried out during periods of relatively low hepatitis activity. Evaluating the characteristics and severity of histological lesions requires expertise outside the scope of clinical virology laboratories. The virologist can, however, be requested to evaluate the virus content of liver tissue, using immunohistochemistry, in situ hybridization, or gene amplification methods. By immunofluorescence, HBsAg is detectable only in the cytoplasm of hepatocytes, whereas HBcAg can found in both the cytoplasm and the nucleus. The presence of HBcAg in the nucleus only is generally considered an indicator of low to moderate hepatocyte damage.

HCV

Subsequent to the development of accurate diagnostic assays for HBV infection, the existence of an additional form of transfusion-transmitted hepatitis, became apparent. Since the diagnosis was one of exclusion, for several years the disease was referred to as non-A, non-B hepatitis, but the causative agent is now known to be HCV in the great majority of cases. Acute HCV infection is often asymptomatic and may remain undiagnosed. When clinical illness is present, it is generally mild. Because acute infection is so benign, the clinical importance of hepatitis C is mainly due to its remarkable ability to persist and to produce chronic and irreversible liver damage. Approximately 80% of infected individuals become chronic virus carriers, with little or no propensity to resolve the infection spontaneously. Chronic hepatitis C tends to be more slowly progressive than chronic hepatitis B, but the long-term consequences are possibly more severe. For this reason and because of the absolute lack of vaccines, hepatitis C is presently the major public health concern associated with hepatitis viruses.

The Virus

Evidence that a specific virus was implicated in the etiology of what was then known as non-A, non-B hepatitis was obtained when, in the late 1970s, it was shown that chimpanzees could be infected with patients' blood. The agent proved hard to characterize and remained elusive until 1989, when its genome was cloned and described in detail. Because HCV shares significant properties with members of the family *Flaviviridae*, it is presently classified within this family in the newly proposed genus *Hepacivirus*. Due to major difficulties encountered in directly characterizing virus morphology, physicochemical properties, and interactions with host cells, most of what we presently understand about HCV structure and replicative strategies (Table 6) is inferred from the organization of its positive-sense single-stranded (ss) RNA genome (Fig. 4) and from presumed similarities to related flaviviruses.

In infected humans, HCV shows a distinct tropism for the liver, but there is growing evidence that it can also replicate in other tissues, albeit much less actively. Indeed, in situ hybridization, immunohistochemistry, and, more recently, antigenome-specific PCR have provided evidence of local virus replication not only in hepatocytes but also in several lymphoid and nonlymphoid tissues, including biliary epithelial cells, bone marrow, spleen, peripheral blood mononuclear cells, pancreas, thyroid, and salivary and adrenal glands (Cho et al., 1996; Laskus et al., 1998a; Sansonno et al., 1998). Essentially similar results have been obtained with chimpanzees, the only animal known to be consistently susceptible to experimental infection. Several primary tissue cultures and established cell lines have been reported to support limited growth of HCV in vitro, but the amounts

TABLE 6 Structure and replication of HCV; salient features

Structure

Round 55- to 65-nm-diameter virus particle with density ranging between 1.14 and 1.16 g/ml

Icosahedral 30- to 35-nm-diameter capsid, or core, composed of numerous copies of a single protein (p21)

Envelope with short projections composed of two virus-coded glycoproteins (E1 or gp31 and E2 or gp68-72) and cell-derived lipids

Linear ssRNA (approximately 9.5 kb) with messenger polarity

Most of the genome is a single large ORF encoding a 3,011- to 3,033-amino-acid precursor from which approximately 10 structural and nonstructural proteins are formed (Fig. 4)

The two extremities of the genome, the 5' UTR (324 to 341 nucleotides long) and the 3' UTR, have no known coding functions but possess defined secondary structures and exert essential regulatory functions in genome replication and expression

Replication

Poorly understood

Virus entry into cell probably mediated by E2 binding to receptor(s): both the receptor for serum low-density lipoproteins and an extracellular loop of CD81 have been implicated in attachment, but their roles in entry are uncertain

Release of the viral genome into the cytoplasm, where the entire process seems to take place

The viral ORF is translated into the polyprotein precursor; translation is cap independent and is controlled by an internal ribosome entry site located in the 5' UTR

Co- and posttranslationally, the polyprotein is cleaved by viral and possibly cellular proteases to generate the three major structural proteins and several nonstructural proteins with catalytic (including an RNA-dependent RNA polymerase, two proteases, and an NTPase-helicase) and regulatory functions

Synthesis of antigenomic minus-strand RNA, which serves as a template for progeny

Envelope probably acquired by budding into cytoplasmic vesicles

No overt cytopathic effects; mechanisms of release from cells unknown

of virus produced are too small to be of much utility (Lau, 1998; Morrica et al., 1999). Thus, despite recent advances (Lohman et al., 1999), we still lack a reliable in vitro infection system. Infectivity is readily destroyed by ether, chloroform, β-propriolactone, and formalin, and the virus is relatively unstable at room temperature; however, complete inactivation of infected plasma requires heating it to 100°C for 5 min.

HCV isolates show a considerable degree of genetic heterogeneity. Variability is unevenly distributed across the genome and is especially high in the regions encoding the envelope glycoproteins, while the noncoding terminal 5' untranslated region (UTR), a portion of the 3' UTR, and the region coding for the capsid protein are the most highly conserved. Maximum heterogeneity occurs in a 30-amino-acid stretch (hypervariable region 1 [HVR-1]) of the envelope glycoprotein E2, believed to represent a key target for protective antiviral immune responses. Based on their genetic diversity, HCV isolates have been grouped into six distinct genotypes, designated types 1 through 6, and

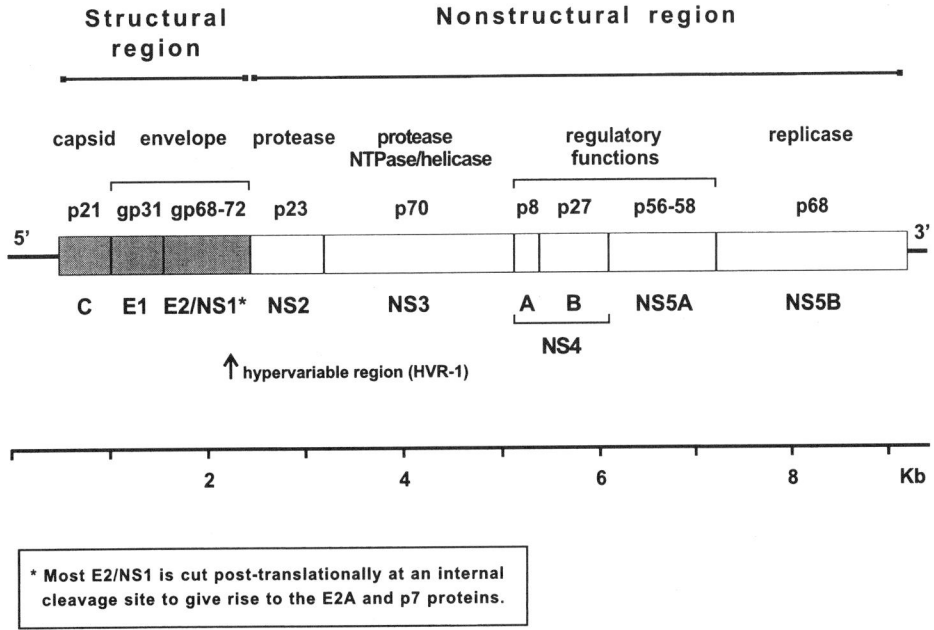

FIGURE 4 Organization of the HCV genome and encoded proteins (Table 6). C, capsid (core); E, envelope; NS, nonstructural.

several dozen subtypes, designated by lowercase letters (Simmonds, 1995). Sequence analysis of the NS5B region of the viral genome has been found to be especially informative for genotyping purposes (Simmonds et al., 1993). In this region, isolates differing by 12% or less are considered to belong to the same subtype, and those differing by 28% or less are considered to belong to the same genotype. Since there are no true in vitro serum neutralization assays available, the significance of genetic heterogeneity in terms of cross-protective immunity is presently unknown. The observation that hosts infected with a given genotype can be superinfected with another isolate of the same or different genotype casts doubt on whether distinct serotypes of HCV exist and have practical importance.

More comprehensive reviews of HCV structure and biology can be found in Major and Feinstone, 1997, and Bendinelli et al., 1999.

Dynamics of HCV Replication and Antiviral Immune Responses in Infected Patients

The clinical course and outcomes of HCV infection can be extremely varied and unpredictable. However, since the viral events that characterize the infection are rather stereotyped, the only useful distinction from the standpoint of the clinical virology laboratory is that between acute and chronic infections.

Acute Infections

Information about acute HCV infection is mainly derived from posttransfusion cases (Fig. 5A). Under these circumstances, the virus becomes detectable in the circulation in 1 to 3 weeks, although it probably reaches the liver much earlier (in experimentally infected chimpanzees, this organ becomes virus positive several days earlier than blood). As discussed below, many community-acquired forms of hepatitis C occur in patients with no known risk of parenteral infection. Since it is likely that in these cases infection is initiated by small doses of virus gaining access to the body through the mucous membranes, it is also likely that the virus takes longer to become detectable in blood. The extrahepatic sites of virus amplification, if any, in these early stages, and the routes used to reach the liver are unknown. In any case, viremia peaks at titers of 10^8 genomes/ml or greater a few weeks later.

Persons with acute infection are either asymptomatic (60 to 70%) or present with generally mild jaundice (20 to 30%) or nonspecific manifestations (10 to 20%) after an incubation period of 4 to 20 weeks. ALT elevations may exhibit different patterns (single or multiple peaks, plateaus, etc.) but infrequently exceed 1,000 U/liter, reflecting generally moderate liver damage. Fulminant hepatic failure is rare and usually occurs in conjunction with other hepatotropic noxae.

There are few reports of the viral events occurring during the acute stages of HCV infection being studied systematically using recently developed methods which accurately quantitate viral loads. However, it is well established that in most acutely infected patients, the initial phase of florid virus replication is followed, within months, by a gradual decline of viremia (Naito et al., 1994). Symptoms also disappear, and ALT concentrations progressively normalize. However, this does not mean that the infection will be cleared. In fact, complete resolution of primary HCV infection is observed in less than 15 to 25% of acutely infected persons. Of note, no clinical or epidemiological features can

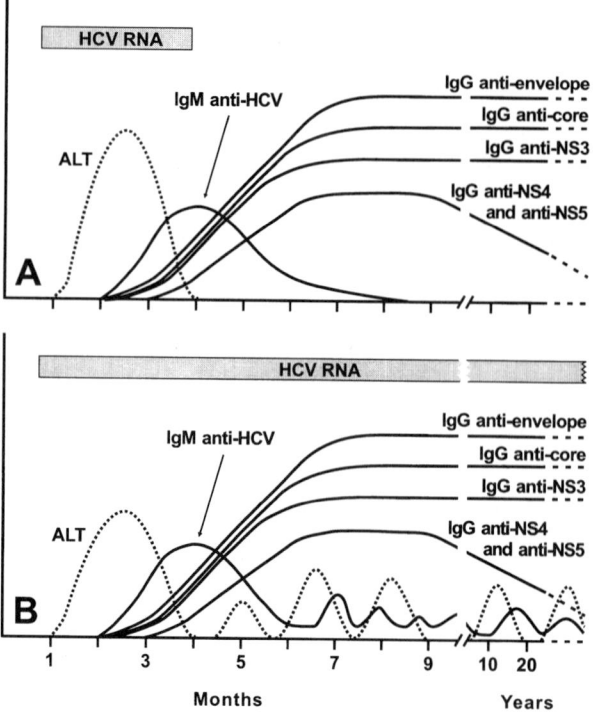

FIGURE 5 Typical courses of HCV infection. (A) Acute self-limited infection. (B) Infection that does not resolve and becomes chronic (the great majority of cases). While the virological and immunological events depicted are seen in virtually all patients, clinical symptoms develop in a minority. The shaded bars represent the time periods during which viral RNA is demonstrable in the circulation.

be used to predict whether the infection will resolve or become persistent.

The modulation of HCV viremia observed in the course of acute infections is believed to result essentially from the host's immune response to the replicating virus (Yoshimura et al., 1994). Both humoral and cell-mediated immune responses are vigorous but tend to develop more slowly than in most other systemic viral infections. In particular, antibodies measurable by commercially available kits usually do not become demonstrable until 8 to 9 weeks after the initiation of infection, and occasionally the preserological phase can be much longer (up to 9 months in some patients). It should be noted, however, that our present understanding of anti-HCV immune responses is based mainly on the use of recombinant and synthetic antigens derived from the structural and nonstructural regions of the viral genome, while the development of functional assays using infectious virus as the antigen has been hindered by the lack of practical methods for growing the virus in cultured cells. For example, there are no serum neutralization tests available, although surrogate methods have been proposed which measure the ability of antibodies to block HCV entry into cells (Rosa et al., 1996). It is not clear, therefore, whether failure to eradicate the virus results from an inherent resistance of the virus to antibody-mediated neutralization, from delayed or defective immune responses, or from both.

Reportedly, the titers of antibodies to oligopeptides representing the N terminus of the HVR-1 domain of E2, which is believed to contain important neutralization epitopes, were found predominantly in subjects with self-limited acute infections (Zibert et al., 1997). Moreover, it has been reported that self-limited acute infections are characterized by robust T helper (Th) cell functions and by the prevalence of T lymphocytes with a type Th0 or Th1 cytokine profile, whereas infections that do not resolve exhibit less pronounced antiviral lymphoproliferative responses and cytokine profiles of the Th2 type (Missale et al., 1996; Diepolder et al., 1997; Tsai et al., 1997). Nevertheless, there are no markers, immunological or of other type, that can be used to reliably predict whether acute HCV infection will be cleared or persist (Beld et al., 1999; Ray et al., 1999).

The first antiviral antibodies detected in infected patients are usually of the IgM class, despite difficulties that still exist in their routine demonstration and measurement (see below). IgG antibodies develop later and with kinetics that vary depending on the viral antigen considered: those to the envelope, core, and NS3 antigens usually develop earlier than antibodies to NS4 and NS5. No differences have been detected between the antibody responses to envelope epitopes in patients who resolve the infection and those who evolve toward chronicity (Zibert et al., 1999).

Chronic Infections

As discussed above, the great majority of HCV-infected individuals become virus carriers indefinitely (Fig. 5B). After the acute phase, patients who have not resolved the infection are usually symptom free, with ALT levels in the normal range or only slightly elevated, for at least 1 or 2 decades. However, with duration of infection, in most chronically infected persons ALT levels become intermittently or persistently elevated and liver function tends to progressively deteriorate as a result of accumulating hepatocellular damage. Other subjects maintain constantly normal ALT values but nevertheless are viremia positive, and a liver biopsy may reveal considerable damage, ranging from persistent or active hepatitis to cirrhosis. Still others do not appear to develop clinically significant histopathological changes.

It is generally accepted that patients with chronic HCV infection, even if they show an indolent course, should be evaluated for severity of hepatic damage and possible treatment and that accurate prognosis requires long-term follow-up with periodical measurement of ALT levels. Initial reports suggested that in chronically infected patients viremia could fluctuate markedly and even disappear for variably protracted periods of time, as a presumed result of the opposing forces of virus replicative activity in the liver and host immune defenses. However, more recent longitudinal studies using accurate quantitative assays indicate that, in the absence of antiviral treatments, viremia levels may vary between 10^3 and 10^8 genomes/ml (usual range, 10^5 to 10^6 genomes/ml) in different patients but remain essentially stable over time in each individual patient, with variations in titer that usually do not exceed 1 log unit, and that intermittent viremia is uncommon (Yoshimura et al., 1997; Gordon et al., 1998). Essentially confirming these results, the mean difference between maximum and minimum HCV RNA values (monthly tests) was found to be 0.7 log units in patients with constantly normal ALT and 1.3 log units in patients with fluctuating ALT (Pontisso et al., 1999). Spontaneous clearance of viremia is very rare in chronically infected individuals. Recently, interesting data on the kinetics of HCV replication during chronic infection have been obtained by sequentially measuring the viremia levels of responder patients given IFN-α at doses sufficient to markedly reduce virus production and release from infected cells. The half-life of virions in plasma was estimated to be 2.7 h, while the number of virions produced and cleared per day was on the order of 10^{12} (Neumann et al., 1998).

The spectrum of antiviral antibodies found in chronically infected patients is approximately the same as that observed in advanced acute infections. However, antiviral IgM is detected in fewer patients and generally in a discontinuous fashion, and anti-NS4 and anti-NS5 antibodies may slowly decline considerably and occasionally even become undetectable after years of persistence. By testing peripheral blood or intrahepatic lymphocytes, Th and CTL reactivities to several HCV antigens are readily detected in most chronically infected subjects; however, attempts to correlate these immune parameters with the progression of the infectious course and the severity of liver lesions have so far been disappointing. An important contribution of cell-mediated immunity to both containment of infection and generation of hepatocellular injury, however, is likely (Chisari, 1997; Cooper et al., 1999). Patients with clinically stable chronic infections also exhibit high titers of antibodies that block virus attachment to cells (Ishii et al., 1998).

In summary, HCV paradoxically appears to be more immunogenic but less prone to immunologic control than HBV. Although the ability of HCV to undergo prompt genetic variation is considered important, the reasons behind its outstanding capacity to escape immune control and persist are essentially unknown (Bassett et al., 1999; Beld et al., 1999; Cramp et al., 1999; Ray et al., 1999). The other flaviviruses do not normally establish persistent infections.

Epidemiology

HCV is present worldwide, with prevalence rates that range, in the general population, between 1% in Western Europe and North America and 20% in certain African countries, even though most infected people are unaware of their infections because they are clinically healthy. In developed countries, the highest prevalence is found in the 20- to 49-year-old age group, in males, and in low socioeconomic classes (Centers for Disease Control and Prevention, 1998; Alter et al., 1999). In the categories at risk for parenteral infection, HCV prevalence is significantly higher than in the general population. For example, it can be as high as 60% in hemophiliacs who received commercial clotting factors before these blood derivatives were treated to inactivate viruses, and it usually ranges between 50 and 90% in injecting drug users. Posttransfusion infections have dropped dramatically in recent years, thanks to donor testing for anti-HCV antibody and, prior to the development of HCV-specific assays, for surrogate markers (ALT levels and anti-HBV). The risk is now calculated to be on the order of 1 per 1,000 units transfused. Prior to the implementation of virucidal treatments, intravenous immune globulins could also transmit the infection. Renal patients requiring dialysis are also frequently infected (Pereira and Levy, 1997), but with the exception of these patients, descriptions of nosocomial infections are scarce. Individuals occupationally exposed to blood are also at increased risk of infection. Needle stick accidents and other types of percutaneous or permucous exposure represent important modes of transmission. Moreover, iatrogenic transmission is believed to have contributed substantially to HCV diffusion in Egypt and possibly other countries.

The HCV infections that occur in people with no identifiable parenteral risk are called sporadic, cryptogenetic, or community acquired. Understanding the sources and modes of contagion in these people is difficult, in part probably due to hesitance to disclose previous risk behaviors. There is evidence that HCV may be shed with genital secretions and saliva. Both mother-to-child and venereal transmissions are known to occur but are considerably less efficient than for HBV, unless expecting mothers and sexual partners have especially high viremia levels due to acute infection, concomitant HIV infection, or other reasons. Approximately 5% of infants born to infected mothers are virus positive. Transmission to household contacts is also infrequent. In a significant proportion of community-acquired infections, no apparent occasion of exposure to the virus can be recognized. There are no indications that HCV or other blood-borne hepatitis viruses can be spread by insect vectors.

Investigations have shown that genotypes 1 to 3 and their subtypes are ubiquitous, but their relative prevalence rates vary in different geographical regions. Subtype 1a is predominant in North America, and subtype 1b is predominant in Western Europe and Japan. Genotypes 4 to 6 appear to be less common and restricted to specific regions (types 4 and 5 to Africa and type 6 to Southeast Asia), although further studies are needed to obtain a comprehensive map of HCV genotype distributions. Recently, subtype 2c, which was considered a rare genotype, was found to represent the second most frequent genotype in Italian patients with community-acquired infections (Maggi et al., 1997b). Interestingly, genotype distributions in Europe have been found to vary with patients' age, probably reflecting temporally distinct, independent epidemic waves. Thus, subtype 1b has been found to be responsible for over 80% of infections in persons 60 years old or more but for fewer than 30% of the cases observed in people less than 30 years old. Differences in genotype distribution have also been observed, depending on the mode of infection. Thus, hemophiliacs tend to be infected with genotypes that are predominant in the areas from which the contaminated blood used for clotting factor production derived, rather than with those most prevalent in their own countries (Kinoshita et al., 1993; Pistello et al., 1994).

Prevention and Treatment

Except for the lack of means for active immunoprophylaxis, primary prevention of HCV infection relies on measures similar to those discussed for HBV. Those directed at interrupting virus transmission through iatrogenic or accidental exposure to infected blood are of primary importance. Indeed, serological screening of donors has greatly reduced posttransfusion and posttransplant HCV infections, and it is now debated whether donors should also be tested for viral RNA to exclude those in the preserological phase of infection (Roth et al., 1999). Blood derivatives are now also safe due to improved blood supply and virucidal treatments (Pistello et al., 1991; Alter, 1995). Secondary prevention measures are also analogous to those outlined for HBV. Detailed guidelines on these matters have been published recently (Centers for Disease Control and Prevention, 1998).

The efficacy of postexposure prophylaxis with immune globulins is probably only marginal. Attempts to develop vaccines against HCV are under way. Pilot experiments have given some encouraging results (Abrignani and Rosa, 1998); however, a vaccine against HCV still seems a long way off. The great genetic diversity of HCV, the lack of efficient in vitro culture systems, the scarcity of experimental hosts, and the poor resistance to reexposure to the virus demonstrated by infected humans and chimpanzees herald formidable obstacles.

The therapeutic armamentarium presently available for hepatitis C is limited to IFN-α given alone or in combination therapies still under evaluation. It is generally agreed that chronic hepatitis C patients with persistently elevated ALT, high viremia levels, and liver biopsy showing portal and bridging fibrosis and significant necroinflammatory changes should be treated, because their risk of developing cirrhosis is greater. There is also wide consensus that patients with persistently normal ALT and those with advanced cirrhosis should be managed on an individual basis or in the context of clinical trials. However, precisely which patients should receive antiviral treatment and which should not is still an open matter.

The goals of therapy are a sustained virological (virus undetectable in blood) and biochemical (normalization of ALT for at least 12 months) response, but neither has proved easy to achieve. Prolonged treatment with IFN-α has been shown to reduce biochemical abnormalities in approximately 40 to 50% of chronic infections and to ameliorate liver lesions in a large proportion of responder patients, but clinical benefits are often transient, and in about half the initial responders, symptoms recur after cessation of treatment. Complete virus eradication is even less common. Three million units given subcutaneously three times weekly for 12 to 18 months has been demonstrated to result in sustained clearance of HCV from plasma in approximately 20% of patients, but treatment doses and regimens vary widely (Hoofnagle, 1998; Camma et al., 1999). At least theoretically, using higher dosages might have appreciable advantages (Neumann et al., 1998). The role of anti-IFN antibodies in determining the limited efficacy of IFN-α therapy is under examination. IFN-α treatment has been found to also be of some benefit in acute infections (Camma et al., 1996).

The therapeutic problems posed by nonresponder and relapsing patients are also the subject of intensive clinical investigation and debate. The strategies considered include more drastic IFN-α regimens, different types of IFN (consensus IFN and IFN-β have been used less extensively but appear to be effective), and combination therapies. Recently, the combination of IFN-α and ribavirin has provided significantly better results than IFN-α alone. Attempts to develop more innovative therapeutic approaches are also under way, including antisense oligonucleotides and ribozymes. It is also expected that a better understanding of HCV enzymes will lead to the design of new effective drugs (Ahmed and Keeffe, 1999).

Currently, HCV-associated cirrhosis and HCC are the indication for liver transplantation in at least one-fourth of cases. Following transplantation, nearly all transplant recipients have recurrent, IFN-resistant HCV viremia and signs of hepatitis in the grafted organ. The consequences for long-term graft survival have yet to be fully evaluated, but the risk of cirrhosis appears to be significant (Feray et al., 1999).

Indicators Predictive of Beneficial Response to Antiviral Treatments

The limited efficacy of IFN therapy must be weighed against its considerable side effects and cost. Therefore, criteria that might be used to predict the likelihood of a sustained response are being actively sought. Existing data indicate that several variables influence the outcome of treatment, but none is 100% predictive. Old age, race, high viremia titers, and the presence of hepatic fibrosis are considered dependable predictors of poor responsiveness. The viral genotype has also been shown to affect the outcome of IFN treatment, since rates of response are higher against genotypes 2 and 3 than against genotypes 1 (especially subtype 1b) and 4 (Sherlock, 1995). Whether this reflects genotype-specific differences in resistance, higher initial virus loads, or more advanced disease due to longer duration of infections caused by certain genotypes is not clear (Hoofnagle, 1998). The sensitivities of other genotypes to IFN are still to be determined. Comparative genome analysis of genotype 1b isolates obtained prior to IFN treatment from responder and nonresponder Asian patients has also suggested that responsiveness correlates with the amino acid composition of a small segment of the NS5A protein (Enomoto et al., 1996), but this has not been confirmed with American and European patients (Rispeter et al., 1998). An elevated complexity of HCV quasispecies has also generally, albeit not invariably, correlated with poor responses and high relapse rates (Shindo et al., 1996; Pawlotsky et al., 1998a; Lopez-Labrador et al., 1999). Attempts are also being made to predict the long-term outcome of therapies from the changes in viremia levels brought about by the early doses of treatment. In recent studies of patients treated with IFN-α, those with an early normalization of ALT levels (Davis et al., 1994), an early disappearance of viremia (Wiley et al., 1998; McHutchison et al., 1999), or a fall in viremia greater than 3 log units during the first 4 weeks of treatment (Zeuzem et al., 1998) exhibited the best long-term responses. In a similar study, patients who had not resolved the infection after a first cycle of IFN therapy responded better to a second cycle if they had shown a reduction in viremia of at least 2 log units during the first cycle (Tong et al., 1998). Attempts to correlate response to therapy with the expression of IFN receptor genes in the liver are also under way (Morita et al., 1999).

Laboratory Diagnosis and Follow-Up

Demonstration of HCV infection is obtained by examining serum or plasma with a combination of immunological and molecular assays (Fig. 6). The latter methods are also of considerable utility for correct clinical management of infected patients, because they can be used to quantitatively assess HCV replicative activity.

Tests for Antiviral Antibodies

In acute HCV infections, the appearance of viremia precedes that of biochemical indicators of liver damage by 3 to 10 weeks and that of antiviral antibodies by several more weeks. Thus, in the early phase when antiviral antibodies are not manifest (the window period), infection can be demonstrated solely by testing for viremia. However, the laboratory is rarely requested to perform a diagnosis of HCV infection this early in infection. In most instances, the patient to be tested is under clinical examination for symptoms suggestive of ongoing hepatitis or, more often, has already been found to have altered ALT levels during routine blood testing performed for unrelated reasons. Because

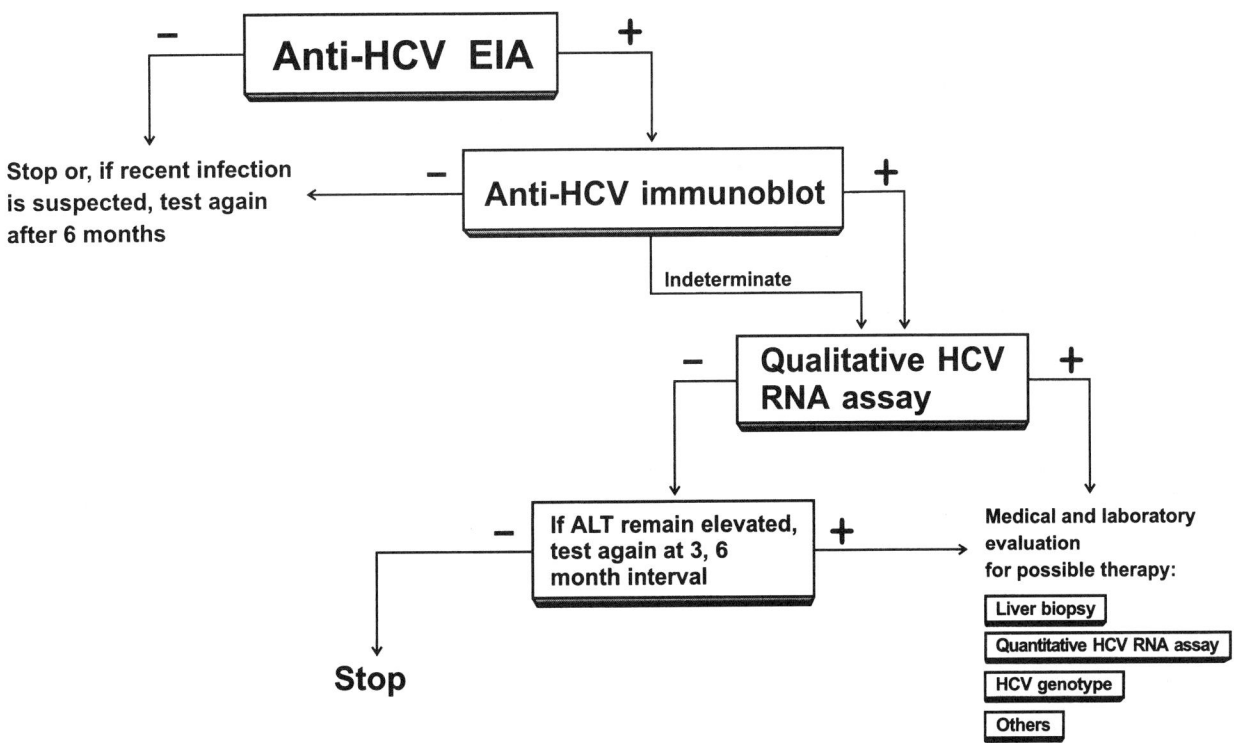

FIGURE 6 Algorithm for the laboratory diagnosis of HCV infection.

by the time ALT are elevated most patients possess readily detectable antiviral antibodies, the first approach to the diagnosis of HCV infection is usually via serology.

Commercial assays for demonstrating anti-HCV antibodies have been redesigned several times to shorten the duration of the preserological (window) period and increase sensitivity and specificity, thus meeting the needs of blood banks as well as of clinical virology laboratories. The third-generation kits presently in use consist of various formats of EIA and immunoblot assays prepared with carefully selected recombinant and synthetic antigens derived from conserved domains of the core, NS3, NS4, and NS5 regions of the viral genome. Compared to the kits containing fewer and less well characterized antigens that were commercialized soon after HCV was cloned, the present kits have permitted a reduction in the duration of the window phase of 4 to 6 weeks. As demonstrated in the screening of hundreds of thousands of blood donations, the specificity of current EIA is generally excellent, and false-positive results are very limited (due to hypergammaglobulinemia, aged sera, rheumatoid factors, and unexplained causes). Nonetheless, especially when persons with normal ALT levels are being tested, it is correct to verify positive EIA results with supplemental tests, such as immunoblot assays that permit a better definition of the specificities of reacting antibodies. With these assays, sera are considered positive when they react with epitopes derived from at least two viral regions and indeterminate when they are reactive with a single region. Particularly in individuals at low risk for HCV infection, indeterminate immunoblot results may be indicative of a false positivity. However, it is important to keep in mind that, since anti-NS4 and -NS5 antibodies tend to be produced later than other antiviral antibodies and may wane following spontaneous or posttreatment resolution of infection, the absence of these reactivities may be indicative of a recent infection still in the process of seroconversion or of a resolved infection. Indeterminate immunoblot results are frequently also observed in immunocompromised patients whose reactivity is often restricted to core and NS3 antigens.

Even with the optimized serological assays presently in use, no consistent correlation has been observed between the presence or titer of IgG to specific epitopes and the stage or other parameters of infection. In this regard, the search for virus-specific IgM is poorly informative, since this antibody is detected in 50 to 90% of acute infections versus as many as 50 to 70% of chronic infections. Thus, although in some studies IgM levels have been seen to correlate with ALT or viremia levels and the presence of IgM has been suggested to be predictive of a poor responsiveness to therapy, it is uncertain whether assays for these antibodies will one day become truly useful for staging the infection and identifying periods of particularly intense viral activity. Furthermore, the sensitivity and specificity performance of presently available commercial assays for anti-HCV IgM (generally based on viral core antigens, although other antigens are under evaluation) is not completely satisfactory. Since the avidity of anti-HCV IgG has been shown to increase significantly with the duration of infection, tests that measure this parameter have also been proposed but have not become routine.

Detection and Measurement of Viremia

The patient's serological profile as determined by current assays does not differentiate between ongoing and resolved infections, and there are no cell culture systems for demon-strating the presence of HCV. Furthermore, immunological detection of the core antigen in plasma has been reported only recently (Kobayashi et al., 1998) and its clinical utility is presently under evaluation (Aoyagi et al., 1999; Komatsu and Takasaki, 1999). It is, therefore, not surprising that the molecular assays that detect and quantitate the copies of viral RNA in plasma or serum have so rapidly acquired a pivotal role in the diagnosis and management of HCV infection despite their recent introduction and relatively poor standardization. Both commercial and in-house assays exploiting different gene amplification and detection principles can be used. The specificities, sensitivities, and hands-on requirements of these methods are continuously being improved. Presently, the lower limits of detection of commercial kits range around 50 to 1,000 genomes/ml, and intra- and interassay variabilities have been markedly reduced. However, it is advisable to repeat the tests several times in order to minimize false-negative results and identify patients who might have intermittent viremia, as is often seen during the course of spontaneous resolution or therapy. Also, the possibility of false-positive results demands that rigorous quality assurance controls be in place in the performing laboratory.

Due to the high genetic variability of HCV, correct selection of the viral genome segment targeted is critical for satisfactory performance of gene amplification assays. Despite being the most conserved portion of the genome, even the 5′ UTR region presents sufficient diversity between genotypes to affect assay sensitivity. This explains why the first versions of commercial kits for viremia measurement underestimated the titers of genotypes 2 and 3 by approximately 1 log unit (Hawkins et al., 1997). The versions currently marketed have been remodeled to reduce this shortcoming, although they probably will need further balancing (Berger et al., 1998; Mellor et al., 1999).

Testing the patient's plasma for the viral genome is useful in a number of clinical situations (Table 5); however, presently its most frequent use is for assessing whether infection has been cleared, either spontaneously or as a result of therapy. In a recent study, a sustained virological response to IFN therapy could be determined with 97 and 99% certainty at 4 weeks and at 3 to 4 months after the completion of therapy, respectively (Shiratori et al., 1999). The best noninvasive approach for monitoring whether and how patients are responding to antivirals in the course of treatment is the use of quantitative tests together with measurement of ALT levels. A positive correlation between HCV RNA levels in plasma and virus replication in the liver has been reported (Agnello et al., 1998; Negro et al., 1999); however, it is generally recognized that viral loads are of limited relevance for prognosis (Chemello et al., 1994). It is generally recommended that quantitative assays not be used to confirm infections or establish whether they have resolved; however, it is likely that with the development and commercialization of increasingly sensitive and reliable kits, this limitation will soon be overcome. Also, it is important to note that at present the results of different assays and even different versions of the same assay are not comparable and that there are no conversion tables; therefore, each patient should be continuously monitored by using the same type of assay.

HCV Genotyping

Numerous studies have retrospectively or prospectively investigated the existence of possible associations between specific HCV genotypes and clinical and virological features of the infections produced, but in general, the conclusions have been negative or controversial (Poynard et al., 1997; Simmonds, 1995). Thus, presently the only correlation widely accepted is that with responsiveness to antivirals (see above). However, since this is a critical issue, the laboratory is often requested to characterize the genotype of a patient's isolate.

Sequence analysis of the viral genome is considered the "gold standard" for genotyping HCV as well as other viruses, and it has also been shown that sequencing short segments is sufficient for practical purposes, if the segment analyzed is chosen with care (Bukh et al., 1995; Robertson et al., 1998; Pistello et al., 1999). Because sequencing is expensive, time-consuming, and available only in select laboratories, several easier alternative methodologies have been proposed, some of which are commercially available. In general, these methods exploit one of two approaches. In one approach (serotyping), the genotype is determined indirectly by probing the reactivity of the patient's antiviral antibodies against genotype-specific synthetic peptides deduced from the core and NS4 regions of the viral genome. A significant limitation of this assay is that it does not distinguish among subtypes, which considerably reduces its clinical usefulness. In the other approach, the genotype is studied directly by characterizing the amplicons obtained from appropriate domains of the viral genome using restriction fragment length polymorphism, binding to genotype-specific probes, or other techniques. All of these "rapid" methods have advantages and disadvantages in terms of sensitivity, specificity, and hands-on time, and furthermore, they need to be evaluated against the genotypes prevalent in the area where they are used (Vatteroni et al., 1997).

HCV Quasispecies Analysis

Similar to many other viruses, especially those with an RNA genome, HCV is present in infected hosts in the form of a variably complex mix of genetic variants, generally known as quasispecies. Since the quasispecies nature of viruses has been invoked to explain many aspects of their natural history (Domingo and Holland, 1997; Maggi et al., 1997a, 1999b; Ray et al., 1999), it is possible that the quasispecies characteristics of HCV contribute to the determination of clinically relevant features of the infections produced. As mentioned above, the complexity and temporal evolution of the HCV quasispecies have been proposed as parameters useful in making decisions about the therapeutic management of patients. However, at present determining the composition of viral quasispecies is still essentially regarded as a research tool. The methods used (cloning and sequencing, single-strand polymorphism analysis, etc.) are cumbersome and, since they are prone to significant artifacts, should be performed in laboratories with skilled personnel.

Liver Biopsy

The indications and utility of liver biopsy in hepatitis C are essentially the same as those discussed for hepatitis B (Perrillo, 1997). Presently, no test or aggregate of tests permits an evaluation of disease activity in the liver and of outcome. Liver biopsy remains the gold standard for assessing the severity of liver injury. Contrary to most previous reports, recent in situ hybridization studies using riboprobes for both positive- and negative-strand RNAs have shown that during chronic infections HCV may be replicating in the majority of hepatocytes without apparent tissue damage (Agnello et al., 1998).

HDV (DELTA VIRUS)

HDV, also known as delta virus, is a defective viroidlike agent that replicates productively only in cells infected with HBV and for this reason is found only in people infected with this virus. Especially when superimposed on a preexisting HBV infection, acute HDV infections are often associated with an appreciable aggravation of the underlying liver disease and with increased occurrence of fulminant hepatitis, while the manifestations associated with chronic HDV infections may be either more severe or, more frequently, not substantially different than those caused by HBV alone.

The Virus

HDV was first demonstrated in the late 1970s, when a nuclear antigen (delta or HD antigen [HDAg]) that had previously been identified in the hepatocytes of a subset of HBV-infected patients was recognized as belonging to a separate, albeit HBV-dependent, viral entity (Rizzetto et al., 1980). The virus has a number of interesting and unusual properties that make it unique among known animal viruses (Table 7), including an extraordinary structural and functional, but not phylogenetic, similarity of its tiny negative-sense ssRNA genome with those of viroids and analogous infectious RNA agents of plants. The gene products HBV supplies to HDV are essential structural components, namely, the envelope glycoproteins representing HBsAg, which form the outer coat of the satellite as well as that of the helper virus. This complementation is, therefore, essential for maturation, cell-to-cell spread, and transmissibility of progeny HDV virions. HDAg is the sole HDV-encoded protein consistently found in viral particles and infected cells, and it exists in two molecular species (Fig. 7) that together represent the nucleoprotein and, in addition, exert opposing regulatory functions on HDV RNA replication (Table 7). HDV is genetically unrelated to any other animal virus, including HBV. For this reason, it is presently classified in a separate, free-standing genus, *Deltavirus*, created ad hoc.

Experimental propagation of HDV is readily obtained in chimpanzees infected with HBV (where HDV appears to induce more severe hepatitis than do other human hepatitis viruses) and in eastern woodchucks infected by the corresponding hepadnavirus (woodchuck hepatitis virus), which provides the required outer coat as effectively as HBV. These animal models mimic HDV infections in humans and exhibit similar liver lesions, while in mice HDV undergoes an incomplete replication due to the host's inability to support HBV growth. Unlike HBV, extrahepatic replication of HDV has never been detected in infected humans and chimpanzees. Probably due to the dependence on HBV, which cannot infect cultured cell lines, there are no in vitro culture systems of practical utility for HDV, although limited growth occurs in primary hepatocytes coinfected with HBV. Other cell types can be transfected with cloned DNA containing portions of the HDV and HBV genomes and are used for investigating specific aspects of HDV genome replication. Initially, it was believed that direct virus-induced cytopathic effects might be more important in HDV than in other hepatitis virus infections; however, the mechanisms of hepatocellular damage by HDV are now thought to be essentially immunopathogenic, similar to those caused by HBV and HCV (Smedile and Verme, 1999). Interestingly, transgenic mice expressing HDAg in their livers showed no obvious hepatocellular damage (Guilhot et al., 1994).

The genetic variation observed in HDV is considerable, since independent isolates may exhibit divergences of up to

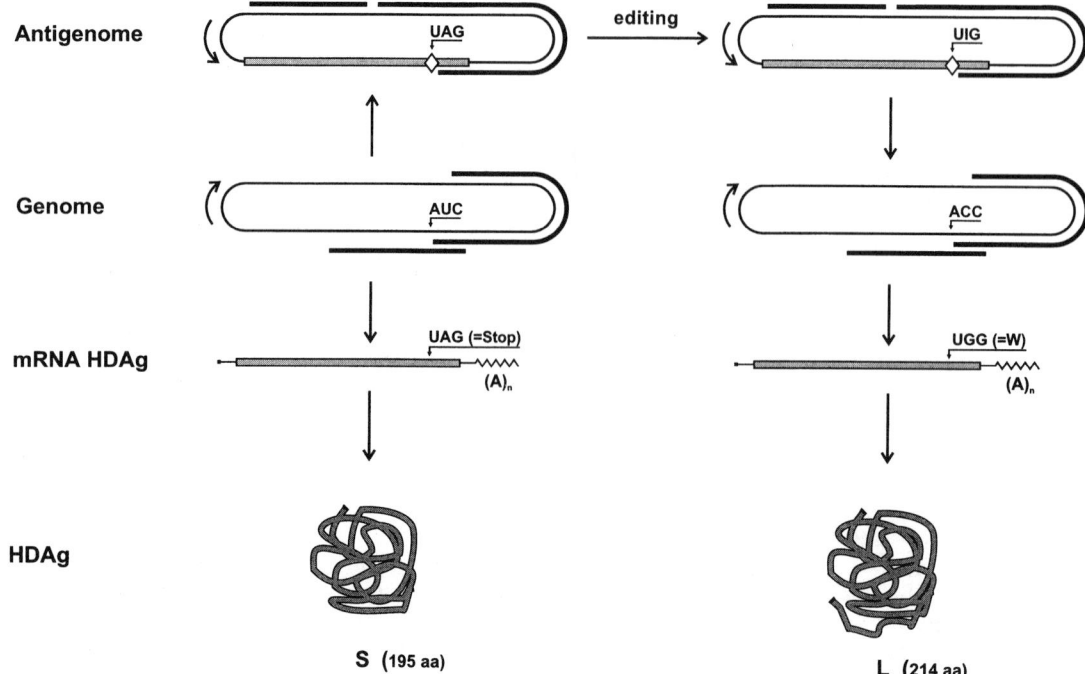

FIGURE 7 Organization of the HDV genome and encoded proteins. The ellipses represent the genomic and antigenomic circular viral RNA. The shaded segment of the antigenome represents the only consistently expressed ORF, while the external thick lines represent predicted but unexpressed ORFs. A variable proportion of antigenomes is modified by host cell enzymes at a specific nucleotide position (the editing site [diamond]) contained in the functional ORF. This editing leads to the replacement, in the HDAg mRNA, of an amber stop codon with a tryptophan codon (W) and results in the readthrough of 19 additional amino acids. By this mechanism, variable proportions of S-HDAg and L-HDAg are produced. The two forms of HDAg have substantially distinct functions (Table 7). The extent of editing has been reported to be related to the course of infection in experimental animals (Yang et al., 1995). aa, amino acids.

40%. Based upon sequence similarities and phylogenetic analysis of a semiconserved segment of the HDAg-encoding ORF, three major genotypes (designated I, II, and III) and several subtypes (designated with letters) have been distinguished. While these genotypes have different geographical distributions (see below), it is uncertain whether they also contribute to the determination of differences in phenotypic properties that may affect the severity of infection. A virus variant lacking one of the two forms of HDAg has been reported in an outbreak of fulminant hepatitis (Tang et al., 1993). Some variation is also observed within individual patients and may result from the quasispecies effect and, occasionally, from the occurrence of mixed infections (Casey, 1996; Wu et al., 1999).

HDV is rapidly inactivated by nonionic detergents and lipid solvents, but there are few reports dealing with its resistance to inactivants. More extensive coverage of HDV structure and replication can be found in reviews by Lai (1995), Casey (1996), Hadziyannis (1997), and Karayiannis (1998).

Dynamics of HDV Replication and Antiviral Immune Responses in Infected Patients

As defined in humans and experimentally infected chimpanzees, the course of virological, immunological, and clinical events differs substantially depending on the HBV status of the host at the time of exposure to HDV (Fig. 8).

Coinfections

The coinfection pattern occurs when an HBV-naive individual becomes simultaneously infected with HDV and HBV as a result of exposure to a mixed inoculum. Coinfection has generally been studied in adults living in developed countries (typically, drug addicts and other subjects exposed percutaneously to infected blood), in whom the helper HBV infection usually resolves in a few months. Therefore, under these circumstances, HDV infection is also self-limited in greater than 95% of cases.

In the early phases postcoinfection, the two viruses replicate acutely, but since the release of progeny HDV virions is limited by the availability of HBsAg, HDV generally becomes detectable in the circulation later than HBV. Thus, initially, the profile of infection markers is that typical of an acute HBV infection, but soon thereafter, generally shortly before the appearance of clinical symptoms and the rise of anti-HBc IgM, HDV starts to circulate in the plasma and then progressively peaks, with a delay of approximately 1 week relative to the HBV peak. Concomitant with this HDV viremia peak, a marked reduction in the titers of circulating HBV occurs that is believed to stem from inhibition of HBV expression possibly mediated by the HDAg protein (Wu et al., 1991). In coinfected patients, the immune responses to HBV are similar in dynamics and strength to those seen in patients singly infected with this virus, while those to HDV are usually weak and often restricted to IgM. Anti-HD antibodies develop slowly,

TABLE 7 Structure and replication of HDV: salient features

Structure

    Spherical, slightly pleiomorphic 36-nm-diameter virions

    Spherical 18-nm-diameter nucleocapsid consisting of viral RNA and numerous copies of the nucleoprotein HDAg

    HDAg exists in two versions with the same sequence except that the larger form (L-HDAg) has an additional 19 amino acids at the carboxy terminus compared to the short (195-amino-acid) form (S-HDAg) and is partly phosphorylated

    Envelope composed of cell-derived lipids and the same three proteins forming the surface antigen of the helper virus (HBV)

    Circular, negative-sense 1.7-kb ssRNA, which folds into an unbranched rod-shaped structure due to extensive (70%) intramolecular complementarity and base pairing

    The viral RNA possesses ribozyme activity, which autocatalytically cuts itself at a specific cleavage site, and self-ligating activity

    Several ORFs are present in both the genomic and antigenomic strands, but only one of the latter strands is consistently expressed and codes for the HDAg

Replication

    Poorly understood

    Entry into cells probably mediated by the HBsAg envelope and unknown receptors

    RNA replication takes place in the cell nucleus, presumably by a double rolling-circle mechanism involving cellular RNA polymerase II followed by autocatalytic cleavages and by ligations, and leads to the production of both genomic and antigenomic RNAs, including linear monomeric and oligomeric species as well as circular monomeric species

    Synthesis of S-HDAg and L-HDAg from the same ORF (Fig. 7)

    Synthesis of viral RNA regulated by the two forms of HDAg: S-HDAg transactivates while L-HDAg down-regulates RNA replication

    With the assistance of HBsAg,[a] progeny genomes and HDAg assemble in the nucleus to form the virion

    Release probably occurs through modalities similar to those used by HBV

    No overt cytopathic effects produced by HDV itself

[a]Whereas HDAg is mainly a nuclear protein, HBsAg is essentially cytoplasmic. How they come to interact remains essentially unknown.

remain at low titer, and may be lost rapidly, so they can escape detection unless tested for repeatedly. Both viruses are generally cleared in a few months. In fact, although HDV-HBV coinfections tend to have a more severe course than infection by HBV alone, progression to chronicity remains an infrequent outcome, observed in 1 to 3% of cases. After recovery, anti-HD antibodies generally disappear within months or a few years.

The clinical outcome of HDV-HBV coinfections is generally benign. When present, the symptoms of acute hepatitis develop after an incubation period of 5 to 10 weeks and, probably depending on the relative titers of the two viruses in the infecting inoculum, present a monophasic or biphasic course. In the latter case, the first bout of manifestations is generally attributable to HBV and the second to HDV.

Superinfections

The superinfection pattern is observed when HDV affects persons who are already chronically infected with HBV, and it results in persistent HDV infection in more than 70% of cases. Since the HBsAg necessary for productive HDV replication is already available in the hepatocytes, the course of infection is more rapid and florid than in coinfections. HDV starts to circulate in the plasma within a couple of weeks and reaches titers of up to 10^{11} infectious doses/ml. Concomitantly, a suppression of HBV viremia occurs that may be so pronounced transiently as to cause the HBsAg to be temporarily undetectable.

As judged from IgM and IgG antibody development, during superinfections the immune responses to HDV mount more rapidly than in coinfections; however, their contribution to the natural history of HDV infection is poorly understood. As expected from the internal location of the antigen recognized, anti-HD antibodies appear unable to neutralize the virus and probably play no role in protection, whereas anti-HBs antibodies are believed to be protective against HDV as well as against HBV. On the other hand, cell-mediated immune responses to HDV are just beginning to be investigated. The limited evidence available suggests that sensitized lymphocytes contribute to both amplification of liver injury and control of HDV replication (Nisini et al., 1997). This is in line with findings showing that HDV viremia is particularly elevated in patients with advanced HIV infection (Roingeard et al., 1992). Whatever the significance of anti-HD immune responses, HDV persistence is characterized by the continuous presence of virus in the plasma and liver, although at lower levels than in the acute phase (Simpson et al., 1994), in spite of persisting high titers of anti-HD IgG and IgM. In fact, the anti-HD IgM disappears only in rare patients who eventually also succeed in resolving the infection.

The changes of clinical status brought about by HDV superinfections depend on the patients' clinical conditions. In asymptomatic HBV carriers, HDV may cause changes typical of a generally severe acute hepatitis that usually develop after a shorter incubation period (3 to 5 weeks) than in coinfections. On the other hand, in patients with symptomatic chronic hepatitis B, HDV superinfection generally produces an exacerbation of the underlying disease. Rapidly progressing subacute forms of hepatitis have also been described, and the reported overall rates of mortality have varied between 2 and 20%. Once HDV infection has become chronic, the clinical evolution is frequently toward an active chronic hepatitis of considerable gravity and an accelerated or enhanced tendency to cirrhosis but no evident increased risk of HCC (Huo et al., 1996). In a recent community-based cross-longitudinal study, 62% of HDV RNA-positive HBsAg carriers had high levels of serum ALT versus 9% of those who were HDV RNA negative (Sakugawa et al., 1997).

Epidemiology

Similar to HBV, the only known natural host of HDV is humans. Seroprevalence studies have shown that HDV is present worldwide in susceptible (i.e., HBV-infected) indi-

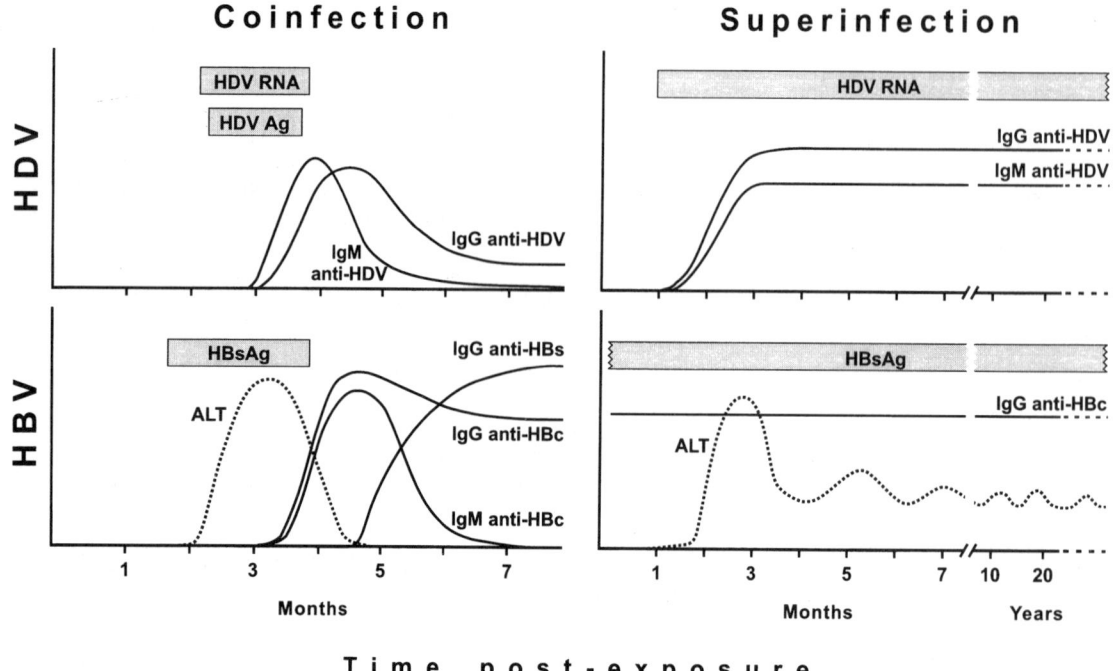

FIGURE 8 Typical courses of HDV infection. The major markers of the helper HBV infection are also shown. (A) Typical coinfection. (B) Typical superinfection. The shaded bars represent the time periods during which the markers indicated are demonstrable in the circulation. HDAg detectability is based on EIA methods.

viduals but at rates that vary markedly in different geographic regions. In general, it has been found that HDV infects higher proportions of HBV carriers in those areas where HBV is most prevalent, but this correlation is a loose one. For example, HDV endemicity is low in many parts of Asia where HBV is widespread. The rates for HDV-seropositive HBV carriers are 20 to 60% in South America and Central and East Asia, 5 to 20% in the Mediterranean basin, Eastern Europe, and West Africa, and 0.5 to 2% in North America and northern Europe. In low-prevalence countries, HDV is virtually restricted to illicit injecting drug abusers, among whom it has been seen to spread rapidly, while penetrance among other categories of HBV carriers is moderate.

Except for parenteral exposure to HDV-containing blood, the routes and modes of transmission are poorly understood. In particular, the transmission routes responsible for the high endemicity of certain areas are unknown, although it is widely assumed that they are the same as for HBV and are favored by poor socioeconomic conditions and poor hygiene. Sexual transmission is inefficient but has been documented in spouses of infected persons as well as among homosexual men and prostitutes, who have also been found to harbor mixed HDV genotypes as a likely result of multiple exposures (Wu et al., 1995, 1999). Perinatal mother-to-child transmission has also been observed. However, the cases in which no obvious risk factor can be recognized are numerous.

Most HDV isolates throughout the world belong to genotype I. The other two genotypes appear to be geographically restricted. Genotype II has been detected in Japan and Taiwan, and genotype III has been found in northern South America, where the induced disease is particularly severe (Shakil et al., 1997).

Prevention and Treatment

Potentially, all HBV carriers are at risk for contracting HDV infection. These individuals should be counseled about the additional health risk represented by HDV superinfection and how to prevent it. Unfortunately, the populations at higher risk are also the most difficult to reach with educational campaigns. In HBV-free individuals, all of the practices that have proven useful in preventing HBV infection, primarily accurate control of blood supply and anti-HBV vaccination, have proved valuable for control of HDV spread as well. Recently, a decrease in HDV endemicity has been noted in several countries and has been attributed to the interruption of HBV transmission resulting from extensive vaccination and other anti-HBV preventive measures (Karayiannis, 1998).

Since HDV replication does not persist after HBsAg clearance, the treatments that lead to HBV elimination also eradicate HDV infection (Battegay et al., 1994). However, IFN-α and other current treatments are at best transiently beneficial in only a small fraction of patients (Malaguernera et al., 1996). In addition, there is evidence that HDV infection reduces the efficacy of anti-HBV treatments (Lau et al., 1999). Since better therapeutic options are needed, attempts to develop drugs that may selectively control HDV replication are under way (Glenn et al., 1998). Speculation that in patients treated with orthotopic liver transplantation HDV might not require concomitant HBV replication has not been substantiated (Smedile et al., 1998).

Laboratory Diagnosis and Follow-Up

Performing HDV diagnostic tests in HBV-uninfected individuals is of no value. By contrast, the possible coexistence of an HDV infection is an important consideration in the management of all HBsAg-positive subjects, primarily if they belong to risk groups or live in countries where HDV endemicity is high. Suspicion should especially arise when the liver disease worsens or persists in spite of a negative test for HBeAg.

Similar to HBV, laboratory diagnosis of HDV depends extensively on the use of immunoassays that demonstrate the presence of HDAg and anti-HDV antibodies in serum. Molecular methods for demonstrating the presence of viral RNA are also extremely useful, although not as widely used as for hepatitis B and C. As a matter of fact, where HDV endemicity is low, the entire set of diagnostic tests for this virus is usually available only in a few specialized laboratories. A key to the interpretation of diagnostic tests for HDV is provided in Table 8.

Tests for Viral Antigen and RNA

The presence of viral antigen and RNA markers in serum is indicative of ongoing active HDV replication and has been found to correlate with significant liver inflammation and necrosis. EIA for HDAg demonstrate the antigen during the preserological stage of acute infection but then become negative because the patient's anti-HD antibodies compete with those used to capture the antigen in these assays. Western blot methods for HDAg detection suffer less of this limitation and have the added advantage of separately visualizing the S-HDAg and the L-HDAg, but they are not commercially available and are labor-intensive. In coinfections with moderate or no clinical involvement, the HDAg often remains undetectable because of limited virus replication, and the infection may be signaled only by the appearance of anti-HD. The HDAg also is almost always undetectable by EIA in chronic infections.

The approaches used for HDV RNA detection include dot and slot hybridization assays with labeled probes, which have a lower limit of sensitivity (approximately 10^5 to 10^6 genomes/ml), and gene amplification methods, such as reverse transcription-PCR, which are 10,000-fold more sensitive. Due to higher sensitivity and lack of interference by the patient's anti-HD response, these assays significantly extend the duration of viremia detectability in coinfections (Fig. 8) and are positive in the great majority of chronically infected patients (Huang et al., 1998). The assays can also

be rendered semiquantitative by testing serial dilutions or quantitative by more complex procedures, thus providing a tool for patient follow-up in the course of therapy. A summary of the clinical utility of molecular tests for HDV is provided in Table 5.

Tests for Antiviral Antibodies

Anti-HD IgM assays have an important role in the diagnosis of both acute and chronic infections. When an acute coinfection is suspected, they should be repeated several times before an HDV infection is excluded because antibody responses to HDV are generally feeble (see above). In chronic infections, IgM detection is considered useful for long-term prognosis because it helps in distinguishing those infections with highly active virus replication and ongoing liver disease (high-titer IgM) from those with a more favorable course (low-titer IgM), although this correlation is far from absolute (Huang et al., 1998). Moreover, in a recent study, the behavior of anti-HD IgM has been proposed as the best predictor of sustained responsiveness to IFN because these antibodies were seen to decline in patients who responded to therapy with a long-term normalization of ALT but not in those who did not respond or responded only transiently. Based on evidence obtained from liver transplant patients, in the same study anti-HD IgM reactivity was found to correlate with HDV-induced hepatocellular damage more closely than with HDAg production (Borghesio et al., 1998).

The tests for total anti-HD antibodies essentially measure those of the IgG class. In coinfections, these antibodies become detectable 3 to 8 weeks after clinical onset and can be missed due to their brief persistence. Anti-HD IgG is found both in chronic active infections and in subjects who have cleared the virus, two situations that can be distinguished by testing for HDV RNA. In chronic infections, high titers of these antibodies are considered an indicator of poor prognosis.

Establishing whether an HDV infection results from coinfection or superinfection also has importance for patient management and prognosis. A useful parameter is the presence of high titers of anti-HBc IgM, which signals a recent HBV infection and is therefore strongly indicative of an ongoing coinfection (Table 8).

TABLE 8 Interpretation of test results for diagnosis of HDV infection

Results[a]							Interpretation
HDV				HBV			
RNA	HDAg	Anti-HD IgM	Total anti-HD	HBsAg	Anti-HBs	Anti-HBc IgM	
+	+	−	−	+	−	+	Acute coinfection; early phase
+	−	+	+	+	−	+	Acute coinfection; late phase
+	±	±	±	+	−	±[b]	Acute superinfection
+	−	+	+	+	−	±[b]	Chronic infection
−	−	−	+	−	+	−	Past infection
−	−	−	−	−	−	−	No current infection

[a]+, positive; −, negative; ±, may be either positive or negative.
[b] Sensitive tests can be positive but at low titer.

HDV Genotyping

Characterizing the genotype of HDV is of limited utility in clinical practice. It is generally performed by sequencing or, more simply, by restriction fragment length polymorphism analysis of amplicons derived from the HDAg-encoding ORF (Wu et al., 1999).

Liver Biopsy

Direct examination of liver tissue condition is important for the management of chronic HDV-HBV infections. The necroinflammatory liver lesions are similar to those in chronic HBV infection alone but generally are more severe. Histopathology can be complemented with the immunochemical demonstration of HDAg or quantitative evaluation of HDV genomes in liver tissue.

ADDITIONAL VIRUSES UNDER CONSIDERATION FOR POSSIBLE HEPATOPATHOGENESIS

As mentioned in the introduction, several novel transfusion-transmitted viruses are presently under scrutiny for a possible role in the genesis of acute and chronic hepatitis of cryptic etiology. These include HGV, TTV, and SEN-V.

HGV or GB Type C Virus

Several years after its definitive identification, HGV or GB type C virus is still a disease orphan virus. In spite of early enthusiasm that led to one of its present names, its role as an agent of hepatitis is still undecided and there are no other clinical manifestations clearly associated with it. It is, instead, well established that the virus is widespread in humans and can persist and circulate in the blood of infected individuals for long periods of time in the absence of detectable signs or symptoms of disease.

The Virus

In the 1960s, tamarins and marmosets inoculated with serum from an American physician were found to harbor a filterable agent that produced liver damage in these hosts. Many years later, samples from these monkeys were reexamined using molecular approaches similar to those that had permitted HCV identification, and this reexamination resulted in the identification of three distinct but related viral genomes, one of which (GB type C) was shown to rep-

resent a human virus (Simons et al., 1995). While these studies were in progress, an independent group (Linnen et al., 1996) had retrieved a virus from monkeys inoculated with serum from another individual with acute hepatitis and called it hepatitis G virus. By comparing the genomes, GB type C virus and HGV were soon recognized as different isolates of the same virus (approximately 95% identity at the amino acid level), but the two names still coexist in the literature, although the latter is probably used more commonly.

HGV is a single-stranded, positive-sense RNA virus that is presently classified within the family *Flaviviridae*. Its structure and replication characteristics are poorly understood and mostly inferred from those of related viruses, such as HCV (Fig. 9 and Table 9). It should be noted, however, that the amino acid sequence homology with HCV is only 30%. A salient distinguishing feature of HGV is the presence of a grossly truncated gene for a capsid protein and even its absence in some isolates (Shimizu et al., 1999). Antibodies to a synthetic peptide deduced from the putative nucleocapsid protein, however, have been shown to correlate with infection (Xiang et al., 1998). In infected hosts, the virus has been detected at low titers in numerous tissues, including liver, bone marrow, spleen, and peripheral blood mononuclear cells (Laskus et al., 1998b; Radkowski et al., 1999); however, the sites of actual virus replication are still poorly characterized. Even the marked hepatotropism, which was initially considered an important feature of HGV, has recently been questioned based on low virus content and undetectability of antigenomic viral RNA in the liver (Kobayashi et al., 1999; Laras et al., 1999). Several primate species have been found to be susceptible to HGV replication, but the consequences of infection vary: thus, tamarins show clearly evident liver damage, while chimpanzees develop persistent infections but no symptoms. There are few reports of attempts to grow the virus in tissue culture; however, limited replication has been described in primary cultures of stimulated peripheral blood mononuclear cells and human hepatocytes and in several cell lines of hepatic and lymphoid origin.

Efforts to group HGV isolates into genotypes have been problematical. Unlike what is observed with HCV, where multiple short genomic segments have proved useful for these purposes, the classification and phylogenetic analysis of HGV has required examination of complete genomes or of polymorphisms found in the 5′ UTR (Robertson et al.,

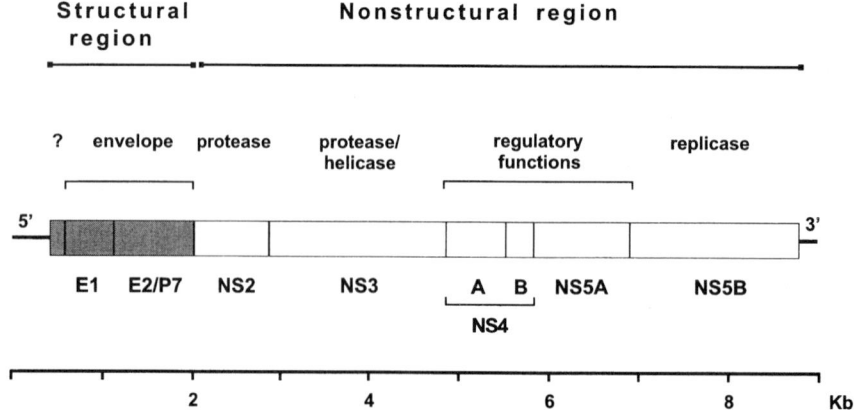

FIGURE 9 Organization of the HGV genome and encoded proteins (Table 9). ?, capsid protein is either grossly truncated or absent in different isolates; E, envelope; NS, nonstructural.

TABLE 9 Structure and replication of HGV; known and inferred properties

Structure
  Spherical, 40- to 60-nm-diameter virions
  Small or absent capsid protein
  Pericapsid or envelope composed of two or three virus-coded proteins
  Linear, positive-sense, 9.4-kb ssRNA
  Genome organized in structural (E1, E2, possibly core, and p7) and nonstructural (NS2, NS3, NS4, and NS5) coding regions and two
    highly conserved noncoding regions (UTRs) located at each terminus; an expressed ORF is also present in the antigenome
  Internal ribosome entry site in the 5′ UTR
  No hypervariable domains detected in the genome

Replication
  Enzymatic activities apparently coded: two proteases, helicase, RNA polymerase
  A minus RNA strand complementary to the viral genome can be detected in infected cells
  Cap-independent synthesis of a polyprotein
  Polyprotein cleaved by viral and possibly cellular proteases

1998; Cong et al., 1999). This has led to the recognition of at least four genetic groups that diverge by a maximum of 12% over the entire sequence and appear to have uneven geographic distributions (Naito et al., 1999). More detailed descriptions of HGV can be found in recent reviews (Karayiannis et al., 1998; Kiyosawa and Tanaka, 1999).

Dynamics of HGV Replication and Antiviral Immune Responses in Infected Individuals

The natural history of HGV in infected subjects is still poorly understood. Follow-up studies of patients who have acquired the infection via transfused blood or dialysis have clearly shown that HGV infection can either resolve spontaneously in a matter of weeks or become chronic. In most patients, the virus is found by PCR in serum by 2 weeks postexposure, where it remains readily detectable throughout the duration of the acute infection. Patients who resolve the infection develop high titers of antibodies to the envelope protein E2, and this is usually followed, within a few weeks, by the disappearance of HGV from the circulation and an apparently complete eradication of the infection (Dille et al., 1997; Feucht et al., 1997; Hwang et al., 1999). Most likely, anti-E2 antibodies are important for virus control, as suggested by the observation that patients who possess these antibodies do not seem to experience HGV reinfections following liver transplantation (Hassoba et al., 1998). However, they probably fail to neutralize and act in concert with other hitherto-undefined antiviral effectors, as indicated by reactivations of infection observed in anti-E2-positive patients and other findings (Yamada-Osaki et al., 1999). A role for CTL and other effectors of cell-mediated immunity in infection control has been hypothesized but remains to be investigated.

In subjects who become chronic virus carriers, anti-E2 antibodies remain undetectable and the virus is demonstrable in the circulation continuously, or sometimes intermittently, for months or years (Alter, 1995; Alter et al., 1997a, 1997b; Lefrére et al., 1999b). Whether failure to develop an effective antibody response in these subjects results from some deficit of their immune systems or from an inherent ability of HGV to dodge immune surveillance is presently unknown, although the observation that chronic infections are especially frequent in immunocompromised individuals seems to support the first possibility. It should be noted, however, that the precise proportion of acutely infected individuals who fail to clear the virus and become chronic carriers also needs to be determined.

Recent data indicate that HGV circulates in plasma at average titers higher than those exhibited by HCV (Lefrére et al., 1999a), but the rates of virus production and clearance are unknown. Sequential studies have shown that the rate of mutation exhibited by HGV in chronically infected patients is low (less than 10^{-6} base substitutions per site per year), remains relatively constant over time, and is similar throughout the entire viral sequence (Suzuki et al., 1999). The absence in the HGV genome of domains especially prone to variations has led to speculation that the strategies this virus uses to persist in its host differ from those of viruses that present hypervariable regions in their genomes, such as HCV and HIV. Interestingly, HGV is present in infected hosts in the form of quasispecies but, probably due to its considerable genetic stability, master sequences appear to remain unchanged for long intervals (Pickering et al., 1997; Kato et al., 1998).

Initially, HGV infection was associated with acute and chronic liver pathologies, including biliary duct inflammation and fulminant hepatitis occurring in patients with no other known etiology (Heringlake et al., 1996). However, an unequivocal demonstration that the virus is indeed implicated in these pathologies is lacking. On the contrary, there are numerous reports of acute and chronic HGV infections, acquired via contaminated transfusions or otherwise, in individuals with no biochemical evidence of liver damage. Moreover, in the few patients infected with HGV alone who present elevations of aminotransferases, these are usually modest and temporally unrelated to viremia fluctuations. Even in patients infected with HBV or HCV, concomitant infection with HGV does not seem to increase the severity of the clinical course and liver histopathology or affect the outcome of therapies (Hayashi et al., 1999). It seems likely, therefore, that the hepatitis tag was premature. In fact, consensus appears to be growing that HGV is essentially a harmless virus capable of producing significant pathology solely under certain conditions. According to some studies, in HIV-infected patients concomitant HGV viremia might actually represent a favorable prognostic element (Lefrére et al., 1999b).

Epidemiology

For some time, searching for the viral genome has been the only tool for studying the epidemiology of HGV. The results have shown that active infection is present worldwide, with prevalence rates in the general population varying between less than 1% and 4% in different countries (Minton, 1998).

With the recent development of tests for anti-HGV antibodies (see below), it has been found that 3 to 20% of healthy people have antibodies, thus indicating that the virus is more widespread than previously appreciated. In particular, rates of HGV seropositivity have ranged between 2 and 8% in Asia and North America, between 10 and 15% in Europe, and around 20% in South Africa and South America (Dille et al., 1997; Ross et al., 1998).

The routes of host-to-host transmission used by HGV have been deduced mainly from the prevalence of infection in different categories. Thus, the high prevalence rates detected in injecting drug abusers (33 to 75%) and in patients at risk for parenteral infection in general (6 to 26%) have demonstrated that the infection is frequently blood borne. On the other hand, the high frequency of HGV infection in HCV- and HBV-infected patients (10 to 19%) has suggested that these infections disseminate through similar routes. Mother-to-child transmission has been documented, although the precise modes of virus spread are unknown (Lin et al., 1998). Sexual transmission is considered likely (Feucht et al., 1999), but its efficiency appears to be low (Tan et al., 1999). The saliva and semen of infected individuals have been shown to contain small concentrations of virus. Despite these data, our present understanding of the modalities of HGV transmission provides no clues as to why this virus is more widespread than HBV and HCV. The possibility that HGV is spread by vectors has been considered (Salvey et al., 1998).

Prevention and Treatment

In theory, the strategies for prevention of HGV infection should proceed along lines similar to those used in prevention of HBV and HCV infection. Whether blood donations should be screened for HGV genomes and/or antibody is, however, a matter of debate. At present, the cons are judged to outweigh the pros, but it is likely that, with the development of more standardized and reliable HGV diagnostic methods and increasing concern for blood supply safety, the matter will be reconsidered. There are no drugs with proven efficacy for HGV infection, and what we know about HGV pathogenicity would not justify specific clinical trials. In patients given IFN-α for HBV or HCV infection, this treatment has been seen to frequently abate concomitant HGV viremia, but the effect was generally transient (Enomoto et al., 1998; Pawlotsky et al., 1998b; Tanaka et al., 1999). Ribavirin and other compounds are also under evaluation in doubly infected patients.

Laboratory Diagnosis

Diagnosis of HGV infection is usually carried out by testing serum for the viral genome and/or antiviral antibodies. It should be noted, however, that the conditions for which performing such tests is deemed clinically useful are limited. In fact, at present, the only patients considered worth testing for HGV are those with presumed viral, acute, or chronic liver pathologies that are now tagged as non-A to E. As a consequence, only a few laboratories offer the assays.

The presence of HGV in serum is usually evaluated by reverse transcription-PCR. Several protocols have been suggested, and, judging from reported data, those using the 5′ UTR as an amplification target seem to perform better than others. However, more standardized methods are clearly needed (Kunkel et al., 1998). Methods for quantitating the viral genomes in serum have also been described but appear to be of little clinical utility.

The development of assays for detecting anti-HGV antibodies has been slowed by initial difficulties in identifying B epitopes of the virus endowed with sufficient immunoreactivity. However, at least four distinct strongly reactive epitopes have been detected in the NS3, NS4, and NS5A products. Since these epitopes appear essentially conformational, the procedure used for their production is important for antigenicity. Satisfactory results have recently been obtained using EIA based on recombinant E2 protein expressed in CHO cells (Dille et al., 1997; Tacke et al., 1997). Despite considerable improvements, currently available EIA still yield a small but significant proportion of false-positive and false-negative results. Proposed confirmatory tests include a line probe assay based on recombinant structural and nonstructural proteins (Schroter et al., 1999), a radioimmunoprecipitation assay using E2 protein expressed in BHK-21 cells (Pilot-Matias et al., 1996), and a sandwich EIA using E2 protein expressed in CHO cells (Lou et al., 1997). However, none of these assays has been extensively evaluated. As summarized in Table 10, the uncertainties existing in the interpretation of available HGV diagnostic test results are still considerable. In part, this is also due to the fact that attempts to develop sensitive and specific assays for anti-HGV IgM have so far been frustrating. It is likely that existing problems will be solved in the near future, especially if this elusive virus is proven to have clinical relevance.

TTV

The novel TTV has been identified by representational difference analysis of the nucleic acids present in sequential sera from a Japanese patient with posttransfusion non-A-to-G hepatitis (Nishizawa et al., 1997). Although characterization has just started, what is emerging seems to indicate that it possesses a number of interesting properties. The medical significance remains to be determined. Early indications led to suggestions that TTV might be endowed with a marked hepatotropism and could represent a major cause of those forms of hepatitis of presumed viral etiology that

TABLE 10 Interpretation of diagnostic test results for HGV infection

Results[a]		Interpretation
Anti-E2 antibody	HGV RNA	
−	+	Acute or chronic infection
+	+	Resolving infection, reactivation, or chronic infection
+	−	Past infection
−	−	No infection

[a] +, positive; −, negative.

could not be attributed to known viruses. Indeed, TTV was initially demonstrated in patients with hepatitis, including fulminant forms, and was reported to be present in the liver at concentrations 10 to 100 times higher than in plasma and, in one patient, even in the absence of detectable viremia (Yamamoto et al., 1998). Ongoing studies, however, have raised considerable doubt about this disease association and tend also to exclude the possibility that TTV coinfection represents a determinant of increased liver injury in patients infected with other hepatitis viruses. Attempts to correlate TTV infection with other cryptogenetic pathologies of presumed viral etiology, such as systemic lupus erythematosus, psoriasis, and rheumatoid arthritis, have also yielded negative results (Maggi et al., 1999a). In fact, suggestions have been put forward that TTV might be essentially apathogenic (Simmonds et al., 1999). As discussed below, the prevalence in the general population of persistent productive TTV infection is high, and this circumstance can render identification of possible etiological associations extremely arduous.

The Virus

Based on the few properties presently understood, it has been suggested that TTV might be a parvovirus, the first member of the family *Circoviridae* (prototype, the chicken anemia virus) shown to infect humans, or even a member of a new family. Its nucleic acid is a circular ssDNA of approximately 3.8 kb and negative polarity. The genome (Fig. 10) is organized in two large partially overlapping ORFs with a coding capacity for 770 amino acids (structural region?) and 202 amino acids (nonstructural?), respectively, and several smaller ORFs. Of the actual virion, we know only the approximate size (30 to 50 nm), the buoyant density in CsCl, and that it lacks an external lipid envelope. The location and modalities of replication are entirely unknown. There have been no reported attempts to grow the virus in tissue culture, while recently it has been reported that it can be transmitted to chimpanzees in the absence of biochemical or histological evidence of hepatitis (Mushahwar et al., 1999). Although TTV has been detected by PCR in the liver and peripheral blood mononuclear cells (Okamoto et

al., 1999), there is no evidence that these represent sites where the virus does indeed replicate.

Early studies had suggested the existence of two major genotypes of TTV, designated 1 and 2 and differing by over 30% in a short segment of ORF1, and four subtypes (Okamoto et al., 1998a). More recent studies, however, have indicated that there may be other genotypes (Takayama et al., 1999). Further studies should investigate whether TTV genotypes differ in geographical distribution or in other respects, as observed for other viruses.

Dynamics of TTV Replication and Antiviral Immune Responses in Infected Individuals

It is well established that TTV can persist for protracted periods of time in infected hosts. In one study, patients were found to carry the virus for up to 22 years (Matsumoto et al., 1999). A few cases of self-limited infections have also been described (Tsuda et al., 1999). By endpoint dilution and real-time PCR, the titers of TTV found in the plasma of infected patients have been shown to range around 10^3 to 10^8 genome copies/ml and to be essentially stable over periods of time up to 5 years (M. Bendinelli, unpublished data; Maggi et al., 1999a). Most of the patients examined so far by quantitative methods were presumably persistently infected. It is not known whether viremia levels vary depending on the time elapsed from initial infection.

Immune responses to TTV are just beginning to be examined. By immunoprecipitation, anti-TTV IgG was detected in one of six viremic subjects and in 11 of 38 nonviremic individuals. Moreover, in two patients with posttransfusion hepatitis, free antibodies developed at the time they cleared the virus from the blood (Tsuda et al., 1999). Using an EIA employing synthetic peptides deduced from ORF1, antiviral IgG has been detected in a proportion of both viremic and nonviremic subjects (Pistello et al., in press). Furthermore, recent data indicate that most circulating TTV is bound to IgG (Nishizawa et al., 1999).

Epidemiology

There are already several studies that investigate the prevalence of TTV infection in the general population and in selected groups of patients. The emerging data are quite diverse; however, it is not known whether this diversity stems from differences in the geographic origin and other variables of the population examined or rather reflects differences in the sensitivities of the PCR methods used for TTV detection. The general impression, however, is that TTV is ubiquitous. Reports have indicated carrier rates in the general population ranging between 1 and 13% in the United States and northern Europe, 7 to 83% in African countries, 12% in Japan, 36% in Thailand, and 62% in Brazil. However, using one set of primers that improved PCR sensitivity by 10 to 100 times, the prevalence of viremia in the general population of Japan has been reported to be as high as 92% (Takahashi et al., 1998), thus emphasizing that more standardized detection methods are needed to establish the true prevalence of TTV wordwide. Seroprevalence rates also appear to be high (Tsuda et al., 1999; Pistello et al., in press).

Reports also indicate that individuals at risk for parenteral infection may show enhanced rates of active infection relative to the general population examined in the corresponding studies. These include hemophiliacs treated with blood concentrates prior to 1986 (57 to 73%); patients on hemodialysis (46 to 53%), polytransfused (18 to

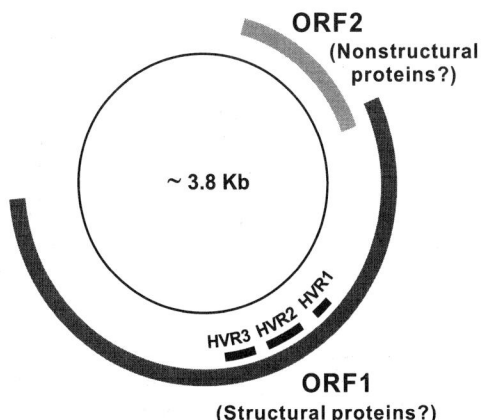

FIGURE 10 Organization of the TTV genome. The shaded arcs represent the two major ORFs identified. Several smaller ORFs also exist. The solid bars marked HVR-1 to -3 indicate three HVRs recently recognized (Nishizawa et al., 1999).

55%), or with thalassemia (19 to 93%); and injecting drug abusers (19 to 40%). Although from these studies it would seem that parenteral exposure is important for TTV spread, other routes of transmission appear possible, as suggested by the detection of virus in stools, saliva, and breast milk (Okamoto et al., 1998b; Ross et al., 1999) and by the very high infection rates detected in healthy persons with no parenteral risk, at least in some studies. Sexual transmission, however, seems to be inefficient (MacDonald et al., 1999). Examining 660 Italian patients affected by various pathologies, we found infection rates that varied between 28 and 73% in different groups and an overall rate of 50%. In this study, TTV infection was as frequent in children under 4 years of age as in older age groups, thus suggesting that infection occurs early in life (Davidson et al., 1999; Maggi et al., 1999a). The possibility that TTV is transmitted across the placenta is still debated (Morrica et al., 2000). It is also worth noting that TTV-like sequences have been detected in farm animals as well as in nonhuman primates, suggesting further sources of human infection (Leary et al., 1999).

Prevention and Treatment

There are no studies of how TTV infection could be prevented. If the high prevalence rates detected in some studies are confirmed, effective prevention may prove extremely hard to achieve, even in the context of blood and organ donation. There also are few studies designed to specifically investigate TTV sensitivity to antiviral treatments (Chayama et al., 1999). However, TTV viremia levels were found to remain unchanged or to only transiently decline in most patients given the standard INF-α treatments used for control of HCV infection, regardless of whether levels of the latter virus declined (Maggi et al., 1999a).

Laboratory Diagnosis

At present, there are no tests for TTV-specific antigens or antibodies that have been proven useful for diagnosis of infection. Thus, the only approach available is demonstrating the virus in plasma and possibly other clinical specimens by nucleic acid amplification techniques. The methods described include several formats of PCR and have generally used primers derived from ORF1 or from the noncoding regions of the viral genome. The variations in detection sensitivity encountered using different sets of primers have already been mentioned and may at least in part reflect the genetic diversity of the virus (Okamoto et al., 1998a; Leary et al., 1999). Another major problem is the selection of patients to test for TTV infection. At present, chronic infection has not been firmly associated with any clinical manifestations; thus, it seems reasonable to limit the tests to patients for whom all the other possible causes of infection have been excluded. Efforts should also be made to identify patients undergoing acute infections, in an attempt to identify possible clinical manifestations produced by first encounters with the virus. Genotyping of TTV has generally been carried out by sequence analysis; however, a method based on restriction fragment length polymorphism analysis has been described (Tanaka et al., 1998), which in our hands correctly characterized over 90% of 157 isolates examined (Maggi et al., 1999a). Cases of patients harboring more than one genotype (up to five) have been reported and appear to be a frequent occurrence, primarily in individuals at high risk for blood-borne infections (Takayama et al., 1999).

SEN Virus

According to the only published report (Anonymous, 1999), this virus, discovered recently in the blood of an injecting drug abuser, is frequently detected in HIV-, HBV-, and HCV-infected individuals and is especially prevalent in patients with chronic posttransfusion non-A-to-G hepatitis. Full disclosure of its distinguishing features in peer review publications is expected soon.

FUTURE DIRECTIONS

From a time when clinical virology essentially consisted of confirming or excluding a clinical diagnosis of infection, we have rapidly progressed to the present state of the art, in which an increasing number of viral infections also can be systematically monitored to evaluate how they evolve and respond to antiviral therapies. Blood-borne hepatitis viruses, which have always represented important challenges for clinical virologists as well as for physicians, are among the few viruses that have paved the way in this direction. For HBV, methods for patient follow-up have been available for a long time, thanks to early realization that infected persons circulate abundant viral antigen in their plasma and develop specific subsets of antiviral antibodies in an orderly fashion that may be used for staging the infection. The subsequent development of reliable molecular methods for quantitative evaluation of viral genomes has only provided a further and more sensitive approach to the follow-up of hepatitis B and D. When HCV was discovered, the era of gene amplification techniques was at its radiant dawn, and it is therefore not surprising that molecular methods have found such an important place in the diagnostics and follow-up of this infection. Future research in the diagnostics for HBV, HCV, and HDV infections should be directed toward the development of improved virological or immunological indicators that can be exploited for optimum therapeutic management of patients. These include indicators that may permit identification of those acutely infected patients that will evolve toward chronicity, indicators for establishing as early as possible which persistently infected patients are likely to show serious disease with passing time, and indicators for assessing the severity and progression of liver injury without having to resort to liver biopsy. Since hepatocellular damage by these viruses does not depend solely on the extent of viral replication, viremia levels are only partially informative in this regard. The diagnostics of drug resistance is already useful for the management of HBV-infected patients. It is an easy prediction that, as a result of ongoing attempts to develop truly effective antivirals, clinicians will soon have at their disposal a larger choice of therapeutic options. Methods that can guide the selection and monitoring of therapies will, therefore, become increasingly important. Since those presently available are too cumbersome for extensive routine application, easier-to-perform and more flexible methodologies for demonstrating virus resistance to antivirals need to be urgently developed and validated.

HGV, TTV, and especially SEN-V have been discovered too recently for a full appreciation of their disease-inducing potential. However, from what we are learning about HGV and TTV, it seems that their pathogenicity is limited. If this is confirmed, it appears likely that these viruses will be found to produce disease, possibly including liver pathology, only under special circumstances. Since HGV and TTV are widespread and produce long-lasting infections, simple detection of their presence will say very little about the pathological role exerted in a patient. For HGV and TTV, therefore, it will also be important for methods to be developed and made available in the near future for staging the infection and establishing the causative link with the pathology under scrutiny.

REFERENCES

Abrignani, S., and D. Rosa. 1998. Perspectives for a hepatitis C virus vaccine. *Clin. Diagn. Virol.* **10:**181–185.

Agnello, V., G. Abel, G. B. Knight, and E. Muchmore. 1998. Detection of widespread hepatocyte infection in chronic hepatitis C. *Hepatology* **28:**573–584.

Ahmed, A., and E. B. Keeffe. 1999. Treatment strategies for chronic hepatitis C: update since the 1997 National Institutes of Health Consensus Development Conference. *J. Gastroenterol. Hepatol.* **14:**S12–S18.

Allen, M. I., J. Gauthier, M. DesLauriers, E. J. Bourne, K. M. Carrick, F. Baldanti, L. L. Ross, M. W. Lutz, and L. D. Condreay. 1999. Two sensitive PCR-based methods for detection of hepatitis B virus variants associated with reduced susceptibility to lamivudine. *J. Clin. Microbiol.* **37:**3338–3347.

Alter, H. J. 1995. To C or not to C: these are the questions. *Blood* **85:**1681–1695.

Alter, H. J., Y. Nakatsuji, J. Melpolder, J. Wages, R. Wesley, J. W. K. Shih, and J. P. Kim. 1997a. The incidence of transfusion-associated hepatitis G virus infection and its relation to liver disease. *N. Engl. J. Med.* **336:**747–754.

Alter, M. J., M. Gallagher, T. T. Morris, L. A. Moyer, E. L. Meeks, K. Krawczynski, J. P. Kim, and H. S. Margolis. 1997b. Acute non-A-E hepatitis in the United States and the role of hepatitis G virus infection. *N. Engl. J. Med.* **336:**741–746.

Alter, M. J., D. Kruszon-Moran, O. V. Nainan, G. M. McQuillan, L. A. Moyer, R. A. Kaslow, and H. S. Margolis. 1999. The prevalence of hepatitis C virus in the United States, 1988 through 1994. *N. Engl. J. Med.* **341:**556–562.

Altfeld, M., J. K. Rockstroh, M. Addo, B. Kupfer, I. Pult, H. Will, and U. Spengler. 1998. Reactivation of hepatitis B in a long-term anti-HBs-positive patient with AIDS following lamivudine withdrawal. *J. Hepatol.* **29:**306–309.

Angus, P. W. 1997. Review: hepatitis B and liver transplantation. *J. Gastroenterol. Hepatol.* **12:**217–223.

Anonymous. 1999. SEN-V, a new blood-borne pathogenic hepatitis virus is discovered. *Antivir. Agents Bull.* **12:**227–228.

Aoyagi, K., C. Ohue, K. Iida, T. Kimura, E. Tanaka, K. Kiyosawa, and S. Yagi. 1999. Development of a simple and highly sensitive enzyme immunoassay for hepatitis C virus core antigen. *J. Clin. Microbiol.* **37:**1802–1808.

Assad, S., and A. Francis. 1999. Over a decade of experience with a yeast recombinant hepatitis B vaccine. *Vaccine* **18:**57–67.

Bassett, S. E., D. L. Thomas, K. M. Brasky, and R. E. Lanford. 1999. Viral persistence, antibody to E1 and E2, and hypervariable region 1 sequence stability in hepatitis C virus-inoculated chimpanzees. *J. Virol.* **73:**1118–1126.

Battegay, M., L. H. Simpson, J. H. Hoofnagle, R. Sallie, and A. M. Di Bisceglie. 1994. Elimination of hepatitis delta virus infection after loss of hepatitis B surface antigen in patients with chronic delta hepatitis. *J. Med. Virol.* **44:**389–392.

Beld, M., M. Penning, M. van Putten, V. Lukashov, A. van den Hoek, M. McMorrow, and J. Goudsmit. 1999. Quantitative antibody response to structural (core) and nonstructural (NS3, NS4, and NS5) hepatitis C virus proteins among seroconverting injecting drug users: impact of epitope variation and relationship to detection of HCV RNA in blood. *Hepatology* **29:**1288–1298.

Bendinelli, M., M. L. Vatteroni, F. Maggi, and M. Pistello. 1999. Hepatitis C virus, p. 65–127. *In* S. Specter (ed.), *Viral Hepatitis. Diagnosis, Therapy and Prevention.* Humana Press, Totowa, N.J.

Berger, A., J. Braner, H. W. Doerr, and B. Weber. 1998. Quantification of viral load: clinical relevance for human immunodeficiency virus, hepatitis B virus and hepatitis C virus infection. *Intervirology* **41:**24–34.

Bond, W. W., M. S. Favero, N. J. Petersen, C. R. Gravelle, J. W. Ebert, and J. E. Maynard. 1981. Survival of hepatitis B virus after drying and storage for one week. *Lancet* **i:**550–551.

Borghesio, E., F. Rosina, A. Smedile, M. Lagget, M. G. Niro, G. Marinucci, and M. Rizzetto. 1998. Serum immunoglobulin M antibody to hepatitis D as a surrogate marker of hepatitis D in interferon-treated patients and in patients who underwent liver transplantation. *Hepatology* **27:**873–876.

Brunetto, M. R., U. A. Rodriguez, and F. Bonino. 1999. Hepatitis B mutants. *Intervirology* **42:**69–80.

Bukh, J., R. H. Miller, and R. H. Purcell. 1995. Genetic heterogeneity of hepatitis C virus: quasispecies and genotypes. *Semin. Liver Dis.* **15:**41–63.

Camma, C., P. Almasio, and A. Craxi. 1996. Interferon as treatment for acute hepatitis C. A meta-analysis. *Dig. Dis. Sci.* **41:**1248–1255.

Camma, C., M. Giunta, G. Pinzello, A. Morabito, P. Verderio, and L. Pagliaro. 1999. Chronic hepatitis C and interferon alpha: conventional and cumulative meta-analyses of randomized controlled trials. *Am. J. Gastroenterol.* **94:**581–595.

Casey, J. L. 1996. Hepatitis delta virus. Genetics and pathogenesis. *Clin. Lab. Med.* **16:**451–464.

Centers for Disease Control and Prevention. 1998. Recommendations for prevention and control of hepatitis C (HCV) infection and HCV-related chronic disease. *Morb. Mortal. Wkly. Rep.* **47:**1–39.

Centers for Disease Control and Prevention. 1999. Update: recommendations to prevent hepatitis B virus transmission—United States. *Morb. Mortal. Wkly. Rep.* **48:**33–34.

Chan, H. L., M. Hussain, and A. S. Lok. 1999. Different hepatitis B virus genotypes are associated with different mutations in the core promoter and precore regions during hepatitis B e antigen seroconversion. *Hepatology* **29:**976–984.

Chang, K.-M., and F. V. Chisari. 1999. Immuno-pathogenesis of hepatitis B virus infection. *Clin. Liver Dis.* **3:**221–239.

Chang, M. H. 1998. Chronic hepatitis virus infection in children. *J. Gastroenterol. Hepatol.* **13:**541–548.

Chang, M. H., C.-J. Chen, M.-S. Lai, H. M Hsu, T. C. Wu, M. S. Kong, D. C. Liang, W. Y. Shau, and D. S. Chen. 1997. Universal hepatitis B vaccination in Taiwan and the incidence of hepatocellular carcinoma in children. *N. Engl. J. Med.* **336:**1855–1859.

Chayama, K., M. Kobayashi, A. Tsubota, M. Kobayashi, Y. Arase, Y. Suzuki, S. Saitoh, N. Murashima, K. Ikeda, K. Okamoto, M. Hashimoto, M. Matsuda, H. Koike, M. Kobayashi, and H. Kumada. 1999. Susceptibility of TT virus to interferon therapy. *J. Gen. Virol.* **80:**631–634.

Chemello, L. A., A. Alberti, K. Rose, and P. Simmonds. 1994. Hepatitis C serotype and response to interferon therapy. *N. Engl. J. Med.* **330:**143.

Chisari, F. V. 1997. Cytotoxic T cells and viral hepatitis. *J. Clin. Investig.* **99:**1472–1477.

Cho, S. W., S. G. Hwang, D. C. Han, S. Y. Jin, M. S. Lee, C. S. Shim, D. W. Lee, and H. B. Lee. 1996. In situ detection of hepatitis C virus RNA in liver tissue using a digoxigenin-labeled probe created during a polymerase chain reaction. *J. Med. Virol.* **48:**227–233.

Coleman, P. F., Y. C. Chen, and I. K. Mushahwar. 1999. Immunoassay detection of hepatitis B surface antigen mutants. *J. Med. Virol.* **59:**19–34.

Cong, M., M. W. Fried, S. Lambert, E. N. Lopareva, M. Zhan, F. H. Pujol, S. P. Thyagarajan, K. S. Byun, H. A. Fields, and Y. E. Kudyakov. 1999. Sequence heterogeneity within three different regions of the hepatitis G virus genome. *Virology* **255:**250–259.

Cooper, S., A. L. Erickson, E. J. Adams, J. Kansopon, A. J. Weiner, D. Y. Chien, M. Houghton, P. Parham, and C. M. Walker. 1999. Analysis of a successful immune response against hepatitis C virus. *Immunity* **10:**439–449.

Cramp, M. E., P. Carucci, S. Rossol, S. Chokshi, G. Maertens, R. Williams, and N. V. Naoumov. 1999. Hepatitis C virus (HCV) specific immune responses in anti-HCV positive patients without hepatitis C viraemia. *Gut* **44:**424–429.

Davidson, F., D. MacDonald, J. L. Mokili, L. E. Prescott, S. Graham, and P. Simmonds. 1999. Early acquisition of TT virus (TTV) in an area endemic for TTV infection. *J. Infect. Dis.* **179:**1070–1076.

Davis, G. L., K. Lindsay, J. Albrecht, H. C. Bodenheimer, Jr., L. A. Balart, R. P. Perrillo, J. L. Dienstag, C. Tamburro, E. R. Schiff, W. Carey, J. Payne, I. M. Jacobson, D. H. van Theil, J. Lefkowitch, B. Sanghvi, and The Hepatitis Interventional Therapy Group. 1994. Clinical predictors of response to recombinant interferon-α treatment in patients with chronic non-A, non-B hepatitis (hepatitis C). *J. Viral Hepat.* **1:**55–63.

Dhedin, N., C. Douvin, M. Kuentz, M. F. Saint Marc, O. Reman, C. Rieux, F. Bernardin, F. Norol, C. Cordonnier, D. Bobin, J. M. Metreau, and J. P. Vernant. 1998. Reverse seroconversion of hepatitis B after allogeneic bone marrow transplantation: a retrospective study of 37 patients with pretransplant anti-HBs and anti-HBc. *Transplantation* **66:**616–619.

Dienstag, J. L., E. R. Shiff, T. L. Wright, R. P. Perrillo, H. W. Hann, Z. Goodman, L. Crowther, L. D. Condreay, M. Woessner, M. Rubin, and N. A. Brown. 1999. Lamivudine as initial treatment of chronic hepatitis B in the United States. *N. Engl. J. Med.* **341:**1256–1263.

Diepolder, H. M., J.-T. Gerlach, R. Zachoval, R. M. Hoffmann, M.-C. Jung, E. A. Wierenga, S. Scholz, T. Santantonio, M. Houghton, S. Southwood, A. Sette, and G. R. Pape. 1997. Immunodominant CD4+ T-cell epitope within nonstructural protein 3 in acute hepatitis C virus infection. *J. Virol.* **71:**6011–6019.

Dille, B. J., T. K. Surowy, R. A. Gutierrez, P. F. Coleman, M. F. Knigge, R. J. Carrik, R. D. Aach, F. B. Hollinger, C. E. Stevens, L. H. Barbosa, G. J. Nemo, J. W. Mosley, G. J. Dawson, and I. K. Mushahwar. 1997. An ELISA for detection of antibodies to the E2 protein of GB virus C. *J. Infect. Dis.* **175:**458–461.

Dodson, S. F., C. A. Bonham, D. A. Geller, T. V. Cacciarelli, J. Rakela, and J. J. Fung. 1999. Prevention of de novo hepatitis B infection in recipients of hepatic allografts from anti-HBc positive donors. *Transplantation* **68:**1058–1061.

Domingo, E., and J. J. Holland. 1997. RNA virus mutations and fitness for survival. *Annu. Rev. Microbiol.* **51:**151–178.

Enomoto, M., S. Nishiguchi, K. Fukuda, T. Kuroki, M. Tanaka, S. Otani, M. Ogami, and T. Monna. 1998. Characteristics of patients with hepatitis C virus with and without GB virus C/hepatitis G virus coinfection and efficacy of interferon alfa. *Hepatology* **27:**1388–1393.

Enomoto, N., I. Sakuma, Y. Asahina, M. Kurosaki, T. Murakami, C. Yamamoto, Y. Ogura, N. Izumi, F. Marumo,

and C. Sato. 1996. Mutations in the nonstructural protein 5A gene and response to interferon in patients with chronic hepatitis C virus 1b infection. *N. Engl. J. Med.* **334:**77–81.

Feray, C., L. Caccamo, G. J. Alexander, B. Ducot, J. Gugenheim, T. Casanovas, C. Lonaz, M. Gigou, P. Burra, L. Barkholt, R. Esteban, T. Bizzolon, J. Lerut, A. Minello-Franza, P. H. Bernard, K. Nachbaur, D. Botta-Fridlund, H. Bismuth, S. W. Schalm, and D. Samuel. 1999. European collaborative study on factors influencing outcome after liver transplantation for hepatitis C. *Gastroenterology* **117:**619–625.

Feucht, H. H., B. Zollner, S. Polywka, B. Knodler, M. Schroter, H. Nolte, and R. Laufs. 1997. Distribution of hepatitis G viremia and antibody response to recombinant proteins with special regard to risk factors in 709 patients. *Hepatology* **26:**491–494.

Feucht, H. H., M. Schroter, B. Zollner, S. Polywka, and R. Laufs. 1999. Age-dependent acquisition of hepatitis G virus/CB virus C in a nonrisk population: detection of the virus by antibodies. *J. Clin. Microbiol.* **37:**1294–1297.

Ganem, D. 1996. *Hepadnaviridae* and their replication, p. 2703–2737. *In* B. N. Fields, D. M. Knipe, and P. M. Howley (ed.), *Fields Virology*, 3rd ed. Lippincott-Raven, Philadelphia, Pa.

Glenn, J. S., J. C. Marsters, Jr., and H. B. Greenberg. 1998. Use of a prenylation inhibitor as a novel antiviral agent. *J. Virol.* **72:**9303–9306.

Gordon, S. C., P. J. Dailey, A. L. Silverman, B. A. Khan, V. P. Kodali, and J. C. Wilber. 1998. Sequential serum hepatitis C viral RNA levels longitudinally assessed by branched DNA signal amplification. *Hepatology* **28:**1702–1706.

Guilhot, S., S.-N. Huang, Y. P. Xia, N. La Monica, M. M. C. Lai, and F. V. Chisari. 1994. Expression of hepatitis delta virus large and small antigens in transgenic mice. *J. Virol.* **68:**1052–1058.

Hadziyannis, S. J. 1997. Review: hepatitis delta. *J. Gastroenterol. Hepatol.* **12:**289–298.

Hassoba, H. M., M. G. Pessoa, N. A. Terrault, N. J. Lewis, M. Hayden, J. C. Hunt, X. Qiu, S. C. Lou, and T. L. Wright. 1998. Antienvelope antibodies are protective against GBV-C reinfection: evidence from the liver transplant model. *J. Med. Virol.* **56:**253–258.

Hawkins, A., F. Davidson, and P. Simmonds. 1997. Comparison of plasma virus loads among individuals infected with hepatitis C virus (HCV) genotypes 1, 2, and 3 by Quantiplex HCV RNA assay and in-house limiting dilution method. *J. Clin. Microbiol.* **35:**187–192.

Hayashi, J., K. Ueno, Y. Kawakami, Y. Kishihara, I. Ariyama, N. Furuyo, Y. Sawayama, Y. Etoh, and S. Kashiwagi. 1999. Clinical course of chronic hepatitis C virus infection is not influenced by concurrent hepatitis G virus infection. *Dig. Dis. Sci.* **44:**618–623.

Heringlake, S., S. Osterkamp, C. Trautwein, H. L. Tillman, K. Boker, S. Muerhoff, G. Mushahwar, G. Hunsmann, and M. P. Manns. 1996. Association between fulminant hepatic failure and a strain of GBV virus C. *Lancet* **348:**1626–1629.

Hollinger, F. B. 1996. Hepatits B virus, p. 2739–2807. *In* B. N. Fields, D. M. Knipe, and P. M. Howley (ed.), *Fields Virology*, 3rd ed. Lippincott-Raven, Philadelphia, Pa.

Hoofnagle, J. H. 1998. Therapy of viral hepatitis. *Digestion* **59:**563–578.

Huang, Y. H., J. C. Wu, W. Y. Sheng, T. I. Huo, F. Y. Chang, and S. D. Lee. 1998. Diagnostic value of anti-hepatitis D virus (HDV) antibodies revisited: a study of total and IgM anti-HDV compared with detection of HDV-RNA by polymerase chain reaction. *J. Gastroenterol.* **33:**512–516.

Huo, T. I., J. C. Wu, C. R Lai, C. L. Lu, W. Y. Sheng, and S. D. Lee. 1996. Comparison of clinico-pathological features in hepatitis B virus-associated hepatocellular carcinoma with or without hepatitis D virus superinfection. *J. Hepatol.* **25**:439–444.

Hwang, S. J., S. D. Lee, R. H. Hu, C. Y. Chan, L. Lai, R. L. Loi, and M. J. Tang. 1996. Comparison of three different hybridization assays in the quantitative measurement of serum hepatitis B virus DNA. *J. Virol. Methods* **62**:123–129.

Hwang, S. J., R. H. Lu, C. Y. Chan, F. Y. Chang, and S. D. Lee. 1999. Detection of antibodies to E2-protein of GB virus-C/hepatitis G virus in patients with acute posttransfusion hepatitis. *J. Med. Virol.* **57**:85–89.

Idilman, R., N. De Maria, A. Colantoni, and D. H. van Thiel. 1998. Pathogenesis of hepatitis B and C-induced hepatocellular carcinoma. *J. Viral Hepat.* **5**:285–299.

Ishii, K., D. Rosa, Y. Watanabe, T. Katayama, H. Harada, C. Wyatt, K. Kiyosawa, H. Aizaki, Y. Matsura, M. Houghton, S. Abrignani, and T. Miyamura. 1998. High titers of antibodies inhibiting the binding of envelope to human cells correlate with natural resolution of chronic hepatitis C. *Hepatology* **28**:1117–1120.

Jannsen, H. L., G. Gerken, V. Carreno, P. Marcellino, N. V. Naumov, A. Craxi, H. Ring-Larsen, G. Kitis, J. van Hattun, R. A. de Vries, P. P. Michielsen, F. J. ten Kate, W. C. Hop, R. A. Heijtink, P. Honkoop, and S. W. Schalm. 1999. Interferon alpha for chronic hepatitis B infection: increased efficacy of prolonged treatment. The European Concerted Action on Viral Hepatitis (EUROHEP). *Hepatology* **30**:238–243.

Jurinke, C., B. Zollner, H. H. Feucht, D. van den Boom, A. Jacob, S. Polywska, R. Laufs, and H. Koster. 1998. Application of nested PCR and mass spectroscopy for DNA-based virus detection: HBV-DNA detected in the majority of isolated anti-HBc positive sera. *Genet. Anal.* **14**:97–102.

Karayiannis, P. 1998. Hepatitis D virus. *Rev. Med. Virol.* **8**:13–24.

Karayiannis, P., J. Pickering, R. Zampino, and H. C. Thomas. 1998. Natural history and molecular biology of hepatitis G virus/GB virus C. *Clin. Diagn. Virol.* **15**:103–111.

Kato, T., M. Mizokami, T. Nakano, E. Orito, K. Ohba, Y. Kondo, Y. Tanaka, R. Ueda, M. Mukaide, K. Fujita, K. Yasuda, and S. Iino. 1998. Heterogeneity in E2 region of GBV-C/hepatitis G virus and hepatitis C virus. *J. Med. Virol.* **55**:109–117.

Kinoshita, T., K. Miyake, H. Okamoto, and S. Mishiro. 1993. Imported hepatitis C virus genotype in Japanese hemophiliacs. *J. Infect. Dis.* **168**:249–250.

Kiyosawa, K., and E. Tanaka. 1999. GB virus C/hepatitis G virus. *Intervirology* **42**:185–195.

Kobayashi, M., E. Tanaka, A. Matsumoto, K. Yoshizawa, H. Imai, T. Sodeyama, and K. Kiyosawa. 1998. Clinical application of hepatitis C virus core protein in early diagnosis of acute hepatitis C. *J. Gastroenterol.* **33**:508–511.

Kobayashi, M., E. Tanaka, J. Nakayama, C. Furuwatari, T. Katsuyama, S. Kawasaki, and K. Kiyosawa. 1999. Detection of GB virus C/hepatitis G virus genome in peripheral blood mononuclear cells and liver tissue. *J. Med. Virol.* **57**:114–121.

Komatsu, F., and K. Takasaki. 1999. Determination of serum hepatitis C virus (HCV) core protein using a novel approach for quantitative evaluation of HCV viraemia in anti-HCV-positive patients. *Liver* **19**:375–380.

Kunkel, U., M. Hohne, T. Berg, U. Hopf, A. S. Kekule, G. Frosner, G. Pauli, and E. Schreier. 1998. Quality control study on the performance of GB virus C/hepatitis G virus PCR. *J. Hepatol.* **28**:978–984.

Lai, M. M. C. 1995. The molecular biology of hepatitis delta virus. *Annu. Rev. Biochem.* **64**:259–286.

Laras, A., G. Zacharakis, and S. J. Hadziyannis. 1999. Absence of negative strand of GBV-C/HGV RNA from the liver. *J. Hepatol.* **30**:383–388.

Laskus, T., M. Radkowski, L. F. Wang, H. Vargas, and J. Rakela. 1998a. Search of hepatitis C virus extrahepatic replication sites in patients with acquired immunodeficiency syndrome: specific detection of negative-strand viral RNA in various tissue. *Hepatology* **28**:1398–1401.

Laskus, T., M. Radkowski, L. F. Wang, H. Vargas, and J. Rakela. 1998b. Detection of hepatitis G virus replication sites by using highly strand-specific Tth-based reverse transcriptase PCR. *J. Virol.* **72**:3072–3075.

Lau, D. T., E. Doo, Y. Park, D. E. Kleiner, P. Schmid, M. C. Kuhns, and J. H. Hoofnagle. 1999. Lamivudine for chronic delta hepatitis. *Hepatology* **30**:546–549.

Lau, J. Y.-N. (ed.). 1998. *Hepatitis C Protocols.* Humana Press, Totowa, N.J.

Leary, T. P., J. C. Erker, M. L. Chalmers, S. M. Desai, and I. K. Mushawar. 1999. Improved detection system for TT virus reveals high prevalence in humans, non-human primates and farm animals. *J. Gen. Virol.* **80**:2115–2120.

Lee, W. M. 1997. Hepatitis B virus infection. *N. Engl. J. Med.* **337**:1733–1745.

Lefrère, J. J., C. Ferec, F. Roudot-Thoraval, P. Loiseau, J. F. Cantaloube, P. Biagini, M. Mariotti, G. LeGac, and B. Mercier. 1999a. GBV/hepatitis G virus (HGV) RNA load in immunodeficient individuals and in immunocompetent individuals. *J. Med. Virol.* **59**:32–37.

Lefrère, J. J., F. Roudot-Thoraval, L. Morand-Joubert, J. C. Petit, J. Lerable, M. Thauvin, and M. Mariotti. 1999b. Carriage of GB virus C/hepatitis G virus RNA is associated with a slower immunologic, virologic, and clinical progression of human immunodeficiency virus disease in coinfected persons. *J. Infect. Dis.* **179**:783–789.

Liaw, Y. F. 1997. Current therapeutic trends in therapy for chronic viral hepatitis. *J. Gastroenterol. Hepatol.* **12**:S346–S353.

Lin, H. H., J. H. Kao, K. Y. Yeh, D. P. Liu, M. H. Chang, P. J. Chen, and D. S. Chen. 1998. Mother-to-infant transmission of GB virus C/hepatitis G virus: the role of high-titered maternal viremia and mode of delivery. *J. Infect. Dis.* **177**:1202–1206.

Lindh, M., J. E. Gonzales, G. Norkrans, and P. Horal. 1998. Genotyping of hepatitis B virus by restriction pattern analysis of a pre-S amplicon. *J. Virol. Methods* **72**:163–174.

Lindh, M., C. Hannoun, A. P. Dhillon, G. Norkrans, and P. Horal. 1999. Core promoter mutations and genotypes in relation to viral replication and liver damage in East Asian hepatitis B virus carriers. *J. Infect. Dis.* **179**:775–782.

Ling, R., and T. J. Harrison. 1999. Functional analysis of mutations conferring lamivudine resistance on hepatitis B virus. *J. Gen. Virol.* **80**:601–606.

Linnen, J., J. Wages, Jr., Z. Y. Zhang-Keck, K. E. Fry, K. Z. Krawczynski, H. Alter, E. Koonin, M. Gallagher, M. Alter, S. Hadziyannis, P. Karayiannis, K. Fung, Y. Nakatsuji, J. W. K. Shih, L. Young, M. Piatak, Jr., C. Hoover, J. Fernandez, S. Chen, J. C. Zou, T. Morris, K. C. Hyams, S. Ismay, J. D. Lifson, G. Hess, S. H. K. Foung, H. Thomas, D. Bradley, H. Margolis, and J. P. Kim. 1996. Molecular cloning and disease association of hepatitis G virus: a transfusion-transmissible agent. *Science* **271**:505–508.

Locarnini, S. A. 1998. Hepatitis B virus surface antigen and polymerase gene variants: potential virological and clinical significance. *Hepatology* **27**:294–297.

Lohman, V., F. Korner, J. Koch, U. Herian, L. Theilmann, and R. Bartenschlager. 1999. Replication of subgenomic hepatitis C virus RNAs in a hepatoma cell line. *Science* **285:**110–113.

London, W. T., and A. E. Evans. 1996. The epidemiology of hepatitis viruses B, C, and D. *Clin. Lab. Med.* **16:**251–271.

Lopez-Labrador, F. X., S. Ampurdanes, M. Gimenez-Barcons, M. Guilera, J. Costa, M. T. Jimenez de Anta, J. M. Sanchez-Tapias, J. Rodes, and J. C. Saiz. 1999. Relationship of the genomic complexity of hepatitis C virus with liver disease severity and response to interferon in patients with chronic HCV genotype 1b infection. *Hepatology* **29:**897–903.

Loriot, M. A., P. Marcellin, F. Walker, N. Boyer, C. Degott, I. Randrianatoavina, J. P. Benhamou, and S. Erlinger. 1997. Persistence of hepatitis B virus DNA in serum and liver from patients with chronic hepatitis after loss of HBsAg. *J. Hepatol.* **27:**251–258.

Lou, S., X. Qiu, G. Tegtmeier, S. Leitza, J. Brackett, K. Cousineau, A. Varma, H. Seballos, S. Kundu, S. Kuemmerle, and J. C. Hunt. 1997. Immunoassays to study prevalence of antibody against GB virus C in blood donors. *J. Virol. Methods* **68:**45–55.

MacDonald, D. M., G. R. Scott, D. Clutterbuck, and P. Simmonds. 1999. Infrequent detection of TT virus infection in intravenous drug users, prostitutes, and homosexual men. *J. Infect. Dis.* **179:**686–689.

Maggi, F., C. Fornai, A. Morrica, F. Casula, M. L. Vatteroni, S. Marchi, P. Ciccorossi, L. Riente, M. Pistello, and M. Bendinelli. 1999a. High prevalence of TT virus viremia in Italian patients regardless of age, clinical diagnosis, and previous interferon treatment. *J. Infect. Dis.* **180:**838–842.

Maggi, F., C. Fornai, A. Morrica, M. L. Vatteroni, M. Giorgi, S. Marchi, P. Ciccorossi, M. Bendinelli, and M. Pistello. 1999b. Divergent evolution of hepatitis C virus in the liver and peripheral blood mononuclear cells of infected patients. *J. Med. Virol.* **57:**57–63.

Maggi, F., C. Fornai, M. L. Vatteroni, M. Giorgi, A. Morrica, M. Pistello, G. Cammarota, S. Marchi, P. Ciccorossi, A. Bionda, and M. Bendinelli. 1997a. Differences in hepatitis C virus quasispecies composition between liver, peripheral blood mononuclear cells and plasma. *J. Gen. Virol.* **78:**1521–1525.

Maggi, F., M. L. Vatteroni, C. Fornai, A. Morrica, M. Giorgi, M. Bendinelli, and M. Pistello. 1997b. Subtype 2c of hepatitis C virus is highly prevalent in Italy and is heterogeneous in the NS5A region. *J. Clin. Microbiol.* **35:**161–164.

Major, M. E., and S. M. Feinstone. 1997. The molecular virology of hepatitis C. *Hepatology* **25:**1527–1538.

Malaguernera, M., S. Restuccia, G. Pistone, P. Ruello, I. Giugno, and B. A. Trovato. 1996. A meta-analysis of interferon-alpha treatment of hepatitis D virus infection. *Pharmacotherapy* **16:**609–614.

Maruyama, T., A. McLachlan, S. Iino, K. Koike, K. Kurokawa, and D. Milich. 1993. The serology of hepatitis B infection revisited. *J. Clin. Investig.* **91:**2586–2595.

Mason, A. L., L. Xu, L. Guo, M. Kuhns, and R. P. Perrillo. 1998. Molecular basis for persistent hepatitis B virus infection in the liver after clearance of serum hepatitis B surface antigen. *Hepatology* **27:**1736–1742.

Matsumoto, A., A. E. T. Yeo, W. K. Shih, E. Tanaka, K. Kiyosawa, and H. J. Alter. 1999. Transfusion-associated TT virus infection and its relationship to liver disease. *Hepatology* **30:**283–288.

McHutchison, J. G., L. M. Blatt, R. Ponnudurai, K. Goodarzi, J. Russell, and A. Conrad. 1999. Ultracentrifugation and concentration of a large volume of serum for HCV RNA during treatment may predict sustained and relapse response in chronic HCV infection. *J. Med. Virol.* **57:**341–355.

Mellor, J., A. Hawkins, and P. Simmonds. 1999. Genotype dependence of hepatitis C virus load measurement in commercially available quantitative assays. *J. Clin. Microbiol.* **37:**2525–2532.

Michalak, T. I., I. U. Pardoe, C. S. Coffin, N. D. Churchill, D. S. Freake, P. Smith, and C. L. Trelegan. 1999. Occult lifelong persistence of infectious hepadnavirus and residual liver inflammation in woodchucks convalescent from acute viral hepatitis. *Hepatology* **29:**928–938.

Minton, J. 1998. Transfusion-associated hepatitis G virus infection. *Rev. Med. Microbiol.* **9:**207–215.

Missale, G., R. Bertoni, V. Lamonaca, A. Valli, M. Massari, C. Mori, M. G. Rumi, M. Houghton, F. Fiaccadori, and C. Ferrari. 1996. Different clinical behaviors of acute hepatitis virus C infection are associated with different vigor of the anti-viral cell-mediated immune response. *J. Clin. Investig.* **98:**706–714.

Morita, K., K. Tanaka, S. Saito, T. Kitamura, T. Kiba, T. Fujii, K. Numata, and H. Sekihara. 1999. Expression of interferon receptor genes in the liver as a predictor of interferon response in patients with chronic hepatitis C. *J. Med. Virol.* **58:**359–365.

Morrica, A., M. Giorgi, F. Maggi, C. Fornai, M. L. Vatteroni, S. Marchi, A. Ricchiuti, G. Antonelli, M. Pistello, and M. Bendinelli. 1999. Susceptibility of human and non-human cell lines to HCV infection as determined by the centrifugation-facilitated method. *J. Virol. Methods* **77:**207–215.

Morrica, A., F. Maggi, M. L. Vatteroni, C. Fornai, M. Pistello, P. Ciccorossi, E. Grassi, A. Gennazzani, and M. Bendinelli. 2000. TT virus: evidence for transplacental transmission. *J. Infect. Dis.* **181:**803–804.

Mushahwar, I. K., J. C. Erker, A. S. Muerhoff, T. P. Leary, J. N. Simons, L. G. Birkenmeyer, M. L. Chalmers, T. J. Pilot-Matias, and S. M. Dexai. 1999. Molecular and biophysical characterization of TT virus: evidence for a new virus family infecting humans. *Proc. Natl. Acad. Sci. USA* **96:**3177–3182.

Naito, H., K. M. Win, and K. Abe. 1999. Identification of a novel genotype of hepatitis G virus in Southeast Asia. *J. Clin. Microbiol.* **37:**1217–1220.

Naito, M., N. Hayashi, H. Hagiwara, K. Katayama, A. Kasahara, H. Fusamoto, M. Kato, M. Masuzawa, and T. Kamada. 1994. Serial quantitative analysis of serum hepatitis C virus RNA level in patients with acute and chronic hepatitis C. *J. Hepatol.* **20:**755–759.

Negro, F., K. Krawczynski, R. Quadri, L. Rubbia-Brandt, M. Mondelli, J. P. Zarski, and A. Hadengue. 1999. Detection of genomic- and minus-strand of hepatitis C virus RNA in the liver of chronic hepatitis C patients by strand-specific semiquantitative reverse transcriptase polymerase chain reaction. *Hepatology* **29:**536–542.

Neumann, A. U., N. P. Lam, H. Dahari, D. R. Gretch, T. E. Wiley, T. J. Layden, and A. S. Perelson. 1998. Hepatitis C viral dynamics in vivo and the antiviral efficacy of interferon-α therapy. *Science* **282:**103–107.

Ngui, S. L., R. Hallett, and C. G. Teo. 1999. Natural and iatrogenic variation in hepatitis B virus. *Rev. Med. Virol.* **9:**183–209.

Niitsuma, H., M. Ishii, M. Miura, K. Kobayashi, and T. Toyota. 1997. Low level hepatitis B viremia detected by polymerase chain reaction accompanies the absence of HBe antigenemia and hepatitis in hepatitis B virus carriers. *Am. J. Gastroenterol.* **92:**119–123.

Nishizawa, T., H. Okamoto, K. Konishi, H. Yoshizawa, Y. Miyakawa, and M. Mayumi. 1997. A novel DNA virus (TTV) associated with elevated transaminase levels in posttransfusion hepatitis of unknown etiology. *Biochem. Biophys. Res. Commun.* **241**:92–97.

Nishizawa, T., H. Okamoto, F. Tsuda, T. Aikawa, Y. Sugai, K. Konishi, Y. Akahane, M. Ukita, T. Tanaka, Y. Miyakawa, and M. Mayumi. 1999. Quasispecies of TT virus (TTV) with sequence divergence in hypervariable regions of the capsid protein in chronic TTV infection. *J. Virol.* **73**:9604–9608.

Nisini, R., M. Paroli, D. Accapezzato, F. Bonino, F. Rosina, T. Santantonio, F. Sallusto, A. Amoroso, M. Houghton, and V. Barnaba. 1997. Human CD4+ T-cell response to hepatitis delta virus: identification of multiple epitopes and characterization of T-helper cytokine profiles. *J. Virol.* **71**:2241–2251.

Nowak, M. A., S. Bonhoeffer, A. M. Hill, R. Boehme, H. C. Thomas, and H. McDade. 1996. Viral dynamics in hepatitis B virus infection. *Proc. Natl. Acad. Sci. USA* **93**:4398–4402.

Okamoto, H., T. Nishizawa, N. Kato, M. Ukita, H. Ikeda, H. Iizuka, Y. Miyakawa, and M. Mayumi. 1998a. Molecular cloning and characterization of a novel DNA virus (TTV) associated with posttransfusion hepatitis of unknown etiology. *Hepatol. Res.* **10**:1–16.

Okamoto, H., Y. Akahane, M. Ukita, M. Fukuda, F. Tsuda, Y. Miyakawa, and M. Mayumi. 1998b. Fecal excretion of a nonenveloped DNA virus (TTV) associated with posttransfusion of non-A-G hepatitis. *J. Med. Virol.* **56**:128–132.

Okamoto, H., N. Kato, H. Iizuka, F. Tsuda, Y. Miyakawa, and M. Mayumi. 1999. Distinct genotypes of a nonenveloped DNA virus associated with posttransfusion non-A to G hepatitis (TT virus) in plasma and peripheral blood mononuclear cells. *J. Med. Virol.* **57**:252–258.

Ono-Nita, S. K., N. Kato, Y. Shiratori, K. H. Lau, H. Yoshida, F. J. Carrilho, and M. Omata. 1999. Susceptibility of lamivudine-resistant hepatitis B to other reverse transcriptase inhibitors. *J. Clin. Investig.* **103**:1635–1640.

Pawlotsky, J.-M., M. Pellerin, M. Bouvier, F. Roudot-Thoraval, G. Germanidis, A. Bastie, F. Darthuy, J. Rémiré, C.-J. Soussy, and D. Dhumeaux. 1998a. Genetic complexity of the hypervariable region 1 (HVR1) of hepatitis C virus (HCV): influence on the characteristics of the infection and responses to interferon alfa therapy in patients with chronic hepatitis C. *J. Med. Virol.* **54**:256–264.

Pawlotsky, J. M., F. Roudot-Thoraval, A. S. Muerhoff, M. Pellerin, G. Germanidis, S. M. Desai, A. Bastie, F. Darthuy, J. Rémiré, E. S. Zafrani, C. J. Soussy, I. K. Mushahwar, and D. Dhumeaux. 1998b. GB virus C (GBV-C) infection in patients with chronic hepatitis C. Influence on liver disease and on hepatitis virus behaviour: effect of interferon alfa therapy. *J. Med. Virol.* **54**:26–37.

Pereira, B. J. G., and A. S. Levy. 1997. Hepatitis C virus infection in dialysis and renal transplantation. *Kidney Int.* **51**:981–999.

Perrillo, R. P. 1997. The role of liver biopsy in hepatitis C. *Hepatology* **26**:57S–61S.

Pessoa, M. G., and T. L. Wright. 1999. Update on clinical trials in the treatment of hepatitis B. *J. Gastroenterol. Hepatol.* **14**:S6–S11.

Pickering, J. M., H. C. Thomas, and P. Karayiannis. 1997. Genetic diversity between hepatitis G virus isolates: analysis of nucleotide variation in the NS-3 and putative "core" peptide genes. *J. Gen. Virol.* **78**:53–60.

Pilot-Matias, T. J., A. S. Muerhoff, J. N. Simons, T. P. Leary, S. L. Buijk, M. L. Chalmers, J. C. Erker, G. J. Dawson, S. M. Desai, and I. K. Mushahwar. 1996. Identification of antigenic regions in the GB hepatitis viruses GBV-A, GBV-B, and GBV-C. *J. Med. Virol.* **48**:329–338.

Pistello, M., L. Ceccherini-Nelli, N. Cecconi, M. Bendinelli, and F. Panicucci. 1991. Hepatitis C virus seroprevalence in Italian hemophiliacs injected with virus-inactivated concentrates: five-year follow-up and correlation with antibodies to other viruses. *J. Med. Virol.* **33**:43–46.

Pistello, M., F. Maggi, M. L. Vatteroni, N. Cecconi, F. Panicucci, G. P. Bresci, L. Gambardella, M. Taddei, A. Bionda, M. Tuoni, and M. Bendinelli. 1994. Prevalence of hepatitis C virus genotypes in Italy. *J. Clin. Microbiol.* **32**:232–234.

Pistello, M., F. Maggi, C. Fornai, A. Leonildi, A. Morrica, M. L. Vatteroni, and M. Bendinelli. 1999. Classification of hepatitis C virus type 2 isolates by phylogenetic analysis of core and NS5 regions. *J. Clin. Microbiol.* **37**:2116–2117.

Pistello, M., A. Morrica, F. Maggi, M. L. Vatteroni, G. Freer, C. Fornai, F. Casula, S. Marchi, P. Ciccorossi, P. Rovero, and M. Bendinelli. TT virus levels in the plasma of infected individuals with different hepatic and extrahepatic pathologies. *J. Med. Virol.*, in press.

Pontisso, P., G. Bellati, M. Brunetto, L. Chemello, G. Colloredo, R. Di Stefano, M. Nicoletti, M. G. Rumi, M. G. Ruvoletto, R. Soffredini, L. M. Valenza, and G. Colucci. 1999. Hepatitis C virus RNA profiles in chronically infected individuals: do they relate to disease activity? *Hepatology* **29**:585–589.

Poynard, T., P. Bedossa, and P. Opolon. 1997. Natural history of liver fibrosis progression in patients with chronic hepatitis C. *Lancet* **349**:825–832.

Radkowski, M., L. F. Wang, J. Cianciaria, J. Rakela, and T. Laskus. 1999. Analysis of hepatitis G virus/GB virus C quasispecies and replication sites in human subjects. *Biochem. Biophys. Res. Commun.* **258**:296–299.

Ranki, M., H. M. Schatzl, R. Zachoval, M. Uusi-Oukari, and P. Lehtovaara. 1995. Quantification of hepatitis B virus DNA over a wide range from serum for studying viral replicative activity in response to treatment and in recurrent infection. *Hepatology* **21**:1492–1499.

Ray, S. C., Y. M. Wang, O. Laeyendecker, J. R. Ticehurst, S. A. Villano, and D. L. Thomas. 1999. Acute hepatitis C virus structural gene sequences as predictors of persistent viremia: hypervariable region 1 as a decoy. *J. Virol.* **73**:2938–2946.

Rispeter, K., M. Lu, A. Zibert, M. Wiese, J. M. de Oliviera, and M. Roggendorf. 1998. The "interferon sensitivity determining region" of hepatitis C virus is a stable sequence element. *J. Hepatol.* **29**:352–361.

Rizzetto, M., M. G. Canese, J. L. Gerin, W. T. London, D. L. Sly, and R. H. Purcell. 1980. Transmission of the hepatitis B virus-associated delta antigen to chimpanzees. *J. Infect. Dis.* **141**:590–602.

Robertson, B., G. Myers, C. Howard, T. Brettin, J. Bukh, B. Gaschen, T. Gojobori, G. Maertens, M. Mizokami, O. Nainan, S. Netesov, K. Nishioka, P. Shin-i, P. Simmonds, D. Smith, L. Stuyver, and A. Weiner. 1998. Classification, nomenclature, and database development for hepatitis C virus (HCV) and related viruses: proposals for standardization. *Arch. Virol.* **143**:2493–2503.

Roingeard, P., F. Dubois, P. Marcellin, J. Bernuau, S. Bonduelle, J. P. Benhamou, and A. Goudeau. 1992. Persistent delta antigenemia in chronic delta hepatitis and its relation with human immunodeficiency virus infection. *J. Med. Virol.* **38**:191–194.

Rosa, D., S. Campagnoli, C. Moretto, E. Guenzi, L. Cousens, M. Chin, C. Dong, A. J. Weiner, J. Y. N. Lau, Q.-L. Choo, D. Chien, P. Pileri, M. Houghton, and S. Abrignani. 1996. A quantitative test to estimate neutralizing antibodies to the hepatitis C virus: cytofluorometric assessment of envelope glycoprotein 2 binding to target cells. *Proc. Natl. Acad. Sci. USA* **93**:1759–1763.

Ross, R. S., S. Viazov, U. Schmitt, S. Schmolke, M. Tache, B. Ofenloch-Haehnle, M. Holtmann, N. Muller, G. Da Villa, C. F. Yoshida, J. M. Oliveira, A. Szabo, N. Paladi, J. P. Kruppenbacher, T. Philipp, and M. Roggendorf. 1998. Distinct prevalence of antibodies to the E2 protein of GB virus C/hepatitis G virus in different parts of the world. *J. Med. Virol.* **54:**103–106.

Ross, R. S., S. Viazov, V. Runde, U. W. Schaefer, and M. Roggendorf. 1999. Detection of TT virus DNA in specimens other than blood. *J. Clin. Virol.* **13:**181–184.

Roth, W. K., M. Weber, and E. Seifried. 1999. Feasibility and efficacy of routine PCR screening of blood donations for hepatitis C virus, hepatitis B virus, and HIV-1 in a blood-bank setting. *Lancet* **353:**359–363.

Sakugawa, H., H. Nakasone, Y. Kawakami, F. Adaniya, T. Mizushima, T. Nakayoshi, F. Kinjo, A. Saito, H. Zukeran, Y. Miyagi, S. Yakabi, M. Taira, M. Kinoshita, and Y. Yamakawa. 1997. Determination of hepatitis delta virus (HDV)-RNA in asymptomatic cases of HDV infection. *Am. J. Gastroenterol.* **92:**2232–2236.

Salvey, L. A., C. A. Hyland, L. Mison, N. Solomon, and E. J. Gowans. 1998. Is there evidence for vector transmission of GBV-C? *Lancet* **351:**1104.

Sansonno, D., C. Lotesoriere, V. Cornacchiulo, M. Fanelli, P. Gatti, G. Iodice, V. Racanelli, and F. Dammaco. 1998. Hepatitis C virus infection involved CD34(+) hematopoietic progenitor cells in hepatitis C virus chronic carriers. *Blood* **92:**3328–3337.

Schroter, M., H. H. Feucht, P. Schafer, B. Zollner, and R. Laufs. 1999. GB virus C/hepatitis G virus infection in hemodialysis patients: determination of seroprevalence by a four-antigen recombinant immunoblot assay. *J. Med. Virol.* **57:**230–234.

Shakil, A. O., S. Hadziyannis, J. H. Hoofnagle, A. M. Di Bisceglie, J. L. Gerin, and J. L. Casey. 1997. Geographic distribution and genetic variability of hepatitis delta virus genotype I. *J. Virol.* **71:**5408–5414.

Sherlock, S. 1995. Introduction: combination antiviral therapy in chronic hepatitis C. *J. Hepatol.* **23**(Suppl. 2):1–2.

Shimizu, Y. K., M. Hijikata, T. Kiyohara, Y. Kitamura, and H. Yoshikura. 1999. Replication of GB virus C (hepatitis G virus) in interferon-resistant Daudi cells. *J. Virol.* **73:**8411–8414.

Shindo, M., K. Hamada, S. Koya, K. Arai, Y. Sokawa, and T. Okuno. 1996. The clinical significance of changes in genetic heterogeneity of the hypervariable region 1 in chronic hepatitis C with interferon therapy. *Hepatology* **24:**1018–1023.

Shiratori, Y., N. Kato, H. Yoshida, F. Imazeki, K. Okano, O. Yokosuka, and M. Omata. 1999. How soon can a virological sustained response be determined after withdrawl of interferon therapy in chronic hepatitis C? *J. Gastroenterol. Hepatol.* **14:**79–84.

Simmonds, P. 1995. Variability of hepatitis C virus. *Hepatology* **21:**570–583.

Simmonds, P., E. C. Holmes, T.-A. Cha, S.-W. Chan, F. McOmish, B. Irvine, E. Beall, and P. L. Yap. 1993. Classification of hepatitis C virus into six major genotypes and a series of subtypes by phylogenetic analysis of the NS-5 region. *J. Gen. Virol.* **74:**2391–2399.

Simmonds, P., L. E. Prescott, C. Logue, F. Davidson, A. E. Thomas, and C. A. Ludlam. 1999. TT virus. Part of the normal human flora? *J. Infect. Dis.* **180:**1748–1749.

Simons, J. N., T. P Leary, G. J. Dawson, T. J. Pilot-Matias, A. S. Muerhoff, G. G. Schlauder, S. M. Desai, and I. K.

Mushahwar. 1995. Isolation of novel virus-like sequences associated with human hepatitis. *Nat. Med.* **1:**564–569.

Simpson, L. H., M. Battegay, J. M Hoofnagle, J. G. Waggoner, and A. M. Di Bisceglie. 1994. Hepatitis delta virus RNA in serum of patients with chronic delta hepatitis. *Dig. Dis. Sci.* **39:**2650–2655.

Smedile, A., and G. Verme. 1999. Hepatitis D virus, p. 129–150. *In* S. Specter (ed.), *Viral Hepatitis. Diagnosis, Therapy, and Prevention.* Humana Press, Totowa, N.J.

Smedile, A., J. L. Casey, P. J. Cote, M. Durazzo, B. Lavezzo, R. H. Purcell, M. Rizzetto, and J. L. Gerin. 1998. Hepatitis D viremia following orthotopic liver transplantation involves a typical HDV virion with a hepatitis B surface antigen envelope. *Hepatology* **27:**1723–1729.

Specter, S., M., Bendinelli, and H. Friedman (ed.). 1998. *Rapid Detection of Infectious Agents*, p. 216. Plenum Press, New York, N.Y.

Suzuki, Y., K. Katayama, S. Fukushi, T. Kageyama, A. Oya, H. Okamura, Y. Tanaka, M. Mizokami, and T. Gojobori. 1999. Slow evolutionary rate of GB virus C/hepatitis G virus. *J. Mol. Evol.* **48:**383–389.

Tacke, M., K. Kiyosawa, K. Stark, V. Schlueter, B. Ofenloch-Haehnle, G. Hess, and A. M. Engel. 1997. Detection of antibodies to a putative hepatitis G virus envelope protein. *Lancet* **349:**318–320.

Takahashi, K., H. Hoshino, Y. Ohta, N. Yoshida, and S. Mishiro. 1998. Very high prevalence of TT virus (TTV) infection in general population of Japan revealed by a new set of PCR primers. *Hepatol. Res.* **12:**233–239.

Takayama, S., S. Yamazaki, S. Matsuo, and S. Sugii. 1999. Multiple infection of TT virus (TTV) with different genotypes in Japanese hemophiliacs. *Biochem. Biophys. Res. Commun.* **256:**208–211.

Tan, D., A. Matsumoto, C. Conry-Cantilena, J. C. Melpolder, J. W. Shih, M. Lauther, G. Hess, J. W. Gibble, P. M. Ness, and H. J. Alter. 1999. Analysis of hepatitis G virus (HGV), antibody to HGV envelope protein, and risk factors for blood donors coinfected with HGV and hepatitis C virus. *J. Infect. Dis.* **179:**1055–1061.

Tanaka, T., G. Hess, V. Schlueter, D. Zdunek, S. Tanaka, and M. Kohara. 1999. Correlation of interferon treatment response with GBV-C/HGV genomic RNA and anti-envelope 2 protein antibody. *J. Med. Virol.* **57:**370–375.

Tanaka, Y., M. Mizokami, E. Orito, T. Ohno, T. Nakano, T. Kato, H. Kato, M. Mukaide, Y. M. Park, B. S. Kim, and R. Ueda. 1998. New genotypes of TT virus (TTV) and a genotyping assay based on restriction fragment length polymorphism. *FEBS Lett.* **437:**201–206.

Tang, J. R., O. Hantz, L. Vitvitski, J. P. Lamelin, R. Parana, L. Cova, J. L. Lesbordes, and C. Trepo. 1993. Discovery of a novel point mutation changing the HDAg expression of a hepatitis delta virus isolate from Central African Republic. *J. Gen. Virol.* **74:**1827–1835.

Tong, M. J., L. M. Blatt, L. T. Tong, K. Sayadzadech, and A. Conrad. 1998. Long-term retreatment in chronic hepatitis C patients who were non-responders to initial course of interferon-alpha 2b. *J. Viral Hepat.* **5:**323–331.

Tsai, S.-L., Y.-F. Liaw, M.-H. Chen, C.-Y. Huang, and G. C. Kuo. 1997. Detection of type2-like T-helper cells in hepatitis C virus infection: implication for hepatitis C virus chronicity. *Hepatology* **25:**449–458.

Tsuda, F., H. Okamoto, M. Ukita, T. Tanaka, Y. Akahane, K. Konishi, H. Yoshizawa, Y. Miyakawa, and M. Mayumi. 1999. Determination of antibodies to TT virus (TTV) and application to blood donors and patients with post-transfusion non-A to G hepatitis in Japan. *J. Virol. Methods* **77:**199–206.

Vajro, P., F. Migliaro, A. Fontanella, and G. Orso. 1998. Interferon: a meta-analysis of published studies in pediatric chronic hepatitis B. *Acta Gastroenterol. Belg.* **61:**219–223.

Vatteroni, M. L., F. Maggi, A. Morrica, C. Fornai, M. Giorgi, M. Pistello, and M. Bendinelli. 1997. Comparative evaluation of five rapid methods for identifying subtype 1b and 2c of hepatitis C virus isolates. *J. Virol. Methods* **66:**187–194.

Vyas, G. N., and T. S. B. Yen. 1999. Hepatitis B virus: biology, pathogenesis, epidemiology, clinical description, and diagnosis, p. 35–63. *In* S. Specter (ed.), *Viral Hepatitis. Diagnosis, Therapy, and Prevention.* Humana Press, Totowa, N.J.

Wiley, T. E., L. Briedi, N. Lam, and T. J. Layden. 1998. Early HCV RNA values after interferon predict response. *Dig. Dis. Sci.* **43:**2169–2172.

Wu, J. C., P. J. Chen, M. Y. P. Kuo, S. D. Lee, D. S. Chen, and L. P. Ting. 1991. Production of hepatitis delta virus and suppression of helper hepatitis B virus in a human hepatoma cell line. *J. Virol.* **65:**1099–1104.

Wu, J. C., C. M. Chen, I. J. Sheen, S. D. Lee, H. M. Tzeng, and K. B. Choo. 1995. Evidence of transmission of hepatitis D virus to spouses from sequence analysis of the viral genome. *J. Virol.* **69:**7593–7600.

Wu, J. C., I. A. Huang, Y. H. Huang, J. Y. Chen, and I. J. Sheen. 1999. Mixed genotypes infection with hepatitis D virus. *J. Med. Virol.* **57:**64–67.

Xiang, J., D. Klinzman, J. McLinden, W. N. Schmidt, D. R. LaBrecque, R. Gish, and J. T. Stapleton. 1998. Characterization of hepatitis G virus (GB-C virus) particles: evidence for a nucleocapsid and expression of sequences upstream of the E1 protein. *J. Virol.* **72:**2738–2744.

Yamada-Osaki, M., R. Sumazaki, M. Tsuchida, K. Koike, T. Fukushima, and A. Matsui. 1999. Persistence and clinical outcome of hepatitis G virus infection in pediatric bone marrow transplant recipients and children treated for hematological malignancy. *Blood* **93:**721–727.

Yamamoto, T., K. Kajino, M. Ogawa, I. Gotoh, S. Matsuoka, K. Suzuki, M. Moriyama, H. Okubo, M. Kudo, Y. Arakawa, and O. Hino. 1998. Hepatocellular carcinomas infected with the novel TT DNA virus lack viral integration. *Biochem. Biophys. Res. Commun.* **251:**339–343.

Yang A., P. Karayiannis, H. Thomas, and J. Monjardino. 1995. Editing efficiency of hepatitis delta virus RNA is related to the course of infection in woodchucks. *J. Gen. Virol.* **76:**3071–3078.

Yoshimura, E., J. Hayashi, Y. Tani, M. Ohmiya, K. Nakashima, H. Ikematsu, N. Kinukawa, Y. Maeda, and S. Kashiwagi. 1994. Inverse correlation between the titer of antibody to hepatitis C virus and the degree of hepatitis C viremia. *J. Infect.* **29:**147–155.

Yoshimura, E., J. Hayashi, K. Ueno, Y. Kishihara, K. Yamaji, Y. Etoh, and S. Kashiwagi. 1997. No significant changes in levels of hepatitis C virus (HCV) RNA by competitive polymerase chain reaction in blood samples from patients with chronic HCV infection. *Dig. Dis. Sci.* **42:**772–777.

Yotsuyanagi, H., K. Yasuda, S. Iino, K. Moriya, Y. Shintani, H. Fujie, T. Tsutsumi, S. Kimura, and K. Koike. 1998. Persistent viremia after recovery from self-limited acute hepatitis B. *Hepatology* **27:**1377–1382.

Zaaijer, H. L., F. ter Borg, H. T. M. Cuypers, M. C. A. H. Hermus, and P. N. Lelie. 1994. Comparison of methods for detection of hepatitis B virus DNA. *J. Clin. Microbiol.* **32:**2088–2091.

Zeuzem, S., R. A. de Man, P. Honkoop, W. K. Roth, S. W. Schalm, and J. M. Schmidt. 1997. Dynamics of hepatitis B virus infection in vivo. *J. Hepatol.* **27:**431–436.

Zeuzem, S., J.-H. Lee, A. Franke, B. Rüster, O. Prümmer, G. Herrmann, and W. K. Roth. 1998. Quantification of the initial decline of serum hepatitis C virus RNA and response to interferon alfa. *Hepatology* **27:**1149–1156.

Zibert, A., W. Krass, H. Meisel, G. Jung. and M. Roggendorf. 1997. Epitope mapping of antibodies directed against hypervariable region 1 in acute self-limiting and chronic infection due to hepatitis C virus. *J. Virol.* **71:**4123–4127.

Zibert, A., W. Kraas, R. S. Ross, H. Meisel, S. Lechner, G. Jung, and M. Roggendorf. 1999. Immunodominant B-cell domains of hepatitis C virus envelope proteins E1 and E2 identified during early and late time points of infection. *J. Hepatol.* **30:**177–184.

Zuckerman, A. J., and J. N. Zuckerman. 1999. Molecular epidemiology of hepatitis B virus mutants. *J. Med. Virol.* **58:**193–195.

Rabies

CHARLES V. TRIMARCHI

28

Rabies is a viral infection of the central nervous system (CNS) of mammals caused by a neurotropic virus. The disease is maintained in host populations of canines and other terrestrial carnivores and bats. Transmission is through infectious saliva, transferred by bites of clinically rabid animals to other susceptible hosts. The disease is characterized by a long and variable incubation period, followed by an acute, progressive encephalitis culminating in the death of nearly every infected animal.

Modern methods now permit the prevention of rabies infection in humans and animals by pre- or postexposure vaccination. Elimination of rabies has been achieved in canine populations over large areas by widespread dog vaccination programs combined with dog control activities. The disease has even been controlled in some wildlife populations by campaigns to vaccinate the main vector through distribution of baits containing vaccines that induce effective immunity by ingestion. However, the disease still exacts a horrible toll in terms of human and domestic-animal mortality in large areas of the world, particularly where resource limitations prevent the implementation of preventive measures. It is also associated with enormous opportunity costs where extraordinarily expensive programs must be initiated and maintained to protect humans and domestic animals.

Although there are numerous early historical references to rabies in wildlife throughout the world, the zoonotic potential of the disease always was and remains exemplified by its presence in domestic dogs. Where rabies has been controlled in dogs, human mortality from the disease has promptly become uncommon. However, because human populations are not natural hosts of rabies virus, the disease in humans is incidental to the reservoir of the disease in wild and domestic animals. Consequently, a more appropriate measure of the impact of rabies on public health in the areas where canine rabies has been controlled is the estimate of the prevalence of the disease in animal populations and the expense involved in preventing transmission to humans (Smith and Seidel, 1993). Expenditures for preventive rabies treatments for humans and domestic animals in the United States exceed $100 million annually (Uhaa et al., 1992).

HISTORY

Several authors have produced exhaustive and fascinating summaries of the history of rabies and humans (Theodor-

ides, 1986; Steele and Fernandez, 1991; Baer et al., 1996). Rabies has been one of the most widely and consistently feared zoonotic diseases since the earliest recorded human history. The transmission of rabies to humans by the bite of mad dogs was included in the Eschunna Code of ancient Mesopotamia in the 23rd century B.C. Greek mythology included a god whose task it was to counteract the effects of rabies. Aristotle's *Natural History of Animals*, written in the fourth century B.C., described a "madness" in dogs and went on to say, "this causes them to become very irritable and all mammals they bite become diseased." Note that by the philosopher's time it was already observed that rabies was a disease of mammals, transmitted by the bite of the afflicted animal, and resulting in behavioral changes, including aggression.

The earliest methods of treatment for the prevention of rabies included submersion of the patient in water, cautery of the wound with heat or caustics, wound treatment according to the pharmacological ideas of the time, and systemic treatment with an astounding variety of substances of animal and herbal origin, including the "hair of the dog that bit you." Subsequent developments in the prevention and diagnosis of rabies have either led or paralleled human accomplishments in the sciences of infectious diseases. Georg Zinke in 1804 discussed the "agent of rabies" and described experiments that demonstrated the path of the virus from saliva to wound to saliva (Wilkinson, 1988). During the period 1881 to 1884, Louis Pasteur described the involvement of the CNS and salivary glands in the disease and discussed attenuation of the virus and the theoretical basis of immunizing injections. In 1885 he reported the landmark first human postexposure vaccination of Joseph Meister, and by 1896 he reported on the treatment of 350 exposed persons, of whom only one developed rabies.

Successful canine vaccination campaigns became a reality with the contributions of Koprowski and Cox in 1948 to the development of modified-virus vaccines by serial chicken embryo passage (Koprowski and Cox, 1948). The severe reactions that too frequently accompanied exposure to nervous-tissue-derived human vaccines were avoided with the development of cell culture vaccines after Wiktor and Koprowski adapted fixed rabies virus to human diploid cell cultures in 1965 (Wiktor and Koprowski, 1965). Even with potent vaccines, an efficacy approaching 100% for postexposure treatment (PET) of humans became regularly achievable only when it was recognized by Habel in 1945 that immune serum given at the site of the bite was also critical to survival (Habel, 1945).

The diagnosis of rabies entered the modern era when Negri in 1903 discovered the intracytoplasmic inclusions in large nerve cells of rabid animals, pathognomonic for rabies, that would bear his name and be the basis for rabies postmortem diagnosis for more than half a century (Negri, 1903). The development of the mouse inoculation test in 1935 by Webster and Dawson (Webster and Dawson, 1935) introduced a reliable diagnostic and research virus isolation tool, upgraded dramatically in speed beginning in 1978 by the application of in vitro cell culture techniques (Smith et al., 1978). The application of the direct fluorescent antibody test (FAT) to rabies examination by Goldwasser and Kissling in 1958 added modern sensitivity, specificity, and rapidity to rabies laboratory science (Goldwasser and Kissling, 1958). The application of monoclonal antibody methods to rabies virology (Wiktor et al., 1980) revolutionized the study of rabies and rabies-related viral antigenic relationships.

BIOLOGY OF THE VIRUS

Rabies is the prototype virus of the genus *Lyssavirus* of the order *Mononegavirales*, family *Rhabdoviridae* (Wunner et al., 1995). The members of the order *Mononegavirales* are characterized by a nonsegmented, negative-stranded RNA genome, encapsulated tightly into ribonucleocapsid structures. The members of the family *Rhabdoviridae* are classified as a group based on similar conical or bullet-shaped appearance by electron microscopy. The host range for the genus is highly diversified, including plants, arthropods, fish, and mammals. Previously known simply as the rabies and rabies-related virus group (Shope et al., 1970), the genus *Lyssavirus* is presently composed of genotype 1, classical rabies virus, and six other genotypes that are closely related antigenically and genetically and that cause a clinical disease indistinguishable from rabies. Genotype 1 includes the majority of field viruses of global distribution in terrestrial mammals and in insectivorous and hematophagous bats of the western hemisphere, as well as the laboratory and vaccine strains (World Health Organization [WHO], 1994). The distribution of the nonrabies lyssaviruses is restricted to the Old World. They include genotype 2, Lagos bat virus, isolated from African bats; genotype 3, Mokola virus, isolated from African rodents; genotype 4, Duvenhage virus, isolated from African bats; genotypes 5 and 6, European bat lyssaviruses 1 and 2, isolated from European bats (Kissi et al., 1995); and genotype 7, Australian bat lyssavirus, isolated from Australian bats (Hooper et al., 1997). Lyssaviruses are serologically distinct from other rhabdoviruses (Shope and Tesh, 1987).

The rigid structure of the rabies virion measures approximately 180 by 75 nm (Hummeler et al., 1967; Vernon et al., 1972). The virus is hemispherical at one end and usually planar at the other end, where it buds off last from the surface membrane of an infected cell. The particle contains a helical nucleocapsid surrounded by a lipid bilayer envelope. Spikelike surface projections protrude 10 nm from the outer surface of the lipid bilayer. Disruption of the virus discloses five proteins. The internal ribonucleoprotein complex (RNP) contains the viral RNA associated with three internal proteins: a large (190-kDa) transcriptase, or L protein; a 55-kDa nucleoprotein, N; and a 38-kDa nonstructural phosphorylated protein, NS. The viral envelope is composed of a 26-kDa matrix protein, M, an envelope sheath

consisting of lipids derived from the host cell plasma membrane, and the surface spikes formed by a 67-kDa glycoprotein, G (Wunner et al., 1988).

The rabies virus genome consists of a single-stranded, nonsegmented RNA molecule of negative sense polarity. It is approximately 1,200 nucleotides in length and has a molecular mass of approximately 4.6×10^6 kDa. The viral RNA is transcribed into five polyadenylated, monocistronic mRNA species, corresponding to the five viral proteins. The negative polarity of the rabies genome prevents direct translation into viral proteins, requiring an autonomous transcription step facilitated by the RNA polymerase (Tordo, 1996).

The RNP functions in the transcription and replication of the virion. Accumulations of RNP constitute the intracytoplasmic inclusions in infected cells, which have diagnostic importance because they can be detected by direct observation with histologic methods and by antigen detection methods employing N protein-specific antibodies. The G protein is the viral antigen that induces the production of virus-neutralizing antibodies, conferring immunity against exposure (Wiktor et al., 1984). The induction of antibody and the conferred immunity are dependent upon the intact secondary and tertiary structures of the G protein (Koprowski , 1991). The development of PCR gene amplification and direct nucleotide sequence analysis (Sacramento et al., 1991) has substantially accelerated progress toward the understanding of the structure-function relationships of the various elements of the rabies virus (Tordo, 1996).

Rabies virus does not generally cause host cell destruction. It is synthesized in the cytoplasm of infected cells and is released by budding through cell membranes (Murphy, 1986). The virus is somewhat resistant to air drying and freeze-thaw cycles and is relatively stable at pHs of 5 to 10. However, it is labile to pasteurization temperatures, UV light and lipid solvents, ethanol, iodine disinfectants, and quaternary ammonium compounds (Kaplan, 1996).

PATHOGENESIS AND PATHOLOGY

All mammals are susceptible to rabies virus infection. With very rare exceptions, rabies virus infection terminates in the death of each infected individual. Rabies virus has evolved a pathogenesis within the individual animal that facilitates the maintenance of the virus in the true host, the reservoir species population. Maintenance of the virus in the reservoir population by direct host-to-host transmission is dependent on simultaneous infection of the brain and salivary glands: it is the impact on behavior resulting from infection of the limbic system that induces biting behavior, and infection of the salivary gland tissue results in infectious doses of virus in the saliva to serve as an infectious inoculum for bite transmission. This pathogenic pattern has permitted the entrenchment of the virus in host populations and the continued risk to humans of exposure (Murphy, 1986).

Empirical and laboratory evidence accumulated over centuries related to rabies cycles vectored by domestic and wild species supports the conclusion that rabies is transmitted by entrance of the virus through bites of rabid animals. However, there also has been apparent nonbite transmission to humans and animals by introduction of the virus through inhalation of infectious aerosols in a very unusual bat cave environment in the southwestern United States (Constantine, 1962). The conclusion that the two associated human cases in that setting were definitely the result of exposure to airborne virus may need reconsideration in light of current observations that bat bites capable of rabies transmission may be associated with limited injury (Rupprecht, 1996). Rabies transmission to humans may also have occurred by aerosol in two laboratory accidents (Winkler et al., 1973; Center for Disease Control, 1977). Transmission by direct contamination of mucous membranes by saliva has also been reported (Afshar, 1979). Infection by ingestion of infected tissues was reported in dogs feeding on infected fox carcasses in the Arctic (Mansel, 1951) and numerous laboratory studies (Charlton, 1988). Human-to-human transmission has been reported by bite and mucous membrane exposure (Fekadu et al., 1996) and as a result of corneal transplant from rabies victims in eight occurrences worldwide (Centers for Disease Control and Prevention [CDC], 1999a).

When a reservoir host animal becomes infected following exposure from the bite of an infected animal, the virus may invade peripheral nerves or nerve endings directly or may first be "amplified" by invasion of striated muscle cells prior to infection of the nerve endings (Charlton, 1988). It is not clear if the infection of myocytes at the site of exposure is an essential aspect of the pathogenesis of rabies or how this growth contributes to long incubation periods. The early events of viral replication and muscle and nerve cell infection at the site of exposure occur without substantial stimulation of the immune system. The binding of rabies virus to the host cell is mediated by the viral G protein. The rabies receptor is complex and may vary with cell type. The putative receptor in muscle cells is the nicotinic acetylcholine receptor (Lentz et al., 1984; Baer and Lentz, 1991), but there is evidence that the membrane oligosaccharides and lipoprotein components of neurons may also serve as rabies virus receptors (Tsiang, 1988). After entering sensory or motor nerve endings, its genome progresses centripetally transneuronally, only by retrograde axoplasmic flow, to the CNS (Kristensson and Olsson, 1973) and similarly within the CNS from first-order to second-order neurons (Kelly and Strick, 1997). The nervous-tissue pathway to the CNS and the lack of viremia may limit exposure to the immune system, explaining the lack of an early antibody response (Krebs et al., 1995).

Virus replication in the CNS occurs mainly in neurons, with extensive distribution in the brain and spinal cord. Recognizable clinical signs of rabies generally do not appear until several replication cycles have occurred in the brain (Kaplan, 1985). Centrifugal spread occurs simultaneously via anterograde axoplasmic flow from the CNS to peripheral nerves and to some nonnervous tissues, including, most importantly, the salivary glands. This accounts for the appearance of rabies virus in some tissues and fluids up to a few days before the recognized onset of rabies symptoms (Charlton, 1988). Although infectious doses of virus in the saliva of vectors is paramount for the maintenance of the virus in host populations, virus presence in saliva may be sporadic during and just prior to the clinical period (Constantine, 1967; Fekadu et al., 1982). Terminally, rabies antigen may be demonstrated in many tissues, including the buccal, nasal, and intestinal mucosa; urinary bladder; epidermis; cornea; lungs; kidneys; adrenal medulla; and brown fat (Debbie and Trimarchi, 1970). Virus in these nonsalivary tissues has not been shown to be responsible for host-to-host transmission of epidemiological significance.

Studies with animal models have demonstrated that immune mechanisms are involved in the neuropathogenesis of rabies. There have been numerous observations that animals and humans that die despite some exposure to rabies vaccine prior to or immediately after exposure to rabies often succumb with a shorter incubation period than that of naive individuals (WHO, 1984). In an in vitro model using a mouse macrophage cell line, the presence of rabies-neutralizing antibodies in concentrations below protective levels actually enhanced the ability of rabies virus to infect these cells (King et al., 1984). Immunosuppression has been demonstrated to have a sparing effect in some situations (Smith et al., 1982). Furthermore, virus replication may suppress production of cellular neuropeptides and neurotransmitters, leading to functional CNS failure and the fatal outcome of rabies infection (Fu, 1997).

The classical rabies pathogenesis described above results in a variable but long incubation period, typically 10 days to several months but rarely of many years' duration (Smith et al., 1991). The incubation period is followed by an acute, undelayed progression of disease following onset through the classical rabies encephalitic stages, culminating in death. Other observed pathogenic patterns exist, but they are extremely uncommon in naturally occurring disease. These rare and atypical pathologies may include shorter or longer incubation periods (less than 10 days and up to many years), a prolonged clinical period, variations in excretion of the virus, recovery with or without chronic disability, and a carrier state of virus shedding without clinical manifestations. Although infection with an Ethiopian strain of rabies virus was demonstrated to result in some dogs that shed virus in their saliva for long periods without clinical manifestations (Fekadu et al., 1981), no convincing evidence of such a carrier state exists in naturally occurring rabies in North America (Charlton, 1988).

CLINICAL RABIES

Following the variable incubation period, the disease in humans is marked by a brief prodromal period of several days' duration, with complaints of nonspecific symptoms, including malaise, anorexia, fatigue, headache, and fever. Characteristically, during this period there is pain and parathesia, or "tingling," at the site of exposure, which are usually the first rabies-specific symptoms (Bernard, 1986). Behavioral manifestations may include apprehension, anxiety, irritability, and insomnia. Following the prodromal period, patients develop a rapidly progressive neurologic course, with a range of symptoms that may include disorientation, hallucinations, paralysis, nuchal rigidity, aerophobia, pharyngeal spasms, hydrophobia, hypersalivation, dysphagia, focal or generalized seizures, cardiac and respiratory arrhythmias, and hypertension, leading to coma and death (Matyas et al., 1999). A review of 32 human rabies deaths in the United States from 1980 to 1996 identified agitation and confusion, hypersalivation, hydrophobia or aerophobia, and limb pain and weakness as the most commonly observed signs of clinical rabies. The cases had a median clinical period of 19 days (range, 7 to 28 days). In 12 of the 32 cases, the disease was only diagnosed postmortem (Noah et al., 1998). In the absence of intensive care and secondary-support therapies, death usually comes in human rabies cases within 7 days, generally from respiratory failure (Gode et al., 1976). In patients receiving intensive care, the disease eventually severely affects nearly every major organ system, and death occurs as a result of the cessation of cerebral and cardiovascular activity (Fishbein, 1991). There have been just four well-documented human survivals of clinical rabies worldwide, with outcomes from full recovery without sequelae, or with some partial paralysis, to major residual neurologic impairment. Three of the four patients had a history of some pre- or postexposure vaccination.

Rabies in vector animal species is similar in most respects to the disease in humans, but the prodromal period often is followed by a stage of excitation with or without aggression, either followed by or intermixed with periods of lethargy and depression. Progressive paralysis, which often begins referable to the site of exposure, may first be recognized in the posterior limbs or larynx. Paralysis of the throat may result in uncharacteristic vocalizations and accumulation of copious and stringy saliva from the mouth. During the excitative period, animals may exhibit self-mutilation and heightened and inappropriate sexual behaviors. Livestock also may demonstrate aggressive and heightened sexual behavior, but facial and pharyngeal paralysis, hypersalivation, bellowing, straining, and posterior paralysis leading to a "dog sitting" posture and then recumbency are more common (Debbie and Trimarchi, 1992).

EPIZOOTIOLOGY OF ANIMAL RABIES

Rabies is maintained in nature as cycles in bats, wild terrestrial carnivores, and domestic canine populations that serve as reservoirs and vectors of the disease. Specific variants of the virus are associated with each geographically and temporally defined wildlife cycle. Each vector is highly susceptible to the variant that has adapted to that population and is capable of transmission to conspecifics because of coincident aggressive behavior and infectious virus titers in the saliva. A sustained outbreak is also dependent upon an adequate vector population density and a host natural history that provides adequate opportunity for interspecific interactions within the characteristic clinical period. Rabies distribution in animals can be discussed in three general categories: domestic-canine rabies; terrestrial-, sylvatic-wildlife rabies; and rabies in bats.

Rabies virus maintained in domestic-dog populations accounts for 95% of all animal rabies cases reported globally (Fekadu, 1991) and still accounts for most of the zoonotic impact of the disease: 90% of the human exposures to rabies and 99% of the human rabies deaths worldwide are attributed to this cycle (Smith and Seidel, 1993). Although the development of potent vaccines in combination with stray dog control programs has been proven effective in extinguishing dog rabies epizootics, dog rabies is still epizootic in most countries of Asia, Africa, and South America. Worldwide, there are an estimated 60,000 human deaths each year from dog-transmitted rabies and 10,000,000 postexposure vaccination regimens administered as a consequence of this problem. In 1996 dog rabies was still a threat in 87 countries, with an estimated 2.4 billion humans at risk (WHO, 1997). Europe and North America controlled rabies in domestic dogs by stray animal control and widespread vaccination programs during the 1950s and 1960s so that it now accounts for less than 5% of animal rabies cases in those regions. Canine rabies is often associated with urban rabies cycles. The genetic similarity of dog rabies variants found throughout Africa and the Americas confirms the common origin by introduction of infected dogs during colonization by Europeans (Smith and Seidel, 1993).

Rabies in terrestrial wildlife is also present nearly globally. The disease in red foxes has been prevalent in subarctic and northern parts of North America, in subarctic Asia, and in central and eastern Europe. The raccoon dog and gray wolf are reservoirs in northern Eurasia (Wandeler et al., 1994). In Africa, the dog rabies variant has established enzootic rabies in wolf, jackal, and wild-dog populations. In South Africa, a distinct strain of rabies is maintained in the yellow mongoose (Cleveland, 1998). Surveillance and reporting in Asia is sporadic, but the disease is present in foxes, jackals, and wolves (Kaplan, 1985). In South America distinct rabies variants exist in some terrestrial wildlife, but less is known about these because of the importance there of dog rabies (Krebs et al., 1995). Rabies is present in the Indian mongoose in Grenada, Puerto Rico, Cuba, and the Dominican Republic in the Caribbean (Everard and Everard, 1985).

In North America, rabies in terrestrial species is maintained in geographically defined outbreaks with a single antigenically or genetically distinctive variant. The variant is maintained by intraspecific disease transmission within the population of a single predominant vector species, after which the outbreak is named. Infection of other species in the area occurs, but these "spillover" cases have only rarely established sustained intraspecific transmission in another species. The disease may persist in the vector population for decades, and the geographic area affected can grow, diminish, or shift gradually or rapidly (Smith, 1996). The establishment of a rabies epizootic in a new area in a susceptible population can also be the result of human translocation of wildlife, exemplified by the introduction of raccoon rabies into the mid-Atlantic states in 1977 (Jenkins et al., 1998).

Raccoon rabies was first recognized in Florida in the 1940s and became endemic in the Southeast. This intense outbreak presently affects areas of the 15 eastern seaboard states as well as Vermont, West Virginia, Alabama, and Ohio. The epizootic has been characterized by a rapid northeasterly spread, very large numbers of rabid raccoons in newly affected areas, and a subsequent cyclic nature, without ever completely dissipating in areas once affected. Spillover to an exceptionally diverse group of other mammalian domestic and wild species, most commonly by far to striped skunks, has occurred from this outbreak. However, no human rabies cases from this variant have yet been recognized.

Skunk-vectored rabies enzootics exist across a broad region of the continent, including most of the central states and provinces, and in California. The viruses associated with this outbreak are actually composed of three genetically distinct lineages, maintained in the north central, south central, and Pacific coast areas (Orciari, 1995). Where rabies exists in other North American terrestrial vectors, there is significant spillover into skunks. The skunk cases in those areas result from infection with the outbreak-associated variant, and the distribution in skunks is limited to those areas where the disease exists in the primary vector.

Rabies exists in red foxes in Alaska, parts of northern New England, and the Canadian provinces of Ontario and Quebec. The variants are similar in all of the regions, as these outbreaks are vestiges of the southward spread of Arctic fox rabies that became established in the red fox and swept across northern areas of the continent during the 1950s (MacInnes, 1988). Two different variants of rabies virus persist in gray fox populations of west Texas and in Arizona (Smith, 1989).

A recent emergence of dog- and coyote-vectored rabies occurred in southernmost Texas, adjacent to a longstanding focus of canine rabies at the Texas-Mexico border. Beginning in 1988, the outbreak expanded northward to encompass most of south Texas. By 1999, oral rabies vaccination efforts had nearly extinguished this outbreak.

Spillover to other terrestrial mammal species, particularly to skunks and foxes, exists in each of these areas of rabies endemicity. Sporadic cases are also seen in these areas in groundhogs, bobcats, coyotes, beavers, otters, opossums, deer, and rabbits. Spillover to unvaccinated domestic animals, including cats, dogs, ferrets, cattle, horses, sheep, goats, and swine, also occurs. Rabies in large rodents, particularly woodchucks and beaver, is reported in areas of terrestrial rabies, especially in the midst of the raccoon rabies outbreak. Rabies is uncommon in other rodents and lagomorphs but has occasionally been confirmed in squirrels, rats, and domestic rabbits and even more rarely in prairie dogs, porcupines, chipmunks, and mice (Childs et al., 1997).

Rabies in bats and rabies in terrestrial mammals are independent cycles. The rabies variants isolated from bats are antigenically and genetically distinct from those associated with terrestrial rabies (Smith, 1989). In bats, classical rabies virus infections (genotype 1 lyssavirus) are limited in distribution to the western hemisphere, where these mammals are of major importance as reservoirs and transmitters of rabies virus infection. Vampire bats are found throughout the more tropical areas of Latin America, and cattle losses from vampire bat-transmitted rabies range from 30,000 annually in Mexico alone (Brass, 1994) to more than 1 million in all of Latin America (Fenner et al., 1993). Human mortality from vampire bat-transmitted rabies continues to be reported, with recent reporting of 19 deaths in Brazil in 1990 and 1991 (Schneider et al., 1996) and 29 deaths in two Peruvian jungle communities (Lopez et al., 1992).

Rabies is widespread in insectivorous bats of North America, both geographically and by species. The disease has been confirmed in 49 states and in most of the 39 bat species indigenous to the United States. In 1997, the 958 laboratory-confirmed rabid bats from 46 states were the greatest total since 1984 and accounted for 11.3% of all U.S. animal rabies cases for the year. Six states reported rabies in bats in 1997, but not in terrestrial species (Krebs et al., 1998). Most laboratory-confirmed rabid bats in the nation are not identified to the species level, but among those that are, most cases occur in the Mexican freetailed bat and the big brown bat, both widely distributed common commensal species. Like rabies in terrestrial mammals, virus transmission in bats is mainly intraspecific, and distinct rabies variants can be identified antigenically or genetically for specific bat species (Smith, 1996). The analysis of bat rabies variants has been hampered by a failure to identify the species of most bats tested and by submission biases that result in most bats that are tested being of the common commensal species. Nevertheless, associations of particular variants with the major colonial and migratory species have been described (Smith et al., 1995).

Occasional interspecific transmission of these variants is recognized among bats and, less frequently, to terrestrial mammals. Although terrestrial mammals infected with bat virus variants can have virus in their salivary glands during clinical disease (Trimarchi et al., 1986), there has been no evidence that infection with a bat virus variant has established an outbreak in a terrestrial species. While estimates of disease prevalence in samples from random collection of asymptomatic bats at roosts are generally less than 1%, rabies positivity rates among bats submitted to public health laboratories for testing range from 5 to 15% (Brass, 1994). In New York State from 1981 to 1998, 4% (1,052) of 26,250 bats submitted for testing after an encounter with a person or a domestic animal were rabid (Trimarchi, 1999). There is an obvious strong bias toward sick and injured bats in these public health samples. Nevertheless, these rates are relevant for public health decisions because they accurately reflect the likelihood of rabies infection in individual bats encountered under common circumstance by people and pets.

The closely related European bat lyssaviruses 1 and 2 (genotype 5 and 6 lyssaviruses), the Australian bat lyssavirus (genotype 7 lyssavirus), and the less closely related Lagos bat virus (genotype 2 lyssavirus) and Duvenhage virus (genotype 4 lyssavirus) of Africa broaden the distribution of rabies-related encephalitis associated with bats. Current information on their prevalence, vectors, and distribution is based on limited surveillance to date. These viruses are capable of producing rabieslike encephalitis in humans or other mammals and may eventually prove to be of greater epidemiological importance as we learn more about their natural history.

EPIDEMIOLOGY OF HUMAN RABIES

Worldwide, data regarding human rabies are largely unreliable. Underestimates are common because human rabies in developing areas is largely a rural problem, and many rural cases go undiagnosed or unreported (Nicholson, 1994). Human rabies is still common in developing, tropical countries of Africa, Asia, and Latin America, with an estimated 35,000 to 60,000 deaths annually worldwide. Because most of these deaths are due to rabid-dog bites and a lack of, or inadequate, rabies postexposure prophylaxis, the mortality from the disease is largely a problem of poverty and lack of access to health care (Meslin et al., 1994).

In the United States, human rabies deaths became uncommon following control of dog-vectored rabies in the early 1950s. There were 99 cases during the 1950s, 15 in the 1960s, 23 in the 1970s, 10 in the 1980s, and 27 from 1990 through 1998. Of the 37 deaths since 1980, 32.4% (12) of the cases were acquired through dog bites occurring outside the United States, 5.4% (2) were a variant associated with indigenous domestic dogs, 2.7% (1) were a variant associated with skunks, and 59.4% (22) were a variant associated with insectivorous bats. In only 22% (8) of the cases was there a definite history of a bite (Noah et al., 1998).

LABORATORY DIAGNOSIS

The rabies laboratory has long played a prominent role in rabies control. The laboratory diagnosis of rabies is most often the result of the postmortem examination of animals that have potentially exposed a human to the disease by a bite or other transdermal contact with saliva or neural tissue. Because the modern rabies laboratory can provide reliable results on the day of receipt of the specimen, the physician's decision to provide or withhold rabies treatment following a bite from a suspect animal commonly is based on the laboratory tests performed on the animal's brain. It is this function that dictates the uniquely high standards of sensitivity and specificity required of these tests, as a false-negative result could easily lead to a patient's death. Alternatively, if a delay in testing for rabies is unavoidable, it may be appropriate to initiate rabies treatment. Subsequent negative results from a reliable laboratory would justify terminating postexposure treatment.

Examination for evidence of rabies infection is also performed on samples from rabies-suspect animals that have potentially exposed domestic animals to infection and in the differential diagnosis of encephalitis in domestic animals, even in the absence of human or animal exposure. The rabies laboratory supports surveillance for the disease in wildlife to aid in the proper allocation and targeting of rabies control programs, such as the efforts to vaccinate wildlife with oral baits. The combined data generated by examination for all reasons are used to define epizootiologic patterns useful for animal bite management when the offending animal is not available for observation or testing.

Not all biting animals need to be killed and tested. Because of laboratory data and empirical knowledge of virus shedding periods prior to onset of signs of the disease, dogs, cats, and ferrets can be confined and observed for signs of rabies for 10 days following a bite to rule out rabies transmission (see "Human Rabies Prevention" below). This sometimes is extended to other domestic species on a case-by-case basis.

Wild animals, especially bats, foxes, skunks, and raccoons, that have bitten or otherwise potentially exposed a human to infection should be tested immediately. Because rabies is a disease affecting mammals, it is never necessary to test arthropods, amphibians, reptiles, or birds. In the United States, small rodents, including mice, rats, and squirrels, are essentially free of rabies, and therefore routine examination is not required; exceptions include rodents involved in unprovoked attacks in areas where rabies is endemic and larger rodents, such as woodchucks, muskrats, and beaver (Childs et al., 1997).

DIAGNOSIS OF RABIES IN ANIMALS

Rabies diagnosis can be achieved with 100% sensitivity only by the postmortem examination of brain tissue. Throughout most of the prolonged incubation period of rabies, there is no reliable means to rule out infection—there is no rise in circulating antibody titer, and neither rabies virus, its antigens, nor rabies RNA can be reliably identified. This is due to the limited and unpredictable distribution of the virus and its proteins during the retrograde axoplasmal movement from the site of the exposure to the CNS during rabies pathogenesis. Rabies virus, its antigens, and rabies RNA do not move centrifugally away from the CNS during the incubation period (Charlton, 1988). However, modern methods can identify the presence of rabies virus in the brain of a rabid animal that dies or is euthanized during, or up to several days before, the onset of the clinical signs of the disease. Therefore, after an exposure, when a decision is made to sacrifice and test the animal for rabies infection, it is never necessary to delay testing for further development of the disease to achieve a reliable diagnosis. Most importantly, centrifugal spread of the virus from the CNS to the salivary glands (and therefore a potentially infectious bite) does not precede the appearance of demonstrable rabies antigen in the brain (Charlton, 1988). Therefore, a negative result of brain examination by acceptable methods assures that the bite of the animal could not have caused an exposure to rabies.

COLLECTING, PREPARING, AND SUBMITTING RABIES SPECIMENS

The animal species, nature of the exposure, variant of rabies virus, and time and cause of death may affect the terminal distribution of rabies virus and its antigens in the brain. Generally, the brain stem, the cerebellum, and the hippocampi constitute the best diagnostic samples, and areas of each tissue are examined to provide a reliable negative

report. Therefore, the intact head is the preferred diagnostic sample for the postmortem diagnosis of rabies in animals. An animal can be euthanized for rabies examination with barbiturate or nonbarbiturate injectables or gases or by other humane means that do not damage the brain. The specimen must be preserved by refrigeration immediately and kept at refrigeration temperature (preferably in wet ice) until it arrives at the laboratory. Should refrigeration not be possible, freezing is an acceptable but less desirable alternative. A single freezing and thawing will not prevent reliable diagnosis, but freezing will make the dissection more difficult and may delay the test. Repeated freeze-thaw cycles can damage the specimen. The head should be removed from the neck before the first vertebrae with caution to avoid contaminating injury or the creation of infectious aerosols. Those capturing suspected rabid animals or handling and decapitating the carcass should receive rabies preexposure immunization (CDC, 1999a). Bats and other very small animals should be submitted intact to avoid damage to the animal's CNS during decapitation. For large livestock species, including cattle and horses, sample portions of the brainstem and cerebellum can be removed through the foramen magnum by a veterinarian (Debbie and Trimarchi, 1992).

The specimen should be directly and immediately transported to the laboratory. Specimens can also be shipped by a prompt parcel delivery service, if properly packaged and labeled. Some state laboratories provide standard rabies specimen containers for shipping heads to the laboratory. Alternatively, the specimen may be sealed in two heavy (4-mil) plastic bags, individually sealed by knotting. The double-bagged head should then be placed in another bag, similarly sealed, and also containing several hard-frozen gel refrigerant packs. This in turn must be contained in an inner Styrofoam box within a wax-treated outer cardboard box. An envelope attached to the outside of the container should contain a fully completed standard rabies specimen history form, if available. If no form is available, provide all the significant information, including the names, addresses, and telephone numbers of the owner of the animal, the complainant, and all humans in contact with the specimen. Also include information on the clinical observations, date of death or means of euthanasia, exact location of capture, and information on the person or agency to receive the report. Generally, reports of rabies-positive specimens are made immediately by telephone. Reporting practices vary widely, however, and the submitter should ascertain the local practice.

DISSECTION AND SAMPLE PREPARATION

When the entire head is received at the laboratory for examination, the flesh is removed from the cranium and an anterior and two lateral cuts are made in the cranium by chisel or saw to permit the calvarium to be reflected posteriorly. After removal of the meninges, the left and right horns of the hippocampus, the cerebellum, and the brainstem are removed. An alternative method, useful for surveillance-only examinations, employs the removal of a core of brain tissue by the insertion of a soda straw or similar hollow tube into the foramen magnum and advancing forward to capture samples of the brainstem, cerebellum, and hippocampus that can then be forced out of the straw and used for slide and suspension preparation (Barrat, 1996). Touch impressions or slip smears of each tissue are made on microscope slides immediately. Mainly for research purposes for examination of nonneural tissues, frozen sections of tissue can be prepared using a cryostat. A 10% suspension of a mixture of the diagnostic tissues is also made in suitable diluents for animal inoculation or cell culture virus isolation. A sample of each brain tissue is saved at $-70°C$ for further testing.

DIRECT IMMUNOFLUORESCENCE FOR RABIES ANTIGEN

With achievable sensitivity and specificity approaching 100% and a routine turnaround time of less than a day, the direct FAT is the preferred procedure for the diagnosis of rabies. Glass microscope slides containing slip smears or impression slides of brain tissue, frozen sections of other tissues, or monolayers of tissue culture cells are fixed for 1 to 4 h in acetone at $-20°C$ to assure permeability of the cells to the diagnostic antibodies and to aid in tissue adherence to the slide. An alternative microwave fixation process has been described (Davis et al., 1998). After being air dried, the slides are flooded with the diagnostic immunofluorescence reagent. These reagents are made by the conjugation of fluorescein isothiocyanate (FITC) to purified immunoglobulin G specific for rabies nucleocapsid protein. The antibodies can be extracted and purified from antiserum of rabies-hyperimmune hamsters, rabbits, or horses. The development of mouse monoclonal antibody technology has permitted the production of highly specific and uniform diagnostic reagents employing a cocktail of these antibodies specific for different epitopes on the rabies antigen (Wiktor et al., 1980). The conjugated reagent may also contain a counterstain, such as Evans blue (R. J. Rudd and C. V. Trimarchi, *Abstr. 8th Annu. Rabies Am. Conf.*, Kingston, Ontario, Canada, 1997). After a 30-min incubation at 37°C in a moist incubator, the diagnostic reagent is thoroughly washed off in multiple saline baths and a distilled water rinse. The FITC-conjugated antibodies remain attached to the brain film only where rabies nucleocapsid protein is present. After being air dried, the slides are examined with a light microscope fitted with a mercury vapor or xenon lamp and appropriate excitation and barrier filters to show FITC-labeled structures as yellow-green fluorescent objects against a dark background. Rabies-specific staining in brain tissue appears as characteristic round or oval intracytoplasmic inclusions that are most prominent in the large neurons of the cerebellum, hippocampus, and brainstem (Trimarchi and Debbie, 1991).

Rabies immunofluorescence testing of fresh brain tissue can be comparable in sensitivity or superior to isolation procedures. A true indication of the value and reliability of the FAT is the absence of any known human rabies cases in the United States when rabies postexposure treatment was withheld based on immunofluorescence-negative laboratory reports. Important factors in the avoidance of false-negative results in rabies diagnosis include strict adherence to a uniform methodology, the use of proper scientific controls for all procedures, diagnostic conjugate quality and proper dilution, and optimization of lamp and microscope performance. The specificity of the procedure can also approach 100% agreement with virus isolation when procedures and practice ensure avoidance of cross contamination and conjugate-diluent and wash salines and staining conditions are optimized to eliminate nonspecific fluorescence. The laboratory can use selected application of tests for specificity of observed fluorescence, including paired staining with virus-absorbed and sham-absorbed conjugates and staining with FITC-labeled immunoglobulin G conjugate specific for another antigen.

The diagnosis of rabies in animals is not a Clinical Laboratory Improvement Amendment 88 analysis. Therefore, proficiency testing is not mandatory. Voluntary rabies proficiency testing programs were conducted periodically from 1973 to 1992 and each year since 1994. The performance of rabies diagnostic laboratories enrolled in proficiency testing in the United States has been excellent when panels of rabies-positive and -negative test slides were evaluated by the direct FAT. Excellent consensus has been observed among participants for strong positive and negative test samples. Discrepancies have mainly occurred with very weakly positive slides (J. Powell, *Abstr. 8th Annu. Rabies Am. Conf.*, Kingston, Ontario, Canada, 1997). One of the most important factors for efficacy of the rabies FAT is the laboratory's recruitment and retention of properly trained and experienced microscopists (Hanlon et al., 1999).

A critical requirement for sensitive rabies diagnosis by immunofluorescence is a good-quality brain sample. Brain tissue exposed to chemical fixative, repeated freeze-thaw cycles, or elevated temperatures may result in denatured or masked rabies antigens, hampering recognition by the diagnostic reagents. Decomposition will affect the sensitivity of all rabies diagnostic procedures. Immunofluorescence tests may remain positive for a period after isolation of virus is no longer possible (Rudd and Trimarchi, 1989). Evidence of rabies infection based on FAT of decomposed or mutilated tissue fragments may support a valid rabies-positive report, confirmable by isolation, immunohistochemistry, or molecular methods. One of the most difficult diagnostic decisions confronting the rabies laboratorian is to determine at what stage of decomposition it is no longer possible to issue a reliable negative rabies report. Certainly, once the CNS has become foul smelling and green, with some liquefaction, negative results are not reliable. Occasionally, even specimens with the appearance of only early decomposition on gross inspection may result in slides with each microscopic field so overgrown with autofluorescing decomposition bacteria that reliable examination is not possible. Mutilation of the tissue by trauma to the extent that the necessary sample regions of cerebellum, brain stem, and hippocampus are not available or are unrecognizable also precludes a reliable negative result. Generally, it is not possible for laboratory personnel, based solely on a verbal description, to determine before the dissection whether a decomposed or mutilated specimen is testable. Consequently, unless it is clear that the carcass is in the last stages of decomposition, or it is mutilated to the extent that no recognizable CNS exists, it is wise to submit compromised specimens to the laboratory for evaluation of suitability.

OTHER METHODS OF ANTIGEN DETECTION

Antigenic sites recognized by anti-nucleocapsid diagnostic reagents in the standard FAT may be masked by bonds created when CNS tissues are fixed by exposure to formalin. As a result, the sensitivity of the procedure is greatly reduced for fixed tissues. Advances have recently increased the sensitivity of direct and indirect immunofluorescence and immunohistochemical procedures applied to fixed tissue. Digestion of the tissue with enzymes such as proteinase prior to staining can expose the antigenic sites (Warner et al., 1997). Use of avidin-biotin complex amplification and high-affinity monoclonal antibodies for the first label have improved the performance of these procedures so that they may be approaching the reliability of FAT on fresh tissues (Hamir et al., 1995). Further evaluation is necessary before negative results for chemically fixed tissues can be used for public health decisions. When applicable, confirmatory testing of fixed tissues by another sensitive technique, such as reverse transcription (RT)-PCR, would be appropriate.

Rapid and simple diagnosis of some viral infections can be achieved by using enzyme-linked immunosorbent assays (ELISA). A method has been developed to demonstrate the presence of rabies antigen, and recent methods utilizing avidin-biotin amplification show promise of improved sensitivity, approaching that of FAT and virus isolation (Bourhy and Perrin, 1996). They offer the advantages of automated reading and good reliability for partially decomposed tissues, offering applicability under field conditions. While these techniques can be used as a backup to FAT examination and for epizootiologic surveillance testing, they are not widely used in U.S. public health laboratories.

HISTOLOGIC EXAMINATION

Negri bodies are intracytoplasmic, acidophilic inclusion bodies that can be demonstrated best in the Purkinje cells of the cerebella and the pyramidal cells of the hippocampis (Perl and Good, 1991) of many rabies-infected animals by microscopic examination of tissue stained with basic fuchsin and methylene blue (Tierkel and Antanasiu, 1996) or hematoxylin and eosin (Lepine and Atanasiu, 1996). The presence, distribution, and size of Negri bodies is related to the species of animal, variant of rabies virus, and duration of the clinical period before death or euthanasia. Demonstration of these pathognomonic inclusions is very specific and, coupled with evidence of encephalitic inflammatory response, may provide a reliable positive rabies diagnosis when reported by an experienced pathologist. However, the sensitivity of the method is poor, with numerous histologic surveys indicating that 25% or more of rabid animals have no demonstrable Negri bodies, severely limiting the value of this diagnostic method for medical decisions (Perl and Good, 1991).

VIRUS ISOLATION

The dire consequences of false-negative results in cases of human exposure support a continued practice of utilizing rabies virus isolation as a backup, confirmatory procedure, despite the proven sensitivity and reliability of the FAT. It may be employed as a general quality control procedure or applied only in instances of bites to humans from highly suspect animals. In either case, it serves to sustain confidence in the reliability of the FAT results and to relieve the microscopist of the full burden of responsibility. The propagation of virus in the laboratory is also a critical component for identification of virus variants and in the production of diagnostic reagents. In vitro and in vivo methods are both widely employed.

The mouse inoculation test (MIT) was introduced for diagnostic purposes in 1935. It is a very sensitive and reproducible procedure. When used as a confirmatory procedure for FAT results, portions of the same tissues that are used for the microscopic examination are ground into a suspension, employing a diluent of physiologic salt solution containing serum and antibiotic supplements. The suspension is inoculated intracerebrally into weanling Swiss albino mice. The mice are carefully observed daily for 30 days for evidence of rabies infection. Mice that develop illness during the obser-

vation period are immediately euthanized, and the brains are examined by FAT. A valuable attribute of the MIT is its ability to detect small quantities of rabies virus even in very weakly positive specimens. It can also be applied to mutilated and decomposed samples. Its weakness lies in the typical 7- to 18-day period between inoculation and recognized illness in the mice. The limitation of the procedure results from the possibility that if treatment were withheld following a bite due to a false-negative FAT, detection by this backup method would occur after a period that would cause great concern about vaccine failure (Rudd and Trimarchi, 1980). The period can be shortened by the use of neonates and with daily sacrifice and FAT examination of numerous individual animals. However, this greatly increases the labor-intensive nature of the procedure and can be prohibitive in a laboratory performing routine diagnoses on large numbers of specimens.

The delay associated with the demonstration of virus by the MIT can be avoided by the isolation and identification of rabies virus on continuous cell culture (Rudd and Trimarchi, 1989). Tissue from the diagnostic regions of the brain of the suspect animal is ground into a suspension in a cell culture medium as the diluent. The suspension is incubated for one to several days after inoculation on a monolayer of cells of a continuous cell line selected for its susceptibility to infection with rabies virus, generally a mouse neuroblastoma cell line. The sensitivity of the procedure can be increased with the treatment of the cells with DEAE-dextran (Kaplan et al., 1967). The cell monolayers are then washed, fixed in cold acetone or a formalin-methanol mix, and examined by FAT. If infectious rabies virus is present in the brain tissue, characteristic intracytoplasmic inclusions of rabies antigen will be observed in fluorescent foci in the cells. The sensitivity of the procedure is comparable to that of the FAT and the MIT (Webster and Casey, 1996). Because results are available within a few days of receipt of the specimen, it serves as a much better means of confirming negative FAT results, as a false-negative FAT would be recognized in a period permitting timely initiation of PET.

MOLECULAR METHODS

Direct molecular methods for the diagnosis of rabies employ probes for the presence of existing rabies virus RNA in tissue samples by using dot or blot hybridization assays. These methods generally lack sufficient sensitivity to show the presence of rabies RNA among the total RNAs of the sample, except in the more heavily infected samples (Tordo et al., 1995).

RT-PCR is a more sensitive and useful technique. RNA extracted directly from infected brain material is reverse transcribed to cDNA, which is then amplified by PCR. RT-PCR requires primers, or short synthesized oligonucleotide sequences derived from conserved regions of the rabies virus genome, usually from the nucleoprotein, or N protein, gene. The optimal sensitivity is achieved with a nested PCR, in which a second round of amplification is performed on the initial PCR product using primers internal to the original primers (Kamolvarin et al., 1993). The products of PCR amplification of the rabies genome can be detected and analyzed in numerous ways, including direct visualization in agarose gels stained with ethidium bromide following electrophoresis. Indirect detection can be done using DNA probes revealed by radioactive labels, digoxigenin immunologic detection, or enzymatic revelation with indicators such as alkaline phosphatase.

Molecular methods have important applications in the rabies laboratory: in the antemortem diagnosis of human rabies (Noah et al., 1998), as a sensitive backup test to FAT, and, when paired with restriction fragment length polymorphism or with nucleotide sequencing analysis, in the epizootiologic investigations of rabies (Nadin-Davis et al., 1996). Currently, its use in routine rabies diagnosis is limited. For a number of reasons, the FAT is unlikely to be replaced by RT-PCR. The FAT for rabies antigen in brain tissue is rapid, sensitive, specific, easy to perform, and relatively inexpensive. In the United States, no human case of rabies has ever been attributed to contact with an animal found negative for rabies by FAT of brain material. Tests for the sensitivity and specificity of FAT by comparison to virus isolation approach 100% (Smith, 1999). While nested PCR can be 100 to 1,000 times more sensitive than FAT, this extreme sensitivity brings with it a need for extraordinary quality control practices to avoid false-positive results. Small fragments of RNA generated by tissue processing during necropsy, or transferred from sample to sample during RNA extractions, could generate false-positive results during the 100,000 or more public health rabies examinations conducted annually in laboratories throughout the United States. Furthermore, primer selection and the identification of truly universal primer sets for rabies and rabies-related viruses may still be a barrier to withholding rabies treatment based solely on negative PCR results.

DIAGNOSIS OF HUMAN RABIES

Despite the dire prognosis in rabies infection, testing should be done in all cases of acute, progressive human encephalitis of unknown etiology, even in the absence of a history of bite exposure (CDC, 1997). As a result of the efficacy of modern postexposure treatment regimens, human rabies cases in the United States and other developed nations are no longer commonly associated with vaccine failure. Also, in these regions, rabies prophylaxis is provided in all cases of known exposure, and even in most cases of suspected exposure, to rabies. Therefore, human cases are most often identified in the absence of a clear history of suspicious animal bite or other exposure. In numerous recent human rabies cases in the United States, the disease was not suspected or diagnosed during the clinical illness and was only recognized postmortem, and sometimes after a lengthy delay (Noah et al., 1998).

Antemortem diagnosis is a valuable tool to permit early identification and postexposure treatment of family and health care staff potentially exposed by contact with the patient's saliva. It also aids in patient management and allows efforts to prepare the family for the invariably fatal outcome of the disease. Postmortem diagnosis of rabies in cases of fatal encephalitis of unknown etiology is critical to gain greater knowledge of the prevalence of rabies encephalitis in humans, the frequency of failure of pre- and postexposure vaccination, and the probable vectors and variants that pose the greatest risk to human health. The recognition of each human rabies case is very important in identification of the highest-risk exposure routes, vectors, and rabies virus variants—information essential to the development of effective rabies control and exposure management protocols.

Although brain biopsy would be the most sensitive antemortem diagnostic method, the risks associated with the procedure make its use uncommon. There are numerous less invasive intravitam tests that can confirm rabies infection.

However, rabies virus, its antigens, and rabies RNA do not move centrifugally away from the CNS during the incubation period and move only slowly during the clinical period (Charlton, 1988). Similarly, humoral antibody responses do not occur during the incubation period and are generally not demonstrable until the second week of illness. Antemortem diagnosis is therefore attempted by the analysis of numerous tissues by several methods, searching for rabies-specific antibody in the serum or cerebrospinal fluid (CSF) or for viral antigen, live virus, or viral RNA in body fluids (saliva), peripheral nerves (skin biopsy), or epithelial cells (corneal impression). Antigen detection can best be accomplished by FAT performed on a full-thickness skin biopsy specimen taken from the nape of the neck and including several hair follicles (Smith, 1999). FAT can also be used to demonstrate rabies antigen in corneal-impression slides (Zaidman and Billingsly, 1998). It is recommended that these samples be taken by an ophthalmologist because of the risk of corneal abrasion. Virus isolation by MIT or cell culture inoculation can be applied to saliva and CSF. Antibody assay for this purpose can be performed by neutralization test, ELISA, or indirect FAT on serum and CSF. Nested RT-PCR is applied to saliva, skin biopsy specimens, corneal impressions, or CSF. Demonstration of rabies antigen by FAT in any tissue, rabies RNA in saliva, or rabies antibody in serum and CSF or isolation of rabies virus from any tissue is confirmatory of rabies virus infection. Antibody in serum alone may not be indicative of clinical rabies in a patient with a known or unclear history of rabies vaccination or very recent treatment with rabies immune globulin (RIG).

The samples for antibody assay are 1 ml or more of CSF and serum, submitted in a plastic tube or vial. The skin biopsy specimen can be submitted on a gauze sponge moistened with sterile physiologic saline and sealed in a small plastic container. Corneal-impression slides should be submitted in a plastic slide container with the surface of the slide containing the impression clearly marked. A 1-ml sample of frank saliva should be collected in a plastic sputum jar. Alternatively, a buccal swab can be taken and submitted immersed in a tube containing 1 ml of sterile saline. All of these samples can be stored at −70°C and shipped on dry ice. Postmortem testing methods for human rabies are similar to those described for animals. If the patient dies and an autopsy is performed, the ideal specimens for postmortem diagnosis are 1-cm³ samples of unfixed cerebellum, brain stem, and hippocampus preserved by refrigeration.

The laboratory should be contacted immediately when human rabies is suspected. If original samples collected in the first week of symptoms do not disclose evidence of rabies infection, it must be understood that this does not conclusively rule out rabies and that repeat samples may be necessary—antemortem tests may remain negative well into the clinical period. Review of 32 human rabies deaths in the United States from 1980 to 1996 disclosed that in 12 of the cases rabies was not suspected until after death. Of the remaining 20 cases, antemortem evidence of rabies was found by one or more tests in 18 cases. Antibody to rabies was detected in 10 of the 18 cases, virus was isolated from saliva in 9 of 15 cases, and RNA was detected by PCR in each of the 10 patients tested. Antigen was demonstrated by FAT in skin in 10 of 15 patients tested and in corneal impressions in 2 of 7 patients tested, and antigen was present in brain biopsy specimens in all three patients examined in that manner (Noah et al., 1998).

ANTEMORTEM DIAGNOSIS OF ANIMAL RABIES

It is possible to apply the methods described for antemortem diagnosis of human rabies to suspect animals as well. Skin biopsy has been demonstrated to be particularly sensitive when applied to biopsy specimens taken from the snouts of terrestrial carnivores, which permits examination of the innervation of the tactile hairs (Blenden et al., 1983). However, the same limitations apply to antemortem diagnosis in animals: since all tests can remain negative well into or throughout the clinical period, negative results do not rule out rabies. Rabies cannot be reliably diagnosed by current methods during most of the incubation period—claims that molecular testing of the saliva of a biting animal can be used for decisions on bite management are false, as rabies virus may be shed in saliva sporadically prior to and during the clinical period (Charlton, 1988). Therefore, these tests are of little value for public health decisions and should never be a substitute when circumstances require 10-day observation or euthanasia and examination of brain tissue.

RABIES VIRUS VARIANT TYPING

Methods that characterize the antigenic and genetic attributes of isolates of rabies virus and the rabies-related members of the genus *Lyssaviruses* now enable the laboratory to identify the virus variants responsible for epizootics as well as individual cases of rabies in animals and humans. Distinctive differences in the variants responsible for the major terrestrial outbreaks worldwide, the numerous bat rabies variants, the laboratory strains of rabies virus, and the rabies-related lyssaviruses are distinguishable by these means. Rabies variant identification yields a greater knowledge of the epizootiologic relationships of virus and vectors, allowing the development of more effective animal contact guidelines and rabies control strategies. Reaction patterns in indirect FATs, employing panels of monoclonal antibodies specific for unique viral nucleocapsid epitopes, permit such discrimination (Smith et al., 1986). The immunofluorescence assays can be performed on brain tissue of the original subject or mouse- or cell-culture-passaged virus. Genetic analysis permits more precise detail of the evolutionary relatedness of isolates, investigation of the spatial and temporal changes that may occur, and, particularly, the measure of similarity among virus isolates. This is accomplished by the extraction, transcription, and amplification of the RNA of an isolate by RT-PCR and subsequent sequence analysis of the cDNA nucleotide or deduced amino acid sequence for the entire or partial nucleocapsid or glycoprotein genes. Using computer algorithms to perform pairwise comparisons, estimates of genetic identity can be calculated and expressed as percent homology among isolates (Smith et al., 1992).

RABIES ANTIBODY ASSAY

Assays to demonstrate and quantify rabies antibody in sera of humans or animals serve numerous functions in rabies diagnosis and control. Serologic testing for rabies-specific antibody titer is performed on human serum to determine the response to pre- and postexposure vaccination and to determine the timing of booster vaccinations to maintain a rabies-immune status. Evidence of rabies antibody in serum and CSF is used in the antemortem diagnosis of rabies in humans. Antibody detection in human and animal vacci-

nees is used in vaccine efficacy trials and in evaluation of field wildlife vaccination campaigns. Long quarantines for cats and dogs entering rabies-free countries can be reduced if an immunologic response can be confirmed by this test (WHO, 1992).

Because these antibody tests are generally employed to measure immune status, the most widely used assays measure neutralizing antibodies. Constant virus, varying serum dilution neutralization assays are most commonly employed. Dilutions of heat-inactivated serum are combined with a standardized amount of rabies virus and then incubated for 1 h at 37°C. Mouse or cell culture inoculation is performed after incubation to demonstrate residual virus after incubation. The mouse inoculation test developed in 1935 (Webster and Dawson, 1935) is a very reproducible method, still employed in some laboratories, and is used as the standard to evaluate other procedures.

The commonly used cell culture virus neutralization techniques are those in which residual virus is demonstrated by immunofluorescence in the inoculated cell monolayers. The most widely employed test for human postvaccinal titer determination is the rabies fluorescent focus inhibition test (RFFIT) (Smith et al., 1996). This technique determines the serum neutralization endpoint titer by a mathematical interpolation of the number of microscopic fields containing fluorescent foci at serum dilutions of 1:5 and 1:50. While results may be given as reciprocals of the calculated endpoint dilution, they are often expressed in terms of international units of neutralizing activity, determined for each test by the titration of a standard reference serum. Alternative in vitro methods that actually determine the last dilution of the patient's serum that neutralizes the virus challenge in a manner similar to the standard mouse test also are utilized and generally report titers in international units (Trimarchi et al., 1996). One of these tests, the fluorescent-antibody virus neutralization test, has been selected by the Office International des Epizooties as the standard of examination of animal sera for evidence of successful vaccination prior to movement into rabies-free areas (Office International des Epizooties, 1996). A recent comparison of the sensitivities of the fluorescent-antibody virus neutralization test and the RFFIT in one laboratory concluded that they are comparable if quality control measures are stringent, particularly assurance that a well-defined challenge virus, obtained from the same source, was utilized (Briggs et al., 1998).

Other serologic tests are employed for rabies antibody assay, particularly for research and vaccine evaluation procedures. High degrees of sensitivity and specificity are reported with an ELISA directed against whole virus for the measurement of neutralizing antibodies (Savy and Atanasiu, 1978). The rabies G protein should serve as the immunosorbent for the ELISA. A competitive ELISA, employing a neutralizing monoclonal antibody, reportedly achieves a high degree of sensitivity, freedom from the need for species-specific intermediate antibodies, and measurement of neutralizing antibody (Elmgren and Wandeler, 1996).

RABIES CONTROL

If the control of rabies is defined as elimination of human mortality from the disease, then it can be achieved by success with the following strategies, individually or in combination: elimination of the virus in animal populations, elimination of human exposure to infected animals, prevention of human infection by prior or postexposure vaccination, or development of an efficacious cure for clinical disease. Because rabies virus has demonstrated adaptability to a wide variety of host populations, eradication of the virus, as has been achieved for smallpox, may not be a realistic goal. Control in domestic-dog populations and some wildlife vectors in geographically defined areas has proven to be achievable by stray dog control coupled with vaccination programs and by oral rabies vaccination campaigns for wildlife. However, resource limitations, an increasing movement of domestic and wild animals, and an expanding role and distribution of rabies and rabies-related viruses in bats threaten to make even regional elimination unlikely for most of the world. Modern cell culture vaccines and human RIG (HRIG) and purified RIG have made safe and highly efficacious pre- and postexposure immunization a reality, and yet resource limitations prevent application of these methods for much of the world's population (Meslin et al., 1994).

In North America, Europe, and other developed areas, great reduction in human mortality has been achieved by the virtual elimination of dog-vectored rabies outbreaks by vaccination programs (often compulsory) and leash and stray dog measures. Further reduction has accrued from the development and availability of modern biologics for preexposure treatment and PET, as human cases almost never are associated with vaccination failures in these areas. Other components of the rabies control effort include education of the public and health care professionals regarding exposure avoidance, the proper management of potential exposure to rabies, including animal confinement and observation, prompt and accurate rabies diagnosis, and accessibility to prompt and proper PET or preexposure vaccination when warranted.

Wildlife rabies control by vector population reduction has only rarely proven to be effective, and the North American experience has not been encouraging (MacInnes, 1988). However, reduction or elimination of wildlife rabies epizootics has proven to be achievable in some situations by wildlife vaccination. Oral rabies vaccination has controlled fox rabies in large areas of Europe and Ontario. A modified live-virus vaccine has been used in Canada, and efforts in Europe have used baits containing either modified live virus or a recombinant, vaccinia-vectored vaccine. The baits have been distributed by hand, helicopter, or fixed-wing aircraft in one or two campaigns per year in programs that have been continued for many years. The modified live-virus vaccines used in Canadian and European baiting programs do not vaccinate the most common terrestrial rabies vectors of the United States (raccoons and skunks) well by the oral route (Rupprecht et al., 1989). A recombinant vaccinia-rabies glycoprotein (V-RG) has proven safe and efficacious in laboratory and field trials (Rupprecht et al., 1993) and was licensed by the U.S. Department of Agriculture in 1997 for use in oral rabies vaccination programs for raccoons conducted by state and federal agencies. It also has been used with apparent success to control coyote vectored rabies in south Texas (G. Fearneyhough, *Proc. IX Int. Meet. Res. Adv. Rabies Control Am.*, abstr. 34, p. 961, 1998).

MANAGING DOMESTIC-ANIMAL EXPOSURES TO RABIES

Preexposure rabies vaccination of domestic animals is presently achievable with a variety of highly immunogenic and efficacious cell culture vaccines. Those licensed for use in the United States are now all killed-virus vaccines and are listed along with the terms of administration annually in the *Compendium of Animal Rabies Control* prepared by the National Association of State Public Health Veterinarians (National Association of State Public Health Veterinarians, 1999). Rabies vaccines are presently licensed for use in dogs, cats, cattle, horses, sheep, and ferrets. An animal is currently vaccinated 1 month after primary vaccination, which establishes a current vaccination status for the remainder of 1 year. A booster dose is required at 1 year after primary vaccination to continue vaccinated status, and that booster and subsequent doses may provide vaccinated status for up to 3 years, depending on the vaccine and species. Animals exposed to rabies that are not current on rabies vaccination status should immediately be euthanized. If the owner is unwilling to have this done, the animal must be strictly isolated for 6 months. Animals that are currently vaccinated should be revaccinated immediately following exposure to rabies.

HUMAN RABIES PREVENTION

The prompt management of potential human exposure to rabies is a critical component of human rabies prevention. All potential rabies exposures should be evaluated for rabies risk based upon the nature of the contact, the species of animals involved, the circumstances of the incident (e.g., behavior of offending animal or provoked or unprovoked encounter), and, the current local status of animal rabies and the adequacy of surveillance. The rabies vaccination status of the animal should not be used to rule out the need for further consideration of rabies transmission. Not all biting animals need to be killed and tested. A dog, cat, or ferret that has bitten a person but is wanted by the owner and is not demonstrating signs of rabies infection can be confined and observed daily for 10 days. Because of knowledge regarding virus-shedding patterns in the days preceding onset of rabies-specific signs (Niezgoda et al., 1997), survival of the animal without rabies onset for 10 days after the bite rules out the need for postexposure prophylaxis. If, however, the animal dies or signs of rabies develop during the observation period, it must be immediately euthanized if necessary and examined. Similarly, when a 10-day confinement of the offending animal is not possible (if it is symptomatic or has died) or is inappropriate (the offending animal is a wild or exotic species or hybrid), the animal must be humanely euthanized if required and tested.

The Advisory Committee on Immunization Practices (ACIP) of the U.S. Department of Health and Human Services says that exposure to rabies occurs when infectious virus is introduced into bite wounds or open cuts in the skin or onto mucous membranes. Any penetration of the skin by the teeth, regardless of the location of the wound, must be considered a bite exposure. Nonbite exposures have occasionally led to rabies infection (Afshar, 1979), with exposure to infectious aerosols or corneal transplants from rabies victims carrying the greatest risk. Direct contamination of an open wound (abrasion or scratch) or mucous membrane with saliva or other potentially infectious material (e.g., neural tissue) from a rabid animal is also considered rabies exposure. Contact with blood, urine, or feces (including bat guano), or merely petting a rabid animal, is generally not an indication of exposure. Rabies virus in saliva on environmental surfaces is quite labile; therefore, if the material on a surface is dry, it generally can be considered noninfectious (CDC, 1999a). Generally, bites or other contact with a rabid animal occurring more than 10 days prior to the recognized onset of signs of rabies are not considered potential rabies exposure.

From 1990 through 1998 there were 22 deaths from indigenously acquired human rabies in the United States. Despite the existence of numerous and widespread terrestrial rabies outbreaks in the country during the period, a surprising 91% (20) of these cases were caused by bat rabies variants (CDC, 1999b). In only one of these cases was there a confirmed bite exposure. In many others there had been an indoor bat encounter under circumstances where a bat bite, with its limited injury in comparison to bites of terrestrial carnivores, may have gone unrecognized. As a result, the ACIP has developed specific language regarding bat encounters and rabies treatment: "Rabies postexposure prophylaxis is recommended for all persons with bite, scratch, or mucous membrane exposure to a bat, unless the bat is available for testing and is negative for evidence of rabies. Postexposure prophylaxis might be appropriate even if a bite, scratch, or mucous membrane exposure is not apparent when there is reasonable probability that such exposure might have occurred" (CDC, 1999a). One must be careful not to assume that merely being in close proximity to, or in the same room with, a bat constitutes an exposure. Particular concern must be directed to those situations in which contact was possible but where there was a reasonable probability a bite may have gone unnoticed (Debbie and Trimarchi, 1998). Examples of scenarios justifying consideration of rabies exposure include a sleeping person awakening to find a bat in the room or an adult witnessing a bat in the room with a previously unattended young child or mentally handicapped person.

Because it is very often difficult to accurately reconstruct details immediately following an indoor bat encounter and to evaluate the likelihood of a reasonable probability of exposure, it is prudent to recommend the capture and retention for testing of bats involved in incidents with the potential for human exposure. As rabies positivity rates among insectivorous bats encountered by the general public typically are 3 to 5% (Childs et al., 1994), more than 90% of the bats encountered in potential-exposure circumstance will test negative for the disease, eliminating the need for further difficult investigations or decisions and avoidable PET. This practice will also help ensure that PET is provided to those who have actually had contact with the relatively small proportion of encountered bats that are rabid (Debbie and Trimarchi, 1998).

HUMAN RABIES PROPHYLAXIS

The relatively long incubation period in rabies infection permits efficacious PET, and modern biologics have afforded the potential for 100% success with proper wound treatment and prompt and appropriately administered immune globulin and a course of vaccine. An immediate and thorough washing of bite wounds with soap and water and a viricidal agent are valuable measures for the prevention of rabies (Griego et al., 1995). With the exception of patients with previous immunization, rabies PET should always include passive immunization with HRIG to neutralize virus at the site of exposure and active immunization with vaccine to produce neutralizing antibody that develops in 7 to 10 days after vaccination is initiated. Active immunization also triggers a cell-mediated response that is critical to the success of PET. The RIG is administered only once, at the beginning of the prophylaxis, but if there is a delay in administering the RIG it can be given through the seventh day after the administration of the first dose of vaccine. The recommended RIG dose is 20 IU per kg of body weight. If anatomically feasible, the full dose should be infiltrated into and around the wounds. Any remaining volume or, for treatments for exposures with no recognizable wounds, the entire dose, should be injected intramuscularly at one or more sites distant from the site of vaccine inoculation. Three immunologically comparable vaccines grown in three differently derived cell culture systems are currently licensed for use in the United States: a human diploid cell vaccine (HDCV), a fetal rhesus lung diploid cell culture vaccine, and a purified chicken embryo cell vaccine. All are packaged for and administered as a 1-ml intramuscular dose for pre- or postexposure administration. The HDCV is also approved and marketed as a 0.1-ml intradermal dose, solely for use in preexposure vaccination. For PET, any of the three vaccines can be used in a five-dose course, with 1-ml intramuscular injections given in the deltoid area (or for small children the anterolateral aspect of the thigh) on days 0, 3, 7, 14, and 28, commencing as soon as possible after the exposure is known. PET for previously vaccinated individuals is composed solely of two doses of vaccine administered as a 1.0-ml intramuscular injection in the deltoid region on days 0 and 3, commencing as soon as possible after the exposure is known. RIG is not administered to previously vaccinated patients. For this purpose, previously vaccinated status applies only to individuals who have received one of the recommended pre- or postexposure regimens of the currently licensed vaccines or who received another vaccine or regimen and had a documented adequate neutralizing antibody titer (0.5 IU or greater or complete neutralization at 1:5 on an RFFIT).

Although several other efficacious cell culture vaccines are widely available outside the United States, nervous-tissue vaccines and immune serum of equine origin are still employed in some areas. A purified RIG of equine origin has been used effectively in developing countries (Wilde et al., 1989). Several alternative PET schedules are employed outside the United States, including multisite regimens that are accelerated by administration of more than one dose on the first day of treatment by either intramuscular or intradermal inoculation. These regimens may induce early antibody response, which is beneficial where RIG is not available, and can reduce the amount of scarce vaccine required (Dreesen, 1997).

Preexposure rabies vaccination is available for persons at risk of rabies exposure, such as veterinarians, animal control officers, animal handlers, rabies laboratory workers, others whose activities bring them into contact with a rabies vector species, and certain travelers. Preexposure vaccination does not preclude the need for PET following a known exposure but eliminates the need for RIG and reduces the number of vaccine doses to two. Furthermore, a previously vaccinated individual might be protected from inapparent exposures or when PET is unavoidably delayed (CDC, 1999a). The primary preexposure regimen consists of three 1-ml intramuscular or 0.1-ml intradermal injections given one per day on days 0, 7, and 21 or 28. Only the HDCV is currently licensed and packaged for intradermal use. The intradermal product is not licensed for postexposure use in the United States and should not be administered concurrently with antimalarial prophylactic drugs (Bernard et al., 1985). Depending on the risk category, booster vaccinations may be recommended when rabies-neutralizing antibody titer is less than 0.5 IU or when there is less than complete neutralization at a 1:5 serum dilution by RFFIT.

Routine determination of an adequate serum rabies-neutralizing antibody titer following primary pre- or postexposure vaccination is not required unless the person is immunosuppressed. Persons at the highest risk of inapparent exposure, such as those working in rabies research or vaccine production laboratories, should have a titer determination every 6 months. Others at frequent risk, such as those working with mammals in areas where rabies is enzootic, should be serologically tested at 2-year intervals. A single booster vaccination is recommended when the titer is less than 0.5 IU or when there is complete neutralization at 1:5 on the RFFIT. The ACIP also recommends that because certain travelers and animal workers in areas with low animal rabies rates are in an infrequent-exposure category, they do not require routine preexposure booster doses following primary vaccination (CDC, 1999a).

Reactions following vaccination with the currently licensed cell culture vaccines and administration of HRIG are less serious and less common than with previously available biologics. Mild local reactions, such as pain, erythema, and itching are not uncommon, having been reported in 30 to 74% of vaccinees (Noah et al, 1996). Mild systemic reactions, including headache, nausea, abdominal pain, muscle aches, and dizziness, are reported in 5 to 40% of recipients. Guillain-Barré syndrome-like neurologic illness and other CNS and peripheral nervous system disorders temporally associated with HDCV administration have not been linked in a causal relationship with the vaccination (Tornatore and Richert, 1990). A delayed (2 to 21 days postinoculation) immune-complex-like reaction was reported among 6% of patients receiving booster doses of HDCV. The reaction included hives, sometimes accompanied by arthralgia, arthritis, angioedema, nausea, vomiting, fever, and malaise, but the reaction has not been life threatening (Dreesen et al., 1986). No fetal abnormalities have been associated with rabies vaccination, and pregnancy is not a contraindication (Chutivongse et al., 1995).

RESEARCH DIRECTIONS

At its simplest, there are two remaining ultimate goals of rabies research: the development of a cure for the disease after symptoms develop and the eradication of the virus. Although eradication would eliminate the need for a cure, worldwide elimination of the virus may be virtually unattainable because of the myriad of variants of rabies and rabies-related viruses that cycle in numerous and changing terrestrial and chiropteran vectors.

We now have effective means to combat rabies transmission to humans. Rapid and reliable postmortem diagnosis of rabies is readily achievable. Highly effective and safe biologics exist for the prevention of rabies in humans and domestic animals by pre- and postexposure vaccination. Efficacious dog vaccines and stray dog control programs have proven to effectively reduce human rabies deaths to rare events by eliminating the dog as a significant vector of the disease, as has occurred across North America and Europe. Encouraging evidence is mounting for the control of rabies outbreaks in wildlife over large areas by oral rabies vaccination strategies employing vaccine-laden baits distributed in the environment. Sadly, resource limitations, and to a lesser degree politics and societal resistance, have prevented successful implementation of these strategies in the areas of the world accommodating the vast majority of the world's population. As a result, human mortality from rabies infection is still commonplace in all but a relatively small number of developed nations. Nervous tissue vaccines, which are asso-

ciated with much higher rates of postvaccinal adverse events that far outweigh their one advantage of low cost, still constitute 25% of all human rabies vaccine produced worldwide (Dreesen and Hanlon, 1998). The efforts to muster the resources required to effect affordable rabies control programs worldwide can benefit from economic analysis to make optimal resource allocation decisions (Meltzer and Rupprecht, 1998).

Sufficient resources will remain unattainable for a successful cosmopolitan rabies control program if the present costs to produce, store, and deliver therapeutic biologics are not drastically reduced. Consequently, a priority research imperative for rabies control is the development of safe, effective, inexpensive, and mass-producible products for active and passive immunization. Recombinant vaccine has already been instrumental in rabies control, as a vaccinia-rabies glycoprotein recombinant has been used in oral rabies vaccination programs in Europe and the United States (Brochier et al., 1996). The vaccination of free-ranging wildlife and dogs could be greatly simplified with nonpathogenic vector viruses that are transmitted horizontally by host-to-host spread of infection or vertically in the host population by transgenic technology. The transgenic approach has already been utilized to produce tomatoes expressing rabies virus G protein (McGarvey et al., 1995).

Recombinant-vaccine technology offers hope of multivalent childhood vaccines protecting against numerous infectious diseases, and it has been suggested that this could be incorporated into childhood vaccination programs in areas of high rabies endemicity where postexposure vaccination is not readily accessible (Dreesen, 1997). An avian poxvirus construct expressing rabies virus glycoprotein has been demonstrated to effectively produce rabies immunity (Fries et al., 1996). Multivalent recombinant vaccines for the immunization of animals are also in development (Hu et al., 1997). Another possible way to reduce the cost of production and increase stability of the immunizing product, as well as possibly to confer lifelong duration of immunity from infant vaccination, is the development and use of DNA vaccines (Butts et al., 1998). This technology utilizes the observations that introduction of a plasmid encoding a reporter gene into an animal results in the expression of that gene in vivo, the DNA persists for long periods, and the protein produced by the plasmid can induce an immune response (Babiuk et al., 1999). Rabies DNA vaccines have been developed, and preliminary efficacy trials in nonhuman primates have demonstrated a capability to induce protective antibody (Lodmell et al., 1998).

Although evidence is convincing that infiltration of the exposure site with neutralizing antibody is necessary for optimal protection by PET, availability prevents the consistent use of passive protection with PET throughout much of the world. HRIG is used exclusively in the United States, but a purified immune globulin of equine origin is available that is used widely elsewhere and has been well tolerated. The use of monoclonal antibody technology may permit the production of an affordable and consistent alternative to avoid the costs and acute shortages associated with antiserum-derived immunoglobulins. Monoclonal antibodies of human and mouse origin have been shown to possess significant protective activity when evaluated in a postexposure prophylaxis model in hamsters (C. A. Hanlon, C. DeMattos, C. DeMattos, J. Shaddock, and C. E. Rupprecht, *Proc. IX Int. Meet. Res. Adv. Rabies Control Am.*, abstr. 113, p. 970, 1998).

The use of antibody therapies, antiviral compounds, and interferon and its precursors have shown some promise as a cure for rabies in vitro, but they all have proven futile when utilized in vivo during clinical rabies. Reversal of the disease process once symptoms begin will be very difficult without a greater knowledge of the pathogenesis and pathological mechanisms of the infection at the host and cellular levels. One interesting proposed strategy is the use of antisense oligodeoxynucleotides complementary to rabies virus genomic RNA. If rabies pathogenesis is due to virus replication in neural cells, leading to suppression of host gene expression, it has been hypothesized that exposure to antisense sequences might reverse the disease process (Fu, 1997). Preliminary evidence in vitro suggests that one such oligodeoxynucleotide can almost completely block rabies virus infection of cells and inhibit cell-to-cell spread of rabies virus in mouse neuroblastoma cell culture (Fu et al., 1996).

In the United States, 22 of 37 human rabies cases since 1980 resulted from infection with bat rabies variants. In only two of these cases was there evidence of a bat bite. In many of the remaining cases there was some physical contact with a bat without an evident bite. In several others there had been one or more indoor encounters with a bat with no contact recognized. Bat teeth are small and sharp, and a bat bite may not draw blood or be noticed (C. E. Rupprecht, abstr., *Symp. Bat Rabies*, U.S. Animal Health Association, Little Rock, Ark., 1996). These observations have prompted the changes in the guidelines for the management of potential rabies exposures following encounters with a bat. What may be even more puzzling than the cryptic nature of the recent human rabies cases is the identification in the vast majority of these cases of a rabies variant associated only with two species of bats that are very rarely encountered by humans. In 16 of the 22 bat rabies-related human deaths in the United States since 1980, the variant has been identified by antigenic or genetic typing as that associated with silver-haired and pipistrelle bats. These species are noncommensal bats that are very rarely seen and even more rarely involved in recognized contact with the public (Childs et al., 1994). Preliminary investigations have suggested that the silver-haired–pipistrelle bat rabies variant may possess special growth characteristics related to temperature and nonneural cell adaptation with possible implications for transmission by superficial intradermal exposures (Moromoto et al., 1996). These data invite further investigation of the cryptic nature of the transmission of rabies to humans in the United States and of the preponderance of bat rabies, and particularly this variant.

The tremendous reduction in the risk of PET-associated adverse reactions realized with modern cell culture vaccines and HRIG resulted in a dramatic increase in the number of PETs in the United States in the early 1980s. The increase in rabies in densely populated areas of the mid-Atlantic and northeastern states due to the raccoon rabies epizootic and the expansion of treatment criteria to include certain non-bite exposures have accounted for enormous increases in the numbers of rabies treatments in those areas (Wyatt et al., 1999). The recommendation to consider PET in bat encounters where there is a reasonable probability that a bite could have occurred but gone unnoticed (CDC, 1999a) will further escalate shortages of biologics and the annual costs for human vaccination in this country. Applied research on the epidemiology of human rabies PET is required to verify the appropriateness of current treatment algorithms. This is particularly true in light of realistic risk assessments and the consideration of the opportunity costs of the expenditures that would be required to try to bring the already low rabies mortality rate in the United States to zero (Rupprecht et al., 1996).

I gratefully acknowledge the careful reading of and valuable suggestions to the text by Jean Smith, Rabies Laboratory, CDC, Atlanta, Ga.

REFERENCES

Afshar, A. A. 1979. A review of non-bite transmission of rabies virus infection. *Br. Vet. J.* **135:**142–148.

Babiuk, L. A., J. Lewis, S. Van den Hurk, and R. Braun. 1999. DNA immunization: present and future, p. 163–180. *In* R. D. Schultz (ed.), *Veterinary Vaccines and Diagnostics.* Academic Press, New York, N.Y.

Baer, G. M., and T. L. Lentz. 1991. Rabies pathogenesis to the central nervous system, p. 106–120. *In* G. M. Baer (ed.), *The Natural History of Rabies*, 2nd ed. CRC Press, Boca Raton, Fla.

Baer, G. M., J. Neville, and G. S. Turner. 1996. *Rabbis and Rabies: A Pictorial History of Rabies Through the Ages.* Laboratorios Baer, Condessa, Mexico.

Barrat, J. 1996. Simple technique for the collection and shipment of brain specimens for rabies diagnosis, p. 425–432. *In* F. X. Meslin, M. M. Kaplan, and H. Koprowski (ed.), *Laboratory Techniques in Rabies*, 4th ed. World Health Organization, Geneva, Switzerland.

Bernard, K. W. 1986. Clinical rabies in humans, p. 43–48. *In* D. B. Fishbein, L. A. Sawyer, and W. G. Winkler (ed.), *Rabies Concepts for Professionals*, 2nd ed. Merieux Institute, Miami, Fla.

Bernard, K. W., D. B. Fishbein, and K. D. Miller. 1985. Preexposure rabies immunization with human diploid-cell vaccine: decreased antibody responses in persons immunized in developing countries. *Am. J. Trop. Med. Hyg.* **34:**633–647.

Blenden, D. C., J. F. Bell, A. T. Tsao, and J. V. Umoh. 1983. Immunofluorescent examination of the skin of rabies-infected animals as a means of early detection of rabies virus antigen. *J. Clin. Microbiol.* **18:**631–636.

Bourhy, H., and P. Perrin. 1996. Rapid rabies enzyme immunodiagnosis (RREID) for rabies antigen detection, p. 105-113. *In* F. X. Meslin, M. M. Kaplan, and H. Koprowski (ed.), *Laboratory Techniques in Rabies*, 4th ed. World Health Organization, Geneva, Switzerland.

Bourhy, H., B. Kissi, and N. Tordo. 1993. Molecular diversity of the Lyssavirus genus. *Virology* **194:**70–81.

Brass, D. A. 1994. Prevalence and distribution of rabies in insectivorous bats, p. 131–150. *In* D. A. Brass (ed.), *Rabies in Bats.* Livia Press, Ridgefield, Conn.

Briggs, D. J., J. S. Smith, F. L. Mueller, J. Schwenke, R. D. Davis, C. R. Gordon, K. Schweitzer, L. A. Orciari, P. A. Yager, and C. E. Rupprecht. 1998. A comparison of two serological methods for detecting the immune response after rabies vaccination in dogs and cats being exported to rabies-free areas. *Biologicals* **26:**347–355.

Brochier, B., M. F. A. Aubert, P. P. Pastoret, E. Masson, J. Schon, M. Lombard, G. Chappuis, B. Languet, and P. Desmettre. 1996. Field use of a vaccinia-rabies recombinant vaccine for the control of sylvatic rabies in Europe and North America. *Rev. Sci. Tech. Off. Int. Epizoot.* **15:**947–970.

Butts, C., I. Zubkoff, D. S. Robbins, S. Cao, and M. Sarzotti. 1998. DNA immunization of infants: potential and limitations. *Vaccine* **16:**1444–1449.

Center for Disease Control. 1977. Rabies in a laboratory worker—New York. *Morb. Mortal. Wkly. Rep.* **26:**183–184.

Centers for Disease Control and Prevention. 1997. Human rabies—Montana and Washington, 1997. *Morb. Mortal. Wkly. Rep.* **46:**770–774.

Centers for Disease Control and Prevention. 1999a. Human rabies prevention—United States, 1999: recommendations of the Advisory Committee on Immunization Practices (ACIP). *Morb. Mortal. Wkly. Rep.* **48**(RR-1):1–21.

Centers for Disease Control and Prevention. 1999b. Human rabies in Virginia, 1998. *Morb. Mortal. Wkly. Rep.* **48**:95–97.

Charlton, K. M. 1988. The pathogenesis of rabies, p. 101–150. *In* J. B. Campbell and K. M. Charlton (ed.), *Rabies.* Academic Press, Norwell, Mass.

Childs, J. E., C. V. Trimarchi, and J. W. Krebs. 1994. The epidemiology of bat rabies in New York State. *Epidemiol. Infect.* **113**:501–511.

Childs, J. E., L. Colby, J. W. Krebs, T. Strine, M. Feller, D. Noah, C. Drenzek, J. S. Smith, and C. E. Rupprecht. 1997. Surveillance and spatiotemporal associations of rabies in rodents and lagomorphs in the United States, 1985–1994. *J. Wildl. Dis.* **33**:20–27.

Chutivongse, S., H. Wilde, M. Benjavongkulchai, P. Chomchey, and S. Punthawong. 1995. Postexposure rabies vaccination during pregnancy: effect on 202 women and their infants. *Clin Infect. Dis.* **20**:818–820.

Cleveland, S. 1998. Epidemiology and control of rabies: the growing problem of rabies in Africa. *Trans. R. Soc. Trop. Med. Hyg.* **92**:131–134.

Constantine, D.G. 1962. Rabies transmission by nonbite route. *Public Health Rep.* **77**:287–289.

Constantine, D. G. 1967. Bat rabies in the southwestern United States. *Public Health Rep.* **82**:867–888.

Davis, C., S. Neill, and P. Raj. 1998. Microwave fixation of rabies specimens for fluorescent antibody testing. *J. Virol. Methods* **68**:177–182.

Debbie, J. G., and C. V. Trimarchi. 1970. Pantropism of rabies virus in free-ranging rabid red fox *Vulpes fulva. J. Wildl. Dis.* **6**:500–506.

Debbie, J. G., and C. V. Trimarchi. 1992. Rabies, p. 116–120. *In* A. E. Castro and W. P. Heuschele (ed.), *Veterinary Diagnostic Virology.* Mosby Year Book, Boston, Mass.

Debbie, J. G., and C. V. Trimarchi. 1998. Prophylaxis for suspected exposure to bat rabies. *Lancet* **350**:1790–1791.

Dreesen, D. W. 1997. A global review of rabies vaccines for human use. *Vaccine* **15**:S2–S6.

Dreesen, D. W., and C. A. Hanlon. 1998. Current recommendations for the prophylaxis and treatment of rabies. *Drugs* **56**:801–809.

Dreesen, D. W., K. W. Bernard, R. A. Parker, A. J. Deutsch, and J. Brown. 1986. Immune complex-like disease in 23 persons following a booster dose of rabies human diploid cell vaccine. *Vaccine* **4**:45–49.

Elmgren, L. D., and A. I. Wandeler. 1996. Competitive ELISA for the detection of rabies virus-neutralizing antibodies, p. 200–208. *In* F. X. Meslin, M. M. Kaplan, and H. Koprowski (ed.), *Laboratory Techniques in Rabies,* 4th ed. World Health Organization, Geneva, Switzerland.

Everard, C. O., and J. D. Everard. 1985. Mongoose rabies in Grenada, p. 43–70. *In* P. J. Bacon (ed.), *Population Dynamics of Rabies in Wildlife.* Academic Press, New York, N.Y.

Fekadu, M. 1991. Canine rabies, p. 367–378. *In* G. M. Baer (ed.), *The Natural History of Rabies,* 2nd ed. CRC Press, Boca Raton, Fla.

Fekadu, M., J. H. Shaddock, and G. M. Baer. 1981. Intermittent excretion of rabies virus in the saliva of a dog two and six months after it had recovered from experimental rabies. *Am. J. Trop. Med. Hyg.* **30**:1113–1115.

Fekadu, M., J. H. Shaddock, and G. M. Baer. 1982. Excretion of rabies virus in the saliva of dogs. *J. Infect. Dis.* **145**:715–719.

Fekadu, M., T. Endeshaw, A. Wondimagegnehu, Y. Bogale, T. Teshager, and J. G. Olsen. 1996. Possible human-to-human transmission of rabies in Ethiopia. *Ethiop. Med. J.* **34**:123–127.

Fenner, F. J., E. O. Gibbs, F. A. Murphy, R. Rott, M. J. Studdert, and D. O. White. 1993. *Veterinary Virology,* 2nd ed. Academic Press, Toronto, Canada.

Fishbein, D. B. 1991. Rabies in humans, p. 519–549. *In* G. M. Baer (ed.), *The Natural History of Rabies,* 2nd ed. CRC Press, Boca Raton, Fla.

Fries, L. F., J. Tartaglia, J. Taylor, E. K. Kaufman, B. Meignier, E. Paoletti, and S. Plotkin. 1996. Human safety and immunogenicity of a canarypox-rabies glycoprotein recombinant vaccine: an alternative poxvirus vector system. *Vaccine* **14**:428–434.

Fu, Z. F. 1997. Rabies and rabies research: past, present and future. *Vaccine* **15**:S20–S24

Fu, Z. F., E. Wickstrom, M. Jiang, S. Corisdoe, B. Dietzschold, and H. Koprowski. 1996. Inhibition of rabies virus infection by an oligodeoxynucleotide complimentary to rabies virus genomic RNA. *Antisense Nucleic Acid Drug Dev.* **6**:87–93.

Gode, G. R., A. V. Raju, T. S. Jayalakshmi, H. L. Kaul, and N. K. Bide. 1976. Intensive care in rabies therapy: clinical observations. *Lancet* **ii**:6.

Goldwasser, R. A., and R. E. Kissling. 1958. Fluorescent antibody staining of street and fixed rabies virus antigens. *Proc. Soc. Exp. Biol. Med.* **98**:219–223.

Griego, R. D., T. Rosen, I. F. Orengo, and J. E. Wolf. 1995. Dog, cat and human rabies: a review. *J. Am. Acad. Dermatol.* **33**:1019–1029.

Habel, K. 1945. Seroprophylaxis in experimental rabies. *Public Health Rep.* **60**:545–551.

Hamir, A. N., Z. F. Moser, F. Fu, B. Dietzschold, and C. E. Rupprecht. 1995. Immunohistochemical test for rabies: identification of a diagnostically superior monoclonal antibody. *Vet. Rec.* **136**:295–296.

Hanlon, C., J. Smith, G. Anderson, C. V. Trimarchi, and D. Schnurr. 1999. Laboratory diagnosis of rabies: report of the National Working Group on Prevention and Control of Rabies. *J. Am. Vet. Med. Assoc.* **215**:1444–1446.

Hooper, P. T., R. A. Lunt, A. R. Gould, H. Samaratunga, A. D. Hyatt, L. J. Gleeson, B. J. Rodwell, C. E. Rupprecht, J. S. Smith, and P. K. Murray. 1997. A new lyssavirus—the first endemic rabies-related virus recognized in Australia. *Bull. Inst. Pasteur* **95**:209–218.

Hu, L., C. Ngichabe, C. Trimarchi, J. Esposito, and F. Scott. 1997. Raccoon poxvirus live recombinant feline panleukopenia virus VP2 and rabies virus glycoprotein bivalent vaccine. *Vaccine* **15**:1466–1472.

Hummeler, K., H. Koprowski, and T. J. Wiktor. 1967. Structure and development of rabies virus in tissue culture. *Nature* **294**:275–278.

Jenkins, S. R., B. D. Perry, and W. G. Winkler. 1998. Ecology and epidemiology of raccoon rabies. *Rev. Infect. Dis.* **10**:S620–S625.

Kamolvarin, N., T. Tirawatnpong, R. Rattanasiwamoke, S. Tirawatnpong, T. Panpanich, and T. Hemachudha. 1993. Diagnosis of rabies by polymerase chain reaction with nested primers. *J. Infect. Dis.* **167**:207–210.

Kaplan, C. 1985. Rabies: a worldwide disease, p. 1–21. *In* P. J. Bacon (ed.), *Population Dynamics of Rabies in Wildlife.* Academic Press, New York, N.Y.

Kaplan, M. M. 1996. Safety precautions in handling rabies virus, p. 3–8. *In* F. X. Meslin, M. M. Kaplan, and H. Koprowski (ed.), *Laboratory Techniques in Rabies,* 4th ed. World Health Organization, Geneva, Switzerland.

Kaplan, M. M., T. J. Wiktor, R. F. Maes, J. B. Campbell, and H. Koprowski. 1967. Effect of polyions on the infectivity of rabies virus in tissue culture: construction of a single-cycle growth curve. *J. Virol.* **1:**145–151.

Kelly, R. M., and P. L. Strick. 1997. Retrograde transport of rabies virus through the cerebello-thalamocortical circuits of primates. *Soc. Neurosci. Abstr.* **23:**1828.

King, A. A., J. J. Sands, and J. S. Porterfield. 1984. Antibody-mediated enhancement of rabies virus infection in a mouse macrophage cell line (P388DI). *J. Gen. Virol.* **65:**1091–1093.

Kissi, B., N. Tordo, and H. Bourhy. 1995. Genetic polymorphism in the rabies virus nucleoprotein gene. *Virology* **209:**526–537.

Koprowski, H. 1991. The virus: overview. *In* G. M. Baer (ed.), *The Natural History of Rabies*, 2nd ed. CRC Press, Boca Raton, Fla.

Koprowski, H., and H. R. Cox. 1948. Studies on chick embryo adapted rabies virus. I. Culture characteristics and pathogenicity. *J. Immunol.* **60:**533–539.

Krebs, J. W., M. L. Wilson, and J. E. Childs. 1995. Rabies-epidemiology, prevention and future research. *J. Mammal.* **76:**681–694.

Krebs, J. W., J. S. Smith, C. E. Rupprecht, and J. E. Childs. 1998. Rabies surveillance in the United States during 1997. *J. Am. Vet. Med. Assoc.* **213:**1713–1728.

Kristensson, D., and Y. Olsson. 1973. *Prog. Neurobiol.* **1:**85–106.

Lentz, T. L., P. T. Wilson, E. Hawrot, and D. W. Speicher. 1984. Amino acid sequence similarity between rabies virus glycoprotein and snake venom curare mimetic neurotoxins. *Science* **226:**847–848.

Lepine, P., and P. Atanasiu. 1996. Histopathological diagnosis, p. 66–79. *In* F. X. Meslin, M. M. Kaplan, and H. Koprowski (ed.), *Laboratory Techniques in Rabies*, 4th ed. World Health Organization, Geneva, Switzerland.

Lodmell, D. L., N. B. Ray, M. J. Parnell, L. C. Ewalt, C. A. Hanlon, J. H. Shaddock, D. S. Sanderlin, and C. E. Rupprecht. 1998. DNA immunization protects nonhuman primates against rabies virus. *Nat. Med.* **4:**949–952.

Lopez, A., P. Miranda, E. Tejada, and D. Fishbein. 1992. Outbreak of human rabies in the Peruvian jungle. *Lancet* **339:**1408–1411.

MacInnes, C. D. 1988. Control of wildlife rabies, p. 381–406. *In* J. B. Campbell and K. M. Charlton (ed.), *Rabies*. Kluwer Academic Publishers, Boston, Mass.

Mansel, G. A. 1951. *Royal Canadian Mounted Police Report, "G" Division, Port Harrison, Quebec.*

Matyas, B. T., M. W. McGuill, and A. DeMaria. 1999. Reemergence of rabies in the US. *Infect. Med.* **16:**129–138.

McGarvey, P. B., J. Hammond, M. M. Dienelt, D. C. Hooper, F. H. Michaels, Z. F. Fu, B. Dietzschold, and H. Koprowski. 1995. Expression of the rabies virus glycoprotein in transgenic tomatoes. *Bio/Technology* **13:**1484–1487.

Meltzer, M. I., and C. E. Rupprecht. 1998. A review of the economics of the prevention and control of rabies. Part I. global impact and rabies in humans. *Pharmacoeconomics* **14:**365–383.

Meslin, F. X., D. B. Fishbein, and H. C. Matter. 1994. Rationale and prospects for rabies elimination in developing countries. *Curr. Top. Microbiol. Immunol.* **187:**1–26.

Moromoto, K., M. Patel, S. Corisdeo, D. C. Hooper, Z. F. Fu, and C. E. Rupprecht. 1996. Characterization of a unique variant of bat rabies responsible for newly emerging human cases in the United States. *Proc. Natl. Acad. Sci USA* **93:**5653–5658.

Murphy, F. A. 1986. The rabies virus and pathogenesis of the disease, p. 11–16. *In* D. B. Fishbein, L. A. Sawyer, and W. G. Winkler (ed.), *Rabies Concepts for Professionals*, 2nd ed. Merieux Institute, Miami, Fla.

Nadin-Davis, S. A., W. Huang, and A. I. Wandeler. 1996. The design of strain-specific polymerase chain reactions for discrimination of the raccoon rabies virus strain from indigenous rabies viruses of Ontario. *J. Virol. Methods* **57:**141–156.

National Association of State Public Health Veterinarians. 1999. *Compendium of Animal Rabies Control, 1999.* National Association of State Public Health Veterinarians, Richmond, Va.

Negri, A. 1903. Contributo allo studio della etiologia della rabia. *Boll. Soc. Med. Chir. Pavia* **3:**88–95.

Nicholson, K. G. 1994. Rabies, p. 595–620. *In* A. J. Zuckerman, J. E. Banatvala, and J. R. Pattison (ed.), *Principals and Practice of Clinical Virology*, 3rd ed. John Wiley & Sons, New York, N.Y.

Niezgoda, M., D. J. Briggs, J. Shaddock, D. W. Dreesen, and C. E. Rupprecht. 1997. Pathogenesis of experimentally induced rabies in domestic ferrets. *Am. J. Vet. Res.* **58:**1327–1331.

Noah, D. L., M. G. Smith, J. C. Gottardt, J. W. Krebs, D. Green, and J. E. Childs. 1996. Mass human exposure to rabies in New Hampshire: exposures, treatment, and cost. *Am. J. Public Health* **86:**1149–1151.

Noah, D. L., C. L. Drenzek, J. S. Smith, J. W. Krebs, L. Orciari, J. Shaddock, D. Sanderlin, S. Whitfield, M. Fekadu, J. G. Olson, C. E. Rupprecht, and J. E. Childs. 1998. Epidemiology of human rabies in the United States, 1980 to 1996. *Ann. Intern. Med.* **128:**922–930.

Office International des Epizooties. 1996. Rabies, p. 211–213. *In OIE Manual of Standards for Diagnostic Tests and Vaccines*, 3rd ed. Office International des Epizooties, Paris, France.

Orciari, L. A. 1995. Genetic analysis of rabies virus isolates from skunks in the United States. M.S. thesis. University of Georgia, Athens.

Perl, D. P., and P. F. Good. 1991. The pathology of rabies in the central nervous system, p. 163–190. *In* G. M. Baer (ed.), *The Natural History of Rabies*, 2nd ed. CRC Press, Boca Raton, Fla.

Rudd, R. J., and C. V. Trimarchi. 1980. Tissue culture technique for routine isolation of street strain rabies virus. *J. Clin. Microbiol.* **12:**590–593.

Rudd, R. J., and C. V. Trimarchi. 1989. The development and evaluation of an in vitro virus isolation procedure as a replacement for the mouse inoculation test in rabies diagnosis. *J. Clin. Microbiol.* **27:**2522–2528.

Rupprecht, C. E., B. Dietzschold, J. H. Cox, and L. G. Schneider. 1989. Oral vaccination of raccoons (*Procyon lotor*) with an attenuated (SAD-B19) rabies vaccine. *J. Wildl. Dis.* **25:**548–554.

Rupprecht, C. E., C. A. Hanlon, M. Niezgoda, J. R. Buchanan, D. Diehl, and H. Koprowski. 1993. Recombinant rabies vaccine: efficacy assessment in free ranging animals. *Onderstepoort J. Vet. Res.* **60:**463–468.

Rupprecht, C. E., J. S. Smith, J. Krebs, M. Niezgoda, and J. E. Childs. 1996. Current issues in rabies prevention in the United States. *Public Health Rep.* **111:**400–407.

Sacramento, D., H. Bourhy, and N. Tordo. 1991. PCR technique as an alternative method for rabies diagnosis and molecular epidemiology of rabies virus. *Mol. Cell. Probes* **6:**229–240.

Savy, V., and P. Atanasiu. 1978. Rapid immunoenzymatic technique for titration of rabies antibodies IgG and IgM. *Dev. Biol. Stand.* **40:**247–253.

Schneider, M. C., C. Santos-Burgoa, J. Aron, B. Munoz, S. Ruiz-Velazco, and W. Uieda. 1996. Potential force of infection of human rabies transmitted by vampire bats in the Amazonian Region of Brazil. *Am. J. Trop. Med. Hyg.* **55:**680–684.

Shope, R. E., and R. B. Tesh. 1987. Lyssaviruses, p. 509–534. *In* R. R. Wagner (ed.), *The Rhabdoviruses.* Plenum Press, New York, N.Y.

Shope, R. E., F. A. Murphy, A. K. Harrison, O. R. Causey, G. E. Kemp, D. I. Simpson, and D. L. Moore. 1970. Two African viruses serologically and morphologically related to rabies virus. *J. Virol.* **6:**690–692.

Smith, A. L., G. H. Tignor, R. W. Emmons, and J. D. Woodie. 1978. Isolation of field rabies virus strains in CER and murine neuroblastoma cell cultures. *Intervirol.* **9:**359–361.

Smith, J. S. 1989. Rabies virus epitopic variation: use in ecologic studies. *Adv. Virus Res.* **36:**215–253.

Smith, J. S. 1996. New aspects of rabies with emphasis on epidemiology, diagnosis, and prevention of the disease in the United States. *Clin. Microbiol. Rev.* **9:**166–176.

Smith, J. S. 1999. Rabies virus, p. 1099–1106. *In* P. R. Murray, E. J. Baron, M. A. Pfaller, F. C. Tenover, and R. H. Yolken (ed.), *Manual of Clinical Microbiology,* 7th ed. ASM Press, Washington, D.C.

Smith, J. S., and D. H. Seidel. 1993. Rabies: a new look at an old disease, p. 82–106. *In* J. L. Melnick (ed.), *Prog. Med. Virol.* Karger, Basel, Switzerland.

Smith, J. S., C. L. McClelland, F. L. Reid, and G. M. Baer. 1982. Dual role of the immune response in street rabies virus infection of mice. *Infect. Immun.* **35:**213–217.

Smith J. S., F. L. Reid-Sanden, L. F. Roumillat, C. V. Trimarchi, K. Clark, G. M. Baer, and W. G. Winkler. 1986. Demonstration of antigen variation among rabies virus isolates by using monoclonal antibodies to nucleocapsid proteins. *J. Clin. Microbiol.* **24:**573–580.

Smith, J. S., D. B. Fishbein, C. E. Rupprecht, and K. Clark. 1991. Unexplained rabies in three immigrants in the United States. *N. Engl. J. Med.* **324:**205–211.

Smith, J. S., L. A. Orciari, P. A.Yager, H. D. Seidel, and C. K. Warner. 1992. Epidemiologic and historical relationships among 87 rabies virus isolates as determined by limited sequence analysis. *J. Infect. Dis.* **166:**296–307.

Smith, J. S., L. A. Orciari, and P. A. Yager. 1995. Molecular epidemiology of rabies in the United States. *Semin. Virol.* **6:**387–400.

Smith, J. S., P. A. Yager, and G. M. Baer. 1996. A rapid fluorescent focus inhibition test (RFFIT) for determining rabies virus-neutralizing antibody, p. 181–192. *In* F. X. Meslin, M. M. Kaplan, and H. Koprowski (ed.), *Laboratory Techniques in Rabies,* 4th ed. World Health Organization, Geneva, Switzerland.

Steele, J. H., and P. J. Fernandez. 1991. History of rabies and global aspects, p. 1–24. *In* G. M. Baer (ed.), *The Natural History of Rabies,* 2nd ed. CRC Press, Boca Raton, Fla.

Theodorides, J. 1986. *Histoire de la Rage.* Mason, Paris, France.

Tierkel, E. S., and P. Atanasiu. 1996. Rapid microscopic examination for Negri bodies and preparation of specimens for biological tests, p. 55–65. *In* F. X. Meslin, M. M. Kaplan, and H. Koprowski (ed.), *Laboratory Techniques in Rabies,* 4th ed. World Health Organization, Geneva, Switzerland.

Tordo, N. 1996. Characteristics and molecular biology of rabies virus, p. 28–52. *In* F. X. Meslin, M. M. Kaplan, and H. Koprowski (ed.), *Laboratory Techniques in Rabies,* 4th ed. World Health Organization, Geneva, Switzerland.

Tordo, N., H. Bourhy, and D. Sacramento. 1995. PCR technology for *Lyssavirus* diagnosis, p. 125–145. *In* J. Clewly (ed.),

The Polymerase Chain Reaction (PCR) for Human Viral Diagnosis. CRC Press, London, United Kingdom.

Tornatore, C. S., and J. R. Richert. 1990. CNS demyelination associated with diploid cell rabies vaccine. *Lancet* **335:**1346–1347.

Trimarchi, C. V. 1999. Bat rabies from a national and state perspective, p. 883–884. *In Convention Notes.* American Veterinary Medical Association, New Orleans, La.

Trimarchi, C. V., and J. G. Debbie. 1991. The fluorescent antibody in rabies, p. 220–233. *In* G. M. Baer (ed.), *The Natural History of Rabies,* 2nd ed. CRC Press, Boca Raton, Fla.

Trimarchi, C. V., R. J. Rudd, and M. K. Abelseth. 1986. Experimentally induced rabies in four cats inoculated with a rabies virus isolated from a bat. *Am. J. Vet. Res.* **47:**777–780.

Trimarchi, C. V., R. J. Rudd, and M. Safford. 1996. An in vitro virus neutralization test for rabies antibody, p. 193–199. *In* F. X. Meslin, M. M. Kaplan, and H. Koprowski (ed.), *Laboratory Techniques in Rabies,* 4th ed. World Health Organization, Geneva, Switzerland.

Tsiang, H. 1988. Interactions of rabies virus and host cells, p. 67–100. *In* J. Campbell and K. M. Charlton (ed.), *Rabies.* Kluwer Academic Publishers, Boston, Mass.

Uhaa, I. J., V. N. Dato, F. E. Sorhage, J. W. Beckley, D. E. Roscoe, R. D. Gorsky, and D. B. Fishbein. 1992. Benefits and costs of using an orally absorbed vaccine to control rabies in raccoons. *J. Am. Vet. Med. Assoc.* **201:**1873–1882.

Vernon, S. K., A. R. Neurath, and B. A. Rubin. 1972. Electron microscopic studies on the structure of rabies virus. *J. Ultrastruct. Res.* **41:**29–42.

Wandeler, A. I., S. A. Nadin-Davis, R. R. Tinline, and C. E. Rupprecht. 1994. Rabies epidemiology: some ecological and evolutionary perspectives, p. 297–324. *In* C. E. Rupprecht, B. Dietzschold, and H. Koprowski (ed.), *Lyssaviruses.* Springer-Verlag, Berlin, Germany.

Warner, C. K., S. G. Whitfield, M. Fekadu, and H. Ho. 1997. Procedures for reproducible detection of rabies virus antigen, mRNA and genome in situ in formalin fixed tissues. *J. Virol. Methods* **67:**5–12.

Webster, L. T., and J. R. Dawson. 1935. Early diagnosis of rabies by mouse inoculation. Measurement of humoral immunity to rabies by mouse protection test. *Proc. Soc. Exp. Biol. Med.* **32:**570–573.

Webster, W. A., and G. A. Casey. 1996. Virus isolation in neuroblastoma cell culture, p. 96–104. *In* F. X. Meslin, M. M. Kaplan, and H. Koprowski (ed.), *Laboratory Techniques in Rabies,* 4th ed. World Health Organization, Geneva, Switzerland.

Wiktor, T. J., and H. Koprowski. 1965. Successful immunization of primates with rabies vaccine prepared in human diploid cell strain WI38. *Proc. Soc. Exp. Biol. Med.* **118:**1069–1073.

Wiktor, T. J., A. Flamand, and H. Koprowski. 1980. Use of monoclonal antibodies in diagnosis of rabies virus infection and differentiation of rabies and rabies-related viruses. *J. Virol. Methods* **1:**33–46.

Wiktor, T. J., R. I. Macfarlan, K. J. Reagan, B. Dietzschold, P. J. Curtis, W. H. Wunner, M. P. Kieny, R. Lathe, J. P. Lecocq, M. Mackett, B. Moss, and H. Koprowski. 1984. Protection from rabies by a vaccinia virus recombinant containing the rabies virus glycoprotein gene. *Proc. Natl. Acad. Sci. USA* **81:**7194–7198.

Wilde, H., P. Chomchey, P. Puyaratabandhu, and S. Chutivongse. 1989. Purified equine rabies immune globulin: a safe and affordable alternative to rabies immune globulin. *Bull. W.H.O.* **67:**731–736.

Wilkinson, L. 1988. Understanding the nature of rabies: an historical perspective, p. 1–23. *In* J. B. Campbell and K. M. Charlton (ed.), *Rabies*. Kluwer Academic Publishers, Boston, Mass.

Winkler, W. G., T. R. Fashinell, L. Leffingwell, P. Howard, and P. Conomy. 1973. Airborne rabies transmission in a laboratory worker. *JAMA* **226:**1219–1221.

World Health Organization. 1984. *WHO Expert Committee on Rabies, 7th Report.* WHO Technical Report Series, no. 709. World Health Organization, Geneva, Switzerland.

World Health Organization. 1992. *WHO Expert Committee on Rabies, 8th Report.* WHO Technical Report Series, no. 824. World Health Organization, Geneva, Switzerland.

World Health Organization. 1994. *Workshop on Genetic and Antigenic Molecular Epidemiology of Lyssaviruses*, p. 1–14. World Health Organization, Geneva, Switzerland.

World Health Organization. 1997. *World Health Report 1996.* World Health Organization, Geneva, Switzerland.

Wunner, W. H., J. K. Larson, B. Dietzschold, and C. L. Smith. 1988. The molecular biology of rabies viruses. *Rev. Infect. Dis.* **10:**S771–S784.

Wunner, W. H., C. H. Calisher, R. G. Dietzgen, R. G. Jackson, A. O. Kitajima, M. F. Lafon, J. C. Leong, S. T. Nichol, D. Peters, J. S. Smith, and P. J. Walker. 1995. *Rhabdoviridae*, p. 275–280. *In* Classification and Nomenclature of Viruses, Sixth Report of the International Committee on Taxonomy of Viruses. Springer-Verlag, New York, N.Y.

Wyatt, J. D., W. H. Barker, N. M. Bennett, and C. A. Hanlon. 1999. Human rabies postexposure prophylaxis during a raccoon rabies epizootic in New York, 1993 and 1994. *Emerg. Infect. Dis.* **5:**1–11.

Zaidman, G. W., and A. Billingsly. 1998. Corneal impression test for the diagnosis of acute rabies encephalitis. *Ophthalmology* **105:**249–251.

Arboviruses

JOHN T. ROEHRIG

29

LABORATORY PROCEDURES FOR DETECTING VIRUSES

Introduction

The term arbovirus is a contraction of "arthropod-borne virus" and has no phylogenetic or classification significance. This term describes the mechanism by which these viruses are transmitted and maintained in nature: through the bite of a hematophagous arthropod. Most medically important arboviruses are transmitted by either mosquitoes or ticks. In the United States alone, representatives from at least five virus families can be transmitted by biting arthropods (Table 1). We will focus on only the medically important arboviruses in this chapter.

Because the viruses are transmitted by arthropods, arboviral disease usually manifests itself during the warmer months in the temperate climates of the world. Arboviral disease can, however, be contracted in the winter months in milder climates, and disease transmission can occur year-round in the tropics. During the milder times of the year, or depending on the patient's travel history, testing for arboviruses should be included in the laboratory diagnosis of cases compatible with arboviral infections.

There are 535 arboviruses listed in the *International Catalogue of Arboviruses* (Karabatsos, 1985), but most have not been associated with human disease. Continued encroachment on the world's tropical rainforests, however, coupled with rapid transport of humans and animals, makes arbovirus infections emerging and reemerging infectious diseases. This observation means that new arboviruses may be associated with human diseases or known arboviruses may cause outbreaks in previous or new locales. Identification of these agents will, by definition, be difficult, with the medical and veterinary community depending on specialty reference laboratories capable of working with and identifying these biosafety level 3 and 4 pathogens (Centers for Disease Control and Prevention [CDC], 1993). The World Health Organization (WHO) sponsors a laboratory network of WHO Collaborating Centers, distributed throughout the world, which specialize in diagnosing arboviral diseases. It is likely that a clinical sample from an arbovirus infection will arrive at one of these laboratories for diagnosis or confirmation.

History

Yellow fever (YF) epidemics probably occurred as early as 1648 in the Yucatan Peninsula of Mexico. *Aedes aegypti* mosquitoes, which are the urban vectors of YF, also transmit dengue (DEN) virus, the cause of DEN fever. DEN fever outbreaks occurred quite frequently in the southern United States until the 1920s, when populations of the vector mosquito were controlled. Both DEN fever and YF continue to occur in tropical America and Africa, even though an effective YF vaccine exists. This inability to control YF despite the availability of an effective vaccine reflects the poor economic conditions of the countries where YF is endemic, where this vaccine is still too expensive for general use (Monath, 1991). DEN virus also causes dengue hemorrhagic fever and dengue shock syndrome (DHF-DSS), which currently occur as major, lethal epidemics in children in Southeast Asia and appeared for the first time in the New World in Cuba in 1981 (Kouri et al., 1983; Guzman et al., 1984, 1990).

The primary clinical manifestation of life-threatening arboviral disease in North America has been encephalitis. Three mosquito-borne viruses that cause human encephalitis were discovered during the 1930s. Western equine encephalitis (WEE) virus was isolated in 1930 from horses (Meyer et al., 1931) and in 1938 was associated with encephalitis in humans in California. It now occurs infrequently in the irrigated farmland of the western United States and Canada. Eastern equine encephalitis (EEE) was isolated in 1933 from horses (TenBroeck and Merrill, 1933). It was subsequently isolated from humans in 1938. Currently, EEE has a distribution throughout most of the eastern half of the United States. The most important cause of epidemic encephalitis in the United States is St. Louis encephalitis (SLE). The first outbreak of SLE occurred in 1933 in St. Louis, Mo., with 1,095 reported cases (Cumming, 1935). The last major SLE epidemic was in 1975, with 1,815 reported cases. Endemic (rural) SLE may occur each year in much of the western United States (Monath and Tsai, 1987; Tsai et al., 1987b; Reisen et al., 1990, 1992a, 1992b; Reisen and Chiles, 1997). It has been hypothesized that major urban SLE outbreaks occur every 7 to 10 years; however, this no longer appears to be the case. The reduction in the incidence of SLE may be due to life-style modifications, such as the use of air conditioning and television. Focal outbreaks can occur each year; however, many times they are localized in the poorer urban areas in scattered locations, such as Chicago, Philadelphia, Houston, and New Orleans.

The third major equine encephalitides is Venezuelan equine encephalitis (VEE). VEE viruses either are associated with major epidemics or are maintained in nature in enzootic cycles. The varieties of VEE viruses associated with these differing epidemiological presentations can be separated both serologically and through genetic analysis. VEE virus has caused major human epidemics periodically throughout Central and South America since the 1930s, the most recent in 1995 in Colombia and Venezuela (Kinney et al., 1989, 1998; Sneider et al., 1993; Weaver et al., 1996; Rivas et al., 1997). It is now believed that the earliest VEE epidemics were caused by incompletely inactivated vaccines (Sneider et al., 1993; Weaver et al., 1999). VEE epidemics are now caused by epidemic strains of VEE virus thought to have evolved from naturally occurring enzootic VEE viruses (Rico-Hesse et al., 1995; Powers et al., 1997; Kinney et al., 1998). The reasoning for this is derived partly from the inability to isolate epidemic VEE viruses during interepidemic periods.

Detailed reviews of all of these viruses as well as a currently emerging encephalitis caused by the California (CAL) serogroup virus, LaCrosse (LAC) encephalitis, are recommended for further study (Calisher and Thompson, 1983; Monath, 1988, 1996; Trent et al., 1989; Tsai and Monath, 1996; McJunkin et al., 1998).

Methods Used

General Considerations

Laboratory diagnosis of arboviral infections has traditionally been based upon serological identification of antiviral antibodies and/or isolation of virus. While the classical serological assays, hemagglutination inhibition (HI), complement fixation (CF), and neutralization (NT) of virus infectivity, have been replaced by enzyme-linked immunosorbent assay (ELISA), each of the earlier tests still has applicability. The timing after infection of certain viral infections can sometimes be ascertained with the CF test. While the laboratorian can readily distinguish between virus families (e.g., flaviviruses and togaviruses), within an individual family, many of the viruses are so closely related antigenically that only the virus NT test can differentiate infections.

TABLE 1 Medically important arboviruses in the United States

Virus family	Pathogen	Related viruses
Togaviridae	EEE virus	HJ virus
	WEE virus	
	VEE virus	
Flaviviridae	SLE virus	None
	POW virus	
	DEN virus	
Bunyaviridae	CAL serogroup viruses	Many
	LAC virus	
	CAL encephalitis virus	
	SSH virus	
	Jamestown Canyon virus	
	Cache Valley virus	
Reoviridae	CTF virus	None
Rhabdoviridae	VSV	Rabies virus

While the expensive technique of virus isolation by inoculation into susceptible cell culture is losing ground to more rapid assays, like PCR and antigen detection ELISA, the former approach is still useful. For example, alphaviruses replicate in common continuous cell cultures, like Vero or BHK-21 cells, often demonstrating virus-specific cytopathic effects within 24 h. These virus-infected cells can then be used to identify the infecting agent by indirect immunofluorescence assay (IFA) using well-characterized virus-specific murine monoclonal antibodies (MAbs). Very few PCR assays developed for arboviruses have been critically and completely analyzed to the extent that they now function as simple and reproducible laboratory tests. The caveat with all newer assays that detect only viral protein or nucleic acid is that they are unable to produce replicating virus usable for future serological or genetic analysis.

Basic Principles

A historical analysis of arboviral infections investigated at the WHO Collaborating Center for arboviruses in the Division of Vector-Borne Infectious Diseases, CDC, identified 24 viruses that cause enough disease to warrant their inclusion in routine diagnostic virology testing panels. These panels include representatives from all virus families and can be organized by their geographic distribution. The decision of which virus panel is used can be based upon the patient's location and travel history (Fig. 1). These viruses are not the only arboviral agents responsible for disease, but the virus panels should detect the majority of arboviral infections. Regardless of the assay employed, confirmation of arboviral infection requires acute- and convalescent-phase serum samples that yield a demonstrable increase in antiviral-antibody activity. The introduction of ELISA protocols that measure virus-specific immunoglobulin M (IgM), especially when applied to acute-phase serum or cerebrospinal fluid (CSF) samples, yields good approximations of recent infections when the timing of the specimen is appropriate. For some arboviruses, however, IgM reactivity can be measured weeks after the onset of disease. For most arboviruses, serological cross-reactivity with related viruses increases as the infection progresses. Because of this and the close antigenic relatedness of many of the agents within the same virus family, IgG ELISA assays are often not very specific. Still, it is a simple matter to differentiate viruses from different families (e.g., togaviruses from flaviviruses). Because epitopes that elicit virus-neutralizing antibody are under the most severe immunologic pressure, these epitopes are usually the most virus specific. Consequently, the virus NT assay demonstrates a fair amount of serological specificity, even with convalescent-phase serum samples.

Applications

Serology for Antibody Testing

HI

The HI test utilizes the ability of antiviral antibody to block the capacity of the virus to agglutinate erythrocytes (Clarke and Casals, 1958). This was the first technique used to characterize arboviruses. The HI test successfully differentiated togaviruses (group A arboviruses, primarily alphaviruses) from flaviviruses (group B arboviruses) long before modern biochemical techniques confirmed this observation. Many laboratories still utilize the HI test, although with the advent of ELISA, it is being replaced.

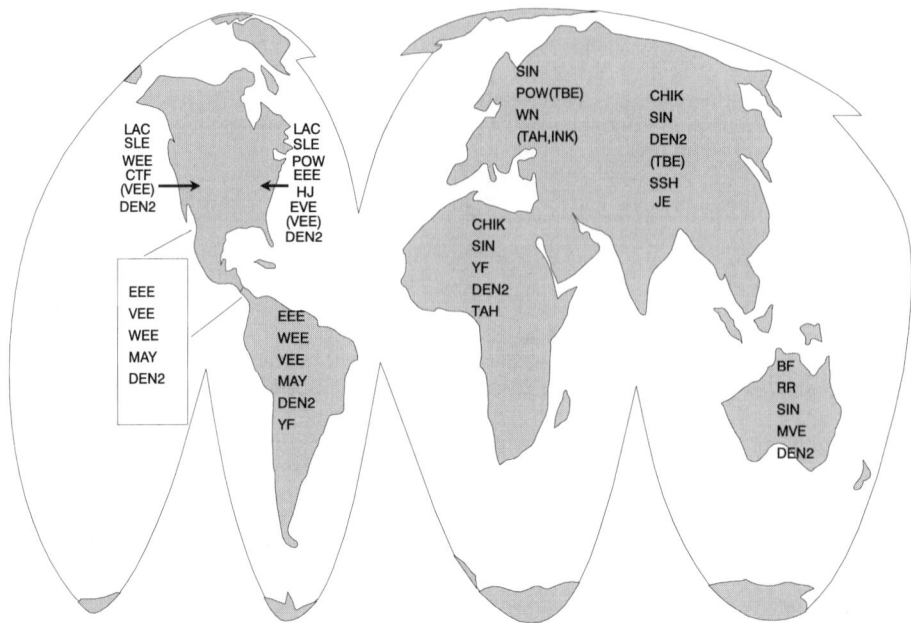

FIGURE 1 Antigen panels for arboviral testing based upon geographic distribution and prevalence. EVE, Everglades VEE; MAY, Mayaro; SIN, Sindbis; WN, West Nile encephalitis; TAH, Tahyna; INK, Inkoo; CHIK, chikungunya; BF, Barmah Forest; RR, Ross River.

The HI test requires the tedious preparation of hemagglutination buffers, continual test standardization, and the routine availability of gander erythrocytes. As the disease progresses, virus cross-reactivity in the HI test also increases. It is not uncommon for convalescent-phase serum samples to react with two or more virus antigens within the same virus family, making even a fourfold or higher serum HI titer rise between acute- and convalescent-phase serum samples difficult to interpret.

CF

The CF test measures the ability of the antiviral antibody to fix complement in the presence of virus antigen. It is more difficult to maintain proper quality control of the CF test than the HI test. Consequently, it is used only in special situations, such as when attempting to determine the time after infection that an individual serum sample was taken (Monath et al., 1980). Because CF antibody appears later in infection but has a shorter half-life (around 2 years), this test has been used as an indicator of more recent primary infection. With the advent of the IgM ELISA, allowing direct measurement of the early IgM antibody, CF tests are no longer as useful. The CF test may, however, indicate a recent infection if the serum sample is taken after the IgM antibody has waned.

PRNT

The plaque reduction neutralization test (PRNT) is a contradiction among assays used to diagnose arboviral infection. It is by far the most expensive and problematic test to perform, but it is still the only serological assay able to differentiate between infection by two closely antigenically related viruses. The subtlety of the PRNT is based upon the plaquing requirements of various arboviruses. Plaquing of arboviruses is usually performed in a variety of continuous mammalian cell lines. The most common of

these are Vero, BHK-21, and CER cells. Both plaque size and morphology might differ, depending on the cell type used. Time to plaque formation also varies. It may take 7 to 10 days for flavivirus plaques to form, while alphaviruses usually plaque in 24 to 48 h. To perform the PRNT, a virus seed of known titer must be available. Since many arboviruses lose titer upon freeze-thawing, it is best to have multiple aliquots of the virus seed. A constant amount of virus (50 to 100 PFU) is mixed individually with dilutions of the serum being tested. Following plating on cells, the plaques are visualized by adding a solution of the vital dye neutral red. The number of plaques in an individual plate is then divided by the starting number of virions to calculate percent neutralization. Typically, the PRNT is interpreted at a 70% PRNT titer, that is, the last dilution of serum that inhibits 70% of the total added plaques.

MAC-ELISA

Currently, the ELISA is used to measure either IgM or IgG individually. As with other infections, IgM titers usually signify recent virus infection. While many IgM protocols have been designed over the years, the most appropriate protocol for measuring IgM is the IgM capture ELISA (MAC-ELISA) (Westaway et al., 1974; Heinz et al., 1981; Burke and Nisalak, 1982; Jamnback et al., 1982; Monath et al., 1984; Bundo and Igarashi, 1985; Burke et al., 1985a, 1985b; Calisher et al., 1985a, 1985c, 1986a, 1986b, 1986c; Carter et al., 1985; Dykers et al., 1985; Besselaar et al., 1989; Cardosa et al., 1992; Sahu et al., 1994; Kittigul et al., 1998). This approach minimizes the interference of the higher-avidity IgG with IgM binding to antigen and consequently is more sensitive than the indirect-ELISA format for IgM (Heinz et al., 1981). The capture design also permits using antigens from a variety of sources, including those that normally have too much irrelevant protein for direct coating of plates. In the MAC-ELISA, human antiviral antibody is first captured in a 96-well ELISA plate by precoated commercial anti-human IgM antibody. The virus specificity of this captured IgM is determined by reacting individual wells with different virus antigens. The captured virus antigen is then detected with an antiviral antibody. The most efficient MAC-ELISA design uses broadly cross-reactive murine MAbs, conjugated to an enzyme as antiviral antigen detector molecules. Three of these MAbs—2A2C-3 (a broad alphavirus reactor), 6B6C-1 (a broad flavivirus reactor), and 10G-4 (a broad bunyavirus reactor)—are currently used to identify viral antigens from these three virus families (Table 2) (Roehrig et al., 1983, 1990a; Ludwig et al., 1991). Since the absorbance recorded is dependent upon the amount of antiviral antibody in the sample (provided that antigen is in excess), this ELISA can be run first at a single screening dilution (e.g., 1:400). The results from the MAC-ELISA are usually interpreted by dividing the absorbance of the test sample on antigen (P) by the absorbence of a negative control serum on antigen (N). In our laboratory, P/N ratios of >2.0 are considered positive, with the caveat that P/N values between 2 and 3 are often false positives. In this case, another serological assay (e.g., PRNT) should be performed to confirm equivocal results. Alternatively, a convalescent-phase serum can be tested. The antibody titer in this specimen should have increased from that in the acute-phase specimen. The MAC-ELISA is capable of distinguishing between infections caused by the medically important alphaviruses (EEE, WEE, and VEE viruses). Commercial IgM enzyme immunoassay kits are now being produced. While these kits may simplify testing procedures, they have yet to be critically evaluated by an unbiased source.

IgG ELISA

Standard indirect IgG ELISA can be used with arboviruses (Frazier and Shope, 1979; Roehrig, 1982). The problem with this approach is the wide variety of agents causing these diseases. It is simply too difficult and time-consuming to prepare pure virus antigen for even the limited subset of arboviruses used in antigen panels. To circumvent this problem, an indirect IgG ELISA has been developed in which virus antigen is captured in wells with broadly cross-reactive murine MAbs for each of the virus families (Table 2). The wells are first coated with these MAbs, and then the appropriate viral antigen is added. After viral antigen has been captured, this IgG assay functions like any other indirect ELISA. Patient serum samples are added, and the binding of antiviral antibody is detected with a commercial anti-species antibody conjugated to enzyme. While antigen is typically prepared as virus-infected mouse brain that has been processed to remove nonspecific inhibitors, virus-infected cell culture fluids can also be used. The latter antigen is typically lower in activity. There are commercial ELISA kits available; however, their reliability has yet to be conclusively proven.

Indirect IFA

One of the oldest commercial assays for antibody to SLE, WEE, EEE, and LAC viruses is based upon endpoint titration of sera by IFA. The kit is used by many public health and commercial laboratories. Since this is an indirect format, the problem of IgG competition for IgM binding occurs in the IgM IFA. The IFA lacks both the sensitivity and quantitative characteristics of the ELISA. For this rea-

son, serological diagnosis based upon IFA titrations is not preferred.

Virus Isolation and Identification

Three approaches are currently used to identify virus in complex solutions. The oldest method is to inoculate specimens into susceptible cell cultures, wait for virus-specific cytopathic effects, and then identify the virus isolate by a complex serological testing scheme. More recent techniques use antiviral antibody to capture virus antigen from a solution to be identified later with antiviral antibody. While both of these assays rely on viral proteins for the identification process, the advent of PCR assays has now established genome identification as one of the primary tests for virus identification. The evolution of rapid genome sequencing and the accumulation of large numbers of virus gene sequences have allowed PCR identification to evolve into a precise virus identification procedure. For those laboratories that do not have PCR capability, the development of virus-specific MAbs has improved the classic techniques of virus isolation and serological identification to the extent that these approaches are still options for the clinical virology laboratory.

IFA identification

Previously, virus NT assays were necessary to differentiate closely related viruses, such as flaviviruses. There now exist MAb reagents capable of identifying a specific virus by IFA (Table 2) (Roehrig, 1986, 1990; Heinz and Roehrig, 1990; Roehrig and Bolin, 1997). There are also MAbs capable of identifying virus complexes and even larger virus groups (e.g., all alphaviruses or all flaviviruses). While these MAbs have replaced virus-grouping antisera prepared by the National Institutes of Health (NIH), the NIH grouping serum samples are still quite useful in characterizing those arboviruses for which few MAbs are available (e.g., bunyaviruses). Because of the short supply of the NIH grouping sera, the reagents are usually available only to reference laboratories, whereas the MAb reagents are available to all public health laboratories and can identify all domestic medically important arboviruses.

Antigen capture ELISA

Because of the high avidity and precise specificity of MAbs, these reagents are currently being incorporated into antigen capture ELISA protocols (Hildreth et al., 1982, 1984; Beaty et al., 1983; Hildreth and Beaty, 1983; Kuno et al., 1985; Monath et al., 1986; Scott and Olson, 1986; Tsai et al., 1987a, 1988; Gajanana et al., 1995; Brown et al., 1996). In these assays, viral proteins are immobilized on a solid phase by an antiviral MAb. The captured antigen is then detected by using an antiviral antibody conjugated to enzyme. For simplicity, the detecting MAbs are usually broadly cross-reactive, such as the flavivirus MAb 6B6C-1. This approach reduces the number of enzyme conjugates necessary for virus identification. These protocols are currently formulated in ELISA format, but some are being redesigned for commercial use as dipstick assays. The dipstick assays will be particularly useful in mosquito surveillance efforts, where smaller numbers of specimens are routinely tested and the test is more amenable to field analyses. Antigen capture ELISA has been developed for EEE, WEE, SLE, LAC, and DEN viruses. While there have been no pro-

TABLE 2 MAbs useful in arbovirus identification

Virus and MAb	Virus specificity	ELISA			IFA	IHC[b]	Reference[c]
		MAC	IgG	Ag capture[d]			
Alphaviruses							
VEE virus							
1A2B-10	Wild-type VEE	−	−	D	+	+	Roehrig et al., 1991
5B4D-6	TC-83 VEE	−	−	D	+	+	Roehrig et al., 1982
1A3A-5	1AB,1C, 2	−	−	D	+	+	Roehrig and Mathews, 1985
1A4D-1	1AB, 1C, 1D	−	−	D	+	+	Roehrig and Mathews, 1985
1A1B-9	1D, 1E, 1F	−	−	D	+	+	Rico-Hesse et al., 1988
1A3A-9	All Subtype 1	−	−	D	+	+	Roehrig and Mathews, 1985
1A3B-7	All VEE complex	−	−	D	+	+	Roehrig and Mathews, 1985
WEE virus							
2B1C-6	WEE	−	−	D	+	+	Hunt and Roehrig, 1985
2A3D-5	WEE complex	−	−	C	+	+	Hunt and Roehrig, 1985
2D4-1	HJ	−	−	D	+	+	Karabatsos et al., 1988
2A2C-3	All alphaviruses	+	−	D	+	+	Karabatsos et al., 1988
EEE virus							
1B5C-3	NA EEE	−	−	D	+	+	Roehrig et al., 1990a
1B1C-4	EEE complex	−	−	D	+	+	Roehrig et al., 1990a
1A4B-6	All alphaviruses	−	+	C	+	+	Roehrig et al., 1990a
Flaviviruses							
SLE virus							
6B5A-2	SLE	−	−	D	+	+	Roehrig et al., 1983
4A4C-4	SLE	−	−	C	+	+	Roehrig et al., 1983
6B6C-1	All flaviviruses	+	−	D	+	+	Roehrig et al., 1983
JE virus							
JE314H52	JE	−	−	D	+	+	Unpublished data
6B4A-10	JE complex	−	−	C	+	+	Guirakhoo et al., 1992
6A4D-1	JE, MVE	−	−	D	+	+	Guirakhoo et al., 1992
MVE virus							
4B6C-2	MVE	−	−	D	+	+	Hawkes et al., 1988
YF virus							
5E3	YF	−	−	D	+	+	Schlesinger et al., 1983
2D12	YF	−	−	D	+	+	Schlesinger et al., 1983
864	Vaccine YF	−	−	D	+	+	Gould et al., 1985
117	Wild-type YF	−	−	D	+	+	Gould et al., 1989
DEN virus							
D2-1F1-3	DEN1	−	−	D	+	+	Unpublished data
3H5-1-21	DEN2	−	−	D	+	+	Henchal et al., 1985
D6-8A1-12	DEN3	−	−	D	+	+	Unpublished data
1H10-6-7	DEN4	−	−	D	+	+	Henchal et al., 1982
4G2	All flaviviruses	−	+	C	+	+	Henchal et al., 1982
Bunyaviruses							
LAC virus							
807-18	LA	−	−	D	+	+	Gonzalez-Scarano et al., 1982
10G4	CAL group	+	+	C and D	+	+	Ludwig et al., 1991

[a]+, useful; −, not useful.
[b]Useful for immunohistochemistry (IHC).
[c]Reference where MAb was first described. The publication lists all important biological characteristics of the MAb.
[d]Used as capture (C) or detector (D) antibodies. Ag, antigen.
[e]North American (NA) EEE viruses only.

tocols published for VEE virus antigen detection, the MAbs are available and development of this assay should not be far off. MAbs useful in antigen capture ELISA, as either capture or detector antibodies, are included in Table 2.

PCR and genome sequencing

PCR has opened a whole new vista in virus identification. There are a number of arbovirus PCR assays; however, many of them have not been rigorously tested. The extreme sensitivity of the PCR assay is attractive; however, it is this very sensitivity that compounds the problems of inadvertent introduction of nucleic acid that may result in false-positive reactions. The power of the PCR has been increased by the advent of rapid genome-sequencing techniques. By sequencing amplimers and comparing these sequences to known viral sequences, one can not only rapidly identify the unknown virus but in many cases its geographic source. PCR assays have been developed for many medically important arboviruses (Deubel et al., 1990, 1993; Eldadah et al., 1991; Chen et al., 1992; Howe et al., 1992a, 1992b; Campbell et al., 1993; Chow et al., 1993; Fulop et al., 1993; Porter et al., 1993; Chan et al., 1994; Chang et al., 1994; Wasieloski et al., 1994; Campbell and Huang, 1995, 1996; Kuno et al., 1996; Nawrocki et al., 1996; Figueiredo et al., 1997a, 1997b; Johnson et al., 1997; Meiyu et al., 1997; Pfeffer et al., 1997; Attoui et al., 1998a; Brightwell et al., 1998; Paranjpe and Banerjee, 1998).

Advantages and Disadvantages and Tips

Incorporation of MAb reagents into both serological assays and virus identification procedures has led to a new level of test standardization among diagnostic laboratories (Roehrig et al., 1998b). These readily reproducible and highly defined reagents continue to improve the rapidity, sensitivity and specificity of all diagnostic procedures. Similarly, the exquisite sensitivity of the PCR has created a paradigm shift in how infectious agents are handled and identified. High-containment viruses can be handled safely after they have been subjected to nucleic acid extraction techniques. Since the enhanced sensitivity of the PCR may lead to false-positive results, a diagnosis should not be based solely on a positive PCR result but should be confirmed by a diagnostic serological assay. Even though PCR and antigen detection ELISA are rapid and sensitive techniques, it is still useful to actually isolate a virus. Without having a virus in hand, future analyses will be impossible.

While the sensitivity of the newer assays can be spectacular, false positives can still occur. In the MAC-ELISA, the majority of equivocal results occur when P/N ratios are between 2.0 and 3.0. In these instances, it is still necessary to confirm these results by an alternative serological assay. In the antigen capture ELISA, the inhibition control is often not run, making the capture results uninterpretable (Roehrig et al., 1998b). Similarly, the classical approach of having paired serum samples is also still useful. Even though MAC-ELISA appears to be an excellent way to determine current infections, many times serum samples are taken so early after onset that even the IgM antibody titer is not yet measurable. Both IgM and IgG antibodies will usually be found in convalescent-phase serum samples of these individuals (Roehrig et al., 1998b).

Finally, the best approach to identifying and limiting arbovirus outbreaks is through good disease surveillance. This surveillance may involve sampling the mosquito vectors or animal reservoirs. Whichever approach is taken,

since arboviral diseases know no political boundaries, communication of arbovirus activities to agencies like CDC is imperative to formulate national strategy for disease intervention. CDC is also available to confirm laboratory testing for all laboratories that have possible arbovirus activity, especially those that have little experience with these diseases and, therefore, utilize commercial laboratories for their arboviral testing. It must be remembered that there are no national certification processes for these commercial laboratories, and consequently, the results may not be completely accurate.

Future and Conclusions

There is a continuing need for improvement in laboratory testing for arboviruses. A serological binding assay to identify the virus specificity of an antiflavivirus antibody response would be quite useful. Even though sequence analysis can identify many unique regions among and between flaviviruses, the conformational dependence of many flavivirus epitopes dictates that sophisticated modeling and structure-function analysis will be needed before these new antigens can be made. New MAbs capable of identifying many arboviruses (especially the medically important bunyaviruses) are also needed. Better and more rigorous testing of new PCR assays is necessary before they can be used routinely in the diagnostic laboratory. Standardization of diagnostic techniques would greatly improve lab-to-lab reproducibility. As new assays are developed, the pharmaceutical industry must take the lead in commercializing these tests and ensuring their validity.

With the exception of human immunodeficiency virus, arboviral infections can be considered the most important emerging or reemerging viral diseases. There are three reasons for this. First, as the world's population continues to grow, humans will continue to encroach on the habitats of these zoonotic viral infections. This encroachment, usually by nonimmune individuals, will result in epidemics of completely new viral diseases or reemergence of quiescent diseases. Second, as new agents are introduced into the human population, their ability to expand to new areas is facilitated by rapid transportation. An individual infected with DEN virus in Southeast Asia can be back in the United States before symptoms occur. These events could lead to new epidemics in completely new areas, where both physicians and laboratory personnel are unfamiliar with symptoms, diagnosis, and control measures. Finally, there are very few approaches to prevention of these diseases, since neither vaccines nor therapeutic pharmaceuticals exist. With this in mind, the physician and the diagnostic virology laboratory must consider arboviral diseases whenever symptoms, timing, exposure to insects, or travel history indicate possible arbovirus infections.

VIRAL PATHOGENS

Introduction

While there are a few exceptions that will be noted later, arboviruses are usually associated with two major disease syndromes—encephalitis and hemorrhagic fevers. The case incidence can vary from hundreds of thousands, as with DEN, to a handful, as with the tick-transmitted Powassan (POW) virus, which has caused only 21 reported cases of human encephalitis in the United States and Canada since it was first isolated in 1958. The severity of symptoms of arboviral infections can also vary. Most cases of arboviral encephalitis are subclinical; however, infection with EEE can result in death or severe lifelong neurological deficits. The continental United States has no indigenous arboviruses that cause hemorrhagic fever; however, travelers are at reasonable risk from infection with YF and DEN viruses. Fortunately for the physician, many of the arboviral infections are caused by closely related viruses, so their ecologies, entomologies, and epidemiologies are very similar, regardless of the continent on which the exposure occurred. For example, the recent outbreak of West Nile (WN) encephalitis in Bucharest, Romania, had many features in common with urban outbreaks of SLE in the United States (Tsai et al., 1998; Han et al., 1999). A summary of the biochemical characteristics of the major families of arboviruses is shown in Table 3.

Biology

Alphaviruses

The genus *Alphavirus* in the family *Togaviridae* contains many members that cause disease throughout the world. These viruses can cause classic encephalitis (WEE, EEE, and VEE viruses) or more disseminated disease (chikungunya, o'nyong-nyong, Semliki Forest, Sindbis, Ross River, Barmah Forest, and Mayaro viruses). While these viruses cause a variety of symptoms, their basic biologies are identical. An excellent and very comprehensive review of alphavirus molecular biology has been recently published (Strauss and Strauss, 1994). Alphaviruses are small (60- to 70-nm-diameter) viruses with a membrane-derived envelope surrounding an icosahedral nucleocapsid. The nucleocapsid encloses one positive-sense single-stranded (ss) RNA molecule of about 12 kDa. The genome encodes four nonstructural proteins (nsp1 to nsp4), the capsid (C) protein, and two virus envelope (E) glycoproteins, E1 and E2. Little is known about the early events in alphavirus replication. Recent evidence implicates laminin as the alphavirus receptor protein for some cell types (Strauss et al., 1994; Ludwig et al., 1996). Following attachment and endocytosis, the alphavirus must undergo an acid-catalyzed conformational change in its surface glycoproteins that initiates fusion of the viral envelope with the membrane of the endocytic vesicle. This fusion process releases the capsid into the cytoplasm and initiates RNA synthesis. During replication, the structural proteins (C, E1, and E2) are synthesized from a subgenomic mRNA of about one-third of the total genome. This allows for abundant synthesis of the structural proteins for inclusion in progeny virions. Progeny viruses bud through cellular membranes that have been modified by the addition of the E1 and E2 glycoproteins to release infectious virions (Strauss et al., 1995).

Alphaviruses typically kill infected tissue culture cells within 24 to 48 h. Cell death has recently been shown to occur through apoptosis (Levine et al., 1994, 1996; Ubol et al., 1994; Despres et al., 1995; Lewis et al., 1996; Griffin and Hardwick, 1997). Alphaviruses are also extremely efficient at shutting down host cell synthesis. Alphaviruses grow well in a number of continuous cell lines, such as Vero, BHK-21, and the mosquito cell line C6/36, any of which are acceptable for virus isolation and subsequent characterization protocols. In fact, by using inoculation procedures and IFA typing with MAbs, alphaviruses can usually be identified in 24 to 48 h.

The E2 protein appears to be the virion protein associated with attachment to susceptible cells. Preincubation of virus with neutralizing anti-E2 monoclonal antibodies alone will block virus attachment to cells (Roehrig et al., 1982, 1988; Roehrig and Mathews, 1985). Because of its ability to elicit neutralizing antibody, the E2 protein is under pressure from the immune response, which results in greater sequence divergence than for other alphavirus proteins. It is, therefore, this protein that is most responsible for the specificity of the PRNT in sera from infected individuals. The E1 glycoprotein appears to mediate cell membrane fusion and contains alphavirus group-reactive epitopes (Schmaljohn et al., 1983; Boggs et al., 1989). These E1 epitopes serve as the targets for the broadly reactive detector MAbs used in the diagnostic ELISA protocols (MAbs 2A2C-3 and 1A4B-6) (Roehrig et al., 1982, 1990a; Hunt and Roehrig, 1985).

Flaviviruses

The family *Flaviviridae* is the family of viruses that is responsible for most arboviral disease. The family includes the DEN, YF, Japanese encephalitis (JE), SLE, and tick-borne encephalitis (TBE) viruses. Other medically important flaviviruses are WN, Murray Valley encephalitis (MVE), and POW viruses. Flaviviruses are small (40 to 60 nm diameter) and are composed of an icosahedral nucleocapsid surrounded by a membrane-derived envelope. Similar to alphaviruses, the nucleocapsid encloses one positive-sense ssRNA molecule of about 10 to 11 kDa. This genome encodes three structural proteins, C, premembrane (prM), and E, and seven nonstructural proteins, NS1, NS2a, NS2b, NS3, NS4a, NS4b, and NS5 (Rice et al., 1985). Flavivirus attachment and entry are similar to those of alphaviruses, requiring an acid-catalyzed conformational shift in the E

TABLE 3 Characteristics of the families of common arboviruses

Characteristic	Value[a]				
	Togaviridae	*Flaviviridae*	*Bunyaviridae*	*Reoviridae*	*Rhabdoviridae*
Size (nm)	60–70	40–60	80–120	60–80	50–100 by 100–400
Morphology	Spherical	Spherical	Spherical	Spherical	Bullet shaped
Nucleic acid	ssRNA	ssRNA	ssRNA (3 segments)	dsRNA (12 segments)	ssRNA
Polarity of nucleic acid	Positive	Positive	Negative	Negative	Negative
Enveloped	Yes	Yes	Yes	No	Yes
Structural M proteins	E1, E2, C	E, C, M	L, G1, G2, N, NS_M, NS_s	?	L, G, N, P,
Nucleocapsid symmetry	Icosahedral	Icosahedral	Helical	Icosahedral	Helical
No. of nucleocapsids	1	1	3	?	1
RNA polymerase	No	No	Yes	Yes	Yes

[a] ?, unknown; ds, double stranded.

glycoprotein to effect membrane fusion and release of capsid into the cytoplasm (Roehrig et al., 1990b; Guirakhoo et al., 1991, 1992, 1993). Unlike alphaviruses, flaviviruses do not have a subgenomic RNA from which the structural proteins are derived. During maturation, the prM protein is cleaved by a furin-like cellular enzyme to M protein, which along with the E glycoprotein is found in the virion envelope (Stadler et al., 1997). Virus attachment and membrane fusion are both mediated by the E glycoprotein (Guirakhoo et al., 1989; Mandl et al., 1989; Roehrig et al., 1990b). The 2-Å crystal structure was recently solved for the amino-terminal 400-amino-acid fragment of the TBE virus E glycoprotein (Rey et al., 1995). The three-dimensional structure confirmed much of the biology of the E glycoprotein. For a more extensive review of the flavivirus antigenic structure and function, there are a number of good sources (Heinz, 1986; Heinz and Roehrig, 1990; Heinz and Mandl, 1993; Roehrig, 1997). Unlike alphaviruses, flaviviruses do not shut down host cell synthesis. In general, flaviviruses infect a number of continuous cell types, but are more selective and take longer to grow. Many flaviviruses may require up to 7 days for adequate antigen expression. Most of the mosquito-borne flaviviruses grow well in the mosquito cell line C6/36.

The E glycoprotein elicits virus-neutralizing antibody, so this protein is subjected to immune pressure and, as a result, is responsible for eliciting virus-specific antibody (Peiris et al., 1982; Kimura-Kuroda and Yasui, 1983; Roehrig et al., 1983, 1998a; Hawkes et al., 1988; Barrett et al., 1990). This protein also contains epitopes that are flavivirus cross-reactive (Gentry et al., 1982; Henchal et al., 1982, 1985; Hawkes et al., 1988; Roehrig et al., 1983, 1998a). These cross-reactive epitopes serve as the targets for the broadly reactive detector MAbs used in the diagnostic-ELISA protocols (MAbs 6B6C-1 and 4G2) (Roehrig et al., 1983; Henchal et al., 1985).

Bunyaviruses

The family *Bunyaviridae* contains the most vector-borne viruses, only a few of which have been consistently associated with human disease. For the United States, the California serogroup viruses, primarily LAC encephalitis virus, are the most important pathogens. Other bunyaviruses associated with human disease are the Cache Valley, Jamestown Canyon, Snowshoe hare (SSH), Tahyna, and Inkoo viruses. The bunyaviruses are larger than either alphaviruses or flaviviruses, about 80 to 120 nm in size. The virion contains a tripartite genome with three negative-sense ssRNA segments enclosed in helical nucleocapsids (Obijeski et al., 1976b). The L genome segment encodes the L polymerase, the M genome segment encodes the NS_M protein and the two surface glycoproteins G1 and G2, and the S genome segment encodes the nucleocapsid (N) and NS_S proteins (Obijeski et al., 1976a; Gentsch et al., 1977; Bishop et al., 1980, 1982). Because of their tripartite genome, there is a potential that bunyaviruses undergo genetic reassortment in nature, similar to orthomyxoviruses (Gentsch et al., 1977, 1979; Bishop et al., 1978; Bishop, 1979; Bishop and Beaty, 1988; Baldridge et al., 1989; Chandler et al., 1990, 1991; Urquidi and Bishop, 1992).

Coltiviruses

Little is known about the molecular biology of coltiviruses, which are members of the family *Reoviridae*. The virion is naked (60 to 80 nm diameter) and carries 12 negative-sense double-stranded RNA segments within its nucleocapsid (Knudson, 1981). The specific structure of the virion and its structural proteins have not been defined (Attoui et al., 1997, 1998b). The coding assignments are just now being determined, and the functions of the encoded proteins have also not been well defined. Because of the multiple genomic segments, coltiviruses, like bunyaviruses, might be able to undergo genetic reassortment in nature, but this has not been demonstrated (Karabatsos et al., 1987).

Rhabdoviruses

Rhabdoviruses are larger viruses (50 to 100 nm by 100 to 400 nm) and have a characteristic bullet shape. Rabies virus is the rhabdovirus of most public health significance; however, the prototype virus, vesicular stomatitis virus (VSV), has recently been associated with outbreaks in horses and cattle. This virus is included here because of its possible transmission to animal handlers, although such transmission has been rare. Because of its similarities to rabies virus, the molecular biology of VSV has been intensively studied, and for detailed reviews, the reader is referred elsewhere (Dietzschold et al., 1996).

Pathogenesis

Arboviruses gain entry through the skin by the bite of an infected arthropod; however, some are capable of being transmitted by aerosol in the laboratory setting. While knowledge of the initial events of infection is superficial, evidence is accumulating that early interactions of virus, cells, and mosquito saliva might play some role in the outcome of infection (Zeidner et al., 1999). The mosquito saliva enters the dermis and at times enters the small capillaries directly when the mosquito's proboscis penetrates the vessel. It is presumed that the virus replicates initially in the dermal tissues, including the capillary endothelium, although it is also possible that virus is transported directly in the blood to primary target organs. Replication also occurs in the regional lymph nodes, and from there the blood is seeded, inducing a secondary viremia, which in turn carries virus to infect muscle and connective-tissue cells. This viremia is often of very high titer and is accompanied by fever, leukopenia, and malaise. It is during this viremic phase that an arthropod may feed and become infected. The period between infection and viremia (the intrinsic incubation period) is usually short, from 1 to 3 days. Viremia may last 2 to 5 days. Colorado tick fever viremia is of much longer duration because immature erythrocytes are infected and virus remains in the blood cells for 2 to 6 weeks.

The vast majority of human arboviral infections are either asymptomatic or self-limited febrile illnesses. Antibody is produced, and it complexes with and neutralizes circulating virus. The process is accompanied by complete recovery and leads to the lifelong presence of antibody. Occasionally, however, an infected person develops encephalitis. The mechanism of entry of virus into the central nervous system (CNS) is not completely understood, nor is it understood why one person develops encephalitis and another apparently similar individual does not. Virus may reach the brain by seeding of cerebral capillaries during viremia and then by direct invasion of the brain parenchyma through the capillary walls. Alternatively, certain neural cells, such as the olfactory neurons, are exposed directly to circulating blood; viremia may seed these nerve endings, and the virus may pass directly to the olfactory lobe of the brain (Monath et al., 1983). Regardless of the mechanism, it is important to note that the process of seeding the brain and productive infection of brain cells takes time. By the time the patient presents with encephalitis, serum antibody is usually detectable, as is antibody in the CSF. At this stage of infection, viremia has ceased and diagnosis is made by serological assay.

Epidemiology

EEE

EEE occurs throughout the eastern part of the United States. There have been 172 confirmed cases of EEE in the United States since 1964 (Fig. 2). Epidemics of EEE are rare, but a few human cases occur every summer and fall. Equine epizootics can also occur in regions as far north as Canada. The virus is maintained in nature by an enzootic cycle involving birds and a variety of mosquito species. The swampy environments necessary for the EEE vector mosquitoes usually limit the dissemination of this disease. *Culiseta melanura* is the main mosquito infecting birds; human and equine infections are associated with *Aedes sollicitans* and *Aedes vexans* in temperate regions and with *Aedes taeniorhynchus*, *Culex taeniopus*, and *Culex nigripalpus* in the tropics. A related alphavirus, Highlands J (HJ), also occurs in the eastern United States. While it is not frequently associated with human disease, cross-reactions with the virus can confuse laboratory diagnosis.

Inapparent cases of EEE are rare. Onset is abrupt, with high fever, headache, and vomiting followed by drowsiness, coma, and severe convulsions. On examination, there is neck stiffness, spasticity, and, in infants, bulging fontanelles. Death may occur within 3 to 5 days of onset. Sequelae are common (30%), including convulsions, paral-

ysis, and mental retardation. The case/fatality ratio for EEE can reach as high as 30%.

WEE

While the last major United States epidemic of WEE occurred in the 1970s, WEE remains an important cause of encephalitis in North America with 639 confirmed cases since 1964 (Fig. 2) (Reeves, 1987). The enzootic cycle involves passerine birds (in which the infection is inapparent) and culicine mosquitoes, principally *Culex*. Human cases are first seen in June or July in the Northern Hemisphere, but the mechanism of overwintering of the virus is unknown. Children, especially those under 1 year old, are affected more severely than adults and may be left with permanent brain damage, which is also seen in about 5% of adults. The mortality is about 25%. Strains of WEE virus appear to be relatively homogeneous by oligonucleotide fingerprinting and are clearly different from the serologically related HJ virus (Trent and Grant, 1980; Hunt and Roehrig, 1985; Karabatsos et al., 1988).

VEE

VEE virus was isolated in Venezuela in 1938 from the brain of a horse; like the EEE and WEE viruses, it causes encephalitis in members of the family Equidae and humans. The enzootic cycle of the VEE virus is still incompletely understood, but it appears to involve a variety of rodents rather than avian species, which are the hosts of the EEE

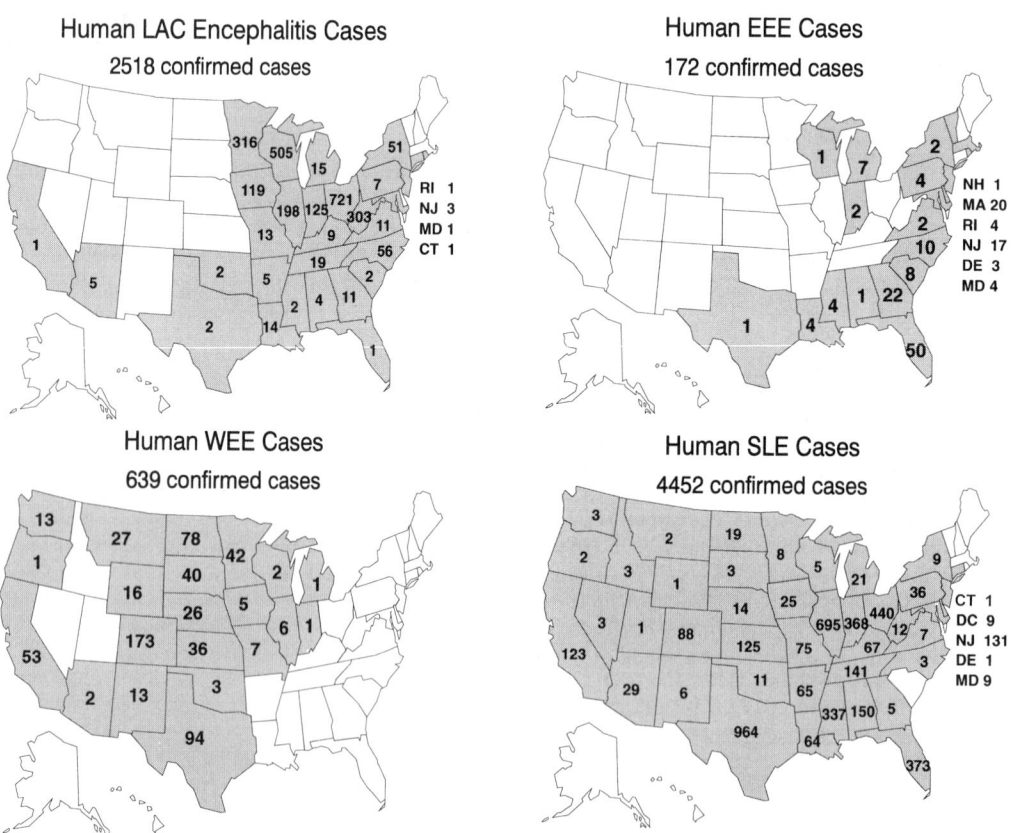

FIGURE 2 Distribution and incidence of cases of human arbovirus infections in the United States from 1964 to 1998.

and WEE viruses. Infection of humans is less severe than with the other two alphaviruses, and fatalities are rare. Adults usually develop only an influenzalike illness, and overt encephalitis is usually confined to children. Six antigenic subtypes of VEE viruses (1 to 6) are now recognized, with subtype 1 subdivided into five variants, 1AB to 1F (Calisher et al., 1980, 1985b; Kinney et al., 1983). Only subtype 1AB and 1C viruses have been associated with major epidemics and epizootics (Powers et al., 1997). The other VEE virus subtypes are involved in enzootic VEE virus transmission (Oberste et al., 1996, 1998a, 1998b; Watts et al., 1997, 1998).

A major VEE epizootic spread through Central America to reach Texas in 1971, where it was controlled by a massive equine vaccination program, using the live attenuated TC-83 vaccine. Over 200,000 horses died in the outbreak, and there were several thousand human infections. It is now believed that that epidemic was caused by poorly inactivated vaccine (Sneider et al., 1993; Weaver et al., 1999). The most recent outbreak of epizootic VEE occurred in 1995 in Colombia and Venezuela (Weaver et al., 1996; Rivas et al., 1997).

SLE

SLE virus is the most important mosquito-borne viral cause of epidemic encephalitis in the United States (Monath and Tsai, 1987; Tsai et al., 1987b; Marfin et al., 1993; McCaig et al., 1994). It can also be found throughout the Western Hemisphere. The last major U.S. SLE epidemic occurred in the late 1970s, when thousands of individuals in the Midwest were infected. The average yearly number of cases of SLE is 193; however, only about 1% of infections are clinically apparent. The CDC has registered 4,452 confirmed human SLE cases since 1964 (Fig. 2). The overall case/fatality ratio is 5 to 15%. Clinical SLE infections have an age-dependent distribution, with the elderly at highest risk. SLE virus is maintained in nature in a virus-bird-virus cycle. Mosquito vectors in the eastern part of the United States are usually *Culex pipiens pipiens* or *Culex pipiens quinquefasciatus*. *Culex tarsalis* is the primary SLE virus vector in the West. Anecdotal evidence indicates that the eastern form of SLE is symptomatically more severe than the western form. The reasons for this are not known. While SLE is seasonal in temperate areas, year-round transmission can occur in milder climatic areas, such as Florida.

YF

YF is believed to have originated in Africa; the first recorded outbreak was in Barbados in 1647. This was followed by innumerable epidemics in the West Indies, Central and South America, and the eastern United States as far north as New York and by introductions through seaports in more temperate regions in the Western Hemisphere. YF virus was the first virus associated with mosquito transmission and the first flavivirus for which an effective vaccine was developed. Control of the *Aedes aegypti* mosquito almost completely eradicated urban YF. However, the disease persisted sporadically in rural areas, as a consequence of a sylvan cycle involving monkeys and forest-dwelling mosquitoes, e.g., *Haemagogus* and *Sabethes* spp. in South America, *Aedes africanus* in East Africa, and a variety of *Aedes* spp. in West Africa.

The disease varies from an inapparent infection to a fulminating disease, terminating in death. After an incubation period of 3 to 6 days, the illness begins suddenly with fever, rigors, headache, and backache. The patient is intensely ill

and may suffer from from nausea and vomiting. A tendency to bleeding may be seen early on. This stage of active congestion is followed quickly by one of stasis. The facial edema and flushing are replaced by a dusky pallor, the gums become swollen and bleed easily, and there is a pronounced hemorrhagic tendency with black vomit, melaena, and ecchymoses. Death, when it occurs, is usually within 6 to 7 days of onset and is rare after 10 days of illness. The jaundice, which gives the disease its name, is generally apparent only in convalescing patients. Mortality can range from 5 to 50%. At autopsy, the organs most affected are the liver, spleen, kidneys, and heart.

DEN

There are four DEN virus serotypes (1 to 4), all of which are endemic throughout the tropics, particularly in Asia, the Caribbean, the Pacific, and some areas of West Africa. DEN is currently the most important arboviral disease, with hundreds of thousands of cases occurring each year and millions of people at risk. In many areas, several types of DEN virus cocirculate, and successive epidemics may occur, caused by different serotypes, because cross protection between DEN virus types in humans lasts only a short time.

DEN is endemic in tropical areas where *Stegomyia* mosquitoes are constantly active. *A. aegypti* is the most important vector, particularly in urban areas, but other *Stegomyia* spp. play a role in rural areas of Asia and the Pacific Islands. These include *Aedes albopictus*, *Aedes polynesiensis*, and *Aedes scutellaris*. There is some evidence that monkeys are involved in maintenance of the virus, but there is no proof of a vertebrate maintenance host other than humans.

The clinical picture of classic DEN fever usually affects adults and older children. There is an incubation period of 5 to 8 days followed by the sudden onset of fever (which is often biphasic), severe headache, chills, and generalized myalgia. A maculopapular rash generally appears on the trunk between the third and fifth day of illness and later spreads to the face and extremities. The illness generally lasts for about 10 days, after which recovery is usually complete, although convalescence may be protracted.

DHF-DSS

In Southeast Asia, DHF-DSS occurs almost entirely in children (Halstead, 1988). More recently, DHF has been reported in Cuba (Guzman et al., 1984, 1987, 1988) and in Brazil (Nogueira et al., 1989); in both of these outbreaks, substantial numbers of adults were affected.

JE

JE is widespread throughout Asia, from the maritime provinces of the former USSR to South India and Sri Lanka (Umenai et al., 1985). Epidemics occur in late summer in temperate regions, but the infection is endemic in many tropical areas. Culicine mosquitoes breeding in rice fields are the main vectors, transmitting virus to humans from water birds and pigs, which act as amplifying hosts. The onset of symptoms is usually sudden and may progress to frank encephalitis. The mortality in most outbreaks is less than 10% but has exceeded 30%.

TBE

TBE is caused by two variants of the same flavivirus: central European encephalitis (CEE) and Russian spring and summer encephalitis (RSSE). While these two viruses are serologically closely related, the diseases they cause differ in severity. RSSE is the more severe infection, carrying a mortality of up to 25% in some outbreaks, whereas that in CEE seldom exceeds 5%. RSSE is transmitted by *Ixodes persulcatus* ticks, while *Ixodes ricinus* transmits CEE. CEE can occur in enzootic foci extending from Scandinavia in the north to Greece and Yugoslavia in the south. Males, particularly those who spend large amounts of time in the forests, are at greatest risk from TBE infection. Infection can also be acquired by ingestion of infected cow or goat milk. The infection ranges from mild, influenza-type illness or a benign, aseptic meningitis to fatal meningoencephalitis. Fever is often biphasic, and there may be severe headache and neck rigidity, with transient pareses of the limbs or shoulder girdle or, less commonly, of the respiratory musculature. A few patients are left with residual flaccid paralysis (Ackermann et al., 1986).

POW

POW virus is a rare cause of acute viral CNS disease in Canada and the United States, but it is also present in Russia, where it has been recovered from mosquitoes, ticks, and humans. It was first isolated in Canada in 1958 and has since caused 21 cases of encephalitis in Canada and the eastern United States. Patients who recover may have residual neurological problems. In addition to isolations from humans, the virus has been recovered from ticks (*Ixodes marxi*, *Ixodes cookei*, and *Dermacentor andersoni*) and from the tissues of a spotted skunk (*Spilogale putorius*) (Johnson, 1987).

CAL Serogroup Encephalitis

CAL encephalitis virus was isolated in 1943 from *Aedes* mosquitoes in California and was later associated serologically with three pediatric encephalitis cases in California. Not until 1964, however, was the full significance of the CAL serogroup viruses realized. In that year, a virus closely related to CAL encephalitis virus was isolated from the stored brain of a child who had died in 1960 in LaCrosse, Wis. Starting in the early 1960s, the LAC virus has been associated in the United States with about 30 to 140 cases of CAL serogroup encephalitis per year, and LAC encephalitis is currently the most prevalent arboviral encephalitis in the United States. There have been 2,518 confirmed cases of LAC encephalitis in the United States since 1964 (Fig. 2). Recently, disease caused by LAC encephalitis virus has been identified in areas outside of its classical range of the upper Midwest. LAC encephalitis cases have now been identified in West Virginia, Virginia, Kentucky, Tennessee, North Carolina, and Alabama, indicating that either this virus is emerging into new territory or it has always been present in these areas and increased surveillance has led to the recognition of disease (McJunkin et al., 1998). Two other closely related CAL serogroup viruses, SSH and James Canyon viruses, have been etiologically associated with a small number of encephalitis cases in the United States and Canada since 1980 (Artsob et al., 1980, 1982, 1986; Grimstad et al., 1982).

Most cases of LAC encephalitis are subclinical. Typically, clinical cases of LAC encephalitis occur in children under the age of 15 years. While infection with LAC virus can progress to frank encephalitis, LAC encephalitis is rarely fatal. LAC encephalitis is an endemic disease associated with hardwood forests. The primary mosquito vector is *Aedes triseriatus*, and the virus is maintained in nature in a mosquito-rodent (usually ground squirrels or chipmunks) cycle. An unusual feature of this virus is its ability to be transferred from mother to offspring by a mechanism known as transovarial transmission in mosquitoes. This mechanism of transmission assists the virus in establishing enzootic foci of infection.

CTF

Colorado tick fever (CTF) is prevalent in the western mountain region of the United States and was initially confused with Rocky Mountain spotted fever until the virus was isolated in 1944 (Florio et al., 1944). There have been 864 confirmed cases of CTF in the United States since 1964, with 551 of the cases occurring in Colorado. It is most common in campers, hikers, and other persons coming in contact with the tick vector, *D. andersoni* (Burgdorfer, 1977; Bowen et al., 1981; Lane et al., 1982; Eads and Smith, 1983; Emmons, 1985, 1988; McLean et al., 1989). Typical symptoms are diphasic fever, muscle aches, malaise, and, occasionally, hemorrhagic or CNS complications in children.

VSV

It is not clear whether VSV is an arbovirus in the classical sense or whether insects are purely mechanical vectors of the disease. The main symptoms of VSV in animals are vesicles in the mouths of the infected animals. There is no good evidence that VSV routinely infects humans; and if it does, most of the infections are subclinical. The disease presents as a mild fever. Infection of humans with VSV must be associated with contact with previously infected animals.

Diagnosis

Inclusion of arboviruses as possible etiologic agents of infection in a laboratory differential diagnosis can initially be based on three considerations: case location, time of year the case occurred, and the patient's travel history. A history of mosquito or tick bites is also useful; however, it is usually difficult to document accurately. Domestic arboviral infection has traditionally occurred in the late summer or fall as the numbers of insect vectors increase and enough time has passed for virus amplification in mammalian hosts to allow for transmission to humans. To decide which antigens to use in a laboratory test for arboviruses, the physician must be aware of any unusual travel prior to the onset of symptoms. If a patient lives in California and has not recently visited the eastern part of the United States, the chance of an EEE infection is nil; however, infection with either WEE or SLE is a possibility. Similarly, if a patient presents with symptoms consistent with DEN, and traveled to Puerto Rico 6 months before, the chance that the current infection is DEN would be small, even though DEN is endemic in Puerto Rico.

Another confounding issue in the diagnosis of encephalitis is the multitude of agents that can cause similar symptoms and the low frequency of arboviral encephalitis. Herpesviruses and enteroviruses also cause encephalitis. Enterovirus encephalitis occurring in the summer months may be confused with arboviral encephalitis. Fortunately, the age distribution of enteroviral encephalitis cases is usually sufficiently different from arboviral encephalitides that arboviruses can be ruled out.

The clinical laboratory findings and histopathology of arboviral encephalitis are often not helpful in arriving at an etiologic diagnosis. A definitive diagnosis can be made only in the diagnostic virology laboratory. The histopathology is characterized by perivascular cuffing, neuronal chromatolysis, cell shrinkage, and neuronophagia. EEE brain lesions are unusually necrotizing and are associated with high lympho-cyte counts and modestly elevated protein levels in the CSF.

Prevention and Therapy

The most effective ways to prevent or contain an arboviral outbreak are through vaccination and chemical control of the arthropod vector. Only YF has a currently licensed vaccine that is readily available to the general public. The YF vaccine is a live, attenuated vaccine that has a long, successful track record and should be recommended for anyone with planned travel into areas where YF is endemic. Both inactivated and live attenuated vaccines for JE virus have been developed and successfully used in Asia (Hennessy et al., 1996; Liu et al., 1997). The killed vaccine has been approved for use in the United States. A killed vaccine for CEE virus has been in use in Europe for a number of years, but it is not approved for use in the United States. There are a number of veterinary vaccines available for the equine encephalitides, and there are also experimental vaccines that can be used, with appropriate approval, to protect laboratory personnel from EEE, WEE, and VEE virus infection. These human vaccines, developed by the U.S. Department of Defense, have investigational new drug status and possess characteristics that will always limit their use by the general public.

Interrupting the virus transmission cycle by reducing human exposure to the arthropod vector remains the most common approach to intervening in an arboviral outbreak. The reduction of human exposure can be accomplished in one of three ways. Since mosquitoes require water in which to breed, source reduction of mosquito breeding sites can reduce the risk of human infection. Source reduction may be either drastic (e.g., draining swamps) or more subtle (e.g., removing items that collect water, such as discarded tires or cans). A second and relatively easy approach to reducing human exposure is using insect repellents or reducing time spent outside during the time that mosquitoes are most active. Many of the arboviral mosquito vectors are most active at dusk. Reducing or modifying times of outside activity during the early evening will, therefore, reduce the chance of human infection. The third approach to reducing human exposure is applying either adulticides to reduce the number of biting mosquitoes or larvicides to reduce future mosquito generations. Either of these techniques is temporary, and frequent reapplications may be required. As society becomes more sensitized to the general application of chemicals, arthropod control through insecticide treatment becomes more difficult. Many of the modern insecticides have been shown to be environmentally and physiologically safe to use and should still be considered where control efforts need to be applied to larger areas. However, with frequent administration of insecticides, mosquito populations can acquire resistence.

As with most viral infections, few therapeutic treatments are available for arboviral infections. Treatment is usually only supportive. A recent study of LAC encephalitis did demonstrate the possible effectiveness of treating patients with ribavirin (McJunkin et al., 1997). This was the first study of its kind and will require confirmation in subsequent analyses.

Unusual Features, Insights, Future, and Conclusions

Because of the unique and complex ecology of arboviruses, prevention and control of arboviral disease are difficult. Control of the mosquito vector or reduction of mosquito breeding sites is usually only a temporary solution. Because of cost, even development of a vaccine as effective as the YF 17D vaccine has not completely controlled this important arboviral disease. What then, are our options to reduce the incidence of these diseases? Vaccine development for those viruses that cause a significant number of infections (e.g., DEN) is finally being pursued by the private sector. The goal will be to produce these products at a cost that will not prohibit their use in disadvantaged target populations.

For the "orphan" arboviral diseases (e.g., SLE and EEE), there is little hope that these markets will be lucrative enough for commercial vaccine development. Other approaches must be pursued. One possibility would be the development of protective immune globulin to be used prophylactically in the face of an expanding epidemic. Progress in producing human MAbs, or humanization of protective murine MAbs, may lead to a cheap source of readily available products capable of aborting viral infection. In animal models, it appears that preexisting neutralizing antibody may be sufficient to abort infection (Mathews and Roehrig, 1982, 1984; Boere et al., 1983; Roehrig and Mathews, 1985; Schlesinger et al., 1985; Kaufman et al., 1987). It is not quite as clear whether these reagents could be used therapeutically as well. What is known is that through adequate disease surveillance, epidemics could be detected very early, and the at-risk population could be identified and protected by these reagents.

As with most viral infections, it is not easy to produce therapeutic agents. The use of ribavirin for LAC infections appears to be promising. Since most medically important arboviruses are positive-stranded RNA viruses, there are few such compounds available. As with vaccines, because the potential market for these therapeutic agents may be small, it will be difficult to entice the pharmaceutical companies to develop therapeutics specifically for arboviruses. For flaviviruses, research on and development of therapeutic drugs for hepatitis C may be directly applicable to the mosquito- and tick-borne members of this family. It is clear, however, that therapeutic drugs for most of these viruses are not on the immediate horizon. Because of these factors, arboviral diseases will continue to affect humankind for the foreseeable future.

REFERENCES

Ackermann, R., K. Kruger, M. Roggendorf, B. Rehse-Kupper, M. Mortter, M. Schneider, and I. Vukadinovic. 1986. Spread of early-summer meningoencephalitis in the Federal Republic of Germany. *Dtsch. Med. Wochenschr.* **111:**927–933. (In German.)

Artsob, H., L. Spence, C. Th'ng, and R. West. 1980. Serological survey for human arbovirus infections in the province of Quebec. *Can. J. Public Health* **71:**341–346.

Artsob, H., L. Spence, G. Surgeoner, B. Helson, J. Thorsen, L. Grant, and C. Th'ng. 1982. Snowshoe hare virus activity in Southern Ontario. *Can. J. Public Health* **73:**345–349.

Artsob, H., L. Spence, C. Th'ng, V. Lampotang, D. Johnston, C. MacInnes, F. Matejka, D. Voigt, and I. Watt. 1986. Arbovirus infections in several Ontario mammals, 1975–1980. *Can. J. Vet. Res.* **50:**42–46.

Attoui, H., P. De Micco, and X. de Lamballerie. 1997. Complete nucleotide sequence of Colorado tick fever virus segments M6, S1 and S2. *J. Gen. Virol.* **78:**2895–2899.

Attoui, H., F. Billoir, J. M. Bruey, P. de Micco, and X. de Lamballerie. 1998a. Serologic and molecular diagnosis of Colorado tick fever viral infections. *Am. J. Trop. Med. Hyg.* **59:**763–768.

Attoui, H., R. N. Charrel, F. Billoir, J. F. Cantaloube, P. de Micco, and X. de Lamballerie. 1998b. Comparative sequence analysis of American, European and Asian isolates of viruses in the genus Coltivirus. *J. Gen. Virol.* **79:**2481–2489.

Baldridge, G. D., B. J. Beaty, and M. J. Hewlett. 1989. Genomic stability of La Crosse virus during vertical and horizontal transmission. *Arch. Virol.* **108:**89–99.

Barrett, A. D., J. H. Mathews, B. R. Miller, A. R. Medlen, T. N. Ledger, and J. T. Roehrig. 1990. Identification of monoclonal antibodies that distinguish between 17D-204 and other strains of yellow fever virus. *J. Gen. Virol.* **71:**13–18.

Beaty, B. J., T. L. Jamnback, S. W. Hildreth, and K. L. Brown. 1983. Rapid diagnosis of La Crosse virus infections: evaluation of serologic and antigen detection techniques for the clinically relevant diagnosis of La Crosse encephalitis. *Prog. Clin. Biol. Res.* **123:**293–302.

Besselaar, T. G., N. K. Blackburn, and N. Aldridge. 1989. Comparison of an antibody-capture IgM enzyme-linked immunosorbent assay with IgM-indirect immunofluorescence for the diagnosis of acute Sindbis and West Nile infections. *J. Virol. Methods* **25:**337–345.

Bishop, D. H. 1979. Genetic potential of bunyaviruses. *Curr. Top. Microbiol. Immunol.* **86:**1–33.

Bishop, D. H., and B. J. Beaty. 1988. Molecular and biochemical studies of the evolution, infection and transmission of insect bunyaviruses. *Philos. Trans. R. Soc. Lond. B* **321:**463–483.

Bishop, D. H., J. R. Gentsch, and L. H. el-Said. 1978. Genetic capabilities of bunyaviruses. *J. Egypt Public Health Assoc.* **53:**217–234.

Bishop, D. H., C. H. Calisher, J. Casals, M. P. Chumakov, S. Y. Gaidamovich, C. Hannoun, D. K. Lvov, I. D. Marshall, N. Oker-Blom, R. F. Pettersson, J. S. Porterfield, P. K. Russell, R. E. Shope, and E. G. Westaway. 1980. Bunyaviridae. *Intervirology* **14:**125–143.

Bishop, D. H., K. G. Gould, H. Akashi, and C. M. Clerx-van Haaster. 1982. The complete sequence and coding content of snowshoe hare bunyavirus small (S) viral RNA species. *Nucleic Acids Res.* **10:**3703–3713.

Boere, W. A., B. J. Benaissa-Trouw, M. Harmsen, C. A. Kraaijeveld, and H. Snippe. 1983. Neutralizing and non-neutralizing monoclonal antibodies to the E2 glycoprotein of Semliki Forest virus can protect mice from lethal encephalitis. *J. Gen. Virol.* **64:**1405–1408.

Boggs, W. M., C. S. Hahn, E. G. Strauss, J. H. Strauss, and D. E. Griffin. 1989. Low pH-dependent Sindbis virus-induced fusion of BHK cells: differences between strains correlate with amino acid changes in the E1 glycoprotein. *Virology* **169:**485–488.

Bowen, G. S., R. G. McLean, R. B. Shriner, D. B. Francy, K. S. Pokorny, J. M. Trimble, R. A. Bolin, A. M. Barnes, C. H. Calisher, and D. J. Muth. 1981. The ecology of Colorado tick fever in Rocky Mountain National Park in 1974. II. Infection in small mammals. *Am. J. Trop. Med. Hyg.* **30:**490–496.

Brightwell, G., J. M. Brown, and D. M. Coates. 1998. Genetic targets for the detection and identification of Venezuelan equine encephalitis viruses. *Arch. Virol.* **143:**731–742.

Brown, J. M., D. M. Coates, and R. J. Phillpotts. 1996. Evaluation of monoclonal antibodies for generic detection of flaviviruses by ELISA. *J. Virol. Methods* **62:**143–151.

Bundo, K., and A. Igarashi. 1985. Antibody-capture ELISA for detection of immunoglobulin M antibodies in sera from Japanese encephalitis and dengue hemorrhagic fever patients. *J. Virol. Methods* **11:**15–22.

Burgdorfer, W. 1977. Tick-borne diseases in the United States: Rocky Mountain spotted fever and Colorado tick fever. A review. *Acta Trop.* **34:**103–126.

Burke, D. S., and A. Nisalak. 1982. Detection of Japanese encephalitis virus immunoglobulin M antibodies in serum by antibody capture radioimmunoassay. *J. Clin. Microbiol.* **15:**353–361.

Burke, D. S., A. Nisalak, M. A. Ussery, T. Laorakpongse, and S. Chantavibul. 1985a. Kinetics of IgM and IgG responses to Japanese encephalitis virus in human serum and cerebrospinal fluid. *J. Infect. Dis.* **151:**1093–1099.

Burke, D. S., M. Tingpalapong, M. R. Elwell, P. S. Paul, and R. A. Van Deusen. 1985b. Japanese encephalitis virus immunoglobulin M antibodies in porcine sera. *Am. J. Vet. Res.* **46:**2054–2057.

Calisher, C. H., R. E. Shope, W. Brandt, J. Casals, N. Karabatsos, F. A. Murphy, R. B. Tesh, and M. E. Wiebe. 1980. Proposed antigenic classification of registered arboviruses. I. Togaviridae, Alphavirus. *Intervirology.* **14:**229–232.

Calisher, C. H., and W. H. Thompson. 1983. *California Serogroup Viruses,* vol. 123. Alan R. Liss, New York, N.Y.

Calisher, C. H., O. Meurman, M. Brummer-Korvenkontio, P. E. Halonen, and D. J. Muth. 1985a. Sensitive enzyme immunoassay for detecting immunoglobulin M antibodies to Sindbis virus and further evidence that Pogosta disease is caused by a western equine encephalitis complex virus. *J. Clin. Microbiol.* **22:**566–571.

Calisher, C. H., T. P. Monath, C. J. Mitchell, M. S. Sabattini, C. B. Cropp, J. Kerschner, A. R. Hunt, and J. S. Lazuick. 1985b. Arbovirus investigations in Argentina, 1977–1980. III. Identification and characterization of viruses isolated, including new subtypes of western and Venezuelan equine encephalitis viruses and four new bunyaviruses (Las Maloyas, Resistencia, Barranqueras, and Antequera). *Am. J. Trop. Med. Hyg.* **34:**956–965.

Calisher, C. H., J. D. Poland, S. B. Calisher, and L. A. Warmoth. 1985c. Diagnosis of Colorado tick fever virus infection by enzyme immunoassays for immunoglobulin M and G antibodies. *J. Clin. Microbiol.* **22:**84–88.

Calisher, C. H., V. P. Berardi, D. J. Muth, and E. E. Buff. 1986a. Specificity of immunoglobulin M and G antibody responses in humans infected with eastern and western equine encephalitis viruses: application to rapid serodiagnosis. *J. Clin. Microbiol.* **23:**369–372.

Calisher, C. H., A. O. el-Kafrawi, M. I. Al-Deen Mahmud, A. P. Travassos da Rosa, C. R. Bartz, M. Brummer-Korvenkontio, S. Haksohusodo, and W. Suharyono. 1986b. Complex-specific immunoglobulin M antibody patterns in humans infected with alphaviruses. *J. Clin. Microbiol.* **23:**155–159.

Calisher, C. H., C. I. Pretzman, D. J. Muth, M. A. Parsons, and E. D. Peterson. 1986c. Serodiagnosis of La Crosse virus infections in humans by detection of immunoglobulin M class antibodies. *J. Clin. Microbiol.* **23:**667–671.

Campbell, J., L. Iacono-Connors, S. Walz, and W. Schultz. 1993. The Langat model for tick-borne encephalitis virus. Specific detection by RT-PCR. *J. Virol. Methods* **44:**235–240.

Campbell, W. P., and C. Huang. 1995. Detection of California serogroup viruses using universal primers and reverse transcription-polymerase chain reaction. *J. Virol. Methods* **53:**55–61.

Campbell, W. P., and C. Huang. 1996. Detection of California serogroup Bunyaviruses in tissue culture and mosquito pools by PCR. *J. Virol. Methods* **57:**175–179.

Cardosa, M. J., P. H. Tio, S. Nimmannitya, A. Nisalak, and B. Innis. 1992. IgM capture ELISA for detection of IgM antibodies to dengue virus: comparison of 2 formats using hemagglutinins and cell culture derived antigens. *Southeast Asian J. Trop. Med. Public Health* **23**:726–729.

Carter, I. W., L. D. Smythe, J. R. Fraser, N. D. Stallman, and M. J. Cloonan. 1985. Detection of Ross River virus immunoglobulin M antibodies by enzyme-linked immunosorbent assay using antibody class capture and comparison with other methods. *Pathology* **17**:503–508.

Centers for Disease Control and Prevention. 1993. *Biosafety in Microbiological and Biological Laboratories*, 3rd ed., p. 177. Public Health Service, U.S. Department of Health and Human Services, Washington, D.C.

Chan, S. Y., I. Kautner, and S. K. Lam. 1994. Detection and serotyping of dengue viruses by PCR: a simple, rapid method for the isolation of viral RNA from infected mosquito larvae. *Southeast Asian J. Trop. Med. Public Health* **25**:258–261.

Chandler, L. J., B. J. Beaty, G. D. Baldridge, D. H. Bishop, and M. J. Hewlett. 1990. Heterologous reassortment of bunyaviruses in Aedes triseriatus mosquitoes and transovarial and oral transmission of newly evolved genotypes. *J. Gen. Virol.* **71**:1045–1050.

Chandler, L. J., G. Hogge, M. Endres, D. R. Jacoby, N. Nathanson, and B. J. Beaty. 1991. Reassortment of La Crosse and Tahyna bunyaviruses in Aedes triseriatus mosquitoes. *Virus Res.* **20**:181–191.

Chang, G. J., D. W. Trent, A. V. Vorndam, E. Vergne, R. M. Kinney, and C. J. Mitchell. 1994. An integrated target sequence and signal amplification assay, reverse transcriptase-PCR–enzyme-linked immunosorbent assay, to detect and characterize flaviviruses. *J. Clin. Microbiol.* **32**:477–483.

Chen, H. S., H. Y. Guo, H. Y. Chen, and Y. K. Liag. 1992. Amplification of dengue 2 virus ribonucleic acid sequence using the polymerase chain reaction. *Southeast Asian J. Trop. Med. Public Health* **23**:30–36.

Chow, V. T., C. L. Seah, and Y. C. Chan. 1993. Use of NS3 consensus primers for the polymerase chain reaction amplification and sequencing of dengue viruses and other flaviviruses. *Arch. Virol.* **133**:157–70.

Clarke, D. H., and J. Casals. 1958. Techniques for hemagglutination and hemagglutination-inhibition with arthropod-borne viruses. *Am. J. Trop. Med. Hyg.* **7**:561–572.

Cumming, H. S. 1935. *Report on the St. Louis Outbreak of Encephalitis*. Public Health Service, U.S. Treasury Department, Washington, D.C.

Despres, P., J. W. Griffin, and D. E. Griffin. 1995. Effects of anti-E2 monoclonal antibody on Sindbis virus replication in AT3 cells expressing bcl-2. *J. Virol.* **69**:7006–7014.

Deubel, V., M. Laille, J. P. Hugnot, E. Chungue, J. L. Guesdon, M. T. Drouet, S. Bassot, and D. Chevrier. 1990. Identification of dengue sequences by genomic amplification: rapid diagnosis of dengue virus serotypes in peripheral blood. *J. Virol. Methods* **30**:41–54.

Deubel, V., R. M. Nogueira, M. T. Drouet, H. Zeller, J. M. Reynes, and D. Q. Ha. 1993. Direct sequencing of genomic cDNA fragments amplified by the polymerase chain reaction for molecular epidemiology of dengue-2 viruses. *Arch. Virol.* **129**:197–210.

Dietzschold, B., C. E. Rupprecht, Z. F. Fu, and H. Koprowski. 1996. Rhabdoviruses, p. 1137–1159. *In* B. N. Fields, D. M. Knipe, and P. M. Howley (ed.), *Fields Virology*. Lippincott Raven Publishers, Philadelphia, Pa.

Dykers, T. I., K. L. Brown, C. B. Gundersen, and B. J. Beaty. 1985. Rapid diagnosis of LaCrosse encephalitis: detection of specific immunoglobulin M in cerebrospinal fluid. *J. Clin. Microbiol.* **22**:740–744.

Eads, R. B., and G. C. Smith. 1983. Seasonal activity and Colorado tick fever virus infection rates in Rocky Mountain wood ticks, Dermacentor andersoni (Acari: Ixodidae), in north-central Colorado, USA. *J. Med. Entomol.* **20**:49–55.

Eldadah, Z. A., D. M. Asher, M. S. Godec, K. L. Pomeroy, L. G. Goldfarb, S. M. Feinstone, H. Levitan, C. J. Gibbs, and D. C. Gajdusek. 1991. Detection of flaviviruses by reverse-transcriptase polymerase chain reaction. *J. Med. Virol.* **33**:260–267.

Emmons, R. W. 1985. An overview of Colorado tick fever. *Prog. Clin. Biol. Res.* **178**:47–52.

Emmons, R. W. 1988. Ecology of Colorado tick fever. *Annu. Rev. Microbiol.* **42**:49–64.

Figueiredo, L. T., W. C. Batista, and A. Igarashi. 1997a. Detection and identification of dengue virus isolates from Brazil by a simplified reverse transcription-polymerase chain reaction (RT-PCR) method. *Rev. Inst. Med. Trop. Sao Paulo* **39**:79–83.

Figueiredo, L. T., W. C. Batista, and A. Igarashi. 1997b. A simple reverse transcription-polymerase chain reaction for dengue type 2 virus identification. *Mem. Inst. Oswaldo Cruz* **92**:395–398.

Florio, L., M. O. Stewart, and E. R. Mugrage. 1944. The experimental transmission of Colorado tick fever. *J. Exp. Med.* **80**:165–188.

Frazier, C. L., and R. E. Shope. 1979. Detection of antibodies to alphaviruses by enzyme-linked immunosorbent assay. *J. Clin. Microbiol.* **10**:583–585.

Fulop, L., A. D. Barrett, R. Phillpotts, K. Martin, D. Leslie, and R. W. Titball. 1993. Rapid identification of flaviviruses based on conserved NS5 gene sequences. *J. Virol. Methods* **44**:179–188.

Gajanana, A., R. Rajendran, V. Thenmozhi, P. P. Samuel, T. F. Tsai, and R. Reuben. 1995. Comparative evaluation of bioassay and ELISA for detection of Japanese encephalitis virus in field collected mosquitoes. *Southeast Asian J. Trop. Med. Public Health* **26**:91–97.

Gentry, M. K., E. A. Henchal, J. M. McCown, W. E. Brandt, and J. M. Dalrymple. 1982. Identification of distinct antigenic determinants on dengue-2 virus using monoclonal antibodies. *Am. J. Trop. Med. Hyg.* **31**:548–555.

Gentsch, J., L. R. Wynne, J. P. Clewley, R. E. Shope, and D. H. Bishop. 1977. Formation of recombinants between snowshoe hare and La Crosse bunyaviruses. *J. Virol.* **24**:893–902.

Gentsch, J. R., G. Robeson, and D. H. Bishop. 1979. Recombination between snowshoe hare and La Crosse bunyaviruses. *J. Virol.* **31**:707–717.

Gonzalez-Scarano, F., R. E. Shope, C. E. Calisher, and N. Nathanson. 1982. Characterization of monoclonal antibodies against the G1 and N proteins of LaCrosse and Tahyna, two California serogroup bunyaviruses. *Virology* **120**:42–53.

Gould, E. A., A. Buckley, N. Cammack, A. D. Barrett, J. C. Clegg, R. Ishak, and M. G. Varma. 1985. Examination of the immunological relationships between flaviviruses using yellow fever virus monoclonal antibodies. *J. Gen. Virol.* **66**:1369–1382.

Gould, E. A., A. Buckley, P. A. Cane, S. Higgs, and N. Cammack. 1989. Use of a monoclonal antibody specific for wild-type yellow fever virus to identify a wild-type antigenic variant in 17D vaccine pools. *J. Gen. Virol.* **70**:1889–1894.

Griffin, D. E., and J. M. Hardwick. 1997. Regulators of apoptosis on the road to persistent alphavirus infection. *Annu. Rev. Microbiol.* **51**:565–592.

Grimstad, P. R., C. L. Shabino, C. H. Calisher, and R. J. Waldman. 1982. A case of encephalitis in a human associated with a serologic rise to Jamestown Canyon virus. *Am. J. Trop. Med. Hyg.* **31:**1238–1244.

Guirakhoo, F., F. X. Heinz, and C. Kunz. 1989. Epitope model of tick-borne encephalitis virus envelope glycoprotein E: analysis of structural properties, role of carbohydrate side chain, and conformational changes occurring at acidic pH. *Virology* **169:**90–99.

Guirakhoo, F., F. X. Heinz, C. W. Mandl, H. Holzmann, and C. Kunz. 1991. Fusion activity of flaviviruses: comparison of mature and immature (prM-containing) tick-borne encephalitis virions. *J. Gen. Virol.* **72:**1323–1329.

Guirakhoo, F., R. A. Bolin, and J. T. Roehrig. 1992. The Murray Valley encephalitis virus prM protein confers acid resistance to virus particles and alters the expression of epitopes within the R2 domain of E glycoprotein. *Virology* **191:**921–931.

Guirakhoo, F., A. R. Hunt, J. G. Lewis, and J. T. Roehrig. 1993. Selection and partial characterization of dengue 2 virus mutants that induce fusion at elevated pH. *Virology* **194:**219–223.

Guzman, M. G., G. Kouri, L. Morier, M. Soler, and A. Fernandez. 1984. A study of fatal hemorrhagic dengue cases in Cuba, 1981. *Bull. Pan. Am. Health Org.* **18:**213–220.

Guzman, M. G., G. Kouri, E. Martinez, J. Bravo, R. Riveron, M. Soler, S. Vazquez, and L. Morier. 1987. Clinical and serologic study of Cuban children with dengue hemorrhagic fever/dengue shock syndrome (DHF/DSS). *Bull. Pan. Am. Health Org.* **21:**270–279.

Guzman, M. G., G. Kouri, J. Bravo, M. Soler, L. Morier, S. Vazquez, A. Diaz, R. Fernandez, A. Ruiz, A. Ramos, et al. 1988. Dengue in Cuba: history of an epidemic. *Rev. Cubana Med. Trop.* **40:**29–49. (In Spanish.)

Guzman, M. G., G. P. Kouri, J. Bravo, M. Soler, S. Vazquez, and L. Morier. 1990. Dengue hemorrhagic fever in Cuba, 1981: a retrospective seroepidemiologic study. *Am. J. Trop. Med. Hyg.* **42:**179–184.

Halstead, S. B. 1988. Pathogenesis of dengue: challenges to molecular biology. *Science* **239:**476–481.

Han, L. L., F. Popovici, J. P. Alexander, Jr., V. Laurentia, L. A. Tengelsen, C. Cernescu, H. E. Gary, Jr., N. Ion-Nedelcu, G. L. Campbell, and T. F. Tsai. 1999. Risk factors for West Nile virus infection and meningoencephalitis, Romania, 1996. *J. Infect. Dis.* **179:**230–233.

Hawkes, R. A., J. T. Roehrig, A. R. Hunt, and G. A. Moore. 1988. Antigenic structure of the Murray Valley encephalitis virus E glycoprotein. *J. Gen. Virol.* **69:**1105–1109.

Heinz, F. X. 1986. Epitope mapping of flavivirus glycoproteins. *Adv. Virus Res.* **31:**103–168.

Heinz, F. X., and C. W. Mandl. 1993. The molecular biology of tick-borne encephalitis virus. *APMIS* **101:**735–745.

Heinz, F. X., and J. T. Roehrig. 1990. Immunochemistry of flaviviruses, p. 289–305. *In* M. H. V. van Regenmortel and H. Neurath (ed.), *Immunochemistry of Viruses. The Basis for Serodiagnosis and Vaccines.* Elsevier, Amsterdam, The Netherlands.

Heinz, F. X., M. Roggendorf, H. Hofmann, C. Kunz, and F. Deinhardt. 1981. Comparison of two different enzyme immunoassays for detection of immunoglobulin M antibodies against tick-borne encephalitis virus in serum and cerebrospinal fluid. *J. Clin. Microbiol.* **14:**141–146.

Henchal, E. A., M. K. Gentry, J. M. McCown, and W. E. Brandt. 1982. Dengue virus-specific and flavivirus group determinants identified with monoclonal antibodies by indirect immunofluorescence. *Am. J. Trop. Med. Hyg.* **31:**830–836.

Henchal, E. A., J. M. McCown, D. S. Burke, M. C. Seguin, and W. E. Brandt. 1985. Epitopic analysis of antigenic deter-

minants on the surface of dengue-2 virions using monoclonal antibodies. *Am. J. Trop. Med. Hyg.* **34:**162–169.

Hennessy, S., Z. Liu, T. F. Tsai, B. L. Strom, C. M. Wan, H. L. Liu, T. X. Wu, H. J. Yu, Q. M. Liu, N. Karabatsos, W. B. Bilker, and S. B. Halstead. 1996. Effectiveness of live-attenuated Japanese encephalitis vaccine (SA14-14-2): a case-control study. *Lancet* **347:**1583–1586.

Hildreth, S. W., and B. J. Beaty. 1983. Application of enzyme immunoassays (EIA) for the detection of La Crosse viral antigens in mosquitoes. *Prog. Clin. Biol. Res.* **123:**303–312.

Hildreth, S. W., B. J. Beaty, J. M. Meegan, C. L. Frazier, and R. E. Shope. 1982. Detection of La Crosse arbovirus antigen in mosquito pools: application of chromogenic and fluorogenic enzyme immunoassay systems. *J. Clin. Microbiol.* **15:**879–884.

Hildreth, S. W., B. J. Beaty, H. K. Maxfield, R. F. Gilfillan, and B. J. Rosenau. 1984. Detection of eastern equine encephalomyelitis virus and Highlands J virus antigens within mosquito pools by enzyme immunoassay (EIA). II. Retrospective field test of the EIA. *Am. J. Trop. Med. Hyg.* **33:**973–980.

Howe, D. K., M. H. Vodkin, R. J. Novak, C. J. Mitchell, and G. L. McLaughlin. 1992a. Detection of St. Louis encephalitis virus in mosquitoes by use of the polymerase chain reaction. *J. Am. Mosq. Control Assoc.* **8:**333–335.

Howe, D. K., M. H. Vodkin, R. J. Novak, R. E. Shope, and G. L. McLaughlin. 1992b. Use of the polymerase chain reaction for the sensitive detection of St. Louis encephalitis viral RNA. *J. Virol. Methods* **36:**101–110.

Hunt, A. R., and J. T. Roehrig. 1985. Biochemical and biological characteristics of epitopes on the E1 glycoprotein of western equine encephalitis virus. *Virology* **142:**334–346.

Jamnback, T. L., B. J. Beaty, K. L. Hildreth, K. L. Brown, and C. B. Gundersen. 1982. Capture immunoglobulin M system for rapid diagnosis of La Crosse (California encephalitis) virus infections. *J. Clin. Microbiol.* **16:**557–580.

Johnson, A. J., N. Karabatsos, and R. S. Lanciotti. 1997. Detection of Colorado tick fever virus by using reverse transcriptase PCR and application of the technique in laboratory diagnosis. *J. Clin. Microbiol.* **35:**1203–1208.

Johnson, H. N. 1987. Isolation of Powassan virus from a spotted skunk in California. *J. Wildl. Dis.* **23:**152–153.

Karabatsos, N. 1985. *International Catalogue of Arboviruses.* American Society for Tropical Medicine and Hygiene, San Antonio, Tex.

Karabatsos, N., J. D. Poland, R. W. Emmons, J. H. Mathews, C. H. Calisher, and K. L. Wolff. 1987. Antigenic variants of Colorado tick fever virus. *J. Gen. Virol.* **68:**1463–1469.

Karabatsos, N., A. L. Lewis, C. H. Calisher, A. R. Hunt, and J. T. Roehrig. 1988. Identification of Highlands J virus from a Florida horse. *Am. J. Trop. Med. Hyg.* **39:**603–606.

Kaufman, B. M., P. L. Summers, D. R. Dubois, and K. H. Eckels. 1987. Monoclonal antibodies against dengue 2 virus E-glycoprotein protect mice against lethal dengue infection. *Am. J. Trop. Med. Hyg.* **36:**427–434.

Kimura-Kuroda, J., and K. Yasui. 1983. Topographical analysis of antigenic determinants on envelope glycoprotein V3 (E) of Japanese encephalitis virus, using monoclonal antibodies. *J. Virol.* **45:**124–132.

Kinney, R. M., D. W. Trent, and J. K. France. 1983. Comparative immunological and biochemical analyses of viruses in the Venezuelan equine encephalitis complex. *J. Gen. Virol.* **64:**135–147.

Kinney, R. M., B. J. Johnson, J. B. Welch, K. R. Tsuchiya, and D. W. Trent. 1989. The full-length nucleotide sequences of the virulent Trinidad donkey strain of Venezuelan equine encephalitis virus and its attenuated vaccine derivative, strain TC-83. *Virology* **170**:19–30.

Kinney, R. M., M. Pfeffer, K. R. Tsuchiya, G. J. Chang, and J. T. Roehrig. 1998. Nucleotide sequences of the 26S mRNAs of the viruses defining the Venezuelan equine encephalitis antigenic complex. *Am. J. Trop. Med. Hyg.* **59**:952–964.

Kittigul, L., S. Suthachana, C. Kittigul, and V. Pengruan-grojanachai. 1998. Immunoglobulin M-capture biotin-streptavidin enzyme-linked immunosorbent assay for detection of antibodies to dengue viruses. *Am. J. Trop. Med. Hyg.* **59**:352–356.

Knudson, D. L. 1981. Genome of Colorado tick fever virus. *Virology* **112**:361–364.

Kouri, G., P. Mas, M. G. Guzman, M. Soler, A. Goyenechea, and L. Morier. 1983. Dengue hemorrhagic fever in Cuba, 1981: rapid diagnosis of the etiologic agent. *Bull. Pan. Am. Health Organ.* **17**:126–132.

Kuno, G., D. J. Gubler, and N. S. Santiago de Weil. 1985. Antigen capture ELISA for the identification of dengue viruses. *J. Virol. Methods* **12**:93–103.

Kuno, G., C. J. Mitchell, G. J. Chang, and G. C. Smith. 1996. Detecting bunyaviruses of the Bunyamwera and California serogroups by a PCR technique. *J. Clin. Microbiol.* **34**:1184–1188.

Lane, R. S., R. W. Emmons, V. Devlin, D. V. Dondero, and B. C. Nelson. 1982. Survey for evidence of Colorado tick fever virus outside of the known endemic area in California. *Am. J. Trop. Med. Hyg.* **31**:837–843.

Levine, B., J. M. Hardwick, and D. E. Griffin. 1994. Persistence of alphaviruses in vertebrate hosts. *Trends Microbiol.* **2**:25–28.

Levine, B., J. E. Goldman, H. H. Jiang, D. E. Griffin, and J. M. Hardwick. 1996. Bcl-2 protects mice against fatal alphavirus encephalitis. *Proc. Natl. Acad. Sci. USA* **93**:4810–4815.

Lewis, J., S. L. Wesselingh, D. E. Griffin, and J. M. Hardwick. 1996. Alphavirus-induced apoptosis in mouse brains correlates with neurovirulence. *J. Virol.* **70**:1828–1835.

Liu, Z. L., S. Hennessy, B. L. Strom, T. F. Tsai, C. M. Wan, S. C. Tang, C. F. Xiang, W. B. Bilker, X. P. Pan, Y. J. Yao, Z. W. Xu, and S. B. Halstead. 1997. Short-term safety of live attenuated Japanese encephalitis vaccine (SA14-14-2): results of a randomized trial with 26,239 subjects. *J. Infect. Dis.* **176**:1366–1369.

Ludwig, G. V., B. A. Israel, B. M. Christensen, T. M. Yuill, and K. T. Schultz. 1991. Monoclonal antibodies directed against the envelope glycoproteins of La Crosse virus. *Microb. Pathog.* **11**:411–421.

Ludwig, G. V., J. P. Kondig, and J. F. Smith. 1996. A putative receptor for Venezuelan equine encephalitis virus from mosquito cells. *J. Virol.* **70**:5592–5599.

Mandl, C. W., F. Guirakhoo, H. Holzmann, F. X. Heinz, and C. Kunz. 1989. Antigenic structure of the flavivirus envelope protein E at the molecular level, using tick-borne encephalitis virus as a model. *J. Virol.* **63**:564–571.

Marfin, A. A., D. M. Bleed, J. P. Lofgren, A. C. Olin, H. M. Savage, G. C. Smith, P. S. Moore, N. Karabatsos, and T. F. Tsai. 1993. Epidemiologic aspects of a St. Louis encephalitis epidemic in Jefferson County Arkansas, 1991. *Am. J. Trop. Med. Hyg.* **49**:30–37.

Mathews, J. H., and J. T. Roehrig. 1982. Determination of the protective epitopes on the glycoproteins of Venezuelan equine encephalomyelitis virus by passive transfer of monoclonal antibodies. *J. Immunol.* **129**:2763–2767.

Mathews, J. H., and J. T. Roehrig. 1984. Elucidation of the topography and determination of the protective epitopes on the E glycoprotein of Saint Louis encephalitis virus by passive transfer with monoclonal antibodies. *J. Immunol.* **132**:1533–1537.

McCaig, L. F., H. T. Janowski, R. A. Gunn, and T. F. Tsai. 1994. Epidemiologic aspects of a St. Louis encephalitis outbreak in Fort Walton Beach, Florida in 1980. *Am. J. Trop. Med. Hyg.* **50**:387–391.

McJunkin, J. E., R. Khan, E. C. de los Reyes, D. L. Parsons, L. L. Minnich, R. G. Ashley, and T. F. Tsai. 1997. Treatment of severe La Crosse encephalitis with intravenous ribavirin following diagnosis by brain biopsy. *Pediatrics* **99**:261–267.

McJunkin, J. E., R. R. Khan, and T. F. Tsai. 1998. California-La Crosse encephalitis. *Infect. Dis. Clin. N. Am.* **12**:83–93.

McLean, R. G., R. B. Shriner, K. S. Pokorny, and G. S. Bowen. 1989. The ecology of Colorado tick fever in Rocky Mountain National Park in 1974. III. Habitats supporting the virus. *Am. J. Trop. Med. Hyg.* **40**:86–93.

Meiyu, F., C. Huosheng, C. Cuihua, T. Xiaodong, J. Lianhua, P. Yifei, C. Weijun, and G. Huiyu. 1997. Detection of flaviviruses by reverse transcriptase-polymerase chain reaction with the universal primer set. *Microbiol. Immunol.* **41**:209–213.

Meyer, K. F., C. M. Haring, and B. Howitt. 1931. The etiology of epizootic encephalomyelitis in horses in the San Joaquin Valley. *Science* **74**:227–228.

Monath, T. P. 1988. *Arboviruses: Epidemiology and Ecology*, vol. 1. CRC Press, Boca Raton, Fla.

Monath, T. P. 1991. Yellow fever: Victor, Victoria? Conqueror, conquest? Epidemics and research in the last forty years and prospects for the future. *Am. J. Trop. Med. Hyg.* **45**:1–43.

Monath, T. P. 1996. Flaviviruses, p. 1133–1185. *In* D. Richman, R. Whitley, and F. Hayden (ed.), *Clinical Virology*. Churchill-Livingstone, New York, N.Y.

Monath, T. P., and T. F. Tsai. 1987. St. Louis encephalitis: lessons from the last decade. *Am. J. Trop. Med. Hyg.* **37**:40S–59S.

Monath, T. P., R. B. Craven, D. J. Muth, C. J. Trautt, C. H. Calisher, and S. A. Fitzgerald. 1980. Limitations of the complement-fixation test for distinguishing naturally acquired from vaccine-induced yellow fever infection in flavivirus-hyperendemic areas. *Am. J. Trop. Med. Hyg.* **29**:624–634.

Monath, T. P., C. B. Cropp, and A. K. Harrison. 1983. Mode of entry of a neurotropic arbovirus into the central nervous system. Reinvestigation of an old controversy. *Lab. Investig.* **48**:399–410.

Monath, T. P., R. R. Nystrom, R. E. Bailey, C. H. Calisher, and D. J. Muth. 1984. Immunoglobulin M antibody capture enzyme-linked immunosorbent assay for diagnosis of St. Louis encephalitis. *J. Clin. Microbiol.* **20**:784–790.

Monath, T. P., J. R. Wands, L. J. Hill, M. K. Gentry, and D. J. Gubler. 1986. Multisite monoclonal immunoassay for dengue viruses: detection of viraemic human sera and interference by heterologous antibody. *J. Gen. Virol.* **67**:639–650.

Nawrocki, S. J., Y. H. Randle, M. H. Vodkin, J. P. Siegel, and R. J. Novak. 1996. Evaluation of a reverse transcriptase-polymerase chain reaction assay for detecting St. Louis encephalitis virus using field-collected mosquitoes (Diptera: Culicidae). *J. Med. Entomol.* **33**:123–127.

Nogueira, R. M., H. G. Schatzmayr, M. P. Miagostovich, M. F. Farias, and J. D. Farias Filho. 1989. Virological study of a dengue type 1 epidemic at Rio de Janeiro. *Mem. Inst. Oswaldo Cruz* **83**:219–225.

Oberste, M. S., M. D. Parker, and J. F. Smith. 1996. Complete sequence of Venezuelan equine encephalitis virus subtype IE reveals conserved and hypervariable domains within the C terminus of nsP3. *Virology* **219**:314–320.

Oberste, M. S., M. Fraire, R. Navarro, C. Zepeda, M. L. Zarate, G. V. Ludwig, J. F. Kondig, S. C. Weaver, J. F. Smith, and R. Rico-Hesse. 1998a. Association of Venezuelan equine encephalitis virus subtype IE with two equine epizootics in Mexico. *Am. J. Trop. Med. Hyg.* **59:**100–107.

Oberste, M. S., S. C. Weaver, D. M. Watts, and J. F. Smith. 1998b. Identification and genetic analysis of Panama-genotype Venezuelan equine encephalitis virus subtype ID in Peru. *Am. J. Trop. Med. Hyg.* **58:**41–46.

Obijeski, J. F., D. H. Bishop, F. A. Murphy, and E. L. Palmer. 1976a. Structural proteins of La Crosse virus. *J. Virol.* **19:**985–997.

Obijeski, J. F., D. H. Bishop, E. L. Palmer, and F. A. Murphy. 1976b. Segmented genome and nucleocapsid of La Crosse virus. *J. Virol.* **20:**664–675.

Paranjpe, S., and K. Banerjee. 1998. Detection of Japanese encephalitis virus by reverse transcription/polymerase chain reaction. *Acta Virol.* **42:**5–11.

Peiris, J. S., J. S. Porterfield, and J. T. Roehrig. 1982. Monoclonal antibodies against the flavivirus West Nile. *J. Gen. Virol.* **58:**283–289.

Pfeffer, M., B. Proebster, R. M. Kinney, and O. R. Kaaden. 1997. Genus-specific detection of alphaviruses by a semi-nested reverse transcription-polymerase chain reaction. *Am. J. Trop. Med. Hyg.* **57:**709–718.

Porter, K. R., P. L. Summers, D. Dubois, B. Puri, W. Nelson, E. Henchal, J. J. Oprandy, and C. G. Hayes. 1993. Detection of West Nile virus by the polymerase chain reaction and analysis of nucleotide sequence variation. *Am. J. Trop. Med. Hyg.* **48:**440–446.

Powers, A. M., M. S. Oberste, A. C. Brault, R. Rico-Hesse, S. M. Schmura, J. F. Smith, W. Kang, W. P. Sweeney, and S. C. Weaver. 1997. Repeated emergence of epidemic/epizootic Venezuelan equine encephalitis from a single genotype of enzootic subtype ID virus. *J. Virol.* **71:**6697–6705.

Reeves, W. C. 1987. The discovery decade of arbovirus research in western North America, 1940–1949. *Am. J. Trop. Med. Hyg.* **37:**94S–100S.

Reisen, W. K., and R. E. Chiles. 1997. Prevalence of antibodies to western equine encephalomyelitis and St. Louis encephalitis viruses in residents of California exposed to sporadic and consistent enzootic transmission. *Am. J. Trop. Med. Hyg.* **57:**526–529.

Reisen, W. K., J. L. Hardy, W. C. Reeves, S. B. Presser, M. M. Milby, and R. P. Meyer. 1990. Persistence of mosquito-borne viruses in Kern County, California, 1983–1988. *Am. J. Trop. Med. Hyg.* **43:**419–437.

Reisen, W. K., J. L. Hardy, S. B. Presser, M. M. Milby, R. P. Meyer, S. L. Durso, M. J. Wargo, and E. Gordon. 1992a. Mosquito and arbovirus ecology in southeastern California, 1986–1990. *J. Med. Entomol.* **29:**512–524.

Reisen, W. K., M. M. Milby, S. B. Presser, and J. L. Hardy. 1992b. Ecology of mosquitoes and St. Louis encephalitis virus in the Los Angeles Basin of California, 1987–1990. *J. Med. Entomol.* **29:**582–598.

Rey, F. A., F. X. Heinz, C. Mandl, C. Kunz, and S. C. Harrison. 1995. The envelope glycoprotein from tick-borne encephalitis virus at 2 Å resolution. *Nature* **375:**291–298.

Rice, C. M., E. M. Lenches, S. R. Eddy, S. J. Shin, R. L. Sheets, and J. H. Strauss. 1985. Nucleotide sequence of yellow fever virus: implications for flavivirus gene expression and evolution. *Science* **229:**726–733.

Rico-Hesse, R., J. T. Roehrig, and R. W. Dickerman. 1988. Monoclonal antibodies define antigenic variation in the ID

variety of Venezuelan equine encephalitis virus. *Am. J. Trop. Med. Hyg.* **38:**187–194.

Rico-Hesse, R., S. C. Weaver, J. de Siger, G. Medina, and R. A. Salas. 1995. Emergence of a new epidemic/epizootic Venezuelan equine encephalitis virus in South America. *Proc. Natl. Acad. Sci. USA* **92:**5278–5281.

Rivas, F., L. A. Diaz, V. M. Cardenas, E. Daza, L. Bruzon, A. Alcala, O. De la Hoz, F. M. Caceres, G. Aristizabal, J. W. Martinez, D. Revelo, F. De la Hoz, J. Boshell, T. Camacho, L. Calderon, V. A. Olano, L. I. Villarreal, D. Roselli, G. Alvarez, G. Ludwig, and T. Tsai. 1997. Epidemic Venezuelan equine encephalitis in La Guajira, Colombia, 1995. *J. Infect. Dis.* **175:**828–832.

Roehrig, J. T. 1982. Development of an enzyme-linked immunosorbent assay for the identification of arthropod-borne togavirus antibodies. *J. Gen. Virol.* **63:**237–240.

Roehrig, J. T. 1986. The use of monoclonal antibodies in studies of the structural proteins of alphaviruses and flaviviruses, p. 251–278. *In* S. Schlesinger and M. J. Schlesinger (ed.), *The Viruses: The Togaviridae and Flaviviridae.* Plenum Press, New York, N.Y.

Roehrig, J. T. 1990. Immunochemistry of togaviruses, p. 271–288. *In* M. H. V. van Regenmortel and A. R. Neurath (ed.), *Immunochemistry of Viruses. The Basis for Serodiagnosis and Vaccines.* Elsevier, Amsterdam, The Netherlands.

Roehrig, J. T. 1997. Immunochemistry of dengue viruses, p. 199–219. *In* D. J. Gubler and G. Kuno (ed.), *Dengue and Dengue Hemorrhagic Fever.* CAB International, New York, N.Y.

Roehrig, J. T., and R. A. Bolin. 1997. Monoclonal antibodies capable of distinguishing epizootic from enzootic varieties of subtype 1 Venezuelan equine encephalitis viruses in a rapid indirect immunofluorescence assay. *J. Clin. Microbiol.* **35:**1887–1890.

Roehrig, J. T., and J. H. Mathews. 1985. The neutralization site on the E2 glycoprotein of Venezuelan equine encephalomyelitis (TC-83) virus is composed of multiple conformationally stable epitopes. *Virology* **142:**347–356.

Roehrig, J. T., J. W. Day, and R. M. Kinney. 1982. Antigenic analysis of the surface glycoproteins of a Venezuelan equine encephalomyelitis virus (TC-83) using monoclonal antibodies. *Virology* **118:**269–278.

Roehrig, J. T., J. H. Mathews, and D. W. Trent. 1983. Identification of epitopes on the E glycoprotein of Saint Louis encephalitis virus using monoclonal antibodies. *Virology* **128:**118–126.

Roehrig, J. T., A. R. Hunt, R. M. Kinney, and J. H. Mathews. 1988. In vitro mechanisms of monoclonal antibody neutralization of alphaviruses. *Virology* **165:**66–73.

Roehrig, J. T., A. R. Hunt, G. J. Chang, B. Sheik, R. A. Bolin, T. F. Tsai, and D. W. Trent. 1990a. Identification of monoclonal antibodies capable of differentiating antigenic varieties of eastern equine encephalitis viruses. *Am. J. Trop. Med. Hyg.* **42:**394–398.

Roehrig, J. T., A. J. Johnson, A. R. Hunt, R. A. Bolin, and M. C. Chu. 1990b. Antibodies to dengue 2 virus E-glycoprotein synthetic peptides identify antigenic conformation. *Virology* **177:**668–675.

Roehrig, J. T., R. A. Bolin, A. R. Hunt, and T. M. Woodward. 1991. Use of a new synthetic-peptide-derived monoclonal antibody to differentiate between vaccine and wild-type Venezuelan equine encephalomyelitis viruses. *J. Clin. Microbiol.* **29:**630–631.

Roehrig, J. T., R. A. Bolin, and R. G. Kelly. 1998a. Monoclonal antibody mapping of the envelope glycoprotein of the dengue 2 virus, Jamaica. *Virology* **246:**317–328.

Roehrig, J. T., T. M. Brown, A. J. Johnson, N. Karabatsos, D. M. Martin, C. J. Mitchell, and R. Nasci. 1998b. Alphaviruses, p. 7–18. *In* J. R. Stephenson and A. Warnes (ed.), *Methods in Molecular Biology: Diagnostic Virology Protocols.* Humana Press, Totowa, N.J.

Sahu, S. P., A. D. Alstad, D. D. Pedersen, and J. E. Pearson. 1994. Diagnosis of eastern equine encephalomyelitis virus infection in horses by immunoglobulin M and G capture enzyme-linked immunosorbent assay. *J. Vet. Diagn. Investig.* **6:**34–38.

Schlesinger, J. J., M. W. Brandriss, and T. P. Monath. 1983. Monoclonal antibodies distinguish between wild and vaccine strains of yellow fever virus by neutralization, hemagglutination inhibition, and immune precipitation of the virus envelope protein. *Virology* **125:**8–17.

Schlesinger, J. J., M. W. Brandriss, and E. E. Walsh. 1985. Protection against 17D yellow fever encephalitis in mice by passive transfer of monoclonal antibodies to the nonstructural glycoprotein gp48 and by active immunization with gp48. *J. Immunol.* **135:**2805–2809.

Schmaljohn, A. L., K. M. Kokubun, and G. A. Cole. 1983. Protective monoclonal antibodies define maturational and pH-dependent antigenic changes in Sindbis virus E1 glycoprotein. *Virology* **130:**144–154.

Scott, T. W., and J. G. Olson. 1986. Detection of eastern equine encephalomyelitis viral antigen in avian blood by enzyme immunoassay: a laboratory study. *Am. J. Trop. Med. Hyg.* **35:**611–618.

Sneider, J. M., R. M. Kinney, K. R. Tsuchiya, and D. W. Trent. 1993. Molecular evidence that epizootic Venezuelan equine encephalitis (VEE) I-AB viruses are not evolutionary derivatives of enzootic VEE subtype I-E or II viruses. *J. Gen. Virol.* **74:**519–523.

Stadler, K., S. L. Allison, J. Schalich, and F. X. Heinz. 1997. Proteolytic activation of tick-borne encephalitis virus by furin. *J. Virol.* **71:**8475–8481.

Strauss, J. H., and E. G. Strauss. 1994. The alphaviruses: gene expression, replication, and evolution. *Microbiol. Rev.* **58:**491–562.

Strauss, J. H., K. S. Wang, A. L. Schmaljohn, R. J. Kuhn, and E. G. Strauss. 1994. Host-cell receptors for Sindbis virus. *Arch. Virol. Suppl.* **9:**473–484.

Strauss, J. H., E. G. Strauss, and R. J. Kuhn. 1995. Budding of alphaviruses. *Trends Microbiol.* **3:**346–350.

TenBroeck, C., and M. Merrill. 1933. A serological difference between eastern and western equine encephalomyelitis virus. *Proc. Exp. Biol. Med.* **31:**217–220.

Trent, D. W., and J. A. Grant. 1980. A comparison of New World alphaviruses in the western equine encephalomyelitis complex by immunochemical and oligonucleotide fingerprint techniques. *J. Gen. Virol.* **47:**261–282.

Trent, D. W., J. T. Roehrig, and T. F. Tsai. 1989. Alphaviruses, p. 1–41. *In* D. H. Gilden and H. Lipton (ed.), *Clinical and Molecular Aspects of Neurotropic Virus Infections.* Kluwer Academic Publishers, Boston, Mass.

Tsai, T., and T. P. Monath. 1996. Alphaviruses, p. 1217–1225. *In* D. D. Richman, R. J. Whitley, and F. G. Hayden (ed.), *Clinical Virology.* Churchill-Livingstone, New York, N.Y.

Tsai, T. F., R. A. Bolin, M. Montoya, R. E. Bailey, D. B. Francy, M. Jozan, and J. T. Roehrig. 1987a. Detection of St. Louis encephalitis virus antigen in mosquitoes by capture enzyme immunoassay. *J. Clin. Microbiol.* **25:**370–376.

Tsai, T. F., W. B. Cobb, R. A. Bolin, N. J. Gilman, G. C. Smith, R. E. Bailey, J. D. Poland, J. J. Doran, J. K. Emerson, and K. J. Lampert. 1987b. Epidemiologic aspects of a St. Louis encephalitis outbreak in Mesa County, Colorado. *Am. J. Epidemiol.* **126:**460–473.

Tsai, T. F., C. M. Happ, R. A. Bolin, M. Montoya, E. Campos, D. B. Francy, R. A. Hawkes, and J. T. Roehrig. 1988. Stability of St. Louis encephalitis viral antigen detected by enzyme immunoassay in infected mosquitoes. *J. Clin. Microbiol.* **26:**2620–2625.

Tsai, T. F., F. Popovici, C. Cernescu, G. L. Campbell, and N. I. Nedelcu. 1998. West Nile encephalitis epidemic in southeastern Romania. *Lancet* **352:**767–771.

Ubol, S., P. C. Tucker, D. E. Griffin, and J. M. Hardwick. 1994. Neurovirulent strains of Alphavirus induce apoptosis in bcl-2-expressing cells: role of a single amino acid change in the E2 glycoprotein. *Proc. Natl. Acad. Sci. USA* **91:**5202–5206.

Umenai, T., R. Krzysko, T. A. Bektimirov, and F. A. Assaad. 1985. Japanese encephalitis: current worldwide status. *Bull. W. H. O.* **63:**625–631.

Urquidi, V., and D. H. Bishop. 1992. Non-random reassortment between the tripartite RNA genomes of La Crosse and snowshoe hare viruses. *J. Gen. Virol.* **73:**2255–2265.

Wasieloski, L. P. J., A. Rayms-Keller, L. A. Curtis, C. D. Blair, and B. J. Beaty. 1994. Reverse transcription-PCR detection of LaCrosse virus in mosquitoes and comparison with enzyme immunoassay and virus isolation. *J. Clin. Microbiol.* **32:**2076–2080.

Watts, D. M., V. Lavera, J. Callahan, C. Rossi, M. S. Oberste, J. T. Roehrig, C. B. Cropp, N. Karabatsos, J. F. Smith, D. J. Gubler, M. T. Wooster, W. M. Nelson, and C. G. Hayes. 1997. Venezuelan equine encephalitis and Oropouche virus infections among Peruvian army troops in the Amazon region of Peru. *Am. J. Trop. Med. Hyg.* **56:**661–667.

Watts, D. M., J. Callahan, C. Rossi, M. S. Oberste, J. T. Roehrig, M. T. Wooster, J. F. Smith, C. B. Cropp, E. M. Gentrau, N. Karabatsos, D. Gubler, and C. G. Hayes. 1998. Venezuelan equine encephalitis febrile cases among humans in the Peruvian Amazon River region. *Am. J. Trop. Med. Hyg.* **58:**35–40.

Weaver, S. C., R. Salas, R. Rico-Hesse, G. V. Ludwig, M. S. Oberste, J. Boshell, and R. B. Tesh. 1996. Re-emergence of epidemic Venezuelan equine encephalomyelitis in South America. VEE Study Group. *Lancet* **348:**436–440.

Weaver, S. C., M. Pfeffer, K. Marriott, W. Kang, and R. M. Kinney. 1999. Genetic evidence for the origins of Venezuelan equine encephalitis virus subtype 1AB outbreaks. *Am. J. Trop. Med. Hyg.* **60:**441–448.

Westaway, E. G., A. J. Della-Porta, and B. M. Reedman. 1974. Specificity of IgM and IgG antibodies after challenge with antigenically related togaviruses. *J. Immunol.* **112:**656–663.

Zeidner, N. S., S. Higgs, C. M. Happ, B. J. Beaty, and B. R. Miller. 1999. Mosquito feeding modulates Th1 and Th2 cytokines in flavivirus susceptible mice: an effect mimicked by injection of sialokinins, but not demonstrated in flavivirus resistant mice. *Parasite Immunol.* **21:**35–44.

Papovaviruses

KEERTI V. SHAH

30

Papovaviruses are small, naked, icosahedral viruses that have a double-stranded DNA genome and that multiply in the nucleus. The family *Papovaviridae* consists of two subfamilies, the papillomaviruses (wart viruses) and polyomaviruses. Viruses of both subfamilies are widely distributed in nature (Table 1). Humans are hosts to more than 100 different human papillomaviruses (HPVs) (designated HPV type 1 [HPV-1], HPV-2, etc.) and to the polyomaviruses BK virus (BKV) and JC virus (JCV). Viruses of each subfamily have a common evolutionary origin. All members of a subfamily are immunologically related and share some nucleotide sequences, but there is no evidence of a relationship between the papillomaviruses and the polyomaviruses.

Papillomaviruses are larger than polyomaviruses (a virion diameter of 55 nm compared with 45 nm) and have a larger genome (8×10^3 compared with 5×10^3 bp). All of the genetic information in papillomaviruses is located on one strand. In contrast, the genetic information in polyomaviruses is about equally divided between the two strands. Viruses of the two subfamilies also differ biologically. Papillomaviruses infect surface epithelia and produce warts at the site of multiplication on the skin or the mucous membrane. On the other hand, polyomaviruses, after initial multiplication at the site of entry in the respiratory or the gastrointestinal tract, reach internal organs (the kidneys, lungs, and brain) following viremia. Although viruses of both subfamilies can transform cells and produce tumors experimentally, only papillomaviruses are etiologically associated with naturally occurring tumors.

HPV

The infectious nature of human warts has been suspected for many centuries (Rowson and Mahy, 1967). Their viral etiology was established in 1907 by experimental transmission of warts from person to person by inoculation of a cell-free extract of wart tissue. The virus was visualized in the 1950s soon after electron microscopy came into general use and, on the basis of morphological similarities and the nuclear site of multiplication, was grouped with polyomavirus of mice and vacuolating agent (Simian virus 40 [SV40]) of rhesus monkeys to form the papova (papilloma, polyoma, and vacuolating agent) group (Melnick, 1962). Warts have characteristic histopathological features and

have been recognized at many different sites in humans (skin, genital tract, respiratory tract, and oral cavity) and in many mammalian species. However, papillomaviruses still cannot be grown in culture. The existence of a large number of distinct HPVs became evident only after the development of recombinant DNA technology, which permitted the cloning of viral genomes from different sites and the comparison of the nuclear sequences of these genomes (Orth et al., 1977; Gissmann et al., 1982).

Characteristics of the Virus

Papillomaviruses are classified on the basis of species of origin (human, bovine, etc.) and the degree of genetic relatedness with other papillomaviruses infecting the same species. New types are defined by the extent of sequence variation from known types in specific regions of the genome. A compilation and analysis of nucleic acid and amino acid sequences of papillomaviruses is available from the database maintained by the Los Alamos National Laboratory, N. Mex. (Los Alamos National Laboratory, 1998).

The virion is nonenveloped and has a diameter of 55 nm, icosahedral symmetry, and 72 capsomers. The viral genome is a double-stranded, circular DNA molecule with 8×10^3 bp and a molecular mass of 5.2×10^6 Da. Complete nucleotide sequences are known for many HPV types. All of the open reading frames (ORFs) in papillomavirus DNA are located on only one of the two strands, indicating that only one strand carries the genetic information. Detailed physical maps have been constructed for almost all of the HPV genomes. The viral genome is divided into an early region that contains eight ORFs (E1 to E8) and a late region that has two ORFs (L1 and L2).

The functions of the papillomavirus ORFs are listed in Table 2. The viral capsid is made up of two structural proteins, L1 and L2. L1 is the major structural protein of the virus. It mediates viral attachment to susceptible cells and the immunologic responses to viral infection. The L1 protein produced in yeast or in baculovirus vector self-assembles as a viruslike particle that is conformationally similar to the authentic virion (Schiller and Roden, 1995). The proposed HPV-16-based preventive vaccines now in phase 1 and 2 trials are based on the use of viruslike particles (either L1 alone or L1 plus L2) as the immunogens.

The early proteins E1 and E2 are viral regulatory proteins involved in viral DNA replication and viral transcription. E4 is a late protein expressed in terminally differentiated cells and is found in association with viral capsids. E6 and E7 are the oncoproteins of HPVs that are responsible for the immortalization and transformation of keratinocytes. The E6 and E7 proteins exert their effects, in part, by complexing with and inactivating, respectively, the tumor suppressor proteins p53 and pRb (Howley, 1996). The E6 and E7 proteins are invariably expressed in cells of HPV-associated cervical cancer and are the targets for HPV-based immunotherapy protocols, which aim to destroy established cervical cancers (Wu, 1994).

Pathogenesis and Disease Potential

HPVs infect only epithelia of skin and mucous membranes. The virus probably infects cells of the basal layer of the epithelium, which undergo proliferation and form the wart. Histologically, a wart is localized epithelial hyperplasia with a defined boundary and an intact basement membrane. All layers of the normal epithelium are represented in the wart. The prickle cell layer is irregularly thickened, the granular layer contains foci of koilocytotic cells, and the cornified layer displays hyperkeratosis. The viral capsid antigen and virus particles are found only in the nuclei of cells of the differentiated, nondividing, superficial layers of the wart. In the infected cell, the multiple copies of the viral genome are present in an unintegrated state.

Warts and other papillomavirus-related lesions vary widely in appearance, morphology, site of occurrence, and pathogenic potential. The spectrum of HPV infection ranges from completely subclinical infection, to transient, barely noticed, self-limiting, benign infections, to malignancies of the skin or the genital tract. Many factors determine the clinical significance of papillomavirus infection, as described below.

Location of the Lesion

The importance of the location of the lesion is best exemplified by laryngeal papilloma. Although the tumors are benign, they may cause life-threatening respiratory obstruction because of their location on the vocal cords.

TABLE 1 Natural hosts of papovaviruses

Host	Papillomavirus(es)	Polyomavirus(es)
Humans	HPV-1 to -83+	BKV and JCV
Monkeys	Rhesus papillomavirus	SV 40 of macaques; simian agent 12 of baboons; lymphotrophic papovavirus of African green monkeys
Cattle	Bovine papillomavirus types 1 to 6	Bovine polyomavirus
Rabbits	Cottontail rabbit papillomavirus	Rabbit kidney vacuolating virus
Rodents	*Mastomys natalensis* papillomavirus	Polyomavirus and K virus of mice; hamster papovavirus
Birds	Chaffinch papillomavirus	Budgerigar fledgling disease virus
Other	Horse, dog, sheep, European elk, and deer papillomaviruses	

TABLE 2 Papillomavirus ORFs and their functions and products

ORF[a]	Function or product
L1	L1 protein is the major structural protein; proposed immunogen for preventive vaccines
L2	L2 protein is a minor capsid protein
E1	Initiation of viral DNA replication
E2	Regulation of viral transcription
E4	Expressed late; disrupts cytokeratin and aids in virus release
E5	Interacts with growth factors; oncogenic for bovine papillomavirus
E6	Transforming protein; targets p53 degradation
E7	Transforming protein; complexes with retinoblastoma protein

[a]ORFs E3 and E8 have no known functions.

Genotype of the Virus

There is a strong correlation between the genotype of the infecting virus and the morphology and site of the lesion. For example, almost all flat warts of the skin yield HPV-3 or HPV-10. Most deep plantar warts are caused by HPV-1, and common warts are caused by HPV-2. Virus types HPV-6 and HPV-11 are recovered from most of the genital warts (condylomas). Oncogenic potential is also correlated with the viral genotype. In the genital tract, HPV-16 and HPV-18 are strongly associated with malignancies and HPV-6 and HPV-11 are associated with benign warts. In the rare dermatologic disorder epidermodysplasia verruciformis (EV), lesions caused by HPV-5, HPV-8, and HPV-13 have a greater tendency to convert to malignancy than lesions caused by several other virus types.

Host Factors

Warts tend to increase in size and numbers in conditions associated with immunologic impairment, especially T-lymphocyte deficiency. The immunologic impairment may be subtle, as in pregnancy, or gross, as in organ transplant recipients, patients receiving anticancer therapy, and AIDS patients.

Papillomavirus infection is acquired in a variety of ways: through skin abrasions (skin warts), by sexual intercourse (genital warts), during passage through an infected birth canal (juvenile-onset laryngeal papilloma), and probably in other ways (e.g., papillomas of the oral cavity by autoinoculation or by oral sex).

Clinical Illnesses Related to HPVs

The HPVs naturally fall into two groups, cutaneous HPVs and mucosal HPVs. The viruses are site specific. Cutaneous HPVs are seldom encountered in the genital tract, and genital HPVs are rarely found on the skin. The reservoir for all of the mucosal HPVs is the genital tract, with two exceptions: HPV-13 and HPV-32 are viruses of the oral cavity associated with a condition called focal epithelial hyperplasia, which is prevalent largely in some aboriginal populations. Genital HPVs are also recovered from other mucosal sites, especially the aerodigestive tract. HPV-6 and HPV-11, which are responsible for condylomas in the genital tract, may be transmitted intrapartum from an infected mother to the child and produce juvenile-onset recurrent respiratory papillomatosis (laryngeal papilloma). The HPV-associated illnesses and the most common types of virus responsible for these conditions are listed in Table 3.

Cutaneous HPVs

Cutaneous Warts

There are many morphological types of warts, and each type may have preferred locations on the skin (Bunney, 1992). Specific HPV types are associated with different morphological types of lesions (Croissant et al., 1985). Common warts (caused by HPV-2 and HPV-4) are found on the hands and generally occur as multiple warts. The warts are characteristically dome shaped, with numerous conical projections (papillomatosis) that give their surfaces a velvety appearance. Deep plantar warts (on the bottom surface of the foot; caused by HPV-1) generally occur singly and have a highly thickened corneal layer (hyperkeratosis). Flat warts (with little or no papillomatosis; caused by HPV-3 and HPV-10) almost always occur as multiple warts and are found most often on the arms and face and around the knees. The threadlike filiform warts occur most often on the face and neck.

Skin warts are transmitted by direct contact with an infected individual or indirectly by contact with contaminated objects. The incubation period is difficult to estimate but may be as short as 1 week or as long as several months. As a rule, warts in an otherwise healthy individual are few and small, but a large number of warts may develop in immunodeficient individuals or in apparently normal persons. Most warts regress within 2 years, probably as a result of cell-mediated immune responses. Treatment or excision of one wart often results in regression of the remaining warts. This may result from a "triggering" of an immune response due to immune-competent cells which come in contact with antigens that are released as a result of treatment.

Warts are most prevalent in children and young adults. At any time, as many as 10% of school children may have warts at some site. It is not known if the reduced prevalence in the older population represents acquired immunity, reduced exposure, or both. The incidence of warts in the general population is believed to be increasing. Recreational activity in which bare skin may be exposed to virus-contaminated objects (for example, swimming in communally used pools) increases the risk of acquiring warts, especially plantar warts (Bunney, 1992).

EV

EV is a rare, lifelong disease in which a patient is unable to resolve the wart virus infection (Jablonska and Majewski, 1972). Most patients exhibit defects of cell-mediated immunity. The disease probably has a genetic basis. Patients frequently give a history of parental consanguinity and, despite the rarity of the disease, multiple cases occur in some families. It is postulated that EV patients have an inherited immunologic defect as a result of homozygosity for a rare recessive autosomal gene. The nature of the presumed genetic defect is not known.

The onset of EV occurs in infancy or childhood. The patient develops multiple, disseminated, polymorphic wart-like lesions that tend to become confluent. The warts are of two clinical types: flat warts and red or reddish-brown macular plaques resembling pityriasis versicolor. The warts contain abundant amounts of virus particles, viral antigen, and viral DNA. The flat warts of EV patients yield HPV-3 or -10, the same genotypes that are recovered from flat warts of healthy individuals. However, a bewildering variety of viral genotypes are recovered from the macular plaques of EV patients (Jablonska and Majewski, 1972). These EV genotypes have not been recovered from skin warts, so it is not clear how EV patients acquire these infections. Recent reports suggest that these EV-related viruses are widely seeded in normal skin epithelium (Boxman et al., 1997).

In about 33% of the cases, multiple foci of malignant transformation arise in the reddish-brown plaques, especially in lesions occurring in areas exposed to sunlight. Histologically, the tumors may be in situ (bowenoid) or invasive squamous cell carcinomas. The tumors grow slowly and are generally nonmetastasizing. The malignant cells contain multiple copies of episomal viral DNA (HPV-5, -8, or -14) but no viral particles or capsid antigen. HPV DNA is also recovered from metastatic tumor cells.

The carcinomas occurring in EV patients illustrate how several factors working in concert result in papillomavirus-induced malignancy. Viruses of specific genotypes infecting an immunologically impaired host produce malignant transformation in lesions that are exposed to sunlight.

Skin Cancers

Numerous HPV types, many of them novel types, have been recovered from nonmelanoma skin cancers in renal transplant recipients as well as in healthy individuals. It is not yet clear if they contribute to the development of these cancers (Kawashima et al., 1990; Berkhout et al., 1995).

Mucosal HPVs

About 40 HPV types infect the genital tract. Genital HPV infections are the most prevalent sexually transmitted infections, with prevalence as high as 40 to 45% in sexually active young women (Schneider and Koutsky, 1992). A large majority of the HPV-positive women have no cytological abnormalities. The infections are of 6 to 24 months duration. The prevalence decreases markedly with increasing age.

TABLE 3 Clinical associations of HPVs

Disease location and type	Predominant HPV type
Skin	
Deep plantar wart	HPV-1
Common wart	HPV-2, -4
Flat wart	HPV-3, -10
EV macular plaques	HPV-5, -8, etc.
Mucosa	
Genital warts	HPV-6, -11
Cervical cancer	
High risk	HPV-16, -18, -31, -45
Intermediate risk	HPV-33, -35, -39, -51, -52, -56, -58, -59, -68
Low risk	HPV-6, -11, -42, -43, -44, etc.
Vulvar cancer	HPV-16
Penile cancer	HPV-16
Oropharyngeal cancer	HPV-16
Respiratory papillomas	HPV-6, -11
Focal epithelial hyperplasia of the oral cavity	HPV-13, -32

Genital Warts (Anogenital Warts, Condyloma, and Genital Papilloma)

Papillomavirus infection of the genital tract occurs predominantly in young adults and in sexually promiscuous populations. Genital warts are the most frequent clinical manifestation of HPV infections. The incidence of genital warts has increased in recent years. In the United States, an estimated 946,000 individuals consulted private physicians for genital warts in 1981 compared with an estimated 169,000 in 1966 (Centers for Disease Control, 1992). The number of comparable consultations for genital herpes in 1981 was 295,000, or about 31% of that for condyloma. In the United Kingdom, the annual incidence of genital warts per 100,000 population rose from about 30 in 1971 to 50 in 1978. In sexually transmitted disease clinics, genital warts account for about 4% of patient visits, compared with 24% of visits accounted for by gonorrhea; however, in a population-based study in Rochester, Minn., the incidence rate for genital warts was about one-half that for gonorrhea (Chuang et al., 1984).

The incubation period for condylomas is estimated to be between 3 weeks and 8 months, with an average of 2.8 months (Oriel, 1971). About 66% of the sexual partners of condyloma patients develop the disease. Condylomas may be papillary (condyloma acuminatum) or flat (condyloma planum). The most frequent sites for papillary (or exophytic) condylomas are the penis, around the anus, and on the perineum in the male and the vaginal introitus, the vulva, the perineum, and around the anus in the female. On the cervix, flat condylomas are far more frequent than papillary condylomas (Meisels et al., 1982). The flat lesion on the cervix was not recognized to be due to papillomavirus infection until the late 1970s. It is now known to be a common clinical manifestation of genital HPV infection in the female. The lesion is generally seen only by a colposcopic examination and is confirmed by cytology and histopathology.

In a large number of infected individuals, condylomas occur at more than one site in the genital tract. Condylomas may increase in number and size during pregnancy and regress after delivery. Immunosuppressed populations—for example, patients with AIDS—have a high prevalence of condylomas. The closely related HPV-6 and -11 are responsible for a large majority of the condylomas (Gissmann et al., 1983). Many genital warts regress with time, but some may persist for long periods. They may cause local irritation and itching, become infected, and cause severe physical and psychological difficulties for the patient if they enlarge in size or increase in numbers. The presence of condylomas during pregnancy is a risk factor for the transmission of HPV from mother to newborn during birth and for the consequent development of respiratory papilloma in the offspring.

Respiratory Papilloma

Respiratory papilloma is a chronic, rare, and recurrent disease in which benign viral papillomas in the respiratory tract may become life threatening because of their location (Shah and Howley, 1996). The vocal cords in the larynx are the site most often affected, although the disease may occur at other locations (e.g., the trachea) without laryngeal involvement. The most common presenting symptom is hoarseness of voice or change of voice. The papillomas may produce respiratory distress and obstruction, especially in children. The disease tends to recur following surgical removal of the papilloma, and patients may require frequent operations, sometimes as often as every 2 to 4 weeks. Surgery may lead to dissemination of disease to other sites, for example, to the lungs. Malignant conversion of papilloma is rare; it may be associated with a history of previous radiation therapy but may also occur in the course of a long-term chronic papillomatosis.

The highest risk of onset of respiratory papilloma is under the age of 5 years. About 33 to 50% of the cases occur by that age, and the onset of illness in about 33% of the cases occurs in adult life (Mounts and Shah, 1984). The viral types recovered from both juvenile- and adult-onset disease are HPV-6 and -11, the viruses that are responsible for genital warts (Gissmann et al., 1977; Mounts et al., 1982; Pfister, 1990). The transmission of virus in juvenile-onset cases probably occurs during the process of birth in the course of fetal passage through an infected birth canal. Mothers of patients with laryngeal papilloma frequently have a history of genital warts during pregnancy. The risk of acquiring laryngeal papilloma for children born to mothers with active genital papillomavirus infection is crudely estimated to be between 1 in 30 and 1 in 100. Cesarean delivery prior to rupture of the membrane very likely reduces the risk of virus transmission. The transmission of virus in adult-onset disease does not occur intrapartum. It is suspected that in adults, the infection is acquired by oral contact with infected genitalia.

Warts at Other Sites

Several morphological types of warts occur in the oral cavity. They have been described as common warts, flat warts, condylomas, or respiratory papillomas on the basis of their clinical and histologic features. HPV-6 and HPV-11 are the viruses recovered most frequently from these lesions. A clinically well-defined entity, focal epithelial hyperplasia, has been described in the oral mucosa. The condition occurs with high frequency in American Indians in North and South America, but it has also been seen in other races and in many parts of the world. Clinically, there are discrete, multiple, elevated nodules on the oral mucosa (lips, buccal mucosa, and tongue), which may persist for many years and have the histologic appearance of warts. These lesions are associated with HPV types 13 and 32 (Pfister, 1990).

Cervical Cancer

Worldwide, about 500,000 new cases of cervical cancer occur each year (Peto, 1986). The incidence of cervical cancer is high in developing countries, where it is the most common female malignancy and accounts for about 24% of all female cancers. In the United States, there are about 14,000 new cases of cervical cancer and about 4,800 deaths, annually.

Nearly all cervical cancers originate in the squamo-columnar transformation zone of the cervix, where columnar cells from the endocervix form a junction with the stratified epithelium of the vagina. A progressive spectrum of abnormalities, classified as low-grade and high-grade squamous intraepithelial neoplasia, precedes invasive cancer. Most of these abnormalities resolve on their own, and only a small proportion progress to invasive cancer. The time interval between early cytological abnormalities and cervical cancer may span 10 to 20 years. During this interval, cytological abnormalities are readily detected by Pap smear examination and can be successfully treated.

HPVs and Cervical Cancer

The evidence for an etiologic link between HPV infections and cervical cancer is conclusive (International Agency for Research on Cancer, 1995). Almost all cervical cancers in all parts of the world are caused by HPV infections (Bosch et al., 1995; Walboomers et al., 1999). This evidence is briefly summarized below.

Cancer of the cervix has the epidemiological characteristics of a sexually transmitted disease. The wide differences in cervical cancer incidence in the world are fully explained by taking into account HPV endemicity, sexual practices of women and of their male partners, and standards of health care (i.e., access to Pap smear screening and to treatment of preinvasive disease) (Skegg et al., 1982; Bosch et al., 1994).

HPV DNA sequences are found in a large majority of invasive cancers as well as in the entire clinical spectrum of precursor lesions of invasive cancers. There is a preferential association of some HPV types with invasive cancer (Table 3). While all HPV types found in the genital tract are associated with some form of mild dysplasia and/or subclinical infections, HPV-16, HPV-18, HPV-31, and HPV-45 predominate in invasive cancers. The viral genome is found in the cancer cells themselves and is present in both the primary and metastatic tumors. The viral genome is extrachromosomal in most dysplasias but is integrated in all HPV-18-associated, and many HPV-16-associated, invasive cancers (Cullen et al., 1991). The linearization of the viral genome prior to integration occurs most frequently by a break in the El-E2 region, leading to the deregulation, and increased transcription, of the viral genes E6 and E7. The E6 and E7 genes, which code for the transforming functions of HPV, are always expressed in HPV-associated cancers. Experimental data indicate that the E6 and E7 proteins of the high-risk HPVs exert their oncogenic effect by complexing with and inactivating the tumor suppressor proteins p53 and pRb105, respectively (Scheffner et al., 1991).

HPV-16 is the virus most predominantly associated with cervical cancers in worldwide studies and accounts for about 50% of the cancers. HPV-18 is associated with 10 to 20% of cervical cancers and is not distributed uniformly in different geographic areas. HPV-18 and HPV-16 infections account for about equal proportions of adenocarcinoma of the cervix.

Cancers at Other Sites

HPV infections are associated with cancers at other genital sites, for example, the vulva, the penis, and the anal canal. They also appear to be responsible for a subset of oropharyngeal cancers (Snijders et al., 1992).

Diagnosis

Clinically, a papillary wart is seldom misdiagnosed as something else, but other dermatologic conditions (e.g., molluscum contagiosum, plantar corn, or skin tags) may be mistaken for warts. Histologic examination of the tissue generally establishes the diagnosis of a wart but does not assist in identification of the genotype of the infecting virus. No serologic tests are available for virus identification.

Tests for Viral Capsid Antigen

A broadly cross-reactive genus-specific antiserum is available which is capable of recognizing capsid antigen of all HPV and animal papillomavirus by immunoperoxidase or immunofluorescence tests (Jenson et al., 1980). Tests can be performed on sections of routinely collected, formalin-fixed, paraffin-processed tissues, as well as on exfoliated cells. The viral antigen is present in the nuclei of cells of the superficial layers of the epithelium. For detection of virus, an immunologic test for viral capsid antigen is considerably more sensitive than demonstration of virus particles by electron microscopy. However, the antigen is not detectable in at least 25% of histologically confirmed warts. In antigen-positive tissues, the number of cells displaying antigen is variable, ranging from only one or two cells to a large number of cells in the section. Only a proportion of cytologically affected cells exhibit antigen. Warts at different sites differ markedly with respect to their yield of virus particles and patterns of antigen distribution. Virus particles and antigen are abundant in some plantar and common warts but are scarce in genital tract and laryngeal papillomas.

The capsid antigen assay has several limitations. Productive infection ceases as the lesion progresses toward malignancy, so the assay is negative in HPV-associated cancers. Also, the assay, when positive, indicates that there is productive infection with an HPV but does not identify the genotype of the infecting HPV.

HPV Identification by DNA-Based Assays

HPVs can be specifically identified only by nucleic acid-based assays because the viruses cannot be grown in culture and type-specific immunologic reagents are not available. Tests in common use are (i) PCR-based assays in which a segment of the HPV genome is first amplified and then identified by hybridization, by sequencing, or by its restriction fragment pattern; (ii) hybridization of unamplified tissue DNA with viral probes by Hybrid Capture or by Southern hybridization; and (iii) in situ hybridization of tissue sections to localize the viral genome to specific cells.

The PCR assays are most frequently used for epidemiological investigations. Hybrid Capture is used for both epidemiological and clinical studies. The in situ hybridization assay is most useful in studies in which it is important to localize the virus to the tumor cells.

PCR-Based Assays

PCR-based assays have high analytic sensitivity and can detect as few as 10 to 100 copies of the HPV genome. They require very small amounts of the specimen. The most widely used assays to identify genital HPVs employ a one-step treatment of the clinical specimens and a single amplification reaction with consensus primers which are capable of amplifying most of the genital-tract HPVs. The PCR products are then identified with a large number of type-specific probes, by sequencing the PCR products, or by an analysis of the restriction fragment patterns of the PCR products.

The two most widely used PCR assays both amplify a segment of the L1 ORF. The MY09-MY11-HMB01 primers are degenerate, and they amplify a 450-bp segment of the L1 ORF of HPVs (Manos et al., 1989; Hildesheim et al., 1994). The PCR products are identified in a dot blot format by hybridization with individual biotinylated type-specific HPV probes employing Amersham's enhanced chemiluminescence system. A simpler variation of this assay has been published recently in which amplification is performed with biotinylated primers and the PCR products are identified by reaction with immobilized probes on a strip (Gravitt et al., 1998). This line blot assay requires a single hybridization assay. The GP5+ and GP6+ primers, which are also widely used, are consensus, nondegenerate primers which amplify a 140-bp region of the L1 gene (de Roda Husman et al., 1995; Jacobs et al., 1995).

Hybrid Capture Assays

The Hybrid Capture II assay (Digene, Inc., Beltsville, Md.) is performed in a 96-well microtiter format on unamplified tissue DNA (Lorincz, 1996). The sensitivity of the assay is enhanced by signal amplification. The tissue DNAs are screened separately with each of two pools of full-length RNA probes: probe A consists of the low-risk HPV types HPV-6, -11, -41, -42, and -43, and probe B consists of the 13 high-risk and intermediate-risk HPVs listed in Table 3. The single-stranded DNA in denatured specimens hybridizes with the RNA probe. The DNA-RNA hybrids are captured and immobilized with a hybrid-specific antibody on the bottom of the well and detected in an enzyme-linked immunos or bent assay (ELISA)-like format with the use of a chemiluminescent compound. The virus in the specimen is identified as belonging to a low-risk or high-risk group but is not identified as a specific type. The test allows for a degree of quantitation of the viral DNA in the specimens (Cox et al., 1995). The Hybrid Capture II system is available commercially and is approved by the U.S. Food and Drug Administration.

Southern Hybridization

The PCR-based assays, as well as Hybrid Capture, can be performed on crude extracts of specimen DNA. Southern hybridization is performed on cellular DNA purified through a series of phenol-chloroform extractions. The technique is labor-intensive and requires significant amounts of purified cellular DNA. It is now seldom used for HPV diagnosis in clinical or epidemiological investigations.

In Situ Hybridization

In situ hybridization is the only method which permits localization of the viral genome to a specific cell type, e.g., tumor cells. Tests have been devised to detect viral DNA, viral transcripts, or expression of specific viral genes (Durst et al., 1992; Stoler et al., 1992).

Prevention and Treatment

Most skin warts and genital warts regress spontaneously. The patient seeks treatment for cosmetic reasons, pain, discomfort, and disability depending on the location and size of the warts. The most difficult problems for therapy are posed by children with recurrent laryngeal papilloma, patients with EV, pregnant women with genital warts, and warts in immunocompromised individuals. There is no "one-time" treatment for all warts. The therapies in use include application of caustic agents, such as podophyllin and salicylic acid; cryotherapy; surgical removal; antimetabolites, such as 5-fluorouracil applied in a cream or a solution; and treatment with interferon. Both laryngeal papillomas and genital warts are reported to respond to interferon therapy, but recurrence after cessation of therapy is not uncommon. Preinvasive cervical lesions are readily treated by a variety of procedures (cryotherapy, laser, loop electrocautery excision procedure, etc.).

There are high hopes that anti-HPV-based strategies may help in the prevention and control of cervical cancers. The diagnosis of infection with high-risk HPVs may help select women who need close monitoring (Cox et al.,

1995). HPV assays have the potential for use in cervical cancer screening as an adjunct to, or as a replacement for, Pap smears (Meijer et al., 1997). Prophylactic HPV vaccines based on viruslike particles are in phase 1 and 2 trials; if effective, they would prevent HPV infections. Therapeutic HPV vaccines, aimed at destroying tumor cells by targeting the viral E6 and E7 proteins present in these cells, are also in development.

HUMAN POLYOMAVIRUSES

The first conclusive evidence of human infection with polyomaviruses was obtained in the mid-1960s, when polyomavirus particles were consistently demonstrated by electron microscopy in the enlarged nuclei of oligodendrocytes in the affected areas of brains of patients with progressive multifocal leukoencephalopathy (PML) (Zu Rhein, 1969). In 1971, JCV, the causative agent of PML, was isolated from a brain of a patient with PML in primary human fetal glial cell cultures. In the same year, another polyomavirus, BKV, was isolated in Vero cell cultures from the urine of a renal-transplant recipient (Padgett and Walker, 1976; Gardner, 1977). Subsequent studies have shown that following clinically inapparent primary infection in childhood, both JCV and BKV persist in the kidney and in B lymphocytes in the peripheral blood. The viruses are reactivated in a variety of conditions that impair cell-mediated immune responses. Almost all of the pathological effects of BKV and JCV infections occur in immunodeficient individuals.

Between 1955 and 1961, millions of people were inadvertently exposed to SV40, an oncogenic polyomavirus of Asian macaques, which had contaminated inactivated (Salk) poliovirus vaccines and experimental live poliovirus vaccines prepared from virus pools grown in primary rhesus monkey kidney cultures (Shah and Nathanson, 1976).

Characteristics of the Virus

The polyomavirus virion is nonenveloped and has a diameter of 44 nm, icosahedral symmetry, and 72 capsomers. The viral genome is a double-stranded, circular DNA molecule with 5×10^3 bp and a molecular mass of 3.2×10^6 Da. Each of the two DNA strands carries about 50% of the genetic information. Complete nucleotide sequences, as well as detailed physical and physiologic maps, are known for both the JCV and BKV genomes (Cole, 1996). There is extensive nucleotide sequence homology between BKV and JCV throughout their genomes, with the highest conservation in the late region, which codes for the capsid proteins.

Both JCV and BKV hemagglutinate human erythrocytes. The capsid consists of three virus-specified proteins (VP1, VP2, and VP3) and three cellular histones (VP4, VP5, and VP6). The major capsid protein, VP1, accounts for more than 70% of the virion mass and has a molecular mass of 39 to 44 kDa. Cells infected with or transformed by the viruses express T antigens, which are coded by the early regions of the genomes and are not part of the viral capsid. T antigens are required for the initiation of viral DNA synthesis and for the initiation and maintenance of virus-induced transformation.

Despite the extensive nucleotide sequence homology between the two genomes, JCV and BKV can be readily distinguished from one another by immunologic and DNA hybridization tests. Antibodies to the two viruses in human sera display minimal or no cross-reactivity in neutralization, hemagglutination inhibition, or ELISAs. Both JCV and

BKV share with other polyomaviruses a genus-specific immunologic determinant(s) that is physically located on VP1 but internal to the virion surface (Shah, 1996). Antibodies against the genus-specific determinant(s), prepared by immunization with disrupted capsids, react with all human and animal polyomaviruses.

Both JCV and BKV transform cells in tissue culture and are oncogenic in laboratory animals. JCV transforms hamster brain cells and human amnion cells, whereas BKV transforms cells of hamster, rat, rabbit, monkey, and mouse origin. Both viruses are oncogenic for newborn hamsters. JCV also produces cerebral neoplasm in owl and squirrel monkeys, and provides the only model of a primate central nervous system tumor caused by a virus (London et al., 1978, 1983).

Pathogenesis and Disease Potential

Primary infections with JCV and BKV occur in childhood. Infection with JCV is acquired at a later age than BKV infection. In the United States, 50% of children develop antibodies to JCV by the age of 10 to 14 years and to BKV by 3 to 4 years. Infection in healthy children is most often subclinical. Serologic studies suggest that primary BKV infection may be associated with mild upper respiratory disease, but BKV has not been isolated from respiratory secretions (Goutsmit et al., 1982). Tonsillar tissue may be the site of primary JCV infection (Monaco et al., 1996).

The viruses persist in the kidney following primary infection. Viral genomes can be demonstrated in cadaver kidney tissues. It is likely that after multiplication at the site of entry, the viruses reach the kidney by a process of viremia, and infection of the kidney is associated with transient viruria. However, JCV and BKV have been recovered very rarely from blood, urine, or any other site in healthy, immunocompetent children.

Nearly all significant illnesses, as described below, due to BKV and JCV occur in immunocompromised hosts, mainly as a result of reactivation of viruses latent in the kidney. Conditions in which viruses are reactivated include pregnancy, diabetes, organ transplantation, antitumor therapy, and AIDS and other immunodeficiency diseases. Unchecked virus multiplication after primary infection of immunodeficient individuals may also lead to pathological consequences.

PML

PML is a rare, fatal, subacute demyelinating disease of the central nervous system that results from JCV infection of oligodendrocytes in the brain (Padgett and Walker, 1976; Johnson, 1982). It occurs as a complication of a wide variety of conditions associated with T-cell deficiencies. These conditions include lymphoproliferative disorders, such as Hodgkin's disease, chronic lymphocytic leukemia, and lymphosarcoma; chronic diseases, such as sarcoidosis and tuberculosis; primary immunodeficiency diseases; prolonged immunosuppressive therapy, as, for example, in renal-transplant recipients and patients with rheumatoid arthritis, systemic lupus erythematosus, and myositis; and AIDS. Most cases of PML occur in middle age or later life, but the disease is being increasingly identified at earlier ages, e.g., in children with primary immunodeficiency diseases, in renal-transplant recipients, and in AIDS patients. Cases of PML in older patients are most likely the result of reactivation of latent JCV. In the younger patient, it is possible that unchecked primary JCV infection may lead to PML. The

AIDS pandemic has brought about a marked increase in the number of PML cases, from less than 0.2 per million persons in 1984 to 3.3 per million persons in 1994 (Holman et al., 1998). Between 1991 and 1994, 89% of PML deaths were among AIDS patients. PML is diagnosed in 1 to 2% of AIDS-associated deaths and in 3.8% of AIDS patients who have neurological abnormalities.

PML has unique pathological features. The affected area of the brain contains foci of demyelination, which have at their edges enlarged oligodendrocytes. The nuclei of the oligodendrocytes are two to three times their normal size and basophilic, and they may contain basophilic or eosinophilic inclusion bodies. Most lesions also have bizarre, giant astrocytes with hyperchromatic pleomorphic nuclei. Inflammation is minimal or absent. Neurons are unaffected. The characteristics of these lesions and the occurrence of PML in immunodeficient individuals led Richardson (1961) to propose that the key event in the pathogenesis of PML was infection of oligodendrocytes with a common virus when immune defenses were impaired. Demyelination was a result of destruction of oligodendrocytes, which are normally responsible for the formation and maintenance of myelin sheaths. The nuclei of affected oligodendrocytes contain abundant numbers of JCV particles.

Clinically, PML has an insidious onset and may occur at any time in the course of the underlying illness. The signs and symptoms point to a multifocal involvement of the brain. Impaired speech and vision and mental deterioration are common early features of the disease. The patient remains afebrile, and headache is uncommon. The cerebrospinal fluid remains normal. As a rule, the disease is progressive, resulting in death within 3 to 6 months after onset. Paralysis of limbs, cortical blindness, and sensory abnormalities occur in later stages. A few patients may survive for years, with stabilization of the condition and even apparent remission. A longer survival time is thought to be associated with a more marked inflammatory response in the brain.

The diagnosis of PML can be conclusively established by pathological examination of a biopsy or specimen postmortem. Macroscopically, the brain shows foci of demyelination that may vary widely in size and may become confluent and necrotic in the advanced stages of disease. The lesions are most frequent in the subcortical white matter. The cerebrum is almost always affected. Microscopically, the presence of enlarged oligodendrocyte nuclei around the foci of demyelination is diagnostic. These altered nuclei contain abundant amounts of JCV particles, antigen, and DNA. JCV particles or antigen are not found in healthy brains or in nondiseased areas of PML brains. Small amounts of viral DNA may be recovered from extraneural sites, such as the kidney, liver, lymph node, and spleen (Grinnell et al., 1983).

In recent years, noninvasive techniques, such as computerized axial tomography scans or nuclear magnetic resonance imaging of the brain, have provided effective means for the diagnosis of PML.

Renal- and Bone Marrow Transplant Recipients

About 33 to 50% of renal- and bone marrow transplant recipients excrete JCV, BKV, or both in urine in the posttransplantation period. The duration of viruria varies from a few days to several months. BKV infection is more common than JCV infection, and most infections are due to reactivation of latent viruses. The frequency of infection is higher in recipients with a history of diabetes (Hogan et al., 1980).

The risk of BKV infection is increased if a kidney from a seropositive donor is transplanted into a seronegative recipient (Andrews et al., 1983). The infections may sometimes lead to severe pathological consequences, e.g., loss of renal function or rejection of allografts. Some cases of ureteral obstruction, an uncommon and late complication in renal transplantation, have been ascribed to JCV and BKV infections. In addition, BKV reactivation appears to be responsible for hemorrhagic cystitis in bone marrow transplant recipients (Arthur et al., 1986).

Primary Immunodeficiency Diseases

BKV has been isolated from the urine of patients with primary immunodeficiency diseases. A fatal result of BKV infection has been reported. A 6-year-old boy with hyperimmunoglobulin M deficiency developed massive BKV viruria, tubulointerstitial nephritis with viral inclusions in the lesions, and irreversible renal failure (Rosen et al., 1983).

Pregnancy

Both JCV and BKV are reactivated in some women during normal pregnancy. In a prospective study, cytopathology in cells obtained from urine sediment suggested JCV and BKV infections in 3.2% of pregnant women. This was most frequently observed in the last trimester of pregnancy (Coleman et al., 1980). In another study, 16% of the women showed a rise in antibody to one or the other virus during pregnancy. All the infections were reactivations of latent viruses in antibody-positive individuals (Andrews et al., 1983). It has been reported that fetal sera may have BKV-specific immunoglobulin M, indicating transplacental transmission of the virus; these observations have not been confirmed.

Role in Human Malignancies

JCV and BKV are oncogenic for laboratory animals, and they transform cultured cells. These viruses, as well as SV40, therefore, have been investigated for their roles in human malignancies (Barbanti-Brodano et al., 1998; Butel and Lednicky, 1999). Tumors of the nervous system have received special attention. Occasionally, multifocal astrocytomas seem to arise in lesions of PML demyelination, suggesting that the tumors arose in these lesions. There are reports of finding BKV, JCV, and SV40 genomes in nervous system tumors and in mesotheliomas and osteosarcomas, but a reproducible and consistent etiologic association of JCV, BKV, or SV40 with any human malignancy has not been demonstrated (Shah, 2000).

Diagnosis

Evidence of multifocal brain disease in an immunocompromised individual suggests the possibility of PML. The unique histopathological features of PML are seen by light microscopy. Except for two cases in 1971 in which SV40 was isolated from brains of PML patients, all other cases have yielded JCV. The virus can be specifically identified by an immunoperoxidase test of frozen sections or paraffin sections of the affected tissue using a monospecific anti-JCV serum. Alternatively, the viral genome in the lesion can be identified by hybridization of the total DNA extracted from the affected tissue (Grinnell et al., 1983), by in situ hybridization of paraffin sections with a JCV probe, or by

detection of JCV by PCR technology (Arthur et al., 1989). Serologic studies are not helpful in the diagnosis of PML. Anti-JCV antibodies are present at the onset of the disease, and they do not show any marked increase as the disease progresses. Viral antibodies are not detected in the cerebrospinal fluid.

In conditions other than PML, cytomorphology of the urinary tract epithelial cells suggestive of virus excretion in the urine is often the first indication of virus infection. Virus-infected cells are enlarged, and their nuclei contain single, large, basophilic inclusions that may occupy the whole nucleus (Coleman, 1975). There are no cytoplasmic inclusions. Differential diagnosis includes cells infected with cytomegalovirus and sometimes urothelial cancer cells. Cytomegalovirus-infected cells may have both nuclear and cytoplasmic inclusions; the nuclear inclusions are small and surrounded by a clear peripheral zone (halo) and are either basophilic or eosinophilic. The malignant-cell nucleus has rough-textured chromatin in contrast to the structureless inclusion in the virus-infected nucleus.

The cytological abnormalities are not always present or clear-cut during viruria, and these abnormalities do not distinguish between JCV and BKV infections. JCV grows best in primary human fetal glial cells, and BKV grows best in primary human embryonic kidney or human diploid fibroblast cells. Both viruses also grow in primary urothelial cell cultures collected from infant urine. Isolation of virus in tissue culture is inefficient, and it may take several weeks before a specific diagnosis is possible. Human fetal glial cells or urinary tract epithelial cells are difficult to obtain. An ELISA for detection of viral antigens in urine and DNA dot blot hybridization assays for identification of viral genomes in cells from the urinary tract offer the prospect of rapid diagnosis of JCV and BKV. However, the most effective method for diagnosing BKV and JCV employs PCR technology (Arthur et al., 1989).

Treatment

Attempts to treat PML have not been successful, although some remissions have been reported with the use of nucleic acid base analogs, adenine arabinoside, and cytosine arabinoside. It would be useful, when possible, to reduce or discontinue immunosuppressive therapy. No attempts have been made to treat urinary tract infections.

REFERENCES

Andrews, C., R. Daniel, and K. Shah. 1983. Serologic studies of papovavirus infections in pregnant women and renal transplant recipients, p. 133–141. In J. L. Sever and D. L. Madden (ed.), Polyomaviruses and Human Neurological Diseases. Alan R. Liss, New York, N.Y.

Arthur, R. R., K. V. Shah, S. J. Baust, G. W. Santos, and R. Saran. 1986. Association of Bk viruria with hemorrhagic cystitis in recipients of bone marrow transplants. N. Engl. J. Med. 315:230–234.

Arthur, R. R., S. Dagostin, and K. V. Shah. 1989. Detection of BKV and JCV in urine and brain tissue by the polymerase chain reaction. J. Clin. Microbiol. 27:1174–1179.

Barbanti-Brodano, G., F. Martini, M. De Mattei, L. Lazzarin, A. Corallini, and M. Tognon. 1998. BK and JC human polyomaviruses and simian virus 40: natural history of infection in humans, experimental oncogenicity, and association with human tumors. Adv. Virus Res. 50:69–99.

Berkhout, R. J. M., L. M. Tieben, H. L. Smits, J. N. Bavinck, B. J. Vermeer, and J. ter Schegget. 1995. Nested

PCR approach for detection and typing of epidermodysplasia verruciformis-associated human papillomavirus types in cutaneous cancers from renal transplant recipients. *J. Clin. Microbiol.* **33:**690–695.

Bosch, F. X., N. Munoz, S. de Sanjosé, E. Guerrero, A. M. Ghaffari, J. Kaldor, X. Castellsague, and K. V. Shah. 1994. Importance of human papillomavirus endemicity in the incidence of cervical cancer: an extension of the hypothesis on sexual behavior. *Cancer Epidemiol. Biomarkers Prev.* **3:**375–379.

Bosch, F. X., M. M. Manos, N. Muñoz, M. Sherman, A. M. Jansen, J. Peto, M. H. Schiffman, V. Moreno, R. Kurman, K. V. Shah, and the IBSCC Study Group. 1995. Prevalence of human papillomavirus in cervical cancer: a worldwide perspective. *J. Natl. Cancer Inst.* **87:**796–802.

Boxman, I. L. A., R. J. M. Berkhout, L. H. C. Mulder, M. D. Wolkers, J. N. Bouwes Bavinck, B. J. Vermeer, and J. ter Schegget. 1997. Detection of human papillomavirus DNA in plucked hairs from renal transplant recipients and healthy volunteers. *J. Investig. Dermatol.* **108:**712–715.

Bunney, M. 1992. *Viral Warts: Their Biology and Treatment*, 2nd ed. Oxford University Press, Oxford, United Kingdom.

Butel, J. S., and J. A. Lednicky. 1999. Cell and molecular biology of simian virus 40: implications for human infection and disease. *J. Natl. Cancer Inst.* **91:**119–134.

Centers for Disease Control. 1992. HIV infection and expanded surveillance case definition for AIDS among adolescents and adults. *Morb. Mortal. Wkly. Rep.* **41:**1–13.

Chuang, T. Y., H. O. Perry, L. T. Kurland, and D. M. Ilstrup. 1984. Condylomata acuminatum in Rochester, Minn., 1950–1978. *Arch. Dermatol.* **120:**469–483.

Cole, C. N. 1996. Polyomavirinae: the viruses and their replication, p. 1997–2025. *In* B. N. Fields, D. M. Knipe, and P. M. Howley (ed.), *Virology.* Lippincott-Raven, Philadelphia, Pa.

Coleman, D., M. Wolfendale, R. Daniel, N. Dhanjal, S. Gardner, P. Gibson, and A. Field. 1980. A prospective study of human polyomavirus infection in pregnancy. *J. Infect. Dis.* **142:**1–8.

Coleman, D. V. 1975. The cytodiagnosis of human polyomavirus infection. *Acta Cytol.* **19:**93–96.

Cox, J. T., A. T. Lorincz, M. H. Schiffman, M. E. Sherman, A. Cullen, and R. J. Kurman. 1995. Human papillomavirus testing by hybrid capture appears to be useful in triaging women with a cytologic diagnosis of atypical squamous cells of undetermined significance. *Am. J. Obstet. Gynecol.* **172:**946–954.

Croissant, O., F. Breitburd, and G. Orth. 1985. Specificity of cytopathic effect of cutaneous human papillomaviruses. *Clin. Dermatol.* **3:**43–55.

Cullen, A. P., R. Reid, M. Campion, and A. Lorincz. 1991. Analysis of the physical state of different human papillomavirus DNAs in intraepithelial and invasive cervical neoplasia. *J. Virol.* **65:**606–612.

de Roda Husman, A. M., J. M. M. Walboomers, A. J. C. van den Brule, C. J. L. M. Meijer, and P. J. F. Snijders. 1995. The use of general primers GP5 and GP6 elongated at their 3′ ends with adjacent highly conserved sequences improves human papillomavirus detection by PCR. *J. Gen. Virol.* **76:**1057–1062.

Durst, M., D. Glitz, A. Schneider, and H. zur Hausen. 1992. Human papillomavirus type 16 (HPV16) gene expression and DNA replication in cervical neoplasia: analysis by in situ hybridization. *Virology* **189:**132–140.

Gardner, S. 1977. The new human papovaviruses: their nature and significance, p. 93–115. *In* A. P. Waterson (ed.), *Recent Advances in Clinical Virology.* Livingstone, New York, N.Y.

Gissmann, L., H. Pfister, and H. zur Hausen. 1977. Human papillomaviruses: characterization of four different isolates. *Virology* **76:**569–580.

Gissmann, L., V. Diehl, H. J. Schultz-Loulon, and H. zur Hausen. 1982. Molecular cloning and characterization of human papillomavirus DNA derived from a laryngeal papilloma. *J. Virol.* **44:**393–400.

Gissmann, L., L. Wolnik, H. Ikenberg, U. Koldovsky, H. Schnurch, and H. zur Hausen. 1983. Human papillomavirus types 6 and 11 DNA sequences in genital and laryngeal papillomas and in some cervical cancers. *Proc. Natl. Acad. Sci. USA.* **80:**560–563.

Goutsmit, J., P. Wertheim-van Dillen, A. van Strein, and J. van der Noordaa. 1982. The role of BK virus in acute respiratory tract disease and the presence of BKV DNA in tonsils. *J. Med. Virol.* **10:**91–99.

Gravitt, P. E., C. L. Peyton, R. J. Apple, and C. M. Wheeler. 1998. Genotyping of 27 human papillomavirus types by using L1 consensus PCR products by a single-hybridization, reverse line blot detection method. *J. Clin. Microbiol.* **36:**3020–3027.

Grinnell, B., B. Padgett, and D. Walker. 1983. Distribution of nonintegrated DNA from JC papovavirus in organs of patients with progressive multifocal encephalopathy. *J. Infect. Dis.* **147:**669–675.

Hildesheim, A., M. H. Schiffman, P. Gravitt, A. G. Glass, C. E. Greer, T. Zhang, D. R. Scott, B. B. Rush, P. Lawler, M. E. Sherman, et al. 1994. Persistence of type-specific human papillomavirus infection among cytologically normal women in Portland, Oregon. *J. Infect. Dis.* **169:**235–240.

Hogan, T., E. Borden, J. McBain, B. Padgett, and D. Walker. 1980. Human polyomavirus infections with JC virus and BK virus in renal transplant patients. *Ann. Intern. Med.* **92:**373–378.

Holman, R. C., R. J. Torok, E. D. Belay, R. S. Janssen, and L. B. Schonberger. 1998. Progressive multifocal leukoencephalopathy in the United States, 1979–1994: increased mortality associated with HIV infection. *Neuroepidemiology* **17:**303–309.

Howley, P. M. 1996. Papillomavirinae: the viruses and their replication, p. 2045–2076. *In* B. N. Fields, D. M. Knipe, and P. M. Howley (ed.), *Virology.* Lippincott-Raven, Philadelphia, Pa.

International Agency for Research on Cancer. 1995. *IARC Monograph on the Evaluation of Carcinogenic Risks to Humans,* vol. 64. *Human Papillomaviruses.* International Agency for Research on Cancer, Lyon, France.

Jablonska, S., and S. Majewski. 1972. Epidermodysplasia verruciformis: immunological and clinical aspects, p. 157–175. *In* H. zur Hausen (ed.), *Human Pathogenic Papillomaviruses.* Springer Verlag, Heidelberg, Germany.

Jacobs, M. V., A. M. de Roda Husman, A. J. C. van den Brule, P. J. F. Snijders, C. J. L. M. Meijer, and J. M. M. Walboomers. 1995. Group specific differentiation between high- and low-risk human papillomavirus genotypes by general primer-mediated polymerase chain reaction and two cocktails of oligonucleotide probes. *J. Clin. Microbiol.* **33:**901–905.

Jenson, A., J. Rosenthal, W. D. Lancaster, C. Olson, and K. V. Shah. 1980. Immunological relatedness of papilloma viruses from different species. *J. Natl. Cancer Inst.* **64:**495–500.

Johnson, R. T. 1982. *Viral Infections of the Nervous System,* p. 241–247. Raven Press, New York, N.Y.

Kawashima, M., M. Favre, S. Obalek, S. Jablonska, and G. Orth. 1990. Premalignant lesions and cancers of the skin in the general population: evaluation of the role of human papillomaviruses. *J. Investig. Dermatol.* **95:**537–542.

London, W. T., S. A. Houff, D. L. Madden, D. A. Fuccillo, M. Gravell, W. C. Wallen, A. E. Palmer, J. L. Sever, B. L. Padgett, D. L. Walker, G. M. Zu Rhein, and T. Ohashi. 1978. Brain tumors in owl monkeys inoculated with a human polyomavirus (JC virus). *Science* **201**:1249.

London, W. T., S. A. Houff, P. E. McKeever, W. C. Wallen, J. L. Sever, B. L. Padgett, and A. D. L. Walker. 1983. Viral-induced astrocytomas in squirrel monkeys. *Prog. Clin. Biol. Res.* **105**:227–237.

Lorincz, A. 1996. Hybrid Capture™ method for detection of human papillomavirus DNA in clinical specimens. *Papillomavirus Rep.* **7**:1–5.

Los Alamos National Laboratory. 1998. *Human Papillomaviruses 1998: A Compilation And Analysis of Nucleic Acid and Amino Acid Sequences.* Los Alamos National Laboratory, Los Alamos, N. Mex.

Manos, M. M., Y. Ting, D. K. Wright, A. J. Lewis, T. R. Broker, and S. M. Wolinsky. 1989. The use of polymerase chain reaction amplification for the detection of genital human papillomaviruses. *Cancer Cells* **7**:209–214.

Meijer, C. J. L. M., L. Rozendaal, J. C. van der Linden, T. J. M. Helmerhorst, F. J. Voorhorst, and J. M. M. Walboomers. 1997. Human papillomavirus testing for primary cervical cancer screening, p. 338–347. *In* E. Franco and J. Monsenego (ed.), *New Developments in Cervical Cancer Screening and Prevention.* Blackwell Science, Oxford, United Kingdom.

Meisels, A., C. Morin, and M. Casas-Cordero. 1982. Human papillomavirus infection of the uterine cervix. *Int. J. Gynecol. Pathol.* **1**:75–94.

Melnick, J. 1962. Papovavirus group. *Science* **135**:1128–1130.

Monaco, M., W. Atwood, M. Gravell, C. Tornatore, and E. Major. 1996. JC virus infection of hematopoietic progenitor cells, primary B lymphocytes, and tonsillar stromal cells: implications for viral latency. *J. Virol.* **70**:7004–7012.

Mounts, P., and K. V. Shah. 1984. Respiratory papillomatosis: etiological relation to genital tract papillomaviruses. *Prog. Med. Virol.* **29**:90–114.

Mounts, P., K. V. Shah, and H. Kashima. 1982. Viral etiology of juvenile- and adult-onset squamous papilloma of the larynx. *Proc. Natl. Acad. Sci. USA* **79**:5425–5429.

Oriel, J. 1971. Natural history of genital warts. *Br. J. Vener. Dis.* **47**:1–13.

Orth, G., M. Favre, and O. Croissant. 1977. Characterization of a new type of human papillomavirus that causes skin warts. *J. Virol.* **24**:108–120.

Padgett, B., and D. Walker. 1976. New human papovaviruses. *Prog. Med. Virol.* **22**:1–35.

Peto, R. 1986. Introduction: geographic patterns and trends, p. 3–15. *In* R. Peto and H. zur Hausen (ed.), *Viral Etiology of Cervical Cancer.* Cold Spring Harbor Laboratory, Cold Spring Harbor, N.Y.

Pfister, H. 1990. General introduction to papillomaviruses, p. 1–9. *In* H. Pfister (ed.), *Papillomaviruses and Human Cancer.* CRC Press, Boca Raton, Fla.

Richardson, E. 1961. Progressive multifocal leukoencephalopathy. *N. Engl. J. Med.* **265**:315–316.

Rosen, S., W. Harmon, A. Krensky, P. Edelson, B. Padgett, B. Grinnell, M. Rubino, and D. Walker. 1983. Tubulo-interstitial nephritis associated with polyomavirus (BK type) infection. *N. Engl. J. Med.* **308**:1192–1196.

Rowson, K. E. K., and B. W. J. Mahy. 1967. Human papova (wart) virus. *Bacteriol. Rev.* **31**:110–131.

Scheffner, M., K. Munge, J. Byrne, and P. Howley. 1991. The state of the p53 and retinoblastoma genes in human cervical carcinoma cell lines. *Proc. Natl. Acad. Sci. USA* **88**:5523–5527.

Schiller, J. T., and R. B. S. Roden. 1995. Papillomavirus-like particles. *Papillomavirus Rep.* **6**:121–128.

Schneider, A., and L. A. Koutsky. 1992. Natural history and epidemiologic features of genital HPV infection, p. 25–52. *In* N. Munoz, F. X. Bosch, K. V. Shah, and A. Meheus (ed.), *The Epidemiology of Human Papillomavirus and Cervical Cancer.* International Agency for Research on Cancer, Lyon, France.

Shah, K. V. 1996. Polyomaviruses, p. 2027–2043. *In* B. N. Fields, D. M. Knipe, and P. M. Howley (ed.), *Virology.* Lippincott-Raven, Philadelphia, Pa.

Shah, K. V. 2000. Polyoma Viruses (JC virus, BK virus and simian virus 40) and human cancer, p. 461–474. *In* J. J. Goedert (ed.), *Infectious Causes of Cancer: Targets for Intervention.* Humana Press, Inc., Totowa, N.J.

Shah, K. V., and P. M. Howley. 1996. Papillomaviruses, p. 2077–2109. *In* B. N. Fields, D. M. Knipe, and P. M. Howley (ed.), *Virology.* Lippincott-Raven, Philadelphia, Pa.

Shah, K. V., and N. Nathanson. 1976. Human exposure to SV40: review and comment. *Am. J. Epidemiol.* **103**:1–12.

Skegg, D. C. G., P. A. Corwin, and C. Paul. 1982. Importance of the male factor in cancer of the cervix. *Lancet* **ii**:581–583.

Snijders, P. J. F., F. V. Cromme, A. van den Brule, H. F. J. Schrijnemakers, G. B. Snow, and C. J. L. M. Meijer. 1992. Prevalence and expression of human papillomavirus in tonsillar carcinomas, indicating a possible viral etiology. *Int. J. Cancer* **51**:845–850.

Stoler, M. H., C. R. Rhodes, A. Whitbeck, S. M. Wolinsky, L. T. Chow, and T. R. Broker. 1992. Human papillomavirus type 16 and 18 gene expression in cervical neoplasias. *Hum. Pathol.* **23**:117–128.

Walboomers, J. M. M., M. V. Jacobs, M. M. Manos, F. X. Bosch, J. A. Kummer, K. V. Shah, P. J. F. Snijders, J. Peto, C. J. L. M. Meijer, and N. Muñoz. 1999. Human papillomavirus is a necessary cause of invasive cervical cancer worldwide. *J. Pathol.* **189**:12–19.

Wu, T. C. 1994. Immunology of the human papilloma virus in relation to cancer. *Curr. Opin. Immunol.* **6**:746–754.

Zu Rhein, G. 1969. Association of papova virions with a human demyelinating disease (progressive multifocal leukoencephalopathy). *Prog. Med. Virol.* **11**:185–247.

Herpes Simplex Viruses

LAURE AURELIAN

31

The word "herpes" has been in the medical vocabulary for at least 25 centuries. In Greek it means "to creep" and was originally used to define "an animal that goes on all fours" (*herpeton*). In the Hippocratic corpus it was used to describe an assortment of cutaneous lesions, including descriptions compatible with herpes simplex or herpes zoster lesions. In 1736, Jean Astruc classified the condition as a sexually transmitted disease and noted its frequency among homosexuals. In the early 19th century, six clinical entities were delineated, including facial and genital herpes, but the disease was not considered communicable, probably because of the idiosyncratic appearance of symptoms in conjunction with disparate, well-defined febrile illnesses. It took 100 years to demonstrate that material derived from human herpetic lesions that was passed through a filter with pores small enough to retain bacteria produces a serially transmissible keratoconjunctivitis in rabbits (Luger and Lauda, 1921). In 1929, Goodpasture concluded that the virus travels along the axonal route from peripheral sites of infection and resides in a latent state within the nerve cells of the ganglion. In the 1930s, it became evident that the manifestation is gingivostomatitis in seronegative subjects and recurrent "fever blisters" in those who are seropositive. Now it is widely known that herpesvirus causes recurrent facial lesions commonly identified as fever blisters and sexually transmitted recurrent genital lesions. Accordingly, the word "herpes" usually refers to the herpes simplex viruses, types 1 (HSV-1) and 2 (HSV-2).

CHARACTERISTICS AND CLASSIFICATION

HSV is a member of the family *Herpesviridae*. It belongs to the subfamily *Alphaherpesvirinae* and the genus *Simplexvirus*. Criteria for classification in the *Alphaherpesvirinae* are virion architecture, variable host range, relatively short reproductive cycle, rapid spread in culture, efficient cell destruction, and the capacity to establish latent infection in sensory ganglia. HSV shares four biological properties with other *Herpesviridae*: (i) DNA codes for enzymes involved in nucleic acid metabolism, (ii) DNA synthesis and capsid assembly occur in the nucleus, (iii) production of infectious progeny is accompanied by cell death, and (iv) viruses remain latent in their natural hosts. The HSV virion is 150 to 200 nm in diameter. The electron-dense core has a diameter of approximately 77.5 nm. It contains the 152-kbp, linear, double-stranded DNA. It is surrounded by the capsid, a 100- to 110-nm-diameter protein shell with icosadeltahedral symmetry composed of 162 capsomeres, each with a hole along the long axis. The capsid is surrounded by the tegument, an electron-dense, amorphous material that is asymmetrically distributed and may appear to be fibrous upon negative staining. The outer covering of the virion is the lipid bilayer envelope, acquired from cell membranes. The virion is decorated with spikes, approximately 8 nm long, containing HSV-specified glycoproteins (Fig. 1).

HSV virions contain at least 30 known proteins, which have been given serial numbers. Approximately half of the HSV-1 structural proteins are in the tegument, and some are involved in initiation of the replicative cycle. This includes UL36 (also known as VP1/2), which is associated with the release of viral DNA from incoming capsids; UL41 (also known as vhs), which is responsible for decreased mRNA stability and inhibition of host cell translation; UL48 (also known as Vmw65, VP16 or α-TIF), which is responsible for the *trans*-activation of immediate-early (IE) genes; UL13, which has protein kinase (PK) activity; RR1, the large subunit of the viral ribonucleotide reductase (RR) that also has PK activity and is required for optimal expression of HSV-2 IE genes (Smith and Aurelian, 1997); IE proteins ICP4 and ICP0, which regulate the expression of the other viral genes (Yao and Courtney, 1991); and UL17 gene products, which are required for cleavage and packaging of viral DNA (Salmon et al., 1998). Eleven virion proteins are glycosylated (gB, gC, gD, gE, gG, gH, gI, gJ, gK, gL, and gM). HSV glycoproteins contain N-linked high-mannose and O-linked and complex heterogeneous glycans. Virions also contain polyamines (reviewed by Roizman and Sears [1996]).

HSV Capsids

There are three types of capsids. Type A capsids, found primarily in the nucleus, lack DNA and are never enveloped. Type B capsids are intranuclear and contain DNA. Type C capsids contain DNA and are enveloped. Type A capsids consist of four proteins, VP5 (encoded by the UL19 gene), VP19C (UL38), VP23 (UL18), and the scaffolding protein. VP5, the major capsid protein, is a component of both pentameric and hexameric capsomeres. It interacts with VP19C and VP23. During capsid assembly, one or two VP5 molecules bind to one to six molecules of scaffolding pro-

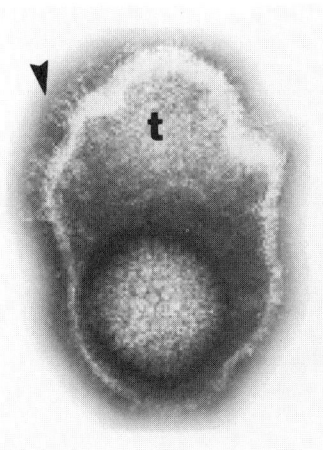

FIGURE 1 Electron micrograph of a negatively stained HSV virion reveals a capsid composed of capsomeres surrounded by the tegument (t), which in turn is surrounded by the electron-dense envelope decorated with spikes (arrowhead).

tein to form an internal core. They are cross-linked by triplexes consisting of V19C and VP23. UL26 encodes a protease that cleaves itself and gives rise to capsid proteins VP21 and VP24. It also cleaves the UL26.5-encoded protein pre-VP22a and gives rise to capsid protein VP22a. VP22a (also known as ICP35) is the primary scaffolding protein in HSV-1-infected cells. VP21 can also serve as an effective scaffolding protein. VP22a exits the capsid at the time that DNA enters. It is not found in the mature virion (Thomsen et al., 1995; Newcomb et al., 1999). At least seven genes (UL6, UL15, UL17, UL25, UL28, UL32, and UL33) encode the proteins that are required for cleavage and packaging of HSV-1 DNA (Salmon and Baines, 1998). The UL16 gene product copurifies with HSV-2 type C capsids and is involved in HSV-2 DNA cleavage and packaging (Oshima et al., 1998).

HSV DNA and Unique Features of the Genetic Information

DNA extracted from virions contains ribonucleotides, nicks, and gaps and has a G+C content of 68% for HSV-1 and 69% for HSV-2. The DNA consists of two covalently linked components containing unique sequences (U_L and U_S) that are bracketed by inverted repeats; 15 kb of DNA sequence represent inverted repeats of terminal regions inserted between the U_L and U_S domains. The U_L and U_S components can invert relative to each other, such that DNA extracted from virions or from infected cells consists of four equimolar populations differing in relative orientations. Genes are contained within both the unique and repeat sequences. They are given serial numbers following the designation U_L or U_S (reviewed in Roizman and Sears, 1996). The HSV-1 genome has been sequenced entirely, predicting the existence of 56 open reading frames (ORFs) in U_L and 12 in U_S (McGeoch et al., 1988). However, additional ORFs were identified, and HSV-1 is known to express at least 84 proteins. Thirty-eight HSV-1 ORFs cannot be deleted without ablating the capacity of the virus to replicate in cell culture (reviewed by Roizman and Sears [1996]). Infected cells also contain transcripts from genome

domains not known to specify proteins. These include the latency-associated transcripts (LATs), which have been implicated in latency regulation, and an RNA (ori_S RNA) derived by transcription of two of the three origins of viral DNA synthesis which map in the inverted repeats. ORFs which map in the inverted repeats (viz., $\gamma_1 34.5$, ORF P) are present in two copies per viral genome. Also, three sets of genes are located antisense to each other. ORFs O and P map antisense to the $\gamma_1 34.5$ gene, UL43.5 maps antisense to UL43, and UL27.5 maps antisense to the gene encoding gB. The expression of antisense genes is mutually exclusive (Carter et al., 1996; Chang et al., 1998), suggesting that they may be involved in various aspects of HSV pathogenesis. Because the C terminus of $\gamma_1 34.5$ may have originated from a cellular gene that inhibits stress-induced apoptosis (viz., MyD116), it has been suggested that $\gamma_1 34.5$ has a cell-type-dependent anti-apoptotic function (Chou and Roizman, 1994). Indeed, $\gamma_1 34.5$ binds protein phosphatase 1α and directs it to dephosphorylate the α subunit of the translation initiation factor eIF-2, thereby allowing unabated protein synthesis (He et al., 1997). Randall and Roizman (1997) suggested that ORF P is one of several genes that are involved in the establishment of HSV-1 latency. Yeast two-hybrid studies indicated that ORF P interacts with a component of the ASF/SF2 splicing factor and colocalizes with spliceosomes (Bruni and Roizman, 1996), but how this interaction relates to latency establishment is unclear. Less is known about HSV-2 proteins and their role in virus growth.

Intertypic Variation and Intratypic Polymorphism

The overall homology of HSV-1 and HSV-2 DNA is 50%. The genetic maps are largely colinear, but the two viruses differ in restriction endonuclease cleavage sites and the apparent size of viral proteins. Genome regions have very high (80 to 90%) or relatively low (35%) homology. Regions of limited homology include the first one-third of the RR1 gene, which codes for a serine/threonine PK activity (RR1 PK) (Chung et al., 1989; Cooper et al., 1995; Nelson et al., 1996; Peng et al., 1996) and glycoprotein gG, which is used in serotype-specific antibody assays. Antigenic and biologic markers that differentiate between HSV-1 and HSV-2 (Table 1) suggest that the two viruses have evolved different regulatory functions and are not as similar as originally assumed.

Intratypic polymorphism includes base substitutions that add or eliminate restriction endonuclease cleavage sites and/or change an amino acid. Restriction endonuclease site polymorphism is present in isolates from epidemiologically distinct sites and subjects, a finding used in studies of HSV transmission. Based on the presence or absence of 225 restriction endonuclease sites scattered throughout the HSV-1 genome, the nucleotide diversity (average differences per nucleotide) in isolates from the United States was estimated at 0.0046 (Sakaoka et al., 1994). Nucleotide substitutions in genes that are targets of antiviral agents (viz., thymidine kinase [TK]) are particularly relevant in immunocompromised subjects with persistent or recurrent HSV infection, for whom drug resistance is a major problem.

REPLICATIVE CYCLE

To understand the multipotentiality of HSV and its ability to cause a wide spectrum of disease in infected humans, a basic knowledge of its replicative cycle is required. In pro-

TABLE 1 Properties of HSV serotypes

Property	HSV-1	HSV-2
Site of infection	Primarily nongenital	Primarily genital[a]
Site of recurrence	Primarily trigeminal	Primarily sacral
Transmission	Primarily nonsexual	Primarily sexual
Biochemical properties		
% G + C content of viral DNA	67	69
Homology of viral DNA		Approximately 50%
Electrophoretic mobility of viral proteins		Some differences
Antigenic properties		Mostly cross-reactive with intratypic variation; some type-specific determinants
Biologic properties		
Pock size in CAM[b]	Small	Large
Plaque size in chick embryo cells	None or small	Yes or large
CPE in other cells	Tight adhesion of rounded cells	Loose aggregation propensity for syncytium formation
Growth cycle		
Titer in PRK cells	6×10^7	8×10^5
Particle-to-PFU ratio in PRK cells	36	2,000
% Enveloped virus in PRK cells[c]	38	6.8
Microtubules[d]	No	Yes
Sensitivity to:		
Temp (37°C) (log loss/h)	0.07	0.27
IUdR or BUdR	Sensitive	Relatively insensitive
IFN	Relatively resistant	Sensitive
Neurotropism	More	Less
Virulence ($\log_{10}$ PFU/LD$_{50}$)[e]	4.0	0.5

[a]A significant increase in the frequency of genital HSV-1 infection in the last decade has been seen.
[b]CAM, chorioallantoic membrane of fertilized eggs.
[c]Electron microscopy analysis reveals a high proportion of nonenveloped virus particles in HSV-1-infected cells.
[d]The formation of unique microtubule structures in HSV-2-infected cells has been described in electron microscopy studies.
[e]Done in mice. Large variations among HSV-2 isolates have been seen.

ductively infected (permissive) cells, the HSV-1 growth cycle is 8 to 16 h long, depending on the cell type. The HSV-2 replicative cycle is somewhat longer.

Attachment and Penetration

Attachment to and penetration of the cell membrane are the initial stages of the replicative cycle. The first interaction is with glycosaminoglycans (GAGs) that are widely distributed on the cell surface, preferentially heparan sulfate. This interaction facilitates the binding of the virus to the cellular receptor (Shieh et al., 1992). However, HSV-1 and HSV-2 glycoproteins interact with different affinities for different structural features of heparin, potentially influencing cell tropism (Herold et al., 1996). Expression cloning assays devised for isolating plasmids that encode cell surface proteins that mediate HSV-1 entry into penetration-resistant cells have identified a protein designated "herpesvirus entry protein" (HveA). It binds both HSV-1

and HSV-2 and is a new member of the human tumor necrosis factor receptor family. Additional receptors for HSV-1 and HSV-2 include poliovirus receptor-related proteins 1 and 2 (PRR1, PRR2), which have been redesignated HVeC and HVeB, respectively, and the herpesvirus immunoglobulin-like receptor (HIgR), which appears to be an alternative splice variant of PRR1 (Cocchi et al., 1998; Geraghty et al., 1998). Five envelope glycoproteins are necessary and sufficient to induce membrane fusion. Glycoprotein gC (and in some cases also gB) binds to GAGs. Glycoprotein gD binds the HVe receptors (Krummenacher et al., 1998). Glycoproteins gH and gL are also involved in fusion of the virion envelope to the plasma membrane, which is required for intracellular penetration (Roop et al., 1993). The redundancy resulting from attachment to different cell surface receptors, mediated by various viral glycoproteins, is probably responsible for the broad-host-range spectrum of HSV. Upon entry into the cell, the capsids are transported to the nuclear pores, a transport that is likely mediated by the cellular cytoskeleton. Release of viral DNA into the nucleoplasm may require a virus function. Viral DNA is transcribed in the nucleus, and all of the viral proteins are synthesized in the cytoplasm.

Regulation of HSV Gene Expression

HSV proteins form several kinetic groups that are coordinately regulated and sequentially ordered in a cascade fashion. However, the rate of synthesis and protein abundance may vary for each protein. The IE or α genes are the first to be expressed. The rate of IE protein synthesis reaches maximal levels at 2 to 4 h postinfection (p.i.), but the proteins continue to accumulate at nonuniform rates until late in infection. IE genes are defined by (i) the presence of the sequence TAATGARAT in one or several copies within several hundred base pairs upstream of the cap site, (ii) expression in the absence of other viral protein synthesis, and (iii) enhancement of expression by VP16, a protein located in the virion tegument. VP16 forms a multiprotein DNA complex with cellular proteins oct-1 and HCF. It first associates with HCF, a ubiquitous nuclear protein required for progression through the G_1 phase of the cell cycle (Wilson et al., 1997). This is followed by association with oct-1 on the TAATGARAT motif (Lai and Herr, 1997). Only five proteins (ICP0, ICP4, ICP22, ICP27, and ICP47) were originally classified as IE proteins. With the exception of ICP47, they have regulatory functions required for the synthesis of subsequent protein groups. Functional ICP4 is required for the synthesis of all other viral proteins (DeLuca and Schaffer, 1985; O'Hare and Hayward, 1985). ICP27 sometimes functions with ICP4 to initiate expression of the next class of proteins (Samaniego et al., 1995). ICP22 is required for late gene expression (Rice et al., 1995) and HSV-1 growth (Smith et al., 1986), and ICP0 is a promiscuous *trans*-activator (Everett et al., 1999). Viruses lacking ICP4, ICP27, and ICP22 are growth defective (Wu et al., 1996). ICP47 appears to shield HSV from cytotoxic T-cell (CTL) activity. It is a competitive inhibitor of peptide binding to the transporter associated with antigen presentation (Hill et al., 1995; Jugovic et al., 1998).

The early (E), or β, proteins include the β1 (viz., ICP8, the major DNA binding protein) and β2 (viz., TK and DNA polymerase) groups. Their synthesis reaches peak rates at 5 to 7 h p.i., correlates with the onset of DNA replication, and requires the presence of competent IE genes, specifically ICP4. RR1, originally classified as a β gene, was recently shown to fulfill all the criteria for IE gene classification. It has the TAATGARAT promoter sequence, its expression is independent of IE genes ICP4, ICP27, and ICP22, and it is activated by the oct-1/VP16 complex (Wymer et al., 1989; Desai et al., 1993; Zhu and Aurelian, 1997). RR1 PK has a major regulatory function in HSV-2 growth, as evidenced by the finding that it is required for optimal expression of the IE genes ICP4 and ICP27 (Smith et al., 1998).

The late (L), or γ, genes form a continuum differing in their kinetics and dependence on viral DNA synthesis. For example, glycoproteins gB and gD are synthesized relatively early in infection and are minimally affected by inhibitors of viral DNA synthesis, while glycoprotein gC is made late in infection and its expression is inhibited in the presence of inhibitors of viral DNA synthesis (reviewed by Roizman and Sears [1996]).

Most viral proteins are posttranslationally modified. Processing includes cleavage, phosphorylation, sulfation, glyco-sylation, myristylation, ADP-ribosylation, and nucleotidylation. For example, ICP4, ICP0, ICP22, and ICP27 are nucleotidylated, and this requires a viral β or γ function (Blaho et al., 1993). However, with the exception of some glycoproteins (viz., gD), the extent to which processing is required for protein function and virus growth is still unclear. HSV specifies at least three PKs, i.e., US3, UL13, and RR1 PK (Chung et al., 1989; Purves et al., 1992; Cooper et al., 1995; Daikoku et al., 1997). Studies of appropriate HSV-1 mutants indicated that UL13 is involved in the phosphorylation of the IE protein ICP22 (Purves et al., 1993). It also complexes with glycoprotein gE and mediates phosphorylation of the viral Fc receptor (Ng et al., 1998). US3 phosphorylates a membrane protein encoded by UL34 (Purves et al., 1992). HSV-2 RR1 PK phosphorylates the small RR subunit and cellular proteins rasGAP and hSOS, which are involved in Ras signaling activated by RR1 PK (Smith et al., 1994; Hunter et al., 1995b). RR1 PK is required for HSV-2 growth (Smith et al., 1992, 1998).

HSV DNA Synthesis

HSV DNA synthesis occurs in the nucleus. It is first detected at 3 h p.i. but continues for at least 12 h, with the bulk of viral DNA being made relatively late in infection (Igarashi et al., 1993). At that time, at least, DNA replicates by a rolling circle mechanism (Jacob et al., 1979). Proteins that are involved in the production of precursors for DNA synthesis include (i) TK, which phosphorylates purine pentosides and a variety of nucleoside analogs that are not efficiently phosphorylated by cellular kinases and which is the target of antiviral agents such as acyclovir and ganciclovir; (ii) RR, which consists of the RR1 and RR2 subunits (Frame et al., 1985) and functions to reduce ribonucleotides to deoxyribonucleotides, creating the pool of substrates for DNA synthesis (RR activity is required for virus growth in cells that are not actively dividing, such as neurons [Goldstein and Weller, 1988]); (iii) uracil DNA glycosylase, which functions in DNA repair and proofreading; and (iv) dUTPase, which hydrolyzes dUTP to dUMP (Caradonna and Cheng, 1981). Alkaline nuclease, coded by UL12, repairs nicks and gaps in replicating DNA and processes replication intermediates into a form suitable for encapsidation (Goldstein and Weller, 1998).

The HSV genome contains three origins of viral DNA replication, defined operationally as sequences needed for amplification of a DNA fragment in cells coinfected with a helper virus. Two of these map in the U_S component (*ori$_s$*), the third in the middle of the U_L component (*ori$_L$*). Transfection experiments using a plasmid that contains an origin of DNA synthesis indicate that origin-initiated DNA synthesis requires seven genes: UL30, which specifies the viral DNA polymerase; UL29, which specifies the single-strand-specific DNA-binding protein ICP8; UL9, which specifies a protein that binds three sites at or near the origin of viral DNA synthesis; UL42, which specifies a protein that confers processivity on viral DNA polymerase; and three additional proteins, encoded by UL5, UL8, and UL52, that form a complex that functions as a primase, and in the presence of ICP8, as a helicase (Parris et al., 1988; Crute et al., 1989; Hernandez and Lehman, 1990; Gao and Knipe, 1991; Reddig et al., 1994; Falkenberg et al., 1997). However, none of the origins is uniquely required for viral DNA replication (Igarashi et al., 1993). Since DNA replicates as a rolling circle, there is no need for origin-based de novo initiation of synthesis each time a viral DNA copy is made.

Virion Assembly

Newly synthesized viral DNA is processed and packaged into preformed capsids in the nucleus (Fig. 2). Processing involves amplification of some sequences and cleavage of viral DNA from concatemers that lack free ends. Seven viral proteins (UL6, UL15, UL17, UL25, UL28, UL32, and UL33) are responsible for cleavage-packaging events (Salmon and Baines, 1998). Isomerization is associated with DNA replication, cleavage, and packaging and results from the inversion of the U_L and U_S components relative to each other. DNA extracted from a plaque generated by a single virus particle, presumed to be in one arrangement, contains all four isomers of HSV DNA in equimolar concentrations. The physiologic function of inversion is unclear, and genomes frozen in one orientation by deletion of internal inverted repeats are viable (Jenkins and Roizman, 1986).

Virion Envelopment and Egress

Empty capsids are only rarely enveloped, and those containing DNA fragments less than the genome length are generally retained in the nucleus (Church et al., 1998). Early electron microscopy studies showing thickening of nuclear membrane patches next to DNA containing assembled capsids (Fig. 2) were interpreted to mean that (i) the virion envelope is acquired at the inner nuclear membrane and (ii) maturation of envelope glycoproteins occurs as vesicles containing these virions move through the secretory pathway. Consistent with this interpretation, two tegument proteins were found in the nucleus, close to putative sites of capsid assembly termed "assemblons" (Ward et al., 1996). However, other tegument proteins colocalized in the cytoplasm (Elliott et al., 1995) and the tegument was shown to assemble independently of the capsids (Rixon et al., 1992).

Studies of glycoprotein localization and processing and the presence of numerous unenveloped capsids in the infected cell cytoplasm have led to the hypothesis that capsids are enveloped at the inner lamella of the nuclear membrane and become deenveloped at its outer lamella and reenveloped in the endoplasmic reticulum (Fig. 3). They are released in the extracellular environment by envelopment at the plasma membrane or by fusion of vesicles carrying enveloped virus at the plasma membrane (Browne et al., 1996). Studies of an HSV-1 mutant in which the VP22 tegument protein was replaced with green fluorescent protein, the presence of which was monitored by time lapse confocal microscopy, suggested that tegument proteins are acquired downstream of capsid translocation through the nuclear membrane, with final envelopment occurring at a cytoplasmic location further along the exocytic pathway to the cell periphery (Elliott and O'Hare, 1999). Glycoprotein gK is involved in the translocation of infectious virions from the cytoplasm to the extracellular spaces (Jayachandra et al., 1997), and structural protein VP22 is responsible for intercellular HSV-1 transport via a nonclassical Golgi-independent mechanism (Elliott and O'Hare, 1997).

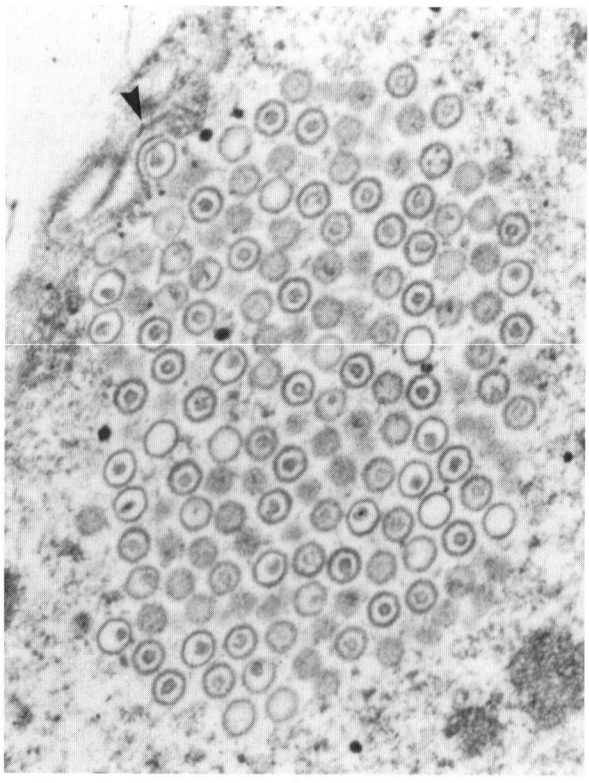

FIGURE 2 Electron micrograph of an HSV-2-infected cell nucleus reveals capsids in different stages of assembly and DNA incorporation. A thickened nuclear membrane is next to a DNA-containing capsid (arrowhead).

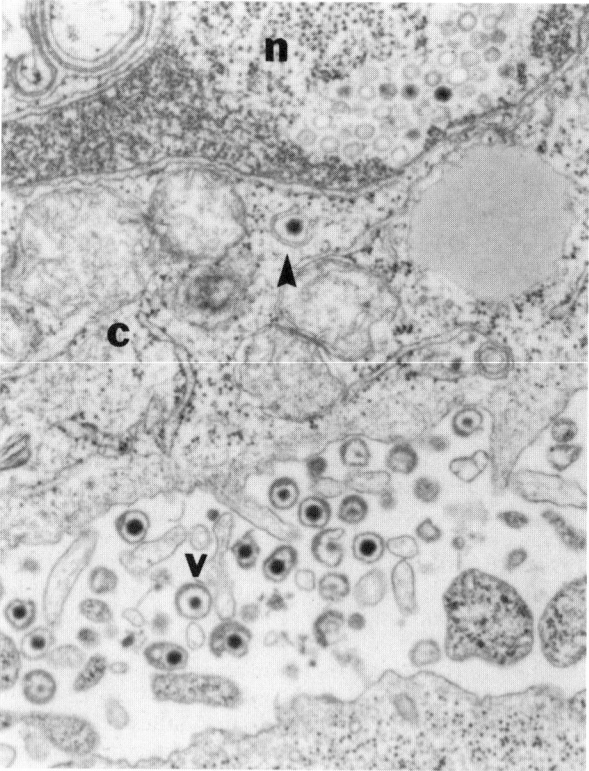

FIGURE 3 Electron micrograph of an HSV-2-infected cell reveals intranuclear (n) capsids, chromatin margination, and apparent reduplication of the nuclear membrane. Virus envelopment in the endoplasmic reticulum (arrowhead) is evident in the cytoplasm (c). Extracellular virus (v) is also shown.

Virus-Induced Alterations of the Infected Cell

The outcome of productive infection is cell death. The host cell macromolecular metabolism is shut off early in infection. The first stage is mediated by the virion tegument protein vhs (encoded by UL41), which causes degradation of host mRNA (Read et al., 1993). HSV-2 strains generally shut off the host more rapidly than do HSV-1 strains, a property associated with three regions of the HSV-2 vhs protein (Everly and Read, 1997). The second stage requires de novo synthesis of viral proteins (Nishioka and Silverstein, 1977). The earliest manifestation of host cytopathic effects is an enlarged, peripherally displaced nucleolus. Concurrently, there is chromatin margination (Fig. 3) and the nucleus becomes multilobed. A basophilic Feulgen-positive inclusion body that contains viral DNA is seen early in infection; an eosinophilic intranuclear inclusion body (type A) devoid of virus material develops late in infection. Cellular membranes are altered, giving the impression of reduplication late in infection (Fig. 3). Cell surface changes include the acquisition of a receptor for the Fc domain of immunoglobulin (a function assigned to a complex of glycoproteins gE and gI [Johnson et al., 1988]); a receptor for C3b, a fragment of the third component of complement (a function assigned to glycoprotein gC [Friedman et al., 1986]); and antigenic determinants that are major targets of the immune response to the virus. The gE-gI glycoprotein complex interacts with components of cell junctions and may be involved in intercellular virus spread (Dingwell and Johnson, 1998). Most HSV-1 and HSV-2 strains cause cells to round up and attach to each other. Some strains cause cell fusion (polykaryocytosis) that is cell type specific or independent. Mutations that confer this phenotype map in at least five loci in the HSV-1 genome, within gB, gK, gL, UL24, and UL20 (Bond and Person, 1984; Pogue-Geile and Spear, 1987; Hutchinson et al., 1993; Roop et al., 1993). Polykaryocytosis is the cell alteration used in the cytologic diagnosis of HSV infections.

Apoptosis is a sequentially ordered response to intracellular and extracellular factors that culminates in cell death. It has been suggested that HSV-1 triggers apoptosis at multiple metabolic checkpoints and in turn has evolved mechanisms to block apoptosis at each point, in a cell-type-dependent manner (Galvan and Roizman, 1998). HSV proteins that interact with cellular proteins could trigger or block signaling pathways that lead to apoptosis. For example, HSV-1 ICP0 binds to and stabilizes a ubiquitin-specific protease and the elongation factor 1δ (Kawaguchi et al., 1997; Everett et al., 1999), and ICP27 binds to spliceosomes and blocks the maturation of spliced mRNAs (Roizman and Sears, 1996). ICP22 affects the phosphorylation of RNA polymerase II (Rice et al., 1995), and HSV-2 RR1 PK binds rasGAP and hSOS and phosphorylates rasGAP (Smith et al., 1994).

LATENCY

Latency is defined as virus persistence in the infected host in a repressed state that is compatible with cell survival. It is often followed by subsequent episodes of virus reactivation and clinical symptoms. Latency prompts the following questions: (i) what tissues harbor latent infection, (ii) how is latency established and maintained, (iii) what mechanisms underlie virus reactivation, and (iv) what mechanisms regulate the development of clinical recurrences? Most of the available information regarding these questions has been obtained from studies using animal models. These include HSV-1 and HSV-2 models (mouse cutaneous, ocular, and vaginal) that do not sustain recurrent disease but in which virus can be reactivated by various stimuli and models (guinea pig cutaneous and vaginal HSV-2 and rabbit ocular HSV-1) that sustain spontaneous recurrence.

What Tissues Harbor Latent Virus?

Infectious virus can only rarely be recovered from sensory or autonomic nervous system ganglia from animal models and humans (obtained postmortem). However, maintenance and growth of neural cells in tissue culture results in the production of infectious virus and subsequent permissive infection of susceptible cells, a process called cocultivation. Using this process, trigeminal and sacral dorsal root (S2 to S4) ganglia were identified as the most common sites for latent HSV-1 and HSV-2 infections, respectively. Studies of the temporal appearance of infectious virus in tissues from infected mice suggest that latency establishment is preceded by virus replication at the inoculation site in epidermal and dermal cells (reviewed by Stevens [1989]). In mice, the higher the virus input and the better it replicates at the periphery, the higher the number of latently infected neurons. In guinea pigs, the frequency of recurrences is proportional to the input virus titers. In humans, the relationship is to the severity of the primary infection, which presumably reflects virus titer (Leib et al., 1989a, 1989b; Sawtell, 1997; Lekstrom-Himes et al., 1998). However, mutants that cannot replicate in any cell in vivo can establish and maintain latency, suggesting that replication at the periphery is not essential for latency establishment (Katz et al., 1990). Studies from our (unpublished) and other (Buchman et al., 1979) laboratories indicate that virus isolates obtained at various times from a single patient can have different restriction endonuclease patterns, indicating that ganglia can be colonized by multiple HSV strains.

Unenveloped capsids are transported from the periphery to the nerve-cell bodies in the ganglia innervating the site of infection by retrograde axonal transport, involving microtubules (Kristensson et al., 1986). In humans, the time between peripheral infection and virus spread to the ganglia is unknown. In mice and guinea pigs, virus is recovered from ganglia within 1 to 2 days after peripheral inoculation. In animal models, there is a short period of virus replication in the ganglia that, for HSV-1, involves UL24 (Jacobson et al., 1998). Because UL24 overlaps the TK gene, older findings that TK is required for ganglionic replication and latency may actually reflect the contribution of UL24 rather than TK. A conserved region in ICP4 that contains 13 nearly contiguous serine residues may also be critical for HSV-1 replication in trigeminal ganglia (Bates et al., 1998). It is unclear whether ganglionic replication occurs in humans. By 2 to 4 weeks, at most, after resolution of the primary disease, latency is established and virus can no longer be recovered from ganglionic homogenates.

HSV DNA persistence at peripheral sites has been demonstrated by PCR, but its role in disease pathogenesis is unclear. HSV-1 DNA has been seen in the cornea, both in humans and in experimental animals. In mice, it has been found in the epithelial cells of the cornea, conjunctiva, and eyelid for up to 4 months following infection. A spatial and temporal relationship has been demonstrated between DNA presence and corneal lesions, suggesting that it may reflect active persistent infection rather than bona fide latency (Maggs et al., 1998). In humans, HSV DNA was found in skin from herpes-associated erythema multiforme (HAEM) patients for 3 to 5 months after lesion resolution. However, virus was not reactivated, even by cocultivation (Imafuku et al., 1997).

How Is Latency Established and Maintained?

In latently infected neurons, viral DNA is a circular episome (Mellerick and Fraser, 1987). Quantitative PCR studies have shown that the HSV-1 genome copy numbers in mouse trigeminal ganglia range from <10 to >1,000 (Katz et al., 1990). The higher the titer of input virus, the higher the number of neurons that have high numbers of viral genome copies (Sawtell, 1997). It has been estimated that in human trigeminal ganglia there are 0.01 to 0.4 HSV genome equivalents per cell. Because neurons constitute only 10% of the cells in neural tissues, the number was calculated to be 20 copies per latently infected neuron (reviewed by Stevens [1989]).

The mechanisms by which latency is established and maintained are still unclear. Mutations that decrease the efficiency of virus replication (viz., TK-negative HSV-1 mutants) have a negative impact on the ability of the virus to establish latency and reactivate (Hay et al., 1995). ICP0 was implicated in the establishment of HSV-1 latency and its reactivation (Leib et al., 1989b; Wilcox et al., 1997), and VP16 was implicated in HSV-1 replication in trigeminal ganglia and latency establishment (Tal-Singer et al., 1999). LAT is the only transcript in latently infected sensory neurons. It maps to the long terminal repeat region of the HSV genome and is largely confined to the nucleus. Since it is transcribed from the DNA strand opposite that encoding the major *trans*-activator ICP0, LAT was originally hypothesized to play a role in latency establishment and maintenance by inhibiting ICP0 expression. However, its function in latency establishment could not be definitively established. Thus, the absence of LAT does not allow ICP0 expression, LAT mutants establish latency (Javier et al., 1988), and two ORFs (ORF O and ORF P), which are read antisense to LAT on the HSV genome, may have contributed to the phenotype of the mutants used to examine LAT's function.

How Does the Virus Reactivate?

In mice, virus reactivation is induced by physical trauma, iontophoresis or epinephrine, transient hyperthermia, corneal scarification, or cadmium injection. In the guinea pigs and rabbits, reactivation is spontaneous. It is unknown whether some HSV strains reactivate more frequently than others. Spontaneous reactivation in humans is induced by exposure to local or systemic stimuli. Local stimuli include injury to tissues innervated by latently infected neurons (viz., by sunlight [UV]). Systemic stimuli include fever due to bacterial or viral infections, menstruation, physical or emotional stress, and hormonal imbalance. They can cause simultaneous reactivation of latent virus in the trigeminal (HSV-1) and sacral (HSV-2) ganglia. The genome copy number in latently infected ganglia (which is determined by the rate of virus replication at the site of infection) is an important parameter for HSV reactivation (Maggioncalda et al., 1996; Sawtell, 1997; Lekstrom-Himes et al., 1998).

Studies of appropriate mutants have suggested that TK is required for reactivation rather than for latency establishment (Coen et al., 1989). Supporting this interpretation, the frequency of reactivation was decreased by TK

inhibitors (Watkins et al., 1998). Some studies of HSV-1 LAT mutants have suggested that reactivation requires LAT (Perng et al., 1996; Drolet et al., 1998), while others have suggested that LAT is not required (Ho and Mocarski, 1988; Leib et al., 1989a; Block et al., 1990; Deshmane et al., 1993). The major LAT transcript (50 to 90% of the total, depending on virus strain or experimental model) is a 2-kb stable intron derived from a primary transcript of approximately 8.3 kb. Another transcript (10 to 50% of the total) is a spliced derivative of the former transcript and is about 1.5 kb. The 1.5-kb LAT transcript appears to be required for HSV-1 reactivation in the rabbit model. Since this transcript does not overlap other known HSV-1 genes, it does not function via an antisense mechanism and may involve a direct DNA or RNA mechanism (Fraser et al., 1992; Farrell et al., 1993; Drolet et al., 1998). More recently, however, it was shown that deregulated expression of an ORF contained within the 2-kb LAT greatly enhances virus growth presumably by substituting for deficiencies in IE gene expression by a protein-dependent effect (Thomas et al., 1999). An intertypic recombinant in which a 2.8-kb region from HSV-1 LAT was substituted for the respective HSV-2 sequence was reactivated with an HSV-1 phenotype, suggesting that LAT is involved in the site-specific reactivation phenotypes of HSV-1 and HSV-2 (Yoshikawa et al., 1996). However, when used at a 10-fold-higher input, genital HSV-1, recurred as frequently as HSV-2, suggesting that regional specificity is also determined by the quantity of latent DNA (Lekstrom-Himes et al., 1998). LAT is not involved, or at best is an inefficient and weak determinant of spontaneous HSV-2 reactivation, in guinea pigs (Bourne et al., 1994; Wang et al., 1997).

An HSV-2 mutant deleted in RR1 PK has a decreased ability to replicate in the mouse and guinea pig models, as well as to establish latency and reactivate. This presumably reflects the role of RR1 PK in HSV-2 replication (Smith et al., 1992, 1998). However, the RR1 PK mutant was not reactivated from the rare ganglia in which latency was established (DNA positive), suggesting that RR1 PK is directly involved in HSV-2 reactivation (Aurelian et al., 1999; Wachsman et al., unpublished data). We proposed that RR1 is an early response to latency reactivating stimuli (Aurelian, 1998; Smith et al., 1998), because the stimuli upregulate AP-1 transcription factors (Valyi-Nagy et al., 1991) and RR1 is the only viral promoter that contains AP-1 *cis*-response elements and responds to AP-1 with basal expression (independent of VP16, ICP0, or ICP4) (Wymer et al., 1989; Zhu and Aurelian, 1997). RR1 is required for virus replication in neuronal cells (Jacobson et al., 1989; Idowu et al., 1992). It also provides the PK activity that is required for optimal IE gene expression and limited viral DNA synthesis, both of which are critical to virus reactivation (Kosz-Vnenchak et al., 1993). An amplification loop (increased AP-1) is generated by the ability of RR1 PK to activate the Ras signaling pathway (Smith et al., 1994; Hunter et al., 1995b). Reactivating stimuli also cause nuclear localization of a cellular protein (host cell factor) that is required for VP16-mediated IE gene activation (Kristie et al., 1999). The outcome is initiation of the lytic cascade and production of infectious virus. Consistent with this interpretation, RR1 and TK transcripts are detected before IE transcripts during reactivation (Tal-Singer et al., 1997). An alternative, but not mutually exclusive, interpretation is that reactivating stimuli down-regulate or functionally impair cellular proteins (viz., octamer protein oct-

2) that inhibit viral gene expression in latently infected ganglia (Latchman, 1994).

What Determines Disease Causation?

Reactivated virus is transported to the body surface, presumably by reverse axonal transport. However, only some reactivation episodes result in clinical disease. Factors responsible for the development of clinically apparent recurrent lesions are still poorly understood. They are likely to include the efficiency of virus clearing by host factors and the resulting titers of reactivated virus. In guinea pigs, recurrent HSV-2 lesions have been linked to decreased T-cell responses (Iwasaka et al., 1983). Similarly, longitudinal studies of human subjects with HSV-1 or HSV-2 infections have indicated that clinical lesions are associated with unimpeded virus replication resulting from transient down-regulation of HSV-specific T-cell responses (reviewed by Aurelian [1989]). Decreased HSV-specific T-cell responses were first seen during prodrome, at 1 to 2 days before lesion onset (Sheridan et al., 1982, 1985; Sheridan and Aurelian, 1983). The decrease may reflect a shift in the balance of HSV-specific T-helper cells in favor of the type 2 (Th2) population, which decreases production of Th1-specified cytokines (viz., gamma interferon [IFN-γ]) that contribute to HSV containment. Indeed, peripheral blood lymphocytes (PBL) from patients with UV-induced HSV recurrence evidenced decreased levels of HSV-specific T-cell proliferation and increased levels of the Th2 cytokine interleukin 6 (IL-6). Proliferation was restored by the addition of IL-6 antibody, indicating that IL-6 is involved in the immune down-regulation associated with UV-induced recurrent disease (Miura et al., 1994). Th2-type cytokines may also function in ganglionic reactivation, since IL-6 production by satellite cells is increased in latently infected ganglia from UV-treated mice, while the Th1 lymphokines IFN-γ and IL-2 are decreased (Shimeld et al., 1999). Prostaglandin may be involved in T-cell down-regulation, as suggested by the finding of increased PGE$_2$ levels in skin from mice treated with recurrence-inducing stimuli but not in skin from mice treated with a stimulus that reactivates ganglionic virus but fails to produce recurrent disease (Hayashi and Aurelian, 1986; Yasumoto et al., 1987). In humans, the frequency of recurrence has been significantly reduced by prolonged treatment with the prostaglandin inhibitor indomethacin (Wachsman et al., 1990).

PATHOGENESIS

HSV infects neonates, children, and adults and can produce a wide spectrum of diseases (Table 2). The primary (or initial) infection is generally in the form of a localized lesion. The incubation period is 1 to 26 days (median, 6 to 8 days). The severity varies among individuals, ranging from asymptomatic to severe systemic illness that can be fatal. Factors that can influence severity are age, sex, genetic or hormonal factors, immune competence, associated illnesses, and virulence of the infecting virus strain. Recurrent skin disease, the hallmark of HSV pathogenesis, is usually localized to the site of the primary lesion, or close to it. Nearly all people with a clinically recognized primary HSV-2 infection and 55% of those with clinical HSV-1 infection develop at least one recurrent episode within 1 year p.i. With the exception of severely immunocompromised patients, recurrent herpes is relatively benign, producing fewer and smaller lesions that heal faster than those seen in the primary

TABLE 2 Diseases associated with or caused by HSV

Stomatitis
Herpes labialis
Genital lesions
Anal and/or rectal lesions
Atypical and hyperproliferative lesions
Primary herpetic dermatitis
Eczema herpeticum
Traumatic herpes
Herpetic withlow
Herpes gladiatorum
Persistent mucocutaneous lesions
Acute herpetic rhinitis
Keratoconjunctivitis
Keratitis
Chorioretinitis
Neonatal herpes
Meningitis
Mollaret's meningitis
Encephalitis
Bell's palsy
Progressive dementia syndrome
Monofocal epilepsy
Necrotizing myelitis
Pharyngitis
Hepatitis
Pneumonitis
Urethritis
Cystitis
Monoarticular arthritis
Adrenal necrosis
Atherosclerosis
Autonomic system dysfunction
Peptic ulcer
HAEM

infection. Systemic symptoms are rare, and the duration of virus shedding is shorter. A prodrome often signals recurrence; it is characterized by a tingling sensation at the site at which the lesions will appear, preceding them by a few hours to 2 days. This sensation may be accompanied by radiating radicular pain. Virus reactivation evidences regional specificity, with HSV-1 reactivating most frequently from trigeminal ganglia and HSV-2 from sacral ganglia (Lafferty et al., 1985).

Initial HSV-1 disease is generally gingivostomatitis, a serious infection of the gums, tongue, mouth, lip, facial area, and pharynx that is seen primarily in children from 1 to 3 years old and is often accompanied by high fever, malaise, myalgias, swollen gums, irritability, inability to eat, and cervical lymphadenopathy. Reactivated HSV-1 is associated with the development of mucosal ulcerations or with lesions at the mucocutaneous junction of the lip, presenting as small vesicles that last 4 to 7 days and are often referred to as herpes labialis, facialis, febrilis, cold sores, or fever blisters. Both HSV-1 and HSV-2 were isolated from the posterior pharynx in 11% of cases of pharyngitis. Concomitant lesions of the tongue, buccal mucosa, and gingiva were seen in one-third of cases. Acute HSV-1 rhinitis is a primary infection of the nose recognized by the appearance of tiny vesicles in the nostrils usually associated with fever and enlarged cervical lymph nodes. Other HSV-1 skin diseases include eczema herpeticum (usually a manifestation of an infection in which the skin is the portal of entry) and traumatic herpes (resulting from traumatic breaks due to burns or abrasions in the normal skin of a susceptible child). Herpetic whitlow and herpes gladiatorum are occupational hazards (of dentists and hospital personnel and of wrestlers, respectively) resulting from infections of broken skin (often on fingers) that has been in contact with virus on another individual. HSV-1 infections of the eye can lead to keratoconjunctivitis, which, in its most serious form, can cause blindness. HSV keratitis can also be acquired after penetrating keratoplasty (incidence is 1.2 cases per 1,000 person years), occurring in the first 2 years after transplantation (Remeijer et al., 1997). The disease has an immunopathology component that involves HSV-specific CD4$^+$ T-helper type 1 (Th1) cells (Niemialtowski and Rouse, 1992). Chorioretinitis is a manifestation of disseminated HSV infection that may occur in neonates or in patients with AIDS. The eye can be the only site of HSV involvement in neonates. HSV-1 may be excreted in the saliva or tears from asymptomatic individuals. Serologic studies suggest that most childhood HSV-1 infections are asymptomatic but that they may be followed by subsequent clinically symptomatic recurrent disease.

The genitourinary tract is the primary site of HSV-2 infection. In females, the infection is manifested by vesicles on the mucous membranes of the labia and the vagina. Cervical involvement is common and it may contribute to the development of abnormal Papanicolaou smears. Severe forms result in ulcers that cover the entire area surrounding the vulva. Symptoms of primary infection include itching, pain, and lymphadenopathy. Systemic symptoms are more common in women than in men. They often accompany the appearance of primary lesions and include fever, headache, photophobia, malaise, and generalized myalgias. As a rule they are not seen in recurrent disease. HSV infections of the genitourinary tract in women are easily mistaken for more common urinary tract infections, as the major symptoms—dysuria, urinary retention, urgency and frequency, pain, and discharge—are similar. Therefore, when a patient presents with these symptoms, particularly in the absence of typical HSV lesions in the vulvar, vaginal, or cervical regions, cultures should be taken for both HSV and common gram-negative organisms. In males, common sites of HSV-2 infection are the shaft of the penis, the prepuce, and the glans penis. Urethritis is probably the main local extension, accompanied by a watery discharge often resulting in dysuria. Symptomatic urethritis is rare in recurrent disease, but virus often can be cultured from the urethra. Anal and rectal infections occur in both sexes. They are particularly seen in women and in homosexual men as a result of anal intercourse. Atypical genital herpes is often described in immunocompromised patients and can present as large hyperkeratotic ulcers. Acyclovir-resistant strains are occasionally isolated from such ulcers (Tyring et al., 1998). In AIDS patients, HSV can produce persistent mucocutaneous disease. A newly described HSV-2 presentation takes the form of large hyperproliferative lesions that may be caused by acyclovir-resistant virus and are difficult to treat (Beasley et al., 1997). HAEM is a severe inflammatory disease that can follow recurrent as well as primary HSV-1 or HSV-2 episodes. Disease is associated with HSV DNA fragments, but lesion development is spatially and temporally associated with gene expression. It has an immunopathology component that involves HSV-specific CD4$^+$ V$_\beta$2 T cells and production of Th1 cytokines, most notably IFN-γ

(Imafuku et al., 1997; Aurelian et al., 1998; Kokuba et al., 1998a, 1999). It is still unclear how HSV DNA fragments are transported to the peripheral site of HAEM development.

HSV-2 infection is associated with acquisition of human immunodeficiency virus (HIV-1) and human T-cell lymphotropic virus infections (Holmberg et al., 1988; Aurelian, 1990; Keet et al., 1990; Hook et al., 1992). The findings that HSV (i) activates HIV-1 gene expression in macrophages that harbor latent HIV (Feng et al., 1993) and (ii) persistently infects PBL and CD4$^+$ T-cell lines which, upon superinfection with HIV, produce HIV pseudotyped by HSV-1 envelope (Calistri et al., 1999), may be relevant to AIDS pathogenesis.

The central nervous system (CNS) is also a target of HSV infection, and when this happens the disease is usually severe. HSV encephalitis is the most commonly reported viral CNS infection, accounting for 10 to 20% of all cases. In children and young adults, encephalitis usually results from a generalized primary infection and is generally due to HSV-1. HSV-2 is involved in approximately 4 to 6% of cases. Virus presumably enters the CNS through neurotropic spread by way of the olfactory bulb and can be isolated from other organs in addition to the CNS. In adults, encephalitis is often preceded by cutaneous lesions, and there is a level of uncertainty as to the contribution of reinfection with exogenous virus versus reactivation of endogenous virus. Because it is difficult to differentiate HSV encephalitis from other viral encephalitides and focal CNS diseases, PCR is rapidly becoming the method of choice for its diagnosis.

HSV-1 and HSV-2 are also associated with mild or atypical encephalitis and are the most commonly recognized causes of acute sporadic encephalitis in the United States, often associated with immunocompromise (Fodor et al., 1998). HSV-2 is a common cause of aseptic meningitis (Read and Kurtz, 1999). It also causes benign recurrent meningitis or Mollaret's meningitis (Gignoux et al., 1998). HSV-2-induced meningitis preceded by a common cystitis was reported in a patient with no history of primary or recurrent genital infection, suggesting that HSV-2 infections are underdiagnosed (Venot et al., 1998). HSV reactivation from geniculate ganglia is the most important cause of Bell's palsy, an idiopathic peripheral facial paralysis of sudden onset that accounts for >50% of all cases of facial paralysis (Schirm and Mulkens, 1997). HSV DNA was found in monofocal epilepsy seizure epicenters (Sanders et al., 1997), and HSV-1 and HSV-2 were associated with a form of acute necrotizing myelitis, generally with an ascending pattern (Nakajima et al., 1998). Necrotizing HSV encephalitis was implicated in progressive dementia syndrome (Zachhuber et al., 1995). HSV DNA was also found in gastrointestinal sensory neurons, suggesting that HSV is involved in recurrent gastrointestinal disorders and peptic ulcer (Lohr et al., 1990). A high proportion (45%) of biopsies from coronary arteries of patients undergoing coronary artery bypass grafting were positive for HSV-2, but not HSV-1, antigen and evidenced inflammatory changes, suggesting that HSV-2 infection is associated with atherosclerosis (Raza-Ahmad et al., 1995). Semen can also contain HSV, especially in primary infection but also in asymptomatic subjects. HSV antigen and DNA have been seen in spermatozoa (Color plate 3) (see color insert) and may be involved in infertility (Aurelian, unpublished data; Kotronias and Kapranos, 1998).

Approximately 75 to 80% of neonatal HSV-2 infections are acquired by passage through an infected birth canal, including patients with asymptomatic shedding (Brown et al., 1991). However, congenitally acquired infections have been described, mostly in association with primary HSV-2 infection of the mother during pregnancy. In utero infection results from transplacental or ascending infection (Gagnon, 1968). Postnatal infection as a consequence of nursing on an infected breast has also been documented (Dunkle et al., 1979). Neonatal HSV infections may remain localized to the site of infection (viz., skin, eye, mouth), extend to the CNS, or disseminate to multiple organs. Neonates have the highest frequency of visceral and CNS involvement of all HSV-infected patients. Skin lesions are the most commonly recognized features of the disease, but in at least 70% of untreated cases disseminated disease is also seen. Factors influencing the extent of HSV disease in the neonate include maternal and host immunity, virus load, site of inoculation, and virus strain. The most significant predictor of mortality and morbidity in neonates is disease severity. Babies with the most severe disease (defined by level of consciousness or development of pneumonia) have the worst outcome.

Among recipients of bone marrow transplants, HSV pneumonitis appears to account for 6 to 8% of cases of interstitial pneumonia. Mortality due to HSV pneumonia in immunosuppressed patients is above 80%. Early diagnosis is essential for initiation of antiviral chemotherapy. HSV has been isolated from 40% of patients with acute respiratory distress syndrome. A high incidence of pulmonary HSV infection associated with adult respiratory distress syndrome also has been described in burn patients (Byers et al., 1996). Other uncommonly reported complications of HSV include monoarticular arthritis, adrenal necrosis, idiopathic thrombocytopenia, and glomerulonephritis. In immunocompromised patients, generalized HSV with involvement of adrenal glands, pancreas, small and large intestine, and bone marrow is reported. Severe generalized disease that is occasionally responsible for herpetic hepatitis is also described.

EPIDEMIOLOGY

HSV infections are found worldwide. Transmission occurs during close personal contact with active ulcerative skin lesions or virus-positive secretions from asymptomatic subjects. Primary HSV-1 infection usually occurs before 5 years of age and is generally asymptomatic. Only 10 to 15% of primary infections produce clinical disease. Virus shedding in the saliva is a significant reservoir for transmission. HSV-1 can be isolated from the mouth for 7 to 10 days after primary infection. It is uncommon in children younger than 6 months old. It has been reported in 20% of children from 7 months to 2 years old and in 18% of those from 3 to 14 years old. Shedding decreases with advancing age, with salivary excretion reported in 2 to 9% of asymptomatic adults. There is no seasonal variation in the incidence of infection. Geographic location, socioeconomic status, and age influence the frequency of HSV infection (reviewed by Whitley [1996]). More recent serosurveys suggest that presumed middle-class individuals in industrialized societies become infected later in life, with prevalences of 40 and 60% in the second and third decades of life, respectively. However, only 5 of 18 countries surveyed in the National Health and Nutrition Examination Survey (NHANES) had an HSV-1 antibody prevalence less than 70% (Nahmias et al., 1990).

Clustered outbreaks of HSV stomatitis have been reported in orphanages (Hale et al., 1953) and hospitals (Pugh et al., 1955).

The risk of HSV-2 infection is correlated with markers of sexual promiscuity, including early age of first intercourse, large numbers of sexual partners, history of other sexually transmitted diseases (STDs), and increasing age. The highest prevalence of HSV-2 antibodies (75 to 98%) is seen in female prostitutes and homosexual men. Women are about 45% more likely than men to be infected with HSV-2. The risk of susceptible females for contracting HSV-2 from infected males is 80% after a single contact. The probability of infection with HSV-2 is less than 10% for U.S. women having one partner. It increases to 40%, 62%, and greater than 80% as the number of partners increases to 2 to 10, 11 to 50, and more than 50, respectively. For heterosexual men, these risks are 20, 35, and 70% for each of the three risk groups, respectively. For homosexual men, the risks are greater than 60 and 90% for those with 11 to 50 and more than 50 partners, respectively (Nahmias et al., 1990).

Serosurveys using type-specific antibody assays have indicated an increase of greater than 30% in the prevalence of HSV-2 infections over the past 2 decades. Of particular concern is the finding that HSV-2 infection occurs at a younger age, with seroprevalence having quintupled in white teenagers and doubled in young adults in their 20s over the past 2 decades (Fleming et al., 1997). In the most recent NHANES (NHANES III), 15% of white men and 20% of white women in the United States were HSV-2 positive. Among black persons, 35% of men and 55% of women were HSV-2 positive. The number of individuals with genital HSV ranges between 40 and 60 million. In STD clinics in the United States, United Kingdom, and Australia, the HSV-2 seroprevalence is as high as 50 to 66% (Bassett et al., 1994). The frequency of HSV-1 isolation from genital lesions has also increased significantly during the last 10 years, particularly among people with first-episode infections (Slomka et al., 1998). It is now estimated that 25 million U.S. adults have genital HSV-1 infections resulting from self-inoculation or from increased use of oral sexual practices. HSV-1 and HSV-2 can colonize and reactivate in the same anatomic region (Aurelian et al., 1990; Sucato et al., 1998).

HSV-2 infections are often asymptomatic (Aurelian and Kessler, 1985; Strand et al., 1986; Barton et al., 1987; Aurelian et al., 1990; Wald et al., 1997). The frequency of subclinical shedding varies among individuals and over time within an individual. It is up to three times more frequent during the first 3 to 6 months after primary infection than later. In established infections, subclinical HSV-2 shedding from the genital tract has been estimated to occur on 1 to 6% of days (Barton et al., 1987). Relatively common shedding of HSV from perianal and rectal sites has also been described to occur in men who have sex with men (Krone et al., 1998). Other undiagnosed persons have symptoms that are not recognized as being caused by HSV. Atypical manifestations include vulvar, perianal, and penile fissures, localized erythema, and back pain without genital lesions. All of these groups are at risk of transmitting virus. In a prospective study of couples where one had genital HSV-2 and the other did not, transmission occurred at an annualized rate of 10%; 70% of the transmission events were during times when the source partner was asymptomatic (Aurelian et al., 1990; Mertz et al., 1992).

The estimated incidence of HSV encephalitis is 2.3 cases per million persons per year. The age distribution is biphasic, with one peak at 5 to 30 years of age and a second peak at above 50 years of age. HSV-1 accounts for 95% of cases. HSV-2 was isolated from the cerebrospinal fluid (CSF) in 0.5 to 3% of patients with aseptic meningitis.

The major risk to the fetus is maternal primary or initial genital infection with HSV, the rate of which is 0.5 to 10% per year (Brown et al., 1991). Infection acquired near the time of labor is associated with neonatal herpes and perinatal morbidity. Recurrent infection is the most common form of infection during gestation. The incidence of HSV-2 shedding at delivery is 0.01 to 0.6%, irrespective of past history and time of gestation. The incidence of cervical shedding in pregnant women with asymptomatic infection averages 3% (range, 0.2 to 7.4%). The frequency of shedding does not vary by trimester during gestation. Most women whose children are infected (60 to 80%) are asymptomatic at the time of delivery. Using stringent identification criteria (virus isolation within 48 h of life), approximately 30 babies have been identified in the world literature with symptomatic congenital HSV disease. This frequently results from ascending infection in women who have had prolonged rupture of membranes before delivery (reviewed by Whitley [1996]). HSV has also been reported in a mother and neonate with herpes (pemphigoid) gestationis, a rare autoimmune disease with extensive cutaneous involvement that appears during pregnancy and in the immediate postpartum (Chen et al., 1999).

DIAGNOSIS

Virus Culture

Sensitivity, specificity, speed, and availability determine the choice of a diagnostic test for HSV infection (Table 3). The presence of viable virus in a clinical specimen as determined by virus isolation in culture is the most definitive evidence of HSV infection, particularly within the context of identifying a reservoir for virus transmission. HSV can be isolated from oral and genital lesions, ocular samples, throat swabs and washings, bronchioalveolar washings, biopsies from brain and other tissues, and CSF. Cytopathic effect (CPE) can be seen within 18 h p.i., but it generally takes 2 to 3 days (depending on the sensitivity of the cells and the amount of virus inoculated). Cultures are routinely kept for 14 days. This is a major limitation when rapid diagnosis is essential, such as in encephalitis or systemic diseases where therapy must be initiated early.

When a patient presents with skin vesicles, they should be punctured and the vesicular fluid taken on a swab. The base of the lesion is also swabbed, and both samples are used to inoculate cell cultures. For best results, specimens should be inoculated within 1 h after collection. Otherwise, they should be put in transport medium and refrigerated or frozen (at −70°C) in order to preserve the specimen. The transport medium routinely used in our laboratory is Eagle's minimal essential medium with Earle's salts and L-glutamine supplemented with 0.8% bovine serum albumin fraction V, 10 mM HEPES buffer, and antibiotics (100 μg of penicillin and streptomycin per ml, 5 μg of amphotericin per ml). Primary cells are generally more sensitive than established lines. The most sensitive are primary rabbit kidney (PRK) and human embryo cells. Commonly used cell lines are human lung fibroblasts (MRC-5) and human lung carcinoma cells (A549) (McCarter and Robinson, 1997).

HSV causes a characteristic CPE consisting of rounded refractile cells. If hematoxylin-eosin stain is applied, the cells exhibit eosinophilic intranuclear inclusions. Polykaryocytosis depends on the virus strain and the cell type. Results may be confounded by the presence in the specimens of other viruses that cause a similar CPE, such as adenovirus, a common isolate in ocular specimens. To address this problem, at least two cell types that differ in the ability to support the growth of these viruses should be used routinely for virus culture. We use MRC-5 and A549 cells. HSV grows in MRC-5 cells; adenovirus does not. In A549 cells, adenovirus CPE is first seen on day 4 or 5; that of HSV is seen on days 1 through 3. Ultimately, however, virus identification can be done only using specific antisera in an appropriate immunologic test. The best virus culture results are obtained with specimens collected when the lesion is still in the vesicular or early ulcerative stage (days 1 through 3 postonset), at which time mean virus titers are $\log_{10}$ PFU $\pm$ SEM = 2.5 $\pm$ 0.5. By day 4 postonset, virus titers are often much lower ($\log_{10}$ PFU $\pm$ SEM = 1.3 $\pm$ 1), and the probability of successful virus isolation is greatly reduced.

Antigen Detection

Because the quantity of virus present in clinical specimens is variable (depending on lesion status and collection procedure), antigen detection may provide an alternative to virus culture which has the ability to identify nonviable virus. The basic principle is that antigens in the lesion interact with the added HSV antibody, producing a complex that may be detected by a variety of methods, such as immunofluorescence (IF) or immunoperoxidase (IP) staining, enzyme immunoassay (EIA) (viz., ELISA), or radioimmunoassay (RIA). Recently available monoclonal antibodies to HSV proteins have improved the specificity of these assays and provided the necessary tools for differentiation between the two HSV serotypes. Antigen detection assays are rapid and relatively inexpensive, but there is still uncertainty as to their sensitivities and specificities.

For direct examination, the base of the vesicle is scraped and the scrapings are placed in a drop of phosphate-buffered saline (PBS) on slides. After drying, the cells are fixed with acetone, washed with PBS, and stained (30 min, 37°C) with HSV antibody conjugated to a detector molecule such as fluorescein isothiocyanate (FITC). If monoclonal antibodies are used, fixation may have to be in ethanol or methanol, since some epitopes are acetone sensitive. Detection of fluorescence indicates HSV presence. The indirect assay uses unconjugated HSV antibody followed by the appropriate FITC-conjugated anti-immunoglobulin (30 min, 37°C). It can also utilize the high affinity of biotin for avidin, an egg white protein. Anti-HSV antibody conjugated to biotin is added to the culture, followed by the addition of avidin-fluorescein conjugate. In a direct comparison, we found that the sensitivities of the IF assay were 70 and 53.8% of that of virus culture when specimens were collected from primary and recurrent lesions, respectively. This correlates with mean titers of total recoverable virus ($\log_{10}$ PFU $\pm$ SEM) of 4.0 $\pm$ 1.0 and 2.5 $\pm$ 0.5, respectively, in primary and recurrent lesions. The indirect assay is more sensitive, but it can be less specific. Specificity may be a problem particularly in presence of vesicle fluid, mucus, or other protein-binding secretions. Antigen detection is an effective method of diagnosing ocular HSV infection. The cornea is anesthetized, and corneal epithelium is scraped with a sterile scalpel blade and transferred to sterile PBS on glass slides. The blade is washed in transport medium and a

TABLE 3 Methods for diagnosis of HSV infections

Type of infection	Method	Advantage(s)	Disadvantage(s)	Cost	General availability
Skin, mucocutaneous	Isolation	High sensitivity, specificity	Slow; complex	High	Variable
	Cytology (Tzanck)	Rapid, simple	Poor to fair sensitivity, specificity	Low	Yes
	Antigen				
	IF, IP	Good sensitivity; rapid, simple	Poor to fair specificity	Low	Yes
	RIA	Good sensitivity	Inconvenient	High	Variable
	EIA	Good sensitivity, specificity; rapid, simple	Variable results	Low	Yes
	Shell vial amplification	Better sensitivity, specificity than direct	Slower than direct	Low/high	Yes
	DNA				
	PCR	Highest sensitivity; rapid	Potential specificity problems	Medium	Variable
Keratitis	Isolation	High sensitivity, specificity	Slow, complex	High	Variable
	Antigen				
	IF; IP	Good sensitivity; rapid	Poor to fair specificity	Low	Yes
Encephalitis (biopsy or CSF)	Isolation	High sensitivity, specificity	Slow, complex	High	Variable
	Antigen				
	IF; IP	Good specificity	Not completely reliable	Low	Yes
	DNA				
	PCR	Highest sensitivity; rapid	Potential specificity problems	Medium	Variable
Previous infection	Serology				
	Increased titers	Good sensitivity, specificity	Only primary infection	Medium	Variable
	Serotype	Good specificity, sensitivity	Variable results	Medium	Variable

swab of the affected area is placed in virus transport medium. Standard procedures for the indirect IF test are followed to obtain a preliminary diagnosis while awaiting confirmation from virus culture. In the IP procedure, the anti-immunoglobulin is conjugated to the enzyme horseradish peroxidase and the sample is treated with diamino-benzidine, a substrate for the enzyme. If HSV is in the sample, it binds the HSV antibody, resulting in an enzyme reaction that is characterized by reddish brown areas. The IP and IF assays are equally sensitive and specific in direct comparison, but IP has an advantage in that it requires an ordinary light microscope and the slides can be kept indefinitely. The sensitivities of the antigen detection procedures are increased by amplification through short-term (16- to 24-h) growth in tissue culture, a procedure also referred to as shell vial. Specimens are inoculated onto cell cultures on coverslips in shell vials. Low-speed centrifugation (700 × g, 40 min, room temperature) of inoculated shell vials increases virus adsorption onto the cell monolayers, thereby enhancing the sensitivity of the assay (Gleaves et al., 1985). Known negative and HSV-infected (positive) cell cultures are used as controls. A modification of the shell vial assay (suspension-infection) combines the specimen with suspended culture cells before allowing them to grow (Rich and Johnson, 1998). In a direct comparison, the sensitivities of the direct IF assay relative to the shell vial assay were 75.8% for HSV-1 and 93.1% for HSV-2 (Reina et al., 1997). The sensitivities of the shell vial assays can be identical to that of virus culture (Gleaves et al., 1985; Espy and Smith, 1988) or significantly lower (66.2%) (Seal et al., 1991). However, the advantage of the shell vial is its relative rapidity (16 to 24 h, as compared to 4 to 14 days for culture).

EIA is another antigen detection assay. It is rapid, versatile, and adaptable to automation. The test specimen is directly attached to the microtiter well or bound to antibody-coated wells. It is treated with HSV antibody and antiglobulin that is conjugated to an enzyme, such as alkaline phosphatase. Addition of substrate results in the formation of a soluble colored reaction product that can be read quantitatively in a spectrophotometer. Variables that affect the results include the quality of the HSV antibody, the buffer, the conditions used to bind the specimen to the plates, and the plastic used to make the plates. Optimal results are obtained with polyvinyl plates exposed to the specimen in PBS (or distilled H_2O) for 2 h and washed with PBS-Tween. Irreproducible results and/or higher background levels are observed with polystyrene plates and when carbonate buffer, 1% glutaraldehyde, or Ca^{2+}/Mg^{2+}-free PBS is used and incubation is overnight. Restandardization must be performed for each new lot of microtiter plates, as the plates vary in the ability to bind antigen. When performed under optimal conditions, EIA has a false-negative level of 6.2% and a false-positive level of 8.3% compared to virus culture (Aurelian, 1982).

Tissue culture amplification increases assay sensitivity. Compared to virus culture, the direct EIA showed 97% sensitivity in nongenital specimens and 88% sensitivity in genital specimens. The culture-amplified EIA (4 days) showed 95% sensitivity in all samples (Kok et al., 1998). A recent adaptation is the enzyme-linked virus-inducible system (ELVIS), which employs a genetically altered BHK cell line that allows for overnight HSV detection and simultaneous serotyping. The specimen is incubated overnight on shell vials containing these cells and stained with galactopyranoside. Blue cells indicate that HSV is present; fluorescence indicates HSV-2. In the absence of fluorescence, cells are stained with FITC-labeled HSV-1 antibody for HSV-1 confirmation. In direct comparison with shell vials of primary rabbit kidney cells that were IF stained on day 4 p.i., ELVIS detected 42 of 43 specimens that were positive by shell vial, and there were only 3 specimens in which the number of blue cells was too small to interpret (Turchek and Huang, 1999).

Immunoassays can also use radioisotope (^{125}I) as the detector molecule (RIA). The disadvantages are that radioactive agents are relatively unstable, there are working hazards, precautions are required, and expensive equipment is needed. However, RIA has the advantage that "background" enzymes do not lower specificity, as may occur in EIA. Spot blotting is another antigen detection assay in which the test specimens are subjected to appropriate cell lysis and placed on a membrane that captures HSV antigen, if present. Following thorough washing to remove excess material, the membrane is treated with peroxidase-conjugated HSV antibody and a dye solution to generate a colorimetric reaction that is read by comparison to membranes containing positive and negative controls.

DNA Technology

In situ hybridization with radioactively labeled or biotinylated DNA probes can be used as a detection method for samples inoculated onto shell vials. It may be as sensitive and specific as antigen detection or virus culture (Espy and Smith, 1988). DNA restriction endonuclease analysis is used in the identification of common-source outbreaks of HSV infections. It is based on the ability of restriction endonucleases to cleave DNA at specific sites, generating cleavage products the numbers and sizes of which differ among HSV isolates (Buchman et al., 1979). Cleavage products are separated by slab gel electrophoresis and visualized by chemical staining or by autoradiography if the viral DNA was labeled with ^{32}P while grown in culture.

PCR is an extremely sensitive and relatively simple assay that has been rapidly replacing standard HSV detection protocols. Commonly used primers are directed to HSV DNA polymerase, UL42, gB, gD, or TK sequences, and they target different sequences in these genes. Their sensitivities (viz., the number of detected HSV DNA copies) often vary by 10- to 1,000-fold. Restriction digestion of PCR products can be used for direct HSV typing. Assay modifications include (i) nested PCR, in which amplification is done twice, the second time with primers internal to those used in the first set (this is likely to increase sensitivity), and (ii) multiplex PCR, in which coamplification is done simultaneously with primers for various agents. For example, multiplex PCR has been described for HSV-1, HSV-2, and cytomegalovirus by Cassinoti and Siegel (1998) and for HSV-2, *Haemophilus ducreyi*, and *Treponema pallidum* by Beyrer et al. (1998). However, the sensitivity of some primer sets can be artifactually decreased in coamplification, and this is generally difficult to control.

PCR has been directly compared to virus culture and antigen detection by EIA. HSV-2 was detected in 76 of 194 (39%) genital swabs by EIA, 93 of 194 (48%) genital swabs by virus isolation, and 115 of 194 (59%) genital swabs by PCR. Comparison of the three methods was as follows: for cell culture versus PCR, sensitivity was 80.9% and specificity was 100%; for EIA versus PCR, sensitivity was 65.2% and specificity was 98.7%; and for EIA versus cell culture, sensitivity was 80.7% and specificity was 99%. EIA was less effective in detecting HSV among recurrent than first-episode patients in comparison to culture and PCR. The data indicate that PCR is more sensitive than antigen detection and virus culture (Slomka et al., 1998). Accordingly, PCR is more effective than virus culture in diagnosing an old, crusted lesion, in which virus titers are low. Indeed, in a study of 246 patients attending an STD clinic, only 59% of HSV-positive samples were correctly diagnosed by the clinicians, and 11% had an atypical appearance. Thirty-one samples were PCR positive and culture negative, and these were generally from older, crusted lesions. However, PCR was negative in 27 patients with clinically diagnosed HSV, including 2 with vesicular and 15 with ulcerative lesions (Safrin et al., 1997), suggesting that PCR results must be interpreted with caution.

False-positive PCR results due to specimen contamination can be a major problem in a clinical setting that processes large numbers of specimens daily (Fomsgaard et al., 1998). To control for such problems, negative specimens must be interspersed between the clinical specimens and the specificity of the amplicants must be confirmed by hybridization with specific (internal amplicant sequence) and unrelated (negative-control) probes. Hybridization with the appropriate probe also increases the sensitivity of the assay, but it is relatively cumbersome since it uses radioactivity and its turnaround is 2 to 4 days. As an alternative to hybridization, various commercial assays use colorimetric microtiter detection. The sensitivity and specificity of one such assay were, respectively, 63 and 100% compared to hybridization. Two other commercial assays had 98.2 and 100% sensitivity and 100 and 96.9% specificity (Tang et al., 1998).

PCR is rapidly becoming the method of choice for diagnosis of HSV encephalitis or meningitis (Dennet et al., 1996; Mitchell et al., 1997). This is particularly true in the case of CSF samples from patients treated with antiviral drugs, in which the amounts of HSV DNA can range between <20 and 20,000 copies/ml of CSF obtained within 4 to 5 days after therapy onset. In a two-center study in which PCR was used on CSF specimens from 43 patients suspected of HSV encephalitis, combined clinical radiological and laboratory results were positive in 6 of 43 (14%) patients. Discrepant PCR results between the two centers were obtained in eight cases (18%) and consisted of five false positives and three false negatives. The two centers had used different methods to prepare the DNA and different nested PCR protocols. DNA purification, albeit more cumbersome, improved data concordance (Hirsch and Bossart, 1999). Because of the wide variability in HSV DNA levels in CSF specimens from patients with encephalitis, a quantitative PCR assay may be helpful in establishing prognosis and in monitoring patient response to antiviral chemotherapy. One such assay is a competitive PCR. Here, specimens are coamplified with an internal standard, and the ratio of the amount of HSV to the amount of internal standard is compared to the ratios on a standard curve in which the same internal standard is coamplified with known amounts of HSV DNA. The assay can measure 10^0 to 10^6 copies of HSV DNA (Hobson et al., 1997; Domingues et al., 1998). It has been recommended that PCR be performed together with an immunoglobulin G IgG capture EIA for HSV intrathecal antibody production in order to span the disease cycle from symptom onset to weeks after hospitalization (Fomsgaard et al., 1998).

The widespread use of PCR has raised questions about the clinical significance of DNA detection within the context of disease causation and transmission. For example, it is unclear whether the presence of HSV DNA in epilepsy seizure centers (Sanders et al., 1997) or gastrointestinal sensory neurons (Lohr et al., 1990) means that HSV causes monofocal epilepsy and peptic ulcer, respectively. Similarly, the proportion of HSV DNA-positive secretions from women in labor has been reported to be about eight times higher by PCR than previously reported by culture methods (Cone et al., 1994). However, the frequencies of HSV transmission to the infant and neonatal disease were not increased. Because the likelihood of infection depends on input virus load, a DNA-positive, culture-negative specimen may not be infectious. Alternatively, such specimens may contain DNA fragments rather than the full-length HSV genome (Imafuku et al., 1997). To ensure that DNA positivity in a culture-negative specimen reflects the presence of low virus titers, PCR must be done with a set of primers for HSV genes that span the entire genome and are equally sensitive.

Cytology

Polykaryocytosis is detected by cytologic examination of tissue scrapings (Zanck assay). Cytology is somewhat useful for screening, but its insensitivity and its inability to distinguish between HSV serotypes argues against its use at a time when other, equally rapid tests are available. In direct comparison, virus isolation detected HSV in 84% of cases compared with 57% by indirect IF or IP and only 40% by cytologic examination.

Serology

Serodiagnosis of a primary HSV infection in a patient presenting with clinically visible lesions is based on the demonstration of a significant rise (≥4-fold) in serum anti-

body levels over the course of the illness. This is done by taking paired serum samples. The first (acute-phase) sample is taken early in the course of disease, preferably before 10 days, which is the time required to develop antibodies. The second (convalescent-phase) sample is taken 2 to 3 weeks later. Many factors may confound the results of serodiagnosis. Because of the considerable cross-reactivity between the two HSV serotypes, a type-specific assay must be used in order to detect HSV-2 antibodies in patients who already have high titers of antibodies to HSV-1, and vice versa. Furthermore, patients with recurrent disease do not evidence demonstrable increases in antibody levels. Therefore, serology is not commonly used in virus diagnosis. When it is done, its value is limited to (i) documentation of primary HSV infection in which a clear-cut rise in antibody titers is definitive or (ii) providing evidence of prior infection that may constitute an indication for prophylactic therapy (viz., in immunocompromised patients).

Type-specific serodiagnostic techniques, which are currently becoming available, are useful in evaluating past HSV-2 infections from a transmission or prevalence standpoint. Distinguishing between HSV types may also be of some value in predicting recurrences, since genital HSV-2 is more likely to recur than genital HSV-1. Several type-specific serologic assays using glycoprotein G from HSV-1 (gG-1) and HSV-2 (gG-2), which are only 35% homologous, have been developed. They include the immunodot and immunoblot strip methods, Western blotting with extracts from HSV-infected cells, immunoblotting with baculovirus-expressed gG, and EIA with gG purified on immunoaffinity columns. Three commercially available assays for gG-2 antibodies (one strip immunoblot and two indirect EIAs) were used to study 1,250 serum samples. The sensitivities of the assays were 99.2, 99.7, and 89.9%, and their specificities were 97.1, 96.7, and 99.3%, respectively, but a qualitative agreement between the three assays was found in only 86.4% of the sera (Groen et al., 1998). In cohort studies, 6.1 to 21.2% of tested specimen sets shifted from seropositive to seronegative over time, an inaccuracy that was most pronounced for HSV-2. Day-to-day variability and inconsistent color intensity, which do not occur in EIAs that are not serotype specific, were interpreted to indicate limited sensitivity (Schmid et al., 1999). However, since gG-2 is 35% homologous to gG-1, variability may actually reflect a level of cross-reactivity. EIAs with peptide antigens that are unique to HSV-1 or HSV-2 could overcome this problem, as suggested by the high sensitivity and specificity of an assay using a gG-2 peptide (Marsden et al., 1998). Concern with the gG assays is that they may miss antibodies to other HSV proteins, thereby giving rise to false-negative results. Indeed, gG-1-negative sera recognized type-specific peptides in the HSV-1 glycoprotein C (Ackermann et al., 1998). It may well be that the reliability of type-specific serologic assays will be improved only by using mixtures of type-specific antigens. Assays using HSV-1 and HSV-2 unique peptides from viral proteins other than gG (viz., RR1 PK), are under development (Aurelian et al., unpublished data).

IMMUNITY AND VACCINE DEVELOPMENT

HSV infection is accompanied by the development of virus-specific humoral and cell-mediated immune responses that are primarily directed against HSV-type common antigenic determinants. HSV antibody is likely to be involved in protection from secondary infection, but HSV recurrences occur in the presence of high titers of neutralizing antibody. Experimental studies, including adoptive transfer of HSV-specific T cells, indicate that protective immunity is primarily mediated by CD4[+] T cells, mainly of the Th1 type (Sethi et al., 1983; Aurelian et al., 1991; Wachsman et al., 1992; Manickan et al., 1995; Milligan et al., 1998). Both in the guinea pig model and in humans, recurrent disease has been linked to down-regulation of T-cell responses, primarily of the Th1 type, resulting from an imbalance between the Th1- and Th2-type responses (Sheridan et al., 1982, 1985; Iwasaka et al., 1983; Sheridan and Aurelian, 1983; Miura et al., 1994). This suggests that recurrences can be interrupted by conditions which bias towards a Th1 phenotype, such as the use of appropriate immunomodulators (viz., IL-12) and/or vaccination of HSV-infected subjects with the appropriate recombinant vaccines (therapeutic vaccination). Indeed, increased production of Th2-type cytokines (viz., IL-6, IL-10) has been associated with suppression of cellular immune responses and with UV-induced recurrent disease, both in animal models and humans (Hayashi and Aurelian, 1986; Yasumoto et al., 1987; Miura et al., 1994; Sin et al., 1999). Thermally injured mice, which are deficient in the production of the Th1 cytokine IFN-γ and the generation of CTLs (a typical type 1 effector T cell) but evidence increased production of type 2 cytokines and the generation of CD8[+] type 2 T helper cells (Th2), are 100 times more susceptible to HSV-1 infection than are untreated mice. Administration of IL-12 to thermally injured mice triggers events that bias HSV-specific immunity to a Th1 response and inhibit type 2 cytokine production while improving resistance against HSV-1 infection (Matsuo et al., 1996). By the same token, CD4[+] Th1 responses contribute to HSV keratitis (Niematowski and Rouse, 1992) and HAEM (Aurelian et al., 1998; Kokuba et al., 1998a). IL-12 administration in this environment causes an immunopathologic HSV disease (Kanangat et al., 1996).

Several HSV vaccine strategies have been developed during the last 10 to 15 years, most of them predicated on the generation of systemic cell-mediated immunity. Subunit proteins prepared from virus-infected cells or by recombinant methods have been shown to reduce genital disease in animal models when given in combination with adjuvant (Meignier et al., 1987). However, they have failed to induce protective immunity in humans (Burke, 1993). Recombinant HSV-1 viruses expressing HSV-2 glycoproteins D, E, G, and part of I were studied in guinea pigs and owl monkeys and in human clinical trials, with limited success (Meignier et al., 1988). A human adenovirus recombinant expressing HSV-2 gB was tested in mice (McDermott et al., 1984), and vaccinia virus recombinants expressing HSV-1 or HSV-2 gD were tested in mice and guinea pigs (Wachsman et al., 1987, 1989a, 1989b; Aurelian et al., 1991). Major lessons taught by these studies include the following: (i) glycosylation-related epitopes are essential in the induction of protective immunity by viral glycoproteins, (ii) protection from fatal HSV disease and clearance of high-dose HSV-2 from the skin are mainly a function of HSV-specific Th1 responses, and (iii) development of CD4[+] Th1-mediated protective immunity depends on "relevant" antigen presentation, which is predicated on the specific recombinant vector used in immunization. Studies of antigen-coding plasmid DNA, including gD DNA, demonstrated the ability of DNA immunization to induce protective immunity in certain animal models, but the induced immunity was

often incomplete (Ghiasi et al., 1995). Administration of IL-12 cDNA together with a gD DNA vaccine was shown to drive HSV-specific CD4[+] Th1 responses, resulting in reduced HSV-2 morbidity and mortality (Sin et al., 1999). In situ transfection of mouse cornea with an IFN-α1-containing DNA plasmid protected from ocular HSV-1 challenge by reducing virus replication in the eye and trigeminal ganglia, suggesting that ectopic cytokine expression may be therapeutically effective (Noisakran et al., 1999).

Recombinant viruses that are genetically compromised for growth have also been studied for the ability to induce protective immunity when used as live virus vaccines. Guinea pigs immunized with an ICP8-deleted HSV-2 mutant had less severe skin lesions and reduced episodes of recurrent disease upon HSV-2 challenge (Da Costa et al., 1997). An HSV-2 mutant deleted in both copies of the γ[1]34.5 gene and the UL55 and UL56 ORFs reduced HSV-2 lesion development and severity in guinea pigs (Spector et al., 1998), and an HSV-2 mutant deleted in the gH gene reduced HSV-2 replication and protected from HSV-2-induced disease (McLean et al., 1994). These recombinants appear to be more effective than the other preparations in reducing HSV-2 replication and preventing disease due to HSV-2 challenge, presumably because they present a wide array of antigenic targets, thereby inducing a higher, broader, and more long-lasting immune response. Their disadvantages are that (i) they retain RR1 PK, which causes neoplastic transformation of human cells (Smith et al., 1994; Aurelian, 1995, 1998a; Hunter et al., 1995b), and (ii) there is the possibility of homologous recombination with sequences in the complementing cell line used for their growth, leading to functional reconstruction of the deleted virus gene and subsequent full virulence (McLean et al., 1994). We have shown that the RR1 PK oncogene is required for HSV-2 growth in cultured cells (Smith et al., 1992, 1998) and in animal models (Aurelian et al., 1999). Mice and guinea pigs immunized with an HSV-2 mutant that lacks RR1 PK are protected from HSV-2 challenge, likely mediated by the induction of the relevant HSV-specific T-cell response (Aurelian et al., 1999). The mutant is defective in latency establishment and therefore in reactivation. In guinea pigs, immunization prevents HSV-2 recurrences, also when given therapeutically after HSV-2 infection (therapeutically) (Table 4). Furthermore, the mutant grows in noncomplementing cells, thereby avoiding potential reconstruction of the deleted gene (Aurelian et al., 1999; Wachsman et al., unpublished data).

THERAPY

The ideal HSV therapy would both reduce disease severity and prevent latency establishment. In reality, there is no known agent that can prevent latency, presumably because it is already established by the time symptomatic disease appears. Barrier prevention during intercourse and foreplay is, therefore, recommended for HSV-2. The major goals for the treatment of primary HSV disease are to reduce (i) the time to complete healing of the lesions and resolution of clinical symptoms, (ii) the likelihood of complications and, should they occur, their severity, and (iii) the time of virus shedding and therefore the likelihood of transmission. Agents used to treat HSV are listed in Table 5. While some of these have little scientific basis, they often remain popular because clinical manifestations are to some degree influenced by a placebo effect. Research is focused on the devel-

TABLE 4 RR1 PK mutant has therapeutic HSV-2 vaccine activity in guinea pigs[a]

Recurrent cutaneous disease	Mock vaccinated	Vaccinated
No. of animals with lesions/total (%)	11/14 (79)	1/15 (7)
Day of 1st recurrence	21.9 ± 1.4	32
Duration (days)	10.4 ± 2.2	3
Peak lesion score	2.5 ± 0.2	2
Peak footpad width (cm)	0.56 ± 0.07	0.29
No. of recurrences/animal	2.4 ± 0.4	1

[a]Guinea pigs were infected with HSV-2 (3×10^6 PFU) and mock vaccinated (with PBS) or vaccinated with RR1 PK mutant (10^6 PFU) on days 7 and 17 after HSV-2 infection. Animals were monitored for 40 days for the development of recurrent HSV-2 cutaneous disease. Data are expressed as means ± SEM.

opment of wide spectrum microbicidal agents that can be applied vaginally to prevent infection with HSV-2 as well as other sexually transmitted agents (Kokuba et al., 1998b).

Specific antiviral agents are listed in Table 6. All are targeted at the viral DNA polymerase and block viral DNA replication. They are nucleoside analogues, structurally related to endogenous pyrimidine or purine bases that specifically inhibit the activity of viral enzymes and function as false substrates for DNA synthesis. The synthetic purine nucleoside analogue, acyclovir (ACV) is licensed for the treatment of primary genital HSV infections. At therapeutic doses, ACV interferes with viral DNA synthesis while leaving cell DNA synthesis largely unaffected. HSV TK, which has physicochemical and biologic properties that are different from those of the cell enzyme, mistakenly recognizes ACV as an endogenous nucleoside and phosphorylates it. By contrast, the cellular TK phosphorylates ACV at a rate 10^6-fold slower. ACV-monophosphate is phosphorylated twice as much by host cell enzymes. ACV-triphosphate, the active form of the drug, serves as a competitive

inhibitor of viral DNA polymerase, which recognizes it much more readily than does the host cell DNA polymerase. In general, the active metabolite is recognized by HSV DNA polymerase as an alternative substrate and is incorporated into the growing viral DNA chain, slowing down replication or causing termination of chain growth. As DNA synthesis is required for ACV to function, the drug cannot destroy latent virus.

ACV is available in topical, intravenous, and oral formulations. Oral ACV speeds the healing and resolution of symptoms of primary and recurrent episodes of genital HSV infection and reduces virus shedding, also in immunocompromised patients (Whitley and Gnann, 1992). Intravenous administration of 10 mg/kg of body weight every 8 h for at least 10 days is still the standard therapy for severe HSV infections (viz., encephalitis). Intravenous and oral ACV also prevent reactivation of HSV in seropositive patients undergoing immunosuppressive chemotherapy. Routine use for recurrent episodes of disease, especially for mild episodes, is not recommended. Long-term daily suppressive therapy with oral ACV (two to five 200-mg capsules) suppresses recurrences in approximately 71% of patients with a history of frequent recurrences but does not eliminate ganglionic virus, and recurrences return after ACV is discontinued. A study of 279 immunocompetent patients with a history of frequently recurring genital HSV who stopped suppressive therapy after 6 years indicated that 85.8% had at least one recurrence and 75% had at least two recurrences in the subsequent year. Significantly, paired pre- and posttherapy virus isolates from 13 of these patients showed no trend toward resistance development (Fife et al., 1994). The incidence of resistance among immunocompetent hosts is 3 to 5%. However, resistance is increasingly common among immunocompromised patients, where resistant

TABLE 5 Agents that have been used to treat HSV

Topical Inactivators
    Ether/chloroform
    Photodynamic inactivation
    Laser treatment
    Microbicidal agents

Immunotherapy
    Unrelated vaccines (smallpox, BCG)
    HSV vaccines
    Isoprinosine
    Levamisole
    IFN
    IFN inducers

Others
    Cotrimoxazole
    Griseofulvin
    Corticosteroid creams
    Lithium
    Chlorpromazine
    DMSO
    Glutaraldehyde
    Lysine
    Zinc
    Salicylic acid
    Vitamins B_{12}, C, E

TABLE 6 Antiviral agents used to treat HSV

Agent	Commercial name
Acyclovir (ACV)	Zovirax; generic drug also available
Valacyclovir	Valtrex
Famciclovir	Famvir
Penciclovir	Denavir
Cidofovir	Vistide; generic drug also available
Phosphonoformatic acid (PFA)	Foscarnet
AraA	Vidarabine
Trifluorothymidine (TFT)	

strains can cause problematic mucocutaneous infections. This is generally the result of deficient ACV phosphorylation due to alterations in the viral TK, including loss or reduced levels of activity or altered substrate specificity. Sequence analysis indicated that 46.7% of the resistant isolates had frameshift mutations in homopolymer runs of Gs or Cs, while 53.3% had point mutations in conserved or nonconserved regions (Gaudreau et al., 1998). Mutations in HSV DNA polymerase have also been reported.

Valacyclovir, the oral prodrug of ACV, has increased bioavailability but is not significantly more effective. Famciclovir is the oral prodrug of penciclovir, the active antiviral compound. It is approved for genital HSV infections. Its mechanism of action is similar to that of ACV. Cross-resistance for ACV is with famciclovir and valacyclovir but not Vidarabine or Foscarnet. Vidarabine is available in parenteral and ophthalmic preparations for the treatment of HSV encephalitis, neonatal herpes, and herpetic keratitis. Because it is phosphorylated by cellular enzymes, it has been used as alternative therapy in ACV-resistant infections. However, it is no longer a drug of choice, mainly because of its relatively low potency, rapid degradation (by deamination), and poor aqueous solubility. Foscarnet is a pyrophosphate analogue. It inhibits viral DNA polymerase directly and does not need to be activated intracellularly. It competes with pyrophosphate for the pyrophosphate binding site of DNA polymerase. Foscarnet is a noncompetitive inhibitor of DNA polymerase. It is used for the treatment of ACV-resistant HSV. It is available for intravenous delivery and is relatively toxic. Significant resistance to Foscarnet may develop in immunosuppressed patients, where it may be beneficial to use combination therapy with Foscarnet and ACV (Safrin et al., 1994). Foscarnet has recently been developed for oral delivery (Pechere et al., 1998). For therapy of ACV- and/or Foscarnet-resistant mucocutaneous HSV, topical trifluorothymidine (TFT) and topical or intravenous cidofovir have yielded encouraging results. The biochemical basis for their effectiveness is that cidofovir inhibits DNA polymerase in competition with dCTP and that TFT is efficiently phosphorylated by cellular enzymes. Cidofovir is currently under U.S. Food and Drug Administration review for topical treatment of refractory HSV (Martinez and Luks-Golger, 1997). TFT and cidofovir are under development for ocular infections (Romanowski et al., 1999).

Ongoing research is focused on the development of agents that may be effective against ACV-resistant strains, such as compounds that inhibit viral RR (viz., peptidomimetic inhibitor BILD 1633) (Duan et al., 1998) or DNA polymerase (viz., H2G) (Neyts et al., 1998). A class of non-nucleoside antiviral agents (viz., SCH 43478) that have selective activity against HSV-2 and function at the level of IE gene trans-activation (Albin et al., 1997) and antisense oligonucleotides that target IE genes or their trans-activation (Smith et al., 1986; Gao et al., 1990; Kulka et al., 1993, 1994) have also been described. The clinical significance of the antisense approach is that it could overcome the problem of drug resistance. However, improved delivery methods must be developed.

UNIQUE PROPERTIES

A unique property of HSV-2 is its ability to cause neoplastic transformation. Transformation is mediated by RR1 PK, which is a strong oncogene, both in rodent and human cells. Indeed, human cells transformed by RR1 PK have a cloning efficiency in agarose that is comparable to or greater than (3- to 14-fold) that of other oncogenes, including src, myc, fps, erbB, fos, and sis (Falcone et al., 1985; Lee et al., 1993; Yang et al., 1994). Cells transformed by RR1 PK cause tumors in animals, and tumor formation is inhibited by RR1 PK inhibition (Aurelian, 1998). RR1 PK is a novel member of the serine/threonine growth factor receptor family and is located on its own branch of the phylogenetic tree (Hunter et al., 1995a). Anchorage-independent growth requires functional PK activity and is not evidenced by cells infected with HSV-2 mutants that lack or are deficient in RR1 PK or are transfected with plasmids in which the PK activity has been abrogated through site-directed mutations, such as in the ATP-binding site or the transmembrane domain. The mechanism of transformation by RR1 PK has been elucidated and shown to involve activation of the Ras signaling pathway (Smith et al., 1992, 1994; Hunter et al., 1995b). There is relatively little homology between the HSV-2 and HSV-1 RR1 PKs. Most notable are the absence of an ATP-binding Lys residue and a nonfunctional transmembrane domain in HSV-1 RR1 PK, which probably explain its failure to cause cellular transformation (Chung et al., 1989; Cooper et al., 1995; Hunter et al., 1995a). However, HSV-1 is mutagenic and causes chromosome breakage (reviewed by Aurelian [1995]).

HSV VECTORS FOR GENE THERAPY

Vectors based on HSV-1 show promise for gene transfer into mammalian cells because of their wide host range, efficient infection, and ability to deliver genes to nondividing cells. They are particularly promising for gene therapy of neurological disorders. Research is focused on a number of approaches. One of these is using amplicons that are plasmid vectors that are unable to replicate on their own but contain HSV-1 sequences that, in the presence of helper virus, support DNA replication and packaging into virus particles. Amplicon vectors that express tyrosine hydroxylase have been used in a rat model of Parkinson's disease and shown to cause long-term biochemical and behavioral recovery (Geller et al., 1997). The problem with this approach is the toxicity associated with the helper virus used to package the amplicon. Ongoing research is designed to develop nonpathogenic and efficient packaging systems (Stavropoulos and Stratdhee, 1998). Another approach is using attenuated viruses to deliver the transgene. Single gene mutants of HSV-1 deleted or altered in the gene that codes for TK, RR1, or $\gamma_1 34.5$ were injected intratumorally and shown to kill malignant glioma and meningioma cells and other nervous system tumors (Markert et al., 1993; Boviatsis et al., 1994; McMenamin et al., 1998). A recently described multiple mutant in $\gamma_1 34.5$ and RR1 loci that is avirulent upon intracerebral inoculation was shown to kill human glioma and malignant meningioma cells, decrease tumor growth, and prolong survival in tumor-bearing nude mice (Yazaki et al., 1995). Another approach, commonly termed "suicide gene therapy," is based on transduction of HSV TK by adenovirus or retrovirus vectors in conjunction with treatment with ganciclovir (GCV), an inhibitor of DNA replication that is preferentially activated by HSV TK (Ramesh et al., 1999). This approach involves a "bystander effect," whereby the number of cells killed significantly exceeds the number of cells transduced with the foreign gene. The approach has been used in various tumor models to achieve local control and prolong survival as well as to reduce spontaneous metastatic cancer (Hall et al.,

1998). Combined transduction with interleukin genes could also increase the efficacy of the HSV TK-GCV system (Okada et al., 1999); this is an area of ongoing research interest.

REFERENCES

Ackermann, G., F. Ackermann, H. J. Eggers, U. Wieland, and J. E. Kühn. 1998. Mapping of linear antigenic determinants on glycoprotein C of herpes simplex virus type 1 and type 2 recognized by human serum immunoglobulin G antibodies. *J. Med. Virol.* **55:**281–287.

Albin, R., R. Chase, C. Risano, M. Lieberman, E. Ferrari, A. Skelton, P. Buontempo, S. Cox, J. DeMartino, J. Wright-Minogue, G. Jirau-Lucca, J. Kelly, A. Afonso, A. D. Kwong, E. J. Rozhon, and J. F. O'Connell. 1997. SCH 43478 and analogs: in vitro activity and in vivo efficacy of novel agents for herpesvirus type 2. *Antivir. Res.* **35:**139–146.

Ashley, R. L., and A. Wald. 1999. Genital herpes: review of the epidemic and potential use of type-specific serology. *Clin. Microbiol. Rev.* **12:**1–8.

Aurelian, L. 1982. Herpes simplex virus diagnosis: antigen detection by ELISA and flow microfluorometry. *Diagn. Gynecol. Obstet.* **4:**375–388.

Aurelian, L. 1989. Herpes simplex, p. 73–100. *In* S. Specter, M. Bendinelli, and H. Friedman (ed.), *Virus-Induced Immunosuppression.* Plenum Publishing Corp., New York, N.Y.

Aurelian, L. (ed.). 1990. *Herpesviruses, the Immune System and AIDS.* Kluwer Academic Publishers, Boston, Mass.

Aurelian, L. 1995. Transformation and mutagenic effects induced by herpes simplex virus types 1 and 2, p. 253–280. *In* G. Barbanti-Brodano, M. Bendinelli, and H. Friedman (ed.), *DNA Tumor Viruses: Oncogenic Mechanisms.* Plenum Publishing Corp., New York, N.Y.

Aurelian, L. 1998. Herpes simplex virus type 2: unique biological properties include neoplastic potential mediated by the PK domain of the large subunit of ribonucleotide reductase. *Front. Biosci.* **3:**d237–d249. [Online.] http://bioscience.org/1998/v3/d/aurelian/list.htm.

Aurelian, L., and I. I. Kessler. 1985. Subclinical herpes virus infections of the genital tract are commonly associated with viral shedding. *Cervix* **3:**235–248.

Aurelian, L., M. Wachsman, and J. W. Burnett. 1990. Clinical and subclinical HSV infection resulting from exposure to asymptomatic patients. *Br. J. Dermatol.* **122:**117–119.

Aurelian, L., C. C. Smith, M. Wachsman, and E. Paoletti. 1991. Immune responses to herpes simplex virus in guinea pigs (footpad model) and mice immunized with vaccinia virus recombinants containing herpes simplex virus glycoprotein D. *Rev. Infect. Dis.* **13:**S924–S934.

Aurelian, L., H. Kokuba, and J. W. Burnett. 1998. Understanding the pathogenesis of HSV-associated erythema multiforme. *Dermatology* **197:**219–222.

Aurelian, L., H. Kokuba, and C. C. Smith. 1999. Vaccine potential of a herpes simplex virus type 2 mutant deleted in the PK domain of the large subunit of ribonucleotide reductase (ICP10). *Vaccine* **17:**1951–1963.

Barton, S. E., J. M. Davis, V. W. Moss, A. S. Tyms, and P. E. Munday. 1987. Asymptomatic shedding and subsequent transmission of genital herpes simplex virus. *Genitourin. Med.* **63:**102–105.

Bassett, I., B. Donovan, N. J. Bodsworth, P. R. Field, D. W. Ho, S. Jeansson, and A. L. Cunningham. 1994. Herpes simplex virus type 2 infection of heterosexual men attending a sexual health centre. *Med. J. Aust.* **160:**697–700.

Bates, P. A., and N. A. DeLuca. 1998. The polyserine tract of herpes simplex virus ICP4 is required for normal viral gene expression and growth in murine trigeminal ganglia. *J. Virol.* **72:**7115–7124.

Beasley, K. L., G. E. Cooley, G. F. Kao, M. H. Lowitt, J. W. Burnett, and L. Aurelian. 1997. Herpes simplex vegetans: atypical genital herpes infection in a patient with common variable immunodeficiency. *J. Am. Acad. Dermatol.* **37:**860–863.

Beyrer, C., K. Jitwatcharanan, C. Natpratan, R. Kaewvichit, K. E. Nelson, C. Y. Chen, J. B. Weiss, and S. A. Morse. 1998. Molecular methods for the diagnosis of genital ulcer disease in a sexually transmitted disease clinic population in northern Thailand: predominance of herpes simplex virus infection. *J. Infect. Dis.* **178:**243–246.

Blaho, J. A., C. Mitchell, and B. Roizman. 1993. Guanylylation and adenylylation of the α regulatory proteins of herpes simplex virus require a viral β or γ function. *J. Virol.* **67:**3891–3900.

Block, T. M., J. G. Spivack, I. Steiner, S. Deshmane, M. T. McIntosh, R. P. Lirette, and N. W. Fraser. 1990. A herpes simplex virus type 1 latency-associated transcript mutant reactivates with normal kinetics from latent infection. *J. Virol.* **64:**3417–3426. (Erratum, **64:**4603.)

Bond, V. C., and S. Person. 1984. Fine structure physical map locations of alterations that affect cell fusion in herpes simplex virus type 1. *Virology* **132:**368–376.

Bourne, N., L. R. Stanberry, B. L. Connelly, J. Kurawadwala, S. E. Straus, and P. R. Krause. 1994. Quantity of latency-associated transcript produced by herpes simplex virus is not predictive of the frequency of experimental recurrent genital herpes. *J. Infect. Dis.* **169:**1084–1087.

Boviatsis, E. J., J. M. Scharf, M. Chase, K. Harrington, N. W. Kowall, X. O. Breakefield, and E. A. Chiocca. 1994. Antitumor activity and reporter gene transfer into rat brain neoplasms inoculated with herpes simplex virus vectors defective in thymidine kinase or ribonucleotide reductase. *Gene Ther.* **1:**323–331.

Brown, Z. A., J. Benedetti, R. Ashley, S. Burchett, S. Selke, S. Berry, L. A. Vontver, and L. Corey. 1991. Neonatal herpes simplex virus infection in relation to asymptomatic maternal infection at the time of labor. *N. Engl. J. Med.* **324:**1247–1252.

Browne, H., S. Bell, T. Minson, and D. W. Wilson. 1996. An endoplasmic reticulum-retained herpes simplex virus glycoprotein H is absent from secreted virions: evidence for reenvelopment during egress. *J. Virol.* **70:**4311–4316.

Bruni, R., and B. Roizman. 1996. Open reading frame P—a herpes simplex virus gene repressed during productive infection encodes a protein that binds a splicing factor and reduces synthesis of viral proteins made from spliced mRNA. *Proc. Natl. Acad. Sci. USA* **93:**10423–10427.

Buchman, T. G., B. Roizman, and A. J. Nahmias. 1979. Demonstration of exogenous genital reinfection with herpes simplex virus type 2 by restriction endonuclease fingerprinting of viral DNA. *J. Infect. Dis.* **140:**295–304.

Burke, R. L. 1993. Current developments in herpes simplex virus vaccines. *Semin. Virol.* **4:**187–197.

Byers, R. J., P. S. Hasleton, A. Quigley, C. Dennett, P. E. Klapper, G. M. Cleator, and E. B. Faragher. 1996. Pulmonary herpes simplex in burns patients. *Eur. Respir. J.* **9:**2313–2317.

Calistri, A., C. Parolin, M. Pizzato, P. Calvi, I. Giaretta, and G. Palù. 1999. Herpes simplex virus chronically infected human T lymphocytes are susceptible to HIV-1 superinfection and support HIV-1 pseudotyping. *J. Acquir. Immune Defic. Syndr. Hum. Retrovirol.* **21:**90–98.

Caradonna, S. J., and Y. C. Cheng. 1981. Induction of uracil-DNA glycosylase and dUTP nucleotidohydrolase activity in herpes simplex virus-infected human cells. *J. Biol. Chem.* **256:**9834–9837.

Carter, K. L., P. L. Ward, and B. Roizman. 1996. Characterization of the products of the U(L)43 gene of herpes simplex virus 1: potential implications for regulation of gene expression by antisense transcription. *J. Virol.* **70:**7663–7668.

Cassinotti, P., and G. Siegl. 1998. A nested-PCR assay for the simultaneous amplification of HSV-1, HSV-2, and HCMV genomes in patients with presumed herpetic CNS infections. *J. Virol. Methods* **71:**105–114.

Chang, Y. E., L. Menotti, F. Filatov, G. Campadelli-Fiume, and B. Roizman. 1998. U$_L$27.5 is a novel γ_2 gene antisense to the herpes simplex virus 1 gene encoding glycoprotein B. *J. Virol.* **72:**6056–6064.

Chen, S. H., K. Chopra, T. Y. Evans, S. S. Raimer, M. L. Levy, and S. K. Tyring. 1999. Herpes gestationis in a mother and child. *J. Am. Acad. Dermatol.* **40:**847–849.

Chou, J., and B. Roizman. 1994. Herpes simplex virus 1 g$_1$34.5 gene function, which blocks the host response to infection, maps in the homologous domain of the genes expressed during growth arrest and DNA damage. *Proc. Natl. Acad. Sci. USA* **91:**5247–5251.

Chung, T. D., J. P. Wymer, C. C. Smith, M. Kulka, and L. Aurelian. 1989. Protein kinase activity associated with the large subunit of herpes simplex virus type 2 ribonucleotide reductase (ICP10). *J. Virol.* **63:**3389–3398.

Church, G. A., A. Dasgupta, and D. W. Wilson. 1998. Herpes simplex virus DNA packaging without measurable DNA synthesis. *J. Virol.* **72:**2745–2751.

Cocchi, F., M. Lopez, L. Menotti, M. Aoubala, P. Dubreuil, and G. Campadelli-Fiume. 1998. The V domain of herpesvirus Ig-like receptor (HIgR) contains a major functional region in herpes simplex virus-1 entry into cells and interacts physically with the viral glycoprotein D. *Proc. Natl. Acad. Sci. USA* **95:**15700–15705.

Coen, D. M., M. Kosz-Vnenchak, J. G. Jacobson, D. A. Leib, C. L. Bogard, P. A. Schaffer, K. L. Tyler, and D. M. Knipe. 1989. Thymidine kinase-negative herpes simplex virus mutants establish latency in mouse trigeminal ganglia but do not reactivate. *Proc. Natl. Acad. Sci. USA* **86:**4736–4740.

Cone, R. W., A. C. Hobson, Z. Brown, R. Ashley, S. Berry, C. Winter, and L. Corey. 1994. Frequent detection of genital herpes simplex virus DNA by polymerase chain reaction among pregnant women. *JAMA* **272:**792–796.

Cooper, J., J. B. Clements and J. Conner. 1995. Characterization of the novel protein kinase activity present in the R1 subunit of herpes simplex virus ribonucleotide reductase. *J. Virol.* **69:**4979–4985.

Crute, J. J., T. Tsurumi, L. A. Zhu, S. K. Weller, P. D. Olivo, M. D. Challberg, E. S. Mocarski, and I. R. Lehman. 1989. Herpes simplex virus 1 helicase-primase: a complex of three herpes-encoded gene products. *Proc. Natl. Acad. Sci. USA* **86:**2186–2189.

Da Costa, X. J., N. Bourne, L. R. Stanberry, and D. M. Knipe. 1997. Construction and characterization of a replication-defective Herpes simplex virus 2 ICP8 mutant strain and its use in immunization studies in a guinea pig model of genital disease. *Virology* **232:**1–12.

Daikoku, T., S. Shibata, F. Goshima, S. Oshima, T. Tsurumi, H. Yamada, Y. Yamashita, and Y. Nishiyama. 1997. Purification and characterization of the protein kinase encoded by UL13 gene of herpes simplex virus type 2. *Virology* **235:**82–93.

DeLuca, N. A., and P. A. Schaffer. 1985. Activation of immediate early and late promoters by temperature-sensitive

and wild-type forms of herpes simplex virus type 1 protein ICP4. *Mol. Cell. Biol.* **5:**1997–2008.

Dennett, C., P. E. Klapper, and G. M. Cleator. 1996. Polymerase chain reaction in the investigation of relapse following herpes simplex encephalitis. *J. Med. Virol.* **48:**129–132.

Desai, P., R. Ramakrishnan, Z. W. Lin, B. Osak, J. C. Glorioso, and M. Levine. 1993. The RR1 gene of herpes simplex virus type 1 is uniquely *trans* activated by ICP0 during infection. *J. Virol.* **67:**6125–6135.

Deshmane, S. L., M. Nicosia, T. Valyi-Nagy, L. T. Feldman, A. Dillner, and N. W. Fraser. 1993. An HSV-1 mutant lacking the LAT TATA element reactivates normally in explant cocultivation. *Virology* **196:**868–872.

Dingwell, K. S., and D. C. Johnson. 1998. The herpes simplex virus gE-gI complex facilitates cell-to-cell spread and binds to components of cell junctions. *J. Virol.* **72:**8933–8942.

Domingues, R. B., F. D. Lakeman, M. S. Mayo, and R. J. Whitley. 1998. Application of competitive PCR to cerebrospinal fluid samples from patients with herpes simplex encephalitis. *J. Clin. Microbiol.* **36:**2229–2234.

Drolet, B. S., G. C. Perng, J. Cohen, S. M. Slanina, A. Yukht, A. B. Nesburn, and S. L. Wechsler. 1998. The region of the herpes simplex virus type 1 LAT gene involved in spontaneous reactivation does not encode a functional protein. *Virology* **242:**221–232.

Duan, J., M. Liuzzi, W. Paris, M. Lambert, C. Lawetz, N. Moss, J. Jaramillo, J. Gauthier, R. Déziel, and M. G. Cordingley. 1998. Antiviral activity of a selective ribonucleotide reductase inhibitor against acyclovir-resistant herpes simplex virus type 1 in vivo. *Antimicrob. Agents Chemother.* **42:**1629–1635.

Dunkle, L. M., R. R. Schmidt, and D. M. O'Connor. 1979. Neonatal herpes simplex infection possibly acquired via maternal breast milk. *Pediatrics* **63:**250–251.

Elliott, G., and P. O'Hare. 1997. Intercellular trafficking and protein delivery by a herpesvirus structural protein. *Cell* **88:**223–233.

Elliott, G., and P. O'Hare. 1999. Live-cell analysis of a green fluorescent protein-tagged herpes simplex virus infection. *J. Virol.* **73:**4110–4119.

Elliott, G., G. Mouzatikisi, and P. O'Hare. 1995. VP16 interacts via its activation domain with VP22, a tegument protein of herpes simplex virus, and is relocated to a novel macromolecular assembly in coexpressing cells. *J. Virol.* **69:**7932–7941.

Espy, M. J., and T. F. Smith. 1988. Detection of herpes simplex virus in conventional tube cell cultures and in shell vials with a DNA probe kit and monoclonal antibodies. *J. Clin. Microbiol.* **26:**22–24.

Everett, R. D., M. Meredith, and A. Orr. 1999. The ability of herpes simplex virus type 1 immediate-early protein Vmw110 to bind to a ubiquitin-specific protease contributes to its roles in the activation of gene expression and stimulation of virus replication. *J. Virol.* **73:**417–426.

Everly, D. N. J., and G. S. Read. 1997. Mutational analysis of the virion host shutoff gene (UL41) of herpes simplex virus (HSV): characterization of HSV type 1 (HSV-1)/HSV-2 chimeras. *J. Virol.* **71:**7157–7166.

Falcone, G., F. Tato, and S. Alema. 1985. Distinctive effects of the viral oncogenes *myc*, *erb*, *fps* and *src* on the differentiation program of quail myogenetic cells. *Proc. Natl. Acad. Sci. USA* **82:**426–430.

Falkenberg, M., D. A. Bushnell, P. Elias, and I. R. Lehman. 1997. The UL8 subunit of the heterotrimeric herpes simplex virus type 1 helicase-primase is required for the unwinding of single strand DNA-binding protein (ICP8)-coated DNA substrates. *J. Biol. Chem.* **272:**22766–22770.

Farrell, M. J., J. M. Hill, T. P. Margolis, J. G. Stevens, E. K. Wagner, and L. T. Feldman. 1993. The herpes simplex virus type 1 reactivation function lies outside the latency-associated transcript open reading frame ORF-2. *J. Virol.* **67:**3653–3655.

Feng, C. P., M. Kulka, and L. Aurelian. 1993. NF-KB binding proteins induced by HSV-1 infection of U937 cells are not involved in activation of human immunodeficiency virus. *Virology* **192:**491–500.

Fife, K. H., C. S. Crumpacker, G. J. Mertz, E. L. Hill, G. S. Boone, and the Acyclovir Study Group. 1994. Recurrence and resistance patterns of herpes simplex virus following cessation of or ≥6 years of chronic suppression with acyclovir. *J. Infect. Dis.* **169:**1338–1341.

Fleming, D. T., G. M. McQuillan, R. E. Johnson, A. J. Nahmias, S. O. Aral, F. K. Lee, and M. E. St. Louis. 1997. Herpes simplex virus type 2 in the United States, 1976 to 1994. *N. Engl. J. Med.* **337:**1105–1111.

Fodor, P. A., M. J. Levin, A. Weinberg, E. Sandberg, J. Sylman, and K. L. Tyler. 1998. Atypical herpes simplex virus encephalitis diagnosed by PCR amplification of viral DNA from CSF. *Neurology* **51:**554–559.

Fomsgaard, A., N. Kirkby, I. P. Jensen, and B. F. Vestergaard. 1998. Routine diagnosis of herpes simplex virus (HSV) encephalitis by an internal DNA controlled HSV PCR and an IgG-capture assay for intrathecal synthesis of HSV antibodies. *Clin. Diagn. Virol.* **9:**45–56.

Frame, M. C., H. S. Marsden, and B. M. Dutia. 1985. The ribonucleotide reductase induced by herpes simplex virus type 1 involves minimally a complex of two polypeptides (136K and 38K). *J. Gen. Virol.* **66:**1581–1587.

Fraser, N. W., T. M. Block, and J. G. Spivack. 1992. The latency-associated transcripts of herpes simplex virus: RNA in search of function. *Virology* **191:**1–8.

Friedman, H. M., J. C. Glorioso, G. H. Cohen, J. C. Hastings, S. L. Harris, and R. J. Eisenberg. 1986. Binding of complement component C3b to glycoprotein C of herpes simplex virus type 1: mapping of gC-binding sites and demonstration of conserved C3b binding in low-passage clinical isolates. *J. Virol.* **60:**470–475.

Gagnon, R. A. 1968. Transplacental inoculation of fatal herpes simplex in the newborn. Report of 2 cases. *Obstet. Gynecol.* **31:**682–684.

Galvan, V., and B. Roizman. 1998. Herpes simplex virus 1 induces and blocks apoptosis at multiple steps during infection and protects cells from exogenous inducers in a cell-type-dependent manner. *Proc. Natl. Acad. Sci. USA* **95:**3931–3936.

Gao, M., and D. M. Knipe. 1991. Potential role for herpes simplex virus ICP8 DNA replication protein in stimulation of late gene expression. *J. Virol.* **65:**2666–2675.

Gao, W. Y., J. W. Jaroszewski, J. S. Cohen, and Y. C. Cheng. 1990. Mechanisms of inhibition of herpes simplex virus type 2 growth by 28-mer phosphorothioate oligodeoxycytidine. *J. Biol. Chem.* **265:**20172–20178.

Gaudreau, A., E. Hill, H. H. J. Balfour, A. Erice, and G. Boivin. 1998. Phenotypic and genotypic characterization of acyclovir-resistant herpes simplex viruses from immunocompromised patients. *J. Infect. Dis.* **178:**297–303.

Geller, A. I., L. Yu, Y. Wang, and C. Fraefel. 1997. Helper virus-free herpes simplex virus-1 plasmid vectors for gene therapy of Parkinson's disease and other neurological disorders. *Exp. Neurol.* **144:**98–102.

Geraghty, R. J., C. Krummenacher, G. H. Cohen, R. J. Eisenberg, and P. G. Spear. 1998. Entry of alphaherpesviruses mediated by poliovirus receptor-related protein 1 and poliovirus receptor. *Science* **280:**1618–1620.

Ghiasi, H., S. Cai, S. Slanina, A. B. Nesburn, and S. L. Wechsler. 1995. Vaccination of mice with herpes simplex virus type 1 glycoprotein D DNA produces low levels of protection against lethal HSV-1 challenge. *Antivir. Res.* **28:**147–157.

Gignoux, L., P. Ryvlin, F. Najioullah, and F. Mauguière. 1998. Recurrent Mollaret's meningitis of herpetic origin. *Presse Med.* **27:**1470–1472. (In French.)

Gleaves, C. A., D. J. Wilson, A. D. Wold, and T. F. Smith. 1985. Detection and serotyping of herpes simplex virus in MRC-5 cells by use of centrifugation and monoclonal antibodies 16 h postinoculation. *J. Clin. Microbiol.* **21:**29–32.

Goldstein, D. J., and S. K. Weller. 1988. Factor(s) present in herpes simplex virus type-1 infected cells can compensate for the loss of the large subunit of the viral ribonucleotide reductase: characterization of an ICP6 deletion mutant. *Virology* **166:**41–51.

Goldstein, J. N., and S. K. Weller. 1998. In vitro processing of herpes simplex virus type 1 DNA replication intermediates by the viral alkaline nuclease, UL12. *J. Virol.* **72:**8772–8781.

Groen, J., G. Van Dijk, H. G. Niesters, W. I. Van Der Meijden, and A. D. Osterhaus. 1998. Comparison of two enzyme-linked immunosorbent assays and one rapid immunoblot assay for detection of herpes simplex virus type 2-specific antibodies in serum. *J. Clin. Microbiol.* **36:**845–847.

Hale, B. D., R. C. Reindtorff, L. C. Walker, and A. N. Roberts. 1953. Epidemic herpetic stomatitis in an orphanage nursery. *JAMA* **183:**1068–1072.

Hall, S. J., M. A. Sanford, G. Atkinson, and S. H. Chen. 1998. Induction of potent antitumor natural killer cell activity by herpes simplex virus-thymidine kinase and ganciclovir therapy in an orthotopic mouse model of prostate cancer. *Cancer Res.* **58:**3221–3225.

Hay, K. A., A. Gaydos, and R. B. Tenser. 1995. The role of herpes simplex thymidine kinase expression in neurovirulence and latency in newborn vs. adult mice. *J. Neuroimmunol.* **61:**41–52.

Hayashi, Y., and L. Aurelian. 1986. Immunity to herpes simplex virus type 2: viral antigen-presenting capacity of epidermal cells and its impairment by ultraviolet B irradiation. *J. Immunol.* **136:**1087–1092.

He, B., M. Gross, and B. Roizman. 1997. The $\gamma_1 34.5$ protein of herpes simplex virus 1 complexes with protein phosphatase 1α to dephosphorylate the α subunit of the eukaryotic translation initiation factor 2 and preclude the shutoff of protein synthesis by double-stranded RNA-activated protein kinase. *Proc. Natl. Acad. Sci. USA* **94:**843–848.

Hernandez, T. R., and I. R. Lehman. 1990. Functional interaction between the herpes simplex-1 DNA polymerase and UL42 protein. *J. Biol. Chem.* **265:**11227–11232.

Herold, B. C., S. I. Gerber, B. J. Belval, A. M. Siston, and N. Shulman. 1996. Differences in the susceptibility of herpes simplex virus types 1 and 2 to modified heparin compounds suggest serotype differences in viral entry. *J. Virol.* **70:**3461–3469.

Hill, A., P. Jugovic, I. York, G. Russ, J. Bennink, J. Yewdell, H. Ploegh, and D. Johnson. 1995. Herpes simplex virus turns off the TAP to evade host immunity. *Nature* **375:**411–415.

Hirsch, H. H., and W. Bossart. 1999. Two-centre study comparing DNA preparation and PCR amplification protocols for herpes simplex virus detection in cerebrospinal fluids of patients with suspected herpes simplex encephalitis. *J. Med. Virol.* **57:**31–35.

Ho, D. Y., and E. S. Mocarski. 1989. Herpes simplex virus latent RNA (LAT) is not required for latent infection in the mouse. *J. Virol.* **86:**7596–7600.

Hobson, A., A. Wald, N. Wright, and L. Corey. 1997. Evaluation of a quantitative competitive PCR assay for measuring herpes simplex virus DNA content in genital tract secretions. *J. Clin. Microbiol.* **35:**548–552.

Holmberg, S. D., J. A. Stewart, A. R. Gerber, R. H. Byers, F. K. Lee, P. M. O'Malley, and A. J. Nahmias. 1988. Prior herpes simplex virus type 2 infection as a risk factor for HIV infection. *JAMA* **259:**1048–1050.

Hook, E. W. D., R. O. Cannon, A. J. Nahmias, F. F. Lee, C. H. J. Campbell, D. Glasser, and T. C. Quinn. 1992. Herpes simplex virus infection as a risk factor for human immunodeficiency virus infection in heterosexuals. *J. Infect. Dis.* **165:**251–255.

Hunter, J. C. R., C. C. Smith, and L. Aurelian. 1995a. The HSV-2 LA-1 oncoprotein is a member of a novel family of serine/threonine receptor kinases. *Int. J. Oncol.* **7:**515–522.

Hunter, J. C. R., C. C. Smith, D. Bose, M. Kulka, R. Broderick, and L. Aurelian. 1995b. Intracellular internalization and signaling pathways triggered by the large subunit of HSV-2 ribonucleotide ductase (ICP10) *Virology* **210:**345–360.

Hutchinson, L., F. L. Graham, W. Cai, C. Debroy, S. Person, and D. C. Johnson. 1993. Herpes simplex virus (HSV) glycoproteins B and K inhibit cell fusion induced by HSV syncytial mutants. *Virology* **196:**514–531.

Idowu, A. D., E. B. Fraser-Smith, K. L. Poffenberger, and R. C. Herman. 1992. Deletion of the herpes simplex virus type 1 ribonucleotide reductase gene alters virulence and latency in vivo. *Antivir. Res.* **17:**145–156.

Igarashi, K., R. Fawl, R. J. Roller, and B. Roizman. 1993. Construction and properties of a recombinant herpes simplex virus 1 lacking both S-component origins of DNA synthesis *J. Virol.* **67:**2123–2132.

Imafuku, S., H. Kokuba, L. Aurelian, and J. Burnett. 1997. Expression of herpes simplex virus DNA fragments located in epidermal keratinocytes and germinative cells is associated with the development of erythema multiforme lesions. *J. Investig. Dermatol.* **109:**550–556.

Iwasaka, T., J. F. Sheridan, and L. Aurelian. 1983. Immunity to herpes simplex virus type 2: recurrent lesions are associated with the induction of suppressor cells and soluble suppressor factors. *Infect. Immun.* **42:**955–964.

Jacob, R. J., L. S. Morse, and B. Roizman. 1979. Anatomy of herpes simplex virus DNA. XII. Accumulation of head-to-tail concatemers in nuclei of infected cells and their role in the generation of the four isomeric arrangements of viral DNA. *J. Virol.* **29:**448–457.

Jacobson, J. G., S. H. Chen, W. J. Cook, M. F. Kramer, and D. M. Coen. 1998. Importance of the herpes simplex virus UL24 gene for productive ganglionic infection in mice. *Virology* **242:**161–169.

Jacobson, J. G., D. A. Leib, D. J. Goldstein, C. L. Bogard, P. A. Schaffer, S. K. Weller, and D. M. Coen. 1989. A herpes simplex virus ribonucleotide reductase deletion mutant is defective for productive acute and reactivatable latent infections of mice and for replication in mouse cells. *Virology* **173:**276–283.

Javier, R. T., J. G. Stevens, V. B. Dissette, and E. K. Wagner. 1988. A herpes simplex virus transcript abundant in latently infected neurons is dispensable for establishment of the latent state. *Virology* **166:**254–257.

Jayachandra, S., A. Baghian, and K. G. Kousoulas. 1997. Herpes simplex virus type 1 glycoprotein K is not essential for infectious virus production in actively replicating cells but is required for efficient envelopment and translocation of infectious virions from the cytoplasm to the extracellular space. *J. Virol.* **71:**5012–5024.

Jenkins, F. J., and B. Roizman. 1986. Herpes simplex virus 1 recombinants with noninverting genomes frozen in different isomeric arrangements are capable of independent replication. *J. Virol.* **59:**494–499.

Johnson, D. C., M. C. Frame, M. W. Ligas, A. M. Cross, and N. D. Stow. 1988. Herpes simplex virus immunoglobulin G Fc receptor activity depends on a complex of two viral glycoproteins, gE and gI. *J. Virol.* **62:**1347–1354.

Jugovic, P., A. M. Hill, R. Tomazin, H. Ploegh, and D. C. Johnson. 1998. Inhibition of major histocompatibility complex class I antigen presentation in pig and primate cells by herpes simplex virus type 1 and 2 ICP47. *J. Virol.* **72:**5076–5084.

Kanangat, S., J. Thomas, S. Gangappa, J. S. Babu, and B. T. Rouse. 1996. Herpes simplex virus type 1-mediated up-regulation of IL-12 (p40) mRNA expression. Implications in immunopathogenesis and protection. *J. Immunol.* **156:**1110–1116.

Katz, J. P., E. T. Bodin, and D. M. Coen. 1990. Quantitative polymerase chain reaction analysis of herpes simplex virus DNA in ganglia of mice infected with replication-incompetent mutants. *J. Virol.* **64:**4288–4295.

Kawaguchi, Y., R. Bruni, and B. Roizman. 1997. Interaction of herpes simplex virus 1 α regulatory protein ICP0 with elongation factor 1δ: ICP0 affects translational machinery. *J. Virol.* **71:**1019–1024.

Keet, I. P., F. K. Lee, G. J. van Griensven, J. M. Lange, A. Nahmias, and R. A. Coutinho. 1990. Herpes simplex virus type 2 and other genital ulcerative infections as a risk factor for HIV-1 acquisition. *Genitourin. Med.* **66:**330–333.

Kok, T., L. Mickan, and S. Schepetiuk. 1998. Rapid detection, culture-amplification and typing of herpes simplex viruses by enzyme immunoassay in clinical samples. *Clin. Diagn. Virol.* **10:**67–74.

Kokuba, H., S. Imafuku, S. Huang, L. Aurelian, and J. Burnett. 1998a. Erythema multiforme lesions are associated with expression of a herpes simplex virus (HSV) gene and qualitative alterations in the HSV-specific T-cell response. *Br. J. Dermatol.* **138:**952–964.

Kokuba, H., L. Aurelian, and A. R. Neurath. 1998b. 3-Hydroxyphthaloyl beta-lactoglobulin. IV. Antiviral activity in the mouse model of genital herpesvirus infection. *Antivir. Chem. Chemother.* **9:**353–357.

Kokuba, H., L. Aurelian, and J. W. Burnett. 1999. Herpes simplex virus gene expression in erythema multiforme lesional skin correlates with the presence of IFN-alpha, TGF-beta, and p21[waf]. *J. Investig. Dermatol.* **113:**808–815.

Kosz-Vnenchak, M., J. Jacobson, D. M. Coen, and D. M. Knipe. 1993. Evidence for a novel regulatory pathway for herpes simplex virus gene expression in trigeminal ganglion neurons. *J. Virol.* **67:**5383–5393.

Kotronias, D., and N. Kapranos. 1998. Detection of herpes simplex virus DNA in human spermatozoa by in situ hybridization technique. *In Vivo* **12:**391–394.

Kristensson, K., E. Lycke, M. Roytta, B. Svennerholm, and A. Vahlne. 1986. Neuritic transport of herpes simplex virus in rat sensory neurons in vitro. Effects of substances interacting with microtubular function and axonal flow [nocodazole, taxol and erythro-9-3-(2-hydroxynonyl)adenine]. *J. Gen. Virol.* **67:**2023–2028.

Kristie, T. M., J. L. Vogel, and A. E. Sears. 1999. Nuclear localization of the C1 factor (host cell factor) in sensory neurons correlates with reactivation of herpes simplex virus from latency. *Proc. Natl. Acad. Sci. USA* **96:**1229–1233.

Krone, M. R., S. R. Tabet, M. Paradise, A. Wald, L. Corey, and C. L. Celum. 1998. Herpes simplex virus shedding among human immunodeficiency virus-negative men who have sex with men: site and frequency of shedding. *J. Infect. Dis.* **178:**978–982.

Krummenacher, C., A. V. Nicola, J. C. Whitbeck, H. Lou, W. Hou, J. D. Lambris, R. J. Geraghty, P. G. Spear, G. H. Cohen, and R. J. Eisenberg. 1998. Herpes simplex virus glycoprotein D can bind to poliovirus receptor-related protein 1 or herpesvirus entry mediator, two structurally unrelated mediators of virus entry. *J. Virol.* **72:**7064–7074.

Kulka, M., C. C. Smith, J. Levis, R. Fishelevich, J. C. Hunter, C. D. Cushman, P. S. Miller, P. O. Ts'o, and L. Aurelian. 1994. Synergistic antiviral activities of oligonucleoside methylphosphonates complementary to herpes simplex virus type 1 immediate-early mRNAs 4, 5, and 1. *Antimicrob. Agents Chemother.* **38:**675–680.

Kulka, M., M. Wachsman, S. Miura, R. Fishelevich, P. S. Miller, P. O. Ts'o, and L. Aurelian. 1993. Antiviral effect of oligo(nucleoside methylphosphonates) complementary to the herpes simplex virus type 1 immediate early mRNAs 4 and 5. *Antivir. Res.* **20:**115–130.

Lai, J. S., and W. Herr. 1997. Interdigitated residues within a small region of VP16 interact with Oct-1, HCF, and DNA. *Mol. Cell. Biol.* **17:**3937–3946.

Lafferty, W. M., C. Remington, A. Winter, A. Fahnlander, and L. Corey. 1985. Natural history of concomitant pharyngeal and genital HSV infection: influence on viral type and anatomic site on recurrence rates. *Clin. Res.* **33:**408A.

Latchman, D. S. 1994. Herpes simplex virus latency and immediate early gene repression by the cellular octamer binding protein Oct-2, p. 238–252. *In* Y. Becker and G. Darai (ed.), *Pathogenicity of Human Herpesviruses Due to Specific Pathogenicity Genes.* Springer-Verlag, Berlin, Germany.

Lee, M. -S., J. -H. Yang, Z. Salehi, P. Arnstein, L. S. Chen, G. Jay, and J. S. Rhim. 1993. Neoplastic transformation of a human keratinocyte cell line by the v-*fos* oncogene. *Oncogene* **8:**387–393.

Leib, D. A., C. L. Bogard, M. Kosz-Vnenchak, K. A. Hicks, D. M. Coen, D. M. Knipe, and P. A. Schaffer. 1989a. A deletion mutant of the latency-associated transcript of herpes simplex virus type 1 reactivates from the latent state with reduced frequency. *J. Virol.* **63:**2893–2900.

Leib, D. A., D. M. Coen, C. L. Bogard, K. A. Hicks, D. R. Yager, D. M. Knipe, K. L. Tyler, and P. A. Schaffer. 1989b. Immediate-early regulatory gene mutants define different stages in the establishment and reactivation of herpes simplex virus latency. *J. Virol.* **63:**759–768.

Lekstrom-Himes, J. A., L. Pesnicak, and S. E. Straus. 1998. The quantity of latent viral DNA correlates with the relative rates at which herpes simplex virus types 1 and 2 cause recurrent genital herpes outbreaks. *J. Virol.* **72:**2760–2764.

Lohr, J. M., J. A. Nelson, and M. B. A. Oldstone. 1990. Is herpes simplex virus associated with peptic ulcer disease? *J. Virol.* **64:**2168–2174.

Luger, A., and E. Lauda. 1921. Transmissibility of herpetic keratitis in man to the cornea of rabbit. *Wien Klin. Wochenschr.* **34:**132.

Maggioncalda, J., A. Mehta, Y. H. Su, N. W. Fraser, and T. M. Block. 1996. Correlation between herpes simplex virus type 1 rate of reactivation from latent infection and the number of infected neurons in trigeminal ganglia. *Virology* **225:**72–81.

Maggs, D. J., E. Chang, M. P. Nasisse, and W. J. Mitchell. 1998. Persistence of herpes simplex virus type 1 DNA in chronic conjunctival and eyelid lesions of mice. *J. Virol.* **72:**9166–9172.

Manickan, E., R. J. Rouse, Z. Yu, W. S. Wire, and B. T. Rouse. 1995. Genetic immunization against herpes simplex virus. Protection is mediated by CD4+ T lymphocytes. *J. Immunol.* **155:**259–265.

Markert, J. M., A. Malick, D. M. Coen, and R. L. Martuza. 1993. Reduction and elimination of encephalitis in an experimental glioma therapy model with attenuated herpes simplex mutants that retain susceptibility to acyclovir. *Neurosurgery* **32:**597–603.

Marsden, H. S., K. MacAulay, J. Murray, and I. W. Smith. 1998. Identification of an immunodominant sequential epitope in glycoprotein G of herpes simplex virus type 2 that is useful for serotype-specific diagnosis. *J. Med. Virol.* **56:**79–84.

Martinez, C. M., and D. B. Luks-Golger. 1997. Cidofovir use in acyclovir-resistant herpes infection. *Ann. Pharmacother.* **31:**1519–1521.

Matsuo, R., M. Kobayashi, D. N. Herndon, R. B. Pollard, and F. Suzuki. 1996. Interleukin-12 protects thermally injured mice from herpes simplex virus type 1 infection. *J. Leukoc. Biol.* **59:**623–630.

McCarter, Y. S., and A. Robinson. 1997. Comparison of MRC-5 and primary rabbit kidney cells for the detection of herpes simplex virus. *Arch. Pathol. Lab. Med.* **121:**122–124.

McDermott, M. R., J. R. Smiley, P. Leslie, J. Brais, H. E. Rudzroga, and J. Bienenstock. 1984. Immunity in the female genital tract after intravaginal vaccination of mice with an attenuated strain of herpes simplex virus type 2. *J. Virol.* **51:**747–773.

McGeoch, D. J., M. A. Dalrymple, A. J. Davison, A. Dolan, M. C. Frame, D. McNab, L. J. Perry, J. E. Scott, and P. Taylor. 1988. The complete DNA sequence of the long unique region in the genome of herpes simplex virus type 1. *J. Gen. Virol.* **69:**1531–1574.

McLean, C. S., M. Erturk, R. Jennings, D. Ni Challanain, A. C. Minson, L. Duncan, M. E. G. Boursnell, and S. C. Inglis. 1994. Protective vaccination against primary and recurrent disease caused by herpes simplex virus (HSV) type 2 using a genetically disabled HSV-1. *J. Infect. Dis.* **170:**1100–1109.

McMenamin, M. M., A. P. Byrnes, F. G. Pike, H. M. Charlton, R. S. Coffin, D. S. Latchman, and M. J. Wood. 1998. Potential and limitations of a gamma 34.5 mutant of herpes simplex 1 as a gene therapy vector in the CNS. *Gene Ther.* **5:**594–604.

Meignier, B., T. M. Jourdier, B. Norrild, L. Pereira, and B. Roizman. 1987. Immunization of experimental animals with reconstituted glycoprotein mixtures of herpes simplex virus 1 and 2: protection against challenge wth virulent virus. *J. Infect. Dis.* **155:**921–930.

Meignier, B., R. Longnecker, P. Mavromara-Nazos, A. E. Sears, and B. Roizman. 1988. Virulence and establishment of latency by genetically engineered deletion mutants of herpes simplex virus 1. *Virology* **162:**251–254.

Mellerick, D. M., and N. W. Fraser. 1987. Physical state of the latent herpes simplex virus genome in a mouse model system: evidence suggesting an episomal state. *Virology* **158:**265–275.

Mertz, G. J., J. Benedetti, R. Ashley, S. A. Selke, and L. Corey. 1992. Risk factors for the sexual transmission of genital herpes. *Ann. Intern. Med.* **116:**197–202.

Milligan, G. N., D. I. Bernstein, and N. Bourne. 1998. T lymphocytes are required for protection of the vaginal mucosae and sensory ganglia of immune mice against reinfection with herpes simplex virus type 2. *J. Immunol.* **160:**6093–6100.

Mitchell, P. S., M. J. Espy, T. F. Smith, D. R. Toal, P. N. Rys, E. F. Berbari, D. R. Osmon, and D. H. Persing. 1997. Laboratory diagnosis of central nervous system infections with herpes simplex virus by PCR performed with cerebrospinal fluid specimens. *J. Clin. Microbiol.* **35:**2873–2877.

Miura, S., M. Kulka, C. C. Smith, S. Imafuku, J. W. Burnett, and L. Aurelian. 1994. Cutaneous ultraviolet radiation (UVR) inhibits herpes simplex virus induced lymphoproliferation in latently infected subjects. *Clin. Immunol. Immunopathol.* **72:**62–69.

Nahmias, A. J., F. K. Lee, and S. Beckman-Nahmias. 1990. Sero-epidemiological and -sociological patterns of herpes simplex virus infection in the world. *Scand. J. Infect. Dis. Suppl.* **69:**19–36.

Nakajima, H., D. Furutama, F. Kimura, K. Shinoda, N. Ohsawa, T. Nakagawa, A. Shimizu, and H. Shoji. 1998. Herpes simplex virus myelitis: clinical manifestations and diagnosis by the polymerase chain reaction method. *Eur. Neurol.* **39:**163–167.

Nelson, J., J. Zhu, C. C. Smith, M. Kulka, and L. Aurelian. 1996. ATP and SH3 binding sites in the protein kinase of the large of herpes simplex virus ribonucleotide reductase (ICP10). *J. Biol. Chem.* **271:**17021–17027.

Newcomb, W. W., F. L. Homa, D. R. Thomsen, B. L. Trus, N. Cheng, A. Steven, F. Booy, and J. C. Brown. 1999. Assembly of the herpes simplex virus procapsid from purified components and identification of small complexes containing the major capsid and scaffolding proteins. *J. Virol.* **73:**4239–4250.

Neyts, J., G. Andrei, and E. De Clercq. 1998. The antiherpesvirus activity of H2G [(R)-9-[4-hydroxy-2-(hydroxymethyl)butyl]guanine] is markedly enhanced by the novel immunosuppressive agent mycophenolate mofetil. *Antimicrob. Agents Chemother.* **42:**3285–3289.

Ng, T. I., W. O. Ogle, and B. Roizman. 1998. UL13 protein kinase of herpes simplex virus 1 complexes with glycoprotein E and mediates the phosphorylation of the viral Fc receptor: glycoproteins E and I. *Virology* **241:**37–48.

Niemialtowski, M. G., and B. T. Rouse. 1992. Predominance of Th1 cells in ocular tissues during herpetic stromal keratitis. *J. Immunol.* **149:**3035–3039.

Nishioka, Y., and S. Silverstein. 1977. Degradation of cellular mRNA during infection by herpes simplex virus. *Proc. Natl. Acad. Sci. USA* **74:**2370–2374.

Noisakran, S., I. L. Campbell, and D. J. Carr. 1999. Ectopic expression of DNA encoding IFN-alpha 1 in the cornea protects mice from herpes simplex virus type 1-induced encephalitis. *J. Immunol.* **162:**4184–4190.

O'Hare, P., and G. S. Hayward. 1985. Evidence for a direct role for both the 175,000- and 110,000-molecular-weight immediate-early proteins of herpes simplex virus in the transactivation of delayed-early promoters. *J. Virol.* **53:**751–760.

Okada, H., K. M. Giezeman-Smits, H. Tahara, J. Attanucci, W. K. Fellows, M. T. Lotze, W. H. Chambers, and M. E. Bozik. 1999. Effective cytokine gene therapy against an intracranial glioma using a retrovirally transduced IL-4 plus HSVtk tumor vaccine. *Gene Ther.* **6:**219–226.

Oshima, S., T. Daikoku, S. Shibata, H. Yamada, F. Goshima, and Y. Nishiyama. 1998. Characterization of the UL16 gene product of herpes simplex virus type 2. *Arch. Virol.* **143:**863–880.

Parris, D. S., A. Cross, L. Haarr, A. Orr, M. C. Frame, M. Murphy, D. J. McGeoch, and H. S. Marsden. 1988. Identification of the gene encoding the 65-kilodalton DNA-binding protein of herpes simplex virus type 1. *J. Virol.* **62:**818–825.

Pechère, M., W. Wunderli, L. Trellu-Toutous, M. Harms, J. H. Saura, and J. Krischer. 1998. Treatment of acyclovir-resistant herpetic ulceration with topical foscarnet and antiviral sensitivity analysis. *Dermatology* **197:**278–280.

Peng, T., J. C. R. Hunter, and J. W. Nelson. 1996. The novel protein kinase of the RR1 subunit of herpes simplex virus has autophosphorylation and transphosphorylation activity that differs in its ATP requirements for HSV-1 and HSV-2. *Virology* **216:**184–196.

Perng, G. C., H. Ghiasai, S. M. Slanina, A. B. Nesburn, and S. L. Wechsler. 1996. The spontaneous reactivation function of the herpes simplex virus type 1 LAT gene resides completely within the first 1.5 kilobases of the 8.3-kilobase primary transcript. *J. Virol.* **70:**976–984.

Pogue-Geile, K. L., and P. G. Spear. 1987. The single base pair substitution responsible for the Syn phenotype of herpes simplex virus type 1, strain MP. *Virology* **157:**67–74.

Pugh, R. C. B., J. A. Dudgeon, and M. Bodia. 1955. Kaposi's varicelliform eruption (eczema herpeticum) with typical visceral necrosis. *J. Pathol. Bacteriol.* **69:**67–70.

Purves, F. C., W. O. Ogle, and B. Roizman. 1993. Processing of the herpes simplex virus regulatory protein alpha 22 mediated by the UL13 protein kinase determines the accumulation of a subset of alpha and gamma mRNAs and proteins in infected cells. *Proc. Natl. Acad. Sci. USA* **90:**6701–6705.

Purves, F. C., D. Spector, and B. Roizman. 1992. U_L34, the target of the herpes simplex virus U_S3 protein kinase, is a membrane protein which in its unphosphorylated state associates with novel phosphoproteins. *J. Virol.* **66:**4295–4303.

Ramesh, R., A. Munshi, A. J. Marrogi, and S. M. Freeman. 1999. Enhancement of tumor killing using a combination of tumor immunization and HSV-tk suicide gene therapy. *Int. J. Cancer* **80:**380–386.

Randall, G., and B. Roizman. 1997. Transcription of the derepressed open reading frame P of herpes simplex virus 1 precludes the expression of the antisense $\gamma_1 34.5$ gene and may account for the attenuation of the mutant virus. *J. Virol.* **71:**7750–7757.

Raza-Ahmad, A., G. A. Klassen, D. A. Murphy, J. A. Sullivan, C. E. Kinley, R. W. Landymore, and J. R. Wood. 1995. Evidence of type 2 herpes simplex infection in human coronary arteries at the time of coronary artery bypass surgery. *Can. J. Cardiol.* **11:**1025–1029.

Read, G. S., B. M. Karr, and K. Knight. 1993. Isolation of a herpes simplex virus type 1 mutant with a deletion in the virion host shutoff gene and identification of multiple forms of the vhs (UL41) polypeptide. *J. Virol.* **67:**7149–7160.

Read, S. J., and J. B. Kurtz. 1999. Laboratory diagnosis of common viral infections of the central nervous system by using a single multiplex PCR screening assay. *J. Clin. Microbiol.* **37:**1352–1355.

Reddig, P. J., L. A. Grinstead, S. J. Monahan, P. A. Johnson, and D. S. Parris. 1994. The essential in vivo function of the herpes simplex virus UL42 protein correlates with its ability to stimulate the viral DNA polymerase in vitro. *Virology* **200:**447–456.

Reina, J., J. Saurina, V. Fernandez-Baca, M. Munar, and I. Blanco. 1997. Evaluation of a direct immunofluorescence cytospin assay for the detection of herpes simplex virus in clinical samples. *Eur. J. Clin. Microbiol. Infect. Dis.* **16:**851–854.

Remeijer, L., P. Doornenbal, A. J. Geerards, W. A. Rijneveld, and W. H. Beekhuis. 1997. Newly acquired herpes simplex virus keratitis after penetrating keratoplasty. *Ophthalmology* **104:**648–652.

Rice, S. A., M. C. Long, V. Lam, P. A. Schaffer, and C. Spencer. 1995. Herpes simplex virus immediate-early protein ICP22 is required for viral modification of host RNA polymerase II and establishment of the normal viral transcription program. *J. Virol.* **69:**5550–5559.

Rich, T., and F. B. Johnson. 1998. Performance of six cell lines in the suspension-infection test used for the detection of herpes simplex virus. *Diagn. Microbiol. Infect. Dis.* **32:**81–84.

Rixon, F. J., C. Addison, and J. McLauchlan. 1992. Assembly of enveloped tegument structures (L particles) can occur independently of virion maturation in herpes simplex virus type 1 infected cells. *J. Gen. Virol.* **73:**277–284.

Roizman, B., and A. E. Sears. 1996. Herpes simplex viruses and their replication, p. 2231–2295. *In* B. N. Fields, D. M. Knipe, and P. M. Howley. (ed.), *Virology*, 3rd ed. Lippincott-Raven Publishers, Philadelphia, Pa.

Romanowski, E. G., S. P. Bartels, and Y. J. Gordon. 1999. Comparative antiviral efficacies of cidofovir, trifluridine, and acyclovir in the HSV-1 rabbit keratitis model. *Investig. Ophthalmol. Vis. Sci.* **40:**378–384.

Roop, C., L. Hutchinson, and D. C. Johnson. 1993. A mutant herpes simplex virus type 1 unable to express glycoprotein L cannot enter cells, and its particles lack glycoprotein H. *J. Virol.* **67:**2285–2297.

Safrin, S., S. Kemmerly, B. Plotkin, T. Smith, N. Weissbach, D. De Veranez, L. D. Phan, and D. Cohn. 1994. Foscarnet-resistant herpes simplex virus infection in patients with AIDS. *J. Infect. Dis.* **169:**193–196.

Safrin, S., H. Shaw, G. Bolan, J. Cuan, and C. S. Chiang. 1997. Comparison of virus culture and the polymerase chain reaction for diagnosis of mucocutaneous herpes simplex virus infection. *Sex. Transm. Dis.* **24:**176–180.

Sakaoka, H., K. Kurita, Y. Iida, S. Takada, K. Umene, Y. T. Kim, C. S. Ren, and A. J. Nahmias. 1994. Quantitative analysis of genomic polymorphism of herpes simplex virus type 1 strains from six countries: studies of molecular evolution and molecular epidemiology of the virus. *J. Gen. Virol.* **75:**513–527.

Salmon, B., and J. D. Baines. 1998. Herpes simplex virus DNA cleavage and packaging: association of multiple forms of U_L15-encoded proteins with B capsids requires at least the U_L6, U_L17, and U_L28 genes. *J. Virol.* **72:**3045–3050.

Salmon, B., C. Cunningham, A. J. Davison, W. J. Harris, and J. D. Baines. 1998. The herpes simplex virus type 1 U_L17 gene encodes virion tegument proteins that are required for cleavage and packaging of viral DNA. *J. Virol.* **72:**3779–3788.

Samaniego, L. A., A. L. Webb, and N. A. De Luca. 1995. Functional interactions between herpes simplex virus immediate early proteins during infection: gene expression as a consequence of ICP27 and different domains of ICP4. *J. Virol.* **69:**5705–5715.

Sanders, V. J., S. L. Felisan, A. E. Waddell, A. J. Conrad, P. Schmid, B. E. Swartz, M. Kaufman, G. O. Walsh, A. A. De Salles, and W. W. Tourtellotte. 1997. Presence of herpes simplex DNA in surgical tissue from human epileptic seizure foci detected by polymerase chain reaction: preliminary study. *Arch. Neurol.* **54:**954–960.

Sawtell, N. M. 1997. Comprehensive quantification of herpes simplex virus latency at the single-cell level. *J. Virol.* **71:**5423–5431.

Schirm, J., and P. S. Mulkens. 1997. Bell's palsy and herpes simplex virus. *APMIS* **105:**815–823.

Schmid, D. S., D. R. Brown, R. Nisenbaum, R. L. Burke, D. Alexander, R. Ashley, P. E. Pellett, and W. C. Reeves. 1999. Limits in reliability of glycoprotein G-based type-specific sero-

logic assays for herpes simplex virus types 1 and 2. *J. Clin. Microbiol.* **37:**376–379.

Seal, L. A., P. S. Toyama, K. M. Fleet, K. S. Lerud, S. R. Heth, A. J. Moorman, J. C. Woods, and R. B. Hill. 1991. Comparison of standard culture methods, a shell vial assay, and a DNA probe for the detection of herpes simplex virus. *J. Clin. Microbiol.* **29:**650–652.

Sethi, K. K., Y. Omata, and K. E. Schneweis. 1983. Protection of mice from fatal herpes simplex virus type 1 infection by adoptive transfer of cloned virus-specific and H-2-restricted cytotoxic T lymphocytes. *J. Gen. Virol.* **64:**443–447.

Sheridan, J. F., and L. Aurelian. 1983. Immunity to herpes simplex virus type 2. V. Risk of recurrent disease following primary infection: modulation of T cell subsets and lymphokine (LIF) production. *Diagn. Immunol.* **1:**246–256.

Sheridan, J. F., M. Beck, L. Aurelian, and M. Radowsky. 1985. Immunity to herpes simplex virus. Virus reactivation modulates lymphokine (NK enhancing) activity. *J. Infect. Dis.* **152:**449–456.

Sheridan, J. F., A. D. Donnenberg, L. Aurelian, and D. J. Elpern. 1982. Immunity to herpes simplex virus type 2. IV. Impaired lymphokine production during recrudescence correlates with an imbalance in T lymphocyte subsets. *J. Immunol.* **129:**326–331.

Shieh, M. T., D. WuDunn, R. I. Montgomery, J. D. Esko, and P. G. Spear. 1992. Cell surface receptors for herpes simplex virus are heparan sulfate proteoglycans. *J. Cell Biol.* **116:**1273–1281.

Shimeld, C., D. L. Easty, and T. J. Hill. 1999. Reactivation of herpes simplex virus type 1 in the mouse trigeminal ganglion: an in vivo study of virus antigen and cytokines. *J. Virol.* **73:**1767–1773.

Sin, J. I., J. J. Kim, R. L. Arnold, K. E. Shroff, D. McCallus, C. Pachuk, S. P. McElhiney, M. W. Wolf, S. J. Pompa-de Bruin, T. J. Higgins, R. B. Ciccarelli, and D. B. Weiner. 1999. IL-12 gene as a DNA vaccine adjuvant in a herpes mouse model: IL-12 enhances Th1-type CD4+ T cell-mediated protective immunity against herpes simplex virus-2 challenge. *J. Immunol.* **162:**2912–2921.

Slomka, M. J., L. Emery, P. E. Munday, M. Moulsdale, and D. W. Brown. 1998. A comparison of PCR with virus isolation and direct antigen detection for diagnosis and typing of genital herpes. *J. Med. Virol.* **55:**177–183.

Smith, C. C., and L. Aurelian. 1997. The large subunit of herpes simplex virus type 2 ribonucleotide reductase (ICP10) is associated with the virion tegument and has PK activity. *Virology* **234:**235–242.

Smith, C. C., L. Aurelian, M. P. Reddy, P. S. Miller, and P. O. P. Ts'o. 1986. Antiviral effect of an oligonucleoside methylphosphonate complementary to the splice junction of herpes simplex virus type 1 immediate early pre-mRNA 4 and 5. *Proc. Natl. Acad. Sci. USA* **83:**2787–2791.

Smith, C. C., M. Kulka, J. P. Wymer, T. D. Chung, and L. Aurelian. 1992. Expression of the large subunit of herpes simplex virus type 2 ribonucleotide reductase (ICP10) is required for virus growth and neoplastic transformation. *J. Gen. Virol.* **73:**1417–1428.

Smith, C. C., J. H. Luo, J. C. R. Hunter, J. V. Ordonez, and L. Aurelian. 1994. The transmembrane domain of the large subunit of HSV-2 ribonucleotide reductase (ICP10) is required for the transformation-related signaling pathways that involve *ras* activation. *Virology* **200:**598–612.

Smith, C. C., T. Peng, M. Kulka, and L. Aurelian. 1998. The PK domain of the large subunit of herpes simplex virus type 2 ribonucleotide reductase (ICP10) is involved in IE gene transcription and virus growth. *J. Virol.* **72:**9131–9141.

Spector, F. C., E. R. Kern, J. Palmer, R. Kaiwar, T.-A. Cha, P. Brown, and R. R. Spaete. 1998. Evaluation of a live attenuated recombinant virus RAV 9395 as a herpes simplex virus type 2 vaccine in guinea pigs. *J. Infect. Dis.* **177:**1143–1154.

Stavropoulos, T. A., and C. A. Strathdee. 1998. An enhanced packaging system for helper-dependent herpes simplex virus vectors. *J. Virol.* **72:**7137–7143.

Stevens, J. G. 1989. Human herpesviruses: a consideration of the latent state. *Microbiol. Rev.* **53:**318–332.

Strand, A., A. Vahlne, B. Svennerholm, J. Wallin, and E. Lycke. 1986. Asymptomatic virus shedding in men with genital herpes infection. *Scand. J. Infect. Dis.* **18:**195–197.

Sucato, G., A. Wald, E. Wakabayashi, J. Vieira, and L. Corey. 1998. Evidence of latency and reactivation of both herpes simplex virus (HSV)-1 and HSV-2 in the genital region. *J. Infect. Dis.* **177:**1069–1072.

Tal-Singer, R., T. M. Lasner, W. Podrzucki, A. Skokotas, J. J. Leary, S. L. Berger, and N. W. Fraser. 1997. Gene expression during reactivation of herpes simplex virus type 1 from latency in the peripheral nervous system is different from that during lytic infection of tissue cultures. *J. Virol.* **71:**5268–5276.

Tal-Singer, R., R. Pichyangkura, E. Chung, T. M. Lasner, B. P. Randazzo, J. Q. Trojanowski, N. W. Razer, and S. J. Triezenberg. 1999. The transcriptional activation domain of VP16 is required for efficient infection and establishment of latency by HSV-1 in the mouse peripheral and central nervous system. *Virology* **259:**20–33.

Tang, Y. W., P. N. Rys, B. J. Rutledge, P. S. Mitchell, T. F. Smith, and D. H. Persing. 1998. Comparative evaluation of colorimetric microtiter plate systems for detection of herpes simplex virus in cerebrospinal fluid. *J. Clin. Microbiol.* **36:**2714–2717.

Thomas, S. K., G. Gough, D. S. Latchman, and R. S. Coffin. 1999. Herpes simplex virus latency-associated transcript encodes a protein which greatly enhances virus growth, can compensate for deficiencies in immediate-early gene expression, and is likely to function during reactivation from virus latency. *J. Virol.* **73:**6618–6625.

Thomsen, D. R., W. W. Newcomb, J. C. Brown, and F. L. Homa. 1995. Assembly of the herpes simplex virus capsid: requirement for the carboxyl-terminal twenty-five amino acids of the proteins encoded by the UL26 and UL26.5 genes. *J. Virol.* **69:**3690–3703.

Turchek, B. M., and Y. T. Huang. 1999. Evaluation of ELVIS HSV ID/Typing System for the detection and typing of herpes simplex virus from clinical specimens. *J. Clin. Virol.* **12:**65–69.

Tyring, S. K., S. S. Carlton, and T. Evans. 1998. Herpes. Atypical clinical manifestations. *Dermatol. Clin.* **16:**783–788.

Valyi-Nagy, T., S. Deshmane, A. Dillner, and N. W. Fraser. 1991. Induction of cellular transcription factors in trigeminal ganglia of mice by corneal scarification, herpes simplex virus type 1 infection, and explantation of trigeminal ganglia. *J. Virol.* **65:**4142–4152.

Venot, C., A. Beby, A. Bourgoin, G. Giraudeau, B. Becq-Giraudon, and G. Agius. 1998. Genital recurrent infection occurring 6 months after meningitis due to the same herpes simplex virus type 2 (HSV-2) strain evidenced by restriction endonuclease analysis. *J. Infect.* **36:**233–235.

Wachsman, M., L. Aurelian, and J. W. Burnett. 1990. The prophylactic use of cyclooxygenase inhibitors in recurrent herpes simplex infections. *Br. J. Dermatol.* **123:**375–380.

Wachsman, M., L. Aurelian, C. C. Smith, B. R. Lipinskas, M. E. Perkus, and E. Paoletti. 1987. Vaccinia virus recombinants expressing herpes simplex virus glycoprotein D protect guinea pigs from primary and recurrent HSV-2 cutaneous disease. *J. Infect. Dis.* **155:**1188–1189.

Wachsman, M., L. Aurelian, C. C. Smith, M. E. Perkus, and E. Paoletti. 1989a. Regulation of expression of herpes simplex virus (HSV) glycoprotein D in vaccinia recombinants affects their ability to protect from cutaneous HSV-2 disease. *J. Infect. Dis.* **159:**625–634.

Wachsman, M., J. H. Luo, L. Aurelian, and E. Paoletti. 1992. Protection from herpes simplex virus type 2 is associated with T cells involved in delayed type hypersensitivity that recoginize glycosylation-related epitopes on glycoprotein D. *Vaccine* **10:**447–454.

Wachsman, M., J. H. Luo, L. Aurelian, M. E. Perkus, and E. Paoletti. 1989b. Antigen-presenting capacity of epidermal cells infected with vaccinia virus recombinants containing the herpes simplex virus glycoprotein D, and protective immunity. *J. Gen. Virol.* **70:**2513–2520.

Wald, A., L. Corey, R. Cone, A. Hobson, G. Davis, and J. Zeh. 1997. Frequent genital herpes simplex virus 2 shedding in immunocompetent women. Effect of acyclovir treatment. *J. Clin. Investig.* **99:**1092–1097.

Wang, K., L. Pesnicak, and S. E. Straus. 1997. Mutations in the 5′ end of the herpes simplex virus type 2 latency-associated transcript (LAT) promoter affect LAT expression in vivo but not the rate of spontaneous reactivation of genital herpes. *J. Virol.* **71:**7903–7910.

Ward, P. L., W. O. Ogle, and B. Roizman. 1996. Assemblons: nuclear structures defined by aggregation of immature capsids and some tegument proteins of herpes simplex virus 1. *J. Virol.* **70:**4623–4631.

Watkins, A. M., P. J. Dunford, A. M. Moffatt, P. Wong-Kai-In, M. J. Holland, D. S. Pole, G. M. Thomas, J. Martin, N. A. Roberts, and M. J. Mulqueen. 1998. Inhibition of virus-encoded thymidine kinase suppresses herpes simplex virus replication in vitro and in vivo. *Antivir. Chem. Chemother.* **9:**9–18.

Whitley, R. J. 1996. Herpes simplex viruses and their replication, p. 2297–2342. *In* B. N. Fields, D. M. Knipe, and P. M. Howley (ed.), *Virology*, 3rd ed. Lippincott-Raven Publishers, Philadelphia, Pa.

Whitley, R. J., and J. W. Gnann, Jr. 1992. Acyclovir: a decade later. *N. Engl. J. Med.* **327:**782–799. (Erratum, **328:**671).

Wilcox, C. L., R. L. Smith, R. D. Everett, and D. Mysofski. 1997. The herpes simplex virus type 1 immediate-early protein ICP0 is necessary for the efficient establishment of latent infection. *J. Virol.* **71:**6777–6785.

Wilson, A. C., R. N. Freiman, H. Goto, T. Nishimoto, and W. Herr. 1997. VP16 targets an amino-terminal domain of HCF involved in cell cycle progression. *Mol. Cell. Biol.* **17:**6139–6146.

Wu, N., S. C. Watkins, P. A. Schaffer, and N. A. DeLuca. 1996. Prolonged gene expression and cell survival after infection by a herpes simplex virus mutant defective in the immediate-early genes encoding ICP4, ICP27, and ICP22. *J. Virol.* **70:**6358–6369.

Wymer, J. P., T. D. Chung, Y.-N. Chang, G. S. Hayward, and L. Aurelian 1989. Identification of immediate-early-type cis-response elements in the promoter for the ribonucleotide veductase large subunit from herpes simplex virus type 2. *J. Virol.* **63:**2773–2784.

Yang, D., S. K. Kohler, V. M. Maher, and J. J. McCormick. 1994. v-sis oncogene-induced transformation of human fibroblasts into cells capable of forming benign tumors. *Carcinogenesis* **15:**2167–2175.

Yao, F., and R. J. Courtney. 1991. Association of a major transcriptional regulatory protein, ICP4, of herpes simplex virus type 1 with the plasma membrane of virus-infected cells. *J. Virol.* **65:**1516–1524.

Yasumoto, S., Y. Hayashi, and L. Aurelian. 1987. Immunity to herpes simplex virus type 2: suppression of virus-induced immune responses in UVB irradiated mice. *J. Immunol.* **139:**2788–2793.

Yazaki, T., H. J. Manz, S. D. Rabkin, and R. L. Martuzza. 1995. Treatment of human malignant meningiomas by G207, a replication-competent multimutated herpes simplex virus 1. *Cancer Res.* **55:**4752–4756.

Yoshikawa, T., J. M. Hill, L. R. Stanberry, N. Bourne, J. F. Kurawadwala, and P. R. Krause. 1996. The characteristic site-specific reactivation phenotypes of HSV-1 and HSV-2 depend upon the latency-associated transcript region. *J. Exp. Med.* **184:**659–664.

Zachhuber, C., F. Leblhuber, K. Jellinger, C. Bancher, G. P. Tilz, and L. Binder. 1995. Necrotizing herpes simplex encephalitis as the cause of a progressive dementia syndrome. *Deutsche Med. Wochenschr.* **120:**1278–1282. (In German.)

Zhu, J., and L. Aurelian. 1997. AP-1 *cis*-response elements are involved in basal expression and Vmw110 transactivation of the large subunit of herpes simplex virus types 2 ribonucleotide reductase (ICP10). *Virology* **231:**301–312.

Cytomegalovirus, Varicella-Zoster Virus, and Epstein-Barr Virus

KIRSTEN ST. GEORGE, DAVID T. ROWE, AND CHARLES R. RINALDO, JR.

32

CMV

History

Cytomegalovirus (CMV) was first noticed by Ribbert in 1881, when he saw large "protozoan-like" cells in the kidney of a stillborn infant (Ribbert, 1904). Similar reports appeared sporadically for some years, with Goodpasture and Talbot coining the term "cytomegalia" (Goodpasture and Talbot, 1921). Von Glahn and Pappenheimer (1925) published the first description of a case in an adult and concluded that the inclusions were caused by a virus of the herpes group. Despite numerous reports of apparent cytomegalic inclusion disease (CID), the frequency of infection was not realized until the 1932 report of Farber and Wolbach. These authors found intranuclear and cytoplasmic inclusions in the submaxillary glands of 14% of cases in a series of postmortem examinations of infants who had died of various causes (Farber and Wolbach, 1932). The term "generalized CID" was suggested in 1950 by Wyatt et al. (Wyatt et al., 1950) for the lethal congenital infection. Soon thereafter Fetterman reported these same enlarged cells with intranuclear inclusions in the urine of a 3-day-old premature infant with jaundice, purpura, hepatosplenomegaly, and intracerebral calcifications (Fetterman, 1952). The first report of the visualization of CMV by electron microscopy appeared in 1953, when cytomegalic inclusion cells from an infant's pancreas were viewed and particles were observed in both the cytoplasm and the clear halo around the inclusions (Minder, 1953). The virus was finally isolated in three different centers independently (Rowe et al., 1956; Smith, 1956; Weller et al., 1957).

The spectrum of disease known to be caused by CMV has broadened enormously since these early studies. A CMV mononucleosis syndrome in immunologically competent adults was first described in 1965 (Klemola and Kaariainen, 1965). However, it is in patients with impaired immune function that the virus now commonly produces a wide range of potentially serious and even life-threatening disease manifestations. Most notably, transplant recipients, deliberately immune suppressed to prevent graft rejection, and patients with AIDS, who become more severely immunosuppressed than transplant recipients, can develop a host of CMV syndromes (Ho, 1977; Jacobson and Mills, 1988; Ives, 1997; Fishman and Rubin, 1998). Enormous advances have also been made in recent decades in our understanding of the biology and molecular biology of CMV, the immunology and pathogenesis of CMV infections, and the prevention and treatment of CMV disease.

Biology and Pathogenesis

To distinguish it from the CMVs in other animals, human CMV is often abbreviated HCMV. In this chapter however, unless otherwise specified, CMV refers only to human CMV. A member of the *Betaherpesvirinae* subfamily of the *Herpesviridae* family, CMV (human herpesvirus type 5) has the typical structure of a herpesvirus, indistinguishable from the others by electron microscopy. The viral DNA is double stranded and linear and, in "wild-type" CMV, approximately 245,000 bp in length. It is protected by an icosahedral nucleocapsid composed of 162 hexagonal capsomeres, which is surrounded by an amorphous tegument and a loose lipid envelope. The capsid proteins include a major and a minor capsid protein and an assemblin protein which is cleaved to produce a protease enzyme involved in particle maturation. The tegument is composed largely of phosphoproteins, a number of which are immunogenic and/or DNA-binding proteins involved in the initial stages of gene expression. The virus-coded envelope proteins include those with herpes simplex virus (HSV) envelope protein homology, such as gB and gH, from which their commonly used names were derived. Similar to other enveloped viruses, CMV is readily inactivated by heat, organic solvents, and dessication. While there are many genetic variants, termed strains, the genetic differences do not allow subclassification into genotypes. And while infection with one strain produces cross-reactive immunity to others, it does not necessarily provide protection either from superinfection by an additional strain or from disease, particularly in immunosuppressed hosts.

The genome is divided into the unique short (U_s) and unique long (U_L) regions, both of which are bounded by repeat sequences. Both regions can be inverted, giving CMV the ability to form four isomers, all of which are produced in equal amounts in culture. The most extensively studied CMV genome is that of the laboratory strain AD169. This strain has been completely sequenced and has a little over 200 predicted proteins (Chee et al., 1990). Its genomic size has recently been reassessed as being 230,283 bp (Dargan et al., 1997; Mocarski et al., 1997), with a 15-kb

410

deletion of at least 19 genes in the U$_s$ region compared with clinical isolates and some other laboratory strains (Cha et al., 1996).

CMV replication in vitro can be accomplished in fibroblasts, macrophages, endothelial cells, epithelial cells, trophoblasts, and smooth muscle cells (Smith, 1986; Tugizov et al., 1996; Woodroffe et al., 1997; Fish et al., 1998; Halwachs-Baumann et al., 1998; Söderberg-Nauclér et al., 1998). Interestingly, the virus can attach to a wide variety of mammalian cells (Nowlin et al., 1991), and in vivo, infected epithelial cells are found in the kidney, liver, bile ducts, salivary gland, gut epithelium, lung, and pancreas, as well as in endothelial cells. Furthermore, CMV replication has been described in polymorphonuclear leukocytes, monocytes, and T and B lymphocytes (Meyer-Konig et al., 1995). The range of in vitro cell types able to sustain CMV replication is therefore probably a reflection of the requirement for maintaining differentiated cells for CMV replication in culture and possibly of the transient expression of receptors (Jarvis et al., 1999). In vitro studies with human hematopoietic progenitor cells have shown that CMV can persist in precursor cells, but activation of extensive gene expression and recovery of infectious viral progeny correlated with terminal differentiation of the cells (Zhuravskaya et al., 1997). Monocytes/macrophages and endothelial cells are important in latent and acute CMV infections (Taylor-Wiedeman et al., 1991). Both are involved in the spread of the virus, with endothelial cells providing the interface between infected circulating macrophages and organ tissues and possibly mediating lytic and host inflammatory damage to infected organs (Waldman et al., 1998). Sensitive PCR assays applied to monocyte-enriched samples from peripheral blood have detected CMV DNA in all seropositive blood donors and 55% of seronegative donors (Larsson et al., 1998b). In vitro experiments with aortic macrovascular and brain-derived microvascular endothelial cells have found that CMV replication takes place in both, with similar percentages of infected cells and similar amounts of progeny virus. However, the amount of intracellular virus was much greater in microvascular cells and resulted in rapid cell lysis whereas the macrovascular cells maintained nonlytic infection with continuous release of virus (Fish et al., 1998). Epithelial cells within organs are a major target of CMV infection, and proposed mechanisms of disease include ischemia, direct cytopathic effect (CPE), and immune-system-mediated tissue damage.

Despite the broad host range of human CMV for cell attachment, its replication is extremely species specific, both in vivo and in vitro. While primate CMVs have been shown to replicate readily in human fibroblasts in vitro and possibly in vivo (Michaels et al., 1997; M. G. Michaels, F. J. Jenkins, K. St. George, M. A. Nalesnik, T. E. Starzl, and C. R. Rinaldo, *Abstr. 39th Intersci. Conf. Antimicrob. Agents Chemother.*, abstr. 2080, 1999), the replication of human CMV in primate fibroblasts has not been demonstrated.

CMV exhibits the typical herpesvirus replication cycle, with initial attachment via as-yet-unidentified cell receptors, although heparan sulfate and CD13 have been implicated (Soderberg et al., 1993). Following fusion of the viral envelope with the cell membrane, the nucleocapsid of CMV enters the cell and presumably travels to the nucleus. Some of the tegument phosphoproteins are also translocated to the nucleus, while others remain in the cytoplasm. The viral DNA moves from the nucleocapsid into the nucleus, and an alpha-beta-gamma sequence of gene expression follows. The immediate-early, or alpha, gene

proteins, primarily transactivators, appear in the infected cell first, followed by the early, or beta, gene proteins, primarily nonstructural glycoproteins. Rolling-circle replication of the viral DNA occurs next and requires at least 11 virus-coded proteins (Pari and Anders, 1993). Replication is initiated at the *cis*-acting origin of replication, *ori*Lyt, near the center of the U$_L$ region of the genome (Anders and McCue, 1996). During this time, the production of late, or gamma, gene proteins, primarily structural components of the progeny virions, also takes place. Appropriate packaging of DNA is achieved in the nucleus by a "headful" mechanism, with cleavage of concatemers occurring when no more DNA can be inserted into the capsid. The virus is then believed to move through the nuclear membrane via an enveloping–de-enveloping process, with the final envelope of the progeny virus probably being acquired from the cytoplasmic membrane (Gibson, 1996; Zini et al., 1999). Despite its reputation as a relatively slow-growing virus in vitro, recent evidence suggests that CMV is far more dynamic in vivo than previously suspected, with an average doubling time or half-life in peripheral blood of approximately 1 day (Emery et al., 1999a).

The major immediate-early (MIE) promoter of CMV controls production of the immediate-early gene products, and its sequence and function are highly conserved (Sorg and Stamminger, 1998). Activation of this promoter is essential for initiation of the replicative cycle and for reactivation of CMV from the latent state. Interestingly prostaglandins, tumor necrosis factor alpha, interleukin-1β (IL-1β), IL-6, and IL-10 all upregulate the MIE promoter (Kline et al., 1998). Since these are all products of activated macrophages, acute inflammation has been proposed as one potential trigger for CMV reactivation. Furthermore, MIE gene products have been shown to enhance IL-6 production, which suggests a role for CMV in producing inflammatory states (Iwamoto and Konicek, 1997). CMV also has self-regulating proteins, such as the pIRS1 products, which down-regulate expression of the immediate-early proteins (Romanowski and Shenk, 1997). The immediate-early genes regulate viral and cellular gene expression throughout the replicative cycle. The MIE gene region contains a complex promoter, and differentially spliced transcripts generate a series of proteins (Stenberg, 1996). Two main transcription regions, commonly referred to as IE1 and IE2 (Stinsky et al., 1983), contain the respective coding information for the two most well-known MIE proteins, p72 and p86 (Stinsky et al., 1983; Pizzorno et al., 1991). These are the most abundant of the MIE proteins during infection of human fibroblasts (Stenberg et al., 1989) and, in addition to their DNA-binding abilities, interact with cellular transcription factors.

The CMV genome is believed to encode approximately 65 glycoproteins (Chee et al., 1990; Cha et al., 1996) with a wide variety of functions. Glycoprotein B (gB or gpUL55) is the most abundant protein in the CMV envelope. All clinical isolates express gB, and serologic responses to gB can be detected in infected individuals. Specific monoclonal antibodies against gB can inhibit virus-cell binding (Ohizumi et al., 1992), and studies suggest the transmembrane region of the protein is responsible for fusion (Tugizov et al., 1994). Glycoprotein H (gH or gpUL75) is the next most abundant component of the envelope and has also been proposed to play a role in virus-cell fusion (Keay and Baldwin, 1991). Its covalent association with glycoprotein L (gL or gpUL115) is believed to be essential for its intracellular transport (Kaye et al., 1992; Spaete et al., 1993). The integral membrane protein, gM (gpUL100), is very hydrophobic and extremely insoluble. It contains many membrane-spanning domains (Lehner et al., 1989) and can be detected in virus particles.

There is evidence that cell and anatomical-site tropism may vary across different gBs and other subtypes of the virus (Meyer-Konig et al., 1998c; Verbraak et al., 1998) and that gB types have different disease associations (Bongarts et al., 1996; Hebart et al., 1997; Torok-Storb et al., 1997; Woo et al., 1997). Other studies, however, have failed to find any significant associations between gB type and disease (Vogelberg et al., 1996; Chern et al., 1998; Peek et al., 1998). These studies are complicated by the fact that there is variable predominance of gB type across different patient populations (Rasmussen et al., 1997a; Woo et al., 1997; Zipeto et al., 1998), geographic locations (Zipeto et al., 1998), and race (Chou and Dennison, 1991; Fries et al., 1994; Wada et al., 1997). Furthermore, there is intragenic variability of gB type in clinical isolates, with different sites within the gB gene from a single isolate giving different typing results (Chou, 1992a; Meyer-König et al., 1998b).

Infection with CMV results in lifelong latency of the virus with possible reactivations at various times caused by a variety of triggers. Monocytes and granulocyte-macrophage progenitor cells in the bone marrow and peripheral blood may be important sites of latency (Minton et al., 1994). CMV reactivation from granulocyte-macrophage precursors can be induced by cocultivation with permissive cells or by cytokine treatment (Kondo et al., 1994; Hahn et al., 1998), and reactivation from mononuclear cells can be induced with dendritic cell markers by allogeneic stimulation (Söderberg-Nauclér et al., 1997). Latency-associated transcripts have been identified, including sense and antisense transcripts of the IE1-IE2 region (Kondo and Mocarski, 1995), and antibodies to proteins encoded by them have been detected in the sera of blood donors (Kondo et al., 1996). Studies using PCR in situ hybridization have shown that while the genomic copy number in experimental latency models is similar to that seen in vivo, the percentage of cells latently infected is very low (0.004 to 0.01%) following natural infections (Slobedman and Mocarski, 1999). Evidence from studies of humans (Hendrix et al., 1997) and animal models (Collins et al., 1993) of CMV, however, strongly suggest that the virus can achieve latency in a more widespread fashion throughout the body.

An increasing number of CMV gene products are being identified which modulate host cell functions associated with transcription control, signaling, and protein synthesis. Host cell protein synthesis is shut off immediately following contact with the cell, whether or not the cell is permissive for replication and even if the virus has been UV inactivated. This function appears to be mediated by the tegument protein pp65 (UL83) (Bodaghi et al., 1995). CMV infection also inhibits cell cycle progression and alters the expression of cell cyclins (Jault et al., 1995), although the pattern and extent of disruption is dependent upon the cell replication phase at which CMV enters the host cell (Salvant et al., 1998). More recently, CD10 and CD13 have been shown to be down-regulated by CMV (Phillips et al., 1998), while the MIE gene product IE2 p86 transactivates the p53 gene, resulting in a five- to sixfold increase in p53 protein levels in infected cells (Muganda et al., 1998). The IE proteins are also involved in the up-regulation of NF-κB (Yurochko et al., 1997). CMV proteins have been shown to form complexes with cell proteins, such as that between IE2

p86 and the retinoblastoma protein (Rb) (Sommer et al., 1994), which may further enhance its repertoire of cellular effects. CMV has also been shown to disrupt the normal balance of cellular lipids in cultured human cells (Abrahamsen et al., 1996) and to block the induction of apoptosis (Zhu et al., 1995).

Both infectious CMV and UV-inactivated virus have been shown to transform various cells in vitro (Albrecht and Rapp, 1973; Geder et al., 1976; Smiley et al., 1988) with activation of proto-oncogenes after CMV infection (Boldogh et al., 1990, 1991). Three morphological transforming regions (mtr) of the CMV genome have been identified. The first, mtrI, is not retained in transformed cells and led to the "hit-and-run" mechanism theory (Nelson et al., 1984). Subsequently, the mtrII and mtrIII regions were identified; retention of the mtrII region is required for maintenance of the transformed phenotype (El-Beik et al., 1986). The mtrII gene is now considered the CMV oncogene, and mtrII protein is expressed in transformed cells. The mtrII protein binds to the cellular tumor suppressors p53 and Rb. While p53 is a multifunctional protein, the binding site for the mtrII protein is in a domain critical for transcriptional activation (Muralidhar et al., 1996). Additionally, although IE72 and IE86 have not been shown to have transforming potential in vitro, they also interact with p53 and Rb and have been suggested to exert hit-and-run mutations in cellular genes (Shen et al., 1997b).

Propagation of CMV in appropriate cell cultures produces progeny virus particles, dense bodies, and noninfectious enveloped particles. Dense bodies are composed largely of tegument proteins with a particularly high percentage of pp65. They are surrounded by a lipid envelope indistinguishable from that of a virus particle, contain no capsid or DNA, and move into permissive cells with exactly the same kinetics as infectious particles (Roby and Gibson, 1986; Topilko and Michelson, 1994). Noninfectious enveloped particles contain an empty capsid and tegument surrounded by the lipid envelope (Roby and Gibson, 1986).

Several envelope glycoproteins elicit strong humoral responses, including the production of virus-neutralizing antibodies. Most of the virus-neutralizing activity in the sera of naturally infected individuals is from antibodies against gB (Britt et al., 1990; Gonczol et al., 1991; Liu et al., 1991; Marshall et al., 1992), and virus-neutralizing activity is associated with high-affinity anti-gB antibodies rather than low-affinity antibodies (Boppana and Britt, 1995). Antigenic domain 1 (AD-1) was the first antibody-binding site identified on gB (Utz et al., 1989). More than 50% of the gB-specific antibody response in serum is directed against AD-1 (Kneiss et al., 1991), and point mutations within this region tend to eliminate the antigenicity of the whole peptide (Wagner et al., 1992). Furthermore, this tends to be one of the most highly conserved regions of gB, which suggests a critical role in the structure or function of the protein. Antibodies to gH can also be detected following natural infection. While measurements of their prevalence are variable, assays using full-length protein detect them in almost 100% of seropositive people (Rasmussen et al., 1991; Boppana et al., 1995), and in some individuals they account for the majority of the virus-neutralizing activity of serum (Urban et al., 1996).

Cell-mediated immune responses play a major role in the control of CMV disease in infected hosts. Major histocompatibility complex (MHC)-restricted CD8+ cytotoxic T-lymphocyte (CTL) responses and proliferative CD4+

responses have been described for gB (Riddell et al., 1991; Utz et al., 1992; Liu et al., 1993; Beninga et al., 1995; He et al., 1995), and CD4$^+$-lymphocyte proliferative responses, but not MHC-restricted CD8$^+$-CTL activity, have been reported for gH. Cell-mediated responses to gM, IE1 and -2, pp71, and gpUL18 have also been reported (Beninga et al., 1995; He et al., 1995). However, the major CTL response to CMV is that to the lower matrix tegument protein, pp65 (Wills et al., 1996). The response to individual pp65 epitopes is highly focused, and CTL clones undergo extensive expansion (Weekes et al., 1999). The presence of CMV-specific cytotoxic CD8$^+$ T cells correlates with recovery from acute CMV infection (Quinnan et al., 1982) and prevention of disease in bone marrow transplant recipients (Krause et al., 1997; Reusser et al., 1997; Podlech et al., 1998). Moreover, CMV disease can be prevented in bone marrow transplant recipients by the injection of CMV-specific CD8$^+$-T-cell clones (Walter et al., 1995).

CMV has been shown to suppress immune function in numerous ways. Lymphocytes from patients with CMV mononucleosis were found to have reduced responses to certain mitogens (Rinaldo et al., 1977), possibly mediated by suppressor cells (Rinaldo et al., 1980). Furthermore, there were reduced proliferative responses to other viruses to which the patients had prior immunity, and proliferative responses to CMV did not achieve levels comparable with those of CMV-immune donors until several months after the acute illness (Levin et al., 1979).

MHC class I and II expression on antigen-presenting cells is crucial for cell-mediated immune responses. Protein products from several genes in the U$_S$ region of CMV down-regulate class I and class II expression. US2 and US11 dislocate class I heavy chains from the endoplasmic reticulum to the cytosol (Jones et al., 1995; Wiertz et al., 1996; Jones and Sun, 1997), US3 facilitates their retention in the endoplasmic reticulum (Jones et al., 1996), and US6 inhibits TAP translocation of the peptides (Ahn et al., 1997; Hengel et al., 1997; Lehner et al., 1997). CMV also inhibits gamma interferon-induced class II expression (Sedmak et al., 1994) downstream of the activation site of STAT1 (le Roy et al., 1999), and US2 causes the degradation of two essential proteins in the MHC class II pathway (Tomazin et al., 1999). This inhibition reduces the presentation of CMV antigens and therefore recognition by T-cell clones, allowing escape from the cell-mediated immune response. Prior to the expression of these genes in the replication cycle of CMV however, there is abundant production of the IE72 protein that would be expected to cause a CTL response. It appears that to avoid this, the pp65 matrix protein phosphorylates the IE, and phosphorylated IE fails to activate CTL (Gilbert et al., 1996). Additionally, cells that have lost cell surface expression of MHC class I are normally destroyed by NK cells. However, the UL18 gene of CMV encodes a class I homolog which may inhibit this NK function (Reyburn et al., 1997), although this has been disputed (Leong et al., 1998). Additionally, there is strong evidence that immediate-early or early gene products of CMV down-regulate the expression of cell surface lymphocyte function-associated antigen-3, which also reduces susceptibility to NK lysis (Fletcher et al., 1998).

There are numerous interactions between CMV-coded proteins and chemokines. At least three CMV genes, US27, US28, and UL33, encode proteins with CC chemokine receptor homology. The US28 gene product, for example, can bind beta chemokines (Neote et al., 1993), an event

which is followed by a rise in intracellular calcium (Gao and Murphy, 1994) and intense sequestering of extracellular chemokines (Bodaghi et al., 1998).

Epidemiology and Disease Manifestations

CMV has a ubiquitous worldwide distribution and is nonseasonal. Seroprevalences range from 30 to 100%, depending on age, geographic location, and ethnic and socioeconomic background. In general, the seroprevalence in developing countries is close to 100% and infections tend to be acquired early in life. Populations in developed countries tend to exhibit lower seroprevalence rates, and infections are more commonly acquired during adolescence and adulthood, with higher rates of acquisition in lower socioeconomic groups. This classic picture of CMV distribution in developed countries may, however, be changing due to the high level of CMV infection in daycare facilities and their increased use among the higher socioeconomic groups in these countries (Stagno and Cloud, 1994).

Human CMV cannot be studied directly in laboratory animals due to its species specificity. CMVs have been found in numerous animals, including guinea pigs, mice, rats, hamsters, moles, dogs, horses, and a variety of nonhuman primates (Ho, 1991). These animal models, particularly mouse and guinea pig CMV, have been used to study various aspects of CMV infection and disease. Small-animal models can be useful for the preclinical evaluation of new drugs, with mouse CMV drug susceptibilities being generally closer to human CMV than those of guinea pig CMV. However, the disease manifestations in some models may not resemble those of human CMV. For example, while rabbits can be intraocularly inoculated with CMV, there is no disease despite active viral replication. Nevertheless, preclinical evaluations with these models to assess efficacy, toxicity, pharmacokinetics, plasma levels, adsorption, and tissue distribution can be helpful. Murine CMV has been used extensively for immune evasion studies (Wiertz et al., 1997), and guinea pig CMV has been used as a model for congenital CMV disease (Griffith et al., 1985). Primate models have generally not proven useful due to their high cost and the fact that the very high seroprevalence levels make the study of primary infection and disease difficult.

CMV disease is the most common congenital and perinatal viral infection throughout the world (Sessions and Taeusch, 1991). Mother-to-baby transmission can occur during pregnancy, during birth, or through breast milk. Breast milk secretion of CMV peaks about a month after delivery, and viral titers are higher in breast milk whey than in milk cells (Numazaki, 1997). The precise mechanism of transmission from mother to fetus is unknown. Infected maternal cells may pass through breaks in the usual maternofetal barrier, or virus may directly infect the syncytiotrophoblast and be transmitted to the underlying fetal placental cells. Interestingly, villous syncytiotrophoblasts have been recently shown to be permissive for CMV replication in vitro (Hemmings et al., 1998). Horizontal transmission occurs via blood contact or close physical contact. Child-to-child and child-to-staff transmission occurs in playgroups and daycare settings, and infected children can transmit CMV to their parents (Pass et al., 1987, 1990). Thus, there is an increased risk of primary CMV infection to seronegative mothers in a second or subsequent pregnancy when previous children are in daycare (Stagno and Cloud, 1994), and in fact, point source acquisition of multiple cases of congenital CMV from daycare settings has been demonstrated (Murph et al., 1998). Interestingly, health care workers in pediatric settings are not at increased risk for acquiring CMV (Dworsky et al., 1983a).

The risk of intrauterine infection during primary infection in the mother is approximately 40%, much higher than in reactivated infections (Fowler et al., 1992). Higher cell-mediated and humoral immune responses have been associated with a reduced risk of transmission to the fetus (Boppana et al., 1993). Estimates of the percentage of infections likely to result in severe symptomatic disease at birth vary from 7 to 15% (Pass et al., 1980; Conboy et al., 1987), and 30 to 60% of infants born with inapparent or mild CMV infection go on to develop adverse consequences later in life (Stagno et al., 1977; Pass et al., 1980). A primary infection during the first half of gestation is likely to result in a more severe outcome, although virus transmission is not necessarily more likely during this time than later in pregnancy (Stagno et al., 1986; Conboy et al., 1987). In developed countries, a serious outcome is more likely in babies of mothers from upper socioeconomic groups, as their infections are more likely to be primary (Stagno et al., 1982). While the majority of infections will result in no apparent disease, up to 2.5% of asymptomatic babies infected in utero, as well as virtually all symptomatic cases, will secrete the virus for prolonged periods after birth. The presence of disease, or the risk of developing disease, has been shown to be associated with high viral titers in neonatal urine although this association declines to insignificance by 3 months of age (Stagno et al., 1975).

Like other herpesviruses, CMV persists in a latent form following primary infection and is transmissible during either primary infection, reactivation, or superinfection with a different strain. Infections with multiple strains are common in both immunocompetent (Meyer-König et al., 1998a) and immunosuppressed (Spector et al., 1984) patient populations. Following a primary infection, the virus may be shed for months or, particularly in the case of perinatal infections, years in numerous body fluids, including blood, urine, oropharyngeal secretions, breast milk, cervicovaginal secretions, and semen. Virus is commonly reactivated and shed from these sites during reactivations, facilitating transmission. However, reactivations in immunocompetent people are almost invariably asymptomatic. CMV is also sexually transmissible (Handsfield et al., 1985), with high seroprevalence rates among patients of sexually transmitted disease clinics (Shen et al., 1994) and male homosexuals (Mintz et al., 1983). The virus can also be transmitted via blood transfusion and transplantation. Since virus cannot usually be isolated from donated blood and posttransfusion CMV infection occurs in about 5% of seronegative recipients of seropositive blood, CMV presumably exists in a latent state before transfusion. CMV infection and disease following solid-organ transplantation is most commonly caused by virus from the donated organ (Ho et al., 1975; Grundy et al., 1988), whereas that following bone marrow transplantation is commonly derived from the recipient (Winston et al., 1985). Furthermore, while seronegative recipients of solid organs from seropositive donors are at high risk for posttransplant CMV disease, bone marrow transplant recipients who are seropositive and/or in receipt of bone marrow or peripheral stem cells from a seropositive donor are at high risk for CMV disease.

Symptomatic congenital CMV infection, or CID, may present with any combination of jaundice, petechial rash, hepatosplenomegaly, thrombocytopenic purpura, intrauterine growth retardation, myocarditis, microcephaly, mental retardation, spastic diplegia, pneumonitis, chorioretinitis, optic atrophy, and blindness. Death, psychomotor retardation, mental retardation, or neurologic abnormalities, dental defects, and sensorineural hearing loss are also common sequelae (Pass et al., 1980). Importantly, 10 to 25% of apparently asymptomatic congenital infections at birth result in psychomotor and/or intellectual disabilities later in life, and approximately 15% of those infected will develop hearing defects. However, computerized tomography scans of congenitally infected newborns can assist with the prognosis of adverse neurodevelopmental outcomes (Boppana et al., 1997), and neurodevelopmental or intellectual impairment are likely to be clinically evident by 12 months of age (Ivarsson et al., 1997). CMV infection may also be acquired from cervicovaginal secretions during birth or afterwards from breast milk (Stagno et al., 1980). This perinatal or neonatal transmission, however, rarely results in clinical illness, with the exception of occasional cases of pneumonitis (Brasfield et al., 1987), although transmission through breast milk to very low birth weight premature infants may be associated with a more serious outcome (Dworsky et al., 1983b; Vochem et al., 1998). Neither congenital nor perinatal CMV infection appears to adversely affect subsequent humoral immune responses to other antigens (Reynolds et al., 1978).

The majority of CMV infections in the postperinatal years in immunocompetent hosts are either asymptomatic or associated with a relatively mild self-limiting illness. The mononucleosis syndrome most commonly seen usually presents with headache, fever, lymphocytosis, malaise, lethargy, and myalgia for a few weeks. Mild hepatitis is a common complication, while pneumonitis and central nervous system (CNS) involvement in these patients is rare, and pharyngitis, lymphadenopathy, and splenomegaly are less common than in Epstein-Barr virus (EBV) mononucleosis (Klemola et al., 1970). Though relatively rare, severe CMV infections in immunocompetent adults have a poor prognosis without the early initiation of therapy (Eddleston et al., 1997). Postnatal primary CMV infection can be followed by Guillain-Barré syndrome (Boucquey et al., 1991), and CMV has been implicated in some other diseases. The worldwide distribution and high seroprevalence of the virus make etiological associations difficult. However, evidence suggests a role in malignancies (Spector and Spector, 1984; Doniger et al., 1999), insulin-dependent diabetes (Pac et al., 1988), and atherosclerosis (Persoons et al., 1994). CMV has been detected in the arterial walls of atherosclerosis patients (Hendrix et al., 1990; Melnick et al., 1994; Sorlie et al., 1994), but results of studies of the association between CMV seropositivity and coronary artery disease in immunocompetent people are conflicting (Adam et al., 1987; Zhou et al., 1996; Adler et al., 1998b; Blum et al., 1998; Rothenbacher et al., 1999). Transplant-associated arteriosclerosis, however, is the main cause of death in heart transplant recipients 12 months after transplantation (Gao et al., 1988; Olivari et al., 1989) and is increased by CMV infection (Grattan et al., 1989; McDonald et al., 1989; Loebe et al., 1990). Though the role of the virus is not clearly understood, CMV-mediated vascular damage in immunocompetent and immunosuppressed patients may be via enhancement of inflammatory responses (Vossen et al., 1996).

Significant morbidity and mortality are seen in CMV infections in immunosuppressed patient populations, notably solid-organ and cell transplant recipients, cancer patients, and human immunodeficiency virus (HIV)-

positive patients. These may be the result of either primary infection, reactivation of latent virus, or superinfection with a different strain, and patients have been shown to be symptomatically infected by more than one strain (Verbraak et al., 1998). Symptoms may present as a variety of syndromes or disease states ranging from mild fever to multiorgan damage. In immunosuppressed hosts, the virus is capable of producing a wide variety of diseases, including thrombocytopenia, retinitis, pneumonitis, gastroenteritis, hepatitis, and neurologic disease. Patients commonly experience spiking fevers, malaise, lethargy, myalgia, or arthralgia, as well as symptoms associated with specific organ damage. While there are numerous reports of the association between viral load and likelihood of progression to disease in both transplant recipients (van den Berg et al., 1991; Koskinen et al., 1993b; Niubò et al., 1996; Stéphan et al., 1997; Zaia et al., 1997; Evans et al., 1998; Mendez et al., 1998b) and HIV-positive patients (Rasmussen et al., 1997b; Spector et al., 1998, 1999; Walmsley et al., 1998), the relationship is not absolute (Bek et al., 1996; Tufail et al., 1999) and may be dependent on the pathogenic mechanism of disease at different anatomical sites. In bone marrow transplant recipients, for example, the most serious manifestation of CMV infection is pneumonia. However, this may be preceded by relatively low viral loads. The pathogenesis is thought to be immunologically mediated (Grundy et al., 1987), a theory supported by the rarity of CMV pneumonitis in both bone marrow transplant recipients prior to engraftment and AIDS patients, despite the presence of CMV in their lungs (Bower et al., 1990; Miles et al., 1990). Alternatively, high viral loads may damage cells and disrupt organ function due to direct toxicity or virus-mediated cell lysis.

CMV disease in AIDS patients is associated with an increased risk of death, and risk factors for CMV disease in these patients include older age, number of years since first diagnosis of AIDS, female gender, male homosexuality, CD4$^+$ lymphocyte counts of less than 50 per μl, zidovudine or *Pneumocystis carinii* prophylaxis prior to the development of AIDS, toxoplasmosis, and neurologic disease (Monforte et al., 1997; Saillour et al., 1998). Although less frequent in pediatric HIV disease patients, CMV infection is also a major contributor to morbidity and mortality in this patient group (Kitchen et al., 1997). While rare in transplant recipients, retinitis is the most common CMV disease manifestation in AIDS patients, possibly as a result of their extended periods of high levels of CMV viremia. The next most common sites of CMV disease in AIDS patients are the gastrointestinal tract and CNS (Ives, 1997; Saillour et al., 1997). Polyradiculomyelopathy, while an uncommon CMV manifestation in AIDS patients, is virtually unheard of in transplant recipients. CMV may interact with HIV to increase its pathogenicity. While still a controversial subject, this is supported by cross-transactivation of the HIV long terminal repeat by CMV proteins (Davis et al., 1987; Ghazal et al., 1991), coinfection of cells and organs by the two viruses (Nelson et al., 1988; Webster et al., 1995), and the increased risk of AIDS among male homosexuals with persistent CMV (Detels et al., 1994). Furthermore, CMV has been shown to code for a protein (US28) that mediates HIV-1 infection by acting as a coreceptor on cell surfaces (Pleskoff et al., 1997). It is important to note that there has been a marked decrease in the frequency of CMV disease in AIDS patients since the introduction of retrovirus protease inhibitor drugs and combination anti-HIV therapies (Brodt et al., 1997; Gerna et al., 1998d).

CMV disease in transplant recipients is associated with increased cost and length of hospital stay (McCarthy et al., 1993; Falagas et al., 1997a), is commonly accompanied by neutropenia and systemic immune suppression, and places the patient at increased risk for secondary bacterial or fungal infections (Chatterjee et al., 1978; Falagas et al., 1996; George et al., 1997). Organ disease is commonly most severe in the transplanted organ. Thus, for example, CMV pneumonitis is more common in lung or heart-lung recipients, and CMV has been associated with accelerated vascular damage and atherosclerosis after cardiac transplantation (Koskinen et al., 1993a; Dummer et al., 1994) and with hepatitis C reactivation and cirrhosis after liver transplantation (Rosen et al., 1997). CMV infection has also been associated with an increased risk of graft rejection in solid-organ recipients (Grattan et al., 1989; Keenan et al., 1991; Reinke et al., 1994; Lautenschlager et al., 1997), failed or delayed engraftment in bone marrow transplant recipients (Apperley et al., 1989), increased incidence or severity of graft-versus-host disease (Bostrom et al., 1988), and increased risk of posttransplantation lymphoproliferative disorder (Ho et al., 1985; Mañez et al., 1997). The role of CMV in vanishing-bile-duct syndrome in hepatic allografts, bronchiolitis obliterans in lung allografts, and glomerulopathy in renal allografts is still controversial (Fishman and Rubin, 1998). The determination of CMV serologic status of both donor and recipient prior to transplantation is important for identifying those patients at highest risk for severe CMV disease. Seronegative recipients of solid organs from seropositive donors and bone marrow transplant recipients where either the recipient or donor is seropositive are at highest risk of serious CMV disease manifestations. Patients undergoing treatment for graft rejection are also at increased risk of CMV disease (Hibberd et al., 1992; Bailey et al., 1993), and indeed a reduction in immunosuppressive therapy, in the absence of any signs of rejection, may be attempted in some situations to control CMV disease. CMV disease typically occurs 30 to 100 days after transplantation. CMV disease prior to engraftment in bone marrow transplant recipients, while uncommon, is associated with the presence of other opportunistic infections and has a high fatality rate (Limaye et al., 1997).

Diagnosis

CMV is fairly readily recognized histologically by its formation of typical "owl's eye" intranuclear inclusion bodies in infected cells. They have been observed in numerous tissues, including kidney, liver, lung, and gastrointestinal tract, but are less evident in the brain. Confirmation of CMV by histological assessment of biopsy or autopsy tissue is a routine procedure in the event of organ disease and may be aided by the use of immunohistological procedures (Hackman et al., 1985). Cytological assessment of cells, such as those from bronchoalveolar lavage (BAL) specimens, can also be performed, but the sensitivity is low when compared to that of virus culture (Paradis et al., 1988). Nucleic acid hybridization for the detection of CMV nucleic acid in tissue sections and cell smears has also been attempted for the detection and diagnosis of CMV. In some studies, the sensitivity of culture was found to be higher (Chou and Merigan, 1983; Gleaves et al., 1989), while others found them to be equivalent (Myerson et al., 1984). While the sensitivity of detection of nucleic acid in tissue sections is increased by the use of in situ PCR, this has not gained widespread popularity and has been overshadowed by the development of PCR assays for blood, body fluids, and tissue homogenates rather than tissue sections. CMV can also be visualized by electron microscopy, but its appearance is indistinguishable from that of other herpesviruses and the technique is not generally used for clinical diagnostic purposes.

Diagnosis of CID has classically been made by the isolation of CMV from urine or saliva within 1 week of birth. Isolation after this time will not differentiate congenital from perinatal infection. Recently, however, it has been suggested that PCR detection of CMV DNA in placental tissue may be a reliable diagnostic tool (Ozono et al., 1997). Diagnosis and prognosis of primary CMV infection during pregnancy remains a difficult issue. Transmission to the fetus in utero cannot be predicted by the presence of immunoglobulin M (IgM) antibodies in maternal serum, since not all primary infections result in a detectable IgM response and IgM responses are often seen during reactivations. In fact, less than 10% of women with detectable CMV IgM in the peripheral blood deliver congenitally infected infants (Lazzarotto et al., 1998a). Studies of the maternal peripheral blood levels of pp65 antigenemia, viremia, and leuko-DNAemia, suggest that PCR detection of CMV DNA in maternal blood leukocytes 1 month after onset of infection is the most reliable for the detection of infection. However, quantitation did not correlate with outcome for the fetus in any of the assays (Revello et al., 1998b). Reported sensitivities of shell vial culture or PCR for CMV DNA performed on amniotic fluid for the diagnosis of congenital infection range from about 70 to 100%, with culture giving lower sensitivity but higher positive predictive value (Lazzarotto et al., 1998a; Revello et al., 1998a). Repeat sampling may be necessary, however, to ensure a high negative predictive value (Revello et al., 1998a), and the repeated amniocentesis carries an associated increased risk of spontaneous abortion.

The majority of commercially available assays for CMV IgG antibodies use either viral lysate preparations or semipurified viral proteins as antibody-binding targets and have similar sensitivities. Those for detecting IgM antibodies, however, show more variation in sensitivity, depending on the antigen preparation used. In general, the necessity for detection of structural and nonstructural viral proteins in order to achieve optimal sensitivity renders those IgM assays which use viral lysates more sensitive than those that use semipurified viral-protein preparations. However, certain pooled recombinant-antigen preparations specifically designed for the purpose have also been found to be good (Landini et al., 1990; Vornhagen et al., 1994; Greijer et al., 1999) and avoid problems such as patient antibody reactivity with cellular antigens, which may produce false-positive results (Revello et al., 1986). While the detection of IgG antibodies provides confirmation of previous exposure to CMV and implies the presence of latent virus, it does not indicate that the individual is protected from reactivation of the latent virus or superinfection with a different strain. The clinical utility of CMV IgG assays is therefore quite limited. There are instances where it is appropriate to demonstrate seroconversion, particularly in pregnant women with a suspected primary infection. However, even there, more recent work would suggest that additional data from tests of antibody avidity may help clarify the prognosis (Grangeot-Keros et al., 1997; Lazzarotto et al., 1997). The detection of CMV IgM in this setting is not helpful, as less than 10% of IgM-positive pregnant women subsequently deliver a congenitally infected infant (Lazzarotto et al., 1998a). In organ transplant recipients, the delayed development of high-avidity antibody is correlated with delayed

virus clearance and an associated more risky prognosis (Lutz et al., 1994; Lazzarotto et al., 1998b). Serology, however, is generally not recommended for immunocompromised patients. They may be unable to mount a detectable humoral response or may have circulating IgG antibodies from transfusions or immunotherapy treatments. Likewise, IgM detection is not recommended for immunosuppressed patients and, since CMV IgM responses can be raised even in response to reactivations, the general utility is limited.

CMV is usually isolated for diagnostic purposes by culture in human fibroblasts. However, the appearance of CMV CPE, while typically requiring 1 to 2 weeks of incubation, can take as long as a month. An excellent advancement on this was made with the development of centrifugation-amplified culture combined with pre-CPE antigen detection. Generally known as the shell vial assay (Gleaves et al., 1984), this technique provides a result within 24 to 48 h of inoculation and can be applied to a variety of specimens. The usefulness of surveillance cultures of urine or saliva for immunosuppressed patients is controversial but is generally considered unreliable due to a lack of consistent disease association (Reinke et al., 1994; Falagas et al., 1997c; Shinkai et al., 1997). Isolation in culture, from saliva or urine specimens taken within the first week of birth, is, however, the preferred method for diagnosis of CID. While isolation of CMV from peripheral blood strongly supports a diagnosis of CMV mononucleosis, isolation from urine may be due to asymptomatic shedding and purely coincidental. Blood culture is also recommended for symptomatic pregnant women, although a negative test result does not guarantee a safe outcome for the fetus. Culture of CMV from BAL fluid for the diagnosis of pneumonia tends to have high negative but low positive predictive value (Mann et al., 1997; Uberti-Foppa et al., 1998).

Direct antigen detection for CMV diagnosis has been applied primarily to specimens of peripheral blood, cerebrospinal fluid (CSF), and BAL fluid. Direct detection of early and late antigens with pooled antibody reagents can be used to detect CMV in the cellular component of BAL specimens. The sensitivity of this procedure, however, is low, although its specificity and positive predictive value for CMV pneumonitis are probably quite high (Martin and Smith, 1986; Gleaves and Meyers, 1989). Detection of the CMV pp65 antigen by immunofluorescence on cells from the CSF has been found to be a rapid and sensitive technique for confirming CMV polyradiculomyelopathy (Revello et al., 1994; Flood et al., 1997), seen almost exclusively in AIDS patients. This condition produces very high leukocyte counts in the CSF, rendering cell smears relatively easy to prepare, and a high percentage of the leukocytes are strongly antigen positive. The detection of the same antigen in peripheral blood leukocytes (van der Bij et al., 1988) and endothelial cells (Gerna et al., 1998a), the CMV antigenemia assay, is generally considered to be superior in sensitivity to CMV culture (Boland et al., 1990; Brumback et al., 1997) and has been extensively used for confirmation of CMV infection, prognosis, and response to therapy (Baldanti et al., 1998a; van den Berg et al., 1989; Gerna et al., 1991; Boeckh et al., 1992; Francisci et al., 1997; Lipson et al., 1998). Its utility for all three is well documented, although controversy continues over appropriate cutoff points for the initiation of antiviral therapy. This is complicated by the fact that the appropriate level for treatment is variable, depending on the patient population, and, as for any test, the viral load coped with by the immune system varies from patient to patient. The most reasonable approach is to determine the likelihood of progression to disease based on data from extensive clinical trials (Baldanti et al., 1998a; Boeckh and Boivin, 1998; Gerna et al., 1998c). The maximum time between specimen collection and processing for the antigenemia assay before specimen deterioration reaches an unacceptable level is controversial (Gerna et al., 1992b; Niubò et al., 1994; Landry et al., 1995; F. J. Michalski, Letter, *J. Clin. Microbiol.* **31:**400, 1993). While shorter than that for PCR assays for CMV DNA (Schäfer et al., 1997), it may be increased by the use of leukocyte stabilization reagents (Bush and Sluchak-Carlsen, 1998). Additionally, the assay is somewhat time-consuming and does not lend itself to automation. While the pp65 antigen is localized mostly to neutrophils, antigen may also be seen in monocytes and endothelial cells, infection of the latter being generally associated with more advanced disease (Percivalle et al., 1993; Salzberger et al., 1997). The utility of the assay is hampered by specimens with low white cell counts, although this tends to be problematic only in bone marrow transplant recipients before engraftment. Since CMV disease prior to engraftment is a rare event, the clinical importance of this is probably low.

There are now numerous tests available for the detection of CMV nucleic acid. Amplification assays, primarily PCR, have been successfully applied to CSF specimens for the detection and diagnosis of CMV encephalitis (Atkins et al., 1994; Shinkai and Spector, 1995; Arribas et al., 1995; Vogel et al., 1996) and are the tests of choice for this disease. They have also been applied to many other specimens, notably those from tissue biopsies (Kjellström et al., 1997; Kusne et al., 1997) and BAL (Fajac et al., 1997; Hansen et al., 1997; de la Hoz et al., 1998) with varying results. The largest body of work, however, is on their utility for the detection of CMV in peripheral blood. Generalizations are difficult, since the sensitivities and specificities of the assays are infinitely variable due to differences in nucleic acid extraction methods, genomic targets, primer sequences, cycling conditions, and detection systems (Chou, 1992b; Levy et al., 1996; Levy et al., 1996; Hamprecht et al., 1997; Mendez et al., 1998a; Zweygberg-Wirgart et al., 1998). There are now a number of commercially available assays with target or signal amplification systems which detect either CMV DNA or RNA. For the detection of CMV DNA for clinical purposes, most leading authorities in the field agree that quantitation is important (van der Meer et al., 1996). As with antigenemia testing, the appropriate cutoff for initiating treatment varies across patient groups and therefore a simple plus-minus answer is overly simplistic (Schäfer and Laufs, 1996). Also as with antigenemia assays, higher viral loads as measured by nucleic acid assays have been found to correlate with an increased likelihood of disease (Gerna et al., 1998b; Mendez et al., 1998b; Aitken et al., 1999), and nucleic acid quantitation can be used to track response to therapy (Toyoda et al., 1997; Roberts et al., 1998). Development of CMV DNA positivity in blood early rather than later posttransplantation, has also been shown to correlate with a higher risk of the development of CMV disease (Manez et al., 1996). PCR assays for CMV DNA in blood need not be at the most sensitive end of the scale relative to the ultimate capabilities of PCR, since when applied to that degree, CMV PCR detects clinically irrelevant CMV nucleic acid from latent genetic material (Stanier et al., 1989; Bevan et al., 1991; Delgado et al., 1992; Weber et al., 1994; Zhang et al., 1995; Lo et al., 1997). In contrast, this issue is not of concern in the PCR testing of CSF, where high levels of sensitivity are desirable. Specimen deteriora-

tion with time after sample collection is also not as problematic with PCR assays as other tests for CMV (Roberts et al., 1997). The detection of CMV DNA in plasma rather than leukocytes or whole blood has been suggested in some studies to correlate with likelihood of progression to disease (Spector et al., 1992; Boeckh et al., 1997). Similarly, there have been reports of good prognostic power with the detection of mRNA transcripts by either reverse transcription (RT)-PCR or nucleic acid sequence-based amplification, particularly if late rather than early CMV gene transcripts are targeted (Gozlan et al., 1993, 1996; Nelson et al., 1996; Blok et al., 1998; Lam et al., 1998; Gerna et al., 1999). Virtually all nucleic acid amplification assays are susceptible to inhibition. The degree to which this is problematic is dependent on the type of specimen and the way it is processed. The problem, however, does occur frequently enough for the inclusion of an amplification control to be considered important in diagnostic testing.

With the development of potent antiviral drugs for CMV has come the emergence of antiviral-drug-resistant strains and the need for laboratory tests with which to detect them (Erice, 1999). All tests fall into the two broad categories of phenotypic and genotypic assays. Phenotypic assays, which give measurements of the percentage of viral growth inhibited by a drug, are generally either plaque reduction or yield reduction assays. Results from either are usually reported in terms of the 50% inhibitory concentration level, referring to the drug concentration required to inhibit virus production by 50%. The classic phenotypic test for antiviral sensitivity in CMV, the plaque reduction assay (Plummer and Benyesh-Melnick, 1964), is quite time-consuming and difficult to read and requires relatively high titers of virus to perform. More simple, rapid adaptations have therefore been developed (Telenti and Smith, 1989; Gerna et al., 1992a). In plaque reduction assays there is a decrease in plaque size as well as plaque count with increasing concentration of drug, although only the decrease in plaque count is assessed. This potential inaccuracy in the measurement of virus inhibitory power is overcome by the use of yield reduction assays, where the total progeny virus is measured rather than counting the number of plaques. A number of methods are available for progeny measurement, including plaque titration or more rapid techniques, such as viral nucleic acid hybridization (Dankner et al., 1990). However, there is always concern that the in vitro passaging required to generate sufficient virus for phenotypic testing results in the generation of a virus population with characteristics altered from those in vivo. In particular, a resistant strain representing a minority of the virus population in vivo may be overgrown by wild-type virus in vitro during passaging and not detected by subsequent phenotypic assays. Genotypic assays, in addition to giving precise information regarding the actual resistance mutation(s) present, can frequently be performed directly on patient specimens or on PCR-amplified genetic material, which more closely resembles the in vivo picture. Numerous rapid molecular assays for CMV drug resistance mutations have now been reported (Chou et al., 1995; Prix et al., 1999), but all of them detect only a handful of mutations, generally in the UL97 gene and associated with ganciclovir (GCV) resistance exclusively. Those which target the most common mutations will probably detect about 70 to 80% of the GCV-resistant mutants in clinical practice. For 100% detection, regions of the relevant genes must be sequenced. Fortunately, in the UL97 gene, these are localized to a region of approximately 450 bp. In the polymerase gene, however, the location of resistance mutations for GCV, foscarnet, and cidofovir are spread over large areas of the gene and detection requires more extensive sequencing (Erice, 1999).

Therapy

Antiviral drugs are available for both the treatment and prevention of CMV disease. While more commonly used for HSV infection, acyclovir (ACV) (Zovirax; Glaxo Wellcome) has been shown to be phosphorylated by the UL97 kinase of CMV (Zimmerman et al., 1997; Talarico et al., 1999), which explains its suppression of CMV disease in transplant recipients (Meyers et al., 1988; Balfour et al., 1989; Prentice et al., 1994) and AIDS patients (Sha et al., 1991) and the therapeutic benefit of valacyclovir, the valyl ester prodrug, for CMV in AIDS patients (Feinberg et al., 1998; Griffiths et al., 1998; Emery et al., 1999b). A nucleoside analog, ACV is monophosphorylated by viral enzymes (Elion et al., 1977; Fyfe et al., 1978) and subsequently converted to the triphosphate form by cellular enzymes. In its triphosphate form, ACV selectively inhibits viral polymerases (Elion et al., 1977; Cheng et al., 1983) and, once incorporated into replicating DNA chains, results in chain termination (McQuirt and Furman., 1982; Cheng et al., 1983). ACV, however, is not as potent in its anti-CMV activity as other anti-CMV drugs and, even when added to treatment regimens with more powerful agents in transplant recipients, does not afford any additional benefit (Green et al., 1997; Boeckh et al., 1998).

GCV(Cytovene; Roche), also a nucleoside analog and also phosphorylated by the UL97 kinase of CMV (Sullivan et al., 1992), is used for the treatment of CMV disease in preference to ACV. Its ability to inhibit CMV replication is stronger than that of ACV due to the higher levels of the triphosphate active form which are generated in CMV-infected cells. While its CMV-inhibitory powers are greater that those of ACV, it has considerably more toxicity problems, notably bone marrow suppression (Shepp et al., 1985). Phosphorylated GCV is also incorporated into growing DNA chains but does not cause complete chain growth termination (Hamzeh et al., 1990).

Few trials of GVC treatment for symptomatic congenital CMV disease have been performed, but there is some evidence of clinical benefit (Nigro et al., 1994; Halwachs et al., 1995; Stronati et al., 1995). While excretion of CMV decreases during treatment, it returns to near-pretreatment levels upon cessation of therapy (Fan-Harvard et al., 1989; Hocher et al., 1990; Reigstad et al., 1992; Attard-Montalto et al., 1993). Short-term hearing improvement has been observed in 16% of treated babies (Whitley et al., 1997), but effects on long-term clinical outcome may not be substantial (Reigstad et al., 1992).

Numerous studies with a variety of intravenous prophylactic regimens have found GCV to be successful in reducing the occurrence and severity of CMV disease in bone marrow transplant recipients (Goodrich et al., 1993; Winston et al., 1993; Atkinson et al., 1998), solid-organ transplant recipients (Winston et al., 1995; Badley et al., 1997; Peddi et al., 1997; Seu et al., 1997; Shen et al., 1997a), and AIDS patients, as well as for treating established disease (Gudnason et al., 1989; Spector et al., 1993; Gerna et al., 1997). In solid-organ transplant recipients, however, prophylactic regimens are commonly unsuccessful in the patients at highest risk for CMV disease (Merigan et al.,

1992; Martin et al., 1994), and in bone marrow transplant recipients prophylactic therapy did not result in an overall reduction in mortality, as its CMV disease-reducing benefits were countered by its bone marrow toxicity (Goodrich et al., 1993; Winston et al., 1993). Preemptive therapy with intravenous GCV, initiated following the detection of CMV infection but prior to the development of symptomatic disease, has also been successful in reducing the incidence and severity of CMV disease in bone marrow transplant recipients (Vlieger et al., 1992; Einsele et al., 1995; Koehler et al., 1997) and solid-organ transplant recipients (Hibberd et al., 1995; Grossi et al., 1996; Akposso et al., 1997; Brennan et al., 1997a; Egan et al., 1998; Yang et al., 1998) and avoids unnecessary antiviral prophylaxis for many patients. When long-term outcomes were assessed in bone marrow transplant recipients, however, prophylaxis and preemptive therapy were found to result in similar survival rates (Boeckh et al., 1996). Detection of CMV infection for successful preemptive therapy requires the use of more sensitive tests than culture, since a significant proportion of patients will develop CMV disease prior to or concurrent with the shedding of culturable virus (Goodrich et al., 1991; Einsele et al., 1995). Additionally, early treatment of CMV disease in bone marrow transplant recipients has been associated with an increased incidence of "late CMV disease" occurring more than 100 days posttransplant (Li et al., 1994).

More recent trials of oral GCV have been successful, particularly with prophylactic therapy, in both transplant recipients (Ahsan et al., 1997; Brennan et al., 1997b; Gane et al., 1997; Flechner et al., 1998; Nankivell et al., 1998; Speich et al., 1999) and AIDS patients (Drew et al., 1995; Danner and Matheron, 1996; Spector et al., 1996; Brosgart et al., 1998; The Oral Ganciclovir European and Australian Study Group, 1995). However, the very low bioavailability of the oral form of the drug (Anderson et al., 1995; Spector et al., 1995) has raised concerns regarding the potential for an increase in the emergence of drug-resistant virus (Baldanti et al., 1998b; Drew et al., 1999). The prodrug of GCV, valganciclovir, in contrast has high bioavailability when administered orally (Jung and Dorr, 1999) and shows promise in recent clinical trials as an effective alternative. As with intravenously administered GCV, oral GCV doses need to be adjusted in patients with renal failure (Pescovitz et al., 1998). GCV can also be administered locally to the eye via an intravitreal implant for the treatment of CMV retinitis (Musch et al., 1997). While effective, the device must be surgically replaced every several months and does not provide protection against CMV disease at other anatomical sites.

Foscarnet (Foscavir; Astra) is a pyrophosphate analog that inhibits polymerase function and, since it enters cells in its active form, requires no modifications by viral or cellular enzymes before producing its antiviral effect. As with GCV, foscarnet exhibits considerably more toxicity problems than ACV, notably electrolyte imbalance (Jacobson et al., 1991) and nephrotoxicity (Deray et al., 1989), the latter being alleviated by prehydration (Katlama et al., 1992; Taburet et al., 1992). It cannot be administered orally because of the resultant low bioavailability and gastrointestinal tract toxicity (Sjovall et al., 1988; Noormohamed et al., 1998). Its use for the treatment of CMV disease (Palestine et al., 1991; Aschan et al., 1992) is usually as the second drug of choice in the event of GCV failure. When used, it has been found to be effective in the treatment of established disease (Dietrich et al., 1997; Gerna et al., 1997) and in both prophylactic (Ippoliti et al., 1997) and preemptive therapy regimens (Moretti et al., 1998), and it shows better penetration of the blood-brain barrier than GCV (Hengge et al., 1993).

Another nucleoside analog, cidofovir (Vistide; Gilead), is phosphorylated by cellular enzymes and inhibits the viral DNA polymerase of CMV. It is attractive due to its long intracellular half-life, which enables treatment with weekly injections, but it is also nephrotoxic (Hitchcock et al., 1996; Polis et al., 1996) and requires probenecid and intravenous saline rehydration to minimize the risk of kidney damage (Lea and Bryson, 1996) as well as dosage adjustments for patients with impaired renal function. Cidofovir has been approved for the treatment of retinitis, where it has been shown to be of clinical benefit (Lalezari et al., 1997).

Formivirsen (Vitravene; Novartis), an antisense oligonucleotide that inhibits CMV replication, has been approved for the local treatment of CMV retinitis by intravitreal injection. It has been approved in America and Europe for use in newly diagnosed or advanced disease where other therapies are considered unsuitable or have been ineffective (Marwick, 1998) and offers the clinical benefit of absence of cross-resistance with other anti-CMV drugs.

CMV immune globulin (CMVIG) (CytoGam; MedImmune) preparations have also been used in the prevention and treatment of CMV disease. Although results of clinical trials are variable, overall, prophylactic CMVIG appears to reduce the incidence of CMV disease and death in both bone marrow and solid-organ transplant recipients (Snydman et al., 1987; Bass et al., 1993; Wittes et al., 1996; Falagas et al., 1997b; Flynn et al., 1997; Ruutu et al., 1997).

A benzimidazole riboside, 1263W94 (Glaxo Wellcome), has shown great potential as an anti-CMV therapeutic agent in clinical studies. Exhibiting negligible side effects, good bioavailability, and no in vivo emergence of resistant strains during preliminary trials, it produced a dramatic decrease in viral titer in clinical specimens from treated patients (Lalezari et al., 1998). Therapy with a CMV protease inhibitor has also been found to be effective (Casado et al., 1998), but like 1263W94, it is not licensed for use in the United States. At the time of this writing, further trials of these and other new potential anti-CMV agents have slowed following the dramatic decrease of CMV disease in AIDS patients since the introduction of retrovirus protease inhibitors and combination anti-HIV therapies (Brodt et al., 1997; Gerna et al., 1998d).

Combination therapies have also been used for the prevention and treatment of CMV disease. In general, the use of combined intravenous therapy or combined intravenous and oral therapy with more than one drug is reserved for serious infections not responding to a single drug, since these techniques tend to result in a high incidence of severe toxicity (Anders et al., 1998; Jacobson et al., 1999). However, there have been a number of encouraging studies with combinations of antiviral drugs and CMV IG (Nicol et al., 1993) and with anti-CMV drugs administered locally and systemically at the same time (Martin et al., 1999).

Prolonged antiviral therapy for CMV disease carries the risk of emergence of resistant strains. Drug-resistant CMV, first reported clinically in AIDS patients (Erice et al., 1989), has now also been reported in both bone marrow (Drobyski et al., 1991; Slavin et al., 1993) and solid-organ transplant recipients (Lurain et al., 1996; Alain et al., 1997) and in patients with other immunodeficiency disorders

(Wolf et al., 1998). The appearance of resistant CMV is more common in AIDS patients, probably as a result of the more extended periods of continual antiviral treatment. Importantly, resistant strains show no reduction in virulence from that of drug-sensitive strains. In the event of suspected emergence of drug-resistant CMV, rapid and sensitive detection is essential for appropriate adjustments to therapy. Sensitive monitoring of the viral load will often generate evidence of recurring disease in the face of therapy before the onset of symptoms and may in itself be sufficient evidence for changing treatment strategy. Since lack of response to therapy may be caused by problems such as reduced drug penetration to certain tissues, the confirmation of the presence of drug-resistant virus can only be achieved with phenotypic-susceptibility assays or the genotypic detection of resistance mutations. While data are limited, it appears that some resistance mutations may be biologically stable and remain in the in vivo virus population following the cessation of therapy while others disappear in the absence of drug pressure (Wolf et al., 1995; Erice et al., 1997). Mutations conferring GCV resistance may occur in either the UL97 gene or the polymerase (UL54) gene, although those in the UL97 gene are more common. Mutations conferring foscarnet or cidofovir resistance occur in the polymerase gene. Furthermore, some UL54 mutations are capable of producing resistance to more than one drug, and in general, strains with mutations in both the UL97 and UL54 genes are highly resistant to GCV (Erice, 1999).

Prevention

Given the worldwide ubiquitous distribution of CMV and its ease of transmission via breast milk and sexual contact, general public health prevention measures to curb its spread would appear to be somewhat futile. Similarly, regardless of basic measures such as rigorous attention to hygienic practices, it is unlikely that the high incidence of CMV infection in daycare settings can be reduced. Widespread application of safer sexual practices, advertised primarily for stemming the HIV infection rate, may assist in reducing the incidence of infection among sexually active people with multiple partners. However, it is unlikely that the overall prevalence of CMV infection will decrease, and indeed it may well increase. A more important and realistic goal is the prevention of disease, rather than infection. In some patient populations, however, the risk of serious disease from a primary infection is so high that prevention of infection remains important.

The reduction of CMV transmission via blood transfusions has classically been attempted by using blood from seronegative donors for patients at high risk for developing disease. This approach is not absolute, as there are seronegative individuals who harbor latent virus detectable by PCR (Larsson et al., 1998b). The use of leukocyte filters, which may remove CMV from blood prior to transfusion, has been proposed as a possible alternative (Smith et al., 1993; Bowden et al., 1995; Narvios et al., 1998). More recently, CD13-positive mononuclear cells have been identified as the peripheral blood site of productive CMV infection (Larsson et al., 1998a), raising the possibility of specific cell elimination prior to transfusion or bone marrow transplantation for the reduction of virus transmission. In solid-organ transplantation settings, attempts are made to reduce the risk of CMV disease by using organs from seropositive donors only for transplantation to seropositive recipients.

In practice, this is frequently not possible due to the scarcity of organ donors relative to patients awaiting transplant. In xenotransplantation settings, the use of CMV-seronegative donor animals is recommended following the recent finding of transmission of primate CMV to a human xenotransplant recipient (Michaels et al., *39th ICAAC*, 1999).

In addition to prophylactic treatments with antiviral drugs, discussed in the section on therapy above, adoptive immunotherapy has been proposed as an alternative for the prevention of CMV disease in transplant recipients. The infusion of ex vivo-expanded CMV-specific donor-derived CD8$^+$-CTL clones has been shown to be well tolerated and effective in reducing the incidence of CMV disease in high-risk bone marrow transplant recipients (Riddell et al., 1992, 1994; Walter et al., 1995), although cytotoxic activity declines in patients deficient in CD4$^+$ T helper cells.

Since primary CMV infection in pregnant women poses the greatest risk of congenital CMV in the infant, maternal immunization prior to pregnancy may protect newborns from the disease. In an effort to decrease the incidence of congenital CMV infection, suggested recipients of a CMV vaccine include toddlers, preteen children, and young unmarried mothers (Pass, 1996). Likewise, primary CMV infections in solid-organ transplant recipients carry the greatest risk of serious manifestations, and vaccination prior to transplantation may moderate the severity of subsequent CMV disease. Studies with live attenuated CMV Towne vaccine showed that the majority of male, female, and pediatric recipients develop cell-mediated immune responses which persist for 6 months and dose-dependent antibody titers comparable to those seen following natural infections (Adler et al., 1998a). Importantly, seronegative recipients of seropositive kidneys who received this vaccine not only developed humoral and cellular immunity but were protected from severe CMV disease (Adler, 1996). Combined-vaccine strategies have also been suggested with recombinant canarypox live vaccine expressing CMV gB used to prime antibody responses prior to vaccination with the attenuated Towne vaccine (Adler et al., 1999).

Circulating levels of anti-gB antibody have been shown to be inversely proportional to systemic viral load in HIV-infected patients (Alberola et al., 1998), and high titers of glycoprotein-specific antibodies correlate with the absence of viral DNA in the blood of bone marrow transplant recipients (Schoppel et al., 1998). This suggests a role for anti-glycoprotein antibodies in the prevention of CMV disease or modulation of its progression. The relatively poor immunogenicity of viral glycoproteins requires the addition of powerful adjuvants to purified gB vaccines. Preliminary trials with one such mixture have shown promising results but may require as many as four doses to generate persistent neutralizing-antibody responses (Pass et al., 1999).

VZV

History

Varicella-zoster virus (VZV) was first definitively recognized by isolation in cell culture in 1952 (Weller and Stoddard, 1952). The alphaherpesvirus (human herpesvirus type 3) was previously recognized by its characteristic giant-cell pathology related to the common disease syndromes of childhood varicella, or chickenpox, and adult-onset zoster, or shingles. The advent of an effective vaccine licensed in the United States in 1995 has greatly enhanced control of this sometimes severe or fatal viral infection.

Biology and Pathogenesis

Our knowledge of the molecular biology of VZV has greatly expanded in recent years. The virion consists of an icosahedral nucleocapsid core approximately 100 nm in diameter surrounding linear, double-stranded DNA (Cohen and Straus, 1996). The nucleocapsid is surrounded by a tegument that includes immediate-early proteins involved in transactivation of viral DNA. The tegument is in turn coated by a lipid bilayer membrane that is derived by budding of the nucleocapsid through various membranes of the host cell. The envelope contains virus-encoded glycoprotein spikes about 8 nm long that project from its surface. The intact virion is about 180 to 200 nm in diameter.

Since VZV is an enveloped virus, its infectivity is degraded by standard physicochemical treatment, including organic solvents, heating at 56°C for 30 min, and cryo-preservation for prolonged periods above −70°C.

The VZV genome of 125 kbp is relatively small for a herpesvirus. The genome has been sequenced, and it contains approximately 69 open reading frames (Davison and Scott, 1986). Most of these are homologs of genes of HSV and code for similar regulatory and structural proteins (Kinchington and Cohen, in press). Like that of HSV, the genome consists of unique long and short segments with terminal inverted-repeat regions, but it is predominantly found as two isomers rather than the four of HSV. The replication cycle of VZV for the most part follows the typical course of other herpesviruses. The virion attaches via its glycoproteins to heparan sulfate proteoglycan and as-yet-undefined receptors on the cell membrane, fusing the membranes of the virus and cell. The viral nucleocapsid then enters the cell cytoplasm and is transported to the nucleus. There, a series of immediate-early, early, and late viral genes are transcribed to form viral nonstructural proteins required for viral DNA function and structural proteins that coat newly replicated viral DNA to form nucleocapsids. Characteristic inclusion bodies are formed in the nucleus by excess viral proteins and nucleocapsids, but they also include cell proteins, such as heat shock protein 70 (Ohgitani et al., 1998). The nucleocapsids bud out of the nucleus and thereby become coated with a lipid bilayer, although this viral envelope can be acquired by budding of nucleocapsids from other cellular organelles after release from the nucleus.

As an alphaherpesvirus, VZV has the general properties of variable host range, short replication cycle, rapid dissemination in cell culture, CPE, and ability to establish latency primarily in sensory ganglia (Roizman et al., 1981). Its highly restricted host range is of particular interest because it has greatly limited studies of the pathogenesis of the virus and hindered development of appropriate treatment and vaccines. Thus, VZV replicates efficiently in vitro only in a limited variety and number of cells of human, monkey, rat, and guinea pig origin. There is no really versatile animal model for VZV infection, although some latency models have been developed in rats and guinea pigs.

There is minimal genetic variability among epidemiologically distinct isolates of VZV. There is only one serotype, and there are no distinct antigenic or clinical subtypes of VZV that could account for differences in virulence and pathogenicity.

VZV replicates in vitro in several types of human diploid cell lines, particularly fibroblasts (e.g., HF [fetal or neonatal foreskin cells], MRC-5 [a variant of HeLa continuous cervical carcinoma cells], HEL [human embryonic lung cells], PHK [primary human kidney cells], and A549 [continuous human lung carcinoma cells]) and B lymphoblastoid cell lines, and in primary cell cultures derived from CNS tissues, including astrocytes and ganglia (Gershon et al., 1995). The virus also replicates in certain nonhuman cells, including primary rhesus monkey kidney cells and guinea pig embryo fibroblasts. The virus is highly cell associated, with little extracellular virus detected and most of the virions being replication defective. The low release of virus into the extracellular environment has been related to the interaction of newly formed virions with mannose 6-phosphate receptors, with subsequent diversion of these to lysosomes in the cell (Gabel et al., 1989). This is seemingly in contrast to the extensive CPE, including giant-cell formation and apoptosis (Sadzot-Delvaux et al., 1995), induced by VZV replication.

VZV growth in vivo is characterized by tropism for respiratory epithelial cells during primary infection followed by viremic spread to lymphoreticular cells in the liver and spleen (Arvin, 1996). This is followed by a secondary viremia with transportation to cutaneous epidermal cells by infected lymphocytes and formation of characteristic vesicular skin lesions. Other host tropisms include the lung alveoli, liver parenchyma, and CNS tissues, which can result in significant CPE. This can lead to severe, disseminated disease, as well as encephalitis. It is apparent that in the intact host, virus is not highly restricted to infected cells as it is in vitro. This is particularly problematic in immunosuppressed individuals, where immune control of the virus is compromised and unable to quell reactivation of latent virus.

Replication of VZV results in cellular alterations that may affect the ability of the host to control the viral infection. Notably, the virus encodes a glycoprotein, gE, on the cell surface that resembles the Fc γRII receptor (Litwin and Grose, 1992). As with many other herpesviruses (Rinaldo, 1994; Ploegh, 1998), VZV down-regulates MHC class I expression on infected cells (Cohen, 1998). These properties may allow VZV to evade host antibody and CD8$^+$-T-cell-mediated responses, respectively.

The hallmark property of VZV replication in vivo is establishment of a permanent latent state in several types of sensory neuronal cells, including trigeminal, thoracic, vestibular, spiral, and geniculate ganglia (Mahalingam et al., 1990; Lungu et al., 1995; Dueland, 1996; Steiner, 1996; Furuta et al., 1997; Kennedy et al., 1998). Nonneuronal cells proximate to neuronal cells may also harbor latent VZV (Croen and Straus, 1991). The lack of an appropriate animal model of VZV latency and reactivation has hampered research in this area. As discussed in detail by Kinchington (1999), the latent state of VZV is unique among herpesviruses in that it expresses different RNA transcripts and is present in unusual anatomic and cellular sites compared to other herpesviruses, particularly HSV. The latent state of the virus may be controlled by several viral regulatory proteins, particularly ORF63 (Sadzot-Delvaux et al., 1997), a putative immediate-early protein of VZV. Latency may be maintained by localization of these regulatory proteins in the cytoplasm of latently infected neurons rather than their usual location in the nucleus during lytic replication of VZV (Lungu, et al., 1998). In addition, it has been suggested that host CTL reactivity to latency proteins, such as ORF63, inhibits reactivation of latent virus (Sadzot-Delvaux, et al., 1997). The mechanisms of reactivation of latent VZV have not been elucidated but presumably involve modulation of these regulatory proteins.

Reactivation of latent VZV probably occurs throughout the life of the host after primary infection but is controlled by immune mechanisms and normally remains subclinical.

Indeed, VZV can be detected by PCR amplification of viral DNA in the blood of older, asymptomatic individuals (Devlin et al., 1992; Schunemann et al., 1998) as well as immunocompromised, asymptomatic patients (Wilson et al., 1992; Mainka et al., 1998). Reactivation of latent VZV in dorsal root ganglia occurs primarily in elderly persons, presumably due to a progressively lower T-cell immune response to the virus during older age. This results in either a subclinical outcome or a clinical syndrome, including a chronically painful zoster without rash (zoster sine herpete) or overt zoster with a classic, thoracic dermatoform vesicular rash (shingles). These clinical conditions can be much more severe and life-threatening in immunocompromised hosts, such as organ and tissue transplant recipients, cancer patients, and persons infected with HIV type 1 (HIV-1).

Primary infection induces a humoral immune response to VZV, characterized by circulating IgG, IgM, and IgA that include neutralizing antibodies directed against viral glycoproteins and regulatory proteins (Arvin, 1996). The levels of IgM antibodies subside within a few months, whereas IgG antibodies persist for the life of the host and may be of importance in prevention of reinfection and reactivation of latent virus. T-cell immunity is considered to be of paramount significance in control of both primary and reactivated VZV infections. In particular, development of both CD8$^+$ HLA class I-restricted-CTL reactivity and HLA class II-restricted CD4$^+$ helper and CTL responses specific for VZV structural and regulatory proteins has been associated with control of primary viremia and latent virus. Lower levels of anti-VZV T-cell responses are associated with increased risk for severe primary and reactivated infections.

Epidemiology and Disease Manifestations

VZV infection is present throughout the world, with greater prevalence of varicella during inclement months due to the spread of virus indoors to close contacts by respiratory secretions (Whitley, 1995; Arvin, 1996; Wharton, 1996). Prior to the advent of a prophylactic vaccine, over 90% of adults in the United States were immune to VZV. The prevalence is lower in less-developed countries, particularly in tropical regions. The virus infection is spread by respiratory secretions and is therefore almost ubiquitous in children, resulting in a self-limiting, predominantly symptomatic illness that rarely recurs. Prodromal symptoms of fever and malaise occur during the prolonged, 2-week incubation period prior to outbreak of the classic maculopapular rash that progresses to a vesicular rash. The rash occurs primarily on the scalp and trunk. These vesicular lesions crust over within 1 to 2 days but do not fully heal for up to 3 weeks. Complications are uncommon in immunocompetent hosts but can include opportunistic bacterial infections of the skin lesions, thrombocytopenia, and varicella pneumonia and encephalitis. Additionally, congenital infection can be disseminated and quite severe, as can neonatal infections if the infant has a seronegative mother and therefore does not have maternal antibodies.

Immunocompromised patients, particularly children with HIV-1 infection or underlying cancers such as leukemia, may have a more severe varicella syndrome (Gershon et al., 1997; Kavaliotis et al., 1998). This is characterized by greater numbers of lesions and systemic infection of visceral tissues, with high mortality. This is a major problem in children with immunosuppression related to bone marrow transplantation (Maltezou et al., 1998), which is now being more effectively controlled through prophylactic vaccination.

Herpes zoster occurs in about 20% of nonimmunocompromised persons who have had primary VZV infection. It is a manifestation of reactivated latent VZV infection occurring predominantly in immunocompetent persons over 45 years of age. The pathology usually follows an anatomic route around the torso along the dorsal ganglia, as virus spreads cell to cell along the neurons to epithelial cells in the skin. This results in vesicular lesions similar to those of varicella. The lesions are controlled by host immunity, particularly anti-VZV CTL, and ultimately resolve within several weeks. A significant common complication of zoster is postherpetic neuralgia (Gilden, 1994; Schmader, 1998). This is a chronic, debilitating pain that begins approximately 1 month after onset of skin lesions and can persist long after the resolution of the overt signs and symptoms of zoster. The pain may be due to destruction of neurons by VZV infection. The incidence of this syndrome increases with older age. Other rare but severe complications of zoster in immunocompetent individuals include encephalitis, bullous erythema multiforme, Stevens-Johnson syndrome, Ramsey-Hunt syndrome, and ophthalmic disease (herpes zoster opthalmicus), including optic neuritis, perivasculitis and atypical necrotizing retinopathy (Garweg and Bohnke, 1997; Weismann et al., 1998; Terada et al., 1998; Murakami et el, 1998; Wenkel et al., 1998).

Immunocompromised individuals, such as adults with HIV-1 infection and certain cancers, are at risk for development of serious and sometimes life-threatening zoster (Weller, 1983; Gnann and Whitley, 1991). The incidence of zoster in HIV-1-infected men is approximately 15 times greater than in HIV-1-seronegative men (Buchbinder et al., 1992). As with varicella, the number of lesions is higher and the lesions last longer in these immunocompromised patients. Complications include pneumonia, encephalitis, and necrotizing retinopathy (Cohen and Grossman, 1989; Miller et al., 1997).

Notably, the incidence and severity of opportunistic infections in HIV-1-infected persons has significantly declined since 1997 with the advent of potent antiretroviral drug therapy (Jacobson and French, 1998). Although data are as yet incomplete, the incidence of zoster in such persons should decline along with other opportunistic infections. This is related to the low burden of HIV-1 infection in patients treated with combinations of HIV-1 protease inhibitors and reverse transcriptase inhibitors, with consequent enhancement of antimicrobial T-cell reactivity (Powderly et al., 1998).

Diagnosis

The importance of laboratory determination of VZV infection is highly dependent on the nature of the clinical illness. Thus, laboratory diagnosis of VZV infection is not recommended for the conventional, uncomplicated varicella or zoster syndromes in otherwise healthy children and adults. This is based on recognition of the classic signs and symptoms of varicella, the backdrop of an endemic of varicella in the geographic location, or known exposure to someone with varicella and the lack of indicators of underlying immune deficiency. This also is prudent management of health care costs.

Nevertheless, there are bona fide circumstances where a laboratory-based diagnosis of varicella is appropriate. These are first, the presence of immunosuppressive conditions in the patient, such as cancer and HIV-1 infection; second,

the occurrence of varicella in a neonate, a presumed immune person, or a person who has been vaccinated; third, zoster in a young person; and fourth, occurrence of significant clinical complications of VZV infection or of unusual syndromes that have been related to VZV infection. In these cases, a laboratory diagnosis of VZV infection can be essential to determining the clinical prognosis and making therapeutic decisions, i.e., when to start, stop, or change antiviral therapy.

A traditional method for determining VZV infection is to identify intranuclear inclusions and giant cells in scrapings from the base of a fresh vesicular lesion that have been stained with either Tzanck, Giemsa, or hematoxylin-and-eosin preparations (Woods and Walker, 1996). Such histological diagnoses are complicated, however, by the identical inclusion bodies formed in HSV-infected cells. This is of major significance in determining disease prognosis and proper antiviral-drug treatment regimens. Staining of cell smears with colloidal gold complexed to anti-VZV IgG has also been used to identify VZV directly in cell smears by electron microscopy (Vreeswijk et al., 1988).

The conventional method for diagnosing VZV infection has been identification of characteristic CPE in monolayer cultures of human fibroblasts grown on the side walls of glass tubes (Gershon et al., 1995). The cell lines of choice for isolation of VZV are the human diploid cell lines HF, MRC-5, and A549. This procedure, however, is fraught with problems, including the need for fresh vesicle fluids, the relatively long time to positivity (usually 1 week), and the loss of the cell cultures to microbial contamination. An improvement in the culture technique that is now the standard method in most diagnostic laboratories is inoculation of specimens into fibroblast cultures (e.g., MRC-5) that have been grown on glass coverslips in shell vials. The coverslips are fixed with acetone after 3 and 6 days of culture and stained with fluorescein isothiocyanate-conjugated monoclonal IgG antibody specific for VZV glycoprotein (Gleaves et al., 1988; West et al., 1988). Positive cells have cytoplasmic, apple green fluorescence. This procedure is more sensitive and usually more rapid for identification of VZV than standard tube cultures.

A method for even more timely diagnosis of VZV infection is direct staining of vesicular-cell smears with fluorescent monoclonal antibodies to VZV, gp98-gp62 (Gleaves et al., 1988; Coffin and Hodinka, 1995; Dahl et al., 1997). This procedure can be done within hours of obtaining the specimen. The method is highly dependent on the quality of the specimen, as too few epithelial cells is a common problem. This procedure is more sensitive than conventional tube cell culture (97.5 versus 49.4%) (Coffin and Hodinka, 1995) and is less costly, particularly in labor. Also, molecular detection of VZV DNA by in situ hybridization and PCR have been reported to be sensitive and specific methods for diagnosis of VZV infections in vesicle fluids and skin biopsies (Annunziato et al., 1996; Beards et al., 1998). Rapid differential diagnosis of zoster and HSV lesions, which is important for prognosis and treatment, has been done using PCR (Rubben et al., 1997). These molecular procedures may in time replace antigen detection methods for diagnosis of VZV infections if they can be made more cost-effective for the routine laboratory.

Detection of VZV infection in CSF directly by PCR and hybridization and indirectly by measurement of anti-VZV antibodies can be of value in diagnosis and treatment of varicella and zoster encephalitis (Puchhammer-Stockl et al., 1991; Burke et al., 1997; Koskiniemi et al., 1997; Gilden et al., 1998). A problem is that PCR also detects VZV DNA in CSF of persons with subclinical virus reactivation (Cinque et al., 1997). However, detection of VZV DNA in oropharyngeal swabs of patients with zoster sine herpete (Furuta et al., 1997) and in aqueous and vitreous humors of patients with retinitis (Danise et al., 1997; Knox et al., 1998) has been proposed as a new standard for diagnosis of these syndromes. Additionally, PCR has been used to detect VZV in amniotic fluids of mothers with incident varicella in early pregnancy (Mouly et al., 1997). This may be of benefit in diagnosing and possibly treating congenital varicella.

Serologic assays for detection of anti-VZV IgG and IgM antibodies have been in use for many years (Arvin, 1996). They are of very limited value in diagnosis due to the need for matched acute and convalescent or chronic samples, long turnaround, indirect measure of infection, and false positives for IgM. The determination of VZV antibody is important, however, when there is a question regarding the immune status of a person and particularly in immunosuppressed patients.

Therapy

Drug prophylaxis for VZV infection has improved dramatically during the last 20 years. There are now three licensed antiviral drugs effective against VZV infection (Arvin, 1996; Wutzler, 1997). These have replaced alpha interferon (Merigan et al., 1978) and vidarabine (Shepp et al., 1986) for treatment of VZV. ACV is recommended for VZV infections in persons with underlying immunodeficiencies, such as HIV-1 infection, neoplasia, and organ or tissue transplantation; neonates; and persons with severe clinical complications. The drug has greatly improved the prognosis of VZV infection in such immunocompromised children and adults but does not eliminate latent virus infection. ACV is used for both acute varicella and zoster, particularly in immunocompromised hosts. There are several variations in dosage recommended based on the disease syndrome, especially in immunocompromised patients with chronic, recurrent disease (Arvin, 1996). The drug may be most effective in quelling serious diseases, such as retinitis in immunocompromised patients, when given in combination with other anti-VZV drugs (Moorthy et al., 1997).

ACV has also been used successfully as prophylaxis for varicella in at-risk, exposed persons (Lin et al., 1997; USPHS/IDSA Prevention of Opportunistic Infections Working Group, 1997; Ogilvie, 1998; Yoshikawa et al., 1998). Resistance of VZV to ACG can occur during chronic administration through mutations in the viral thymidine kinase gene (Jacobson et al., 1990; Reusser, 1996; Fillet et al., 1998), and requires a switch to another antiviral regimen (Reusser et al., 1996).

Famciclovir (Famvir; SmithKline Beecham) is the oral prodrug form of penciclovir that has greater bioavailability against VZV infection after oral administration (Cirelli et al., 1996; Balzarini et al., 1998). Famciclovir is an acyclic guanine derivative that is rapidly transformed into the active antiviral form of penciclovir. The drug is phosphorylated by viral thymidine kinase to penciclovir monophosphate and then by cellular kinases to the triphosphate form of penciclovir. It inhibits the VZV DNA polymerase by competing with deoxyguanosine triphosphate. The drug enhances healing of acute zoster and decreases the pain of postherpetic neuralgia.

Valacyclovir (Valtrex; Glaxo Wellcome) is the hydrochloride salt of the L-valyl ester of acyclovir (Balzarini et al., 1998). This prodrug is rapidly converted to ACG and has greater bioavailability. It decreases the pain of postherpetic neuralgia.

Foscarnet, or phosphonoformic acid, is an organic analog of inorganic pyrophosphate. It inhibits herpesvirus replication at the pyrophosphate binding site of the viral DNA polymerase (Reusser, 1996). It is effective for treatment of ACV-resistant VZV. Foscarnet-resistant VZV has been reported, with the likely mutations conferring this being in the viral DNA polymerase (Visse et al., 1998).

VZV immune globulin (VZIG) (American Red Cross) is a high-titer, anti-VZV antibody preparation used as postexposure prophylaxis to prevent VZV infection in certain patients at high risk of serious disease (USPHS/IDSA Prevention of Opportunistic Infections Working Group, 1997; Desjardin and Snydman, 1998).

The recent crystallization of protease of VZV (Qiu et al., 1997) offers a new target for the therapeutic intervention of VZV infection. Postulated active sites of the viral protease may be exploited to develop highly specific and inhibitory antiviral agents.

Prevention

In 1995, the first human herpesvirus vaccine, Varivax (Oka; Merck) was licensed for use in the United States (Arvin and Gershon, 1996; Jennings et al., 1998). This vaccine was originally isolated by Takahashi and colleagues in the mid-1970s from a child with varicella and then passaged in human and guinea pig cells to achieve a live attenuated preparation (Takahashi et al., 1974). Trials in the United States indicated that the Merck vaccine is safe, induces a long-lasting T-cell immunity to VZV (Gershon et al., 1989; Zerboni et al., 1998), is highly effective in preventing primary varicella in healthy children (Weibel et al., 1984), and decreases the incidence of zoster in leukemic children (Lawrence et al., 1988; Hardy et al., 1991). The vaccine is now recommended for use for prevention of varicella in the United States in children 1 year of age or older and in all susceptible healthcare workers (Lyznicki et al., 1998).

The Oka-Merck vaccine has also been used as a therapeutic regimen to prevent or at least alleviate the severity of complications of VZV infection (Oxman, 1995). This is based on the correlation of decreased VZV-specific T-cell immunity with increased risk of zoster (Arvin, 1996). As proof of efficacy of this approach, cellular and humoral immune responses to VZV have been boosted by vaccination of elderly patients with the Oka vaccine (Berger et al., 1998; Levin et al., 1998). The risk for zoster appears to be decreased in leukemic children either by exposure to natural varicella or by vaccination with the Oka vaccine (Gershon et al., 1996). Additionally, administration of a heat-killed VZV vaccine has reduced the severity of zoster in leukemic children after bone marrow transplantation (Redman et al., 1997).

There are concerns about the Oka-Merck vaccine, however. The vaccine can cause disease, particularly in immunosuppressed persons (Larussa et al., 1998). The vaccine strain has also been transmitted to healthy household contacts, resulting in overt skin lesions (Larussa et al., 1997). Breakthrough infections have been noted in Japan (Takayama et al., 1997). It is important that evidence so far suggests that, even though the vaccine establishes latency, it does not commonly reactivate and cause zoster (Arvin, 1996).

EBV

Historical Perspective

EBV was discovered over 3 decades ago during a search for the etiology of Burkitt's lymphoma (BL) affecting children in sub-Saharan Africa. The high incidence of lymphomas was initially described by Denis Burkitt, who suspected the involvement of an infectious, possibly arthropod-borne, agent (Burkitt, 1958, 1963). Cells obtained by Burkitt from the tumors of Ugandan children were sent via air freight to the laboratory of Epstein and Barr at Middlesex Hospital in London. After surviving for several days at ambient temperatures, some of the specimens contained viable lymphoblasts which grew in culture and harbored virus particles detected by electron microscopy (Epstein et al., 1964). Cultures of BL cells making virus were then found to contain antigens detectable with sera from BL patients (Henle and Henle, 1966). This provided the basis for a seroepidemiological study that led to the discovery that the Epstein-Barr virus was the causative agent of infectious mononucleosis (IM) (Henle et al., 1968b; Niederman et al., 1970). Ironically, the first clue to the connection between EBV and IM came from serum specimens obtained from a research assistant in the laboratory of Walter and Gertrude Henle, who had been conducting the seroepidemiological surveys. Sera obtained from the assistant before a classic episode of IM were negative, while sera obtained immediately after IM were positive (Henle et al., 1968a).

Initial efforts to grow the virus by cocultivation of irradiated BL cells with human neonatal lymphocytes led to the next surprising discovery (Henle et al., 1967; Miller et al., 1969). Instead of causing cytopathology, the virus induced the neonatal cells to proliferate indefinitely in vitro. The efficiency of the process was so high that a limiting-dilution endpoint assay for immortalization of B cells became the standard assay for measuring the titer of biologically active virus in a culture supernatant. Research over the last 3 decades has focused on the immortalizing phenotype of EBV principally because this activity has been proposed as the mechanism underlying the pathogenesis of the lymphoproliferative diseases with which the virus has come to be associated.

Biology and Pathogenesis

EBV (human herpesvirus type 4) is the prototype for the *Lymphocryptovirus* subgroup of the subfamily *Gammaherpesvirinae* (Roizman, 1996). The strict tropism for B lymphocytes is based upon the ability of the gp340 major viral envelope glycoprotein to bind the CD21 surface marker which acts as the virus receptor (Nemerow et al., 1985). In vitro virus-mediated cross-linking of CD21 serves to trigger an initial activating signal that leads to blast transformation, homotypic cell-cell adhesion, expression of B-cell activation antigens, and production of cytokines (Gordon et al., 1986; Tanner et al., 1987, 1996). Within 3 to 4 h of attachment, the linear 172-kb DNA genome is delivered to the nucleus and coated with host cell histones. All the biological and molecular analyses of the events following the infection of B cells suggest that the virus is predisposed to initiate the immortalization program. Virus-driven lymphoproliferation is characterized by the expression of at least 10 viral genes. Six Epstein-Barr nuclear antigens (EBNAs) have been identified (EBNA1, EBNA2, EBNA3A, EBNA3B, EBNA3C, and EBNA-LP), along with two latent membrane proteins (LMP1 and LMP2) and two short RNAs (EBER1 and EBER2) with extensive putative secondary structure (Kieff, 1996). Of these genes, six (EBNA1, EBNA2, EBNA3A, EBNA3C, EBNA-LP, and LMP1 genes) have been implicated directly in the process of immortalization by inhibition with antisense oligonucleotides (Roth et al., 1994) or by knockout recombinational analyses (Hammerschmidt and Sugden, 1989; Cohen et al., 1991; Khanna et al., 1992; Tomkinson and Kieff, 1992; Kaye et al., 1993; Tomkinson et al., 1993a, 1993b).

In many respects, EBV-immortalized cells look and behave like B cells that are stimulated to proliferate by the cross-linking of cell surface CD40 and the addition of IL-4 to the culture medium (Gordon et al., 1986). In addition to CD19 and CD20 cell surface markers, immortalized cells express CD23, CD30, CD39, CD70, B7, CD58, and CD54 (Gerber et al., 1972; Gordon et al., 1984, 1986; Greenspan et al., 1985, 1990; Gratama et al., 1988, 1994; Hamilton-Dutoit et al., 1991). The density and serum dependence are suggestive of autocrine feedback loops operating to sustain and enhance growth (Gordon et al., 1984). All cells in a cultured cell line express the viral immortalization proteins and maintain the viral genome as a multicopy episome (Sugden and Mark, 1977; Sugden et al., 1979). Despite the obvious alterations that accompany virus immortalization of B cells, and the potential disease implications, this virus-cell interaction has been described as "latent" because most of the cells in a lymphoblastoid cell line produce no detectable virus. For most virus-immortalized cell lines, chemical or physical agents that induce B cells to differentiate towards a plasma cell phenotype will trigger the virus into a cell-lytic, virion-productive phase.

The characteristics of latently infected cells found in the blood of human hosts are actually quite different from immortalized cells. The existence of latently infected cells in the blood and lymphoid tissues has been known for decades principally because explanted tissues rich in lymphocytes spontaneously produced lymphoblastoid cell lines with the features of EBV-immortalized B cells (Pope and Scott, 1968; Nilsson et al., 1971; Rocchi et al., 1977; Rickinson and Epstein, 1985; Yao et al., 1985; Lewin et al., 1987). That latent infection occurs in the lymphoid compartment was originally demonstrated by two key pieces of evidence. (i) During prolonged ACG treatment, virus shedding in the oropharynx ceases; however, the frequency of establishment of spontaneous immortalized B-cell lines from the peripheral blood remains unchanged. (ii) Analyses of EBV strains recovered from bone marrow transplant recipients indicates that existing latent infection can be eradicated from seropositive recipients during replacement of their hematopoietic cell lineages with bone marrow from seronegative donors (Gratama et al., 1988; Yao et al., 1989). The presence and distribution of EBV-positive cells in lymphoid tissues has been examined by in situ hybridization with EBER probes (Howe and Steitz, 1986; Weiss and Movahed, 1989; Karlin et al., 1990; Strickler et al., 1993; Anagnostopoulos et al., 1995; Reynolds et al., 1995; Niedobitek et al., 1997a, 1997b). In one study, 4 of 12 normal lymph nodes contained small EBER-positive cells localized to extrafollicular areas at a frequency of approximately 1 to 10/0.5 cm^2 (Niedobitek et al., 1997a). This was contrasted with acute-stage-mononucleosis lymph nodes, where the distribution was similar but the frequency was >1,000/0.5 cm^2 and large positive cells were mixed in with the small positive cells.

The development of PCR has allowed the detection of viral DNA from latently infected cells directly and sensitively. Pioneering work in this area used PCR to detect viral DNA in tissue specimens obtained at autopsy from patients with no indication of EBV-related disease (Cheung and Dosch, 1991). Viral genomes were readily detectable in the parotid gland (7 of 15), submandibular gland (8 of 10), nasopharynx (8 of 10), tonsil (8 of 10), larynx (5 of 6), lung (5 of 9), cervical lymph node (7 of 10), mediastinal lymph node (7 of 10), abdominal lymph node (4 of 10), spleen (6 of 10), and kidney (4 of 10). Less frequent detection was reported for the liver (1 of 10), pancreas (1 of 4), ovary (1 of 5), uterine cervix (1 of 4), and testis (1 of 3). The frequency of positive cells in these tissues was not investigated. Detection and quantitation of latently infected B cells in the peripheral blood of healthy carriers has also been studied (Thorley-Lawson and Israelsohn, 1987; Tosato et al., 1990; Miyashita et al., 1995, 1997). In one recent study, the frequency of EBV-positive cells in nine carriers ranged from 23 to 625/10^7 B cells, with a mean of 125 (Miyashita et al., 1997). For a given individual, the viral load was relatively stable over a period of at least 2 years. The number of genomes per cell was estimated to be less than five. By presorting peripheral blood lymphocytes based upon cell surface markers, infected cells were detected among the CD19$^+$, IgD$^-$, CD23$^-$, CD80$^-$, and Ki67$^-$ population, consistent with the view that latency was in resting nonactivated memory B cells (Miyashita et al., 1997; Babcock et al., 1998). Recent estimates comparing PCR-measured frequency of latently infected cells directly to the spontaneous-outgrowth frequency revealed that spontaneous reactivation was fivefold less than expected (Babcock et al., 1998). This suggests that in vitro reactivation from latency may not be a very efficient process and/or that the latently infected population may be heterogeneous, containing a large fraction of cells incapable of producing virus.

Another application of the PCR techniques is for the detection of RNA expression (RT-PCR). Several groups have reported results of RT-PCR assays on RNA taken directly from peripheral blood lymphocytes of healthy carriers (Qu and Rowe, 1992; Tierney et al., 1994; Chen et al., 1995; Miyashita et al., 1997). The different studies agree on some important points. First, the messages for key proteins involved in immortalization are not detected (e.g., LMP1 and EBNA2). Second, all of the studies detected LMP2a expression (16 of 18 carriers in total). Third, all of the studies detected EBNA transcription unit promoter activity, although there was no consensus on which promoter was active or whether any EBNA genes were expressed. In the most recent reports, LMP2a RNA was readily detectable, while EBNA messages were only occasionally detected (Miyashita et al., 1997; Babcock et al., 1998). Latency, therefore, most likely occurs in resting B cells, where very few copies (<5) of the viral episome express the LMP2a message (Miyashita et al., 1997).

Epidemiology

The virus is transmitted primarily in the saliva, and early studies demonstrated the presence of biologically active virus in the salivary secretions of 15 to 20% of healthy asymptomatic carriers (Gerber et al., 1972; Chang et al., 1973; Evans, 1982; Strauch et al., 1987). Although examples of transmission through blood transfusion and along with the donor organ during transplantation have been documented, these are considered to be atypical routes of transmission. The age of acquisition varies depending upon geographic location, being much earlier in underdeveloped nations (>90% of preschoolers) than in Western Europe and America (30 to 40% of preschoolers). Socioeconomic status within developed nations also seems to influence transmission, as a greater percentage of poor children are seropositive in age-matched studies (Evans, 1982).

Two strains of EBV have been recognized (EBV-1 and EBV-2), and seroepidemiological and virus isolation studies suggest that most people in Europe and North America are infected with the EBV-1 strain (Yao et al., 1991; Gratama et al., 1994; Rickinson and Kieff, 1996). A much smaller percentage (<10%) is infected with the EBV-2 virus. Double infections have recently been described on the basis of extremely sensitive PCR assays of oropharyngeal secretions. In these studies, 10% of normal donors were shedding both types of viral DNA (Sixbey et al., 1989). Infections occur even in remote isolated tribes in the Amazon basin and New Guinea highlands. Viruses recovered from these sites show nucleic acid polymorphisms that genetically distance them from both EBV-1 and EBV-2, suggesting that the virus can persist and evolve along with even very small groups of humans (Lung et al., 1990, 1991; Abdul-Hamid et al., 1992).

IM is not a reportable disease, so incidence statistics are not maintained. In one study, the reported rate in Connecticut was 48 cases/10^5 population/year for the general population, while a rate of 840 cases/10^5 population/year was recorded among college students, the group at greatest risk (Miller, 1990). No true epidemics of EBV infection have been described, although occasional clusters of cases have been reported. Recent advances in molecular typing of viruses have allowed transmission of virus from donor to recipient in bone marrow and solid-organ transplantation to be documented.

Clinical Manifestations

IM

Most infections with EBV occur in childhood and are clinically inapparent (Miller, 1990; Rickinson and Kieff, 1996). Symptomatic infection in adolescents and adults is infrequent but can cause a recognizable clinical entity (IM). The incubation period is between 30 and 50 days, followed by sore throat, headache, malaise, and fatigue. Fever lasts about 10 days. General lymphadenopathy involving the cervical, axillary, epitrochlear, inguinal, mediastinal, and mesenteric nodes is observed. Splenomegaly occurs in approximately 50% of cases, and hepatosplenomegaly occurs in approximately 10% of cases. Since EBV replication can be detected at high rates in the oropharynx during the symptomatic phase, trials with oral ACG have been conducted (Andersson et al., 1987). While the drug did block virus shedding in the oropharynx, it did not alter the disease course or change the load of virus-infected cells in the blood (Andersson et al., 1987; Yao et al., 1989). Thus, it has been suggested that the disease symptoms are not caused by virus replication. An immunopathological etiology for the symptoms is suspected because they are precisely correlated with the period of atypical lymphocytosis caused by the exaggerated T-cell response and with the heavy infiltration of affected organs by these reactive T cells. The final consequence of EBV infection is a lifelong carrier state characterized by episodic asymptomatic shedding of virus into saliva, persistent low IgG antibody titers to EBNA1, a high frequency of circulating CTL precursors to EBV-infected B cells, and a corps of latently infected B cells that escape immunosurveillance (Miller, 1990; Rickinson and Kieff, 1996).

LD

EBV infection in the immunocompromised host is accompanied by the risk of developing lymphoproliferative disease (LD). Posttransplantation LDs (PT-LDs) occur in approximately 1% of renal transplants (Nalesnik et al., 1988; Wilkinson et al., 1989; Preiksaitis et al., 1992; Cockfield et al., 1993; Opelz and Henderson, 1993; Leblond et al., 1995), 2% of liver transplants (Nalesnik et al., 1988; Opelz et al., 1993; Morgan and Superina, 1994), 6% of pancreas transplants (Nalesnik et al., 1988; Wilkinson et al., 1989; Stratta et al., 1994; Davis et al., 1995; Lumbreras et al., 1995), 2 to 10% of heart transplants (Swinnen et al., 1990; Preiksaitis et al., 1992; Chen et al., 1993; Opelz et al., 1993; Leblond et al., 1994), 4 to 10% of heart-lung and lung transplants (Nalesnik et al., 1988; Armitage et al., 1991; Preiksaitis et al., 1992; Leblond et al., 1995; Aris et al., 1996; Wood et al., 1996), 20% of small-bowel transplants (Kocoshis, 1994), and 23% of multivisceral and 2% of bone marrow transplants (Deeg et al., 1984; Kocoshis, 1994; Todo et al., 1995). Recent estimates indicate that 87% of PT-LDs are of B-cell origin, 12.5% are of T-cell origin, and 0.5% are of null-cell origin. For the B-cell PT-LDs, the clinical and pathological features vary across a spectrum from polyclonal B-cell hyperplasia to monoclonal immunoblastic lymphoma (Nalesnik, 1998). The current view is that the disease is progressive. Early stages have the characteristics of nonspecific reactive hyperplasia but can quickly evolve into a more diffuse polymorphic hyperplasia and finally into a disease with lesions resembling immunoblastic lymphoma or multiple myeloma (Frezzera et al., 1981; Nalesnik et al., 1988; Knowles et al., 1995; Harris et al., 1997). Pathologists have had difficulty delineating specific stages by inspection of diseased-tissue biopsy specimens, and this has led to an inability to reliably determine the clinical behavior of the disease based on histopathological assessment. Complicating the diagnosis is the observation that, simultaneously, different lesions within the same patient may present the features of different stages of disease.

HIV-infected individuals are also at risk for developing LDs, which have been divided morphologically into two broad categories, the BL-like lymphomas and the immunoblastic lymphomas (Hamilton-Dutoit et al., 1991). The BL-like lymphomas generally appear early in the progression of HIV disease towards AIDS, while the immunoblastic lymphomas are more likely to be diagnosed during the severely immunocompromised end stages of AIDS. The latter tumors are similar in most respects to the LDs in transplant recipients with the exception that primary CNS presentation (rare in transplant recipients) is common in AIDS lymphomas.

OHL

Another disease manifestation of EBV infection that is a consequence of immunosuppression is oral hairy leuko-plakia (OHL), characterized by a thickening of the epithelium in the lateral borders of the tongue (Greenspan et al., 1985). Histopathologically, there is a greatly disturbed basal-to-apical organization of the squamous epithelium that is correlated with the presence of EBV in the affected cells. The lytic-cycle marker BZLF1 is present, but expression is restricted to the apical regions of the lesion. This kind of evidence has been used to suggest that during the normal course of virus shedding, the virus infects epithelial cells and replicates in the terminally differentiated cells in a manner analogous to the replication of papillomaviruses (Sixbey et al., 1984). Because OHL is primarily a lytic-virus lesion, it responds well to antiviral therapy (Resnick et al., 1988; Greenspan, 1990).

Virus-Associated Malignancies

The endemic African form of BL described by Burkitt shows a nearly 100% correlation with EBV. All the cells of the tumor are virus DNA and EBNA1 positive (Neri et al., 1991). Histologically, all these tumors are dominated by a homogeneous population of small, noncleaved malignant B cells interspersed with occasional macrophages that give the specimen its characteristic "starry sky" appearance (Magrath, 1990). The nonendemic form of BL has only a 30 to 40% association with EBV. The current view of the pathogenesis of BL is that it is a multistep process involving environmental factors (that enhance germinal-center activity), genetic alterations (the deregulation of the c-myc oncogene through an 8:14 chromosome translocation), EBV infection, and other factors and events not as well characterized. Thus, EBV infection contributes to but does not cause BL (Rickinson and Kieff, 1996).

The other well-documented EBV-associated tumor is nasopharyngeal carcinoma (NPC). The link between EBV and NPC was first suggested by elevated anti-EBV antibody titers in NPC patients (Old et al., 1966; Henle et al., 1970a). The presence of virus DNA in all cases of World Health Organization type II and III undifferentiated NPC and the monoclonal nature of the EBV episome in each case has suggested a strong causal association (Raab-Traub and Flynn, 1986). The exact mechanism by which the virus contributes to NPC, a disease with marked demographic patterns of incidence, is not known. As with BL, the contribution of other factors, both genetic and environmental, suggests a multistep process in which the virus is just one player.

Diagnosis

The clinical differential diagnosis of EBV is difficult, especially in a complicated case or before the full-blown syndrome is evident. Mononucleosis must be distinguished from a variety of other infectious diseases that cause fever and inflammation of the upper respiratory tract. Serologic tests for the presence of antibodies to EBV are the principal means of confirming a diagnosis. The seroepidemiology of EBV infection was accomplished with a series of fluorescence assays developed during the first decade of research on the virus (Henle et al., 1968a, 1970b; Reedman and Klein, 1973). These measure three classical antigens: EBNA, EA (early antigen), and VCA (viral capsid antigen). A number of enzyme-linked immunosorbent assay-based assays and other easy-to-read antibody tests based on specific peptide epitopes have been introduced (Pearson and Luka, 1986; Henle et al., 1987; Hille et al., 1993; Van Grunsven et al., 1994). None has gained widespread acceptance, and all need to be interpreted in terms of the original fluorescence assays, since virtually all of the information available on the primary antibody response to EBV infection relied on these tests.

The diagram in Fig. 1 summarizes the kinetics of appearance and persistence of antibodies to EBV following primary infection of an immunocompetent host. Some tests detect IgM antibodies to VCA that increase very soon after infection and are usually substantial by the time symptoms appear. IgG anti-EA and IgG anti-VCA titers peak during the acute phase of IM, lagging slightly behind the atypical lymphocytosis and correlating well with the period of maximum virus shedding. The anti-EA responses drop faster and become undetectable during convalescence. Persistently elevated anti-EA responses have been interpreted as evidence of active virus infection. The appearance of an anti-EBNA response is delayed and usually occurs late in infection during the convalescent phase. Patients with severe chronic active EBV infection, fatal mononucleosis, or a disseminated polyclonal LD often fail to develop any detectable anti-EBNA response (Henle and Henle, 1979; Rickinson and Kieff, 1996).

In addition to EBV-specific antibody responses, IM patients develop a nonspecific rise in the levels of serum immunoglobulins and the appearance of so-called heterophile antibodies, mostly of the IgM class, with autoreactive specificities. These antibodies appear very early in the course of infection and may be the products of virus-infected cells. They have the property of being able to agglutinate or lyse sheep, goat, horse, and bovine red blood cells and form the basis of the Paul-Bunnell or "spot" test for IM. The test may not be positive at first and may have to be repeated after 3 to 4 weeks, and it may remain negative in young children even as EBV-specific serology confirms that an infection is due to EBV.

The definitive diagnosis of an LD requires histological examination of a tissue biopsy specimen, and in situ hybridization with an EBV-specific EBER RNA probe unambiguously defines the involvement of virus-infected cells (Nalesnik, 1998). Because of the immunosuppressed state of the patient and the use of prophylactic pooled immunoglobulin therapy, EBV serology is unreliable. Recently, several semiquantitative and quantitative PCR tests for the detection of viral DNA in the peripheral blood have been developed (Riddler et al., 1994; Savoie et al., 1994; Kenagy et al., 1995; Rooney et al., 1995b; Rowe et al., 1997). These tests have shown that the circulating viral load can be used as a diagnostic marker for LD. Current efforts are being directed towards determining whether preemptive steps can avert a PT-LD when a rising viral load is detected early after transplantation.

Therapy

For IM, the use of the antiviral compounds ACV and GCV, as discussed above, is not effective. Therapy is essentially palliative. Bed rest is helpful and may be necessary in severe cases. Analgesics and antipyretics can provide relief from sore throat pain and fever. Patients with enlarged spleens should avoid contact sports for 3 to 6 months following recovery. Approximately 30% of IM patients experience a secondary pharyngeal infection with group A streptococci that may require antibiotic treatment. Ampicillin is not recommended because it usually causes a rash in IM patients (Miller, 1990).

For LDs, the picture is more complex. The disease is actually being caused by the underlying immunosuppression. If the immunosuppression can be reduced, for instance, in PT-LD in transplant recipients, then recovering T-cell immunity can eliminate the virus-driven LD. A number of clinical studies have reported that transplant recipients given antiviral agents, principally as prophylaxis for posttransplantation CMV infection, have had fewer PT-LDs than expected from historical experience with similar recipients (Stratta et al., 1992; Davis et al., 1995; Lumbreras et al., 1995; Winston et al., 1995; Badley et al., 1997; Darenkov et al., 1997; Gane et al., 1997; Keay et al., 1998). This body of evidence has been balanced by an equal number of reports showing that prophylactic antiviral therapy has not prevented PT-LD in nonrandomized trials (Shapiro et al., 1988; Zutter et al., 1988; Walker et al., 1995; Aris et al., 1996; Haque et al., 1996; Newell et al., 1996; Wood et al., 1996). Some of the studies have supplemented antivirals with various doses of intravenous immunoglobulin, while others have not included any immunoglobulin. The value of supplemental immunoglobulin in prophylaxis for EBV infections has not been determined.

Recently, experimental therapies employing adoptive transfer of donor-derived EBV-specific CTLs to allogeneic bone marrow transplant recipients have been tried (Heslop et al., 1994, 1996; Papadopoulos et al., 1994; Rooney et al., 1995a, 1995b; Bonini et al., 1996). In one series, five patients with PT-LD were treated with infusions of donor peripheral blood mononuclear cells, achieving complete and sustained remissions (Papadopoulos et al., 1994). The further development of strategies for growing EBV-specific CD8+ HLA class I-restricted CTLs in vitro will soon permit expanded trials in solid-organ transplant recipients and other immunocompromised patients. These adoptive immunotherapeutic strategies should also prove useful in treating EBV-associated malignancies.

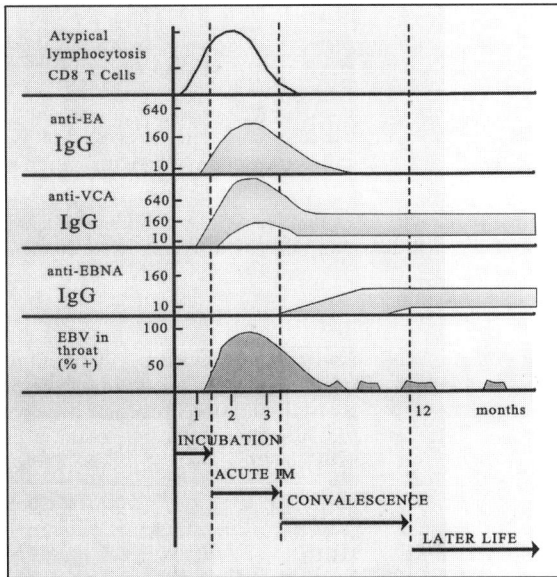

FIGURE 1 Antibody responses to EBV antigens during the course of IM are plotted relative to the period of atypical lymphocytosis and virus shedding. Reciprocal titers are shown for the virus-specific IgG class of antibody detected by immunofluorescence assays for the classical EA, VCA, and EBNA.

REFERENCES

Abdul-Hamid, M., J. J. Chen, N. Constantine, M. Massoud, and N. Raab-Traub. 1992. EBV strain variation: geographical distribution and relation to disease state. *Virology* **190:**168–175.

Abrahamsen, L. H., M. J. Clay, J. M. Lyle, J. M. Zink, L. J. Fredrikson, A. J. DeSiervo, and M. A. Jerkofsky. 1996. The effects of cytomegalovirus infection on polar lipids and neutral lipids in cultured human cells. *Intervirology* **39:**223–229.

Adam, E., J. L. Melnick, J. L. Probtsfield, B. L. Petrie, J. Burek, K. R. Bailey, C. H. McCollum, and M. E. DeBakey. 1987. High levels of cytomegalovirus antibody in patients requiring vascular surgery for atherosclerosis. *Lancet* **ii:**291–293.

Adler, S. P. 1996. Current prospects for immunization against cytomegaloviral disease. *Infect. Agents Dis.* **5:**29–35.

Adler, S. P., S. H. Hempfling, S. E. Starr, S. A. Plotkin, and S. Riddell. 1998a. Safety and immunogenicity of the Towne strain cytomegalovirus vaccine. *Pediatr. Infect. Dis. J.* **17:**200–206.

Adler, S. P., J. Kyun Her, J. Ben Wang, and G. W. Vetrovec. 1998b. Prior infection with cytomegalovirus is not a major risk factor for angiographically demonstrated coronary artery atherosclerosis. *J. Infect. Dis.* **177:**209–212.

Adler, S. P., S. A. Plotkin, E. Gonczol, M. Cadoz, C. Meric, J. B. Wang, P. Dellamonica, A. M. Best, J. Zahradnik, S. Pincus, K. Berencsi, W. I. Cox, and Z. Gyulai. 1999. A canarypox vector expressing cytomegalovirus (CMV) glycoprotein B primes for antibody responses to a live attenuated CMV vaccine (Towne). *J. Infect. Dis.* **180:**843–846.

Ahn, K., A. Gruhler, B. Galocha, T. R. Jones, E. J. Wiertz, H. L. Ploegh, P. A. Peterson, Y. Yang, and K. Fruh. 1997. The ER-luminal domain of the HCMV glycoprotein US6 inhibits peptide translocation by TAP. *Immunity* **6:**613–621.

Ahsan, N., M. J. Holman, and H. C. Yang. 1997. Efficacy of oral ganciclovir in prevention of cytomegalovirus infection in post-kidney transplant patients. *Clin. Transplant.* **11:**633–639.

Aitken, C., W. Barrett-Muir, C. Millar, K. Templeton, J. Thomas, F. Sheridan, D. Jeffries, M. Yaqoob, and J. Breuer. 1999. Use of molecular assays in diagnosis and monitoring of cytomegalovirus disease following renal transplantation. *J. Clin. Microbiol.* **37:**2804–2807.

Akposso, K., E. Rondeau, J.-P. Haymann, M.-N. Peraldi, C. Marlin, and J.-D. Sraer. 1997. Long-term prognosis of renal transplantation after preemptive treatment of cytomegalovirus infection. *Transplantation* **63:**974–976.

Alain, S., P. Honderlick, D. Grenet, M. Stern, C. Vadam, M. Sanson-Le Pors, and M. Mazeron. 1997. Failure of ganciclovir treatment associated with selection of a ganciclovir-resistant cytomegalovirus strain in a lung transplant recipient. *Transplantation* **63:**1533–1536.

Alberola, J., V. Dominguez, L. Cardenoso, J. Lopez-Aldeguer, M. Blanes, F. Estelles, C. Ricart, A. Pastor, R. Igual, and D. Navarro. 1998. Antibody response to human cytomegalovirus (HCMV) glycoprotein B (gB) in AIDS patients with HCMV end-organ disease. *J. Med. Virol.* **55:**272–280.

Albrecht, T., and F. Rapp. 1973. Malignant transformation of hamster embryo fibroblasts exposed to ultraviolet-irradiated human cytomegalovirus. *Virology* **55:**53–61.

Anagnostopoulos, I., H. Stein, C. Kreschel, and M. Hummel. 1995. Morphology, immunophenotype, and distribution of latently and/or productively infected cells in acute infectious mononucleosis: implications for the inter-individual infection route of EBV. *Blood* **85:**744–750.

Anders, D. G., and L. A. McCue. 1996. The human cytomegalovirus genes and proteins required for DNA synthesis. *Intervirology* **39:**378–388.

Anders, H. J., N. Weiss, J. A. Bogner, and F. D. Goebel. 1998. Ganciclovir and foscarnet efficacy in AIDS-related CMV polyradiculopathy. *J. Infect.* **36:**29–33.

Anderson, R. D., K. G. Griffy, D. Jung, A. Dorr, J. D. Hulse, and R. B. Smith. 1995. Ganciclovir absolute bioavailability and steady-state pharmacokinetics after oral administration of two 3000mg/d dosing regimens in human immunodeficiency virus-seropositive patients. *Clin. Ther.* **17:**425–432.

Andersson, J., B. Skoldenberg, I. Emberg, S. Britten, W. Henle, and U. Anderson. 1987. Acyclovir treatment in primary Epstein-Barr virus infection. A double-blind placebo-controlled study. *Scand. J. Infect. Dis. Suppl.* **48:**107–115.

Annunziato, P., O. Lungu, A. Gershon, D. N. Silvers, P. Larussa, and S. J. Silverstein. 1996. *In situ* hybridization detection of varicella zoster virus in paraffin-embedded skin biopsy samples. *Clin. Diagn. Virol.* **7:**69–76.

Apperley, J. F., C. Dowding, J. Hibbin, J. Buiter, E. Matutes, P. J. Sissons, M. Gordon, and J. M. Goldman. 1989. The effect of cytomegalovirus on hemopoiesis: in vitro evidence for selective infection of marrow stromal cells. *Exp. Hematol.* **17:**38–45.

Aris, R. M., D. M. Maia, I. P. Neuringer, K. Gott, S. Kiley, K. Gertis, and J. Handy. 1996. Post-transplantation lymphoproliferative disorder in the Epstein-Barr virus-naive lung transplant recipient. *Am. J. Respir. Crit. Care Med.* **154:**1712–1717.

Armitage, J. M., R. L. Kormos, R. S. Stuart, F. J. Fricker, B. P. Griffith, M. Nalesnik, R. L. Hardesty, and J. S. Dummer. 1991. Posttransplant lymphoproliferative disease in thoracic organ transplant patients: ten years of cyclosporine-based immunosuppression. *J. Heart Lung Transplant.* **10:**877–886.

Arribas, J. R., D. B. Clifford, C. J. Fichtenbaum, D. L. Commins, W. G. Powderly, and G. A. Storch. 1995. Level of cytomegalovirus (CMV) DNA in cerebrospinal fluid of subjects with AIDS and CMV infection of the central nervous system. *J. Infect. Dis.* **172:**527–531.

Arvin, A. M. 1996. Varicella-zoster virus. *Clin. Microbiol. Rev.* **9:**361–381.

Arvin, A. M., and A. A. Gershon. 1996. Live attenuated varicella vaccine. *Annu. Rev. Microbiol.* **50:**59–100.

Aschan, J., O. Ringden, P. Ljungman, B. Lonnqvist, and S. Ohlman. 1992. Foscarnet for treatment of cytomegalovirus infections in bone marrow transplant recipients. *Scand. J. Infect. Dis.* **24:**143–150.

Atkins, J. T., G. J. Demmler, W. D. Williamson, J. M. McDonald, A. S. Istas, and G. J. Buffone. 1994. Polymerase chain reaction to detect cytomegalovirus DNA in the cerebrospinal fluid of neonates with congenital infection. *J. Infect. Dis.* **169:**1334–1337.

Atkinson, K., I. Nivison-Smith, A. Dodds, A. Concannon, S. Milliken, and K. Downs. 1998. A comparison of the pattern of interstitial pneumonitis following allogeneic bone marrow transplantation before and after the introduction of prophylactic ganciclovir therapy in 1989. *Bone Marrow Transplant.* **21:**691–695.

Attard-Montalto, S. P., M. C. English, L. Stimmler, and G. J. Snodgrass. 1993. Ganciclovir treatment of congenital cytomegalovirus infection: a report of two cases. *Scand. J. Infect. Dis.* **25:**385–388.

Babcock, G., L. Decker, M. Volk, and D. Thorley-Lawson. 1998. EBV persistence in memory B cells in vivo. *Immunity* **9:**395–404.

Badley, A. D., E. C. Seaberg, M. K. Porayko, R. H. Wiesner, M. R. Keating, M. P. Wilhelm, R. C. Walker, R. Patel, W. F. Marshall, M. DeBernardi, R. Zetterman, J. L. Steers, and

C. V. Paya. 1997. Prophylaxis of cytomegalovirus infection in liver transplantation: a randomized trial comparing a combination of ganciclovir and acyclovir to acyclovir. NIDDK Liver Transplantation Database. *Transplantation* **64**:66–73.

Bailey, T. C., W. G. Powderly, G. A. Storch, S. B. Miller, J. D. Dunkel, R. S. Woodward, E. Spitznagel, D. W. Hanto, and W. C. Dunagan. 1993. Symptomatic cytomegalovirus infection in renal transplant recipients given either Minnesota antilymphoblast globulin (MALG) or OKT3 for rejection prophylaxis. *Am. J. Kidney. Dis.* **21**:196–201.

Baldanti, F., M. G. Revello, E. Percivalle, and G. Gerna. 1998a. Use of the human cytomegalovirus (HCMV) antigenemia assay for diagnosis and monitoring of HCMV infections and detection of antiviral drug resistance in the immunocompromised. *J. Clin. Virol.* **11**:51–60.

Baldanti, F., L. Simoncini, A. Sarasini, M. Zavattoni, P. Grossi, M. G. Revello, and G. Gerna. 1998b. Ganciclovir resistance as a result of oral ganciclovir in a heart transplant recipient with multiple human cytomegalovirus strains in blood. *Transplantation* **66**:324–329.

Balfour, H. H., Jr., B. A. Chace, J. T. Stapleton, R. L. Simmons, and D. S. Fryd. 1989. A randomized, placebo-controlled trial of oral acyclovir for the prevention of cytomegalovirus disease in recipients of renal allografts. *N. Engl. J. Med.* **320**:1381–1387.

Balzarini, J., L. Naesens, and E. Declercq. 1998. New antivirals—mechanism of action and resistance development. *Curr. Opin. Microbiol.* **1**:535–546.

Bass, E. B., N. R. Powe, S. N. Goodman, S. L. Graziano, R. I. Griffiths, T. S. Kickler, and J. R. Wingard. 1993. Efficacy of immune globulin in preventing complications of bone marrow transplantation: a meta-analysis. *Bone Marrow Transplant.* **12**:273–282.

Beards, G., C. Graham, and D. Pillay. 1998. Investigation of vesicular rashes for HSV and VZV by PCR. *J. Med. Virol.* **54**:155–157.

Bek, B., M. Boeckh, J. Lepenies, B. Bieniek, K. Arasteh, W. Heise, K.-M. Deppermann, G. Bornhöft, M. Stöffler-Meilicke, I. Schuller, and G. Höffken. 1996. High-level sensitivity of quantitative pp65 cytomegalovirus (CMV) antigenemia assay for the diagnosis of CMV disease in AIDS patients and follow-up. *J. Clin. Microbiol.* **34**:457–459.

Beninga, J., B. Kropff, and M. Mach. 1995. Comparative analysis of fourteen individual human cytomegalovirus proteins for helper T cell response. *J. Gen. Virol.* **76**:153–160.

Berger, R., E. Trannoy, G. Hollander, F. Bailleux, C. Rudin, and H. Creusvaux. 1998. A dose-response study of a live attenuated varicella-zoster virus (OKA strain) vaccine administered to adults 55 years of age and older. *J. Infect. Dis.* **178**:S99–S103.

Bevan, I. S., R. A. Daw, P. J. Day, F. A. Ala, and M. R. Walker. 1991. Polymerase chain reaction for detection of human cytomegalovirus infection in a blood donor population. *Br. J. Haematol.* **78**:94–99.

Blok, M. J., V. J. Goossens, S. J. Vanherle, B. Top, N. Tacken, J. M. Middeldorp, M. H. Christiaans, J. P. van Hooff, and C. A. Bruggeman. 1998. Diagnostic value of monitoring human cytomegalovirus late pp67 mRNA expression in renal-allograft recipients by nucleic acid sequence-based amplification. *J. Clin. Microbiol.* **36**:1341–1346.

Blum, A., M. Giladi, M. Weinberg, G. Kaplan, H. Pasternack, S. Laniado, and H. Miller. 1998. High anti-cytomegalovirus (CMV) IgG antibody titer is associated with coronary artery disease and may predict post-coronary balloon angioplasty restenosis. *Am. J. Cardiol.* **81**:866–868.

Bodaghi, B., P. DalMonte, L. Picard, C. Bessia, and S. Michelson. 1995. Human cytomegalovirus protein pp65 (ppUL83) plays a role in inhibition of host cell protein synthesis. *Scand. J. Infect. Dis. Suppl.* **99**:41–42.

Bodaghi, B., T. R. Jones, D. Zipeto, C. Vita, L. Sun, L. Laurent, F. Arenzana-Seisdedos, J.-L. Virelizier, and S. Michelson. 1998. Chemokine sequestration by viral chemoreceptors as a novel viral escape strategy: withdrawal of chemokines from the environment of cytomegalovirus-infected cells. *J. Exp. Med.* **188**:855–866.

Boeckh, M., and G. Boivin. 1998. Quantitation of cytomegalovirus: methodologic aspects and clinical applications. *Clin. Microbiol. Rev.* **11**:533–554.

Boeckh, M., R. A. Bowden, J. A. Goodrich, M. Pettinger, and J. D. Meyers. 1992. Cytomegalovirus antigen detection in peripheral blood leukocytes after allogeneic marrow transplantation. *Blood* **80**:1358–1364.

Boeckh, M., T. A. Gooley, D. Myerson, T. Cunningham, G. Schoch, and R. A. Bowden. 1996. Cytomegalovirus pp65 antigenemia-guided early treatment with ganciclovir versus ganciclovir at engraftment after allogeneic marrow transplantation: a randomized double-blind study. *Blood* **88**:4063–4071.

Boeckh, M., G. M. Gallez-Hawkins, D. Myerson, J. A. Zaia, and R. A. Bowden. 1997. Plasma polymerase chain reaction for cytomegalovirus DNA after allogeneic marrow transplantation: comparison with polymerase chain reaction using peripheral blood leukocytes, pp65 antigenemia, and viral culture. *Transplantation* **64**:108–113.

Boeckh, M., T. A. Gooley, and R. A. Bowden. 1998. Effect of high-dose acyclovir in survival in allogeneic marrow transplant recipients who received ganciclovir at engraftment or for cytomegalovirus pp65 antigenemia. *J. Infect. Dis.* **178**:1153–1157.

Boland, G. J., G. C. de Gast, R. J. Hené, G. Jambroes, R. Donckerwolcke, T. H. The, and G. C. Mudde. 1990. Early detection of active cytomegalovirus (CMV) infection after heart and kidney transplantation by testing for immediate early antigenemia and influence of cellular immunity on the occurrence of CMV infection. *J. Clin. Microbiol.* **28**:2069–2075.

Boldogh, L., S. AbuBakar, and T. Albrecht. 1990. Activation of proto-oncogenes: an immediate early event in human cytomegalovirus infection. *Science* **247**:561–564.

Boldogh, L., S. AbuBakar, C. Z. Deng, and T. Albrecht. 1991. Transcriptional activation of cellular oncogenes *fos, jun,* and *myc* by human cytomegalovirus. *J. Virol.* **65**:1568–1571.

Bongarts, A., D. von Laer, C. Vogelberg, K. Ebert, J. van Lunzen, J. Garweg, P. Vaith, F. T. Hufert, O. Haller, and U. Meyer-Konig. 1996. Glycoprotein B genotype of human cytomegalovirus: distribution in HIV-infected patients. *Scand. J. Infect. Dis.* **28**:447–449.

Bonini, C., S. Berzeletti, C. Traversari, E. Zappone, P. Servida, S. Rossini, and C. Bordicnon. 1996. A pilot study of HSV-TK gene transfer into donor peripheral blood lymphocytes for controlled BVL after allo BMT. *Blood* **88**:487–497.

Boppana, S. B., and W. J. Britt. 1995. Antiviral antibody responses and intrauterine transmission after primary maternal cytomegalovirus infection. *J. Infect. Dis.* **171**:1115–1121.

Boppana, S. B., R. F. Pass, and W. J. Britt. 1993. Virus-specific antibody responses in mothers and their newborn infants with asymptomatic congenital cytomegalovirus infections. *J. Infect. Dis.* **167**:72–77.

Boppana, S. B., M. A. Polis, A. A. Kramer, W. J. Britt, and S. Koenig. 1995. Virus specific antibody responses to human cytomegalovirus (HCMV) in human immunodeficiency virus type 1-infected individuals with HCMV retinitis. *J. Infect. Dis.* **171:**182–185.

Boppana, S. B., K. B. Fowler, Y. Vaid, G. Hedlund, S. Stagno, W. J. Britt, and R. F. Pass. 1997. Neuroradiographic findings in the newborn period and long-term outcome in children with symptomatic congenital cytomegalovirus infection. *Pediatrics* **99:**409–414.

Bostrom, L., O. Ringden, B. Sundberg, A. Linde, J. Tollemar, and B. Nilsson. 1988. Pretransplant herpesvirus serology and acute graft-versus-host disease. *Transplantation* **46:**548–552.

Boucquey, D., C. J. Sindic, M. Lamy, M. Delmee, J. P. Tomasi, and E. C. Laterre. 1991. Clinical and serological studies in a series of 45 patients with Guillain-Barre syndrome. *J. Neurol. Sci.* **104:**56–63.

Bowden, R. A., S. J. Slichter, M. Sayers, D. Weisdorf, M. Cays, G. Schoch, M. Banaji, R. Haake, K. Welk, L. Fisher, J. McCullough, and W. Miller. 1995. A comparison of filtered leukocyte-reduced and cytomegalovirus seronegative blood products for the prevention of transfusion-associated CMV infection after marrow transplant. *Blood* **86:**3598–3603.

Bower, M., S. E. Barton, M. R. Nelson, J. Bobby, D. Smith, M. Youle, and B. G. Gazzard. 1990. The significance of the detection of cytomegalovirus in the bronchoalveolar lavage fluid in AIDS patients with pneumonia. *AIDS* **4:**317–320.

Brasfield, D. M., S. Stagno, R. J. Whitley, G. Cloud, G. Cassell, and R. E. Tiller. 1987. Infant pneumonitis associated with cytomegalovirus, *Chlamydia, Pneumocystis,* and *Ureaplasma:* follow-up. *Pediatrics* **79:**76–83.

Brennan, D. C., K. A. Garlock, B. A. Lippmann, R. S. Buller, M. Gaudreault-Keener, J. A. Lowell, S. B. Miller, S. Shenoy, T. K. Howard, and G. A. Storch. 1997a. Control of cytomegalovirus-associated morbidity in renal transplant patients using intensive monitoring and either preemptive or deferred therapy. *J. Am. Soc. Nephrol.* **8:**118–125.

Brennan, D. C., K. A. Garlock, G. G. Singer, M. A. Schnitzler, B. J. Lippmann, R. S. Buller, M. Gaudreault-Keener, J. A. Lowell, S. Shenoy, T. K. Howard, and G. A. Storch. 1997b. Prophylactic oral ganciclovir compared with deferred therapy for control of cytomegalovirus in renal transplant recipients. *Transplantation* **64:**1843–1846.

Britt, W. J., L. Vugler, E. J. Butfiloski, and E. B. Stephens. 1990. Cell surface expression of human cytomegalovirus (HCMV) gp55-116 (gB): use of HCMV-vaccinia recombinant virus infected cells in analysis of the human neutralizing antibody response. *J. Virol.* **64:**1079–1085.

Brodt, H. R., B. S. Kamps, P. Gute, B. Knupp, S. Staszewski, and E. B. Helm. 1997. Changing incidence of AIDS-defining illnesses in the era of antiretroviral combination therapy. *AIDS* **11:**1731–1738.

Brosgart, C. L., T. A. Louis, D. W. Hillman, C. P. Craig, B. Alston, E. Fisher, D. I. Abrams, R. L. Luskin-Hawk, J. H. Sampson, D. J. Ward, M. A. Thompson, and R. A. Torres. 1998. A randomized, placebo-controlled trial of the safety and efficacy of oral ganciclovir for prophylaxis of cytomegalovirus disease in HIV-infected individuals. *AIDS* **12:**269–277.

Brumback, B. G., S. N. Bolejack, M. V. Morris, C. Mohla, and T. E. Schutzbank. 1997. Comparison of culture and the antigenemia assay for detection of cytomegalovirus in blood specimens submitted to a reference laboratory. *J. Clin. Microbiol.* **35:**1819–1821.

Buchbinder, S. P., M. H. Katz, N. A. Hessol, J. Y. Liu, P. M. O'Malley, R. Underwood, and S. D. Holmberg. 1992. Herpes zoster and human immunodeficiency virus infection. *J. Infect. Dis.* **166:**1153–1156.

Burke, D. G., R. C. Kalayjian, V. R. Vann, S. A. Madreperla, H. E. Shick, and D. G. B. Leonard. 1997. Polymerase chain reaction detection and clinical significance of varicella-zoster virus in cerebrospinal fluid from human immunodeficiency virus-infected patients. *J. Infect. Dis.* **176:**1080–1084.

Burkitt, D. 1958. A sarcoma involving the jaws in African children. *Br. J. Surg.* **46:**218–223.

Burkitt, D. 1963. Determining the climatic limitations of a children s cancer common in Africa. *Br. Med. J.* **2:**1019–1023.

Bush, C. E., and J. A. Sluchak-Carlsen. 1998. Evaluation of a leukocyte stabilization reagent for use in the cytomegalovirus pp65 antigenemia assay. *J. Clin. Microbiol.* **36:**3410–3411.

Casado, J. L., M. J. Perez-Elias, P. Marti-Belda, A. Antela, M. Suarez, E. Ciancas, B. Frutos, M. D. Perez, and A. Guerrero. 1998. Improved outcome of cytomegalovirus retinitis in AIDS patients after introduction of protease inhibitors. *J. Acquir. Immune. Defic. Syndr. Hum. Retrovirol.* **19:**130–134.

Cha, T. A., E. Tom, G. W. Kemble, G. M. Duke, E. S. Mocarski, and R. R. Spaete. 1996. Human cytomegalovirus clinical isolates carry at least 19 genes not found in laboratory strains. *J. Virol.* **70:**78–83.

Chang, R. S., J. P. Lewis, and C.F. Abildgaard. 1973. Prevalence of oropharyngeal excretors of leukocyte-transforming agents among a human population. *N. Engl. J. Med.* **289:**1325–1329.

Chatterjee, S. N., M. Fiala, J. Weiner, J. A. Stewart, B. Stacey, and N. Warmer. 1978. Primary cytomegalovirus and opportunistic infections. Incidence in renal transplant recipients. *JAMA* **240:**2446–2449.

Chee, M. S., A. T. Bankier, S. Beck, R. Bohni, C. M. Brown, R. Cerny, T. Horsnell, C. A. Hutchison, T. Kouzarides, J. A. Martignetti, E. Preddie, S. C. Satchwell, P. Tomlinson, K. M. Weston, and B. G. Barrell. 1990. Analysis of the protein-coding content of the sequence of human cytomegalovirus strain AD169. *Curr. Top. Microbiol. Immunol.* **154:**125–170.

Chen, F., J. Zou, L. diRenzo, G. Winberg, L. Hu, G. Kleim, and I. Ernberg. 1995. A subpopulation of normal B cells latently infected with Epstein-Barr virus resembles Burkitt lymphoma cells in expressing EBNA1 but not EBNA2 or LMP1. *J. Virol.* **69:**3752–3758.

Chen, J. M. B., M. L. Barr, A. Chadburn, G. Frizzera, F. A. Schenkel, R. R. Sciacca, D. S. Reison, L. J. Addonizio, E. A. Rose, D. M. Knowles, and R. E. Michler. 1993. Management of lymphoproliferative disorders after cardiac transplantation. *Ann. Thorac. Surg.* **56:**527–538.

Cheng, Y. C., S. P. Grill, G. E. Dutschman, K. Nakayama, and K. F. Bastow. 1983. Metabolism of 9-(1,3-dihydroxy-2-propoxymethyl)guanine, a new anti-herpesvirus compound, in herpes simplex virus-infected cells. *J. Biol. Chem.* **258:**12460–12464.

Chern, K. C., D. B. Chandler, D. F. Martin, B. D. Kuppermann, R. A. Wolitz, and T. P. Margolis. 1998. Glycoprotein B subtyping of cytomegalovirus (CMV) in the vitreous of patients with AIDS and CMV retinitis. *J. Infect. Dis.* **178:**1149–1153.

Cheung, R., and H. Dosch. 1991. The tyrosine kinase lck is critically involved in the growth transformation of human B lymphocytes. *J. Biol. Chem.* **266:**8667–8670.

Chou, S. 1992a. Comparative analysis of sequence variation in gp116 and gp55 components of glycoprotein B of human cytomegalovirus. *Virology* **188:**388–390.

Chou, S. 1992b. Effect of interstrain variation on diagnostic DNA amplification of the cytomegalovirus major immediate-early gene region. *J. Clin. Microbiol.* **30:**2307–2310.

Chou, S., and T. C. Merigan. 1983. Rapid detection and quantitation of human cytomegalovirus in urine through DNA hybridization. *N. Engl. J. Med.* **308:**921–925.

Chou, S., A. Erice, M. C. Jordan, G. M. Vercellotti, K. R. Michels, C. L. Talarico, S. C. Stanat, and K. K. Biron. 1995. Analysis of the UL97 phosphotransferase coding sequence in clinical cytomegalovirus isolates and identification of mutations conferring ganciclovir resistance. *J. Infect. Dis.* **171:**576–583.

Chou, S. W., and K. M. Dennison. 1991. Analysis of interstrain variation in cytomegalovirus glycoprotein B sequences encoding neutralization-related epitopes. *J. Infect. Dis.* **163:**1229–1234.

Cinque, P., S. Bossolasco, L. Vago, C. Fornara, S. Lipari, S. Racca, A. Lazzarin, and A. Linde. 1997. Varicella-zoster virus (VZV) DNA in cerebrospinal fluid of patients infected with human immunodeficiency virus-VZV disease of the central nervous system or subclinical reactivation of VZV infection. *Clin. Infect. Dis.* **25:**634–639.

Cirelli, R., K. Herne, M. McCrary, P. Lee, and S. K. Tyring. 1996. Famciclovir: review of clinical efficacy and safety. *Antiviral Res.* **29:**141–151.

Cockfield, S. M., J. K. Preiksaitis, L. D. Jewell, and N. A. Parfrey. 1993. Post-transplant lymphoproliferative disorder in renal allograft recipients. *Transplantation* **56:**88–96.

Coffin, S. E., and R. L. Hodinka. 1995. Utility of direct immunofluorescence and virus culture for detection of varicella-zoster virus in skin lesions. *J. Clin. Microbiol.* **33:**2792–2795.

Cohen, J., F. Wang, and E. Kieff. 1991. Epstein-Barr virus nuclear protein 2 mutations define essential domains for transformation and transactivation. *J. Virol.* **65:**2545–2554.

Cohen, J. I. 1998. Infection of cells with varicella-zoster virus down-regulates surface expression of class I major histocompatibility complex antigens. *J. Infect. Dis.* **177:**1390–1393.

Cohen, J. I., and S. E. Straus. 1996. Varicella-zoster virus and its replication, p. 2525–2545. *In* B. N. Fields, D. M. Knipe, and P. M. Howley (ed.), *Virology*, 3rd ed. Lippincott-Raven Publishers, Philadelphia, Pa.

Cohen, P. R., and M. E. Grossman. 1989. Clinical features of human immunodeficiency virus-associated disseminated herpes zoster virus infection—a review of the literature. *Clin. Exp. Dermatol.* **14:**273–276.

Collins, T., C. Pomeroy, and M. C. Jordan. 1993. Detection of latent cytomegalovirus DNA in diverse organs of mice. *J. Infect. Dis.* **168:**725–729.

Conboy, T. J., R. F. Pass, S. Stagno, C. A. Alford, G. J. Myers, W. J. Britt, F. P. McCollister, M. N. Summers, C. E. McFarland, and T. J. Boll. 1987. Early clinical manifestations and intellectual outcome in children with symptomatic congenital cytomegalovirus infection. *J. Pediatr.* **111:**343–348.

Croen, K. D., and S. E. Straus. 1991. Varicella-zoster virus latency. *Annu. Rev. Microbiol.* **45:**265–282.

Dahl, H., J. Marcoccia, and A. Linde. 1997. Antigen detection: the method of choice in comparison with virus isolation and serology for laboratory diagnosis of herpes zoster in human immunodeficiency virus-infected patients. *J. Clin. Microbiol.* **35:**347–349.

Danise, A., P. Cinque, S. Vergani, M. Candino, S. Racca, A. Debona, R. Novati, A. Castagna, and A. Lazzarin. 1997. Use of polymerase chain reaction assays of aqueous humor in the differential diagnosis of retinitis in patients infected with human immunodeficiency virus. *Clin. Infect. Dis.* **24:**1100–1106.

Dankner, W. M., D. Scholl, S. C. Stanat, M. Martin, R. L. Sonke, and S. A. Spector. 1990. Rapid antiviral DNA-DNA hybridization assay for human cytomegalovirus. *J. Virol. Methods* **28:**293–298.

Danner, S. A., and S. Matheron. 1996. Cytomegalovirus retinitis in AIDS patients: a comparative study of intravenous and oral ganciclovir as maintenance therapy. *AIDS* **10** (Suppl. 4):S7–S11.

Darenkov, I., M. A. Marcarelli, G. P. Basadonna, M. L. Friedman, K. M. Lorber, J. G. Howe, J. Crouch, M. J. Bia, A. S. Kliger, and M. I. Lorber. 1997. Reduced incidence of Epstein-Barr virus-associated posttransplant lymphoproliferative disorder using preemptive antiviral therapy. *Transplantation* **64:**848–852.

Dargan, D. J., F. E. Jamieson, J. Maclean, A. Dolan, C. Addison, and D. J. McGeoch. 1997. The published DNA sequence of human cytomegalovirus strain AD169 lacks 929 base pairs affecting genes UL42 and UL43. *J. Virol.* **71:**9833–9836.

Davis, C. L., K. L. Harrison, J. P. McVicar, P. J. Forg, M. P. Bronner, and C. L. Marsh. 1995. Antiviral prophylaxis and the Epstein Barr virus-related post-transplant lymphoproliferative disorder. *Clin. Transplant.* **9:**53–59.

Davis, M. G., S. C. Kenney, J. Kamine, J. S. Pagano, and E. S. Huang. 1987. Immediate early gene region of human cytomegalovirus trans-activates the promoter of human immunodeficiency virus. *Proc. Natl. Acad. Sci. USA* **84:**8642–8646.

Davison, A. J., and J. E. Scott. 1986. The complete DNA sequence of varicella-zoster virus. *J. Gen. Virol.* **67:**1759–1816.

Deeg, H. J., J. Sanders, P. Martin, A. Fefer, P. Neiman, J. Singer, R. Storb, and E. D. Thomas. 1984. Secondary malignancies after marrow transplantation. *Exp. Hematol.* **12:**660–666.

de la Hoz, R. E., S. K. Byrne, S. Hayashi, C. Sherlock, D. Cook, and J. C. Hogg. 1998. Diagnosis of cytomegalovirus infection in HIV-infected patients with respiratory disease. *Clin. Diagn. Virol.* **10:**1–7.

Delgado, R., C. Lumbreras, C. Alba, M. A. Pedraza, J. R. Otero, R. Gómez, E. Moreno, A. R. Noriega, and C. Payá. 1992. Low predictive value of polymerase chain reaction for diagnosis of cytomegalovirus disease in liver transplant recipients. *J. Clin. Microbiol.* **30:**1876–1878.

Deray, G., F. Martinez, C. Katlama, B. Levaltier, H. Beaufils, M. Danis, M. Rozenheim, A. Baumelou, E. Dohin, M. Gentilini, and C. Jacobs. 1989. Foscarnet nephrotoxicity: mechanism, incidence and prevention. *Am. J. Nephrol.* **9:**316–320.

Desjardin, J. A., and D. R. Snydman. 1998. Antiviral immunotherapy—a review of current status. *Biodrugs* **9:**487–507.

Detels, R., C. T. Leach, K. Hennessey, Z. Liu, B. R. Visscher, J. D. Cherry, and J. V. Giorgi. 1994. Persistent cytomegalovirus infection of semen increases risk of AIDS. *J. Infect. Dis.* **169:**766–768.

Devlin, M. E., D. H. Gilden, R. Mahalingam, A. N. Dueland, and R. Cohrs. 1992. Peripheral blood mononuclear cells of the elderly contain varicella-zoster virus DNA. *J. Infect. Dis.* **165:**619–622.

Dieterich, D. T., M. A. Poles, E. A. Lew, S. Martin-Munley, J. Johnson, D. Nix, and M. J. Faust. 1997. Treatment of gastrointestinal cytomegalovirus infection with twice-daily foscarnet: a pilot study of safety, efficacy, and pharmacokinetics in patients with AIDS. *Antimicrob. Agents Chemother.* **41:**1226–1230.

Doniger, J., S. Muralidhar, and L. J. Rosenthal. 1999. Human cytomegalovirus and human herpesvirus 6 genes that transform and transactivate. *Clin. Microbiol. Rev.* **12:**367–382.

Drew, W. L., D. Ives, J. P. Lalezari, C. Crumpacker, S. E. Follansbee, S. A. Spector, C. A. Benson, D. N. Friedberg, L. Hubbard, M. J. Stempien, A. Shadman, and W. Buhles. 1995. Oral ganciclovir as maintenance treatment for cytomegalovirus retinitis in patients with AIDS. *N. Engl. J. Med.* **333:**615–620.

Drew, W. L., M. J. Stempien, J. Andrews, A. Shadman, S. J. Tan, R. Miner, and W. Buhles. 1999. Cytomegalovirus

(CMV) resistance in patients with CMV retinitis and AIDS treated with oral or intravenous ganciclovir. *J. Infect. Dis.* **179:**1352–1355.

Drobyski, W. R., K. K. Knox, D. R. Carrigan, and R. C. Ash. 1991. Foscarnet therapy for ganciclovir-resistant cytomegalovirus in marrow transplantation. *Transplantation* **52:**155–157.

Dueland, A. N. 1996. Latency and reactivation of varicella zoster virus infections. *Scand. J. Infect. Dis.* Suppl. **100:**46–50.

Dummer, S., A. Lee, M. K. Breinig, R. Kormos, M. Ho, and B. Griffith. 1994. Investigation of cytomegalovirus infection as a risk factor for coronary atherosclerosis in the explanted hearts of patients undergoing heart transplantation. *J. Med. Virol.* **44:**305–309.

Dworsky, M., M. Yow, S. Stagno, R. F. Pass, and C. Alford. 1983b. Cytomegalovirus infection of breast milk and transmission in infancy. *Pediatrics* **72:**295–299.

Dworsky, M. E., K. Welch, G. Cassady, and S. Stagno. 1983a. Occupational risk for primary cytomegalovirus infection among pediatric health-care workers. *N. Engl. J. Med.* **309:**950–953.

Eddleston, M., S. Peacock, M. Juniper, and D. A. Warrell. 1997. Severe cytomegalovirus infection in immunocompetent patients. *Clin. Infect. Dis.* **24:**52–56.

Egan, J. J., J. Lomax, L. Barber, S. S. Lok, R. Martyszczuk, N. Yonan, A. Fox, A. K. Deiraniya, A. J. Turner, and A. A. Woodcock. 1998. Preemptive treatment for the prevention of cytomegalovirus disease in lung and heart transplant recipients. *Transplantation* **65:**747–752.

Einsele, H., G. Ehninger, H. Hebart, K. M. Wittkowski, U. Schuler, G. Jahn, P. Mackes, M. Herter, T. Klingebiel, J. Loffler, S. Wagner, and C. A. Muller. 1995. Polymerase chain reaction monitoring reduces the incidence of cytomegalovirus disease and the duration and side effects of antiviral therapy after bone marrow transplantation. *Blood* **86:**2815–2820.

El-Beik, T., A. Razzaque, R. Jariwalla, R. L. Cihlar, and L. J. Rosenthal. 1986. Multiple transforming regions of human cytomegalovirus DNA. *J. Virol.* **60:**645–652.

Elion, G. B., P. A. Furman, J. A. Fyfe, P. de Miranda, L. Beauchamp, and H. J. Schaeffer. 1977. Selectivity of action of an anti-herpetic agent, 9-(2-hydroxyethoxymethyl)guanine. *Proc. Natl. Acad. Sci. USA* **74:**5716–5720.

Emery, V. C., A. V. Cope, E. F. Bowen, D. Gor, and P. D. Griffiths. 1999a. The dynamics of human cytomegalovirus replication in vivo. *J. Exp. Med.* **190:**177–182.

Emery, V. C., C. Sabin, J. E. Feinberg, M. Grywacz, S. Knight, and P. D. Griffiths. 1999b. Quantitative effects of valacyclovir on the replication of cytomegalovirus (CMV) in persons with advanced human immunodeficiency virus disease: baseline CMV load dictates time to disease and survival. *J. Infect. Dis.* **180:**695–701.

Epstein, M. A., B. G. Achong, and Y. M. Barr. 1964. Virus particles in cultured lymphoblasts from Burkitt's lymphoma. *Lancet* **i:**702–703.

Erice, A., S. Chou, K. K. Biron, S. C. Stanat, H. H. Balfour, Jr., and M. C. Jordan. 1989. Progressive disease due to ganciclovir-resistant cytomegalovirus in immunocompromised patients. *N. Engl. J. Med.* **320:**289–293.

Erice, A., C. Gil-Roda, J. Pérez, H. H. Balfour, Jr., K. J. Sannerud, M. N. Hanson, G. Boivin, and S. Chou. 1997. Antiviral susceptibilities and analysis of UL97 and DNA polymerase sequences of clinical cytomegalovirus isolates from immunocompromised patients. *J. Infect. Dis.* **175:**1087–1092.

Erice, E. 1999. Resistance of human cytomegalovirus to antiviral drugs. *Clin. Microbiol. Rev.* **12:**286–297.

Evans, A. S. 1982. The transmission of EB viral infections, p. 211–225. *In* J. Hooks and G. Jordan (ed.), *Viral Infections in Oral Medicine.* Elsevier North Holland, New York, N.Y.

Evans, P. C., A. Soin, T. G. Wreghitt, and G. J. Alexander. 1998. Qualitative and semiquantitative polymerase chain reaction testing for cytomegalovirus DNA in serum allows prediction of CMV related disease in liver transplant recipients. *J. Clin. Pathol.* **51:**914–921.

Fajac, A., F. Stéphan, A. Ibrahim, E. Gautier, J. F. Bernaudin, and J. L. Pico. 1997. Value of cytomegalovirus detection by PCR in bronchoalveolar lavage routinely performed in asymptomatic bone marrow recipients. *Bone Marrow Transplant.* **20:**581–585.

Falagas, M. E., M. Arbo, R. Ruthazer, J. L. Griffith, B. G. Werner, R. Rohrer, R. Freeman, W. D. Lewis, and D. R. Snydman. 1997a. Cytomegalovirus disease is associated with increased cost and hospital length of stay among orthotopic liver transplant recipients. *Transplantation* **63:**1595–1601.

Falagas, M. E., D. R. Snydman, J. Griffith, B. G. Werner, and the Boston Center for Liver Transplantation CMVIG Study Group. 1996. Exposure to cytomegalovirus from the donated organ is a risk factor for bacteremia in orthotopic liver transplant recipients. *Clin. Infect. Dis.* **23:**468–474.

Falagas, M. E., D. R. Snydman, R. Ruthazer, J. Griffith, B. G. Werner, R. Freeman, R. Rohrer, and the Boston Center for Liver Transplantation CMVIG Study Group. 1997b. Cytomegalovirus immune globulin (CMVIG) prophylaxis is associated with increased survival after orthotopic liver transplantation. *Clin. Transplant.* **11:**432–437.

Falagas, M. E., D. R. Snydman, R. Ruthazer, B. G. Werner, J. Griffith, and the Boston Center for Liver Transplantation CMVIG Study Group. 1997c. Surveillance cultures of blood, urine, and throat specimens are not valuable for predicting cytomegalovirus disease in liver transplant recipients. *Clin. Infect. Dis.* **24:**824–829.

Fan-Harvard, P., M. C. Nahata, and M. T. Brady. 1989. Ganciclovir—a review of pharmacology, therapeutic efficacy and potential use for treatment of congenital cytomegalovirus pneumonia. *J. Clin. Pharmacol. Ther.* **14:**329–340.

Farber, S., and S. G. Wolbach. 1932. Intranuclear and cytoplasmic inclusions ("protozoan-like bodies") in the salivary glands and other organs of infants. *Am. J. Pathol.* **8:**123–135.

Feinberg, J. E., S. Hurwitz, D. Cooper, F. R. Sattler, R. R. MacGregor, W. Powderly, G. N. Holland, P. D. Griffiths, R. B. Pollard, M. Youle, M. J. Gill, F. J. Holland, M. E. Power, S. Owens, D. Coakley, J. Fry, and M. A. Jacobson. 1998. A randomized, double-blind trial of valacyclovir prophylaxis for cytomegalovirus disease in patients with advanced human immunodeficiency virus infection. *J. Infect. Dis.* **177:**48–56.

Fetterman, G. H. 1952. A new laboratory aid in the clinical diagnosis of inclusion disease of infancy. *Am. J. Clin. Pathol.* **22:**424–425.

Fillet, A. M., B. Dumont, E. Caumes, B. Visse, H. Agut, F. Bricaire, and J. M. Huraux. 1998. Acyclovir-resistant varicella-zoster virus—phenotypic and genetic characterization. *J. Med. Virol.* **55:**250–254.

Fish, K. N., C. Soderberg-Naucler, L. K. Mills, S. Stenglein, and J. Nelson. 1998. Human cytomegalovirus persistently infects aortic endothelial cells. *J. Virol.* **72:**5661–5668.

Fishman, J. A., and R. H. Rubin. 1998. Infection in organ-transplant recipients. *N. Engl. J. Med.* **338:**1741–1751.

Flechner, S. M., R. K. Avery, R. Fisher, B. A. Mastroianni, D. A. Papajcik, K. J. O'Malley, M. Goormastic, D. A. Goldfarb, C. S. Modlin, and A. C. Novick. 1998. A randomized prospective controlled trial of oral acyclovir versus oral ganciclovir for cytomegalovirus prophylaxis in high-risk kidney transplant recipients. *Transplantation* **66:**1682–1688.

Fletcher, J. M., H. G. Prentice, and J. E. Grundy. 1998. Natural killer cell lysis of cytomegalovirus (CMV)-infected cells correlates with virally induced changes in cell surface lymphocyte function-associated antigen-3 (LFA-3) expression and not with the CMV-induced down-regulation of cell surface Class I HLA. *J. Immunol.* **161:**2365–2374.

Flood, J., W. L. Drew, R. Miner, D. Jekic-McMullen, L.-P. Shen, J. Kolberg, J. Garvey, S. Follansbee, and M. Poscher. 1997. Diagnosis of cytomegalovirus (CMV) polyradiculopathy and documentation of in vivo anti-CMV activity in cerebrospinal fluid by using branched DNA signal amplification and antigen assays. *J. Infect. Dis.* **176:**348–352.

Flynn, J. T., B. A. Kaiser, S. S. Long, S. L. Schulman, A. Deforest, M. S. Polinsky, and H. J. Baluarte. 1997. Intravenous immunoglobulin prophylaxis of cytomegalovirus infection in pediatric renal transplant recipients. *Am. J. Nephrol.* **17:**146–152.

Fowler, K. B., S. Stagno, R. F. Pass, W. J. Britt, T. J. Boll, and C. A. Alford. 1992. The outcome of congenital cytomegalovirus infection in relation to maternal antibody status. *N. Engl. J. Med.* **326:**663–667.

Francisci, D., A. Tosti, F. Baldelli, G. Stagni, and S. Pauluzzi. 1997. The pp65 antigenemia test as a predictor of cytomegalovirus-induced end-organ disease in patients with AIDS. *AIDS* **11:**1341–1345.

Frezzera, G., D. W. Hanto, K. J. Gajl-Peczalska, J. Rosai, R. W. McKenna, R. K. Sibley, K. P. Holahan, and L. L. Lindquist. 1981. Polymorphic diffuse B cell hyperplasias and lymphomas in transplant recipients. *Cancer Res.* **41:**4262–4270.

Fries, B. C., S. Chou, M. Boeckh, and B. Torok-Storb. 1994. Frequency distribution of cytomegalovirus envelope glycoprotein genotypes in bone marrow transplant recipients. *J. Infect. Dis.* **169:**769–774.

Furuta, Y., S. Fukuda, S. Suzuki, T. Takasu, Y. Inuyama, and K. Nagashima. 1997. Detection of varicella-zoster virus DNA in patients with acute peripheral facial palsy by the polymerase chain reaction, and its use for early diagnosis of zoster sine herpete. *J. Med. Virol.* **52:**316–319.

Fyfe, J. A., P. M. Keller, P. A. Furman, R. L. Miller, and G. B. Elion. 1978. Thymidine kinase from herpes simplex virus phosphorylates the new antiviral compound 9-(2-hydroxyethoxymethyl)guanine. *J. Biol. Chem.* **253:**8721–8727.

Gabel, C. A., L. Dubey, S. P. Steinberg, D. Sherman, M. D. Gershon, and A. A. Gershon. 1989. Varicella-zoster virus glycoprotein oligosaccharides are phosphorylated during post-translational maturation. *J. Virol.* **63:**4264–4276.

Gane, E., F. Saliba, G. J. C. Valdecasas, J. O'Grady, M. D. Pescovitz, S. Lyman, and C. A. Robinson. 1997. Randomised trial of efficacy and safety of oral ganciclovir in the prevention of cytomegalovirus disease in liver-transplant recipients. *Lancet* **350:**1729–1733.

Gao, J. L., and P. M. Murphy. 1994. Human cytomegalovirus open reading frame US28 encodes a functional beta chemokine receptor. *J. Biol. Chem.* **269:**28539–28542.

Gao, S. Z., J. S. Schroeder, S. Hunt, and E. B. Stinson. 1988. Retransplantation for severe accelerated coronary artery disease in heart transplant recipients. *Am. J. Cardiol.* **62:**876–881.

Garweg, J., and M. Bohnke. 1997. Varicella-zoster virus is strongly associated with atypical necrotizing herpetic retinopathies. *Clin. Infect. Dis.* **24:**603–608.

Geder, L., R. Lausch, F. O'Neill, and F. Rapp. 1976. Oncogenic transformation of human embryo lung cells by human cytomegalovirus. *Science* **192:**1134–1137.

George, M. J., D. R. Snydman, B. G. Werner, J. Griffith, M. E. Falagas, N. N. Dougherty, and R. H. Rubin. 1997. The independent role of cytomegalovirus as a risk factor for invasive fungal disease in orthotopic liver transplant recipients. *Am. J. Med.* **103:**106–113.

Gerber, P., M. Nonoyama, S. Lucas, E. Perlin, and L. I. Goldstein. 1972. Oral excretion of Epstein-Barr virus by healthy subjects and patients with infectious mononucleosis. *Lancet* **ii:**988–989.

Gerna, G., D. Zipeto, M. Parea, M. G. Revello, E. Silini, E. Percivalle, M. Zavattoni, P. Grossi, and G. Milanesi. 1991. Monitoring of human cytomegalovirus infections and ganciclovir treatment in heart transplant recipients by determination of viremia, antigenemia and DNAemia. *J. Infect. Dis.* **164:**488–498.

Gerna, G., F. Baldanti, M. Zavattoni, A. Sarasini, E. Percivalle, and M. G. Revello. 1992a. Monitoring of ganciclovir sensitivity of human cytomegalovirus strains coinfecting blood of an AIDS patient by an immediate-early antigen plaque assay. *Antivir. Res.* **19:**333–345.

Gerna, G., M. G. Revello, E. Percivalle, and F. Morini. 1992b. Comparison of different immunostaining techniques and monoclonal antibodies to the lower matrix phospho protein (pp65) for optimal quantitation of human cytomegalovirus antigenemia. *J. Clin. Microbiol.* **30:**1232–1237.

Gerna, G., A. Sarasini, F. Baldanti, E. Percivalle, D. Zella, M. G. Revello, and the Italian Foscarnet GID Study Group. 1997. Quantitative systemic and local evaluation of the antiviral effect of ganciclovir and foscarnet induction treatment in human cytomegalovirus gastrointestinal disease of patients with AIDS. *Antivir. Res.* **34:**39–50.

Gerna, G., M. Zavattoni, F. Baldanti, M. Furione, L. Chezzi, M. G. Revello, and E. Percivalle. 1998a. Circulating cytomegalic endothelial cells are associated with high human cytomegalovirus (HCMV) load in AIDS patients with late-stage disseminated HCMV disease. *J. Med. Virol.* **55:**64–74.

Gerna, G., M. Zavattoni, F. Baldanti, A. Sarasini, L. Chezzi, P. Grossi, and M. G. Revello. 1998b. Human cytomegalovirus (HCMV) leukodnaemia correlates more closely with clinical symptoms than antigenemia and viremia in heart and heart-lung transplant recipients with primary HCMV infection. *Transplantation* **65:**1378–1385.

Gerna, G., E. Percivalle, F. Baldanti, A. Sarasini, M. Zavattoni, M. Furione, M. Torsellini, and M. G. Revello. 1998c. Diagnostic significance and clinical impact of quantitative assays for diagnosis of human cytomegalovirus infection/disease in immunocompromised patients. *New Microbiol.* **21:**293–308.

Gerna, G., A. D. Monforte, M. Zavattoni, A. Sarasini, L. Testa, and M. G. Revello. 1998d. Sharp drop in the prevalence of human cytomegalovirus leuko-DNAemia in HIV-infected patients following highly active antiretroviral therapy. *AIDS* **12:**118–120.

Gerna, G., F. Baldanti, J. M. Middeldorp, M. Furione, M. Zavattoni, D. Lilleri, and M. G. Revello. 1999. Clinical significance of expression of human cytomegalovirus pp67 late transcript in heart, lung, and bone marrow transplant recipients as determined by nucleic acid sequence-based amplification. *J. Clin. Microbiol.* **37:**902–911.

Gershon, A. A., S. P. Steinberg, and the Varicella Vaccine Collaborative Study Group of the National Institute of Allergy and Infectious Diseases. 1989. Persistence of immunity to varicella in children with leukemia immunized with live attenuated varicella vaccine. *N. Engl. J. Med.* **320:**892–897.

Gershon, A. A., P. LaRussa, and S. P. Steinberg. 1995. Varicella-zoster virus, p. 895–904. *In* P. R. Murray, E. J. Baron, M. A. Pfaller, F. C. Tenover, and R. H. Yolken (ed.), *Manual of Clinical Microbiology,* 6th ed. American Society for Microbiology, Washington, D.C.

Gershon, A. A., P. LaRussa, S. Steinberg, N. Mervish, S. H. Lo, and P. Meier. 1996. The protective effect of immunologic boosting against zoster: an analysis in leukemic children who were vaccinated against chickenpox. *J. Infect. Dis.* **173:**450–453.

Gershon, A. A., N. Mervish, P. Larussa, S. Steinberg, S. H. Lo, D. Hodes, S. Fikrig, V. Bonagura, and S. Bakshi. 1997. Varicella-zoster virus infection in children with underlying human immunodeficiency virus infection. *J. Infect. Dis.* **176:**1496–1500.

Ghazal, P., J. Young, E. Giulietti, C. DeMattei, J. Garcia, R. Gaynor, R. M. Stenberg, and J. A. Nelson. 1991. A discrete *cis* element in the human immunodeficiency virus long terminal repeat mediates synergistic *trans* activation by cytomegalovirus immediate early proteins. *J. Virol.* **65:**6735–6742.

Gibson, W. 1996. Structure and assembly of the virion. *Intervirology* **39:**389–400.

Gilbert, M. J., S. R. Riddell, B. Plachter, and P. D. Greenberg. 1996. Cytomegalovirus selectively blocks antigen processing of its immediate-early gene product. *Nature* **384:**720–722.

Gilden, D. H. 1994. Herpes zoster with postherpetic neuralgia—persisting pain and frustration. *N. Engl. J. Med.* **330:**932–933.

Gilden, D. H., J. L. Bennett, B. K. Kleinschmidtdemasters, D. D. Song, A. S. Yee, and I. Steiner. 1998. The value of cerebrospinal fluid antiviral antibody in the diagnosis of neurologic disease produced by varicella zoster virus. *J. Neurol. Sci.* **159:**140–144.

Gleaves, C. A., and J. D. Meyers. 1989. Rapid detection of cytomegalovirus in bronchoalveolar lavage specimens from marrow transplant patients: evaluation of a direct fluorescein-conjugated monoclonal antibody reagent. *J. Virol. Methods* **26:**345–350.

Gleaves, C. A., T. F. Smith, E. A. Shuster, and G. R. Pearson. 1984. Rapid detection of cytomegalovirus in MRC-5 cells inoculated with urine specimens by using low-speed centrifugation and monoclonal antibody to an early antigen. *J. Clin. Microbiol.* **19:**917–919.

Gleaves, C. A., C. F. Lee, C. I. Bustamante, and J. D. Meyers. 1988. Use of murine monoclonal antibodies for laboratory diagnosis of varicella-zoster virus infection. *J. Clin. Microbiol.* **26:**1623–1625.

Gleaves, C. A., D. Myerson, R. A. Bowden, R. C. Hackman, and J. D. Meyers. 1989. Direct detection of cytomegalovirus from bronchoalveolar lavage samples by using a rapid in situ DNA hybridization assay. *J. Clin. Microbiol.* **27:**2429–2432.

Gnann, J. W., and R. J. Whitley. 1991. Natural history and treatment of varicella-zoster in high-risk populations. *J. Hosp. Infect.* **18:**317–329.

Gonczol, E., C. de Taisne, G. Hirka, K. Berensci, W. C. Lin, E. Paoletti, and S. Plotkin. 1991. High expression of human cytomegalovirus (HCMV)-gB protein in cells infected with a vaccinia-gB recombinant: the importance of the gB protein in HCMV immunity. *Vaccine* **9:**631–637.

Goodpasture, E. W., and F. B. Talbot. 1921. Concerning the nature of "protozoan-like" cells in certain lesions of infancy. *Am. J. Dis. Child.* **21:**415–425.

Goodrich, J. M., M. Mori, C. A. Gleaves, C. Du Mond, M. Cays, D. F. Ebeling, W. C. Buhles, B. DeArmand, and J. D. Meyers. 1991. Early treatment with ganciclovir to prevent cytomegalovirus disease after allogeneic bone marrow transplantation. *N. Engl. J. Med.* **325:**1601–1607.

Goodrich, J. M., R. A. Bowden, L. Fisher, C. Keller, G. Schoch, and J. D. Meyers. 1993. Ganciclovir prophylaxis to prevent cytomegalovirus disease after allogeneic marrow transplant. *Ann. Intern. Med.* **118:**173–178.

Gordon, J., S. Ley, M. Melamed, P. Aman, and N. Hughes-Jones. 1984. Soluble factor requirements for the autostimulatory growth of B lymphoblasts immortalized by Epstein-Barr virus. *J. Exp. Med.* **159:**1554–1559.

Gordon, J., L. Walker, G. Guy, G. Brown, M. Rowe, and A. Rickinson. 1986. Control of human B-lymphocyte replication. II. Transforming Epstein-Barr virus exploits three distinct viral signals to undermine three separate control points in B-cell growth. *Immunology* **58:**591–595.

Gozlan, J., J. M. Salord, C. Chouaïd, C. Duvivier, O. Picard, M. C. Mejohas, and J. C. Petit. 1993. Human cytomegalovirus (HCMV) late mRNA detection in peripheral blood of AIDS patients: diagnostic value for HCMV disease compared with those of viral culture and HCMV DNA detection. *J. Clin. Microbiol.* **31:**1943–1945.

Gozlan, J., J. P. Laporte, S. Lesage, M. Labopin, A. Najman, N. C. Gorin, and J. C. Petit. 1996. Monitoring of cytomegalovirus infection and disease in bone marrow recipients by reverse transcription-PCR and blood and urine culture. *J. Clin. Microbiol.* **34:**2085–2088.

Grangeot-Keros, L., M. J. Mayaux, P. Lebon, F. Freymuth, G. Eugene, R. Stricker, and E. Dussaix. 1997. Value of cytomegalovirus (CMV) IgG avidity index for the diagnosis of primary CMV infection in pregnant women. *J. Infect. Dis.* **175:**944–946.

Gratama, J., M. Oosterveer, F. Zwaan, J. Lepoutre, G. Klein, and I. Ernberg. 1988. Eradication of Epstein-Barr virus by allogeneic marrow transplantation: implication for the site of viral latency. *Proc. Natl. Acad. Sci. USA* **85:**8693–8699.

Gratama, J. W., M. A. P. Oosterveer, and W. Weiner. 1994. Detection of multiple 'ebnotypes' in individual EBV carriers following lymphocyte transformation by virus derived from the peripheral blood and oropharynx. *J. Gen. Virol.* **75:**85–94.

Grattan, M. T., C. E. Moreno-Cabral, V. A. Starnes, P. E. Oyer, E. B. Stinson, and N. E. Shumway. 1989. Cytomegalovirus infection is associated with cardiac allograft rejection and atherosclerosis. *JAMA* **261:**3561–3566.

Green, M., M. Kaufmann, J. Wilson, and J. Reyes. 1997. Comparison of intravenous ganciclovir followed by oral acyclovir with intravenous ganciclovir alone for prevention of cytomegalovirus and Epstein-Barr virus disease after liver transplantation in children. *Clin. Infect. Dis.* **25:**1344–1349.

Greenspan, D., Y. G. DeSouza, M. A. Conant, H. Hollander, S. K. Chapman, E. T. Lennette, V. Petersen, and J. S. Greenspan. 1990. Efficacy of desiclovir in the treatment of Epstein-Barr virus infection in oral hairy leukoplakia. *J. Acquir. Immune Defic. Syndr.* **3:**571–578.

Greenspan, J. S., D. Greenspan, E. T. Lennett, D. I. Abrams, M. A. Conant, V. Petersen, and U. K. Freese. 1985. Replication of Epstein-Barr virus within the epithelial cells of oral "hairy" leukoplakia, an AIDS-associated lesion. *N. Engl. J. Med.* **313:**1564–1571.

Greijer, A. E., J. M. G. van der Crommert, S. J. C. Stevens, and J. M. Middeldorp. 1999. Molecular fine-specificity analysis of antibody responses to human cytomegalovirus and design of novel synthetic-peptide-based serodiagnostic assays. *J. Clin. Microbiol.* **37:**179–188.

Griffith, B. P., S. R. McCormick, C. K. Fong, J. T. Lavallee, H. L. Lucia, and E. Goff. 1985. The placenta as a site of cytomegalovirus infection in guinea pigs. *J. Virol.* **55:**402–409.

Griffiths, P. D., J. E. Feinberg, J. Fry, C. Sabin, L. Dix, G. Gor, A. Ansari, and V. C. Emery. 1998. The effect of valacyclovir on cytomegalovirus viremia and viruria detected by polymerase chain reaction in patients with advanced human immunodeficiency virus disease. *J. Infect. Dis.* **177:**57–64.

Grossi, P., S. Kusne, C. Rinaldo, K. St. George, M. Magnane, J. Rakela, J. Fung, and T. E. Starzl. 1996. Guidance of ganciclovir therapy with pp65 antigenemia in cytomegalovirus-free recipients of livers from seropositive donors. *Transplantation* **61:**1659–1660.

Grundy, J. E., J. D. Shanley, and P. D. Griffiths. 1987. Is cytomegalovirus interstitial pneumonitis in transplant recipients an immunopathological condition? *Lancet* **ii:**996–999.

Grundy, J. E., S. F. Lui, M. Super, N. J. Berry, P. Sweny, O. N. Fernando, J. Moorhead, and P. D. Griffiths. 1988. Symptomatic cytomegalovirus infection in seropositive kidney recipients: reinfection with donor virus rather than reactivation of recipient virus. *Lancet* **ii:**132–135.

Gudnason, T., K. K. Belani, and H. H. Balfour, Jr. 1989. Ganciclovir treatment of cytomegalovirus disease in immunocompromised children. *Pediatr. Infect. Dis. J.* **8:**436–440.

Hackman, R. C., D. Myerson, J. D. Meyers, H. M. Shulman, G. E. Sale, L. C. Goldstein, M. Rastetter, N. Fluornoy, and E. D. Thomas. 1985. Rapid diagnosis of cytomegaloviral pneumonia by tissue immunofluorescence with a murine monoclonal antibody. *J. Infect. Dis.* **151:**325–329.

Hahn, G., R. Jores, and E. S. Mocarski. 1998. Cytomegalovirus remains latent in a common precursor of dendritic and myeloid cells. *Proc. Natl. Acad. Sci. USA* **95:**3937–3942.

Halwachs, G., J. Kutschera, A. Tiran, H. Rosegger, H. Engele, M. Danda, H. Lackner, U. Maurer, and M. Wilders-Truschnig. 1995. Antiviral treatment of congenitally infected children with a positive cytomegalovirus polymerase chain reaction in the cerebrospinal fluid. *Scand. J. Infect. Dis. Suppl.* **99:**89–90.

Halwachs-Baumann, G., M. Wilders-Truschnig, G. Desoye, T. Hahn, L. Kiesel, K. Klingel, P. Rieger, G. Jahn, and C. Sinzger. 1998. Human trophoblast cells are permissive to the complete replicative cycle of human cytomegalovirus. *J. Virol.* **72:**7598–7602.

Hamilton-Dutoit, S. J., G. Pallesen, M. B. Franzmann, J. Karkov, F. Black, P. Skinhoj, and C. Pedersen. 1991. AIDS-related lymphoma: histopathology, immunophenotype and association with Epstein-Barr virus as demonstrated by *in situ* nucleic acid hybridization. *Am. J. Pathol.* **138:**149–163.

Hammerschmidt, W., and B. Sugden. 1989. Genetic analysis of immortalizing functions of Epstein-Barr virus in human B lymphocytes. *Nature* **340:**393–397.

Hamprecht, K., E. Mikeler, and G. Jahn. 1997. Semi-quantitative detection of cytomegalovirus DNA from native serum and plasma by nested PCR—influence of DNA extraction procedures. *J. Virol. Methods* **69:**125–135.

Hamzeh, F. M., P. S. Lietman, W. Gibson, and G. S. Hayward. 1990. Identification of the lytic origin of DNA replication in human cytomegalovirus by a novel approach utilizing ganciclovir-induced chain termination. *J. Virol.* **64:**6184–6195.

Handsfield, H. H., S. H. Chandler, V. A. Caine, J. D. Meyers, L. Corey, E. Medeiros, and J. K. McDougall. 1985. Cytomegalovirus infection in sex partners: evidence for sexual transmission. *J. Infect. Dis.* **151:**344–348.

Hansen, K. K., J. Vestbo, T. Benfield, J. D. Lundgren, and L. R. Mathiesen. 1997. Rapid detection of cytomegalovirus in bronchoalveolar lavage fluid and serum samples by polymerase chain reaction: correlation of virus isolation and clinical outcome for patients with human immunodeficiency virus infection. *Clin. Infect. Dis.* **24:**878–883.

Haque, T., J. A. Thomas, K. I. Falk, R. Parratt, B. J. Hunt, M. Yacoub, and D. H. Crawford. 1996. Transmission of donor Epstein-Barr virus (EBV) in transplanted organs causes lymphoproliferative disease in EBV-seronegative recipients. *J. Gen. Virol.* **77:**1169–1172.

Hardy, I., A. A. Gershon, S. P. Steinberg, P. LaRussa, and the Varicella Vaccine Collaborative Study Group. 1991. The incidence of zoster after immunization with live attenuated varicella vaccine. A study in children with leukemia. *N. Engl. J. Med.* **325:**1545–1550.

Harris N. L., J. A. Ferry, and S. H. Swerdlow. 1997. Post-transplant lymphoproliferative disorders (PTLD): summary of the Society for Haematology Workshop. *Semin. Diagn. Pathol.* **14:**8–13.

He, H., C. R. Rinaldo, Jr., and P. A. Morel. 1995. T cell proliferative responses to five human cytomegalovirus proteins in healthy seropositive individuals: implications for vaccine development. *J. Gen. Virol.* **76:**1603–1610.

Hebart, H., M. Greif, H. Krause, L. Kanz, G. Jahn, C. A. Muller, and H. Einsele. 1997. Interstrain variation of immediate early DNA sequences and glycoprotein B genotypes in cytomegalovirus clinical isolates. *Med. Microbiol. Immunol.* **186:**135–138.

Hemmings, D. G., R. Kilani, C. Nykiforuk, J. Preiksaitis, and L. J. Guilbert. 1998. Permissive cytomegalovirus infection of primary villous term and first trimester trophoblasts. *J. Virol.* **72:**4970–4979.

Hendrix, M. G., M. M. Salimans, C. P. van Boven, and C. A. Bruggeman. 1990. High prevalence of latently present cytomegalovirus in arterial walls of patients suffering from grade III atherosclerosis. *Am. J. Pathol.* **136:**23–28.

Hendrix, R. M., M. Wagenaar, R. L. Slobbe, and C. A. Bruggeman. 1997. Widespread presence of cytomegalovirus DNA in tissues of healthy trauma victims. *J. Clin. Pathol.* **50:**59–63.

Hengel, H., J. O. Koopmann, T. Flohr, W. Muranyi, E. Goulmy, G. J. Hammerling, U. H. Koszinowski, and F. Momburg. 1997. A viral ER-resident glycoprotein inactivates the MHC-encoded peptide transporter. *Immunity* **6:**623–632.

Hengge, U. R., N. H. Brockmeyer, R. Malessa, U. Ravens, and M. Goos. 1993. Foscarnet penetrates the blood-brain barrier: rationale for therapy of cytomegalovirus encephalitis. *Antimicrob. Agents Chemother.* **37:**1010–1014.

Henle, G., and W. Henle. 1966. Immunofluorescence in cells derived from Burkitt's lymphoma. *J. Bacteriol.* **91:**1248–1256.

Henle, G., and W. Henle. 1979. The virus as the etiologic agent of infectious mononucleosis, p. 297–320. *In* M. A. Epstein and B. G. Achong (ed.), *The Epstein-Barr Virus.* Springer-Verlag, Berlin, Germany.

Henle, G., W. Henle, and V. Diehl. 1968a. Immunofluorescence in cells derived from Burkitt's lymphoma. *J. Bacteriol.* **91:**1248–1256.

Henle, G., W. Henle, and V. Diehl. 1968b. Relation of Burkitt's tumor-associated herpes-type virus to infectious mononucleosis. *Proc. Natl. Acad. Sci. USA* **59:**94–101.

Henle, G., W. Henle, H. G. Ho, P. Burtin, Y. Cachin, P. Clifford, A. De Schryver, G. De The, V. Diehl, and G. Klein. 1970a. Antibodies to Epstein-Barr virus in nasopharyngeal carcinoma, other head and neck neoplasms and control groups. *J. Natl. Cancer Inst.* **44:**225–231.

Henle, W., V. Diehl, G. Kohn, H. zur Hausen, and G. Henle. 1967. Herpes type-virus and chromosome marker in normal leukocytes after growth with irradiated Burkitt cells. *Science* **157:**1064–1065.

Henle, W., G. Henle, B. A. Zajac, G. Pearson, R. Waubke, and M. Scriba. 1970b. Differential reactivity of human serums with early antigens induced by Epstein-Barr virus. *Science* **169:**188–190.

Henle, W., G. Henle, J. Andersson, I. Ernberg, G. Klein, C. A. Horwitz, G. Marklund, L. Rymo, C. Wellinder, and S. E. Straus. 1987. Antibody responses to Epstein-Barr virus-determined nuclear antigen (EBNA)-1 and EBNA 2 in acute and chronic Epstein-Barr virus infection. *Proc. Natl. Acad. Sci. USA* **84:**570–574.

Heslop, H. E., M. K. Brenner, and C. M. Rooney. 1994. Donor T cells to treat EBV-associated lymphoma. *N. Engl. J. Med.* **331**:679–684.

Heslop, H. E., C. Y. C. Ng, C. Li, C. A. Smith, S. K. Lofton, R. A. Krance, M. K. Brenner, and C. M. Rooney. 1996. Longterm restoration of immunity against EBV infection by adoptive transfer of gene-modified T lymphocytes. *Nat. Med.* **2**:551–558.

Hibberd, P. L., N. E. Tolkoff-Rubin, A. B. Cosimi, R. T. Schooley, D. Isaacson, M. Doran, A. Delvecchio, F. L. Delmonico, H. Auchincloss, Jr., and R. H. Rubin. 1992. Symptomatic cytomegalovirus disease in cytomegalovirus antibody seropositive renal transplant recipients treated with OKT3. *Transplantation* **53**:68–72.

Hibberd, P. L., N. E. Tolkoff-Rubin, D. Conti, F. Stuart, J. R. Thistlethwaite, J. F. Neylan, D. R. Snydman, D. Freeman, M. I. Lorber, and R. H. Rubin. 1995. Preemptive ganciclovir therapy to prevent cytomegalovirus disease in antibody positive renal transplant recipients. *Ann. Intern. Med.* **123**:18–26.

Hille, A., K. K. Klein, S. Baumler, F. A. Grasser, and N. Mueller-Lantzsch. 1993. Expression of Epstein-Barr virus nuclear antigen 1, 2A and 2B in the baculovirus expression system: serological evaluation of human antibodies to these proteins. *J. Med. Virol.* **39**:233–241.

Hitchcock, M. J. M., H. S. Jaffe, J. C. Martin, and R. J. Stagg. 1996. Cidofovir, a new agent with potent anti-herpesvirus activity. *Antimicrob. Agents Chemother.* **7**:115–127.

Ho, M. 1977. Virus infections after transplantation in man. *Arch. Virol.* **55**:1–24.

Ho, M. 1991. Nonhuman cytomegaloviruses, p. 319–326. *In* M. Ho (ed.), *Cytomegalovirus Biology and Infection*, 2nd ed. Plenum Medical Book Company, New York, N.Y.

Ho, M., S. Suwansirikul, J. N. Dowling, L. A. Youngblood, and J. A. Armstrong. 1975. The transplanted kidney as a source of cytomegalovirus infections. *N. Engl. J. Med.* **293**:1109–1112.

Ho, M., G. Miller, R. W. Atchison, M. K. Breinig, J. S. Dummer, W. Andiman, T. E. Starzl, R. Eastman, B. P. Griffith, R. L. Hardesty, H. T. Bahnson, T. R. Hakala, and J. T. Rosenthal. 1985. Epstein-Barr virus infections and DNA hybridization studies in posttransplantation lymphoma and lymphoproliferative lesions: the role of primary infection. *J. Infect. Dis.* **152**:876–886.

Hocher, J. R., L. N. Cook, G. Adams, and G. P. F. Rabalais. 1990. Ganciclovir therapy of congenital cytomegalovirus pneumonia. *Pediatr. Infect. Dis. J.* **9**:743–745.

Howe, J., and J. Steitz. 1986. Localization of Epstein-Barr virus encoded small RNAs by in situ hybridization. *Proc. Natl. Acad. Sci. USA* **83**:9006–9010.

Ippoliti, C., A. Morgan, D. Waarkentin, K. van Besien, R. Mehra, I. Khouri, S. Giralt, J. Gajewski, R. Champlin, B. Andersson, and D. Przepiorka. 1997. Foscarnet for prevention of cytomegalovirus infection in allogeneic marrow transplant recipients unable to receive ganciclovir. *Bone Marrow Transplant.* **20**:491–495.

Ivarsson, S.-A., B. Lernmark, and L. Svanberg. 1997. Ten-year clinical, developmental, and intellectual follow-up of children with congenital cytomegalovirus infection without neurologic symptoms at one year of age. *Pediatrics* **99**:800–803.

Ives, D. V. 1997. Cytomegalovirus disease in AIDS. *AIDS* **11**:1791–1797.

Iwamoto, G. K., and S. A. Konicek. 1997. Cytomegalovirus immediate early genes upregulate interleukin-6 gene expression. *J. Investig. Med.* **45**:175–182.

Jacobson, M. A., and M. French. 1998. Altered natural history of AIDS-related opportunistic infections in the era of potent combination antiretroviral therapy. *AIDS* **S157–S163**.

Jacobson, M. A., and J. Mills. 1988. Serious cytomegalovirus disease in the acquired immunodeficiency syndrome (AIDS). Clinical findings, diagnosis, and treatment. *Ann. Intern. Med.* **108**:585–594.

Jacobson, M. A., T. G. Berger, S. Fikrig, P. Becherer, J. W. Moohr, S. C. Stanat, and K. K. Biron. 1990. Acyclovir-resistant varicella zoster virus infection after chronic oral acyclovir therapy in patients with the acquired immunodeficiency syndrome (AIDS). *Ann. Intern. Med.* **112**:187–191.

Jacobson, M. A., J. G. Gambertoglio, F. T. Aweeka, D. M. Causey, and A. A. Portale. 1991. Foscarnet-induced hypocalcemia and effects of foscarnet on calcium metabolism. *J. Clin. Endocrinol. Metab.* **72**:1130–1135.

Jacobson, M. A., S. Wilson, H. Stanley, C. Holtzer, J. Cherrington, and S. Safrin. 1999. Phase I study of combination therapy with intravenous cidofovir and oral ganciclovir for cytomegalovirus retinitis in patients with AIDS. *Clin. Infect. Dis.* **28**:528–533.

Jarvis, M. A., C. E. Wang, H. L. Meyers, P. P. Smith, C. L. Corless, G. J. Henderson, J. Vieira, W. J. Britt, and J. A. Nelson. 1999. Human cytomegalovirus infection of Caco-2 cells occurs at the basolateral membrane and is differentiation state dependent. *J. Virol.* **73**:4552–4560.

Jault, F. M., J. M. Jault, F. Ruchti, E. A. Fortunato, C. Clarke, J. Corbeil, D. D. Richman, and D. H. Spector. 1995. Cytomegalovirus infection induces high levels of cyclins, phosphorylated Rb, and p53, leading to cell cycle arrest. *J. Virol.* **69**:6697–6704.

Jennings, R., T. Green, and G. R. Kinghorn. 1998. Herpesvirus vaccines—an update. *Biodrugs* **10**:257–264.

Jones, T. R., and L. Sun. 1997. Human cytomegalovirus US2 destabilizes major histocompatibility complex class I heavy chains. *J. Virol.* **71**:2970–2979.

Jones, T. R., L. K. Hanson, L. Sun, J. S. Slater, R. M. Stenberg, and A. E. Campbell. 1995. Multiple independent loci within the human cytomegalovirus unique short region downregulate expression of major histocompatibility complex class I heavy chains. *J. Virol.* **69**:4830–4841.

Jones, T. R., E. J. Wiertz, L. Sun, K. N. Fish, J. A. Nelson, and H. L. Ploegh. 1996. Human cytomegalovirus US3 impairs transport and maturation of major histocompatibility complex class I heavy chains. *Proc. Natl. Acad. Sci. USA* **93**:11327–11333.

Jung, D., and A. Dorr. 1999. Single-dose pharmacokinetics of valganciclovir in HIV- and CMV-seropositive subjects. *J. Clin. Pharmacol.* **39**:800–804.

Karlin, S., B. Blaisdell, and G. Schachtel. 1990. Contrasts in codon usage of latent versus productive genes of Epstein-Barr virus: data and hypotheses. *J. Virol.* **64**:4264–4273.

Katlama, C., E. Dohin, E. Caumes, I. Cochereau-Massin, C. Brancon, M. Robinet, O. Rogeaux, R. Dahan, and M. Gentilini. 1992. Foscarnet induction therapy for cytomegalovirus retinitis in AIDS: comparison of twice-daily and three-times-daily regimes. *J. Acquir. Immune. Defic. Syndr.* **5** (Suppl. 1):S18–S24.

Kavaliotis, J., I. Loukou, M. Trachana, N. Gombakis, H. Tsagaropoulou-Stigga, and D. Koliouskas. 1998. Outbreak of varicella in a pediatric oncology unit. *Med. Pediatr. Oncol.* **31**:166–169.

Kaye, J. F., U. A. Gompels, and A. C. Minson. 1992. Glycoprotein H of human cytomegalovirus (HCMV) forms a stable complex with the HCMV UL115 gene product. *J. Gen Virol.* **73**:2693–2698.

Kaye, K., K. Izumi, and E. Kieff. 1993. Epstein-Barr virus latent membrane protein 1 is essential for B-lymphocyte growth transformation. *Proc. Natl. Acad. Sci. USA* **90:**9150–9154.

Keay, S., and B. Baldwin. 1991. Anti-idiotype antibodies that mimic gp86 of human cytomegalovirus inhibit viral fusion but not attachment. *J. Virol.* **65:**5124–5128.

Keay, S., D. Oldach, A. Wiland, D. Klassen, E. Schweitzer, L. V. Abruzzo, D. Kumar, and S. Bartlett. 1998. Post-transplant lymphoproliferative disorder associated with OKT3 and decreased antiviral prophylaxis in pancreas transplant recipients. *Clin. Infect. Dis.* **26:**596–600.

Keenan, R. J., M. E. Lega, J. S. Dummer, I. L. Paradis, J. H. Dauber, H. Rabinowich, S. A. Yousem, R. L. Hardesty, B. P. Griffith, R. J. Duquesnoy, and A. Zeevi. 1991. Cytomegalovirus serologic status and postoperative infection correlated with risk of developing chronic rejection after pulmonary transplantation. *Transplantation* **51:**433–438.

Kenagy, D. N., Y. Schlesinger, K. Weck, J. N. Ritter, M. M. Gaudreault-Keener, and G. A. Storch. 1995. Epstein-Barr virus DNA in peripheral blood leukocytes of patients with posttransplant lymphoproliferative disease. *Transplantation* **60:**547–554.

Kennedy, P. G. E., E. Grinfeld, and J. W. Gow. 1998. Latent varicella-zoster virus is located predominantly in neurons in human trigeminal ganglia. *Proc. Natl. Acad. Sci. USA* **95:**4658–4662.

Khanna, R., R. Slade, L. Poulsen, D. Moss, S. Burrows, J. Nicholls, and J. Burrows. 1992. Evolutionary dynamics of genetic variation in Epstein-Barr virus isolates of diverse geographical origins. Evidence for immune-pressure-independent genetic drift. *J. Virol.* **71:**8340–8346.

Kieff, E. 1996. Herpesviridae, p. 2343–2396. *In* B. N. Fields, D. M. Knipe, and P. M. Howley (ed.), *Virology,* 3rd ed. Raven Press, New York, N.Y.

Kinchington, P. R. 1999. Latency of varicella zoster virus: a persistently perplexing state. *Front. Biosci.* **4:**1–13.

Kinchington, P. R., and J. I. Cohen. Varicella zoster virus proteins. *In* A. Gershon and A. Arvin (ed.), *Varicella-Zoster Virus,* in press.

Kitchen, B. J., H. D. Engler, V. J. Gill, D. Marshall, S. M. Steinberg, P. A. Pizzo, and B. U. Mueller. 1997. Cytomegalovirus infection in children with human immunodeficiency virus infection. *Pediatr. Infect. Dis. J.* **16:**358–363.

Kjellström, C., T. Bergström, G. Martensson, A. Ricksten, F. Nilsson, S. Olofsson, and V. P. Collins. 1997. Relation between polymerase chain reaction findings and morphological changes during cytomegalovirus infection in transplanted lung. *Diagn. Mol. Pathol.* **6:**267–276.

Klemola, E., and L. Kaariainen. 1965. Cytomegalovirus as a possible cause of a disease resembling infectious mononucleosis. *Br. Med. J.* **ii:**1099–1102.

Klemola, E., R. Von Essen, G. Henle, and W. Henle. 1970. Infectious-mononucleosis-like disease with negative heterophil agglutination test. Clinical features in relation to Epstein-Barr virus and cytomegalovirus antibodies. *J. Infect. Dis.* **121:**608–614.

Kline, J. N., G. M. Hunninghake, B. He, M. M. Monick, and G. W. Hunninghake. 1998. Synergistic activation of the human cytomegalovirus major immediate early promoter by prostaglandin E2 and cytokines. *Exp. Lung. Res.* **24:**3–14.

Kneiss, N., M. Mach, J. Fay, and W. J. Britt. 1991. Distribution of linear antigenic sites on glycoprotein gp55 of human cytomegalovirus. *J. Virol.* **65:**138–146.

Knowles, D. M., E. Cesarman, A. Chadburn, G. Frizzera, J. Chen, E. A. Rose, and R. E. Michler. 1995. Correlative morphology and molecular genetic analysis demonstrates

three distinct categories of lymphoproliferative disorders. *Blood* **85:**552–562.

Knox, C. M., D. Chandler, G. A. Short, and T. P. Margolis. 1998. Polymerase chain reaction-based assays of vitreous samples for the diagnosis of viral retinitis—use in diagnostic dilemmas. *Ophthalmology* **105:**37–44.

Kocoshis, S. A. 1994. Small bowel transplantation in infants and children. *Pediatr. Gastroenterol.* **23:**727–742.

Koehler, M., K. St. George, G. D. Ehrlich, J. Mirro, S. M. Neudorf, and C. Rinaldo. 1997. Prevention of CMV disease in allogeneic BMT recipients by CMV antigenemia-guided preemptive therapy. *J. Pediatr. Hematol./Oncol.* **19:**43–47.

Kondo, K., and E. S. Mocarski. 1995. Cytomegalovirus latency and latency-specific transcription in hematopoietic progenitors. *Scand. J. Infect. Dis. Suppl.* **99:**63–67.

Kondo, K., H. Kaneshima, and E. S. Mocarski. 1994. Human cytomegalovirus latent infection of granulocyte-macrophage progenitors. *Proc. Natl. Acad. Sci. USA* **91:**11879–11883.

Kondo, K., J. Xu, and E. S. Mocarski. 1996. Human cytomegalovirus latent gene expression in granulocyte-macrophage progenitors in culture and in seropositive individuals. *Proc. Natl. Acad. Sci. USA* **93:**11137–11142.

Koskinen, P. K., M. S. Nieminen, L. A. Krogerus, K. B. Lemstrom, S. P. Mattila, P. J. Hayry, and I. T. Lautenschlager. 1993a. Cytomegalovirus infection and accelerated cardiac allograft vasculopathy in human cardiac allografts. *J. Heart Lung Transplant.* **12:**724–729.

Koskinen, P. K., M. S. Nieminen, S. P. Mattila, P. J. Häyry, and I. T. Lautenschlager. 1993b. The correlation between symptomatic CMV infection and CMV antigenemia in heart allograft recipients. *Transplantation* **55:**547–551.

Koskiniemi, M., L. Mannonen, A. Kallio, and A. Vaheri. 1997. Luminometric microplate hybridization for detection of varicella-zoster virus PCR product from cerebrospinal fluid. *J. Virol. Methods* **63:**71–79.

Krause, H., H. Hebart, G. Jahn, C. A. Müller, and H. Einsele. 1997. Screening for CMV-specific T cell proliferation to identify patients at risk of developing late onset CMV disease. *Bone Marrow Transplant.* **19:**1111–1116.

Kusne, S., R. Manez, B. L. Frye, K. St. George, K. Abu-Elmagd, J. Tabasco-Menguillon, J. J. Fung, S. Todo, C. Rinaldo, and G. D. Ehrlich. 1997. Use of DNA amplification for diagnosis of cytomegalovirus enteritis after intestinal transplantation. *Gastroenterology* **112:**1121–1128.

Lalezari, J. P., J. A. Aberg, L. H. Wang, R. C. Miner, M. B. Wire, M. A. Jacobson, and W. L. Drew. 1998. In vivo anti-CMV activity safety and pharmacokinetics of oral 1263W94 in HIV-infected subjects with asymptomatic CMV shedding. *In Fifth Conference on Retroviruses and Opportunistic Infections,* abstr. 762, Chicago, Ill.

Lalezari, J. P., R. J. Stagg, B. D. Kuppermann, G. N. Holland, F. Kramer, D. V. Ives, M. Youle, M. R. Robinson, W. L. Drew, and H. S. Jaffe. 1997. Intravenous cidofovir for peripheral CMV retinitis in patients with AIDS. A randomized, controlled trial. *Ann. Intern. Med.* **126:**257–263.

Lam, K. M., N. Oldenburg, M. A. Khan, V. Gaylore, G. W. Mikhail, P. D. Strouhal, J. M. Middeldorp, N. Banner, and M. Yacoub. 1998. Significance of reverse transcription polymerase chain reaction in the detection of human cytomegalovirus gene transcripts in thoracic organ transplant recipients. *J. Heart Lung Transplant.* **17:**555–565.

Landini, M. P., M. X. Guan, G. Jahn, W. Lindenmaier, M. Mach, A. Ripalti, A. Necker, T. Lazzarotto, and B. Plachter. 1990. Large-scale screening of human sera with cytomegalovirus recombinant antigens. *J. Clin. Microbiol.* **28:**1375–1379.

Landry, M. L., D. Ferguson, S. Cohen, K. Huber, and P. Wetherill. 1995. Effect of delayed specimen processing on cytomegalovirus antigenemia test results. *J. Clin. Microbiol.* **33:**257–259.

Larsson, S., C. Söderberg-Nauclér, and E. Möller. 1998a. Productive cytomegalovirus (CMV) infection exclusively in CD13-positive peripheral blood mononuclear cells from CMV-infected individuals. *Transplantation* **65:**411–415.

Larsson, S., C. Söderberg-Nauclér, F. Z. Wang, and E. Möller. 1998b. Cytomegalovirus DNA can be detected in peripheral blood mononuclear cells from all seropositive and most seronegative healthy blood donors over time. *Transfusion* **38:**271–278.

Larussa, P., S. Steinberg, F. Meurice, and A. Gershon. 1997. Transmission of vaccine strain varicella-zoster virus from a healthy adult with vaccine-associated rash to susceptible household contacts. *J. Infect. Dis.* **176:**1072–1075.

Larussa, P., S. Steinberg, A. Arvin, D. Dwyer, M. Burgess, M. Menegus, K. Rekrut, K. Yamanishi, and A. Gershon. 1998. Polymerase chain reaction and restriction fragment length polymorphism analysis of varicella-zoster virus isolates from the United States and other parts of the world. *J. Infect. Dis.* **178:**S64–S66.

Lautenschlager, I., K. Höckerstedt, H. Jalanko, R. Loginov, K. Salmela, E. Taskinen, and J. Ahonen. 1997. Persistent cytomegalovirus in liver allografts with chronic rejection. *Hepatology* **25:**190–194.

Lawrence, R., A. A. Gershon, R. Holzman, S. P. Steinberg, and the NIAID Varicella Vaccine Collaborative Study Group. 1988. The risk of zoster after varicella vaccination in children with leukemia. *N. Engl. J. Med.* **318:**543–548.

Lazzarotto, T., P. Spezzacatena, P. Pradelli, D. A. Abate, S. Varani, and M. P. Landini. 1997. Avidity of immunoglobulin G directed against human cytomegalovirus during primary and secondary infections in immunocompetent and immunocompromised subjects. *Clin. Diagn. Lab. Immunol.* **4:**469–473.

Lazzarotto, T., B. Guerra, P. Spezzacatena, S. Varani, L. Gabrielli, P. Pradelli, F. Rumpianesi, C. Banzi, L. Bovicelli, and M. P. Landini. 1998a. Prenatal diagnosis of congenital cytomegalovirus infection. *J. Clin. Microbiol.* **36:**3540–3544.

Lazzarotto, T., S. Varani, P. Spezzacatena, P. Pradelli, L. Potena, A. Lombardi, V. Ghisetti, L. Gabrielli, D. A. Abate, C. Magelli, and M. P. Landini. 1998b. Delayed acquisition of high-avidity anti-cytomegalovirus antibody is correlated with prolonged antigenemia in solid organ transplant recipients. *J. Infect. Dis.* **178:**1145–1149.

Lea, A. P. and H. M. Bryson. 1996. Cidofovir. *Drugs* **52:**225–230.

Leblond, V., L. Sutton, R. Dorent, F. Davi, M. O. Bitker, J. Gabarre, F. Charlotte, J. J. Ghoussoub, C. Fourcade, and A. Fischer. 1995. Lymphoproliferative disorders after organ transplantation: a report of 24 cases observed in a single center. *J. Clin. Oncol.* **13:**961–968.

Lehner, P. J., J. T. Karttunen, G. W. Wilkinson, and P. Cresswell. 1997. The human cytomegalovirus US6 glycoprotein inhibits transporter associated with antigen processing-dependent peptide translocation. *Proc. Natl. Acad. Sci. USA* **94:**6904–6909.

Lehner, R., H. Meyer, and M. Mach. 1989. Identification and characterization of a human cytomegalovirus gene coding for a membrane protein that is conserved among human herpesviruses. *J. Virol.* **63:**3792–3800.

Leong, C. C., T. L. Chapman, P. J. Bjorkman, D. Formankova, E. S. Mocarski, J. H. Phillips, and L. L. Lanier. 1998. Modulation of natural killer cell cytotoxicity in human cytomegalovirus infection: the role of endogenous class I major histocompatability complex and a viral class I homolog. *J. Exp. Med.* **187:**1681–1687.

le Roy, E., A. Mühlethaler-Mottet, C. Davrinche, B. Mach, and J.-L. Davignon. 1999. Escape of human cytomegalovirus from HLA-DR-restricted CD4⁺-T-cell response is mediated by repression of gamma interferon-induced class II transactivator expression. *J. Virol.* **73:**6582–6589.

Levin, M. J., C. R. Rinaldo, Jr., P. L. Leary, J. A. Zaia, and M. S. Hirsch. 1979. Immune response to herpesvirus antigens in adults with acute cytomegaloviral mononucleosis. *J. Infect. Dis.* **140:**851–857.

Levin, M. J., D. Barber, E. Goldblatt, M. Jones, B. Lafleur, C. Chan, D. Stinson, G. O. Zerbe, and A. R. Hayward. 1998. Use of a live attenuated varicella vaccine to boost varicella-specific immune responses in seropositive people 55 years of age and older—duration of booster effect. *J. Infect. Dis.* **178:**S109–S112.

Levy, R., F. Najioullah, D. Thouvenot, S. Bosshard, M. Aymard, and B. Lina. 1996. Evaluation and comparison of PCR and hybridization methods for rapid detection of cytomegalovirus in clinical samples. *J. Virol. Methods* **62:**103–111.

Lewin, N., P. Aman, M. Masucci, G. Klein, B. Oderg, H. Strander, W. Henle, and G. Henle. 1987. Characterization of EBV carrying B cell populations in healthy seropositive individuals with regard to density, release of transforming virus and spontaneous outgrowth. *Int. J. Cancer* **39:**472–476.

Li, C. R., P. D. Greenberg, M. J. Gilbert, J. M. Goodrich, and S. R. Riddell. 1994. HLA-restricted cytomegalovirus (CMV)-specific T-cell responses after allogeneic bone marrow transplant: correlation with CMV disease and effect of ganciclovir prophylaxis. *Blood* **83:**1971–1979.

Limaye, A. P., R. A. Bowden, D. Myerson, and M. Boeckh. 1997. Cytomegalovirus disease occurring before engraftment in marrow transplant recipients. *Clin. Infect. Dis.* **24:**830–835.

Lin, T. Y., Y. C. Huang, H. C. Ning, and C. Hsueh. 1997. Oral acyclovir prophylaxis of varicella after intimate contact. *Pediatr. Infect. Dis. J.* **16:**1162–1165.

Lipson, S. M., M. E. Match, A. L. Toro, M. H. Kaplan, and D. H. Shepp. 1998. Application of a standardized cytomegalovirus antigenemia assay in the management of patients with AIDS. *Diagn. Microbiol. Infect. Dis.* **32:**75–79.

Litwin, V., and C. Grose. 1992. Herpesviral Fc receptors and their relationship to the human Fc receptors. *Immunol. Res.* **11:**226–238.

Liu, Y. N., A. Klaus, B. Kari, M. F. Stinski, J. Edkhardt, and R. C. Gehrz. 1991. The N-terminal 513 amino acids of the envelope glycoprotein gB of human cytomegalovirus stimulates both B- and T-cell immune responses in humans. *J. Virol.* **65:**1644–1648.

Liu, Y. N., J. Curtsinger, P. R. Donahue, A. Klaus, G. Optiz, J. Cooper, R. W. Karr, F. H. Bach, and R. C. Gehrz. 1993. Molecular analysis of the immune response to human cytomegalovirus glycoprotein B. I. Mapping of HLA-restricted helper T-cell epitopes on gp93. *J. Gen. Virol.* **74:**2207–2214.

Lo, C. Y., K. N. Ho, K. Y. Yuen, S. L. Lui, F. K. Li, T. M. Chan, W. K. Lo, and I. K. P. Cheng. 1997. Diagnosing cytomegalovirus disease in CMV seropositive renal allograft recipients: a comparison between the detection of CMV DNAemia by polymerase chain reaction and antigenemia by CMV pp65 assay. *Clin. Transplant.* **11:**286–293.

Loebe, M., S. Schuler, O. Zais, H. Warnecke, E. Fleck, and R. Hetzer. 1990. Role of cytomegalovirus infection in the development of coronary artery disease in the transplanted heart. *J. Heart Transplant.* **9:**707–711.

Lumbreras, C., I. Fernandez, J. Velosa, S. Munn, S. Sterioff, and C. V. Paya. 1995. Infectious complications following pancreatic transplantation: incidence, microbiological and clinical characteristics, and outcome. *Clin. Infect. Dis.* **20**:514.

Lung, M. L., R. S. Chang, M. L. Huang, H. Y. Guo, D. Choy, J. Sham, S. Y. Tsao, P. Cheng, and M. H. Ng. 1990. Epstein-Barr virus genotypes associated with nasopharyngeal carcinoma in Southern China. *Virology* **177**:44–53.

Lung, M. L., W. P. Lam, J. Sham, D. Choy, Z. Yong-Sheng, H. Y. Guo, and M. H. Ng. 1991. Detection and prevalence of the Af@ variant of Epstein-Barr virus in Southern China. *Virology* **185**:67–71.

Lungu, O., P. W. Annunziato, A. Gershon, S. M. Staugaitis, D. Josefson, P. LaRussa, and S. J. Silverstein. 1995. Reactivated and latent varicella-zoster virus in human dorsal root ganglia. *Proc. Natl. Acad. Sci. USA* **92**:10980–10984.

Lungu, O., C. A. Panagiotidis, P. W. Annunziato, A. A. Gershon, and S. J. Silverstein. 1998. Aberrant intracellular localization of varicella-zoster virus regulatory proteins during latency. *Proc. Natl. Acad. Sci. USA* **95**:7080–7085.

Lurain, N. S., H. C. Ammons, K. S. Kapell, V. V. Yeldandi, E. R. Garrity, and J. P. O'Keefe. 1996. Molecular analysis of human cytomegalovirus strains from two lung transplant recipients with the same donor. *Transplantation* **62**:497–502.

Lutz, E., K. N. Ward, and J. J. Gray. 1994. Maturation of antibody avidity after primary human cytomegalovirus infection is delayed in immunosuppressed solid organ transplant patients. *J. Med. Virol.* **44**:317–322.

Lyznicki, J. M., R. J. Bezman, and M. Genel. 1998. Report of the Council on Scientific Affairs American Medical Association—immunization of healthcare workers with varicella vaccine. *Infect. Control Hosp. Epidemiol.* **19**:348–353.

Magrath, I. 1990. The pathogenesis of Burkitt's lymphoma. *Adv. Cancer Res.* **55**:133–269.

Mahalingam, R., M. Wellish, W. Wolf, A. N. Dueland, R. Cohrs, A. Vafai, and D. Gilden. 1990. Latent varicella-zoster viral DNA in human trigeminal and thoracic ganglia. *N. Engl. J. Med.* **323**:627–631.

Mainka, C., B. Fuss, H. Geiger, H. Hofelmayr, and M. H. Wolff. 1998. Characterization of viremia at different stages of varicella-zoster virus infection. *J. Med. Virol.* **56**:91–98.

Maltezou, H. C., D. Petropoulos, M. Gardner, D. Abisaid, E. C. Mantzouranis, K. V. I. Roston, and K. W. Chan. 1998. Varicella-zoster virus infection in children with hematopoietic stem cell transplant. *Int. J. Pediatr. Hematol. Oncol.* **5**:345–351.

Manez, R., S. Kusne, C. Rinaldo, J. M. Aguado, K. St. George, P. Grossi, B. Frye, J. J. Fung, and G. Ehrlich. 1996. Time to detection of cytomegalovirus (CMV) DNA in blood leukocytes is a predictor for the development of CMV disease in CMV-seronegative recipients of allografts from CMV-seropositive donors following liver transplantation. *J. Infect. Dis.* **173**:1072–1076.

Mañez, R., M. C. Breinig, P. Linden, J. Wilson, J. Torre-Cisneros, S. Kusne, S. Dummer, and M. Ho. 1997. Posttransplant lymphoproliferative disease in primary Epstein-Barr virus infection after liver transplantation: the role of cytomegalovirus disease. *J. Infect. Dis.* **176**:1462–1467.

Mann, M., J. H. Shelhamer, H. Masur, V. J. Gill, W. Travis, D. Solomon, J. Manischewitz, F. Stock, H. C. Lane, and F. P. Ognibene. 1997. Lack of clinical utility of bronchoalveolar lavage cultures for cytomegalovirus in HIV infection. *Am. J. Respir. Crit. Care Med.* **155**:1723–1728.

Marshall, D. and C. Sample. 1995. EBV nuclear antigen 3c is a transcriptional regulator. *J. Virol.* **69**:3624–3630.

Marshall, G. S., G. P. Rabalais, G. G. Stout, and S. L. Waldeyer. 1992. Antibodies to recombinant-derived glycoprotein B after natural human cytomegalovirus infection correlate with neutralizing activity. *J. Infect. Dis.* **165**:381–384.

Martin, D. F., B. D. Kuppermann, R. A. Wolitz, A. G. Palestine, H. Li, C. A. Robinson, and the Roche Ganciclovir Study Group. 1999. Oral ganciclovir for patients with cytomegalovirus retinitis treated with a ganciclovir implant. *N. Engl. J. Med.* **340**:1063–1070.

Martin, M., R. Manez, P. Linden, D. Estores, J. Torre-Cisneros, S. Kusne, L. Ondick, R. Ptachcinski, W. Irish, D. Kisor, I. Felser, C. Rinaldo, A. Stieber, J. Fung, M. Ho, R. Simmons, and T. Starzl. 1994. A prospective randomized trial comparing ganciclovir-high dose acyclovir to high dose acyclovir for prevention of cytomegalovirus disease in adult liver transplant recipients. *Transplantation* **58**:779–785.

Martin, W. J., II, and T. F. Smith. 1986. Rapid detection of cytomegalovirus in bronchoalveolar lavage specimens by a monoclonal antibody method. *J. Clin. Microbiol.* **23**:1006–1008.

Marwick, C. 1998. First "antisense" drug will treat CMV retinitis. *JAMA* **280**:871.

McCarthy, J. M., M. A. Karim, H. Krueger, and P. A. Keown. 1993. The cost impact of cytomegalovirus in renal transplant recipients. *Transplantation* **55**:1277–1282.

McDonald, K., T. S. Rector, E. A. Braulin, S. H. Kubo, and M. T. Olivari. 1989. Association of coronary artery disease in cardiac transplant recipients with cytomegalovirus infection. *Am. J. Cardiol.* **64**:359–362.

McQuirt, P. V., and P. A. Furman. 1982. Acyclovir inhibition of viral DNA chain elongation in herpes simplex virus-infected cells. *Am. J. Med.* **73**(Suppl. 1A):67–71.

Melnick, J. L., C. Hu, J. Burek, E. Adam, and M. E. BeBakey. 1994. Cytomegalovirus DNA in arterial walls of patients with atherosclerosis. *J. Med. Virol.* **42**:170–174.

Mendez, J., M. Espy, T. F. Smith, J. Wilson, R. Wiesner, and C. V. Paya. 1998b. Clinical significance of viral load in the diagnosis of cytomegalovirus disease after liver transplantation. *Transplantation* **65**:1477–1481.

Mendez, J. C., M. J. Espy, T. F. Smith, J. A. Wilson, and C. V. Paya. 1998a. Evaluation of PCR primers for early diagnosis of cytomegalovirus infection following liver transplantation. *J. Clin. Microbiol.* **36**:526–530.

Merigan, T. C., K. H. Rand, R. B. Pollard, P. A. Abdallah, G. W. Jordan, and R. P. Fried. 1978. Human leukocyte interferon for the treatment of herpes zoster in patients with cancer. *N. Engl. J. Med.* **298**:981–987.

Merigan, T. C., D. G. Renlund, S. Keay, M. R. Bristow, V. Starnes, J. B. O'Connell, S. Resta, D. Dunn, P. Gamberg, R. M. Ratkovec, W. E. Richenbacher, R. C. Millar, C. DuMond, B. DeAmond, V. Sullivan, T. Cheney, W. Buhles, and E. B. Stinson. 1992. A controlled trial of ganciclovir to prevent cytomegalovirus disease after heart transplantation. *N. Engl. J. Med.* **326**:1182–1186.

Meyer-Konig, U., A. Serr, D. von Laer, G. Kirste, C. Wolff, O. Haller, D. Neumann-Haefelin, and F. T. Hufert. 1995. Human cytomegalovirus immediate early and late transcripts in peripheral blood leukocytes: diagnostic value in renal transplant recipients. *J. Infect. Dis.* **171**:705–709.

Meyer-König, U., K. Ebert, B. Schrage, S. Pollak, and F. T. Hufert. 1998a. Simultaneous infection of healthy people with multiple human cytomegalovirus strains. *Lancet* **352**:1280–1281.

Meyer-König, U., M. Haberland, D. von Laer, O. Haller, and F. T. Hufert. 1998b. Intragenic variability of human cytomegalovirus glycoprotein B in clinical strains. *J. Infect. Dis.* **177**:1162–1169.

Meyer-Konig, U., C. Vogelberg, A. Bongarts, D. Kampa, R. Delbruck, G. Wolff-Vorbeck, G. Kirste, M. Haberland, F. T. Hufert, and D. von Laer. 1998c. Glycoprotein B genotype correlates with cell tropism in vivo of human cytomegalovirus infection. J. Med. Virol. 55:75–81.

Meyers, J. D., E. C. Reed, D. H. Shepp, M. Thornquist, P. S. Dandliker, C. A. Vicary, N. Flournoy, L. E. Kirk, J. H. Kersey, E. D. Thomas, and H. H. Balfour, Jr. 1988. Acyclovir for prevention of cytomegalovirus infection and disease after allogeneic marrow transplantation. N. Engl. J. Med. 318:70–75.

Michaels, M. G., D. J. Alcendor, K. St. George, C. R. Rinaldo, Jr., G. D. Ehrlich, M. J. Becich, and G. S. Hayward. 1997. Distinguishing baboon cytomegalovirus from human cytomegalovirus: importance for xenotransplantation. J. Infect. Dis. 176:1476–1483.

Miles, P. R., R. P. Baughman, and C. C. Linneman. 1990. Cytomegalovirus in the bronchoalveolar lavage fluid of patients with AIDS. Chest 97:1072–1076.

Miller, G., J. F. Enders, H. Lisco, and H. I. Kohn. 1969. Establishment of lines from normal human blood leukocytes by co-cultivation with a leukocyte line derived from a leukemic child. Proc. Soc. Exp. Biol. Med. 132:247–252.

Miller, I. G. 1990. Epstein-Barr Virus, p. 2397–2436. In B. N. Fields, D. M. Knipe, and P. M. Howley (ed.), Virology, 2nd ed. Lippincott-Raven Publishers, Philadelphia, Pa.

Miller, R. F., N. S. Brink, J. Cartledge, Y. Sharvell, and P. Frith. 1997. Necrotising herpetic retinopathy in patients with advanced HIV disease. Genitourin. Med. 73:462–466.

Minder, W. H. 1953. Die Aetiologie der Cytomegalia infantum. Schweiz. Med. Wochenschr. 83:1180–1182.

Minton, E. J., C. Tysoe, J. H. Sinclair, and J. G. P. Sissons. 1994. Human cytomegalovirus infection of the monocyte/macrophage lineage in bone marrow. J. Virol. 68:4017–4021.

Mintz, L., W. L. Drew, R. C. Miner, and E. H. Braff. 1983. Cytomegalovirus infections in homosexual men. An epidemiological study. Ann. Intern. Med. 99:326–329.

Miyashita, E., B. Yang, K. Lam, D. Crawford, and D. Thorley-Lawson. 1995. A novel form of Epstein-Barr virus latency in normal B cells in vivo. Cell 80:593–601.

Miyashita, E., B. Yang, G. Babcock, and D. Thorley-Lawson. 1997. Identification of the site of Epstein-Barr virus persistence in vivo as a resting B cell. J. Virol. 71:4882–4891.

Mocarski, E. S., M. N. Prichard, C. S. Tan, and J. M. Brown. 1997. Reassessing the organization of the UL42-UL43 region of the human cytomegalovirus strain AD169 genome. Virology 239:169–175.

Monforte, A. D., F. Mainini, L. Testa, L. Vago, L. Balotta, M. Nebuloni, S. Antinori, T. Bini, and M. Moroni. 1997. Predictors of cytomegalovirus disease, natural history and autopsy findings in a cohort of patients with AIDS. AIDS 11:517–524.

Moorthy, R. S., D. V. Weinberg, S. A. Teich, B. B. Berger, J. T. Minturn, S. Kumar, N. A. Rao, S. M. Fowell, I. A. Loose, and L. M. Jampol. 1997. Management of varicella-zoster virus retinitis in AIDS. Br. J. Ophthalmol. 81:189–194.

Moretti, S., P. Zikos, M. T. van Lint, E. Tedone, D. Occhini, F. Gualandi, T. Lamparelli, N. Mordini, G. Berisso, S. Bregante, B. Bruno, and A. Bacigalupo. 1998. Foscarnet vs ganciclovir for cytomegalovirus (CMV) antigenemia after allogeneic hemopoietic stem cell transplantation (HSCT): a randomized study. Bone Marrow Transplant. 22:175–180.

Morgan, G., and R. A. Superina. 1994. Lymphoproliferative disease after pediatric liver transplantation. J. Pediatr. Surg. 29:1192–1196.

Mouly, F., V. Mirlesse, J. F. Meritet, F. Rozenberg, M. H. Poissonier, P. Lebon, and F. Daffos. 1997. Prenatal diagnosis of fetal varicella-zoster virus infection with polymerase chain reaction of amniotic fluid in 107 cases. Am. J. Obstet. Gynecol. 177:894–898.

Muganda, P., R. Carrasco, and Q. Qian. 1998. The human cytomegalovirus IE2 86 kDa protein elevates p53 levels and transactivates the p53 promoter in human fibroblasts. Cell Mol. Biol. 44:321–331.

Murakami, S., Y. Nakashiro, M. Mizobuchi, N. Hato, N. Honda, and K. Gyo. 1998. Varicella-zoster virus distribution in Ramsay-Hunt-syndrome revealed by polymerase chain reaction. Acta Oto-Laryngol. 118:145–149.

Muralidhar, S., J. Doniger, E. Mendelson, J. C. Araujo, F. Kashanchi, N. Azumi, J. M. Brady, and L. J. Rosenthal. 1996. Human cytomegalovirus mtrII oncoprotein binds to p53 and down-regulates p53 activated transcription. J. Virol. 70:8691–8700.

Murph, J. R., I. E. Souza, J. D. Dawson, P. Benson, S. J. Petheram, D. Pfab, A. Gregg, M. E. O'Neill, B. Zimmerman, and J. F. Bale, Jr. 1998. Epidemiology of congenital cytomegalovirus infection: maternal risk factors and molecular analysis of cytomegalovirus strains. Am. J. Epidemiol. 147:940–947.

Musch, D. C., D. F. Martin, J. F. Gordon, M. D. Davis, and B. D. Kuppermann. 1997. Treatment of cytomegalovirus retinitis with a sustained-release ganciclovir implant. N. Engl. J. Med. 337:83–90.

Myerson, D., R. C. Hackman, J. A. Nelson, D. C. Ward, and J. K. McDougall. 1984. Widespread presence of histologically occult cytomegalovirus. Hum. Pathol. 15:430–439.

Nalesnik, M. 1998. Clinical and pathological features of post transplant lymphoproliferative disorders. Springer Semin. Immunopathol. 20:325–342.

Nalesnik, M., R. Jaffe, T. E. Starzl, A.J. Demetris, K. Porter, J. A. Burnham, L. Makowka, M. Ho, and J. Locker. 1988. The pathology of post-transplant lymphoproliferative disease occurring in the setting of Cyclosporine A-prednisone immunosuppression. Am. J. Pathol. 133:173–183.

Nankivell, B. J., M. A. Malour, G. R. Russ, P. Barclay, A. R. Glanville, R. D. M. Allen, P. J. O'Connell, and J. R. Chapman. 1998. Maintenance therapy with oral ganciclovir after treatment of cytomegalovirus infection. Clin. Transplant. 12:270–273.

Narvios, A. B., D. Przepiorka, J. Tarrand, K. W. Chan, R. Champlin, and B. Lichtiger. 1998. Transfusion support using filtered unscreened blood products for cytomegalovirus-negative allogeneic marrow transplant recipients. Bone Marrow Transplant. 22:575–577.

Nelson, J. A., B. Fleckenstein, G. Jahn, D. A. Galloway, and J. K. McDougall. 1984. Structure of the transforming region of human cytomegalovirus AD169. J. Virol. 49:109–115.

Nelson, J. A., C. Reynolds-Kohler, M. B. Oldstone, and C. A. Wiley. 1988. HIV and HCMV coinfect brain cells in patients with AIDS. Virology 165:286–290.

Nelson, P. N., B. K. Rawal, Y. S. Boriskin, K. E. Mathers, R. L. Powles, H. M. Steel, Y. S. Tryhorn, P. D. Butcher, and J. C. Booth. 1996. A polymerase chain reaction to detect a spliced late transcript of human cytomegalovirus in the blood of bone marrow transplant recipients. J. Virol. Methods 56:139–148.

Nemerow, G., R. Wolfert, M. McNaughton, and N. Cooper. 1985. Identification and characterization of the Epstein-Barr virus receptor on human B lymphocytes and its relationship to the C3d complement receptor (CR2). J. Virol. 55:347–351.

Neote, K., D. DiGregorio, J. Y. Mak, R. Horuk, and T. J. Schall. 1993. Molecular cloning, functional expression, and signaling characteristics of a C-C chemokine receptor. *Cell* **72**:415–425.

Neri, A., F. Barriga, G. Ighirami, D. M. Knowles, J. Neequaye, I. T. Magrath, and R. Dalla-Favera. 1991. Epstein-Barr virus infection precedes clonal expansion in Burkitt's and acquired immunodeficiency syndrome-associated lymphoma. *Blood* **77**:1092–1095.

Newell, K. A., E. M. Alonso, P. F. Whitington, D. S. Bruce, J. M. Millis, J. B. Piper, E. S. Woodle, S. M. Kelly, H. Koeppen, J. Hart, C. M. Rubin, and J. R. Thistlethwaite, Jr. 1996. Posttransplant lymphoproliferative disease in pediatric liver transplantation. *Transplantation* **62**:370–375.

Nicol, D. L., A. S. MacDonald, P. Belitsky, S. Lee, A. D. Cohen, H. Bitter-Suermann, J. Lowen, and A. Whalen. 1993. Reduction by combination prophylactic therapy with CMV hyperimmune globulin and acyclovir of the risk of primary CMV disease in renal transplant recipients. *Transplantation* **55**:841–846.

Niederman, J. C., A. S. Evans, M. S. Subrahmangan, and R. W. McCollum. 1970 Prevalence, incidence, and persistence of Epstein-Barr virus antibody in young adults. *N. Engl. J. Med.* **282**:361–365.

Niedobitek, G., E. Kremmer, H. Herbst, L. Whitehead, C. Dawson, E. Niedobitek, C. von Ostarr, N. Rooney, F. Grasser, and L. Young. 1997a. Immunohistochemical detection of the Epstein-Barr virus-encoded latent membrane protein 2A in Hodgkins disease and infectious mononucleosis. *Blood* **90**:1664–1672.

Niedobitek, G., L. Young, D. Wright, L. Whitehead, and H. Herbst. 1997b. EBV infection in infectious mononucleosis: virus latency, replication and phenotype of EBV infected cells. *J. Pathol.* **182**:151–159.

Nigro, G., H. Scholz, and U. Bartmann. 1994. Ganciclovir therapy for symptomatic congenital cytomegalovirus infection in infants: a two-regimen experience. *J. Pediatr.* **124**:318–322.

Nilsson, K., G. Klein, W. Henle, and G. Henle. 1971. The establishment of lymphoblastoid cell lines from adult and from foetal human lymphoid tissue and its dependence on EBV. *Int. J. Cancer* **8**:443–450.

Niubò, J., J. L. Pérez, A. Carvajal, C. Ardanuy, and R. Martín. 1994. Effect of delayed processing of blood samples on performance of cytomegalovirus antigenemia assay. *J. Clin. Microbiol.* **32**:1119–1120.

Niubò, J., J. L. Pérez, J. T. Martínez-Lacasa, A. García, J. Roca, J. Fabregat, S. Gil-Vernet, and R. Martín. 1996. Association of quantitative cytomegalovirus antigenemia with symptomatic infection in solid organ transplant patients. *Diagn. Microbiol. Infect. Dis.* **24**:19–24.

Noormohamed, F. H., M. S. Youle, C. J. Higgs, S. Martin-Munley, B. G. Gazzard, and A. F. Lant. 1998. Pharmacokinetics and absolute bioavailability of oral foscarnet in human immunodeficiency virus-seropositive patients. *Antimicrob. Agents Chemother.* **42**:293–297.

Nowlin, D. M., N. R. Cooper, and T. Compton. 1991. Expression of a human cytomegalovirus receptor correlates with infectibility of cells. *J. Virol.* **65**:3114–3121.

Numazaki, K. 1997. Human cytomegalovirus infection of breast milk. *FEMS Immunol. Med. Microbiol.* **18**:91–98.

Ogilvie, M. M. 1998. Antiviral prophylaxis and treatment in chickenpox—a review prepared for the UK advisory group on chickenpox on behalf of the British Society for the Study of Infection. *J. Infect.* **36**:31–38.

Ohgitani, E., K. Kobayashi, K. Takeshita, and J. Imanishi. 1998. Induced expression and localization to nuclear-inclusion bodies of hsp70 in varicella-zoster virus-infected human diploid fibroblasts. *Microbiol. Immunol.* **42**:755–760.

Ohizumi, Y., H. Suzuki, Y. Matsumoto, Y. Masuho, and Y. Numazaki. 1992. Neutralizing mechanisms of two human monoclonal antibodies against human cytomegalovirus glycoprotein 130/55. *J. Gen. Virol.* **73**:2705–2707.

Old, L. J., P. Clifford, E. A. Boyse, H. F. Oettgen, E. de Harven, G. Geering, and B. Williamson. 1966. Precipitating antibody in human serum to an antigen present in cultured Burkitt's lymphoma cells. *Proc. Natl. Acad. Sci. USA* **56**:1699–1704.

Olivari, M. T., D. C. Homans, R. F. Wilson, S. H. Kubo, and W. S. Ring. 1989. Coronary artery disease in cardiac transplant patients receiving triple-drug immunosuppressive therapy. *Circulation* **80**:111–115.

Opelz, G., and R. Henderson. 1993. Incidence of non-Hodgkin lymphoma in kidney and heart transplant recipients. *Lancet* **342**:1514–1516.

The Oral Ganciclovir European and Australian Cooperative Study Group. 1995. Intravenous versus oral ganciclovir: European/Australian comparative study of efficacy and safety in the prevention of cytomegalovirus retinitis recurrence in patients with AIDS. *AIDS* **9**:471–477.

Oxman, M. N. 1995. Immunization to reduce the frequency and severity of herpes zoster and its complications. *Neurology* **45**:S41–S46.

Ozono, K., S. Mushiake, T. Takeshima, and M. Nakayama. 1997. Diagnosis of congenital cytomegalovirus infection by examination of placenta: application of polymerase chain reaction and in situ hybridization. *Pediatr. Pathol. Lab. Med.* **17**:249–258.

Pac, C. Y., H.-M. Eun, R. G. McArthur, and J. W. Yoon. 1988. Association of cytomegalovirus infection with autoimmune type 1 diabetes. *Lancet* **i**:1–4.

Palestine, A. G., M. A. Polis, M. D. de Smet, B. F. Baird, J. Falloon, J. A. Kovacs, R. T. Davey, J. J. Zurlo, K. M. Zunich, M. Davis, L. Hubbard, R. Brothers, F. L. Ferris, E. Chew, J. L. Davis, B. I. Rubin, S. D. Mellow, J. A. Metcalf, J. Manischewitz, J. R. Minor, R. B. Nussenblatt, H. Masur, and H. C. Lane. 1991. A randomized, controlled trial of foscarnet in the treatment of cytomegalovirus retinitis in patients with AIDS. *Ann. Intern. Med.* **115**:665–673.

Papadopoulos, E. B., M. Ladanyi, and D. Emanuel. 1994. Infusions of donor leukocytes as treatment of EBV associated lymphoproliferative disorders complicating allogeneic bone marrow transplantation. *N. Engl. J. Med.* **330**:1185–1189.

Paradis, I. L., W. F. Grgurich, J. S. Dummer, A. Dekker, and J. H. Dauber. 1988. Rapid detection of cytomegalovirus pneumonia from lung lavage cells. *Am. Rev. Respir. Dis.* **138**:697–702.

Pari, G. S., and D. G. Anders. 1993. Eleven loci encoding *trans*-acting factors are required for transient complementation of human cytomegalovirus oriLyt-dependent DNA replication. *J. Virol.* **67**:6979–6988.

Pass, R. F. 1996. Immunization strategy for prevention of congenital cytomegalovirus infection. *Infect. Agents Dis.* **5**:240–244.

Pass, R. F., A. M. Duliege, S. Boppana, R. Sekulovich, S. Percell, W. Britt, and R.L. Burke. 1999. A subunit cytomegalovirus vaccine based on recombinant envelope glycoprotein B and a new adjuvant. *J. Infect. Dis.* **180**:970–975.

Pass, R. F., E. A. Little, S. Stagno, W. J. Britt, and C. A. Alford. 1987. Young children as a probable source of maternal and congenital cytomegalovirus infection. *N. Engl. J. Med.* **316**:1366–1370.

Pass, R. F., S. Stagno, G. J. Myers, and C. A. Alford. 1980. Outcome of symptomatic congenital cytomegalovirus infection: results of a long-term longitudinal follow-up. *Pediatrics* **66**:758–762.

Pass, R. F., C. Hutto, M. D. Lyon, and G. Cloud. 1990. Increased rate of cytomegalovirus infection among day care center workers. *Pediatr. Infect. Dis. J.* **9:**465–470.

Pearson, G. R., and J. Luka. 1986. Characterization of the virus-determined antigens, p. 47–73. *In* M. A. Epstein and B. G. Achong (ed.), *The Epstein-Barr Virus: Recent Advances.* Heinemann, London, United Kingdom.

Peddi, V. M., S. Hariharan, T. J. Schroeder, and M. R. First. 1997. Impact of ganciclovir prophylaxis on cytomegalovirus infection in recipients of cadaveric renal allografts. *Nephron* **76:**49–55.

Peek, R., F. Verbraak, M. Bruinenberg, A. van der Lelij, G. van den Horn, and A. Kijlstra. 1998. Cytomegalovirus glycoprotein B genotyping in ocular fluids and blood of AIDS patients with cytomegalovirus retinitis. *Investig. Ophthalmol. Vis. Sci.* **39:**1183–1187.

Percivalle, E., M. G. Revello, L. Vago, F. Morini, and G. Gerna. 1993. Circulating endothelial giant cells permissive for human cytomegalovirus (HCMV) are detected in disseminated HCMV infections with organ involvement. *J. Clin. Investig.* **92:**663–670.

Persoons, M. C. J., M. J. A. P. Daemen, J. H. Bruning, and C. A. Bruggeman. 1994. Active cytomegalovirus infection of arterial smooth muscle cells in immunocompromised rats: a clue to herpesvirus-associated atherogenesis? *Circ. Res.* **75:**214–220.

Pescovitz, M. D., T. L. Pruett, T. Gonwa, B. Brook, R. McGory, K. Wicker, K. Griffy, C. A. Robinson, and D. Jung. 1998. Oral ganciclovir dosing in transplant recipients and dialysis patients based on renal function. *Transplantation* **66:**1104–1107.

Phillips, A. J., P. Tomasec, E. C. Wang, G. W. Wilkinson, and L. K. Borysiewicz. 1998. Human cytomegalovirus infection downregulates expression of the cellular aminopeptidases CD10 and CD13. *Virology* **250:**350–358.

Pizzorno, M. C., M. A. Mullen, Y. N. Chang, and G. S. Hayward. 1991. The functionally active IE2 immediate early regulatory protein of human cytomegalovirus is an 80-kilodalton polypeptide that contains two distinct activator domains and a duplicated nuclear localization signal. *J. Virol.* **65:**3839–3852.

Pleskoff, O., C. Treboute, A. Brelot, N. Heveker, M. Seman, and M. Alizon. 1997. Identification of a chemokine receptor encoded by human cytomegalovirus as a cofactor for HIV-1 entry. *Science* **276:**1874–1878.

Ploegh, H. L. 1998. Viral strategies of immune evasion. *Science* **280:**248–253.

Plummer, G., and M. Benyesh-Melnick. 1964. A plaque reduction neutralization test for human cytomegalovirus. *Proc. Soc. Exp. Biol. Med.* **117:**145–150.

Podlech, J., R. Holtappels, N. Wirtz, H. P. Steffens, and M. J. Reddehase. 1998. Reconstitution of CD8 T cells is essential for the prevention of multiple-organ cytomegalovirus histopathology after bone marrow transplantation. *J. Gen. Virol.* **79:**2099–2104.

Polis, M. A., K. M. Spooner, B. F. Baird, J. F. Manischewitz, H. S. Jaffe, P. E. Fisher, J. Falloon, R. T. Davey, Jr., J. A. Kovacs, R. E. Walker, S. M. Whitcup, R. B. Nussenblatt, H. C. Lasne, and H. Masur. 1996. Anticytomegaloviral activity and safety of cidofovir in patients with human immunodeficiency virus infection and cytomegalovirus viruria. *Antimicrob. Agents Chemother.* **39:**882–886.

Pope, J., and W. Scott. 1968. Transformation of foetal human leukocytes in vitro by filtrates of a human leukemia cell line containing herpes-like virus. *Int. J. Cancer* **3:**857–866.

Powderly, W. G., A. Landay, and M. M. Lederman. 1998. Recovery of the immune system with antiretroviral therapy—the end of opportunism. *JAMA* **280:**72–77.

Preiksaitis, J. D., F. Diaz-Mitoma, F. Mirzayans, S. Roberts, and D. L. J. Tyrrell. 1992. Quantitative oropharyngeal Epstein-Barr virus shedding in renal and cardiac transplant recipients: relationship to immunosuppressive therapy, serologic responses, and the risk of posttransplant lymphoproliferative disorder. *J. Infect. Dis.* **166:**986–994.

Prentice, H. G., E. Gluckman, R. L. Powles, P. Ljungman, N. Milpied, J. M. Fernandez-Ranada, F. Mandelli, P. Kho, L. Kennedy, and A. R. Bell. 1994. Impact of long-term acyclovir on cytomegalovirus infection and survival after allogeneic bone marrow transplantation: European Acyclovir for CMV Prophylaxis Study Group. *Lancet* **343:**749–753.

Prix, L., K. Hamprecht, B. Holzhuter, R. Handgretinger, T. Klingebiel, and G. Jahn. 1999. Comprehensive restriction analysis of the UL97 region allows early detection of ganciclovir-resistant human cytomegalovirus in an immunocompromised child. *J. Infect. Dis.* **180:**491–495.

Puchhammer-Stockl, E., T. Popow-Kraupp, F. X. Heinz, C. W. Mandl, and C. Kunz. 1991. Detection of varicella-zoster virus DNA by polymerase chain reaction in the cerebrospinal fluid of patients suffering from neurological complications associated with chicken pox or herpes zoster. *J. Clin. Microbiol.* **29:**1513–1516.

Qiu, X. Y., C. A. Janson, J. S. Culp, B. Richardson, C. Debouck, W. W. Smith, and S. S. Abdelmeduid. 1997. Crystal structure of varicella-zoster virus protease. *Proc. Natl. Acad. Sci. USA* **94:**2874–2879.

Qu, L., and D. Rowe. 1992. Epstein-Barr virus latent gene expression in uncultured peripheral blood lymphocytes. *J. Virol.* **66:**3715–3724.

Quinnan, G. V., Jr., N. Kirmani, A. H. Rook, J. F. Manischewitz, L. Jackson, G. Moreschi, G. W. Santos, R. Saral, and W. H. Burns. 1982. Cytotoxic T cells in cytomegalovirus infection: HLA-restricted T-lymphocyte and non-T-lymphocyte cytotoxic responses correlate with recovery from cytomegalovirus infection in bone-marrow-transplant recipients. *N. Engl. J. Med.* **307:**7–13.

Raab-Traub, N., and K. Flynn. 1986. The structure of the terminal of the Epstein-Barr virus as a marker of clonal cellular proliferation. *Cell* **47:**883–889.

Rasmussen, L., C. Matkin, R. Spaete, C. Pachl, and T. C. Merigan. 1991. Antibody response to human cytomegalovirus glycoproteins gB and gH after natural infection in humans. *J. Infect. Dis.* **164:**835–842.

Rasmussen, L., C. Hong, D. Zipeto, S. Morris, D. Sherman, S. Chou, R. Miner, W. L. Drew, R. Wolitz, A. Dowling, A. Warford, and T. C. Merigan. 1997a. Cytomegalovirus gB genotype distribution differs in human immunodeficiency virus-infected patients and immunocompromised allograft recipients. *J. Infect. Dis.* **175:**179–184.

Rasmussen, L., D. Zipeto, R. A. Wolitz, A. Dowling, B. Efron, and T. C. Merigan. 1997b. Risk for retinitis in patients with AIDS can be assessed by quantitation of threshold levels of cytomegalovirus DNA burden in blood. *J. Infect. Dis.* **176:**1146–1155.

Redman, R. L., S. Nader, L. Zerboni, C. Liu, R. M. Wong, B. W. Brown, and A. M. Arvin. 1997. Early reconstitution of immunity and decreased severity of herpes zoster in bone marrow transplant recipients immunized with inactivated varicella vaccine. *J. Infect. Dis.* **176:**578–585.

Reedman, B. M., and G. Klein. 1973. Cellular localisation of an Epstein-Barr virus (EBV)-associated complement-fixing antigen in producer and non-producer lymphoblastoid cell lines. *Int. J. Cancer* **11:**599–620.

Reigstad, H., R. Bjerknes, T. Markestad, and H. Myrmel. 1992. Ganciclovir therapy of congenital cytomegalovirus disease. *Acta Paediatr.* **81:**707–708.

Reinke, P., E. Fietze, S. Ode-Hakim, S. Prosch, J. Lippert, R. Ewert, and H. D. Volk. 1994. Late-acute renal allograft rejection and symptomless cytomegalovirus infection. *Lancet* **344:**1737–1738.

Resnick, L., J. S. Herbst, D. V. Ablashi, S. Atherton, B. Frank, L. Rosen, and S. N. Horwitz. 1988. Regression of oral hairy leukoplakia after orally administered acyclovir therapy. *JAMA* **259:**384–388.

Reusser, P. 1996. Herpesvirus resistance to antiviral drugs: a review of the mechanisms, clinical importance and therapeutic options. *J. Hosp. Infect.* **33:**235–248.

Reusser, P., C. Cordonnier, H. Einsele, D. Engelhard, D. Link, A. Locasciulli, and P. Ljungman. 1996. European survey of herpesvirus resistance to antiviral drugs in bone marrow transplant recipients. Infectious Diseases Working Party of the European Group for Blood and Marrow Transplanttion (EBMT). *Bone Marrow Transplant.* **17:**813–817.

Reusser, P., R. Attenhofer, H. Hebart, C. Helg, B. Chapuis, and H. Einsele. 1997. Cytomegalovirus-specific T-cell immunity in recipients of autologous peripheral blood stem cell or bone marrow transplants. *Blood* **89:**3873–3879.

Revello, M. G., E. Percivalle, and G. Gerna. 1986. Immunoglobulin M to the membrane of uninfected fibroblasts in primary human cytomegalovirus infections. *Microbiology* **9:**127–138.

Revello, M. G., E. Percivalle, A. Sarasini, F. Baldanti, M. Furione, and G. Gerna. 1994. Diagnosis of human cytomegalovirus infection of the nervous system by pp65 detection in polymorphonuclear leukocytes of cerebrospinal fluid from AIDS patients. *J. Infect. Dis.* **170:**1275–1279.

Revello, M. G., A. Sarasini, M. Zavattoni, F. Baldanti, and G. Gerna. 1998a. Improved prenatal diagnosis of congenital human cytomegalovirus infection by a modified nested polymerase chain reaction. *J. Med. Virol.* **56:**99–103.

Revello, M. G., M. Zavattoni, A. Sarasini, E. Percivalle, L. Simoncini, and G. Gerna. 1998b. Human cytomegalovirus in blood of immunocompetent persons during primary infection: prognostic implications for pregnancy. *J. Infect. Dis.* **177:**1170–1175.

Reyburn, H. T., O. Mandelboim, M. Valés-Gómez, D. M. Davis, L. Pazmany, and J. L. Strominger. 1997. The class I MHC homologue of human cytomegalovirus inhibits attack by natural killer cells. *Nature* **386:**514–517.

Reynolds, D., M. Gulley, and P. Banks. 1995. New characterization of infectious mononucleosis and a phenotypic comparison with Hodgkin's disease. *Am. J. Pathol.* **146:**379–388.

Reynolds, D. W., S. Stagno, K. L. Herrman, and C. A. Alford. 1978. Antibody response to live virus vaccines in congenital and neonatal cytomegalovirus infections. *J. Pediatr.* **92:**738–742.

Ribbert, H. 1904. Ueber proozoanartige Zellen in der Niere eines syphilitischen Neugeborenen und in der Parotis von Kindern. *Zentbl. Allg. Pathol.* **15:**945–948.

Rickinson, A. B., and M. Epstein. 1985. A re-examination of the Epstein-Barr virus carrier state in healthy seropositive individuals. *Int. J. Cancer* **35:**35–42.

Rickinson, A. B., and E. Kieff. 1996. Epstein Barr Virus, p. 2397–2436. *In* B. N. Fields, D. M. Knipe, and P. M. Howley (ed.), *Virology*, 3rd ed. Lippincott-Raven Publishers, Philadelphia, Pa.

Riddell, S. R., M. Rabin, A. P. Geballe, W. J. Britt, and P. D. Greenburg. 1991. Class I MHC-restricted cytotoxic T lymphocyte recognition of cells infected with human cytomegalovirus does not require endogenous viral gene expression. *J. Immunol.* **146:**2795–2804.

Riddell, S. R., K. S. Watanabe, J. M. Goodrich, C. R. Li, M. E. Agha, and P. D. Greenberg. 1992. Restoration of viral immunity in immunodeficient humans by the adoptive transfer of T cells. *Science* **257:**238–241.

Riddell, S. R., B. A. Walter, M. J. Gilbert, and P. D. Greenberg. 1994. Selective reconstitution of CD8+ cytotoxic T lymphocyte responses in immunodeficient bone marrow transplant recipients by the adoptive transfer of T cell clones. *Bone Marrow Transplant.* **14**(Suppl. 4)**:**S78–S84.

Riddler, S. A., M. D. Breinig, and J. L. C. McKnight. 1994. Increased levels of circulating Epstein-Barr virus (EBV)-infected lymphocytes and decreased EBV nuclear antigen antibody responses are associated with the development of posttransplant lymphoproliferative disease in solid-organ transplant recipients. *Blood* **84:**972–984.

Rinaldo, C. R., Jr. 1994. Modulation of major histocompatibility complex antigen expression by viral infection. *Am. J. Pathol.* **144:**637–650.

Rinaldo, C. R., Jr., P. H. Black, and M. S. Hirsch. 1977. Interaction of cytomegalovirus with leukocytes from patients with mononucleosis due to cytomegalovirus. *J. Infect. Dis.* **136:**667–678.

Rinaldo, C. R., Jr., W. P. Carney, B. S. Richter, P. H. Black, and M. S. Hirsch. 1980. Mechanisms of immunosuppression in cytomegaloviral mononucleosis. *J. Infect. Dis.* **141:**488–495.

Roberts, T. C., R. S. Buller, M. Gaudreault-Keener, K. E. Sternhell, K. Garlock, G. G. Singer, D. C. Brennan, and G. A. Storch. 1997. Effects of storage temperature and time on qualitative and quantitative detection of cytomegalovirus in blood specimens by shell vial culture and PCR. *J. Clin. Microbiol.* **35:**2224–2228.

Roberts, T. C., D. C. Brennan, R. S. Buller, M. Gaudreault-Keener, M. A. Schnitzler, K. E. Sternhell, K. A. Garlock, G. G. Singer, and G. A. Storch. 1998. Quantitative polymerase chain reaction to predict occurrence of symptomatic cytomegalovirus infection and assess response to ganciclovir therapy in renal transplant recipients. *J. Infect. Dis.* **178:**626–635.

Roby, C., and W. Gibson. 1986. Characterization of phosphoproteins and protein kinase activity of virions, noninfectious enveloped particles, and dense bodies of human cytomegalovirus. *J. Virol.* **59:**714–727.

Rocchi, G., A. DeFelici, G. Ragona, and A. Heinz. 1977. Quantitative evaluation of Epstein-Barr virus-infected mononuclear peripheral blood leukocytes in infectious mononucleosis. *N. Engl. J. Med.* **296:**132–134.

Roizman, B. 1996. Herpesviridae, p. 2221–2230. *In* B. N. Fields, D. M. Knipe, and P. M. Howley (ed.), *Virology*, 3rd ed. Lippincott-Raven Publishers, Philadelphia, Pa.

Roizman, B., L. E. Carmichael, F. Deinhardt, G. de-The, A. J. Nahmias, W. Plowright, F. Rapp, P. Sheldrick, M. Takahashi, and K. Wolf. 1981. Herpesviridae. Definition, provisional nomenclature, and taxonomy. The Herpesvirus Study Group, the International Committee on Taxonomy of Viruses. *Intervirology* **16:**201–217.

Romanowski, M. J., and T. Shenk. 1997. Characterization of the human cytomegalovirus irs1 and trs1 genes: a second immediate-early transcription unit within irs1 whose product antagonizes transcriptional activation. *J. Virol.* **71:**1485–1496.

Rooney, C. M., S. K. Loftin, M. S. Holladay, M. K. Brenner, R. A. Krance, and H. E. Heslop. 1995a. Early identification of Epstein-Barr virus associated posttransplantation lymphoproliferative disease. *Br. J. Haematol.* **89:**98–103.

Rooney, C. M., C. A. Smith, C. Y. C. Ng, S. Loftin, C. Li, R. A. Krance, M. K. Brenner, and H. E. Heslop. 1995b. Use of gene-modified virus-specific T lymphocytes to control EBV related lymphoma. *Lancet* **345:**9–15.

Rosen, H. R., S. Chou, C. L. Corless, D. R. Gretch, K. D. Flora, A. Boudousquie, S. L. Orloff, J. M. Rabkin, and K. G. Benner. 1997. Cytomegalovirus viremia: risk factor for allograft cirrhosis after liver transplantation for hepatitis C. *Transplantation* **64:**721–726.

Roth, G., T. Curiel, and J. Lacy. 1994. Epstein-Barr viral nuclear antigen 1 antisense oligodeoxynucleotide inhibits proliferation of Epstein-Barr virus-immortalized B cells. *Blood* **84:**582–587.

Rothenbacher, D., A. Hoffmeister, G. Bode, P. Wanner, W. Koenig, and H. Brenner. 1999. Cytomegalovirus infection and coronary heart disease: results of a German case-control study. *J. Infect. Dis.* **179:**690–692.

Rowe, D. T., L. Qu, J. Reyes, N. Jabbour, E. Yunis, P. Putnam, S. Todo, and M. Green. 1997. Use of quantitative-competitive PCR to measure EBV genome load in lymphoproliferative disorders. *J. Clin. Microbiol.* **35:**1612–1615.

Rowe, W. P., J. W. Hartley, S. Waterman, H. C. Turner, and R. J. Huebner. 1956. Cytopathogenic agent resembling human salivary gland virus recovered from tissue cultures of human adenoids. *Proc. Soc. Exp. Biol. Med.* **92:**418–424.

Rubben, A., J. M. Baron, and E. I. Grussendorfconen. 1997. Routine detection of herpes simplex virus and varicella zoster virus by polymerase chain reaction reveals that initial herpes zoster is frequently misdiagnosed as herpes simplex. *Br. J. Dermatol.* **137:**259–261.

Ruutu, T., P. Ljungman, L. Brinch, S. Lenhoff, B. Lönnqvist, O. Ringdén, P. Ruutu, L. Volin, D. Albrechtsen, B. Sallerfors, F. Ebeling, and G. Myllylä. 1997. No prevention of cytomegalovirus infection by anti-cytomegalovirus hyperimmune globulin in seronegative bone marrow transplant recipients. *Bone Marrow Transplant.* **19:**233–236.

Sadzot-Delvaux, C., P. Thonard, S. Schoonbroodt, J. Piette, and B. Rentier. 1995. Varicella-zoster virus induces apoptosis in cell culture. *J. Gen. Virol.* **76:**2875–2879.

Sadzot-Delvaux, C., P. R. Kinchington, S. Debrus, B. Rentier, and A. M. Arvin. 1997. Recognition of the latency-associated immediate early protein IE63 of varicella-zoster virus by human memory T lymphocytes. *J. Immunol.* **159:**2802–2806.

Saillour, F., N. Bernard, J.-M. Ragnaud, P. Morlat, Y. Blanchard, E. Monlun, E. Labouyrie, G. Chene, and F. Dabis. 1997. Incidence of cytomegalovirus disease in the Aquitaine cohort of HIV-infected patients: a retrospective survey, 1987–1993. *J. Infect.* **35:**155–161.

Salvant, B. S., E. A. Fortunato, and D. H. Spector. 1998. Cell cycle dysregulation by human cytomegalovirus: influence of the cell cycle phase at the time of infection and effects on cyclin transcription. *J. Virol.* **72:**3729–3741.

Saillour, F., N. Bernard, L. Dequae-Merchadou, C. Marimoutou, V. Journot, and F. Dabis. 1998. Predictive factors of occurrence of cytomegalovirus disease and impact on survival in the Aquitaine cohort in France, 1985 to 1994. *J. Acquir. Immune Defic. Syndr. Hum. Retrovirol.* **17:**171–178.

Salzberger, B., D. Myerson, and M. Boeckh. 1997. Circulating cytomegalovirus (CMV)-infected endothelial cells in marrow transplant patients with CMV disease and CMV infection. *J. Infect. Dis.* **176:**778–781.

Savoie, A., C. Perpete. L. Carpentier, J. Joncas, and C. Alfieri. 1994. Direct correlation between the load of Epstein-Barr virus-infected lymphocytes in the peripheral blood of pediatric transplant patients and the risk of lymphoproliferative disease. *Blood* **83:**2712–2722.

Schäfer, P., and R. Laufs. 1996. Experience with quantitative PCR for the management of HCMV disease. *Intervirology* **39:**204–212.

Schäfer, P., W. Tenschert, K. Gutensohn, and R. Laufs. 1997. Minimal effect of delayed sample processing on results of quantitative PCR for cytomegalovirus DNA in leukocytes compared to results of an antigenemia assay. *J. Clin. Microbiol.* **35:**741–744.

Schmader, K. 1998. Postherpetic neuralgia in immunocompetent elderly people. *Vaccine* **16:**1768–1770.

Schoppel, K., C. Schmidt, H. Einsele, H. Hebart, and M. Mach. 1998. Kinetics of the antibody response against human cytomegalovirus-specific proteins in allogeneic bone marrow transplant recipients. *J. Infect. Dis.* **178:**1233–1243.

Schunemann, S., C. Mainka, and M. H. Wolff. 1998. Subclinical reactivation of varicella-zoster virus in immunocompromised and immunocompetent individuals. *Intervirology* **41:**98–102.

Sedmak, D. D., A. M. Guglielmo, D. A. Knight, D. J. Birmingham, E. H. Huang, and W. J. Waldman. 1994. Cytomegalovirus inhibits major histocompatability class II expression on infected endothelial cells. *Am. J. Pathol.* **144:**683–692.

Sessions, C. F., and H. W. Taeusch. 1991. Viral infections of the newborn. *In* H. W. Taeusch, A. R. Ballard, and M. E. Avery (ed.), *Diseases of the Newborn.* W. B. Saunders, Philadelphia, Pa.

Seu, P., D. J. Winston, C. D. Holt, F. Kaldas, and R. W. Busuttil. 1997. Long-term ganciclovir prophylaxis for successful prevention of primary cytomegalovirus (CMV) disease in CMV-seronegative liver transplant recipients with CMV-seropositive donors. *Transplantation* **64:**1614–1617.

Sha, B. E., C. A. Benson, T. A. Deutsch, P. A. Urbanski, J. P. Phair, and H. A. Kessler. 1991. Suppression of cytomegalovirus retinitis in persons with AIDS with high-dose intravenous acyclovir. *J. Infect. Dis.* **164:**777–780.

Shapiro, R. S., K. McClain, G. Frizzera, K. J. Gajl-Peczalska, J. H. Kersey, B. R. Blazar, D. C. Arthur, D. F. Patton, J. S. Greenberg, B. Burke, N. K. C. Ramsey, P. McGlave, and A. H. Filipovich. 1988. Epstein-Barr virus associated B cell lymphoproliferative disorders following bone marrow transplantation. *Blood* **71:**1234–1243.

Shen, C. Y., S. F. Chang, H. J. Lin, H. N. Ho, T. S. Yeh, S. L. Yang, E. S. Huang, and C. W. Wu. 1994. Cervical cytomegalovirus infection in prostitutes and in women attending a sexually transmitted disease clinic. *J. Med. Virol.* **43:**362–366.

Shen, G. K., E. J. Alfrey, C. L. Knoppel, D. C. Dafoe, and J. D. Scandling. 1997a. Eradication of cytomegalovirus reactivation disease using high-dose acyclovir and targeted intravenous ganciclovir in kidney and kidney/pancreas transplantation. *Transplantation* **64:**931–933.

Shen, Y., H. Zhu, and T. Shenk. 1997b. Human cytomegalovirus IE1 and IE2 proteins are mutagenic and mediate "hit and run" oncogenic transformation in cooperation with the adenovirus E1A proteins. *Proc. Natl. Acad. Sci. USA* **94:**3341–3345.

Shepp, D. H., P. S. Dandliker, P. de Miranda, T. C. Burnette, D. M. Cederberg, L. E. Kirk, and J. D. Meyers. 1985. Activity of 9-[2-hydroxy-1-(hydroxymethyl)ethoxymethyl]guanine in the treatment of cytomegalovirus pneumonia. *Ann. Intern. Med.* **103:**368–373.

Shepp, D. H., P. S. Dandliker, and J. D. Meyers. 1986. Treatment of varicella-zoster virus infection in severely immunocompromised patients. *N. Engl. J. Med.* **314:**208–212.

Shinkai, M., and S. A. Spector. 1995. Quantitation of human cytomegalovirus (HCMV) in cerebrospinal fluid by competitive PCR in AIDS patients with different HCMV central nervous system diseases. *Scand. J. Infect. Dis.* **27:**559–561.

Shinkai, M., S. A. Bozzette, W. Powderly, P. Frame, and S. A. Spector. 1997. Utility of urine and leukocyte cultures and plasma DNA polymerase chain reaction for identification of AIDS patients at risk for developing human cytomegalovirus disease. *J. Infect. Dis.* **175:**302–308.

Sixbey, J. W., J. G. Nedrud, N. Raab-Traub, R. A. Hanes, and J. S. Pagano. 1984. Epstein-Barr virus replication in oropharyngeal epithelial cells. *N. Engl. J. Med.* **310:**1225–1230.

Sixbey, J. W., P. Shirley, P. J. Chesney, D. M. Buntin, and L. Resnick. 1989. Detection of a second widespread strain of Epstein-Barr virus. *Lancet* **ii:**761–765.

Sjovall, J., A. Karlson, S. Ogenstad, E. Sandstrom, and M. Saarimaki. 1988. Pharmacokinetics and absorption of foscarnet after intravenous and oral administration to patients with human immunodeficiency virus. *Clin. Pharmacol. Ther.* **44:**65–73.

Slavin, M. A., R. R. Bindra, C. A. Gleaves, M. B. Pettinger, and R. A. Bowden. 1993. Ganciclovir sensitivity of cytomegalovirus at diagnosis and during treatment of cytomegalovirus pneumonia in marrow transplant recipients. *Antimicrob. Agents Chemother.* **37:**1360–1363.

Slobedman, B., and E. S. Mocarski. 1999. Quantitative analysis of latent human cytomegalovirus. *J. Virol.* **73:**4806–4812.

Smiley, M. L., E. Mar, and E. Huang. 1988. Cytomegalovirus infection and viral-induced transformation of human endothelial cells. *J. Med. Virol.* **25:**213–226.

Smith, J. D. 1986. Human cytomegalovirus: demonstration of permissive epithelial cells and nonpermissive fibroblastic cells in a survey of human cell lines. *J. Virol.* **60:**583–588.

Smith, K. L., T. Cobain, and R. A. Dunstan. 1993. Removal of cytomegalovirus DNA from donor blood by filtration. *Br. J. Haematol.* **83:**640–642.

Smith, M. G. 1956. Propagation in tissue cultures of a cytopathogenic virus from human salivary gland virus (SGV) disease. *Proc. Soc. Exp. Biol. Med.* **92:**424–430.

Snydman, D. R., B. G. Werner, B. Heinze-Lacey, V. P. Berardi, N. L. Tilney, R. L. Kirkman, E. L. Milford, S. I. Cho, H. L. Bush, Jr., A. S. Levey, T. B. Strom, C. B. Carpenter, R. H. Levey, W. E. Harmon, C. E. Zimmerman II, M. E. Shapiro, T. Steinman, F. LoGerfo, B. Idelson, G. P. J. Schroter, M. Levin, J. McIver, J. Leszczynski, and G. F. Grady. 1987. Use of cytomegalovirus immune globulin to prevent cytomegalovirus disease in renal transplant recipients. *N. Engl. J. Med.* **317:**1049–1054.

Soderberg, C., S. Larsson, S. Bergstedt-Lindqvist, and E. Moller. 1993. Definition of a subset of human peripheral blood mononuclear cells that are permissive to human cytomegalovirus infection. *J. Virol.* **67:**3166–3175.

Söderberg-Nauclér, C., K. N. Fish, and J. A. Nelson. 1997. Reactivation of latent human cytomegalovirus by allogeneic stimulation of blood cells from healthy donors. *Cell* **91:**119–126.

Söderberg-Nauclér, C., K. N. Fish, and J. A. Nelson. 1998. Growth of human cytomegalovirus in primary macrophages. *Methods Companion Methods Enzymol.* **16:**126–138.

Sommer, M. H., A. L. Scully, and D. H. Spector. 1994. Transactivation by the human cytomegalovirus IE2 86-kilodalton protein requires a domain that binds to both the TATA box-binding protein and the retinoblastoma protein. *J. Virol.* **68:**6223–6231.

Sorg, G., and T. Stamminger. 1998. Strong conservation of the constitutive activity of the IE1/2 transcriptional control region in wild type strains of human cytomegalovirus. *J. Gen. Virol.* **79:**3039–3047.

Sorlie, P. D., E. Adam, S. L. Melnick, A. Folsom, T. Skelton, L. E. Chambless, R. Barnes, and J. L. Melnick. 1994.

Cytomegalovirus/herpesvirus and carotid atherosclerosis: the ARIC study. *J. Med. Virol.* **42:**33–37.

Spaete, R. R., K. Perot, P. I. Scott, J. A. Nelson, M. F. Stinski, and C. Pachl. 1993. Coexpression of truncated human cytomegalovirus gH with the UL115 gene product or the truncated human fibroblast growth factor receptor results in transport of gH to the cell surface. *Virology* **193:**853–861.

Spector, D. H., and S. A. Spector. 1984. The oncogenic potential of human cytomegalovirus. *Prog. Med. Virol.* **29:**45–89.

Spector, S. A., K. K. Hirata, and T. R. Newman. 1984. Identification of multiple cytomegalovirus strains in homosexual men with acquired immunodeficiency syndrome. *J. Infect. Dis.* **150:**953–956.

Spector, S. A., R. Merrill, D. Wolf, and W. M. Dankner. 1992. Detection of human cytomegalovirus in plasma of AIDS patients during acute visceral disease by DNA amplification. *J. Clin. Microbiol.* **30:**2359–2365.

Spector, S. A., T. Weingeist, R. B. Pollard, D. T. Dieterich, T. Samo, C. A. Benson, D. F. Busch, W. R. Freeman, P. Montague, H. J. Kaplan, L. Kellerman, M. Crager, M. DeArmond, W. Buhles, J. Feinberg, AIDS Clinical Trials Group, and Cytomegalovirus Cooperative Study Group. 1993. A randomized, controlled study of intravenous ganciclovir therapy for cytomegalovirus peripheral retinitis in patients with AIDS. *J. Infect. Dis.* **168:**557–563.

Spector, S. A., B. F. Busch, S. Follansbee, K. Squires, J. P. Lalezari, M. A. Jacobson, J. D. Connor, D. Jung, A. Shadman, B. Mastre, W. Buhles, W. L. Drew, AIDS Clinical Trials Group, and Cytomegalovirus Cooperative Study Group. 1995. Pharmacokinetic, safety, and antiviral profiles of oral ganciclovir in persons infected with human immunodeficiency virus: a phase I/II study. *J. Infect. Dis.* **171:**1431–1437.

Spector, S. A., G. F. McKinley, J. P. Lalezari, T. Samo, R. Andruczk, S. Follansbee, P. D. Sparti, D. V. Havlir, G. Simpson, W. Buhles, R. Wong, and M. J. Stempien. 1996. Oral ganciclovir for the prevention of cytomegalovirus disease in persons with AIDS. *N. Engl. J. Med.* **334:**1491–1497.

Spector, S. A., R. Wong, K. Hsia, M. Pilcher, and M. J. Stempien. 1998. Plasma cytomegalovirus (CMV) DNA load predicts CMV disease and survival in AIDS patients. *J. Clin. Investig.* **101:**497–502.

Spector, S. A., K. Hsia, M. Crager, M. Pilcher, S. Cabral, and M. J. Stempien. 1999. Cytomegalovirus (CMV) DNA load is an independent predictor of CMV disease and survival in advanced AIDS. *J. Virol.* **73:**7027–7030.

Speich, R., R. Thurnheer, A. Gaspert, W. Weder, and A. Boehler. 1999. Efficacy and cost effectiveness of oral ganciclovir in the prevention of cytomegalovirus disease after lung transplantation. *Transplantation* **67:**315–320.

Stagno, S., and G. A. Cloud. 1994. Working parents: the impact of day care and breastfeeding on cytomegalovirus infections in offspring. *Proc. Natl. Acad. Sci. USA* **91:**2384–2389.

Stagno, S., D. W. Reynolds, A. Tsiantos, D. A. Fuccillo, W. Long, and C. A. Alford. 1975. Comparative serial virologic and serologic studies of symptomatic and subclinical congenitally and natally acquired cytomegalovirus infections. *J. Infect. Dis.* **132:**568–577.

Stagno, S., D. W. Reynolds, C. S. Amos, A. J. Dahle, F. P. McCollister, I. Mohindra, R. Ermocilla, and C. A. Alford. 1977. Auditory and visual defects resulting from symptomatic and subclinical congenital cytomegaloviral and toxoplasma infections. *Pediatrics* **59:**669–678.

Stagno, S., D. Reynolds, R. Pass, and C. Alford. 1980. Breast milk and the risk of cytomegalovirus infection. *N. Engl. J. Med.* **302:**1073–1076.

Stagno, S., R. F. Pass, M. E. Dworsky, R. E. Henderson, E. G. Moore, P. D. Walton, and C. A. Alford. 1982. Congenital cytomegalovirus infection: the relative importance of primary and recurrent maternal infection. *N. Engl. J. Med.* **306:**945–949.

Stagno, S., R. F. Pass, G. Cloud, W. J. Britt, R. E. Henderson, P. D. Walton, D. A. Veren, F. Page, and C. A. Alford. 1986. Primary cytomegalovirus infection in pregnancy. Incidence, transmission to fetus, and clinical outcome. *JAMA* **256:**1904–1908.

Stanier, P., D. L. Taylor, A. D. Kitchen, N. Wales, Y. Tryhorn, and A. S. Tyms. 1989. Persistence of cytomegalovirus in mononuclear cells in peripheral blood from blood donors. *Br. Med. J.* **299:**897–898.

Steiner, I. 1996. Human herpesviruses latent infection in the nervous system. *Immunol. Rev.* **152:**157–173.

Stenberg, R. M. 1996. The human cytomegalovirus major immediate-early gene. *Intervirology* **39:**343–349.

Stenberg, R. M., A. S. Depto, J. Fortney, and J. A. Nelson. 1989. Regulated expression of early and late RNA and protein from the human cytomegalovirus immediate-early gene region. *J. Virol.* **63:**2699–2708.

Stéphan, F., A. Fajac, D. Grenet, P. Honderlick, S. Ricci, I. Frachon, S. Friard, I. Caubarrere, J.-F. Bernaudin, and M. Stern. 1997. Predictive value of cytomegalovirus DNA detection by polymerase chain reaction in blood and bronchoalveolar lavage in lung transplant patients. *Transplantation* **63:**1430–1435.

Stinski, M. F., D. R. Thomsen, R. M. Stenberg, and L. C. Goldstein. 1983. Organization and expression of the immediate early genes of human cytomegalovirus. *J. Virol.* **46:**1–14.

Stratta, R. J., M. S. Shaefer, K. A. Cushing, R. S. Markin, E. C. Reed, A. N. Langnas, T. J. Pillen, and B. W. Shaw, Jr. 1992. A randomized prospective trial of acyclovir and immune globulin prophylaxis in liver transplant recipients receiving OKT3 therapy. *Arch. Surg.* **127:**55–63.

Stratta, R. J., R. J. Taylor, J. S. Bynon, J. A. Lowell, M. S. Cattral, K. Frisbie, S. Miller, S. J. Radio, and D. C. Brennan. 1994. Viral prophylaxis in combined pancreas-kidney transplant recipients. *Transplantation* **57:**506–512.

Strauch, B., N. Siegel, L. Andrews, and G. Miller. 1987. Oropharyngeal excretion of Epstein-Barr virus by renal transplant recipients and other patients treated with immunosuppressive drugs. *Lancet* **i:**234–237.

Strickler, J., F. Fedeli, C. Horwitz, C. Copenhaver, and G. Frizzeri. 1993. Infectious mononucleosis in lymphoid tissue. Histopathology, in situ hybridization, and differential diagnosis. *Arch. Pathol. Lab. Med.* **117:**269–278.

Stronati, M., M. G. Revello, R. M. Cerbo, M. Furione, G. Rondini, and G. Gerna. 1995. Ganciclovir therapy of congenital human cytomegalovirus hepatitis. *Acta Paediatr.* **84:**340–341.

Sugden, B., and W. Mark. 1977. Clonal transformation of adult human leukocytes by Epstein-Barr virus. *J. Virol.* **23:**503–508.

Sugden, B., M. Phelps, and J. Domoradzki. 1979. Epstein-Barr virus DNA is amplified in transformed lymphocytes. *J. Virol.* **31:**590–595.

Sullivan, V., C. L. Talarico, S. C. Stanat, M. Davis, D. M. Coen, and K. K. Biron. 1992. A protein kinase homologue controls phosphorylation of ganciclovir in human cytomegalovirus-infected cells. *Nature* **358:**162–164.

Swinnen, L. J., M. R. Costanzo-Nordin, S. G. Fisher, E. J. O'Sullivan, M. R. Johnson, A. L. Heroux, G. J. Dizikes, R. Pifarre, and R. I. Fisher. 1990. Increased incidence of lymphoproliferative disorder after immunosuppression with the monoclonal antibody OKT3 in cardiac-transplant recipients. *N. Engl. J. Med.* **323:**1723–1728.

Taburet, A.-M., C. Katlama, C. Blanshard, G. Gorza, D. Gazzard, E. Dohin, B. G. Gazzard, C. Frostegard, and E. Singlas. 1992. Pharmacokinetics of foscarnet and distribution to cerebrospinal fluid after intravenous infusion in patients with human immunodeficiency virus infection. *Antimicrob. Agents Chemother.* **36:**1821–1824.

Takahashi, M., T. Otsuka, Y. Okuno, Y. Asano, and T. Yazaki. 1974. Live vaccine used to prevent the spread of varicella in children in hospital. *Lancet* **ii:**1288–1290.

Takayama, N., M. Minamitani, and M. Takayama. 1997. High incidence of breakthrough varicella observed in healthy Japanese children immunized with live attenuated varicella vaccine (OKA strain). *Acta Paediatr. Jpn.* **39:**663–668.

Talarico, C. L., T. C. Burnette, W. H. Miller, S. L. Smith, M. G. Davis, S. C. Stanat, T. I. Ng, Z. He, D. M. Coen, B. Roizman, and K. K. Biron. 1999. Acyclovir is phosphorylated by the human cytomegalovirus UL97 protein. *Antimicrob. Agents Chemother.* **43:**1941–1946.

Tanner, J., J. Weis, D. Fearon, Y. Whang, and E. Kieff. 1987. Epstein-Barr virus gp350/220 binding to the B lymphocyte C3d receptor mediates adsorption, capping, and endocytosis. *Cell* **50:**203–213.

Tanner, J., C. Alfieri, T. Chatila, and F. Diaz-Mitoma. 1996. Induction of interleukin-6 after stimulation of human B-cell CD21 by Epstein-Barr virus glycoproteins gp350 and gp220. *J. Virol.* **70:**570–575.

Taylor-Wiedeman, J., J. G. Sissons, L. K. Borysiewicz, and J. H. Sinclair. 1991. Monocytes are a major site of persistence of human cytomegalovirus in peripheral blood mononuclear cells. *J. Gen. Virol.* **72:**2059–2064.

Telenti, A., and T. F. Smith. 1989. Screening with a shell vial assay for antiviral activity against cytomegalovirus. *Diagn. Microbiol. Infect. Dis.* **12:**5–8.

Terada, K., T. Niizuma, S. Kawano, N. Kataoka, T. Akisada, and Y. Orita. 1998. Detection of varicella-zoster virus DNA in peripheral mononuclear cells from patients with Ramsay-Hunt-syndrome or zoster sine herpete. *J. Med. Virol.* **56:**359–363.

Thorley-Lawson, D., and E. Israelsohn. 1987. Generation of specific cytotoxic T cells with a fragment of the Epstein-Barr virus-encoded p63/latent membrane protein. *Proc. Natl. Acad. Sci. USA* **84:**5384–5388.

Tierney, R., N. Steven, L. Young, and A. Rickinson. 1994. Epstein-Barr virus latency in blood mononuclear cells: analysis of viral gene transcription during primary infection and in the carrier state. *J. Virol.* **68:**7374–7385.

Todo, S., A. Tzakis, K. Abu-Elmagd, J. Reyes, H. Furukawa, B. Nour, J. Fung, A. Demetris, and T. E. Starzl. 1995. Abdominal multivisceral transplantation. *Transplantation* **59:**234–240.

Tomazin, R., J. Boname, N. R. Hegde, D. M. Lewinsohn, Y. Altschuler, T. R. Jones, P. Cresswell, J. A. Nelson, S. R. Riddell, and D. C. Johnson. 1999. Cytomegalovirus US2 destroys two components of the MHC class II pathway, preventing recognition by CD4+ T cells. *Nat. Med.* **5:**1039–1043.

Tomkinson, B., and E. Kieff. 1992. Use of second-site homologous recombination to demonstrate that Epstein-Barr virus nuclear protein 3B is not important for lymphocyte infection or growth transformation in vitro. *J. Virol.* **66:**2893–2903.

Tomkinson, B., E. Robertson, and E. Kieff. 1993a. Epstein-Barr virus nuclear proteins EBNA-3A and EBNA-3C are essential for B-lymphocyte growth transformation. *J. Virol.* **67:**2014–2205.

Tomkinson, B., E. Robertson, R. Yalamanchili, R. Long-necker, and E. Kieff. 1993b. Epstein-Barr virus recombinants from overlapping cosmid fragments. *J. Virol.* **67:**7298–7306.

Topilko, A., and S. Michelson. 1994. Hyperimmediate entry of human cytomegalovirus virions and dense bodies into human fibroblasts. *Res. Virol.* **145:**75–82.

Torok-Storb, B., M. Boeckh, C. Hoy, W. Leisenring, D. Myerson, and T. Gooley. 1997. Association of specific cytomegalovirus genotypes with death from myelosuppression after marrow transplantation. *Blood* **90:**2097–2102.

Tosato, G., J. Tanner, K. Jones, M. Revel, and S. Pike. 1990. Identification of interleukin-6 as an autocrine growth factor for Epstein-Barr virus-immortalized B cells. *J. Virol.* **64:**3033–3041.

Toyoda, M., J. B. Carlos, O. A. Galera, K. Galfayan, X. Zhang, Z. Sun, L. S. C. Czer, and S. C. Jordan. 1997. Correlation of cytomegalovirus DNA levels with response to antiviral therapy in cardiac and renal allograft recipients. *Transplantation* **63:**957–963.

Tufail, A., A. A. Moe, M. J. Miller, E. A. Wagar, D. A. Bruckner, and G. N. Holland. 1999. Quantitative cytomegalovirus DNA level in the blood and its relationship to cytomegalovirus retinitis in patients with acquired immune deficiency syndrome. *Ophthalmology* **106:**133–141.

Tugizov, S., D. Navarro, P. Paz, Y. Wang, I. Qadri, and L. Pereira. 1994. Function of human cytomegalovirus glyco-protein B: syncytium formation in cells constitutively expressing gB is blocked by virus-neutralizing antibodies. *Virology* **201:**263–276.

Tugizov, S., E. Maidji, and L. Pereira. 1996. Role of apical and basolateral membranes in replication of human cytomegalovirus in polarized retinal pigment epithelial cells. *J. Gen. Virol.* **77:**61–74.

Uberti-Foppa, C., F. Lillo, M. R. Terreni, A. Puglisi, M. Guffanti, M. Gianotti, and A. Lazzarin. 1998. Cytomegalo-virus pneumonia in AIDS patients: value of cytomegalovirus culture from BAL fluid and correlation with lung disease. *Chest* **113:**919–923.

Urban, M., M. Klein, W. J. Britt, E. Hassfurther, and M. Mach. 1996. Glycoprotein H of human cytomegalovirus is a major antigen for the neutralizing humoral immune response. *J. Gen. Virol.* **77:**1537–1547.

USPHS/IDSA Prevention of Opportunistic Infections Working Group. 1997. 1997 USPHS/IDSA guidelines for the prevention of opportunistic infections in persons infected with human immunodeficiency virus: disease-specific recommendations. *Clin. Infect. Dis.* **25:**S313–S335.

Utz, U., W. J. Britt, L. G. Vugler, and M. Mach. 1989. Identification of a neutralizing epitope on the glycoprotein gp58 of human cytomegalovirus. *J. Virol.* **63:**1995–2001.

Utz, U., S. Koenig, J. E. Coligan, and W. E. Biddison. 1992. Presentation of three different viral peptides, HTLV-1 tax, HCMV gB, and influenza virus M1, is determined by common structural features of the HLA-A2.1 molecule. *J. Immunol.* **149:**214–221.

van den Berg, A. P., I. J. Klompmaker, E. B. Haagsma, A. Scholten-Sampson, C. M. A. Bijleveld, J. Schirm, M. van der Giessen, M. J. H. Slooff, and T. H. The. 1991. Antigenemia in the diagnosis and monitoring of active cytomegalovirus infection after liver transplantation. *J. Infect. Dis.* **164:**265–270.

van den Berg, A. P., W. van der Bij, W. J. van Son, J. Anema, M. van der Giessen, J. Schirm, A. M. Tegzess, and T. H. The. 1989. Cytomegalovirus antigenemia as a useful marker of symptomatic cytomegalovirus infection after renal transplantation—a report of 130 consecutive patients. *Transplantation* **48:**991–995.

van der Bij, W., J. Schirm, R. Torensma, W. J. van Son, A. M. Tegzess, and T. H. The. 1988. Comparison between viremia and antigenemia for detection of cytomegalovirus in blood. *J. Clin. Microbiol.* **26:**2531–2535.

van der Meer, J. T., W. L. Drew, R. A. Bowden, G. J. Galasso, P. D. Griffiths, D. A. Jabs, C. Katlama, S. A. Spector, and R. J. Whitley. 1996. Summary of the international consensus symposium on advances in the diagnosis, treatment and prophylaxis of cytomegalovirus infection. *Antivir. Res.* **32:**119–140.

Van Grunsven, W. M. J., W. J. M. Spaan, and J. M. Middle-dorp. 1994. Localization and diagnostic application of immunodominant domains of the BFRF3-encoded Epstein-Barr virus capsid protein. *J. Infect. Dis.* **170:**13–19.

Verbraak, F. D., M. Bruinenberg, G. J. van den Horn, C. Meenken, A. van der Lelij, C. B. Hoyng, A. Kijlstra, and R. Peek. 1998. Cytomegalovirus (CMV) strain differences between the eye and blood in AIDS patients with CMV retinitis. *AIDS* **12:**713–718.

Visse, B., B. Dumond, J. M. Huraux, and A. M. Fillet. 1998. Single amino acid change in DNA polymerase is associated with foscarnet resistance in a varicella-zoster virus strain recovered from a patient with AIDS. *J. Infect. Dis.* **178:**S55–S57.

Vlieger, A. M., G. J. Boland, N. M. Jiwa, R. A. de Weger, R. Willemze, G. C. de Gast, and J. H. Falkenburg. 1992. Cytomegalovirus antigenemia assay or PCR can be used to monitor ganciclovir treatment in bone marrow transplant recipients. *Bone Marrow Transplant.* **9:**247–253.

Vochem, M., K. Hamprecht, G. Jahn, and C. P. Speer. 1998. Transmission of cytomegalovirus to preterm infants through breast milk. *Pediatr. Infect. Dis. J.* **17:**53–58.

Vogel, J. U., J. Cinatl, A. Lux, B. Weber, A. J. Driesel, and H. W. Doerr. 1996. New PCR assay for rapid and quantitative detection of human cytomegalovirus in cerebrospinal fluid. *J. Clin. Microbiol.* **34:**482–483.

Vogelberg, C., U. Meyer-Konig, F. T. Hufert, G. Kirste, and D. von Laer. 1996. Human cytomegalovirus glycoprotein B genotypes in renal transplant recipients. *J. Med. Virol.* **50:**31–34.

Von Glahn, W. C., and A. M. Pappenheimer. 1925. Intranuclear inclusions in visceral disease. *Am. J. Pathol.* **1:**445–465.

Vornhagen, R., B. Plachter, W. Hinderer, T. H. The, J. Van Zanten, L. Matter, C. A. Schmidt, H. H. Sonneborn, and G. Jahn. 1994. Early serodiagnosis of acute cytomegalovirus infection by enzyme-linked immunosorbent assay using recombinant antigens. *J. Clin. Microbiol.* **32:**981–986.

Vossen, R. C. R. M., M. C. E. van Dam-Mieras, and C. A. Bruggeman. 1996. Cytomegalovirus infection and vessel wall pathology. *Intervirology* **39:**213–221.

Vreeswijk, J., E. Folkers, F. Wagenaar, and J. G. Kapsenberg. 1988. The use of colloidal gold immunoelectron microscopy to diagnose varicella-zoster virus (VZV) infections by rapid discrimination between VZV, HSV-1 and HSV-2. *J. Virol. Methods* **22:**255–271.

Wada, K., S. Mizuno, K. Kato, T. Kamiya, and K. Ozawa. 1997. Cytomegalovirus glycoprotein B sequence variation among Japanese bone marrow transplant recipients. *Intervirology* **40:**215–219.

Wagner, B., B. Kropff, H. Kalbacher, W. Britt, V. A. Sundqvist, L. Ostberg, and M. Mach. 1992. A continuous sequence of more than 70 amino acids is essential for antibody binding to the dominant antigenic site of glycoprotein gp58 of human cytomegalovirus. *J. Virol.* **66:**5290–5297.

Waldman, W. J., D. A. Knight, and P. W. Adams. 1998. Cytolytic activity against allogeneic human endothelia: resistance of cytomegalovirus-infected cells and virally activated lysis of uninfected cells. *Transplantation* **66:**67–77.

Walker, R. C., W. F. Marshall, J. G. Strickler, R. H. Wiesner, J. A. Velosa, T. M. Haberman, C. G. McGregor, and C. V. Paya. 1995. Pretransplantation assessment of the risk of lymphoproliferative disorder. *Clin. Infect. Dis.* **20:**1346–1353.

Walmsley, S., K. O'Rourke, C. Mortimer, A. Rachlis, I. Fong, and T. Mazzulli. 1998. Predictive value of cytomegalovirus (CMV) antigenemia and Digene hybrid capture DNA assays for CMV disease in human immunodeficiency virus-infected patients. *Clin. Infect. Dis.* **27:**573–581.

Walter, E. A., P. D. Greenberg, M. J. Gilbert, R. J. Finch, K. S. Watanabe, E. D. Thomas, and S. R. Riddell. 1995. Reconstitution of cellular immunity against cytomegalovirus in recipients of allogeneic bone marrow by transfer of T-cell clones from the donor. *N. Engl. J. Med.* **333:**1038–1044.

Weber, B., U. Nestler, W. Ernst, H. Rabenau, J. Braner, A. Birkenbach, E.-H. Scheuermann, W. Schoeppe, and H. W. Doerr. 1994. Low correlation of human cytomegalovirus DNA amplification by polymerase chain reaction with cytomegalovirus disease in organ transplant recipients. *J. Med. Virol.* **43:**187–193.

Webster, A., J. E. McLaughlin, M. A. Johnson, V. C. Emery, and P. D. Griffiths. 1995. Use of the polymerase chain reaction to detect genomes of human immunodeficiency virus and cytomegalovirus in post-mortem tissues. *J. Med. Virol.* **47:**23–28.

Weekes, M. P., M. R. Wills, K. Mynard, A. J. Carmichael, and J. G. P. Sissons. 1999. The memory cytotoxic T-lymphocyte (CTL) response to human cytomegalovirus infection contains individual peptide-specific CTL clones that have undergone extensive expansion in vivo. *J. Virol.* **73:**2099–2108.

Weibel, R. E., B. J. Neff, B. J. Kuter, H. A. Guess, C. A. Rothenberger, A. J. Fitzgerald, K. A. Connor, A. A. McLean, M. R. Hilleman, E. B. Buynak, and E. M. Scolnick. 1984. Live attenuated varicella virus vaccine. Efficacy trial in healthy children. *N. Engl. J. Med.* **310:**1409–1415.

Weismann, K., C. S. Petersen, C. W. Blichmann, N. H. Nielsen, and B. M. Hultberg. 1998. Bullous erythema multiforme following herpes zoster and varicella-zoster virus infection. *J. Eur. Acad. Dermatol. Venereol.* **11:**147–150.

Weiss, L., and L. Movahed. 1989. In situ demonstration of Epstein-Barr viral genomes in viral-associated B cell lymphoproliferations. *Am. J. Pathol.* **134:**651–659.

Weller, T. H. 1983. Varicella and herpes zoster. Changing concepts of the natural history, control, and importance of a not-so-benign virus. *N. Engl. J. Med.* **309:**1434–1440.

Weller, T. H., and M. B. Stoddard. 1952. Intranuclear inclusion bodies in cultures of human tissue inoculated with varicella vesicle fluid. *J. Immunol.* **68:**311–319.

Weller, T. H., J. C. Macauley, J. M. Craig, and P. Wirth. 1957. Isolation of intranuclear inclusion producing agents from infants with illnesses resembling cytomegalic inclusion disease. *Proc. Soc. Exp. Biol. Med.* **94:**4–12.

Wenkel, H., V. Rummelt, B. Fleckenstein, and G. O. H. Naumann. 1998. Detection of varicella zoster virus DNA and viral antigen in human eyes after herpes zoster ophthalmicus. *Ophthalmology* **105:**1323–1330.

West, P. G., B. Aldrich, R. Hartwig, and G. J. Haller. 1988. Increased detection rate for varicella-zoster virus with combination of two techniques. *J. Clin. Microbiol.* **26:**2680–2681.

Wharton, M. 1996. The epidemiology of varicella-zoster virus infections. *Infect. Dis. Clin. N. Am.* **10:**571–581.

Whitley, R. J. 1995. Varicella-zoster virus. *In* G. L. Mandell, J. E. Bennett, and R. Dolin (ed.), *Principles and Practice of Infectious Diseases*, 4th ed. Churchill Livingstone, New York, N.Y.

Whitley, R. J., G. Cloud, W. Gruber, G. A. Storch, G. J. Demmler, R. F. Jacobs, W. Dankner, S. A. Spector, S. Starr, R. F. Pass, S. Stagno, W. J. Britt, C. Alford, Jr., S. Soong, X. J. Zhou, L. Sherrill, J. M. Fitzgerald, and J. P. Sommadossi. 1997. Ganciclovir treatment of symptomatic congenital cytomegalovirus infection: results of a phase II study. *J. Infect. Dis.* **175:**1080–1086.

Wiertz, E., A. Hill, D. Tortorella, and H. Ploegh. 1997. Cytomegaloviruses use multiple mechanisms to elude the host immune response. *Immunol. Lett.* **57:**213–216.

Wiertz, E. J. H. J., T. R. Jones, L. Sun, M. Bogyo, H. J. Geuze, and H. L. Ploegh. 1996. The human cytomegalovirus US11 gene product dislocates MHC class I heavy chains from the endoplasmic reticulum to the cytosol. *Cell* **84:**769–779.

Wilkinson, A. H., J. L. Smith, L. G. Hunsicker, J. Tobacman, D. P. Kapelanski, M. Johnson, F. H. Wright, D. M. Behrendt, and R. J. Corry. 1989. Increased frequency of posttransplant lymphomas in patients treated with cyclosporine, azathioprine, and prednisone. *Transplantation* **47:**293–296.

Wills, M. R., A. J. Carmichael, K. Mynard, X. Jin, M. P. Weekes, B. Plachter, and J. G. Sissons. 1996. The human cytotoxic T-lymphocyte (CTL) response to cytomegalovirus is dominated by structural protein pp65: frequency, specificity, and T-cell receptor usage of pp65-specific CTL. *J. Virol.* **70:**7569–7579.

Wilson, A., M. Sharp, C. M. Koropchak, S. F. Ting, and A. M. Arvin. 1992. Subclinical varicella-zoster virus viremia, herpes zoster, and T lymphocyte immunity to varicella-zoster viral antigens after bone marrow transplantation. *J. Infect. Dis.* **165:**119–126.

Winston, D. J., E. S. Huang, M. J. Miller, C. H. Lin, W. G. Ho, R. P. Gale, and R. E. Champlin. 1985. Molecular epidemiology of cytomegalovirus infections associated with bone marrow transplantation. *Ann. Intern. Med.* **102:**16–20.

Winston, D. J., W. G. Ho, K. Bartoni, C. Du Mond, D. F. Ebeling, and W. C. Buhles. 1993. Ganciclovir prophylaxis of cytomegalovirus infection and disease in allogeneic bone marrow transplant recipients: results of a placebo-controlled double-blind trial. *Ann. Intern. Med.* **118:**179–184.

Winston, D. J., D. Wirin, A. Shaked, and R. W. Busuttil. 1995. Randomized comparison of ganciclovir and high-dose acyclovir for long-term cytomegalovirus prophylaxis in liver transplant recipients. *Lancet* **346:**69–74.

Wittes, J. T., A. Kelly, and K. M. Plante. 1996. Meta-analysis of CMVIG studies for the prevention and treatment of CMV infection in transplant patients. *Transplant. Proc.* **28:**(Suppl. 2):17–24.

Wolf, D. G., D. J. Lee, and S. A. Spector. 1995. Detection of human cytomegalovirus mutations associated with ganciclovir resistance in cerebrospinal fluid of AIDS patients with central nervous system disease. *Antimicrob. Agents Chemother.* **39:**2552–2554.

Wolf, D. G., I. Yaniv, A. Honigman, I. Kassis, T. Schonfeld, and S. Ashkenazi. 1998. Early emergence of ganciclovir-resistant human cytomegalovirus strains in children with primary combined immunodeficiency. *J. Infect. Dis.* **178:**535–538.

Woo, P. C. Y., C. Y. Lo, S. K. F. Lo, H. Siau, J. S. M. Peiris, S. S. Y. Wong, W. K. Luk, T. M. Chan, W. W. Lim, and K. Y. Yuen. 1997. Distinct genotypic distributions of cytomegalovirus (CMV) envelope glycoprotein in bone marrow and renal transplant recipients with CMV disease. *Clin. Diagn. Lab. Immunol.* **4:**515–518.

Wood, B. L., D. Sabath, V. C. Broudy, and G. Raghu. 1996. The recipient origin of posttransplant lymphoproliferative disorders in pulmonary transplant patients. *Cancer* **78:**2223–2228.

Woodroffe, S. B., J. Hamilton, and H. M. Garnett. 1997. Comparison of the infectivity of the laboratory strain AD169 and a clinical isolate of human cytomegalovirus to human smooth muscle cells. *J. Virol. Methods* **63:**181–191.

Woods, G. L., and D. H. Walker. 1996. Detection of infection or infectious agents by use of cytologic and histologic stains. *Clin. Microbiol. Rev.* **9:**382–404.

Wutzler, P. 1997. Antiviral therapy of herpes simplex and varicella-zoster virus infections. *Intervirology* **40:**343–356.

Wyatt, J. P., J. Saxton, R. S. Lee, and H. Pinkerton. 1950. Generalized cytomegalic inclusion disease. *J. Pediatr.* **36:**271–294.

Yang, C. W., Y. O. Kim, Y. S. Kim, S. Y. Kim, I. S. Moon, H. J. Ahn, Y. B. Koh, and B. K. Bang. 1998. Clinical course of cytomegalovirus (CMV) viremia with and without ganciclovir treatment in CMV-seropositive kidney transplant recipients. Longitudinal follow-up of CMV pp65 antigenemia assay. *Am. J. Nephrol.* **18:**373–378.

Yao, Q., A. Rickinson, and M. Epstein. 1985. A re-examination of the Epstein-Barr virus carrier state in healthy seropositive individuals. *Int. J. Cancer* **35:**35–42.

Yao, Q. P., M. Rowe, B. Martin, L. Young, and A. B. Rickinson. 1991. The EBV carrier state: dominance of a single growth-transforming isolate in the blood and oropharynx of healthy carriers. *J. Gen. Virol.* **72:**1579–1590.

Yao, Q. Y., P. Ogan, M. Rowe, M. Wood, and A. B. Rickinson. 1989. The Epstein-Barr virus-infected B cells persist in the circulation of acyclovir-treated virus carriers. *Int. J. Cancer* **43:**67–71.

Yoshikawa, T., S. Suga, T. Kozawa, S. Kawaguchi, and Y. Asano. 1998. Persistence of protective immunity after postexposure prophylaxis of varicella with oral acyclovir in the family setting. *Arch. Dis. Child.* **78:**61–63.

Yurochko, A. D., M. W. Mayo, E. E. Poma, A. S. Baldwin, Jr., and E. S. Huang. 1997. Induction of the transcription factor Sp1 during human cytomegalovirus infection mediates upregulation of the p65 and p105/p50 NF-κB promoters. *J. Virol.* **71:**4638–4648.

Zaia, J. A., G. M. Gallez-Hawkins, B. R. Tegtmeier, A. ter Veer, X. Li, J. C. Niland, and S. J. Forman. 1997. Late cytomegalovirus disease in marrow transplantation is predicted by virus load in plasma. *J. Infect. Dis.* **176:**782–785.

Zerboni, L., S. Nader, K. Aoki, and A. M. Arvin. 1998. Analysis of the persistence of humoral and cellular immunity in children and adults immunized with varicella vaccine. *J. Infect. Dis.* **177:**1701–1704.

Zhang, L. J., P. Hanff, C. Rutherford, W. H. Churchill, and C. S. Crumpacker. 1995. Detection of human cytomegalovirus DNA, RNA, and antibody in normal donor blood. *J. Infect. Dis.* **171:**1002–1006.

Zhou, Y. F., M. B. Leon, M. A. Waclawiw, J. J. Popma, Z. X. Yu, T. Finkel, and S. E. Epstein. 1996. Association between prior cytomegalovirus infection and the risk of restenosis after coronary atherectomy. *N. Engl. J. Med.* **335:**624–630.

Zhu, H., Y. Shen, and T. Shenk. 1995. Human cytomegalovirus IE1 and IE2 proteins block apoptosis. *J. Virol.* **69:**7960–7970.

Zhuravskaya, T., J. P. Maciejewski, D. M. Netski, E. Bruening, F. R. Mackintosh, and S. St. Jeor. 1997. Spread of human cytomegalovirus (HCMV) after infection of human hematopoietic progenitor cells. *Blood* **90:**2482–2491.

Zimmerman, A., D. Michel, I. Pavic, W. Hampl, A. Lüske, J. Neyts, E. De Clercq, and T. Mertens. 1997. Phosphorylation of acyclovir, ganciclovir, penciclovir and S2242 by the cytomegalovirus UL97 protein: a quantitative analysis using recombinant vaccinia viruses. *Antivir. Res.* **36:**35–42.

Zini, N., M. C. Battista, S. Santi, M. Riccio, G. Bergamini, M. P. Landini, and N. M. Maraldi. 1999. The novel structural protein of human cytomegalovirus, pUL25, is localized in the viral tegument. *J. Virol.* **73:**6073–6075.

Zipeto, D., C. Hong, G. Gerna, M. Zavattoni, D. Katzenstein, T. C. Merigan, and L. Rasmussen. 1998. Geographic and demographic differences in the frequency of human cytomegalovirus gB genotypes 1-4 in immunocompromised patients. *AIDS Res. Hum. Retrovir.* **14:**533–536.

Zurlo, J. J., D. O'Neill, M. A. Polis, J. Manischewitz, R. Yarchoan, M. Baseler, H. C. Lane, and H. Masur. 1993. Lack of clinical utility of cytomegalovirus blood and urine cultures in patients with HIV infection. *Ann. Intern. Med.* **118:**12–17.

Zutter, M. M., P. J. Martin, G. E. Sale, H. M. Shulman, L. Fisher, E. D. Thomas, and D. M. Durnam. 1988. Epstein-Barr virus lymphoproliferation after bone marrow transplantation. *Blood* **72:**520.

Zweygberg-Wirgart, B., M. Brytting, A. Linde, B. Wahren, and L. Grillner. 1998. Sequence variation within three important cytomegalovirus gene regions in isolates from four different patient populations. *J. Clin. Microbiol.* **36:**3662–3669.

Human Herpesviruses 6, 7, and 8

PHILIP E. PELLETT AND SHEILA C. DOLLARD

33

Human herpesvirus 6 variants A and B (HHV-6A and HHV-6B), HHV-7, and HHV-8 (also known as Kaposi's sarcoma [KS]-associated herpesvirus) were discovered between 1986 and 1994 (Salahuddin et al., 1986; Frenkel et al., 1990; Chang et al., 1994). These are not new viruses in an evolutionary sense; all evidence suggests that they have been part of human biology since near its inception. That they were discovered only recently is a measure of both their subtle interaction with immunocompetent hosts and the relative youth of virology. Their discovery can be attributed to the conflux of increasing populations of immunocompromised patients and the development of new methods for lymphocyte propagation and molecular identification of infectious agents.

HHV-6B is the major etiologic agent of roseola (also known as roseola infantum, exanthem subitum, or sixth disease). HHV-7 accounts for a subset of roseola cases. Both HHV-6 variants and HHV-7 have been associated with disease in immunocompromised organ transplant recipients. HHV-8 is the etiologic agent of KS and has been closely associated with multicentric Castleman's disease (MCD) and a form of body cavity lymphomas known as primary effusion lymphomas (PEL). Each of the viruses has also been associated with a number of other diseases for which proof of an etiologic role is incomplete or has been refuted. An example of an intriguing, but unproven, association is between the HHV-6 variants and multiple sclerosis. These issues will be discussed in more detail below. Prevalences and disease associations are summarized in Tables 1 and 2.

As for all herpesviruses, an important aspect of the biology of these viruses is that they establish lifelong infections of their hosts. These persistent infections are maintained through a combination of establishing latent (nonproductive) infections in some cells and intermittent or persistent lytic infections in other cells or tissues (Table 2). Latent virus can be reactivated to a lytic state. Virus produced during the lytic state can then reseed both the latent and persistent repositories as well as provide infectious material for transmission. Reactivation events are normally subclinical but can be associated with disease, particularly in immunocompromised patients. Issues relating to persistence of HHV-6, HHV-7, and HHV-8 were reviewed recently (Meng and Pellett, 1999).

These viruses present diagnostic challenges for the clinician and the laboratorian. It is likely that much will change on these fronts over the next few years, with further development of both rapid diagnosis and therapy that will work best in conjunction with careful monitoring of viral activity.

HHV-6A AND HHV-6B

Discovery and Identification of the Variants

Salahuddin and colleagues discovered HHV-6 in lymphocytes of immunocompromised patients with lymphoproliferative disorders (Salahuddin et al., 1986). It soon became apparent that isolates of the virus were segregated in two distinct groups based on their in vitro cell tropisms, antigenicity, nucleotide sequences, and epidemiology (Ablashi et al., 1991, 1993; Aubin et al., 1991; Schirmer et al., 1991; reviewed in Braun et al., 1997, and Campadelli-Fiume et al., 1993). The variant groups are designated HHV-6A and HHV-6B. While distinct, HHV-6A and HHV-6B are more closely related than any other pair of recognized herpesvirus species. Many aspects of their biology overlap, and they are sufficiently antigenically cross-reactive that variant-specific serologic assays are not available. Nonetheless, because they are clearly distinct, we will discuss them separately to the extent possible. When discussing studies where distinctions were not made, we will refer to them collectively as HHV-6.

Biology

Structure

HHV-6A and HHV-6B share the morphology that is common to all herpesviruses: an electron-dense core containing the viral genome contained in a capsid, which is surrounded by a layer known as the tegument, all of which is enclosed in a lipid envelope that is studded with virus-encoded glycoproteins and integral membrane proteins. Virions are 150 to 200 nm in diameter and have a characteristic appearance by electron microscopy (Fig. 1A) (Biberfeld et al., 1987; Yoshida et al., 1989). HHV-6 maturation appears to proceed by a pathway that includes (i) acquisition of tegument in nuclear structures called tegusomes, (ii) envelopment of tegumented capsids at inner nuclear membranes that lack the full complement of viral glycoproteins, (iii) deenvelopment into the cytoplasm, (iv) reenvelopment at annulate lamellae or cis-Golgi cisterna membranes that are studded with viral glycoproteins, (v) progressive maturation of viri-

TABLE 1 Prevalences of recently discovered herpesviruses

Virus and area	Prevalence (%)	
	Children (3–12 yr)	Adults
HHV-6A	?[a]	>50
HHV-6B	>80	>90
HHV-7	70	60–95
HHV-8		
KS nonendemic	0–5	0–10
KS endemic	10–30	30–>50

[a]?, unknown.

on glycoproteins during passage through the Golgi apparatus, and (vi) virion release from vesicles at the cell surface or during cell lysis (Nii et al., 1990; Roffman et al., 1990; Cardinali et al., 1998; Torrisi et al., 1999).

Genetic Content

HHV-6 genomes are approximately 160 kb in length and include approximately 100 genes (Gompels et al., 1995; Dominguez et al., 1999; Isegawa et al., 1999). The genomic termini are bounded by arrays of the sequence element, (TAACCC)n; similar sequences are present at the telomeres of mammalian chromosomes. Different groups of genes are conserved at each taxonomic level, proceeding from the level of family to subfamily, genus, and species. About 40 genes are homologs of genes that are shared across the family *Herpesviridae*; these are referred to as "herpesvirus core genes." Most of these genes encode structural proteins or proteins that are involved in lytic-cycle genome replication. Some genes are represented in all members of the subfamily *Betaherpesvirinae* (including human cytomegalovirus [HCMV]), and a handful are specific to the genus *Roseolovirus* (the HHV-6 variants and HHV-7). One gene (U94) that is in common to HHV-6A and HHV-6B but is not present in HHV-7 is a homolog of the gene encoding the important parvovirus nonstructural protein *ns1*, or *rep* (Thomson et al., 1991). Between the variants, amino acid sequence identities generally range from 94 to 99% in the genes that are conserved across the family *Herpesviridae* and from 70 to 90% in other genes (Dominguez et al., 1999; Isegawa et al., 1999). Several putative genes are specific to each of the HHV-6 variants.

Taxonomy

As described above, HHV-6A and HHV-6B are classified in the genus *Roseolovirus* of the subfamily *Betaherpesvirinae*. Based on the definition of herpesvirus species put forth by the International Committee for Taxonomy of Viruses, the variants should be formally recognized as distinct species (Braun et al., 1997; Campadelli-Fiume et al., 1999; Dominguez et al., 1999).

Growth Cycle

The HHV-6 variants can infect a variety of cell types in vitro but grow best in primary umbilical cord blood lymphocytes. Several strains have been adapted for reliable replication in continuous lymphocyte cell lines. HHV-6A isolates grow in HSB-2 and JJHAN cells, and HHV-6B variants grow best in Molt-3 and MT-4 cells (Ablashi et al., 1991). Pathological studies indicate that the in vivo host cell range is much wider and includes cells of lymphoid, epithelial, and neuronal origin (Table 2) (reviewed in Braun et al., 1997). CD4[+] cells are frequent targets of infection (Takahashi et al., 1989), although CD4 itself is not the receptor (Lusso et al., 1989). Interestingly, HHV-6A infection can induce CD4 expression on cells that were previously CD4[−], conferring de novo susceptibility to human immunodeficiency virus (HIV) infection in culture (Lusso et al., 1991, 1995). HHV-6-infected lymphocytes become greatly enlarged and refractile in 3 to 5 days (Ablashi et al., 1988; Black et al., 1989). Some cells become multinucleated. Host cell protein synthesis continues in the face of infection (Black et al., 1992), although host DNA replication is shut down (Di Luca et al., 1990). In general, infectious yields are low, with titers normally ranging from 10^3 to 10^5/ml.

Epidemiology

Due to the lack of variant-specific serologic assays, it is difficult to describe the distribution and prevalence of the variants individually. It is clear that in many populations, the combined seroprevalence exceeds 90% and can approach 100% (Table 1) (reviewed in Braun et al., 1997). Based on studies of children that combined serology and variant-specific PCR assays, HHV-6B infections occur most often before the age of 2 and nearly everyone who is HHV-6 seropositive is likely to have been infected with this variant (Pruksananonda et al., 1992; Hall et al., 1994). HHV-6A prevalence is more difficult to pin down, but PCR studies of lung and skin biopsy specimens suggest that it may exceed 50% in the United States and Italy (Cone et al., 1996; Di Luca et al., 1996). The age of acquisition for HHV-6A has not been identified. The extent to which prior infection by one variant protects from or modifies the course of infection by the other variant is not known.

Transmission

Saliva is the probable primary route of transmission for HHV-6B (Mukai et al., 1994; Di Luca et al., 1995; Tanaka-Taya et al., 1996). Viral DNA has been detected in various

TABLE 2 Tissue distribution of and diseases associated with HHV-6A, HHV-6B, HHV-7, and HHV-8

Virus	Predominant tissue tropism in vivo	Principal clinical syndrome(s)
HHV-6A	CD4[+] lymphocytes, skin, brain, monocytes, lung	AIDS- and transplant-associated opportunistic infection
HHV-6B	CD4[+] lymphocytes, monocytes, salivary glands, lymph node, brain, lung	Febrile illness, roseola, posttransplantation disease
HHV-7	CD4[+] lymphocytes, salivary glands	Subset of roseola, CMV-associated posttransplantation disease
HHV-8	Diseased tissue associated with KS, PEL, MCD; CD19[+] PBMC	KS, MCD, PEL

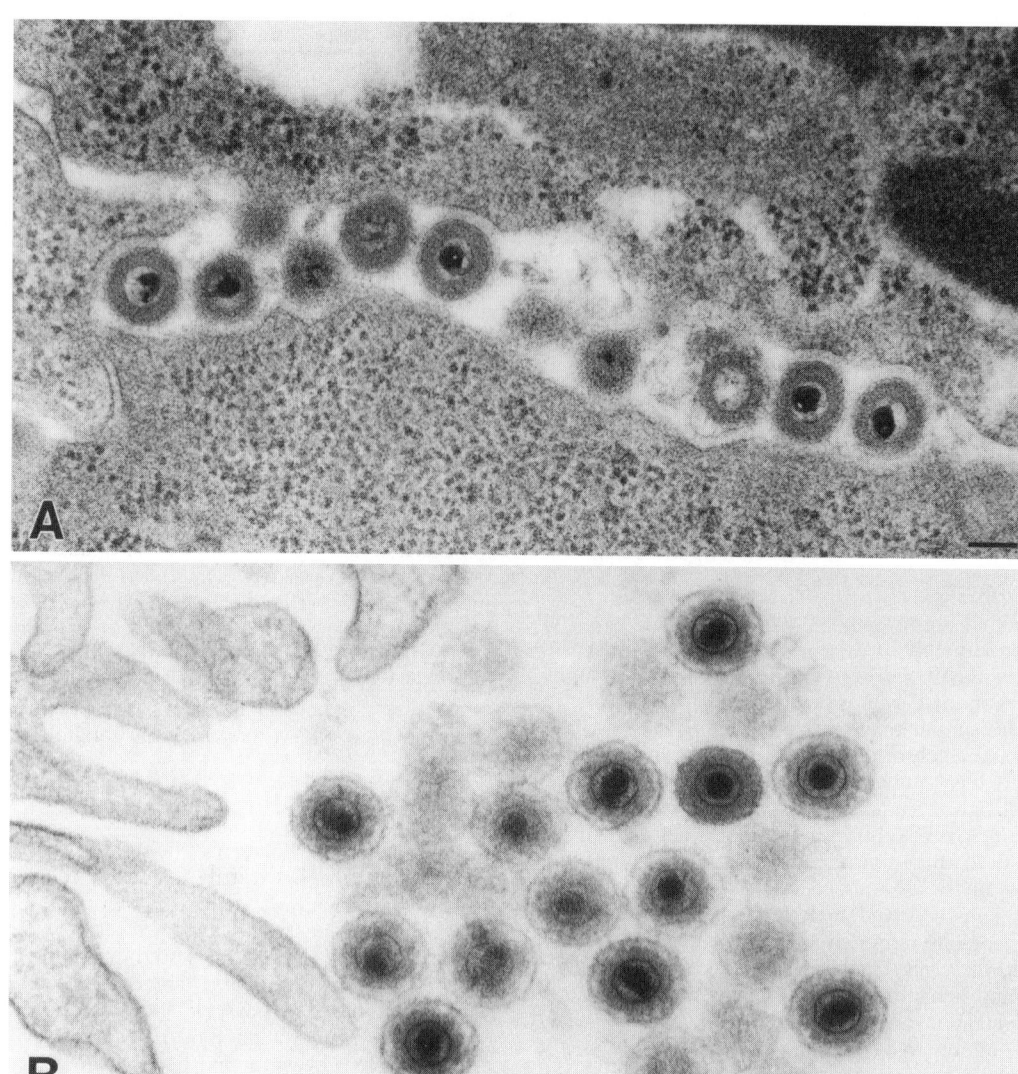

FIGURE 1 HHV-6B (A) and HHV-7 (B) virions grown in cord blood lymphocytes. The virus particles are extracellular and mature. Culture material was provided by Jodi Black and Carlos Lopez, and electron microscopy and photography were by Cynthia Goldsmith, Centers for Disease Control and Prevention. Bars = 100 nm.

percentages of cervical secretions (Leach et al., 1994; Okuno et al., 1995), cord blood (Adams et al., 1998), and tissues from spontaneously aborted fetuses (Aubin et al., 1992). This suggests that congenital or sexual transmission is possible, although it is not likely to be frequent. HHV-6A has been detected infrequently in saliva but more commonly in skin (Di Luca et al., 1996), suggesting that the virus might be transmitted by direct contact. This route is also possible for HHV-6B.

In addition to transmission via an infectious process, HHV-6 can be transmitted genetically via a genome integrated near the telomere of human chromosome 1 (Daibata et al., 1998). In other work, several patients were identified whose peripheral blood mononuclear cells (PBMC) harbor HHV-6 genomes integrated into the short arm of chromosome 17 band p3 at levels that allow their detection by Southern blotting in PBMC (Torelli et al., 1995); for one patient, the site of integration has been mapped to near the telomere of chromosome 17 (Morris et al., 1999). It is not known whether these cases represent germ line events. Integration in the vicinity of telomeres may be related to the presence of telomerelike sequences at HHV-6 genomic termini.

Pathogenicity and Clinical Aspects

The significant clinical aspects of HHV-6A and HHV-6B infections can be broken down by age (children and adults) and immune status. Disease associated with primary infection normally occurs early in life, and reactivated latent virus (or very rare primary infections) is associated with disease in immunocompromised organ transplant recipients. Over the years, numerous diseases have been associated with HHV-6 infections (Table 3). For roseola, the evidence is robust and incontrovertible. For some other diseases, such as infections following organ transplantation, the available data are compelling and consistent with a pathogenic role,

TABLE 3 Spectrum of HHV-6 disease associations

Age group	Associated disease	
	Proven or highly probable	Proposed[a]
Children	Roseola Febrile illness Febrile convulsions Posttransplantation disease Meningitis and encephalitis	Hepatosplenomegaly Thrombocytopenia
Adults	Lymphadenopathy Meningitis and encephalitis Mononucleosis-like disease Posttransplantation disease	AIDS-associated disease Acute lymphoblastic leukemia MS Non-Hodgkin's lymphoma S100[+] chronic lymphoproliferative disease Spontaneous abortion

[a]Etiology not proven because data are either lacking or conflicting.

although some elements of formal proof are missing. Multiple sclerosis (MS) is an example of a disease for which intriguing and tantalizing observations of a possible etiologic association have not been reliably reproduced and other aspects of etiologic proof are incomplete. In some cases, proof of an etiologic association is difficult to come by because of the scarcity of a disease. The proposed associations of HHV-6 with histiocytic necrotizing lymphadenitis and Sjögren's syndrome fell by the wayside after more complete studies were done.

HHV-6 is a commensal inhabitant of the central nervous system, with over 80% of adult brains harboring the virus (Luppi et al., 1994), and it is typically present in a latent form in circulating lymphocytes (Jarrett et al., 1990; Cone et al., 1993; Cuende et al., 1994; Di Luca et al., 1994; Wilborn et al., 1994b); CD34[+] hematopoietic progenitor cells and their differentiated progeny, including monocytes and macrophages (Kondo et al., 1991; Katsafanas et al., 1996; Luppi et al., 1999); and PBMC-derived dendritic cells (Asada et al., 1999) (Table 2). The only viral transcript identified in latently infected lymphocytes spans the U94 gene (Rotola et al., 1998).

Primary Infection

HHV-6B is an important pathogen in young children. Approximately 30% of children in the United States experience roseola, which is normally a mild, self-limiting disease characterized by 1 to 3 days of fever (39 to 40°C) followed by a rash that lasts 1 to 3 days. A transient immunoglobulin M (IgM) response develops quickly, followed within 2 weeks by a sustained IgG response of increasing antibody avidity (Ueda et al., 1989; Balachandra et al., 1991; Ward et al., 1993a) (Fig. 2). HHV-6B can be detected by culture most efficiently during the febrile phase that precedes the rash in roseola (Asano et al., 1991a; Okada et al., 1993). Viral DNA can be readily detected in blood by PCR for over a month (Pruksananonda et al., 1992).

Primary infection may also present as fever without rash or rash without fever (Asano et al., 1989; Suga et al., 1989), or it may be asymptomatic. It is not uncommon for children to experience more severe symptoms in association with primary HHV-6B infection, including fever of ≥39.5°C, inflamed tympanic membranes, gastrointestinal and respiratory distress, throbmocytopenia, intussusception, and

febrile seizures (Table 3) (Asano et al., 1991b; Pruksananonda et al., 1992; Hall et al., 1994; Yoshikawa et al., 1998). By some estimates, HHV-6 infections account for 20 to 40% of febrile admissions to pediatric emergency departments for children 6 months to 2 years old (Pruksananonda et al., 1992; Portolani et al., 1993; Hall et al., 1994). Papers by Asano, Hall, and colleagues are particularly informative on these issues and are highly recommended (Asano et al., 1991a, 1994; Pruksananonda et al., 1992; Suga et al., 1993; Caserta et al., 1994; Hall et al., 1994).

Disseminated, sometimes fatal, HHV-6 infections have been described in children, as well as associations with cases of idiopathic thrombocytic purpura and hemophagocytic syndrome (reviewed in Braun et al., 1997, and Hoang et al., 1999). Although primary HHV-6 infections normally occur following the waning of maternal antibody at around 6 months of age, infections can occur in the first weeks of life.

A rapid, point-of-care, diagnostic assay for HHV-6 infection may reduce unnecessary use of antibiotics and hospital admissions. It is likely that some children with high fever due to primary HHV-6 infection are treated with antibiotics. Furthermore, a roseola rash may appear following antibiotic therapy, which could be mistakenly diagnosed as drug allergy, unnecessarily precluding future use of the drug.

Primary infection in adults is rare but can be severe. It has been associated with lymphadenopathy, hepatospleno-

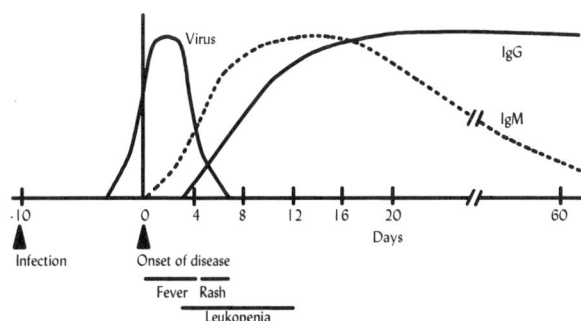

FIGURE 2 Immune response to HHV-6 showing temporal relationship between viral load, clinical signs and symptoms of infection (roseola), and production of specific IgM and IgG antibodies.

megaly, and other symptoms similar to mononucleosis (reviewed in Braun et al., 1997).

Disease in Immunocompromised Patients

Because these viruses establish lifelong latent infections, viral activity in immunocompromised patients can occur without new virus exposure.

HHV-6 activity in organ transplant recipients has been identified by virus culture, PCR, serology, and immunohistochemistry. About 90% of the HHV-6 that has been detected following organ transplantation has been HHV-6B; HHV-6A has seldom been detected in blood, while both variants have been detected in the lungs. Disease associations include febrile rash episodes that resemble graft-versus-host disease, sinusitis, pneumonitis, graft suppression and rejection, encephalitis, and possibly increased risk of "CMV disease" (reviewed in Braun et al., 1997).

The degree of HHV-6 activity and extent of association with disease in transplant recipients has varied widely from study to study. There are a number of major variables: recipient age, immunosuppressive regimen and other aspects of case management, material assayed (circulating lymphocytes, tissue biopsies, or serum), assay methods (serology, quantitative versus qualitative PCR, or variant identification), interactions with other viruses (especially HCMV and HHV-7), and study design (small, intensively studied cohort or retrospective analysis of materials collected from patients who fit a particular profile, such as having idiopathic pneumonitis). Over the years, investigators at some centers have consistently charted a more benign course of posttransplantcation HHV-6 activity than have others (Carrigan and Knox, 1994; Cone et al., 1999). It will be interesting to ultimately learn which variables led to the differences. In the meantime, we consider it likely that HHV-6 causes disease in at least some transplant recipients.

For renal transplants, HHV-6 activity has been most frequently observed in patients who received anti-lymphocyte therapy (anti-CD3 monoclonal antibody or anti-lymphocyte globulin) as part of their immunosuppressive regimen. This is possibly related to the enhancing effect of anti-CD3 antibodies on viral growth that has been observed in vitro (Kikuta et al., 1990; Roffman and Frenkel, 1991). HHV-6 antigens are much more likely to be detected in allografts during acute rather than chronic rejection.

In bone marrow transplant (BMT) recipients, HHV-6 has been associated with sinusitis, graft-versus-host disease and other exanthems, pneumonitis, idiopathic febrile episodes, marrow suppression, and delayed granulocyte and platelet engraftment (reviewed in Braun et al., 1997). In a recent comparison of lymphotropic herpesvirus activity following allogeneic peripheral blood stem cell transplantation or BMT (Maeda et al., 1999), HHV-6 was detected more frequently in BMT recipients, usually 3 to 4 weeks following the transplantation. Platelet engraftment was significantly delayed in patients from either group who were HHV-6 positive at any time during the first 4 weeks. In another recent report, Cone et al. (1999) studied HCMV-seronegative allogeneic BMT recipients; 90% of the patients had substantial increases in blood HHV-6B levels consistent with virus reactivation, one patient had an apparent primary infection accompanied by rash, and another had a reactivation-associated rash. Other than these two mild and self-limited bouts, no other clinical episodes in these patients were linked with HHV-6 activity,

suggesting that in the absence of HCMV activity, HHV-6 is less pathogenic.

HHV-6 has been hypothesized to have a role in AIDS based on several in vitro studies. HHV-6 and HIV can up- or downregulate each other's growth (depending on conditions), HHV-6 gene products can induce expression from the HIV long terminal repeat, and HHV-6 can induce or upregulate CD4 expression, increasing the pool of cells susceptible to HIV infection (reviewed in Braun et al., 1997). Late in AIDS, HHV-6 DNA and antigens are frequently disseminated into many tissues. Important questions include whether the in vitro growth-modulating mechanisms are active in vivo and whether HHV-6 dissemination late in AIDS is a cause or effect of disease. A recent study provided evidence for in vivo synergism between HIV and HHV-6B by showing that when both viruses are present in tissues, their quantities are elevated by approximately 1 $\log_{10}$ unit (Emery et al., 1999). Given that progression of HIV disease is related to HIV levels, it is possible that this virologic synergism translates to an effect on disease progression.

Two important observations were made in a study of HHV-6 in children vertically infected with HIV (Kositanont et al., 1999). First, HHV-6B infection was acquired more slowly by the HIV-infected children (no HHV-6A infections were detected). Children 4 to 12 months old with low CD4+-cell ratios (<15%) were less likely to acquire HHV-6, suggesting that the lower rate of HHV-6 acquisition was due to target cell depletion. Differences in caregiver contact that might also influence the transmission rate were not studied. Second, HIV disease progressed in all children who acquired HHV-6 infections but not in the HIV-infected children who remained free of HHV-6. This situation warrants further study and consideration of trials of antiviral intervention.

Other Disease Associations

HHV-6 infections have been associated with numerous other diseases (Table 3) (reviewed in Braun et al., 1997); several warrant mention here. Some studies have described elevated levels of HHV-6 activity or antibodies in patients with chronic fatigue syndrome, but the evidence for an etiologic association is not convincing. In a small percentage of encephalitis cases, HHV-6 is the only infectious agent detected (McCullers et al., 1995; Kolski et al., 1998). In an interesting report, of 22 herpes simplex encephalitis cases, HHV-6B was simultaneously detected in 3. Two of the 3 dually infected patients died, in comparison with only 1 of 19 with only herpes simplex virus infections (Tang et al., 1997). This suggests a pathogenic synergism between herpes simplex virus and HHV-6 in encephalitis.

HHV-6 DNA can transform primary cells to a tumorigenic phenotype, and the transforming activity has been mapped to at least two genes. Many malignancies have been examined for HHV-6 associations, but no conclusive linkages have been identified (reviewed in Braun et al., 1997). The most plausible associations are with some cases of oral cancer (at least in Malaysia), the nodular-sclerosis type of Hodgkin's disease, and S100-positive chronic lymphoproliferative disease. As described below, HHV-8 is clearly the etiologic agent of KS. HHV-6B and HHV-7 antigens have been detected in KS lesions and localized to cells of the monocyte/macrophage lineage (Bovenzi et al., 1993; Kempf et al., 1997), although their roles in KS pathogenesis remain to be defined.

Several lines of evidence suggest a possible association between HHV-6 and MS. Evidence for the plausibility of this association includes the presence of the virus in most brains; the ability of the virus to replicate (albeit inefficiently) in primary astrocytes, neurons, and oligodendrocytes; and the association of the virus with neurologic symptoms. More directly, in some studies HHV-6 serologic titers were higher in MS patients than in the control population, including patients with other neurologic diseases, and HHV-6 DNA was detected in cerebrospinal fluid or serum from more MS patients than controls (Sola et al., 1993; Wilborn et al., 1994a; Sanders et al., 1996; Soldan et al., 1997; Ablashi et al., 1999). Most intriguingly, HHV-6 antigens were detected in MS plaque regions by immunohistochemistry, with a distribution different than that seen in healthy brains (Challoner et al., 1995). Unfortunately, many of these results have not been replicated (Enbom et al., 1997; Martin et al., 1997; Fillet et al., 1998), and the linkage between HHV-6 and MS remains unproven. Collaborative studies will be required among the several laboratories working in this area to ascertain whether the differences in results are due to technical differences, experimental error, or differences in the populations being studied. Ultimate proof of a specific etiology for MS, such as HHV-6, may require therapeutic trials that are coupled with refined diagnostic methods that would allow proof of the intended effect. For MS, this is particularly difficult because of the relative inaccessibility of the affected tissue. In addition, it seems likely that if HHV-6 is etiologically involved in MS, therapy will be most efficacious early in the progression of the disease. This will complicate both identification of candidates for therapy and measurement of positive outcomes.

Diagnosis

For an experienced pediatrician, classic roseola can be reliably diagnosed on a clinical basis, and no laboratory confirmation is needed. Other presentations of primary HHV-6 infection would most reliably be diagnosed with the aid of laboratory testing, although it is seldom done other than in research settings. In two studies, 40% of clinical diagnoses of measles or rubella in children between the ages of 3 months and 2 years that were not supported by laboratory results were associated with HHV-6 activity (Black et al., 1996a; Tait et al., 1996). HHV-6 activity can be monitored by culture, PCR, or serology. Specific organ involvement

may be best addressed by immunohistochemistry. Commercially available reagents and assays include monoclonal antibodies that react with either or both variants, serologic reagents and assays, and PCR primers.

Virus Detection

HHV-6 viremia seldom occurs other than during periods of frank viral activity, such as during the febrile phase of roseola and during reactivation events in organ transplant recipients. PCR from serum or plasma is useful because it allows discrimination between active infections in which virus is being released (overt viremia) and latent infections of lymphocytes that are likely to be of little clinical relevance (Secchiero et al., 1995a, 1995b; Chiu et al., 1998; Suga et al., 1998). Culture can be accomplished by purification of patient lymphocytes followed by stimulation with phytohemagglutinin and growth in the presence of interleukin 2 (IL-2) (Black et al., 1989). Some laboratories culture patient lymphocytes alone, while others cocultivate the activated lymphocytes with similarly purified and activated human umbilical cord blood lymphocytes. Because the cytopathic effect in HHV-6-infected cells is indistinguishable from that induced by other viruses, including HHV-7 and HIV, culture isolation must be confirmed. This is most easily accomplished through the use of a monoclonal antibody, some of which allow variant discrimination (Table 4). Methods for variant discrimination are summarized in Table 4.

Numerous PCR primer and probe sets have been described, including nested (Cone et al., 1996), nonnested (Aubin et al., 1994; Yamamoto et al., 1994), nonisotopic (Osiowy et al., 1998), quantitative (Secchiero et al., 1995b; Clark et al., 1996), and multiplexed (Kidd et al., 1998; Minjolle et al., 1999). Several of these, as well as other primer sets, also provide HHV-6 variant discrimination (Table 4). Given the recent wider availability of rapid PCR cycling and real-time fluorescence detection equipment, the prospect of having clinically accessible methods for rapid diagnosis of febrile children and monitoring organ transplant recipients is on the horizon.

Antibody Tests

IgM seroconversion occurs 5 to 7 days after the onset of symptoms associated with primary infection, peaks at 2 to 3 weeks, and wanes by 2 months. The IgG response becomes detectable within 7 to 10 days and is sustained in most cases

TABLE 4 References for analytical methods that distinguish HHV-6A from HHV-6B[a]

Monoclonal antibody reactivity	Restriction endonuclease profiles	Nucleotide sequence analysis	PCR product analysis	Reference
+				Chandran et al., 1992; Campadelli-Fiume et al., 1993; Foa-Tomasi et al., 1995
+	+			Ablashi et al., 1991; Schirmer et al., 1991
+		+		Aubin et al., 1993; Pellett et al., 1993
+	+		+	Dewhurst et al., 1992
		+		Gompels et al., 1993
		+	+	Chou and Marousek, 1994; Yamamoto et al., 1994
			+	Drobyski et al., 1993; Aubin et al., 1994; Cone et al., 1996; Gautheret et al., 1996; Kidd et al., 1998; Wilborn et al., 1998

[a]+, method discussed.

for life (Fig. 2). A host of HHV-6 antibody assays have been developed in every standard format, including immunofluorescent assays (IFA), microtiter-dish-based enzyme immunoassays (EIA), and immunoblots (Black et al., 1989, 1996b; Robert et al., 1990; Couillard et al., 1992; Coyle et al., 1992; Chokephaibulkit et al., 1997). Several of these assays are available commercially.

HHV-6 serology is of limited use. Given the high prevalence of HHV-6 infections, a single positive result simply places a patient in the vast group of HHV-6-seropositive individuals. No diagnostically useful titer has been identified. Serologic diagnosis of primary infection and of reactivation events requires paired acute and convalescent specimens; fourfold antibody titer increases are interpreted as evidence for viral activity. Obtaining appropriately paired specimens is neither fast nor convenient; the relevant clinical event may have passed by the time antibody levels rise. In addition to increased titers, primary infections can be discriminated from reactivation by antibody avidity, which increases with time after infection (Ward et al., 1993a, 1993b), and it is easily assessed by inclusion of urea in the primary antibody incubation step.

No serologic assay is variant specific. In addition to the extensive antigenic cross-reactivity between the HHV-6 variants, the variants have sufficient cross-reactivity with HHV-7 to lead to occasional false-positive results unless antigen adsorptions are done or an immunoblot assay that is based on non-cross-reactive antigens is used (Black et al., 1996b). Cross-reaction with HCMV has also been seen in some assays (Adler et al., 1993; Chokephaibulkit et al., 1997).

IgM assays offer some promise in the area of rapid diagnosis, but their use has not been thoroughly studied. One commercially available assay is an IgM assay that uses a purified viral protein, p38/41, as the antigen (Patnaik et al., 1995). Groups of MS patients and chronic fatigue syndrome patients have had mean titers higher than those of controls in this assay, but more study will be required to define its diagnostic utility.

Therapy

In vitro, HHV-6 isolates are sensitive to ganciclovir and foscarnet but not acyclovir (reviewed in Braun et al., 1997). No clinical trials have been conducted to ascertain the clinical utility of antiviral therapy for any HHV-6-associated disease. As suggested by Braun et al., three situations that might warrant such trials include cases of encephalitis, idiopathic pneumonitis in organ transplant recipients, and MS (Braun et al., 1997).

HHV-7

HHV-7 was discovered by Frenkel and colleagues in circulating PBMC obtained from a healthy adult (Frenkel et al., 1990). It was subsequently identified as a commensal inhabitant of saliva, with 75% of saliva specimens from adults carrying cell-free infectious virus (Wyatt and Frenkel, 1992; Black et al., 1993; Hidaka et al., 1993). HHV-7 is closely related to the HHV-6 variants and shares many of their molecular and biological properties (reviewed in Black and Pellett, 1999). HHV-7 is highly and widely prevalent, with infection normally being acquired early in life (Tables 1 and 2). Two disease states associated with the virus are of particular importance: a subset of roseola cases and CMV disease

in organ transplant recipients. It is likely that additional disease associations will be identified.

Biology

As described above for HHV-6, HHV-7 virion structure is similar to that of other herpesviruses (Fig. 1B) (Frenkel and Roffman, 1996; Black et al., 1997; Klussmann et al., 1997). HHV-7 genomes are more compact than HHV-6 genomes, with a length of approximately 145 kb (Nicholas, 1996; Megaw et al., 1998). As for the HHV-6 variants, HHV-7 genomes are bounded by sequences that resemble mammalian telomeres. Nearly every HHV-7 gene has a counterpart in HHV-6. As mentioned above, a significant difference is that both HHV-6 variants contain a homolog of the parvovirus *rep*, or *ns1*, gene, but HHV-7 does not. Amino acid sequence identity between HHV-7 and the HHV-6 variants ranges from 22 to 75%, with most genes having identities of 40 to 60%.

The HHV-7 growth cycle has many similarities with those of the HHV-6 variants. It grows best in activated primary cord blood lymphocytes and has been adapted for growth in the SupT1 T-cell line (Berneman et al., 1992; Black et al., 1997; Cermelli et al., 1997; Ablashi et al., 1998). The cytopathic effect it induces is indistinguishable from HHV-6 cytopathic effect, and it also induces host cell protein synthesis (Black et al., 1997). Cell death in HHV-7-infected cultures can be due to either virus-induced apoptosis or lysis (Secchiero et al., 1997). Unlike HHV-6, HHV-7 uses CD4 as its cellular receptor, as does HIV (Lusso et al., 1994; Yasukawa et al., 1995). Under appropriate conditions in vitro, HIV and HHV-7 can inhibit each other's growth by receptor competition (Lusso et al., 1994).

Cocultivation with HHV-7 of fresh peripheral blood lymphocytes that harbor latent HHV-6B can result in reactivation of HHV-6B to a lytic state (Katsafanas et al., 1996). Consistent with this, DNA levels for both viruses were elevated in HIV disease patients when HHV-6 and HHV-7 were simultaneously present in tissues (Emery et al., 1999). The mechanism of this activity is not known.

Epidemiology

The prevalence of HHV-7 in adults ranges globally from approximately 60 to 90% (reviewed in Black and Pellett, 1999). Most infections are acquired early in life, generally following primary HHV-6 infection. A subset of HHV-6 and HHV-7 antigens are cross-reactive (Yasukawa et al., 1993; Foa-Tomasi et al., 1994; Black et al., 1996b), which raises the question of whether prior infection with one virus affects infection by the other. Infectious HHV-7 is present in the saliva of most seropositive individuals; thus, it is likely that saliva is the primary route of transmission. Because HHV-7 seems to be much more readily available for transmission, it is a mystery why HHV-6B is normally acquired earlier.

Pathogenicity and Clinical Aspects

It is likely that the full spectrum of HHV-7-associated disease has not been defined. To date, primary infection with HHV-7 has been associated with a subset (<10%) of roseola cases and disease similar to that commonly associated with HCMV infections in organ transplant recipients. The prevalence of HHV-7 reaches 60 to 70% by the age of 2. Since herpesvirus-related disease is often more severe the later primary infection occurs, it is possible that HHV-7 is associated with idiopathic fevers and/or rash illnesses in older children and young adults (Black and Pellett, 1999).

Primary Infection

The virologic aspects of primary HHV-7 infection in young children who experience roseola symptoms are similar to those seen for HHV-6B: viremia, clearance, establishment of latent infections in peripheral blood lymphocytes, and development of IgG antibodies (Tanaka et al., 1994; Ueda et al., 1994; Torigoe et al., 1995). Neurologic complications, such as febrile convulsions, have been described for nearly half of the small number of primary infections that have been described (Torigoe et al., 1995, 1996; Clark et al., 1997; Caserta et al., 1998; van den Berg et al., 1999). It is not clear whether this is a general property of HHV-7 primary infections or is due to the cases reported representing the more severe and less common end of the spectrum.

Disease in Immunocompromised Patients

HHV-7 activity and association with disease have been studied in renal, bone marrow, and allogeneic peripheral blood stem cell transplant recipients (Yalcin et al., 1994; Osman et al., 1996; Wang et al., 1996; Chan et al., 1997; Maeda et al., 1999). A general observation has been that HHV-7 activity peaks near the beginning of the second month after transplantation, which precedes the usual onset of HCMV activity by 2 to 4 weeks. HHV-7 activity frequently occurs in association with CMV disease. CMV disease is characterized as any of the following: (i) idiopathic febrile illness with leukopenia or a fall in platelet count in the presence or absence of transaminitis plus laboratory evidence of HCMV activity, (ii) clinical pneumonitis plus detection of HCMV, or (iii) gastrointestinal symptoms plus histologic evidence of HCMV activity (Osman et al., 1996). The relative risk for CMV disease is highest when both HHV-7 and HCMV are concurrently active. Although increased HHV-7 activity is commonly seen in transplant recipients, in one study of BMT and allogeneic peripheral blood stem cell transplant recipients, HHV-7 activity was not above a baseline intended to discriminate latent from active infection (Maeda et al., 1999).

Because HHV-7 and HIV both use CD4 as their cellular receptor, it is possible that the viruses interact in vivo, but there is no evidence in support of this hypothesis (Emery et al., 1999).

Pityriasis Rosea

Pityriasis rosea is a well-known rash illness that is characterized by the initial appearance of a "herald patch," followed several days later by crops of smaller lesions scattered over the body. Waves of additional lesions appear at approximately 3-day intervals for several weeks. The disease can recur, and herpesviruses have been suspected as possible causes.

In one study, HHV-7 was cultured from the peripheral blood lymphocytes of pityriasis rosea patients and was detected by PCR in lesions (Drago et al., 1997). Serology was not done to ascertain whether the HHV-7 activity was due to primary or reactivated infection. Other investigators have been unable to confirm the association (Kempf et al., 1999). The absence of independent confirmation or addi-

tional supporting information indicates that it is unlikely the virus causes the disease.

Diagnosis

As for HHV-6, the high prevalence of HHV-7 makes it impossible to interpret single positive serologic results. Paired acute- and convalescent-phase specimens can help clarify temporal associations between viral activity and disease. The circumstance that most clearly warrants monitoring HHV-7 activity is organ transplantation. Quantitative PCR or PCR from serum is likely to provide the most useful evidence for viral activity, but their use for this purpose is an area of research, and specific guidelines for their application are not available. Because of the labor and other costs associated with screening for multiple agents, multiplexed PCR assays that allow screening for several agents simultaneously are likely to come into wider use for monitoring organ transplant recipients (Kidd et al., 1998; Minjolle et al., 1999).

Virus Detection

Methods available for detecting HHV-7 include culture, monoclonal antibodies, and PCR. Culture methods for HHV-7 are identical to those used for HHV-6: culture of activated lymphocytes in the presence of IL-2 or coculture of such lymphocytes with additional purified and activated lymphocytes, preferably from human umbilical cord blood (Secchiero et al., 1994; Black et al., 1997). Monoclonal antibodies useful for culture confirmation are commercially available. Several PCR systems for HHV-7 have been described, including quantitative methods (Berneman et al., 1992; Secchiero et al., 1995b; Wilborn et al., 1995; Kidd et al., 1996).

Antibody Tests

IFA, EIA, and immunoblot assays have been described for detecting HHV-7 antibodies (Wyatt et al., 1991; Clark et al., 1993; Torigoe et al., 1995; Black et al., 1996b; Huang et al., 1997). False-positive results can be obtained in the IFA and EIA due to cross-reactivity with HHV-6. Immunoblot assays using antigens specific for each of the viruses do not cross-react (Black et al., 1996b; Foa-Tomasi et al., 1996), but reagents for such assays are not readily available. The only HHV-7 serodiagnostic reagents available commercially are slides for IFA.

Therapy

In vitro, HHV-7 is susceptible to cidofovir and phosphonoformic acid, less sensitive to ganciclovir than HHV-6 or HCMV, and insensitive to acyclovir and its relatives (Black et al., 1997; Takahashi et al., 1997; Yoshida et al., 1998). Antiviral therapy might be of value for treatment of HHV-7-associated CMV disease in organ transplant recipients, but there have been no clinical trials, and thus, specific recommendations cannot be made.

HHV-8

HHV-8 was discovered as the result of a directed effort to identify the etiologic agent of KS. Chang and colleagues used representational-difference analysis to identify genetic material in KS lesions that was not present in unaffected skin (Chang et al., 1994). The process resulted in the identification of a small genomic segment of what turned out to be a novel herpesvirus that was named KS-associated herpesvirus and is now recognized as HHV-8.

Biology

HHV-8 virions have characteristic herpesvirus morphology (Fig. 3). Based on sequence and biological properties, the virus is classified in the *Rhadinovirus* genus of the subfamily *Gammaherpesvirinae* (Moore et al., 1996a). Its closest HHV relative is Epstein-Barr virus (EBV). The replicative form of the HHV-8 genome is approximately 160 kb long and contains over 80 genes (Russo et al., 1996). In addition to the set of conserved herpesvirus core genes, HHV-8 contains genes that have homologs only in other gammaherpesviruses or rhadinoviruses. Most herpesviruses have a small number of genes that are of obvious host derivation, such as the thymidylate synthetase genes that appear to have been acquired by different herpesviruses independently at different points in their evolution. HHV-8 stands out in that it has 12 genes that have cellular homologs, including genes whose products stimulate cell division, inhibit apoptosis, and modulate immune function (Neipel et al., 1997). Four HHV-8 genes have independent growth transforming properties. Collectively, this set of growth and immune-regulatory genes is clearly involved in HHV-8 pathogenesis, although the regulatory networks are only partially defined at present (reviewed in Neipel and Fleckenstein, 1999).

A gene of particular interest is K1, which maps at the extreme left end of the HHV-8 genome. It is a transforming gene whose product is a membrane-associated glycoprotein with signaling properties (Lee et al., 1998). The nucleotide and amino acid sequences of K1 are remarkably divergent among viral strains collected around the world and from various HHV-8-associated diseases; amino acid sequences encoded by K1 can differ by over 30% (Cook et al., 1999; Hayward, 1999; Meng et al., 1999). This is in contrast to sequence variation elsewhere in the genome, where it seldom exceeds 1%. Although K1 sequences from different sources are highly divergent, they are constant in different tissues and over time in the same individual and thus are of use in HHV-8 molecular epidemiology. K1 analysis has led to the identification of four major genotypes that reflect global patterns of human migration. This provides substantial evidence that HHV-8 is not a recent introduction in humans but has been a part of human biology on an evolutionary time scale.

In addition to the four K1 genotypes, Hayward and colleagues have identified two major and highly divergent alleles of the K15 gene (Hayward, 1999), which maps at the far right end of the genome. In contrast to K1, the K15 alleles are highly conserved from strain to strain. K15 is homologous to the LMP-2 gene of EBV, which has a role in raising the switching threshold between latent and lytic infections. The pathological significance of the K15 allelic variation remains to be determined.

Although there have been many attempts to identify cell systems for HHV-8 primary isolation and propagation, none has been identified that allows efficient replication of fresh isolates from KS lesions (Renne et al., 1998, and references therein). The virus does grow in continuous cell lines derived from PEL, and PEL-derived virus can be propagated in dermal endothelial cells that have been immortalized with the E6 and E7 genes of human papillomavirus type 16 (Moses et al., 1999). PEL cell lines frequently also harbor EBV, but several that contain only HHV-8 have been iden-

tified. These lines are in wide use for laboratory studies of HHV-8 molecular and cellular biology, as well as for generation of serodiagnostic reagents (reviewed in Cesarman and Knowles, 1999). HHV-8 is resident in every cell in PEL cell lines. Under normal growth conditions for PEL cell lines, the virus is latent in most cells but spontaneously replicates lytically in about 1% of cells. In response to treatment with a phorbol ester (tetradecanoyl phorbol acetate) or sodium butyrate, lytic replication can be induced in 10 to 30% of cells. This system is analogous to the culture systems that have been used for many years to study EBV replication. Viral gene expression in latently infected cells is limited to a handful of latent genes, including orf73 (latent nuclear antigen), orf72 (v-cyclin), and K12 (kaposin). K12 encodes highly transcribed RNA species that are present at $>10^5$ copies per cell. Another small RNA, T1.1, or *nut-1*, is expressed at very high copy numbers in cells that are in the lytic state (Zhong et al., 1996). The high copy numbers of T0.7 and T1.1 transcripts, coupled with their being markers for latent and lytic replication, respectively, make these transcripts ideal targets for in situ hybridization studies (Staskus et al., 1999).

Epidemiology

HHV-8 is distributed worldwide, but its prevalence varies widely, corresponding to the geographic and demographic distribution of KS. Thus, in many African countries and portions of Italy, KS is endemic and HHV-8 seroprevalence exceeds 60%; in the United States and the United Kingdom, where KS is rare, HHV-8 seroprevalence in the general population is less than 10% (0 to 2% in some studies) (reviewed in Schulz, 1999). These values must be considered as minimum estimates because serologic assays for HHV-8 currently miss 5 to 15% of KS patients, who by definition are infected with the virus; it is possible that some people with KS do not continuously produce measurable HHV-8 antibody. In countries where KS is endemic, HHV-8 seroprevalence is distributed across the population in the absence of identified behavioral risk factors. In countries where KS is rare, HHV-8 seroprevalence is highest in KS patients, next highest in HIV-positive patients who are men who have sex with men (MSM), followed by HIV-negative MSM, with few healthy blood donors testing positive (see Fig. 4).

Even in areas where KS is endemic and HHV-8 prevalence is high, only a small fraction of HHV-8-infected individuals develop the disease, indicating that in addition to HHV-8, cofactors are needed for KS to develop. It is interesting that in western Africa, where KS is relatively uncommon, HHV-8 is approximately as prevalent as in Uganda, where KS is the most common malignancy. In The Gambia, a country of western Africa, HIV-1-infected patients have greater than 10-fold-higher risk of KS than do those infected with HIV-2 (Ariyoshi et al., 1998). This suggests that a factor associated with HIV-1 infection specifically increases the risk of KS.

Transmission

The epidemiology of HHV-8 suggests that there are at least two major routes of transmission, one that leads to infection of children born to HHV-8-seropositive mothers in Africa (Bourboulia et al., 1998; Mayama et al., 1998; Gessain et al., 1999) and another that involves sexual transmission, predominately among MSM (Martin et al., 1998; Renwick et al., 1998; Blackbourn et al., 1999; Grulich et al., 1999).

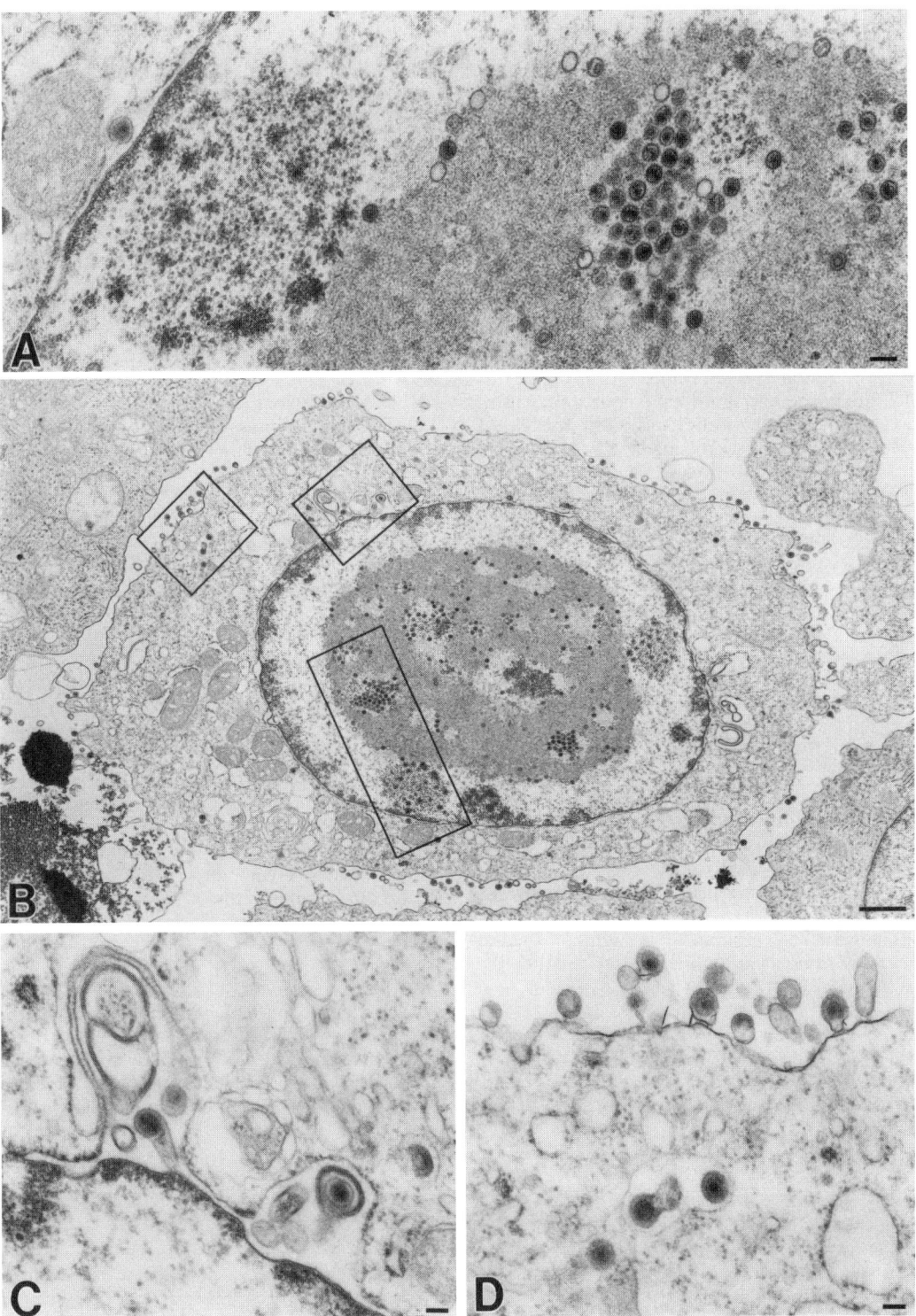

FIGURE 3 HHV-8 virions in BCBL-1 cells incubated with 0.3 mM sodium butyrate for 5 days. The boxed areas in panel B are shown at higher magnification in panels A, C, and D. (B) Induced cells show signs of lytic infection, including condensed and marginated chromatin and numerous nuclear inclusions that contain maturing virus particles (A). Virions in membrane-bound vesicles associated with the nuclear (C) and cytoplasmic (D) membranes are shown. Culture material was provided by Margaret Offermann (Emory University) and Jodi Black (Centers for Disease Control and Prevention [CDC]) (Yu et al., 1999). Electron microscopy and photography were by Cynthia Goldsmith, CDC. Bar = 1 μm (B) or 100 nm (A, C, and D).

HHV-8 DNA has been detected in saliva, nasal secretions, and semen of male AIDS patients and conceivably can be transmitted via any of them (Blackbourn and Levy, 1997; Koelle et al., 1997; Blackbourn et al., 1998; Pellett et al., 1999). Several reports have identified linkages between HHV-8 transmission and oral-anal contact, but this is not conclusive, nor is there a convincing linkage to any specific sexual practice. In addition to its presence in bodily fluids from males, HHV-8 DNA has been detected in cervicovaginal specimens of HHV-8-seropositive women at rates approximately equivalent to the rate of detection of the DNA in semen from HIV-positive men (on the order of 10%) (Calabro et al., 1999; Whitby et al., 1999). Further studies will be required to determine whether this is a source for virus transmission.

HHV-8 was identified in the blood of an apparently healthy blood donor following multiple donations, raising the question of whether the virus can be transmitted via blood or blood products (Blackbourn et al., 1997). In a small study of blood and blood product recipients, 10 (71%) of 14 recipients of blood from donors who were positive for both HIV and HHV-8 became HIV infected, while none seroconverted for HHV-8 (Operskalski et al., 1997). This suggests that the risk of HHV-8 transmission via blood is substantially lower than for HIV. Larger studies are underway to more precisely define the risk or lack thereof.

Although most cases of posttransplantation KS appear to be due to preexisting infection, virus transmission via the transplanted organ that resulted in KS has been described (Parravicini et al., 1997). Appropriate donor-recipient matching should be considered, especially in regions where KS is endemic.

Pathogenicity and Clinical Aspects

Primary Infection

Little is known about primary HHV-8 infection. In one report from Uganda, 4 (8%) of 53 children with their first febrile episode were PCR positive for the virus (Kasolo et al., 1997), but the study was too preliminary to allow the conclusion that the virus caused the fevers.

KS

KS is a neoplasm that is characterized by reddish-brown plaque or nodular lesions originating in endothelial tissues. Lesions can occur on the extremities, trunk, and internal organs. Groups of lesions can expand and coalesce into large tumor masses. Pathologically, KS lesions are characterized by the presence of cells with a spindlelike appearance, extensive networks of vascular slits, extravasated red blood cells, and purplish pigmentation from hemosiderin. Clinical diagnosis is normally confirmed by pathological examination of a lesion biopsy specimen.

The four major forms of KS are summarized in Table 5. All forms of the disease are more common in men than in women, but recent trends show that the gap in some forms of KS is narrowing, particularly in Africa. In one study, the clinical course of KS was more aggressive in women (Nasti et al., 1999). KS is the most common malignancy in many parts of Africa and is the most common malignancy in AIDS patients in the United States, with 12.5% of male AIDS patients in the United States having KS as their AIDS-defining illness and a total of 24% of male patients experiencing KS during the course of AIDS (Jones et al., 1999).

The accumulated epidemiological, pathological, and molecular data make it clear that HHV-8 is the etiologic agent of all forms of KS. HHV-8 DNA, transcripts, and proteins are present in the spindle cells of KS lesions, including the earliest stages of lesion development (Boshoff et al., 1995; Li et al., 1996; Staskus et al., 1997, 1999; Dupin et al., 1999). Most HHV-8-containing cells express latent antigens or transcripts, while lytic replication takes place in a small percentage of cells.

KS is seldom tightly linked to primary infection with HHV-8. HHV-8 seroconversion and the presence of HHV-8 DNA in blood can precede KS by many months or years (Moore et al., 1996b). In one study, patients who were seropositive for HHV-8 before acquiring HIV had KS less frequently than did patients who acquired HHV-8 after HIV, indicating that KS is less likely to develop in the face of a more complete immune response (Melbye et al., 1998). The triggers that lead to development of KS in HHV-8-positive individuals have not been identified.

PEL

Tissues from many malignancies have been examined for the presence of HHV-8 DNA. The most striking association was with a class of non-Hodgkin's lymphomas known as body cavity-based lymphoma (Cesarman et al., 1995). Such tumors occur as nonsolid tumors in body cavities.

TABLE 5 Epidemiological forms of KS

Form	Population	Associated with immune impairment	Male/female ratio
Classical	Older individuals of Mediterranean or Eastern European Jewish ancestry	No	5–10:1
African endemic	All age groups in equatorial Africa	No	2–5:1 (adults) 1.8:1 (children)
AIDS related	HIV type 1-infected individuals	Yes	
	Overall	Yes	50:1
	IVDU[a] or heterosexual acquisition	Yes	2.5:1[b]
Transplant related	Iatrogenically immune suppressed transplant recipients	Yes	2:1

[a]IVDU, intravenous drug user.
[b]Jones et al., 1999.

TABLE 6   Common properties of PEL[a]

Lymphomatous effusion without contiguous tumor mass in pleural, peritoneal, and/or pericardial cavity
Seldom disseminated beyond cavity of origin
Cell morphology bridges large-cell immunoblastic lymphoma and anaplastic large-cell lymphoma
Expression of CD45 and activation-associated antigens
Infrequent expression of B-cell-associated antigens
Clonal immunoglobulin gene rearrangements indicating B-cell origin
Infrequent c-*myc* rearrangements or bcl-2, ras, and p53 alterations
Frequently coinfected with EBV
HHV-8 DNA in every tumor cell

[a]Based on Cesarman and Knowles, 1999.

Because of several distinguishing properties, HHV-8-associated body cavity lymphomas are now referred to as PEL. PEL are rare and occur more frequently in HIV-infected patients. PEL properties are summarized in Table 6. As mentioned above, much of what we know of the molecular biology of HHV-8 has been learned from cell lines derived from these tumors. Given the several mechanisms by which HHV-8 can transform cells, it is likely that the virus is the direct cause of PEL, although considering how uncommon PEL are, it is clear that other factors are involved in their pathogenesis.

MCD

MCD is a multicentric lymph node angiofollicular hyperplasia characterized by vascular proliferation in germinal centers. It occurs primarily in AIDS patients, sometimes linked with KS or non-Hodgkin's lymphoma. HHV-8 is present in the hyperplastic regions of MCD lesions in a subset of patients with MCD. Three rapidly progressing and severe MCD cases were identified over a brief interval in one hospital in patients who had recently begun highly active antiretroviral therapy (Zietz et al., 1999), raising the possibility that some aspect of the antiretroviral therapy promoted the disease. Thus far, there have been no indications that this represents anything other than a sporadic outbreak.

Other Disease Associations

Numerous diseases have been studied for possible HHV-8 associations. In some cases, the initial observations have been refuted, while others await further study.

One possible association that received a lot of attention was that between HHV-8 and multiple myeloma. The association was plausible because multiple myeloma is characterized by high IL-6 levels and HHV-8 encodes an IL-6 homolog in addition to its transforming genes. In the initial study, bone marrow dendritic cells from all 15 multiple-myeloma patients studied were HHV-8 positive by PCR (Rettig et al., 1997). Unfortunately, subsequent PCR, serologic, and epidemiological studies from many laboratories failed to corroborate the initial observations, and the association now seems unlikely (reviewed and discussed in Tarte et al., 1999, and Berenson and Vescio, 1999).

Similar stories of initial positive associations that were not confirmed have been told for sarcoidosis, cutaneous T-cell lymphoma, and angiosarcoma (reviewed in Schulz, 1999). Positive associations (sometimes case reports) for which follow-up studies have not been reported include cerebral B-cell lymphoma (Luppi et al., 1996), encephalitis (Said et al., 1997), and premalignant Bowen's disease and malignant squamous cell carcinoma in Japan (Inagi et al., 1996).

Diagnosis

Presently, HHV-8 diagnostics are primarily used to study epidemiology and disease associations in research settings. Areas in which they are likely to enter clinical practice in the near future include aiding the diagnosis of KS, prediction of KS development, donor-recipient matching in organ transplantation, and monitoring the effects of therapy directed at the virus.

Virus Detection

Systems for inefficient, usually short-term culture of primary isolates of HHV-8 have been described, but their utility is limited (Vieira et al., 1997; Renne et al., 1998). Virus detection can be accomplished by using PCR or in situ methods. Numerous PCR primer sets have been described. The most widely used, and still one of the most sensitive sets, is that described in the paper reporting the discovery of the virus (Chang et al., 1994). In addition, nested primer sets (Moore et al., 1996b), sets for use in the Taqman system (Kennedy et al., 1997), and quantitative-competitive PCR sets (Lock et al., 1997) have been described. Sensitive methods for in situ hybridization are based on the high-copy-number transcripts that are specific for either latent or lytically infected cells (Staskus et al., 1999). Monoclonal antibodies specific for HHV-8 have been described (Chan et al., 1998; Dupin et al., 1999; Sadler et al., 1999; Zoeteweij et al., 1999). They have not been widely used for diagnosis, but this is likely to change soon.

Antibody Tests

The development of serologic assays for HHV-8 has been a very active area of research. Assays come in two main varieties: those that target latent antigens and those that target lytic antigens. In addition, assays come in a variety of formats, including IFA (Gao et al., 1996; Lennette et al., 1996; Smith et al., 1997; Whitby et al., 1998), EIA (Simpson et al., 1996; Chatlynne et al., 1998; Pau et al., 1998), and immunoblot (Zhu et al., 1999).

Assay sensitivity has been difficult to define in absolute terms relative to infection. Because KS patients are infected with HHV-8, KS patient serum is normally used as the serologic "gold standard." Most HHV-8 serologic assays identify

70 to 90% of KS patients as positive (reviewed in Schulz, 1999, and Chatlynne and Ablashi, 1999). At first glance, this lack of sensitivity in detecting KS cases can be taken as an indication that the assays are inherently insensitive in detecting HHV-8 antibodies. However, KS patients who are seropositive frequently have very high antibody titers. In addition, as described below, the major HHV-8 antigens have been identified, and the various assays are sufficiently refined so that there is no reason to suppose that they are inherently less sensitive in detecting antibody than similar assays for other agents. It is likely that the generally low antibody titers in seropositive healthy individuals and in a subset of KS patients are a reflection of the strategy used by the virus to establish long-term latent infections in a manner that in some individuals does not lead to a substantial detectable immune response. This might be accomplished by the virus through its choice of target tissue(s), its mode of replication, and its ability to modulate the immune response. Although the serologic response in some virus-positive individuals is below the limit of detection (Meng et al., 2000), it is important to remember that the immune response is sufficient to control the infection; cellular immunity is likely to play a role (Osman et al., 1999).

Under circumstances where immune control is diminished, such as in organ transplant recipients or in HIV-infected individuals, viral activity can become unchecked, leading to the development of KS. The high-titer immune response that develops in most KS patients is reminiscent of the response to EBV antigens seen in patients with nasopharyngeal carcinoma.

Assay sensitivity cannot be fully evaluated without considering assay specificity. As cutoffs for discriminating positives from negatives are lowered in an attempt to increase sensitivity, specificity can be compromised. In addition, the possibility of serologic cross-reaction with tissues or other agents must be considered.

Current IFA are based either on the latent nuclear antigen or on cytoplasmic antigens that are produced in cells induced to a lytic state by treatment with phorbol ester or sodium butyrate. There is no hint that the latent assays are anything but highly specific, but their sensitivity is questionable because in many studies less than 90% of KS patients are positive. An anti-human IgG monoclonal antibody has been used to increase sensitivity in the lytic antigen IFA (mIFA) (Lennette et al., 1996). The mIFA is more sensitive at detecting HHV-8 in KS patients, but questions have been raised about its specificity because over 20% of healthy donor specimens have been found positive in the absence of corroboration from other assays, such as PCR. Although cross-reactivity with EBV antibodies has been hypothesized and detected (Andre et al., 1997), experimental evidence indicates that such cross-reactivity is not significant (Lennette et al., 1996). The more important source of false-positive results appears to be related to reactions with cellular debris resulting from chemical induction and the absence of otherwise equivalent negative control cells that have been chemically treated but lack the virus. The possible problems are exacerbated at high serum concentrations, such as in the study described in the initial publication of the mIFA, in which a serum dilution of 1:10 was used (Lennette et al., 1996).

Few investigators accept positive results at 1:10 serum dilutions as being specific. A more widely accepted cutoff is reactivity at a dilution of 1:40 in the presence of Evans' blue counterstain (Zhu et al., 1999), although some investigators prefer even higher dilutions, such as 1:100 or 1:160. In our

experience, many specimens with titers of only 1:10 or 1:20 in the mIFA are corroborated by reactivity in other assays, including peptide-based EIA and PCR of PBMC from the same individual (Spira et al., 2000). Nonetheless, other specimens that were positive at such high serum concentrations cannot be similarly corroborated and are possible false positives; thus, we evaluate such specimens as equivocal.

The differences in the sensitivities of the latent and lytic assays are clearest in studies of populations of blood donors or others who are at low risk for KS. In the United States and the United Kingdom, seroprevalences of 0 to 3% have been found in these populations by latent IFA, while a value of 20% was observed by mIFA at a 1:10 serum dilution and 8% by lytic IFA at a 1:40 serum dilution (reviewed in Schulz, 1999).

To date, three major antigens and their genes have been identified: the virion glycoprotein encoded by open reading frame K8.1 (Chandran et al., 1998; Raab et al., 1998), a capsid protein encoded by open reading frame 65 (Fig. 4) (Simpson et al., 1996), and the major latent nuclear antigen encoded by open reading frame 73 (Rainbow et al., 1997). Recent developments in HHV-8 serology include the use of multiple synthetic or purified recombinant viral antigens to improve sensitivity (Zhu et al., 1999; Spira et al., 2000); these assays have been implemented in EIA and immunoblot formats.

Prevention and Therapy

HHV-8 infections acquired after HIV seroconversion are more likely to lead to KS (Renwick et al., 1998); thus, although the routes of transmission have not been unambiguously defined, safe sex practices, such as the use of latex condoms during every act of sexual intercourse, are recommended to reduce the risk of exposure (Centers for Disease Control and Prevention, 1999).

Traditionally, KS has been treated with cytotoxic cancer therapies. The identification of HHV-8 as the cause of the disease offers the possibility of specific therapy directed at the agent. In vitro, HHV-8 lytic replication is extraordinarily sensitive to cidofovir and is also sensitive to ganciclovir and foscarnet (Kedes and Ganem, 1997; Medveczky et al., 1997). Some, but not all, retrospective studies indicated a possible therapeutic benefit from ganciclovir or foscarnet (Costagliola and Mary-Krause, 1995; Jones et al., 1995; Glesby et al., 1996; Mocroft et al., 1996). In addition, in a placebo-controlled, double-blind trial of ganciclovir in AIDS patients with HCMV retinitis, the risk of developing KS was reduced by 75 and 93% in patients who received oral or intravenous ganciclovir, respectively (Martin et al., 1999). Although the benefits of ganciclovir or foscarnet therapy have shown up in large and long-term studies, only a marginal positive effect of foscarnet on KS was seen in one small treatment study (Cordero et al., 1997), and neither drug affected the HHV-8 load in PBMC in another small series (Boivin et al., 1999). Thus, while some of these results offer hope, further studies will be needed before therapeutic recommendations can be made (Cordero et al., 1997; Centers for Disease Control and Prevention, 1999; Gnann et al., 2000). In addition, it is clear that the most effective therapy for KS in immunocompromised patients is immune system rejuvenation. Thus, KS in organ transplant recipients frequently disappears upon cessation of immune-suppressive therapy, KS incidence has diminished markedly in the face of combination therapy for HIV infections (Jones et al., 1999), and in a small series, KS in HIV-infect-

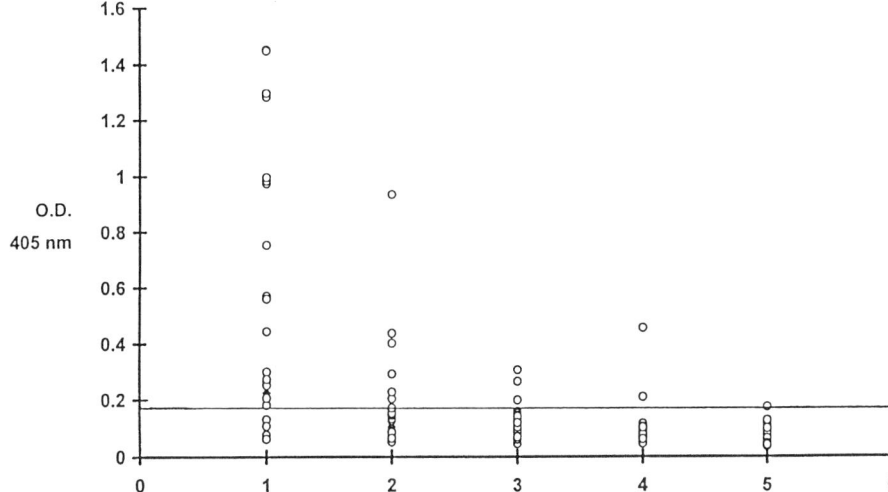

FIGURE 4 Relative HHV-8 antibody reactivity measured by enzyme-linked immunosorbent assay with recombinant orf65 protein among various groups: 1, AIDS KS; 2, HIV-infected homosexual men without KS; 3, United Kingdom blood donors; 4, intravenous drug users; 5, hemophiliacs. O.D., optical density. Figure courtesy of Thomas Schulz, The University of Liverpool, and The Lancet Ltd. (Simpson et al., 1996).

ed patients improved markedly in response to retonovir (Conant et al., 1997).

CONCLUSIONS

The viruses described here have the shared property of being innocuous in most healthy hosts; only HHV-6B has a frequent association with disease during primary infection. Little is known of the mechanisms of latency employed by these viruses; unraveling these mechanisms will reveal much about how host defenses can be compromised or avoided. The relative apathogenicity of the HHV-6 variants and HHV-7 makes them candidates for vector development. HHV-8 is unlike any other herpesvirus in the restriction of its high-prevalence natural host range to specific geographic areas or behaviors; the basis of this restriction remains an enigma.

Rapid diagnostics for HHV-6 and HHV-7 are needed to more accurately evaluate and appropriately treat febrile children. Refined diagnostics for HHV-8 are likely to help predict KS development and will enable monitoring of antiviral therapy. The role of these viruses in posttransplantation disease remains to be fully defined, and areas of uncertain etiologic involvement, such as between HHV-6 and MS, need to be resolved.

REFERENCES

Ablashi, D., H. Agut, Z. Berneman, G. Campadelli-Fiume, D. Carrigan, L. Ceccerini-Nelli, B. Chandran, S. Chou, H. Collandre, R. Cone, T. Dambaugh, S. Dewhurst, D. DiLuca, L. Foà-Tomasi, B. Fleckenstein, N. Frenkel, R. Gallo, U. Gompels, C. Hall, M. Jones, G. Lawrence, M. Martin, L. Montagnier, F. Neipel, J. Nicholas, P. Pellett, A. Razzaque, G. Torrelli, B. Thomson, S. Salahuddin, L. Wyatt, and K. Yamanishi. 1993. Human herpesvirus-6 strain groups: a nomenclature. *Arch. Virol.* **129:**363–366.

Ablashi, D. V., S. F. Josephs, C. Buchbinder, K. Hellman, S. Nakamura, T. Llana, P. Lusso, M. Kaplan, J. Dahlberg, S.

Memon, F. Imam, K. L. Ablashi, P. D. Markham, B. Kramarsky, G. R. F. Krueger, P. Biberfeld, F. Wong-Staal, S. Z. Salahuddin, and R. C. Gallo. 1988. Human B-lymphotropic virus (human herpesvirus-6). *J. Virol. Methods* **21:**29–48.

Ablashi, D. V., N. Balachandran, S. F. Josephs, C. L. Hung, G. R. F. Krueger, B. Kramarsky, S. Z. Salahuddin, and R. C. Gallo. 1991. Genomic polymorphism, growth properties, and immunologic variations in human herpesvirus-6 isolates. *Virology* **184:**545–552.

Ablashi, D. V., M. Handy, J. Bernbaum, L. G. Chatlynne, W. Lapps, B. Kramarsky, Z. N. Berneman, A. L. Komaroff, and J. E. Whitman. 1998. Propagation and characterization of human herpesvirus-7 (HHV-7) isolates in a continuous T-lymphoblastoid cell line (SupT1). *J. Virol. Methods* **73:**123–140.

Ablashi, D. V., W. Lapps, M. Kaplan, J. E. Whitman, J. R. Richert, and G. R. Pearson. 1999. Human herpesvirus-6 (HHV-6) infection in multiple sclerosis: a preliminary report. *Mult. Scler.* **4:**490–496.

Adams, O., C. Krempe, G. Kogler, P. Wernet, and A. Scheid. 1998. Congenital infections with human herpesvirus 6. *J. Infect. Dis.* **178:**544–546.

Adler, S. P., M. McVoy, S. Chou, S. Hempfling, K. Yamanishi, and W. Britt. 1993. Antibodies induced by a primary cytomegalovirus infection react with human herpesvirus 6 proteins. *J. Infect. Dis.* **168:**1119–1126.

Andre, S., O. Schatz, J. R. Bogner, H. Zeichhardt, M. Stoffler-Meilicke, H. U. Jahn, R. Ullrich, A. K. Sonntag, R. Kehm, and J. Haas. 1997. Detection of antibodies against viral capsid proteins of human herpesvirus 8 in AIDS-associated Kaposi's sarcoma. *J. Mol. Med.* **75:**145–152.

Ariyoshi, K., M. Schim van der Loeff, P. Cook, D. Whitby, T. Corrah, S. Jaffar, F. Cham, S. Sabally, D. O'Donovan, R. A. Weiss, T. F. Schulz, and H. Whittle. 1998. Kaposi's sarcoma in the Gambia, West Africa is less frequent in human immunodeficiency virus type 2 than in human immunodeficiency virus type 1 infection despite a high prevalence of human herpesvirus 8. *J. Hum. Virol.* **1:**193–199.

Asada, H., V. Klaus-Kovtun, H. Golding, S. I. Katz, and A. Blauvelt. 1999. Human herpesvirus 6 infects dendritic cells and suppresses human immunodeficiency virus type 1 replication in coinfected cultures. *J. Virol.* **73**:4019–4028.

Asano, Y., S. Suga, T. Yoshikawa, A. Urisu, and T. Yazaki. 1989. Human herpesvirus type 6 infection (exanthem subitum) without fever. *J. Pediatr.* **115**:264–265.

Asano, Y., T. Nakashima, T. Yoshikawa, S. Suga, and T. Yazaki. 1991a. Severity of human herpesvirus-6 viremia and clinical findings in infants with exanthem subitum. *J. Pediatr.* **118**:891–895.

Asano, Y., T. Yoshikawa, S. Suga, T. Hata, T. Yamazaki, and T. Yazaki. 1991b. Simultaneous occurrence of human herpesvirus 6 infection and intussusception in three infants. *Pediatr. Infect. Dis. J.* **10**:335–337.

Asano, Y., T. Yoshikawa, S. Suga, I. Kobayashi, T. Nakashima, T. Yazaki, Y. Kajita, and T. Ozaki. 1994. Clinical features of infants with primary human herpesvirus 6 infection (exanthem subitum, roseola infantum). *Pediatrics* **93**:104–108.

Aubin, J. T., H. Collandre, D. Candotti, D. Ingrand, C. Rouzioux, M. Burgard, S. Richard, J. M. Huraux, and H. Agut. 1991. Several groups among human herpesvirus 6 strains can be distinguished by Southern blotting and polymerase chain reaction. *J. Clin. Microbiol.* **29**:367–372.

Aubin, J. T., L. Poirel, H. Agut, J. M. Huraux, C. Bignozzi, Y. Brossard, N. Mulliez, J. Roume, F. Lecuru, and R. Taurelle. 1992. Intrauterine transmission of human herpesvirus 6. *Lancet* **340**:482–483.

Aubin, J. T., H. Agut, H. Collandre, K. Yamanishi, B. Chandran, L. Montagnier, and J. M. Huraux. 1993. Antigenic and genetic differentiation of the two putative types of human herpes virus 6. *J. Virol. Methods* **41**:223–234.

Aubin, J. T., L. Poirel, C. Robert, J. M. Huraux, and H. Agut. 1994. Identification of human herpesvirus 6 variants A and B by amplimer hybridization with variant-specific oligonucleotides and amplification with variant-specific primers. *J. Clin. Microbiol.* **32**:2434–2440.

Balachandra, K., P. Bowonkiratikachorn, B. Poovijit, A. Thattiyaphong, C. Jayavasu, C. Wasi, M. Takahashi, and K. Yamanishi. 1991. Human herpesvirus 6 (HHV-6) infection and exanthem subitum in Thailand. *Acta Paediatr. Jpn.* **33**:434–439.

Berenson, J. R., and R. A. Vescio. 1999. HHV-8 is present in multiple myeloma patients. *Blood* **93**:3157–3159.

Berneman, Z. N., D. V. Ablashi, G. Li, M. Eger-Fletcher, M. S. Reitz, Jr., C. L. Hung, I. Brus, A. L. Komaroff, and R. C. Gallo. 1992. Human herpesvirus 7 is a T-lymphotropic virus and is related to, but significantly different from, human herpesvirus 6 and human cytomegalovirus. *Proc. Natl. Acad. Sci. USA* **89**:10552–10556.

Biberfeld, P., B. Kramarsky, S. Z. Salahuddin, and R. C. Gallo. 1987. Ultrastructural characterization of a new human B-lymphotropic DNA virus (HBLV) isolated from patients with lymphoproliferative disease. *J. Natl. Cancer Inst.* **79**:933–941.

Black, J. B., and P. E. Pellett. 1999. Human herpesvirus 7. *Rev. Med. Virol.* **9**:245–262.

Black, J. B., K. C. Sanderlin, C. S. Goldsmith, H. E. Gary, C. Lopez, and P. E. Pellett. 1989. Growth properties of human herpesvirus-6 strain Z29. *J. Virol. Methods* **26**:133–145.

Black, J. B., C. Lopez, and P. E. Pellett. 1992. Induction of host cell protein synthesis by human herpesvirus 6. *Virus Res.* **22**:13–23.

Black, J. B., N. Inoue, K. Kite-Powell, S. Zaki, and P. E. Pellett. 1993. Frequent isolation of human herpesvirus 7 from saliva. *Virus Res.* **29**:91–98.

Black, J. B., E. Durigon, K. Kite-Powell, L. de Souza, S. P. Curli, A. M. S. Afonso, M. Theobaldo, and P. E. Pellett. 1996a. Seroconversion to human herpesvirus 6 and human herpesvirus 7 among Brazilian children with clinical diagnoses of measles or rubella. *Clin. Infect. Dis.* **23**:1156–1158.

Black, J. B., T. F. Schwarz, J. L. Patton, K. Kite-Powell, P. E. Pellett, S. Wiersbitzky, R. Bruns, C. Muller, G. Jager, and J. Stewart. 1996b. Evaluation of immunoassays for detection of antibodies to human herpesvirus 7. *Clin. Diagn. Lab. Immunol.* **3**:79–83.

Black, J. B., D. A. Burns, C. S. Goldsmith, P. M. Feorino, K. Kite-Powell, R. F. Schinazi, P. W. Krug, and P. E. Pellett. 1997. Biologic properties of human herpesvirus 7 strain SB. *Virus Res.* **52**:25–41.

Blackbourn, D. J., and J. A. Levy. 1997. Human herpesvirus 8 in semen and prostate. *AIDS* **11**:249–250.

Blackbourn, D. J., J. Ambroziak, E. Lennette, M. Adams, B. Ramachandran, and J. A. Levy. 1997. Infectious human herpesvirus 8 in a healthy North American blood donor. *Lancet* **349**:609–611.

Blackbourn, D. J., E. T. Lennette, J. Ambroziak, D. V. Mourich, and J. A. Levy. 1998. Human herpesvirus 8 detection in nasal secretions and saliva. *J. Infect. Dis.* **177**:213–216.

Blackbourn, D. J., D. Osmond, J. A. Levy, and E. T. Lennette. 1999. Increased human herpesvirus 8 seroprevalence in young homosexual men who have multiple sex contacts with different partners. *J. Infect. Dis.* **179**:237–239.

Boivin, G., A. Gaudreau, E. Toma, R. Lalonde, J. P. Routy, G. Murray, J. Handfield, and M. G. Bergeron. 1999. Human herpesvirus 8 DNA load in leukocytes of human immunodeficiency virus-infected subjects: correlation with the presence of Kaposi's sarcoma and response to anticytomegalovirus therapy. *Antimicrob. Agents Chemother.* **43**:377–380.

Boshoff, C., T. F. Schultz, M. M. Kennedy, A. K. Graham, C. Fisher, A. Thomas, J. O'McGee, R. A. Weiss, and J. J. O'Leary. 1995. Kaposi's sarcoma-associated herpesvirus infects endothelial and spindle cells. *Nat. Med.* **1**:1274–1278.

Bourboulia, D., D. Whitby, C. Boshoff, R. Newton, V. Beral, H. Carrara, A. Lane, and F. Sitas. 1998. Serologic evidence for mother-to-child transmission of Kaposi sarcoma-associated herpesvirus infection. *JAMA* **280**:31–32.

Bovenzi, P., P. Mirandola, P. Secchiero, R. Strumia, E. Cassai, and D. Di Luca. 1993. Human herpesvirus 6 (variant A) in Kaposi's sarcoma. *Lancet* **341**:1288–1289.

Braun, D. K., G. Dominguez, and P. E. Pellett. 1997. Human herpesvirus 6. *Clin. Microbiol. Rev.* **10**:521–567.

Calabro, M. L., J. R. Fiore, A. Favero, A. Lepera, A. Saracino, G. Angarano, T. F. Schulz, and L. Chieco-Bianchi. 1999. Detection of human herpesvirus 8 in cervicovaginal secretions and seroprevalence in human immunodeficiency virus type 1-seropositive and -seronegative women. *J. Infect. Dis.* **179**:1534–1537.

Campadelli-Fiume, G., S. Guerrini, X. Liu, and L. Foà-Tomasi. 1993. Monoclonal antibodies to glycoprotein B differentiate human herpesvirus 6 into two clusters, variants A and B. *J. Gen. Virol.* **74**:2257–2262.

Campadelli-Fiume, G., P. Mirandola, and L. Menotti. 1999. Human herpesvirus 6: an emerging pathogen. *Emerg. Infect. Dis.* **5**:353–366.

Cardinali, G., M. Gentile, M. Cirone, C. Zompetta, L. Frati, A. Faggioni, and M. R. Torrisi. 1998. Viral glycoproteins accumulate in newly formed annulate lamellae following infection of lymphoid cells by human herpesvirus 6. *J. Virol.* **72:**9738–9746.

Carrigan, D. R., and K. K. Knox. 1994. Human herpesvirus 6 (HHV-6) isolation from bone marrow: HHV-6-associated bone marrow suppression in bone marrow transplant patients. *Blood* **84:**3307–3310.

Caserta, M. T., C. B. Hall, K. Schnabel, K. McIntyre, C. Long, M. Costanzo, S. Dewhurst, R. Insel, and L. G. Epstein. 1994. Neuroinvasion and persistence of human herpesvirus 6 in children. *J. Infect. Dis.* **170:**1586–1589.

Caserta, M. T., C. B. Hall, K. Schnabel, C. E. Long, and N. D'Heron. 1998. Primary human herpesvirus 7 infection: a comparison of human herpesvirus 7 and human herpesvirus 6 infections in children. *J. Pediatr.* **133:**386–389.

Centers for Disease Control and Prevention. 1999. 1999 USPHS/IDSA guidelines for the prevention of opportunistic infections in persons infected with human immunodeficiency virus. *Morb. Mortal. Wkly. Rep.* **48**(RR-10):1–66.

Cermelli, C., P. Pietrosemoli, M. Meacci, M. Pecorari, A. M. Sabbatini, B. Colombari, and M. Portolani. 1997. SupT-1: a cell system suitable for an efficient propagation of both HHV-7 and HHV-6 variants A and B. *New Microbiol.* **20:**187–196.

Cesarman, E., and D. M. Knowles. 1999. The role of Kaposi's sarcoma-associated herpesvirus (KSHV/HHV-8) in lymphoproliferative diseases. *Semin. Cancer Biol.* **9:**165–174.

Cesarman, E., Y. Chang, P. S. Moore, J. W. Said, and D. M. Knowles. 1995. Kaposi's sarcoma-associated herpesvirus-like DNA sequences in AIDS-related body cavity-based lymphomas. *N. Engl. J. Med.* **332:**1186–1191.

Challoner, P. B., K. T. Smith, J. D. Parker, D. L. MacLeod, S. N. Coulter, T. M. Rose, E. R. Schultz, J. L. Bennett, R. L. Garber, M. Chang, P. A. Schad, P. M. Stewart, R. C. Nowinski, J. P. Brown, and G. C. Burmer. 1995. Plaque-associated expression of human herpesvirus 6 in multiple sclerosis. *Proc. Natl. Acad. Sci. USA* **92:**7440–7444.

Chan, P. K., J. S. Peiris, K. Y. Yuen, R. H. Liang, Y. L. Lau, F. E. Chen, S. K. Lo, C. Y. Cheung, T. K. Chan, and M. H. Ng. 1997. Human herpesvirus-6 and human herpesvirus-7 infections in bone marrow transplant recipients. *J. Med. Virol.* **53:**295–305.

Chan, S. R., C. Bloomer, and B. Chandran. 1998. Identification and characterization of human herpesvirus-8 lytic cycle-associated ORF 59 protein and the encoding cDNA by monoclonal antibody. *Virology* **240:**118–126.

Chandran, B., S. Tirawatnapong, B. Pfeiffer, and D. V. Ablashi. 1992. Antigenic relationships among human herpesvirus-6 isolates. *J. Med. Virol.* **37:**247–254.

Chandran, B., M. S. Smith, D. M. Koelle, L. Corey, R. Horvat, and E. Goldstein. 1998. Reactivities of human sera with human herpesvirus-8-infected BCBL-1 cells and identification of HHV-8-specific proteins and glycoproteins and the encoding cDNAs. *Virology* **243:**208–217.

Chang, Y., E. Cesarman, M. S. Pessin, F. Lee, J. Culpepper, D. M. Knowles, and P. S. Moore. 1994. Identification of herpesvirus-like DNA sequences in AIDS-associated Kaposi's sarcoma. *Science* **266:**1865–1869.

Chatlynne, L. G., and D. V. Ablashi. 1999. Seroepidemiology of Kaposi's sarcoma-associated herpesvirus (KSHV). *Semin. Cancer Biol.* **9:**175–185.

Chatlynne, L. G., W. Lapps, M. Handy, Y. Q. Huang, R. Masood, A. S. Hamilton, J. W. Said, H. P. Koeffler, M. H. Kaplan, A. Friedman-Kien, P. S. Gill, J. E. Whitman, and D. V. Ablashi. 1998. Detection and titration of human herpesvirus-8-specific antibodies in sera from blood donors, acquired immunodeficiency syndrome patients, and Kaposi's sarcoma patients using a whole virus enzyme-linked immunosorbent assay. *Blood* **92:**53–58.

Chiu, S. S., C. Y. Cheung, C. Y. Tse, and M. Peiris. 1998. Early diagnosis of primary human herpesvirus 6 infection in childhood: serology, polymerase chain reaction, and virus load. *J. Infect. Dis.* **178:**1250–1256.

Chokephaibulkit, K., P. A. Brunell, V. Vimal, C. Long, K. Schnabel, and C. B. Hall. 1997. Enzyme-linked immunosorbent assay for detection of antibody to human herpesvirus 6. *Clin. Diagn. Lab. Immunol.* **4:**687–691.

Chou, S., and G. I. Marousek. 1994. Analysis of interstrain variation in a putative immediate-early region of human herpesvirus 6 DNA and definition of variant-specific sequences. *Virology* **198:**370–376.

Clark, D. A., M. L. Freeland, L. K. Mackie, R. F. Jarrett, and D. E. Onions. 1993. Prevalence of antibody to human herpesvirus 7 by age. *J. Infect. Dis.* **168:**251–252.

Clark, D. A., M. Ait-Khaled, A. C. Wheeler, I. M. Kidd, J. E. McLaughlin, M. A. Johnson, P. D. Griffiths, and V. C. Emery. 1996. Quantification of human herpesvirus 6 in immunocompetent persons and post-mortem tissues from AIDS patients by PCR. *J. Gen. Virol.* **77:**2271–2275.

Clark, D. A., I. M. Kidd, K. E. Collingham, M. Tarlow, T. Aueni, A. Riordan, P. D. Griffiths, V. C. Emery, and D. Pillay. 1997. Diagnosis of primary human herpesvirus 6 and 7 infections in febrile infants by polymerase chain reaction. *Arch. Dis. Child.* **77:**42–45.

Conant, M. A., K. M. Opp, D. Poretz, and R. G. Mills. 1997. Reduction of Kaposi's sarcoma lesions following treatment of AIDS with ritonavir. *AIDS* **11:**1300–1301.

Cone, R. W., M. L. Huang, R. Ashley, and L. Corey. 1993. Human herpesvirus 6 DNA in peripheral blood cells and saliva from immunocompetent individuals. *J. Clin. Microbiol.* **31:**1262–1267.

Cone, R. W., M.-L. Huang, R. C. Hackman, and L. Corey. 1996. Coinfection with human herpesvirus 6 variants A and B in lung tissue. *J. Clin. Microbiol.* **34:**877–881.

Cone, R. W., M. L. Huang, L. Corey, J. Zeh, R. Ashley, and R. Bowden. 1999. Human herpesvirus 6 infections after bone marrow transplantation: clinical and virologic manifestations. *J. Infect. Dis.* **179:**311–318.

Cook, P. M., D. Whitby, M.-L. Calabro, M. Luppi, D. N. Kakoola, H. Hjalgrim, K. Ariyoshi, B. Ensoli, A. J. Davison, T. F. Schulz, and the International Collaborative Group. 1999. Variability and evolution of Kaposi's sarcoma-associated herpesvirus in Europe and Africa. *AIDS* **13:**1165–1176.

Cordero, E., L. F. Lopez-Cortes, P. Viciana, A. Alarcon, and J. Pachon. 1997. Foscarnet and AIDS-associated Kaposi's sarcoma. *AIDS* **11:**1787–1788.

Costagliola, D., and M. Mary-Krause. 1995. Can antiviral agents decrease the occurrence of Kaposi's sarcoma? *Lancet* **346:**578.

Couillard, M., J. R. Joly, L. Deschenes, and G. Richer. 1992. Evaluation of variables in immunofluorescence procedures for the detection of antibodies against human herpesvirus 6 (HHV-6). *Diagn. Microbiol. Infect. Dis.* **15:**313–320.

Coyle, P. V., M. Briggs, R. S. Tedder, and J. D. Fox. 1992. Comparison of three immunoassays for the detection of anti-HHV6. *J. Virol. Methods* **38**:283–295.

Cuende, J. I., J. Ruiz, M. P. Civeira, and J. Prieto. 1994. High prevalence of HHV-6 DNA in peripheral blood mononuclear cells of healthy individuals detected by nested-PCR. *J. Med. Virol.* **43**:115–118.

Daibata, M., T. Taguchi, T. Sawada, H. Taguchi, and I. Miyoshi. 1998. Chromosomal transmission of human herpesvirus 6 DNA in acute lymphoblastic leukaemia. *Lancet* **352**:543–544.

Dewhurst, S., B. Chandran, K. McIntyre, K. Schnabel, and C. B. Hall. 1992. Phenotypic and genetic polymorphisms among human herpesvirus-6 isolates from North American infants. *Virology* **190**:490–493.

Di Luca, D., G. Katsafanas, E. C. Schirmer, N. Balachandran, and N. Frenkel. 1990. The replication of viral and cellular DNA in human herpesvirus 6-infected cells. *Virology* **175**:199–210.

Di Luca, D., R. Dolcetti, P. Mirandola, V. De Re, P. Secchiero, A. Carbone, M. Boiocchi, and E. Cassai. 1994. Human herpesvirus 6: a survey of presence and variant distribution in normal peripheral lymphocytes and lymphoproliferative disorders. *J. Infect. Dis.* **170**:211–215.

Di Luca, D., P. Mirandola, T. Ravaioli, R. Dolcetti, A. Frigatti, P. Bovenzi, L. Sighinolfi, P. Monini, and E. Cassai. 1995. Human herpesviruses 6 and 7 in salivary glands and shedding in saliva of healthy and human immunodeficiency virus positive individuals. *J. Med. Virol.* **45**:462–468.

Di Luca, D., P. Mirandola, T. Ravaioli, B. Bigoni, and E. Cassai. 1996. Distribution of HHV-6 variants in human tissues. *Infect. Agents Dis.* **5**:203–214.

Dominguez, G., T. R. Dambaugh, F. R. Stamey, S. Dewhurst, N. Inoue, and P. E. Pellett. 1999. Human herpesvirus 6B genome sequence: coding content and comparison with human herpesvirus 6A. *J. Virol.* **73**:8040–8052.

Drago, F., E. Ranieri, F. Malaguti, E. Losi, and A. Rebora. 1997. Human herpesvirus 7 in pityriasis rosea. *Lancet* **349**:1367–1368.

Drobyski, W. R., M. Eberle, D. Majewski, and L. A. Baxter-Lowe. 1993. Prevalence of human herpesvirus 6 variant A and B infections in bone marrow transplant recipients as determined by polymerase chain reaction and sequence-specific oligonucleotide probe hybridization. *J. Clin. Microbiol.* **31**:1515–1520.

Dupin, N., C. Fisher, P. Kellam, S. Ariad, M. Tulliez, N. Franck, E. van Marck, D. Salmon, I. Gorin, J. P. Escande, R. A. Weiss, K. Alitalo, and C. Boshoff. 1999. Distribution of human herpesvirus-8 latently infected cells in Kaposi's sarcoma, multicentric Castleman's disease, and primary effusion lymphoma. *Proc. Natl. Acad. Sci. USA* **96**:4546–4551.

Emery, V. C., M. C. Atkins, E. F. Bowen, D. A. Clark, M. A. Johnson, I. M. Kidd, J. E. McLaughlin, A. N. Phillips, P. M. Strappe, and P. D. Griffiths. 1999. Interactions between beta-herpesviruses and human immunodeficiency virus in vivo: evidence for increased human immunodeficiency viral load in the presence of human herpesvirus 6. *J. Med. Virol.* **57**:278–282.

Enbom, M., C. Martin, S. Fredrikson, L. Jagdahl, H. Dahl, and A. Linde. 1997. Intrathecal antibody production to lymphotropic herpesviruses in patients with multiple sclerosis. *Neurol. Infect. Immunol.* **2**:107–111.

Fillet, A. M., P. Lozeron, H. Agut, O. Lyon-Caen, and R. Liblau. 1998. HHV-6 and multiple sclerosis. *Nat. Med.* **4**:537.

Foà-Tomasi, L., E. Avitabile, L. Ke, and G. Campadelli-Fiume. 1994. Polyvalent and monoclonal antibodies identify major immunogenic proteins specific for human herpesvirus 7-infected cells and have weak cross-reactivity with human herpesvirus 6. *J. Gen. Virol.* **75**:2719–2727.

Foà-Tomasi, L., E. Avitabile, and G. Campadelli-Fiume. 1995. Selection of a monoclonal antibody specific for variant B human herpesvirus 6-infected mononuclear cells. *J. Virol. Methods* **51**:289–296.

Foà-Tomasi, L., M. P. Fiorilli, E. Avitabile, and G. Campadelli-Fiume. 1996. Identification of an 85 kDa phosphoprotein as an immunodominant protein specific for human herpesvirus 7-infected cells. *J. Gen. Virol.* **77**:511–518.

Frenkel, N., and E. Roffman. 1996. Human herpesvirus 7, p. 2609–2622. *In* B. N. Fields, D. M. Knipe, P. M. Howley, R. M. Channock, J. L. Melnick, T. P. Monath, B. Roizman, and S. E. Straus (ed.), *Fields Virology*, 3rd ed., vol. 2. Lippincott-Raven, Philadelphia, Pa.

Frenkel, N., E. C. Schirmer, L. S. Wyatt, G. Katsafanas, E. Roffman, R. M. Danovich, and C. H. June. 1990. Isolation of a new herpesvirus from CD4+ T cells. *Proc. Natl. Acad. Sci. USA* **87**:748–752.

Gao, S.-J., L. Kingsley, M. Li, W. Zheng, C. Parravicini, J. Ziegler, R. Newton, C. R. Rinaldo, A. Saah, J. Phair, R. Detels, Y. Chang, and P. S. Moore. 1996. KSHV antibodies among Americans, Italians and Ugandans with and without Kaposi's sarcoma. *Nat. Med.* **2**:925–928.

Gautheret, A., J. T. Aubin, L. Poirel, S. Chopineau, H. Agut, and E. Dussaix. 1996. Detection and variant identification of HHV-6 by a non-radioactive hybridization microplate assay for amplimer detection. *J. Virol. Methods* **58**:33–40.

Gessain, A., P. Mauclere, M. van Beveren, S. Plancoulaine, A. Ayouba, J. L. Essame-Oyono, P. M. Martin, and G. de The. 1999. Human herpesvirus 8 primary infection occurs during childhood in Cameroon, Central Africa. *Int. J. Cancer* **81**:189–192.

Glesby, M. J., D. R. Hoover, S. Weng, N. M. H. Graham, J. P. Phair, R. Detels, M. Ho, and A. J. Saah. 1996. Use of antiherpes drugs and the risk of Kaposi's sarcoma: data from the Multicenter AIDS Cohort Study. *J. Infect. Dis.* **173**:1477–1480.

Gnann, J. W., P. E. Pellett, and H. W. Jaffe. 2000. Human herpesvirus 8 (HHV-8) and Kaposi's sarcoma in persons infected with human immunodeficiency virus. *Clin. Infect. Dis.* **30**:S72–S76.

Gompels, U. A., D. R. Carrigan, A. L. Carss, and J. Arno. 1993. Two groups of human herpesvirus 6 identified by sequence analyses of laboratory strains and variants from Hodgkin's lymphoma and bone marrow transplant patients. *J. Gen. Virol.* **74**:613–622.

Gompels, U. A., J. Nicholas, G. Lawrence, M. Jones, B. J. Thomson, M. E. Martin, S. Efstathiou, M. Craxton, and H. A. Macaulay. 1995. The DNA sequence of human herpesvirus-6: structure, coding content, and genome evolution. *Virology* **209**:29–51.

Grulich, A. E., S. J. Olsen, K. Luo, O. Hendry, P. Cunningham, D. A. Cooper, S. J. Gao, Y. Chang, P. S. Moore, and J. M. Kaldor. 1999. Kaposi's sarcoma-associated herpesvirus: a sexually transmissible infection? *J. Acquir. Immune. Defic. Syndr.* **20**:387–393.

Hall, C. B., C. E. Long, K. C. Schnabel, M. T. Caserta, K. M. McIntyre, M. A. Costanzo, A. Knott, S. Dewhurst, R. A. Insel, and L. G. Epstein. 1994. Human herpesvirus-6 infection in children. A prospective study of complications and reactivation. *N. Engl. J. Med.* **331**:432–438.

Hayward, G. S. 1999. KSHV strains: the origins and global spread of the virus. *Semin. Cancer Biol.* **9:**187–199.

Hidaka, Y., Y. Liu, M. Yamamoto, R. Mori, C. Miyazaki, K. Kusuhara, K. Okada, and K. Ueda. 1993. Frequent isolation of human herpesvirus 7 from saliva samples. *J. Med. Virol.* **40:**343–346.

Hoang, M. P., K. F. Ross, D. B. Dawson, R. H. Scheuermann, and B. B. Rogers. 1999. Human herpesvirus-6 and sudden death in infancy: report of a case and review of the literature. *J. Forensic Sci.* **44:**432–437.

Huang, L. M., C. Y. Lee, M. Y. Liu, and P. I. Lee. 1997. Primary infections of human herpesvirus-7 and herpesvirus-6: a comparative, longitudinal study up to 6 years of age. *Acta Paediatr.* **86:**604–608.

Inagi, R., H. Kosuge, S. Nishimoto, K. Yoshikawa, and K. Yamanishi. 1996. Kaposi's sarcoma-associated herpesvirus (KSHV) sequences in premalignant and malignant skin tumors. *Arch. Virol.* **141:**2217–2223.

Isegawa, Y., T. Mukai, K. Nakano, M. Kagawa, J. Chen, Y. Mori, T. Sunagawa, K. Kawanishi, J. Sashihara, A. Hata, P. Zou, H. Kosuge, and K. Yamanishi. 1999. Comparison of the complete DNA sequences of human herpesvirus 6 variants A and B. *J. Virol.* **73:**8053–8063.

Jarrett, R. F., D. A. Clark, S. F. Josephs, and D. E. Onions. 1990. Detection of human herpesvirus-6 DNA in peripheral blood and saliva. *J. Med. Virol.* **32:**73–76.

Jones, J. L., D. L. Hanson, S. Y. Chu, J. W. Ward, and H. W. Jaffe. 1995. AIDS-associated Kaposi's sarcoma. *Science* **267:**1078–1079.

Jones, J. L., D. L. Hanson, M. S. Dworkin, D. L. Alderton, P. L. Fleming, J. E. Kaplan, and J. Ward. 1999. Surveillance for AIDS-defining opportunistic illness, 1992–1997. *Morb. Mortal. Wkly. Rep.* **48(SS-2):**1–22.

Kasolo, F. C., E. Mpabalwani, and U. A. Gompels. 1997. Infection with AIDS-related herpesviruses in human immunodeficiency virus-negative infants and endemic childhood Kaposi's sarcoma in Africa. *J. Gen. Virol.* **78:**847–855.

Katsafanas, G. C., E. C. Schirmer, L. S. Wyatt, and N. Frenkel. 1996. In vitro activation of human herpesviruses 6 and 7 from latency. *Proc. Natl. Acad. Sci. USA* **93:**9788–9792.

Kedes, D. H., and D. Ganem. 1997. Sensitivity of Kaposi's sarcoma-associated herpesvirus replication to antiviral drugs. Implications for potential therapy. *J. Clin. Investig.* **99:**2082–2086.

Kempf, W., V. Adams, N. Wey, R. Moos, M. Schmid, E. Avitabile, and G. Campadelli-Fiume. 1997. CD68+ cells of monocyte/macrophage lineage in the environment of AIDS-associated and classic-sporadic Kaposi sarcoma are singly or doubly infected with human herpesviruses 7 and 6B. *Proc. Natl. Acad. Sci. USA* **94:**7600–7605.

Kempf, W., V. Adams, M. Kelinhans, G. Burg, F. O. Nestle, and G. Campadelli-Fiume. 1999. Pityriasis rosea is not associated with human herpesvirus 7. *Arch. Dermatol.* **135:**1070–1072.

Kennedy, M. M., S. B. Lucas, R. R. Jones, D. D. Howells, S. J. Picton, E. E. Hanks, J. O. McGee, and J. J. O'Leary. 1997. HHV8 and Kaposi's sarcoma: a time cohort study. *Mol. Pathol.* **50:**96–100. (Erratum, **50:**167.) *Mol. Pathol.* **50:**96–100.

Kidd, I. M., D. A. Clark, M. Ait-Khaled, P. D. Griffiths, and V. C. Emery. 1996. Measurement of human herpesvirus 7 load in peripheral blood and saliva of healthy subjects by quantitative polymerase chain reaction. *J. Infect. Dis.* **174:**396–401.

Kidd, I. M., D. A. Clark, J. A. Bremner, D. Pillay, P. D. Griffiths, and V. C. Emery. 1998. A multiplex PCR assay for the simultaneous detection of human herpesvirus 6 and human herpesvirus 7, with typing of HHV-6 by enzyme cleavage of PCR products. *J. Virol. Methods* **70:**29–36.

Kikuta, H., H. Lu, K. Tomizawa, and S. Matsumoto. 1990. Enhancement of human herpesvirus 6 replication in adult human lymphocytes by monoclonal antibody to CD3. *J. Infect. Dis.* **161:**1085–1087.

Klussmann, J. P., E. Krueger, T. Sloots, Z. Berneman, G. Arnold, and G. R. Krueger. 1997. Ultrastructural study of human herpesvirus-7 replication in tissue culture. *Virchows Arch.* **430:**417–426.

Koelle, D. M., M. L. Huang, B. Chandran, J. Vieira, M. Piepkorn, and L. Corey. 1997. Frequent detection of Kaposi's sarcoma-associated herpesvirus (human herpesvirus 8) DNA in saliva of human immunodeficiency virus-infected men: clinical and immunologic correlates. *J. Infect. Dis.* **176:**94–102.

Kolski, H., E. L. Ford-Jones, S. Richardson, M. Petric, S. Nelson, F. Jamieson, S. Blaser, R. Gold, H. Otsubo, H. Heurter, and D. MacGregor. 1998. Etiology of acute childhood encephalitis at The Hospital for Sick Children, Toronto, 1994–1995. *Clin. Infect. Dis.* **26:**398–409.

Kondo, K., T. Kondo, T. Okuno, M. Takahashi, and K. Yamanishi. 1991. Latent human herpesvirus 6 infection of human monocytes/macrophages. *J. Gen. Virol.* **72:**1401–1408.

Kositanont, U., C. Wasi, N. Wanprapar, P. Bowonkiratikachorn, K. Chokephaibulkit, S. Chearskul, K. Chimabutra, R. Sutthent, S. Foongladda, R. Inagi, T. Kurata, and K. Yamanishi. 1999. Primary infection of human herpesvirus 6 in children with vertical infection of human immunodeficiency virus type 1. *J. Infect. Dis.* **180:**50–55.

Leach, C. T., E. R. Newton, S. McParlin, and H. B. Jenson. 1994. Human herpesvirus 6 infection of the female genital tract. *J. Infect. Dis.* **169:**1281–1283.

Lee, H., R. Veazey, K. Williams, M. Li, J. Guo, F. Neipel, B. Fleckenstein, A. Lackner, R. C. Desrosiers, and J. U. Jung. 1998. Deregulation of cell growth by the K1 gene of Kaposi's sarcoma-associated herpesvirus. *Nat. Med.* **4:**435–440.

Lennette, E. T., D. J. Blackbourn, and J. A. Levy. 1996. Antibodies to human herpesvirus type 8 in the general population and in Kaposi's sarcoma patients. *Lancet* **348:**858–861.

Li, J. J., Y. Q. Huang, C. J. Cockerell, and A. E. Friedman-Kien. 1996. Localization of human herpes-like virus type 8 in vascular endothelial cells and perivascular spindle-shaped cells of Kaposi's sarcoma lesions by in situ hybridization. *Am. J. Pathol.* **148:**1741–1748.

Lock, M. J., P. D. Griffiths, and V. C. Emery. 1997. Development of a quantitative competitive polymerase chain reaction for human herpesvirus 8. *J. Virol. Methods* **64:**19–26.

Luppi, M., P. Barozzi, A. Maiorana, R. Marasca, and G. Torelli. 1994. Human herpesvirus 6 infection in normal human brain tissue. *J. Infect. Dis.* **169:**943–944.

Luppi, M., P. Barozzi, R. Marasca, M. Savarino, and G. Torelli. 1996. HHV-8-associated primary cerebral B-cell lymphoma in HIV-negative patient after long-term steroids. *Lancet* **347:**980.

Luppi, M., P. Barozzi, C. Morris, A. Maiorana, R. Garber, G. Bonacorsi, A. Donelli, R. Marasca, A. Tabilio, and G. Torelli. 1999. Human herpesvirus 6 latently infects early bone marrow progenitors in vivo. *J. Virol.* **73:**754–759.

Lusso, P., R. C. Gallo, S. E. DeRocco, and P. D. Markham. 1989. CD4 is not the membrane receptor for HHV-6. *Lancet* **i:**730.

Lusso, P., A. De Maria, M. Malnati, F. Lori, S. E. DeRocco, M. Baseler, and R. C. Gallo. 1991. Induction of CD4 and susceptibility to HIV-1 infection in human CD8+ T lymphocytes by human herpesvirus 6. *Nature* **349:**533–535.

Lusso, P., P. Secchiero, R. W. Crowley, A. Garzino-Demo, Z. N. Berneman, and R. C. Gallo. 1994. CD4 is a critical component of the receptor for human herpesvirus 7: interference with human immunodeficiency virus. *Proc. Natl. Acad. Sci. USA* **91:**3872–3876.

Lusso, P., A. Garzino-Demo, R. W. Crowley, and M. S. Malnati. 1995. Infection of γ/δ T lymphocytes by human herpesvirus 6: transcriptional induction of CD4 and susceptibility to HIV infection. *J. Exp. Med.* **181:**1303–1310.

Maeda, Y., T. Teshima, M. Yamada, K. Shinagawa, S. Nakao, Y. Ohno, K. Kojima, M. Hara, K. Nagafuji, S. Hayashi, S. Fukuda, H. Sawada, K. Matsue, K. Takenaka, F. Ishimaru, K. Ikeda, K. Niiya, and M. Harada. 1999. Monitoring of human herpesviruses after allogeneic peripheral blood stem cell transplantation and bone marrow transplantation. *Br. J. Haematol.* **105:**295–302.

Martin, C., M. Enbom, M. Soderstrom, S. Fredrikson, H. Dahl, J. Lycke, T. Bergstrom, and A. Linde. 1997. Absence of seven human herpesviruses, including HHV-6, by polymerase chain reaction in CSF and blood from patients with multiple sclerosis and optic neuritis. *Acta Neurol. Scand.* **95:**280–283.

Martin, D. F., B. D. Kuppermann, R. A. Wolitz, A. G. Palestine, H. Li, C. A. Robinson, and the Roche Ganciclovir Study Group. 1999. Oral ganciclovir for patients with cytomegalovirus retinitis treated with a ganciclovir implant. *N. Engl. J. Med.* **340:**1063–1070.

Martin, J. N., D. E. Ganem, D. H. Osmond, K. A. Page-Shafer, D. Macrae, and D. H. Kedes. 1998. Sexual transmission and the natural history of human herpesvirus 8 infection. *N. Engl. J. Med.* **338:**948–954.

Mayama, S., L. E. Cuevas, J. Sheldon, O. H. Omar, D. H. Smith, P. Okong, B. Silvel, C. A. Hart, and T. F. Schulz. 1998. Prevalence and transmission of Kaposi's sarcoma-associated herpesvirus (human herpesvirus 8) in Ugandan children and adolescents. *Int. J. Cancer* **77:**817–820.

McCullers, J. A., F. D. Lakeman, and R. J. Whitley. 1995. Human herpesvirus 6 is associated with focal encephalitis. *Clin. Infect. Dis.* **21:**571–576.

Medveczky, M. M., E. Horvath, T. Lund, and P. G. Medveczky. 1997. "In vitro" antiviral drug sensitivity of the Kaposi's sarcoma-associated herpesvirus. *AIDS* **11:**1327–1332.

Megaw, A. G., D. Rapaport, B. Avidor, N. Frenkel, and A. J. Davison. 1998. The DNA sequence of the RK strain of human herpesvirus 7. *Virology* **244:**119–132.

Melbye, M., P. M. Cook, H. Hjalgrim, K. Begtrup, G. R. Simpson, R. J. Biggar, P. Ebbesen, and T. F. Schulz. 1998. Risk factors for Kaposi's-sarcoma-associated herpesvirus (KSHV/HHV-8) seropositivity in a cohort of homosexual men, 1981–1996. *Int. J. Cancer* **77:**543–548.

Meng, Y.-X., and P. E. Pellett. 1999. Herpesviruses 6, 7, and 8, p. 269–296. *In* R. Ahmed and I. Chen (ed.), *Persistent Viral Infections*. John Wiley and Sons, Ltd., Sussex, United Kingdom.

Meng, Y.-X., T. J. Spira, G. J. Bhat, C. J. Birch, J. D. Druce, B. R. Edlin, R. Edwards, C. Gunthel, R. Newton, F. R. Stamey, C. Wood, and P. E. Pellett. 1999. Individuals from North America, Australasia, and Africa are infected with four different genotypes of human herpesvirus 8. *Virology* **261:**106–119.

Meng, Y.-X., L. L. Lam, K. Kite-Powell, F. R. Stamey, C.-P. Pau, P. E. Pellett, and T. J. Spira. 2000. Human immunodeficiency virus-seropositive individual with persistent human herpesvirus 8 infection for >11 years without development of Kaposi's sarcoma. *Clin. Infect. Dis.* **30:**221–222.

Minjolle, S., C. Michelet, I. Jusselin, M. Joannes, F. Cartier, and R. Colimon. 1999. Amplification of the six major human herpesviruses from cerebrospinal fluid by a single PCR. *J. Clin. Microbiol.* **37:**950–953.

Mocroft, A., M. Youle, B. Gazzard, J. Morcinek, R. Halai, and A. N. Phillips. 1996. Anti-herpesvirus treatment and risk of Kaposi's sarcoma in HIV infection. *AIDS* **10:**1101–1105.

Moore, P. S., S.-J. Gao, G. Dominguez, E. Cesarman, O. Lungu, D. M. Knowles, R. Garber, P. E. Pellett, D. J. McGeoch, and Y. Chang. 1996a. Primary characterization of a herpesvirus agent associated with Kaposi's sarcoma. *J. Virol.* **70:**549–558.

Moore, P. S., L. A. Kingsley, S. D. Holmberg, T. Spira, P. Gupta, D. R. Hoover, J. P. Parry, L. J. Conley, H. W. Jaffe, and Y. Chang. 1996b. Kaposi's sarcoma-associated herpesvirus infection prior to onset of Kaposi's sarcoma. *AIDS* **10:**175–180.

Morris, C., M. Luppi, M. McDonald, P. Barozzi, and G. Torelli. 1999. Fine mapping of an apparently targeted latent human herpesvirus type 6 integration site in chromosome band 17p13.3. *J. Med. Virol.* **58:**69–75.

Moses, A. V., K. N. Fish, R. Ruhl, R. Smith, P. P. Smith, J. G. Strussenberg, L. Zhu, B. Chandran, and J. A. Nelson. 1999. Long-term infection and transformation of dermal microvascular endothelial cells by human herpesvirus 8. *J. Virol.* **73:**6892–6902.

Mukai, T., T. Yamamoto, T. Kondo, K. Kondo, T. Okuno, H. Kosuge, and K. Yamanishi. 1994. Molecular epidemiological studies of human herpesvirus 6 in families. *J. Med. Virol.* **42:**224–227.

Nasti, G., D. Serraino, A. Ridolfo, A. Antinori, G. Rizzardini, C. Zeroli, L. Nigro, M. Tavio, E. Vaccher, and U. Tirelli. 1999. AIDS-associated Kaposi's sarcoma is more aggressive in women: a study of 54 patients. *J. AIDS Hum. Retrovirol.* **20:**337–341.

Neipel, F., and B. Fleckenstein. 1999. The role of HHV-8 in Kaposi's sarcoma. *Semin. Cancer Biol.* **9:**151–164.

Neipel, F., J. C. Albrecht, and B. Fleckenstein. 1997. Cell-homologous genes in the Kaposi's sarcoma-associated rhadinovirus human herpesvirus 8: determinants of its pathogenicity? *J. Virol.* **71:**4187–4192.

Nicholas, J. 1996. Determination and analysis of the complete nucleotide sequence of human herpesvirus 7. *J. Virol.* **70:**5975–5989.

Nii, S., M. Yoshida, F. Uno, T. Kurata, K. Ikuta, and K. Yamanishi. 1990. Replication of human herpesvirus 6 (HHV-6): morphological aspects. *Adv. Exp. Med. Biol.* **278:**19–28.

Okada, K., K. Ueda, K. Kusuhara, C. Miyazaki, K. Tokugawa, M. Hirose, and K. Yamanishi. 1993. Exanthema subitum and human herpesvirus 6 infection: clinical observations in fifty-seven cases. *Pediatr. Infect. Dis. J.* **12:**204–208.

Okuno, T., H. Oishi, K. Hayashi, M. Nonogaki, K. Tanaka, and K. Yamanishi. 1995. Human herpesviruses 6 and 7 in cervixes of pregnant women. *J. Clin. Microbiol.* **33:**1968–1970.

Operskalski, E. A., M. P. Busch, J. W. Mosley, and D. H. Kedes. 1997. Blood donations and viruses. *Lancet* **349:**1327.

Osiowy, C., I. Prud'homme, M. Monette, and S. Zou. 1998. Detection of human herpesvirus 6 DNA in serum by a microplate PCR-hybridization assay. *J. Clin. Microbiol.* **36:**68–72.

Osman, H. K. E., J. S. M. Peiris, C. E. Taylor, P. Warwicker, R. F. Jarrett, and C. R. Madeley. 1996. "Cytomegalovirus disease" in renal allograft recipients: is human herpesvirus 7 a cofactor for disease progression? *J. Med. Virol.* **48:**295–301.

Osman, M., T. Kubo, J. Gill, F. Neipel, M. Becker, G. Smith, R. Weiss, B. Gazzard, C. Boshoff, and F. Gotch. 1999. Identification of human herpesvirus 8-specific cytotoxic T-cell responses. *J. Virol.* **73:**6136–6140.

Parravicini, C., S. J. Olsen, M. Capra, F. Poli, G. Sirchia, S. J. Gao, E. Berti, A. Nocera, E. Rossi, G. Bestetti, M. Pizzuto, M. Galli, M. Moroni, P. S. Moore, and M. Corbellino. 1997. Risk of Kaposi's sarcoma-associated herpes virus transmission from donor allografts among Italian posttransplant Kaposi's sarcoma patients. *Blood* **90:**2826–2829.

Patnaik, M., A. L. Komaroff, E. Conley, E. A. Ojo-Amaize, and J. B. Peter. 1995. Prevalence of IgM antibodies to human herpesvirus 6 early antigen (p41/38) in patients with chronic fatigue syndrome. *J. Infect. Dis.* **172:**1264–1267.

Pau, C. P., L. L. Lam, T. J. Spira, J. B. Black, J. A. Stewart, P. E. Pellett, and R. A. Respess. 1998. Mapping and serodiagnostic application of a dominant epitope within the human herpesvirus 8 ORF 65-encoded protein. *J. Clin. Microbiol.* **36:**1574–1577.

Pellett, P. E., D. Sanchez-Martinez, G. Dominguez, J. B. Black, E. Anton, C. Greenamoyer, and T. R. Dambaugh. 1993. A strongly immunoreactive virion protein of human herpesvirus 6 variant B strain Z29: identification and characterization of the gene and mapping of a variant-specific monoclonal antibody reactive epitope. *Virology* **195:**521–531.

Pellett, P. E., T. J. Spira, O. Bagasra, C. Boshoff, L. Corey, L. de Lellis, M. L. Huang, J. C. Lin, S. Matthews, P. Monini, P. Rimessi, C. Sosa, C. Wood, and J. A. Stewart. 1999. Multicenter comparison of PCR assays for detection of human herpesvirus 8 DNA in semen. *J. Clin. Microbiol.* **37:** 1298–1301.

Portolani, M., C. Cermelli, A. Moroni, M. F. Bertolani, D. Di Luca, E. Cassai, and A. M. Sabbatini. 1993. Human herpesvirus-6 infections in infants admitted to hospital. *J. Med. Virol.* **39:**146–151.

Pruksananonda, P., C. B. Hall, R. A. Insel, K. McIntyre, P. E. Pellett, C. E. Long, K. C. Schnabel, P. H. Pincus, F. R. Stamey, T. R. Dambaugh, and J. A. Stewart. 1992. Primary human herpesvirus 6 infection in young children. *N. Engl. J. Med.* **326:**1445–1450.

Raab, M. S., J. C. Albrecht, A. Birkmann, S. Yaguboglu, D. Lang, B. Fleckenstein, and F. Neipel. 1998. The immunogenic glycoprotein gp35-37 of human herpesvirus 8 is encoded by open reading frame K8.1. *J. Virol.* **72:**6725–6731.

Rainbow, L., G. M. Platt, G. R. Simpson, R. Sarid, S. J. Gao, H. Stoiber, C. S. Herrington, P. S. Moore, and T. F. Schulz. 1997. The 222- to 234-kilodalton latent nuclear protein (LNA) of Kaposi's sarcoma-associated herpesvirus (human herpesvirus 8) is encoded by orf73 and is a component of the latency-associated nuclear antigen. *J. Virol.* **71:**5915–5921.

Renne, R., D. Blackbourn, D. Whitby, J. Levy, and D. Ganem. 1998. Limited transmission of Kaposi's sarcoma-associated herpesvirus in cultured cells. *J. Virol.* **72:**5182–5188.

Renwick, N., T. Halaby, G. J. Weverling, N. Dukers, G. R. Simpson, R. A. Coutinho, J. Lange, T. F. Schulz, and J. Goudsmit. 1998. Seroconversion for human herpesvirus 8 during HIV infection is highly predictive of Kaposi's sarcoma. *AIDS* **12:**2481–2488.

Rettig, M. B., H. J. Ma, R. A. Vescio, M. Pold, G. Schiller, D. Belson, A. Savage, C. Nishikubo, C. Wu, J. Fraser, J. W. Said, and J. R. Berenson. 1997. Kaposi's sarcoma-associated herpesvirus infection of bone marrow dendritic cells from multiple myeloma patients. *Science* **276:**1851–1854.

Robert, C., H. Agut, J. T. Aubin, H. Collandre, D. Ingrand, A. Devillechabrolle, P. LeHoang, and J. M. Huraux. 1990. Detection of antibodies to human herpesvirus-6 using immunofluorescence assay. *Res. Virol.* **141:**545–555.

Roffman, E., and N. Frenkel. 1991. Replication of human herpesvirus-6 in thymocytes activated by anti-CD3 antibody. *J. Infect. Dis.* **164:**617–618.

Roffman, E., J. P. Albert, J. P. Goff, and N. Frenkel. 1990. Putative site for the acquisition of human herpesvirus 6 virion tegument. *J. Virol.* **64:**6308–6313.

Rotola, A., T. Ravaioli, A. Gonelli, S. Dewhurst, E. Cassai, and D. Di Luca. 1998. U94 of human herpesvirus 6 is expressed in latently infected peripheral blood mononuclear cells and blocks viral gene expression in transformed lymphocytes in culture. *Proc. Natl. Acad. Sci. USA* **95:**13911–13916.

Russo, J. J., R. A. Bohenzky, M. C. Chien, J. Chen, M. Yan, D. Maddalena, J. P. Parry, D. Peruzzi, I. S. Edelman, Y. Chang, and P. S. Moore. 1996. Nucleotide sequence of the Kaposi sarcoma-associated herpesvirus (HHV8). *Proc. Natl. Acad. Sci. USA* **93:**14862–14867.

Sadler, R., L. Wu, B. Forghani, R. Renne, W. Zhong, B. Herndier, and D. Ganem. 1999. A complex translational program generates multiple novel proteins from the latently expressed kaposin (K12) locus of Kaposi's sarcoma-associated herpesvirus. *J. Virol.* **73:**5722–5730.

Said, J. W., T. Tasaka, S. de Vos, and H. P. Koeffler. 1997. Kaposi's sarcoma-associated herpesvirus/human herpesvirus type 8 encephalitis in HIV-positive and -negative individuals. *AIDS* **11:**1119–1122.

Salahuddin, S. Z., D. V. Ablashi, P. D. Markham, S. F. Josephs, S. Sturzenegger, M. Kaplan, G. Halligan, P. Biberfeld, F. Wong-Staal, B. Kramarsky, and R. C. Gallo. 1986. Isolation of a new virus, HBLV, in patients with lymphoproliferative disorders. *Science* **234:**596–601.

Sanders, V. J., S. Felisan, A. Waddell, and W. W. Tourtellotte. 1996. Detection of herpesviridae in postmortem multiple sclerosis brain tissue and controls by polymerase chain reaction. *J. Neurovirol.* **2:**249–258.

Schirmer, E. C., L. S. Wyatt, K. Yamanishi, W. J. Rodriguez, and N. Frenkel. 1991. Differentiation between two distinct classes of viruses now classified as human herpesvirus 6. *Proc. Natl. Acad. Sci. USA* **88:**5922–5926.

Schulz, T. F. 1999. Epidemiology of Kaposi's sarcoma-associated herpesvirus/human herpesvirus 8. *Adv. Cancer Res.* **76:**121–160.

Secchiero, P., Z. N. Berneman, R. C. Gallo, and P. Lusso. 1994. Biological and molecular characterisitics of human herpesvirus 7: in vitro growth optimization and development of a syncytia inhibition test. *Virology* **202:**506–512.

Secchiero, P., D. R. Carrigan, Y. Asano, L. Benedetti, R. W. Crowley, A. L. Komaroff, R. C. Gallo, and P. Lusso. 1995a. Detection of human herpesvirus 6 in plasma of children with primary infection and immunosuppressed patients by polymerase chain reaction. *J. Infect. Dis.* **171:**273–280.

Secchiero, P., D. Zella, R. W. Crowley, R. C. Gallo, and P. Lusso. 1995b. Quantitative PCR for human herpesviruses 6 and 7. *J. Clin. Microbiol.* **33:**2124–2130.

Secchiero, P., L. Flamand, D. Gilbellini, E. Falcieri, I. Robuffo, S. Capitani, R. C. Gallo, and G. Zauli. 1997. Human herpesvirus 7 induces CD4+ T-cell death by two distinct mechanisms: necrotic lysis in productively infected cells and apoptosis in uninfected or nonproductively infected cells. *Blood* **90:**4502–4512.

Simpson, G. R., T. F. Schulz, D. Whitby, P. M. Cook, C. Boshoff, L. Rainbow, M. R. Howard, S.-J. Gao, R. A. Bohenzky, P. Simmonds, C. Lee, A. de Ruiter, A. Hatzakis, R. S. Tedder, I. V. D. Weller, R. A. Weiss, and P. S. Moore. 1996. Prevalence of Kaposi's sarcoma associated herpesvirus infection measured by antibodies to recombinant capsid protein and latent immunofluorescence antigen. *Lancet* **348:**1133–1138.

Smith, M. S., C. Bloomer, R. Horvat, E. Goldstein, J. M. Casparian, and B. Chandran. 1997. Detection of human herpesvirus 8 DNA in Kaposi's sarcoma lesions and peripheral blood of human immunodeficiency virus-positive patients and correlation with serologic measurements. *J. Infect. Dis.* **176:**84–93.

Sola, P., E. Merelli, R. Marasca, M. Poggi, M. Luppi, M. Montorsi, and G. Torelli. 1993. Human herpesvirus 6 and multiple sclerosis: survey of anti-HHV-6 antibodies by immunofluorescence analysis and of viral sequences by polymerase chain reaction. *J. Neurol. Neurosurg. Psychiatry* **56:**917–919.

Soldan, S. S., R. Berti, N. Salem, P. Secchiero, L. Flamand, P. A. Calabresi, M. B. Brennan, H. W. Maloni, H. F. McFarland, H. C. Lin, M. Patnaik, and S. Jacobson. 1997. Association of human herpes virus 6 (HHV-6) with multiple sclerosis: increased IgM response to HHV-6 early antigen and detection of serum HHV-6 DNA. *Nat. Med.* **3:**1394–1397.

Spira, T. J., L. Lam, S. C. Dollard, Y.-X. Meng, C. P. Pau, J. B. Black, D. Burns, B. Cooper, M. Hamid, J. Huong, K. Kite-Powell, and P. E. Pellett. 2000. Comparison of serologic assays and PCR for diagnosis of human herpesvirus 8 infection. *J. Clin. Microbiol.* **38:**2174–2180.

Staskus, K. A., W. Zhong, K. Gebhard, B. Herndier, H. Wang, R. Renne, J. Beneke, J. Pudney, D. J. Anderson, D. Ganem, and A. T. Haase. 1997. Kaposi's sarcoma-associated herpesvirus gene expression in endothelial (spindle) tumor cells. *J. Virol.* **71:**715–719.

Staskus, K. A., R. Sun, G. Miller, P. Racz, A. Jaslowski, C. Metroka, H. Brett-Smith, and A. T. Haase. 1999. Cellular tropism and viral interleukin-6 expression distinguish human herpesvirus 8 involvement in Kaposi's sarcoma, primary effusion lymphoma, and multicentric Castleman's disease. *J. Virol.* **73:**4181–4187.

Suga, S., T. Yoshikawa, Y. Asano, T. Yazaki, and S. Hirata. 1989. Human herpesvirus-6 infection (exanthem subitum) without rash. *Pediatrics* **83:**1003–1006.

Suga, S., T. Yoshikawa, Y. Asano, T. Kozawa, T. Nakashima, I. Kobayashi, T. Yazaki, H. Yamamoto, Y. Kajita, T. Ozaki, et al. 1993. Clinical and virological analyses of 21 infants with exanthem subitum (roseola infantum) and central nervous system complications. *Ann. Neurol.* **33:**597–603.

Suga, S., T. Yoshikawa, Y. Kajita, T. Ozaki, and Y. Asano. 1998. Prospective study of persistence and excretion of human herpesvirus-6 in patients with exanthem subitum and their parents. *Pediatrics* **102:**900–904.

Tait, D. R., K. N. Ward, D. W. G. Brown, and E. Miller. 1996. Measles and rubella misdiagnosed in infants as exanthem subitum (roseola infantum). *Br. Med. J.* **312:**101–102.

Takahashi, K., S. Sonoda, K. Higashi, T. Kondo, H. Takahashi, M. Takahashi, and K. Yamanishi. 1989. Predominant CD4 T-lymphocyte tropism of human herpesvirus 6-related virus. *J. Virol.* **63:**3161–3163.

Takahashi, K., M. Suzuki, Y. Iwata, S. Shigeta, K. Yamanishi, and E. DeClercq. 1997. Selective activity of various nucleoside and nucleotide analogues against human herpesvirus 6 and 7. *Antivir. Chem. Chemother.* **8:**24–31.

Tanaka, K., T. Kondo, S. Torigoe, S. Okada, T. Mukai, and K. Yamanishi. 1994. Human herpesvirus 7: another causal agent for roseola (exanthem subitum). *J. Pediatr.* **125:**1–5.

Tanaka-Taya, K., T. Kondo, T. Mukai, H. Miyoshi, Y. Yamamoto, S. Okada, and K. Yamanishi. 1996. Seroepidemiological study of human herpesvirus-6 and -7 in children of different ages and detection of these two viruses in throat swabs by polymerase chain reaction. *J. Med. Virol.* **48:**88–94.

Tang, Y. W., M. J. Espy, D. H. Persing, and T. F. Smith. 1997. Molecular evidence and clinical significance of herpesvirus coinfection in the central nervous system. *J. Clin. Microbiol.* **35:**2869–2872.

Tarte, K., Y. Chang, and B. Klein. 1999. Kaposi's sarcoma-associated herpesvirus and multiple myeloma: lack of criteria for causality. *Blood* **93:**3159–3163.

Thomson, B. J., S. Efstathiou, and R. W. Honess. 1991. Acquisition of the human adeno-associated virus type-2 rep gene by human herpesvirus type-6. *Nature* **351:**78–80.

Torelli, G., P. Barozzi, R. Marasca, P. Cocconcelli, E. Merelli, L. Ceccherini-Nelli, S. Ferrari, and M. Luppi. 1995. Targeted integration of human herpesvirus 6 in the p arm of chromosome 17 of human peripheral blood mononuclear cells in vivo. *J. Med. Virol.* **46:**178–188.

Torigoe, S., T. Kumamoto, W. Koide, K. Taya, and K. Yamanishi. 1995. Clinical manifestations associated with human herpesvirus 7 infection. *Arch. Dis. Child.* **72:**518–519.

Torigoe, S., W. Koide, M. Yamada, E. Miyashiro, K. Tanaka-Taya, and K. Yamanishi. 1996. Human herpesvirus 7 infection associated with central nervous system manifestations. *J. Pediatr.* **129:**301–305.

Torrisi, M. R., M. Gentile, G. Cardinali, M. Cirone, C. Zompetta, L. V. Lotti, L. Frati, and A. Faggioni. 1999. Intracellular transport and maturation pathway of human herpesvirus 6. *Virology* **257:**460–471.

Ueda, K., K. Kusuhara, M. Hirose, K. Okada, C. Miyazaki, K. Tokugawa, M. Nakayama, and K. Yamanishi. 1989. Exanthem subitum and antibody to human herpesvirus-6. *J. Infect. Dis.* **159:**750–752.

Ueda, K., K. Kusuhara, K. Okada, C. Miyazaki, Y. Hidaka, K. Tokugawa, and K. Yamanishi. 1994. Primary human herpesvirus 7 infection and exanthema subitum. *Pediatr. Infect. Dis. J.* **13:**167–168.

van den Berg, J. S., J. H. van Zeijl, J. J. Rotteveel, W. J. Melchers, F. J. Gabreels, and J. M. Galama. 1999. Neuroinvasion by human herpesvirus type 7 in a case of exanthem subitum with severe neurologic manifestations. *Neurology* **52:**1077–1079.

Vieira, J., M. L. Huang, D. M. Koelle, and L. Corey. 1997. Transmissible Kaposi's sarcoma-associated herpesvirus (human herpesvirus 8) in saliva of men with a history of Kaposi's sarcoma. *J. Virol.* **71:**7083–7087.

Wang, F.-Z., H. Dahl, A. Linde, M. Brytting, A. Ehrnst, and P. Ljungman. 1996. Lymphotropic herpesviruses in allogenic bone marrow transplantation. *Blood* **88:**3615–3620.

Ward, K. N., J. J. Gray, M. W. Fotheringham, and M. J. Sheldon. 1993a. IgG antibodies to human herpesvirus-6 in young children: changes in avidity of antibody correlate with time after infection. *J. Med. Virol.* **39:**131–138.

Ward, K. N., J. J. Gray, M. E. Joslin, and M. J. Sheldon. 1993b. Avidity of IgG antibodies to human herpesvirus-6 distinguishes primary from recurrent infection in organ transplant recipients and excludes cross-reactivity with other herpesviruses. *J. Med. Virol.* **39:**44–49.

Whitby, D., M. Luppi, P. Barozzi, C. Boshoff, R. A. Weiss, and G. Torelli. 1998. Human herpesvirus 8 seroprevalence in blood donors and lymphoma patients from different regions of Italy. *J. Natl. Cancer Inst.* **90:**395–397.

Whitby, D., N. A. Smith, S. Matthews, S. O'Shea, C. A. Sabin, R. Kulasegaram, C. Boshoff, R. A. Weiss, A. de Ruiter, and J. M. Best. 1999. Human herpesvirus 8: seroepidemiology among women and detection in the genital tract of seropositive women. *J. Infect. Dis.* **179:**234–236.

Wilborn, F., C. A. Schmidt, V. Brinkmann, K. Jendroska, H. Oettle, and W. Siegert. 1994a. A potential role for human herpesvirus type 6 in nervous system disease. *J. Neuroimmunol.* **49:**213–214.

Wilborn, F., C. A. Schmidt, R. Zimmermann, V. Brinkmann, F. Neipel, and W. Siegert. 1994b. Detection of herpesvirus type 6 by polymerase chain reaction in blood donors: random tests and prospective longitudinal studies. *Br. J. Haematol.* **88:**187–192.

Wilborn, F., C. A. Schmidt, F. Lorenz, R. Peng, H. Gelderblom, D. Huhn, and W. Siegert. 1995. Human herpesvirus type 7 in blood donors: detection by the polymerase chain reaction. *J. Med. Virol.* **47:**65–69.

Wilborn, F., T. Binder, F. Neipel, R. Blasczyk, and W. Siegert. 1998. Human herpesvirus type 6 variants identified by single-strand conformation polymorphism analysis. *J. Virol. Methods* **73:**21–29.

Wyatt, L. S., and N. Frenkel. 1992. Human herpesvirus 7 is a constitutive inhabitant of adult human saliva. *J. Virol.* **66:**3206–3209.

Wyatt, L. S., W. J. Rodriguez, N. Balachandran, and N. Frenkel. 1991. Human herpesvirus 7: antigenic properties and prevalence in children and adults. *J. Virol.* **65:**6260–6265.

Yalcin, S., T. Karpuzoglu, G. Suleymanlar, G. Mutlu, T. Mukai, T. Yamamoto, Y. Isegawa, and K. Yamanishi. 1994. Human herpesvirus 6 and human herpesvirus 7 infections in renal transplant recipients and healthy adults in Turkey. *Arch. Virol.* **136:**183–190.

Yamamoto, T., T. Mukai, K. Kondo, and K. Yamanishi. 1994. Variation of DNA sequence in immediate-early gene of human herpesvirus 6 and variant identification by PCR. *J. Clin. Microbiol.* **32:**473–476.

Yasukawa, M., Y. Yakushijin, M. Furukawa, and S. Fujita. 1993. Specificity analysis of human CD4+ T-cell clones directed against human herpesvirus 6 (HHV-6), HHV-7, and human cytomegalovirus. *J. Virol.* **67:**6259–6264.

Yasukawa, M., Y. Inoue, E. Sada, Y. Yakushijin, M. Furukawa, and S. Fujita. 1995. CD4 down-modulation by ganglioside and phorbol ester inhibits human herpesvirus 7 infection. *J. Gen. Virol.* **76:**2381–2385.

Yoshida, M., F. Uno, Z. L. Bai, M. Yamada, S. Nii, T. Sata, T. Kurata, K. Yamanishi, and M. Takahashi. 1989. Electron microscopic study of a herpes-type virus isolated from an infant with exanthem subitum. *Microbiol. Immunol.* **33:**147–154.

Yoshida, M., M. Yamada, T. Tsukazaki, S. Chatterjee, F. D. Lakeman, S. Nii, and R. J. Whitley. 1998. Comparison of antiviral compounds against human herpesvirus 6 and 7. *Antivir. Res.* **40:**73–74.

Yoshikawa, T., M. Morooka, S. Suga, Y. Niinomi, T. Kaneko, K. Shinoda, Y. Muraki, K. Takahashi, N. Sugaya, and Y. Asano. 1998. Five cases of thrombocytopenia induced by primary human herpesvirus 6 infection. *Acta Paediatr. Jpn.* **40:**278–281.

Yu, Y., J. B. Black, C. S. Goldsmith, P. J. Browning, K. Bhalla, and M. K. Offermann. 1999. Induction of human herpesvirus-8 DNA replication and transcription by butyrate and TPA in BCBL-1 cells. *J. Gen. Virol.* **80:**83–90.

Zhong, W., H. Wang, B. Herndier, and D. Ganem. 1996. Restricted expression of Kaposi sarcoma-associated herpesvirus (human herpesvirus 8) genes in Kaposi sarcoma. *Proc. Natl. Acad. Sci. USA* **93:**6641–6646.

Zhu, L., R. Wang, A. Sweat, E. Goldstein, R. Horvat, and B. Chandran. 1999. Comparison of human sera reactivities in immunoblots with recombinant human herpesvirus (HHV)-8 proteins associated with the latent (ORF73) and lytic (ORFs 65, K8.1A, and K8.1B) replicative cycles and in immunofluorescence assays with HHV-8-infected BCBL-1 cells. *Virology* **256:**381–392.

Zietz, C., J. R. Bogner, F. D. Goebel, and U. Lohrs. 1999. An unusual cluster of cases of Castleman's disease during highly active antiretroviral therapy for AIDS. *N. Engl. J. Med.* **340:**1923–1924.

Zoeteweij, J. P., S. T. Eyes, J. M. Orenstein, T. Kawamura, L. Wu, B. Chandran, B. Forghani, and A. Blauvelt. 1999. Identification and rapid quantification of early- and late-lytic human herpesvirus 8 infection in single cells by flow cytometric analysis: characterization of antiherpesvirus agents. *J. Virol.* **73:**5894–5902.

Poxviruses*

R. MARK L. BULLER

34

Historically, poxvirus infections were diagnosed by a combination of approaches. The ability to distinguish among orthopoxvirus species was important, since variola and monkeypox viruses presented with an identical clinical picture but differed in their disease severity for, and transmissibility between, humans. Since orthopoxvirus species are indistinguishable antigenically by routine serologic methods or structurally by electron microscopy, laboratory diagnosis usually involved propagation of the isolate on the chorioallantoic membranes of 12-day-old chicken embryos. Diagnosis of parapoxvirus and yatapoxvirus infections was accomplished by electron microscopy of lesion material in combination with clinical and epidemiological findings. Finally, the presence of molluscum bodies in a squash preparation of lesion material or a stained wax section of a biopsy specimen was considered pathognomonic for molluscum contagiosum virus (MCV) infection.

BIOLOGY OF HUMAN POXVIRUS PATHOGENS

Classification

The earliest classification of poxviruses was based on criteria which considered disease signs or symptoms and gross pathology. This resulted in the grouping of animal diseases which were characterized by pocks on the skin, including such distinct diseases as smallpox (variola virus), chickenpox (varicella-zoster virus), and syphilis, or large pox (the spirochete *Treponema pallidum*). This group of diseases was culled with the application of a more stringent criterion based on morphological characterization of virions and cytoplasmic inclusion bodies. Poxvirus virions, or elementary bodies, were observed by a succession of early investigators: Chaveau in 1868, Buist in 1886, Calmette and Guérin in 1901, Prowazek in 1905, and Paschen in 1906. The infectious nature of the virion was finally elucidated by Ledingham (1931).

Under the auspices of a Poxvirus Subcommittee created at the Sixth International Congress for Microbiology in 1953, Fenner and Burnet wrote a review which summarized the characteristics of the poxvirus group (Fenner and Bur-

net, 1957). This article remains the basis for the classification of poxviruses. The family *Poxviridae* is divided into two subfamilies: *Entomopoxvirinae* (poxviruses of the insects) and *Chordopoxvirinae* (poxviruses of the vertebrates). The vertebrate poxviruses share a group-specific nucleoprotein (NP) antigen and were further subclassified into genera by comparing cross protection in animal studies and cross neutralization of infectivity in tissue culture and through the analysis of genetic polymorphisms in genomic DNA from virions (Table 1). The latter is now the method of choice for classification of poxvirus isolates. In general, members of the same genera have similar morphologies and biological properties.

Virion Morphology

Poxviruses are the largest of all animal viruses and can be visualized by light microscopy, although the details of the virion structure remain obscure. With the advent of high-resolution electron microscopy, the morphology of the virion structure began to be revealed during the 1940s and 1950s (Dawson and McFarlane, 1948; Peters, 1956). Poxvirus virions appear to be oval or brick-shaped structures about 200 to 400 nm long with axial ratios of 1.2 to 1.7 (Fig. 1A). Two forms of the nonenveloped virion have been identified in clinical specimens: the intact M (mulberry) form is found mainly in the vesicular fluid, and the C (capsule) form, which appears to be a degenerated M form, is associated with dried scabs.

Virion Structure and Composition

The virion contains a noninfectious, linear, double-stranded-DNA genome that can vary from 130 to 300 kbp, depending on the poxvirus species (see Moss, 1996, for a review). The virion has more than 100 polypeptides arranged in three distinct structures (core, lateral body, and membrane envelope [Fig. 1B]) as determined by electron microscopy of virions subjected to thin sectioning, cryosectioning, and/or negative-staining procedures. The membrane envelope that surrounds the virion can be in at least two forms, which differentiate the intracellular and extracellular viruses from each other. The former is doubly wrapped in the intermediate compartment and remains cell associated, whereas the latter receives an additional double membrane on passage through the Golgi apparatus, losing one of the two following fusion with the plasma membrane on exiting the cell (Fig. 2).

---

*This chapter contains information presented by the author in section 8.7 of *Infectious Diseases* (D. Armstrong and J. Cohen [ed.], Mosby, London, England, 1999).

Poxvirus Life Cycle

A Single Cycle of Virus Replication

The replication cycle of poxviruses occurs in the cytoplasm of the cell and varies greatly in duration. It takes between 37 and 75 h from the time of infection to obtain maximum levels of progeny Yaba monkey tumor virus infectivity in a continuous monkey cell line (CV-1). This contrasts with a vaccinia virus replication cycle of 12 to 24 h in another continuous monkey cell line, BS-C1. The majority of information concerning the replication of poxviruses has been obtained by using vaccinia virus infections of tissue culture cell lines, which is depicted in Fig. 2.

Poxvirus Replication and Spread in the Host

Poxviruses infect the host mainly through the cornified epithelium of the skin or the mucosa of the respiratory tract. Infection via the skin is probably through microscopic abrasions, which allow access of the virus to the epidermal or dermal layer. The orthopoxviruses (except for variola and monkeypox viruses), parapoxviruses, yatapoxviruses, and MCV infect their hosts in this manner. Epidemiological evidence strongly suggests that variola virus was transmitted in excretions from the mouth or nose (and not from scab material). The exact area of the respiratory tract that is initially infected, however, remains obscure (Fenner et al., 1988). It is likely that human monkeypox can infect via the oropharynx or nasopharynx, through abrasions of the skin, or even possibly through the oral cavity (Jezek and Fenner, 1988).

Following infection, poxviruses replicate locally in epidermal cells, causing changes in the cellular structure which can be detected with commonly used histochemical stains, such as eosin and hematoxylin. Differential reactivity of these stains reveals the presence of at least three virus-specific staining patterns in infected cells. Areas of basophilic staining are referred to as B-type or Guarnieri inclusion bodies and indicate sites of virus DNA synthesis (Guarnieri, 1892). Additionally ectromelia (mousepox) virus- and cowpox virus-infected epithelial cells contained Marchal (Marchal, 1930) and Downie (Downie, 1939) bodies, respectively, which are acidophilic in character and are now referred to as A-type inclusion bodies. A-type inclusion bodies usually contain virions embedded in a proteinaceous matrix that is composed mainly of a single polypeptide, which in the case of cowpox virus has a mass of ~160 kDa. Although A-type inclusion bodies are not a property of all poxviruses, when they are present they are thought to provide a means of ensuring survival of virus in the environment. This hypothesis is supported by the apparent dependence of insect poxviruses on inclusion structures (called spherules) for dissemination and survival in nature (Bergoin and Dales, 1971). MCV infection has its own unique intracytoplasmic, eosinophilic, granular inclusion called the molluscum body, which increases in size throughout the virus life cycle until the keratinocyte is devoid of any intracellular structure except virions.

Virus that has replicated at the site of infection spreads to the draining lymph node via the lymphatics, and possibly in infected cells. At this point poxvirus pathogenesis follows one of two courses. Some viruses cause a localized, self-limited infection with little spread from the original site of inoculation; this is the case with MCV, Shope fibroma virus, or Yaba monkey tumor virus. Others cause a fulminant, systemic infection characterized by a generalized rash and a high mortality rate, as with variola virus, human monkeypox virus, or ectromelia virus infections of mice. In certain cases, both of these disease patterns can be caused by the same virus but in different host species. Classic examples are infections by myxoma virus or cowpox virus. In humans and cows, cowpox virus causes a localized pustular skin lesion, whereas generalized infections with pustular rashes are observed in many species from the family *Felidae*. In tapeti (wild rabbit), myxoma virus produces localized benign fibromas, whereas the infection of the domestic (European) rabbit results in a generalized infection with a high mortality rate.

In systemic infections (based on the mousepox model [Fenner, 1948]), virus passes out of the draining lymph node via the efferent lymphatics and enters the venous system through the thoracic ducts. This primary viremia results in the seeding of virus into the liver, spleen, and/or other organs, with the target organs and/or cells differing among the poxviruses. For example, ectromelia virus replicates to high levels in the fibroreticular cells but not the lymphocytes of the reticuloendothelial system, whereas myxoma virus replicates well in the lymphocytes (McFadden, 1988; Karupiah et al., 1993). Further virus replication in these internal organs liberates large quantities of virus into the bloodstream, and this secondary viremia results in productive infections of kidneys, lungs, intestines, skin, and/or other organs. Infection of the skin can give rise to the classic rash, which in variola and human monkeypox is a single crop of discrete maculopapular lesions.

Transmission of virus from an index case can originate from the primary site of infection, which must be the case in

TABLE 1 Poxviruses of the vertebrates

Genus	Species
Orthopoxvirus	Camelpox, cowpox, ectromelia, monkeypox, raccoon pox, skunkpox, taterapox, Uasin Gishu, vaccinia,[a] variola, and vole pox viruses
Parapoxvirus	Auzduk, chamois contagious ecthyma, orf,[a] parapox of seal, parapox of deer, pseudocowpox, and bovine papular stomatitis viruses
Avipoxvirus	Canarypox, fowlpox,[a] juncopox, mynahpox, pigeonpox, psittacinepox, quailpox, sparrowpox, starlingpox, and turkeypox viruses
Capripoxvirus	Goatpox, lumpy skin disease, and sheep pox[a] viruses
Leporipoxvirus	Hare fibroma, myxoma,[a] rabbit (Shope) fibroma, and squirrel fibroma viruses
Suipoxvirus	Swinepox virus[a]
Molluscipoxvirus	MCV[a]
Yatapoxvirus	Yaba monkey tumor virus and tanapox[a] virus

[a]Prototypical member.

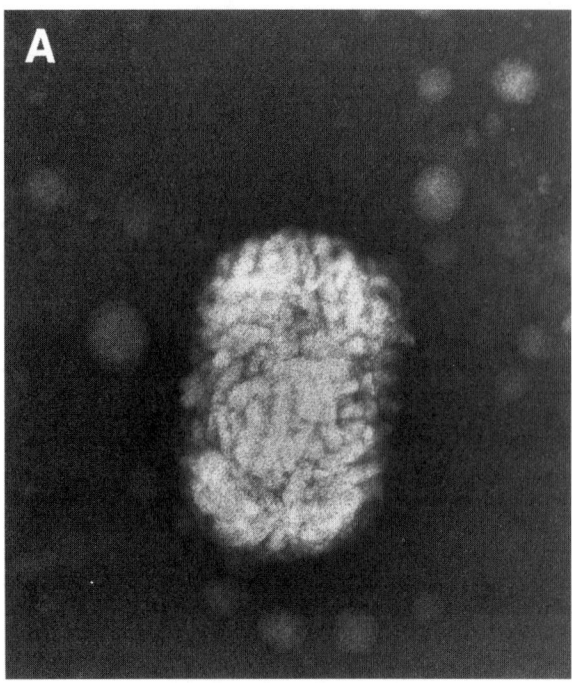

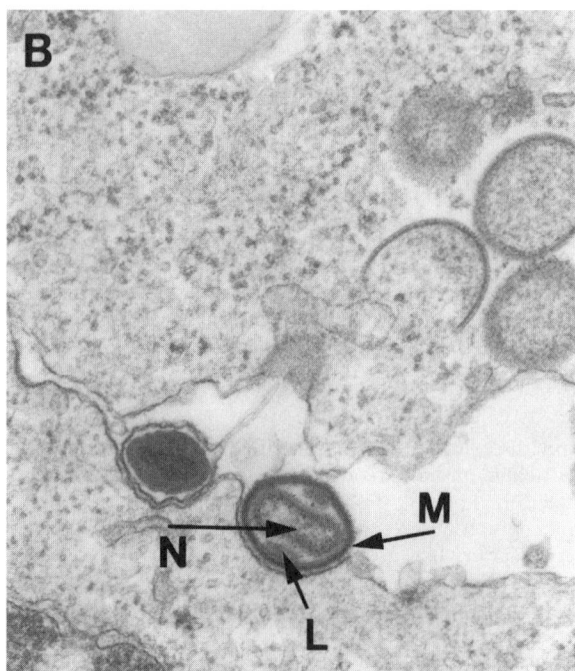

FIGURE 1 Morphology and structure of a poxvirus virion. (A) Electron micrograph of a negatively stained M form of an MCV virion. Note the textured surface. Magnification, ×120,000. (B) Electron micrograph of a thin section of a cowpox virus virion. N, nucleosome; L, lateral body; M, membrane. Note the immature forms of the virus in various stages of morphogenesis in the upper portion of the photograph. Magnification, ×120,000.

localized self-limited infections such as molluscum contagiosum, or from virus produced in the epithelia as a result of the secondary viremia. In the case of smallpox, variola virus was transmitted from lesions in the mouth, nose, and pharynx into the nasal and oropharyngeal secretions during the first week of the rash.

Recovery from poxvirus infections involves both nonspecific mechanisms (interferon and possibly complement and natural killer cells) and learned responses (delayed-type hypersensitivity, cytotoxic T lymphocytes, and specific antibody alone and in combination with natural killer cells [antibody-dependent cell cytotoxicity]) (Buller and Palumbo, 1991). Of these host responses, cell-mediated T-cell responses are thought to be the most important, although the complete repertoire of the host defense armamentarium would likely be required for recovery from infection with highly virulent poxviruses, such as variola. On recovery from infection, poxvirus immunity from disease can be life-long, as in the case of variola virus, or transient, as with MCV.

POXVIRUSES PATHOGENIC FOR HUMANS

Ten poxviruses have been documented to infect humans (Table 2). Except for the "extinct" variola virus and MCV, the poxvirus diseases are zoonoses. With rare exceptions, these zoonotic poxviruses fail to establish a human chain of transmission. Most human poxvirus infections occur through minor abrasions in the skin. The parapoxvirus orf virus and MCV cause the most frequent poxvirus infections worldwide, with the incidence of molluscum contagiosum on the rise, especially as an opportunistic infection of late-stage AIDS patients. Indeed, as populations of immunosuppressed patients increase, the frequency of all human poxvirus-induced disease may rise. Individuals with atopic

dermatitis may be predisposed to poxvirus infections, such as molluscum contagiosum, orf, or cowpox.

MCV

Pathogenesis

Transmission

The incubation period of molluscum contagiosum ranges from weeks to months. It is observed in children and adults, with spread within the latter group governed in part by sexual practices. Nonsexual transmission is a consequence of infection by direct contact or through fomites. Case histories have suggested transmission from surgeons' fingers, swimming pools, bath towels in gymnasiums, contact between wrestlers, and tattooing (Postlethwaite, 1970). Transmission between persons in the absence of fomites requires fairly close contact. Lesions can be commonly observed on opposing surfaces, and the virus can be further spread on the person by autoinoculation.

Lesion Histopathology

Epidermis. MCV has one of the narrowest cell tropisms of any virus, replicating only in the keratinocyte of the human epidermis (Buller et al., 1995). MCV replicates during the 9 to 15 days it takes the infected keratinocyte to reach the stratum granulosum, with effects of the virus life cycle detectable initially in the lower layers of the stratum spinosum and becoming more prominent with progression of virus replication. As the virus-infected cell approaches the skin surface, the accumulation of progeny virions in a granular matrix in the cytoplasm forces the cell organelles, including the nucleus, to the periphery of the cell, further contributing to an overall increase in the size of the cell.

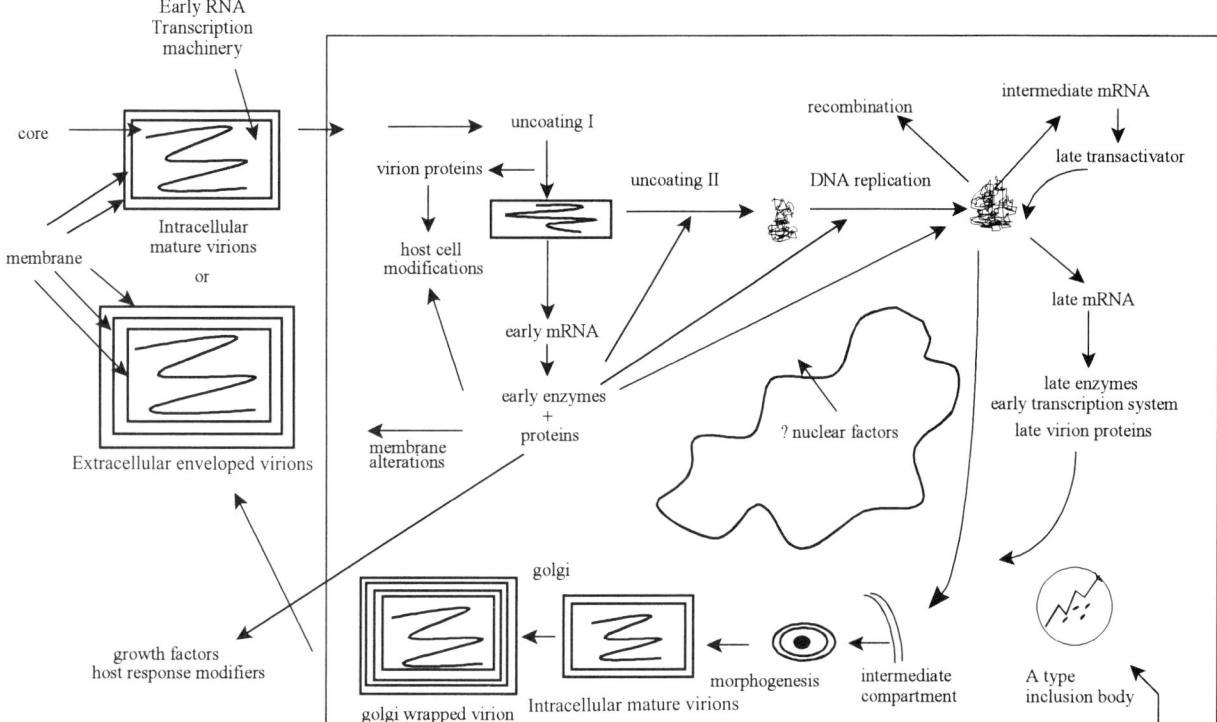

FIGURE 2 The cytoplasmic poxvirus replication cycle. Poxvirus virions containing early RNA transcription machinery attach to and fuse with the plasma membrane, releasing the cores into the cytoplasm (uncoating I). Early genes are expressed that code for a variety of functions which modify the host cell for optimal virus replication, attenuate the host responses to infection, and mediate virus synthetic processes. Following further uncoating (II), the virus genome is replicated via concatemers, late transcription factors are expressed from intermediate genes, and late-gene RNA is synthesized. Late genes encode the early transcription system, enzymes, and structural proteins necessary for virion assembly, which commences with the formation of membrane structures in the intermediate compartment and the packaging of resolved unit-length genomic DNA. The intracellular mature virion has two membranes derived from the intermediate compartment and may remain in the cytoplasm or become occluded in the A-type inclusion body or become wrapped by a further two membranes in the Golgi apparatus and be exported from the cell with the loss of one membrane (extracellular enveloped virions). The extracellular enveloped virions may remain attached to the exterior cell surface (cell-associated enveloped virions) or can be released into the extracellular milieu. The extracellular virus is important for virus spread and pathogenesis in animals. Nuclear factors may also be involved in the cytoplasmic poxvirus replication cycle. This replication scheme is based on the study of the prototypic poxvirus vaccinia virus (Moss, 1996). Other poxvirus species differ from this model mainly in the types of growth factors and host response modifiers encoded by the virus and the amounts of extracellular enveloped virus produced. Adapted from Moss, 1996.

Using light microscopy, stained cells are seen as hyaline acidophilic masses; these are referred to as molluscum or Henderson-Patterson bodies and are pathognomonic for disease (Fig. 3). Higher magnification of the molluscum body reveals a cell entirely filled with virions (Fig. 4). As a consequence of the hypertrophy of the infected cells and hyperplasia of the basal cells, the molluscum lesion extends above the adjacent skin as a tumor and projects into the dermis without breaking the basement membrane.

Dermis. In lesions in which the basement membrane is intact, the dermis is usually uninvolved.

Epidemiology

Geographical Range
MCV has a worldwide distribution but is more prevalent in the tropics. Restriction endonuclease analysis of genomic DNA from MCV isolates has revealed the existence of at least four virus subtypes. There is one recent study that suggests the distribution of subtypes can vary geographically (Nakamura et al., 1995).

Prevalence and Incidence
Nonsexually transmitted molluscum contagiosum is more prevalent in the tropics than in Europe. For example, molluscum contagiosum was diagnosed in 1.2% of outpatients in Aberdeen between 1956 and 1963, the mean age of infection was between 10 and 12 years, and spread within households and schools was infrequent. On the other hand, in Fiji in 1966, 4.5% of the population of a village had the disease, mean age of infection was between 2 and 3 years, and 25% of households harbored more than one case (Postlethwaite, 1970; Porter et al., 1992). In England between 1971 and 1985 there was a 400% increase in cases of genital molluscum contagiosum, with the majority of the cases in the 15- to 24-year-old age group and with affected women being younger than affected men (Porter et al., 1992). In the United States

TABLE 2 Poxviruses pathogenic for humans[a]

Genus	Species[b]	Reservoir	Geographic distribution	Disease
Molluscipoxvirus	MCV	Humans[c]	Worldwide	Single or multiple skin nodules
Parapoxvirus	Orf virus (contagious pustular dermatitis virus or contagious ecthyma virus)	Sheep,[c] goats	Worldwide	Localized skin lesions
	Bovine papular stomatitis virus	Cattle[c]	Worldwide	Localized skin lesions
	Pseudocowpox virus (milker's nodule virus or paravaccinia virus)	Cattle[c]	Worldwide	Localized skin lesions
Orthopoxvirus	Cowpox virus	Rodents,[c] cats, cows, zoo animals	Europe, western Asia	Localized pustular lesions
	Monkeypox virus	Squirrels,[c] monkeys	Western and central Africa	Generalized disease
	Vaccinia virus	Natural host unknown (buffaloes[d])	India, laboratory, vaccine	Localized pustular lesions
	Variola major, variola minor viruses (alastrim, amaas, or kaffir virus)	Eradicated from humans in 1977	Worldwide	Generalized disease
Yatapoxvirus	Tanapox virus	Rodents?	Eastern and central Africa	Localized nodular skin lesions
	Yaba monkey tumor virus	Monkeys?	Western Africa	Localized nodular skin lesions

[a]Adapted from Fenner, 1996.
[b]Alternative names are in parentheses.
[c]Natural host.
[d]During the smallpox eradication program, water buffalo in India were infected with the local vaccine strain of vaccinia virus.

between 1966 and 1983 there was a 10-fold increase in cases in patients aged 25 to 29 years (Porter et al., 1992). Molluscum contagiosum is a common and sometimes severely disfiguring opportunistic infection of between 5 and 18% of HIV-infected patients, especially those with severely depressed CD4[+]-T-cell numbers (Schwartz and Myskowski, 1992).

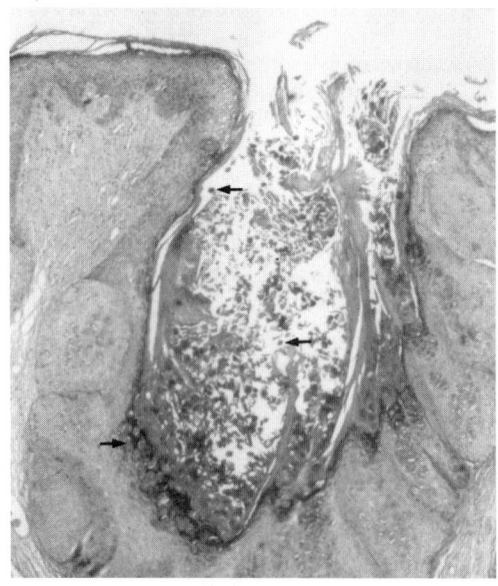

FIGURE 3 Histologic section of a molluscum contagiosum lesion. A hematoxylin- and eosin-stained wax section of a skin biopsy specimen with hyperkeratosis and acanthosis of the epidermis is shown. Note that the hyperplasia associated with the lesion causes severe invagination of the epidermis without loss of integrity of the basal layer. The arrows indicate molluscum bodies. Magnification, ×100.

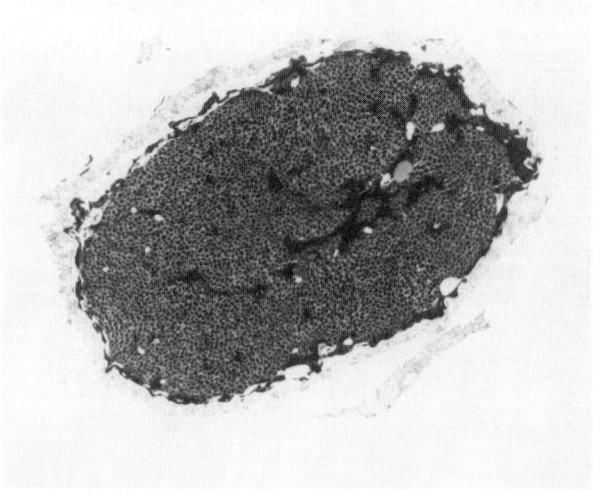

FIGURE 4 Electron micrograph of a thin section of an MCV-infected cell, or molluscum body. All of the cellular organelles are unrecognizable, having been pushed to the periphery of the cell by the masses of virions. Magnification, ×3,000.

Diagnosis

Clinical Signs, Symptoms, and Severity

Molluscum contagiosum was first recognized in 1817. It presents as small clusters of lesions in immunocompetent individuals. There are no signs or symptoms associated with the disease. The lesions are generally painless, appearing on the trunk and limbs (except the palms and soles) in the nonsexually transmitted disease. In children, it can also be fairly common in the skin of the eyelids, with solitary or multiple lesions, and can be complicated by chronic follicular conjunctivitis and later by a superficial punctate keratitis (Al-Hazzaas and Hidayat, 1993). There may be an associated erythema 1 to 11 months after the appearance of the lesion with no correlation to a history of allergy or eczema (DeOreo et al., 1956). As a sexually transmitted disease in teenagers and adults, the lesions are mostly on the lower abdominal wall, pubis, inner thighs, and genitalia. Lesions can persist for as little as 2 weeks or as long as 2 years. Reinfections can be common. As yet there is no solid correlation of virion DNA type with specific pathology or location (i.e., genital versus nongenital) of the lesions (Thompson et al., 1990).

In immunocompromised individuals (especially the HIV-infected population), the infection is not self-limiting and more and larger lesions are observed, especially on the face, neck, scalp, and upper body, with multiple adjacent lesions sometimes becoming confluent. Molluscum contagiosum can be considered a cutaneous marker of severe immunodeficiency.

Gross Lesion Pathology

In immunocompetent patients, the MCV lesions begin as pimples and become umbilicated, epidermal, flesh-colored, raised nodules 2 to 5 mm in diameter. A semisolid caseous material can be expressed from the center of the lesion; it is rich in molluscum bodies and is probably responsible for disease transmission. Rarely, the disease will present as a large lesion (>5 mm in diameter) called a giant molluscum (Fig. 5). Giant molluscum lesions have been reported more fre-

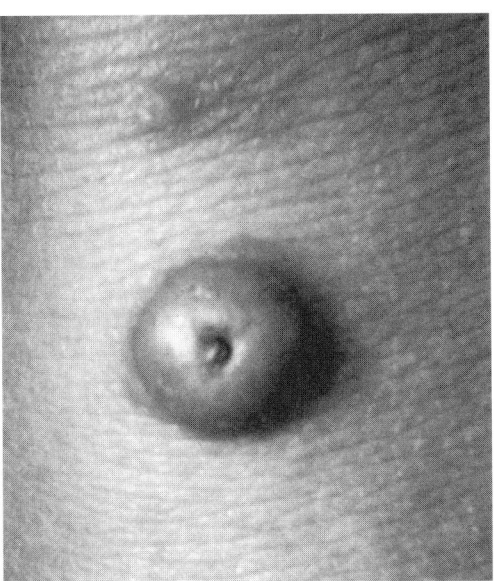

FIGURE 5 Molluscum contagiosum lesions. A giant molluscum lesion is shown next to a more typical lesion. Note the umbilicated center. Courtesy of J. Burnett.

quently in severely immunodeficient HIV-infected individuals (Schwartz and Myskowski, 1992; Izu et al., 1994), although they were observed in New Guinea before the introduction of HIV and therefore are not solely a consequence of immunodeficiency.

Diagnosis and Differential Diagnosis

The diagnosis is usually made clinically, based on the gross appearance of the lesions and their chronic nature. Confirmation is easily obtained by hematoxylin and eosin staining of a biopsy specimen or by a squash preparation of material expressed from the lesion (Fig. 3).

Molluscum contagiosum (especially giant molluscum) can be confused with a number of other disorders, such as keratoacanthoma, warty dyskeratoma, syringomas, hidrocystomas, basal cell epithelioma, trichoepithelioma, ectopic sebaceous glands, giant condylomata acuminata, chalazion, sebaceous cysts, verrucae, milia, lid abscess, or granuloma on the eyelids (Janniger and Schwartz, 1993; Itin and Gilli, 1994; O'Neil, 1995). In immunodeficient patients, disseminated cutaneous cryptococcosis and histoplasmosis can resemble molluscum contagiosum (Janniger and Schwartz, 1993). An inflamed molluscum lesion without the association of typical lesions can be mistaken for a bacterial infection.

Parapoxviruses

The parapoxviruses orf virus, bovine stomatitis virus, and pseudocowpox virus generally cause occupational infections of humans, with orf infections being the most common. Because of the clinical similarity of the diseases caused by these agents, they have been referred to collectively as farmyard pox diseases. Wildlife (skinning animals such as deer and reindeer) have also been sources of orf virus infection. The majority of human parapoxvirus infections probably go unreported, as many sheep farmers and rural physicians are aware of the diseases and make a diagnosis based solely on clinical findings. No human-to-human transmission of parapoxvirus infections has been reported.

Pathogenesis

Transmission

Direct transmission of orf virus has been observed as a consequence of bottle feeding lambs and from animal bites to the hand and contact with sheep and goat products during slaughter. Fomites including splinters, barbed wire, or farmyard surfaces, such soil, feeding troughs, or barn beams, have been implicated as sources of virus.

Pseudocowpox virus from lesions on teats of cows is a major source of virus for milkers' nodule of the hand.

Bovine papular stomatitis virus infection of humans generally occurs from lesions confined to the mouth, tongue, lips, or nares and occasionally from the teats of infected cattle.

Lesion Histopathology

The histopathological features of human orf and pseudocowpox virus lesions are indistinguishable and are similar to the human lesions caused by bovine papular stomatitis virus. For this reason, and since orf infections are the most common human parapoxvirus infections, only the histopathological features of human orf will be presented, as observed by Groves et al. (1991).

Epidermis. The most striking change in the epidermis is hyperplasia, in which strands of epidermal keratinocytes penetrate into the dermis. Generally a mild-to-moderate degree of acanthosis is detected, and parakeratosis is a common feature. Cytoplasmic vacuolation, nuclear vacuolation, and deeply eosinophilic, homogeneous cytoplasmic inclusion bodies often surrounded by a pale halo are also characteristic of the infection. An intense infiltrate of lymphocytes, polymorphs, or eosinophils frequently involves the epidermis.

Dermis. A dense, predominantly lymphohistiocytic inflammatory cell infiltrate is present in all cases. Also, there is marked edema both vertically and horizontally that may contribute to the overall papillomatous appearance. The most striking feature of the dermis is the massive capillary proliferation and dilation. The former is thought to be the result of the action of a virus-encoded growth factor which has homology with mammalian vascular endothelial growth factor (Lyttle et al., 1994).

Epidemiology

Geographical Range

Orf in sheep and goat populations has been reported in Canada, the United States, Europe, Japan, New Zealand, and Africa. Pseudocowpox virus is maintained in European-derived dairy herds in all parts of the world. Bovine papular stomatitis virus is similarly distributed but is found more often associated with beef cattle rather than dairy animals.

Prevalence and Incidence

In a 1-year New Zealand study, 500 meat workers out of a population of 20,000 at risk were infected with orf virus, with the highest risk (4%) of infection for those involved in the initial stunning, killing, and hanging of the sheep (Robinson and Petersen, 1983). Serologic surveys of orf virus-infected sheep and goat herds yielded orf antibody prevalences of up to 90%. The high seroprevalence of orf antibody in herds is likely due to the highly stable nature of the orf virion which contaminates the pasture.

In one study, pseudocowpox virus has been found to be endemic in small herds tested in West Dorset, England (Nagington et al., 1965). Pseudocowpox and bovine papular stomatitis viruses are probably endemic in all European-derived dairy herds.

Diagnosis

Clinical Signs, Symptoms, and Severity

The clinical presentation of orf is usually 3 to 4 weeks postinfection. The disease involves the appearance of single or multiple nodules (diameters of 6 to 27 mm [Groves et al., 1991]) which are sometimes painful, usually on the hands and less frequently on the head or neck. Orf virus infection can also be associated with a low-grade fever, swelling of the lymph nodes, and/or erythema multiforme bullosum. Resolution of the disease occurs over a period of 4 to 6 weeks, usually without complication; however, autoinoculation of the eye may lead to serious sequelae, and enlarged lesions can arise in humans suffering from immunosuppressive conditions, burns, or atopic dermatitis (Robinson and Lyttle, 1992). Lesion healing can be complicated by bullous pemphigoid (Murphy and Ralfs, 1996), and reinfections have been documented (Becker, 1940).

Pseudocowpox virus lesions usually appear on the hands and are relatively painless but may itch. The draining lymph node may be enlarged. The nodules are gradually absorbed and disappear in 4 to 6 weeks (Carson and Kerr, 1967).

In bovine papular stomatitis, lesions occur on the hands, diminish after 14 days, and are no longer evident 3 to 4 weeks after onset (Carson and Kerr, 1967).

Gross Lesion Pathology

The orf lesion characteristically goes through a maculopapular target stage in which a red center is surrounded by a white ring of cells which is surrounded by a red halo of inflammation as shown in Fig. 6; however, patients usually present later, when the lesion is at the granulomatous or papillomatous stages 3 to 4 weeks following the initial infection. It takes the lesion at least 4 to 7 weeks to heal (Robinson and Lyttle, 1992).

In pseudocowpox virus infection, milkers' nodules are first observed as round cherry-red papules; these develop into purple, smooth nodules up to 2 cm in diameter and may be umbilicated. The lesions rarely ulcerate (Becker, 1940).

The lesions of bovine papular stomatitis appear as circumscribed wartlike nodules that gradually enlarge until they are 3 to 8 mm in diameter (Carson and Kerr, 1967).

Diagnosis and Differential Diagnosis

Diagnosis of parapoxvirus infection is done by clinical (lesion morphology) and epidemiological (recent contact with cattle or sheep) evidence and electron microscopy of negatively stained lesion material (presence of ovoid particles [Groves et al., 1991]).

Without knowing the animal source of the infection, orf cannot be differentiated from milkers' nodule based on clinical findings, histology, or electron microscopy (i.e., disease acquired from sheep is orf and disease from cattle is milkers' nodule or possibly bovine papular stomatitis [Groves et al., 1991]).

Atypical giant orf lesions in patients who are immunocompromised or suffering from burns or atopic dermatitis may be confused with pyogenic granuloma (Robinson and Lyttle, 1992; Tan et al., 1991).

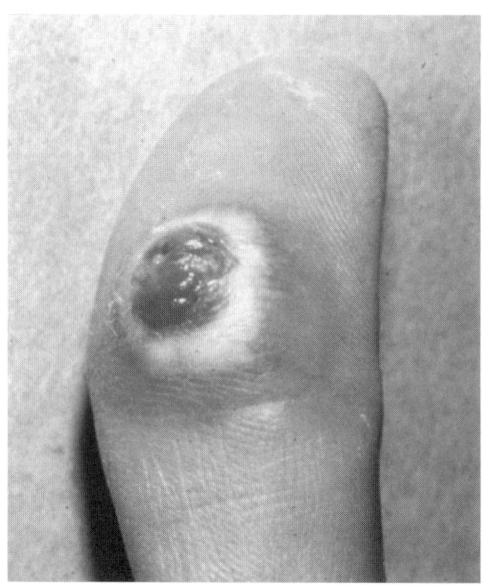

FIGURE 6 A typical orf lesion at the target stage of development. Courtesy of Andrew Mercer.

Orthopoxviruses

Four orthopoxviruses have been shown to cause disease in humans. The most notorious of the diseases was smallpox, which is believed to have originated around 6,000 B.C. in Mesopotamia or Egypt (Fenner et al., 1988). By 1967, two clinically distinct types of smallpox, caused by variola major and variola minor viruses, were endemic in 33 countries. Variola major virus circulated in Asia and caused the most severe disease, with case fatality rates between 5 and 40%. Variola minor virus was thought to have evolved from variola major in the 19th century and was associated with case fatality rates between 0.1 and 2%. With the global eradication of smallpox in 1979, variola virus no longer circulates in human populations (Fenner et al., 1988). At this time, all known stocks of variola virus are held in two World Health Organization (WHO) collaborating centers, the Centers for Disease Control and Prevention (CDC) in Atlanta, Ga., and the State Center of Virology and Biotechnology (VECTOR) in Kotsovo, Russia. Although the WHO Committee on Orthopoxvirus Infections has recommended the destruction of the remaining stocks of variola virus, this course of action has been delayed pending the assessment of future scientific needs for live variola virus.

The eradication of smallpox was made possible by the availability of a live vaccine and at least three characteristics of variola virus: one stable serotype, virus transmission after a long incubation period and prodrome, and no animal reservoir of the virus. The live vaccine was another orthopoxvirus, vaccinia virus, which did not provide sterilizing immunity but did protect vaccinees from severe disease following a subsequent infection with variola virus. Vaccinia virus, however, was not completely avirulent and caused a large number of vaccine-related complications (accidental infection, generalized vaccinia, eczema vaccinatum, erythema multiforme, progressive vaccinia, and postvaccinal encephalitis, with 1,253 cases/million primary vaccinations [Buller and Palumbo, 1991]). With the eradication of variola virus from human populations, the need for continued immunization was diminished, and by March 1983, WHO reported that >96% of the 160 member states had officially discontinued immunization. Presently, vaccinia virus infection of humans is limited to the controlled immunization of personnel that handle the virus in the laboratory and the military. A vaccinia virus-like virus called buffalopox also circulates in water buffalo in India.

Buffalopox was documented in the 1960s and 1970s in herds of milk buffalos and dairy cattle in Asia and Africa (Kolhapure, 1997). The disease is transmitted to humans by direct contact with infected buffalo; lesions are confined to the fingers, hands, and face. No generalized human infection has been documented. Limited person-to-person transmission may occur. Biological data and limited DNA analyses of isolates from an outbreak in India in 1985 suggest that buffalopox virus may be derived from vaccinia virus vaccine strains that were transmitted to livestock during the smallpox immunization era (Pathak et al., 1974).

Presently, only monkeypox and cowpox viruses cause significant human infections. Cowpox virus is not the virus from which vaccinia virus was derived, as commonly stated; rather it is a distinct species of orthopoxvirus.

Pathogenesis

Transmission

The exact mode of transmission of monkeypox virus from an animal source to humans is not known but may be via the oropharynx or nasopharynx or through abrasions of the skin or oral cavity. Person-to-person transmission (like that of the eradicated smallpox) is believed to be via the upper respiratory tract, with virus released in oropharyngeal secretions of patients who have a rash (Jezek and Fenner, 1988). Unlike smallpox, monkeypox person-to-person transmission is very inefficient and rarely exceeds 3 generations (Jezek and Fenner, 1988).

Cowpox is usually acquired by direct introduction of the virus from an animal source into minor abrasions in the skin; however, 30% of infections show no known risk factor (Baxby et al., 1994).

Lesion Histopathology

Orthopoxvirus lesions are characterized by epidermal hyperplasia, with infected cells becoming swollen and vacuolated and undergoing "ballooning degeneration." The cells contain the irregular, faint B-type inclusion body found in all poxvirus-infected cells. Cowpox virus skin lesions from animals contained A-type inclusion bodies in epidermal cells, sebaceous glands, and endothelial cells; however, they were not observed in the few human cowpox lesions examined (Baxby et al., 1994).

Epidemiology

Geographical Range

Monkeypox virus is found in the tropical rain forests of countries in western and central Africa, most notably in the Democratic Republic of the Congo (formerly Zaire). The reservoir of monkeypox virus in nature is most likely the African arboreal squirrels (*Funisciurus* and *Heliosciurus* spp.) and perhaps monkeys (Richardson and Dumbell, 1994).

Cowpox virus is endemic in Europe and some western states of the former Soviet Union. Rodents (voles, wood mice, and rats) have been implicated as reservoirs of cowpox virus, with cows, zoo animals, and cats as incidental hosts (Baxby et al., 1994).

Prevalence and Incidence

A WHO intensive surveillance for human monkeypox between 1981 and 1986 in Zaire confirmed 65 cases annually, with the greatest risk of infection to inhabitants of small villages within 100 m of tropical rain forests (Jezek and Fenner, 1988). Recently, monkeypox has reemerged on a greater scale than previously seen. Between February 1996 and October 1997, 250 cases of human monkeypox were verified in a population of 500,000. Approximately 75% of the cases appeared to be due to human-to-human transmission, although the secondary attack rate was ~8% among unvaccinated contacts, which is similar to the historical value (World Health Organization, 1996a, 1996b; Centers for Disease Control and Prevention, 1997a, 1997b; Mukinda, 1997). The reemergence of monkeypox may be due to waning immunity following cessation of the smallpox immunization program or increased encroachment of larger populations into the primary habitat of the animal reservoir of the virus.

The domestic cat is responsible for the majority of human cowpox, although human infections have been acquired from cows and rodents. Between 1969 and 1993 there were approximately 45 human cowpox cases in Britain, three published case histories from Germany, and two each from Belgium, Sweden, and France (Baxby et al., 1994).

Diagnosis

Clinical Signs, Symptoms, and Severity

Approximately 12 days after infection with monkeypox virus, fever and headache occur. This is followed 1 to 3 days later by a rash and generalized lymphadenopathy. The rash (the number of lesions is variable) appears first on the face and generally has a centrifugal distribution (Fig. 7). The illness lasts 2 to 4 weeks, depending on its severity. The case fatality rate is approximately 12% (Jezek and Fenner, 1988).

With cowpox virus infection, a lesion, usually solitary, appears on the hands or face; this can be extremely painful, and the patient can present with systemic symptoms, including pyrexia, malaise, lethargy, sore throat, and local lymphadenopathy. Complete recovery takes between 3 and 8 weeks. Person-to-person transmission has not been reported. Complications can include ocular or generalized infections; the latter occurs in patients with atopic dermatitis, allergic asthma, or atopic eczema and in one case was associated with death (Baxby et al., 1994).

Gross Lesion Pathology

The monkeypox virus skin lesions begin as macules which rapidly progress to pustules. At about 8 or 9 days after the onset of the rash, the pustules become umbilicated and dry up; by 14 to 16 days after the onset of the rash, a crust has formed. Most skin lesions are about 0.5 cm in diameter (Jezek and Fenner, 1988).

The cowpox lesion appears as an inflamed macule and progresses through an increasingly hemorrhagic vesicle stage to a pustule that ulcerates and crusts over by the end of the second week, becoming a deep-seated, hard black eschar 1 to 3 cm in diameter (Fig. 8) (Baxby et al., 1994).

Diagnosis and Differential Diagnosis

The diagnosis of monkeypox requires clinical (rash), epidemiological (equatorial Africa), and laboratory (brick-shaped virions in scab material) findings. Although the rash with associated lymphadenopathy is usually pathognomonic, the sporadic nature of the disease contributes to failure to arrive at an accurate diagnosis solely on clinical grounds (Jezek and Fenner, 1988).

Cowpox diagnosis is rarely made based on clinical findings (lesion morphology and systemic illness) and usually requires laboratory results (brick-shaped virions in scab material). Cowpox should be considered in patients who have had contact with cats and who present between July and October with a painful hemorrhagic vesicle or black eschar, with or without erythema, accompanied by lymphadenopathy and a systemic illness (Baxby et al., 1994).

Monkeypox can be confused with a number of other conditions that result in a rash:

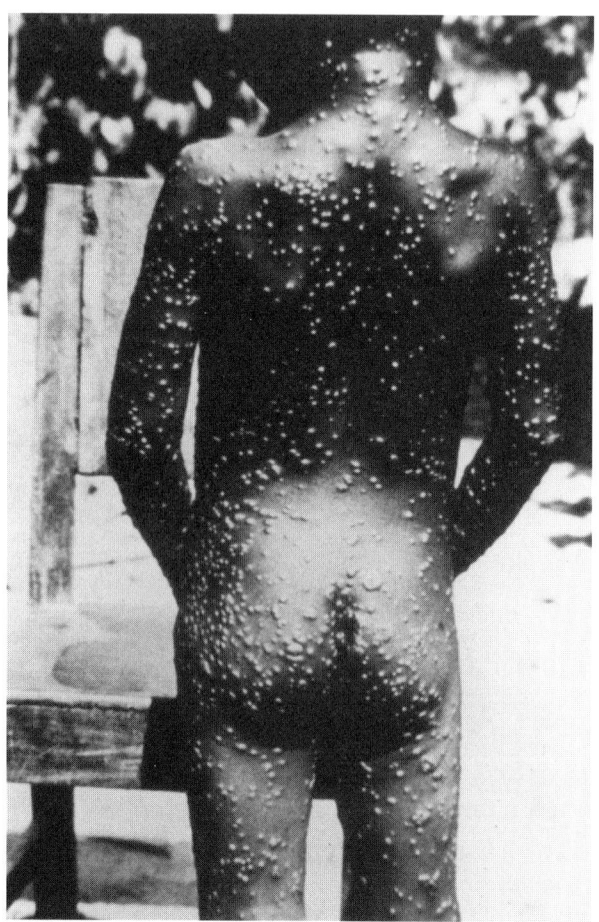

FIGURE 7 Monkeypox rash. A 7-year-old Zairian girl is shown 2 days after the onset of the rash. Courtesy of M. Szczeniowski.

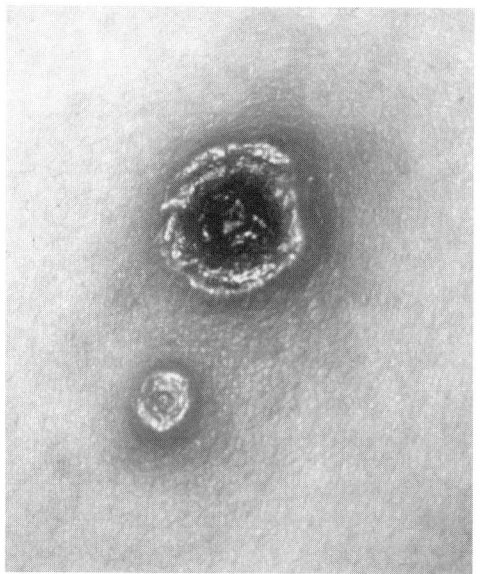

FIGURE 8 Primary and secondary lesions of cowpox. The primary lesion is at the early eschar stage (probably 2 to 3 weeks after infection), whereas the secondary lesion (below) is at the early vesicular stage. Provided by M. White with permission from Baxby et al., 1994.

- chickenpox, although its varicella-zoster lesions are more superficial, appear in crops, and have a centripetal distribution;
- tanapox, except that the tanapox lesions evolve slowly and are nodular, large, and without pustulation; and
- syphilis, although the secondary rash of syphilis does not evolve past the papular stage (Jezek and Fenner, 1988).

Generalized cowpox can be misdiagnosed as eczema herpeticum, whereas localized cowpox is most frequently misdiagnosed as one of the following:

- orf or milkers' nodules, although the parapoxvirus lesion is clinically distinct and is usually not painful and there are often no systemic-disease symptoms;
- herpesvirus reactivation, even though herpes lesions are not usually hemorrhagic or erythematous and the scab is not so deep-seated and is of lighter color;
- anthrax, although the anthrax lesion is painless and rapidly progresses to the eschar stage (5 to 6 days) (Baxby et al., 1994).

Yatapoxviruses

The genus *Yatapoxvirus* has three members, tanapoxvirus, Yaba-like disease virus, and Yaba monkey tumor virus, which are serologically related (Rouhandeh, 1988). Yaba-like disease virus and tanapox virus, but not Yaba monkey tumor virus, are closely related, as indicated by DNA restriction endonuclease analysis of genomic DNA (Ropp et al., 1999). Yaba-like disease virus and Yaba monkey tumor virus have been isolated only from animal handlers, whereas tanapox virus has been found in nature.

Pathogenesis

Transmission
Tanapox virus infection may occur via scratches or possibly via arthropod vectors. Yaba monkey tumor virus is a very rare infection of animal handlers at nonhuman primate facilities. There is no evidence for human-to-human transmission with either virus, and autoinoculation of virus to other areas of the body is not common.

Lesion Histopathology
Little is known about the pathology of yatapoxviruses except from the study of Yaba monkey tumor virus in nonhuman primate models (Niven et al., 1961).

Epidemiology

Geographical Range
Tanapox is endemic in equatorial Africa. The animal reservoir is not known.

Yaba monkey tumor virus appeared in primate colonies but has yet to be seen in nature.

Prevalence and Incidence
In the town of Lisa (population, 70,000) in northern Zaire, 264 laboratory-confirmed tanapox cases were observed between 1979 and 1983 (Jezek et al., 1985).

There have been no reported human cases of Yaba monkey tumor poxvirus in over 2 decades.

Diagnosis

Clinical Signs, Symptoms, and Severity
In most patients infected by the tanapox virus, fever (38 to 39°C) commenced 1 to 2 days prior to skin eruptions and was frequently accompanied by severe headache, backache, and prostration. In most patients, only a single lesion was observed, which developed on parts of the body not usually covered by clothing. Multiple lesions, when observed, ranged from 2 to 10. Regional lymph nodes became enlarged with lesion development. The lesions usually disappeared spontaneously within 6 weeks, unless there was a secondary infection (Jezek et al., 1985).

Gross Lesion Pathology
By the end of the first week, the lesion was greater than 10 mm in diameter, with a large erythematous areola several centimeters wide surrounded by edematous skin. Some lesions developed into large nodules, but most ulcerated without pus. The maximum diameter of the lesion was reached in the second week (Jezek et al., 1985).

Diagnosis and Differential Diagnosis
Diagnosis of tanapox is made by a combination of clinical (lesion character and number), epidemiological (equatorial Africa), and laboratory (enveloped brick-shaped virions in lesion material) findings (Jezek et al., 1985). Patients with multiple lesions can be misdiagnosed as having monkeypox (see above).

COLLECTION, HANDLING, AND STORAGE OF SPECIMENS

In the United States, poxvirus diagnostic specimens should be evaluated by commercial diagnostic laboratories or local government health laboratories unless smallpox is suspected. A suspected case of smallpox must be immediately reported to the nearest government health department. Also, pending diagnosis, the patient should be sequestered with appropriate medical care and contact with persons outside the immediate family should be minimized. If the health department review cannot exclude a diagnosis of smallpox, then the case must be immediately reported to the WHO Collaborating Center for Smallpox and Other Poxvirus Infections, Poxvirus Section, CDC, Atlanta, GA 30333 (Ropp et al., 1999).

The appropriate biosafety level precautions must be taken for the handling, transport, and processing of infected lesion material (Nakano, 1979; Ropp et al., 1999). Skin lesions are the specimen of choice for diagnosis of a poxvirus infection. Virions are usually present in this material and can remain viable even after several weeks of storage with or without refrigeration. Because lesion material may be analyzed by several different approaches, it is important that sufficient specimen be collected for submission. For suspected orthopoxvirus, parapoxvirus, and yatapoxvirus infections, two to four biopsy specimens should be adequate and may be acquired from lesions at the macular-papular, vesicular-pustular, and/or crusting stage. During the vesicular-pustular stage, the fluid (including cells from the base of the lesion, where the virus is often found in high concentration) can be collected in capillary tubes, on dry swabs, or as thick droplets on glass slides (Nakano, 1979; Ropp et al., 1999). Carrier medium should not be added to any of the specimens as it may interfere with electron microscopic analysis. The specimens can be stored for a short time at 4°C; however, −20 or −70°C is preferable for longer-term storage.

International, U.S., and/or local shipping regulations must be followed when preparing for shipment of the specimen. The specimen should be initially packaged in a rigid receptacle matched to the size of the specimen with a leakproof cap; the cap should be further wrapped with parafilm to prevent either the contamination of the exterior of the shipping container or the pollution of the specimen from "melted" conventional or dry ice. To prevent possible cross contamination among samples, each sample should have its own primary container. This primary container should be packaged in a crush-resistant outer container. The specimens should be shipped on wet ice if possible.

DIAGNOSTIC METHODS

Identification and differentiation of poxvirus species have been carried out using a variety of analytical approaches, including virus growth, histology, electron microscopy, and serology; however, analysis of lesion material for poxvirus DNA by PCR appears to be the technique of the future.

Virus Growth

Traditionally the chorioallantoic membrane of the 12-day-old chicken embryo has been used to culture the orthopoxviruses variola, monkeypox, vaccinia, cowpox, and buffalopox viruses, since all of these viruses grow well and can be differentiated from one another by their characteristic pock (lesion) morphology; however, this approach does not support the growth of parapoxviruses and yatapoxviruses. Tissue culture is another approach used to cultivate poxviruses. Orthopoxviruses produce cytopathic effect in Vero cells, and the parapoxviruses orf virus and pseudocowpox virus can be propagated in cells of ovine and bovine origin, respectively. Although tanapox virus can be grown in LLC-MK$_2$ cells, the process is rather inefficient, as only 20 isolations could be made from 145 specimens in which tanapox virus virions could be detected by electron microscopy (Nakano, 1979). MCV has yet to be propagated by either of these approaches. Because virus growth assays are laborious, expensive, and slow and can be relatively insensitive, they will likely be supplanted by PCR analysis of virion DNA from lesion material.

Histology

Histological examination of poxvirus lesions can be informative with both parapoxvirus and MCV infections. Parapoxvirus lesions exhibit an epidermis which is hyperplastic, with strands of epidermal keratinocytes penetrating into the dermis. The molluscum contagiosum lesion has pear-shaped lobules filled with intracytoplasmic Henderson-Paterson or molluscum bodies. In practice, histology is usually employed only for confirmation of MCV infections. This can be carried out via wax sections or a squash preparation. For the squash preparation, the keratotic dome-shaped molluscum lesion is placed on a regular slide under a coverslip or second slide. The lesion is flattened and can be examined by light microscopy directly or after being stained with Wright's or methylene blue stain. Round to ovoid molluscum bodies up to 37 by 27 μm are diagnostic of molluscum contagiosum. Alternatively, a biopsy specimen is fixed in 10% formal saline and submitted for wax embedding, sectioning at 5 μm, and staining with hematoxylin and eosin. Microscopic examination should provide a field of view similar to Fig. 3.

Electron Microscopy

Electron microscopy has been the most dependable and rapid method available for diagnosis of a poxvirus infection, when the results are considered along with clinical findings. Scab material or fluid from the vesicle should be processed by standard methods for negative-staining electron mi-

croscopy (Nakano, 1979). When poxviruses are viewed in this manner, the virions can be divided into two morphologically distinct groups (Fig. 1). One group, represented by vaccinia virus, is brick shaped and includes the variola, cowpox, vaccinia, monkeypox, and tanapox viruses and MCV (size range, 140 to 230 by 210 to 380 nm) (Fig. 1A). The second group, represented by orf virus, is ovoid and elongated and includes the orf and milker's nodule viruses (size range, 120 to 160 by 250 to 310 nm). Clinically, chickenpox is the disease most often confused with monkeypox (and smallpox prior to its eradication), whereas electron microscopy reveals the respective virions to be quite distinct. Parapoxvirus and yatapoxvirus infections can also be confidently diagnosed by electron microscopy with the availability of clinical and epidemiological findings. The reliability of electron microscopy for detection of poxvirus virions is high; in the case of variola virus, the CDC rated electron microscopy and chicken embryo chorioallantoic membrane as 98.6 and 89% accurate, respectively (Boulter et al., 1961). Similarly, parapoxvirus diagnosis by electron microscopy was more reliable than tissue culture (Nakano, 1979). It is felt that the reliability of electron microscopy could be close to 100% provided sufficient specimen is made available.

Serology

If virus lesion material is not available, then serology is an alternative approach for establishing the etiology of the disease. Historically, a large number of serologic assays have been described for measuring exposure to orthopoxviruses, parapoxviruses, and MCV, but at present only neutralization tests, Western blotting, and enzyme-linked immunosorbent assay are in common use, and then only as research tools or on a case-by-case basis at CDC (Nakano, 1979; Mukinda, 1997; Wantanabe, 1998; Ropp et al., 1999). The routine use of serologic assays in the diagnosis of poxvirus infections is hampered for several reasons. First, the existence of antibodies generated more than 20 years earlier following immunization with vaccinia virus during the smallpox eradication program hinders the specific detection of antibodies against other orthopoxviruses, unless one employs species-specific antigen (Ropp et al., 1999). Presently, there are no sensitive, specific, and reliable serologic tests to retrospectively differentiate among orthopoxvirus infections. Also, following orf, and milker's nodule virus and MCV infections, antibodies have not always been detected or have persisted only transiently (Porter et al., 1992; Robinson and Lyttle, 1992; Watanabe et al., 1998; Ropp et al., 1999). Finally, disease diagnosis usually requires a fourfold rise in titer between sera drawn at the acute and convalescent phases, an inconvenience that not all patients will tolerate when confronted with a non-life-threatening illness.

DNA Analysis

The determination of genomic DNA endonuclease cleavage profiles or nucleotide sequences has become the most definitive method for identification of poxvirus species; however, these approaches usually require some method for amplification of the virus from lesion material prior to analysis. This is not the case with PCR DNA analysis. Using primer pairs from the gene encoding the major A-type inclusion body protein (Meyer et al., 1994) or the hemagglutinin (Ropp et al., 1995), PCR analysis can distinguish among the known orthopoxvirus species. Importantly,

PCR analysis can be used directly on nucleic acid extracted from scabs or dried vesicle fluid (Ropp et al., 1995), and detailed protocols have recently been published (Meyer et al., 1995; Ropp et al., 1999). PCR analysis has the potential to provide the most reliable, rapid, and cost-effective tool for the detection of poxvirus nucleic acid in lesion material, although at this time PCR protocols have yet to be described for parapoxvirus or yatapoxvirus DNA. Although this approach looks exceedingly promising, its efficacy will become more apparent with experience gained from analysis of additional specimens acquired from different sources and maintained under various conditions.

Evaluation of Laboratory Diagnostic Methods

Recently, PCR has been used in the rapid identification of orthopoxvirus species, and this approach could readily be adapted to the diagnosis of other poxvirus infections should the need arise (Meyer et al., 1995). The approach has the potential to give a definitive diagnosis based on one assay without the need for additional clinical or epidemiological information. The ability to detect virus DNA directly from lesion material without amplification through either tissue culture or chicken embryo chorioallantoic membranes makes PCR the method of choice for diagnosing orthopoxvirus infections. With parapoxvirus and yatapoxvirus infections, virus propagation is rarely carried out as part of the laboratory analysis; instead, the diagnosis is accomplished by electron microscopy of lesion material in combination with clinical and epidemiological findings. In the near term, the diagnosis of molluscum contagiosum will not be affected by the development of the PCR approach (even though a protocol has been described [Nunez et al., 1996]), since squash preparations and wax sections are both reliable and inexpensive.

Poxvirus Diagnostics in the Future

PCR should be the diagnostic assay of the future for poxvirus infections once instruments are available which automate DNA preparation from specimens, carry out the PCR amplification reactions, and are coupled to a detection system. Even the diagnosis of molluscum contagiosum would probably be done in this manner. The development of PCR-based diagnosis may benefit from the funding initiatives directed at developing detection systems for pathogens considered potential biological weapons. Even now, portable PCR machines have been designed for use in the field, without the need for a sophisticated laboratory environment (Baker, 1996).

Conclusions

Traditionally poxvirus infections have been analyzed by a combination of laboratory tests which required the consideration of clinical and epidemiological findings prior to a definitive diagnosis. In the future, PCR technology alone has the potential to provide a diagnosis with reduced expenditure of time and money.

PREVENTION AND THERAPY

Except for MCV, most poxvirus infections are zoonoses that fail to establish a human chain of transmission, thus mitigating the need for vaccine strategies. Also, since MCV causes a benign, self-limiting disease in immunocompetent patients, there is a similar lack of urgency for development of prevention strategies.

The management of orthopoxvirus, parapoxvirus, and yatapoxvirus infections is supportive. There are no systemic or topical chemotherapeutic agents commercially available, although vaccinia-immune gammaglobulin has been useful in the treatment of two patients who had cowpox and a number of human vaccinia virus infections. Prevention of secondary bacterial infections through the use of antibiotic ointments is an option. In the case of cowpox, steroids are contraindicated and may exacerbate the illness (Baxby et al., 1994).

Because of the chronic nature of molluscum contagiosum, one can consider curettage and cryotherapy of lesions. With children, pretreatment of lesions with lidocaine and prilocaine for between 15 and 60 min resulted in a decrease in the reported pain on lesion removal with a comedo extractor or curette (de Waard-van der Spek et al., 1990). With cryotherapy, lesions are treated for 6 to 30 s with a cotton-tipped applicator dipped in liquid nitrogen, and the treatment is repeated at 3-week intervals as needed (Bardenstein and Elmets, 1995). Cryotherapy is relatively painless, cost-effective, and yields good cosmetic results. With patients infected with human immunodeficiency virus (HIV), this treatment approach has the added advantage of mitigating the risk of disease transmission to medical personnel. Removal of giant molluscum lesions, but not regular lesions, usually results in scar formation.

WILL SMALLPOX OR A SMALLPOXLIKE DISEASE APPEAR IN THE FUTURE?

Smallpox was one of the most feared diseases known. With the 9 December 1979 certification of global eradication of smallpox, the need for preventive immunization with the vaccinia virus vaccine ceased, and a growing percentage of the world's population now lacks protective immunity to this deadly disease. Thus, the world's population is at risk for reemergence of variola virus or the evolution of a new poxvirus that can efficiently spread by aerosol transmission and is associated with high mortality. The reemergence of variola virus could arise from at least three sources. Since variola virus has no known animal reservoir, the only source of the virus in nature would be from the "graveyards" of the world. Persistence of infectious variola virus in the soils or waters of temperate or tropical climates is considered unlikely, leaving only the remote possibility of preservation of virus in corpses locked in permafrost (Fenner et al., 1988). A second possible source of variola virus is from accidental infections from laboratory stocks. This possibility is also considered remote, as over the last 5 decades the facilities and procedures used to handle variola virus have been upgraded substantially, while the numbers of laboratories carrying out variola virus-related research has declined dramatically. At this time, all known stocks of variola virus are held in two WHO collaborating centers. The most likely source for the reintroduction of variola virus into the human population is from clandestine stocks held by a terrorist group or a rogue nation. Although many pathogens other than variola virus would seem to be better candidates for bioterrorism or biological warfare, the individual(s) that would perpetrate such an act may not be concerned with the scientific basis of such an argument.

Perhaps a more likely scenario is the evolution of a poxvirus which efficiently infects and is transmitted in human populations with associated severe mortality and morbidity. Based on our knowledge of poxvirus biology, the orthopoxvirus monkeypox would be the most likely candidate. Presently it infects humans, causing a disease that is clinically indistinguishable from smallpox but without the same level of mortality and transmissibility. Although poxviruses have a relatively low mutation frequency compared to RNA viruses, the availability of a growing human population lacking immunity provides the opportunity for more frequent monkeypox infections of humans in central Africa, which could allow the selection of a virus more virulent and transmissible for humans.

If indeed smallpox or a smallpoxlike disease appeared in the future, does the United States have the resources in place to quickly contain an outbreak? A small outbreak, probably; a large outbreak, probably not. We have three inextricably linked problems. We lack adequate levels of the live vaccinia virus vaccine. Also, the presence of a large immunodeficient population makes rapid, large-scale immunizations with a live virus vaccine problematic. Finally, vaccinia immune globulin, which is required for vaccine testing, is not currently available.

Wyeth Laboratories is currently the only licensed producer of the live attenuated vaccinia virus smallpox vaccine. Since 1983, Wyeth has stopped its distribution to civilians, and the CDC is the only nonmilitary source; however, this vaccine is approximately 20 years old and is showing signs of loss of potency. Even if the vaccine still retained its potency and sufficient doses had been stockpiled, its use in a widespread immunization program would be ill advised due to the increase in the general population of persons with an underlying immunodeficiency due to infection with HIV or immunosuppressive treatments for cancer or tissue transplantation. Vaccinia virus immunizations of individuals with impaired cellular immunity results in progressive vaccinia, a severe complication that is almost uniformly fatal if left untreated. Recently, two military personnel were immunized with the vaccinia virus prior to being identified as HIV positive, and they subsequently developed a disseminated vaccinia virus infection (Centers for Disease Control, 1982; Redfield et al., 1987). As there is no therapeutic antiviral (although research continues), the only treatment for progressive vaccinia is vaccinia immune globulin, which is not always efficacious. At the time of writing, vaccinia immune globulin is not available. Since accessible stocks of vaccinia immune globulin have been a prerequisite for immunizing humans with the vaccinia virus vaccine, its lack has curtailed the routine immunization of scientists and laboratory personnel who use vaccinia virus as a tool in their research. Similarly, new inductees into the military presumably will not receive their immunizations as a safety measure in the event of exposure to variola virus during bioterrorism or biowarfare. Finally, the testing of new lots of live vaccinia virus vaccine could be delayed.

CONCLUSIONS

Ten poxviruses have been documented to infect humans (Table 2). Except for MCV, these poxvirus diseases are zoonoses. With rare exceptions, these zoonotic poxviruses fail to establish a human chain of transmission, and where human transmission is documented, as in the case of monkeypox, it usually ceases at 3 generations. Most human poxvirus infections occur through minor abrasions in the skin. The parapoxvirus orf virus and MCV cause the most frequent poxvirus infections worldwide; however, the incidence of monkeypox in sub-Saharan Africa may be on the rise. Unless an individual is immunocompromised, the zoonotic poxvirus infections usually resolve themselves

uneventfully; however, monkeypox is the exception, with mortality rates of ~12%. Due to the sporadic and/or benign nature of the infections, health care professionals have not developed prevention strategies for human poxvirus infections. As might be expected, specific treatment modalities are focused mainly on molluscum contagiosum, which is relatively common among children, adults, and severely immunocompromised HIV-positive patients.

REFERENCES

Al-Hazzaas, A. F., and A. A. Hidayat. 1993. Molluscum contagiosum of the eyelid and infraorbital margin—a clinicopathologic study with light and electron microscopic observations. *J. Pediatr. Ophthalmol. Strabismus* **30:**58–59.

Baker, C. S., F. Capriano, and S. R. Palumbi. 1996. Molecular genetic identification of whale and dolphin products from commercial markets in Korea and Japan. *Mol. Ecol.* **5:**671–685.

Bardenstein, D. S., and C. Elmets. 1995. Hyperfocal cryotherapy of multiple molluscum contagiosum lesions in patients with the acquired immune deficiency syndrome. *Ophthalmology* **102:**1031–1034.

Baxby, D., M. Bennett, and B. Getty. 1994. Human cowpox 1969–93: a review based on 54 cases. *Br. J. Dermatol.* **131:**598–607.

Becker, F. T. 1940. Milker's nodules. *JAMA* **115:**2140.

Bergoin, M., and S. Dales. 1971. Comparative observations on poxviruses of invertebrates and vertebrates, p. 171–203. *In* K. Maramorosch and E. Kurstak (ed.), *Comparative Virology.* Academic Press, Inc., New York, N.Y.

Boulter, E. A., J. C. N. Westwood, and H. B. Maber. 1961. Value of serotherapy in a virus disease (rabbitpox). *Lancet* **ii:**1012–1015.

Buller, R. M. L., and G. K. Palumbo. 1991. Poxvirus pathogenesis. *Microbiol. Rev.* **55:**80–122.

Buller, R. M. L., J. Burnett, W. Chen, and J. Kreider. 1995. Replication of molluscum contagiosum virus. *Virology* **213:**655–659.

Carson, C. A., and K. M. Kerr. 1967. Bovine papular stomatitis with apparent transmission to man. *J. Am. Vet. Med. Assoc.* **151:**183–187.

Centers for Disease Control. 1982. Epidemiologic notes and reports: Disseminated vaccinia infection in a college student in Tennessee. *Morb. Mortal. Wkly. Rep.* **31:**682–683.

Centers for Disease Control and Prevention. 1997a. Human monkeypox—Kasai Oriental, Democratic Republic of Congo, February 1996–October 1997. *Morb. Mortal. Wkly. Rep.* **46:**1168–1171.

Centers for Disease Control and Prevention. 1997b. Human monkeypox—Kasai Oriental, Zaire, 1996–1997. *Morb. Mortal. Wkly. Rep.* **46:**304–307.

Dawson, I. M., and A. S. McFarlane. 1948. Structure of an animal virus. *Nature* **161:**464–466.

DeOreo, G. A., H. H. Johnson, Jr., and G. W. Binkley. 1956. An eczematous reaction associated with molluscum contagiosum. *Arch. Dermatol.* **74:**344–348.

de Waard-van der Spek, F. B., A .P. Oranje, S. Lillieborg, C. J. Hop, and E. Stolz. 1990. Treatment of molluscum contagiosum using a lidocaine/prilocaine cream (EMLA) for analgesia. *J. Am. Acad. Dermatol.* **23:**685–688.

Downie, A. W. 1939. A study of the lesions produced experimentally by cowpox virus. *J. Pathol.* **48:**361–378.

Fenner, F. 1948. The pathogenesis of the acute exanthems. An interpretation based on experimental investigations with mousepox (infectious ectromelia of mice). *Lancet* ii:915–930.

Fenner, F. 1996. Poxviruses, p. 2673–2702. *In* B. N. Fields, D. M. Knipe, and P. M. Howley (ed.), *Virology*, 3rd ed. Lippincott-Raven Publishers, Philadelphia, Pa.

Fenner, F., and F. M. Burnet. 1957. A short description of the poxvirus group (vaccinia and related viruses). *Virology* **4**:305–314.

Fenner, F., D. A. Henderson, I. Arita, Z. Jezek, and I. D. Ladnyi. 1988. *Smallpox and Its Eradication.* World Health Organization, Geneva, Switzerland.

Groves, R. W., E. Wilson-Jones, and D. M. MacDonald. 1991. Human orf and milkers' nodule: a clinicopathologic study. *J. Am. Acad. Dermatol.* **25**:706–711.

Guarnieri, G. 1892. Richerche sulla patogenesi ed etiologia dell'infezione vaccinia e variolosa. *Arch. Sci. Med.* **16**:243–247.

Itin, P. H., and L. Gilli. 1994. Molluscum contagiosum mimicking sebaceous nevus of Jadassohn, ecthyma and giant condylomata acuminata in HIV-infected patients. *Dermatology* **189**:396–398.

Izu, R., D. Manzano, J. Gardeazabal, and J. L. Diaz-Perez. 1994. Giant molluscum contagiosum presenting as a tumor in an HIV-infected patient. *Int. J. Dermatol.* **33**:266–267.

Janniger, C. K., and R. Schwartz. 1993. Molluscum contagiosum in children. *Cutis* **52**:194–196.

Jezek, Z., and F. Fenner. 1988. Human monkeypox, p. 81–110. *In* J. L. Melnick (ed.), *Monographs in Virology.* Karger, Basel, Switzerland.

Jezek, Z., I. Arita, M. Szczeniowski, K. M. Paluku, K. Ruti, and J. H. Nakano. 1985. Human tanapox in Zaire: clinical and epidemiological observations on cases confirmed by laboratory studies. *Bull. W.H.O.* **63**:1027–1035.

Karupiah, G., T. N. Fredrickson, K. L. Holmes, L. H. Khairallah, and R. M. Buller. 1993. Importance of interferons in recovery from mousepox. *J. Virol.* **67**:4214–4226.

Kolhapure, R. M., R. P. Deolankar, C. D. Tupe, C. G. Rant, A. Basu, B. M. Dame, S. D. Pawar, M. V. Joshi, V. S. Padbidri, M. K. Goverdhan, and K. Banerjeck. 1997. Investigation of buffalopox outbreaks in Maharashtra State during 1992–1996. *Indian J. Med. Res.* **106**:441–446.

Ledingham, J. C. G. 1931. The aetiological importance of the elementary bodies in vaccinia and fowl-pox. *Lancet* ii:525–526.

Lyttle, D. J., K. M. Fraser, A. A. Mercer, and A. J. Robinson. 1994. Homologs of vascular endothelial growth factor are encoded by the poxvirus orf virus. *J. Virol.* **68**:84–92.

Marchal, J. 1930. Infectious ectromelia. A hitherto undescribed virus disease of mice. *J. Pathol.* **33**:713–728.

McFadden, G. 1988. Poxvirus of rabbits, p. 37–62. *In* G. Darai (ed.), *Virus Diseases in Laboratory and Captive Animals.* Martinus Nijhoff, Boston, Mass.

Meyer, H. P., M. Pfeffer, and H. J. Rziha. 1994. Sequence alterations within and downstream of the A-type inclusion protein genes allow differentiation of Orthopoxvirus species by polymerase chain reaction. *J. Gen. Virol.* **75**:1975–1981.

Meyer, H. R., S. L. Ropp, and J. J. Esposito. 1995. *In* J. S. A. Warnes (ed.), *Methods in Molecular Biology: Diagnostic Virology Protocols*, p. 1–25. Humana Press, Totowa, N.J.

Moss, B. 1996. Poxviridae: the viruses and their replication, p. 2637–2671. *In* B. N. Fields, D. M. Knipe, and P. M. Howley (ed.), *Virology*, 3rd ed. Lippincott-Raven Publishers, Philadelphia, Pa.

Mukinda, V. B. K., G. Mwema, M. Kilundu, D. L. Heymann, A. S. Khan, and J. J. Esposito. 1997. Re-emergence of human monkeypox in Zaire in 1996. *Lancet* **349**:1449–1450.

Murphy, J. K., and I. G. Ralfs. 1996. Bullous pemphigoid complicating human orf. *Br. J. Dermatol.* **134**:929–930.

Nagington, J., G. H. Tee, and J. W. Smith. 1965. Milker's nodule virus infections in Dorset and their similarity to Orf. *Nature* **208**:505–507.

Nakamura, J., Y. Muraki, M. Yamada, Y. Hatano, and S. Nii. 1995. Analysis of molluscum contagiosum virus genomes isolated in Japan. *J. Med. Virol.* **46**:339–348.

Nakano, J. H. 1979. Poxviruses, p. 257–308. *In* E. H. Lennette and N. J. Schmidt (ed.), *Diagnostic Procedures for Viral, Rickettsial and Chlamydial Infections.* American Public Health Association, Washington, D.C.

Niven, J. S. F., J. A. Armstrong, C. H. Andrewes, H. G. Pereira, and R. C. Valentine. 1961. Subcutaneous "growths" in monkeys produced by a poxvirus. *J. Pathol. Bacteriol.* **81**:1–14.

Nunez, A. F., J. M. Funes, M. Agromayor, M. Moratilla, A. J. Vargas, J. L. Lopes-Estabaranz, M. Esteben, and A. Martin-Gallardo. 1996. Detection and typing of molluscum contagiosum virus in skin lesions by using a simple lysis method and polymerase chain reaction. *J. Med. Virol.* **50**:342–349.

O'Neil, C. 1995. Pearly penile papules on the shaft. *Arch. Dermatol.* **131**:491–492.

Pathak, P. N., G. V. Rao, and W. A. Tompkins. 1974. In vitro cellular immunity to unrelated pathogens in chickens infected with fowlpox virus. *Infect. Immun.* **10**:34–41.

Peters, D. 1956. Morphology of resting vaccinia virus. *Nature* **178**:1453–1455.

Porter, C. D., N. W. Blake, J. J. Cream, and L. C. Archard. 1992. Molluscum contagiosum virus. *Mol. Cell Biol. Hum. Dis. Ser.* **1**:233–257.

Postlethwaite, R. 1970. Molluscum contagiosum. *Arch. Environ. Health* **21**:432–452.

Redfield, R. R., D. C. Wright, W. D. James, T. S. Jones, C. Brown, and D. S. Burke. 1987. Disseminated vaccinia in a military recruit with human immunodeficiency virus (HIV) disease. *N. Engl. J. Med.* **316**:673–676.

Richardson, M., and K. Dumbell. 1994. Comparisons of monkeypox viruses from animal and human infections in Zaire. *Trop. Geogr. Med.* **46**:327–329.

Robinson, A. J., and D. J. Lyttle. 1992. Parapoxviruses: their biology and potential as recombinant vaccines, p. 285–327. *In* M. M. Binns and G. L. Smith. (ed.), *Recombinant Poxviruses.* CRC Press, Inc., Boca Raton, Fla.

Robinson, A. J., and G. V. Petersen. 1983. Orf virus infection of workers in the meat industry. *N. Z. Med. J.* **96**:81–83.

Ropp, S. L., Q. Jin, J. C. Knight, R. F. Massung, and J. J. Esposito. 1995. PCR strategy for identification and differentiation of smallpox and other orthopoxviruses. *J. Clin. Microbiol.* **33**:2069–2076.

Ropp, S. L., J. J. Esposito, V. N. Loparev, and G. J. Palumbo. 1999. Poxviruses infecting humans, p. 1131–1138. *In* P. Murray (ed.), *Manual of Clinical Microbiology.* American Society for Microbiology, Washington, D.C.

Rouhandeh, H. 1988. Yaba virus, p. 1–5. *In* G. Darai (ed.), *Virus Diseases in Laboratory and Captive Animals.* Martinus Nijhoff, Boston, Mass.

Schwartz, J. J., and P. L. Myskowski. 1992. Molluscum contagiosum in patients with human immunodeficiency virus infection. *J. Am. Acad. Dermatol.* **27**:583–588.

Tan, S. T., G. B. Blake, and S. Chambers. 1991. Recurrent orf in an immunocompromised host. *Br. J. Plastic Surg.* **44:**465–467.

Thompson, C. H., R. T. De Zwart-Steffe, and I. M. Biggs. 1990. Molecular epidemiology of Australian isolates of molluscum contagiosum. *J. Med. Virol.* **32:**1–9.

Watanabe, T., S. Morikawa, K. Suzuki, T. Miyamura, K. Tamake, and Y. Ueda. 1998. Two major antigenic polypeptides of molluscum contagiosum virus. *J. Infect. Dis.* **177:**284–292.

World Health Organization. 1996a. Monkeypox, Zaire. *Wkly. Epidemiol. Rec.* **72:**101–108.

World Health Organization. 1996b. Monkeypox, Zaire. *Wkly. Epidemiol. Rec.* **71:**326

Parvoviruses

STANLEY J. NAIDES

35

The family *Parvoviridae* is composed of small, nonenveloped, single-stranded DNA viruses. Parvo derives from the Latin "parvus," meaning small. The family is composed of two subfamilies, *Parvovirinae* and *Densovirinae*. *Parvovirinae* comprises three genera: *Parvovirus, Erythrovirus*, and *Dependovirus*. The genus *Parvovirus* includes those members of the family that infect nonerythroid vertebrate host cells and replicate autonomously. The genus, *Erythrovirus* consists of those members of the family that infect erythroid mammalian host cells and are autonomous in their ability to replicate. Previously, the prototype erythrovirus, B19, was classified with the parvoviruses. In the current *Sixth Report of the International Committee on Taxonomy of Viruses*, B19 and related primate viruses were classified separately in recognition of their tropism for erythroid precursors and the ability of B19 to cause erythematous rashes, hence *Erythrovirus*. The genus *Dependovirus* consists of those members of the family that require the presence of a helper virus, such as adenovirus or herpesvirus, in order to replicate. The dependoviruses include the adeno-associated viruses. The subfamily *Densovirinae* is composed of those autonomously replicating insect viruses previously known as densonucleosis viruses because of characteristic nuclear changes in the host cell occurring during infection.

In 1975, Yvonne Cossart and her colleagues reported the discovery of parvoviruslike particles in human serum originally screened for hepatitis B virus surface antigen (HBsAg). In a screen of 3,219 sera received in a routine clinical laboratory, 3 sera were found to be positive for HBsAg by electrophoresis but not by more sensitive reverse passive hemagglutination or radioimmunoassay (RIA) (Vandervelde et al., 1974). Two of the three sera were positive for HBsAg by electron microscopy. One of the three was included in a control panel of sera for HBsAg tests. This serum was labeled no. 19 in panel B. Cossart and her colleagues realized that the sera that were positive in electrophoresis assays for HBsAg, but negative in the more sensitive tests, contained a new viral antigen. The antibody source for the detection of HBsAg by the electrophoresis tests was human serum that contained antibodies against HBsAg as well as antibodies against the new antigen. The antibody source used for detection of HBsAg in the reverse passive hemagglutination test and RIA was hyperimmune antisera raised in animals by using a purified preparation of HBsAg. Thus, these animal antisera did not contain anti-

bodies against the new antigen. The nonidentity of the new antigen to HBsAg was demonstrated in Ouchterlony gel diffusion tests (Fig. 1) (Cossart et al., 1975). Electron microscopy of the serum containing the new antigen revealed spherical particles and empty shells typical of the family *Parvoviridae*. On cesium chloride gradients, the antigen banded at a buoyant density of 1.36 to 1.40, also typical of the family *Parvoviridae*. A human parvovirus had not been described previously. In deference to the serum identifier of the initial positive serum, the new virus was named B19. The antigen was next identified in sera from nine healthy blood donors, one patient with acute hepatitis, and one recent renal transplant recipient. Convalescent-phase sera from four of these individuals demonstrated loss of virus with concomitant seroconversion. While Cossart et al. were unable to associate the presence of virus with a specific illness, initial studies found 30% of adults had antibodies against this virus. Cossart et al. noted the similarity of their new antigen to particles in feces described in patients with acute gastroenteritis (Paver et al., 1973; Pattison, 1988).

CHARACTERISTICS OF PARVOVIRUS B19

Parvovirus B19 is a nonenveloped icosahedral virus which by electron microscopy may appear spherical. The particle measures approximately 23 nm in diameter, although a range of particle diameters from 20.5 to 25 nm has been described. The parvovirus B19 genome is a 5.6-kb, single-stranded DNA molecule characterized by imperfect palindromes at both the 3' and 5' ends. The palindromic ends form terminal hairpin loops (Summers et al., 1983; Cotmore and Tattersall, 1984; Astell and Blundell, 1989). While most autonomous parvoviruses possess unique sequences at each terminus, B19 differs in that its termini are inverted terminal repeats (Deiss et al., 1990). In this respect, B19 resembles adeno-associated virus of the sister genus *Dependovirus* (Berns and Hauswirth, 1984). As a result of this structure, B19 is able to package either a plus- or a minus-sense DNA strand in a given virion, yielding progeny viral stocks with approximately equal numbers of plus- and minus-sense genomes. A single copy of the genome is encapsidated.

B19 replication follows a modified rolling-hairpin model of replication characteristic of the autonomous parvoviruses (Tattersall and Ward, 1976). The imperfect palindrome at

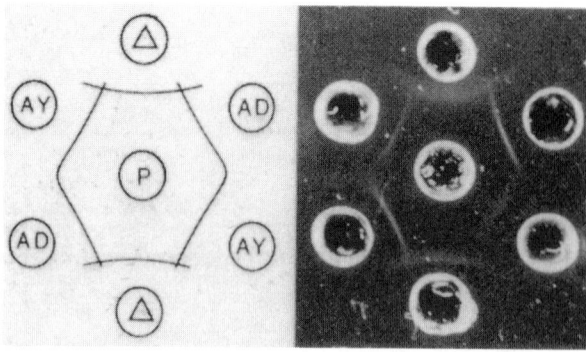

FIGURE 1 Ouchterlony gel diffusion assay showing newly discovered B19 antigen (Δ) nonidentity with HBsAg subtypes ad and ay, using anti-B19 antibody-positive antiserum P (from Cossart et al., 1975, with permission).

the 3' end of the molecule forms a region of double-stranded DNA that primes 3' elongation during replication. As the 3' complementary strand (C_1) elongates, it unfolds the 5' end of the parental strand (V_1). This represents the monomeric replicative form. The 3' end of C_1, which is now complementary to the parental 5' V_1 hairpin, is able to fold back on itself and continue elongation. As C_1 elongates, a complementary strand (V_2) with the same sequence as the parental virus (V_1) is made. As the 3'-end elongation continues, a second complementary strand (C_2) is synthesized. The resulting form is the dimeric replicative form. The viral nonstructural protein NS1 serves as the "nickase" which reduces the replicative forms to progeny virus. Nickase reduction results in two distinct configurations of the distal 375-nucleotide palindromes, which are inverted complements of each other. These alternate configurations of the terminal hairpins have been found in all parvoviruses analyzed so far and are referred to as "flip" and "flop" (Astell, 1990; Carter et al., 1990). While parvovirus B19 is autonomous in that it does not require a helper virus, productive replication is still restricted to rapidly dividing cells, where cellular factors required for viral replication are found in the nuclei. Parvovirus B19, like other parvoviruses, replicates and assembles in the cell nucleus.

The virus has been sequenced, and while sequence variation between isolates has been identified, there is no association with clinical manifestations (Shade et al., 1986; Blundell et al., 1987; Erdman et al., 1996; Johansen et al, 1998; Takahashi et al., 1999). The virus employs a somewhat simple coding strategy. Interposed between the palindromes, a single strong promoter at map unit 6 initiates transcription of the left-hand nonstructural-protein gene region, as well as the right-hand structural-gene region (Blundell et al., 1987). The nonstructural protein NS1 has a mass of approximately 74,000 Da and is encoded between nucleotides 435 and 2,448. NS1, as demonstrated for homologous nonstructural proteins in other parvoviruses, has a DNA binding domain, an ATP binding site, and helicase activity, thereby providing the nickase activity necessary to reduce replicative forms to progeny virus (Willwand et al., 1997; Corbau et al., 1999; Dettwiler et al., 1999). NS1 may also play a role in the assembly of viral DNA into mature viral capsids. Cotmore and Tattersall (1989) identified NS1 on the external surfaces of mature B19 viral particles covalently linked to virion DNA.

The structural proteins VP1 and VP2 are encoded by nucleotides 2444 to 4786 and 3125 to 4786 and have masses of 84,000 and 58,000 Da, respectively (Cotmore et al., 1986; Ozawa and Young, 1987). Both structural proteins are encoded in the same open reading frame, but VP2 results from an alternatively spliced transcript. mRNA for VP2 is initiated at an ATG codon at nucleotides 3125 to 3127 (Ozawa et al., 1987a). Digestion of B19 isolates with panels of restriction endonucleases has revealed strain variation and sequence drift, but the variation is limited and is not associated with consistent strain patterns, serological variation, or differences in clinical syndromes, as noted above (Mori et al., 1987; Erdman et al., 1996; Johansen et al, 1998; Takahashi et al., 1999).

PATHOGENESIS OF VIRUS INFECTION

Much of the early knowledge of the natural history of B19 infection was based upon studies of experimental infection in normal volunteers conducted by Mary Anderson and her colleagues at the Common Cold Research Unit, Harvard Hospital, Salisbury, England (Anderson et al., 1985a). Plasma containing parvovirus B19 was obtained from a healthy blood donor. The presence of other infectious agents was ruled out by inoculation of samples of this plasma in an array of bacterial and viral screening tests. The plasma was diluted in Hanks' buffered saline containing 0.2% bovine serum albumin, and 0.5 ml of the diluted plasma was inoculated into each nostril of a volunteer. Volunteers with preexistent anti-B19 antibodies did not develop viremia. Virus was first detected 6 days after inoculation in previously seronegative individuals. Peak titers of virus were reached 8 and 9 days postinoculation. Viremia was present for up to 7 days. Virus was detected in nasal washes and gargle specimens between days 7 and 11 postinoculation, during the same period as the viremia. Virus was not detected in urine or fecal specimens from any of the volunteers, except for one woman whose urine was contaminated with menstrual blood. High-titer IgM antibody to B19 developed during the second week after inoculation. Anti-B19 immunoglobulin G (IgG) antibody began to develop at the end of the second week or early in the third week after inoculation. In one individual with a trace amount of preexisting anti-B19 IgG, increased levels of IgG antibody were detected at the time of a small IgM antibody response 9 to 10 days after inoculation. None of the volunteers with a significant anti-B19 IgG antibody level prior to inoculation developed a subsequent IgM antibody response.

The clinical illness associated with experimental B19 infection was biphasic. In some individuals the viremia was asymptomatic, but in others it was associated with a flulike illness consisting of malaise, myalgia, and/or transient fever. Several individuals experienced headache, pruritis, and chills. The second phase of the illness began towards the end of the second week postinoculation and was characterized by rash, arthralgia, and arthritis.

Hematological alterations associated with B19 infection were noted by Anderson and her colleagues. There was an absolute areticulocytosis from the period of peak viremia until several days after the onset of the antibody response. The hemoglobin fell during the week after onset of areticulocytosis, but the decrease in hemoglobin was not clinically significant. Neutropenia was detected as early as day 8 postinoculation and persisted through the second week following infection. Lymphocyte and platelet counts were transiently depressed during the second week postinoculation. No significant trends were observed in the monocyte, basophil, or eosinophil counts. The incubation period between inoculation and onset of symptoms in natural infections can be inferred to range from 6 to 18 days based upon the volunteer study and epidemiological studies of B19 outbreaks; however, studies of some outbreaks have suggested that the incubation period may be as long as 28 days. Individuals are infectious during the period of viremia. Since many patients will not present until the onset of either rash or joint symptoms, most patients will no longer be infectious at the time of presentation. Onset of rash, polyarthralgia, or polyarthritis are temporally associated with development of an anti-B19 antibody response which, as we have seen, results in clearance of viremia and cessation of virus shedding (Joseph, 1986). Anti-B19 IgM antibody may be present for up to 2 months but declines thereafter. Anti-B19 IgG antibody response is long-lived. Anti-B19 IgG antibody seroprevalence in the adult population ranges between 40 and 60%. Acquisition of IgG antibody to B19, suggesting infection, is accelerated after 5 years of age, when many children first enter school (Anderson et al., 1986).

CLINICAL MANIFESTATIONS OF PARVOVIRUS B19 INFECTION

After Cossart's report, parvovirus B19 was a virus in search of a disease. While screening serum in a clinical laboratory for evidence of parvovirus B19 infection, Pattison and his colleagues observed six patients with sickle cell disease in aplastic crisis who had evidence of a recent B19 infection (Pattison et al., 1981). Aplastic crisis in sickle cell disease was first described in 1950 as an acute fall in hemoglobin associated with cessation of new erythrocyte formation, or areticulocytosis. Following areticulocytosis lasting approximately 7 to 10 days, a brisk reticulocyte rebound occurred with eventual return of hemoglobin to baseline levels. While the etiology of aplastic crisis in sickle cell anemia and other hemolytic anemias was not known, an infectious agent was considered likely because the crisis usually occurred only once in any given individual and occasionally occurred in outbreaks within chronic hemolytic anemia clinic patient populations. Shortly after Pattison's observation, Serjeant and his colleagues confirmed that B19 was the cause of an epidemic of aplastic crisis in Jamaica between 1979 and 1981 (Serjeant et al., 1981). Subsequent studies demonstrated that B19 infection may cause aplastic crisis in chronic hemolytic anemia regardless of the anemia's underlying etiology. In short order, B19-induced aplastic crisis was also found to occur in individuals with hereditary spherocytosis, alpha and beta thalassemias, pyruvate kinase deficiency, glucose-6-phosphate dehydrogenase deficiency, pyrimidine 5'-nucleotidase deficiency, hereditary stomatocytosis, autoimmune hemolytic anemia, and hereditary erythrocytic multinuclearity associated with a positive acidified (HAMS) test (Pattison et al., 1981; Duncan et al., 1983; Rao et al., 1983; Davidson et al., 1984; Evans et al., 1984; Green et al., 1984; Kelleher et al., 1984; Bertrand et al., 1985; Summerfield and Wyatt, 1985; Tsukada et al., 1985; Lefrere et al., 1986a, 1986b, 1986c, 1986d, 1986f; Saarinen et al., 1986; Takahashi et al., 1986; West et al., 1986; Hanada et al., 1988; Rappaport et al., 1989; Rechavi et al., 1989; Goldman et al., 1990; Mabin and Chowdhury, 1990). All of these patients shared the need for brisk reticulocytosis to maintain their hemoglobin levels in the face of shortened erythrocyte survival. During the period of areticulocytosis, individuals usually required transfusion support.

Examination of the bone marrow revealed growth arrest at the giant pronormoblast stage of erythrocyte development. Marginal intranuclear inclusions were sometimes seen and represented accumulated B19 virus (Burton, 1986; Caul et al., 1988; Knisely et al., 1988).

Since the end of the 19th century, when pediatricians enumerated the rash illnesses of childhood, the fifth rash, erythema infectiosum, or fifth disease, has been a well-described clinical entity. In May 1983, an outbreak of a rash illness in a primary school in North London, England, which had typical features of erythema infectiosum was described and was confirmed to be due to B19 infection (Cohen et al., 1983a; Anderson et al., 1984). The majority of the children had the typical rash of classic erythema infectiosum, which presents with bright red "slapped cheeks." The rash may also be seen on the torso and extremities (Color plate 4) (see color insert). The exanthem is frequently lacy or reticular in pattern but may be macular, maculopapular, or occasionally vesicular or hemorrhagic. It is pruritic in approximately half the cases. In the majority of children, the rash may recur after initial resolution. Recurrence is usually precipitated by sun exposure, a hot bath, or physical activity. Recrudescence may occur for weeks or months following the acute infection, but during episodes of rash recrudescence, children are not infectious. While infection in children may be asymptomatic, when symptoms do occur they tend to be mild. These include sore throat, headache, fever, cough, anorexia, vomiting, diarrhea, and arthralgia (Brandrup and Larsen, 1976; Cramp and Armstrong, 1976; Lauer et al., 1976; Shneerson et al., 1980; Anderson et al., 1983; Andrews et al., 1984; Clarke, 1984; Okabe et al., 1984; Mynott, 1985; Nunoue et al., 1985; Plummer et al., 1985; Shiraishi et al., 1985; Chorba et al., 1986; van Elsacker-Niele and Anderson, 1987; Mansfield, 1988; Naides et al., 1988). At the time of presentation, children with a rash usually have anti-B19 IgM antibodies (Shiraishi et al., 1985).

Erythema infectiosum may also be seen in adults not previously infected. In adults, the rash tends to be more subtle and the bright red "slapped cheeks" are absent. When adults are symptomatic, they tend to have a more severe flulike illness in which polyarthralgia and joint swelling are more prominent. For example, Ager and his colleagues observed adult involvement in an outbreak of erythema infectiosum in Port Angeles, Washington, in 1961 and 1962. Subjects were identified by the presence of a typical rash. Only 5.1% of children under 10 years of age had joint pain, and 2.8% had joint swelling. In children infected during their adolescent years, joint pain and swelling occurred in 11.5 and 5.3%, respectively. However, in the 20-year-old or older age group, joint pain occurred in 77.2% and joint swelling occurred in 59.6% (Ager et al., 1966).

White and his colleagues demonstrated that B19 could cause a chronic rheumatoidlike arthropathy (White et al., 1985). Sera were collected from individuals presenting at an "early synovitis" clinic at the Royal National Hospital for the Rheumatic Diseases, Bath, England, beginning in mid-1979. Primary care physicians were invited to refer patients to the clinic as soon as possible after the onset of acute joint symptoms. Traditionally, referral would wait until it became clear that joint symptoms had become chronic, usually a period of 3 months or more. Sera from 153 patients with early synovitis were retrospectively tested when tests for parvovirus B19 became available. Nineteen patients had evidence of a recent B19 infection. Although 49 of the cohort were men, all 19 patients with evidence of a recent

B19 infection were women. Eighteen of the infected patients presented with acute, moderately severe symmetric polyarthritis that usually started in the hands or knees and within 24 to 48 h involved wrists, ankles, feet, elbows, and shoulders. The cervical spine was involved in two cases, and the lumbosacral spine was involved in one case. In three cases, the upper limbs alone were affected. All patients complained of joint pain, stiffness, and variable swelling. The acute polyarthropathy usually improved within 2 weeks but completely resolved in only two cases. In 17 cases, symptoms persisted for more than 2 months, and in three cases, they persisted for more than 4 years. Thirteen of the 19 B19-infected patients reported an influenzalike illness with malaise, fever, gastrointestinal symptoms, and/or rash. Two-thirds of the patients had episodic flares but remained symptomatic between flares. One-third had episodic flares but were symptom free between flares. There was no associated long-term joint damage (White et al., 1985). The distribution of joint involvement and its symmetry may suggest a diagnosis of rheumatoid arthritis (Reid et al., 1985; White et al., 1985; Woolf et al., 1989; Naides et al., 1990). Many patients experience morning stiffness. About half of the patients with chronic B19 arthropathy meet the criteria of the American Rheumatism Association for a diagnosis of rheumatoid arthritis (Arnett et al., 1988; Naides et al., 1990; A. J. Silman, editorial, *Br. J. Rheumatol.* **27**:341–343, 1988). Patients may have a transient expression of autoantibodies during acute infection, including rheumatoid factor, anti-DNA antibodies, and anti-lymphocyte antibodies (Luzzi et al., 1985; Semble et al., 1987; Naides and Field, 1988; Sasaki et al., 1989; Soloninka et al., 1989). While initial reports suggested that chronic B19 arthropathy was associated with the major histocompatibility antigen HLA DR4, as is seen in classic erosive rheumatoid arthritis, subsequent studies have demonstrated no increased association with DR4 (Klouda et al., 1986; Woolf et al., 1987). In apparently immune-competent individuals with chronic B19 arthropathy, B19 DNA has been found in bone marrow aspirates and synovium (Foto et al., 1993; Naides et al., 1991). B19 DNA has also been detected using sensitive nested PCR techniques in synovium from healthy military recruits undergoing arthroscopy for trauma, suggesting that B19 may persist in a latent state (Söderlund et al., 1997). While B19 may present in a pattern similar to that of rheumatoid arthritis, it has generally not been thought to cause classic rheumatoid-factor-positive rheumatoid arthritis (Naides, 1998). A report suggesting that B19 upregulates production of tumor necrosis factor alpha and interleukin 6, two cytokines central to inflammation in the joint in rheumatoid arthritis, remains unconfirmed (Takahashi et al., 1998).

During a regional outbreak of erythema infectiosum in Scotland, six pregnant women were found to have serologically documented B19 infection. Four delivered normal-term infants. Two aborted grossly hydropic fetuses with anemia during the second trimester. There was evidence of fetal leukoerythroblastic reaction with eosinophilic changes in hematopoietic cell nuclei, hepatitis, and excess iron pigment in the liver. Hybridization with radiolabeled B19 DNA probes demonstrated B19 DNA in fetal liver, heart, thymus, kidney, adrenal, and placental tissues (Anand et al., 1987). Additional observations have confirmed the association of maternal B19 infection with fetal hydrops (Brown et al., 1984; Knott et al., 1984; Mortimer et al., 1985a; Bond et al., 1986; Gray et al., 1986; Anand et al., 1987; Carrington et al., 1987; Clewley et al., 1987; Mat-

sunaga et al., 1987; Woernle et al., 1987; Anderson and Hurwitz, 1988; van Elsacker-Niele et al., 1989; Knisely, 1990; Salimans, 1990).

The fetus is similar to an individual with chronic hemolytic anemia in that red cell survival is only 45 to 70 days and fetal red cell mass increases 34-fold during the second trimester (Gray et al., 1987). The infected fetus develops B19-induced aplastic crisis, resulting in a high-output cardiac failure with hydrops evidenced as soft tissue edema, ascites, pleural effusions, and, in some cases, polyhydramnios on ultrasound examination. A rise in maternal serum alpha-fetoprotein during the first trimester and early second trimester may herald ultrasound evidence of fetal hydrops (Carrington et al., 1987). Until a report by Woernle and his colleagues in 1987, fetal B19 infection was thought to be uniformly fatal. However, Woernle reported four pregnant women with anti-B19 IgM-positive serology, one of whom delivered a stillborn hydropic fetus whose tissues were positive for B19 DNA by nucleic acid hybridization. While the other three IgM-positive women gave birth to healthy offspring, one of the neonates had anti-B19 IgM antibody-positive cord serum. A second apparently healthy neonate was anti-B19 IgM negative but IgG positive; the anti-B19 IgG antibody persisted in the infant's serum at 9 months of age, confirming that it was of fetal origin, consistent with an in utero infection (Woernle et al., 1987). While microophthalmia with abnormal lens development has been reported in an abortus, congenital anomalies have not been a common feature of B19 infection (Weiland et al., 1987; Kinney et al., 1988). However, case reports suggest that occasional abortuses may show evidence of developmental anomalies and survivors of B19 infection in utero may demonstrate evidence of a congenital syndrome characterized by anemia, thrombocytopenia, and cardiac and hepatic dysfunction (S. J. Naides, G. Cuthbertson, J. C. Murray, and J. T. Stapleton, *Program Abstr. 28th Intersci. Conf. Antimicrob. Agents Chemother.*, abstr. 199, 1988). Viral cardiomyopathy as a mechanism of fetal hydrops has been reported (Naides and Weiner, 1989). Hepatic dysfunction in neonates as well as adult B19 infection has been described (Naides, 1987; Naides et al., *28th ICAAC*; Metzman et al., 1989). Whether developmental anomalies represent a direct viral effect or the indirect effect of severe illness during gestation remains to be determined. Fetuses have now been successfully treated for anemia by in utero transfusion with excellent salvage and outcome (Peters and Nicolaides, 1990; Soothill, 1990; Sahakian et al., 1991).

Patients with congenital or acquired immune deficiency, including Nezelof's syndrome, prior chemotherapy for lymphoproliferative disorders, immunosuppressive therapy for transplantation, or AIDS, may fail to clear B19 viremia (Kurtzman et al., 1987, 1988, 1989b; Graeve et al., 1989; Young et al., 1989; Frickhofen and Young, 1989, 1990; Chrystie et al., 1990; de Mayolo and Temple, 1990; Frickhofen et al., 1990; Rao et al., 1990; Naides et al., 1993). In the immune-competent host, the IgM antibody response may last 2 months or more. Anti-B19 IgM antibody and acute-phase IgG antibody (less than 1 week postinoculation) recognize determinants on VP2. In convalescent-phase serum, anti-B19 IgG antibody recognizes determinants on the VP1 structural protein (Kurtzman et al., 1989b). B19 VP1 and VP2 are products of alternate transcription of the same open reading frame, and VP1 contains an additional 227 N-terminal amino acids not present in VP2 (Shade et al., 1986). VP1 therefore contains unique determinants not present in the truncated form represented by VP2. These determinants may be in the unique nonoverlapping N-terminal region or may represent conformational differences in the sequences shared by the two proteins. Western blot analysis of sera from individuals with congenital immune deficiency, prior chemotherapy, or AIDS demonstrated the absence of convalescent-phase anti-B19 IgG antibodies directed against VP1. These sera were unable to neutralize B19 virus in bone marrow in vitro culture systems. While this work suggests that neutralizing determinants are unique to VP1 (since antibodies from immune-deficient patients recognized VP2), recent work with synthetic peptides suggests that neutralizing determinants may also be found on VP2 (Kurtzman et al., 1989a; Sato et al., 1991a, 1991b). Efforts to use recombinant empty capsids containing VP1 and VP2 as a vaccine are underway (Bansal et al., 1993).

B19-neutralizing activity is found in commercially available pooled immunoglobulin, since the seroprevalence of anti-B19 IgG antibodies in the adult population is approximately 50% (Anderson et al., 1986; Frickhofen et al., 1990; Naides et al., 1993). Immunocompromised individuals may fail to mount a neutralizing antibody response to B19, allowing virus to persist in the bone marrow and to cause chronic or intermittent suppression of one or more hematopoietic lineages. For example, B19 is a major cause of red cell aplasia in individuals with human immunodeficiency virus type 1 infection. Intravenous immunoglobulin may be employed to treat B19-associated bone marrow suppression and B19 persistence (Frickhofen et al., 1990). Intramuscular immunoglobulin therapy may be effective in the treatment of concurrent B19 infection in AIDS patients refusing intravenous immunoglobulin (Naides et al., 1993). However, immunoglobulin therapy may not be universally successful in clearing B19 persistence, and retreatment may be necessary (Bowman et al., 1990).

Infection control measures in the outpatient setting are limited to avoiding exposure for high-risk groups. Unfortunately, most infections result from exposure to index cases during the period of viremia during which the index case either is asymptomatic or has nonspecific influenzalike symptoms. In the hospitalized patient, infection control measures are important in avoiding exposure of hospital personnel and patients at risk. Outbreaks of B19 infection among hospital staff have been documented (Bell et al., 1989). Patients with fifth disease should be isolated with secretion precautions for 24 h after the onset of rash, arthralgia, or arthritis. Secretion precautions require wearing gowns and gloves when handling body fluids or secretions, e.g., saliva, nasal aspirates, urine, stool, or blood. Patients who are likely to be viremic are isolated with secretion precautions. However, patients with hemolytic anemia, pregnancy, or immune compromise should not be permitted to share a room with a potentially viremic patient. Isolation of viremic patients should be continued for the duration of the illness and in most cases for the entire period of hospitalization. Mothers of infected newborns should be placed in isolation under secretion precautions for the duration of the hospital stay. Mothers may visit neonates in the nursery but must follow infection control procedures that include careful hand washing and avoidance of contact with other infants. Household contacts of viremic patients should be isolated with secretion precautions from day 7 until day 18 after contact. Throughout the period of hospitalization, employees, visitors, and patients who are preg-

nant should be separated from persons who are likely to be viremic (including those with initial fifth disease rash) or who are household contacts of such a patient (Naides, 1989). While these measures have been universally accepted for infection control, some centers have also recommended respiratory precautions for viremic patients (Anderson et al., 1989). Respiratory precautions require separation of the patient in a single room and wearing of a mask by staff during all patient contacts.

Special care should be taken when handling B19 in the laboratory. Laboratory acquisition of infection has been suggested, with the most likely source being aerosolization during centrifugation, resuspension of viral pellets, and the washing stages of immunoassays (Cohen et al., 1988). Exposure of B19 virus to UV light decreases infectivity, but it also reduces the antigenicity of virus preparations used in diagnostic assays. It would appear prudent to survey laboratory personnel for their serological status and to caution individuals who are pregnant or immunocompromised or who have hemolytic anemia about working directly with B19 virus or in situations that could lead to their direct exposure to the virus.

PREFERRED SITES FOR VIRUS ISOLATION

Virus is most easily isolated from serum obtained from individuals with aplastic crisis or fifth disease, the latter prior to development of rash or arthritis, in high titer (10^{11} or more particles/ml). Virus may also be isolated from sera of individuals with persistent infection in the setting of immunocompromise, but viral titers tend to be lower (10^5 to 10^6 particles/ml). Virus may be found in bone marrow aspirates during the period of areticulocytosis in individuals with aplastic crisis or in those with persistent virus infection. In immunodeficient individuals with persistent infection and in chronic B19 arthropathy, B19 may be detected in bone marrow even in the absence of detectable viremia (Rao et al., 1990; Foto et al., 1993). In fetal infection, virus has been isolated from cord serum, ascites, or amniotic fluid (Naides and Weiner, 1989). Small, round, parvoviruslike particles have been found in stools from individuals with gastroenteritis, but B19 has not been isolated from such specimens nor has it been reported in stools from individuals with known B19 infection. Rather, the small, round, parvoviruslike particles found in stool may be a related human parvovirus with sequence homology similar to that of B19 (Oliver and Phillips, 1988; Turton et al., 1990). Detection of B19 DNA in urine has been reported in one case of neonatal B19 infection (Naides et al., *28th ICAAC*).

VIRUS STABILITY

B19 is not affected by ether, chloroform, 0.25% sodium deoxycholate, RNase, micrococcal DNase, potassium iodide, or heating at 45°C for 30 min. It is readily inactivated when heated at 56°C for 5 min, by treatment with 1 mg of proteinase K/ml, or by treatment with 0.05 N NaOH. Prior treatment of B19 with either 0.05 N HCl or glycine-HCl, pH 2.8, is about 75% effective in inhibiting B19 growth in bone marrow culture (Young et al., 1984). B19 may be stored frozen at −85°C for indefinite periods without loss of virion infectivity. Special handling is not required, although repetitive freezing and thawing may result in some loss of virus titer.

B19 may survive the dry-heat processing of factor VIII and IX concentrates at 80°C for 72 h (Lyon et al., 1989). Solvent-detergent inactivation of factor VIII concentrate with tri-(n-butyl) phosphate detergents will not inactivate B19, and terminal dry heating of these preparations at 100°C for 10 to 30 min or more has been recommended (Rubinstein and Rubinstein, 1990). Despite these efforts, there remains concern that B19 may be transmitted via pooled blood products (Saldanha and Minor, 1996; Erdman et al., 1997).

PROPAGATION OF B19 IN VITRO

Routine culture of parvovirus B19 in vitro is not available. All continuous cell lines tested failed to support B19 growth, including erythroleukemic cell lines such as K562 and HEL. B19 was first grown in primary bone marrow suspension cultures supplemented with erythropoietin (Ozawa et al., 1986). Optimally, the bone marrow should be derived from individuals with hemolytic anemia, e.g., sickle cell anemia, in whom the erythroid precursor pool is increased. Amplification of virus and the death of cells as a consequence of viral replication limits bone marrow culture as a source of virus production. The input-to-output ratio of virus is only approximately 1:50 for culture supernatants and 1:200 for total cultures (Ozawa et al., 1987b). Propagation of B19 in primary fetal liver cell culture and human cord blood culture have been reported, but as with bone marrow cultures, these are not efficient systems (Brown et al., 1991; Srivastava et al., 1992). Two megakaryocytic leukemic cells lines, UT-7 and MB-2, will support long-term virus growth, but these systems are inefficient as well (Komatsu et al., 1991; Shimomura et al., 1992; Munshi et al., 1993; Takahashi et al., 1989). The difficulty in culturing B19 has led investigators to seek a readily renewable antigen source through recombinant DNA technology.

PRESENCE OF B19 IN VARIOUS TISSUES

In the immunocompromised patient, B19 may be found continuously or intermittently in serum during periods of viremia. Virus may also be found in the bone marrow of these individuals during this period, presumably representing persistent infection. In fetal infections, virus may be found in serum, amniotic fluid, and fetal ascites, as well as an array of body tissues, including liver, spleen, kidney, thymus, bone marrow, heart, and placenta. The period after maternal infection during which B19 can be recovered from fetal tissues is not entirely clear and may depend on the severity of infection, the gestational age of the fetus at the time of infection, and whether transfusion therapy for the fetus is attempted. In apparently immune-competent individuals with chronic B19 arthropathy, B19 DNA has been found in bone marrow aspirates and synovium (Naides et al., 1991; Foto et al., 1993). Whether mature virus may be isolated from these tissues remains to be determined.

LABORATORY DIAGNOSIS OF B19 INFECTION

A number of approaches have been used to diagnose parvovirus B19 infection.

Electron Microscopy

Electron microscopy provides morphological identification of the virus. Standard touch preparations that are negatively stained are made by allowing a liquid sample to dry on an electron microscopy grid that was previously coated with a thin layer of plastic. After the sample has dried, it is stained with phosphotungstic acid or uranyl acetate, both electron-dense materials that accumulate around the particle to give a base relief appearance, or a negative image, and then viewed. In the absence of endogenous antibody to B19, incubation of the sample with specific antiserum or monoclonal antibody to B19 may cause aggregation of viral particles, which can be visualized by electron microscopic examination (Cohen, 1988).

A combined pseudoreplica-immunochemical staining technique offers the advantage of preserving the morphology while providing a specific serologic diagnosis (Naides and Weiner, 1989). This approach may be useful even in the presence of endogenous B19 antibodies that cause virus to aggregate, since antigenic sites for binding exogenous anti-B19 antibodies may still be available. Virus in various body fluids may be examined by this technique. A 25-μl sample is allowed to be absorbed into an agarose block, leaving the viral particles on the surface. The agarose is layered or coated with plastic which, after hardening, is floated off the agarose, inverted, and applied to a support grid. The pseudoreplica may then be negatively stained with phosphotungstic acid or uranyl acetate. Immunoelectron microscopy may be performed on pseudoreplicated samples prior to negative staining. The samples for immunoelectron microscopy are applied to nickle support grids as described above and then incubated ("blocked") with goat serum diluted 1:10 in 0.1 M phosphate-buffered saline (PBS) with 0.1% glycine before incubation with anti-B19 monoclonal antibody. We use 162-2B, an anti-B19 mouse monoclonal antibody of IgM isotype developed by Anderson and his colleagues at the Centers for Disease Control (Anderson et al., 1986). The samples are washed and then incubated with a polyclonal goat anti-mouse IgM (μ-chain-specific) antibody conjugated to colloidal gold to enable visualization of the antibody by electron microscopy. The samples are then negatively stained with uranyl acetate and carbon coated (Fig. 2) (Fritz et al., 1992). Observation of viral particles with specific colloidal-gold-conjugated antibody allows species identification. B19 virus may be difficult to distinguish from enteroviruses on the basis of morphology alone. B19 particles may exclude uranyl acetate or phosphotungstic acid, giving them a "full" appearance, or stain may enter the capsid, giving the appearance of an "empty" shell.

Immunoassay

Both RIA and enzyme-linked immunosorbent assay (ELISA) have been used to detect B19 antigen and specific antibody to B19. In antigen capture assays, an anti-human IgM or IgG antibody is allowed to coat a solid phase. In the second step, a serum known to contain either anti-B19 IgM or IgG antibody is then incubated on the plate. Excess antibody is removed by washing. In the third step, the test serum is added to allow capture of B19 antigen should any be present. Captured antigen is detected by a mouse monoclonal antibody to viral antigen followed by an anti-mouse antibody labeled with either ^{125}I (RIA) or peroxidase (ELISA). Cohen and his colleagues first described the antibody capture assay for anti-B19 IgM antibody, in which the serum to be tested for anti-B19 antibody is added in the second step and a serum known to contain B19 virus is substituted in the third step (Cohen et al., 1983b). Detection of the captured antigen indicates the presence of antibodies to B19 virus in the patient's serum. As an alternative to wild-type virus, recombinant empty capsids have been used as a source of antigen.

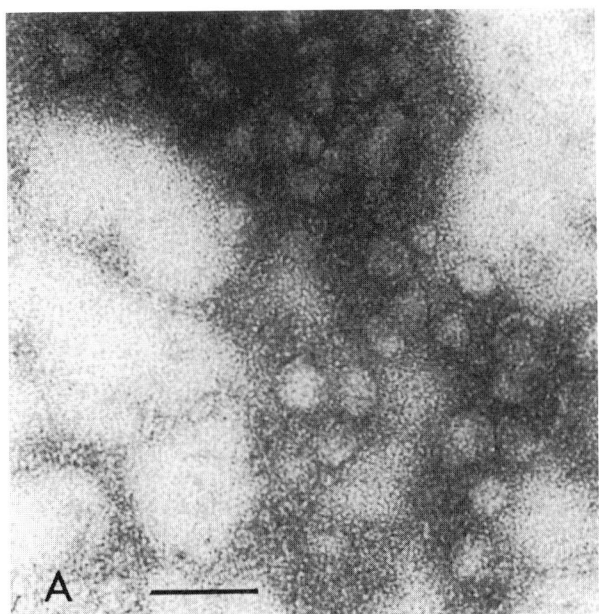

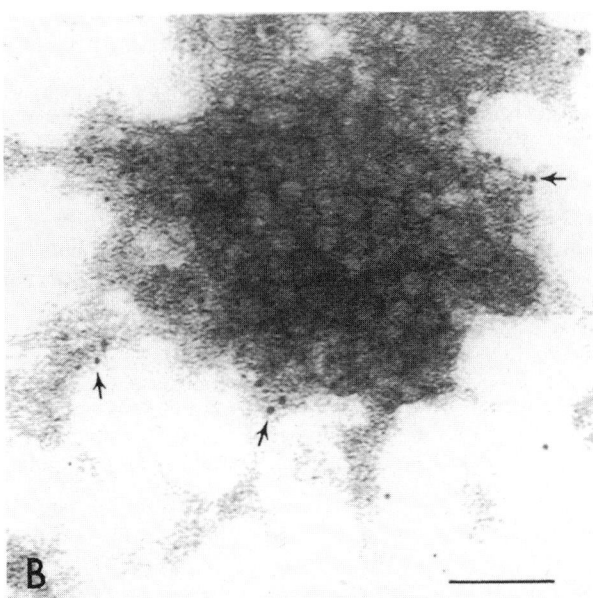

FIGURE 2 (A) Fetal ascites from a hydropic fetus showing viral particles without label, prepared by a pseudoreplica technique and negatively stained with uranyl acetate. Original magnification, ×100,000; image magnification, ×280,000; bar, 50 nm. (B) The fetal ascites prepared by pseudoreplica technique and identified as B19 by indirect labeling with colloidal gold (arrow) before negative staining with uranyl acetate. Original magnification, ×50,000; image magnification, ×140,000; bar, 100 nm. (From Naides and Weiner, 1989, with permission.)

We employ an antibody capture ELISA for detection of anti-B19 IgM and IgG antibodies developed by Anderson and his colleagues at the Centers for Disease Control. Microtiter plates designed for use in ELISA are coated with goat antihuman IgM or IgG antibody diluted in carbonate-

bicarbonate buffer (Anderson et al., 1986). After overnight incubation, the plates are washed with PBS–0.15% Tween 20, and serum to be tested for antibody is added. After 1.5 h at 35°C, the plates are washed with PBS–0.15% Tween 20, and then serum containing high-titer B19 virus that contains no endogenous antibody, or recombinant empty capsid, is added to the test wells. A control antigen-negative serum is added to parallel wells. After overnight incubation, the plates are washed, and monoclonal antibody 162-2B (a mouse IgM) is added to the wells. The monoclonal antibody is diluted in PBS with 0.5% gelatin and 0.15% Tween 20 for the IgG serology, and in PBS with 0.5% gelatin, 0.15% Tween 20, and 2% normal goat serum for the IgM test. After washing, a peroxidase-conjugated goat anti-mouse IgM antibody is added. After incubation for 1 h at 35°C, the wells are washed, and then 3,3′,5,5′-tetramethylbenzidine dihydrochloride color developer is added to each well in the presence of hydrogen peroxide, and the reaction is stopped with 25 μl of 2 M H_2SO_4. A serum is considered positive when the value of the patient's serum sample incubated with virus (a serum from a viremic patient) is greater than 3 standard deviations above the mean value of controls (all patients' serum samples in an assay incubated with normal serum), and the difference between the values from wells with and without virus for a given test serum is greater than 3 standard deviations (Bell et al., 1989). The presence of B19 virus in a prospective antigen source is detected by antigen capture ELISA and confirmed by direct DNA hybridization methods or by PCR with B19-specific primers, followed by hybridization of the amplified product with B19-specific probes (Naides et al., 1990). Anderson and his colleagues have demonstrated that the antibody capture ELISA is highly sensitive and specific (Anderson et al., 1986). Early reports of cross-reactivity between anti-B19 and anti-rubella antibodies were based on counterimmunoelectrophoresis techniques that have been replaced by RIA and ELISA methods. In the RIA and ELISA methods, cross-reactivity has not been a problem (Cohen and Shirley, 1985; Kurtz and Anderson, 1985; Cohen and Supran, 1987).

In response to the dependence on B19 viremic serum as a source of antigen, a number of recombinant antigens have been developed for B19 testing. While these are not yet widely available, they provide useful diagnostic and research reagents. B19 VP1 and VP2 proteins have been expressed in Chinese hamster ovary cells transfected with a B19 plasmid construct. The VP1 and VP2 self-assembled into empty capsids (Kajigaya et al., 1989). Transfected cell lysates were useful as an antigen source. VP1 and VP2 were coexpressed in the baculovirus system as well (Brown et al., 1990). B19 synthetic peptides based upon sequence analysis of the viral capsid gene have also been employed as an antigen source. Synthetic peptides provide approximately 90% sensitivity and specificity in comparing serologic titers using native virus (Fridell et al., 1991). B19 antigen has also been prepared as fusion proteins expressed in *Escherichia coli* (Morinet et al., 1989). Recombinant B19 proteins have been the basis for diagnostic kits available in Europe. A kit for B19 serologies has recently been approved by the Food and Drug Administration for diagnostic testing in the United States.

Nucleic Acid Probes

B19 DNA may be detected by hybridization with cDNA probes, riboprobes (synthetic RNA), or synthetic oligonucleotide probes. B19-specific probes have been used to identify virus by in situ hybridization (Hassam et al., 1990; Schwarz et al., 1991). Anderson and her colleagues and Clewley first reported detection of B19 viral DNA using molecularly cloned probes that were ^{32}P labeled. A virus-containing sample was blotted onto nitrocellulose filters using a dot blot manifold. It was then baked and hybridized with virus-specific probe (Anderson et al., 1985b; Clewley, 1985).

Since B19 encapsidates both positive- and negative-sense strands in equal numbers, B19 DNA forms double-stranded molecules on extraction and purification (Clewley, 1984). Clewley purified double-stranded B19 DNA and then treated it with nuclease S1 to remove hairpin palindromic loops by cutting at exposed single-stranded regions in the termini. A nearly full-length genomic fragment was cloned by homopolymeric tailing after addition of cytidylic acid tails using terminal transferase. An insertion site was produced by adding guanylic acid tails to the cloning vector pBR322 after its linearization by restriction at the PstI site (Clewley, 1985).

Nonradioactive labels for probes may be used for safety and a long shelf life. For example, Mori and her colleagues used a biotin-labeled DNA probe and streptavidin-alkaline phosphatase conjugate (Mori et al., 1989). A number of investigators have used digoxigenin-labeled probes for detection of B19 (Azzi et al., 1990; Zerbini et al., 1990).

PCR

PCR offers exquisite sensitivity and the ability to detect B19 DNA in an array of clinical specimens. Primers directed against sequences in both the nonstructural protein and the viral capsid protein genes have been used. PCR is highly sensitive, as demonstrated by Clewley, who reported that 60 out of 95 anti-B19 IgM-positive serum samples were found to be positive by PCR while only 1 was positive by dot blot hybridization. PCR was 99% specific in that only one sample in a control panel of 100 sera from individuals with other infections was positive by PCR as well as by dot blot hybridization. He concluded that the high rate of detection of B19 DNA by PCR represents a low rate of decay of viral DNA after peak viremia and is not a clinically significant phenomenon (Clewley, 1989). Koch and Adler, using oligonucleotide primers and detection of amplification products on agarose gels, found PCR to be 10,000 times more sensitive than dot blot hybridization. Southern analysis of amplified product using a radiolabeled oligonucleotide probe complementary to a sequence between the primers is typically 10^3 times more sensitive than dot blot hybridization. PCR has detected B19 DNA in urine, amniotic fluid, pleural fluid, ascites, and leukocyte extracts (Koch and Adler, 1990). PCR has also been used to detect B19 in fetal and adult tissues (Clewley, 1989; Salimans et al., 1989; Foto et al., 1993). PCR has been successfully utilized to diagnose persistent B19 infection in immunocompromised patients in whom small amounts of B19 DNA in serum may be detected (Frickhofen and Young, 1990; Naides et al., 1993). The utility of attributing clinical syndromes to B19 infection after detection of B19 DNA in tissues by sensitive PCR methods has been questioned. Söderlund and colleagues (1997) demonstrated B19 DNA in synovial tissue in healthy military recruits undergoing arthroscopy for traumatic knee injuries, using a sensitive PCR-based method.

UNUSUAL CLINICAL FEATURES

A number of uncommon dermatological manifestations of B19 infection have been reported. A vesiculopustular eruption has been seen in adult patients with fifth disease. It is unusual because it combines the features of morbilliform and vesiculopustular lesions. Subepidermal edema, ballooning necrosis of the dermis, and a lymphohistiocytic infiltrate may be noted. The infiltrate may contain unusual binucle-

ate giant cells. The vesicle itself may contain a neutrophilic infiltrate. Extravasation of erythrocytes into the dermis eventually gives the vesiculopustules a hemorrhagic appearance (Naides et al., 1988). Purpura may be seen in the absence of thrombocytopenia (Lefrere et al., 1985; Mortimer et al., 1985b; Shiraishi et al., 1989). Some patients may have purpura as a result of thrombocytopenia (Kilbourne et al., 1987; Lefrere et al., 1989). The clinical appearance may suggest Henoch-Schönlein purpura (Lefrere et al., 1985, 1986e). A "gloves and socks" syndrome of acral erythema with sharp demarcation at the wrists and ankles has been described, although this is not specific to B19 infection (Etienne and Harms, 1996; Stone and Murph, 1993; Drago et al., 1997).

B19 infection may present as isolated neutropenia, thrombocytopenia, anemia, or idiopathic thrombocytopenic purpura (Anderson and Cohen, 1987; Van Elsacker-Niele et al., 1996). Finger and toe paresthesias, progressive arm weakness, encephalopathy, and aseptic meningitis have all been reported (Denning et al., 1987; Faden et al., 1990, 1992; Walsh et al., 1988; Suzuki et al., 1995; Umene and Nunoue, 1995). Carpal tunnel syndrome may occur as a result of wrist swelling (Samii et al., 1996). B19 infection may trigger fibromyalgia, a muscle pain and fatigue syndrome, but this appears to represent a nonspecific viral trigger (Berg et al., 1993). Self-limited benign acute lymphadenopathy, hemophagocytic syndrome, and hemophagocytic syndrome in association with lymphadenopathy resembling necrotizing lymphadenitis (Kikuchi's disease) have also been reported (Tsuda et al., 1993; Muir et al., 1992; Shirono and Tsuda, 1995; Watanabe et al., 1994).

In addition, transiently abnormal liver enzymes have been observed in healthy adults with acute B19 arthropathy, and in neonates surviving anemia and nonimmune hydrops secondary to B19 infection in utero. B19 infection has been associated with acute hepatitis and some cases of non-A, non-B, non-C acute fulminant liver failure with or without associated aplastic anemia (Naides, 1987; Karetnyi et al., 1999; Tsuda, 1993; Yoto et al., 1996; Langnas et al., 1995; Longo et al., 1998). There is some question as to whether B19 can precipitate cutaneous vasculitis, polyarteritis nodosa, or Wegener's granulomatosis (Corman and Dolson, 1992; Corman and Staud, 1995; Finkel et al., 1994; Leruez-Ville et al., 1994; Nikkari et al., 1994, 1997). A role for B19 infection in juvenile rheumatoid arthritis has been suggested, but the arthritis is monoarticular or pauci-articular, in contrast to the symmetric polyarthritis seen in adults, suggesting either that the disease presents differently in children or that age is a confounding factor in that juvenile rheumatoid arthritis occurs at a time when children acquire B19 infection (Nocton et al., 1993).

Simpson and his colleagues reported a small parvovirus-like agent isolated from synovial tissue of a patient with severe rheumatoid arthritis after coculture with the continuous cell line WI38 and then passage through suckling mouse brain. The virus, designated RA-1, elicited a syndrome in neonatal mice that included neurological disturbances, dwarfism, alopecia, blepharitis, and thoracic spine curvature (Simpson et al., 1984). However, the presence of anti-RA-1 antibodies has not been associated with any clinical syndrome in humans. Anti-B19 antibodies do not react with RA-1 isolates (Stierle et al., 1987). When various clinical specimens from more than 100 human donors, including healthy subjects and patients with rheumatic disorders, were tested for RA-1 sequences by PCR amplification and Southern blot analysis, all donors tested positive for RA-1-related DNA sequences. Therefore, RA-1 is distinct from

B19. Reports of fecal parvoviruslike agents described above suggest that human parvoviruses other than B19 exist. However, RA-1 has not been associated with any clinical syndrome to date, nor is it entirely clear that it is a human virus.

PREVENTION OF AND THERAPY FOR DISEASES DUE TO B19

Parvovirus B19 is ubiquitous. It is therefore difficult to prevent exposure. Community contacts are frequently asymptomatic. However, avoidance of potentially viremic individuals by those at high risk for complications of B19 infection remains a prudent course.

There is no specific antiviral therapy for parvovirus B19. In general, management of patients is symptomatic and supportive. Patients with aplastic crisis usually require transfusion support during the period of areticulocytosis. Adults with chronic B19 arthropathy benefit from nonsteroidal anti-inflammatory drugs which control symptoms of pain and swelling. We have attempted a short course of oral prednisone at 5 mg daily in one patient with B19 arthropathy without apparent benefit (Naides et al., 1990). When fetal infection is accompanied by severe anemia, transfusion support may be required for fetal survival. Fetuses treated in this manner have survived to term and have been born without apparent long-term sequelae (Peters and Nicolaides, 1990; Soothill, 1990; Sahakian et al., 1991). We reported one case of B19-induced fetal hydrops without significant anemia, the hydrops in this fetus being due to a viral cardiomyopathy. Digitalization of the fetus in utero resulted in resolution of ascites and other stigmata of congestive heart failure, but the fetus died despite the good response (Naides and Weiner, 1989). As described above, individuals with immunodeficiency may develop persistent B19 infection because they lack the ability to produce anti-B19 neutralizing antibodies. Infusion of commercial immune serum globulin, or immunoglobulin, has been used successfully to clear persistent infection in patients with congenital immune deficiency, hematological malignancy, or AIDS (Kurtzman et al., 1988, 1989b; Frickhofen et al., 1990). Intravenous immunoglobulin is given at a dose of 0.4 g/kg of body weight daily for 5 or 10 days. Therapy results in the clearing of B19 viremia and improvement in bone marrow suppression (Frickhofen et al., 1990). As an alternative to the intravenous route, we have given an AIDS patient with chronic anemia and a monthly transfusion requirement immune serum globulin intramuscularly with good initial response. Prior to treatment, reticulocytes were undetectable and the anemia did not respond to erythropoietin. After initial intramuscular immunoglobulin injection, the reticulocyte count increased and the transfusion requirement decreased (Naides et al., 1993). While intramuscular immune serum globulin has the advantage of ease of administration and significantly decreased costs, the efficacy of the intramuscular route compared to intravenous immune serum globulin administration remains to be determined. Specific antiviral chemotherapy for B19 disease has not been evaluated.

REFERENCES

Ager, A. E., T. D. Y. Chin, and J. D. Poland. 1966. Epidemic erythema infectiosum. *N. Engl. J. Med.* **275**:1326–1331.

Anand, A., E. S. Gray, T. Brown, J. P. Clewley, and B. J. Cohen. 1987. Human parvovirus infection in pregnancy and hydrops fetalis. *N. Engl. J. Med.* **316**:183–186.

Anderson, L. J., and E. S. Hurwitz. 1988. Human parvovirus B19 and pregnancy. *Clin. Perinatol.* **15**:273–286.

Anderson, L. J., R. A. Tsou, T. L. Chorba, H. Wulff, P. Tattersall, and P. P. Mortimer. 1986. Detection of antibodies and antigens of human parvovirus B19 by enzyme-linked immunosorbent assay. *J. Clin. Microbiol.* **24**:522–526.

Anderson, M. J., and B. J. Cohen. 1987. Human parvovirus B19 infections in United Kingdom 1984–86. *Lancet* i:738–739.

Anderson, M. J., S. E. Jones, S. P. Fisher Hoch, E. Lewis, S. M. Hall, C. L. Bartlett, B. J. Cohen, P. P. Mortimer, and M. S. Pereira. 1983. Human parvovirus, the cause of erythema infectiosum (fifth disease)? *Lancet* i:1378.

Anderson, M. J., E. Lewis, I. M. Kidd, S. M. Hall, and B. J. Cohen. 1984. An outbreak of erythema infectiosum associated with human parvovirus infection. *J. Hyg.* **93**:85–93.

Anderson, M. J., P. G. Higgins, L. R. Davis, J. S. Willman, S. E. Jones, I. M. Kidd, J. R. Pattison, and D. A. Tyrrell. 1985a. Experimental parvoviral infection in humans. *J. Infect. Dis.* **152**:257–265.

Anderson, M. J., S. E. Jones, and A. C. Minson. 1985b. Diagnosis of human parvovirus infection by dot-blot hybridization using cloned viral DNA. *J. Med. Virol.* **15**:163–172.

Anderson, M. J., S. M. Hall, and G. R. Serjeant. 1989. Risks associated with human parvovirus B19 infection. *Morb. Mortal. Wkly. Rep.* **38**:81–97.

Andrews, M., R. W. Y. Martin, A. R. Duff, H. D. Greig, and S. A. H. Frost. 1984. Fifth disease: report of an outbreak. *J. R. Coll. Gen. Pract.* **34**:573–574.

Arnett, F. C., S. M. Edworthy, D. A. Bloch, D. J. McShane, J. F. Fries, N. S. Cooper, L. A. Healey, S. R. Kaplan, M. H. Liang, H. S. Luthra, T. A. Medsger, Jr., D. M. Mitchell, D. H. Neustadt, R. S. Pinals, J. G. Schaller, J. T. Sharp, R. L. Wilder, and G. G. Hunder. 1988. The American Rheumatism Association 1987 revised criteria for the classification of rheumatoid arthritis. *Arthritis Rheum.* **31**:315–324.

Astell, C. R. 1990. Terminal hairpins of parvovirus genomes and their role in DNA replication, p. 59–79. *In* P. Tijssen (ed.), *Handbook of Parvoviruses*, vol. I. CRC Press, Inc., Boca Raton, Fla.

Astell, C. R., and M. C. Blundell. 1989. Sequence of the right hand terminal palindrome of the human B19 parvovirus genome has the potential to form a "stem plus arms" structure. *Nucleic Acids Res.* **17**:5857.

Azzi, A., K. Zakrzewska, G. Gentilomi, M. Musiani, and M. Zerbini. 1990. Detection of B19 parvovirus infections by a dot-blot hybridization assay using a digoxigenin-labeled probe. *J. Virol. Methods* **27**:125–133.

Bansal, G. P., J. A. Hatfield, F. E. Dunn, A. A. Kramer, F. Brady, C. H. Riggin, M. S. Collett, K. Yoshimoto, S. Kajigaya, and N. S. Young. 1993. Candidate recombinant vaccine for human B19 parvovirus. *J. Infect. Dis.* **167**:1034–1044.

Bell, L. M., S. J. Naides, P. Stoffman R. L. Hodinka, and S. A. Plotkin. 1989. Human parvovirus B 19 infection among hospital staff members after contact with infected patients. *N. Engl. J. Med.* **321**:485–491.

Berg, A. M., S. J. Naides, and R. W. Simms. 1993. Established fibromyalgia syndrome and parvovirus B19 infection. *J. Rheumatol.* **20**:1941–1943.

Berns, K. I., and W. W. Hauswirth. 1984. Adeno-associated virus DNA structure and replication, p. 1–31. *In* K. I. Berns (ed.), *The Parvoviruses*. Plenum Press, New York, N.Y.

Bertrand, Y., J. J. Lefrere, G. Leverger, A. M. Courouce, C. Feo, M. Clark, G. Schaison, and J. P. Soulier. 1985. Autoimmune haemolytic anaemia revealed by human parvovirus linked erythroblastopenia. *Lancet* ii:382–383.

Blundell, M. C., C. Beard, and C. R. Astel. 1987. In vitro identification of a B19 parvovirus promoter. *Virology* **157**:534–538.

Bond, P. R., E. Q. Caul, J. Usher, B. J. Cohen, J. P. Clewley, and A. M. Field. 1986. Intrauterine infection with human parvovirus. *Lancet* **i**:448–449.

Bowman, C. A., B. J. Cohen, D. R. Norfolk, and C. J. Lacey. 1990. Red cell aplasia associated with human parvovirus B19 and HIV infection: failure to respond clinically to intravenous immunoglobulin. *AIDS* **4**:1038–1039.

Brandrup, F., and P. O. Larsen. 1976. Erythema infectiosum (fifth disease). *Br. Med. J.* **ii**:47–48.

Brown, C. S., M. J. A. W. M. van Bussel, A. L. M. Wassenaar, A. W. van Elsacker-Niele, H. T. Weiland, and M. M. M. Salimans. 1990. An immunofluorescence assay for the detection of parvovirus B19 IgG and IgM antibodies based on recombinant viral antigen. *J. Virol. Methods* **29**:53–62.

Brown, K. E., J. Mori, B. J. Cohen, and A. M. Field. 1991. In vitro propagation of parvovirus B19 in primary foetal liver culture. *J. Gen. Virol.* **72**:741–745.

Brown, T., A. Anand, L. D. Ritchie, J. P. Clewley, and T. M. Reid. 1984. Intrauterine parvovirus infection associated with hydrops fetalis. *Lancet.* **ii**:1033–1034.

Burton, P. A. 1986. Intranuclear inclusions in marrow of hydropic fetus due to parvovirus infection. *Lancet* **ii**:1155.

Carrington, D., D. H. Gilmore, M. J. Whittle, D. Aitken, A. A. Gibson, W. J. Patrick, T. Brown, E. O. Caul, A. M. Field, and J. P. Clewley. 1987. Maternal serum alpha-fetoprotein—a marker of fetal aplastic crisis during intrauterine human parvovirus infection. *Lancet* **i**:433–435.

Carter, B. J., E. Mendelson, and J. P. Trempe. 1990. AAV DNA replication, integration, and genetics, p. 169–226. *In* P. Tijssen (ed.), *Handbook of Parvoviruses*, vol. I. CRC Press, Inc., Boca Raton, Fla.

Caul, E. O., M. J. Usher, and P. A. Burton. 1988. Intrauterine infection with human parvovirus B19: a light and electron microscopy study. *J. Med. Virol.* **24**:55–66.

Chorba, T., P. Coccia, R. C. Holman, P. Tattersall, L. J. Anderson, J. Sudman, N. S. Young, E. Kurczynski, U. M. Saarinen, and R. Moir. 1986. The role of parvovirus B 19 in aplastic crisis and erythema infectiosum (fifth disease). *J. Infect. Dis.* **154**:383–393.

Chrystie, I. L., J. D. Almeida, and J. Welch. 1990. Electron microscopic detection of human parvovirus (B19) in a patient with HIV infection. *J. Med. Virol.* **30**:249–252.

Clarke, H. C. 1984. Erythema infectiosum: an epidemic with a probable posterythema phase. *Can. Med. Assoc. J.* **130**:603–604.

Clewley, J. P. 1984. Biochemical characterization of a human parvovirus. *J. Gen. Virol.* **65**:241–245.

Clewley, J. P. 1985. Detection of human parvovirus using a molecularly cloned probe. *J. Med. Virol.* **15**:173–181.

Clewley, J. P. 1989. Polymerase chain reaction assay of parvovirus B19 DNA in clinical specimens. *J. Clin. Microbiol.* **27**:2647–2651.

Clewley, J. P., B. J. Cohen, and A. M. Field. 1987. Detection of parvovirus B19 DNA, antigen, and particles in the human fetus. *J. Med. Virol.* **23**:367–376.

Cohen, B. J. 1988. Laboratory tests for the diagnosis of infection with B19 virus, p. 69–83. *In* J. R. Pattison (ed.), *Parvoviruses and Human Disease*. CRC Press, Inc., Boca Raton, Fla.

Cohen, B. J., and J. A. Shirley. 1985. Dual infection with rubella and human parvovirus. *Lancet* **ii**:662–663.

Cohen, B. J., and E. M. Supran. 1987. IgM serology for rubella and human parvovirus B19. *Lancet* **i**:393.

Cohen, B. J., P. P. Mortimer, and M. S. Pereira. 1983a. Human parvovirus, the cause of erythema infectiosum (fifth disease)? *Lancet* **i**:1378.

Cohen, B. J., P. P. Mortimer, and M. S. Pereira. 1983b. Diagnostic assays with monoclonal antibodies for the human serum parvovirus-like virus (SPLV). *J. Hyg.* **91**:113–130.

Cohen, B. J., A. M. Courouce, T. F. Schwarz, K. Okochi, and G. J. Kurtzman. 1988. Laboratory infection with parvovirus B19. *J. Clin. Pathol.* **41**:1027–1028.

Corbau, R., N. Salom, J. Rommelaere, and J. P. Nuesch. 1999. Phosphorylation of the viral nonstructural protein NS1 during MVMp infection of A9 cells. *Virology* **5**:402–415.

Corman, L. C., and R. Staud. 1995. Association of Wegener's granulomatosis with parvovirus B19 infection: comment on the concise communication by Nikkari et al. *Arthritis Rheum.* **38**:1174–1175.

Corman, L. C., and D. J. Dolson. 1992. Polyarteritis nodosa and parvovirus B19 infection. *Lancet* **339**:491.

Cossart, Y. E., A. M. Field, B. Cant, and D. Widdows. 1975. Parvovirus-like particles in human sera. *Lancet* **i**:72–73.

Cotmore, S. F., and P. Tattersall. 1984. Characterization and molecular cloning of a human parvovirus genome. *Science* **226**:1161–1165.

Cotmore, S. F., and P. Tattersall. 1989. A genome-linked copy of the NS-1 polypeptide is located on the outside of infectious parvovirus particles. *J. Virol.* **63**:3902–3911.

Cotmore, S. F., V. C. McKie, L. J. Anderson, C. R. Astell, and P. Tattersall. 1986. Identification of the major structural and non-structural proteins encoded by human parvovirus B19 and mapping of their genes by procaryotic expression of isolated genomic fragments. *J. Virol.* **60**:548–557.

Cramp, H. E., and B. D. J. Armstrong. 1976. Erythema infectiosum: an outbreak of "slapped cheek" disease in north Devon. *Br. Med. J.* **i**:885–886.

Davidson, R. J., T. Brown, and D. Wiseman. 1984. Human parvovirus infection and aplastic crisis in hereditary spherocytosis. *J. Infect.* **9**:298–300.

Deiss, V., J. D. Tratschin, M. Weitz, and G. Siegl. 1990. Cloning of the human parvovirus B19 genome and structural analysis of its palindromic termini. *Virology* **175**:247–254.

de Mayolo, J. A., and J. D. Temple. 1990. Pure red cell aplasia due to parvovirus B19 infection in a man with HIV infection. *South. Med. J.* **83**:1480–1481.

Denning, D. W., A. Amos, P. Rudge, and B. J. Cohen. 1987. Neuralgic amyotrophy due to parvovirus infection. *J. Neurol. Neurosurg. Psychiatry* **50**:641–642.

Dettwiler, S., J. Rommelaere, and J. P. Nuesch. 1999. DNA unwinding functions of minute virus of mice NS1 protein are modulated specifically by the lambda isoform protein kinase C. *J. Virol.* **73**:7410–7420.

Drago, F., A. Parodi, and A. Rebora. 1997. Gloves-and-socks syndrome in a patient with Epstein-Barr virus infection. *Dermatology* **194**:374.

Duncan, J. R., C. B. Potter, M. D. Cappellini, J. B. Kurtz, M. J. Anderson, and D. J. Weatherall. 1983. Aplastic crisis due to parvovirus infection in pyruvate kinase deficiency. *Lancet* **ii**:14–16.

Erdman, D. D., E. L. Durigon, Q. Y. Wang, and L. J. Anderson. 1996. Genetic diversity of human parvovirus B19: sequence analysis of the VP1/VP2 gene from multiple isolates. *J. Gen. Virol.* **77**:2767–2774.

Erdman, D. D., B. C. Anderson, T. J. Torok, T. H. Finkel, and L. J. Anderson. 1997. Possible transmission of parvovirus B19 from intravenous immune globulin. *J. Med. Virol.* **53**:233–236.

Etienne, A., and M. Harms. 1996. Cutaneous manifestations of parvovirus B19 infection. *Presse Med.* **25:**1162–1165.

Evans, J. P., M. A. Rossiter, T. O. Kumaran, G. W. Marsh, and P. P. Mortimer. 1984. Human parvovirus aplasia: case due to cross infection in a ward. *Br. Med. J.* **iii:**681.

Faden, H., G. W. Gary, Jr., and M. Korman. 1990. Numbness and tingling of fingers associated with parvovirus B19 infection. *J. Infect. Dis.* **161:**354–355.

Faden, H., G. W. Gary, Jr., and L. J. Anderson. 1992. Numbness and tingling of fingers associated with parvovirus B19 infection. *Clin. Infect. Dis.* **15:**595–597.

Feder, H. M., Jr. 1994. Fifth disease. *N. Engl. J. Med.* **331:** 1062.

Finkel, T. H., T. J. Torok, P. J. Ferguson, E. L. Durigon, S. R. Zaki, D. Y. M. Leung, R. J. Harbeck, E. W. Gelfand, F. T. Saulsbury, J. R. Hollister, and L. J. Anderson. 1994. Chronic parvovirus B19 infection and systemic necrotizing vasculitis: opportunistic infection or aetiological agent? *Lancet* **343:**1255–1258.

Foto, F., K. G. Saag, L. L. Scharosch, E. J. Howard, and S. J. Naides. 1993. Parvovirus B19-specific DNA in bone marrow from B19 arthropathy patients: evidence for B19 virus persistence. *J. Infect. Dis.* **167:**744–748.

Frickhofen, N., and N. S. Young. 1989. Persistent parvovirus B19 infections in humans. *Microb. Pathog.* **7:**319–327.

Frickhofen, N., and N. S. Young. 1990. Polymerase chain reaction for detection of parvovirus B19 in immunodeficient patients with anemia. *Behring Inst. Mitt.* **85:**46–54.

Frickhofen, N., J. L. Abkowitz, M. Safford, J. M. Berry, J. Antunez de Mayolo, A. Astrow, R. Cohen, I. Halperin, L. King, and D. Mintzer. 1990. Persistent B19 parvovirus infection in patients infected with human immunodeficiency virus type 1 (HIV-1): a treatable cause of anemia in AIDS. *Ann. Intern. Med.* **113:**926–933.

Fridell, E., B. J. Cohen, and B. Wahren. 1991. Evaluation of a synthetic-peptide enzyme-linked immunosorbent assay for immunoglobulin M to human parvovirus B19. *J. Clin. Microbiol.* **29:**1376–1381.

Fritz, B., K. Moore, and S. J. Naides. 1992. A combined pseudoreplica immunochemical technique for research and diagnostic virology. *J. Microsc. Res. Techn.* **21:**59–64.

Goldman, F., H. Rotbart, K. Gutierrez, and D. Ambruso. 1990. Parvovirus-associated aplastic crisis in a patient with red blood cell glucose-6-phosphate dehydrogenase deficiency. *Pediatr. Infect. Dis. J.* **9:**593–594.

Graeve, J. L., P. A. de Alarcon, and S. J. Naides. 1989. Parvovirus B19 infection in patients receiving cancer chemotherapy: the expanding spectrum of disease. *Am. J. Pediatr. Hematol. Oncol.* **11:**441–444.

Gray, E. S., A. Anand, and T. Brown. 1986. Parvovirus infections in pregnancy. *Lancet* **i:**208.

Gray, E. S., R. J. Davidson, and A. Anand. 1987. Human parvovirus and fetal anaemia. *Lancet* **i:**1144.

Green, D. H., A. J. Bellingham, and M. J. Anderson. 1984. Parvovirus infection in a family associated with aplastic crisis in an affected sibling pair with hereditary spherocytosis. *J. Clin. Pathol.* **37:**1144–1146.

Hanada, T., K. Koike, T. Takeya, T. Nagasawa, Y. Matsunaga, and H. Takita. 1988. Human parvovirus B19-induced transient pancytopenia in a child with hereditary spherocytosis. *Br. J. Haematol.* **70:**113–115.

Hassam, S., J. Briner, J. D. Tratschin, G. Siegl, and P. U. Heitz. 1990. In situ hybridization for the detection of human parvovirus B19 nucleic acid sequences in paraffin-embedded specimens. *Virchows Arch. B.* **59:**257–261.

Johansen, J. N., L. S. Christensen, K. Zakrzewska, K. Carlsen, A. Hornsleth, and A. Azzi. 1998. Typing of European strains of parvovirus B19 by restriction endonuclease analyses and sequencing: identification of evolutionary lineages and evidence of recombination of markers from different lineages. *Virus Res.* **53:**215–223.

Joseph, P. R. 1986. Incubation period of fifth disease. *Lancet* **i:**1390–1391.

Kajigaya, S., T. Shimada, S. Fujita, and N. S. Young. 1989. A genetically engineered cell line that produces empty capsids of B19 (human) parvovirus. *Proc. Natl. Acad. Sci. USA* **86:**7601–7605.

Karetnyi, Y. V., P. R. Beck, R. S. Markin, A. N. Langnas, and S. J. Naides. 1999. Human parvovirus B19 infection in acute fulminant liver failure. *Arch. Virol.* **144:**1713–1724.

Kelleher, J. F., Jr., N. L. Luban, B. J. Cohen, and P. P. Mortimer. 1984. Human serum parvovirus as the cause of aplastic crisis in sickle cell disease. *Am. J. Dis. Child.* **138:**401–403.

Kilbourne, E. D., C. P. Cerini, M. W. Khan, J. W. Mitchell, Jr., and P. L. Ogra. 1987. Immunologic response to the influenza virus neuraminidase is influenced by prior experience with the associated viral hemagglutinin. *J. Immunol.* **138:**3010–3013.

Kinney, J. S., L. J. Anderson, E. Farrar, R. A. Strikas, M. L. Kumar, R. M. Kuegman, J. L. Sever, E. S. Hurwitz, and R. K. Sikes. 1988. Risk of adverse outcomes of pregnancy after human parvovirus B19 infection. *J. Infect. Dis.* **157:**663–667.

Klouda, P. T., S. A. Corbin, B. A. Bradley, B. J. Cohen, and A. D. Woolf. 1986. HLA and acute arthritis following human parvovirus infection. *Tissue Antigens* **28:**318–319.

Knisely, A. S. 1990. Parvovirus B19 infection in the fetus. *Lancet* **336:**443.

Knisely, A. S., P. A. O'Shea, P. McMillan, D. B. Singer, and M. S. Magid. 1988. Electron microscopic identification of parvovirus virions in erythroid-line cells in fatal hydrops fetalis. *Pediatr. Pathol.* **8:**163–170.

Knott, P. D., G. A. Weiply, and M. J. Anderson. 1984. Serologically proved intrauterine infection with parvovirus. *Br. Med. J.* **iii:**1660.

Koch, W. C., and S. P. Adler. 1990. Detection of human parvovirus B19 DNA by using the polymerase chain reaction. *J. Clin. Microbiol.* **28:**65–69.

Komatsu, N., H. Nakauchi, A. Miwa, T. Ishihara, M. Eguchi, M. Moroi, M. Okada, Y. Sato, H. Wada, Y. Yawata, T. Suda, and Y. Miura. 1991. Establishment and characterization of a human leukemic cell line with megakaryocytic features: dependency on granulocyte macrophage colony-stimulating factor, interleukin 3, or erythropoietin for growth and survival. *Cancer Res.* **51:**341–348.

Kurtz, J. B., and M. J. Anderson. 1985. Cross-reactions in rubella and parvovirus specific IgM tests. *Lancet* **ii:**1356.

Kurtzman, G., N. Frickhofen, J. Kimball, D. W. Jenkins, A. W. Nienhuis, and N. S. Young. 1989a. Pure red-cell aplasia of 10 years duration due to persistent parvovirus B19 infection and its cure with immunoglobulin therapy. *N. Engl. J. Med.* **321:**519–523.

Kurtzman, G. J., K. Ozawa, B. Cohen, G. Hanson, R. Oseas, and N. S. Young. 1987. Chronic bone marrow failure due to persistent B19 parvovirus infection. *N. Engl. J. Med.* **317:** 287–294.

Kurtzman, G. J., B. Cohen, P. Meyers, A. Amunullah, and N. S. Young. 1988. Persistent B19 parvovirus infection as a cause of severe chronic anaemia in children with acute lymphocytic leukaemia. *Lancet* **ii:**1159–1162.

Kurtzman, G. J., B. J. Cohen, A. M. Field, R. Oseas, R. M. Blaese, and N. S. Young. 1989b. Immune response to B19 parvovirus and an antibody defect in persistent viral infection. *J. Clin. Investig.* **84:**1114–1123.

Langnas, A. N., R. S. Markin, M. S. Cattral, and S. J. Naides. 1995. Parvovirus B19 as a possible causative agent of fulminant liver failure and associated aplastic anemia. *Hepatology* **22:**1661–1665.

Lauer, B. A., J. N. MacCormack, and C. Wilfert. 1976. Erythema infectiosum: an elementary school outbreak. *Am. J. Dis. Child.* **130:**252–254.

Lefrere, J. J., A. M. Courouce, J. Y. Muller, M. Clark, and J. P. Soulier. 1985. Human parvovirus and purpura. *Lancet* **i:**730.

Lefrere, J. J., A. M. Courouce, Y. Bertrand, R. Girot, and J. P. Soulier. 1986a. Human parvovirus and aplastic crisis in chronic hemolytic anemias: a study of 24 observations. *Am. J. Hematol.* **23:**271–275.

Lefrere, J. J., A. M. Courouce, C. Boucheix, C. Chomienne, A. Bernadou, and J. P. Soulier. 1986b. Aplastic crisis and erythema infectiosum (fifth disease) revealing a hereditary spherocytosis in a familial human parvovirus infection. *Nouv. Rev. Fr. Hematol.* **28:**7–9.

Lefrere, J. J., A. M. Courouce, R. Girot, Y. Bertrand, and J. P. Soulier. 1986c. Six cases of hereditary spherocytosis revealed by human parvovirus infection. *Br. J. Haematol.* **62:**653–658.

Lefrere, J. J., A. M. Courouce, R. Girot, and P. Cornu. 1986d. Human parvovirus and thalassaemia. *J. Infect.* **13:**45–49.

Lefrere, J. J., A. M. Courouce, J. P. Soulier, M. P. Cordier, M. C. Guesne Girault, C. Polonovski, and A. Bensman. 1986e. Henoch-Schonlein purpura and human parvovirus infection. *Pediatrics* **78:**183–184.

Lefrere, J. J., R. Girot, A. M. Courouce, M. Maier Redelsperger, and P. Cornu. 1986f. Familial human parvovirus infection associated with anemia in siblings with heterozygous beta-thalassemia. *J. Infect. Dis.* **153:**977–979.

Lefrere, J. J., A. M. Courouce, and C. Kaplan. 1989. Parvovirus and idiopathic thrombocytopenic purpura. *Lancet* **i:**279.

Leruez-Ville, M., A. Lauge, F. Morinet, L. Guillevin, and P. Deny. 1994. Polyarteritis nodosa and parvovirus B19. *Lancet* **344:**263–264.

Longo, G., M. Luppi, M. Bertesi, L. Ferrara, G. Torelli, and G. Emilia. 1998. Still's disease, severe thrombocytopenia, and acute hepatitis associated with acute parvovirus B19 infection. *Clin. Infect. Dis.* **26:**994–995.

Luzzi, G. A., J. B. Kurtz, and H. Chapel. 1985. Human parvovirus arthropathy and rheumatoid factor. *Lancet* **i:**1218.

Lyon, D. J., C. S. Chapman, C. Martin, K. E. Brown, J. P. Clewley, A. J. Flower, and V. E. Mitchell. 1989. Symptomatic parvovirus B19 infection and heat-treated factor IX concentrate. *Lancet* **i:**1085.

Mabin, D. C., and V. Chowdhury. 1990. Aplastic crisis caused by human parvovirus in two patients with hereditary stomatocytosis. *Br. J. Haematol.* **76:**153–154.

Mansfield, F. 1988. Erythema infectiosum. Slapped face disease. *Aust. Fam. Physician* **17:**737–738.

Matsunaga, Y., T. Matsukura, S. Yamazaki, M. Sugase, and R. Izumi. 1987. Hydrops fetalis caused by intrauterine human parvovirus infection. *Jpn. J. Med. Sci. Biol.* **40:**165–169.

Metzman, R., A. Anand, P. A. DeGiuho, and A. S. Knisely. 1989. Hepatic disease associated with intrauterine parvovirus B19 infection in a newborn premature infant. *J. Pediatr. Gastroenterol. Nutr.* **9:**112–114.

Mori, J., P. Beattie, D. W. Melton, B. J. Cohen, and J. P. Clewley. 1987. Structure and mapping of the DNA of human parvovirus B19. *J. Gen. Virol.* **68:**2797–2806.

Mori, J., A. M. Field, J. P. Clewley, and B. J. Cohen. 1989. Dot blot hybridization assay of B19 virus DNA in clinical specimens. *J. Clin. Microbiol.* **27:**459–464.

Morinet, F., L. D'Auriol, J. D. Tratschin, and F. Galibert. 1989. Expression of the human parvovirus B19 protein fused to protein A in *Escherichia coli*: recognition by IgM and IgG antibodies in human sera. *J. Gen. Virol.* **70:**3091–3097.

Mortimer, P. P., B. J. Cohen, M. M. Buckley, J. E. Cradock Watson, M. K. Ridehalgh, F. Burkhardt, and U. Schilt. 1985a. Human parvovirus and the fetus. *Lancet* **ii:**1012.

Mortimer, P. P., B. J. Cohen, M. A. Rossiter, S. M. Fairhead, and A. F. M. S. Rahman. 1985b. Human parvovirus and purpura. *Lancet* **i:**730–731.

Muir, K., W. T. A. Todd, W. H. Watson, and E. Fitzsimons. 1992. Viral-associated haemophagocytosis with parvovirus-B19-related pancytopenia. *Lancet* **339:**1139–1140.

Munshi, N. C., S. Zhou, M. J. Woody, D. A. Morgan, and A. Srivastava. 1993. Successful replication of parvovirus B19 in the human megakaryocytic leukemia cell line MB-02. *J. Virol.* **67:**562–566.

Mynott, M. J. 1985. An epidemic of erythema infectiosum in a school. *Practitioner* **229:**767–768.

Naides, S. J. 1987. Transient liver enzyme abnormalities in acute human parvovirus (HPV) infection. *Clin. Res.* **35:**859A.

Naides, S. J. 1989. Infection control measures for human parvovirus B19 in the hospital setting. *Infect. Control Hosp. Epidemiol.* **10:**326–329.

Naides, S. J. 1998. Rheumatic manifestations of parvovirus B19 infection. *Rheum. Dis. Clin. N. Am.* **24:**375–401.

Naides, S. J., and E. H. Field. 1988. Transient rheumatoid factor positivity in acute human parvovirus B19 infection. *Arch. Intern. Med.* **148:**2587–2589.

Naides, S. J., and C. P. Weiner. 1989. Antenatal diagnosis and palliative treatment of nonimmune hydrops fetalis secondary to fetal parvovirus B19 infection. *Prenat. Diagn.* **9:**105–114.

Naides, S. J., W. Piette, L. A. Veach, and Z. Argenyi. 1988. Human parvovirus B19-induced vesiculopustular skin eruption. *Am. J. Med.* **84:**968–972.

Naides, S. J., L. L. Scharosch, F. Foto, and E. J. Howard. 1990. Rheumatologic manifestations of human parvovirus B19 infection in adults. Initial two-year clinical experience. *Arthritis Rheum.* **33:**1297–1309.

Naides, S. J., F. Foto, J. L. Marsh, L. L. Scharosch, and E. J. Howard. 1991. Synovial tissue analysis in patients with chronic parvovirus B19 arthropathy. *Clin. Res.* **39:**733A.

Naides, S. J., E. J. Howard N. S. Swack, and J. T. Stapleton. 1993. Parvovirus B19 infection in human immunodeficiency virus type 1 infected persons failing or intolerant to Zidovudine therapy. *J. Infect. Dis.* **168:**101–105.

Nikkari, S., J. Mertsola, H. Korvenranta, R. Vainionpää, and P. Toivanen. 1994. Wegener's granulomatosis and parvovirus B19 infection. *Arthritis Rheum.* **37:**1707–1708.

Nikkari, S., R. Vainionpaa, P. Toivanen, W. L. Gross, N. Mistry, E. Csernok, W. Szpirt, B. Baslund, and A. Wiik. 1997. Association of Wegener's granulomatosis with parvovirus B19 infection: comment on the concise communication by Nikkari et al., reply. *Arthritis Rheum.* **38:**1175.

Nocton, J. J., L. C. Miller, L. B. Tucker, and J. G. Schaller. 1993. Human parvovirus B19-associated arthritis in children. *J. Pediatr.* **122:**186–190.

Nunoue, T., K. Okochi, P. P. Mortimer, and B. J. Cohen. 1985. Human parvovirus (B19) and erythema infectiosum. *J. Pediatr.* **107**:38–40.

Okabe, N., S. Koboyashi, O. Tatsuzawa, and P. P. Mortimer. 1984. Detection of antibodies to human parvovirus in erythema infectiosum (fifth disease). *Arch. Dis. Child.* **59**:1016–1019.

Oliver, A. R., and A. D. Phillips. 1988. An electron microscopical investigation of faecal small round viruses. *J. Med. Virol.* **24**:211–218.

Ozawa, K., and N. Young. 1987. Characterization of capsid and noncapsid proteins of B19 parvovirus propagated in human erythroid bone marrow cell cultures. *J. Virol.* **61**:2627–2630.

Ozawa, K., G. Kurtzman, and N. Young. 1986. Replication of the B19 parvovirus in human bone marrow cell cultures. *Science* **233**:883–886.

Ozawa, K., J. Ayub, Y. S. Hao, G. Kurtzman, T. Shimada, and N. Young. 1987a. Novel transcription map for the B19 (human) pathogenic parvovirus. *J. Virol.* **61**:2395–2406.

Ozawa, K., G. Kurtzman, and N. Young. 1987b. Productive infection by B19 parvovirus of human erythroid bone marrow cells in vitro. *Blood* **70**:384–391.

Pattison, J. R. (ed.). 1988. *Parvoviruses and Human Disease*, p. 1–4. CRC Press, Inc., Boca Raton, Fla.

Pattison, J. R., S. E. Jones, J. Hodgson, L. R. Davis, J. M. White, C. E. Stroud, and L. Murtaza. 1981. Parvovirus infections and hypoplastic crisis in sickle-cell anaemia. *Lancet* **i**:664–665.

Paver, W. K., E. O. Caul, C. R. Ashley, and S. K. Clarke. 1973. A small virus in human faeces. *Lancet* **i**:237–239.

Peters, M. T., and K. H. Nicolaides. 1990. Cordocentesis for the diagnosis and treatment of human fetal parvovirus infection. *Obstet. Gynecol.* **75**:501–504.

Plummer, F. A., G. W. Hammond, K. Forward, L. Sekia, L. M. Thompson, S. E. Jones, I. M. Kidd, and M. J. Anderson. 1985. An erythema infectiosum-like illness caused by human parvovirus infection. *N. Engl. J. Med.* **313**:74–79.

Rao, K. R., A. R. Patel, M. J. Anderson, J. Hodgson, S. E. Jones, and J. R. Pattison. 1983. Infection with parvovirus-like virus and aplastic crisis in chronic hemolytic anemia. *Ann. Intern. Med.* **98**:930–932.

Rao, S. P., S. T. Miller, and B. J. Cohen. 1990. Severe anemia due to B19 parvovirus infection in children with acute leukemia in remission. *Am. J. Pediatr. Hematol. Oncol.* **12**:194–197.

Rappaport, E. S., G. Quick, D. Ransom, B. Helbert, and L. S. Frankel. 1989. Aplastic crisis in occult hereditary spherocytosis caused by human parvovirus (HPV B19). *South. Med. J.* **2**:247–251.

Rechavi, G., A. Vonsover, Y. Manor, F. Mileguir, O. Shpilberg, G. Kende, F. Brok Simoni, M. Mandel, T. Gotlieb Stematski, and I. Ben Bassat. 1989. Aplastic crisis due to human B19 parvovirus infection in red cell pyrimidine-5′-nucleotidase deficiency. *Acta Haematol.* **82**:46–49.

Reid, D. M., T. M. Reid, T. Brown, J. A. Rennie, and C. J. Eastmond. 1985. Human parvovirus-associated arthritis: a clinical and laboratory description. *Lancet* **i**:422–425.

Rubinstein, A. I., and D. B. Rubinstein. 1990. Inability of solvent-detergent (S-D) treated factor VIII concentrate to inactivate parvoviruses and non-lipid enveloped non-A, non-B hepatitis virus in factor VIII concentrate: advantages to using sterilizing 100° C dry heat treatment. *Am. J. Hematol.* **35**:142.

Saarinen, U. M., T. L. Chorba, P. Tattersall, N. S. Young, L. J. Anderson, E. Palmer, and P. F. Coccia. 1986. Human parvovirus B19-induced epidemic acute red cell aplasia in patients with hereditary hemolytic anemia. *Blood* **67**:1411–1417.

Sahakian, V., C. P. Weiner, S. J. Naides, R. A. Williamson, and L. L. Scharosch. 1991. Intrauterine transfusion treatment of nonimmune hydrops fetalis secondary to human parvovirus B19 infection. *Am. J. Obstet. Gynecol.* **164**:1090–1091.

Saldanha, J., and P. Minor. 1996. Detection of human parvovirus B19 DNA in plasma pools and blood products derived from these pools: implications for efficiency and consistency of removal of B19 DNA during manufacture. *Br. J. Haematol.* **93**:714–719.

Salimans, M. M. 1990. Detection of human parvovirus B19 DNA by dot-hybridization and the polymerase chain reaction: applications for diagnosis of infections. *Behring Inst. Mitt.* **85**:39–45.

Salimans, M. M., F. M. van de Rijke, A. K. Raap, and A. M. van Elsacker Niele. 1989. Detection of parvovirus B19 DNA in fetal tissues by in situ hybridization and polymerase chain reaction. *J. Clin. Pathol.* **42**:525–530.

Samii, K., P. Cassinotti, J. de Freudenreich, Y. Gallopin, D. Le Fort, and H. Stalder. 1996. Acute bilateral carpal tunnel syndrome associated with human parvovirus B19 infection. *Clin. Infect. Dis.* **22**:162–164.

Sasaki, T., Y. Takahashi, K. Yoshinaga, K. Sugamura, and H. Shiraishi. 1989. An association between human parvovirus B-19 infection and autoantibody production. *J. Rheumatol.* **16**:708–709.

Sato, H., J. Hirata, M. Furukawa, N. Kuroda, H. Shiraki, Y. Maeda, and K. Okochi. 1991a. Identification of the region including the epitope for a monoclonal antibody which can neutralize human parvovirus B19. *J. Virol.* **65**:1667–1672.

Sato, H., J. Hirata, N. Kuroda, H. Shiraki, Y. Maeda, and K. Okochi. 1991b. Identification and mapping of neutralizing epitopes of human parvovirus B 19 by using human antibodies. *J. Virol.* **65**:5485–5490.

Schwarz, T. F., A. Nerlich, B. Hottentrager, G. Jager, I. Wiest, S. Kantimm, H. Roggendorf, M. Schultz, K.-P. Gloning, T. Schramm, W. Holzgreve, and M. Roggendorf. 1991. Parvovirus B 19 infection of the fetus: histology and in situ hybridization. *Am. J. Clin. Pathol.* **96**:121–126.

Semble, E. L., C. A. Agudelo, and P. S. Pegram. 1987. Human parvovirus B19 arthropathy in two adults after contact with childhood erythema infectiosum. *Am. J. Med.* **83**:560–562.

Serjeant, G. R., J. M. Topley, K. Mason, B. E. Serjeant, J. R. Pattison, S. E. Jones, and R. Mohamed. 1981. Outbreak of aplastic crises in sickle cell anaemia associated with parvovirus-like agent. *Lancet* **ii**:595–597.

Shade, R. O., M. C. Blundell, S. F. Cotmore, P. Tattersall, and C. R. Astell. 1986. Nucleotide sequence and genome organization of human parvovirus B19 isolated from the serum of a child during aplastic crisis. *J. Virol.* **58**:921–936.

Shimomura, S., N. Komatsu, N. Frickhofen, S. Anderson, S. Kajigaya, and N. S. Young. 1992. First continuous propagation of B19 parvovirus in a cell line. *Blood* **79**:18–24.

Shiraishi, H., D. Wong, R. H. Purcell, R. Shirachi, E. Kumasaka, and Y. Numazaki. 1985. Antibody to human parvovirus in outbreak of erythema infectiosum in Japan. *Lancet* **i**:982–983.

Shiraishi, H., K. Umetsu, H. Yamamoto, Y. Hatakeyama, N. Yaegashi, and K. Sugamura. 1989. Human parvovirus (HPV B19) infection with purpura. *Microbiol. Immunol.* **33**:369–372.

Shirono, K., and H. Tsuda. 1995. Parvovirus B19-associated haemophagocytic syndrome in healthy adults. *Br. J. Haematol.* **89**:923–926.

Shneerson, J. M., P. P. Mortimer, and E. M. Vandervelde. 1980. Febrile illness due to a parvovirus. *Br. Med. J.* **280:**1580.

Simpson, R. W., L. McGinty, L. Simon, C. A. Smith, C. W. Godzeski, and R. J. Boyd. 1984. Association of parvoviruses with rheumatoid arthritis of humans. *Science* **223:**1425–1428.

Söderlund, M., R. von Essen, J. Haapasaari, U. Kiistala, O. Kiviluoto, and K. Hedman. 1997. Persistence of parvovirus B19 DNA in synovial membranes of young patients with and without chronic arthropathy. *Lancet* **349:**1063–1065.

Soloninka, C. A., M. J. Anderson, and C. A. Laskin. 1989. Anti-DNA and antilymphocyte antibodies during acute infection with human parvovirus B19. *J. Rheumatol.* **16:**777–781.

Soothill, P. 1990. Intrauterine blood transfusion for non-immune hydrops fetalis due to parvovirus B19 infection. *Lancet* **336:**121–122.

Srivastava, C. H., S. Z. Zhou, N. C. Munshi, and A. Srivastava. 1992. Parvovirus B19 replication in human umbilical cord blood cells. *Virology* **189:**456–461.

Stierle, G., K. A. Brown, S. G. Rainsford, C. A. Smith, D. Hamerman, H. E. Stierle, and D. C. Dumonde. 1987. Parvovirus associated antigen in the synovial membrane of patients with rheumatoid arthritis. *Ann. Rheum. Dis.* **46:**219–223.

Stone, M. S., and J. R. Murph. 1993. Papular-purpuric gloves and socks syndrome: a characteristic viral exanthem. *Pediatrics* **92:**864–865.

Summerfield, G. P., and G. P. Wyatt. 1985. Human parvovirus infection revealing hereditary spherocytosis. *Lancet* **ii:**1070.

Summers, J., S. E. Jones, and M. J. Anderson. 1983. Characterization of the genome of the agent of erythrocyte aplasia permits its classification as a human parvovirus. *J. Gen. Virol.* **64:**2527–2532.

Suzuki, N., S. Terada, and M. Inoue. 1995. Neonatal meningitis with human parvovirus B19 infection. *Arch. Dis. Child. Fetal Neonatal* **73:**F196–F197.

Takahashi, M., T. Koike, Y. Moriyama, A. Shibata, R. Koike, M. Sanada, and T. Tsukada. 1986. Inhibition of erythropoiesis by human parvovirus-containing serum from a patient with hereditary spherocytosis in aplastic crisis. *Scand. J. Haematol.* **37:**118–124.

Takahashi, N., N. Takada, T. Hashimoto, and T. Okamoto. 1999. Genetic heterogeneity of the immunogenic viral capsid protein region of human parvovirus B19 isolates obtained from an outbreak in a pediatric ward. *FEBS Lett.* **450:**289–293.

Takahashi, T., K. Ozawa, K. Mitani, K. Miyazono, S. Asano, and F. Takaku. 1989. B19 parvovirus replicates in erythroid leukemic cells in vitro. *J. Infect. Dis.* **160:**548–549.

Takahashi, Y., C. Murai, S. Shibata, Y. Munakata, T. Ishii, K. Ishii, T. Saitoh, T. Sawai, K. Sugamura, and T. Sasaki. 1998. Human parvovirus B19 as a causative agent for rheumatoid arthritis. *Proc. Natl. Acad. Sci. USA* **95:**8227–8232.

Tattersall, P., and D. C. Ward. 1976. Rolling hairpin model for replication of parvovirus and linear chromosomal DNA. *Nature* **263:**106–109.

Tsuda, H. 1993. Liver dysfunction caused by parvovirus B19. *Am. J. Gastroenterol.* **88:**1463.

Tsuda, H., Y. Maeda, and K. Nakagawa. 1993. Parvovirus B19-related lymphadenopathy. *Br. J. Haematol.* **85:**631–632.

Tsukada, T., T. Koike, R. Koike, M. Sanada, M. Takahashi, A. Shibata, and T. Nunoue. 1985. Epidemic of aplastic crisis in patients with hereditary spherocytosis in Japan. *Lancet* **i:**1401.

Turton, J., H. Appleton, and J. P. Clewley. 1990. Similarities in nucleotide sequence between serum and faecal human parvovirus DNA. *Epidemiol. Infect.* **105:**197–201.

Umene, K., and T. Nunoue. 1995. A new genome type of human parvovirus B19 present in sera of patients with encephalopathy. *J. Gen. Virol.* **76:**2645–2651.

Vandervelde, E. M., C. Goffin, B. Megson, N. Mahmood, A. Porter, and Y. E. Cossart. 1974. User's guide to some new tests for hepatitis-B antigen. *Lancet* **ii:**1066.

van Elsacker-Niele, A. W., and M. J. Anderson. 1987. First picture of erythema infectiosum? *Lancet* **i:**229.

van Elsacker-Niele, A. M., M. M. Salimans, H. T. Weiland, C. Vermey Keers, M. J. Anderson, and J. Versteeg. 1989. Fetal pathology in human parvovirus B19 infection. *Br. J. Obstet. Gynaecol.* **96:**768–775.

Van Elsacker-Niele, A. M. W., H. T. Weiland, A. C. M. Kroes, and M. C. Kappers-Klunne. 1996. Parvovirus B19 infection and idiopathic thrombocytopenic purpura. *Ann. Hematol.* **72:**141–144.

Walsh, K. J., R. D. Armstrong, and A. M. Turner. 1988. Brachial plexus neuropathy associated with human parvovirus infection. *Br. Med. J.* **296:**896.

Watanabe, M., Y. Shimamoto, M. Yamaguchi, S. Inada, S. Miyazaki, and H. Sato. 1994. Viral-associated haemophagocytosis and elevated serum TNF-a with parvovirus-B19-related pancytopenia in patients with hereditary spherocytosis. *Clin. Lab. Haematol.* **16:**179–182.

Weiland, H. T., C. Vermey Keers, M. M. Salimans, G. J. Fleuren, R. A. Verwey, and M. J. Anderson. 1987. Parvovirus B19 associated with fetal abnormality. *Lancet* **i:**682–683.

West, N. C., R. E. Meigh, M. Mackie, and M. J. Anderson. 1986. Parvovirus infection associated with aplastic crisis in a patient with HEMPAS. *J. Clin. Pathol.* **39:**1019–1020.

White, D. G., A. D. Woolf, P. P. Mortimer, B. J. Cohen, D. R. Blake, and P. A Bacon. 1985. Human parvovirus arthropathy. *Lancet* **i:**419–421.

Willwand, K., A. Q. Baldauf, L. Deleu, E. Mumtsidu, E. Costello, P. Beard, and J. Rommelaere. 1997. The minute virus of mice (MVM) nonstructural protein NS1 induces nicking of MVM DNA at a unique site of the right-end telomere in both hairpin and duplex conformation in vitro. *J. Gen. Virol.* **78:**2647–2655.

Woernle, C. H., L. J. Anderson, P. Tattersall, and J. M. Davison. 1987. Human parvovirus B19 infection during pregnancy. *J. Infect. Dis.* **156:**17–20.

Woolf, A. D., G. V. Campion, P. T. Klouda, A. Chiswick, B. J. Cohen, and P. A. Dieppe. 1987. HLA and the manifestations of human parvovirus B19 infection. *Arthritis Rheum.* **30:**S52.

Woolf, A. D., G. V. Campion, A. Chishick, S. Wise, B. J. Cohen, P. T. Kiouda, O. Caul, and P. A. Dieppe. 1989. Clinical manifestations of human parvovirus B19 in adults. *Arch. Intern. Med.* **149:**1153–1156.

Yoto, Y., T. Kudoh, K. Haseyama, N. Suzuki, and S. Chiba. 1996. Human parvovirus B19 infection associated with acute hepatitis. *Lancet* **347:**868–869.

Young, N. S., P. P. Mortimer, J. G. Moore, and R. K. Humphries. 1984. Characterization of a virus that causes transient aplastic crisis. *J. Clin. Investig.* **73:**224–230.

Young, N. S., B. Baranski, and O. Kurtzman. 1989. The immune system as mediator of virus-associated bone marrow failure: B19 parvovirus and Epstein-Barr virus. *Ann. N. Y. Acad. Sci.* **554:**75–80.

Zerbini, M., M. Musiani, S. Venturoli, O. Gallinella, D. Gibellini, O. Gentilomi, and M. La Placa. 1990. Rapid screening for B19 parvovirus DNA in clinical specimens with a digoxigenin-labeled DNA hybridization probe. *J. Clin. Microbiol.* **28:**2496–2499.

Measles, Mumps, and Rubella

WILLIAM J. BELLINI AND JOHN L. SEVER

36

MEASLES VIRUS

Humans are the only known natural host for measles (rubeola) virus. Enders and Peebles in 1954 first reported the successful isolation of measles virus in human and rhesus monkey kidney tissue cultures (Enders and Peebles, 1954). At that time there were more than 400,000 cases of measles reported each year in the United States. However, since virtually all children would acquire measles, the true number probably exceeded 3 million cases per year. In 1963 both an inactivated and a live attenuated vaccine (Schwartz, 1962) were licensed for use in the United States. The killed vaccine eventually proved less effective, and children who received this material were at risk of developing an atypical severe form of the disease when subsequently exposed to live measles virus. In 1967, the inactivated vaccine was discontinued. A live, further-attenuated vaccine (Moraten strain) was licensed in 1968 and is the vaccine used currently in the United States. Initially, vaccine was administered to children over 9 months of age, but it became apparent that the young infants did not mount an adequate immune response. It is now recommended that vaccination be initiated at age 12 to 15 months and that a second dose be administered at 4 to 6 years of age or by 11 to 12 years of age combined with mumps and rubella vaccines (MMR).

Characteristics of Virus

Measles virus is an RNA virus and is classified as a member of the paramyxovirus group. The measles virion is spherical, with a diameter ranging from 120 to 250 nm (Hall and Martin, 1974). It has an envelope composed of glycoproteins and lipids and bears short surface projections. The envelope encloses an elongated helical nucleocapsid in which protein units are arranged in a spiral around the nucleic acid.

Clinical Aspects

Measles is spread through direct contact with infected droplets originating from a cough or sneeze or with contaminated fomites. It is a highly contagious, acute biphasic disease with a prominent prodrome preceding the exanthemic phase. Susceptible persons intimately exposed to a measles patient have a 99% chance of acquiring the disease. Prior to the use of vaccines, more than 90% of the population had measles before the age of 10 years. After an incubation period of 9 to 11 days, there is an initial 3- to 4-day prodromal period characterized by fever, cough, coryza, and conjunc-

tivitis. The incubation period in adults may last up to 3 weeks. Fever occurs 24 h or less before other symptoms appear, and these increase in severity, reaching a peak with the appearance of the rash on the fourth or fifth day.

Bluish-white lesions with a red halo (spots) appear on the buccal or labial mucosa in 50 to 90% of cases, 2 to 3 days after the onset of the prodrome. These lesions are small, irregular red spots with a bluish-white speck in the center; they are located on the inner lip or opposite the lower molars and are pathognomonic for measles (Koplik, 1896). They may be few in number early in the prodrome; however, they increase rapidly to spread over the entire surface of the mucous membranes. A lesion somewhat similar in appearance to Koplik spots has been reported with ECHO-9 (coxsackie A23) and coxsackie A16 and A9 virus infections. The rash is first evident behind the ears or on the forehead. The lesions are red macules 1 to 2 mm in diameter which become maculopapules over the next 3 days. By the end of the second day, the trunk and upper extremities are covered with rash, and by the third day, the lower extremities are affected. The rash resolves in the same sequence, lasting approximately 6 days. The lesions turn brown and persist for 7 to 10 days and are followed by a fine desquamation.

The most frequent complication of measles involves infections of the lower respiratory tract. Croup, bronchitis, bronchiolitis, and, rarely, giant-cell interstitial pneumonia may occur. Otitis media is a common bacterial complication of measles. Prior to the advent of antibiotics, these complications contributed to a high number of fatalities and significant morbidity. Excluding pneumonia and otitis media, the most frequent serious complication of measles is postinfectious encephalitis. It occurs in 0.1 to 0.2% of measles patients during any stage of the illness, although it is most common 2 to 7 days after the onset of the exanthem. Death has occurred in 1 or 2 out of every 1,000 reported cases in the United States. Other complications include thrombocytopenic purpura, appendicitis, myocarditis, and mesenteric lymphadenitis (Gershon and Krugman, 1979).

Subacute sclerosing panencephalitis (SSPE), also called Dawson's encephalitis, is a late or "slow-virus" complication of measles. The incidence of SSPE was approximately 1 per 100,000 to 1,000,000 cases, but after the advent of the measles vaccine, there was a dramatic decrease in the frequency of the disease. SSPE is a progressive, invariably fatal encephalopathy characterized by personality changes, mental deterioration, involuntary movements, muscular rigidity, and death. It usually begins 4 to 17 years after the patient has recovered from measles. Measles virus was successfully isolated from brain and lymphoid tissues of SSPE patients (Horta-Barbosa et al., 1969, 1971). Extensive use of measles vaccine has almost eliminated SSPE from the United States (Bloch et al., 1985).

Transplacental infections have been associated with some effects on the fetus. There is an apparent increased frequency of premature labor, low birth weight, abortions, and stillbirths (Eberhart-Phillips et al., 1993). The teratogenic potential of gestational measles has neither been proved nor refuted (Fuccillo and Sever, 1973; South and Alford, 1980).

Atypical measles occurred in some children previously vaccinated with killed measles virus vaccines when they became infected with wild measles (Fulginiti et al., 1967). Fever, a prodromal period, and subsequent rash characterized the disease. During the prodrome, some patients experienced malaise, myalgia, headache, nausea, and vomiting. Symptoms usually lasted for 2 to 3 days, and frequently individuals had a sore throat, conjunctivitis, and photophobia along with nonproductive cough and pneumonia. Chest X rays often showed patchy infiltrates. The rash produced was different from that of typical measles. It could be a mixture of macules, papules, vesicles, and pustules. Frequently, there was a petechial component, which began at the distal extremities and was concentrated on the hands, wrists, ankles, and feet and then progressed centrally toward the trunk. Koplik spots were not reported, and the face was rarely involved. Edema often occurred in the extremities. The appearance of atypical measles could be confused with that of Rocky Mountain spotted fever.

Among immunocompromised patients measles can be severe and prolonged. This is a particular risk for patients with certain leukemias, lymphomas, or human immunodeficiency virus infection (Markowitz et al., 1988).

Laboratory Diagnosis

Overview

While the diagnosis of a case of measles can be made based on observation of clinical signs and symptoms, expedient laboratory confirmation is extremely important because of possible confusion with other rash-causing illnesses. The incidence of measles has decreased in many areas to the extent that medical personnel may be unfamiliar with the clinical presentation of the disease. Moreover, milder forms of measles have been reported to occur in previously vaccinated individuals (Edmonson et al., 1990), and in many instances these patients do not meet the clinical case definition. Because of the highly infectious nature of measles virus, it is important that laboratory confirmation of suspected cases be completed as quickly as possible so that control measures can be initiated. Therefore, the preferred laboratory diagnostic procedure is the detection of measles-specific immunoglobulin M (IgM) antibody in a single serum sample obtained during the acute phase of the disease (Tuokko, 1984). There are a number of sensitive and specific commercial enzyme immunoassy (EIA) kits available which are relatively easy to perform and give results within 4 h. IgM can also be detected by using indirect immunofluorescent antibody (IFA) assays, but this method is not as sensitive as EIA.

Other serologic methods include hemagglutination inhibition (HI) and the plaque neutralization test (PNT), but these tests are time-consuming and require that both acute- and convalescent-phase serum samples be available for testing. At present it is desirable to obtain the necessary specimens when the patient is first seen, since in many regions of the world convalescent-phase serum specimens may never be obtained. Finally, diagnosis can be made by detecting viral antigens or RNA using either IFA or reverse transcriptase (RT)-PCR. However, IFA and RT-PCR are not as sensitive as detection of IgM (W. J. Bellini, unpublished data), and these methods require special laboratory techniques, which may not be available in most areas.

Serologic Diagnosis

Traditional antibody tests, such as HI, the PNT, and EIA, have been used extensively in the serologic diagnosis of measles. However, because of the availability of sensitive and specific commercial kits, EIA has become the most widely used test format. Commercial EIAs can also be used to measure measles-specific IgM and IgG responses and therefore have particular importance in diagnosis as well as in measles control programs. Some of the available kits were found to have sensitivities and specificities that compared favorably with plaque reduction neutralization (Ratnam et al., 1995). Most laboratories are already equipped to run EIAs and have the necessary equipment and expertise.

Standard EIAs

Production of measles antigen for EIA has traditionally been done using tissue culture-adapted strains (e.g., Edmonston strain) grown in Vero cells. Measles is a highly cell-associated virus, so infected cells must be disrupted by freeze-thaw or sonication before the clarified cell lysates are used as antigens. More recently, measles antigens have been produced by recombinant-DNA expression systems, such as baculovirus (Summers and Smith, 1986). EIAs using the baculovirus-expressed nucleoprotein (MVN) in both capture and indirect formats had high levels of sensitivity and specificity compared to those of the PNT and other commercial EIAs (Hummel et al., 1992). The baculovirus system has an added advantage in that the yields of antigen possible in the insect cell culture system far exceed those of virus growing in mammalian cells. Also, the necessity of handling large amounts of a virus that is pathogenic for humans is avoided.

For detecting IgG, indirect EIA has been successful (Erdman et al., 1991). For this indirect test, either whole virus antigen diluted in 0.05 M bicarbonate buffer (pH 9.5) or recombinant antigen diluted in phosphate-buffered saline (PBS) is placed in polystyrene microtiter plates. Serum specimens are diluted in PBS containing 4% normal goat serum (plus 4% *Spodoptera frugiptera* [Sf9] cell lysate for recombinant antigen), and 0.05% EDTA is then added to the washed plates. Bound antibody can be detected with standard commercial reagents, such as goat anti-human IgG conjugated to either alkaline phosphatase or biotin. The assays are developed with the appropriate substrate, and the plates can be read either by eye or with the aid of a spectrophotometer.

For the detection of measles-specific IgM in single, acute-phase serum samples, commercial kits based on both indirect and IgM capture formats have been used (Erdman et al., 1991; Mayo et al., 1991). Though IgM capture is generally regarded as the more sensitive format, some of the commercial indirect-EIA kits had sensitivities and specificities that approached those of the capture format (Arista et al., 1995).

Although the configurations of the commercial measles IgM kits vary, the following description will provide an example of a measles IgM capture EIA. Microtiter plates are coated with goat anti-human IgM diluted in 0.01 M PBS for 1 h at 37°C. The plates are washed, and diluted serum specimens are added to consecutive wells and incubated for 1 h at 37°C. After the plates are washed, baculovirus-expressed MVN and Sf9 negative control antigen are added to duplicate wells for each specimen and incubated for 1 h at 37°C. The plates are washed, and biotinylated monoclonal antibody to MVN is added. Following another incubation for 1 h at 37°C, the plates are washed and streptavidin-peroxidase is added. Following a 20-min incubation period at 37°C, the plates are washed and a 3,3′,5,5′-tetramethylbenzidine substrate solution is added. After 15 min at room temperature, the color reaction is stopped by acidification, and the color intensity is determined spectrophotometrically.

Interpretation of EIA results. The occurrence of clinically defined measles in individuals years after receiving vaccine has been attributed to primary vaccine failure due to insufficient primary antigenic stimulation, as well as secondary vaccine failure due to a putative loss of protective antibody or waning immunity. The use of sensitive EIAs for detection of measles IgM antibody has provided a better description of primary and secondary antibody responses following primary and secondary vaccination and natural measles virus infection. The findings of these studies indicated that (i) an IgM response follows primary measles vaccination or measles infection in the immunologically naive, (ii) an IgM response is absent on revaccination of those previously immunized, and (iii) an IgM response may follow clinical measles virus infection independent of prior immunization status. There is also evidence that patient contacts who have a history of natural infection or vaccination and have a resident IgG response may also develop a secondary IgG or an IgM response to currently circulating virus (Muller et al., 1996; Helfand et al., 1998). Close contacts may show few, if any,

of the clinical signs of measles infection. Table 1 summarizes the possible interpretations of EIA results.

The time at which the serum specimen is collected may affect the results of the capture IgM EIA. It was possible to detect MVN-specific IgM in 77% of serum specimens taken from measles patients during the first 72 h after rash onset, while 100% of specimens taken 4 to 11 days after rash onset contained detectable IgM. Therefore, some false-negative reactions may occur if samples are taken within 72 h after rash onset, and a second serum sample may be required (Helfand et al., 1997).

New EIAs

A new and potentially promising development has been the use of EIAs to measure the avidity of IgG antibodies to measles (Tuokko, 1995; de Souza, et al., 1997; Narita et al., 1997). As the immune response matures, low-avidity antibodies are replaced with high-avidity antibodies. These avidity differences can be detected by using protein denaturants, typically 8 M urea, in the washing step of the indirect EIA for measles IgG. An avidity index is then calculated by comparing the optical densities obtained with and without urea in the wash buffer. These tests have been able to successfully differentiate between primary and secondary responses to vaccination and to natural infection.

The HI Test

Because of the availability of EIAs, the HI test is now more useful as a research tool than as a diagnostic test. The HI test measures antibodies to the measles hemagglutinin and has a sensitivity equivalent to that of EIA. The HI test can be performed in a few hours, but it requires special reagents and paired serum samples. Other limitations of the HI test are the requirement for fresh vervet monkey red blood cells (RBC), the difficulty of producing sufficient antigen, and the possible presence of nonspecific inhibitors of hemagglutination (HA) in serum.

Serum should be inactivated at 56°C for 30 min before being tested. Nonspecific inhibitors may be removed by absorption of the serum with an equal volume of 25% (wt/vol) kaolin in PBS.

RBC should be obtained from monkeys lacking measles antibody, and RBC from some animals demonstrate HA

TABLE 1 Interpretation of measles EIA results

IgM result[a]	IgG result[a]	Previous infection history	Current infection	Comments
+	+ or −	Not vaccinated; no history of measles	Recent first MMR	Seroconvert[b]
+	+ or −	Not vaccinated; no history of measles	Wild-type measles	Seroconvert[b]; classic measles
+	+ or −	Previously vaccinated; primary vaccine failure	Recent second MMR	Seroconvert[b]
−	+	Previously vaccinated; IgG+	Recent second MMR	IgG level may stay same or rise
+	+	Previously vaccinated; IgG+	Wild-type measles	May have few or no symptoms[c]
+	+	Recently vaccinated	Exposed to wild-type measles	Cannot distinguish if vaccine or wild type; evaluate on epidemiological grounds[d]
+ or −	+	Distant history of measles	Wild-type measles	May have few or no symptoms[c]; if clinically compatible, may have been misdiagnosed initially

[a]+, positive; −, negative.
[b]IgG response depends on timing of specimen collection.
[c]If so, do not consider contagious unless clinical presentation is consistent with measles.
[d]If IgM negative, this is helpful in ruling out wild-type-measles infection.

better than those from other animals. It is necessary to pre-select donor animals for a strong positive HA with the measles antigen. RBC can be stored in Alsever's solution at 4°C for up to 1 week. On the day of the test, RBC should be washed three times in PBS and resuspended in PBS at 0.5%.

Viral antigen is propagated in Vero, BSC-1, or dog kidney cells or lung fibroblasts. When the cytopathic effect (CPE) approaches 100%, the cultures are frozen and thawed three times. Tween 80 is added to a concentration of 0.125%, and the cultures are mixed at 4°C and then centrifuged at 1,000 × g for 20 min. The supernatant is saved as an antigen. The choice of the measles strain to use as an antigen is important, since many wild-type viruses do not agglutinate RBC. More consistent results will be obtained with the Edmonston strain obtained from the American Type Culture Collection.

The titer of the hemagglutinin is determined as follows. Twofold dilutions of antigen in 0.025 ml of PBS are added to V-bottom microtiter plates. To each well, 0.025 ml of 0.5% RBC is added. The plate is shaken briefly and incubated at 37°C for 30 min to 1 h. The highest dilution of antigen giving complete HA is the endpoint and contains 1 HA unit.

To perform the test, serial twofold dilutions of serum specimens in 0.025 ml of PBS are added to microtiter wells. An equal volume of antigen diluted in PBS to contain 4 HA units is added, and the mixtures are incubated for 1 h at 37°C. Then 0.05 ml of 0.5% RBC suspension is added, and the mixtures are shaken and allowed to settle at 37°C for 1 to 2 h. Titers are recorded as the highest serum dilution resulting in complete HI.

The PNT

The PNT, which measures neutralizing antibodies that are directed against the surface glycoproteins of measles virus, is more sensitive than HI or EIA (Albrecht et al., 1981). Since functional antibodies are being detected, the PNT provides the best serologic correlate for the assessment of immune protection. However, the PNT is not suitable for routine serologic diagnosis because it is very labor-intensive, requires paired serum samples, and takes 5 to 7 days to perform.

In the PNT, anti-measles antibody in serum combines with and neutralizes measles virus, preventing it from infecting a cell monolayer and forming a plaque under the overlay. The endpoint for the test is the highest dilution of serum which will reduce the number of plaques by 50%. Serum dilutions are made in 96-well microtiter trays, proceeding either in twofold or fourfold steps (depending on the expected titer of the serum). Once the dilutions of serum samples are made, an equal volume (120 μl) of a dilution of virus containing 25 to 35 PFU is added to each well and incubated for 2.5 h at 36°C. After incubation, 100 μl of the serum-virus mixture is added to each of two 16-mm-diameter tissue culture plates containing Vero cell monolayers. These trays are then incubated for 1 h at 36°C. After the incubation, the inoculum is removed by aspiration and the monolayers are covered with overlay medium consisting of either 2% carboxymethylcellulose in Leibovitz-15 medium or 1% agarose in Eagle's medium. The trays are then incubated for 5 days at 36°C. On day 4, the trays are stained with a solution of neutral red in cell culture medium. On day 5, the overlay is removed and the plaques are counted. The plaque counts in the two wells representing a given serum dilution are averaged.

Other Serologic Methods

Attention has also focused on the development of serologic assays that could be used in developing countries where measles continues to be a major public health problem. Such assays must be simple to perform in the field, be inexpensive, require little equipment or refrigeration, be suitable for large numbers of samples, and be able to detect measles-specific IgM antibodies. While several tests are under development, none is commercially available at present.

Direct Examination of Specimens for Viruses and Viral Antigens

Virus Isolation

Although other cell lines have traditionally been used to grow measles virus, an Epstein-Barr virus-transformed B lymphoblastoid cell line, B95a, is the preferred cell line for primary isolation of measles virus (Kobune et al., 1990). These cells are up to 10,000 times more sensitive for isolation of measles virus from clinical specimens than other cell lines, such as Vero and primary monkey kidney cells. B95a cells are relatively easy to maintain in the laboratory, and the measles CPE is readily observed. However, laboratorians should note that this cell line does produce Epstein-Barr virus and should be handled as infectious material at all times.

B95-8 cells are available from the American Type Culture Collection (CRL 1612). When cultured in Dulbecco's modified minimum essential medium supplemented with 100 U of penicillin/ml, 100 μg of streptomycin/ml, and fetal bovine serum (FBS), these cells will adhere to the surface of the culture vessel, and the adherent cells are referred to as B95a. Cell growth is sustained in medium containing 8 to 10% FBS. FBS is used at a 2% concentration for cell maintenance during virus isolation. Cell stocks can be prepared using standard cryoprotection medium.

B95a cells should be at 85% confluency when inoculated with specimens suspected of containing measles virus. After inoculation with clinical material, the cells should be maintained in Dulbecco's modified minimum essential medium plus 2% FBS and antibiotics. Change the medium every 3 to 5 days, and passage the cells by splitting them at 1:2 every 5 to 7 days. Check for viral CPE (syncytium formation) daily. When CPE is visible, continue to feed the cells until the CPE becomes extensive. When CPE is visible over at least 75% of the monolayer, virus stocks can be prepared by freezing the cells at −70°C. It is advisable to perform three or four blind passages before discontinuing efforts for virus isolation.

Cytologic Examination

Intranuclear and intracytoplasmic inclusions and giant cells are characteristic CPEs for cells infected with measles virus. Cytologic examination of various tissue specimens and secretions for these Warthin-Finkeldey giant cells can be used as a diagnostic procedure. Secretions are obtained by aspiration of mucus from the nose or by swabbing the nasal mucosa with a sterile cotton-tipped applicator. Slides can be stained with either Wright stain or hematoxylin and eosin. Tissue samples may be fixed in 10% formalin, embedded in paraffin, sectioned, and then stained with hematoxylin and eosin stain (Gershon and Krugman, 1979). Staining of tissue specimens with monoclonal antibodies to the measles nucleoprotein has been used for the diagnosis of giant-cell pneumonia and measles inclusion body encephalitis (Zaki and Bellini, 1997).

Immunofluorescence

Detection of measles virus can be achieved by using an IFA assay to examine clinical specimens as well as B95a cell cultures infected with clinical material. The standard assay uses a commercially available monoclonal antibody to the nucleoprotein of measles virus and fluorescein-conjugated goat anti-mouse antiserum. Nasal secretions should be centrifuged at 800 × g to pellet the cells. The cell pellets are then washed several times with sterile saline before being applied to a glass slide and fixed in cold 80% acetone for 10 min at −20°C (Minnich et al., 1991; Smaron et al., 1991).

RT-PCR

RT-PCR has been used in research settings to detect measles virus RNA in clinical specimens and infected cells (Nakayama et al., 1995; Rota et al., 1995). While serologic testing is best for routine diagnosis of measles infections, these assays do not provide any information about the virus. Therefore, RT-PCR can be considered for diagnostic use in special cases where cell culture is not a practical alternative and/or when genetic characterization of the virus is required. In this regard, RT-PCR has been particularly useful for the diagnosis of measles inclusion body encephalitis, SSPE, and giant-cell pneumonia. Sequence data obtained from the PCR products can sometimes identify the source of the virus and differentiate between vaccine and wild-types strains of measles (Bellini and Rota, 1998).

RT-PCR techniques have had the greatest impact on molecular-surveillance activities. Measles RNA is easily amplified from RNA extracted from infected cell cultures, and it has been possible to detect measles RNA in nasal, urinary, and sometimes serum samples by RT-PCR even when virus isolation has been unsuccessful (Jin et al., 1997).

Control and Prevention

Individuals having an illness compatible with a diagnosis of measles should be cared for in such a way that contact with other people is minimal. The communicability of measles virus is extremely high. Therefore, any susceptible individuals who have had direct face-to-face contact with the infectious individual should obtain prophylactic treatment. Risk, other than with face-to-face contact, is very low, and therefore postexposure prophylaxis is unnecessary. Measles vaccination may provide protection if given within 72 h of exposure (American Academy of Pediatrics, 1997). The Immunization Practices Advisory Committee supports readmission to school of all previously unimmunized children immediately following vaccination (Centers for Disease Control, 1989). Immune globulin, given within 6 days of exposure, can prevent or modify measles virus infection. It is indicated for susceptible close contacts of measles patients, particularly if they are less than 1 year old. If immune globulin is used for a child of this age, measles vaccine should be given 5 or 6 months later provided that the child is then at least 12 months old.

After a further-attenuated variant of the Edmonston B vaccine was introduced in 1968, the reported cases of measles took a dramatic downward turn. In 1960, the cumulative total number of cases was 399,852 from week 1 to week 35. In 1970, the total was 39,365; in 1981, it was 2,562; in 1982, it was 1,188; and in 1983, for the same period, the total number of cases was 1,194 (Centers for Disease Control, 1982). There was hope that 1983 would be the year in which measles would be eliminated from the United States, but this goal was not accomplished. In fact, reported cases increased every year until 1986, when there were 6,282 cases. A small decrease in total cases was reported for 1987 and 1988, but total cases rose again during 1989 to 16,236 cases (Centers for Disease Control, 1990). Since 1992, fewer than 1,000 cases have been reported each year, and more recently there have been fewer than 100 cases per year in the United States.

In 1993, the Childhood Immunization Initiative called for the elimination from the United States by 1996 of indigenous transmission of measles, rubella, congenital rubella syndrome, and three other childhood diseases. The Pan American Health Organization adopted the goal of eliminating measles throughout the Americas by 2000. Epidemiological and laboratory data suggest that the transmission of indigenous measles was interrupted in the United States in 1993 (Watson et al., 1998).

MUMPS VIRUS

Mumps virus infection was probably first described around the fifth century B.C. by Hippocrates. The name "mumps" is thought to be derived from the mumbling speech of patients afflicted with this disease. Johnson and Goodpasture identified the etiologic agent as a virus in 1934 (Johnson and Goodpasture, 1934). The virus was first isolated in the amniotic cavity of a chicken embryo in 1945 (Habel, 1945). Buynak and Hilleman, in 1966, developed the first successful live attenuated vaccine by passage of the virus in chicken embryo cell cultures (Buynak and Hilleman, 1966).

Characteristics of Virus

Mumps virus is a member of the paramyxovirus group. The virus particles range from 85 to 300 nm in diameter (Cantell, 1961). The virus has a single strand of RNA, contains a nucleoprotein core, and has an outer envelope. The envelope contains a hemagglutinin, neuraminidase, and a hemolysin.

Clinical Aspects

Humans are the only known host and reservoir of the mumps virus. The infection can be either clinically apparent or subclinical. Infection is endemic worldwide, usually affecting the 6- to 10-year-old age group; it occurs predominantly in the spring. About 30 to 40% of mumps infections produces parotitis, 15 to 20% are asymptomatic, and 50% are associated with nonspecific or respiratory symptoms. The average incubation period is 16 to 18 days. Parotitis occurs more frequently in children (Philip et al., 1959). The parotitis is sudden and may not be preceded by any prodromal symptoms. Swelling of the glands reaches a maximum after 48 h, and they usually remain swollen for a period of 7 to 10 days. There may be little or no increase in body temperature. Approximately 20 to 30% of postpubital men acquiring mumps develop epididymo-orchitis between 1 and 2 weeks following the parotitis. Sterility is not a common sequela of infection, since only 1 to 12% of the cases are bilateral.

Another complication of mumps virus infection is meningoencephalitis, which has an incidence of 5 to 10%. Encephalitis is one central nervous system complication, but mumps virus infection has been linked to other, rare central nervous system complications, such as transverse myelitis, cerebellar ataxia, poliomyelitislike syndrome, and Guillain-Barré syndrome. About 5% of adult females with mumps may develop oophoritis. Other complications, such as pancreatitis, thyroiditis, neuritis, inflammation of the eye, and inner-ear infection can be encountered. There have been reports of diabetes mellitus being associated with mumps, but at present information is inconclusive (Sultz et al., 1975; Ratzmann et al., 1984). Other reports that intrauterine mumps can lead to endocardial fibroelastosis have not been confirmed (St. Geme et al., 1966).

Seroepidemiological surveys have indicated that 80 to 90% of adults have evidence of prior mumps infection. Mumps is transmitted by saliva containing the virus either by direct transfer, by air-suspended droplets, or by recently contaminated fomites. Approximately 85% of susceptible contacts can become infected when first exposed, and 25 to 40% of the infections may be asymptomatic. The virus is thought to multiply in the upper respiratory tract, then invade the bloodstream, and finally affect the salivary glands and other organs. About 18 days elapse between the time of exposure and the first detectable enlargement of the salivary glands. The incubation period may range from 14 to 24 days. The period of communicability can be from 7 days before the salivary gland involvement until 9 days thereafter. The virus is also excreted in the urine for as long as 14 days after the onset of illness.

Laboratory Diagnosis

Clinical diagnosis of mumps infection can be made reliably when typical parotitis is evident at the time of patient examination. However, since parotitis may be caused by other viral and nonviral diseases or conditions, and as much as 20% of mumps infections can be asymptomatic, diagnosis by viral isolation or serologic techniques is preferable. Moreover, the incidence of mumps in the United States has decreased by more than 99% with the advent of MMRII vaccine, and the disease no longer exhibits seasonality (Centers for Disease Control and Prevention, 1998). The laboratory criteria for the diagnosis of mumps are (i) isolation of the mumps virus from clinical specimens, (ii) a significant rise between acute- and convalescent-phase titers in serum, or (iii) a positive IgM response in an EIA, when only a single specimen is available.

Virus isolation from saliva, blood, urine, or cerebrospinal fluid (CSF) confirms the presence of recent mumps infection. While primary monkey kidney cell cultures are likely the most sensitive cells, availability, risk of infection with herpes B virus, and animal rights issues have resulted in the use of continuous cell lines. Continuous cell lines, such as HeLa and Vero, are currently the cells of choice, but growth of virus has been noted to occur in primary human cell cultures as well (Hopps and Parkman, 1979). Although mumps virus infection results in characteristic CPE consisting of large syncytia, some isolates vary in the intensity and frequency of the CPE and thus must be confirmed by IFA staining (Swierkosz, 1995).

Mumps virus is stable for several days at 4°C. The stability increases with decreasing temperature, particularly if 1 to 2% serum or protein is added to the storage medium. Mumps virus may be stored for months or longer at −70°C. Throat swabs, urine, and CSF specimens should be inoculated onto susceptible cells within a few hours following sample collection.

Conventional Serology

Many serologic assays have been used to detect mumps antibodies. HI, complement fixation (CF), and neutralization tests (NT) are standard methods for mumps virus and supply reproducible results. The serologic confirmation of mumps using any of these methods requires a fourfold difference in serial dilution endpoint titration of acute- and convalescent-phase serum specimens collected 2 to 3 weeks apart. All three methods primarily measure IgG, and the interpretation of the results can be confounded by cross-reactivity with parainfluenza antibodies. These assays have been described in detail elsewhere (Hopps and Parkman, 1979; Black, 1996) and are not likely to be used routinely in the clinical laboratory. Nevertheless, standard reagents to perform each of these assays, including CF and NT, remain commercially available. Due to their enhanced sensitivity, high throughput, relative ease, and low cost, EIAs are the serologic assays of choice (Tuokko, 1984; Glickmann et al., 1986; Berbers et al., 1993).

EIAs are widely available commercially. These assays use a partially purified mumps antigen bound to a solid support, such as the wells of a microtiter plate. IgM assays formatted in this way must provide a means to remove potentially interfering mumps IgG from the assay. These indirect IgM assays are also vulnerable to the presence of rheumatoid factor in serum as a source of false-positive reactions. Capture IgM formats that would circumvent the interference of rheumatoid factor were not commercially available at the time of writing. Of considerable interest is the use of oral fluid rather than serum in the determination of immune status for measles, mumps, and rubella. The IgG EIA results indicated that oral fluids were a good surrogate for serum and yielded 94% sensitivity and specificity relative to serum specimens in the indirect EIA (Thieme et al., 1994).

Other Methods

The hemolysis-in-gel method has been used primarily as an alternative to CF and approximates the sensitivity of neutralization (Grillner and Blomberg, 1976). The method requires coating erythrocytes with mumps antigen and suspending them in agarose. Heat-inactivated serum specimens are applied to precut wells and allowed to diffuse into the agarose for 24 h. Following the addition of complement, the zone of lysis around the well is measured and compared to standards of known antibody titers to mumps. The method has practical disadvantages, since coated erythrocytes and viral antigen must be freshly prepared, as no commercial source of the reagents is available. Another test for immune status (IgG) is the dot immunobinding assay (Condorelli and Ziegler, 1993). Filter paper soaked in a dilution of patient's serum is applied to mumps antigen dotted onto nitrocellulose. Following 30 min of incubation, the filter paper is removed and the nitrocellulose sheet is washed thoroughly. Next, an enzyme-linked anti-human antibody is added. Finally, the assay is developed with an appropriate chromogen substrate, and the color change is assessed relative to known positive and negative serum specimens. The assay was determined to provide results similar to those of an EIA run in parallel.

Control and Prevention

In the United States, the mumps vaccine was introduced in 1967 and was recommended for routine use in 1977. Between 1968 and 1995, the number of cases decreased from 185,691 to 906 annually. The enforcement of state vaccination laws requiring students to be vaccinated before entering school has been a major factor in reducing the incidence of the disease (Chaiken et al., 1987). Mumps vaccine is given along with measles and rubella vaccines (MMR) at 12 to 15 months of age and again at 4 to 6 years or by 11 to 12 years of age.

Mumps vaccine has not been found to be effective in preventing infection if given after exposure. Mumps immune globulin is of no value, and it is no longer available in the United States.

RUBELLA

Background
Rubella (German or 3-day measles) was first described by German authors in the 18th century and was accepted by an International Congress of Medicine in London as a disease independent of measles and scarlet fever in 1881 (Cooper, 1985). It is usually a mild contagious viral disease causing a fine maculopapular rash, which begins on the face and spreads to the abdomen and the extremities. Rubella would be of little medical importance were it not for the profound defects rubella virus (RV) infection can cause in the unborn child. More than 75% of infants born to mothers infected with RV during the first trimester of pregnancy are affected. The spectrum of defects includes sensorineural hearing loss, cardiovascular abnormalities, cataracts, and meningoencephalitis and is collectively known as congenital rubella syndrome (CRS). The association between rubella and birth defects was first recognized by N. McAlister Gregg in 1941, 60 years after the recognition of rubella as a distinct disease (Gregg, 1941). Gregg's insightful discovery occurred when all birth defects were thought to be inherited and rubella was considered nothing more than a mild early-childhood disease. Rubella virus is one of the clearest examples of a potent infectious teratogen (Webster, 1998).

Prior to the availability of vaccines against rubella, epidemics occurred every 6 to 9 years in the United States, and larger epidemics occurred at intervals of up to 30 years (Horstmann, 1990). The mechanism(s) by which major epidemics gained force is not clear, since smaller epidemics occurred frequently and only one serotype of rubella virus exists. The last major epidemic in the United States was in 1964–1965, resulting in about 11,000 fetal deaths or induced terminations of pregnancy and 20,000 children born with CRS (Orenstein et al., 1984).

The virus is difficult to grow in tissue culture, and it was first isolated in 1962 (Parkman et al., 1962; Weller and Neva, 1962). Clinical specimens containing RV do not produce easily observed changes in standard tissue culture cells, but clever assays for the detection of RV in clinical specimens, based on RV's interference with the replication of lytic enteroviruses, proved to be very useful (Schiff and Sever, 1966). Attenuation was shown to be possible by virus passage in tissue culture, and in 1969, only 7 years after RV was isolated, the first live-virus vaccine was licensed.

In the United States, immunization against RV has been consistently encouraged, and the Immunization Practices Advisory Committee now suggests two doses of RV-containing vaccine beginning at age 12 to 15 months and completed before school entry at age 4 to 5. Cases of CRS in the United States have been dramatically reduced (Centers for Disease Control and Prevention, 1997a), but the worldwide problem is far from over. There remain an estimated 20,000 CRS cases annually in the Americas (Hinman et al., 1998), and there are probably more than 200,000 CRS cases each year worldwide.

Description of Virus and Viral Replication
RV is a 60- to 70-nm-diameter particle consisting of a core particle surrounded by a lipid envelope (Banatvala and Best, 1998). The envelope contains two viral glycoproteins, E1 and E2, and the core contains the viral capsid protein (C) and an infectious single-stranded RNA genome of about 10,000 nucleotides. The 5′ two-thirds of the genome serves as an mRNA for a polyprotein, which is cleaved into nonstructural proteins necessary for virus replication. Structural proteins (E1, E2, and C) are cleaved from a second polyprotein translated from a subgenomic mRNA produced during viral replication. The subgenomic RNA has the nucleotide sequence of the 3′ one-third of the genome. New virions are produced when core particles bud through membranes containing E1 and E2.

RV is a member of the family *Togaviridae* and is the only member of the genus *Rubivirus*. Closely related viruses, Sindbis virus, Venezuelan equine encephalitis virus, and Semliki Forest virus, are classified in the genus *Alphavirus*. Most togaviruses have invertebrate hosts, but RV is known to infect only humans.

Clinical Aspects—Postnatal Rubella
Postnatal rubella is usually a mild disease requiring little treatment. The virus can be isolated from throat and nasopharyngeal specimens and, with more difficulty, from blood, urine, and even skin. Virus is present in the upper respiratory tract from about 7 to 21 days postexposure and may be found in blood and urine before about day 16. Communicability is greatest between about day 14 and day 19 postexposure and is thought to occur from exposure to respiratory secretions.

The symptoms of postnatal rubella and the approximate duration in days postexposure of each include fine maculopapular rash, days 16 to 18; postauricular and suboccipital lymphadenopathy, days 7 to 28; arthralgia, mostly among postpubertal females, days 19 to 28; and low-grade fever, days 14 to 21. About 60% of postpubertal females have joint involvement, which may be very painful and may, in rare cases, last a month. The virus has been reported to replicate in synovial-membrane cell culture (Grayzel and Beck, 1971). Rare complications of postnatal rubella include thrombocytopenia (about 1 in 3,000 cases) and postrubella encephalitis (about 1 in 6,000 cases). Postnatal rubella may be asymptomatic, an important fact when considering potential transmission to the fetus.

Clinical Aspects—Congenital Rubella
The clinical course of rubella in pregnant women is similar to that in nonpregnant women. Rubella shortly before pregnancy (with the rash developing within 11 days of the last menstrual period) is not likely to result in an infected fetus. However, the probability of transmission to the fetus increases quickly, and when the maternal rash occurs 3 weeks after the last menstrual period, most fetuses become infected (Enders et al., 1988). RV infection, even subclinical infection, occurring during the first 16 to 20 weeks of gestation is usually transmitted to the fetus (more than 75% of the time). The likelihood of defects declines during the first 20 weeks of gestation. About 60% of confirmed maternal infections during the first month of pregnancy result in significant birth defects (Rakowsky and Sever, 1998). Spontaneous abortion occurs in about 20% of infections acquired in the first 8 weeks of pregnancy.

During maternal viremia, 7 through 16 days postexposure, the placenta may be infected, particularly endothelial cells, and the resulting damage likely allows the virus to cross the placenta (Webster, 1998). In the first trimester, almost any fetal organ (heart, brain, and primary lens cells) can be infected. Fetal spread likely occurs via the vascular system, where fetal endothelial cells are damaged. Neonates with CRS usually have RV-specific IgM but remain virus positive.

When maternal rubella occurs after 16 weeks of gestation, the more severe defects are unlikely, but hearing impairment may still occur (Katow, 1999). Maternal rubella after the first trimester often leads to infection of fetal tissue, but fetal damage is limited, presumably because organogenesis is complete, transfer of maternal IgG to the fetus is higher, and a fetal immune response is present (Rakowsky and Sever, 1998).

Late manifestations of CRS have been recognized. These are disabilities that do not make their appearance until years after birth. One of the first disabilities to be found was insulin-dependent diabetes mellitus (Menser et al., 1967; Plotkin and Kay, 1970). Additional studies revealed a 20% incidence of latent or overt diabetes in a study involving 50 older subjects from the 1964 rubella epidemic (Forrest et al., 1971). All of the patients had a prenatal history of rubella before the 16th week of gestation. All were deaf, and most had other rubella-associated defects of the eyes and heart. In a follow-up study, 40% of these CRS patients had developed evidence of overt or latent diabetes (Menser et al., 1974). Other endocrine disorders have been seen in small numbers of survivors of congenital rubella. They include hypothyroidism (Ziring et al., 1977; P. R. Ziring, B. A. Fedun, and L. A. Cooper, Letter, *J. Pediatr.* **88:**1002, 1975), hyperthyroidism (Floret et al., 1980), hypoadrenalism (Ziring et al., 1977), and a growth hormone deficiency (Preece et al., 1977).

Ocular consequences of CRS are observed during and after the neonatal period. One study described 13 patients having glaucoma 3 to 22 years after birth. All had cataracts early in life, which either had been removed surgically or had resolved spontaneously. Other groups of patients were found to have keratic precipitates without other evidence of acute ocular inflammation (Boger, 1980, 1981).

Another late-appearing disability associated with congenital rubella is bilateral hearing loss. There has been one report of a child with CRS who had numerous audiograms, developed normal speech patterns, and attended to everyday activities until age 10, when signs of progressive deafness began to appear (Desmond et al., 1978).

The last disability associated with CRS is progressive rubella panencephalitis (PRP) (Townsend et al., 1976). This is a slow-virus manifestation of rubella that is similar to SSPE due to measles. PRP usually appears during the second decade of life. Progressive deterioration of intellectual and motor function occurs, with dementia close to the time of death. There is an intense immune response against rubella antigens, and high titers of rubella antibody are present in both serum and CSF (Weller et al., 1964). Virus has been recovered from the brain by rescue techniques (Wolinsky, 1978). The pathological findings are similar to those of SSPE but without the inclusions and with perivascular deposits (Rosenberg et al., 1981). PRP was reported in two patients with postnatal rubella infection and in nine cases of CRS (Townsend et al., 1975; Weil et al., 1975; Wolinsky, 1978; P. Lebon and G. Lyon, Letter, *Lancet* **ii:**468, 1974). No correlation can be made between the occurrence of PRP and the presence of rubella-associated defects or the severity of neonatal infection.

The pathology produced with congenital rubella appears to result from a chronic viral infection with an inhibition of cell multiplication at critical points in organogenesis. This causes the hypoplastic organ development and other char-acteristic structural defects seen with this disease (Rawls and Melnick, 1966). The immune response may also contribute to permanent damage in the developing child either by an impaired immunity or by inflicting damage through inflammatory mechanisms (Fuccillo et al., 1974; Rosenberg et al., 1981).

Clinical Diagnosis—Postnatal Rubella

Postauricular or suboccipital lymphadenopathy and a fine maculopapular rash is characteristic of postnatal rubella, but infection with enteroviruses and adenoviruses and mild cases of measles and scarlet fever can have similar symptoms. Clinical diagnosis is further complicated because postnatal rubella is a rare illness in some countries, leaving many physicians with little experience with rubella cases. Therefore, laboratory confirmation of RV infection is often necessary for an accurate diagnosis.

Clinical Diagnosis—Congenital Rubella

As with postnatal rubella, the clinical symptoms characteristic of CRS are not unique, and laboratory confirmation of RV infection is necessary. Particular emphasis on laboratory confirmation is necessary in the United States, where CRS is rare.

In order to identify probable CRS cases and for surveillance purposes, the Centers for Disease Control and Prevention has adopted the following procedure (Centers for Disease Control and Prevention, 1997b). CRS defects are divided into two groups. Group A is cataracts and congenital glaucoma, congenital heart disease (usually patent ductus arteriosus or peripheral pulmonary artery stenosis), loss of hearing, and pigmentary retinopathy. Group B is purpura, splenomegaly, jaundice, microcephaly, mental retardation, meningoencephalitis, and radiolucent bone disease. Two symptoms from group A or one from A and one from B indicate a probable CRS case. The most common defects seen in CRS are hearing loss, cataracts, retinopathy, and congenital heart disease. Laboratory evidence of congenital RV infection confirms a probable CRS case.

Detailed clinical descriptions of CRS are available and are useful in diagnosis (Rakowsky and Sever, 1998). A history of maternal vaccination and any exposures of a seronegative mother to RV from areas of endemicity during gestation are important pieces of information to gather when determining an index of suspicion for rubella.

Laboratory Diagnosis

Laboratory diagnosis of both postnatal and congenital RV infections is made using serologic and/or virus culture techniques. Throat swabs are the usual specimens for virus culture. The timing of serum collections is important, particularly for postnatal rubella cases (see below). Antibodies specific for RV antigens can be detected by HI assays, NT, EIA, or latex agglutination methods. Most serologic assays necessary for laboratory diagnosis are available commercially (Frey, in press).

Serum positive for RV-specific IgM is diagnostic for recent RV infection. An IgM capture assay is usually preferred because of fewer difficulties, such as false positives due to rheumatoid antibodies, but indirect EIAs which avoid such difficulties are acceptable. An indirect EIA for RV-specific IgG is usually used (e.g., during prenatal screening for immunity to rubella).

Adapted rubella virus strains produce CPE on a variety of cells. Virus from clinical specimens typically does not produce CPE, and this limits the routine use of virus culture techniques to document RV infection. However, RV in clinical specimens does replicate in a variety of cell lines, including Vero, RK13, BHK-21, and GMK-AH-1, and viral replication can be detected by IFA or RT-PCR. The most common virus culture assay currently used is IFA on infected Vero cells.

Postnatal Rubella

In postnatal infections, IgM is detectable a few days after the rash appears and persists for about 4 weeks. If a serum specimen taken a few days after the appearance of rash is negative for RV-specific IgM, testing should be repeated with serum taken about 5 days later. If acute- and convalescent-phase sera are available, a fourfold rise in RV-specific IgG is diagnostic for RV infection; such sera should be taken within 7 days of the appearance of rash and 17 to 21 days after the appearance of rash, respectively. Since virus is usually shed in throat or nasopharyngeal specimens for only about 5 days after the onset of rash, many patients will present when virus shedding is low. This fact, plus the difficulty of virus culture techniques, limits the value of isolation as a method for diagnosis of postnatal rubella.

Congenital Rubella

The utility of laboratory techniques is different for CRS and postnatal rubella. The presence of IgM in infants less than 6 months of age is diagnostic of in utero infection, but IgM declines between 6 and 12 months after birth. Of course, positive virus culture in specimens from a newborn is diagnostic of congenital infection, and virus can be shed for up to 1 year and up to 3 years in some specimens, such as cataractous lens aspirates. Thus, virus culture is a more commonly used technique in CRS diagnosis than in postnatal rubella diagnosis. About half of the CRS cases in the United States are confirmed by RV-specific IgM, and half are confirmed by RV culture. The presence of maternal RV-specific IgG limits the use of RV-specific IgG in CRS cases. Infant IgG levels that are high or increasing in the first year of life, when maternal IgG declines, are consistent with congenital RV infection.

Reinfection

Reinfection may present a difficult situation diagnostically, since RV is shed and IgM antibodies may again be produced. Differentiation between primary infection and reinfection is sometimes important in the pregnant woman, because reinfection has a lower probability of fetal damage (Banatvala and Best, 1998). Reinfection is usually asymptomatic, but this does not differentiate it from asymptomatic primary infection, which may be transmitted to the fetus. Reinfection does produce higher-avidity IgG than recent primary infection, but laboratory measurement of IgG avidity is difficult and is usually not available. Differentiation between primary infection and reinfection may require an accurate history of exposure, information about previous tests for rubella IgG, and vaccination history.

Control and Prevention

Rubella is poorly controlled in most of the world. The percentage of seronegative women of childbearing age varies depending on previous epidemics. Rubella levels, as a result of epidemics and other factors, is sometimes above 25% (Cutts et al., 1997). During rubella outbreaks, lack of immunity in women of childbearing age typically results in one to two CRS cases per 1,000 births (Cutts et al., 1997).

The means to control rubella is immunization. Safe and effective attenuated live-virus vaccines have been developed by serial passage in tissue culture. The RA27/3 vaccine in use in the United States since 1979, a component of the MMR vaccine, was passaged 25 times in WI38 cells. It produces a more wild-type immune response than the previously used vaccine, HPV77DE5, and induces 95% seroconversions of susceptibles. It produces the symptoms of mild rubella in 10 to 15% of vaccinees, occasionally even producing a rash. Although virus is shed from the upper respiratory tract, transmission of vaccine virus is very rare. Joint symptoms occur in about 60% of postpubertal women, and rarely, these develop into arthritis. Although pregnancy is a contraindication for vaccination, hundreds of seronegative women have been inadvertently vaccinated during the first trimester, and no fetal defects have been observed (Enders, 1985).

Control of rubella through vaccination in the United States has resulted in CRS cases falling from highs of 20,000 per year to about 10 per year (Centers for Disease Control and Prevention, 1997a). A total of less than 500 cases of postnatal rubella per year is now typically reported to the Centers for Disease Control and Prevention. Most postnatal rubella in the United States occurs in populations with close contacts with areas of endemicity, highlighting the need for international efforts to control rubella (Hinman et al., 1998). Outbreaks still occur in the United States and they are controlled by identifying contacts. If previous immunization cannot be documented or serologic evidence of immunity cannot be obtained, the contact is offered vaccine (with consideration for contraindication for vaccination).

The first dose of rubella vaccine should be given at 12 to 15 months of age combined with measles and mumps vaccines (MMR). The second dose of MMR vaccine should be given at 4 to 6 years of age or by 11 to 12 years of age (Centers For Disease Control and Prevention, 1998). Prenatal serologic screening of women is indicated for all women who do have acceptable evidence of immunity based on documented administration of rubella vaccines or laboratory evidence of immunity. Women who do not have serologic evidence of rubella immunity or documented rubella vaccination should be vaccinated with MMR vaccine after termination of the pregnancy. They should be counseled to avoid conception for 3 months after vaccination.

Prenatal testing for immunity to rubella and counseling of seronegative pregnant women to avoid contact with rubella patients is an important part of CRS prevention. Postpartum vaccination of seronegative mothers is recommended. The most commonly missed opportunity for prevention of CRS in the United States is postpartum vaccination (Schluter et al., 1998).

Provided sufficient laboratory capacity exists, enhanced surveillance can be implemented by referring cases for laboratory confirmation with less specific clinical criteria than those required for a probable CRS case (e.g., one group A symptom). Enhanced surveillance is useful in elimination efforts when total CRS is low and every CRS case needs to be identified.

REFERENCES

Albrecht, P., K. Herrmann, and G. R. Burns. 1981. Role of virus strain in conventional and enhanced measles plaque neutralization test. *J. Virol. Methods* **3:**251–260.

American Academy of Pediatrics. 1997. *1997 Red Book. Report of the Committee on Infectious Diseases*, 24th ed. American Academy of Pediatrics, Elk Grove Village, Ill.

Arista, S., D. Ferraro, A. Cascio, E. Vizzi, and R. di Stefano. 1995. Detection of IgM antibodies specific for measles virus by capture and indirect enzyme immunoassays. *Res. Virol.* **146:**225–232.

Banatvala, J. E., and J. M. Best. 1998. Rubella. *In* L. Collier, A. Balows, and M. Sussman (ed.), *Microbiology and Microbial Infections,* vol. 1. Oxford University Press, Inc., New York, N.Y.

Bellini, W. J., and P. A. Rota. 1998. Genetic diversity of wild-type measles viruses: implications for global measles elimination programs. *Emerg. Infect. Dis.* **4:**1–7.

Berbers, G. A. M., A. H. J. O. Marzec, M. Bastmeijer, P. G. M. van Gageldonk, and A. D. Plantinga. 1993. Blocking ELISA for detection of mumps and antibodies in human sera. *J. Virol. Methods* **42:**155–168.

Black, F. L. 1996. Measles and mumps, p. 688–692. *In* N. R. Rose, E. Conway de Macario, J. D. Folds, H. C. Lane, and R. M. Nakamura (ed.), *Manual of Clinical Laboratory Immunology,* 6th ed. American Society for Microbiology, Washington, D.C.

Bloch, A. B., W. A. Orenstein, H. C. Stetler, S. G. Wassilak, R. W. Amler, K. J. Bart, C. D. Kirby, and A. R. Hinman. 1985. Health impact of measles vaccination in the United States. *Pediatrics* **76:**524–532.

Boger, W. P., III. 1980. Late ocular complications in congenital rubella syndrome. *Opthalmology* **87:**1244–1252.

Boger, W. P., III. 1981. Spontaneous absorption of the lens in the congenital rubella syndrome. *Am. J. Ophthalmol.* **99:**433–434.

Buynak, E. B., and M. R. Hilleman. 1966. Live attenuated mumps virus vaccine. *Proc. Soc. Exp. Biol. Med.* **123:**768–775.

Cantell, K. 1961. Mumps virus. *Adv. Virus Res.* **8:**123–164.

Centers for Disease Control. 1982. Countdown toward the elimination of measles. *Morb. Mortal. Wkly. Rep.* **31:**447–478.

Centers for Disease Control. 1989. Measles prevention: recommendations of the Immunization Practices Advisory Committee (ACIP). *Morb. Mortal. Wkly. Rep.* **38:**1–17.

Centers for Disease Control. 1990. *Morb. Mortal. Wkly. Rep.* **38:**891.

Centers for Disease Control and Prevention. 1997a. Rubella and congenital rubella syndrome—United States, 1994–1997. *Morb. Mortal. Wkly. Rep.* **46:**350–354.

Centers for Disease Control and Prevention. 1997b. Case definitions for infectious conditions under public health surveillance. *Morb. Mortal. Wkly. Rep.* **46**(RR-10):30.

Centers for Disease Control and Prevention. 1998. Measles, mumps and rubella—vaccine use and strategies for elimination of measles, rubella and congenital rubella syndrome and control of mumps. *Morb. Mortal. Wkly. Rep.* **47**(RR-8):1–57.

Chaiken, B. P., N. M. Williams, S. R. Preblud, W. Parkin, and R. Altman. 1987. The effect of a school entry law on mumps activity in a school district. *JAMA* **257:**2455–2458.

Condorelli, F., and T. Ziegler. 1993. Dot immunoblotting for simultaneous detection of specific immunoglobulin G antibodies to measles virus, mumps virus, and rubella virus. *J. Clin. Microbiol.* **31:**717–719.

Cooper, L. Z. 1985. The history and medical consequences of rubella. *Rev. Infect. Dis.* S2–S10.

Cutts, F. T., S. E. Robertson, J.-L. Diaz-Ortega, and R. Samuel. 1997. Control of rubella and congenital rubella syndrome (CRS) in developing countries. Part 1. Burden of disease from CRS. *Bull. W. H. O.* **75:**55–68.

Desmond, M. M., E. S. Fisher, A. L. Vorderman, H. G. Schaffer, L. P. Andrew, T. E. Zion, and F. I. Catlin. 1978. The longitudinal course of congenital rubella encephalitis in non-retarded children. *J. Pediatr.* **93:**584–591.

de Souza, V. A., C. S. Pannuti, L. M. Sumita, and H. F. de Andrade, Jr. 1997. Enzyme-linked immunosorbent assay-IgG antibody avidity test for single sample serologic evaluation of measles vaccines. *J. Med. Virol.* **52:**275–279.

Eberhart-Phillips, J.E., P. D. Frederick, R. C. Baron, and L. Mascola. 1993. Measles in pregnancy: a descriptive study of 58 cases. *Obstet. Gynecol.* **82:**797–801.

Edmonson, M., D. Addiss, J. McPherson, J. Berg, S. Circo, and J. Davis. 1990. Mild measles and secondary vaccine failure during a sustained outbreak in a highly vaccinated population. *JAMA* **163:**2467–2471.

Enders, G. 1985. Rubella antibody titres in vaccinated and non-vaccinated women and results of vaccination during pregnancy. *Rev. Infect. Dis.* **7:**S103–S112.

Enders, G., U. Nikerl-Pacher, E. Miller, and J. E. Cradock-Watson. 1988. Outcome of confirmed periconceptional maternal rubella. *Lancet* **i:**1445–1447.

Enders, J. F., and T. C. Peebles. 1954. Propagation in tissue cultures of cytopathogenic agents from patients with measles 1954. *Proc. Soc. Exp. Biol. Med.* **86:**277–286.

Erdman, D. D., L. J. Anderson, D. R. Adams, J. A. Stewart, L. E. Markowitz, and W. J. Bellini. 1991. Evaluation of monoclonal antibody-based capture enzyme immunoassays for detection of specific antibodies to measles virus. *J. Clin. Microbiol.* **29:**1466–1471.

Floret, D., D. Rosenberg, G. N. Hage, and P. Monnet. 1980. Case report: hyperthyroidism, diabetes mellitus and the congenital rubella syndrome. *Acta Paediatr. Scand.* **69:**250–261.

Forrest, J. M., M. A. Menser, and J. A. Burgess. 1971. High frequency of diabetes mellitus in young adults with congenital rubella. *Lancet* **ii:**332–334.

Frey, T. K. Rubella. *In* R. Nakamura and J. Folds (ed.), *Clinical Diagnostic Immunology,* in press. Blackwell Science, Malden, Mass.

Fuccillo, D. A., and J. L. Sever. 1973. Viral teratology. *Bacteriol. Rev.* **37:**19–31.

Fuccillo, D. A., R. W. Steele, S. A. Henson, M. M. Vincent, J. B. Hardy, and J. A. Bellanti. 1974. Impaired cellular immunity to rubella virus in congenital rubella. *Infect. Immun.* **9:**81–84.

Fulginiti, V. A., J. J. Eller, A. W. Downie, and C. H. Kempe. 1967. Altered reactivity to measles virus. *JAMA* **202:**1075–1080.

Gershon, A., and S. Krugman. 1979. Measles virus, p. 665–693. *In* H. Lennette, J. Schmidt (ed.), *Diagnostic Procedures for Viral, Rickettsial, and Chlamydial Infections.* American Public Health Association, Washington, D.C.

Glickmann, G., M. Pedersen, and C. H. Mordhurst. 1986. Detection of specific immunoglobulin M to mumps virus in serum and cerebrospinal fluid samples from patients with acute mumps infection, using an antibody-capture enzyme immunoassay. *Acta Pathol. Microbiol. Immunol. Scand. Sect. C* **94:**145–156.

Grayzel, A. I., and C. Beck. 1971. The growth of vaccine strain of rubella virus in cultured human synovial cells. *Proc. Soc. Exp. Biol. Med.* **136:**496–498.

Gregg, N. 1941. Congenital cataract following german measles in the mother. *Trans. Ophthalmol. Soc. Aust.* **3:**35–46.

Grillner, L., and J. Blomberg. 1976. Hemolysis-in-gel and neutralization tests for determination of antibodies to mumps virus. *J. Clin. Microbiol.* **4:**11–15.

Habel, K. 1945. Cultivation of mumps virus in the developing chick embryo and its application to studies of immunity to mumps in man. *Public Health Rep.* **60:**201–212.

Hall, W. W., and S. J. Martin. 1974. The biochemical and biological characteristics of the surface components of measles virus. *J. Gen. Virol.* **22:**363–374.

Helfand, R. F., J. L. Heath, L. J. Anderson, E. F. Maes, D. Guris, and W. J. Bellini. 1997. Diagnosis of measles with an IgM capture EIA: the optimal timing of specimen collection after rash onset. *J. Infect. Dis.* **175:**195–199.

Helfand, R. F., D. K. Kim, H. E. Gary, G. L. Edwards, G. P. Bisson, M. J. Papania, J. L. Heath, D. L. Schaff, W. J. Bellini, S. C. Redd, and L. J. Anderson. 1998. Nonclassic measles infections in an immune population exposed to measles during a college bus trip. *J. Med. Virol.* **56:**337–341.

Hinman, A. R., B. S. Hersh, and C. A. de Quadros. 1998. Rational use of rubella vaccine for prevention of congenital rubella syndrome in the Americas. *Rev. Panam. Salud Pública* **4:**156–160.

Hopps, H. E., and P. D. Parkman. 1979. *In* E. H. Lennette and N. J. Schmidt (ed.), *Diagnostic Procedures for Viral, Rickettsial, and Chlamydial Infections.* American Public Health Association, Washington, D.C.

Horstmann, D. M. 1990. Rubella, p. 617–630. *In* A. S. Evans (ed.), *Viral Infections of Humans: Epidemiology and Control*, 3rd ed. Plenum Medical Book Company, New York, N.Y.

Horta-Barbosa, L., D. A. Fuccillo, J. L. Sever, and W. Zeman. 1969. Subacute sclerosing panencephalitis: isolation of measles virus from a brain biopsy. *Nature* **221:**974.

Horta-Barbosa, L., R. Hamilton, B. Wiltig, D. A. Fuccillo, J. L. Sever, and M. L. Vernon. 1971. Subacute sclerosing panencephalitis: isolation of suppressed measles virus from lymph node biopsies. *Science* **173:**840–841.

Hummel, K. B., D. D. Erdman, J. L. Heath, and W. J. Bellini. 1992. Baculovirus expression of the nucleoprotein gene of measles virus and utility of the recombinant protein in diagnostic enzyme immunoassays. *J. Clin. Microbiol.* **30:**2874–2880.

Jin, L., D. W. G. Brown, M. E. B. Ramsay, P. A. Rota, and W. J. Bellini. 1997. The diversity of measles virus in the UK, 1992–1995. *J. Gen. Virol.* **78:**1287–1294.

Johnson, S. C., and E. W. Goodpasture. 1934. An investigation of the etiology of mumps. *J. Exp. Med.* **59:**1–19.

Katow, S. 1999. Rubella virus genome diagnosis during pregnancy and mechanism of congenital rubella. *Intervirology* **41:**163–169.

Kobune, F., H. Sakata, and A. Sugiura. 1990. Marmoset lymphoblastoid cells as a sensitive host for isolation of measles virus. *J. Virol.* **64:**700–705.

Koplik, H. 1896. The diagnosis of the invasion of measles from a study of the exanthemata as it appears on the buccal mucus membrane. *Arch. Pediatr.* **13:**918–922.

Markowitz, L. E., F. W. Chandler, E. O. Roldan, M. J. Saldana, K. C. Roach, S. S. Hutchins, S. R. Preblud, C. D. Mitchell, and G. B. Scott. 1988. Fatal measles pneumonia without rash in a child with AIDS. *J. Infect. Dis.* **158:**480–483.

Mayo, D. R., T. Brennan, D. P. Corimer, J. Hadler, and P. Lamb. 1991. Evaluation of a commercial measles virus immunoglobulin M enzyme immunoassay. *J. Clin. Microbiol.* **29:**2865–2867.

Menser, M. A., L. Dods, and J. D. Harley. 1967. A twenty-five year follow-up of congenital rubella. *Lancet* **ii:**1347–1350.

Menser, M. A., J. M. Forrest, M. C. Honeyman, and J. A. Burgess. 1974. Diabetes, HLA antigens, and congenital rubella. *Lancet* **ii:**1058–1059.

Minnich, L. L, F. Goodenough, and C. G. Ray. 1991. Use of immunofluorescence to identify measles virus infections. *J. Clin. Microbiol.* **29:**1148–1150.

Muller, C. P., S. Huiss, and F. Schneider. 1996. Secondary immune response in parents of children with recent measles. *Lancet*, **ii:**1379–1380.

Nakayama, T., T. Mori, S. Yamaguchi, S. Sonoda, S. Asamura, R. Yamashita, Y. Takeuchi, and T. Urano. 1995. Detection of measles virus genome directly from clinical samples by reverse transcriptase-polymerase chain reaction and genetic variability. *Virus Res.* **35:**1–16.

Narita, M., S. Yamada, Y. Matsuzono, O. Itakura, T. Togashi, and H. Kikuta. 1997. Measles virus-specific immunoglobulin G subclass response in serum and cerebrospinal fluid. *Clin. Diagn. Virol.* **8:**233–239.

Orenstein, W. A., K. J. Bart, A. R. Hinman, S. R. Preblud, W. L. Greaves, S. W. Doster, H. C. Stetler, and B. Sirotkin. 1984. The opportunity and obligation to eliminate rubella from the United States. *JAMA* **251:**1988–1994.

Parkman, P. D., E. L. Buescher, and M. S. Artenstein. 1962. Recovery of rubella virus from Army recruits. *Proc. Soc. Exp. Biol. Med.* **111:**225–230.

Philip, R. N., K. R. Reinhard, and D. B. Lackman. 1959. Observations on a mumps epidemic in a "virgin" population. *Am. J. Hyg.* **69:**91–95.

Plotkin, S. A., and R. Kaye. 1970. Diabetes melitus and congenital rubella. *Pediatrics* **46:**650–651.

Preece, M. A., P. J. Kearney, and W. C. Marshall. 1977. Growth-hormone deficiency in congenital rubella. *Lancet* **ii:**842–844.

Rakowsky, A., and J. L. Sever. 1998. Rubella (german measles), p. 1301–1311. *In* S. L. Gorbach, J. G. Bartlett, and N. R. Blacklow (ed.), *Infectious Diseases*, 2nd ed. W. B. Saunders Company, Philadelphia, Pa.

Ratnam, S., V. Gadag, R. West, J. Burris, E. Oates, F. Stead, and N. Bouilianne. 1995. Comparison of commercial enzyme immunoassay kits with plaque reduction neutralization test for detection of measles virus antibody. *J. Clin. Microbiol.* **33:**811–815.

Ratzmann, K. P., J. Strese, S. Witt, H. Berling, H. Keilacker, and D. Michaelis. 1984. Mumps infection and insulin-dependent diabetes mellitus (IDDM). *Diabetes Care* **7:**170–173.

Rawls, W. E., and J. L. Melnick. 1966. Rubella virus carrier cultures derived from congenitally infected infants. *J. Exp. Med.* **123:**795–816.

Rosenberg, H. S., E. H. Oppenheimer, and J. R. Esterly. 1981. Congenital rubella syndrome: the late effects and their relation to early lesions. *Perspect. Pediatr. Pathol.* **6:**183–202.

Rota, P. A., A. S. Khan, E. Durigon, T. Yuran, Y. S. Villamarzo, and W. J. Bellini. 1995. Detection of measles virus RNA in urine specimens from vaccine recipients. *J. Clin. Microbiol.* **33:**2485–2488.

Schiff, G. M., and J. L. Sever. 1966. Rubella: recent laboratory and clinical advances. *Prog. Med. Virol.* **8:**30–61.

Schluter, W. W., S. E. Reef, S. C. Redd, and C. A. Dykewicz. 1998. Changing epidemiology of congenital rubella syndrome in the United States. *J. Infect. Dis.* **178:**36–41.

Schwartz, A. J. F. 1962. Preliminary tests of a highly attenuated measles vaccine. *Am. J. Dis. Child.* **103:**386–389.

Smaron, M. F., E. Saxon, L. Wood, C. McCarthy, and J. A. Morello. 1991. Diagnosis of measles by fluorescent antibody and culture of nasopharyngeal secretions. *J. Virol. Methods* **33:**223–229.

South, M. A., and C. A. Alford. 1980. The immunology of chronic intrauterine infections, p. 702–714. *In* R. Stiehm and V. A. Fulginiti (ed.), *Immunologic Disorders in Infants and Children*, 2nd ed. W. B. Saunders, Philadelphia, Pa.

St. Geme, J. W., Jr., G. R. Noren, and P. Adams, Jr. 1966. Proposed embryopathic relation between mumps virus and primary endocardial fibroelastosis. *N. Engl. J. Med.* **275:**339–347.

Sultz, H. A., B. A. Hart, M. Zielezny, and E. R. Schlesinger. 1975. Is mumps virus an etiologic factor in juvenile diabetes mellitus? *J. Pediatr.* **86:**654–656.

Summers, M. D., and G. E. Smith. 1986. *A Manual of Methods for Baculovirus Vectors and Insect Cell Culture Procedures.* Bulletin no. 1555. Texas Agricultural Experiment Station, College Station, Tex.

Swierkosz, E. M. 1995. Mumps virus, p. 963–967. *In* P. R. Murray, E. J. Baron, M. A. Pfaller, F. C. Tenover, and R. H. Yolken, (ed.), *Manual of Clinical Microbiology*, 6th ed. American Society for Microbiology, Washington, D.C.

Thieme, T., S. Piacentini, S. Davidson, and K. Steingart. 1994. Determination of measles, mumps, and rubella immunization status using oral fluid samples. *JAMA* **272:**219–221.

Townsend, J. J., J. R. Baringer, J. A. Wolinsky, N. Malamud, J. P. Medick, H. S. Panirch, R. A. T. Scott, L. S. Oshiro, and N. E. Cremer. 1975. Progressive rubella panencephalitis. *N. Engl. J. Med.* **292:**990–993.

Townsend, J. J., W. G. Stroop, J. R. Baringer, J. H. Wolinsky, J. H. McKerrow, and B. O. Berg. 1976. Neuropathy of progressive rubella panencephalitis after childhood rubella. *Neurology* **32:**185–190.

Tuokko, H. 1984. Comparison of nonspecific reactivity in indirect and reverse immunoassays for measles and mumps immunoglobulin M antibodies. *J. Clin. Microbiol.* **20:**972–976.

Tuokko, H. 1995. Detection of acute measles infections by indirect and mu-capture enzyme immunoassays for immunoglobulin M antibodies and measles immunoglobulin G antibody avidity enzyme immunoassay. *J. Med. Virol.* **45:**306–311.

Watson, J. C., S. C. Redd, P. H. Rhodes, and S. C. Hadler. 1998. The interruption of transmission of indigenous measles in the United States during 1993. *Pediatr. Infect. Dis. J.* **17:**363–366.

Webster, W. S. 1998. Teratogen update: congenital rubella. *Teratology* **58:**13–23.

Weil, M. L., H. H. Itabashi, N. E. Cremer, L. S. Oshiro, E. H. Lennette, and L. Carnay. 1975. Chronic progressive panencephalitis due to rubella virus simulating subacute sclerosing panencephalitis. *N. Engl. J. Med.* **292:**994–998.

Weller, T. H., and F. A. Neva. 1962. Propagation in tissue culture of cytopathic agents from patients with rubella-like illness. *Proc. Soc. Exp. Biol. Med.* **111:**215–225.

Weller, T. H., C. A. Alford, and F. A. Neva. 1964. Retrospective diagnosis by serological means of congenitally acquired rubella infection. *N. Engl. J. Med.* **270:**1039–1041.

Wolinsky, J. S. 1978. Progressive rubella panencephalitis, p. 331–341. *In* P. J. Vinken and G. W. Bruyn (ed.), *Handbook of Clinical Neurology*, vol. 34. North-Holland, Amsterdam, The Netherlands.

Zaki, S., and W. J. Bellini. 1997. Measles virus, p. 233–245. *In* D. H. Connor, F. W. Chandler, D. A. Schwartz, D. J. Manz, and E. E. Lack, (ed.) *Pathology of Infectious Diseases*, vol 1. Appleton and Lange, Stamford, Conn.

Ziring, P. R., G. Gallo, M. Finegold, E. Buimovici-Klein, and P. Ogra. 1977. Chronic lymphocytic thyroiditis: identification of rubella virus antigen in the thyroid of a child with congenital rubella. *J. Pediatr.* **90:**419–420.

Human Retroviruses

JÖRG SCHÜPBACH AND ROBERT C. GALLO

37

Retroviruses are enveloped positive-strand RNA viruses with diameters of about 80 to 120 nm. The infectious particles (i.e., virions) contain two copies of a single-stranded genomic RNA about 7 to 11 kb long. The RNA is surrounded by viral enzymes and structural proteins forming a nucleocapsid and a matrix shell, to the outside of which a lipid envelope derived from the host cell membrane is attached. Inserted into the membrane are viral glycoprotein oligomers which mediate adsorption to receptors located on suitable host cells and penetration of the cell membrane.

For a long time, retroviruses were well known as the causative agents of leukemias, lymphomas, other cancers, or chronic inflammations in various animal species, while evidence for a pathogenic role in humans was missing. Advances in cell culture techniques, particularly the possibility of growing T cells with interleukin 2 (IL-2) (Morgan et al., 1976; Ruscetti et al., 1977), and tests for the retroviral enzyme reverse transcriptase (RT) first led to the isolation of the human T-cell leukemia virus type 1 (HTLV-1) (Poiesz et al., 1981). HTLV-1 was soon identified as the causative agent of adult T-cell leukemia or lymphoma (ATLL), a rapidly progressing cancer of CD4$^+$ lymphocytes first described in the southeastern parts of Japan (Takatsuki et al., 1977). Knowledge gained from studying HTLV-1 was fundamental for the detection of other human retroviruses, first the related HTLV-2 (Kalyanaraman et al., 1982). Soon thereafter, human immunodeficiency virus type 1 (HIV-1) was discovered (Barre-Sinoussi et al., 1983) and rapidly identified as the causative agent of the newly recognized AIDS (Gallo et al., 1984; Levy et al., 1984; Popovic et al., 1984; Sarngadharan et al., 1984; Schupbach et al., 1984a). Two years later, another AIDS-causing virus, HIV-2, was found (Clavel et al., 1986a). Meanwhile, in addition to a few sporadic cases in which foamy retroviruses (Anonymous, 1997; Heneine et al., 1998) or simian immunodeficiency virus (SIV) (Khabbaz et al., 1994) was found in humans as a result of direct cross-species nosocomial transmission from monkeys, two additional novel exogenous retroviral agents have been identified. One is a type D retrovirus closely related to Mason-Pfizer monkey virus or simian retrovirus type 1, which was isolated from two B-cell lymphomas and from bone marrow of a single HIV-1-positive patient with AIDS (Bohannon et al., 1991; Ford et al., 1992). Another virus, human retrovirus 5, was originally isolated from the salivary gland of a patient with Sjögren's

syndrome (Griffiths et al., 1997) and was subsequently also detected in other autoimmune disorders (Griffiths et al., 1999). An overview of the currently known exogenous human retroviruses and the diseases associated with them is shown in Table 1. In addition, the question of whether endogenous retroviruses might contribute to human autoimmune disease has been raised in connection with Sjögren's syndrome (Garry et al., 1990), multiple sclerosis (Perron et al., 1997), and insulin-dependent diabetes (Conrad et al., 1997).

HIVs

Biology

HIV-1 and HIV-2 are the causative agents of AIDS. AIDS is the end stage of a protracted pathogenic process in which the immune system of an infected person and its ability to control infections or malignant proliferative disorders are progressively destroyed.

The HIVs are members of the genus *Lentivirinae* of the family *Retroviridae*. HIV-1 is the more aggressive virus and is responsible for the AIDS pandemic. HIV-2, identified in 1986 (Clavel et al., 1986a, 1986b), is much less pathogenic than HIV-1. The rates of heterosexual and mother-to-child transmission of HIV-2 are low, and latency dominates the clinical picture. Rarely, however, the virus may cause AIDS (D'Aquila, 1996; R. Marlink, Editorial, *AIDS* **10**:689–699, 1996).

Primate Lentiviruses

A group of related viruses, the SIVs, infect various species of Old World monkeys and the chimpanzee (Fig. 1). They may have arisen several million years ago in *Cercopithecus* monkeys. These primate lentiviruses are categorized into five major lineages. Lineage 1 contains the various isolates of HIV-1, which are subclassified into three groups, M (main), O (outlier), and a new group, N, recently identified in Cameroonians (Simon et al., 1998). From the phylogenetic tree in Fig. 1, it is evident that the various group M isolates (e.g., HIV-1/LAI) are more closely related to two isolates from the chimpanzee (*Pan troglodytes*), SIVcpz-GAB1 and SIVcpzUS, which were both isolated from the subspecies *Pan troglodytes troglodytes*, than to isolates of HIV-1 group O (e.g., HIV-1/ANT70) or to another chim-

TABLE 1 Overview of retroviruses isolated from humans

Virus	Affiliation	Disease association
HIV-1, HIV-2	Subfamily *Lentivirinae*; primate lentiviruses	AIDS and related conditions
HTLV-1	HTLV-bovine leukemia virus group	ATLL
		HAM-TSP
		HAU
HTLV-2	HTLV-bovine leukemia virus group	Unclear; rare forms of T-cell lymphoproliferative disorders, HAM-TSP (?)
Human foamy virus[a]	Subfamily *Spumavirinae*	Nosocomial infection with no known disease association
SIV	Subfamily *Lentivirinae*; primate lentiviruses	Nosocomial infection with too-short observation time
Type D retrovirus isolated from a B-cell Burkitt-type lymphoma of a patient with AIDS[b]	D-type virus group	Unclear; single-case observation
Human retrovirus 5	Related to simian type D retroviruses, rodent intracisternal type A particles, and murine mammary tumor virus	Autoimmune disorders?

[a]Anonymous, 1997.
[b]Hoffman et al., 1985; Levy et al., 1985.

panzee isolate, SIVcpzANT, which was isolated from a different subspecies, *Pan troglodytes schweinfurthii*. These data indicate that the HIV-1 epidemic is the result of zoonotic virus transmissions from *P. t. troglodytes* to humans. All HIV-1 isolates, including the M, O, and N groups, appear to be derived from the same chimpanzee subspecies, *P. t. troglodytes*. Sequence comparisons further suggest that the HIV-1 N sequence is a mosaic generated by an ancestral recombination of HIV-1- and SIVcpzUS-related sequences in a chimpanzee host (Gao et al., 1999). Analysis of HIV-1 sequences from a Congolese serum sample dating back to 1958 further shows that the diversication of all group M subtypes has occurred very recently, i.e., within the past 50 years (Zhu et al., 1998).

Lineage 2 contains the different isolates of HIV-2, which are closely related to viruses infecting sooty mangabeys (SIV$_{sm}$). This virus has also been naturally transmitted to macaques. An HIV-2 strain (HIV-2/ST) that differs less from its simian counterparts than it does from other human viruses has been identified. This, along with the fact that SIV infections in their natural simian hosts do not cause disease, has led to the conclusion that the HIV-2 epidemic is also the result of simian-to-human cross-species transmission that has occurred on several different occasions (reviewed in Sharp et al., 1994).

HIV Groups and Subtypes

The extraordinary variability of HIV, due to rapid mutation and recombination, has led to the development and geographical distribution of various distinctive groups and subtypes of viruses. HIV-1 group M is divided into subtypes A to J. Isolates of group O differ as much from each other as do viruses from different subtypes of group M, but their limited number has so far precluded definition of distinct subtypes. Group N viruses have been isolated from only two individuals (Simon et al., 1998). In HIV-2, five subtypes (A to E) have been defined. HIV-1 subtype B viruses were reponsible for the first epidemic in North America, Europe, Australia,

and the metropolitan centers of other continents in the late 1970s and early 1980s. Subtype E is most prevalent in Thailand and Southeast Asia. Subtype F has been found in Brazil and Romania, and G is prevalent in Russia. Almost all HIV-1 subtypes are found in sub-Saharan Africa, with subtypes A, C, and D being the most prevalent in that continent. C is also common in India. Viruses of the M group, which are responsible for the AIDS pandemic, are rapidly spreading worldwide, whereas group O viruses are rarely isolated and virtually exclusively restricted to persons originating from Cameroon, Gabon, and Equatorial Guinea (Sharp et al., 1994). Recent investigations indicate that 25 to 30% of novel HIV-1 infections in various European countries are now due to non-B subtypes (Barin et al., 1997, 1998; Böni et al., 1999).

Full-length-genome analysis indicates that HIV-1 group M subtypes A, B, C, and D are largely, if not entirely, distinguishable throughout the genome and show no clear evidence of intersubtype recombination. In contrast, all available sequences of subtypes E (from both Southeast Asia and Central Africa) and G are recombinant with subtype A; the nonrecombinant parental isolates of these subtypes have not yet been identified. Subtypes E and G, and some A-D recombinant HIV-1 isolates, have retained the cytoplasmic domain of gp41 from subtype A. Subtype I is a complex recombinant comprising sequences of A, G, and I (Gao et al., 1998). Evidence for superinfection and recombination has also been documented for HIV-2 (Sharp et al., 1994).

The Retroviral Replication Cycle

An overview of the retroviral replication cycle is given in Fig. 2A. For infection of a host cell, the HIV virion binds to a virus receptor, the CD4 molecule, located on the membrane (Dalgleish et al., 1984; Klatzmann et al., 1984). Interaction with a coreceptor is required before fusion of the viral and host cell membranes occurs, followed by membrane penetration and virus entry, release of the viral core (uncoating), and a progressive dissolution of the viral core, which

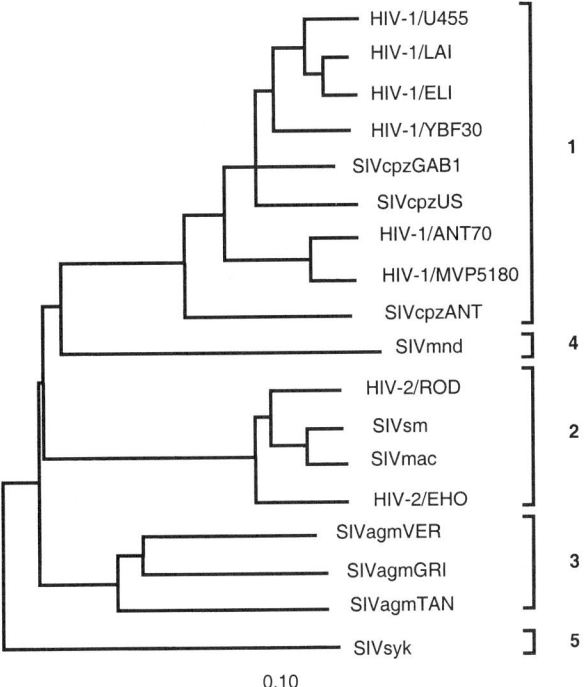

FIGURE 1 Phylogenetic tree of primate lentiviruses, derived from Pol sequences. The numbers 1 to 5 indicate five major lineages. HIV-1/U455 is a group M, subtype A isolate. ELI is group M subtype D, and LAI is group M subtype B. ANT70 and MVP5180 represent group O, and YBF30 is group N. ROD and EHO represent different subtypes of HIV-2. SIVcpzGAB1 and SIVcpzUS are chimpanzee isolates from the subspecies *P. t. troglodytes*, while SIVcpzANT is an isolate from *P. t. schweinfurthii.* mnd, mandrill; agm, African green monkey; syk, Sykes' monkey; sm, sooty mangabey. HIV-1 isolates are derived from SIV chimpanzee isolates, and HIV-2 is derived from SIVsm. The bar at the bottom denotes genomic diversity. Combined and simplified from Sharp et al., 1994, and Gao et al., 1999.

leads to the preintegration complex that enters the nucleus. Like all retroviruses, HIV particles contain a characteristic enzyme, reverse transcriptase (RT). The enzyme is cleaved, and thereby activated, from a precursor by the action of another retroviral enzyme, the viral protease (PR). RT possesses three distinct enzymatic functions. It acts as an RNA-dependent DNA polymerase, the RT activity in the strict sense of the word; an RNase H; and a DNA-dependent DNA polymerase. After infection of a host cell and the step of uncoating, these different RT functions serve in turn to synthesize a cDNA of the viral RNA, to degrade the RNA from the cDNA-RNA heteroduplex, and to duplicate the cDNA strand (Fig. 2B). Regulatory sequences present at both ends of the viral RNA (R-U5 at the 5′ end and U3-R at the 3′ end) are thereby complemented and partially duplicated in a way that yields the so-called long terminal repeats (LTR). These contain U3-R-U5 and are located at both ends of the viral double-stranded DNA (Fig. 2B). This double-stranded DNA, which is still associated with some viral proteins in the preintegration complex, migrates into the nucleus, where it is integrated into the host cell genome by a

third specific retroviral enzyme, the integrase (IN). The integrated retroviral DNA genome is called the provirus.

Genome and Transcripts

The genomic organization of the HIV-1 and HIV-2 proviruses is shown in Fig. 2C. Like all retroviruses, HIVs possess the open reading frames (ORFs) *gag* and *env* for structural proteins, namely, the precursor proteins of the viral capsid and envelope, respectively, and *pol* for the enzymes. Additional overlapping ORFs code for the *trans*-acting transcriptional activator (Tat) and the regulator of viral expression (Rev), which are both essential for virus replication. Furthermore, both HIV types contain ORFs for several accessory or auxiliary proteins, including, in the case of HIV-1, Vif, Vpr, Vpu, and Nef and, in the case of HIV-2, Vif, Vpx, Vpr, and Nef.

The proviral DNA is transcribed by the cellular RNA polymerase II into a polyadenylated unspliced mRNA. A variety of qualitatively different mRNAs (the major ones are shown in Fig. 2D) have been found in HIV-1-infected cells. They are divided into early and late transcripts. Early transcripts are completely, or at least doubly, spliced and code for the accessory proteins Tat, Rev, and Nef, which, in a complex interplay with cellular regulators, coordinate the production of structural proteins necessary for particle assembly. Late transcripts are singly spliced, or remain entirely unspliced, and code for Vpu, Env, Vif, Vpr, or Gag and Pol. As for all retroviruses, the unspliced viral mRNA also serves as the encapsidated genomic RNA. Late transcripts are only exported to the cytoplasm in the presence of Rev.

Structural Proteins and Particle Composition

Figure 3A gives an overview of the different viral proteins, and Fig. 3B shows their localization in the mature particle. In addition to the proteins shown in the figure, the nucleocapsid also contains Vpr (or Vpx in the case of HIV-2 [*vpx* is a duplicate of *vpr*]) at molar quantities equivalent to those of Gag proteins. Small amounts of Nef (p27) are also present. Furthermore, several host cell proteins also assemble in the virions. These include cyclophilins, which act as molecular chaperons for the capsid protein p24, and cell membrane proteins, like the β-2 microglobulin, HLA-α, and HLA-β chains, and HLA-DR. Particles are assembled at the cell membrane and are, in a still immature and noninfectious form, released by budding. For full maturation into infectious particles, the viral Gag and Gag-Pol precursor proteins must be cleaved into the different subunit proteins by the PR.

Virus Entry

The CD4 molecule is located on CD4$^+$ T lymphocytes, macrophages, and dendritic or other antigen-presenting cells. Interaction of the CD4-binding domain of the HIV surface glycoprotein, gp120, with CD4 induces a conformational change, whereupon the complex of CD4 and gp120 interacts with a coreceptor of the family of seven-transmembrane chemokine receptors (Alkhatib et al., 1996; Deng et al., 1996; Dragic et al., 1996; Feng et al., 1996), after which the fusion domain of the transmembrane protein gp41 inserts itself into the cell membrane, leading to the fusion of the viral and cellular membranes. These coreceptors are G-protein-coupled signaling receptors which bind the chemokines that are involved in controlling the activation and migration of various leukocytes to a site of infection.

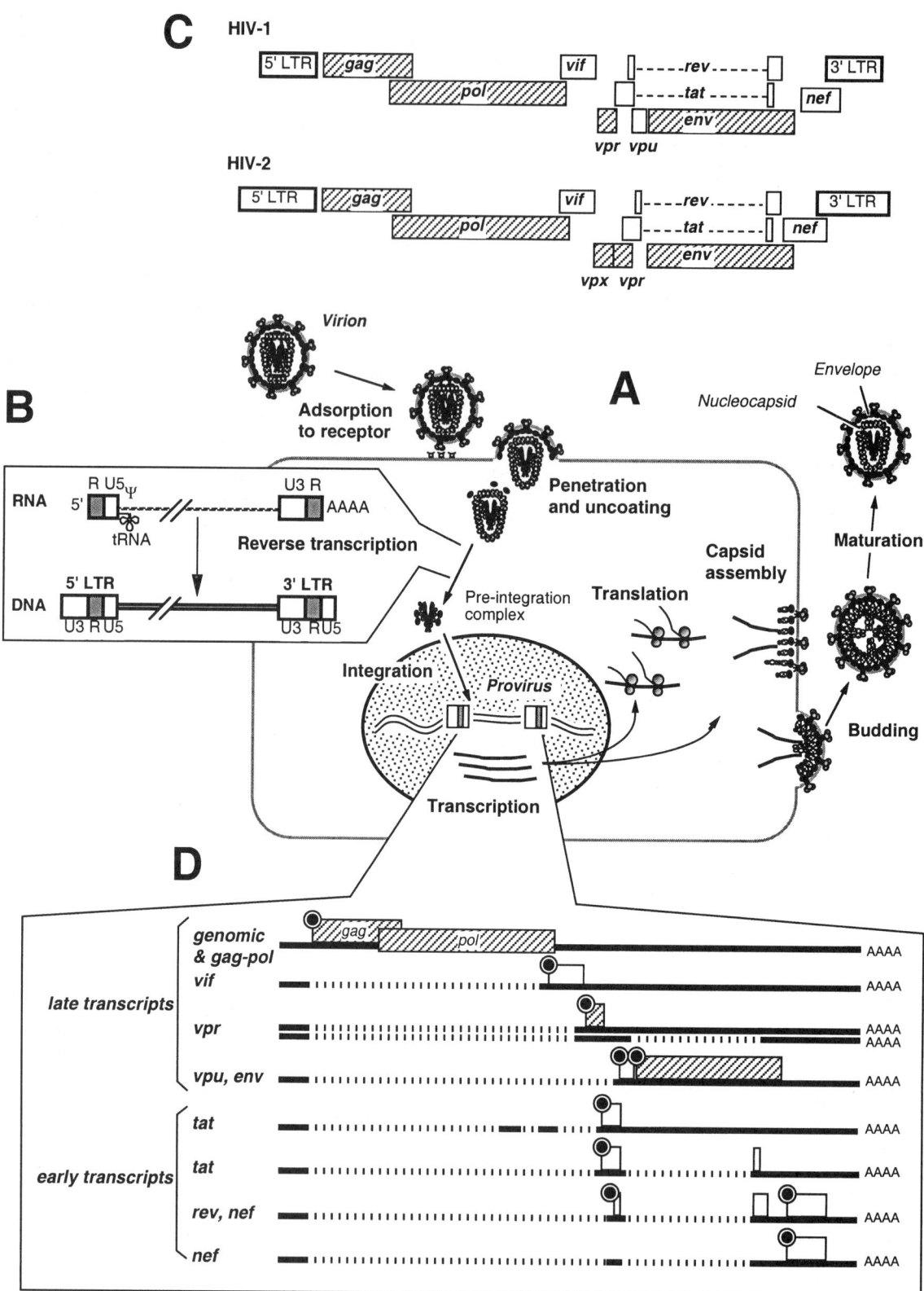

FIGURE 2 HIV replication cycle. (A) Overview. (B) Reverse transcription. The retroviral genome contained in virions consists of RNA. Its characteristic features include terminal repeats (R), U5 (5′ untranslated), U3 (3′ untranslated), 3′ polyadenylation, a binding site for a tRNA which serves as the primer for reverse transcription, and the encapsidation signal Ψ. During reverse transcription, the viral RNA is reverse transcribed into double-stranded DNA, and terminal sequences are partially duplicated in a way that leads to LTR composed of U3-R-U5. (C) Genomic organization of HIV-1 and HIV-2. The hatched boxes denote ORFs for proteins contained in particles. (D) Early and late viral transcripts found in HIV-1-infected cells. Drawing extended from Schüpbach, 1999.

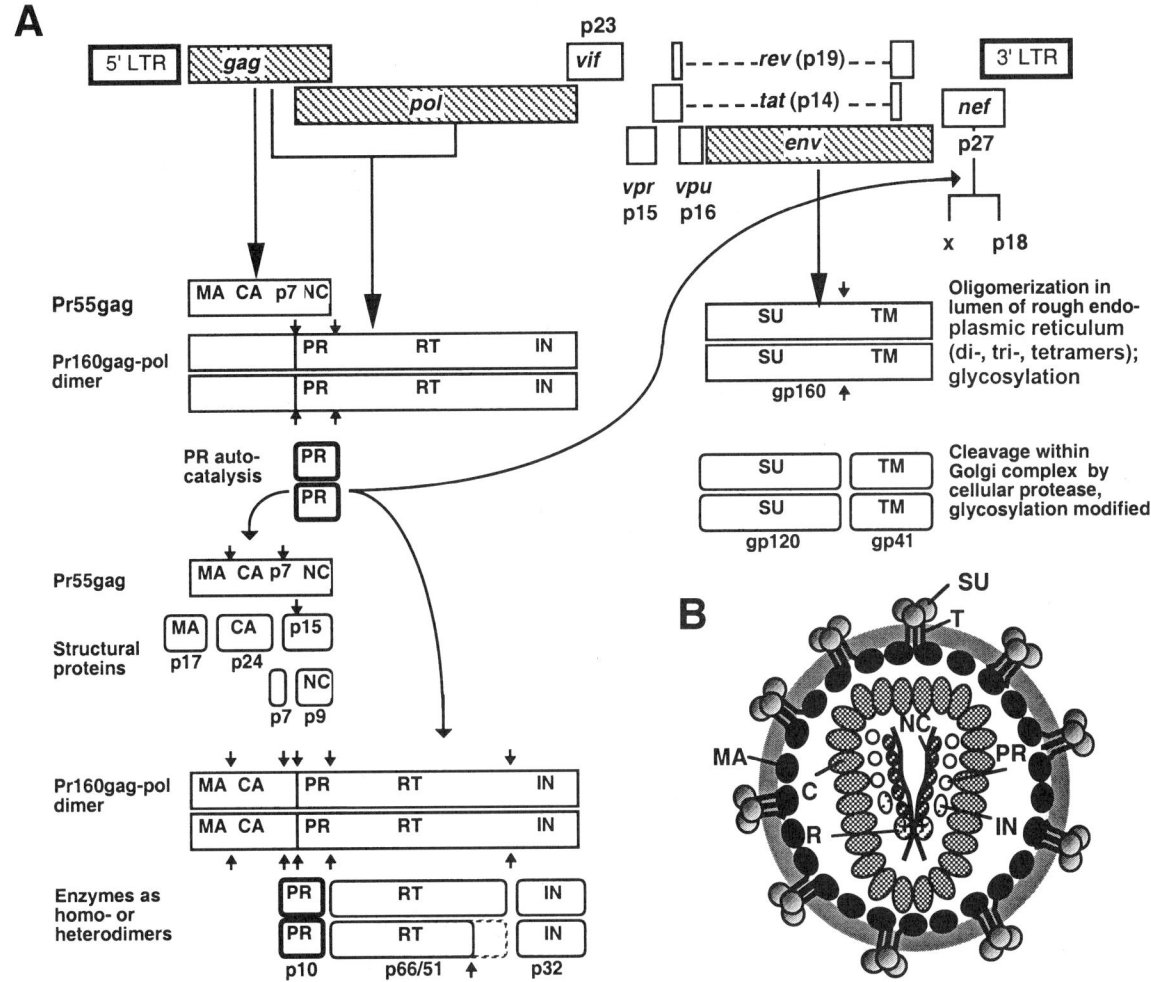

FIGURE 3 Translational products of HIV-1 and particle composition. (A) Translation. The open boxes in the genome representation on top denote ORFs of the accessory proteins Tat, Rev, Nef, Vif, Vpr, and Vpu which are translated into proteins of final size (note that a subset of Nef which is incorparated into particles is further cleaved by PR). The hatched boxes denote ORFs translated into precursor proteins. The products of the *gag, pol,* and *env* genes are synthesized as polyprotein precursors. The principal Gag precursor, Pr55*gag*, is cleaved by the viral protease (PR or p10) into the matrix (MA) protein p17, the capsid (CA) protein p24, and a C-terminal protein, p15, which is subsequently cleaved into p7 and the nucleocapsid (NC) protein p9. Cleavage of Pr160*gag-pol*, which is produced by ribosomal frameshifting at the *gag-pol* junction, yields PR, RT, and the IN. All three enzymes remain dimerized after cleavage. RT first forms a homodimer, p66-p66, which is subsequently modified into the heterodimer p66-p51. The Env precursor gp160 is glycosylated in the Golgi system, oligomerizes into dimers and trimers, and is cleaved by a cellular protease into the SU protein gp120 and the smaller TM protein gp41. The small arrows indicate protease cleavage sites. (B) Localization of viral proteins in mature virions. Reprinted from Schüpbach, 1999.

Langerhans cells or monocytes/macrophages exhibit the CCR5 chemokine receptor, which permits infection by the M (macrophage)-tropic viruses of the non-syncytium-inducing (NSI) phenotype (now called R5 viruses [Berger et al., 1998]) usually present in early infection (Alkhatib et al., 1996; Deng et al., 1996; Dragic et al., 1996). This interaction is inhibited by the natural ligands of CCR5, the β-chemokines RANTES, MIP-1α, and MIP-1β, whose identifiction as HIV suppressor factors led to the discovery of these long-sought coreceptors (Cocchi et al., 1995). In contrast, the coreceptor on CD4+-T-cell lines is the CXCR4 molecule, which binds only T-cell-tropic, syncytium-inducing isolates (now called X4 viruses [Berger et al., 1998]) and whose natural ligand is the stroma-derived factor SDF-1 (Bleul et al., 1996; Oberlin et al., 1996). Primary CD4+ T cells exhibit both coreceptors, CCR5 and CXCR4, and thus can be infected by both M- and T-tropic viruses.

Not all cells infectable with HIV (for example, astrocytes) express CD4. In these cells, galactosylceramide has been described as an alternative virus receptor for gp120. M cells of the enteric mucosa, which may be involved in rectal infections, are also negative for CD4 but positive for galactosylceramide (Harouse et al., 1991; McAlarney et al., 1994). Similarly, intestinal or vaginal epithelial cells, spermatozoa, and oligodendrocytes, all negative for CD4, may be infectable via a galactosylceramide or related glycolipid receptor (Albright et al., 1996; Strizki et al., 1996; Delézay et al., 1997).

Sequence Diversity as a Result of RT Errors and Genomic Recombination

Retroviral RTs do not possess a 3′–5′ exonuclease activity for proofreading and thus have a much higher misincorporation rate than cellular DNA polymerases. For HIV-1 RT this rate is particularly high, ranging from 1/1,700 to 1/4,000 nucleotides. Additional errors may occur during transcription, since the RNA polymerase II also does not proofread. For the 9.5-kb HIV genome, the in vivo error rate is estimated to be on the order of one to three misincorporations per replication cycle (Coffin, 1992). Another mechanism that contributes to sequence diversity is genomic recombination. The prerequisite for recombination of two retroviral genomes is coinfection of a cell with both viruses and encapsidation of both viral RNAs in the same particle (heterozygosity). Recombination then occurs during reverse transcription, probably while the minus-strand DNA is synthesized, and is well documented in several genetic subtypes of HIV-1 (see above). Recombination may have played a key role in the recent evolution of HIV-1, and the geographic intermixing of subtypes, which is increasing, is likely to foster the emergence of an even greater variety of recombinant strains.

Sequence diversity is manifested not only on the level of the pandemic but also in the infected individual, where it is generated. Moreover, the rapidity with which virus replicates is an important factor contributing to the rapid accumulation of virus variants. Selective-pressure factors, such as the local availability of host cell receptors or coreceptors, cellular or humoral antiviral immune responses, or antiretroviral drugs, may then act upon this pool of variant viruses, inhibiting growth of some variants and favoring replication of others that exhibit a better-suited phenotype. The outgrowth of such a group of viruses under selection pressure is called a quasispecies (Wain-Hobson, 1992). The many quasispecies in each patient evolve both in time and space. It is estimated that the sequence variability in an infected person increases by about 1% per year. In a given patient, different quasispecies are present at different sites in the body, for example, in Langerhans cells of different skin patches (Sala et al., 1994) or individual microdissected splenic white pulps (Cheynier et al., 1994). Quasispecies distinct from those of the rest of the body are also found in the brain or the genital tract (Zhu et al., 1996).

Regulation of HIV Expression

Host cell activation induces transcription of the viral genes from the promoter located in the U3 region of the 5′ LTR (Fig. 4, inset). In addition to the core promoter, which contains attachment sites for the multisubunit RNA polymerase II transcription complex TFIID and the transcription factors SP-1 and LBP-1 (leader binding protein 1), U3 also contains an upstream enhancer region with sites for the cellular enhancer factor NF-κB (whose liberation from an inhibitor and migration from the cytoplasm into the nucleus is induced during cell activation) and, still further upstream, a modulatory region for interaction with various cellular modifiers of transcription. Depending on the host cell type, one or the other of these modifiers may bind to this region, thus influencing the local environment of the promoter and thereby positively or negatively modifying the level of transcription.

Viral transcription is initiated at the 5′ end of the upstream R sequence (Fig. 4). In the absence of the Tat transactivator protein, very few full-length transcripts are produced and, in the absence of the Rev protein, most of these few will be multiply spliced before they are exported to the cytoplasm. Such multiply spliced early transcripts serve for translation of the accessory proteins Nef, Tat, and Rev. Nef has the ability to activate signal transduction pathways, namely, the NF-κB system, thereby further enhancing viral transcription (Lee et al., 1995; Saksela et al., 1995). Tat migrates back into the nucleus and binds to a bulge of a stable hairpin structure, the Tat-responsive element (TRE), which is localized on newly synthesized viral transcripts (Fig. 4, inset). The TRE-bound Tat then recruits a number of cellular proteins that also bind to TAR, with a final effect of greatly enhancing the processivity of RNA polymerase II. This results in the production of numerous full-length transcripts (reviewed in Cullen, 1998). Rev also migrates back into the nucleus, where it binds to a complicated stem-loop structure, the Rev-responsive element, which is located in the middle of the env gene and is present on only unspliced or singly spliced mRNA. Multiple Rev molecules then polymerize along those first deposited, thus efficiently shielding such RNAs from further splicing. In addition, Rev mediates the transport of viral RNAs from the nucleus to the cytoplasm. In addition to these early proteins, one of the late proteins, Vpr, may also have an indirect effect on transcription: it arrests the host cell in the G² phase, thus preventing it from progressing to mitosis. Since the HIV-1 LTR promoter appears to be more active in G²-arrested cells, Vpr may thereby exert a stimulatory effect on virus expression (reviewed in Cullen, 1998).

Since cell activation induces HIV expression, virus production is also under the influence of a variety of cytokines that regulate the immune system (Fig. 5). Low levels of cytokines are continually secreted even when the immune system appears quiescent and are increased in response to antigen. HIV replicates more efficiently in activated cells, and virus levels consistently increase when the immune systems of HIV-infected persons are activated, e.g., by opportunistic pathogens or by immunization with antigens, such as influenza or tetanus toxoid. Such immune activation is characterized by cellular activation and expression of the proinflammatory cytokines tumor necrosis factor alpha (TNF-α), IL-1β, and IL-6. These agents strongly promote HIV replication. Other cytokines, such as transforming growth factor β (TGF-β), IL-10, and IL-16, downregulate virus replication, in part by downregulating the expression of proinflammatory cytokines. HIV replication is also inhibited by the chemokines RANTES, MIP-1α, MIP-1β, or SDF-1, which bind to the CCR5 or the CXCR4 coreceptor, thus interfering with virus entry (reviewed in Fauci, 1996). However, since these chemokines may also induce signal transduction, their effects are not restricted to inhibition of viral replication. For example, RANTES may block entry of M-tropic virus into CD4⁺ T cells by interaction with CCR5 but induce activation of the cell by CCR5-mediated signaling, thus increasing CXCR4-mediated infection of the cell by T-tropic viruses (Weissman et al., 1997).

Not only is HIV activation influenced by the immune system, but the immune system is correspondingly activated by HIV expression. Chronic HIV infection leads to a constant production of viral antigens. In addition to the virus's

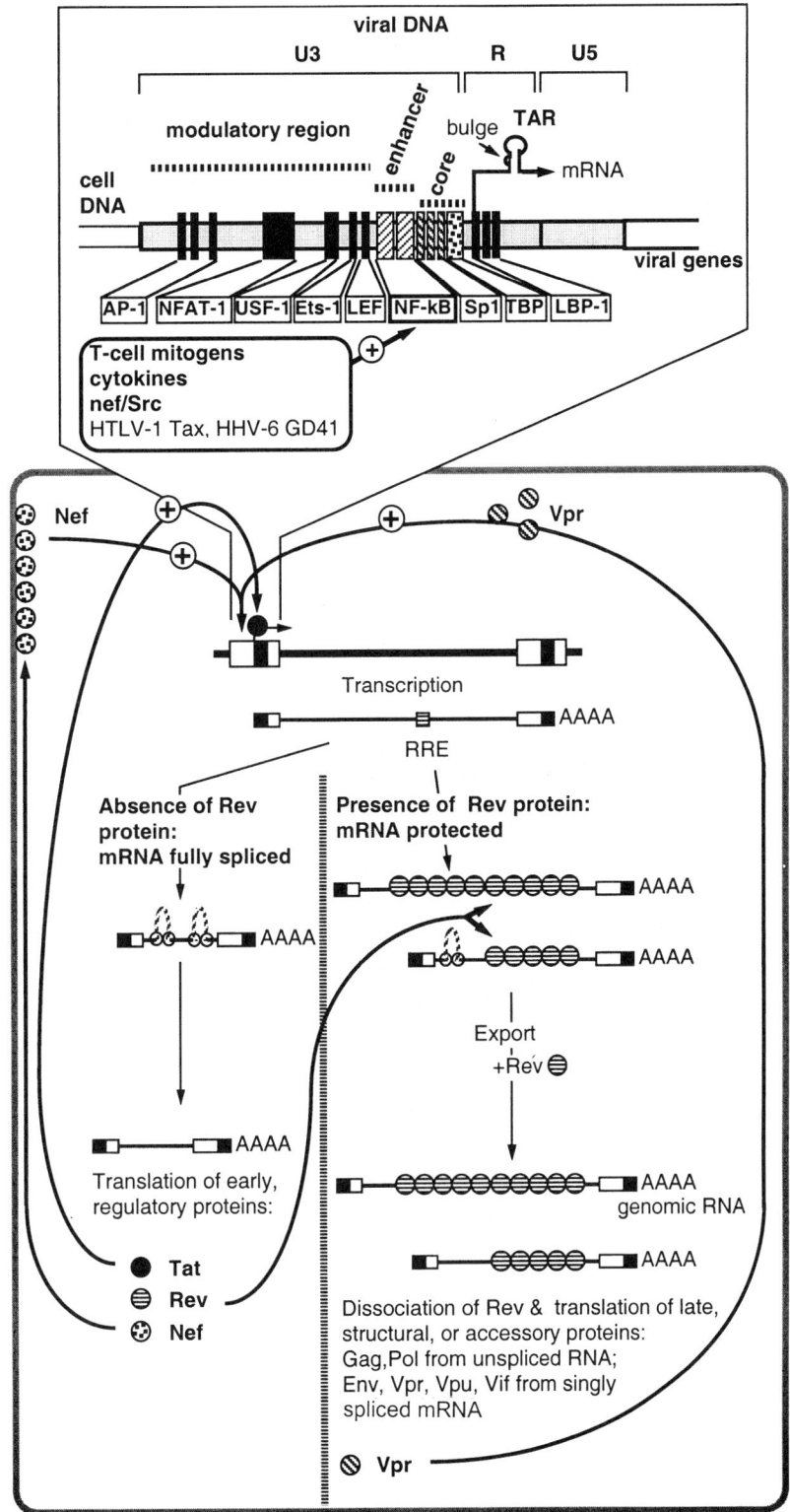

FIGURE 4 Regulation of HIV expression. Transcription is induced by the core promoter in U3 of the viral LTR (enhanced inset). The LTR's enhancer region contains binding sites for NF-κB, which is indirectly activated by Nef, mitogens, cytokines, and some *trans*-acting proteins of other viruses. Transcription is further modified by the modulatory region, which binds various cellular transcription factors, shown in boxes. Transcription is initiated at the 5′ boundary of the R region. In the absence of Tat, most transcripts are terminated prematurely and very few make it into full-length, polyadenylated mRNA, which is the precursor of all viral mRNAs (main figure). In the absence of Rev, full-length mRNA is multiply spliced, exported, and translated into the early proteins Tat, Rev, and Nef. Nef leads to activation of signal transduction by NF-κB and to enhanced basal transcription. Tat migrates back into the nucleus, where it binds to the TRE (TAR) in the R region of the viral mRNA (inset), which stimulates transcription initiation. Subsequent binding of Tat to the moving RNA polymerase II enhances the processivity of this enzyme, which results in a large increase in full-length viral mRNAs. The Rev protein also migrates into the nucleus and polymerizes on the viral mRNA, protecting it from splicing. In addition, Rev promotes the export of these viral RNAs. Consequently, Rev promotes the production of late-phase structural proteins and viral enzymes and of genomic RNA for particle assembly. An additional upregulatory effect is exerted by Vpr.

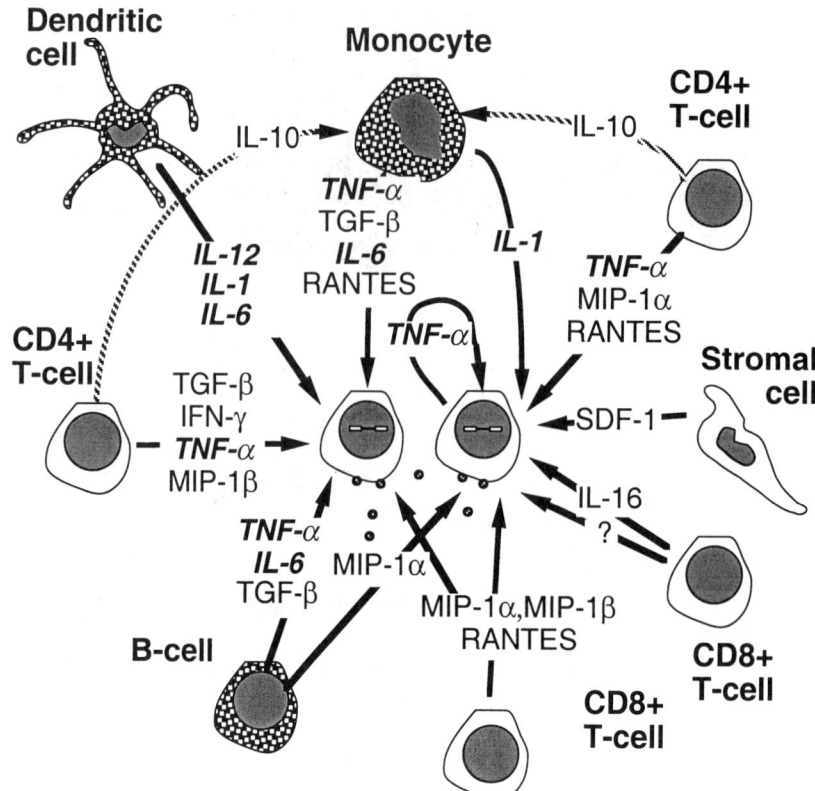

FIGURE 5 Cytokine regulation of HIV-1 expression. The proinflammatory cytokines TNF-α, IL-1β, and IL-6 (in boldface italics) strongly promote HIV replication. TGF-β, IL-10, and IL-16 downregulate virus replication, in part by downregulating the expression of proinflammatory cytokines. HIV replication is also inhibited by the chemokines RANTES, MIP-1α, MIP-1β, and SDF-1, which interfere with viral entry. Reprinted from Schüpbach, 1999.

effect on cells with specificities for the various viral antigens, gp120 is capable of activating CD4+ cells nonspecifically by binding to the CD4 receptor (Kornfeld et al., 1988). The effect of this permanent stimulation is a chronic hyperactivation of the immune system (reviewed in Fauci, 1993). Efficient antiretroviral combination therapy by blocking new host cell infection decreases the levels of viral proteins, thus resulting in a normalization of the immune system's state of activation (Autran et al., 1997). However, this has the disadvantage of keeping the virus in proviral latency, in which it can be attacked neither by the immune system nor by the therapy. It is therefore unlikely that HIV infection can be eliminated by antiretroviral drugs alone.

Pathogenesis of HIV Infection

Establishment of Infection

HIV is predominantly transmitted by unprotected sexual intercourse, connatally from mother to child, postnatally by breast feeding, or by parenteral inoculation. The efficacy of sexual transmission is enhanced by concomitant genital infections that increase not only the number of infected cells in the genital secretions of the virus transmitter but, when present in the recipient, also the number of susceptible target cells. Moreover, local infection enhances immune activation, which in turn favors virus replication (reviewed in Vernazza et al., 1999).

In the globally most frequent mode of transmission, women are infected vaginally by unprotected sexual intercourse. In addition to sperm, which do not transmit infection, semen contains macrophages, lymphocytes, and cell-free virus in seminal fluid as potential sources of HIV transmission. Studies with SIV in macaques showed that free virus is able to penetrate the untraumatized cervicovaginal epithelium and that the first cells infected are dendritic cells of the Langerhans type located within the epithelium or in the subjacent lamina propria (Spira et al., 1996). In humans, however, sequence comparisons suggest that it is not the free virus of the seminal plasma but more likely macrophage-associated R5 virus that is usually transmitted; it has also been shown that cervical epithelial cells are infectable by cell-associated virus in vitro (Tan and Phillips, 1996). Virus released from epithelial cells may then infect dendritic cells or macrophages in the second line. Male genital infection may be mediated by Langerhans cells present at relatively high density in the penile foreskin. A higher risk of infection in uncircumcised men has been reported in several studies (Moses et al., 1996). Monocytes/macrophages present in the urethra might serve as alternative virus targets.

In rectal infection, a possible primary target of the virus is the intestinal M cells, which are present at high frequency in the rectal mucosa, while dendritic cells are absent. M cells transport particulate materials through the intestinal epithelium and deliver them to the gut-associated lymphoid tissues located underneath (Amerongen et al., 1991; Hussain and Lehner, 1995; Schneider et al., 1996). Alternatively, virus might infect intestinal epithelial cells by using galactosylceramide and CXCR4 (Delézay et al., 1997). CXCR4 could favor infection by virulent T-tropic X4 viruses, which might partially explain why receptive anal intercourse is associated with a faster progression to AIDS (Phair et al., 1992). Peroral infection is the route in mother-to-child transmission by breastfeeding and has been implicated in cases of transmission where the only risk factor was receptive oral intercourse (Lifson et al., 1990; Schacker et al., 1996; M. M. Berrey and T. Shea, Letter, *J. Acquir. Immune Defic. Syndr. Hum. Retrovirol.* **14:**475, 1997). The likely targets of such infection are the nasopharyngeal lymphoid tissues. Studies of nontraumatic inoculation of neonate macaques with SIV have demonstrated an efficiency of oral infection that was 6,000-fold higher than that of rectal inoculation (Baba et al., 1996). In parenteral infections, it is assumed that the likely primary target cells of intravenously inoculated virus are the circulating CD4+ T cells or monocytes/macrophages, while the targets would be Langerhans cells or monocytes/macrophages in the case of intracutaneous or percutaneous lesions.

Propagation of Infection and Generalization

Langerhans cells are bone marrow-derived professional antigen-presenting cells that are positive for CD4, Fcγ receptors, and HLA class II antigens and reside in many different tissues, including skin and all mucosal epithelia except those of the urethra and the rectum (Hussain and Lehner, 1995). Upon contact with antigen, they become activated and migrate through the afferent lymphatics to the regional lymph nodes in order to present the antigen to T cells (Fig. 6). HIV-infected Langerhans cells by themselves appear not to replicate HIV well. However, if CD4+ T cells are added to Langerhans cells in vitro, a very active virus replication is observed (Cameron et al., 1992; Pope et al., 1995; Weissman et al., 1995; Zoeteweij and Blauvelt, 1998). Massive virus production was also observed in multinucleated giant cells that exhibited dendritic cell markers

and that were found in the mucosal surface of the nasopharyngeal adenoids of HIV-positive individuals (Frankel et al., 1996).

These observations, in conjunction with the SIV studies with rhesus monkeys (Spira et al., 1996), suggest that during the close contact of dendritic cells and CD4+ T cells that is required for antigen presentation the virus is transmitted to the CD4+ T cells, whence a very active virus production is initiated in conjugates of both cell types, or even a fusion product of the two. Such encounters leading to bursts of virus production may occur all along the route dendritic cells take to reach the draining lymph nodes, i.e., the submucosal tissues of the viral entry port, the afferent lymphatics, and, finally, the mantle zone of the lymph node follicles, where the CD4+ T cells reside. The availability of densely packed CD4+ T cells in the absence of an efficient immune response results in large-scale virus production. In consequence, free virus particles, as well as virus-infected cells, leave the lymph node by the efferent lymphatics to infect lymph node stations further downstream and to enter the blood, which leads to generalized infection of all lymphoid tissues, as well as the central nervous system (CNS). Infection of the latter is probably mediated by infected cells of the monocyte/macrophage lineage, which are known to cross the blood-brain barrier. The SIV model has shown that this propagation is very rapid: infection of Langerhans cells in the lamina propria of the vagina can be detected within 2 days, as well as infection of regional lymph nodes, and 5 days after inoculation the virus can be demonstrated in the plasma (Spira et al., 1996). In humans, initial plasma viremia is estimated to occur between 4 and 11 days after infection (Niu et al., 1993).

Acute Phase and Chronicity

Figure 7 summarizes the virological and immunological courses of acute and chronic HIV infection. Hematogenic generalization, probably by infected cells, leads to infection of all lymphoid tissues in the body. Replication of HIV within the lymphatics—which comprise about 98% of the

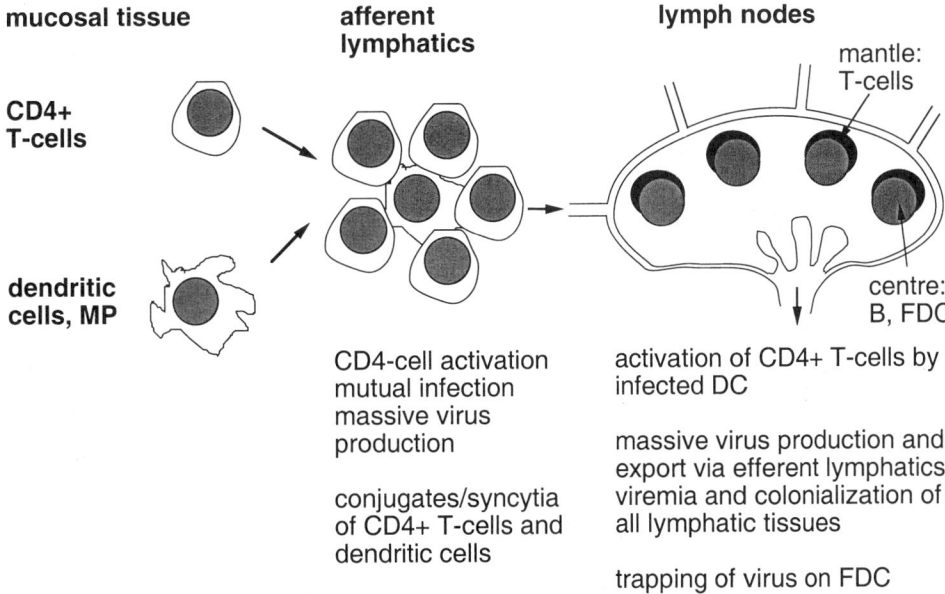

FIGURE 6 Propagation of HIV from the entry port to the lymphatics and the blood stream. Reprinted from Schüpbach, 1999.

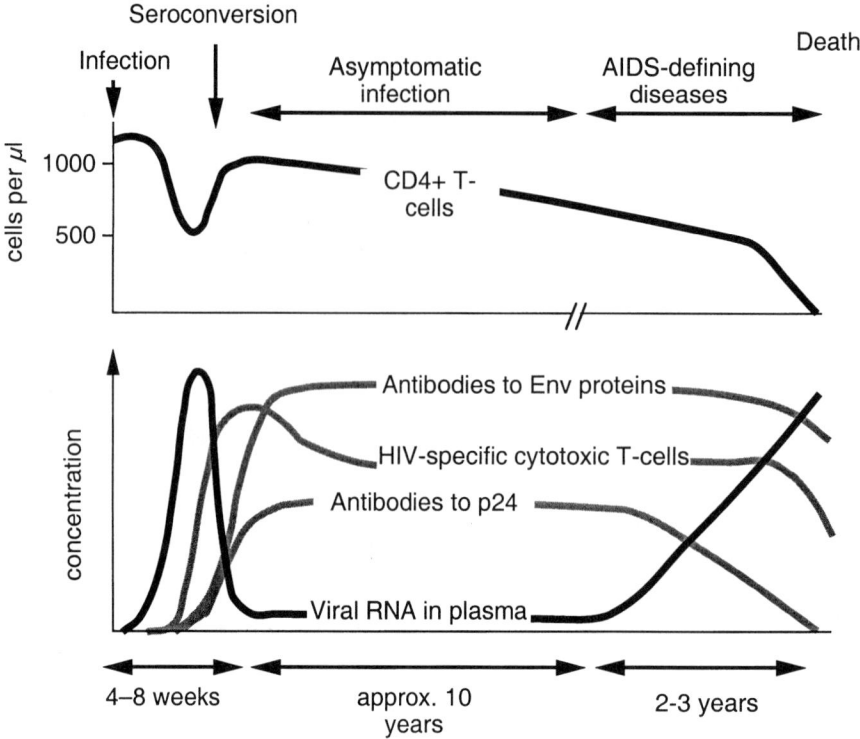

FIGURE 7 Virological and immunological parameters in the typical course of HIV infection.

total lymphocytes in the body—causes, in the absence of a specific immune response, a rapid increase in production and release of viral particles and virus-infected cells. In the blood, this is manifested as a concomitant burst in concentrations of cell-free or cell-associated infectious virus, particle-associated viral RNA, p24 antigen, and cell-associated DNA (Clark et al., 1991; Daar et al., 1991; Graziosi et al., 1993; Piatak et al., 1993; Koup et al., 1994). Concentrations of viral RNA in plasma may vary widely from tens of thousands to more than 1 million copies per ml (Schacker et al., 1998). The earliest virus population observed following HIV transmission is most frequently of the R5 (NSI) phenotype and genotypically very homogenous (Zhu et al., 1993, 1996; Delwart et al., 1994), even after exposure to an inoculum of mixed R5-X4 phenotype (Cornelissen et al., 1995). Transmission of X4 viruses, however, has also been demonstrated in about 15% of those infected (Roos et al., 1992).

Lymph node biopsy performed close to acute infection shows germinal center hyperplasia with a hyperplastic-reticulum network of follicular dendritic cells (FDC), with abundance of B cells in the mantle zone and of CD4+ T cells and interdigitating cells in the paracortex. Increased expression of IL-1, IL-2, IL-6, IL-10, TNF-α, and gamma interferon (IFN-γ) is present (Gray et al., 1996). In the blood, severe primary HIV infection is characterized by an initial lymphopenia followed by CD8+ lymphocytosis and inversion of the CD4/CD8 ratio. Subsequently, the CD8+-cell count gradually returns to normal, whereas the CD4/CD8 ratio remains inverted because of a relatively low number of CD4+ lymphocytes. Primary infection is followed by a prolonged and severe cellular hyporesponsive-

ness to both mitogens and antigen (Pedersen et al., 1990; Sinicco et al., 1990).

Virus levels decrease with the onset of the antiviral immune response, namely, the production of HIV-specific cytotoxic T lymphocytes (CTL) (Koup et al., 1994; Pantaleo et al., 1994; Connick et al., 1996) (Fig. 7). During this initial cellular response, as much as 6% of the CD8+ cells may represent HIV-specific CTL (Yang et al., 1996; Borrow et al., 1997), and the magnitude and persistence of the CD8+-cell response is associated with disease progression (Roos et al., 1992). Studies of SIV-infected macaques in which CD8+ T cells were temporarily ablated by infusion of a CD8-specific monoclonal antibody have also demonstrated the importance of these cells in lowering the viral load in both primary and chronic infection (Jin et al., 1999; Schmitz et al., 1999). Moreover, anti-virus antibodies that bind to virus particles and to which complement is fixed may, after seroconversion, increase virus retention on FDC that carry complement receptors at high density and thus can retain large quantities of complexed infectious virions (Embretson et al., 1993; Pantaleo et al., 1993a; Heath et al., 1995). In agreement with this, the viral-RNA load in early HIV infection is high in lymphoid tissues but low in plasma (Pantaleo et al., 1998).

After the initial peak, concentrations of virus in the blood are, at least in some patients, stabilized on individually different levels, the so-called "set point" or "inflection point," which is strongly associated with disease outcome (Jurriaans et al., 1994; Henrard et al., 1995; Mellors et al., 1996; Schacker et al., 1998). The set point is the equilibrium resulting from the interplay of viral, host cell, and immunological factors and is usually reached within a few months to 1 year of infection (Kaufmann et al., 1998b; Schacker et al., 1998). Viral titers in plasma subsequently change little or increase only slowly for a long time, which coincides with clinical latency. During this time, the CD4$^+$ T-cell count decreases continuously at an individually different but constant rate. A marked increase in the level of viral RNA in plasma is seen only in advanced immunodeficiency, when the CD4$^+$ T-cell count has dropped to below 200/μl. This is seen as a final complete breakdown of the mechanisms that previously maintained at least some control of virus replication. Frequently, the final increase is preceded by an emergence of syncytium-inducing X4 viruses (Schellekens et al., 1992; Koot et al., 1993).

Several factors that influence the outcome in the chronic phase are already effective during the acute phase. Generally, adult patients presenting with an acute retroviral syndrome, and particularly those with a severe one, progress more rapidly than those with asymptomatic primary infection (Pedersen et al., 1989; Sinicco et al., 1990, 1993; Keet et al., 1993; Schacker et al., 1998). The importance of viral factors is suggested by the observation that receptive anal intercourse with many partners (Phair et al., 1992) or infection by a patient with advanced disease (Carre et al., 1996) is associated with rapid progression. These conditions have an increased likelihood of infection by X4 viruses shown to be associated with rapid CD4$^+$ T-cell decline (Roos et al., 1992). The higher pathogenicity of X4 viruses is also evident from the demonstration that they block specific antibody responses in normal human lymphoid tissues infected in vitro, an effect not seen with NSI viruses (Glushakova et al., 1998). Rarely, attenuated virus strains with *nef* or *vpr* deletions are found in long-term nonprogressors (Deacon et al., 1995; Huang et al., 1995; Kirchhoff et al., 1995; Wang et al., 1996). A truncated *nef* gene was also associated with a slower progression in HIV-2 infections (Switzer et al., 1998).

The importance of immunological factors in disease outcome is shown by the demonstration that a more broadly directed CTL response in early infection and at later stages is associated with an ensuing lower viral load and slower disease progression (Greenough et al., 1997; Pantaleo et al., 1997). The importance of CD8$^+$ T cells in the control of virus was also evident in monkeys depleted of these cells prior to experimental infection with SIV: after the initial peak, high levels of virus were maintained, followed by a rapid fatal course, while animals treated with a control antibody developed an efficient CTL response that resulted in a massive viral-load reduction (Schmitz et al., 1999). In chronically infected animals, CD8$^+$ T-cell ablation led to a 3- to 4-log-unit increase in plasma virus concentrations that returned to undetectable levels when CD8$^+$ cells returned (Jin et al., 1999; Schmitz et al., 1999). As is known from other chronic infections, a strong CD4$^+$ T-helper cell response is also of crucial importance for the control of HIV infection. Except for the long-term nonprogressors, strong T-helper cell responses cannot be found in most HIV-infected patients, and there is a negative correlation between plasma viremia and HIV-specific CD4$^+$ T-helper cells. It has also been shown that efficient antiretroviral treatment at the earliest possible point in the infection can induce HIV-specific CD4$^+$-T-cell responses as strong as those seen in long-term nonprogressors (Rosenberg et al., 1997). This is compatible with the hypothesis that without such treatment these CD4$^+$ T cells are attacked and eliminated early in the course of infection, when they become activated in response to HIV exposure (Rosenberg et al., 1999). Further-

more, a number of major histocompatibility complex-encoded genes have been identified which, alone or in combination, may positively or negatively affect disease progression (reviewed in Haynes et al., 1996).

The importance of cellular factors in disease outcome is illustrated by the demonstration that resistance to HIV infection or disease progression may be controlled on the level of the viral coreceptor. Individuals homozygous for a 32-bp deletion of the CCR5 gene (CKR5–Δ32/Δ32), which leads to truncation and functional loss of CCR5 coreceptors are, despite a few reported exceptions, strongly protected against HIV-1 infection. Moreover, progression to AIDS in heterozygous (CKR5–wt/Δ32) HIV-infected individuals is postponed by 2 to 4 years (Dean et al., 1996). Another HIV chemokine receptor mutation, CCR2–64I, while not involved in resistance to infection, affects disease progression. HIV-1-infected individuals carrying the CCR2–64I allele progressed to AIDS 2 to 4 years later than individuals homozygous for this allele. Taken together, the data shows that in about 38 to 45% of patients with rapid progression to AIDS (within 3 years), this can be attributed to their CCR2–+/+ or CCR5–+/+ genotypes, whereas about 30% of long-term survivors free of AIDS after at least 16 years of infection had either a CCR2–wt/64I or CKR5–wt/Δ32 genotype (Smith et al., 1997).

The common denominator of these prognostically relevant factors is the extent to which the virus replicates, or is permitted to replicate, in the host. Defects in coreceptors, a vigorous antiviral immune response, or an attenuated virus strain all contribute to restrict viral replication. On the other hand, immune activation, e.g., by chronic infections such as tuberculosis, enhances viral production. The resulting viral load after the establishment of equilibrium (i.e., the set point) is thus the most important prognostic factor.

During the asymptomatic phase of the disease, virus production in the lymphatics continues steadily (Biberfeld et al., 1986; Cameron et al., 1987; Tenner-Racz et al., 1988; Embretson et al., 1993; Pantaleo et al., 1993a). Within the lymphoid tissues, virus is activated and released from latently infected CD4$^+$ T cells that are predominantly located in the mantle zone of follicles; however, macrophages may also be an important source of infectious virus (Orenstein et al., 1997). Released virus interacts with Env-specific antibodies, resulting in immune complex formation. These then pass through the FDC network, where the complexes are trapped. Trapped virus remains infectious even in the presence of neutralizing antibodies and may thus lead to the infection of T cells that migrate through the lymph node (Heath et al., 1995).

Virus Replication Dynamics

The development of effective antiretroviral drugs capable of rendering most of the virus produced by virus-expressing cells noninfectious has permitted determination of virus replication dynamics in various host cell compartments (Fig. 8). More than 99% of virus production is thought to be maintained by productively infected lymphocytes in which the interval between infection, virus production, and cell death is very short, with a half-life of about 1 day. The half-life of virus in the plasma of patients has been estimated at about 6 h. In order to keep the virus concentration at equilibrium, at least 10^{10} virus particles must be released per day. Experiments with macaques that were infused with a large bolus of SIV indicated a plasma half-life on the order of minutes, with only 1 to 10% of the virions cleared into

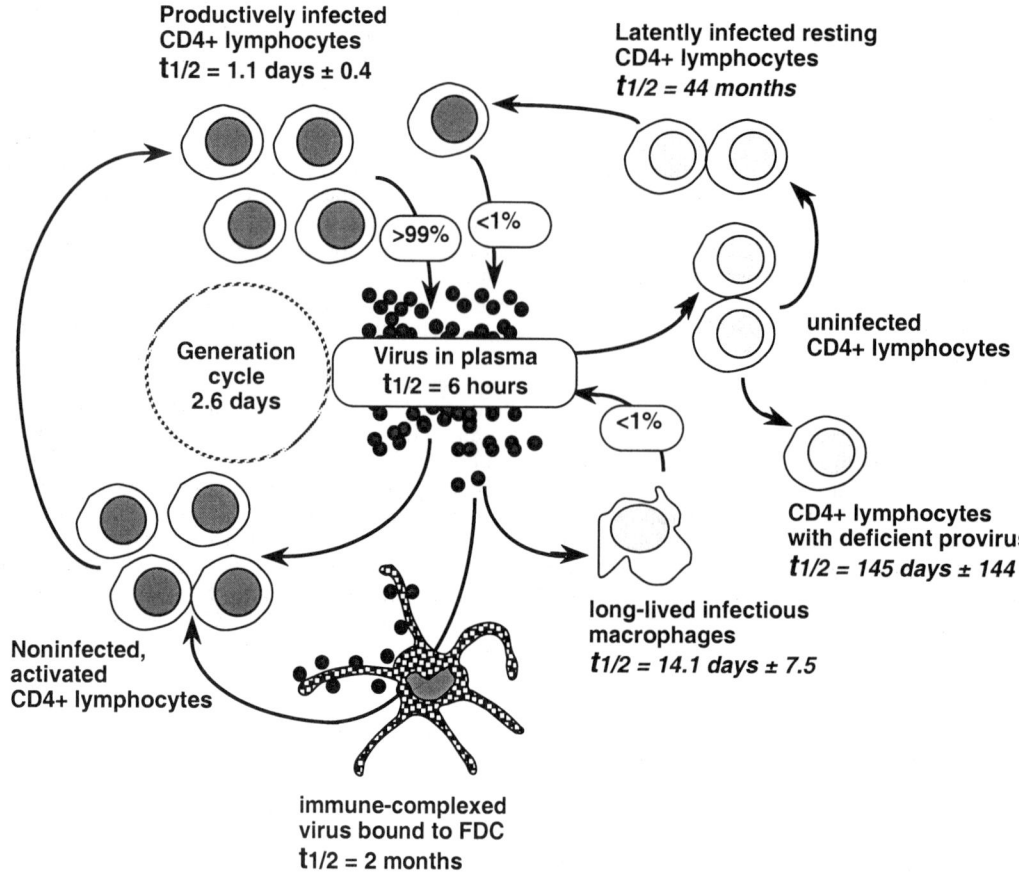

FIGURE 8 Kinetics of HIV production and elimination in vivo (based on Perelson et al., 1996, 1997; Finzi et al., 1999; and Zhang et al., 1999a, 1999b).

lymph nodes, spleen, lungs, and liver but not to other tissues examined (Zhang et al., 1999a). The length of a replication cycle of acutely producing CD4$^+$ T cells has been estimated at 2.6 days. Thirty-five million CD4$^+$ T cells are lost daily from the blood and must be replaced; in the whole body, this amounts to about 2×10^9 cells per day (Ho et al., 1995; Wei et al., 1995; Perelson et al., 1996, 1997).

In addition to the bulk virus production in acutely infected CD4$^+$ cells, latently infected CD4$^+$ cells and longer-lived cells, such as macrophages or Langerhans cells, also contribute to virus production, but their share in untreated patients is on the order of 1%. These proportions change, however, when efficient antiretroviral treatment blocks the acute replication cycle. In this phase, the longevity of these cell types, with half-lives on the order of weeks to several years in the case of resting CD4$^+$ T cells, necessitates continuation of aggressive therapy for many years, if not for life, if ways to activate the virus from such cells cannot be found (Finzi et al., 1997, 1999; Wong et al., 1997).

HIV-Associated Diseases

Acute Retroviral Syndrome

The first clinical manifestations may appear a few days to a few weeks after infection with a transient condition, the acute retroviral syndrome. It is found in 50 to 70% of infected patients and is characterized by signs of immune activation and multisystem dysfunction. Patients frequently seek medical help for a flulike or infectious mononucleosis–like disease with fever, generalized lymphadenopathy, sore throat, arthralgia, myalgia, fatigue, rash, and/or weight loss. The rash consists of a maculopapular exanthem, especially of the trunk, with occasional transition to a papulovesicular appearance. Lesions of the oral mucosa, often aphthous, may also occur. Occasionally, diarrhea, pancreatitis, mild liver disturbance, bacterial sepsis, thrombocytopenia, epiglottitis, lymphocytic alveolitis, or self-limiting neurological disorders, like meningitis, encephalitis, polyneuropathy, or myelopathy, may be found. Opportunistic infections that are normally seen only in advanced immunodeficiency, such as esophageal candidiasis, may be present occasionally. These symptoms typically resolve within 5 to 30 days.

Clinical Latency

Acute infection is followed by a long period of disease-free clinical latency. Clinical latency is, however, not paralleled by viral latency, as demonstrated by frequently high concentrations of HIV particles in plasma and, even more so, in lymphoid tissues (Tenner-Racz et al., 1986; Cameron et al., 1987; Embretson et al., 1993; Pantaleo et al., 1993a).

AIDS

The relentless production of HIV proteins, maintained by continuous viral replication in productively infected cells, and the ensuing elimination of host cells over many years finally lead to the destruction of the immune system, which is clinically manifested by the occurrence of opportunistic infections and tumors (Table 1). In addition, infection of the CNS may lead to distinct HIV-associated diseases, including the HIV-associated dementia complex, vacuolar myelopathy, and sensory neuropathy (Price, 1996).

Several disease classifications have been introduced since the first description of AIDS in 1981 (Gottlieb et al., 1981). The currently used Centers for Disease Control and Prevention (CDC) 1993 classification (Table 2) is based on a combination of clinical and CD4$^+$-T-cell count categories that define nine mutually exclusive stages (Anonymous, 1993a).

AIDS Incubation Time

The median time to AIDS in untreated adult patients has been estimated at 10 to 11 years and is similar for homosexual men and drug addicts. Incubation time varies considerably. In the 5 to 10% of patients who are rapid progressors, AIDS develops within 2 to 3 years after infection. On the other end of the spectrum, 5 to 10% of patients, the nonprogressors, are free of symptoms after 7 to 10 years and have stable CD4$^+$ T-cell counts, although these are significantly lower than those of uninfected controls. Data from large prospective cohort studies suggest that about 13% of HIV-positive homosexual men will remain AIDS free for 20 years after seroconversion (Munoz et al., 1995). In maternally transmitted pediatric infection, disease progression appears to follow a bimodal distribution, with a subgroup of children progressing rapidly to AIDS at a median age of approximately 5 months and 20% of infected children developing AIDS within 12 months (Downs et al., 1995). The mean time from birth to a stage of severe symptoms is estimated at 6.3 to 6.6 years (Barnhart et al., 1996; Pliner et al., 1996), and the mean time to death is estimated at 6.3 to 9.4 years (Barnhart et al., 1996). Based on 2,148 perinatally infected children, the mean duration of the symptom-free interval (stage N of the CDC 1994 pediatric classification [Caldwell et al., 1994]) was 10 months, that of stage A (mild signs or symptoms) was 4 months, that of stage B (moderate signs or symptoms) was 65 months, and that of stage C (severe signs or symptoms) was 34 months. Though children usually develop moderate symptoms by the second year of life, they may not progress further for a long time (Barnhart et al., 1996).

Epidemiology

The predominant modes of HIV transmission are unprotected sexual intercourse, connatal transmission from mother to child (one-third by intrauterine transmission and the rest during delivery), and postnatal transmission by breast-feeding or by parenteral inoculation. On average, a single exposure by sexual intercourse with an infected individual is associated with a 0.01 to 1.0% risk of infection (about 0.01% in the case of receptive vaginal intercourse and about 1% in the case of receptive anal intercourse). The risk for the insertive partner is smaller, though not minimal. The average risk is 0.5 to 1% for one-time injecting drug use, 12 to 50% for connatal mother-to-child transmission, 12% for breast-feeding, 90% for a contaminated blood transfusion, and 0.1 to 1.0% for nosocomial transmission by needlesticks, etc. (reviewed in Levy, 1997). In general, the risk is proportional to the viral load. Thus, it is considerably higher if the virus donor is in the acute phase of the infection or in an advanced stage of disease (Fig. 7) and it is reduced by efficient antiretroviral therapy. The virus is not transmitted through casual contact in a household setting, and there is no evidence for transmission by nonhuman vectors.

Diagnosis of HIV Infection

The two principal questions in HIV diagnostics are whether a person is infected and, if infected, how actively the virus is replicating. The susceptibility of a patient's virus to antiretroviral drugs may establish itself as another question of eminent practical importance.

Diagnostic Procedures in HIV Infection

HIV infection can be detected by a variety of tests. Virus can be directly detected by assays for the various virus components. These include proteins, especially p24, which can be specifically assayed by immunological tests; RT, whose enzymatic activity can be detected by functional tests; and viral DNA or RNA, which can be identified by molecular tests. Most frequently, however, HIV infection is diagnosed by tests that assess whether an individual's immune system has produced an HIV-specific immune response. Since retroviruses are known to establish infections that persist for life, demonstration of an HIV-specific immune response, if it is strong and directed against various viral antigens, can be trusted to truly reflect an ongoing infection. Testing for HIV-specific antibodies is therefore still the main tool for the diagnosis of HIV, at least in adults. In contrast, virus component tests are the assays of choice in infants.

As probably in no other infection, the diagnosis of HIV infection relies on commercially available test kits. Strong competition among the manufacturers and strict evaluation and control by the regulatory authorities of different countries have led to the availability of a large number of excellent commercial diagnostic products that are of high sensitivity and specificity, are well standardized, and provide a continuously high-quality standard of diagnostics in the

TABLE 2 CDC 1993 classification of HIV infections

CD4$^+$ cell category	Clinical category		
	A[a]	B[b]	C[c]
1 ($\geq$500/μl; $\geq$29% of lymphocyte count)	A1	B1	C1
2 (200–499/μl; 14–28% of lymphocyte count)	A2	B2	C2
3 (<200/μl; <14% of lymphocyte count)	A3	B3	C3

[a]Asymptomatic, acute (primary) HIV or PGL.
[b]Symptomatic; not category A or C conditions (see Table 3).
[c]AIDS indicator conditions (see Table 4).

TABLE 3 Signs and conditions defining HIV clinical category B

Symptomatic conditions in an HIV-infected adolescent or adult that are not included among conditions listed in clinical category C and that meet at least one of the following criteria: (i) attributed to HIV infection or indicative of a defect in cell-mediated immunity or (ii) considered by physicians to have a clinical course or to require management that is complicated by HIV infection. Examples include the following:

    Bacillary angiomatosis
    Candidiasis, oropharyngeal (thrush)
    Candidiasis, vulvovaginal; persistent, frequent, or poorly responsive to therapy
    Cervical dysplasia (moderate or severe) or cervical carcinoma in situ
    Constitutional symptoms, such as fever (38.5°C) or diarrhea for >1 month
    Hairy leukoplakia, oral
    Herpes zoster (shingles) involving at least two distinct episodes or >1 dermatome
    Idiopathic thrombocytopenic purpura
    Listeriosis
    Pelvic inflammatory disease, particularly if complicated by tubo-ovarian abscess
    Peripheral neuropathy

user laboratory. These kits are frequently better and yield more consistent results than research procedures developed in the diagnostic laboratory. Good commercial tests are therefore strongly recommended. Using unregistered tests for screening or for certain types of supplemental testing is even illegal in many countries.

Only very general descriptions of procedures will be given in the following sections, as commercial test kits all contain detailed step-by-step instructions. For procedures that are not commercially available, the reader is asked to consult the literature. The intent is to guide the reader through the multitude of available procedures and to discuss their points of strength and weakness.

Screening Tests

HIV-specific antibodies are produced within a few weeks after infection. The time to positivity (i.e., to seroconversion) in antibody tests may depend on the infecting virus, the infectious dose, the transmission mode, and the sensitivity of the antibody assay.

In a study based on first-generation screening assays (see below), seroconversion was estimated to occur after an average of 45 days after infection; with 95% certainty, the window period for 90% of individuals was less than 20 weeks (Petersen et al., 1994). The usefulness of other tests in reducing the average window period has since been estimated as follows: contemporary third-generation anti-HIV-

TABLE 4 Signs and conditions defining HIV clinical category C (AIDS indicator diseases)

Candidiasis of bronchi, trachea, or lungs
Candidiasis, esophageal
Cervical cancer, invasive[a]
Coccidioisomycosis, disseminated or extrapulmonary
Cryptococcosis, extrapulmonary
Cryptosporidiosis, chronic intestinal (>1 month's duration)
Cytomegalovirus disease (other than liver, spleen, or nodes)
Cytomegalovirus retinitis (with loss of vision)
Encephalopathy, HIV-related
Herpes simplex: chronic ulcer(s) (>1 month's duration) or bronchitis, pneumonitis, or esophagitis
Histoplasmosis, disseminated or extrapulmonary
Isosporiasis, chronic intestinal (>1 month's duration)
Kaposi's sarcoma
Lymphoma, Burkitt's (or equivalent term)
Lymphoma, immunoblastic (or equivalent term)
Lymphoma, primary, of brain
Mycobacterium avium complex or *Mycobacterium kansasii*, disseminated or extrapulmonary
Mycobacterium tuberculosis, any site (pulmonary[a] or extrapulmonary)
Mycobacterium, other species or unidentified species, disseminated or extrapulmonary
P. carinii pneumonia
Pneumonia, recurrent[a]
Progressive multifocal leukoencephalopathy
Salmonella septicemia, recurrent
Toxoplasmosis of brain
Wasting syndrome due to HIV

[a]Added in the 1993 expansion of the AIDS surveillance case definition.

1 and -2 enzyme immunoassays, −20.3 days (95% confidence interval [CI], 8.0 to 32.5); p24 antigen and PCR for proviral DNA, −26.4 days (95% CI, 12.6 to 38.7); PCR for viral RNA in plasma, −31.0 days (95% CI, 16.7 to 45.3) (Busch et al., 1995). With modern third-generation antibody-screening assays, half of the infected individuals should become antibody positive within 3 weeks after infection. Most of the other half should become positive within 2 months, but 5% still seroconvert more than 6 months after infection. In these, the use of tests for RNA, DNA, or p24 reduces this period only insignificantly, by 1 to 2 weeks (reviewed in Busch and Satten, 1997).

There are numerous commercial HIV tests for screening, and it may be difficult to recognize the advantages and disadvantages of a particular test based on the information given by the manufacturer. An overview of different test formats and their properties is given in Fig. 9.

The most important kit formats used for HIV antibody screening are the indirect binding assay, the antibody capture assay, and the double-antigen sandwich assay. Indirect binding assays comprise the so-called first-generation enzyme-linked immunosorbent assays (ELISA), which are based on purified viral lysate, and "second-generation" tests based on recombinant antigen, usually representing Gag and transmembrane (TM) protein. First-generation indirect binding assays are not restricted to ELISA but also include immunofluorescence and Western blotting (WB). Line immunoassays (LIA) that use recombinant proteins may be considered second-generation indirect binding assays. Antibody capture assays usually employ recombinant proteins; their principle is that of an indirect binding assay reversed. Double-antigen sandwich assays (frequently also called third-generation assays) usually employ recombinant antigen. Particle agglutination, which may be considered a variant of a double-antigen sandwich assay, is based on viral lysate. Although all of these test formats detect antibodies, they vary in their precise diagnostic questions and answers. Indirect binding assays and antibody capture assays verify, by binding the sample's HIV-specific antibodies to an immunoglobulin-specific reagent, that a sample component reactive in such a test is in fact an immunoglobulin. In contrast, the identity of a component reactive in a double-antigen sandwich assay is not assessed and remains unknown— the only information provided is that it is capable of linking solid-phase HIV antigen with liquid-phase tracer antigen.

The different kit formats are affected differently in certain diagnostic problem situations. One typical problem is antigenic variation. The virus with which a patient is infected may exhibit proteins that differ considerably from the antigens used in the test. Consequently, the patient's antibodies may not bind well to the test kit's antigens, and if they are present at low titer, a false-negative result may be generated. This type of problem was first recognized after the discovery of HIV-2, when it was found that antibodies induced by HIV-2 were not well recognized by screening kits based solely on HIV-1 antigens. This led to the inclusion of HIV-2 components, usually of the TM protein, in the kits. A similar problem was recognized when group O viruses were discovered (De Leys et al., 1990; Gurtler et al., 1994). Double-antigen sandwich assays are most affected by antigen variation, because in order to generate a positive result, antibodies must bind at least two antigen molecules. Double binding is unlikely if the kit antigen and the patient's antibodies do not fit. In indirect binding assays and antibody capture assays, antibodies need only bind a single antigen molecule in order to generate a signal. The sensitiv-

ity of double-antigen sandwich assays, however, was relatively easily improved by the inclusion of recombinant group O antigens in the solid and/or liquid phase.

Another typical problem is early seroconversion. Antibodies in this phase are restricted to a few viral proteins (usually envelope protein and p24), they are of low titer and low affinity, and the dominating immunoglobulin isotypes are immunoglobulin M (IgM) and possibly IgA. In addition, these antibodies may be partially saturated with HIV antigen, which is usually present at high concentrations in the early phase. In this situation, it is important that the test provide a high concentration of the antigen that is best recognized. This goal is more easily achieved with recombinant proteins than with viral lysate. Furthermore, the test must select for the few HIV-specific antibodies present in the bulk immunoglobulin; this is impossible with antibody capture assays that bind immunoglobulins of all antigenic specificities. In addition, the test should detect IgM because, in the presence of antigenemia, its pentameric structure with a total of 10 antigen binding sites, is most likely to have several sites remaining accessible. The clear winner in this situation is the double-antigen sandwich assay: it initially selects for HIV-specific antibodies (binding to solid phase), and it does not discriminate against non-IgG isotypes. The first test of this principle which went into diagnostic use was the particle agglutination assay, which was introduced in the mid-1980s, i.e., long before third-generation ELISA were developed. This test, which uses viral lysate for the antigen, performs remarkably well in seroconversion panels and, due to the broad spectrum of antigens present in the viral lysate, has in addition a broad detection range for antigenic variation (Constantine et al., 1994; Vercauteren et al., 1995).

The practical relevance of these considerations is shown when the performances of different kits with seroconversion panels are compared. Among 23 different commercial kits whose performances on at least 15 different commercially available seroconversion panels were compared by the Swiss Federal Office of Public Health, the 17 double-antigen sandwich assays were the most sensitive and occupied ranks 1 to 15, 17, and 18. The four indirect binding assays occupied ranks 16, 19, 21, and 22, and the two antibody capture assays ranked 20th and 23rd (unpublished data). Seroconversion panel comparisons also demonstrate the inferior sensitivity of immunofluorescence tests and WB, which rank together with other indirect antibody binding assays. Assessment of the performance in seroconversion panels followed by revocation of approval for the 10 to 20% of least sensitive kits is one of the most powerful instruments of regulatory agencies to guarantee continuous further technical improvement of diagnostic tests for HIV (Schupbach, 1996).

Combined tests for antibodies and antigen. Several companies now offer kits that detect both antibodies and antigen. In seroconversion panel analysis, these kits now rank first among all screening tests even if they are based on the antibody capture format. The average gain in time to detection compared with third-generation kits is 4 to 5 days (Gurtler et al., 1998; Weber et al., 1998).

In conclusion, for antibody screening in the laboratory one should use kits that detect antibodies to both HIV-1 and HIV-2, have a documented good sensitivity to group M and O infections, and perform well in seroconversion panels. Reactive samples or those with results in a borderline zone must be retested and, if repeatedly reactive, must be submitted to confirmatory or supplemental testing.

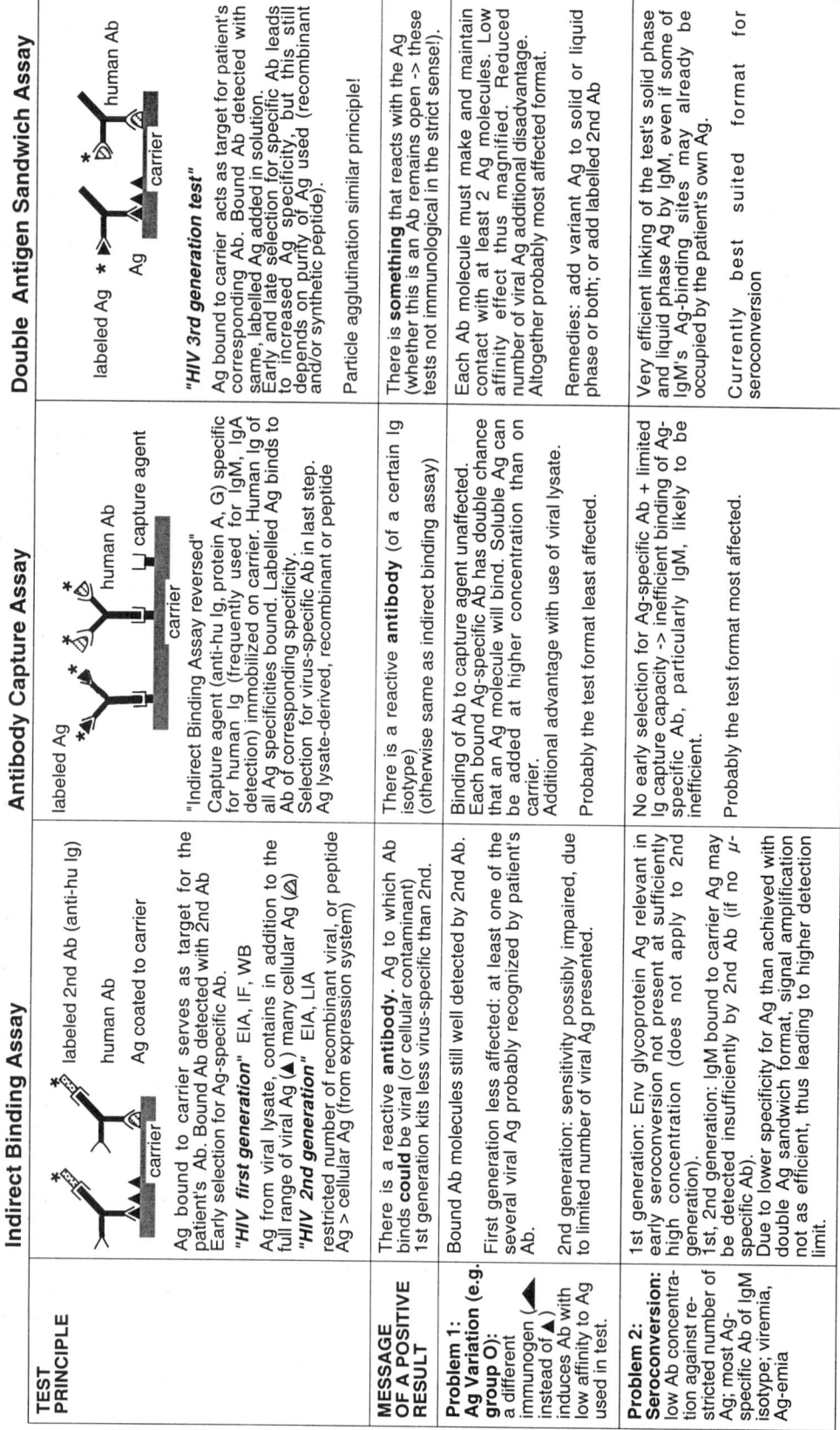

	Indirect Binding Assay	Antibody Capture Assay	Double Antigen Sandwich Assay
TEST PRINCIPLE	Ag bound to carrier serves as target for the patient's Ab. Bound Ab detected with 2nd Ab. Early selection for Ag-specific Ab. "HIV first generation" EIA, IF, WB. Ag from viral lysate, contains in addition to the full range of viral Ag (▲) many cellular Ag (Ⓐ). "HIV 2nd generation" EIA, LIA. restricted number of recombinant viral, or peptide Ag > cellular Ag (from expression system).	"Indirect Binding Assay reversed" Capture agent (anti-hu Ig, protein A, G) specific for human Ig (frequently used for IgM, IgA detection) immobilized on carrier. Human Ig of all Ag specificities bound. Labelled Ag binds to Ab of corresponding specificity. Selection for virus-specific Ab in last step. Ag lysate-derived, recombinant or peptide	"HIV 3rd generation test" Ag bound to carrier acts as target for patient's corresponding Ab. Bound Ab detected with same, labelled Ag added in solution. Early and late selection for specific Ab leads to increased Ag specificity, but this still depends on purity of Ag used (recombinant and/or synthetic peptide). Particle agglutination similar principle!
MESSAGE OF A POSITIVE RESULT	There is a **reactive antibody.** Ag to which Ab binds **could** be viral (or cellular contaminant) 1st generation kits less virus-specific than 2nd.	There is a reactive **antibody** (of a certain Ig isotype) (otherwise same as indirect binding assay)	There is **something** that reacts with the Ag (whether this is an Ab remains open -> these tests not immunological in the strict sense!).
Problem 1: Ag Variation (e.g. group O): a different immunogen (▲) instead of (▲) induces Ab with low affinity to Ag used in test.	Bound Ab molecules still well detected by 2nd Ab. First generation less affected: at least one of the several viral Ag probably recognized by patient's Ab. 2nd generation: sensitivity possibly impaired, due to limited number of viral Ag presented.	Binding of Ab to capture agent unaffected. Each bound Ag-specific Ab has double chance that an Ag molecule will bind. Soluble Ag can be added at higher concentration than on carrier. Additional advantage with use of viral lysate. Probably the test format least affected.	Each Ab molecule must make and maintain contact with at least 2 Ag molecules. Low affinity effect thus magnified. Reduced number of viral Ag additional disadvantage. Altogether probably most affected format. Remedies: add variant Ag to solid or liquid phase or both; or add labelled 2nd Ab
Problem 2: Seroconversion: low Ab concentration against restricted number of Ag; most Ag-specific Ab of IgM isotype; viremia, Ag-emia	1st generation: Env glycoprotein Ag relevant in early seroconversion not present at sufficiently high concentration (does not apply to 2nd generation). 1st, 2nd generation: IgM bound to carrier Ag may be detected insufficiently by 2nd Ab (if no μ-specific Ab). Due to lower specificity for Ag than achieved with double Ag sandwich format, signal amplification not as efficient, thus leading to higher detection limit.	No early selection for Ag-specific Ab + limited Ig capture capacity -> inefficient binding of Ag-specific Ab, particularly IgM, likely to be inefficient. Probably the test format most affected.	Very efficient linking of the test's solid phase and liquid phase Ag by IgM, even if some of IgM's Ag-binding sites may already be occupied by the patient's own Ag. Currently best suited format for seroconversion

FIGURE 9 Kit design and test performance of HIV screening tests. Synopsis of the most frequently used test formats, their principles, the meaning of positive results, and performance in two typical problem situations. Ag, antigen; Ab, antibody; Ig, immunoglobulin; EIA, enzyme immunoassay; Slightly modified from Schupbach, 1996, with kind permission from *Clinical and Diagnostic Virology*.

Rapid tests and home testing. Rapid tests are assays that can be performed with minimal or no laboratory equipment and yield results within 30 min. Such tests may be useful in certain situations, e.g., in assessing the risk of HIV transmission in needle stick injuries and similar exposures to possibly HIV-contaminated materials, in organ donations, or whenever a laboratory test result may not be available quickly. Rapid tests may be of different formats, including double-antigen sandwich, indirect binding, immunoglobulin capture, agglutination, or chromatographic assay. Some of them have good sensitivity and specificity. In more extensive seroconversion panel comparisons, the sensitivities of most of these tests, however, seem somewhat inferior to the most recent generation of ELISA-based tests (Malone et al., 1993; Samdal et al., 1996; Kuun et al., 1997; Vallari et al., 1998). Therefore, the use of such tests should normally be restricted to emergency situations.

Many persons infected with HIV are not tested until they develop symptoms. Up to one-third of patients receive their HIV diagnosis within 2 months of progression to AIDS. The hope that such individuals would get tested earlier has led to the development of new testing strategies, some of which have already been approved by the U.S. Food and Drug Administration (FDA). Already approved are home test collection systems, where kits may be ordered by phone and are delivered by express courier. Blood is collected on filter paper by finger pricking and sent to a designated laboratory for screening. If the screening is reactive, supplemental tests can be performed on the same sample. Results, as well as personal counseling, are available by calling an information phone number. Such testing systems have very good sensitivity and specificity, since testing remains in the hands of professionals. Collecting a sufficiently large specimen may be the biggest problem, which may affect 7 to 10% of the users. As an alternative, testing systems for other fluids, namely, oral fluids or urine, have also been approved by the FDA, despite the fact that their sensitivities cannot be evaluated by seroconversion panels. The observation that HIV-exposed seronegative individuals considered uninfected have a high frequency of HIV-reactive IgA in urine and that 8% of even supposedly unexposed controls exhibit such reactions presents certain problems regarding the specificity of such tests (Mazzoli et al., 1997). It can also be expected that more problems would be encountered with true home testing kits in which all steps of the test have to be performed by the user. Such kits have therefore not yet been approved (Brodie and Sax, 1997).

Supplemental Testing—Antibody

WB was introduced into HIV testing in 1984 (Sarngadharan et al., 1984; Schupbach et al., 1984a), was proposed for systematic confirmation of reactive ELISA results in 1985 (Schupbach et al., 1985), and has remained a principal confirmatory tool used worldwide. It has become clear that in contrast to other tests that have been continuously improved, WB has remained limited to first-generation kits which have some well-known flaws. Their sensitivities in seroconversion panels are clearly inferior to those of double-antigen sandwich tests, and they rank together with first-generation indirect antibody binding ELISA. Furthermore, they are prone to detect cross-reactive antibodies induced by agents different from HIV, which results in a high rate of indeterminate results.

At least one improvement, the use of recombinant proteins for the production of the strips, has been realized by a few manufacturers. When recombinant proteins are used

instead of viral lysate, strips can be produced as LIA in which the purified antigens are applied in separate bands to the membranes. Some of these tests have better sensitivity and specificity than conventional WB and can discriminate between HIV-1 and HIV-2 infection. They still have a high rate of indeterminate results (Pollet et al., 1991). This shows that indeterminate WB reactions are caused by antibodies that cross-react with HIV proteins rather than with cellular contaminants of the viral lysate. In another product, recombinant Env TM protein of HIV-2 has been added to strips otherwise produced with viral lysate in order to improve detection of HIV-2 antibodies. While this format does improve detection of HIV-2 infections, a separate WB for HIV-2 is still needed to confirm the reaction.

WB is more prone to problems with carryover contamination than screening assays. The use of the convenient multichannel troughs for the incubation of the strips with the test sample presents a certain risk. Contamination with minute volumes of a strongly positive serum may lead to faint Env bands, even when the dilution is up to 1 millionfold. While intra-assay contamination can be ruled out by repeating the assay in an isolated test chamber, repeat testing will not identify contamination of the specimen tube. Touching the wet inner side of a specimen tube lid with the gloved fingers may carry enough material to the lid of a subsequently opened tube to result in faint WB reactivity to Env antigens. The probability of such events depends on the proportion of strongly positive sera among the specimens tested by a laboratory and on how many times a specimen tube is opened. In order to minimize this risk, handling and testing samples from known HIV-positive patients together with diagnostic samples should be avoided, and gloves that have become contaminated with specimen must be changed immediately. It should also be recognized that samples with initial borderline results carry an increased risk of contamination, since these tubes are opened repeatedly for supplemental testing. This results in a higher cumulative risk of contamination. Alarm bells should ring when a sample with borderline or low positive results in screening is faintly reactive in WB. It may be an early seroconversion sample, but it need not be. The contamination problem is a strong reason why WB interpretation should follow the most stringent, and not the most sensitive, guidelines.

Indeterminate WB results have been described for patients with autoimmune disorders, in particular systemic lupus erythematosus, after infection with certain viruses, including herpes simplex virus type 1 or cytomegalovirus, or after vaccination against influenza or rabies virus. For the latter, epitopes related to HIV have been implicated. Such information, however, is of little practical value, and the origin of indeterminate WB reactions in individual patients usually remains obscure.

In spite of all these flaws, a Western blot with a "full-house" pattern of reactive antibodies probably remains the most convincing laboratory evidence for an HIV infection. In addition, WB may offer some information on whether the infection is caused by HIV-1 or HIV-2 (especially if the strips contain recombinant HIV-2 protein in addition to the viral lysate of HIV-1) and on the clinical stage of the infection (Schupbach et al., 1985). When reactive bands are few and their intensities are low, interpretation is hazardous, and a diagnosis must not be based on WB alone.

Western blot interpretation guidelines. In an attempt to render WB more sensitive, the Association of State and Territorial Public Health Laboratory Directors and the

CDC (ASTPHLD-CDC) issued interpretation recommendations that specify antibody reaction (of undescribed intensity) to any two of three antigen bands, including gp120-gp160 (considered as one antigen), gp41, and p24 (Anonymous, 1989). Since most of the gp160 and gp120 bands on Western blots are not due to the Env precursor or the surface (SU) protein but instead represent tetramers or trimers of gp41 (Pinter et al., 1989), confirmed positivity established on reaction with gp120-gp160 and gp41 can be based on reaction with a single protein, TM, alone and is thus inherently unsafe. The same is true for the very similar recommendations by the Consortium for Retrovirus Serology Standardization (CRSS), the only difference being that p24 may be replaced by p31 (*pol*). In the absence of other bands, this combination is, however, extremely rare, and interpretation by CRSS is usually the same as that by ASTPHLD-CDC. Similarly, the World Health Organization (WHO) recommendation specifies any two among gp160, gp120, and gp41 (Anonymous, 1990). This means practically that TM-reactive antibodies must be present at a concentration sufficient for detection of not only the strongest but also the second-strongest TM bands. The strongest TM band is usually the largest antigen, i.e., gp160, which migrates least in the sodium dodecyl sulfate-polyacrylamide gel electrophoresis (SDS-PAGE)-based separation of proteins and thus is the sharpest and best-detected band. Depending on the manufacturer, either gp120 or gp41 may be the second-strongest band. Due to varying degrees of glycosylation, gp41 migrates in SDS-PAGE as a very diffuse band. Reactive antibodies thus generate a signal that is much less well recognized than if the same antibodies bind to the sharp gp160 band.

Due to the propensity of WB to detect cross-reactive antibodies, a combination of Env and p24 bands is also not sufficiently stringent for confirmation. WB analysis of 100 screening-negative but otherwise unselected Swiss blood donors showed isolated reaction with p24 in 9% and with gp120-gp160 in 3%. The likelihood of a chance combination leading to "confirmed positivity" in a healthy donor would thus be 0.09×0.03, or 0.0027, i.e., 1 out of 370 (Schupbach et al., 1990). We and others (Healey and Bolton, 1993) have observed several cases that satisfied the ASTPHLD-CDC criteria for WB positivity and kept this pattern essentially unchanged over years but were negative in long-term follow-up in all direct tests for HIV components, including virus culture, PCR for HIV DNA and RNA, and an antigen test of a sensitivity comparable to that of PCR (see below). Representative Western blots of a recent such case are shown in Fig. 10. In another such case, very low concentrations of particle-associated RT activity were detected in plasma, and the activity was shown to be transmissible to fresh leukocytes in culture, thus suggesting an infectious agent undetectable in HIV-specific virus component tests (Schupbach et al., 1996b).

More-stringent interpretation criteria have been issued by the American Red Cross (ARC) and by the FDA. ARC specifies at least one band each from Env, Gag, and Pol but does not specify the identities of these bands. The most stringent—but least sensitive—recommendation is that of the FDA, which specifies reactions with p24, p31, and Env.

In our opinion, only the most stringent interpretation guidelines are to be applied if a diagnosis of HIV infection is established using WB as the only supplemental test. If some true cases of HIV infection are WB indeterminate by the FDA guidelines, this is of no concern as long as their WB patterns are suggestive of HIV infection. This is always the case if ASTPHLD-CDC interpretation would render these

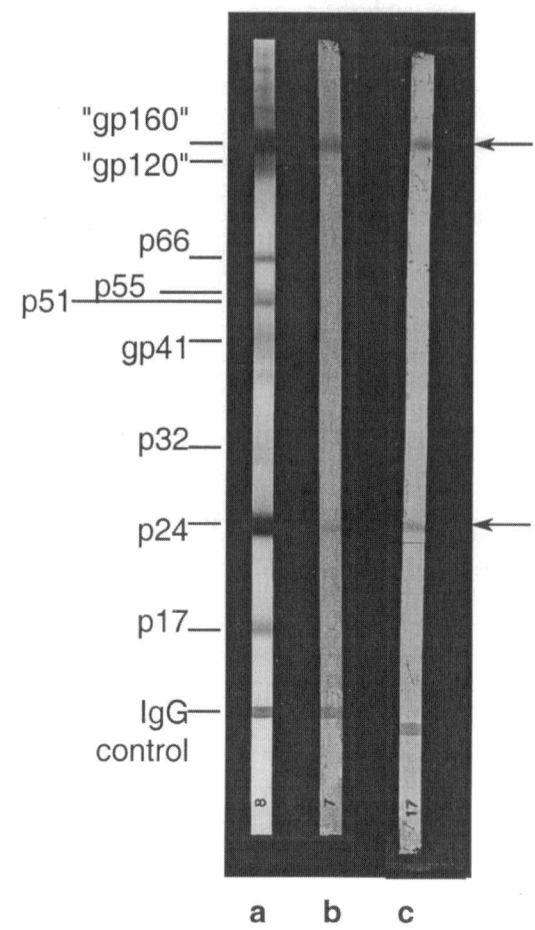

FIGURE 10 Example of false-positive HIV-1 Western blot interpretation according to ASTPHLD-CDC or CRSS guidelines. Lane (a), weakly positive control. Lane (b), sample from a healthy individual exhibiting weak reactions with gp160, gp120 (very weak), and p24. This sample was taken 3.5 months after an initial sample with the same pattern (not shown). Lane (c), sample from the same individual taken 1 month after the initial sample. P24 antigen with signal amplification-boosted ELISA was negative in all three plasma samples; PCR for viral DNA was negative in PBMC from the samples in lanes b and c, and RNA was negative in the plasma in lane b. Culture with PBMC depleted of CD8[+] T cells from the sample in lane b was negative for p24 antigen and RT by PERT assay; this test was also negative with the plasmas in lanes b and c. Reprinted from Schüpbach, 1999.

patterns positive. In all such cases, a safe diagnosis can be established based on supplemental tests for viral components. One also has to take into account the fact that the ASTPHLD-CDC criteria were established based on a single commercial product, the Du Pont WB kit. Meanwhile, other valuable kits are available whose protein compositions may differ. Guidelines established for one particular kit and with one particular sample cohort cannot be applied to other kits or populations from other geographical regions without careful reexamination of their validity.

Another attempt to make WB more sensitive in early infection is its use for detection of IgM. However, such testing lacks specificity. As detailed below, Gag-reactive antibodies of IgM, IgA, and IgG isotypes are frequently detected by WB in sera of infants born to HIV-negative mothers. Therefore, many of these reactions appear to be due to common agents unrelated to HIV.

In order to avoid the practical and financial problems associated with WB, the WHO has recommended alternative test strategies based on the use of at least two different screening tests (Sato et al., 1994). Large studies have shown that such testing algorithms may yield results that are at least equivalent to those of the conventional testing algorithm, in which all samples repeatedly reactive in a screening assay are submitted to WB testing. In one algorithm, reactive samples are submitted to another screening assay and only those samples with discrepant results are submitted to WB testing. As an alternative, initial screening is performed with two different tests and those with discrepancies are submitted to supplemental testing. Supplemental testing by WB can also be replaced without loss in sensitivity or specificity by a third screening assay (Laleman et al., 1991; van der Groen et al., 1991; Urassa et al., 1992; A. Ittiravivongs, S. Likanonsakul, T. D. Mastro, S. Tansuphasawadikul, N. Young, T. Naiwatunakul, D. Kitayaporn, and K. Limpakarnjanarat, Letter, *J. Acquir. Immune Defic. Syndr. Hum. Retrovirol.* **13**:296–297, 1996; Nkengasong et al., 1999).

Given the further development of screening tests since the time these studies were done, such testing algorithms have become even more attractive. In particular, initial screening with two tests (which should ideally combine a double-antigen sandwich assay with either an indirect binding assay or an antibody capture assay) is bound to increase the sensitivity of the procedure still more. However, as mentioned above, this strategy does not establish whether an infection is due to HIV-1 or HIV-2 (see below). This question is of importance regarding the validity of results of sequence-based viral-load determination or the need for antiretroviral treatment. WB or, where available, an LIA or dot immunoassay will therefore frequently be done anyway. There are also some rapid tests that can discriminate between HIV-1 and HIV-2 infections.

Supplemental Testing—Virus Components

In many cases, supplemental testing requires the use of tests for virus components. This applies to samples from patients presenting with primary HIV infection as well as to those with indeterminate results in antibody testing. In addition, diagnosis of pediatric HIV infection is best established based on virus component tests. Virus components that can be assayed include the p24 antigen, viral DNA or RNA, and RT activity. In addition, the capability of the virus to replicate can be assessed by virus isolation in cell culture. This requires an HIV-specific virus component test for identification of virus released into the supernatant, usually an antigen test.

Commercially available diagnostic tests for these components include p24 antigen assays that are available from several companies. A commercial kit is available for PCR detection of HIV-1 DNA. Amplification-based tests for viral RNA are available from several manufacturers and currently include the PCR for reverse-transcribed RNA, the nucleic acid sequence based amplification (NASBA), and a signal-amplification procedure involving branched DNA (bDNA) probes.

Antigen tests. Antigen tests for p24 are easy to perform and are diagnostically valuable in early infection, when antigen is usually present at high titers in the peripheral blood while HIV-specific antibodies may still be undetectable or present at concentrations too low to permit a confirmed positive diagnosis. The test principle consists of binding the p24 antigen present in a sample to p24-specific, usually monoclonal, "capture" antibodies, with which a solid support has been coated. Unbound sample components are washed away, and bound antigen is reacted with another p24-specific "tracer" antibody to which is conjugated an enzyme (horseradish peroxidase or alkaline phosphatase) capable of signal generation when combined with a suitable substrate (Fig. 11A). For confirmation of a reactive result, the sample must be subjected to a neutralization assay. This means that the antigen test is repeated in the presence of HIV-specific antibodies. These bind the antigen in immune complexes, thus preventing its detection in the test (Fig. 11B). The most advanced antigen test that has recently been approved by the regulatory agencies of several European countries detects not only groups M and O of HIV-1 very efficiently, but also HIV-2.

Antigen p24 measurement is frequently confronted with two problems. One is the presence of p24-specific antibodies, which—as in the neutralization assay—complex the antigen, thus causing underdetection or false-negative results (Fig. 11B). This problem is exemplified in the course of early HIV infection, during which the increase of antibody concentrations in serum or plasma leads to immune complexating of the antigen and a rapid decrease of its detectability. This may lead to a situation in which WB may still be inconclusive and the antigen test may be negative. The other problem is the presence of immunoglobulin-specific, rheumatoid-factor-like antibodies that may bridge the capture and the tracer antibodies of an antigen test and thus cause overdetection or false-positive results (Fig. 11C). This type of problem is present when in the neutralization test the addition of HIV-specific antibodies to the test sample does not result in a higher degree of signal reduction than the addition of normal control antibodies.

Solutions to these problems that have been proposed and introduced into routine testing were primarily aimed at improving detection of immune-complexed antigen. Acidification or base treatment leads to a significant, though incomplete, release of antigen, thus increasing the propor-

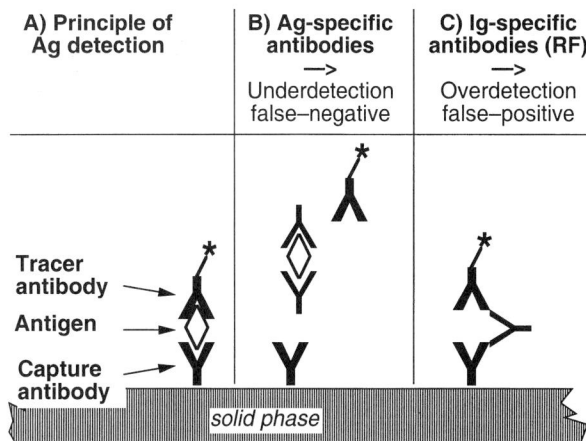

FIGURE 11 Principle of antigen (Ag) testing and interference by antigen-specific or immunoglobulin (Ig)-specific antibodies. For descriptions, refer to the text. Reprinted from Schüpbach, 1999.

tion of positives significantly. These treatments will, however, also release rheumatoid factors from preformed immunoglobulin–anti-immunoglobulin complexes, thus aggravating the problem of false positivity (M. Gutierrez, A. Vallejo, and V. Soriano, Letter, *Vox Sang.* **68:**132–133, 1995). A third problem of antigen assays is their insufficient analytical and diagnostic sensitivity, which renders the practical usefulness of these tests rather low when compared to molecular tests for viral RNA or DNA.

Recently, modifications at several steps of the antigen detection procedure have resulted in considerable improvements, however (Fig. 12A). First, boiling diluted serum for 5 min prior to antigen testing destroys the antigen binding capacity of all antibodies, thus irreversibly eliminating both of the above interferences while leaving the antigen in a condition permitting its detection by some, though not all, commercial assays (Schupbach and Boni, 1993). Moreover, use of plasma instead of serum and

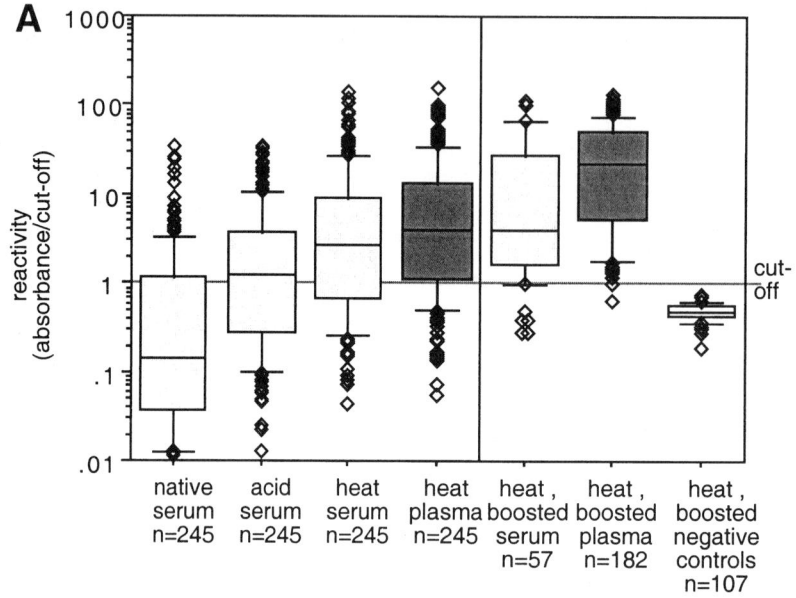

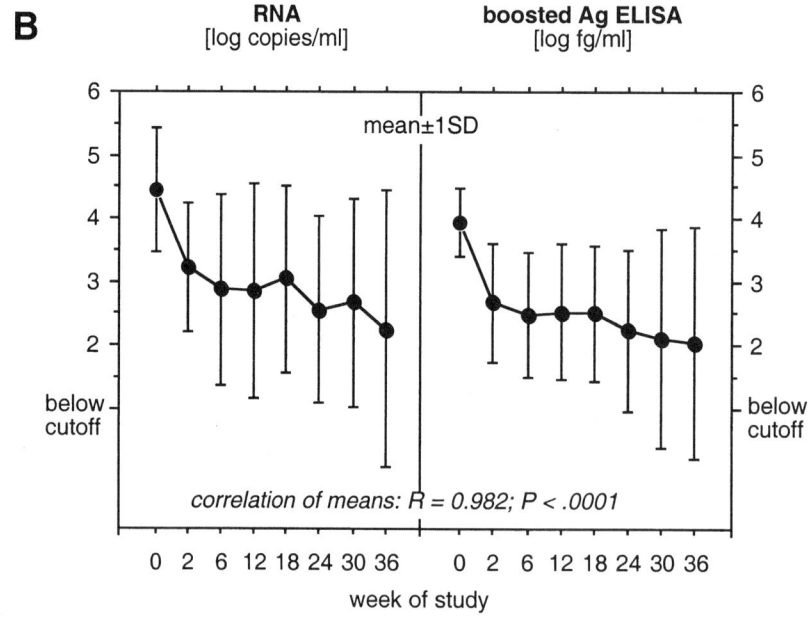

FIGURE 12 Improvements in p24 antigen detection and use of a signal amplification-boosted p24 antigen (Ag) ELISA of heat-denatured plasma for antiretroviral-treatment monitoring. (A) Improvements achieved by various procedures involving the same cohort of serum or plasma samples from 245 HIV-positive patients at different stages of the infection. (B) Comparison of viral load testing by commercial reverse transcription-PCR and signal amplification-boosted antigen ELISA. SD, standard deviation. Reprinted from Schüpbach, 1999.

boosting the antigen assay with a simple signal amplification step rendered antigen detection in adult HIV-1 positive patients as sensitive as a commercial reverse transcription-PCR kit for quantification of viral RNA, whose detection limit is about 200 to 400 copies/ml. This variant of the antigen assay is thus of interest as a highly sensitive and specific, yet inexpensive, diagnostic tool for both pediatric and adult HIV-1 infection (Lyamuya et al., 1996; Schupbach et al., 1996a; Nadal et al., 1999). The same procedure can also be used as a quantitative test for anti-retroviral-treatment monitoring (see below).

Tests for viral RNA or DNA. Three different techniques for the sensitive detection of HIV RNA or DNA are presently available. In PCR, double-stranded DNA is denatured, a pair of HIV-specific primers is annealed to the separated viral-DNA strands, and these primers are extended by a heat-resistent DNA-dependent DNA polymerase (*Taq* polymerase). This procedure is continued for 30 to 40 cycles, each of which comprises a high-temperature denaturation, a low-temperature primer annealing, and an intermediate-temperature primer extension (DNA synthesis) element (Kwok et al., 1987). If the starting material for PCR is RNA, a cDNA must first be generated by reverse transcription. This cDNA can then be amplified by the regular procedure (Byrne et al., 1988). In the current commercial kit (Amplicor HIV-1 Monitor version 1.5; Roche Diagnostics) for HIV-1 group M genomic RNA, a 155-bp sequence from the *gag* gene is amplified and the double-stranded DNA product is analyzed by ELISA.

In NASBA (NucliSens; Organon Teknika), RNA is amplified in an isothermal multienzymatic procedure mediated by the enzymatic effects of RNA-dependent DNA polymerase (RT), RNase H, DNA-dependent DNA polymerase, and DNA-dependent RNA polymerase. This procedure thus mimics the retroviral nucleic acid replication cycle (Fig. 2). The product of this procedure is RNA (Kievits et al., 1991). A sequence from *gag* similar to that of the reverse transcription-PCR kit is amplified.

In the bDNA method (Quantiplex, Chiron Diagnostics), viral RNA is captured on a solid surface by immobilized specific capture probes. The captured RNA is then reacted with "connector" probes, which hybridize with one end to a series of short sequences of the *pol* region of the RNA and with the other end mediate fixation of bDNA detector probes. These bDNAs are then reacted with still other bDNAs that hybridize to the first. Enzyme-labeled tracer probes are finally hybridized to all the branches, and the analysis is based on chemiluminescence (Urdea et al., 1993). The procedure has been compared to a Christmas tree: the more branches it has, the more candles can be attached and the brighter is the light that emerges from a single tree. The principal difference between the bDNA method and PCR or NASBA is the lack of amplification of viral sequences and, in consequence, of a carryover problem. What is called signal amplification is in fact a mere signal accumulation in which interacting molecular probes added to the reaction at a predetermined high concentration are hybridized to captured viral RNA in an ordered process that results in the deposition of consecutive probe layers. The procedure is in fact comparable to an indirect antibody binding assay, where a primary antigen-specific antibody binds to the antigen and is then detected by secondary antibodies with specificity for the first. With this system, a greater degree of precision is achieved than with

nucleic acid amplification techniques, whose outcomes depend on the efficacy of primer annealing in each cycle. Another important advantage of the bDNA method is the broad specificity, which is due to the use of about 40 different probes that cover most of the *pol* gene, the most conserved retroviral gene. The drawback of the method is its relatively high detection level, which must be compensated for by large specimen volumes (2 ml of plasma).

These nucleic acid detection kits are primarily designed for the quantification of viral RNA. Depending on the detection level, some of them may, under certain conditions, be suitable for diagnostic (qualitative) purposes, but caution must be exercised, since false-positive results have been reported (C. de Mendoza, A. Holguin, and V. Soriano, Letter, *AIDS* **12**:2076–2077, 1998; Rich et al., 1999). For diagnostic questions, it is therefore better to test for viral DNA. For HIV-1 DNA, there is one commercial PCR kit (Roche Diagnostics). Many diagnostic laboratories, including ours, have developed their own PCR methods, in particular for qualitative detection of viral DNA or RNA. Detailed step-by-step instructions for our own diagnostic procedures, which are capable of detecting a single DNA copy, even without using nested-PCR protocols (which carries a high risk of carryover contamination [see below]), have been published (Boni, 1996). For RNA quantification, the use of commercial kits is preferable, due to a better comparability of results obtained in different laboratories.

The availability of commercial kits for the detection of HIV-1 RNA or DNA by sequence or signal amplification has rendered these tests attractive for laboratories with no background in molecular biology. The fact that under optimal circumstances a single gene copy can be detected by some of these tests has created a relatively uncritical attitude, resulting in a degree of trust not yet justified in several respects. More than antigen tests, molecular tests are sensitive to sequence variation, resulting in false-negative results if inappropriate, i.e., not well conserved, sequences are used. Carryover contamination, whose main source is amplified DNA but which may also originate from other specimens, can also be a problem, in particular with PCR, and commercial products or systems must therefore have built-in carryover protection devices. In addition, precautionary measures are a must for all laboratories performing such tests (Boni, 1996).

Important factors in molecular testing are sample handling and sample preparation. Particle-associated RNA in plasma has been claimed to be very unstable, demanding special expensive plasma preparation tubes from a certain manufacturer and immediate separation of plasma from the cellular pellet before the sample is shipped to the laboratory that performs the test. Independent investigators have not been able to confirm these claims. Their work indicates that HIV-1 RNA levels are stable (variance, less than 0.3 log units) up to 3 days after collection when stored either at room temperature or at 4°C as cell-free plasma in the EDTA-plasma preparation tubes or even as EDTA-anticoagulated whole blood in regular tubes. Comparison of paired HIV-1-positive plasma and serum specimens revealed that RNA quantitation was 20 to 65% lower in serum than in plasma. EDTA plasma is thus the preferred specimen for these assays and provides the highest levels of RNA. EDTA-plasma that is prepared and frozen within 8 h of collection can thus be trusted to represent a good material for these tests. It can be thawed and frozen up to three times before RNA levels decrease significantly (Pachl et al., 1993; Sebire et al., 1996; C. C. Ginocchio, Y. P. Wang, M.

H. Kaplan, G. Mulligan, M. Cronin, D. Carroll, and J. Romano, *Proc. 4th Conf. Retrovir. Opportune Infect.*, p. 624, 1997).

The maximal sensitivity of molecular tests is limited by sample size and sequence variation. The detection limit of PCR is a single DNA molecule. However, 1 µg of genomic DNA, which contains the DNA of approximately 150,000 cells, corresponds to the number of peripheral blood mononuclear cells (PBMC) contained in only about 75 µl of blood. Consequently, even with a detection limit of one provirus/µg of DNA, roughly 70,000 infected cells must be present in 5,000 ml of blood for PCR analysis to be positive. A detection limit of 10 copies per reaction is more realistic because of the Poisson distribution, which means, for example, that not every sample with a nominal concentration of one copy/sample indeed contains such a copy. Similarly, the sensitivity for the detection of particle-associated HIV RNA is limited by the efficiency of the reverse-transcription step and by the volume of analyzed plasma. In order to achieve higher sensitivity, sample input is now frequently increased to milliliter plasma volumes (Schockmel et al., 1997). Claims of a detection limit of five RNA copies/ml, or even less, when testing a 1-ml plasma specimen should be met with reservations, but quantification limits of 50 copies/ml are realistic for "ultrasensitive procedures." Loss of sensitivity may also be due to sequence divergence of HIV. Attempts to develop a unique pair of primers that allow PCR amplification from all HIV isolates have so far failed. Subtypes A, E, F, and G were underdetected by version 1 of the Roche Amplicor HIV-1 Monitor reverse transcription-PCR test, and A and G are underdetected by the Nuclisense kit (NASBA). Virus loads were thus systematically underestimated in patients infected with these subtypes (Coste et al., 1996; Simons et al., 1997; Debyser et al., 1998; Parekh et al., 1999). The addition of "mix-in" primers and the use of modified primers in the HIV-1 Monitor version 1.5 appear, however, to have considerably increased the recognition of the different group M subtypes, while group O isolates and HIV-2 remain undetectable by all commercial tests (Parekh et al., 1999; C. Loveday, H. Devereux, A. Burke, L. Dann, A. Phillips, and M. Johnson, *Abstr. 12th World AIDS Conf.*, abstr. 42167, 1998; V. Michael, G. Anderegg, F. Campbell, and A. F. Amor, *Abstr. 12th World AIDS Conf.*, abstr. 42155, 1998; K. Triques, J. Coste, J. L. Perret, J. Reynes, S. Herman, J. Spadoro, and M. Peters, *Abstr. 12th World AIDS Conf.*, abstr. 41145, 1998).

The application of PCR to the detection of HIV at the screening level is currently precluded by the low level of automation, the lack of a pair of primers with the capability to amplify DNA from all HIV isolates, and the occurrence of infected cells below the detection limit in some individuals. PCR is very helpful when serology fails to provide an answer. This applies in particular to specimens with borderline reactivity in screening assays or incomplete patterns on confirmatory WB, specimens from individuals with suspected acute-phase disease (although an antigen test would suffice in most instances), and babies of HIV-infected mothers.

Virus isolation. For virus isolation, leukocytes are separated from anticoagulated blood samples by Ficoll centrifugation and cocultured with phytohemagglutinin-stimulated leukocytes from healthy blood donors. The culture supernatants are periodically assayed for viral antigen. Cultures usually become positive within 2 weeks, but culture times of up to 60 days are sometimes required

(Ho et al., 1989; Burgard et al., 1992). The procedure has a sensitivity of about 90% over all stages; the success rate is lower with samples from asymptomatic patients. A significant increase in efficiency of virus isolation is achieved by more complicated procedures combining concentration of possibly infected cells; removal of cells that might interfere with virus replication, such as $CD8^+$ cells; and activation of infected lymphocytes by phytohemagglutinin or antibody to CD3 and CD28. Such procedures enabled virus isolation from the lymph node cells of all patients whose viral RNA in plasma had become undetectable under antiretroviral treatment (Finzi et al., 1997; Wong et al., 1997).

Despite these recent improvements in sensitivity, virus isolation remains a time-consuming and costly diagnostic tool for HIV infection. Since PCR for viral DNA or RNA, as well as the signal amplification-boosted p24 antigen test, have diagnostic sensitivities of more than 96% and yield results within a few hours, we consider virus isolation unsuitable for the routine diagnosis of both adult and pediatric HIV infections. We still do virus cultures, but only with samples for which the presence of indeterminate results is also suggestive of infection with a retrovirus different from HIV. Virus culture, of course, remains essential for many research questions.

PERT assay. Particle-associated RT is a unique marker of possibly infectious retrovirus particles. Measuring particle-associated RT activity with a sensitive procedure can therefore be used as a general screening test for retroviruses. A family of novel tests collectively named PERT assays (for product-enhanced RT) combine the broad detection range of RT tests with the high sensitivity of nucleic acid amplification procedures. PERT assays are based on the selective enhancement, by PCR or one of the various other nucleic acid amplification methods, of the cDNA product synthesized by the test sample's RT activity from an RNA template. A detailed overview of these procedures and detailed step-by-step instructions for a PCR-based PERT assay used in our laboratory have been published (Boni and Schüpbach, 1997; Schüpbach and Böni, 1998). Assays based on this principle were also established by others (Silver et al., 1993; Heneine et al., 1995).

The PERT assay is 10^6 to 10^7 times more sensitive than a conventional RT test. The assay detects a variety of different retroviruses or RTs at a concentration corresponding to 1 nU of RT activity in the case of murine leukemia virus, which corresponds to 2.1×10^2 enzyme molecules, or 3 to 11 particles (Pyra et al., 1994). The PERT assay's sensitivity in fact rivals that of PCR for viral RNA, as demonstrated in HIV-1 infection. Testing of serial dilutions of HIV-1 by PERT assay and by PCR for viral RNA indicated a detection level of five particles. With a 100-µl plasma sample, all of 30 untreated patients at different stages of HIV-1 infection were positive. In HIV-1 seroconversion panels, the PERT assay detected more positives than any other method, including PCR for viral RNA. Importantly, the test's high sensitivity is not compromised by low specificity; the prevalence of elevated RT activity among 160 unselected blood donors was only 1.9% (Boni et al., 1996b). Since the source of RT in these cases is unknown, no precise figure for the PERT assay's specificity can be given.

In conjunction with HIV diagnosis, the PERT assay is useful when a low $CD4^+$-lymphocyte count or indeterminate serological results, for example, a WB positive according to ASTPHLD-CDC but indeterminate by FDA guidelines, suggest the presence of an HIV infection, possibly with an agent that is not well detected with current HIV-specific amplification methods. A negative result in the PERT assay renders an HIV infection very unlikely. If the PERT assay is positive but HIV-specific PCR involving different genomic regions remains negative, identification and characterization of the suspected retroviral agent can be achieved with other recently developed procedures (Conrad, et al., 1997; Weissmahr et al., 1997). The PERT assay is also useful for quantification of HIV-1, particularly of group O, and of HIV-2, neither of which is detectable by commercial kits.

Diagnosis of Pediatric HIV Infection

Diagnosis of HIV infection in babies born to HIV-positive mothers is complicated by the presence of HIV-specific IgG antibodies of maternal origin. IgG concentrations in term-born babies are as high as in their mothers. Since the half-life of IgG is about 3 weeks, HIV-specific maternal antibodies disappear slowly, and traces may remain detectable by WB in some uninfected infants well beyond the 15 months given as an upper limit by most authors. We have observed traces of Env-reactive IgG in uninfected children at up to 22 months of age. These antibodies disappeared thereafter, and none of the children has since shown any sign of infection (Schupbach et al., 1994a).

Since IgM and IgA isotypes are not subject to transplacental transport, their presence in the infant has been explored as a diagnostic marker. Detailed studies showed that Env-specific IgA antibodies have good specificity, except in cord blood and during the first few days of life, when they are detectable in a high proportion of uninfected children, probably due to transplacental microtransfusion (Petit et al., 1995). After the newborn period, the specificity of Env-gp160-specific IgA was excellent. Diagnostic sensitivity in the first few months of life, however, is too low, and a reliable diagnosis based on IgA reaction with gp160 is only possible after the first half year. IgM antibodies are predominantly directed against p24 and are of low specificity, since they are also produced in uninfected infants born to HIV-positive mothers and are even present in about 40% of control infants born to uninfected mothers. A further complication is the presence of IgG-reactive rheumatoid-factor-like IgM or IgA, which may also lead to false-positive results and which therefore requires quantitative removal of the IgG present in a sample, for example, by absorption with protein G-Sepharose (Schupbach et al., 1994a). Considering all this, serological diagnosis based on HIV-specific IgA or IgM cannot be recommended.

HIV infection in maternally exposed infants is best diagnosed with virus component tests. In pediatric HIV infection, virus components are, on average, present at higher concentrations than in adult infection (Schupbach et al., 1994b). Although virus culture is still regarded by some as a diagnostic "gold standard" for pediatric HIV infection, PCR for viral DNA or RNA is clearly more sensitive. Approximately one-third of maternally transmitted infections, probably those representing transmission in utero, can be detected within the first 10 days of life. In a few cases, PCR for DNA from PBMC may still be negative when PCR for viral RNA in plasma is already positive. The remainder, assumed to have become infected at birth, become PCR positive within the next 2 months. In Switzerland, babies with negative HIV results at birth are usually not retested before the age of 1 to 2 months. Using two different primer pairs and procedures published in detail (Boni, 1996) the prospective sensitivity of DNA PCR in samples from infants older than 10 days has been 100% in our laboratory (Nadal et al., 1999).

The same sensitivity as with PCR, however, is also achieved with improved tests for p24 antigen (see above). Heat-mediated immune complex dissociation of samples and the use of a regular unboosted antigen ELISA with a detection limit of about 2 pg/ml resulted in an overall diagnostic sensitivity of 96% (Schupbach et al., 1994b). The combination of using plasma as a sample, heat-mediated immune complex dissociation, and signal amplification-boosted p24 antigen ELISA with a detection limit of about 0.5 pg/ml increased the prospective sensitivity in infants older than 10 days to 100%, and in newborns less than 10 days old, p24 detection was actually the most sensitive test (Table 5). In a prospective study over 4 years, we have not seen a single sample positive by PCR for viral DNA or RNA which would have been negative by the signal-boosted antigen assay (Nadal et al., 1999). The high sensitivity of this antigen detection procedure has also been confirmed by others in a study of African samples (Lyamuya et al., 1996). Used in combination, PCR and this antigen test permit independent detection of HIV infection in a single sample. Infection can then be immediately verified with a freshly drawn sample. A safe diagnosis of HIV infection can thus be established during the first few weeks and months of life.

Regarding specificity, very low levels of p24 (<1 pg/ml) may be present in the first days of life in infants who do not

TABLE 5 Prospective diagnostic sensitivities of HIV-1 detection methods in pediatric infection

Age of patient	Sensitivity[a]			
	Antigen neutralized	In-house PCR (viral DNA)	In-house PCR (viral RNA)	HIV-1 Monitor (viral RNA)
≤10 days	6/12 (50)	5/12 (42)	3/7 (43)	Not done[b]
11 days–3 months	10/10	8/8	7/7	6/6
>3–6 months	19/19	12/12	12/12	9/9
>6 months	191/191	66/66	26/26	120/120
>10 days (total)	220/220 (100)	86/86 (100)	45/45 (100)	135/135 (100)

[a]n positive/n tested samples (percent).
[b]A sample that was positive in the antigen assay but negative by in-house PCR for viral DNA and for RNA was also negative by the ultrasensitive HIV-1 Monitor version 1.5, which is capable of detecting 50 copies/ml.

become infected. In one instance, this antigen was fully neutralizable and could be confirmed and neutralized in a second sample obtained a few weeks later, though its concentration was about 10-fold reduced. It is unclear whether such cases represent protein that is immune complexed with maternal IgG and actively transported through the placenta, result from transplacental microtransfusion of maternal blood, or are the manifestation of a transient infection (Rowland-Jones and McMichael, 1995). It is important that a diagnosis of HIV infection in an infant not be based exclusively on antigen results from samples obtained in the first week of life.

Diagnosis of HIV-2 Infection

Screening tests, at least in several European countries, must demonstrate good sensitivity to HIV-2 on at least 200 different samples in order to receive approval by regulatory agencies. Detection of HIV-2 infection at the level of screening is therefore not a problem. In contrast, determination of whether reactivity in screening is due to infection with HIV-1, HIV-2, or both may sometimes be difficult. On HIV-1 Western blots, sera from HIV-2-infected individuals frequently have strong reactions with Gag and Pol proteins compared to the reactions to Env. In particular, they may present with an unusually strong p32 (integrase) band. A suspected HIV-2 infection is further supported by reaction with the recombinant HIV-2 TM protein present on the products of some manufacturers. We have, however, occasionally seen a sample that did not react with this band, although HIV-2 infection was subsequently confirmed. The presence of isolated Gag bands on

HIV-1 Western blots or of strong reactivity against p24 and Gag precursors (p55, p43, and p39) is not an indication of HIV-2 infection. Similarly, the presence of faint Gag-reactive bands on HIV-1 blots is usually not an indication for an HIV-2 Western blot.

Results of HIV-1 WB suggestive of HIV-2 infection should be confirmed by a Western blot produced from HIV-2 lysate or by tests employing specific recombinant proteins, e.g., an LIA or a rapid test that differentiates between the two. If reactions to both HIV-1 and HIV-2 proteins are present at similar intensities, diagnostic PCR for both viruses will be necessary. HIV-2-infected asymptomatic individuals, i.e., the overwhelming majority, have much lower viral loads than those infected with HIV-1. Virus culture for HIV-2 has only rarely been positive in our hands, despite repeated attempts to isolate the virus from the half dozen asymptomatic patients we have seen, even when CD8+ T cells were removed prior to culturing and the highly sensitive PERT assay was used for monitoring of RT production.

Diagnostic Algorithms and General Considerations

A synopsis of the use of the various methods in HIV diagnostic testing of adults is shown in Fig. 13. It is based on the consideration that the combination of two simple and relatively inexpensive procedures, antibody screening and antigen testing, will detect not only virtually all untreated infections that can be diagnosed with molecular-amplification methods but also most of those that are missed by these molecular tests due to sequence variability or too low numbers of viral-genome copies in the specimen.

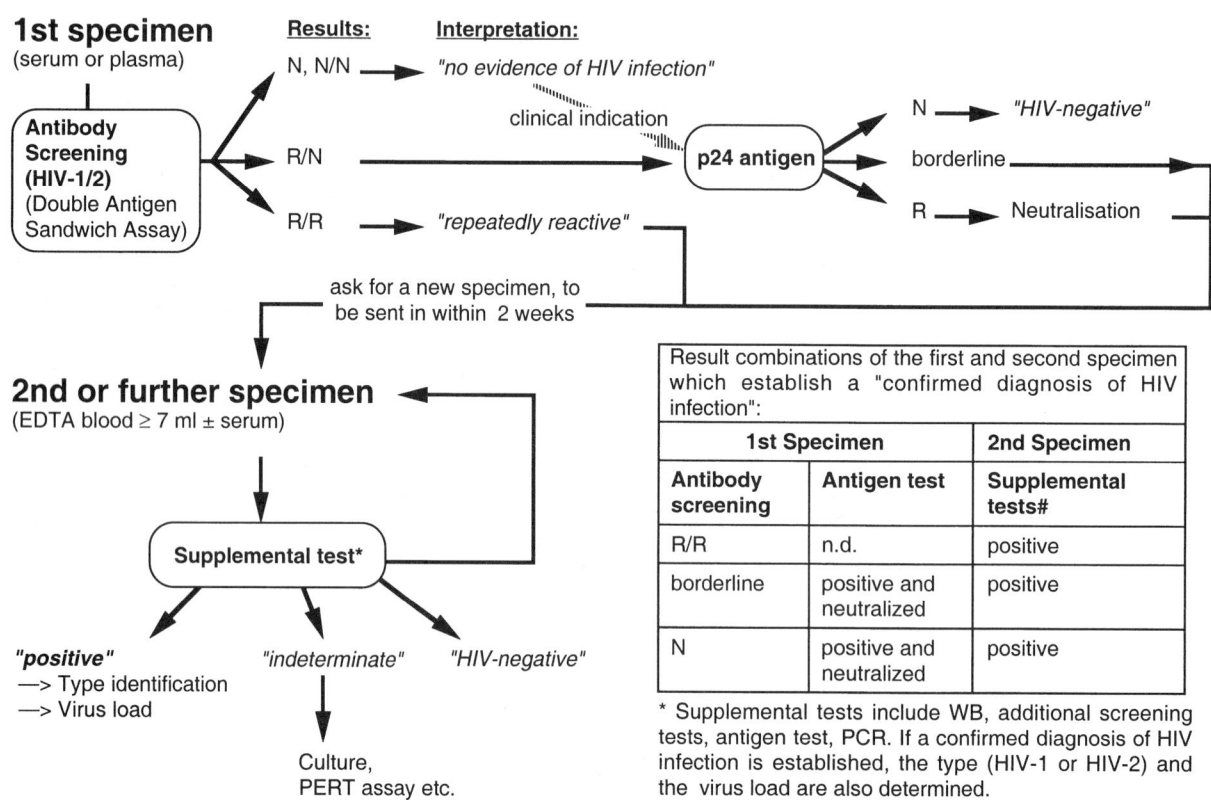

FIGURE 13 Algorithms for diagnosis of HIV infections in adults. Reprinted from Schüpbach, 1999.

It must be emphasized that HIV infection must never be established on a single sample, even if reactivity in screening has been confirmed by at least one supplemental test. The possibility that an error, for example, a sample mix-up or contamination, might lead to a false-positive diagnosis dictates verification of all positive results with a second, freshly drawn sample. If the first sample presented with a low reactive result in screening and required different supplemental tests for confirmation, the supplemental tests should also be done on the second sample unless it is now clearly positive. Demonstration that the second sample has the same low reactivity in screening is not sufficient verification.

Also, indeterminate results of different methods never add up to a positive result. For example, a borderline screening test plus a borderline Western blot with a pattern positive by ASTPHLD-CDC criteria but indeterminate by ARC or FDA criteria plus a reactive antigen test which, however, cannot be confirmed by neutralization are not sufficient for a positive diagnosis. Testing must continue until clear-cut positive results are demonstrated.

When testing the first specimen, samples reactive in a screening test should be retested in a second, different screening assay. A more costly alternative is screening all samples by two different tests, preferably performed in different runs, which increases the sensitivity in the early stage of infection and provides maximal protection against human error, which is the most relevant source of false-positive or false-negative diagnosis. Use of two different screening test kits is less necessary if testing systems featuring automated positive sample identification from the bar-coded original test tube are used. Due to the high specificities of current test kits, a specimen reactive in two different screening tests is very unlikely to be a false positive, and the tentative diagnosis "HIV positive" can be verified with a freshly drawn sample without a need for supplemental testing of the initial sample. Antigen screening should be performed liberally on all samples from patients with possible exposure to HIV, clinical signs of an acute retroviral syndrome, or discordant results of antibody screening or whenever absorbance values are elevated. If antibody screening and antigen testing do not yield a clear result, a fresh sample, ideally an EDTA-anticoagulated blood sample, is required. This second sample, which permits all the different types of tests mentioned above, should be taken without delay. There is no rationale for a 3-month delay to diagnose a primary HIV infection if it is suggested by the results of the first sample.

The assays performed on the second specimen are chosen according to the results from the first sample. A confirmatory laboratory should be able to perform a variety of supplemental tests in order to establish a "confirmed diagnosis of HIV infection." A third positive screening assay, a positive WB according to FDA interpretation guidelines, a positive and neutralized antigen assay, or PCR for viral RNA or DNA may all be used alone or in combination to establish this diagnosis. Tests performed with the second sample should also establish the type of virus (HIV-1 or HIV-2) and the viral load, because this information is needed in order to assess whether an antiretroviral treatment should be given.

Treatment Monitoring

Viral-Load Determination

Viral-load determination is instrumental in several aspects of the clinical management of HIV infection. First, in early infection at the set point, it serves to assess the like-

ly course the infection will take. Based on the risk of progression, which depends on the viral load, appropriate treatment decisions can be made. The short-term efficacy of a specific antiretroviral treatment can then be assessed by measuring the reduction in virus concentrations achieved within the first 2 to 4 weeks after treatment initiation. Long-term efficacy is likely when virus levels continuously decrease below the level of detection and remain undetectable by the most sensitive assays.

Prognostic value. Higher HIV RNA levels correlate with lower baseline CD4$^+$ T-cell counts, a more rapid decline in CD4$^+$ T-cell counts, and more rapid disease progression. Patients with more than 100,000 copies/ml of plasma within 6 months of seroconversion were 10-fold more likely to progress to AIDS over 5 years than those with fewer copies. Maintenance of <10,000 copies/ml in early HIV infection is associated with a decreased risk of progression to AIDS. However, in patients with more advanced disease, a low RNA count does not protect from progression; up to 30% of patients with <10,000 copies/ml progressed (Coombs et al., 1996; Mellors et al., 1996; Saag et al., 1996; Welles et al., 1996; O'Brien et al., 1997). Patients with advanced disease can present with high or low viral-RNA titers.

As mentioned above, PCR, NASBA, and bDNA methods are valid procedures for viral-load determinations. In the plasma of most untreated patients, viral RNA is detectable at all stages of disease. If RNA is undetectable in an untreated patient, this may reflect a very low viral load, as seen in long-term nonprogressors. However, the negative result may also be due to a virus subtype not well detected by the assay. For example, group O viruses are detected by none of the existing commercial RNA measurement procedures. If quantitative tests of untreated patients, especially if they show signs of disease progresssion, are negative, one should check the likelihood of an infection with an alternative subtype (A, E, F, or G) and verify that a procedure with good sensitivity to all the various subtypes was employed. While the bDNA method initially appeared to be superior to other procedures in this respect, improvements realized in version 1.5 of the Roche HIV-1 Monitor apparently result in a significantly improved detection of subtypes A, E, F, and G (Loveday et al., *Abstr. 12th World AIDS Conf.*; Michael et al., *Abstr. 12th World AIDS Conf.*; Triques et al., *Abstr. 12th World AIDS Conf.*; Parekh et al., 1999). In addition, in the ultrasensitive procedure in which a larger sample volume (200 instead of 40 μl) is assessed, the limit of quantification has been lowered to about 50 copies/ml. The analytical sensitivities of the NASBA and the bDNA procedures have also been improved, but only extended side-by-side comparisons of these tests can show whether sequence variability is no longer a problem. A negative result in other virus component tests less affected by sequence variation supports a low load. Such alternative methods include in particular the PERT assay, which, being a functional test for RT activity, is completely independent of viral sequence.

Contrary to the general perception, p24 may also serve as a valid prognostic marker. In a study of 169 patients representing all stages of chronic infection, signal amplification-boosted p24 ELISA of heat-denatured plasma yielded results that were significantly better correlated with the ensuing CD4$^+$ T-cell decline and survival over a median observation period of 2.7 years (range, 0.1 to 4.9), while

RNA prevailed as a marker of progression to clinically defined AIDS (Ledergerber et al., 2000).

Treatment monitoring. Treatment monitoring assesses the short- and long-term treatment-induced reductions of the baseline viral load. Effective antiretroviral therapy significantly decreases HIV RNA levels in plasma within 1 week of initiation. A lack of significant decrease within this period indicates absence of effect. Historically, zidovudine monotherapy resulted in a median 0.7-log-unit decrease in RNA levels in plasma within 2 weeks, which usually returned to baseline by 6 months, indicating the outgrowth of a zidovudine-resistant viral quasispecies. Combinations of two nucleoside analog inhibitors of RT (NRTIs) led to more prominent reductions (-1.5 log units), which often persisted for more than 1 year. Combination of two RT inhibitors with a protease inhibitor (PI) or a nonnucleoside RT inhibitor (NNRTI), so-called triple therapies, or "highly active antiretroviral therapies" (HAART), may result in sustained reduction of the plasma viral RNA below the detection limit of even the ultrasensitive procedures, which may last for prolonged periods, especially when initiated in previously untreated patients who do not yet harbor drug-resistant virus mutants. Similar results may also be achievable with combinations of two NRTIs with hydroxyurea or of three NRTIs. Treatment of patients in early disease with such combinations has resulted in a decrease in viral load that led to persistent nondetectability of viral RNA in plasma for more than 3 years, a slow decrease in PBMC-associated proviral DNA, and even decreasing titers of HIV-specific antibodies. Analysis of lymphoid tissues shows, however, that virus-expressing cells, not to mention provirus, are still detectable after years, and replication-competent virus can also be isolated from lymphod tissues of such patients (Finzi et al., 1997, 1999; Wong et al., 1997). Interruption of the treatment regimen for very brief periods may thus lead within a few days to the reappearance of virus in plasma to levels seen before initiation of the therapy (Neumann et al., 1999).

Concentrations of viral RNA in plasma are thus only imperfectly correlated with the situation in the lymphoid tissues, which harbor 98% of the body's lymphocytes. Nevertheless, since the lymphoid tisuess have, in the efferent lymphatics, an outlet to the peripheral blood, virus concentrations in plasma should, except in recently infected individuals (Pantaleo et al., 1998), more or less reflect the lymphoid tissue situation. The rapid decline of viral RNA in plasma observed after initiation of efficient antiretroviral combination therapy is in fact paralleled by a similar decline of viral RNA in lymphoid tissue (Cavert et al., 1997). Specifically, in tonsil tissue there was a rapid drop in mononuclear cells acutely producing virus, with a half-life of 0.9 days, which is comparable to the estimated half-life of 1.1 days of acutely infected CD4+ T cells in the blood (Ho et al., 1995; Wei et al., 1995; Perelson et al., 1996, 1997). Viral RNA bound to FDC followed the decline with an initial half-life of 1.7 days (versus 0.6 days for virus in plasma). A phase of slower decay, with half-lives of 15 days for expressing cells and 14 days for FDC-bound virus, followed from day 2 onward. After 6 months of treatment, there were still infected cells and low levels of virus expression in a majority of patients (Cavert et al., 1997); cell-associated viral DNA and mRNA levels reached a plateau after about 500 days of treatment, after which no further decrease was observed (Furtado et al., 1999). Ongoing low-level replication after more than 18 months of aggressive treatment has been confirmed by the demonstration of unintegrated circular forms of viral DNA or of sequence evolution (Chun et al., 1997; Furtado et al., 1999; Zhang et al., 1999b).

Viral load studies under antiretroviral treatment have so far focused on effects in plasma and the lymphatics. Little is known about the impact of these regimens on HIV infection of the CNS, and viral kinetics in the CNS have not been established.

RNA-based tests are currently viewed as the only feasible methods for viral-load determination. These tests are expensive, however, which restricts their use. Again, the signal amplification-boosted p24 antigen test may represent a valuable alternative. Viral load monitoring in many patients can be done at considerably less expense with this assay (see above). In a study of 127 plasma samples taken at regular intervals from 23 patients with CD4+-cell counts of less than 50/ml who received a defined antiretroviral treatment, the viral load was comparatively assessed by antigen assay and commercial quantitative reverse transcription-PCR (detection limit, 200 copies/ml). Overall detection during a median observation time of 25 weeks amounted to 75.6% for antigen and 73.6% for reverse transcription-PCR. The antigen detection limit was 0.2 pg/ml. Antigen was detectable in all 23 baseline samples, while RNA was undetectable in one. The antigen and RNA levels in 79 samples positive for both markers were correlated ($R = 0.714$; $P < 0.0001$). Average changes in levels of p24 and RNA at eight time points were correlated ($R = 0.982$; $P < 0.0001$) (Fig. 10B). In individual patients, the two parameters behaved similarly, in certain cases virtually identically. In conclusion, the performance of this antigen detection procedure was comparable to that of commercial quantitative reverse transcription-PCR, thus providing a simple, high-throughput alternative in monitoring the efficacy of antiretroviral treatment (Boni et al., 1997). Similar findings were observed in pediatric HIV infection (Nadal et al., 1999).

HIV Drug Resistance Testing

The Achilles heel of any antimicrobial chemotherapy is the development of resistance; in the case of the antiretroviral drugs used against HIV, this is caused by mutations of the genes targeted by these drugs, namely, the RT or the PR gene. Mutations conferring resistance, however, are not restricted to the enzyme genes themselves, as shown by certain mutations in the proteolytic processing sites encoded by the *gag* gene which are targeted by the viral PR (Carrillo et al., 1998). Since many resistance mutations affect different drugs, resulting in cross-resistance, a detailed knowledge of them would be valuable for the design of treatment in individual patients. This is particularly true in view of the fact that viruses with resistance mutations may be transmitted and thus be present in patients prior to any antiretroviral treatment (Imrie et al., 1997). The frequency of such transmissions in the United States and Western Europe is estimated at 10 to 15%; transmission of multidrug-resistant viruses has also been observed (reviewed in Rodriguez-Rosado et al., 1999).

There are two types of resistance mutations. Primary mutations directly reduce the susceptibility of an enzyme to an antiretroviral drug, are relatively specific for each drug, and appear soon after treatment initiation (Table 6). They permit the mutant to replicate in the presence of the drug, but replicative capacity is usually impaired due to a

TABLE 6 Primary mutations associated with resistance to antiretroviral drugs[a]

Antiretroviral family	Drug	Primary mutations (codon[s])
Nucleoside analogs	Zidovudine	41, 70, 215
	Didanosine	74
	Zalcitabine	69
	Lamivudine	184
	Stavudine	75 (?)
	Abacavir	184
Nonnucleoside analogs	Nevirapine	103, 181
	Delavirdine	103, 181
	Efavirenz	103
PI	Ritonavir	82
	Indinavir	46, 82
	Saquinavir	48, 90
	Nelfinavir	30
	Amprenavir	50

[a]Reprinted from *AIDS* (Rodriguez-Rosado et al., 1999), with permission from the publisher.

decreased activity of the mutated enzyme. Thus, when treatment with the failing drug is continued, virus strains with secondary mutations compensating for the enzymatic impairment are usually selected over time. Most of these secondary (or compensatory) mutations do not further increase resistance to the drug but restore the replicative capacity of the mutant virus. While there is little overlapping among primary mutations, many of the compensatory mutations are shared among drugs of the same family: NRTIs, NNRTIs, and PIs. Failing drug regimens should therefore be switched as soon as they are recognized, in order to prevent secondary mutations.

Not all apparent resistance to antiretroviral treatment is due to viral mutation and drug-induced selection. Viruses of group O and HIV-2 are intrinsically resistant to NNRTIs, since their RTs do not bind these drugs, which were developed based on the B subtype prevalent in Western countries. For some drugs, such as stavudine, cellular phosphokinases are needed to activate the drug to its triphosphate form intracellularly. Previous treatment with zidovudine appears to decrease this enzyme activity, thus impairing the intracellular concentrations of active drug. Frequently, however, the reasons for a failing viral response to antiretroviral treatment are a lack of adherence to treatment, impaired intestinal drug absorbance, pharmacokinetic interactions, or continuing viral replication at sanctuary sites at which drug concentrations are inadequate. These points should be evaluated before a resistance analysis is considered.

Drug resistance testing might be envisaged in the following situations: (i) in individuals with acute primary HIV infection in whom an efficient early therapy has been shown to improve the outcome (Rosenberg et al., 1997; Kahn and Walker, 1998) but where there is also a sizeable chance of infection with a drug-resistant mutant; (ii) in drug-naive HIV-infected pregnant women, in order to minimize the risk of vertical transmission; and (iii) to choose alternative regimens for pretreated patients with therapeutic failure, especially for those heavily pretreated.

In the area of methods, there are genotypic and phenotypic assays. Genotypic tests examine the population of viral genomes in a test sample for the presence of mutations known to confer resistance. Phenotypic assays measure how sensitive to a drug a population of viruses or their enzyme(s) targeted by the drug are; the result of this assay is given as a 50 or 90% inhibitory concentration.

All genotypic assays use PCR for amplifying the population of viral sequences which code for the viral enzymes, RT and PR, that are targeted by the antiretroviral therapy. The analysis of the amplified sequences for known resistance mutations can then be done by two different methods. One uses reverse hybridization of the amplified material to sequence probes immobilized on a carrier surface. This principle is used in the line probe assay, in which a limited number of mutations are assessed by hybridizing the amplified sequence to probes which discriminate between wild-type and mutant sequences and are immobilized on a Western-blot-like strip. On a more general scale, this principle is also utilized by chip-based systems, such as the Affimetrix assay. The alternative method of product analysis is by automated cyclic sequence analysis of the target region, e.g., with the Perkin-Elmer–Applied Biosystems method.

Phenotypic assays are performed with viruses isolated from patients, after growing them in a suitable target cell culture. For better standardization, however, phenotypic resistance analysis is preferentially performed by a recombinant-virus approach. In this method, the RT or PR sequences are amplified from the patient's sample and are inserted into a cloned defective virus backbone that lacks the viral enzyme to be investigated. The insertion renders the virus competent for replication in a given cell line, and the recombinant viruses will exhibit standardized replication properties for all gene products except for PR or RT. Regarding susceptibility to antiretrovirals, the recombinants will represent the patient's own virus population. Two such systems are currently available (Virco and ViroLogic).

A last and simple method based on the PERT assay directly measures to what degree the RT activity of HIV particles contained in a plasma sample is inhibited by antiretroviral drugs (Schüpbach and Böni, 1998). This approach seems to be applicable to both NRTIs and NNRTIs (Garcia Lerma et al., 1999).

A problem of all resistance assays is their limited capability to detect a minority of resistant mutants against a background of wild-type sequences. Even under optimal conditions, resistant sequences usually need to be present in at least 25 to 30% of the HIV genomes in order to be reliably detected. Under practical conditions, as assessed in recent round trials in which panels with standardized mixtures of wild-type and mutant sequences of a cloned virus were assessed by genotyping laboratories around the world, results were far from optimal, indicating, as in the early days of PCR, the urgent need for standardization of these procedures and quality control (Schuurman et al., 1999). Despite these technical problems and the difficulty of result interpretation, there is now both retrospective and prospective evidence in support of the clincal utility of resistance testing (Durant et al., 1999).

Prevention and Treatment of HIV Infection

Retrovirus Stability and Safety Issues

Although the consequences of an infection with HIV are probably more severe than for HTLV, the HIVs and HTLVs are both classified as biological safety hazard agents of moderate risk (class 3). Since the physical compositions of the two viruses are similar, the following information derived from investigations of HIV-1 can largely be applied to the HTLVs. Since transmission of the HTLVs requires direct cell-to-cell contact, one is even more on the safe side with the recommended safety measures.

The risk of laboratory-acquired infection with these viruses stems primarily from contamination of the hands and mucous membranes of the eyes, nose, and mouth by infectious blood and other body fluids. There is no evidence that HIV or HTLV is transmitted by the airborne route. Strict adherence to the safety precautions, outlined in more detail in a handy booklet from the WHO (Anonymous, 1991), is paramount in preventing nosocomial infections. Good gloves and a protective laboratory gown should always be worn, and the eyes should be protected from spills. Disposable unbreakable plasticware should be used, never glassware or other sharp or breakable objects. When sharp instruments cannot be avoided, the nondominant hand should be protected with a stainless steel chain mail or Kevlar glove.

Spills or contaminations of laboratory surfaces must be decontaminated immediately. Whenever possible, a type 2 laminar flow biological safety cabinet should be used when handling any patient sample that might contain infectious virus. Use of a laminar flow biological safety cabinet is mandatory for virus isolation work. Centrifuges, including those used in laboratories that perform only serological testing, should be equipped with sealed centrifuge buckets. Research work with HIV and virus production require a designated biosafety level 3 laboratory with restricted access, a ducted exhaust-ventilation system that maintains negative air pressure and creates a directional airflow, and other safety measures.

HIVs, HTLVs, and other retroviruses are rapidly inactivated by detergents and disinfectants that are effective against other enveloped viruses. Otherwise, HIV at least is relatively stable. At autopsy, HIV has been isolated up to 16.5 days postmortem from various tissues (Douceron et al., 1993). Suspensions of the virus in protein-containing fluids or dried preparations are also relatively stable (Tjotta et al., 1991). At the optimum pH of 7.1, the half-life of the virus ranged from approximately 24 h at 37°C to no significant loss over 6 months at −75°C. Drying the virus on a glass surface or freezing caused a 5- to 12-fold and a 4- to 5-fold decrease in activity, respectively. The dried preparations, however, were about as stable as when stored in a buffered solution (Tjotta et al., 1991). In another study, 1 log unit of inactivation in culture fluid, seawater, sewage, and dechlorinated tap water (all sterile and kept at 16°C in the dark) required 1.3, 1.6, 2.9, and 1.8 days, respectively. After the

first 4 days, the inactivation became even slower (1 log unit of inactivation after 4.3, 2.6, 5.7, and 4.6 days, respectively). HIV was more stable than herpes simplex virus but less stable than poliovirus (Sattar and Springthorpe, 1991). These data are not meant to suggest that HIV transmission might occur by exposure to water, for which there is absolutely no basis. They should, however, make clear that caution when working with HIV and rapid disinfection of spills and contaminations are important.

The standard disinfectant recommended for contaminated surfaces is a hypochlorite solution with a concentration of 0.5% available chlorine (5 g/liter; 5,000 ppm). When working with HIV cultures and virus preparations, a higher concentration of 1% available chlorine is recommended (Anonymous, 1991; Van Bueren et al., 1995). Fresh 2% solutions of alkaline glutaraldehyde are effective, but care should be taken that they are not too dilute or have not become stale when used for disinfecting HIV associated with organic matter. A solution of iodine and detergent (2% Jodopax) will remove all detectable HIV-1 activity. In contrast, 70% industrial methylated spirit or 70% ethanol is ineffective in inactivating dried protein-rich spills of cell-free or cell-associated HIV within a reasonable time; complete inactivation requires up to 20 min (Tjotta et al., 1991; van Bueren et al., 1994).

The risk of HIV infection following percutaneous needle stick exposure to HIV-contaminated blood is estimated to be between 0.13 and 0.5%. The risk of infection by such accidents depends on the depth of the penetration (the relative risk [RR] of percutaneous lesions is 16.1), whether there is visible contamination of the penetrating object with blood (RR, 5.2), whether the object was used for an intravenous or intra-arterial injection (RR, 5.1), and the disease stage (i.e., the viral load) of the index patient (RR of AIDS, 6.4) (Anonymous, 1995).

Needle stick or other puncture wounds, cuts, and skin contaminated by spills or splashes of specimen material should be thoroughly washed with soap and water and disinfected with a nonirritating disinfectant (cells of the immune system should not be attracted!). Bleeding from any wound should be encouraged. In case of a percutaneous penetration, an antiretroviral prophylactic treatment should be started immediately. Postexposure monotherapy with azidothymidine lowered the risk of infection by 80% (Anonymous, 1995). Presently, the recommended postexposure chemoprophylaxis for HIV consists of a triple combination that includes two RT inhibitors and an efficient PI. This should be given for at least 2 weeks but not longer than 4 weeks. Chemoprophylaxis is not recommended when intact skin is exposed to blood or blood-containing body fluids.

Treatment of HIV Infection

Due to the high variability of HIV and the generation of drug-resistant mutants, successful long-lasting suppression of virus replication can only be achieved with combination treatment, frequently also called HAART. Such regimens are usually composed of at least three drugs, currently selected from the NRTI, NNRTI, and PI groups (Montaner et al., 1998; Hirschel and Opravil, 1999). NRTIs function as nucleoside triphophates for RT-mediated cDNA synthesis and act as chain terminators; this group includes drugs such as zidovudine, didanosine, zalcitabine, stavudine, lamivudine, and abacavir. In contrast, NNRTIs bind directly to the RT, thereby blocking its active site either directly or indirectly; examples include nevirapine, delavirdine, and efavirenz. PIs block the active site of the viral PR, thereby inhibiting the processing of the Gag-Pol and Gag precursor proteins; this group includes saquinavir, indinavir, ritonavir, and nelfinavir. Agents that block the third viral enzyme, the IN, are in clinical evaluation. In addition, the antimetabolite hydroxyurea (HU), which inhibits the cellular enzyme ribonucleoside diphosphate reductase, thus lowering the intracellular concentrations of dATP in particular, is used (Lori et al., 1994). When HU is given in combination with didanosine and stavudine, it leads to an increased incorporation of these chain terminators into the nascent cDNA. A further advantage is the absence of resistance, since this is a cellular enzyme.

A triple combination frequently contains a PI and two NRTIs, but there are also efficient combinations that are composed of NNRTIs and NRTIs or of NRTIs alone, with or without HU. As a rule, the clinical benefit of antiretroviral therapy depends on the degree to which viral replication is suppressed. Thus, patients with a reduction of the plasma viral RNA load below the detection limit of ultrasensitive RNA assays have the highest chance of a long-lasting response to treatment. However, a rising $CD4^+$ T-cell count may also be seen in some patients treated with PI-containing regimens whose virus titers did not respond well (Kaufmann et al., 1998a).

HAART entered clinical trials in 1995, and within a short time, their general use outside clinical studies as well markedly reduced both HIV-associated morbidity and mortality (Egger et al., 1997; Palella et al., 1998). In sporadic cases, antiretroviral treatment may even have led to a situation in which the virus was well under control even after discontinuation of the treatment, as manifested by a very low viral load combined with high concentrations of HIV-specific CTLs (J. Lisziewicz, H. Jessen, D. Finzi, R. F. Siliciano, and F. Lori, Letter, *Lancet* **352**:199–200, 1998). Normally, however, treatment discontinuation leads to an immediate rebound of the virus, and treatment thus must be given for many years, if not for life (Finzi et al., 1999; Zhang et al., 1999). Since HAART interrupts the positive rebound of viral proteins on virus expression through the activation of the immune system, it also drives the virus into proviral latency. Combination of HAART with measures that lead to a controlled activation of HIV-infected cells, e.g., by IL-2 or gp120, or possibly even short interruptions of antiretroviral treatment may serve to renew the expression of viral proteins and to stimulate the antiviral effector systems of the immune system, but the results of such clinical trials are not yet available.

HTLVs

Biology

PTLVs, HTLV Types and Subtypes, and Genomic Variability

The HTLVs belong to the genus-level HTLV-bovine leukemia virus group, whose members are characterized by particles with a central spherical core and a complex genome. Similarly to the HIVs, which are closely related to the SIVs, the HTLVs are closely related to a group of simian viruses called primate T-lymphotropic retroviruses (PTLVs). The PTLV type I (PTLV-I) group includes, in addition to HTLV-1, the related simian T-cell leukemia or lymphoma virus type 1 (STLV-1), which is highly prevalent in many different Old World monkey species. The PTLV-II group includes HTLV-2 and the related STLV-2, a single isolate from a New World monkey. The PTLV-L group is represented by a single isolate from a baboon of the species *Papio hamadryas* wild born in Eritrea, and the STLV-PP group, which is more closely related to HTLV-2 than to HTLV-1, but still is quite distinct from HTLV-2, contains a pygmy chimpanzee (*Pan paniscus*) isolate. Phylogenetic analysis of these various retroviruses indicates that the HTLVs were transmitted from monkey to human on several different occasions in the ancient past (reviewed in Gessain and de The, 1996a).

Three major genetic lineages have been identified for HTLV-1 (Melanesian, Central African, and cosmopolitan), and three have been identified for HTLV-2 (A, B, and C). The cosmopolitan lineage of HTLV-1 includes four subtypes, A (transcontinental), B (Japanese), C (West African), and D (North African), that are indigenous to all continents except Australia/Melanesia (Yamashita et al., 1996). In addition to viruses of the cosmopolitan group, Africa also harbors the Central African subtype. In the Americas, the majority of isolates from endemic infections among Amerindian peoples of North and South America belong to subtype A and were probably introduced by ancient immigration from Asia across the Bering strait. In addition, subtype C was brought to the Caribbean through the slave trade (Van Dooren et al., 1998), although most Caribbean isolates are also of subtype A (Yamashita et al., 1996). In addition to subtype A, which is found in indigenous populations living on Hokkaido, the northernmost island, and Okinawa in the extreme south, Japan also harbors the B subtype, which is prevalent in the southwest. The Melanesian lineage is found in the indigenous peoples living on Papua New Guinea and the Solomon Islands, as well as in Australian aborigines. The isolates of this group are highly divergent among themselves and quite distinct from the other two lineages. The nucleotide sequence differs from that of the cosmopolitan lineage by 6 to 7.5% for the transcontinental subtype, 2 to 4.5% for the Central African subtype, 1 to 2% for the West African subtype, and 0.5 to 2% for the Japanese subtype. The different HTLV-1 lineages appear to be derived from at least three different ancient events of interspecies transmission from monkey to human, one thought to have given rise to the Melanesian subtype several thousand years ago and two others that occurred in Africa, giving rise to the Central African and Cosmopolitan lineages. Less than 1% divergence is present among different isolates of one subtype, and the intraperson variability is less than 0.5% (Wattel et al., 1996; reviewed in Gessain et al., 1985).

HTLV-2 differs from HTLV-1 overall by 40% of its coding sequence. The sequence homology is low in MA (59%), SU (63%), and Rex (61%) but higher in CA (85%) and *tax* (78%). Subtypes A and B differ by about 4 to 6%, depending on the region analyzed, and the strains within each subtype differ by up to 0.4%. Subtype A (representing the prototype isolate MO) is prevalent among intravenous drug addicts but has also been found to be endemic among some Amerindian peoples of South America. Subtype B is endemic in Amerindian tribes scattered throughout North, Central, and South America. A third subtype (C) was identified in urban Brazilian and Indian populations (Ferreira et al., 1997). HTLV-2 was considered a retrovirus of the New World, possibly brought in as early as 40,000 years ago by migration across the Bering land bridge. However, HTLV-2 isolates were also more recently identified in some remote populations of Central Africa, including Pygmies, and appear to represent an additional subtype D. An HTLV-2 D isolate from the Efe Bambuti Pygmy tribe differed by 2.4% in *tax-rex* to 10.7% in the LTR from both subtypes HTLV-2 A and HTLV-2 B. Phylogenetic analysis indicated that the Pygmy HTLV-2 Efe2 strain had the longest independent evolution and suggested an African origin of HTLV-2 (P. Goubau, J. Desmyter, J. Ghosquiere, and B. Kaseruka, Letter, *Nature* **359**:201, 1992; Lal et al., 1994; P. Goubau, A. Vandamme, K. Beuselinck, and J. Desmyter, Letter, *J. Acquir. Immune Defic. Syndr.* **12**:208–209, 1996; Vandamme et al., 1998).

Genome, Transcripts, and Particularities of Viral Replication

The HTLV genome contains an additional coding region downstream of the *gag*, *pol*, and *env* genes originally called the X region for its then-unknown function (Fig. 14). The sizes of the different proteins are indicated in the figure. The X region codes for the regulatory proteins Tax and Rex, which are both essential for viral replication and exert functions analogous to those of Tat and Rev of HIV, and a number of additional proteins encoded by overlapping ORFs whose functions are largely unknown (reviewed by Gessain and de The, 1996b; Ferreira et al., 1997).

Like HIV Tat, but through an entirely different mechanism, the transactivator protein Tax (40 kDa in HTLV-1 and 37 kDa in HTLV-2) enhances the LTR-mediated transcription of viral genes. The interaction with the LTR is indirect and involves a number of cellular transcription factors, like NF-κB, CREB/ATF-1 (cyclic-AMP-reponsive element binding protein/activating transcription factor 1), SRF (serum-reactive factor), and others whose normal function is the regulation of cellular genes but which may also bind to the HTLV LTR, resulting in enhanced viral transcription. Through the same and other mechanisms, Tax is capable of activating a number of cellular oncogenes

and transcription factors, like c-*fos*, c-*myc*, and *egr-1*; growth factors or cellular receptors, such as the alpha chain of the IL-2 receptor (IL-2Rα); a variety of interleukins, including IL-1α and β and IL-2, -3, -4, -5, -6, -8, -9 and -15; c-*sis* (platelet-derived growth factor), granulocyte-macrophage colony-stimulating factor (GM-CSF), TGF-β, gamma interferon (IFN-γ), TNF-α and -β, parathyreotropic hormone-related protein, vimentin, and major histocompatibility complex class I. Induction of mitosis is thus an important effect of Tax (reviewed in Buckle et al., 1996). In addition, Tax downregulates the DNA repair enzyme β polymerase, which results in an impaired base excision repair of oxidative damage to DNA (Philpott and Buehring, 1999). These interactions are important for pathogenesis (reviewed in Taylor et al., 1996; Yoshida, 1996; Ferreira et al., 1997; Uchiyama, 1997).

The Rex protein (p27 for HTLV-1; p26 for HTLV-2) is functionally similar to the HIV Rev protein. The phosphoprotein is located in the nucleus and nucleolus and controls the expression of virion proteins on the posttranscriptional level by increasing the number of unspliced or singly spliced viral RNAs that are exported from the nucleus. Like Rev, Rex forms homopolymeric complexes along the viral RNA, but at a site (the Rex-responsive element) that is located in the U3-R region at the 3' end of the viral RNA. Thus, unlike Rev, Rex binds to all viral mRNAs, and the selective effect on incompletely spliced RNAs involves additional elements (reviewed in Rabson and Graves, 1997).

The replication of the HTLVs follows the same general principles as that of the HIVs (Fig. 2), though with some important differences: The HTLVs are characterized by a remarkable genetic stability and high concentrations of virus-infected cells in the absence of a detectable plasma viremia. In some infected individuals, particularly in those developing neurological disease, up to one-fifth of PBMC

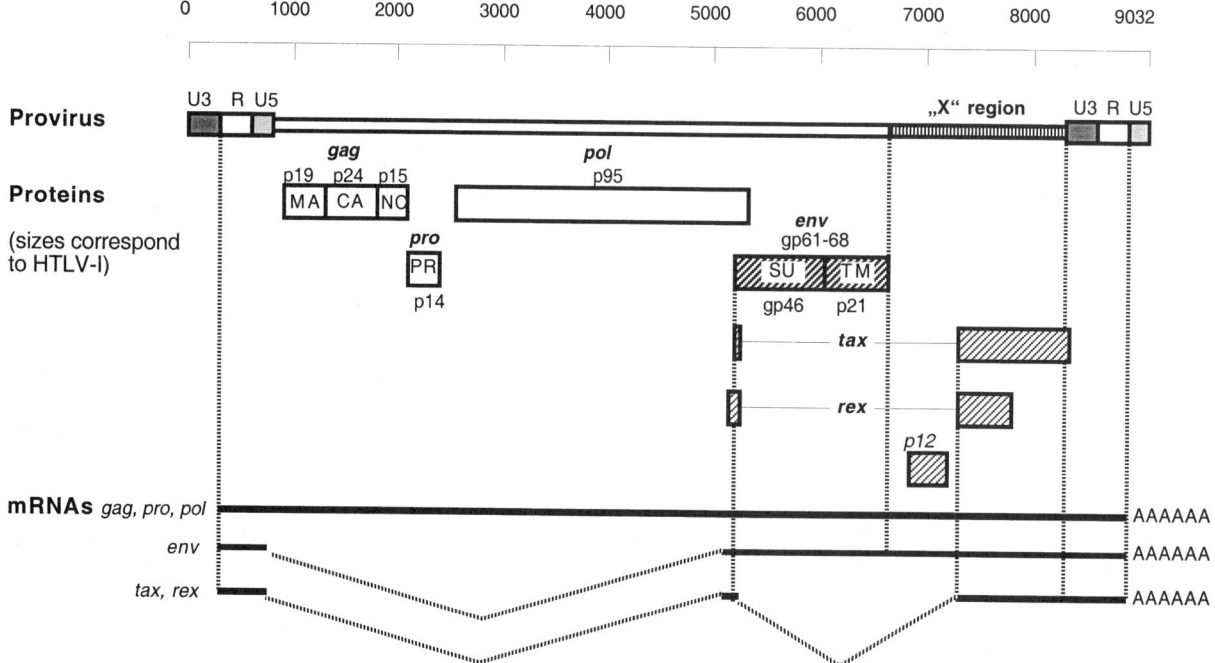

FIGURE 14 Genome, proteins, and transcripts of HTLV-1. Simplified from Cann and Chen, 1996, and Ferreira et al., 1997.

may carry the HTLV provirus. The virus replicates essentially by mitosis, whereby the provirus is duplicated, as can be demonstrated by PCR amplification of the conserved HTLV-1 integration sites. This is true of all stages of HTLV infection. These differences in the mode of replication are due to different effects of Tax with respect to Rex compared with those of Tat and Rev of HIV. In addition to its activation of viral transcription, Tax, but not Tat, has a mitogenic effect. Both Rex and Rev have a negative effect on their own expression. Regarding the viral transactivators, Tat is less dependent on Rev than Tax is on Rex, owing to a form of Tat that is translated from a fully spliced, Rev-independent RNA (Fig. 2D). In contrast, Tax is exclusively translated from the same mRNA as Rex (see Fig. 14). Thus, activation of HTLV transcription can never lead to the same sustained high level of virus production seen with HIV (Wattel et al., 1996).

HTLV-Associated Disease and Pathogenesis

HTLV-I-Associated Diseases

Most individuals infected with HTLV-1 remain disease-free carriers throughout their lifetimes. In 2 to 6% of infected individuals, however, chronic disease may develop, usually after a long incubation time. Three characteristic disease entities have been etiologically linked with the infection: (i) ATLL, (ii) HTLV-associated myelopathy/tropical spastic paraparesis (HAM/TSP), and (iii) HTLV-associated uveitis (HAU) (Table 1). In addition, HTLV-1 infection may be etiologically associated with an arthropathy (HTLV-1-associated arthropathy [HAAP]) and a bronchopneumopathy. Other manifestations, including polymyositis (Caldwell et al., 1996; Higuchi et al., 1996), Sjögren's syndrome (Kompoliti et al., 1996; Osato et al., 1996), Graves' disease (especially in combination with uveitis and/or arthropathy) (Kawai et al., 1995), lymphadenitis and infectious dermatitis (La Grenade et al., 1996; Manns et al., 1999) have also been described.

ATLL

In the mid-1970s, ATLL was recognized as a new disease in Japan based on characteristic clinical features, origin of the patients from a distinct geographic region, and a CD4$^+$-T-cell phenotype of the leukemic cells (Takatsuki et al., 1977; Uchiyama et al., 1977). ATLL develops after a long latency in a small fraction (2 to 6% lifetime risk) of chronically infected individuals, who normally are infected shortly after birth (Tajima, 1990; Tajima and Cartier, 1995). The mean age at disease onset is 55, and the male-to-female ratio is 1.4:1 (Uchiyama et al., 1977; Tajima, 1990; Shimoyama, 1991). ATLL is currently divided into four clinical subgroups: acute, chronic, smoldering, and lymphoma types. The classification is based on the frequency of abnormal T lymphocytes in the blood, serum levels of lactic acid dehydrogenase (LDH), presence or absence of tumor infiltrates in various organs, and the clinical course (Yamaguchi et al., 1983; Kawano et al., 1985; Shimoyama, 1991). More than 50% of ATLL cases are of the acute type, 20% are of the lymphoma type, 20% are of the chronic type, and about 5% are of the smoldering type.

Characteristically, acute ATLL presents with general malaise, fever, cough, dyspnea, abdominal fullness, thirst, and drowsiness. The examination reveals lymph node enlargement, hepatosplenomegaly, and jaundice. Abnormal laboratory findings include a marked leukocytosis, hypercalcemia, high serum LDH, and high levels of a soluble form of IL-2Rα. Blood smears show characteristic leukemic cells

with deeply convoluted or lobulated nuclei (Takatsuki et al., 1977; Shimoyama, 1991) (Fig. 15). Chronic ATLL is characterized by milder symptoms and signs and a more protracted clinical course. Patients with smoldering ATLL have only a few leukemic cells in their blood and frequently present with skin lesions, such as papules, nodules, and erythema. Lymph node enlargement and splenomegaly in these patients are minimal, and serum LDH is normal to slightly elevated (Yamaguchi et al., 1983). In lymphoma-type ATLL, the predominant finding is lymph node enlargement.

Major complications of ATLL include hypercalcemia and serious opportunistic infections by bacteria, fungi, protozoa, and viruses. As in AIDS, *Pneumocystis carinii* pneumonia, aspergillosis or candidiasis, and cytomegalovirus pneumonia are common (Shimoyama, 1991). The prognosis of ATLL is poor, with a median survival time of 6 months for the acute type, 10 months for the lymphoma type, and 2 years for the chronic type. Four-year survival rates are only 5% for both the acute and lymphoma types, 27% for the chronic type, and 63% for smoldering ATLL (Shimoyama, 1991).

The pathogenesis of ATLL involves a number of factors and several disease steps. HTLV-1 replicates in a variety of cells, including immature bone marrow cells, macrophages, fibroblasts, and endothelial cells, but it transforms only CD4$^+$ T cells. HTLV-1 does not act through a viral oncogene or through activation of a particular cellular oncogene. Nevertheless, HTLV-1 integration in the leukemic cells of an individual patient is monoclonal, but at random, thus indicating that ATLL arises from a single virus-infected clone. Increased spontaneous proliferation is observed when T cells from HTLV-1 infected individuals are cultured in vitro. With time, the proliferation becomes independent of exogenously added IL-2. Freshly isolated T cells of HTLV-1 infected individuals also exhibit an activated phenotype characterized by high expression of IL-2Rα, HLA-DR, and ICAM-1. T-cell activation, as evidenced by these features, may be caused by several different effects induced by viral protein(s). The HTLV-1 SU protein (gp46) exerts a mitogenic effect on resting T cells that is independent of antigen presentation by accessory cells (reviewed in Green and Chen, 1994). Tax transactivates a number of cellular proteins, in particular the IL-2Rα chain, which is constitutively expressed in all HTLV-1-infected cells and whose expression is regulated by NF-κB. In resting T cells NF-κB is localized in the cytoplasm and bound to an inhibitor. Tax apparently displaces this inhibitor, which allows NF-κB to migrate into the nucleus and to activate the promoter of the IL-2Rα gene. The IL-2Rα chain is the high-affinity IL-2 binding component of the IL-2R complex, which contains in addition the IL-2Rβ and IL-2Rγ chains. The binding of IL-2 to the IL-2Rα chain causes a heterodimerization of the IL-2Rβ and IL-2Rγ chains, which induces a pathway leading to cell signaling. Signaling may be potentiated by binding of the X-region-encoded p12 to the IL-2Rβ and IL-2Rγ chains. Tax also transactivates the GM-CSF gene. Increased release of GM-CSF may activate macrophages. IL-1 and other cytokines secreted by these cells may further stimulate T-cell replication.

Together, these mechanisms lead to the development of a permanent IL-2-dependent proliferation of CD4$^+$ T cells with polyclonally integrated provirus that is characteristic of all HTLV-1-infected individuals. Tax-mediated inappropriate expression of cellular oncogenes, such as c-*fos*, *egr-1*, and *egr-2*, may also contribute to leukemogenesis. Tax-mediated effects, however, appear to be no longer relevant in leukemic cells, since *tax* mRNA is undetectable in most circulating

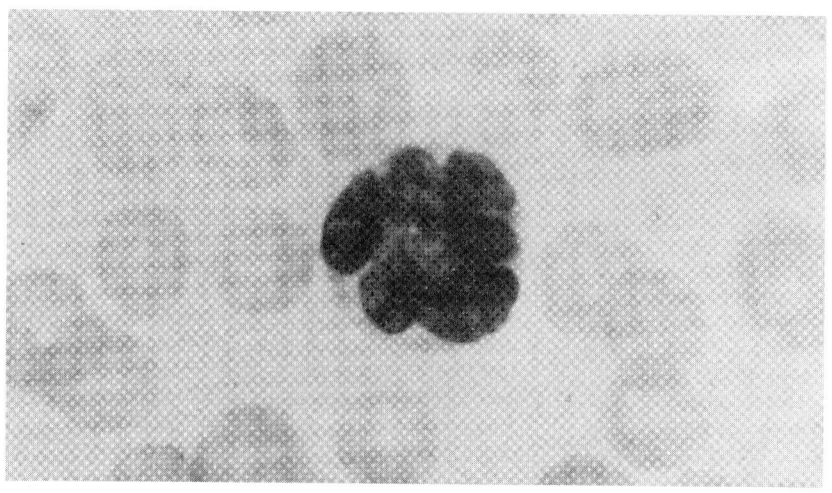

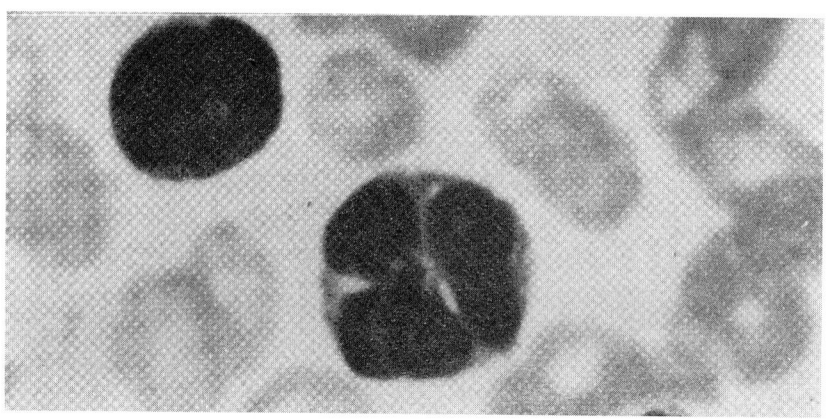

FIGURE 15 Morphology of typical neoplastic cells from the peripheral blood of an ATLL patient. Reprinted from Di Marzo Veronese et al., 1992.

ATLL cells. The next step, IL-2-independent proliferation of a small subset of HTLV-1-infected cells, as seen in smoldering ATLL, would require additional mutational changes, which may be facilitated by an impaired DNA repair resulting from downregulation of the repair enzyme β-polymerase by Tax (Philpott and Buehring, 1999). This phase may then progress from a stage of controllable monoclonal-T-cell growth to one of uncontrollable growth. Finally, full-blown ATLL would require additional genetic changes, like chromosomal translocations or mutations of the p53 tumor suppressor gene (reviewed in Franchini, 1995; Buckle et al., 1996; Yoshida, 1996; Ferreira et al., 1997; Uchiyama, 1997).

HAM-TSP

Following identification of HTLV-1 as the etiologic agent of ATLL, a serological association with HTLV-1 was independently reported for patients from the Caribbean who suffered from TSP (Gessain et al., 1985) and for patients from Japan who presented with a myelopathy (Gessain et al., 1985; M. K. Osame, K. Usuku, S. Izumo, N.

Ljichin, H. Amitani, A. Iguta, M. Matsumoto, and M. Tara, Letter, *Lancet* **i:**1031–1032, 1986). Comparative studies subsequently demonstrated that the viruses found in these diseases were genetically indistinguishable from the HTLV-1 strains found in ATLL (M. Yoshida, M. Osame, K. Usuku, M. Matsumoto, and A. Igata, Letter, *Lancet* **i:**1085–1086, 1987). Rarely, HAM-TSP and ATLL are seen together in the same patient (C. Bartholomew, F. Cleghorn, W. Charles, P. Ratan, L. Roberts, K. Maharaj, N. Jankey, H. Daisley, B. Hanchard, and W. Blattner, Letter, *Lancet* **ii:**99–100, 1986; Kawai et al., 1989). A paraparesis can be experimentally induced in rats neonatally infected with HTLV-1. About 25% of the animals infected with the Japanese prototype cell line MT-2, which was established from a patient with ATLL, developed paraparesis (Kira et al., 1997).

The lifetime risk of developing HAM-TSP among seropositive individuals varies from less than 0.1% in Japan to 1.7 to 7% reported from Africa, the Caribbean, and the United States (reviewed in Taylor, 1998). The mean age at onset is 30 to 40 years, and the male-to-female ratio is about

1:3. The incubation time extends from years to decades, but may occasionally be as short as 18 weeks (Osame et al., 1990). Disease onset is usually slow; patients often have had symptoms for years before the diagnosis is established.

Clinically, patients with HAM-TSP show spasticity of the lower limbs with hyperreflexia and a positive Babinski sign, paraparesis or paraplegia, urinary or fecal incontinence, impotence, and sensory disturbances with poorly defined thoracic sensory levels (Osame et al., Letter, 1986; Levin and Jacobson, 1997). The pathology reveals a myelopathy that primarily affects the pyramidal tracts and, to a lesser extent, the sensory system. The disease is gradually progressive with no spontaneous remissions. The diagnosis is corroborated by laboratory evidence of an infection with HTLV-1. Analysis of the cerebrospinal fluid (CSF) shows mild protein increase, modest lymphocytic pleocytosis with lymphocytes that may have lobulated nuclei, and positivity for HTLV-1 specific antibodies. Atypical lymphocytes can also be seen in the blood. Magnetic resonance imaging may show diffuse lesions of the white matter and the paraventricular region of the brain. In contrast to ATLL, HAM-TSP has a much better prognosis; the mean time to death is usually more than 10 years. Major causes of death are other complications of HTLV-1 infection, namely, infections and cancers (Gessain et al., 1985; Osame et al., Letter, 1986; Jacobson et al., 1988).

The neuropathological findings in HAM-TSP are closely correlated with the clinical and neuroradiological data (reviewed in Levin and Jacobson, 1997). The spinal cord lesions—characteristically located on the thoracic level—show extensive mononuclear infiltration in the early phase, consisting of a mixed population of CD4$^+$ T cells, CD8$^+$ T cells, B lymphocytes, and foamy macrophages. With transition to the chronic phase (>5 years), the infiltrates decrease in cell number and are dominated by CD8$^+$ T lymphocytes and, to a lesser degree, foamy macrophages, while CD4$^+$ T cells are rare and B cells are virtually absent. The spinal cord in the chronic phase is atrophic, and there is progressive thickening and fibrosis of the leptomeninges and blood vessels. Symmetrical, widespread loss of myelin and axonal dystrophy of the lateral columns is seen, particularly of the corticospinal tracts. Damage is most severe in the thoracic and lumbar regions, although the cervical region and the brainstem may also be affected.

Characteristically, the HTLV-1 proviral load in the blood of HAM-TSP patients is high: 2 to 20 proviral copies per 100 PBMC are seen in HAM-TSP compared to 0.04 to 8 copies in asymptomatic carriers (Kubota et al., 1993), and the prevalence of the disease rises exponentially with the log (proviral load) once the proviral load exceeds 1% of the PBMC (Nagai et al., 1998). The presence of HTLV-1 infected cells in the affected regions of the CNS has been documented in several studies (reviewed in Levin and Jacobson, 1997). Virus-infected CD4$^+$ T cells probably carry the virus to the CNS (Richardson et al., 1990). In the early phase of HAM-TSP, while CD4$^+$ T cells in the inflammatory infiltrates are numerous, HTLV-1 DNA sequences may be colocalized with these infiltrates. With progression into the chronic phase, this association may become less clear (Kubota et al., 1994) and may be absent in chronic lesions from which CD4$^+$ T cells have completely disappeared (Kira et al., 1992; Ohara et al., 1992). In situ hybridization and in situ PCR studies have shown that the HTLV-1 RNA in an acute case of HAM-TSP was localized to infiltrating CD4$^+$ T cells (Moritoyo et al., 1996), but it was localized to astrocytes in the chronic phase of the disease (Lehky et al., 1995).

The pathogenesis of HAM-TSP is still unclear, and several models have been proposed. The first involves elimination of HTLV-1-infected cells by CD8$^+$ CTL that invade the CNS and are predominantly directed against Tax (reviewed in Levin and Jacobson, 1997; Taylor, 1998; Bangham et al., 1999). Tax-specific CTL precursors are indeed present at high frequency in HAM-TSP patients (Shida et al., 1987; Jacobson et al., 1990), but this could also merely be in response to the high proviral load usually found in them. Tax-specific CTL are also easily detectable in asymptomatic carriers, which renders the role of CTL in pathogenicity unlikely (Daenke et al., 1996). More recently, it has been demonstrated that CTL exert a positive selection pressure upon the *tax* gene, as evidenced by the occurrence of *tax* CTL escape mutants. There is evidence for an even higher degree of selection in asymptomatic carriers than in patients with HAM-TSP, suggesting that a vigorous Tax-specific CTL response may actually protect against the disease (Niewiesk and Bangham, 1996). This is particularly likely to be the case if an HLA-A2 allele is present, which has been found associated with a high CTL response and especially efficient killing of HTLV-1-infected cells: bearers of the HLA-A2 allele are significantly more frequent among healthy carriers of the virus than among HAM-TSP patients and have a lower provirus load (Bangham et al., 1999; Jeffery et al., 1999; Wodarz et al., 1999). The second model of pathogenesis involves an HTLV-1-induced autoimmune response against a CNS antigen (reviewed by Oger and Dekaban, 1995; Kompoliti et al., 1996). In accordance with this model, a CD4$^+$-T-cell clone that is directed against an epitope present in normal spinal cord tissue has been isolated from a patient with HAM-TSP (Nagai et al., 1996). In the third model, Tax-specific CTL invading the CNS and killing virus-infected cells would release proinflammatory cytokines that are cytotoxic to nearby glial cells (Biddison et al., 1997). Such a "bystander effect" appears to be the currently favored model of pathogenesis (Bangham et al., 1999).

HAU

In 1992, a uveitis of HTLV-1-infected patients of otherwise unexplained etiology was proposed as a separate disease entity (Mochizuki et al., 1994; M. Mochizuki, T. Watanabe, K. Yamaguchi, K. Tajima, K. Yoshimura, S. Nakashima, M. Shirao, S. Araki, N. Miyata, S. Mori, et al., Letter, *J. Infect. Dis.* **166**:943–944, 1992). These patients present with blurred vision with acute or subacute onset, preserved visual acuity in most instances, iritis, vitreous opacities, retinal vasculitis with exudates, and hemorrhages. The ages of patients are similar to those of patients with with HAM-TSP. The condition usually responds well to ocular or oral administration of corticosteroids, although recurrence is observed frequently.

Proviral DNA of HTLV-1 can be demonstrated in about 60% of T-cell clones established from the intraocular fluid of such patients, which suggests that the intraocular presence of HTLV-1-infected T lymphocytes is due to positive selection and is not the result of a mere coincidence. These T cells are activated and release a number of inflammatory cytokines, which are deemed responsible for the disease, since they can be abrogated by corticosteroid therapy (Sagawa et al., 1995). The condition frequently seems to be associated with a history of Graves' disease (Yamaguchi et al., 1994; Watanabe et al., 1997).

HAAP

A chronic arthropathy developing in HTLV-1-infected individuals was first described in 1989 (K. Nishioka, I. Maruyama, K. Sato, I. Kitajima, Y. Nakajima, and M. Osame, Letter, *Lancet* i:441, 1989). HTLV-1-specific antibodies can be shown in synovial fluids of the affected joints, and proviral DNA can be demonstrated in the synovial tissues and in synovial fluid lymphocytes (Kitajima et al., 1991).

Evidence that the arthropathy of HTLV-1-infected persons is not due to a mere coincidence with a disease that is frequent in most countries includes the demonstration that mice that are transgenic for the HTLV-1 *env*-pX region develop chronic inflammatory lesions that are manifested by swelling of the ankle combined with redness and/or swelling of the footpad near the ankle. Histologically, these lesions closely resemble the changes observed in human rheumatoid arthritis (Iwakura et al., 1991). Autoimmunity appears to play a major role in the disease development of this model (Iwakura et al., 1995). A similar, immune-mediated disease can also be induced by the infection of neonatal rats with certain isolates of HTLV-1 (Kira et al., 1997).

HTLV-1-Associated Bronchopneumopathy

Pulmonary complications are more frequent in patients with ATLL and lymphoma than in patients with other hematological malignancies. Pulmonary complications of ATLL are present in more than 90% of patients and include, in addition to infiltration by leukemic cells, opportunistic infections due to *P. carinii*, cytomegalovirus, fungi, or tubercle bacilli. Patients with HAM-TSP or HAU also have frequent pulmonary complications characterized by T-lymphocytic alveolitis in the absence of leukemic cells or opportunistic pathogens, and a similar pulmonary involvement is also detectable in asymptomatic carriers. Characteristically, respiratory symptoms and chest radiographic abnormalities are rarely found in these patients, although 60 to 80% have T-lymphocytic bronchiolitis, alveolitis, or interstitial pneumonia. Chronic sinusitis is also frequent, especially in patients with diffuse panbronchiolitis. Infiltrates consist of CD3+, CD8+, or CD4+ T lymphocytes. For pathogenesis, mechanisms similar to those discussed for HAM-TSP are proposed, namely, a CTL effect and an inflammatory effect exerted by release of cytokines by activated cells (reviewed in Sugimoto et al., 1998).

Possible, Less Likely, and Unproven Associations of HTLV with Other Diseases

In a large study of the most common hematological diseases in Europe (n = 730, plus 210 controls), HTLV-1 infection was reliably detected (by serology backed up by PCR for *tax* and *pol* in ≥3 samples for each positive patient) in 11 of 67 (17%) patients with a myelodysplastic syndrome (a neoplasia of myeloid cells), 1 of 26 patients with T-cell non-Hodgkins lymphoma, and 1 of 1 patient with T-ALL (Karlic et al., 1997). In contrast, claims that cutaneous T-cell lymphomas (CTCL) are associated with a presence of the HTLV-1 *tax* gene in the absence of antibodies to structural proteins (Manca et al., 1994; Pancake et al., 1995, 1996) could not be confirmed by other groups even when CTCL patients from regions where HTLV-1 is endemic or from the same country from which positive reports originated were analyzed (Boni et al., 1996a; Bazarbachi et al., 1997; Kikuchi et al., 1997; Wood et al., 1997). One of the groups reporting a Tax association of CTCL reported that relatives of Tax-positive CTCL patients were also Tax positive and that the prevalence of HTLV-1 infection in drug addicts, and even in blood donors, as detected by serology using a recombinant Tax antigen and PCR far exceeded the prevalence based on conventional serology. These reports, however, have not been substantiated by other groups and are in our view suggestive of technical problems affecting the specificity of the tests rather than of true geographical differences in the etiologies of these malignancies (Zucker-Franklin et al., 1997; Zucker-Franklin and Pancake, 1998). Similarly, claimed associations of HTLV-1 with multiple sclerosis (Koprowski et al., 1985; Greenberg et al., 1989) or amyotrophic lateral sclerosis (W. K. Engel, Letter, *Ann. Neurol.* 30:431–433, 1991; Ferrante et al., 1995) have not been confirmed (Ehrlich et al., 1991; J. C. Winkelmann, Editorial, *J. Lab. Clin. Med.*, 122:230–231, 1993).

HTLV-2-associated disease. HTLV-2 was originally isolated from a T-cell line (Mo-T) derived from a patient with a T-cell variant of hairy-cell leukemia (Kalyanaraman et al., 1982). Subsequently, the virus was also isolated in a similar case which upon closer examination demonstrated the coexistence of two proliferative processes, namely, a CD8+-T-cell leukemia with monoclonally integrated HTLV-2 and a B-cell hairy-cell leukemia that was negative for integrated HTLV-2 (Rosenblatt et al., 1986). A pathogenetic role of HTLV-2 in these and other malignancies involving CD8+ T cells, in which the virus has occasionally been found, is supported by tropism of the virus for CD8+ T cells in vitro (Ijichi et al., 1992) and the fact that lymphocytes of HTLV-2-infected patients proliferate in vitro in the absence of antigenic stimulation, similar to cells of HTLV-1-infected individuals (Wiktor et al., 1991). Epidemiological studies have confirmed that the typical form of B-cell hairy-cell leukemia is not associated with HTLV-2 (Hjelle et al., 1991). HTLV-2 has also been detected in a few cases of a spastic myelopathy with variable degrees of ataxia. In addition, a neurological syndrome resembling HAM-TSP has been described in isolated cases, sometimes as a coinfection with HIV-1 (reviewed in Ferreira et al., 1997).

Epidemiology and Transmission of the HTLVs

Foci of HTLV-1 infection are found geographically clustered, amounting to several million individuals worldwide. Endemicity is present in southern and northern Japan, the Caribbean, parts of Africa, the Middle East, South America, the Pacific Melanesian Islands, and Papua New Guinea. HTLV is also prevalent in individuals originating from these areas who have immigrated to areas of low prevalence, such as North America or Europe. HTLV-1 prevalence ranges from 3 to 6% in Trinidad, Jamaica, and other Caribbean islands to 30% in rural Miyazaki in southern Japan. Among low-risk populations in the United States and Europe, the prevalence is less than 1%. Within a given population, the seroprevalence increases with age and is twice as high in women as in men. As discussed above in more detail, HTLV-2 is endemic in Amerindian peoples of the Americas. Similar to HIV-1, HTLV-1 and -2 have been introduced more recently among homosexual men and intravenous drug users of the United States and Europe (reviewed in Gessain and de The, 1996a; Manns et al., 1999).

Like the HIVs, the HTLVs are transmitted by sexual intercourse, by mother-child transmission, and by parenteral inoculation. Among sexually transmitted cases, male-to-female transmission predominates over female-to-male transmission owing to the larger number of infected T cells present in semen than in vaginocervical secretions. Both HTLV-1 and HTLV-2 are also transmitted among homosexual men. Mother-to-child transmission occurs in 18 to 30% of infected women, a frequency similar to that of HIV-1 infection, and occurs both connatally and postnatally through breast milk. The latter mode of transmission seems to be more efficient than for HIV-1 and occurs with a time- or dose-dependent frequency. In one study, overall transmission was 16%. It was 5% among infants breast-fed for less than 3 months after birth and 27% among infants breast-fed for 3 months or longer. Of 78 bottle-fed infants, 13% proved to be infected, suggesting connatal transmission (Hirata et al., 1992). In other studies, connatal transmission was much less frequent (3%). HTLV-1-infected cells were detected by PCR in 2.5% of the cord blood samples from HTLV-1-positive pregnancies, but this was not associated with infection if the babies were formula fed (Hino et al., 1996). Transmission by blood products is, in contrast to HIV, strictly cell associated; the virus is not transmitted by plasma or plasma-derived products. Recipients of contaminated blood seroconvert with a 40 to 60% probability, and the median seroconversion time is estimated at 51 days (Manns et al., 1999). Both HTLV-1 and HTLV-2 (especially subtype A) have, like HIV, been spread among intravenous drug users (reviewed by Gessain and de Thé, 1996b; Ferreira et al., 1997).

Diagnosis of HTLV Infections

The principles, tools, and problems of HTLV diagnosis are, with some modifications, the same as those for the diagnosis of HIV (see above). Screening is based on tests for HTLV-specific antibodies by ELISA or particle agglutination tests. Confirmatory tests are based on WB. Since there are many indeterminate Western blots, confirmation must be backed by supplemental tests, such as the direct demonstration of the virus, either by PCR for viral DNA or by the more time-consuming virus isolation evaluated by demonstration of HTLV antigen. Tests for virus components (antigen or viral RNA) in plasma have not been used for diagnostic purposes, although the presence of p24 antigen in circulating immune complexes in sera of ATLL patients or carriers (Schupbach et al., 1984c; Schupbach and Kalyanaraman, 1989) and in the CSF of patients with chronic progressive myelopathy has been described. In the latter group, all seven investigated patients were antigen positive, with concentrations in the range of hundreds of nanograms per milliliter of CSF (Bhagavati et al., 1988).

Screening Tests for HTLV-1 and -2

For HTLV-1 and -2 screening, tests analogous to those reviewed in Fig. 9 for the detection of HIV are used, with the same principal advantages and drawbacks. Since all subtypes of HTLV-1 diverge less than 10% from each other, underdetection of these viruses owing to sequence diversity is not a problem. Since certain regions, for example, the CA region of gag, exhibit a high degree of sequence similarity, there is broad serological cross-reactivity between HTLV-1 and HTLV-2. Due to the low virus particle load, antibody titers may be low. Thus, third-generation tests (double-antigen sandwich assays and particle agglutination) are the test formats that should yield the highest sensitivity. Tests based on ELISA, particle agglutination, or indirect immunofluorescence are available from various commercial companies (Defer et al., 1995; Taylor et al., 1996; Farias de Carvalho et al., 1997; Hale et al., 1997; Vrielink et al., 1999).

Supplemental Tests for HTLV-1 and -2

WB

Serological confirmation of HTLV-1 and -2 infection requires the demonstration of antibodies to both *gag* (p24) and *env* (gp46 and/or gp68 protein) by WB and/or radioimmunoprecipitation assay (RIPA) (Anonymous, 1993b). In our view, WB is preferable to in-house methods of confirmation like RIPA, since the commercial WB kits are better standardized and much easier to run. WB kits are provided by a number of companies. Of particular interest are strips that contain, in addition to the viral proteins derived from viral lysate, recombinant proteins representing TM of HTLV-1 (which, due to high homology, is also detected by antibodies from individuals infected with HTLV-2) and type-specific SU (gp46) of HTLV-1 and -2. This increases the sensitivity, since the concentration of the gp46 Env protein relative to that of the gp68 Env protein on the strips is usually low in kits derived from lysate alone. Based on the pattern of reactivity, it is frequently, though not always, possible to decide whether infection by HTLV-1 or HTLV-2 is present. With such strips, intense reaction with the *gag* proteins p19 and p24 and the *env* recombinant proteins rgp21 and rgp46 of HTLV-1 satisfies the criteria for positivity for HTLV-1 and intense reaction with p24, rgp21, and rgp46 of HTLV-2 satisfies the criteria for positivity for HTLV-2 (Medrano et al., 1997). Still, in many cases WB does not permit an unequivocal diagnosis because there is a high percentage of samples with indeterminate results due to reaction with lysate-derived natural Gag proteins, but also with recombinant Env proteins (Fig. 16). Some of these indeterminates may have very intense reactions with p19 and a variety of larger proteins, such as p26, p28, p32, p36, p45, and p53. Proteins of similar size present in HTLV-1-infected cells were all determined to contain either a p19 or a p24 moiety or both (Schupbach and Kalyanaraman, 1989). Reaction with several of these various proteins thus may signify no more than reaction (though strongly) with an epitope of p19 or p24 alone (Schupbach et al., 1988). Sometimes, such intense *gag* patterns may also be combined with weak reaction to Env rgp21 and/or rgp46[I] or rgp46[II], but even this does not necessarily imply an infection with an HTLV. Weak reactions with recombinant Env proteins in all possible combinations may also be found in the absence of reaction with Gag proteins. Line immunoassays based entirely on recombinant proteins of both HTLV-1 and -2 may sometimes be of some help (Sabino et al., 1999).

In many instances, however, confirmation by PCR and/or virus isolation is necessary. This is particulary true for geographical areas or populations in which these viruses are not endemic. Under such conditions, any suggestive serological result not strongly antibody positive should be subjected to PCR and/or virus isolation. Indeterminate reactions, however, are also frequent in populations at risk for exposure, as in drug addicts (Medrano et al., 1997). On the other hand, serologically indeterminate results may also be present in truly infected individuals (Zehender et al., 1996; Caterino-de-Araujo et al., 1998).

PCR

PCR analysis for HTLV-1 and/or HTLV-2 DNA is necessary for all serologically indeterminate results in which antibody reaction to Env proteins (rgp21, rgp46[I], or rgp46[II]) is present. Antibody reaction with the Gag proteins p19 and/or p24 alone, or rgp21 alone, has been found by PCR not to be associated with HTLV infection (Defer et al.,

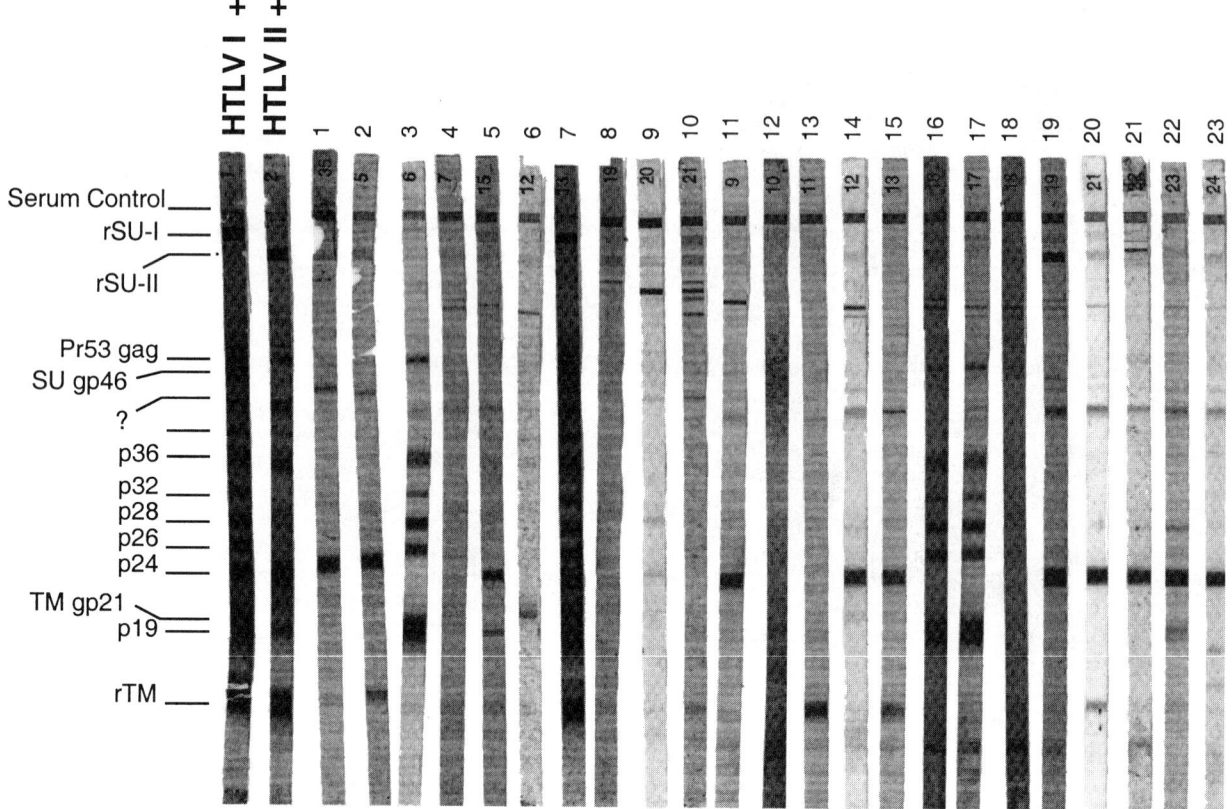

FIGURE 16 Positive and indeterminate HTLV-1 and -2 Western blots. The strips labeled "+" represent positive controls. Reaction with the recombinant surface envelope proteins (rSU) of the respective viruses, the recombinant transmembrane protein (rTM), and Gag proteins p19 and p24 is required. Among samples 1 to 23, only no. 7 was confirmed by PCR (HTLV-1). Note the strong reactions of PCR-negative samples 16 and 17 with p19 and various Gag precursors and the faint reaction, especially of 16, with rSU-I and rSU-II. Note also the patterns with antibodies to p24 and to rSU-I and/or -II.

1995). PCR is performed on Ficoll-purified PBMC and frequently uses a sequence of *tax* that is conserved for both HTLV-1 and HTLV-2 which is amplified by primers designated SK43 and SK44 and detected by probe SK45 (Kwok et al., 1988, 1990). Differentiation of HTLV-1 and HTLV-2 in samples positive for these oligonucleotide sequences is then achieved by amplification of a region in *pol* by primers SK110 and SK111 followed by hybridization of probe SK112, which is specific for HTLV-1- or HTLV-2-specific SK188. This system is also available as a commercial kit (Amplicor HTLV 1-2 PCR test; Roche Diagnostic Systems).

RIPA

RIPA is a complex procedure that is difficult to standardize; it is suitable for a research laboratory but is not recommended for a diagnostic laboratory. In brief, HTLV-1 or HTLV-2 producer cell lines are metabolically labeled with [^{35}S]cysteine. Lysate of cells (which contains relatively high concentrations of the Env precursor gp61-gp68 as well as of Tax, both not present in viral lysate) is incubated with a small volume of a patient's serum. Immune complexes are then immobilized on protein A and resolved by SDS-PAGE followed by autoradiography at −70°C (Defer et al., 1995).

Virus Isolation

In comparison to PCR, virus isolation is more time-consuming and probably overall more expensive for the laboratory. However, the procedure is still justified in view of a possible presence of unknown retroviruses serologically related to HTLV that might be responsible for some of the many serologically indeterminate results. Virus isolation should thus be evaluated not only with HTLV-specific reagents (antigen assay or immunofluorescence), but also by the PERT assay for particle-associated RT (see "Supplemental Testing—Virus Components" in the section on HIV above). For virus isolation, Ficoll-purified PBMC are cocultured with phytohemagglutinin-preactivated normal PBMC or cord blood leukocytes in an IL-2 containing medium. The supernatant is analyzed twice weekly for RT by PERT assay. In cases of HTLV infection, the PERT assay usually becomes positive within a few days, while detection of RT by conventional assays may take several weeks. When the PERT assay has become positive, tests for HTLV p24 using an antigen assay can be performed. Differentiation of HTLV-1 and HTLV-2 growing in culture is probably best achieved by PCR. RT positivity combined with nonreactivity in PCR for *tax* in such a situation would suggest the presence of a novel virus serologically related to HTLV. The use of generic PCR primers and probes capable of detecting all the various primate T-lymphotropic retroviruses may be helpful in such a situation (Vandamme et al., 1997).

Treatment of HTLV Infection

In contrast to HIV, infection by HTLV is not treated, as long as it is asymptomatic. Since the virus appears to replicate by mitosis of infected cells rather than by production of infectious particles, a benefit from typical blockers of the retroviral replication cycle, such as inhibitors of RT or PR, remains doubtful. Among a variety of protocols, prednisolone, danazol (an attenuated androgen), and IFN-α have proven effective in HAM-TSP, although the effect may be transient in many cases (reviewed by Taylor, 1998). For ATLL, multiple chemotherapy combinations have been tried with disappointing results. Although complete remission could be induced in 43% of patients treated with an eight-drug combination, there was no benefit for the overall survival time, and the median remained unchanged at 8 months. Surprisingly, a combination of IFN-α and zidovudine seems to be able to induce complete remission in some patients, even some who had not responded to classical combination chemotherapy. Since ATLL is a monoclonal disease that does not depend on viral replication, it is unlikely that zidovudine in this case acts by its RT-inhibitory effect; an antineoplastic mechanism is more likely. Another rationale for therapy now being evaluated in clinical trials involves ^{90}Y-conjugated monoclonal antibody to the IL-2Rα chain that is highly expressed on all ATLL cells. By this or by the use of toxin-conjugated IL-2, ATLL cells can be relatively selectively targeted and eliminated. Other therapeutic approaches have included the use of IFN-α, and -γ, but the overall effect is disappointing. ATLL remains a disease with poor prognosis, but prevention of the disease is promising. Refraining from breast-feeding and the introduction of HTLV testing of all blood donors have led to a significant decline of the carrier rate among the younger generation in Japan (reviewed in Takatsuki et al., 1996; Ferreira et al., 1997; Pawson et al., 1998).

REFERENCES

Albright, A. V., J. Strizki, J. M. Harouse, E. Lavi, M. O'Connor, and F. Gonzalez-Scarano. 1996. HIV-1 infection of cultured human adult oligodendrocytes. *Virology* **217:**211–219.

Alkhatib, G., C. Combadiere, C. C. Broder, Y. Feng, P. E. Kennedy, P. M. Murphy, and E. A. Berger. 1996. CC CKR5: a RANTES, MIP-1alpha, MIP-1beta receptor as a fusion cofactor for macrophage-tropic HIV-1. *Science* **272:**1955–1958.

Amerongen, H. M., R. Weltzin, C. M. Farnet, P. Michetti, W. A. Haseltine, and M. R. Neutra. 1991. Transepithelial transport of HIV-1 by intestinal M cells: a mechanism for transmission of AIDS. *J. Acquir. Immune Defic. Syndr.* **4:**760–765.

Anonymous. 1989. Interpretation and use of the western blot assay for serodiagnosis of human immunodeficiency virus type 1 infections. *Morb. Mortal. Wkly. Rep.* **38:**1–7.

Anonymous. 1990. Acquired immunodeficiency syndrome (AIDS). Proposed WHO criteria for interpreting results from western blot assays for HIV-1, HIV-2, and HTLV-I/HTLV-II. *Wkly. Epidemiol. Rec.* **65:**281–283.

Anonymous. 1991. *Biosafety Guidelines for Diagnostic and Research Laboratories Working with HIV.* World Health Organization, Geneva, Switzerland.

Anonymous. 1993a. Centers for Disease Control and Prevention and the USPHS Working Group: guidelines for counselling persons infected with human T-lymphotropic virus type I (HTLV-I) and type II (HTLV-II). *Ann. Intern. Med.* **118:**448–454.

Anonymous. 1993b. From the Centers for Disease Control and Prevention. False-positive serologic tests for human T-cell lymphotropic virus type I among blood donors following influenza vaccination, 1992. *JAMA* **269:**2076–2078.

Anonymous. 1995. Case-control study of HIV seroconversion in health-care workers after percutaneous exposure to HIV-infected blood—France, United Kingdom, and United States, January 1988–August 1994. *Morb. Mortal. Wkly. Rep.* **44:**929–933.

Anonymous. 1997. Nonhuman primate spumavirus infections among persons with occupational exposure—United States, 1996. *Morb. Mortal. Wkly. Rep.* **46:**129–131.

Autran, B., G. Carcelain, T. S. Li, C. Blanc, D. Mathez, R. Tubiana, C. Katlama, P. Debre, and J. Leibowitch. 1997. Positive effects of combined antiretroviral therapy on CD4+ T cell homeostasis and function in advanced HIV disease. *Science* **277:**112–116.

Baba, T. W., A. M. Trichel, L. An, V. Liska, L. N. Martin, M. Murphey-Corb, and R. M. Ruprecht. 1996. Infection and AIDS in adult macaques after nontraumatic oral exposure to cell-free SIV. *Science* **272:**1486–1489.

Bangham, C. R., S. E. Hall, K. J. Jeffery, A. M. Vine, A. Witkover, M. A. Nowak, D. Wodarz, K. Usuku, and M. Osame. 1999. Genetic control and dynamics of the cellular immune response to the human T-cell leukaemia virus, HTLV-I. *Phil. Trans. R. Soc. Lond. B* **354:**691–700.

Barin, F., A. M. Courouce, J. Pillonel, L. Buzelay, J. Baudelot, C. Botte, S. Chamaret, F. Durand, M. H. Elghouzzi, J. M. Lemaire, P. Maisonneuve, M. Maniez, P. Moncharmont, L. Noel, M. L. North, Y. Piquet, C. Rouzioux, W. Smilovici, and C. Tirtaine. 1997. Increasing diversity of HIV-1(M) Serotypes in French blood donors over a 10-year period (1985–1995). *AIDS* **11:**1503–1508.

Barnhart, H. X., M. B. Caldwell, P. Thomas, L. Mascola, I. Ortiz, H. W. Hsu, J. Schulte, R. Parrott, Y. Maldonado, and R. Byers. 1996. Natural history of human immunodeficiency virus disease in perinatally infected children: an analysis from the Pediatric Spectrum of Disease Project. *Pediatrics* **97:**710–716.

Barre-Sinoussi, F., J. C. Chermann, F. Rey, M. T. Nugeyre, S. Chamaret, J. Gruest, C. Dauguet, C. Axler-Blin, F. Vezinet-Brun, C. Rouzioux, W. Rozenbaum, and L. Montagnier. 1983. Isolation of a T-lymphotropic retrovirus from a patient at risk for acquired immune deficiency syndrome (AIDS). *Science* **220:**868–871.

Bazarbachi, A., V. Soriano, R. Pawson, A. Vallejo, T. Moudgil, E. Matutes, J. Peries, A. Molina, H. de The, T. F. Schulz, D. Catovsky, and P. S. Gill. 1997. Mycosis fungoides and Sezary syndrome are not associated with HTLV-I infection: an international study. *Br. J. Haematol.* **98:**927–933.

Berger, E. A., R. W. Doms, E. M. Fenyo, B. T. M. Korber, D. R. Littman, J. P. Moore, Q. J. Sattentau, H. Schuitemaker, J. Sodroski, and R. A. Weiss. 1998. A new classification for HIV-1. *Nature* **391:**240.

Bhagavati, S., G. Ehrlich, R. W. Kula, S. Kwok, J. Sninsky, V. Udani, and B. J. Poiesz. 1988. Detection of human T-cell lymphoma/leukemia virus type I DNA and antigen in spinal fluid and blood of patients with chronic progressive myelopathy. *N. Engl. J. Med.* **318:**1141–1147.

Biberfeld, P., K. J. Chayt, L. M. Marselle, G. Biberfeld, R. C. Gallo, and M. E. Harper. 1986. HTLV-III expression in infected lymph nodes and relevance to pathogenesis of lymphadenopathy. *Am. J. Pathol.* **125:**436–442.

Biddison, W. E., R. Kubota, T. Kawanishi, D. D. Taub, W. W. Cruikshank, D. M. Center, E. W. Connor, U. Utz, and S. Jacobson. 1997. Human T cell leukemia virus type I (HTLV-I)-specific CD8+ CTL clones from patients with HTLV-I-associated neurologic disease secrete proinflammatory cytokines, chemokines, and matrix metalloproteinase. *J. Immunol.* **159:**2018–2025.

Bleul, C. C., M. Farzan, H. Choe, C. Parolin, I. Clark-Lewis, J. Sodroski, and T. A. Springer. 1996. The lymphocyte chemoattractant SDF-1 is a ligand for LESTR/fusin and blocks HIV-1 entry. *Nature* **382:**829–833.

Bohannon, R. C., L. A. Donehower, and R. J. Ford. 1991. Isolation of a type D retrovirus from B-cell lymphomas of a patient with AIDS. *J. Virol.* **65:**5663–5672.

Boni, J. 1996. PCR detection of HIV. *Methods Mol. Biol.* **50:**93–107.

Boni, J., and J. Schupbach. 1997. Reverse transcriptase assay based on product enhancement for drug susceptibility assays. *In* D. Kinchington and R. F. Schinazi (ed.), *Antiviral Evaluation.* Humana Press, Totowa, N.J.

Boni, J., H. Pyra, and J. Schupbach. 1996a. Sensitive detection and quantification of particle-associated reverse transcriptase in plasma of HIV-1-infected individuals by the product-enhanced reverse transcriptase (PERT) assay. *J. Med. Virol.* **49:**23–28.

Boni, J., M. Opravil, Z. Tomasik, M. Rothen, K. Bisset, P. Grob, R. Lüthy, and J. Schupbach. 1997. Simple monitoring of antiretroviral therapy with a signal-amplification-boosted HIV-1 P24 antigen assay with heat-denatured plasma. *AIDS* **11:**F47–F52.

Boni, J., H. Pyra, M. Gebhardt, L. Perrin, P. Bürgisser, L. Matter, W. Fierz, P. Erb, J. C. Piffaretti, E. Minder, P. Grob, J. J. Burckhardt, M. Zwahlen, J. Huder, and J. Schüpbach. 1999. High frequency of non-B subtypes in newly diagnosed HIV-1 infections in Switzerland. *J. Acquir. Immune Defic. Syndr. Hum. Retrovirol.* **22:**174–179.

Boni, R., A. Davis-Daneshfar, G. Burg, D. Fuchs, and G. S. Wood. 1996b. No detection of HTLV-I proviral DNA in lesional skin biopsies from Swiss and German patients with cutaneous T-cell lymphoma. *Br. J. Dermatol.* **134:**282–284.

Borrow, P., H. Lewicki, X. Wei, M. S. Horwitz, N. Peffer, H. Meyers, J. A. Nelson, J. E. Gairin, B. H. Hahn, M. B. Oldstone, and G. M. Shaw. 1997. Antiviral pressure exerted by HIV-1-specific cytotoxic T lymphocytes (CTLs) during primary infection demonstrated by rapid selection of CTL escape virus. *Nat. Med.* **3:**205–211.

Brodie, S., and P. Sax. 1997. Novel approaches to HIV antibody testing. *AIDS Clin. Care* **9:**1–6.

Buckle, G. J., D. A. Hafler, and P. Hollsberg. 1996. HTLV-I-induced T-cell activation. *J. Acquir. Immune Defic. Syndr. Hum. Retrovirol.* **13:**S107–S113.

Burgard, M., M. J. Mayaux, S. Blanche, A. Ferroni, M. L. Guihard-Moscato, M. C. Allemon, N. Ciraru-Vigneron, G. Firtion, C. Floch, F. Guillot, et al. 1992. The use of viral culture and p24 antigen testing to diagnose human immunodeficiency virus infection in neonates. The HIV Infection in Newborns French Collaborative Study Group. *N. Engl. J. Med.* **327:**1192–1197.

Busch, M. P., and G. A. Satten. 1997. Time course of viremia and antibody seroconversion following human immunodeficiency virus exposure. *Am. J. Med.* **102:**117–124. (Discussion, 125–126).

Busch, M. P., L. L. Lee, G. A. Satten, D. R. Henrard, H. Farzadegan, K. E. Nelson, S. Read, R. Y. Dodd, and L. R. Petersen. 1995. Time course of detection of viral and serologic markers preceding human immunodeficiency virus type 1 sero-

conversion: implications for screening of blood and tissue donors. *Transfusion* **35:**91–97.

Byrne, B. C., J. J. Li, J. Sninsky, and B. J. Poiesz. 1988. Detection of HIV-1 RNA sequences by in vitro DNA amplification. *Nucleic Acids Res.* **16:**4165.

Caldwell, C. J., W. Y. Barrett, J. Breuer, S. F. Farmer, and M. Swash. 1996. HTLV-1 polymyositis. *Neuromuscul. Disord.* **6:**151–154.

Caldwell, M. B., M. J. Oxtoby, R. J. Simonds, M. L. Lindegren, and M. F. Rogers. 1994. 1994 Revised classification system for human immunodeficiency virus infection in children less than 13 years of age. *Morb. Mortal. Wkly. Rep.* **43:**1–10.

Cameron, P. U., R. L. Dawkins, J. A. Armstrong, and E. Bonifacio. 1987. Western blot profiles, lymph node ultrastructure and viral expression in HIV-infected patients: a correlative study. *Clin. Exp. Immunol.* **68:**465–478.

Cameron, P. U., P. S. Freudenthal, J. M. Barker, S. Gezelter, K. Inaba, and R. M. Steinman. 1992. Dendritic cells exposed to human immunodeficiency virus type-1 transmit a vigorous cytopathic infection to CD4+ T cells. *Science* **257:**383–387. (Erratum, **257:**1848.)

Cann, A. L., and I. S. Y. Chen. 1996. Human T-cell leukemia virus types I and II, p. 1849–1880. *In* B. N. Fields, D. M. Knipe, and P. M. Howley (ed.), *Virology*, 3rd ed. Lippincott-Raven Publishers, Philadelphia, Pa.

Carre, N., L. Meyer, F. Boufassa, C. Deveau, A. Persoz, C. Rouzioux, and J. L. Vilde. 1996. High risk of HIV disease progression after infection through a sexual partner with AIDS. *AIDS* **10:**77–80.

Carrillo, A., K. D. Stewart, H. L. Sham, D. W. Norbeck, W. E. Kohlbrenner, J. M. Leonard, D. J. Kempf, and A. Molla. 1998. In vitro selection and characterization of human immunodeficiency virus type 1 variants with increased resistance to ABT-378, a novel protease inhibitor. *J. Virol.* **72:**7532–7541.

Caterino-de-Araujo, A., E. de los Santos-Fortuna, M. C. Meleiro, J. Suleiman, M. L. Calabro, A. Favero, A. De Rossi, and L. Chieco-Bianchi. 1998. Sensitivity of two enzyme-linked immunosorbent assay tests in relation to western blot in detecting human T-cell lymphotropic virus types I and II infection among HIV-1 infected patients from Sao Paulo, Brazil. *Diagn. Microbiol. Infect. Dis.* **30:**173–182.

Cavert, W., D. W. Notermans, K. Staskus, S. W. Wietgrefe, M. Zupancic, K. Gebhard, K. Henry, Z. Q. Zhang, R. Mills, H. McDade, J. Goudsmit, S. A. Danner, and A. T. Haase. 1997. Kinetics of response in lymphoid tissues to antiretroviral therapy of HIV-1 infection. *Science* **276:**960–964.

Cheynier, R., S. Henrichwark, F. Hadida, E. Pelletier, E. Oksenhendler, B. Autran, and S. Wain-Hobson. 1994. HIV and T cell expansion in splenic white pulps is accompanied by infiltration of HIV-specific cytotoxic T lymphocytes. *Cell* **78:**373–387.

Chun, T. W., L. Carruth, D. Finzi, X. Shen, J. A. DiGiuseppe, H. Taylor, M. Hermankova, K. Chadwick, J. Margolick, T. C. Quinn, Y. H. Kuo, R. Brookmeyer, M. A. Zeiger, P. Barditch-Crovo, and R. F. Siliciano. 1997. Quantification of latent tissue reservoirs and total body viral load in HIV-1 infection. *Nature* **387:**183–188.

Clark, S. J., M. S. Saag, W. D. Decker, S. Campbell-Hill, J. L. Roberson, P. J. Veldkamp, J. C. Kappes, B. H. Hahn, and G. M. Shaw. 1991. High titers of cytopathic virus in plasma of patients with symptomatic primary HIV-1 infection. *N. Engl. J. Med.* **324:**954–960.

Clavel, F., D. Guetard, F. Brun-Vezinet, S. Chamaret, M. A. Rey, M. O. Santos-Ferreira, A. G. Laurent, C. Dauguet, C. Katlama, C. Rouzioux, et al. 1986a. Isolation of a new human retrovirus from West African patients with AIDS. *Science* 233:343–346.

Clavel, F., M. Guyader, D. Guetard, M. Salle, L. Montagnier, and M. Alizon. 1986b. Molecular cloning and polymorphism of the human immune deficiency virus type 2. *Nature* 324:691–695.

Cocchi, F., A. L. De Vico, A. Garzino-Demo, S. K. Arya, R. C. Gallo, and P. Lusso. 1995. Identification of RANTES, MIP-1 alpha, and MIP-1 beta as the major HIV-suppressive factors produced by CD8+ T cells. *Science* 270:1811–1815.

Coffin, J. M. 1992. Genetic diversity and evolution of retroviruses. *Curr. Top. Microbiol. Immunol.* 176:143–164.

Connick, E., D. G. Marr, X. Q. Zhang, S. J. Clark, M. S. Saag, R. T. Schooley, and T. J. Curiel. 1996. HIV-specific cellular and humoral immune responses in primary HIV infection. *AIDS Res. Hum. Retrovir.* 12:1129–1140.

Conrad, B., R. N. Weissmahr, J. Boni, R. Arcari., J. Schupbach, and B. Mach. 1997. A human endogenous retroviral superantigen as candidate autoimmune gene in type I diabetes. *Cell* 90:303–313.

Constantine, N. T., G. van der Groen, E. M. Belsey, and H. Tamashiro. 1994. Sensitivity of HIV-antibody assays determined by seroconversion panels. *AIDS* 8:1715–1720.

Coombs, R. W., S. L. Welles, C. Hooper, P. S. Reichelderfer, R. T. D'Aquila, A. J. Japour, V. A. Johnson, D. R. Kuritzkes, D. D. Richman, S. Kwok, J. Todd, J. B. Jackson, V. De Gruttola, C. S. Crumpacker, and J. Kahn. 1996. Association of plasma human immunodeficiency virus type 1 RNA level with risk of clinical progression in patients with advanced infection. AIDS Clinical Trials Group (ACTG) 116B/117 Study Team. ACTG Virology Committee Resistance and HIV-1 RNA Working Groups. *J. Infect. Dis.* 174:704–712.

Cornelissen, M., G. Mulder-Kampinga, J. Veenstra, F. Zorgdrager, C. Kuiken, S. Hartman, J. Dekker, L. van der Hoek, C. Sol, R. Coutinho, et al. 1995. Syncytium-inducing (SI) phenotype suppression at seroconversion after intramuscular inoculation of a non-syncytium-inducing/SI phenotypically mixed human immunodeficiency virus population. *J. Virol.* 69:1810–1818.

Coste, J., B. Montes, J. Reynes, M. Peeters, C. Segarra, J. P. Vendrell, E. Delaporte, and M. Segondy. 1996. Comparative evaluation of three assays for the quantitation of human immunodeficiency virus type 1 RNA in plasma. *J. Med. Virol.* 50:293–302.

Cullen, B. R. 1998. HIV-1 auxiliary proteins: making connections in a dying cell. *Cell* 93:685–692.

Daar, E. S., T. Moudgil, R. D. Meyer, and D. D. Ho. 1991. Transient high levels of viremia in patients with primary human immunodeficiency virus type 1 infection. *N. Engl. J. Med.* 324:961–964.

Daenke, S., A. G. Kermode, S. E. Hall, G. Taylor, J. Weber, S. Nightingale, and C. R. Bangham. 1996. High activated and memory cytotoxic T-cell responses to HTLV-1 in healthy carriers and patients with tropical spastic paraparesis. *Virology* 217:139–146.

Dalgleish, A. G., P. C. Beverley, P. R. Clapham, D. H. Crawford, M. F. Greaves, and R. A. Weiss. 1984. The CD4 (T4) antigen is an essential component of the receptor for the AIDS retrovirus. *Nature* 312:763–767.

D'Aquila, R. T. 1996. Human immunodeficiency virus type 2: human biology of the other AIDS virus. *Curr. Clin. Top. Infect. Dis.* 16:84–101.

Deacon, N. J., A. Tsykin, A. Solomon, K. Smith, M. Ludford-Menting, D. J. Hooker, D. A. McPhee, A. L. Greenway, A. Ellett, C. Chatfield, et al. 1995. Genomic structure of an attenuated quasispecies of HIV-1 from a blood transfusion donor and recipients. *Science* 270:988–991.

Dean, M., M. Carrington, C. Winkler, G. A. Huttley, M. W. Smith, R. Allikmets, J. J. Goedert, S. P. Buchbinder, E. Vittinghoff, E. Gomperts, S. Donfield, D. Vlahov, R. Kaslow, A. Saah, C. Rinaldo, R. Detels, and S. J. O'Brien. 1996. Genetic restriction of HIV-1 infection and progression to AIDS by a deletion allele of the CKR5 structural gene. Hemophilia Growth and Development Study, Multicenter AIDS Cohort Study, Multicenter Hemophilia Cohort Study, San Francisco City Cohort, ALIVE Study. *Science* 273:1856–1862.

Debyser, Z., E. Vanwijngaerden, K. Vanlaethem, K. Beuselinck, M. Reynders, E. Declercq, J. Desmyter, and A. M. Vandamme. 1998. Failure to quantify viral load with two of the three commercial methods in a pregnant woman harboring an HIV type 1 subtype G strain. *AIDS Res. Hum. Retrovir.* 14:453–459.

Defer, C., J. Coste, F. Descamps, S. Voisin, J. M. Lemaire, M. Maniez, and A. M. Courouce. 1995. Contribution of polymerase chain reaction and radioimmunoprecipitation assay in the confirmation of human T-lymphotropic virus infection in French blood donors. Retrovirus Study Group of the French Society of Blood Transfusion. *Transfusion* 35:596–600.

Delézay, O., N. Koch, N. Yahi, D. Hammache, C. Tourres, C. Tamalet, and J. Fantini. 1997. Co-expression of CXCR4/fusin and galactosylceramide in the human intestinal epithelial cell line HT-29. *AIDS* 11:1311–1318.

Delwart, E. L., H. W. Sheppard, B. D. Walker, J. Goudsmit, and J. I. Mullins. 1994. Human immunodeficiency virus type 1 evolution in vivo tracked by DNA heteroduplex mobility assays. *J. Virol.* 68:6672–6683.

De Leys, R., B. Vanderborght, M. Vanden Haesevelde, L. Heyndrickx, A. van Geel, C. Wauters, R. Bernaerts, E. Saman, P. Nijs, B. Willems, et al. 1990. Isolation and partial characterization of an unusual human immunodeficiency retrovirus from two persons of west-central African origin. *J. Virol.* 64:1207–1216.

Deng, H., R. Liu, W. Ellmeier, S. Choe, D. Unutmaz, M. Burkhart, P. Di Marzio, S. Marmon, R. E. Sutton, C. M. Hill, C. B. Davis, S. C. Peiper, T. J. Schall, D. R. Littman, and N. R. Landau. 1996. Identification of a major co-receptor for primary isolates of HIV-1. *Nature* 381:661–666.

Di Marzo Veronese, F., P. Lusso, J. Schüpbach, and R. C. Gallo. 1992. Human retroviruses, p. 585–625. *In* S. Specter and G. Lance (ed.), *Clinical Virology Manual*, 2nd ed. Elsevier Science Publishers, New York, N.Y.

Douceron, H., L. Deforges, R. Gherardi, A. Sobel, and P. Chariot. 1993. Long-lasting postmortem viability of human immunodeficiency virus: a potential risk in forensic medicine practice. *Forensic Sci. Int.* 60:61–66.

Downs, A. M., G. Salamina, and R. A. Ancelle-Park. 1995. Incubation period of vertically acquired AIDS in Europe before widespread use of prophylactic therapies. *J. Acquir. Immune Defic. Syndr. Hum. Retrovirol.* 9:297–304.

Dragic, T., V. Litwin, G. P. Allaway, S. R. Martin, Y. Huang, K. A. Nagashima, C. Cayanan, P. J. Maddon, R. A. Koup, J. P. Moore, and W. A. Paxton. 1996. HIV-1 entry into CD4+ cells is mediated by the chemokine receptor CC-CKR-5. *Nature* 381:667–673.

Durant, J., P. Clevenbergh, P. Halfon, P. Delgiudice, S. Porsin, P. Simonet, N. Montagne, C. A. B. Boucher, J. M. Schapiro, and P. Dellamonica. 1999. Drug-resistance genotyping in HIV-1 therapy: the VIRADAPT randomised controlled trial. *Lancet* 353:2195–2199.

Egger, M., B. Hirschel, P. Francioli, P. Sudre, M. Wirz, M. Flepp, M. Rickenbach, R. Malinverni, P. Vernazza, and M. Battegay. 1997. Impact of new antiretroviral combination therapies in HIV infected patients in Switzerland: prospective multicentre study. Swiss HIV Cohort Study. *Br. Med. J.* **315:**1194–1199.

Ehrlich, G. D., J. B. Glaser, V. Bryz-Gornia, J. Maese, T. A. Waldmann, B. J. Poiesz, and S. J. Greenberg. 1991. Multiple sclerosis, retroviruses, and PCR. The HTLV-MS Working Group. *Neurology* **41:**335–343.

Embretson, J., M. Zupancic, J. L. Ribas, A. Burke, P. Racz, K. Tenner-Racz, and A. T. Haase. 1993. Massive covert infection of helper T lymphocytes and macrophages by HIV during the incubation period of AIDS. *Nature* **362:**359–362.

Farias de Carvalho, S. M., M. S. Pombo de Oliveira, L. C. Thuler, M. Rios, R. C. Coelho, L. C. Rubim, E. M. Silva, A. M. Reis, and D. Catovsky. 1997. HTLV-I and HTLV-II infections in hematologic disorder patients, cancer patients, and healthy individuals from Rio de Janeiro, Brazil. *J. Acquir. Immune Defic. Syndr.* **15:**238–242.

Fauci, A. S. 1993. Multifactorial nature of human immunodeficiency virus disease: implications for therapy. *Science* **262:**1011–1018.

Fauci, A. S. 1996. Host factors and the pathogenesis of HIV-induced disease. *Nature* **384:**529–534.

Feng, Y., C. C. Broder, P. E. Kennedy, and E. A. Berger. 1996. HIV-1 entry cofactor: functional cDNA cloning of a seven-transmembrane, G protein-coupled receptor. *Science* **272:**872–877.

Ferrante, P., M. E. Westarp, R. Mancuso, S. Puricelli, M. P. Westarp, M. Mini, D. Caputo, and M. R. Zuffolato. 1995. HTLV tax-rex DNA and antibodies in idiopathic amyotrophic lateral sclerosis. *J. Neurol. Sci.* **129:**140–144.

Ferreira, O. C., Jr., V. Planelles, and J. D. Rosenblatt. 1997. Human T-cell leukemia viruses: epidemiology, biology, and pathogenesis. *Blood Rev.* **11:**91–104.

Finzi, D., M. Hermankova, T. Pierson, L. M. Carruth, C. Buck, R. E. Chaisson, T. C. Quinn, K. Chadwick, J. Margolick, R. Brookmeyer, J. Gallant, M. Markowitz, D. D. Ho, D. D. Richman, and R. F. Siliciano. 1997. Identification of a reservoir for HIV-1 in patients on highly active antiretroviral therapy. *Science* **278:**1295–1300.

Finzi, D., J. Blankson, J. D. Siliciano, J. B. Margolick, K. Chadwick, T. Pierson, K. Smith, J. Lisziewicz, F. Lori, C. Flexner, T. C. Quinn, R. E. Chaisson, E. Rosenberg, B. Walker, S. Gange, J. Gallant, and R. F. Siliciano. 1999. Latent infection of CD4+ T cells provides a mechanism for lifelong persistence of HIV-1, even in patients on effective combination therapy. *Nat. Med.* **5:**512–517.

Ford, R. J., L. A. Donehower, and R. C. Bohannon. 1992. Studies on a type D retrovirus isolated from an AIDS patient lymphoma. *AIDS Res. Hum. Retrovir.* **8:**742–751.

Franchini, G. 1995. Molecular mechanisms of human T-cell leukemia/lymphotropic virus type I infection. *Blood* **86:**3619–3639.

Frankel, S. S., B. M. Wenig, A. P. Burke, P. Mannan, L. D. Thompson, S. L. Abbondanzo, A. M. Nelson, M. Pope, and R. M. Steinman. 1996. Replication of HIV-1 in dendritic cell-derived syncytia at the mucosal surface of the adenoid. *Science* **272:**115–117.

Furtado, M. R., D. S. Callaway, J. P. Phair, K. J. Kunstman, J. L. Stanton, C. A. Macken, A. S. Perelson, and S. M. Wolinsky. 1999. Persistence of HIV-1 transcription in peripheral-blood mononuclear cells in patients receiving potent antiretroviral therapy. *N. Engl. J. Med.* **340:**1614–1622.

Gallo, R. C., S. Z. Salahuddin, M. Popovic, G. M. Shearer, M. Kaplan, B. F. Haynes, T. J. Palker, R. Redfield, J. Oleske, B. Safai, et al. 1984. Frequent detection and isolation of cyto-

pathic retroviruses (HTLV-III) from patients with AIDS and at risk for AIDS. *Science* **224:**500–503.

Gao, F., D. L. Robertson, C. D. Carruthers, Y. Li, E. Bailes, L. G. Kostrikis, M. O. Salminen, F. Bibollet-Ruche, M. Peeters, D. D. Ho, G. M. Shaw, P. M. Sharp, and B. H. Hahn. 1998. An isolate of human immunodeficiency virus type 1 originally classified as subtype I represents a complex mosaic comprising three different group M subtypes (A, G, and I). *J. Virol.* **72:**10234–10241.

Gao, F., E. Bailes, D. L. Robertson, Y. L. Chen, C. M. Rodenburg, S. F. Michael, L. B. Cummins, L. O. Arthur, M. Peeters, G. M. Shaw, P. M. Sharp, and B. H. Hahn. 1999. Origin of HIV-1 in the chimpanzee Pan troglodytes troglodytes. *Nature* **397:**436–441.

Garcia Lerma, J., R. F. Schinazi, A. S. Juodawlkis, V. Soriano, Y. Lin, K. Tatti, D. Rimland, T. M. Folks, and W. Heneine. 1999. A rapid non-culture-based assay for clinical monitoring of phenotypic resistance of human immunodeficiency virus type 1 to lamivudine (3TC). *Antimicrob. Agents Chemother.* **43:**264–270.

Garry, R. F., C. D. Fermin, D. J. Hart, S. S. Alexander, L. A. Donehower, and H. Luo-Zhang. 1990. Detection of a human intracisternal A-type retroviral particle antigenically related to HIV. *Science* **250:**1127–1129.

Gessain, A., and G. de The. 1996a. Geographic and molecular epidemiology of primate T lymphotropic retroviruses: HTLV-I, HTLV-II, STLV-I, STLV-PP, and PTLV-L. *Adv. Virus Res.* **47:**377–426.

Gessain, A., and G. de The. 1996b. Geographic and molecular epidemiology of primate T lymphotropic retroviruses: HTLV-I, HTLV-II, STLV-I, STLV-PP, and PTLV-L. *Adv. Virus Res.* **47:**377–426.

Gessain, A., F. Barin, J. C. Vernant, O. Gout, L. Maurs, A. Calender, and G. de The. 1985. Antibodies to human T-lymphotropic virus type-I in patients with tropical spastic paraparesis. *Lancet* **ii:**407–410.

Glushakova, S., J. C. Grivel, W. Fitzgerald, A. Sylwester, J. Zimmerberg, and L. B. Margolis. 1998. Evidence for the HIV-1 phenotype switch as a causal factor in acquired immunodeficiency. *Nat. Med.* **4:**346–349.

Gottlieb, G. J., A. Ragaz, J. V. Vogel, A. Friedman-Kien, A. M. Rywlin, E. A. Weiner, and A. B. Ackerman. 1981. A preliminary communication on extensively disseminated Kaposi's sarcoma in young homosexual men. *Am. J. Dermatopathol.* **3:**111–114.

Gray, C. M., L. Morris, J. Murray, J. Keeton, S. Shalekoff, S. F. Lyons, P. Sonnenberg and D. J. Martin. 1996. Identification of cell subsets expressing intracytoplasmic cytokines within HIV-1-infected lymph nodes. *AIDS* **10:**1467–1475.

Graziosi, C., G. Pantaleo, L. Butini, J. F. Demarest, M. S. Saag, G. M. Shaw, and A. S. Fauci. 1993. Kinetics of human immunodeficiency virus type 1 (HIV-1) DNA and RNA synthesis during primary HIV-1 infection. *Proc. Natl. Acad. Sci. USA* **90:**6405–6409.

Green, P. L., and I. S. Y. Chen. 1994. Molecular features of the human T-cell leukemia virus. Mechanisms of transformation and leukemogenicity, p. 277–311. *In* J. A. Levy (ed.), *The Retroviridae,* vol. 3. Plenum Press, New York, N.Y.

Greenberg, S. J., G. D. Ehrlich, M. A. Abbott, B. J. Hurwitz, T. A. Waldmann, and B. J. Poiesz. 1989. Detection of sequences homologous to human retroviral DNA in multiple sclerosis by gene amplification. *Proc. Natl. Acad. Sci. USA* **86:**2878–2882.

Greenough, T. C., D. B. Brettler, M. Somasundaran, D. L. Panicali, and J. L. Sullivan. 1997. Human immunodeficiency virus type 1-specific cytotoxic T lymphocytes (CTL), virus load, and CD4 T cell loss: evidence support-ing a protective role for CTL in vivo. *J. Infect. Dis.* **176:**118–125.

Griffiths, D. J., P. J. Venables, R. A. Weiss, and M. T. Boyd. 1997. A novel exogenous retrovirus sequence identified in humans. *J. Virol.* **71:**2866–2872.

Griffiths, D. J., S. P. Cooke, C. Herve, S. P. Rigby, E. Mallon, A. Hajeer, M. Lock, V. Emery, P. Taylor, P. Pantelidis, C. B. Bunker, R. du Bois, R. A. Weiss, and P. J. Venables. 1999. Detection of human retrovirus 5 in patients with arthritis and systemic lupus erythematosus. *Arthritis Rheum.* **42:**448–454.

Gurtler, L., A. Muhlbacher, U. Michl, H. Hofmann, G. G. Paggi, V. Bossi, R. Thorstensson, R. G. Villaescusa, A. Eiras, J. M. Hernandez, W. Melchior, F. Donie, and B. Weber. 1998. Reduction of the diagnostic window with a new combined p24 antigen and human immunodeficiency virus antibody screening assay. *J. Virol. Methods* **75:**27–38.

Gurtler, L. G., P. H. Hauser, J. Eberle, A. von Brunn, S. Knapp, L. Zekeng, J. M. Tsague, and L. Kaptue. 1994. A new subtype of human immunodeficiency virus type 1 (MVP-5180) from Cameroon. *J. Virol.* **68:**1581–1585.

Hale, A., T. Leung, S. Sivasubramaniam, J. Kenny, and S. Sutherland. 1997. Prevalence of antibodies to HTLV in antenatal clinic attenders in south east London. *J. Med. Virol.* **52:**326–329.

Harouse, J. M., S. Bhat, S. L. Spitalnik, M. Laughlin, K. Stefano, D. H. Silberberg, and F. Gonzalez-Scarano. 1991. Inhibition of entry of HIV-1 in neural cell lines by antibodies against galactosyl ceramide. *Science* **253:**320–323.

Haynes, B. F., G. Pantaleo, and A. S. Fauci. 1996. Toward an understanding of the correlates of protective immunity to HIV infection. *Science* **271:**324–328.

Healey, D. S., and W. V. Bolton. 1993. Apparent HIV-1 glycoprotein reactivity on western blot in uninfected blood donors. *AIDS* **7:**655–658.

Heath, S. L., J. G. Tew, J. G. Tew, A. K. Szakal, and G. F. Burton. 1995. Follicular dendritic cells and human immunodeficiency virus infectivity. *Nature* **377:**740–744.

Heneine, W., S. Yamamoto, W. M. Switzer, T. J. Spira, and T. M. Folks. 1995. Detection of reverse transcriptase by a highly sensitive assay in sera from persons infected with human immunodeficiency virus type 1. *J. Infect. Dis.* **171:**1210–1216.

Heneine, W., W. M. Switzer, P. Sandstrom, J. Brown, S. Vedapuri, C. A. Schable, A. S. Khan, N. W. Lerche, M. Schweizer, D. Neumann-Haefelin, L. E. Chapman, and T. M. Folks. 1998. Identification of a human population infected with simian foamy viruses. *Nat. Med.* **4:**403–407.

Henrard, D. R., J. F. Phillips, L. R. Muenz, W. A. Blattner, D. Wiesner, M. E. Eyster, and J. J. Goedert. 1995. Natural history of HIV-1 cell-free viremia. *JAMA* **274:**554–558.

Higuchi, I., K. Hashimoto, E. Matsuoka, R. Rosales, M. Nakagawa, K. Arimura, S. Izumo, and M. Osame. 1996. The main HTLV-I-harboring cells in the muscles of viral carriers with polymyositis are not macrophages but CD4+ lymphocytes. *Acta Neuropathol.* **92:**358–361.

Hino, S., S. Katamine, H. Miyata, Y. Tsuji, T. Yamabe, and T. Miyamoto. 1996. Primary prevention of HTLV-I in Japan. *J. Acquir. Immune Defic. Syndr.* **13:**S199–S203.

Hirata, M., J. Hayashi, A. Noguchi, K. Nakashima, W. Kajiyama, S. Kashiwagi, and T. Sawada. 1992. The effects of breastfeeding and presence of antibody to p40tax protein of human T cell lymphotropic virus type-I on mother to child transmission. *Int. J. Epidemiol.* **21:**989–994.

Hirschel, B., and M. Opravil. 1999. The year in review: antiretroviral treatment. *AIDS* **13**(Suppl. A)**:**S177–S187.

Hjelle, B., R. Mills, S. Swenson, G. Mertz, C. Key, and S. Allen. 1991. Incidence of hairy cell leukemia, mycosis fungoides, and chronic lymphocytic leukemia in first known HTLV-II-endemic population. *J. Infect. Dis.* **163:**435–440.

Ho, D. D., T. Moudgil, and M. Alam. 1989. Quantitation of human immunodeficiency virus type 1 in the blood of infected persons. *N. Engl. J. Med.* **321:**1621–1625.

Ho, D. D., A. U. Neumann, A. S. Perelson, W. Chen, J. M. Leonard, and M. Markowitz. 1995. Rapid turnover of plasma virions and CD4 lymphocytes in HIV-1 infection. *Nature* **373:**123–126.

Hoffman, A. D., B. Banapour, and J. A. Levy. 1985. Characterization of the AIDS-associated retrovirus reverse transcriptase and optimal conditions for its detection in virions. *Virology* **147:**326–335.

Huang, Y., L. Zhang, and D. D. Ho. 1995. Biological characterization of *nef* in long-term survivors of human immunodeficiency virus type 1 infection. *J. Virol.* **69:**8142–8146.

Hussain, L. A., and T. Lehner. 1995. Comparative investigation of Langerhans' cells and potential receptors for HIV in oral, genitourinary and rectal epithelia. *Immunology* **85:**475–484.

Ijichi, S., M. B. Ramundo, H. Takahashi, and W. W. Hall. 1992. In vivo cellular tropism of human T cell leukemia virus type II (HTLV-II). *J. Exp. Med.* **176:**293–296.

Imrie, A., A. Beveridge, W. Genn, J. Vizzard, and D. A. Cooper. 1997. Transmission of human immunodeficiency virus type 1 resistant to nevirapine and zidovudine. Sydney Primary HIV Infection Study Group. *J. Infect. Dis.* **175:**1502–1506.

Iwakura, Y., M. Tosu, E. Yoshida, M. Takiguchi, K. Sato, I. Kitajima, K. Nishioka, K. Yamamoto, T. Takeda, M. Hatanaka, et al. 1991. Induction of inflammatory arthropathy resembling rheumatoid arthritis in mice transgenic for HTLV-I. *Science* **253:**1026–1028.

Iwakura, Y., S. Saijo, Y. Kioka, J. Nakayama-Yamada, K. Itagaki, M. Tosu, M. Asano, Y. Kanai, and K. Kakimoto. 1995. Autoimmunity induction by human T cell leukemia virus type 1 in transgenic mice that develop chronic inflammatory arthropathy resembling rheumatoid arthritis in humans. *J. Immunol.* **155:**1588–1598.

Jacobson, S., C. S. Raine, E. S. Mingioli, and D. E. McFarlin. 1988. Isolation of an HTLV-1-like retrovirus from patients with tropical spastic paraparesis. *Nature* **331:**540–543.

Jacobson, S., H. Shida, D. E. McFarlin, A. S. Fauci, and S. Koenig. 1990. Circulating CD8+ cytotoxic T lymphocytes specific for HTLV-I pX in patients with HTLV-I associated neurological disease. *Nature* **348:**245–248.

Jeffery, K. J., K. Usuku, S. E. Hall, W. Matsumoto, G. P. Taylor, J. Procter, M. Bunce, G. S. Ogg, K. I. Welsh, J. N. Weber, A. L. Lloyd, M. A. Nowak, M. Nagai, D. Kodama, S. Izumo, M. Osame, and C. R. Bangham. 1999. HLA alleles determine human T-lymphotropic virus-I (HTLV-I) proviral load and the risk of HTLV-I-associated myelopathy. *Proc. Natl. Acad. Sci. USA* **96:**3848–3853.

Jin, X., D. E. Bauer, S. E. Tuttleton, S. Lewin, A. Gettie, J. Blanchard, C. E. Irwin, J. T. Safrit, J. Mittler, L. Weinberger, L. G. Kostrikis, L. Q. Zhang, A. S. Perelson, and D. D. Ho. 1999. Dramatic rise in plasma viremia after CD8(+) T cell depletion in simian immunodeficiency virus-infected macaques. *J. Exp. Med.* **189:**991–998.

Jurriaans, S., B. Van Gemen, G. J. Weverling, D. Van Strijp, P. Nara, R. Coutinho, M. Koot, H. Schuitemaker, and J. Goudsmit. 1994. The natural history of HIV-1 infection: virus load and virus phenotype independent determinants of clinical course? *Virology* **204**:223–233.

Kahn, J. O., and B. D. Walker. 1998. Acute human immunodeficiency virus type 1 infection. *N. Engl. J. Med.* **339**:33–39.

Kalyanaraman, V. S., M. G. Sarngadharan, M. Robert-Guroff, I. Miyoshi, D. Golde, and R. C. Gallo. 1982. A new subtype of human T-cell leukemia virus (HTLV-II) associated with a T-cell variant of hairy cell leukemia. *Science* **218**:571–573.

Karlic, H., M. Mostl, H. Mucke, B. Pavlova, M. Pfeilstocker, and R. Heinz. 1997. Association of human T-cell leukemia virus and myelodysplastic syndrome in a central European population. *Cancer Res.* **57**:4718–4721.

Kaufmann, D., G. Pantaleo, P. Sudre and A. Telenti. 1998a. Cd4-cell count in HIV-1-infected individuals remaining viraemic with highly active antiretroviral therapy (HAART). *Lancet* **351**:723–724.

Kaufmann, G. R., P. Cunningham, A. D. Kelleher, J. Zaunders, A. Carr, J. Vizzard, M. Law, and D. A. Cooper. 1998b. Patterns of viral dynamics during primary human immunodeficiency virus type 1 infection. The Sydney Primary HIV Infection Study Group. *J. Infect. Dis.* **178**:1812–1815.

Kawai, H., Y. Nishida, M. Takagi, S. Nakamura, and S. Saito. 1989. HTLV-I associated myelopathy (HAM) with adult T-cell leukemia (ATL). *Rinsho Shinkeigaku* **29**:588–592. (In Japanese.)

Kawai, H., K. Yokoi, M. Akaike, M. Kunishige, M. Abe, Y. Tanouchi, H. Mine, Y. Mimura, and S. Saito. 1995. Graves' disease in HTLV-I carriers. *J. Mol. Med.* **73**:85–88.

Kawano, F., K. Yamaguchi, H. Nishimura, H. Tsuda, and K. Takatsuki. 1985. Variation in the clinical courses of adult T-cell leukemia. *Cancer* **55**:851–856.

Keet, I. P., P. Krijnen, M. Koot, J. M. Lange, F. Miedema, J. Goudsmit, and R. A. Coutinho. 1993. Predictors of rapid progression to AIDS in HIV-1 seroconverters. *AIDS* **7**:51–57.

Khabbaz, R. F., W. Heneine, J. R. George, B. Parekh, T. Rowe, T. Woods, W. M. Switzer, H. M. McClure, M. Murphey-Corb, and T. M. Folks. 1994. Brief report: infection of a laboratory worker with simian immunodeficiency virus. *N. Engl. J. Med.* **330**:172–177.

Kievits, T., B. van Gemen, D. van Strijp, R. Schukkink, M. Dircks, H. Adriaanse, L. Malek, R. Sooknanan, and P. Lens. 1991. NASBA isothermal enzymatic in vitro nucleic acid amplification optimized for the diagnosis of HIV-1 infection. *J. Virol. Methods* **35**:273–286.

Kikuchi, A., T. Nishikawa, Y. Ikeda, and K. Yamaguchi. 1997. Absence of human T-lymphotropic virus type I in Japanese patients with cutaneous T-cell lymphoma. *Blood* **89**:1529–1532.

Kira, J., Y. Itoyama, N. Koyanagi, J. Tateishi, M. Kishikawa, S. Akizuki, I. Kobayashi, N. Toki, K. Sueishi, H. Sato, et al. 1992. Presence of HTLV-I proviral DNA in central nervous system of patients with HTLV-I-associated myelopathy. *Ann. Neurol.* **31**:39–45.

Kira, J., T. Hamada, Y. Kawano, M. Okayama, and K. Yamasaki. 1997. An association of human T-cell lymphotropic virus type I infection with vascular dementia. *Acta Neurol. Scand.* **96**:305–309.

Kirchhoff, F., T. C. Greenough, D. B. Brettler, J. L. Sullivan, and R. C. Desrosiers. 1995. Brief report: absence of intact nef sequences in a long-term survivor with nonprogressive HIV-1 infection. *N. Engl. J. Med.* **332**:228–232.

Kitajima, I., K. Yamamoto, K. Sato, Y. Nakajima, T. Nakajima, I. Maruyama, M. Osame, and K. Nishioka. 1991. Detection of human T cell lymphotropic virus type I proviral DNA and its gene expression in synovial cells in chronic inflammatory arthropathy. *J. Clin. Investig.* **88**:1315–1322.

Klatzmann, D., E. Champagne, S. Chamaret, J. Gruest, D. Guetard, T. Hercend, J. C. Gluckman, and L. Montagnier. 1984. T-lymphocyte T4 molecule behaves as the receptor for human retrovirus LAV. *Nature* **312**:767–768.

Kompoliti, A., B. Gage, L. Sharma, and J. C. Daniels. 1996. Human T-cell lymphotropic virus type 1-associated myelopathy, Sjogren syndrome, and lymphocytic pneumonitis. *Arch. Neurol.* **53**:940–942.

Koot, M., I. P. Keet, A. H. Vos, R. E. de Goede, M. T. Roos, R. A. Coutinho, F. Miedema, P. T. Schellekens, and M. Tersmette. 1993. Prognostic value of HIV-1 syncytium-inducing phenotype for rate of CD4+ cell depletion and progression to AIDS. *Ann. Intern. Med.* **118**:681–688.

Koprowski, H., E. C. DeFreitas, M. E. Harper, M. Sandberg-Wollwheim, W. A. Sheremata, M. Robert-Guroff, C. W. Saxinger, M. B. Feinberg, F. Wong-Staal, and R. C. Gallo. 1985. Multiple sclerosis and human T-cell lymphotropic retroviruses. *Nature* **318**:154–160.

Kornfeld, H., W. W. Cruikshank, S. W. Pyle, J. S. Berman, and D. M. Center. 1988. Lymphocyte activation by HIV-1 envelope glycoprotein. *Nature* **335**:445–448.

Koup, R. A., J. T. Safrit, Y. Cao, C. A. Andrews, G. McLeod, W. Borkowsky, C. Farthing, and D. D. Ho. 1994. Temporal association of cellular immune responses with the initial control of viremia in primary human immunodeficiency virus type 1 syndrome. *J. Virol.* **68**:4650–4655.

Kubota, R., T. Fujiyoshi, S. Izumo, S. Yashiki, I. Maruyama, M. Osame, and S. Sonoda. 1993. Fluctuation of HTLV-I proviral DNA in peripheral blood mononuclear cells of HTLV-I-associated myelopathy. *J Neuroimmunol.* **42**:147–154.

Kubota, R., F. Umehara, S. Izumo, S. Ijichi, K. Matsumuro, S. Yashiki, T. Fujiyoshi, S. Sonoda, and M. Osame. 1994. HTLV-I proviral DNA amount correlates with infiltrating CD4+ lymphocytes in the spinal cord from patients with HTLV-I-associated myelopathy. *J. Neuroimmunol.* **53**:23–29.

Kuun, E., M. Brashaw, and A. D. P. Heyns. 1997. Sensitivity and specificity of standard and rapid HIV-antibody tests evaluated by seroconversion and non-seroconversion low-titre panels. *Vox Sang.* **72**:11–15.

Kwok, S., J. J. Lipka, N. McKinney, D. E. Kellogg, B. Poiesz, S. K. Foung, and J. J. Sninsky. 1990. Low incidence of HTLV infections in random blood donors with indeterminate western blot patterns. *Transfusion* **30**:491–494.

Kwok, S., D. Kellogg, G. Ehrlich, B. Poiesz, S. Bhagavati, and J. J. Sninsky. 1988. Characterization of a sequence of human T cell leukemia virus type I from a patient with chronic progressive myelopathy. *J. Infect. Dis.* **158**:1193–1197.

Kwok, S., D. H. Mack, K. B. Mullis, B. Poiesz, G. Ehrlich, D. Blair, A. Friedman-Kien, and J. J. Sninsky. 1987. Identification of human immunodeficiency virus sequences by using in vitro enzymatic amplification and oligomer cleavage detection. *J. Virol.* **61**:1690–1694.

La Grenade, L., R. A. Schwartz, and C. K. Janniger. 1996. Childhood dermatitis in the tropics: with special emphasis on infective dermatitis, a marker for infection with human T-cell leukemia virus-I. *Cutis* **58**:115–118.

Lal, R. B., S. M. Owen, J. Mingle, P. H. Levine, and A. Manns. 1994. Presence of human T lymphotropic virus types I and II in Ghana, west Africa. *AIDS Res. Hum. Retrovir.* **10**:1747–1750.

Laleman, G., M. Kambale, I. Van Kerckhoven, N. Kapila, M. Konde, U. Selemani, P. Piot, and G. van der Groen. 1991. A simplified and less expensive strategy for confirming anti HIV-1 screening results in a diagnostic laboratory in Lubumbashi, Zaire. *Ann. Soc. Belg. Med. Trop.* **71**:287–294.

Ledergerber, B., M. Flepp, J. Boni, Z. Tomasik, R. W. Cone, R. Lüthy, and J. Schupbach. 2000. Human immunodeficiency virus type 1 p24 concentratioin measured by boosted ELISA of heat-denatured plasma correlates with decline in CD4 cells, progression to AIDS, and survival: comparison with viral RNA measurement.. *J. Infect. Dis.* **181:**1280–1288.

Lee, C. H., B. Leung, M. A. Lemmon, J. Zheng, D. Cowburn, J. Kuriyan, and K. Saksela. 1995. A single amino acid in the SH3 domain of Hck determines its high affinity and specificity in binding to HIV-1 Nef protein. *EMBO J.* **14:**5006–5015.

Lehky, T. J., C. H. Fox, S. Koenig, M. C. Levin, N. Flerlage, S. Izumo, E. Sato, C. S. Raine, M. Osame, and S. Jacobson. 1995. Detection of human T-lymphotropic virus type I (HTLV-I) tax RNA in the central nervous system of HTLV-I-associated myelopathy/tropical spastic paraparesis patients by in situ hybridization. *Ann. Neurol.* **37:**167–175.

Levin, M. C., and S. Jacobson. 1997. HTLV-I associated myelopathy/tropical spastic paraparesis (HAM/TSP): a chronic progressive neurologic disease associated with immunologically mediated damage to the central nervous system. *J. Neurovirol.* **3:**126–140.

Levy, J. A., A. D. Hoffman, S. M. Kramer, J. A. Landis, J. M. Shimabukuro, and L. S. Oshiro. 1984. Isolation of lymphocytopathic retroviruses from San Francisco patients with AIDS. *Science* **225:**840–842.

Levy, J. A., J. Shimabukuro, T. McHugh, C. Casavant, D. Stites, and L. Oshiro. 1985. AIDS-associated retroviruses (ARV) can productively infect other cells besides human T helper cells. *Virology* **147:**441–448.

Levy, J. E. 1997. *HIV and the Pathogenesis of AIDS*, 2nd ed. ASM Press, Washington, D.C.

Lifson, A. R., P. M. O'Malley, N. A. Hessol, S. P. Buchbinder, L. Cannon, and G. W. Rutherford. 1990. HIV seroconversion in two homosexual men after receptive oral intercourse with ejaculation: implications for counseling concerning safe sexual practices. *Am. J. Public Health* **80:**1509–1511.

Lori, F., A. Malykh, A. Cara, D. Sun, J. N. Weinstein, J. Lisziewicz, and R. C. Gallo. 1994. Hydroxyurea as an inhibitor of human immunodeficiency virus-type 1 replication. *Science* **266:**801–805.

Lyamuya, E., U. Bredberg-Raden, A. Massawe, E. Urassa, G. Kawo, G. Msemo, T. Kazimoto, A. Ostborn, K. Karlsson, F. Mhalu, and G. Biberfeld. 1996. Performance of a modified HIV-1 p24 antigen assay for early diagnosis of HIV-1 infection in infants and prediction of mother-to-infant transmission of HIV-1 in Dar es Salaam, Tanzania. *J. Acquir. Immune Defic. Syndr. Hum. Retrovirol.* **12:**421–426.

Malone, J. D., E. S. Smith, J. Sheffield, D. Bigelow, K. C. Hyams, S. G. Beardsley, R. S. Lewis, and C. R. Roberts. 1993. Comparative evaluation of six rapid serological tests for HIV-1 antibody. *J. Acquir. Immune Defic. Syndr.* **6:**115–119.

Manca, N., E. Piacentini, M. Gelmi, P. Calzavara, M. A. Manganoni, A. Glukhov, F. Gargiulo, M. De Francesco, F. Pirali, G. De Panfilis, et al. 1994. Persistence of human T cell lymphotropic virus type 1 (HTLV-1) sequences in peripheral blood mononuclear cells from patients with mycosis fungoides. *J. Exp. Med.* **180:**1973–1978. (Erratum, **181:**441, 1995.)

Manns, A., M. Hisada, and L. La Grenade. 1999. Human T-lymphotropic virus type I infection. *Lancet* **353:**1951–1958.

Mazzoli, S., D. Trabattoni, S. Caputo, S. Piconi, C. Ble, F. Meacci, S. Ruzzante, A. Salvi, F. Semplici, R. Longhi, M. Fusi, N. Tofani, M. Biasin, M. Villa, F. Mazzotta, and M. Clerici. 1997. HIV-specific mucosal and cellular immunity in HIV-seronegative partners of HIV-seropositive individuals. *Nat. Med.* **3:**1250–1257.

McAlarney, T., S. Apostolski, S. Lederman, and N. Latov. 1994. Characteristics of HIV-1 gp120 glycoprotein binding to glycolipids. *J. Neurosci. Res.* **37:**453–460.

Medrano, F. J., V. Soriano, E. J. Calderon, C. Rey, M. Gutierrez, R. Bravo, M. Leal, J. Gonzalez-Lahoz, and E. Lissen. 1997. Significance of indeterminate reactivity to human T-cell lymphotropic virus in western blot analysis of individuals at risk. *Eur. J. Clin. Microbiol. Infect. Dis.* **16:**249–252.

Mellors, J. W., C. R. Rinaldo, Jr., P. Gupta, R. M. White, J. A. Todd, and L. A. Kingsley. 1996. Prognosis in HIV-1 infection predicted by the quantity of virus in plasma. *Science* **272:**1167–1170.

Mochizuki, M., K. Tajima, T. Watanabe, and K. Yamaguchi. 1994. Human T lymphotropic virus type 1 uveitis. *Br. J. Ophthalmol.* **78:**149–154.

Montaner, J. S. G., R. Hogg, J. Raboud, R. Harrigan, and M. O'Shaughnessy. 1998. Antiretroviral treatment in 1998. *Lancet* **352:**1919–1922.

Morgan, D. A., F. W. Ruscetti, and R. Gallo. 1976. Selective in vitro growth of T lymphocytes from normal human bone marrows. *Science* **193:**1007–1008.

Moritoyo, T., T. A. Reinhart, H. Moritoyo, E. Sato, S. Izumo, M. Osame, and A. T. Haase. 1996. Human T-lymphotropic virus type I-associated myelopathy and tax gene expression in CD4+ T lymphocytes. *Ann. Neurol.* **40:**84–90.

Moses, S., F. A. Plummer, J. E. Bradley, A. Ndinya, N. J. Nagelkerke, and A. R. Ronald. 1996. Association between lack of male circumcision and risk for HIV infection: review of the epidemiological evidence. *Int. Conf. AIDS* **11:**452.

Munoz, A., A. J. Kirby, Y. D. He, J. B. Margolick, B. R. Visscher, C. R. Rinaldo, R. A. Kaslow, and J. P. Phair. 1995. Long-term survivors with HIV-1 infection: incubation period and longitudinal patterns of CD4+ lymphocytes. *J. Acquir. Immune Defic. Syndr. Hum. Retrovirol.* **8:**496–505.

Nadal, D., J. Böni, C. Kind, O. Varnier, F. Steiner, Z. Tomasik, and J. Schüpbach. 1999. Prospective evaluation of amplification-boosted ELISA for heat-denatured p24 antigen for diagnosis and monitoring of pediatric HIV-1 infection. *J. Infect. Dis.* **180:**1089–1095.

Nagai, M., S. Yashiki, T. Fujiyoshi, C. Fujiyama, B. Kitze, S. Izumo, M. Osame, and S. Sonoda. 1996. Characterization of a unique T-cell clone established from a patient with HAM/TSP which recognized HTLV-I-infected T-cell antigens as well as spinal cord tissue antigens. *J. Neuroimmunol.* **65:**97–105.

Nagai, M., K. Usuku, W. Matsumoto, D. Kodama, N. Takenouchi, T. Moritoyo, S. Hashiguchi, M. Ichinose, C. R. M. Bangham, S. Izumo, and M. Osame. 1998. Analysis of HTLV-I proviral load in 202 HAM/TSP patients and 243 asymptomatic HTLV-I carriers: high proviral load strongly predisposes to HAM/TSP. *J. Neurovirol.* **4:**586–593.

Neumann, A. U., R. Tubiana, V. Calvez, C. Robert, T. S. Li, H. Agut, B. Autran, and C. Katlama. 1999. HIV-1 rebound during interruption of highly active antiretroviral therapy has no deleterious effect on reinitiated treatment. *AIDS* **13:**677–683.

Niewiesk, S., and C. R. Bangham. 1996. Evolution in a chronic RNA virus infection: selection on HTLV-I tax protein differs between healthy carriers and patients with tropical spastic paraparesis. *J. Mol. Evol.* **42:**452–458.

Niu, M. T., J. A. Jermano, P. Reichelderfer, and S. M. Schnittman. 1993. Summary of the National Institutes of Health workshop on primary human immunodeficiency virus type 1 infection. *AIDS Res. Hum. Retrovir.* **9:**913–924.

Nkengasong, J. N., C. Maurice, S. Koblavi, M. Kalou, D. Yavo, M. Maran, C. Bile, K. N'Guessan, J. Kouadio, and

S. Bony. 1999. Evaluation of HIV serial and parallel serologic testing algorithms in Abidjan, Cote d'Ivoire. *AIDS* **13:**109–117.

Oberlin, E., A. Amara, F. Bachelerie, C. Bessia, J. L. Virelizier, F. Arenzana-Seisdedos, O. Schwartz, J. M. Heard, I. Clark-Lewis, D. F. Legler, M. Loetscher, M. Baggiolini, and B. Moser. 1996. The CXC chemokine SDF-1 is the ligand for LESTR/fusin and prevents infection by T-cell-line-adapted HIV-1. *Nature* **382:**833–835.

O'Brien, W. A., P. M. Hartigan, E. S. Daar, M. S. Simberkoff, and J. D. Hamilton. 1997. Changes in plasma HIV RNA levels and CD4+ lymphocyte counts predict both response to antiretroviral therapy and therapeutic failure. VA Cooperative Study Group on AIDS. *Ann. Intern. Med.* **126:**939–945.

Oger, J., and G. Dekaban. 1995. HTLV-I associated myelopathy: a case of viral-induced auto-immunity. *Autoimmunity* **21:**151–159.

Ohara, Y., Y. Iwasaki, S. Izumo, I. Kobayashi, and A. Yoshioka. 1992. Search for human T-cell leukemia virus type I (HTLV-I) proviral sequences by polymerase chain reaction in the central nervous system tissue of HTLV-I-associated myelopathy. *Arch. Virol.* **124:**31–43.

Orenstein, J. M., C. Fox, and S. M. Wahl. 1997. Macrophages as a source of HIV during opportunistic infections. *Science* **276:**1857–1861.

Osame, M., R. Janssen, H. Kubota, H. Nishitani, A. Igata, S. Nagataki, M. Mori, I. Goto, H. Shimabukuro, R. Khabbaz, et al. 1990. Nationwide survey of HTLV-I-associated myelopathy in Japan: association with blood transfusion. *Ann. Neurol.* **28:**50–56.

Osame, M., K. Usuku, S. Izumo, N. Ijichi, H. Amitani, A. Igata, M. Matsumoto, and M. Tara. 1986. HTLV-I associated myelopathy, a new clinical entity. *Lancet* **i:**1031–1032.

Osato, M., K. Yamaguchi, S. Tamiya, H. Yamasaki, T. Okubo, H. Suzushima, N. Asou, K. Sakata, M. Kawakita, and K. Takatsuki. 1996. A human T-cell lymphotropic virus type-I carrier with chronic renal failure, aplastic anemia, myelopathy, uveitis, Sjogren's syndrome and panniculitis. *Intern. Med.* **35:**742–745.

Pachl, C., M. Saxer, T. Elbeik, M. Stempien, S. J. Fong, D. Kern, P. Sheridan, B. Hoo, D. Besemer, R. Kokka, et al. 1993. Quantitation of HIV-1 RNA in plasma using a branched DNA (bDNA) signal amplification assay: evaluation of specimen collection and stability. *Natl. Conf. Hum. Retrovir. Relat. Infect.* **16:**110.

Palella, F. J., Jr., K. M. Delaney, A. C. Moorman, M. O. Loveless, J. Fuhrer, G. A. Satten, D. J. Aschman, and S. D. Holmberg. 1998. Declining morbidity and mortality among patients with advanced human immunodeficiency virus infection. HIV Outpatient Study Investigators. *N. Engl. J. Med.* **338:**853–860.

Pancake, B. A., D. Zucker-Franklin, and E. E. Coutavas. 1995. The cutaneous T cell lymphoma, mycosis fungoides, is a human T cell lymphotropic virus-associated disease. A study of 50 patients. *J. Clin. Investig.* **95:**547–554.

Pancake, B. A., E. H. Wassef, and D. Zucker-Franklin. 1996. Demonstration of antibodies to human T-cell lymphotropic virus-I tax in patients with the cutaneous T-cell lymphoma, mycosis fungoides, who are seronegative for antibodies to the structural proteins of the virus. *Blood* **88:**3004–3009.

Pantaleo, G., C. Graziosi, J. F. Demarest, L. Butini, M. Montroni, C. H. Fox, J. M. Orenstein, D. P. Kotler, and A. S. Fauci. 1993a. HIV infection is active and progressive in lymphoid tissue during the clinically latent stage of disease. *Nature* **362:**355–358.

Pantaleo, G., C. Graziosi, and A. S. Fauci. 1993b. New concepts in the immunopathogenesis of human immunodeficiency virus infection. *N. Engl. J. Med.* **328:**327–335.

Pantaleo, G., J. F. Demarest, H. Soudeyns, C. Graziosi, F. Denis, J. W. Adelsberger, P. Borrow, M. S. Saag, G. M. Shaw, R. P. Sekaly, et al. 1994. Major expansion of CD8+ T cells with a predominant V beta usage during the primary immune response to HIV. *Nature* **370:**463–467.

Pantaleo, G., J. F. Demarest, T. Schacker, M. Vaccarezza, O. J. Cohen, M. Daucher, C. Graziosi, S. S. Schnittman, T. C. Quinn, G. M. Shaw, L. Perrin, G. Tambussi, A. Lazzarin, R. P. Sekaly, H. Soudeyns, L. Corey, and A. S. Fauci. 1997. The qualitative nature of the primary immune response to HIV infection is a prognosticator of disease progression independent of the initial level of plasma viremia. *Proc. Natl. Acad. Sci. USA* **94:**254–258.

Pantaleo, G., O. J. Cohen, T. Schacker, M. Vaccarezza, C. Graziosi, G. P. Rizzardi, J. Kahn, C. H. Fox, S. M. Schnittman, D. H. Schwartz, L. Corey, and A. S. Fauci. 1998. Evolutionary pattern of human immunodeficiency virus (HIV) replication and distribution in lymph nodes following primary infection—implications for antiviral therapy. *Nat. Med.* **4:**341–345.

Parekh, B., S. Phillips, T. C. Granade, J. Baggs, D. J. Hu, and R. Respess. 1999. Impact of HIV type 1 subtype variation on viral RNA quantitation. *AIDS Res. Hum. Retrovir.* **15:**133–142.

Pawson, R., G. J. Mufti, and A. Pagliuca. 1998. Management of adult T-cell leukaemia/lymphoma. *Br. J. Haematol.* **100:**453–458.

Pedersen, C., B. O. Lindhardt, B. L. Jensen, E. Lauritzen, J. Gerstoft, E. Dickmeiss, J. Gaub, E. Scheibel, and T. Karlsmark. 1989. Clinical course of primary HIV infection: consequences for subsequent course of infection. *Br. Med. J.* **299:**154–157.

Pedersen, C., E. Dickmeiss, J. Gaub, L. P. Ryder, P. Platz, B. O. Lindhardt, and J. D. Lundgren. 1990. T-cell subset alterations and lymphocyte responsiveness to mitogens and antigen during severe primary infection with HIV: a case series of seven consecutive HIV seroconverters. *AIDS* **4:**523–526.

Perelson, A. S., A. U. Neumann, M. Markowitz, J. M. Leonard, and D. D. Ho. 1996. HIV-1 dynamics in vivo: virion clearance rate, infected cell life-span, and viral generation time. *Science* **271:**1582–1586.

Perelson, A. S., P. Essunger, Y. Z. Cao, M. Vesanen, A. Hurley, K. Saksela, M. Markowitz, and D. D. Ho. 1997. Decay characteristics of HIV-1-infected compartments during combination therapy. *Nature* **387:**188–191.

Perron, H., J. A. Garson, F. Bedin, F. Beseme, G. Paranhos-Baccala, F. Komurian-Pradel, F. Mallet, P. W. Tuke, C. Voisset, J. L. Blond, B. Lalande, J. M. Seigneurin, and B. Mandrand. 1997. Molecular identification of a novel retrovirus repeatedly isolated from patients with multiple sclerosis. The Collaborative Research Group on Multiple Sclerosis. *Proc. Natl. Acad. Sci. USA* **94:**7583–7588.

Petersen, L. R., G. A. Satten, R. Dodd, M. Busch, S. Kleinman, A. Grindon, and B. Lenes. 1994. Duration of time from onset of human immunodeficiency virus type 1 infectiousness to development of detectable antibody. The HIV Seroconversion Study Group. *Transfusion* **34:**283–289.

Petit, T., E. Gluckman, E. Carosella, Y. Brossard, O. Brison, and G. Socie. 1995. A highly sensitive polymerase chain reaction method reveals the ubiquitous presence of maternal cells in human umbilical cord blood. *Exp. Hematol.* **23:**1601–1605.

Phair, J., L. Jacobson, R. Detels, C. Rinaldo, A. Saah, L. Schrager, and A. Munoz. 1992. Acquired immune deficiency syndrome occurring within 5 years of infection with human immunodeficiency virus type-1: the Multicenter AIDS Cohort Study. *J. Acquir. Immune Defic. Syndr.* **5:**490–496.

Philpott, S. M., and G. C. Buehring. 1999. Defective DNA repair in cells with human T-cell leukemia bovine leukemia viruses: role of tax gene. *J. Natl. Cancer Inst.* **91:**933–942.

Piatak, M., Jr., M. S. Saag, L. C. Yang, S. J. Clark, J. C. Kappes, K. C. Luk, B. H. Hahn, G. M. Shaw, and J. D. Lifson. 1993. High levels of HIV-1 in plasma during all stages of infection determined by competitive PCR. *Science* **259:**1749–1754.

Pinter, A., W. J. Honnen, S. A. Tilley, C. Bona, H. Zaghouani, M. K. Gorny, and S. Zolla-Pazner. 1989. Oligomeric structure of gp41, the transmembrane protein of human immunodeficiency virus type 1. *J. Virol.* **63:**2674–2679.

Pliner, V., J. Weedon, and P. Thomas. 1996. Estimation of long term survival to AIDS in perinatally infected children. *Int. Conf. AIDS* **11:**3473.

Poiesz, B. J., F. W. Ruscetti, M. S. Reitz, V. S. Kalyanaraman, and R. C. Gallo. 1981. Isolation of a new type C retrovirus (HTLV) in primary uncultured cells of a patient with Sezary T-cell leukaemia. *Nature* **294:**268–271.

Pollet, D. E., E. L. Saman, D. C. Peeters, H. M. Warmenbol, L. M. Heyndrickx, C. J. Wouters, G. Beelaert, G. van der Groen, and H. Van Heuverswyn. 1991. Confirmation and differentiation of antibodies to human immunodeficiency virus 1 and 2 with a strip-based assay including recombinant antigens and synthetic peptides. *Clin. Chem.* **37:**1700–1707.

Pope, M., S. Gezelter, N. Gallo, L. Hoffman, and R. M. Steinman. 1995. Low levels of HIV-1 infection in cutaneous dendritic cells promote extensive viral replication upon binding to memory CD4+ T cells. *J. Exp. Med.* **182:**2045–2056.

Popovic, M., M. G. Sarngadharan, E. Read, and R. C. Gallo. 1984. Detection, isolation, and continuous production of cytopathic retroviruses (HTLV-III) from patients with AIDS and pre-AIDS. *Science* **224:**497–500.

Price, R. W. 1996. Neurological complications of HIV infection. *Lancet* **348:**445–452.

Pyra, H., J. Boni, and J. Schupbach. 1994. Ultrasensitive retrovirus detection by a reverse transcriptase assay based on product enhancement. *Proc. Natl. Acad. Sci. USA* **91:**1544–1548.

Rabson, A. B., and B. J. Graves. 1997. Synthesis and processing of viral RNA, p. 205–262. *In* J. M. Coffin, S. H. Hughes, and H. E. Varmus (ed.), *Retroviruses*. Cold Spring Harbor Laboratory Press, Cold Spring Harbor, N.Y.

Rich, J. D., N. A. Merriman, E. Mylonakis, T. C. Greenough, T. P. Flanigan, B. J. Mady, and C. C. Carpenter. 1999. Misdiagnosis of HIV infection by HIV-1 plasma viral load testing: a case series. *Ann. Intern. Med.* **130:**37–39.

Richardson, J. H., A. J. Edwards, J. K. Cruickshank, P. Rudge, and A. G. Dalgleish. 1990. In vivo cellular tropism of human T-cell leukemia virus type 1. *J. Virol.* **64:**5682–5687.

Rodriguez-Rosado, R., C. Briones, and V. Soriano. 1999. Introduction of HIV drug-resistance testing in clinical practice. *AIDS* **13:**1007–1014.

Roos, M. T., J. M. Lange, R. E. de Goede, R. A. Coutinho, P. T. Schellekens, F. Miedema, and M. Tersmette. 1992. Viral phenotype and immune response in primary human immunodeficiency virus type 1 infection. *J. Infect. Dis.* **165:**427–432.

Rosenberg, E. S., J. M. Billingsley, A. M. Caliendo, S. L. Boswell, P. E. Sax, S. A. Kalams, and B. D. Walker. 1997. Vigorous HIV-1-specific CD4(+) T cell responses associated with control of viremia. *Science* **278:**1447–1450.

Rosenberg, E. S., L. LaRosa, T. Flynn, G. Robbins, and B. D. Walker. 1999. Characterization of HIV-1-specific T-helper cells in acute and chronic infection. *Immunol. Lett.* **66:**89–93.

Rosenblatt, J. D., D. W. Golde, W. Wachsman, J. V. Giorgi, A. Jacobs, G. M. Schmidt, S. Quan, J. C. Gasson, and I. S. Chen. 1986. A second isolate of HTLV-II associated with atypical hairy-cell leukemia. *N. Engl. J. Med.* **315:**372–377.

Rowland-Jones, S. L., and A. McMichael. 1995. Immune responses in HIV-exposed seronegatives: have they repelled the virus? *Curr. Opin. Immunol.* **7:**448–455.

Ruscetti, F. W., D. A. Morgan, and R. C. Gallo. 1977. Functional and morphologic characterization of human T cells continuously grown in vitro. *J. Immunol.* **119:**131–138.

Saag, M. S., M. Holodniy, D. R. Kuritzkes, W. A. O'Brien, R. Coombs, M. E. Poscher, D. M. Jacobsen, G. M. Shaw, D. D. Richman, and P. A. Volberding. 1996. HIV viral load markers in clinical practice. *Nat. Med.* **2:**625–629.

Sabino, E. C., M. Zrein, C. P. Taborda, M. M. Otani, G. Ribeiro-Dos-Santos, and A. Saez-Alquezar. 1999. Evaluation of the INNO-LIA HTLV I/II assay for confirmation of human T-cell leukemia virus-reactive sera in blood bank donations. *J. Clin. Microbiol.* **37:**1324–1328.

Sagawa, K., M. Mochizuki, K. Masuoka, K. Katagiri, T. Katayama, T. Maeda, A. Tanimoto, S. Sugita, T. Watanabe, and K. Itoh. 1995. Immunopathological mechanisms of human T cell lymphotropic virus type 1 (HTLV-I) uveitis. Detection of HTLV-I-infected T cells in the eye and their constitutive cytokine production. *J. Clin. Investig.* **95:**852–858.

Saksela, K., G. Cheng, and D. Baltimore. 1995. Proline-rich (PxxP) motifs in HIV-1 Nef bind to SH3 domains of a subset of Src kinases and are required for the enhanced growth of Nef+ viruses but not for down-regulation of CD4. *EMBO J.* **14:**484–491.

Sala, M., G. Zambruno, J. P. Vartanian, A. Marconi, U. Bertazzoni, and S. Wain-Hobson. 1994. Spatial discontinuities in human immunodeficiency virus type 1 quasispecies derived from epidermal Langerhans cells of a patient with AIDS and evidence for double infection. *J. Virol.* **68:**5280–5283.

Samdal, H. H., B. G. Gutigard, D. Labay, S. I. Wiik, K. Skaug, and A. G. Skar. 1996. Comparison of the sensitivity of four rapid assays for the detection of antibodies to HIV-1/HIV-2 during seroconversion. *Clin. Diagn. Virol.* **7:**55–61.

Sarngadharan, M. G., M. Popovic, L. Bruch, J. Schupbach, and R. C. Gallo. 1984. Antibodies reactive with human T-lymphotropic retroviruses (HTLV-III) in the serum of patients with AIDS. *Science* **224:**506–508.

Sato, P. A., W. J. Maskill, H. Tamashiro, and D. L. Heymann. 1994. Strategies for laboratory HIV testing: an examination of alternative approaches not requiring Western blot. *Bull. W. H. O.* **72:**129–134.

Sattar, S. A., and V. S. Springthorpe. 1991. Survival and disinfectant inactivation of the human immunodeficiency virus: a critical review. *Rev. Infect. Dis.* **13:**430–447.

Schacker, T., A. C. Collier, J. Hughes, T. Shea, and L. Corey. 1996. Clinical and epidemiologic features of primary HIV infection. *Ann. Intern. Med.* **125:**257–264.

Schacker, T. W., J. P. Hughes, T. Shea, R. W. Coombs, and L. Corey. 1998. Biological and virologic characteristics of primary HIV infection. *Ann. Intern. Med.* **128:**613–620.

Schellekens, P. T., M. Tersmette, M. T. Roos, R. P. Keet, F. de Wolf, R. A. Coutinho, and F. Miedema. 1992. Biphasic rate of CD4+ cell count decline during progression to AIDS correlates with HIV-1 phenotype. *AIDS* **6:**665–669.

Schmitz, J. E., M. J. Kuroda, S. Santra, V. G. Sasseville, M. A. Simon, M. A. Lifton, P. Racz, K. Tenner-Racz, M. Dalesandro, B. J. Scallon, J. Ghrayeb, M. A. Forman, D. C. Montefiori, E. P. Rieber, N. L. Letvin, and K. A. Reimann. 1999.

Control of viremia in simian immunodeficiency virus infection by CD8(+) lymphocytes. *Science* **283:**857–860.

Schneider, T., R. Ullrich, and M. Zeitz. 1996. The immunologic aspects of human immunodeficiency virus infection in the gastrointestinal tract. *Semin. Gastrointest. Dis.* **7:**19–29.

Schockmel, G. A., S. Yerly, and L. Perrin. 1997. Detection of low HIV-1 RNA levels in plasma. *J. AIDS Hum. Retrovirol.* **14:**179–183.

Schupbach, J. 1996. Licencing of diagnostic HIV kits—to keep a sensible standpoint in an area of permanent revolution. *Clin. Diagn. Virol.* **5:**137–146.

Schüpbach, J. 1999. Human imunodeficiency viruses, p. 847–870. *In* E. J. Baron, P. R. Murray, M. A. Pfaller, F. C. Tenover, and R. H. Yolken (ed.), *Manual of Clinical Microbiology*, 7th ed. ASM Press, Washington, D.C.

Schupbach, J., and J. Boni. 1993. Quantitative and sensitive detection of immune-complexed and free HIV antigen after boiling of serum. *J. Virol. Methods* **43:**247–256. (Erratum, **45:**245, 1993.)

Schüpbach, J., and J. Böni. September 1998. Process for the detection of reverse transcriptase. U.S. patent 5,807,669.

Schupbach, J., and V. S. Kalyanaraman. 1989. Detection of high concentrations of HTLV-1 p24 and a novel gag precursor, p45, in serum immune complexes of a healthy seropositive individual. *Int. J. Cancer.* **44:**90–94.

Schupbach, J., M. Popovic, R. V. Gilden, M. A. Gonda, M. G. Sarngadharan, and R. C. Gallo. 1984a. Serological analysis of a subgroup of human T-lymphotropic retroviruses (HTLV-III) associated with AIDS. *Science* **224:**503–505.

Schupbach, J., M. G. Sarngadharan, and R. C. Gallo. 1984b. Antigens on HTLV-infected cells recognized by leukemia and AIDS sera are related to HTLV viral glycoprotein. *Science* **224:**607–610.

Schupbach, J., V. S. Kalyanaraman, M. G. Sarngadharan, P. A. Bunn, D. W. Blayney, and R. C. Gallo. 1994. Demonstration of viral antigen p24 in circulating immune complexes of two patients with human T-cell leukaemia/lymphoma virus (HTLV) positive lymphoma. *Lancet* **i:**302–305.

Schupbach, J., O. Haller, M. Vogt, R. Luthy, H. Joller, O. Oelz, M. Popovic, M. G. Sarngadharan, and R. C. Gallo. 1985. Antibodies to HTLV-III in Swiss patients with AIDS and pre-AIDS and in groups at risk for AIDS. *N. Engl. J. Med.* **312:**265–270.

Schupbach, J., A. Baumgartner, and Z. Tomasik. 1988. HTLV-1 in Switzerland: low prevalence of specific antibodies in HIV risk groups, high prevalence of cross-reactive antibodies in normal blood donors. *Int. J. Cancer* **42:**857–862.

Schupbach, J., Z. Tomasik, and J. Boni. 1990. Wide divergence of different HIV Western Blot (WB) interpretation standards in low-positive sera. *Int. Conf. AIDS* **6:**215.

Schupbach, J., Z. Tomasik, J. Jendis, J. Boni, R. Seger, and C. Kind. 1994a. IgG, IgM, and IgA response to HIV in infants born to HIV-1 infected mothers. Swiss Neonatal HIV Study Group. *J. AIDS* **7:**421–427.

Schupbach, J., J. Boni, Z. Tomasik, J. Jendis, R. Seger, and C. Kind. 1994b. Sensitive detection and early prognostic significance of p24 antigen in heat-denatured plasma of human immunodeficiency virus type 1-infected infants. Swiss Neonatal HIV Study Group. *J. Infect. Dis.* **170:**318–324.

Schupbach, J., M. Flepp, D. Pontelli, Z. Tomasik, R. Luthy, and J. Boni. 1996a. Heat-mediated immune complex dissociation and enzyme-linked immunosorbent assay signal amplification render p24 antigen detection in plasma as sensitive as HIV-1 RNA detection by polymerase chain reaction. *AIDS* **10:**1085–1090.

Schupbach, J., H. Pyra, J. Jendis, Z. Tomasik, and J. Boni. 1996b. Isolation of an in vitro transmissible agent with reverse transcriptase activity from a blood donor with a borderline-positive HIV-1 serology for more than five years. *Clin. Diagn. Virol.* **5:**197–203.

Schuurman, R., L. Demeter, P. Reichelderfer, J. Tijnagel, T. de Groot, and C. Boucher. 1999. Worldwide evaluation of DNA sequencing approaches for identification of drug resistance mutations in the human immunodeficiency virus type 1 reverse transcriptase. *J. Clin. Microbiol.* **37:**2291–2296.

Sebire, K., K. McGavin, S. Land, and C. Birch. 1996. Stability of HIV RNA in blood specimens. *Annu. Conf. Australas. Soc. HIV Med.* **8:**145.

Sharp, P. M., D. L. Robertson, D. L. Gao, and B. H. Hahn. 1994. Origins and diversity of human immunodeficiency viruses. *AIDS* **8:**S27–S42.

Shida, H., T. Tochikura, T. Sato, T. Konno, K. Hirayoshi, M. Seki, Y. Ito, M. Hatanaka, Y. Hinuma, M. Sugimoto, et al. 1987. Effect of the recombinant vaccinia viruses that express HTLV-I envelope gene on HTLV-I infection. *EMBO J.* **6:**3379–3384.

Shimoyama, M. 1991. Diagnostic criteria and classification of clinical subtypes of adult T-cell leukaemia-lymphoma. A report from the Lymphoma Study Group (1984–87). *Br. J. Haematol.* **79:**428–437.

Silver, J., T. Maudru, K. Fujita, and R. Repaske. 1993. An RT-PCR assay for the enzyme activity of reverse transcriptase capable of detecting single virions. *Nucleic Acids Res.* **21:**3593–3594.

Simon, F., P. Mauclere, P. Roques, I. Loussertajaka, M. C. Mullertrutwin, S. Saragosti, M. C. Georgescourbot, F. Barre Sinoussi, and F. Brunvezinet. 1998. Identification of a new human immunodeficiency virus type 1 distinct from group M and group O. *Nat. Med.* **4:**1032–1037.

Simons, P., G. Muyldermans, P. Lacor, G. Zissis, and S. Lauwers. 1997. False-negative HIV viral load in AIDS patients. *AIDS* **11:**1783–1784.

Sinicco, A., G. Palestro, P. Caramello, D. Giacobbi, G. Giuliani, G. Paggi, M. Sciandra, and P. Gioannini. 1990. Acute HIV-1 infection: clinical and biological study of 12 patients. *J. Acquir. Immune Defic. Syndr.* **3:**260–265.

Sinicco, A., R. Fora, M. Sciandra, A. Lucchini, P. Caramello, and P. Gioannini. 1993. Risk of developing AIDS after primary acute HIV-1 infection. *J. Acquir. Immune Defic. Syndr.* **6:**575–581.

Smith, M. W., M. Dean, M. Carrington, C. Winkler, G. A. Huttley, D. A. Lomb, J. J. Goedert, T. R. Obrien, L. P. Jacobson, R. Kaslow, S. Buchbinder, E. Vittinghoff, D. Vlahov, K. Hoots, M. W. Hilgartner, and S. J. Obrien. 1997. Contrasting genetic influence of CCR2 and CCR5 variants on HIV-1 infection and disease progression. *Science* **277:**959–965.

Spira, A. I., P. A. Marx, B. K. Patterson, J. Mahoney, R. A. Koup, S. M. Wolinsky, and D. D. Ho. 1996. Cellular targets of infection and route of viral dissemination after an intravaginal inoculation of simian immunodeficiency virus into rhesus macaques. *J. Exp. Med.* **183:**215–225.

Strizki, J. M., A. V. Albright, H. Sheng, M. O'Connor, L. Perrin, and F. Gonzalez-Scarano. 1996. Infection of primary human microglia and monocyte-derived macrophages with human immunodeficiency virus type 1 isolates: evidence of differential tropism. *J. Virol.* **70:**7654–7662.

Sugimoto, M., M. Kitaichi, A. Ikeda, S. Nagai, and T. Izumi. 1998. Chronic bronchioloalveolitis associated with human T-cell lymphotrophic virus type I infection. *Curr. Opin. Pulm. Med.* **4:**98–102.

Switzer, W. M., S. Wiktor, V. Soriano, A. Silva-Graca, K. Mansinho, I. M. Coulibaly, E. Ekpini, A. E. Greenberg, T. M. Folks, and W. Heneine. 1998. Evidence of Nef truncation in human immunodeficiency virus type 2 infection. *J. Infect. Dis.* **177:**65–71.

Tajima, K. 1990. The 4th nation-wide study of adult T-cell leukemia/lymphoma (ATL) in Japan: estimates of risk of ATL and its geographical and clinical features. The T- and B-cell Malignancy Study Group. *Int. J. Cancer* **45:**237–243.

Tajima, K., and L. Cartier. 1995. Epidemiological features of HTLV-I and adult T cell leukemia. *Intervirology* **38:**238–246.

Takatsuki, K., J. Uchiyama, K. Sagawa, and J. Yodi. 1977. Adult T-cell leukemia in Japan, p. 73–77. *In* S. Senos, F. Takaku, and S. Irino (ed.), *Topics in Hematology*. Excerpta Medica, Amsterdam, The Netherlands.

Takatsuki, K., M. Matsuoka, and K. Yamaguchi. 1996. Adult T-cell leukemia in Japan. *J. Acquir. Immune Defic. Syndr. Hum. Retrovirol.* **13:**S15–S19.

Tan, X., and D. M. Phillips. 1996. Cell-mediated infection of cervix derived epithelial cells with primary isolates of human immunodeficiency virus. *Arch. Virol.* **141:**1177–1189.

Taylor, G. P. 1998. Pathogenesis and treatment of HTLV-I associated myelopathy. *Sex. Transm. Infect* **74:**316–322.

Taylor, M. B., S. P. Parker, H. H. Crewe-Brown, J. McIntyre, and W. D. Cubitt. 1996. Seroepidemiology of HTLV-I in relation to that of HIV-1 in the Gauteng region, South Africa, using dried blood spots on filter papers. *Epidemiol. Infect.* **117:**343–348.

Tenner-Racz, K., P. Racz, M. Bofill, A. Schulz-Meyer, M. Dietrich, P. Kern, J. Weber, A. J. Pinching, F. Veronese-Dimarzo, M. Popovic, et al. 1986. HTLV-III/LAV viral antigens in lymph nodes of homosexual men with persistent generalized lymphadenopathy and AIDS. *Am. J. Pathol.* **123:**9–15.

Tenner-Racz, K., P. Racz, H. Schmidt, M. Dietrich, P. Kern, A. Louie, S. Gartner, and M. Popovic. 1988. Immunohistochemical, electron microscopic and in situ hybridization evidence for the involvement of lymphatics in the spread of HIV-1. *AIDS* **2:**299–309.

Tjotta, E., O. Hungnes, and B. Grinde. 1991. Survival of HIV-1 activity after disinfection, temperature and pH changes, or drying. *J. Med. Virol.* **35:**223–227.

Uchiyama, T. 1997. Human T cell leukemia virus type I (HTLV-I) and human diseases. *Annu. Rev. Immunol.* **15:**15–37.

Uchiyama, T., J. Yodoi, K. Sagawa, K. Takatsuki, and H. Uchino. 1977. Adult T-cell leukemia: clinical and hematologic features of 16 cases. *Blood* **50:**481–492.

Urassa, W. K., U. Bredberg-Raden, E. Mbena, K. Palsson, E. Minja, R. A. Lema, K. Pallangyo, F. S. Mhalu, and G. Biberfeld. 1992. Alternative confirmatory strategies in HIV-1 antibody testing. *J. Acquir. Immune Defic. Syndr.* **5:**170–176.

Urdea, M. S., J. C. Wilber, T. Yeghiazarian, J. A. Todd, D. G. Kern, S. J. Fong, D. Besemer, B. Hoo, P. J. Sheridan, R. Kokka, et al. 1993. Direct and quantitative detection of HIV-1 RNA in human plasma with a branched DNA signal amplification assay. *AIDS* **7:**S11–S14.

Vallari, A. S., R. K. Hickman, J. R. Hackett, C. A. Brennan, V. A. Varitek, and S. G. Devare. 1998. Rapid assay for simultaneous detection and differentiation of immunoglobulin G antibodies to human immunodeficiency virus type 1 (HIV-1) group M, HIV-1 group O, and HIV-2. *J. Clin Microbiol.* **36:**3657–3661.

van Bueren, J., D. P. Larkin, and R. A. Simpson. 1994. Inactivation of human immunodeficiency virus type 1 by alcohols. *J. Hosp. Infect.* **28:**137–148.

van Bueren, J., R. A. Simpson, H. Salman, H. D. Farrelly, and B. D. Cookson. 1995. Inactivation of HIV-1 by chemical disinfectants: sodium hypochlorite. *Epidemiol. Infect.* **115:**567–579.

Vandamme, A. M., K. Van Laethem, H. F. Liu, M. Van Brussel, E. Delaporte, C. M. de Castro Costa, C. Fleischer, G. Taylor, U. Bertazzoni, J. Desmyter, and P. Goubau. 1997. Use of a generic polymerase chain reaction assay detecting human T-lymphotropic virus (HTLV) types I, II and divergent simian strains in the evaluation of individuals with indeterminate HTLV serology. *J. Med. Virol.* **52:**1–7.

Vandamme, A. M., M. Salemi, M. Van Brussel, H. F. Liu, K. Van Laethem, M. Van Ranst, L. Michels, J. Desmyter, and P. Goubau. 1998. African origin of human T-lymphotropic virus type 2 (HTLV-2) supported by a potential new HTLV-2d subtype in Congolese Bambuti Efe Pygmies. *J. Virol.* **72:**4327–4340.

van der Groen, G., I. Van Kerckhoven, G. Vercauteren, and P. Piot. 1991. Simplified and less expensive confirmatory HIV testing. *Bull. W.H.O.* **69:**747–752.

Van Dooren, S., E. Gotuzzo, M. Salemi, D. Watts, E. Audenaert, S. Duwe, H. Ellerbrok, R. Grassmann, E. Hagelberg, J. Desmyter, and A. M. Vandamme. 1998. Evidence for a post-Columbian introduction of human T-cell lymphotropic virus in Latin America. *J. Gen. Virol.* **79:**2695–2708.

Vercauteren, G., G. Beelaert, and G. van der Groen. 1995. Evaluation of an agglutination HIV-1 + 2 antibody assay. *J. Virol. Methods* **51:**1–8.

Vernazza, P. L., J. J. Eron, S. A. Fiscus, and M. S. Cohen. 1999. Sexual transmission of HIV: infectiousness and prevention. *AIDS* **13:**155–166.

Vrielink, H., H. W. Reesink, M. R. Habibuw, M. Schuller, C. B. M. van der Meer, and P. N. Lelie. 1999. Comparison of four HTLV-I and HTLV-I plus II ELISAs. *Vox Sang.* **76:**187–191.

Wain-Hobson, S. 1992. Human immunodeficiency virus type 1 quasispecies in vivo and ex vivo. *Curr. Top. Microbiol. Immunol.* **176:**181–193.

Wang, B., Y. C. Ge, P. Palasanthiran, S. H. Xiang, J. Ziegler, D. E. Dwyer, C. Randle, D. Dowton, A. Cunningham, and N. K. Saksena. 1996. Gene defects clustered at the C-terminus of the vpr gene of HIV-1 in long-term nonprogressing mother and child pair: in vivo evolution of vpr quasispecies in blood and plasma. *Virology* **223:**224–232.

Watanabe, T., M. Mochizuki, and K. Yamaguchi. 1997. HTLV-1 uveitis (HU). *Leukemia* **3:**582–584.

Wattel, E., M. Cavrois, A. Gessain, and S. Wain-Hobson. 1996. Clonal expansion of infected cells: a way of life for HTLV-I. *J. Acquir. Immune Defic. Syndr. Hum. Retrovirol.* **13:**S92–S99.

Weber, B., E. H. M. Fall, A. Berger, and H. W. Doerr. 1998. Reduction of diagnostic window by new fourth-generation human immunodeficiency virus screening assays. *J. Clin. Microbiol.* **36:**2235–2239.

Wei, X., S. K. Ghosh, M. E. Taylor, V. A. Johnson, E. A. Emini, P. Deutsch, J. D. Lifson, S. Bonhoeffer, M. A. Nowak, B. H. Hahn, et al. 1995. Viral dynamics in human immunodeficiency virus type 1 infection. *Nature* **373:**117–122.

Weissmahr, R. N., J. Schupbach, and J. Boni. 1997. Reverse transcriptase activity in chicken embryo fibroblast culture supernatants is associated with particles containing endogenous avian retrovirus EAV-0 RNA. *J. Virol.* **71:**3005–3012.

Weissman, D., Y. Li, J. Ananworanich, L. J. Zhou, J. Adelsberger, T. F. Tedder, M. Baseler, and A. S. Fauci. 1995. Three populations of cells with dendritic morphology exist in peripheral blood, only one of which is infectable with human immunodeficiency virus type 1. *Proc. Natl. Acad. Sci. USA* **92:**826–830.

Weissman, D., R. L. Rabin, J. Arthos, A. Rubbert, M. Dybul, R. Swofford, S. Venkatesan, J. M. Farber, and A. S. Fauci. 1997. Macrophage-tropic Hiv and Siv envelope proteins induce a signal through the Ccr5 chemokine receptor. *Nautre* **389:**981–985.

Welles, S. L., J. B. Jackson, B. Yen-Lieberman, L. Demeter, A. J. Japour, L. M. Smeaton, V. A. Johnson, D. R. Kuritzkes, R. T. D'Aquila, P. A. Reichelderfer, D. D. Richman, R. Reichman, M. Fischl, R. Dolin, R. W. Coombs, J. O. Kahn, C. McLaren, J. Todd, S. Kwok, and C. S. Crumpacker. 1996. Prognostic value of plasma human immunodeficiency virus type 1 (HIV-1) RNA levels in patients with advanced HIV-1 disease and with little or no prior zidovudine therapy. AIDS Clinical Trials Group Protocol 116A/116B/117 Team. *J. Infect. Dis.* **174:**696–703.

Wiktor, S. Z., S. Jacobson, S. H. Weiss, G. M. Shaw, J. S. Reuben, V. J. Shorty, D. E. McFarlin, and W. A. Blattner. 1991. Spontaneous lymphocyte proliferation in HTLV-II infection. *Lancet* **337:**327–328.

Wodarz, D., M. A. Nowak, and C. R. Bangham. 1999. The dynamics of HTLV-I and the CTL response. *Immunol. Today* **20:**220–227.

Wong, J. K., M. Hezareh, H. F. Gunthard, D. V. Havlir, C. C. Ignacio, C. A. Spina, and D. D. Richman. 1997. Recovery of replication-competent HIV despite prolonged suppression of plasma viremia. *Science* **278:**1291–1295.

Wood, G. S., J. M. Schaffer, R. Boni, R. Dummer, G. Burg, M. Takeshita, and M. Kikuchi. 1997. No evidence of HTLV-I proviral integration in lymphoproliferative disorders associated with cutaneous T-cell lymphoma. *Am. J. Pathol.* **150:**667–673.

Yamaguchi, K., H. Nishimura, H. Kohrogi, M. Jono, Y. Miyamoto, and K. Takatsuki. 1983. A proposal for smoldering adult T-cell leukemia: a clinicopathologic study of five cases. *Blood* **62:**758–766.

Yamaguchi, K., M. Mochizuki, T. Watanabe, K. Yoshimura, M. Shirao, S. Araki, N. Miyata, S. Mori, T. Kiyokawa, and K. Takatsuki. 1994. Human T lymphotropic virus type 1 uveitis after Graves' disease. *Br. J. Ophthalmol.* **78:**163–166.

Yamashita, M., E. Ido, T. Miura, and M. Hayami. 1996. Molecular epidemiology of HTLV-I in the world. *J. Acquir. Immune Defic. Syndr. Hum. Retrovirol.* **13:**S124–S131.

Yang, O. O., S. A. Kalams, M. Rosenzweig, A. Trocha, N. Jones, M. Koziel, B. D. Walker, and R. P. Johnson. 1996. Efficient lysis of human immunodeficiency virus type 1-infected cells by cytotoxic T lymphocytes. *J. Virol.* **70:**5799–5806.

Yoshida, M. 1996. Molecular biology of HTLV-I: recent progress. *J. Acquir. Immune Defic. Syndr. Hum. Retrovirol.* **13:**S63–S68.

Zehender, G., M. Girotto, C. De Maddalena, G. Francisco, M. Moroni and M. Galli. 1996. HTLV infection in ELISA-negative blood donors. *AIDS Res. Hum. Retrovir.* **12:**737–740.

Zhang, L. Q., P. J. Dailey, T. He, A. Gettie, S. Bonhoeffer, A. S. Perelson, and D. D. Ho. 1999a. Rapid clearance of simian immunodeficiency virus particles from plasma of rhesus macaques. *J. Virol.* **73:**855–860.

Zhang, L. Q., B. Ramratnam, K. Tenner-Racz, Y. X. He, M. Vesanen, S. Lewin, A. Talal, P. Racz, A. S. Perelson, B. T. Korber, M. Markowitz, and D. D. Ho. 1999b. Quantifying residual HIV-1 replication in patients receiving combination antiretroviral therapy. *N. Engl. J. Med.* **340:**1605–1613.

Zhu, T., H. Mo, N. Wang, D. S. Nam, Y. Cao, R. A. Koup, and D. D. Ho. 1993. Genotypic and phenotypic characterization of HIV-1 patients with primary infection. *Science* **261:**1179–1181.

Zhu, T., N. Wang, A. Carr, D. S. Nam, R. Moor-Jankowski, D. A. Cooper, and D. D. Ho. 1996. Genetic characterization of human immunodeficiency virus type 1 in blood and genital secretions: evidence for viral compartmentalization and selection during sexual transmission. *J. Virol.* **70:**3098–3107.

Zhu, T. F., B. T. Korber, A. J. Nahmias, E. Hooper, P. M. Sharp, and D. D. Ho. 1998. An African HIV-1 sequence from 1959 and implications for the origin of the epidemic. *Nature* **391:**594–597.

Zoeteweij, J. P., and A. Blauvelt. 1998. HIV-dendritic cell interactions promote efficient viral infection of T cells. *J. Biomed. Sci.* **5:**253–259.

Zucker-Franklin, D., and B. A. Pancake. 1998. Human T-cell lymphotropic virus type 1 tax among American blood donors. *Clin. Diagn. Lab. Immunol.* **5:**831–835.

Zucker-Franklin, D., B. A. Pancake, M. Marmor, and P. M. Legler. 1997. Reexamination of human T cell lymphotropic virus (HTLV-I/II) prevalence. *Proc. Natl. Acad. Sci. USA* **94:**6403–6407.

Chlamydiae

JULIUS SCHACHTER

38

Although members of the genus *Chlamydia* cause a number of human diseases, in clinical virology the diagnosis most often requested is that of localized *Chlamydia trachomatis* lower genital tract infections (Schachter and Stamm, 1999). The lymphogranuloma venereum (LGV) serovars of *C. trachomatis* cause a more invasive systemic sexually transmitted disease which is relatively uncommon in the United States. *C. psittaci*, another species, is a very common organism in domestic mammals and is virtually ubiquitous in the avian kingdom. It affects humans only as a zoonosis, and this diagnosis is usually established serologically (Schachter and Stamm, 1999). Serology is also usually used to diagnose *C. pneumoniae* infection. This organism is a common cause of human respiratory disease (Grayston et al., 1990). Knowledge of *C. pneumoniae* infections is still in its infancy, with research exploring a broader clinical spectrum. It is uncertain how much pressure will be placed on diagnostic laboratories to test for *C. pneumoniae* infections.

During their growth, chlamydiae produce characteristic intracytoplasmic inclusions which were first seen in the epithelia of experimentally infected nonhuman primates (Halberstaedter and von Prowazek, 1907). By 1910, similar studies using Giemsa staining of epithelial-cell scrapings had demonstrated that *C. trachomatis* could infect the conjunctivae of adults and newborns, as well as cervical or urethral epithelial cells (Lindner, 1910).

The trachoma agent was first isolated by T'ang and colleagues in China in 1957 (T'ang et al., 1957). The interest in sexually transmitted chlamydial infections was renewed by the reports of Jones and colleagues, which documented recovery of the organism from the urethras and cervixes of adults in England (Jones, 1964). The introduction of tissue culture procedures for the isolation of *C. trachomatis* increased the clinical relevance of its detection, as the earlier yolk sac procedures used for isolation of the organism were very time-consuming (Gordon and Quan, 1965). The original cell culture isolation procedure involved irradiation of McCoy cells. This has now been supplanted by treatment of the tissue culture cells with cycloheximide (Ripa and Mardh, 1977). Culture diagnosis can be made in 2 to 7 days after processing of the specimen. Nonculture methods, such as direct fluorescent antibody (DFA) techniques and enzyme immunoassays (EIAs), made diagnosis of chlamydial infections widely available (Stamm, 1988). These procedures are being replaced by more sensitive nucleic acid amplification procedures (Schachter, 1998).

The importance of *C. psittaci* as a human pathogen was made apparent in a pandemic in 1929–1930 (Schachter and Dawson, 1978). Human psittacosis was recognized as a severe respiratory disease, which was life threatening until the introduction of the tetracyclines. More recently, *C. pneumoniae*, an organism which was isolated during trachoma surveys and recognized to be an atypical chlamydia when compared to trachoma strains, was found to be an important cause of human respiratory disease (Grayston et al., 1990). Current research is exploring the possible etiologic role of this organism in coronary artery disease (Saikku et al., 1988).

BIOLOGY

The chlamydiae are among the more common pathogens found throughout the animal kingdom (Storz, 1971; Page, 1972; Schachter, 1978; Schachter and Dawson, 1978). They are nonmotile, gram-negative, obligate intracellular bacteria. Their unique developmental cycle differentiates them from all other microorganisms (Moulder et al., 1984). They replicate within the cytoplasm of host cells, forming characteristic membrane-bound inclusions that can be seen by light microscopy. They differ from the viruses in possessing both RNA and DNA and cell walls that are quite similar in structure to those of gram-negative bacteria. However, they lack peptidoglycan; their structural integrity depends on disulfide binding of outer membrane proteins. They are susceptible to many broad-spectrum antibiotics, possess a number of enzymes, and have a restricted metabolic capacity. None of these metabolic reactions results in the production of energy. Thus, they have been considered energy parasites that use the ATP produced by the host cell for their own metabolic requirements (Moulder, 1966). Many aspects of chlamydial molecular biology are not well understood. The recent sequencing of the *C. trachomatis* genome should provide researchers with many relevant leads (Stephens et al., 1998).

Taxonomy

Chlamydiae are presently placed in their own order, *Chlamydiales*, family *Chlamydiaceae*, with one genus, *Chlamydia* (Moulder, 1984). There are four species, *C. trachomatis*, *C. psittaci*, *C. pecorum*, and *C. pneumoniae*. *C. trachomatis* includes the organisms causing trachoma, inclusion conjunctivitis, LGV, and other sexually transmitted infections and some strains producing pneumonia in rodents. *C. trachomatis*

561

strains are sensitive to sulfonamides. They produce a glycogenlike material while replicating within the cytoplasmic inclusion vacuole, which stains with iodine. *C. psittaci* strains infect many avian species and mammals, producing the diseases psittacosis, ornithosis, feline pneumonitis, bovine abortion, etc. (Storz, 1971; Page, 1972). They are resistant to the action of sulfonamides and produce inclusions which do not stain with iodine. *C. pecorum* is quite similar to *C. psittaci*. It can be differentiated by monoclonal antibody or DNA homology testing and is a pathogen of ruminants (Fukushi et al., 1992). *C. pneumoniae* has characteristics similar to those of *C. psittaci* but shows little DNA relatedness to the other species. The elementary bodies (EBs) of *C. pneumoniae* appear to be pear shaped rather than round, as are the EBs of other *Chlamydia* species (Grayston et al., 1989).

Growth Cycle

Chlamydiae are ingested by susceptible host cells (Byrne and Moulder, 1978). Following attachment at specific sites on the surface of the cell, the EB enters the cell through a process similar to receptor-mediated endocytosis and resides in an endosome, where the entire growth cycle is completed. The chlamydiae prevent phagosome-lysosome fusion. The inclusion membrane is modified by insertion of chlamydia-specific antigens (Rockey et al., 1995). Once the EB (diameter, 0.25 to 0.35 μm) has entered the cell, it reorganizes into a reticulate body (RB), which is larger (0.5 to 1 μm) and contains more RNA. After approximately 8 h, the RB begins dividing by binary fission. Approximately 18 to 24 h after infection, these RBs begin to become EBs by a reorganization or condensation process that is poorly understood. The EBs are then released to initiate another cycle of infection. The EBs are specifically adapted for extracellular survival and are the infectious form, whereas the intracellular metabolically active and replicating form, the RB, does not survive well outside the host cell and is adapted for an intracellular milieu. Thus, the defining characteristic of chlamydiae is the unique growth cycle that involves alternation between two highly specialized morphological forms.

Immunology

The chlamydiae possess group (genus)-specific, species-specific, and type-specific antigens. Most of these are located within the cell wall, but the precise structural relationships are not known. The major outer membrane protein (MOMP) contains species-, subspecies-, and serovar-specific antigens (Caldwell and Schachter, 1982; Stephens et al., 1982). The group antigen, shared by all members of the genus, appears to be a lipopolysaccharide (LPS) with a ketodeoxyoctanoic acid as the reactive moiety (Dhir et al., 1971). It is similar to the LPSs of some gram-negative bacteria (Nurminen et al., 1983). Specific antigens of *C. psittaci* strains can be demonstrated by neutralization tests (Banks et al., 1970). The specific antigens of *C. trachomatis* are best recognized by a microimmunofluorescence (micro-IF) technique (Wang and Grayston, 1970), although these antigens are also associated with a toxic factor (large numbers of viable chlamydiae may kill mice in less than 24 h after intravenous inoculation). The MOMP molecule has been cloned and sequenced, and species- and serovar-specific antigens have been identified within its variable regions (Stephens et al., 1988).

PATHOGENESIS

C. trachomatis is almost exclusively a human pathogen (Grayston and Wang, 1975; Schachter, 1978). Serovars within this species cause trachoma (serovars A, B, Ba, and C have been associated with endemic trachoma, the most common form of preventable blindness), inclusion conjunctivitis, and LGV (serovars L1, L2, and L3). Where sexual transmission of *C. trachomatis* strains other than LGV has been studied, serovars D through K have been found to be the major identifiable cause of nongonococcal urethritis in men and may also cause epididymitis. Proctitis may occur in either sex. In women, cervicitis is a common result of chlamydial infection, and acute salpingitis may occur. The agent in the cervix may be transmitted to the neonate as it passes through the infected birth canal; eye disease, inclusion conjunctivitis of the newborn, and a characteristic chlamydial pneumonia of infants may develop (Beem and Saxon, 1977). Vaginal infection and enteric infection in neonates have been recognized.

The trachoma biovar is essentially a parasite of squamocolumnar epithelial cells; the LGV biovar is more invasive and will involve lymphoid cells. Typical of the genus, *C. trachomatis* strains are capable of causing chronic and clinically inapparent infections. Because their growth cycle is approximately 48 to 72 h long, the incubation periods are relatively long, generally 1 to 3 weeks. *C. trachomatis* causes cell death as a result of its replicative cycle and thus is capable of producing cell damage whenever it persists. However, because there are no toxic effects demonstrated nor is there sufficient cell death as a result of replication, it is likely that the majority of the disease manifestations are due to immunopathological mechanisms or nonspecific host responses to the organism or its byproducts. Genus-specific proteins can be found in extracts of EBs, and one such protein has been implicated as a potential sensitizing antigen capable of inducing hypersensitivity reactions in the eye and skin in previously infected hosts (Watkins et al., 1986). This protein has been identified as a heat shock protein (HSP60) which shares antigenic epitopes with similar proteins of other bacteria (Morrison et al., 1989).

The prevailing theory is that repeat infections sensitize the host, and local delayed hypersensitivity reactions to HSP60 (at the site of infection) cause damage. Persistent or recurrent infections are associated with development of complications of simple epithelial infections. A common endpoint of these late consequences is scarring of mucous membranes. The genital complications include pelvic inflammatory disease (PID) and its late consequences of infertility and ectopic pregnancy (Hillis, 1997), while ocular infections may lead to blinding trachoma (Schachter and Dawson, 1978). High levels of antibody to HSP60 are associated with tubal-factor infertility, ectopic pregnancy (Toye et al., 1993; Wagar et al., 1990), and scarring trachoma (Peeling et al., 1998).

In the absence of therapy, chlamydial infections may persist for years, although symptoms usually abate. The pathogenic mechanisms of *C. pneumoniae* have yet to be elucidated. The same may be said for *C. psittaci*, except that this agent infects cells very efficiently, and disease may reflect cytopathic effects.

EPIDEMIOLOGY

C. trachomatis Infections

Trachoma

The World Health Organization estimates that approximately 6 million people have been blinded by trachoma and that hundreds of millions live in areas of endemicity (Thylefors et al., 1995). The areas of the world where trachoma is hyperendemic are in North Africa and sub-Saharan Africa, the Middle East, the drier regions of the Indian subcontinent, and Southeast Asia. Foci of trachoma persist in Australia, the South Pacific, and Latin America.

In areas of hyperendemicity, the prevalence of trachoma is essentially 100% by the second or third year of life (Dawson et al., 1976). Active disease is most common in young children, who are the reservoir of the disease. By adult life, active infection is infrequent, but the sequelae of the disease result in blindness. In such areas, trachoma constitutes the major cause of blindness. Worldwide, trachoma is considered the most common cause of preventable blindness.

Genital Infection

C. trachomatis is the most common sexually transmitted bacterial pathogen. Estimates of the annual incidence are approximately 90 million infections worldwide, with 3 to 4 million occurring in the United States (World Health Organization, 1996; Centers for Disease Control and Prevention, 1993). *C. trachomatis*, like other genital pathogens, has been found with higher frequency among individuals who are younger, more often nonwhite, unmarried, and of lower socioeconomic status. In the United States, most of the estimated 2 million cases of acute urethritis are nongonococcal, and *C. trachomatis* causes 30 to 50% of these cases (Holmes et al., 1975).

Epididymitis, a severe complication of urethritis, occurs in approximately 500,000 men per year in the United States, and roughly half of these cases are caused by *C. trachomatis*. *C. trachomatis* is recovered from 20 to 30% of men with gonorrhea. Postgonococcal urethritis is usually a chlamydial infection following gonococcal infection, as the concomitant infection with *C. trachomatis* does not respond to the treatment for gonorrhea. Chlamydial and gonococcal infections also occur together in women (35 to 45% double infection is not uncommon), and inadequate treatment may be followed by salpingitis. This condition, often called PID, is the most important consequence of genital chlamydial infection. Late consequences of PID include tubal-factor infertility and ectopic pregnancy, and chlamydial PID may cause more than 30% of these conditions. A silent salpingitis may occur and still produce sufficient tubal damage to cause the late consequences.

C. trachomatis infections may be persistent and asymptomatic ("latent" or "silent" infection) for prolonged periods. For example, infants have been documented to shed organisms from their conjunctivae for more than 2 years postdelivery. Similarly, women with cervical infection have been culture positive for over 15 months (McCormack, 1979).

Approximately 20 to 30% of infants exposed to *C. trachomatis* in the birth canal will develop conjunctivitis, and 10 to 15% will develop pneumonia (Schachter et al., 1986; Centers for Disease Control and Prevention, 1993). Conjunctivitis develops after 5 to 19 days of life. Infant pneumonia develops from 2 weeks to 4 months of age. *C. trachomatis* has been estimated to cause 20 to 30% of pneumonia cases in infants under 6 months of age.

C. psittaci Infection (Psittacosis)

Psittacosis is a relatively rare disease in the United States. In 1994, 38 cases were reported (0.2/1,000,000 population) (Centers for Disease Control and Prevention, 1994). Most of the patients had exposure to psittacine birds. Virtually all avian species are potential reservoirs. Meyer listed 130 species known to be such sources in 1967, and the list could no doubt be lengthened (Meyer, 1967). The risk of exposure is greatly increased in those occupations where handling birds or their carcasses is common.

C. pneumoniae Infection

Studies in many countries found 20 to 80% of adults with antibody to *C. pneumoniae* and in most countries the prevalence is 50% or more (Grayston et al., 1986). Antibody is uncommon under the age of 5 years, and then age-specific prevalence increases rapidly from 5 to 20 years. The prevalence then increases slowly but steadily throughout life. In Seattle, men over 60 have a prevalence of 70%. Rates are higher in men than in women after the age of 20. There is some loss of antibody noted 3 to 5 years after the first infection, so the high rates of antibody after 70 suggest that virtually everyone is infected at some time in their lives. Reinfections can occur. There is no evidence for other than human-to-human transmission of *C. pneumoniae* infection.

Epidemics have been studied in military recruits in Finland and in one adjacent civilian community (Kleemola et al., 1988). Attack rates varied from 60 to 80 per 1,000 men. Transmission was slow, with epidemics lasting as long as 6 months.

LABORATORY DIAGNOSIS

Collection of Specimens

For cytological studies, impression smears of involved tissues or scrapings of involved epithelial-cell sites should be appropriately fixed (methanol may be used for immunofluorescent and Giemsa stains). It is imperative that samples be collected from the involved epithelial-cell sites by vigorous swabbing or scraping. This is also true for the isolation attempts. Purulent discharges are inadequate and should be cleaned from the site prior to sampling.

For most *C. trachomatis* infections of humans, the involved mucous membranes should be vigorously swabbed or sampled by scraping. Thus, the conjunctiva for trachoma inclusion conjunctivitis, the anterior urethra (several centimeters into the male urethra) for urethritis, or the cervix (within the endocervical canal) for cervicitis would be sampled. As these strains appear to infect only columnar or squamocolumnar cells, cervical specimens must be collected at the transitional zone or within the os. The organism can also infect the urethra of the female, and it may improve recovery rates if another sample is collected from the urethra and sent to the laboratory for testing in the same tube with the cervical sample. For women with salpingitis, the samples may be collected by needle aspiration of the involved fallopian tube, or endometrial specimens may yield the agent. Rectal mucosa, nasopharynx, and throat may also be sampled. For infants with pneumonia, swabs may be collected from the posterior nasopharynx or the throat.

Isolation

The recommended procedures for primary isolation of chlamydiae use cultured cells for the growth of the organism. The most common technique involves inoculation of clinical specimens by centrifugation into cycloheximide-treated McCoy cells (Ripa and Mardh, 1977). Some workers have found human cells to be more sensitive for isolation of both *C. trachomatis* and *C. pneumoniae*. At present, HL or HEp2 cells appear to be the cell line of choice for isolation of *C. pneumoniae* (Cles and Stamm, 1990; Hyman et al., 1995). *C. psittaci* will grow in cell culture, in yolk sac, and in mice by a variety of routes of inoculation. These organisms present a threat to laboratory workers, and laboratory infections have been common.

Nonculture Methods

There are a number of ways to detect chlamydiae directly in clinical specimens; some of the procedures have been used for many years. For example, direct microscopic examination for organisms in impression smears of infected avian tissues is still a useful procedure. A Macchiavello stain or modifications, such as the Gimenez stain, may be useful in detecting EBs in infected avian species or infected animals, but not humans (Schachter and Dawson, 1978). Direct light microscopy of human specimens is less often used, as modern techniques have replaced the Giemsa stain. Both EIA and DFA procedures are commercially available. Newer procedures, first DNA hybridization and more recently nucleic acid amplification tests, are finding extensive use in diagnosing chlamydial infection. These nonculture assays are not used where diagnosis may be required for legal purposes (such as instances of sexual abuse or rape) because of limitations of specificity.

Nonculture tests do offer a number of advantages. They are less technically demanding than the culture methods and thus are more widely available. However, because all nonculture tests have some false-positive reactions, they must be used with care in low-prevalence settings. The DFA procedures are best used for evaluating relatively small numbers of specimens, as they can be relatively time-consuming and require expensive microscopes and trained microscopists. They will, however, provide laboratories with the assurance that an adequate specimen was collected. The same is not true for the EIAs or nucleic acid assays (nor for that matter for culture). The EIAs offer the advantage of allowing batch processing of specimens.

Nucleic Acid Technology

Direct methods of detection of chlamydial DNA have not been more sensitive than the antigen detection methods. However, the introduction of amplified nucleic acid technology, such as PCR, ligase chain reaction, and transcription-mediated amplification, in the mid-1990s revolutionized chlamydial diagnostics (Jaschek, 1993; Schachter et al., 1994; Crotchfelt et al., 1998). These tests are highly specific and are more sensitive than any other diagnostic tests, including culture. The chlamydial genes can be detected in epithelial-cell specimens, in male and female urine, and in vaginal swabs (Pate et al., 1998; Stary et al., 1998). The potential for widespread use of noninvasive screening tests is an important option offered by this technology. As a group, nucleic acid amplification tests are clearly the tests of choice for diagnosis of C. trachomatis infection.

Fluorescent-Antibody Technique

Fluorescein-conjugated monoclonal antibodies are commercially available and are routinely used in some laboratories to identify chlamydial inclusions in infected cell cultures (Stamm et al., 1983). These antibodies may be used directly on clinical specimens as well (Tam et al., 1984). Species-specific antibodies to C. trachomatis are recommended for staining clinical specimens suspected of containing the organisms. Genus-specific antibodies may be used for infected cell cultures, but they are less suited for use in direct detection in clinical specimens because of the irregular staining of the LPS in the EB. C. pneumoniae-specific monoclonal antibodies appear to be better able to detect the organism in infected cell cultures, as the infectivity of the organism appears to be relatively low. The DFA technique is somewhat less sensitive than isolation in culture but is faster and less expensive and may represent an alternative method of diagnosing chlamydial infections in settings where cell culture isolation is not available and nucleic acid amplification tests are too expensive.

EIA

Most EIAs use polyclonal or monoclonal antibodies to detect chlamydial LPS in clinical specimens. This is a soluble antigen and is more readily detected in soluble-phase assays than is MOMP. The advantages and disadvantages of the test have been discussed previously. These tests are also generally less sensitive than cell culture performed under ideal settings but, as with the fluorescent-antibody procedures, there are advantages: specimens do not have to be maintained on ice and can be collected from clinic settings distant from the processing laboratory. The introduction of blocking assays for confirmatory purposes improves the specificity of the EIAs to the point where they can theoretically be used for screening in low-prevalence settings. The EIAs are easy to perform and suitable for testing many specimens in high-volume laboratories. However, the sensitivity is still somewhat lower than that of culture. Except for the lower cost per test, there is no reason for routine use of EIAs. Nucleic acid amplification tests offer far superior performance.

Serodiagnosis

The most widely used serological test for diagnosing chlamydial infections is the genus-specific complement fixation (CF) test. It is useful in diagnosing psittacosis, in which paired sera often show fourfold or greater increases in titer. About one-third of patients with other evidence of C. pneumoniae infection will also show rising antibody titers in the CF test. Single-point titers greater than 1:64 may support a diagnosis of LGV. With LGV, it is difficult to demonstrate rising antibody titers, since the nature of the disease results in the patient being seen by the physician after the acute stage. Any titer above 1:16 is considered significant evidence of exposure to chlamydiae. There are some commercially available serological tests based on measurement of antibodies reactant to chlamydial inclusions in cell culture or EIA with chlamydial antigens. These tests are often described as "specific" by the manufacturers. However, the inclusions contain LPS, and genus-specific cross-reaction will occur. At this writing, none of these tests are recommended for routine use. None has a high predictive value in diagnosis of uncomplicated lower genital tract infection. The CF test is not useful in diagnosing trachoma inclusion conjunctivitis or the related genital tract infections, and it plays no role in diagnosing neonatal chlamydial infections.

The micro-IF method is a much more sensitive procedure for measuring antichlamydial antibodies. It may be used in diagnosing psittacosis and C. pneumoniae infection, in which paired sera show rising immunoglobulin G (IgG) titers (and often IgM antibody). With LGV, it is again difficult to demonstrate rising titers, but single-point titers in active cases usually have relatively high levels of IgM (>32) and IgG (2,048) antibodies. Trachoma, inclusion conjunctivitis, and the genital tract infections may be diagnosed by the micro-IF technique if appropriately timed paired acute- and convalescent-phase sera are obtained. However, it is often difficult to demonstrate rising antibody titers, particularly in sexually active populations, since many of these individuals are seen for chronic or repeat infections.

Serology is particularly useful in diagnosing chlamydial pneumonia in neonates. In this case, high levels of IgM antibody are regularly found in association with disease (Schachter et al., 1982). IgG antibodies are less useful because the infants are being seen at a time when they have considerable levels of circulating maternal IgG, since all of these infections are acquired from the infected mother, who is seropositive. It takes between 6 and 9 months for maternal antichlamydial antibodies to disappear from the infant. Older infants may be tested for determination of the prevalence of chlamydial infection without fear of confounding the effects of maternal antibody. Infants with inclusion conjunctivitis or respiratory-tract carriage of chlamydiae without pneumonia usually have very low levels of IgM antibodies. Thus, a single titer of 32 or greater may support the diagnosis of chlamydial pneumonia.

With *C. pneumoniae* infection, seroconversion may take more than 4 weeks, thus requiring a delay in collection of convalescent sera. However, in a consensus statement, researchers in the field felt that seroconversion was the most reliable indicator of a recent *C. pneumoniae* infection (File et al., 1997).

PREVENTION AND TREATMENT

Although there has been considerable effort to develop a chlamydial vaccine, at this writing there is no effective vaccine for any human chlamydial disease. Prevention is based on treatment of those infected to prevent transmission and complications.

C. psittaci Infection (Psittacosis)

Control of psittacosis depends on control of avian sources of infection. The 1929 pandemic of psittacosis was stopped by banning shipment or importation of psittacine birds. An effective method for controlling psittacosis in parakeets or other seed-eating birds has been developed—a chlortetracycline-impregnated seed. Prophylaxis for other birds is given in the form of a tetracycline-containing mash. Current requirements for quarantine of imported birds are for 30 days of treatment.

For treating human cases, the antibiotic of choice is tetracycline in a dose (for adults) of at least 250 mg four times a day continued for at least 3 weeks to avoid relapse. Severely ill patients may need measures for cardiovascular and respiratory support. Erythromycin (500 mg four times a day orally) is an alternative therapy.

C. trachomatis Infections

Trachoma

There is growing evidence that community-wide treatment with oral azithromycin may help control blinding trachoma (Schachter et al., 1999). Standard treatment of cases in areas of endemicity has been based on long-term (6 weeks) daily topical application of 1% tetracycline ophthalmic ointment.

Genital Infection

Control of these infections will be based on screening and treatment programs. In addition to reducing the prevalence of infection, these programs prevent expensive complications and have been shown to be cost-effective. Thus, screening and treatment for pregnant women will prevent postpartum endometritis and pneumonia in infants (Schachter et al., 1986). Screening and treatment in a family planning clinic will prevent PID (Scholes et al., 1996).

Tetracyclines have been the mainstay of therapy (Centers for Disease Control and Prevention, 1998). Tetracycline hydrochloride or oxytetracycline (500 mg 4 times daily), minocycline (100 mg twice daily), or doxycycline (100 mg twice daily) are equally effective oral treatments for uncomplicated infection in men and women. The treatment of choice is a single 1-g oral dose of azithromycin, which has been shown to be as effective as the 7-day doxycycline regimen (Martin et al., 1992). In parallel evaluations, success rates of greater than 96% for uncomplicated genital chlamydial infection were obtained with both regimens. Of the quinolones, only ofloxacin (400 mg daily for 7 days) is recommended (Ridgway, 1995). Erythromycin (500 mg 4 times daily for 7 days) is recommended for pregnant women, with amoxicillin (500 mg 3 times daily for 7 days) as an alternative (Crombleholme, 1990). In practice, azithromycin will see increasing use in the treatment of chlamydial infections in pregnancy.

C. pneumoniae Infections

Grayston and colleagues found that both tetracycline and erythromycin are effective against *C. pneumoniae* in vitro; they recommended either drug in a dosage of 2 g daily for 10 to 14 days or 1 to 1.5 g daily for 21 days but stated that relapses may occur even after this intensive treatment (Grayston et al., 1989). The role of azithromycin is under study.

INSIGHTS AND THE FUTURE

The next few years could provide important movement in controlling chlamydial infections and their complications. The potential for control of blinding trachoma is exciting. The importance of the finding that first-catch urine from males and females or vaginal swabs are suitable specimens for diagnosis of genital infection using nucleic acid amplification tests cannot be overstated. Exquisitely sensitive and specific noninvasive diagnostic tests will provide the basis for public health screening programs. Together with single-dose treatment, there is a chance for major successes.

There are still major needs. These include rapid tests to diagnose genital infection in order to speed up treatment of the infected (Pate et al., 1998), better diagnostics for *C. pneumoniae*, more information on the real clinical spectrum of *C. pneumoniae* infection, and, of course, more information on chlamydial biology, especially on protective immunity.

REFERENCES

Banks, J., B. Eddie, M. Sung, N. Sugg, J. Schachter, and K. F. Meyer. 1970. Plaque reduction technique for demonstrating neutralizing antibodies for Chlamydia. *Infect. Immun.* **2:**443–447.

Beem, M. O., and E. M. Saxon. 1977. Respiratory-tract colonization and a distinctive pneumonia syndrome in infants infected with *Chlamydia trachomatis*. *N. Engl. J. Med.* **296:**306–310.

Byrne, G. I., and J. W. Moulder. 1978. Parasite-specified phagocytosis of *Chlamydia psittaci* and *Chlamydia trachomatis* by L and HeLa cells. *Infect. Immun.* **19:**598–606.

Caldwell, H. D., and J. Schachter. 1982. Antigenic analysis of the major outer membrane protein of *Chlamydia* spp. *Infect. Immun.* **35:**1024–1031.

Centers for Disease Control and Prevention. 1993. Recommendation for the prevention and management of *Chlamydia trachomatis* infections, 1993. *Morb. Mortal Wkly. Rep.* **42:**(RR-12):1–39.

Centers for Disease Control and Prevention. 1994. Summary of notifiable diseases, United States, 1993. *Morb. Mortal Wkly. Rep.* **42:**1–73.

Centers for Disease Control and Prevention. 1998. 1998 Guidelines for treatment of sexually transmitted diseases. *Morb. Mortal Wkly. Rep.* **47:**1–111.

Cles, L. D., and W. E. Stamm. 1990. Use of HL cells for improved isolation and passage of *Chlamydia pneumoniae. J. Clin. Microbiol.* **28:**938–940.

Crombleholme, W. R., J. Schachter, M. Grossman, D. V. Landers, and R. L. Sweet. 1990. Amoxicillin therapy for *Chlamydia trachomatis* in pregnancy. *Obstet. Gynecol.* **75:**752–756.

Crotchfelt, K. A., B. Pare, C. Gaydos, and T. C. Quinn. 1998. Detection of *Chlamydia trachomatis* by the Gen-Probe AMPLIFIED Chlamydia trachomatis Assay (AMP CT) in urine specimens from men and women and endocervical specimens from women. *J. Clin. Microbiol.* **36:**391–394.

Dawson, C. R., T. Daghfous, M. Messadi, I. Hoshiwara, and J. Schachter. 1976. Severe endemic trachoma in Tunisia. *Br. J. Ophthalmol.* **60:**245–252.

Dhir, S. P., G. E. Kenny, and J. T. Grayston. 1971. Characterization of the group antigen of *Chlamydia trachomatis. Infect. Immun.* **4:**725–730.

File, T. M., Jr., J. G. Bartlett, G. H. Cassell, C. A. Gaydos, J. T. Grayston, M. R. Hammerschlag, R. B. Jones, J. B. Kahn, T. J. Marrie, J. A. Ramirez, P. Saikku, J. Schachter, H. R. Schumacher, W. E. Stamm, C. W. Stratton, and V. L. Yu. 1997. The importance of *Chlamydia pneumoniae* as a pathogen: the 1996 consensus conference on *Chlamydia pneumoniae* infections. *Infect. Dis. Clin. Pract.* **6**(Suppl. 2):S28–S31.

Fukushi, H., and K. Hirai. 1992. Proposal of *Chlamydia pecorum* sp. nov. for *Chlamydia* strains derived from ruminants. *Int. J. Syst. Bacteriol.* **42:**306–308.

Gordon, F. B., and A. L. Quan. 1965. Isolation of the trachoma agent in cell culture. *Proc. Soc. Exp. Biol. Med.* **118:**354.

Grayston, J. T., and S. Wang. 1975. New knowledge of chlamydiae and the diseases they cause. *J. Infect. Dis.* **132:**87–105.

Grayston, J. T., C. C. Kuo, S. P. Wang, and J. Altman. 1986. A new *Chlamydia psittaci* strain, TWAR, isolated in acute respiratory tract infections. *N. Engl. J. Med.* **315:**161–168.

Grayston, J. T., C.-C. Kuo, L. A. Campbell, and S.-P. Wang. 1989. *Chlamydia pneumoniae* sp. nov. for *Chlamydia* sp. strain TWAR. *Int. J. Syst. Bacteriol.* **39:**88–90.

Grayston, J. T., L. A. Campbell, C. C. Kuo, C. H. Mordhorst, P. Saikku, D. Thom, and S. P. Wang. 1990. A new respiratory tract pathogen: *Chlamydia pneumoniae* strain TWAR. *J. Infect. Dis.* **161:**618–625.

Halberstaedter, L., and S. von Prowazek. 1907. Über Zelleinschliesse parasitärer Natur beim Trachom. *Arb. Kais. Gesund.* **26:**44–47.

Hillis, S. D., L. M. Owens, P. A. Marchbanks, L. F. Amsterdam, and W. R. MacKenzie. 1997. Recurrent chlamydial infections increase the risks of hospitalization for ectopic pregnancy and pelvic inflammatory disease. *Am. J. Obstet. Gynecol.* **176:**103–107.

Holmes, K. K., H. H. Handsfield, S. P. Wang, B. B. Wentworth, M. Turck, J. B. Anderson, and E. R. Alexander. 1975. Etiology of nongonococcal urethritis. *N. Engl. J. Med.* **292:**1199–1205.

Hyman, C. L., P. M. Roblin, C. A. Gaydos, T. C. Quinn, J. Schachter, and M. R. Hammerschlag. 1995. Prevalence of asymptomatic nasopharyngeal carriage of *Chlamydia pneumoniae* in subjectively healthy adults: assessment by polymerase chain reaction-enzyme immunoassay and culture. *Clin. Infect. Dis.* **20:**1174–1178.

Jaschek, G., C. A. Gaydos, L. E. Welsh, and T. C. Quinn. 1993. Direct detection of *Chlamydia trachomatis* in urine specimens from symptomatic and asymptomatic men by using a rapid polymerase chain reaction assay. *J. Clin. Microbiol.* **31:**1209–1212.

Jones, B. R. 1964. Ocular syndromes of TRIC virus infection and their possible genital significance. *Br. J. Vener. Dis.* **40:**3–18.

Kleemola, M., P. Saikku, R. Visakorpi, S. P. Wang, and J. T. Grayston. 1988. Epidemics of pneumonia caused by TWAR, a new *Chlamydia* organism, in military trainees in Finland. *J. Infect. Dis.* **157:**230–236.

Lindner, K. 1910. Ätiologie der gonokokkenfreien Urethritis. *Wien. Klin. Wochenschr.* **23:**283–284.

Martin, D. H., T. F. Mroczkowski, Z. A. Dalu, J. McCarty, R. B. Jones, S. J. Hopkins, and R. B. Johnson. 1992. A controlled trial of a single dose of azithromycin for the treatment of chlamydial urethritis and cervicitis. *N. Engl. J. Med.* **327:**921–925.

McCormack, W. M., S. Alpert, D. E. McComb, R. L. Nichols, D. Z. Semine, and S. H. Zinner. 1979. Fifteenmonth follow-up study of women infected with *Chlamydia trachomatis. N. Engl. J. Med.* **300:**123–125.

Meyer, K. F. 1967. The host spectrum of psittacosis-lymphogranuloma venereum (PL) agents. *Am. J. Ophthalmol.* **63** (Suppl.):1225–1246.

Morrison, R. P., R. J. Belland, K. Lyng, and H. D. Caldwell. 1989. Chlamydial disease pathogenesis. The 57-kD chlamydial hypersensitivity antigen is a stress response protein. *J. Exp. Med.* **170:**1271–1283.

Moulder, J. W. 1966. The relation of the psittacosis group (Chlamydiae) to bacteria and viruses. *Annu. Rev. Microbiol.* **20:**107–130.

Moulder, J. W. 1984. Order *Chlamydiales* and family *Chlamydiaceae*, p. 729–739. *In* N. R. Krieg (ed.), *Bergey's Manual of Systematic Bacteriology*, vol 1. Williams and Wilkins, Baltimore, Md.

Nurminen, M., M. Leinonen, P. Saikku, and P. H. Makela. 1983. The genus-specific antigen of Chlamydia: resemblance to the lipopolysaccharide of enteric bacteria. *Science* **220:**1279–1281.

Page, L. A. 1972. Chlamydiosis (ornithosis), p. 414–417. *In* M. S. Hofstad (ed.), *Diseases of Poultry.* Iowa State University Press, Ames.

Pate, M. S., P. B. Dixon, K. Hardy, M. Crosby, and E. W. Hook III. 1998. Evaluation of the Biostar Chlamydia OIA assay with specimens from women attending a sexually transmitted disease clinic. *J. Clin. Microbiol.* **36:**2183–2186.

Peeling, R. W., R. L. Bailey, D. J. Conway, M. J. Holland, A. E. Campbell, O. Jallow, H. C. Whittle, and D. C. Mabey. 1998. Antibody response to the 60-kDa chlamydial heat-shock protein is associated with scarring trachoma. *J. Infect. Dis.* **177:**256–259.

Ridgway, G. L. 1995. Quinolones in sexually transmitted diseases. Global experience. *Drugs* **49** (Suppl. 2):115–122.

Ripa, K. T., and P. A. Mardh. 1977. New simplified culture technique for *Chlamydia trachomatis*, p. 323–327. *In* K. K. Holmes and D. Hobson (ed.), *Non-Gonococcal Urethritis and Related Infections.* American Society for Microbiology, Washington, D.C.

Rockey, D. D., R. A. Heinzen, and T. Hackstadt. 1995. Cloning and characterization of a *Chlamydia psittaci* gene coding for a protein localized in the inclusion membrane of infected cells. *Mol. Microbiol.* **15:**617–626.

Saikku, P., M. Leinonen, K. Mattila, M. R. Ekman, M. S. Nieminen, P. H. Makela, J. K. Huttunen, and V. Valtonen.

1988. Serological evidence of an association of a novel *Chlamydia*, TWAR, with chronic coronary heart disease and acute myocardial infarction. *Lancet* **ii:**983–986.

Schachter, J. 1978. Chlamydial infections. *N. Engl. J. Med.* **298:**428–435, 490–495, 540–549.

Schachter, J. 1998. Diagnosis of human chlamydial infections, p. 577–586. *In* R. S. Stephens, G. I. Byrne, G. Christiansen, I. N. Clarke, J. T. Grayston, R. G. Rank, G. L. Ridgway, P. Saikku, J. Schachter, and W. E. Stamm (ed.), *Chlamydial Infections: Proceedings of the Ninth International Symposium on Human Chlamydial Infection.* International Chlamydia Symposium, San Francisco, Calif.

Schachter, J., and C. R. Dawson. 1978. *Human Chlamydial Infections.* PSG Publishing Co., Littleton, Mass.

Schachter, J., and W. E. Stamm. 1999. Chlamydia, p. 795–806. *In* P. R. Murray, E. J. Baron, M. A. Pfaller, F. C. Tenover, and R. H. Yolken (ed.), *Manual of Clinical Microbiology,* 7th ed. American Society for Microbiology, Washington, D.C.

Schachter, J., M. Grossman, and P. H. Azimi. 1982. Serology of *Chlamydia trachomatis* in infants. *J. Infect. Dis.* **146:**530–535.

Schachter, J., M. Grossman, R. L. Sweet, J. Holt, C. Jordan, and E. Bishop. 1986. Prospective study of perinatal transmission of *Chlamydia trachomatis.* *JAMA* **255:**3374–3377.

Schachter, J., W. E. Stamm, T. C. Quinn, W. W. Andrews, J. D. Burczak, and H. H. Lee. 1994. Ligase chain reaction to detect *Chlamydia trachomatis* infection of the cervix. *J. Clin. Microbiol.* **32:**2540–2543.

Schachter, J., S. K. West, D. Mabey, C. R. Dawson, L. Bobo, R. Bailey, S. Vitale, T. C. Quinn, A. Sheta, S. Sallam, H. Mkocha, D. Mabey, and H. Faal. 1999. Azithromycin in control of trachoma. *Lancet* **354:**630–635.

Scholes, D., A. Stergachis, F. E. Heidrich, H. Andrilla, K. K. Holmes, and W. E. Stamm. 1996. Prevention of pelvic inflammatory disease by screening for cervical chlamydial infection. *N. Engl. J. Med.* **334:**1362–1366.

Stamm, W. E. 1988. Diagnosis of *Chlamydia trachomatis* genitourinary infections. *Ann. Intern. Med.* **108:**710–717.

Stamm, W. E., M. Tam, M. Koester, and L. Cles. 1983. Detection of *Chlamydia trachomatis* inclusions in McCoy cell cultures with fluorescein-conjugated monoclonal antibodies. *J. Clin. Microbiol.* **17:**666–668.

Stary, A., E. Schuh, M. Kerschbaumer, B. Gotz, and H. Lee. 1998. Performance of transcription-mediated amplification and ligase chain reaction assays for detection of chlamydial infection in urogenital samples obtained by invasive and noninvasive methods. *J. Clin. Microbiol.* **36:**2666–2670.

Stephens, R. S., M. R. Tam, C. C. Kuo, and R. C. Nowinski. 1982. Monoclonal antibodies to *Chlamydia trachomatis:* antibody specificities and antigen characterization. *J. Immunol.* **128:**1083–1089.

Stephens, R. S., E. A. Wagar, and G. K. Schoolnik. 1988. High-resolution mapping of serovar-specific and common antigenic determinants of the major outer membrane protein of *Chlamydia trachomatis.* *J. Exp. Med.* **167:**817–831.

Stephens, R. S., S. Kalman, C. Lammel, J. Fan, R. Marathe, L. Aravind, W. Mitchell, L. Olinger, R. L. Tatusov, Q. Zhao, E. V. Koonin, and R. W. Davis. 1998. Genome sequence of an obligate intracellular pathogen of humans: *Chlamydia trachomatis.* *Science* **282:**754–759.

Storz, J. 1971. *Chlamydia and Chlamydia-Induced Diseases.* Charles C Thomas, Springfield, Ill.

Tam, M. R., W. E. Stamm, H. H. Handsfield, R. Stephens, C. C. Kuo, K. K. Holmes, K. Ditzenberger, M. Krieger, and R. C. Nowinski. 1984. Culture-independent diagnosis of *Chlamydia trachomatis* using monoclonal antibodies. *N. Engl. J. Med.* **310:**1146–1150.

T'ang, F.-F., H.-L. Chang, Y.-T. Huang, and K.-C. Wang. 1957. Trachoma virus in chick embryo. *Natl. Med. J. China.* **43:**81.

Thylefors, B., A. D. Negrel, R. Pararajasegaram, and K. Y. Dadzie. 1995. Global data on blindness. *Bull. W. H. O.* **73:**115–121.

Toye, B., C. Laferriere, P. Claman, P. Jessamine, and R. Peeling. 1993. Association between antibody to the chlamydial heat-shock protein and tubal infertility. *J. Infect. Dis.* **168:**1236–1240.

Wagar, E. A., J. Schachter, P. Bavoil, and R. S. Stephens. 1990. Differential human serologic response to two 60,000 molecular weight *Chlamydia trachomatis* antigens. *J. Infect. Dis.* **162:**922–927.

Wang, S. P., and J. T. Grayston. 1970. Immunologic relationship between genital TRIC, lymphogranuloma venereum, and related organisms in a new microtiter indirect immunofluorescence test. *Am. J. Ophthalmol.* **70:**367–374.

Watkins, N. G., W. J. Hadlow, A. B. Moos, and H. D. Caldwell. 1986. Ocular delayed hypersensitivity: a pathogenetic mechanism of chlamydial-conjunctivitis in guinea pigs. *Proc. Natl. Acad. Sci. USA* **83:**7480–7484.

World Health Organization. 1996. *Global Prevalence and Incidence of Selected Curable Sexually Transmitted Diseases: Overview and Estimates.* World Health Organization, Geneva, Switzerland.

Rodent-Borne Viruses

BRIAN HJELLE

39

Rodents are hosts of several pathogens that afflict humans (zoonotic diseases). Rodent-borne agents range from bacterial and rickettsial microbes, such as the plague bacillus (*Yersinia pestis*), *Borrelia* spp., and *Ehrlichia* spp., to parasites (*Babesia microti*). Small mammals, including rodents, can play a role in the sylvatic cycles of some arboviruses, but rodents serve as the primary reservoirs of two major groups of medically important viruses: those of the family *Arenaviridae* and those of the genus *Hantavirus*, family *Bunyaviridae*. Unlike many zoonotic pathogens, hantaviruses and arenaviruses do not require or use an arthropod vector to maintain infection in reservoir populations or to infect humans.

Viruses of the families *Bunyaviridae* and *Arenaviridae* exhibit no cross-reactivity at the antigenic level and no significant nucleotide homology, but the similarities between members of the genus *Hantavirus* and those of the family *Arenaviridae* are striking nonetheless. Both have segmented, negative-sense, single-stranded RNA genomes (Schmaljohn, 1996; Southern, 1996). Both are found in rodent populations in a very focal distribution and confine themselves predominantly to one or a few closely related species within the family *Muridae*. The same rodent host may harbor both a hantavirus and an arenavirus. The list of recognized species is expanding rapidly for both virus groups (Schmaljohn and Hjelle, 1997). For both, closely related virus species can exist sympatrically in association with specific reservoir species without influencing one another's evolution (Rawlings et al., 1996). Transmission to humans, and possibly also among the reservoir rodents, is believed to occur through contaminated aerosols of rodent urine, feces, or saliva (Tsai, 1987). There are similarities in the diseases caused by arenaviruses and hantaviruses as well. Such diseases can all be regarded as within the clinical continuum of hemorrhagic fevers and frequently involve thrombocytopenia, coagulopathy, shock, and capillary leak syndrome. Members of both virus groups are susceptible to the antiviral drug ribavirin in vitro and in vivo.

There are certainly important differences as well. Arenaviruses are frequently transmitted vertically in the animal reservoir. Vertical transmission does not appear to occur with hantaviruses. While both hantaviruses and arenaviruses coevolve with their predominant rodent host, the coevolution of virus and host is somewhat less tight for arenaviruses than hantaviruses (Bowen et al., 1997; Schmaljohn and Hjelle, 1997). Arenavirus disease often

includes a profound neurologic component, and hepatitis is common. Hantaviruses can cause elevation of serum transaminases, but rarely is it severe, and neurologic disease is absent. In both the patient and the rodent host, antibody and neutralizing antibody responses elicited by arenaviruses are weak and arise late in infection, whereas hantaviruses elicit early and strong antibody responses (Hjelle et al., 1997; M. Bharadwaj and B. Hjelle, unpublished data). Treatment of some arenavirus diseases can include passive immunization using immune plasma, but the efficacy of such treatment in hantavirus infections is not known.

Research on pathogenesis is currently on a much higher technical plane for arenaviruses than for hantaviruses. The main reason is that lymphocytic choriomeningitis (LCM) virus naturally infects the laboratory mouse *Mus musculus* and has been the subject of years of scrutiny as a model for the study of viral pathogenesis and immune responses (Oldstone et al., 1985). Numerous animal disease models are available for arenaviruses. Fundamental studies of the immune responses to antigens, dissection of the role of the major histocompatibility complex genes in antigen presentation, the description of cytotoxic T lymphocytes (CTL), and other basic immunology studies made early and critical uses of the LCM virus model. Antigenic epitopes responsible for viral clearance by CTL have been mapped, as have escape mutants. Both viral and host determinants of persistence, clearance, and pathogenesis of several types of diseases have been identified. Transgenic mice have been extensively employed to evaluate host determinants of immune response, clearance, and pathogenesis.

Some hantaviruses can be coaxed to elicit at least transient infections in hamsters or in their native rodent hosts, but no physiologically relevant disease model is available. Very little is known about the mechanisms of pathogenesis or clearance. Hantaviruses are much harder to isolate from natural sources than are arenaviruses. Once an isolate has completed its adaptation to tissue culture, such as in Vero E6 cells, it will generally propagate consistently (Schmaljohn, 1996).

Ten hantavirus species and eight arenavirus species are known to be pathogenic in humans (Table 1). For both virus groups, Old World forms cause human diseases that are related to, but distinguishable from, those caused by New World forms.

BIOLOGY

Structure and Physical Properties

Arenavirus virions are spherical enveloped structures that generally measure between 90 and 120 nm in diameter (Burns and Buchmeier, 1993). T-shaped glycoprotein spikes approximately 7 to 10 nm long can be seen surrounding the surface membrane. The arenavirus particle is disrupted by detergent or solvents. When exposed to a pH of less than 5.5 or greater than 8.5, the hydrophilic head becomes detached from the remainder of the spike, and the virus can no longer bind to the cell surface receptor (Di Simione et al., 1994).

The hantavirus virion is a sphere approximately 80 to 110 nm in diameter. The envelope consists of a highly structured lipid bilayer with two transmembrane glycoproteins, G1 and G2. There are three helical nucleocapsid core structures. Hantaan (HTN) virus is inactivated by treatment at pH 5 or with 0.1% deoxycholate, ether, acetone, or 70% ethanol or by storage at 37°C (White et al., 1982).

Genome and Antigens

The single-stranded RNA genome of arenaviruses is composed of two segments, large (L) and small (S), which measure approximately 7,200 and 3,400 nucleotides (nt), respectively. Each 3′ terminus has a 19-nt region of conserved bases that may hybridize to its inverted complement on the opposite end of the segment. Both intramolecular and intermolecular complexes may be formed through hybridization of the terminal inverted repeats. The L segment encodes the viral RNA-dependent RNA polymerase (RDRP, or L protein) and a second protein, Z, that is presumed to serve a regulatory or structural role (Singh et al., 1987; Iapalucci et al., 1989a, 1989b; Salvato and Shimomaye, 1989; Salvato et al., 1989). The S segment encodes the external glycoproteins GP-1 and GP-2, as well as the core nucleoprotein antigen NP (Auperin et al., 1984; Southern et al., 1987). GP-1 and GP-2 are produced by posttranslational cleavage of a precursor glycoprotein, GP-C, by a trypsinlike protease.

Both the L and S genomic segments of arenaviruses employ an ambisense coding strategy. The mRNA for the NP antigen is complementary to the genomic RNA, whereas GP-C is translated from a genomic-sense mRNA. Similarly, the Z protein is synthesized from a genomic-sense mRNA, and the L protein is synthesized from an antigenomic-sense mRNA. GP-C is approximately 500 amino acids (aa) in length, with five or six N-linked glycosylation sites in GP-1 and two in GP-2. GP-1 and GP-2 are each homotetramers. GP-2 is an integral transmembrane glycoprotein and is needed for acid-dependent membrane fusion (Di Simione et al., 1994), whereas GP-1 is held in place through noncovalent interactions with GP-2 and interacts with the cell surface receptor (Borrow and Oldstone, 1992; Cao et al., 1998). NP is approximately 560 aa in length, L protein is about 2,200 aa long, and Z is about 90 aa long. NP is also subject to posttranslational modification via phosphorylation, a modification that is more abundant in persistently infected cells (Bruns et al., 1986).

The hantavirus genome is composed of three minussense RNA segments approximately 6,500 (L segment), 3,700 (middle [M] segment), or 1,700 to 2,100 (S segment) nt long. There are short regions (22 to 23 nt) of conserved bases on the termini that are imperfectly complementary to one another within a segment, which allows each segment to form a panhandle structure through hybridization of the terminal inverted-repeat sequences. The L segment encodes a 2,150-aa RDRP. The M segment encodes an ~1,140-residue envelope glycoprotein precursor. Through the cotranslational proteolytic processing of the envelope precursor in the endoplasmic reticulum, G1 (652 aa) and G2 (488 aa) are produced (Schmaljohn, 1996). The S segment encodes an RNA-binding nucleocapsid (N) core antigen, which makes up the majority of the viral core. As with the other bunyaviruses, the genomic RNAs of hantaviruses are not polyadenylated, nor are the corresponding mRNAs. One unconfirmed report has contended that the M segment of Sin Nombre (SN) virus is indeed polyadenylated (Hutchinson et al., 1996). As with other bunyaviruses, cap snatching is used to stabilize hantavirus RNAs (Patterson and Kolakofsky, 1984; Schmaljohn, 1996).

The hantavirus envelope consists of a lattice of G1-G2 heterodimers in a lipid bilayer. The amino terminus of each transmembrane glycoprotein is on the external surface of the membrane, and the carboxyl terminus is on the inner leaflet. The assembly of bunyavirus envelopes occurs in the cis- or medial-Golgi apparatus, although assembly of SN and Black Creek Canal viruses may occur at the plasma membrane instead (Goldsmith et al., 1995; Ravkov et al., 1997). There are three conserved N-linked glycosylation sites in all hantavirus G1 antigens and one in the G2 antigens.

PATHOGENESIS

Arenaviruses readily infect a wide variety of mammals, and a large number of experimental models have been developed that have employed laboratory mice, guinea pigs, hamsters, and primates. Several of the animal models have strong similarities to the human diseases caused by arenaviruses (Peters et al., 1996). Thus, the models have helped improve the understanding of the mechanisms of pathogenesis of arenaviruses in humans. Unlike hantaviruses, arenaviruses can also have deleterious effects on their reservoir hosts (Webb et al., 1975; Childs and Peters, 1993). The nature of these effects depends upon the age of the rodent, the dose of virus, the viral serotype, the route of infection, and probably also the genetic background of the individual animal.

Arenaviruses are shed from their infected reservoir hosts in the urine, saliva, nasal secretions, and feces (Childs and Peters, 1993). In the human, infection is acquired through the inhalation of virus-contaminated aerosols. There is an early involvement of macrophages, and early viral replication occurs in hilar lymph nodes (Kenyon et al., 1992). Arenaviruses exert an important immunosuppressive effect in animal models and in humans through necrosis of follicles and destruction of antigen-presenting cells in the lymph nodes and spleen of the host (Mims and Tosolini, 1969; Gonzalez et al., 1980). Prominent cellular destruction in vivo is a reflection of the immune response and is probably not due to the cytopathogenicity of the virus itself. While many arenaviruses will form plaques in monolayers, cytopathic effects are generally not seen.

During natural infections, arenaviruses frequently elicit only late, low-titer antibody responses, but they evoke profound T-cell responses, involving both T helper cells and CTL. The CTL responses, ironically, are thought to be essential both for immunopathogenesis and for clearance and recovery in the host (Lehmann-Grube et al., 1993). In both the mouse models and the human host, LCM virus antigens are thought to elicit virus-specific CD8[+] T cells that are necessary and sufficient to cause neurologic disease

TABLE 1 Pathogenic hantaviruses and arenaviruses

Virus	Distribution	Host	Disease	Comments
Hantavirus	Asia, especially China, Korea, and Russia	Striped field mouse (*Apodemus agrarius*)	HFRS, severe	Prototype hantavirus, 5–10% case-fatality ratio; 50,000–100,000 cases/yr in China
Seoul virus	Worldwide; most disease occurs in Asia	Commensal rats (*Rattus* spp.)	HFRS, mild to moderate	Primary cause of HFRS in urban China
Puumala virus	Western Europe	Bank vole (*Clethrionomys glareolus*)	HFRS, mild ("nephropathia epidemica")	Fatalities very rare despite thousands of cases in Europe
Dobrava-Belgrade virus	Balkan states, central Europe	Yellow-necked field mouse (*A. flavicollis*); *A. agrarius*	HFRS, severe form	Increasingly recognized as far north as northern Germany
SN (Four Corners) virus	Widespread in United States and Canada, especially in western regions	Deer mouse (*Peromyscus maniculatus*)	HCPS	Prototype pathogenic New World hantavirus
New York virus	Long Island, New York	White-footed mouse (*Peromyscus leucopus*)	HCPS	On mainland, *P. leucopus* carries SN and not NY virus
Bayou virus	Eastern Texas and Louisiana; Georgia	Rice rat (*Oryzomys palustris*)	HCPS	Possibly greater renal involvement than with SN virus
Black Creek Canal virus	Southern Florida	Cotton rat (*Sigmodon hispidus*)	HCPS	Etiologic role in HCPS not definitively established; possibly increased renal disease
Andes virus	Andes cordillera, especially Argentine and Chilean Patagonia	Long-tailed pygmy rice rat (colilargo) (*Oligoryzomys longicaudatus*)	HCPS	Transmissible person to person; probably synonymous with Oran, Lechiguanas, and other *Oligoryzomys*-borne viruses of Argentina
Laguna Negra virus	Paraguay	Vesper mouse (*Calomys laucha*); possibly *Oligoryzomys* spp. as well	HCPS	Synonymous with previously described Rio Mamoré virus of *Oligoryzomys microtis*?
Arenaviruses				
LCM virus	Worldwide	House mouse (*M. musculus*)	Choriomeningitis	Prototype arenavirus
Lassa virus	Africa	Multimammate rats (*Mastomys* spp.)	Lassa fever (shock, hemorrhage, meningitis)	Susceptible to ribavirin
Flexal virus	Brazil	*Oryzomys* spp.	Two symptomatic laboratory infections	
Junin virus	Argentina	*Calomys musculinis*	Argentine hemorrhagic fever	Treated with convalescent-phase plasma of patients who have recovered from disease; ribavirin
MAC virus	Bolivia	*C. callosus*[a]	Bolivian hemorrhagic fever	Now rare and subject to control by intensive trapping
Guanarito virus	Venezuela	*Zygodontomys brevicauda*	Venezuelan hemorrhagic fever	Host originally misidentified as *S. alstoni*
Sabiá virus	Brazil	Unknown	Hemorrhagic fever	Three cases known; two via laboratory infection
Tacaribe virus	Trinidad	*Artibeus* sp. bats?	Febrile syndrome	One laboratory infection with mild central nervous system symptoms

(Byrne et al., 1984). The CTL attack and lyse the infected cells in the central nervous system, with disruption of the blood-brain barrier. In experimental models in which CD8$^+$ cells are depleted, CD4$^+$ cells may induce the same disease, but the relevance of this finding to human disease is uncertain (Muller et al., 1992).

Arenavirus diseases of humans range from asymptomatic seroconversion through a grippelike syndrome (fever, prostration, headache, nausea, and vomiting) to meningitis and hemorrhagic fever. The incubation period for Lassa fever is about 5 to 21 days (Peters et al., 1996). In LCM virus infection, leukopenia and thrombocytopenia, as well as elevated liver enzymes, may be seen, although in many infections there are no hematologic abnormalities (Baum et al., 1966; Biggar et al., 1975; Fisher-Hoch et al., 1988). Hepatitis, as well as hemorrhage, are much more common and severe with Lassa fever and with the South American hemorrhagic fevers than with LCM virus infection.

Capillary leak and widespread organ involvement accompanied by shock, adult respiratory distress syndrome, and/or hemorrhage are responsible for most deaths due to Lassa virus and the South American hemorrhagic fever viruses. The pathological findings at necropsy, which can include petechiae and hemorrhage as well as hepatocellular and splenic necrosis, are generally not severe enough to explain the death of the patient (Child et al., 1967; Elsner et al., 1973; Walker et al., 1982). The case fatality ratios range from a few percent to as much as 34% for Bolivian hemorrhagic fever. Fetal loss is common when women become infected with Lassa virus while pregnant. All arenaviruses have some propensity to affect the human nervous system. Aseptic meningitis, coma, and seizures may be seen acutely, while Lassa virus and possibly Machupo (MAC) virus can produce nerve deafness (Cummins et al., 1990). Arenavirus infections provoke high systemic alpha interferon and tumor necrosis factor alpha responses. The presence of such mediators may partially explain the clinical features of arenavirus diseases (Levis et al., 1984; Marta et al., 1999).

Hantaviruses probably enter via the respiratory tract on particles that have been inhaled and then make contact with cells bearing the β3 integrin entry molecule (Gavrilovskaya et al., 1998). The virus is presumably taken up into regional lymph nodes, where it undergoes primary replication. After hematogenous dissemination, it invades vascular endothelium throughout the body, where it undergoes further cycles of replication and secondary viremia. In necropsy samples, viral antigen is detectable in endothelial cells, particularly those lining small capillaries and venules, especially in the lung, kidney, liver, lymphoid tissue, and heart (Kim et al., 1993; Zaki et al., 1995; Green et al., 1998). Some antigen is detectable in the macrophage compartment in a small minority of hantavirus cardiopulmonary syndrome (HCPS) cases. Renal tubular epithelial cells express antigen in hemorrhagic fever with renal syndrome (HFRS) (Kim et al., 1993; Groen et al., 1996). The prodromal symptoms, such as headache, fever, myalgias, and chills, begin after an incubation period of approximately 7 to 21 days and probably coincide with secondary viremia.

The prodromal stage of illness usually lasts from 2 to 8 days; 2 to 5 days is most typical. For both HFRS and HCPS, the prodromal symptoms become more severe in the 24 to 48 h immediately preceding the onset of capillary leak. Nausea, vomiting, and/or diarrhea often ensue. Facial flushing and petechiae may develop early in the course of HFRS, whereas for HCPS, a cough may herald the onset of pulmonary disease. Immediately before the onset of capillary leak syndrome, enlarged and highly activated lymphocytes (immunoblasts) become detectable in the peripheral blood. During this time, viral antigen, expressed in the vascular endothelium, may lead to the recruitment of virus-specific CTLs to the lungs and kidneys. Autopsy studies verify the presence of immunoblasts in the interstitium of the lungs, the lymphoid organs, and the liver (Hjelle et al., 1995b; Nolte et al., 1995).

There is abundant evidence that hantavirus disease is predominantly caused by immunologic injury rather than direct viral lytic attack. The defining lesion for both HFRS and HCPS is capillary leak syndrome. For HFRS, capillary leak occurs retroperitoneally, whereas for HCPS it is confined to the pulmonary bed (Hjelle et al., 1995b). The breakdown at the level of vascular endothelial integrity is not visible microscopically but is strictly functional in nature (Hjelle et al., 1995a; Mertz et al., in press). The functional breakdown does not allow the egress of formed blood elements, such as red blood cells, but instead only protein is allowed to cross into the interstitium. The lack of visible lysis of endothelial cells that express hantavirus antigens can be taken as evidence against direct destruction of such cells by either virus or virus-specific CTL.

By the time symptoms appear, all patients have hantavirus-specific immunoglobulin M (IgM) antibodies against the viral N antigen. Given a diagnostic test of adequate sensitivity, essentially all have specific IgG antibodies as well (Jenison et al., 1994; Hjelle et al., 1997). Viral RNA titers are decreasing by the time of admission (Terajima et al., 1999). HLA B*35-restricted CTL directed against viral N antigen in HCPS, as well as HLA A1- and B51-restricted CTL against N in HFRS, are detectable in the blood (Van Epps et al., 1999).

Immunoblasts (Fig. 1) are believed to play a central role in the pathogenesis of HCPS, and probably in HFRS as well. In the blood, they commonly make up more than 20% of lymphocytes and are felt to be the same as those detected in the tissues at autopsy. Flow cytometric studies and in situ immunologic stains performed on autopsy samples verify that immunoblasts, despite their morphologic resemblance to activated plasma cells, are predominantly T cells (Nolte et al., 1995; Zaki et al., 1995). CD8$^+$ cells predominate over CD4$^+$ cells. Macrophages are also found in the infiltrates. While there is at present no evidence that CTL activity per se is responsible for the capillary leak syndrome, there is evidence for profoundly increased cytokine activity in the pulmonary bed in HCPS (Mori et al., 1999). Taken together, these findings suggest a model in which viral antigens are presented by highly activated vascular endothelial cells, resulting in recruitment of T cells to the parenchyma. T cells and resident histocytes, as well as endothelial cells themselves, may establish a milieu in which proinflammatory cytokines induce a reversible breakdown of the barrier function of the intracellular junctions at the vascular endothelium (Fig. 2) (Puri and Rosenberg, 1989).

EPIDEMIOLOGY

For both hantaviruses and arenaviruses, reservoir rodents develop chronic infections that result in transient or periodic virus shedding in urine, feces, and saliva. Infection is considered to be lifelong despite the continued presence of high titers of neutralizing and nonneutralizing antibodies in hantavirus-infected rodents.

The relationships between arenaviruses and their rodent hosts are complex, and relatively little is known about the

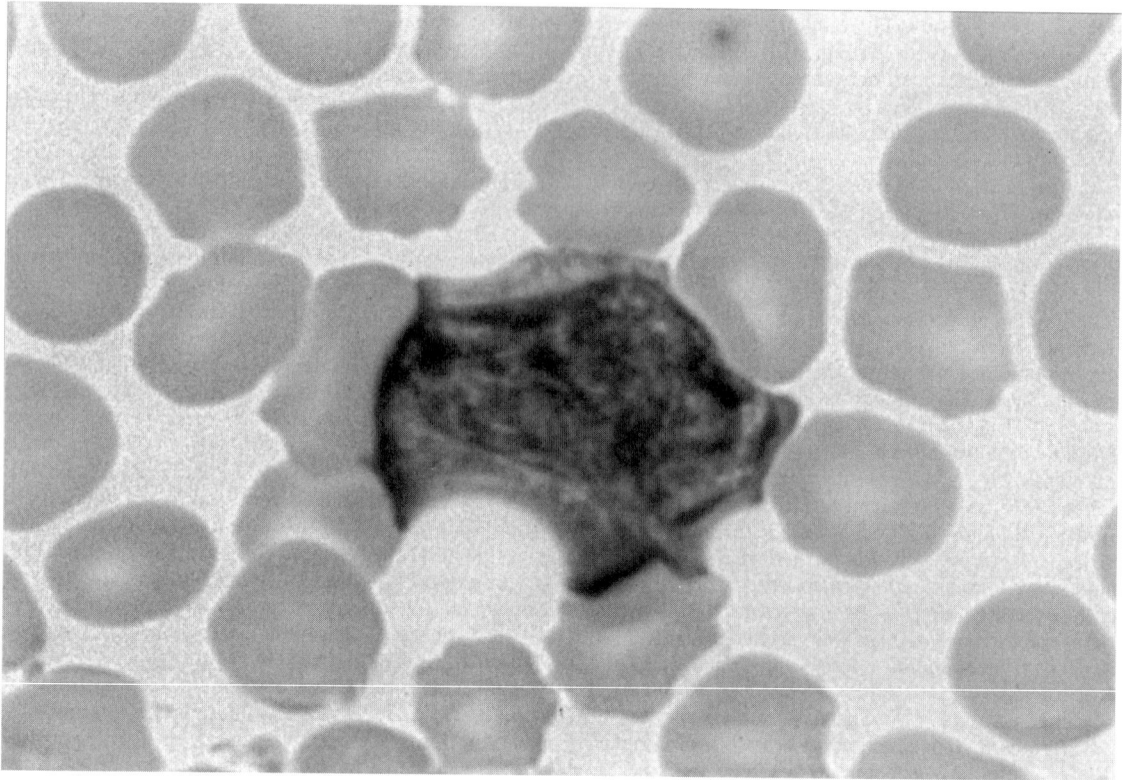

FIGURE 1 Peripheral blood smear (oil immersion) showing a typical immunoblast from a patient with acute HCPS. Note the large size (~16 μm diameter), immature chromatin, and basophilic cytoplasm. Magnification, ×100. Photo courtesy of M. K. Foucar.

factors that promote or inhibit arenavirus infection in wild rodent populations. Both horizontal transmission, which may be by the venereal route, and vertical transmission may occur. Horizontal transmission to immunocompetent adult animals may result in clearance, whereas neonatal infections often result in chronic viremia and shedding. When *Calomys* sp. mice more than 9 days of age are experimentally infected with MAC virus, some develop chronic viremia in the same manner as do neonates and some mount a strong antibody response and have minimal or no viremia. After a recent venereal infection, vertical transmission of MAC virus via milk results in chronic viremia in the pups. Dams can also transmit virus transovarially or in utero; if the dam has been viremic before becoming pregnant, fetal death may ensue (Childs and Peters, 1993). Fighting, including biting, is felt to be an important route for transmission of LCM virus in *M. musculus* (Skinner and Knight, 1973).

Several different epidemiological patterns of human arenavirus infection are known. LCM virus infection has a worldwide distribution, with seroprevalences ranging from approximately 1 to 10% (Peters et al., 1996). Outbreaks of LCM virus infection have occurred among people working in animal colonies or with infected animal cells or tissues (Baum et al., 1966; Biggar et al., 1975). By comparison, Lassa fever and the South American hemorrhagic fevers occur in sporadic or epidemic form in regions in which the reservoir rodent occurs. For the South American hemorrhagic fevers in particular, it is apparent that disease outbreaks are localized to discrete, small portions of the ranges of the reservoir rodents. While Venezuelan hemorrhagic fever has thus far remained confined to a small portion of the range of *Zygodontomys brevicauda*, the area of endemici-

ty for Argentine hemorrhagic fever has increased steadily in the years since its initial description (Maiztegui et al., 1986). Both Lassa virus and MAC virus may be transmitted interpersonally, such as in hospital settings (White, 1972; Peters et al., 1974; Centers for Disease Control and Prevention, 1994). Probable male-to-female transmission, most likely sexual, has been noted for several arenaviruses.

For hantaviruses, virtually all, if not all, infected reservoir animals develop detectable antibody responses. Seroprevalences range from about 3 to 50%, with higher prevalences in the United States generally recorded in western states (Hjelle et al., 1995a; Mills et al., 1997; Bennett et al., 1999). The reservoir host is not adversely affected by the infection. The mechanisms by which rodents transmit the virus to one another are largely unknown. Horizontal transmission is clearly the predominant, if not the exclusive, means of transmission. For many hantaviruses, males have higher seroprevalence than females. Hantaviruses may infect rodents other than their primary reservoir hosts, especially during epizootics. Such infections also appear to be persistent, at least when the secondary host is closely related to the predominant host species (see, for example, Song et al., 1995; Rawlings et al., 1996).

Humans acquire hantavirus infection by exposure to contaminated rodent excreta. Overwhelmingly, the evidence is that indoor exposure in poorly ventilated spaces is much riskier than outdoor activities, such as hiking, planting crops, or gardening. Vacuuming, sweeping, shaking rugs, and dusting are among the activities that aerosolize particulates that are then inhaled via the respiratory route. It is quite common for patients to report recent sighting of living rodents or other evidence of very recent rodent activity;

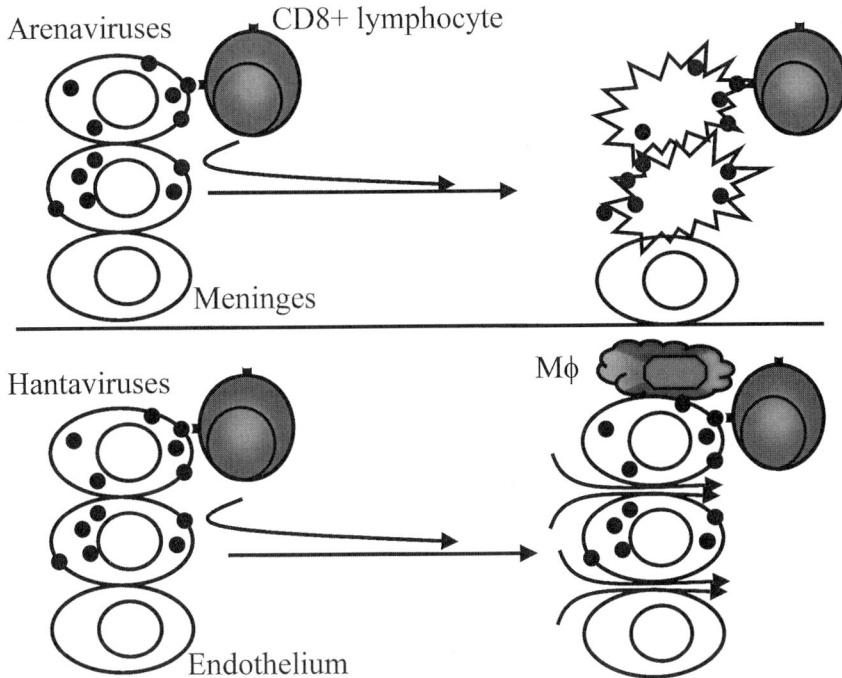

FIGURE 2 In this highly simplified diagram, the most important of the presumed immunopathological mechanisms of arenavirus infection (top) is contrasted with that of hantavirus infection (bottom). After recruitment by virus-infected cells, virus-specific CD8+ cells may engage in a direct cytolytic attack and destroy the infected cells, as seen in experimental arenavirus infections. Alternatively, they may help establish a milieu that results in a functional, transient defect in the barrier function among endothelial cells, resulting in transudation of plasma from the vascular space into the interstitium. Most likely the CD8+ and/or CD4+ lymphocyte must collaborate with resident tissue macrophages (Mφ) to bring about the defect in the endothelial barrier function.

thus, one may suspect that only recently produced rodent excreta are highly infectious to humans. Most exposure occurs in peridomestic settings, but recreational and occupational exposures are being recognized with increasing frequency.

Hantavirus infections occur somewhat more frequently in men than women; male/female infection ratios range from 1:1 to more than 4:1 (Niklasson and LeDuc, 1987; Niklasson et al., 1987; Ruo et al., 1994; Centers for Disease Control and Prevention, 1998; Ferrer et al., 1998). Hantaviruses rarely infect children under 12, except for Andes virus in Argentina and Chile (Pini et al., 1998). Some have suggested that children are less likely than adults to develop severe disease when infected with hantaviruses, but too few data are available on pediatric infections to render a decisive opinion. All racial and ethnic groups are susceptible to hantavirus infection and disease. In the United States, native Indian groups are overrepresented among HCPS patients. The overrepresentation of Indians can easily be attributed to their more frequent involvement in rural, agrarian lifestyles compared to members of the majority culture.

Interpersonal transmission of Andes hantavirus is a well-established phenomenon in Argentina and possibly has occurred in Chile as well (Enria et al., 1996; Padula et al., 1998; M. Ferres, personal communication). Both nosocomial and intrahousehold transmission have been documented, although not at equal levels of rigor. The mechanism of interpersonal transmission is not established and is certainly not obvious. No such transmission has been demonstrated for any other hantavirus. Clustered cases of human infec-

tion have been noted for North American hantaviruses, but common-source exposures to rodent-infested houses, buildings, and vehicles have been judged to be more likely in each such instance (Wells et al., 1997; unpublished data of the author).

DIAGNOSIS

Diagnosis of acute infection with arenaviruses is often problematic, because specific antibody responses to arenavirus infections are often delayed until days to weeks after the illness has run its course. IgM and IgG antibodies may be detected by indirect immunofluorescence assay using infected cells as a substrate or by enzyme-linked immunofluorescence assay, Western blotting, or other immunoblot assay using recombinant antigens or viral lysates. Neutralizing-antibody responses may be detected but are not diagnostically useful in the acute-care setting. Assays for circulating viral antigen, which use sandwich immunoassays, are sensitive for diagnosis of acute Lassa fever, at least in fatal cases (Peters et al., 1996). Unfortunately, such tests are not widely available in the acute-care setting. Virus isolation is possible either in susceptible mice or in tissue culture, but it is not practical due to its inconsistent sensitivity and the high biohazard (biosafety level 4) associated with propagation of all of the natural human arenavirus pathogens other than LCM virus. For these reasons, tests for viral genetic material in inactivated patient blood have been increasingly favored in the diagnosis of acute arenavirus infection (Lunkenheimer et al., 1990;

Lozano et al., 1995). Unfortunately, such technology is not readily available in many of the remote, impoverished regions of the world, such as western Africa or northern Bolivia, where arenavirus diseases occur. In the absence of a sensitive "gold standard" for detection of infection, it was noted that Junin virus RNA could be detected in some febrile patients who did not seroconvert to Junin virus, suggesting the possibility that Junin virus may cause infection without seroconversion in some cases.

By comparison to arenaviruses, specific serologic diagnosis of acute hantavirus infection is straightforward. Given a test of sufficient sensitivity, hantavirus antibodies of both the IgM and IgG classes are readily detectable in virtually all patients from the onset of symptoms (Jenison et al., 1994; Hjelle et al., 1994a, 1997). Tests using recombinant antigens that have been affixed to a membrane (Western blotting or strip immunoblot assay) appear to be more sensitive than enzyme-linked immunosorbent assays, especially in the IgG format (unpublished data of the author; Hjelle et al., 1997). For the more severe hantavirus diseases, such as HCPS and the Asian forms of HFRS, viral RNA is readily detectable by reverse transcription-PCR, using peripheral blood mononuclear cells as the source of RNA (Hjelle et al., 1994c). Detection of Puumala virus RNA by this method is somewhat more difficult (Plyusnin et al., 1997). Hantavirus antigens may also be readily detected in paraffin-embedded tissues of patients at necropsy, but tissue biopsy is not necessary to diagnose infection in living patients

(Zaki et al., 1995; Green et al., 1998). Viral RNA is readily detectable by reverse transcription-PCR using frozen tissues and also may be detected in paraffin-embedded tissues (Nichol et al., 1993; Hjelle et al., 1994b).

Specific diagnostic techniques are generally available only in regional laboratories. Even given the several-hour turnaround times for antibody tests offered in some laboratories, shipping of samples may impose an obligatory 24 to 30 h delay in diagnosis. Such delay may be unacceptable for diagnosis of HCPS and some cases of HFRS. For HCPS, clinical algorithms may be used to ascertain more quickly whether the disease is progressing as expected for HCPS. Thrombocytopenia is nearly universal upon presentation, even at the earliest stages. Thrombocytopenia progresses very rapidly, often with declines of 50,000/μl or more in as little as 12 h. Serial platelet counts may be monitored pending the results of specific antibody tests. Manual blood smears may be evaluated for the presence of immature, nonactivated granulocytes and the appearance of immunoblasts (Fig. 1). Such approaches may be equally useful for some forms of HFRS, but laboratory abnormalities in milder diseases such as HFRS may progress less decisively. Given evidence that the disease is progressing as expected for HCPS or severe HFRS, it is advisable that the patient be transported to a facility with advanced tertiary-care capabilities. Abnormalities that are judged to be especially sensitive and specific for early recognition of HCPS (and, to a lesser extent, for HFRS) are outlined in Table 2.

TABLE 2 Clinical and laboratory findings especially helpful in clinical diagnosis of HCPS and HFRS

Characteristic	Sensitivity	Specificity	Comments
Prodrome of fever, headache, myalgias with or without nausea and vomiting	100%	Low	Nausea and vomiting are common but not universal. Muscle aches may be severe, even limiting, and tend to involve lower muscle groups in thighs and buttocks.
Cough	Moderate	Low	Predicts pulmonary edema.
Elevated lactate dehydrogenase; liver function test abnormalities	High	Modest	AST and ALT elevation may occur relatively late in disease. Lactate dehydrogenase elevation occurs early but is not always profound.
Left shift	High; nearly 100%	Low	Not "bandemia" but must include granulocytes at myelocyte or earlier stage to be specific. Granulocytes are rarely highly activated.
Thrombocytopenia	100%, but is, on rare occasions, delayed for 1–2 days	Moderate	A key laboratory abnormality that can often decisively exclude HCPS when absent. Sensitivity with some forms of HFRS is less impressive.
Elevated serum lactate level	Modest	Modest	Predicts severe disease.
Flushing, petechiae	Moderate (HFRS only)	Moderate	Common for HFRS caused by HTN virus. Petechiae may occur in palate, pharynx, or conjunctivae.
Conjunctival injection, photophobia, ophthalmic pain, glaucoma	Moderate (HFRS only)	Moderate	HFRS only.
Renal insufficiency, proteinuria	Moderate (HFRS >> HCPS)	Moderate in HFRS, low in HPCS	Common in HTN virus-associated HFRS; much less so in Puumala virus HFRS or HCPS.
Bilateral interstitial infiltrates	100% in HCPS; low in HFRS	Moderate	Some SN virus infections are now recognized without pulmonary involvement. HTN virus HFRS may result in interstitial infiltrates in up to 20% of cases.

PREVENTION AND THERAPY

Prevention

One may reduce one's risk of exposure to hantaviruses and arenaviruses by reducing contact with wild rodents. By keeping cooking and eating areas clean, containing garbage, sealing holes in buildings, and clearing trash from the outside of buildings, one may make the home less attractive to rodents. Trapping or poisoning may be used to reduce rodent infestations.

Indoor spaces that have become infested with rodents represent a significant risk, and their cleaning must be handled carefully. The space should be ventilated in advance. Rodent droppings should be thoroughly wetted with detergent or 10% hypochloride solution before cleaning. Even at that point, every effort should be made to avoid aerosolizing dust, such as by avoiding the use of leaky handheld vacuum cleaners or the brisk sweeping of dust in closed spaces. Using liquids extensively during cleaning helps prevent any residual live virus from becoming airborne. One should wear gloves and, when available, HEPA masks during cleaning activities. Rodent carcasses should be disposed of in sealed plastic bags after spraying them with detergent or bleach. Insecticide spray should be applied to the rodent carcass in areas in which plague occurs.

Therapy

Arenaviruses, such as Lassa virus and Junin virus, are quite sensitive to the broad-spectrum antiviral drug ribavirin in vitro and in vivo, and intravenous ribavirin has been used during outbreaks of arenavirus disease (McCormick et al., 1986; Enria and Maiztegui, 1994). Unfortunately, ribavirin is not always available. Convalescent-phase plasma obtained from patients who have recovered from Junin virus contains neutralizing antibodies. Depending upon the neutralizing-antibody titer, as little as 2 to 3 U of convalescent-phase plasma may reduce the mortality of Junin virus infection from 20 to 30% to 1 to 2%. A minority of patients treated with convalescent plasma may return with a late neurologic syndrome, but the neurologic syndrome is usually self-limited, and full recovery is the rule.

Treatment of hantavirus infections is largely supportive, although ribavirin has assumed an important place in the treatment of HFRS due to HTN virus in China (Huggins et al., 1991). For HCPS, early placement of the patient in the intensive-care unit is thought to be critical in reducing mortality. For HFRS, there is a larger spectrum of disease severity. Nephropathia epidemica due to Puumala virus, the predominant European form of HFRS, is usually mild and can often be treated in the outpatient setting. Physicians in the region of endemicity often quickly recognize HFRS due to HTN virus, a much more severe disease, during the prodrome period. Such early recognition allows timely intervention with ribavirin.

A controlled trial of ribavirin for HCPS is under way in the United States and Canada, but no information is available about its efficacy. An uncontrolled trial of ribavirin was undertaken briefly during the 1993–1994 outbreak, but the trial was not conducted in such a way as to allow any determination of its efficacy, so it had to be abandoned (Mertz et al., 1997). For patients with severe cardiopulmonary manifestations that are predictive of demise (a cardiac index of <2 and serum lactate at >4 mg/dl), extracorporeal membrane oxygenation has been used and has prevented death in several gravely ill patients.

FUTURE

Preventing morbidity and mortality from arenavirus and hantavirus infections is a daunting task, fraught with many logistical and economic difficulties. However, many important inroads have been made. While Lassa fever is a significant cause of morbidity and mortality in western Africa, preventing and treating Lassa fever is rendered very difficult by the limited economic resources available to the affected populations. Bolivian hemorrhagic fever has been controlled by rapid and aggressive rodent-trapping programs, and Argentine hemorrhagic fever is potentially controllable by the Candid 1 vaccine (Maiztegui et al., 1998). The mortality of HFRS in China has been reduced, but by no means to a comfortable level, by use of ribavirin and advances in early recognition of the syndrome. While education has undoubtedly had important but unquantifiable effects, less progress has been made in the control of Venezuelan hemorrhagic fever, nephropathia epidemica, or HCPS. For HCPS, which affects the wealthiest democracies of North and South America, education has been inadequate to significantly reduce the mortality of the disease, whether in outbreak or sporadic forms.

For Lassa fever, Bolivian and Venezuelan hemorrhagic fevers, HFRS, and HCPS, vaccines represent the most probable route toward control of morbidity and mortality. Commercial interest in vaccine development is slight, however, either because the market is considered to be too small or because the affected populations do not have the economic resources to pay for the vaccine. It is nevertheless important for scientists to proceed apace with developing such vaccines and demonstrating their efficacy in animal and human systems so that their existence can be highlighted in the commercial and political arenas in which their ultimate fate will be decided.

CONCLUSIONS

In addition to their considerable economic importance in destruction of crops and food stores, rodents carry a variety of diseases affecting humans. The viral diseases carried by rodents include the hemorrhagic fevers caused by arenaviruses (family *Arenaviridae*) and hantaviruses (family *Bunyaviridae*). These disparate virus groups, related at a most distant level phylogenetically, have many fascinating similarities to one another and equally fascinating differences. By striking rural, often impoverished populations with severe diseases, frequently in epidemic form, they present unique challenges to the world's public health infrastructure. Several inroads have been made, but there is still much to be done.

This work was supported by Public Health Service grants RO1 AI36336 and RO1 AI41692 and by the Defense Advanced Research Projects Agency.

REFERENCES

Auperin, D. D., V. Romanowski, M. Galinski, and D. H. L. Bishop. 1984. Sequencing studies of Pichinde arenavirus S RNA indicate a novel coding strategy: an ambisense viral S RNA. *J. Virol.* **52:**897–904.

Baum, S. G., A. M. Lewis, Jr., W. P. Rowe, and R. J. Huebner. 1966. Epidemic nonmeningitic lymphocytic-choriomeningitis-virus infection. An outbreak in a population of laboratory personnel. *N. Engl. J. Med.* **274:**934–936.

Bennett, S. G., J. P. Webb, Jr., M. B. Madon, J. E. Childs, J. E. Ksiazek, T. G. Torrez-Martinez, and B. Hjelle. 1999.

Hantavirus (Bunyaviridae) infections in rodents from Orange and San Diego Counties, California. *Am. J. Trop. Med. Hyg.* **60:**75–84.

Biggar, R. J., J. P. Woodall, P. D. Walter, and G. E. Haughie. 1975. Lymphocytic choriomeningitis outbreak associated with pet hamsters. Fifty-seven cases from New York State. *JAMA* **232:**494–500.

Borrow, P., and M. B. Oldstone. 1992. Characterization of lymphocytic choriomeningitis virus-binding protein(s): a candidate cellular receptor for the virus. *J. Virol.* **66:**7270–7281.

Bowen, M. D., C. J. Peters, and S. T. Nichol. 1997. Phylogenetic analysis of the Arenaviridae: patterns of virus evolution and evidence for cospeciation between arenaviruses and their rodent hosts. *Mol. Phylogenet. Evol.* **8:**301–316.

Bruns, M., W. Zeller, H. Rohdewohld, and F. Lehmann-Grube. 1986. Lymphocytic choriomeningitis virus. IX. Properties of the nucleocapsid. *Virology* **151:**77–85.

Burns, J. W., and M. J. Buchmeier. 1993. Glycoproteins of arenaviruses, p. 17–35. *In* M. S. Salvato (ed.), *The Arenaviridae.* Plenum Press, New York, N.Y.

Byrne, J. A., R. Ahmed, and M. B. Oldstone. 1984. Biology of cloned cytotoxic T lymphocytes specific for lymphocytic choriomeningitis virus. I. Generation and recognition of virus strains and H-2b mutants. *J. Immunol.* **133:**433–439.

Cao, W., M. D. Henry, P. Borrow, H. Yamada, J. H. Elder, E. V. Ravkov, S. T. Nichol, R. W. Compans, K. P. Campbell, and M. B. Oldstone. 1998. Identification of alpha-dystroglycan as a receptor for lymphocytic choriomeningitis virus and Lassa fever virus. *Science* **282:**2079–2081.

Centers for Disease Control and Prevention. 1994. Bolivian hemorrhagic fever—El Beni Department, Bolivia, 1994. *Morb. Mortal. Wkly. Rep.* **43:**943–946.

Centers for Disease Control and Prevention. 1998. Hantavirus pulmonary syndrome—Colorado and New Mexico 1998. *Morb. Mortal. Wkly. Rep.* **47:**449–452.

Child, P. L., R. B. MacKenzie, L. R. Valverde, and K. M. Johnson. 1967. Bolivian hemorrhagic fever. A pathologic description. *Arch. Pathol.* **83:**434–445.

Childs, J. E., and C. J. Peters. 1993. Ecology and epidemiology of arenaviruses and their hosts, p. 331–385. *In* M. S. Salvato (ed.), *The Arenaviridae.* Plenum Press, New York, N.Y.

Cummins, D., J. B. McCormick, D. Bennett, J. A. Samba, B. Farrar, S. J. Machin, and S. P. Fisher-Hoch. 1990. Acute sensorineural deafness in Lassa fever. *JAMA.* **264:**2093–2096.

Di Simione, C., M. A. Zandonatti, and M. J. Buchmeier. 1994. Acidic pH triggers LMCV membrane fusion activity and conformational change in the glycoprotein spike. *Virology* **198:**455–465.

Elsner, B., E. Schwarz, O. G. Mando, J. Maiztegui, and A. Vilches. 1973. Pathology of 12 fatal cases of Argentine hemorrhagic fever. *Am. J. Trop. Med. Hyg.* **22:**229–236.

Enria, D. A., and J. I. Maiztegui. 1994. Antiviral treatment of Argentine hemorrhagic fever. *Antiviral Res.* **23:**23–31.

Enria, D., P. Padula, E. L. Segura, N. Pini, A. Edelstein, C. Riva Posse, and M. C. Weissenbacher. 1996. Hantavirus pulmonary syndrome in Argentina: possibility of person to person transmission. *Medicina* (Buenos Aires) **56:**709–711.

Ferrer, J. F., C. B. Jonsson, E. Esteban, D. Galligan, M. A. Basombrio, M. Peralta-Ramos, M. Bharadwaj, N. Torrez-Martinez, J. Callahan, A. Segovia, and B. Hjelle. 1998. High prevalence of hantavirus infection in Indian communities of the Paraguayan and Argentinean Gran Chaco. *Am. J. Trop. Med. Hyg.* **59:**438–444.

Fisher-Hoch, S., J. B. McCormick, D. Sasso, and R. B. Craven. 1988. Hematologic dysfunction in Lassa fever. *J. Med. Virol.* **26:**127–135.

Gavrilovskaya, I. N., M. Shepley, R. Shaw, M. H. Ginsberg, and E. R. Mackow. 1998. Beta 3 integrins mediate the cellular entry of hantaviruses that cause respiratory failure. *Proc. Natl. Acad. Sci. USA* **95:**7074–7079.

Goldsmith, C. S., L. H. Elliott, C. J. Peters, and S. R. Zaki. 1995. Ultrastructural characteristics of Sin Nombre virus, causative agent of hantavirus pulmonary syndrome. *Arch. Virol.* **140:**2107–2122.

Gonzalez, P.H., P. M. Cossio, R. Arana, J. L. Maiztegui, and R. P. Laguens. 1980. Lymphatic tissue in Argentine hemorrhagic fever. Pathologic features. *Arch. Pathol. Lab. Med.* **104:**250–254.

Green, W., R. Feddersen, O. Yousef, M. Behr, K. Smith, J. Nestler, S. Jenison, T. Yamada, and B. Hjelle. 1998. Tissue distribution of hantavirus antigen in naturally infected humans and deer mice. *J. Infect. Dis.* **177:**1696–1700.

Groen, J., J. A. Bruijn, M. N. Gerding, J. G. Jordans, A. W. Moll van Charante, and A. D. Osterhaus. 1996. Hantavirus antigen detection in kidney biopsies from patients with nephropathia epidemica. *Clin. Nephrol.* **46:**379–383.

Hjelle, B., F. Chavez-Giles, N. Torrez-Martinez, T. Yamada, J. Sarisky, M. Ascher, and S. Jenison. 1994a. Dominant glycoprotein epitope of Four Corners hantavirus is conserved across a wide geographical area. *J. Gen. Virol.* **75:**2881–2888.

Hjelle, B., S. Jenison, N. Torrez-Martinez, T. Yamada, K. Nolte, R. Zumwalt, K. MacInnes, and G. Myers. 1994b. A novel hantavirus associated with an outbreak of fatal respiratory disease in the southwestern United States: evolutionary relationships to known hantaviruses. *J. Virol.* **68:**592–596.

Hjelle, B., C. F. Spiropoulou, N. Torrez-Martinez, S. Morzunov, C. J. Peters, and S. T. Nichol. 1994c. Detection of Muerto Canyon virus RNA in peripheral blood mononuclear cells from patients with hantavirus pulmonary syndrome. *J. Infect. Dis.* **170:**1013–1017.

Hjelle, B., B. Anderson, N. Torrez-Martinez, W. Song, W. F. Gannon, and T. L. Yates. 1995a. Prevalence and geographic genetic variation of hantaviruses of New World harvest mice (*Reithrodontomys*): identification of a divergent genotype from a Costa Rican *Reithrodontomys mexicanus*. *Virology* **207:**452–459.

Hjelle, B., S. A. Jenison, D. E. Goade, W. B. Green, R. M. Feddersen, and A. A. Scott. 1995b. Hantaviruses: clinical, microbiologic, and epidemiologic aspects. *Crit. Rev. Clin. Lab. Sci.* **32:**469–508.

Hjelle, B., S. Jenison, N. Torrez-Martinez, B. Herring, S. Quan, A. Polito, S. Pichuantes, T. Yamada, C. Morris, F. Elgh, H. W. Lee, H. Artsob, and R. Dinello. 1997. Rapid and specific detection of Sin Nombre virus antibodies in patients with hantavirus pulmonary syndrome by a strip immunoblot assay suitable for field diagnosis. *J. Clin. Microbiol.* **35:**600–608.

Huggins, J. W., C. M. Hsiang, T. M. Cosgriff, M. Y. Guang, J. L. Smith, Z. O. Wu, J. W. LeDuc, Z. M. Zheng, J. M. Meegan, Q. N. Wang, D. D. Oland, X. E. Gui, P. H. Gibbs, G. H. Yuan, and T. M. Zhang. 1991. Prospective, double-blind, concurrent, placebo-controlled clinical trial of intravenous ribavirin therapy of hemorrhagic fever with renal syndrome. *J. Infect. Dis.* **164:**1119–1127.

Hutchinson, K. L., C. J. Peters, and S. T. Nichol. 1996. Sin Nombre virus mRNA synthesis. *Virology* **224:**139–149.

Iapalucci, S., R. Lopez, O. Rey, N. Lopez, M.-T. Franze-Fernandez, G. N. Cohen, M. Lucero, A. Ochoa, and M. M. Zakin. 1989a. Tacaribe virus L gene encodes a protein of 2210 amino acid residues. *Virology* **170:**40–47.

Iapalucci, S., N. Lopez, O. Rey, M. M. Zakin, G. N. Cohen, and M.-T. Franze-Fernandez. 1989b. The 5′ region of Tacaribe virus L RNA encodes a protein with a potential metal binding domain. *Virology* **173**:357–361.

Jenison, S., T. Yamada, C. Morris, B. Anderson, N. Torrez-Martinez, N. Keller, and B. Hjelle. 1994. Characterization of human antibody responses to four corners hantavirus infections among patients with hantavirus pulmonary syndrome. *J. Virol.* **68**:3000–3006.

Kenyon, R. H., K. T. McKee, Jr., P. M. Zack, M. K. Rippy, A. P. Vogel, C. York, J. Meegan, C. Crabbs, and C. J. Peters. 1992. Aerosol infection of rhesus macaques with Junin virus. *Intervirology* **33**:23–31.

Kim, S., E. T. Kang, Y. G. Kim, J. S. Han, J. S. Lee, Y. I. Kim, W. C. Hall, J. M. Dalrymple, and C. J. Peters. 1993. Localization of Hantaan viral envelope glycoproteins by monoclonal antibodies in renal tissues from patients with Korean hemorrhagic fever H. *Am. J. Clin. Pathol.* **100**:398–403.

Lehmann-Grube, F., J. Lohler, O. Utermohlen, and C. Gegin. 1993. Antiviral immune responses of lymphocytic choriomeningitis virus-infected mice lacking CD8+ T lymphocytes because of disruption of the beta 2-microglobulin gene. *J. Virol.* **67**:332–339.

Levis, S. C., M. C. Saavedra, C. Ceccoli, E. Falcoff, M. R. Feuillade, D. A. Enria, J. I. Maiztegui, and R. Falcoff. 1984. Endogenous interferon in Argentine hemorrhagic fever. *J. Infect. Dis.* **149**:428–433.

Lozano, M. E., D. Enria, J. L. Maiztegui, O. Grau, and V. Romanowski. 1995. Rapid diagnosis of Argentine hemorrhagic fever by reverse transcriptase PCR-based assay. *J. Clin. Microbiol.* **33**:1327–1332.

Lunkenheimer, K., F. T. Hufert, and H. Schmitz. 1990. Detection of Lassa virus RNA in specimens from patients with Lassa fever by using the polymerase chain reaction. *J. Clin. Microbiol.* **28**:2689–2692.

Maiztegui, J., M. Feuillade, and A. Briggiler. 1986. Progressive extension of the endemic area and changing incidence of Argentine hemorrhagic fever. *Med. Microbiol. Immunol.* **175**:149–152.

Maiztegui, J. I., K. T. McKee, Jr., J. G. Barrera Oro, L. H. Harrison, P. H. Gibbs, M. R. Feuillade, D. A. Enria, A. M. Briggiler, S. C. Levis, A. M. Ambrosio, N. A. Halsey, C. J. Peters, J. L. Maiztegui, K. T. McKee, Jr., J. G. Barrera Oro, L. H. Harrison, P. H. Gibbs, M. R. Feuillade, D. A. Enria, A. M. Briggiler, S. C. Levis, A. M. Ambrosio, N. A. Halsey, and C. J. Peters. 1998. Protective efficacy of a live attenuated vaccine against Argentine hemorrhagic fever. AHF Study Group. *J. Infect. Dis.* **177**:277–283.

Marta, R. F., V. S. Montero, C. E. Hack, A. Sturk, J. L. Maiztegui, and F. C. Molinas. 1999. Proinflammatory cytokines and elastase-alpha-1-antitrypsin in Argentine hemorrhagic fever. *Am. J. Trop. Med. Hyg.* **60**:85–89.

McCormick, J. B., I. J. King, P. A. Webb, C. L. Scribner, R. B. Craven, K. M. Johnson, L. H. Elliot, and R. Belmont-Williams. 1986. Lassa fever. Effective therapy with ribavirin. *N. Engl. J. Med.* **314**:20–26.

Mertz, G. J., B. L. Hjelle, and R. T. Bryan. 1997. Hantavirus infection. *Adv. Intern. Med.* **42**:369–421.

Mertz, G. J., B. Hjelle, T. M. Williams, and F. T. Koster. Host responses in the hantavirus cardiopulmonary syndrome. In J. F. Saluzzo and B. Dodet (ed.), *Factors in the Emergence and Control of Rodent-Borne Viral Diseases*, in press. Elsevier, Paris, France.

Mills, J. N., T. G. Ksiazek, B. A. Ellis, P. E. Rollin, S. T. Nichol, T. L. Yates, W. L. Gannon, C. E. Levy, D. M.

Engelthaler, T. Davis, D. T. Tanda, J. W. Frampton, C. R. Nichols, C. J. Peters, and J. E. Childs. 1997. Patterns of association with host and habitat: antibody reactive with Sin Nombre virus in small mammals in the major biotic communities of the southwestern United States. *Am. J. Trop. Med. Hyg.* **56**:273–284.

Mims, C. A., and F. A. Tosolini. 1969. Pathogenesis of lesions in lymphoid tissue of mice infected with lymphocytic choriomeningitis (LCM) virus. *Br. J. Exp. Pathol.* **50**:584–592.

Mori, M., A. L. Rothman, I. Kurane, J. M. Montoya, K. B. Nolte, J. E. Norman, D. C. Waite, F. T. Koster, and F. A. Ennis. 1999. High levels of cytokine-producing cells in the lung tissues of patients with fatal hantavirus pulmonary syndrome. *J. Infect. Dis.* **179**:295–302.

Muller, D., B. H. Koller, J. L. Whitton, K. E. LaPan, K. K. Brigman, and J. A. Frelinger. 1992. LCMV-specific, class II-restricted cytotoxic T cells in beta 2-microglobulin-deficient mice. *Science* **255**:1576–1578.

Nichol, S. T., C. F. Spiropoulou, S. Morzunov, P. E. Rollin, T. G. Ksiazek, H. Feldmann, A. Sanchez, J. Childs, S. Zaki, and C. J. Peters. 1993. Genetic identification of a hantavirus associated with an outbreak of acute respiratory illness. *Science* **262**:914–917.

Niklasson, B., and J. W. LeDuc. 1987. Epidemiology of nephropathia epidemica in Sweden. *J. Infect. Dis.* **155**:269–276.

Niklasson, B., J. Leduc, K. Nystrom, and L. Nyman. 1987. Nephropathia epidemica: incidence of clinical cases and antibody prevalence in an endemic area of Sweden. *Epidemiol. Infect.* **99**:559–562.

Nolte, K. B., R. M. Feddersen, K. Foucar, S. R. Zaki, F. T. Koster, D. Madar, T. L. Merlin, P. J. McFeeley, E. T. Umland, and R. E. Zumwalt. 1995. Hantavirus pulmonary syndrome in the United States: a pathological description of a disease caused by a new agent. *Hum. Pathol.* **26**:110–120.

Oldstone, M. B., R. Ahmed, J. Byrne, M. J. Buchmeier, Y. Riviere, and P. Southern. 1985. Virus and immune responses: lymphocytic choriomeningitis virus as a prototype model of viral pathogenesis. *Br. Med. Bull.* **41**:70–74.

Padula, P. J., A. Edelstein, S. D. Miguel, N. M. Lopez, C. M. Rossi, and R. D. Rabinovich. 1998. Hantavirus pulmonary syndrome outbreak in Argentina: molecular evidence for person-to-person transmission of Andes virus. *Virology* **241**:323–330.

Patterson, J. L., and D. Kolakofsky. 1984. Characterization of La Crosse virus small-genome transcripts. *J. Virol.* **49**:680–685.

Peters, C. J., R. W. Kuehne, R. R. Mercado, R. H. Le Bow, R. O. Spertzel, and P. A. Webb. 1974. Hemorrhagic fever in Cochabamba, Bolivia, 1971. *Am. J. Epidemiol.* **99**:425–433.

Peters, C. J., M. Buchmeier, P. Rollin, and T. G. Ksiazek. 1996. Arenaviruses, p. 1521–1551. In B. N. Fields, D. M. Knipe, P. M. Howley, et al. (ed.), *Virology*, 3rd ed. Lippincott-Raven Publishers, Philadelphia, Pa.

Pini, N. C., A. Resa, G. del Jesus Laime, G. Lecot, T. G. Ksiazek, S. Levis, and D. A. Enria. 1998. Hantavirus infection in children in Argentina. *Emerg. Infect. Dis.* **4**:85–87.

Plyusnin, A., J. Horling, M. Kanerva, J. Mustonen, Y. Cheng, J. Partanen, O. Vapalahti, S. K. Kukkonen, J. Niemimaa, H. Henttonen, B. Niklasson, A. Lundkvist, and A. Vaheri. 1997. Puumala hantavirus genome in patients with nephropathia epidemica: correlation of PCR positivity with HLA haplotype and link to viral sequences in local rodents. *J. Clin. Microbiol.* **35**:1090–1096.

Puri, R. K., and S. A. Rosenberg. 1989. Combined effects of interferon α and interleukin 2 on the induction of a vascular leak syndrome in mice. *Cancer Immunol. Immunother.* **28**:267–274.

Ravkov, E. V., S. T. Nichol, and R. W. Compans. 1997. Polarized entry and release in epithelial cells of Black Creek Canal virus, a New World hantavirus. *J. Virol.* **71**:1147–1154.

Rawlings, J. A., N. Torrez-Martinez, S. U. Neill, G. M. Moore, B. N. Hicks, S. Pichuantes, A. Nguyen, M. Bharadwaj, and B. Hjelle. 1996. Cocirculation of multiple hantaviruses in Texas, with characterization of the S genome of a previously-undescribed virus of cotton rats (*Sigmodon hispidus*). *Am. J. Trop. Med. Hyg.* **55**:672–679.

Ruo, S. L., Y. L. Li, Z. Tong, Q. R. Ma, Z. L. Liu, Y. W. Tang, K. L. Ye, J. B. McCormick, S. P. Fisher-Hoch, and Z. Y. Xu. 1994. Retrospective and prospective studies of hemorrhagic fever with renal syndrome in rural China. *J. Infect. Dis.* **170**:527–534.

Salvato, M., E. Shimomaye, and M. B. Oldstone. 1989. The primary structure of the lymphocytic choriomeningitis virus L gene encodes a putative RNA polymerase. *Virology* **169**:377–384.

Salvato, M. S., and E. M. Shimomaye. 1989. The completed sequence of lymphocytic choriomeningitis virus reveals a unique RNA structure and a gene for a zinc finger protein. *Virology* **173**:1–10.

Schmaljohn, C., and B. Hjelle. 1997. Hantaviruses: a global disease problem. *Emerg. Infect. Dis.* **3**:95–104.

Schmaljohn, C. S. 1996. Bunyaviridae: the viruses and their replication, p. 1447–1471. *In* B. N. Fields, D. M. Knipe, P. M. Howley, et al. (ed.), *Virology*, 3rd ed. Lippincott-Raven Publishers, Philadelphia, Pa.

Singh, M. K., F. V. Fuller-Pace, M. J. Buchmeier, and P. J. Southern. 1987. Analysis of the genomic L RNA segment from lymphocytic choriomeningitis virus. *Virology* **161**:448–456.

Skinner, H. H., and E. H. Knight. 1973. Natural routes for post-natal transmission of murine lymphocytic choriomeningitis. *Lab. Anim.* **7**:171–184.

Song, W., N. Torrez-Martinez, W. Irwin, J. Harrison, R. Davis, M. Ascher, M. Jay, and B. Hjelle. 1995. Isla Vista virus: a genetically novel hantavirus of the California vole *Microtus californicus*. *J. Gen. Virol.* **76**:3195–3199.

Southern, P. J. 1996. Arenaviridae: the viruses and their replication, p. 1505–1519. *In* B. N. Fields, D. M. Knipe, P. M. Howley, et al. (ed.), *Virology*, 3rd ed. Lippincott-Raven Publishers, Philadelphia, Pa.

Southern, P. J., M. K. Singh, Y. Riviere, D. R. Jacoby, M. J. Buchmeier, and M. B. Oldstone. 1987. Molecular characterization of the genomic S RNA segment from lymphocytic choriomeningitis virus. *Virology* **157**:145–155.

Terajima, M., J. D. Hendershot III, H. Kariwa, F. Koster, B. Hjelle, D. Goade, C. DeFronzo, and F. A. Ennis. 1999. High levels of viremia in patients with the hantavirus cardiopulmonary syndrome. *J. Infect. Dis.* **180**:2030–2034.

Tsai, T. F. 1987. Hemorrhagic fever with renal syndrome: mode of transmission to humans. *Lab. Anim. Sci.* **37**:428–430.

Van Epps, H. L., C. S. Schmaljohn, and F. A. Ennis. 1999. Human memory cytotoxic T-lymphocyte (CTL) responses to Hantaan virus infection: identification of virus-specific and cross-reactive CD8($^+$) CTL epitopes on nucleocapsid protein. *J. Virol.* **73**:5301–5308.

Walker, D. H., J. B. McCormick, K. M. Johnson, P. A. Webb, G. Komba-Kono, L. H. Elliott, and J. J. Gardner. 1982. Pathologic and virologic study of fatal Lassa fever in man. *Am. J. Pathol.* **107**:349–356.

Webb, P. A., G. Justines, and K. M. Johnson. 1975. Infection of wild and laboratory animals with Machupo and Latino viruses. *Bull. W. H. O.* **52**:493–499.

Wells, R. M., J. Young, R. J. Williams, L. R. Armstrong, K. Busico, A. S. Khan, T. G. Ksiazek, P. E. Rollin, S. R. Zaki, S. T. Nichol, and C. J. Peters. 1997. Hantavirus transmission in the United States. *Emerg. Infect. Dis.* **3**:361–365.

White, H. A. 1972. Lassa fever. A study of 23 hospital cases. *Trans. R. Soc. Trop. Med. Hyg.* **66**:390–401.

White, J. D., F. G. Shirey, G. R. French, J. W. Huggins, O. M. Brand, and H. W. Lee. 1982. Hantaan virus, aetiological agent of Korean haemorrhagic fever, has Bunyaviridae-like morphology. *Lancet* **i**:768–771.

Zaki, S. R., P. W. Greer, L. M. Coffield, C. S. Goldsmith, K. B. Nolte, K. Foucar, R. M. Feddersen, R. E. Zumwalt, G. L. Miller, A. S. Khan, P. E. Rollin, T. G. Ksiazek, S. T. Nichol, B. W. J. Mahy, and C. J. Peters. 1995. Hantavirus pulmonary syndrome. Pathogenesis of an emerging infectious disease. *Am. J. Pathol.* **146**:552–579.

APPENDIXES: Reference Laboratories

Virology Services Offered by the Federal Reference Laboratories of the Centers for Disease Control and Prevention*

BRIAN W. J. MAHY

APPENDIX
1

A strong collaborative effort among local, state, and federal laboratories provides the foundation for a successful nation-wide program for the surveillance, prevention, and control of infectious diseases. Laboratories at each level have distinct responsibilities. As the state public health laboratories provide reference and disease surveillance (RDS) at the state level, so too do the Centers for Disease Control and Prevention (CDC) provide RDS at the national level. The effectiveness and timeliness with which each laboratory fulfills its responsibilities greatly influence the success of the nationwide program.

Providing RDS in microbiology, hematology, histopathology, and immunology at CDC is the responsibility of the National Center for Infectious Diseases (NCID). These program activities, performed in collaboration with state and other qualified laboratories, constitute an important segment of NCID's mission. All RDS specimens, with proper justification and a completed request form, must be submitted to NCID by or through the state public health laboratory with the knowledge and consent of the state laboratory director or his/her designee. Within the limitations outlined in the following sections, NCID will provide RDS on the following:

- Cultures, serum, or cerebrospinal fluid (CSF) samples, transudates, exudates, tissues, or histologic specimens from patients suspected of having an unusual infectious disease and/or other kinds of specimens (e.g., vectors, foods, liquids) that aid in the diagnosis of life-threatening, unusual, or exotic infectious diseases
- Cultures or serum specimens obtained from patients who have infectious diseases that occur only sporadically or from patients who are involved in outbreaks of diseases caused by organisms for which satisfactory diagnostic reagents are not commercially or widely available
- Organisms suspected of being unusual pathogens or that are associated with hospital-acquired infections
- Specimens forwarded to NCID for confirmation of quality assurance for test performance

- Serum specimens or cultures that are clinically important and are sent to NCID for confirmation because the results in state laboratories were atypical, aberrant, or difficult to interpret or because difficulties were encountered with the reagents used
- Arthropod and vertebrate specimens necessary for confirmation of zoonotic diseases

In assigning reference priorities, NCID places strong emphasis on the quality of both the specimen and the accompanying information. Prior consultation on urgent or unusual specimens enables NCID to be more responsive and efficient. Protecting the nation's health through an effective RDS program requires teamwork, cooperation, and good communication.

NCID ORGANIZATION

Listed below are the different kinds of laboratories of the National Center for Infectious Diseases. Virology laboratories are located in the NCID divisions or programs as shown below. The staff members to contact for consultation are listed by organization in Table 1.

Bacteriology Laboratories
Division of Bacterial and Mycotic Diseases (DBMD)
Hospital Infections Program (HIP)
Division of AIDS, Sexually Transmitted Diseases and TB Laboratory Research (DASTLR)
Division of Vector-Borne Infectious Diseases (DVBID)
Arctic Investigations Program (AIP)

Mycology Laboratories
Division of Bacterial and Mycotic Diseases (DBMD)

Parasitology Laboratories
Division of Parasitic Diseases (DPD)

Rickettsial Laboratories
Division of Viral and Rickettsial Diseases (DVRD)

Virology Laboratories
Division of Vector-Borne Infectious Diseases (DVBID) (includes arbovirus laboratories)

*Material in this appendix is reprinted with modifications from *Reference and Disease Surveillance* (February 1986), prepared by the Centers for Disease Control, Atlanta, Ga., with permission.

581

TABLE 1 Laboratories of the NCID that handle virus specimens

NCID division or program and staff	Location	Office phone no.
Vector-Borne Infectious Diseases Director, Duane J. Gubler	Ft. Collins, Colo.	(970) 221-6428
Arbovirus Diseases Branch Chief, John T. Roehrig	Ft. Collins, Colo.	(970) 221-6442
Dengue Branch Chief, Gary G. Clark	San Juan, P.R.	(787) 766-5181

Viral and Rickettsial Diseases
    Acting Director, James W. LeDuc
    Deputy Director, Rima F. Khabbaz
    Assistant Director for Public Health,
       Lawrence B. Schonberger
    Assistant Director for Blood Safety,
       Mary E. Chamberland
	Atlanta, Ga.
	Atlanta, Ga.
	Atlanta, Ga.
	
	Atlanta, Ga.
		(404) 639-3574
		(404) 639-2775
		(404) 639-3091
		
		(404) 639-4350

Infectious Disease Pathology Activity Chief, Sherif R. Zaki	Atlanta, Ga.	(404) 639-3134
Hepatitis Branch Chief, Harold S. Margolis	Atlanta, Ga.	(404) 639-2339
Influenza Branch Chief, Nancy J. Cox	Atlanta, Ga.	(404) 639-3591
Respiratory and Enteric Viruses Branch Chief, Larry J. Anderson	Atlanta, Ga.	(404) 639-3596
Special Pathogens Branch Acting Chief, Thomas Ksiazek	Atlanta, Ga.	(404) 639-1511
Viral Exanthems and Herpesvirus Branch Chief, William C. Reeves	Atlanta, Ga.	(404) 639-1338
Viral and Rickettsial Zoonoses Branch Chief, James Childs	Atlanta, Ga.	(404) 639-1075
AIDS, STD and TB Laboratory Research Director, Harold W. Jaffe	Atlanta, Ga.	(404) 639-4581
HIV Immunology and Diagnostics Branch Chief, Steve McDougal	Atlanta, Ga.	(404) 693-3434
HIV/AIDS and Retrovirology Branch Chief, Thomas M. Folks	Atlanta, Ga.	(404) 639-1024
Arctic Investigations Program Director, Jay C. Butler	Anchorage, Alaska	(907) 729-3404

Division of Viral and Rickettsial Diseases (DVRD)

Division of AIDS, Sexually Transmitted Diseases and TB Laboratory Research (DASTLR)

Arctic Investigations Program (AIP)

REFERENCE AND DISEASE SURVEILLANCE

Tables 2 and 3 list the virology RDS activities of NCID. These tables are organized by disease category and alphabet-ized for easy reference. Use Table 3 and the footnoted instructions to obtain the shortest possible response time from NCID laboratories.

RDS activities are not undertaken for diseases for which satisfactory reagents or tests are commercially available or for which there is no public health need. The deputy direc-tor of NCID [Joseph McDade, phone, (404) 639-3967], is available to consult with state health department laboratory directors about firms that produce reagents or offer inter-state diagnostic services.

TABLE 2 Hepatitis testing available at CDC

Disease[a]	Center	Tests Available[b]		Type of analyte			Specimen requested
		Analyte	Format	Antibody	Antigen	DNA/RNA	
Hepatitis A	DVRD	Anti-HAV	EIA	X			Serum or plasma
	DVRD	IgM anti-HAV	EIA	X			Serum or plasma
Hepatitis B	DVRD	HBsAg	RIA, EIA		X		Serum or plasma
	DVRD	Total anti-HBs (qualitative)	EIA	X			Serum or plasma
	DVRD	Total anti-HBs (quantitative)	EIA	X			Serum or plasma
	DVRD	Total anti-HBc	RIA, EIA	X			Serum or plasma
	DVRD	IgM anti-HBc	EIA	X			Serum or plasma
	DVRD	HBeAg	EIA		X		Serum or plasma
	DVRD	Total anti-HBe	EIA	X			Serum or plasma
	DVRD	HBV DNA (quantitative)	PCR			X	Serum
Hepatitis C	DVRD	Anti-HCV	EIA, RIBA	X			Serum or plasma
	DVRD	HCV RNA (qualitative)	RT-PCR			X	Serum or plasma[c]
	DVRD	HCV RNA (quantitative)	RT-PCR			X	Serum or plasma[c]
Hepatitis D	DVRD	Total anti-HDV	EIA	X			Serum or plasma
Hepatitis E	DVRD	IgG anti-HEV	EIA	X			Serum or plasma
	DVRD	IgM anti-HEV	EIA	X			Serum or plasma

[a]Prior consultation and arrangements with DVRD are required before submission of specimens; indicate on the request form that such arrangements have been made.
[b]Abbreviations: HAV, hepatitis A virus; EIA, enzyme immunoassay; Ig, immunoglobulin; RIA, radioimmunoassay; RT, reverse transcription.
[c]Heparinized plasma is not suitable for RT-PCR.

REQUIREMENTS FOR ALL SPECIMENS

General Information

The information provided below is intended as a general guideline for those who wish to send specimens to CDC for evaluation. Although this appendix is comprehensive, special circumstances may arise and the user is encouraged to consult appropriate staff within NCID in such instances.

1. Etiologic agents should be cultivated and shipped in a medium that protects and ensures the viability of the microorganisms during transit.

2. Only pure cultures of etiologic agents should be sent; mixed cultures cannot be accepted without written justification.

3. Optimum containers for shipping different groups of etiologic agents vary depending on the agent and the distance involved in shipment. In all instances, however, the primary container should be of a durable material that, when properly packaged, is leakproof and can withstand the temperature and pressure variations likely to occur in the air and on the ground during shipment to CDC.

4. Serum specimens for serologic testing should be aseptically separated from whole blood. Contaminated serum specimens are unsuited for almost all purposes. Paired serum specimens are preferred and, in many cases, required. (See specific requirements for various etiologic agents.) Generally, the first specimen should be obtained and refrigerated as soon after the onset of illness as possible. The second specimen should be collected 2 to 4 weeks later. The optimal interval for collecting the second serum specimen varies with different infectious agents.

5. When whole blood is sent for isolation of particular viral agents, the blood should be kept cold but not frozen prior to shipment and shipped on wet ice (not dry ice); however, whole blood submitted for isolation of suspected rickettsial agents must be packed in dry ice and shipped frozen.

6. Slides with tissue sections, blood films, or smears of clinical material should be dry, free of immersion oil, properly labeled, and carefully packed in a slide container. The slide container must be wrapped in absorbent material and placed in a sealed, leakproof container.

Instructions for Identifying Specimens

1. Identify individual specimen tubes or vials by encircling them with adhesive tape that contains a typed or printed patient identification number and other identifying information. Do not use ballpoint pens, wax, or other writing instruments that tend to smear.

2. Print patient's identification number, type of specimen, and date of specimen collection on the specimen label.

Instructions for Packing Specimens

1. Package specimens properly in accordance with the provisions of subparagraph 72.3(a) of Title 42, Code of Federal Regulations (*Federal Register,* 20 August 1980), to protect both the material while in transit and the personnel who handle the packages.

- Never mail clinical specimens or cultures in glass or plastic Petri plates or similar containers!
- Never enclose dry ice in hermetically sealed containers!
- Enclose patient history and specimen information using CDC form 50.34, which can be obtained directly from your state health department.

2. Avoid delay at CDC by addressing shipments as follows:

Data and Specimen Handling Section
Bldg. 4, Rm. B-35, Mail Stop G12
National Center for Infectious Diseases
Centers for Disease Control and Prevention
Atlanta, GA 30333

TABLE 3 Viral diseases for which testing is available at CDC[a,b]

Disease or agent	Organizational unit	Serological tests available	Isolation specimens requested	Antigen detection tests available	Nucleic acid detection tests available
Arboviruses					
California encephalitis (LaCrosse)	DVBID	EIA, HI, CF, NT	Brain, mosquitoes, CSF, serum	IIF, EIA[h]	RT-PCR
Colorado tick fever	DVBID	EIA, CF, NT	Blood (unfrozen), ticks	IIF, EIA[h]	RT-PCR
Dengue 1–4	DVBID	EIA, HI, CF, NT	Serum, organs, mosquitoes	IIF, EIA[h]	RT-PCR
Eastern equine encephalitis	DVBID	EIA, HI, CF, NT	Brain, CSF, serum, mosquitoes	IIF, EIA[h]	RT-PCR
Japanese encephalitis	DVBID	EIA, HI, CF, NT	Brain, CSF, serum, mosquitoes	IIF, EIA[h]	RT-PCR
Murray Valley encephalitis	DVBID	EIA, HI, CF, NT	Brain, CSF, serum, mosquitoes	IIF, EIA[h]	RT-PCR
Powassan	DVBID	EIA, HI, CF, NT	Brain, CSF, serum, mosquitoes	IIF, EIA[h]	RT-PCR
St. Louis encephalitis	DVBID	EIA, HI, CF, NT	Brain, CSF, serum, mosquitoes	IIF, EIA[h]	RT-PCR, dot blot
Venezuelan equine encephalitis	DVBID	EIA, HI, CF, NT	Brain, CSF, serum, mosquitoes	IIF, EIA[h]	RT-PCR
Western equine encephalitis	DVBID	EIA, HI, CF, NT	Brain, CSF, serum, mosquitoes	IIF, EIA[h]	RT-PCR
Yellow fever	DVBID	EIA, HI, CF, NT	Serum, organs, mosquitoes	IIF, EIA[h]	RT-PCR
Other arboviruses	DVBID	EIA, HI, CF, NT	Serum, arthropods	IIF, EIA[h]	RT-PCR
Congo-Crimean hemorrhagic fever[d]	DVRD	IgM capture EIA	Serum, liver, spleen		RT-PCR
Tick-borne encephalitis[d]	DVRD	EIA[h], HI, CF, NT	Brain, CSF, serum, ticks	IIF	
Hantaviruses	DVRD	EIA	NA		RT-PCR
Arenaviruses					
Junin[d]	DVRD	EIA	Serum, liver, spleen		RT-PCR
Lassa[d]	DVRD	EIA	Serum, liver, spleen	IIF	RT-PCR
Lymphocytic choriomeningitis	DVRD	EIA	Serum, CSF		RT-PCR
Machupo[d]	DVRD	EIA	Serum, liver, spleen		RT-PCR
Enteroviruses					
Coxsackieviruses[e]	DVRD	NT[f]	Brain, CSF, throat swab, stool, nasopharyngeal swab		RT-PCR
Echoviruses[e]	DVRD	NT[f]	Brain, CSF, throat swab, stool, nasopharyngeal swab		RT-PCR
Enteroviruses[e]	DVRD	NT[f]	Brain, CSF, throat swab, stool, nasopharyngeal swab		RT-PCR
Polioviruses 1–3[e]	DVRD	NT[f,m]	Brain, CSF, throat swab, stool, nasopharyngeal swab		RT-PCR
Respiratory viruses					
Adenoviruses[e]	DVRD	EIA, NT[f]	Nasopharyngeal aspirate/swab, throat swab, eye swab, urine	IIF, EIA	PCR
Coronaviruses[e]	DVRD			IIF, EIA	RT-PCR
Influenza[g]	DVRD	HI, NT		IIF, EIA	
Measles[h]	DVRD	EIA, NT	Throat, swab, urine, saliva	IIF, EIA	RT-PCR
Mumps[h]	DVRD	EIA	Throat swab, lymphocytes, urine	IIF	RT-PCR
Parainfluenza[e]	DVRD	EIA	Nasopharyngeal aspirate/swab	IIF, EIA	RT-PCR
Parvovirus, human[e]	DVRD	EIA		EIA	PCR
Respiratory syncytial virus[e]	DVRD	EIA	Nasopharyngeal aspirate/swab[c]	IIF, EIA	RT-PCR
Rhinovirus[e]	DVRD	EIA	Nasopharyngeal aspirate/swab		RT-PCR

(Continued on next page)

TABLE 3 (Continued)

Herpesviruses					
Cytomegalovirus[e]	DVRD	EIA	Urine (unfrozen), throat swab	EIA	
Herpes simplex[e]	DVRD	EIA	Vesicular fluid, brain	IIF, EIA	
Infectious mononucleosis (Epstein-Barr)[e]	DVRD	IIF, OCH			
Varicella-zoster	DVRD	IIF, EIA	Vesicular fluid, scabs[j]		RT-PCR
Miscellaneous viruses					
Rubella	DVRD	EIA	Throat swab		RT-PCR
Orf-paravaccinia[j]	DVRD	IIF			
Vaccinia[j]	DVRD	HI, IIF, EIA	Vesicular fluid, scabs, brain, saliva[h]		RT-PCR
Marburg and Ebola[d]	DVRD	EIA	Serum, liver, spleen		RT-PCR
Hendra[d]	DVRD	EIA			RT-PCR
Nipah[d]	DVRD	EIA			RT-PCR
Rift Valley fever[d]	DVRD	EIA			RT-PCR
Rabies	DVRD	RFFIT, IIF	Saliva, brain or skin biopsy		
Retroviruses					
HIV-1[l]	DASTLR	EIA, WIB	Blood	EIA	RT-PCR
HIV-2[l]	DASTLR	EIA, WIB	Blood	EIA	RT-PCR
HTLV-1 and -2	DASTLR	EIA, WIB	Blood	EIA	RT-PCR
Gastroenteritis viruses					
Rotavirus, human[h]	DVRD	EIA	Stool	EIA	RT-PCR
Norwalk[h]	DVRD	EIA	Stool	EIA	RT-PCR
Adenovirus[h] 40/41	DVRD	EIA, NT	Stool	EIA	RT-PCR
Astrovirus[h]	DVRD	EIA	Stool	EIA	RT-PCR

[a] The absence of a notation in any column indicates that the service is not offered for that virus. Abbreviations: CF, complement fixation; CSF, cerebrospinal fluid; DFA, direct fluorescent-antibody test; EIA, enzyme immunoassay; HI, hemagglutination inhibition; IIF, indirect immunofluorescence; NA, not applicable; NT, neutralization; OCH, ox cell hemolysin; PPT, precipitation test; RFFIT, rapid fluorescent focus inhibition test; RIP, radioimmunoprecipitation; RIA, radioimmunoassay; WIB, Western immunoblot.

[b] Other tests or tests for agents not listed may be made by prior consultation and arrangement. All serum specimens for serologic tests should be sterile and shipped frozen or on wet ice. Serum specimens may be sent without refrigeration if delivery is assured within 48 h. In most instances, paired (acute- and convalescent-phase serum specimens are required for serologic diagnosis. Specimens for viral isolation should be shipped frozen on dry ice if they will not arrive within 48 h of collection; if delivery to CDC will take less than 48 h, specimens for viral isolation should be shipped chilled by cold packs or wet ice.

[c] Send specimens on wet ice or cold packs; do not freeze.

[d] These viruses are among the most highly pathogenic viruses known. Infections (and deaths) may occur in medical and supportive personnel through close contact with patients and their exudates. Contact SPB, DVRD, NCID, CDC, for special instructions before taking, packing, and shipping specimens suspected of containing these agents. Do not submit these specimens for testing without making prior arrangements.

[e] Because of the ubiquitous nature of these viruses, primary diagnosis by serologic testing or virus isolation cannot routinely be offered. These services may be provided by prior consultation and arrangement for outbreaks or cases of unusual public health significance.

[f] Neutralization tests are performed by prior arrangement only when virus is isolated from patient.

[g] Because of the epidemic nature of the virus, primary diagnosis by serologic testing or virus isolation cannot routinely be offered. These services may be provided by individual agreement for outbreaks or cases of unusual public health significance. Reference antigenic analysis of representative or unusual isolates ("strain comparison") is organized and conducted yearly by the World Health Organization (WHO) Collaborating Center for Influenza at CDC. Isolates should be sent directly to that center (bldg. 7, room 112). The completed WHO data form, not CDC form 50.34, is required. The WHO data form may be obtained from the Influenza Branch, DVRD, NCID, CDC.

[h] Prior consultation and arrangements with the appropriate branch are required before submission of specimens; indicate on the request form that such arrangements have been made.

[i] Immune status testing is performed only by prior consultation and arrangement; indicate on the request form that such arrangements have been made.

[j] For information on collecting, packing, and shipping specimens from a person with suspected smallpox, contact the Viral Exanthems and Herpesvirus Branch, DVRD, NCID, CDC.

[k] Laboratory support and collaboration for HTLV-1 and HTLV-2, other non-AIDS retroviruses, and diseases associated with these viruses (adult T-cell leukemia/lymphoma, tropical spastic paraparesis, or HTLV-1-associated myelopathy) are available through prior arrangements with the HIV/AIDS and Retrovirology Branch, DASTLR, NCID, CDC.

[l] Laboratory support and collaboration for HIV-1 and HIV-2, and for AIDS, are available through prior arrangement with the HIV Immunology and Diagnostics Branch, DASTLR, NCID, CDC.

[m] Neutralization tests are preformed by prior arrangement to confirm immunity in at-risk laboratory workers or in cases of paralytic disease compatible with poliomyelitis.

3. Whenever possible, time the shipment to arrive at CDC at the beginning or middle of the work week, *not* just before or on a weekend or holiday.

SPECIAL REQUIREMENTS FOR VIRAL DISEASES

Considerable research and development in the rapid diagnosis of viral diseases, together with the recognition of many new viral agents responsible for infectious disease, have prompted NCID to encourage state laboratories to request consultation and reference testing for those diseases and agents listed in Tables 2 and 3, as well as for new agents that may not be included.

The Division of Viral and Rickettsial Diseases (DVRD) and the Division of AIDS, Sexually Transmitted Diseases and TB Laboratory Research (DASTLR) provide confirmation of serologic and virologic diagnosis when deemed essential by a state laboratory (after that laboratory has consulted the DVRD specialty laboratory or DASTLR). Highest priority is given to the diagnosis of a life-threatening condition or to an outbreak of illness of public health importance. Infection with a highly pathogenic virus, such as Lassa or Marburg, is a major health hazard not only for the individual patient but also for the medical and support personnel involved. In such cases, prompt contact with the Special Pathogens Branch (SPB), DVRD, is important. Laboratory support is also available to evaluate problem specimens and to study complications of viral vaccines. The viral disease programs of the Division of Vector-Borne Infectious Diseases (DVBID) are concerned with surveillance of arthropod-borne viral diseases in the United States and offer clinical and public health laboratory support in (i) primary or confirmatory diagnosis of arboviral infections, (ii) identification of arbovirus isolates, and (iii) serologic and virologic surveillance of arthropods and vertebrates.

Clinical specimens from persons or animals with suspected arbovirus diseases must be accompanied by the following items of clinical and epidemiologic history: dates and places of travel; yellow fever, Japanese encephalitis, or other relevant immunizations (Western, Eastern, and Venezuelan equine encephalitis virus vaccines in horses and avians); and dates of onset and specimen collection.

ADDITIONAL SUGGESTIONS FOR VIRAL SPECIMENS

General

When shipping specimens in tubes or vials, use only tight-fitting soft rubber stoppers or leak-proof screw caps; seal well with waterproof tape. Avoid direct contact between specimen containers and dry ice to prevent breakage. To prevent thawing of specimens in the event of transit delays, use enough dry ice to last 48 h beyond the expected arrival time.

Specimens for Isolation of Etiologic Agents

Select specimens during the acute, febrile phase of illness. Depending on the disease suspected, the specimen submitted may be nasal or throat washings, sputum, urine, feces, CSF, skin scrapings, aspirate from lesions, blood, serum, or various tissues. Throat, nasal, or rectal swabs must be immersed in an appropriate virus transport fluid to prevent drying. Blood, CSF, and tissues should be handled aseptical-

ly. Autopsy tissues from several different organs should be placed in separate containers, not pooled. Most specimens deliverable to the laboratory within 24 h of collection may be sent chilled by using ice packs to maintain the initial condition. If more than 24 h are required, the specimens should be frozen and kept frozen during shipment. (Specimens suspected of containing cytomegalovirus should not be frozen but sent chilled.) Consult the appropriate laboratory if you are uncertain about the optimum method for shipping a given specimen. When shipment of frozen specimens is not possible, some types of specimens can be preserved in buffered glycerin (50% glycerin–50% phosphate-buffered saline [pH 7.2]), but partial loss of virus generally occurs (complete rickettsial and chlamydial loss occurs). Consultation is required before buffered glycerin is used for transport of specific virus specimens.

Specimens for Serologic Testing

Generally, serum may be shipped unfrozen if it is drawn aseptically and delivery within 48 h is assured. Never add a preservative to serum that will be used in serologic tests.

Arthropod Specimens (Virus Isolation)

Arthropod specimens must be collected alive, killed by freezing or exposure to ether vapor, sealed in ampules, and shipped on dry ice. Cyanide or chloroform must not be used, because these chemicals inactivate viruses and rickettsiae.

REQUIREMENTS FOR SPECIMENS FOR DVRD AND DVBID

Sera and Other Body Fluids

Paired sera (acute and convalescent phase) are preferred for serologic diagnosis, but in some cases a presumptive diagnosis is possible if only a single convalescent-phase serum specimen is available. Consult with the respective laboratory if you have questions in this regard. The DVBID requests CSF specimens from patients with encephalitis. Newly introduced antibody capture tests of antibody in CSF offer a rapid and sensitive means of specific diagnosis. Detection of viral antigen in other body fluids is under investigation, and other fluids may be accepted for testing if approved. Pre-arrangements for the latter tests should be made with DVBID. In cases of suspected viral hemorrhagic fever, the SPB, DVRD, and the DVBID require acute-phase sera both for immediate serologic testing, which may be diagnostic, and for virus isolation. The handling and transport of such specimens must be discussed in advance with either the SPB, DVRD, or the DVBID, as appropriate.

Tissues

Biopsy or autopsy specimens from patients with suspected arboviral disease or viral hemorrhagic fever may be submitted for viral isolation and/or examination by immunofluorescence for viral antigen. Tissues submitted for immunofluorescence testing should be embedded in appropriate medium for frozen sections or sent cold for processing by DVBID or DVRD. Formalin-fixed tissues may be suitable for immunohistochemical examination in some instances. Prearrangements for these procedures are required.

Isolates

Viral isolates referred to DVBID or DVRD for identification and/or characterization should be accompanied by as complete an isolation history as possible. The source of the isolate, date of isolation, host system used for isolation, success of reisolation and passage level, and titer of the material being sent should be given. If available, information concerning sensitivity to lipid solvents, spectrum of sensitive host systems, presence or absence of hemagglutinin activity, and antigenic relationship to other established viruses also should be provided. Prearrangements for these studies must be made with the appropriate laboratory.

Arthropod and Avian Specimens

In the event of a known or presumed arthropod-borne viral encephalitis outbreak, arthropods may be submitted for virus isolation. Contact with appropriate DVBID personnel is required to arrange for testing prior to submission of specimens. Specimens should be carefully placed in lipless glass vials, tightly stoppered, and shipped on dry ice. The date(s), method(s), and location(s) of collection should accompany the specimens. Sera from wild birds may also be submitted for antibody tests under similar conditions. Sera should be collected from abundant passerine bird species from the potential outbreak area. Birds may be bled from the jugular vein (0.2 ml) and the blood specimen added to 0.9 ml of diluent (buffered saline, pH 7.4 to 7.8), preferably with a protein stabilizer such as fetal bovine serum (5 to 10%) or bovine serum albumin (0.5%). Specimens should be centrifuged at low speed to remove red cells; the supernatant fluid should then be removed, frozen, and shipped on dry ice. As is required for arthropods, prearrangements for testing of avian sera must be secured.

REFERENCE AND DISEASE SURVEILLANCE REQUEST FORM

Instructions for Completing the Request Form for Reference and Disease Surveillance Support (CDC 50.34 REV. 11-90)

The CDC form is a combined request, specimen information, patient history, and result report form. The upper third of the front side of the form must contain the required information for identification of the specimen that is used to start a computer record on each specimen, to track a specimen while being tested in NCID laboratories, and to direct the report to the proper addressee. All of the information requested on the upper front third of the form must be provided. The back of the form is used to provide NCID laboratories with essential information about the specimen, the patient, and the assistance requested. The required information for this portion of the form depends on the type of specimen submitted and the assistance requested. The results of tests performed at CDC can be affixed to the lower portion of the front of the form (once the information has been entered into the computer), and this portion constitutes the report that is returned to the state health department laboratory or other authorized sender. Detailed instructions for completing the form are as follows.

Front of Form (Fig. 1)

Justification

The justification must be completed and signed by a state health department laboratory representative before a specimen can be accepted by CDC.

Name, address, and phone number of requesting physician or organization

Print or type the name, address, and telephone number of the physician, microbiologist, or organization from which the specimen originated. (Include the person or institution to contact if additional information is needed and to whom the final report will be forwarded by the state laboratory.)

State health department laboratory address

Print or stamp the address of the state health department laboratory or agency sending the specimen. (This will ensure that reports are directed through the correct state health department.)

State health department number

Print the state health department laboratory number assigned to the specimen, if any (used as a cross-reference for specimen identification).

Date sent to CDC

Print date (month, day, and year) that the specimen is shipped to CDC.

Hospital number

For those specimens originating from a hospital, print the patient's hospital number. (This is not required by CDC, but it is helpful information for the hospital record office in matching the report to the patient's record.)

Name

Print last name, first name, and middle initial of the patient or other equally appropriate specimen identification (required to track and locate specimens).

Birthdate

Print the date (month, day, and year) of the patient's birth and age in years (or months if an infant).

Sex

Print "M" (male) or "F" (female).

Clinical diagnosis

Print the patient's clinical diagnosis; if none has been made, indicate why assistance is requested (e.g., possible outbreak, exotic isolate, or possible disease).

Associated illness

Print the patient's associated or underlying illness, such as cancer, arthritis, hypertension, immunocompromise, or enter the patient's major symptoms.

Date of onset

Print the date (month, day, and year) that illness started. (This date is critical for the interpretation of serologic results; if uncertain, give an approximate date.)

Fatal?

Check one box only. (This element has epidemiologic significance.)

Justification must be completed by State health department laboratory before specimen can be accepted by CDC. Please check the first applicable statement and when appropriate complete the statement with the *.

1. **Disease suspected to be of public health importance. Specimen is:**
   (a) ☐ from an outbreak.
   (b) ☐ from uncommon or exotic disease.
   (c) ☐ an isolate that cannot be identified, is atypical, shows multiple antibiotic resistance, or from a normally sterile site(s).
   (d) ☐ from a disease for which reliable diagnostic reagents or expertise are unavailable in State.
2. ☐ Ongoing collaborative CDC/State project.
3. ☐ Confirmation of results requested for quality assurance.

*Prior arrangement for testing has been made. Please bring to the attention of:

(Name) _____

Name, Address and Phone Number of Physician or Organization:

Completed by:

Date:_____

STATE HEALTH DEPARTMENT LABORATORY ADDRESS:

STATE HEALTH DEPT. NO.:

DATE SENT TO CDC: Month Day Year

PATIENT IDENTIFICATION Hospital No.:

NAME: Last (18-37)   First (38-47)   Middle Initial (48)

BIRTHDATE: (49-54) Month Day Year SEX: (55)

CLINICAL DIAGNOSIS: (56-57)

ASSOCIATED ILLNESS: (58-59)

DATE OF ONSET (Mo. Da. Yr.) (60-65) FATAL? (66) ☐YES ☐NO

(For CDC Use Only) CDC NUMBER

UNIT FY 3-4) NUMBER (5-10) SUF (11)

DATE RECEIVED(12-17) Month Day Year

REVERSE SIDE OF THIS FORM MUST BE COMPLETED   Type Specimen

THIS FORM MUST BE EITHER PRINTED OR TYPED

PLEASE PREPARE A SEPARATE FORM FOR EACH SPECIMEN

D.A.S.H.

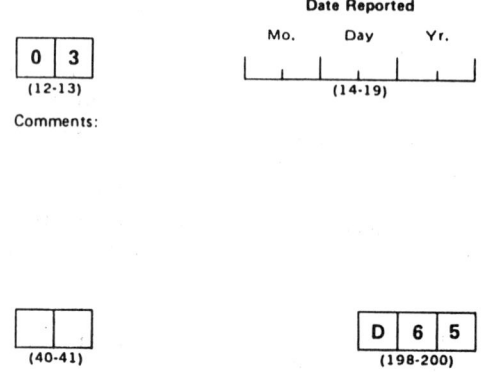

Date Reported
Mo. Day Yr.

| 0 | 3 |
(12-13)

(14-19)

Comments:

| | |
(40-41)

| D | 6 | 5 |
(198-200)

DEPARTMENT OF HEALTH AND HUMAN SERVICES
Public Health Service
Centers for Disease Control
Center for Infectious Diseases
Atlanta, Georgia 30333

CDC
CENTERS FOR DISEASE CONTROL
AND PREVENTION

CDC 50.34 REV. 11-90
(Formerly 3.203)

FIGURE 1   Front of CDC request form for reference and disease surveillance support.

LABORATORY EXAMINATION(S) REQUESTED (31-36)
☐ **AN**timicrobial Susceptibility   ☐ **ID**entification   ☐ **SE**rology (Specify Test) _____
☐ **HI**stology   ☐ **IS**olation   ☐ **OT**her (Specify) _____

CATEGORY OF AGENT SUSPECTED (37)   ☐ Bacterial   ☐ Viral   ☐ Fungal   ☐ Rickettsial   ☐ Parasitic   ☐ Other (Specify) _____

SPECIFIC AGENT SUSPECTED: _____ (38-40) [   ] **OTHER ORGANISM(S) FOUND:** _____ (41-46) [   ]

ISOLATION ATTEMPTED? (47) ☐ Yes ☐ No   **NO. TIMES ISOLATED (48-49)** _____ **NO. TIMES PASSED (50-51)** _____
SPECIMEN SUBMITTED IS (52): ☐ Original Material   ☐ Pure Isolate   ☐ Mixed Isolate

DATE SPECIMEN TAKEN (53-58)
Mo.   Da.   Year
| | | | | | |

ORIGIN (59-60)
☐ **FO**od   ☐ **AN**imal (specify) _____
☐ **HU**man   ☐ **SO**il   ☐ **OT**her (Specify) _____

SOURCE OF SPECIMEN (61-62): ☐ **BL**ood   ☐ **GA**stric   ☐ **SE**rum   ☐ **SP**utum   ☐ **UR**ine   ☐ **CSF**   ☐ **HA**ir   ☐ **SK**in   ☐ **ST**ool   ☐ **TH**roat
☐ **WO**und (Site) _____   ☐ **TI**ssue (Specify) _____
☐ **EX**udate (Site) _____   ☐ **OT**her (Specify) _____

SUBMITTED ON (63-64): ☐ **ME**dium (Specify) _____   ☐ **EG**g   ☐ **TI**ssue Culture (Type) _____
☐ **AN**imal (Specify) _____   ☐ **OT**her (Specify) _____

SERUM INFORMATION:
Mo.   Da.   Yr.
(65-72) ☐ **AC**ute | | | |
(73-80) ☐ **CO**nvalescent | | | |
(81-88) ☐ **S3** | | | |
(89-96) ☐ **S4** | | | |
(97-104) ☐ **S5** | | | |

IMMUNIZATIONS:
Mo.   Yr.
_____ (105-110)
_____ (111-116)
_____ (117-122)
_____ (123-128)

TREATMENT: Drugs Used ☐ None (129)
Date Begun   Date Completed
Mo. Da. Yr.   Mo. Da. Yr.
_____ (130-143)
_____ (144-157)
_____ (158-171)

EPIDEMIOLOGICAL DATA: (172-173)
☐ **S**ingle Case   ☐ **SP**oradic   ☐ **CO**ntact   ☐ **EP**idemic   ☐ **CA**rrier
Family Illness _____ (174-175)
Community Illness _____ (176-177)
Travel and Residence (Location)
Mo.   Yr.
☐ **F**oreign _____ (178-183)
☐ **USA** _____ (184-189)
Animal Contacts (Species) _____ (190-191)
Arthropod Contacts: (192) ☐ None ☐ Exposure Only ☐ Bite
Type of Arthropod _____ (193-194)
Suspected Source of Infection _____ (195-196)

PREVIOUS LABORATORY RESULTS/OTHER CLINICAL INFORMATION:
(Information supplied should be related to this case and/or specimen (s) and relative to the test (s) requested.

CLINICAL TEST RESULTS: (12-13) [0][2]
Sputum and Histological Findings _____
Blood Counts _____   Urine Exams _____
Date
Type Skin Tests Performed   Mo. Da. Yr.   Strength   Pos.   Neg.
_____ (14-21) — — — (22)
_____ (23-30) — — — (31)
_____ (32-39) — — — (40)

SIGNS AND SYMPTOMS
(48-49) ☐ **FE**ver
Maximum Temperature: _____ (50-53)
Duration: _____ Days (54-55)
(56-57) ☐ **CH**ills
RASH:
(58-59) ☐ **MA**culopapular
(60-61) ☐ **HE**morrhagic
(62-63) ☐ **VE**sicular
(64-65) ☐ Erythema Nodosum
(66-67) ☐ Erythema Marginatum
(68-69) ☐ **OT**her _____
RESPIRATORY:
(70-71) ☐ **RH**initis
(72-73) ☐ **PU**lmonary
(74-75) ☐ **PH**aryngitis
(76-77) ☐ **CA**lcifications
(78-79) ☐ **PN**eumonia (type) _____
(80-81) ☐ **OT**her _____
CARDIOVASCULAR:
(82-83) ☐ **MY**ocarditis
(84-85) ☐ **PE**ricarditis
(86-87) ☐ **EN**docarditis
(88-89) ☐ **OT**her _____
GASTROINTESTINAL:
(90-91) ☐ **D**iarrhea
(92-93) ☐ **BL**ood
(94-95) ☐ **MU**cous
(96-97) ☐ **CO**nstipation
(98-99) ☐ **AB**dominal pain
(100-101) ☐ **VO**miting
(102-103) ☐ **OT**her _____

CENTRAL NERVOUS SYSTEM:
(104-105) ☐ **HE**adache
(106-107) ☐ **ME**ningismus
(108-109) ☐ **MI**crocephalus
(110-111) ☐ **HY**drocephalus
(112-113) ☐ **SE**izures
(114-115) ☐ **CE**rebral Calcification
(116-117) ☐ **CH**orea
(118-119) ☐ **PA**ralysis
(120-121) ☐ **OT**her _____
MISCELLANEOUS:
(122-123) ☐ **JA**undice
(124-125) ☐ **MY**algia
(126-127) ☐ **PL**eurodynia
(128-129) ☐ **CO**njunctivitis
(130-131) ☐ **CH**orioretinitis
(132-133) ☐ **SP**lenomegaly
(134-135) ☐ **HE**patomegaly
(136-137) ☐ Liver Abscess
(138-139) ☐ **LY**mphadenopathy
(140-141) ☐ **MU**cous Membrane Lesions
(142-143) ☐ **OT**her _____
STATE OF ILLNESS:
(144-145) ☐ **SY**mptomatic
(146-147) ☐ **AS**ymptomatic
(148-149) ☐ **SU**bacute
(150-151) ☐ **CH**ronic
(152-153) ☐ **DI**sseminated
(154-155) ☐ **LO**calized
(156-157) ☐ **IN**traintestinal
(158-159) ☐ **EX**traintestinal
(160-161) ☐ **OT**her _____

FOR CDC USE ONLY [0][1] (12-13)   No. Specimens: (16-20) _____   No. Tests: (21-25) _____

TYPE SERVICE: (14-15)
01-Reference
02-Epid. Aid
03-Proficiency Testing
04-Special Projects
| | | | - Other

LOCATION CODE: (26-27)
AR Argentina
AS Australia
AU Austria
BC Bermuda
BE Belgium
BH British Honduras
BL Bolivia
BR Brazil
CA Canada
CB Cambodia
CI Chile

CM Cameroon
CO Colombia
CS Costa Rica
CY Cyprus
DR Dominican Rep.
EC Ecuador
ES El Salvador
ET Ethiopia
FR France
GE Germany
GQ Guam

GT Guatemala
HA Haiti
HO Honduras
IN India
IS Israel
IT Italy
IV Ivory Coast
JM Jamaica
MX Mexico
MY Malaysia
NI Nigeria

NU Nicaragua
NZ New Zealand
PA Paraguay
PE Peru
PK Pakistan
PL Poland
PN Panama
PP New Guinea
RP Philippines
RQ Puerto Rico
SL Sierra Leone
SP Spain

SZ Switzerland
TD Trinidad-Tobago
TH Thailand
TW Taiwan
UK United Kingdom
UR Soviet Union
UY Uruguay
VE Venezuela
VN Vietnam
VQ Virgin Islands
| | | Other _____

SPECIMEN SUBMITTED BY: (28-30)
100-Health Dept.
200-CDC Clinic
205-Proficiency Testing
225-CDC Non-clinic
301-Army
302-Navy
303-Air Force
307-V.A. Hosp.
310-U.S.D.A.

323-Indian Hosp.
325-NIH
400-Foreign
402-Peace Corps.
550-University
606-Physician/Clinic
| | | | - Other

CDC 50.34 REV. 11-90 (BACK)
(Formerly 3.203)

CDC NUMBER | UNIT | FY | NUMBER | SUF.

*U.S.GPO:1996-734-363

FIGURE 2 Back of CDC request form for reference and disease surveillance support.

Type of specimen

Print the type of specimen (e.g., serum, CSF, fungus culture) in the shaded area.

Exercise good judgment in determining the relevance of these items. Paired sera are required for serologic diagnosis of viral and bacterial diseases, and a single serum specimen is required for mycotic and parasitic diseases and for syphilis serology (congenital syphilis excepted). In all instances, the date(s) of collection of serum specimens must be provided. Immunization history is required when such information relates to the kind of serologic analysis requested (e.g., required for polio or measles). Information on treatment, such as administration of immune serum or globulin or antibiotics, is often of great benefit in serologic testing or for identifying reference cultures. As much relevant epidemiologic data as can be obtained should be provided. A history of travel and animal or arthropod contacts are required for those RDS requests for which this kind of information is clearly necessary. If any required item of information is not available after efforts have been made to obtain it, please so indicate.

Back of Form (Fig. 2)

Previous laboratory results and/or other clinical information

Include any additional information related to a case or specimen that might be helpful to the laboratorian in determining or selecting appropriate laboratory tests or in interpreting results. The types of specimens usually sent to CDC laboratories are serum specimens, reference cultures, or clinical specimens. To assist state health department laboratories and others in obtaining the request form information that NCID requires, Table 4 is provided as a guide for each of the three types of specimens.

TABLE 4 Required and useful information for form CDC 50.34 Rev. 11-90 (back)

Specimen type	Information[a]	
	Required	Useful
Serum specimens	Laboratory examination(s) requested Specific agent suspected Serum information Immunizations Treatment Epidemiological data Previous laboratory results	Other clinical information Signs, symptoms
Reference cultures	Laboratory examination(s) requested Category of agent suspected Specific agent suspected Specimen submitted[b] Origin (of specimen) Source of specimen Submitted on (medium type) Previous laboratory results Biochemical reactions[c]	Isolation attempted? Date specimen taken No. of times isolated Other clinical information Clinical test results Signs and symptoms Other organism(s) found Epidemiological data Treatment
Clinical specimens	Laboratory examination(s) requested Category of agent suspected Specific agent suspected Specimen submitted is Date specimen taken Source of specimen Epidemiological data Previous laboratory results	Other clinical information Clinical test results Signs and symptoms

[a]As stated on form CDC 50.34 Rev. 11-90 (back).
[b]Original material, pure isolate, or mixed isolate.
[c]May be attached as a separate sheet.

State Public Health Laboratory Virology Services[*]

RICHARD L. HODINKA AND STEPHEN A. YOUNG

APPENDIX
2

State public health laboratories, much like the federal laboratories at the Centers for Disease Control and Prevention (CDC), are charged with providing laboratory diagnosis of viral infections when local services are not available. Thus, most states do not encourage routine use of the state virology laboratories as a primary diagnostic laboratory, preferring to defer to another laboratory service. The availability of only limited diagnostic services by state laboratories is often necessitated by limited financial support. Therefore, private hospital, commercial, or local public health laboratories are often used for routine primary diagnostic virology services. The state virus laboratory should be utilized for diagnostic problems that go beyond the scope or capability of local laboratories, especially for viruses for which statewide surveillance is performed (e.g., influenza viruses, arboviruses) as well as for viruses of epidemiologic significance. Although many state laboratories accept specimens for routine primary isolation and/or detection, their budgets may be adversely affected if the laboratories are confronted with large numbers of specimens. This may impinge unfavorably on their ability to perform a key function as a center to collate and disseminate information regarding viral diseases to the CDC.

Within this context, we have listed some of the functions and viral diagnostic services available in state and U.S. territorial public health laboratories. The information was provided through (i) the directors of the various laboratories, (ii) an April 1998 profile survey distributed to the 56 U.S. state and territorial public health laboratories, and (iii) the 28th edition of the Consolidated Annual Report, which provides quantitative information about public health laboratory services identified in 15 workload areas. The survey and report were tabulated by the Association of Public Health Laboratories (APHL) (formerly the Association of State and Territorial Public Health Laboratory Directors), 1211 Connecticut Ave., N.W., Suite 608, Washington, DC 20036; phone, (202) 822-5227; fax, (202) 887-5098; website: www.aphl.org.

SUBMISSION OF SPECIMENS

All states and U.S. territories have laboratories that accept specimens for the diagnosis of viral diseases (Table 1). The submission of specimens to the state laboratory can be directly from the requesting physician or institution or via local public health laboratories. Each laboratory has its own set of requirements for the processing, shipping, and type of specimens that are acceptable for the detection and/or identification of particular viruses. In most cases, these are described in detail in the written and electronic literature provided by the appropriate state authority. In many states, conditions for specimen submission are similar; however, specific requirements are imposed by other states. Anyone who desires to submit a specimen to their state clinical virology laboratory should contact the laboratory head to determine the requirements for submitting clinical material. Some generally accepted requirements include:

1. A good clinical history with a listing of the patient's name and age, date of specimen collection, date of onset of illness, type of specimen and collection site, major clinical symptoms, relevant immunization history, virus(es) for which specimens are to be tested, and physician's name, address, and telephone number.

2. Serologic testing for antiviral antibodies generally requires simultaneous submission of acute- and convalescent-phase sera, except for special screening studies (e.g., measles, mumps, rubella, or varicella-zoster virus immune status).

3. Neonatal serum should be accompanied by a maternal serum sample.

Many state laboratories request that, during an epidemic, only a limited number of specimens be submitted for determination of the causative agents, rather than specimens from each patient seen in the course of the epidemic. Shipping instructions often include conditions for handling and packaging specimens as well as prepayment of shipping costs. Many states do not charge a fee for testing; however, some states may have minimal fees. Published fee schedules are available from most state laboratories that charge for their services. Submission of specimens may be limited, e.g., Georgia, Maryland, and New York have requirements that limit submission of specimens by physicians and/or patients who reside in their states. Several states limit specimen submission to licensed physicians, whereas others also accept

---

[*]This appendix contains information presented in chapter 35, by Steven Specter and Gerald Lancz, in the second edition of this manual.

specimens from veterinarians and other legitimate public health services, including hospitals, state agencies, and public or community health laboratories.

Specimens for the detection of rabies virus often have additional requirements for handling, shipping, and clinical history. Many states provide specific instructions as well as a Rabies Investigation Report Form to accompany such specimens.

SCOPE OF SERVICES

An overview of the services available in each state laboratory is provided in Table 1. The diversity of services offered by different states is apparent. These range from laboratories that offer only serology for select viruses to those that provide extensive serology, isolation, and direct detection services for many viruses. There are laboratories that (i) exclude class IV agents (e.g., lassa fever virus); (ii) exclude viruses that could be easily tested for in hospital or private laboratories (e.g., herpes simplex viruses); and (iii) include only viruses that have epidemiologic importance (e.g., influenza viruses, arboviruses). A listing of such specific services by state has been avoided here because these services no doubt change periodically. Again, it is recommended that you refer to your state laboratory to determine the extent of services offered. Many state laboratories now publish a Client Services Guide and periodically distribute newsletters with updated information. The addresses, phone and fax numbers, and websites for the state laboratories, as of August 1998, are provided in Table 2. Approximately 70% (39 of 56) of the state and territorial laboratories have their own websites.

Most laboratories send specimens to the CDC if they do not handle them on site. This may be limited to class IV specimens or may cover a broad range of viral agents. Also listed in Table 1 are some special services offered by the laboratories, including participation in national or international surveillance programs for arboviruses and influenza viruses; serological screening for human immunodeficiency virus (HIV) and hepatitis A, B, and C viruses; testing for *Chlamydia trachomatis;* and detection services for rabies virus, which usually results in rapid reporting in suspected rabies exposure cases. However, a service that is considered to be special by one state laboratory may be part of the normal services offered by another state laboratory. Many laboratories also provide surveillance and typing of enteroviruses as well as surveillance for respiratory (e.g., adenoviruses, parainfluenza virus types 1, 2, and 3, and respiratory syncytial virus) and enteric (e.g., rotavirus) viruses. Depending on the particular state laboratory, testing for these agents may involve viral isolation and/or direct detection methods such as immunofluorescence, enzyme immunoassays, electron microscopy, or nucleic acid amplification. Of particular interest, 31 state public health laboratories report performing nucleic acid amplification and 5 state that they do electron microscopy as part of their current technology. Molecular amplification assays are being used by participating state laboratories for the qualitative detection of HIV, hepatitis C virus, hantavirus (Sin Nombre), norwalk viruses, and *C. trachomatis* from infected individuals and to detect arboviruses, enteroviruses, and herpes-group viruses from cerebrospinal fluid of patients with central nervous

system disease. Quantitative monitoring of HIV and/or hepatitis C virus RNA also is offered in some laboratories. The number and types of viral serologic assays performed varies between state laboratories but may include either immunoglobulin G (IgG) or IgM assays for measles virus, rubella virus, mumps virus, parvovirus B19, cytomegalovirus, herpes simplex virus, varicella-zoster virus, hantavirus, influenza and parainfluenza viruses, respiratory syncytial virus, enteroviruses, dengue viruses, and human T-cell leukemia virus types 1 and 2. A total of 71% (30 of 42) of the responding state public health laboratories have the capacity to handle viral agents classified as biosafety level 3 pathogens.

Regulation of licensure of virology laboratories by state laboratories is still a common practice. Approximately 79% (33 of 42) of the state laboratories that responded to the April 1998 APHL survey indicated that they were involved in regulating primary laboratories. Additionally, some state laboratories indicated that they are not involved in the regulation of primary laboratory licensure but that another state agency performs this function. In some states, there are no practicing primary virology laboratories that require regulation. It would seem that some regulation is desirable in all states to ensure that standard, accepted practices are used to obtain reliable results. The establishment of some level of regulation within states would promote this standardization.

TURNAROUND TIME FOR REPORTING OF RESULTS

The bane of viral diagnosis by state laboratories in the past has been the long turnaround time from submission of a specimen by the physician until a report is returned to that physician. Frequently, this was a matter of months in all but emergency cases, as with exposure to rabies virus. Today, the turnaround time for diagnosis of many viral diseases is no longer a significant problem. Most laboratories send out a written report (35 of 35 respondents) upon identification of a virus; in some states, reports are also transmitted by fax (26 of 35 respondents) and/or electronically via computer (20 of 35 respondents). No responding laboratories provide reports by telephone (0 of 28 respondents).

Virtually all laboratories indicate that the length of time for reporting results is variable and dependent on the type of testing to be performed, as well as whether a specimen is positive or negative. Serologic results are frequently reported from within a few days to 1 week of receipt of the acute- and convalescent-phase sera; however, a few laboratories indicate this may take as long as 2 to 4 weeks. Testing for immune status against a particular virus is reported from a few days to 4 weeks after receipt of a single serum sample. Positive isolation of many viruses is reported within 72 h, whereas some isolation (as in the case of cytomegalovirus or varicella-zoster virus) may take as long as 2 to 4 weeks. Reports on specimens that are negative for virus isolation may be sent out as soon as 2 weeks or not until 6 weeks after receipt of the specimen.

We thank the directors of the various state laboratories and APHL for supplying the information used in this appendix.

TABLE 1 Virology services available in state and territorial public health laboratories[a]

State or territory	Viral serology[b]	Viral isolation	Refer specimens to CDC[c]	Regulate primary labs in state	Arbovirus surveillance	Influenza surveillance	Rabies detection	HIV serology	Hepatitis serology (A/B/C)	Chlamydia testing
Alabama	+	+[d]	+	+[e]	+	+	+	+	No/+/No	+
Alaska	+	+	No	+	+	+	+	+	+/+/+	+
American Samoa	NI	NI	NI	NI	NI	NI	NI	NI	NI	NI
Arizona	+	+	+	+	+	NI	+	+	+/+/No	+
Arkansas	+	No	+	+	+	No	+	+	No/+/No	+
California	+	+	+	+	+	No	+	+	+/+/+	+
Colorado	+	No	+	+	+	No	+	+	No/+/No	+
Connecticut	+	+	+	No	No	+	+	+	No/+/No	+
Delaware	+	+	+	+	+	+	+	+	+/+/No	+
District of Columbia	+	No	+	+[e]	No	No	+	+	+/+/+	+
Florida	+	+	+	+[e]	+	+	+	+	+/+/No	+
Georgia	+	+	+	+[e]	No	+	+	+	+/+/No	+
Guam	+	No	+	No	NI	NI	NI	NI	NI	NI
Hawaii	+	+	+	+	No	+	+	+	No/+/No	+
Idaho	+	+	+	+	No	+	+	+	+/+/+	+
Illinois	+	+	+	+	+	No	+	+	No/+/No	+
Indiana	+	+	+	No	+	+	+	+	+/+/+	+
Iowa	+	+	+	No	+	+	+	+	+/+/+	+
Kansas	+	+	+	+	No	+	No	+	No/+/No	+
Kentucky	+	+	+	+	No	+	+	+	+/+/No	+
Louisiana	+	+[f]	+	+	+	+	+	+	No/+/+	+
Maine	+	+	+	+	No	+	+	+	No/+/+	+
Maryland	+	+	+	+	+	+	+	+	+/+/+	+
Massachusetts	+	+	+	+	+	+	+	+	No/No/+	+
Michigan	+	+	+	No	No	+	+	+	+/+/+	+
Minnesota	+	+	+	No	No	No	+	+	No/No/No	+
Mississippi	+	No	+	+	+	No	+	+	+/+/+	+
Missouri	+	+	+	No	No	+	+	+	+/+/+	+
Montana	+	+	+	No	No	+	No	No	+/+/+	+
Nebraska	+	+	+	No	No	+	No	+	+/+/+	+
Nevada	+	No	+	No	No	No	No	No	No/+/No	No
New Hampshire	+	+	+	+[e]	+	+	+	+	+/+/+	+
New Jersey	+	+	+	+[e]	+	+	+	+	+/+/+	+
New Mexico	+	+	+	+	+	+	+	+	+/+/+	+
New York	+	+	+	+	+	+	+	+	+/+/+	+
North Carolina	+	+	+	+	+	+	+	+	+/No/No	+
North Dakota	+	+	+	+	+	+	+	+	+/+/+	+
Northern Mariana Islands	No	No	+	No	No	No	No	No	No/+/No	No
Ohio	+	+	+	No	+	+	+	+	No/+/No	+

(Continued on next page)

TABLE 1 (Continued)

State								
Oklahoma	+	+	No	No	+	+	+/+/No	+
Oregon	+	+	+	No	+	+	+/+/+	+
Pennsylvania	+	+	+	No	+	+	No/No/No	+
Puerto Rico	+	+	+	+	No	+	+/+/+	+
Rhode Island	No	+	No	No	+	+	No/No/No	+
South Carolina	+	+	No	No	+	+	+/+/+	+
South Dakota	+	+	+	+	+	+	+/+/No	+
Tennessee	+	+	+[e]	No	+	+	No/No/No	+
Texas	+	+	+	+	+	+	+/+/+	+
Utah	+[f]	+	+	+	+	+	+/+/No	+
Vermont	No	+	No	No	No	+	No/+/+	+
Virgin Islands	No	+	No	No	+	No	+/+/+	+
Virginia	+	+	+	+	+	+	+/+/+	+
Washington	+[f]	+	No	No	+	+	No/No/No	+
West Virginia	+	+	+	No	+	+	No/+/No	+
Wisconsin	+	+	+[e]	+	+	+	+/+/+	+
Wyoming	No	+	+[e]	No	No	No	+/+/No	+

[a]Information provided by individual laboratory directors and from an April 1998 profile survey and the 28th edition of the Consolidated Annual Report, tabulated by APHL. NI, no information provided; +, yes.

[b]May include either IgG or IgM assays for measles, rubella, mumps, parvovirus B19, cytomegalovirus, herpes simplex virus, varicella-zoster virus, hantavirus, influenza and parainfluenza viruses, respiratory syncytial virus, enteroviruses, dengue viruses, and human T-cell leukemia virus types 1 and 2.

[c]Specimens submitted to CDC refer to those not normally tested in the state laboratories.

[d]Respiratory viruses only.

[e]Regulation by state authorities but not through the public health laboratory.

[f]Influenza viruses only.

TABLE 2 State and territorial public health laboratories

ALABAMA
Director
Bureau of Clinical Laboratories
State Department of Public Health
8140 University Dr.
Montgomery, AL 36130-3017
Phone: (334) 260-3400
Fax: (334) 244-5083
Website:
www.alapubhealth.org/admin/report/
pages/lab.htm

ALASKA
Chief
Section of Laboratories
State Public Health Laboratory
527 East 4th Ave. Suite 7
Anchorage, AK 99501
Phone: (907) 269-7941
Fax: (907) 269-7939
Website: www.hss.state.ak.us/labs/

AMERICAN SAMOA
Director
Department of Health Services
Government of American Samoa
LBJ Tropical Medical Center
Pago Pago, American Samoa 96799
Phone: (684) 633-1222
Fax: not available
Website: not available

ARIZONA
Director
Bureau of State Laboratory Services
Arizona Department of Health Services
1520 West Adams St.
Phoenix, AZ 85007-2698
Phone: (602) 542-1194
Fax: (602) 542-0760
Website: www.hs.state.az.us/lab/lab.html

ARKANSAS
Director
Public Health Laboratories
Arkansas Department of Health
4815 West Markham St.
Little Rock, AR 72205-3867
Phone: (501) 661-2191
Fax: (501) 661-2310
Website:
health.state.ar.us/htm/labfacts.htm

CALIFORNIA
Director
Laboratory Science
California State Department of Health
Services
2151 Berkeley Way, Room 703
Berkeley, CA 94704
Phone: (510) 540-2408
Fax: (510) 540-3075
Website:
www.dhs.ca.gov/ps/ls/lsindex.htm

COLORADO
Director
Laboratory and Radiation Services Division
Colorado Department of Public Health
8100 Lowry Blvd.
Denver, CO 80220-6928
Phone: (303) 692-3090
Fax: (303) 344-9989
Website:
www.cdphe.state.co.us/lr/lrhom.asp

CONNECTICUT
Bureau Chief
Connecticut Department of Public
Health and Addiction Services
10 Clinton St.
Hartford, CT 06106
Phone: (860) 509-8500
Fax: (860) 509-8697
Website:
www.state.ct.us/dph/BASS/LabHP.htm

DELAWARE
Director
Delaware Public Health Laboratory
30 Sunnyside Rd.
Smyrna, DE 19977
Phone: (302) 653-2870
Fax: (302) 653-2877
Website: not available

DISTRICT OF COLUMBIA
Director
Bureau of Laboratories
Department of Human Services
300 Indiana Ave. NW, Room 6154
Washington, DC 20001
Phone: (202) 727-0557
Fax: (202) 727-0582
Website: not available

FLORIDA
Chief
Department of Health
Bureau of Laboratories
1217 Pearl St.
Jacksonville, FL 32202
Phone: (904) 791-1550
Fax: (904) 791-1567
Website: www.doh.state.fl.us/

GEORGIA
Director
Georgia State Laboratory of Public
Health
Department of Human Resources
1749 Clairmont Rd.
Decatur, GA 30033-4050
Phone: (404) 327-7900
Fax: (404) 327-7919
Website: www.ph.dhr.state.ga.us/org/
laboratory.htm

GUAM
Director
Guam Public Health Laboratory
P.O. Box 2816
Agana, GU 96910
Phone: (671) 734-7164
Fax: (671) 734-5910
Website: not available

HAWAII
Chief
Laboratories Division
State of Hawaii Department of Health
2725 Waimano Home Rd., 3rd Floor
Pearl City, HI 96782
Phone: (808) 453-6652
Fax: (808) 453-6662
Website: not available

IDAHO
Chief
Bureau of Laboratories
Department of Health and Welfare
2220 Old Penitentiary Rd.
Boise, ID 83712
Phone: (208) 334-2235
Fax: (208) 334-2382
Website: www2.state.id.us/phd1/lab/
labinfo.html

ILLINOIS
Chief
Division of Laboratories
Illinois Department of Public Health
825 N. Rutledge
Springfield, IL 62794-9453
Phone: (217) 782-6562
Fax: (217) 524-7924
Website: www.idph.state.il.us/about/
laboratories/index.htm

INDIANA
Director
Laboratory Resource Center
Indiana State Department of Health
1330 West Michigan St.
Indianapolis, IN 46206-1964
Phone: (317) 233-8006
Fax: (317) 383-6868
Website: www.state.in.us/isdh/labs/
labsvcfr.htm

IOWA
Director
University Hygienic Laboratory
University of Iowa
Iowa City, IA 52242
Phone: (319) 335-4500
Fax: (319) 335-4600
Website: www.uhl.uiowa.edu/

(Continued on next page)

TABLE 2 State and territorial public health laboratories *(Continued)*

KANSAS
Director
Division of Health and Environmental
    Laboratories
Kansas Department of Health and
    Environment
Forbes Building, #740
Topeka, KS 66620
Phone: (913) 296-1619
Fax: (913) 296-1641
Website: www.kdhe.state.ks.us/labs/

KENTUCKY
Director
Division of Laboratory Services
Department for Health Services
100 Sower Blvd., Suite 204
Frankfort, KY 40601
Phone: (502) 564-4446
Fax: (502) 564-7019
Website: not available

LOUISIANA
Director
Division of Laboratories
Office of Public Health
Louisiana State Department of Health
    and Hospitals
325 Loyola Ave., 7th Floor
New Orleans, LA 70112
Phone: (504) 568-5375
Fax: (504) 568-5393
Website: www.dhh.state.la.us/
    oph/labs.htm

MAINE
Chief
Laboratory Operations
Health and Environmental Testing
    Laboratory
Department of Human Services
221 State St.—Station No. 12
Augusta, ME 04333
Phone: (207) 287-2727
Fax: (207) 287-6832
Website: janus.state.me.us/dhs/etl/
    homepage.htm

MARYLAND
Director
Laboratories Administration
State Department of Health and Mental
    Hygiene
201 W. Preston St.
Baltimore, MD 21201
Phone: (410) 767-6100
Fax: (410) 333-5403
Website: not available

MASSACHUSETTS
Director
State Laboratory Institute
Department of Public Health
305 South St.
Boston, MA 02130-3597
Phone: (617) 983-6201
Fax: (617) 983-6210
Website: www.state.ma.us/dph/bls/

MICHIGAN
Laboratory Director
Michigan Department of Community
    Health
3500 Martin Luther King Blvd.
Building 44
Lansing, MI 48909
Phone: (517) 335-8063
Fax: (517) 335-8051
Website: www.mdch.state.mi.us/pha/bofl/

MINNESOTA
Director
Public Health Laboratory
Minnesota Department of Health
717 Delaware St., SE
Minneapolis, MN 55414
Phone: (612) 676-5331
Fax: (612) 676-5514
Website: www.health.state.mn.us/
    divs/phl/phl.html

MISSISSIPPI
Director
State Public Health Laboratory
Mississippi Department of Health
2423 N. State St.
Jackson, MS 39216
Phone: (601) 576-7582
Fax: (601) 354-6124
Website: not available

MISSOURI
Director
State Public Health Laboratory
Missouri Department of Health
307 W. McCarty St.
Jefferson City, MO 65102
Phone: (573) 751-3334
Fax: (573) 751-7219
Website: www.health.state.mo.us/About
    theDepartment/DS4.html#SPHL

MONTANA
Director
Public Health Laboratory
State Department of Health and Envi-
    ronmental Services
Cogswell Building
Helena, MT 59604-6489
Phone: (406) 444-2642
Fax: (406) 444-1802
Website: www.dphhs.state.mt.us/
    hpsd/pubheal/labserv/phlab/
    index.htm

NEBRASKA
Director
Public Health Laboratory
UNMC
600 42nd St.
Omaha, NE 68198
Phone: (402) 559-4116
Fax: (402) 559-4077
Website: not available

NEVADA
Director
Nevada State Laboratory
University of Nevada School of Medicine
1600 N. Virginia St.
Reno, NV 89503-1738
Phone: (775) 688-1335
Fax: (775) 688-1460
Website: not available

NEW HAMPSHIRE
Director
Public Health Laboratory
Division of Public Health Services
State Laboratory Building
6 Hazen Dr.
Concord, NH 03301
Phone: (603) 271-4657
Fax: (603) 271-4783
Website: not available

NEW JERSEY
Commissioner
New Jersey Department of Health and
    Senior Services
Division of Public Health
P.O. Box 360
Trenton, NJ 08625
Phone: (609) 984-2201
Fax: (609) 984-9601
Website: www.state.nj.us/health/phel/
    phelorg.htm

(Continued on next page)

TABLE 2 *(Continued)*

NEW MEXICO
Director
Scientific Laboratory Division
700 Camino de Salud, N.E.
Albuquerque, NM 87106
Phone: (505) 841-2500
Fax: (505) 841-2543
Website: sld.state.nm.us

NEW YORK
Director
Division of Laboratory Quality
    Certification
Wadsworth Center for Laboratories and
    Research
Empire State Plaza
P.O. Box 509
Albany, NY 12201-0509
Phone: (518) 474-7885
Fax: (518) 474-3439
Website: www.wadsworth.org

NORTH CAROLINA
Director
Division of Laboratory Services
306 N. Wilmington St.
Raleigh, NC 27601
Phone: (919) 733-7834
Fax: (919) 733-8695
Website: www.schs.state.nc.us/NCSLPH/

NORTH DAKOTA
Director
Division of Microbiology
State Department of Health
1205 Ave. A West
Bismarck, ND 58501
Phone: (701) 328-5262
Fax: (701) 328-5270
Website: not available

NORTHERN MARIANA ISLANDS
Secretary of Health
Department of Public Health
Commonwealth Health Center
P.O. Box 409 CK
Saipan, MP 96950
Phone: (670) 234-8950
Fax: (670) 234-8930
Website: not available

OHIO
Chief
State Department of Health
1571 Perry St.
Columbus, OH 43201
Phone: (614) 466-2278
Fax: (614) 752-9863
Website: www.odh.state.oh.us/directory/
    directory-f.htm

OKLAHOMA
Chief
Public Health Laboratory Services
State Department of Health
1000 N.E. 10th St.
Oklahoma City, OK 73124
Phone: (405) 271-5070
Fax: (405) 271-4850
Website: www.health.state.ok.us/
    program/phl/index.html

OREGON
Director
Oregon Health Division
Center for Public Health Laboratories
1717 S.W. 10th Ave.
Portland, OR 97201
Phone: (503) 229-5296
Fax: (503) 229-5682
Website: www.ohd.hr.state.or.us/phl/
    welcome.htm

PENNSYLVANIA
Director
Bureau of Laboratories
Pennsylvania Department of Health
Pickering Way and Welsh Pool Rd.
Lionville, PA 19353
Phone: (610) 363-8500
Fax: (610) 436-3346
Website: not available

PUERTO RICO
Director
Institute of Health Laboratories
Department of Health
Commonwealth of Puerto Rico
Building A—Call Box 70184
San Juan, PR 00936-8184
Phone: (787) 274-7817
Fax: (787) 759-6210
Website: not available

RHODE ISLAND
Executive Director
Environmental Health
Rhode Island Department of Health
    Laboratories
3 Capitol Hill, Room 209
Providence, RI 02908
Phone: (401) 222-3118
Fax: (401) 222-6953
Website: not available

SOUTH CAROLINA
Director
Bureau of Laboratories
Department of Health and
    Environmental Control
P.O. Box 2202
8231 Parklane Rd.
Columbia, SC 29202
Phone: (803) 935-7042
Fax: (803) 935-7357
Website: www.state.sc.us/dhec/LAB/
    labgenrl.htm

SOUTH DAKOTA
Director
State Health Laboratory
615 East Fourth St.
Pierre, SD 57501
Phone: (605) 773-4757
Fax: (605) 773-6129
Website: www.state.sd.us/doh/lab/
    lindex.htm

TENNESSEE
Director
Laboratory Services
Tennessee Department of Health
630 Hart Ln.
Nashville, TN 37247-0801
Phone: (615) 262-6300
Fax: (615) 262-6393
Website: www.state.tn.us/health/lab/

TEXAS
Chief
Bureau of Laboratories
Texas Department of Health
1100 West 49th St.
Austin, TX 78756
Phone: (512) 458-7318
Fax: (512) 458-7294
Website: www.tdh.state.tx.us/lab/

UTAH
Director
Utah State Health Laboratory
46 Medical Dr., Room 207
Salt Lake City, UT 84113-1105
Phone: (801) 584-8450
Fax: (801) 584-8486
Website: hlunix.hl.state.ut.us/els/

VERMONT
Director
Health Surveillance
Vermont Department of Health
    Laboratory
P.O. Box 70
108 Cherry St.
Burlington, VT 05402-0070
Phone: (802) 863-7246
Fax: (802) 865-7701
Website: www.state.vt.us/health/
    healthsu.htm

VIRGIN ISLANDS
Director
Public Health Laboratory
St. Thomas Hospital
St. Thomas, VI 00802
Phone: (340) 776-8311
Fax: (340) 776-0610
Website: not available

(Continued on next page)

TABLE 2 State and territorial public health laboratories (*Continued*)

VIRGINIA
Deputy Director
Division of Consolidated Laboratory
  Services
Department of General Services
Commonwealth of Virginia—Box 1877
One North 14th St., Room 231
Richmond, VA 23219
Phone: (804) 786-7905
Fax: (804) 371-7973
Website: www.dgs.state.va.us/dcls/

WASHINGTON
Assistant Secretary
Public Health Laboratories
1610 N.E. 150th St.
Seattle, WA 98155-7224
Phone: (206) 361-2800
Fax: (206) 361-2904
Website:
  www.doh.wa.gov/phl/default.htm

WEST VIRGINIA
Director
Office of Laboratory Services
State of West Virginia/DHHR
167 11th Ave.
South Charleston, WV 25303-1137
Phone: (304) 558-3530
Fax: (304) 558-2006
Website: www.wvdhhr.org/bph/lab.htm

WISCONSIN
Director
State Laboratory of Hygiene
William D. Stovall Building
465 Henry Mall
Madison, WI 53706
Phone: (608) 262-3911
Fax: (608) 262-3257
Website: www.slh.wisc.edu/

WYOMING
Director
Wyoming State Public Health Laboratory
Division of Preventive Medicine
517 Hathaway Building
Cheyenne, WY 82002
Phone: (307) 777-7431
Fax: (307) 777-6422
Website: wdhfs.state.wy.us/lab/

Laboratories Offering Viral Diagnostic Services*

RICHARD L. HODINKA AND STEPHEN A. YOUNG

APPENDIX
3

This appendix identifies laboratories that offer viral diagnostic services; these include hospital, university, and commercial laboratories (Table 1). We have not included public health laboratories, because the services they provide are generally specialized and not intended for primary diagnosis of viral illnesses but for epidemiology. Because diagnostic services offered by any individual laboratory may change, we have not attempted to indicate the extent of services available. Additionally, laboratories that provide viral diagnostic services may open, close, or discontinue services at any time. This must be considered when reference is made to the information in Table 1 for possible submission of specimens and utilization of testing services.

The information presented was supplied by individual clinical virologists and state public health laboratories, who provided names of laboratories with which they were familiar. In this regard, the editors are especially indebted to Stephen Racioppi, Micro Test, Inc., Lilburn, Ga., and Dan McGill, BioWhittaker, Inc., Walkersville, Md. For some states, the state laboratory personnel were unable to supply such a list and it was not possible for us to identify individuals with knowledge of viral diagnostic services. Thus, the lack of a list of services available in a particular state may reflect either our inability to identify those laboratories or that laboratory services are not offered in that state.

Some commercial laboratories have a national network for processing clinical specimens. These laboratories fill the void when there is no local laboratory that performs the needed services (Table 2).

The extent of services available from viral diagnostic laboratories varies widely. Some laboratories offer viral diagnostic and serology services for virtually all common human pathogenic viruses, while others perform only a limited number of tests (e.g., herpes simplex virus isolation or rubella virus serology). Individuals should contact the laboratory(ies) listed in their locale to determine how their needs can best be served.

---

*This appendix contains information presented in chapter 36, by Steven Specter and Gerald Lancz, in the second edition of this manual.

TABLE 1 Laboratories that perform viral diagnosis

ALABAMA

Birmingham
Alabama Reference Laboratories
Birmingham Branch Laboratories
Children's High Tower Medical Center
LabCorp
Medical Laboratory Associates
University of Alabama Medical Center

Mobile
University of South Alabama Medical
Center

Montgomery
Alabama Reference Laboratories

ALASKA

Anchorage
Providence Hospital

ARIZONA

Phoenix
Bolin Laboratories, Inc.
Cigna Health Plan of Arizona
Consultants Medical Laboratory
Good Samaritan Regional Medical Center
LabCorp
National Health Laboratories, Inc.
Phoenix Baptist Hospital
Sonora Laboratory Sciences
St. Joseph's Hospital
St. Luke's Hospital

Scottsdale
Blood Systems of Arizona

Tempe
Sonora-Quest Laboratories

Tucson
American Red Cross
Kino Community Hospital
Tucson Medical Center
University of Arizona Medical Center
Yuma Regional Medical Center

ARKANSAS

Little Rock
Arkansas Children's Hospital
University of Arkansas Medical School

CALIFORNIA

Bakersfield
Physicians Automated Laboratory

Berkeley
Kaiser Permanente Regional Laboratory
Virolab, Inc.

Burbank
Health Line Clinical Laboratories

Cerritos
Health Care Clinical Laboratories

Cypress
MRL Reference Laboratory

Emeryville
Virolab, Inc.

Fresno
Valley Children's Hospital

Garden Grove
West Cliff Laboratories

Glendale
Pathology Clinical Laboratories

Greenbrae
Marin General Hospital

Huntington Park
Apex Laboratories

La Jolla
Scripps Clinic

Long Beach
Long Beach Memorial Medical Center
Medical Reference Laboratory

Los Angeles
Cedars Sinai Medical Center
Children's Hospital of Los Angeles
CLMG, Inc.
Kaiser Permanente
Pathology Associates Laboratory
Specialty Laboratories, Inc.
UCLA Clinical Laboratories
USC/LAC Medical Center

Newberg Park
Reference Laboratory

North Hollywood
Southern California Kaiser Permanente

Northridge
Dimensions Medical

Orange
St. Joseph Hospital
University of California, Irvine

Pasadena
Immunology Consultants
Marist Laboratories
Montgomery Watson Laboratories

Pico Rivera
California Laboratory Network, Inc.

Roseville
LabCorp

Sacramento
Physician's Clinical Laboratory

San Diego
Children's Hospital
UCSD Medical Center
Veterans Affairs Medical Center

San Francisco
California Pacific Medical Center
Children's Hospital
Davies Medical Center
Mt. Zion Hospital
San Francisco General Hospital
University of California
Virologic, Inc.

San Jose
Columbia San Jose Medical Center
Unilabs

San Juan Capistrano
Quest Diagnostics, Inc.

San Luis Obispo
Sierra Vista Hospital

Santa Ana
HCA Laboratory

Santa Barbara
Santa Barbara Cottage Hospital

Santa Monica
Specialty Laboratories, Inc.

Stanford
Stanford University Medical Center

Tarzana
Unilabs

Torrance
Harbor-UCLA Medical Center

Van Nuys
Consolidated Laboratory Services
Quest Diagnostics, Inc.

COLORADO

Denver
Children's Hospital
Clinical Laboratories of Colorado
Immunological Associates of Denver
St. Joseph's Hospital
University of Colorado Health Sciences
Center

Glenwood Springs
Valley View Hospital

Grand Junction
St. Mary's Hospital

Greeley
ARUP—North Colorado Medical Center

CONNECTICUT

Bridgeport
Quest Diagnostics, Inc.

Danbury
Danbury Hospital

Farmington
University of Connecticut Medical Center

Hartford
Hartford Hospital
St. Francis Hospital

New Britain
BBI-North American Clinical Laboratories

New Haven
Yale-New Haven Hospital

West Haven
Veterans Affairs Medical Center

DELAWARE

Newark
Christiana Hospital

New Castle
ARUP—Med Labs, Inc.

(Continued on next page)

TABLE 1 *(Continued)*

Wilmington
Delaware Medical Laboratories
Medical Center of Delaware
Wilmington Hospital

DISTRICT OF COLUMBIA
Children's Hospital National Medical
 Center
Columbia Hospital for Women
Georgetown University Hospital
George Washington University Hospital
Providence Laboratory Associates
Veterans Affairs Hospital
Walter Reed Army Medical Center

FLORIDA
Bay Pines
Veterans Affairs Medical Center
Clearwater
Morton Plant Mease Health Care
Fort Lauderdale
Holy Cross Hospital
Fort Myers
Diagnostic Services, Inc.
Gainesville
Shands Teaching Hospital at the
 University of Florida
Jacksonville
Baptist Medical Center
Mayo Clinic
St. Vincent's Hospital
Lakeland
Lakeland Regional Medical Center
Melbourne
Holmes Regional Medical Center
Miami
Bascom Palmer Eye Institute
Cedars Healthcare Group, Ltd.
Children's Hospital
Jackson Memorial Hospital
Mercy Hospital
Sekot Laboratories, Inc.
University of Miami
Naples
Diagnostic Services, Inc.
Orlando
Orlando Regional Medical Center
Palm Beach
St. Mary's Hospital
Rockledge
Westhoff Reference Laboratories
Sarasota
Sarasota Memorial Hospital
St. Petersburg
All Children's Hospital
LabCorp
Tampa
H. Lee Moffitt Cancer Center
James A. Haley VA Hospital

Quest Diagnostics, Inc.
St. Joseph's Hospital
Tampa General Hospital

GEORGIA
Athens
Athens Regional Medical Center
N.E. Georgia Reference Lab., Inc.
Atlanta
Egleston Children's Hospital
Emory University Laboratory
Grady Memorial Hospital
Northside Hospital
Augusta
Humana Hospital
Medical College of Georgia
St. Joseph's Hospital
Ft. Gordon
Pathology Laboratory
Marietta
Promina Kennestone Hospital
Tucker
Quest Diagnostics, Inc.

HAWAII
Honolulu
Clinical Laboratories of Hawaii
Diagnostic Lab Services
Kaiser Permanente Medical Center
Tripler Army Medical Center

IDAHO
Boise
St. Luke's Hospital

ILLINOIS
Arlington Heights
Northwestern Community Hospital
Chicago
Children's Memorial Hospital
Columbus Hospital
Cook County Hospital
Illinois Masonic Medical Center Hospital
Michael Reese Hospital
Mt. Sinai Hospital
Northwestern Memorial Hospital
Rush-Presbyterian St. Luke's Hospital
VA West Side Memorial Center
University of Chicago Hospital
University of Illinois Medical Center
Elmhurst
LabCorp
Lutheran General Hospital
Memorial Hospital
Evanston
Evanston Hospital
Great Lakes
Naval Hospital
Litchfield
Swedish American Hospital

Maywood
Loyola University Medical Center
Oak Park
West Suburban Hospital Medical Center
Park Ridge
Victoria Clinic Reference Laboratory
Peoria
Mobilab, Inc.
Rockford
Rockford School of Medicine
Springfield
Memorial Medical Center
St. John's Hospital

INDIANA
Indianapolis
Clarian Health
Community Hospital
Indiana Medical Center
Mid America Clinical Laboratories
Wishard Memorial Hospital
Lafayette
Lafayette Home Hospital
Muncie
Pathologists Associated
South Bend
Notre Dame University
South Bend Medical Foundation
Valparaiso
Porter Memorial Hospital

IOWA
Cedar Rapids
Weland Clinical Laboratory
Des Moines
Iowa Methodist Medical Center
Mercy Hospital
Iowa City
University of Iowa Hospital and Clinics
Sioux City
Marian Health Center
St. Luke's Hospital

KANSAS
Kansas City
Kansas University Medical Center
Prairie Village
St. Luke's Hospital
Topeka
St. Francis Hospital
Wichita
Via Christi Regional Medical Center
Consolidated Biological Laboratory
St. Francis Hospital
Wesley Medical Center

(Continued on next page)

TABLE 1 Laboratories that perform viral diagnosis (*Continued*)

KENTUCKY

Lexington
University of Kentucky Medical Center
Veterans Affairs Medical Center

Louisville
Alliant Health System
American Red Cross
Humana Hospital—University of Louisville
Jewish Hospital
Kosair Children's Hospital
National Health Laboratory
NKC, Inc.
Suburban Hospital
University Louisville Pediatric Virology
    Laboratory
ViroMed Laboratory

Madisonville
Regional Medical Center

Owensboro
Western Kentucky Regional Blood Center

LOUISIANA

Baton Rouge
Louisiana Reference Laboratory

Jefferson
Alton Oschner Clinic

Lake Charles
Pathology Laboratory

New Orleans
Charity Hospital of Louisiana
Children's Hospital
LSU Medical Center
Tulane University Medical Center

Shreveport
Christus Health System
LSU Medical Center

MAINE

Bangor
Affiliated Laboratory, Inc.

Portland
Maine Medical Center
Osteopathic Hospital of Maine

MARYLAND

Baltimore
Greater Baltimore Medical Center
Johns Hopkins Bayview Medical Center
Johns Hopkins Hospital
Maryland Medical Laboratories
Quest Diagnostics, Inc.
Sinai Hospital
University of Maryland Medical Center
Bethesda Naval Medical Command, NIH

Fort Meade
Kimbrough Hospital

Hagerstown
Hagerstown Medical Laboratory, Inc.
Washington County Hospital

Rockville
NIH Clinical Center
Providence Lab Associates

Walkersville
BioWhittaker Laboratory

MASSACHUSETTS

Boston
Beth Israel Hospital
Boston Medical Center
Brigham and Women's Hospital
Channing Laboratories
Children's Hospital
City of Boston Hospital
Dana-Farber Cancer Institute
Lahey Hitchcock Clinic Medical Center
Massachusetts General Hospital
New England Deaconess Hospital
Tufts-New England Medical Center

Brookline
New England Medical Center

Cambridge
Mount Auburn Hospital
Quest Diagnostics, Inc.

Hyannis
Cap Code Hospital

Pittsfield
Berkshire Medical Center

Salem
Salem Hospital

Springfield
Pro Med
Valley Medical Laboratories

Taunton
Morton Hospital

West Springfield
Baystate Medical Center

Wilmington
New England Pathology

Worcester
University of Massachusetts Medical Center

MICHIGAN

Ann Arbor
University of Michigan Hospital
Warde Medical Laboratory

Detroit
Children's Hospital
Detroit Medical Center
Henry Ford Hospital
St. John Hospital

Grand Rapids
Butterworth Hospital
Continental Clinical Biochemical
St. Mary's Hospital

Kalamazoo
Borgess Medical Center
Bronson Methodist Hospital

Lansing
Edward Sparrow Hospital
Ingham Regional Medical Center

Livonia
LabCorp

Royal Oaks
William Beaumont Hospital

Southfield
Providence Hospital

Traverse City
Munson Medical Center

MINNESOTA

Andover
Hennepin County Medical Center

Duluth
St. Mary's Medical Center

Minneapolis
Children's Hospitals & Clinics
Hennepin County Metropolitan Medical
    Center
Lufkin Laboratories
University of Minnesota Hospital
Veterans Hospital

Minnetanka
ViroMed Laboratories, Inc.

Rochester
Mayo Clinic

St. Cloud
North Central Laboratories

St. Paul
Children's Hospital
Ramsey Medical Center

MISSISSIPPI

Jackson
University of Mississippi Medical Center

Meridian
Medical Pathology Laboratories

MISSOURI

Columbia
Boyce and Bynum Laboratories
Veterans Affairs Medical Center

Kansas City
Children's Mercy Hospital
North Kansas City Memorial Hospital
Physician's Reference Laboratory
St. Luke's Hospital

Springfield
Lester Cox Medical Center
St. John's Health System

St. Louis
Cardinal Glennon Children's Hospital
Barnes-Jewish Hospital
Medical Pediatrics
Quest Diagnostics, Inc.
St. Louis Children's Hospital

(*Continued on next page*)

TABLE 1 *(Continued)*

St. Louis University Medical Center
Washington University Medical School

MONTANA
Bozeman
Bozeman Hospital
Missoula
Community Hospital

NEBRASKA
Hastings
Mary Lanning Memorial Hospital
Lincoln
Lincoln General Hospital
Omaha
AMI/St. Joseph's Hospital
Children's Memorial Hospital
Physicians Laboratory
University of Nebraska Medical Center

NEVADA
Las Vegas
Associated Pathologists Laboratories
Sunrise Hospital Laboratory
Reno
Associated Pathology Laboratories
Sierra Nevada Laboratory
Washoe Medical Center

NEW HAMPSHIRE
Lebanon
Dartmouth-Hitchcock Medical Center
Manchester
New Hampshire Medical Laboratories
Rochester
Pathology Laboratories, Inc.

NEW JERSEY
Camden
Cooper Hospital University Medical Center
Edison
Edison Medical Laboratories
Elmwood Park
Bio-Reference Laboratory
Englewood
Englewood Hospital Association
Hackensack
Hackensack Medical Center
Livingston
Saint Barnabas Medical Center
Long Branch
Monmouth Medical Center
Morris Plains
Tri Delta Diagnostics, Inc.
Newark
St. Michael's Hospital
University Hospital—UMDNJ
New Brunswick
Middlesex Hospital
Robert Wood Johnson Medical School

Raritan
LabCorp
Teterboro
Quest Diagnostics, Inc.

NEW MEXICO
Albuquerque
Lovelace Medical Center
SED Medical Laboratories
Tri Core Reference Laboratories
Farmington
San Juan Regional Medical Center

NEW YORK
Albany
Albany Medical Center
Wadsworth Center for Labs and Research
Baldwinsville
Radisson Clinical Laboratories
Binghamton
Binghamton General Hospital
Bronx
Montifiore Medical Center
Veterans Affairs Medical Center
Brooklyn
King's County Hospital Center
Quentin Medical Laboratory, Inc.
SUNY Health Science Center
Buffalo
Children's Hospital of Buffalo
Erie County Medical Center
East Meadow
Nassau County Medical Center
Floral Park
Queens Hospital Center
Hauppauge
Sunrise Medical Laboratory
Johnson City
United Health Services
Kingston
Kingston City Community Laboratory
Manhasset
North Shore University Hospital
New Hartford
Centrex Clinical Laboratories
New Hyde Park
Long Island Jewish Medical Center
New York City
Bellevue Hospital
Beth Israel Medical Center
Columbia-Presbyterian Medical Center
Cornell University Medical Center
Memorial Sloan-Kettering Cancer Center
Mt. Sinai Hospital
New York Hospital
New York University Medical Center
St. Luke's Roosevelt Hospital
Veterans Affairs Medical Center
Plattsburgh
Champlain Valley Physicians

Rochester
Genesee Hospital
Rochester General Hospital
University of Rochester-Strong Memorial
  Hospital
Staten Island
IBR Consolidated Clinical Laboratory
Staten Island University Hospital
Stony Brook
University Hospital at Stony Brook
Syracuse
Crouse-Irving Memorial Hospital
SUNY Health Science Center at Syracuse
Valley Cottage
ARUP—Medilabs, Inc.
White Plains
White Plains Hospital Medical Center

NORTH CAROLINA
Asheville
Memorial Mission Hospital
St. Joseph's Health System
Burlington
LabCorp
Chapel Hill
University of North Carolina Hospitals
Charlotte
Carolina Medical Center
Charlotte Mecklenburg Hospital
Charlotte Memorial Hospital
Presbyterian Hospital
Durham
Duke University Medical Center
Veterans Affairs Medical Center
Greensboro
Moses Cone Memorial Hospital
Westly Long Hospital
Greenville
East Carolina University Virology
  Laboratory
Pitt County Memorial Hospital
Raleigh
Rex Hospital, Inc.
Winston-Salem
North Carolina Baptist Hospital

NORTH DAKOTA
None identified

OHIO
Akron
Akron City Hospital
Children's Hospital Medical Center
Beachwood
Mt. Sinai Medical Center
Canton
Aultman Hospital
Timkin Mercy Medical Center
Cincinnati
Children's Hospital Medical Center

(Continued on next page)

TABLE 1 Laboratories that perform viral diagnosis (*Continued*)

Health Alliance Laboratory Services
Jewish Hospital of Cincinnati
The Christ Hospital
University of Cincinnati Hospital

Cleveland
Case Western Reserve University Hospital
Cleveland Clinic Foundation
Cleveland Metropolitan General Hospital
Metrohealth Medical Center
Mt. Sinai Healthcare System
St. Luke's Hospital
St. Vincent's Medical Center
University Hospitals of Cleveland

Columbus
Children's Hospital
Doctor's Hospital
Grant/Riverside Methodist Hospital
Ohio State University Hospital

Dayton
Children's Hospital Medical Center
Veterans Affairs Medical Center
Wright State University

Dublin
LabCorp

Kettering
Kettering Medical Center

Miamisburg
Diagnostic Virology Service

Moraine
Compunet Clinical Laboratories

Parma
Mt. Sinai Reference Laboratory

Rootstown
North East Ohio Universities

Springfield
Mercy Medical Center
Springfield Community Hospital

Steubenville
Ohio Valley Hospital

Toledo
Medical College of Ohio
St. Vincent Mercy Medical Center
Toledo Hospital

Youngstown
Northside Medical Center
Western Reserve Healthcare System

OKLAHOMA

Oklahoma City
Children's Hospital of Oklahoma
HCA Presbyterian Hospital
Integris Baptist Medical Center
Medical Arts Laboratory
Northwest Laboratories
Oklahoma Medical Center
St. Anthony Hospital

Tulsa
St. Francis Hospital
St. John's Hospital
Vironostics Laboratories, Inc.

OREGON

Bend
St. Charles Medical Center

Clackamas
Kaiser Regional Laboratory

Coos Bay
Coastal Medical Laboratory, Inc.

Corvallis
Good Samaritan Hospital Laboratory

Eugene
Oregon Medical Laboratories

Gresham
Mt. Hood Medical Center

Medford
Rouge Valley Medical Center

Pendleton
Interpath Laboratory, Inc.

Portland
Emmanuel Hospital/Metro Laboratory
Good Samaritan Hospital Legacy
    Laboratories
Kaiser Permanente Oregon Regional
    Laboratories
Providence Medical Center

Roseburg
Douglas Community Hospital

The Dalles
Mid Columbia Medical Center

PENNSYLVANIA

Abington
Abington Memorial Hospital

Allentown
Health Network Laboratories

Bethlehem
St. Luke's Hospital

Camp Hill
Holy Spirit Hospital

Chester
Crozer Chester Medical Center
Sacred Heart Hospital

Danville
Geisinger Medical Center

Erie
Associated Clinical Laboratory
Clinical Pathology Institute
Quest Diagnostics, Inc.

Harrisburg
Pinnacle Health at Harrisburg Hospital

Hershey
Hershey Medical Center

Horsham
Quest Diagnostics, Inc.

Langhorne
Delaware Valley Medical Center

Latrobe
Latrobe Area Hospital

Norristown
Quest Diagnostics, Inc.

Paoli
Main Line Clinical Laboratories

Philadelphia
Albert Einstein Medical Center
Ayer Clinical Laboratory/Pennsylvania
    Hospital
Children's Hospital of Philadelphia
Hospital of the Medical College of
    Pennsylvania
Jeanes Hospital
Lankenau Hospital
St. Christopher's Hospital for Children
Temple University Hospital
Thomas Jefferson University Hospital
University of Pennsylvania Medical Center

Pittsburgh
Allegheny General Hospital
Children's Hospital of Pittsburgh
Clinical Pathology Facility
Eye and Ear Hospital
Fulton Medical Laboratory
Magee Women's Hospital
Med-Chek Laboratories
Mercy Hospital
Presbyterian University Hospital
St. Francis Hospital
University of Pittsburgh Medical Center
    Health System
Western Pennsylvania Hospital

Plymouth Meeting
Dekalb Laboratories

Sayre
Robert Packer Hospital

State College
Centre Community Hospital

Throop
Clinical Laboratories, Inc.

Trevose
Quest Diagnostics, Inc.

West Chester
Chester County Hospital

York
York Hospital

RHODE ISLAND

Providence
Brown University Medical Center
Rhode Island Hospital
Women & Infants Hospital

SOUTH CAROLINA

Charleston
Medical University of South Carolina
Roper Hospital

Florence
McLeod Regional Medical Center

Greenville
Greenville Memorial Hospital

(*Continued on next page*)

TABLE 1 *(Continued)*

SOUTH DAKOTA
Sioux Falls
Veterans Affairs Medical Center
Vermillion
University of South Dakota Virology
  Laboratory

TENNESSEE
Chattanooga
Allied Laboratories of Chattanooga
Johnson City
East Tennessee State Medical School
Knoxville
East Tennessee Children's Hospital
University of Tennessee
Memphis
Baptist Regional Laboratories
LeBonheur Children's Hospital
Regional Medical Center
St. Francis Memorial Hospital
St. Jude's Children's Research Hospital
Nashville
Baptist Hospital
Meharry Medical College
Quest Diagnostics, Inc.
St. Thomas Hospital
Vanderbilt University Medical Center

TEXAS
Abilene
Hendrick Medical Center
Austin
Austin Clinical Pathology Laboratories
Austin Regional Clinic
Seton Medical Center
Bastrop
University of Texas/M. D. Anderson
  Cancer Center
Corpus Christi
Driscoll Children's Hospital
Dallas
Baylor University Medical Center
LabCorp
National Pathology Laboratory
Parkland Memorial Hospital
Quest Diagnostics, Inc.
University of Texas Southwest Medical
  Center
Fort Sam Houston
Brooke Army Medical Center
Fort Worth
Cook-Fort Worth Children's Hospital
Harris Medical Laboratory
Galveston
Shriners Burn Institute
University of Texas Medical Branch
Houston
Baylor College of Medicine
Bentaur General Hospital
Christus St. Joseph's Hospital

Hermann Hospital
Influenza Research Center/Baylor College
  of Medicine
LabCorp
Methodist Hospital
Microbiology Specialists, Inc.
St. Luke's Episcopal Hospital
Hurst
Allied Clinical Laboratory
Lubbock
Southwest Clinical Laboratory
St. Mary's Hospital of the Plains
Texas Tech University Health Science
  Center
Richardson
Viral Diagnostics, Inc.
San Antonio
Brooks Air Force Base
LabCorp
Medical Center Hospital/University of Texas
Oak Hills Pathology Laboratories
Severans Reference Laboratory
Southwest Texas Methodist Hospital
University Health Systems
Virus Reference Laboratories
Temple
Scott and White Clinic
Texarkana
Doctors Diagnostic Laboratories
Victoria
Lyster Reference Laboratory

UTAH
Salt Lake City
Association of Regional and University
  Pathologists (ARUP), Inc.
Primary Children's Medical Center
University of Utah Medical Center

VERMONT
Burlington
Medical Center Hospital of Vermont

VIRGINIA
Chantilly
American Medical Laboratories
Charlottesville
University of Virginia Medical Center
Herndon
LabCorp
McLean
United Medical Laboratories, Inc.
Norfolk
DePaul Medical Center
Medical College of Hampton Roads
Portsmouth
Naval Medical Center
Richmond
Division of Consolidated Lab Services
Medical College of Virginia

Virginia Commonwealth University
Salem
Veterans Administration Medical Center
Vienna
National Health Laboratories

WASHINGTON
Camas
Southwest Washington Medical Center
Everett
Phoenix Central Laboratory
Kent
LabCorp
Richland
Kadlec Medical Center
Seattle
Children's Hospital Medical Center
Fred Hutchinson Cancer Research Center
Laboratory of Pathology for Seattle, Inc.
Northwest Hospital
Quest Diagnostics, Inc.
University of Washington
Virginia Mason Clinic
Spokane
Deaconess Hospital
Pathology Associates Medical Laboratories
Sacred Heart Medical Center

WEST VIRGINIA
Charleston
Charleston Area Medical Center
Charleston Memorial Hospital
Huntington
Marshall University School of Medicine
Morgantown
West Virginia University Hospital

WISCONSIN
Brookfield
Franciscan Shared Laboratory—Elmbrook
Eau Claire
Sacred Heart Hospital
Green Bay
Bellin Memorial Hospital
St. Vincent Hospital
LaCrosse
Gunderson Lutheran Hospital
Madison
University of Wisconsin
Marshfield
Marshfield Clinic
St. Joseph's Hospital
Milwaukee
Aurora Health Care
Children's Hospital of Wisconsin
Herpesvirus Diagnostics, Inc.
Medical College of Wisconsin
Milwaukee Medical Clinic
Saint Mary's Hospital
United Region Medical Services

(Continued on next page)

TABLE 1 Laboratories that perform viral diagnosis *(Continued)*

Watertown	WYOMING
Watertown Memorial Hospital	**Jackson**
Wauwatosa	Intermountain Virology Laboratory
Children's Hospital	

TABLE 2 Commercial laboratories that offer viral diagnostic services nationally

Laboratory	Headquarters
American Medical Laboratories	Chantilly, Va.
ARUP Laboratories	Salt Lake City, Utah
BBI-North American Clinical Laboratories	New Britain, Conn.
LabCorp	Burlington, N.C.
Mayo Medical Laboratories	Rochester, Minn.
MRL Reference Laboratory	Cypress, Calif.
Quest Laboratories, Inc.	Teterboro, N.J.
Specialty Laboratories, Inc.	Santa Monica, Calif.
ViroMed Laboratories, Inc.	Minneapolis, Minn.

Author Index

Subject Index

609